# TEXTBOOK OF ANATOMY

# TEXTBOOK OF ANATOMY

FOURTH EDITION

**W. Henry Hollinshead, Ph.D.**

Professor Emeritus of Anatomy,
Mayo Graduate School of Medicine (University of Minnesota),
Rochester, Minnesota;
Visiting Professor Emeritus of Anatomy,
University of North Carolina School of Medicine,
Chapel Hill, North Carolina

**Cornelius Rosse, M.D., D.Sc.**

Professor and Chairman, Department of Biological Structure,
University of Washington School of Medicine,
Seattle, Washington

**HARPER & ROW, PUBLISHERS**
**Philadelphia**

Cambridge
New York
Hagerstown
San Francisco

London
Mexico City
São Paulo
Sydney

1817

The authors and publisher have exerted every effort to ensure that drug selection and dosage set forth in this text are in accord with current recommendations and practice at the time of publication. However, in view of ongoing research, changes in government regulations, and the constant flow of information relating to drug therapy and drug reactions, the reader is urged to check the package insert for each drug for any change in indications and dosage and for added warnings and precautions. This is particularly important when the recommended agent is a new or infrequently employed drug.

Acquisitions Editor: William Burgower
Sponsoring Editor: Richard Winters
Manuscript Editor: Mary K. Smith
Indexer: Eleanor Kuljian
Art Director: Maria S. Karkucinski
Designer: Patrick Turner
Production Supervisor: Kathleen P. Dunn
Production Assistant: Carol Florence
Compositor: Bi-Comp, Inc.
Printer/Binder: R. R. Donnelley & Sons

Fourth Edition



1 3 5 6 4 2

**Library of Congress Cataloging in Publication Data**

Hollinshead, W. Henry (William Henry), 1906–
Textbook of anatomy.

Includes bibliographies and index.
1. Anatomy, Human. I. Rosse, Cornelius. II. Title.
[DNLM: 1. Anatomy. QS 4 H741t]
QM23.2.H57 1985 611 84-25308
ISBN 0-06-141263-5

# PREFACE

The unprecedented expansion of biomedical knowledge during recent decades has inevitably demanded an evaluation of curricula in the health-related professions. Consequently, the relatively stable period in medical education that prevailed during the middle third of this century was followed by experimentation with curricular content and the design of new strategies and resources for learning. One of the noteworthy features of this process was the active involvement of students, not only as experimental material but as critical monitors of experimental results.

Three main conclusions may be drawn at this stage from the curricular reforms of the recent past: 1) Experimentation and evaluation must continue, and flexibility must be built into the health-related professional curricula, anticipating not only the continued expansion of the biomedical sciences, but also the changes in the educational background of future student populations and in the needs of the health professions. 2) Students are capable of and motivated for independent learning. Appropriate learning resources must be developed for this purpose, and curricular structure must place the responsibility for much of the learning on the student to foster an attitude for continued independent learning beyond graduation. 3) Curricular content must not be defined by the sets of facts and skills on which the everyday practice of the profession relies. Unless the use of such facts and skills is based on a thorough and broad *understanding* of human biology, the practitioner will be restricted to the routine and conventional, degrading health care delivery from the professional to the trade level. Preparation for the practice of a *profession* demands education rather than mere training. *Education* is a protracted developmental process, resulting in a qualitative change in the learner's ability to integrate and use diverse types of information to handle not only routine but also novel and unfamiliar situations. In the process of education, the learner inevitably encounters many facts and concepts that will not surface in the everyday practice of the profession; yet, without such facts and concepts, an understanding of human biology could not be attained.

The fourth edition of this *Textbook of Anatomy* was prepared with the above considerations in mind. Anatomy is basic to education in the health-related professions. Although the day-to-day practice of different disciplines and specialities requires a widely different knowledge base of anatomical *facts*, a general understanding of the structure of the human body and an appreciation of functional anatomy is a requisite for the interpretation of many symptoms and essentially all physical signs. The fourth edition of this book is no more "clinical," however, than its predecessors have been. Its purpose remains unchanged from those of previous editions: It presents to the student of medicine and dentistry, and the graduate student of anatomy, an account of anatomical facts and concepts necessary to appreciate the structural organization of the human body at an intellectually satisfying level of complexity, thus providing a foundation on which to build when more detailed and specialized knowledge becomes necessary. To enhance the understanding of anatomy, developmental considerations have been included liberally without an attempt to present a comprehensive account of embryology. Contrary to the opinions of many, the study of anatomy need not consist of the memorization of long lists of names; rather, it should rely on the visualization of parts and regions of the body in three dimensions based on an understanding of how these relationships have come about and why they exist. The embryology as well as the abundant clinical examples in the book serve the same purpose: they are included to aid the understanding of normal anatomy and make its study more meaningful.

Curricular reforms have led to a tendency to reduce the presentation of anatomy in many recent textbooks to the so-called essential facts. Apart from the intractable difficulty of defining what is essential and for what purpose, such an approach divests anatomy of the attributes it possesses as one of the oldest of all scientific disciplines. Of more serious concern is the undesirable influence such an approach has on students in the health professions: it encourages them to learn unsupported facts parrot-fashion and to accept didactic statements unquestioningly. Because anatomy is usually the first discipline that students can relate directly to the practice of their chosen profession, it is important that teachers and authors of anatomical textbooks capitalize on this opportunity to present the subject as a discipline of science that relies on the scientific method and thus stimulate the development of an inquiring, analytical attitude in the student. The concrete subject matter of the discipline lends itself well for this purpose. Therefore there are many references in this book to the methods of obtaining anatomical data and also to the gaps of knowledge that still exist in specific areas. The lists of recommended readings at the end of each chapter have been selected to focus the students' attention on anatomy as an active and evolving biomedical science to which they may contribute during their undergraduate and postgraduate careers.

The fourth edition conforms in general to the previous editions. Following an introductory section in Part I, eight brief chapters give a general account of the systems of the body in Part II with the emphasis on continuities in a particular system from one region of the body to another. The function of each system is discussed as a whole. Parts III to IX encompass the remaining 24 chapters, which deal with the regions of the body.

The number of chapters in the fourth edition has been increased from 28 to 34 owing to the reorganization of the parts concerned with the thorax and abdomen. Both the thorax and abdomen now are introduced by a general chapter followed by three additional chapters, in conformity with the organization of those parts of the book dealing with the upper and lower limbs, and with the head and neck. Parts VI to VIII, dealing with the thorax, abdomen, pelvis, and perineum, have been entirely rewritten. Because of the advent and importance of computer-assisted tomography, transverse sections have been introduced in these regions. Developmental considerations also have been substantially augmented. The objective was to provide an understanding of how the topographical arrangements come about developmentally in the thorax, abdomen, pelvis, and perineum without having to refer to a textbook of embryology. As a rule, embryology texts deal with the development of organ systems without emphasizing topographical relationships. In addition to these main parts of the book, sections of Chapters 1, 3, and 7 also have been rewritten.

The appearance of the fourth edition is noticeably different from that of its predecessors because of the extensive use of color. Color has been added to nearly 200 black and white figures from the third edition, and many old illustrations have been modified. There are 113 entirely new figures in the fourth edition, including many x-rays obtained by a variety of radiographic techniques.

In this edition, terminology conforms as much as possible to *Nomina Anatomica,* 4th edition, 1977. Some useful terms not included in this edition of the N.A. will be found in the text. Others appear in the Glossary of Synonyms and Eponyms. With few exceptions, the use of eponyms has been avoided in the text.

It is our hope that the changes made in the fourth edition of this *Textbook of Anatomy* respond in a constructive manner to the demands of contemporary educational needs in the health-related professions. By aiming to provide the conceptual and functional basis for understanding the human body at the macroscopic level of organization, we hope to stimulate the student to acquire the ability to think in anatomical terms. The presentation of the material will, we hope, lend itself to independent study of the subject as well as to use in conjunction with structured courses. The degree of our success in realizing our intentions can be gauged only by the criticism that we invite from our students and colleagues who select this book as an aid to learning and teaching anatomy.

W. HENRY HOLLINSHEAD, PH.D.

CORNELIUS ROSSE, M.D., D.SC.

# ACKNOWLEDGMENTS

We wish to record our thanks to a number of our colleagues and associates who have generously contributed to the fourth edition of this book.

The sections of Chapter 21 dealing with the normal and abnormal development of the heart were written in close collaboration with Dr. Lore Tenckhoff, cardiologist and Director of Cardiac Ultrasound, Children's Orthopedic Hospital and Medical Center, Seattle, Washington; and Clinical Professor of Pediatrics and Radiology at the University of Washington. One of us (C.R.) wishes to place on record an indebtedness to her for the help she provided during the writing of the chapter and for the rewarding collaboration in the classroom over a number of years. The numerous radiographs that appear throughout the text of Parts VI to VIII without specific credit have all been provided by members of the faculty of the Department of Radiology, University of Washington. We are particularly grateful to Dr. Melvin M. Figley, Professor and Chairman Emeritus of this Department, who has not only provided many of the x-rays but has coordinated the assembly of radiographic illustrations for this textbook. The helpfulness and the high professional standards of this Department deserve the highest praise and our boundless gratitude. We would like to thank especially Drs. Lawrence, Mack, Leon A. Phillips, and Charles A. Rohrmann for their generosity and cooperation.

Charlotte P. G. Kaiser drew essentially all the new illustrations, and it is a particular pleasure to place on record our thanks for her patience, dedication, and fine work. She is also responsible for the modification of existing figures and for the arduous task of working out the color separations of existing and new drawings that give such a different appearance to this edition. Without the help of Susan A. Marett the preparation of this new edition would probably never have been accomplished. She has served as an invaluable resource in many respects including text editing, literature search, bibliography, liaison with collaborators and publishers, and as a most valuable general assistant responsible for all administrative duties related to the revision of the book. We are also grateful to Doris E. Ringer for the preparation of the manuscript and for the numerous other ways in which she has contributed to the completion of this project.

We wish to thank Mr. Vincent P. Destro of the Section of Medical Graphics, Mayo Clinic, for his cooperation in providing for us numerous illustrations owned by the Mayo Clinic that have been published either in this textbook or in *Anatomy for Surgeons*. Likewise, we thank W. B. Saunders for permitting us to use figures from *Functional Anatomy of the Limbs and Back*. Our publishers have been patient during the protracted preparation of this new edition and we thank them for the cooperation they have given us.

# CONTENTS

# TEXTBOOK OF ANATOMY

# PART I

# INTRODUCTION

# 1

# THE STUDY OF ANATOMY

## Content and Subdivisions of the Subject

The word "anatomy" is derived from Greek roots that mean "to cut up" or "to dissect." The study of human anatomy in its early stages was adequately defined by this term, for anatomy dealt only with structures that could be displayed by dissection and that were visible to the naked eye, what we now call "gross anatomy." Although essentially a morphological science, anatomy was never purely that. Even in the earliest writings there were speculations regarding the importance of the various parts and how they worked. Thus, a consideration of the use to which a part is put and how it fulfills its functions has always been a part of anatomy. Without considering function, anatomic study would be analogous to learning the names and arrangement of all the parts of an automobile engine and having no concept of what the engine does or how it works. Although there are a few exceptions, all structures have a function associated with them that is quite obvious, even to a lay person. Function is dependent on structure, whether it is gross, microscopic, or molecular.

Physiology, a discipline primarily concerned with the study of function in biological systems, became separated from anatomy as a science in its own right as methods of investigating structure and function became increasingly complex. The division between anatomy and physiology, however, can never be as sharp as intimated by their separate names and the different academic units concerned with their study.

It is difficult to make a sharp distinction between anatomy and physiology, and it is even more difficult to do so between the various subdivisions of anatomy. These subdivisions fall conveniently into four general spheres: gross anatomy, neuroanatomy, microscopic anatomy or histology, and developmental anatomy or embryology. Such subdivisions are purely arbitrary and for the convenience of instruction only. It is important, furthermore, to appreciate that none of the subdivisions can be understood without some knowledge of the others.

In short, the study of anatomy is concerned primarily with structure, ranging from the molecular to the macroscopic, and properly includes a consideration of the formation and functional importance of all parts of the organism. Human anatomy, therefore, is a sector of special interest within the field of human biology. Just as it is true in medicine that treatment should be directed at the patient and not at a specific disease, so it is true of anatomy, that it should concern itself with all the aspects of human biology that are necessary for an understanding of the function of the living body. The field of anatomy is an enormous, expanding area that no one ever completely masters. Different specialties in the health professions concern themselves in greater or lesser depth with certain aspects of this large body of knowledge. To apply this knowledge, it is necessary, however, to have a basic understanding of the structure of the human body as a whole.

Gross anatomy itself is sometimes divided according to methods of approach. Systemic anatomy attempts to treat the body according to systems: skeletal, vascular, and so forth. Regional anatomy deals with several systems located in a particular region of the body. Practical or surgical anatomy emphasizes certain features that are of particular importance to the practitioners of medicine and surgery. All three of these approaches necessarily deal with the same basic subject matter. The student of gross anatomy needs to combine all three of these approaches. A general understanding of the systems of the body aids in an appreciation of the more detailed regional anatomy and of the interrelationships of parts and regions. The regional approach is commonly used in the dissection room and is also most useful to the physician and surgeon. Consideration of the functional and clinical aspects of anatomy provides a better background upon which to build clinical knowledge. This functional approach aids learning by emphasizing the importance of what may seem to be a mass of unrelated minutiae.

## Contents of This Text

Any one-volume textbook of anatomy necessarily omits many details; there is enough available information to fill volumes. Therefore, this book does not contain enough detail to meet all the possible present and future requirements of the student. Other sources of information will be needed. Fortunately, however, it is possible to recognize a fairly large body of information as being essential to any thorough study of gross anatomy, and that is what this book attempts to present.

Even though gross anatomy is an old science, interpretations of anatomic facts sometimes change rapidly, and there are some points about which there are long-standing differences of opinion among anatomists. Ideally, the student should be made aware of such differences of opinion. However, divergent views can properly be presented only in detail, and this is not possible in a short text. Occasionally, therefore, views presented in this book will not agree with those of the student's instructors, because some necessarily represent only one opinion.

Certain aspects of gross anatomy are understandable only when there is a background of knowledge in neuroanatomy, histology, and embryology. Yet the students are rarely well versed in these fields at the time they are introduced to gross anatomy. For this reason, certain fundamentals that belong more properly to another field, such as a discussion of the histology of a nerve or the explanation of the development of a part, are introduced at appropriate places in the text. The aim is to elucidate the gross anatomy and not to give a well-rounded picture of any other field. It is assumed that the student has texts in those fields and will turn to them as necessary for further information.

Although some of the systemic anatomy presented in this text is acquired by many students in courses in comparative anatomy, parts of it will be new to most students. It is suggested, therefore, that the student read pertinent chapters of the systemic anatomy as early as possible in the course. Chapter 2 (Anatomic Terminology) should, of course, be understood from the very beginning. Because one cannot dissect anywhere without encountering skin, fascia, muscles, nerves, and vessels, Chapter 3 (The Connective Tissues), that part of Chapter 4 (The Nervous System) dealing with nerves, Chapter 5 (The Circulatory System), Chapter 6 (The Muscular System), and Chapter 10 (The Skin and Its Appendages) contain background information useful from the very first part of the dissection. The parts of Chapter 4 dealing with the autonomic and central nervous systems can be postponed, if desired, until dissection involves some parts of these, and so can Chapters 7 (The Digestive and Respiratory Systems), 8 (The Urogenital System), and 9 (The Endocrine System). The references in the regional parts of the text to the more general discussions are intended to save repetition, where possible, and yet bring the student back from study of specifics to one of general facts and theories, thus helping to build up understanding as well as factual knowledge.

In the discussion of regional parts, some of the material appears in small print. This is material that usually cannot be verified in the dissecting laboratory, or it is of an explanatory nature that amplifies the main part of the text. Since it includes discussion of functional and developmental aspects and clinical applications, it should not be regarded as of lesser importance; indeed, much of the material in small print is necessary to an understanding of the presentation as a whole.

In the discussion, structures are described, on the whole, in the order in which they can be found during dissection. Because this type of presentation does not allow for complete description of nerves and vessels, only parts of which are usually visible at any one stage of dissection, summaries of the nerves and vessels follow most of the descriptions of the regions. These are intended for review of the nerves and vessels themselves, but they can also be used in the laboratory as a review of the entire region. Similarly, introductory chapters on the upper and lower limbs should be reread after dissection, both for brief reviews of these regions as a whole and for help in systematizing the detailed knowledge that the student has gained.

## Hints on Studying

Anatomy is a visual science. It is essential for the student to engage the "mind's eye." The time spent on repeated reading of the text will be less effective than the time spent thinking about the subject and visualizing a structure and its relationship to other parts or regions. It is well to read the text before going to the dissection room, to become familiar with the region, to learn what to look for, and to gain some understanding of the functional implications of the parts to be dissected. Even at this stage it is useful to try to visualize what is being read, and it is an added advantage to be able to see it with the book or atlas closed. The text should be consulted again to fill in specific gaps rather than rereading a complete section. Dissection is usually performed with the aid of special guides, and in performing the dissection, the student should anticipate the structures that are already in the mind's eye. This way, dissection will be a challenging experience and its reward will be discovery. Knowledge gained from the textbook and atlases will be reinforced and expanded by dissection. Before embarking on the next area of study, the student should again, by way of a summary, recall visually the dissected parts and return to the textbook to reinforce the known concepts and to pick up additional details that may have been overlooked in the first reading.

It is helpful to meet with fellow students periodically to discuss the material and quiz each other on what has been learned. Being able to explain anatomy is the best proof for its thorough understanding.

# 2

# ANATOMIC TERMINOLOGY

Anatomy is a descriptive science. Therefore, its terminology is mainly descriptive; it is also logical. An accurate use of anatomic terms in anatomy is essential to avoid confusion. There is an international nomenclature, the basis of which is Latin. Despite the use of Latin, however, many of the meanings of the words are immediately obvious. For those that are not, the student must learn the meaning. In many instances, it is common and accepted practice to translate the Latin into the vernacular of the particular country concerned, but it is largely a matter of custom as to which terms are so translated and which are rendered in their original Latin form.

Over the history of anatomy, many structures have acquired multiple names. It was estimated in the latter part of the 19th century that there were some 50,000 anatomic terms in use but that these actually applied only to some 5000 to 6000 structures. The German Anatomical Society studied the question of multiplicity in terms and, in 1895, in Basle, adopted a standard nomenclature that came to be known as the Basle Nomina Anatomica, or the B.N.A. This was generally adopted in Germany and in the English-speaking countries. In succeeding years, however, as it became obvious that there were some inaccuracies and infelicities in the B.N.A., several countries undertook to revise it, and synonyms again began to be common. A thorough revision of the B.N.A. was finally undertaken by the International Congress of Anatomists and was approved at a meeting in Paris in 1955. The list of terms there adopted is known as the Nomina Anatomica, or N.A., and is now the official terminology.

Nomina Anatomica retains the B.N.A. principles of eliminating synonyms and trying to make terms precisely descriptive. The International Nomenclature Committee from time to time revises N.A.

In this text, the 1977 edition of N.A. is followed, usually in the anglicized form. The original Latin terminology is given where there might be cause for confusion. Although there should no longer be any necessity to learn synonyms, the student cannot avoid hearing and seeing some. Synonyms and alternate terminology will be encountered in the literature and in clinical practice, because many instructors of anatomy learned the B.N.A. and clinical instructors often have had no opportunity or need to keep their terminology current. For this reason, some of the more common synonyms, usually B.N.A. terms, are included in the text, and there is also appended a short, incomplete glossary of terms frequently encountered in the clinical literature.

Anatomic terms can be divided into two general categories: names of parts of the body and descriptive terms that include adjectives and verbs. Not only the names but also the adjectives (*e.g.*, proximal, distal) and the verbs (*e.g.*, flex, extend) must be used precisely to avoid misunderstanding.

## Parts of the Body

The body is divided into three parts, or regions: the head and neck, the trunk, and the limbs. The trunk, in turn, is divided into the thorax, abdomen, and perineum. The limbs, formerly called **appendages,** now are called **membra,** but this is generally translated into **limbs** rather than members. Names of parts of the limbs can best be considered later, but it might be noted that "arm" and "leg," although popularly used as synonyms for the upper and lower limbs (superior and inferior members), are not anatomic synonyms for these limbs; the arm (brachium) is that part of the upper limb between shoulder and elbow, and the leg (crus) is that part of the lower limb between knee and ankle.

To understand even introductory anatomy, it is necessary to know the meaning of some Latin terms. The head is the **caput,** hence the term **capitis** (of the head). The skull is the **cranium,** hence the adjective **cranial; cephalic** is used in the same sense as cranial, but the prefix **cephalo-** is used as a combining form (*e.g.*, cephalothoracic, relating to the head and chest). There are two words that mean neck, **collum** and **cervix,** and from these terms are derived the possessive forms, **colli** and **cervical.** The back of the neck is the **nucha,** but its derivative, **nuchal,** is more commonly used for structures relating to the neck.

## Terms of Position, Planes, Direction, and Movement

Terms of position, planes, direction, and movement must be used in reference to the anatomic position and must be used consistently to avoid confusion. The **anatomic position** is defined as the erect position with the arms at the sides and palms of the hands facing forward.

The **median plane** bisects the body into right and left halves. Any plane parallel with the median plane is a **sagittal plane. Medial** describes a position nearer to the median plane, and **lateral** describes a position farther from it. A **coronal plane** is at right angles to the sagittal plane and bisects the body into **anterior** and **posterior** portions. The position of a structure nearer to the front of the body is described as **anterior** and that nearer to the back of the body as **posterior.** The terms **ventral** and **dorsal** may be used synonymously with anterior and posterior, but properly should apply only to nerves. A **transverse plane** is horizontal and bisects the body into upper and lower parts. **Superior** means lying above, and **inferior,** lying below. Terms such as **internal** and **external, superficial** and **deep,** are readily understandable and refer to relationships to the surface of the body. **Proximal** and **distal,** meaning closer to and farther from, are general terms, but in reference to the limbs, they are used to designate structures nearer to or farther from the attachment of the limb to the trunk.

In describing the limbs, all of the above terms may be used. However, to avoid confusion, it is customary to describe the limbs in terms of the positions of their paired bones. Thus, in the upper limb, the radius is the lateral bone of the forearm and the ulna is the medial one, so **radial** refers to the thumb side and **ulnar** to the little-finger side. In the leg, the tibia is the medial bone and the fibula is the lateral one, so **tibial** refers to the big-toe side and **fibular** to the little-toe side. Also, the adjective used in describing structures of the anterior surface of the hand is **palmar** rather than anterior; similarly, **plantar** is used in describing structures of the sole of the foot. The other surface of both hand and foot is known as the **dorsum.**

In the embryo, some additional terms are used: Structures close to the head or movements toward the head are described as **cranial** or **rostral** (rostrum meaning beak) rather than superior; structures closer to the tail or movements toward the tail are described as **caudal** (cauda meaning tail, since the embryo possesses a tail) rather than inferior.

Terms describing movement also are a necessary part of anatomic terminology. For purposes of clarity, they sometimes need to be qualified, and at some joints, movements are complex enough to demand special description and definition. Terms generally used are flexion and extension, abduction and adduction, rotation, and circumduction.

**Flexion** means bending in a direction that approximates the ventral surfaces of the body. **Extension** is the reverse of flexion and usually straightens a part by movement toward the original dorsal surface of the body. These terms need to be qualified when describing movements of the foot. Because in the anatomic position the foot makes a 90° angle with the tibia, its original ventral surface, the sole, faces downward and its original dorsal surface faces upward. Therefore, flexion of the foot is spoken of as **plantarflexion,** a movement that increases the 90° angle; extension of the foot is referred to as **dorsiflexion,** a movement that decreases the same angle.

**Abduction** means to draw away from and **adduction,** the reverse, drawing toward. These terms are used in reference to the median plane of the body except when they pertain to movements of the digits. Abduction of the trunk is the same as bending it laterally, and its adduction restores it to the vertical position. Abduction of the limbs lifts them away from the body, and adduction approximates them. The reference line for movement of the digits is the middle finger or the second toe. Spreading the digits is abduction, and drawing them together is adduction.

**Rotation** is a movement of a part around its long axis. When the anterior surface rotates laterally, the movement is called **lateral rotation,** and rotation of the anterior surface medially is **medial rotation.** Movements that are reminiscent of rotation take place in the forearm and in the foot. These movements receive the names of **supination** and **pronation** of the hand and **inversion** and **eversion** of the foot. In the anatomic position, the forearm and the hand are supinated. When the dorsum of the hand is turned forward without rotation of the upper arm, the hand and the forearm are pronated. With the elbow bent, the palm of the supinated hand faces the ceiling; in the same position at the elbow, the palm of the pronated hand faces the floor. Inversion is the movement that turns the sole of the foot inward or medially, and eversion turns the sole outward or laterally.

Some joints permit several movements to take place. When a part is moved successively through flexion, abduction, extension, and adduction, it circumscribes a cone of space, with its distal point drawing a circle. Such a movement is called **circumduction.**

## Abbreviations

Most abbreviations used in anatomic texts and figures are clear if one knows the words being abbreviated; for instance, "flex. poll. long.," when referring to a muscle, adequately identifies the flexor pollicis longus muscle. There are, however, a few commonly used abbreviations with which the student needs to be familiar: a. for artery or the Latin *arteria,* aa. for arteries or *arteriae;* v. for vein or *vena,* vv. for veins or *venae;* n. for nerve or *nervus,* nn. for nerves or *nervi;* and m. for muscle or *musculus,* mm. for muscles or *musculi.*

In referring to vertebrae or spinal nerves, which are designated both by region and by number within the region (*e.g.,* the 6th cervical or the 8th thoracic), it is often convenient to use shorter forms: **C** for cervical, **T** for thoracic, **L** for lumbar, and **S** for sacral. Thus, the third nerve and vertebra of each of the regions may be identified as C3, T3, L3, and S3. In older literature, dorsal (D) is synonymous with thoracic; however, since vertebrae in all regions are located dorsally in the body, this use of the term should be abandoned.

# PART II

# SYSTEMIC ANATOMY

# 3

# THE CONNECTIVE TISSUES

The term "connective tissue" is used to include a variety of tissues that, despite their common derivation from embryonic mesenchyme, assume distinctive gross and microscopic features and subserve a variety of functions. As the name implies, the universal function of connective tissue is to hold together more specialized tissues, welding them into organs and retaining them in their proper anatomic relationships. Connective tissue also circumscribes the various morphological and functional units of organs and supports nerves and blood vessels as they course between different parts of the body. In its specialized forms, connective tissue supports the entire body and retains the shape of many of its parts. These **specialized connective tissues** include **bone** and **cartilage,** in which connective tissue cells (osteocytes and chondrocytes) are encased in a solid or semisolid matrix. **Ordinary connective tissue** is distinguished by a gellike matrix, in which the cells (fibroblasts and fibrocytes) are dispersed among numerous pliable fibers that lend tensile strength or elasticity to the tissue. Under the term "ordinary connective tissue" are usually included **loose connective tissue,** which functions as packing material between organs, and **dense connective tissue,** which is organized into such structures as tendons and ligaments.

Other tissues that develop from embryonic mesenchyme include **blood** and **lymph,** and these are also sometimes classified among the connective tissues. For the purposes of this book, blood and lymph are not considered as connective tissue. This chapter deals only with the various types of ordinary connective tissue and its two specialized forms, bone and cartilage.

The ordinary connective tissues, as well as bone and cartilage, are characterized by the dispersion of cells in an **extracellular matrix.** The different physical and biochemical properties of the matrix essentially determine the type of connective tissue. The matrix of ordinary connective tissue consists of various types of protein molecules that are either assembled into fibers or exist in solution. The matrix of bone is rendered rigid by the deposition of calcium salts, whereas cartilage matrix is semisolid owing to polymerization and cross-linking of some of its constituents.

The connective tissue cells are responsible for synthesizing the matrix, including the proteins that make up the various types of fibers, and the amorphous ground substance. The **fibers** are made of either collagen or elastin. When collagen fibers predominate, one speaks of *fibrous tissue*. Fibrous connective tissue with a high degree of organization is found in ligaments and tendons, whereas in the dermis and in scar tissue, the fibers are not regularly organized. *Elastic tissue,* in which elastic fibers predominate, is found in the wall of large, distensible arteries and also in some ligaments. When loose connective tissue is extensively infiltrated by fat cells, it is spoken of as *adipose tissue.* Loose connective tissue is also readily distended by extracellular fluid, which causes swelling (edema) of various parts and organs.

All connective tissues are dynamic in that there is a constant turnover of their constituent molecules, and they are readily reorganized in their structure in response to various stimuli and mechanical stresses. Various types of connective tissue may change radically even in an adult body, and new connective tissue may arise. This is most dramatically demonstrated in the remodeling of bone in healing fractures or the reorganization of scar tissue.

## GROSS ANATOMIC FORMS OF CONNECTIVE TISSUE

### CONNECTIVE TISSUE PROPER

Ordinary connective tissue is recognized by the naked eye as loose connective tissue, fascia, ligaments, tendons, and fat.

#### Loose Connective Tissue

Macroscopically, loose connective tissue, or *areolar tissue,* is a fine, cobweblike packing material that fills the interstices between organs and serves as padding. It may be seen, for instance, when the fiber bundles in a muscle like the biceps are pulled apart. The same type of tissue is found between the dermis and an underlying structure (muscle or bone) over parts of the body that are devoid of subcutaneous fat (the dorsum of the hand). Areolar tissue creates a plane of cleavage along which tissues or organs may be pulled away from one another, and it also aids the sliding of these structures upon each other.

There may be large numbers of fat cells among the connective tissue fibers, in which case the tissue is usually referred to as **fatty (adipose) tissue** rather than as loose connective tissue. Accumulations of fat in fat-storing cells form the visible depots of fat in the body. Such adipose tissue appears yellow to the naked eye. Body fat under the skin is known to gross anatomists as **superficial fascia.** In Latin terminology, superficial fascia is referred to as *tela subcutanea* or *paniculus adiposus.* Superficial fascia serves as padding to fill out the body contours and to conserve body heat. Over the palms of the hands and soles of the feet, the fat is arranged in rather tight lobules to absorb pressure.

Microscopically, loose connective tissue is a framework of interwoven collagen fibers intermingled with occasional elastic fibers. The large interfibrillar spaces are filled with fluid and also contain the cells found in connective tissue (*e.g.,* fibroblasts, phagocytes, migratory elements of the blood). This basic structure also exists in adipose tissue, but most of the spaces are occupied by the distended fat cells. Adipose tissue that is not under the skin is similar in structure to superficial fascia but is known simply as fat.

#### Fascia

The term "fascia" is rather loosely applied in anatomy and in surgery. When used without qualification, it most often denotes a readily visible connective tissue membrane that consists of a thin layer of dense connective tissue without obvious organization of its fibers. The exception to the definition is superficial fascia, which is not a membrane but a padding. Most fascias are arranged in sheets or tubes and form a more or less obvious connective tissue layer between or around structures. For instance, all muscles, nerves, and blood vessels, as well as most organs, are encased in such connective tissue coverings. Some fascias are dense, rather strong sheets; others are flimsy. No exact division exists between the indistinct, flimsy layer of fascia and loose connective tissue.

Descriptions of fascias tend to be confusing, because fascias are simply more obvious layers of the general connective tissue packing of the body; all connective tissue in the body is continuous with all other connective tissue. Thus, in one sense, a fascia has no beginning and no end, and any description of fascias is necessarily somewhat arbitrary. For instance, custom and convenience usually determine whether a fascia attaching to a bone is considered to continue across the bone or to end there, replaced beyond the bone by a differently named fascia. Similarly, when a fascia splits into two or more layers, custom and convenience largely determine whether the two layers are separately named.

The description of a fascia as forming a dissectable layer implies that the connective tissue on each side of it, binding it to surrounding structures, is looser in texture and more irregularly arranged. However, there is no standard as to how skillful the dissector must be or how sharp the scalpel, and no absolute criterion determines what "looser" and "firmer" mean. Thus, arguments as to the existence of a certain fascia may be largely a question of semantics. Surgically useful fascias are those that are strong enough to hold sutures, but the fact that a connective tissue layer is that strong is not an invariable justification for giving it a name of its own.

Fascias may actually be a part of the wall of an organ, as, for instance, the outer connective tissue layer of blood vessels, the *adventitia*. The connective tissue layer that forms the surface of a voluntary muscle, the *perimysium*, is also called the fascia of the muscle. Structures lying in or traversing loose connective tissue tend to have a condensation of fascia on or around them. For example, the visceral structures in the neck and pelvis and the structures in the loose connective tissue of the posterior abdominal wall have fascial layers or tubes associated with them.

Microscopically, fascia is distinguished from loose connective tissue by the greater amount of collagen fibers. The majority of these fibers may be more organized than in loose connective tissue, but are much more irregular than collagen fibers in tendons or aponeuroses.

In addition to the numerous and variable fascias, there are two rather distinct fascial systems that are of special importance to the anatomist and surgeon. The internal fascia lines the thoracic and abdominal cavities, and the external or investing fascia lies deep to the tela subcutanea and is usually referred to as the deep fascia.

**Internal Fascia.** The internal fascia that lines the thoracic cavity is called *endothoracic fascia*, and the fascia that lines the abdominal cavity is the *endoabdominal fascia*. Both these internal fascias form a barely discernable lining of the thoracic and abdominal cavities, and their chief function is to affix the parietal layer of the serous sacs, the pleura in the thorax and the peritoneum in the abdomen, to the inner aspect of the body wall. The terms "endothoracic" and "endoabdominal fascia" do not usually include the loose connective tissue that fills the spaces between the thoracic and abdominal organs. Such loose connective tissue is particularly abundant in the pelvis, and here the packing material is often spoken of as *endopelvic fascia*. The endopelvic fascia is a continuation of the endoabdominal fascia into the pelvic cavity.

Specializations exist in the internal fascias that are named in their own right. For instance, the suprapleural membrane is a thickening in the endothoracic fascia as it roofs over the superior aperture of the thorax and covers the apices of the lungs. Internal fascia may also be named according to the muscles with which it is in contact (*e.g.*, transversalis fascia, diaphragmatic fascia, psoas fascia, superior fascia of the levator ani). In most instances, these named fascias covering the muscles are not distinguishable from each other and are synonymous with the endoabdominal fascia. This is particularly important to remember in connection with the transversalis fascia, the name most often used when speaking of the endoabdominal fascia. Only over the psoas is the endoabdominal fascia sufficiently thickened to make it anatomically distinct from the rest of the endoabdominal fascia. Likewise, specializations exist in endopelvic fascia. These include membranous condensations of endopelvic fascia over some of the muscles that form the walls of the pelvic cavity (fascia of the obturator internus) and rather poorly defined bands of dense connective tissue that support the pelvic organs and are called ligaments (*e.g.*, pubocervical, uterosacral).

The inner layers of endoabdominal fascia may become laden with fat, which can be separated from the more membranous layer adjacent to the abdominal wall muscles. In some areas, such adipose tissue is specially named (*e.g.*, around the kidneys, perirenal fat); otherwise, it is spoken of, especially by radiologists, as *preperitoneal fat*.

**Deep Fascia.** After removal of the superficial fascia (*tela subcutanea*) by blunt dissection, the deep fascia appears in most locations as a thin grayish layer on the surface of the muscles, separable from them only by sharp dissection. Where one muscle overlies another, the deep fascia between them serves as packing, but upon dissection, it is typically split into two layers, one for each muscle. The ease of splitting and the clarity of the two fascial layers depend upon the density of the fascia adjacent to the muscles' surfaces and the amount of looser connective tissue between the muscles. Where muscles attach to bone, the deep fascia becomes continuous with the periosteum of the bone.

In the neck and in the limbs, the deep fascia is a tough fibrous layer that surrounds the entire part and from its deep surface sends septa among the muscles. In this deep fascia, the fibers have two prevailing directions, longitudinal and transverse, so that they are interwoven into a membrane. In some locations, such as the back of the arm, the deep fascia is fused to the surface of the muscle; in other locations, such as the front of the arm, it forms a loosely fitting envelope around the muscles. In some locations, also, the deep fascia and its septa give rise to muscles, so that it is more tendinous than fascial in nature (with a majority of fibers running in one direction rather than criss-crossed). Where the deep fascia and its septa meet subcutaneous bone (*e.g.*, around the joints of the limbs) they blend with the periosteum.

In the limbs, the chief septa of the deep fascia are medial and lateral ones, which reach the bone and separate the originally ventral (anterior or flexor) muscles from the originally dorsal (posterior or extensor) ones. In the neck, the septa are more complicated and are usually described as forming, in addition to sheaths for various muscles, several distinct layers in the front of the neck. At the levels of the wrist and ankle joints, the deep fascia is strengthened by special transverse fibers. These thickened portions hold the tendons close to the joints across which they pass, and are known as *retinacula*.

## Ligaments

The term "ligament" (N.A.: *ligamentum;* plural, ligamenta; meaning a binding together) is used so loosely in anatomy that it conveys no clear idea of structure. However, there are in general two kinds, those that connect viscera to each other or to the body wall and those that connect one bone to another. The two groups have nothing in common, save that they both "connect" something to something else. Tendons also connect, but they always connect voluntary muscle to something else.

**Visceral Ligaments.** It is in this group that the greatest diversity lies. Some visceral ligaments, such as those connected to the stomach, are actually only parts of a mesentery (a thin sheet of connective tissue, with epithelial—mesothelial—surfaces, that conducts blood vessels, lymph vessels, and nerves to a viscus). Others, such as those between liver and diaphragm, are modified mesenteries in which the connective tissue is much increased and actually holds two parts closely together. Some are fibromuscular remains of parts that functioned in the fetus (for instance, the medial umbilical ligaments are formed from the umbilical arteries). Others, such as some of those in the pelvis, are merely the connective tissue padding about blood vessels and nerves. Still others contain some smooth muscle or are formed largely by it.

Some authorities have insisted that a true visceral ligament must contain smooth muscle and that those that do not are false ligaments, but the distinction is not generally followed. In the first place, there is increasing evidence that some of the supposedly "true" ligaments actually contain no smooth muscle other than that of the blood vessels in them. Second, a ligament containing a smooth muscle bundle is often called a muscle rather than a ligament, or even both. For instance, the rectococcygeus muscle, a band connecting the bowel to the coccyx, is called a ligament only when it is fibrous rather than muscular, and the suspensory ligament and suspensory muscle of the duodenum are parts of the same structure.

**Skeletal Ligaments.** Skeletal ligaments are more or less distinct bands of connective tissue that bind together two bones or bony parts. Across joints that have cavities, they usually blend with the fibrous wall of the joint cavity and are perhaps best regarded in this instance as being special thickenings of the fibrous capsule. Indeed, because of their close association with the capsules of joints, some of them were once known as capsular ligaments.

Since skeletal ligaments must withstand pull at the joints they cross, the majority of their fibers run in the same direction. Most are composed almost entirely of closely packed collagen fibers and therefore allow little stretch. However, some ligaments, notably the ligamenta flava of the vertebral column, are composed almost entirely of elastic fibers. They stretch with movement in one direction, shorten with movement in the opposite direction, and, therefore, tend to remain taut rather than to become lax and double up. Ligaments composed of elastic tissue are, in the fresh condition, yellowish (elastic fibers are sometimes called "yellow elastic fibers"). Those composed primarily of collagenous tissue are white, and some of the heavier ones resemble tendons; however, ligaments are usually arranged in more loosely packed parallel bundles, and they usually do not have quite the glistening, shiny white appearance of tendons. Heavier ligaments do resemble tendons though, and a few so-called ligaments really are tendons. For example, the tendon at the front of the knee attaches to the tibia, but is so interrupted by the kneecap (patella) that its lower segment is called the patellar ligament.

### Tendons

Tendons are actually parts of muscles and are best discussed in detail along with other aspects of muscles. However, they deserve mention with fibrous connective tissue because they are one type of very dense collagenous tissue. Even more than in most ligaments, their fibers run in one direction, so that they are able to withstand great pull. Because the densely packed parallel bundles of a tendon are held together by only a few cross-fibers, it can be fairly easily shredded. The predominance of parallel bundles gives a tendon its white, shiny appearance.

A tendon is defined as the tissue that attaches voluntary muscle to another structure; thus, one end of a tendon is always attached to muscle. At the end away from the muscle, tendon fibers blend with the fibrous connective tissue of the structure to which they attach—usually bone, where they are continuous both with the fibrous outer covering, the periosteum, and with the fibers that form a part of the substance of the bone. In some instances, they blend with the dense connective tissue that forms the deep layer of the skin (the dermis or corium).

Tendons, although thinner than the muscle

to which they belong, are usually of the same general form (*i.e.*, broad and flat when they are a part of broad and flat muscles, cordlike when they are a part of long slender muscles). The long rounded tendons running into the hand and foot are colloquially called "leaders," but have no scientific name to distinguish them. Very broad, flat tendons are known as **aponeuroses** and because of their breadth and thinness resemble dense fascias. The characteristic difference is that an aponeurosis, like other tendons, is composed of predominantly parallel collagenous bundles, whereas a fascia has interwoven fibers; however, there are intergrades between these two types of fibrous membranes, and in some cases custom more than anything else determines which a membrane is called.

Tendons usually have special protection where they work across bone or across other tendons at such angles that friction and fraying may occur. Usually, this protection is simply the special lubrication offered by a bursa or tendon sheath. However, in some locations cartilage or bone (sesamoid bones) develops within tendons at points of special friction—the fibroblasts that form connective tissue fibers can also form cartilage or bone, even in the adult, although it is not known exactly what conditions determine which they form.

Since tendons, fascia, and ligaments all have the same fundamental structure, they tend to blend with each other when they are close together. A fascia may contain tendon fibers, and a ligament may be reinforced by a tendon that blends with it. Indeed, tendons and ligaments are so similar that some ligaments are apparently tendons that have lost their muscle fibers.

## Tissue Spaces, Fascial Spaces, Bursae, and Tendon Sheaths

**Tissue spaces** filled with extracellular fluid surround or are adjacent to most living cells. Extracellular fluid serves as the medium for interchange between the cells and the blood and is drained by lymphatics or directly into the venous system. The largest tissue spaces occur in loose connective tissue, where the interstices between fibers are spacious and the cellular elements are few. For this reason, fluid collects easily in the subcutaneous tissue if interference with the lymphatic drainage of a part occurs, if local block or cardiac failure impedes venous drainage, or if excess body fluid collects as a result of inadequate excretion of fluid by the kidneys or because of certain nutritional deficiencies. This excess accumulation of fluid, known as edema (dropsy), may occur in tissue spaces at any site and may seriously interfere with the function of an organ or part through pressure, restriction of mobility, or reduction of nutrients or oxygen reaching the cells.

**Fascial spaces** are potential spaces filled with loose connective tissue between or among more dense layers of connective tissue. Fascial spaces are usually described as being areas of loose connective tissue that lie between muscles or are bounded by connective tissue layers dense enough to be called fascia. Because of the looseness of their connective tissue and their distensible walls, fluid such as pus or blood can accumulate in these crevices and convert them into "spaces" of considerable size.

Fluid or air accumulating in a fascial space may spread into other fascial spaces with which it is continuous. This spread is determined by the arrangement of the loose connective tissue and the surrounding fascial layers. It seems definite that the spread of air and noninfected fluid is determined largely by the arrangement of the looser connective tissue, but clinical opinion differs sharply in regard to the importance of fascial spaces in the spread of infectious processes. One opinion is that infections do so spread, following the path of least resistance. The contrasting opinion is that the swelling and the proliferation of connective tissue produced by infection tend to encapsulate (form a wall around) the process and that this prevents spread through fascial spaces. Since both phenomena have been documented by clinical series, it may be that the nature of the infection, rather than the anatomy of the fascial spaces, is often the determining factor.

**Bursae** (bursa meaning purse) are closed connective tissue sacs filled with fluid and with a slippery inner surface. They develop in response to friction between tendon and bone, ligament, or other tendons, or between bone and skin. Most deep-lying bursae develop before birth, but subcutaneous bursae may appear in adult life over prominences exposed to unusual friction. Certain bursae that develop in close relation to joint cavities frequently fuse with these, so that what was once a bursa becomes an extraarticular extension or recess of the joint cavity.

There are some 50 named and fairly constant bursae in man and many other, usually small, inconstant ones. The majority of both groups are subtendinous or submuscular, but a few are subfascial and some rather constant ones are subcutaneous.

Bursae are typically flattened and essentially collapsed spaces, containing only enough fluid to moisten their walls. When they become infected or are injured, the inflammation (bursitis) causes swelling. Since deep bursae lie between a muscle and a bone or between two muscles, any movement that involves contraction of an adjoining muscle will produce pressure on the inflamed bursa and hence will be painful. Subcutaneous bursae, between bony prominences or muscles and tendons and the skin, become visibly swollen and tender to the touch when they are inflamed.

**Tendon sheaths** are similar to bursae in their fundamental structure, but rather than being simple connective tissue pockets, they are complex tubes wrapped completely around tendons. Tendon sheaths develop like bursae (*i.e.*, the connective tissue splits and forms a cavity), but their structure can be most easily understood if a tendon sheath is thought of as first developing as a bursa on one side of the tendon and then extending around the tendon (Fig. 3-1). One layer of the tendon sheath is closely applied to the surface of the tendon and is usually referred to as the visceral layer. The outer layer, surrounding the cavity, is known as the parietal layer. These two smooth, glistening layers, for the most part separated from each other by a film of fluid, are continuous with each other at the ends of the tendon sheath and are often also continuous with each other through a mesotendon (Fig. 3-2). Mesotendons vary greatly, for they may persist throughout the length of a tendon sheath, may in part disappear or become much fenestrated, or may be completely

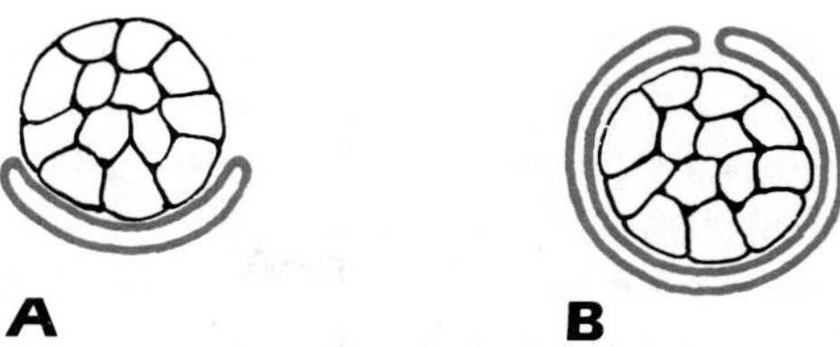


**FIG. 3-1.** Diagram indicating the similarity between bursae and tendon sheaths; a bursa underlying a tendon, **A**, can be converted into a tendon sheath, **B**, simply by extension around the tendon. Synovial membrane is represented by a red line.

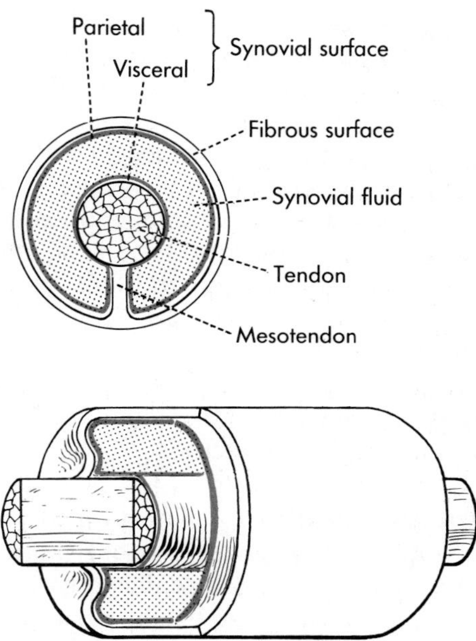


**FIG. 3-2.** Diagrams of a tendon sheath. Cut edges of synovial membrane are represented by red lines. Above is a cross section, with a mesotendon also included. Below is a view from the side with a large segment removed and parts cut away at various levels on the left side.

absent. Mesotendons allow blood vessels to reach the tendon as it courses through the sheath rather than to have to enter only at the ends of the tendon sheath. The small and rather constant remains of mesotendons in connection with the flexor tendons of the fingers and toes are known as vincula.

## CARTILAGE

In contrast to fibrous connective tissue, which is adapted particularly to withstand pull, cartilage (like bone) is so formed as to withstand both pressure and pull: although it contains connective tissue fibers, the interstices between the fibers are filled not with fluid but with a substance that gives cartilage its solid feel and appearance. However, this matrix (intercellular substance) of cartilage is softer than that of bone, so cartilage possesses a springiness that bone does not. The matrix contains no cavities except those occupied by the cells, so blood vessels that invade cartilage must do so by destroying it. This occurs normally during growth of the body and abnormally in varying disease conditions. Otherwise, cartilage is avascular.

**General Distribution.** Almost all the bones of the body first take form as cartilage. The cartilage is destroyed and replaced by bone in the prenatal or early postnatal period. However, in many bones, growth in length depends upon continued growth of cartilage, so small pieces of cartilage (epiphyseal cartilages or plates) typically persist in bones until the person has ceased to grow in height. Cartilage also persists, normally throughout adult life, at the ends of bones that enter into freely movable (synovial) joints; since cartilage has a smoother surface than bone, its persistence on the end of a bone, where it is called articular cartilage, allows the joint to move with less friction than would bony surfaces. Presumably, also, cartilage's greater springiness allows the articular cartilages to absorb some of the shock delivered to the joints during movement.

Distribution of cartilage in the adult is limited mostly to articular surfaces of bones and to certain locations where springiness in the skeleton is an advantage. The cartilaginous support of the ear, for example, and of the lower part of the nose allows bending, whereas bone in the same location would fracture. In the same way, the persistence of the anterior ends of the ribs as cartilages contributes greatly to the elasticity of the thoracic wall. Cartilaginous rather than bony support of the air passage to the lungs is also advantageous, since this passage undergoes constant movement with breathing.

Calcification of (deposition of calcium salts in) cartilage occurs regularly in cartilage that is to be replaced by bone. It also may occur, with advancing age, in the permanent cartilages, which, as might be expected, reduces their resiliency and renders them brittle.

**Types of Cartilage.** Although all cartilage contains connective tissue fibers, the fibers may not be easily demonstrable or, if they are, may be either collagen or elastic; thus, cartilage is divisible into three types: hyaline cartilage, fibrocartilage, and elastic cartilage.

**Hyaline cartilage** is glassy in appearance (the meaning of hyaline) because its felted collagen fibers are masked by the homogeneous ground substance (matrix) deposited about the fibers. **Fibrocartilage** also contains collagen fibers, but the fibers are in heavier bundles and there is less ground substance; fibrocartilage seems to be a developmental stage between fibrous tissue and hyaline cartilage and may persist throughout life, become transformed into hyaline cartilage, or return again to dense fibrous connective tissue. **Elastic cartilage** has elastic rather than collagen fibers embedded in its matrix.

Hyaline cartilage in the adult is largely confined to articular cartilages, to the cartilages of the ribs, and to the rings and plaques that support the trachea and its branches.

The greatest concentration of fibrocartilage is in disks that lie between adjacent vertebrae (intervertebral disks), but fibrocartilage or very dense fibrous connective tissue—so-called fibrocartilages are usually truly cartilaginous only where they bear weight—also exists in some freely movable joints, notably those of the lower jaw, the ends of the clavicle, and the knee joint, where it completely or partially subdivides the joint into two cavities. Fibrocartilage is especially capable of absorbing shocks, an important function of the intervertebral disks. Within joint cavities, the function of disks seems to be to assist in gliding movements and to aid in the spread of the lubricating fluid in the joint.

Elastic cartilage primarily occurs in the skeleton of the external ear and the lower part of the nose.

**Growth and Repair.** After its rapid growth during developmental stages, cartilage grows little. Damaged articular cartilage does not regenerate.

Grafts of cartilage, whittled to the proper shape and size, are sometimes used in plastic repair of the face. Grafts of living cartilage, usually obtained from a rib cartilage of the person on whom the plastic operation is being done, typically live but neither grow nor decrease in size. Grafts of preserved cartilage, on the other hand, are gradually resorbed.

## BONES

Bone (*os*) is a connective tissue in which the cells are arranged at regular intervals and in which the intervening matrix hides the numerous collagen fibers that lie in it. This matrix is impregnated with calcium salts, mostly in a complex form of calcium phosphate and calcium hydroxide belonging to the apatite series of minerals. Unlike cartilage, bone contains blood vessels. Thicker pieces of bone are regularly organized around the blood vessels to form the lamellated branching series of tubes known as *haversian canals*.

Bone, because it is formed by connective tissue cells, can be deposited in almost any location where there is fibrous connective tissue—for example, in a muscle. However, the function of bone is to give support and protection, so bony tissue is normally organized into units of definite shape and form. It is these organized units that are called bones.

Excluding the water content of some 25% to 30%, adult bone consists of about 30% to 40% collagen and about 60% to 70% mineral deposit. Collagen fibers give the bone tensile strength (resistance to being pulled apart), and calcium salts give it compressive strength (resistance to being compressed or crumbled). In general, bones of children have an excess of collagen as compared to mineral matter and are therefore more easily bent. In bones of elderly people, bone becomes less compact, and the calcium salts are often reduced, so that the bones become brittle and fragile. Although there are no constant values for the strength of bone, good adult bone is said to have a tensile strength of about 12,000 to more than 17,000 $lb/in^2$ (that of copper is 28,000 lb, that of white oak along the grain is about 12,500) and a compressive strength of about 18,000 to almost 25,000 (that of copper is about 42,000, that of granite 15,000, and that of white oak along the grain only 7000).

### Shape and Structure of Bones

Bones vary greatly in shape and size. It is possible by careful study to learn to distinguish even the smaller bones from each other and to state from which side of the body the bone was taken. Despite their diversity, however, it is obvious that some bones, such as the larger ones of the limbs and the bones of the fingers and toes, can be classified as **long bones;** others, particularly some of those of the skull, can be classified as **flat bones;** and still others, such as those of the wrist and ankle, are so irregular in shape that they fall into neither category.

Regardless of their shape, bones have the same general structure (Fig. 3-3). The outer surface (the **cortex**) is dense and hard, composed of bony layers closely packed together; it may be thick or thin. At the articular surfaces of freely movable joints, the cortex is covered by a layer of hyaline cartilage (articular cartilage). The space within the cortex of a flat or irregularly shaped bone, and that at the

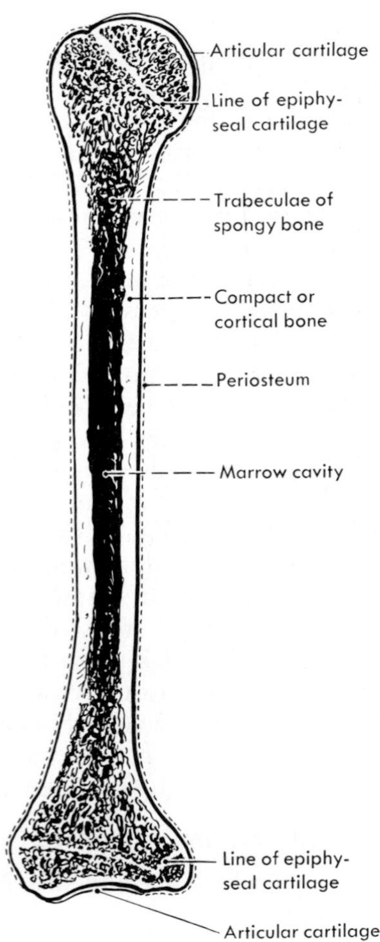


**FIG. 3-3.** Structure of a typical long bone.

ends of a long bone, is occupied by **trabeculae** (beams) of bone, individually weak but together forming a strong bracing system. Because the trabeculation gives a spongy appearance, the bone is called **spongy bone.** The interstices between trabeculae are filled with marrow that is concerned with forming blood cells; from its color derives the name "red marrow." The major part of a long bone (the body or shaft) has a particularly heavy cortex that is tubelike and surrounds a large marrow cavity that is not traversed by bony trabeculae. The marrow here, although potentially capable of forming blood cells, is, in the normal adult, a storage place for fat, hence its name of "yellow marrow."

Bones are covered except at the cartilaginous articular surfaces by a connective tissue membrane, the **periosteum.** The outer layer of the periosteum is fibrous, like a strong fascia, and contains periosteal blood vessels. The inner layer is looser connective tissue and contains **fibroblasts** that can, even in the adult, proliferate and change into osteoblasts for the reconstruction of fractured bone. The surface of the bone abutting on the marrow cavity is also lined by a layer of connective tissue, the **endosteum,** thinner than the periosteum but likewise containing cells capable of forming bone; thus, both endosteum and periosteum normally participate in the healing of a fracture. Blood vessels of the periosteum communicate with those of the bone, but periosteum is much less vascular than many other soft tissues, such as muscle; hence, when it is necessary to remove part of a bone, a more nearly bloodless approach can usually be obtained by subperiosteal resection rather than by cutting away muscles adjacent to the bone.

Periosteum is bound to bone by fibers that continue into the bone to become fibers of the bone itself, but the density of these fibers varies. In many locations periosteum can be stripped from bone with little difficulty, but it is always firmly bound to the bone at the places of insertion of tendons, for whereas some tendon fibers blend with the periosteum, the majority continue through the periosteum to become continuous with the fibers of the bone.

As bones grow, they tend to be modeled by tensile and compressive forces. A well-known example of this occurs in the femur, where the trabeculae of the head and neck (upper end) are arranged in interlocking arches that correspond nicely with the calculated lines of stress evoked by the weight of the upper part of the body (Figs. 3-4 and 3-5). The tube of cortical bone that forms the body of the femur is more efficient, weight for weight, in withstanding the strain placed upon it than is any other structural arrangement of the body. The trabeculae of the distal end of the femur, which differ in arrangement from the upper end, are arranged to withstand the strain at this location. As a corollary of this, a bone subjected to

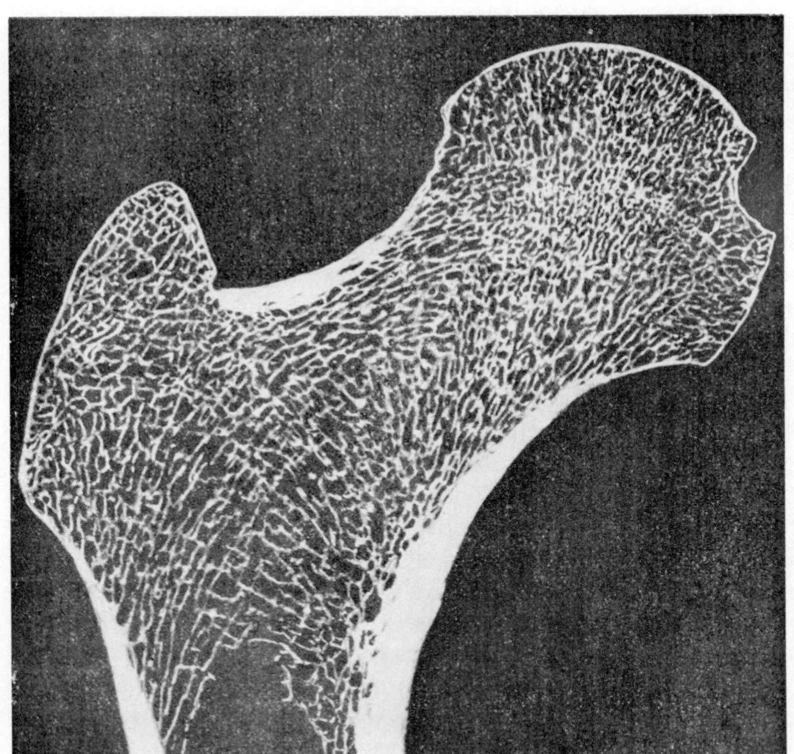

**FIG. 3-4.** A thin coronal section of the upper end of the femur to show the trabecular structure. Note the correspondence between the trabeculae here and the lines of tension and compression in Figure 3-5. (Koch JC: Am J Anat 21:177, 1917)

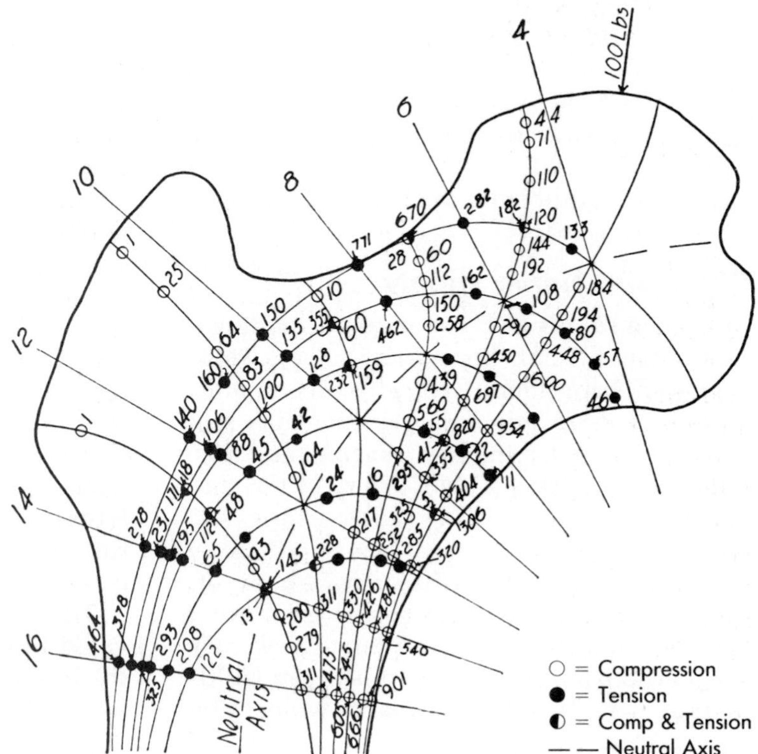


**FIG. 3-5.** Koch's calculation of the maximal tensile and compressive stresses in the upper end of the femur when there is a load of 100 lb on the femoral head. The numbered lines extending across the femur represent the levels of the cross sections that he studied for the bony architecture. (Koch JC: Am J Anat 21:177, 1917)

unusual strain as a result of change in direction of the forces applied to it will gradually undergo resorption and rebuilding of its structure to fit the new pattern of forces acting upon it; this occurs, for instance, from weight bearing after an improperly set fracture.

The pull of muscles upon a bone also participates in modeling it. Most of the projections of bones, whether called ridges, crests, tubercles, or whatever, are results of this modeling by muscles. It is because of this that the skeleton of a male is typically better marked than that of a female and that more heavily muscled bodies, regardless of sex, have more clearly marked bones than do less-muscled ones. Since most of the finer external modeling of bones is produced by soft tissues, study of a dry bone without consideration of the muscles, and sometimes the nerves and vessels related to it, may lead to rote memorization rather than understanding.

After the growth period, and in the absence of repair consequent to a fracture or other damage, bone is relatively quiescent; however, its life depends upon the vascular system within it, and death follows too great an interruption of its vascular supply. Moreover, bone is a great storehouse of calcium and phosphorus, and these substances are constantly being incorporated into the structure of the bone or released from the bone into the bloodstream in equilibrium with the needs of the entire body.

## Blood and Nerve Supply

The **blood supply** to bones varies according to their shape: Small bones may have a single artery entering them (typically accompanied by two or more veins); larger ones may have several, often irregularly spaced; and long bones tend to have a chief artery of the shaft and a variable number of smaller arteries supplying each end.

Because the artery to the shaft, or body, of a long bone is usually the largest, it is known as the **nutrient artery.** The nutrient artery enters the bone during early development, and as the bone grows, the **nutrient canal** in which it lies usually has its external end carried toward the faster growing end. In most long bones, growth in length of the bone occurs much more at one end than at the other; thus, the slant of the canal from surface to marrow cavity is commonly toward the end that has grown less rapidly, which allows one to determine the relative growth of the two ends. The nutrient artery or arteries supply most of the marrow and cortex, their branches running in haversian canals of the cortical bone.

The ends of long bones typically have several arteries entering them and several veins leaving, so that numerous vascular foramina are usually visible close to the ends of dried bones. The vessels to the ends of long bones supply both the marrow and the bone.

**Lymphatics** occur in the periosteum of bone, and it is said that some accompany blood vessels into the bone.

**Nerve fibers** to bone are not numerous, but some do accompany blood vessels and some occur in the periosteum. They apparently consist of both afferent (sensory) and sympathetic efferent fibers, the latter representing the motor nerve supply to the blood vessels. This nerve supply apparently does not play any important part in controlling the growth or repair of bone. Presumably, the afferent fibers within bone are primarily concerned with pain, which may be the first indication of a tumor within the bone. Most of the pain associated with fracture of a bone, however, undoubtedly arises from surrounding soft tissues.

## The Bony Skeleton

The adult bony skeleton (Fig. 3-6) consists of approximately 206 bones. This number is subject to some variation, for supernumerary bones are fairly common, especially in the hands and feet. The number of bones in the fetus and infant is greater because some separate bony elements subsequently fuse together.

The bony skeleton can conveniently be divided into axial and appendicular skeletons, the latter being the skeleton of the limbs, the former that of the head and trunk.

**Axial Skeleton.** The bones of the **skull** form the uppermost part of the axial skeleton. These bones can, in turn, be divided into two groups: those that protect and support the brain and the organs of sight, hearing, and balance; and those of the jaws and face. The remainder of the axial skeleton tends to be arranged segmentally (metamerically). The **vertebral column,** or backbone, is composed of about 32 or 33 segments that remain as separate elements throughout the neck and most of the trunk but fuse together at the level of the hip bones to form the sacrum and, below this, the coccyx. The **ribs,** extending laterally from the vertebral column, are strictly metameric. Although reduced in man to a normal occurrence in the thoracic region, elsewhere they are vestigial and form parts of the vertebrae. The **sternum** (breastbone), connecting the ventral ends of many of the ribs, is metameric in its development; its segments so fuse that in the adult there are only three parts.

Although there are regular differences in the vertebrae in different regions, they are all built upon the same fundamental pattern. The seven vertebrae of the neck are cervical vertebrae. They are followed by 12 thoracic vertebrae, which have the ribs articulated with them. These are followed normally by five lumbar vertebrae that lie in the small of the back. The next five vertebrae fuse together, leaving signs of their fusion, to form the sacrum, which transmits the weight of the upper part of the body to the hip bones. Because the original segments can still be counted, it is customary to refer to, for instance, the second sacral segment or vertebra, even though the vertebrae are not separate. The remaining vertebrae are vestigial and together form the coccyx.

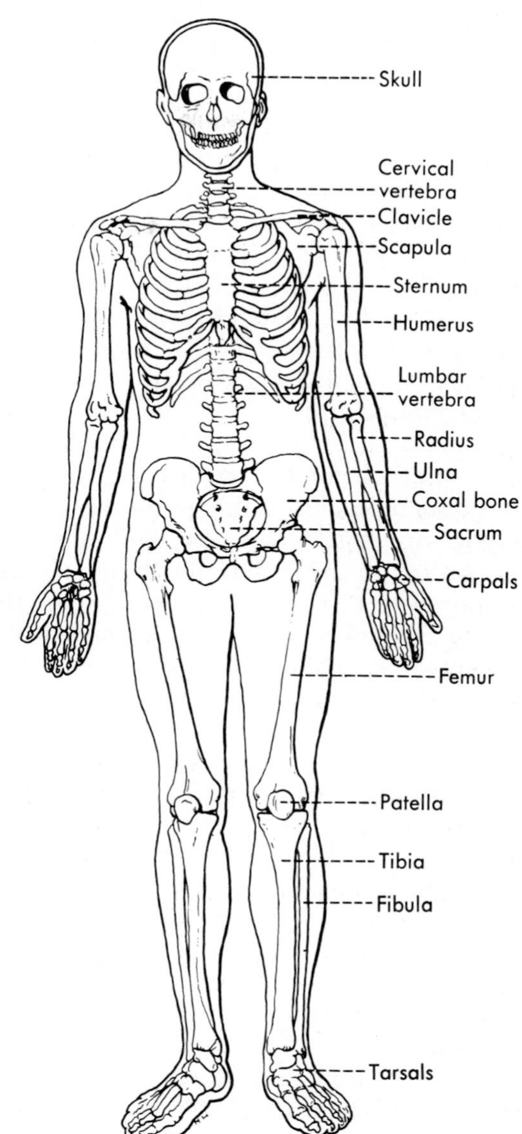


**FIG. 3-6.** The human skeleton, anterior view.

**Appendicular Skeleton.** Of the appendicular skeleton (so termed from the common use of "appendage" to designate a limb), the bones of the two upper limbs are essentially mirror images of each other, as are those of the two lower limbs. In addition, the bones of the upper and lower limbs are analogous. Those of each limb can be divided into two chief parts: a **girdle,** closely associated with the trunk, and the skeleton of the **free limb,** the part distal to the shoulder joint or hip joint.

The upper-limb girdle is also called the **pectoral girdle** (pectus meaning breast) and consists of two bones, the **clavicle** or collar bone and the **scapula** or shoulder blade. The clavicle is united to the axial skeleton by a movable joint, but the scapula is united to the axial skeleton by muscles only, permitting the shoulder to move freely within the limitations imposed by the attachment of the clavicle. Phylogenetically, the girdle of the upper limb consisted of three elements, but in man one of these has disappeared completely, the second persists only as a process (coracoid process) of the scapula, and the third alone persists as a large element, almost the entire scapula; the clavicle is a phylogenetically new bone.

The lower-limb girdle is also called the **pelvic girdle** (pelvis meaning basin), because in their normal position, the two girdles and the lower part of the vertebral column form a basin. In contrast to the pectoral girdle, the three bones on each side, the **pubis, ilium,** and **ischium,** are fused together and are also firmly united to the vertebral column, so that they can better transmit the weight of the trunk to the lower limbs. These three fused bones constitute the **os coxae** or hip bone, formerly called the innominate bone (bone without a name).

The upper and lower limbs are much more similar in regard to the **skeleton of the free limb** than they are in regard to the girdle. In each limb, a single bone forms the proximal segment of the skeleton of the limb: the **humerus** in the arm and the **femur** in the thigh. Further, each bone articulates with its girdle through a ball-and-socket joint, a type of joint that allows more free movement in all directions than any other kind of joint. In the next segment of the limbs there are two bones: the **radius** and the **ulna** in the forearm, the tibia and the **fibula** in the leg. Although the elbow points backward and the knee forward, the joints here are somewhat similar in that they are both modified hinge joints, allowing movement primarily only backward and forward. At the elbow, an additional joint allows rotation of the forearm so that the palm of the horizontally held hand can turn down or up: pronation and supination. The leg and foot are so fixed that the sole of the foot remains constantly down, so that when leg and foot are to be compared with forearm and hand, the hand must be pronated; then the thumb and the big toe and the little finger and the little toe correspond, as do the radius to the tibia and the ulna to the fibula.

The wrist and ankle each consists of a number of bones. Those of the wrist, the **carpals,** contribute very little to the length of the palm of the hand, but the bones of the ankle, the **tarsals,** contribute almost half the length of the sole of the foot.

Five bones, the **metacarpals,** form the major part of the palm of the hand; the skeleton of the anterior part of the foot likewise consists of five bones, the **metatarsals.** Each bone that forms a segment of a digit is known as a **phalanx,** without regard to whether it is in the foot or the hand. There are two phalanges each for the thumb and the big toe and three phalanges for each of the other digits of both limbs. (It might be noted in passing that "digits" is a neutral term, referring to either the fingers and thumb or the toes.)

**Sesamoid bones** (so called because the smallest ones resemble in size the seed of the sesame plant) are bones that develop within tendons or ligaments. They occur regularly in some locations, with varying frequency in others. The pisiform bone, one of the bones of the wrist, is usually said to be a sesamoid bone that interrupts the tendon of a flexor muscle of the wrist. The patella or kneecap, the largest sesamoid bone of the body, interrupts the great extensor tendon (quadriceps femoris) on the front of the knee. In addition to these constant sesamoids, there are usually two at the base of the thumb and two more at the base of the big toe; all of these are much larger than implied by the term "sesamoid." Others may occur in ligaments of the toes and fingers, or in certain tendons at points of increased friction.

Sesamoid bones develop first as sesamoid cartilages, which are not opaque to x-rays and can therefore be identified only by dissection. Even when they are partly or entirely bone, they may be difficult to demonstrate roentgenographically because of the greater size and density of the bones against which they lie.

## Formation and Growth of Bones

The first stage in the formation of any bone is

a condensation of **mesenchyme,** the loose embryonic cellular connective tissue, largely derived from mesoderm, that gives rise to most of the tissues of the body except those few derived from ectoderm and endoderm. The cells in the mesenchymal condensation may then give rise to bone in one of two different ways. A limited number of bones, primarily the bones of the roof of the skull, the lower jaw, and probably the clavicles, are formed by the direct transition of mesenchyme into bony tissue. Phylogenetically, bones of this type are believed to be derived from the dermis of the skin (this is also a derivative of mesenchyme) and are often termed "derm bones"; however, because the mesenchymal condensation forms a rather tough membrane in which the bone develops, they are also known as **membrane bones.** Most bones, however, including the majority of those of the skull, the entire vertebral column, the ribs and sternum, and all the bones of the limbs except the clavicles, are first formed in cartilage. The mesenchymal forerunner produces a rough model of the bone in cartilage. As the cartilage grows, it is invaded by blood vessels and accompanying cells that destroy the cartilage and replace it with bone. Bones of this type are known as **cartilage bones.** It should be emphasized that cartilage is never actually transformed into bone; rather, whenever bone replaces cartilage, the cartilage is destroyed and bone is deposited in its place.

**Growth of a membrane bone** is essentially by a process of accretion on the surfaces and edges of the bone. This is illustrated by the gradual closure of the fontanels, the soft spots of the child's skull, postnatally. The fontanels are simply areas of membrane between edges and corners of certain bones, and the bones grow into the membrane and close the fontanels by simple addition of bone along these edges. At the same time that bone is added to the outside of a membrane bone, there is absorption and reorganization of the inner part. The marrow cavity grows along with the growth of the bone as a whole, preventing the bone from becoming essentially a solid sheet, which would be much heavier than a hollow bone but little stronger.

**Formation and growth of cartilage bones** are somewhat more complicated than those of membrane bones. Briefly, the cartilage representing a bone is invaded, usually at about its center, by blood vessels that grow in as the cartilage is destroyed. Osteoblasts brought in by the mesenchyme along the blood vessels give rise to bone, representing the first **center of ossification** of the bone. As the center of ossification becomes larger, bone is also laid down by the adjacent periosteum. Growth in the diameter of a cartilage bone therefore occurs in essentially the same fashion as does that of membrane bone, by the laying down of successive concentric layers by the periosteum. Constant remodeling of the marrow cavity ensures an adequate balance between lightness and strength.

In some cartilage bones, notably the small ones of the wrist and most of those of the ankle, the rapidly growing cartilage is eventually replaced entirely by bone spreading from the original center of ossification. The vertebrae and the long bones of the limbs, however, develop more than one center of ossification (Figs. 3-7 and 3-8); these additional centers are known as **epiphyses.** Epiphyseal centers of ossification appear for the most part after birth and enlarge rapidly. During the entire growth period they are separated from the chief center of ossification, which forms the **diaphysis,** or body, by a plate of cartilage, the **epiphyseal cartilage.** The epiphyseal cartilages are responsible for almost all the growth in length of a long bone. Once they have disappeared, growth in length ceases entirely, although growth in diameter, through the activity of the periosteum, can continue.

A large long bone has at least two epiphyses, one at each end. Smaller "long" bones, such as those of the fingers and toes, typically have an epiphysis at only one end. There may be separate epiphyseal centers of ossification for projections of bone developed in connection with the attachments of muscles (*e.g.*, the trochanters of the femur, Chap. 17); these are sometimes known as traction epiphyses. Finally, the epiphysis forming the coracoid process of the scapula represents a still different type of epiphysis, an atavistic one, for phylo-

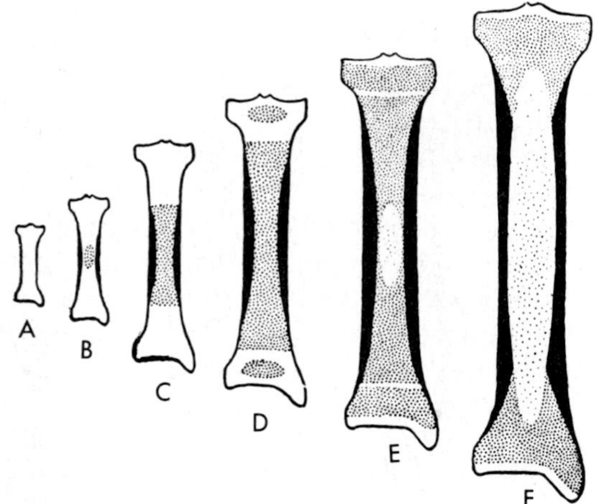


**FIG. 3-7.** Schema of the ossification and growth of a long bone. White represents cartilage; stipple, spongy (endochondral) bone; black, compact (perichondral) bone. **A** is the cartilaginous stage; in **B** and **C** both endochondral and perichondral bone appear and increase; in **D** the epiphyseal centers have appeared; in **E** the epiphyses have reached their full growth; and in **F** they have joined the diaphysis. In the last two stages the marrow cavity (light stipple) appears and spreads through resorption of spongy bone. (Arey LB: Developmental Anatomy, 6th ed. Philadelphia, WB Saunders, 1954)

genetically it is a separate bone and not a part of the scapula.

Growth of the body (shaft) of the bone is responsible for most of the increase in length and occurs entirely from the epiphyseal cartilages at the ends of the bone. The epiphyseal cartilages proliferate cartilage toward the diaphysis or body, and this cartilage is then replaced by bone. Once the cartilage of which they are first formed has been replaced by bone, growth of the epiphyses occurs from the articular cartilages persisting on their surfaces; the epiphyseal cartilages do not contribute to growth of the epiphyses. As growth of the epiphyseal cartilages slows toward the end of puberty, the rate of ossification begins to approach and then exceeds the rate of proliferation of cartilage, so that the entire epiphyseal cartilage is replaced by bone and growth in length ceases. In general, centers of ossification appear earlier in the female than they do in the male, and the final disappearance of the epiphyseal cartilages, with cessation of growth, usually occurs about 3 years earlier in the female.

In the long bones that have two epiphyses, growth usually occurs much faster at one epiphyseal cartilage than at the other. The faster growing end is constant for any one bone but may be different for two adjacent bones. In the femur, the faster growing epiphyseal cartilage is the lower one, but in the adjacent tibia, it is the upper one.

Interference with the growth of an epiphyseal cartilage may occur as a result of disease or of hormonal imbalance. Poliomyelitis has been an especially frequent cause of improper growth of a limb, although the exact mechanism by which this disease slows growth of bones is not known. Since disparity in the length of the two lower limbs may be a considerable handicap, and poliomyelitis may affect only one limb, two methods of preventing too great a disparity have been used: the epiphyseal cartilage is removed from the faster growing limb, thus stopping growth entirely, or the epiphysis and diaphysis are stapled together so that the growing epiphyseal cartilage is brought under greater and greater pressure, resulting in slower and slower growth.

## Fracture and Repair

Fracture of a bone may result from a crushing injury, but in the case of long bones it is more often due to a bending of the bone. Since the tensile strength of bone is less than its compressive strength, fracture of a long bone regularly begins on the outer edge of the curve into which it is bent. The heavier fibrous and lesser mineral content of young growing bones leads to irregular splitting fractures resembling those produced when a green stick is broken, for which reason fractures in young people are frequently referred to as greenstick fractures. In old age, the lessened fibrous and mineral content of bones may lead to such fragility that they can no longer carry out their usual tasks. Fracture of the neck of the femur, a common fracture of the aged, is now thought to result not always from a fall, as once believed, but sometimes from weakening

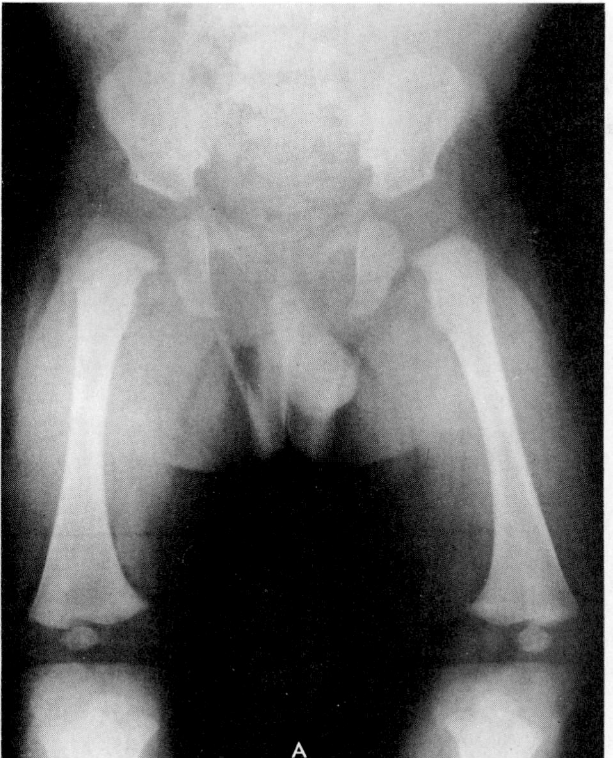


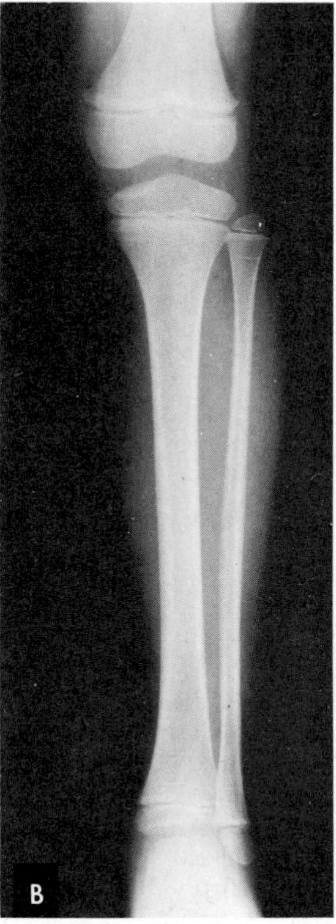


**FIG. 3-8.** Roentgenograms of stages in the ossification of bone. Cartilage casts little more shadow than soft tissues and is not visible; bone, because of its mineral content, casts a shadow of varying intensity. In **A,** the pelvis, thighs, and knees of an infant 4 months old, the bony elements that form the pelvis are united by cartilage and hence seem to be disconnected. The upper end or head of the femur, the bone of the thigh, is represented entirely by cartilage and hence is also not visible. At the knee there is an oval center of ossification for the lower end of the femur and a similar one for the upper end of the tibia, the larger bone of the leg, but the upper end of the fibula is still formed entirely by cartilage. In **B,** the knee and leg of a child 5 years old, the lower end of the femur and both ends of the tibia and fibula are well developed in bone; they appear to be separated from the diaphyses because the epiphyseal cartilages are still present.

of the neck to the extent that it can no longer sustain the weight of the body; the sudden fracture is then thought to produce the fall, rather than the fall producing the fracture.

A fracture may be merely a crack extending partly into the bone, or it may involve complete interruption of the bone. In the latter event, the ends of the bone fragments may be jammed together (impacted) or widely separated from each other. The direction in which the fragments are displaced may depend entirely upon the direction of the force causing the fracture but frequently is determined in part by the actions of muscles that cross the fractured area. Overriding of the ends of the fragments of a fractured bone, as a result of the pull of the muscles crossing the fracture line, is common, and there may be wide displacement if muscles pull the proximal fragment in one direction and the distal fragment in another (Fig. 3-9). The sharp, displaced ends of fractured bones sometimes penetrate blood vessels or nerves and may protrude through the skin (compound fracture). Pain from displacement of the ends of a fractured bone produces reflex or automatic muscular

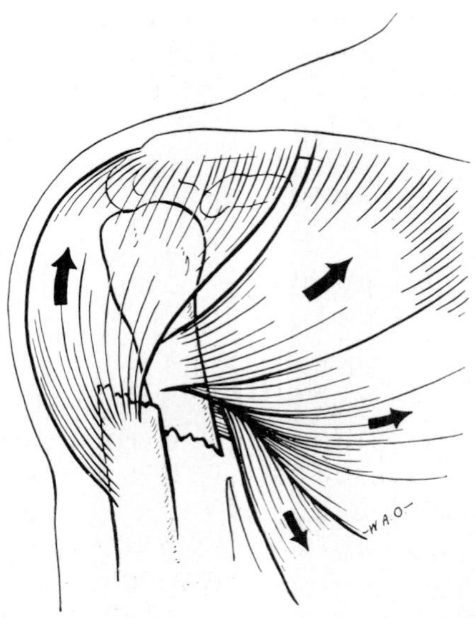

FIG. 3-9. Shortening and displacement of a fractured bone (humerus) as a result of muscle pull. (Bickel WH: In Morris GM (ed): The Cyclopedia of Medicine, Surgery, Specialties, vol 5, p 843. Philadelphia, FA Davis, 1956)

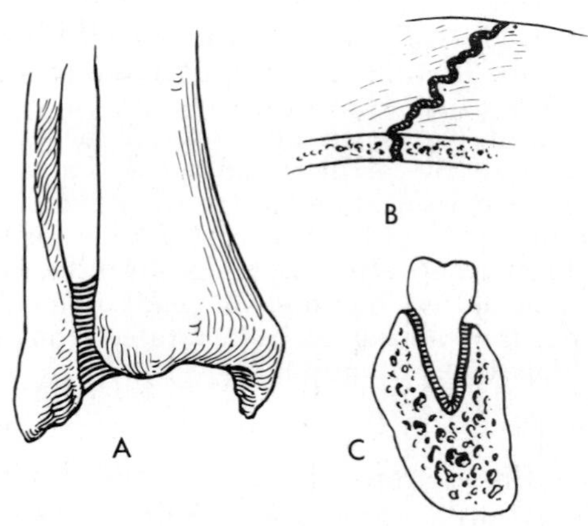


FIG. 3-10. Types of fibrous joints. **A** is the tibiofibular syndesmosis, **B**, a suture of the skull, **C**, a gomphosis, the attachment of a tooth.

spasm. Knowledge of the anatomy of the involved muscles aids in determining the forces and directions of this spasm, which must be overcome in reducing the fracture.

Repair of a fracture takes place primarily through the combined activity of the periosteum and the endosteum, both of which contain cells (osteoblasts) capable of forming bone. Best repair of bone occurs when the ends can be put in close apposition and held that way, making careful reduction of the fracture and fixation of the bone fragments through splints or other methods a necessity.

Bone grafts are sometimes used in orthopaedic surgery. The grafted bone itself apparently does not survive, but does serve as a scaffold upon which new bone is formed to unite firmly the edges of the bone already present. In addition, grafts apparently stimulate the formation of new bone.

## JOINTS

Joints, or articulations, are structures that unite two or more bones. In addition to holding skeletal pieces together, the construction of joints also determines the degree and type of movement possible between them. Whether the chief function of the joint is to provide a stable union or a free range of movement is primarily influenced by the type of tissue the joint is constructed of. In **fibrous joints,** the tissue that unites the articulating bones is fibrous tissue; in **cartilaginous joints,** it is cartilage. Both types of joints are immovable or slightly movable (*synarthroses; amphiarthroses*). Freely movable joints (*diarthroses*) are made of a fibrous capsule that encloses a synovial cavity between the articulating bones; hence, they are called **synovial joints.**

### Fibrous Joints

Fibrous joints are subdivided into several different categories, depending upon the form of the joint (Fig. 3-10).

In a **syndesmosis** (syndesmo meaning connective tissue) the two bones entering into the joint are typically separated from each other by a considerable space but united by fibrous connective tissue (a ligament) that bridges this space. This connective tissue is usually largely collagenous, so that little movement is possible.

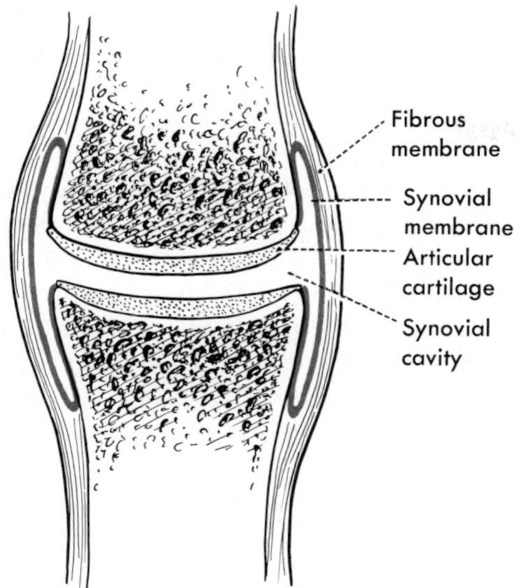


**FIG. 3-11.** Basic structure of a synovial joint. Synovial membrane is indicated in red.

Another type of fibrous junction is that in which the bones are closely adjacent, often firmly interlocking along a wavy line and united by a small amount of dense connective tissue. This type of union, found especially between the flat bones of the skull, is known as a **suture** (suture meaning to sew).

A third type of fibrous union, like the suture in that practically no movement is provided between the two parts, takes the form of a peg fitting into a hole and held there by connective tissue. The attachment of the roots of the teeth to the bone of the jaws is an example of this type, called a **gomphosis** (gomphos meaning nail).

## Cartilaginous Joints

Cartilaginous joints are those in which the union of the bones is by means of cartilage. Two subvarieties are described, the symphysis, in which the cartilage is fibrocartilage, and the synchondrosis, in which hyaline cartilage is present.

In a **symphysis** (meaning grown together) the binding element is fibrocartilage, which allows a limited but nevertheless important amount of movement between the two articulating bones. Probably the most important examples of symphyses are the unions between individual bones of the vertebral column (intervertebral disks, Chap. 14); these symphyses bind the bones firmly together and yet at the same time allow the column to bend.

The most common type of **synchondrosis** (united by cartilage) is that of developing bone, where the bony epiphysis and bony body of a single bone are united by an epiphyseal cartilage. The great majority of synchondroses, therefore, disappear upon full development of the skeleton, although a few persist in connection with the skull and the sternum. Synchondroses typically allow no movement. The manubriosternal synchrondrosis (between two parts of the sternum) is an exception, but it becomes a symphysis because most of the original hyaline cartilage is converted into fibrocartilage.

## Synovial Joints

The synovial joint (synovia refers to the fluid in the joint, which is somewhat like the white of an egg or ovum) is characterized by the presence of a joint cavity. The articulating bones of the joint are separated by this cavity but united by a surrounding articular capsule. The articular capsule consists of two parts, an outer fibrous membrane in which specific thickenings (ligaments) may be visible, and an inner synovial membrane that directly lines the joint cavity except where this cavity is bordered by the hyaline cartilage of the bearing surfaces of the bones (Fig. 3-11). The **fibrous membrane** and any ligaments and muscles that cross the joint bind the bones together. The fibrous membrane is typically composed of collagen fibers, with only a small amount of elastic tissue. Most of the associated ligaments of synovial joints are also collagenous. The fibrous membranes or ligaments uniting the small bones of the middle ear cavity are an exception to this, as they consist largely of elastic tissue.

The **synovial membrane** is for the most part closely adherent to the fibrous membrane, but is always attached to the edge of the articular cartilage; thus, in those joints in which the fibrous membrane attaches some distance from the edges of the articular cartilage, the synovial membrane leaves the fibrous one to be reflected back along the periosteum to the edge of the cartilage. The synovial membrane is a condensation of connective tissue around a joint space and thus has no complete cellular lining (mesothelium) comparable to that found lining the major serous cavities (pleural, pericardial, and peritoneal). Its inner surface is especially rich in cells and capillaries, however, and it is commonly held that both of these contribute to the formation of the synovial fluid (synovia). Folds of synovial membrane frequently project into a joint and probably help to lubricate it. In some locations, a synovial membrane may lie on a relatively thick pad of fat, and the two together move into and out of the joint cavity as the joint is moved back and forth.

The **nerve supply** to a synovial joint is usually derived from several different nerves, a general rule being that each nerve supplying muscles acting across a joint gives at least one branch to that joint. Some of the nerves follow blood vessels and are apparently vasomotor. Some ending in the capsule and on blood vessels are thought to be associated with sensations of pain. Still others in the capsule are probably responsible for the appreciation of movement and position at a joint. The articular cartilage of a joint has neither nerve nor blood supply. It is nourished by blood vessels in the adjacent bone and, especially in the adult, by the synovial fluid.

As already noted, some joints are completely or partially subdivided by **fibrocartilaginous disks** or *menisci* (see descriptions of the sternoclavicular and knee joints).

**Types.** Synovial joints are called either simple or composite according to whether the surfaces of two bones or more than two enter into the joint. Beyond this, classification is based upon the shapes of the articulating surfaces (Fig. 3-12), and these are so varied that no completely satisfactory classification is possible. Recent terminology names seven types: plane, spheroid or cotyloid, condylar, ellipsoid, trochoid, sellar, and hinge (ginglymus). As various joints are studied, the category into which many of them fit will become obvious, as will also the fact that some joints do not fit exactly into any category.

A **plane joint** allows only a limited amount of sliding between two almost flat surfaces (Fig. 3-12A). In all other synovial joints the articulating surfaces are reciprocally curved so that the convexity of one surface fits into the concavity of the other.

In **spheroid (cotyloid) joints,** a convexly rounded head fits into a concavity (Fig. 3-12B and C). Cotylica means a cup and therefore implies the deeper socket, but cotyloid has been used both as synonymous with spheroid and to distinguish the deeper socket. Commonly, this type is called a "ball-and-socket" joint. A ball-and-socket joint allows rather free movement in all directions, including rotation of a bone around its long axis unless ligaments or muscles are so placed as to make this impossible.

A **condylar joint** is essentially a modified ball-and-socket joint, in which a shallow ellipsoidal socket accommodates a "ball" that is not truly round and whose articular surface extends relatively little onto the sides (see Fig. 3-12D). Rotation and abduction–adduction in a condylar joint are limited, but other movements are similar to those of ball-and-socket joints.

An **ellipsoid joint,** although sometimes considered synonymous with a condylar joint, differs from it in having both an ellipsoidal socket and an ellipsoidal ball that resembles a football (Fig. 3-12E). It allows flexion, extension, abduction, and adduction, but no rotation.

A **trochoid joint** is a pivot joint, in which one bone serves as a pin and the other, with its ligaments, forms a circle enclosing the pins (Fig. 3-12F). The chief movement at this joint is rotation.

A **sellar joint** (saddle joint) has complexly curved surfaces resembling those of a saddle (Fig. 3-12G). The surface of one bone is concave in one direction and convex in the other, and that of the second is reciprocally curved

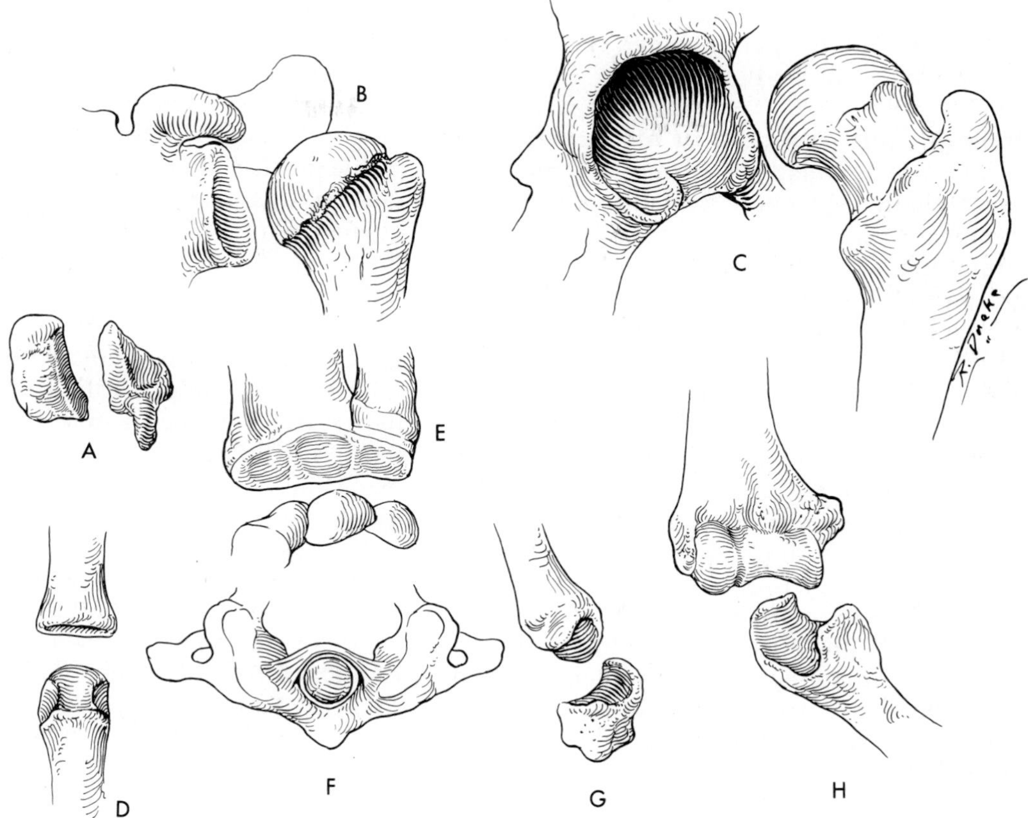


**FIG. 3-12.** Major types of synovial joints. **A,** plane, between two carpal bones; **B** and **C,** spheroid or cotyloid, the shoulder and hip joints; **D,** condylar, a metacarpophalangeal or knuckle joint; **E,** ellipsoid, the radiocarpal joint; **F,** trochoid, the middle atlantoaxial joint (between the first two cervical vertebrae); **G,** sellar, between the trapezium, a carpal bone, and the metacarpal of the thumb; and **H,** ginglymus or hinge, between the humerus and the ulna at the elbow.

so that its convexity and concavity fit respectively the concavity and convexity of the first. This joint allows less free movements of flexion–extension and abduction–adduction than does the ball-and-socket type, and rotation is especially limited.

A **ginglymus** or hinge joint has its surfaces so arranged that movement can take place only through one axis, like the hinge on a door, and this axis is placed transversely so that movement is limited to flexion and extension (Fig. 3-12H).

In the preceding brief discussion, the implication has been that the two surfaces entering into a simple joint are usually reciprocally curved so that they are congruent. In some joints, however, the curves of the convexity of one bone and the concavity of the other differ, and complete congruency apparently exists at no joint, even though the curves may be similar or the planes almost flat. This is said to be in accord with an engineering principle that large weight-bearing surfaces are difficult or impossible to lubricate but that a slight difference in the curvatures of two moving parts does allow proper lubrication, for the contact is then more nearly point-to-point than surface-to-surface. In some of the joints of the lower limb, the curvatures are such that congruency is least during flexion, which facilitates movement, and greater in extension, creating a more stable joint that can best support the weight of the body.

**Strength and Freedom of Movement.** In addition to ligaments, the bones of a joint are held together through a combination of a number of other factors: the soft tissue stretching across the joint, including the skin and various fascial layers but in particular the muscles and tendons; atmospheric pressure; and, for many joints but not all, gravity. In the case of the hip joint and the knee joint, for instance, weight bearing (gravity) obviously tends to force the joint surfaces closer together. In contrast, in the case of the shoulder joint, gravity (the weight of the upper limb) tends to pull the humerus out of the socket, but this tendency is opposed by atmospheric pressure and by a ligament and certain muscles and tendons crossing the joint.

The type and amount of movement allowed at any joint are primarily determined by the shape and extent of the joint surfaces, but other factors also enter in. To a certain extent, maximal strength and freedom of mobility are incompatible, and joints typically represent compromises in which strength is somewhat sacrificed for mobility or mobility is somewhat sacrificed for strength. The shoulder and hip joints, both ball-and-socket joints, afford a good comparison in this regard. The socket of the shoulder joint is shallow and provided with few ligaments to strengthen it. In consequence, it allows free movement but has to depend for its strength upon the muscles surrounding it. In contrast, the hip joint is intrinsically strong, for the socket is deep and the joint capsule is reinforced by heavy ligaments. Because of the depth of the socket and the checking action of the ligaments, movement of the thigh at the hip joint is much more limited than is movement of the arm at the shoulder joint, but, on the other hand, the hip joint is better suited to support the weight thrust upon it.

Joint movement may be checked in several ways: by the apposition of soft tissue; by the limitations of the articular surface and, therefore, impingement of bone against bone; by muscles, acting essentially as ligaments; and by ligaments. Ligaments that check movements often do so as the joint reaches the most stable position, and as a consequence of being taut in this most stable position, they frequently play an important role in stabilizing the joint. Examples of this are seen in the tightening of ligaments at the knee and hip joints when these are extended. In both cases, the line of gravity is so related to the extended joint that the weight falling upon it tends to hyperextend it; the ligaments resist hyperextension, so that the joints are maintained in extension by weight bearing and the ligaments, with little or no help from the muscles.

Ligaments also sometimes guide movements at joints. Certain ligaments at the base of the thumb, for instance, are in part responsible for the fact that when the thumb is brought across the palm, it always rotates so that its pad tends to face the pads of the fingers.

There are, therefore, a number of factors that tend to limit or modify movement at joints, and different joints vary somewhat as to which factor is the most important. However, generally speaking, the tension of opposing muscles tends to be the most important limiting factor in most movements. Similarly, the strength of the muscles crossing a joint is often an important factor in the strength of that joint.

## GENERAL REFERENCES AND RECOMMENDED READINGS

BLACK BM: The prenatal incidence, structure and development of some human synovial bursae. Anat. Rec 60:333, 1934

BOURNE GH (ed): The Biochemistry and Physiology of Bone. New York, Academic Press, 1956

CLARK WE LeG: The Tissues of the Body: An Introduction to the Study of Anatomy, 4th ed. Oxford, Clarendon Press, 1958

COGGESHALL HC, WARREN CF, BAUER W: The cytology of normal human synovial fluid. Anat. Rec 77:129, 1940

DAVIES DV, EDWARDS DAW: The blood supply of the synovial membrane and intraarticular structures. Ann R Coll Surg Engl 2:142, 1948

DODD RM, SIGEL B, DUNN MR: Localization of new cell formation in tendon healing by tritiated thymidine autoradiography. Surg Gynecol Obstet 122:805, 1966

EVANS FG: Stress and Strain in Bones: Their Relation to Fractures and Osteogenesis. Springfield, Ill, Charles C Thomas, 1957

FLECKER H: Time of appearance and fusion of ossification centers as observed by roentgenographic methods. Am J Roentgenol 47:97, 1942

FRANCIS CC: The appearance of centers of ossification from 6 to 15 years. Am J Phys Anthropol 27:127, 1940

FRANCIS CC, WERLE PP, BEHM A: The appearance of centers of ossification from birth to 5 years. Am J Phys Anthropol 24:273, 1939

FRAZER JES: Anatomy of the Human Skeleton, 5th ed. Revised by Breathnach AS. London, J & A Churchill, 1958

GARDNER E: Physiology of movable joints. Physiol Rev 30:127, 1950

GARDNER E: Physiology of joints. Instr Course Lect 10:251, 1953

KELLGREN JH, SAMUEL EP: The sensitivity and innervation of the articular capsule. J. Bone Joint Surg (Br) 32:84, 1950

KELLY PJ: Anatomy, physiology, and pathology of the blood supply of bones. J Bone Joint Surg (Am) 50:766, 1968

KOCH JC: The laws of bone architecture. Am J Anat 21:177, 1917

MALL FP: On ossification centers in human embryos less than one hundred days old. Am J Anat 5:432, 1906

MC CLEAN FC: The ultrastructure and function of bone. Science 127:451, 1958

MILLER MR, KASAHARA M: Observations on the innervation of human long bones. Anat Rec 145:13, 1963

NOBACK CR: The developmental anatomy of the human osseous skeleton during the embryonic, fetal and circumnatal periods. Anat Rec 88:91, 1944

NOBACK CR, ROBERTSON GG: Sequences of appearance of ossification centers in the human skeleton during the first five prenatal months. Am J Anat 89:1, 1951

PAYTON CG: The growth in length of the long bones in the madder-fed pig. J Anat 66:414, 1932

PAYTON CG: The growth of the epiphyses of the long bones in the madder-fed pig. J Anat 67:371, 1933

SALTER N: Methods of measurement of muscle and joint function. J Bone Joint Surg (Br) 37:474, 1955

SIGURDSON LA: The structure and function of articular synovial membranes. J Bone Joint Surg 28:603, 1930

STEARNS ML: Studies on the development of connective tissue in transparent chambers in the rabbit's ear. Am J Anat 66:133, 1940

# 4

# THE NERVOUS SYSTEM

The province of this book does not include detailed discussion of the structure and function of the nervous system, because neuroanatomy and neurophysiology are specialized subjects that have an enormous literature of their own. However, some understanding of basic structure and function is necessary to an adequate appreciation of the gross anatomy of the nervous system.

## THE NEURON

The unit of structure of the nervous system is the nerve cell or **neuron** (Fig. 4-1), which consists of a *cell body* (perikaryon) and processes that extend for a variable distance beyond the cell body proper.

**Dendrites** (dendron meaning tree), of which there are usually several to a single nerve cell, are characterized structurally by a branching, treelike form. They are sometimes defined as cell processes that conduct toward the cell body. In some instances, the dendrite provides the only contact between a nerve cell and the cells that send impulses to it. In other instances, the majority of the synapses are at the base of the dendrites or between them on the cell body itself. Still other loci of synapses have been described. A dendrite seems to be largely a simple expansion of the cell body, and the cytoplasm of the larger dendrites is structurally similar to that of the cell body. They apparently aid in the nutrition of the cell through the greatly expanded surface that they offer.

In contrast to dendrites, **axons** are relatively unbranched over most of their length, although they typically branch repeatedly close to their endings. The cytoplasm of the axon is modified from that of the cell body, for, except for neurofibrils, it contains none of the formed elements included in the body of the cell. The largest axons measure about 18 $\mu$ in diameter, and the smallest ones measure less than 1 $\mu$; yet, in length, many axons can be measured at

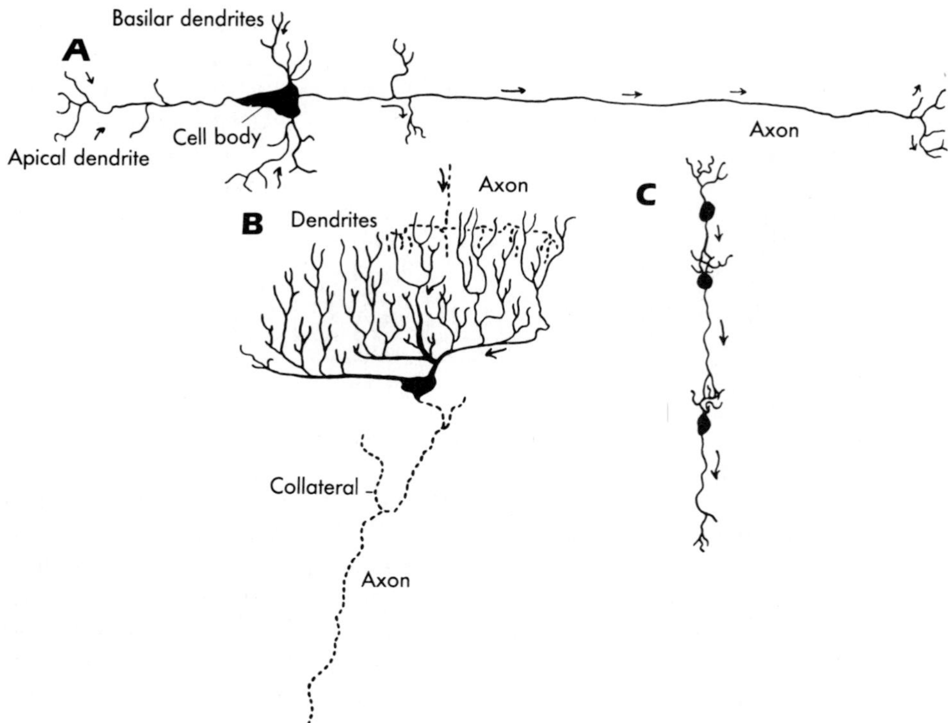


**FIG. 4-1.** Some types of neurons. **A** and **B** are large cells from the brain, with different arrangements of dendrites. In **B** an axon making contact with the dendrites of the cell is also shown. In **C** is shown the manner in which small cells can be linked to form a chain of neurons. The arrows on the figures indicate the direction of conduction. (Mettler FA: Neuroanatomy, 2nd ed. St Louis, CV Mosby, 1948)

least in millimeters, and the longest ones in centimeters or even in feet. Almost all information concerning the conduction of the nerve impulse has been derived from the study of axons, and the term "nerve fibers" refers to axons, not dendrites.

The majority of nerve cells have a single axon. However, the cells that form the sensory ganglia of nerves have two processes that are structurally axons (usually arising from a single stem) and no dendrites. In this one, but very striking, instance, structural axons conduct toward the cell body rather than away from it.

The usual concept of the transmission of the nerve impulse is that conduction along the nerve fiber involves a shift in ions and is therefore essentially electrical, but that conduction across the synapse or from a nerve ending to the end organ is by means of a humoral agent. The concept of humoral agents as transmitters of the nerve impulse is especially well established for the autonomic system, where it has been shown that either norepinephrine (adrenaline) or acetylcholine may be the humoral agent. Acetylcholine is also the humoral agent that acts at the ending of a nerve fiber or voluntary muscle. Both acetylcholine and norepinephrine are apparently among the humoral agents acting at synapses in the central nervous system. Those fibers that liberate norepinephrine at their endings are known as **adrenergic fibers,** and those that liberate acetylcholine are known as **cholinergic fibers.**

Basically, neurons simply link a **receptor** and an **effector,** but only in certain primitive invertebrates does a single nerve cell form such a direct link. In higher animals, neurons are linked into chains or units of varying complexity, so that even the simplest reaction resulting from a stimulus must be mediated through two or more neurons, and most reactions involve great numbers of such cells. Some neurons are in the direct line of transmission from receptor organ to effector organ, but others are parts of circuits that interlock with this direct line of transmission and through their activity vary the response resulting from the stimulus.

Because the speed of conduction of the nerve impulse along a nerve fiber (axon) can be measured very precisely (it varies from about 120 m/sec in fibers of largest diameter to about 0.3 m/sec in the smallest), and because each synapse produces a definite slowing in the passage of the nerve impulse, the number of neurons involved in a given reaction can, in fairly simple instances, be calculated from the time elapsing between stimulus and response.

### Some Basic Groupings of Neurons

Nerve cell bodies are not scattered at random through other tissues, even though there are relatively few tissues into which nerve fibers do not penetrate. The greatest accumulation of cell bodies and nerve fibers is represented by the brain and spinal cord, together known as the central nervous system; nerve fibers and cell bodies outside the brain and spinal cord constitute the peripheral nervous system. The distinction is a gross anatomic one only. The central nervous system is dependent upon the peripheral nervous system both for reception of stimuli and for transmission to the effector organ, and the connections between the afferent and efferent parts of the peripheral nervous system occur in the central nervous system. Also, many of the cell bodies whose fibers form the peripheral nervous system lie themselves within the central nervous system.

Within the central nervous system, accumulations of cell bodies are known as **gray matter** because of their color in the fresh condition. Accumulations of axons are known as **white matter** because many axons have a fatty sheath (myelin) that appears white in the fresh condition. Accumulations of nerve cells outside the central nervous system are known as **ganglia** (ganglion meaning swelling), and bundles of nerve fibers outside the central nervous system form the nerves. There are two types of ganglia: sensory ganglia on the spinal and cranial nerves, and motor (autonomic) ganglia, which lie more peripherally.

Cell bodies and nerve fibers in the central nervous system are enmeshed in **neuroglia,** a connective tissue peculiar to the central nervous system. True neuroglia differs from connective tissue elsewhere not only in its form but also in its origin, for it is derived from ectoderm rather than mesoderm. However, a certain amount of ordinary mesodermal connective tissue is brought into the central nervous system by the blood vessels that penetrate it. The cells and fibers of the ganglia and nerves of the peripheral nervous system are bound together by the usual type of fibrous connective tissue, although some cellular elements within ganglia and nerves are apparently derived, like the neuroglia, from ectoderm.

## DEVELOPMENT OF THE NERVOUS SYSTEM

The first evidence of the nervous system appears in the embryo at about 19 days as a longitudinal thickening of the ectoderm along the dorsal midline; this is called the **neural plate.** Very quickly the edges of this plate become raised above the surrounding ectoderm to form **neural folds,** which arch toward each other. During the fourth week of development these folds meet to form the **neural tube,** which separates from and sinks deep to the general ectoderm to become the central nervous system (Fig. 4-2). The enlarged cephalic end of the tube becomes the brain, and the narrow caudal portion becomes the spinal cord. Cells at the border of the neural folds, where they were continuous with the general ectoderm, separate from both the neural tube and the ectoderm and are the **neural crests.**

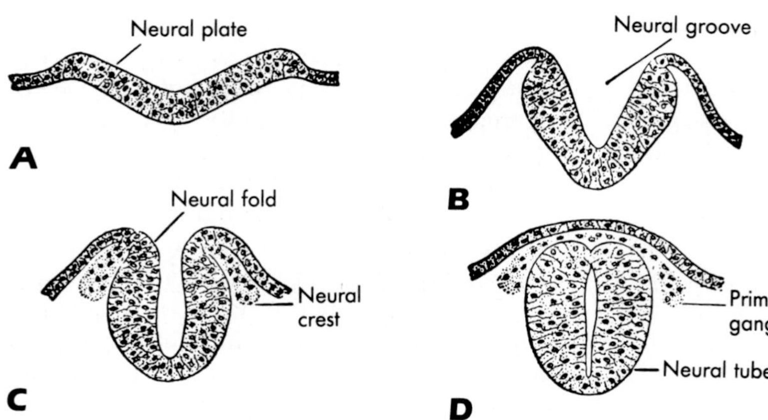


**FIG. 4-2.** Early development of the spinal cord, in cross section. (Arey LB: Developmental Anatomy, 6th ed. Philadelphia, WB Saunders, 1954)

Eventually, these crests form a series of segmentally arranged cell groups on either side of the spinal cord, the forerunners of the spinal (sensory) ganglia. More irregularly arranged cell groups alongside the brain, also representing neural crest, are the forerunners of most of the cranial sensory ganglia. Some of the cranial ganglia seem to arise directly from the surface ectoderm lying close to it after the neural tube is closed.

## Peripheral Nervous System

Most of the neural crest cells that remain in the spinal and cranial sensory ganglia form neurons, but others apparently form nonneuronal supporting cells, which, as flattened "capsular cells," surround the nerve cell bodies of the sensory ganglia. Other neural crest cells migrate away from the sensory ganglia, and their fate is more difficult to determine. These migratory neural crest cells are said by most authors to give rise to the neurons of the autonomic ganglia (and hence the glandular cells of the suprarenal medulla, known to originate from the same cells as these ganglia) and to flattened cells (sheath cells, cells of Schwann) that surround peripheral nerve fibers.

The transformation of a neural crest cell into a *neuroblast,* or young nerve cell, is simply a matter of a cell giving rise to sprouts that grow out as nerve processes. The cells of the spinal and cranial ganglia, destined to become **sensory nerve cells,** develop only two processes, in contrast to the number that develop from many other neuroblasts. Both processes are alike, being axons or true nerve fibers anatomically. One grows out to the periphery, for instance, to the skin, as a sensory (afferent) fiber of a nerve; the other grows into the neural tube, making connections with cells there. With other similar fibers, the latter forms the sensory (afferent) root of the nerve. Thus, a single nerve cell conducts from the peripheral sense organ to the central nervous system. Eventually most nerve cells of the spinal and cranial ganglia become unipolar; the peripherally and centrally oriented axons are brought together, apparently by differential growth of the cell body, until they arise from a common stem. The ganglion cells connected with the internal ear, however, retain their primitive bipolar form.

The fibers that form the **motor roots** of the spinal and cranial nerves arise not from neural crest cells but as processes of neuroblasts that lie within the central nervous system. These become multipolar cells, but their numerous dendrites remain within the central nervous system, whereas their single axons grow out to join the sensory part of the nerve peripheral to the sensory ganglion. Once they have mingled with the sensory fibers, some motor fibers grow out to striated skeletal muscle, to end there, whereas others leave the main nerve and go to autonomic ganglia, to end on the nerve cells forming the ganglia.

**Myelin,** a fatty layer formed around many

nerve fibers, is probably developed through the activity of both the nerve axons and the adjacent sheath cells. The connective tissue that binds the nerve fibers of peripheral nerves together is derived from mesodermal mesenchyme, as is most connective tissue.

The migratory cells that give rise to the **autonomic ganglia** assume various positions in the body, but most accumulate either along the anterolateral aspect of the vertebral column to form the paired sympathetic trunks or in rather large masses around the origins of the great vessels to the viscera to form the prevertebral or collateral ganglia of the sympathetic system. Others form the relatively small and irregularly placed autonomic ganglia of the head, and still others migrate into the wall of the digestive tract to form tiny autonomic ganglia or isolated autonomic cells. The neuroblasts of the autonomic ganglia develop short dendrites and a single axon that grows out to smooth muscle, cardiac muscle, or glands. These cells are essentially similar in appearance to many of the multipolar cells in the central nervous system.

### Central Nervous System

Many of the cells of the primitive neural tube become transformed, after a period of proliferation, into neuroblasts by the growth of processes. Neuroblasts, in turn, are transformed into true nerve cells. Other cells in the neural tube are transformed into neuroglia (supporting) cells, and a limited number remain as rather primitive epithelial cells, ependymal cells, that form the lining of the cavity of the neural tube.

The cell bodies of the neurons of the central nervous system tend to remain grouped close to the central canal, but they send their longer fibers into the periphery of the tube where they grow up and down or across it to form tracts. Thus, the gray matter of the central nervous system tends to be centrally located and the white matter, peripherally located. Exceptions to the central location of gray matter are the thin layers (cortices) on the surfaces of the cerebellum and cerebrum (parts of the brain). Here, although some gray matter does remain centrally located, many cells migrate to the periphery and cover the white matter. Although most neurons of the central nervous system lie entirely within that system, certain ones, as already noted, have only their cell bodies and dendrites within the system, sending their axons into the motor roots of nerves.

Many of the nerve fibers within the central nervous system eventually acquire myelin, just as many peripheral nerve fibers do. In the central nervous system, myelin is added around nerve fibers at about the time they begin to function; thus, the fibers that begin to function first acquire their myelin first. Not all axons within the central nervous system, however, become myelinated, and myelin is never deposited around nerve cell bodies or around dendrites.

The further development of the neural tube into brain and spinal cord cannot be entered into here. However, the concept of a tubular central nervous system requires emphasis. The central nervous system originates as a tube, and in spite of marked developmental changes, it remains essentially a tube. The adult brain, although showing pronounced thickening of its walls, outpouchings, and bendings, still contains a continuous central cavity, the larger subdivisions of which are called ventricles. The spinal cord of the adult is also a tube, but its walls are so thick and its lumen so tiny that the latter is of no functional import; indeed, the central canal of the spinal cord of the adult is often blocked by cellular debris.

The central nervous system becomes surrounded by saclike membranes, the **meninges,** of which the outer one, the dura mater, is definitely derived from mesoderm. The more delicate inner ones, the pia mater and arachnoid, are usually described as being derived in part or entirely from neural crest cells.

## NERVES

Nerves are bundles of nerve fibers that lie outside the central nervous system, and most nerves contain both motor and sensory fibers. The nerves that make their exit through the skull are known as **cranial nerves.** Twelve pairs of these are described and named as being typical of all mammals. With the exception of a part of one that arises from the cervical part of the spinal cord and runs upward to make its exit from the skull, they are all attached to the brain.

Those nerves that make their exit below the skull and between vertebrae are **spinal nerves.** There are 31 pairs of these, arising from the spinal cord throughout its length.

All peripheral nerves, whether spinal or cranial, are very similar in structure, and the features common to all can therefore be discussed together. The differences between spinal and cranial nerves (chiefly, the functional types of fibers that they contain and the way in which they arise from the central nervous system) do not affect their basic anatomy.

## Structure of Peripheral Nerves

**Connective Tissue.** After a nerve, whether spinal or cranial, has taken form and left the shelter of the membranes and bones surrounding the spinal cord and brain, its constituent nerve fibers or axons become bound together, somewhat like the individual wires in a cable, by connective tissue. This is largely collagenous, but contains some elastic fibers and, in some nerves, an appreciable amount of fat. The connective tissue of a large nerve is described as epineurium, perineurium, and endoneurium (Fig. 4-3). The **epineurium** surrounds the entire nerve and holds it loosely to the connective tissue through which it runs. It also sends septa into the nerve that divide the nerve fibers into bundles (*fasciculi* or *funiculi*) of varying size. The **perineurium** consists of both connective tissue and a squamous epithelium. It surrounds each fasciculus and branches with every branching of that, eventually enveloping every nerve fiber in its own perineurial tube. The **endoneurium** is a delicate layer of connective tissue around each nerve fiber, lying therefore within the perineurium.

The connective tissue of a nerve allows a certain amount of stretch without damage to the nerve fibers. The fibers tend to be wavy, and by the time the slack in them is taken up, the connective tissue fibers are also in tension and resist further stretch. The maximal

**FIG. 4-3.** Cross section of part of the human median nerve. (Bremer JL, Weatherford HL: A Textbook of Histology, Philadelphia, P Blakiston & Son Co, 1944)

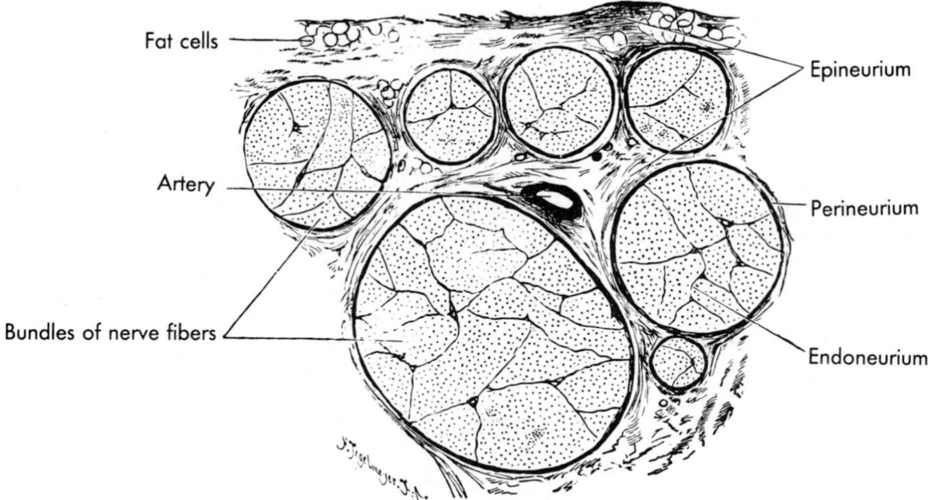

stretch that can be imposed upon any given length of nerve without appreciable damage is approximately 10%, although nerves that contain fat allow a little more. Nerves so placed that they are most likely to be stretched during normal movements typically contain more fat than do other nerves.

**Anastomoses and Branches.** When a nerve branches, the branching is a separation of nerve bundles, not a branching of axons. (Some axons do branch before they near their ending, but this is not common, nor does it usually occur at a grossly visible nerve branching.) Similarly, when two nerves join, their fibers do not unite but simply intermingle and become surrounded by a common epineurium.

The connective tissue of a nerve is so tough, as compared to the nerve fibers themselves, that it generally cannot be removed without tearing the fibers. However, when two nerves join, it is sometimes possible to separate them for a little distance beyond their point of juncture because they may not interchange fibers immediately and their epineuria may fuse only superficially instead of blending decisively. Thus, nerves may seem to join higher or lower than usual, depending on the density of the connective tissue first connecting them. The bundles of such nerves do not remain separate long, however, but divide and rejoin each other repeatedly, so that a little beyond the junction no one bundle contains fibers from one nerve only.

The same type of branching and regrouping of bundles, forming an intraneural plexus, occurs throughout the length of all the larger nerves, whether or not they are formed by the union of other nerves. The intraneural plexus is a sorting and resorting of sensory and motor fibers, or fibers from different segments, which finally leads to the isolation of a particular bundle of nerve fibers. This bundle then leaves the main nerve as a branch. The fibers destined for a given branch are grouped together a variable distance above the level at which they leave the nerve and are merely held to the remainder of the nerve by enveloping connective tissue. Depending, therefore, upon the development of the surrounding connective tissue, a branch may leave a nerve either higher or lower than its usual level of exit. Variations in the level of origin of nerve branches, largely the result of variations in the surrounding connective tissue, are very commonly seen.

**Myelin and Neurolemma.** Nerve fibers greater than about 2 $\mu$ in diameter typically have a fatty sheath, the **myelin sheath,** about them. Those of less than a micron typically have no sheath and hence are termed "unmyelinated fibers." Since large and small nerve fibers occur both in the central nervous system and in peripheral nerves, peripheral fibers may or may not be myelinated. All peripheral nerve fibers, however, have related to them a type of cell, the sheath or **Schwann cell,** also called neurolemma, that is not found, at least in the same form, in the central nervous system. This is an elongated cell that has groups of unmyelinated fibers embedded in it or is wrapped many times around one myelinated fiber. Further, myelinated peripheral nerve fibers show segmentation of the myelin. The constrictions separating successive segments of myelin are called **nodes of Ranvier,** and typically there is one sheath cell to each internodal segment. Outside each sheath cell is the delicate connective tissue of the endoneurium.

**Blood Supply.** Nerves receive twigs from adjacent blood vessels along their course. The twigs to large nerves are of macroscopic size, somewhat irregularly placed, and their branches tend to join together (anastomose) to form a longitudinally running major trunk that supplies the nerve, giving off subsidiary branches that may also run longitudinally within the nerve. Opinions differ as to whether damage to a nerve from pressure or stretch is due to interference with its vascular supply or to direct trauma to nerve fibers.

## Degeneration and Regeneration of Nerves

There is no regeneration of the bodies of nerve cells either inside or outside the central nervous system, for once a neuroblast differentiates into a nerve cell it loses its capacity to divide. Loss of nerve cells, whether in the brain or in the peripheral nervous system, is, therefore, irreparable. Clinical recovery from an apparent loss results only from one of two processes: either the nerve cells concerned were not actually destroyed but merely injured enough to produce temporary arrest of their function, or other, uninjured, cells were capable of taking over, to a reasonable extent, the functions of the destroyed cells.

In contrast, nerve fibers do regenerate after injury, but only in the peripheral nervous system. Nerve fibers within the central nervous system, although they may grow very slightly, apparently never regenerate to form functional connections again. Similarly, interruption of the sensory roots of spinal and cranial nerves is not followed by functional regeneration, since the fibers must enter and grow within the central nervous system to make functional connections. Consequently, section of sensory nerve roots, sometimes carried out for relief of intractable pain or for other reasons, can be expected to produce permanent results.

Presumably, the absence of sheath cells and endoneurium is the primary factor responsible for lack of regeneration in the central nervous system, for regeneration of nerve fibers seems to follow the same general principles that govern and guide the early outgrowth of the fibers in the embryo. The most important principle is that the outgrowing fiber must have a structure that it can grow along if it is to reach an end organ. Regenerating peripheral nerves, but not fibers in the central nervous system, have a guiding thread in the cells and connective tissue fibers immediately around the axon.

**Degeneration.** When an axon in a peripheral nerve is interrupted, all of that axon distal to the level of interruption—that is, the part that is no longer connected to the nerve cell body—degenerates, and if the axon is myelinated, its myelin sheath degenerates also. For a time, degenerating axons and myelin can be stained and recognized microscopically, at least to a certain extent, in both the central and peripheral nervous system. Later, however, phagocytic cells remove the remains of both axon and myelin. In the peripheral nervous system, degeneration does not at first involve the sheath cells nor the thin tubes of connective tissue around these cells and the axon. So a freshly degenerated peripheral nerve still has the same connective tissue framework that it had before degeneration. It is said, however, that the perineurial epithelium, which eventually surrounds each nerve fiber as a perineurial tube, is the most important element remaining after the nerve fibers have degenerated.

Axons can be injured severely enough that they cease to function and yet not severely enough to produce interruption of anatomic continuity. In this event, neither degeneration nor regeneration occurs, although the symptoms (in peripheral nerves, typically loss of sensation and paralysis of skeletal muscle) may be identical with complete anatomic interruption of the nerve fibers. Recovery in this case is merely a question of regaining function in anatomically intact fibers. In minor cases this may be only a matter of minutes, in major ones a matter of months. Recovery of function of severely injured nerve fibers may be impossible to distinguish clinically from degeneration and regeneration of anatomically interrupted fibers.

**Regeneration.** After complete interruption of a peripheral nerve fiber, its proximal end, still connected to the cell body, passes through a period of relative inactivity. This may include some degeneration of its most distal part but is apparently primarily a preparation for regeneration. During this period, the sheath cells are usually described as multiplying and, by their apposition, forming tiny threads within the endoneurial tubes. These tubes, partly filled by sheath cells, maintain approximately their original sizes, although some proliferation of connective tissue occurs within the nerve (and if the onset of regeneration is too long delayed, this proliferation leads to collapse of the tubes, with consequent difficulty in regeneration).

Actual regeneration begins with the sprouting of several very fine fibrils—there may be dozens, but usually only three or four—from the distal end of the intact part of each axon, each fibril having an ameboid growing tip. These tips either seek out or encounter by chance the ends of nearby endoneurial tubes and attempt to grow down them (as the ameboid tip of a nerve fiber in tissue culture will grow along a fine cobweb). Of the regenerating axon tips that enter a single endoneurial tube, some are quickly outstripped in their growth by others and tend to degenerate. Of the others, the one that first reaches the end organ and establishes a functional connection with it persists as the single nerve fiber within the endoneurial tube, while the others degenerate.

Since regenerating nerve fibers must reach endoneurial tubes if they are to grow any great distance, it is important that tubes and ameboid tips be close together when regeneration begins. If there has been no break in the continuity of the nerve as a whole, but only in the nerve fibers, the endoneurial tubes distal to the level of interruption are still continuous with those surrounding the regenerating axon. Each nerve fiber then tends to regenerate along the same pathway that it originally followed and to reach the same end organ with which it was originally connected. Obviously, therefore, a lesion of a nerve that does not interrupt the continuity of its connective tissue offers essentially ideal conditions for regeneration, and regeneration may be expected to occur both as swiftly and as completely as possible.

If the nerve as a whole has been interrupted, bringing the cut ends together and holding them in that position is the nearest approximation to continuity that can be achieved. The most important problem is to prevent scar tissue from forming between the adjacent ends, which would block the growth of axons into the peripheral part of the nerve. When the nerve ends are so far separated that they cannot be approximated without too much tension, a nerve graft, employing appropriate lengths of a cutaneous nerve from the same person, must be used as a bridge to connect the two ends if regeneration is to be obtained. Regenerating axons will not grow over any appreciable distance, even through normal tissue, to reach the proximal end of a degenerated nerve because a majority tend to be diverted by the intervening connective tissue fibers.

For perfect regeneration, every regenerating axon would seemingly have to make connection with exactly the same end organ with which it was once connected. This may be reasonably well achieved where there has been no interruption in the continuity of the connective tissue framework of the nerve. In any other circumstance, it seems inevitable that some growing axons would regenerate down the wrong endoneurial tubes and, hence, terminate in contact with end organs with which they were not previously associated. If the end organ is of the same type as that originally reached by the fiber, functional regeneration can be expected. For example, recovery is probable if an axon that once supplied muscle regenerates to muscle. The fiber is capable of forming functional connections with muscle cells that it did not originally supply, and the central nervous system is capable of readjusting its signals and sending them to a given muscle by a nerve different from the one that has previously conducted impulses to that muscle. This ability to readjust is sometimes purposely used by surgeons. For instance, in a case of irreparable damage to a proximal part of the facial nerve, with resulting paralysis of the face, the accessory nerve (to muscles of the shoulder) can be sectioned and its proximal end sutured to the distal part of the facial nerve. After the regenerating accessory fibers have reached the facial muscles, the patient obtains fairly adequate control over these muscles, although it requires, at first, thinking of shrugging his shoulders when he wants to move his face.

However, if a sensory fiber regenerates down an endoneurial tube that brings it in contact with muscle fibers, it cannot form motor endings on those muscles as would a motor nerve fiber. Neither can a fiber originally connecting with muscle form functional sensory endings in the skin. Regeneration of fibers down the wrong pathways may be disastrous, leading to no functional regeneration at all. Because of this danger, the surgeon, in suturing a nerve, takes particular care not to twist either end, for he hopes to bring together nerve bundles that were originally in continuity with each other.

Although nerve fibers grow very rapidly (some of them at the rate of more than 3 mm/day), the great distances over which they sometimes must regenerate, which may be measurable in feet when the injury is close to the spinal cord, make the functional regeneration of nerve fibers a slow process. As a general rule, the farther a regenerating axon tip gets from the nerve cell body, the more slowly it grows. For example, an axon that starts regenerating at the rate of 3 mm/day may drop its rate to 0.5 mm/day or less as it reaches a more peripheral location. Additional time at the beginning and the end of the process of anatomic regeneration must be allowed for in clinical practice. The first period is necessary for reorganization of the axon before it begins to regenerate, and the second, for functional maturation of the fiber after it has reached the end organ. The complete process may occupy periods of from several months to a year or more.

## SPINAL NERVES

The 31 pairs of spinal nerves are named and numbered according to the regions of the vertebral column with which they are associated. Since they are all attached to the spinal cord, and this is enclosed in the vertebral canal formed by the vertebrae, they must leave the canal to reach the periphery, and they do this in a regular manner. Of the eight **cervical nerves,** the first pair make their exit between the skull and the 1st cervical vertebra; the next six emerge between the seven cervical vertebrae; and the 8th or last cervical nerve emerges between the 7th cervical and the 1st thoracic vertebra. Except in the cervical region, where the nerves outnumber the vertebrae because the uppermost one lies above the 1st cervical vertebra, the nerves are named according to the vertebra *below which* each passes. Thus, there are 12 (pairs of) **thoracic nerves,** the first leaving the vertebral canal below the pedicles of the 1st thoracic vertebra, the 12th leaving below the 12th thoracic (and above the 1st lumbar) vertebra; there are five **lumbar nerves,** the 1st leaving below the 1st lumbar vertebra, the 5th below the 5th lumbar; and there are five **sacral nerves,** the first four of which make their exits through foramina in the sacrum (representing the intervertebral spaces of other parts of the column), while the 5th leaves below the sacrum. There is only one **coccygeal nerve,** running below the 1st coccygeal segment; the other rudimentary coccygeal segments, the remains of a tail, have no nerves associated with them.

As they leave the spinal cord, spinal nerves are metamerically (segmentally) arranged, and the distribution of their fibers to skin and muscle is referred to as the segmental distribution of the spinal nerves. Many branches of the spinal nerves do not, however, run separately to the muscles and skin that they innervate but, rather, join or exchange branches with other spinal nerves. This mingling and interchange (a plexus) gives rise to new nerves containing fibers from several spinal nerves. Thus, many peripheral nerves are plurisegmental; their distribution is usually referred to as the distribution of peripheral nerves. Although these nerves are of spinal origin, they do not represent individual spinal nerves, and peripheral and spinal (segmental) nerves are sometimes contrasted on that basis.

**Composition.** Each spinal nerve is typically attached to the spinal cord by a dorsal and a

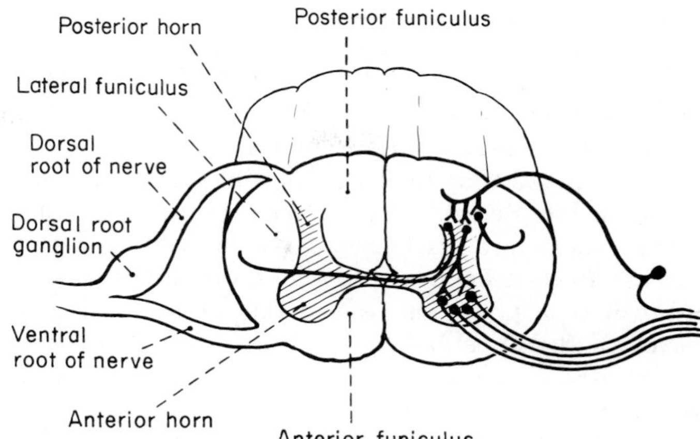


**FIG. 4-4.** Diagram of a short section of spinal cord. A cell body of the dorsal root ganglion is shown on the right with its central fiber entering the spinal cord; the cell bodies in the gray matter include those that send fibers to neighboring cell bodies, those that send fibers into the white matter, and the neurons whose fibers form the ventral roots of the spinal nerves.

ventral root (Fig. 4-4). The fibers forming the dorsal root are afferent, or sensory, and arise from a collection of nerve cell bodies, the **spinal or dorsal root ganglion,** that lies at the distal end of the dorsal root. Each body gives rise to a single axon that soon bifurcates. One branch extends peripherally in the spinal nerve to the area of sensory distribution of that nerve. The other branch extends centrally to form the dorsal root and enter the spinal cord. The fibers of the dorsal root branch after they enter the spinal cord. Some branches of a single fiber end around nerve cells at the level of entrance, whereas others run for varying distances up and down to end around cells at other levels.

The fibers of the **ventral roots** of spinal nerves are efferent, or motor, and are derived from cell bodies that lie in the spinal cord. Dorsal and ventral roots come together just distal to the spinal ganglion, and sensory and motor fibers thoroughly intermingle to form a "mixed" spinal nerve. (Although "afferent," implying conduction to the spinal cord, is actually a better term than "sensory," implying conduction that affects consciousness, "sensory" is more easily understood and is commonly used in the general sense of afferent. The term "efferent," implying conduction from—in this case, the spinal cord—has little advantage over "motor." "Mixed" refers to a nerve containing both sensory and motor fibers.)

The ventral roots of all the thoracic, the upper two or three lumbar, and about the 2nd to 4th sacral nerves contain *two* types of motor fibers. As in all the ventral roots, large fibers (with their cell bodies inside the spinal cord) run without interruption, sometimes over a distance of several feet, to end on skeletal muscle fibers. There are also, in these particular roots, small fibers that leave the spinal nerve at some point to end in collections of nerve cells called autonomic ganglia. These ganglia, in turn, send their fibers to smooth muscle, cardiac muscle, or glands. From the functional standpoint, therefore, the motor fibers of the ventral roots of many nerves can be classified as **somatic efferent,** because skeletal muscle has to do with the body as a whole and autonomic, or general **visceral efferent,** because the involuntary structures innervated by them are usually viscera.

Similarly, of the sensory or afferent fibers in the **dorsal roots,** some go to skin or to sensory endings of skeletal muscle, and others go to viscera. Hence, afferent fibers, like motor fibers, can be classified as **somatic** or **visceral** (in spinal nerves they are termed general somatic afferent or general visceral afferent).

Since the dorsal roots of spinal nerves consist entirely of afferent fibers, and the ventral roots of motor fibers, either sensory or motor fibers alone can be interrupted by sectioning the roots close to the spinal cord. (Differential section of somatic and visceral fibers in the roots is not possible, although visceral fibers, motor and sensory, alone can be interrupted by cutting certain branches of the nerves.) Once the two roots have united to form the mixed nerve, the sensory and motor fibers become so mixed that their differential section is impossible. Because the two roots come together just before the nerve leaves the vertebral canal, their exposure for section requires removal of a posterior part of one or more vertebrae (laminectomy).

**Branches.** Each mixed spinal nerve formed by the union of dorsal and ventral roots divides fairly quickly into two major branches or **rami** (ramus meaning branch), dorsal and ventral (Fig. 4-5). The dorsal rami turn sharply around the vertebral column to supply the muscles most closely associated with the dorsal aspect of the column (true muscles of the back), usually continuing also to skin of the back. The ventral branches supply the anterolateral skin and muscles of the neck and trunk and all the skin and muscles of the limbs.

Close to the point at which its two roots join, each spinal nerve usually gives off a small branch (meningeal) that runs back toward the spinal cord and supplies tissue within the vertebral canal. The ventral ramus of each spinal nerve typically receives, close to its origin, a small nerve from a ganglion of the sympathetic trunk (a part of the autonomic nervous system). A little beyond this, the thoracic and upper lumbar nerves that contain autonomic fibers in their ventral roots send these fibers to the sympathetic trunk by a second small nerve. These connections between the spinal nerves and the sympathetic trunk are the rami communicantes. Since they consist exclusively of visceral fibers, it is at this level that operative section of sensory and motor visceral fibers avoids damage to somatic fibers.

**Distribution of "Typical" Nerve.** The **dorsal rami** of all the spinal nerves are essentially similar and, in contrast to many ventral rami, remain for the most part separate from each other. They are distributed in a segmental fashion to the musculature and skin of the back, each nerve supplying a region centered below the center of distribution of the nerve arising above it. Soon after its origin, a dorsal ramus typically divides into two branches, both of which supply the vertebral muscula-

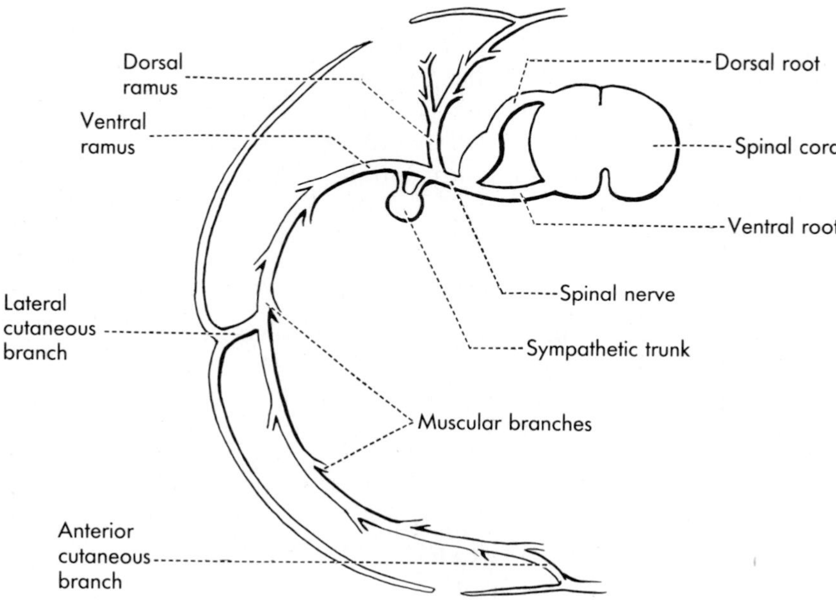


**FIG. 4-5.** Formation and branches of a thoracic or "typical" spinal nerve.

ture, but only one of which usually continues to the skin.

The distribution of many of the **ventral rami** of the spinal nerves is complicated by the interchange of branches or by formation of plexuses (see the following section) in which the identities of the contributing nerves are lost. The ventral branches of the thoracic nerves, however, are largely kept apart by the ribs and can often be traced by dissection to the muscles and skin that each innervates. Because of the simplicity of their ventral branches, the thoracic nerves can be conveniently described as "typical" spinal nerves. Part of the ventral branch of the 1st thoracic nerve, and the ventral branches of the next ten, are also known as *intercostal nerves,* and that of the 12th as the *subcostal nerve,* because of their relation to the ribs (costae).

Most nerves tend to run caudally as they leave the vertebral canal. Beyond this point the slant of the intercostal (and subcostal) nerves, as they run laterally and forward, is determined by the slant of the ribs. Although, because of this slant, the anterior branches of approximately the lower six thoracic nerves continue from the thoracic to the abdominal wall, they undergo no fundamental change in muscular relationships or in branches as compared to upper ones.

An intercostal nerve, then, runs laterally, caudally, and forward between layers of muscle and gives off twigs that supply this muscle. When it reaches a position roughly in a vertical line from the axilla (armpit), it gives off a lateral cutaneous branch that perforates the overlying muscle and, as it approaches the skin, divides into anterior and posterior subbranches. These run anteriorly and posteriorly, respectively, in the connective tissue underlying the skin and, through their twigs to the skin, supply a strip on the anterolateral wall of the trunk, extending backward to skin innervated by dorsal rami and forward to skin innervated by another cutaneous branch of the same intercostal nerve.

After giving off its lateral cutaneous branch, the intercostal nerve continues forward between muscles of the body wall, whether of the thorax or abdomen, and supplies these muscles as far as they extend toward the anterior (ventral) midline. A little lateral to the midline, the nerve gives off its second cutaneous branch, an anterior cutaneous one, that runs toward the skin and subdivides into branches that supply skin medially as far as the midline and laterally as far as the skin innervated by the lateral cutaneous branch of that same nerve.

From this description it should be plain that a "typical" spinal nerve innervates, through its dorsal branch, dorsally (posteriorly) lying muscle and skin and, through its ventral branch, ventrally (anteriorly) lying muscle and skin. Moreover, if an intercostal nerve is taken to be a typical ventral branch, such a branch innervates muscle and skin lying at its own segmental level. Complications are introduced by overlap in cutaneous innervation through adjacent segmental nerves, to be discussed later, but the greatest complications are introduced by the growth of the limbs (in the development of which the segmental origin of the muscles and skin cannot be observed) and by the formation of plexuses in which the segmental ventral branches are lost.

**Nerve Plexuses.** As noted, the ventral rami of most spinal nerves, except those of the thoracic region, join together in complex intertwinings and interchangings of fibers that are known as plexuses. Three chief plexuses are formed: the **cervical plexus,** formed by the ventral rami of the upper cervical nerves; the **brachial plexus,** formed by the ventral rami of the lower cervical nerves and most of the ventral ramus of the 1st thoracic nerve; and the **lumbosacral plexus,** formed by the ventral rami of most of the lumbar and sacral nerves. The cervical plexus supplies most of the anterior muscles of the neck, some skin of the head and much of that of the neck, and skin on the upper portion of the shoulder and thorax; the brachial plexus supplies the musculature and skin of the upper limb; and the lumbosacral plexus supplies the musculature and skin of the lower limb.

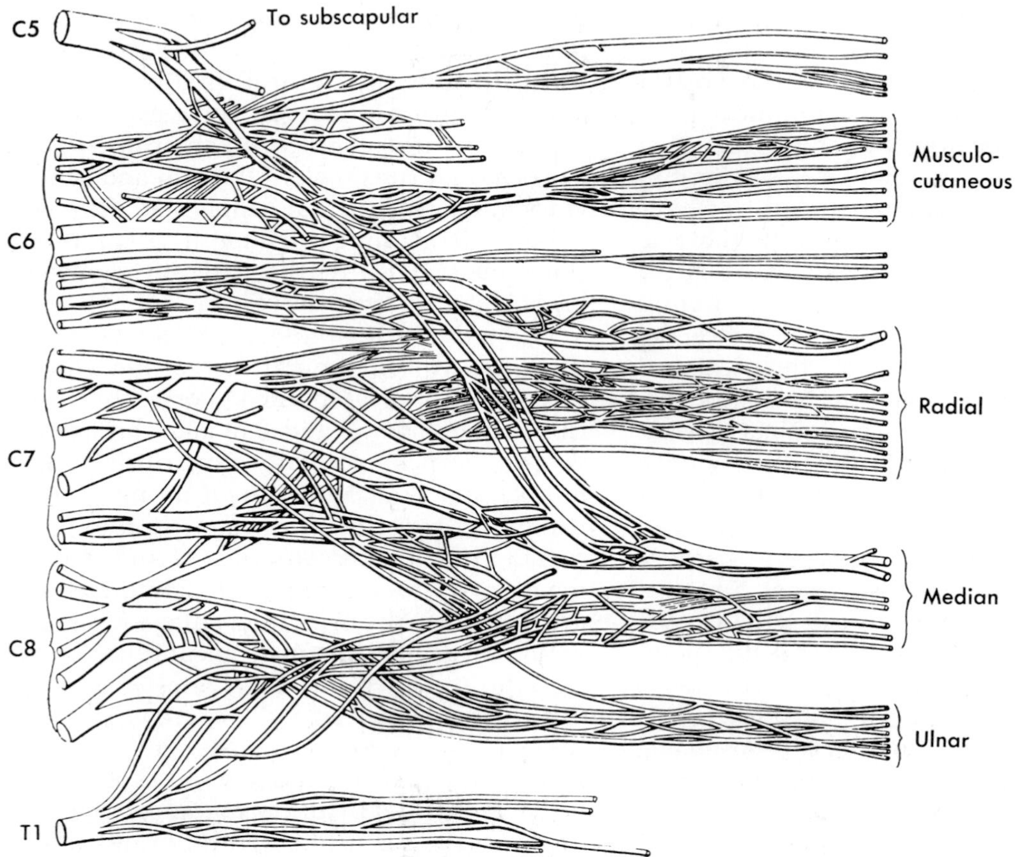


**FIG. 4-6.** Interchange of fiber bundles in a plexus. This is the brachial plexus in which the large fiber bundles have been dissected into smaller ones following maceration of the connective tissue. Note that the fiber bundles arising from the ventral rami of spinal nerves (labeled on the left) join, divide, and rejoin each other in such a fashion that the peripheral nerves (labeled on the right) contain fibers from more than one spinal nerve, and note that it is impossible to follow individual fibers through the plexus. (Kerr AT: Am J Anat 23:285, 1918)

The nerves entering into a plexus lose their identity by interchanging fibers with each other, and the nerves emerging from the plexus, termed the branches of the plexus, do not at all correspond to the nerves entering it; rather, almost every branch of a plexus contains fibers from two or more spinal nerves (Fig. 4-6). The branches of a plexus therefore must be named differently from the spinal nerves contributing to it and are commonly named from their position (for instance, ulnar) or their distribution (for instance, medial antebrachial cutaneous). Where the ventral branches of the spinal nerves come together to form a plexus, much of the interchange of fibers is hidden by the connective tissue that wraps all peripheral nerves; hence, it is often difficult to determine exactly which spinal nerves contribute to which branches of the plexus. Knowledge concerning this has been gained both from careful dissections of plexuses after removal of the surrounding connective tissue and from clinical studies. The composition of the major branches of the plexuses is fairly accurately known, as are the major variations. The matter becomes more difficult when small branches of the plexus are concerned, or when small branches arise from a large branch of the plexus some distance from the plexus.

Since a spinal nerve contributing to a plexus usually has some of its fibers continuing into several branches of the plexus, and a branch of a plexus is typically derived from several spinal nerves, it becomes important to distinguish between the distribution of a spinal nerve (segmental distribution) and that of a branch of a plexus (peripheral nerve distribution). The segmental distribution of the nerves to the back and to the thoracic and abdominal wall is largely identical with the peripheral nerve distribution. Since these branches do not usually enter into plexuses, their segmental distribution can be fairly well determined by simple dissection. In the limbs, in contrast, the branches of the plexuses form the peripheral nerves. In consequence, it is the distribution of the peripheral nerves that is determined by dissection of the limbs (by tracing a branch of a plexus to certain muscles and to a certain area of skin). The segmental innervation of the muscles and the segmental innervation to the skin (dermatomal distribution) must be determined by other means.

## Peripheral Nerve and Segmental (Spinal Nerve) Innervation

As just noted, the distribution of peripheral nerves to both muscle and skin can be determined with some exactness in the dissecting room. Except on the trunk, however, the segmental innervation of muscle and skin is obscured by the plexuses and must be investigated by other methods.

**Muscle.** The **peripheral nerve innervation** of muscle is the easiest of all innervations to determine in the dissecting room, because nerves to muscles are usually large enough to be dissectible with no great difficulty, most of their finer branching occurs within the muscle, and only rarely do branches supplying one muscle continue through to an adjacent muscle. A statement of the peripheral nerve innervation of a muscle is regularly a part of the description of the muscle. Fortunately, since knowledge of segmental innervation of muscle is limited, knowledge of the exact peripheral, rather than segmental, innervation of muscles is commonly of greater clinical use. This is because most injuries to nerves involve some part of their long peripheral course rather than the nerve roots or short segmental parts of the nerves proximal to the plexuses. The peripheral innervation of muscle is constant enough for one to predict with reasonable certainty the paralysis that will follow a lesion (injury) of a given nerve at a given level or to diagnose rather accurately the location of a nerve lesion from the paralysis found on examination of the patient.

The common variations of nerves to muscles are the number of branches that a muscle receives from a given nerve and the level at which these branches leave the nerve. Some muscles regularly receive a single nerve branch, some two or more, but variations in the number of branches received by a single muscle are frequently encountered. This probably means only that in some cases all the fibers to the muscle arise from the nerve so close together that they have a common connective tissue sheath, and branching occurs largely or entirely within the muscle, whereas in other cases what could be branches of a single muscular nerve arise separately from the main nerve stem. The other variation, that of level of origin, is probably also largely a matter of the surrounding epineurium: nerve fibers for a given muscle are usually segregated in the main nerve some distance (from as little as 1 mm to 15 cm or more) above the level at which the fibers leave the nerve, and the external origin of a muscular branch depends on how soon these segregated fibers leave the connective tissue wrapping common to the entire nerve. Awareness of the possibility of variation is necessary in clinical practice. Sometimes, also, knowledge of the specific variations concerning a given muscle or muscle group is useful, although so rarely that it is safe to depend upon reference to the literature rather than memorization.

In contrast to the case with peripheral innervations, determination of the **segmental innervation** of muscles of the limbs by anatomic means is very inaccurate, for it is often largely a matter of exclusion. If, for instance, a given branch of the brachial plexus is known to contain fibers from only the 5th, 6th, and 7th cervical nerves, any muscle innervated by this branch must have a segmental innervation no greater than from C5 to C7, and the approximate segmental innervation of certain muscles can be deduced. However, it does not necessarily follow that every muscle innervated by that branch of the plexus receives fibers from all three spinal nerves; one could be innervated by C5 and C6, another by C6 and C7. Nor can it be stated that the two or more segments supplying a muscle contribute equally to its innervation. Not even an approximation can be made by anatomic means for many muscles. Some nerves contain fibers derived from almost all the spinal nerves entering a plexus, yet any one of their muscular branches usually contains fibers from only two or three of these nerves. In these instances it is impossible, because of the intraneural plexuses in the larger nerves, to trace specific spinal nerve fibers through an appreciable length of nerve. For many muscles, therefore, only clinical methods are available to determine their segmental innervations.

**Skin.** The general **peripheral nerve innervation** of skin can be determined reasonably satisfactorily by dissection, but details cannot because the finer branches of the nerves, running first in the subcutaneous tissue and then in the dense dermis (connective tissue layer) of the skin, necessarily become, at some point, too small to trace. More accurate assessment of the distribution of peripheral nerves to skin therefore must be made by clinical means. When this is done, by mapping the loss of sensation produced by complete interruption of one or several cutaneous nerves, it becomes obvious that a certain amount of overlap exists between any two adjacent nerves. "Overlap" means that two or more nerves exclusively supply an area rather than only one. Further, not all sensations are lost over the same area, so the different functional elements in a nerve do not have exactly the same distribution.

The overlap between nerves varies somewhat with the peripheral nerves under consideration. The general rule for the segmental nerves to the skin of the trunk is that the area of distribution of one nerve is completely overlapped by the nerves lying above and below it in the series. Thus, section of a single intercostal nerve does not result in any area of complete loss of sensation. The upper part of the area of distribution of the sectioned nerve is supplied also by fibers from the nerve above, the lower part by fibers from the nerve below. At most, section of a single intercostal nerve should give an area of diminished sensation, but the change in sensation is usually so slight that the patient is not conscious of it. Typically, interruption of two or more successive nerves is required to produce total loss of sensation of an area on the trunk.

The situation is somewhat different in the limbs, because the overlap between peripheral nerves there is less extensive. Complete interruption of a single peripheral nerve typically produces changes in sensation that are appreciated by the patient. The area of sensory loss is regularly less than is the area of distribution of the nerve determined or estimated by dissection, because of the overlap by adjacent nerves. Thus, successive plotting on the skin of the area of sensory loss resulting after interruption of each one of the nerves to skin of the upper limb separately produces a chart (Fig. 4-7) in which each area of sensory loss is bordered by a more or less wide area to which no single innervation can be assigned. This is the area of overlap between two or more contiguous nerves, where loss of one produces insufficient change in sensation to be detected reliably. Further, because the fibers concerned with appreciation of pain and temperature in two adjacent nerves overlap more than do the fibers for touch, each area of distribution of a peripheral nerve, as determined by loss of sensation after its injury, is divisible into two zones: a central zone in

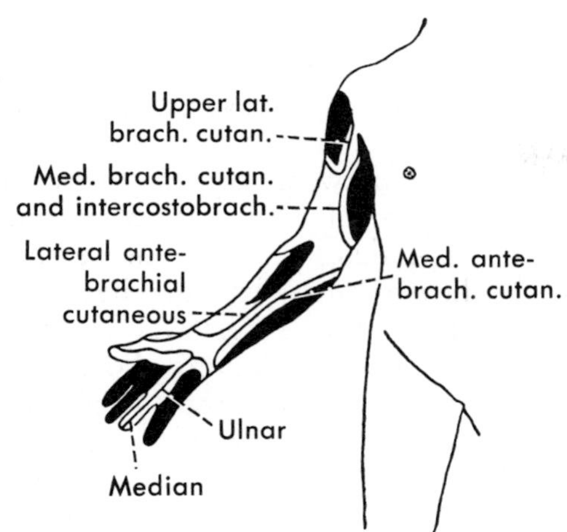


**FIG. 4-7.** Overlap in the distribution of peripheral cutaneous nerves, as indicated by the average sensory loss after separate interruption of the individual cutaneous nerves to the front of the upper limb. The black areas are those of complete sensory loss; the outlined areas adjacent to them are those in which sense of touch is lost but pain and temperature sense are preserved in part, because of the overlap from adjacent nerves; the remaining areas are those of almost total sensory overlap between adjoining nerves. (After Lewandowsky and Foerster. Woltman HW, Kernohan JW: In Baker AB (ed): Clinical Neurology, vol 3. New York, Paul B Hoeber, 1955)

which all cutaneous sensation is lost, and a peripheral zone in which touch is lost but pain and temperature can still be appreciated because this zone is partially supplied by pain and temperature fibers from adjacent nerves. Because of the overlap, a lesion of two or more adjacent nerves necessarily produces a sensory loss over an area greater than the sum of the areas of loss encountered in lesions of these nerves separately.

Although clinical findings thus modify somewhat the conclusions that might be drawn from purely anatomic ones, the latter are a safe guide to what may be expected clinically, provided one remembers that the area of cutaneous sensory loss produced by interruption of a nerve is always less than the area of anatomic distribution of the nerve and that the same peripheral nerve may have a greater or lesser distribution than normal in different people.

**Segmental innervation** of the skin conforms to a **dermatome,** an area of skin supplied by a single spinal (segmental) nerve. Determination of the segmental innervation of the skin of the trunk is a relatively easy matter, since the peripheral nerves are themselves segmental. The several clinical charts of the dermatomes here are in general agreement with each other and with anatomic findings. Determination of dermatomes of the limbs is almost entirely a clinical matter, on the other hand, because of the impossibility of tracing spinal nerve fibers through the plexus, the larger nerves, and their cutaneous branches.

There are two dermatomal charts in use today, and, as will be apparent from a comparison of Figure 4-8 with Figures 4-9 and 4-10, they differ considerably in detail. In Figure 4-8, dermatomes are shown extending as continuous strips from the base of the limb toward or to the most distal part, whereas in Figures 4-9 and 4-10 some dermatomes are shown as being restricted to the distal part of the limb and having no connection with its base. The former has a particularly regular and logical pattern that appeals to anatomists, but clinicians still disagree as to which is the more accurate representation of their findings.

Actually, either chart is compatible with what is known of the embryologic development of the limbs. Since each limb arises as an outgrowth from the body wall, the skin along its original cranial border (thumb or big-toe side) would be expected to be innervated by a higher segmental nerve than the skin along its originally caudal border, and the skin between these two borders on both surfaces might be expected to be innervated in regular sequence by the intermediate nerves entering the limb. Both dermatomal charts shown here follow this principle. Both also indicate that on the original ventral surface, some of the segmental nerves have their distribution entirely in the distal part of the limb, so that along the ventral surface of the proximal part, the nerves originally at the borders of the limb—in the case of the upper limb, C5 or C6 for the upper or preaxial bor-

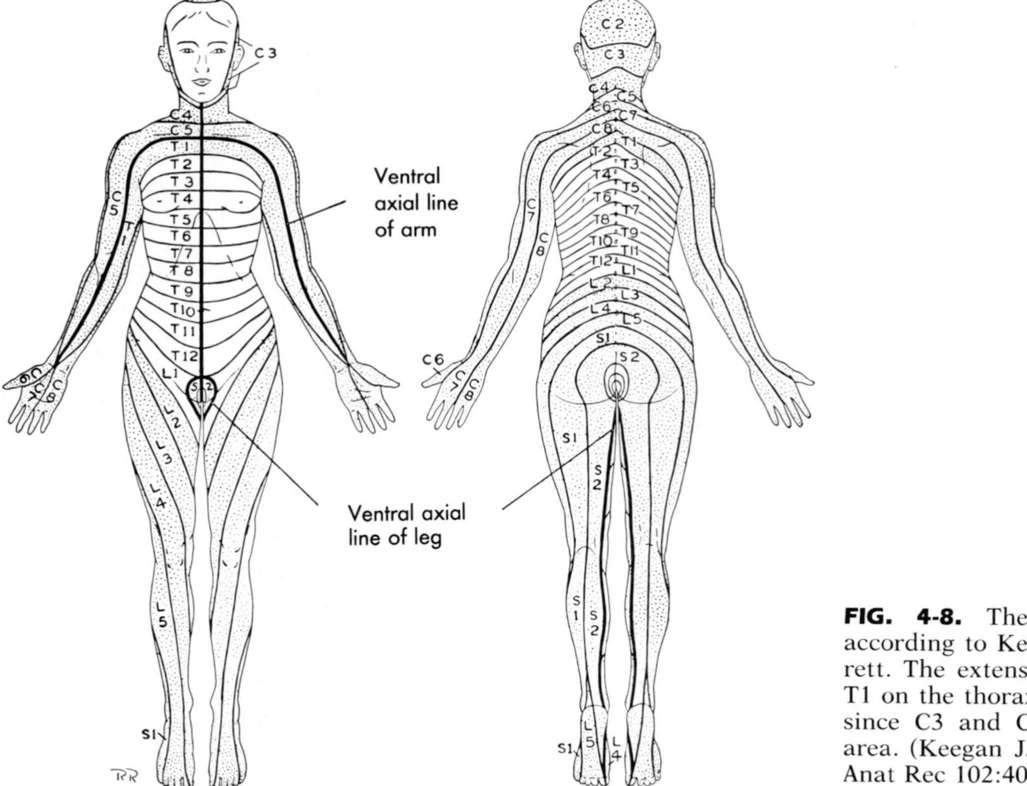


**FIG. 4-8.** The dermatomes according to Keegan and Garrett. The extension of C5 and T1 on the thorax is a mistake, since C3 and C4 supply this area. (Keegan JJ, Garrett FD: Anat Rec 102:409, 1948)

der, T1 or T1 and T2 for the lower—are immediately adjacent to each other. Lines along which originally widely separated dermatomes have met are known as axial lines. All dermatomal schemata show a ventral axial line, but not all agree on the existence of a dorsal one.

## CRANIAL NERVES

The 12 pairs of cranial nerves differ so much from each other that the origin, composition, and distribution of each must be described individually. The present section is only a brief summary of features that should be understood before these nerves are considered in detail. The nerves are not only named but numbered. The names and numbers of the cranial nerves are as follows:

**I** Olfactory
**II** Optic
**III** Oculomotor
**IV** Trochlear
**V** Trigeminal
**VI** Abducens
**VII** Facial
**VIII** Vestibulocochlear
**IX** Glossopharyngeal
**X** Vagus
**XI** Accessory
**XII** Hypoglossal

Common practice allows use of either the name or the number, whichever is more convenient, in referring to these nerves. The 8th nerve, however, is more commonly referred to by its number than name. A good practice in writing, not always adhered to now, is to use Roman numerals for the cranial nerves, thus clearly distinguishing them from spinal nerves.

All the functional elements of spinal nerves are present in the 12 cranial nerves, although not in each nerve. These are general somatic afferent and general visceral afferent fibers, distributed to skin or other somatic

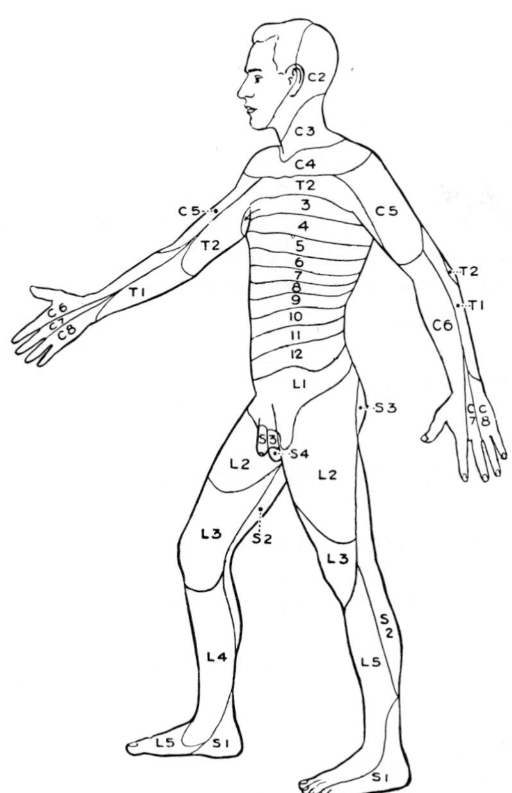


**FIG. 4-9.** Anterolateral view of the dermatomes according to Foerster. (After Foerster, Haymaker W, Woodhall B: Peripheral Nerve Injuries, 2nd ed. Philadelphia, WB Saunders, 1953)

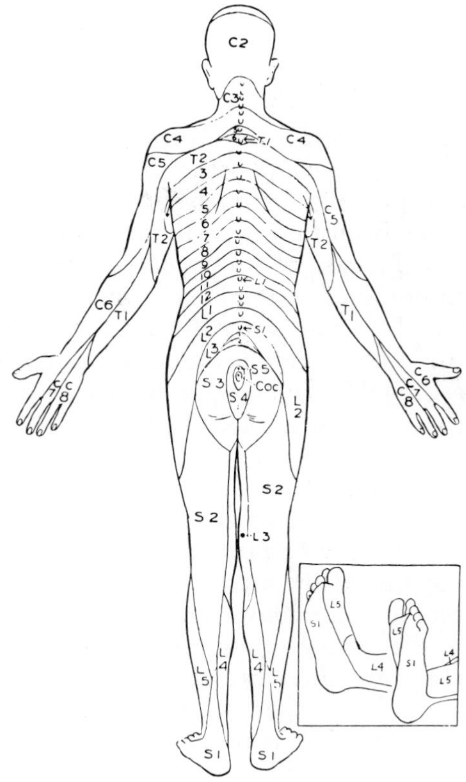


**FIG. 4-10.** Posterior view of the dermatomes according to Foerster. (After Foerster. Haymaker W, Woodhall B: Peripheral Nerve Injuries, 2nd ed. Philadelphia, WB Saunders, 1953)

structures and to the viscera, respectively, and somatic motor and general visceral motor (autonomic) fibers, distributed respectively to skeletal muscle derived from somites and to smooth muscle or glands by way of autonomic ganglia. Cranial nerves also contain elements that are not found in spinal nerves. Only in the head and neck is there muscle that develops from the wall of the pharynx (branchiomeric muscle, p. 102), and the cranial nerves originally connected with the wall of the pharynx are therefore the only nerves containing voluntary motor fibers to this type of voluntary muscle. [Branchiomeric muscles are not distinguishable histologically from the usual type of skeletal muscle (somatic), but their nerve fibers are sometimes called special visceral motor because the muscle they supply is not the usual smooth muscle that develops from the wall of the gut.] Further, some cranial nerves serve types of sensory organs that occur only in the head, and the fibers of these nerves are usually known as special sensory ones. If they are concerned primarily with the gut, as taste fibers are, for instance, they are known as special visceral sensory (afferent) fibers, but if they are connected with the body as a whole, as are the fibers from the eye, they are known as special somatic sensory fibers. No single cranial nerve contains all seven types of fibers that are found in all 12 of these nerves combined.

Although they differ from spinal nerves in the specific types of fibers that they contain and are connected to the brain rather than to the spinal cord, most of the cranial nerves are essentially similar in structure to spinal nerves. The cell bodies for the motor fibers of cranial nerves, whether voluntary motor or autonomic, are located within the central nervous system; the sensory cell bodies form sensory ganglia that are located on the nerve outside the central nervous system; and the connective tissue and other structures of a peripheral part of a cranial nerve are indistinguishable from that of a spinal nerve.

As in the case of spinal nerves, the voluntary motor fibers extend to the voluntary muscle that they innervate; the autonomic fibers synapse in autonomic ganglia rather than proceeding directly to smooth muscle or other effectors; and the afferent fibers, as they enter the nervous system, tend to branch and make multiple connections with cells there. Perhaps the biggest difference is in terminology. Cranial nerves may be purely motor, purely sensory, or mixed; in mixed nerves the sensory and motor fibers may or may not join proximal to the sensory ganglion. Dorsal and ventral roots are therefore not seen, although a nerve may have separate sensory and motor roots that are the equivalent of those of the spinal nerves. Similarly, the ganglia of the cranial nerves are the equivalent of the spinal or dorsal root ganglia but are individually named. The sensory fibers between the ganglion and the brain will not regenerate functionally after they have been interrupted, just as dorsal root fibers will not. Also, cranial nerves may send to or receive from certain autonomic ganglia branches that are comparable to the rami communicantes that unite spinal nerves to other autonomic ganglia.

The individual cranial nerves are briefly summarized in Chapter 32.

## THE AUTONOMIC SYSTEM

Reference to the autonomic nervous system has necessarily been made in preceding discussions, for some of the cells of this system lie within the central nervous system and send their fibers through the ventral roots of many spinal nerves and the motor parts of certain cranial nerves. It should be clear, therefore, that although the autonomic nervous system is usually defined as a distinct portion of the peripheral nervous system, it is anatomically distinct only in part and, like

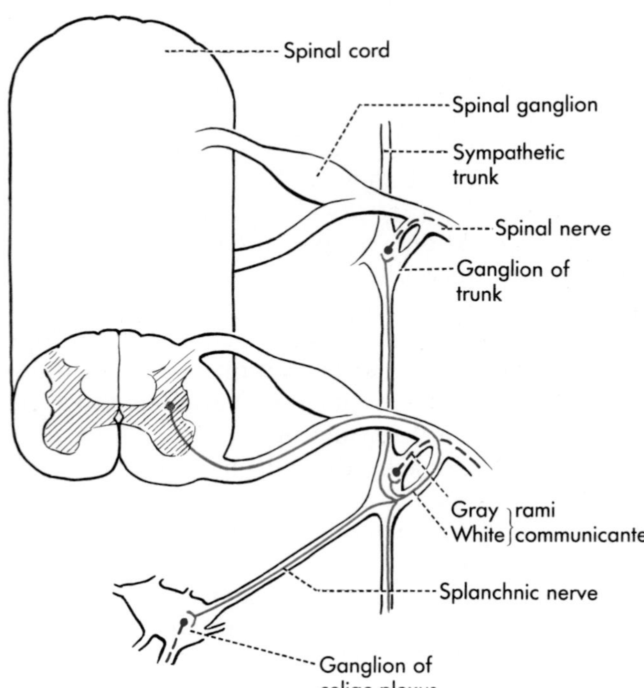


**FIG. 4-11.** Diagram of a short section of the sympathetic nervous system. Preganglionic fibers (represented by the solid red line) are shown leaving the spinal cord by way of the ventral root of a nerve, entering the sympathetic trunk through the white ramus communicans, and thereafter taking various courses, including leaving the trunk by way of a splanchnic nerve to make connection with a more peripherally placed ganglion. Postganglionic fibers are shown as blue lines. The preganglionic fiber does not branch in the ganglion of the sympathetic trunk; rather, there are separate preganglionic axons that synapse, ascend, descend, or form a splanchnic nerve.

other portions of the peripheral nervous system, lies in part within the central nervous system. It is really a functional rather than an anatomic system.

In the central nervous system, the autonomic system is played upon by both **somatic** and **visceral afferent fibers** (as is the voluntary motor part of the nervous system), and the synapses concerned are located within the central nervous system. Also, the cells of the autonomic system that send their fibers out through cranial or spinal nerves are played upon by certain higher centers, just as are the voluntary motor cells of the brain and spinal cord. It is customary to speak of autonomic centers and pathways within the central nervous system—meaning, by this, centers and pathways that act primarily upon and through the autonomic system.

The one anatomic feature that most clearly distinguishes the autonomic system from the peripheral voluntary motor system (to voluntary muscle) is that autonomic fibers that leave the central nervous system do not terminate on an end organ as do voluntary motor fibers, but terminate in **autonomic ganglia** (Figs. 4-11 and 4-12). The ganglion cells then give rise to fibers that proceed to the end organ. Thus, two neurons are regularly required to transmit a nerve impulse from the central

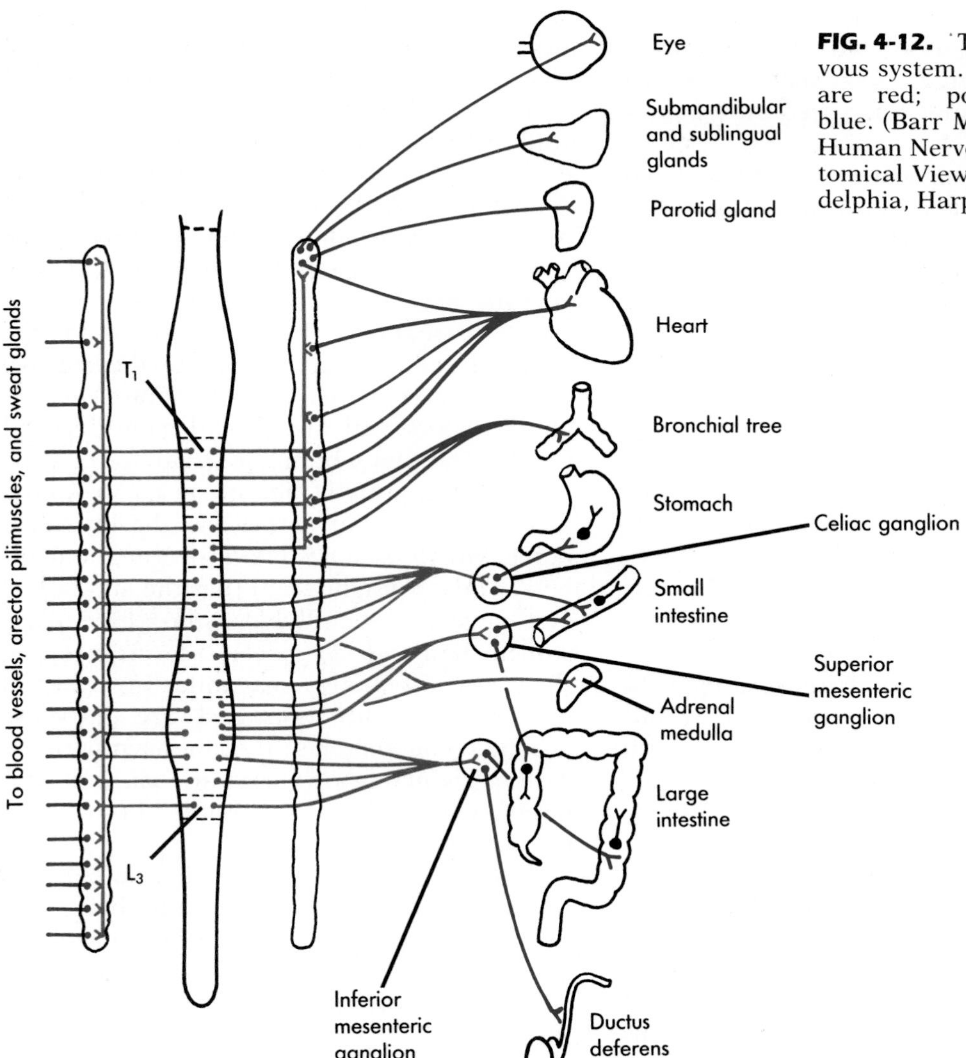


**FIG. 4-12.** The sympathetic nervous system. Preganglionic fibers are red; postganglionic fibers blue. (Barr ML, Kiernan JA: The Human Nervous System: An Anatomical Viewpoint, 4th ed. Philadelphia, Harper & Row, 1983)

nervous system to an end organ over the autonomic system, instead of the single neuron employed in the case of skeletal muscle.

Of the two neurons transmitting an autonomic impulse, one has its cell body located within the central nervous system and sends its axon through the ventral root of a spinal nerve, or the motor portion of a cranial nerve, to an autonomic ganglion; the second has its cell body in an autonomic ganglion and sends its axon to the end organ. The neuron whose cell body is in the central nervous system is known as a **preganglionic neuron,** and its axon is a preganglionic axon until it ends in an autonomic ganglion, regardless of how many ganglia it may traverse before it ends. Similarly, the neuron with its cell body located in a ganglion is known as a **postganglionic neuron,** and it is postganglionic axons that reach end organs innervated by the autonomic system.

The autonomic nervous system differs from the voluntary motor system in that it supplies smooth muscle, cardiac muscle, and certain glands, structures over which we ordinarily exercise no volitional control. Older names for the autonomic nervous system, such as the visceral nervous system, the involuntary nervous system, or the vegetative nervous system, also convey this idea of lack of volitional control. Control of the activity of the bladder and of the bowel, both of which have smooth muscle in their walls and are therefore innervated by the autonomic system, is usually said to be an exception to the rule; however, there is increasing evidence that voluntary control of these organs depends not upon direct control of the smooth muscle by volitional control of the autonomic system, but upon control of skeletal muscle that in turn affects the emptying or retention of contents by these organs.

## Subdivisions

Two fundamental parts of the autonomic system are described. These differ in certain aspects of both their anatomy and physiology but are similar in that both are composed of preganglionic and postganglionic neurons and control involuntary structures in the body. The **sympathetic system** consists of all the preganglionic neurons that send their fibers through the ventral roots of the thoracic and upper lumbar nerves, and of the ganglia (postganglionic neurons) and their branches with which these preganglionic axons make connection (Fig. 4-12). The **parasympathetic system** consists of all the preganglionic neurons that send their fibers through cranial nerves or through sacral nerves and the postganglionic neurons with which these preganglionic neurons are connected (Fig. 4-13).

If one examines the roots of all the nerves, cranial and spinal, for the presence of preganglionic fibers, it becomes obvious that these fibers emerge only through certain nerves and can be classified according to their level of emergence: a group of preganglionic fibers that emerge through the 3rd, 7th, 9th, and 10th cranial nerves; a group that emerge through the ventral roots of all the thoracic nerves and the first two or three lumbar nerves; and a group that emerge through the ventral roots of the 2nd and 3rd, or 2nd, 3rd, and 4th sacral nerves. The cranial group (outflow or stream) is thus separated from the thoracolumbar group not only by the last two cranial nerves, which do not contain preganglionic fibers, but also by all eight cervical nerves. The thoracolumbar stream or outflow is usually separated from the sacral one by the lower three lumbar nerves and the 1st sacral nerve, which do not contain preganglionic fibers. Thus, it might seem that the autonomic system should be divided into three parts rather than two. However, the cranial and the sacral parts are very similar in regard to anatomy and function and are therefore grouped together as the parasympathetic system, which shows certain constant differences from the sympathetic (thoracolumbar) system.

A clinically important distinction between the sympathetic and parasympathetic systems is that they differ in their pharmacology. Certain drugs will produce the effects of parasympathetic stimulation, others will produce the effects of sympathetic stimulation, and

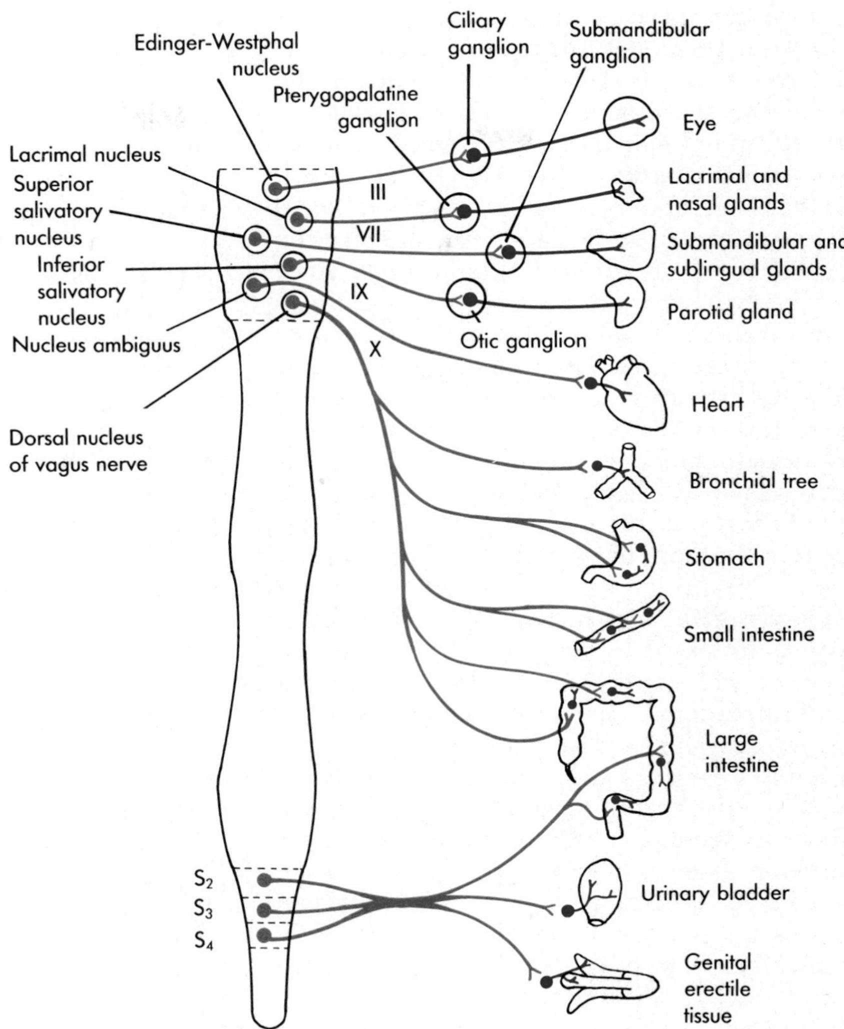


**FIG. 4-13.** The parasympathetic nervous system. Preganglionic fibers are red; postganglionic fibers blue. (Barr ML, Kiernan JA: The Human Nervous System: An Anatomical Viewpoint, 4th ed. Philadelphia, Harper & Row, 1983)

still others will paralyze differentially one of the two systems. These different effects depend on differences in the nature of sympathetic and parasympathetic fibers: although apparently all preganglionic fibers, sympathetic and parasympathetic, are cholinergic, almost all sympathetic postganglionic fibers (those to most sweat glands and some blood vessels being the exception) are adrenergic; postganglionic parasympathetic fibers, in contrast, are cholinergic.

## Sympathetic System

The sympathetic nervous system is, in some ways, more complicated and is more widely distributed than the parasympathetic nervous system, but it is, at the same time, more nearly segmental and therefore lends itself more easily to a general rather than a detailed description.

**White Rami Communicantes.** The cell bodies of the preganglionic neurons of the sympathetic system are located in the spinal cord at the level of origin of all the thoracic and of the first two lumbar nerves. Occasionally they send fibers through the 8th cervical nerve or through the 3rd lumbar nerve. (The cell bodies form a cell group, the intermediolateral, in the spinal cord.) The prepanglionic axons traverse the ventral roots of the nerves of which they are a part, never the dorsal roots, and become a component of the mixed nerve at the level at which dorsal and ventral roots join. They run a short distance farther peripherally, into the ventral branch of the nerve, and then, joined by sensory fibers that are going to the same viscera, leave the spinal nerves as rami communicantes. Since most of the preganglionic autonomic fibers are myelinated, as are many of the sensory fibers to the viscera, a ramus communicans leaving the spinal nerve is called a white ramus (Fig. 4-11). The white rami communicantes are relatively short, because soon after they leave the spinal nerves they reach the sympathetic trunk, located on the lateral aspect of the front of the vertebral column.

**Sympathetic Trunks.** The sympathetic trunks (often called sympathetic chains) are paired, each consisting of a series of ganglia (accumulations of nerve cells) connected by intervening fibers. The ganglia appear somewhat like beads strung some distance apart on a string. Each trunk extends from the level of the 1st cervical vertebra to the tip of the sacrum. They are never strictly segmental, one for each spinal nerve, in the adult. Instead of eight cervical ganglia, there are typically only three, sometimes four. There are more commonly 11 thoracic ganglia rather than 12. There may be five or more lumbar ganglia, but four is most common, and there are commonly less than five ganglia in the sacrococcygeal region.

Since the preganglionic fibers of the sympathetic system usually leave the spinal cord only through the 12 thoracic and the first two lumbar nerves, these nerves alone have white rami communicantes connecting them to the sympathetic trunk; white rami communicantes, therefore, join only the thoracic and upper lumbar portions of the trunk. However, the preganglionic fibers in the rami communicantes do not necessarily stop in the first ganglion of the trunk that they reach. In fact, at most levels, the majority do not stop but take alternative courses, turning up the trunk to end in ganglia above the level they entered, turning down the trunk to end in ganglia below this level, or simply traversing the trunk and proceeding forward toward the viscera that they ultimately innervate (Figs. 4-11 and 4-14). The parts of the trunk between ganglia are formed by the nerve fibers, preganglionic and accompanying afferent ones, that ascend or descend in the trunk. Above the uppermost white ramus, in the cervical part of the trunk, the fibers are almost entirely ascending. Below the lowest white ramus, in the lower lumbar and sacral parts of the trunk, they are descending. In between, the trunk contains

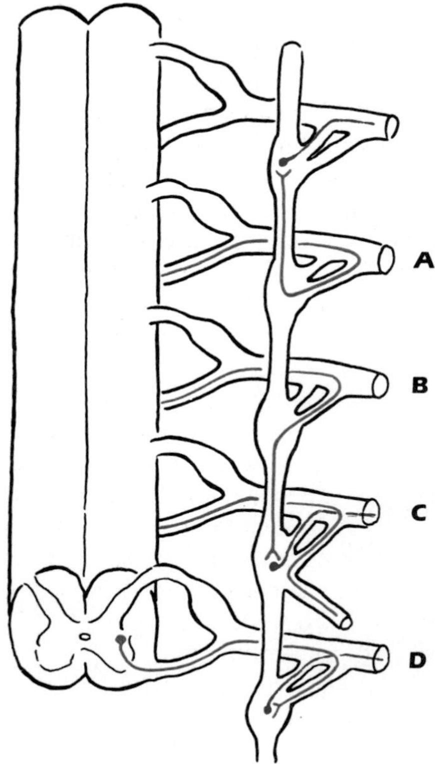


**FIG. 4-14.** Various courses that preganglionic fibers (solid red lines) may take through the thoracic sympathetic trunks. Blue lines represent postganglionic fibers. In **A** the preganglionic fiber ascends to synapse at a higher level in the trunk; in **B** it descends to a lower level; in **C** it traverses the trunk, leaving it as a visceral (splanchnic) branch; and in **D** it ends in the ganglion at its own segmental level. The preganglionic fibers in any one nerve may do all of these.

fibers going in both directions. It is, of course, only through such a spread of preganglionic fibers from the thoracic and upper lumbar regions that the cervical, the lower lumbar, and the sacral ganglia can receive impulses from the central nervous system, since they receive no white rami communicantes.

The sympathetic ganglia are dependent for their function upon their synaptic relation to the preganglionic fibers. They can be cut off from their connections with the central nervous system, hence, paralyzed, by cutting those particular thoracic or lumbar white rami communicantes that are known to contribute to them, by transecting the sympathetic trunk just above the level of the 1st thoracic nerve (the highest normal level of entrance of sympathetic fibers into the trunk) in the case of the cervical ganglia, or by transecting the lumbar trunk a little below the second or third lumbar level (the lowest level of entrance of sympathetic fibers into the trunk) in the case of the lower lumbar and sacral ganglia. (To prevent regeneration, a part of the trunk is customarily removed—an operation called "chain ganglionectomy"—instead of simply cutting nerve fibers.)

**Gray Rami Communicantes.** The postganglionic neurons whose cell bodies form the ganglia of the sympathetic trunk typically send their axons back to the spinal nerves. These axons form gray rami communicantes, so called because the axons are for the most part unmyelinated. Each spinal nerve receives at least one gray ramus communicans, in contrast to the limited occurrence of white rami, and therefore the peripheral branches of all spinal nerves contain sympathetic fibers. Once the postganglionic fibers from the sympathetic trunk have entered a spinal nerve by way of a gray ramus communicans, they mingle with the voluntary motor and afferent fibers of the nerve and are distributed through the branches of the nerve to blood vessels in the skin or elsewhere and to the smooth muscle and sweat glands of the skin.

Through the gray rami, postganglionic fibers reach the neck, body wall, and limbs. Since the sympathetic trunk does not extend into the head or come in close contact with the cranial nerves, the fibers to the head arise from the uppermost ganglion (superior cervical) of the trunk and largely follow arteries to the head. These branches (plexuses) along the arteries are the functional equivalent of gray rami.

**Splanchnic (Visceral) Nerves and Ganglia.** As noted earlier, a preganglionic fiber reaching the sympathetic trunk by means of a white ramus need not necessarily terminate within the trunk but may instead pass through it and proceed toward the viscera. Many preganglionic fibers do merely pass through the thoracic and lumbar portions of the trunk and form nerves that are generally known as visceral branches or as splanchnic nerves. These consist primarily of preganglionic fibers that have passed through the trunk without synapse and of accompanying afferent fibers. Many also contain a small proportion of postganglionic fibers, derived from the cells in the ganglia of the trunk—although most of these cells send their fibers to the body wall and limbs, not all of them do so—and a few contain mostly postganglionic fibers.

The preganglionic fibers of the visceral or splanchnic nerves do not end directly upon the organ that they innervate, for the law of structure of the autonomic system is that two neurons, a preganglionic and a postganglionic, must be involved. The largest splanchnic branches, for instance, reach the abdomen, and there the branches from the two sides form heavy nerve plexuses along the abdominal aorta (the chief artery in the abdomen) and especially around the bases of the major branches of that vessel. These plexuses, the best known of which is probably the celiac plexus, contain ganglion cells, and it is on these ganglion cells that the preganglionic fibers end. The postganglionic fibers arising from these ganglion cells tend to follow blood vessels to the viscera, where they end on the smooth muscle and perhaps the glands of these viscera.

In the celiac plexus, the cell bodies of the

postganglionic nerve cells are grouped together into large ganglia, but in some of the other plexuses, the ganglia are small and scattered. All ganglia, whether grossly identifiable or not, are often grouped together as the prevertebral or collateral ganglia, since they tend to lie in front of the vertebral column. A general rule, to which there are exceptions, is that a preganglionic sympathetic fiber controlling smooth muscle in the body wall or limb synapses in the sympathetic trunk, whereas a preganglionic sympathetic fiber controlling smooth muscle of the viscera passes through the sympathetic trunk to synapse in a prevertebral ganglion.

**Summary of Anatomy of Sympathetic System.** An understanding of the fundamental anatomy of the sympathetic system is absolutely essential if the rather complex and detailed anatomy encountered in the dissecting room, and the rationale of surgical procedures on this system, is to be appreciated. A brief summary of the features may be helpful.

Preganglionic fibers of the sympathetic system leave the spinal cord through the ventral roots of all 12 thoracic and of the upper 2 lumbar nerves and, in turn, leave these nerves as white rami communicantes to join the sympathetic trunk. Some preganglionic fibers end within the trunk, many running up or down within it for some distance. Others simply traverse the trunk and form visceral or splanchnic nerves. The postganglionic neurons whose cell bodies form the ganglia of the trunk send their axons, for the most part, into all the spinal nerves, forming gray rami communicantes through which they are distributed to visceral structures of the body wall and limbs (Fig. 4-14) or, from the upper end of the trunks, to the head along blood vessels. The preganglionic fibers that traverse the trunk and leave it as visceral branches or splanchnic nerves end in collateral ganglia situated in general in plexuses about the great vessels. The postganglionic fibers derived from these ganglia then follow blood vessels to the various viscera and end there on smooth muscle or glands.

### Parasympathetic System

The fundamentals of organization of the parasympathetic nervous system can be much more briefly described. As is true of the sympathetic nervous system, the parasympathetic system is organized upon the basis of preganglionic and postganglionic neurons. The preganglionic neurons constitute parts of the nuclei of certain cranial nerves or lie in the sacral segments of the cord. Their axons emerge either as a component of a cranial nerve or as a part of the ventral root of a sacral nerve. These preganglionic axons usually leave the nerve with which they are associated and end in ganglia composed of postganglionic neurons. (They are not described as forming rami communicantes, since this term is applied mostly to the sympathetic system).

In contrast to the almost regular spacing of the ganglia of the sympathetic trunk, the ganglia of the parasympathetic system are widely scattered and irregularly spaced, and many of them are of microscopic size. They also differ from sympathetic ganglia by being located very close to the organs that they innervate, or even within the walls of these organs.

The four pairs of **cranial parasympathetic ganglia** that innervate structures in the head (Fig. 4-13) are located very close to the organs that they innervate. Certain smooth muscle in the eye is innervated by a ganglion lying just behind the eyeball. The salivary gland in front of the ear is innervated by a ganglion lying just deep to it, and the two salivary glands lying close to the lower border of the lower jaw are innervated by a ganglion that also lies in this position. The mucous membrane and blood vessels of the nose are innervated by a ganglion lying just outside the lateral wall of the nose. This is in marked contrast to the sympathetic innervation of these structures, which is from a ganglion situated not in the head but in the upper part of the neck.

In similar fashion, the small ganglia that innervate the heart and lungs lie immediately adjacent to, on, and in those organs. Most of the ganglia that innervate the digestive tract lie in the walls of that tract, where they are called collectively **enteric ganglia** or, sometimes, terminal ganglia. The cells that innervate the urinary bladder lie very close to or in that organ.

The cranial ganglia are associated with the cranial part of the parasympathetic outflow; one receives its preganglionic fibers from the 3rd cranial nerve, two receive theirs from the 7th, and one receives its from the 9th, as described in more detail in Chapter 32. The ganglia supplying the heart and lungs and most of the enteric ganglia also receive their preganglionic fibers from the cranial outflow. The nerve that supplies them is the 10th cranial nerve or vagus (wandering) nerve, so called because it, unlike other cranial nerves, extends into the thorax and abdomen. When the vagus reaches the abdomen, its fibers mingle with sympathetic fibers that are also going to the digestive tract. It is not known exactly how far the vagus extends caudally. However, the enteric ganglia of a lower part of the digestive tract and the ganglia supplying the urinary bladder receive their preganglionic fibers from the sacral parasympathetic outflow, fibers in about the 2nd to 4th sacral nerves, which leave these nerves to run to the bowel and bladder.

The parasympathetic system does not, in general, employ the same nerves or the same ganglia as the sympathetic system and is for the most part an anatomically discrete system. Only where the two systems converge on their way to the same part do their fibers become mixed.

**Resumé of Distribution.** Since the parasympathetic nervous system does not send fibers back to spinal nerves, its distribution is limited. In contrast to the sympathetic system, which is distributed to all major parts of the body, the distribution of the parasympathetic system is confined to the head and neck and to some of the viscera of the trunk.

## Functions of Autonomic System

Volumes have been written concerning the functions of the sympathetic and parasympathetic systems and the type and extent of control that they exert over specific organs. Although both systems control involuntary structures (smooth and cardiac muscle, glands), they have different effects that can best be brought out by contrasting them.

**Contrast of Sympathetic and Parasympathetic.** The most succinct summary of the functions of the sympathetic and parasympathetic systems is that the sympathetic system is primarily an emergency one, preparing the body so that the individual can flee or fight, whichever seems wisest, when faced with danger; the parasympathetic system, in contrast, is primarily a homeostatic one, tending to promote quiet and orderly processes of the body.

The **sympathetic system,** for example, dilates the pupil of the eye, thus providing a wider field of vision and a better opportunity to see a dangerous object approaching from the side. It speeds up the rate and increases the strength of the heart beat. It constricts the blood vessels in the skin and those of the digestive tract, allowing blood to be diverted from these structures for circulation to the voluntary muscles. It tends also to slow down the peristaltic action of the digestive tract, a function appropriate to the decreased circulation to that part. Through its action on the suprarenal (adrenal) medulla, the sympathetic system also produces the liberation of epinephrine (adrenaline) into the bloodstream. This epinephrine has, in general, the same effect on smooth muscle and cardiac muscle as does stimulation by sympathetic fibers.

The **parasympathetic system,** on the other hand, constricts the pupil of the eye, protecting the retina by preventing excess light from reaching it. It also adjusts the lens so that near objects are seen clearly (accommodation). The parasympathetic system decreases the rate and strength of the heart beat, conserving energy, and it acts on the digestive tract to promote the orderly processes of digestion, governing, at least in part, the secretion of digestive juices and tending to increase peristalsis. Finally, the parasympathetic system is responsible for the autonomic phase of emptying the urinary bladder and the rectum, ridding the body of wastes.

From this it may be gathered that the sympathetic and parasympathetic nervous systems often have somewhat opposing actions. Indeed, their different functions afford a reason, in addition to the differences in anatomy, for subdividing the autonomic system into these two parts. However, it should not be concluded that every organ has both a sympathetic and a parasympathetic innervation and that the two systems are constantly competing with each other. Although many phases of autonomic activity are not yet clearly understood, a considerable mass of smooth muscle in the body is certainly not innervated by both systems, and in those organs that are, there is a coordination of activity rather than competition.

**Differences in Distribution.** Most blood vessels in the limbs, body wall, and digestive tract do not have a parasympathetic innervation. Although constricted by sympathetic activity, they are either dilated passively by the pressure of the blood as the contraction of their muscle is eased or, in some instances apparently, dilated actively by less numerous vasodilator fibers of the sympathetic system; the parasympathetic system produces vasodilation primarily in the head and in the genital organs. A few abdominal viscera apparently have only a sympathetic innervation, and whether most of the smooth muscle of the urinary bladder has a sympathetic innervation, as well as a parasympathetic one, has long been subject to argument.

In such instances as these cited, with the possible exception of the last, there obviously can be no competition between the two systems. The degree of contraction or relaxation of the smooth muscle concerned would have to depend purely upon the degree of activity, or lack of activity, in the single system supplying it.

**Coordinated Action.** Control of the size of the pupil of the eye is probably one of the clearest examples of the differing but coordinated actions of the two systems on a single part. The parasympathetic system constricts the pupil by stimulating a circular band of smooth muscle, the sphincter pupillae, to contract. The sympathetic system dilates the pupil by stimulating a different muscle, the dilator pupillae, to contract. It should be noted that the two systems are described as acting on different smooth muscles. Actually, although these muscles have opposing ("antagonistic") actions, they no more compete in adjusting the size of the pupil than two antagonistic muscles at a joint compete in moving the joint. Rather, as one muscle contracts, the other smoothly relaxes.

The heart definitely has a double innervation, sympathetic and parasympathetic, but again these systems do not normally compete with each other. In most mammals, depriving the heart of its parasympathetic innervation produces an increase in the rate and strength of the heart beat, but depriving it of its sympathetic innervation produces no immediate change, indicating that under normal circumstances the parasympathetic system tends to conserve the resources of the body by maintaining cardiac contraction at the lowest effective level. The sympathetic system takes over only in an "emergency" (which may be no more than standing up suddenly from a chair or walking or running up a flight of stairs). At the same time that the sympathetic system takes over, the parasympathetic system ceases to control the heart.

## THE CENTRAL NERVOUS SYSTEM

The spinal cord and the brain compose the central nervous system (Fig. 4-15). Between them, they give rise to all the motor fibers in the spinal and cranial nerves and receive all the sensory ones. They consist of enormous numbers of neurons. One cerebral cortex, only a part of one side of the brain, has been estimated to contain about 7 billion nerve cells. The nerve cell bodies are usually grouped together into masses that are collectively called gray matter, whereas the longer fibers tend

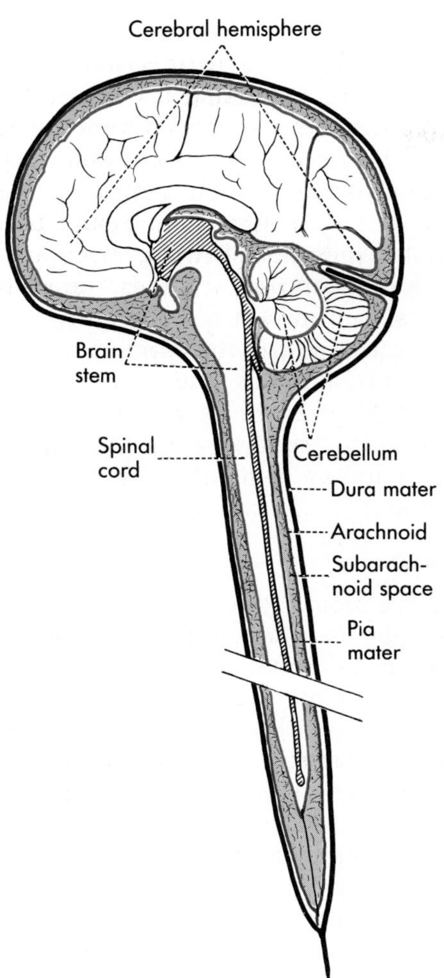


**FIG. 4-15.** Diagram of the central nervous system and its meninges in a sagittal section. The spinal cord is much foreshortened.

to be grouped as the white matter. The white matter of the central nervous system, composed of axons, serves to connect various groups of gray matter and therefore consists of fibers running in many different directions. A general rule is that there are two-way connections between two connected masses of gray matter, so that the activity of each group influences that of the other.

## Spinal Cord

The spinal cord (medulla spinalis, or "marrow of the spine") is the simplest part of the central nervous system and can be thought of as a modified tube with enormously thickened walls. The lumen of the tube is called the central canal. The gray matter in its thick walls is arranged around the central canal, and the white matter largely surrounds the gray matter.

The spinal cord is discussed in more detail in the section on the back, and only a few general aspects need be mentioned here. In cross section (Fig. 4-4), the uneven grouping of the gray matter around the central canal is obvious. The projections on each side of the midline are known as the **posterior** and **anterior horns.** In general, the posterior horns receive the incoming sensory fibers composing the dorsal roots of the spinal nerves and help to disperse their nerve impulses to other parts of the cord and to the brain. In the anterior horns are large cells that give rise to the fibers in the ventral roots of spinal nerves that go to skeletal muscle; thus, the anterior horn is largely motor to such muscle. In some parts of the cord, there is a small **lateral horn** that contains the cell bodies of the preganglionic autonomic neurons that also leave through ventral roots.

The white matter is composed of fibers that run between parts of the spinal cord, from the spinal cord to parts of the brain, or from parts of the brain to the gray matter of the cord. Through it, therefore, impulses entering over spinal nerves are disseminated as necessary through the spinal cord and forwarded to the higher levels of activity in the brain. Also through it various parts of the brain influence the activity of the voluntary motor and autonomic cells located in the spinal cord. Fibers that have the same function tend to be grouped together in the spinal cord, and such groupings, although not necessarily visible ones, constitute the tracts of the cord. The major tracts are sensory, ascending toward the brain, and motor, descending to various levels of the spinal cord.

## Brain

The brain (encephalon meaning in the head), the enlarged upper end of the central nervous system lodged within the skull, is enormously complicated. Except for general features of its gross anatomy, it is properly studied in the neuroanatomy laboratory and described in neuroanatomy texts. A further brief discussion occurs in this text in the section on the head and neck.

A part of the brain, the **brain stem,** is a direct upward continuation of the spinal cord and, although somewhat more complex than the spinal cord, resembles it in general. In the brain stem, gray and white matter are more intermixed than in the spinal cord. In some places, the surface is formed of white matter as it is in the cord, and in others, gray matter comes to the surface. The gray matter that corresponds most closely to the horns of the spinal cord tends to be fairly close to the modified central canal (ventricles where it is expanded) of the brain stem. Much of this gray matter sends motor fibers into, or receives sensory fibers from, cranial nerves through cell groups that are called the nuclei of the cranial nerves. The white matter of the brain stem is, like that of the cord, largely collected into tracts. Some of these tracts are continuous with sensory and motor tracts in the spinal cord, whereas others originate from gray matter at the brain stem level, or end there.

Surmounting the brain stem, and originating as outgrowths from it, are the **cerebellum** (little brain) and the **cerebrum** (brain). The cerebrum is composed of paired cerebral hemispheres. These are distinguished from the brain stem by their size and by their structure. Although both contain deep-lying masses of gray matter, nuclei, each also has an extensive convoluted surface composed of gray matter, known as **pallium** (cloak or mantle) or **cortex** (bark, therefore outer layer). Here nerve cells are spread out in a relatively thin layer that allows exit and ingress of the incredibly large number of fibers that form the underlying white matter. Through the white matter, the cortices of both the cerebellum and the cerebrum are connected to lower-lying centers of the brain stem and spinal cord and to each other. Various parts of the cerebral cortex of the same and opposite sides also have extensive connections with each other.

The cerebellum is particularly concerned with helping to guide, through lower centers, the activity of voluntary muscle. The cerebrum is literally "the brain" in the colloquial sense of the word and is concerned not only with voluntary muscle (it initiates and helps control voluntary movements) but also with sensations, judgments, interpretations, emotions, and all those activities grouped together as mental. Most sensory impulses to the cerebrum cross to the opposite side before they reach it, and most impulses to voluntary muscle cross after they leave it, so that each cerebral cortex mediates, for the most part, sensation and movement on the opposite side of the body.

## Reflexes

Most of the motor activity induced by the nervous system is reflex activity. A reflex is an involuntary or unwilled reaction to a stimulus. As such, it may be very simple, involving only a small segment of the central nervous system, or it may be exceedingly complicated. In general, the simplest reflexes occur through the spinal nerves and the spinal cord, although the same basic types of reflexes occur through cranial nerves and the brain stem. Far more complex reflexes also occur through the brain, however, and those at the level of the cerebral cortex may be difficult to distinguish from voluntary activity.

**Spinal Reflexes.** Although the motor cells in the lateral and anterior horns of the spinal cord are greatly influenced by the motor pathways derived from higher centers, the basic initiation and control of their activity is local. This is accomplished through afferent impulses brought in by the sensory divisions of the spinal nerves. These impulses reach the motor cells either directly or after synapse with intercalated (connecting) neurons in the gray matter of the cord. The basic function of the nervous system is to connect a receptor to an effector organ, and in man, the simplest pathway from a sense organ to a motor one is through this connection between sensory and motor fibers of the nerves. A response that involves only the spinal nerves and their connections in the cord, rather than pathways to centers in the brain, is known as a spinal reflex.

Perhaps one of the simplest spinal reflexes is the tendon- or muscle-stretch reflex, a rather simple muscular contraction exemplified by the well-known knee jerk. In this reflex, the stimulus is a sudden stretching of the tendon and muscle, brought about by tapping the tendon. The sensory endings in the tendon and muscle are stimulated by this stretch and send impulses along the spinal nerve into the spinal cord. Some branches of the ingoing sensory fibers make almost immediate synaptic connection with the neurons of the anterior horn, and these deliver back to the muscle concerned, through the same nerve, the impulse to respond to the stretch by a sudden contraction. The simplest spinal reflex, therefore, involves only two or three sets of neurons, the afferent or sensory neurons, the efferent or motor neurons, and in some cases an intercalated neuron that makes the connection between the afferent and efferent neurons (Fig. 4-16). In such an instance, since both ingoing and outgoing paths of the reflex are along the same nerve or nerves, little spread up or down the spinal cord is necessary. In other instances, however, such as withdrawal of a limb because of a painful stimulus to the foot, the reflex not only spreads up and down the cord to involve all the centers contributing to the stimulated limb but also spreads to the other side so that the limb or limbs there will support the extra weight.

Although spinal reflexes are normally enhanced or diminished by impulses coming from higher centers, they are organized reactions even in the absence of such impulses. Indeed, most reflexes seem to be decidedly purposeful, for they are actions that are usually essentially the same as would consciously be carried out after there was awareness of the stimulus (*e.g.*, jerking the hand away from a painful stimulus or increasing the contraction of the extensor muscles at the knee when their stretching gives notice that the knee is about to buckle beneath the weight upon it). They are thus stereotyped, but otherwise fairly adequate, responses to stimuli that commonly impinge upon the body. As such, they govern most fundamental neuromuscular activities, whether of voluntary or involuntary muscle. Most conscious or truly voluntary motor activity is simply a series of adjustments of the basic reflexes that have to do with flexion and extension of a limb or maintaining a group of muscles in sufficient contraction to withstand the effect of gravity. The fact that many stimuli that are adequate to produce a reflex response are also adequate to affect higher centers in the brain accounts for the modification of many of these reflexes. The brain can combine the information just received from one stimulus with information received simultaneously from other sources or stored as a result of previous experience.

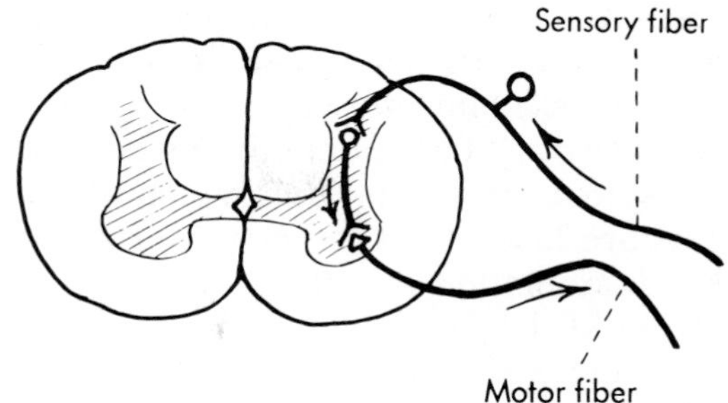


**FIG. 4-16.** The basic anatomy of a spinal reflex. The neuron in the spinal cord connecting the sensory and motor neurons is an intercalary neuron.

The most easily elicited spinal reflexes are protective in function, designed to remove the affected part of the body from a source of pain or to alleviate the pain as much as possible if it arises within the body. Spinal reflexes in response to pain may take precedence over voluntary action if the pain is severe enough. All the muscles around a very painful joint, for instance, may go into such marked spasmodic contraction (to prevent movement of the joint and exacerbation of the pain) that the joint cannot be voluntarily moved. In the same way, the abdominal muscles over an affected viscus may contract to protect from pressure from the outside, and if the pain becomes severe enough, "doubling up" to relieve the intraabdominal pressure may ensue.

Although reflexes that involve skeletal

muscle are the most obvious, it should be remembered that all autonomic activity is reflex. As for skeletal muscle, the simplest reflexes involving the autonomic system occur at the spinal level, while more complicated reflexes are mediated by various parts of the brain. Examination of reflexes (mostly those involving skeletal muscle) demonstrates the integrity of, or an interference with, the reflex pathway in the spinal nerves and the cord. It also gives information concerning the integrity of various other pathways in the cord, and even of centers in the brain, since reflexes are usually modified by impulses descending through the white matter of the cord.

### Meninges and Cerebrospinal Fluid

The meninges are connective tissue wrappings of the central nervous system that intervene between it and the surrounding bone and, with the fluid that they enclose, cushion the nervous system. There are three layers, each continuous from brain to spinal cord.

The **dura mater,** the tough outermost layer of the meninges, lines the skull as it surrounds the brain and forms a simple tube around the spinal cord. Immediately internal to the dura mater, and forming a similarly shaped sac, is the thin **arachnoid membrane** (arachnoidea). The arachnoid membrane is separated from the innermost meninx, the **pia mater,** by a fluid-filled space. The pia mater is also a thin layer, but unlike the arachnoid, it is closely applied to the brain and spinal cord. The space between the arachnoid and pia is the **subarachnoid space,** and the fluid that fills the space is called cerebrospinal fluid.

The colorless **cerebrospinal fluid** is formed largely in the cavities (ventricles) of the brain and escapes from them to fill the subarachnoid space. It is constantly being formed in the ventricles and constantly leaving the subarachnoid space to enter the venous system.

The average quantity of cerebrospinal fluid necessary to fill the ventricles and the subarachnoid space has been estimated at about 135 ml, and the total quantity produced per day at about 500 ml, so interference with circulation and absorption can lead to a rapid increase in volume, and hence in pressure. Large increases in the amount of cerebrospinal fluid (hydrocephalus, or "water on the brain") may exert sufficient pressure to damage the brain severely.

Pressure of the cerebrospinal fluid can be measured by inserting hollow needles through the skull and certain parts of the brain into the ventricles, or into the larger parts of the subarachnoid space that lie in the lower part of the skull, or in the lumbar part of the vertebral column. A similar technique also may be used to withdraw fluid for examination, to introduce medicines into the space, or to withdraw some fluid and replace it with air or radiopaque medium (a substance that resists penetration by roentgen rays) so that a ventricle or the subarachnoid space can be studied by roentgenograms.

Simple observation of a sample will show whether the cerebrospinal fluid is tinged with blood, and microscopic and chemical examination may reveal other changes. The fluid normally contains sugar, chlorides, protein, nonprotein nitrogen, and urea, as well as some 3 to 5 mononuclear cells per cubic millimeter. Changes in the amounts and proportions of these constituents are often of particular diagnostic value.

## GENERAL REFERENCES AND RECOMMENDED READINGS

ADAMS WE: The blood supply of nerves: Historical review. J Anat 76:323, 1942

CANNON WB: The argument for chemical mediation of nerve impulses. Science 90:521, 1939

DUNCAN D: A relation between axone diameter and myelination determined by measurement of myelinated spinal root fibers. J Comp Neurol 60:437, 1934

FOERSTER O: The dermatomes in man. Brain 56:1, 1933

HAMMOND WS, YNTEMA CL: Depletions in the thoracolumbar sympathetic system following removal of neural crest in the chick. J Comp Neurol 86:237, 1947

HARRISON RG: Outgrowth of nerve fibers as a mode of protoplasmic movement. J Exp Zool 9:787, 1910

HEINBECKER P, BISHOP GH, O'LEARY JL: Functional and histologic studies of somatic and autonomic nerves of man. Arch Neurol Psychiatry 35:1233, 1936

KEEGAN JJ, GARRETT FD: The segmental distribution of the cutaneous nerves in the limbs of man. Anat Rec 102:409, 1948

KUNTZ A: The Autonomic Nervous System, 4th ed. Philadelphia, Lea & Febiger, 1953

LAST RJ: Innervation of the limbs. J Bone Joint Surg (Br) 31: 452, 1949

LELE PP, WEDDELL G: The relationship between neurohistology and corneal sensibility. Brain 79:119, 1956

MITCHELL GAG: Anatomy of the Autonomic Neurons System. Edinburgh, Livingstone, 1953

MOYER EK, KIMMEL DL, WINEBORNE LW: Regeneration of sensory spinal nerve roots in young and in senile rats. J Comp Neurol 98:283, 1953

O'CONNELL JEA: The intraneural plexus and its significance. J Anat 70:468, 1936

PEARSON AA, ECKHARDT AL: Observations on the gray and white rami communicantes in human embryos. Anat Rec 138:115, 1960

RANSON SW: Degeneration and regeneration of nerve fibers. J Comp Neurol 22:487, 1912

SHANTHAVEERAPPA TR, BOURNE GH: The effects of transection of the nerve trunk on the perineural epithelium with special reference to its role in nerve degeneration and regeneration. Anat Rec 150:35, 1964

SHAWE GDH: On the number of branches formed by regenerating nerve fibers. Br J Surg 42:474, 1955

SHERRINGTON CS: The Integrative Action of the Nervous System, 2nd ed. New Haven, Yale University Press, 1947

SUNDERLAND S: Rate of regeneration in human peripheral nerves: Analysis of the interval between injury and onset of recovery. Arch Neurol Psychiatry 58:251, 1947

SUNDERLAND S: Factors influencing the course of regeneration and the quality of the recovery after nerve suture. Brain 75:19, 1952

SUNDERLAND S: The connective tissues of peripheral nerves. Brain 88:841, 1966

WHITE JC, SMITHWICK RH, SIMEONE FA: The Autonomic Nervous System: Anatomy, Physiology, and Surgical Application, 3rd ed. New York, Macmillan, 1952

WOOLLARD HH, WEDDELL G, HARPMAN JA: Observations on the neurohistological basis of cutaneous pain. J Anat 74:413, 1940

# 5

# THE CIRCULATORY SYSTEM

The circulatory or vascular system (vas meaning vessel) is divisible into two major parts, the **blood vascular** and the **lymphatic systems.** These two communicate, for the lymphatic system empties into the blood vascular system. The blood vascular system consists of the heart, which pumps blood into the arteries; these terminate in capillaries, from which blood is returned to the heart by veins. The lymphatic system consists of lymphatic capillaries that begin blindly in the tissues and collect tissue fluid. The lymphatic vessels convey this fluid, called lymph, toward the base of the neck, where they empty it into veins. The lymphatic system also includes collections of lymphatic tissue that have to do with the immune responses of the body.

Roentgenographic examination of various parts of the circulatory system in the living person has, in recent years, become a useful diagnostic tool. The technique, known generally as angiography, consists of injecting a small amount of radiopaque material into an appropriate vessel and taking roentgenograms that will show the passage of the material through the region to be examined. It has been used to study the heart and great vessels (cardioangiography), various arteries (arteriography), veins (venography), and lymphatics and lymph nodes (lymphangiography or lymphography).

## GENERAL STRUCTURE

### VESSELS

Since blood vessels and lymphatics are built upon the same fundamental plan, their essential structure can be discussed together. The classification of blood vessels is into **arteries,** vessels that conduct blood away from the heart; **veins,** vessels that conduct blood toward the heart; and **capillaries,** very small vessels distributed through the tissues of the body to bring the blood into close contact with the cells of the tissues and connect the arteries to the veins. **Lymphatics** are divisible, more simply, into lymphatic capillaries and larger lymphatic vessels.

Since an artery conducts blood from the heart, its wall must be sufficiently strong to withstand the sudden thrust imposed upon it with every beat of the heart. The walls of arteries are, therefore, thicker and stronger than those of veins of similar diameter. Lymphatics have the thinnest walls of all and are particularly difficult to recognize in dissection because of their generally small size and because their contents are colorless.

All blood vessels and lymphatics have a lining of flattened cells known as **endothelial cells.** The integrity of this layer is essential to normal blood flow. If it is injured, blood cells begin to stick at the point of injury and to build up a blood clot. In vessels larger than capillary size, the endothelial cells are supported by a thin layer of connective tissue. The endothelium plus its supporting connective tissue constitutes the **intima** (tunica intima). In blood vessels of the size seen at dissection, the next layer wrapped around the intima is the **media** (tunica media), which consists of a layer of smooth muscle and elastic tissue. The third and outer coat of such a blood vessel is composed primarily of loose connective tissue, the **adventitia** (tunica adventitia). The adventitia binds vessels loosely to the connective tissue in which they run and contains nerves and small blood vessels that supply the wall of the vessel. Larger lymphatics have a similar structure.

The structural differences among arteries, veins, and lymphatics are primarily the degree of development of the media and adventitia, especially the former. For any given size, arteries tend to have a much stronger and thicker media, veins a more poorly developed one, and lymphatics the most poorly developed of all. In both veins and lymphatics, the smooth muscle of the media may be so poorly developed that the connective tissue of the typical three layers blends together with no clear distinction between the layers. Also, the walls of arteries contain fairly prominent elastic tissue, whereas those of veins and lymphatics do not.

## Arteries

Arteries are generally divisible into **elastic** and **muscular types,** although the media of most arteries contains some of both types of tissue. Elastic tissue allows the wall of an artery to be distended by the sudden thrust of blood from the heart and then to contract again, which helps force the blood forward, with no initiation of energy by the wall of the vessel. To move the same amount of blood, the thrust—the work of the heart—would have to be much greater if the arteries were inelastic. The large arteries near the heart typically contain a great deal more elastic tissue than do the smaller, more distal ones; in general, the greater the pressure in an artery, the more elastic tissue there is. For example, the pressure in the great trunk (pulmonary trunk) going to the lungs is much less than that in the great trunk (aorta) carrying blood to the body as a whole; therefore, the wall of the aorta contains far more elastic tissue than the pulmonary trunk. Elastic tissue serves the double purpose of cushioning the sudden rise of pressure induced by the heart beat and of smoothing, by its automatic recoil, what would, in an inelastic system, be a sudden drop in pressure. In hardening of the arteries (arteriosclerosis), in which the elasticity is interfered with by the deposition of fatty and, eventually, calcified material in the intima and sometimes the media, the blood pressure undergoes abnormally great fluctuation with each cardiac cycle (beginning of one heart beat to beginning of the next).

In the larger arteries, particularly the aorta, most of the strength of the wall is provided by elastic tissue. If this undergoes degenerative changes or is destroyed by disease, the arterial wall may bulge, either in one spot or all around. Such a widening of the arterial lumen is an *aneurysm* and always involves the danger that the weakened wall may burst, with almost immediate death ensuing if the vessel is large.

As branches of the aorta are traced distally,

the relative amount of elastic tissue becomes less and the relative amount of smooth muscle more. The elastic tissue of the peripheral arteries tends to smooth the pressure and lessen the velocity of the blood. The smooth muscle, controlled by the autonomic nervous system, can contract or relax to vary the caliber of the vessel and hence the blood flow through it. Even rather large vessels, for instance the chief arteries of the arm and thigh, contain sufficient smooth muscle to reduce to a dangerous level the blood flow through them if they are incited to maximal contraction (blood flow in arteries drops off much more rapidly with constriction than it would in inelastic tubes of the same size and will cease entirely while there is still pressure in the artery). The smallest arteries, **arterioles,** have a media composed almost entirely of smooth muscle and are particularly contractile. It is through the activity of the arterioles that blood is shunted from one place to another, and they are the most important governors of the peripheral resistance to blood flow. If too many arterioles are simultaneously relaxed, blood pressure drops drastically; if too many are contracted, or so diseased that they cannot relax, high blood pressure ensues.

**Anastomoses and End Arteries.** Most parts of the body receive branches from more than one artery, and where two or more arteries supply the same territory, they usually connect with each other (anastomose). The number and size of anastomoses vary considerably with the region or organ (and often with the individual). In many locations, there are one or more macroscopic (dissectible) anastomoses and numerous finer ones. An anastomosis between vessels may be by small terminal branches that result from repeated divisions, or two vessels of considerable size may simply join end to end (inosculate) with such little diminution in caliber that it is impossible to tell at what point the two vessels meet.

Anastomoses between arteries provide alternate ways in which blood can reach a given tissue or organ, and, therefore, their number and size (ability to conduct blood) are critical when the blood supply from one or more sources is cut off. Usually a region is supplied more particularly by one artery than by others, and the anastomotic alternative paths by which blood can reach the field of distribution of the chief artery are known as the **collateral circulation.** Since blood vessels enlarge as more and more blood tries to go through them, collateral circulation is best developed by slow occlusion of the main artery. This ability to gradually enlarge does no good in instances of sudden occlusion unless the collateral circulation is capable of dilating rapidly enough to carry sufficient blood to keep the tissue alive. This depends, in general, upon the total caliber of the collateral circulation. (Collateral circulation can sometimes, but not always, be effectively increased by a sympathectomy that paralyzes the smooth muscle). Knowledge of the anatomy of the collateral circulation to a part is essential if the reasons for death or survival of a part (and of a patient) following an arterial occlusion are to be understood. During surgical procedures, the anatomy of the collateral circulation may also be of critical importance; some vessels can be ligated at any level with impunity, others can never be ligated without damage because there is insufficient collateral circulation, and still others must be ligated at a level calculated to spare the greatest possible amount of collateral circulation if further injury is not to be done.

Until recently, there has been no method of averting gangrene (massive death of tissue), or sometimes death of the patient, when an artery without adequate collateral circulation was occluded by a disease process or by ligation. With new techniques of suturing end to end and of grafting (arteries, veins, and synthetic material, such as Teflon, in the form of tubes, have all been used as grafts), continuity of the vessel after removing the injured segment can often be restored, thus preventing or minimizing the damage that would otherwise ensue. The technique is especially useful in handling injuries of the aorta and its larger branches, including the major vessels of the limbs. It is of course not applicable to very small arteries or to those not surgically accessible.

Just as anastomotic channels occur between arteries to a region, so may they also occur between arteries that supply an organ. This is true, for instance, of the vessels in the wall of the stomach, which anastomose so freely with each other that the surgeon does not hesitate to ligate any of the supplying vessels. However, other organs, among them the most vital—the brain, the heart, the liver, the kidneys—are supplied by arteries that, beyond a certain point, usually within the organ, have no anastomoses with each other, or such small ones as to be totally inadequate. Arteries that do not anastomose are known as **end arteries.** Occlusion of an end artery interrupts the blood supply to a whole segment of the organ, producing necrosis (death) of that segment; the area of necrosis is known as an **infarct.** Results of infarction necessarily vary according to the size and location of the infarct. True end arteries occur in the brain, kidney, and liver, but infarctions in the brain are more injurious for a given size because uninjured tissue of the kidney and liver is usually functionally adequate, whereas there are many parts of the brain for which no other part can substitute. The heart does not have true end arteries in its wall, but anastomoses are normally so small that most of the vessels are functional end arteries. The results of ischemia (decreased blood supply) to the heart vary from very mild to fatal heart attacks, the final outcome depending, in large part, upon the size of the infarct.

## Capillaries, Sinusoids, and Arteriovenous Anastomoses

Most arterioles empty into capillaries, tiny channels (little larger than the diameter of a red blood cell) with endothelial walls through which the blood is in close contact with the tissues. Because of the pressure of the blood, a certain amount of its fluid escapes through the endothelial wall of capillaries to become tissue fluid, thus providing the liquid environment in which all living cells need to be suspended. Through interchange between the blood plasma and the tissue fluid, oxygen, carbon dioxide, and nutrients carried by the bloodstream and metabolic wastes produced by the cells can pass back and forth between the bloodstream and the cells.

Although capillary walls apparently contain contractile elements, circulation through any given capillary bed is determined primarily by the arterioles that feed it. In many tissues, capillaries form a dense interconnecting network, fed by a number of arterioles, but bone and tendon have a relatively poor blood supply, cartilage has no capillary network, and stratified epithelium contains no blood vessels at all. The density of the capillary bed seems to be related primarily to the functional activity of the organ. In some organs, as was intimated in the discussion of end arteries, a large capillary bed receives its blood entirely from arterioles that are all branches of the same artery.

Sinusoids, found in a few locations (for instance, the liver and spleen) are somewhat like capillaries of unusually large diameter. Instead of the usual lining endothelial cells, however, the walls of sinusoids are largely composed of special phagocytic cells (in the liver, Kupffer cells). Sinusoids form a part of the reticuloendothelial system, a defense mechanism concerned with both phagocytosis (engulfment of particulate matter) and the formation of antibodies.

Although capillaries are the usual connection between arteries and veins, there also are sometimes larger connections, arteriovenous anastomoses. Arteriovenous anastomoses are essentially arterioles with especially contractile walls that open directly into venules. Through their contraction and relaxation, they can play an important part in the circulation to the tissues, since their relaxation diverts blood directly into the venous system that would otherwise pass through the capillary network. Arteriovenous anastomoses have been found in many locations, but the exact way in which they fit into the physiology of the bloodstream is not yet understood. Arteriovenous anastomoses that are normal structures are not to be confused with abnormal communications between arteries and veins. These are properly known as arteriovenous aneurysms or fistulae. In certain locations, arteriovenous anastomoses sometimes give rise to painful tumors, known as "glomus tumors."

### Veins and Sinuses

The smallest veins, **venules,** are formed by the junction of capillaries and consist only of an intima and a thin layer of connective tissue, the adventitia. Larger and larger veins are formed by the junction of smaller veins, and in veins of medium size, a media appears, containing smooth muscle in variable amount but never to the degree that arteries do. In the larger veins close to the heart (where the blood pressure is very low), there may be no media, and the adventitia often has longitudinally arranged smooth muscle.

Veins anastomose much more freely than do arteries and, therefore, afford more abundant collateral circulation on the venous side. Occlusion of veins is, for this reason, rarely a cause of necrosis, although it may be if the occluded vein is large and there are only small anastomotic channels, or if a considerable length of vein is occluded (*e.g.*, by a thrombus) and many collateral channels are shut off. Occlusion of even large veins commonly leads only to temporary edema (accumulation of fluid in the tissues), which disappears as the collateral circulation enlarges.

**Sinuses** (vascular sinuses; there are also air sinuses, and certain pathological openings are likewise called sinuses) are veins with very thin walls for their diameter. In the best example, the intracranial sinuses, the true wall of the vein is nothing but endothelium. The endothelium is supported by the heavy connective tissue surrounding the brain (the dura mater) in which the sinuses lie, so this actually replaces the usual media and adventitia of the venous channel.

**Valves.** The two chief factors in directing the flow of blood in veins toward the heart are the slight positive or even the negative pressure obtaining in the thorax and the massaging action of muscles on veins. In the limbs, the effectiveness of the massage, which might under certain circumstances drive blood toward the periphery rather than toward the heart, is ensured by the presence of valves in the veins. Venous valves are infoldings of the intima, so arranged that they allow free passage of blood toward the heart but impede or prevent passage of blood in the reverse direction. Venous valves are often bicuspid (Fig. 5-1), the elements of a pair meeting along their edges in the middle of the vessel, but

**FIG. 5-1.** Longitudinal section of a valved vein. **ti** is the tunica intima, **tm** the media, and **ta** the adventitia; **va** is one of the cusps of a bicuspid valve, **si** the valve sinus. (Kampmeier OF, Birch CLaF: Am J Anat 38:451, 1927)

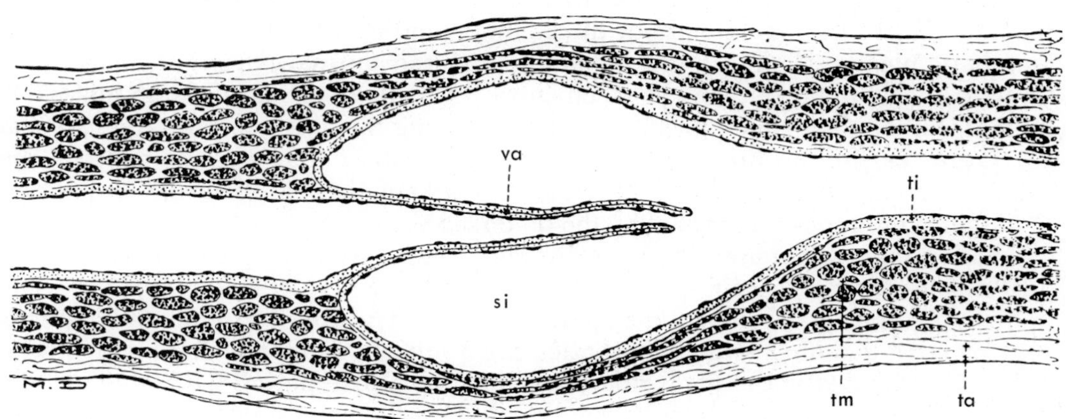

sometimes they have only one cusp, or they may have three. Inefficiency of the valves in the veins connecting the deep and superficial veins of the lower limb allows the pumping action of the muscles to force blood from the deep venous system into the superficial veins, thus overloading and dilating them. Dilated, tortuous veins are known as varicose veins.

There are few or no functionally efficient valves in the veins of the abdomen, but under the squeezing action of the abdominal muscles and the diaphragm, blood flow into the region of lesser pressure provided by the thorax is far easier than backflow into the valved veins of the lower extremity, so valves are not really needed here. In the same way, the veins of the head and neck do not, for the most part, need valves to resist backflow, since flow toward the heart is assisted both by the decreased pressure and by gravity; there are few valves in this part of the venous system.

## Lymphatics and Lymph Nodes

Lymphatic capillaries are essentially similar to blood capillaries in that they have only an endothelial wall, through which water and certain larger molecules can pass freely. They differ in that they begin blindly. Lymph, derived from tissue fluid, enters the lymphatic capillaries by passing through the capillary wall. In some tissues, lymphatic capillaries are sparse. In some, such as the central nervous system, they apparently do not exist at all. In still other locations, for instance the dermis of the skin, they may form plexuses so dense that it is almost impossible to inject substances into the tissue without filling the lymphatic plexus. The rapid spread obtained through the lymphatics is used when intradermal, rather than subcutaneous, injections are given, as, for instance, for prophylaxis against typhoid fever.

Larger lymphatics are formed by the union of smaller ones, and these lymphatics resemble veins in structure except that they have an even more poorly developed media than a vein of corresponding size. Lymphatics are regularly provided with valves, like those of the veins of the limbs, to ensure flow of lymph away from the tissues and toward the venous system. They tend to anastomose freely.

In their course toward the venous system, many lymphatics empty into lymph nodes, instead of joining others directly. **Lymph nodes** are collections of lymphocytes (one type of white blood cell) held together by connective tissue and permeated by lymphatic channels. Each lymph node typically receives a number of lymphatic vessels (termed "afferent," since they are carrying lymph to the node), and the lymph from all these circulates through the lymph channels of the node, leaving it usually by a single efferent vessel. Lymph may pass for some distance through larger and larger lymphatic channels without passing through a lymph node, or it may pass successively through a number of nodes. Typically, all lymph has passed through several lymph nodes before it is returned to the venous system.

If a lymphatic vessel bears carcinoma (cancer) cells that have invaded it and become broken off so that they are floating in its stream, the filtering action of the lymph node tends to retain these cells within the node, rather than to allow them to pass on into the bloodstream; hence, those cancers that spread through the lymphatic system tend to migrate (metastasize) first to lymph nodes, where they typically grow at the expense of the node, gradually destroying it. Since they do not migrate farther until they are well established in the first node that they reach, cancer that has begun to spread can sometimes be eradicated entirely by removing, in addition to the original lesion, the lymph nodes that first receive the drainage from the region of the lesion (called "primary lymph nodes") or by exposing the nodes to radiation that can inhibit or destroy that type of cancer. Thus, detailed knowledge of the usual lymphatic drainage of an organ and of the lymph nodes into which this drainage passes may be of great importance.

The filtering action of lymph nodes, particularly dramatic where it involves cancer cells, is also apparent when the lymph nodes about the lung of a city dweller are examined. The carbon particles breathed in from the smoke in city air are deposited not only in the connective tissue of the lung (taken there by phagocytes) but also in the lymph nodes draining the lung, so that these nodes may be black and very gritty in texture.

Another function of lymph nodes, albeit less dramatic, is to produce lymphocytes, the second most common type of white blood cell. Lymph nodes are responsible for adding most of the lymphocytes to the bloodstream, although the spleen also contributes. Lymph nodes also participate in the production of antibodies.

The lymphatic vessels act with blood capillaries and veins to remove tissue fluid that leaks out from the arterial side of the capillaries; therefore, obstruction of either the venous or the lymphatic drainage of a part may produce excessive accumulation of tissue fluid, **edema.** In the digestive tract, lymphatics also carry out a function that the blood capillaries apparently cannot perform: They receive almost all of the fat absorbed by the digestive system. In addition to the lymphocytes and the fat that the lymphatic return to the venous system contains, lymph has a high protein content. Loss of large amounts of lymph through rupture of a major lymphatic vessel produces serious deficiencies in the protein and salt content of the blood.

## Blood Supply and Innervation of Vessels

Since larger blood vessels and lymphatics have relatively thick walls, the tissue of these walls cannot receive adequate nutrition from the blood contained within the vessel itself; thus, blood vessels that are approximately a millimeter or more in diameter receive vessels, *vasa vasorum,* from adjacent small blood vessels. These lie in the adventitia and form a capillary network there, but do not pass very far into the media in arteries; in veins they are said to go sometimes as far as the intima. Lymphatics even smaller than a millimeter in diameter have a blood supply.

Little is known concerning the innervation of lymphatics. As might be expected, the more muscular arteries have a better innervation than do the less muscular arteries and veins. Except in a few locations, the motor innervation of the blood vessels is entirely by the sympathetic nervous system, and the usual effect of activity of this system is to cause contraction of the circular smooth muscle of the vessel, hence constriction of its diameter. The arterioles, with their almost entirely muscular walls, have a particularly rich innervation, and it is through these vessels, primarily, that peripheral blood flow is regulated by the sympathetic system.

Arteries and veins also receive afferent fibers. Many of these fibers are concerned with pain. In certain locations also, especially on the great veins and arteries close to the heart and on the chief arteries (internal carotids) to the brain, there are areas of afferent endings that react to changes of the blood pressure within the vessel and that reflexly affect the autonomic system so as to raise or lower this pressure. The rate and strength of the heart beat is controlled in part, for example, by impulses originating within the walls of the vascular system, as is the constriction and relaxation of arterioles.

**Source of Nerve Fibers.** It is common for blood vessels to be accompanied by a nerve plexus, sometimes macroscopic, sometimes microscopic and embedded in the adventitia; but apparently only in the case of the blood vessels to the thoracic and abdominal viscera and to the head, where the nerves accompany the blood vessels primarily because they are both going to the same organs, does a plexus beginning at the base of a vessel extend throughout the length of that vessel. It is of particular importance to recognize that in the limbs (frequently the site of vascular disturbances that can be alleviated, in greater or lesser degree, by sympathectomy) the nerve plexus along a chief artery and its branches is not one continuous plexus but a series of interlocking plexuses that are fed at more or less regular intervals by nerve branches derived directly from neighboring peripheral nerves. Interruption of the nerve plexus along, for instance, the chief artery to the lower limb, the external iliac, will not abolish the nerve plexus on the femoral or popliteal arteries (the direct continuation of the external iliac into the limb) nor those on their branches. In consequence, a periarterial sympathectomy (stripping the plexus from around a part of an artery), or separation of a length of artery from all neighboring nerves so as to sever the branches of the nerves to the artery, produces only a local denervation of the vessel. If sympathetic denervation of an entire limb is desired, this can be accomplished only by interrupting the sympathetic nervous system before it has joined the major nerves of the limb. Although the sympathetic fibers travel in these major nerves, it is obviously impractical to interrupt them by sectioning the nerves, since these are also the source of the voluntary motor activity and sensation in the limb.

## HEART

The heart (cor, cardia) begins its embryological development as a contractile tube. In spite of the fact that it departs markedly from its original size and shape and, in the adult, is completely divided into right and left sides, it can still be likened to a double-barreled tube, receiving blood at one end (the **atria**) from veins and pumping it out at the other end (the **ventricles**) into the arteries. Since the heart must pump blood in only one direction, it is provided with valves to ensure against backflow from a region of higher to one of lower pressure.

The intimal lining of the heart, continuous with the intima of the vessels connecting to it, is known as the **endocardium** and does not differ essentially from the intima of blood vessels. It also forms the valves that lie between atria and ventricles and at the bases of the two great arterial trunks leaving the heart.

The muscular part of the heart, the **myocardium,** is equivalent to the media of a blood vessel. It is a special type of muscle (cardiac muscle) found only in the heart and the great vessels as they attach to it. Although cardiac muscle is striated like voluntary muscle, it differs in all other respects.

The musculature of the atria is thin, since the atria work at a low pressure and have only to drive blood into the relaxed ventricles. The musculature of both ventricles is much thicker than that of the atria, and the musculature of the left ventricle is much thicker than that of the right since the former must pump blood all over the body whereas the latter must pump only to the lungs.

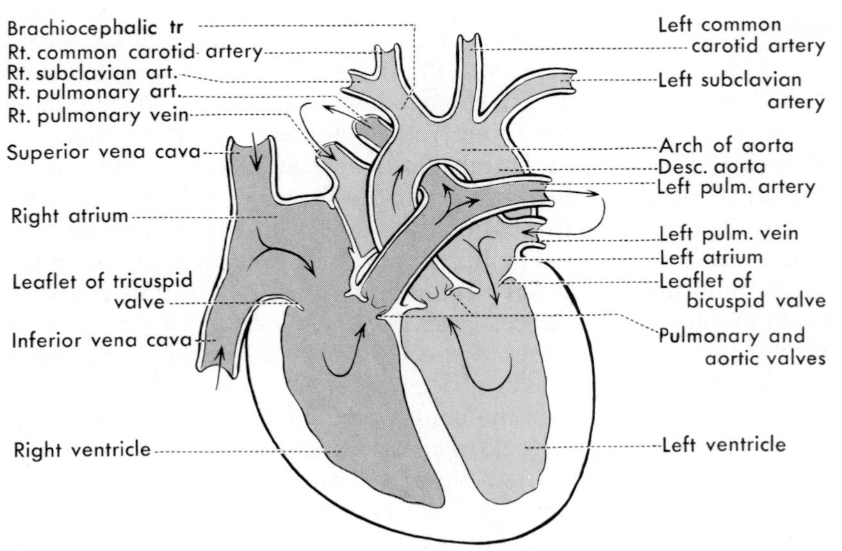


**FIG. 5-2.** Diagram of the heart and the great vessels connected to it. The right side of the heart and its connecting vessels are blue; the left side of the heart and its connecting vessels are pink. Arrows indicate the direction of blood flow.

The outermost layer of the heart is the **epicardium** (visceral layer of the serous pericardium), whose outer surface is a single layer of mesothelial (pavement epithelial) cells continuous with the serous (inner) surface of the pericardium.

The **pericardium** is a closed sac surrounding the heart. Its outer wall, the fibrous pericardium, is a tough, fibrous, loose-fitting sac. It is lined with mesothelium (the parietal layer of serous pericardium) that is reflected onto the heart around the great vessels leaving it and along the posterior aspect of the heart where that organ is fused to the pericardium. In these locations the serous layer continues onto the heart as the epicardium or visceral lamina of the serous pericardium.

The gross morphology of the heart is described later. This hollow, muscular organ, roughly the size of one's fist, is, in the adult, normally completely divided into sides, through which the blood circulates separately, entering and leaving one side and then returning to the heart to enter and leave the other side (Fig. 5-2). The great veins carrying blood to the heart enter the thin-walled atria; the veins from all the body except the lungs enter the right atrium, and those from the lungs enter the left atrium. Each atrium opens into the ventricle of its side. The opening is protected by a valve (atrioventricular valve) so arranged that whereas blood can flow freely from atrium to ventricle, the valve closes as blood starts to flow in the reverse direction. Each ventricle gives rise to a large arterial trunk, the right ventricle giving rise to the pulmonary trunk (to the lungs), the left ventricle to the aorta (distributing blood to the rest of the body).

The musculature of the heart is highly vascular and is fed by two arteries, the **coronary arteries,** that arise from the aorta just after it leaves the heart. Obstruction of a major branch of a coronary artery, especially one to a ventricle, is the common cause of a cardiac stroke (myocardial infarct) or sudden heart failure. Too great a narrowing of a major branch of an artery, or occlusion of minor branches, produces ischemia of the cardiac muscle. The pain produced by this is known as *angina pectoris*. For the most part, the cardiac veins that collect the blood delivered to the myocardium by the coronary arteries parallel these arteries, but all the larger ones empty into a single vein, the coronary sinus, that returns the blood to the right atrium.

The heart also has a nerve supply that contains both afferent fibers (including ones of pain) and motor fibers (Chap. 21). The motor nerve supply is a purely regulatory one, however. It acts upon the heart to modify (*i.e.*, increase or decrease) the rate and strength of the heart beat according to the needs of the body.

## CIRCULATION OF THE BLOOD

The anatomy of the circulation of the blood through the heart is very simple. In contrast, the factors governing the peripheral circulation are very complicated and even yet not completely understood. Understanding of a few basic phenomena will, however, serve as an introduction to the subject.

### Circulation Through the Heart

Since the heart is divided into right and left sides, there are two circulations through it (Figs. 5-2 through 5-4). Both, of course, occur simultaneously and are equal in volume, but it is convenient to trace a given quantity of blood from the time it enters one side of the heart until it returns again to that side.

Blood from the head and neck, the upper limbs, and the thoracic wall enters the right atrium by the superior vena cava; blood from the abdomen, pelvis, and lower limbs enters this atrium by the inferior vena cava; and blood that has circulated to the cardiac muscle enters the right atrium through the coronary sinus. The right atrium thus receives blood that has passed through capillaries in the tissues of the body, lost much of its oxygen, and picked up much of the carbon dioxide in the tissues. As the blood runs into the right atrium, some of it continues on into the

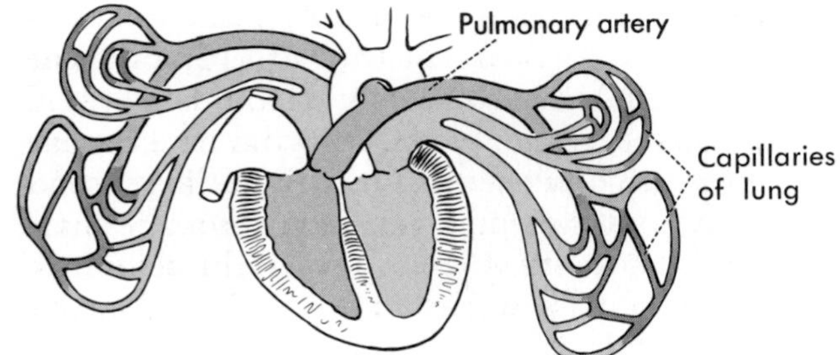


**FIG. 5-3.** Diagram of the pulmonary circulation. The right ventricle and the pulmonary arterial system are blue; the pulmonary veins and the left atrium are pink.

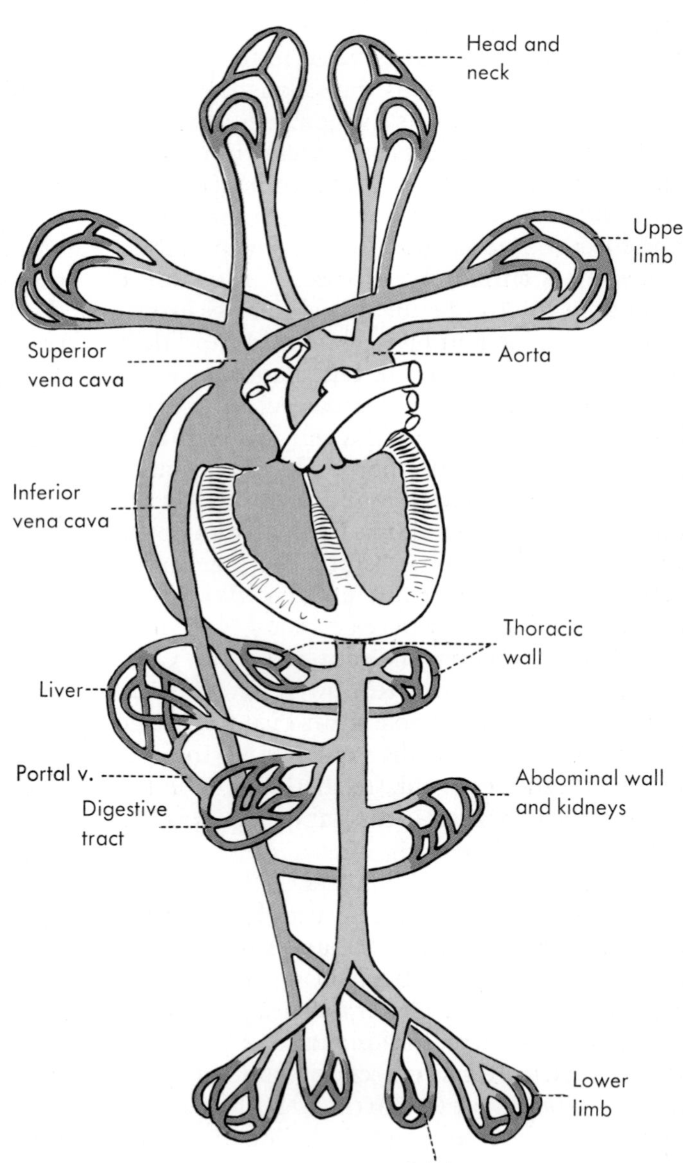


**FIG. 5-4.** Diagram of the systemic circulation. The left ventricle, the aorta, and the other arteries are pink; the veins, including the portal, are blue; the purple connections between vessels represent the circulatory pathways in the various parts and organs indicated.

relaxed right ventricle. Contraction of the right atrium injects more blood into this ventricle. At this moment, the right ventricle begins to contract and to force blood out through the pulmonary trunk to the lungs. Backward passage into the relaxing atrium is normally prevented by the right atrioventricular valve, which is closed by the rising intraventricular pressure. Similarly, the blood forced by the right ventricle into the pulmonary trunk (so called because it is a common trunk for the paired pulmonary arteries) is prevented from returning to the relaxing ventricle by the pulmonary valve at the level of origin of the pulmonary trunk from the ventricle.

After passing through the pulmonary trunk and its branches, and through the capillaries and smaller veins of the lungs, the blood sent to the lungs by the right ventricle is returned to the left atrium by pulmonary veins. As in the case of the right side of the heart, contraction of the left atrium injects additional blood into the left ventricle. Contraction of the ventricle closes the left atrioventricular valve and forces blood out through the aorta. Closure of the aortic valve at the base of the aorta prevents regurgitation of blood into the relaxing left ventricle. The blood is then delivered through the branches of the aorta to the tissues of the body, where it loses oxygen, picks up carbon dioxide, and returns to the right side of the heart to begin its double cycle all over again.

Since the blood is constantly circulating both to the lungs and to the tissues of the body, it is obviously important that both sides of the heart handle the same amount of blood. If, for instance, the right side fails to send to the lungs all the blood it has received from the left side, it will be overfilled by the next inflow of blood and will have to dilate slightly to receive it. If it then does not beat strongly enough to expel all the blood it has just received plus the residue from the previous beat, it must remain dilated or dilate still more. Normally, dilation of the heart produces a stronger beat, so at some stage of dilation, the beat usually becomes strong enough to restore the balance of circulation (*compensation*). If the heart muscle of the dilated side finally begins to fail (*decompensation*), rapid dilation and more serious or fatal interference with the circulation ensues. Digitalis is useful in minimizing decompensation because it increases the strength of the heart beat and, therefore, the cardiac output. Epinephrine, which mimics the effects of sympathetic stimulation, has the same effect but acts more powerfully, more quickly, and over a shorter period of time than does digitalis; it is, therefore, most useful in emergency situations, whereas digitalis is used in chronic cases of incipient cardiac failure.

## Cardiac Cycle

The cardiac cycle is the series of events occurring during one beat of the heart as it fills and empties. The period of contraction of the heart is **systole,** that of relaxation, **diastole.** Since atria and ventricles contract separately, there is actually an atrial systole and a ventricular systole. Atrial and ventricular diastole overlap. Pressure changes in atrial systole can be measured only by instruments lying within the heart, but ventricular systole can be measured roughly simply by the rapidity and strength of the arterial pulse. Clinically, therefore, the term "systole" usually refers to ventricular systole only, and atrial systole, occurring during ventricular diastole, is usually called a presystolic phenomenon.

The electrocardiogram, based upon the principle that contraction of muscle always involves an electrical change, allows the cardiac cycle to be analyzed in detail and thus aids in diagnosis of minor irregularities and local defects in contraction resulting from injury to the musculature. Normal heart sounds are caused by closing of the valves and contraction of the ventricles. A heart murmur is the sound heard when blood under high pressure runs through a narrowed opening. It is often a sign of leakage of blood from ventricle to atrium, or from pulmonary trunk or aorta to ventricle, as a result of imperfect valvular closure.

Contraction of the two atria typically occurs simultaneously and is closely followed by contraction of the ventricles. Although the heart muscle works more continuously than does any other muscle in the body, its periods of rest are greater than its periods of contraction: in a cycle lasting about 0.8 second, atrial contraction lasts for only about 0.1 second, so that the atrial muscle is resting seven-eighths of the time. Ventricular systole lasts only about 0.3 second, so that even the ventricular musculature rests for a longer period than it contracts, and all the cardiac muscle is relaxed at the same time for about half the length of the cardiac cycle.

## Peripheral Circulation

The force of the heart beat and the pulsatile contraction of the elastic arteries carry the blood into the capillaries, but since there is a steady increase in the extensiveness of the vascular bed (the total area of channels through which the blood can flow) between the heart and the capillaries, there is also a steady decrease in blood pressure. Blood that leaves the heart through the aorta at a pressure of about 120 mm Hg falls in the capillaries to only 10 mm to 20 mm. Further, the elastic recoil of the arteries gradually converts the pulsatile flow from the heart into a steady flow through the capillaries, and the expansion of the vascular bed slows the velocity of blood from an average of about 0.5 m/sec near the heart to about 0.5 mm/sec in the capillaries. These changes in the blood flow allow a better interchange between the capillaries' contents and the tissues and result in delivery of a steady stream of blood into the venous side of the circulation. The flow into the veins is delivered at very low pressure, and therefore, in dependent parts of the body, most notably the lower limbs, the venous return of blood to the heart is made difficult by the effect of gravity. The effect of the valves and of the massage by muscles has already been noted. Blood pressure falls steadily in the veins as they near the heart, often to negative levels so that some of the blood is literally sucked into the heart.

The physiology of blood flow and of the factors that affect it and the blood pressure are too complicated to be discussed here, but a few basic principles can be noted. The source of the arterial blood pressure is, of course, the thrust given the blood at ventricular systole, but the pressure and rapidity of flow are determined by this and the peripheral resistance together, for the heart must beat more strongly to move blood against greater peripheral resistance. If the arterioles, the chief governors of peripheral resistance, are tonically contracted beyond normal, the heart must work harder to maintain an adequate flow against the increased resistance, and blood pressure rises. In such event, the work on the heart can be lightened and the blood pressure lowered by the administration of one of the more effective drugs now available; if, however, the cause of the increased resistance is an arteriosclerotic aorta or irreversible changes in the caliber of the arterioles, such treatment will be useless. The reverse effect occurs in vascular shock, in which, as a result of loss of blood or of pooling of blood in the tissues in consequence of general vasodilation, not enough blood returns to the heart to enable it to maintain normal pressure.

Normally, as blood spreads through the branching arterial system, it encounters some vascular beds that are at the moment more constricted and others that are less constricted, and it therefore flows more freely through the latter since they are pathways of less resistance. Local vasoconstriction and vasodilation can, therefore, bring about decreased or increased blood flow to an organ or part and shifts in rate of blood flow among parts on the basis of their activity and resultant needs, with no alterations of blood pressure. It is important to note that the volume of blood flow through an artery does not follow the rules that hold for an inelastic system of pipes, in which rate can be calculated from fluid pressure and size of the pipe. Rather, in arteries, the rate of flow drops off far more rapidly with vasoconstriction and with decreases in blood pressure than mathematical calculations would indicate. A general lowering of blood pressure combined with vasoconstriction in a given part, or even partial block of a large artery, can result in markedly reduced or even total cessation of flow to a part.

Much of the venous blood pressure is hydrostatic, and venous return to the heart is, therefore, expedited by elevating a part above the level of the heart, a procedure commonly followed in alleviating any swelling caused or contributed to by venous congestion.

## BASIC GROSS ANATOMY

A general comprehension of the gross anatomy of the cardiovascular and lymphatic systems is necessary if the parts studied regionally are to be fitted into their proper places in these systems. In regard to the cardiovascular system, it has already been noted that since arteries conduct blood away from the heart, the largest arteries are those that leave the heart. The large arteries branch and rebranch until finally they end in capillaries. In the same fashion, small veins are formed from capillaries, and larger veins are formed by the union of these. The largest veins are those that empty blood back into the heart.

Many branches of the expanding vascular tree, as one follows the arteries, and many tributaries of the contracting vascular tree, as one follows the veins, are named. Generally speaking, although not always, arteries and veins run together and therefore have the same name. There is no rule as to how small an artery must be to be considered unworthy of a name, but usually the named arteries are those that one can dissect with no great difficulty. Even in vital organs where the detailed blood supply is of particular interest, named arteries are, for the most part, at least a millimeter in diameter. In locations (such as the limbs) where the finer pattern of branching is not particularly important, the named vessels are generally much larger.

Both arteries and veins vary somewhat in pattern from one person to another, as might be expected from their method of development. As a rule, variations in the arterial pattern are less common than those in the venous pattern. Indeed, many variations in the venous pattern are so common that little attention is paid to them. For instance, the detailed pattern of the superficial veins on the back of the hand is so varied that probably no two patterns exactly coincide, even though all usually correspond to the same general plan. Variations in superficial veins in regard to precise pattern, size, and termination or connections are particularly numerous. The deep veins that accompany arteries tend to vary in the same manner as do the arteries they accompany. Major variations in those veins of the abdomen and thorax that do not quite parallel arteries can usually be explained on the basis of their complex embryological development.

Since arterial variations are less common than venous ones, they tend to be more striking. Also, since arteries tend to anastomose less freely than veins, and a misplaced artery is often potentially more dangerous (anomalous superficial arteries have been mistaken for veins, and misplaced deep arteries may exert pressure on a vital structure), variations in the arteries tend to be of more anatomic and clinical interest than variations in veins. The common types of arterial variation are those of size and distribution of branches, origin of a vessel at a higher or lower level than usual, combined origin of vessels that usually arise separately, and origin of a usual branch of a vessel either independently from a parent trunk or from another neighboring vessel. More important and less commonly encountered variations (anomalies) are usually combinations of abnormal origin and course; many of these can be understood by knowledge of the normal developmental history of the artery concerned.

### SUBDIVISIONS OF THE BLOOD VASCULAR SYSTEM

The peripheral vascular system is divided into two parts: the pulmonary circulation, to the lungs, and the systemic circulation, to the rest of the body (the former is sometimes called the lesser circulation). Pulmonary arteries and pulmonary veins go to and from the lungs, and systemic arteries and veins go to and from all the tissues, organs, and organ systems of the body (Figs. 5-3 and 5-4). In addition, a system of veins drains most of the digestive tract and ends in the sinusoids of the liver rather than joining veins going to the heart. This system of veins differs from almost all other veins in that it both begins and ends in capillaries. Because it carries blood to the liver, it is known as the portal system. In some lower animals, it may be recalled, there is also a renal portal system, through which blood beginning in venous capillaries is filtered through capillaries in the kidney before returning to the heart, but in man and other mammals the "hepatic" portal system is the only major portal system; hence, the qualifying word "hepatic" is not used.

## Pulmonary Vessels

The two pulmonary arteries, one to each lung, arise from a single pulmonary trunk that in turn arises from the right ventricle. At its base, the pulmonary trunk is provided with a pulmonary valve consisting of three cusps. The trunk is a short stem lying within (*i.e.*, surrounded by) the pericardial cavity, branching at the uppermost part of this cavity into right and left pulmonary arteries. These go to the right and left lungs.

There is a good deal of variation in the number and arrangement of major branches given off by each pulmonary artery. Within the substance of the lung, the arteries branch and rebranch, as do systemic vessels elsewhere, but in general they follow the air passages. The arterioles of the pulmonary arteries are, in turn, continuous with capillaries that form a very close network in intimate contact with the air sacs (alveoli) within the lung and, hence, are particularly adapted to easy interchange of gases between the air sacs and the bloodstream.

The pulmonary veins, draining the capillary plexus of the lung, are formed by the confluence of smaller veins. Instead of all the veins from the right lung going together to form a single vein, and those of the left lung doing likewise, however, there usually are two right and two left pulmonary veins. Both sets of pulmonary veins empty into the left atrium, so that this atrium normally receives four veins.

Bearing in mind the fact that the right atrium and ventricle receive blood returned from the general tissues of the body, it should be obvious that the pulmonary arteries carry to the lungs blood poor in oxygen but rich in carbon dioxide, and the pulmonary veins return to the heart blood low in carbon dioxide but rich in oxygen.

## Systemic Arteries

The left ventricle, receiving oxygenated blood from the lungs by way of the pulmonary veins and left atrium, gives rise to the *aorta,* which in turn gives rise, directly or indirectly, to all the systemic arteries (Fig. 5-5). As the aorta leaves the ventricle, it is directed upward, but it soon arches to the left and backward to attain a position on the left side of the vertebral column. Because of this arrangement, the aorta is conveniently divided into **ascending aorta, arch** (both of which are short), and **descending aorta.** The descending aorta is, in turn, conveniently described as being thoracic or abdominal, the continuity of the two parts being at the level at which the aorta passes through the diaphragm.

Close to its base and almost under cover of two of the cusps of the aortic valve, the ascending aorta gives off paired coronary arteries that supply the musculature of the heart. The arch of the aorta normally gives off three arterial trunks, the first being the **brachiocephalic trunk** (formerly "innominate artery"); the second, the **left common carotid artery;** and the third, the **left subclavian artery.** After a short course upward, the brachiocephalic trunk divides in the base of the neck into a **right subclavian artery** and a **right common carotid.** The left common carotid and left subclavian parallel each other into the base of the neck.

**Head and Neck.** The two common carotid

arteries, left and right, are similar. Each runs upward in the neck, lateral to the trachea or windpipe (where it can be palpated) and divides at about the upper border of the larynx (Adam's apple) into external and internal carotid arteries. Normally, there are no branches of the common carotid except these terminal ones. The **external carotid** gives off a series of branches to structures in the neck (the neck also obtains a blood supply from branches of the subclavian artery), and both superficial and deep branches to the face and jaws. One branch (superficial temporal) becomes subcutaneous in front of the ear, where its pulse easily can be felt. The **internal carotid** artery, in contrast, gives off no branches in the neck. It runs upward and enters the skull and ends by dividing into several important branches to the brain.

**Upper Limb.** Although differing slightly in origin, the two subclavian arteries are essentially similar. At the base of the neck each gives off branches to the neck, the shoulder, and the thorax; one, the vertebral, runs the length of the neck to enter the skull and add to

**FIG. 5-5.** The chief systemic arteries.

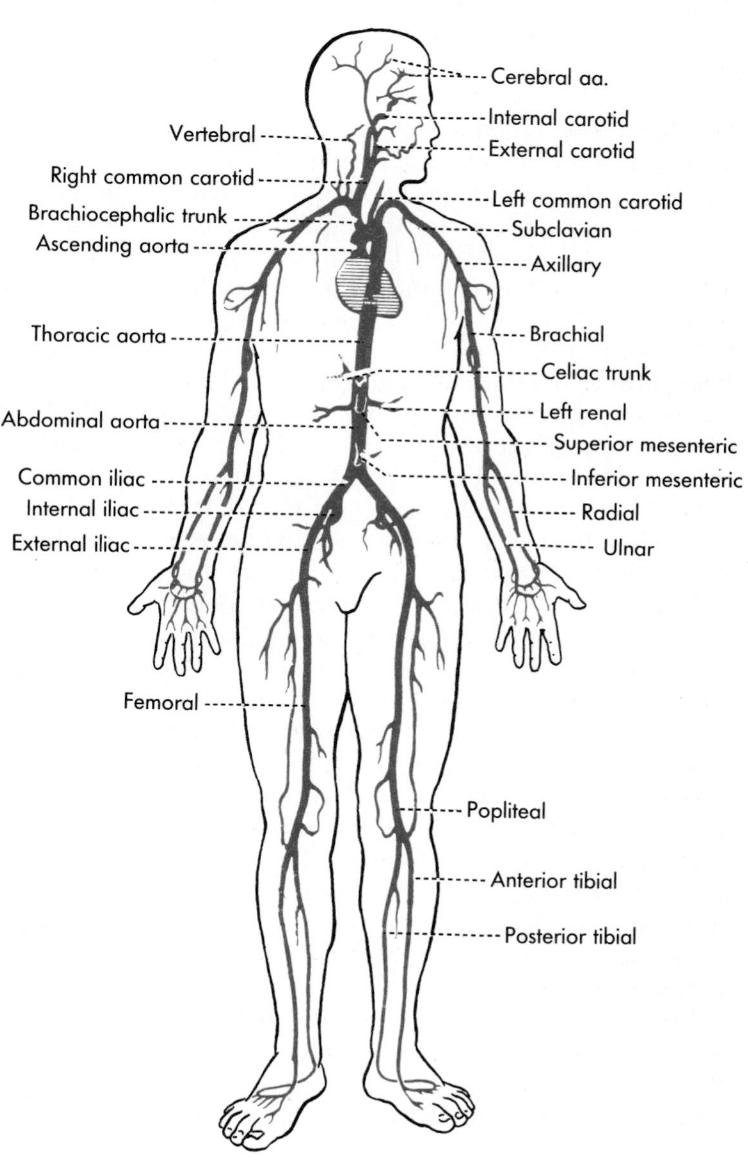

the blood supply of the brain. The subclavian artery leaves the neck by passing across the first rib, between this and the overlying clavicle, into the axilla (armpit), where its name is changed to **axillary artery.**

The axillary artery gives off a number of branches to the shoulder, the uppermost part of the arm, and the thoracic wall. As it leaves the axilla to run down the medial side of the arm, it becomes the **brachial artery.** The brachial artery ends in front of the elbow by dividing into **radial** and **ulnar arteries,** which course downward on the anterior aspect of the forearm in the positions indicated by their names; they end by supplying the hand and digits.

**Thorax.** Since the lungs and heart, the large organs in the thorax, receive their blood supplies from the pulmonary and coronary arteries, respectively, the chief branches of the **thoracic aorta** (thoracic part of the descending aorta) are small, supplying the thoracic wall (intercostal arteries) and the esophagus.

**Abdomen.** The **abdominal aorta** (abdominal part of the descending aorta) is the direct continuation of the thoracic aorta. The name of the vessel simply is changed as it passes through the diaphragm from the thorax into the abdomen. The abdominal aorta gives rise to a number of small paired branches, including ones to the abdominal wall, but its major branches are to the digestive tract and to the kidneys. The three unpaired branches to the digestive tract arise from the front of the aorta. The first is the **celiac trunk,** which, through three major branches, supplies upper abdominal organs (stomach, liver, duodenum and pancreas, spleen); next is the **superior mesenteric artery,** which partly overlaps the distribution of the celiac trunk but supplies especially most of the small intestine and much of the large intestine; last is the **inferior mesenteric artery,** supplying the more caudal part of the large intestine.

The arteries to the kidneys **(renal arteries)** arise from the sides of the aorta and are therefore paired, typically one to each kidney. The aorta ends in the lower part of the abdomen by bifurcating into paired **common iliac arteries.**

**Pelvis and Lower Limb.** The two common iliac arteries, the terminal branches of the aorta, are, like other paired arteries, essentially similar. Each runs downward and laterally and ends by dividing into internal and external iliac arteries.

The **internal iliac artery** (formerly called the hypogastric artery) descends into the pelvis and supplies the pelvic viscera (bladder, uterus and vagina in the female, rectum), sends branches into the buttock to supply muscles there, and sends one branch to the perineum (space between the thighs) to supply the structures there.

The **external iliac** continues, without diminution in size, into the thigh, lying at first anteromedially and in a rather superficial position. As it enters the thigh, its name becomes the **femoral artery.** As the femoral artery passes down the thigh, it comes to lie deep to certain muscles, giving off several named branches that largely end as branches to muscles in the thigh. The femoral artery eventually appears behind the knee (in the space known as the popliteal fossa), and the name of the artery here changes to **popliteal artery.** The popliteal artery in turn ends a little below the knee by dividing into **anterior** and **posterior tibial arteries,** the former passing to and running down the anterior aspect of the leg and into the dorsum of the foot, and the latter running down the posterior aspect of the leg and into the plantar surface of the foot.

## Systemic Veins

With a few exceptions (the most notable of which are the veins of the brain, the subcutaneous veins of the limbs, and, to some extent, the portal vein draining the digestive tract), the veins largely parallel the arteries and are named as they are or are called the *venae comitantes* (accompanying veins) of the arteries. Arteries of moderate size, such as the brachial, radial, and ulnar arteries, typically have a pair of veins accompanying them. The two elements of a pair usually unite by numerous cross channels.

The superficial veins of the limbs form a special system of their own, but unite with the deep veins accompanying the arteries. These veins and their tributaries lie in the subcutaneous tissue and are often visible through the skin. Those in front of the elbow are particularly easy to find and are frequently used for withdrawing blood or giving intravenous medication or feeding. The superficial veins are so valved that blood normally can pass only toward the proximal part of the limb, and their connections with the deep veins are so valved that blood normally can pass only from the superficial into the deep veins.

Finally, the veins draining the major part of the intestinal tract, although in part paralleling the arteries supplying this tract, do not empty into the venous equivalent of the aorta (the **inferior vena cava**) but carry their blood to the liver; they constitute a separate system, the **portal system** of veins. Although the portal system is systemic in comparison with the pulmonary system, for it does carry blood that has lost oxygen by passing through the tissues, it is sometimes not included among the systemic veins; instead, "systemic" is sometimes used in a restricted sense in comparing the two sets of veins in the abdomen. A general diagram of the veins of the trunk and base of the neck is shown in Figure 5-6.

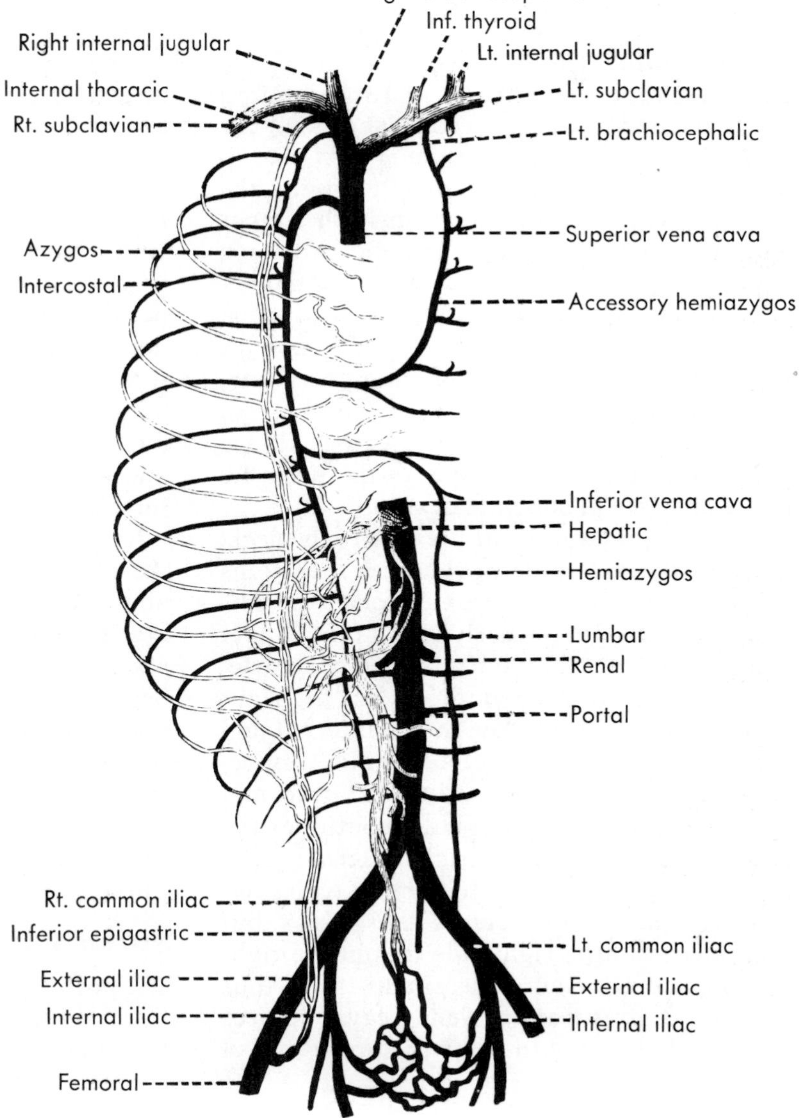


**FIG. 5-6.** Diagram of the systemic venous system. Veins of the anterior wall of the trunk are shown in outline only, and the portal system is lightly shaded. (Henle J: Handbuch der systematischen Anatomie, Bd 3. Braunschweig, Vieweg und Sohn, 1868)

**Head and Neck.** The chief veins draining the head and neck are the **internal jugular veins.** Each begins at the base of the skull, where it receives blood brought to the brain by both the internal carotid and vertebral arteries, and is a large vessel that runs straight down the neck alongside the internal and common carotid arteries. It receives tributaries from the face, jaws, tongue, and structures of the upper part of the neck, and it ends at the base of the neck by joining the **subclavian vein** (from the upper limb). Also, a smaller, superficial **external jugular vein** (jugular meaning neck) helps to drain the scalp and usually connects with the upper part of the internal jugular. Just before it ends in the subclavian vein, the external jugular receives some tributaries from the shoulder.

**Upper Limb.** The deep veins of the upper limb originate in the hand and form paired veins accompanying the radial and ulnar arteries. At the elbow, they unite to form paired **brachial veins** that end above by joining the **axillary vein.** The axillary vein is the upper end of the **basilic vein,** one of the two large superficial veins of the upper limb. As the basilic enters the axilla (armpit), its name changes to axillary. The axillary vein receives the other superficial vein of the upper limb, the **cephalic,** and veins corresponding to branches of the axillary artery. As the axillary vein leaves the arm and enters the neck, its name changes to **subclavian vein.**

The subclavian vein, at the base of the neck, receives the **external jugular** and then joins the **internal jugular.** This union forms the **brachiocephalic vein** (formerly called innominate vein). In contrast to the brachiocephalic arterial trunk, normally found only on the right side, the union of subclavian and internal jugular veins to form a brachiocephalic vein occurs on both sides of the body. After its formation, the right brachiocephalic continues the downward course of the right internal jugular, passing down into the thorax. The left brachiocephalic also enters the thorax but then passes to the right side to join the right brachiocephalic vein. The single trunk thus formed, the **superior vena cava,** passes straight downward to enter the upper end of the right atrium, receiving, just before it does so, the **azygos vein** (the chief venous drainage of the thoracic wall). A number of small veins from the neck and the thorax empty into the brachiocephalic veins.

**Thorax.** Small veins, mostly from the thoracic wall (intercostal veins) and corresponding in general to the thoracic branches of the aorta, empty into a pair of vessels that approximately parallel the thoracic aorta. This pair is the **azygos** ("unpaired") **system,** so named because it is asymmetrical. The veins of the left side empty mostly into the larger vein on the right. The largest vein of the system on the left is called the **hemiazygos vein.** The larger vein on the right is the **azygos vein.** It joins the lowermost part of the superior vena cava.

The cardiac veins, also in the thorax, have already been mentioned. The pulmonary veins are not, of course, part of the systemic venous system.

**Lower Limb and Abdomen.** The deep veins of the lower limb accompany the arteries and bear similar names. The **femoral vein,** paralleling the femoral artery, enters the abdomen, whereupon its name is changed to **external iliac vein.** The external iliac vein is joined by the **internal iliac** from the pelvis and buttock to form the **common iliac vein.** Right and left common iliac veins unite to form the **inferior vena cava.** In its course upward, the inferior vena cava and its tributaries receive vessels corresponding to the branches of the abdominal aorta, except those going to the digestive tract. Finally, just before it leaves the abdomen, the inferior vena cava receives the hepatic veins from the liver, which contain not only the blood from the arteries of the liver but also the blood from the digestive tract brought there by the portal vein. The inferior vena cava ends in the inferior part of the right atrium immediately after passing through the diaphragm.

### Portal System

The portal system of veins drains the gastrointestinal (digestive) tract from the stomach to the upper part of the rectum. The tributaries of this system unite to form the **portal vein,** which instead of returning blood directly to the heart, as do the venae cavae, delivers it to the sinusoids (essentially dilated capillaries) in the liver.

The tributaries of the portal vein generally parallel the arteries going to the digestive tract (and the artery to the spleen) and are similarly named; however, there is no single trunk corresponding to the celiac trunk of the aorta, and the portal vein itself parallels the artery to the liver **(hepatic artery).** The portal vein ends by dividing into branches that enter the substance of the liver and divide repeatedly. Eventually, they empty into the sinusoids of the liver, just as do the hepatic artery and its branches. The small veins opening into the sinusoids thus constitute the termination of the portal system of veins.

Blood from the sinusoids, whether brought in by the hepatic artery or the portal vein, is drained by the **hepatic veins.** These, as already noted, empty into the upper end of the inferior vena cava immediately below the diaphragm.

## LYMPHATIC SYSTEM

The lymphatic system begins in capillaries that begin blindly and drain tissue spaces. Deep lymphatics that drain the muscles and other tissue of the limbs and the body wall are relatively few, but accompany the arteries and veins supplying these parts. Deep lymph nodes are correspondingly scarce in the limbs but typically do occur in a few locations. The chief lymphatic drainage from the limbs and from the body wall is, therefore, by superficial lymphatics.

The lymphatics and lymph nodes of the body are sufficiently varied to make unprofitable even a reasonably brief description of them here. In synopsis, most of the lymphatics of the lower limb converge on lymph nodes located anteriorly and superficially in the uppermost part of the thigh (Fig. 5-7). These then drain upward into the abdomen, where many of the lymphatics and lymph nodes are closely associated with the aorta. The abdominal nodes also receive the lymphatic drainage from the pelvic and abdominal viscera. In the uppermost part of the abdomen, the major lymphatics unite to form the largest lymphatic vessel in the body, the **thoracic duct.** This runs upward along the posterior thoracic wall, receiving lymphatics from that, and ends at the base of the neck on the left side by emptying into the venous system at approximately the angle of junction of the internal jugular and subclavian veins.

The lymphatics of the upper limb converge upon lymph nodes situated around the axillary vessels, and those of the head and neck converge upon lymph nodes grouped especially around the internal jugular veins. Lymphatic trunks from the upper limb and from the head and neck join each other and a trunk from the thorax and empty together, or remain separate and therefore empty separately, into veins at the base of the neck on the right side. On the left side they join the thoracic duct or empty into the venous system close to the ending of that duct.

In addition to the lymphatic vessels and nodes, the lymphatic system includes collections of lymphocytes located under the epithelium of the digestive tract (tonsils, Peyer's patches), the spleen, the thymus, and the bone marrow. Apart from returning tissue fluid from essentially all parts of the body to the venous circulation, the significance of the lymphatic system lies largely in its contribution to the immune responses of the body. The organs of the lymphatic system that are concerned with the production of lymphocytes, the chief immunocompetent cells of the body, are called **primary lymphoid organs.** These include the thymus and the bone marrow. Those lymphoid organs in which the immune response is initiated against antigens are called **secondary lymphoid organs.** These include the lymph nodes, the subepithelial collections of lymphocytes, and the spleen.

The **thymus** is distinct from other lymphoid organs in that instead of being entirely mesodermal in origin, it is derived, in part, from an outgrowth of the embryonic gut (pharyngeal pouch). Its lymphoid elements develop in association with the endodermal epithelium. The **bone marrow** produces all types of blood cells (hematopoietic cells) in addition to lymphocytes.

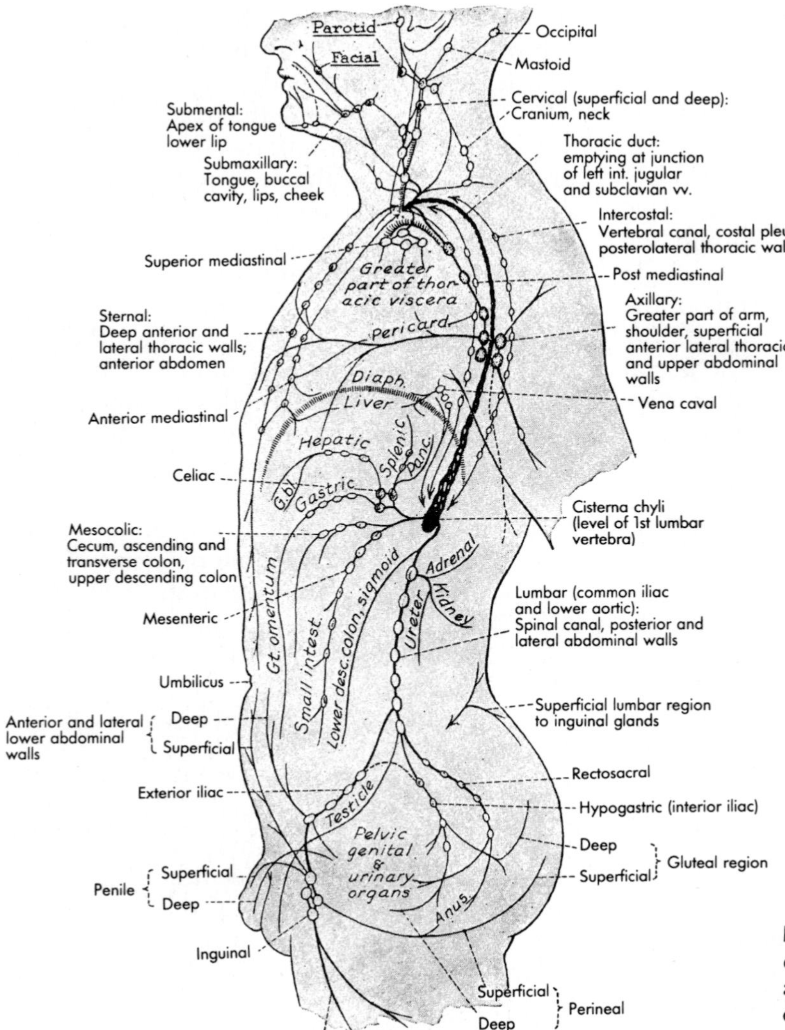


**FIG. 5-7.** Schema of the lymphatic drainage of the body. (Jones T, Shepard WC: Manual of Surgical Anatomy. Philadelphia, WB Saunders, 1945)

Antigens that are picked up in the tissues by the afferent lymph stimulate an immune response first in the lymph nodes; those that are picked up by the blood initiate a response first in the spleen.

**Spleen.** The spleen is enclosed in a part of the mesentery of the stomach, and its parenchyma resembles that of lymph nodes. Lymphatics within the spleen are confined to its capsule and to large trabeculae, so that the lymphatic nodules of the spleen add lymphocytes directly to the bloodstream instead of delivering them first into lymphatic vessels, as the lymph nodes do. Instead of capillaries, the spleen has large sinusoids lined by highly phagocytic cells. The size of the sinusoids allows the spleen to act as a reservoir for red blood cells, which accumulate in it when the splenic circulation is sluggish. The phagocytic walls of the sinusoids are the chief elements concerned with the destruction of red blood cells and the removal of the iron component from them in order that this can be used again to form new cells. If the spleen is removed, these functions are carried out by the bone marrow, lymph nodes, and liver. Under abnormal conditions, the spleen may also begin to produce red cells and myelocytes, normally produced only in the bone marrow.

## SOME DEVELOPMENTAL CONSIDERATIONS

Details of development necessary to an understanding of the variations and anomalies of specific vessels are best noted in connection with those vessels, and only general comment upon the development of the vascular system is needed here. The heart and all vessels of the vascular and lymphatic systems develop from embryonic mesenchyme. Mesenchyme is the versatile, multipotent, embryonic connective tissue formed from *mesoderm* (see Fig. 7-2), the middle germ layer of the embryonic disk. The vasculature of those organs that are primarily derived from *ectoderm* or *endoderm* (the other two germ layers) is also of mesenchymal origin.

### Blood Vessels

All blood vessels develop from capillaries, which grow by sprouting and by coalescing with other capillary spaces to form networks in which certain channels enlarge as a result of a greater amount of blood coursing through them, while others disappear completely or remain as capillaries. This method of development offers opportunity for much variation in the anatomy of the vascular system, and vascular variations are not so much to be wondered at as is the fact that developmental conditions are so relatively constant from one person to another that a basic and prevailing vascular pattern can be recognized in all. Since, from the beginning of the developing circulation, the blood leaving the heart is under greater pressure and flows faster than does blood returning to the heart, it might be expected (in analogy with a river at flood as compared to a slow and winding one) that blood leaving the heart would have the greater tendency to take the shortest and most constant route to a part and that arteries would, therefore, be less variable than veins; indeed, as already noted, this is generally true.

**Abdomen and Thorax.** The first circulations to develop are to the yolk sac and the placenta. Since the yolk sac is nonfunctional in man, this circulation is short-lived, and only the proximal parts of the yolk-sac (*vitelline, omphalomesenteric*) vessels persist as the blood supply to the gut (celiac and mesenteric arteries, portal vein). The numerous, originally paired, yolk-sac arteries become single, unpaired ones, apparently by fusion, and are reduced to three. The veins become reduced to a single pair that are interrupted in their course to the heart by the developing liver and become converted into portal veins; parts of both portal veins contribute to the single definitive portal vein.

Unlike the yolk-sac circulation, that to the placenta enlarges steadily and persists up to the time of birth as the umbilical vessels (the paired vein is reduced to a single one). After birth, the useless parts of the umbilical vessels lose their lumen but remain throughout life as fibrous cords ("ligaments").

The paired lateral branches of the abdominal aorta to the kidneys are remains of much more numerous paired vessels that originally supplied the nephrogenic ridge, especially the evanescent mesonephros, or "middle kidney." Besides the arteries to the kidneys, other vessels of this group persist as paired arteries to the gonads, the suprarenal glands, and the diaphragm.

The inferior vena cava has a particularly complicated developmental history, for all three pairs of veins formed caudal to the heart—the postcardinals (posterior cardinals), subcardinals, and supracardinals—contribute to the development of this unpaired vessel (see Chap. 25). In consequence of this complicated development, major variations of the inferior vena cava are more common than are variations in any other large vein.

The pulmonary veins are a derivative of the venous plexus on the gut, just as the lungs are an outgrowth from the gut. Anomalous openings of one or more pulmonary veins into the systemic venous system are believed to result from improper separation of the two systems.

The pulmonary arteries are derivatives of

the aortic arch system, as are the arch of the aorta and the great vessels that originate from it (Chap. 22).

**Limbs.** The main arterial stems in both upper and lower limbs have a complicated history. In each instance, the first definitive stem is formed by enlargement of one of several vessels feeding the developing limb bud, but in neither case does this stem persist throughout the length of the limb. Rather, a series of branches appear, and first one, then another, takes over the duty of supplying the distal part of the limb, so that the definitive main channels of the adult are composed of portions of several different arteries in series.

The earliest veins of the limbs are superficial and lie especially on the borders of the limb. Most of the channels along the preaxial (radial and tibial, respectively) borders of the limbs atrophy, but those along the postaxial borders tend to persist as the proximal parts of the great veins of the limbs, receiving both superficial veins and the definitive deep veins that develop distally along the arteries.

**Head and Neck.** The veins of the head and neck develop from the paired precardinal (anterior cardinal) veins, which receive the veins from the head and neck and upper limbs just as the posterior cardinal veins originally receive those from the lower limbs and the trunk. The left precardinal vein shunts its blood to the right precardinal and helps to form the superior vena cava.

The arteries of the head and neck have a particularly complicated developmental history. Most of them develop from the aortic arch system, which is similar to the system that supplies the gills in fishes and thus aerates the blood.

### Heart

The heart, in an early stage, is a single tube like the heart of a fish, receiving blood at one end and propelling it from the other into the aortic arch system. Subsequently, it becomes twisted upon itself and some of its original subdivisions disappear. Partitions that appear and separate the heart into right and left sides then produce a four-chambered heart. This differs drastically from the four-chambered heart of a fish: Instead of four chambers in series, there are two right and two left chambers, and, as noted, the blood must circulate twice through the heart in order to become aerated and be returned to the body in general.

### Lymphatics

Lymphatic capillaries originate either by outgrowth from the venous system in certain locations or by the coalescence of blind lymphatic spaces, or both. Like blood vessels, they increase in size according to the flow of lymph through them. In an early stage of development, enlarged lymphatics or lymph sacs are found at the junction of the chief veins of each of the upper and lower limbs with the cardinal veins into which they empty, and two others are found on the posterior abdominal wall. From these sacs, lymphatics grow along blood vessels: those connected with the limbs grow distally into those parts; those connected with the lower limbs also grow centrally and connect with the upper abdominal lymph sac, which gives rise, in turn, to the thoracic duct, which grows to join the sac connected with the left upper limb; the lower abdominal sac grows peripherally to the intestines and centrally to join the thoracic duct; and the sacs connected with the upper limbs either retain their connections to the venous system here (at the base of the neck) or establish new ones, so that all lymph must return to the bloodstream here.

Anomalous openings of the lymphatic system into the venous system (in locations other than the base of the neck), although rare, indicate the close developmental relationship between veins and lymphatics.

## GENERAL REFERENCES AND RECOMMENDED READINGS

ADACHI B: Das Arteriensystem der Japaner. Kyoto, Kenkyusha, 1928

ADACHI B: Das Venensystem der Japaner. Tokyo, Kenkyusha, 1933, 1940

BREMER JL: The earliest blood vessels in man. Am J Anat 16:447, 1914

BURTON AC, YAMADA S: Relation between blood pressure and flow in the human forearm. J Appl Physiol 4:329, 1951

CLARK ER, CLARK EL: Microscopic observations on the growth of blood capillaries in the living mammal. Am J Anat 64:251, 1939

CONGDON ED: Transformation of the aortic arch system during the development of the human embryo. Contrib Embryol 14:47, 1922

DAVIS CL: Development of the human heart from its first appearance to the stage found in embryos of twenty paired somites. Contrib Embryol 19:245, 1927

DORR LD, BRODY MJ: Functional separation of adrenergic and cholinergic fibers to skeletal muscle vessels. Am J Physiol 208:417, 1965

EDWARDS EA: The orientation of venous valves in relation to body surfaces. Anat Rec 64:369, 1936

FRANKLIN KJ: A Monograph on Veins. Springfield, Il, Charles C Thomas, 1937

GOSS CM: The first contractions of the heart in rat embryos. Anat Rec 82:466, 1942

HARVEY W: Movement of the Heart and Blood in Animals: An Anatomical Essay. Franklin, KJ (trans): Springfield, Il, Charles C Thomas, 1957

KAPLAN IW, KARLIN S: Glomus tumor: A report of two cases. Am J Surg 86:192, 1953

KINMONTH JB, SIMEONE FA: Motor innervation of large arteries with particular reference to the lower limb. Br J Surg 39:333, 1952

KRAMER JG, TODD TW: The distribution of the nerves to the arteries of the arm: With a discussion of the clinical values of results. Anat Rec 8:243, 1914

MANN FC, HERRICK JF, ESSEX HE, BALDES EJ: The effect on the blood flow of decreasing the lumen of a blood vessel. Surgery 4:249, 1938

MAXIMENKOV AN: Structural and functional peculiarities in some parts of the venous system. Anat Rec 136:239, 1960

NONIDEZ JF: Identification of the receptor areas in the venae cavae and pulmonary veins which initiate reflex cardiac acceleration (Bainbridge's reflex). Am J Anat 61:203, 1937

POLLEY EH: The innervation of blood vessels in striated muscle and skin. J Comp Neurol 103:253, 1955

QUAIN R: The Anatomy of the Arteries of the Human Body and Its Application to Pathology and Operative Surgery. London, Taylor & Walton, 1844

QUIRING DP: Collateral Circulation: Anatomical Aspects. Philadelphia, Lea & Febiger, 1949

ROUVIÉRE H: Anatomie des lymphatiques de l'homme. Paris, Masson et Cie, 1932

SABIN FR: The origin and development of the lymphatic system. Johns Hopkins Hosp Rep 17:347, 1916

WOOLLARD HH, PHILLIPS R: The distribution of sympathetic fibers in the extremities. J Anat 67:18, 1932

YOFFEY JM, COURTICE FC: Lymphatics, Lymph, and the Lymphomyeloid Complex. London, Academic Press, 1970

# 6

# THE MUSCULAR SYSTEM

There are three types of muscle: cardiac muscle, peculiar to the heart; smooth muscle, found particularly in thin sheets forming part of the wall of hollow organs such as the digestive tract and blood vessels; and skeletal striated muscle. Although different histologically, cardiac and smooth muscle are conveniently grouped together and contrasted with skeletal striated muscle in their physiology. Cardiac muscle markedly, and smooth muscle to some extent, possesses an intrinsic ability to contract, doing so in the absence of impulses from the central nervous system. Both are normally under the control of the involuntary (autonomic) nervous system, and both are therefore sometimes termed "involuntary muscle."

Skeletal muscle, on the other hand, typically contracts only when a nerve impulse reaches it and is directly responsive to volitional control. With the exception of a few named bundles of involuntary muscle, it is the skeletal muscle fibers organized into organs that we refer to as muscles, and it is, of course, muscles that constitute the meat or flesh of those animals that we eat. It is, likewise, skeletal muscles that form the muscular system.

A skeletal muscle, usually referred to simply as "a muscle," is primarily a collection of striated muscle fibers bound together and surrounded by connective tissue, but the tendons by which it is attached and the branches of blood vessels and nerves within it are also a part of the organ.

## Naming of Muscles

Muscles vary much in their shape and size, and in their positions, attachments, and actions. All these features, singly or in combination, have been used in naming them. The trapezius, rhomboid, and deltoid muscles (of the shoulder) are named from their shapes; the latissimus dorsi is named from its size and position (broadest muscle of the back); the interossei of the hand and foot attain their names because of their positions between bones of the hand and foot; and the supraspinatus and infraspinatus are named from their positions above and below the spine of the scapula, respectively. The biceps brachii and the quadriceps femoris are named according to their shape and position; the biceps brachii has two heads of origin and lies in the arm, and the quadriceps femoris has four heads of origin and lies in the thigh. The coracobrachialis is named purely from its attachments to the coracoid process and the arm (brachium). The levator scapulae and supinator muscles are named purely from their actions, lifting the scapula and supinating the forearm, respectively. The two pronator muscles of the forearm are named according to a combination of their actions and shapes; the pronator teres is a rounded muscle that pronates, and the pronator quadratus is a quadrilateral one that pronates. The flexor digitorum superficialis and flexor digitorum profundus are named from their actions and positions, both being flexors of the fingers but one lying superficial to the other. Thus, for the most part, names of muscles are descriptive of some particular feature of the muscle. The student will find it advantageous to attempt to understand why a muscle is so named, since such an understanding will minimize what would otherwise be a task of rote memory.

## Structure

Although some muscles are thin and ribbonlike, some broad and flat, some fan-shaped, and some almost cylindrical with tapering ends, they are all essentially similar in structure. From the standpoint of strength, the most important variation among muscles is not their size but the arrangement of the fiber bundles that form the muscle.

The surface of a muscle is a relatively distinct but thin layer of connective tissue, variously known as its **fascia** or the **epimysium.** In some locations, it is fused to overlying deep fascia and therefore appears to be much thickened. Loose connective tissue or fat often lies between the fascia of one muscle and that of the next so as to afford free play of the muscles across each other. "Cleaning" a muscle in the dissecting laboratory consists of removing adherent dense or loose connective tissue and any fascia that is thick enough to interfere with a clear view of the direction of the muscle fibers. The fascia or epimysium must be removed by sharp dissection because it sends connective tissue septa, together constituting the **perimysium,** into the muscle, subdividing it into smaller bundles. Blood vessels and nerves run in the septa, which branch and rebranch and finally become continuous with a very thin layer of connective tissue, the **endomysium,** that surrounds each individual muscle fiber.

The connective tissue of a muscle is largely collagenous, but some elastic fibers occur among the collagenous ones. The endomysium is composed of the delicate network of collagen fibers usually described as reticular connective tissue. A meshwork of blood capillaries surrounds the individual muscle fibers. Small nerves and lone nerve fibers wind among the muscle fibers and branch to end upon them.

Within a muscle, muscle fibers are bound together into muscle bundles or **fascicles** (fasciculi) by the connective tissue that surrounds them. Within a single fascicle, there may be some muscle fibers that run its entire length (fibers as long as 34 cm have been reported, but most muscle fibers in man are no longer than 10 cm to 15 cm, and many are much shorter); however, within a fascicle, there also will be many muscle fibers that have tapering ends that overlap tapering ends of other fibers and are so bound to them by connective tissue that several fibers in a row can act as a single fiber. Fibers as long as 3.04 cm and as short as 0.14 cm have been found within a single fascicle of a rabbit's muscle.

## Range and Strength of Contraction

The length of fasciculi and their relation to the long axis of the muscle (which is the line along which the muscle pulls), and not the length of individual muscle fibers, determine the range and strength of contraction of a muscle. The statement that an individual muscle fiber can shorten to about half of its relaxed length applies to the fasciculi of a muscle, regardless of whether these fasciculi are made up of short or long fibers and how long the fasciculi are. In practice, therefore, the amount of shortening that a muscle can undergo depends upon the length of the muscle substance between the tendinous, noncontractile parts of the muscle. A fasciculus that is 2 inches (5.0 cm) long can contract until it is about 1 inch long, while one that is only 1 inch long can contract until it is about ½ inch long—the same percentage of contraction, but in one case resulting in a shortening of approximately 1 inch and in the other resulting in shortening of about half that.

Of two muscles with fasciculi parallel to the line of pull, the longer muscle will, therefore, produce a greater amount of movement from any particular attachment than will the shorter muscle. Since the strength of a muscle fiber is proportional to its diameter, it also follows that the strength of two muscles, both of which have parallel fiber bundles, can be compared quite directly by measuring the cross-sectional areas of the two muscles. Muscles with parallel fibers and fiber bundles tend to be flat and thin, since in this way the increase in diameter that the fibers undergo as they contract can be most easily accommodated. In general, therefore, such muscles tend to be relatively weak. Many muscles of this type are, however, rather long from tendon to tendon and have a particularly long range of contraction.

Since the strength of a muscle depends upon the cross-sectional area of its fasciculi, a stronger muscle results if there are more fasciculi. The easiest way to add these without markedly increasing the bulk of the muscle is to have the fasciculi set at an angle to the line of pull. This arrangement of fasciculi does indeed occur in many muscles (Fig. 6-1). In the simplest type, fiber bundles converge obliquely on one side of a tendon to give the muscle somewhat the appearance of half a feather; this type of muscle is known as a

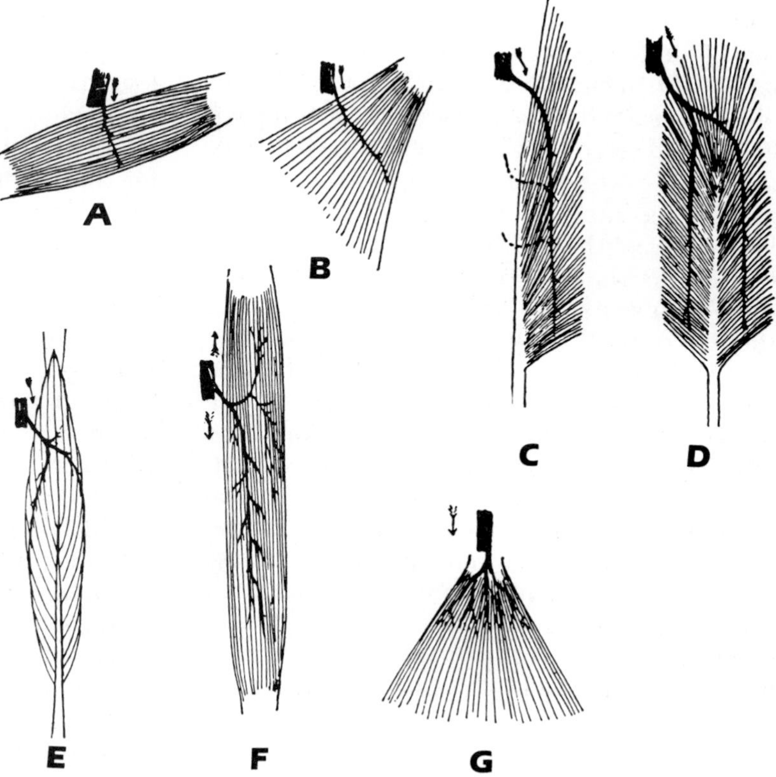


**FIG. 6-1.** Types of nerve distribution in muscles of different fascicular arrangement. **A** and **F** are muscles with parallel muscle bundles, **B** and **G** are fan shaped, **C** is a unipennate muscle, **D** is a bipennate one, and **E** is a fusiform one. (Bardeen CR: Am J Anat 6:259, 1907)

**unipennate muscle.** In a muscle with parallel fibers, the strength of the muscle is proportional to the cross-sectional area of the entire muscle, but in a unipennate muscle the cross-sectional area of the fibers far exceeds the cross-sectional area of the muscle as a whole; hence, a unipennate muscle is much stronger for a given diameter than is a muscle with parallel fibers. In a **bipennate muscle,** two sets of muscle fasciculi converge upon a centrally placed tendon, giving the muscle the appearance of the whole of a feather. Again, the strength of the muscle is determined by the total area of cross section of its fasciculi. Still more complicated forms of muscle occur, for the fiber bundles may arise from a central tendon and curve around to insert on a central tendon, giving rise to a muscle that is usually fusiform in shape and is often described by the rather general term **fusiform muscle;** or a muscle may consist of a number of bipennate parts fused together to give a particularly complex structure **(multipennate).**

Since the range of contraction of a muscle is determined by the length of the fasciculi between tendons, not by the total length of the muscle, this distance is always less in a pennate or fusiform muscle than the total length of the muscle. It thus follows that the range of action and strength of action of a muscle vary inversely; other things being equal, the greater the range of action of a muscle, the less its strength, and the greater its strength, the less its range of action. Although it may seem that a muscle fiber set at an angle to the line of pull would shorten along this line of pull less than a fiber directly in the line of pull, the increase in diameter of the fiber apparently makes up for this difference, so that for practical purposes the maximal shortening of any muscle approximates half the length of its bundles from tendon to tendon, even when these tendons lie within the muscle.

## Leverage

Although the foregoing statements concerning range of action and strength hold true for muscles themselves, the factor of leverage enters into the range and strength of the **movement** produced by muscle. Of two muscles of the same strength crossing and acting upon a joint, the muscle attaching closer to the joint will have less leverage and therefore produce less powerful movement than the muscle attaching farther from the joint. However, at the same time, an angular pull exerted close to the joint will move the distal part of the member through a far greater arc than will a similar angular pull exerted more distally from the joint, so that the muscle inserting closer to the joint will produce the greater range of movement and that inserting more distally will produce the lesser range (Fig. 6-2). Maximal strength and maximal range of movement are therefore incompatible, whether the arrangement of the fasciculi or the leverage of the muscular attachment is considered, because strength and range vary inversely. Both the structure of a muscle and its leverage must be taken into account in estimating its comparative functional strength.

## Attachments and Actions of Muscles

As noted, individual muscle fibers that end within a fasciculus are attached to each other lengthwise by the blending of the endomysium of one fiber with that of the next. Muscle fibers are attached to tendon in the same way, for the endomysium of the individual muscle fibers apparently continues directly into tendon; thus, there apparently is not, as once believed, continuity between muscle fiber and tendon fiber, but continuity between the connective tissue of the muscle and the connective tissue that is the tendon. The tendinous fibers of a muscle may be so short that they are not visible as a gross entity, and the attachment is then frequently described as being *fleshy*.

Most muscles attach to bone or cartilage, and the connective tissue of their tendinous fibers blends with the connective tissue fibers of these structures. Some muscles, such as those of the face, attach to the skin. Muscles of the tongue attach into the mucous membrane of that organ. A few, such as the tensor fasciae latae, attach into fascia. Still others form cir-

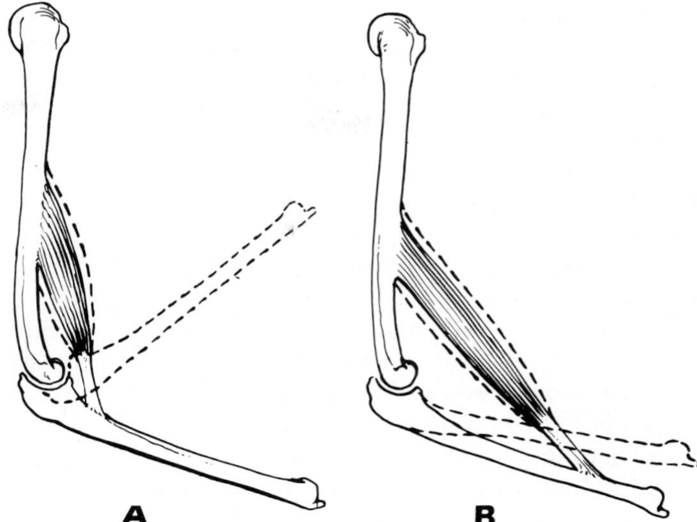


**FIG. 6-2.** Effect of the place of attachment of a muscle on the range of movement. Both muscles are shown as shortening the same amount (difference between length of solid and broken outlines), but that in **A,** attached closer to the joint, moves the lever over a far greater range (difference between solid and broken outlines) than does that of **B.**

cular bands and are termed *sphincter* or orbicular muscles, or attach to another muscle through the intervention of a tendinous *raphe* or intersection. In all instances, regardless of the site of attachment, the method of attachment is the same: a blending of the reticular endomysial and tendinous fibers with the connective tissue of the structure to which the muscle is attached.

Except for sphincteric muscles, which compress and narrow an orifice, muscles typically have attachment at both ends and tend to pull in a straight line between these attachments. For descriptive purposes, these attachments are distinguished, respectively, as the **origin** and **insertion;** the origin is defined as the more fixed attachment, the insertion as the more movable one. In the limbs, the origin is given as proximal, the insertion distal, since distal parts here are usually more movable than proximal ones. In some instances, it is difficult to decide which is the more movable end of a muscle. For example, the rectus abdominis, a muscle on the front of the abdomen, attaches to the sternum and ribs above and to the pelvis below. It acts by moving whichever part happens at the moment to be the more movable. Therefore, some texts give the origin of this muscle as being above, the insertion below, whereas others reverse this.

The **action** of a muscle is the effect of its contraction, which is usually stated in terms of the flexion, extension, or other movement of a part—typically the part that receives its insertion—that contraction produces. The reverse action, in which the insertion end is temporarily more fixed and the end of origin more mobile, is sometimes of particular importance also. For instance, the prominent muscles in the calf of the leg normally act on the heel to raise the body on the toes (plantar flex), but when one stands quietly the more fixed point is the heel, and the muscles then act on the leg to prevent the dorsiflexion (forward bending) at the ankle that would otherwise occur as the result of the weight of the body being centered in front of this joint.

The simplest way of reaching some conclusion regarding the action or actions of a specific muscle is to observe carefully in the dissecting room its origin and insertion, calculate from these the angle of pull, and thus deduce that when the muscle contracts, and the insertion therefore approaches the origin, the action of the muscle is along the line of pull. To be inclusive, the fact that the line of pull may vary somewhat with different positions of the part should also be considered. At best, however, the method of gross observation and deduction, useful though it is, can do no more than indicate what the actions of a muscle *can* be; if accurately done, it will eliminate some actions from the realm of possibility, but it cannot show that the muscle actually has all the actions attributed to it.

Deductions as to the actions of muscles based upon their anatomy can be confirmed in part or modified by the method of electrical stimulation of living individual muscles *in situ*. Again, however, this method simply shows what a muscle can do when it acts as an isolated unit, not what it does under normal circumstances. Few movements are normally produced by the contraction of a single muscle. Therefore, to determine which muscles are involved in a certain action, it is necessary to find out which are contracting and to analyze the possible reasons for the contraction of each. It is possible to detect, by palpation, contraction in superficial muscles, and sometimes in deeper ones when overlying muscles are relaxed or paralyzed. Excellent studies of muscle action have been carried out by this method, particularly when both normal and partially paralyzed subjects were available. Similarly, the technique of electromyography allows permanent records of the contraction of muscles to be made during various actions. Both these methods of determining contraction depend, however, on recognizing with certainty the muscle being investigated and in many instances only prove that a given muscle contracts during a given action, not that it is actually responsible for that action. The latter proof may be difficult to obtain, for in any movement, a number of muscles that do not directly produce the observed action contract during that action.

For any movement, unless it is carried out by the aid of gravity only (in which case muscles producing the opposite action will contract in order to control the movement), the contraction of one or more muscles is necessary. The muscle or muscles that are believed to produce that movement are often called **agonists** or **prime movers.** At the same time that an agonist contracts, other muscles also contract in order to steady or assist the movement. For instance, muscles at the shoulder joint contract in order to fix the arm when the forearm is flexed and extended. The flexors and extensors of the wrist contract in order to stabilize the wrist in a favorable position when the fingers are to be tightly clenched. Such muscles are conveniently known as **fixator muscles** (also as synergists, but this term has come to include *all* muscles that seem to contract in order to aid a certain prime mover, whether as additional agonists or as fixators). Finally, for some movements, **antagonists** (muscles carrying out actions directly the opposite of those of the prime mover) also contract, instead of relaxing, and then, by slow relaxation as the agonist contracts, help to steady the movement. Any muscle is, at various times, an agonist, a fixator, or an antagonist, depending upon the action being considered. Although it is usually easy to decide which muscles are the antagonists in a given movement, it may be difficult to decide when a muscle is acting as a prime mover and when as a fixator.

For these several reasons, the real actions of a number of muscles are not accurately known, and there are a few muscles to which diametrically opposed actions are attributed by different texts. In the present book, the functions given are based, wherever possible, on the combined information obtained from anatomic analysis and the results of muscle stimulation, electromyography, and clinical studies. This combination offers a better clue to function than any one method alone.

## Structure of Tendons

It has already been noted that tendons consist of heavy parallel bundles of collagen fibers. However, while the large bundles composing the tendon are parallel to each other, each large bundle is made up of smaller ones that intertwine, so that the pull of muscle fibers upon it is spread throughout the whole bundle rather than concentrated on individual fibers (Fig. 6-3). In the same way, as a tendon reaches its insertion, the fibers of the bundles fan out and intertwine in such a fashion that pull of any part of the muscle is spread over a considerable part of the insertion, and, as

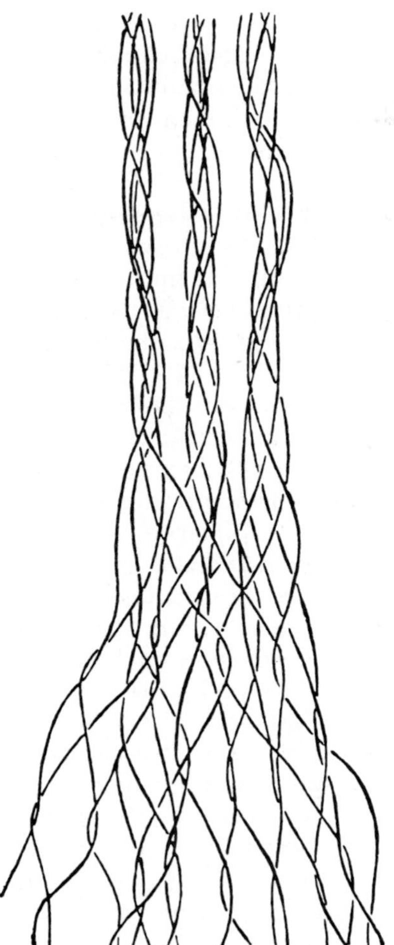

**FIG. 6-3.** Diagram of the lower end of a tendon. The large tendon bundles above, although composed of intertwining subsidiary bundles, parallel each other; as they approach their attachment to bone, in the bottom of the figure, they lose their identity by intertwining with each other. (Mollier G: Morphol Jahrb 79:161, 1937)

movement alters the angle at which the tendon reaches its insertion, there are always some tendon fibers in the straight line of pull.

Tendons are enormously stronger than relaxed muscle for any given cross-sectional area. Tensile strength of human muscle has been said to be about 77 lb/in$^2$ (that is, relaxed muscle will rupture when subjected to that amount of pull), and that of tendon from 8600 to almost 18,000 lb/in.$^2$ Even when maximally contracted, muscle does not approach the strength of tendon; a tendon is stronger than the much larger muscle belly of the muscle of which both are parts. The usual ratio is that a tendon can resist at least twice the pull that the muscle can put upon it. In consequence, when a sudden strain is thrown upon a muscle and its tendon, a tendon that is otherwise normal does not rupture. Instead, the muscle itself will rupture across its belly or at the musculotendinous junction, or the tendon will pull loose at its insertion, often pulling out with it a fragment of the bone to which it is attached. Although tendons ruptured through their middles do occur clinically, especially about the shoulder, this seems to always be a result of previous pathologic weakening of the tendon.

## Development of the Muscular System

Voluntary muscle arises entirely from mesoderm. Some muscles are derived directly from that part of the condensed mesodermal somite known as the *myotome* or muscle plate. Others are derived directly from the looser mesoderm known as mesenchyme, which may or may not be derived from the myotome. Mesodermal cells fuse, becoming multinucleated, and thus are converted into voluntary muscle fibers.

Generally speaking, the musculature connected with the vertebral column originates from the myotomes, and it has been believed that the anterolateral muscles of the trunk are likewise derived by migration from ventral parts of the myotomes. More recently, some workers have claimed, others have denied, that this is true only of those parts nearest the vertebral column and that the more anterior parts originate from the unsegmented mesenchyme of the body wall lateral to the somites. The muscles of the limbs certainly originate largely or entirely from this unsegmented mesenchyme. Some of the muscles of the head (those moving the eyeball and those of the tongue) are derived from rudimentary myotomes occurring in the head region, but many muscles of the head (those concerned with the jaw, the face, the pharynx, and the larynx) arise, like those of the limbs, directly from mesenchyme. The origin of this mesenchyme is known to be from the branchial arches, which are visceral structures since they are a part of the wall of the pharynx. Thus, skeletal striated muscle embryologically is of two types: **somatic** (derived from somites or from mesenchyme of the body wall) and **branchiomeric** (derived from branchial arches—mesenchyme of the pharynx). Histologically and physiologically, however, the two types of voluntary muscle cannot be distinguished.

Individual muscles differentiate from the myotomes as a result of several processes. One is a fusion of adjacent myotomes, so that a muscle may extend for a number of segments instead of being strictly segmental. Another is a tangential splitting of the muscle mass, so that, for instance, several layers of muscles are formed, as on the dorsal aspect of the vertebral column. Here the myotomes have fused superficially to form long muscles, but less fusion has occurred deeply so that the deeper layers of muscles, split off from the superficial layer and each other, become successively shorter as the vertebral column is neared. And, finally, muscle masses, or parts of them, may migrate, as do those forming some of the muscles of the shoulder, or may undergo a change in direction so that they no longer have a direction of pull along the long axis of the body, the original orientation of the myotome. These phenomena are seen in part in muscles of the back and even more in the anterolateral muscles; although the myotomal derivatives are kept apart in the thoracic region by the ribs, they have fused in the abdomen to form muscles extending in general from the thoracic cage to the pelvis. Ventral parts unite to form the rectus muscle, which runs longitudinally between the thoracic cage and the pelvis. More lateral parts split tangentially into three layers, external oblique, internal oblique, and transversus, which in turn have directions of pull different from those of the rectus and different from each other.

The earliest development of muscles derived from mesenchyme is indicated by a condensation of this mesenchyme into premuscle masses. The differentiation of individual muscles from these premuscle masses involves the same processes of splitting and migration that occur in the development of musculature derived from somites.

**Development of Nerve–Muscle Relations.** The nerves grow into premuscle masses early, while the masses are still closely associated with the nervous system. As the muscle masses migrate, the nerves typically maintain their original connection to the muscle masses. Hence, although it is impossible to follow, embryologically, the origin of many muscles derived from mesenchyme, the level of origin of the muscle, regardless of its final position, can, for practical purposes, be determined by ascertaining its innervation. For instance, most of the muscles of mastication are obviously derived from the first or mandibular branchial arch, since they are innervated by the nerve of this arch, which is the mandibular branch of the 5th cranial nerve. Similarly, the latissimus dorsi, although spreading so far caudally as to attain an attachment on the pelvis, is actually derived from mesenchyme originating at the level of the lower cervical region, and its nerve contains fibers from the lower cervical nerves. The fact that most muscles of the limbs receive their nerve supply from two or more spinal nerves is regarded as evidence that these muscles originated from mesoderm associated with two or more segments of the body, even though these muscles cannot be traced to the myotomes themselves. Similarly, variations in the exact segmental innervation of muscles, which are common, are attributed either to unusual segmentation of the body as a whole or to origin of a muscle in part from a slightly higher or lower position than is usual for it.

Since the relationship between developing nerves and muscles is so close and, apparently, so nearly constant, nerves tend to parallel the muscles in their development. Thus, when the developing musculature begins to be subdivided into one part dorsal to the vertebral column and one part ventrolateral to the column, the spinal nerves divide into dorsal and ventral branches for these parts.

The musculature of the limbs is innervated by the ventral branches of spinal nerves, just as are the other ventrolateral muscles. Since the nerves contributing to a limb form plexuses in which the original segmentation of the nerves is lost, it becomes difficult to follow any nerve through the plexus to the muscles that it innervates. However, it is obvious that the branches of the plexus develop with the musculature, for in both upper and lower limbs, the first subdivision of the developing musculature is into anterior (ventral) and posterior (dorsal) parts and the first subdivision of the plexuses of the two limbs is into anterior and posterior parts. The anterior nerves are then distributed to the musculature developing from the anterior premuscle mass, and the posterior nerves are distributed to posteriorly developing musculature. Further differentiation of the nerve proceeds as the muscle masses split into individual muscles, each of these masses carrying with it its own nerve or nerves.

## Innervation and Blood Supply of Muscle and Tendon

Certain aspects of the innervation of muscles have already been mentioned in Chapter 4 in connection with spinal nerves.

**Peripheral Innervation.** From what has been said concerning the close developmental relation between muscle and nerve, it should be clear that the innervation of a muscle is not haphazard, with a muscle receiving nerve fibers from any nerve in its neighborhood. Rather, it is fixed early in development. It is easy to see, in the course of a series of dissections, that a muscle tends to be innervated constantly by branches from a certain major nerve, regardless of what other nerves may be spatially available to it. For instance, either of the two major nerves on the anterior side of the forearm could supply all the muscles there, yet each supplies only certain muscles. Further, although the two nerves are of approximately equal size, one supplies almost all the muscles and the other supplies only a few.

Only a small number of muscles are regularly innervated by two named nerves rather than one, or sometimes by one nerve and sometimes by another; hence, these are easily learned. More confusing is the fact that some muscles, particularly those of the hand, may receive a truly anomalous innervation, that is, an innervation from a nerve that would not be expected to supply them. This apparently results from nerve fibers growing down the wrong one of two nerves that originate and run close together.

Because of its specificity, the peripheral nerve supply of a muscle is just as much a part of that muscle's description as are its origin, insertion, and actions. All are necessary to an appreciation of the muscle's role in the body.

**Segmental Innervation.** Some few muscles (the shortest ones of the back and those between the ribs) actually extend only the length of one segment (from one vertebra to the next or one rib to the next) and are obviously derived from a single original segment of the body. As one would expect, these are supplied by the single spinal nerve belonging to their particular segment.

Certain other muscles (the larger muscles of the back, the anterolateral abdominal muscles) extend over a number of vertebral segments and therefore lie at the levels of a number of the segmental spinal nerves. They, thus, might be expected to be supplied by a number of nerves. Indeed, it happens that these muscles are supplied by nerve branches (dorsal branches for the back muscles, intercostal nerves for the abdominal ones) that maintain the original segmentation of the spinal nerves; hence, these plurisegmental muscles can easily be shown to have a plurisegmental innervation. In general, the muscle is innervated by the nerves of the segments over which it extends, and the longer the muscle, the greater its segmental nerve supply.

Thus, by direct observation, the embryology and gross anatomy of the muscles of the trunk indicate that a muscle is innervated by the spinal (segmental) nerve or nerves associated with the segment or segmental level from which the muscle arose. In contrast, it is not possible in the limbs to establish, embryologically, the segmental origin of most of the muscles, nor, because of the plexuses, is it usually possible to determine, anatomically, the segmental innervation of the individual muscles. Rather, certain assumptions are made in regard to the embryology of the limbs, and the segmental innervation of the muscles is determined largely by clinical means.

The first assumption is that the muscles (and skin) of the limbs are derived from segments corresponding to the spinal nerves contributing to the plexuses of the limbs. This is a reasonable assumption since the limbs are at first relatively broader than they later become and their bases do lie at about the levels of the nerves that grow into them. The second assumption, argued from the knowledge that most limb muscles are supplied by at least two spinal nerves, is that limb muscles are typically plurisegmental in their origin and that, in analogy with the muscles of the trunk, the segmental innervation of a muscle is a clue to its segmental origin. Neither of these assumptions helps to determine the segmental innervations of the muscles of the limbs, but they do rationalize knowledge gained by other means.

Electromyography (recording the electrical changes associated with muscular activity, Fig. 6-4) is probably the most accurate means of assessing the segmental innervation of a muscle, since during the early stages of degeneration of its nerve, a muscle undergoes a functional change (fibrillation) that can be detected by recording electrodes placed in the muscle. Like most clinical methods, this one involves certain difficulties in technique and in the interpretation of results.

Since most movements are carried out by several muscles contracting simultaneously, it is frequently impossible, in clinical practice, to say that a certain muscle is not partici-

**FIG. 6-4.** Electromyograms, or tracings of action potentials of muscle. Electrical activity was detected by a needle electrode inserted into the muscle and recorded on a cathode ray oscillograph. An upward deflection indicates a change of voltage in the negative direction at the electrode. **A** is a recording from a normal resting muscle and shows no electrical activity; **B** is from a normal muscle during voluntary contraction and records the action potentials of motor units close to the electrode; **C** is a recording from denervated muscle showing action potentials of fibrillating muscle fibers. The time signal (wavy line below each recording) is 100 cycles/sec. The amplitudes shown in **B** cannot be directly compared with those in **C**, for the recording conditions were different: The amplitudes in **B** range up to 1 millivolt, whereas those in **C** range up to only 0.1 millivolt. (Courtesy of Dr. E. H. Lambert)

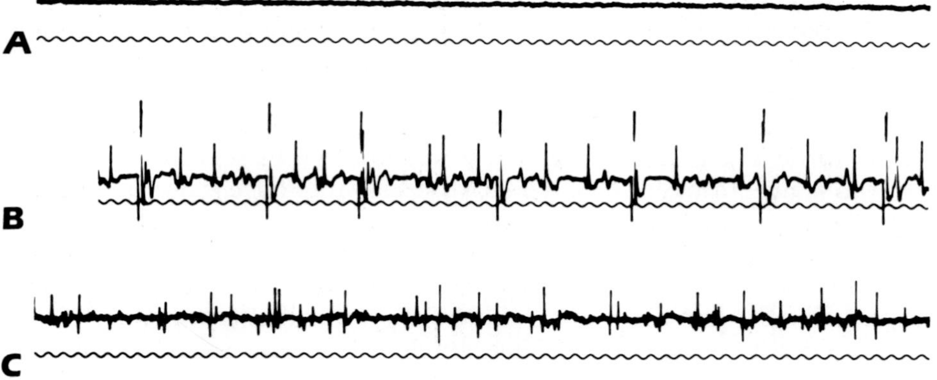

pating unless it is superficially placed and can be palpated with certainty. Further, it is not always known whether a nerve or a group of nerves has been completely or partially interrupted and, therefore, whether a continuing, but weakened, action of a muscle results from an incomplete lesion of all its segmental nerves or from the unimpaired function of segmental nerves not involved in the lesion. Segmental innervations are not accurately known, and tables listing such innervations typically vary to a considerable degree among themselves. How much of this variation is due to actual variation in the innervation and how much to incomplete knowledge is not apparent. However, it is fair to assume that segmental innervations do vary somewhat, depending upon the early relationships between the developing musculature and the spinal nerves. In amphibia, for instance, it is possible to transplant the developing limb into such a position that only some of the nerves normally reaching it will do so, and so that one or more nerves normally coursing past it will grow into the limb and innervate some of its muscles. Variations in the construction of the limb plexuses in man indicate that the relation between the limb bud and the nervous system may vary spontaneously and that in some cases there must be real variations in segmental innervations from one person to another.

Although accumulated data obtained by combined anatomic and clinical means are often sufficiently accurate to be of great value in diagnosis of injuries to the spinal cord or the nerve roots, they as yet do not fit into any regular anatomic pattern: two adjacent muscles apparently may or may not have the same segmental innervation, and muscles placed differently on the limb may have similar segmental innervations. For this reason, segmental innervations are usually looked up in texts, as necessary, rather than memorized. Only two very general "rules" seem of much aid in organizing a concept of a pattern of segmental innervation. One is that the more proximal muscles of a limb, whether on the flexor or extensor surface, tend to be innervated by nerves from higher segments than the more distally placed muscles of the same surface. For example, the muscles on the flexor surface of the arm are innervated primarily by C5 and C6, and those in the palm of the hand are probably innervated by C8 and T1. The other "rule" is that muscles on the front of a limb usually receive their innervation through slightly higher segmental nerves than do muscles correspondingly placed on the back of the limb. For instance, the muscles on the front of the arm that flex the elbow are innervated usually by C5 and C6, and those on the back of the arm that extend the elbow are apparently innervated primarily through C6, C7, and C8. Sufficient exceptions to these "rules," particularly the second, make their application in specific instances questionable.

**Distribution of Nerves within Muscles.** Whether a muscle has a single nerve or several nerves entering it, the distribution of the nerve or nerves within that muscle typically follows a fixed pattern: the nerve branches and rebranches in the connective tissue of the muscle, and the branches follow such courses as to bring them into contact with all the muscle fibers within a muscle. The pattern of distribution of a nerve within a muscle, therefore, depends, in part, upon the shape of the muscle (Fig. 6-1). For instance, in a long, fusiform muscle, the major nerve branches must run longitudinally with the muscle bundles in order to reach all of them, whereas in a short, wide muscle, the major branch or branches may run transverse to the muscle fibers, only the smaller branches running longitudinally to reach muscle fibers at the ends of the muscle.

Every individual muscle fiber eventually receives a nerve fiber, for skeletal muscle contracts only when it is incited to do so by means of a nerve impulse reaching it. The nerve fiber forms, with the substance of the muscle, a structure known as the motor end-plate (neuromuscular junction), at which the impulse delivered by the nerve is transformed into the trigger mechanism that induces contraction of the muscle fiber. Acetylcholine released by the nerve fiber is an essential part of this mechanism, and nerve fibers to skeletal muscle are therefore known as **cholinergic fibers.** Substances that interfere with the action of acetylcholine necessarily produce paralysis (lack of contraction) of the muscle. This is the basis on which curare, the primitive South American poison adapted to clinical use and employed when complete quiescence of muscle is desired, has its effect.

Not all the nerve fibers entering a voluntary muscle are motor ones destined for the muscle fibers. Some are derived from the sympathetic system and innervate blood vessels, although these are markedly smaller in caliber than are the motor fibers to the muscle itself. In addition to motor fibers, a nerve to muscle also contains a large proportion of afferent (conducting toward the spinal cord) fibers, usually said to constitute 40% to 50% of the total number of fibers in the nerve. It is, therefore, not correct to regard a nerve to skeletal muscle as being a purely motor one.

Some of the afferent fibers have to do with the reception of stimuli that give rise to the sensation of pain. Others end in the muscle and its tendons and form endings that are stimulated by the contraction of, or the passive tension on, the muscle and tendon. These endings, being activated by happenings in their immediate environment rather than by stimuli from the outside, are among the types known as **proprioceptive endings.** The best known example is the muscle spindle, a collection of slender muscle fibers enclosed within a delicate connective tissue sheath and having branches of afferent nerve fibers intertwined among the muscle fibers. Proprioceptive endings and fibers in muscle are probably not concerned with conscious appreciation of movement of a part, but are, instead, concerned with reflexes that help maintain the desired or necessary degree of contraction in a muscle. If, for instance, the sensory roots of the nerves contributing to a given muscle or group of muscles are cut, not only are the obvious muscle reflexes—knee jerk, for instance—lost, but the resting muscle also undergoes a diminution in tone (the relatively small amount of contraction demonstrable in muscles that are not completely relaxed). Similarly, deafferentation of an entire limb leads to clumsy and poorly coordinated movements of that limb, which cannot be attributed entirely to loss of conscious sensation in the limb; instead, they are apparently due, in part, to lack of coordination among the muscles as a result of the lack of information delivered to the central nervous system concerning the state of contraction of the various muscles.

**Motor Units.** Although every voluntary muscle fiber must have a nerve fiber ending upon it if it is to grow during developmental stages and to contract after it has matured, there are probably fewer nerve fibers within a nerve entering a muscle than there are muscle fibers within that muscle. Not only do nerves branch within muscles, but individual nerve fibers also branch, so that a single nerve fiber, through multiple branches, ends upon a number of muscle fibers. Since there is no "switch" at the branching of a nerve fiber that will allow a nerve impulse to follow only one of the branches, an impulse along a single nerve fiber normally follows all the branches of that fiber and reaches a number of muscle fibers practically simultaneously, causing all these muscle fibers to contract simultaneously. The group of muscle fibers innervated by a single nerve fiber is known as a **motor unit** and represents the smallest part of the muscle that can be made to undergo isolated contraction. Only in instances in which a muscle has some motor units composed of a single muscle fiber each can one muscle fiber alone contract.

Since the strength of a muscle's contraction depends upon the number of fibers that simultaneously contract (maximal strength implying contraction of all fibers), the motor unit represents the smallest increment of strength within the particular muscle. The general rule is that the finer and more precise the action of the muscle, the smaller its motor units. For instance, muscles of the hip and thigh, which do not need to carry out precise movements, are reported to have motor units ranging from about 150 to perhaps 1600 muscle fibers; those that control the more precise movements of the thumb have much smaller motor units; and those controlling the very precise movements of the eyeball probably have motor units averaging about three muscle fibers.

Under the usual conditions of moving a part or maintaining a position, contraction is automatically rotated among various units so that fatigue is minimal; however, the more powerful the movement demanded, the more motor units must be engaged in that movement. It is by this mechanism of employing more and more motor units that we voluntarily vary the strength of a movement, although the precise mechanisms employed by the central nervous system to do this are not understood. Because we can use different motor units, we can also voluntarily contract only a specific part of a muscle. This becomes particularly important when a single muscle has two opposing actions, as does, for instance, the muscle (deltoid) that lies in the arm just below the tip of the shoulder. The anterior part of this muscle can help in flexing the arm, the posterior part in extending the arm. By choosing which movement it is we desire, we automatically start the nerve circuits that result in only the anterior or the posterior part of the deltoid contracting.

**Denervation, Degeneration, and Regeneration.** When the nerve to a muscle is interrupted, the muscle not only no longer contracts, but the muscle fibers composing it begin to degenerate. This is a slow process in which the muscle fibers gradually become smaller and are eventually largely replaced by connective tissue. Since it is brought on, at least in part, by the lack of contraction of the muscle, electrical stimulation and massage of the muscle are valuable in increasing the blood supply and slowing the degeneration of a denervated muscle. It is generally agreed that the process of degeneration in a muscle as a result of denervation is largely reversible for a period of at least 12 months in man; if nerve fibers succeed in reaching the muscle fibers within this time, the muscle can begin to function again and eventually recover most of its former structure and strength. For a time, the amount of recovery in a denervated muscle probably depends more upon the condition of the nerve to it than upon the extent of degeneration of muscle fibers. However, a permanently denervated muscle is eventually transformed into a fibrous mass in which no muscle fibers are evident.

It already has been stated that a slight twitching of the muscle, known as *fibrillation,* typically occurs in a muscle as and after its nerve degenerates (Fig. 6-4C). Fibrillation is said to involve contraction of only a part of individual muscle fibers, instead of the whole of each participating fiber, and to be the result of a heightened sensitivity to acetylcholine following denervation of the muscle. Since the electrical changes associated with fibrillation, like those associated with normal contraction of a muscle, can be recorded, the electromyogram is useful not only in studying the normal contraction of a muscle but also in determining the extent of denervation of muscles following injury to various specific nerves and in following the course of recovery of function in the muscles.

It was long thought that degenerated muscle fibers are never replaced. Recently, however, it has been established that either the nuclei of degenerated muscle cells or other cells (satellite cells) found among the muscle fibers can form new muscle fibers. The amount of degenerated muscle that can be replaced is small, however. Enlargement (and therefore increased strength) of a muscle, which can result from exercise, is a hypertrophy (increase in size) of the individual muscle fibers already present, not a hyperplasia (increase in the number of cells). Muscle fibers, and hence muscles, become larger with use and smaller with disuse. The degeneration resulting from denervation may be simply an extreme form of disuse atrophy. Since this can go on to the stage of total disappearance of a muscle fiber, however, there eventually comes a time in any denervated muscle when a large proportion or all of its muscle fibers have disappeared. The possibility of the muscle recovering function with reinnervation, then, depends entirely on whether there are any muscle fibers capable of functioning, and the extent of recovery depends on how many such fibers there are and how successfully they can hypertrophy and, thus, substitute greater strength of individual fibers for the irretrievable loss in number of fibers.

**Blood Supply.** As a general rule, an artery and one or two veins accompany the nerve into the muscle, and their larger branches accompany the nerve branches in the connective tissue within the muscle. However, blood vessels do not have the early and permanent connection with developing muscle that nerves do, and muscles that spread over a considerable distance or area frequently gain additional blood supplies from blood vessels in other areas. Thus, the blood supply of a muscle is not as specific as its nerve supply and, in contrast to the latter, the blood supply of individual muscles is seldom stated in descriptions of the muscles.

Each muscle typically has its characteristic pattern of blood supply, which may be a single set of vessels entering at its upper end, a single set entering nearer the middle of the muscle and spreading in both directions, or a series of vessels. A general rule is that the one or several blood vessels that should most obviously supply the muscle, because of the close spatial relationship between it and the vessels, usually do so. Detailed knowledge of the origin of the individual blood vessels to a muscle is rarely of practical significance. Blood vessels outside muscles typically join each other (anastomose) through branches of various sizes, so that if blood cannot reach a given part through one route, it frequently can through some other. Also, since adjacent muscles are usually supplied through the same major blood vessel, occlusion of that blood vessel will be likely to affect them all rather than any single muscle. It is true, however, that within muscles, the pattern of branching and the extent of the anastomosis between vessels vary somewhat from muscle to muscle, and these factors can be correlated, at least in part, with the clinically known fact that an injury to a given blood vessel may affect one muscle of a group much more severely than other muscles of the same group.

Muscular tissue depends upon its blood supply to deliver the oxygen and glucose necessary to its contraction and to dispose of its metabolic wastes. Death of the muscle (gangrene) results from almost complete interruption of its blood supply, and the action of the muscle is interfered with by any drastic lowering of its blood supply even though this is not sufficient to bring about death of the muscle fibers. Clinically, muscle ischemia (reduced blood supply) is marked by pain in the muscles (apparently as a result of accumulation of metabolites), by easy fatigability, and hence, by rapid loss of strength of contraction. Ischemia may be chronic, resulting from permanent narrowing of the vessel supplying the muscle or muscle group, or it may be intermittent or spasmodic, appearing and disappearing as the major blood vessel is narrowed and relaxed.

Whatever the pattern of distribution of the major blood vessels to and within a voluntary muscle, the capillaries uniting the arteries and veins are so arranged among the muscle fibers that each fiber is in intimate contact with one or more capillaries. Muscle is highly vascular, and much oozing of blood can be expected to occur at the cut surface of a living muscle, even though no blood vessel of appreciable size is severed by the cut.

Tendon, in contrast to muscle tissue, has a sparse blood supply, entirely consonant with the fact that tendon is largely a nonliving substance and has low metabolic activity. The vessels of tendons are typically continued into them from the muscle substance, but if tendons are particularly long, there frequently will be not only major vessels continuing from the muscle and usually lying on the surface of the tendon, but other vessels that join the main longitudinal vessels at irregular intervals along the course of the tendon; still others may enter at or close to its insertion into bone. It is not known to what degree changes in the blood supply of a tendon contribute to pathologic weakening of it, but blood supply is of particular importance in the healing of a severed tendon. The increased metabolic processes associated with healing demand an adequate blood supply, and it is well known that tendons that have such a supply heal more quickly, and generally with better end results, than tendons that must depend upon particularly long channels to supply them.

**Lengthening and Shortening of Muscles.** It already has been noted that the extent to which a muscle can shorten is determined by the length of muscle bundles between its tendons. Under normal conditions, therefore, each muscle has muscle fibers and bundles of a length that allow it to produce the maximal desired effect by contracting to about half of the relaxed length of the muscle bundle. The length of the muscle bundles in a given muscle is ordinarily determined, also, by the range of movement that it must allow. The muscle bundles must be long enough to allow a certain amount of movement in the opposite direction through stretching of the bundles without injuring them.

The proper relationship that exists between muscle action and fiber–bundle length is normally achieved during developmental stages but is subject to alteration during postnatal life. Contortionists, who can carry out movements beyond the ranges that most of us can attain, have obviously, by practice, increased the amount of relaxation and contraction obtainable from a muscle. Apparently, this must result from a lengthening of the muscle bundles consequent to their use beyond the normal range. Similarly, muscle bundles of a muscle that is never carried through its range of strength and contraction over a considerable period of time tend to become shorter, so that contraction to about 50% of this constantly shortened length is sufficient to provide the movement demanded. This more or less permanent shortening of a muscle is known as **contracture.**

Any permanent or prolonged shortening of muscles is referred to as contracture, but the type resulting from incomplete use is of particular importance, since it results primarily from improper stretching and limited contraction of a muscle long immobilized (as by a splint) in a shortened position, of a muscle of which the tendon has been cut, or of a group of muscles that are not stretched during movements because the opposing muscles are paralyzed. This type of contracture is for a time reversible, since it is brought about initially by a sustained contraction of the muscle, but if it is maintained over a considerable period, the muscle fibers can no longer be stretched to their original lengths even when not contracting. The shorter length that they assume is just sufficient to bring about the lessened movement demanded of them and is apparently an adaptation to this lessened demand.

Another type of muscular contracture has nothing to do with the action of muscle, but it tends to be the most severe and permanent. This results from degeneration of the muscle fibers and replacement of them by connective tissue, which subsequently shortens as does all connective tissue as it matures.

The danger of muscular contracture in a part immobilized too long or not allowed to undergo its regular arc of motion is an important reason for exercising a part through its full range of movement as early as feasible following injury to that part. In contrast to permanent shortening, which is relatively easy to produce, permanent lengthening of a muscle is difficult to obtain. In consequence, when transferring the attachment of a muscle (so that that muscle can substitute for a paralyzed one) it is necessary to consider whether it is long enough to bring about the desired action.

## GENERAL REFERENCES AND RECOMMENDED READINGS


BASMAJIAN JV: Muscles Alive: Their Functions Revealed by Electromyography, 3rd ed. Baltimore, Williams & Wilkins, 1974

BASMAJIAN JV: Electromyography comes of age: The conscious control of individual motor units in man may be used to improve his physical performance. Science 176:603, 1972

BEEVOR CE: The Croonian Lectures on Muscular Movements and Their Representation in the Nervous System. London, Adlard & Son, 1904

BLOMFIELD LB: Intramuscular vascular patterns in man. Proc R Soc Med 38:617, 1945

BOURNE GH (ed): The Structure and Function of Muscle. New York, Academic Press, 1960

BOWDEN DH, GOYER RA: The size of muscle fibers in infants and children. Arch Pathol 69:188, 1960

BOWDEN REM, GUTTMANN E: Denervation and re-innervation of human voluntary muscle. Brain 67:273, 1944

CLARK WE LeG, BLOMFIELD LB: The efficiency of intramuscular anastomoses, with observations on the regeneration of devascularized muscle. J Anat 79:15, 1945

CRONKITE AE: The tensile strength of human tendons. Anat Rec 64:173, 1936

DUCHENNE GB: Physiology of Motion: Demonstrated by Means of Electrical Stimulation and Clinical Observation and Applied to the Study of Paralysis and Deformities. Kaplan EB (trans-ed): Philadelphia, WB Saunders, 1959

EDWARDS DAW: The blood supply and lymphatic drainage of tendons. J Anat 80:147, 1946

GOSS CM: The attachment of skeletal muscle fibers. Am J Anat 7:259, 1944

GWYN DG, AITKEN JT: The formation of new motor endplates in mammalian skeletal muscle. J Anat 100:111, 1966

HAINES RW: The laws of muscle and tendon growth. J Anat 66:578, 1932

VAN HARREVELD A: Re-innervation of denervated muscle fibers by adjacent functioning motor units. Am J Physiol 144:477, 1945

HUBER GC: On the form and arrangement in fasciculi of striated voluntary muscle fibers. Anat Rec 11:149, 1916

JONES FW: Voluntary muscular movements in cases of nerve lesions. J Anat 54:41, 1920

MACCONAILL MA: The movements of bones and joints: 2. Function of the musculature. J Bone Joint Surg (Br) 31:100, 1949

MCMASTER PE: Tendon and muscle ruptures: Clinical and experimental studies on causes and location of subcutaneous ruptures. J Bone Joint Surg 15:705, 1933

PARRY W: The embryonic origin of the abdominal (ventrolateral) musculature in the albino rat. Am J Anat 122:491, 1968

PATTON NJ, MORTENSEN OA: An electromyographic study of reciprocal activity of muscles. Anat Rec 170:255, 1971

SHERMAN IC: Contractures following experimentally produced peripheral-nerve lesions. J Bone Joint Surg (Am) 30:474, 1948

STEINDLER A: The mechanics of muscular contractures in wrist and fingers. J Bone Joint Surg 14:1, 1932

SAUNDERS RL DE CH, LAWRENCE J, MACIVER DA: Microradiographic studies of the vascular patterns in muscle and skin. In Cosslett JE, Engström A, Pattee H Jr (eds): X-Ray Microscopy and Microradiography, p 539. New York, Academic Press, 1957

WRIGHT WG: Muscle Function. New York, Paul B Hoeber, 1928; Hafner, 1962

# 7

# THE DIGESTIVE AND RESPIRATORY SYSTEMS

The digestive and respiratory systems of mammals are closely connected, for the respiratory system is largely an outgrowth from the digestive system; in adult man, the two share in part a common passageway, the pharynx (Fig. 7-1). It is interesting to note that the respiratory organs of man, the lungs, arise from the region of the pharynx that in fishes is provided with gill slits and gills and, thus, has both respiratory and ingestive functions.

The **digestive system** (*apparatus digestorius*) is essentially a long tube with muscular walls and a glandular epithelial lining. Its upper end is the mouth, and its lower end is the anus; most of it lies coiled within the abdomen. Certain glands of the digestive system have become specialized and are too large to be included in the walls of the tube. These glands have grown out of the tube, assuming the status of independent organs, although still retaining ducts that connect them to the tube. These are the three pairs of salivary glands, whose ducts open into the mouth, and the liver and pancreas, which lie within the abdomen and connect with the small intestine via a ductal system.

The parts of the tubular digestive system are, in order, the mouth, pharynx, esophagus, stomach, small intestine or small bowel (with three subdivisions, the duodenum, jejunum, and ileum), and large intestine or large bowel (including cecum and vermiform appendix, colon, rectum, and anal canal). All these parts between mouth and anus are referred to collectively as the *alimentary canal.*

The **respiratory system** consists of the nose and nasal passages, part of the pharynx, the larynx (voice box), the trachea (windpipe) with its subdivisions, the bronchi, and the lungs.

The lungs and the digestive tract are surrounded by serous sacs, which line the body cavity and, owing to the small amount of serous fluid the sacs contain, facilitate the movements of the lungs and the

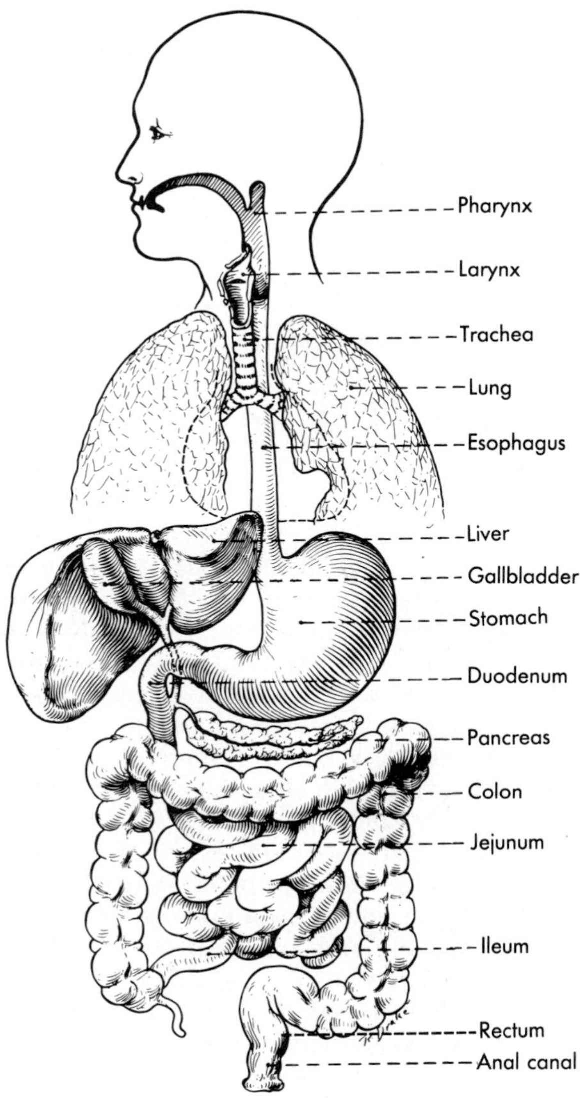


**FIG. 7-1.** Diagram of the digestive and respiratory systems. The nasal cavity, here omitted, opens into the uppermost part of the pharynx.

intestines. In the thoracic cavity, each lung is surrounded by a *pleural sac,* which encloses a pleural cavity; the *peritoneal sac,* with the peritoneal cavity enclosed in it, is associated with the part of the digestive tract located in the abdominal cavity. All these serous sacs, including the one around the heart, which encloses the pericardial cavity, develop from the primitive body cavity of the embryo called the *intraembryonic celom.*

In addition to introducing the digestive and respiratory systems, this chapter also discusses the development of the intraembryonic celom in the three-layered embryo. Much reference will be made in the remainder of this book to both the intraembryonic celom and the tissues that comprise the three-layered embryo.

## THE DIGESTIVE SYSTEM

### General Functional Anatomy

**Mouth and Pharynx.** The mouth (*cavum oris*) is separated from the nasal cavity by the palate. The back end of the palate is movable and capable of coming in contact with the posterior pharyngeal wall so as to cut off the upper (nasal) part of the pharynx from communication with the oral part of the pharynx. The highly mobile tongue largely fills the oral cavity (often referred to as the buccal cavity, although bucca really means cheek). In addition to its function in speech, the tongue is used for positioning food between the teeth so that it can be properly masticated. A number of small glands of the cheeks, lips, tongue, and palate empty their secretion into the oral cavity, as do three large pairs of glands; because they secrete most of the saliva, they are known as *salivary glands.*

The mouth opens into the pharynx through a narrow space, on the lateral walls of which are the tonsils (palatine tonsils). The larynx, the beginning of the respiratory pathway to the lungs, leaves the lower part of the pharynx anteriorly. A little below this, the pharynx narrows and is continued as the tubelike esophagus.

The musculature of the walls of the pharynx is voluntary and promotes the function of swallowing. The movements of the tongue, which tend to pass food back into the pharynx, initiate the act of swallowing and, together with the constriction of the pharyngeal muscles, increase the pressure within the pharynx. During swallowing, the oral and nasal pharynx are temporarily separated by the palate, preventing liquids in the pharynx from flowing into the nose.

**Esophagus and Stomach.** The esophagus is the continuation of the pharynx. It runs down through the neck and thorax and, as soon as it reaches the abdomen, enters the stomach.

The **stomach** (*ventriculus* or *gaster,* hence the adjective "gastric") is a saccular organ, flattened anteroposteriorly and curved along both its upper right and its lower left borders. The esophagus enters the stomach at the left upper end, and the duodenum, the first part of the small intestine, leaves it at its right lower end.

Within the stomach, the digestion of starches, initiated by saliva during mastication, is gradually halted by the effect of the hydrochloric acid secreted there. Excess acid secretion by the stomach is believed to be an important factor in the formation of gastric and duodenal ulcers. Acid-secreting (parietal) cells of the stomach are especially numerous toward its duodenal end. The stomach gradually churns the swallowed food until it becomes liquefied through mixture with the secretions. This liquid material, called *chyme,* is then passed little by little into the duodenum.

**Duodenum, Pancreas, and Liver.** In spite of its short length, the **duodenum** is particularly important because it receives the ducts from the pancreas, liver, and gallbladder. The pancreatic secretions, containing protein-splitting, fat-splitting, and carbohydrate-splitting enzymes, first begin their digestive action on the chyme within the duodenum. The common bile duct delivers bile from the liver and gallbladder to the intestinal contents; this is essential for adequate emulsification of fats, a necessary step in their proper digestion.

The **pancreas** is largely exocrine: pancreatic juice is secreted by the acini (groups of glandular cells that form the secretion) into a duct system that empties into the duodenum. These acini constitute the exocrine portion of the pancreas, which is concerned with digestion. Among the acini are scattered groups of cells that are so isolated from each other that they appear as islands. These pancreatic islets, also known as islets of Langerhans, are the endocrine part of the pancreas (Chap. 9).

The **liver** (*hepar,* hence the adjective "hepatic") is the largest organ of the body, constituting in the average adult about one thirty-sixth of the entire body weight.

The liver is an extremely vascular structure and differs from other glands in that blood flows into it from both the hepatic artery and the portal vein. The **portal vein** is formed by its tributaries, which arise within the walls of the intestinal tract. This large vein, upon entering the liver, breaks up into branches that terminate in sinusoids (dilated capillaries) within the liver. Thus, through this **portal system** of veins, blood containing freshly absorbed products of digestion is brought into intimate contact with liver cells before being returned to the general circulation through the hepatic veins. The detoxifying action of the liver on this blood is essential to life. Although mammals, including man, can survive when only a part of the liver is active, complete removal or destruction of the liver results in death within a few days. In fact, the liver is one of the major centers of chemical activity in the body; the secretion of bile is only one of a number of functions of this organ. Bile pigment, produced from the breakdown of hemoglobin during destruction of red blood cells, is removed from the blood by the liver and secreted as a component of the bile.

**Jejunum and Ileum.** Succeeding the duodenum is the jejunum, followed by the ileum, with no sharp line of division between them. Digestive activity initiated by the pancreatic enzymes in the duodenum is continued in the jejunum and ileum, which are the longest segments of the digestive tract. The small glands in the walls of these organs add their secretion to the chyme. The small intestine, with an enormous mucosal surface provided by folds of its lining and by the villi that project from these folds, is especially adapted for absorption, and most absorption of food from the alimentary tract occurs in the jejunum and ileum.

**Large Intestine.** Rather than opening end to end into the large intestine, the ileum opens into its side a little above a blind lower end;

this blind lower end is the **cecum,** and the **appendix** (vermiform appendix) is a slender projection from it. The ileum joins the large intestine in the lower right side of the abdominal cavity.

The first part of the large intestine above the cecum is the **ascending colon,** which passes up toward the liver, where it bends to the left to pass across the abdominal cavity to become the **transverse colon.** In the left side of the abdominal cavity, close to the stomach and spleen, the large intestine bends again; turning downward, it forms the **descending colon.** Finally, in the lower left part of the abdominal cavity, the colon becomes somewhat tortuous, forming the **sigmoid colon,** which descends into the pelvis. In the pelvis, the colon becomes the **rectum,** which leads to the anus by way of the **anal canal.**

The large intestine absorbs about four-fifths of the water in the intestinal contents that reach it, thus being largely responsible for the dehydration necessary to allow formed stools. It is also an excretory organ for almost all the iron, most of the calcium, and usually about half of the magnesium eliminated by the body; the remaining excretion of these substances is through the urine.

## General Structure

Below the pharynx, the alimentary canal, or digestive tract, remains essentially the same in structure throughout its length. However, there are certain structural variations along the length of the canal that endow particular parts of the tract with specific functions. Basically, the wall of the digestive tract has four layers. There is a thin, outermost layer of mesothelium and connective tissue, the *serosa;* next are the muscular layers (*tunica muscularis*); following is another layer of connective tissue, the *submucosa;* and finally, there is an innermost layer of more delicate connective tissue with an epithelial surface, called the *mucosa*. These layers are discussed from within outward.

The surface epithelium of the **mucosa** (*tunica mucosa*) gives rise to numerous glands that lie in the connective tissue of the mucosa and, in some locations, also invade the submucosa. In the esophagus, subject to possible damage from hot or cold food and from solid particles, the epithelium is stratified squamous and contains few glands; this is also true of the last part of the gut, the anal canal. Elsewhere, the epithelium is columnar and differs in detail from one region to another. In the small intestine, the mucosa forms innumerable tiny, fingerlike processes that project into the lumen and give a mossy appearance and velvety texture to the lining. Each of these tiny processes is a *villus*. Within each villus are blood vessels and a central, blindly ending lymphatic called the *lacteal* (so named because fat, which gives the otherwise clear lymph a milky appearance, is absorbed by these lymphatic capillaries). The villi enormously increase the absorptive surface of the small intestine. Since neither the stomach nor the large intestine contains villi, they are not particularly efficient organs in the process of absorption.

The most important glands of the stomach are those that secrete hydrochloric acid and pepsin, a protein-splitting enzyme. Glands of the small intestine are generally small. Some of them apparently secrete digestive enzymes similar to those secreted by the pancreas. There are also mucus-secreting glands in the small intestine, the mucus serving for lubrication. These become more numerous in the lower part of the small intestine and, in the large intestine, constitute the chief type of gland.

The **submucosa** (*tela submucosa*) is a layer of connective tissue between the mucosal and muscular layers. It is particularly vascular and also contains nerve fibers and postganglionic parasympathetic nerve cells that form a *submucous plexus*. Throughout much of the small intestine, the submucosa and the mucosa together form a series of permanent circular folds. Otherwise, the submucosa acts as padding between the mucosa and the muscular layers, allowing the mucosa to be thrown into temporary folds when the musculature contracts.

Throughout most of the digestive tract, the **musculature** consists of an inner circular and an outer longitudinal layer of smooth muscle.

Smooth muscle is composed of tapering, overlapping cells or fibers bound together by delicate connective tissue. It is called smooth because it shows none of the striations characteristic of cardiac and skeletal muscle. Smooth muscle possesses the ability to contract in the absence of nerve stimuli, although this varies according to the location of the muscle. This autonomous contractility is especially well developed in the muscle of the digestive tract, where peristaltic (rhythmic) movements occur even after complete denervation of the tract. *Peristalsis* is responsible for emptying the stomach and moving the contents of the digestive tract along the length of the intestine. This activity is influenced by the autonomic nervous system. Plexuses of autonomic nerve fibers that contain parasympathetic ganglion cells lie between the longitudinal and circular layers of the muscle (the *myenteric* or intermuscular *plexus*), which is similar to the submucous plexus. The parts of the digestive tract that have voluntary rather than involuntary musculature in their wall are the pharynx and the upper part of the esophagus; the muscle here is striated rather than smooth.

**Serosa** covers the outer surface of the abdominal part of the esophagus, the stomach, the small intestine, and most of the outer surface of the large intestine. The serosa of these organs is, in fact, their *visceral peritoneum.* This layer is present only on those surfaces of organs that are directly adjacent to the peritoneal cavity; for instance, serosa essentially surrounds the muscular layer of most of the small intestine, but on parts of the large intestine occurs only on the anterior surface. The outer surface of the serosa is a mesothelium, whereas its deeper layer, uniting it to the muscular layer, consists of connective tissue. On nonperitonealized surfaces, such as the thoracic part of the esophagus, there is connective tissue only; when this is substantial, it is called the *tunica adventitia.*

## THE RESPIRATORY SYSTEM

The **external nose** and the nasal septum between the two nasal cavities are supported partly by cartilage and partly by bone. The lateral walls of the **nasal cavities** are largely thin bone, on which are prominent ridges, the *conchae* (turbinates), that project medially and downward into the cavity. They are covered by mucous membrane and help to warm and moisten the inspired air. They occupy so much space in the nasal cavity that when they become swollen they may block the air passage completely.

The paired nasal cavities are separated by a median septum, and each opens through a narrow posterior aperture into the nasal part of the pharynx, separated from the mouth by the palate. From the nasal part of the pharynx, inspired air passes into the oral part of the pharynx and then enters the **larynx,** the entrance of which is kept permanently open by supporting cartilages. A little below the entrance, projecting folds in the larynx allow the passageway to be closed for swallowing or holding the breath, or to be narrowed so as to allow a thin column of air to escape between them for phonation. The vocal cords are one pair of laryngeal folds. The prominent anterior cartilage of the larynx (Adam's apple) is the thyroid cartilage, "thyroid" referring to its shape, which is somewhat like that of a shield.

The **trachea,** the continuation of the larynx, is a musculofibrous tube supported by a series of rings of cartilage (trachea meaning "rough") that are incomplete posteriorly. The trachea is thus held permanently open and normally allows easy passage of air. It is prominent in the front of the neck below the larynx; at the base of the neck it disappears into the thorax, where it ends by bifurcating into right and left **bronchi.** Repeated branchings of the bronchi and associated vessels of the lung form the substance of that organ, and the final bronchial branchings, the *alveoli,* provide for intimate contact between the air and blood vessels. (Although "pulmo," from which the adjective "pulmonary" is derived, is the proper name for the lung, common combining forms referring to the lung are "pneumo," and "pneumato," which actually mean gas or air.)

The pulmonary arteries that deliver deoxygenated blood to the lung enter it through its hilum, along with the bronchi. The pulmonary veins leave the lung through its hilum and return the oxygenated blood to the left atrium.

## DEVELOPMENT OF THE CELOM AND THE DIGESTIVE AND RESPIRATORY SYSTEMS

### The Embryonic Disk

At the time when the body cavity makes its first appearance, the embryo consists of a flat disk located between two other cavities: above the disk is the **amniotic cavity;** below it the **yolk sac** (Fig. 7-2A). The amniotic cavity and the yolk sac, with the disk between them, are suspended by the **connecting stalk** in yet another larger cavity called the **extraembryonic celom,** enclosed by the chorion. The orientation of the embryo is indicated by the attachment of the connecting stalk to the caudal region of the disk; the rostral, or cranial, region of the disk is opposite the attachment of the connecting stalk.

The surface of the embryonic disk facing into the amniotic cavity is covered by **ectoderm;** the surface facing into the yolk sac is covered by **endoderm.** On each side of the median axis of the disk, two longitudinal ridges of ectoderm are raised up to form the *neural folds,* the contribution of which to the nervous system has been described in Chapter 4 (Fig. 7-2B). Ectoderm and endoderm are separated from one another everywhere by the third germ layer, the *intraembryonic mesoderm,* except in two small areas: the *oropharyngeal membrane* in the rostral region in front of the neural folds, and the *cloacal membrane,* caudal to the neural folds. These membranes, composed of endoderm and ectoderm, will break down, providing two openings through which the amniotic cavity communicates with the interior of the yolk sac. The gut will be formed from part of the yolk sac.

**Intraembryonic Mesoderm.** The exterior of the amnion, yolk sac, and connecting stalk are covered by *extraembryonic mesoderm;* only intraembryonic mesoderm sandwiched between ectoderm and endoderm contributes to the formation of the body of the embryo. This mesoderm is organized in distinct structures and regions (Fig. 7-2B): (1) Along the central axis of the disk, the mesoderm forms the rod-like **notochord,** around which the vertebral column will be organized. (2) Along each side of the notochord, the so-called *paraxial mesoderm* becomes segmented into discrete units called **somites,** from which the axial skeleton and its musculature are derived (discussed in Chaps. 3 and 6). (3) **Lateral plate mesoderm** spreads as a continuous sheet along the lateral portions of the disk and extends also rostral to the oropharyngeal membrane. (4) On each side, a bar of mesoderm that lies between lateral plate mesoderm and the somites is called **intermediate mesoderm.**

### Formation of the Intraembryonic Celom

The body cavity of the embryo, that is, the intraembryonic celom, is formed by a process of cavitation in lateral plate mesoderm (Fig. 7-2C). The intraembryonic celom splits the lateral plate mesoderm into two laminae: splanchnic mesoderm and somatic meso-

**FIG. 7-2.** Schematic representation of the trilaminar embryonic disk. In **A,** the chorion has been sliced open to show the relationship of the disk to the amniotic cavity, the yolk sac, and the extraembryonic celom. **B** is a view of the ectodermal surface of the embryo seen from the amniotic cavity. A transverse section across the disk and the yolk sac below reveals the components of the intraembryonic mesoderm. In **C,** the location of the intraembryonic celom (black) is shown in lateral plate mesoderm. The left pleuroperitoneal canal is as yet closed; the right one has opened and is in communication with the extraembryonic celom. ▶

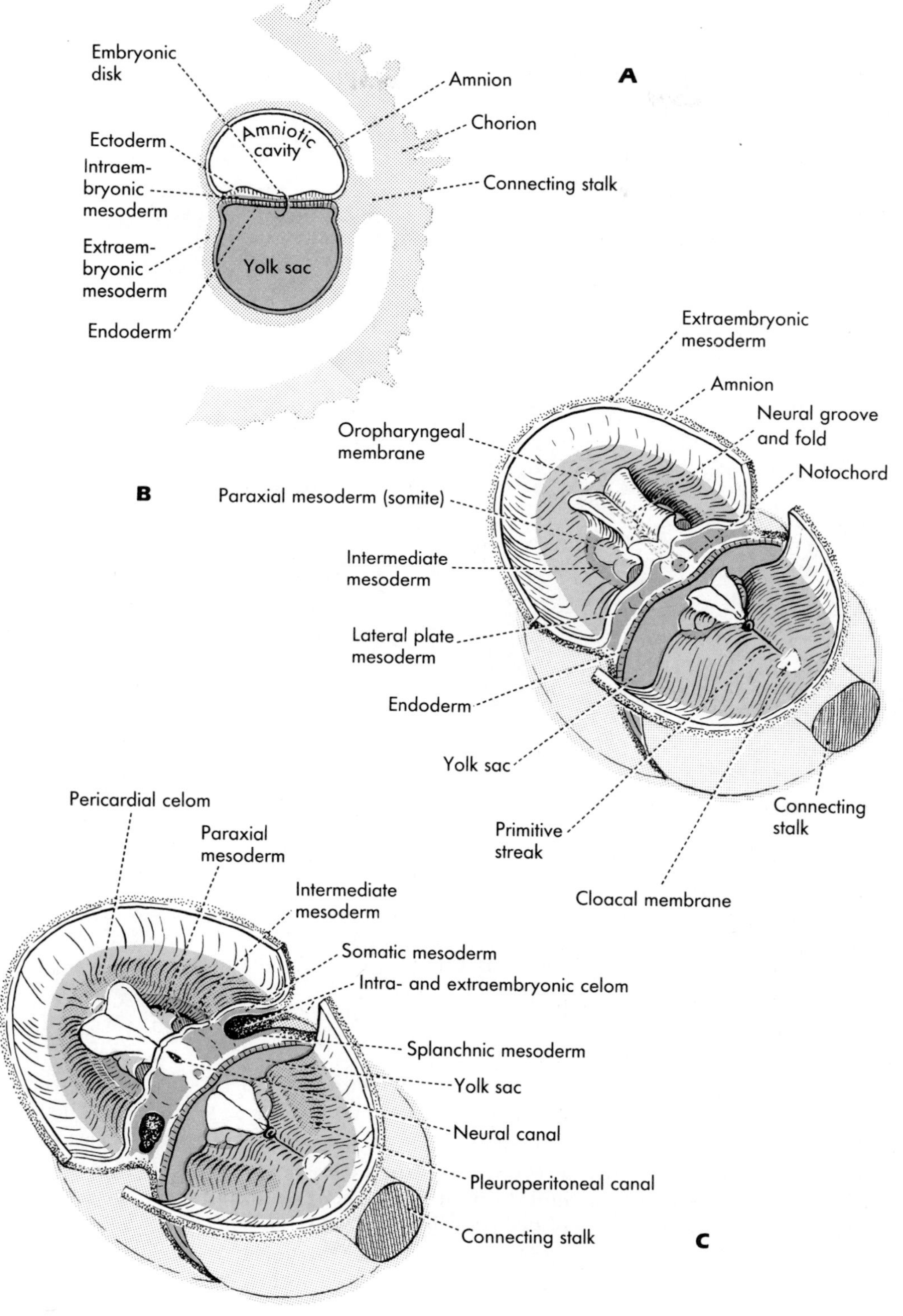
A
Embryonic disk
Amnion
Chorion
Amniotic cavity
Ectoderm
Intraembryonic mesoderm
Connecting stalk
Extraembryonic mesoderm
Yolk sac
Endoderm
B
Extraembryonic mesoderm
Amnion
Neural groove and fold
Oropharyngeal membrane
Notochord
Paraxial mesoderm (somite)
Intermediate mesoderm
Lateral plate mesoderm
Endoderm
Yolk sac
Connecting stalk
Primitive streak
Cloacal membrane
C
Pericardial celom
Paraxial mesoderm
Intermediate mesoderm
Somatic mesoderm
Intra- and extraembryonic celom
Splanchnic mesoderm
Yolk sac
Neural canal
Pleuroperitoneal canal
Connecting stalk

**FIG. 7-2.**

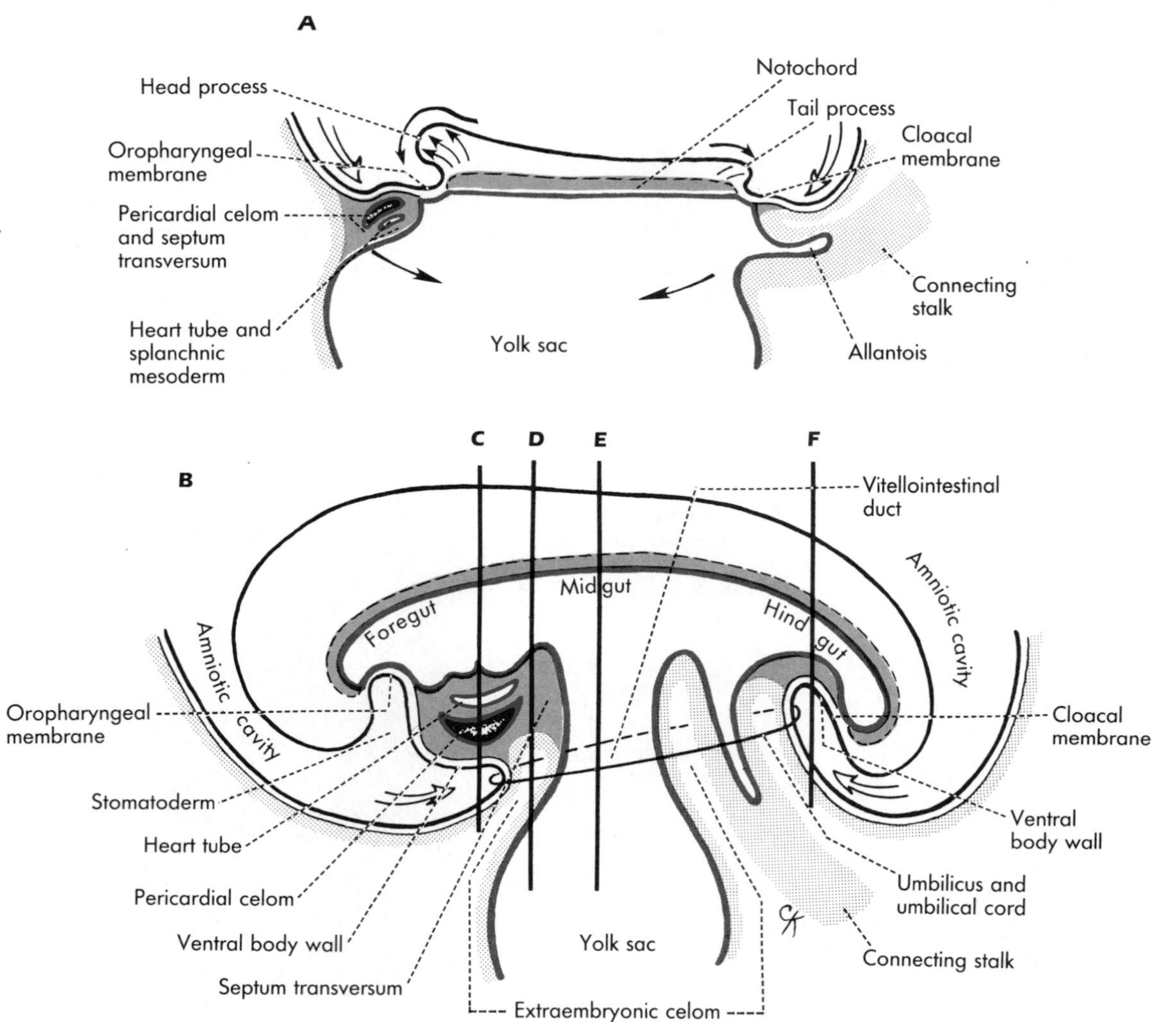


**FIG. 7-3.** Folding of the embryonic disk. **A** and **B** show two successive stages of the folding in sagittal section to illustrate the head fold and tail fold. The positions of transverse sections **C** through **F** are identified in **B. C** is a section through the pericardial portion of the celom; **D,** through the septum transversum; **E,** through the vitellointestinal duct; and **F,** through the hindgut just anterior to the cloacal membrane.

derm. The mesodermal lamina adjacent to endoderm is designated as **splanchnic mesoderm** and gives rise to all smooth muscle, connective tissue, and vasculature of the gut and its derivatives, whereas the epithelial lining of the gut and the parenchymal cells of the organs derived from the gut (liver, pancreas) are formed by endoderm. The mesodermal lamina associated with ectoderm is designated **somatic mesoderm** and, together with the ectoderm, forms most of the tissues of the body wall (Chaps. 19 and 23). The serous pericardial, pleural, and peritoneal sacs develop from those cell layers of these two mesodermal laminae that face into the intraembryonic celom.

The intraembryonic celom extends along each side of the embryonic disk, and these two celomic ducts are connected with each other rostrally but not caudally: a horseshoe-

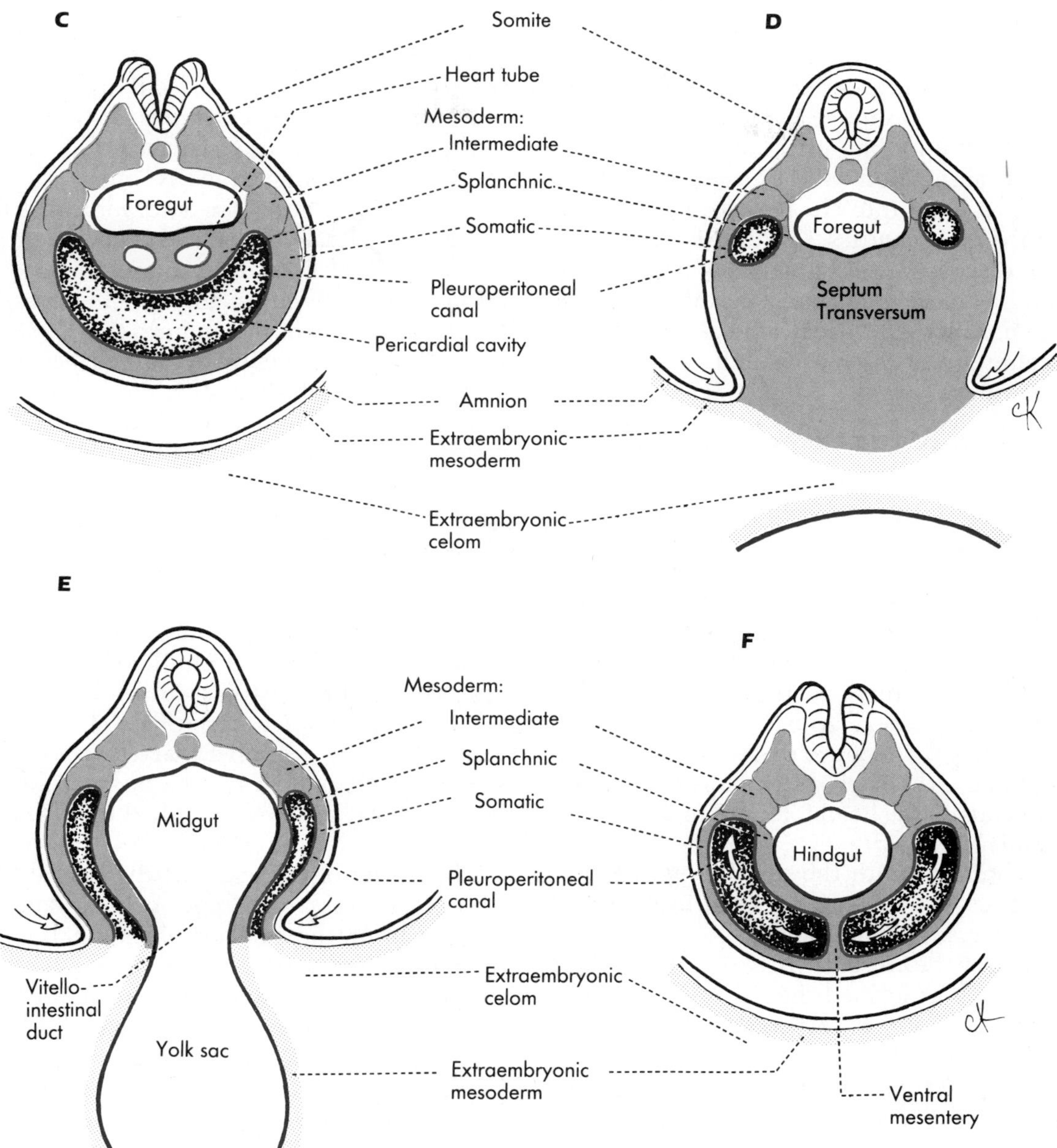


shaped or an inverted **U**-shaped cavity is thus created (Fig. 7-2C). The caudally pointing limbs of the celom, called the *pleuroperitoneal canals,* eventually fuse and give rise to the peritoneal cavity (Chap. 23); the future pleural cavities are situated at the rostral end of the pleuroperitoneal canals, whereas the transverse portion of the celom will become the pericardial cavity (Chaps. 19 and 21).

The medial wall of the pleuroperitoneal canals is formed by intermediate mesoderm; along their lateral wall, somatic and splanchnic laminae of lateral plate mesoderm fuse with each other and with extraembryonic mesoderm (Fig. 7-2C). Thus, at the time of its formation, the intraembryonic celom is a closed cavity; it will, however, soon be breached by the dissolution of a streak of mesoderm in the lateral wall of the pleuroperitoneal canals (Fig. 7-2C). The communication established with the extraembryonic celom admits into the body of the embryo nutrient-rich fluid needed prior to the development of the circulation.

During the folding of the embryonic disk, this communication will be closed and the celom partitioned into the pericardial, pleural, and peritoneal cavities. The division of the celom is discussed in Chapters 19, 23, and 25; the folding of the embryonic disk is dealt with in the next section, because it contributes to the definition of the main subdivisions of the primitive gut.

## Formation of the Foregut, Midgut, and Hindgut

The folding of the flat embryonic disk, which rests on the yolk sac, creates a more or less cylindrical organism. This process entraps portions of the yolk sac within the embryo, which become its alimentary canal or gut and also displaces parts of the celom into a plane that is ventral to the central nervous system and the developing vertebral column of the embryo. There are three components to the folding, which progress more or less simultaneously: the head fold, tail fold, and lateral folds.

Concomitant with the rapid growth of the head process, the so-called **head fold** carries the pericardial portion of the celom ventrally and thereby the cresentic ridge of lateral plate mesoderm that formed the rostral wall of the pericardial cavity sharply indents the yolk sac (Fig. 7-3A and B). The consequences of this head fold are that (1) part of the yolk sac becomes entrapped within the embryo and can henceforth be called the **foregut** (*proenteron*); (2) the pericardial cavity is placed ventral to the foregut (Fig. 7-3C); (3) a ridge of mesoderm is placed transversely across the ventral aspect of the embryo interposed between the pericardial cavity and the yolk sac—this mesoderm is the **septum transversum** (Fig. 7-3D); (4) a ventral body wall covered in ectoderm is created in the ventromedian area of the rostral part of the embryo.

At the caudal end, the **tail fold** likewise indents the yolk sac, entrapping the **hindgut** (*metenteron*) in the embryo, and creates a ventromedian body wall (Fig. 7-3B and F). However, no septum transversum is formed caudal to the yolk sac because a transverse portion of the celom is lacking caudally.

The lateral edges of the embryonic disk also move ventrally as the **lateral folds** indent the lateral wall of the yolk sac and enclose the **midgut** (*mesenteron*) in the now more or less cylindrical body of the embryo (Fig. 7-3E). In this manner, the ventrally curved perimeter of the once disk-shaped embryo constricts around the yolk sac from all directions like a purse string, stopping short around an opening that becomes the umbilicus (Fig. 7-3A and E). A narrow connection persists between the midgut and the extraembryonic yolk sac and is called the *vitellointestinal duct.* Since the amnion remains attached around the embryonic disk during this folding, the amnion will be attached around the umbilicus and will invest the vitellointestinal duct and the connecting stalk, which contains the umbilical veins and the umbilical arteries as these structures enter or leave the body of the embryo, in what can now be called the *umbilical cord.*

The fusion of the edges of the lateral folds with the extraembryonic and splanchnic mesoderm on the exterior of the vitellointestinal duct and with the tail fold closes the communication between the extraembryonic celom and the pleuroperitoneal canals (Fig. 7-3E and F). The definitive anatomy of the **serous sacs** is attained by (1) the separation of the pericardial sac from the pleuroperitoneal canals by the *pleuropericardial membranes* (see Fig. 19-12); (2) expansion of the portion of the pleuroperitoneal canals rostral to the septum transversum to form the bilateral pleural sacs (see Fig. 19-12) and separation of the sacs from the distal portion of the pleuroperitoneal canals by the bilateral *pleuroperitoneal membranes,* which complete the septum transversum dorsally (see Fig. 25-4); (3) formation of the peritoneal sac by expansion of the pleuroperitoneal canals distal to the septum trans-

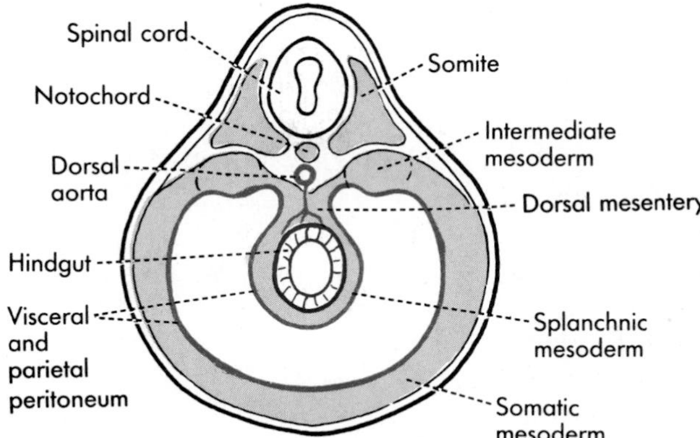


**FIG. 7-4.** The scheme of the peritoneal cavity, derived from developmental stages shown in Figure 7-3.

versum and the fusion of the two canals ventral to the midgut and hindgut so that the gut comes to be suspended by a dorsal mesentery (Fig. 7-4).

### Differentiation of the Primitive Gut

By the breakdown of the oropharyngeal membrane, the foregut establishes communication with the *stomatodeum,* a depression lined by ectoderm from which part of the oral cavity develops (Fig. 7-3B). Similarly, by the breakdown of the cloacal membrane, the hindgut establishes communication with the *proctodeum,* a depression or pit lined by ectoderm from which part of the anal canal develops. The cranial part of the foregut forms the pharynx, in the lateral wall and floor of which a number of ridges and epithelial pouches make their appearance. From the epithelial outgrowths develop a number of glands located in the neck, which lose connection with the pharynx. From one of the ventral diverticula of the pharynx develops the trachea and the lung buds (Chap. 20). This diverticulum retains communication with the pharynx and at their junction develops a specialized sphincter, the larynx, which is also adapted for voice production. The lung buds invaginate the pleural sacs (see Fig. 19-12). The remainder of the foregut becomes the esophagus, the stomach, and part of the duodenum. The liver, gallbladder, and pancreas bud off from the distal end of the foregut and will be joined to the duodenum by their excretory ducts (see Fig. 24-21). From the midgut develop the rest of the small intestine and part of the colon; from the hindgut is formed the rest of the large intestine (see Fig. 23-24).

## GENERAL REFERENCES AND RECOMMENDED READINGS

ANSON BJ, LYMAN RY, LANDER HH: The abdominal viscera in situ: A study of 125 consecutive cadavers. Anat Rec 67:17, 1936

AREY LB, TREMAINE MJ: The muscle content of the lower oesophagus of man. Anat Rec 56:315, 1933

BARON MA: Structure of the intestinal peritoneum in man. Am J Anat 69:439, 1941

CHUDNOFF J, SHAPIRO H: Two cases of complete situs inversus. Anat Rec 74:189, 1939

ELSEN J, AREY LB: On spirality in the intestinal wall. Am J Anat 118:11, 1966

HUNTINGTON GS: The Anatomy of the Human Peritoneum and Abdominal Cavity Considered from the Standpoint of Development and Comparative Anatomy. Philadelphia, Lea Brothers, 1903

JACKSON CM: On the developmental topography of the thoracic and abdominal viscera. Anat Rec 3:361, 1909

JACOBSON LF, NOER RJ: The vascular pattern of the intestinal villi in various laboratory animals and man. Anat Rec 114:85, 1952

MACKLIN CC: X-ray studies on bronchial movements. Am J Anat 35:303, 1925

MOODY RO: The position of the abdominal viscera in healthy, young British and American adults. J Anat 61:223, 1927

TREVES F: Lectures on the anatomy of the intestinal canal and peritoneum in man. Br Med J 1:415; 470; 527; 580, 1885

# 8

# THE UROGENITAL SYSTEM

Urinary and genital organs are closely related both anatomically and developmentally and are therefore grouped together as the urogenital system. In both sexes, the urinary system consists of the paired kidneys, the duct (ureter) leading from each kidney to the urinary bladder situated in the pelvis, the bladder itself, and the channel (the urethra) by which the bladder opens to the outside. The genital or reproductive system in both sexes consists of paired gonads (ovary or testis), a duct system by which the products of the gonads escape to the outside, and the external genitalia that are associated with these ducts. In the male, the ducts from the testis discharge directly to the outside by way of the urethra, which the genital and urinary systems share. In the female, the duct system is largely separate from that of the urinary system, and the paired ducts are in part fused together to form the uterus, in which the fertilized egg develops into a fetus, and the vagina, for internal impregnation of the ovum. Genital and urinary systems in the female share a shallow opening (the vestibule) to the outside.

## URINARY SYSTEM

### Kidney and Ureter

There is no essential difference in the urinary system of the two sexes (Fig. 8-1). Urine is formed in the **kidney** (*ren,* hence the adjective "renal"), in part by a process of filtration from the blood vessels that is dependent upon the pressure of the blood and in part, as this filtrate passes down a long and convoluted tubule, by excretion of additional substances from the filtrate. Beyond their convoluted portions, adjacent tubules join each other to form collecting tubules.

The *cortex,* or outer part of the kidney, is largely concerned with the formation of urine, whereas the *medulla,* or inner part, consists primarily of collecting tubules that serve for transport (cortex and

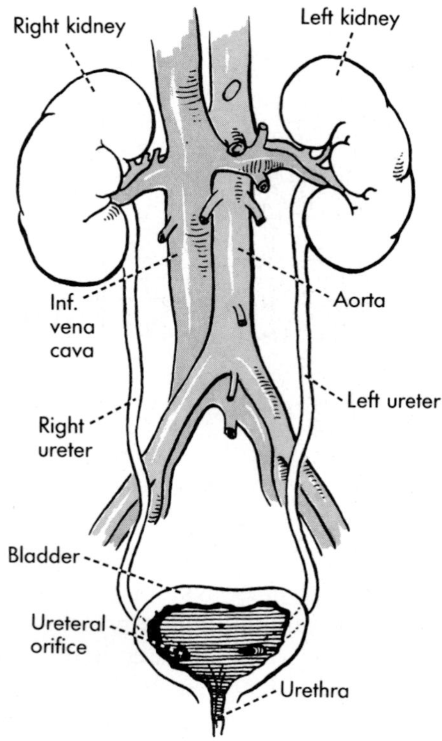


**FIG. 8-1.** Diagram of the urinary system.

medulla mean bark and marrow, respectively). The urine in the collecting tubules is delivered into a single large chamber that is drained by the **ureter,** a thick-walled tube that by peristaltic action conducts urine to the urinary bladder.

Since the kidneys are situated in the upper part of the abdominal cavity, against the posterior abdominal wall, and the urinary bladder is situated in the lowest part of the pelvic cavity, the ureters are some 25 cm to 35 cm long. Any interference with the flow of urine through them endangers the life of the kidney. When a ureter is obstructed, the kidney continues to produce urine and the pressure of this gradually destroys the renal substance, so that eventually the kidney loses its excretory power entirely.

Man readily survives removal of one kidney, but destruction of both kidneys is incompatible with life, for other excretory organs—the skin, the lungs, the colon—cannot take over the functions of the kidney. Urograms (visualization of the urinary passages by roentgenogram) and dye studies offer excellent means of determining not only the condition of the urinary tract and the positions of the kidneys but also the state of function of these important organs.

### Urinary Bladder and Urethra

The **urinary bladder** (*vesica urinaria*) is a highly distensible sac located primarily in the pelvic cavity but rising more or less into the abdomen as it becomes filled with urine. It lies anterior to the rectum in the male and anterior to the uterus and vagina in the female. It receives the two ureters and by the contraction of its smooth muscle discharges the urine to the outside through the urethra.

The **urethra** in the female is short and embedded in the anterior wall of the vagina. Both it and the vagina open between prominent folds, the labia, that are a part of the external genitalia. In the male, the urethra, as it leaves the bladder, is surrounded by the prostate (Fig. 8-2). Enlargement of this gland is a common cause of difficulty in emptying the bladder in the male. After leaving the prostate, the male urethra passes through the thin musculature that forms the floor of the pelvic cavity and enters the penis, continuing to the end of this organ. The ducts of the genital system join the prostatic part of the urethra, so that both urinary and genital systems share the major length of the urethra.

## GENITAL SYSTEM

## MALE GENITAL SYSTEM

The genital system of the male consists of the two **testes** and their ducts, most of the **urethra,** the **prostate** and **seminal vesicles,** and the **external genitalia,** of which the **penis** houses the urethra and the **scrotal sac** houses the testes.

### Development of Testis and Ducts

In their early development, the testes originate very close to a primitive but, in man,

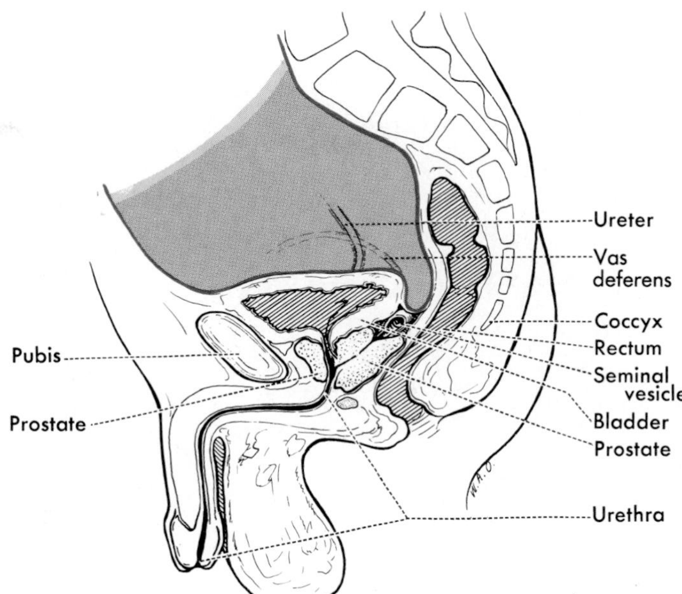


**FIG. 8-2.** Male pelvis in sagittal section, showing a part of the urogenital system. The cut edge of the peritoneum is shown as a red line, and the peritoneal cavity is shaded pink.

evanescent kidney, the *mesonephros,* and they make use of the duct system provided by this kidney. The testis develops a series of *seminiferous tubules* whose walls, after sexual maturity, produce male germ cells, the *spermatozoa.* These tubules, at an early stage of development, become connected to tubules of the mesonephros, which then drain the seminiferous tubules. The mesonephric tubules open into the mesonephric duct, but after this duct becomes connected to the testis by means of the mesonephric tubules, its name is changed to **ductus deferens.** With further development, the lower ends of the mesonephric ducts, or ductus deferentes, open into the urethra.

Close to their lower ends, the ductus deferentes give off blind diverticula that become the **seminal vesicles** (vesicle meaning little bladder), so called because they were once thought to store spermatozoa. They are actually glands that contribute to the *seminal fluid* (ejaculate). The other major gland that contributes to the seminal fluid is the **prostate,** which develops as multiple outgrowths from the urethra and largely surrounds a part of it. The **penis** develops from a swelling just above the opening of the urethra to the outside and by its growth comes to surround the urethra and prolong it.

Although the testis originates in the abdomen, it migrates caudally, passes through a defect in the anterior abdominal wall, and descends into the scrotum, a diverticulum of this wall.

## FEMALE GENITAL SYSTEM

There are many parallels in the development of the male and female genital systems, and homologies of parts of the two systems are easily apparent.

### Development

The **ovary** begins its development, as does the testis, at the level of the mesonephric kidneys; however, unlike the testis, it never separates completely from the celomic epithelium from which it develops. The outer part of the ovary, the *cortex,* persists as the germ-forming layer. In the female, the mesonephric tubules and mesonephric duct never become connected with the ovary, and they largely degenerate. Instead of the ovum's being discharged directly into a tube, therefore, the mature ovum is discharged from the surface of the ovary into the peritoneal cavity and is picked up by a tube that opens into this cavity. The primitive tube is the paramesonephric duct, which

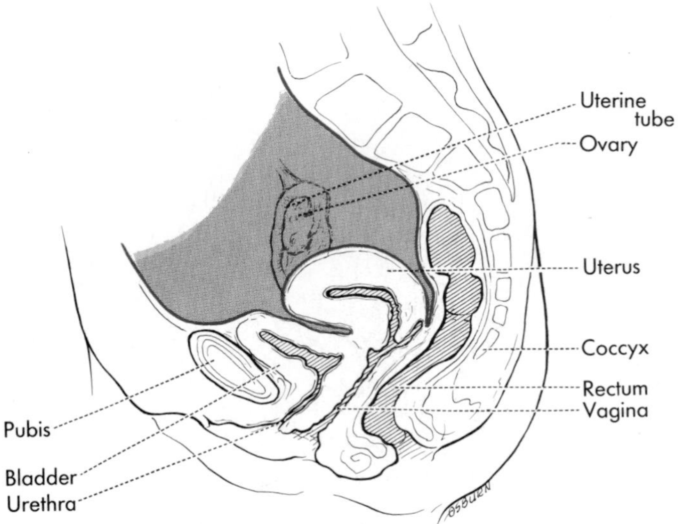


**FIG. 8-3.** Sagittal section of the female pelvis, showing the chief reproductive organs. The cut edge of the peritoneum is shown as a red line, and the peritoneal cavity is shaded pink.

parallels the mesonephric duct and is developed in close relation to it. Unlike the mesonephric duct, the paramesonephric duct opens directly into the peritoneal cavity. As they descend toward the pelvis, the two ducts fuse to form the unpaired uterovaginal canal, and the paired upper parts of the ducts remain as the **uterine tubes.**

The ovaries undergo a descent, for a time paralleling that of the testis, but they and the uterine tubes normally end by taking up a position in the pelvis.

## Anatomy

In the adult female, each **ovary** lies on the lateral wall of the pelvis, close to that end of the uterine tube that opens into the peritoneal cavity. The other ends of the uterine tubes empty into the **uterus** (Fig. 8-3). The uterus is thick walled because its musculature must be heavy to accommodate and then expel the fetus and because its mucosa (*endometrium*) must accommodate the numerous glands whose secretion nourishes the developing egg when it first enters the uterus. The endometrium undergoes a cyclical thickening in preparation for the expected monthly release of the ovum from the ovary. This endometrial hyperplasia occurs as a result of hormones acting upon the uterus. Failure of an egg to be fertilized produces a shift in hormonal balances that leads to rapid endometrial degeneration and its discharge during menstruation.

The uterus empties into the much thinner-walled **vagina** through a narrow lower segment known as the **cervix,** which protrudes into the upper part of the vagina. Since carcinoma of the uterus most often starts at the cervix, examination of the cervix and of material obtained from the vagina by vaginal smear is often useful in its early detection.

The vagina opens through the pelvic floor into the **vestibule,** into which the urethra also opens. The walls of the vestibule, the **labia minora,** are homologues of part of the penis of the male, and the larger folds outside the labia minora, the **labia majora,** are homologues of the scrotal sac of the male. Another part of the penis is represented by the **clitoris,** the tip or glans of which lies anteriorly where the labia minora come together.

## HERMAPHRODITISM

The early development of the male external genitalia is identical with that of the female, and therefore the sex of the fetus cannot at first be determined by examination of the external genitalia. Cessation in development of male genitalia results in their closely resembling those of the female. Overdevelopment of female genitalia produces a condition resembling that of an underdeveloped male. The condition in which testes occur in an individual with apparently female genitalia, or ovaries in a person with apparently male genitalia, is known as *false hermaphroditism.* Male hermaphrodites—individuals with testes but underdeveloped external genitalia—are much more common than are female hermaphrodites. (The terms most frequently used nowadays are "intersex" or "sex intergrade" instead of hermaphrodite.) *True hermaphroditism,* in which an individual has both a testis and an ovary, or a gonad that is a combination of both, is very rare in man.

## GENERAL REFERENCES AND RECOMMENDED READINGS


BOYER CC: The vascular pattern of the renal glomerulus as revealed by plastic reconstruction from serial sections. Anat Rec 125:433, 1956

FARRIS EJ: Human Fertility and Problems of the Male. White Plains, NY, Author's Press, 1950

HALL V: The protoplasmic basis of glomerular ultrafiltration. Am Heart J 54:1, 1957

JONES HW, JR, SCOTT WW: Hermaphroditism, Genital Anomalies and Related Endocrine Disorders. Baltimore, Williams & Wilkins, 1958

MARKEE JE: The morphological and endocrine basis for menstrual bleeding. In Meigs JV, Sturgis SH (ed): Progress in Gynecology, vol II, p 63. New York, Grune & Stratton, 1950

MARKEE JE: Physiology of reproduction. Ann Rev Physiol 13:367, 1951

MOORE RA: The total number of glomeruli in the normal human kidney. Anat Rec 48:153, 1931

PAPANICOLAOU GN: The sexual cycle in the human female as revealed by vaginal smears. Am J Anat 52:519, 1933

PEASE DC, BAKER RF: Electron microscopy of the kidney. Am J Anat 87:349, 1950

SMITH HW: Principles of Renal Physiology. New York, Oxford University Press, 1956

THOMPSON JS, THOMPSON W: Genetics in Medicine, chap 7. Philadelphia, WB Saunders, 1966

YOUNG WC (ed): Sex and Internal Secretions, 3rd ed. Baltimore, Williams & Wilkins, 1961

# 9

# THE ENDOCRINE SYSTEM

The endocrine system is not a gross anatomic system like the systems previously discussed, for it consists of a few widely separated glands that form discrete anatomic organs and of other smaller groups of cells that are located in organs that belong to the digestive and genital systems. From the anatomic standpoint, the endocrine glands are united by one fundamental characteristic: they are glands that do not have ducts but, instead, discharge their secretions (hormones) directly into the bloodstream. A synonym is "ductless glands" (the N.A. term is *glandulae endocrinae*). The cells of endocrine glands are epithelial, as are the cells of exocrine glands. Because of the necessary close relationship to the bloodstream, endocrine tissue is particularly vascular. For instance, the thyroid gland, one of the endocrine glands, has been said to have the greatest blood supply per unit of tissue of any organ in the body.

Although the endocrine glands differ markedly from each other in many anatomic features, and the hormones that they produce differ widely in chemical composition and in physiologic action, they nevertheless form a physiologic system: they all produce hormones that act on other tissues of the body; some of these hormones also act upon other endocrine glands. Many of the fundamental metabolic activities of the body are governed by the endocrine system.

The recognized endocrine glands are the unpaired hypophysis, four parathyroid glands, the unpaired thyroid gland, and the paired suprarenal glands (Fig. 9-1). In addition to these, the ovaries and the testes have endocrine functions; the pancreatic islets are composed of endocrine cells; certain parts of the digestive tract also liberate hormones; and the liver, an organ of many functions, is often said to have endocrine functions. Finally, the pineal body, an outgrowth of the thalamus of the brain, and the thymus, an organ in the upper part of the thorax, are also sometimes described as endocrine organs, although their exact functions are still unknown. Recent evidence seems to indicate endocrine functions for the pineal, but associates the thymus with immune function rather than

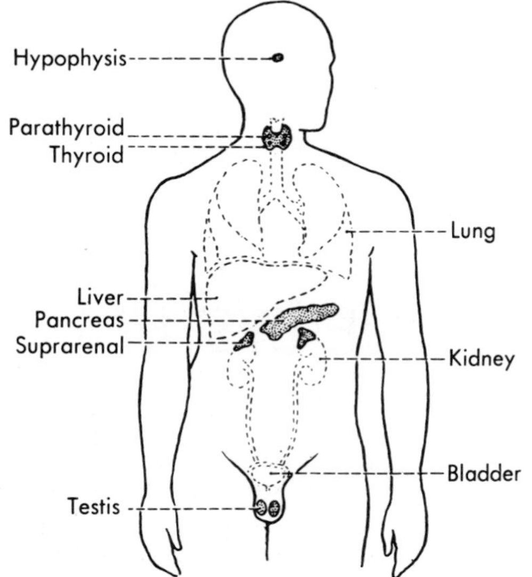


**FIG. 9-1.** Diagram of the chief endocrine glands (stippled). In the female, the ovaries, rather than the testes, would be included. (Baitsell GA: Human Biology. New York, McGraw-Hill, 1940)

with hormones. Since the various endocrine and endocrine-containing organs are described later in this text, according to their regional anatomy, remarks concerning them here will be limited primarily to physiologic considerations.

## Hypophysis

The hypophysis (pituitary gland) is actually two closely associated endocrine organs. The posterior part of the hypophysis (**posterior lobe** or *neurohypophysis*) is an outgrowth from the floor of the hypothalamus of the brain. A larger part is derived from surface ectoderm that in the growth of the mouth is carried back into the roof of the pharynx. When this part separates from the pharyngeal wall, it becomes closely associated with the posterior lobe and forms the important **anterior lobe** and a part, rudimentary in man, known as the pars intermedia.

The hypophysis lies immediately beneath the brain and is connected to it by the hypophyseal stalk, representing the stem by which the posterior lobe grew out from the brain, and by blood vessels around the stalk that arise in the hypothalamus and end in the hypophysis, especially the anterior lobe. The *hypophyseal stalk* consists primarily of nerve fibers that end in the posterior lobe. It is generally accepted that the posterior lobe does not produce the hormones itself that it secretes into the bloodstream but merely releases them into capillaries after they have been formed by nerve cells in the brain and have migrated along the nerve fibers to the posterior lobe. In many animals, the release of hormone is not confined to the posterior lobe, but occurs also throughout the hypophyseal stalk and from the part of the brain that gives rise to the stalk. These parts are, therefore, often included in the neurohypophysis (Fig. 9-2).

The **posterior lobe** apparently secretes two hormones, an antidiuretic hormone (also known as vasopressin) and oxytocin. (Because one hormone may have several different effects, and hormones are difficult to purify, it is often exceedingly difficult to determine whether each of two differing physiologic effects results from the action of a separate hormone or from different parts of a single hormone that may, in the process of attempted purification, be broken away from each other so that two physiologic principles are isolated.) Of the two recognized hormones of the posterior lobe, the antidiuretic hormone, or vasopressin, has to do with controlling the amount of water excreted by the kidney and also produces constriction of arterioles and, therefore, a rise in blood pressure. Oxytocin produces contraction of the smooth muscle of the uterus and also has a lactogenic effect causing sudden expulsion of milk when given to a lactating animal.

The **anterior lobe** of the hypophysis is glandular rather than neural in appearance (hence also called *adenohypophysis*), and the several types of epithelial cells composing it are believed to both elaborate and secrete several different hormones. Although the secretory activity of the anterior lobe is controlled in part by the nervous system, this control is not through nerve fibers, but through substances

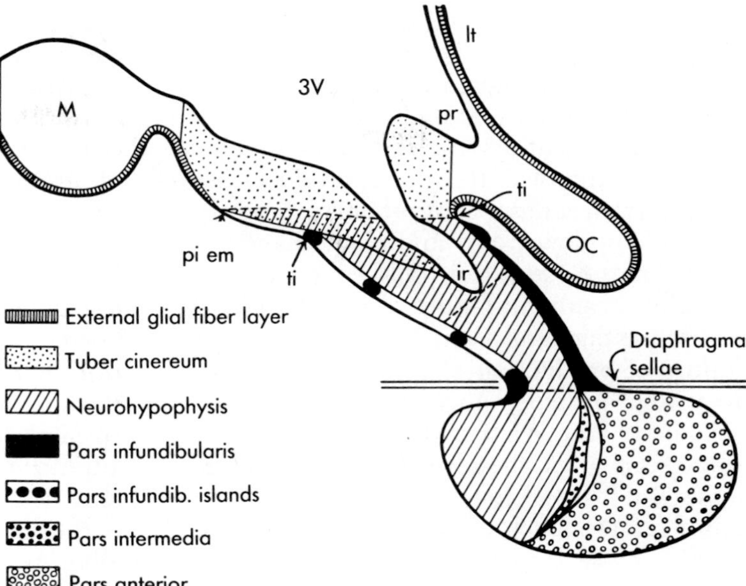


**FIG. 9-2.** The hypophysis of man and the adjacent part of the brain. The diaphragma sellae is a shelf of dura mater that largely separates the hypophysis from the overlying brain. The tuber cinereum is the part of the hypothalamus that gives rise to the stalk of the posterior lobe of the hypophysis; the hypophyseal stalk and a part of the tuber are composed of tissue identical to that of the posterior lobe and are shown as a part of the neurohypophysis; the pars infundibularis (once pars tuberalis) is glandular tissue, somewhat similar to that of the anterior lobe or pars anterior, that extends upward along the hypophyseal stalk. Other labels identify parts of the nervous system: **3V** is the third ventricle; **pr** and **ir** are the optic (preoptic) and infundibular recesses, extensions of the third ventricle over the optic chiasma **(OC)** and into the hypophyseal stalk; **lt** is the lamina terminalis, the most anterior part of the midline portion of the brain; **ti** and **pi em** are markings on the hypothalamus; and **M** is the mamillary body, a part of the hypothalamus. (Haymaker W, Anderson E: In Baker AB (ed): Clinical Neurology, vol 2. New York, Paul B Hoeber, 1955)

released into the blood vessels about the hypophyseal stalk. These begin in the central nervous system and end in the hypophysis (the hypophyseal portal system).

There is no agreement as to exactly how many hormones are secreted by the anterior lobe. The most important active principles of this lobe have to do with growth of the body as a whole, with the development and maintenance of the gonads, and with the maintenance of secretory activity by the thyroid gland and the suprarenal (adrenal) cortex. At least four hormones of the anterior lobe are usually named: a growth hormone, a thyrotropic hormone, an adrenocorticotropic hormone (rather generally known, even to laymen, as ACTH), and one or more gonadotropic hormones. There may be others, but a complete listing is not essential to this discussion. Because of its effects on the endocrine activity of the gonads, the thyroid, and the suprarenal cortex, the anterior lobe of the hypophysis is sometimes known as the master endocrine gland.

Loss of growth hormone, or the fraction of the hormone that has to do with growth, in young animals leads to cessation of growth in body size. Similarly, oversecretion of this hormone produces increased skeletal growth, which may be general and lead to gigantism or, if it occurs at a somewhat later time as a result of a tumor of the gland, produces *acromegaly* (characterized by abnormal enlargement of the lower jaw and of the hands and feet). Lack of secretion of the gonadotropic principle results in failure of secretion by the endocrine portions of the testis and ovary and failure of production of ova or spermatozoa. In young animals, since the development of sexual characteristics is dependent upon these secretions, the secondary sex characteristics fail to develop. A similar lack in the adult female leads to cessation of the reproductive cycle, and in adults of both sexes to disappearance of sexual activity and gradual regression of the sex organs. Lack of thyrotropic hormone greatly diminishes the secretory activity of the thyroid gland and therefore produces a marked drop in metabolic activity. Lack of secretion of adrenocorticotropic hormone lowers the secretory activity of the adrenal cortex and may produce symptoms of adrenal cortical insufficiency.

The *pars intermedia* in lower animals secretes a hormone that acts upon melanophores (pigment cells). Its function in man is unknown.

## Some Hormonal Interrelationships

Although the anterior lobe of the hypophysis has been referred to as the master endocrine gland, it must not be supposed that the secretions of the other endocrine glands have no effect upon each other's activities or upon the activity of the anterior lobe. Interrelationships among the endocrine glands and their hormones are extremely complex, and many facts about them remain to be determined.

One of the more obvious interrelationships between endocrine glands can be seen in the case of the thyroid gland. Although largely governed in its secretory activity by secretion of thyrotropic hormone from the anterior hypophysis, the secretion of the thyroid gland itself affects the metabolic activity of all tissues of the body and, therefore, the activity of the anterior lobe of the hypophysis and of all the other endocrine glands.

Another interendocrine relationship, particularly well known, is the cycle of secretion of adrenocorticotropic hormone by the anterior lobe of the hypophysis and the secretion of certain adrenal cortical hormones by the adrenal cortex. Stated in its simplest form, the anterior lobe secretes adrenocorticotropic hormone in sufficient amounts to cause secretion by the adrenal cortex. However, in turn, the amount of adrenal hormones in the bloodstream determines the rate of secretion of adrenocorticotropic hormone by the anterior lobe. Adrenal cortical hormones, circulating to the anterior lobe, depress the activity of the lobe in secreting adrenocorticotropic hormone, and therefore, when the adrenal hormones have reached an adequate level, the anterior lobe ceases to secrete. Then, as the secretion of adrenal cortical hormones drops off in consequence of lack of stimulation through the anterior lobe's hormones, less circulating hormone reaches the anterior lobe and it begins to secrete once again. Thus, the adrenal cortex, although governed by the anterior lobe, in turn regulates the activity of the anterior lobe in governing it. Exactly similar "feedback" relationships seem to exist between the thyroid gland and the anterior lobe, and the gonads and this lobe.

A clear example of the effect of at least two different hormones on a single organ is seen in regard to skeletal growth. The growth hormone apparently produces its effects by increasing the rate of proliferation of cartilage, which in turn, under normal conditions (thyroid hormone probably being essential for this), increases the rate of ossification and, therefore, the rate of growth of the bones. However, certain sex hormones have a different effect on the epiphyses; they decrease the rate of proliferation and maturation of the cartilage, so that the process of ossification overtakes the growth of new cartilage and the epiphyseal cartilages are totally destroyed. When this occurs, of course, growth ceases. The earlier secretion of larger amounts of sex hormones by the developing female than by the male, evidenced in the earlier sexual maturity of the female, is responsible for the fact that bone growth in females regularly ceases some 1 to 3 years earlier than it does in males. Another feature of this relationship that is of clinical importance is that since sex hormones tend to interfere with growth, they cannot be given indiscriminately to growing children. For instance, certain hormones that may produce descent of an undescended testis before the age of puberty can have a stunting effect upon growth.

### Thyroid Gland

The thyroid gland lies in the neck across and lateral to the trachea. Rather than being arranged in irregular cords permeated by blood vessels, as are the epithelial cells of many endocrine glands, the epithelial cells of the thyroid form vesicles. The outer wall of each vesicle is adjacent to an abundant capillary network, and the hollow within the vesicle serves as a storehouse for thyroid hormone until it is needed by the rest of the body.

The nerves to the thyroid gland are probably purely vasomotor and have nothing directly to do with its secretory activity. This apparently is controlled primarily by the activity of the anterior lobe of the hypophysis.

The hormone of the thyroid gland contains iodine, and a supply of iodine in the diet is necessary to its formation. In the absence of an adequate intake of iodine, the thyroid gland attempts to produce a larger amount of hormone, as if to make up for its deficiency in iodine, and the gland becomes markedly larger. This enlargement of the gland is *simple goiter*, once fairly common in certain areas of Switzerland and the United States, where soil and water are deficient in iodine compounds. Recent techniques, made possible by the availability of radioactive iodine, allow rather precise study of the ability of a thyroid gland to handle iodine.

The chief effects of thyroid hormone are well known. It is essential to normal metabolism, and marked destruction of the thyroid gland in an adult depresses basal metabolism to such an extent that mental activity is interfered with, the pulse rate and temperature of the body fall below normal, and, for reasons not clearly understood, there is a deposition of subcutaneous connective tissue, which causes the face and hands to appear edematous, a condition known as myxedema and readily treated with desiccated thyroid. Absence of adequate secretion by the thyroid gland in infants affects both intelligence and physical growth and, if prolonged, produces idiocy and dwarfism (cretinism). However, treatment with desiccated thyroid begun during the infantile period will usually prevent the otherwise inevitable arrest of physical and mental development.

Quite different from the enlargement of the thyroid gland produced by lack of thyroid hormone is the enlargement occurring, for unknown reasons, in association with markedly increased activity on the part of the gland. In this instance, the enlarged gland secretes more than a normal amount of hormone, with the result that there is an increased metabolic rate, nervousness, and loss of weight (these being symptoms of hyperthyroidism). In connection with this type of enlargement, the eyeballs typically become prominent, so this is known as *exophthalmic* (toxic) *goiter*. Removal of a part of the thyroid gland, thus leaving less glandular tissue to secrete, has been effective in relieving the symptoms of this type of goiter. Removal of more gland than is necessary to achieve the desired results is not serious, since the metabolic activity of the patient can be restored to normal levels by the administration of desiccated thyroid.

### Parathyroids

The parathyroid glands, usually four in number, are closely associated with the thyroid gland (typically lying on its posterior surface) but are different from it both in origin and in physiology. They derive from the walls of the third and fourth branchial pouches and only secondarily become associated with the thyroid gland. Although they are small bodies, the presence of one or more is necessary to health. Surgeons, therefore, so plan their operations on the thyroid that they can be rather confident of not removing all parathyroid tissue.

The hormone of the parathyroid gland exerts control over the calcium content of the blood. The normal constant interchange between blood and bone is only partly independent of parathyroid hormone, and lack of the hormone produces a marked fall in the calcium content of the blood. This, in turn, produces nervous and muscular disturbances and tetany. The opposite effect is produced by oversecretion of parathyroid hormone, typically resulting from a parathyroid tumor; here the blood-calcium level is markedly raised, and since the diet of the individual in regard to calcium usually has not been changed, the calcium is typically withdrawn in abnormal amounts from the bones, with resulting fragility of those elements. A person with an untreated severe parathyroid tumor may break an arm or a rib by simply turning over in bed. Further, the increased blood calcium leads to increased excretion of calcium in the urine, and renal stones are not infrequent accompaniments of a parathyroid tumor. Too much parathyroid secretion produces weakness, diarrhea and vomiting, and, if severe enough, circulatory collapse and death.

## Suprarenal Glands

The suprarenal glands lie behind the peritoneum of the abdominal wall, closely associated with the upper poles of the kidneys, hence their name of suprarenal or adrenal glands. (In man, they are properly called suprarenal glands, while in most animals they are called adrenal glands; however, the latter term is commonly applied to the glands of man.) A suprarenal gland is composed of cords of epithelial cells with abundant capillaries between these cords. Even upon casual examination, it is obvious that there are two distinct types of cells and cell arrangements here. One type of cell forms the outer part of the gland, hence the **cortex,** whereas the other type forms the center of the gland, the **medulla** (Fig. 9-3).

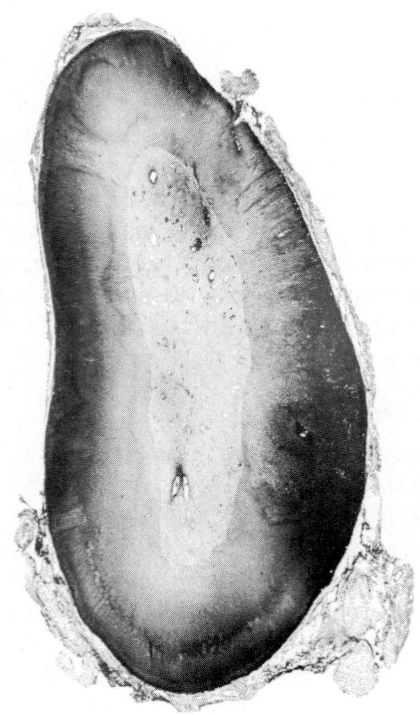

**FIG. 9-3.** Section of the adrenal (suprarenal) gland of a cat, showing cortex and medulla.

The cortex and the medulla of the suprarenal gland are really two separate endocrine glands that have become closely associated. Like the anterior and posterior lobes of the hypophysis, the two parts of the suprarenal have distinct functions and different developmental histories. The **medulla** arises from some of the same cells, usually said to be neural crest cells, that give rise to the great prevertebral sympathetic ganglia, such as the celiac. The **cortex,** however, arises directly by proliferation of the celomic epithelium, in much the same way that the testis and ovary arise. In many lower animals, there is no close relationship between the cortex and medulla, each occurring rather as a series of isolated bodies. Even in man, there are a few isolated collections of cells identical with those of the adrenal medulla, known as *chromaffin tissue* (a term that also includes the medulla), and there may be small nodes of cortical tissue outside the gland that are not associated with medullary tissue.

The functions of the **suprarenal medulla** are believed to be much simpler than those of the cortex. The medulla secretes into the bloodstream a hormone usually referred to as epinephrine or adrenaline, but now known to consist of two forms, *epinephrine* and *norepinephrine*. Both forms act primarily on smooth muscle, producing exactly the effects that stimulation of the sympathetic nervous system does. Epinephrine, for instance, increases both the rate and strength of the heart beat and produces constriction of arterioles and, thus, a rise in the blood pressure. Hence, like activity of the sympathetic nervous system, epinephrine prepares the body for an emergency. Epinephrine dilates the pupil, as does the sympathetic nervous system. As a part of a preparation for an emergency, it also promotes the conversion of glycogen into glucose and the latter's discharge from the liver, so that a greater amount of blood sugar is available to the other tissues of the body. It is because of the hyperglycemia (increased sugar in the blood) produced by the liberation of epinephrine that glycosuria (a spilling over of sugar into the urine) may be found during times of stress and tension in persons who are otherwise normal. The finding of sugar in the urine of a student taking examinations, for instance, is usually attributable to oversecretion by the suprarenal medulla, not to the hypoinsulinism of diabetes.

The suprarenal medulla seems to be the only endocrine gland whose activity is controlled entirely by nervous impulses. The medullary cells have an abundant innervation, derived from the sympathetic nervous system. These fibers, preganglionic ones, end directly on the medullary cells without the interposition of the usual postganglionic cells and fibers—an apparent exception to the rule that in the autonomic system, the preganglionic fibers always end on postganglionic cells, not on effector organs. This is, however, easily reconciled when it is remembered that the medullary cells themselves represent postganglionic nerve cells (having the same origin) and produce the same effects as do postganglionic cells (but over the body in general, through the discharge of epinephrine, rather than in specific locations by sending nerve fibers to these locations). The chromaffin system of the body, of which the suprarenal medulla is the largest component, therefore seems to be an adjunct of the sympathetic nervous system. As is the case with the sympathetic system, there is no evidence that the chromaffin system is essential to life. Destruction of all or almost all of this system in experimental animals is entirely compatible with normal existence in the laboratory, although an animal that had to defend itself or flee in order to save its life would be at a disadvantage.

Chromaffin cell tumors, usually of the suprarenal medulla but sometimes developing from small chromaffin cell groups found associated with the sympathetic trunks in the thorax or abdomen, are one, but not the most common, cause of high blood pressure (hypertension). The hypertension tends to be spasmodic and to reach levels that endanger the life of the patient, so that when the diagnosis is made, the tumor is sought by surgical exploration and removed.

In contrast to the medulla, the **suprarenal cortex** is necessary to life. It has long been known that degeneration of the suprarenal glands (Addison's disease) as a result of tuberculosis or other disease process typically leads to bronzing of the skin, vomiting, and muscular weakness so severe that the individual is incapacitated. The disease typically resulted in death before replacement therapy was available. The exact form in which the secretions of the suprarenal cortex are discharged into the bloodstream is not known. A number of complex steroids have been isolated from the cortex. Some of these, among which cortisone is the most widely known, have been shown to be particularly active physiologically. Cortisone and its related compounds have been used in the treatment of a number of diseases, and it is now possible to treat and maintain in good health persons afflicted with Addison's disease. The importance of the suprarenal cortex in reactions to mental and physical stress has been studied in great detail, increased cortical secretion being apparently the most important element in the "alarm reaction."

Hyperplasia of the cortex or functioning tumors (some tumors apparently do not secrete) lead to an excess of cortical hormones, with resultant overaction on other tissues, including other endocrine organs. Clinical findings in hyperplasia and functioning tumors of the cortex vary greatly, for the secretion may be largely cortisonelike material, sex hormones, or a mixture of both, and the effects of sex hormones vary with the sex of the individual. However, symptoms often include premature puberty in children, sexual changes in adults, hirsutism (abnormal growth of hair) in women, and diabetes mellitus (sugar diabetes); mental changes are not uncommon. Certain of these changes apparently depend upon simple hypersecretion by the gland and are, therefore, amenable to treatment by operative removal of an appropriate amount of a gland. Others are apparently a result of secretion of estrogens or androgens, as a result of malfunction of the cortex, and are usually treated by the administration of cortisone. Tumors of the gland are removed surgically.

## Other Endocrine Tissue

The **endocrine tissue of the pancreas,** the *pancreatic islands* (islets) or islands of Langerhans, consists of small groups of cells scattered among the more numerous acini (the exocrine part of the pancreas that retains its connection to the duct system). The islands, very vascular, consist of two or more types of epitheloid cells, of which one is apparently concerned with the secretion of insulin. *Insulin* is concerned primarily with sugar (glucose) metabolism and is necessary not only for the proper oxidation of sugar but also for its conversion into the form, glycogen, in which it is stored, particularly in the muscles and the liver. (A substance known as glucagon, also produced by the islets, has an opposite effect—it provokes the liver to convert glycogen into glucose.) Too little secretion of insulin leads to accumulation of glucose in the blood, with consequent excretion of it by the kidneys, a part of the syndrome of diabetes mellitus. Improper oxidation of sugar also leads to improper metabolism of fats. In contrast, too much secretion of insulin, which may result from a tumor of one or more pancreatic islands, leads to a dangerous decrease in blood sugar. This can be treated temporarily by supplying more sugar in the diet and can be cured by removing the tumor.

The **endocrine parts of the gonads** are believed to be cells found among or around the sex cells. In the male, they lie among cells of the seminiferous tubules and are called *interstitial* or *Sertoli cells.* In the female, they surround the developing egg and are *follicle cells.* After the egg is liberated, they proliferate, change their character, and become cells of the *corpus luteum.* It is the hormones produced by these cells, which can be grouped together as **sex hormones,** that (with sex hormones liberated by the suprarenals) are responsible for the development of secondary sex characteristics and, in the female, the institution of the menstrual cycle or its cessation with pregnancy.

**Hormones from the digestive tract,** the specific cell of origin being unknown, apparently include a substance released by the duodenal mucosa that causes the gallbladder to discharge bile into the duodenum and another substance that provokes the secretion of enzymes from the exocrine part of the pancreas. One part of the stomach also releases a hormone that provokes the secretion of acid by the stomach.

Whether the **liver** should be classed as an endocrine organ is debatable, but its activity does affect the endocrine balance. Among the activities of the liver is that of inactivating sex hormones (generally classed as *estrogens,* or female sex hormones, and *androgens,* or male sex hormones, both of which are produced in both sexes). With increasing failure of hepatic function, for instance, potent estrogens (which are normally less readily inactivated than androgens) may come to predominate in the bloodstream of a male and produce testicular atrophy, gynecomastia (enlargement of the male breast), and disturbance of the gonadal–hypophyseal relationship.

## GENERAL REFERENCES AND RECOMMENDED READINGS


BENCOSME SA: The histogenesis and cytology of the pancreatic islets in the rabbit. Am J Anat 96:103, 1955

CLARK E: The number of islands of Langerhans in the human pancreas. Anat Anz 43:81, 1913

CROWDER RE: The development of the adrenal gland in man, with special reference to origin and ultimate location of cell types, and evidence in favor of the "cell migration" theory. Contrib Embryol 36:193, 1957

VON EULER US: Hormones of the sympathetic nervous system and the adrenal medulla. Br Med J 1:105, 1951

GARDINER-HILL H (ed): Modern Trends in Endocrinology. New York, Paul B Hoeber, 1958

GREEN JD: The comparative anatomy of the hypophysis, with special reference to its blood supply and innervation. Am J Anat 88:225, 1951

HOLLINSHEAD WH: Anatomy of the endocrine glands. Surg Clin North Am 32:1115, 1952

ISHIKAWA T, KOZUMI K, BROOKS CMCC: Electrical activity recorded from the pituitary stalk of the cat. Am J Physiol 210:427, 1966

LACY PE: Electron microscopic identification of different cell types in the islets of Langerhans of the guinea pig, rat, rabbit and dog. Anat Rec 128:255, 1957

SCHARRER E, SCHARRER B: Hormones produced by neurosecretory cells. Recent Prog Horm Res 10:183, 1954

SWINYARD CA: Growth of the human suprarenal glands. Anat Rec 87:141, 1941

TURNER CD: General Endocrinology, 2nd ed. Philadelphia, WB Saunders, 1955

WELLER GL JR: Development of the thyroid, parathyroid, and thymus glands in man. Contrib Embryol 24:93, 1933

WOLFE JM: Effects of progesterone on the cells of the anterior hypophysis of the rat. Am J Anat 79:199, 1946

# 10

# THE SKIN AND ITS APPENDAGES

The primary function of the skin is to seal off the fluids of the body from the immediate environment of the animal and thus to preserve the fluid environment that cells must have if they are to live. In connection with this protective function, the skin has developed certain appendages that increase its effectiveness: nails as a protection against wear, hair as a protection against too much loss of heat. Because it forms the outside surface of the body, skin also has excretory glands that discharge directly to the outside. It is the medium through which much information concerning the immediate environment of the body is obtained (that is, it is a particularly important sense organ), and it is important in regulating the loss of heat from the body.

The skin (*integument*) is composed of two fundamental layers, an outer epithelial one, the **epidermis,** and an inner one of dense connective tissue, the **dermis** or corium. The term *cutis* sometimes is used to refer to the dermis only, but more properly it refers to the skin as a whole. "Derma" is, however, regularly employed in all combining words (dermatitis, dermatome) in the sense of the skin as a whole. Some of the confusion here may be that the dermis is frequently defined as the true skin. It is this felted layer of dense connective tissue that is treated by tanning to make leather, the epithelial component here being destroyed.

For the most part, the skin is thicker on the dorsal aspect of a part than it is on the ventral aspect, but this condition is reversed in the hand and foot. Here the palmar and plantar surfaces, respectively, are provided with much thicker skin than are their dorsal aspects.

The outer surface of the skin presents a series of delicate creases, intercepting in such a manner as to enclose elongated polyangular spaces and varying in prominence. These minute folds help lend elasticity to the skin and, hence, tend to run in the same direction as do the tension lines of the skin. They are more prominent in loose skin, especially that close to joints. The outer surface of the skin of the palm and sole differs from skin elsewhere in that it is arranged in a series of alternating ridges and grooves (*cristae* and *sulci cutis*), which may be fairly straight lines but toward the tips of the digits become elaborated into loops and whorls. It is the pattern of these that is recorded in fingerprints and footprints. Numerous sweat glands open on the summits of the ridges; hence, fingerprints are left when a smooth object is grasped. The ridges and sulci serve to increase the grip of the hand and foot by increasing friction. The skin of these surfaces is particularly closely bound to deeper structures, this decreased mobility being necessary for a firm grip.

## Epidermis

The epidermis is a layer of stratified squamous epithelium derived directly from the surface ectoderm of the body. Its exposed surface is composed of dead cornified cells that offer effective resistance both to passage of fluid through them and to friction. The thickness of the layer of cornified cells varies markedly. It is thin in such locations as the abdominal wall, but thicker than all other layers of the skin combined on the sole of the foot, especially in the areas most constantly subjected to pressure. Cells from the outermost surface of the cornified layer are constantly being desquamated and replaced by cells that gradually move outward from the deepest layer. The maintenance of the epithelial layer of the skin, therefore, demands a steady mitotic activity on the part of the deepest-lying cells in order to maintain the necessary rate of regeneration of the epidermis.

## Dermis

The dermis immediately adjacent to the epidermis is less dense in texture than elsewhere and contains the terminal capillaries of the skin and the majority of its nerve endings (no blood vessels and only a few nerve fibers penetrate the epidermal layer). On its deep surface, this loose layer, the **papillary layer** (*stratum papillare*), blends with the dense and thicker portion of the dermis, the reticular layer. On its outer surface, the papillary layer gives rise to numerous nipplelike projections, dermal papillae, that fit into conical excavations on the deep surface of the epidermis. It is particularly in the papillae that vascular loops and nerve endings are prominent. The reticular layer of the dermis is composed of densely interwoven connective tissue, largely collagenous but also containing elastic fibers. Its varying thickness is in most locations responsible for differences in the thickness of the skin.

Although the fibers of the dermis run in all directions, in most regions of the skin a greater number of fibers runs in one direction than any other, and this predominant direction tends, in general, to be parallel to the lines along which the skin is folded and stretched during movement. The prevailing direction of these fiber bundles can be determined on the fresh cadaver by inserting a sharp rounded object such as an ice pick into the skin. When this is done, the point separates fiber bundles more than it severs fibers, and the wound left by the instrument becomes slitlike in the direction of the fibers rather than being round. The lines of the skin determined by this method are known as cleavage or tension lines (Fig. 10-1). Cuts made in the direction of these lines produce less disruption of connective tissue bundles than do cuts made across them, and the skin gapes less widely when incisions follow them. Some surgeons do, some do not, consider the prevailing direction of the tension lines in planning surgical incisions. Cutting across the prominent flexion lines at joints almost always results in excess scar formation, however, so incisions are made zigzag across the joint. The incisions above and below the flexion line are connected by an incision almost paralleling the line.

On its deep surface, the dermis is usually connected to the underlying *tela subcutanea*, commonly termed the **superficial fascia.** The loose texture of this provides easy movement of the skin over the underlying structures. In some locations, however, the dermis is bound tightly to underlying deep structures either over a general area, as it is over the tibia in the leg, or over localized areas. The attachment of the skin to the tela subcutanea and deeper structures is through connective tissue bands, the *retinacula cutis*. Where these are locally well developed and attach to firm, deeper-lying tissue, they produce permanent folds and dimples in the skin. The permanent flexure lines in the skin are areas of such firmer anchorage; the skin on either side of them is moved toward these lines during flexion. Certain muscles of the face and neck also attach into the skin, and dimples on the cheek and chin are produced by the action of these muscles on the skin.

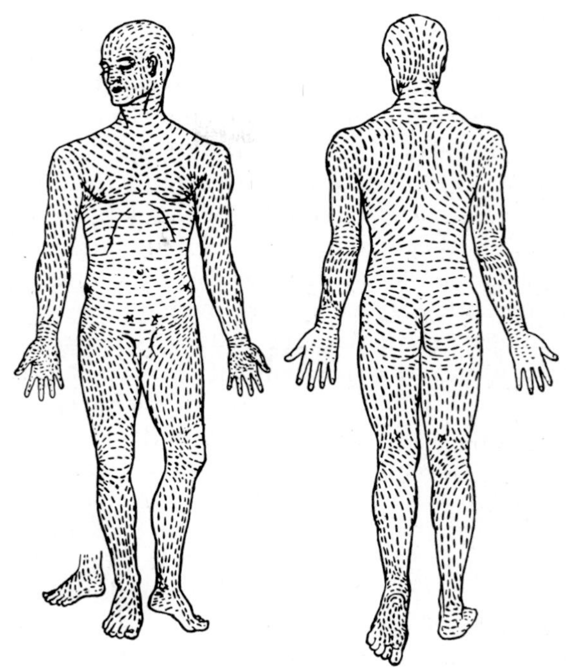

**FIG. 10-1.** Cleavage lines of the skin. (Cox HT: Br J Surg 29:234, 1941)

## Glands

The glands of the skin develop as growths from the epithelial layer; hence, they grow into the dermis with the larger glands growing through the dermis to invade the tela subcutanea. Cutaneous glands fall into two general groups, *sudoriferous* (sweat) and *sebaceous* (oily), but various modifications of these occur in certain regions. The breast is a modified cutaneous gland, of such size that it invades the underlying tela subcutanea. It is described with the pectoral region (Chap. 12).

The more common type of **sweat gland,** sometimes known as an *epicrine* (eccrine) sweat gland, secretes clear sweat that passes through ducts opening directly onto the surface of the skin. The largest sweat glands, found especially in the axilla but also on the areola of the breast and in the genital and circumanal regions, differ from ordinary sweat glands in that they discharge into hair follicles rather than directly onto the surface, and in that the material they discharge is not purely a secretion but consists in part of a portion of the cell cytoplasm. These are called *apocrine glands*. Apocrine glands are primarily responsible for the characteristic odor of sweat. The apocrine sweat glands, like the breast (a modified apocrine sweat gland), enlarge at the time of puberty. The *ceruminous glands*, the secretion of which hardens on exposure to air to form the wax found in the external ear canal, are modified apocrine sweat glands. Stimulation of the sympathetic nervous system provokes a discharge of sweat from sweat glands.

**Sebaceous glands** secrete an oily material, sebum, that coats the hairs. Each hair follicle is provided with from one to six sebaceous glands opening into it. However, sebaceous glands also open directly on the surface of the skin, particularly on the nose (where the openings of the larger ducts, occluded with debris, appear as "blackheads"), in the genital and circumanal region, on the areola and nipple, and in the eyelids (where they are called "tarsal glands").

## Hairs, Nails, and Teeth

**Hairs** (*pili*) are, like the glands of the skin, derived from ectoderm. A glandlike growth of the epidermis into the dermis gives rise to the hair follicle, and the hair, composed of tightly packed cornified cells, grows upward from the base of the follicle through its central aperture. Hairs are constantly being shed and replaced by the growth of new hairs from the same follicle. The earliest hairs of the fetus and newborn, the lanugo, are particularly fine. In parts of the body, hairs only a little coarser than this, usually referred to as down hairs, persist throughout life. In other regions, notably the scalp, the eyebrows and eyelashes, the axilla, the pubic region, and the beard in the male, most of the finer hairs are replaced by coarser ones of succeeding generations, and there is a tendency for the hairs to grow still more coarse with advancing age.

The hairs in certain regions have special names: those of the eyebrows are the supercilia, those of the eyelashes the cilia, those of the beard the barba, the coarse ones in the outer part of the external ear the tragi, those in the outer part of the nasal cavity the vibrissae, those of the scalp the capilli, those of the axilla the hirci, and those of the pubic region the pubes. Few of these terms are actually used very often. Colloquial or descriptive terms (eyebrows, eyelashes, pubic hair) are more commonly used.

Hair follicles typically have associated with them bundles of smooth muscle, the *arrectores pilorum* muscles, that are attached in such fashion to the hair follicle that they tend to erect the usually obliquely lying hairs. At the same time, they exert pressure upon sebaceous glands and provoke the discharge of their secretion into the hair follicle. Contraction of these muscles in many animals serves important functions. For instance, in the cat, contraction of the muscles and raising of the hairs in a cold environment increases the amount of air trapped among the hairs and therefore the insulating value of the hair coat. Also, contraction during anger increases the apparent size of the animal and therefore the possibility of appearing more frightening to an enemy. In man, general contraction of these muscles gives rise to "goose flesh."

**Nails** (*ungues*) are plates of tightly packed, cornified epithelial cells. The thin uncornified epithelium on which the nail plate lies is the *nail bed*. A proximal, thicker part of the nail bed, the *matrix*, is responsible for growth of the nail. The nerves and the blood supply of the nail bed enter it from its deep surface and are derived exclusively from anterior (palmar or plantar) nerves and vessels, never from posterior (dorsal) ones.

**Teeth** (*dentes*) are derived from both epithelial and connective tissue layers of the oral mucous membrane, itself modified skin differing from usual skin in that it does not possess a cornified layer. The hard outer **enamel** of teeth is developed from the epithelium and the bonelike **dentin** from the connective tissue. **Cementum,** modified bone softer than dentin, covers most of the dentin of the root of the tooth and attaches the tooth to the **periodontium** or periodontal membrane (the periosteum that lines the dental alveoli—sockets—of the upper and lower jaws). Nerves and blood vessels running in the bony jaws send branches into the teeth through their apical foramina. The gross anatomy of teeth is briefly described later with the oral cavity (Chap. 31).

## Functional Aspects

The thin, moist skin of animals such as the frog is permeable to a number of substances and is a rather effective organ of respiration, excreting carbon dioxide and absorbing oxygen. A small amount of carbon dioxide is lost through the skin of man, apparently in part dissolved in the sweat, but man's skin is largely impermeable to gases and to water or most liquids, although oily substances are absorbed somewhat better. In contrast to normal skin, skin in which the epidermal layers have been peeled off by blistering allows the ready passage of substances in either watery or oily solution, as do the thinner, moist mucous surfaces such as the conjunctiva and the mucous membrane of the nose that represent modified skin. Most of the **excretion** carried out by the skin is through the sebaceous and the sweat glands. As is well known, salt is a prominent component of sweat, and excessive sweating without adequate intake of salt leads to a deficiency of sodium and chloride ions in the body. Sweat also contains small amounts of protein and urea.

The **protective function** of the skin is best illustrated by the effects of the loss of large areas. For instance, in severe burns of the skin, there is not only the clinical problem of minimizing absorption of toxic substances but also the problems of minimizing infection and loss of fluid, both of which are obvious sequelae to loss of the protective function of the skin.

The **temperature-regulatory function** of the skin is controlled in man by two mechanisms. One depends upon dilation and constriction of the blood vessels of the skin, so that more or less blood is allowed to radiate its heat to the outside. The second is through the evaporation of sweat. Sweat secretion is evoked by a rise in temperature of the blood reaching the central nervous system.

The function of the skin as a **sense organ** depends upon the abundant innervation sent to it by cranial and spinal nerves. The larger nerve trunks to the skin run in the tela subcutanea, and branches leave these to penetrate the skin, where they at first form coarse networks. Branches from these, in turn, form finer and finer networks as they get closer to the epidermis. Certain specialized endings are found in the skin, especially in the dermal papillae; however, a great deal of evidence indicates that there are relatively few specialized sense organs in the skin of man and that the common type of nerve ending in the skin is in connection with the finer networks close to the epidermis. This is apparently true not only for fibers concerned with pain, as has been generally agreed for some time, but also for those concerned with heat and cold and even touch. Hairs add to the sensitivity of a region of skin to touch, since hair follicles have about them a complex plexus of nerve fibers. Deeper lying, encapsulated nerve endings, usually called pacinian corpuscles or Vater-Pacini corpuscles, occur in both the dermis and the tela subcutanea (and also in deep tissues, even within the abdominal cavity) and are generally believed to be associated with sensations of pressure. The concept, once held, that each type of sensation from the skin is mediated by a fiber specific for this type of sensation and also by an anatomically discrete and different nerve ending is no longer widely believed; there are not enough morphologically discrete types of nerve endings to support the concept of *anatomic* specificity of nerve endings.

Gross anatomic aspects of cutaneous innervation, and the concepts of peripheral nerve innervation and segmental innervation or dermatomes, have already been discussed.

## GENERAL REFERENCES AND RECOMMENDED READINGS

COX HT: The cleavage lines of the skin. Br J Surg 29:234, 1941

CUMMINS H, MIDLO C: Finger Prints, Palms and Soles. Philadelphia, Blakiston, 1943

HAMILTON JB (ed): The growth, replacement, and types of hair. Ann NY Acad Sci 53:461, 1951

HUTCHINSON C, KOOP CE: Lines of cleavage in the skin of the newborn infant. Anat Rec 126:299, 1956

LEE MMC: Physical and structural age changes in human skin. Anat Rec 129:473, 1957

MONTAGNA W: The Structure and Function of Skin. New York, Academic Press, 1956

ODLAND GF: The morphology of the attachment between the dermis and the epidermis. Anat Rec 108:399, 1950

TROTTER M: A review of the classification of hair. Am J Phys Anthropol 24:105, 1938

WINKELMANN RK: Nerve Endings in Normal and Pathologic Skin: Contributions to the Anatomy of Sensation. Springfield, Il, Charles C Thomas, 1960

# PART III

# UPPER LIMB

# 11

# THE UPPER LIMB AS A WHOLE

**Development**

The upper limb (*membrum superius*) originates as a swelling, the arm bud, on the lateral body wall of the embryo. As the bud grows, its distal end becomes flattened to form a plate for the hand, and ridges developing upon this plate form the fingers. A constriction above the plate represents the region of the wrist, and one above that, the region of the elbow (Fig. 11-1).

The limb projects at approximately a right angle to the body, making its surfaces and borders easy to recognize. The thumb develops on the border toward the head (the cranial or preaxial border), the little finger develops on the caudal border (the postaxial border), and ventral and dorsal surfaces correspond to the ventral and dorsal (anterior and posterior) surfaces of the trunk of the embryo. With later growth, these relationships are partially distorted by the adduction of the limb to where it almost parallels the trunk instead of projecting at a right angle to it, but even in the adult, the original relations are easily restored by holding the extended and abducted limb horizontally with the palm facing forward: the thumb is then still cranial, the little finger is caudal, the flexor surface on the limb is ventral, and the extensor surface is dorsal.

The developing musculature of the limb divides into ventral and dorsal parts, and this division of the muscles is accompanied by a corresponding division of the nerves growing into the muscles, so that both muscles and nerves of the upper limb can be classified as being either anterior (ventral) or posterior (dorsal). The skeleton at the base of the limb, developing into the shoulder girdle, spreads over the upper part of the trunk. The musculature at the base of the limb also spreads widely, the largest anterior muscles coming to cover much of the anterior thoracic wall, and the largest posterior muscles coming to cover the lateral thoracic wall and most of the structures of the back. This growth of muscles from the base of the limb to an attachment on the axial skeleton provides them with a stable origin through which they can act more powerfully upon the limb.

**FIG. 11-1.** Outline of the development of the upper limb. (After Keith. Coventry MB: In Instructional Course Lectures. The American Academy of Orthopaedic Surgeons, vol 6. Ann Arbor, JW Edwards, 1949)

## Parts and Regions

The chief named parts of the upper limb are the **acromion,** the tip of the shoulder (formed by a bony part of the same name); the **axilla** or armpit; the **brachium** or arm; the **cubitus** or elbow; the **antebrachium** or forearm; the **manus** or hand, hence the term "manual"; the **carpus** or wrist; and the **digits,** which include the fingers and the thumb. The digits are known both by name and by number: the thumb is both the pollex and digit I; digit II is the index finger, digit III the long or middle finger (really middle digit), digit IV the ring finger, and digit V the little finger (digitus minimus).

The described regions of the limb, partly but not entirely corresponding to the above-listed parts, are the deltoid region, the rounded upper part of the limb where it leaves the shoulder—the scapular or shoulder region proper is listed as a region of the back; anterior and posterior brachial regions; anterior and posterior cubital regions, the former presenting a cubital fossa, the hollow in front of the elbow; anterior and posterior antebrachial regions; and the palm and dorsum of the hand.

Because of the relationship of the limb girdle and the muscles at the base of the limb to the trunk, the anterolateral thoracic wall and the superficial region of the back must be included in any study of the upper limb. The anterior part of the thorax is the pectoral region (pectoral refers to the breast or chest) and is conveniently divisible into three regions: the infraclavicular region, on the upper part of the thorax just below the clavicle (collar bone); the mammary region, or region of the breast; and the axillary region, between the base of the limb and the thoracic wall and presenting the depression of the armpit or axillary fossa. On the back, the musculature of the shoulder typically extends from the occipital (posterior) region of the head throughout the entire vertebral (midline) region to the sacral level. Other named regions of the back are the nucha (back of the neck); the scapular region, or region of the shoulder blade; the infrascapular region, below the shoulder blade; and the lumbar region or flank, extending from below the ribs to the level of the bony girdle of the lower limb.

## Surface Anatomy

The **clavicle** (collar bone) lies subcutaneously at the junction of neck and thorax (Fig. 11-2). Its medial end can be palpated without difficulty to its articulation with the sternum (breast bone), and its lateral end to an articulation (the joint not being clearly palpable) with the acromion.

The **sternum** (breast bone) lies subcutaneous in the anterior midline and is palpable throughout its length. Its upper segment is the **manubrium,** which lies in a slightly different plane from the main portion of the bone, the **body.** The slight angle made by the junction of manubrium and body is the **sternal angle.** The third part of the sternum, the **xiphoid process,** is small and lies in the upper part of the depression between the arches of the rib cage on either side (the epigastric region of the abdomen). In the living individual, the xiphoid process is usually slightly movable on the body of the sternum. The concavity of the upper border of the manubrium sterni between the sternal ends of the two clavicles is the jugular notch.

On each side of the sternum, the **ribs** can be felt as they approach the sternum. For further description of these, see Chapter 19. A convenient way of counting the ribs anteriorly is to

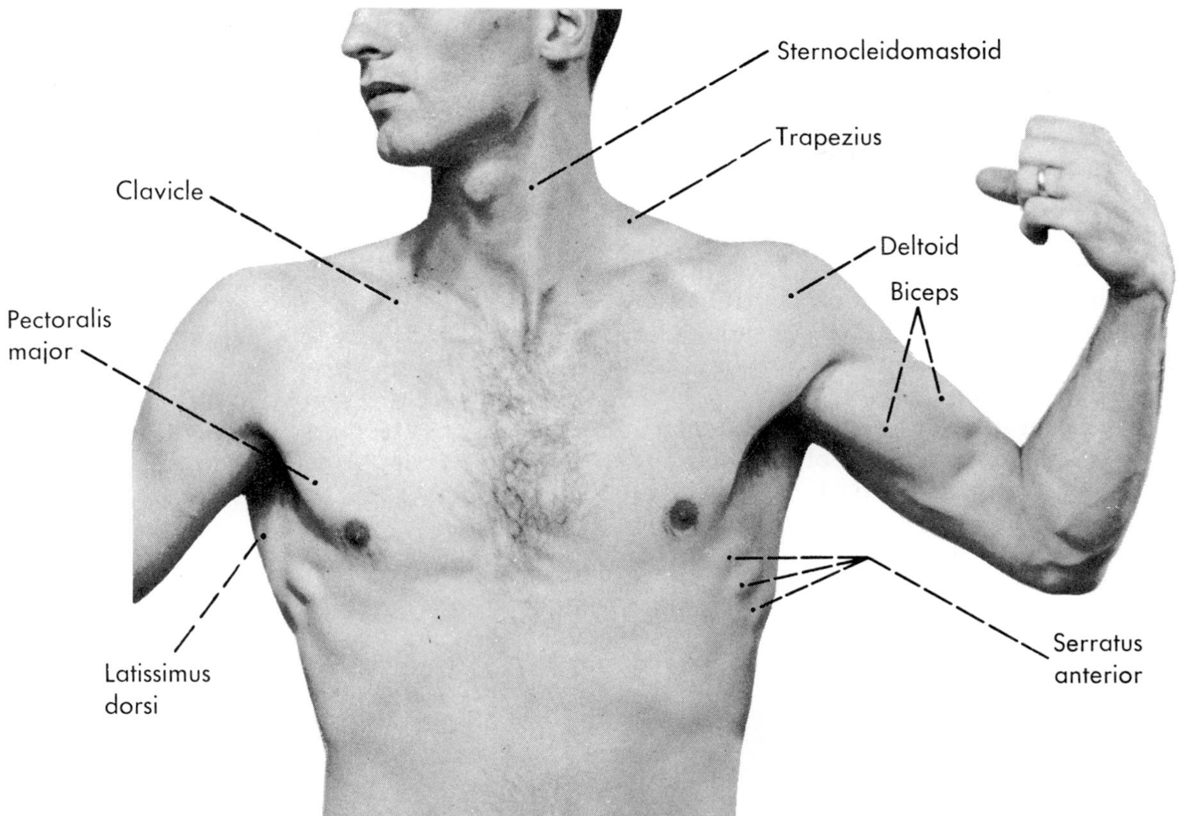


**FIG. 11-2.** Anterior view of the thorax and arm, with many of the associated muscles contracted to make them more prominent.

begin with the 2nd, since the 1st is under and behind the clavicle and cannot be clearly palpated; the 2nd rib nearly always lies at the level of the sternal angle. Most of the region lateral to the sternum and below the infraclavicular region is occupied in the female by the **breast.** In both male and female, both infraclavicular and pectoral regions are occupied by the large **pectoralis major** muscle, which covers most of the anterior thoracic wall and forms the anterior border of the armpit. The nipple in the male is usually situated at the level of the 4th intercostal space (that is, between the 4th and 5th ribs). In the female, the position of the nipple is much more variable, depending primarily upon the shape and size of the breast. Lateral to the pectoralis major muscle, the ribs can be felt fairly easily where they form the lateral portion of the thoracic wall, for they have only a thin covering of muscle over them.

The **armpit** or **axillary fossa** is marked off by anterior and posterior axillary folds. The **axillary line** (midaxillary line) is a convenient point of reference. It is an imaginary line projected vertically downward from the midpoint of the axillary fossa. On occasion, anterior and posterior axillary lines, projected downward from the axillary folds, also are referred to. Another imaginary line on the anterior surface of the thorax, lateral to the anterior midline, is the **midclavicular line,** a vertical line through the midpoint of the clavicle.

The rounded contour of the base of the arm below the acromion is formed by the **deltoid muscle.** Just below the lateral end of the clavicle and medial to the deltoid muscle, another part of the scapula, the **coracoid process,** can be felt. The **acromion** can be followed onto the back, where it is indistinguishably continuous with the **spine of the scapula.** In the living individual, the **medial border** of the scapula, more or less paralleling the posterior midline, can be felt with no difficulty, and it can be clearly seen if the limb is raised forward. Other borders of the scapula are covered by heavier muscles and are more difficult to outline. Most of the muscles of the shoulder are flat, but some of them can be recognized in muscular individuals (Fig. 11-3).

The posterior midline of the **back** is marked by the spinous processes of the vertebrae, which are, for the most part, easily seen and palpated. In most of the neck, however, they lie deeply, and the first spinous process visible at the base of the neck is usually that of the 7th cervical vertebra (also called, because of this feature, the vertebra prominens).

In the **arm** (Figs. 11-2 through 11-4), the fact that the muscles are grouped primarily anteriorly and posteriorly is easily recognized. The bone of the arm, the **humerus,** can be felt fairly clearly between these two muscle masses for much of its length. The grooves on the sides of the **biceps muscle,** the prominent muscle on the front of the arm, are the medial and lateral bicipital grooves (sulci). On the upper part of the medial side of the arm, some of the nerves and vessels descending into the

**FIG. 11-3.** Posterior view of the thorax and arm.

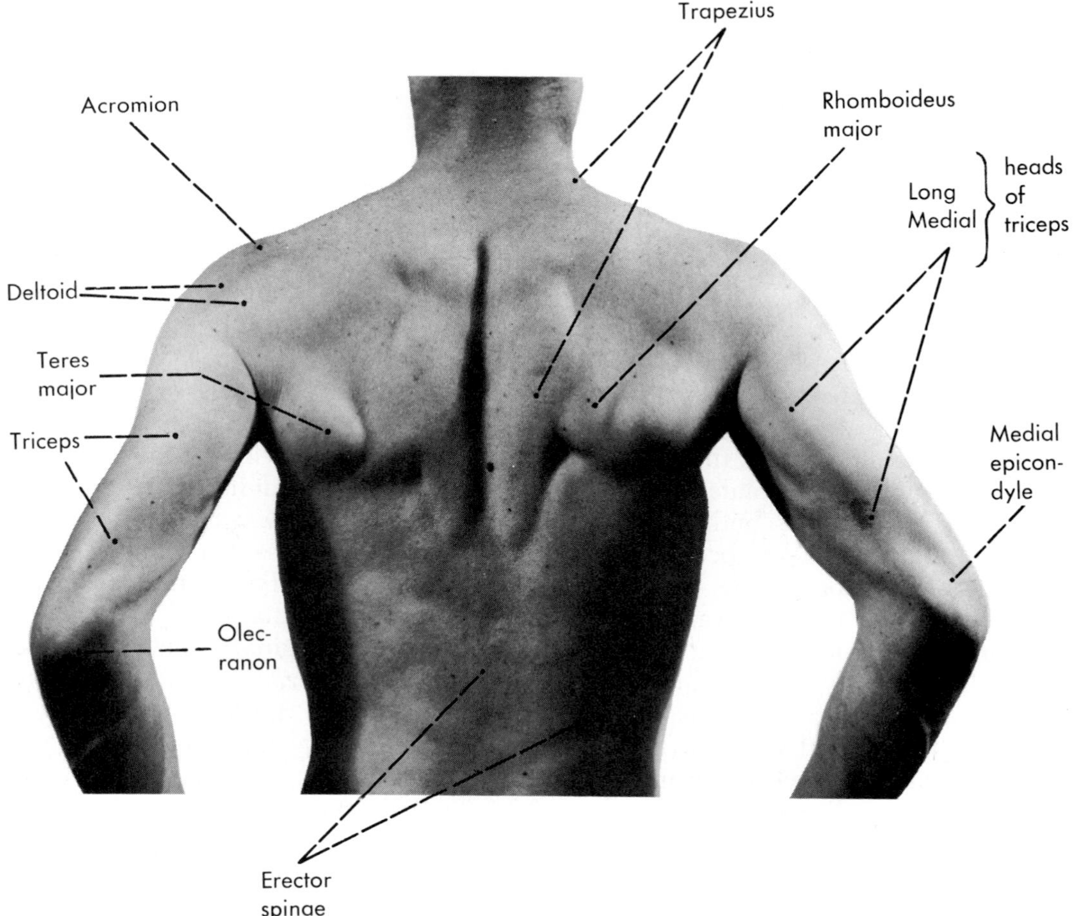

arm from the axilla can be rolled beneath a finger or thumb. At its lower end, the humerus flares out and its prominent projections at the elbow, the **medial** and **lateral epicondyles,** can be palpated without difficulty. The nerve that passes subcutaneously behind the medial epicondyle, giving this part the colloquial name "funny bone," is the **ulnar nerve.** The **cubital fossa,** in front of the elbow, is traversed by the heavy tendon of the biceps muscle.

In the **forearm,** the bony projection behind the elbow is the **olecranon,** the upper end of the **ulna** (one of the bones of the forearm). The ulna is partly subcutaneous throughout its length and can be traced from the elbow to the wrist along the little-finger side of the forearm. The **radius,** the other bone of the forearm, is covered by muscles in its upper part but can be palpated distally as it approaches the wrist. Certain of the carpal (wrist) bones and various tendons crossing the wrist can be recognized, but a description of these and of the surface anatomy of the hand is best reserved for the specific chapter dealing with these parts.

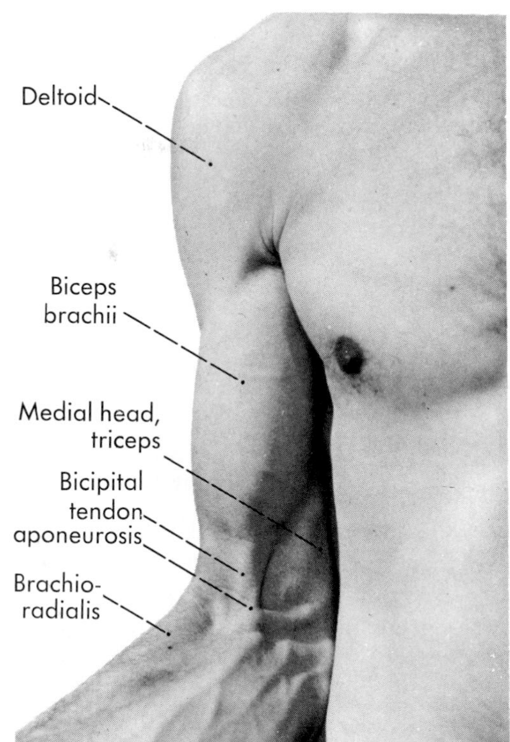


**FIG. 11-4.** Anterior view of the arm and upper part of the forearm.

#### Dermatomes and Cutaneous Nerves

The **3rd** and **4th cervical nerves** supply a limited area of skin over the upper parts of the pectoral region and shoulder, and the **2nd thoracic nerve** usually sends a branch to skin of the medial and upper part of the arm. Otherwise, the skin of the upper limb is supplied by branches from the **brachial plexus,** which is typically formed by the union of branches of the 5th cervical through 1st thoracic nerves, sometimes with the addition of a few fibers from the 4th cervical. The difficulty of tracing segmental nerves through a plexus has already been mentioned and the concepts of dermatomes in the limbs, discussed (Chap. 4). The two commonly used dermatomal charts are those of Foerster and of Keegan (see Figs. 4-8 through 4-10). These do not agree in detail, and some clinicians regard one chart as more applicable, some the other.

The distribution of the specific peripheral cutaneous nerves to the limb (Fig. 11-5) are best described when the various parts are being considered; therefore, only a brief summary is given here. The cutaneous nerves to the arm are small and rather numerous and arise from a number of different larger nerves. The three cutaneous nerves to the forearm all are called **antebrachial cutaneous nerves** and designated according to their distribution as **lateral, medial,** or **posterior.** The antebrachial cutaneous nerves supply some of the skin of the hand close to the wrist, but most of the skin of the hand and fingers is supplied by the three nerves that supply the muscles of the forearm and hand.

## Skeleton

Basic features of bones and joints are described in Chapter 3.

The clavicle and the scapula constitute the **pectoral girdle** or the girdle of the upper limb (Fig. 11-6). The **clavicle** articulates with the axial skeleton by a large synovial joint at its sternal end, which serves as a pivot that allows the shoulder to be moved up and down, forward and backward. In contrast to the clavicle, the **scapula** has no direct attachment to the axial skeleton, for it is attached to the ribs and the vertebral column by muscles only. Its only bony connection to the axial

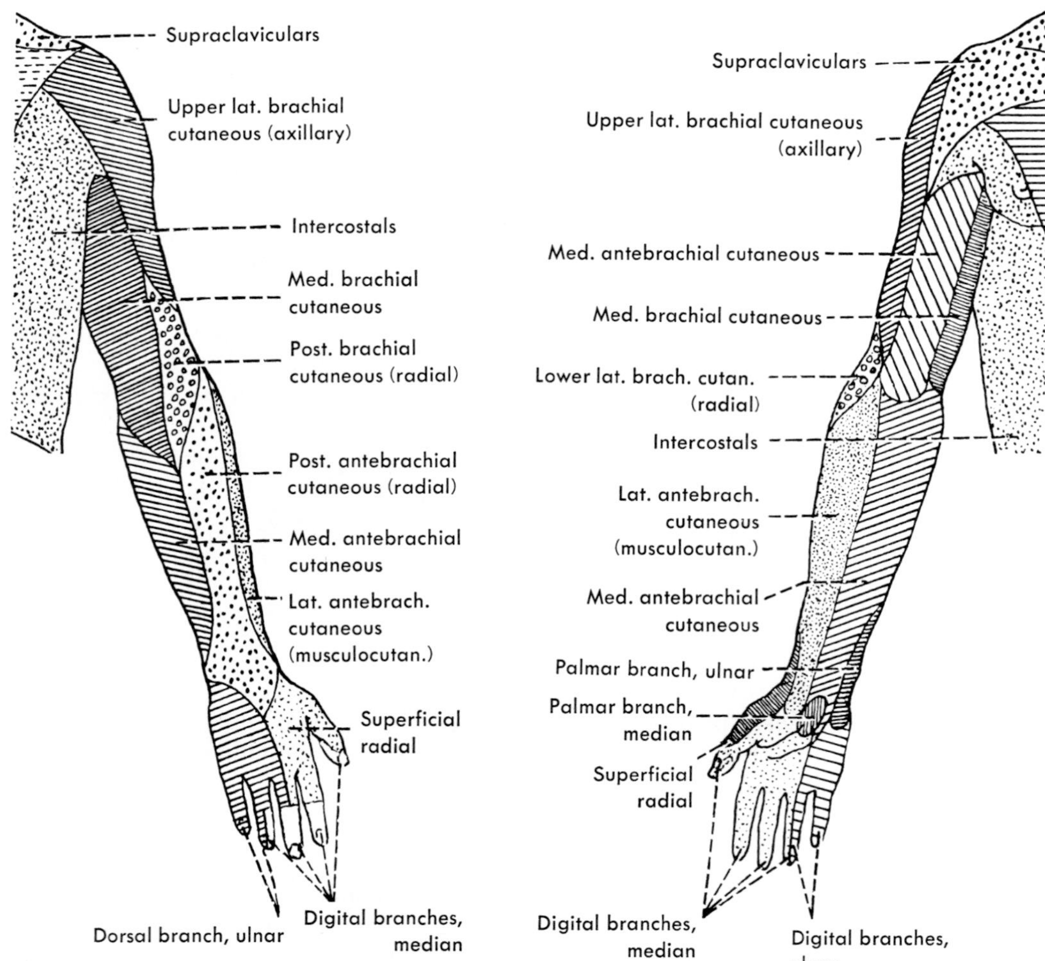


**FIG. 11-5.** Distribution of cutaneous nerves to the upper limb, anterior and posterior views. The distribution of the intercostobrachial, not shown, is largely identical to that of the medial brachial cutaneous. Note that the med. antebrachial cutaneous nerve supplies not only the forearm but a sizable area of skin on the arm. (Flatau E: Neurologische Schemata für die ärtzliche Praxis. Berlin, Springer, 1915)

skeleton is the indirect one through the clavicle. The scapula is roughly triangular. Its lateral corner is expanded and presents a shallow concavity, the **glenoid cavity,** for articulation with the humerus.

The **head of the humerus,** the enlarged upper end, articulates with the scapula. Its body is roughly cylindrical, and at its expanded lower end, between the medial and lateral epicondyles, are surfaces for articulation with the bones of the forearm.

The **radius,** the lateral bone of the forearm (on the thumb side), gives the name radial to this side of the forearm. Its upper end is disklike and allows flexion on the humerus and rotation on the ulna. The movement of turning the palm of the horizontally held hand down and up (pronation and supination) is made possible by the rotatory action of the radius at the elbow. Its expanded lower end articulates with proximal bones of the wrist. The **ulna,** the bone on the little-finger side of the forearm, tapers from its proximal to its distal end. The upper end has a deep notch

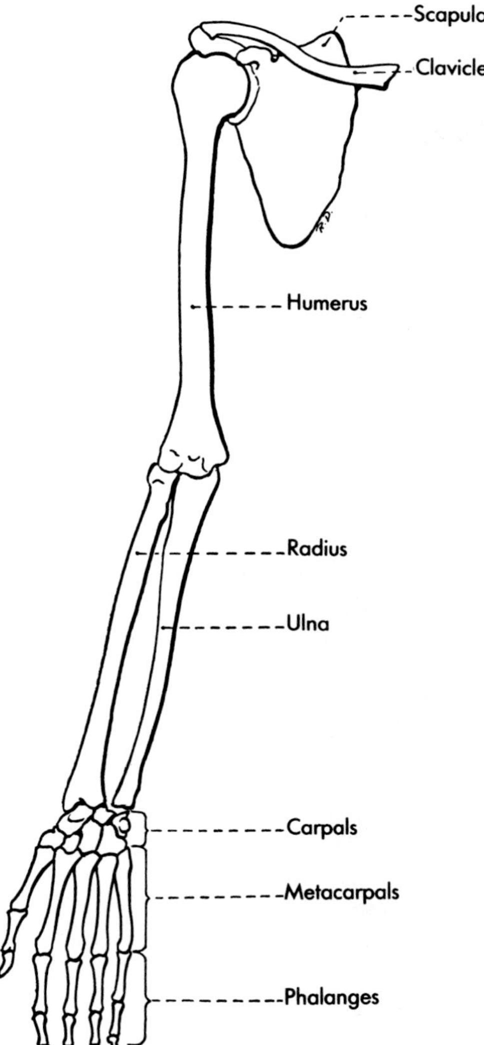


**FIG. 11-6.** Bones of the upper limb.

anteriorly for articulation with the humerus, and the part above this notch, forming the prominence behind the elbow, is the **olecranon.** The small lower end of the ulna articulates with the radius but does not articulate directly with bones of the wrist.

The eight bones of the wrist, the **carpals,** are arranged in two rows, a proximal and a distal. These are followed by five **metacarpals.** The skeleton of the fingers and thumb consists of **phalanges.** The proximal phalanges of the fingers articulate with the metacarpals at the knuckles, and there are three phalanges to each finger and two to the thumb.

**Ossification** of the bones of the upper limb follows the general pattern of ossification already described (Chap. 3); primary centers of ossification form the bodies of the bones. Growth in length occurs from epiphyseal cartilages; the long bones develop one or more secondary or epiphyseal centers that eventually fuse with the body. The latter are more appropriately noted in connection with the description of individual bones. The primary centers of ossification for the upper limb appear over a wide age range: during the fifth or sixth fetal week for the clavicle; during the eighth and ninth weeks for the bodies of the scapula, humerus, radius, ulna, metacarpals, and phalanges; and from shortly after birth up to 10 to 14 years for the various carpals.

## Muscles

Some fundamental aspects of muscle and fascia are discussed in Chapters 6 and 3, respectively.

The muscles of the upper limb (Figs. 11-7 and 11-8) are conveniently divided into pectoral and shoulder muscles, which insert on either the limb girdle or the humerus and have their bellies located on the trunk or on the scapula; muscles of the arm, which have their bellies in the arm and move the forearm at the elbow joint, the humerus at the shoulder joint, or both; muscles of the forearm, which have their bellies in the forearm and act at the elbow, at the wrist, or on the digits; and the intrinsic muscles of the hand, which act upon the digits. These muscles can, in turn, be subdivided into anterior and posterior groups.

Three muscles situated on the anterior thoracic wall represent the anterior muscles at the shoulder. The remaining muscles all belong to the posterior group of muscles. These are more strictly muscles of the shoulder, although the larger ones are topographically muscles of the back.

The anterior muscles of the arm are the original ventral ones and are flexors. The pos-

(*text continues on p. 156*)

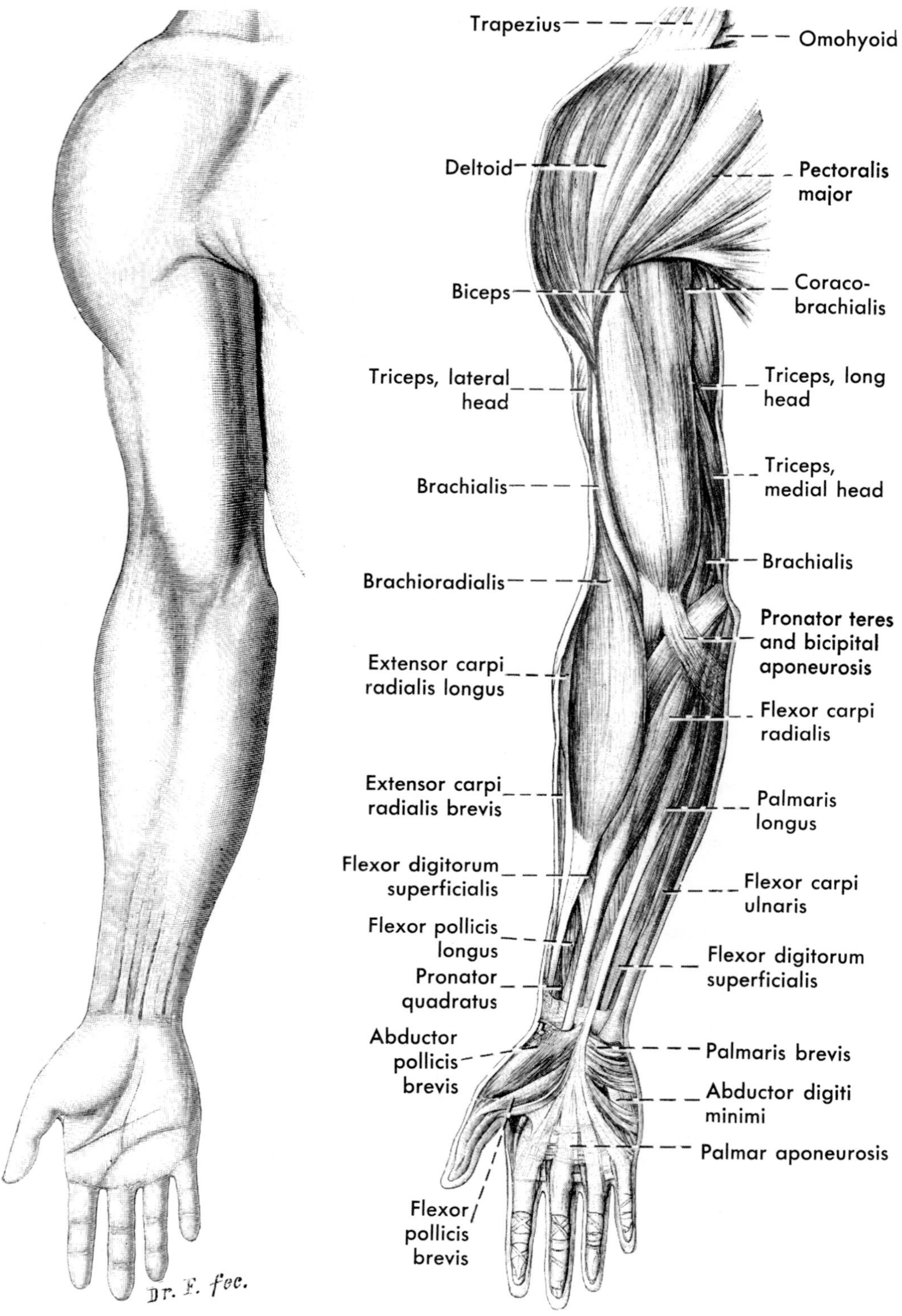


**FIG. 11-7.** Contour of the upper limb and the superficial muscles of the limb from the front. Only small parts of the pectoralis major and the larger shoulder muscles are shown. (Frohse F, Fränkel M: In von Bardeleben K (ed): Handbuch der Anatomie des Menschen, vol 2, sect 2, pt 2. Jena, Fischer, 1908)

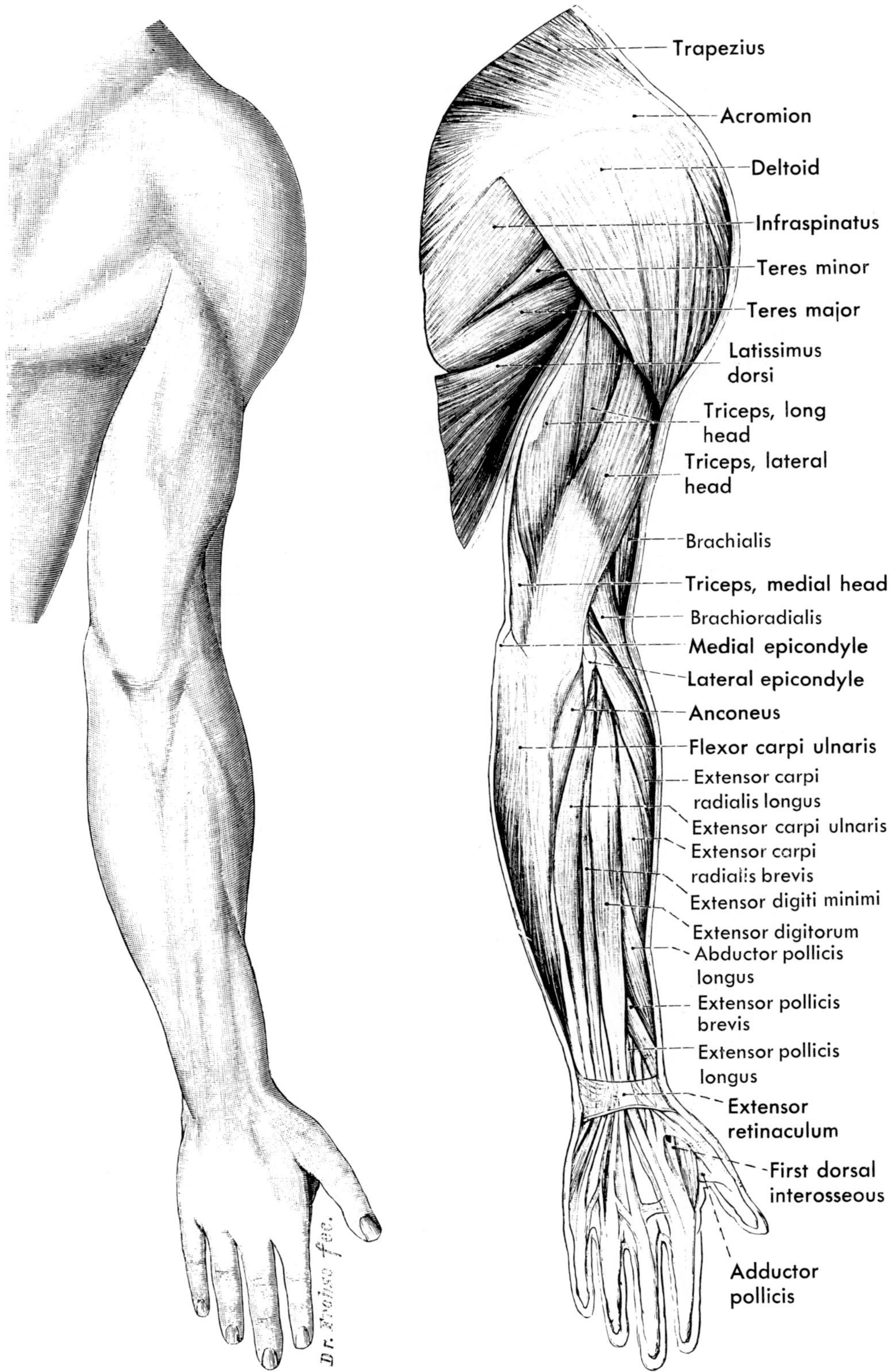


**FIG. 11-8.** Contour of the upper limb, and the musculature, from behind. Only small parts of the trapezius and latissimus dorsi are shown. (Frohse F, Fränkel M: In von Bardeleben K (ed): Handbuch der Anatomie des Menschen, vol 2, sect 2, pt 2. Jena, Fischer, 1908)

terior muscles are the original dorsal ones and are extensors. Of the muscles of the forearm, the anterior group arises chiefly on the ulnar side of the arm or forearm and flexes the wrist and fingers and pronates the forearm. The posterior group arises chiefly on the radial side of the arm (forming the hard knot of muscles on the lateral aspect of the elbow) and on the back of the forearm and extends the wrist and digits and supinates the forearm. The muscles of the hand are normally all anterior ones, arising on the palmar surface.

**Bursae** around the shoulder and **tendon sheaths** at the wrist are described in connection with the region in which they occur.

## Large Nerves

The structure of nerves is discussed in Chapter 4.

The cutaneous nerves of the upper limb have already been briefly mentioned. It was noted that, with minor exceptions, they are all derived from the **brachial plexus** either directly or indirectly. This is true of all the major nerves of the limb, whether cutaneous or not, except for the nerves to a few muscles: the trapezius, a large muscle spread over the upper part of the back, is innervated by the spinal accessory nerve, and a smaller upper muscle, the levator scapulae, is innervated from the cervical plexus.

The brachial plexus is divisible into **anterior** and **posterior parts.** The anterior parts of the plexus supply originally anterior muscles of the limb, and the posterior parts supply posterior muscles. The uppermost anterior branches of the plexus supply the uppermost anterior muscles of the limb, and a series of upper posterior branches from the plexus supply the posteriorly situated shoulder muscles. In addition to two purely cutaneous nerves, there are five mixed nerves continuing into the free limb: the anterior nerves are the musculocutaneous, the median, and the ulnar; the posterior nerves are the axillary and the radial.

The **musculocutaneous nerve** supplies the anterior (flexor) muscles of the arm and continues into the forearm to supply skin only (Fig. 11-9). Both the **median** and the **ulnar nerves** run through the arm without supplying any muscles, giving off their first muscular branches to muscles of the forearm. Between them, these two nerves supply all the anterior muscles of the forearm and all the muscles of the hand. However, the two nerves

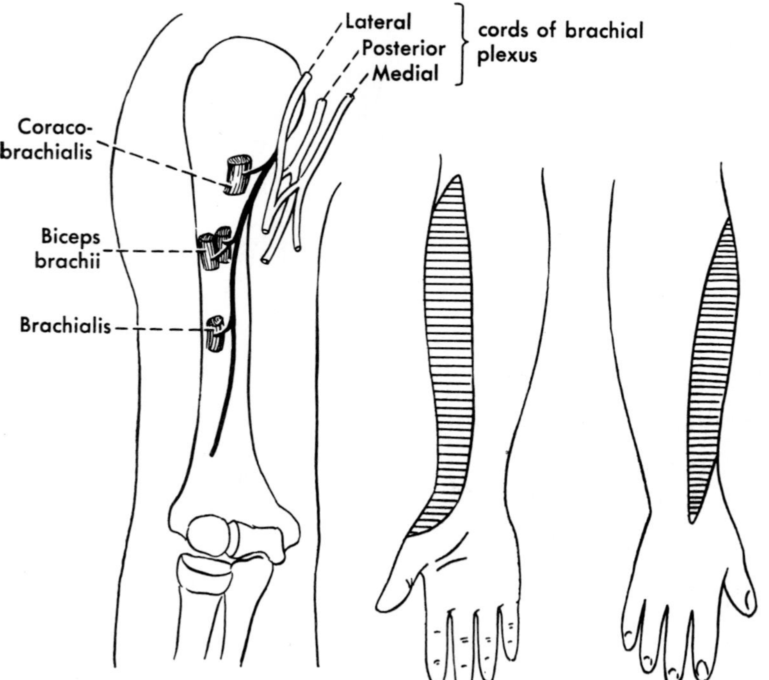


**FIG. 11-9.** Distribution of the musculocutaneous nerve.

do not participate equally in the innervation of the muscles of either the forearm or the hand (Figs. 11-10 and 11-11); the median nerve supplies most of the flexor muscles of the forearm but only a few of those in the hand, whereas the ulnar nerve supplies few muscles in the forearm but most of the muscles of the hand.

Of the two posterior branches of the brachial plexus entering the free limb, the **axillary nerve** is confined in its distribution to the immediate shoulder region, supplying two posterior muscles of the shoulder (Fig. 11-12). The **radial nerve,** therefore, is the only branch of the plexus available for the innervation of the posterior muscles of the free limb; in consequence, it supplies all the posterior muscles of both the arm and forearm.

(*Text continues on p. 161*)

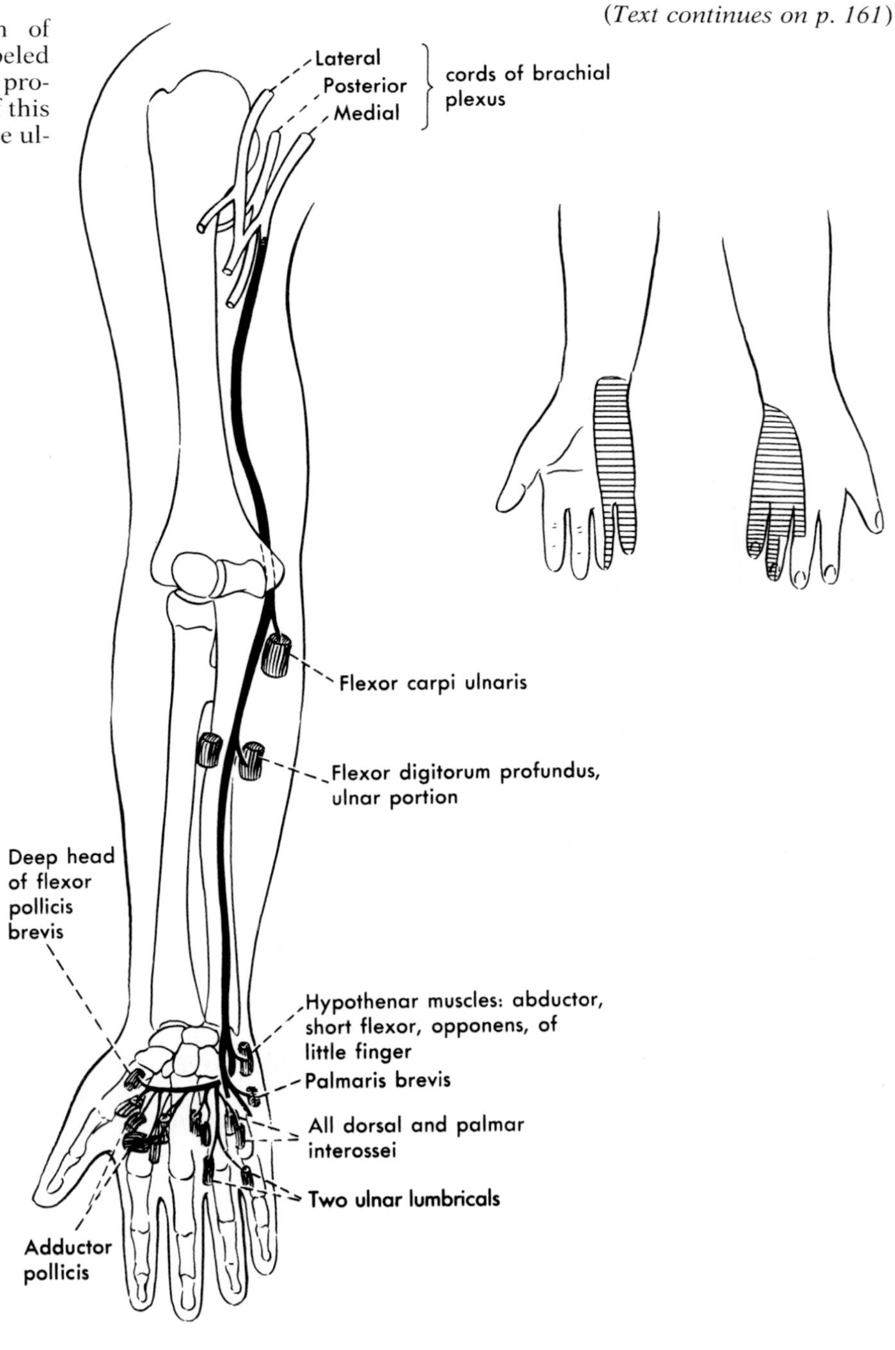

**FIG. 11-10.** Distribution of the ulnar nerve. The unlabeled structure beside the flexor profundus is the radial part of this muscle, not supplied by the ulnar nerve.

Lateral
Posterior
Medial
cords of brachial plexus
Flexor carpi ulnaris
Flexor digitorum profundus, ulnar portion
Deep head of flexor pollicis brevis
Hypothenar muscles: abductor, short flexor, opponens, of little finger
Palmaris brevis
All dorsal and palmar interossei
Two ulnar lumbricals
Adductor pollicis

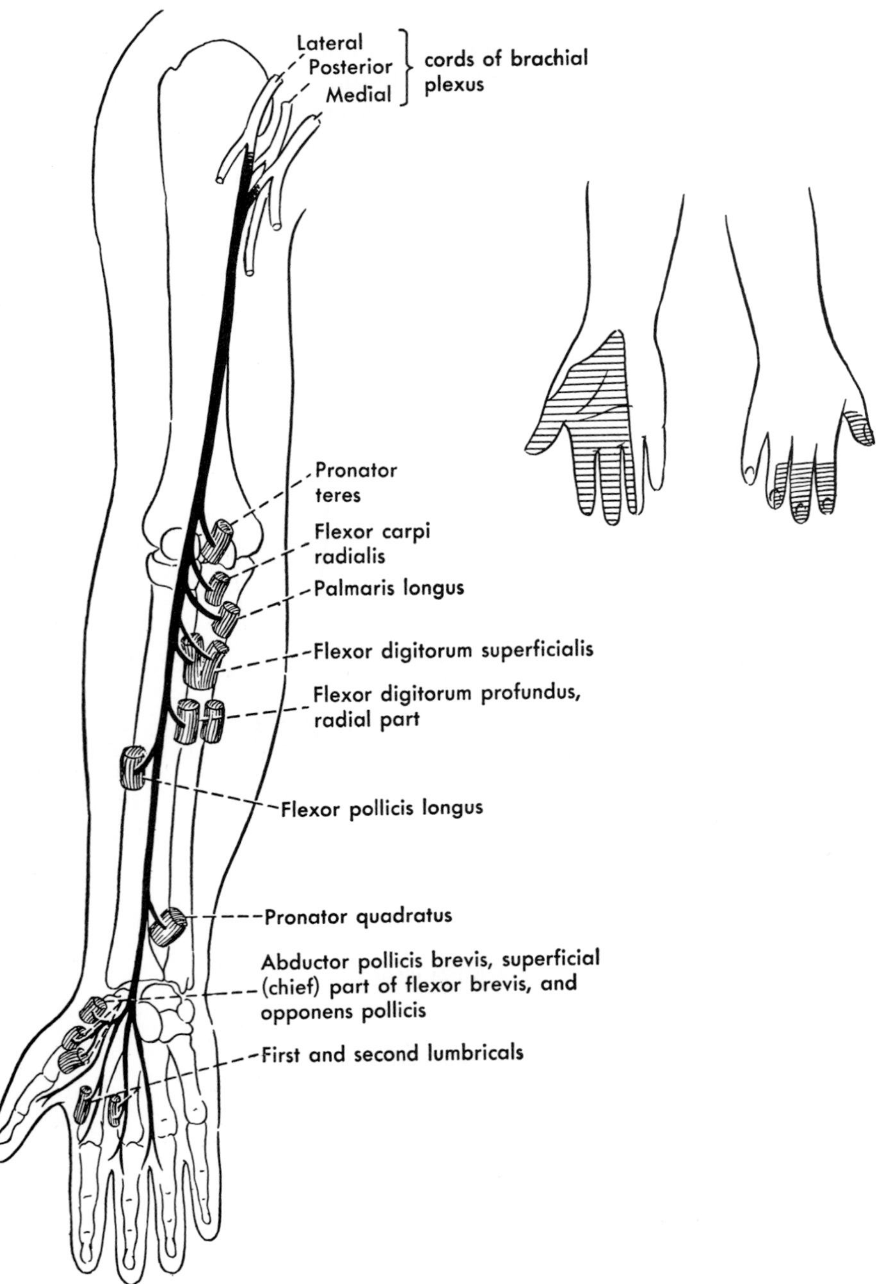


**FIG. 11-11.** Distribution of the median nerve. Here the ulnar part of the flexor digitorum profundus, supplied by the ulnar nerve, is unlabeled. The branches to the flexor digitorum profundus, flexor pollicis longus, and pronator quadratus are given off by the *anterior interosseous* branch of the median nerve, which is not shown separately in this figure.

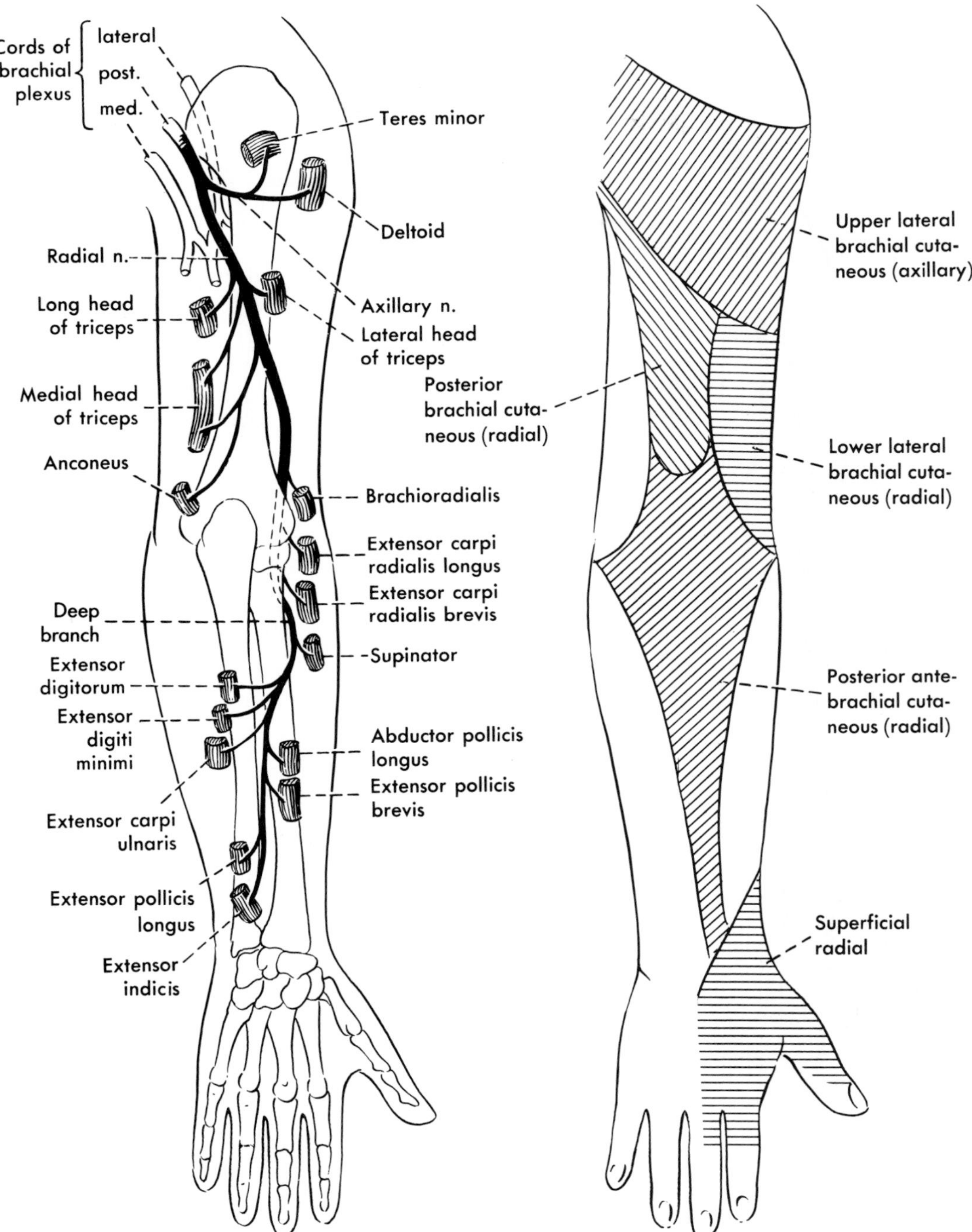


**FIG. 11-12.** Distribution of the axillary and radial nerves. Note that all the muscles supplied below the elbow are innervated by the deep branch of the radial *posterior interosseous nerve;* the course of the superficial radial is not shown.

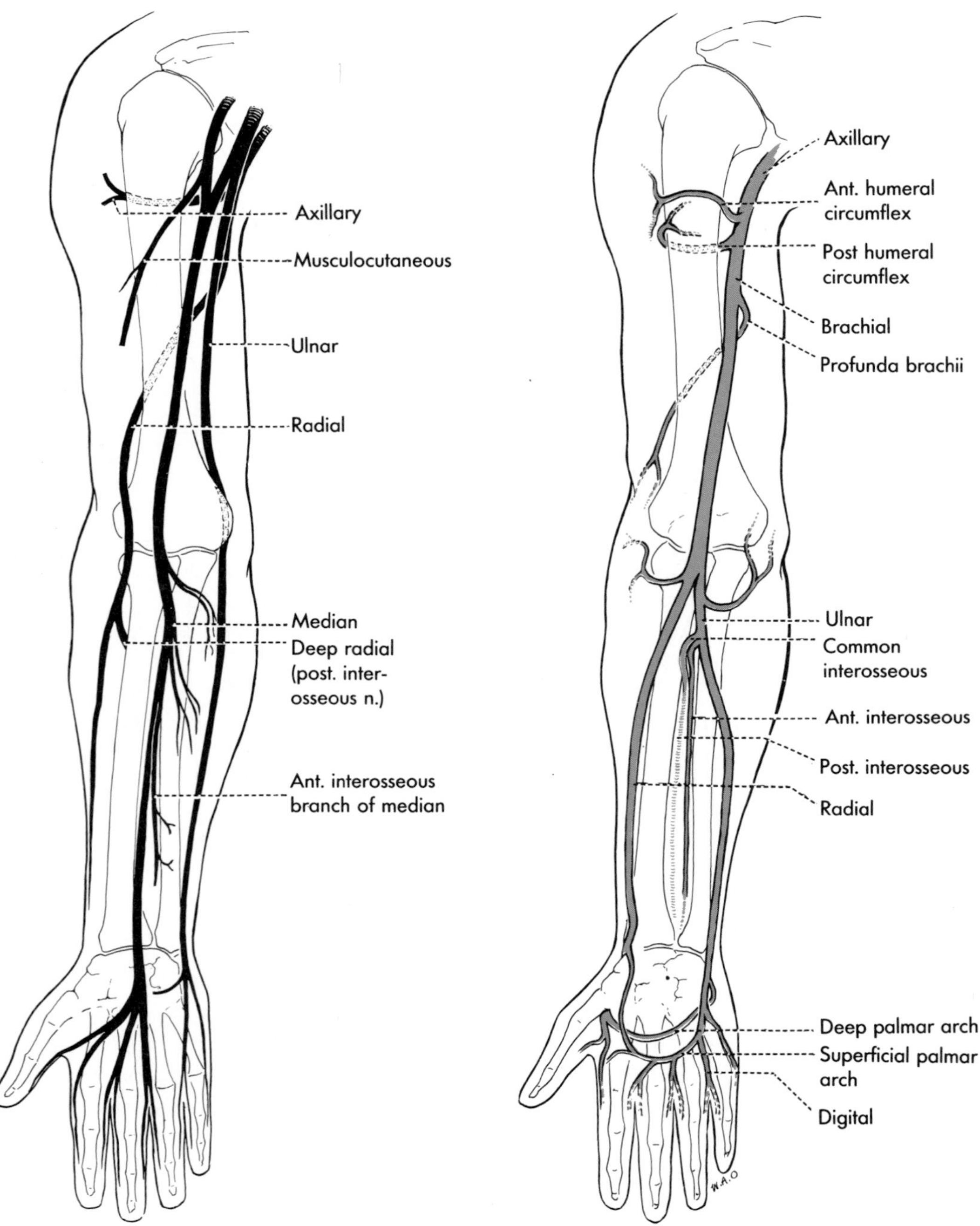


**FIG. 11-13.** The chief nerves and arteries of the upper limb. Note that on the front of the forearm each major artery is paralleled by one of the major nerves.

### Vessels

Basic features and the general pattern of the vascular system are discussed in Chapter 5.

The named arteries of the limb are all deep ones. The arteries are regularly accompanied by one or two deep veins: the larger veins (for instance, brachial veins) bear the same name as the artery; the smaller ones are officially called venae comitantes of the arteries, but in practice they are usually also named after the arteries. In addition to the deep veins, there are relatively large superficial veins of the limb, and these do not accompany arteries.

The **subclavian artery,** lying largely behind the clavicle at the base of the neck and deriving its blood directly or indirectly from the arch of the aorta, is the arterial stem to the upper limb. In its course at the base of the neck, it gives off branches to structures in the neck, to the thoracic wall, and to the shoulder. As it enters the axilla, its name changes to axillary artery (Fig. 11-13). The **axillary artery** gives off branches to the shoulder, arm, and lateral thoracic wall and is continued into the arm as the brachial. The **brachial artery** supplies most of the muscles of the arm and ends in front of the elbow by dividing into the **radial** and **ulnar arteries** that together supply the forearm and hand. The superficial position of the brachial artery on the medial side of the arm allows it to be compressed against the humerus by a blood-pressure cuff, and its superficial position in front of the elbow allows the physician to listen for the return of blood into it as the cuff is released. Similarly, the radial artery is particularly convenient for feeling the pulse, since it lies subcutaneously against the unyielding radius.

The **deep veins** closely parallel the arteries. Much of the venous drainage from the hand is through superficial veins, but paired **radial** and **ulnar veins** accompany the radial and ulnar arteries and, in turn, receive tributaries corresponding to their branches. They unite to form paired **brachial veins** that pass upward along the artery and join the unpaired **axillary vein.** The axillary vein (a continuation of a large superficial vein, the **basilic**) receives tributaries corresponding, in general, to the branches of the axillary artery. It also receives the other superficial vein of the upper limb, the **cephalic vein.** The axillary vein passes between the clavicle and the 1st rib to reach the base of the neck, where its name is changed to subclavian vein. The **subclavian vein** joins or is joined by veins of the neck and returns its blood to the heart by way of the superior vena cava.

The **superficial veins** of the upper limb (Figs. 11-14 and 11-15) begin as networks on the digits and drain mostly onto the dorsal aspect of the hand. The veins leaving the dorsum unite to form two chief channels. Those from the radial side form the **cephalic vein.** Those from the ulnar side form the **basilic vein.**

The **cephalic vein** runs up the radial side of the limb and, after communicating with the basilic vein at the elbow, continues up the lateral side of the anterior surface of the arm. It passes deeply just a little below the clavicle to end in the axillary vein.

The **basilic vein** runs up the ulnar side of the forearm, passing around this border to lie anteromedially at the elbow. It runs superficially in the lower third of the arm, but then penetrates the brachial (deep) fascia and runs deep to this fascia, paralleling the brachial vessels and continuing, as already noted, as the axillary vein.

The **superficial lymphatics** of the upper limb (Figs. 11-16 and 11-17) begin as a dense capillary plexus in the skin of the digits. The lymphatic vessels from the palmar surface tend to run onto the dorsum, as do the superficial veins. Many of the palmar vessels pass around the borders of the digits or hand to join the dorsal vessels. In part because of this, and in part because the loose subcutaneous tissue of the dorsum allows accumulation of fluid, infections of the palm may be evidenced by swelling on the dorsum. The lymphatics from the hand are joined by superficial lymphatics from the forearm, and in the arm, they tend to converge on its anteromedial surface, receiving the lymphatics from the arm and ending in axillary lymph nodes.

**Deep lymphatics** of the upper limb are much less numerous than are the superficial

(*Text continues on p. 164*)

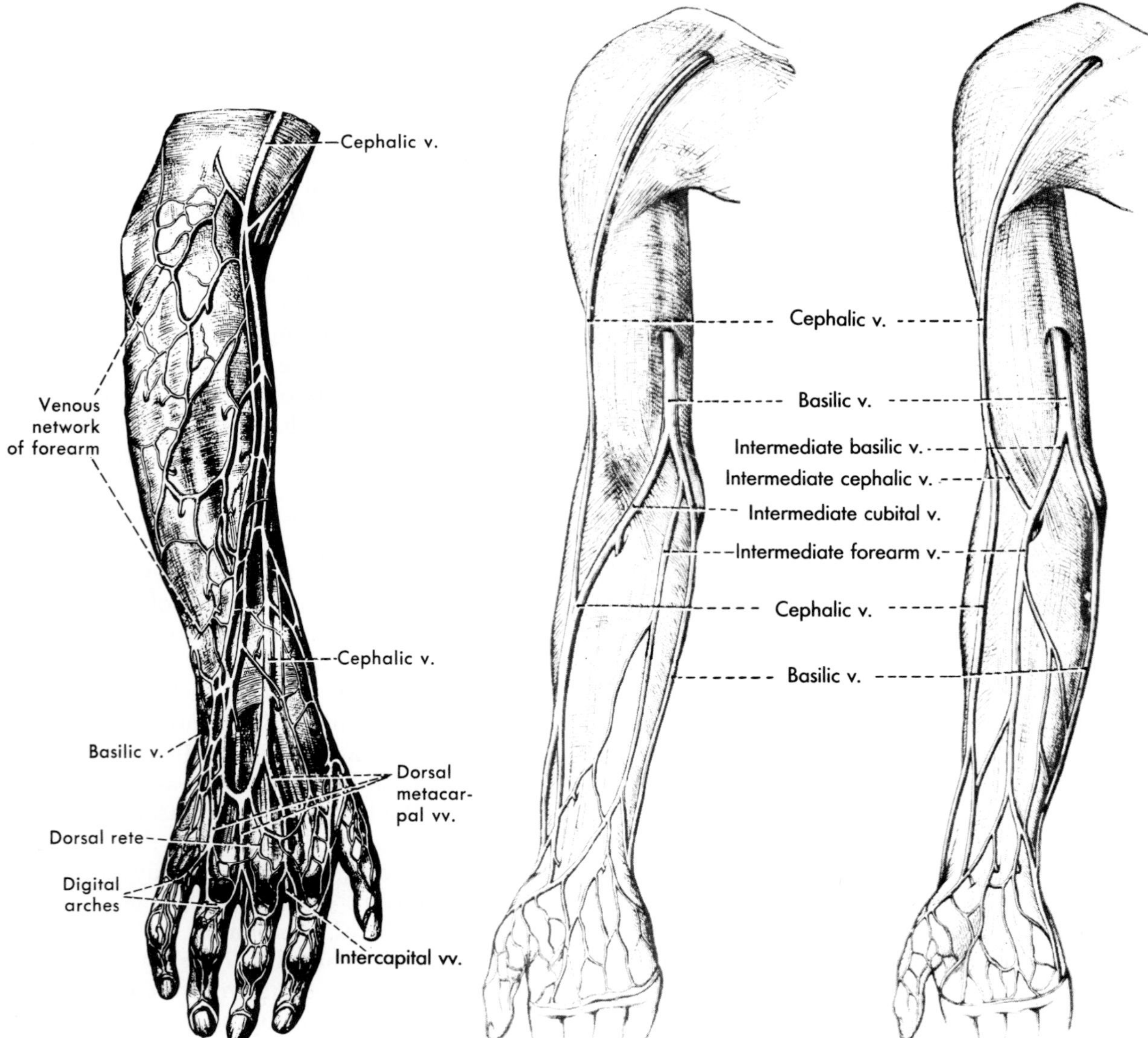


**FIG. 11-14.** Superficial veins of the upper limb, posterolateral view. Above the elbow there are no important venous channels posteriorly. (Toldt C: An Atlas of Human Anatomy, 2nd ed. New York, Macmillan, 1928)

**FIG. 11-15.** Superficial veins of the upper limb, anterior view. Note the difference in arrangement in front of the elbow; the intermediate basilic and intermediate cephalic veins are variants of the intermediate cubital vein. (Toldt C: An Atlas of Human Anatomy, 2nd ed. New York, Macmillan, 1928)

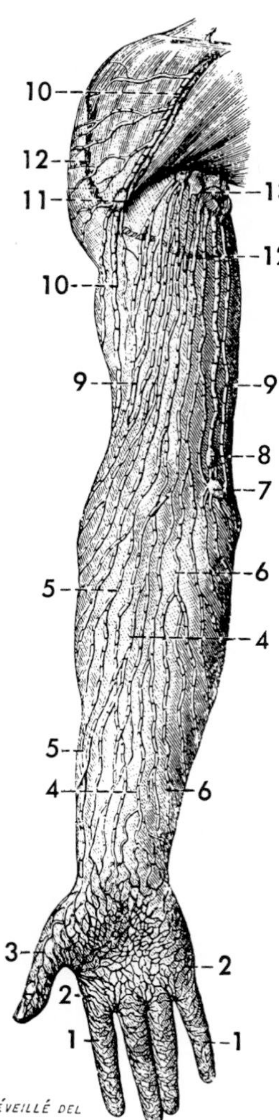


**FIG. 11-16.** Anterior view of the superficial lymphatics of the upper limb. The numbers merely identify various networks and channels, for instance, networks on the fingers, and are of no importance; note, however, the large channels running up to the axillary lymph nodes, **13**, and lymphatics, **10**, following the upper part of the cephalic vein. (After Sappey. Poirier P, Charpy A: Traité d'Anatomie Humaine, 2nd ed, vol 2, fasc 4. Paris, Masson et Cie, 1909)

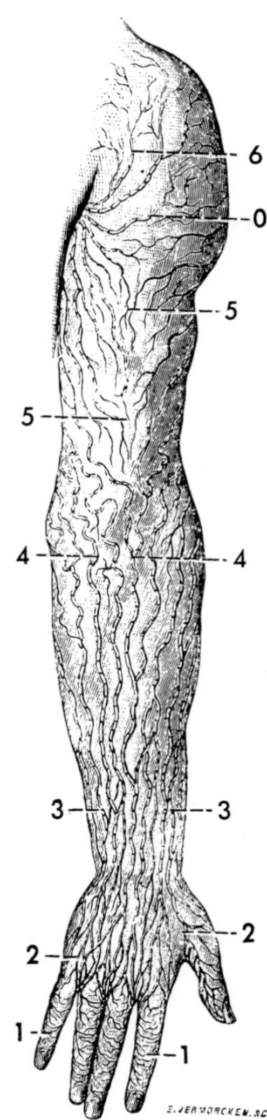


**FIG. 11-17.** Posterior view of the superficial lymphatics of the upper limb. As in the preceding figure, the numbers identifying various channels are mostly unimportant; note, however, at **0, 5, 5,** and **4, 4,** the divergence of the posterior lymphatics: some run medially and some laterally around the limb. (After Sappey. Poirier P, Charpy A: Traité d'Anatomie Humaine, 2nd ed, vol 2, fasc 4. Paris, Masson et Cie, 1909)

lymphatics. They follow the deep arteries and veins, run along with the brachial vessels, and then empty into axillary nodes, as do the superficial lymphatics. A few small and inconstant nodes occur along their course.

## GENERAL REFERENCES AND RECOMMENDED READINGS


FELDMAN MG, KOHAN P, EDELMAN S, JACOBSON JH, II: Lymphangiographic studies in obstructive lymphedema of the upper extremity. Surgery 59:935, 1966

LEWIS WH: The development of the arm in man. Am J Anat 1:145, 1902

LINELL EA: The distribution of nerves in the upper limb, with reference to variabilities and their clinical significance. J Anat 55:79, 1921

O'RAHILLY R: Morphological patterns in limb deficiencies and duplication. Am J Anat 89:135, 1951

ROSSE C: Basic structural plan and functional adaptation in the limbs. In Rosse C, Clawson DK (eds): The Musculoskeletal System in Health and Disease. Hagerstown, Harper & Row, 1980

WHITTAKER CR: The arrangement of the bursae in the superior extremities of the full-time foetus. J Anat Physiol 44:133, 1910

WOOLLARD HH: The development of the principal arterial stems in the forelimb of the pig. Contrib Embryol 14:139, 1922

# 12

# PECTORAL REGION, AXILLA, SHOULDER, AND ARM

Descriptions of the soft parts—fascia, muscles, blood vessels, and nerves—necessarily involve frequent reference to the skeletal system, and, therefore, this system should be borne in mind as other parts are studied. The clavicle, the scapula, and the humerus are of special importance. Only a rudimentary knowledge of the sternum, ribs, vertebral column, and upper ends of the radius and ulna is necessary at this time. Detailed information concerning these bones is given in Chapters 13, 14, and 19, where it can be consulted as necessary. The superficial anatomy, including the bony landmarks that are of importance, has already been described (Chap. 11).

## SKELETON

### PECTORAL GIRDLE

The anteriorly situated clavicle (collarbone) and the posteriorly situated scapula (shoulder blade) constitute the pectoral girdle, the girdle of the upper limb (*cingulum membri superioris*).

#### Clavicle

Viewed from above or below (Figs. 12-1 and 12-2), the clavicle resembles the italic letter *f*, with the concavity of the medial curve directed posteriorly and that of the lateral curve directed anteriorly. Its expanded sternal, or medial, end bears an articular surface for articulation with the manubrium sterni. The entire medial two-thirds is somewhat rounded, but the acromial (lateral) third is flattened superoinferiorly and presents, on its end, an acromial articular surface for the scapula. The clavicle gives attachment to several muscles. Its upper surface shows no special markings. On the infe-

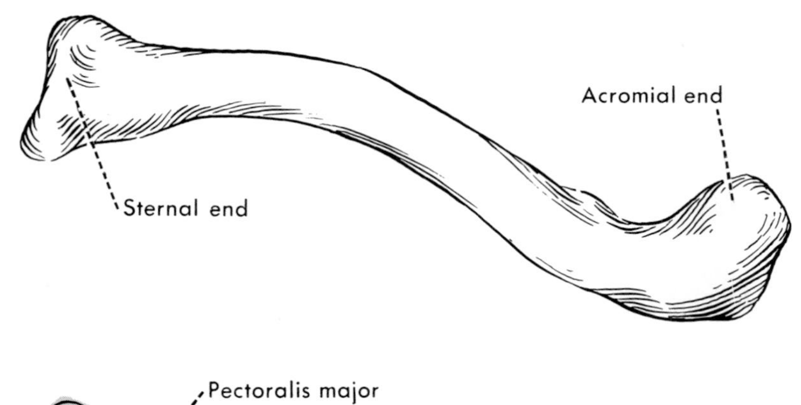


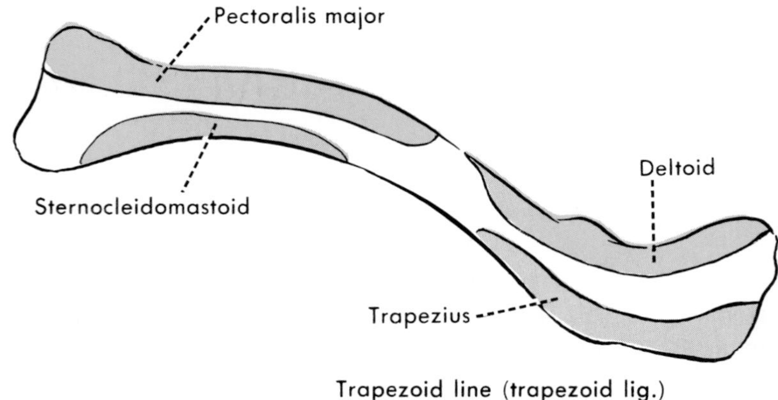


**FIG. 12-1.** The right clavicle from above. Muscle origins are shown in red, insertions in blue.

**FIG. 12-2.** The right clavicle from below.

Trapezoid line (trapezoid lig.)
Conoid tubercle (conoid lig.)
Subclavian groove
Impression for costoclavicular lig.
Acromial articular surface
Sternal articular surface

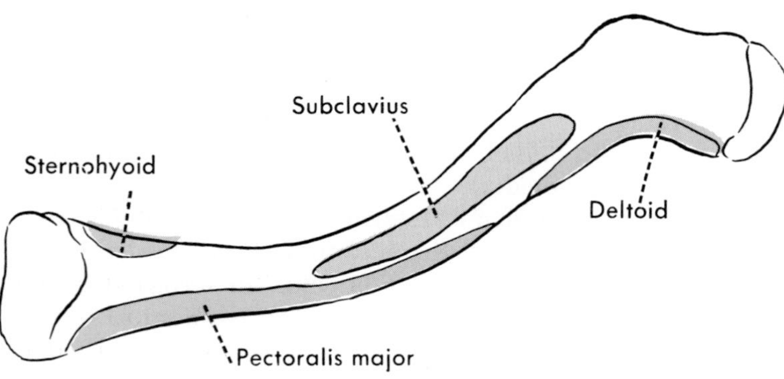

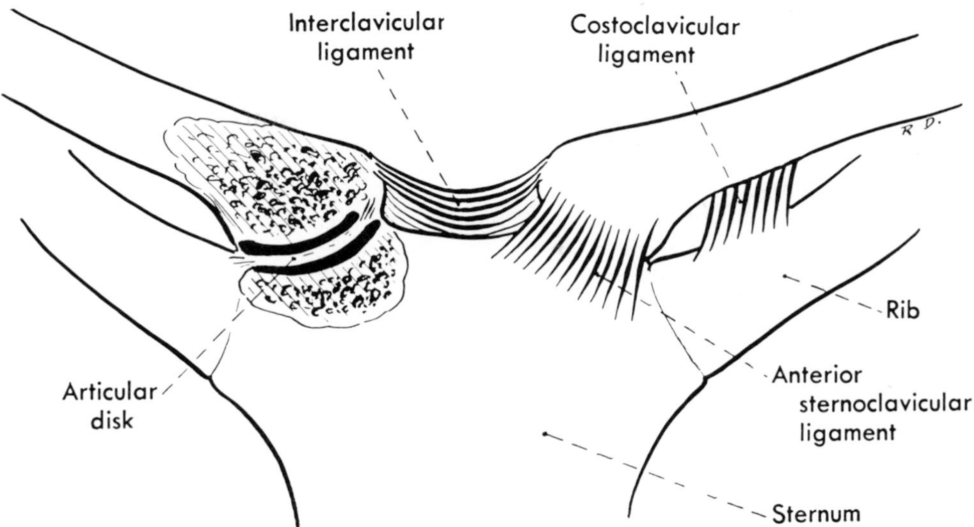


**FIG. 12-3.** The sternoclavicular joint and associated ligaments.

rior surface of the sternal end there is a roughened impression for the attachment of the *costoclavicular ligament,* and the acromial end bears a **conoid tubercle** and a **trapezoid line** marking the attachment of ligaments.

The clavicle is frequently fractured, but commonly heals with no difficulty.

**Ossification** of the clavicle is atypical in several ways. The body is formed from two centers instead of the usual one (failure of union, *clavicular dysostosis,* sometimes occurs). Although it is usually described as a membrane or derm bone, ossification of the clavicle is apparently histologically peculiar, and it subsequently develops epiphyses, typical of cartilage bones. The centers for the body, appearing during the fifth or sixth fetal week, are the first centers to appear. An epiphysis for the sternal end appears between the 18th and 20th years of life and unites with the body about the 25th year. A small epiphysis for the acromial end is said to both appear and unite in the 20th year.

**Articulations.** The synovial joint by which the clavicle articulates with the sternum is the **sternoclavicular joint** (Fig. 12-3). Closely associated with the sternoclavicular joint is the **costoclavicular joint,** a fibrous joint made by the costoclavicular ligament. There are also two lateral joints, a synovial **acromioclavicular joint** and a fibrous **coracoclavicular joint.** The articular surface of the sternal end of the clavicle is much larger than the articular surface of the clavicular notch in the upper corner of the sternum, and this end of the clavicle projects both above and behind the sternum.

The **sternoclavicular joint** typically contains two entirely separate synovial cavities; an *articular disk* attached at its periphery to the joint capsule separates it into a cavity between disk and clavicle and a cavity between disk and sternum. The disk is firmly attached below to the cartilage of the 1st rib at its articulation with the sternum and above to the clavicle. It acts as a hinge upon which the clavicle moves when the shoulder is moved up and down. The lower part of the articular surface of the clavicle swings out and away from the disk as the shoulder moves upward, and back toward the disk as the shoulder moves down. This is the freest movement of the clavicle, but forward and backward movements and rotation about the longitudinal axis also occur. Because of the firm upper attachment of the disk to the clavicle, the two largely move together during these movements.

The fibrous capsule of the sternoclavicular joint is strengthened by anterior and posterior **sternoclavicular ligaments,** of which the posterior is the thicker. The two clavicles are also

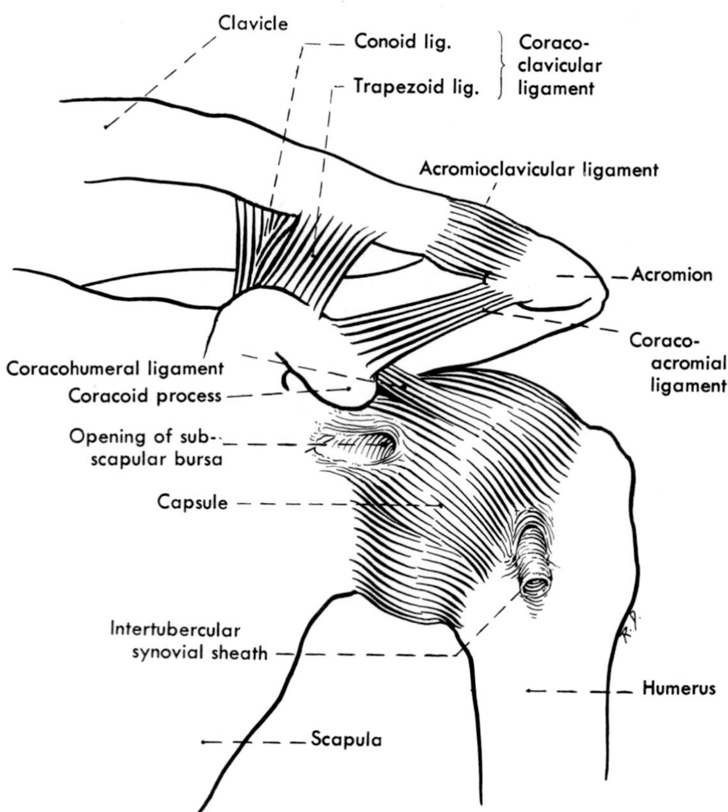


**FIG. 12-4.** Ligaments of the acromial end of the clavicle (and the anterior aspect of the shoulder joint).

united to each other across the upper surface of the sternum by an **interclavicular ligament,** and both this and the sternoclavicular ligaments (whose fibers slant primarily downward and inward) tend to prevent lateral and upward dislocation.

Upward and lateral movements of the sternal end of the clavicle is also resisted by the **costoclavicular ligament** attached to the clavicle and 1st rib. This ligament arises from the cartilage and the distal end of the bone of the 1st rib and passes upward, laterally, and slightly backward to attach to the inferior surface of the clavicle.

The **acromioclavicular joint** (Fig. 12-4) is both weak and small. Small, oval, articular facets on the acromion and the clavicle provide an essentially plane joint surface so that a certain amount of slipping can occur between scapula and clavicle. The articular capsule, known as the **acromioclavicular ligament,** is rather thin and somewhat lax, in order to permit this movement. Within the acromioclavicular joint there usually is an *articular disk* that partly, and sometimes completely, subdivides the cavity.

If the strength of the acromioclavicular joint were dependent upon its ligament, dislocation ("shoulder separation") presumably would be frequent. It is more frequent than dislocation at the sternoclavicular joint, but fracture of the clavicle is far more common than dislocation. The chief bracing ligament at the acromial end of the clavicle is the **coracoclavicular ligament,** the fibrous joint between the clavicle and the coracoid process of the scapula. The coracoclavicular ligament consists of two parts, the trapezoid and the conoid ligaments. The **trapezoid ligament** is attached to the coracoid process for about an inch of its length and runs upward, anteriorly, and somewhat laterally to attach along a similar length on the inferior surface of the clavicle (to the trapezoid line). The **conoid**

**ligament** lies partly behind the trapezoid ligament and is somewhat cone shaped, as its name implies; it is attached to the coracoid process where this makes its acute bend and passes upward and slightly backward, expanding as it does so, to attach to the undersurface of the clavicle at the conoid tubercle, just posteromedial to the trapezoid line. These two ligaments resist independent upward movement of the clavicle or downward movement of the scapula. Between them they also resist independent anteroposterior movement of the clavicle or scapula.

Movement at the acromioclavicular joint, although small, is necessary to movement of the shoulder because the acromial end of the clavicle moves along a curve, the radius of which is determined by the clavicle's length, whereas the scapula, closely following the curve of the thoracic wall, must adjust itself to the varying radii of different parts of this curve. Normally, therefore, the scapula rotates on the clavicle as the shoulder is moved, and interference with this through disease of the joint or ossification of the coracoclavicular ligament tends to make the clavicle and scapula move as a unit and, therefore, limits movements of the shoulder.

The **blood supply of the sternoclavicular joint** is from twigs of vessels that run close to it (the clavicular branch of the *thoracoacromial artery*, the *internal thoracic artery*, and perhaps the *suprascapular*). Its **nerve supply** is similarly from nerves running close to it (the *nerve to the subclavius* muscle and the *medial supraclavicular nerve*). The **blood supply to the acromioclavicular joint** is from a network of small vessels (*acromial rete*) lying superficially over the acromion, contributed to by several arteries of the shoulder. The joint is said to be **innervated** by twigs from the pectoral, suprascapular, and axillary nerves, that is, nerves to shoulder muscles.

## Scapula

The scapula is a rather thin, triangular, flat bone that is strengthened by a prominent ridge of bone, the **scapular spine,** and by a thickened lateral border. Most of it is, however, rather well protected by muscles and by its apposition to the thoracic wall, and most fractures of it, therefore, involve the protruding subcutaneous acromion.

The scapula has a costal surface and a dorsal surface. The **costal surface,** called the *subscapular fossa* (Fig. 12-5), is for the most part slightly concave. The **dorsal surface** (Fig. 12-6) is divided by the scapular spine into an upper *supraspinatous fossa* and a lower *infraspinatous fossa.* The spine continues laterally and upward as the freely projecting **acromion,** and the supraspinatous and infraspinatous fossae communicate lateral to the spine and deep to the acromion. The sharp forward turn of the acromion is marked posterolaterally by a subcutaneous and palpable bony prominence, the *acromial angle*.

The scapula has three **borders:** *medial, lateral,* and *superior*. The **angles** at which these borders meet are the *inferior,* the *lateral,* and the *superior* (the junction of superior and medial borders). The **superior border** of the scapula is marked at about the junction of its medial two-thirds and lateral third by the *scapular notch* (*incisura*); lateral to this, the **coracoid process** projects upward and then bends anterolaterally, where its end is palpable below the clavicle. The *superior transverse scapular ligament* (commonly "transverse scapular ligament") across the notch converts it into a foramen.

The **lateral angle** of the scapula is expanded to form a rough oval whose slightly concave surface is covered by articular cartilage and forms the **glenoid cavity,** which articulates with the head of the humerus (Fig. 12-7). On the upper border of the cavity is the small *supraglenoid tubercle,* and on the lateral border of the scapula immediately below the glenoid cavity is a thickening, the *infraglenoid tubercle.* These tubercles serve for the attachment of the tendons of the long heads of the biceps and triceps muscles, respectively. The narrow part of the scapula medial to the lateral angle is the neck (*collum*).

The scapula receives its **blood supply** from vessels that run in close contact with both its surfaces: primarily, the *subscapular artery* and its *circumflex scapular branch,* and the *suprascapular artery*.

A large number of **muscles** attach to the scapula (Figs. 12-5 and 12-6). The majority attach to its borders or to the projecting coracoid process, acromion, and spine, but all three fossae also give rise to muscles.

In addition to the **superior transverse scapular**

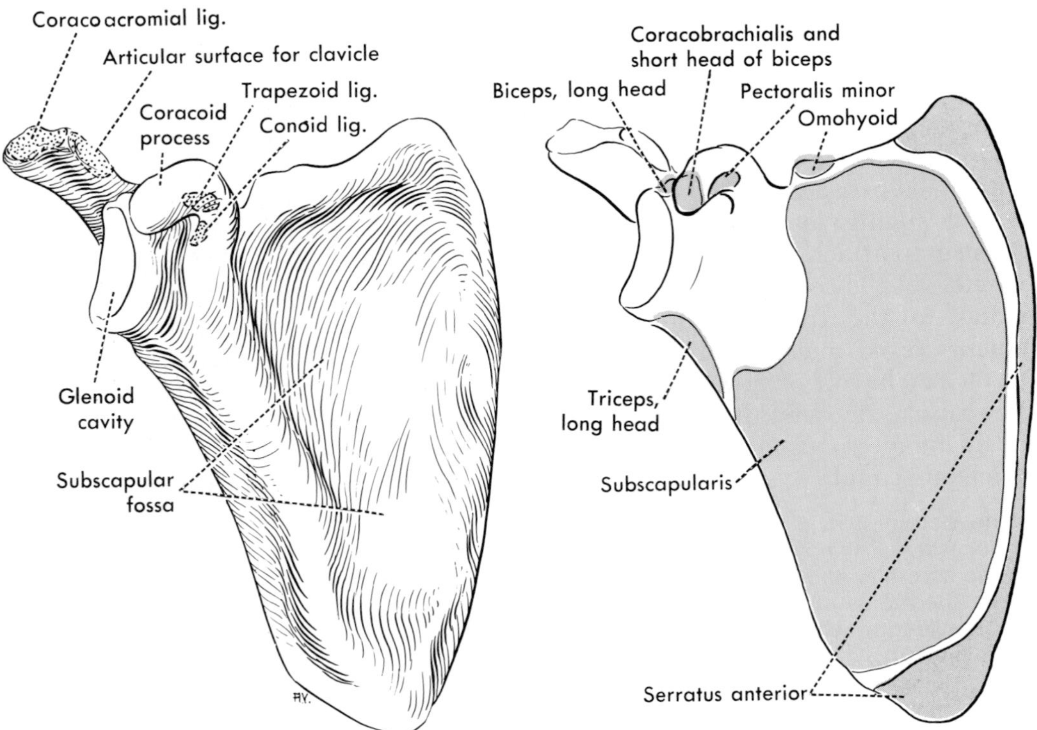


**FIG. 12-5.** Costal surface of the scapula. Muscle origins are shown in red, insertions in blue.

**FIG. 12-6.** Dorsal aspect of the scapula.

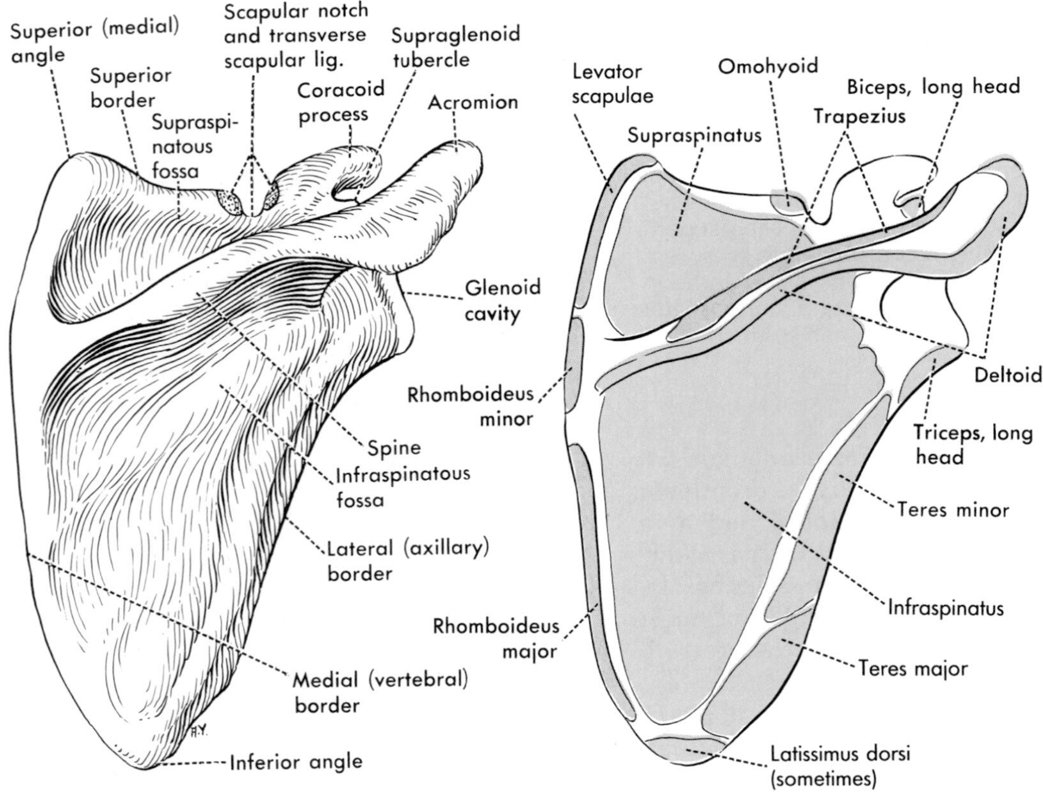

**ligament,** one or more other ligaments connect parts of the scapula; the constant one is the **coracoacromial ligament,** a broad band arising from most of the posterior edge of the coracoid process and passing upward and laterally above the shoulder joint to attach to the free end of the acromion (Fig. 12-4). The **inferior transverse scapular ligament,** when present, converts the notch at the base of the spine into a foramen for the nerve and vessels that continue from the supraspinatous to the infraspinatous fossa.

**Ossification** of the scapula is through a number of centers, of which only the largest secondary (epiphyseal) ones—those for the major parts of the coracoid process and acromion and for the inferior angle of the scapula—are shown in relation to the body in Figure 12-8. Other centers include additional small ones for the coracoid, acromion, medial border, and glenoid cavity.

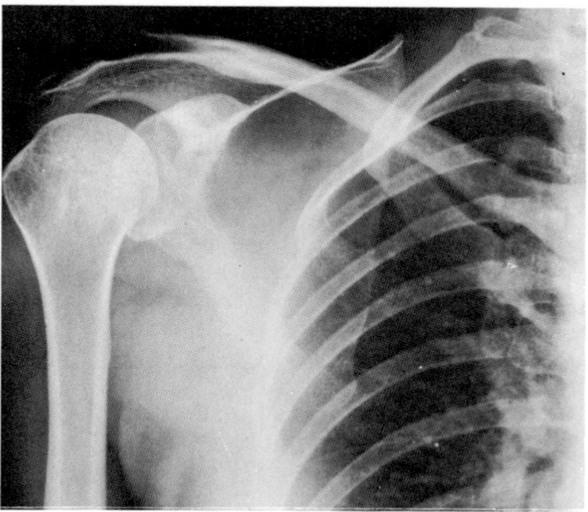

**FIG. 12-7.** Roentgenogram of the shoulder.

## HUMERUS

The humerus has expanded upper and lower ends connected by its shaft, or body (Figs. 12-9 and 12-10). It forms the skeleton of the arm; its upper end articulates with the scapula at the shoulder joint, and its lower end articulates with the radius and the ulna at the elbow joint.

The expanded **upper end** presents three prominences, the head and two tubercles. The **head** is the smooth, rounded, articular surface; the indistinct **anatomic neck** is just distal to the articular cartilage. On the medial side of the front of the humerus below the head, there is a roughened projection, the **lesser tubercle,** and the ridge of bone extending downward from this is the *crest* of the lesser tubercle. The much larger, roughened projection on the lateral side of the upper end of the humerus just below the head is the **greater tubercle,** and the *crest* of the greater tubercle extends downward from it. Tubercles and crests serve for the attachment of muscles. The groove between the two tubercles is the **intertubercular groove** (sulcus), often called the bicipital groove because it houses the tendon of the long head of the biceps muscle. The indefinite area below the tubercles where the bone narrows to continue as the body is the **surgical neck,** so called because fractures here are fairly common.

The **body** of the humerus is roughly cylindrical but is described as having three *borders* and three *surfaces.* The **anterior border** begins above with the crest of the greater tubercle and tends to be indistinct below, but extends to the ridge between the two depressions on the lower end of the front of the humerus (the *coronoid* and *radial fossae*). The **lateral** and **medial borders,** indistinct above, are prominent edges below and here often are called *supracondylar ridges.* Each ends in a thicker **epicondyle.** On the posterior and inferior surface of the medial epicondyle is a smooth

**FIG. 12-8.** Chief epiphyseal centers of the scapula and the upper end of the humerus. The first of two numbers indicates the usual age at the time of appearance, the second at the time of fusion, of the epiphyseal center. (Camp JD, Cilley EIL: Am J Roentgenol 26:905, 1931)

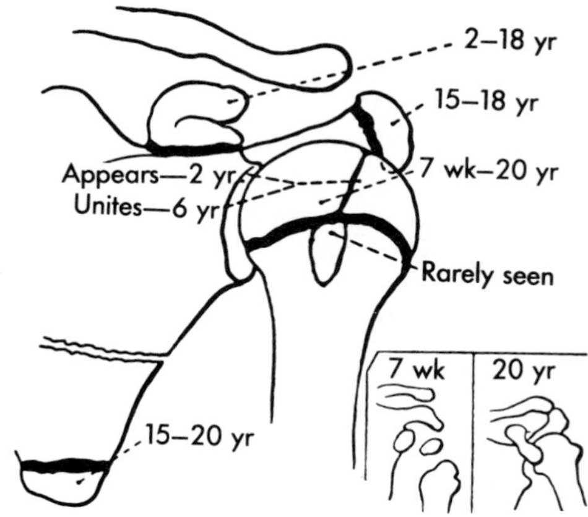

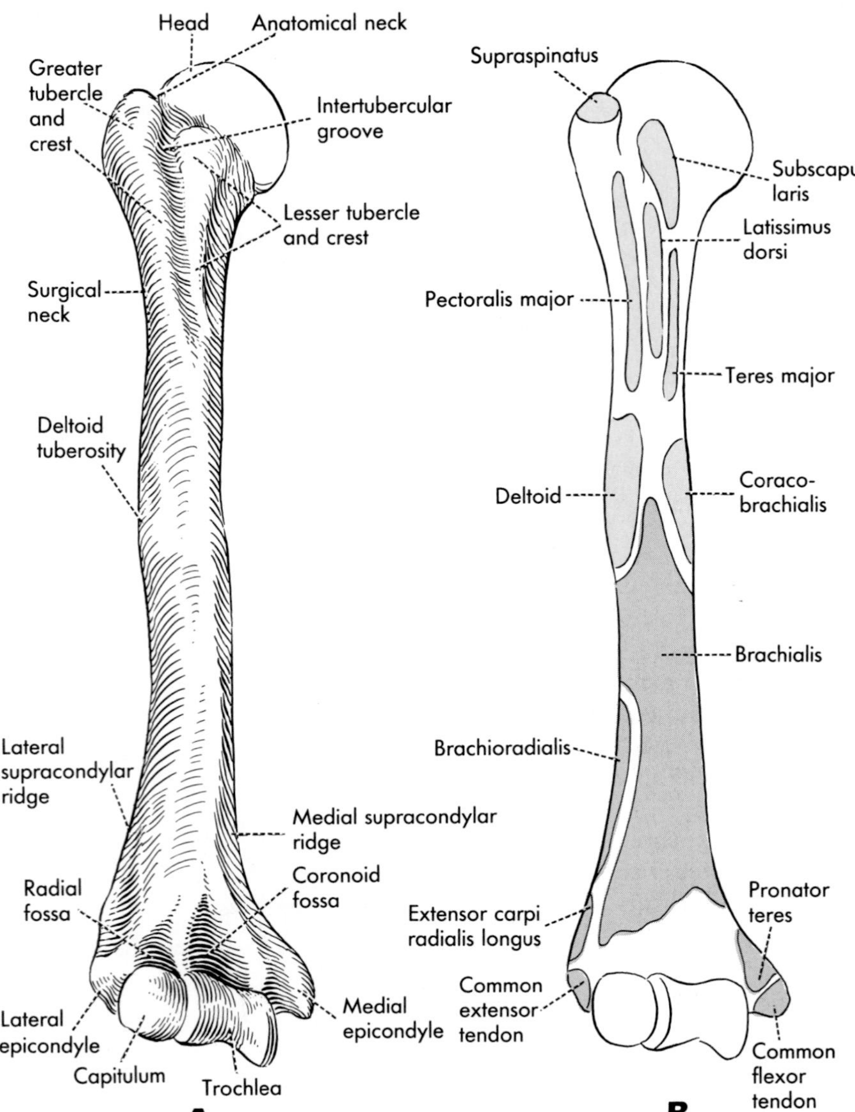


**FIG. 12-9.** Anterior views of the humerus. Muscle origins are shown in red, insertions in blue are shown in **B.**

groove, the *sulcus of the ulnar nerve.* The anterolateral surface of the humerus presents the roughened **deltoid tuberosity** above its middle and tends to blend with the anteromedial surface below. The anteromedial surface has no special markings. The posterior surface is crossed spirally from its medial to its lateral side by a shallow groove, the **sulcus of the radial nerve,** usually marked at the lateral border.

The expanded **lower end** of the humerus between the epicondyles is the **condyle** and consists of articular surfaces and concavities. The slightly convex lateral articular surface, the **capitulum,** is on the anterior and inferior surfaces only, but the more pulleylike medial articular surface, the **trochlea,** extends posteriorly also. On the anterior surface of the humerus above the trochlea is the *coronoid fossa,* and above the capitulum is the *radial fossa.* These fossae receive the coronoid process of the ulna and the head of the radius, respectively, upon flexion of the elbow. On the posterior surface of the condyle is a deep de-

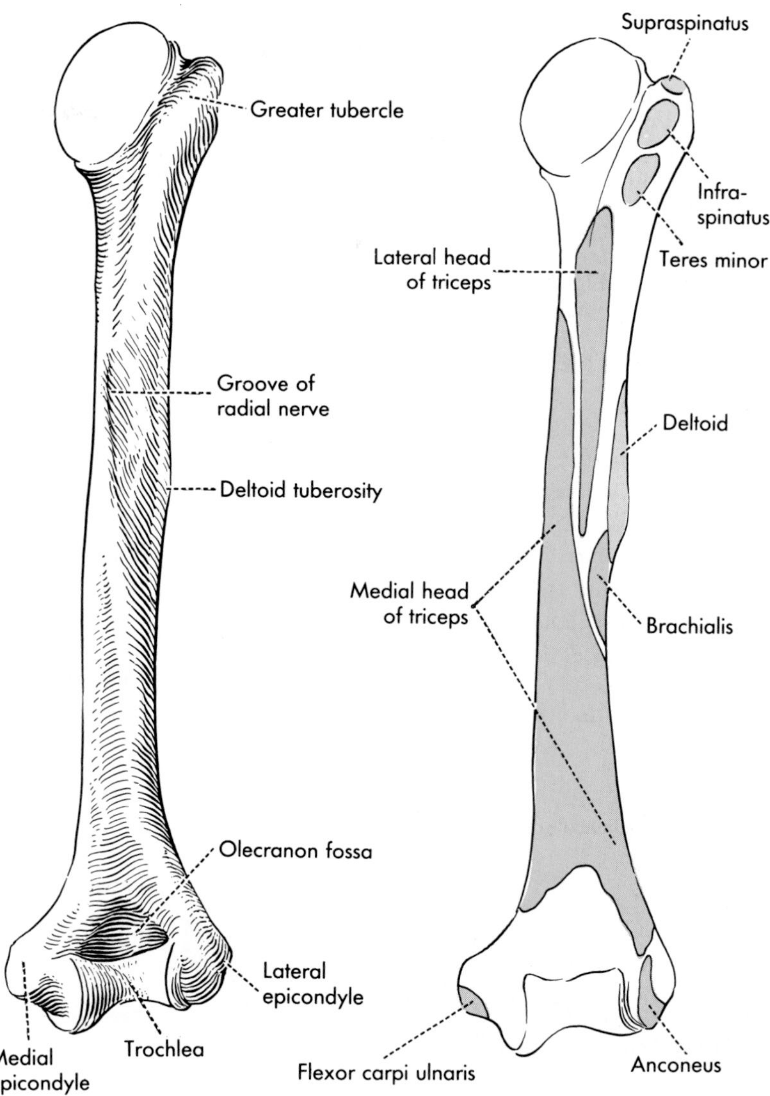


**FIG. 12-10.** Posterior views of the humerus.

pression, the **olecranon fossa,** for the reception of the olecranon of the ulna upon extension of the elbow.

Sometimes, above the medial epicondyle, there is a bony or bony and cartilaginous projection, the **supracondylar process.** It may be a cause of injury to the median nerve as this nerve passes behind it.

The **blood supply** of the humerus at its upper end is from numerous twigs from two vessels encircling its surgical neck. The **nutrient artery** enters the body anteromedially about the middle of the bone if derived from the brachial artery and posteriorly if derived from the profunda brachii, and both nutrient arteries may be present. The lower end is supplied by twigs from the vessels forming the anastomosis about the elbow (Fig. 12-49).

**Ossification** of the body of the humerus, beginning about the eighth or ninth week of embryonic life, is completed before birth. The upper end ossifies through from one to three epiphyseal centers: there is a constant one for the head, usually appearing about the time of birth, and the tubercles either develop from this center or from independent centers that subsequently fuse with the head (Fig. 12-8). Upper end and body fuse at about the age of 17 to 18 years in females and 18 to 21 in males. The lower end of the humerus is typically ossified from four centers, one each for the trochlea, the capitulum, and the two epicondyles (see Fig. 13-8). Not until about 14 years of age in females, 18 in males, are they all fused with each other and the body.

Complete **fractures** of the humerus usually show some overriding and other displacement of the ends as a result of muscular pull; rotatory and angular displacements of the upper fragment are characteristic of fractures of the surgical neck, and angular displacement is characteristic of lower fractures (see Fig. 3-9).

## PECTORAL REGION

The surface anatomy of the pectoral region should be briefly reviewed (Chap. 11). In the male, the contour of this region is formed chiefly by the large **pectoralis major** muscle, which also forms the anterior wall of the **axillary fossa,** and in the female, by the **breast.** The breast is a modified sweat gland that has grown too large to be contained in the skin and therefore has invaded, as do all large cutaneous glands, the underlying *tela subcutanea*. Study of the breast by routine dissection in the laboratory is not particularly rewarding, but the frequent involvement of this organ in the female by carcinoma makes it necessary to understand its fundamental structure.

### Superficial Fascia, Nerves, and Vessels

The **superficial fascia** (*tela subcutanea*) of the pectoral region contains a variable amount of fat. In the male, this fascia tends to be thickest in the region deep to the nipple and areola, and in the female, it forms large masses between and around the glandular tissue of the breast. It contains the superficial vessels and nerves and is continuous with the superficial fascia of the lateral thoracic wall, the abdomen, the arm, and the neck.

The superficial fascia of the uppermost part of the thorax also contains the thin **platysma muscle,** which lies mostly in the superficial fascia of the neck but arises in the thorax and

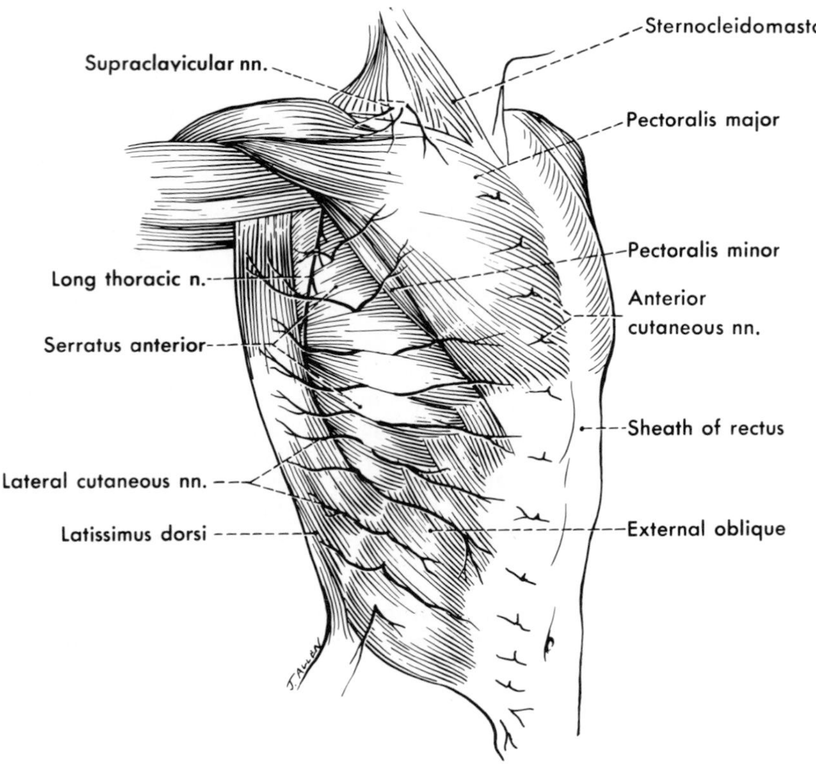


**FIG. 12-11.** Cutaneous nerves of the anterolateral thoracic wall. The long thoracic, also shown because of its superficial position, is the nerve to the serratus anterior muscle.

crosses the clavicle. Skin incisions along the clavicle that go too deeply sever this muscle, as well as the cutaneous nerves that descend from the neck.

The **cutaneous nerve supply** of the pectoral region is through cervical and thoracic nerves (Fig. 12-11). Twigs of the *supraclavicular nerves* (C3 and C4) descend over the clavicle to supply the skin of the thorax over the upper two intercostal spaces; all the rest of the skin on the anterolateral aspects of the thorax is supplied by *branches of intercostal nerves*. The *lateral cutaneous branches* divide into anterior and posterior subbranches that run in the superficial fascia. The anterior subbranches that are distributed to the breast in the female are known also as lateral mammary branches. The pectoral lateral cutaneous nerves are derived from the 3rd through the 6th or 7th intercostal nerves (the lateral cutaneous branch of the 2nd intercostal is distributed to the arm) and emerge between the slips of origin of the serratus anterior, a muscle of the shoulder largely covering the lateral thoracic wall. (The lateral cutaneous branches from the lower five or six intercostal nerves are distributed to skin of the abdomen.)

The *anterior cutaneous branches* of the intercostal nerves are small twigs that in the thoracic region pierce the pectoralis major muscle and spread medially in the superficial fascia to innervate skin as far as the midline and laterally to innervate skin as far as the overlap with the lateral cutaneous branches. The first anterior cutaneous branch is from the 2nd thoracic nerve, for the 1st thoracic nerve has no such branch; the anterior cutaneous branch of the 7th intercostal nerve appears at about the level of the xiphoid process.

The **superficial vessels** of the pectoral region are typically very small, but in the female, some are enlarged to supply the breast. Anteriorly, the *internal thoracic* (formerly internal mammary) *artery* and vein have small *perforating branches* that pierce the pectoralis major with the anterior cutaneous branches of the intercostal nerves, and there are twigs, with no regular arrangement, that reach the pectoral region by passing in the superficial fascia over the lateral edge of the pectoralis major. These are derived from the *lateral thoracic vessels*, which descend on the lateral thoracic wall close to the lateral border of the pectoralis major.

The upper end of the **cephalic vein** lies over the groove between pectoral and deltoid regions and then turns deeply below the clavicle to reach the axilla.

The **lymphatics** of the pectoral region are important primarily as the drainage of the female breast. They are described with that organ.

## BREAST

The male breast needs no description. The female breast (mammary gland, mamma, Fig. 12-12) consists of some 15 to 20 **lobes** of glandular tissue, each of which opens by a separate duct, the **lactiferous duct,** on the tip of the **nipple** (*papilla mammae*). Immediately deep to the **areola** (the pigmented area surrounding the nipple), each lactiferous duct enlarges to form a **lactiferous sinus,** in which milk can accumulate. Beyond the sinus, each duct diminishes in size and branches and rebranches to become continuous with the secretory tissue, the alveoli. The individual lobes vary greatly in size, and typically somewhat less than half of them enlarge to become functional during lactation. Since they radiate from the areola, they tend to be roughly pyramidal in shape, with their apices pointing toward the areola.

The lobes are bound together by fairly dense **connective tissue septa,** derived from the fibrous connective tissue of the tela subcutanea in which the breast develops. Bands of this connective tissue extend from the connective tissue on the anterior aspect of the lobes and attach into the dermis, subdividing the fat that lies superficial to the glandular tissue and that gives the smooth contour to the breast. These connective tissue septa are simply well-developed retinacula cutis, but those over the upper part of the breast are called

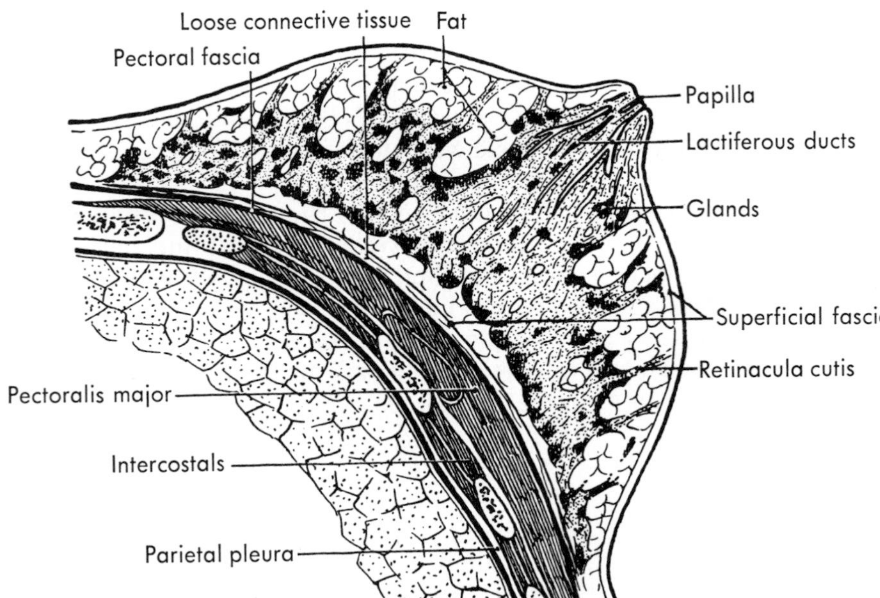


**FIG. 12-12.** Gross structure of the female breast. (Corning HK: Lehrbuch der topographischen Anatomie für Studierende und Ärtze. Munich, JF Bergmann, 1923)

**suspensory ligaments** of the breast. Sometimes carcinoma of the breast calls attention to its presence by so involving the ligaments as to produce a dimpling of the skin through traction upon them. Deep to the glandular tissue of the breast there is usually little fat, but the part of the superficial fascia that rests upon the deep fascia of the pectoral muscle can be separated from it by blunt dissection. Carcinoma of the breast may also make itself apparent by so fixing the breast to the pectoral fascia and muscle that a lessening of mobility becomes obvious.

The glandular tissue always spreads farther than the gross outline of the breast, and there is nearly always an extension into the axilla. Similarly, tissue from one breast may reach or cross the anterior midline or extend downward and medially into the epigastrium (the region below the sternum, between the curves of the two sides of the thoracic cage).

The **blood supply** of the breast is chiefly from superficial vessels. Typically, enlarged lateral branches of the *anterior perforating arteries* (from the internal thoracic) run to the breast as *medial mammary arteries.* They are from any combination of the first four perforating arteries. The *lateral mammary arteries* may be multiple in origin, but commonly are derived from a single branch that rebranches as it approaches the breast, running downward and medially from the axilla over the lateral border of the pectoralis major muscle. They are usually derived from the *lateral thoracic artery.* The superficial veins of the breast form a variable anastomotic pattern and drain along the arterial paths to the breast. The chief venous drainage is typically toward the axilla.

Knowledge of the **lymphatic drainage** of the breast (Fig. 12-13) is of practical importance because of the frequent invasion of lymphatics by carcinoma of this organ. The chief drainage is into axillary lymph nodes, which normally receive more than 75% of the total drainage from the breast and are therefore particularly likely to be involved when there is mammary carcinoma. The largest lymphatics run around the lateral border of the pectoralis major to end in **pectoral nodes** that lie between the pectoralis major and minor and behind the latter muscle. Others usually perforate the pectoralis major and follow the pectoral branch of the thoracoacromial artery to the **apical** (subclavicular) **group of axillary nodes.** The remaining deep lymphatics perforate the pectoralis major and the thoracic wall and empty into **parasternal** (sternal, internal mammary) **nodes** situated along the internal thoracic vessels on the internal surface of the anterior thoracic wall. Finally, the **subcutaneous lymphatics** of the breast anastomose with the **deep lymphatics,** particularly in the neighborhood of the areola, and whereas they account for only a little of the drainage of the breast, they spread across the midline, upward above the clavicle, and downward toward

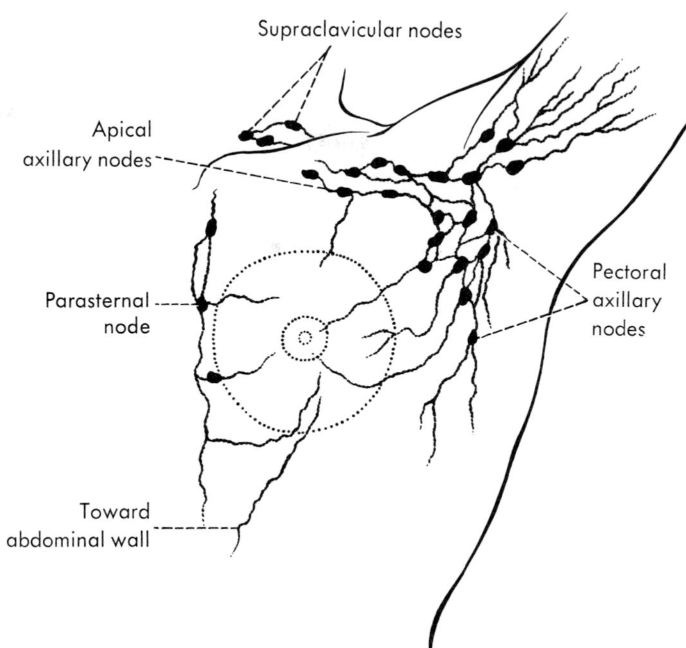


**FIG. 12-13.** Schema of the lymphatic drainage of the breast. (Adapted from Poirier P, Charpy A: Traité d'Anatomie Humaine, 2nd ed, vol 2, fasc 4. Paris Masson et Cie, 1909)

the abdominal wall; thus, they can dissipate carcinoma widely if other lymphatic channels are blocked.

Because of the close relation of the pectoralis major muscle to the lymphatic drainage, and because deep-lying carcinoma may directly invade the muscle, it has been customary to remove the greater part of the pectoralis major and its associated fascia in removal of a breast involved by carcinoma (*radical amputation of the breast*). In this operation, the pectoralis minor and the fascia on the lateral thoracic wall, and as much as possible of the connective tissues in the axilla that may contain lymph nodes, are also removed. Some surgeons also remove parasternal nodes, and a few have removed nodes at the base of the neck that receive drainage from axillary nodes. In recent years, however, there has been much discussion as to whether a radical operation is justified by the results.

### Accessory Mammary Tissue

Accessory breasts or nipples may occur above or below the normal breast in either sex. Commonly, these accessory, or supernumerary, "breasts" consist only of a nipple and an areola, but sometimes true glandular tissue is present also. Supernumerary breasts are usually found along a line extending from the axilla through the normal breast to the groin, this being regarded as the milk line (Fig. 12-14), or the line along which animals with multiple breasts usually develop them. Occasionally, supernumerary breasts are found beyond the usual extent of the milk line, for instance, on the neck or on the vulva.

## MUSCLES AND ASSOCIATED STRUCTURES

**Pectoralis Major.** The pectoralis major is enclosed by **pectoral fascia** that is intimately attached to the muscle. Pectoral fascia on the outer surface of the muscle is thin although fused with the scanty deep fascia of the thorax; that on the deep surface is somewhat thicker, especially in its upper part where it transmits to the muscle branches of the thoracoacromial artery and the lateral pectoral nerve. At the free lateral edge of the muscle, the layers unite and become continuous below with the deep fascia of the abdominal wall, laterally with that of the serratus anterior, and above, through the fascial floor of the axilla, with fascia of the arm and of the latissimus dorsi.

The large, fan-shaped pectoralis major (Figs. 12-11 and 12-15) **arises** from approxi-

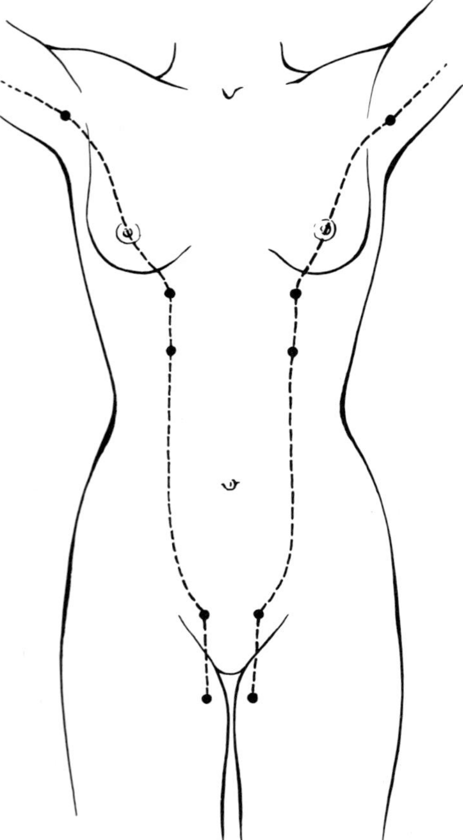

**FIG. 12-14.** The milk line, along which accessory nipples may appear.

mately the medial third of the clavicle, from the lateral part of the entire length of the anterior surface of the manubrium and body of the sternum, and from the cartilages of about the first six ribs (Fig. 12-16). The muscle, therefore, has *clavicular, sternal,* and *costal* parts. Usually there also is a slip from the aponeurosis of the external oblique muscle of the abdomen. As these parts converge toward their **insertion,** the clavicular part, running downward and laterally, blends with the upper sternocostal part, whose fibers pass almost straight laterally, to form an *anterior lamina* to the tendon. The lower portion of the muscle, the fibers of which run laterally and upward, twists as it passes behind the anterior lamina and thus forms a *posterior lamina* in which the fibers of lowest origin insert highest on the humerus. The **bilaminar tendon** passes behind the anterior border of the deltoid muscle, crosses the intertubercular groove, and inserts upon its lateral lip, the crest of the greater tubercle.

Deep to the clavicular head of the pectoralis major, the **lateral pectoral nerve,** supplying the major upper portion of the muscle, and the pectoral branch of the **thoracoacromial artery,** the muscle's chief blood supply, run downward into the muscle. The **medial pectoral nerve,** to a lower part of the muscle, enters the deep surface close to its lateral border after passing through or around the lateral border of the pectoralis minor (Fig. 12-21). The **lateral thoracic artery,** on the thoracic wall behind the lateral border of the muscle; the deltoid branch of the thoracoacromial, lying with the cephalic vein in the groove between the deltoid and the pectoralis major; and the perforating branches of the internal thoracic artery are lesser sources of blood supply to the muscle.

**FIG. 12-15.** The pectoralis major and some associated muscles.

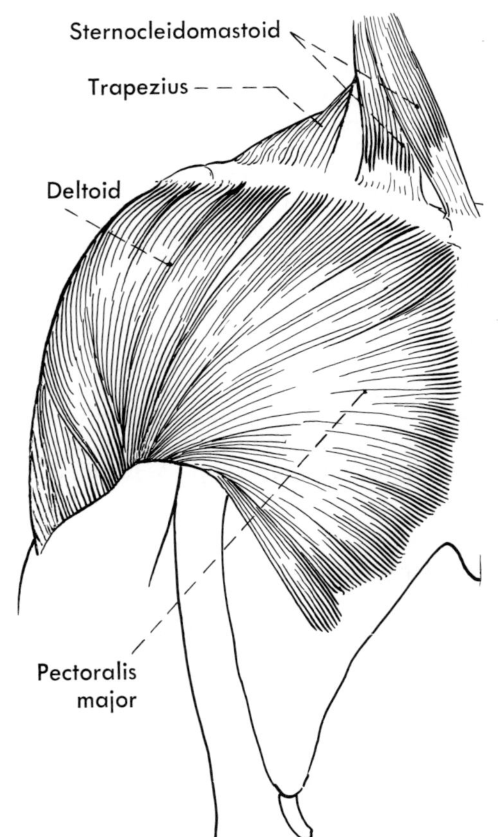

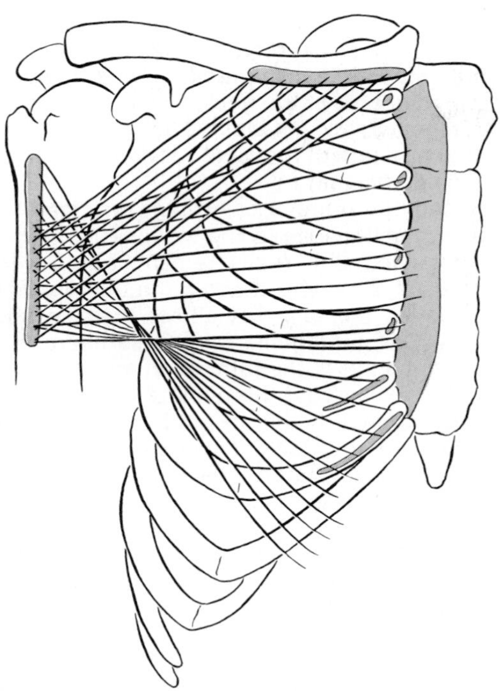

FIG. 12-16. Origin (red) and insertion (blue) of the pectoralis major.

**Pectoralis Minor.** This muscle (Fig. 12-17) is largely covered by the pectoralis major.

The fascia of the pectoralis minor is formed by the splitting of a layer known as the **clavipectoral fascia** (Fig. 12-18). Where this fascia is attached above to the clavicle, it surrounds the subclavius muscle (see the following section), one part passing anterior to the muscle, the other posterior and inferior to it; below and lateral to the muscle the two layers come together to form a single layer that runs downward and laterally to the upper and medial border of the pectoralis minor. Lateral to the subclavius muscle, the fascia attaches to the coracoid process and the 1st rib, as well as to the tendon of insertion of the pectoralis minor, and is particularly strong; this part is sometimes called the *costocoracoid ligament.* The extension to the muscular portion of the pectoralis minor is thinner and is perforated by the thoracoacromial artery and the lateral pectoral nerve. At the origin of the pectoralis minor, the fascia around it blends with fascia over the intercostal muscles. At the inferolateral edge of the muscle, the two layers come together and contribute above to the axillary fascia and below that to fascia on the lateral thoracic wall.

The pectoralis minor typically **arises** from either four or three ribs, usually the 2nd to the 5th or the 3rd to the 5th or 6th. It extends upward and a little laterally to **insert** deep to the deltoid into the medial side of the coracoid process close to its tip.

The **lateral thoracic artery** lies lateral to the pectoralis minor and gives branches to it and to other muscles in its neighborhood. The **medial pectoral nerve** runs downward from the axilla to enter the deep surface of the pectoralis minor and supply the muscle. It either continues through the muscle or sends a branch around its lateral border, to the pectoralis major.

FIG. 12-17. Pectoralis minor and subclavius muscles. Only the lower part of the serratus anterior is shown.

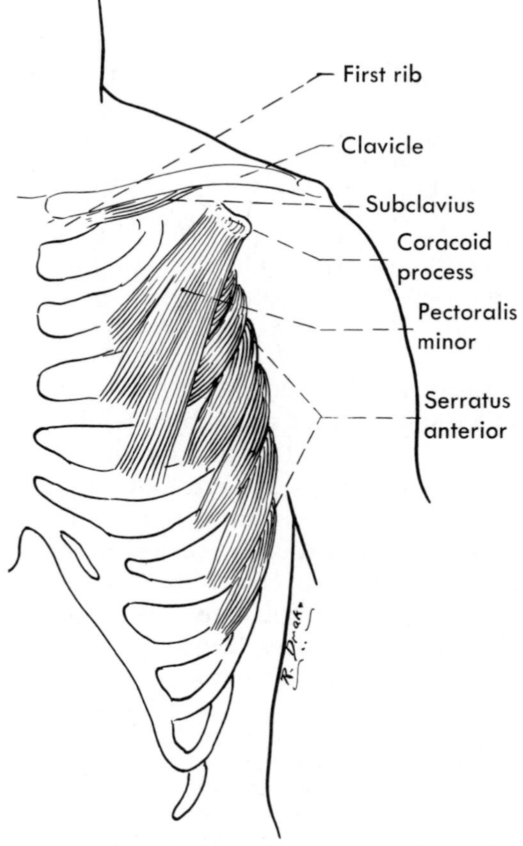

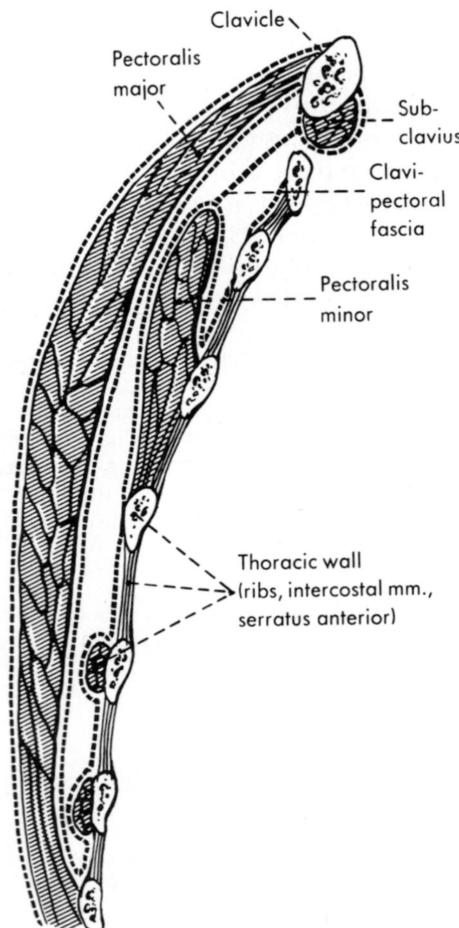


**FIG. 12-18.** Diagram of the clavipectoral and related fascia (broken lines).

**Subclavius.** The subclavius is a small rounded muscle that arises by a tendon from the 1st rib at about the junction of bone and cartilage and passes upward and laterally to insert on the lower surface of the clavicle. It is largely covered by the clavicle and is enclosed between the two layers of the clavipectoral fascia as they leave this bone. Its small nerve enters its deep upper surface and is not visible in a dissection on the thorax.

**Variations, Innervations, and Actions.** Occasional cases of complete or partial absence of the pectoralis major have been reported. Most common is absence of the sternocostal part, with presence of the clavicular part. The pectoralis minor is sometimes absent, sometimes present, when the major is absent.

Truly anomalous slips of muscle are sometimes found in the pectoral region. One such slip is the **sternalis muscle,** a small muscle lying superficial to the medial part of the pectoralis major and having attachments to this muscle, to the sternum, or even to the sheath of the rectus abdominis or to the sternocleidomastoid.

Any one of a number of variably arranged muscle slips may lie close to the axilla. They perhaps represent, like the sternalis, remains of a *panniculus carnosus,* a muscle attaching to the skin of the limb in lower animals that allows them to shake that skin. One arrangement of these fibers, when present in man, is in the form of an arch stretching across the axilla from the pectoralis major to the latissimus dorsi. Whether or not they actually form an arch across the axilla, they are usually grouped together as **axillary arch muscles.** Such a muscle may arise from either the pectoral or the latissimus dorsi muscle, attach into the thoracic fascia or various bony prominences, or run a variable distance down the arm to insert into its fascia.

The **innervation** of the pectoralis major is through both the medial and the lateral pectoral nerves, that of the minor through the medial only. The *medial pectoral nerve* supplies the minor and the lower part of the major with fibers of C8 and T1. The *lateral pectoral nerve* supplies the upper part of the major with fibers from C5 through C7, so that this muscle gets fibers from all the nerves that regularly contribute to the brachial plexus. The *small nerve to the subclavius,* arising from the upper trunk of the brachial plexus, has been reported to have a widely variable segmental origin, but probably usually contains fibers derived from C5 or C6.

The chief **action** of the pectoralis major is to adduct the limb. Because it crosses the front of the humerus to attach to the crest of the greater tubercle, it also medially rotates the limb. The clavicular part assists the anterior part of the deltoid in flexing the arm (drawing it upward and forward) but can do this only until the limb is about horizontal. The lower fibers can help to extend the limb until it is by the side. By pulling down on the humerus, they also can depress and downwardly rotate the scapula (turn the glenoid cavity down). The pectoralis minor pulls the coracoid process downward and thus is a depressor and downward rotator of the scapula. If the scapula is retracted, it can also help to protract it. The action of the subclavius is not really known, but because it runs between the 1st rib and the clavicle, it is usually listed as a depressor of the latter or a levator of the thoracic cage.

**Nerves and Vessels.** The superficial nerves and vessels of the pectoral region already have been described, and the nerves and vessels related to the muscles have been largely described with these. In summary, the only nerves to muscles that appear in this section (although the nerve to the serratus anterior, Fig. 12-11, is only a little more lateral) are the nerves to the pectoral muscles. The nerve to the subclavius has its course entirely in the neck.

The medial and lateral pectoral nerves (Figs. 12-19 and 12-21) were once called "anterior thoracic" nerves. They are named not according to their relative position on the thorax but according to their origins from the medial and lateral cords of the brachial plexus.

The **medial pectoral nerve** passes between the axillary artery and vein to enter the deep surface of the pectoralis minor, supply this, and continue into the pectoralis major. The **lateral pectoral nerve** often has a branch of communication with the medial pectoral nerve before piercing the clavipectoral fascia with the thoracoacromial artery and entering the deep surface of the clavicular head of the pectoralis major.

The two chief arteries of the pectoral region are the thoracoacromial and the lateral thoracic. After its origin from the axillary artery, the **thoracoacromial artery** pierces the clavipectoral fascia and gives rise to four branches: clavicular, pectoral, deltoid, and acromial (Fig. 12-20). The small *clavicular branch* runs toward the sternoclavicular joint, supplying this and a little of the anterior thoracic wall. The *pectoral branch* is the largest branch and runs downward with the lateral pectoral

**FIG. 12-19.** The axilla from the front, after reflection of its anterior wall and removal of most of its contents.

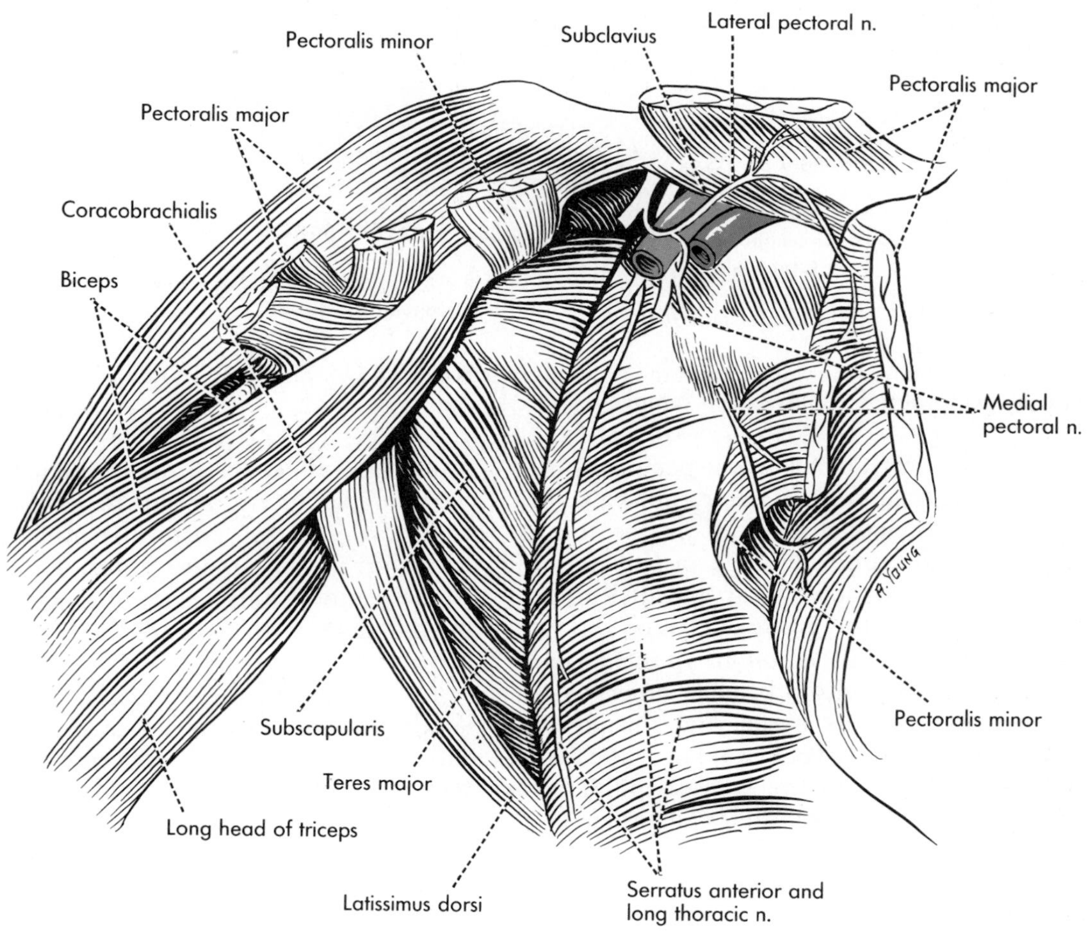

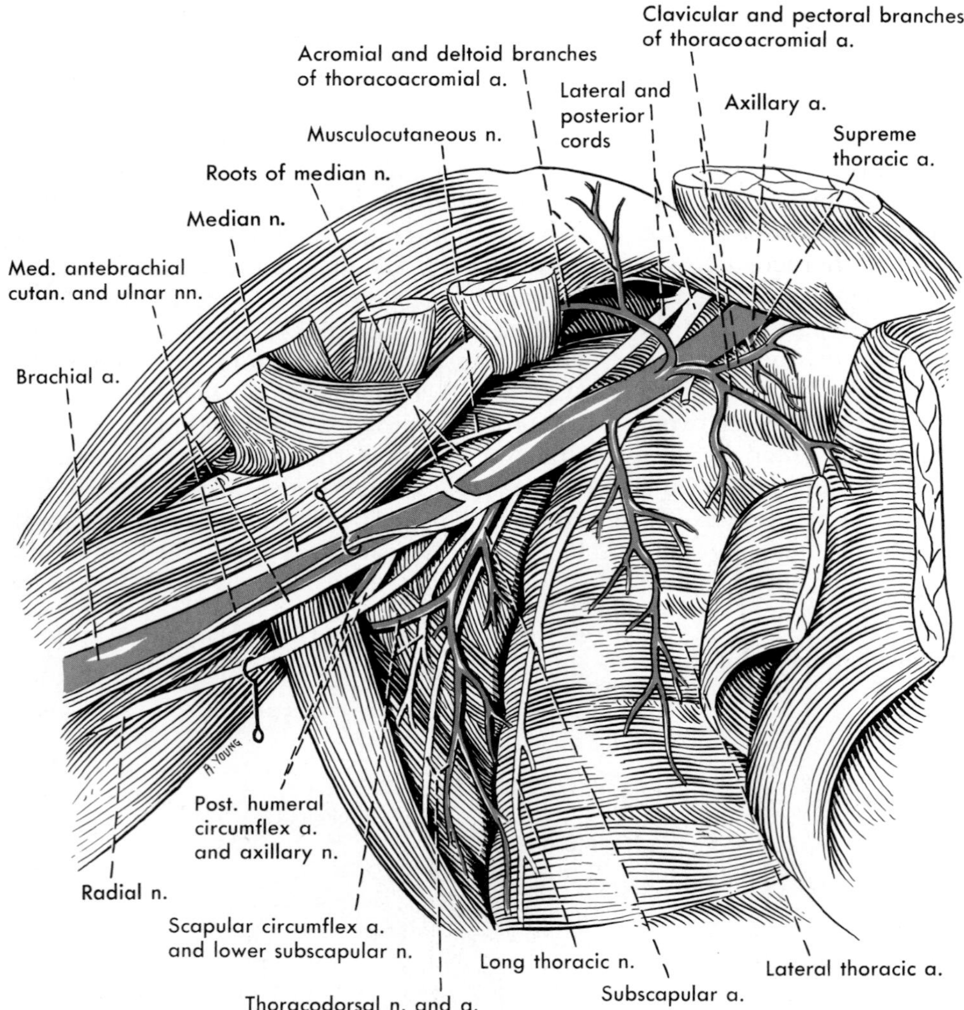


**FIG. 12-20.** The axillary artery and the brachial plexus *in situ*. Lymph nodes, veins, and the pectoral nerves have been removed.

nerve on the deep surface of the pectoralis major. The *deltoid branch* passes forward between the clavicular head of the pectoralis major and the anterior edge of the deltoid muscle and then downward (paralleling the cephalic vein) in the groove between the two muscles. The small *acromial branch,* often arising from the deltoid branch, emerges between the pectoralis major and the deltoid and enters into a network (rete) over the acromion; branches from the suprascapular and posterior humeral circumflex arteries also contribute to the *acromial rete*.

The **lateral thoracic artery** runs down the anterolateral aspect of the thoracic wall for a variable distance at about the level of the lateral border of the pectoralis minor. Besides branches to the serratus anterior muscle and to the pectoral muscles and the breast, it also anastomoses with twigs of the intercostal arteries.

In addition to these vessels, there are the small **perforating arteries** from the internal thoracic and the small **supreme thoracic artery.** The latter arises from the axillary artery and runs medially to a distribution over the 1st or 1st and 2nd intercostal spaces.

The arteries are all accompanied by veins,

which for the most part enter the axillary vein. The **cephalic vein,** which runs between the pectoralis major and the deltoid and then turns deeply between them to join the upper end of the axillary vein, receives veins from these muscles. The **lateral thoracic vein** communicates with the superficial epigastric vein of the lower abdominal wall through the **thoracoepigastric vein.** The thoracoepigastric vein runs in the superficial fascia of the body wall and may become visibly dilated when there is obstruction of the great veins in the abdomen.

## AXILLA

The axilla is a roughly pyramidal compartment between the upper limb and the thoracic wall. Its medial wall is the muscle here covering the thorax, the serratus anterior (Fig. 12-19); its anterior wall is formed by the pectoralis major and minor muscles; its posterior wall is composed largely of the subscapularis muscle, on the costal surface of the scapula, and to a less extent by muscles—teres major and latissimus dorsi—that are closely associated with the lateral border of the scapula; and its lateral wall is a thin strip of the arm between the converging insertions of the muscles of its anterior and posterior walls, therefore essentially the intertubercular groove of the humerus. Its curved floor is fascia continued from the thoracic wall, the pectoral muscles, and the latissimus dorsi to the deep fascia of the arm. It has no roof, for its apex is an aperture (between the subclavius muscle and clavicle, and the 1st rib) that opens into the base of the neck.

The axilla houses the great vessels and nerves of the limb, which are closely grouped together and enclosed in a layer of fascia **(axillary sheath)** brought down by them from the neck. This connective tissue gradually fades out as the nerves and vessels branch. Most of the nerve trunks appearing in the axilla are the lower part of the **brachial plexus** and its major branches and rather closely surround the axillary artery, the continuation of the subclavian artery across the 1st rib. The **axillary artery** is, therefore, conveniently thought of as the central structure of the axilla, with the *axillary vein* closely related to it and the brachial plexus and its branches arranged about it. **Lymph nodes** of the axilla lie in the looser connective tissue that serves as padding about the vital structures and are particularly related to the blood vessels. Larger nodes are usually fairly obvious in dissection, but many smaller ones are removed with the surrounding fat and connective tissue in the process of cleaning the more important structures of the axilla.

### BRACHIAL PLEXUS

Although the upper part of the brachial plexus lies in the neck and cannot be examined until the neck is dissected, an understanding of its method of formation is essential to an appreciation of the axillary portion of the plexus and the importance of the nerves of the upper limb. Relationships in the neck are discussed in Chapter 30.

In brief, the brachial plexus is formed by the union of the ventral rami (see Fig. 4-5) of the 5th, 6th, 7th, and 8th cervical nerves and of most of the ventral ramus of the 1st thoracic nerve. Often, a small part of the 4th cervical nerve also joins the plexus. There may be a small contribution from the 2nd thoracic nerve, but this is thought to be usually autonomic or cutaneous and is generally ignored in descriptions of the plexus. By uniting, dividing, and uniting again, these nerves form, successively, **trunks, anterior** and **posterior divisions,** and **cords** (see Fig. 30-21). The trunks mix the fibers from certain nerves together. The divisions sort out the fibers in the trunks that are destined to go to anterior muscles from those destined for posterior muscles. The cords maintain this sorting, but allow further mixing of nerve fibers from the spinal nerves contributing to the plexus. When the brachial plexus appears in the axilla, this mixing and sorting is complete or almost complete, and it is, therefore, primarily the three cords and their branches, grouped around the axillary artery, that are visible here (Figs. 12-20 and 12-21).

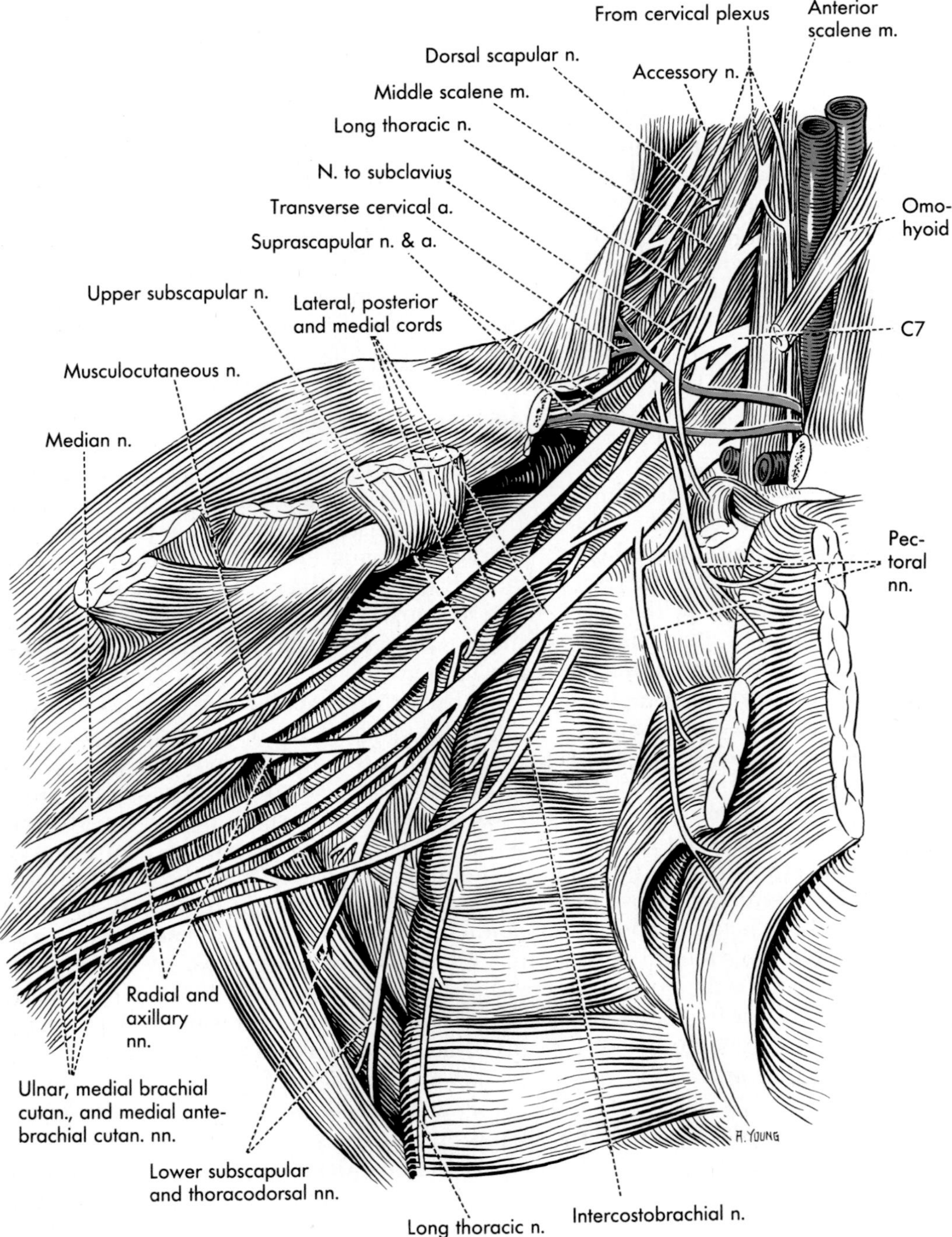


**FIG. 12-21.** The brachial plexus and its branches in a dissection of the axilla and neck. In this specimen, the lateral pectoral nerve arises rather high, above the clavicle instead of from the lateral cord in the axilla.

Of the three **cords,** the lateral and medial represent the anterior divisions of the plexus, and their branches, therefore, go to anterior muscles of the limb. The **lateral cord** contains fibers from C5, C6, and C7, and presumably also from C4, if these join the plexus. The **medial cord** contains fibers from C8 and T1 (and any from T2 that join the plexus). The **posterior cord** is formed by the union of all the posterior divisions and, therefore, receives fibers from all the nerves entering the plexus.

The lateral cord, lying lateral to the axillary artery, and the medial cord, emerging from behind the axillary artery to lie medial to it between it and the axillary vein, are united across the front of the artery by the roots that each sends to the prominent median nerve, which lies anterior or anterolateral to the artery. The posterior cord, lying posterior to the artery, cannot be examined until the artery is displaced. When this is done, it is frequently found that cord formation is completed in the upper part of the axilla. In these cases, the lower trunk of the brachial plexus (from C8 and T1) extends into the axilla before dividing into a small posterior division that completes the posterior cord and a large anterior division that continues as the medial cord.

Some branches of the brachial plexus arise in the neck, but most arise in the axilla, where each cord gives off one or more branches and then ends by dividing into two terminal branches. The distributions of the larger branches are shown schematically in Figures 11-9 through 11-12.

There are only three **branches of the lateral cord,** namely the lateral pectoral nerve and the two terminal branches, the musculocutaneous nerve and the lateral root of the median nerve. The **lateral pectoral nerve** may arise above or below the clavicle, but in the former case it accompanies the lateral cord into the axilla. Its course to the pectoralis major has already been described. The **musculocutaneous nerve** diverges laterally from the lateral cord and leaves the axilla by passing through an upper muscle of the arm (coracobrachialis) to the front of the arm. The **lateral root of the median nerve,** the largest branch, runs downward anterior or anterolateral to the axillary artery to be joined by the medial root and, thus, form the median nerve.

There are typically five **branches of the medial cord:** the medial pectoral, the medial brachial cutaneous, the medial antebrachial cutaneous, and the two terminal branches, namely the ulnar nerve and the medial root of the median nerve. The **medial pectoral nerve** is typically the highest branch of the medial cord. It emerges between the axillary artery and vein, is sometimes connected to the lateral pectoral nerve across the front of the artery, and runs downward to supply both pectoral muscles. The other branches of the medial cord often arise close together. The small **medial brachial cutaneous** may arise with the medial antebrachial cutaneous. It emerges between the axillary artery and vein to be distributed to skin on the medial side of the arm. The larger **medial antebrachial cutaneous** runs downward between the axillary artery and vein and enters the arm in this position. The **ulnar nerve** separates from the medial root of the median nerve and continues downward into the arm posteromedial to the axillary artery and behind the medial antebrachial cutaneous nerve. The **medial root of the median nerve** crosses the front of the axillary artery to join the lateral root from the lateral cord and form the median nerve anterior or anterolateral to the artery. The two roots of the median nerve vary in length, but often form the middle slanting parts of an M (apparent in Fig. 12-21, but not Fig. 12-20) across the front of the axillary artery, with the lateral cord and musculocutaneous nerve forming one upright, the medial cord and ulnar nerve the other.

The **posterior cord** typically gives off three branches in rapid succession before dividing into its two terminal branches. This cord lies behind the axillary artery, on the costal surface of the subscapularis muscle. From it, the **upper subscapular nerve** passes directly into the upper part of the subscapularis muscle. The **thoracodorsal** runs downward, crossing the lateral border of the scapula and the teres major muscle in company with an artery (thoracodorsal), to enter the costal surface of the latissimus dorsi. The **lower subscapular nerve,** shorter than the thoracodorsal, closely parallels it. It gives a branch into the lower part of the subscapularis muscle and continues to end in the teres major, which is attached to the lateral border of the scapula. Finally, the **axillary nerve** turns laterally and posteriorly to disappear with the posterior humeral circumflex vessels between the subscapularis and teres major muscles, whereas the **radial nerve** continues downward behind the axillary artery. The radial nerve may give off one or more small branches before it leaves the axilla.

In addition to these nerves, there are two others in the axilla. One is the **long thoracic nerve,** on the medial wall of the axilla on the surface of the serratus anterior muscle, which it supplies. This nerve arises from the brachial plexus in the neck before the trunks are formed. Of less importance is the **intercostobrachial,** which anastomoses with the medial brachial cutaneous and helps supply skin on the upper medial side of the arm. This nerve, the lateral cutaneous branch of the 2nd intercostal nerve, frequently receives, or is paralleled by, a branch from the 3rd intercostal. It runs across the floor of the axilla to turn down on the medial side of the arm.

### Composition and Variations in Plexus

It has been noted that the 4th cervical nerve may or may not contribute to the brachial plexus. The contribution is usually small, and the distribution of fibers from the 4th cervical nerve through the plexus is usually disregarded in considering segmental innervation of muscles. The plexus may not only receive no 4th cervical but may receive only a part of the ventral ramus of the 5th.

The nerves to the limb receive **sympathetic fibers** through gray rami communicantes (Chap. 4) that connect them with the upper thoracic and lower cervical parts of the sympathetic trunk. The preganglionic fibers to the limb typically enter the trunk through the 2nd thoracic nerve to as low as the 7th to 10th. Most of the synapses apparently occur at the base of the neck, in the 1st thoracic or inferior cervical ganglia. *Sympathectomy* of the upper limb is carried out by sectioning or removing parts of the thoracic sympathetic trunk.

The contribution of the various spinal nerves to the cords and branches of the brachial plexus, as determined by special dissections and by clinical observation, varies somewhat. Of the four branches of the brachial plexus regularly arising in the neck, the dorsal scapular and long thoracic come from the ventral rami of the nerves before they have entered into the plexus, and, therefore, their composition can be determined by careful dissection. The other two, the suprascapular and the nerve to the subclavius, typically arise from the upper trunk of the plexus, and the determination of their composition is, therefore, more difficult. This is especially true of very small nerves and may be the reason why the nerve to the subclavius is reported to have so many variations.

The lateral cord, usually formed by the anterior divisions of the upper and middle trunks, is composed as would be expected. Only in an anomalous plexus can it receive fibers from the 8th cervical or 1st thoracic nerve, for normally there is no communication from these or the lower trunk to the lateral cord. All three branches of the lateral cord commonly receive fibers from all the nerves contributing to it.

The medial cord, typically the continuation of the anterior division of the lower trunk, usually has its expected composition of C8 and T1. The medial cord or its branches can receive fibers from C7 only when there is a contribution from this nerve or the lateral cord to them. This is rare for the medial cord, but common for the ulnar nerve. The communication (the lateral root of the ulnar) usually runs behind the medial root of the median nerve and is often demonstrable only after careful search. The median nerve, formed by branches from the lateral and medial cords, typically receives fibers from all the major nerves entering the plexus.

Although the posterior divisions of all three trunks unite to form the posterior cord, this division of the lower trunk is usually small, and the cord consists mainly of fibers from C5–C7, with a smaller number of fibers from C8 and, in probably less than half of cases, some also from T1. It is often difficult to determine the segmental nerves contributing to the three small branches of the posterior cord, the two subscapular nerves and the thoracodorsal. The axillary nerve fairly frequently arises from the posterior division of the upper trunk or from the combined posterior divisions of upper and middle trunks and can, in these instances, be shown by routine dissection to contain fibers from no more than C5 to C7.

**Gross variations** in the formation of the brachial plexus commonly take the form of higher or lower origin of the branches of the plexus than usual or sometimes of apparent union of parts of the plexus (for instance, of the lateral and medial cords in front of the axillary artery to form an "anterior cord") that are not usually so joined. It has been shown that many of these apparent variations depend primarily upon the connective tissue wrapping of the plexus and are not really variations in its composition or branching. It is often difficult in the usual dissection to know exactly how much connective tissue must be removed to reveal the form of the plexus and at the same time spare nerve fibers that may be binding two nerve trunks together.

### Injuries to Plexus

Wounds in the axilla may produce a great variety of injuries to the brachial plexus, depending upon what structures have been involved. The plexus also can be injured by violent separation of the head and shoulder, such as may occur in a fall from a motorcycle, or may be chronically irritated and injured at the base of the neck.

Injuries resulting from violent separation of the head and shoulder more commonly affect the upper part of the brachial plexus, apparently because the limb is usually by the side when this occurs. This is also a common type of birth injury resulting from a difficult delivery. In other instances, depending upon the position of the arm, the greater traction may be exerted on the lower part of the plexus or may be spread fairly evenly over it. Lesions involving the upper part of the plexus, especially C5 and C6, are sometimes described as resulting in *Erb's* or *Erb-Duchenne paralysis* (affecting especially the shoulder and arm), whereas lesions of the lower part of the plexus, involving C8 and T1, are frequently classified as giving rise to *Klumpke's* or *Klumpke-Dejerine paralysis* (affecting especially the distal part of the limb). Because injuries to the plexus can be so varied, however, there is an increasing tendency to diagnose a lesion as accurately as possible in regard to the specific muscles involved, rather than classifying it loosely into one of these two groups.

Chronic injury to the brachial plexus in the lower part of the neck, often apparently involving also injury to the accompanying artery, has been described under several different terms, depending upon what was thought to be the cause of the injury: "cervical rib syndrome" and "scalenus anticus syndrome" have been the terms most commonly used in the past, but there is now a growing tendency to refer to the condition more generally as *"neurovascular compression,"* or *"thoracic outlet syndrome."* The symptoms in these cases are essentially the same, and point particularly toward involvement of the lower trunk of the plexus at or close to the level at which it crosses the 1st rib.

In some few instances, the lower trunk, in leaving the thorax, must rise higher than usual in order to cross an anomalous cervical rib or a strong fibrous cord attaching such a rib to a normal 1st thoracic rib. In such cases, traction of the trunk over the rib is an obvious cause of injury. This can be corrected by resecting the cervical rib. In the many cases in which no cervical rib is present, the neurovascular compression has been attributed to a variety of causes, including pressure against or by the anterior scalene (which attaches to the 1st rib in front of the neurovascular bundle, Fig. 12-21) and pinching between the 1st rib and the clavicle. Although there probably are several immediate causes, a common precipitating factor is carrying the shoulder abnormally low, so that the neurovascular bundle is dragged against or angulated over some relatively firm structure. In the large majority of instances of neurovascular compression, adequate physical therapy and teaching the patient to carry his shoulder higher will alleviate the symptoms. These symptoms include tingling and numbness, perhaps pain, and often obviously disturbed circulation to the hand, but there is little agreement as to what degree they are caused by irritation or compression of nerves and to what degree caused by vascular involvement.

## BLOOD VESSELS AND LYMPH NODES

The blood vessels of the axilla are the axillary artery and its branches and the axillary vein and its tributaries. These are both intimately related to the brachial plexus, but, as already seen, the axillary artery is especially so. The large lymph nodes are concentrated along the axillary vein.

### Axillary Artery and Branches

The axillary artery is the direct continuation of the subclavian, the change in name occurring as the vessel crosses the 1st rib. It becomes the brachial artery as it leaves the axilla at the lower border of the teres major muscle.

The artery passes behind the pectoralis minor in its course through the axilla and is conveniently described as having three parts: a first part above the muscle, a second part behind, and a third part below the muscle. The

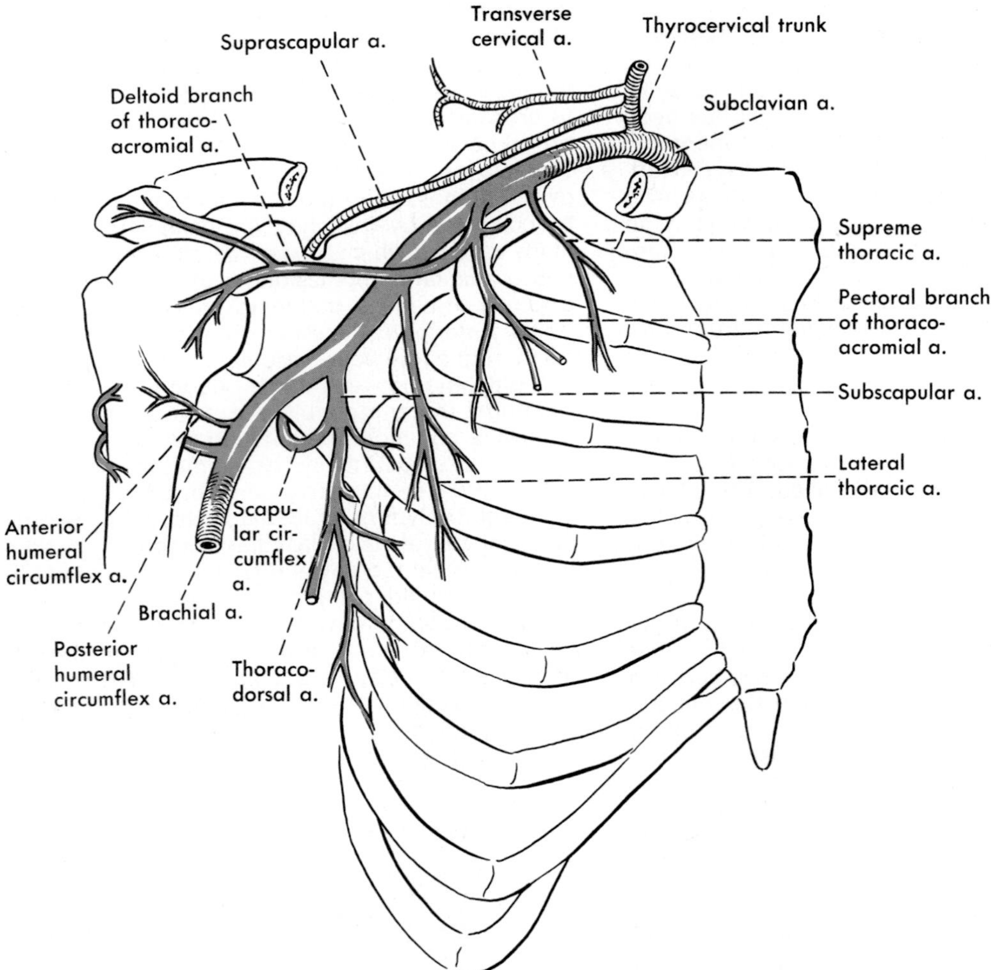


**FIG. 12-22.** Diagram of the axillary artery and its branches; arteries to the shoulder from the thyrocervical trunk are shown uncolored.

artery is paralleled by the axillary vein. The medial cord and its branches lie in part between the artery and the vein. The first part of the artery lies on the upper part of the serratus anterior muscle, with the medial cord behind it and the lateral and posterior cords lateral to and above it. The second and third parts of the artery lie on the subscapularis and teres major muscles, with the three cords of the brachial plexus and their major branches having the relationships already described.

The axillary artery is usually described as giving off six branches (Fig. 12-22), although the number varies because two or more arteries often arise together instead of separately, or two branches of an artery arise separately instead of from the usual common stem. Thus, instead of six, there may be anywhere from 5 to 11 branches. The six named branches of the axillary artery are, in order of origin, the supreme thoracic, the thoracoacromial, the lateral thoracic, the subscapular, and two, anterior and posterior, humeral circumflex arteries. Frequently one or more subscapular rami go directly into the costal surface of the subscapularis muscle.

The **supreme thoracic artery** is a small vessel arising from the first part of the axillary artery and running medially to be distributed over parts of the 1st or 1st and 2nd intercostal spaces. The **thoracoacromial artery** arises from either the first or the second part of the axillary, but passes above the upper border of the pectoralis minor to pierce the clavipectoral fascia and divide into the four branches already described (p. 181). The **lateral thoracic artery** typically arises from the second part of the axillary, but frequently arises instead from the thoracoacromial; it runs downward for a variable distance on the anterolateral thoracic wall.

The **subscapular artery** is usually the largest branch of the axillary and arises from the third part of the artery. From its origin, it typically passes backward and divides into two trunks, of which one, the **scapular circumflex artery,** passes around the lateral border of the scapula between the scapula and the teres major (in the triangular space, Fig. 12-36). The other branch, the **thoracodorsal artery,** continues downward along the posterior axillary wall and is joined in its course by the thoracodorsal nerve (to the latissimus dorsi muscle). The thoracodorsal artery gives branches to muscles of the posterior wall of the axilla and frequently divides into two terminal branches, of which one continues on the latissimus, the other on the lateral thoracic wall. The scapular circumflex and thoracodorsal branches of the subscapular sometimes arise separately from the axillary artery. Occasionally, also, some other artery—more commonly a humeral circumflex—arises with or from the subscapular.

The **anterior and posterior humeral circumflex arteries** are typically the last branches of the axillary artery. They may arise anteriorly and posteriorly, and at the same or different levels, or they may arise by a common stem. The anterior humeral circumflex is small and passes laterally around the front of the surgical neck of the humerus, but the larger posterior humeral circumflex passes around the posterior aspect of the surgical neck of the humerus with the axillary nerve (through the quadrangular space, Figs. 12-34 and 12-36) and runs with the nerve around the humerus under the deltoid muscle, supplying this muscle in particular. Anterior and posterior humeral circumflexes anastomose around the humerus.

A rare but striking **anomaly** of the axillary artery is for it to divide into two branches that proceed down the arm. When this occurs, usually one of the arterial stems in the arm runs more superficially than does the other, and they are, therefore, sometimes distinguished as brachial artery and superficial brachial artery. This apparent doubling of the brachial artery more commonly occurs in the arm than in the axilla and is discussed later.

## Axillary Vein

The axillary vein (Fig. 12-23) is the direct continuation upward of the basilic vein (see Fig. 11-14), a superficial vein that pierces the deep fascia of the arm to lie close to the brachial artery and pass up to the axilla. The basilic vein has its name changed to axillary as it crosses the lower border of the teres major, and the axillary, in turn, has its name changed to subclavian as it crosses the 1st rib. Tributaries of the axillary vein are the brachial veins, the cephalic vein (the other superficial vein of the upper limb), and veins that correspond largely to the branches of the axillary artery, except the veins corresponding to the thoracoacromial artery's branches join the cephalic vein or the upper part of the axillary as several independent veins. The two **brachial veins** typically join the axillary at different levels. The **cephalic vein** reaches the axillary by passing between the deltoid muscle and the clavicular part of the pectoralis major.

## Axillary Lymph Nodes

The lymph nodes of the axilla are embedded in the connective tissue associated with the brachial plexus and the vessels, and surgical removal of them (a part of the procedure of radical amputation of the breast) demands very careful dissection of the axilla. The nodes vary from less than a dozen to about three dozen, and the smaller ones are frequently completely hidden by surrounding connective tissue and fat. Although they form a straggling group with no sharp distinction between members, it is convenient to subdivide them, for purposes of description, into five subgroups (Fig. 12-23).

The **apical nodes** are the highest ones, on the upper part of the axillary vein close to the apex of the axilla. They receive the lymphatic drainage from all the other axillary nodes and give rise to the *subclavian lymphatic trunk* that carries the lymph from the upper limb to the neck. This trunk commonly joins the thoracic duct on the left side and on the right side either enters independently into the venous system or joins lymphatic ducts from the head and neck or from the thorax (see Chap. 30). Some of the lymphatics from the apical group, however, go into lymph nodes in the lower part of the neck (supraclavicular or lower deep cervical); therefore, when axillary lymph nodes are involved by carcinoma, some of the lower deep cervical nodes may be also.

The **central nodes** lie on the axillary vein below the apical nodes, more or less behind the pectoralis minor muscle. They receive lymphatics from the three groups of nodes still to be described and send them to the apical group. Of the remaining three groups, the **lateral nodes** lie on the lower part of the axillary vein and receive the lymphatic drainage from most of the upper limb except the shoulder; the **subscapular nodes** lie on the subscap-

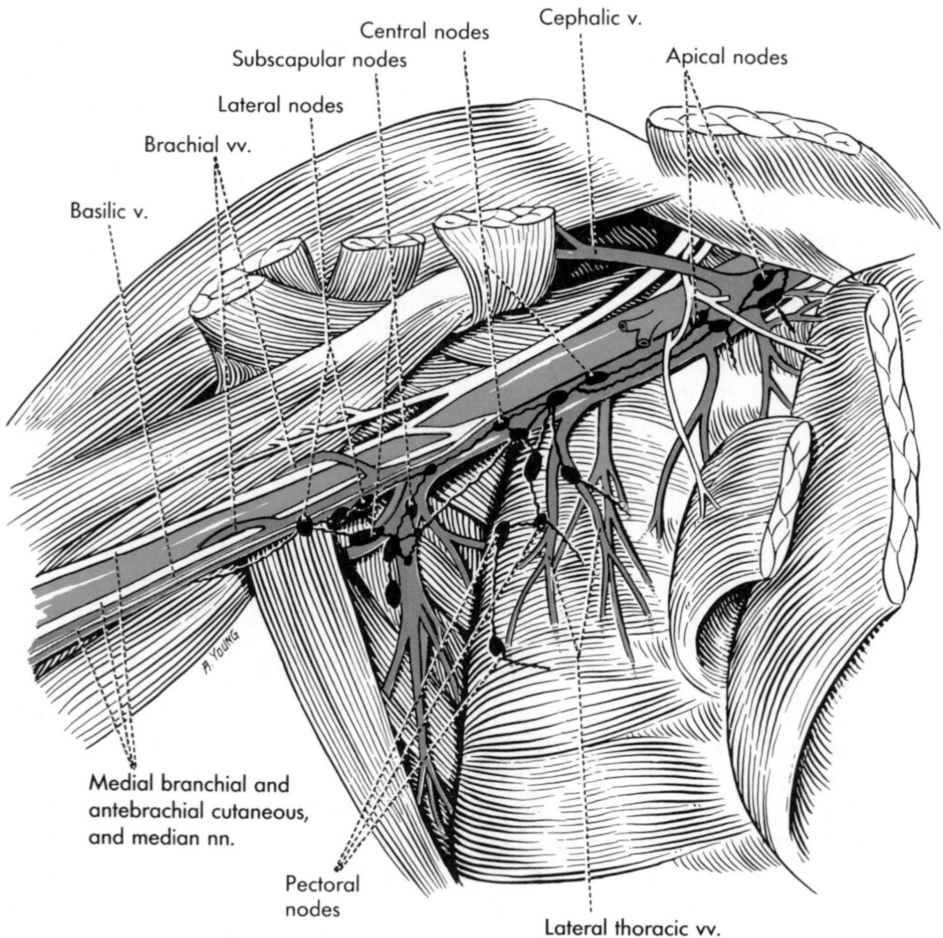


**FIG. 12-23.** Veins and lymph nodes of the axilla.

ular vessels, for the most part on the costal surface of the subscapularis muscle; and the **pectoral nodes** lie largely along the lateral thoracic vessels, close to the lower lateral border of the pectoralis minor, with some often lying between the pectoral muscles. The pectoral nodes receive much of the lateral drainage from the mammary gland as well as other drainage from the anterior thoracic wall and send lymphatics to the apical and central groups.

## SHOULDER

The largest and most superficial muscles of the shoulder, the trapezius and latissimus dorsi, extend from the occipital region of the skull to the iliac crest and the sacrum. The chief points of superficial anatomy to be noted are, therefore, the external occipital protuberance (the prominent posterior midline projection) of the skull, the midline position of the spinous processes of the vertebrae (that of the 7th cervical being usually the highest one that is easily palpable and visible), and the posterior part of the iliac crest, the flared upper edge of the hip (coxal) bone. In the scapular region, the spine of the scapula can be traced to the acromion. The medial border and inferior angle of the scapula can be easily outlined in the living individual, but not necessarily on the cadaver.

### Fascia, Superficial Nerves, and Vessels

The **superficial fascia** is somewhat thick. On its deep surface, it fuses with the deep fascia investing the muscles. Because of this, it is customary to remove the two layers together from the back and shoulder. In so doing, the cutaneous nerves must be watched for as they emerge from the muscles, if they are to be recognized and traced. There may be a subcutaneous bursa in the fascia over the acromion.

The **deep fascia** over the outer surfaces of the trapezius and latissimus dorsi muscles (Fig. 12-25) is thin but strong and intimately bound both to the muscles and to the overlying superficial fascia. At the free borders of these muscles, it joins the fascia on their deep surfaces. In the neck, the single layer formed at the free edge of the trapezius is continued laterally and forward as a strong layer of fascia to the posterior border of the sternocleidomastoid muscle, thereby covering the area known as the posterior triangle of the neck. This fascia is a part of the outer or superficial layer of the cervical fascia (see Chap. 30). At the posterior border of the sternocleidomastoid muscle, the fascia is penetrated by cutaneous nerves (branches of the cervical plexus) that become superficial here.

The fascia on the deep surface of the latissimus dorsi is a distinct layer on the muscular part of the muscle, and as it reaches the lateral free border of the muscle, it fuses with the fascia on the superficial surface. Beyond this border the fascia is continuous with that of adjacent muscles, and in part with the deep fascia of the arm and axilla. Medially, however, the fascia deep to the latissimus loses its identity, fusing with the aponeurosis of origin of the latissimus and helping to form with this and other tendons a thickened fascioaponeurotic layer that covers the true muscles of the back and is known as the posterior layer of the **thoracolumbar fascia.**

The **cutaneous nerves** to the skin of the back and shoulder are largely segmental and consist of dorsal rami of spinal nerves (Chap. 4) for a varying distance on each side of the midline and, more laterally and anteriorly, of *lateral cutaneous branches* of intercostal nerves (Figs. 12-11 and 12-40). The exception to this is the distribution of branches of the cervical plexus to the upper part of the shoulder, over the acromion and lateral end of the clavicle. The nerves here are the intermediate and lateral *supraclavicular nerves,* which descend over the clavicle and acromion to become subcutaneous over the upper part of the deltoid muscle.

The posterior branches of the lateral cutaneous rami of the intercostal nerves run backward and medially in the superficial fascia, crossing the anterior border of the latissimus dorsi, to innervate a variable lateral area of the skin of the back.

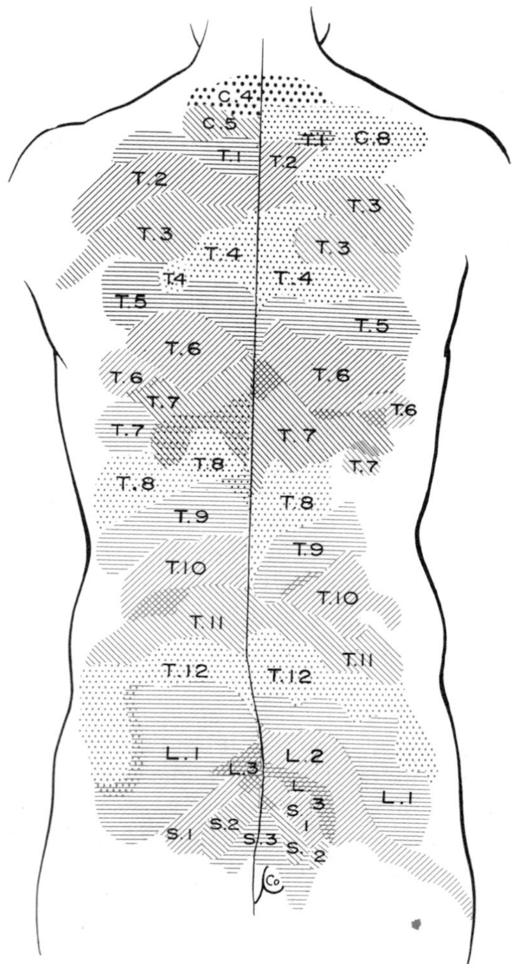


**FIG. 12-24.** Distribution of dorsal rami of spinal nerves to skin of the back. The areas of distribution of the medial branches of the rami are shown in black; those of the lateral branches are shown in red. (Redrawn from Johnston HM: J Anat Physiol 43:80, 1908)

The *dorsal rami* of the spinal nerves become subcutaneous fairly close to the midline of the back. For the most part, as they run posteriorly through the back muscles (deep to the covering shoulder muscles), they supply these muscles and divide into medial and lateral branches. Usually one of these branches ends in the back muscles, whereas the other one continues to the skin. It is the medial branches of the dorsal rami of the cervical and upper six to eight thoracic nerves that become cutaneous; therefore, these emerge from the musculature close to the midline. In contrast, the lateral branches of the lower thoracic and the first three lumbar nerves become cutaneous, emerging along a line distinctly lateral to the line of emergence of the upper cutaneous nerves.

Not all the **dorsal rami of spinal nerves** have cutaneous branches (Fig. 12-24), and there is some variation in distribution. However, in general, the 1st cervical nerve usually has no cutaneous branch. The cutaneous branches of the 2nd and 3rd cervical nerves are distributed upward to the scalp (the large branch of the 2nd, frequently joined by some of the 3rd, forms the **greater occipital nerve,** which pierces the uppermost fibers of the trapezius or appears just lateral to these fibers and is joined by the occipital artery in its course upward; a smaller cutaneous branch of the 3rd lies medial to the greater occipital and is called the **3rd occipital nerve** when it is well developed). The medial branches of the 4th and 5th cervical nerves typically reach skin on the back of the neck, whereas the 6th, 7th, and 8th have no cutaneous branches. Typically, all the dorsal rami of the thoracic nerves have cutaneous branches, medial branches above and lateral branches below. Lateral branches of the upper three lumbar nerves appear just a little above the crest of the ilium to descend over this crest and supply skin of the buttock. The dorsal branches of the last two lumbar nerves do not reach the skin. The lateral branches of the first three sacral nerves communicate with each other and with nerves above and below by simple loops. They also supply skin over part of the buttock. The last two sacral nerves and the coccygeal do not divide into medial and lateral branches, but their posterior divisions do form a series of loops with each other and give off tiny twigs that pass to skin over the lower end of the sacrum and the coccyx.

The **superficial vessels** of the back tend to be segmental like the nerves, for they are primarily from the dorsal rami of the posterior intercostal and lumbar vessels. These dorsal rami also supply the vertebral column and the musculature of the back. Their subcutaneous branches are small. In the cervical region, branches to the back are given off by the longitudinally running vertebral and deep cervical arteries.

## MUSCLES AND ASSOCIATED STRUCTURES

The musculature of the shoulder is conveniently divided into superficial muscles, the **trapezius** and **latissimus dorsi,** that cover most of the other muscles on the back; **deep**

**extrinsic muscles** that, like the superficial muscles, arise from the axial skeleton; and **intrinsic muscles,** arising from the scapula and passing to the humerus. These muscles are supplied by a number of nerves, mostly branches of the brachial plexus. However, the trapezius is supplied by the spinal accessory nerve, and the levator scapulae is supplied by the cervical nerves. Some of the blood vessels of the shoulder arise in the neck, and others are branches of the axillary artery.

## Superficial Muscles

The **trapezius** and the **latissimus dorsi** have their origin along the midline of the back, extending from the external occipital protuberance of the skull to the sacral region (Fig. 12-25). The origin of the trapezius partly covers that of the latissimus. Since both muscles are somewhat triangular, the upper border of the latissimus appears from deep to the trapezius as the two muscles are traced laterally. When the scapula is moved forward, the lower border of the trapezius, the upper border of the latissimus, and the medial border of the scapula form a triangle in which no muscles, except a bit of the lower part of the *rhomboideus major,* intervene between the skin and the ribs with their connecting intercostal muscles. Respiratory sounds picked up by a stethescope placed here are therefore less muffled by intervening tissue than elsewhere on the back.

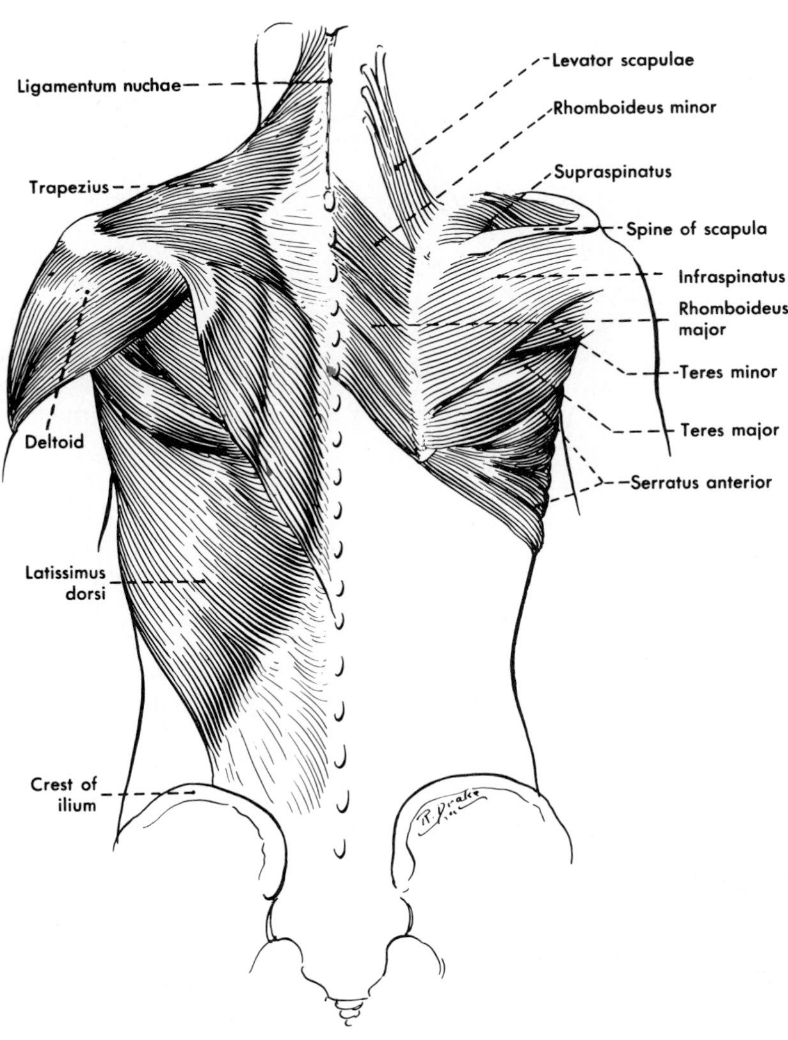


**FIG. 12-25.** Posterior view of muscles of the shoulder; the trapezius, latissimus, and deltoid have all been removed from the right side.

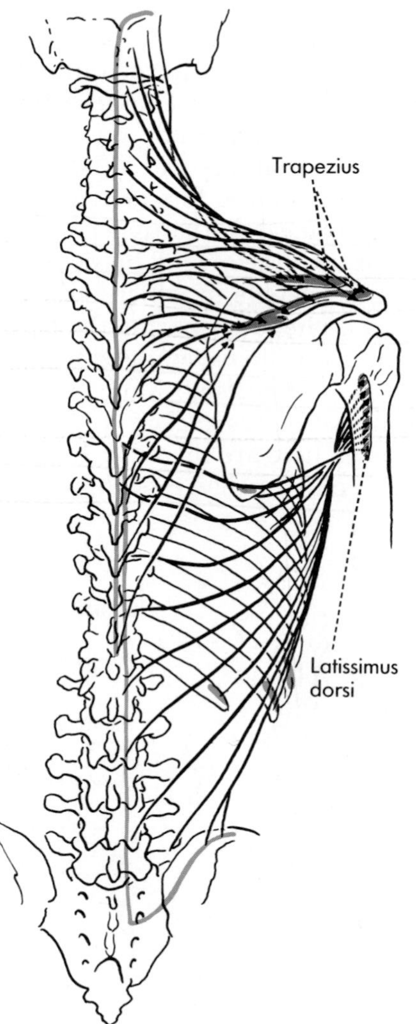


**FIG. 12-26.** Origin (red) and insertion (blue) of the trapezius and the latissimus dorsi.

This triangle is called the *triangle of auscultation.* Just above the crest of the ilium there is typically a small triangle, the *lumbar triangle,* between the lateral border of the latissimus, as it arises from the crest, and the posterior border of the external oblique muscle of the abdomen, as it inserts upon the crest.

**Trapezius.** The uppermost **origin** of the trapezius is usually from a medial part of the *superior nuchal line* on the occipital bone and from the *external occipital protuberance,* but sometimes it fails to reach the skull. Between the skull and the 7th cervical vertebra, the vertebral spinous processes are covered by muscles of the back. Here, the two trapezius muscles are attached to the *ligamentum nuchae.* Except superficially, where the ligamentum nuchae gives attachment to the trapezius and two muscles lying deep to it, this ligament is, in man, a thin fibrous septum that stretches from the external occipital protuberance to the spinous process of the 7th cervical vertebra and extends deeply to attach to the spinous processes of the other cervical vertebrae. At the base of the neck, the trapezius arises from the *spinous process of the 7th cervical vertebra,* and below this, from all the *thoracic vertebral spinous processes* and their connecting supraspinous ligaments (Fig. 12-26). The uppermost fibers run downward and forward to **insert** on approximately the distal third of the *clavicle;* the fibers arising from the lower cervical and upper thoracic region, usually the thickest part of the muscle, pass a little downward and laterally to insert on the *acromion;* and the fibers arising from most of the thoracic region pass laterally and upward to insert along the length of the *spine of the scapula.*

The superficial layer of the cervical fascia extends from the lateral border of the trapezius to the posterior border of the sternocleidomastoid muscle and roofs the posterior triangle of the neck, which lies between these two muscles. Deep to this, another fascial layer covers muscles of the neck and the levator scapulae, a muscle of the shoulder. Between the two layers of fascia are the accessory nerve (to the trapezius), the transverse cervical vessels, and the suprascapular nerve and vessels (Fig. 12-21), all going to the shoulder. The **accessory nerve,** joined by branches from the 3rd and 4th cervical nerves, runs downward on the deep surface of the trapezius with the superficial branch of the **transverse cervical artery.** It gives off branches into the muscle and ends in it. The artery may send a smaller branch upward on the deep surface of the trapezius.

**Latissimus Dorsi.** The latissimus dorsi usually **arises** from the *spinous processes* of the lower six thoracic, the lumbar, and the upper sacral vertebrae; from the posterior part of the *iliac crest;* and from the *lower three or four ribs* by muscular slips that interdigitate here with slips of origin of the external oblique muscle of the abdomen (Fig. 12-26).

The fibers of the latissimus converge toward the axilla, and as the muscle reaches the *teres major* on the lateral border of the scapula, it spirals beneath and anterior to that muscle so that its original anterior surface is directed posteriorly. The muscle fibers end in a flattened tendon that passes with the tendon of the teres major, from which it is partially separated by a bursa (subtendinous bursa of the latissimus), to **insert** into the *crest of the lesser tubercle* and the floor of the intertubercular groove of the humerus. Close to its insertion, the muscle is crossed anteriorly by the vessels and nerves of the axilla and by the coracobrachialis and short head of the biceps, muscles of the arm.

The **thoracodorsal nerve,** the nerve to the latissimus, leaves the axilla on the deep (costal) surface of the muscle (Fig. 12-20) and runs downward on this surface with the **thoracodorsal artery,** a branch of the subscapular.

**Variations, Innervations, and Actions.** The common variation in the trapezius is the height that the muscle reaches in the neck; the two muscles are not necessarily symmetrical in one body. The common variation of the latissimus (the muscle is sometimes associated with axillary arch muscles, is an origin from the inferior angle of the scapula as it crosses this angle.

The **spinal accessory nerve,** a purely motor nerve, supplies the sternocleidomastoid muscle in the neck and then ends in the trapezius. Fibers from the 2nd cervical nerve join it and sometimes continue to the trapezius, which also regularly receives fibers from the 3rd and 4th cervical nerves. In the monkey the cervical nerve fibers going to the trapezius are afferent (sensory, proprioceptive) only, but many clinicians believe that in man they include a variable number of motor fibers. The nerve to the latissimus, the **thoracodorsal nerve,** is derived from the posterior cord of the brachial plexus and brings to the muscle nerve fibers from about C6 to C8, with the contribution from C7 apparently being most important.

The **trapezius** as a whole retracts the scapula. The upper and lower fibers work together, the former pulling upward on the lateral angle, the latter downward on the base of the spine, upwardly rotating the scapula (turning the glenoid cavity up), a movement that normally accompanies abduction of the arm. The heavier part of the muscle inserting on the acromion is particularly important both in upward rotation and in preventing sagging of the lateral angle. No other levator of the scapula inserts on the lateral angle. A regular finding in paralysis of the trapezius as a result of a lesion of the accessory nerve is a downward rotation of the scapula produced by the weight of the arm. The **latissimus** produces a combined movement of adduction, medial rotation, and extension of the arm, as in a swimming stroke. The lower fibers are especially important in preventing upward displacement of the shoulder when the weight of the body tends to produce this, as it does, for example, when an individual hangs by his arms or uses a crutch.

## Deep Extrinsic Muscles

The deep extrinsic muscles are the **levator scapulae,** the **rhomboideus major** and **minor,** and the **serratus anterior** (Fig. 12-25). The first three muscles arise from the vertebral column. The serratus anterior arises anterolaterally from the ribs and largely covers the lateral thoracic wall.

**Levator Scapulae.** The levator scapulae **arises** from the posterior tubercles of the transverse processes of the first three or four cervical vertebrae (Fig. 12-27). It passes downward, laterally, and slightly backward to **insert** into the superior angle of the scapula and the medial border of this bone almost as far as the base of the spine. At its origin, the muscle is covered by the sternocleidomastoid, and in its lower part by the trapezius. Between the two muscles, it forms a part of the floor of the posterior triangle of the neck, and the *accessory nerve* passes from the sternocleidomastoid to the trapezius on its surface. The *dorsal scapular nerve* (to the rhomboids) lies anterior to its lower part and descends around its medial border or pierces the muscle. The superficial and deep branches of the *transverse cervical artery* are separated by it (Fig. 12-35). It receives its **innervation** from deep branches of the cervical plexus that enter it anteriorly close to its origin and may also receive a twig from the dorsal scapular nerve.

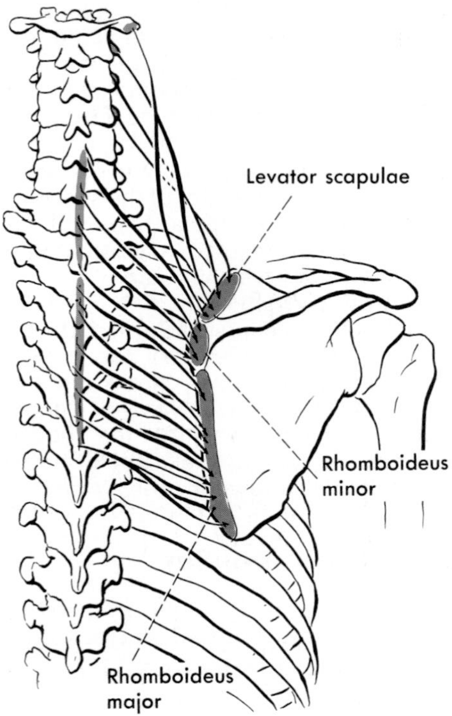


**FIG. 12-27.** Origin (red) and insertion (blue) of the levator scapulae and the rhomboidei.

**Rhomboidei.** The flat rhomboideus minor and rhomboideus major muscles lie deep to the trapezius and extend laterally and downward from the midline of the back to the medial border of the scapula (Figs. 12-25 and 12-27).

The **rhomboideus minor** arises from the lower part of the ligamentum nuchae, from the spinous processes of the 7th cervical and 1st thoracic vertebrae, and from the supraspinous ligament between these. It inserts into the medial border of the scapula at the base of the scapular spine. It may be difficult to separate from the rhomboideus major.

The **rhomboideus major** arises from the spinous processes of the 2nd to 5th thoracic vertebrae and from the intervening supraspinous ligament. It parallels the minor and inserts into the medial border of the scapula from just below the base of the spine to the inferior angle.

On the deep surface of the two rhomboids close to their insertions on the scapula are the *dorsal scapular nerve* and an artery that is either the deep branch of the *transverse cervical* (Fig. 12-35) or a vessel that has a separate origin from the subclavian and is then known as the *dorsal scapular artery*. The nerve, a high branch of the brachial plexus, descends obliquely through the neck and meets the artery deep to the lower end of the levator scapulae muscle and supplies both rhomboidei. The artery and its accompanying vein supply adjacent muscles also and anastomose around the scapula with the suprascapular and subscapular vessels (Fig. 12-36).

**Serratus Anterior.** The serratus anterior covers much of the lateral aspect of the thorax (Fig. 12-11). It **arises** by fleshy slips from the outer surfaces of the upper eight or nine ribs. The first slip is particularly large and arises from the first two ribs and the fascia between them. The muscle then follows the curvature of the thorax to **insert** along the costal aspect of the medial border of the scapula. This attachment is not evenly arranged; the upper four slips insert on most of the medial border, whereas the lower ones converge to insert on the inferior angle. The strongest insertion and the strongest action of the muscle are on this angle.

The upper slips of origin are under the pectoral muscles, but the lower ones are subcutaneous and easily visible in muscular individuals. They interdigitate with origins of the external oblique muscle of the abdomen from these same ribs.

The *long thoracic nerve* descends on the outer surface of the serratus to supply the muscle. The *lateral thoracic artery* approximately parallels the nerve but lies in front of it on the upper part of the thorax. The lower part of the nerve is accompanied by a part of the *thoracodorsal artery* (Fig. 12-20).

**Variations, Innervations, and Actions.** There are no noteworthy common variations in this group of muscles.

The **levator scapulae** is supplied by twigs from the 3rd and 4th cervical nerves. It also receives fibers from C5 if the dorsal scapular nerve helps supply it. The two **rhomboidei** typically receive fibers from C5 only by way of the dorsal scapular nerve. The **serratus anterior** usually receives fibers from C5 through C7 by way of the long thoracic nerve, but the composition of this nerve varies; the root from C6 is usually largest and is the most constant.

The **levator scapulae** elevates the medial angle of the scapula, and the **rhomboidei** elevate the medial border and retract the scapula. If the scapula is not otherwise fixed or moved, all three muscles, in elevating the medial border, turn the glenoid cavity down. The **serratus anterior** protracts (draws forward) the scapula and in so doing, because it follows the curve of the thorax, tends to hold the medial border of the scapula firmly against the thoracic wall. One of the results of a weak serratus is, therefore, an abnormal prominence of the medial border of the scapula when the arms are held horizontally forward; this is referred to as "winging" of the scapula. If, during protraction, the serratus draws the inferior angle forward faster than the remainder of the medial border, the scapula is upwardly rotated. This upward rotation is the most important action of the serratus. It is normally aided by fibers of the trapezius that lift the lateral angle.

### Movements of the Scapula

The muscles moving the scapula are the **extrinsic muscles** attaching either directly to the scapula or to the humerus. In the latter instance, their action upon the scapula is dependent upon the action of intrinsic muscles of the shoulder that hold the head of the humerus in place on the scapula, so that its lateral angle moves with the movements of the humerus.

Movements of the scapula are described as being elevation, depression, protraction (moving it laterally and forward around the thorax), retraction, and upward and downward rotation. In upward rotation, the inferior angle swings laterally and forward so that the glenoid cavity is turned up. In downward rotation, it swings medially and backward so that the glenoid cavity is turned down. Scapular movements usually accompany humeral movements in the same direction, thus increasing the range of movement of the free limb. An accompanying scapular movement is particularly important in abduction of the limb, which is limited to about 90° unless accompanied by upward scapular rotation.

The actions of the various individual muscles already have been described. The following is therefore a summary of the actions that pertain particularly to scapular movements.

The **elevators** of the scapula are the parts of the trapezius inserting laterally on the clavicle and particularly on the acromion (these are called upper, sometimes middle fibers), the levator scapulae, and the two rhomboid muscles.

Gravity alone, especially the weight of the free limb, tends to **depress** the scapula, particularly its lateral angle (thus downwardly rotating the scapula). But active depression of the scapula is brought about by lower fibers of the trapezius, by lower fibers of the pectoralis major, by the pectoralis minor, by lower fibers of the serratus anterior, and by the latissimus dorsi. Of these muscles, the latissimus is the most powerful and the most important. It can resist the upward thrust imposed on the shoulder by the use of a crutch, but the next most powerful muscle, the pectoralis major, cannot do this. The subclavius, if regarded as a depressor of the clavicle, should also be included as a depressor of the scapula.

In order to produce **upward rotation,** a muscle must either advance the inferior angle of the scapula laterally ahead of the rest of the medial border or elevate the lateral angle. The serratus anterior muscle does the former, the acromial fibers of the trapezius the latter; therefore, they are both upward rotators. The strongest movement occurs when they both work together, but in some individuals, the serratus anterior alone can rotate the scapula, whereas the trapezius alone can apparently never produce adequate upward rotation for full abduction of the limb.

**Downward rotators** of the scapula include the levator scapulae and the rhomboideus minor and major, which pull upward on the medial border, and the pectoralis minor, lower fibers of the pectoralis major, and the latissimus dorsi, which pull downward, directly or indirectly, on the lateral angle.

The **protractors** of the scapula are the serratus anterior and the pectoralis minor and major. Of these, the serratus is by far the most important for a powerful movement, such as pushing a heavy object. When it is paralyzed, the shoulder is easily forced back.

**Retractors** are the trapezius, especially its middle portion, the rhomboideus major and minor, and the latissimus dorsi, especially its upper fibers.

It will be noted that many muscles are listed here as belonging in more than one group from the standpoint of action—for example, the trapezius is an elevator, a depressor, and an upward rotator. Further, of two or more muscles in a functional group, there may be actions in addition to the one considered that tend to cancel each other out—as, for instance, the pectoralis minor and major in protraction would tend also to downwardly rotate the scapula, whereas the serratus anterior in protraction tends to upwardly rotate it. Movements of the scapula, like almost all movements, therefore, obviously depend on precise coordination in both time and strength of the contraction of specific parts, according to the movement desired, of several muscles.

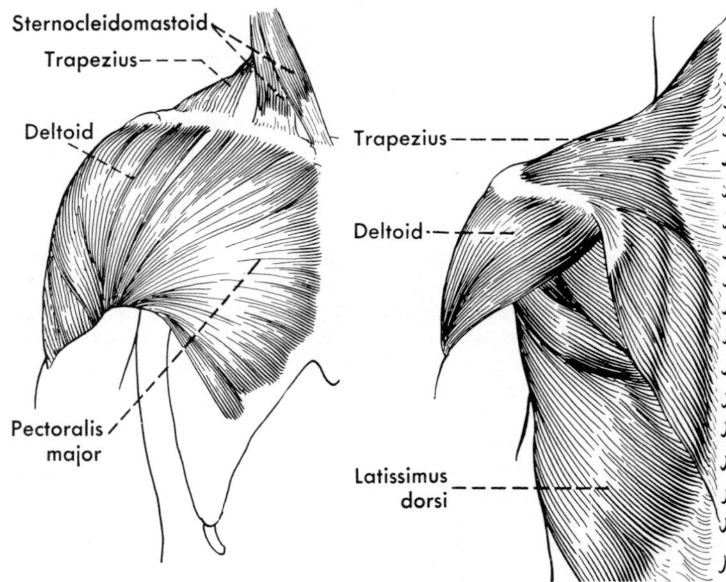


**FIG. 12-28.** The deltoid muscle seen in the anterior view of the right shoulder and in a posterior view of the left.

## Intrinsic Shoulder Muscles

Certain muscles are known as intrinsic shoulder muscles because they both arise and insert on the skeleton of the limb, running from the girdle to the humerus. The superficial muscle of this group is the **deltoid** (Fig. 12-28). The deltoid is particularly prominent as it extends downward from the acromion to the humerus. The other muscles lie deeper, arising from the surfaces or lateral border of the scapula. They are the **supraspinatus,** the **infraspinatus,** the **teres major** and **minor,** and the **subscapularis** (Figs. 12-29 and 12-32). Their attachments are shown diagrammatically in Figures 12-30 and 12-31.

**Deltoid.** The **origin** of the deltoid muscle corresponds approximately to the insertion of the trapezius: from the lateral third of the *clavicle,* from the lateral border of the *acromion,* and from almost all of the *scapular spine.* From this wide origin, the fibers converge to **insert** into the *deltoid tuberosity* on the anterolateral surface of the body of the humerus. Anteriorly and posteriorly, the fibers run approximately parallel to each other, but the more powerful middle part of the muscle has within it *tendinous septa* that serve for the origin and insertion of fibers; the major part of the muscle is therefore multipennate, which makes it particularly strong for its bulk.

**FIG. 12-29.** Intrinsic muscles on the back of the shoulder, and related muscles, after removal of the trapezius and deltoid.

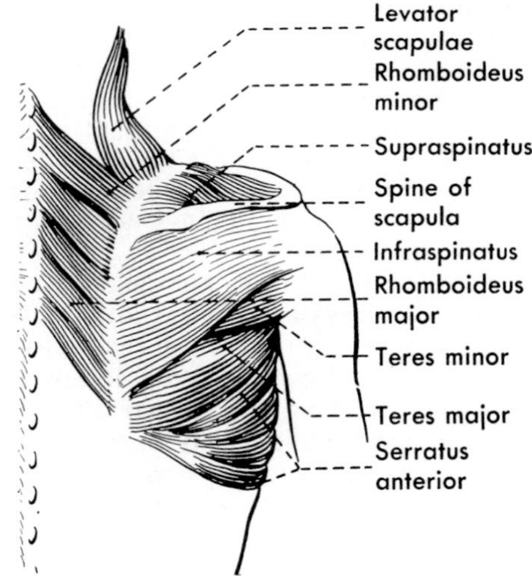


In the groove between the pectoralis major and the anterior border of the deltoid run the *cephalic vein* and the *deltoid branch of the thoracoacromial artery,* which furnish a minor blood supply to the muscle. Around or through the posterior border of the muscle, the *upper lateral brachial cutaneous nerve,* a branch of the axillary, emerges to become

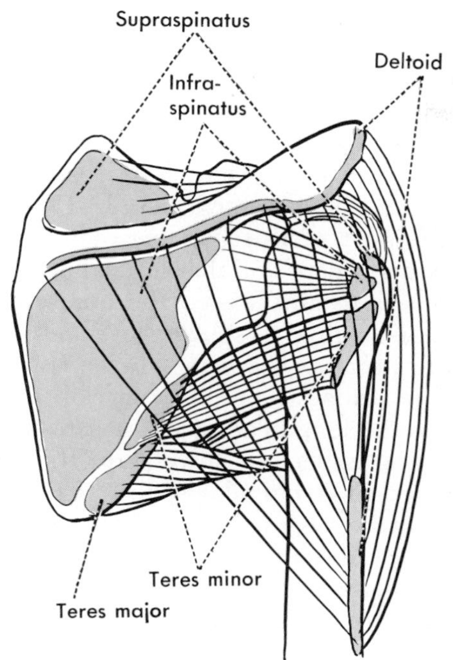


**FIG. 12-30.** Origins (red) and insertions (blue) of the intrinsic muscles of the shoulder, from behind; see also Figure 12-31.

subcutaneous. Deep to the muscle, circling the surgical neck of the humerus and sending branches into the muscle, are the *axillary nerve* and the *posterior humeral circumflex vessels* (Fig. 12-36).

**Supraspinatus.** The supraspinatus muscle arises from most of the bony supraspinatous fossa and from the fascia that covers the muscle. It passes laterally over the top of the shoulder joint to attach to the uppermost of the three facets of the greater tubercle of the humerus. The tendon of this muscle forms the uppermost part of the musculotendinous, or rotator, cuff of the shoulder (Fig. 12-37). It is separated from the overlying acromion, the coracoacromial ligament, and the deltoid muscle by the subacromial and subdeltoid bursae.

The *suprascapular nerve* and *vessels* cross the upper border of the scapula and turn downward between the muscle and the bone (Fig. 12-36); the nerve supplies the muscle, and the vessels supply the muscle and bone. They continue deep to the base of the acromion to run in the infraspinatous fossa.

**Infraspinatus.** The infraspinatus arises from most of the infraspinatous fossa and from the fascia covering the muscle. Its tendon forms an upper posterior part of the musculotendinous, or rotator, cuff, and it inserts into the middle facet on the greater tubercle.

The *suprascapular nerve* and *vessels* continue between the infraspinatus muscle and the scapula, and the nerve ends in the muscle. The *scapular circumflex branch* of the subscapular artery rounds the lateral border of the scapula and runs upward between the bone and the infraspinatus muscle. Here, it and the suprascapular artery anastomose and supply both bone and muscle.

**Teres Minor.** The teres minor arises from about the middle half of the lateral border of the scapula and from septa between it, the infraspinatus, and the teres major. It passes behind the long head of the triceps to insert into the lowest of the three facets of the greater tubercle, thereby forming the lower posterior part of the musculotendinous cuff.

The origin of the muscle is pierced by the

**FIG. 12-31.** Attachments of the intrinsic shoulder muscles, from the front.

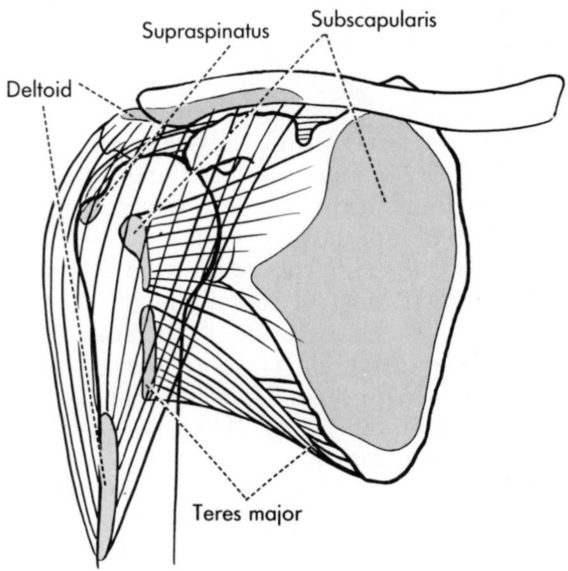

*scapular circumflex artery* as this turns upward in the infraspinatous fossa. Laterally, the *axillary nerve,* as it starts its course around the humerus, appears below the muscle and gives off a branch into it.

**Teres Major.** The teres major arises from about the lower third of the lateral border of the scapula and from intermuscular septa between it and the teres minor, the infraspinatus, and the subscapularis. It passes in front of the long head of the triceps and around the medial side of the humerus with the latissimus dorsi, which spirals around the major so that it lies first posterior, then below, then anterior to this muscle. The teres major attaches to the crest of the lesser tubercle close to the attachment of the latissimus. The bursa of the latissimus dorsi usually lies between the tendons of the latissimus and the teres major and the subtendinous bursa of the teres major between the tendon of the teres major and the humerus.

The anterior (costal) surface of the teres major is crossed by the nerves and vessels continuing from the axilla into the arm (Fig. 12-20). The lower border of the muscle is regarded as the line of division between the axilla and the arm. More medially, the *thoracodorsal branches of the subscapular vessels* supply the teres major as they cross it, and the *lower subscapular nerve* ends in the muscle.

**Subscapularis.** The subscapularis muscle (Fig. 12-32) arises from the costal surface of the scapula, but is usually separated from the neck of that bone by the subscapular bursa. Its broad tendon passes across the front of the shoulder joint, forming the anterior part of the musculotendinous, or rotator, cuff, and inserts into the lesser tubercle of the humerus and its crest. Most of the important nerves and vessels of the axilla lie on the costal surface of the subscapularis. Its tendon of insertion passes behind the coracobrachialis and short head of biceps, as these two muscles arise from the coracoid process.

The lower lateral border of the subscapularis muscle parallels the lateral border of the scapula and is also paralleled by the lower border of the teres minor posteriorly. However, the upper border of the teres major muscle is more nearly horizontal, so between this border, on the one hand, and the lower borders of subscapularis and teres minor, on the other, there is a triangular gap, crossed by the long head of the triceps (Figs. 12-33 and 12-34). The part of this gap between the long head of the triceps and the surgical neck of the humerus is usually called the **quadrangular space.** Through this space the *axillary nerve* and the *posterior humeral circumflex vessels* leave the axilla. The medial part of this gap, between the long head of the triceps and the converging upper and lower borders, is usually called the **triangular space.** The *scapular circumflex vessels* turn around the lower border of the subscapular muscle in this space, but the major parts of the vessels then turn upward through the origin of the teres minor, so that only small twigs appear posteriorly in the space, between teres minor and major.

Branches of the *subscapular artery* enter the costal surface of the subscapularis muscle, and frequently one or more small twigs arising directly from the axillary artery do likewise. While it lies against the muscle, the posterior cord of the brachial plexus gives off the *upper subscapular nerve* into the upper part of the muscle. Farther down it gives off the *lower subscapular nerve* to the lower part of the muscle and to the teres major (Fig. 12-21).

**FIG. 12-32.** The subscapularis muscle.

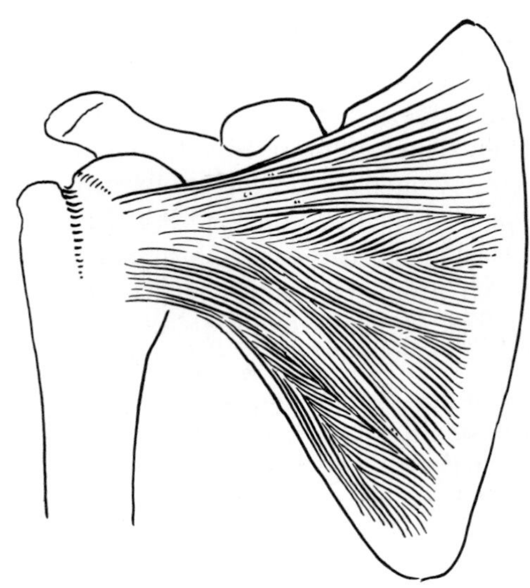

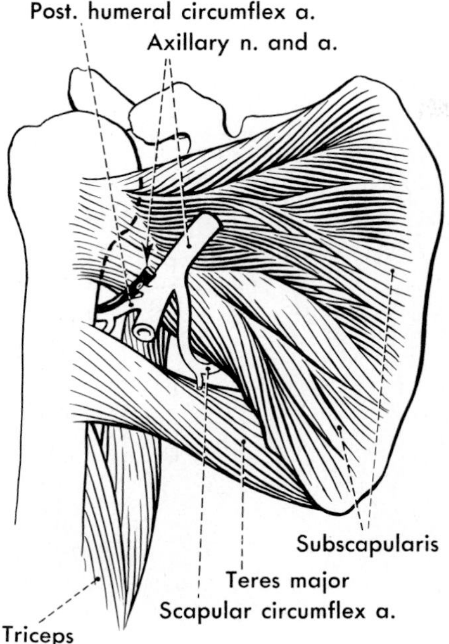


**FIG. 12-33.** The subscapularis muscle and the triangular and quadrangular spaces from the front. The posterior humeral circumflex artery and the axillary nerve transverse the quadrangular space, and the scapular circumflex artery enters the triangular space.

**Variations, Innervations, and Actions.** There are no important variations in the intrinsic shoulder muscles.

The teres minor and deltoid muscles are **innervated** by the axillary nerve, and the supraspinatus and infraspinatus by the suprascapular nerve. The teres major is supplied by the lower subscapular nerve, and the subscapularis by both subscapular nerves. All of these nerves are posterior branches of the brachial plexus and contain fibers derived from C5 and C6.

The **actions** of most of the intrinsic muscles of the shoulder apparently correspond rather well to those that can be deduced from anatomic observations. Most of the muscles have simple actions, but the deltoid, because of its relation to the shoulder joint, is probably the most versatile muscle in the body.

The **supraspinatus** is primarily an *abductor of the arm;* with the arm by the side, it becomes particularly active in supporting the arm against a downward pull. The **infraspinatus** and **teres minor** are both lateral rotators; the latter is also a weak adductor. The teres major is an adductor, medial rotator, and extensor like the latissimus dorsi, but unlike that muscle apparently acts only when the movements are resisted. The **subscapularis** is a medial rotator and by far the most important muscle in this movement. All these muscles except the teres major have another important function, that of keeping the head of the humerus in close contact with the glenoid cavity while the limb is being moved by other muscles.

The most important action of the **deltoid** is to *abduct* the free limb, but since parts of the muscle lie both anterior and posterior to the shoulder joint, these have additional functions: the anterior part can flex and medially rotate the arm, thus helping to move it upward and forward across the thorax, and the posterior part can extend and laterally rotate. Finally, the lower posterior fibers lie below the axis of motion at the shoulder joint when the arm is abducted no more than 45°, and these fibers can adduct from this position. However, it has been said that the primary function of these fibers is probably to prevent the medial rotation produced by the chief adductors, the pectoralis major and latissimus dorsi. Thus, the deltoid can apparently help carry out most of the movements that are possible at the shoulder joint, although it probably does not participate in rotation when only rotation is desired.

**FIG. 12-34.** The triangular and quadrangular spaces from behind.

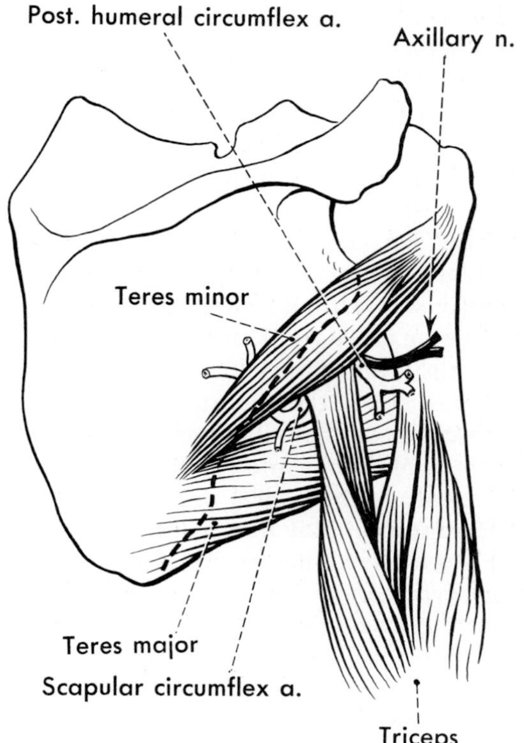

## NERVES AND VESSELS OF SHOULDER

These have been described in the preceding account. All that is necessary here, therefore, is a summary of the deeper nerves and vessels.

### Nerves

The spinal **accessory** emerges from the skull through the jugular foramen, passes downward and backward to the sternocleidomastoid to supply this, and then continues across the posterior triangle of the neck on the outer surface of the levator scapulae to enter the deep surface of the trapezius. The branches from the 3rd and 4th cervical nerves that join the accessory nerve contain sensory fibers for this muscle.

The **nerves to the levator scapulae** arise from the 3rd and 4th cervical nerves and enter the anterior surface of the muscle close to its origin.

The **dorsal scapular nerve** to both rhomboids is typically derived from the 5th cervical nerve close to the intervertebral foramen and runs downward and backward through or across the surface of the middle scalene, parallel to and below the accessory nerve. It then passes medial to or through the levator scapulae (to which it may give a branch) and descends on the deep surface of the rhomboids, parallel to the medial border of the scapula (Fig. 12-35).

The **long thoracic nerve** (Fig. 12-20) is derived usually from the ventral rami of the 5th, 6th, and 7th cervical nerves close to their emergence from the intervertebral foramina. It runs downward and backward through, or with one or more roots in front of or behind, the middle scalene muscle. It soon reaches the uppermost slip of the serratus anterior muscle and descends on the outer surface of this muscle, giving branches into it, for a considerable distance along the lateral thoracic wall.

The **suprascapular nerve** (Fig. 12-36) is derived from the upper trunk of the brachial plexus, typically receiving fibers from the 5th and 6th cervical nerves, and passes downward, laterally, and posteriorly across the lower part of the posterior triangle of the neck to run under cover of the trapezius muscle

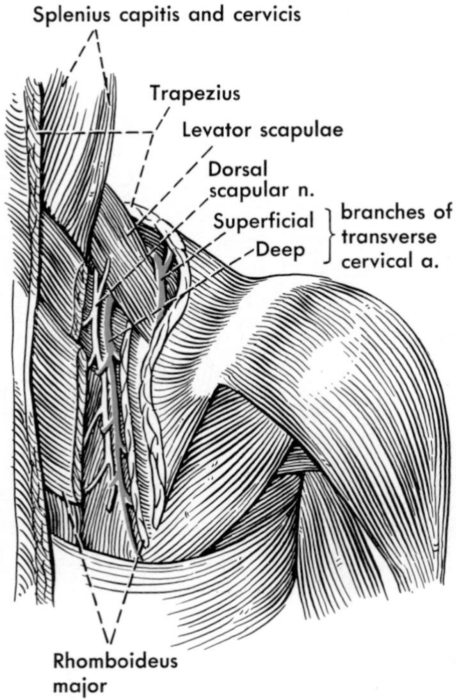


**FIG. 12-35.** The dorsal scapular nerve and branches of the transverse cervical artery.

and, with the suprascapular vein and artery, reach the suprascapular notch. The nerve passes through the notch and descends between the supraspinatus and infraspinatus muscles and the scapula, supplying both muscles.

The **subscapular nerves** and the **thoracodorsal nerve** arise from the posterior cord of the brachial plexus while this is on the costal surface of the subscapularis muscle (Fig. 12-21). Both subscapular nerves usually receive fibers from C5 and C6. The *upper subscapular nerve* passes directly into the upper part of the subscapularis muscle. The *lower subscapular nerve* divides into two branches; one enters the lower part of the subscapularis muscle and the other continues to the teres major.

The thoracodorsal nerve runs downward to the costal surface of the latissimus dorsi. Most of its fibers are derived from C7 with contributions from C6 and C8 and rarely from C5.

The **axillary nerve,** also from the posterior cord, contains fibers primarily from the 5th and 6th cervical nerves. It leaves the axilla by

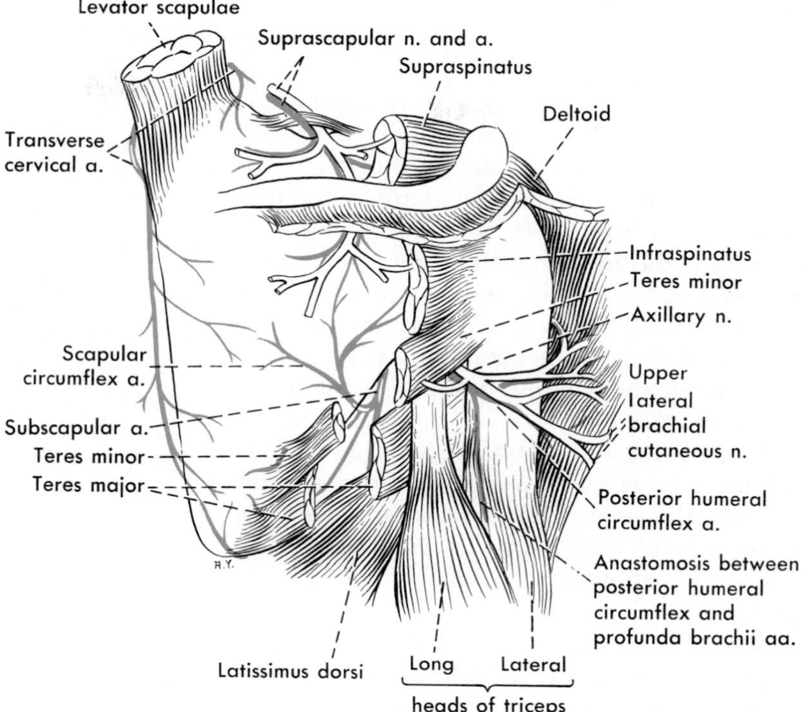


**FIG. 12-36.** Nerves and vessels of the posterior aspect of the shoulder.

passing backward with the posterior humeral circumflex vessels through the quadrangular space. As it passes below the teres minor, it supplies this muscle, and as it reaches the posteromedial aspect of the surgical neck of the humerus (Fig. 12-36), it gives off the *upper lateral brachial cutaneous nerve* to skin overlying the deltoid. It then continues forward around the lateral aspect of the surgical neck of the humerus to supply the entire deltoid muscle.

## Blood Vessels

Veins and arteries of the shoulder parallel each other, except that the cephalic vein, along the anterior border of the deltoid muscle, is accompanied only in its upper part by an artery (the deltoid branch of the thoracoacromial). Also, veins corresponding to parts of the thoracoacromial artery empty into the cephalic vein and the axillary instead of forming a thoracoacromial vein. Other than these two exceptions, the descriptions given here for the arteries hold also for the veins.

As in the case of the nerves, some of the blood vessels to the shoulder arise in the neck, and others arise in the axilla. The arteries arising in the neck are the **transverse cervical** (*transversa colli*) and **suprascapular arteries,** typically both branches of the thyrocervical trunk, a branch of the subclavian artery. These vessels run laterally and then posteriorly across the lower part of the neck, with the transverse cervical lying above the suprascapular (Fig. 12-21). In the back of the neck, the transverse cervical artery passes more medially and, as it reaches the levator scapulae muscle, divides into a *superficial branch* that runs on the deep surface of the trapezius and a *deep branch* that descends on the deep surface of the rhomboids with the dorsal scapular nerve (Fig. 12-35). A common occurrence, which does not affect the distribution of the vessels, is for the two branches of the transverse cervical artery to arise separately, in which case they receive different names (see the account of the thyrocervical trunk in Chap. 30).

The suprascapular artery (Figs. 12-22 and 12-36) is joined by the suprascapular nerve in its course backward toward the scapular notch. It runs above the superior transverse scapular ligament, which separates it from the nerve, but then joins the nerve in ramifying in the supraspinatous and infraspinatous fossae. It anastomoses with the deep branch of the transverse cervical artery and with the scapular circumflex branch of the subscapular artery.

The deltoid branch of the **thoracoacromial artery,** a branch of the axillary, helps to supply the deltoid muscle. The **lateral thoracic artery,** another branch of the axillary, supplies, particularly, an upper part of the serratus anterior.

The **subscapular artery** (Figs. 12-22 and 12-36) arises from the lower part of the axillary artery and soon divides into two branches, the *thoracodorsal* and *scapular circumflex.* The **thoracodorsal artery** continues downward on the subscapularis muscle, supplying this and the teres major, and then usually divides into two subsidiary branches that continue on the costal surface of the latissimus dorsi and the external surface of the serratus anterior. The **scapular circumflex artery** turns dorsally around the lower border of the subscapularis muscle, through the triangular space, and penetrates the origin of the teres minor. There it turns upward between the infraspinatus muscle and the scapula to supply this muscle and anastomose with the suprascapular artery and the deep branch of the transverse cervical.

The **posterior humeral circumflex artery** is the last important artery of the shoulder. It arises from the lower part of the axillary artery and passes posteriorly through the quadrangular space with the axillary nerve. It runs around the surgical neck of the humerus, giving off branches into the deltoid muscle, and ends by anastomosing with the much smaller anterior humeral circumflex artery.

#### Lymphatics

The deep lymphatic vessels of the shoulder in general follow the blood vessels and, therefore, drain largely into axillary lymph nodes. However, some of them follow the transverse cervical and suprascapular vessels and drain into lymph nodes of the lower deep cervical group that lie on these vessels under cover of the trapezius and in the posterior triangle of the neck.

## SHOULDER JOINT

The shoulder joint, formed by the **glenoid cavity of the scapula** and the **head of the humerus,** is completely covered anteriorly, laterally, and posteriorly by the deltoid muscle. Under cover of this muscle, the coracobrachialis muscle and the short head of the biceps (Fig. 12-42), arising together from the coracoid process, are related to the joint anteromedially; the long head of the biceps enters the joint anteriorly, lying in the intertubercular groove; and posteromedially is the long head of the triceps, arising from the infraglenoid tubercle. Still more intimately related to the joint are four muscles that together form a *musculotendinous cuff* over all of the joint except its inferior aspect. Since most of these muscles are also rotators, the musculotendinous cuff is likewise called the **rotator cuff.** The *subscapularis muscle* and *tendon* form the anterior part of the cuff; the *supraspinatus* forms the uppermost part; the *infraspinatus* forms the upper posterior part; and the *teres minor* forms the lower posterior part (Fig. 12-37). The subscapularis inserts upon the lesser tubercle, and the other three muscles insert upon the greater tubercle.

The upper surface of the musculotendinous cuff, particularly the outer surface of the supraspinatus muscle, is separated from the overlying acromion, coracoacromial ligament, and upper part of the deltoid muscle by an important bursa, the **subacromial bursa** (Figs. 12-37 and 12-38). Lateral to this there may be a second bursa, the **subdeltoid,** but the two bursae are frequently fused into one, and a single bursa here is sometimes called subacromial, sometimes subdeltoid.

Calcium deposits in the tendinous floor of the bursa are a not infrequent cause of disability of the shoulder, provoking a *bursitis* that makes any movement of the arm at the shoulder painful.

Tears of one or more tendons of the musculotendinous cuff, either partial or complete, are fairly common, particularly in older people. The supraspinatus tendon is the most frequently torn, but tears of the infraspinatus, the teres minor, and the subscapularis also occur.

When the musculotendinous cuff is stripped away, the **fibrous capsule** of the shoulder joint is found to be very thin, with usually only one ligamentous thickening being visible externally (Fig. 12-4). This is the **coracohumeral ligament,** a band that arises from the lateral edge of the coracoid process and extends over the top of the shoulder joint to attach to the greater tubercle. Its thickened anterior edge is fairly prominent, but the posterior edge fades gradually into the capsule. The coracohumeral ligament apparently prevents downward displacement of the humeral head when the arm is by the side. If there is a heavy downward pull on the arm, the supraspinatus muscle, and to some extent the posterior fibers of the deltoid, will contract to help the ligament. With the arm abducted, the ligament is lax and the strength of the joint depends entirely on the muscles of the musculotendinous cuff.

The fibrous layers and the closely adjacent **synovial layers** of the capsule are attached to the humerus for the most part at the edge of the articular cartilage of the head, but between the two tubercles they send downward a diverticulum, the *intertubercular synovial sheath,* around the tendon of the long head of the biceps. Across the front of this diverticulum is a **transverse humeral ligament** extending between the two tubercles. On the scapula, the capsule blends with the **glenoidal labrum.** This is a fibrocartilaginous ring attached to the rim of the glenoid cavity that increases the cavity's diameter and deepens slightly its concavity. In front of the glenoidal labrum, extending medially to lie between the neck of the scapula and the subscapularis muscle, there is usually an extension of the joint cavity, the **subscapular bursa** or recess. However, a subscapular bursa may have no communication with the joint cavity, or it may have two.

Partial tearing of the labrum from the bony rim is frequently found in older people. Such separation is especially common at the upper part close to the attachment of the long head of the biceps.

When the shoulder joint is opened from behind so that the interior of its anterior surface can be viewed, two or three thickenings can often be seen. These thickenings are known as the **glenohumeral ligaments,** but it is doubtful they are of any importance; any one of the three may be absent or, when present, may be poorly developed.

**FIG. 12-37.** Lateral view of the shoulder joint to show the composition of the musculotendinous or rotator cuff. Synovial membrane is red.

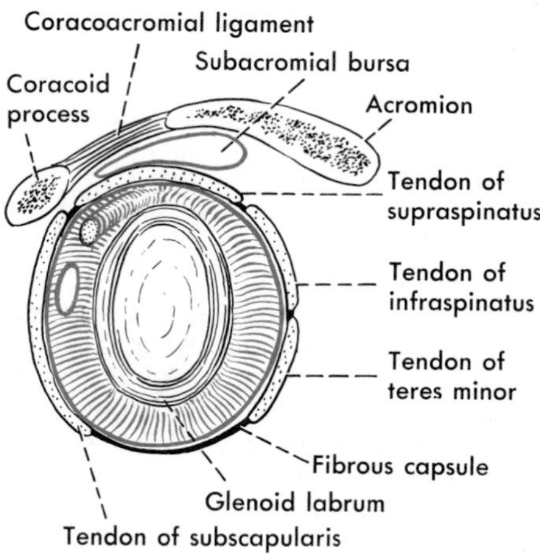


**FIG. 12-38.** Frontal section through the shoulder joint. Synovial membrane is red.

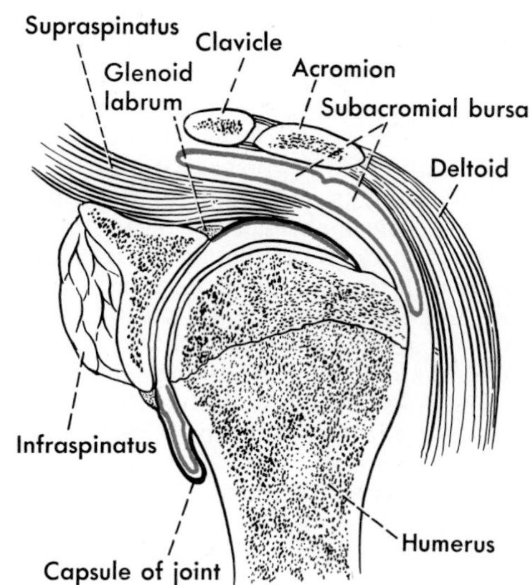

Fraying of the **long head of the biceps** as it passes between the two tubercles within the shoulder joint is fairly frequently seen and may lead to complete *rupture* of the tendon. Dislocation of the biceps tendon medially, over the lesser tubercle, sometimes also occurs.

The **blood supply** to the shoulder joint is primarily by twigs from the *suprascapular, subscapular,* and both *humeral circumflex arteries.* The **nerve supply** is usually from the *axillary, suprascapular,* and *lateral pectoral nerves,* with sometimes a twig from the posterior cord or the radial nerve.

#### Movements at Shoulder Joint

The glenoid cavity typically points somewhat forward. The plane of the scapula in the resting position makes an angle of about 30° with the coronal plane of the body, so humeral movements can be defined in terms of their relation either to the scapula or to the major planes of the body. The latter relation is generally used. Thus, abduction of the arm is moving it away from the side approximately in the coronal plane, and adduction is moving it back to the side. Flexion is a forward movement in the sagittal plane, and extension is the reverse. Rotation, not affected by movement of the scapula, is medial (or internal) when the anterior surface of the humerus turns medially, lateral (or external) when it turns laterally.

Many natural movements at the shoulder joint are actually combinations of some of these. For instance, the frequent movement of flexing the arm and at the same time bringing it upward across the thorax is a combination of flexion, adduction, and medial rotation. Flexion and abduction allow the greatest range of movement, for in either of these the arm can be raised until it is directed vertically above the shoulder.

It must be borne in mind that upward rotation of the scapula is necessary to full abduction of the humerus. Normally, therefore, the trapezius and serratus anterior muscles upwardly rotate the scapula at the same time that the deltoid and supraspinatus are abducting the humerus. The smooth movement of both bones, in which the scapula is said to be rotated 1° to 2° upward for each 2° to 3° of humeral abduction, is referred to as the *scapulohumeral rhythm.* This is perhaps the most striking example of the coordination of scapular and humeral movements, but the two bones regularly move together also in other movements: during flexion of the arm the scapula moves so that the glenoid cavity is pointed more forward; during extension it is retracted so that the cavity points more laterally; and in a downward-reaching movement the scapula rotates downward.

The muscles that move the scapula must also stabilize it so that it affords a firm base upon which the arm can move. Further, some of the longer muscles acting upon the humerus tend to dislocate it from the glenoid cavity. Hence, the muscles most intimately related to the head of the humerus, those forming the musculotendinous cuff, have the important function of maintaining firm contact between scapula and humerus in addition to their participation in movements of the humerus. This is simply another example of the importance of fixator muscles (Chap. 6) in assisting prime movers at many joints.

The muscles producing **flexion** at the shoulder joint are the anterior part of the deltoid, the pectoralis major, the coracobrachialis (a muscle of the arm), and perhaps the biceps brachii (another muscle of the arm). Of these, the anterior part of the deltoid is the most important, capable of bringing about the full range of movement. The clavicular part of the pectoralis major produces a less extensive movement, and the remainder of the pectoralis major flexes only from a hyperextended position. The coracobrachialis adducts much more strongly than it flexes. Both heads of the biceps may contract slightly during flexion at the shoulder, but at most, apparently, it has only a very weak action at the shoulder.

The **extensors** of the arm are usually regarded as being the latissimus dorsi, the posterior part of the deltoid, the sternal part of the pectoralis major, the teres major, and the long head of the triceps (a muscle of the arm). Of these, the latissimus dorsi is especially powerful in extending the arm from a flexed position, and the pectoralis major can carry out extension only from a flexed position. The latissimus helps to hyperextend, but the posterior part of the deltoid moves the humerus farther into extension than does the latissimus. The teres major and the long head of the triceps both help to extend the arm a little beyond the resting position, but apparently contract weakly in pure extension.

There are only two **abductors** of the humerus, the supraspinatus and the deltoid. The supraspinatus may be able to abduct to the horizontal level but often cannot abduct beyond 45° when the deltoid is paralyzed. The deltoid, a much more powerful abductor, regularly can abduct to about the horizontal level. When the arm is laterally rotated, the long head of the biceps passes laterally rather than anteriorly across the shoulder joint and can theoretically help in abduction; however, its contraction in this situation is probably only to protect against hyperextension at the elbow.

The most powerful **adductors** are the pectoralis major, especially the sternocostal part, and the latissimus dorsi, assisted by the teres major. The posterior part of the deltoid may be able to adduct when the arm is about 45° from the side, and when adduction is resisted, the coracobrachialis and the long head of the triceps contract strongly. Contraction of these muscles usually is interpreted as being in part synergistic, to aid the prime movers, and in part a resistance to the downward dislocation of the head of the humerus that the pectoralis and latissimus tend to produce—therefore, in part, contraction of fixator muscles. This double function may also be true of the deltoid.

The chief **lateral rotators** are the infraspinatus and the teres minor. The posterior part of the deltoid also laterally rotates, but apparently only when it is also extending. The chief **medial rotator** is the subscapularis muscle. The teres major and the latissimus dorsi apparently contribute to medial rotation primarily as they extend and adduct, and the anterior part of the deltoid and the pectoralis major contribute primarily as they flex and adduct. The supraspinatus is sometimes listed as a lateral rotator, sometimes as a medial one, but it is said that isolated stimulation of it produces a very slight and weak medial rotation during abduction.

As may be seen from examination of the probable segmental **innervations** of these various muscle groups, the flexors, the abductors, and the lateral rotators are innervated for the most part by the 5th and 6th cervical segments, thus from the upper part of the brachial plexus. The extensors, the adductors, and the medial rotators typically receive some of their nerve supply from the 7th cervical segment or lower, thus from a lower part of the plexus. As a result, an injury to the upper trunk of the plexus or to the 5th and 6th cervical nerves before they join the plexus will paralyze most of the flexors, abductors, and lateral rotators, while sparing in part the muscles of opposite action. A lesion of this type, called *Erb-Duchenne paralysis*, will be accompanied by a typical position of the humerus—somewhat adducted, extended, and medially rotated by the predominant pull of the less-injured muscles. Flexion at the elbow is also particularly weak in this type of paralysis.

## ARM

The humerus, the bone of the arm, has already been described (p. 171). In a study of the muscles, nerves, and vessels of the arm, only a rudimentary knowledge of the elbow joint and of the bones of the forearm is required. The two bones articulating with the humerus at the elbow are the radius and the ulna. These bones are described more fully in Chapter 13 and are illustrated in Figures 13-5 and 13-6.

The upper end of the **radius,** the *head,* is disklike, but slightly concave proximally and covered with articular cartilage. Its upper surface articulates with the *capitulum of the humerus,* whereas the circular articular surface on its margin articulates with the *annular ligament of the radius* (lig. annulare radii) and the *radial notch of the ulna,* which together encircle the head. Thus, the radius can move with the ulna in flexion and extension of the elbow and can rotate on the ulna and humerus within the circle formed by the annular ligament and the radial notch (see Figs. 13-63 and 13-64). Rotation of the radius carries the hand with it and is described as pronation when the horizontally held hand is turned palm downward, supination when it is turned upward. The narrowed portion of the radius immediately succeeding the head is the *neck.* A little below the neck on the ulnar side of the body there is a raised projection, the *radial tuberosity,* that receives the insertion of the biceps brachii muscle.

The **ulna** bears anteriorly, a little below its upper end, a deep *trochlear notch* (incisure). Proximal to this notch, producing the prominent posterior projection at the elbow, is the *olecranon.* The distal border of the notch is the *coronoid process,* which bears at its base the *ulnar tuberosity.* The trochlear notch is continuous on its radial side with a *radial incisure,* which completes the circle around the head of the radius. The trochlear incisure articulates with the *trochlea of the humerus,* thus forming essentially a hinge joint that allows little movement except flexion and extension. The olecranon receives the attachment of the *triceps brachii muscle,* the chief extensor of the arm, and the tuberosity receives the insertion of the *brachialis muscle,* a flexor.

### Fascia, Superficial Vessels, and Nerves

The **superficial fascia** of the arm contains a variable amount of fat, veins, and a number of cutaneous nerves. It is continuous with superficial fascia of the shoulder and forearm.

The **brachial** (deep) **fascia** forms a heavy investing sheath for the arm. It is continuous above with the fascia over the deltoid and pectoral muscles and with the axillary fascia. At

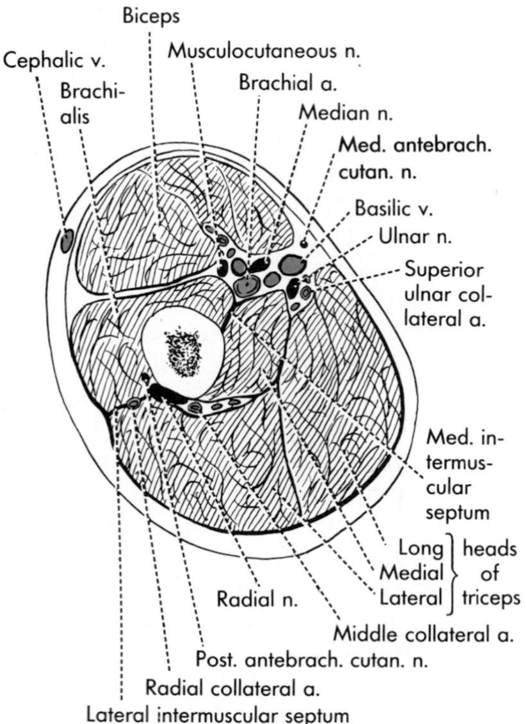


**FIG. 12-39.** Cross section through the middle third of the right arm seen from below. (Redrawn from Eycleshymer AC, Schoemaker DM: A Cross-Section Anatomy. New York, Appleton, 1923)

the elbow, it blends posteriorly and laterally with the subcutaneous projections of bone and anteriorly continues directly into the fascia of the forearm. Between the insertion of the deltoid muscle and the lateral epicondyle, the brachial fascia sends a **lateral intermuscular septum** to the humerus (Fig. 12-39). On the medial side there is a **medial intermuscular septum.** Through these two septa, the major part of the arm is divided into two compartments. The anterior compartment contains the flexor muscles of the arm, and the posterior compartment contains the extensor muscle, the triceps. Over the flexor muscles, the deep fascia is loose, but over the triceps, this fascia is firmly fused to the surface of the muscle.

There are two named superficial veins in the arm, the cephalic and the basilic (see Fig. 11-15). The **cephalic vein** passes upward from the forearm in the superficial fascia. It lies first on the lateral side of the front of the elbow, then in the groove lateral to the biceps, and then in the groove between the deltoid and pectoral muscles, where it passes deeply to join the axillary vein. Its course in the arm is, therefore, entirely superficial. It receives subcutaneous twigs, but no vessels of particular importance.

The **basilic vein** passes from forearm to arm in the superficial fascia on the medial side of the front of the elbow, receiving here one or more communications, the *vena intermediana cubiti,* from the cephalic. At about the junction of the middle and lower thirds of the arm, the basilic vein penetrates the brachial fascia to lie in the anterior compartment of the arm. The basilic vein continues as the axillary vein.

The especially numerous **superficial lymphatics** of the medial side of the arm are mostly a continuation upward of superficial lymphatics from both surfaces of the forearm. They form a series of largely parallel trunks that empty into the lateral (lower) lymph nodes of the axillary group. They are joined by vessels from the back of the arm, some of which run around the lateral aspect of the arm, some around the medial aspect, leaving a lymphatic "divide" along the posterior surface of the arm (see Fig. 11-17). There are typically one or two superficial cubital (supratrochlear) lymph nodes (see 7 and 8 in Fig. 11-16) alongside the basilic vein a little above the medial epicondyle. These receive some of the lymphatics from the ulnar side of the forearm and give rise to lymphatics that run with others along the medial side of the arm. There are said to be numerous communications between the superficial and deep lymphatics of the arm that accompany the brachial vessels, and one or more lymphatics usually follow the cephalic vein to empty into lymph nodes at the apex of the axilla.

The **cutaneous nerves** of the arm are numerous (Figs. 12-40 and 12-41). The *superior lateral brachial cutaneous nerve,* the cutaneous branch of the axillary, supplies largely the skin over the deltoid muscle. Below this, the *inferior lateral brachial cutaneous nerve,* a branch of the radial, emerges close to or with the larger posterior antebrachial cutaneous along the line of the lateral intermuscular septum. It supplies skin of the lateral aspect of the lower part of the arm, whereas the posterior antebrachial proceeds to the forearm. The posterior aspect of the arm is supplied by a branch of the radial, the *posterior brachial cutaneous nerve,* arising in the axilla. The upper medial side of the arm is supplied by the *intercostobrachial nerve,* usually from the 2nd intercostal, and by the *medial brachial cutaneous nerve,* an independent branch of the brachial plexus. The lower medial and anterior sides are supplied by upper branches of the *medial antebrachial cutaneous nerve,* which penetrates the brachial fascia with the basilic vein.

## ANTERIOR MUSCLES AND RELATED STRUCTURES

### Muscles

The anterior muscles of the arm (Figs. 12-42 and 12-44) are only three, the **coracobrachialis,** the **biceps brachii,** and the **brachialis.** Of these, the coracobrachialis crosses the shoulder joint only; the biceps brachii crosses both the shoulder and the elbow joint; and the brachialis crosses the elbow joint only. All three muscles are supplied by the musculocutaneous nerve, from the lateral cord.

**Biceps Brachii.** The biceps brachii has, as its name implies, two heads of origin (Fig. 12-43), a long and a short head. The upper ends of both heads are under cover of the deltoid muscle, but below this the two heads are close together and prominent. The musculocutaneous nerve lies behind the biceps. The brachial artery and veins, the basilic vein, the median nerve, and the medial antebrachial cutaneous nerve all lie medially, somewhat in the groove between the biceps and the brachialis (Fig. 12-45); in the upper part of the arm, the ulnar nerve occupies a similar position, but diverges backward to pass through the medial intermuscular septum and lie in the posterior compartment of the arm.

The **short head of the biceps** has a tendinous origin, fused with the front of the coracobrachialis muscle, from the tip of the coracoid process. The **long head** arises from the supraglenoid tubercle of the scapula by means of a long rounded tendon that passes through the shoulder joint covered by a reflection of synovial membrane. It carries with it a diverticulum of the shoulder joint, the *intertubercular synovial sheath,* as it passes downward in the intertubercular groove. Long and short heads form muscular bellies that unite a little above the elbow. The rounded **tendon** of the muscle passes deeply into the cubital fossa to attach to the tuberosity of the radius. Just before the tendon attaches to the radius, there is a *bicipitoradial bursa* between the two; there also may be an *interosseous cubital bursa* medial to the tendon. The other insertion of the biceps, the **bicipital aponeurosis** (formerly *lacertus fibrosus*), arises from the medial border of the bicipital tendon and from the lower medial portion of the muscle. It has a thickened medial edge that can be plainly felt just above the flexed elbow and passes downward and medially to fuse with the deep fascia over the upper medial side of the forearm and hence reach the ulna.

The *musculocutaneous nerve* sends a branch into the deep surface of each belly. The cutaneous part of the nerve, the *lateral antebrachial cutaneous nerve,* emerges from behind the biceps just lateral to the bicipital tendon.

**Coracobrachialis.** The coracobrachialis muscle arises with the short head of the biceps from the coracoid process and passes downward, separating from the biceps, to insert into the medial border of the body of the humerus about its middle. As the musculocutaneous nerve leaves the medial side of the arm, it does so by penetrating the coracobrachialis muscle, usually sending several twigs into the muscle.

**Brachialis.** The brachialis muscle arises from about the lower half to two-thirds of the front of the body of the humerus, and from both intermuscular septa. It covers the lower part of the front of the humerus and is still broad as it passes across the front of the elbow joint, behind the tendon of the biceps, to insert into the ulnar tuberosity.

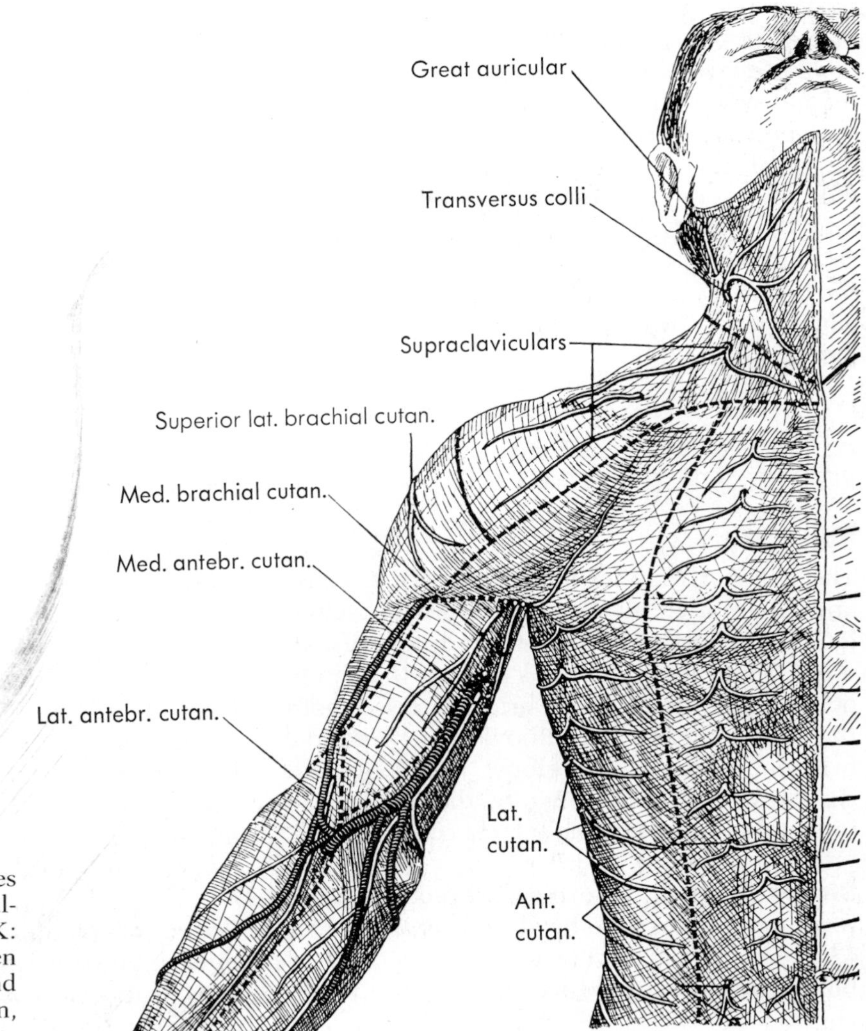


**FIG. 12-40.** Cutaneous nerves of the front of the arm, shoulder, and thorax. (Corning HK: Lehrbuch der topographischen Anatomie für Studierende und Ärtze. Munich, JF Bergmann, 1923)

In the lower part of the arm, the *brachial artery* and its accompanying veins, and the median nerve, pass onto the anterior aspect of the brachialis muscle. They run downward and backward in the cubital fossa posteromedial to the tendon of the biceps (and under cover of the bicipital aponeurosis), with the *median nerve* lying medial to the brachial artery. The *musculocutaneous nerve* lies on the front of the brachialis and sends a branch into the muscle. Behind the lower lateral part of the brachialis, between it and the upper muscles of the extensor forearm group, is the *radial nerve.* It frequently gives off a small twig into the lateral part of the brachialis. This twig is often an articular branch to the elbow joint, but anatomists disagree as to whether it is also sometimes motor to a small part of the brachialis.

**Variations, Innervations, and Actions.** The common variation in the coracobrachialis is its length; occasionally it will almost reach the medial epicondyle. An accessory muscular slip to the bi-

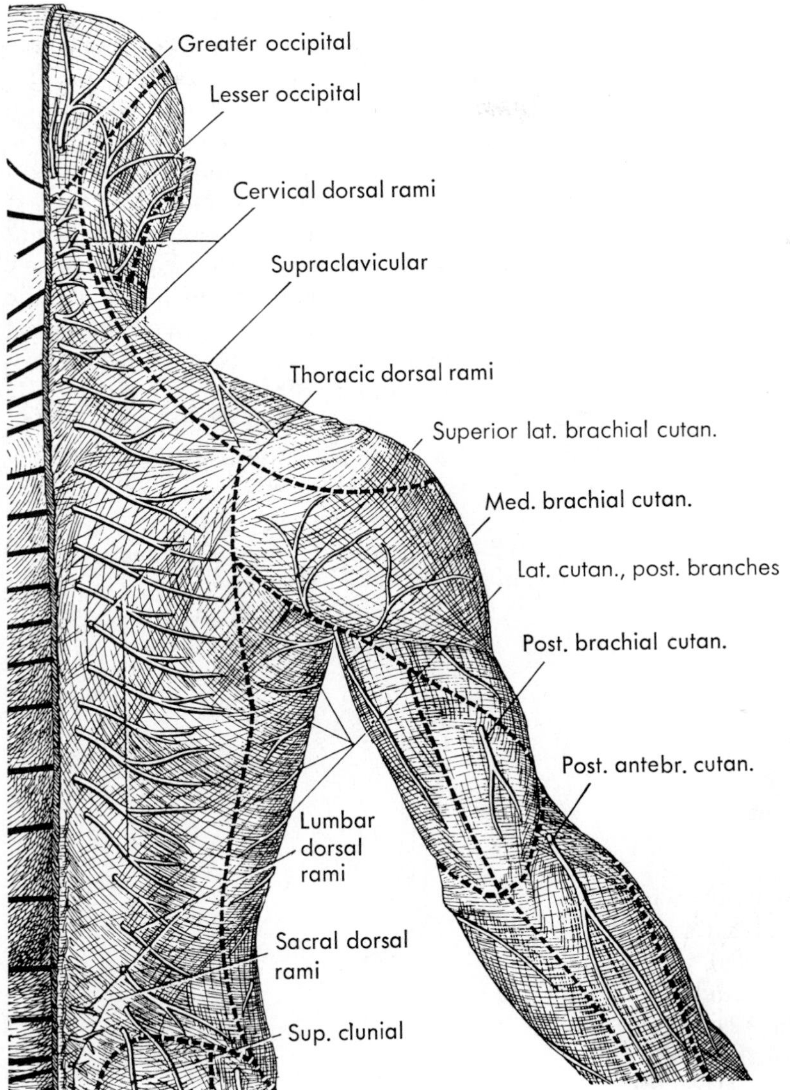


**FIG. 12-41.** Cutaneous nerves of the back of the arm, shoulder, and trunk. (Corning HK: Lehrbuch der topographischen Anatomie für Studierende und Ärtze. Munich, JF Bergmann, 1923)

ceps, usually small, may arise from the humerus or, more rarely, from the pectoralis major.

Through the musculocutaneous nerve, the biceps and brachialis usually receive fibers from C5 and C6, the coracobrachialis from C5 to C7.

The **coracobrachialis** both adducts and flexes the arm, but is a better adductor than flexor. In consequence of its insertion, the **biceps** is both a *flexor* of the forearm and a *supinator.* It necessarily does both at once, however, and therefore typically does not aid materially in supination of the extended forearm or in flexion of the pronated one. Since both heads of the biceps attach to the scapula, the biceps has been regarded also as a flexor at the shoulder, although it participates rather weakly in this action, if at all. The **brachialis** flexes the forearm.

## Nerves and Vessels

**Nerves.** All four major nerves that continue into the arm—the musculocutaneous, the median, the ulnar, and the radial—appear in the anterior compartment of the arm. The musculocutaneous enters the anterior compartment by penetrating the coracobrachialis muscle,

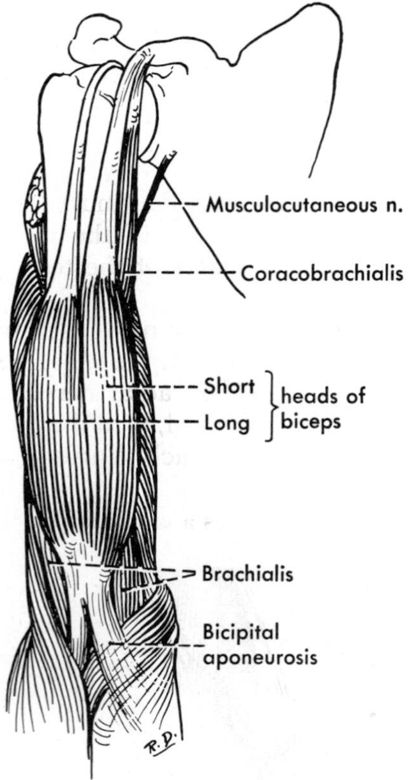


**FIG. 12-42.** Muscles of the front of the arm.

the median and the ulnar run downward with the brachial artery, and the radial enters its lower part by perforating the lateral intermuscular septum to lie against the brachialis.

In summary of these nerves (Fig. 12-45), the **musculocutaneous,** a derivative of the lateral cord, diverges laterally from the rest of the brachial plexus while it is in the axilla, gives off one or more branches to the coracobrachialis muscle, and passes through this muscle to lie between the biceps and brachialis muscles. Here it supplies both muscles and then continues downward to emerge lateral to the tendon of the biceps as the **lateral antebrachial cutaneous nerve.** While it lies in the arm, the musculocutaneous nerve sometimes gives off a branch that joins the median nerve. This usually is interpreted as meaning that fibers that should have run through the lateral root of the median nerve failed to do so but entered, instead, the musculocutaneous and are here rejoining the median nerve. Much less frequently, the musculocutaneous receives a communication from the median nerve.

The **median nerve,** formed on the anterolateral aspect of the axillary artery by the union of its lateral and medial roots (from lateral and medial cords), runs downward on the axillary artery and its continuation, the brachial artery. It gradually crosses the brachial artery anteriorly to lie medial to the artery at the elbow. It gives off no branches to structures of the arm. Its first muscular branch, to the pronator teres (a muscle of the forearm), typically arises at about the level of the medial epicondyle, and a branch to the front of the elbow

**FIG. 12-43.** Origins (red) and insertions (blue) of the anterior muscles of the arm.

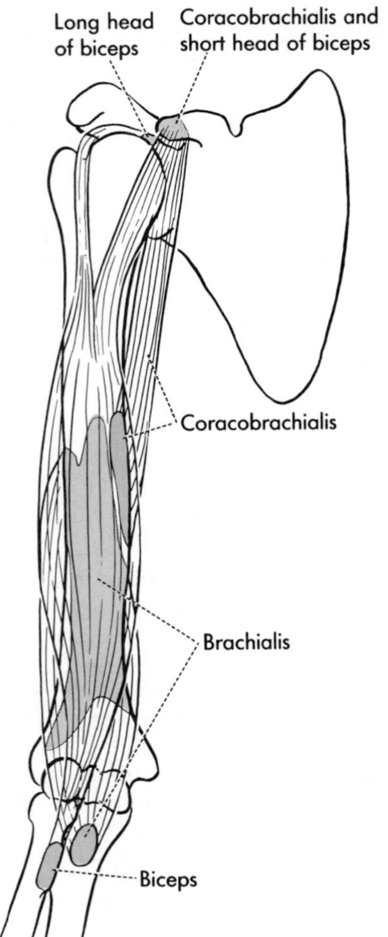

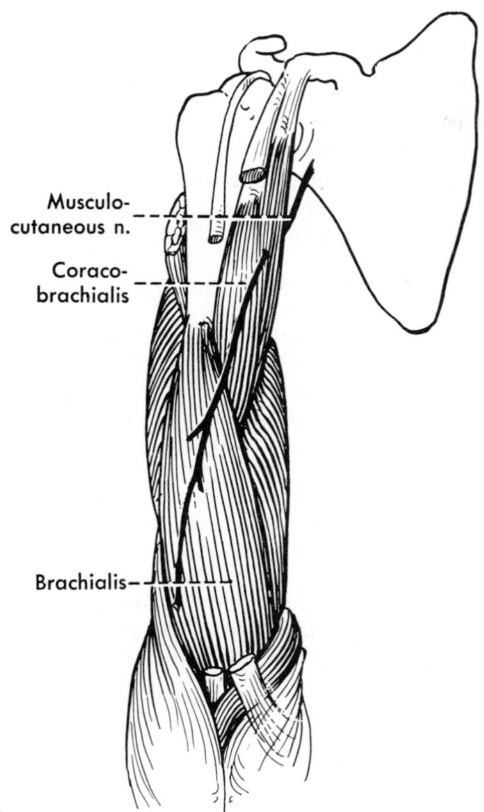


**FIG. 12-44.** Coracobrachialis and brachialis muscles.

joint may also arise here. The nerve and the brachial artery follow the anterior surface of the brachialis muscle into the cubital fossa.

The **ulnar nerve,** from the medial cord of the brachial plexus, runs downward posteromedial to the brachial artery, thus paralleling the median nerve. At about the middle of the arm, it passes through the medial intermuscular septum to lie on the medial aspect of the triceps muscle and pass behind the medial epicondyle into the forearm.

The **radial nerve,** after leaving the axilla, spirals posteriorly around the humerus, but reenters the anterior compartment to pass downward and forward behind the lateral part of the brachialis. Except for the twig from the radial nerve into the brachialis, the branches arising from this part of the radial nerve are to muscles of the extensor forearm group.

In short, therefore, the only nerve that is important to the function of the anterior muscles of the arm is the musculocutaneous nerve.

**Vessels.** The **brachial artery** (Figs. 12-45 and 12-49) is the direct continuation of the axillary artery. The level of the change in name is the lower border of the teres major muscle.

At this level, the median nerve is somewhat anterolateral to the vessel, the ulnar nerve is posteromedial, and the radial nerve is posterolateral. First the radial, then the ulnar, diverge posteriorly. The brachial artery and the

**FIG. 12-45.** Chief nerves and vessels of the arm.

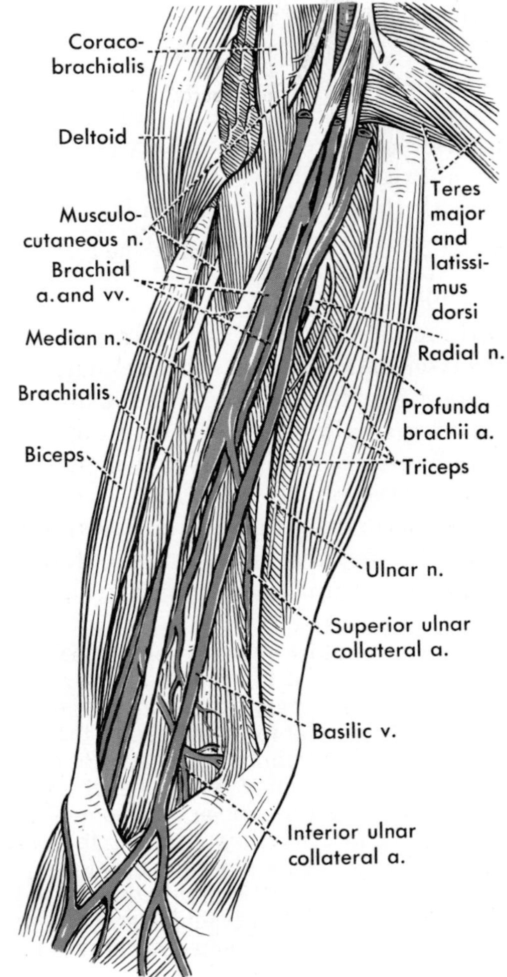

median nerve enter the cubital fossa on the front of the brachialis muscle, where the artery ends by dividing into radial and ulnar arteries.

The first major branch of the brachial artery is the **profunda brachii** (deep brachial), which leaves the artery to pass backward in company with the radial nerve. Its course is described with the structures of the posterior compartment. Below the profunda brachii, the brachial artery gives off **muscular branches,** usually gives rise to a **nutrient artery** that enters the humerus close to the insertion of the coracobrachialis, and gives rise to two other named branches, the superior and inferior ulnar collateral arteries. The **superior ulnar collateral artery** arises about the middle of the arm and passes posteriorly along the medial side of the arm with the ulnar nerve. This artery runs with the nerve behind the medial epicondyle and anastomoses with a recurrent branch from the forearm. The **inferior ulnar collateral artery** arises only a little above the elbow. It typically anastomoses with the superior ulnar collateral and other vessels above the elbow and with a recurrent branch ascending from the forearm (Fig. 12-49).

The brachial artery is typically accompanied by two **brachial veins,** which anastomose with each other freely around the artery and may form a single vein instead of emptying separately into the axillary as they usually do. In the upper two-thirds of the arm, the basilic vein lies superficial to the artery (on its medial side), and the medial antebrachial cutaneous nerve lies between it and the basilic vein.

**Anomalous brachial vessels** sometimes result from a high division of the brachial artery (more rarely of the axillary artery), with two arteries proceeding into the cubital fossa instead of the usual one. In such apparent doubling, whether it occurs at an axillary or a brachial level, one of the two arteries typically lies superficial (anteromedial) to the median nerve. The other, taking the usual course of the brachial, is crossed superficially by the nerve. The deep artery is then regarded as the normal continuation of the brachial, and the more superficially lying artery is named according to its course and branches in the forearm.

Superficial arteries in the arm are relatively common. They have been reported in about 30% of bodies. They are usually unilateral, but sometimes bilateral, and occur in about 18% of limbs. Only when both vessels are followed into the forearm can the proper naming of the superficial vessel be decided upon. It is properly called a *superficial brachial artery* if it bifurcates into radial and ulnar arteries, as the brachial usually does (in this case the arterial stem in the normal position of the brachial is continued as the common interosseous artery).

The artery is properly called a *superficial radial* (sometimes referred to as a high origin of the radial) if it is continued in the forearm as the radial artery. Superficial radial arteries constitute about 75% of all superficial arteries of the arm.

A *superficial ulnar artery,* more common than a superficial brachial artery, is one that continues into the forearm as the ulnar artery (these frequently have abnormal courses here). Occasionally, also, true doubling of the brachial artery is found; the two channels unite in the cubital fossa before the terminal branches arise.

## POSTERIOR MUSCLES AND RELATED STRUCTURES

The important posterior muscle of the arm is the **triceps brachii,** which covers the posterior aspect of the humerus below the deltoid. Associated with the triceps at the elbow is the small **anconeus.**

### Muscles

**Triceps Brachii.** The three heads of origin of the triceps implied by its name are the long head, the lateral head, and the medial head (Figs. 12-46 and 12-47). The **long head** arises by a strong tendon from the infraglenoid tubercle of the scapula and passes downward between the teres minor and the teres major. Here it separates the quadrangular space from the triangular space. The **lateral head** arises from the posterior surface of the body of the humerus above, lateral to the groove for the radial nerve, and from an upper part of the lateral intermuscular septum. The lateral head and the long head soon fuse together to form the superficial part of the triceps muscle. The **medial head** arises from the entire posterior aspect of the humerus below the groove

Then they run downward and laterally around the posterior aspect of the humerus. They first pass posteriorly around the humerus between the humerus and the long head of the triceps and then spiral downward and laterally between the lateral and medial heads, typically resting on upper fibers of the latter, to reach the lateral intermuscular septum. In this course, both nerve and artery give off branches to all three heads of the triceps. As the nerve reaches the lateral border of the triceps muscle, it and a branch of the artery penetrate the lateral intermuscular septum and course downward in front of the elbow

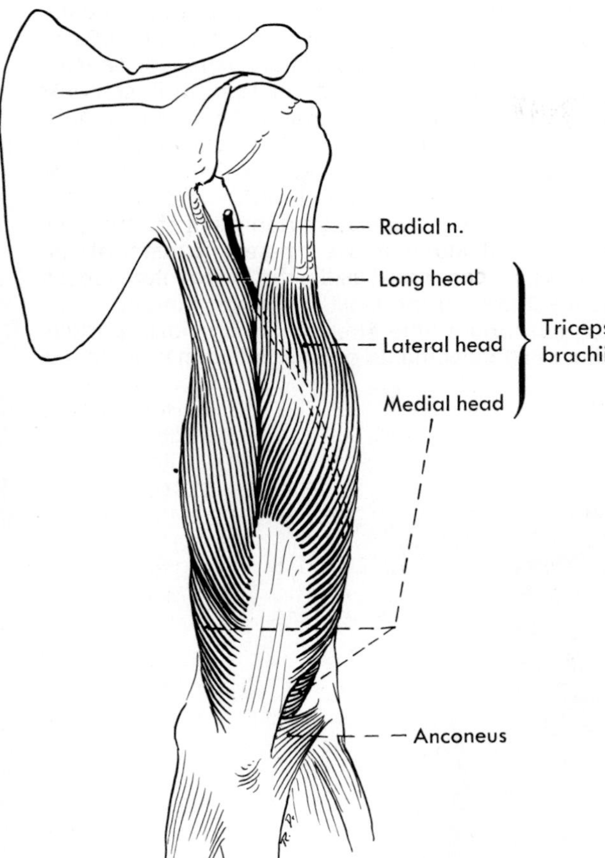


**FIG. 12-46.** Triceps and anconeus muscles.

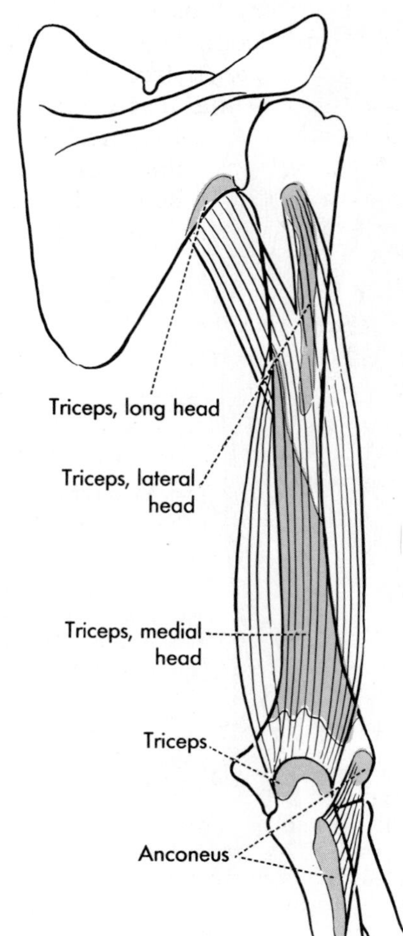


**FIG. 12-47.** Origins (red) and insertions (blue) of the triceps and anconeus muscles.

for the radial nerve, from the medial intermuscular septum, and from the lateral intermuscular septum below the point at which the radial nerve passes through this septum. It attaches to the deep surface of the combined long and lateral heads. From this union there is formed a strong, flattened tendon that passes across the posterior aspect of the elbow joint (it is separated from the elbow joint by the subtendinous bursa of the triceps muscle) to insert into the end of the olecranon. A small slip from the medial head, the *articularis cubiti,* may attach to the capsule of the elbow joint.

The *radial nerve* and the *profunda brachii artery* both give off branches to the triceps while they are on the medial side of the arm.

between the brachialis muscle and upper extensor muscles of the forearm.

**Anconeus.** The anconeus is a small, flat triangular muscle arising from the lateral epicondyle of the humerus and inserting on the lateral side of the olecranon and an adjacent part of the body of the ulna (Figs. 12-46 and 12-47). Its upper border adjoins the lateral portion of the medial head of the triceps, and a twig from the radial nerve descends in this head to end in the anconeus.

**Variations, Innervations, and Actions.** There are no important variations of these two muscles.

Both muscles are **innervated** by the radial nerve. Each head of the triceps receives one or more branches. The segmental innervation of the triceps is usually said to be from about C6 through C8, sometimes higher. The segmental innervation of the anconeus is said to be from C7 and C8.

The triceps is the chief extensor of the forearm (which can also, of course, be easily extended by gravity). Of the three heads, the medial is the most active, the long the least. Although the anconeus can contribute little strength, it is said to participate in all movements of extension. The long head

**FIG. 12-48.** The radial nerve on the posterior side of the arm.

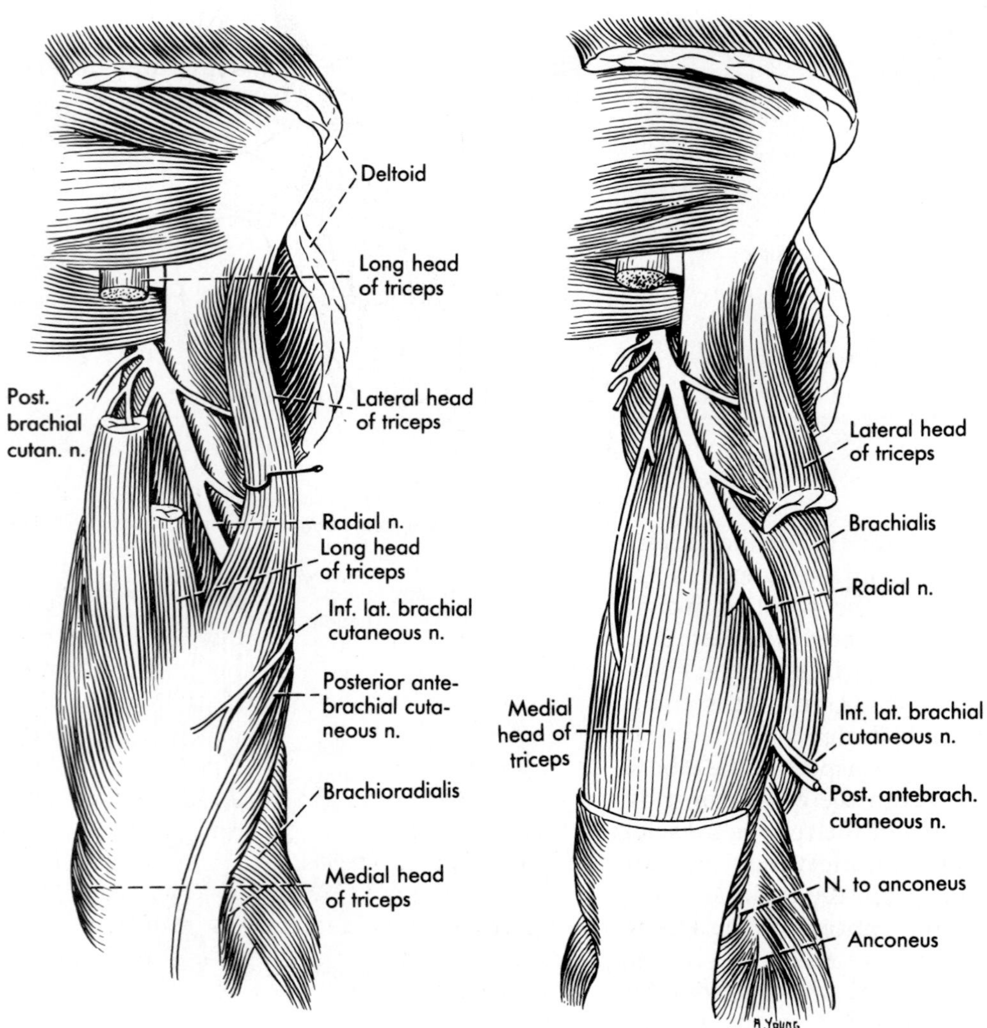

of the triceps, the only one to cross the shoulder joint, probably can assist slightly in adduction and extension of the arm.

## Nerves and Vessels

In summary of the nerves and vessels associated with the posterior muscles of the arm, the association of the **ulnar nerve** is a topographic one only. After piercing the medial intermuscular septum, the ulnar nerve lies on the medial head of the triceps, but gives off no branches into this muscle.

In contrast to the ulnar nerve, the **radial nerve** supplies the posterior muscles of the arm before continuing into the forearm (Fig. 12-48). The radial nerve is the only derivative of the posterior part of the brachial plexus to course downward in the arm and forearm. Before or just after it leaves the axilla, it usually gives off the **posterior brachial cutaneous nerve** and a branch to the long head of the triceps. While it is on the medial side of the arm, it gives off a branch that descends to end in the medial head of the triceps. As it spirals around the humerus between the lateral and medial heads, it supplies the lateral head and gives one or more additional branches into the medial head. After it penetrates the lateral intermuscular septum to lie between the brachialis muscle and the upper extensor muscles of the forearm, it gives off branches to the latter muscles.

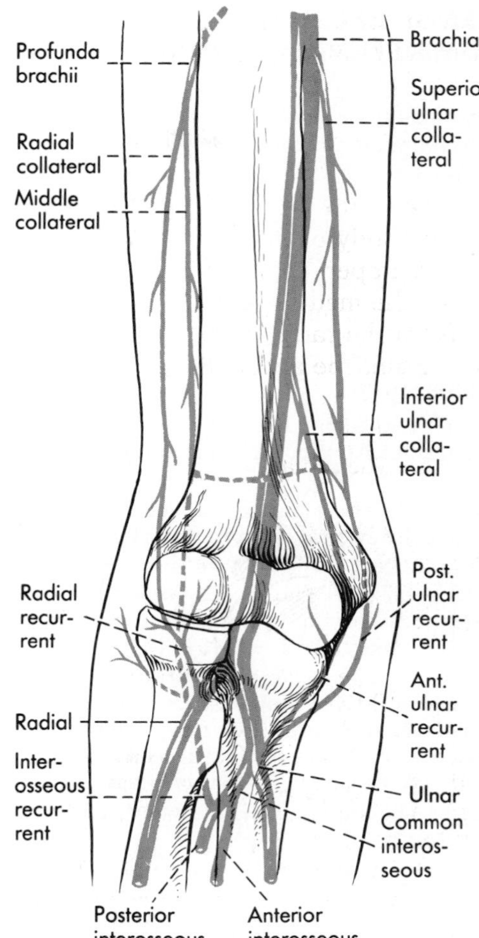


**FIG. 12-49.** Diagram of the lower part of the brachial artery and the anastomoses around the elbow joint.

The triceps receives part of its blood supply from muscular branches arising directly from the brachial artery and from the superior ulnar collateral artery. The **profunda brachii artery** provides muscular branches to the triceps before beginning its course around the humerus in company with the radial nerve. It continues to give off twigs to the muscle as it runs this spiral course. It may give off a nutrient artery to the humerus. Deep to the long head of the triceps it regularly gives rise to a *deltoid branch* that ascends to anastomose with the posterior humeral circumflex artery. This anastomosis accounts for the fact that the profunda brachii sometimes arises from the posterior humeral circumflex, or, more rarely, the circumflex arises from the profunda. The terminal branches of the profunda brachii artery are the *radial* and *middle collateral arteries*, both of which help to form the anastomoses around the elbow (Fig. 12-49). The **radial collateral artery** follows the radial nerve through the lateral intermuscular septum and anastomoses in front of the elbow with the radial recurrent artery. The **middle collateral artery** descends on the triceps, disappears deep to the anconeus, and anastomoses behind the elbow with the interosseous recurrent artery.

## GENERAL REFERENCES AND RECOMMENDED READINGS

ANSON BJ, WRIGHT RR, WOLFER JA: Blood supply of the mammary gland. Surg Gynecol Obstet 69:468, 1939

BASMAJIAN JV, BAZANT FJ: Factors preventing downward dislocation of the adducted shoulder joint: An electromyographic and morphological study. J Bone Joint Surg (Am) 41:1182, 1959

BASMAJIAN JV, LATIF A: Integrated actions and functions of the chief flexors of the elbow: A detailed electromyographic analysis. J Bone Joint Surg (Am) 39:1106, 1957

BROOME HL, BASMAJIAN JV:: The function of the teres major muscle: An electromyographic study. Anat Rec 170:309, 1971

CHARLES CM: On the arrangement of the superficial veins of the cubital fossa in American white and American Negro males. Anat Rec 54:9, 1932

CORBIN KB, HARRISON F: The sensory innervation of the spinal accessory and tongue musculature in the rhesus monkey. Brain 62:191, 1939

DEPALMA AF, CALLERY G, BENNETT GA: Variational anatomy and degenerative lesions of the shoulder joint. In: Instructional Course Lectures. The American Academy of Orthopaedic Surgeons, Vol 6. Ann Arbor, JW Edwards, 1949

GARDNER E: The innervation of the shoulder joint. Anat Rec 102:1, 1948

GREIG HW, ANSON BJ, BUDINGER JM: Variations in the form and attachments of the biceps brachii muscle. Quart Bull, Northwestern Univ M School 26:241, 1952

HALSELL JT, SMITH JR, BENTLAGE CR et al: Lymphatic drainage of the breast demonstrated by vital dye staining and radiography. Ann Surg 162:221, 1965

HUELKE DF: Variation in the origins of the branches of the axillary artery. Anat Rec 135:33, 1959

IP MC, CHANG KSF: A study on the radial supply of the human brachialis muscle. Anat Rec 162:363, 1968

KERR AT: The brachial plexus of nerves in man, the variations in its formation and branches. Am J Anat 23:285, 1918

MCCORMACK LJ, CAULDWELL EW, ANSON BJ: Brachial and antebrachial arterial patterns: A study of 750 extremities. Surg Gynecol Obstet 96:43, 1953

MILLER RA: Observations upon the arrangement of the axillary artery and brachial plexus. Am J Anat 64:143, 1939

PAULY JE, RUSHING JL, SCHEVING LE: An electromyographic study of some muscles crossing the elbow joint. Anat Rec 159:47, 1967

TELFORD ED, MOTTERSHEAD S: Pressure at the cervicobrachial junction: An operative and anatomical study. J Bone Joint Surg (Br) 30:249, 1948

TURNER-WARWICK RT: The lymphatics of the breast. Br J Surg 46:574, 1959

WHITSON RO: Relation of the radial nerve to the shaft of the humerus. J Bone Joint Surg (Am) 36:85, 1954

# 13

# FOREARM AND HAND

There is no truly satisfactory solution in the dissecting room to the problem that the forearm and hand must be considered as a functional unit and yet the hand in itself is so complicated that it deserves special description and dissection. Although tracing structures from the forearm into the hand, if done at all carelessly, may destroy important anatomy of the hand, a concept of the structural and functional continuity seems more important than the acquisition of more detailed knowledge concerning the hand; the two parts are therefore described together here. If, instead of dissecting the two simultaneously, the student is to dissect the forearm and then the hand, it is possible to use the following description by reading first the sections that apply to the forearm and subsequently returning to those that apply to the hand.

## SURFACE AND SKELETAL ANATOMY

The surface anatomy of the forearm has been briefly mentioned. Since much of the surface anatomy of the wrist and hand is skeletal anatomy, it is best discussed in conjunction with the skeleton. In summary of terms, the forearm presents anterior and posterior surfaces (formerly called volar and dorsal) and the hand, palmar and dorsal surfaces. The borders of the forearm are lateral or radial, and medial or ulnar. These terms also are used for the borders of the hand, but sometimes "thenar" is used to describe the thumb side and "hypothenar," the little-finger side.

The skin of the posterior surface of the forearm is thick. That of the dorsum of the hand is thin and freely movable over the underlying tendons and bones. The **ulna,** the medial bone of the forearm, can be palpated throughout its entire length along the posteromedial aspect of the forearm. At the wrist, its lower end forms the rounded prominence just above the dorso-medial border of the hand. Just distal to this, on the ulnar border of the limb, a strong tendon, that of the *extensor carpi ulnaris muscle,* can be felt as it crosses the wrist joint, and just anteromedial to this tendon, the *styloid process* of the ulna can be felt. The **radius,** the lateral bone of the forearm, is largely covered by muscles, but its distal expanded portion can be palpated without difficulty.

The diverging extensor tendons to the fingers can be seen and felt through the skin of the dorsum of the hand and traced to the bases of the digits, but are so closely bound to the dorsal aspect of the fingers that they can be neither felt nor seen much beyond the knuckles. The knuckles are formed by the prominent heads of the **metacarpals,** the bones that are palpable deep to the tendons and that form the major part of the length of the palm of the hand. The 1st metacarpal (that of the thumb) is rather freely movable, the 5th less movable, 2nd and 3rd essentially immovable. When the extended thumb (*pollex*) is separated as widely as possible from the index finger, and the radial side of the wrist is examined, two ridges raised by tendons can be seen converging toward the base of the first metacarpal and enclosing a hollow between them distal to the end of the radius (Fig. 13-1). This hollow is sometimes known as the *anatomic snuff box* because of the use to which it was once put; deep to it, the most distal end of the radius, the *styloid process,* can be palpated. The radial artery passes through this hollow, deep to the tendons, in its course to the hand.

**FIG. 13-1.** The dorsum of the wrist and the "anatomic snuff box."

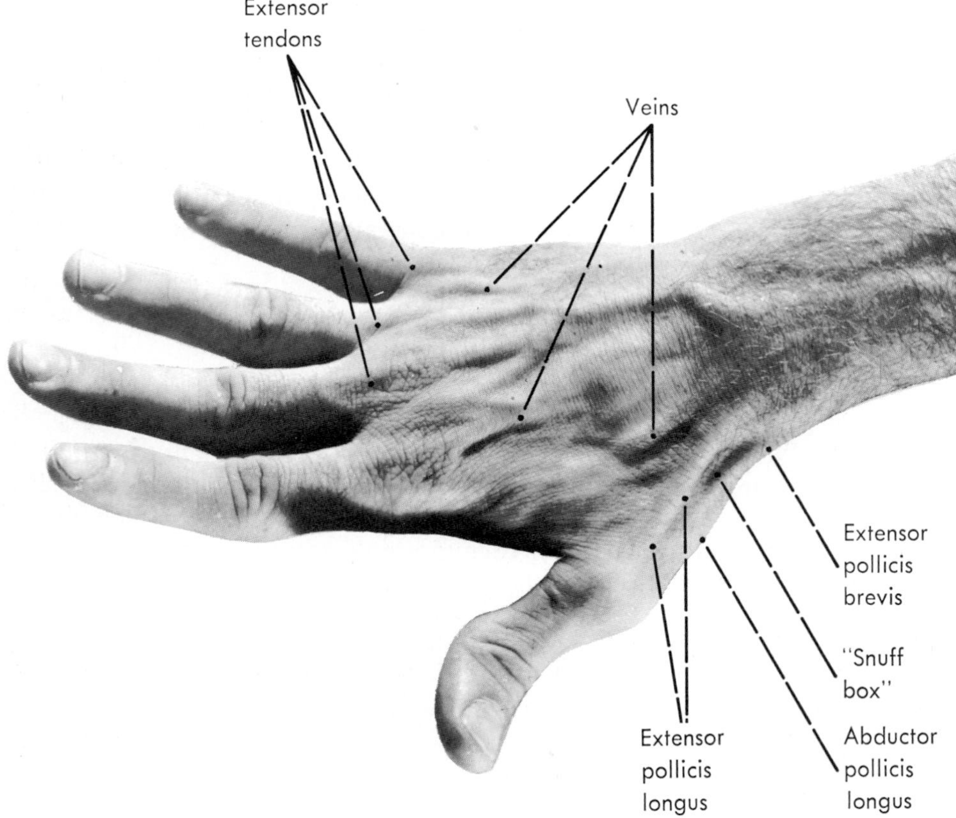

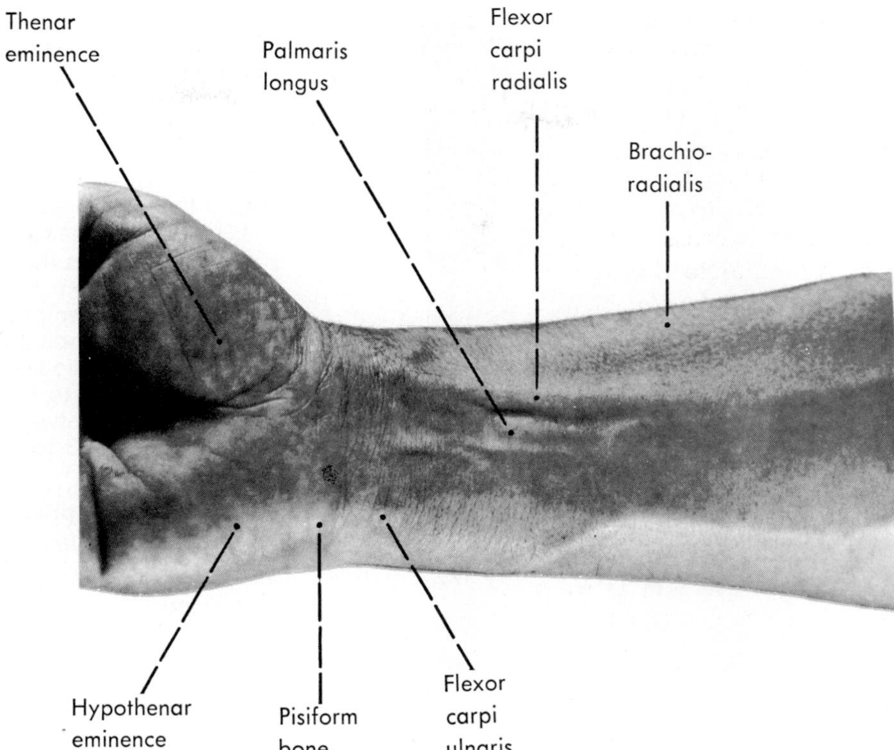


**FIG. 13-2.** The flexor surface of the forearm.

The skin of the anterior surface of the forearm is thin, and through it some of the superficial veins can usually be seen. That of the palmar surface of the hand is much thicker and firmly attached to the underlying deep fascia and bone by rather tough fibrous connective tissue.

The muscular prominence lateral to the cubital fossa and the tendon of the biceps is formed by the *brachioradialis*, which arises from the humerus. The less marked muscular prominence medially is formed by flexor forearm muscles that also arise from the humerus. A little above the wrist, on the ulnar border, the tendon of the *flexor carpi ulnaris* can be easily palpated and often seen (Fig. 13-2). In almost the exact midline of the wrist, the tendon of the *palmaris longus* becomes particularly visible when the wrist is flexed while the fingers are kept extended. This muscle is not always present.

Most of the other tendons palpable at the wrist are flexor tendons of the fingers, but the most radial tendon is that of the *flexor carpi radialis*. The prominence to which this tendon seems to be running is formed by the projections of two closely associated bones, the *scaphoid* and the *trapezium* (Fig. 13-9), difficult to distinguish individually. The rounded contour formed by the muscles of the thumb is the **thenar eminence,** and that formed by little-finger muscles on the ulnar side is the **hypothenar eminence.** The prominence at the base of the hypothenar eminence is formed by the *pisiform bone* and by the hamulus (hook) of the *hamate bone*. These two protruding elements also are difficult to distinguish from each other, but the pisiform lies more proximally and also to the ulnar side of the hook of the hamate.

## RADIUS AND ULNA

The upper ends of the radius and ulna (Figs. 13-3 and 13-4), the bones of the forearm, already have been described in connection with the muscles of the arm (Chap. 12), and their articulations with the humerus and each

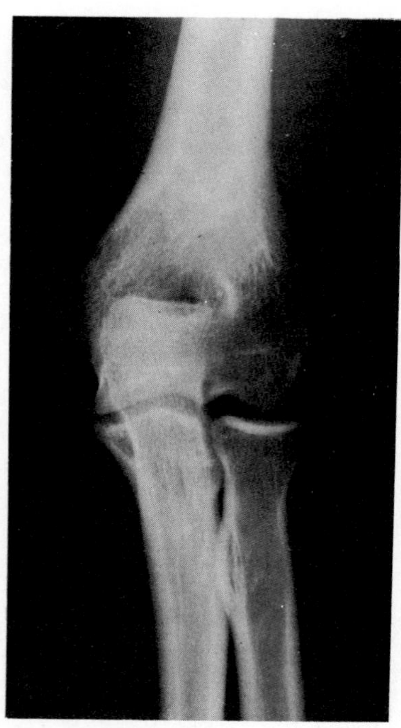

**FIG. 13-3.** Anteroposterior view of the elbow.

other to form the elbow joint are described in more detail later. The articulation between the ulna and the humerus allows primarily only flexion and extension, although a small amount of abduction–adduction can occur during pronation and supination of the forearm. The head of the radius moves on the capitulum of the humerus during flexion and extension, and rotates on the humerus and the ulna during pronation–supination.

The head and neck of the radius are held against the ulna by an annular ligament (Fig. 13-64), and throughout most of their length, the bodies of the two bones are joined by an **antebrachial interosseous membrane.** At the lower end there is a **distal radioulnar synovial joint.** The radius and the ulna receive the insertions of only a few muscles, but they and the intervening interosseous membrane give origin to a great many muscles (Figs. 13-5 and 13-6).

Very rarely, either the radius or the ulna is missing or a distal part is lacking; this is more often true of the radius than of the ulna.

The most common fracture of the forearm is of the lower end of the radius and is known as *Colles' fracture* (Fig. 13-7). This is a complete transverse fracture of the distal inch of the radius in which the distal fragment is displaced posteriorly. It results from forced extension of the hand, usually as the result of trying to ease a fall by the outstretched upper limb, and results in a peculiar deformity that has been given the name of "silver fork deformity."

The radius and the ulna are each **ossified** from three centers, one for the diaphysis and one at each end for the epiphyses. The diaphyseal centers for both bones appear in about the eighth or ninth week of embryonic life, but the epiphyseal centers appear from one to nine years after birth and join the diaphyses between the 14th and 21st years (Figs. 13-8 and 13-15).

**Radius.** The **upper end** of the radius consists of the disklike *head,* the narrower *neck,* and the *radial tuberosity* (Fig. 13-5). Its expanded **distal end** is curved somewhat forward and ends in the slightly concave *carpal articular surface* through which the radius participates in the wrist joint. On the medial side, continu-

**FIG. 13-4.** Lateral view of the elbow.

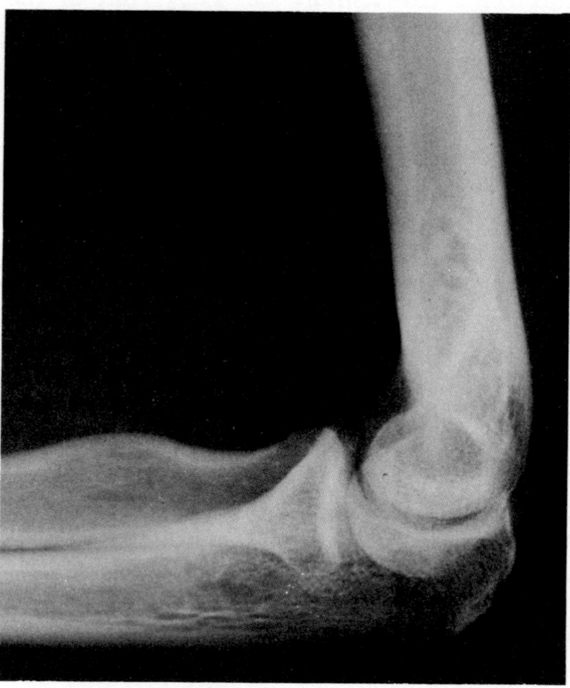

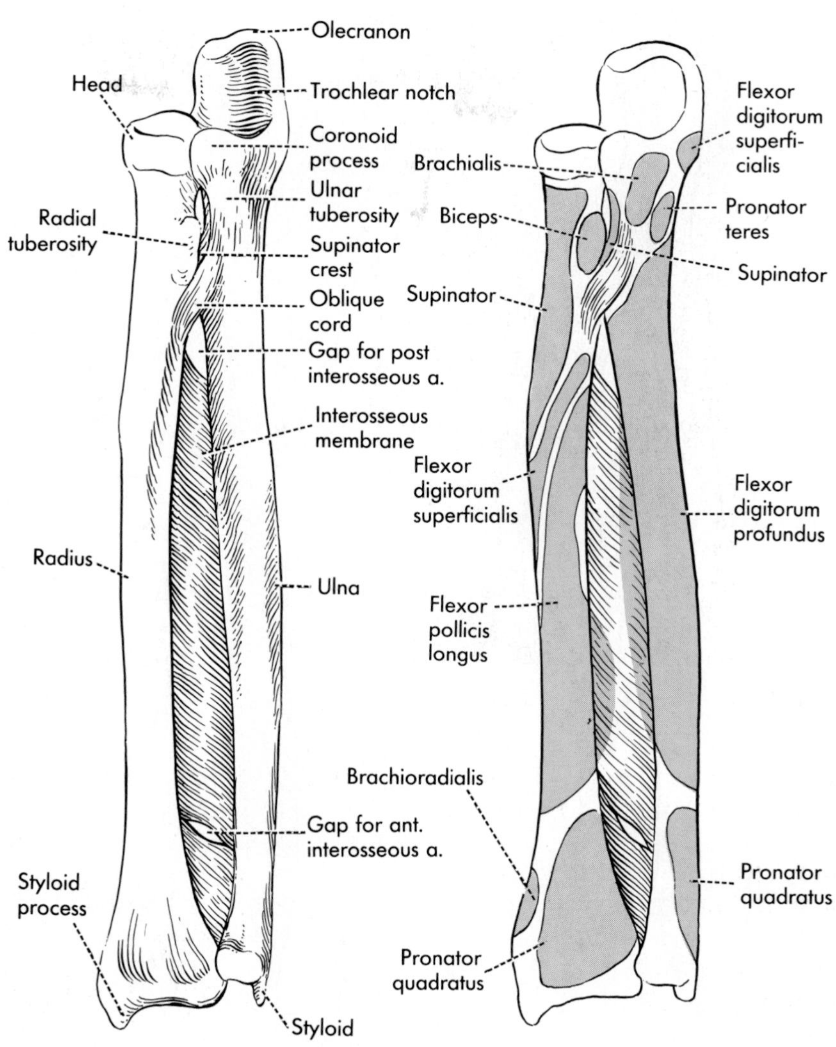


**FIG. 13-5.** Anterior view of the bones of the forearm. Muscle origins are shown in red, insertions in blue.

ous with it, is the *ulnar notch,* which helps form the distal radioulnar joint. Laterally, the bluntly pointed *styloid process* projects downward beyond the carpal articular surface. The **body** of the radius between the radial tuberosity and the distal end is described as having interosseous, anterior, and posterior **borders** and anterior, posterior, and lateral **surfaces.** Of these, only the relatively sharp interosseous (medial) border and the anterior surface are easily defined. The latter lies between the interosseous and the anterior border, which runs obliquely downward from the radial tuberosity to the styloid process; thus, the anterior surface is narrow above, but occupies the whole breadth of the radius below. The lateral surface is lateral to the anterior border and is truly lateral only inferiorly; above, it encircles the radius between the anterior and posterior border of the radial tuberosity. The posterior surface, narrow and mostly medial above, expands and is truly posterior below.

**Ulna.** The large upper or **proximal end** of the ulna has two prominent projections, the *olecranon* and the *coronoid process,* which together bound the *deep trochlear notch* (Fig. 13-5). On the lateral surface of the bone, just

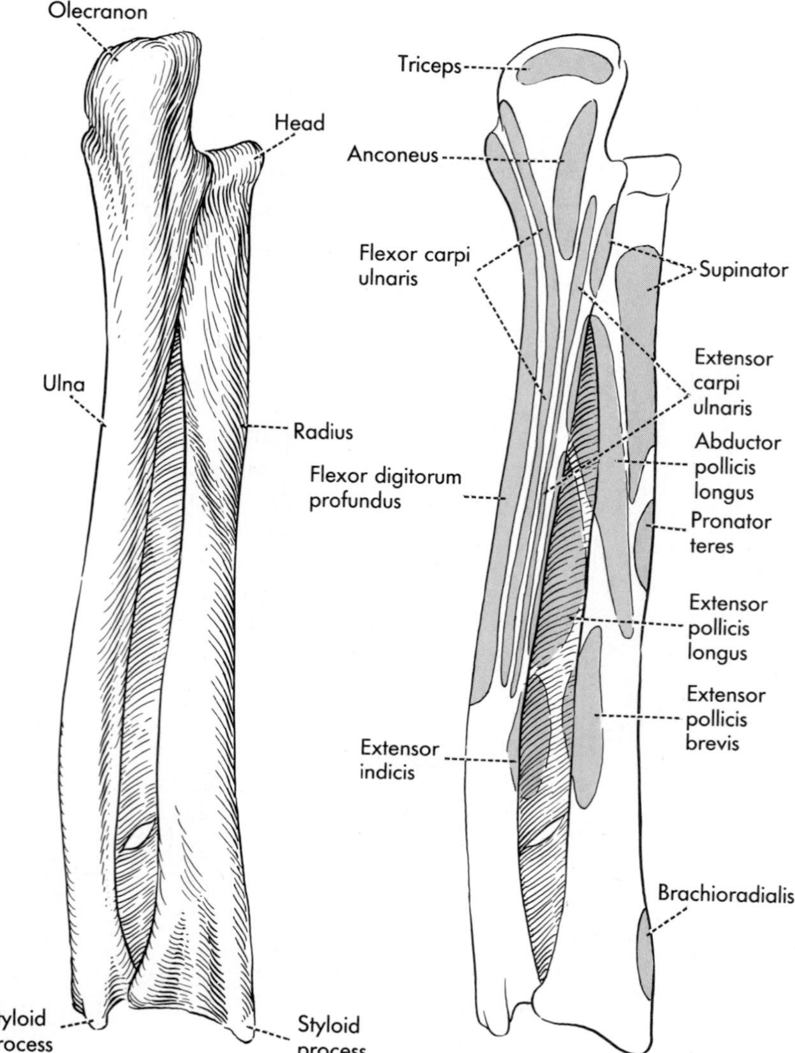


**FIG. 13-6.** Posterior view of the bones of the forearm with muscle origins in red and insertions in blue.

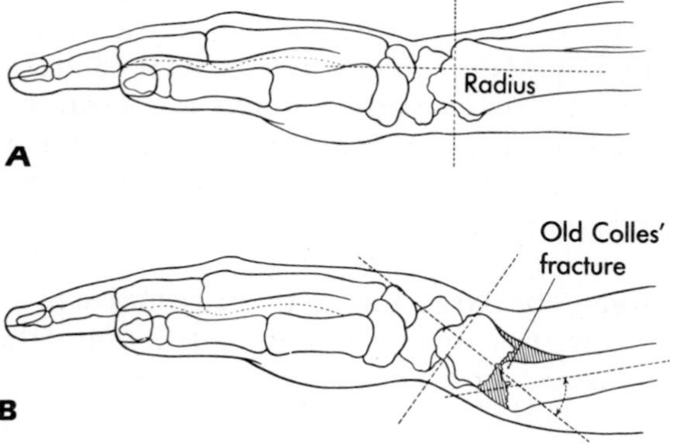


**FIG. 13-7.** Normal alignment of the forearm and hand, **A,** and the angulation ("silver fork deformity") in a Colles' fracture, **B.** The one represented here is an old one, the shaded area indicating the bone formed in healing. (Meyerding HW, Overton LM: Minn Med 18:84, 1935)

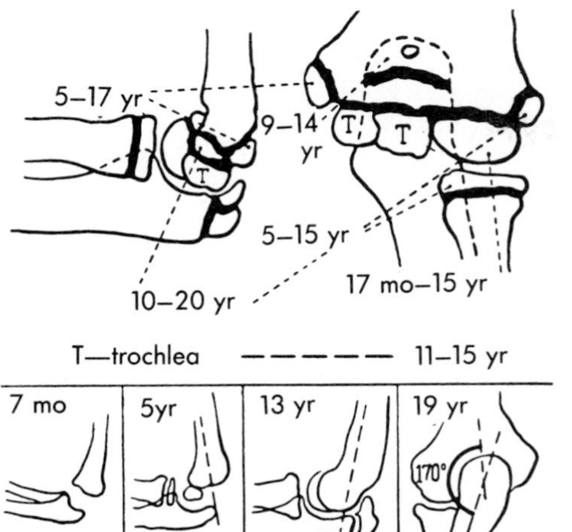


**FIG. 13-8.** Ossification centers in the elbow. For the Epiphyses of the humerus, radius, and ulna, the first of the two numbers indicates the usual age at the time of appearance, the second that at the time of fusion, of the epiphyseal center. The smaller figures represent the appearance of the elbow in roentgenograms at different ages. (Camp JD, Cilley EIL: Am J Roentgenol 26:905, 1931)

distal to the trochlear notch and continuous with it, is the *radial notch* for accommodation of the head of the radius. Anteriorly at the base of the coronoid process is the *ulnar tuberosity.*

The smaller distal end of the ulna, the **head,** consists mostly of a rounded articular circumference, but ends posteriorly in a **styloid process** that is partly separated from the circumference by a shallow groove for the tendon of the extensor carpi ulnaris muscle. The **body** of the ulna, between the coronoid process and the head, is described as having interosseous, anterior, and posterior **borders,** like the radius, and anterior, posterior, and medial **surfaces.** As the bone becomes more rounded distally, the borders tend to fade out and the surfaces to blend. The interosseous (lateral) border is the clearest, but the posterior border can be seen easily (and palpated, since it is subcutaneous) as it extends downward from the posterior surface of the olecranon toward the styloid process. As the posterior border fades out below, it blends with the rounded medial surface, also subcutaneous. The anterior surface (between interosseous and anterior borders) is somewhat shallowly concave above. On the upper part of the "posterior" surface an oblique ridge extends backward and downward from the anterior border of the radial notch. This is the *crest of the supinator muscle,* part of the origin of this muscle.

## SKELETON OF WRIST AND HAND

### Carpals

The skeleton of the carpus, or wrist, is composed of eight carpal bones, arranged in two rows of four each (Figs. 13-9 through 13-12). In the **proximal row,** beginning on the radial side, the *scaphoid,* the *lunate,* and the *triquetral* bones articulate proximally with the radius and the articular disk of the ulna to form the **radiocarpal** or **wrist joint** proper. They articulate transversely with each other and distally with the distal row of carpals. The **distal row,** starting on the radial side, is made up of the *trapezium, trapezoid, capitate,* and *hamate* bones. The *pisiform,* the fourth bone in the proximal row, lies on the anterior surface of the triquetral and articulates with that bone only.

The **scaphoid bone,** so named from its fancied resemblance to a boat, presents proximally a convex surface for articulation with the radius. Distally and medially it has articular surfaces for the three lateral bones of the distal row and for the lunate. Its anterior surface bears a prominent *tubercle,* visible and palpable at the base of the wrist. The convex proximal surface of the **lunate bone** articulates with the radius. Its medial, lateral, and distal surfaces present articular facets for the triquetral, scaphoid, and capitate and hamate bones, respectively. The somewhat pyramidal **triquetral bone** has a proximal surface that is largely nonarticular but bears laterally a small facet for the articular disk at the distal end of the ulna. Laterally it has an articular surface for the lunate, distally one for the ha-

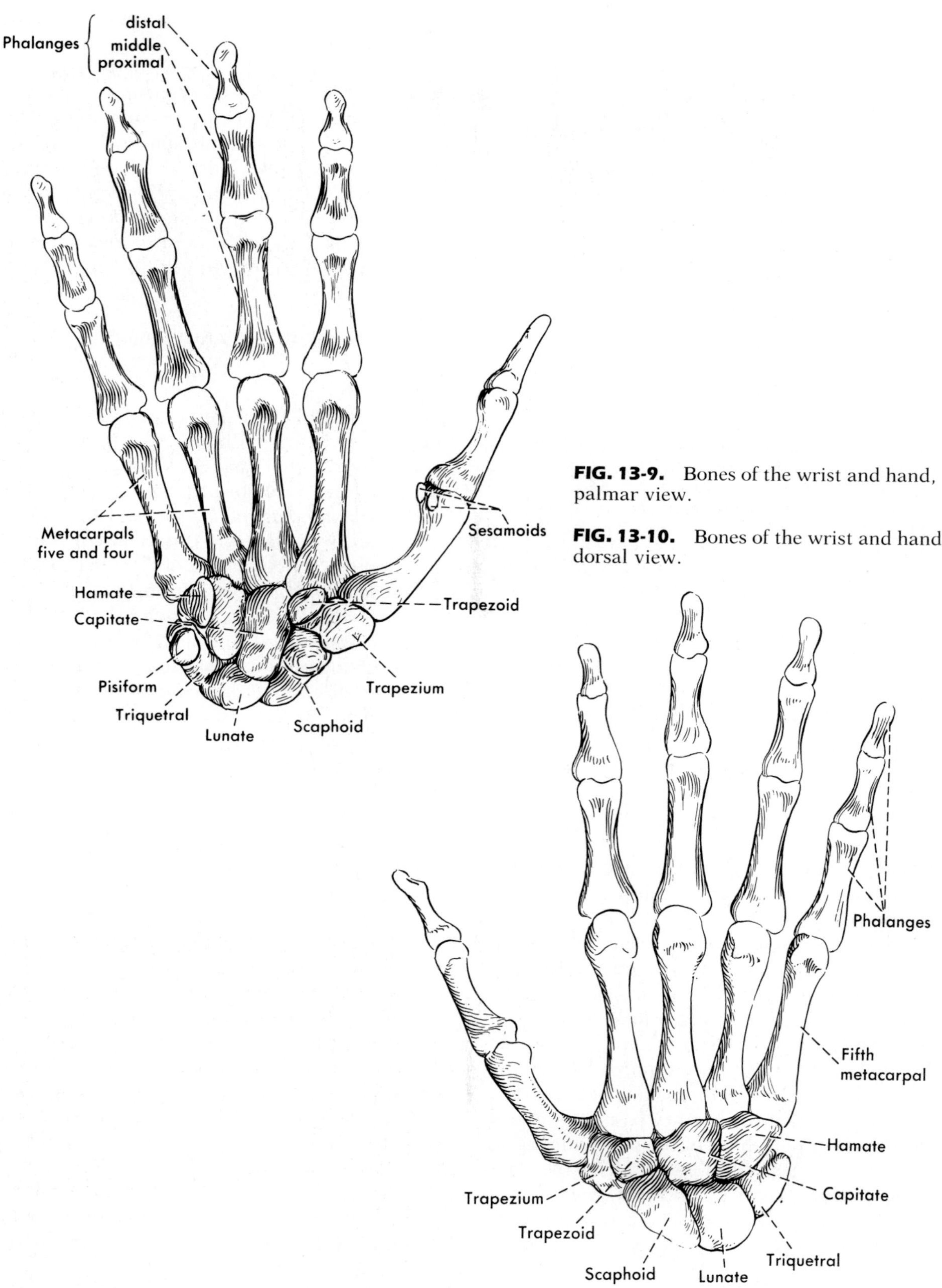


**FIG. 13-9.** Bones of the wrist and hand, palmar view.

**FIG. 13-10.** Bones of the wrist and hand, dorsal view.

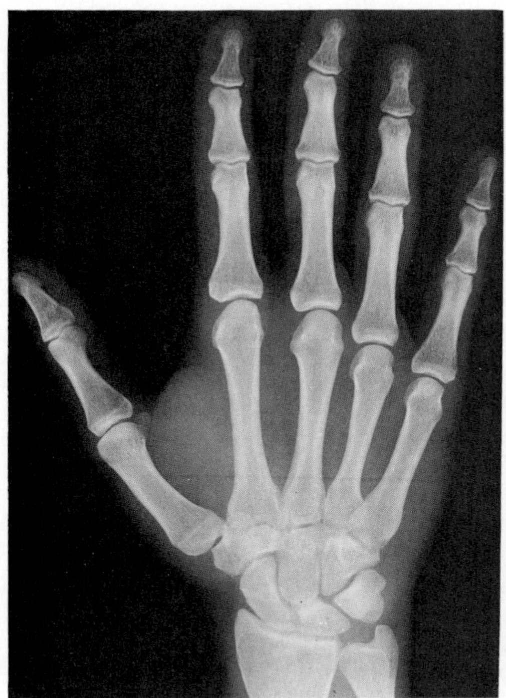

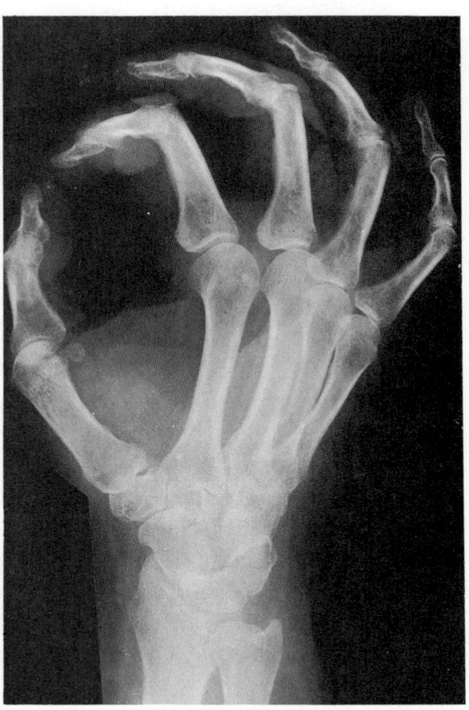

**FIG. 13-11.** Anteroposterior (a.p.) and oblique ("scaphoid") views of two right hands. The carpal bones in the a.p. view can be identified by comparison, if necessary, with Figure 13-10. There is, however, some overlapping between the trapezium and the trapezoid in this view, and only a proximal portion of the pisiform, projecting between the triquetral and the styloid process of the ulna, is plainly visible. The oblique view is used particularly for an examination of the scaphoid bone, the most frequently fractured of the carpals; this view also clearly separates the trapezium and trapezoid. (Courtesy of Dr. David G. Pugh)

**FIG. 13-12.** Roentgenograms of the same hand shown in Figure 13-11, left, in radial and ulnar abduction. Note particularly the different relations of the scaphoid and lunate to the radius, of the scaphoid to the trapezium and trapezoid, and of the hamate, triquetral, and pisiform to each other.

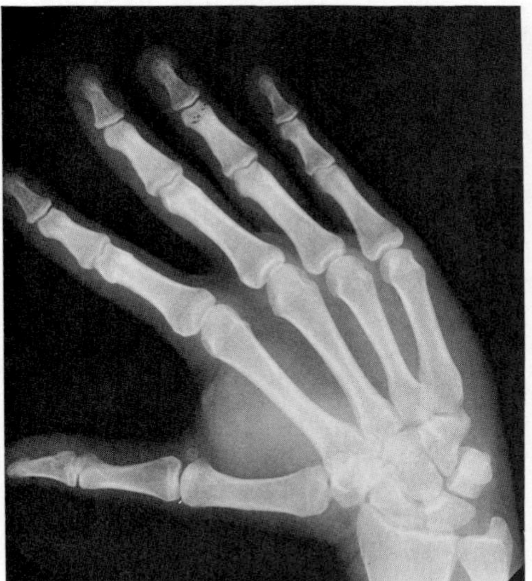

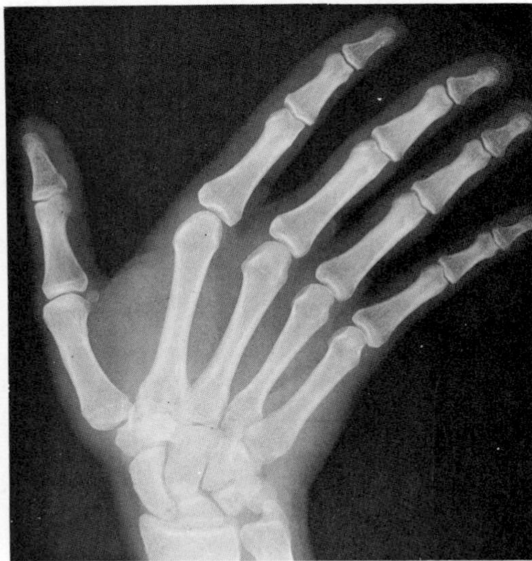

mate, and on the palmar surface, an oval facet for the pisiform bone. The small, somewhat oval, **pisiform bone** (so named because of its resemblance to a pea) bears posteriorly a facet for articulation with the triquetral. The pisiform has long been described as a sesamoid bone, but a more modern study has indicated that it probably is not.

In the distal row, the **capitate** is the largest bone, the **trapezoid** the smallest. Each bone of the distal row presents several articular facets: in general, the proximal surfaces have facets for the proximal row of carpals, the distal surfaces facets for the metacarpals, and the side facets for the adjacent bones in the distal row; the articular surface on the lateral aspect of the **trapezium** is for the first metacarpal. The named parts of these bones are only two, the *tubercle* on the palmar surface of the **trapezium** and the *hamulus* or hooklike projection on the palmar surface of the **hamate** bone.

The carpal articular surfaces and ligaments are such that the proximal row moves fairly freely on the forearm, and the distal row moves on the proximal one. Little movement occurs between the members of a single row. The carpals as a whole have a somewhat convex dorsal surface and a correspondingly concave palmar one. This concavity, the **carpal sulcus** or **groove,** is exaggerated by the projections of the scaphoid and trapezial tubercles on the radial side and by the pisiform bone and the hamate hamulus on the ulnar side. It is converted into a tunnel **(carpal canal)** by the heavy flexor retinaculum that stretches across its front (Fig. 13-39).

**Variations.** More than 20 supposedly different supernumerary bones at the wrist have been described, but few wrists have any. Some of these accessory bones are believed to be sesamoid bones, some to be true accessory carpals (arising from accessory cartilages but not connected with tendons), and some to represent failure of fusion of two centers of ossification arising abnormally within a single carpal element.

Congenital fusion of two or more carpals, although rare, most often involves the lunate and the triquetral bones. This does not interfere with movement of the wrist and can be recognized only roentgenographically.

**Fractures and Dislocations.** Approximately 70% of carpal fractures involve the scaphoid only, usually through its narrow middle part. Since the proximal part of the scaphoid usually derives its blood supply from vessels entering this narrow middle part, a fracture here may deprive it of its blood supply. In this event, the fracture may fail to unite, and the proximal part of the bone may undergo avascular necrosis (death). A hard fall upon the hyperextended hand, the type of accident that also produces Colles' fracture (Fig. 13-7), is the usual mechanism of producing a scaphoid fracture. Forced hyperextension may also produce dislocation of one or more carpals, especially the lunate.

The attachments of muscles to the carpal bones are shown in Figure 13-13; no muscles normally attach to the dorsal aspects of these bones (Fig. 13-14).

## Metacarpals and Phalanges

The five **metacarpals** (Figs. 13-9 to 13-12) form the major part of the skeleton of the palm of the hand. They are essentially similar to each other and are named (by Roman numerals) from the radial side. Each consists of a proximal base, a body, and a head. The **base** of metacarpal I presents a saddle-shaped articular surface for the trapezium, affording the movable carpometacarpal joint characteristic of the thumb. The other metacarpals present flatter facets for articulation with the distal row of carpals and with each other. The third metacarpal has a short styloid process on the dorsal surface of its base.

The metacarpal **bodies** are somewhat triangular in cross section and slightly concave lengthwise on their palmar surface. The convex articular surface of the four medial metacarpal **heads** extends farther onto the palmar than the dorsal aspect but very little onto the sides; that of the thumb extends less onto the palmar surface, and phalangeal flexion here may be more limited. Two small **sesamoid bones,** an ulnar and a radial one, typically lie on the palmar surface of the metacarpophalangeal joint of the thumb. Occasionally there

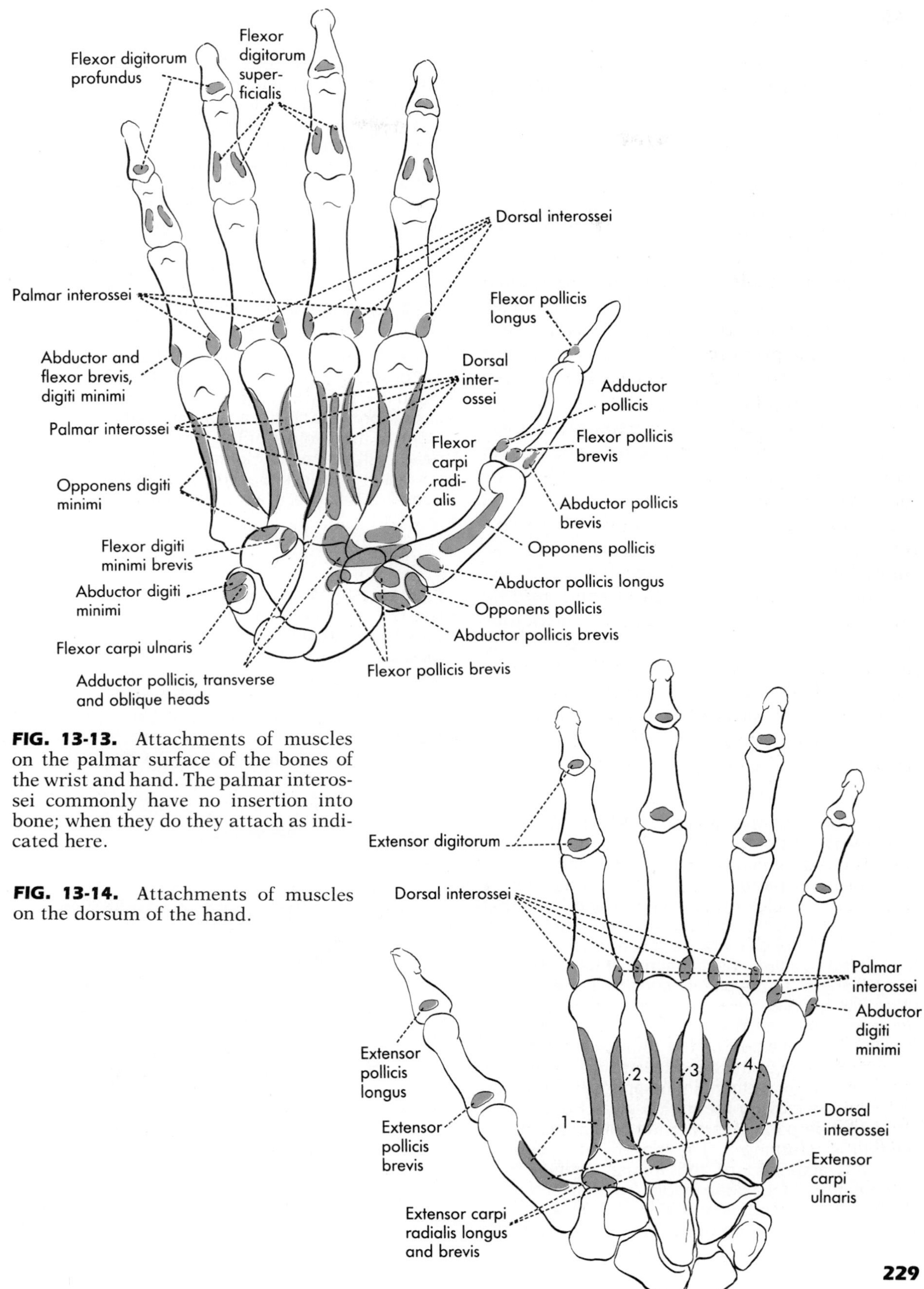


**FIG. 13-13.** Attachments of muscles on the palmar surface of the bones of the wrist and hand. The palmar interossei commonly have no insertion into bone; when they do they attach as indicated here.

**FIG. 13-14.** Attachments of muscles on the dorsum of the hand.

are one or more sesamoids at other metacarpophalangeal joints or at interphalangeal joints.

The **metacarpophalangeal joints** allow flexion (palmar bending) and extension, with a variable degree of hyperextension. They also allow lateral movement (abduction and adduction). Rotation is slight, as it is limited by ligaments that lie on the sides of the joints.

The two **phalanges** of the thumb are named *proximal* and *distal;* the three of each finger are named *proximal, middle,* and *distal.* Each phalanx (the proximal ones are longer and heavier, the more distal ones successively shorter and lighter) consists of three parts, a base, a body, and a head. The expanded *base* is in general shallowly concave for articulation with the preceding metacarpal or phalanx, as the case may be; the *body* is slightly curved palmarward in its long axis, and in cross section, it is somewhat convex dorsally, flat or slightly concave on its palmar surface. The expanded *heads* of the proximal and middle phalanges present pulley-shaped articular surfaces for articulation with the bases of the more distal phalanges. The roughened head of each distal phalanx bears a raised *tuberosity* on its palmar aspect. The interphalangeal joints are strictly hinge joints, allowing only flexion and extension.

**Definitions of Movements of the Thumb.** The movements of the fingers (flexion and extension, abduction—spreading them apart—and adduction) need no particular explanation, but before discussing muscles of the forearm and hand that act upon the thumb, the movements of the thumb should be defined. Because the human thumb in the resting position has its palmar or flexor surface at right angles to the palmar surface of the remainder of the hand and the other four digits, its planes of movement are also at right angles to those of the fingers. Thus, flexion of the thumb is bending it across the palm of the hand toward the ulnar side, extension is a movement in the opposite direction; abduction is moving the thumb away from the index finger perpendicularly to the palmar plane, adduction is returning it. The movement of opposition, characteristic of the human thumb, is a combination of flexion, medial rotation, and adduction that brings the palmar surface of the thumb in contact with the palmar surfaces of the fingers (especially the index finger, as in grasping objects such as a needle). The movement of carrying the thumb away from opposition, a combination of abduction, extension, and lateral rotation, is known as reposition.

The mobile 1st metacarpal participates in all these movements of the thumb, and the carpometacarpal joint is so constituted that the metacarpal always rotates medially during flexion and laterally during extension. This metacarpal movement is necessary for opposition, for only a little rotation can occur at the metacarpophalangeal joint and none at the interphalangeal joint.

**Ossification of the Wrist and Hand.** Each **carpal** ossifies typically from a single center. As indicated in Figure 13-15, some of these bones begin to ossify soon after birth, but others begin much later. Many of the centers of the latter bones usually appear a full year earlier in the female than they do in the male, and in either sex any one of them may appear 2 to 3 years later than usual.

**FIG. 13-15.** Ossification centers in the hand and wrist. For the epiphyses of the radius, ulna, metatarsals, and phalanges, the first of the two numbers indicates the usual age at the time of appearance, the second that at the time of fusion; for the carpals, the numbers represent the range in the time of appearance of their centers. The small figures indicate the appearance of the wrist in roentgenograms at various ages. (Camp JD, Cilley EIL: Am J Roentgenol 26:905, 1931)

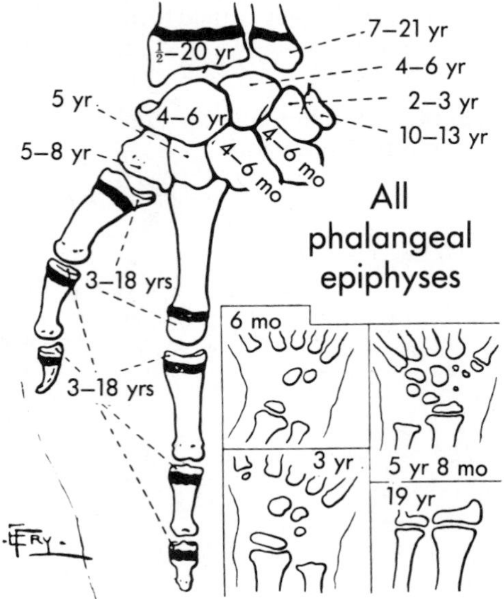

Each **metacarpal** and **phalanx** ossifies from two centers, the one for the body appearing early, approximately in the ninth week of fetal life, and an epiphyseal one appearing usually after the first year of postnatal life. The single epiphysis is at the base of the 1st metacarpal and of each phalanx, but at the heads of the 2nd through 5th metacarpals.

The **sesamoids** of the metacarpophalangeal joint of the thumb usually begin to ossify at or after the age of 18 years.

## FASCIA, SUPERFICIAL NERVES, AND VESSELS

The **superficial fascia** of the forearm is continuous with those of the arm and hand and, over the olecranon, contains a subcutaneous *olecranon bursa*. On the dorsum of the hand, the fascia is thin and loose and allows the easy mobility of the skin. In the palm, it is dense and is traversed by fiber bundles that tightly bind the skin to the underlying deep fascia and bone. The attachments of the heaviest fiber bundles into the skin of the palm are marked by the permanent *flexion lines* of this skin.

On the fingers, superficial and deep fascia blend and have more nearly the characteristics of superficial fascia. There is little fascia on the dorsum of the fingers, but on the palmar aspect it is fairly thick, binds the skin to the underlying tendon sheaths and bone, and transmits the relatively large digital nerves and vessels.

The fascia forms a pad (called *pulp*) of considerable thickness on the distal phalanx. Here the retinacula cutis are so arranged that they tend to guide superficial infections into a deeper position. Infection of the pulp of the terminal phalanx is called a *felon* or *whitlow* and may produce necrosis of bone if it goes deeply enough.

### Vessels

The **superficial veins** of the palm drain largely into the dorsal venous network by passing around the borders of the digits or the hand, or between metacarpal heads as *intercapital veins*. The dorsal network, presenting many different patterns composed of rather large veins, gives rise on its ulnar side to the *basilic vein*, which gradually passes around the ulnar aspect of the forearm to the anterior surface. On its radial side the dorsal network gives rise to the *cephalic vein* that passes around the radial border of the limb to lie on the anterior surface (see Figs. 11-14 and 11-15). A little below the elbow, the cephalic vein gives rise to the *intermediate cubital vein* that passes across the front of the cubital fossa to join the basilic vein (the intermediate cubital is frequently the vein from which blood is drawn). There may be additional connections at this level, and the precise pattern of communiction is variable. Either the cephalic or the intermediate cubital vein typically sends a large connection to the deep veins at the front of the elbow. Fairly frequently a small *intermediate forearm vein* arising from the palm of the hand runs up the anterior surface of the forearm to join either the basilic vein or the intermediate cubital vein.

The **lymphatic drainage** of the palm of the hand is largely into posteriorly placed vessels, just as is the venous drainage. The lymphatics of the hand and forearm are shown in Figures 11-16 and 11-17.

The **superficially placed arteries** of importance are the palmar digital arteries (Fig. 13-16). The *dorsal digital arteries* are tiny. They run subcutaneously, one on each side of each digit, in the scanty connective tissue here and can usually be traced only a little distance along the length of the proximal phalanx. In contrast, the *palmar digital arteries* are large. They appear subcutaneously close to the bases of the fingers and run distally along the palmar aspect of each side of each digit, directly behind the digital nerves. Both nerves and arteries lie far enough to the side of the flexor tendons to be largely protected from pressure when objects are gripped in the hand. The palmar digital arteries give off small branches to the dorsum of the digit as well as to the palmar surface. Each pair on a finger ends in a terminal loop formed by end-to-end anastomosis between the two members. Just proximal to this terminal loop a

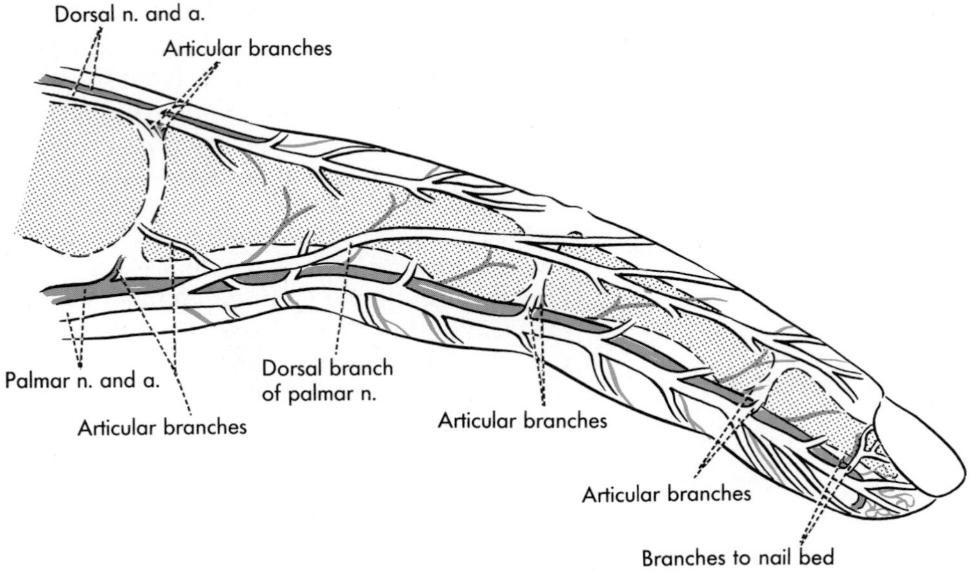


**FIG. 13-16.** Diagram of the digital nerves and vessels on one side of a finger.

large branch is given dorsally on each side to the nail bed, and a series of small subsidiary loops proceed from the loop into the pulp of the distal phalanx.

## Cutaneous Nerves

Of the three **cutaneous nerves to the forearm,** the *lateral antebrachial cutaneous* (continuation of the *musculocutaneous*) is distributed to skin on the anterior and posterior aspects of the radial side of the forearm about as far as the wrist (Figs. 13-17 and 13-18). The *medial antebrachial cutaneous nerve,* an independent branch from the medial cord of the brachial plexus, supplies skin on the anterior and posterior aspects of the ulnar side of the forearm. The *posterior antebrachial cutaneous nerve,* from the radial, supplies the strip down the middle of the posterior surface of the forearm between the distributions of the other two.

The **cutaneous nerves of the hand** are derived from the radial, median, and ulnar nerves. A small *palmar branch of the median nerve* given off a little above the wrist supplies a proximal part of the middle of the palm, and the *palmar branch of the ulnar,* arising in the forearm, supplies a still smaller area of the ulnar side, but the chief innervation to the hand is through digital branches of the three nerves. In the palm (Fig. 13-44), branches of the ulnar and median nerves run distally, give twigs to the palm, and end as *palmar proper digital nerves* that run along each border of a digit in front of the proper digital arteries (Fig. 13-16). Those to the little finger and the ulnar side of the ring finger are typically derived from the ulnar nerve, and the rest from the median nerve. Each gives branches to the skin, the interphalangeal joints, and the nail bed. Except on the thumb, the palmar digital branches of the median nerve (and sometimes one or more of the ulnar) give off branches that run dorsally to supply the skin over the dorsal aspect of the middle and distal phalanges (Figs. 13-16 and 13-19).

The nerves to the dorsum of the hand (Fig. 13-19) usually are derived from the superficial radial nerve and the dorsal branch of the ulnar nerve. The *superficial radial nerve* appears a little above the wrist from under cover of the brachioradialis muscle and runs downward subcutaneously on the radial side of the hand. It supplies this side of the dorsum and gives off *dorsal proper digital nerves* to the thumb and the adjacent one and one-half to

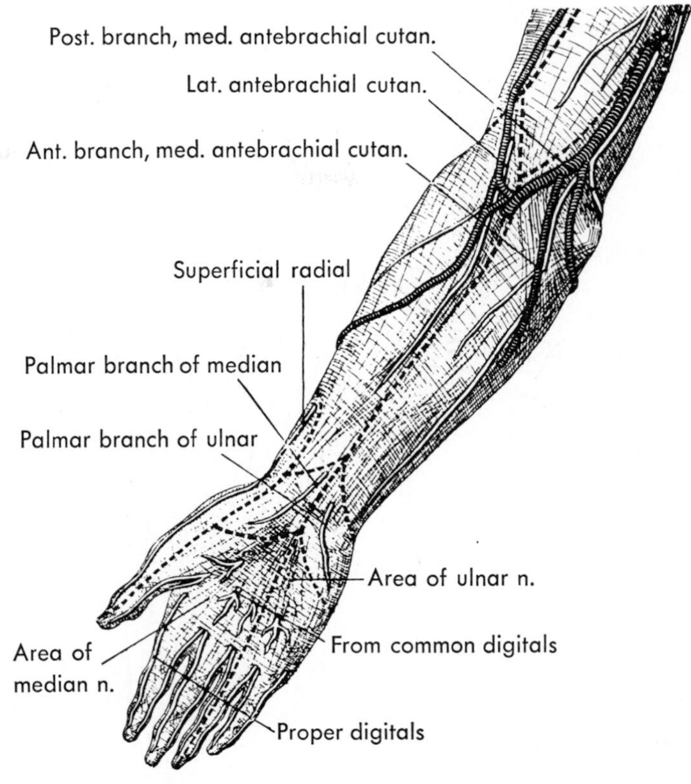


**FIG. 13-17.** Cutaneous nerves of the anterior surface of the forearm and the palm of the hand. (Corning HK: Lehrbuch der topographischen Anatomie für Studierende und Ärtze. Munich, JF Bergmann, 1923)

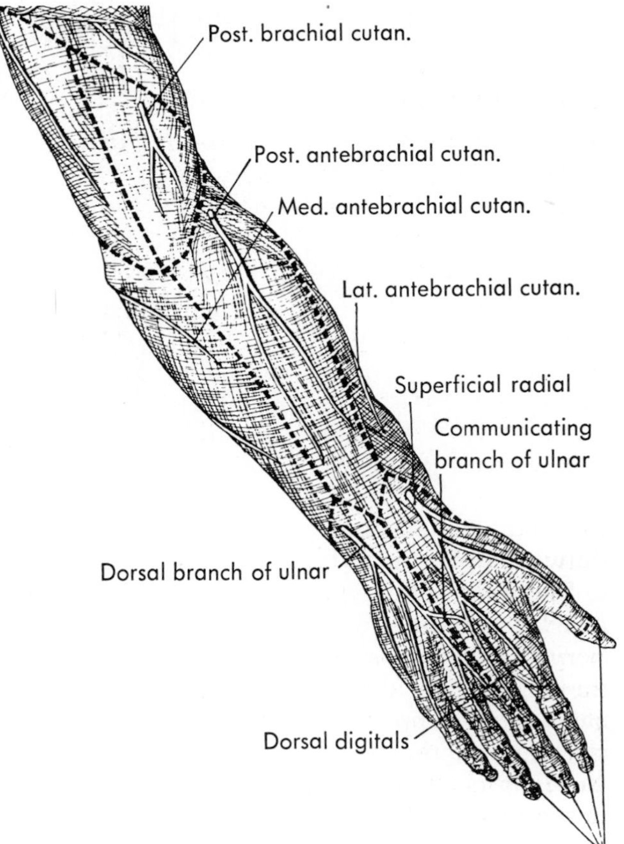


**FIG. 13-18.** Cutaneous nerves of the posterior surface of the forearm and the dorsum of the hand. (Corning HK: Lehrbuch der topographischen Anatomie für Studierende und Ärtze. Munich, JF Bergmann, 1923)

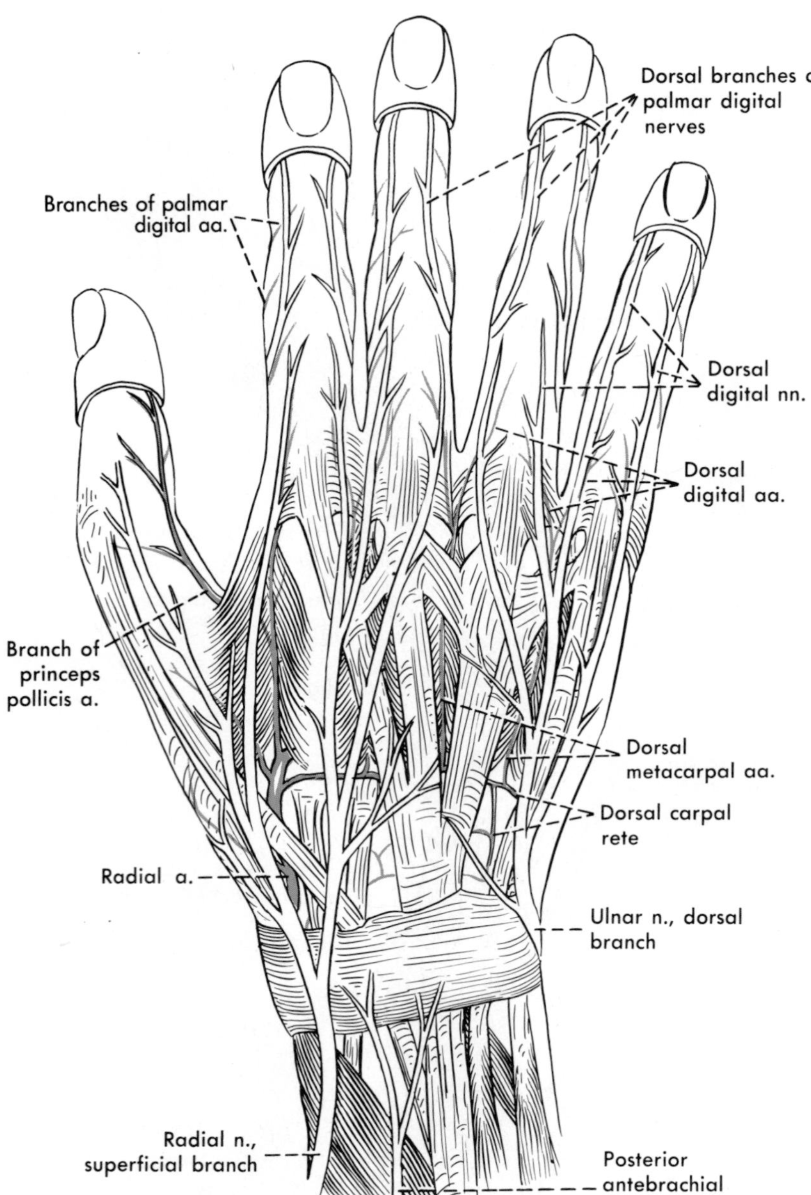


**FIG. 13-19.** Nerves and vessels of the dorsum of the hand.

two and one-half fingers. The branches to the thumb supply skin as far as the nail bed; those to the other digits go little beyond the proximal interphalangeal joints. The *dorsal branch of the ulnar* curves around the ulnar side of the wrist, from deep to the flexor carpi ulnaris, to supply dorsal proper digital branches to the ulnar one and one-half or more digits. There is a good deal of variation in the distribution of these two nerves to the dorsum.

## Deep Fascia

The **antebrachial** (deep forearm) **fascia** encloses both flexor and extensor muscles in a common cylindrical sheath. It is fused to the whole subcutaneous length of the ulna and sends a septum to the radius, thus separating the two groups. In the upper part of the forearm, the flexor and extensor muscles arise from antebrachial fascia and from septa that it sends between them. In the lower anterior

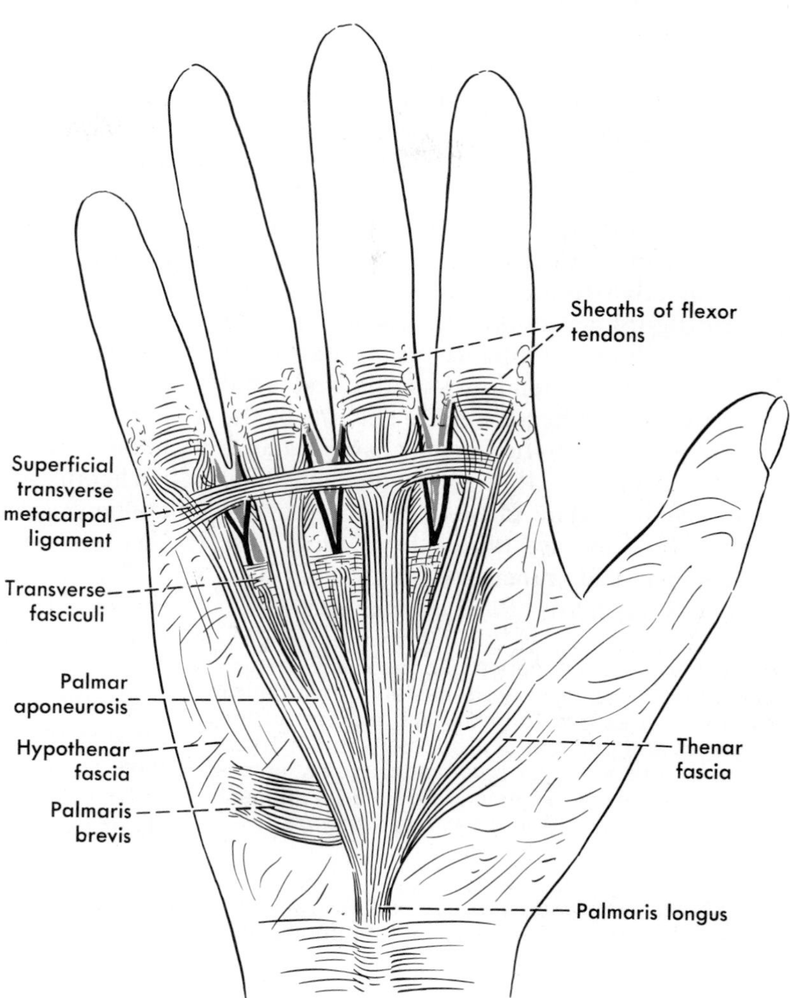


**FIG. 13-20.** The palmar aponeurosis; the digital nerves (black) and digital arteries (red) appear in the connective tissue between the longitudinal fasciculi of the aponeurosis.

part of the forearm, the antebrachial fascia is partially split into two layers, one anterior to the most superficial muscles here (palmaris longus, flexor carpi ulnaris, flexor carpi radialis) and one between these and the other flexors.

The deeper layer is thickened at the wrist by heavy transverse fibers that bridge the carpal groove between the pisiform bone and the hamalus of the hamate on the ulnar side and the scaphoid and trapezium on the radial side and convert the groove into the *carpal canal;* this thickening is the **flexor retinaculum** (Fig. 13-44). The anterior fascial layer shows little thickening and ends mostly by blending with the tendons that lie immediately behind it and with the flexor retinaculum. As it passes from the retinaculum to the anterior surface of the flexor carpi ulnaris and the pisiform bone, however, it runs in front of the ulnar nerve and vessels and leaves distally a gap through which these structures emerge into a superficial position on the hand. The flexor retinaculum, being a thickening of the deep fascia, has no free upper edge. It likewise has no truly free lower edge, for muscles of both thenar and hypothenar eminences arise in part from it, and the heavy palmar aponeurosis is firmly attached to its central portion.

The **palmar aponeurosis** (Fig. 13-20) is the heavy fascia, tendinous in appearance, occupying the center of the palm of the hand. The fascia over the thenar and hypothenar muscles is much less dense, but the aponeurosis consists of heavy fibers that superficially are usually a continuation of the palmaris longus tendon and more deeply arise from what should be the distal edge of the flexor retinaculum. As these fibers run toward the fingers, they become segregated into four slips, one for each finger, and between these slips there appear *transverse fasciculi* that at first bind them together. As the slips near the bases of the fingers, the transverse fasciculi disappear, and in the connective tissue between the slips, the palmar digital nerves and vessels appear. The slips to the fingers typically fuse in part with the digital flexor tendon sheaths and send septa around the flexor tendon sheaths to the metacarpals and to the deep transverse metacarpal ligaments, thus forming a tunnel around each tendon. Close to the bases of the fingers, the slips are united by a much less dense **superficial transverse metacarpal ligament** that lies across their palmar surfaces.

A disease of the palmar aponeurosis, of unknown origin, typically produces abnormal bands of tissue that extend from the aponeurosis to the phalanges and pull one or more digits into marked flexion at the metacarpophalangeal joint, sometimes also flexing or extending interphalangeal joints; this is called *Dupuytren's contracture.*

The **thenar** and **hypothenar fascias** attach to the 1st and 5th metacarpals, respectively, and at the junction of each with the palmar aponeurosis, an **intermuscular septum** runs deeply to attach to these same metacarpals. Thus, the palm of the hand is divided into three **compartments** (Fig. 13-21): the large central compartment behind the palmar aponeurosis contains most of the tendons, nerves, and vessels of the hand; the thenar compartment contains the thenar muscles; and the hypothenar compartment contains the hypothenar muscles. Thin, transversely running muscle bundles, the **palmaris brevis,** arise superficially at about the junction of the palmar aponeurosis and the hypothenar fascia and insert into skin of the ulnar border of the hand.

The fascia of the dorsum of the wrist is thickened by heavy, almost transverse fibers that form the **extensor retinaculum** (Fig. 13-24). This runs obliquely from the distal part of the anterior border of the radius to the styloid process of the ulna and the more ulnar carpals.

Distal to the extensor retinaculum, the **fascia of the dorsum of the hand** is continued out

**FIG. 13-21.** The palmar fascia and the compartments of the hand in a cross section of the hand. Fascial layers of importance are shown in red.

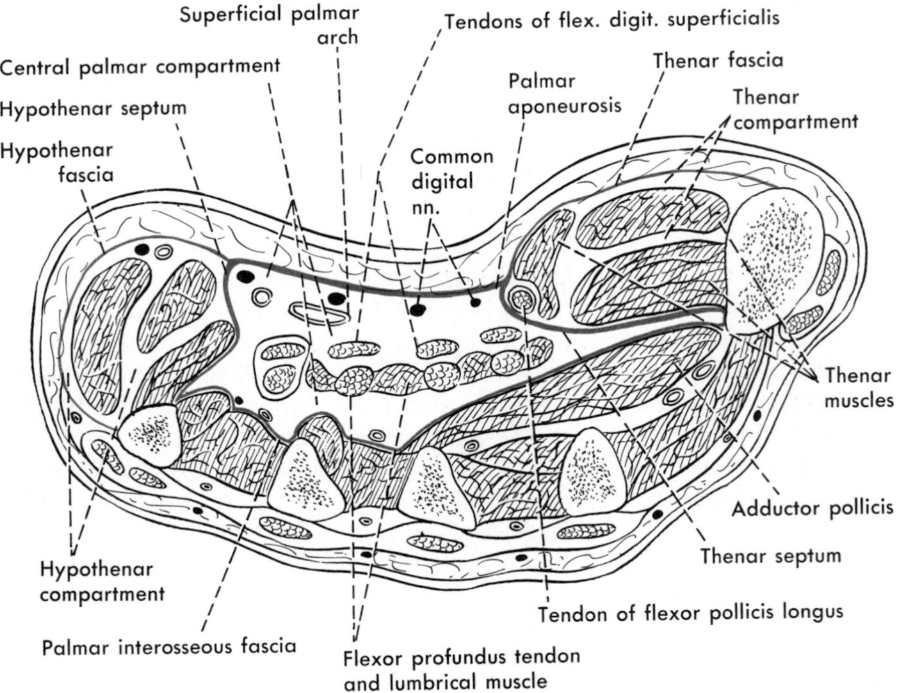

over the extensor tendons as they spread to the fingers. At the bases of the fingers, the fascia fades out by blending with the tissue of the webs of the fingers and with the extensor tendons as they cross the metacarpophalangeal joints.

## POSTERIOR ASPECT

The muscles, nerves, and vessels of the posterior aspect of the forearm and the dorsum of the hand are much simpler than are those of the flexor and palmar aspects. This is particularly true of the hand, where all of the muscles are palmar ones.

The extensor muscles of the forearm are conveniently divisible into two groups, superficial and deep. All the muscles are supplied by the radial nerve or its posterior interosseous branch (see Fig. 11-12). The more radial muscles of the extensor group are supplied by branches of the radial artery, and the others are supplied by the posterior interosseous artery.

### SUPERFICIAL EXTENSOR MUSCLES AND ASSOCIATED STRUCTURES

The superficial extensor muscles (Figs. 13-22 and 13-23) arise from the humerus and from the deep fascia covering them, and one arises also from the ulna. Those that arise from the lateral epicondyle fuse with each other and with intermuscular septa from the antebrachial fascia to form a common extensor tendon attached to the epicondyle.

In the hand, the cutaneous nerves lie superficial to the extensor tendons, but the arteries of the dorsum lie mostly deep to them (Fig. 13-19).

**Extensor Retinaculum.** The extensor retinaculum already has been mentioned as a thickening of the posterior antebrachial fascia at the wrist. As this retinaculum runs obliquely from the lateral border of the radius to the medial border of the ulnar styloid process and the triquetral and pisiform bones, it sends septa to the underlying bones, thus separating the space deep to it into a number of compartments through which run the extensors of the wrist and digits. Within each compartment, the tendon or tendons are protected by a synovial tendon sheath, which begins slightly above the upper end of the retinaculum and extends a variable distance onto the dorsum of the hand. There are typically **six compartments** and therefore six tendon sheaths. In order from the radial side (Fig. 13-24) as they reach the wrist, these compartments and sheaths are for the abductor pollicis longus and the extensor pollicis brevis, the two radial extensors, the extensor pollicis longus, the extensor digitorum and extensor indicis, the extensor digiti minimi, and the extensor carpi ulnaris. A common variation is for the long abductor and short extensor of the thumb to have separate sheaths and compartments, thus increasing the total to seven.

**Brachioradialis.** The brachioradialis, the most anterior member of the extensor group, arises from the anterior surface of the lateral border of the humerus well above the lateral epicondyle, between the triceps and brachialis muscles (Fig. 13-23). It passes in front of the elbow joint, and its long flat tendon inserts on the lateral side of the lower end of the radius. In the lower part of the arm, the radial nerve lies between the brachioradialis and the brachialis and supplies the former before it divides into its superficial and deep branches. In the forearm, the superficial branch of the radial nerve first lies deep to the muscle and then emerges posteriorly to continue into the hand. In the lower part of the forearm, the tendons of this and the following two muscles are crossed superficially by thumb muscles.

**Extensor Carpi Radialis Longus and Brevis.** The **extensor carpi radialis longus** arises from the lateral border of the humerus immediately below the brachioradialis, by which it is partly covered. It passes across the front of the lateral aspect of the elbow joint and tapers to a flat tendon that, deep to the extensor retinaculum, shares a tendon sheath with the extensor carpi radialis brevis and then inserts upon the base of the second metacarpal.

The **extensor carpi radialis brevis** arises from the lateral epicondyle by the *common extensor tendon,* lying at first largely deep to the extensor longus and then on its ulnar side. Its tendon inserts on the base of the third metacarpal.

The *radial nerve* supplies the long radial extensor while its deep branch, the posterior interosseous nerve, supplies the short extensor.

**Extensor Digitorum and Extensor Digiti Minimi.** The **extensor digitorum** arises (Fig. 13-25) from the lateral epicondyle of the humerus by the *common extensor tendon* and lies between the extensor carpi radialis brevis and the extensor carpi ulnaris. It forms a centrally placed superficial mass on the posterior surface of the forearm. On its ulnar side, it gives rise to the extensor digiti minimi muscle, which has its own separate tendon of insertion, and then typically gives rise to four tendons. Muscles of the thumb that arise more deeply become superficial at the radial border of this muscle, spiraling downward and laterally over the extensor muscles already described.

**FIG. 13-22.** Posterior or extensor muscles of the forearm.

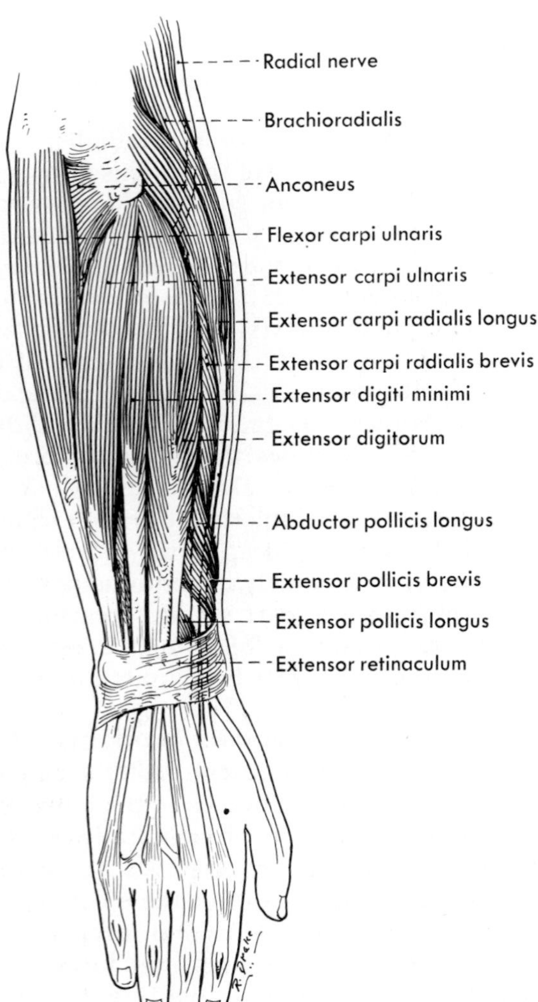


**FIG. 13-23.** Attachments of some of the superficial muscles of the extensor group in the forearm.

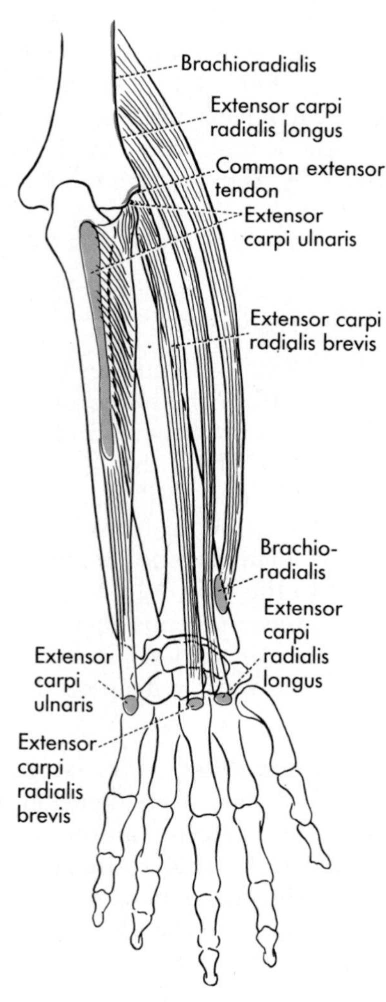

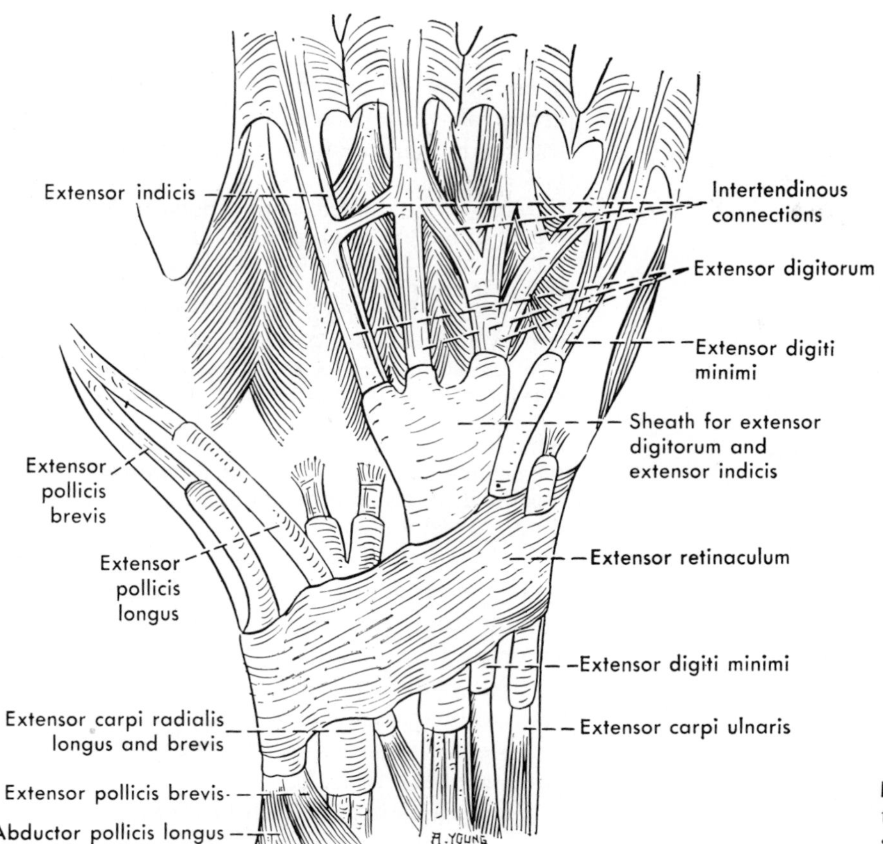


**FIG. 13-24.** Tendons and tendon sheaths on the dorsum of the wrist.

The **four tendons** of the extensor digitorum (there may be only three) pass through a large compartment of the extensor retinaculum and then diverge toward the fingers. As they do so, they exchange fibrous slips, the *intertendinous connections,* between themselves. At the metacarpophalangeal joints, the tendons to the index and little fingers are joined by the tendons of the extensor indicis and extensor digiti minimi, respectively (or if there are only three extensor digitorum tendons, the most ulnar tendon typically sends a slip to the extensor digiti minimi). As the extensor tendons reach the dorsal aspect of the metacarpophalangeal joints, they expand laterally over the sides of the joints, so that in addition to the *central thicker portion* of each tendon, there are thinner *lateral portions* that help to cover the joint. These lateral portions consist in part of fibers that run transversely and hold the tendon in place and in part of fibers from the insertions of the lumbrical and interosseous muscles (pp. 261 and 270). This expansion of the extensor tendon over the metacarpophalangeal joint is referred to by clinicians as the **extensor hood.** Some of its transverse fibers anchor the tendon to the palmar side of the joint.

As the extensor tendon proper continues distally on the proximal phalanx, the more obliquely directed tendinous fibers of the lumbrical and interosseous muscles join it, forming a thin flat tendon often known as the **extensor aponeurosis.** Proximal to the proximal interphalangeal joint, fibers belonging to all the muscles forming the aponeurosis participate in a split that results in three *tendinous bands,* a central one and two lateral ones (Fig. 13-26). The central band ends by inserting on the base of the middle phalanx. The two lateral bands come together on the distal part of the middle phalanx to form a single tendon that crosses the distal interphalangeal joint to insert upon the base of the distal phalanx.

The central and lateral bands at the proximal interphalangeal joint are anchored to

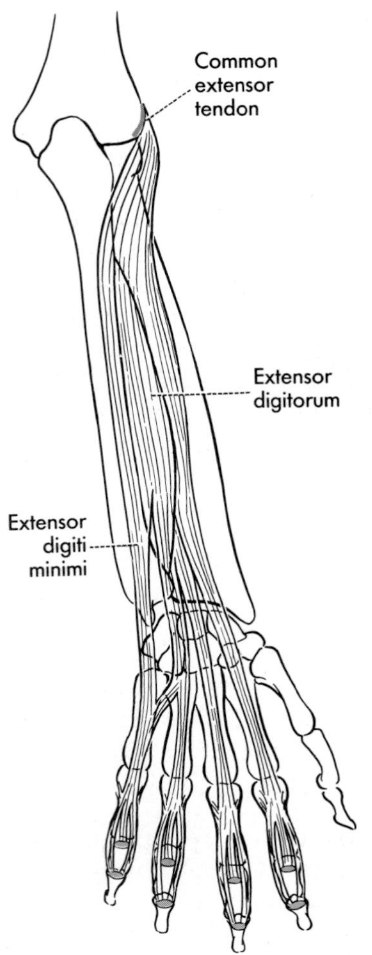


**FIG. 13-25.** Attachments of the extensor digitorum and extensor digiti minimi.

each other and to various elements of the digit by groups of fibers that have been investigated in considerable detail. One of the most important of these, usually called the *oblique cord,* consists of fibers that extend distally and dorsally from the *palmar* aspect of the proximal phalanx to the lateral band. In consequence of this attachment, flexion of the distal phalanx, which pulls the lateral bands distally, so tightens the oblique cords that they produce an equal amount of flexion of the middle phalanx. Similarly, extension of the middle phalanx produces extension of the distal phalanx by pull upon the oblique cords. (These effects were described almost simultaneously by two independent workers and have been generally accepted by surgeons. One group of workers, however, has claimed that the cords merely stabilize the tendon.)

Most of the clinical implications of the attachments of the extensor tendons over the joints cannot be considered here, but it is perhaps easy to see that an injury that detaches the lateral bands from the central band at a proximal interphalangeal joint will allow these bands to slip farther forward than usual during flexion of the finger, and if they slip far enough forward, efforts to extend the joint will result instead in flexion. Similarly, if the central band is torn from its insertion, the lateral bands will take all the pull and as they slip forward will tend to flex the proximal interphalangeal joint but hyperextend the distal one.

The *deep branch of the radial nerve* (also called the *posterior interosseous branch*) emerges from the supinator muscle deep to the extensor digitorum and typically sends several branches into it. The *posterior interosseous artery* also sends branches into the deep surface of the muscle.

The **extensor digiti minimi** (extensor of the little finger) arises from the ulnar aspect of the extensor digitorum. After it separates from this muscle, its tendon passes through a separate compartment and synovial tendon sheath deep to the extensor retinaculum and usually

**FIG. 13-26.** An extensor tendon (extensor aponeurosis) of a finger.

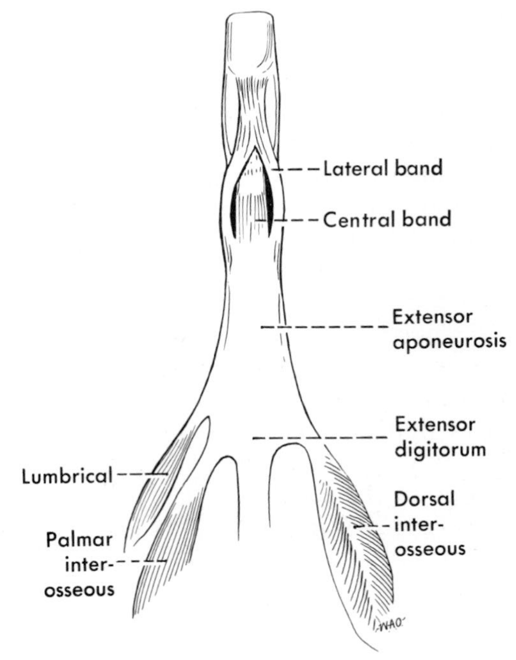

splits into two tendons, of which the radial one joins the extensor digitorum tendon of the little finger or receives an intertendinous connection from the tendon to the ring finger. Attachments on the finger are identical to those already described for the extensor digitorum.

**Extensor Carpi Ulnaris.** The extensor carpi ulnaris arises by two heads (Fig. 13-23), one from the lateral epicondyle through the common extensor tendon, the other from the posterior border of the ulna where an aponeurosis covering it is firmly attached both to the muscle and to the bone. Its tendon passes through a special compartment of the extensor retinaculum and inserts into the medial side of the base of the fifth metacarpal.

**Variations, Innervations, and Actions**

Major **variations** of the superficial extensor muscles are not common. Occasionally, the tendon of a muscle may divide and have an anomalous insertion into a metacarpal, a carpal, or an adjacent tendon. Similarly, there are sometimes aberrant muscle slips uniting adjacent muscles or attaching independently into the wrist or hand.

All these muscles are **innervated** through the *radial nerve,* but the lateral ones (brachioradialis and the long radial extensor) receive their innervation from the radial nerve while this lies on the front of the arm or forearm. The posterior ones (the extensor carpi radialis brevis, digitorum, digiti minimi, and carpi ulnaris) receive their innervation on their deep surfaces through the *deep branch of the radial nerve.* The short radial extensor is supplied in a large proportion of cases by the radial nerve before it divides. It is usually stated that the brachioradialis receives fibers from the 5th and 6th cervical nerves, the radial extensors from about C6 and C7, the extensor digitorum and extensor digiti minimi from C6, C7, and C8, and the ulnar extensor from these or C7 and C8 only.

**Actions.** The brachioradialis, by virtue of its anterior position at the elbow, is primarily a flexor at the elbow. The extensor carpi radialis longus is a weaker flexor at the elbow, and the extensor carpi radialis brevis normally is not a flexor here, but both are extensors at the wrist; both contract in radial abduction at the wrist, a movement often attributed primarily to muscles of the thumb.

The extensor digitorum is an extensor of all three phalanges of each of the four fingers. It also abducts (spreads) the fingers as it extends them. The extensor digiti minimi can both abduct and extend the little finger independently. The extensor carpi ulnaris is an excellent extensor and ulnar abductor at the wrist.

## DEEP EXTENSOR MUSCLES AND ASSOCIATED STRUCTURES

The deep extensor muscles are the supinator, three muscles to the thumb, and the extensor indicis (Fig. 13-27). The supinator is completely covered by overlying muscles. The thumb muscles arise deeply but emerge in the lower part of the forearm from under cover of the extensor digitorum and become superficial as they pass down to the wrist. The extensor indicis occupies a deep position until its tendon reaches the dorsum of the hand.

These muscles are innervated by the deep or posterior interosseous branch of the radial nerve.

**Supinator.** The supinator is a flat muscle that arises from the lateral epicondyle of the humerus, from the radial collateral and anular ligaments, and from the supinator crest of the ulna. It passes downward and laterally around the radius to insert into the lateral surface of the bone (Fig. 13-28), including the part of this surface that is anterior.

The deep branch of the *radial nerve* enters the upper edge of the supinator muscle on the anterior side of the forearm and spirals around the radius to reach the posterior aspect of the forearm within this muscle (Fig. 13-31), typically dividing it into two laminae. In this course, it gives branches into the supinator. The *posterior interosseous artery* lies deep to the supinator and appears at the lower border of the muscle, gives off the *recurrent interosseous artery,* which runs upward across the supinator, and continues downward with the posterior interosseous nerve (Fig. 13-33).

**Abductor Pollicis Longus and Extensor Pollicis Brevis.** The **abductor pollicis longus** (long abductor of the thumb) arises from parts of the posterior surfaces of the ulna and radius and from the intervening interosseous membrane. It emerges from under cover of the extensor digitorum and spirals distally around the radius, crossing the radial extensors in so doing. Its tendon (often double or triple) and that of the closely associated extensor pollicis brevis usually share the same tendon sheath deep to the extensor retinaculum, but even if they have separate sheaths, they form the anterior boundary of the "anatomic snuff box" (Fig. 13-29). The abductor then runs farther palmarward than does the extensor and inserts on the radial (anteriorly directed) surface of the base of the first metacarpal.

The **extensor pollicis brevis** (short extensor of the thumb) arises from the posterior surface of the radius and from the adjacent interosseous membrane below the origin of the long abductor and passes with that muscle's tendon almost to the base of the 1st metacarpal. The short extensor passes along the dorsal aspect of the 1st metacarpal to insert on the base of the proximal phalanx of the thumb, often sending a slip to insert into the long extensor tendon.

The *posterior interosseous artery* and *nerve* run downward across the upper parts of the long abductor and short extensor muscles.

**Extensor Pollicis Longus.** The extensor pollicis longus arises from the posterior surface of the ulna and the adjacent interosseous membrane, appearing from deep to the extensor digitorum a short distance above the wrist. Its tendon passes obliquely through its own compartment deep to the extensor retinaculum. As it crosses the tendons of the

**FIG. 13-27.** Deep muscles of the extensor group in the forearm.

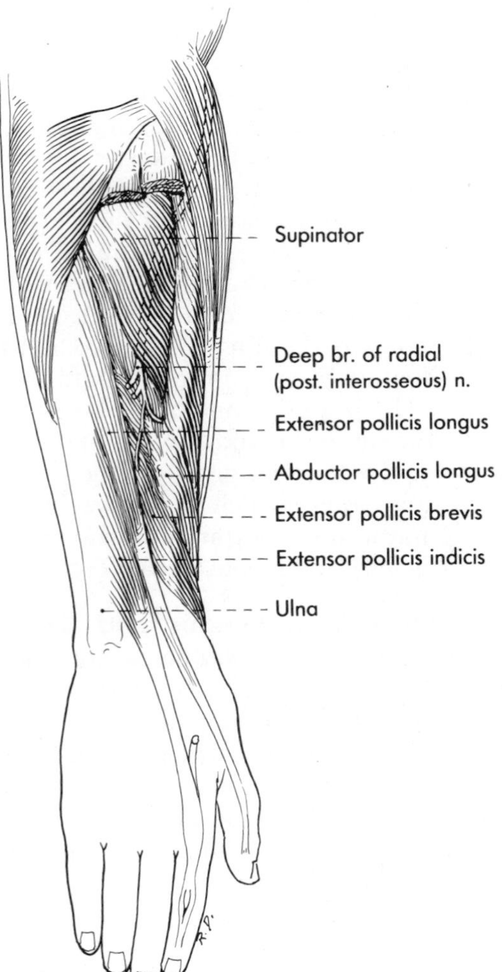


**FIG. 13-28.** Origins (red) and insertions (blue) of the deep group of extensor muscles of the forearm. The insertion of the abductor pollicis longus is too far on the anterior surface to be seen here.

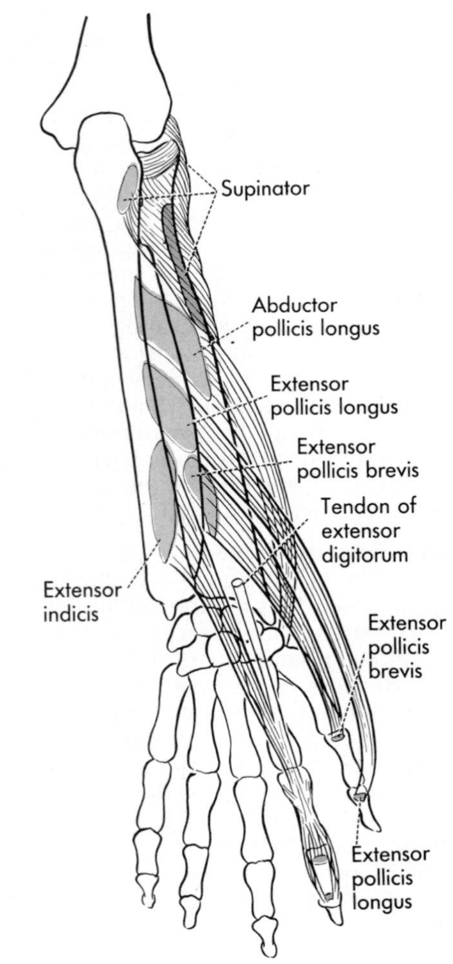

extensor carpi radialis longus and brevis muscles, its tendon sheath typically communicates with that of those muscles.

The extensor pollicis longus tendon forms the posterior border of the anatomic snuff box and then continues along the dorsal aspect of the 1st metacarpal and proximal phalanx to insert on the distal phalanx. Besides a frequent slip from the extensor pollicis brevis, it usually receives a thin tendinous slip on its radial side from the abductor pollicis brevis and one on its ulnar side from the adductor pollicis.

**Extensor Indicis.** The extensor indicis (extensor of the index finger) arises from the posterior surface of the ulna distal to the origin of the long extensor of the thumb and passes downward deep to the extensor digitorum. It crosses the wrist in the tendon sheath of the extensor digitorum, but on the hand, its tendon lies on the ulnar side of that muscle's tendon to the index finger and joins this tendon close to the base of the proximal phalanx.

The *posterior interosseous nerve* runs toward the wrist deep to the extensor indicis and extensor pollicis longus, but the *posterior interosseous artery* crosses them superficially.

#### Variations, Innervations, and Actions

Any of the muscles, but especially those of the thumb, may have slips connecting them or may be so fused as to be inseparable. The most common

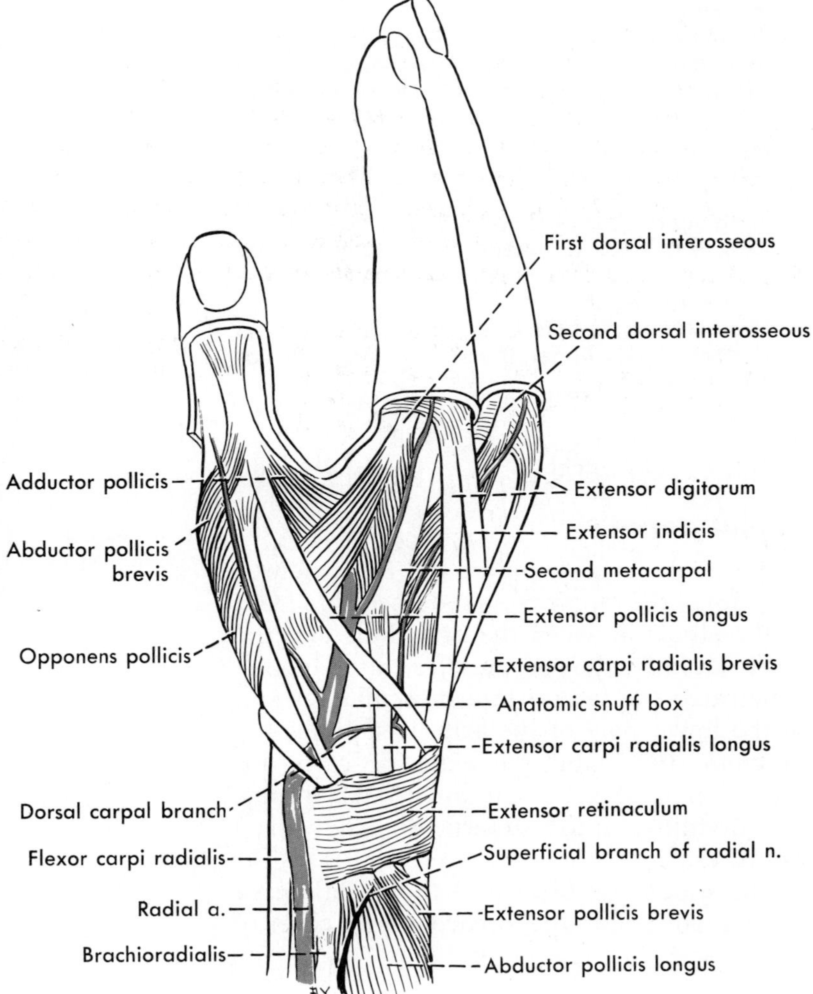


**FIG. 13-29.** Relationships on the radial side of the wrist.

variation is a doubling or tripling of the tendon of the long abductor, reported to occur in more than 60% of limbs; one tendon then usually inserts into the trapezium or into the muscles or fascia of the thenar eminence instead of into the metacarpal.

The **innervation** of all these muscles is from the deep branch of the radial nerve. The supinator is probably innervated primarily by C6, the long abductor and short extensor of the thumb by C6 and C7, and the extensor pollicis longus and extensor indicis by C7 and C8.

**Actions.** The **supinator** participates in all supination, carrying out this function by itself or, when required, with assistance from the biceps brachii.

The **abductor pollicis longus** abducts, extends, and laterally rotates the first metacarpal so as to draw the thumb away from the position of opposition. It and the extensor pollicis brevis together are also radial abductors at the wrist, and because of its anterior insertion, the abductor longus also flexes the wrist.

The **extensor pollicis brevis,** in addition to being a radial abductor at the wrist, is an extensor of the proximal phalanx of the thumb and sometimes of the distal phalanx and is an extensor of the metacarpal, moving this directly radially.

The **extensor pollicis longus** is primarily an extensor of the distal phalanx of the thumb, but in carrying out this action it also extends the proximal phalanx and tends to extend and adduct the first metacarpal. Apparently it can aid somewhat in extension at the wrist.

The **extensor indicis** is an extensor of all the joints of the index finger. It also independently adducts this finger.

## NERVES AND VESSELS

### Radial Nerve

The radial nerve supplies all the extensor muscles of the forearm. As it descends in front of the elbow between the brachialis and the more lateral extensor muscles, after having penetrated the lateral intermuscular septum in the lower part of the arm, it supplies the brachioradialis and the extensor carpi radialis longus (Figs. 13-30 and 13-31). It ends by dividing into superficial and deep branches.

The **superficial branch of the radial nerve** runs downward in the forearm under cover of the brachioradialis and, at about the junction of the middle and lower thirds of the forearm, passes posteriorly around the radius deep to the tendon of the muscle. On the posterior aspect of the forearm, it pierces the deep fascia and continues downward to the dorsum of the

**FIG. 13-30.** Course and muscular branches of the radial nerve in the forearm.

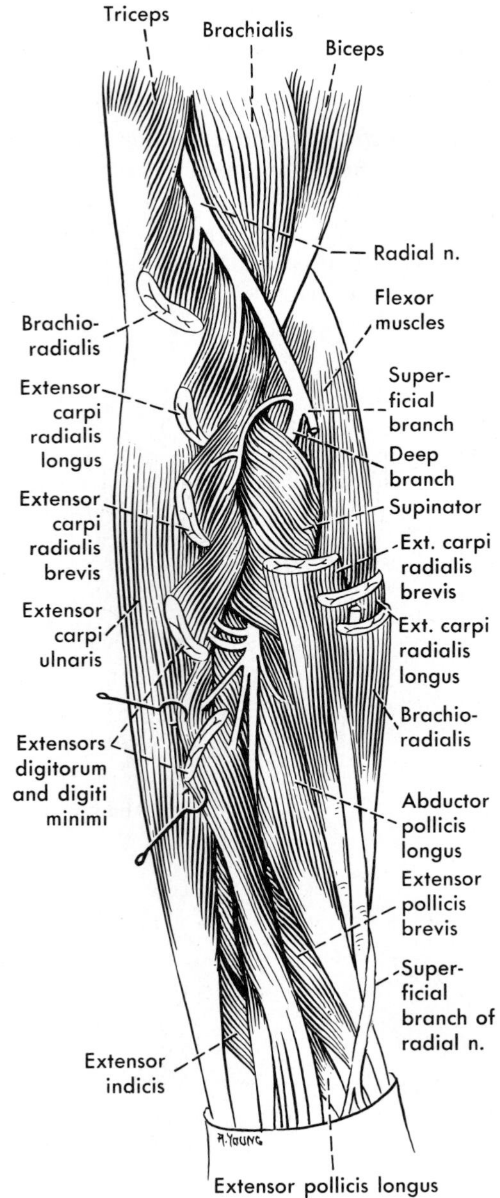

hand (Fig. 13-19). Its branches here have already been described.

The **deep branch of the radial nerve** enters the upper border of the supinator muscle on the anterior aspect of the forearm, gives off twigs into the supinator as it traverses that muscle, and often to the extensor carpi radialis brevis, then it curves through the supinator to the posterior aspect of the forearm. As it emerges from the supinator, it breaks up

**FIG. 13-31.** Further details of the course and branches of the radial nerve after some of the overlying muscles and their nerves have been removed.

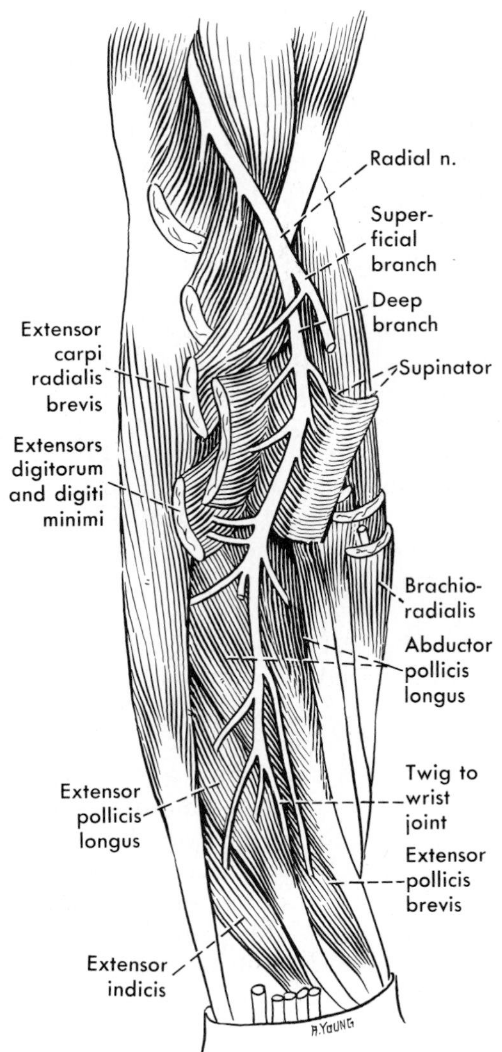


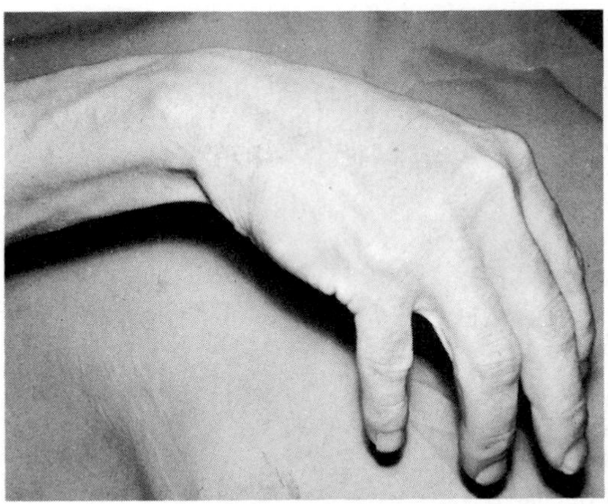

**FIG. 13-32.** Wristdrop, inability to extend the wrist and digits, resulting from a lesion of the radial nerve. (Courtesy of Dr. E. D. Henderson)

into a number of branches that are given off in no regular order, but some of which immediately turn superficially into the overlying extensor digitorum, extensor digiti minimi, and extensor carpi ulnaris muscles; the remaining part, the **posterior interosseous nerve,** runs downward parallel to the posterior interosseous artery to supply all the deeper-lying extensor muscles of the forearm and ends as a twig to the wrist joint. In this course, it passes superficial to the long abductor and short extensor of the thumb, but its terminal branch to the wrist joint passes deep to the extensor pollicis longus and extensor indicis muscles. Because of the numerous branches into which the deep radial nerve breaks up at the lower border of the supinator muscle, surgical repair of the nerve here is difficult.

**Lesions** of the radial nerve in the lower part of the arm may paralyze all the extensor muscles of the forearm, producing a "wristdrop" (flexion of the hand by gravity when the forearm is horizontal) as well as inability to extend the metacarpophalangeal joints of the digits (Fig. 13-32). Because of the shortness of the extensor muscles of the digits, however, a person with wristdrop can passively extend his wrist by tightly clenching his fingers. Lesions of the deep branch may have varying effects, but complete ones abolish extension at the metacarpophalangeal joints of the digits but allow extension of the wrist through the radial extensors.

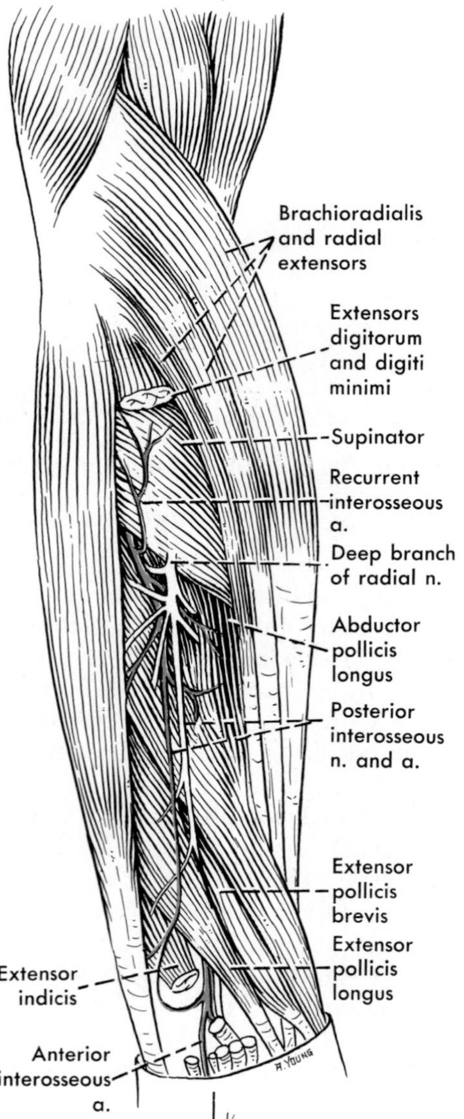


**FIG. 13-33.** Arteries and branches of the radial nerve in the posterior aspect of the forearm.

## Arteries

**Forearm.** The more anterior muscles of the extensor group are closely associated with the **radial artery** (Fig. 13-42), which lies at first under cover of the brachioradialis. Its largest branch to the extensor muscles is the **radial recurrent artery,** which runs upward on the supinator and then between the brachialis and the extensor carpi radialis longus with the radial nerve. It supplies the brachioradialis and the radial carpal extensors and anastomoses with the radial collateral branch of the profunda brachii artery.

The arteries of the posterior aspect of the forearm are two, the posterior interosseous and the posterior terminal branch of the anterior interosseous (Fig. 13-33).

The **posterior interosseous artery,** a branch of the common interosseous, reaches the posterior aspect of the forearm by passing above the upper border of the interosseous membrane and then runs downward on the membrane deep to the supinator muscle. On emerging at the lower border of the muscle, it gives off the **interosseous recurrent artery,** which passes upward on or through the supinator and deep to the anconeus to anastomose behind the lateral epicondyle with the middle collateral branch of the profunda brachii (see Fig. 12-49). Other branches are given off directly to the extensor muscles, and the remaining part of the artery descends across the thumb muscles, where it may anastomose with the posterior branch of the anterior interosseous artery.

The posterior terminal branch of the **anterior interosseous artery** pierces the lower part of the interosseous membrane and runs downward deep to the muscles, giving branches to them. It may anastomose with the posterior interosseous artery and continues downward to participate in the formation of the dorsal carpal rete.

**Hand.** The arteries on the dorsum of the hand are derived from the radial artery and from the dorsal carpal rete. The **radial artery** winds around the wrist deep to the abductor pollicis longus and extensor pollicis brevis tendons, traverses the anatomic snuff box, and passes deep to the extensor pollicis longus tendon to reach the interval between the bases of the 1st and 2nd metacarpals (Figs. 13-29 and 13-34). In this course, it gives off branches to the dorsum of both sides of the thumb, one to the radial side of the index finger, and one or more dorsal carpal branches that run transversely to help form the carpal rete. The radial artery then passes palmarward between the two heads of the first dorsal

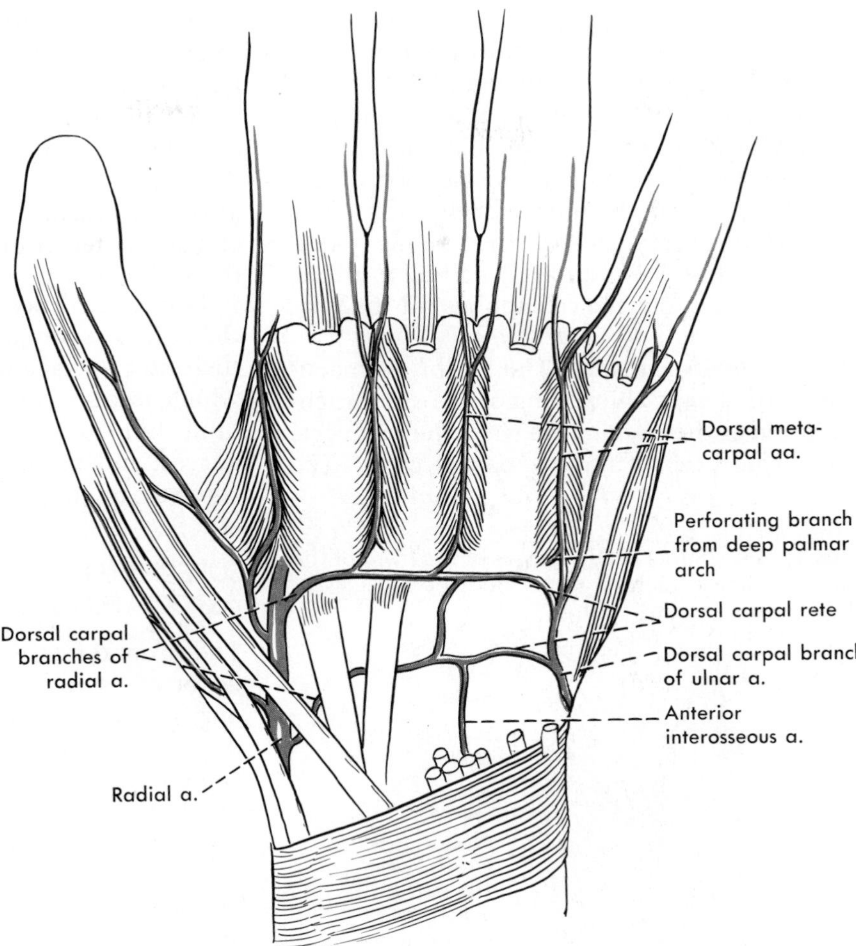


**FIG. 13-34.** Arteries of the dorsum of the hand.

interosseous muscle, dividing as it does so into the *princeps pollicis artery* and the *deep palmar arch.*

The **dorsal carpal branches** of the *radial* and *ulnar arteries* are joined by the terminal posterior branch of the *anterior interosseous artery* to form the **dorsal carpal rete.** This may be a fairly simple arch or a number of anastomosing loops. It gives off three **dorsal metacarpal arteries** distally. As these cross the bases of the 2nd, 3rd, and 4th dorsal interosseous muscles, they receive from the deep palmar arch perforating branches that may be their chief source of blood. They divide close to the heads of the metacarpals into small **dorsal digital arteries** that run a little distance out along the adjacent sides of the two digits with which they are associated. Close to their points of division, the dorsal metacarpal arteries receive perforating branches from the palmar metacarpal ones.

## ANTERIOR ASPECT

The flexor muscles of the forearm are concerned with pronation of the forearm, with flexion and abduction at the wrist, and with flexion of the digits. They are conveniently divided into three groups: a superficial group, an intermediate muscle, and a deep group.

The radial and the ulnar artery run downward in the anterior aspect of the forearm to the palm. The median and ulnar nerves also run down anteriorly and continue into the palm, but the ulnar nerve first sends a cutaneous branch to the dorsum. In the forearm, these nerves are very unequally distributed to the muscles (see Figs. 11-10 and 11-11), for the ulnar nerve supplies only one muscle and part of another, whereas the median nerve supplies all the other anterior muscles of the forearm. In the hand, the muscular distribution is somewhat reversed, for the median nerve typically supplies only four and one-half muscles, whereas the ulnar nerve supplies all the rest.

**Flexor Retinaculum.** The flexor retinaculum is the heavy thickening of the antebrachial fascia at the front of the wrist that converts the carpal groove into the carpal canal. The palmaris longus tendon passes in front of the retinaculum (Fig. 13-35), as do the ulnar nerve and vessels (Fig. 13-44). The median nerve and tendons of the long flexors of the fingers and of the thumb pass behind it, in the carpal canal. Of the two flexors of the wrist, one stops at the wrist and the other runs through a special compartment in the radial attachment of the retinaculum.

Below the flexor retinaculum, the tendon of the palmaris longus becomes continuous with the outer surface of the palmar aponeurosis, which is also continuous with the retinaculum, but most of the important tendons, nerves, and vessels are behind the aponeurosis and are visible only after this is removed.

**FIG. 13-35.** Superficial muscles of the front of the forearm. These include more anterior muscles of the extensor group.

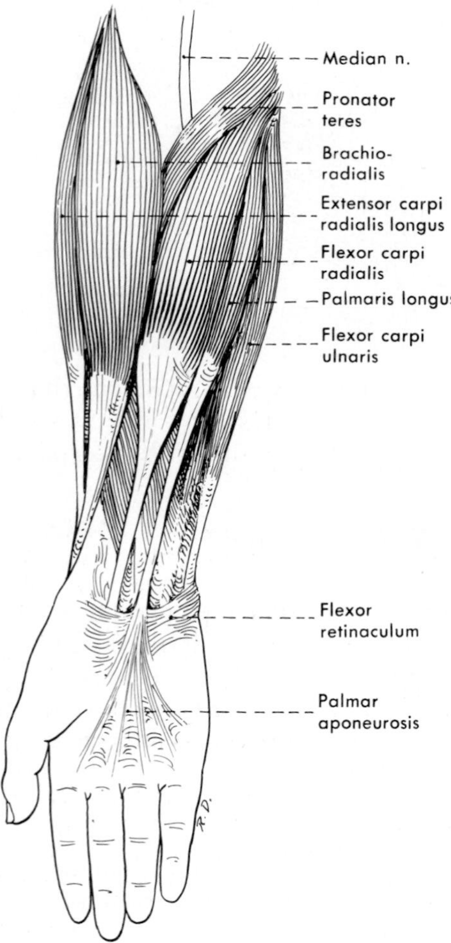


## SUPERFICIAL MUSCLES AND ASSOCIATED STRUCTURES

The pronator teres, the palmaris longus, the flexor carpi radialis, and the flexor carpi ulnaris all arise in part from the medial epicondyle of the humerus through a *common flexor tendon* (Fig. 13-35). The first three are supplied by the median nerve, and the last by the ulnar.

**Pronator Teres.** The pronator teres arises by two heads, a *humeral head* (partly covered by and closely fused to the flexor carpi radialis) from the upper part of the medial epicondyle and an *ulnar head* from the coronoid process of the ulna (Fig. 13-36). As the two heads pass downward and laterally toward the radius, the median nerve passes between them, supplying both, after which they unite and go to an insertion on the lateral side of the radius, at about its middle.

The muscle forms the medial border of the *cubital fossa,* in which the *brachial artery* normally ends. The *ulnar artery* leaves the fossa by passing deep to the pronator teres. The *radial artery* runs downward deep to the brachioradialis muscle and crosses superficial to the pronator close to its insertion.

**Flexor Carpi Radialis.** The flexor carpi radialis arises from the medial epicondyle by the common flexor tendon and also, like all the muscles of the superficial group, from the overlying fascia. It passes downward toward the radial side of the forearm, and at the wrist, its tendon enters a special compartment formed by the splitting of the attachment of the flexor retinaculum to the trapezium. It inserts on the base of the 2nd metacarpal and sometimes sends a slip to the 3rd metacarpal.

As the *median nerve* lies deep to the flexor carpi radialis, it supplies that and the palmaris longus. In the lower third of the forearm, the *radial artery* appears from under cover of the brachioradialis and runs downward in a superficial position lateral to the tendon of the flexor carpi radialis.

**Palmaris Longus.** The palmaris longus arises from the deep fascia, from the medial epicondyle via the common flexor tendon, and from septa between it and adjacent muscles. The muscular belly is typically short and succeeded by a long slender tendon that crosses superficial to the flexor retinaculum and ends by becoming continuous with (inserting into) the superficial fibers of the palmar aponeurosis. Just above the wrist, the tendon lies almost directly in front of the *median nerve,* which is most easily located here through this tendon.

**Flexor Carpi Ulnaris.** The flexor carpi ulnaris arises by *two heads,* one from the medial epicondyle of the humerus by way of the common flexor tendon, the other from the posterior border of about the upper three-fifths of the ulna through an aponeurosis that covers the medial side of the flexor digitorum profundus. They join just below the medial epicondyle, and the *ulnar nerve,* after passing behind the epicondyle, runs between them to lie deep to the muscle and give off branches into it. The muscle inserts on the pisiform bone, usually said to be a sesamoid bone. The continuation of the tendon beyond the pisiform is then said to be represented by the *pisometacarpal* and *pisohamate ligaments.*

Deep to the muscle, the ulnar nerve is joined in its course by the *ulnar artery.* Close to the wrist, the two emerge from under cover of the flexor carpi ulnaris and pass across the front of the flexor retinaculum.

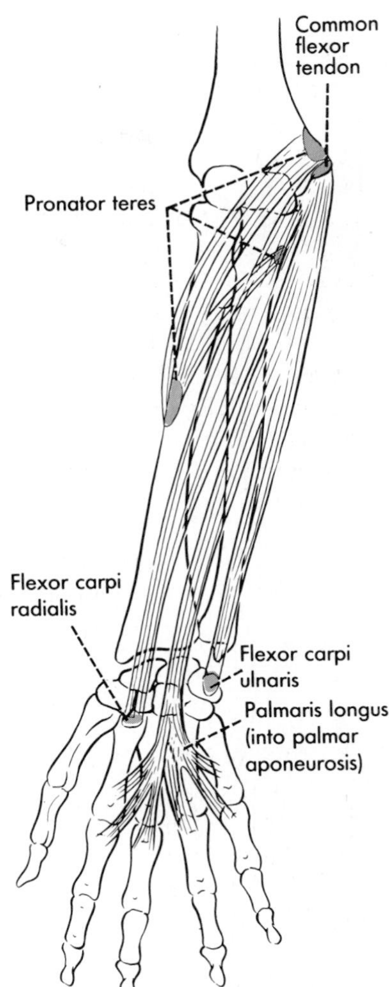


**FIG. 13-36.** Attachments of the superficial layer of flexor muscles; the ulnar origin of the flexor carpi ulnaris, which is on the posterior surface of the ulna, is not visible here.

**Variations, Innervations, and Actions**

The ulnar head of the **pronator teres** has been reported as absent in almost 9% of limbs. The most common variation among this group of muscles, however, is absence of the **palmaris longus,** occurring in some 12% or more of limbs. This muscle also is sometimes completely duplicated or has two tendons or an abnormal origin. Abnormalities of the flexor carpi radialis and flexor carpi ulnaris are apparently rare.

The pronator teres, the flexor carpi radialis, and the palmaris longus are **innervated** by the *median nerve,* and the flexor carpi ulnaris is innervated by the *ulnar.* The number and the exact origin of the nerves to the muscles vary; several branches of the median nerve, for instance, may arise by a common

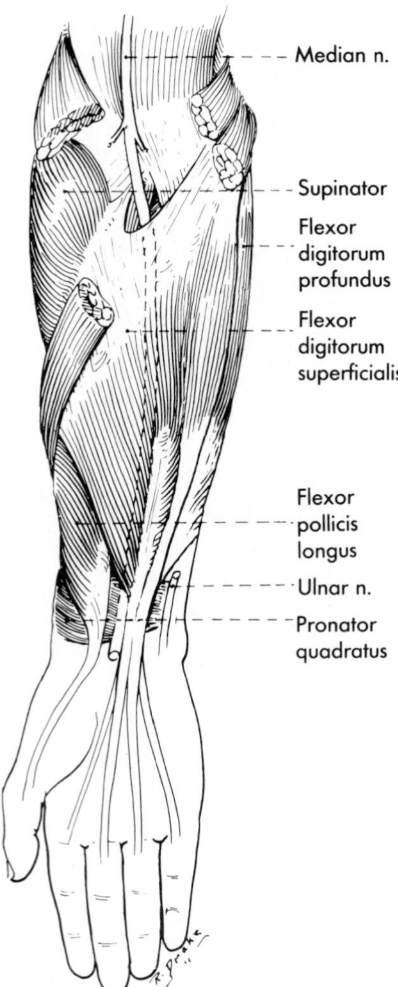


**FIG. 13-37.** The flexor digitorum superficialis, the intermediate layer of anterior muscles of the forearm.

stem or arise separately. The pronator teres and the flexor carpi radialis usually are said to be supplied with fibers from C6 and C7, the palmaris longus with fibers from C7, C8, and T1, and the flexor carpi ulnaris from C8 and T1 and perhaps also fibers from C7 that join the ulnar nerve through its lateral root.

**Actions.** The **pronator teres,** in consequence of its oblique course across the forearm and its insertion into the radius, is primarily a pronator. It is also a weak flexor of the forearm. The **flexor carpi radialis** is primarily a flexor at the wrist. As it flexes, it can aid in pronation, but it does not contract for pronation alone. It is said that it contracts during radial abduction at the wrist, but some authors deny that it actually aids in abduction. The **palmaris longus** has no other function than flexion at the wrist. The **flexor carpi ulnaris** is both a flexor and an ulnar abductor at the wrist.

## INTERMEDIATE MUSCLE AND ASSOCIATED STRUCTURES

The intermediate muscle layer is a single muscle, the **flexor digitorum superficialis** (Figs. 13-37 and 13-38). It arises by two heads and inserts by four tendons on the middle phalanges of the four fingers. It is innervated by the median nerve through branches that

**FIG. 13-38.** Attachments of the flexor digitorum superficialis.

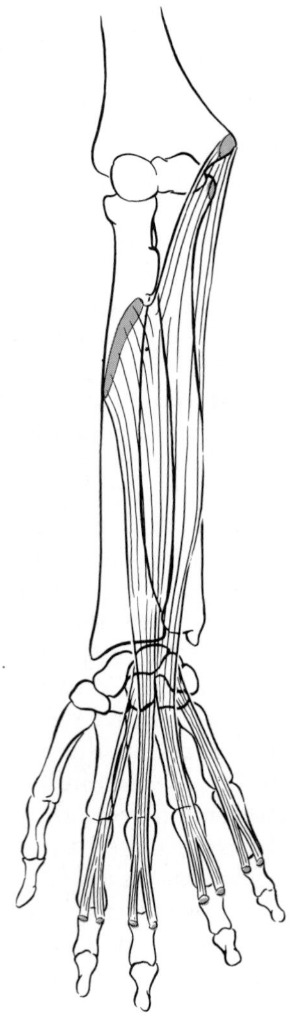

enter its deep surface and supply it with fibers derived from C7 to T1.

Its large *humeroulnar head* arises from the medial epicondyle by way of the common flexor tendon and from the medial border of the base of the coronoid process of the ulna. Its *radial head* arises from the upper part of the anterior border of the radius. The *radial artery* crosses this head anteriorly. The two heads are united high in the forearm by a dense membrane, behind which the *median nerve* and *ulnar artery* run downward. In the lower part of the forearm, the muscle gives rise to four tendons, of which two, to the middle and ring fingers, are placed in front of the two to the index and little fingers. As the tendons approach the wrist in this position, the *median nerve* emerges from behind the muscle on the radial side of the tendons and then passes forward to lie in front of them as nerve and tendons disappear deep to the flexor retinaculum.

As the tendons pass behind the flexor retinaculum, they lie in a large **common synovial tendon sheath** of the flexor muscles that they share with the flexor digitorum profundus. The median nerve lies anterior to this sheath immediately behind the flexor retinaculum, and within the sheath, the superficial flexor tendons lie anterior to the deep flexor tendons (Fig. 13-39). The flexor tendon sheath normally begins a short distance above the upper edge of the flexor retinaculum (Fig. 13-51) and occupies the major part of the carpal canal. On its radial side is the tendon and tendon sheath of the flexor pollicis longus muscle.

In the hand, the superficial branch of the ulnar artery forms the *superficial palmar arch* (Fig. 13-44) immediately in front of the flexor tendons and their sheath. The *digital arteries* derived from the superficial arch, and the *digital branches of the median and ulnar nerves,* run toward the digits anterior to the superfi-

**FIG. 13-39.** Cross section of the wrist showing the flexor retinaculum (red) and the carpal canal behind it.

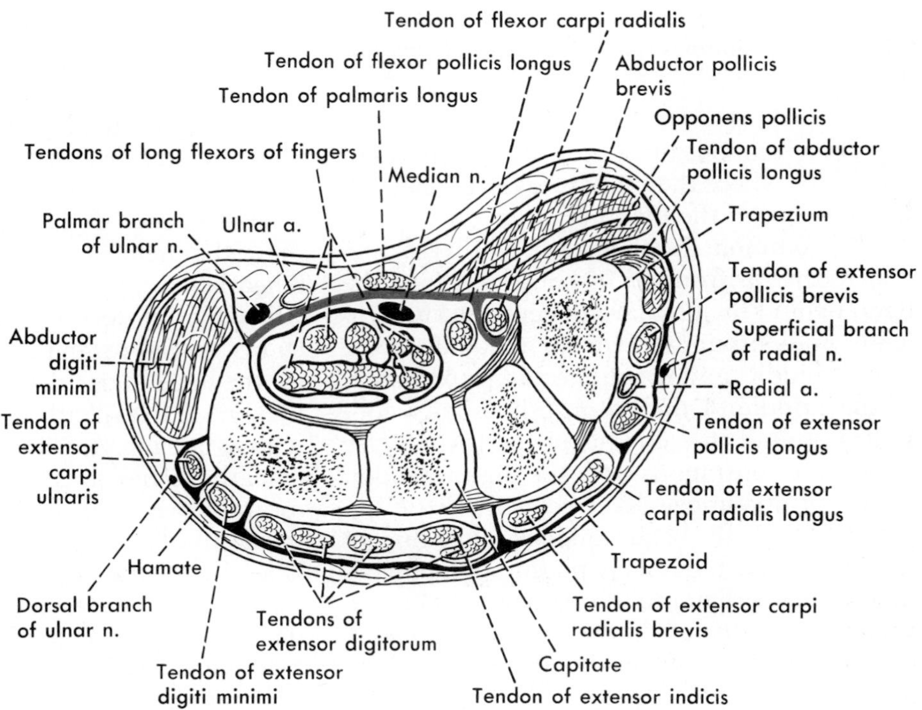

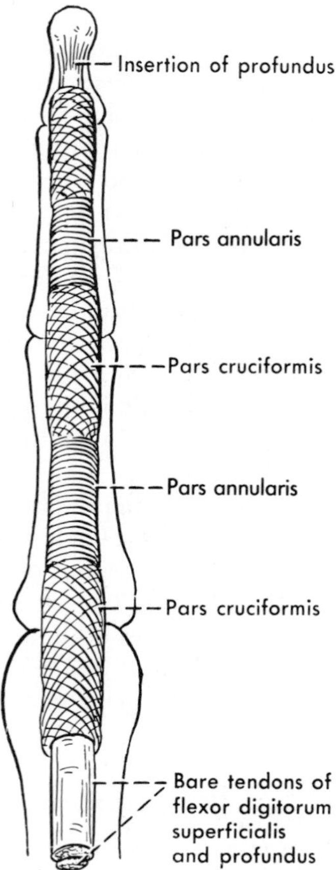


**FIG. 13-40.** Composition of a digital tendon sheath.

cial flexor tendons until they near the bases of the digits, where they sink into the connective tissue between the tendons.

The **common flexor sheath,** with the median nerve in front of it, continues into the hand behind the palmar aponeurosis and typically ends around the flexor tendons to the index, middle, and ring fingers approximately in the middle of the palm, but usually continues around the tendons of the little finger to become continuous with the digital tendon sheath of this finger. Except for the little finger, therefore, the flexor tendons to the fingers (the profundus tendons lying immediately behind the superficial tendons) typically have a variable length of "bare" tendon (unsurrounded by a sheath) in the distal part of the palm.

**Flexor Tendons of Fingers.** As the three radial, bare, tendons of the flexor digitorum superficialis muscle near the metacarpophalangeal joints of the fingers, they and the underlying profundus tendons enter **digital tendon sheaths** on the fingers. These sheaths consist of a rather heavy outer *fibrous portion* and an inner *synovial portion.* The fibrous wall consists of alternating thin and thick portions. In front of the metacarpophalangeal and each interphalangeal joint, fibers run obliquely to form the thin cruciform part of the sheath. Over the bodies of the proximal and middle phalanges, heavier fibers run transversely to form the annular part of the sheath (Fig. 13-40). More recently, the cruciform part in front of the metacarpophalangeal joint has been described by surgeons as being interrupted by a narrow annular part (not shown in the figure) that is attached to the palmar ligaments. The transverse fibers serve as retinacula that keep the long flexor tendons in close contact with the bones and are frequently referred to by clinicians as the "pulleys" of the flexor tendons. The synovial layer of the tendon sheath is a complete tube, but the fibrous layer is an arch across the front of the tendons and is attached to the phalanges and the heavy palmar ligaments that form the front parts of the capsules of the joints. Fibrous and synovial layers of the digital tendon sheaths end distally just short of the insertion of the flexor digitorum profundus tendons on the distal phalanges.

The fibrous digital sheaths end proximally over the metacarpal heads, as do the synovial sheaths, except for that of the little finger. If the usual pattern of the synovial sheaths obtains, therefore, an infection within the sheath of the little finger can travel into the common flexor tendon sheath at the wrist with no difficulty, but infections within the other tendon sheaths either must be retained within them or must break through their walls. If the break is at the weak proximal end of the digital sheath, infections within the sheaths of the index, middle, and ring fingers tend to pass deep to the long flexor tendons into the loose connective tissue here and produce infections in the palmar fascial spaces (Fig. 13-50).

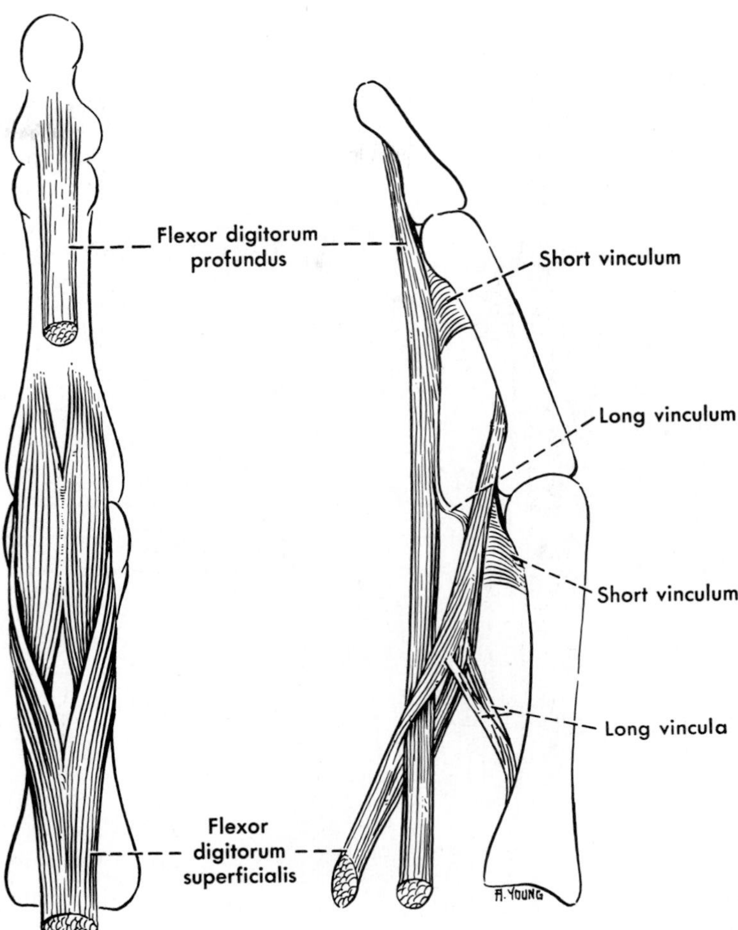


**FIG. 13-41.** The arrangement of the superficial and deep flexor tendons of a digit.

Within the flexor tendon sheaths of the fingers, each tendon of the flexor digitorum superficialis at first lies in front of that of the flexor digitorum profundus, but over the body of the proximal phalanx it becomes deeply grooved and then splits into two parts to allow the profundus to go through (Fig. 13-41). The two parts of the superficialis tendon spiral on each side of the profundus and come together dorsal to it and thus form a tunnel for the profundus tendon. They cross the proximal interphalangeal joint, interchange a few fibers, and insert on each side of the palmar aspect of the base of the middle phalanx. As they come together, but before they reach their insertion, they are held dorsally to the palmar ligaments and the adjacent bone by a triangular *short vinculum.* A slender band, a *long vinculum,* usually attaches each of the two parts of the superficial tendon to the proximal phalanx. Blood vessels reach the tendon by way of the vincula, especially the short one.

Successful surgical repair of the flexor tendons after injury within the digital tendon sheaths has been difficult, mostly because of the difficulty of recreating the sliding mechanism provided by the synovial membranes between the two sets of tendons and their surroundings. Indeed, these parts of the tendons were once termed the "no man's land" of tendon surgery. Traditionally, tendons injured within the digital sheaths were removed; a tendon graft from the palm of the hand to the insertion of the flexor digitorum profoundus was made to replace the profundus tendon, so that the suture lines were close to an adequate blood supply and troublesome adhesions were less likely. The super-

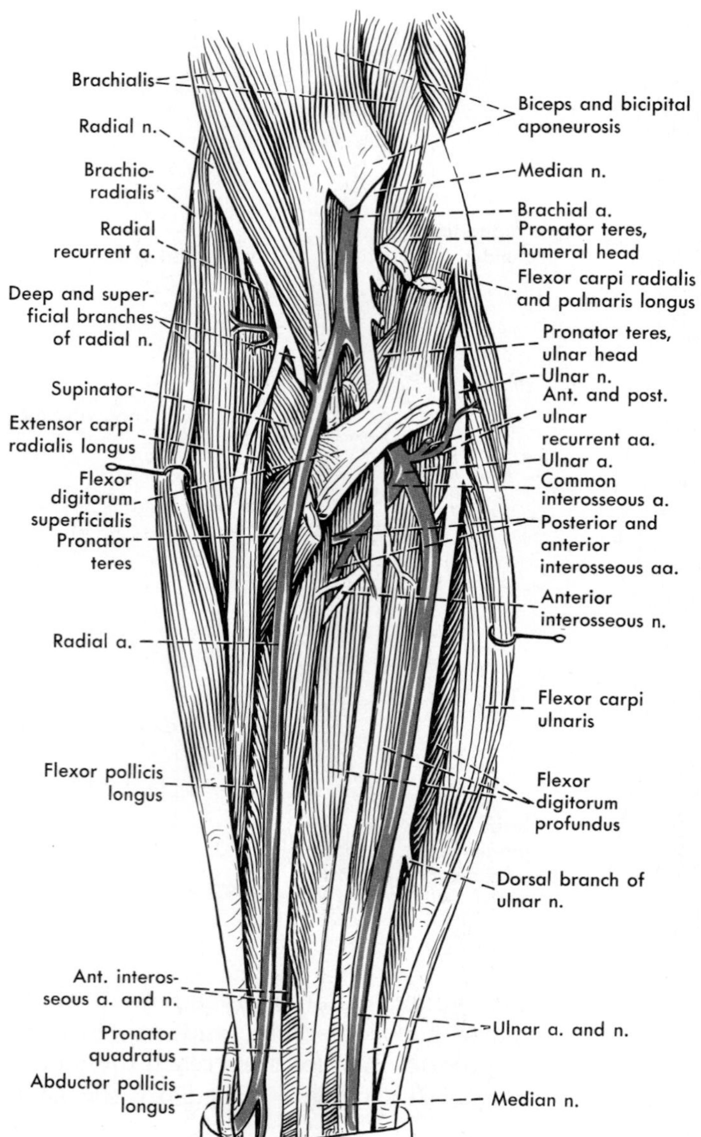


**FIG. 13-42.** Chief nerves and vessels of the anterior aspect of the forearm.

ficial flexor tendon was not replaced, for the deep one can substitute adequately for it in flexing the proximal interphalangeal joint. Improvements in technique have enabled some surgeons to obtain better results by repair of the injured tendon or tendons within the sheath.

## Nerves and Vessels

With the exception of the radial vessels, which run more superficially, all the vessels and nerves of the flexor forearm are closely associated with the deep layer of flexor muscles and can be examined in detail after the flexor digitorum superficialis is reflected or partly reflected.

**Radial Artery.** The radial artery, which is accompanied by two *radial veins,* normally arises (Fig. 13-42) as the lateral terminal branch of the brachial artery in the cubital fossa, on the front of the brachialis muscle. It runs downward and slightly laterally, under cover of the brachioradialis muscle, across the front of the tendon of the biceps, the flexor digitorum superficialis, the insertion of the pronator teres, and the flexor pollicis longus, to appear a little above the wrist between the tendons of the brachioradialis and the flexor carpi radialis. In the middle of the forearm, the *superficial branch of the radial nerve* lies just lateral to it. It leaves the anterior aspect of the forearm by turning laterally around the wrist, deep to extensor tendons of the thumb, thereby attaining the dorsum of the hand (Figs. 13-29 and 13-34). Its course and branches in the hand are described later.

A little below its origin, the radial artery gives off the **radial recurrent artery,** which supplies branches to adjacent muscles and runs upward between the brachialis muscle and the more anterior extensor muscles, following the course of the radial nerve, to anastomose with the radial collateral branch of the profunda brachii artery (see Fig. 12-49). In its further course down the forearm, the radial artery gives rise to unnamed muscular branches, and just before it turns dorsally, it usually gives off a **superficial palmar branch,** which enters the thenar muscles. Its only other named branch in the forearm, the **palmar carpal branch,** is a small vessel that arises at about the same level and passes behind the flexor tendons to help supply the wrist.

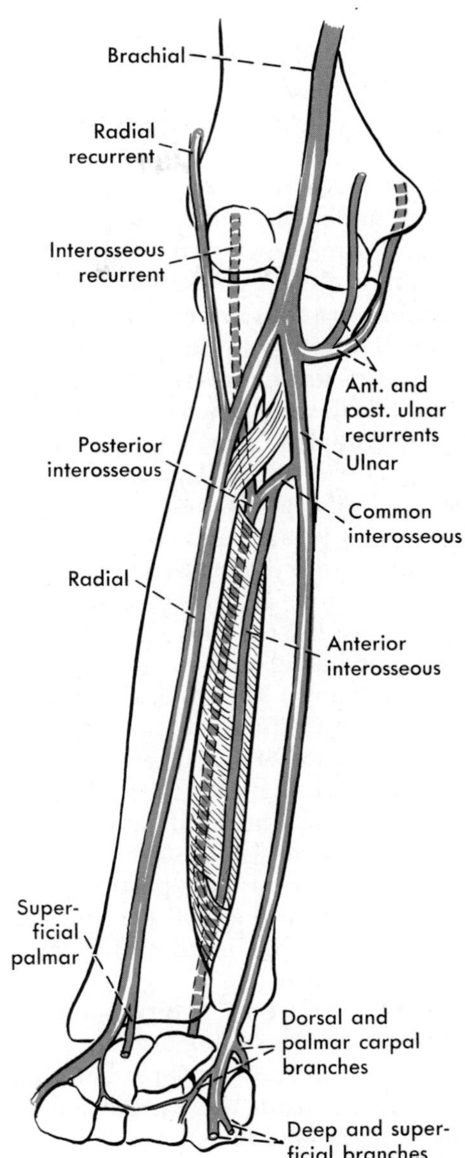


**FIG. 13-43.** Diagram of the chief arteries of the forearm.

**Ulnar Artery.** The ulnar artery, accompanied in the forearm by two *ulnar veins* that join the two radial veins to form the brachial veins, is the larger terminal branch of the brachial artery. It leaves the cubital fossa by passing behind the ulnar head of the pronator teres, by which it is separated from the median nerve. As it passes downward and medially, it first lies medial to the *median nerve* between the flexor digitorum superficialis and the flexor digitorum profundus and then descends with the *ulnar nerve* between the flexor carpi ulnaris and the flexor digitorum profundus (Fig. 13-42). It emerges from behind the flexor carpi ulnaris at the wrist, crosses the anterior surface of the flexor retinaculum, and, as it enters the hand, gives off a small deep branch into the hypothenar muscles and then continues as the superficial palmar arch.

While it lies behind the flexor digitorum superficialis, the ulnar artery gives off the large common interosseous artery and the small ulnar recurrent artery. The **common interosse-**

**ous artery** runs downward and laterally to divide after a short course into anterior and posterior interosseous arteries. The **posterior interosseous artery** immediately passes posteriorly between the flexor digitorum profundus and the flexor pollicis longus to traverse the gap at the upper end of the interosseous membrane and run downward on the posterior side of the limb (Figs. 13-33 and 13-43). The **anterior interosseous artery** gives off a usually small **median artery** to the median nerve and then descends on the interosseous membrane between the flexor digitorum profundus and the flexor pollicis longus, giving off muscular branches as it does so. Occasionally, the median artery, instead of simply supplying the median nerve, is large and continues with the median nerve into the hand, where it joins the superficial palmar arch or gives rise directly to digital vessels. (At one stage of development, the median artery was the chief arterial stem to the hand.) The anterior interosseous ends by passing behind the pronator quadratus (the deepest-lying muscle of the flexor group) and dividing into an anterior and a posterior branch. The small anterior branch continues downward to the front of the wrist, and the posterior branch passes through a hole in the lower part of the interosseous membrane to appear in the lower posterior part of the forearm.

The **ulnar recurrent artery** soon divides into anterior and posterior branches. The anterior branch passes upward between the muscles arising from the medial epicondyle and the brachialis to anastomose with the inferior ulnar collateral artery from the brachial. The posterior branch runs medially and upward to pass behind the medial epicondyle with the ulnar nerve, between the two heads of the flexor carpi ulnaris, and anastomose with the superior ulnar collateral artery (see Fig. 12-49).

Just before it crosses the flexor retinaculum, the ulnar artery gives off a small **palmar carpal branch** that runs behind the flexor tendons to reach the floor of the carpal canal and a **dorsal carpal branch** that runs around the ulnar side of the wrist to join the dorsal carpal rete.

Of the terminal branches of the ulnar artery, given off just after the artery has crossed the front of the flexor retinaculum, the small **deep palmar branch** disappears into the hypothenar muscles. It may or may not join the deep palmar arch. The continuation of the ulnar artery, now called the **superficial palmar arch,** gives off a digital artery to the free border of the little finger and then curves radially across the palm just behind the palmar aponeurosis and in front of the nerves and tendons in the central part of the palm (Fig. 13-44). In this course, it gives off three **palmar common digital arteries** that run distally on the tendons and muscles toward the interspaces between the little and ring, the ring and middle, and the middle and index fingers. They are accompanied by common digital nerves and gradually sink into the connective tissue between the long flexor tendons. Close to the metacarpophalangeal joints, the common digital arteries are joined by palmar metacarpal arteries from the deep palmar arch. They then divide into **proper digital arteries** for the adjacent sides of the fingers between which they lie. The proper digital arteries run out along the fingers behind the proper digital nerves (Fig. 13-16). The two proper digital arteries to the thumb are derived from the *princeps pollicis artery,* a deep branch of the radial that originates as that vessel enters the palm. The proper digital artery to the radial side of the index finger, the *radialis indicis,* has a similar origin.

The superficial palmar arch is usually completed on the radial side by one or more communications that it receives from the radial artery. Most conspicuous, and of frequent occurrence, is the emergence of the **superficial palmar branch of the radial** from among the thumb muscles to join the arch. Even more frequent, apparently, is a communication from the princeps pollicis artery, the radialis indicis, or a common stem that the two may share. Several connections may be present in the same hand.

**Median Nerve.** The median nerve enters the forearm on the front of the brachialis muscle medial to the brachial artery (Fig. 13-42). It leaves the cubital fossa by passing between the two heads of the pronator teres and then passes behind the flexor digitorum superficialis, where it gives off its largest branch to the forearm, the anterior interosseous nerve. A little above the wrist, the median nerve emerges on the radial side of the tendons of the flexor digitorum superficialis and runs forward to lie between these tendons and the flexor retinaculum as nerve and tendons cross the wrist. At approximately the lower border of the flexor retinaculum, the nerve ends by dividing into a muscular branch that enters the thenar muscles and digital branches that supply the thumb and two and one-half fingers.

The median nerve may give off a *twig to the elbow joint* as it crosses that joint, and it gives branches to both heads of the pronator teres

**FIG. 13-44.** Superficial branches of the median and ulnar nerves in the hand, and the superficial palmar arch. In this instance, the median and ulnar nerves anastomose a little distal to the flexor retinaculum.

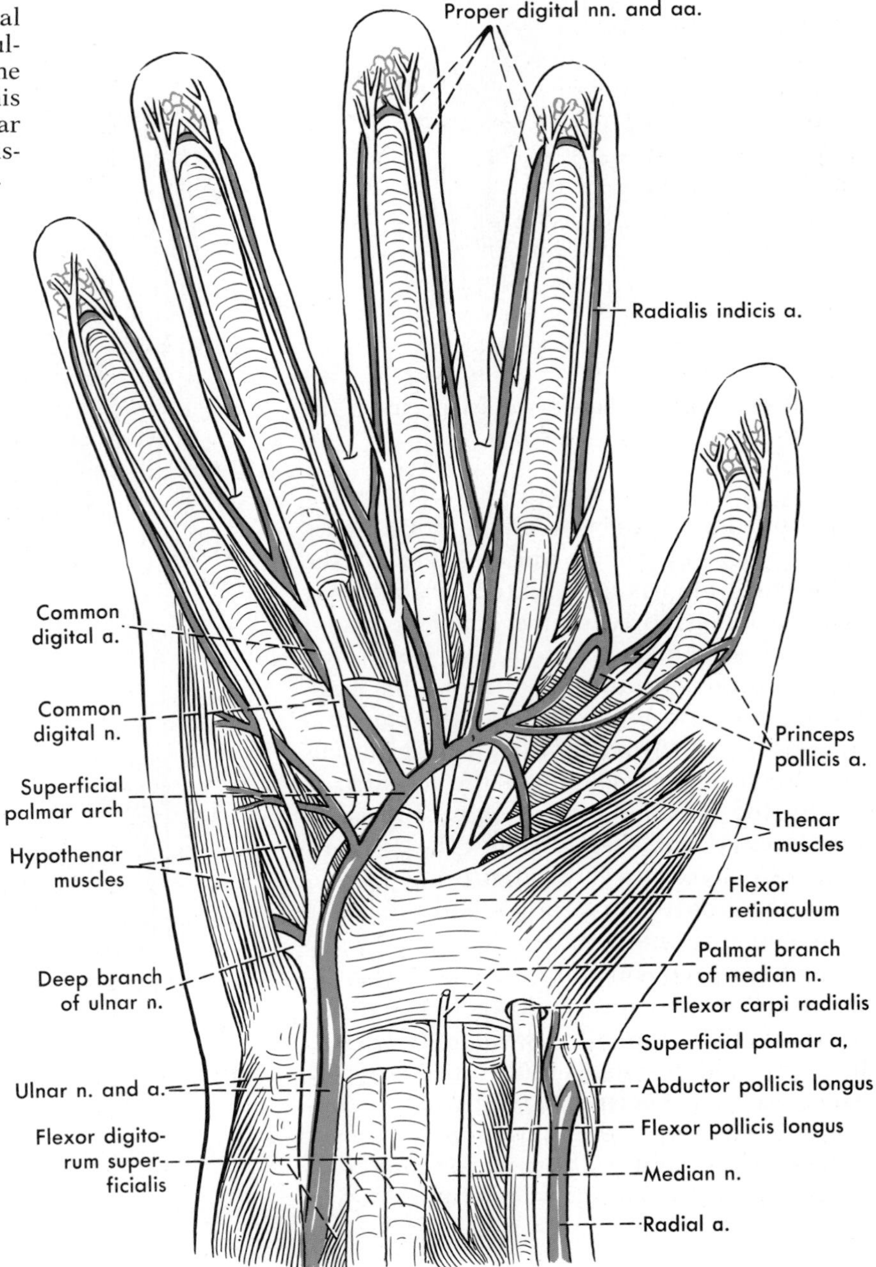

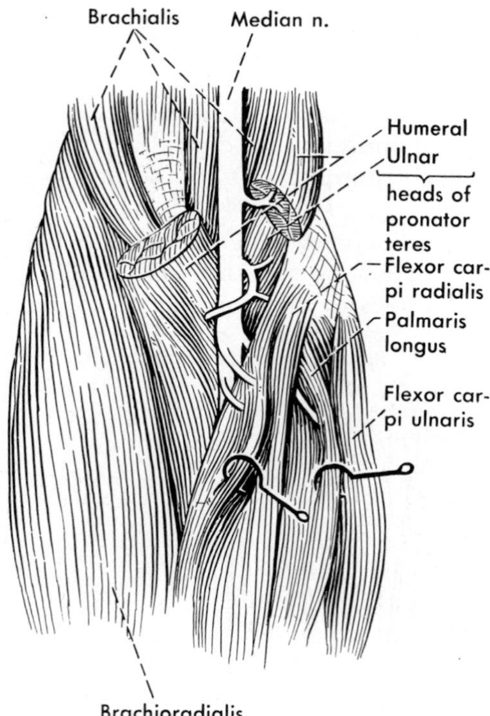


**FIG. 13-45.** The median nerve as it enters the forearm.

before or as it passes between them. **Branches** to the other superficial muscles (except the flexor carpi ulnaris, innervated by the ulnar nerve) arise in no regular order, before and after the nerve passes behind the *flexor digitorum superficialis,* and they supply this muscle, the *flexor carpi radialis,* and the *palmaris longus* (Figs 13-45 and 13-46). The last muscular branch of the median in the forearm is usually the anterior interosseous. The **anterior interosseous nerve** (Fig. 13-47) runs downward across the flexor digitorum profundus to the groove between this and the flexor pollicis longus and then proceeds downward on the interosseous membrane with the anterior interosseous artery. Its branches go to the radial side of the *flexor digitorum profundus,* to the *flexor pollicis longus,* and to the *pronator quadratus,* and a terminal twig usually continues downward to the *wrist joint.*

Below the origin of the anterior interosseous nerve, the median nerve continues downward between the two long flexors of the fingers, and after it has emerged on the radial side of the superficial flexor, it gives rise to a small **palmar branch** that is distributed to skin of the central part of the palm. As the median nerve emerges into the palm, from behind the flexor retinaculum, it divides in a

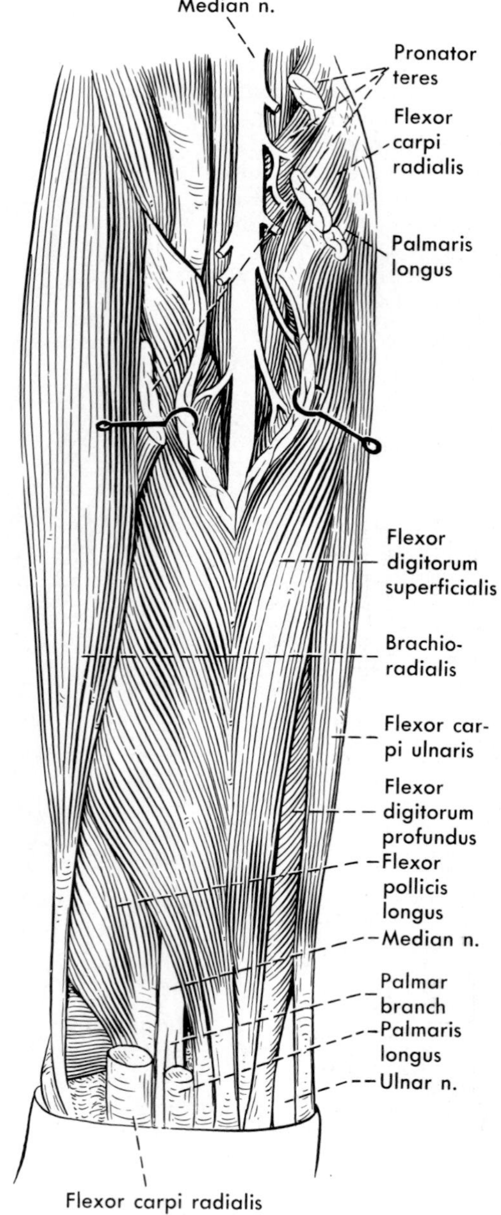


**FIG. 13-46.** Further course of the median nerve.

variable fashion into *digital branches* and a *muscular branch* (Fig. 13-44). The **muscular branch** turns laterally into the thenar muscles, where it is usually said to be distributed to about two and one-half muscles. **Digital branches** are contributed to both sides of the thumb and to the radial side of the index finger, and **common digital nerves** proceed toward the interspaces between the index and middle fingers and the middle and ring fingers, respectively. The latter frequently receives a communication from the ulnar nerve. The digital branches lie at first between the superficial palmar arch and the flexor tendons, but gradually sink into the connective tissue between the tendons. Each common digital nerve divides into *proper digital nerves* for the adjacent sides of each finger; on the digits, the palmar proper digital nerves lie in front of the proper digital arteries, and most of them give off branches that supply distal skin on the dorsum of the finger. While they lie in the palm, the nerve to the radial side of the index finger gives a *branch to* the first *lumbrical muscle* (one of the small muscles associated with the deep flexor tendons), and the common digital nerve to the index and middle fingers gives a branch to the second lumbrical.

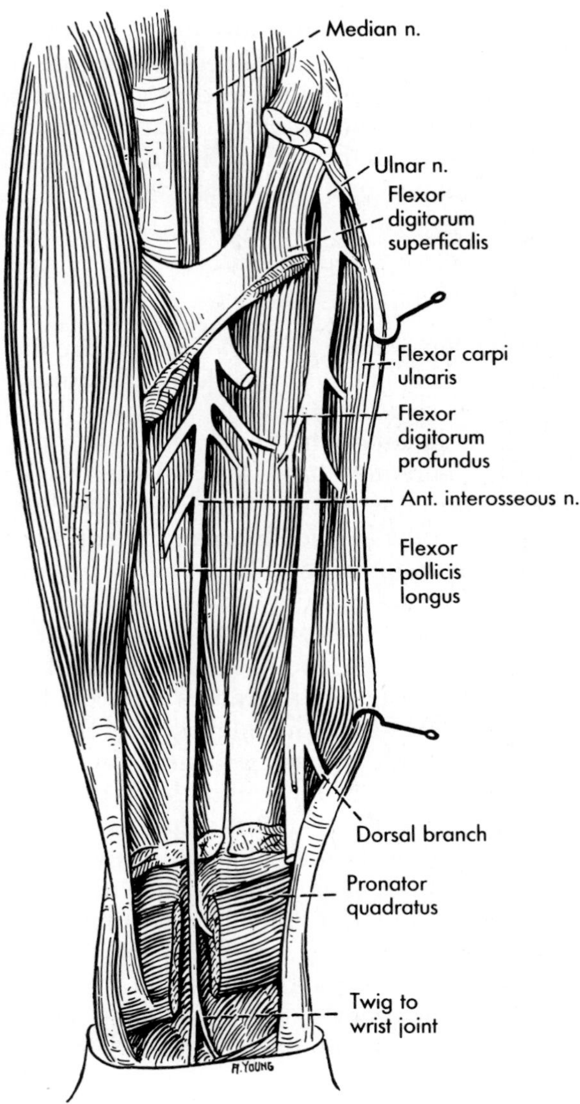


**FIG. 13-47.** The anterior interosseous branch of the median nerve, and the ulnar nerve.

In the forearm, the median nerve fairly often sends a connection to the ulnar nerve or, rarely, receives one from it. Such connections have been shown in some cases to be responsible for anomalous innervation of hand muscles). They are typically long and oblique, arising from either the median nerve or its interosseous branch and slanting downward and medially to join the ulnar. There may be other communications between the two nerves in the hand (for instance, Fig. 13-44), in addition to the one, not always present, between their adjacent common digital branches.

Where the median nerve lies in the carpal canal in front of the long flexor tendons and behind the flexor retinaculum, it is subject to injury **(carpal tunnel syndrome)** as a result of being crushed or caught between these two relatively unyielding structures. Thickening of the synovial tendon sheath is said to be a common cause of such injury.

**Ulnar Nerve.** The ulnar nerve enters the forearm by passing behind the medial epicondyle between the two heads of the flexor carpi ulnaris. It runs downward between the flexor carpi ulnaris and the flexor digitorum profundus, emerges with the ulnar artery from behind the flexor ulnaris just above the wrist, and crosses the anterior surface of the flexor retinaculum close to the pisiform bone, with

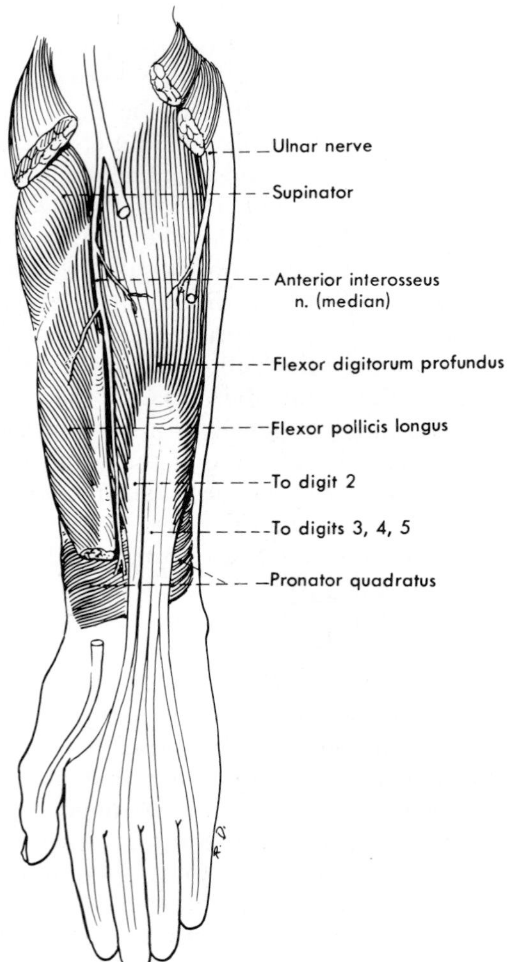


**FIG. 13-48.** The deep layer of anterior forearm muscles.

the ulnar artery on its lateral side. It then divides into superficial and deep branches.

As or just after it enters the forearm, the ulnar nerve supplies a *twig to the elbow* joint and then gives **branches** into the *flexor carpi ulnaris* and the ulnar side of the *flexor digitorum profundus* (Fig. 13-47). A variable distance above the wrist, it gives off a **dorsal branch** that winds around the wrist to give rise to *dorsal digital branches* and a small **palmar branch** that runs down on the ulnar artery to supply skin over the hypothenar eminence.

The **deep branch** of the ulnar nerve arises just distal to the pisiform bone and disappears into the hypothenar muscles. Later, it can be traced through these muscles and across the palm. The **superficial branch** (Fig. 13-44) continues the course of the ulnar nerve, but soon divides into a **proper digital nerve** for the ulnar border of the little finger and a **common digital nerve** for the adjacent sides of the little and ring fingers. The proper digital nerve continues distally on the hypothenar muscles to the little finger, passing deep to the small palmaris brevis muscle and typically innervating that. The common digital nerve pierces the medial intermuscular septum to lie behind the palmar aponeurosis before it runs distally and divides into proper digital nerves. It often gives a communicating branch to the most medial common digital branch of the median nerve.

## DEEP MUSCLE LAYER AND ASSOCIATED STRUCTURES

The deep flexor muscles of the forearm are the flexor pollicis longus, the flexor digitorum profundus, and the pronator quadratus (Fig. 13-48). Associated with the profundus in the hand are four small lumbrical muscles. Some of these muscles are innervated by the median nerve and some by the ulnar, and the profundus is innervated by both.

**Flexor Pollicis Longus.** The flexor pollicis longus arises from about the middle half of the anterior surface of the radius and from the adjacent interosseous membrane (Fig. 13-49). At the wrist, its tendon lies immediately on the radial side of the tendons of the flexor digitorum superficialis and profundus and disappears behind the flexor retinaculum alongside these tendons. It thus lies radially within the carpal canal and is here enclosed in a synovial tendon sheath that accompanies the tendon almost to its insertion on the thumb (Fig. 13-51). Fairly frequently, this sheath communicates with that surrounding the flexor tendons of the fingers.

On the thumb, the **synovial sheath** of the flexor pollicis longus is continued as a fibrous and synovial tendon sheath, similar to those on the fingers; fibrous and synovial layers end together close to the base of the distal phalanx. Within the sheath, the tendon has a prominent short vinculum, but usually no long vinculum.

In the lower part of its course in the forearm, the *radial artery* lies successively on the anterior surfaces of the flexor pollicis longus, the pronator quadratus, and the radius. The *anterior interosseous branch of the median nerve* lies between the flexor pollicis longus and the flexor digitorum profundus and sends branches into both.

**Flexor Digitorum Profundus and Lumbricals.** The flexor digitorum profundus has an extensive origin from about a half to two-thirds of the medial and anterior surfaces of the ulna and from the adjacent interosseous membrane. As it descends deep to the flexor carpi ulnaris and the flexor digitorum superficialis, it typically divides some distance above the wrist into two parts, a smaller radial part that gives rise to a tendon to the index finger and a larger ulnar part that divides at the wrist or in the hand, but may divide earlier, into three tendons for the remaining fingers.

The *ulnar artery and nerve* descend on the surface of this muscle and give branches into it. The ulnar nerve supplies only the ulnar side of the muscle, whereas the anterior interosseous nerve supplies the radial side.

The **tendons** of the flexor digitorum profundus pass into the common flexor tendon sheath behind the superficial flexor tendons, and here the ulnar of the two usual tendons typically divides into three for the three ulnar fingers. As these tendons diverge and leave the tendon sheath, the four small **lumbrical muscles** (Fig. 13-51) arise from them. The *1st lumbrical* arises from the radial side of the tendon of the profundus to the index finger, the *2nd* from the radial side of the tendon to the middle finger or from both the tendons between which it lies, and the *3rd* and *4th lumbricals* from the two profundus tendons between which each lies. Each lumbrical sends its tendon around the radial side of the metacarpophalangeal joint to join the tendon of the extensor digitorum, thereby forming a part of the extensor mechanism of the finger.

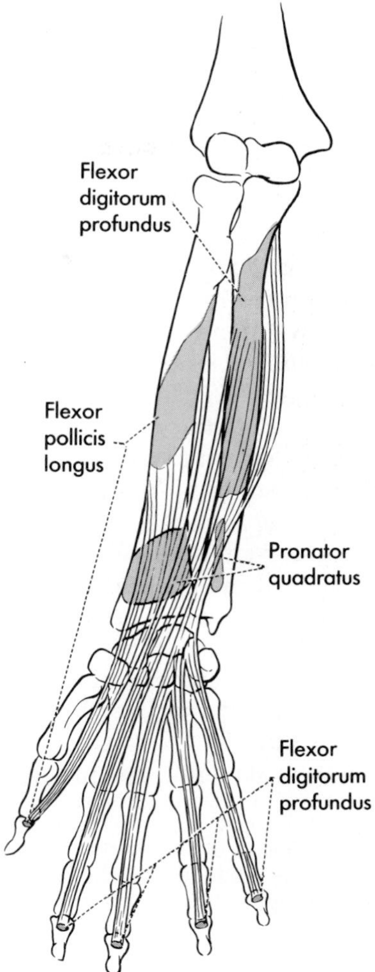


**FIG. 13-49.** Attachments of the deep layer of anterior forearm muscles.

The nerves to the first two lumbricals enter them anteriorly, from digital branches of the median nerve; those to the 3rd and 4th lumbricals usually enter their deep surfaces and are from the deep branch of the ulnar nerve.

The courses of the tendons of the flexor digitorum profundus in the hand necessarily have been largely described in the account of the superficial flexor tendons (p. 252). For a variable distance, each tendon, except that to the little finger, typically has a bare area between the common flexor tendon sheath and the digital tendon sheath. Within the latter, each profundus tendon is at first behind the superficial tendon and then passes through the tunnel formed by the split parts of the superficial tendon to insert on the distal phalanx (Fig. 13-41). Just before it reaches its insertion, the tendon has a prominent *short vinculum;* more proximally, there is usually a *long vinculum.*

**Fascial Spaces of Palm.** Deep to the profundus tendons and the lumbrical muscles in the palm of the hand, there is a good deal of loose connective tissue described as forming palmar fascial spaces. There are some differences of opinion as to the exact anatomy here, but it is usually agreed that the tissue behind the profundus tendons is subdivided into two spaces by a *fascial (oblique) septum* that runs obliquely dorsally and ulnarward from the tendons to the 3rd metacarpal (Fig. 13-50).

The fascial space on the radial side of and behind the oblique septum is called the **thenar space** or sometimes the anterior adductor space, since it lies anterior to the adductor pollicis muscle.

The fascial space medial and anterior to the oblique septum is the **midpalmar space.** Its dorsolateral wall, which separates it from the thenar space, is the oblique septum. Its dorsal wall is fascia over the front of the 3rd and 4th metacarpal bones and the intervening interosseous muscles.

The two spaces extend proximally for a little distance into the carpal canal, where their connective tissue walls come together and are continuous with the connective tissue in the forearm lying deep to the flexor digitorum profundus. Their walls also come together distally, blending with the connective tissue between the flexor tendons in which the common digital nerves and arteries divide to proceed as proper digital ones onto the fingers. Short extensions that continue distally deep to the lumbricals probably act somewhat as bursae between these muscles and the deep transverse metacarpal ligaments.

A completely different concept of the fascial spaces has been offered: all the loose connective tissue of the central compartment between the interosseous fascia and the palmar apponeurosis has been regarded as constitut-

**FIG. 13-50.** Diagram of the palmar fascial spaces: cross section through the palm. The fascial spaces (shaded) are here much exaggerated; the connective tissue in which they develop is shown in red.

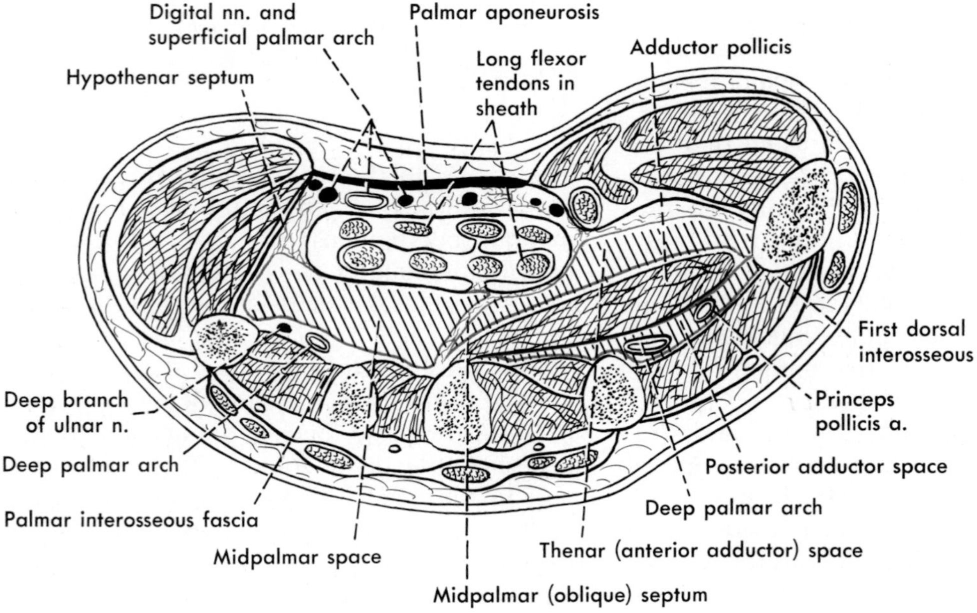

ing a single fascial space in which the long tendons and lumbricals are embedded. Only distally is this compartment divided by the septa that extend around the tendons into eight subdivisions, four for the tendons and four for the lumbrical muscles.

Other fascial spaces sometimes described are the **posterior adductor space,** behind the adductor pollicis, and the **posterior or dorsal interosseous space** (not shaded in Fig. 13-50), between the 1st dorsal interosseous muscle and its covering fascia.

Infections of the fascial spaces of the palm may result from puncture wounds of the hand or from spread of infections from digital tendon sheaths. Apparently, infections rarely spread from one of these fascial spaces into the other. A number of techniques for surgical opening of the spaces have been described, but with the advent of antibiotics, the fascial spaces have become of much less interest than formerly.

**Pronator Quadratus.** The pronator quadratus is a flat muscle lying behind the flexor pollicis longus and the flexor digitorum profundus. It arises from about the distal fourth of the anterior surface of the ulna and runs almost transversely across the front of the forearm to insert into the distal fourth of the anterior surface of the radius.

The anterior interosseous branch of the median nerve usually descends behind the muscle in company with the anterior interosseous artery, supplies the muscle, and then continues downward to the wrist joint.

#### Variations, Innervations, and Actions

Perhaps the most important variation in the deep group of flexor muscles is in the level at which the tendons to the middle, ring, and little fingers become separated from each other. Sometimes this occurs well above the wrist, sometimes close to the lower border of the flexor retinaculum. The level of separation probably accounts for the varying degree of independent movement of the distal phalanges of the fingers. The flexor digitorum profundus may have an accessory origin from the coronoid process, and the flexor pollicis longus often has an accessory origin from either the medial epicondyle or the coronoid process, or both.

The **nerve supply** to the flexor pollicis longus is from the anterior interosseous branch of the median nerve, probably from C7, C8, and T1. Although the ulnar nerve often supplies the ulnar half of the flexor digitorum profundus and the median nerve the radial half, there are variations in the amount of muscle supplied by each. The branch from the median is usually from the anterior interosseous nerve and may supply the muscle with fibers from C7. Both median and ulnar nerves are said to bring into the flexor digitorum profundus fibers from C8 and T1. The nerve to the pronator quadratus from the anterior interosseous branch is usually said to contain fibers from about C7 to T1. The two medial lumbricals, supplied by the deep branch of the ulnar nerve, and the two lateral ones, supplied by the median, probably receive fibers from C8 and T1.

**Actions.** The **flexor pollicis longus** is a flexor of the distal phalanx of the thumb and by continued action, flexes also the proximal phalanx. The **flexor digitorum profundus** primarily flexes the distal interphalangeal joints of the fingers, but in so doing produces an equal amount of flexion at the proximal interphalangeal joints (presumably through the oblique cords of the extensor aponeurosis, p. 240). The flexor digitorum profundus is a poorer flexor at the metacarpophalangeal joints and can help flex the wrist only if the fingers are kept extended. The **lumbricals** are primarily extensors of the interphalangeal joints; secondarily, they flex the metacarpophalangeal joints. The **pronator quadratus** pronates only and takes part in all movements of pronation.

## INTRINSIC MUSCLES OF THE HAND AND ASSOCIATED STRUCTURES

The contents of the central compartment (Fig. 13-21), behind the palmar aponeurosis, have already been largely described. They include the superficial palmar arch and its digital branches, palmar digital branches of the median and ulnar nerves, the long flexors of the fingers, and the lumbrical muscles associated with the deep flexor. The floor of the central palmar compartment, separated from the structures already described by fascial spaces, is composed of the palmar surfaces of some of the metacarpal bones, the interosseous muscles that lie between the metacarpals, and one muscle of the thumb, the adductor pollicis. The central compartment is crossed by the deep branch of the ulnar nerve and the deep palmar arch. These structures can best be examined after the thenar and hypothenar muscles have been dissected.

## Short Muscles of the Thumb

There are four short muscles of the thumb. The abductor pollicis brevis, the flexor pollicis brevis, and the opponens together form the thenar eminence (Fig. 13-51) and are separated from the central compartment of the palm by the septum sent down from the palmar aponeurosis to the 1st metacarpal. The fourth muscle, the adductor pollicis, lies behind and on the ulnar side of this septum and forms the floor of the thenar fascial space. The usual rule, to which there may be many exceptions, is that the branch of the median nerve into the thumb muscles supplies those that form the thenar eminence, whereas the deep branch of the ulnar nerve supplies the adductor and the deep head of the flexor brevis (which adjoins the adductor).

The **abductor pollicis brevis,** the most superficial muscle of the thenar eminence, arises mostly from the distal border of the flexor retinaculum but has some origin from the trapezium (Fig. 13-52). It inserts into the radial side of the base of the proximal phalanx of the thumb, usually giving off a superficial lamina that runs around the radial border of the thumb to attach into the tendon of the extensor pollicis longus.

The **flexor pollicis brevis** is usually de-

**FIG. 13-51.** Muscles of the palm, and the sheaths of the long flexor tendons.

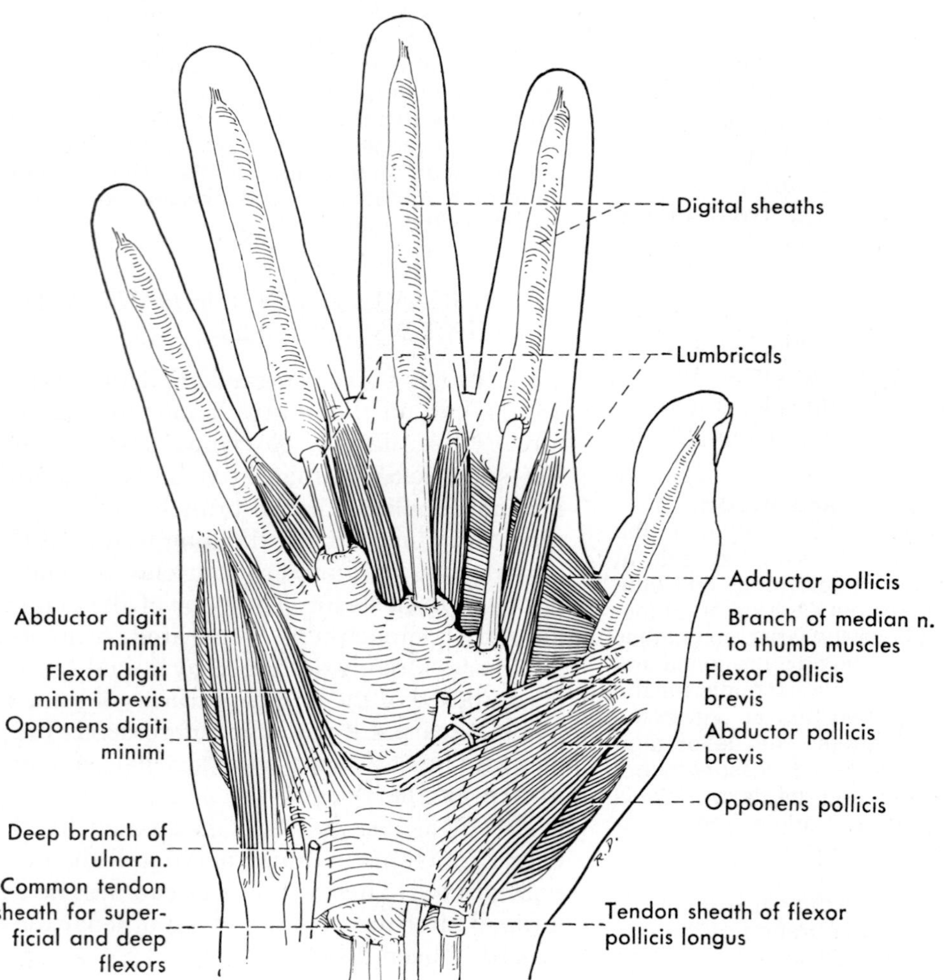

scribed as having two heads, but there is no uniform agreement as to what constitutes the deep head. The *superficial head,* its larger component, arises from the distal portion of the flexor retinaculum and from the trapezium and is partly covered by the abductor. The controversial *deep head* is probably best regarded as consisting of fibers that arise from the floor of the carpal canal (Fig. 13-53) in close association with the oblique head of the adductor pollicis and insert with the superficial head. Such fibers may be missing. When present, they are at first separated from the superficial head by the tendon of the flexor pollicis longus. After the two heads have joined, the muscle inserts on the radial side of the flexor surface of the base of the proximal phalanx, with some attachment to the radial sesamoid of the metacarpophalangeal joint.

The **opponens pollicis,** largely covered by the two preceding muscles, arises like them from the flexor retinaculum and the trapezium and runs obliquely distally and radially; it inserts, however, on the radial side of the body of the first metacarpal (Fig. 13-54).

The **adductor pollicis** regularly arises by two heads (Fig. 13-55), a *transverse head* from the 3rd metacarpal and an *oblique head* from the bases of the 1st, 2nd, and 3rd metacarpals and the ligamentous floor of the carpal canal. Both heads are triangular and join to form a common tendon that has some attachment to the ulnar sesamoid bone of the

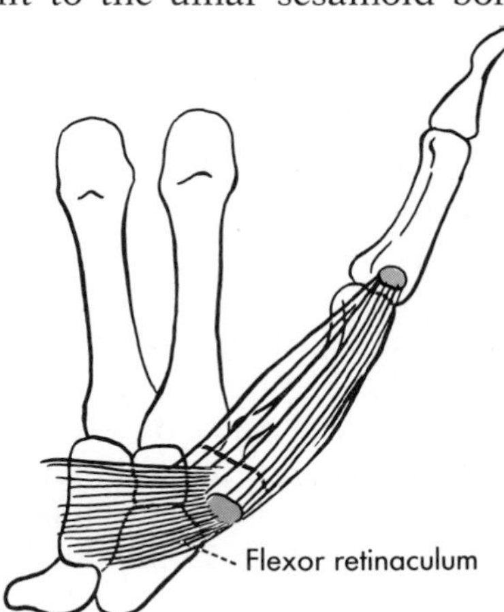


**FIG. 13-52.** Attachments of the abductor pollicis brevis.

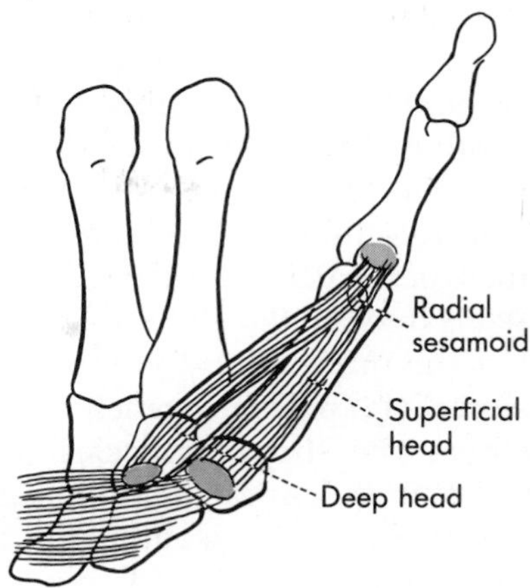


**FIG. 13-53.** Attachments of the flexor pollicis brevis.

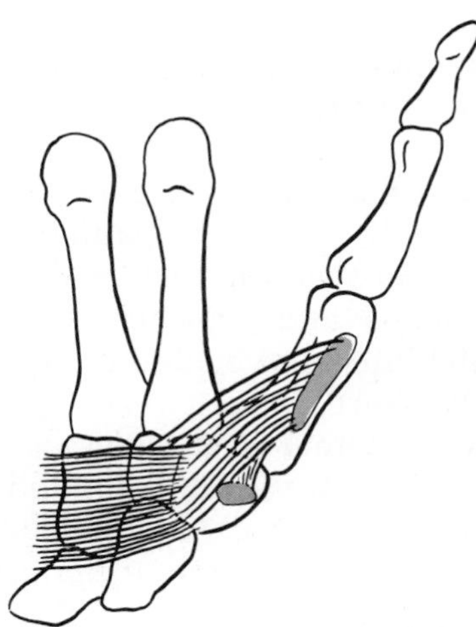

**FIG. 13-54.** Attachments of the opponens pollicis.

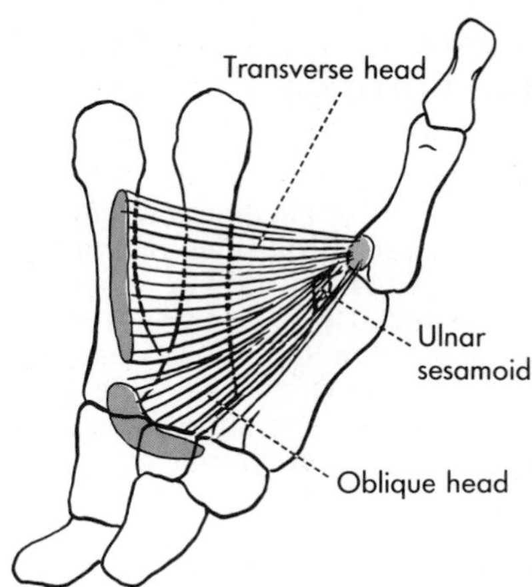


**FIG. 13-55.** Attachments of the adductor pollicis.

metacarpophalangeal joint but inserts primarily into the ulnar side of the base of the proximal phalanx of the thumb. A slip usually passes around the radial side of the thumb to attach to the long extensor tendon. The part of the muscle attaching to the extensor tendon is sometimes described as being a *1st planar interosseous muscle,* thus making four of these instead of the three more commonly listed.

The **superficial palmar branch** of the *radial artery* enters the thenar muscles to supply them. Its communication with the superficial palmar arch, if present, usually runs between the abductor and the opponens muscles.

The last important **muscular branch** of the *median nerve* (twigs to the first two lumbricals come off farther distally) leaves this nerve close to the lower border of the flexor retinaculum (Fig. 13-51) and turns radially into the thenar muscles.

The **deep branch** of the *ulnar nerve,* proceeding radially across the palm of the hand, disappears between the two heads of the adductor pollicis and typically supplies this muscle, the interossei behind it, and the deep head of the short flexor.

The **deep palmar arch** emerges from between the heads of the adductor and runs approximately transversely across the hand. On the ulnar side of the thumb, the *princeps pollicis artery* appears from behind the adductor pollicis, close to the metacarpophalangeal joint, and sends a branch to each side of the thumb. The *radialis indicis artery* also appears here and runs along the radial side of the index finger. It frequently has a connection to the superficial arch.

**Variations.** A certain amount of fusion may occur among the short muscles of the thumb, and doubt as to whether certain muscle fascicles should be included with one or the other muscle has been responsible for a good deal of confusion in regard to the thumb muscles. This is particularly true of the adductor and short flexor, which have been described quite differently by different investigators.

**Innervations.** The adductor and the deep head of the short flexor are typically supplied by the deep branch of the *ulnar nerve,* which brings to them fibers derived from C8 and T1. The other muscles are typically supplied by the *median nerve.* There has been much difference of opinion in regard to their segmental innervation, but evidence is accumulating that it is either C8 or T1 or T1 alone. (See p. 280 for anomalous innervation of thenar muscles.)

**Actions.** The actions of thenar muscles are necessarily discussed in connection with the movements of the thumb (p. 279). In summary, the chief action of each muscle is indicated by its name. However, the abductor can help to flex the proximal phalanx in addition to abducting it and the metacarpal and to extend the distal phalanx (through its attachment to the extensor pollicis longus). The adductor adducts the thumb and flexes the proximal phalanx, but unlike the abductor is reported to be inactive during extension of the distal phalanx. The short flexor is primarily a flexor of the proximal phalanx, and the opponens acts on the metacarpal to produce the chief movement of opposition of the thumb.

## Hypothenar Muscles

The **palmaris brevis** is a small, transversely arranged muscle arising from the ulnar border of the palmar aponeurosis (Fig. 13-20) and inserting into the skin of the ulnar border of the palm and into the pisiform bone. It lies superficial to the hypothenar fascia.

The other three muscles of the hypothenar group lie within the hypothenar compartment; they are thus covered by the hypothenar fascia and are separated from the central palmar compartment by the septum that the hypothenar fascia and the palmar aponeurosis send to the 5th metacarpal (Fig. 13-21). These muscles are the abductor digiti minimi, the most superficial and medial of the group; the flexor digiti minimi brevis; and the opponens digiti minimi (Fig. 13-51). They are all innervated by the deep branch of the ulnar nerve.

The **abductor digiti minimi** arises from the pisiform bone, and often from the adjacent tendon of the flexor carpi ulnaris, and joins the flexor digiti minimi to insert into the medial side of the base of the proximal phalanx of the little finger (Fig. 13-56). It frequently sends a slip of insertion around the ulnar border of the finger to attach into the extensor tendon. The *deep branches of the ulnar nerve* and artery enter the hypothenar muscles between the abductor digiti minimi and the flexor and supply both muscles and the deeper-lying opponens.

The **flexor digiti minimi brevis** arises from the flexor retinaculum and the apex of the hamulus of the hamate bone. It fuses with the abductor to insert with it, but more on the palmar than the medial surface of the proximal phalanx.

The **opponens digiti minimi** is largely covered by the abductor and the flexor. It arises from the distal border of the flexor retinaculum and from the hook of the hamate (Fig. 13-57). Like the opponens of the thumb, it does not cross the metacarpophalangeal joint but inserts on the medial border and palmar surface of the body of the 5th metacarpal.

**Variations.** Fusion of the distal parts of the abductor and short flexor is normal but variable in extent, and there also may be fusion with the opponens. Not infrequently, the short flexor is either absent or so fused with the abductor or the opponens as to be unrecognizable.

**Innervations.** The palmaris brevis is innervated by the ulnar nerve, usually its superficial branch, and the other three muscles are innervated by the deep branch of the ulnar. They all apparently receive fibers from C8 and T1.

**Actions.** The **palmaris brevis** tightens the skin of the hypothenar eminence and may not only increase the effectiveness of the grip but also help to protect the ulnar nerve and vessels from pressure. The **abductor digiti minimi** abducts the little finger and helps to flex the metacarpophalangeal joint (and, if it sends an expansion to the extensor tendon, helps to extend the interphalangeal joints). The **flexor digiti minimi brevis** is primarily a flexor at the metacarpophalangeal joint. The **opponens digiti minimi** carries out the small amount of flexion and lateral rotation allowed by the 5th metacarpal, in so doing cupping the hand so as to deepen the depression of the palm and increasing the power of the grip on tools.

## Interossei and Associated Structures

The interosseous muscles, filling the spaces between the metacarpal bones, are crossed close to their proximal ends by the deep branch of the ulnar nerve and by the deep palmar arterial arch (Fig. 13-58). Branches of

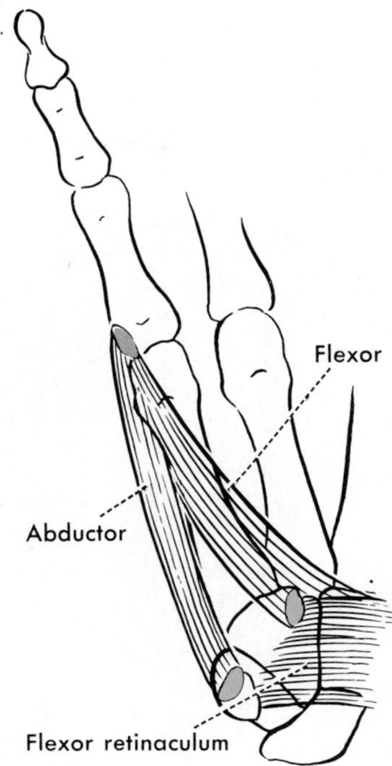


**FIG. 13-56.** Attachments of the abductor digiti minimi and flexor digiti minimi brevis.

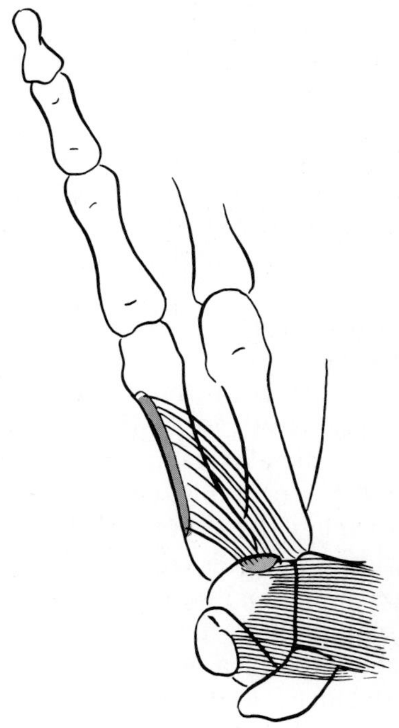

**FIG. 13-57.** Attachments of the opponens digiti minimi.

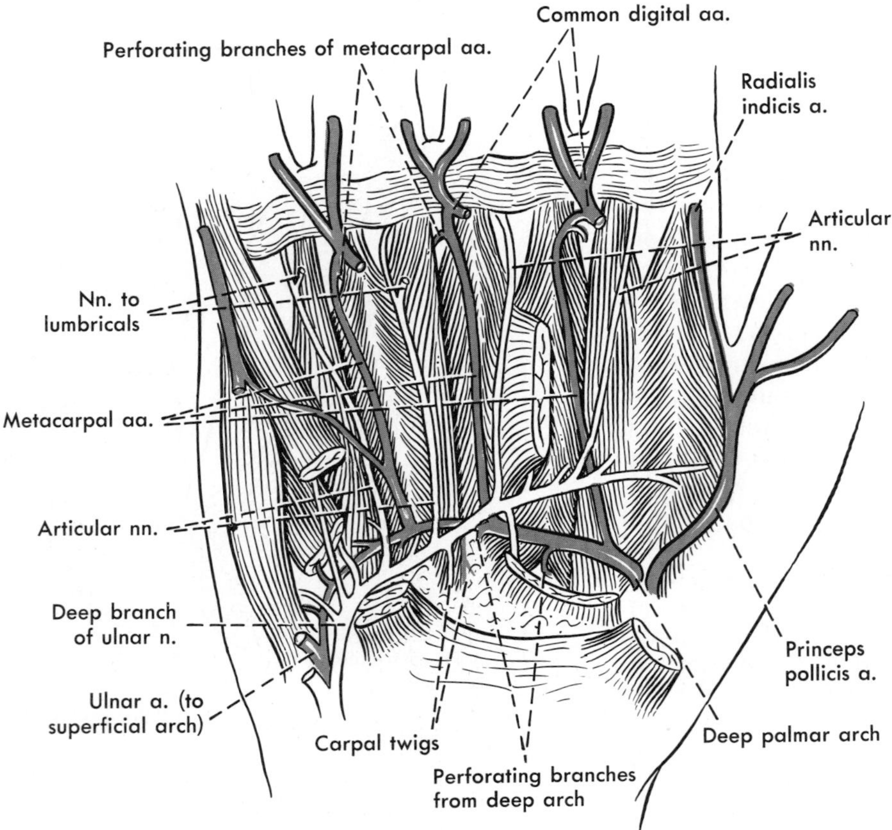


**FIG. 13-58.** The deep palmar arch and the deep branch of the ulnar nerve.

both nerve and arch run distally on their palmar surfaces, and therefore both the nerve and the arch should be examined before the muscles themselves are dissected.

**Vessels and Nerves.** The **deep branch of the ulnar nerve,** one of the terminal branches of the ulnar, arises at about the level of the pisiform bone and passes dorsally between the origins of the abductor digiti minimi and the flexor digiti minimi. It turns radially behind or through the deeper fibers of the opponens and runs across the interossei and the metacarpals approximately parallel to the deep palmar arch. It supplies all three of the *hypothenar muscles* among which it passes and, as it starts across the palm, gives off branches to the most *medial (ulnar) interossei,* to the *4th lumbrical muscle,* and to the metacarpophalangeal joint of the little finger. In its further course, it supplies the *3rd lumbrical,* a variable number of *metacarpophalangeal joints,* and typically the *rest of the interossei.* The branches arise in no regular order and may have separate origins or be clumped together. After passing between the two heads of the *adductor pollicis,* which it supplies, it usually ends in the interossei behind the adductor, but may anastomose with the branch of the median nerve to thumb muscles or extend into the muscles of the thenar eminence.

Behind the adductor pollicis, the **radial artery,** entering the palm from the dorsum, emerges between the two heads of the 1st dorsal interosseous muscle and divides as it does so into the *princeps pollicis artery* and the *deep palmar arch.* The **princeps pollicis** runs distally on the first dorsal interosseous, be-

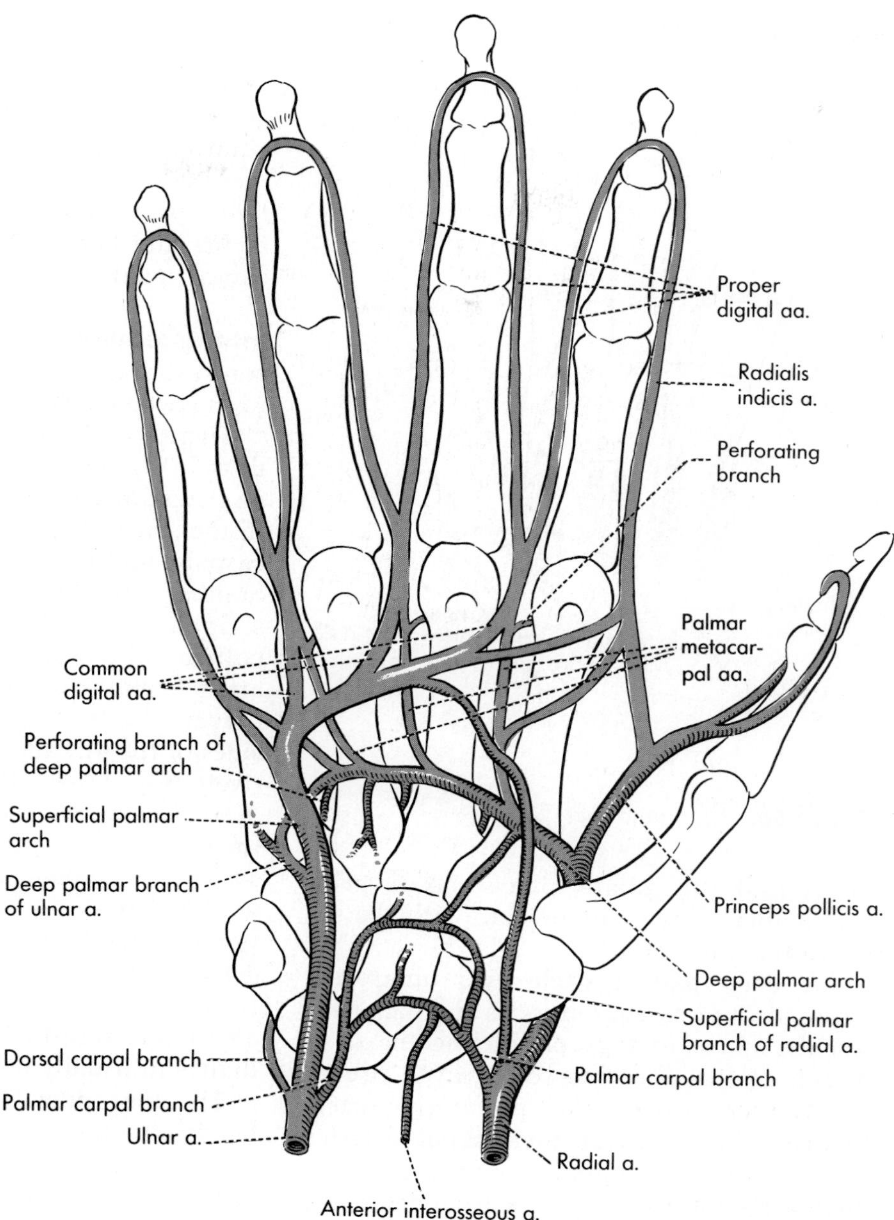


**FIG. 13-59.** Diagram of the arteries of the palm. Many connections exist here between the deep and superficial palmar arteries.

tween that and the adductor, and usually gives off the radialis indicis artery to the radial side of the index finger (this artery may arise directly from the deep palmar arch) and then, as it nears the metacarpophalangeal joint of the thumb, divides into arteries for each side of the thumb. The artery to the radial side of the thumb passes deep to the tendon of the flexor pollicis longus, and on the thumb, the arteries are similar to palmar proper digital arteries (Fig. 13-59). The **radialis indicis** also is the equivalent of a proper digital artery. It runs distally on the 1st dorsal interosseous muscle. Frequently, at the base of the finger, it receives a communication from the 1st palmar metacarpal artery (Fig.

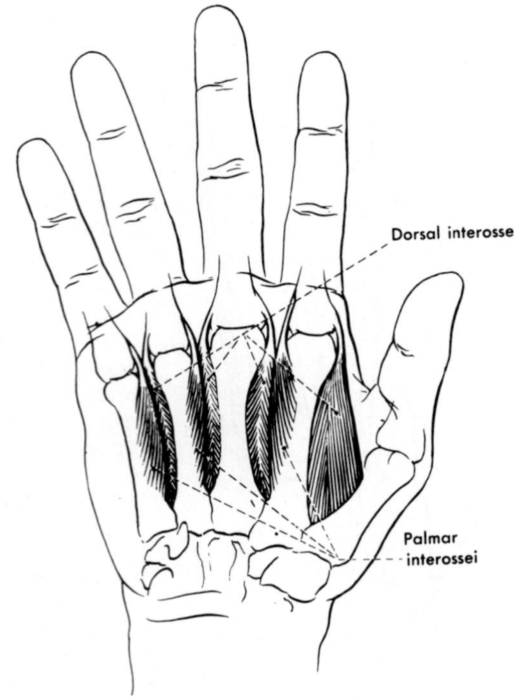


**FIG. 13-60.** The interossei.

13-59). Both the radialis indicis and the princeps pollicis may communicate with the superficial palmar arch.

The **deep palmar arch** runs ulnarward across the interosseous muscles and the metacarpals, approximately paralleling the deep branch of the ulnar nerve. It gives rise to palmar metacarpal and to perforating arteries. The three **palmar metacarpal arteries** pass distally on the interosseous muscles and, close to the level of the metacarpophalangeal joints, give off *perforating branches* that pass dorsally to join the dorsal metacarpal arteries. They then turn palmarward to join the common digital branches of the superficial palmar arch. On the ulnar side of the deep palmar arch, there may be one or more connections to superficial branches of the ulnar artery, and rather frequently the deep branch of the ulnar artery joins the arch.

The **perforating branches** of the deep palmar arch pass straight dorsally between the two heads of each of the 2nd, 3rd, and 4th dorsal interosseous muscles. They carry blood to the dorsal metacarpal arteries and are frequently larger than these arteries at their origin from the dorsal carpal rete. In addition to these branches, the deep palmar arch sends twigs proximally, to help supply the ligaments and bones that form the floor of the carpal canal.

**Interosseous Muscles.** All the interossei, in contrast to the lumbrical muscles, pass dorsal to the deep transverse metacarpal ligaments.

There are two sets of interossei, palmar and dorsal (Fig. 13-60). The **palmar interossei** lie on the palmar surface of the dorsal interossei, in the spaces between the four medial metacarpals, and are therefore typically three in number (some texts describe four; see the description of the adductor pollicis). Of the three palmar interossei here described, the 1st arises from the ulnar side of the 2nd metacarpal and the 2nd and 3rd arise from the radial sides of the 4th and 5th metacarpals, respectively (Fig. 13-61). Each passes dorsal to the deep transverse metacarpal ligament and then farther dorsally to insert into the extensor tendon ("aponeurosis") of the same finger with which its origin is associated. This insertion into the extensor tendon helps to form the extensor hood (Fig. 13-26). In addition to this insertion, any palmar interosseous may also insert into the proximal phalanx of the digit, but usually does not.

The four **dorsal interossei** arise by two heads, one from each of the two metacarpal bones between which each muscle lies (Fig. 13-62). At the base of each muscle there is a small gap between the two heads; the radial artery runs from the dorsum to the palm of the hand in the gap between the heads of the 1st dorsal interosseous, and perforating branches from the deep palmar arch pass between the heads of the other three dorsal interossei. The 1st dorsal interosseous inserts into the base of the proximal phalanx of the index finger, on the radial side of this finger. Rarely, it sends a thin lamina to the extensor tendon. Each of the three remaining interossei usually divides into two parts, a deep one that attaches to the proximal phalanx of the finger on which it inserts and a superficial one that continues dorsally to attach into the extensor tendon.

The dorsal interossei are abductors (spreaders) of the fingers from the midline of the middle finger (a fact easily recalled by abducting the index finger and feeling from the dorsum the contraction of the 1st dorsal interosseous); therefore, the 2nd inserts on the radial side, the 3rd inserts on the ulnar side of the middle digit, to abduct in either direction, and the 4th dorsal interosseous inserts on the ulnar side of the ring finger so that it abducts this finger. (Abduction of the thumb and little finger are carried out by the special musculature of these digits).

**Variations.** The chief variation in the interossei is in regard to the proportion of insertion into bone and into tendon, already mentioned. Generally speaking, it can be expected that the palmar interossei will have their insertions entirely into the extensor tendons, that the first dorsal interosseous will insert entirely into bone, and that the other three dorsal interossei will have a goodly but variable proportion of insertion into both bone and tendon. Truly anomalous arrangements of the interossei are rare.

**Innervations.** The deep branch of the ulnar nerve typically supplies all the interossei with fibers derived from C8 and T1. However, in perhaps 10% of cases, the first dorsal interosseous has an anomalous innervation, being supplied entirely or in part by the median nerve. Thus, injury to the ulnar nerve, although typically paralyzing abduction–adduction of the fingers, may not interfere with abduction of the index finger.

**Actions.** The interossei have two sets of functions: abduction and adduction of the fingers and flexion of the metacarpophalangeal joints with extension of the interphalangeal ones. These are not carried out simultaneously, however, because flexion of the metacarpophalangeal joints adducts the fingers and abduction becomes difficult. The dorsal interossei abduct digits two to four, and the named abductors abduct the thumb and little finger. The palmar interossei adduct digits two, four, and five, and the adductor pollicis adducts the thumb.

All the interossei pass far enough on the flexor aspect of the metacarpophalangeal joints to flex these joints, regardless of whether they insert into bone or into extensor tendon or both. All those that insert into the extensor tendon (therefore typically all except the first dorsal) extend the interphalangeal joints; they do this secondarily, however, and only when the metacarpophalangeal joint is either being flexed or held in flexion.

**FIG. 13-61.** Attachments of the palmar interossei.

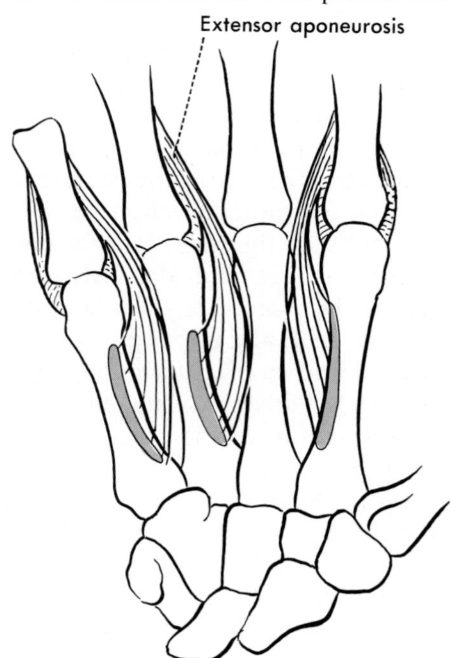


**FIG. 13-62.** Attachments of the dorsal interossei.

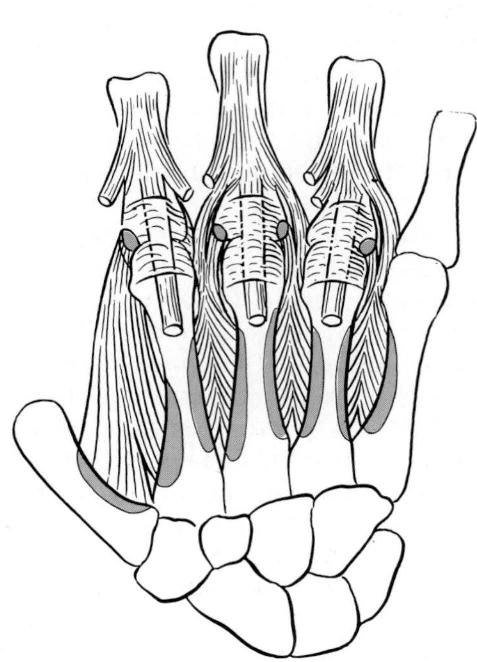

# JOINTS

Some of the muscles of the forearm cross the elbow joint, the wrist joint, or both; others cross, in addition, the joints of the carpus, metacarpus, and digits and are capable of producing movement at all these joints. These are the extrinsic muscles of the hand. The intrinsic muscles of the hand exert their action on the joints of the metacarpophalangeal and interphalangeal joints. The type of motion that results at each joint is influenced by the anatomy of the joint.

## ELBOW

The elbow joint is formed (Figs. 13-3 and 13-4) by the hingelike articulation between the trochlear notch of the ulna and the trochlea of the humerus **(humeroulnar joint);** by the articulation of the shallowly concave upper end of the head of the radius with the capitulum of the humerus **(humeroradial joint);** and by the articulation of the periphery of the head of the radius with the radial notch of the ulna and with the annular ligament **(proximal radioulnar joint).** These articulations are surrounded by a common articular capsule, so there is a single joint cavity here (Fig. 13-63). As is typical of hinge joints, the fibrous membrane of the elbow joint is thin anteriorly and posteriorly but is thickened on the sides by special ligaments. Anteriorly, the capsule is protected by the brachialis, and posteriorly by the triceps.

The fibrous membrane of the **capsule** is attached to the humerus anteriorly above the radial and coronoid fossae and posteriorly above the olecranon fossa, thus well above the articular surfaces. The **synovial membrane** also reaches the bone here, but then turns down over *pads of fat* in the fossae to reach the edges of the articular surfaces. Elsewhere, the fibrous membrane attaches close to the articular surfaces, so that the synovial membrane must run over the bone only a short distance to reach the articular cartilage. Thus, most of the parts within the joint are covered by fat and synovial membrane or by articular cartilage.

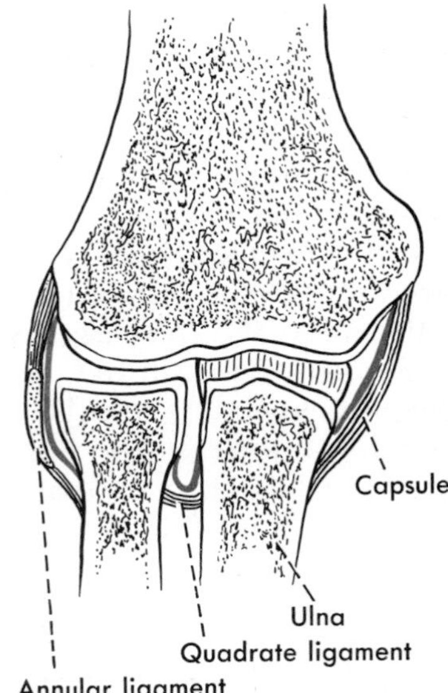


**FIG. 13-63.** Frontal section through the elbow joint; the synovial membrane is red.

That part of the fibrous capsule between ulna and radius, below the neck of the radius and the radial notch, is called the **quadrate ligament.** The other ligaments of the elbow joint are much more distinct. Laterally is the **radial collateral ligament** (Fig. 13-64), which arises from the lateral epicondyle and fans out to attach to the annular ligament of the radius, with only a few fibers reaching the neck of that bone.

The **annular ligament** (the Latin is anulare, rather than annulare) is a band that forms about four-fifths of a circle, the remainder being formed by the radial notch, to the edges of which the annular ligament is attached. The radial collateral ligament also helps to hold this ligament in place. The annular ligament surrounds the head and upper part of the neck of the radius and fits tightly enough around the neck so that distal displacement of the head of the radius cannot occur easily. In some children, displacement takes place without tearing the ligament, and this is usually said to be due to poor development of the radial head.

The medial portion of the fibrous membrane is thickened by a triangular **ulnar collateral ligament** that is attached above to the medial epicondyle and below to the medial edge of the coronoid process, the medial side of the olecranon, and the ridge between the coronoid and olecranon processes (Fig. 13-64). A transverse band of fibers between the olecranon and coronoid processes usually blends with and partly covers this middle portion.

A little below the elbow joint is the **oblique cord,** arising from the tuberosity of the ulna and running obliquely downward and laterally to attach to the radius just distal to the radial tuberosity (Fig. 13-5). The gap between the oblique cord and the upper edge of the interosseous membrane is usually traversed by the posterior interosseous vessels, but sometimes these vessels pass above the cord.

The elbow joint receives twigs from the vessels that form the collateral circulation about the elbow (see Fig. 12-49). Its **nerves** are usually derived from all four nerves that pass across it, the musculocutaneous and the median supplying it anteriorly, the radial and ulnar posteriorly.

Since the ulna cannot be anteriorly dislocated without a concomitant fracture, most **dislocations** at the elbow are posterior. The ulnar nerve is frequently injured by such dislocation. Dislocation of the head of the radius alone is usually anterior. In **fractures** of the humerus close to the elbow, the brachial vessels are sometimes injured. The resulting pressure of blood in the cubital fossa and spasm of collateral vessels may obliterate most of the circulation to the muscles of the forearm.

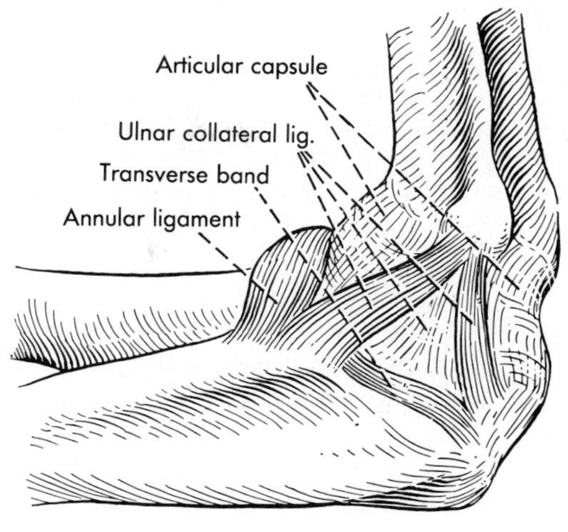


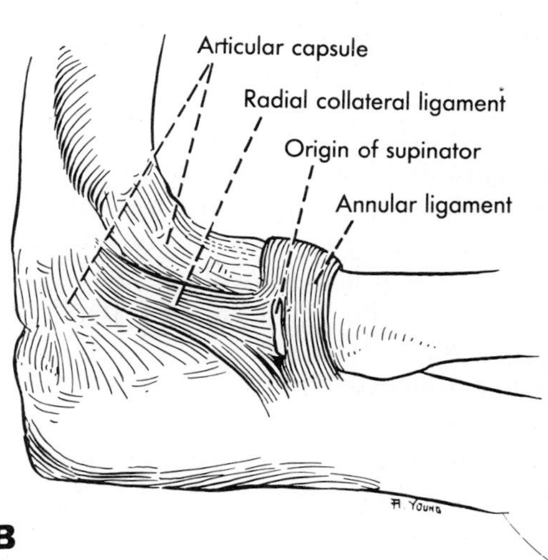


**FIG. 13-64.** Medial, **A,** and lateral, **B,** views of the ligaments of the elbow joint.

#### Movements at Elbow

**Flexion** at the elbow is carried out primarily by the biceps and brachialis muscles and by the brachioradialis. The *brachialis* apparently always contracts for flexion and is primarily responsible for maintaining a flexed position. The *biceps* contracts if the forearm is supinated or if considerable resistance is offered to flexion of the pronated forearm. The *brachioradialis* contracts if the movement is a quick one or if resistance is offered to flexion while the forearm is in a pronated or thumb-up position. The *pronator teres* will also contract against resistance, and probably some of the adjacent muscles of the extensor group, particularly the extensor carpi radialis longus, can also assist. Because of these other flexors, particularly the brachioradialis, fairly good forearm flexion can result against little opposition, even if both the biceps and the brachialis are paralyzed as a result of interference with the musculocutaneous nerve.

**Extension** at the elbow is brought about by the triceps and the anconeus.

**Pronation** is brought about especially by the pronator teres and pronator quadratus muscles. The *pronator quadratus* always participates in pronation, and the *pronator teres* assists, regardless of whether the elbow is flexed or extended, if more power is needed. The flexor carpi radialis and the palmaris longus, although apparently not normally participating in pronation, are usually listed as accessory pronators. The *brachioradialis* may pronate slightly from the extreme position of supination or supinate from an extreme position of pronation; however, according to electromyographic evidence it contracts only when resistance is offered to these movements, and it is not certain that it contributes to them.

**Supination** is carried out primarily by the supinator and by the biceps brachii. The *biceps brachii* supinates only when the forearm is flexed or more power is demanded, but the *supinator* is equally effective whether the forearm is flexed or extended. Accessory supinators are usually said to be the extensor carpi radialis longus and, from the extreme position of pronation, the *brachioradialis.* It should be noted again, however, that the brachioradialis is primarily a flexor and of no real importance in either pronation or supination.

## WRIST

At the wrist, the distal end of the ulna is separated from the wrist joint proper by an *articular disk.* The synovial cavity enclosed between the disk, the distal end of the ulna, and the distal end of the radius is the **distal radioulnar joint.** The synovial membrane of this joint is attached to the articular surfaces of the structures just listed, but extends upward between radius and ulna as the *sacciform recess.* The fibrous capsule is thickened anteriorly, medially, and posteriorly by the ligaments of the wrist that attach to the ulna.

The proximal articular cavity at the wrist is the radiocarpal joint. There are intercarpal joints between the adjacent bones in each row, and there is a midcarpal joint between the bones of the proximal and those of the distal row.

The **radiocarpal joint** lies between the radius and the articular disk of the ulna, proximally, and the scaphoid, lunate, and triquetral bones, distally. The edges of the proximal articular surfaces of the latter three bones are united to each other by *intercarpal interosseous ligaments,* so that the radiocarpal joint typically communicates with no other joint (Fig. 13-65).

Of the **intercarpal joints** of the proximal row of bones, that between the pisiform and triquetral bones is a separate articular cavity, but those between the other three bones are slits that begin proximally at the intercarpal interosseous ligaments and extend distally to open into the midcarpal joint. Similarly, the **midcarpal joint** is continuous with the intercarpal joints between the proximal parts of the distal row of carpals, but is largely sealed off from the distal parts by other *intercarpal interosseous ligaments.* (These ligaments are sometimes defective, that between the trapezium and the trapezoid often being so.)

Distal to the intercarpal interosseous ligaments uniting the bones of the distal row, the intercarpal joints open into the sinuous carpometacarpal joints. The synovial membrane of all these joints is simply reflected from the edges of one articular surface to the edges of

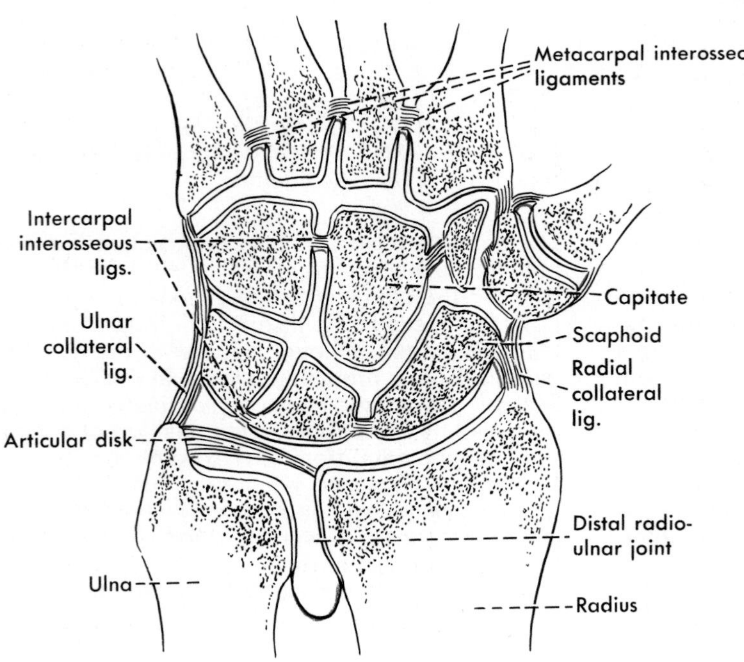


**FIG. 13-65.** Frontal section through the joints at the wrist. The synovial membrane is indicated by red.

the next and, except at the radiocarpal joint, consists of very narrow bits of membrane.

## Ligaments

The external ligaments at the wrist (Figs. 13-66 and 13-67) are thickenings of the fibrous membrane of the articular capsules. On the sides of the wrist, the **radial collateral ligament** extends from the radius to the scaphoid and trapezium and the **ulnar collateral ligament** extends from the ulna to the triquetral and pisiform bones. Anteriorly, the **palmar radiocarpal ligament** runs from the radius to the bones of the proximal row, with some fibers extending also to the capitate. The smaller **palmar ulnocarpal ligament** lies between the palmar radiocarpal ligament and the ulnar collateral ligament. On the dorsum is the large **dorsal radiocarpal ligament.**

Both radiocarpal ligaments slant ulnarward as they pass distally, so that movements of the forearm in either pronation or supination immediately tighten one of the ligaments and assure that the hand is carried with the radius during these movements.

The intercarpal joints are strengthened by a whole series of **palmar intercarpal ligaments,** of which the large ligament extending between proximal and distal rows and centering especially on the capitate is known as the **radiate carpal ligament.** The numerous ligaments on the dorsum are known simply as **dorsal intercarpal ligaments.** Stronger are the more deeply lying **intercarpal interosseous ligaments,** which were included in the description of the intercarpal joints in the preceding section.

Besides being attached to the triquetral bone by an articular capsule, the pisiform has two special ligaments, the **pisohamate** and **pisometacarpal,** that are adequately described by these names. They are often regarded as representing the distal end of the flexor carpi ulnaris, since the pisiform has traditionally been described as a sesamoid bone in the tendon of this muscle.

### Blood Supply and Innervation

The ligaments and bones of the wrist are supplied by twigs from the dorsal carpal rete and, on the palmar side, by palmar carpal branches of the radial and ulnar arteries and twigs from the anterior interosseous artery and the deep palmar arch. The nerve supply is typically from both anterior and posterior interosseous nerves (hence, median and radial nerves) and from the dorsal and the deep palmar branches of the ulnar nerve.

### Movements at Wrist

There is a good deal of variation in the range of movement at the wrist in different persons, and it is difficult to measure with certainty how much movement is produced at the radiocarpal joint and how much among the carpals. However, it is usually agreed that both radiocarpal and midcarpal joints contribute much to palmar flexion and dorsiflexion, whereas in radial abduction most of the movement occurs at the midcarpal joint and in ulnar abduction most is at the radiocarpal joint.

The chief **flexors** of the wrist are the *flexor carpi radialis*, the *flexor carpi ulnaris*, and the *palmaris longus*. The abductor pollicis longus, an accessory flexor at the wrist, is the only flexor not innervated by either the median or the ulnar nerves. The flexor digitorum superficialis and profundus muscles, and the flexor pollicis longus, are effective as wrist flexors only when the digits are kept extended.

The chief **extensors** are the two radial and the single ulnar extensor. The accessory extensors (the several extensors of the digits) can help extend the wrist only when the digits are flexed. All the extensors are innervated by the radial nerve.

The chief **radial abductors** are the *abductor pollicis longus* and *extensor pollicis brevis*. The extensor carpi radialis longus and brevis (both of which have been shown to become active during radial abduction), the flexor carpi radialis, and the extensor pollicis longus are sometimes all listed as aiding in radial abduction, although it is occasionally denied that any of these do so under ordinary conditions. The **ulnar abductors** are the *extensor carpi ulnaris* and the *flexor carpi ulnaris*.

## METACARPAL AND PHALANGEAL JOINTS

It has already been noted that the sinuous **carpometacarpal joint** of the fingers is continuous with the distal parts of the intercarpal joints between the distal row of carpals. This cavity is also continuous with the **intermetacarpal joints** between the bases of the metacarpals of the four fingers. The carpometacarpal joint of the thumb is a separate joint cavity, however, and has its own special ligaments.

The articular capsule of the carpometacar-

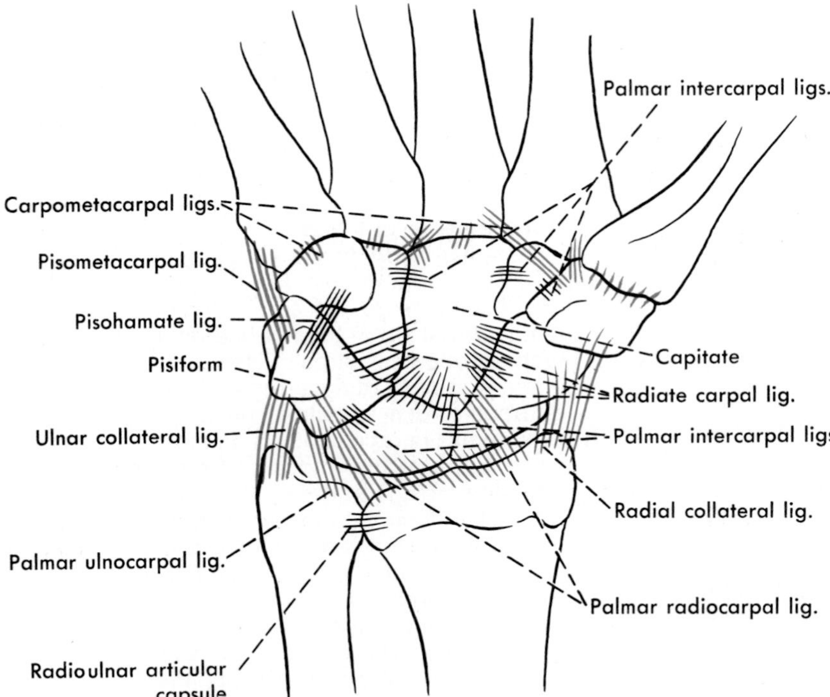


**FIG. 13-66.** Diagram of the palmar ligaments attaching to the carpal bones. Intercarpal ligaments are black, others red.

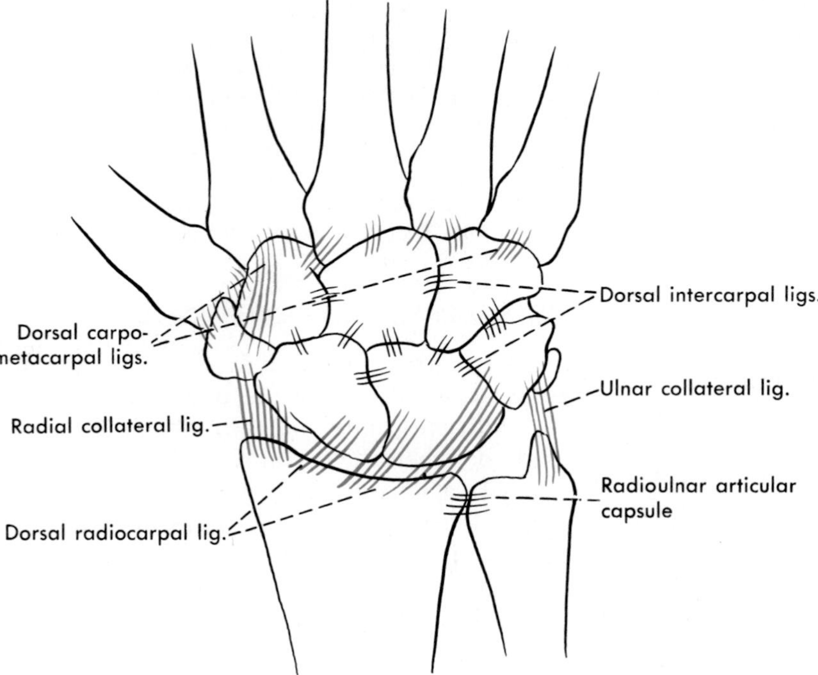


**FIG. 13-67.** Diagram of the dorsal ligaments attaching to the carpal bones. Intercarpal ligaments are black, others red.

pal joint is strengthened by **dorsal** and **palmar carpometacarpal ligaments** arranged approximately as indicated in Figures 13-66 through 13-69, and those of the intermetacarpal joints by **dorsal** and **palmar metacarpal ligaments;** the latter joints are bounded distally by **metacarpal interosseous ligaments.** The heads of the 2nd to 5th metacarpals are bound to each other by heavy **deep transverse metacarpal ligaments.**

The **metacarpophalangeal joints** (Figs. 13-68 through 13-70) are simple. The articular capsule of each joint is reinforced on its palmar surface by a heavy **palmar ligament;** the synovial sheath of the flexor tendons of the digit is fused to the palmar ligament, thus providing a gliding surface for the tendons, and the fibrous parts of the sheath attach to the edges of the ligaments. The deep transverse metacarpal ligaments blend with the palmar ligaments as well as attach to the metacarpals. On each side of a metacarpophalangeal joint is a **collateral ligament** that runs obliquely from a more dorsal attachment on the head of the metacarpal to a more palmar one on the base of the proximal phalanx. They apparently limit abduction when the joint is flexed. On the dorsal aspect, the joint capsule is reinforced by the heavy extensor tendon.

The **interphalangeal articulations** (Fig. 13-70) are all essentially similar to each other and to the metacarpophalangeal articulations of the fingers. The articular capsules are reinforced on their palmar surfaces by **palmar ligaments,** on their sides by **collateral ligaments,** and dorsally by the **extensor tendons;** avulsion of an extensor tendon typically opens up these joints.

**FIG. 13-68.** Diagram of the carpometacarpal, intermetacarpal, and metacarpophalangeal ligaments in a palmar view.

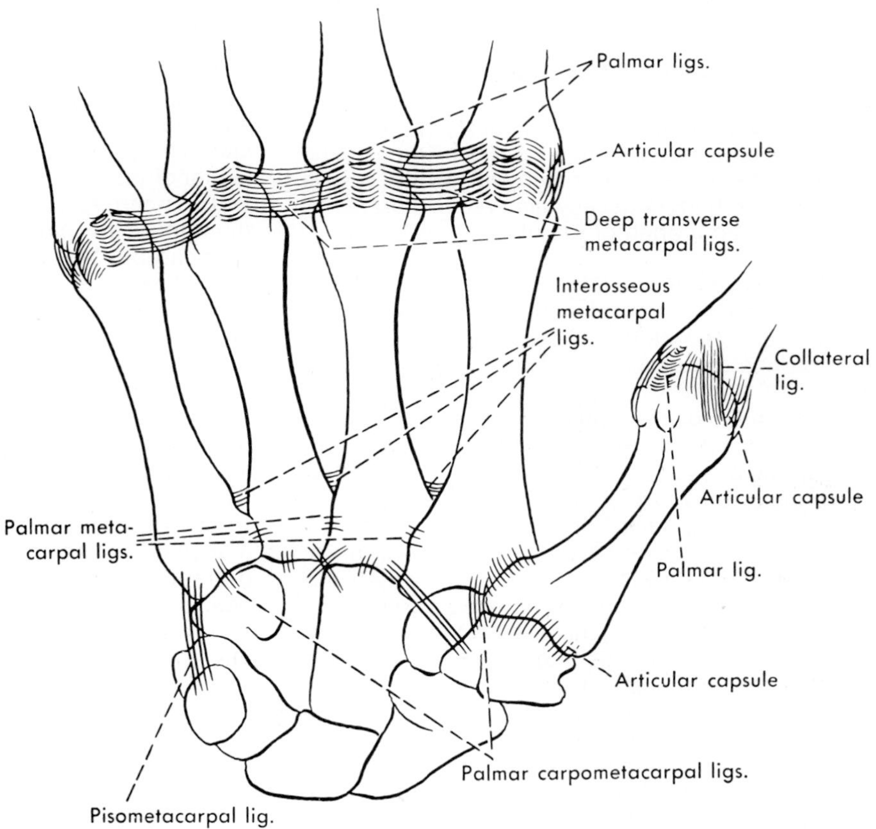

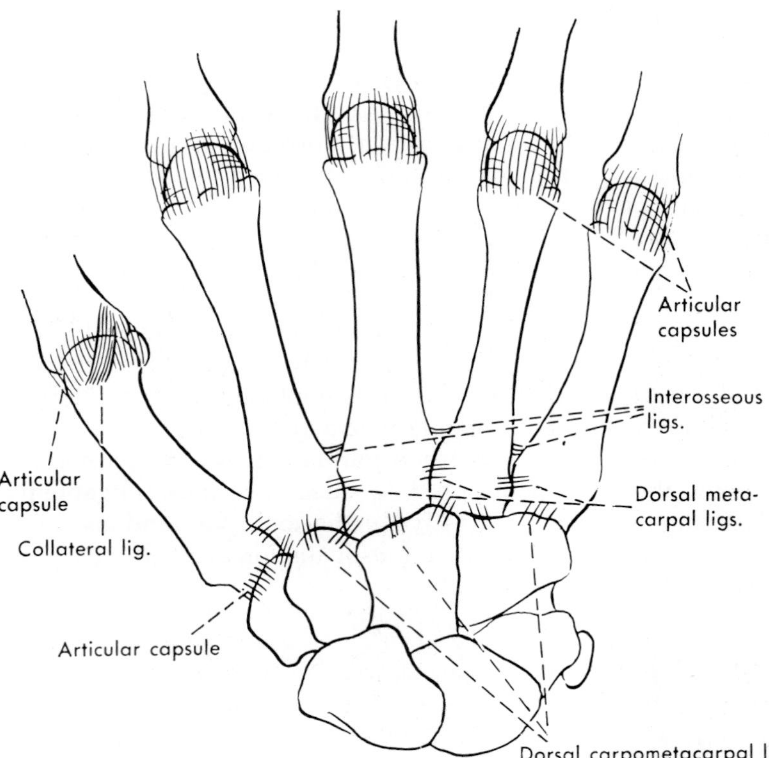


**FIG. 13-69.** Diagram of the carpometacarpal and intermetacarpal ligaments in a dorsal view.

The **blood supply** of the metacarpophalangeal and interphalangeal joints is from the digital arteries. The **nerve supply** to the metacarpophalangeal and interphalangeal joints is from the digital nerves. The metacarpophalangeal joints of the more medial fingers usually also receive branches from the deep branch of the ulnar nerve.

### Movements of the Digits

The thumb is a special instance and needs to be considered separately, but movements of the fingers can be discussed together.

**Movements of the Fingers.** At the *metacarpophalangeal joints* of the fingers, **flexion** can be carried out as a result of continued action of the two *flexor digitorum muscles,* and these muscles acting alone can "make a fist." The lumbricals apparently contract only to extend the interphalangeal joints, although after that they can help flex the metacarpophalangeal ones. The *interosseous muscles* are the primary flexors at the metacarpophalangeal joints. Flexion of the little finger is also assisted by the flexor digiti minimi brevis.

Flexion of the *proximal interphalangeal joints* alone is carried out by the flexor digitorum superficialis, but more commonly the profundus, the only flexor of the distal phalanges, initiates interphalangeal flexion, flexing the middle phalanx at the same time as the distal phalanx.

**Extension** of the *metacarpophalangeal joints* can be carried out only by the *extensor digitorum* and, in the case of index and little fingers, the special extensors of these fingers (which join the extensor digitorum). Extension of the *interphalangeal joints* is carried out by the *extensor digitorum* and the *lumbricals,* aided by the *interossei* only if the metacarpophalangeal joint is held in flexion or is being flexed.

Extension at the wrist decreases the effectiveness of the long extensors of the fingers and at the same time increases that of the long flexors. Flexion at the wrist decreases the action of the finger flexors and, at the same time, exerts passive pull upon the long extensors. Thus, to have good flexion and extension of the fingers, there must be at least one good extensor and one good flexor of the wrist, or the wrist must be otherwise maintained in a position of usefulness—because of the great importance of the grip, one of slight dorsiflexion is best.

The **abductors** of the fingers are the four *dorsal interossei* and both the extensor and the abductor digiti minimi; the extensor digitorum abducts all the fingers as it extends them. The **adductors** of the fingers are the three *palmar interossei,* which move the index, ring, and little fingers toward the middle finger. The extensor indicis can adduct the index finger independently, and the long flexors of the fingers adduct these as they flex them.

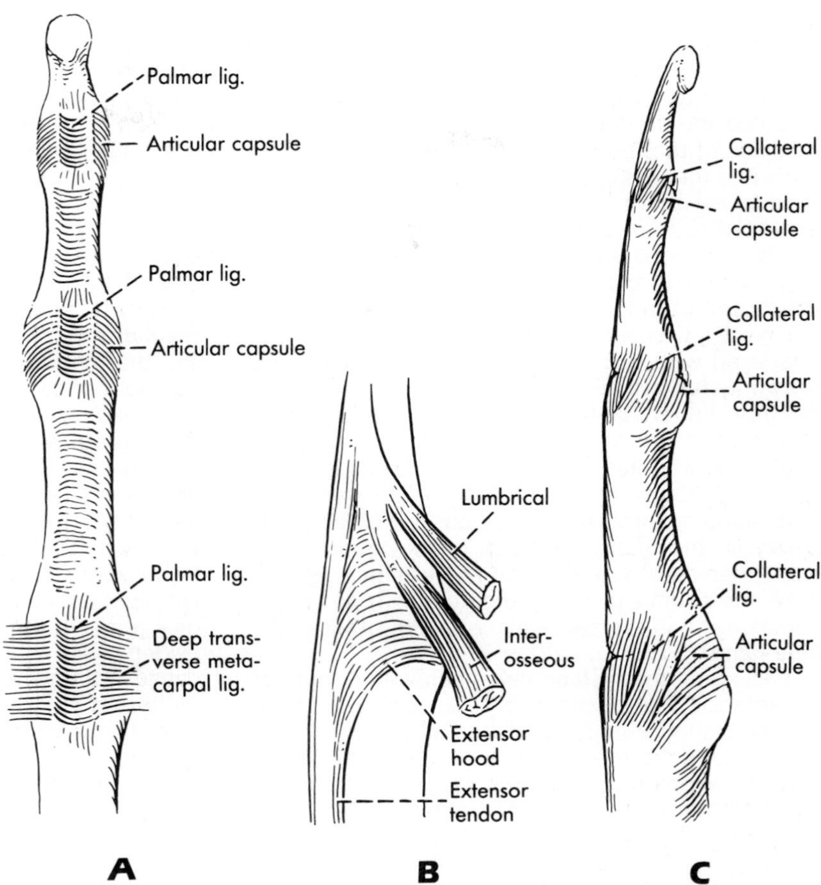


**FIG. 13-70.** Metacarpophalangeal and interphalangeal joints. **A** is a palmar view; **B**, a lateral view of the metacarpophalangeal joint with the extensor tendon in place; **C**, a lateral view of a digit.

**Movements of the Thumb.** These movements (defined on p. 230) are complicated by the fact that the carpometacarpal joint of the thumb is a saddle joint and the metacarpal moves usually with the phalanges. Further, the joint and its ligaments are so arranged that the metacarpal is medially rotated with flexion, laterally rotated with extension. Only a little rotation occurs at the metacarpophalangeal joint, and none at the interphalangeal joint. The chief movements of these joints are flexion and extension, and the phalanges are, for the most part, simply carried passively by the metacarpal during opposition and reposition.

It is difficult to determine exactly which muscles participate in certain movements of the thumb, and there is some disagreement on this subject. However, **abduction** of the metacarpal and, hence, of the thumb as a whole, is brought about primarily by the *abductor pollicis brevis.* The abductor longus is not a pure abductor, but also extends and rotates. The opponens is said to be more active in abduction than in adduction, but its role in both is probably that of stabilizing the metacarpal. **Adduction** is brought about primarily by the *adductor,* but both long and short flexors of the thumb tend to adduct as they flex, and the long extensor tends to adduct as it extends.

**Opposition,** the complex movement of rotation and flexion of the metacarpal that brings the thumb in contact with the finger pads, as in grasping a small object between thumb and forefinger, is especially the function of the *opponens,* the only muscle so arranged as to produce opposition without throwing extra strain on the ligaments of the carpometacarpal joint. The other three short muscles of the thumb are also active during opposition. **Reposition,** movement from opposition, is apparently brought about especially by the *long abductor* and the *long extensor,* and perhaps by the short extensor.

**Flexion** at the *metacarpophalangeal joint* is brought about by the adductor and the short flexor; **extension** is brought about by the short and long extensors. **Flexion** at the *interphalangeal joint* is carried out entirely by the long flexor of the thumb, **extension** by the long extensor of the thumb plus other muscles that send tendons to join its tendon—frequently, therefore, the extensor pollicis brevis and fairly regularly the abductor pollicis brevis.

## INJURIES TO MEDIAN AND ULNAR NERVES

Injuries to the radial nerve have already been mentioned (p. 245, Fig. 13-32).

Since the **median nerve** supplies all the flexor muscles of the forearm except the flexor carpi ulnaris and the ulnar part of the flexor digitorum profundus, an injury to it high in the forearm can be expected to interfere especially with pronation of the forearm and with flexion of the phalanges of digits II and III and of that of the distal phalanx of the thumb (in addition to its effect on muscles of the hand, discussed below). Wrist flexion is less interfered with, since the flexor carpi ulnaris and the abductor pollicis longus are spared. Flexion of the little finger (by the profundus) is rather regularly spared, but that of the ring finger varies according to whether this part of the muscle is innervated by the median or the ulnar nerve. In long-standing median or median and ulnar nerve paralysis, the unopposed pull of the thumb muscles innervated by the radial nerve may gradually draw the thumb dorsally and rotate it, so that it comes to lie in the same plane as the other digits (ape hand).

Lesions of the median nerve below the origin of its anterior interosseous branch are essentially similar in their effects to lesions at the wrist. With such lesions there is loss of sensation in the radial part of the palm and over the more radial fingers, and the thenar muscles may show atrophy and weakness. The terminal phalanx still can be flexed by the flexor pollicis longus, and there may not be as much loss of movement as might be expected, since the muscles innervated by the radial and ulnar nerves are still active. However, unless there is anomalous innervation, opposition and pure abduction (that is, at right angles to the palm) are carried out poorly if at all.

**FIG. 13-71.** Clawing of the fingers as a result of ulnar nerve paralysis. (Courtesy of Dr. E. D. Henderson)

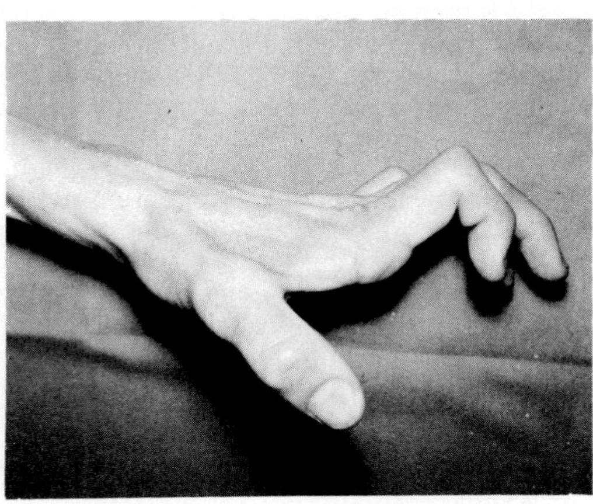

Lesions of the **ulnar nerve,** whether at the elbow or wrist, have their primary effect in the hand. Paralysis of the flexor carpi ulnaris interferes with ulnar deviation at the wrist, and paralysis of the ulnar portion of the flexor profundus abolishes flexion of the distal phalanx of the little finger and often that of the ring finger, but a majority of the muscles of the hand are affected by an ulnar nerve lesion. The chief results are a loss of abduction and adduction of the fingers and of adduction of the thumb, atrophy of the hypothenar eminence, and frequently the production of a "claw hand" (Fig. 13-71). This results when, because of paralysis of the interossei, the metacarpophalangeal joints are drawn by the unopposed extensor digitorum into as much hyperextension as the joint capsules will allow. When there is much hyperextension at these joints, the interphalangeal joints cannot then be extended against the tethering pull of the superficial and deep flexors, so they are flexed. Clawhand is most evident on the ulnar side, since the first two lumbricals oppose hyperextension at the metacarpophalangeal joints but aid extension at the interphalangeal joints of the index and middle digits. It is less evident with a high lesion of the ulnar nerve because of the paralysis of the ulnar part of the flexor digitorum profundus. If the lesion is low, sensory loss may be confined to the palmar side because of sparing of the dorsal branch.

The expected effects of lesions of either the median or the ulnar nerve on muscles of the hand are not always found, for sometimes one or more muscles may have an **anomalous innervation.** As judged by clinical findings, this is particularly true of thumb muscles. It is often difficult to be sure that a given movement of the thumb is or is not carried out by a given muscle, and the percentage of anomalous innervation in thenar muscles may have been reported as much higher than it really is; however, there is clear evidence that anomalous innervation does sometimes occur. Recent electromyographic evidence is said to indicate that the *opponens pollicis* may be supplied by both nerves in approximately one-third of instances and that the *short flexor* of the thumb (apparently the superificial head only) is most commonly supplied by both median and ulnar nerves and in almost a fourth of cases by the ulnar only. Anomalous innervation may be due to the ulnar nerve's spreading farther into the thenar mass than it usually does, but anomalous innervation of hypothenar muscles is apparently never a result of the median nerve sending a direct branch across the palm to the hypothenar muscles; rather, while the ulnar nerve may invade the thenar mass more deeply than usual, or the median nerve may send a branch to the *first dorsal interosseous* muscle, much anomalous innervation seems to be a result of interchange of fibers between the median and ulnar nerves either in the hand or, more commonly, in the forearm or at the brachial plexus.

## GENERAL REFERENCES AND RECOMMENDED READINGS

ATKINSON WB, ELFTMAN H: The carrying angle of the human arm as a secondary sex character. Anat Rec 91:49, 1945

BACKHOUSE KM: The mechanics of normal digital control in the hand and an analysis of the ulnar drift of rheumatoid arthritis. Ann R Coll Surg Engl 43:154, 1968

BASMAJIAN JV, TRAVILL A: Electromyography of the pronator muscles in the forearm. Anat Rec 139:293, 1961

BOJSEN-MØLLER F, SCHMIDT L: The palmar aponeurosis and the central spaces of the hand. J Anat 117:55, 1974

BOSWICK JA JR, STROMBERG WB JR: Isolated injury to the median nerve above the elbow: A review of thirteen cases. J Bone Joint Surg (Am) 47:653, 1967

BUNNELL S: Surgery of the Hand, 3rd ed. Philadelphia, JB Lippincott, 1956

EYLER DL, MARKEE JE: The anatomy and function of the intrinsic musculature of the fingers. J Bone Joint Surg (Am) 36:1, 1954

FORREST WJ, BASMAJIAN JV: Functions of human thenar and hypothenar muscles: An electromyographic study of twenty-five hands. J Bone Joint Surg (Am) 47:1585, 1965

FLYNN JE: Clinical and anatomical investigations of deep fascial space infections of the hand. Am J Surg 55:467, 1942

GARDNER E: The innervation of the elbow joint. Anat Rec 102:161, 1948

GRAY DJ, GARDNER E: The innervation of the joints of the wrist and hand. Anat Rec 151:261, 1965

GREULICH WW, PYLE SI: Radiographic Atlas of Skeletal Development of the Hand and Wrist. Stanford, CA, Stanford University Press, 1950

GRODINSKY M, HOLYOKE EA: The fasciae and fascial spaces of the palm. Anat Rec 79:435, 1941

JAMIESON RW, ANSON BJ: The relation of the median nerve to the heads of origin of the pronator teres muscle: A study of 300 specimens. Q Bull Northwestern Univ M School 26:34, 1952

JONES FW: Voluntary muscular movements in cases of nerve lesions. J Anat 54:41, 1919

KAPLAN EB: Functional and Surgical Anatomy of the Hand, 2nd ed. Philadelphia, JB Lippincott, 1965

LACEY T, II, GOLDSTEIN LA, TOBIN CE: Anatomical and clinical study of the variations in the insertions of the abductor pollicis longus tendon, associated with stenosing tendovaginitis. J Bone Joint Surg (Am) 33:347, 1951

LANDSMEER JMF: The coordination of finger-joint motions. J Bone Joint Surg (Am) 45:1654, 1963

MISRA BD: The arteria mediana. J Anat Soc India 4:48, 1955

MURPHEY F, KIRKLIN JW, FINLAYSON AI: Anomalous innervation of the intrinsic muscles of the hand. Surg Gynecol Obstet 83:15, 1946

O'RAHILLY R: A survey of carpal and tarsal anomalies. J Bone Joint Surg (Am) 35:626, 1953

ROWNTREE T: Anomalous innervation of the hand muscles. J Bone Joint Surg (Br) 31:505, 1949

STOPFORD JSB: The variations in distribution of the cutaneous nerves of the hand and digits. J Anat 53:14, 1918

SUNDERLAND S: The actions of the extensor digitorum communis, interosseous and lumbrical muscles. Am J Anat 77:189, 1945

WRIGHT RD: A detailed study of movement of the wrist joint. J Anat 70:137, 1935

# PART IV

# BACK

# 14

# THE VERTEBRAL COLUMN

Study of the soft tissues of the back is best preceded by knowledge of the vertebral column, a dried specimen of which should be studied prior to dissection. Some of the features of the vertebral column described in this chapter, however, should be checked later during and after the dissection of the soft tissues.

## THE VERTEBRAE

**Composition and Variations.** The vertebral column, often called the *spinal column* or simply *"the spine,"* develops from the somites, the metameric or segmental mesoderm of the embryo, and therefore should itself be metameric. It is, indeed, largely composed of a series of metameric vertebrae (Fig. 14-1), but at the lower end of the column, fusion of adjacent vertebrae to form the sacrum and a part of the coccyx partially obscures the metamerism. The vertebral column in the adult typically consists of seven **cervical,** 12 **thoracic,** and five **lumbar** vertebrae. These are succeeded by the **sacrum,** formed usually by the fusion of five vertebrae, and by the **coccyx,** formed of rudimentary vertebrae, of which the first is commonly separate whereas the succeeding three are fused together. Thus, there are typically 33 vertebrae in the column.

Occasionally there may be 32 or 34 vertebrae instead of the typical 33, but the usual numerical variation in the vertebral column is not an increase or decrease in total number but a variation from the number typical of a certain region. Most important is a lengthening or shortening of the lumbar part of the column as a result of the 1st sacral vertebra failing to fuse with the sacrum and becoming therefore a lumbar vertebra, or as a result of what should be the 5th lumbar vertebra fusing more or less completely to the sacrum.

Studies of skeletons indicate that some 88% to 90% of vertebral columns have a normal number of vertebrae above the sacrum; about 5.5% have some fusion (sacralization) between the 5th lumbar vertebra and the sacrum, but in only one-fifth of these is there such fusion that only four lumbar vertebrae can be recognized; and about 6% have a lengthened thoracolumbar portion, with the 1st sacral vertebra partially or completely separated from the sacrum (lumbarized). Although ribs are typically associated only with thoracic vertebrae, the 7th cervical vertebra may have one (cervical rib). When there is a rib on the vertebra below the 12th thoracic, that rib is usually called a lumbar or "gorilla" rib. There is no correlation between variations in the total number of vertebrae and of ribs.

Increased length of the vertebral column above the sacrum increases the strain put upon the lower part of the lumbar column because of increased leverage. However, persons with either unilateral sacralization or lumbarization tend to have a vertebral column that is more susceptible to damage than is a completely sacralized or lumbarized column, apparently because the articulations between the vertebrae at that level are asymmetrical; both these conditions are regarded as predisposing to backache.

**FIG. 14-1.** The vertebral column from the front and the side.

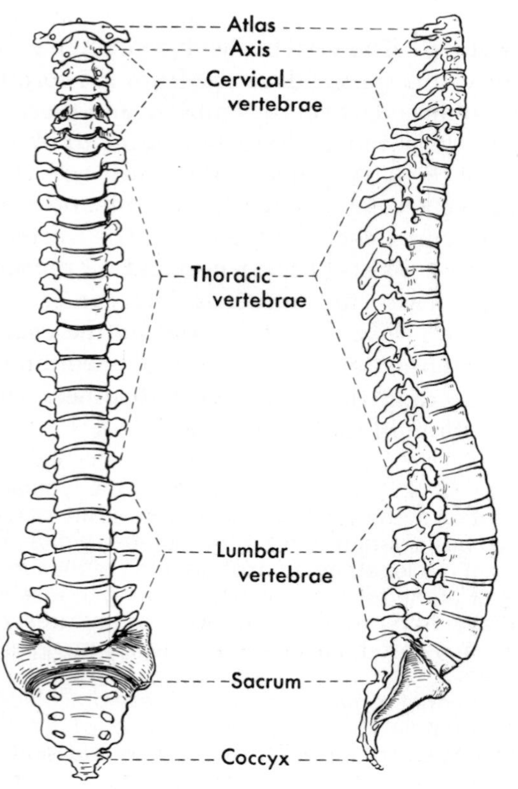


**Typical Vertebra.** The vertebrae vary in size and in other characteristics from one region to another (Fig. 14-6), and to a lesser degree those within a region differ from each other; however, they are all built upon a fundamental plan (Fig. 14-2). The anterior heavier part of the vertebra, resembling a segment of a somewhat ovoid rod, is the **body.** In the dried specimen, the hyaline cartilage that covers most of each end of the body is missing, and the bone appears spongy except at the periphery where epiphyseal rings, fused to the body, cover the spongy bone. Attached posteriorly to the body is the **vertebral arch,** consisting of two *pedicles* and the two *laminae* that unite them. Arch and body together surround the large **vertebral foramen,** and successive vertebral foramina form the **vertebral canal** within which the spinal cord lies. On the arch are projections that serve for attachment of muscles or for articulation with other bones. The typical projections are a **spinous process** that projects posteriorly from the meeting of the laminae; **superior articular processes** (*superior zygapophyses*) that project upward on each side from the upper border of the laminae to form synovial joints with the vertebra above; **inferior articular processes** (*inferior zygapophyses*) that project downward to form synovial joints with the vertebra below; and the laterally projecting **transverse processes** from the region of junction of the laminae and pedicles. Lumbar vertebrae have additional processes.

The laminae of the vertebrae approximate in superoinferior length this dimension of the vertebral bodies, but the pedicles never do; therefore, there are notches, a shallow **superior** and a deep **inferior vertebral incisure,** above and below each pedicle. When vertebrae are fitted together, each two adjacent incisures form an **intervertebral foramen.** Through these foramina the spinal nerves leave the vertebral canal.

In the fresh condition, a typical vertebra articulates with the ones above and below through three types of joints: zygapophyseal, posterolateral **synovial joints** formed by the facets on the articular processes already mentioned; **fibrous joints,** between the laminae

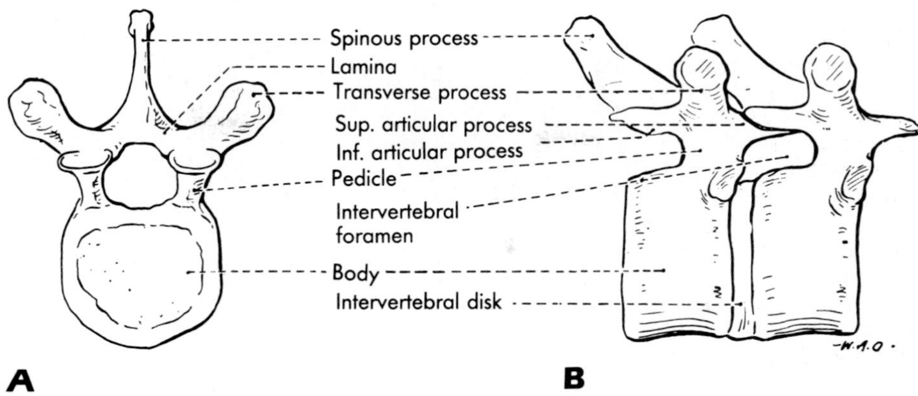


**FIG. 14-2.** Parts of a typical vertebra.

and between the spinous processes; and a **cartilaginous joint** in the form of a fibrocartilaginous *intervertebral disk* that firmly unites the bodies of adjacent vertebrae, yet allows a small amount of movement.

**Regional Characteristics.** Of the seven **cervical vertebrae,** the first two are decidedly different and the 7th cervical vertebra is somewhat modified from a typical one, but a cervical vertebra can be easily recognized by the fact that its transverse process has in it a foramen, the *transverse foramen* (Figs. 14-3, 14-4, and 14-6). This is because the transverse processes of the cervical vertebrae are compound; the posterior root represents a *true transverse process,* but the anterior root, or *costal process,* represents a portion of a rib that has fused both to the body and to the true transverse process. The transverse foramina of about the upper five or six cervical vertebrae transmit the vertebral arteries, vessels that help supply the brain. The transverse processes usually end in anterior and posterior *tubercles,* between which there is a *sulcus for the spinal nerve.* The anterior tubercle of the 6th cervical vertebra is called the *carotid tubercle.*

Cervical vertebrae are also **distinguished** by their small size. Their bodies are about half again as long from side to side as they are anteroposteriorly, their pedicles are rather short, and the articular processes tend to be bulky. The articular surfaces of the superior articular processes face posteriorly and upward, and those of the inferior face anteriorly and downward. Their spinous processes are typically short, but those of the 6th and 7th are much longer; the 3rd through the 6th are usually bifid in white persons, but usually not in blacks; and there is no spinous process on the 1st cervical vertebra.

The **atlas** or 1st cervical vertebra (Figs. 14-3 and 14-5) is peculiar in that it lacks a body and consists chiefly of an *anterior* and a *posterior arch.* Instead of articular processes the atlas has paired *lateral masses* that bear on their upper surfaces *superior articular foveae* for the reception of the occipital condyles of the skull and on their lower surfaces *inferior articular foveae* for articulation with the 2nd cervical vertebra. The anterior arch has anteriorly a small *anterior tubercle* to which muscles attach and posteriorly a facet, the *fovea dentis,* for articulation with the dens of the axis. The spinous process is represented by a small *posterior tubercle.* Along the upper edge of the posterior arch there is a groove, the *sulcus of the vertebral artery,* in which that artery runs after it has traversed the transverse foramen in the lateral mass of the atlas.

The **axis,** the 2nd cervical vertebra (Figs. 14-4 and 14-5), is atypical because it has projecting from the upper part of its body the **dens,** representing the body of the atlas. The dens presents anterior and posterior articular surfaces where synovial joints allow rotation between the skull and atlas on the one hand and the axis on the other. The axis has no real superior articular processes, for its superior

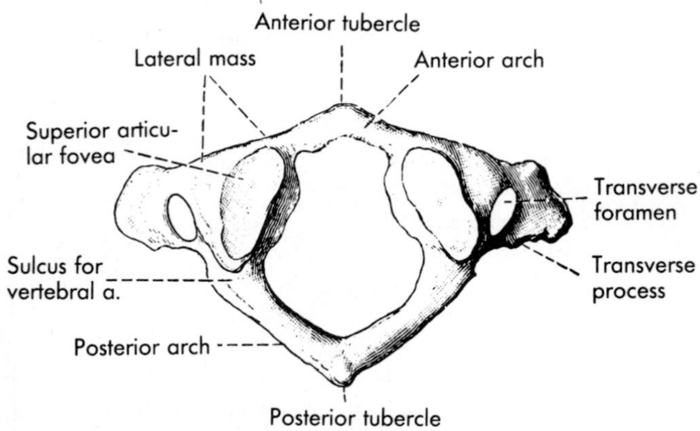


**A**

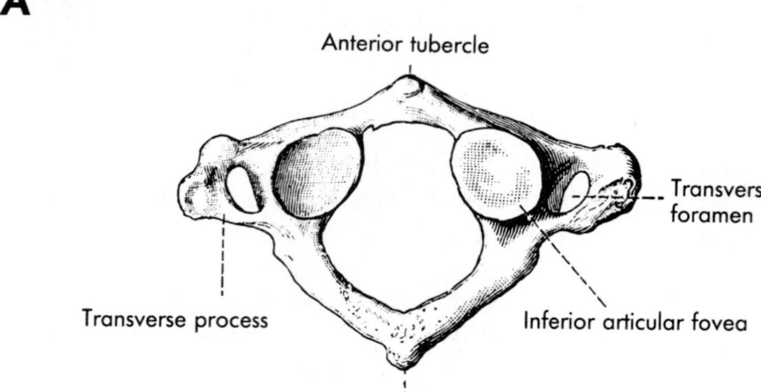


**B**

**FIG. 14-3.** The atlas from above, **A,** and below, **B.** (Disse J: In von Bardeleben K (ed): Handbuch der Anatomie des Menschen, vol 1, sect 1. Jena, Fischer, 1896)



**A**

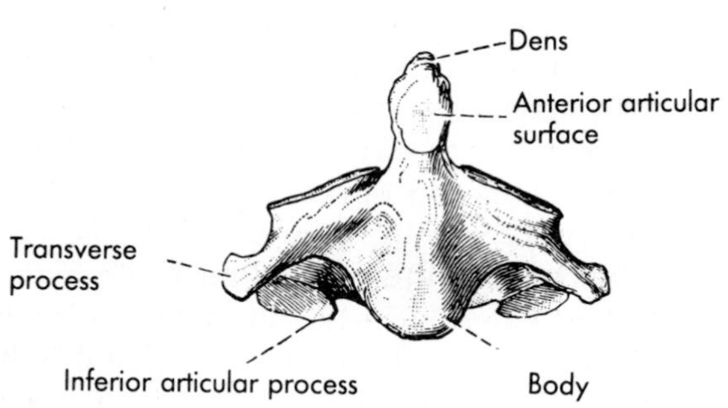


**B**

**FIG. 14-4.** The axis from above, **A,** and the front, **B.** (Disse J: In von Bardeleben K (ed): Handbuch der Anatomie des Menschen, vol 1, sect 1. Jena, Fischer, 1896)

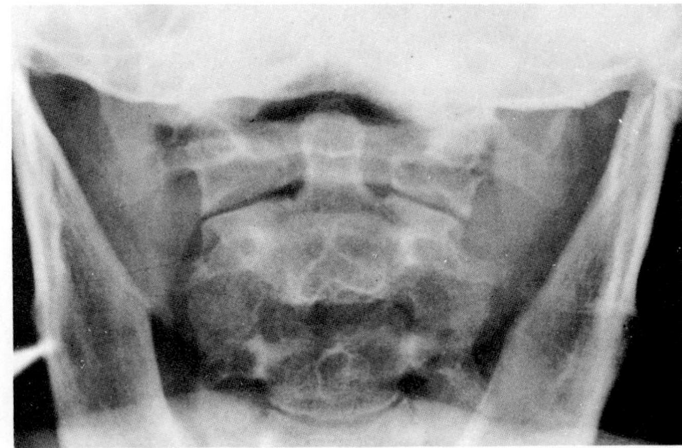

**FIG. 14-5.** Roentgenogram, taken through the open mouth, to show the atlas and axis. The dens shows clearly against the arches of the atlas, and the lateral atlantoaxial joints are also well shown. (Courtesy of Dr. D. G. Pugh)

articular surfaces are placed on the body and pedicles. Because of this placement, the 2nd cervical nerve emerges behind the synovial joint (below this, nerves emerge in front of the joints), and does not groove the transverse process, which therefore has a single tubercle.

The **7th cervical vertebra** has a heavier body and a particularly long spinous process that gives this vertebra its alternative name, the *vertebra prominens*. The transverse foramina are often small, and the anterior or costal element of the transverse process is usually smaller than the posterior one. Occasionally, however, it retains its independence as a rib and is then known as a *cervical rib*.

The bodies of the **thoracic vertebrae** increase in size as they are traced from the 1st to the 12th (Figs. 14-1 and 14-14), with that of the 1st thoracic being only slightly larger than the body of the last cervical, and that of the last thoracic only slightly smaller than the body of the 1st lumbar. The anterior surfaces of the bodies are highly convex, the posterior surfaces slightly concave, from side to side (Fig. 14-6). The *pedicles* are placed toward the upper ends of the bodies, and the inferior vertebral incisures are deep. The *laminae* measure more in a superoinferior direction than do the bodies and somewhat overlap each other. The *spinous processes* are typically long and slender and directed downward enough to overlap at least the succeeding vertebra, but in the lower thoracic region, they become shorter and broader and directed more posteriorly, thus resembling the spinous processes of lumbar vertebrae. The synovial or *zygapophyseal joints* are almost in the coronal plane, for the flat superior articular surfaces face mostly posteriorly and only a little upward and laterally, whereas the inferior surfaces face mostly anteriorly and only a little downward and medially. The long, heavy *transverse processes* project posteriorly and upward as well as laterally; those of the upper ten vertebrae present anterolaterally *articular surfaces* for articulation with the tubercles of the *ribs*.

A **distinguishing feature** of the thoracic vertebrae is that they articulate with ribs and therefore bear **costal foveae** or facets, or parts of them, for these articulations. Most of the thoracic vertebrae articulate with both the head and the tubercle of at least one rib, the articulation with the head of the correspondingly numbered rib being on the upper end of the vertebral body and that with the tubercle being on the transverse process. Most of these vertebrae also articulate with the head of the rib below their own segmental level: the body of the 1st thoracic vertebra has a superior fovea close to its upper end for articulation with the head of the 1st rib and a small inferior fovea close to its lower end for articulation with the upper portion of the head of the 2nd rib; the 2nd through the 8th thoracic vertebrae have approximately a half fovea on each side of their upper borders to accommodate the ribs belonging to their segments and approximately a half fovea below for the following rib; however, the 9th through the 12th ver-

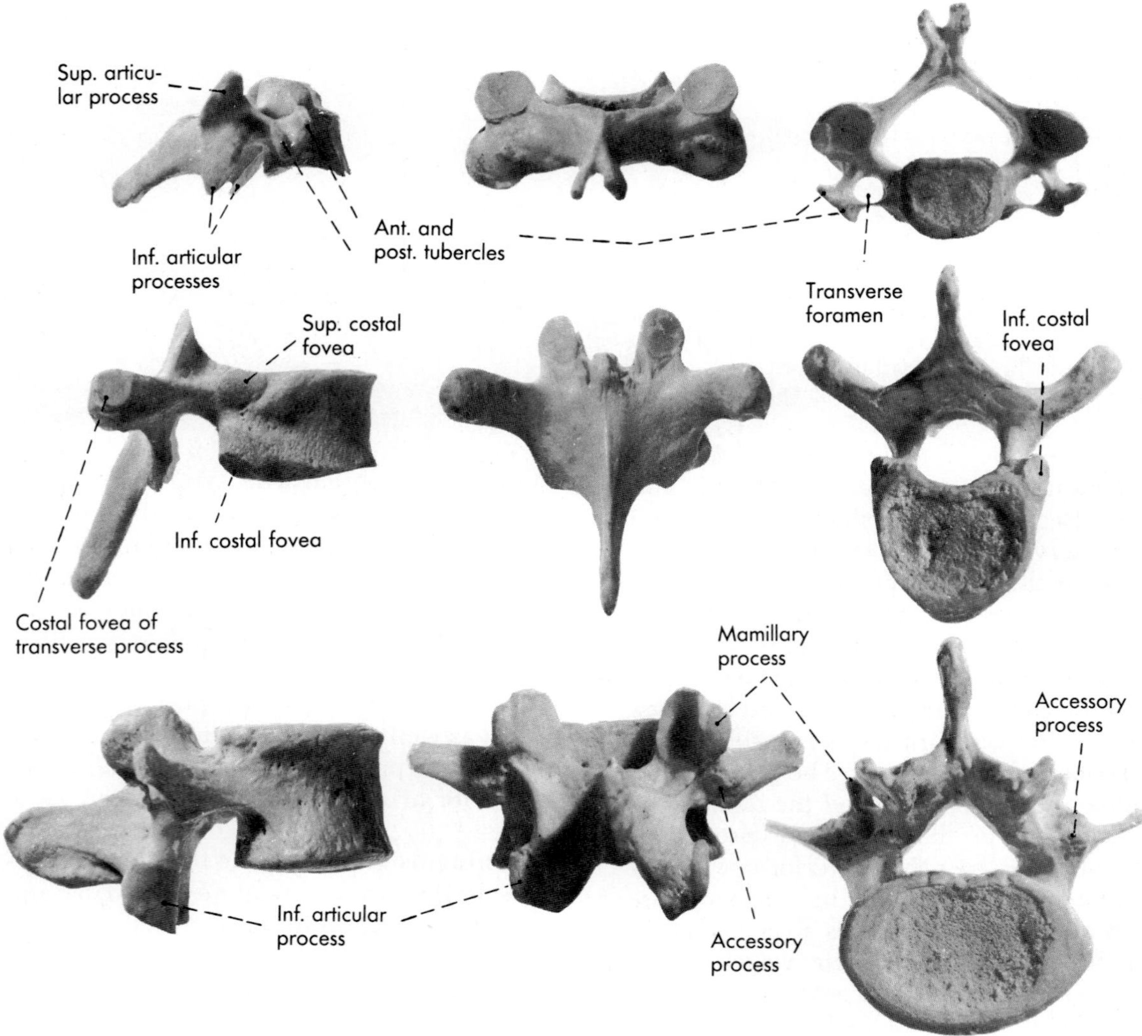


**FIG. 14-6.** Lateral, posterior, and inferior views of the 4th cervical, 7th thoracic, and 4th lumbar vertebrae.

tebrae have single foveae on their upper parts for articulation with the heads of the ribs corresponding to them. Usually the transverse processes of the 11th, and regularly those of the 12th, vertebrae do not articulate with ribs.

The spinous processes of the middle thoracic vertebrae are longest, and differences in the sizes of the vertebrae allow easy distinction between upper and lower ones. The 12th thoracic vertebra can easily be mistaken for a 1st lumbar vertebra if the 12th ribs are lacking and there are therefore no costal foveae on this vertebra; it has a mamillary process as do lumbar vertebrae (see below).

The **lumbar vertebrae** are particularly large and heavy (Figs. 14-6 and 14-15). Their *bodies* are wider transversely than they are deep anteroposteriorly. Those of the lower three vertebrae, particularly that of the 5th, are wedge shaped in lateral view, having a greater vertical height anteriorly than posteriorly. The short and heavy *pedicles* arise from the upper part of the body, so the superior vertebral notch is shallow, the inferior much deeper. The *transverse processes* project somewhat posteriorly and upward as well as laterally. A distal part of each transverse process represents a rudimentary rib that has fused to the process. On the posterior surface of the base of each transverse process there is typically a small tubercle, the *accessory process.* Above and medial to the accessory process, on the posterior surfaces of the superior articular processes, are more marked enlargements, the *mamillary processes.*

In contrast to those of the cervical and thoracic regions, the lumbar *laminae* are regularly shorter vertically than are the bodies. Thus, between any two laminae there is a gap bridged only by ligaments. It is through one of these gaps that lumbar puncture (introducing a needle into the subarachnoid cavity of the lumbar region) is done.

The *spinous processes,* when seen from the side, are broad. When seen from the end they are narrow but expand into an enlarged extremity. The heavy superior articular processes bear the mamillary processes, already mentioned. Their facets are slightly concave, whereas those of the inferior *articular processes* are slightly convex. The synovial joints between the upper lumbar vertebrae for the most part approach the sagittal plane, for the facets of the superior articular processes face largely medially and those of the inferior face largely laterally; however, the inferior articular processes of the 5th lumbar vertebra tend to be less sagittally and more nearly coronally placed, so that they hook over the superior articular processes of the sacrum and help prevent the column from sliding forward on the inclined upper surface of the sacrum. There is thus in the lumbar region a transition from a nearly sagittal to a more nearly coronal position.

The **sacrum** is somewhat triangular in anterior and posterior view (Fig. 14-7). Its central portion consists of the fused bodies of the sacral vertebrae. On its **pelvic surface** are transverse lines or *lineae transversae* (typically four), indicating the regions of fusion of the usual five vertebrae forming it. The wide upper end of the sacrum is the **base.** Projecting upward from it are *superior articular processes* that articulate with the inferior processes of the last lumbar vertebra and face posteriorly or posteromedially. The prominent anterior lip of the base of the sacrum is the *promontory.* On each side of the promontory, the lateral, smooth part of the base is the *ala of the sacrum* (*ala* is wing in Latin). On the pelvic surface are *pelvic sacral foramina* through which the ventral branches of the first four sacral nerves emerge. In an approximately corresponding position on the **dorsal surface** are *dorsal sacral foramina* through which the dorsal branches of the nerves emerge. Each set of sacral foramina opens into an *intervertebral foramen* through which the spinal nerve leaves the sacral portion of the vertebral canal.

In the dorsal midline, the *median sacral* crest represents fused spinous processes. Just medial to the dorsal foramina, a paired series of tubercles, the *intermediate sacral crests,* represent the articular processes of the sacral vertebrae. The **pars lateralis,** between and lateral to the sacral foramina, represents fused transverse and costal processes and presents tubercles, of which only the first may be well developed, that form the *lateral sacral crests.* The smooth articular surface on the lateral aspect of each pars lateralis somewhat resembles the medial side of an ear and is called the *auricular surface.* This forms a synovial joint between the sacrum and the ilium. On the dorsal surface of each lateral mass, lateral to the lateral crest, are roughened raised areas, the *tuberosities* of the sacrum, that receive very heavy *sacroiliac ligaments.* The sacrum is so shaped that the weight of the body, which tends to force the base of the sacrum downward and forward, wedges it even more strongly between the two ilia as it does this.

The lamina of the lowest segment, sometimes more, of the sacrum is incomplete, leaving a gap bridged only by ligaments; this defect is the **sacral hiatus.** *Sacral cornua* lateral to the hiatus represent rudimentary pedicles and articular processes for articulation with the succeeding first coccygeal segment. The apex of the sacrum articulates with the body of the first coccygeal segment.

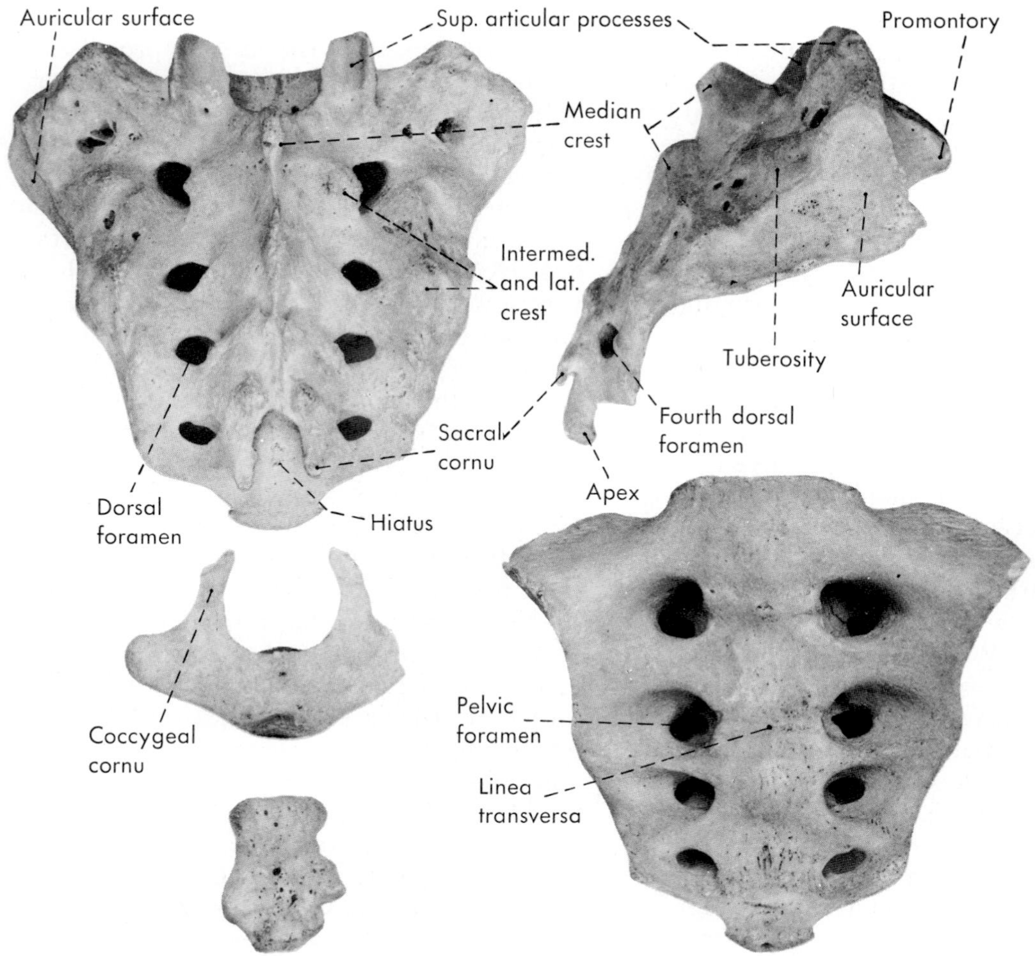


**FIG. 14-7.** Dorsal, lateral, and pelvic surfaces of the sacrum, and a posterior view of the coccyx.

The **coccyx** typically represents four vertebrae (Fig. 14-7), of which the last three are regularly fused together. The first segment, the largest, articulates with the apex of the sacrum through a rudimentary intervertebral disk and has *cornua,* representing pedicles and superior articular processes. *Sacrococcygeal ligaments* unite the coccyx and the sacrum (there is no synovial joint here). This coccygeal vertebra has also short transverse processes. There is commonly a rudimentary fibrocartilaginous or intervertebral disk between the first and second coccygeal segments, allowing some movement at this joint, but the first segment may be fused to the remaining ones. The last three coccygeal segments hardly resemble vertebrae, but do represent the remains of vertebral bodies.

## ARTICULATIONS OF THE VERTEBRAL COLUMN

Special joints exist between the atlas and the skull and the atlas and axis. The other cervical and the thoracic and lumbar vertebrae articulate through joints that are essentially similar in all three regions. As previously noted, these joints are synovial between the articular processes, fibrous between the arches, and cartilaginous (fibrocartilaginous) between the bodies, the last-mentioned being reinforced by special ligaments.

## Synovial Joints

The planes in which the **zygapophyseal** or **synovial joints** of the cervical, thoracic, and lumbar regions lie have already been mentioned briefly in connection with the descriptions of the articular processes and are commented upon again in discussion of the mobility of the vertebral column, since these planes help to determine the movement allowed in each portion of the column. The joints are relatively simple sliding joints, enclosed by the cartilage-covered articular facets of the articular processes and by thin articular capsules that consist as usual of an inner synovial and an outer fibrous membrane.

The **articular capsules** are somewhat lax, to allow the amount of movement demanded at the synovial joints. They do not ordinarily help to check movements of the vertebral column, for these are checked by muscular action, by the heavier ligaments between the arches and between the bodies, by the limitation of movement imposed by the articulations of the bodies with each other, or by impingement of the articular surfaces. The articular capsules are therefore easily strained when excess or abnormal movement occurs, and such strain, as well as arthritis of the joints, are two conditions contributing to low back pain. The synovial joints are **innervated** by twigs from the *dorsal branches of the spinal nerves*, each joint apparently receiving fibers from two spinal nerves.

Because the synovial joints typically lie posterior to the emerging nerves, undue enlargement of a joint as the result of inflammation or bony proliferation may be a source of irritation to the adjacent nerve.

The synovial joints between the atlas and the axis, forming three **atlantoaxial articulations,** are a median and two lateral ones. The *lateral atlantoaxial joints* between the superior facets of the axis and the inferior facets of the atlas are almost flat and more nearly in the transverse plane than are other cervical synovial joints, but are so sloped that rotation between the atlas and the axis produces a slight downward and backward sliding of the atlas on the side toward which rotation is occurring and a slight upward and forward movement on the other side.

The *median atlantoaxial joint* is between the dens of the axis and the anterior arch and

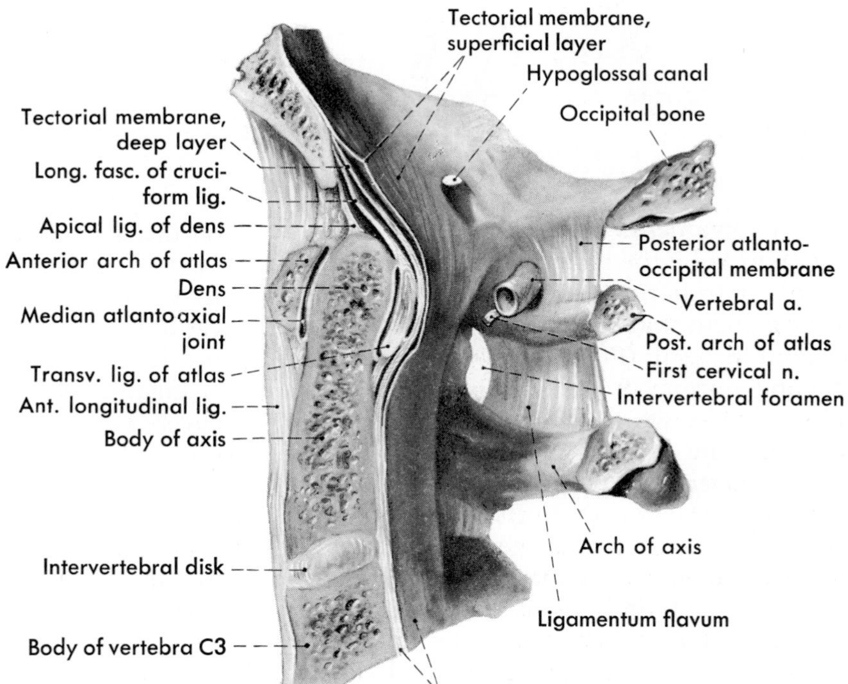


**FIG. 14-8.** Articulations of the axis, atlas, and skull in median longitudinal section. (Redrawn from Fick R: In von Bardeleben K (ed): Handbuch der Anatomie des Menschen, vol 2, sect 1, pt 1. Jena, Fischer, 1904)

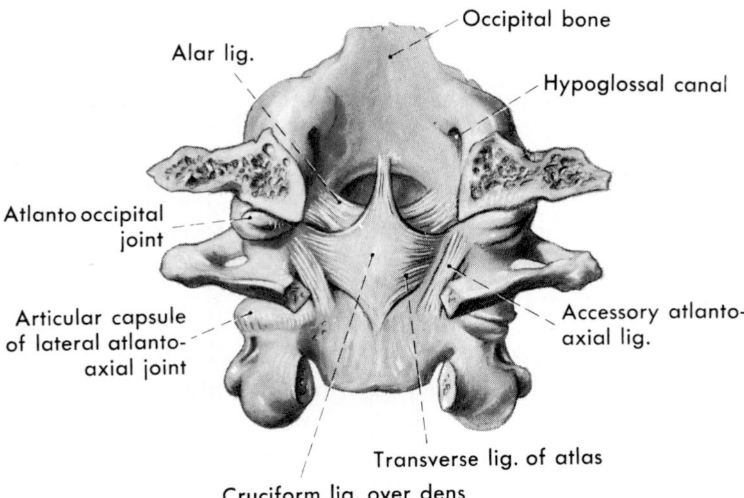


**FIG. 14-9.** Articulations of the axis, atlas, and skull, viewed from behind after removal of the arches, tectorial membrane, and posterior longitudinal ligament. (Redrawn from Fick R: In von Bardeleben K (ed): Handbuch der Anatomie des Menschen, vol 2, sect 1, pt 1. Jena, Fischer, 1904)

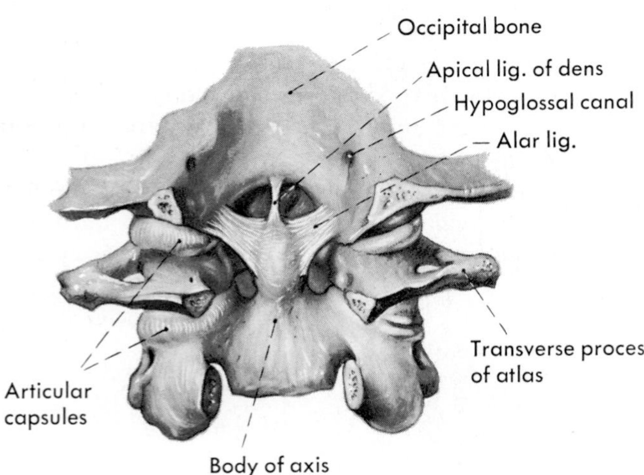


**FIG. 14-10.** Articulations of the axis, atlas, and skull, viewed from behind after removal of the cruciform ligament. (Redrawn from Fick R: In von Bardeleben K (ed): Handbuch der Anatomie des Menschen, vol 2, sect 1, pt 1. Jena, Fischer, 1904)

certain ligaments of the atlas. It is covered posteriorly by the **tectorial membrane,** an upward continuation of the posterior longitudinal ligament that extends from the body of the axis to the skull (Fig. 14-8). Deep to the tectorial membrane in a posterior approach, the dens is held against the anterior arch of the atlas by the **cruciform ligament** (Fig. 14-9). The stronger transverse part of this ligament, attached at both ends to the arch of the atlas and passing posterior to the dens, is the **transverse ligament of the atlas.** Synovial cavities between it and the dens, and between the dens and the anterior arch of the atlas, allow rotation of the skull and the 1st cervical vertebra on the dens. The vertical portion of the ligament (*longitudinal fasciculi*) consists of fibers that run downward to attach to the body of the axis and upward to attach to the occipital bone.

Ligaments attached to the dens, largely covered by the cruciform ligament, are the midline **apical ligament,** between the dens and the occipital bone, and the stronger **alar ligaments,** from the sides of the dens to the occipital bone (Fig. 14-10). The latter ligaments tend to restrict the amount of rotation that can occur upon the dens.

The synovial joints of the **atlantooccipital articulation** are between the occipital condyles of the skull and the superior articular surfaces of the atlas. The convex occipital condyles are more oblong than round, and their anterior ends are closer together than are their posterior ones. The concave articular surfaces of the atlas so fit the condyles that practically no rotational or lateral movement is allowed between the head and the atlas; however, they do allow limited flexion and extension. The other attachments between the atlas and the skull are fibrous ones, the *anterior and posterior atlantooccipital membranes.*

### Posterior Ligaments

The fibrous joints can be divided into those between various parts of the arches and those between the bodies (Figs. 14-11 and 14-12). A **supraspinous ligament** runs over the tips of the spinous processes and blends with the thin **interspinous ligaments** that pass from the lower border of one spinous process to the upper border of the next. In the cervical region, between the spinous process of the 2nd cervical vertebra and that of the 7th where the spinous processes are buried deeply between the heavy muscles on the back of the neck, the supraspinous ligament is represented by a

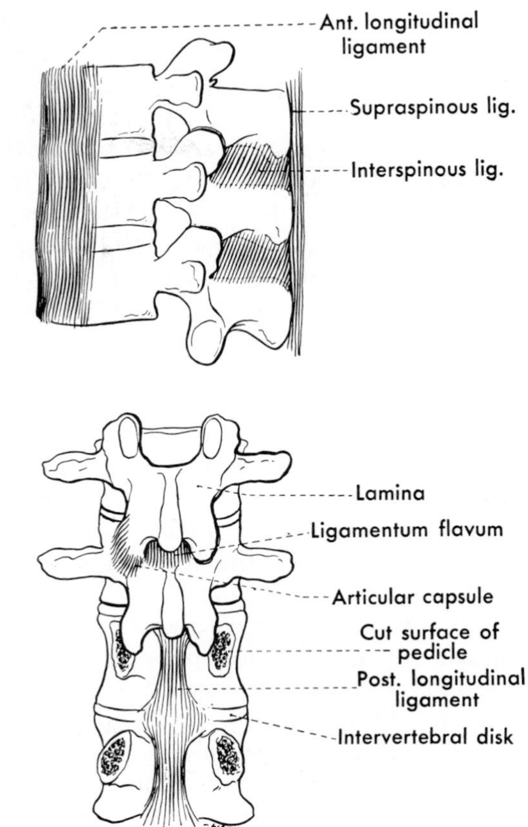


**FIG. 14-11.** Ligaments of the vertebral column.

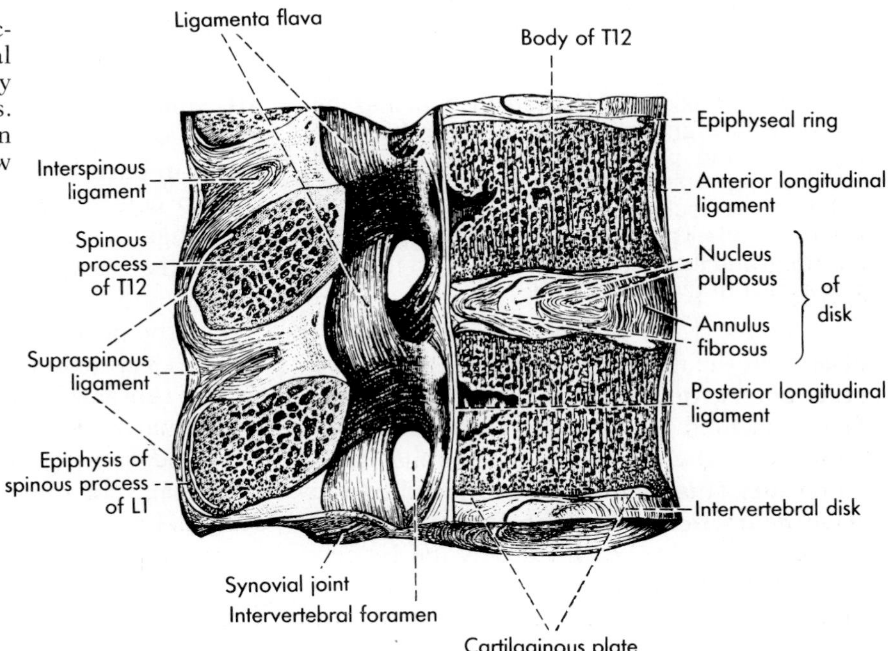


**FIG. 14-12.** Sagittal section of a part of a vertebral column, showing especially the intervertebral disks. (Toldt C: An Atlas of Human Anatomy, 2nd ed. New York, Macmillan, 1928)

thin septum between the musculature of the two sides, the **ligamentum nuchae**. In quadrupeds with heavy heads (for instance, cows), the ligamentum nuchae is a strong and thick band of elastic tissue and aids the muscles in holding up the head. In man, however, the supraspinous and interspinous ligaments and ligamentum nuchae are largely collagenous tissue, of little strength, and in the cervical and lumbar regions, where they should be most important in limiting flexion, the interspinous ligaments are frequently defective or lacking in one or several interspaces, even in young people.

By far the strongest and most important posteriorly placed ligaments are the **ligamenta flava** ("yellow ligaments," because they are composed almost entirely of elastic tissue, which is yellowish). These paired ligaments run between two adjacent laminae and almost entirely fill the space, but are separated in the midline from each other by a narrow slit. Laterally, the ligamenta flava approach close to or blend with fibrous membranes of the synovial joints. Each ligamentum flavum is a flattened band that arises from the anterior surface of the lower edge of one lamina and stretches downward to attach to the upper part of the posterior surface of the succeeding lamina.

Since the laminae are separated farther from each other by flexion, this movement stretches the ligamenta flava and affords more room between laminae for the procedure of lumbar or spinal puncture. Conversely, violent extension (hyperextension) of the neck may carry the cervical laminae so close together that the stretch of the ligamenta flava is lost and they bulge forward against the posterior aspect of the spinal cord.

Between the atlas and the skull, the paired ligamenta flava typical of other interspaces are represented by a single broad and thin membrane, the **posterior atlantooccipital membrane.** This membrane is pierced by the vertebral arteries as they enter the uppermost part of the vertebral canal and by the 1st cervical nerves as they leave this canal.

In addition to the ligaments uniting the arches, there are also **intertransverse ligaments** between succeeding transverse processes. These are best developed in the lumbar region, but probably add little strength to the vertebral column.

## Articulations Between the Bodies

These particularly strong joints are fibrocartilaginous. A fibrocartilaginous intervertebral disk is interposed between the bodies of adjacent vertebrae, and the joints are strengthened anteriorly and posteriorly by longitudinal ligaments.

The **anterior longitudinal ligament** (Fig. 14-11) is a broad band placed on the anterior and anterolateral surfaces of the vertebral bodies and extending from the atlas to the upper part of the pelvic surface of the sacrum. Its deepest fibers extend merely from one vertebra to the next, over the intervertebral disk, and are firmly attached to the margins of the vertebrae. Other fibers extend over two or three, and the most superficial ones over four or five, vertebrae. The fibers are so interlaced as to form a strong band, thickest anteriorly and becoming much thinner laterally. This ligament tends to limit extension of the vertebral column and is especially important in the lumbar region, where the weight of the body tends to increase the normal and permanent dorsiflexion of the lumbar region.

The gap between the skull and the anterior arch of the atlas is bridged by the **anterior atlantooccipital membrane.** This broad membrane, somewhat thicker than the posterior, arises in part from the arch of the atlas and is in part continuous with the anterior longitudinal ligament.

On the posterior surface of the vertebral bodies, and therefore within the vertebral canal, is the **posterior longitudinal ligament.** In the cervical region, this ligament is broad, but in the thoracic and lumbar regions it alternately narrows over the middle of each vertebral body and expands over the ends of the bodies and the intervening disk. The posterior longitudinal ligament consists, like the anterior, of fibers of different lengths. It is anchored firmly to the ends of the bodies, but over their centers relatively loose connective tissue lies between it and the bone, and blood vessels pass from one side to the other, or enter and leave the bone, under cover of this part of the ligament. The central part of the ligament is strong, but its lateral expansions over the intervertebral disks are thinner. The posterior longitudinal ligament tends to check flexion of the vertebral column.

The chief union between vertebral bodies is by the **intervertebral disks** that unite each vertebra to the succeeding one from the 2nd cervical to the lumbosacral junction. They are named and numbered according to the vertebra below which they lie. There is a rudimentary disk at the sacrococcygeal junction, and there may also be one between the first and succeeding segments of the coccyx.

The intervertebral disks are plates of varying thickness that bind two adjacent vertebral bodies together and yet at the same time permit the slight movement between any two vertebrae that, throughout the length of the vertebral column, adds up to surprising mobility of the column as a whole. Each disk is composed of two parts (Fig. 14-12) that blend with each other: there is a firmer outer portion, the *annulus fibrosus,* composed of fibrocartilage, and a softer central portion, the *nucleus pulposus.*

The **annulus fibrosus** is composed of fibrocartilage in which the fibrous element predominates. The fibers run obliquely from one vertebra to the next. They are arranged in concentric rings in which those of each successive ring have a different slant from those of the preceding one. This arrangement gives elasticity to the annulus, for with stretch, the X that the fibers of two adjacent layers make with each other is lengthened and thinned, whereas with compression, it is shortened and broadened. The most peripheral fibers of the annulus insert into the edge of the bone of the vertebral body. The remainder insert into the hyaline cartilage that lies at each end of the disk, covering the cancellous bone of the vertebral body.

The chief component of the **nucleus pulposus** is a mucoid material in which are embedded reticular and collagen fibers. The nucleus contains some 70% to 88% water, the percentage gradually decreasing, on the average, between birth and old age. It is not quite centrally placed, for its center lies somewhat posterior to the center of the body of the vertebra, so that the annulus fibrosus behind the nucleus pulposus is thinner than it is in front of the nucleus.

With its high water content, the nucleus pulposus itself is essentially incompressible; however, its shape can be changed, and this with the compressibility and stretch of the annulus allows the shape of the disk as a whole to be changed, thus permitting movement of one vertebra upon the next. During flexion, for instance, the anterior portion of the disk is compressed and the nucleus pulposus tends to be driven posteriorly and to assume a wedge shape, with its anterior edge thinner. The stretched posterior part of the annulus then normally resists retropulsion of the nucleus. Unfortunately, the annulus fibrosus begins to exhibit degenerative changes, apparently as a result of constant wear and tear upon it, fairly early in life—usually during the third decade—and with increasing age these degenerative changes tend to be more marked, leading to further weakening of the annulus. As a result, depending upon the strains thrown upon it and its state of degeneration, an annulus (most commonly, a lumbar one) may rupture. When it does so, a part of the fibrocartilaginous portion protrudes, and if the tear is complete, a portion of the nucleus pulposus may emerge through the tear. In other cases, a weakened annulus fibrosus may simply bulge without rupturing.

Intervertebral disks may protrude or rupture in any direction, but most common is a posterolateral direction, just lateral to the strong central portion of the posterior longitudinal ligament. This is usually the weakest part of the disk, since the annulus is thinner here and is not supported by other ligaments. Since spinal nerves pass over the posterolateral part of the disk, a protruded or ruptured disk frequently causes irritation of one or more nerves.

The greater the pressure on the nucleus pulposus, the greater pressure it exerts radially to bulge out the annulus, and when the back is used as a lever to lift weight, the load on the lumbar disks becomes great. Thus, measurements of the pressure on the 3rd and 4th lumbar disks of living persons have indicated that leaning forward 20° produces an increase of about 30% in the load, and if weights totalling 44 lb (20 kg) are held in the hands, the load on the disk may range from approximately 500 lb to 750 lb.

The disks contribute about 25% of the length of the vertebral column above the sacrum. Their high water content means that they are subject to dehydration, and enough of this occurs during the course of a day, in consequence of the pressure exerted upon them, to result in the loss of as much as 0.75 inch in height by a man and 0.5 inch by a woman. This dehydration is usually made up by reabsorption of water while the person is lying down, but over a period of years reabsorption does not quite equal water loss, hence the disks gradually thin. This accounts for part of the loss in height that most individuals undergo between young adulthood and old age.

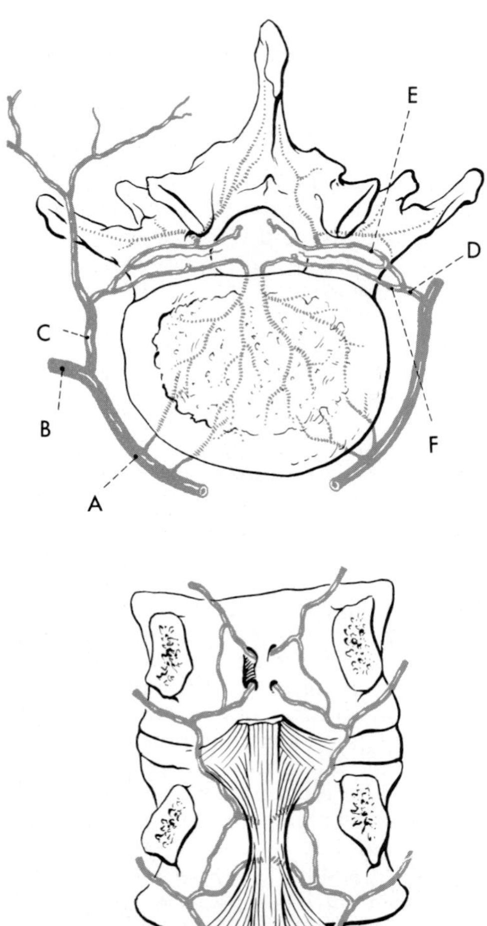


**FIG. 14-13.** Diagram of the blood supply of a vertebra seen from below and from behind with the laminae removed. **A** is a segmental (in this case a lumbar) artery, **B** its ventral continuation, and **C** its dorsal branch; **D** is the spinal branch and **E** and **F** are that branch's dorsal and ventral twigs to tissue of the epidural space and vertebral column. The unlabeled twig between them (a radicular artery) is to the nerve roots and, at some levels, to the spinal cord. In the lower figure only the branches to the vertebral bodies are seen. The cut surfaces of the pedicles are recognizable, and a part of the posterior longitudinal ligament has been removed.

## BLOOD AND NERVE SUPPLY

A typical vertebra is supplied by **segmental arteries** closely associated with it. Because of the method of formation of the vertebrae (p. 304), each thoracic and lumbar vertebra is related at about its middle to paired segmental arteries (intercostal or lumbar) of aortic origin. In the cervical region, similar segmental arteries are branches of the vertebrals, and in the sacral region they are branches of the lateral sacral arteries. Each segmental vessel gives off twigs to the anterolateral aspects of the vertebral body, and larger vessels enter the vertebra from the vertebral canal (Fig. 14-13). The dorsal branches of the segmental arteries, corresponding to dorsal branches of the spinal nerves, are distributed mostly to musculature of the back but give off spinal branches as they pass the intervertebral foramina. These spinal branches enter the vertebral canal and divide into terminal branches that supply the tissue of the epidural space and anastomose above and below with corresponding vessels to form small channels accompanying the venous plexuses here. They supply the dura, nerve roots, and, variably, the spinal cord, and they enter the vertebral arches and bodies to supply them. The vessels to each vertebral body come from segmental arteries entering both above and below the body, so that each body typically receives four arteries. These pass between the posterior longitudinal ligament and the vertebral body and usually penetrate the bone close together or by a common opening.

Much of the **venous drainage** of the vertebrae parallels the arterial supply and enters the *internal vertebral venous plexuses* that surround the spinal cord (see Fig. 15-10). There is also, however, much anterolateral drainage from the vertebrae directly into the segmental veins, and most of the vascular foramina commonly visible on the anterolateral aspect of the vertebral body are for veins.

In the fetus, the **intervertebral disk** has a blood supply reaching it both from its periphery and from the bodies of adjacent vertebrae, but these vessels begin to disappear shortly after birth. Those from each body penetrate the cartilaginous plates on the ends of the body, and sometimes, after they degenerate, the foramina through which they pass do not close. This is thought to allow an intravertebral herniation of the nucleus pulposus, in which a part of the nucleus is forced into the spongy bone of the body and forms a mass that is known as a *"Schmorl body"* or *"intraspongy protrusion."*

The **nerve supply** to the vertebral body has not been carefully investigated, but probably, as in most bone, nerve fibers follow the blood vessels. Evidence that fibers for pain are included is the fact that tumors within a vertebral body often first call attention to their presence by pain. Both the anterior and posterior longitudinal ligaments receive nerve fibers, the latter by way of the *meningeal* (*recurrent*) *branches* of the spinal nerves that leave the nerves just outside the foramina and turn back into the vertebral canal. The meningeal branches divide to ascend and descend and give fibers to the dura, to the adventitia of the blood vessels, to the posterior longitudinal ligament, and to the posterior surface of the annulus fibrosus. In consequence of this innervation, bulging of the intervertebral disk may be a cause of backache.

## VERTEBRAL COLUMN AS A WHOLE

Since the characteristics of individual vertebrae and the joints and ligaments that bind them together have been considered, it is appropriate now to consider the column as a whole.

### Curves

In the early embryo, the vertebral column is C-shaped, as is the trunk of the embryo as a whole, but with succeeding growth the column gradually straightens. In two regions, however, the original curve is never completely abolished. The anterior concavities of the **thoracic** and **sacral curves** are remains of the original or primary curve of the column. By the time of birth, the original anterior concavity of both the cervical and lumbar regions has been abolished, and these curves have begun to reverse themselves. The secondary curve of the cervical portion of the column, with its posterior concavity (Fig. 14-14), becomes established as the infant learns to balance his head and hold it up, and that of the lumbar curve (Fig. 14-15), in the same direction, becomes established as the child learns to sit, stand, and walk. The cervical and lumbar curves are compensatory to the upright habitus of man. The cervical curve acts as a buttress to the weight of the head, thus relieving the posterior muscles of the neck of abnormal strain in holding up the head. Although the thoracic curve tends to put the body at a disadvantage, the lumbar curve, in its turn, again buttresses the weight of the body.

Under normal circumstances, the vertical line of gravity of the body in the erect posture passes behind many of the vertebral bodies of the cervical region but usually through those of the last cervical and first thoracic vertebrae. It passes in front of

**FIG. 14-14.** Lateral roentgenograms of the cervical and thoracic portions of vertebral columns. (Courtesy of Dr. D. G. Pugh)

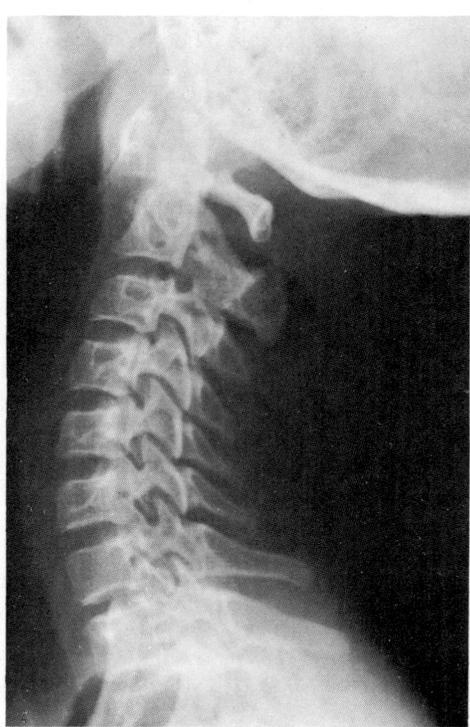

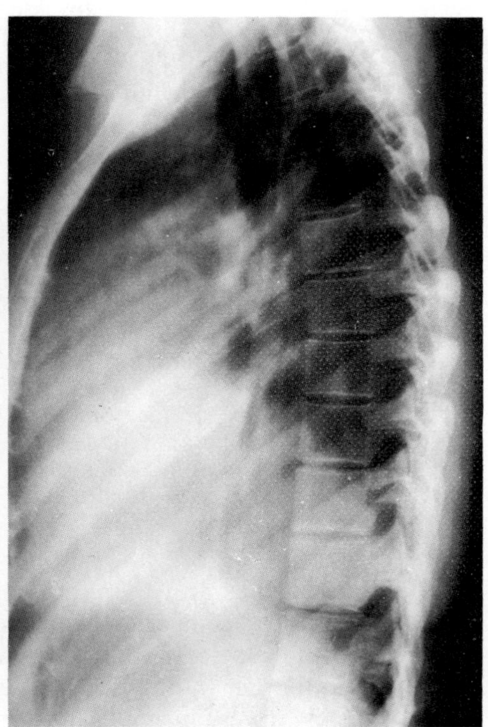

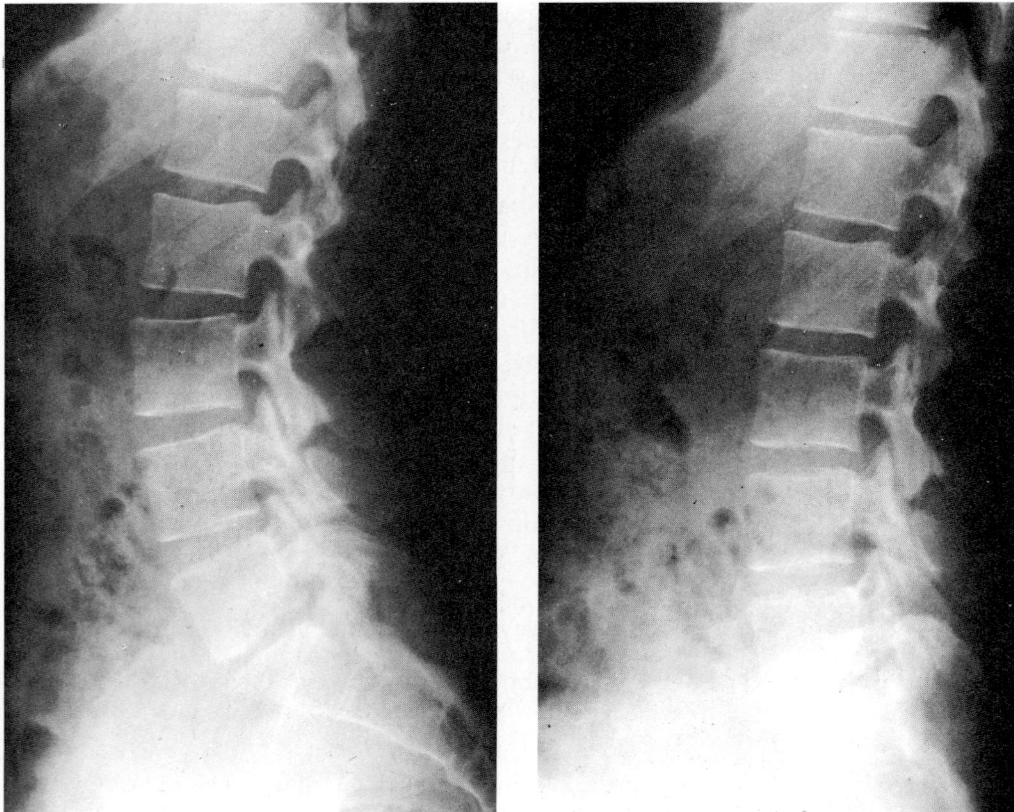

**FIG. 14-15.** Two views of the lumbar curve of the same person. These roentgenograms were both taken with the subject lying on his side. In the view on the left, the thighs were extended at the hips; in the view on the right, they were flexed and in consequence the lumbar curvature was less.

most of the vertebrae of the thoracic region but again through the vertebral bodies at about the thoracolumbar junction. In the lumbar region it passes close to, only a little in front of or behind, the center of the 4th lumbar vertebra. The least expenditure of energy on the part of the musculature of the back is required when the center of gravity is so situated. If one curve is exaggerated so as to throw this center too far from its usual position, an approximately normal position can be restored without additional muscular strain only by altering a curve in the opposite direction.

The cervical, thoracic, and lumbar curves are all gradual ones and therefore relatively strong. The regions of change from one curve to another—the cervicothoracic, thoracolumbar, and lumbosacral junctions—are somewhat sharper curves, in general allow more mobility than occurs within a given region, and are subject to more leverage than are intervertebral junctions within a region. Thus, these junctions are potentially vulnerable to damage, and indeed vertebral fracture is more common at these levels than elsewhere.

The sharpest change in direction and the greatest weight and leverage on the vertebral column come at the **lumbosacral junction** (Fig. 14-16), putting it under particular strain. The sacrum is more oblique to the vertical line of gravity than is any other part of the vertebral column, and its upper surface departs by some 37° to 48° from the horizontal plane. This angle usually is slightly greater in females than in males. There is, thus, a tendency for the thoracolumbar portion of the vertebral column to slide forward on the sacrum. Add to this that abnormalities of the 5th lumbar vertebra are more common than those of other lumbar vertebrae, and the fact that the region of the lumbosacral junction is a common locus of low back pain is not surprising.

The posteriorly concave lumbar curve in part compensates for the obliquity of the sacrum. Where sacral obliquity is more pronounced than normal, therefore, the lumbar curve is also increased, and this lessens the strain on the lumbosacral joint.

The **intervertebral disks** are entirely responsible for the cervical curve, because the vertebral bodies in this region measure slightly less anteriorly than they do posteriorly. In the thoracic region, however, the vertebral bodies are primarily responsible for this curve, and in the lumbar region, the somewhat wedge-shaped disks, thicker anteriorly than posteriorly, are entirely responsible for the upper part of the curve, but both the disks and the vertebrae contribute to the lower part. Note, for instance, the greater anterior than posterior length of the 5th lumbar vertebra and the similar shape of the 5th lumbar (lumbosacral) disk in Figure 14-16.

**Abnormal Curves.** Abnormal curvatures of the vertebral column are designated as kyphosis (hunchback), lordosis (swayback), and scoliosis (a lateral curvature).

In regard to **kyphosis,** the weight of the body obviously tends to increase the thoracic curve, and pronounced kyphosis occurs when some pathologic process destroys or seriously weakens the body of one or more vertebrae. Tuberculosis of the spinal column (Pott's disease) was once a common cause of this condition. Congenital kyphosis can result from defective development of the vertebral bodies. A mild degree of kyphosis is often present in elderly persons.

**Lordosis,** an exaggerated lumbar curve, is usually a compensation for a sacrum that is more oblique than usual. This condition may be congenital or a result of downward rotation of the pelvis because of weakness of the anterior abdominal muscles or a forward shift in the center of gravity. For example, women regularly exhibit lordosis in the later stages of pregnancy; this movement is necessary to restore the line of gravity to an approximately normal position. Any condition that accentuates the wedging of the lumbar vertebrae will produce a permanent lordosis.

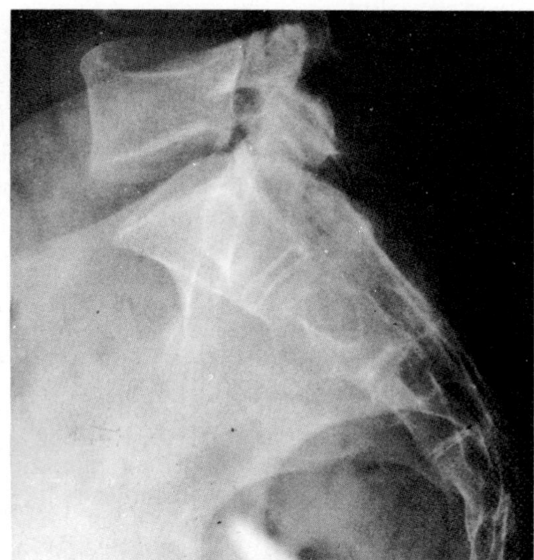

**FIG. 14-16.** Lateral view of the lumbosacral junction and sacrum. (Courtesy of Dr. D. G. Pugh)

**Scoliosis** is a particularly complex curvature. Not only is there a lateral bending but also a rotation of the vertebral column that brings the vertebral bodies toward the convex side of the curve and the spinous processes toward the concavity. Further, one or more compensatory curves may develop following establishment of the primary curve in order that a vertical line of gravity between the head and the pelvis be reestablished. Scoliosis may result from absence of one-half of a vertebral body or from marked paralysis and fibrosis of back muscles, but more often appears to result from some other interference with growth that produces a permanent abnormal wedging of the vertebral bodies. Scoliosis that occurs in a young child and goes untreated usually becomes progressively worse until growth has ceased.

## Mobility and Stability

The remarkable mobility of the vertebral column results primarily from the compressibility and elasticity of the intervertebral disks, which allow movement of the bodies of the vertebrae upon each other.

The amount of movement that can occur in any part of the vertebral column depends in large part upon the ratio between the height of the intervertebral disks and the height of the bony part of the column, but the factors that typically limit movements at joints in general—the checking action of ligaments and muscles, for instance—also play a part. The intervertebral disks will allow limited movement in any direction, and it is primarily the position of the synovial joints that determines the types of movement between any two vertebrae.

In the **cervical region,** the atlantooccipital joint allows primarily flexion and extension, and the atlantoaxial joints allow primarily rotation, but, surprisingly, about the same amount of flexion and extension as does the atlantooccipital joint. In the remainder of this region, all movements are particularly free: the disks are relatively thick, and the anteroposterior and lateral overlap of a vertebra on its neighbors allows flexion and extension (Fig. 14-17), as well as lateral bending. Further, the planes of the synovial joints, which lie almost on the arc of a circle with its center in the vertebral body, permit these movements and also rotation. Because these joints are not exactly on the arc of a circle, however, some flexion to the same side always accompanies rotation in the cervical region.

Movement is most limited in the **thoracic region** because of several factors. The disks are thin here, particularly in the upper and middle portions of the thoracic region (Fig. 14-18), and the attachments of the ribs to the vertebral column and the sternum further circumscribe movement. The almost frontal direction of the synovial joints tends to limit both flexion and extension except in the lower part of the column where the articular surfaces become more nearly sagittally arranged, and the overlapping of adjacent laminae and spinous processes tends to limit extension. Lateral bending is more free, but is limited by the ribs. Rotation also is especially free, but limited by the sternum and ribs. The upper part of the thoracic column is, therefore, particularly limited in its mobility, with the lower part of the column, close to the thoracolumbar junction, being somewhat more mobile.

In the **lumbar region,** the thickness of the disks and the largely sagittal position of the synovial joints in the upper part permit considerable flexion and extension, which are particularly free in this part of the column; the articular facets fit loosely enough together so that lateral flexion is also allowed, but only a little rotation can occur because they lock almost immediately. Usually the last two vertebrae have their lower facets more nearly in the frontal plane, and therefore more rotation can occur in the lower part of the lumbar column.

The muscles acting upon the vertebral column to produce the various movements are discussed later. It is perhaps well to note here, however, that movement of the vertebral column is not brought about exclusively

**FIG. 14-17.** The cervical part of the vertebral column in flexion and in hyperextension (dorsiflexion). (Courtesy of Dr. D. G. Pugh)

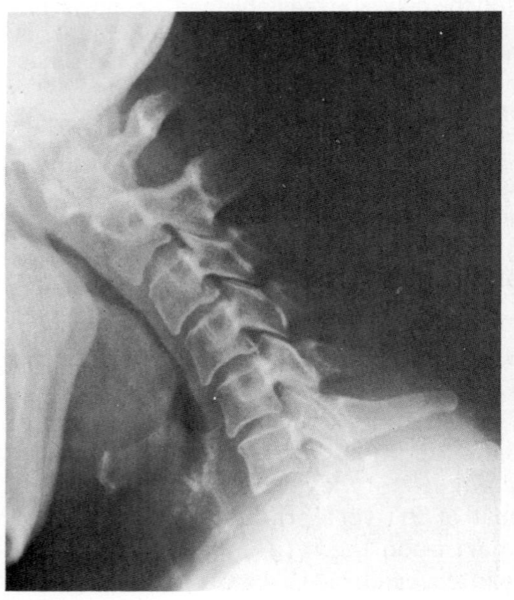

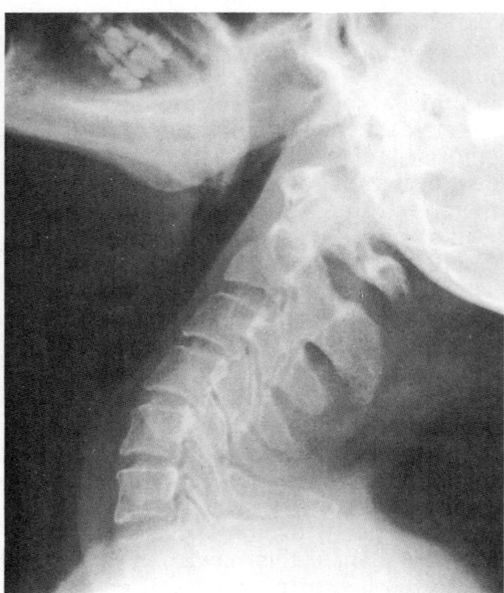

by the musculature attached to it but can result also from gravity and from the action of the muscles of the abdominal wall.

**Stability** of the vertebral column without undue strain upon some part obviously depends upon the presence of normal curves and upon the integrity of the vertebrae and the ligaments that bind them together. Almost any movement of the trunk disturbs the center of gravity, and, therefore, the musculature is particularly important. Further, if the weight of the body is not properly balanced on the vertebral column, undue strain is necessarily thrown upon the muscles. This is undoubtedly a common cause of the low back pain that almost everyone endures from time to time. Abnormal articulations of the vertebral column (particularly frequent at the lumbosacral junction), narrowing of the intervertebral disks as a result of too great dehydration or of herniation of a part of the nucleus pulposus, and abnormal posture may throw great strain upon the synovial joints and the posterior ligaments and give rise to pain. (There seem to be many causes of low back pain: strain of the posterior ligaments; undue strain upon the synovial joints, as just mentioned; muscular strain; perhaps pinching of the synovial membrane of the joints during twisting and lifting movements; and, apparently, herniations of fat and connective tissue through the thoracolumbar fascia, the heavy fascia of the back.) Protrusion of intervertebral disks, with stretching of the posterior fibers of the annulus and of the posterior longitudinal ligament, is apparently also a cause of low back pain, which may precede other evidence of a protruded or ruptured disk (extension of pain down the back of the leg) by many years.

**Fracture and Dislocation.** The most common **fracture** of the column is a crush (compression) fracture of the body of one or more vertebrae, caused by sudden forceful flexion, as in an automobile accident. This type of fracture usually occurs in the region of the thoracolumbar junction, and if it is severe enough to allow acute flexion, it may irreparably damage the spinal cord. In severe crush fractures, the posterior ligaments and parts of the arches may also be torn or fractured, with resulting dislocation of vertebrae. In most crush fractures, however, the strong anterior longitudinal ligament is not injured. Therefore when an injury to the thoracolumbar part of the vertebral column is suspected, it is customary to place the patient in a position of extension in order to keep the anterior longitudinal ligament taut so that any small fragments of the vertebra will be held in place by it. The ligament also acts as a splint against the posterior uninjured parts of the vertebral bodies or arches, tending to prevent dislocation of the vertebra (with possible injury to the spinal cord) while the patient is being handled. The position of extension does not apply in cervical fractures, since these may be caused by either excess flexion or extension.

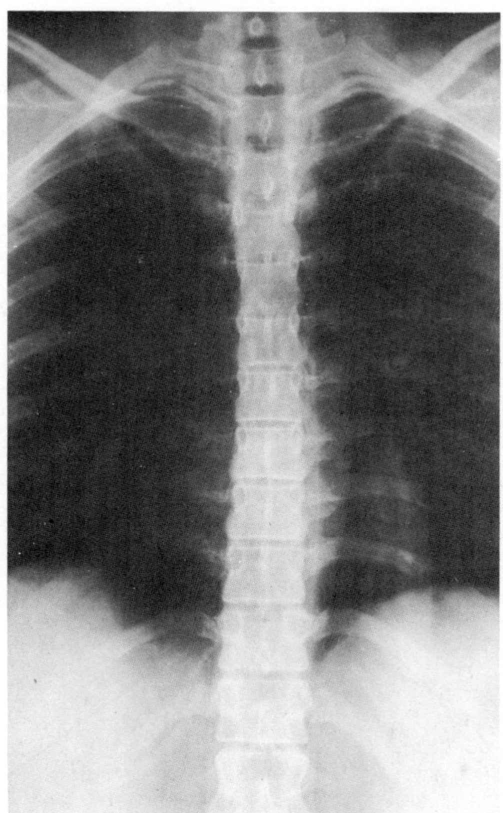

**FIG. 14-18.** Anteroposterior view of the thoracic portion of a vertebral column. As is frequently true, this column is not quite straight, showing a slight scoliosis with the convexity to the left. The curved shadows in the lower part of the figure are produced by the abdominal viscera and the dome of the diaphragm. (Courtesy of Dr. D. G. Pugh)

Fractures of the cervical region, especially those consequent upon the whiplike action that is frequent in automobile accidents, are particularly likely to cause death.

Except in the cervical region, **dislocation** without fracture is impossible because the articular processes interlock. It sometimes happens, however, that the dens is either congenitally absent or has failed to fuse with the axis; in such instances there is gradual dislocation of the head and the atlas upon the axis, since the lateral atlantoaxial articulations are incapable of resisting gradual displacement. Fortunately, the atlas is much larger in circumference than it needs to be to accommodate

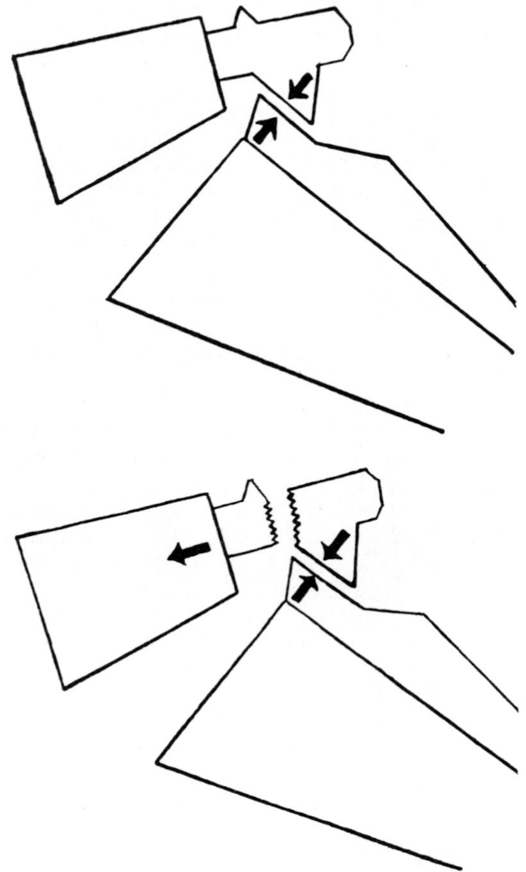

**FIG. 14-19.** Illustration of how the overlapping of the articular processes at the lumbosacral junction prevents anterior displacement of the vertebral column on the sacrum (above), but fails to do so (below) when there is spondylolysis. (Mitchell GAG: J Bone Joint Surg 16:233, 1934)

the spinal cord, and a relatively large amount of displacement can occur before the cord is seriously damaged.

The inferior articular processes of the 5th lumbar vertebra so overlap the superior articular processes of the sacrum (Fig. 14-19) that anterior dislocation of the 5th lumbar vertebra, although favored by the obliquity of the upper end of the sacrum, is impossible under normal circumstances. Occasionally, however, the inferior articular processes are malformed and fail to exert their usual locking action, and downward and forward displacement of the vertebral column on the sacrum results. More commonly, however, such displacement is a result of **spondylolysis.** This term describes a defect in the lamina of a vertebra (usually the 5th lumbar) just caudal to the pedicle. Bilateral defects separate most of the lamina and the inferior articular processes from the pedicles, the body, and the superior articular processes, so that the overlapping of the lumbosacral articular facets holds only the lamina in place (Fig. 14-19). Under these circumstances, there may be a gradual forward displacement of the vertebral column at the lumbosacral disk, perhaps so much as to exert pressure upon the roots of the spinal nerves as they pass onto the upper part of the sacrum. This anterior sliding is known as *spondylolisthesis.*

The cause of spondylolysis is not known, but it apparently develops during the postnatal growth period, for it has not been seen in newborns. The incidence among Americans increases up to about the age of 20, but not thereafter (the incidence varies markedly according to race and sex).

### Development and Growth

The vertebral column originates from mesenchyme derived from the **sclerotome,** the ventromedial part of the segmental mesodermal somites. It is segmental from its beginning. As this mesenchyme collects around the **notochord,** the original longitudinal skeleton of vertebrates, it is at first segmented, as are the myotomes, and the segmental (intersegmental) arteries lie between each two adjacent segments. However, such an arrangement would leave segmental muscles stretching from one end to the other of a single vertebra rather than between vertebrae. The necessary arrangement of alternating and overlapping segments of vertebral column and muscle is obtained by a shift in the segmentation of the vertebrae: each sclerotomal derivative divides into cranial and caudal portions. These portions separate from each other and unite with the caudal and cranial portions, respectively, of the sclerotomal derivatives above and below them (Fig. 14-20). Thus, the vertebrae come to lie adjacent to two myotomes, and the segmental arteries come to lie at the middle of each vertebra instead of at its ends. After the bodies are formed in this fashion, outgrowths extend dorsally around the neural tube to form **neural arches** and ventrolaterally to form **costal processes** that give rise to the transverse processes and ribs.

Between the developing vertebrae, looser mesenchyme accumulates to form the intervertebral disks. As the vertebral body becomes more dense in structure, the part of the notochord within the body gradually disappears (although bits sometimes remain and may later form tumors, *chordomas*). However, pieces of it regularly remain in the center of each developing intervertebral disk and subsequently enlarge to become the nuclei pulposi.

While the mesenchymal arches and costal processes are still growing, **centers of chondrification** (cartilage formation) appear in each costal process and in the right and left sides of each vertebral body. These normally spread until the vertebra is formed largely of cartilage, and as the two sides of the arch meet and come together dorsally during the third prenatal month, the entire vertebra becomes a single piece of cartilage. Before the cartilaginous growth has been completed, **centers of ossification** appear in the bodies of the vertebrae (usually appearing as a single one for each body). A center for each side of the arch also appears, so that a typical vertebra is formed from three primary centers of ossification. At birth, such a vertebra still consists of three bony pieces united by cartilage (but ossification of the coccygeal vertebrae does not appear until after the first year of life). After birth, the two sides of each neural arch unite first. The arches, which include a strip of the body at the base of each pedicle (Fig. 14-21A), then join the vertebral bodies, beginning in the cervical region and proceeding to the sacral one.

As noted in the description of the anatomy of the vertebrae, the costal processes of the cervical and lumbar vertebrae develop into transverse processes that represent both transverse process and rib; those in the thoracic region develop joint cavities and differentiate into separate transverse processes and ribs; and those in the sacral region fuse together to form the lateral parts or masses.

In addition to primary centers of ossification, most of the vertebrae also develop **epiphyses**. At each end of a vertebral body, there is a cartilaginous epiphyseal plate interposed between the bony part of the vertebra and the disk, and at about the time of puberty, a ringlike bony epiphysis appears at the periphery of the cartilage. At about the same time, the cartilaginous tips of the spinous and transverse processes also begin to ossify. Thus, there are typically five **secondary centers of ossification** for each vertebra (Fig. 14-21B and C); however, those vertebrae that have bifid spinous processes have a center for each tip, and the lumbar vertebrae have a center for each mamillary process (Fig. 14-21D). The **epiphyseal rings** join the vertebral bodies between the 18th and 20th years, and the other secondary centers fuse soon thereafter.

Since the body of the **atlas** separates from this bone to fuse with the axis, centers of ossification in the first two cervical vertebrae are different from those of others. There is one center for each half of the posterior arch of the atlas, including the lateral parts that bear the articular surfaces, and a single center for the anterior arch. The centers of ossification for the **axis** include paired centers for the arch and one for the lower part of the body, similar to those of a typical vertebra. Paired centers for the dens, representing the body of the atlas, fuse quickly, so that at birth, the axis consists of four parts. The two parts of the lamina fuse and join the body; the dens is usually firmly united to the body by the age of six years, but the fusion begins superficially and a disk of cartilage may persist deeply for

**FIG. 14-20.** Development of the segmentation of the vertebral column. The original segmentation, marked by the intersegmental arteries and corresponding to that of the myotomes, can still be seen on the left; in the definitive stage, on the right, the caudal half of one segment has fused with the cranial half of the segment below. (After Keyes and Compere. Coventry MG: In Instructional Course Lectures. The American Academy of Orthopaedic Surgeons, vol 6. Ann Arbor, JW Edwards, 1949

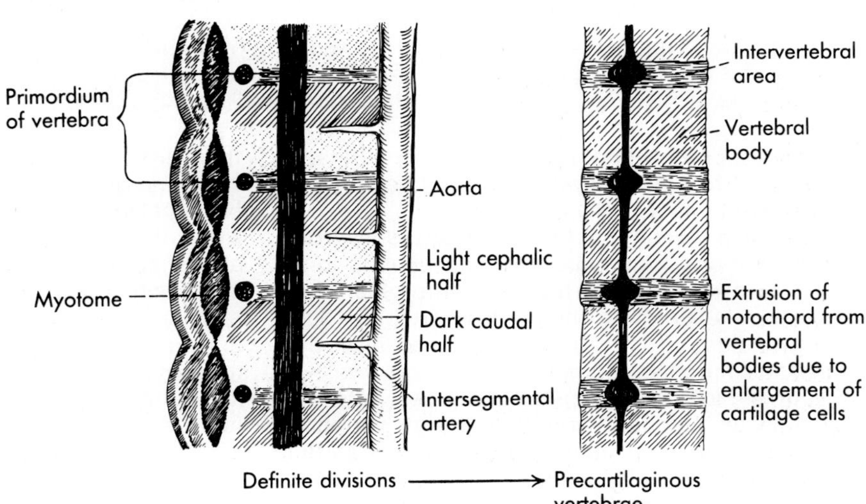

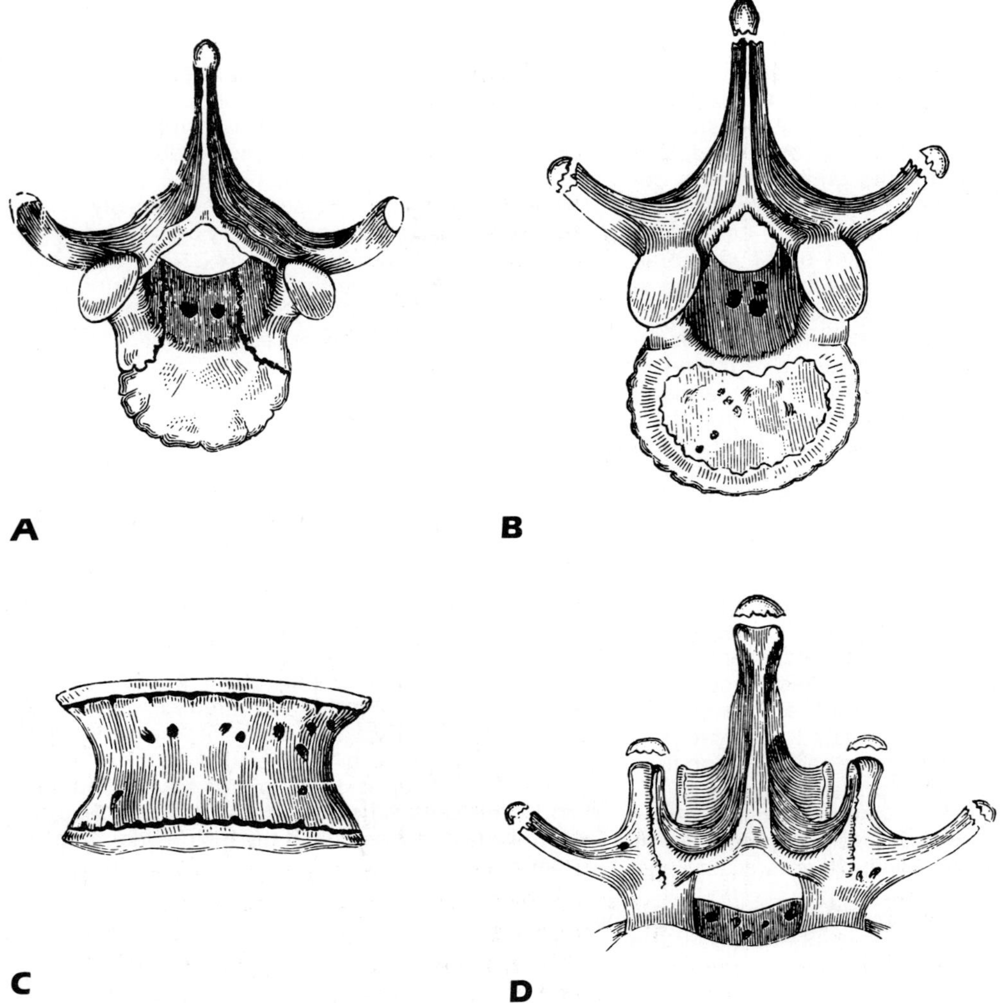


**FIG. 14-21.** Ossification of typical vertebrae. **A** is a thoracic vertebra of a 3-year-old child; the laminae are fused, but the arch has not yet joined the body. **B** shows the epiphyses (separated) of the processes of a thoracic vertebra from a 16-year-old child and the epiphyseal ring on the body. **C** is the body of a lumbar vertebra showing the two epiphyseal rings. **D** is a lumbar vertebra showing the epiphyses not only for the spinous and transverse processes but also for the mamillary processes. (Rauber-Kopsch F: Lehrbuch der Anatomie des Menschen, pt 2. Leipzig, Thieme, 1914)

many years or throughout life. Finally, the cartilaginous tip of the dens develops a bony center that fuses with the rest of the dens.

**Growth in diameter** of the vertebrae results from periosteal bone formation, just as it does in long bones. Growth in length results from proliferation of the cartilages at the ends of the vertebrae, also just as it does in long bones. However, after the vertebral column has completed its growth, the major part of each cartilage persists as a seal between the nucleus pulposus and the spongy vertebral body, and only their peripheries disappear as the ringlike epiphyses at the ends join the bony bodies.

**Spina Bifida.** The most common gross developmental defect of the vertebral column, other than differences in vertebral segmentation in the various regions, is spina bifida (Fig. 14-22). This is brought about by the failure of union of the two sides of an arch and varies from a slight defect to almost complete failure of the arch to appear. The slight defect is known as *spina bifida occulta* and is usually accompanied by no particular symptoms except a greater frequency of low back pain than in the general population. Spina bifida occulta of the last lumbar vertebra has been found in slightly more than 2% of adult columns.

Spina bifida of clinical importance is most common in the 5th lumbar vertebra, but may involve several lumbar vertebrae. Some spina bifida of the sacrum, usually at one or both ends of the sacrum,

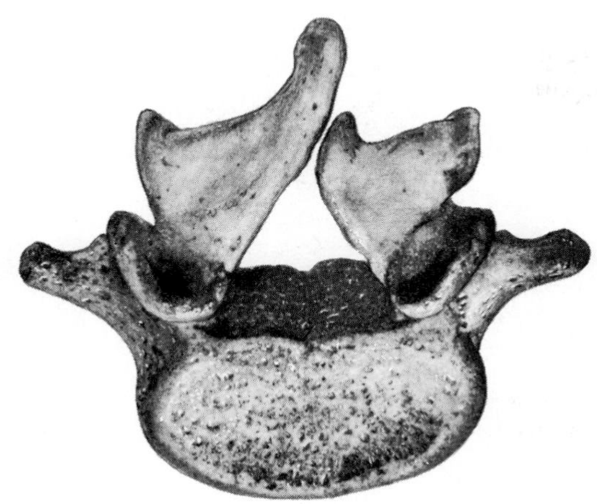

**FIG. 14-22.** Spina bifida occulta of a lumbar vertebra. (Retouched from original. Willis TA: Am J Anat 32:95, 1923)

but sometimes complete, is relatively common but of little clinical importance.

Severe spina bifida of the cervical and thoracolumbar regions is typically complicated by herniation of the meninges and often of the spinal cord. If the meninges alone are involved, the condition is known as *meningocele*, but if the spinal cord also is involved, it is known as *meningomyelocele*. These conditions are usually said to occur in about one of every 1000 births, and most of them are meningomyeloceles. Other congenital anomalies, such as clubfoot and hydrocephalus, are frequently associated with meningomyelocele. Of the children who survive, many show evidence of damage to the nervous system, such as paralysis of the limbs and disturbances of control of the bladder and bowel. Such symptoms may also be associated with spina bifida occulta, but this is much less frequent.

## GENERAL REFERENCES AND RECOMMENDED READINGS


ASMUSSEN E: The weight-carrying function of the human spine. Acta Orthop Scand 29:301, 1960

BARDEEN CR: Numerical vertebral variation in the human adult and embryo. Anat Anz 25:497, 1904

BICK EM, COPEL JW: Longitudinal growth of the human vertebra: Contribution to human osteogeny. J Bone Joint Surg (Am) 32:803, 1950

COVENTRY MB, GHORMLEY RK, KERNOHAN JW: The intervertebral disc: Its microscopic anatomy and pathology: I. Anatomy, development, and physiology. J Bone Joint Surg 27:105, 1945

DAVIS AG: Injuries of the spinal column. In: Instructional Course Lectures. The American Academy of Orthopaedic Surgeons, Vol 6. Ann Arbor, JW Edwards, 1949

FERGUSON WR: Some observations on the circulation in foetal and infant spines. J Bone Joint Surg (AM) 32:640, 1950

HALEY JC, PERRY JH: Protrusions of intervertebral discs: Study of their distribution, characteristics and effects on the nervous system. Am J Surg 80:394, 1950

HOLLINSHEAD WH: Anatomy of the spine: Points of interest to orthopaedic surgeons. J Bone Joint Surg (Am) 47:209, 1965

LANIER RR JR: The presacral vertebrae of American white and Negro males. Am J Phys Anthropol 25:341, 1939

LETTERMAN GS, TROTTER M: Variations of the male sacrum: Their significance in caudal anesthesia. Surg Gynecol Obstet 78:551, 1944

MITCHELL GAG: The lumbosacral junction. J Bone Joint Surg 16:233, 1934

MITCHELL GAG: The significance of lumbosacral transitional vertebrae. Br J Surg 24:147, 1936

MITCHELL PEG, HENDRY NGC, BILLEWICZ WZ: The chemical background of intervertebral disc prolapse. J Bone Joint Surg (Br) 43:141, 1961

NACHEMSON A: The effect of forward leaning on lumbar intradiscal pressure. Acta Orthop Scand 35:314, 1965

PEDERSEN HE, BLUNCK CFJ, GARDNER E: The anatomy of lumbosacral posterior rami and meningeal branches of spinal nerves (sinuvertebral nerves): With an experimental study of their functions. J Bone Joint Surg (Am) 38:377, 1956

ROCHE MB, ROWE GG: The incidence of separate neural arch and coincident bone variations: A summary. J Bone Joint Surg (Am) 34:491, 1952

ROWE GG, ROCHE MB: The etiology of separate neural arch. J Bone Joint Surg (Am) 35:102, 1953

SCANNELL RC: Congenital absence of the odontoid process: A case report. J Bone Joint Surg 27:714, 1945

SCHROCK RD: Congenital abnormalities of the cervicothoracic level. In: Instructional Course Lectures. The American Academy of Orthopaedic Surgeons, Vol 6. Ann Arbor, JW Edwards, 1949

SENSENIG EC: The early development of the human vertebral column. Contrib Embryol 33:21, 1949

VIRGIN WJ: Experimental investigations into the physical properties of the intervertebral disc. J Bone Joint Surg (Br) 33:607, 1951

WILLIS TA: The lumbosacral vertebral column in man, its stability of form and function. Am J Anat 32:95, 1923

WILLIS TA: The thoracicolumbar column in white and Negro stocks. Anat Rec 26:31, 1923

# 15

# SOFT TISSUES OF THE BACK

The innervation of the skin of the back has already been described (Chap. 12, see Fig. 12-24), as have the muscles of the shoulder that extend to the vertebral column and largely cover the true back muscles.

## MUSCLES OF THE BACK

The true musculature of the vertebral column is covered in the back by *muscles of the shoulder girdle* (trapezius, rhomboids, latissimus dorsi) and by *muscles of the ribs* (serratus posterior). The shoulder muscles were discussed in Chapter 12. Before moving to the muscles of the vertebral column, the posterior serratus muscles and associated fascia are discussed.

The muscles of the back are traversed by the **dorsal rami of spinal nerves** and corresponding dorsal branches of segmental vessels. The dorsal rami typically slant inferiorly as they proceed through the back muscles and, therefore, for the most part, supply muscles below the level of origin of the nerve. Each ramus typically divides into a medial and a lateral branch, both of which help to supply muscle. In addition, the medial branch supplies afferent twigs to the synovial joints of the vertebrae and to the posteriorly placed ligaments. One of the two branches, typically the medial one above and the lateral one below, emerges to supply skin of the back. Except for the posterior serratus, this innervation is typical of almost all the back muscles; only a few small lateral muscles are innervated by ventral rami of spinal nerves.

### SERRATUS MUSCLES

In addition to muscles of the shoulder, the vertebral musculature is also partly covered by the serratus posterior muscles of the ribs.

The **serratus posterior superior** (Fig. 15-1) is a flat muscle that arises from the lower part of the ligamentum nuchae and the spi-

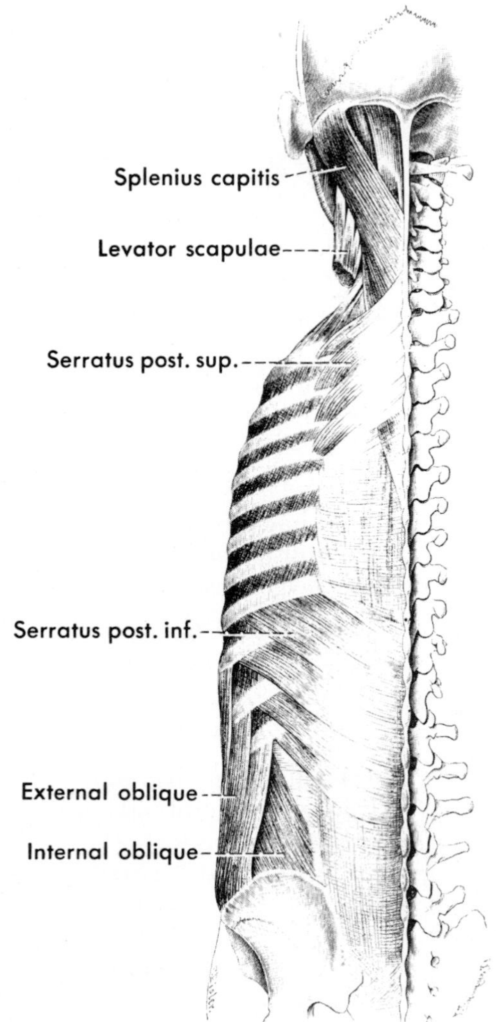


**FIG. 15-1.** Serratus posterior muscles. (Eisler P: In von Bardeleben K (ed): Handbuch der Anatomie des Menschen, vol 2, sect 2, pt 1. Jena, Fischer, 1912)

nous processes of the 7th cervical and first two or three thoracic vertebrae (and the intervening supraspinous ligament). It runs laterally and downward to insert into the upper borders of the 2nd to 4th or 2nd to 5th ribs. The **serratus posterior inferior** arises from the spinous processes of about the lower two thoracic and upper two lumbar vertebrae (and the intervening supraspinous ligament). It runs laterally and upward to insert by three or four muscular slips into the lower borders of the last three or four ribs. The aponeuroses of origin of both these muscles are fused to the underlying fascia covering the back muscles.

The two serratus posterior muscles are inspiratory ones. The superior serratus elevates the upper ribs and the inferior serratus depresses the lower ribs and prevents their being pulled upward by the action of the diaphragm. Both muscles are **innervated** by *intercostal nerves,* the superior typically by the first four intercostals and the inferior by the last four.

**Associated Fascia.** The muscles of the back are covered posteriorly by fascia that separates them more or less completely from overlying structures. They are also in part subdivided by less distinct fascial layers, just as are muscles elsewhere. These layers are, however, too indistinct to be worthy of description.

The covering fascia of the back muscles in the cervical region is part of the deepest layer of fascia (prevertebral) in the front of the neck. It is attached posteriorly to the spinous processes of the lower cervical vertebrae, to the ligamentum nuchae, and to the skull, deep to the attachment of the trapezius.

The fascia covering the muscles of the upper part of the thorax is continuous with that in the neck, but it is thin and transparent except where the aponeurosis of origin of the serratus posterior superior blends with it. It attaches in the midline to the spinous processes of the vertebrae deep to the trapezius muscle. Laterally, at the edge of the back muscles, it blends with the periosteum of the ribs and with the fascia covering the intercostal muscles.

In the lower thoracic and the lumbar region, the fascia is much thicker, for it represents not only fascia but the fused aponeuroses of several muscles. This combination of fascia and aponeuroses is called the **thoracolumbar fascia.** The thoracolumbar fascia lies not only posterior to the musculature of the back but also, below the level of the ribs, anterior to it (Fig. 15-2). As the two deeper layers of the lateral abdominal muscles, the internal oblique and the transversus, approach the

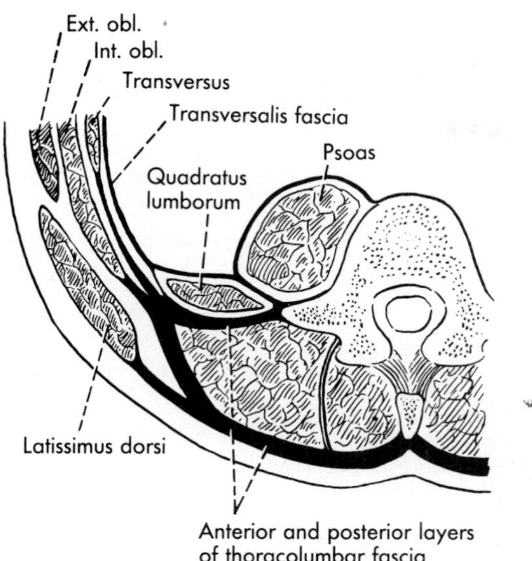


**FIG. 15-2.** Diagram of the thoracolumbar fascia in a cross section at the lumbar level.

vertebral column, their aponeuroses of origin unite and then split so that a part passes posterior to the muscles to attach to the vertebral spinous processes and another part passes anterior to attach to the transverse processes. The part passing anteriorly is known as the *anterior layer* of the thoracolumbar fascia. That passing posteriorly is joined by the aponeuroses of origin of the serratus posterior inferior muscle and the latissimus dorsi to form the *posterior layer* of the thoracolumbar fascia. Thus, the thoracolumbar fascia consists primarily of aponeurosis rather than true fascia.

## POSTERIOR MUSCLES OF THE VERTEBRAL COLUMN

### Structure and General Grouping

The muscles of the back are particularly complex because, although originally strictly segmental and running only from one vertebra to the next, a great deal of fusion between adjacent segmental muscles takes place forming much longer muscles. At the same time, a tangential splitting results in a number of superimposed layers. Further, in contrast to muscles of the limbs, all except the shortest and deepest-lying muscles of the back have multiple origins and insertions, arising from a number of consecutive vertebrae or ribs and inserting above into a number of consecutive vertebrae or ribs. Only those muscles that run just between two vertebrae, or from the vertebral column to the skull, have a single origin or insertion. In most of the muscles, bundles of any one origin fan out to blend with muscle bundles arising a segment or two above and below. From these blended bundles are formed new ones that go to successive insertions on the vertebral columns. Thus, muscle bundles of any one origin are of varying length and diverge toward several insertions, and muscle bundles of any one insertion are also of varying length and converge from several different origins (Fig. 15-3).

There is, perhaps, no entirely satisfactory grouping of the muscles of the back, but a reasonably successful one can be based upon the general direction of the muscle bundles and their approximate lengths. Thus, the longer muscles of the back can be classified into those that arise from the midline and run laterally and upward to their insertion—the *splenius muscles;* those that arise either from the midline or more laterally, but in general run almost longitudinally, with neither a marked outward nor inward slant as they are traced upward—the *erector spinae* group; and muscles that arise laterally but run toward the midline as they are traced up—the *transversospinalis* group. Deep to these muscles are small segmental muscles that run between spinous processes or between transverse processes.

### Splenius Muscles

The splenius muscles are the only two large muscles of the back that arise from the midline and run markedly laterally (Fig. 15-3). They are the most superficial back muscles in the neck, and they received the name splenius (bandage) because they somewhat resemble a bandage wrapped around the neck.

The **splenius capitis** is a broad flat muscle that arises from about the lower half of the ligamentum nuchae and from the spinous processes of the 7th cervical and upper three

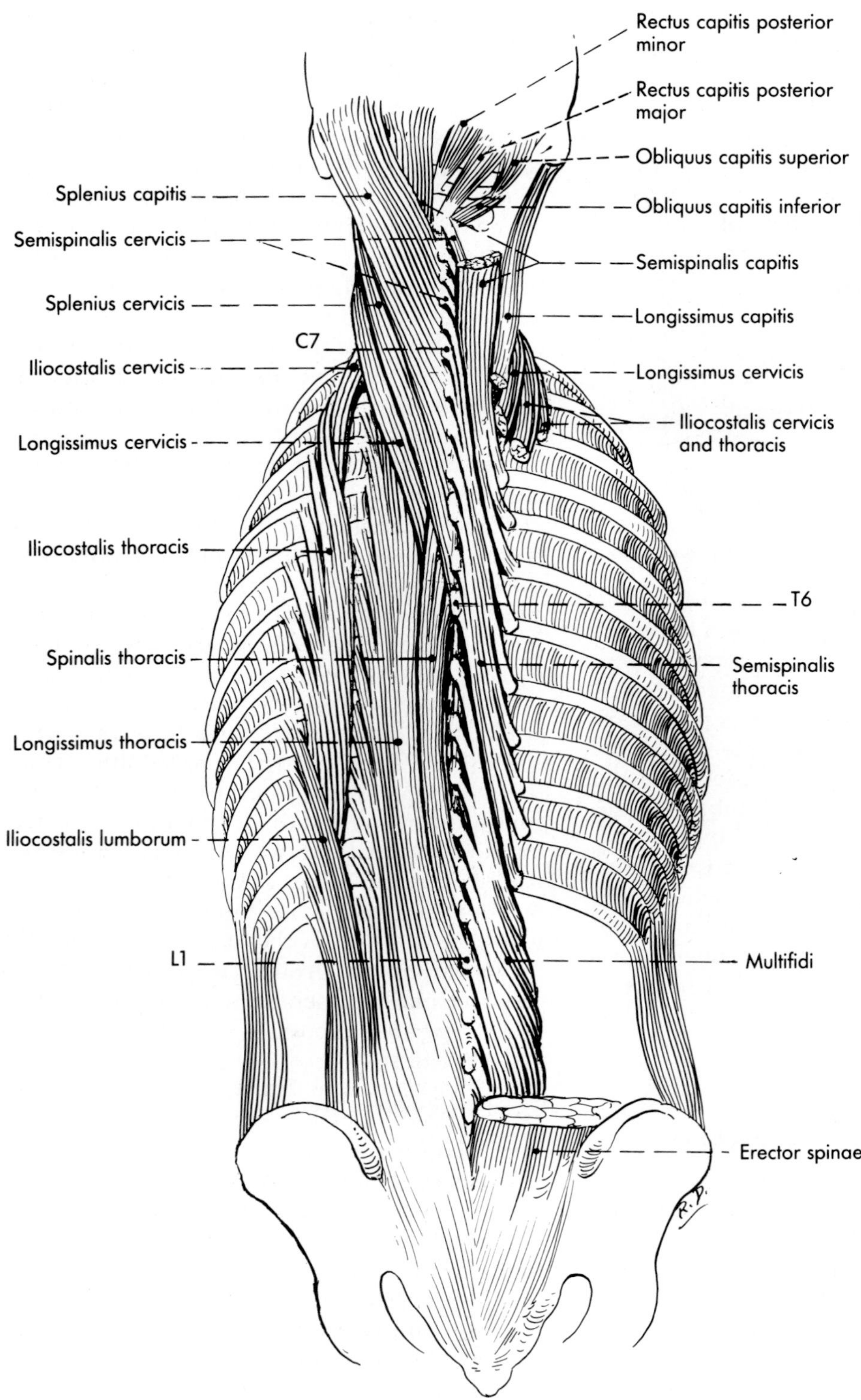


**FIG. 15-3.** General view of the muscles of the back. Many have been removed on the right side to show deeper-lying muscles.

or four thoracic vertebrae. It runs upward and laterally to insert into the skull (occipital bone) just below the lateral part of the superior nuchal line and into the mastoid process of the temporal bone. It is partly covered at its insertion by the sternocleidomastoid muscle and is covered at its origin by the trapezius, but it appears laterally as part of the floor of the posterior triangle of the neck.

The **splenius cervicis** is a narrow muscle arising below the splenius capitis and paralleling its lateral border. It typically arises from the spinous processes of about the 3rd through the 6th thoracic vertebrae and inserts into the transverse processes of the upper two to four cervical vertebrae.

### Erector Spinae

This system, once called the *sacrospinalis,* is the largest muscular mass of the back. Its origin, heavily tendinous superficially, is from the sacrum, the iliac crest, and the spinous processes of most of the lumbar and of the last two thoracic vertebrae. As it becomes largely muscular in the upper lumbar region, it divides into three vertical columns or divisions, each of which in turn is divisible into relatively distinct muscles (Figs. 15-3 and 15-4). The most lateral column or division is the *iliocostalis,* the intermediate division is the *longissimus,* and the medial division is the *spinalis.* The iliocostalis and longissimus each consists of three overlapping parts, a succeeding upper part arising from the same segments on which a lower part inserts. There may or may not be three parts to the spinalis.

**Iliocostalis.** The iliocostalis is divisible into iliocostalis lumborum, iliocostalis thoracis, and iliocostalis cervicis.

The **iliocostalis lumborum,** the lower lateral part of the erector spinae, separates from the remainder of the muscle and divides into six or seven slips of insertion that attach by flattened tendons into the lower borders of the lower six or seven ribs, about as far laterally as their angles. The succeeding **iliocostalis thoracis** arises from the upper borders of the lower six ribs just medial to the insertion of the iliocostalis lumborum and is partly overlapped by it. The iliocostalis thoracis runs almost straight upward to divide into slips that insert into the upper six ribs and sometimes into the transverse process of the 7th cervical vertebra. The **iliocostalis cervicis** arises medial to the iliocostalis thoracis from the angles of approximately the upper six ribs and usually inserts into the transverse processes of the 4th, 5th, and 6th cervical vertebrae.

**Longissimus.** The three muscles forming the longissimus system are the longissimus thoracis, longissimus cervicis, and longissimus capitis.

The **longissimus thoracis** arises as the intermediate part of the erector spinae. It runs almost straight upward to insert both into the lower nine or more ribs and into the tips of the transverse processes associated with these ribs. The **longissimus cervicis** arises medial to the upper end of the longissimus thoracis, from the transverse processes of about the upper four to six thoracic vertebrae. As it runs into the cervical region, it is partly covered by the splenius cervicis and iliocostalis cervicis muscles. It divides into slips that insert into the transverse processes of the 2nd through the 6th cervical vertebrae. The **longissimus capitis** arises medial to the upper end of the longissimus cervicis, partly from the tendons of origin of this muscle and partly from the articular processes of the lower four cervical vertebrae. This muscle runs somewhat more laterally than do the preceding muscles and inserts, under cover of the splenius capitis, into the posterior margin of the mastoid process.

**Spinalis.** The spinalis, the most medial division of the erector spinae, is also the smallest and the most poorly defined. In its best development, there are three muscles in this group: the spinalis thoracis, the spinalis cervicis, and the spinalis capitis. However, the spinalis thoracis is the only constant and relatively independent member of the group.

The **spinalis thoracis** lies medial to the longissimus thoracis, with which it is blended at

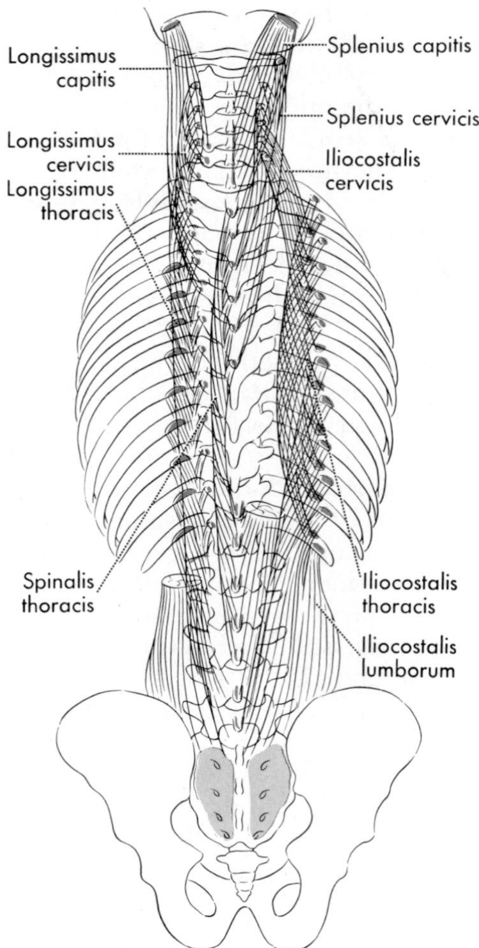


**FIG. 15-4.** Origins (red) and insertions (blue) of the splenius muscles and of the several components of the erector spinae.

its origin. Its independent tendons of origin arise from the spinous processes of about the lower two thoracic and the first two lumbar vertebrae, but the muscle is variable in both its origin and insertion. This slender muscle bundle runs vertically to divide into a number of thin tendons that insert into the spinous processes of some four to eight of the upper thoracic vertebrae. The **spinalis cervicis** is frequently absent or represented by only a few muscle fibers attached to the ligamentum nuchae. When best developed, however, it arises from the lower part of the ligamentum nuchae and from the spinous process of the 7th cervical, or even those of upper thoracic, vertebrae and inserts into the spinous process of the axis and perhaps into those of the 3rd and 4th cervical vertebrae also. The **spinalis capitis** is not a separate muscle, for it is blended laterally with a larger muscle, the *semispinalis capitis*. It arises in part with the tendons of origin of this muscle and inserts with it into the skull, but it may also have some direct origin from upper thoracic spinous processes. If the spinalis capitis is well developed, there is a tendinous intersection (inscription) across the muscle at about the level of the 7th cervical vertebra. It and the semispinalis are crossed by a similar tendinous intersection in the upper part of the neck.

## Transversospinalis Muscles

The transversospinalis muscles lie deep to and are shorter than the erector spinae and, instead of running almost vertically, slant inward from their origins to their insertions. The three muscle groups of this system are the semispinalis, the multifidi, and the rotators (Figs. 15-3 and 15-5). The *semispinalis*, the most superficial group, has the longest muscle bundles. The longest muscle bundles of the deeper-lying *multifidi* are of about the same length as the shortest bundles of the semispinalis. The deepest-lying group, the *rotators*, are no more than two segments long.

**Semispinalis.** The semispinalis consists of three muscles arranged longitudinally in much the same fashion as the muscles forming the divisions of the erector spinae; they are the semispinalis thoracis, cervicis, and capitis. The longest bundles of the semispinalis pass over as many as six vertebrae between their origin and insertion, and the shortest pass over about four.

The **semispinalis thoracis** arises by long slender tendons from the transverse processes of approximately the lower six thoracic vertebrae. It inserts by similar slender tendons into the spinous processes of upper thoracic and lower cervical vertebrae. The **semispinalis cervicis** arises from the transverse processes of the upper five or six thoracic vertebrae and becomes thicker and more muscular as it is traced into the neck. It is largely covered by the semispinalis capitis and inserts by short tendons into the spinous processes of about the 3rd through the 5th cervical vertebrae and by a heavy muscular slip into the spinous process of the axis. The **semispinalis capitis** is a large muscle. It arises by tendons from the tips of the transverse processes of the upper six thoracic vertebrae and the articular processes of the lower three or four cervical ones, extending upward almost vertically as a broad muscle that inserts between the superior and inferior nuchal lines of the occipital bone. The most medial part of this muscle is usually called the spinalis capitis.

**Multifidi.** The multifidi muscles are thickest in the lumbar region and end in the cervical region; however, there is no clear division as they are traced upward, and separate parts are therefore not described. Traditionally, therefore, they have been described as forming a single muscle, the *multifidus*.

In the sacral and lumbar regions, the multifidi are covered by the erector spinae, and in the thoracic and cervical regions by the semispinalis. The fibers of the lumbar part arise from the dorsal surface of the sacrum as low as the 3rd or 4th dorsal sacral foramen, from the medial surface of the posterior superior iliac spine, and from the deep surface of the tendon of origin of the erector spinae, as well as the mamillary processes of the lumbar vertebrae. In the thoracic region, the thinner muscle bundles arise from all the transverse processes, and in the cervical region, the still thinner muscles arise from the articular processes of the lower four cervical vertebrae. From their origin, the fibers extend upward and medially to insert into the sides of the spinous processes of all the vertebrae from the 5th lumbar to the axis.

The muscle bundles of the multifidi are shorter than those of the semispinalis, passing over from two to four vertebrae to reach their insertions. The angles they form with the midline are, therefore, wider than those of the semispinalis.

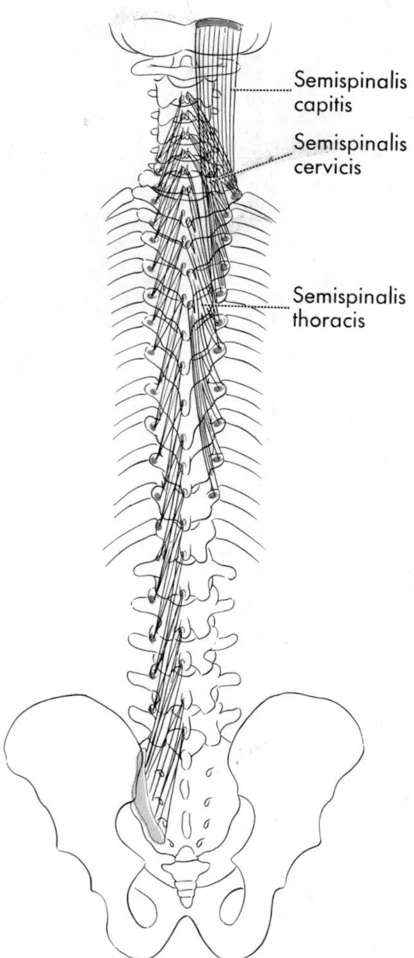


**FIG. 15-5.** Origins and insertions of the semispinalis muscles (right) and the multifidi (left).

**Rotators.** The rotators lie deep to the multifidi and are the deepest members of the transversospinalis system. In contrast to the muscles heretofore considered, each rotator has a single origin and a single insertion (Fig. 15-6 through 15-8). They are divided into two sets, long rotators and short rotators, and occur in the cervical, thoracic, and lumbar regions.

Each **long rotator** arises from the transverse process of one vertebra, passes across the vertebra immediately above, and inserts into the base of the spinous process of the second vertebra above. Each **short rotator** arises from the transverse process of a vertebra but inserts into the base of the spinous process of

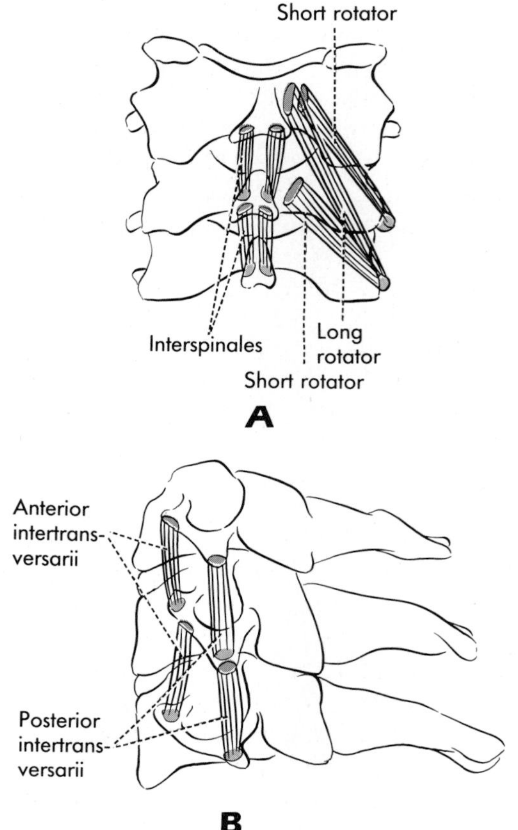


**FIG. 15-6.** Origins (red) and insertions (blue) of the short muscles of the back in the cervical region. **A**, posterior view; **B**, lateral view.

**FIG. 15-7.** Origins and insertions of the short muscles of the back in the thoracic region (right); the levatores costarum are shown on the left.

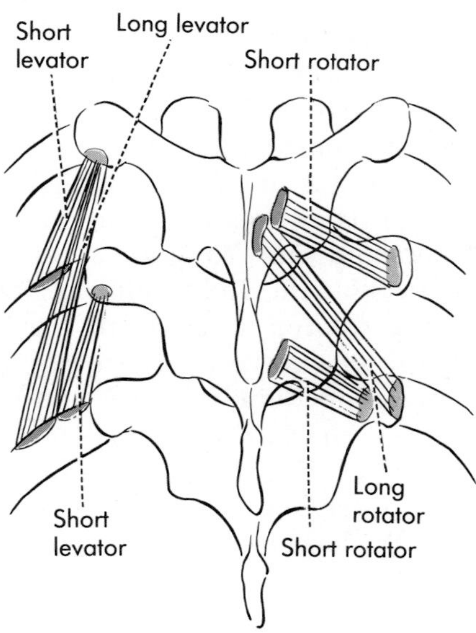


the vertebra immediately above. Thus, the long rotators, which skip one vertebra, have a more oblique course upward than do the short ones, which skip no vertebra but are, instead, strictly segmental.

## Segmental Muscles

The segmental group of muscles extends from one vertebra to the next. The short rotators actually belong to this group, as well as to the transversospinalis system, but are more conveniently considered with the latter. The other strictly segmental muscles are the *interspinales* and the *intertransversarii.*

The **interspinales** (Figs. 15-6 and 15-8) are well developed in the cervical region, where they form six pairs of muscles separated from each other by the thin interspinous ligaments. They stretch between adjacent spinous processes, from that of the 2nd cervical to that of the 1st thoracic. Interspinales are largely lacking in the thoracic region. The lumbar interspinales are commonly well developed and form four pairs that lie between adjacent spinous processes of all five lumbar vertebrae; there may be a pair above or below these.

The **intertransversarii** run vertically between adjacent transverse processes. In the cervical region, there are two intertransversa-

**FIG. 15-8.** Origins and insertions of the short muscles of the back in the lumbar region.

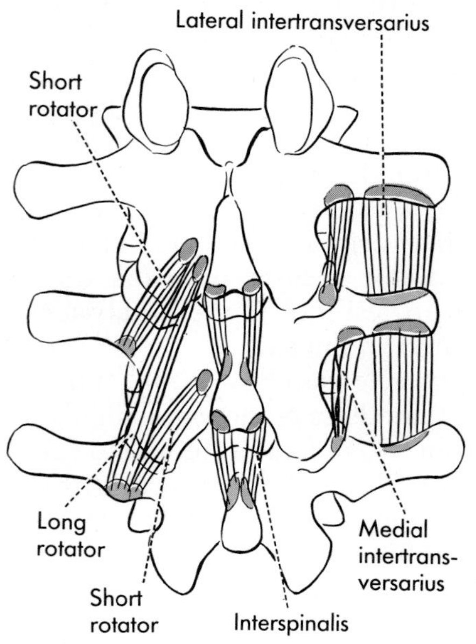

rii on each side. The anterior muscles pass between the anterior tubercles of two adjacent transverse processes, and the posterior muscles pass between the posterior tubercles. There are seven pairs of these. The highest is between the transverse processes of the atlas and axis, the lowest between those of the 7th cervical and 1st thoracic vertebrae.

The intertransversarii are largely lacking in the thoracic region but often are found between the lower three or four thoracic vertebrae. They are frequently subdivided into medial and lateral muscles, like the lumbar intertransversarii. The lateral intertransversarii of the lumbar region pass between the transverse processes of two adjacent vertebrae. The medial intertransversarii arise from the mamillary process of one vertebra and insert on the accessory process of the vertebra above.

## Suboccipital Muscles

A special group of muscles connects the atlas and axis to each other and to the skull. Of these, four are posteriorly located (Fig. 15-9), lying just deep to the semispinalis in the upper part of the neck. Two of these are described as oblique and two as straight muscles.

The **obliquus capitis inferior** arises from the spinous process of the axis and extends laterally and upward to insert upon the transverse process of the atlas. The **obliquus capitis superior** arises from the transverse process of the atlas and passes upward and medially to insert into the occipital bone deep to the insertions of the splenius and semispinalis capitis. The **rectus capitis posterior major** arises from the spinous process of the axis and inserts into the occipital bone a little lateral to the midline. The **rectus capitis posterior minor** arises from the posterior tubercle of the atlas and inserts largely under cover of the major.

Between the obliquus capitis inferior, the obliquus capitis superior, and the rectus capitis posterior major there is a small **suboccipital triangle.** When the muscles are slightly separated here, a part of the posterior arch of the atlas is visible. On the upper border of this arch, the *vertebral artery* runs transversely. This artery has ascended through the transverse foramina of the upper cervical vertebrae and, after passing through that of the atlas, turns medially and posteriorly in the vertebral sulcus on the upper surface of the posterior arch. It then turns forward and penetrates the posterior atlantooccipital membrane. The vertebral artery is an important source of blood supply to the brain.

The dorsal ramus of the 1st cervical nerve **(suboccipital nerve)** emerges into the triangle close to the vertebral artery. It supplies all four of the posterior suboccipital muscles and usually helps supply overlying muscles. The **greater occipital nerve,** representing primarily the dorsal ramus of the 2nd cervical nerve, appears at the lower border of the obliquus capitis inferior and, after giving off a few muscular branches, runs upward across the super-

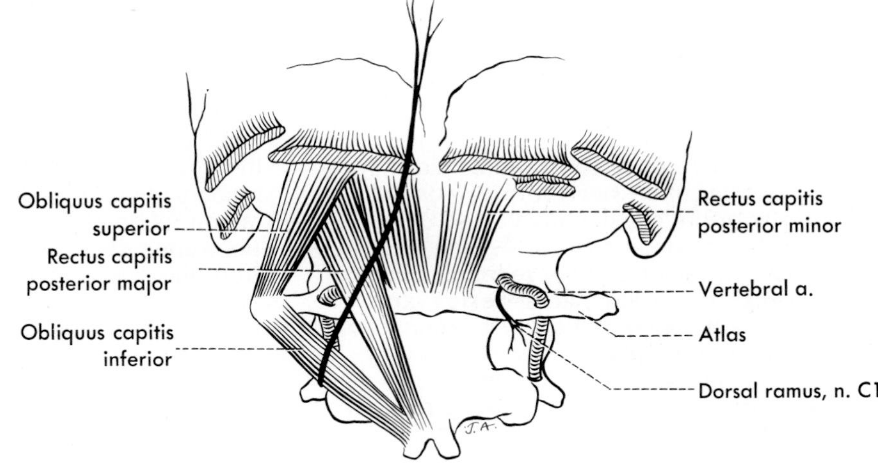


**FIG. 15-9.** Suboccipital muscles and the suboccipital triangle. The nerve on the left crossing the triangle is the greater occipital nerve.

ficial surface of the triangle to reach the scalp, being closely associated here with the occipital artery as this turns upward to the scalp.

### Related Muscles

In addition to the posteriorly placed muscles, a number of other muscles attach to the vertebral column. Most of these are anteriorly or laterally placed and cannot be seen until other regions of the body are dissected. Included among these are the **rectus capitis lateralis** and the **rectus capitis anterior,** short suboccipital muscles placed as their names indicate and closely associated with the **longus capitis** and the **longus colli** on the front of the vertebral column; the three **scalenes,** attached to the anterior tubercles of the cervical portion of the vertebral column; and the **psoas major** and **quadratus lumborum muscles,** associated with the lumbar part of the column. These muscles can be seen only in deep dissections of the neck and abdomen. (See Chaps. 25 and 30.)

The muscles of the shoulder that attach to the vertebral column have already been described, as have the serratus posterior muscles. Another group of muscles closely associated with the vertebral column is the **levatores costarum** (Fig. 15-7). These muscles, divisible like the rotators into short and long ones, arise from transverse processes and insert into ribs.

### Innervation and Actions

**Innervation.** Most of the muscles of the back are innervated by the **dorsal rami of the spinal nerves,** but a few are innervated by ventral rami. The anterior intertransversarii of the cervical region are supplied by ventral rami. The posterior cervical intertransversarii have a double innervation; their more medial parts are supplied by dorsal rami and their lateral parts by ventral rami. In the lumbar region, the medial intertransversarii are supplied by dorsal rami, but the lateral intertransversarii are supplied by ventral rami. The levatores costarum are usually said to be innervated by ventral rami, but one investigator found them innervated by dorsal rami.

**Actions.** The complicated, anatomically overlapping muscles of the back obviously do not have individually discrete functions but act in various combinations of large groups. As previously discussed, the movements of the vertebral column include flexion, extension, lateral flexion, and rotation. Of these movements, flexion is brought about through anteriorly placed muscles or by the action of gravity; in the latter instance, as when one bends over, the posteriorly placed muscles of the back act strongly in order to control the movement that is being carried out primarily by gravity.

All the posterior muscles except the intertransversarii are so placed that it may be assumed they contribute something to extension. Electromyography has indicated that the iliocostalis, longissimus, multifidi, and rotators all become active during flexion (in order to control it) and particularly active again as extension is started. The intertransversarii presumably help only in flexing the column laterally, and the interspinales presumably only in extending the column, but several muscles of the back laterally flex or rotate to one or the other side when they act unilaterally instead of bilaterally. Thus, although the splenius capitis and cervicis help extend the head and neck, they also, when they act unilaterally, flex the head and neck toward the same side and rotate the face toward this side. The erector spinae has been said to initiate lateral flexion and rotation toward the same side, but its activity during these movements also has been interpreted as probably synergistic. It stabilizes the vertebral column against the flexing action of the abdominal muscles, the prime movers in rotation. Once initiated, lateral flexion is controlled by slight activity of all the muscles on the opposite side. Of the muscles of the transversospinalis system, the multifidi apparently participate in controlling flexion to the opposite side, and they and the rotators become active in rotation to the opposite side. The semispinalis would be expected to have similar actions, but one study of the transversospinalis system led the investigators to conclude that the semispinalis act to adjust individual vertebrae rather than move the vertebral column as a whole. Of all the muscles investigated, the rotators alone are said to show almost continuous activity in the relaxed, extended position.

Of the smaller muscles, and those anteriorly placed, the obliquus capitis inferior rotates the atlas and turns the face toward the same side, the obliquus capitis superior extends the head and flexes toward the same side, the rectus capitis major extends the head and rotates toward the same side, and the rectus capitis minor probably only extends but may help to rotate. The laterally and anteriorly placed muscles help to flex, in corresponding directions, the vertebral column and head, but of these, the scalenes act primarily upon the upper two ribs, whereas the psoas major acts primarily upon the lower limb. Finally, the anterolateral abdominal muscles have important actions upon the vertebral column, since they extend from the ribs to the pelvis. The rectus abdominis and the external and internal obliques of the two sides, acting together, are important flexors of the vertebral column, and the internal oblique of one side, perhaps aided by the external oblique of the other, rotates the column toward the side of the internal oblique.

## MENINGES, SPINAL CORD, AND NERVE ROOTS

The meninges and the subarachnoid space have already been briefly discussed, and so has the spinal cord (Chap. 4). The spinal nerves and their general composition and distribution are discussed beginning with page 36. With this material as a background, it is necessary to describe here only the specific anatomy of these parts.

The spinal cord and its surrounding meninges lie in the **vertebral canal,** formed by the successive vertebral foramina. The spinal nerves leave the canal by way of the **intervertebral foramina.** Because the anterior wall of the vertebral canal is formed by the heavy vertebral bodies, study of the spinal cord *in situ* is most easily carried out after a *laminectomy* (removal of laminae) is performed. A laminectomy involving a specific region is also the surgical approach to the spinal cord; the muscles are stripped away from one or more vertebrae to expose the laminae, and as much of a lamina or as many laminae as necessary are then removed. The ligamenta flava, stretching as they do between laminae, are necessarily removed in a laminectomy and should be examined as the spinal cord is exposed by this technique.

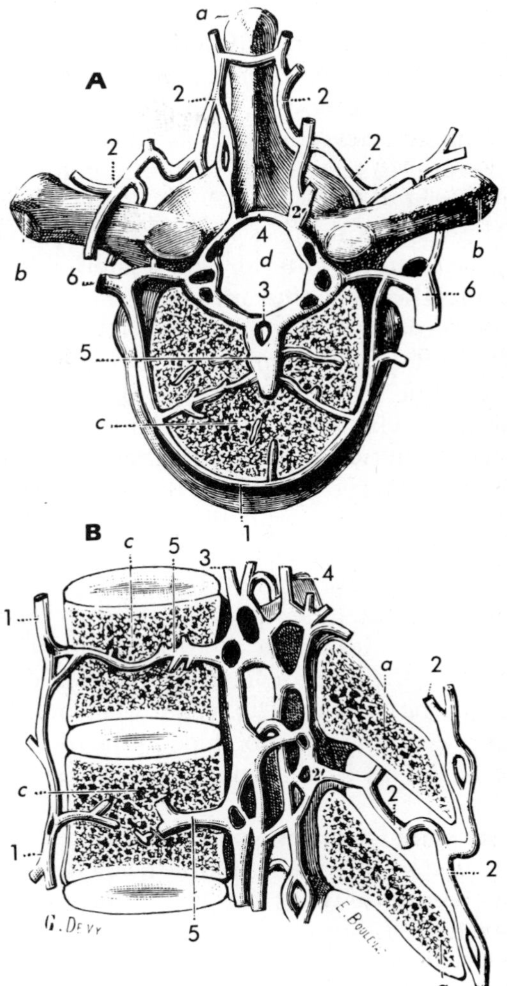


**FIG. 15-10.** The vertebral venous plexuses in a cross section (**A**) and a longitudinal section (**B**) of the vertebral column. **a, b,** and **c** identify the spinous and transverse processes and the body of the vertebra. **1** is the anterior external venous plexus; **2,** the posterior external; **2′**, connections between the posterior internal, **4,** and the posterior external; **3,** the anterior internal plexus; **5,** the major drainage of the vertebral body; **6,** segmental veins and their tributaries. (After Breschet. Testut L: Traité d'Anatomie Humaine, vol 2. Paris, Doin, 1891)

### Epidural Space

The spinal cord and its meninges do not fill the vertebral canal. The space between the walls of this canal and the outer meninx of the cord, the dura mater, is the epidural space, which is filled with fat, connective tissue, and a plexus of veins. This tissue space serves as padding around the spinal cord, affording protection to it in addition to that provided by the cerebrospinal fluid. The **venous plexuses** in the epidural space are particularly large. They receive the chief venous drainage from the vertebrae and connect freely with each other and with the dural venous sinuses that lie within the skull. Since they also drain laterally between vertebrae to connect with veins of the thorax, abdomen, and pelvis, they form an important collateral circulation for the inferior vena cava when intraabdominal pressure is increased and for the dural venous sinuses (which otherwise drain mostly into the large internal jugular veins) when intrathoracic pressure is increased. Because they have no valves, they can also carry blood to the vertebrae or the cranial venous sinuses when pressure conditions are such as to bring about a reverse flow of blood. As a corollary, they sometimes conduct an infection or a metastatic cancer to the vertebral column or the cranial cavity.

The venous plexuses in the epidural space are divided into **anterior** and **posterior internal vertebral venous plexuses** (Fig. 15-10). Each consists of more or less paired longitudinal channels that communicate freely across the midline. Laterally the two plexuses communicate with each other. The connections between the two sides of the anterior internal plexus pass largely between the posterior longitudinal ligament and the vertebral bodies and in so doing receive the major part of the drainage of these bodies. The plexuses drain laterally into segmental veins through the intervertebral foramina and posteriorly between the ligamenta flava into a less dense plexus, the posterior external vertebral venous plexus, that is embedded in the musculature of the back adjacent to the spinous processes.

The **spinal arteries** that enter the vertebral canal and supply the vertebrae (see Fig. 14-13) also give rise to twigs that run in the epidural connective tissue and supply this and the dura. A separate *radicular branch* of each spinal artery supplies the dura around the nerve root and may penetrate the dura to run along the nerve root and become an important part of the blood supply to the spinal cord.

## MENINGES

The meninges surrounding the spinal cord consist of dura mater, arachnoid, and pia mater (Figs. 15-11 and 15-12).

### Dura Mater and Subdural Space

The spinal **dura mater** is a tube of tough collagenous tissue that extends from the foramen magnum to about the middle of the second sacral vertebra and loosely surrounds the spinal cord. At the foramen magnum (the aperture of the skull continuous with the vertebral canal), the *spinal dura* is continuous with the cranial (intracranial) *dura.* Since the cranial dura is fused to the skull, the epidural space about the spinal dura terminates here. The dura ends as a tube at about the middle of the second sacral segment. Below this level it is represented by a slender tough thread, the **filum of the dura mater,** that blends with the posterior longitudinal ligament over the coccyx. The dura is thus anchored above to the skull and below by the filum. In between, its rough outer surface blends with the fibrous connective tissue in the epidural space. Dural sleeves that it sends out around the spinal nerves blend with connective tissue in the intervertebral foramina. The dura is thus partly suspended in the epidural space and partly supported by the epidural connective tissue and vessels.

The inner surface of the dura is smooth. It is largely separated from the adjacent arachnoid by a potential space that contains only enough fluid to moisten its walls. This is the **subdural space.** This space is obliterated caudally where the arachnoid and dura fuse together at the filum of the dura and is obliterated over the nerve roots where the dura and arachnoid become continuous with the connective tissue of the nerves. It is, however, continuous above with the similar subdural space within the cranial cavity.

### Arachnoid, Pia, and Subarachnoid Space

The outer surface of the **arachnoid mater** is rather smooth and forms the inner wall of the subdural space. However, its inner surface gives rise to numerous trabeculae that pass across the underlying **subarachnoid space** to become continuous with the pia mater, which is closely applied to the outer surface of the spinal cord and nerve roots. It is these trabeculae, somewhat resembling a spider web, that have given the arachnoid its name (Arachnida is the name of the zoological class that contains the spiders). Flattened cells that form the inner surface of the arachnoid are continuous around the trabeculae with flattened cells forming the outer surface of the

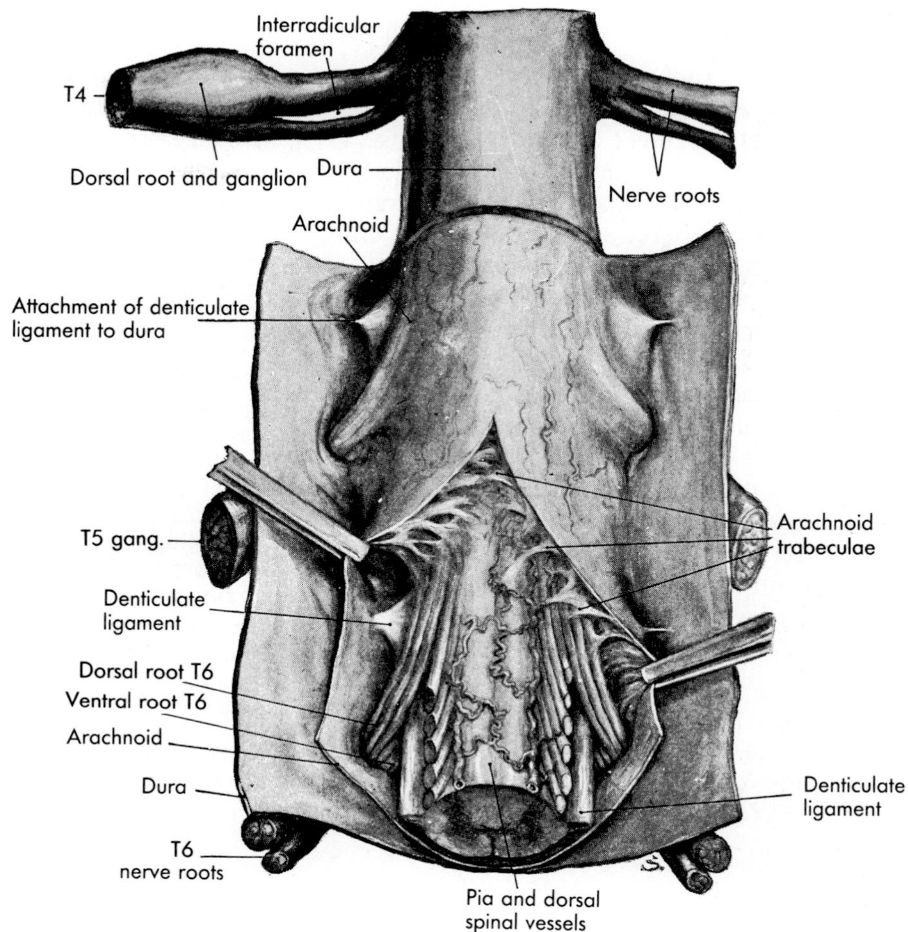


**FIG. 15-11.** Meninges about the cord seen from behind. The dural sac is intact above and reflected below, and the arachnoid has also been opened below to show the arachnoid trabeculae in the subarachnoid space. (Mettler FA: Neuroanatomy, 2nd ed. St Louis, CV Mosby, 1948)

**FIG. 15-12.** Diagram of the meninges and the subarachnoid space in a cross section of the cord. Note their extensions around the nerve roots.

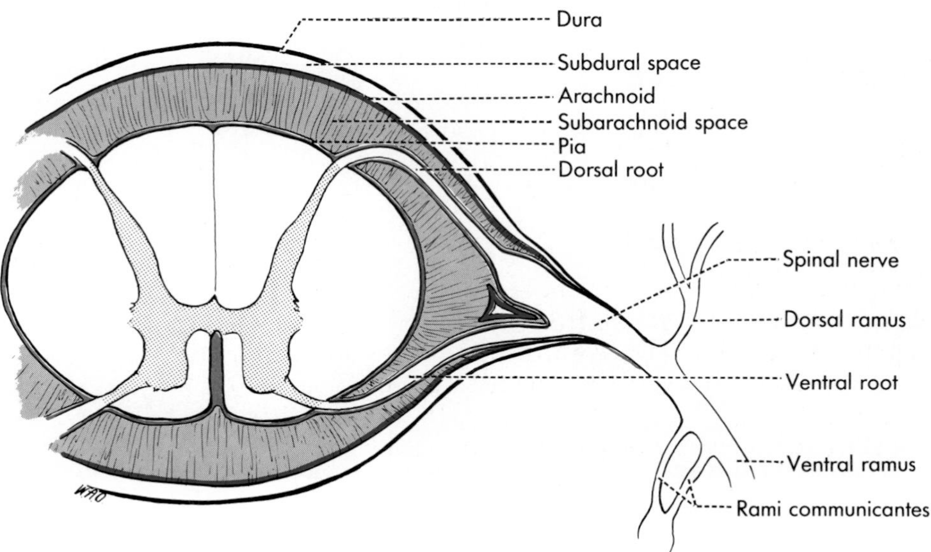

pia. The subarachnoid space is filled with **cerebrospinal fluid,** which thus bathes the cord and helps protect it.

The arachnoid tube in the cervical region is considerably larger than it need be to encircle the spinal cord here, and therefore the subarachnoid space is fairly large. This space, usually more limited in the thoracic region, becomes particularly commodious in the lumbar region. The spinal cord tapers to a point at about the second lumbar level, but the arachnoid sac, like the dura, continues to about the second sacral level. In consequence, the large lower end of the arachnoid sac contains no spinal cord but only caudally running roots of spinal nerves and a considerable amount of cerebrospinal fluid. Since a needle can be inserted here without danger of injuring the spinal cord, *spinal puncture* is usually done between the 3rd and 4th or the 4th and 5th lumbar vertebrae. Spinal puncture allows measurement of the pressure of the cerebrospinal fluid, withdrawal of some of it for bacteriologic and chemical investigations, or introduction of anesthetic, medicinal, or radiopaque material (Fig. 15-13) into the subarachnoid space.

Anesthetic agents introduced into the subarachnoid space mix with the cerebrospinal fluid and bathe the spinal cord and nerve roots, producing *spinal anesthesia*. Sometimes the anesthetic agent is introduced instead into the epidural space where, diffusing through the connective tissue of the space, it comes in contact with the nerve roots only, rather than with the spinal cord itseif. Analgesia so produced is usually known as *caudal analgesia,* because it is used most commonly to produce loss of pain over the lower nerve roots that form the cauda equina. This is most easily done by inserting

**FIG. 15-13.** Roentgenograms of the cervical and lumbar regions of the subarachnoid space, which have been filled with contrast medium. The lateral projections indicate the extension of the subarachnoid space around the nerve roots. (Courtesy of Dr. D. G. Pugh)

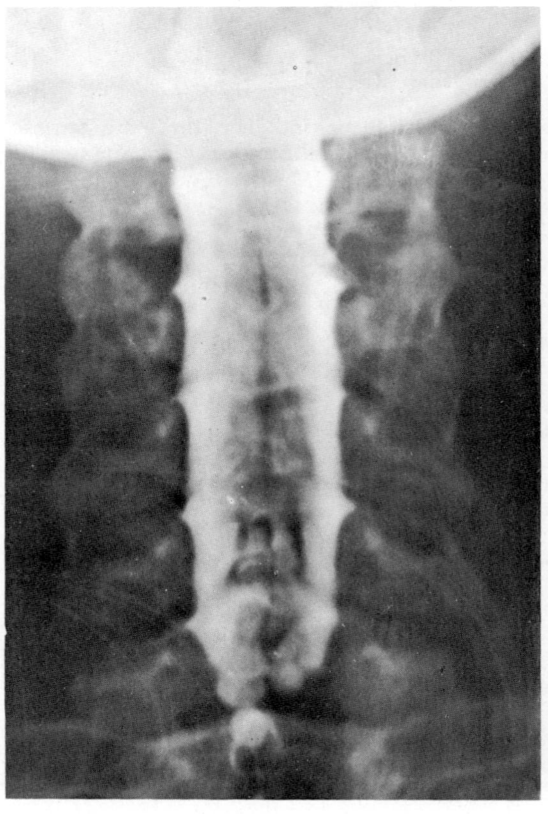

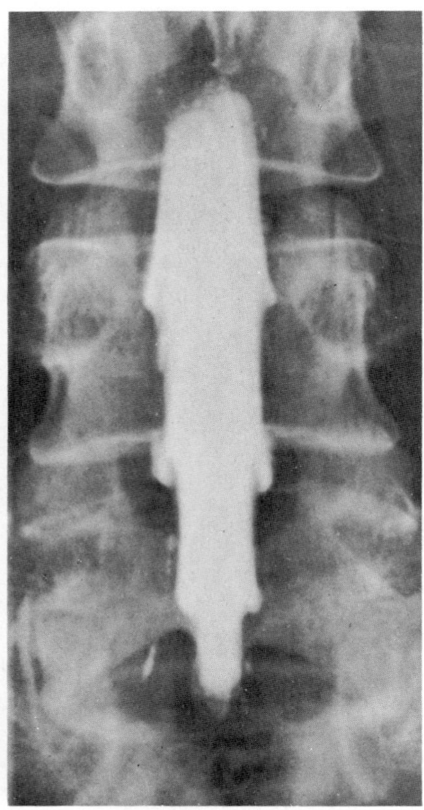

the needle through the sacral hiatus into the epidural space below the ending of the dural sac.

The **pia mater** is closely bound to the spinal cord and to the nerve roots as they cross the subarachnoid space. In certain locations, its connective tissue dips into the spinal cord to form septa here. Arising laterally from it is a band of dense connective tissue, the **denticulate ligament** (Fig. 15-11). The two denticulate ligaments attach the spinal cord to the dura and thus help to suspend the cord within the subarachnoid space. Each ligament presents a serrated lateral border, which has 20 to 21 pointed processes or denticulations that penetrate the arachnoid to become continuous with the dura. Between the denticulations, the denticulate ligament presents a free lateral border. The uppermost denticulation is attached to the dura mater of the skull. Succeeding denticulations are attached between the levels of exit of each adjacent pair of spinal nerves throughout the cervical and thoracic regions of the vertebral column. The last denticulation may attach either above or below the last thoracic nerve.

## SPINAL CORD

Small vessels, particularly veins, are often visible upon the spinal cord (*medulla spinalis*) and nerve roots. Both the arteries and the veins of the spinal cord are primarily longitudinally running vessels that communicate above with cranial vessels; however, these vessels are reinforced at irregular intervals by segmental ones ("radicular vessels") that come in along the nerve roots. The arteries are derived from the spinal arteries; the veins drain into the vertebral venous plexuses. These vessels are important for the spinal cord, since the anterior and posterior spinal arteries are small even at their origins and by themselves can supply only a short upper portion of the cord. The circulation to much of the cord depends upon the vessels coming in along the nerve roots.

While the cord is *in situ,* the **posterior median sulcus,** and the **posterior lateral sulcus** along which successive filaments of the dorsal roots of the spinal nerves emerge, can be recognized with no difficulty. The cord must be rotated before the emergence of the ventral roots from it can be seen. The cervical portion of the cord forms the **cervical enlargement** at the level of emergence of the large nerve roots that contribute to the brachial plexus. The thoracic portion is slenderer and rounder. The lumbrosacral portion enlarges again **(lumbar enlargement)**, but tapers to a point at about the lower end of the 1st lumbar vertebra. The terminal tapered part of the cord is the **conus medullaris,** and the tough thread continued from the conus is the **filum terminale.** Conus and filum terminale are surrounded by longitudinally directed roots (dorsal and ventral) of the more caudal spinal nerves, and these structures together, as they run through the subarachnoid space, are called the **cauda equina.** Since each nerve root turns laterally to leave the vertebral canal through an intervertebral foramen, the elements of the cauda equina become fewer as it is traced inferiorly. Finally, at the lower tapered end of the subarachnoid space, the filum terminale fuses with the filum of the dura. Through these two fila, the caudal end of the spinal cord is anchored to the coccyx.

Although the spinal cord itself is not segmented, the spinal nerves are segmentally arranged. That part of the spinal cord from which a spinal nerve arises is known as a **spinal cord segment.** Since the spinal cord is so much shorter than the vertebral column, it is obvious that a spinal cord segment is typically not as long as is a vertebra; therefore, the farther the cord is traced caudally, the greater discrepancy there is between the positions of the cord segments and the correspondingly numbered vertebrae. This discrepancy begins even in the cervical region. It is much exaggerated in the lower thoracic and upper lumbar regions of the vertebral column, where the lumbar and sacral segments of the cord are very short and a number of them lie at the level of one vertebra. It is this discrepancy between lengths of cord segments and lengths of the vertebrae that accounts for the downward course and increas-

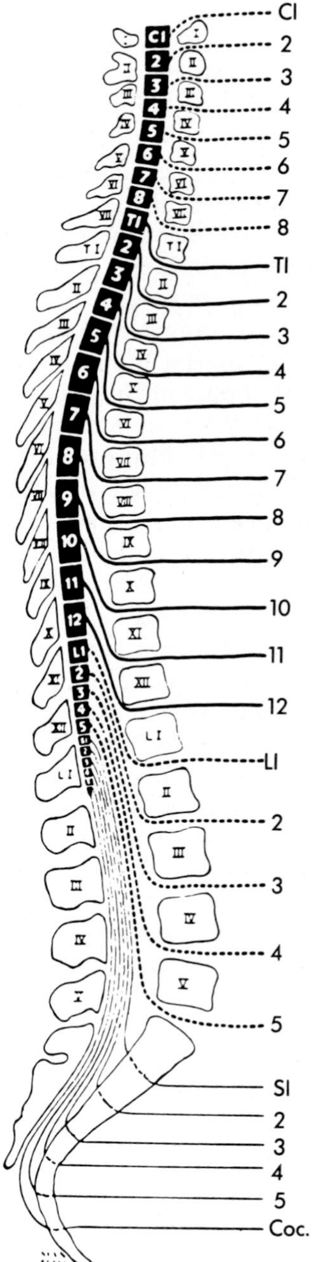


**FIG. 15-14.** Relation of the spinal cord segments and nerve roots to the vertebrae in a sagittal section of the vertebral column. Vertebral spinous processes and bodies bear Roman numerals and spinal segments and nerves, Arabic ones. The relative lengths of the cord segments are clearly shown. (Haymaker W, Woodhall B: Peripheral Nerve Injuries, 2nd ed. Philadelphia, WB Saunders, 1953)

ing lengths of the nerve roots after they leave the spinal cord. The levels of exits of the nerve roots are fixed by their emergence between vertebrae, and, therefore, the upper nerves run obliquely and the lower ones vertically downward, to reach their levels of exit. Figure 15-14 shows the relation between cord segments and vertebrae. These relationships are of particular clinical importance, for a lesion at the level of a given vertebra will usually affect a different segment or segments of the cord. Diagnosis of the level of injury to the cord must be translated into terms of the vertebral level before surgical approach to a lesion is undertaken.

## Finer Anatomy of the Cord

The study of the finer anatomy of the spinal cord can best be carried out in the neuroanatomy laboratory, and only sufficient description to allow appreciation of its general form and function is necessary here.

The spinal cord presents a tiny **central canal,** the remains of the lumen of the tubelike cord of early developmental stages, and very thick walls. Nerve cell bodies grouped around the central canal constitute the *gray matter.* Nerve fibers, many of them myelinated, are grouped outside the gray matter and form the *white matter.*

**Gray Matter.** The nerve cell bodies, plus at least the first parts of the fibers that these cells send out and the termination of fibers that end upon them, compose the gray matter of the spinal cord. In cross section (see Fig. 4-4), the gray matter is seen to be shaped somewhat like a distorted letter H or a butterfly. A narrow band across the center connects the expanded portions in the lateral parts of the spinal cord. Variations in size and shape at different levels occur according to the number of cells located at these levels.

The connecting bar between the two sides is the *gray commissure.* The expanded portions are, in general, divisible into a *posterior* and an *anterior horn,* with, in some locations, a fairly prominent *lateral horn* situated at about the junction of the two. (Although called "horns" when seen in cross section, the horns are actually sections through continuous gray columns; it is therefore correct to speak also of the *gray columns.*)

From a broad functional standpoint, the posterior gray column is primarily concerned with the reception and forwarding of impulses brought in by the peripheral nervous system. The anterior column is generally concerned with sending impulses out along the peripheral nervous system to skeletal muscle. (It is the cells in the anterior column that are particularly affected in poliomyelitis.) The posterior columns tend to be large when a nerve entering at that level has a wide distribution to skin and to sensory endings of muscle and joints, and the anterior columns tend to be large at the level of origin of nerves supplying a large amount of skeletal muscle. The lateral column houses, at certain levels (typically T1 through L2 and S2 through S4), the cell bodies of the autonomic (preganglionic) fibers that leave the central nervous system as parts of the autonomic system.

**White Matter.** The gray matter of the spinal cord is surrounded by large bundles of nerve fibers, the white matter. A *posterior septum* and an *anterior fissure* divide the white matter of the cord into right and left sides. Each side is in turn subdivided, according to convention, into three large subdivisions known as the white columns or the *funiculi*. The *posterior funiculus* lies between the posterior septum and the medial side of the posterior horn and the dorsal nerve root fibers that enter in line with the posterior horn. The *anterior funiculus* lies between the anterior fissure and a line drawn between the anterior horn and the ventral roots of the nerves leaving the cord. The *lateral funiculus* lies largely lateral to the gray matter and between the levels of entrance and exit, respectively, of the dorsal and ventral roots of the spinal nerves.

The *posterior funiculus* consists largely of ascending fibers that are derived from the cells of the (dorsal root) spinal ganglia of the spinal nerves. These fibers enter the spinal cord and send branches into the posterior horn, but the main fiber in many instances turns up to help form the posterior funiculus. The *lateral funiculus* consists in part of ascending fibers, derived from cells in the gray matter of the cord, and in part of descending fibers from a higher level of the nervous system, mostly from various parts of the brain. The *anterior funiculus* also contains some ascending fibers derived from the gray matter of the spinal cord, but contains more descending fibers derived from higher centers. Immediately adjacent to the gray matter in all three columns are short fibers that both arise and end in the gray matter of the cord. Some of these fibers ascend and some descend.

Fibers of similar function tend to be grouped together within the spinal cord. Such a group is known as a *fasciculus* or *tract* of the spinal cord (Fig. 15-15). The posterior funiculus consists of fibers that have to do, for the most part, with muscle and joint sense and other deep sensation, with the exception of pain. Although the posterior funiculus could be fairly regarded as forming a single tract, branches (collaterals) of the ascending fibers of the posterior funiculus descend in several locations in the posterior columns and are named as tracts. Moreover, in the upper part of the spinal cord, a septum tends to separate the fibers of nerves of the lower half of the body from fibers of those of the upper half, so that the posterior funiculus is divisible into two large tracts, a medial and a lateral (*fasciculus gracilis* and *fasciculus cuneatus*). Both of

**FIG. 15-15.** Chief tracts of the spinal cord. The fasciculus proprius (diagonal lines) surrounds the gray matter of the cord (unshaded); otherwise, ascending tracts are shown on the right, descending ones on the left. (Ranson SW, Clark SL: The Anatomy of the Nervous System, 10th ed. Philadelphia, WB Saunders, 1959)

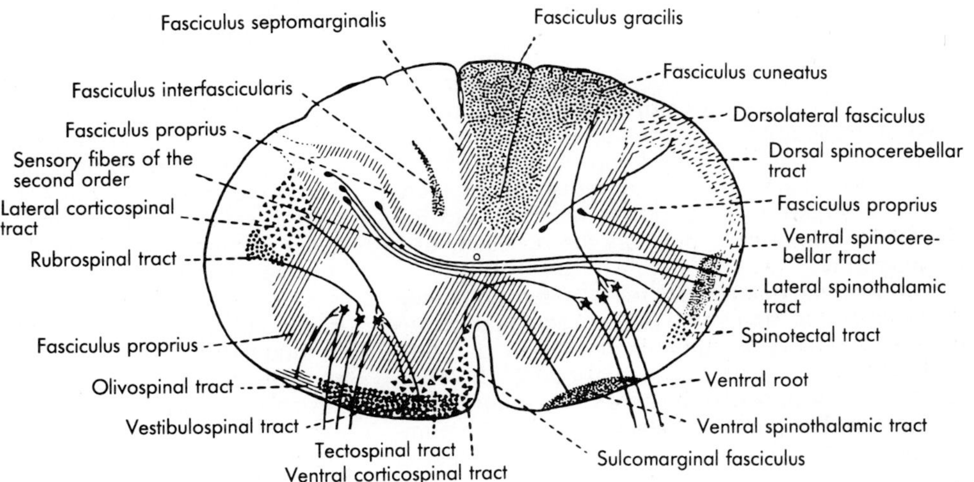

these have the same function but conduct impulses from different parts of the body.

The short fibers adjacent to the gray matter, regardless of what funiculus they are associated with, are called part of the *fasciculus proprius.* This bundle has to do with spinal reflexes involving, as most do, more than one spinal segment.

In the *lateral* and *anterior funiculi,* various tracts of different function are not separated from each other by connective tissue septa, and, indeed, there is considerable overlap among the fibers of different tracts. The definitive positions assigned to the tracts in diagrams of the spinal cord are, therefore, somewhat arbitrary, but do at least indicate the general location of the tract and the greatest concentration of its fibers. Much remains to be learned concerning the function of many of these tracts.

In general, it can be said that the deepest-lying tracts of the lateral funiculi, situated more or less in the concavity between posterior and anterior horns, are the *lateral corticospinal* (*pyramidal*) *tracts.* These tracts arise from the cerebral cortex (highest level of the brain) of one side, cross before they enter the spinal cord, and end in the gray matter, where they influence the activity of those cells of the anterior horn having to do with skeletal muscle. They are responsible in part for initiating voluntary movement. Closely associated with this pathway for voluntary movement are other tracts that also influence movement, but arise from lower centers of the brain. Still other descending fibers in the lateral funiculus control the activity of autonomic cells in the lateral horn of the cord, rather than that of the voluntary motor ones in the anterior horn. Because these descending tracts help to control or initiate motor activity, they are grouped together as *motor tracts.* The more peripheral and ventral parts of the lateral funiculus are composed primarily of ascending fibers, some of them in tracts that end in the cerebellum, others in tracts that go to other parts of the brain, including one that has crossed from the opposite side of the body and has to do with the perception of pain. This latter tract is sometimes cut, unilaterally (on the opposite side from the pain) or bilaterally, for the relief of intractable pain in the operation known as *chordotomy* or *anterolateral chordotomy.* It is the lateral spinothalamic tract. A more recent technique, in which an electrode is inserted into the tract through the skin, obviates the necessity of a laminectomy.

No clear line demarcates the lateral from the anterior funiculus, and some of the tracts that lie primarily in the anterior funiculus may extend in part into the lateral. The more medial part of the *anterior funiculus,* between the anterior median fissure and the anterior horn, consists primarily of a number of small descending (motor) tracts, including one derived from the cerebral cortex and others derived from lower motor centers. The remainder of the anterior funiculus is largely occupied by two tracts, one a descending motor tract derived from subcortical centers and probably having to do largely with maintenance of muscle tone against gravity, the other an ascending crossed tract.

The *functions* of some of the ascending tracts are rather clearly known, especially those that conduct to the level of consciousness—the tracts of the posterior funiculus (see a preceding paragraph), the lateral spinothalamic tract (pain and temperature), and the anterior spinothalamic (touch). However, the exact parts that many of the descending tracts play in the control of voluntary movement are not understood. Cutting various combinations of these tracts as they lie in the spinal cord, for the relief of abnormal muscular contraction or of compulsive involuntary movements, has produced evidence that their function is not a simple one and that there is probably a good deal of overlap in function among the different motor tracts. In general, however, those descending tracts that influence the neurons supplying skeletal muscle either initiate voluntary movement or help distribute and maintain the proper degree of contraction among various muscles to bring about the desired action or maintain a desired posture.

The results of severe local damage to the spinal cord, as from a tumor, fracture–dislocation of a vertebra, or vascular involvement, are disastrous not because of damage to the gray matter but because of damage to the white. Insofar as the gray matter is concerned, damage to it affects primarily only the nerves originating and ending at the level of the damaged gray matter and is, therefore, largely confined to the segment or segments directly affected. It is obviously different for the white matter, however, composed as it is of fibers that should carry impulses across the area of damage. When the spinal cord is badly enough damaged to prevent the conduction of impulses across the damaged region, the temporary and perhaps permanent result is the equivalent of a complete transection of the cord, whether the cord is anatomically interrupted or not. In this condition, no motor, descending, impulses can reach the part of the cord below the level of the lesion, so the parts of the body supplied by this part of the cord are paralyzed—no longer subject to voluntary or even higher reflex control—and spinal reflexes take over. Similarly, ascending impulses concerned with sensation from parts of the body below the level of the lesion cannot reach the brain, so a lower part of the body is devoid of all sensation.

Incomplete transverse lesions of the cord act essentially the same as complete ones, exerting their effects primarily on the white matter. The results of incomplete lesions may differ markedly from one case to another, however, depending upon which tracts are particularly damaged and how severe the damage is. There may, for instance, be muscular weakness with little loss of sensation, dulling but not loss of a certain sensation, or many combinations of sensory or sensory and motor loss.

## NERVE ROOTS

The nerve roots arise as a series of filaments that then blend together to form definitive dorsal and ventral roots. These roots run through the subarachnoid space toward their exits from the vertebral canal. As the nerve roots leave the main dural–arachnoid sac, they carry with them **meningeal sleeves** that surround the two roots of the nerve at least as far as the spinal ganglion (Figs. 15-11 and 15-12). There may be a separate dural sleeve around each root (dorsal and ventral) or the two sleeves may be fused, but there is always a separate arachnoid sheath around each root and within this the subarachnoid cavity surrounds the root.

The dural–arachnoid sleeves in the cervical and thoracic regions project almost straight laterally from the dural–arachnoid sac, but in the lower lumbar region, they run caudally as well as laterally to reach the level of exit of the nerve (Fig. 15-13). A protruding lumbar disk, therefore, may exert pressure upon the dura-clad nerve roots, rather than upon these roots as they lie within the general dural–arachnoid sac. Sometimes just proximal to, sometimes just distal to, the dorsal root ganglion, and in a corresponding position on the ventral root, the dura, arachnoid, and pia become continuous with the connective tissue of the peripheral nerve, and the subarachnoid space is thereby obliterated.

The **dorsal root (spinal) ganglia** regularly lie at the intervertebral foramina through which the nerves make their exit. This means that it is nerve roots, not spinal nerves, that form the cauda equina and that nerve roots vary greatly in length: some typical lengths are 3 mm for the roots of the first cervical nerve, 29 mm for those of the first thoracic, 91 mm for those of the first lumbar, 185 mm for those of the first sacral, and 266 mm for those of the coccygeal. Particularly in the case of the lower nerves, by far the greatest length of the roots lies within the general dural–arachnoid sac, and only a small proportion within the dural–arachnoid sleeve that extends out along the roots.

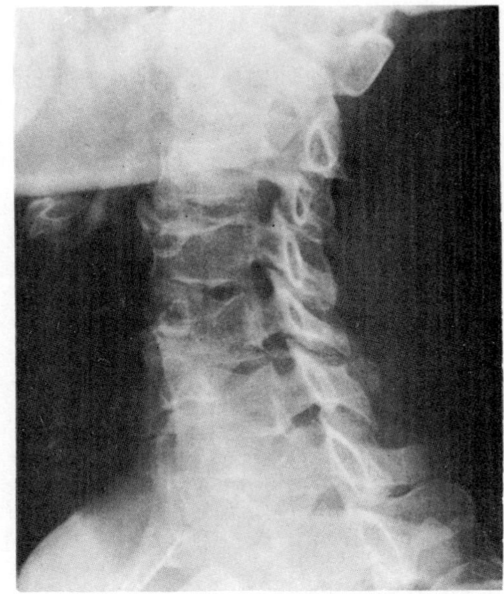

**FIG. 15-16.** Oblique roentgenogram of the cervical region to show intervertebral foramina. (Courtesy of Dr. D. G. Pugh)

### Nerve Roots and Intervertebral Disks

The relationship of the nerve roots and nerves to the intervertebral disks and the intervertebral foramina is of special interest because of the increasing frequency with which injury to the nerves at these levels is recognized.

The **1st cervical nerve** leaves the vertebral canal between the occipital bone and the posterior arch of the atlas, and it is therefore related neither to a bony intervertebral foramen nor to an intervertebral disk. The **2nd cervical nerve** makes its exit between the posterior arch of the atlas and the pedicle of the axis, but again is related to no intervertebral disk nor does it come through a bony intervertebral foramen, because it passes posterior to the lateral atlantoaxial joint, between that and the ligamentum flavum. The **succeeding cervical nerves** make their exits through intervertebral foramina that are bounded posterolaterally by the synovial joints between the articular processes, above and below by the pedicles of the vertebrae, and anteromedially by the bodies of the vertebrae and the intervening intervertebral disks (Fig. 15-16).

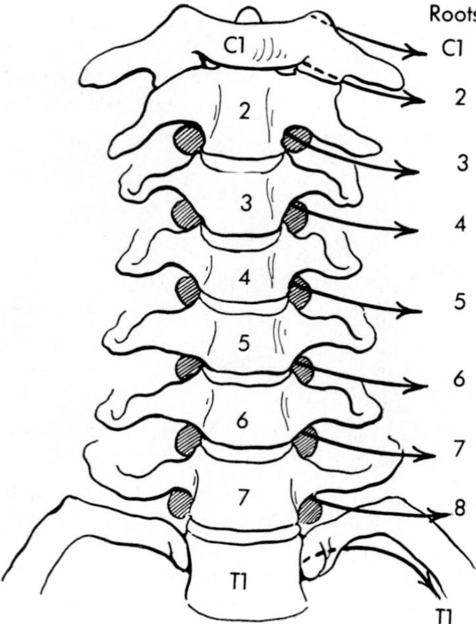


**FIG. 15-17.** Relations of cervical nerves to the vertebrae and intervertebral disks. The nerves emerge above the correspondingly numbered vertebrae, the 8th cervical, of course, emerging above the 8th (1st thoracic) vertebra. (Love JG: In Baker AB (ed): Clinical Neurology, vol 2. New York, Paul B Hoeber, 1955)

The cervical vertebrae so overlap each other laterally and anteriorly that narrowing of a cervical disk results in very little diminution in vertical length of the intervertebral foramina, and these foramina cannot be narrowed enough in the vertical direction to pinch the spinal nerve as it makes its exit. However, the nerves largely fill the anteroposterior diameter of the intervertebral foramina, and therefore anything that tends to narrow this may exert pressure upon the nerve; in consequence, arthritis or swelling of the synovial joints just posterior to the nerves may be a cause of pain along them, and so may a posterolateral bulging of a cervical disk or the bony proliferation from the vertebral edges that sometimes results from marked narrowing of a disk.

A bulging or ruptured cervical disk exerts pressure upon the nerve leaving the intervertebral foramen at its level. However, because the intervertebral disks are numbered according to the vertebra below which each lies (whereas the cervical nerves are numbered according to the vertebra above which each lies, Fig. 15-17), a protrusion of a cervical disk affects the nerve numbered one greater than itself; a protruded 5th cervical disk (the one below the 5th cervical vertebra), for instance, does not affect the 5th cervical nerve but, rather, the 6th one.

In the **thoracic region,** the intervertebral foramina are considerably larger than they need to be to accommodate the relatively small spinal nerves passing through them. The synovial joints lie posterior to the intervertebral foramina, and the pedicles of the two adjacent vertebrae bound the foramina above and below; however, in consequence of the deep inferior vertebral notches, the anterior wall of a thoracic intervertebral foramen is formed largely by the vertebral body, and the intervertebral disk forms only a very small, lowest part of this wall.

The limited mobility of the thoracic portion of the vertebral column, the small size of the disks, and the large intervertebral foramina all tend to protect the thoracic nerves from damage at the level of the disks or the foramina. Disk protrusions

**FIG. 15-18.** View from the vertebral canal of the 3rd and 4th lumbar intervertebral foramina of the right side. Note that the 4th lumbar nerve (dotted lines) does not cross the 4th disk (the lowest one shown here), but does cross the 3rd and can therefore be most easily compressed by that one. (Naffziger HC, Inman V, Saunders JBdeCM: Surg Gynecol Obstet 66:288, 1938. By permission of Surgery, Gynecology & Obstetrics)

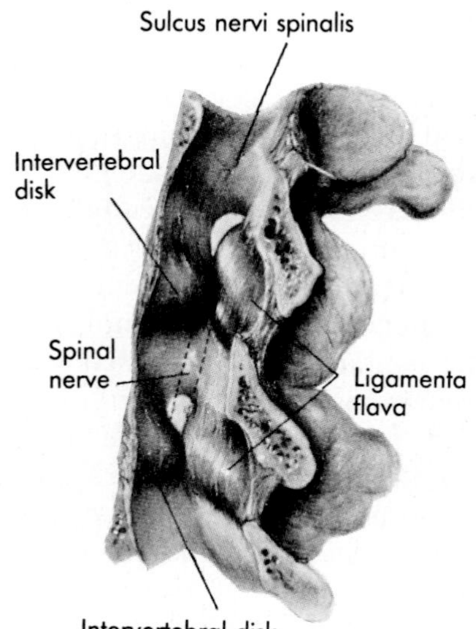

are rare, but if they are posterior, they may damage the spinal cord.

In the **lumbar region,** the intervertebral foramina are particularly long in comparison to their anteroposterior diameter, and the intervertebral notches are so deep that even total removal of a disk would not allow the pedicles above and below the foramen to compress the spinal nerve. However, arthritic changes in the joints, or anything that tends to narrow the anteroposterior diameter of the intervertebral foramen, may cause pain along the nerves.

A posterolateral protrusion or rupture of a lumbar intervertebral disk is relatively common, but the disks lie at the inferior border of the foramen and the nerve makes its exit over the vertebral body above, so unless there is a large mass of protruded disk, the nerve making its exit at the intervertebral space of the disk is not affected. Instead, an intervertebral disk that bulges into the vertebral canal typically comes in contact with the nerve that makes its exit through the foramen below the level of the disk (Fig. 15-18); this nerve lies most laterally within the dural–arachnoid sac, or may have entered its own dural sleeve at the site at which it crosses the bulging disk, and is moreover held laterally by its exit at the foramen just below the disk. Thus, it comes about that a bulging or protruded lumbar disk typically affects the nerve numbered one greater than itself—a lesion of the lumbosacral (5th lumbar) disk affects the 1st sacral nerve, one of the 4th lumbar disk affects the 5th lumbar nerve, and so forth.

The pain resulting from a protruded lumbar disk is due to the nerve being stretched across and pressed upon by the bulge. The muscles of the back usually become spastically contracted when a protrusion is present, and this adds to the pain.

Pain produced by pressure upon the nerve root is usually interpreted as coming from the area of distribution of the sensory fibers of that nerve (this is usually known as "radiation"—extension—of pain down a nerve), and protruded lower lumbar disks, therefore, produce pain in the distribution of the sciatic nerve, the largest nerve arising from the lower lumbar and upper sacral spinal nerves. When the hip is flexed and the knee extended at the same time, this movement exerts traction upon the sciatic nerve and tends to draw the contributing spinal nerves forward against the anterior wall of the vertebral canal, thus against a protruding disk that is present. This maneuver, called the "straight-leg-raising test," is often of value in making the diagnosis of a protruded lumbar disk, for it should reproduce or accentuate the pain. Protrusion or rupture is not always confined to a single disk. Fairly frequently the 5th and 4th disks will be involved simultaneously.

## GENERAL REFERENCES AND RECOMMENDED READINGS


BATSON OV: The function of the vertebral veins and their rôle in the spread of metastases. Ann Surg 112:138, 1940

BATSON OV: The vertebral vein system: Caldwell lecture, 1956. Am J Roentgenol 78:195, 1957

BRIERLEY JB, FIELD EJ: Fate of an intraneural injection as demonstrated by the use of radioactive phosphorus. J Neurol Neurosurg Psychiatry 12:86, 1949

CAVE AJE: The innervation and morphology of the cervical intertransverse muscles. J Anat 71:497, 1937

DUNCAN D, KEYSER LL: Some determinations of the ratio of nerve fibers to nerve cells in the thoracic dorsal roots and ganglia of the cat. J Comp Neurol 64:303, 1936

ECKENHOFF JE: The physiologic significance of the vertebral venous plexus. Surg Gynecol Obstet 131:72, 1970

ELLIOT HC: Cross-sectional diameters and areas of the human spinal cord. Anat Rec 93:287, 1945

ETHELBERG S, RIISHEDE J: Malformation of lumbar spinal roots and sheaths in the causation of low backache and sciatica. J Bone Joint Surg (Br) 34:442, 1952

FRENCH JD, STRAIN WH: Peripheral extension of radiopaque media from the subarachnoid space. Surgery 22:380, 1947

HARMEIER JW: The normal histology of the intradural filum terminale. Arch Neurol Psychiatry 29:308, 1933

HARVEY SC, BURR HS, VANCAMPENHOUT E: Development of the meninges: Further experiments. Arch Neurol Psychiatry 29:683, 1933

HERREN RY, ALEXANDER L: Sulcal and intrinsic blood vessels of human spinal cord. Arch Neurol Psychiatry 41:678, 1939

INGBERT C: An enumeration of the medullated nerve fibers in the dorsal roots of the spinal nerves of man. J Comp Neurol 13:53, 1903

INGBERT CE: An enumeration of the medullated nerve fibers in the ventral roots of the spinal nerves of man. J Comp Neurol 14:209, 1904

JIT I, CHARNALIA VM: The vertebral level of the termination of the spinal cord. J Anat Soc India 8:93, 1959

LANIER VS, McKNIGHT HF, TROTTER M: Caudal analgesia: An experimental and anatomical study. Am J Obstet Gynecol 47:633, 1944

LONG C II, LAWTON EB: Functional significance of spinal cord lesion level. Arch Phys Med Rehabil 36:249, 1955

MORRIS JM, BENNER G, LUCAS DB: An electromyographic study of the intrinsic muscles of the back in man. J Anat 96:509, 1962

REIMANN AG, ANSON BJ: Vertebral level of termination of the spinal cord with report of a case of sacral cord. Anat Rec 88:127, 1944

SUH TH, ALEXANDER L: Vascular system of the human spinal cord. Arch Neurol Psychiatry 41:659, 1939

TARLOV IM: Structure of the nerve root: II. Differentiation of sensory from motor roots; observations on identification of function in roots of mixed cranial nerves. Arch Neurol Psychiatry 37:1338, 1937

# PART V

# LOWER LIMB

# 16

# THE LOWER LIMB AS A WHOLE

The lower limb (*membrum inferius*) is constructed upon the same fundamental plan as the upper limb, and many comparisons between the two are, therefore, possible. Such comparisons are valuable because they lead to a clearer understanding of the fundamental anatomy of both parts and yet emphasize also the specialization of these parts.

Man's abandonment of the quadrupedal habitat has freed the upper limb from weight bearing, and it has become specialized in the direction of greater freedom of movement. At the same time, since the burden upon the lower limb has become greater, it is specialized more in the direction of stability even at the expense of some freedom of movement. (Compare, for instance, the mobility of the fingers with that of the toes.)

**Development**

The lower limb bud appears as a swelling on the side of the body slightly later than does the upper limb bud, but fairly closely parallels it in development. The mesenchymal swelling becomes cylindrical, the distal end becomes flattened as the forerunner of the foot and toes, and constrictions that indicate the regions of the ankle and knee appear. Ridges on the distal flattened portion give rise to the toes.

As in the case of the upper limb, the lower limb bud in an early stage of development projects slightly caudally but soon comes to project at about a right angle to the body. As development proceeds, the big toe or tibial side is directed cranially, while the little toe or fibular side is directed caudally. At this stage, then, the borders and surfaces of the lower limb correspond to those of the upper limb: the cephalic, preaxial, or tibial border, marked by the big toe, corresponds to the radial or thumb border of the upper limb; the caudal, postaxial, or fibular border corresponds to the ulnar or little-finger side of the upper limb; and the ventral and dorsal surfaces, which are flexor and extensor ones, correspond to the ventral and dorsal surfaces of the trunk, just as do those of the upper limb.

In later development, beginning prenatally but not completed until postnatally, the lower limb undergoes a medial rotation, adduction, and extension that change the relative position of its surfaces and borders (Fig. 16-1). Together, these amount to a rotation of almost 180°, so that the original dorsal surface comes to lie mostly anteriorly (ventrally) and laterally, the original ventral surface posteriorly and medially. Thus, the anterolateral side of the thigh and leg, and the dorsum of the foot, represent the original extensor aspect of the limb, even though they now face largely anteriorly. These surfaces, therefore, correspond to the back of the arm and forearm and the dorsum of the hand. Through this rotation, the leg and foot become fixed in a position somewhat corresponding to pronation in the hand, in which position the thumb is medial like the big toe, the little finger lateral like the little toe. In the thigh, greater growth of the original dorsal surface results in this occupying not only the anterior and lateral surfaces but also the lateral half of the posterior surface of the definitive limb (Fig. 16-1D).

As in the upper limb, mesenchymal condensations at the base of the limb bud represent the girdle, and a similar condensation in the center of the projecting bud represents the forerunner of the skeleton here. Cartilage representing the bones is derived from the mesenchymal condensations, for all the bones of the lower limb are cartilage ones. The first center of ossification in the lower limb is in the cartilaginous body of the femur. This center appears during the seventh fetal week. Centers for most of the other long bones appear toward the end of the second fetal month.

**FIG. 16-1.** Rotation of the lower limb. **A** is the limb of a 7-week embryo. **B**, that of an 8-week embryo, seen from the side. **C** shows the limbs at about the end of the third month. **D** shows anterior and posterior views of the adult limb. The original extensor or dorsal surface, directed laterally in **A**, is shaded to differentiate the preaxial part (lined) from the postaxial part (solid black); the original ventral part of the limb, visible in **C** and **D**, is unshaded.

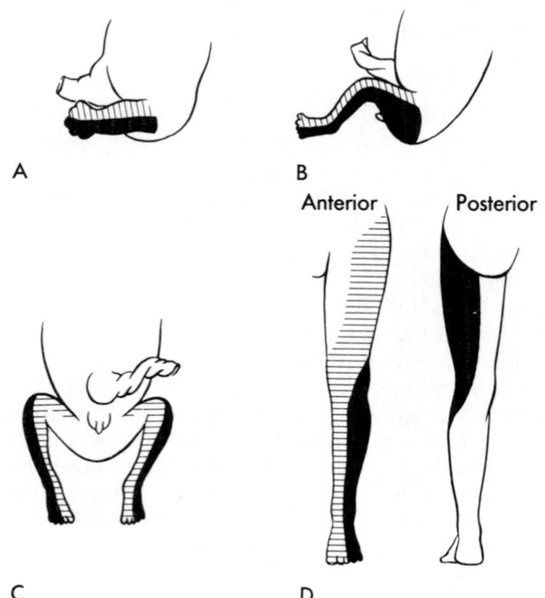


The musculature of the lower limb develops from condensations of mesenchyme around the condensations for the skeleton. As these differentiate, they divide into dorsal and ventral (extensor and flexor) groups, from which individual muscles are differentiated.

The lower limb bud has a wider base than does the upper one and, in consequence, lies at the level of a greater number of spinal nerves than does the upper limb. The nerves at the level of the bud thus form a more complicated plexus, the lumbosacral plexus, into which some eight or more spinal nerves enter; however, this plexus gives rise to nerves that go either to the original ventral part of the limb or to the original dorsal part of the limb and therefore is divisible into ventral and dorsal (anterior and posterior) parts just as is the brachial plexus.

### Parts and Regions

Parts and regions of the lower limb are the **gluteal region** (*natis, clunis,* buttock); the **thigh** (which should be "femur," except that this term designates the bone of the thigh and it is only the adjectival "femoral" that is used to refer to either the thigh or the bone of the thigh); the **knee** (*genu*); the **leg** (*crus*); and the **foot** (*pes*). The knee is divisible into a posterior region, the poples, and an anterior or patellar region. The regions of the leg are anterior (original extensor) and posterior (original flexor); however, *sura* is the calf of the leg and therefore synonymous with the posterior region. The **borders** are *medial* or *tibial,* and *lateral* or *fibular*. The medial and lateral *malleoli* are the subcutaneous bony projections at the ankle. Of the foot, the heel is the *calx,* the region of the ankle is the *tarsus,* and the remaining part of the foot is divided into *metatarsus* and *digits*. The digits are numbered from the big-toe side by Roman numerals; the big toe is usually called the hallux instead of digit I, however, and the little toe the digitus minimus instead of digit V. The borders of the foot are like those of the leg, lateral or fibular and medial or tibial; its surfaces are dorsal (directed cranially) and plantar (the sole of the foot).

### Dermatomes and Cutaneous Nerves

The distinction between dermatomal and peripheral cutaneous distribution of nerves has already been pointed out (Chap. 4), and it was also indicated that of the two commonly used dermatomal charts, those of Foerster (see Figs. 4-9 and 4-10) tend to show the dermatomes as somewhat patchy areas, especially on the anterior surface of the thigh, whereas those of Keegan (see Fig. 4-8) show the dermatomes as long strips that extend from the base of the limb.

The **peripheral cutaneous nerves** of the lower limb are numerous and vary in their derivation; some of those of the thigh, particularly, are derived directly from the lumbosacral plexus (the plexus of the lower limb), whereas others are cutaneous branches of nerves that also supply muscles. Details of these nerves are best described regionally. Figures 16-2 and 16-3 show the general cutaneous distribution to the limb.

**FIG. 16-2.** Distribution of cutaneous nerves to the anterior surface of the lower limb. (Flatau E: Neurologische Schemata für die ärtzliche Praxis. Berlin, Springer, 1915)

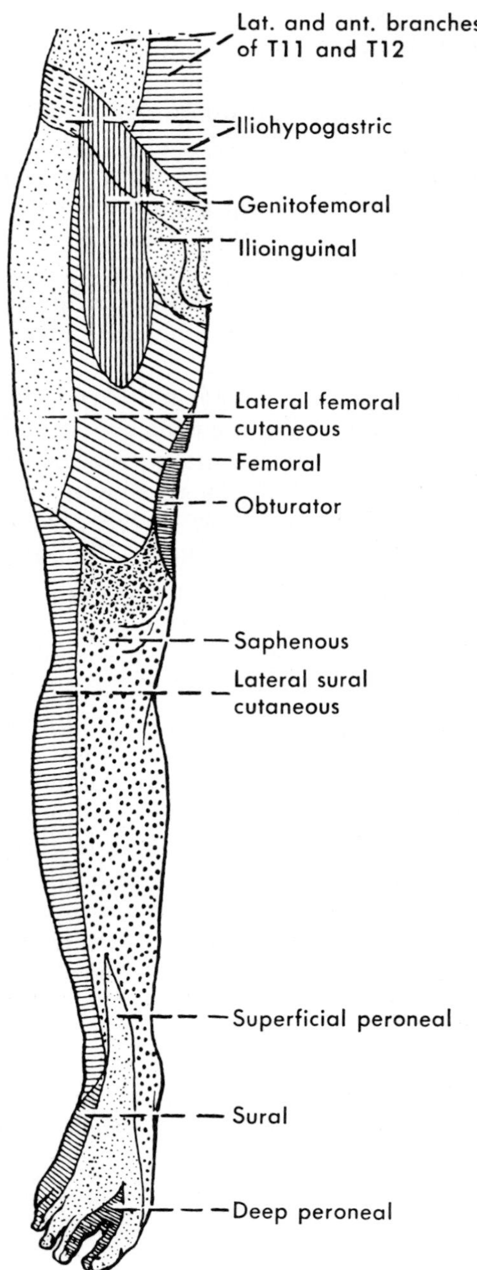


**FIG. 16-3.** Distribution of cutaneous nerves to the posterior surface of the lower limb. (Flatau E: Neurologische Schemata für die ärtzliche Praxis. Berlin, Springer, 1915)

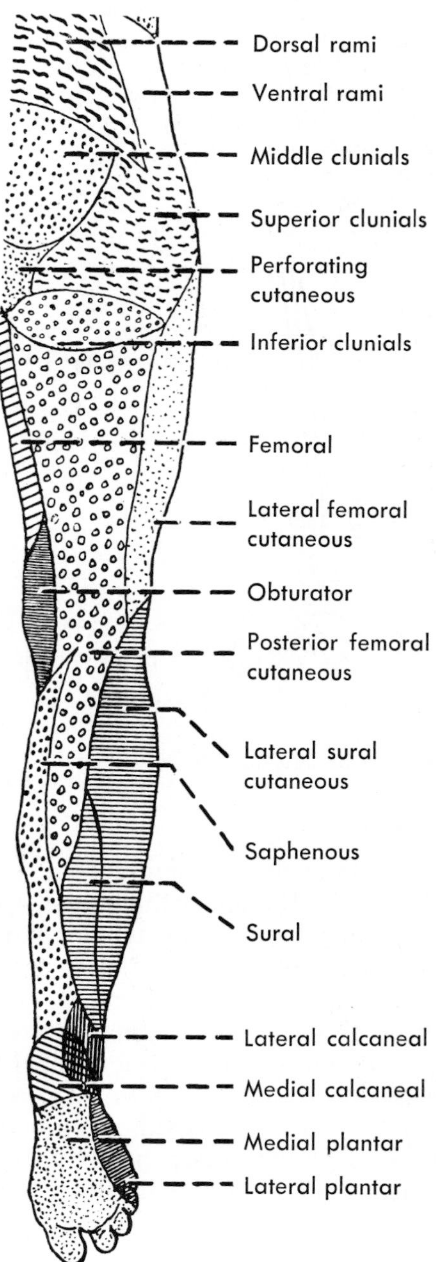

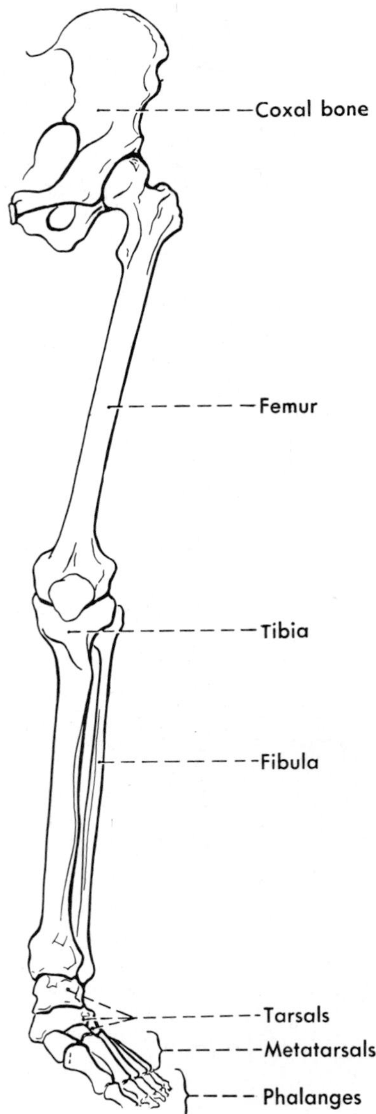


**FIG. 16-4.** Skeleton of the lower limb.

## Skeleton

The skeleton of the lower limb (Fig. 16-4) resembles that of the upper limb in general, but differs in important respects: the **pelvic girdle,** the girdle of the lower limb, is directly and firmly attached to the vertebral column; the three paired bones (ilium, ischium, and pubis) that compose the girdle fuse with each other during development to form a single bone on each side, the **coxal** or **hip bone;** and the two coxal bones are firmly articulated with each other across the midline. These differences can be correlated directly with the importance of the lower limb in supporting the weight of the body.

In accordance with this function, the hip joint is also far stronger structurally than is the shoulder joint, for the upper end of the femur, the head, fits into a deep and cuplike cavity, the *acetabulum,* of the coxal bone. The **femur** in the thigh, corresponding to the humerus of the arm, is the largest bone of the body.

The articulation at the knee joint can be criticized on structural grounds. Although the expanded femoral condyles and the similarly expanded tibial condyles are adapted for the direct downward transmission of weight, the almost flat surface of the tibial condyles and the marked lack of congruity between them and the curved femoral condyles mean that the bony conformation does not resist either anteroposterior or mediolateral displacement; the strength of the knee joint in resisting such movements, therefore, depends upon the ligaments and muscles.

Of the two bones of the leg, the medial tibia and the lateral fibula, the **tibia** supports most of the weight and transmits it from the femur to the ankle. The **fibula** does not enter into the knee joint at all and articulates on the side of the ankle joint, so that although it lends stability to this joint, it transmits only a small part of the total weight across it.

The bones of the ankle and foot bear marked similarities to those of the wrist and hand. The posterior bones of the foot and ankle are called **tarsals,** and although generally much larger than the carpals and arranged in a different fashion, they correspond to them. Succeeding the tarsals are the **metatarsals,** the equivalent of the metacarpals of the hand, and succeeding the metatarsals are the **phalanges** of the digits, much shorter than the corresponding ones of the hand but otherwise resembling them.

## Muscles

Because of the firm articulation of the pelvic girdle to the sacrum (vertebral column), extrinsic muscles comparable to those attach-

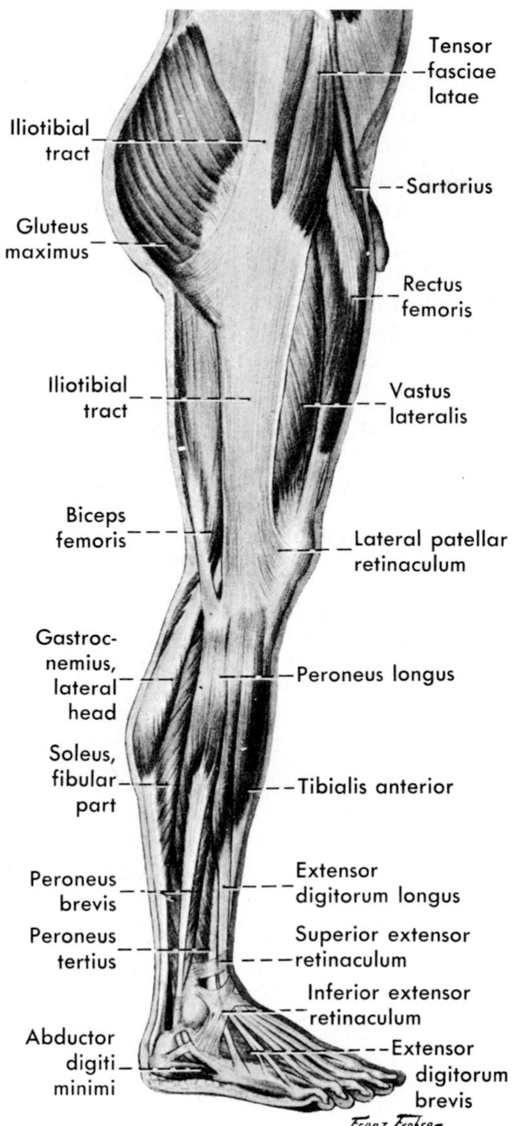


**FIG. 16-5.** Lateral view of muscles of the lower limb. (Frohse F, Fränkel M: In von Bardeleben K (ed): Handbuch der Anatomie des Menschen, vol 2, sect 2, pt 2. Jena, Fischer, 1908)

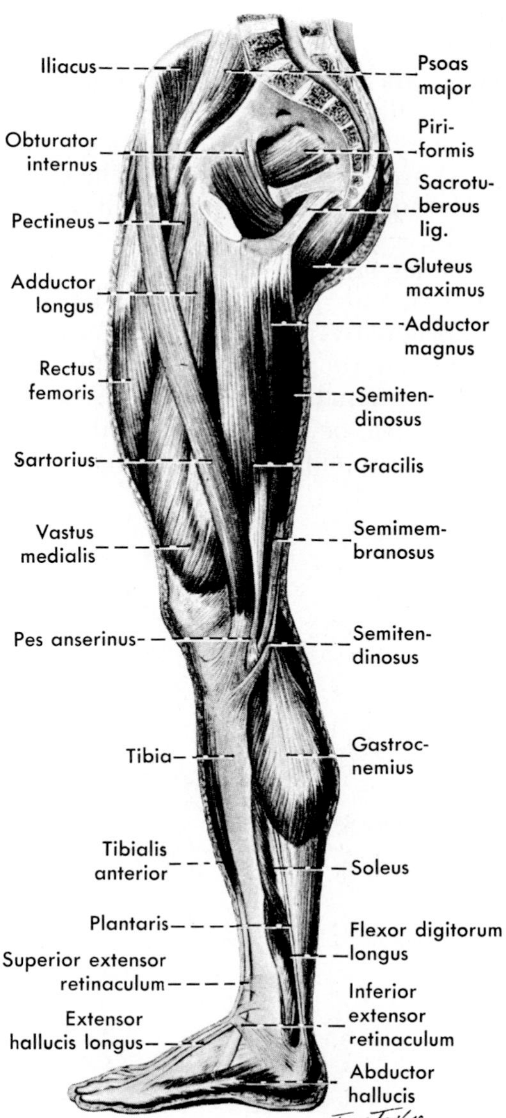


**FIG. 16-6.** Medial view of muscles of the lower limb. (Frohse F, Fränkel M: In von Bardeleben K (ed): Handbuch der Anatomie des Menschen, vol 2, sect 2, pt 2. Jena, Fischer, 1908)

ing the girdle of the upper limb to the vertebral column are sparse. Many muscles do attach to the girdle, but most are either intrinsic limb muscles, moving the free limb, or are abdominal or perineal muscles. Most of the limb musculature is, therefore, intrinsic, arising either from the girdle or from the bones of the free limb.

The musculature of the lower limb (Figs. 16-5 and 16-6) is divisible, as is that of the upper, into original ventral musculature and original dorsal musculature. Muscles of the buttock might be expected from their placement to be dorsal muscles, and the larger and more posterior ones are. However, some of the deeper (more anteriorly lying) small muscles of the buttock, judging by their innervation, are representatives of the ventral muscular group. In the thigh, relations are distorted by the medial rotation that the lower limb undergoes; the original dorsal musculature comes to lie anteriorly and laterally, with only one muscle (the short head of the biceps femoris) still lying posteriorly. The original ventral musculature, therefore, lies largely medially.

In the leg and foot, the muscles of the calf and the plantar surface rather obviously represent original ventral muscles, and those of the anterolateral leg and dorsum of the foot represent original dorsal muscles. Whether in the thigh, leg, or foot, the originally ventral muscles are innervated through ventral branches of the lumbosacral plexus, whereas the originally dorsal muscles are innervated through its dorsal branches. Both divisions of the lumbosacral plexus, the lumbar and sacral plexuses, give off ventral and dorsal branches to the limb.

As is true of muscles of the upper limb, the exact segmental innervation of the muscles of the lower limb is probably not known accurately.

There are a number of **bursae** in the lower limb; the more important and constant ones are mentioned in connection with the regions

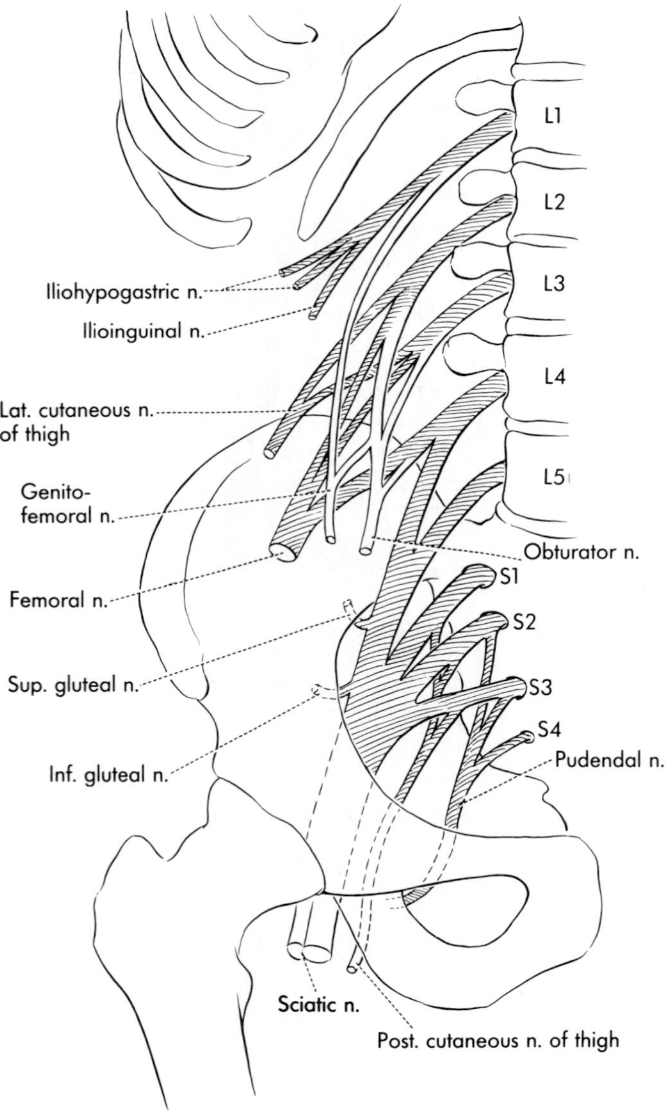


**FIG. 16-7.** Diagram of the lumbosacral plexus. (Sections of Neurology and the Section of Physiology, Mayo Clinic and Mayo Foundation: Clinical Examinations in Neurology, 2nd ed. Philadelphia, WB Saunders, 1963)

in which they occur. **Tendon sheaths** occur at the ankle where the tendons are bound to the underlying bones by retinacula, and there are also plantar digital tendon sheaths on the toes.

## Large Nerves

With the exception of some skin of the buttock, which is supplied by segmental nerves (upper lumbars and upper sacrals), the innervation of the lower limb is entirely through branches of the **lumbosacral plexus** (Fig. 16-7). The distribution of the cutaneous nerves to the lower limb has already been shown, and these nerves are further discussed in the following sections. The nerves to the front of the thigh are from the lumbar plexus. The muscles of the buttock are supplied by several nerves from the sacral plexus. The posterior muscles in the thigh, and all the muscles below the knee, are supplied by the sacral plexus. There are only four large mixed nerves that go to the lower limb; two of them are derived from the lumbar plexus and two from the sacral plexus.

The formation of the **lumbar plexus** must be studied in a deep dissection of the abdomen, and a description of it will be found in Chapter 25. For purposes of the present description, it need only be understood that *the 2nd, 3rd, and 4th lumbar nerves typically contribute to the two major branches of the lumbar plexus.* One of these branches, the **obturator nerve,** is derived from the anterior (ventral) divisions of the plexus and is distributed to the original ventral musculature of the front of the thigh, the anteromedial or adductor group; the distribution of this nerve is shown diagrammatically in Figure 16-8. The other nerve, the **femoral,** is derived from the poste-

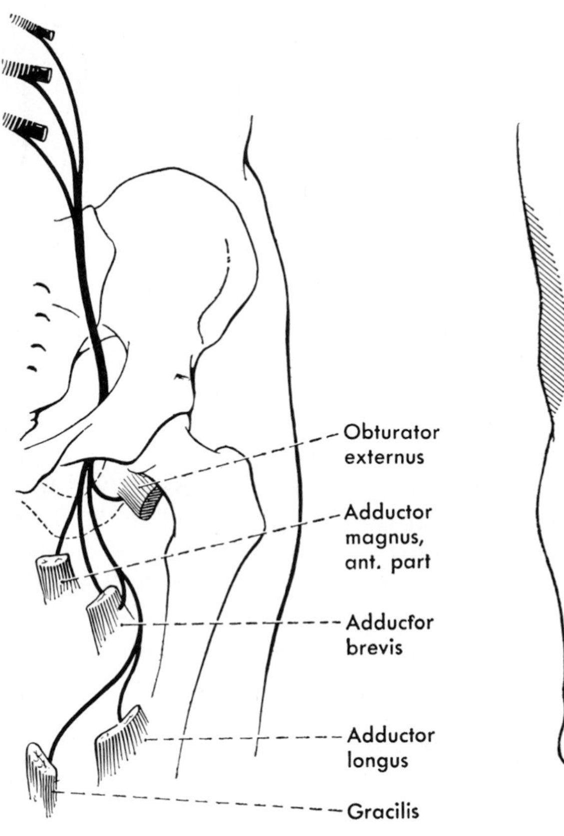


**FIG. 16-8.** Distribution of the obturator nerve. Sometimes the pectineus, shown in Figure 16-9 as being supplied by the femoral nerve, is innervated by the obturator.

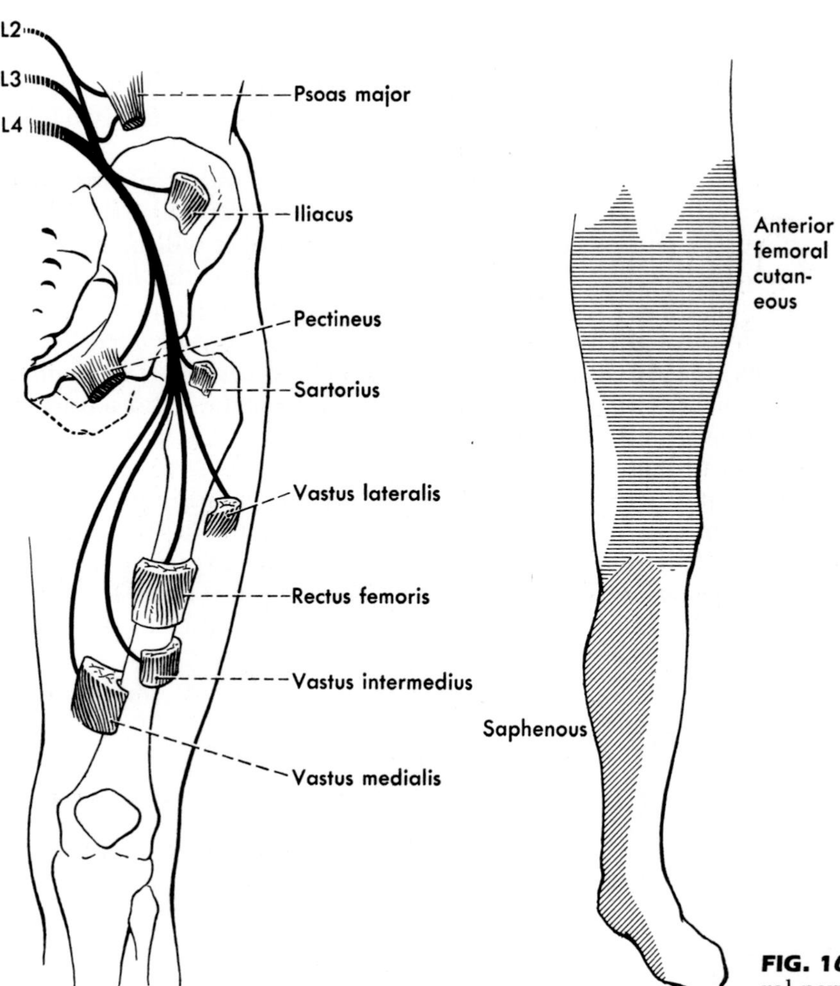


**FIG. 16-9.** Distribution of the femoral nerve.

rior (dorsal) divisions of the plexus and is distributed to originally dorsal muscles of the front of the thigh. Its distribution is shown diagrammatically in Figure 16-9.

The **sacral plexus** is formed in the pelvis and is described with that part; most of its branches enter the buttock, and they are described in the following chapter. Again, for purposes of the present discussion, it is necessary to know only that *this plexus is usually formed by the junction of ventral rami of the 4th and 5th lumbar and the first three or four sacral nerves.* The two large mixed nerves from the sacral plexus, although first bound together by connective tissue to form the **sciatic nerve,** subsequently separate into common peroneal and tibial nerves. The **common peroneal,** the dorsal component, is distributed to original dorsal musculature—one posterolateral muscle of the thigh, the anterolateral muscles of the leg, and the muscles on the dorsum of the foot; its distribution is shown diagrammatically in Figure 16-10. The **tibial nerve,** the ventral component, is distributed to original ventral musculature—the more medial posterior thigh muscles and all the muscles of the calf and of the plantar aspect of the foot; its distribution is shown diagrammatically in Figure 16-11.

(*text continues on p. 343*)

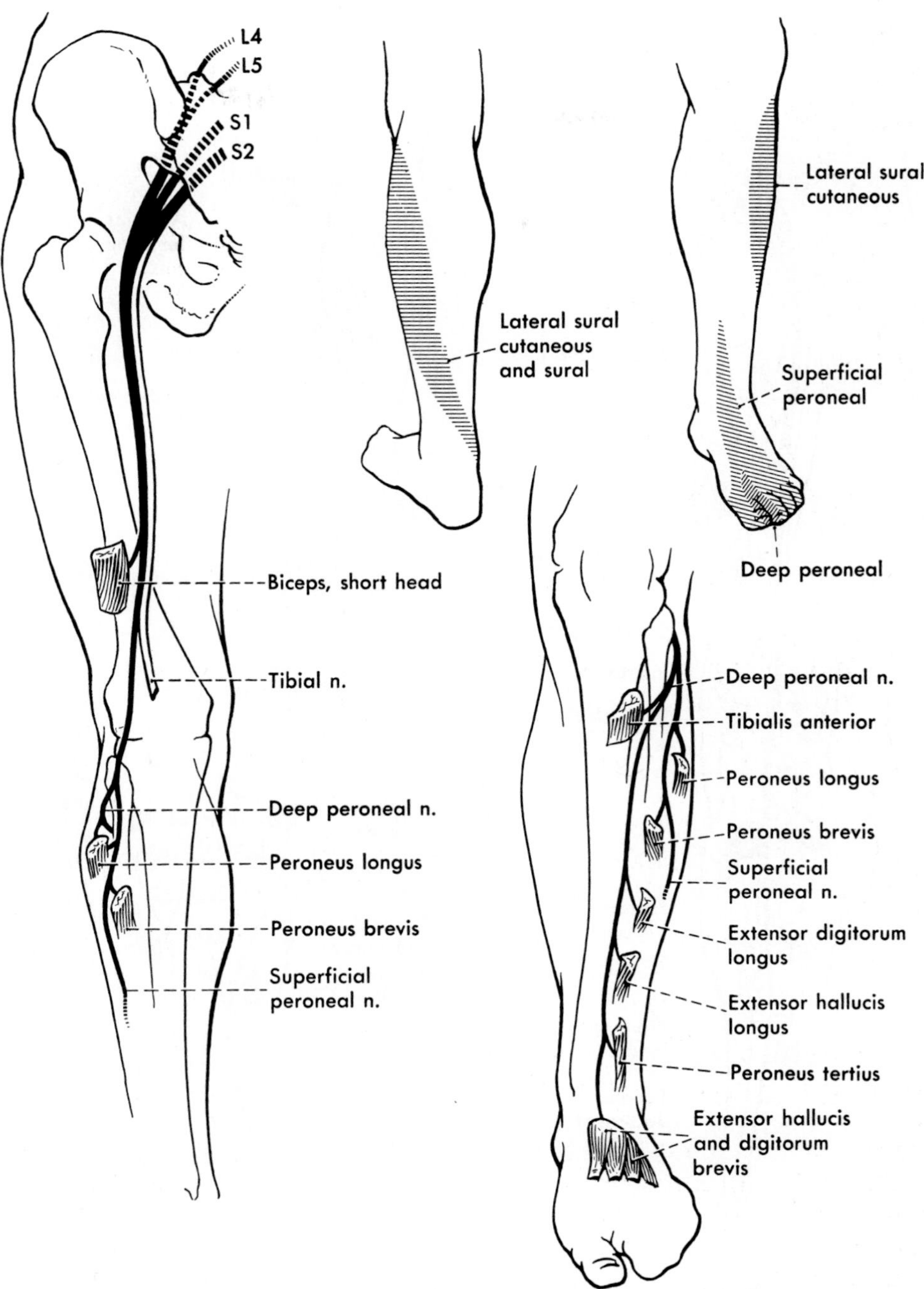


**FIG. 16-10.** Distribution of the peroneal portion of the sciatic nerve.

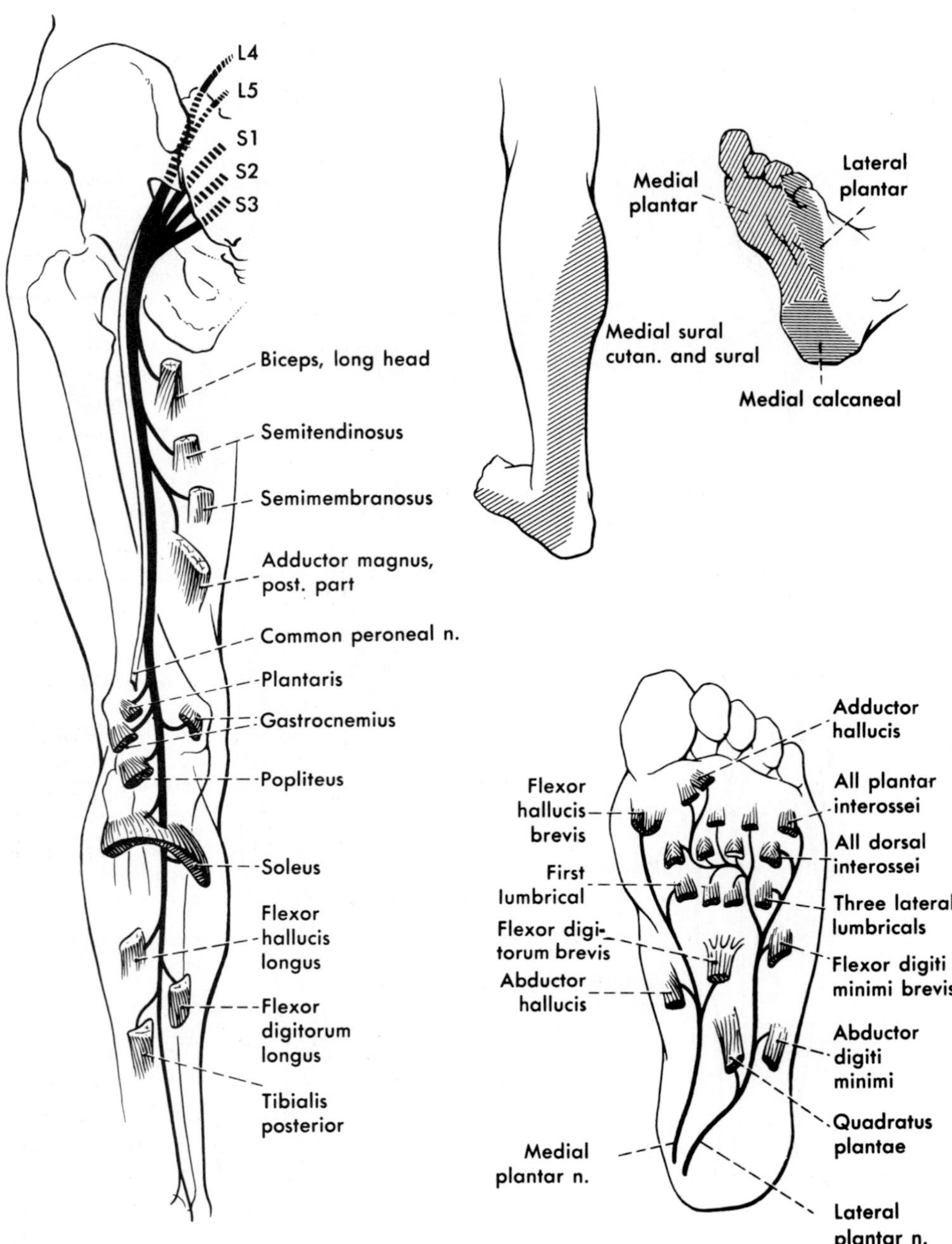


**FIG. 16-11.** Distribution of the tibial portion of the sciatic nerve.

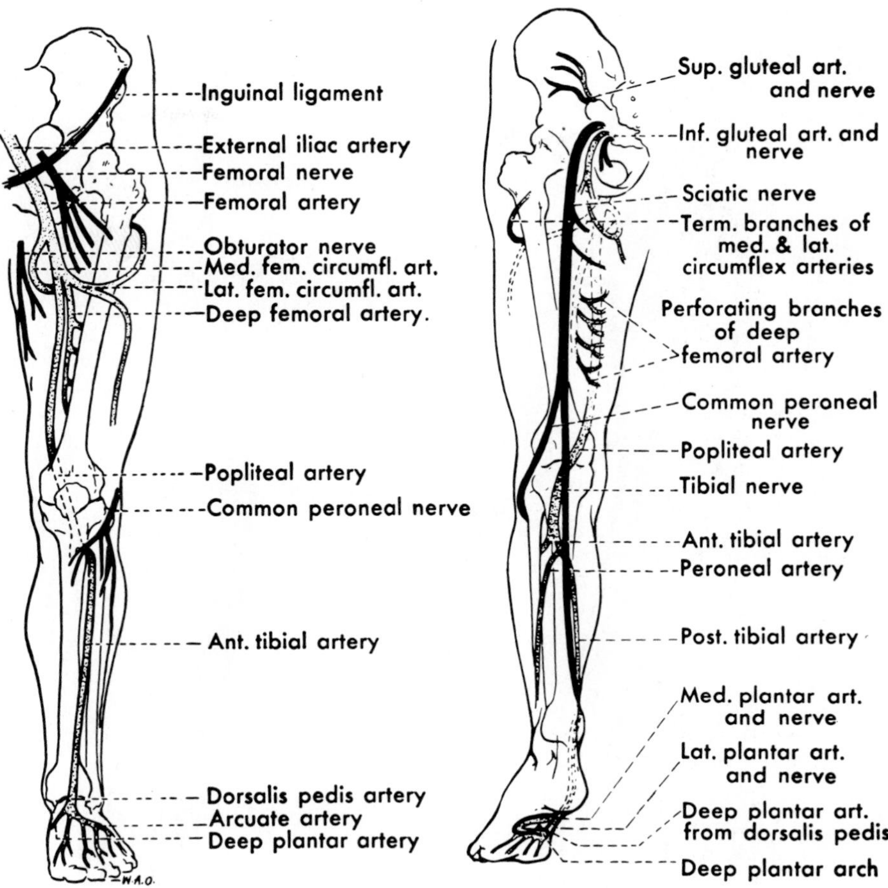


**FIG. 16-12.** Arteries (stippled) and nerves (black) of the lower limb.

## Vessels

Of the three arteries of appreciable size that contribute to the blood supply of the limb, two (the superior and inferior gluteal arteries) pass posteriorly into the buttock and are largely confined to that. The arterial stem distributed to the free limb is the **femoral artery** (Fig. 16-12), the continuation of the external iliac, which derives its blood from the abdominal aorta through the common iliac, a terminal branch of the aorta. The change in name between external iliac and femoral occurs at the level of junction of trunk and thigh (inguinal ligament), and the femoral artery enters and runs down the thigh in an anteromedial position.

A little above the knee, the femoral artery passes to the posterior aspect of the thigh, thus entering the space behind the knee (*popliteal fossa*), and as it does so its name is changed to **popliteal artery.** The popliteal artery ends in the upper part of the leg by dividing into **anterior** and **posterior tibial arteries.** The anterior tibial passes to the front of the leg and runs down onto the dorsum of the foot, and the posterior tibial runs through the calf and into the foot to form plantar vessels there.

Many of the arteries of the lower limb are accompanied by two venae comitantes, but the largest arteries have only single veins paralleling them.

The **superficial veins** of the lower limb are particularly important because they are subject to becoming varicose. A *varicose vein* is a dilated and tortuous one and not only may be

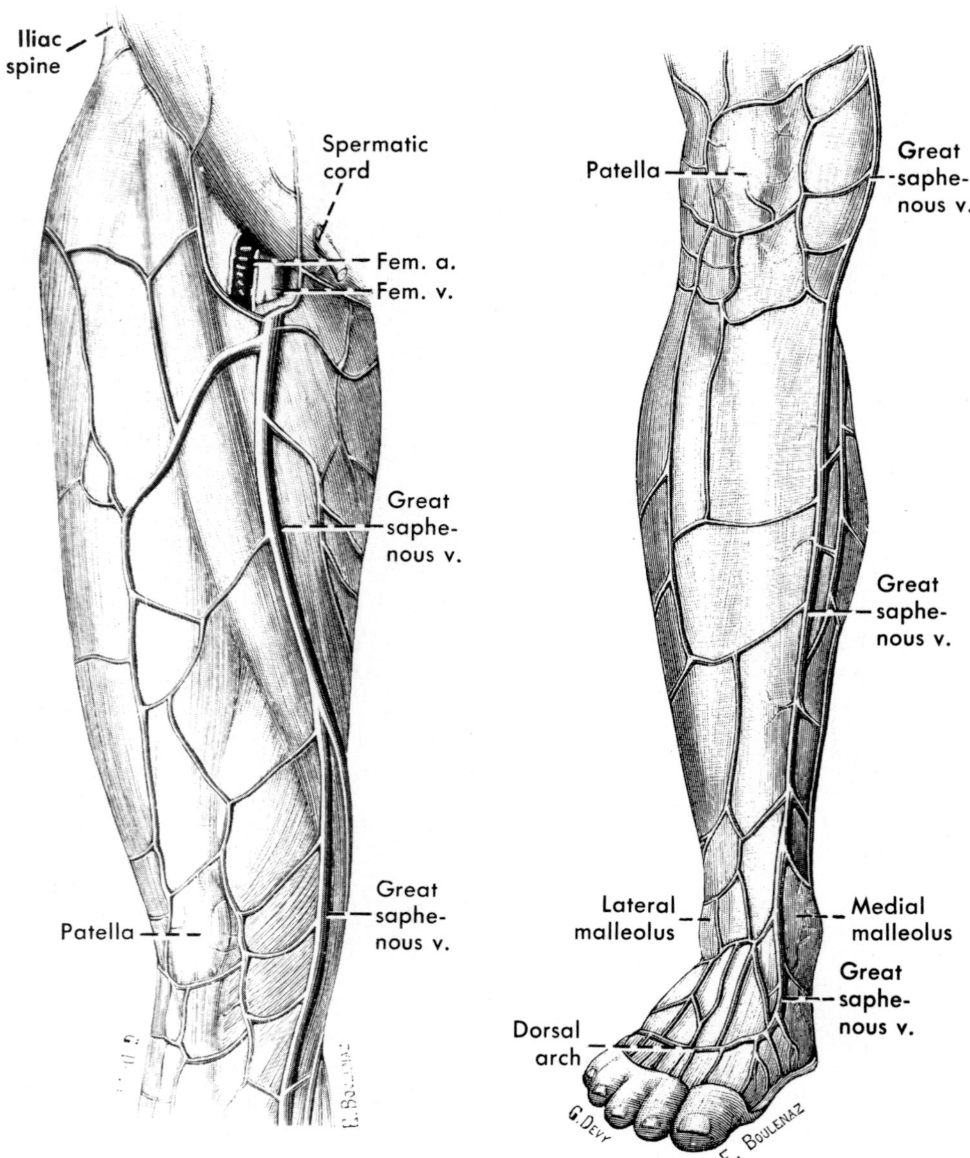


**FIG. 16-13.** Anterior views of the veins of the lower limb. (Testut L: Traité d'Anatomie Humaine, vol 2. Paris, Doin, 1891)

painful but also may lead to nutritive changes in the limb, producing, for instance, ulcers that do not readily heal. The basic cause of varicosities is an overloading of the veins with blood. The superficial veins of the lower limb, which receive neither the support nor the massage that the muscles give the deep veins and yet must, like them, support a column of blood reaching to the level of the heart, are particularly subject to stretch as the result of overfilling.

The precise factors operating to produce varicosities of veins are not understood. A constant finding, however, is that some of the veins that connect the superficial and deep venous systems in the lower limb are defective. These veins, known as **communicating** (perforating) **veins,** or more commonly "perforators," are so valved that they nor-

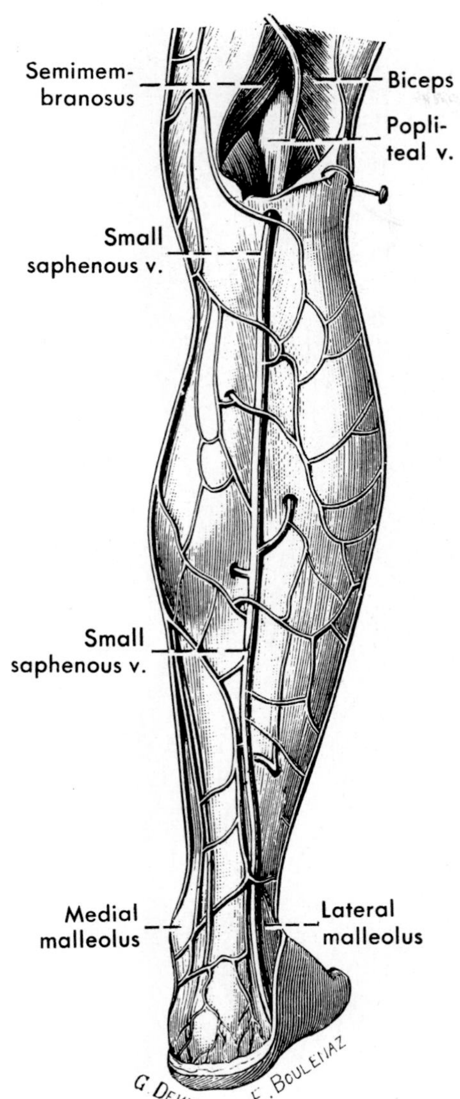


**FIG. 16-14.** The small saphenous vein; as this one ends in the popliteal vein it sends a connection upward toward the great saphenous. (Testut L: Traité d'Anatomie Humaine, vol 2. Paris, Doin, 1891)

mally conduct blood only from the superficial veins into the deep ones. Regularly, however, when the superficial veins of the lower limb are varicose, some (few or many) of the perforating veins can be shown to allow blood to pass from the deep veins into the superficial ones.

The important superficial veins of the lower limb are the great (greater) and small (lesser) saphenous veins. Both veins originate on the dorsum of the foot, and both they and the connections between them give rise to communicating veins that may become "incompetent" (dilated, so that their valves allow blood to run from the deep into the superficial veins).

The **great saphenous vein** originates on the medial border of the foot and runs up the medial side of the leg (Fig. 16-13). It receives communications from the small saphenous vein, so valved that blood can pass normally only from the small to the great saphenous vein. The great saphenous vein crosses the medial side of the knee and then runs upward to end in the femoral vein close to the anterior midline of the thigh.

The **small saphenous vein** (Fig. 16-14) begins along the lateral margin of the foot and passes upward along the posterior aspect of the calf, about its middle. It commonly ends in the popliteal vein.

The **lymphatics** and **lymph nodes** of the lower limb are generally divisible into superficial and deep ones, as are those of the upper limb, and again the superficial lymphatics and nodes are far more important. The **deep lymphatic vessels** follow blood vessels, and there are few deep lymph nodes. In contrast, the **superficial lymphatics** (Figs. 16-15 and 16-16) are numerous and run up all surfaces of the leg. Some of those along the back of the leg tend to converge upon the small saphenous vein and follow it, penetrating the deep fascia with it and ending in deep (popliteal) nodes and vessels, but otherwise the general direction of the lymphatics is upward toward the great saphenous vein.

As they course upward in the thigh, the superficial lymphatics from the lateral side run anteriorly and then medially, and those from the medial side run anteriorly and then laterally, so that they converge toward the path of the great saphenous vein; there is, thus, on the back of the thigh a "lymphatic divide," with lymph on one side running in one direction, that on the other running in another direction. The lymphatics about the great saphenous vein end in a number of prominent superficial inguinal lymph nodes.

Injection of radiopaque material into the

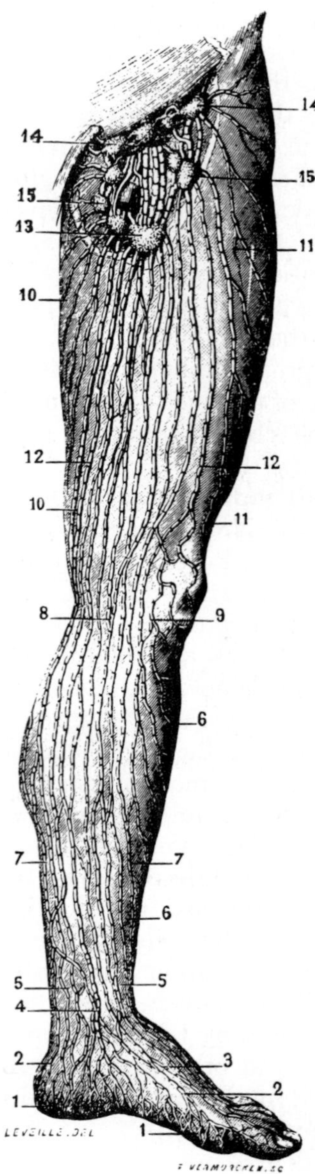


**FIG. 16-15.** Medial view of the superficial lymphatics of the lower limb. The numbers simply call attention to the courses of various lymphatic channels; for instance, **1** identifies lymphatics that course onto the dorsum from the plantar surface of the foot. Superficial inguinal lymph nodes, **13, 14, 15,** are shown at the groin. (Poirier P, Charpy A: Traité d'Anatomie Humaine, 2nd ed, vol 2, fasc 4. Paris, Masson et Cie, 1909)

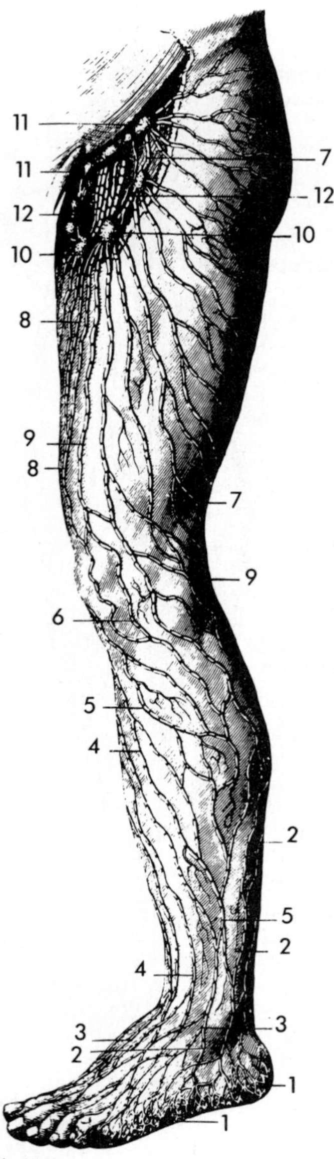


**FIG. 16-16.** Lateral view of the superficial lymphatics of the lower limb. (Poirier P, Charpy A: Traité d'Anatomie Humaine, 2nd ed, vol 2, fasc 4. Paris, Masson et Cie, 1909)

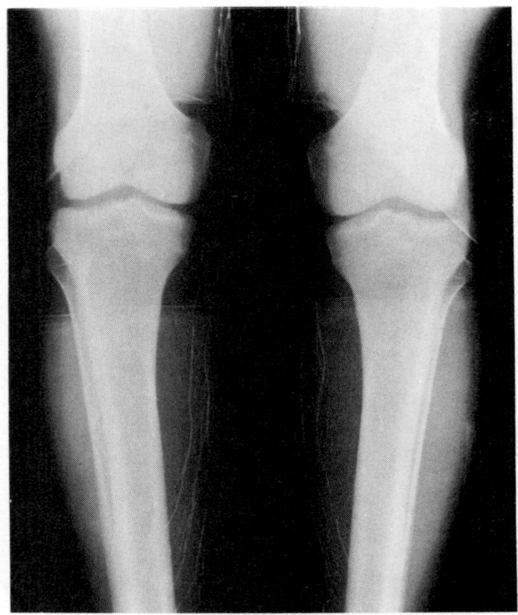

**FIG. 16-17.** Lymphatics of the calf and thigh injected through lymphatics of the feet. (Courtesy of Dr. W. E. Miller)

lymphatics of the lower limb can demonstrate not only lymphatics of the leg and thigh (Fig. 16-17) but also superficial inguinal lymph nodes, iliac and lumbar nodes, and even the thoracic duct.

## GENERAL REFERENCES AND RECOMMENDED READINGS


BARDEEN CR: Studies of the development of the human skeleton: (A) The development of the lumbar, sacral and coccygeal vertebrae. (B) The curves and the proportionate regional lengths of the spinal column during the first three months of embryonic development. (C) The development of the skeleton of the posterior limb. Am J Anat 4:265, 1905

BARDEEN CR: Development and variation of the nerves and the musculature of the inferior extremity and of the neighboring regions of the trunk in man. Am J Anat 6:259, 1907

BARDEEN CR, ELTING AW: A statistical study of the variations in the formation and position of the lumbo-sacral plexus in man. Anat Anz 19:124, 209, 1901

BARDEEN CR, LEWIS WH: Development of the limbs, body-wall and back in man. Am J Anat 1:1, 1901

FLECKER H: Time of appearance and fusion of ossification centers as observed by roentgenographic methods. Am J Roentgenol 47:97, 1942

MULLARKY RE: Termination of the small saphenous vein. Northwest Med 62:878, 1963

O'RAHILLY R: Morphological patterns in limb deficiencies and duplications. Am J Anat 89:135, 1951

PATTERSON AM: The origin and distribution of the nerves to the lower limb. J Anat Physiol 28:84, 169, 1893–1894

SENIOR HD: The development of the arteries of the human lower extremity. Am J Anat 25:55, 1919

# 17

# BUTTOCK, THIGH, AND HIP JOINT

The two coxal bones, the girdles of the lower limbs, articulate almost immovably with the sacrum and each other to form the bony **pelvis** (Figs. 17-1 and 17-2). They and the sacrum move on or with the lumbar part of the vertebral column. The pelvis is capable of tilting movements. A downward tilt (downward rotation) is accompanied by an increase in the lumbar curvature, which is, of course, favored by the fact that the line of gravity is behind most of the bodies of the lumbar column. Upward rotation, resulting in less obliquity of the pelvis, is then brought about by the contraction of the anterolateral abdominal muscles, which pull upward on the anterior part of the pelvis. The pelvis can also be laterally tilted when the weight of the body is supported on one limb. If the two limbs are unequal in length, there will be a permanent lateral tilt. Very much lateral tilting of the pelvis demands a lateral curving of the vertebral column (with its concavity to the high side of the pelvis) if the line of gravity through the body is to be centered over the pelvis once again.

The pelvis serves two purposes: it transmits the weight of the body to the lower limbs and it also forms the lower part of the abdominal cavity (this part being termed the "pelvic cavity" or "pelvis") and houses parts of the abdominopelvic viscera. It, therefore, receives the attachment of abdominal muscles and is provided with special muscles that close its lower end or outlet. Those features of the pelvis that pertain in particular to its role as a part of the body wall are discussed in a later chapter (Chap. 27). Interest here in the pelvis is centered in its relationship to the lower limb.

The muscles of the buttock act primarily across the hip joint only and in general attach to the upper end of the femur, the bone of the thigh. Some of the muscles of the thigh arise from the pelvic girdle, others arise from the femur, and whereas some act exclusively at the hip joint or the knee joint, others are biarticular muscles, crossing both joints.

Nerves and vessels enter the thigh both anteriorly and posteriorly. The anterior artery, the femoral, continues into the leg and foot, but the

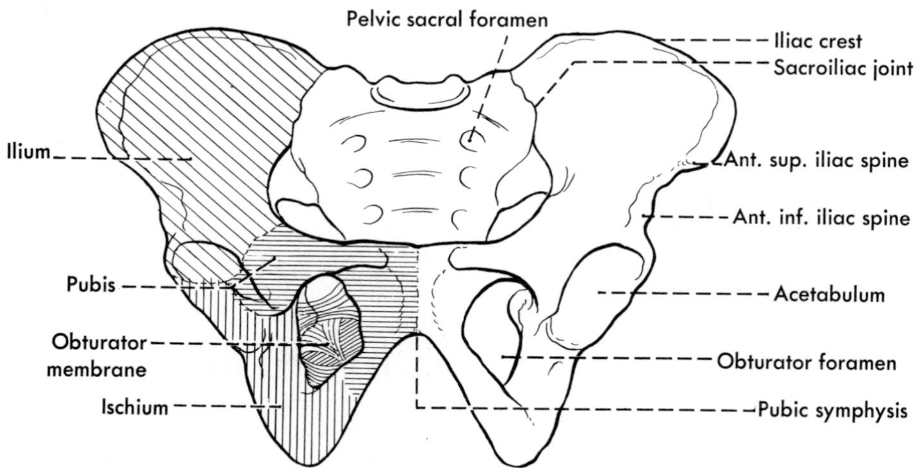


**FIG. 17-1.** The pelvis from the front. On the reader's left the three parts of the coxal bone are differently shaded, and the obturator membrane largely fills the obturator foramen.

anterior nerves, chiefly the femoral and obturator, are distributed almost entirely to the thigh. The reverse is true of the posterior vessels and nerves: the arteries entering the buttock are distributed almost entirely to this part, but the largest posterior nerve, the sciatic, continues through the buttock and the thigh to supply all the muscles of the leg and foot and most of the skin of these parts.

## SKELETON

### PELVIC GIRDLE

Each of the two **coxal** or **hip bones** (formerly known as "innominate bones") consists phylogenetically and developmentally of three separate bones, the ilium, the ischium, and the pubis (Fig. 17-1). In the adult, these bones have so fused that little or no trace of their lines of fusion is left (Figs. 17-2 through 17-4), yet the coxal bone is so complex that it is convenient to refer to the different parts as if they were still separate bones. The **ilium** is the upper, larger part of the os coxae and forms the upper part of the *acetabulum,* the deep, cup-shaped concavity on the lateral aspect of the hip bone. The **pubis** forms the anterior part of the acetabulum and the anteromedial part of the hip bone. The **ischium** forms a posteroinferior part of the acetabulum and the lower posterior part of the hip bone.

The **acetabulum** faces laterally, downward, and forward, and its superior and posterior walls are particularly heavy. Its inferior wall is incomplete, presenting the *acetabular notch,* which leads into the *acetabular fossa,* the rough area in the center of the articular or *lunate surface* (Fig. 17-3).

The somewhat oval **obturator foramen** lies below the acetabulum and is bounded by the pubis and the ischium. Where the inferior ramus of the pubis and the ramus of the ischium meet as they form the lower border of this foramen, there is frequently a line marking their junction. During life, the obturator foramen is closed (obturator means closed) by the **obturator membrane,** except anterosuperiorly; the aperture through which the obturator vessels and nerve leave the pelvis and enter the thigh is the **obturator canal.**

The remaining features of the coxal bone can best be described in terms of the individual bones that compose it. Of these, the **pubis** has a **superior ramus** that enters into the acetabulum, an **inferior ramus** that helps bound the obturator foramen, and, between these, a **body** that presents on its medial side a *symphyseal surface* (Fig. 17-4) at which it is firmly bound by an interpubic disk to the pubis of the other side to form the *pubic symphysis.* The thickened anterior part of the body is the **pubic crest,** which ends laterally at the more prominent **pubic tubercle;** this tubercle is the beginning of a raised line, the **pecten,** which extends upward and laterally along the superior ramus to become continuous with a more or less distinct line on the ilium, the *arcuate line*. On the upper surface of the superior ramus anterior to the pecten, the junction of pubis and ilium is marked by a raised, roughened area, the **iliopubic** (iliopectineal) **eminence;** on the inner surface of this ramus just above the obturator canal, there is a ridge, the *obturator crest,* below which is the *obturator sulcus* leading to the canal.

The parts of the **ischium** are the body and the ramus. The **body** helps form the acetabulum. The **ramus** extends inferiorly and then curves anteriorly to help bound the obturator foramen; the large posteroinferior swelling is the **ischial tuberosity** (tuber ischiadicum), and the smaller and more pointed posterior projection at about the junction of the ramus and body is the **ischial** (ischiadic) **spine.** The concavity between the ischial spine and ischial tuberosity is the **lesser sciatic** (ischiadic) **notch** or incisure; the larger concavity above the ischial spine, involving both the ischium and ilium, is the **greater sciatic notch.**

The **ilium** is described as consisting of a body (*corpus*) and a wing (*ala*). The **body** enters into the formation of the acetabulum and has no named parts. The inner, somewhat concave surface of the wing is the **iliac fossa;** the thin bone of the fossa gives way posteriorly to the thick bone of the sacropelvic sur-

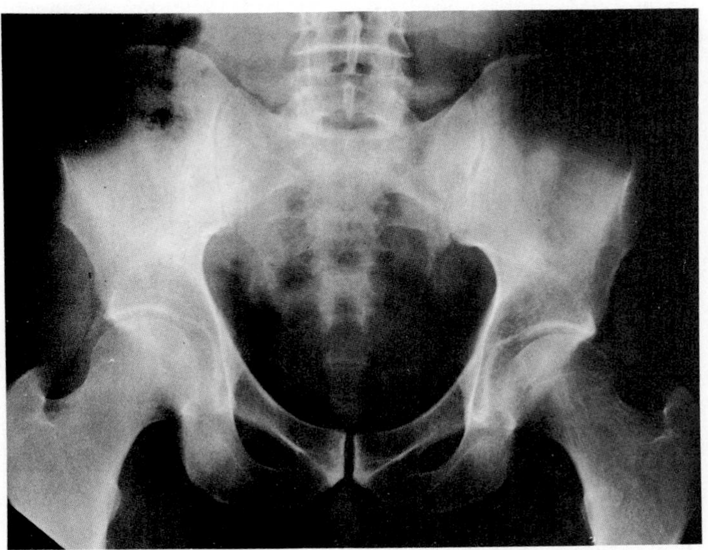

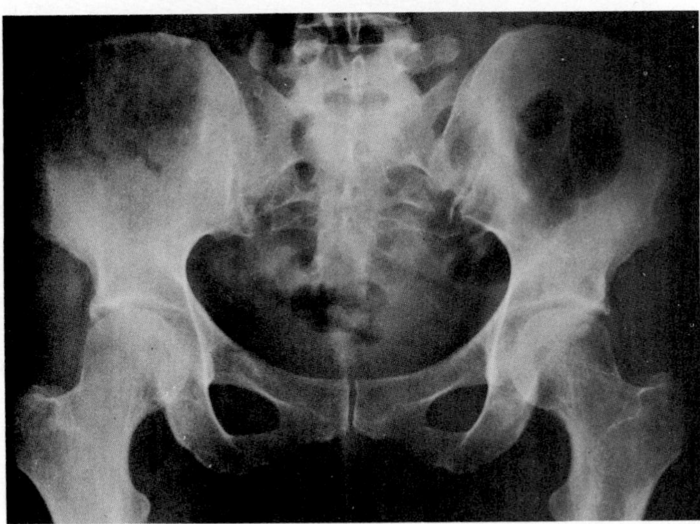

**FIG. 17-2.** Roentgenograms of a male pelvis, above, and a female pelvis, below. (Courtesy of Dr. D. G. Pugh)

face. On the lower part of this surface is the articular facet for the synovial joint between the ilium and the sacrum, called (because of its fancied resemblance to an ear) the **auricular surface;** above this is a rough protuberance, the **iliac tuberosity,** that receives the attachment of heavy ligaments stretching between the sacrum and the ilium.

The upper thickened border of the **wing,** the **iliac crest,** ends anteriorly at the *anterior superior iliac spine* and posteriorly at the *posterior superior iliac spine.* Below the anterior superior iliac spine, on the anterior border of the bone, is the *anterior inferior iliac spine;* below and slightly anterior to the posterior superior iliac spine, at the posterior end of the upper border of the greater sciatic notch, is the *posterior inferior iliac spine.* The **arcuate line** begins at the anterior border of the auricular surface and extends forward and downward, separating the iliac fossa from that part of the bone that helps to form the wall of the true pelvis; it is more or less continuous anteriorly with the pecten of the pubis.

On the lateral surface of the ilium, the only markings of particular importance are the **inferior, anterior,** and **posterior gluteal lines.** These divide the outer surface of the wing of the ilium into three smooth areas from which different gluteal muscles arise.

Because of its situation between the trunk and the free limb, the hip bone receives attachments of muscles of both. The muscular and chief ligamentous attachments are shown in Figures 17-5 and 17-6.

Numerous small **blood vessels** enter and leave the hip bone, particularly its thicker parts where the vascular foramina are usually prominent on the dried bone. The twigs to the bone are from vessels that supply adjacent muscles.

**FIG. 17-3.** Lateral view of the right coxal bone.

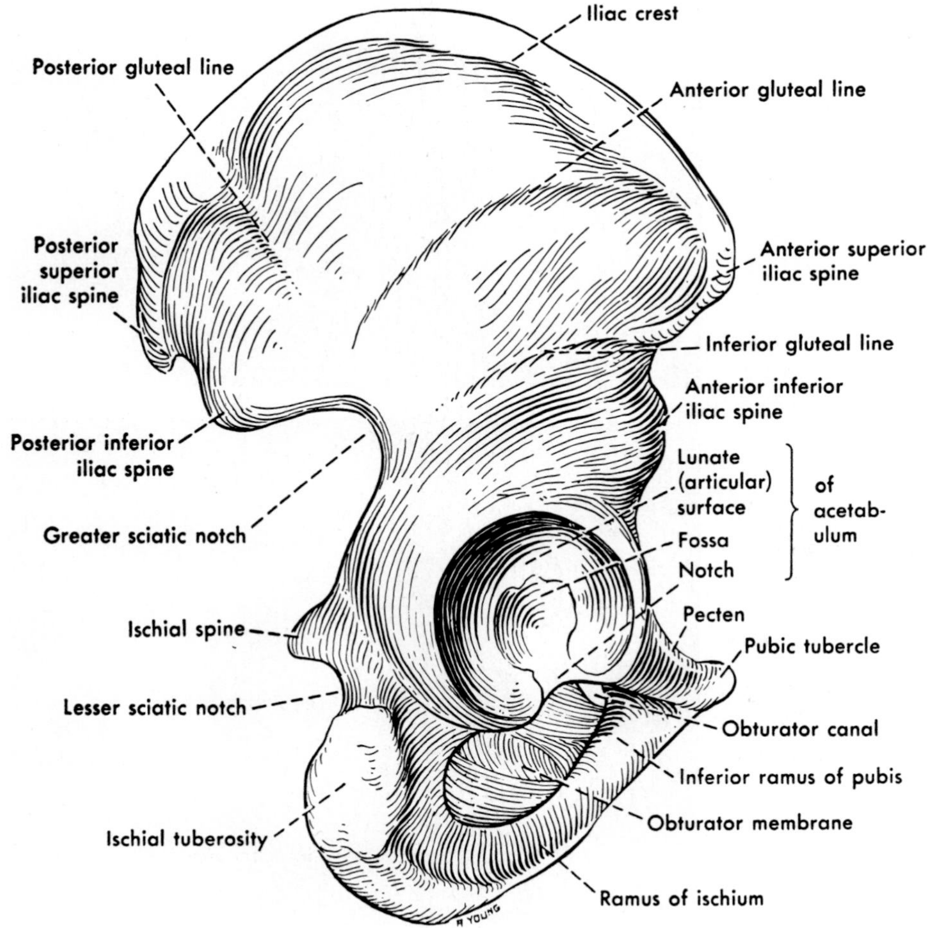

**Fracture** of the hip bone may result from a blow, but the most common cause is compression of the pelvic ring by a crushing force, such as can occur, for instance, in an automobile accident. The fracture normally occurs across a weaker part of the bone and therefore is particularly likely to involve the wing of the ilium or the obturator foramen. If the fracture is such as to allow displacement of a part of the bone, damage to pelvic viscera such as the urinary bladder may occur.

**Ossification** of the coxal bone occurs from three primary centers, one for the ilium, one for the ischium, and one for the pubis. In the infant, these bones are still widely separated by cartilage (Fig. 17-7). The ramus of the ischium and the inferior ramus of the pubis unite at about the seventh or eighth year of life. One or more secondary centers appear in the acetabulum, and the three elements fuse with those centers and each other by about the age of 15 years. The bone is finally completed by other secondary or epiphyseal centers, which include centers for the iliac crest and the ischial tuberosity, and sometimes for the pubis at the symphysis and for the anterior inferior iliac spine (Fig. 17-8). These centers fuse over a considerable range of time, but usually the coxal bone is essentially complete by the age of 20 to 21.

## Pubic Symphysis and Sacroiliac Joint

The pubic symphysis unites the two hip bones to each other, and the sacroiliac joint unites the hip bone to the vertebral column. Both these joints are particularly strong, and upon them depends the stability of the pelvis in supporting the weight of the trunk on the heads of the femurs. The **pubic symphysis** is formed by a fibrocartilaginous *interpubic disk* that unites the symphyseal surfaces of the pubes in the anterior midline. The upper border of the disk blends with the *superior pubic ligament* (over the pubic bodies), and its lower border blends with the *arcuate pubic ligament* (between the two inferior pubic rami).

The **sacroiliac joint** contains a synovial cavity, but is reinforced by heavy ligaments, the

**FIG. 17-4.** Medial view of the right coxal bone.

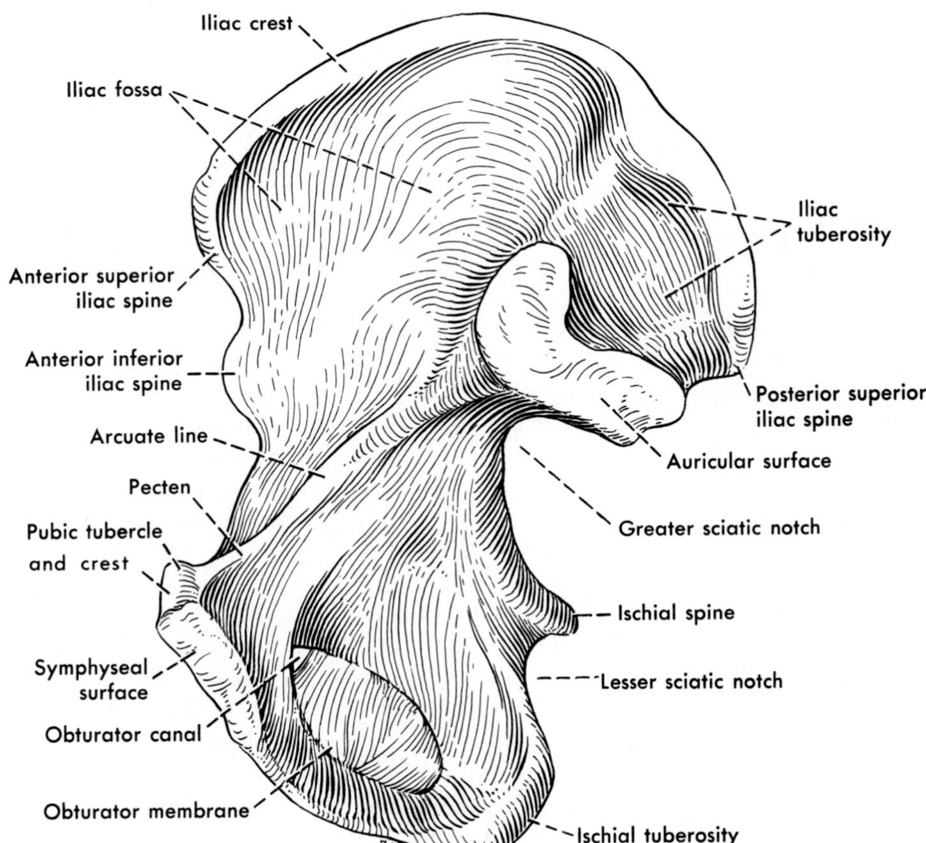

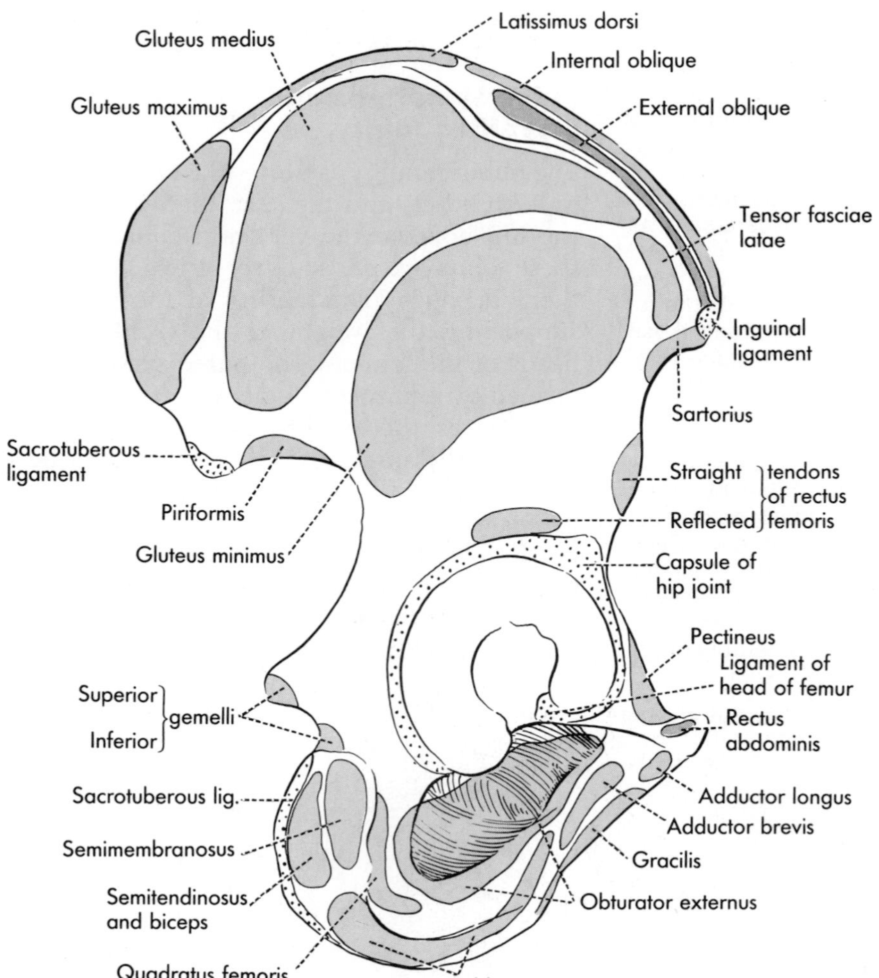


**FIG. 17-5.** Muscular attachments on the lateral surface of the right coxal bone. Origins are red, insertions blue.

strongest of which lie behind the joint cavity. The **sacrotuberous ligament** (Fig. 17-9) is a particularly strong bracing ligament of this joint, attached above to the sacrum and below to the ischial tuberosity. It forms the medial border of both the greater and lesser *sciatic foramina*.

The **sacrospinous ligament,** smaller and more rounded than the sacrotuberous, runs between the sacrum and coccyx and the ischial spine and divides the *greater sciatic foramen* from the *lesser*. These foramina, large on the dried specimen, are filled during life by muscles, nerves, and vessels.

The sacrotuberous and sacrospinous ligaments are so placed as to resist rotation of the sacrum between the coxal bones; the weight on the upper end of the sacrum tends to force it downward and forward, with consequent upward and backward movement of its lower end, and the sacrotuberous and sacrospinous ligaments anchor this lower end. The **iliolumbar ligament,** less important, arises from the transverse process of the 5th lumbar vertebra and attaches to the anterior or pelvic surface of the ilium and sacrum.

The sacroiliac synovial joint (Fig. 17-10) lies between the auricular surfaces of the ilium and sacrum at the levels of about the first three sacral segments. The joint cavity forms only a part of the articulation between sacrum and ilium; the large area posterior to

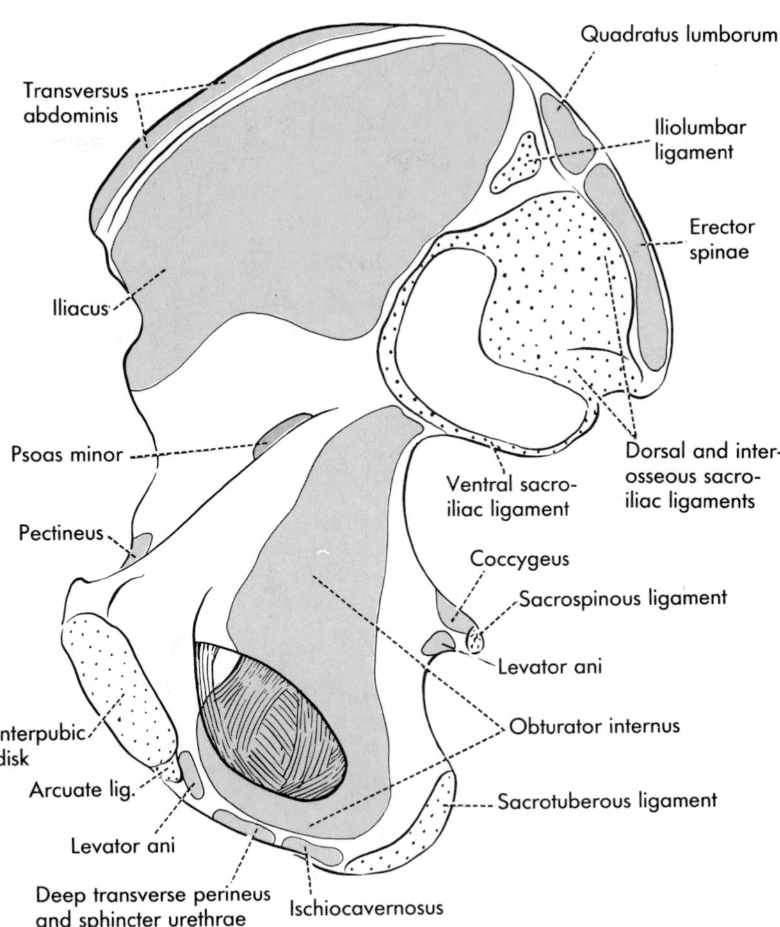


**FIG. 17-6.** Muscular attachments on the medial surface of the right coxal bone.

**FIG. 17-7.** The coxal bones of an infant of 7½ months. The ilium, ischium, and pubis are widely separated by cartilage. The little speck of bone in the acetabulum is the center of ossification of the head of the femur.

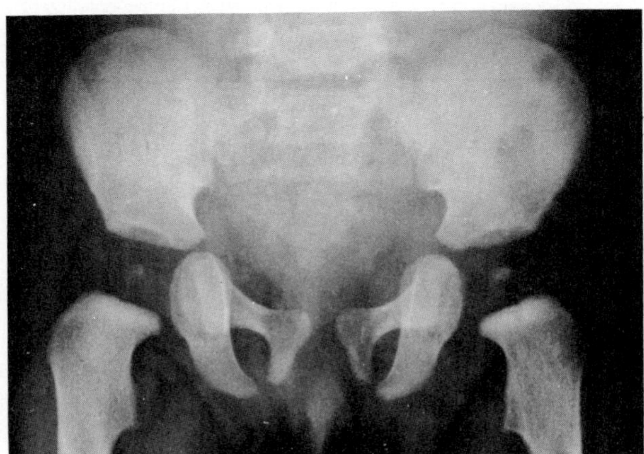

**FIG. 17-8.** Ossification of the coxal bone. **AB** indicates that centers for the bones so marked are present at birth. Two numbers—for instance, **16-25 yr**—indicate respectively the age at which the center so identified appears and the age at which it fuses with the rest of the bone. (Camp JD, Cilley EIL: Am J Roentgenol 26:905, 1931)

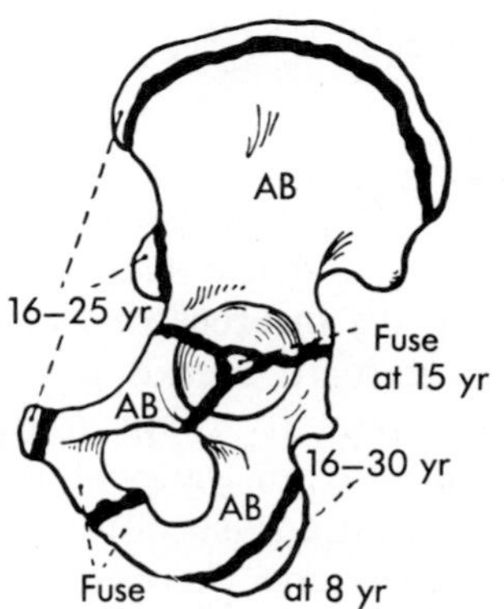

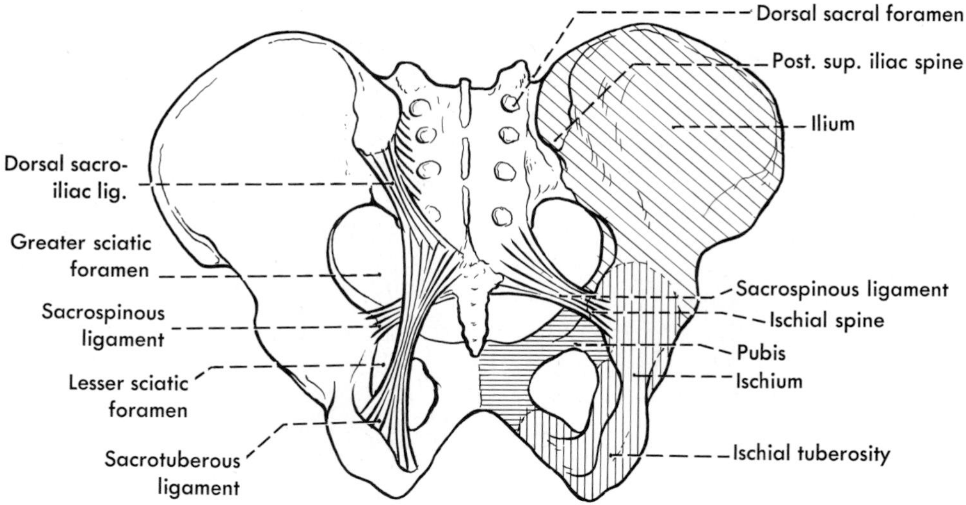


**FIG. 17-9.** The chief bracing ligaments between the sacrum and the coxal bones. In this posterior view the three elements entering into the coxal bone are shaded on the right side.

**FIG. 17-10.** Schematic section through the sacroiliac joint.

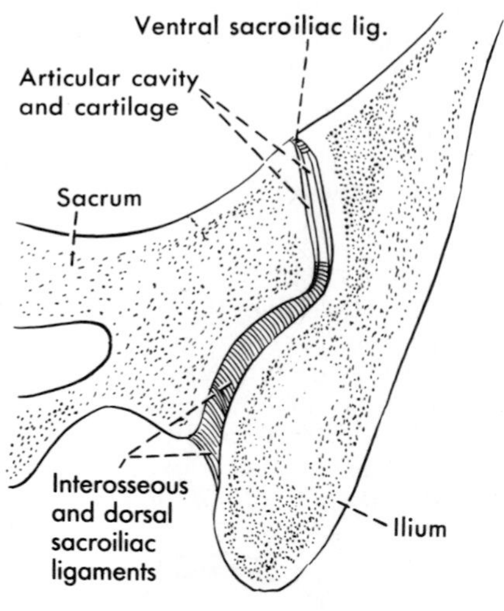


and above the auricular surface, between the iliac tuberosity and the sacrum, is occupied by the very heavy **interosseous sacroiliac ligament.** This ligament blends posteriorly with the more superficially lying *dorsal sacroiliac ligament,* visible from the posterior aspect of the pelvis but otherwise not separable from the interosseous one. Across the front and lower part of the synovial cavity is the much thinner *ventral sacroiliac ligament.* These ligaments, particularly the interosseous one, tend to prevent much movement between sacrum and ilium; further, the articular surfaces of these two bones are so shaped that the weight on the sacrum tends to wedge it firmly between the two ilia, which in turn tends to draw the two ilia closer together because of the pull on the interosseous ligaments. The sacroiliac synovial joint rather regularly shows pathologic changes in adults, and in most males after the age of 50 the joint becomes ankylosed (fused, with the disappearance of the joint cavity).

Because of its sinuous surface and strong ligaments, movement at the sacroiliac joint is limited. However, the hormones of pregnancy soften and loosen the ligaments of the pelvis, allowing them to be more easily stretched, so that mobility at the sacroiliac joint may increase markedly during the latter part of pregnancy.

## FEMUR

The large femur (Figs. 17-11 and 17-12) articulates at its upper end with the hip bone (Fig. 17-2) and at its lower end with both the patella (a sesamoid bone in a tendon) and the tibia (Fig. 17-25). Its **upper end** consists of a *head, neck,* and two *trochanters.* The femur departs markedly from the shape of most long bones because the head and neck sit at an angle to the long axis of the body of the bone. This allows greater mobility at the hip joint but also imposes unusual strains upon the neck of the femur, since the weight is transmitted through a curve. The architecture of the upper end of the femur has been particularly carefully studied, and the trabeculae are said to be arranged in curves that fall nicely along the lines of calculated stress and strain—that is, they accord with engineering principles for bearing such an oblique thrust (see Figs. 3-4 and 3-5).

The ball-like **head** of the femur is covered with cartilage except in a depression, the *fovea capitis,* on its medial side. The fovea serves for the attachment of the ligament of the head. The **neck,** smaller in diameter than the head, is particularly liable to fracture both because of its size and position.

The **greater trochanter** projects above the junction of the neck and the body. On its medial side is a small pit, the *trochanteric fossa,* for the insertion of the obturator externus muscle. The **lesser trochanter** is located posteromedially at the junction of the neck and body. Both trochanters serve for attachment of muscles. On the posterior surface of the femur, the two trochanters are connected by a ridge, the **intertrochanteric crest.** On the anterior surface, the upper part of the **intertrochanteric line** marks the attachment to the femur of the anterior part of the capsule of the hip joint. When the intertrochanteric line is well developed, it can be followed below the lesser trochanter onto the back of the femur. On the posterior surface at about the same level as the lesser trochanter is a roughened **gluteal tuberosity** (for the insertion of the gluteus maximus muscle); sometimes this is raised sufficiently to be termed the *third trochanter.*

The **body** of the femur below the trochanters is approximately cylindrical until it expands at the lower end to form the condyles. Its important marking is the *linea aspera,* the ridge on its posterior aspect. The medial lip of the linea aspera begins above at the *pectineal line* (below the lesser trochanter), and lateral at the gluteal tuberosity. Below, the two lips diverge and leave between them a broad, smooth *popliteal surface.*

The femur ends in rounded **condyles** whose articular surfaces blend together anteriorly to form the *patellar articular surface.* Posteriorly, they are separated by a deep **intercondylar fossa.** This is separated from the popliteal surface by the *intercondylar line.* On the sides of the condyles are roughened **medial** and **lateral epicondyles.** The medial epicondyle is surmounted by a projection, the **adductor tubercle.** The femoral condyles are further discussed in connection with movements at the knee.

The femur gives attachment to many muscles, as indicated in Figures 17-13 and 17-14.

**Ossification** in the body of the femur begins in the seventh or eighth fetal week; a center for the distal end usually appears before birth, one for the head during the first postnatal year, and one for the greater trochanter, and often one for the lesser trochanter, during the fifth to tenth years. The head and the trochanters join the body at about the 16th to 18th year (Fig. 17-15), and the lower end joins a year or so later (see Fig. 18-11).

**Angles of Femur.** The angle between the neck and body of the femur (the *angle of inclination*) averages about 126°. Marked decrease in this angle, that is, a more transverse direction of the neck, can result from weight bearing on a bone that is not capable of standing it. Such an abnormal decrease is known as *coxa vara* (bent hip). It is usually due to rickets, but the same effect can result from disease of the bone or a fracture of the neck that is improperly set.

The head and neck of the femur project also anteriorly, making an angle of about 14° with a transverse line through the centers of the femoral condyles. Finally, the body is not normally exactly vertical, for the two femora converge toward the knees during normal standing. This oblique position partially overcomes the effect of the opposite obliquity of the head and neck and better centers the weight.

**Blood Supply.** The blood supply of the femur consists of multiple arteries entering each end and one or two **nutrient arteries** entering the body. The nutrient arteries are usually derived from upper perforating branches of the *profunda femoris.* They enter the femur close to the linea aspera and run up and down in the marrow cavity. Although they

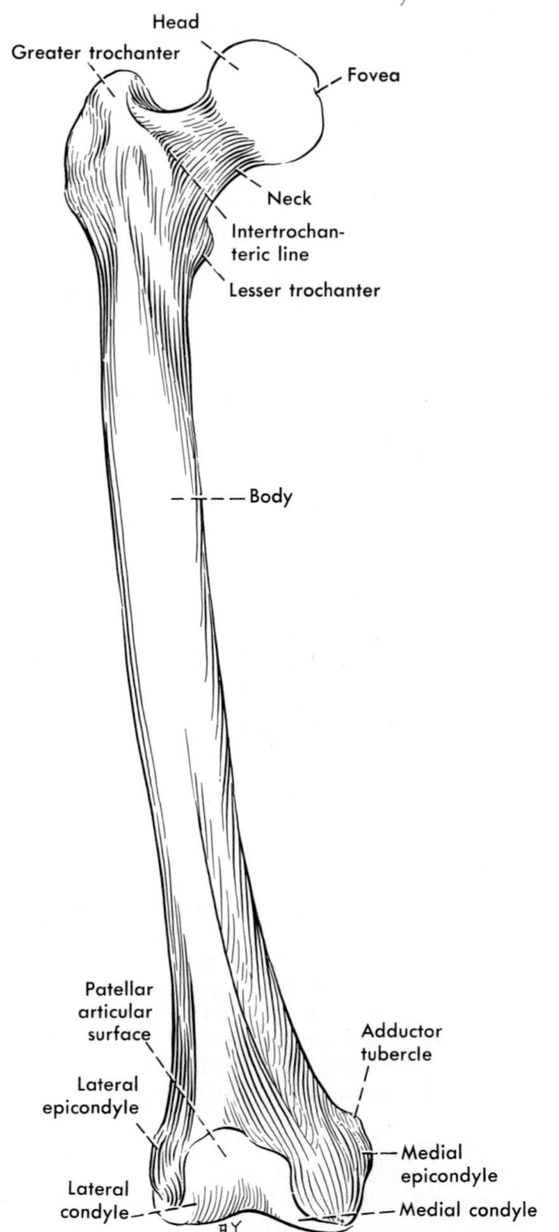


**FIG. 17-11.** Anterior view of the femur.

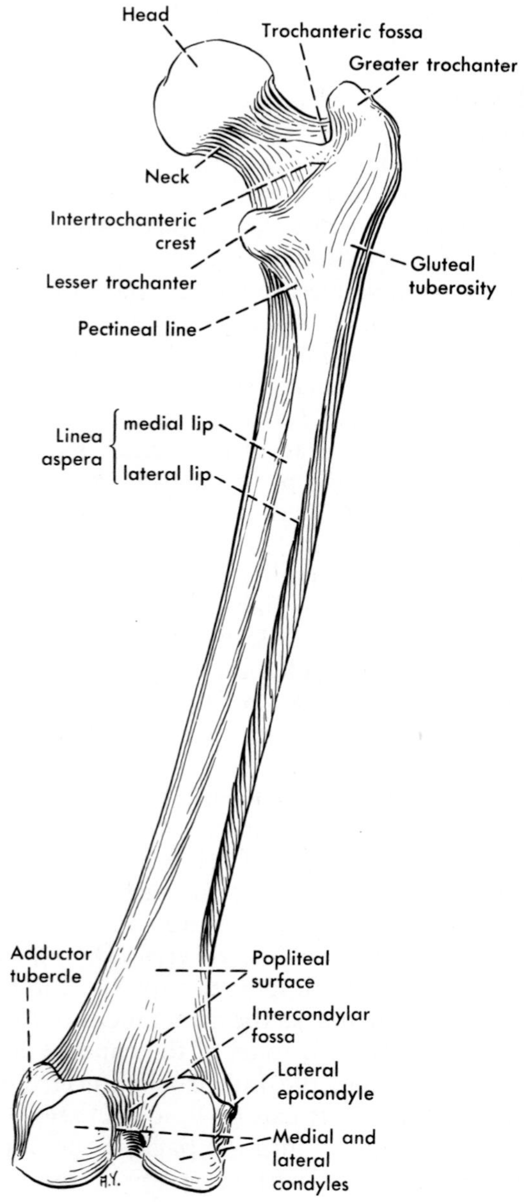


**FIG. 17-12.** Posterior view of the femur.

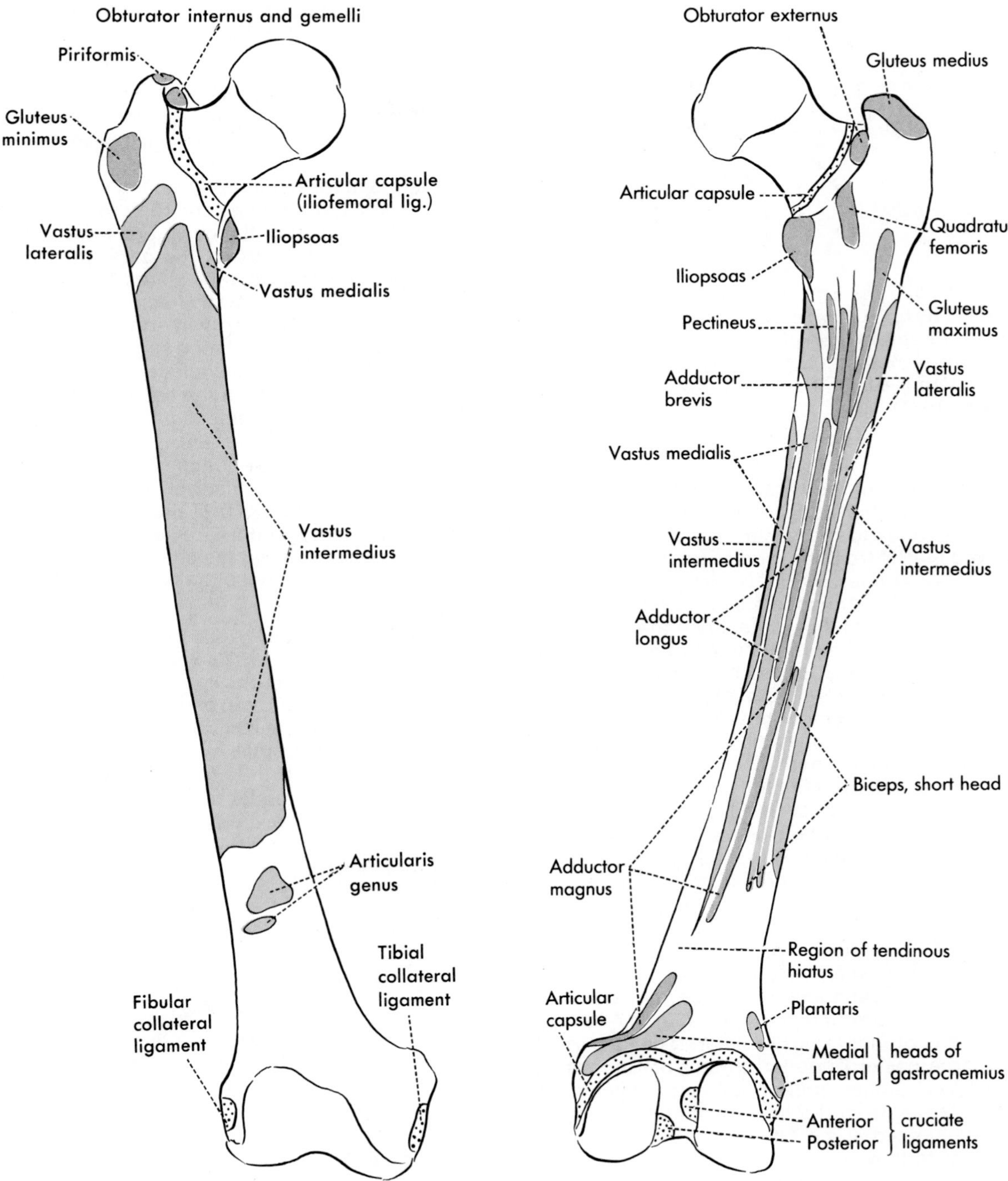


**FIG. 17-13.** Muscular attachments on the anterior surface of the femur.

**FIG. 17-14.** Muscular attachments on the posterior surface of the femur.

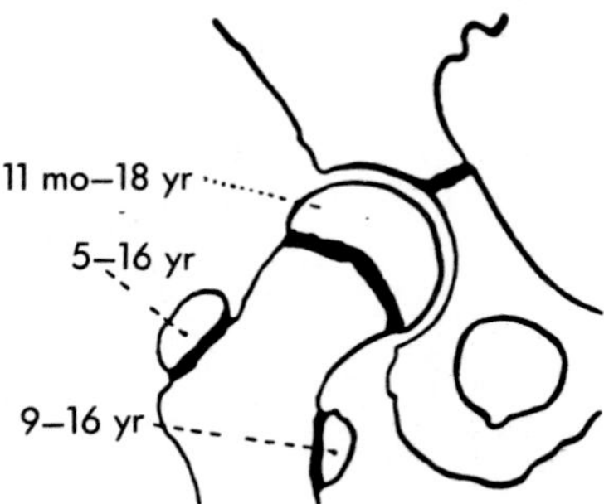


**FIG. 17-15.** Ossification of the upper end of the femur. The first of two numbers indicates the age at which the epiphyseal center can be expected to appear, the second that at which it can be expected to fuse. (Camp JD, Cilley EIL: Am J Roentgenol 26:905, 1931)

anastomose with the vessels at the two ends, the latter are the chief supply here.

The arteries to the upper end of the femur are derived mostly from the *medial and lateral femoral circumflex arteries.* These vessels give numerous branches to the trochanters, which also receive blood from the inferior gluteal and the first perforating branch of the profunda. An **artery of the ligament of the head,** derived from the obturator or the medial femoral circumflex, enters the head through the ligament of the head and supplies a variable amount of bone adjacent to the fovea; otherwise, the head and neck are supplied by branches of the two circumflexes. These branches approach the bone at the level of attachment of the capsule of the hip joint, pierce the capsule, and then run upward along the neck deep to the synovial membrane that is reflected upward around the neck to the cartilage of the head. Because of their relation to the capsule, and because they may raise folds or retinacula of the synovial membrane as they run upward, they have been called both **capsular** and **retinacular arteries,** among other terms. They surround the neck (Fig. 17-16), but three chief groups, a posteroinferior and a particularly large posterosuperior group from the medial femoral circumflex, and an anterior, sometimes a single vessel, from the lateral femoral circumflex, are emphasized in most descriptions.

Because they run upward along the neck, and because the fibrous capsule attaches so low on the neck (Figs. 17-13 and 17-14), the arteries to the neck and head are likely to be interrupted by an intracapsular fracture of the neck. This may leave the head and neck with no blood supply other than that from the artery of the ligament of the head and may result in necrosis of most of the upper fragment and, therefore, failure of the fracture to heal.

Numerous small vessels derived from the arteries forming the anastomoses around the knee joint (see Fig. 18-41) enter the nonarticular surfaces of the condyles.

**Anomalies and Fracture.** Congenital anomalies of the femur are not common. Occasional cases of absence or partial absence have been reported.

Fracture of the femur as a result of violence may occur at any level. A fairly common fracture in elderly people is that of the neck. Fractures outside the capsule, at the base of the neck, present no problems of healing from the standpoint of vascularization, but intracapsular fractures almost always do since they regularly interfere, to a greater or lesser degree, with the blood supply to the upper fragment. Because of the angle through which the weight is transmitted from the head to the body of the femur, accurate alignment of the parts is particularly important. Unless there is impaction (jamming together), the lower fragment tends to be displaced upward by muscular pull. If healing occurs in this position, there will be a shortened functional length of the femur and a varus deformity. In many elderly people, however, the chief problem in fracture of the neck is to keep the patient alive, for the

**FIG. 17-16.** Diagram of the chief arteries to the upper end of the femur. The heavy black lines represent the positions of the epiphyseal cartilages of the femoral head and greater trochanter. **1** is the medial femoral circumflex artery, and its branches **a, b,** and **c** are posteroinferior, posterosuperior, and posterior arteries to the head and neck; **2** is the lateral femoral circumflex artery, and **d** is an anterior branch to the neck; **3** is the artery of the ligament of the head. (After Nussbaum and Funck-Brentano. Mathieu P: Lesions traumatiques du hanche: Fractures du col du femur. In Ombrédanne L, Mathieu P (eds): Traité de Chirurgie Orthopédique, vol 4. Paris, Masson et Cie, 1937)

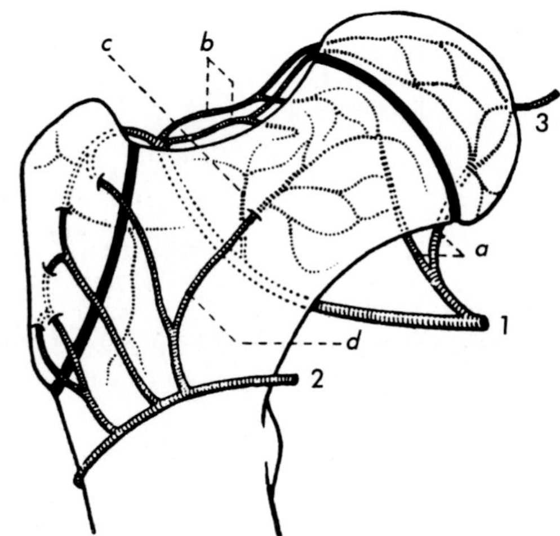

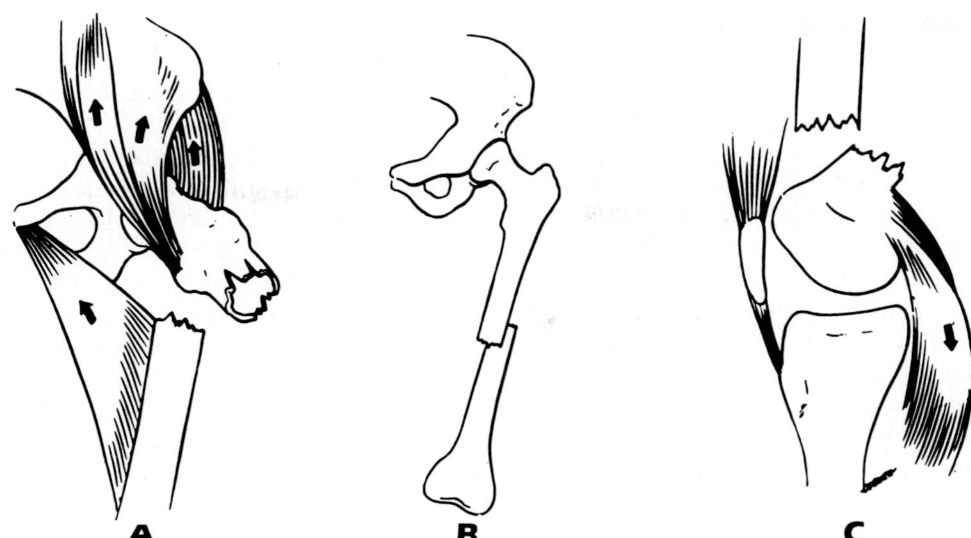


**FIG. 17-17.** Typical displacement in fractures of the upper, middle, and lower parts of the shaft of the femur. In **A**, the upper fragment is flexed, abducted, and laterally rotated; the lower fragment is adducted; and the limb is shortened. In **B**, overriding of the fragments is shown. In **C**, a lateral view, the lower fragment is flexed. (Ivins JC: Modern Medicine Annual, p 222. New York, Harcourt Brace Jovanovich, 1951)

enforced inactivity superimposed upon the already lowered vital capacities of many such persons frequently leads to fatal pneumonia or pulmonary embolism.

Fractures a little below the trochanters are typically accompanied by much displacement of the upper fragment through the pull of the numerous muscles attaching here and by great shortening of the limb as the result of the upward pull of muscles attaching below the line of fracture (Fig. 17-17). Fractures a little above the condyles are typically accompanied by a posterior displacement of the lower fragment by a muscle of the calf (gastrocnemius) that takes its origin here. This displacement is particularly dangerous because it may bring a sharp edge of bone into contact with the large popliteal artery that lies close to the popliteal surface of the bone.

## BUTTOCK

The **superficial fascia** (*tela subcutanea*) of the buttock (*natis* or *clunis*) is a rather thick pad that contains a considerable amount of fat. It is continuous with the superficial fascia of adjoining regions—the back and abdomen, the perineum, and the thigh.

Embedded in the superficial fascia are small vascular twigs, as well as **cutaneous nerves** (Fig. 17-18), that can be most easily found as they come through the deep fascia.

**FIG. 17-18.** Cutaneous nerves of the buttock and posterior aspect of the thigh. (Corning HK: Lehrbuch der topographischen Anatomie für Studierende und Ärtze. Munich, JF Bergmann, 1923)

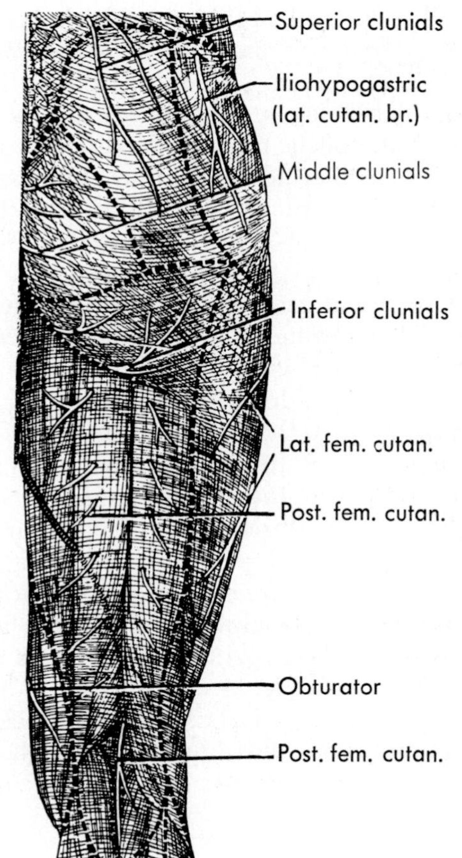

Laterally are the *lateral cutaneous branches of the 12th thoracic* and *iliohypogastric nerves.* These, the branch of the iliohypogastric a little behind that of the 12th thoracic, penetrate the musculature and fascia of the lower part of the lateral abdominal wall and descend over the crest of the ilium toward the region of the greater trochanter. The chief supply of the buttock is through three sets of **clunial nerves:** superior, middle, and inferior. The *superior clunial nerves,* the lateral branches of the dorsal rami of the upper three lumbar nerves, appear close together a little above the iliac crest and descend over the buttock. The *middle clunial nerves,* the lateral branches of the dorsal rami of the first three sacral nerves, extend laterally. The *inferior clunial nerves* curve around the lower border of the gluteus maximus and are derived from the posterior femoral cutaneous nerve.

The **deep fascia** of the gluteal region is attached above to the crest of the ilium and to the sacrum. Over the gluteus maximus, the large superficial muscle of the buttock, it is a thin layer firmly bound both to the muscle and to the superficial fascia. At the lower border of the gluteus maximus, it is continuous with the deep fascia of the thigh (fascia lata), but a thicker layer also curves around the lower border of the muscle and ascends on its deep surface to attach to the underlying sacrotuberous ligament and the ilium. This layer contains the nerves and vessels to the muscle.

Lateral to the gluteus maximus, the two layers of fascia around the muscle come together and continue as a single layer over the gluteus medius, which takes origin in part from this. In this location, also, the fascia is fastened above to the iliac crest, and below the gluteus medius, it blends with the fascia lata. The portion over the gluteus medius constitutes the middle thinner origin of the particularly heavy lateral part of the fascia lata known as the *iliotibial tract.* At the anterolateral border of the gluteus medius, the fascia becomes continuous with the fascia lata once again. Thus, whether it is traced laterally or inferiorly, the deep fascia of the buttock becomes continuous with that of the thigh.

## MUSCLES AND ASSOCIATED STRUCTURES

The large **gluteus maximus** covers most of the important structures in the buttock (Fig. 17-19), but a part of the gluteus medius appears above and lateral to it, and the tensor fasciae latae lies anterior to the gluteus medius.

The gluteus maximus arises (Fig. 17-20) from the outer surface of the ilium behind the posterior gluteal line, from the dorsal surface of the sacrum and coccyx, and from the adjacent sacrotuberous ligament. The largest part of its insertion is into the *iliotibial tract;* approximately the entire upper half of the muscle and the posterior (superficial) half of the remainder have this insertion, whereas only the anterior or deep half of the lower part inserts into the gluteal tuberosity of the femur. There is usually a subcutaneous bursa over the outer surface of this muscle as it passes across the greater trochanter, and a large trochanteric bursa of the gluteus maximus lies between the muscle's tendon and the greater trochanter.

The **inferior gluteal nerve** and **vessels** enter the buttock below the piriformis muscle, penetrate the fascia on the deep surface of the muscle close to its origin, and run laterally between fascia and muscle for some distance before disappearing into the muscle.

Of the structures deep to the gluteus maximus (Fig. 17-19), the **piriformis muscle** occupies a key position. It arises from the pelvic surface of the sacrum, passes through and largely fills the greater sciatic foramen, and inserts into the medial side of the upper end of the greater trochanter (Fig. 17-21). Its nerve supply enters its anterior or pelvic surface inside the pelvis.

Above the piriformis is the gluteus medius, and between the two muscles, the **superior gluteal nerve** and **vessels** enter the buttock. The artery sends its superficial branch into the gluteus maximus, but otherwise the nerve and vessels are not visible until the gluteus medius is reflected, for they turn laterally to run between the gluteus medius and minimus. The other nerves and vessels that enter the buttock from the pelvis, including the in-

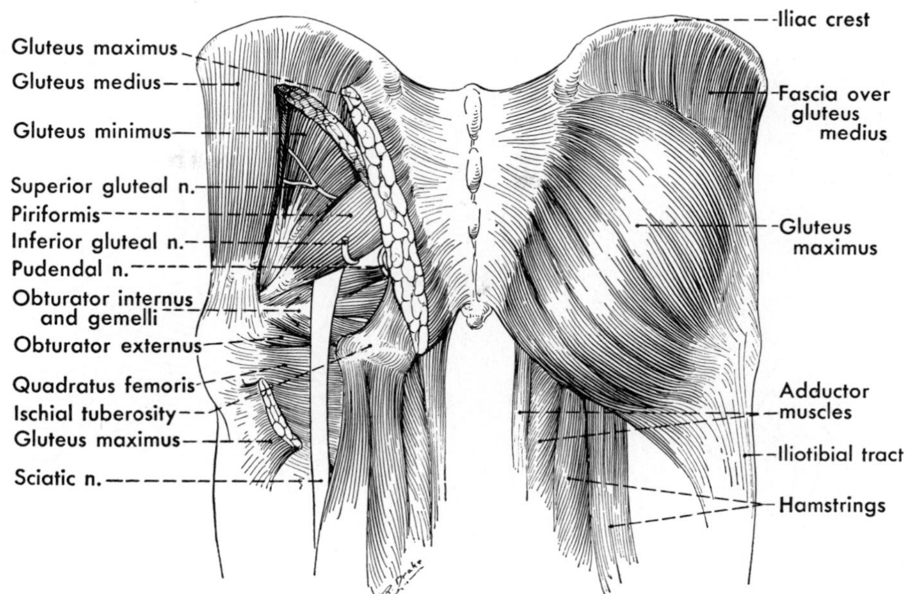


**FIG. 17-19.** Muscles of the buttock. The quadratus femoris has been slightly displaced to show a little of the obturator externus.

**FIG. 17-20.** Attachments of the gluteus maximus.

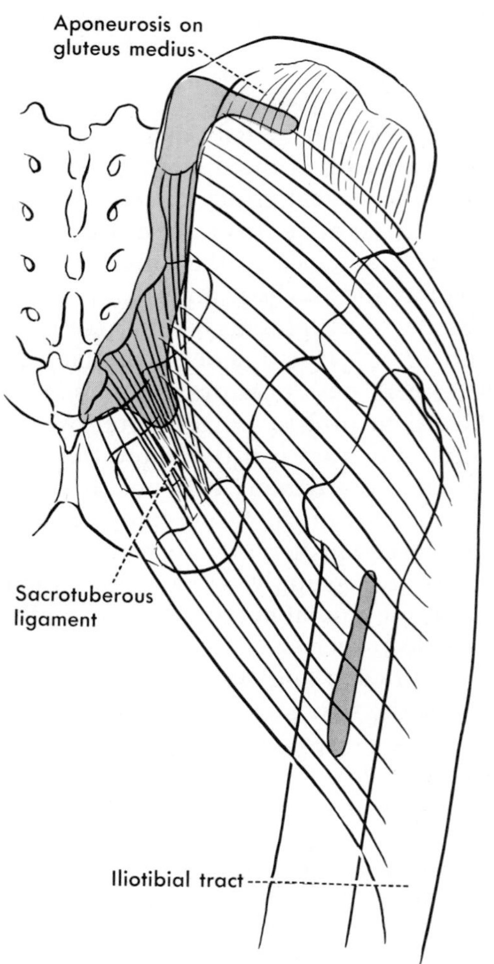


ferior gluteal and internal pudendal nerves and vessels, the large sciatic nerve, and several smaller nerves, normally emerge below the piriformis muscle (Figs. 17-23 and 17-24). The muscles of the buttock lying below the piriformis lie also in front of the sciatic nerve.

The **gluteus medius** arises (Fig. 17-22) from the wing of the ilium between the iliac crest, the anterior gluteal line, and the posterior gluteal line, and also from the deep fascia covering the muscle. The fibers of the gluteus medius converge so that it forms a curved but essentially fan-shaped muscle that inserts into the large posterior part of the upper surface of the greater trochanter.

Deep to the gluteus medius are the *superior gluteal nerve* and the deep branch of the *superior gluteal artery,* running between the gluteus medius and minimus muscles and supplying both. The nerve does not end in the glutei, however, but is continued beyond their anterior edges into the deep surface of the tensor fasciae latae.

The **gluteus minimus** is also fan shaped. It arises from the ilium between the anterior and inferior gluteal lines and inserts in front of the gluteus medius on the upper and anterior surface of the greater trochanter (Fig. 17-21).

The **tensor fasciae latae** arises so far anteri-

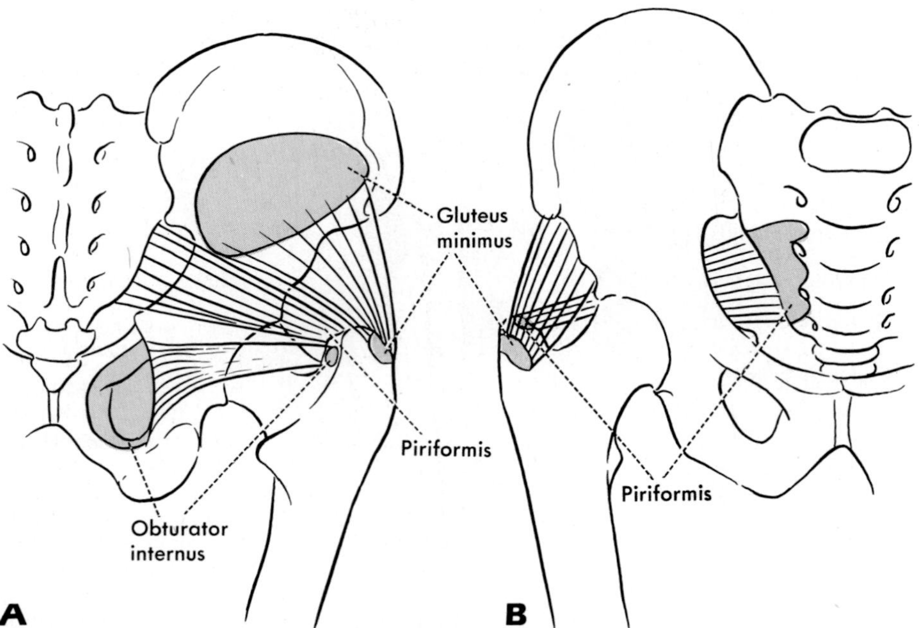


**FIG. 17-21.** Attachments of the piriformis, the gluteus minimus, and the obturator internus. **A,** posterior view; **B,** anterior view.

**FIG. 17-22.** Attachments of the gluteus medius, the gemelli, and the quadratus femoris.

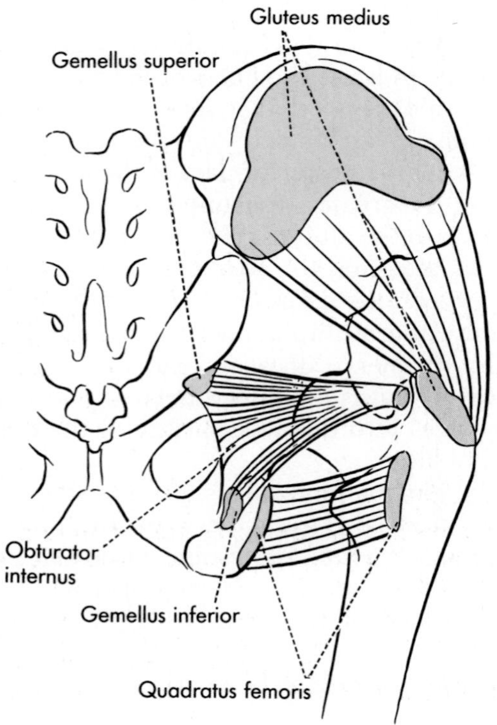


orly that it is most easily dissected with the anterior muscles of the thigh. However, it belongs to the gluteal group, as indicated by its innervation. It arises from the anterior part of the iliac crest close to the anterior superior iliac spine and lies just anterior to the anterior border of the gluteus medius (see Figs. 17-32 and 17-33). It is a short, straplike muscle whose fibers pass downward and somewhat backward to insert into the anterior part of the iliotibial tract a little below the level of the greater trochanter. It is enclosed within the upper part of the fascia lata, and its tendon of insertion helps to form the iliotibial tract. Its nerve supply is the terminal part of the superior gluteal nerve, which is continued into its deep surface.

The following four muscles—the obturator internus, the two gemelli, and the quadratus femoris—are frequently grouped with the piriformis as the short rotators of the buttock, for the principal action of all is lateral rotation.

The **obturator internus** arises on the internal surface of the coxal bone, from the obturator membrane and the bone around that, and converges to a tendon that passes through the lesser sciatic foramen, turning sharply laterally over the smooth bony surface here to insert into the medial surface of the greater trochanter above and in front of the trochanteric fossa (Figs. 17-21 and 17-22). The tendon is separated from the ischium by a large bursa and may be entirely hidden posteriorly by the two gemelli muscles. The gemelli are short muscles that insert into the tendon of the obturator internus. The **superior gemellus** arises from the ischial spine, and the **inferior gemellus** arises from the ischical tuberosity.

Below the obturator internus and the gemelli is the **quadratus femoris,** a short, broad muscle that arises from the ischial tuberosity and inserts on a line extending vertically downward from about the middle of the intertrochanteric crest.

The *sciatic* and *posterior femoral cutaneous nerves* run downward behind the four short rotators (Fig. 17-23). The *nerve to the superior gemellus and the obturator internus* runs medially from in front of these nerves, crosses behind the superior gemellus, and enters the lesser sciatic foramen on the surface of the obturator internus. More medially, the *pudendal nerve* and *internal pudendal vessels* leave the greater sciatic foramen, cross the ischial spine or the sacrospinous ligament, and also enter the lesser sciatic foramen on the surface of the obturator internus. The *nerve to the inferior gemellus and the quadratus femoris* starts downward in front of the sciatic nerve, but passes deep to the gemelli and the obturator internus (Fig. 17-24) and enters the muscles it supplies on their deep surfaces. It also gives off a twig to the posterior aspect of the hip joint. If the quadratus femoris is reflected to follow the nerve, the insertion end of the obturator externus, a member of the anteromedial group of thigh muscles, will be seen passing upward below and behind the capsule of the hip joint to insert into the trochanteric fossa.

Below the quadratus femoris is the uppermost part of the adductor magnus. Branches of the medial femoral circumflex vessels usually appear both above and below the quadratus.

**Variations, Innervations, and Actions.** Important variations in the musculature of the buttock are uncommon. The gemelli typically vary considerably in size, and one of them or the quadratus femoris may be absent. The piriformis muscle may be divided into upper and lower parts by the passage through it of a part or the whole of the sciatic nerve.

**Innervation** of the musculature of the buttock is by five or more nerves, all of them small in relation to the size of the muscle (the gluteus maximus, for instance, probably has one of the largest motor units of any muscle of the body). The **gluteus maximus** is supplied by fibers from the last lumbar and first two sacral nerves through the *inferior gluteal nerve;* the **piriformis** is supplied by twigs on its pelvic surface from the *first two sacral nerves,* or either of these, while these nerves lie on the muscle; the **gluteus medius, gluteus minimus,** and **tensor fasciae latae** are supplied with fibers from the 4th and 5th lumbar and 1st sacral nerves through the *superior gluteal nerve;* the superior gemellus and the **obturator internus** are supplied with fibers from the 5th lumbar and the first two sacrals, or the first three sacral nerves, through the nerve that bears their names; and the inferior gemellus and **quadratus femoris** are supplied, through the nerve that bears their names, with fibers of slightly higher origin, usually from the 4th lumbar through the 1st sacral.

The **action** of the **gluteus maximus** is primarily extension of the hip joint. It is called into play only when the movement must be relatively powerful, as in standing up from a sitting position or in jumping and running. It is a good lateral rotator and adductor. However, when the hip is flexed at a right angle, the lower part of the muscle apparently helps slightly in abduction.

The chief action of both **gluteus medius** and **minimus** is abduction. They not only will abduct the free limb, but, more important, they keep the contralateral side of the pelvis from sagging markedly when the weight is put upon one limb. They are, therefore, particularly important muscles in walking. Both have been said to flex and medially rotate, extend and laterally rotate, but an electromyographic study has apparently indicated that only the minimus is active in flexion and medial rotation, and only the medius in the opposing actions.

The **tensor fasciae latae** is primarily a flexor at the hip, and although it inwardly rotates as it flexes, it apparently is not commonly used for inward rotation alone. It lies too far in front of the joint to be an effective abductor.

The remaining muscles—the piriformis, the obturator internus, the two gemelli, and the quadratus femoris—are primarily lateral rotators of the femur. The piriformis apparently slightly extends and ab-

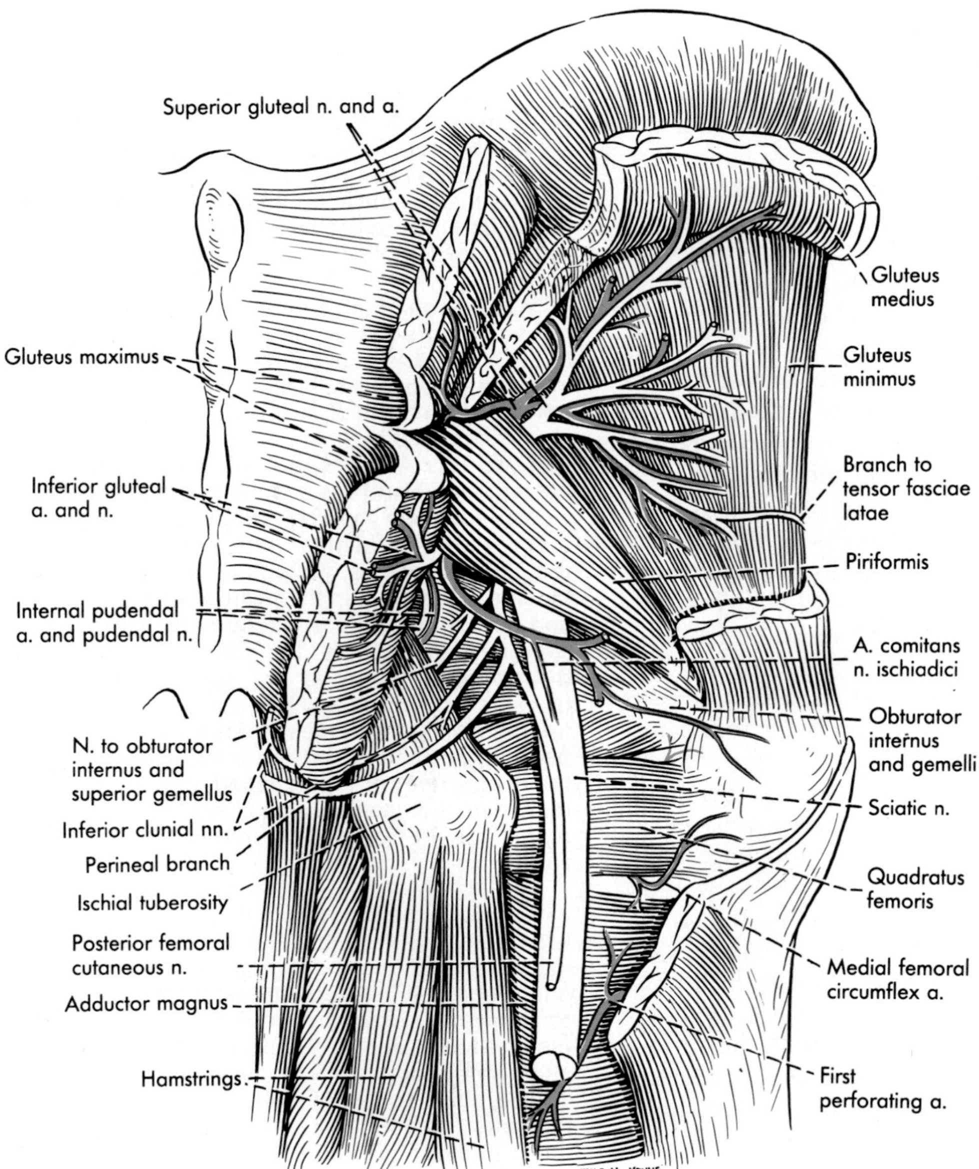


**FIG. 17-23.** Nerves and vessels of the buttock.

ducts as it rotates. The quadratus femoris is usually said to assist slightly in adduction.

## NERVES AND VESSELS

The nerves and blood vessels of the buttock (Figs. 17-23 and 17-24) can be briefly summarized. Those whose distribution is primarily confined to the buttock are the superior and inferior gluteal nerves and vessels and the two small nerves that go to the superior gemellus and obturator internus and to the inferior gemellus and quadratus femoris, respectively. Those that primarily pass through the buttock to another distribution are the internal pudendal vessels and the pudendal, posterior femoral cutaneous, and sciatic nerves.

The **superior gluteal nerve** usually contains fibers from the 4th and 5th lumbar and 1st sacral nerves. The **superior gluteal artery** is a branch of the internal iliac artery, the large artery in the pelvis. Artery and nerve course

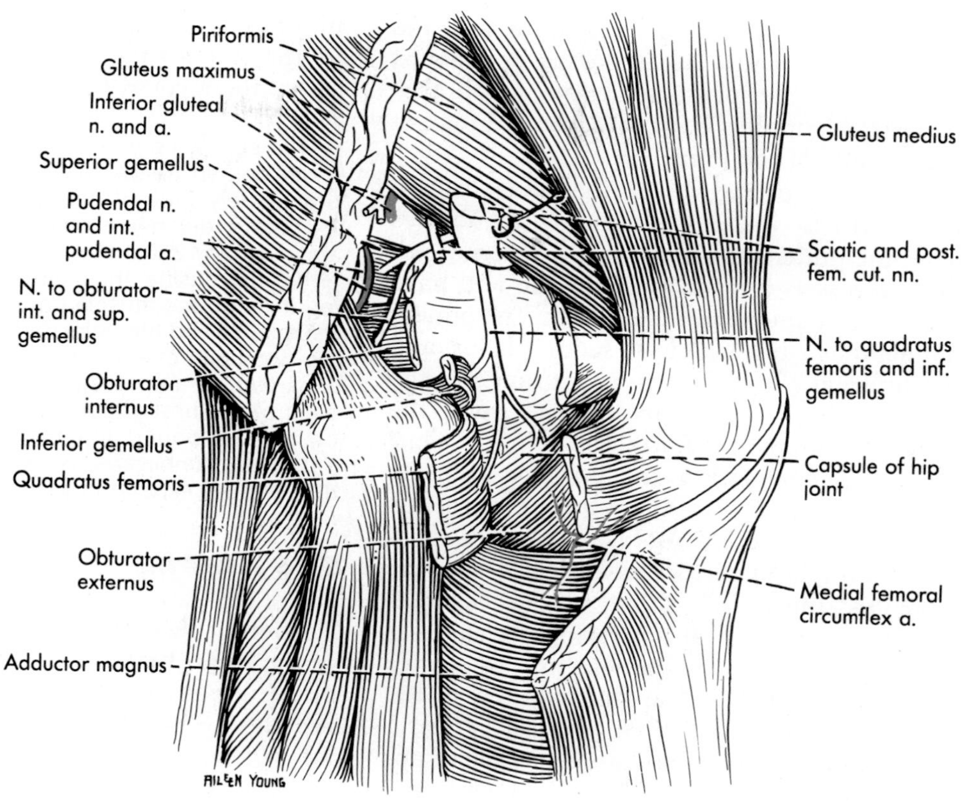


**FIG. 17-24.** Some deeper nerves of the buttock.

together above the piriformis into the buttock, lying between the piriformis below and the gluteus medius and minimus above. The artery sends a superficial branch into the overlying gluteus maximus, and its deep branch and the entire nerve turn upward and laterally between the gluteus medius and minimus. The artery divides into superior and inferior branches. The nerve continues beyond the glutei into the deep surface of the tensor fasciae latae.

The **inferior gluteal nerve** and **vessels** enter the buttock below the piriformis muscle. The nerve usually contains fibers from the last lumbar and first two sacral nerves, and the artery is a branch of the internal iliac artery. Both nerve and artery supply the gluteus maximus. The artery gives off other muscular twigs, sends a named branch, the **arteria comitans nervi ischiadici,** downward on the surface of or into the sciatic nerve, and typically sends an unnamed branch laterally and downward to anastomose with other vessels (medial and lateral femoral circumflexes and the first perforating branch of the profunda) to form an anastomosis that is known by clinicians as the **cruciate** or **crucial anastomosis.**

The remaining two nerves to muscles of the buttock arise anteriorly from the sacral plexus at the upper end of the sciatic nerve. The **nerve to the obturator internus** usually receives fibers from the 5th lumbar and first two sacral nerves. It passes below the piriformis muscle, crosses the posterior surface of the superior gemellus, giving off a twig into that, and then turns medially below the ischial spine to follow the free surface of the obturator internus muscle into the lesser sciatic foramen. The **nerve to the quadratus femoris** is said to receive fibers from the 4th and 5th lumbar and 1st sacral nerves. It also emerges below the piriformis muscle, but runs deep to the superior gemellus and the obturator internus. As it passes downward deep to the inferior gemellus and quadratus femoris, it supplies both muscles and part of the capsule of the joint.

Of the nerves and vessels that pass through the buttock without supplying structures here, the pudendal nerve and the internal pudendal vessels are the most medial and have the shortest course in the buttock. The **pudendal nerve** is a branch of the sacral plexus, and the **internal pudendal artery** a branch of the internal iliac. Nerve, artery, and vein, the nerve lying most medially, the vein most laterally, appear within the buttock for only about the breadth of the superior gemellus muscle, entering the buttock between the piriformis and the superior gemellus, crossing the gemellus, the ischial spine, or the sacrospinous ligament, and disappearing under cover of the sacrotuberous ligament into the lesser sciatic foramen. They are distributed to the perineum.

The **posterior femoral cutaneous nerve** appears at the lower border of the piriformis medial to the sciatic nerve, and while still under cover of the gluteus maximus, it gives off its uppermost branches; these are one or more perineal branches that run superficially around the medial side of the thigh to reach the perineum (area between the thighs) and several inferior clunial nerves that round the lower border of the gluteus maximus to reach skin over part of the buttock. The major part of the posterior femoral cutaneous nerve then runs downward and slightly laterally to lie deep to the fascia lata in almost the exact midline of the thigh.

The **sciatic nerve** is the largest nerve in the body, but actually consists of two nerves, *tibial* and *common peroneal*, that are closely bound together by connective tissue. Normally, it emerges below the piriformis, and it always runs across the posterior surfaces of the obturator internus and gemelli and the quadratus femoris. As it runs down the thigh, it disappears deep to (anterior to) the thigh muscles arising from the ischial tuberosity (hamstring muscles). It gives off no fibers to muscles of the buttock, but its upper branches to the hamstring muscles may arise at or above the level of the ischial tuberosity.

The sciatic nerve has been reported to have its usual course below the piriformis muscle in approximately 85% of cases; in probably less than 1%, the entire nerve comes through the muscle; and in some 12% to 15%, the nerve is divided by the piriformis muscle, the common peroneal part coming through the muscle (most frequently) or above the muscle and the tibial part running below. When the two components of the sciatic nerve are thus separated at their origin, they may simply parallel each other throughout the remainder of the thigh or may be bound together below the piriformis muscle by connective tissue, just as they are when they are not separated at their origins.

## THE THIGH

The femur has already been described. Study of the thigh requires only limited knowledge of the upper end of the tibia and fibula and of the knee joint, parts described in more detail later (Chap. 18).

The **knee joint** is formed between the femoral condyles and the upper end of the tibia, with the *patella* (kneecap) entering into an articulation with the femoral condyles. Although movement at the knee joint is largely limited to flexion and extension, rotation of the tibia upon the femur, or of the femur upon the tibia, occurs. Rotation is freest when the knee is flexed, since some of the ligaments that check rotation are then relaxed.

The broad upper end of the **tibia** is formed mostly by the medial and lateral *tibial condyles,* which present superiorly somewhat oval and almost flat articular surfaces for the femoral condyles. The raised, roughened area on the anterior border of the bone below the condyles is the *tibial tuberosity,* upon which the quadriceps femoris muscle (the large muscle on the front of the thigh) inserts. On the anterior surface of the lateral condyle, there is usually a raised area marking the attachment of the strongest part of the iliotibial tract, and on the posterolateral part of its curved inferior surface is a facet for articulation with the head of the fibula. The only part of the **fibula** concerned with muscles of the thigh is its upper end, the head.

## SUPERFICIAL STRUCTURES

### Patella

The patella is a sesamoid bone, the largest one in the body, developed in the tendon of the quadriceps muscle. It is somewhat triangular, with its base above and its apex below. It interrupts the continuity of the quadriceps tendon so that the lower part of the tendon, extending from the apex of the patella to the tibia, is known as the **patellar ligament.**

The front and sides of the patella are rough for the attachment of tendon fibers and perforated by numerous vascular foramina. The posterior surface of the apex is also rough, but most of the patella's posterior or articular surface is smooth and articulates with the femoral condyles (Fig. 17-25).

The patella affords additional leverage to the quadriceps muscle during the last part of extension at the knee by holding the tendon forward. Removal of it, sometimes necessary when it is particularly badly shattered, is generally reported to result in some weakness and disability of the knee.

### Surface Anatomy

There is relatively little surface anatomy to be observed in the thigh. Anteriorly and above, the bony landmarks are the *anterior superior iliac spine* and the *pubic tubercle;* the flexion crease extending between these two points corresponds approximately to the position of the inguinal ligament and marks the junction of the thigh and the abdominal wall. Posteriorly, the *gluteal fold* delimits the buttock from the thigh. The *ischial tuberosity* is easily palpable. At the knee, the patella, the lower end of the femur, and the upper ends of the tibia and fibula are also palpable. Some of the structures observable at the knee are shown in Figures 17-26 and 17-27.

### Superficial Fascia, Nerves, and Vessels

The **superficial fascia** of the thigh may contain a considerable amount of fat and in its upper part is often clearly divisible into two layers,

**FIG. 17-25.** Roentgenograms of the almost extended and the flexed knee. Note the positions of the patella. (Courtesy of Dr. D. G. Pugh)

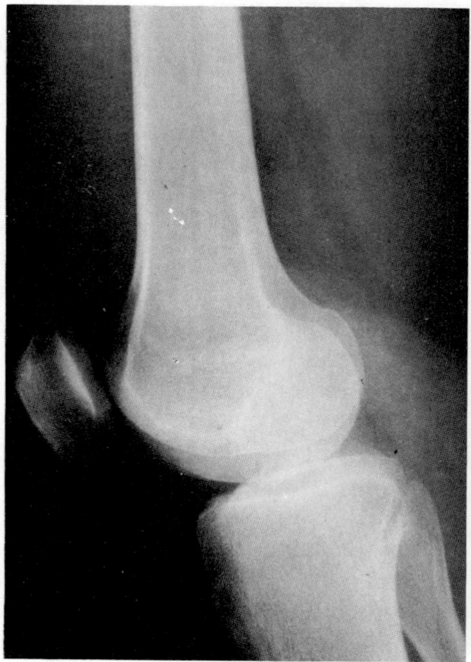

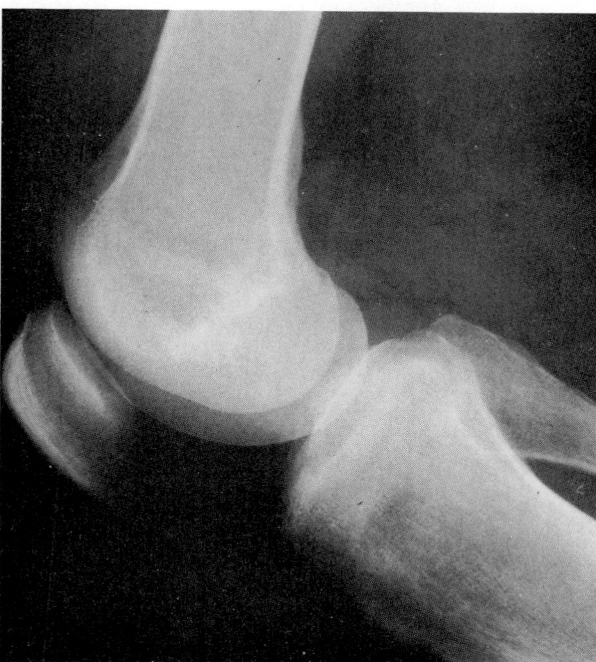

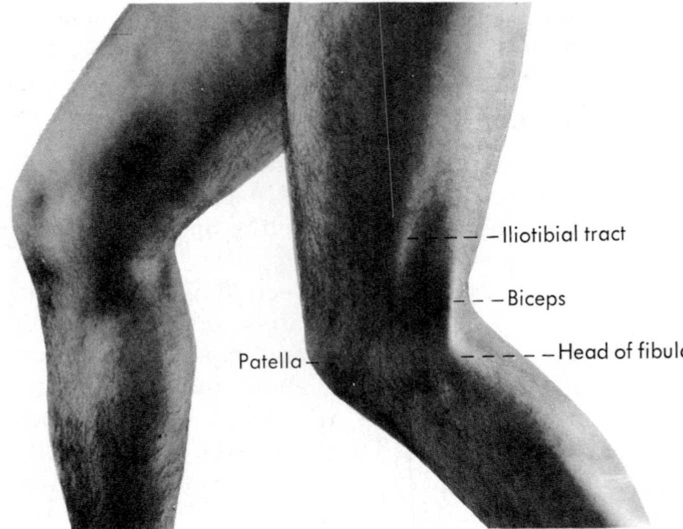


**FIG. 17-26.** Lateral view of structures about the knee.

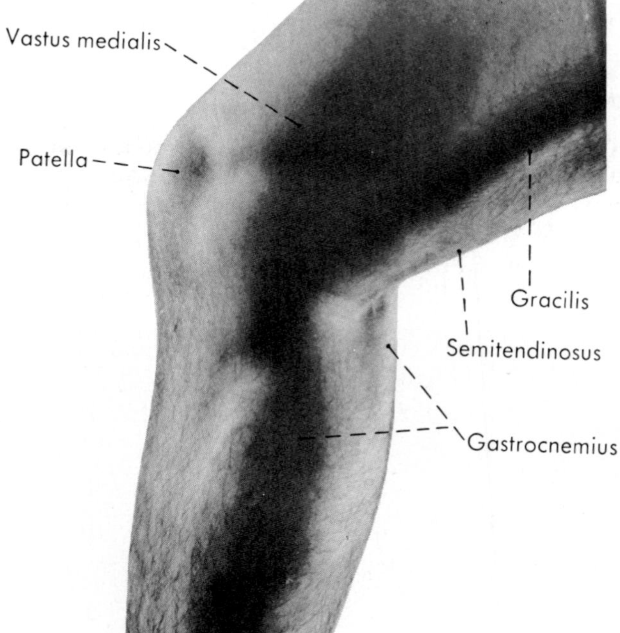


**FIG. 17-27.** Medial view of structures about the knee.

an outer fatty layer and an inner more fibrous one, between which lie the upper end of the great saphenous vein, the superficial inguinal lymph nodes, and the small vessels of this region. The superficial fascia of the thigh is continuous with that of the buttock and the lower part of the abdominal wall. At the knee it loses its fat and blends with the deep fascia.

The chief anterior **cutaneous nerves** are the anterior femoral cutaneous branches of the femoral and the lateral femoral cutaneous nerve (Fig. 17-28), but these are supplemented by the genitofemoral and obturator nerves. The main branches of the **anterior femoral cutaneous nerves** pierce the deep fascia at various levels and supply anterior and anteromedial skin as far as the knee. The **lateral femoral cutaneous** usually appears a little below the anterior superior iliac spine and supplies anterolateral skin as far as the knee. The femoral branch of the **genitofemoral** enters the thigh on the femoral artery and supplies a variable area of skin below the inguinal ligament. The genital branch of the genitofemoral nerve, and the **ilioinguinal nerve,** accompany the spermatic cord or, in the female, the round ligament and are distributed primarily to the scrotum or labia majora rather than to the thigh. The small and variable **cutaneous branch of the obturator** supplies skin on the medial side of the thigh.

The **posterior femoral cutaneous nerve,** after leaving the shelter of the gluteus maximus, descends almost straight down the posterior midline of the thigh immediately deep to the deep fascia (Fig. 17-18). It sends branches through the fascia to skin of the posterior side of the thigh, but the main stem usually does not penetrate the fascia until about the level of the popliteal fossa. Here the nerve does become subcutaneous and typically runs some distance down the posterior aspect of the calf.

The **great saphenous vein** passes upward along the medial side of the thigh to an anteromedial position a little below the inguinal ligament (see Fig. 16-13). It and its tributaries lie in part in, and in part deep to, the superficial fascia. The vein ends a little below the inguinal ligament by turning posteriorly through a defect (the *saphenous hiatus*) in the deep fascia of the thigh and entering the femoral vein.

The tributaries of the great saphenous vein in the thigh are variable, but reach it from both the lateral and medial sides. Small **superficial epigastric, superficial iliac circumflex,** and **external pudendal** veins enter the upper end of the great saphenous, or the femoral vein directly, in various combinations. There may be a large lateral tributary, usually called the **lateral accessory saphenous,** or, less often, a medial accessory saphenous that connects with or represents a small saphenous vein that does not end in the popliteal vein. Perforating or communicating veins that unite the great saphenous to the deep veins of the thigh are few.

Several small **arteries,** branches of the femoral, emerge through the saphenous hiatus or penetrate the fascia lata close to it. They arise in common or separately and are recognizable only through their distributions. The **superficial iliac circumflex artery** runs upward and laterally toward the anterior superior iliac spine; the **superficial epigastric** runs upward and medially toward the umbilicus; and the **external pudendal** runs medially toward the scrotum. (There may be a second external pudendal running deep to the fascia until it is close to the scrotum.)

The superficial **lymphatics** of the thigh converge largely on the upper medial part of the thigh, just below the inguinal ligament, to enter the **superficial inguinal lymph nodes** (Fig. 17-29; see Fig. 16-15). These nodes also receive the lymphatics from the lower part of the abdominal wall, from the buttock (which come

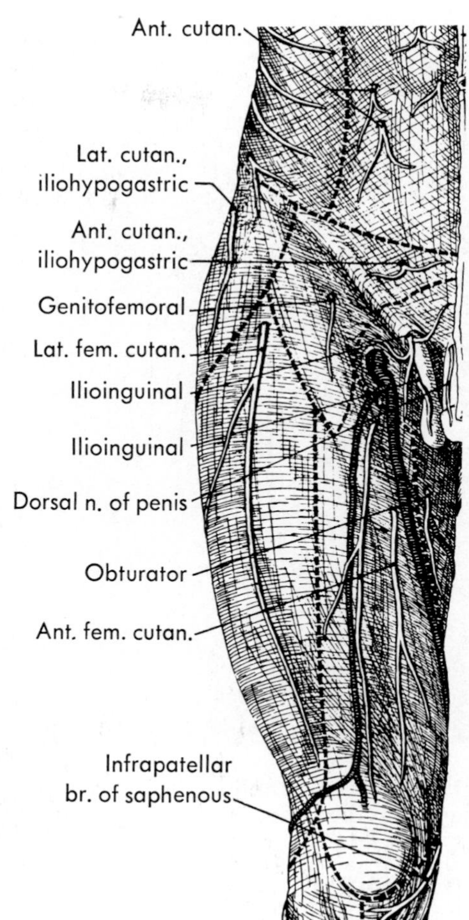


**FIG. 17-28.** Cutaneous nerves of the thigh, anterior view. (Corning HK: Lehrbuch der topographischen Anatomie für Studierende and Ärtze. Munich, JF Bergmann, 1923)

both medially and laterally around the thigh), and from the perineum, including lymphatics from the lower end of the anal canal, lymphatics from the lower end of the vagina in the female, and superficial lymphatics from the penis and scrotum in the male or the labium majus in the female. Thus, carcinoma originating in any part of a wide area may metastasize to the superficial inguinal lymph nodes.

These lymph nodes vary in number and size; apparently there are usually between 12 and 20, although some of them are frequently too small to be recognized in the average dissection. They tend to be arranged in two ill-defined groups, one roughly paralleling the in-

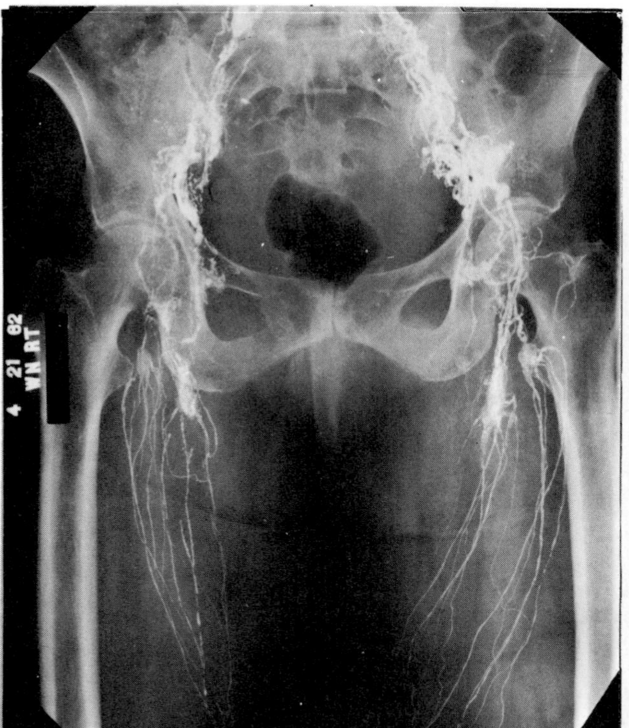


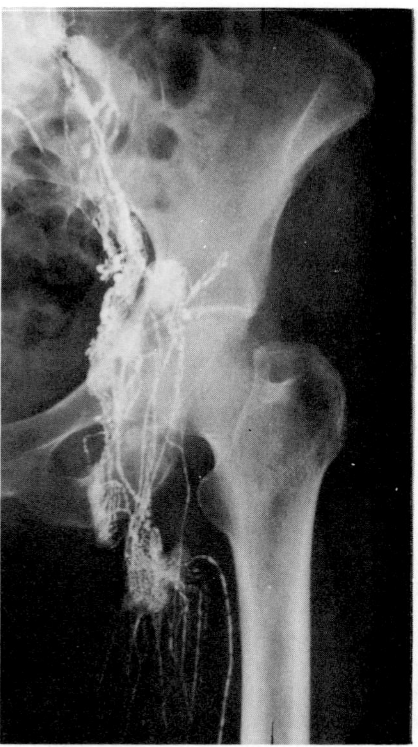

**FIG. 17-29.** Two lymphangiograms showing superficial inguinal nodes and their drainage into iliac nodes. (Courtesy of Dr. W. E. Miller)

guinal ligament, the other roughly paralleling the upper end of the great saphenous vein. The superficial inguinal lymph nodes are drained by lymphatics that run deep to the inguinal ligament, some of them ending in a few deep inguinal lymph nodes, but most passing into the abdomen to end in nodes along the external iliac vessels.

## Deep Fascia

The rather dense deep fascia of the thigh, the **fascia lata,** enwraps the entire thigh, and, in about the distal two-thirds of the thigh, sends **medial** and **lateral intermuscular septa** to the linea aspera of the femur (Fig. 17-34). Above and anteriorly, it is attached to the inguinal ligament. Laterally it is attached to the iliac crest, but splits to go on both sides of the tensor fasciae latae muscle and thus surround this. More posteriorly, still attached above to the iliac crest, it gives origin to fibers of the gluteus medius muscle, and still more posteriorly, it is continuous with the fascia on the superficial and deep surfaces of the gluteus maximus muscle. Medially, it attaches to the body and inferior ramus of the pubis and to the ramus of the ischium. Over the medial (adductor) muscles of the thigh, the fascia lata is rather weak and is attached to the surface of the muscles, but both anteriorly and posteriorly it is stronger and lies rather loosely over the muscles.

The fascia lata becomes particularly strong laterally, where it is reinforced by a middle layer of longitudinally running fibers that are actually tendon fibers derived from the insertions of the tensor fasciae latae and the gluteus maximus muscles. This strong lateral part, the **iliotibial tract,** is thus continuous anteriorly and posteriorly with the tendons of these two muscles. Between them, it passes over the surface of the gluteus medius to reach the iliac crest (Fig. 17-30). Below, it tends to blend, as do other parts of the fascia lata, with the retinacula of the patella and helps form the fibrous capsule of the knee joint, but it can usually be traced down to an attachment on the anterior part of the lateral tibial condyle. For much of its length, the iliotibial tract is anchored to the femur by the lateral intermuscular septum.

In general, the fascia lata loses its identity as it passes across the knee, for it blends with expansions from the tendons at the knee to help form the patellar retinacula and then becomes continuous with the fascia of the leg. Posteriorly, it passes across the popliteal fossa and is directly continuous with the deep fascia of the leg.

Just below the inguinal ligament on the anteromedial side of the thigh, the fascia lata presents a defect, the **saphenous hiatus** (Fig. 17-43), through which the great saphenous vein passes to enter the femoral vein. The part of the superficial fascia covering the hiatus is the **cribriform fascia** (*fascia cribrosa*), so called because it is penetrated by the several small vessels of this region already described and by the efferent lymphatics from the superficial inguinal lymph nodes. The saphenous hiatus has a sharp lateral falciform margin, formed by a fold of the fascia lata. This ends above in a more or less prominent superior cornu that is continuous with the inguinal ligament and a more or less prominent inferior cornu that fades out into the general level of the fascia lata. There is, however, no sharp medial edge to the saphenous hiatus, for on the medial border the fascia lata passes gradually posteriorly to blend with fascia over the pectineus muscle. The femoral artery, vein, and canal descend, enclosed within the femoral sheath, in front of the pectineal fascia and are thus more or less exposed in the saphenous hiatus.

## ANTERIOR AND ANTEROMEDIAL MUSCLES

Closely associated with the anterior muscles of the thigh, and enclosed between two laminae of the fascia lata, is the **tensor fasciae latae** (Figs. 17-32 and 17-33). This muscle has already been described with the gluteal muscles, to which group it belongs in spite of its anterior position.

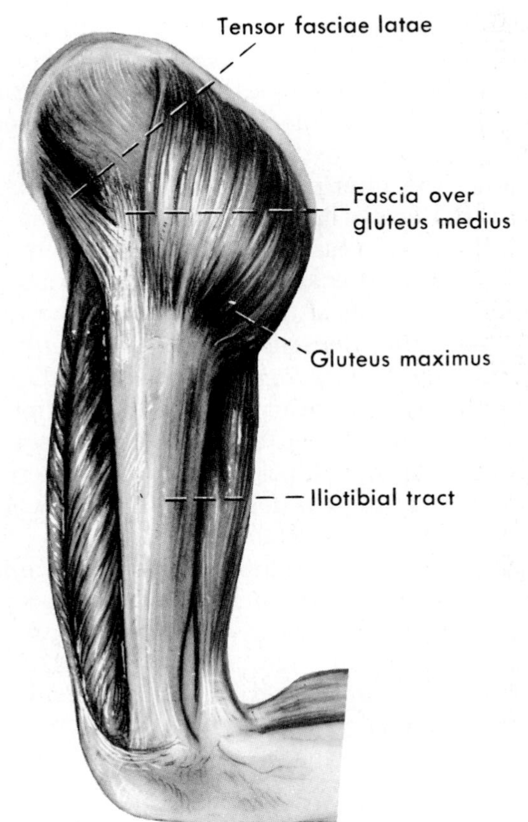


**FIG. 17-30.** The iliotibial tract.

After the fascia lata over the front of the thigh has been removed, the **femoral triangle** and its contents become visible (Fig. 17-31). The upper border of this triangle is the *inguinal ligament*, the lateral border is the *sartorius muscle*, and the medial border is sometimes said to be the medial, sometimes the lateral, border of the *adductor longus muscle*. Within the triangle, the *femoral nerve* is most lateral and shortly after entering the triangle begins to break up into its muscular and cutaneous branches. The *femoral artery and vein*, enclosed within a sheath of connective tissue, the *femoral sheath*, that also contains most medially the essentially empty *femoral canal*, lie in that order medial to the femoral nerve. While it is in the femoral triangle, the femoral artery gives off the small superficial branches already mentioned and its major branch, the *profunda femoris.* The femoral vein, while it lies within the triangle, receives tributaries corresponding to the branches of the artery and also the great saphenous vein.

Since the anterior wall of the femoral triangle is the fascia lata only, the femoral nerve and the chief vessels of the thigh are quite superficial in this area. The floor or posterior wall of the femoral triangle is, from laterally medially, the *iliopsoas muscle,* the *pectineus muscle,* and the *adductor longus.* The adductor brevis may appear between the pectineus and adductor longus to form a small part of this floor. At the apex of the triangle, the femoral vessels disappear deep to the sartorius and enter the *adductor canal.*

There is no sharp subdivision of the anterior and anteromedial muscles of the thigh according to their positions, but on the basis of their innervation, they are conveniently divided into more anterior, developmentally original dorsal muscles, innervated by the femoral nerve, and more medial, originally ventral, muscles innervated by the obturator nerve.

## Anterior (Femoral) Group

The muscles of the anterior group are the sartorius, the quadriceps femoris, the iliopsoas, and usually but not always the pectineus.

The long, straplike **sartorius** (Figs. 17-32 and 17-33) arises from the anterior superior iliac spine and curves downward and medially toward the medial side of the knee. It crosses the joint somewhat behind its axis of motion and turns forward to insert on the upper medial surface of the body of the tibia close to the insertions of two other muscles of the thigh, the gracilis and the semitendinosus. One or more bursae lie deep to the tendon of insertion of the sartorius, separating it from the underlying tendons of the two muscles just mentioned. The tendon of insertion gives off fascial slips that blend below with the fascia of the leg and knee.

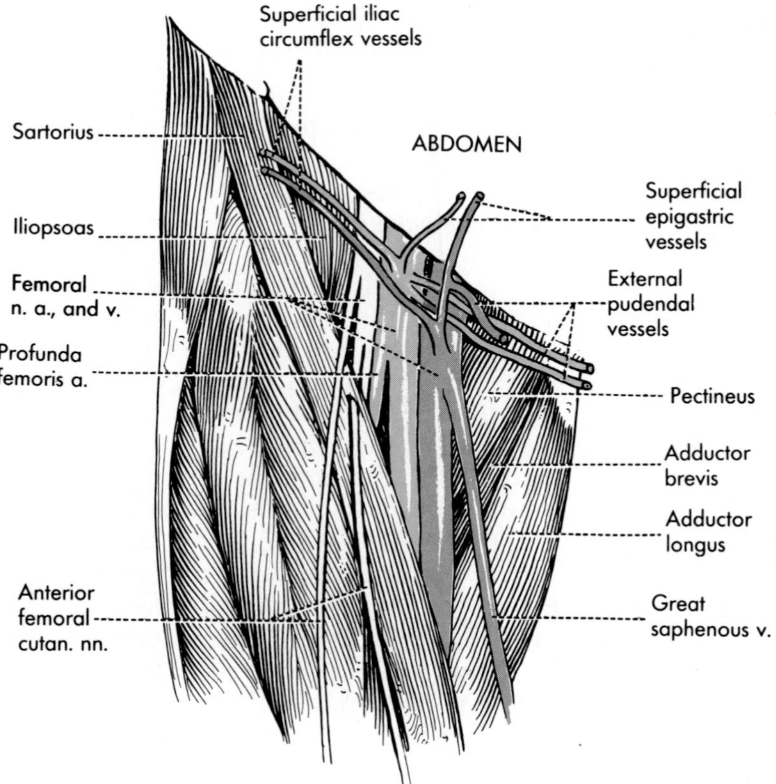


**FIG. 17-31.** The femoral triangle.

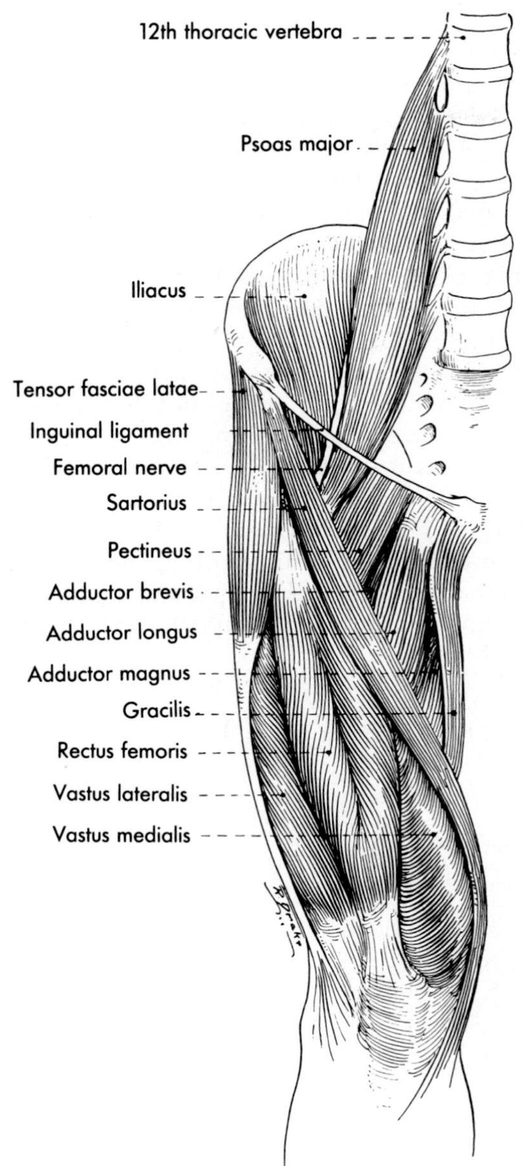


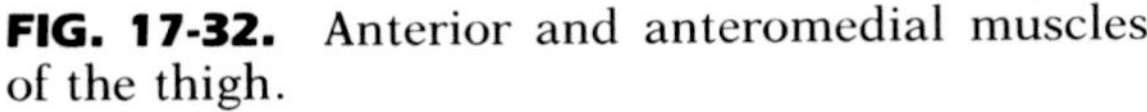

**FIG. 17-32.** Anterior and anteromedial muscles of the thigh.

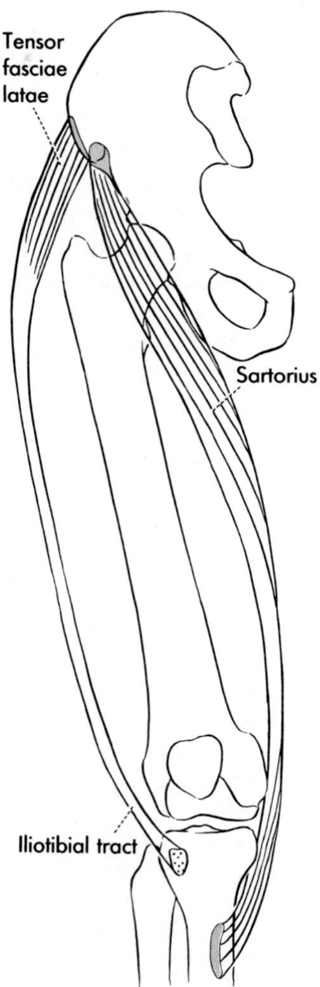


**FIG. 17-33.** Attachments of the sartorius and the tensor fasciae latae.

In its upper part, the sartorius forms the lateral border of the *femoral triangle.* Below the triangle, it lies largely between the anterior and the medial muscles of the thigh and covers the *adductor canal,* which contains the femoral vessels. The sartorius receives two branches from the femoral nerve and is frequently pierced by cutaneous branches of this nerve.

The **adductor canal** is, in cross section, a somewhat triangular space (Fig. 17-34) between the quadriceps muscle and the adductor group, lying deep to the sartorius. It is bounded anterolaterally by the *vastus me-*

*dialis* part of the quadriceps, bounded posteriorly by the *adductor longus and magnus,* and roofed by a heavy layer of fascia that passes between the quadriceps and the adductor muscles. The canal begins above at the apex of the femoral triangle, where the *femoral artery and vein* leave the triangle to run through the canal, accompanied by a branch of the femoral *nerve to the vastus medialis muscle* and by the *saphenous nerve,* a large cutaneous branch of the femoral. The canal ends below at the *tendinous* (adductor) *hiatus,* a gap in the tendon of the adductor magnus through which the femoral artery and vein pass to run downward behind the knee as the popliteal vessels.

The femoral vessels thus run the length of the canal. The nerve to the vastus medialis ends in this muscle. The saphenous nerve leaves the adductor canal a little above the knee and appears at the posterior border of the sartorius muscle to run downward subcutaneously on the medial side of the knee.

The four parts of the **quadriceps femoris** (Figs. 17-32, 17-35, and 17-38) are the rectus femoris, the vastus lateralis, the vastus intermedius, and the vastus medialis. These blend at their insertion, and the three vasti are separable only with difficulty at their origins, since the medial and lateral vasti arise in part from intermuscular septa that they share with the vastus intermedius. In spite of this, the four heads of origin of the muscle are named as if they were separate muscles.

The **rectus femoris,** the anterior member of the group, has a double origin from the ilium: a larger, rounded straight tendon attaches to the anterior inferior iliac spine, and a thinner, flattened reflected tendon, with a lower lacunar border, is given off posteriorly and at-

**FIG. 17-34.** Cross section through the middle third of the left thigh seen from below. (Redrawn from Eycleshymer AC, Schoemaker DM: A Cross-Section Anatomy. New York, Appleton, 1923)

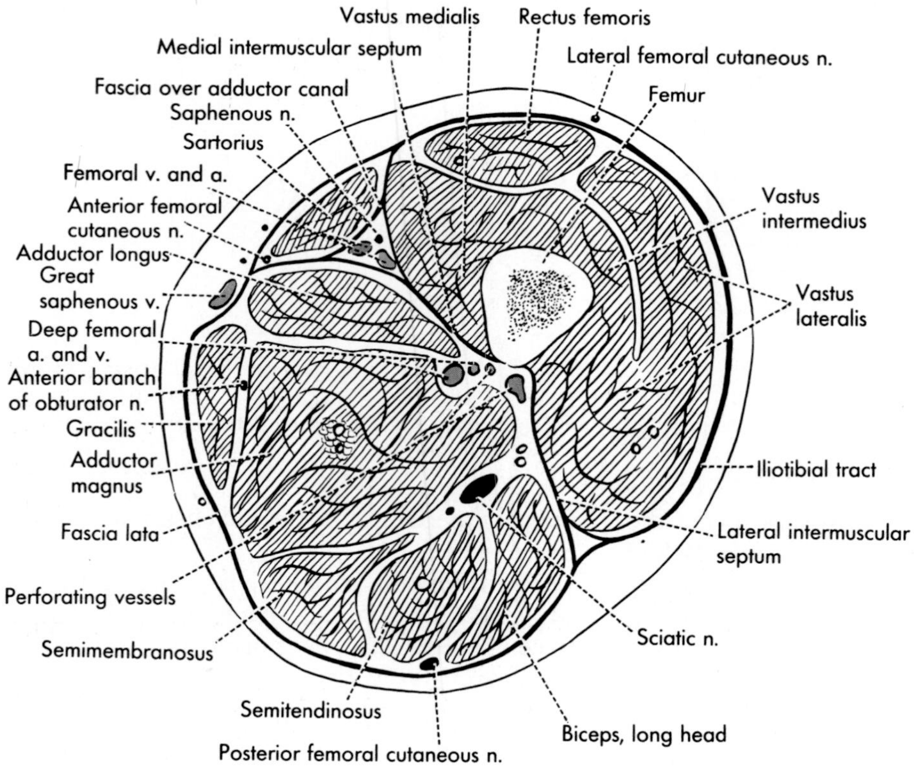

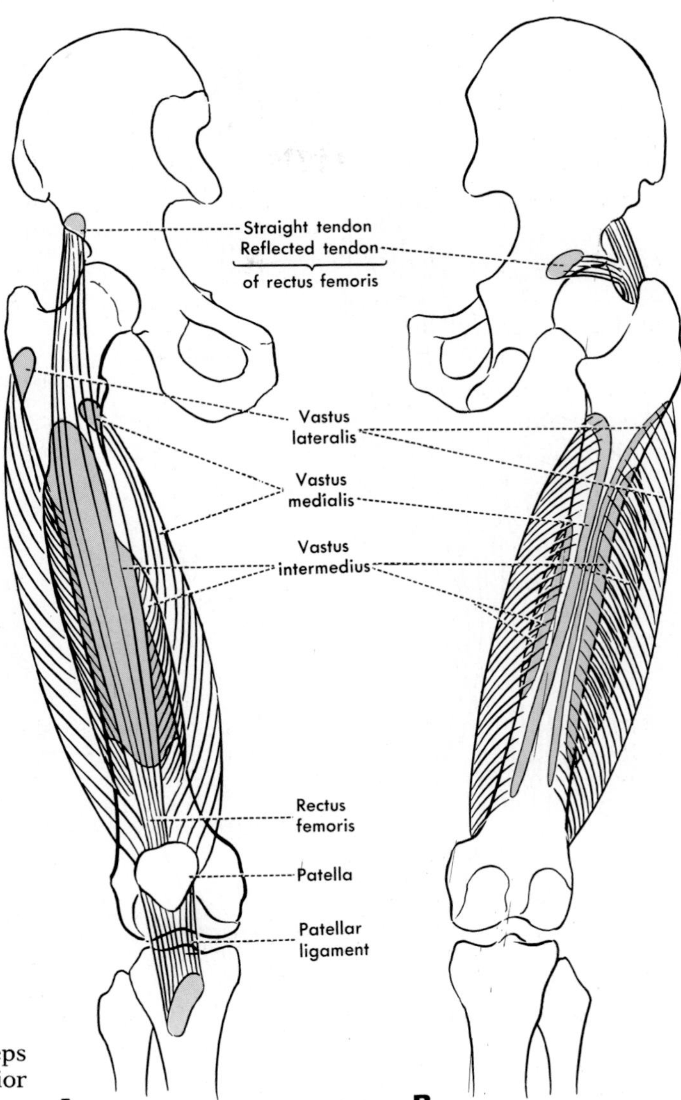


**FIG. 17-35.** Attachments of the quadriceps muscle, anterior and posterior views. **A,** anterior view; **B,** posterior view.

taches to the ilium just above the acetabulum. The tendon of insertion receives, on its deep surface and sides, much of the insertion of the vasti; the combined tendon so formed is the *quadriceps tendon.* This attaches to the upper border of the patella, and the patella is then attached to the tibial tuberosity by a heavy tendon that is known as the *patellar ligament* (really, of course, the lower end of the quadriceps tendon).

Strong fascial expansions, the **patellar retinacula,** attach the sides of the patella and the patellar ligament to the femoral and tibial condyles and help form the capsule of the knee joint. They are derived both from the quadriceps and the fascia lata. In front of the patella, there is usually a *prepatellar subcutaneous bursa,* and in front of the lower end of the patellar ligament, there is usually a *subcutaneous infrapatellar bursa.* Inflammation of these bursae as a result of kneeling (in scrubbing floors) was once known as "housemaid's knee."

The large **vastus lateralis** covers the entire lateral aspect of the thigh and extends onto its anterior and posterior aspects. It is separated

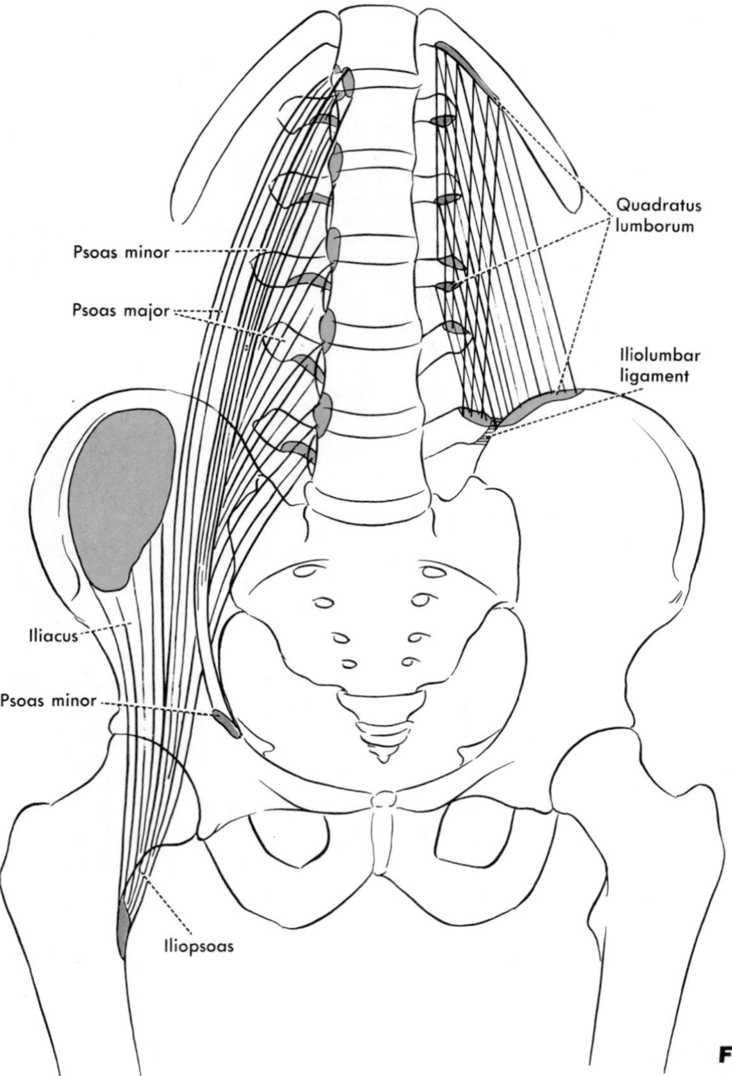


**FIG. 17-36.** Attachments of the iliopsoas and associated muscles.

from most of the lateral surface of the femur by the underlying vastus intermedius. Its origin is partly anterolateral, along the lower border of the greater trochanter, but largely posterior, from the lateral lip of the linea aspera. It also arises from the lateral intermuscular septum and a septum between it and the vastus intermedius. It inserts into the lower part of the tendon of the rectus femoris and the upper and lateral borders of the patella, giving off a tendinous expansion that blends with other connective tissue to help form the lateral *patellar retinaculum*.

The **vastus medialis** is largely kept away from the front and medial side of the femur by the vastus intermedius. Its uppermost fibers arise from the lower anterior portion of the intertrochanteric line, and from this, the origin winds around the medial aspect of the femur to follow the medial lip of the linea aspera. Superficially, the vastus medialis is separated from the adductor muscles by the adductor canal, but posteriorly and deeply both it and the adductor muscles blend with the medial intermuscular septum, which also attaches to the linea aspera, and are therefore closely bound together by this septum. The vastus medialis inserts into the medial border of the quadriceps tendon and into the medial side of the patella, somewhat farther distally on the patella than the vastus lateralis. At the knee, the vastus medialis gives off expansions that help form the medial patellar retinaculum.

The **vastus intermedius** arises from the anterior and lateral surfaces of the body of the femur and from the lateral intermuscular septum. It so covers the major part of the body of the femur that most of the other muscles attaching thereto necessarily do so along the linea aspera. The vastus intermedius is entirely covered by the rectus femoris and the other two vasti and is fused to the latter in its lower part. It inserts into the deep part of the upper border of the patella and into the tendons of the other vasti. Deep to it, as it nears the lower end of the femur, is a small muscle, the *articularis genus,* and a large *suprapatellar bursa* that commonly communicates with the knee joint.

Each muscle of the quadriceps group receives its own nerve supply from the femoral nerve. The nerves to all except the vastus intermedius are usually multiple, and this muscle often receives parts of nerves to the vastus medialis and lateralis in addition to its own nerve.

The **articularis genus** consists of one or more small muscle bundles apparently representing a split-off part of the vastus intermedius. It arises from the femur at about the junction of the upper three-fourths and the lower fourth and inserts into the synovial membrane of the knee joint, usually its upper border. It is supplied by a continuation of a branch of the femoral nerve into the vastus intermedius.

The **iliopsoas,** a muscle of the groin, is composed at its origin of two separate muscles, the iliacus and the psoas major. Since both these muscles arise within the abdomen, their origins must be examined when this part is dissected. As they enter the thigh by passing behind the inguinal ligament, they blend into a single muscle, the iliopsoas, which passes across the front of the hip joint to insert on the lesser trochanter. The femoral nerve lies in the groove of junction of the two muscles.

The **iliacus,** the lateral component of the iliopsoas, is a fan-shaped muscle that arises from the iliac fossa (Fig. 17-36). Its fibers converge to insert into the side of the tendon of the psoas.

The **psoas major,** the medial component of the iliopsoas, arises from the sides of the intervertebral disks and the adjacent ends of the vertebral bodies from the 12th thoracic through the 4th lumbar disk and from the transverse processes of all the lumbar vertebrae.

Within the abdomen, the lumbar plexus takes form in the substance of the psoas major. The muscle receives its innervation by twigs from the lumbar plexus, usually from about the 2nd to the 4th lumbar nerves. The iliacus is innervated by one or more branches from the femoral nerve. As the iliopsoas passes behind the inguinal ligament, it occupies all the space between the lateral part of the ligament and the pelvis (see Fig. 26-17). Its fascia blends anteriorly with the inguinal ligament, but on the medial side of the muscle passes posteriorly, as the *iliopectineal arch,* to attach to the pelvis and blend with fascia over the pectineus muscle, thereby dividing the space behind the inguinal ligament into two parts: the lateral part, occupied by the iliopsoas, is the *lacuna musculorum,* and the medial part is the *lacuna vasorum* (which transmits the great vessels). Deep to the iliopsoas as it crosses the front of the hip joint, there is usually an iliopectineal bursa, which may communicate with the hip joint.

The **pectineus muscle** lies just medial to the iliopsoas and forms a large part of the floor of the femoral triangle. The femoral vessels lie in front of it, and the medial femoral circumflex artery, arising from the femoral or from the profunda femoris, passes backward between it and the iliopsoas. The pectineus muscle arises (Fig. 17-37) from the pecten of the pubis and the bone anterior to the pecten and inserts on the posterior aspect of the femur on the pectineal line (below the lesser trochanter).

In its placement, the pectineus muscle

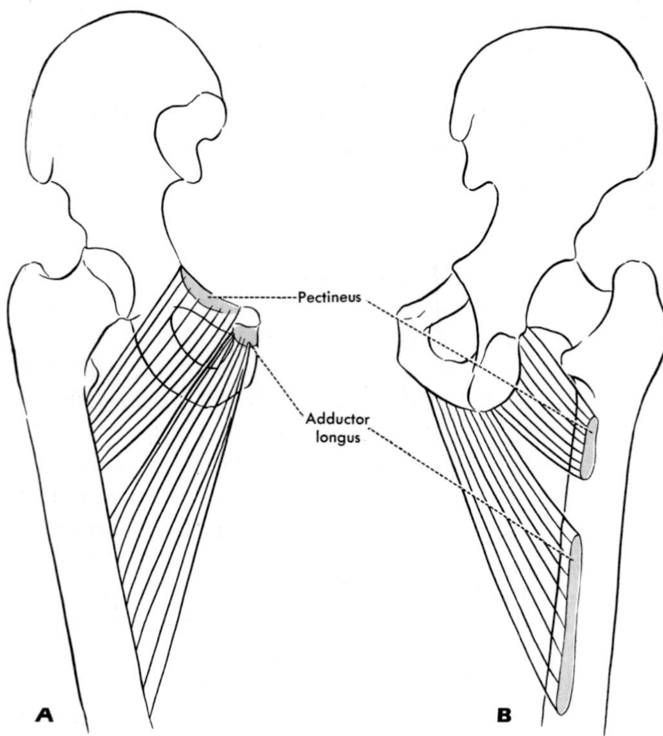


**FIG. 17-37.** Attachments of the pectineus and adductor longus muscles. **A,** anterior view; **B,** posterior view.

seems to be an upper member of the adductor (medial) group of thigh muscles, but its innervation is usually by a branch of the femoral nerve that passes behind the femoral vessels to enter the anterior surface of the muscle. Occasionally, however, the obturator nerve, emerging into the thigh immediately behind the pectineus, partly or entirely supplies the muscle, or, when an accessory obturator nerve is present, the pectineus is innervated by this and the femoral.

**Variations, Innervations, and Actions.** Anomalies of the anterior muscles of the thigh are both rare and of little importance. There may be small aberrant slips of origin or insertion, or the pectineus may be partly fused with the adjacent adductor brevis or adductor longus muscles.

All the muscles of the anterior group are innervated by the **femoral nerve,** with the possible exception, as noted, of a partial or complete innervation of the pectineus by the accessory obturator or obturator. The femoral nerve usually contains fibers from the 2nd through the 4th lumbar nerves. The nerves to the sartorius and to the pectineus are said to contain fibers from the 2nd and 3rd lumbars; those to the quadriceps to contain fibers from the 2nd through the 4th lumbars but with the 3rd and 4th lumbars more important; and the small nerve to the articularis genus to contain fibers from the 3rd and 4th lumbars.

The members of the anterior group of muscles differ considerably in their **actions,** which therefore must be recounted for each muscle individually. In general, the iliopsoas and the pectineus act only at the hip joint; the sartorius and the rectus femoris act at both hip and knee joints (but with opposite actions at the knee joint); and the vasti act only at the knee joint.

The **sartorius,** a rather weak muscle but one with a large range of action, simultaneously flexes the hip and knee and, during flexion at the hip, also slightly rotates the femur laterally. Thus, it helps initiate the action of crossing the legs, a movement from which it receives its name, the "tailor muscle," because of the tailor's habit of sitting crosslegged at his work. Actually, the muscle probably never comes into play when lateral rotation alone is required. Because of its oblique course medially across the front of the thigh, the sartorius also is theoretically an abductor at the hip and often is so listed, but stimulation of it is said to produce no appreciable amount of abduction.

All members of the **quadriceps group,** through

their insertion on the patella and their continuation through the patellar ligament to the tibial tuberosity, extend the leg. The quadriceps is the only muscle that can actively extend at the knee joint, and, therefore, a quadriceps of normal strength is particularly important in stability at this joint. The rectus femoris is most efficient as an extensor at the knee when the hip is extended, but it does not assist in the last 10° to 15° or more of extension, leaving this to the vasti. Since the rectus femoris crosses the front of the hip joint, it is also a flexor at this joint, as well as an extensor at the knee; however, it acts only weakly at the hip, even when the knee is flexed, and still more weakly when the knee is extended. For instance, if extension at the knee is carried out while the hip is sharply flexed, the rectus femoris will not even help to maintain the hip in flexion; instead, extension at the hip results from the passive pull upon the posterior hamstrings, extensors at the hip.

The **articularis genus,** inserting as it does into the suprapatellar bursa of the knee joint, pulls this upward as the knee is extended, thus presumably helping to prevent a redundant fold from getting caught within the knee joint.

The **iliopsoas** is the most powerful flexor of the hip joint, being used in flexing the joint against resistance and in sitting up from a supine position but not in actions requiring little strength. It probably adducts slightly, although it is apparently not normally used as an adductor. It has been listed as both a lateral rotator and a medial rotator, but is apparently active in lateral rotation only.

The **pectineus** is primarily a flexor at the hip joint and works with the tensor fasciae latae in most such movements; electromyographic evidence indicates that it is active also in medial rotation and adduction.

## Anteromedial (Adductor) Group

The anteromedial group of muscles is sometimes known as the obturator group because of its innervation, or as the adductor group because most of the muscles bear that name. This group may include the pectineus, although this is more commonly innervated by the femoral nerve and has, in this book, been described with the anterior muscles. The muscles that regularly belong to this group are the adductor longus, the adductor brevis, the gracilis, the larger anterior part of the adductor magnus (a posterior part really belongs with the hamstrings), and the deep-lying obturator externus.

The **adductor longus** (Figs. 17-32 and 17-37), the most anteriorly placed of the adductor group of muscles, arises by a strong tendon from the front of the body of the pubis just below the pubic tubercle and expands to insert on the linea aspera between the attachments of the vastus medialis and the adductor magnus.

It has already been noted that the adductor longus forms a part of the posterior wall of the *adductor canal,* so that the femoral ("superficial femoral") vessels pass downward on the anterior surface of this muscle (Fig. 17-44). However, the *deep femoral vessels* pass behind the adductor longus, lying on the adductor brevis and the adductor magnus (Fig. 17-45). The anterior branch of the obturator nerve also passes behind the adductor longus and supplies it.

The **adductor brevis,** largely covered by the adductor longus, arises from the body and inferior ramus of the pubis and expands in a triangular fashion as it goes to its insertion on the upper part of the linea aspera (Figs. 17-38 and 17-39). Both the femoral and the deep femoral vessels cross it anteriorly, and so does the anterior branch of the obturator nerve. The posterior branch runs behind it; either branch or both supply the muscle.

The **gracilis** lies superficially along the medial side of the thigh. It arises from the body and inferior ramus of the pubis (Fig. 17-40) and, close to the knee, is covered by the sartorius. Below the knee, its tendon curves forward and expands to insert into the medial surface of the upper end of the tibia below the medial condyle, between the insertions of the sartorius and the semitendinosus (a posterior muscle of the thigh). Between the tendon of the gracilis and the overlying tendon of the sartorius is a bursa, and between the associated tendons of the gracilis, sartorius, and semitendinosus (sometimes referred to as the *pes anserinus* tendon) and the underlying tibial collateral ligament of the knee joint is another bursa, the **anserine bursa.**

The gracilis is innervated through the anterior branch of the obturator nerve.

The **adductor magnus** is a large muscle covered anteriorly by the adductor brevis, adduc-

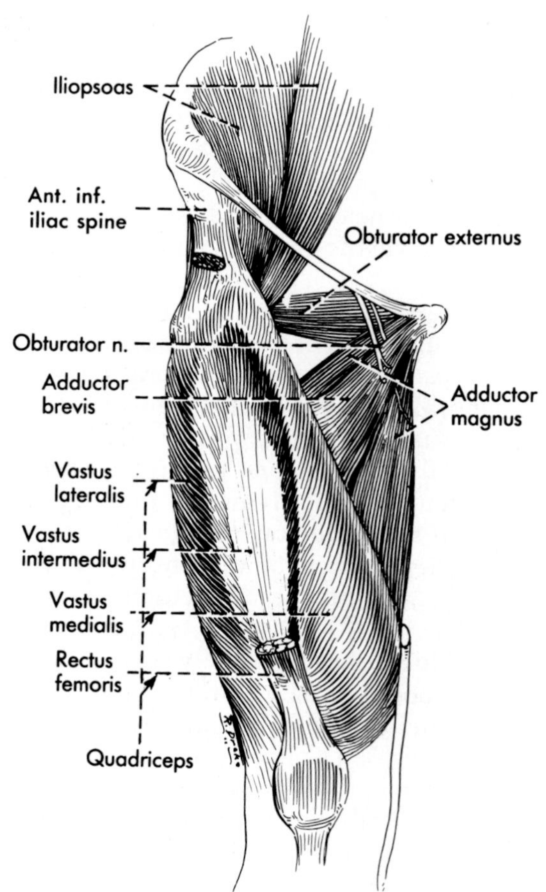


**FIG. 17-38.** Some deeper anterior and anteromedial muscles of the thigh. The pectineus, adductor longus, gracilis, sartorius, tensor fasciae latae, and most of the rectus femoris have been removed.

tor longus, and vastus medialis. It arises from the inferior ramus of the pubis, from the entire length of the ramus of the ischium, and from the ischial tuberosity (Fig. 17-40). From this origin, the muscle spreads out in a rather complicated manner: the most anteriorly arising fibers run almost straight laterally to an insertion on the linea aspera, fibers of succeeding posterior origin run more obliquely downward to the linea aspera, and the most posteriorly arising fibers run almost straight down to insert on the lowest part of the lateral lip of the linea aspera immediately above the adductor tubercle and into that tubercle. In the tendon a little above the adductor tubercle is a gap, the **tendinous** or **adductor hiatus.** This is the lower end of the adductor canal, and through it, the femoral vessels pass to lie against the popliteal surface of the femur, where their name changes to popliteal vessels.

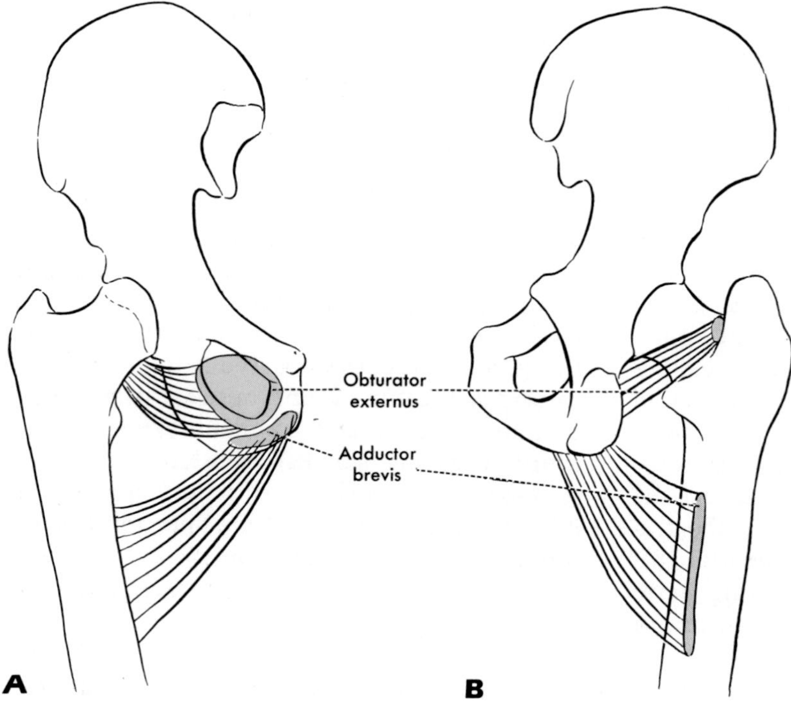


**FIG. 17-39.** Attachments of the adductor brevis and obturator externus. **A,** anterior view; **B,** posterior view.

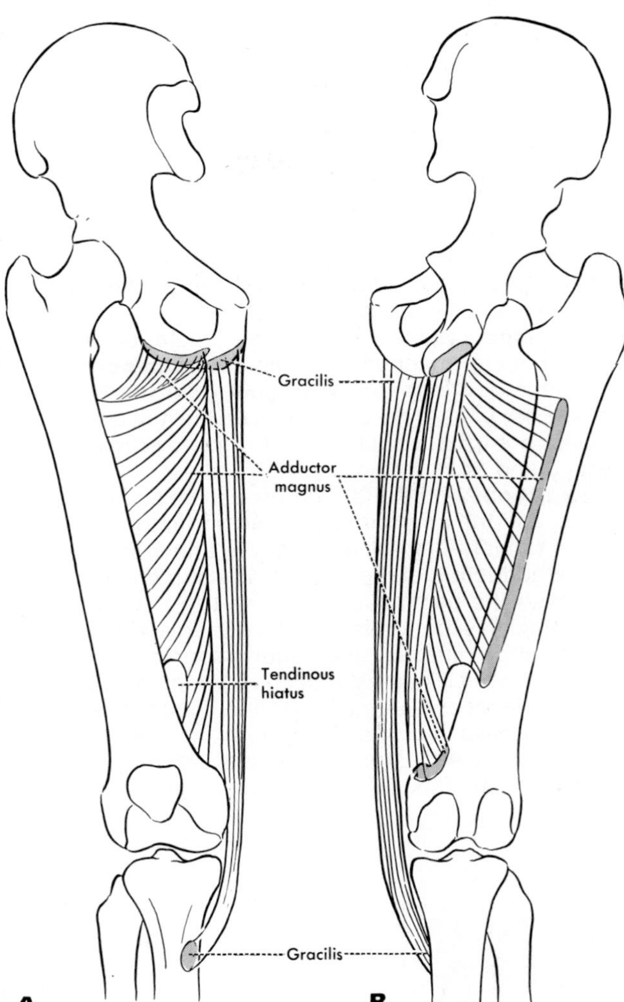


**FIG. 17-40.** Attachments of the adductor magnus and gracilis, **A,** anterior and **B,** posterior views.

The adductor magnus is a composite muscle, apparently consisting originally of *three parts:* the most anteriorly arising and most transverse part; the major anterior oblique part; and the posterior, almost vertical portion that arises particularly from the ischial tuberosity and inserts lowest on the femur. Normally, these parts are so fused together as to be indistinguishable, but the first two parts, forming the anterior portion of the muscle, are *innervated by the posterior branch of the obturator nerve,* whereas the posterior part is innervated by the *tibial nerve* and belongs with the hamstring (posterior) rather than the adductor group of muscles.

The **obturator externus** at its origin lies largely behind the pectineus, but it is also covered in part by the upper ends of all three adductor muscles. It arises from the external surface of the obturator membrane and from the adjacent surface of the pubis and ischium around the obturator foramen (Fig. 17-39). Its fibers converge so that the muscle forms a somewhat twisted cone that runs posteriorly and laterally below the capsule of the hip joint and then upward across the posterior surface of the capsule to insert into the trochanteric fossa.

The *obturator nerve* supplies the obturator externus as it leaves the pelvis. Its anterior branch usually enters the thigh anterior to the obturator externus whereas the posterior branch usually emerges through the muscle.

**Variations, Innervations, and Actions.** The anterior upper part of the adductor magnus may be relatively distinct as an **adductor minimus** muscle. Occasionally there may be some fusion between the adductor brevis and longus, or between these and the pectineus. Real abnormalities of the adductor group of muscles are rare.

Their **innervation,** except for the posterior part of the adductor magnus, is by the **obturator nerve.** This nerve normally receives fibers from the 2nd, 3rd, and 4th lumbar nerves; however, the muscles it supplies are usually said to receive fibers primarily from L3 and L4.

The **actions** of the adductor brevis, the adductor longus, and the anterior part of the adductor magnus are similar. All three muscles are primarily flexors and adductors of the thigh, a function that they share with the pectineus. In contrast to the pectineus, which is a better flexor than it is an adductor, the adductor muscles are better adductors than they are flexors. They have been regarded as both medial rotators and lateral rotators. Stimulation experiments have indicated that both the adductor longus and the adductor magnus laterally rotate as they adduct and flex. Only recently has electromyographic evidence indicated that these muscles act in medial rotation only. This finding might be expected from the results on the pectineus, but it is difficult to understand how two muscles with such similar courses as the pectineus and the iliopsoas rotate in opposite directions. The gracilis strongly adducts the thigh, rotates it medially, and flexes the knee; when the knee is flexed, it medially rotates the leg, and it can flex the hip if the knee is kept extended. The obturator externus is primarily a lateral rotator.

## Nerves and Vessels

The neurovascular structures related to the anteromedial muscles of the thigh are the femoral and the obturator nerves, and the corresponding vessels; the femoral vessels are large and important, whereas the obturator vessels barely appear in the thigh.

**Nerves.** The **femoral nerve** enters the thigh behind the inguinal ligament, lying on the surface of the iliopsoas muscle and immediately lateral to the femoral artery. Only a little below the inguinal ligament, in the upper part of the femoral triangle, the femoral nerve

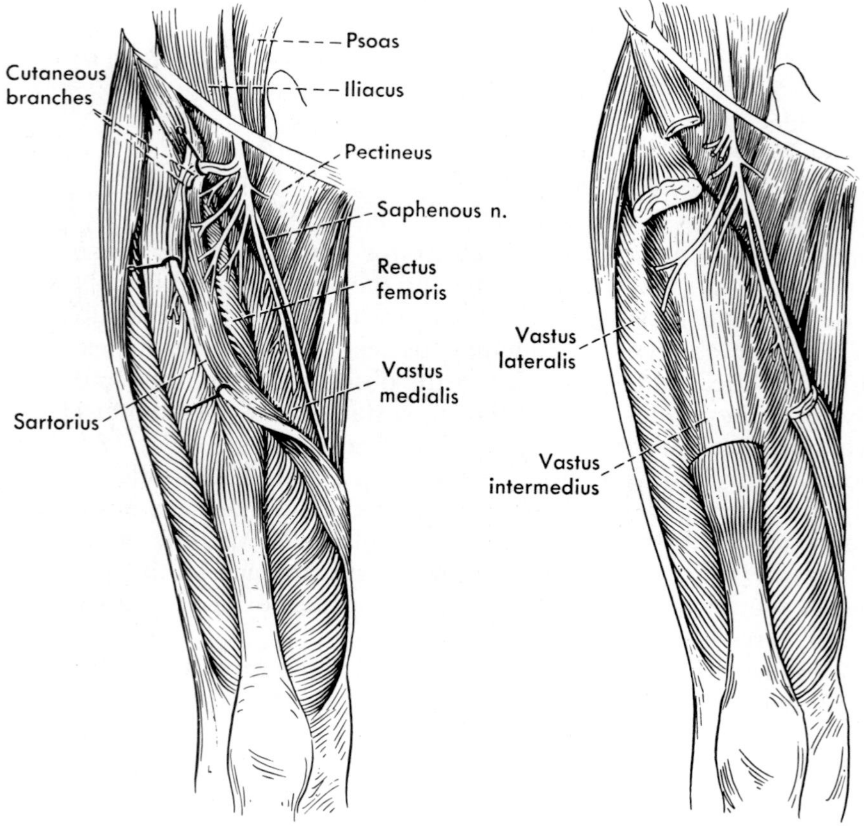


**FIG. 17-41.** Muscular branches of the femoral nerve in the thigh.

breaks up into a number of muscular and cutaneous branches that are given off in no regular order. The lateral femoral circumflex artery runs laterally among its branches.

The femoral nerve's **cutaneous branches** to the thigh are the *anterior femoral cutaneous nerves,* already described. One or more of these may send a branch into the sartorius. The other cutaneous branch of the femoral, the *saphenous nerve,* runs through most of the length of the adductor canal and is distributed to the leg and foot.

The **muscular branches** of the femoral nerve (Fig. 17-41) are the nerve to the *pectineus* (almost always present, for only rarely is this muscle supplied by the obturator nerve alone); two branches, or two sets of branches, to the *sartorius;* usually two branches to the *rectus femoris;* and at least one branch into each of the *three vasti muscles,* with usually some fibers from nerves to the vastus lateralis and the vastus medialis continuing into the vastus intermedius, and a nerve of the vastus intermedius continuing to the articularis genus. *Nerves to the hip joint* may arise independently from the femoral, but frequently arise from its muscular branches. *Nerves to the knee joint* are derived from the nerves to the vasti.

The femoral nerve usually contains fibers from the 2nd, 3rd, and 4th lumbar nerves, but may also receive fibers from the 1st or the 5th lumbar. Apparently many of the fibers from the 2nd lumbar are distributed to skin, and it is said that fibers from the 1st lumbar nerve in the femoral are always so distributed. The chief innervation of the anterior muscles of the thigh is, therefore, through the 3rd and 4th lumbar nerves.

The **obturator nerve,** also from the lumbar plexus, represents the anterior divisions of the same nerves that give rise to the femoral: the 2nd, 3rd, and 4th lumbars. It diverges from the femoral in the substance of the psoas muscle and appears on the medial side of this muscle, whence it passes along the lateral pelvic wall to the obturator canal. It commonly divides into anterior and posterior branches while it is in the canal. Both branches emerge into the thigh behind the pectineus muscle and in front of or through the obturator ex-

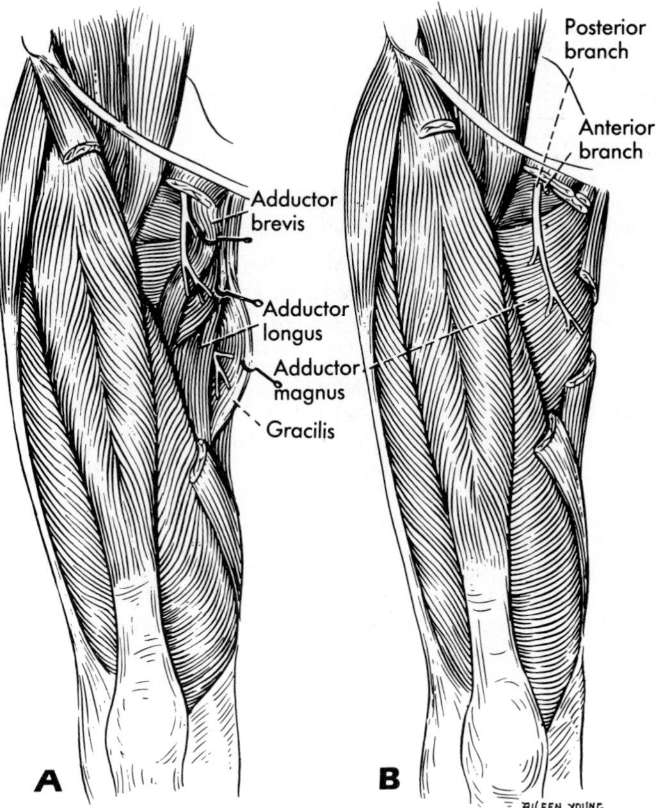


**FIG. 17-42.** Muscular branches of the obturator nerve in the thigh. The distribution of the anterior branch is shown in **A,** and that of the posterior, in **B.**

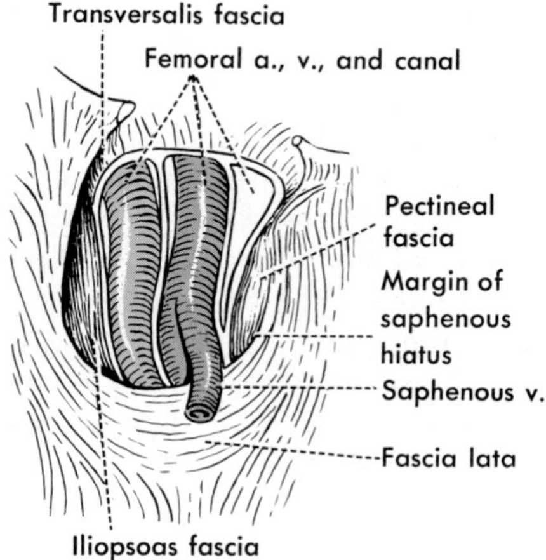


**FIG. 17-43.** The three compartments in the femoral sheath, as seen on the right side opened from the front.

ternus. The *anterior branch* then runs downward and medially in front of the adductor brevis and behind the adductor longus, supplying the longus, sometimes the brevis, and ending in the gracilis. It also gives rise to an *articular branch to the knee joint* that descends along the femoral artery and to the *cutaneous branch* of the obturator to skin on the medial side of the thigh. The *posterior branch* runs behind the adductor brevis and longus, on the anterior surface of the adductor magnus (Fig. 17-42). It may supply the adductor brevis (which may be supplied by either branch or both) and always supplies the anterior part of the adductor magnus. Before or as it enters the thigh, the obturator nerve supplies the obturator externus and gives a branch to the anteromedial part of the *hip joint.*

In about 8% to 10% of sides, there is an **accessory obturator nerve.** It is formed by fibers from the 2nd and 3rd or the 3rd and 4th lumbar nerves, approximately at the level at which the femoral and obturator nerves separate from each other. Unlike the obturator nerve, it passes into the thigh in front of the pubis, on the surface of the pectineus muscle. Its distribution varies somewhat, but it usually supplies at least the pectineus muscle (which is apparently also always supplied by the femoral) and a part of the hip joint. It also commonly communicates with the obturator nerve and, rarely, replaces the obturator nerve in its distribution to some of the muscles usually supplied by this nerve.

**Blood Vessels.** The chief vessels of the thigh are the **femoral artery and vein,** which are continuous at the inguinal ligament with no change in diameter, but with a change in name, with the external iliac vessels and are similarly continuous below with the popliteal vessels. In the femoral triangle, the vein lies medial to the artery, and both are surrounded by a continuation of the fascial layer lining the abdominal cavity, very much as if they had dragged a diverticulum downward as they grew into the thigh. This special fascial investment is the **femoral sheath** (Fig. 17-43). It begins at the level of the inguinal ligament, where it is continuous with the transversalis (internal abdominal) fascia, and ends approximately 3 cm to 4 cm below by becoming continuous with the adventitia of the vessels.

Within the sheath are three compartments, separated by septa that pass between its anterior and posterior walls: the *lateral compartment* contains the femoral artery and the femoral branch of the genitofemoral nerve (which pierces the anterior wall of the sheath to become subcutaneous); the *intermediate compartment* contains the femoral vein; and the *medial compartment,* shorter than the other two, is the **femoral canal,** which contains only a slight amount of loose connective tissue and one or two lymphatic vessels and nodes. The upper end of the femoral canal, adjacent to the peritoneum as this is reflected upward from the posterior onto the anterior abdominal wall, is the **femoral ring.** The femoral canal is funnel shaped, as is the femoral sheath as a whole, but tapers more rapidly to a point than does the femoral sheath and is typically only 1.25 cm to 1.5 cm long. A *femoral hernia* descends through the femoral ring into the femoral canal. As the hernia expands, it, of course, dilates the femoral canal, and it may escape from behind the fascia lata by passing through the saphenous hiatus.

The **femoral artery** (Fig. 17-44) gives off its small superficial branches while it still is enclosed in the femoral sheath, and these vessels, therefore, penetrate the sheath. Similarly, the great saphenous vein and any of the small superficial veins that join the femoral vein directly penetrate the femoral sheath.

In addition to the small superficial arteries (*superficial iliac circumflex, superficial epigastric, external pudendal*—Figs. 17-31 and 17-44), the named branches of the femoral artery are the profunda femoris and the descending genicular, but fairly frequently the femoral may also give rise to the medial or the lateral femoral circumflex artery, normally branches of the profunda. The **profunda femoris artery** typically arises from the posterolateral side of the femoral artery fairly high in the femoral triangle. It and its branches are the chief supply to the muscles of the thigh, although the femoral artery itself gives off occasional small muscular twigs. The deep femoral artery passes posterolaterally and inferiorly and then curves medially to lie behind the femoral artery, at the same time usually giving off its two major branches, the medial and lateral femoral circumflex arteries. Both *femoral cir-*

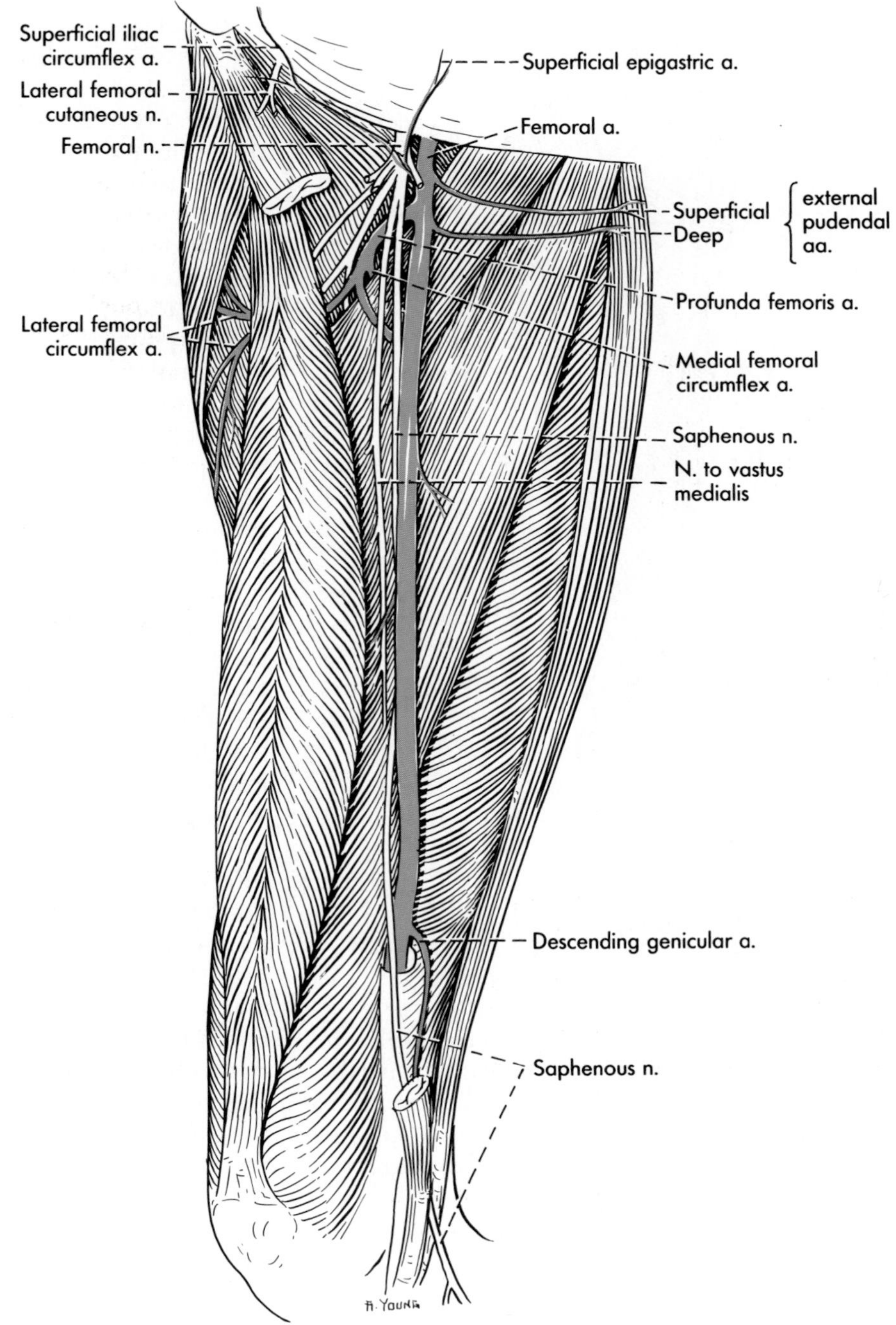


**FIG. 17-44.** The femoral artery in the right thigh after removal of the sartorius muscle and of the fascia covering the adductor canal.

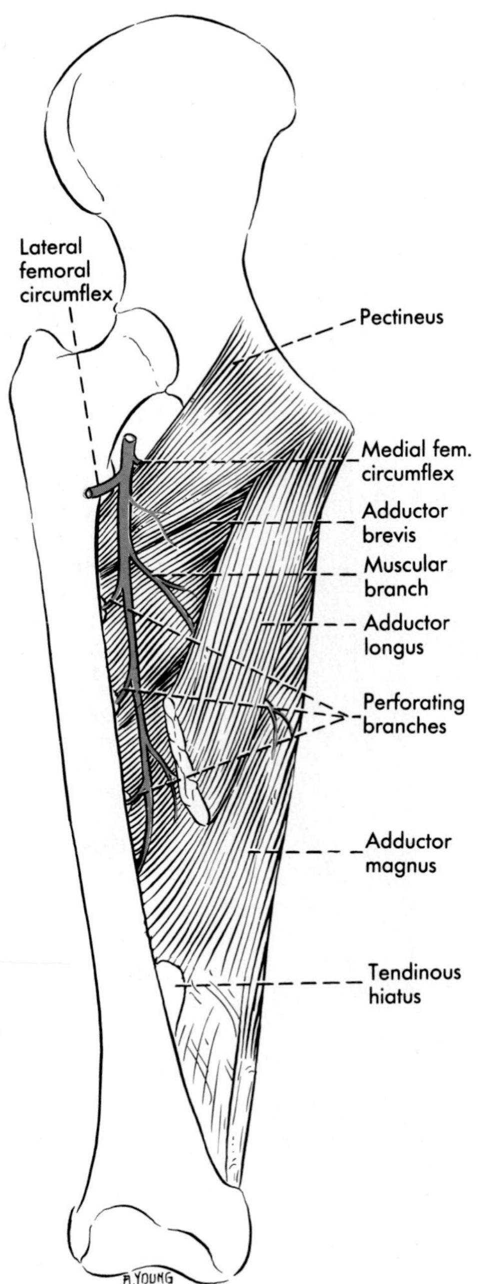


**FIG. 17-45.** The deep femoral artery.

*cumflex veins* typically enter the femoral vein instead of the deep femoral, but otherwise the deep femoral artery and vein are similar. They run down behind the femoral vessels on the surface of the pectineus and adductor brevis muscles, the vein in front of the artery, and then separate from the femoral vessels by passing behind the adductor longus, thus coming to lie directly on the front of the adductor magnus (Fig. 17-45). In addition to branches to the deeper-lying medial muscles, they give rise to perforating vessels that pass through the tendons of insertion of the adductor muscles to become the chief supply of the posterior muscles of the thigh. There are commonly four **perforating branches.** A fairly typical pattern is for the first two perforating vessels to penetrate the adductor brevis and the adductor magnus, and the lower two (the lowest perforating branch being the terminal branch of the profunda) to penetrate the adductor magnus only, below the insertion of the adductor brevis.

The two largest branches of the profunda femoris are the lateral and medial femoral circumflex vessels. The **lateral femoral circumflex artery** typically arises from the lateral side of the upper end of the profunda, but in perhaps some 15% or so of instances it, or a branch of it, arises from the femoral above the profunda. It runs laterally across the front of the iliopsoas muscle and between branches of the femoral nerve, passes behind the sartorius and the rectus femoris muscles, and divides into an *ascending,* a *descending,* and a *transverse branch.* All these branches feed adjacent muscles. The ascending branch runs laterally and upward deep to the rectus femoris and the tensor fasciae latae. It anastomoses with superior gluteal vessels and gives branches to the front of the femur. The descending branch, which sometimes arises separately from the femoral or from the profunda, runs downward behind the rectus femoris and gives branches into both the vastus lateralis and the vastus intermedius muscles. A little above the knee, it anastomoses through the vasti with twigs of the lower perforating arteries and with branches of the genicular vessels and is thus a small contributor to the collateral circulation about the knee joint. The transverse branch of the lateral femoral circumflex disappears into the vastus lateralis. It may emerge through the vastus lateralis to participate in the formation of the so-called *cruciate anastomosis* of the buttock (p. 368).

The **medial femoral circumflex artery** typically arises from the medial or posteromedial side of the profunda (but has an independent origin from the femoral more frequently than does the lateral circumflex) and turns posteriorly between the iliopsoas and pectineus muscles. It gives off muscular branches and usually, close to the inferior border of the hip joint, an *acetabular branch* that anastomoses with the posterior branch of the obturator artery or helps that vessel supply the tissue of the acetabular fossa. The major continuation of the artery is the *deep branch,* which divides in front of the quadratus femoris into an ascending and a transverse branch. The ascending branch appears in the buttock above the quadratus femoris, and the transverse branch appears below it (Fig. 17-23).

The **femoral artery below the profunda** (frequently called by clinicians the *superficial femoral artery,* to distinguish it from the femoral artery above the profunda, sometimes called the *common femoral artery*) runs downward in the adductor canal with the femoral vein behind it, separated from the profunda vessels by the adductor longus muscle. Close to the lower end of the canal, the artery gives off its last named branch, the **descending genicular artery.** This usually divides on the anterior surface of the adductor magnus into articular and saphenous branches. The *articular branch* enters into the anastomosis around the knee, and the *saphenous branch* supplies the lower ends of the muscles on the medial side of the knee and ends by sending twigs to the skin in this area. After giving off the *descending genicular artery,* the femoral artery passes through the tendinous hiatus in the adductor magnus; at this level its name changes to popliteal.

*Thrombosis* of the femoral artery fairly frequently starts at the level of the tendinous hiatus and apparently has its origin as a result of damage to the wall of the vessel by the tendon of the adductor magnus.

The **obturator vessels** play little part in the blood supply of the thigh. The artery typically arises in the pelvis from the internal iliac artery, and the accompanying vein enters the internal iliac vein. As the vessels emerge through the obturator canal, they divide into anterior and posterior branches that, instead of continuing down into the thigh, circle the obturator foramen, supplying particularly, therefore, the obturator externus muscle and the adjacent bone. The posterior branch of the obturator usually gives rise to an *acetabular branch* that enters the acetabular notch and supplies tissue in the acetabular fossa and usually gives rise to the *artery in the ligament of the head of the femur* (this can come from the acetabular branch of the medial femoral circumflex, or from both acetabular arteries).

The relatively few **deep lymphatics** of the thigh drain upward along the femoral artery and end in one to three **deep inguinal lymph nodes** that lie on the medial side of the upper end of the femoral vein. These small nodes, the highest and largest of which often lies at the femoral ring, also receive some connections from the superficial inguinal nodes and vessels from the glans penis or clitoris. The deep nodes drain into nodes along the external iliac artery.

## POSTERIOR MUSCLES AND ASSOCIATED STRUCTURES

The posterior muscles of the thigh are the semitendinosus, the biceps femoris, the semimembranosus, and the posterior part of the adductor magnus (Fig. 17-46). All these muscles, except the short head of the biceps, arise from the ischial tuberosity, and all except the adductor magnus cross the knee joint. As these muscles are traced downward from their origin on the ischial tuberosity, they separate, some distance above the knee, into medial and lateral ones. The lateral muscle is the biceps femoris, and the medial muscles are the semimembranosus and the semitendinosus (and more anteriorly, the adductor magnus).

The **semitendinosus muscle** arises from the ischial tuberosity in common with the long head of the biceps femoris (Fig. 17-47). It passes downward medial to the biceps and posterior to the semimembranosus and diverges medially from the biceps above the knee. Its tendon passes behind the knee joint and then curves forward to insert into the medial aspect of the body of the tibia just poste-

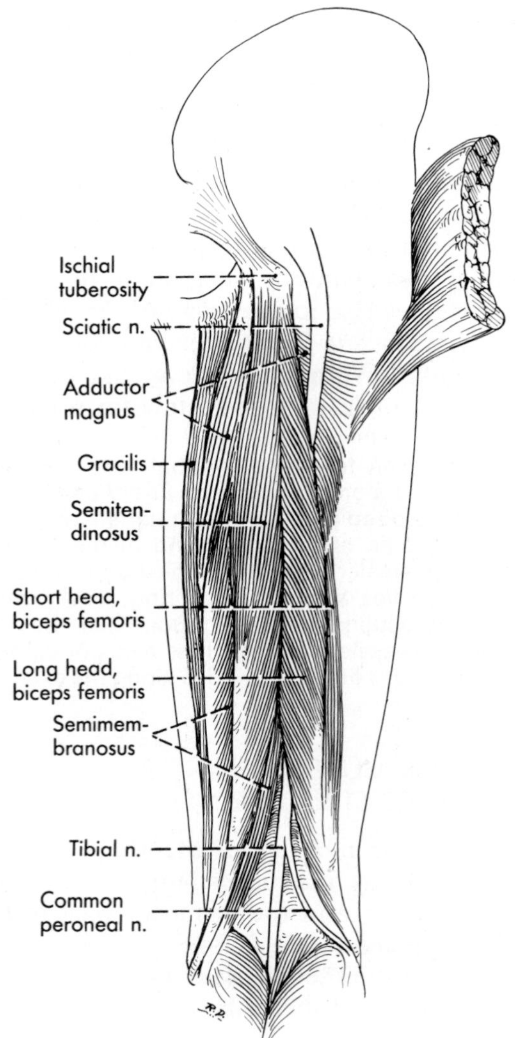


**FIG. 17-46.** Posterior muscles of the right thigh. The gracilis, also shown here, belongs to the adductor group.

rior to the insertion of the gracilis and sartorius, with which it shares the anserine bursa (p. 383). Like most of the tendons around the knee, that of the semitendinosus gives off fibrous expansions that blend with the fascia of the leg.

There are usually two *nerves* to the semitendinosus, one entering its deep surface close to its origin, the other entering it below the tendinous intersection that customarily crosses it. They arise from the *medial* (*tibial*) *side of the sciatic.*

The **biceps femoris** has a long and a short head. The *long head* arises from the ischial tuberosity with the semitendinosus. The *short head* arises from the lateral lip of the linea aspera and the lateral intermuscular septum. The two heads unite in the lower third of the thigh, and their tendon of insertion crosses the posterolateral aspect of the knee joint to insert mostly into the head of the fibula, with some attachment to the lateral tibial condyle; it gives off a fascial expansion to the deep fascia of the leg.

The long head of the biceps receives at least one nerve from the medial, or *tibial, side of the sciatic nerve* and may receive another. The short head receives a branch from the *common peroneal,* or lateral, side of the sciatic nerve.

The **semimembranosus** arises by a particularly long, flat tendon, hence its name, from the posterolateral lower portion of the ischial tuberosity. This broad tendon passes downward in front of the semitendinosus and on the posterior surface of the adductor magnus and, in about the middle of the thigh, gives rise to muscle fibers. These, in turn, give rise to a large rounded tendon that inserts into the posteromedial side of the medial tibial condyle. The tendon gives off a heavy lateral expansion that runs obliquely upward across the posterior aspect of the knee joint, forming a major part of the oblique popliteal ligament (Fig. 17-47; see also Fig. 18-43). It also gives off fascial expansions that reinforce the medial patellar retinaculum and blend with fascia over the popliteus muscle in the upper part of the leg.

A *bursa* usually lies between the tendon of the semimembranosus and the edge of the medial condyle, and part of the bursa deep to the medial head of the gastrocnemius usually separates the semitendinosus tendon from this head. The semimembranosus receives two or more branches from the *tibial part of the sciatic nerve.*

The **adductor magnus** has already been described with other anteromedial muscles of the thigh, with which much of the muscle belongs. It need only be noted here that that part arising from the ischial tuberosity extends almost vertically downward and reaches the lowest attachment of the muscle, the *adductor tubercle.* In its origin, the muscle is, therefore, associated with the posterior muscles. This posterior part is, like them, an extensor at the hip. It receives one or more branches from the *tibial component of the sciatic nerve,* these fairly frequently arising in common with nerves to the semimembranosus.

The **sciatic nerve** leaves the buttock by passing just lateral to the ischial tuberosity and, thereafter, runs downward in the posterior midline, behind the adductor magnus but deep to the other muscles arising from the tuberosity. It may divide into common peroneal and tibial nerves before it reaches the popliteal fossa, but this is of no importance since these nerves do not interchange fibers while they are bound together to form the sciatic.

**Variations, Innervations, and Actions.** Variations in the posterior "hamstring" muscles are not common. There may be a variable amount of fusion among the muscles. Rarely, the short head of the biceps is missing, or there may be an accessory slip of origin or insertion of one of these muscles, more commonly the biceps.

The **innervation** of all the posterior muscles of the thigh is through the sciatic nerve. More specifically, the innervation of all those muscles that arise from the ischial tuberosity, that is, medial to the posterior midline of the thigh, is from the medial component of the sciatic nerve, the tibial nerve. In contrast, the one muscle arising lateral to the posterior midline of the thigh—the short head of the

**FIG. 17-47.** Origins (red) and insertions (blue) of the semitendinosus, biceps, and semimembranosus muscles of the right thigh.

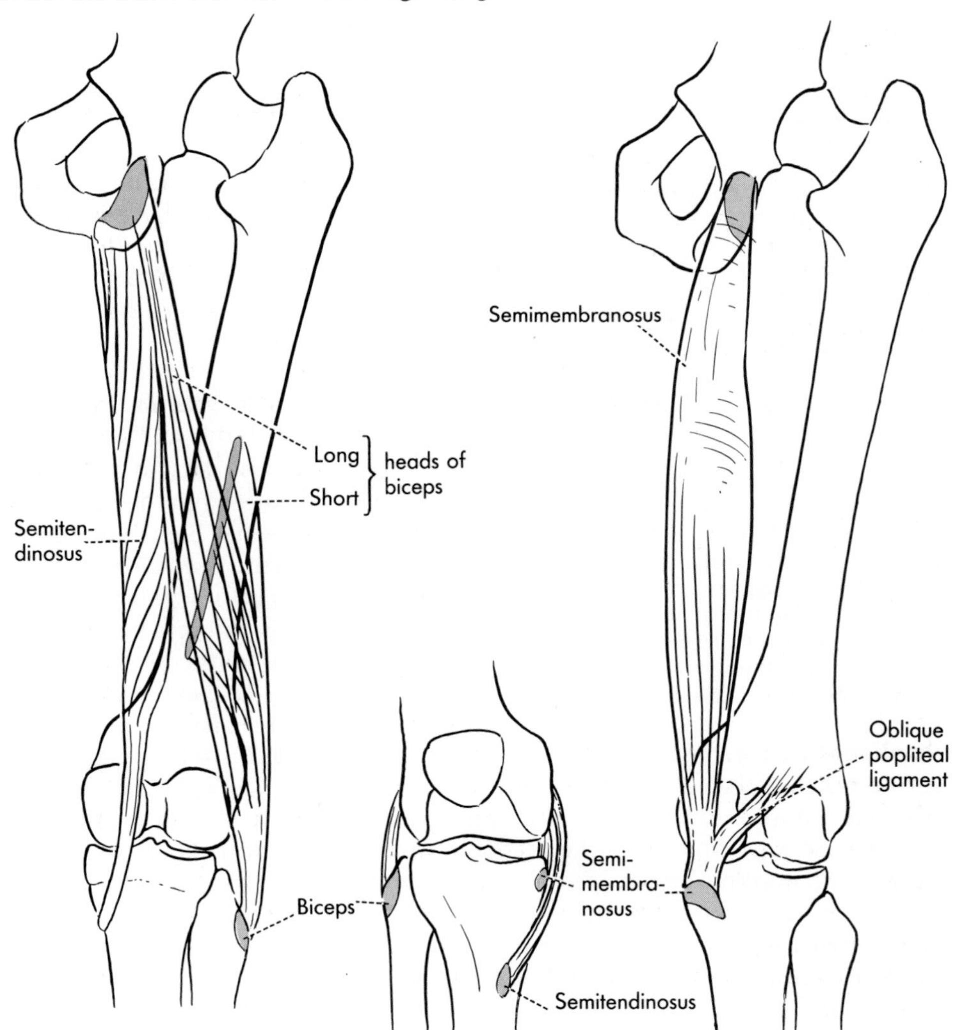

biceps—is innervated from the lateral component of the sciatic nerve, the common peroneal.

The branches of the tibial nerve to the long posterior muscles arise in no regular order. They may originate in various combinations of nerves that then separate to go to different muscles, or even come from a single nerve that leaves the sciatic nerve in the buttock. Usually one or more branches to the upper ends of the muscles leave the sciatic above the level of the ischial tuberosity, and usually, too, each muscle receives more than one nerve. The semitendinosus, the semimembranosus, and the long head of the biceps are said to receive fibers from L5 and S1 or those and S2, the posterior part of the adductor magnus from L4 and L5, and the short head of the biceps from L5 and S1.

In their **actions,** all the muscles arising from the ischial tuberosity assist the gluteus maximus to extend the hip joint, and since they normally maintain more tone than does the gluteus maximus, patients with paralyzed posterior hamstrings tend to fall forward. The semitendinosus, semimembranosus, and long head of the biceps flex the knee, but since the posterior part of the adductor magnus does not, it can extend the thigh at the hip without at the same time producing flexion at the knee—movements that may be incompatible, as, for instance, in rising from a chair. If the leg is flexed, the semitendinosus, and perhaps the semimembranosus, will medially rotate the leg, and both heads of the biceps will laterally rotate it. The short head of the biceps can act only at the knee; it is a more efficient flexor at the knee than is the long head, for the long head apparently relaxes during the last half of flexion, whereas the short head continues to contract. The posterior part of the adductor magnus adducts and also internally rotates the thigh.

It is a common experience that the semitendinosus, the semimembranosus, and the long head of the biceps are not sufficiently long to allow free simultaneous flexion at the hip and extension at the knee, such as are involved in bending over with the knees straight to touch the floor, or in a high kick; sharp flexion at the hip produces flexion at the knee as a result of the resistance of the hamstrings, and extension at the knee produces extension at the hip for the same reason.

**Popliteal Fossa.** Between the diverging biceps on the one hand and the semimembranosus and semitendinosus on the other, and deep to the covering fascia lata, is the upper part of the **popliteal fossa.** The lower borders of the popliteal fossa are formed by the most superficial muscle of the calf, the gastrocnemius, the two heads of which arise from the medial and lateral epicondyles of the femur and converge to a union in the upper part of the calf.

In the loose connective tissue and fat of the popliteal fossa are the lower end of the sciatic nerve and its branches, and the popliteal vessels and their branches and tributaries (Fig. 17-49). In the upper part of the fossa, the sciatic nerve lies posterolateral to the popliteal vessels. The popliteal vein is next, and the popliteal artery is most anterior, lying directly on the popliteal surface of the femur. In the fossa, the common peroneal branch of the sciatic nerve diverges laterally to pass around the lateral side of the leg, but the larger tibial branch descends almost straight down through the fossa. Tibial nerve and popliteal vessels then disappear together deep to the converging heads of the gastrocnemius muscle.

## Nerves and Vessels

After the sciatic nerve leaves the buttock, it runs its entire course in the posterior aspect of the thigh. There is no artery following a similar course, for the chief blood supply to the posterior aspect of the thigh is through branches of the anteriorly placed femoral artery.

**Sciatic Nerve and Branches.** As already noted, the **sciatic nerve** is really two separate nerves that are united only by a common sheath of connective tissue. Of these two components, the tibial nerve is largely medial but a little anterior and the common peroneal nerve is largely lateral but a little posterior.

As the sciatic nerve leaves the buttock lateral to the ischial tuberosity, it disappears deep to the combined origin of the long head of the biceps and the semitendinosus muscle (Fig. 17-48) to proceed downward to the popliteal fossa. All its branches to the thigh except one are from its medial side, from the tibial division. These go to all the muscles arising from the ischial tuberosity. The single lateral branch, from the common peroneal portion of the nerve, usually arises about the middle of the thigh and supplies the short head of the biceps. In or close to the upper end of the pop-

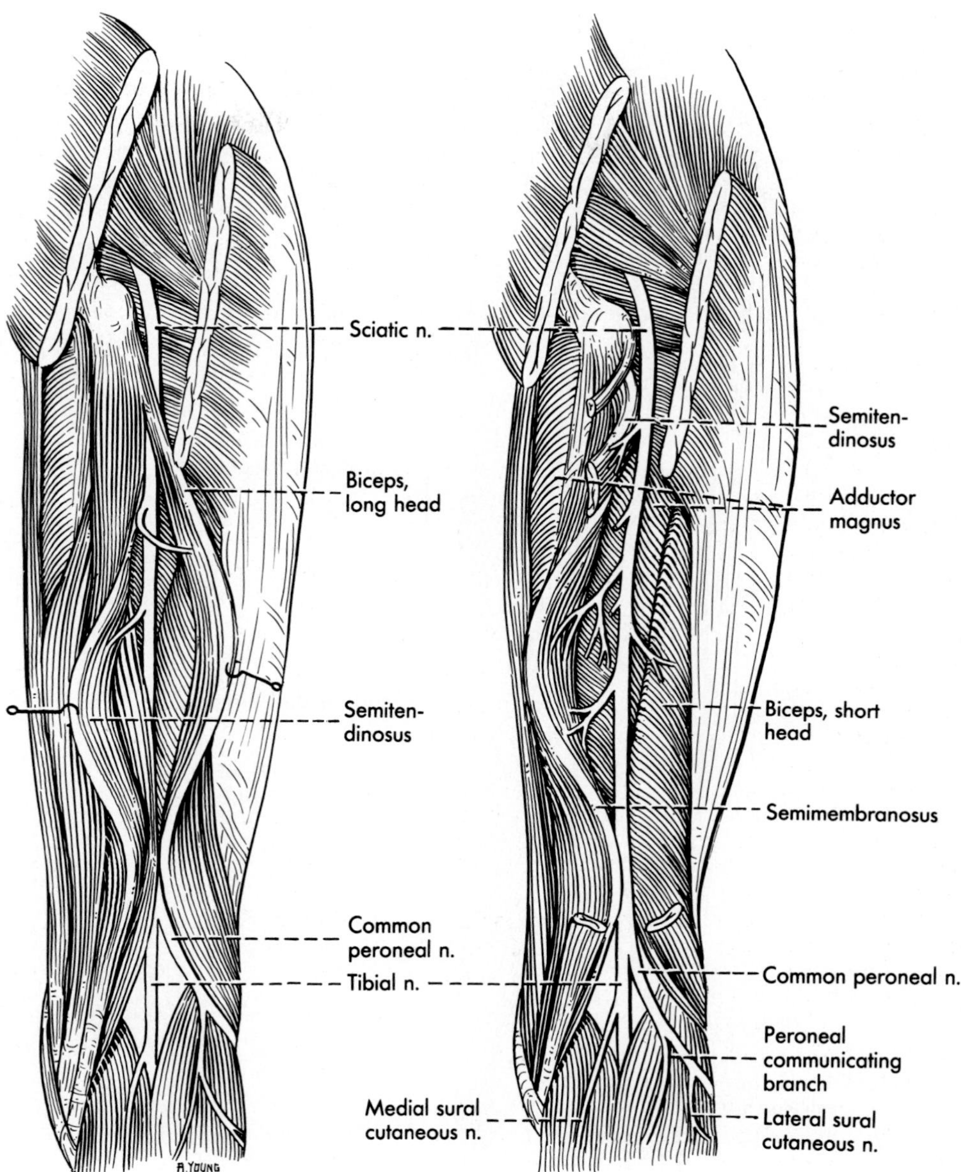


**FIG. 17-48.** The sciatic nerve in the thigh; note that its only lateral branch is to the short head of the biceps.

liteal fossa, the two components of the sciatic nerve separate.

The **common peroneal nerve** diverges laterally and follows approximately the lower edge of the biceps toward the lateral aspect of the leg (thereby crossing superficially the lateral head of the gastrocnemius muscle). Before or soon after it leaves the popliteal fossa, the common peroneal typically gives off two **branches,** a *peroneal communicating branch* and the *lateral sural cutaneous nerve.*

The **tibial nerve** continues almost straight down the middle of the popliteal fossa, gradually passing posterior to the popliteal vessels. The tibial nerve or one of its lower muscular branches gives off a twig that follows the popliteal artery to the knee joint, and in the popliteal fossa it usually gives rise to one cutaneous

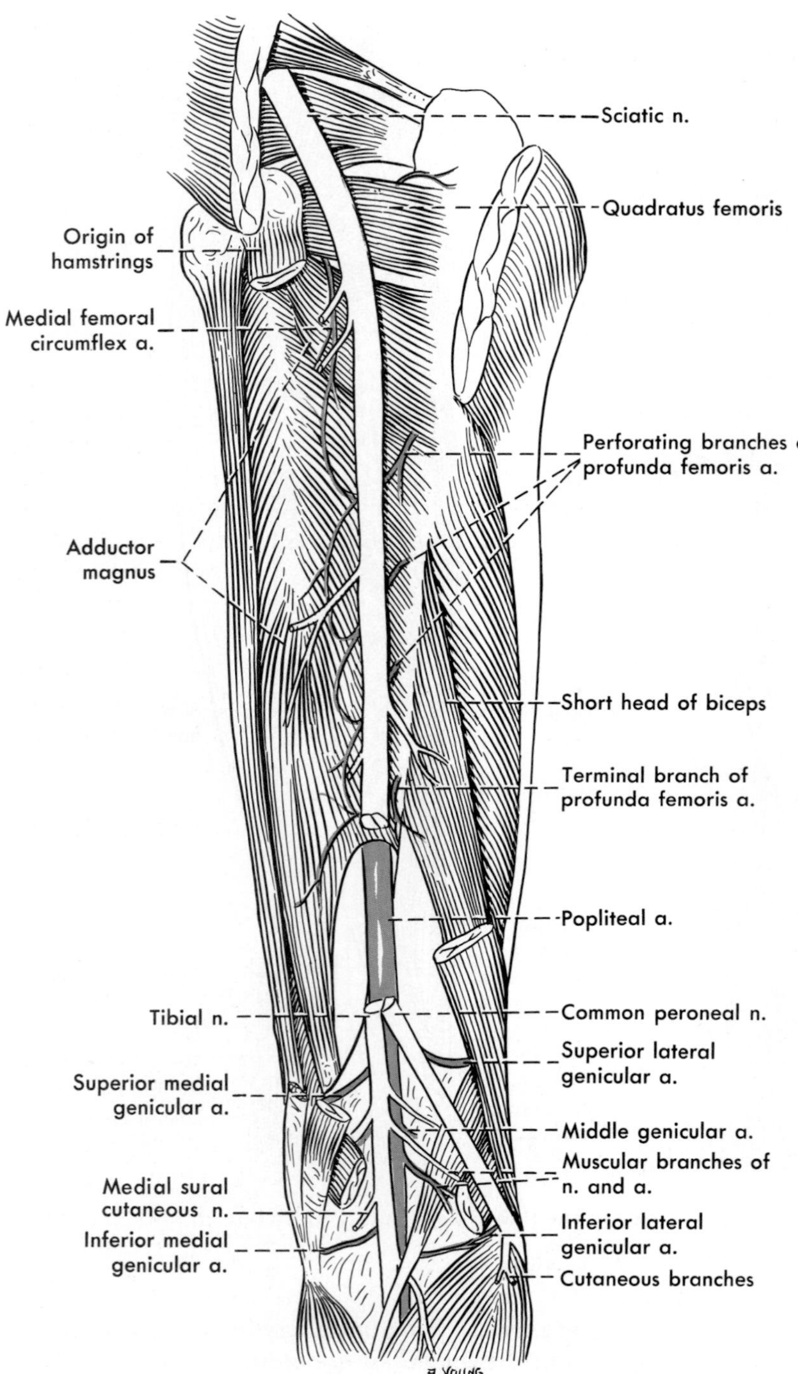


**FIG. 17-49.** Nerves and arteries of the posterior side of the right thigh. Not all the branches of the sciatic nerve are shown; a section has been cut from it, and it has been moved somewhat medially, in order to demonstrate the vessels better. The popliteal vein, not shown, lies between the artery and the sciatic nerve.

and two or three muscular branches. The cutaneous branch is the **medial sural cutaneous nerve,** which runs downward across the union of the heads of the gastrocnemius muscle. The **muscular branches** go to the *plantaris muscle,* a muscle of the leg arising just above the lateral head of the gastrocnemius muscle, and to the *gastrocnemius.*

**Blood Vessels.** The *perforating branches of the profunda femoris vessels,* usually four in number including the terminal branch, appear on the posterior aspect of the thigh by penetrating the adductor magnus muscle. These blood vessels form the chief blood supply to the posterior muscles (Fig. 17-49). However, the *inferior gluteal artery* gives twigs into the upper ends of the muscles attaching to the ischial tuberosity, and the *transverse branch of the medial femoral circumflex* artery appears between the quadratus femoris and the upper border of the adductor magnus. The *popliteal vessels,* the continuation of the femorals, reach the popliteal fossa by passing through the adductor (tendinous) hiatus and run downward, the artery in front of the vein, to disappear between the heads of the gastrocnemius. The popliteal vein usually receives the small saphenous vein, but otherwise its tributaries and the branches of the artery are best seen during dissection of the calf. They are described in Chapter 18.

## THE HIP JOINT

The hip joint is formed by the head of the femur and the deep cuplike cavity of the acetabulum of the coxal bone (Figs. 17-2 and 17-50). Like the glenoid cavity at the shoulder, the acetabulum has a fibrocartilaginous **labrum** attached to its margin. Rather than increasing the diameter of the socket, however, as the glenoidal labrum of the shoulder does, the acetabular labrum increases its depth. The head of the femur fits so deeply into the acetabulum that acetabulum and labrum combined receive more than a hemisphere, and, therefore, the femoral head cannot be removed from the joint without stretching or tearing the labrum.

The continuation of the labrum across the

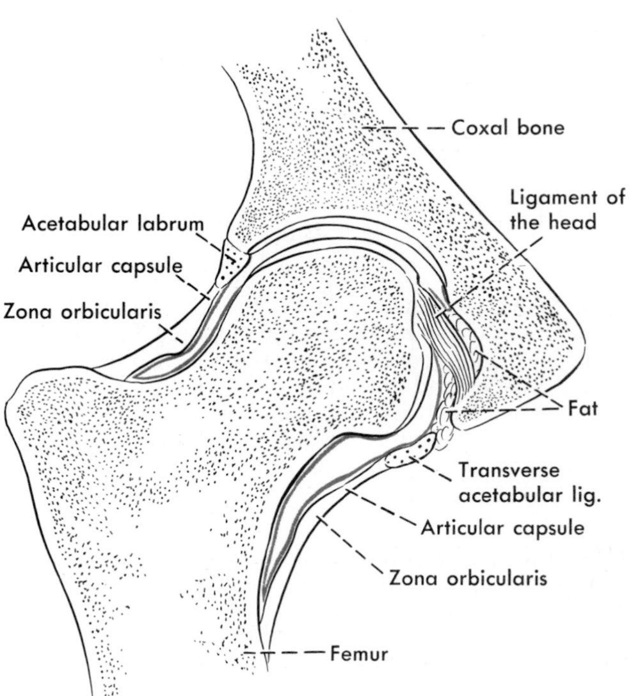


**FIG. 17-50.** A frontal section through the hip joint. Synovial membrane is red.

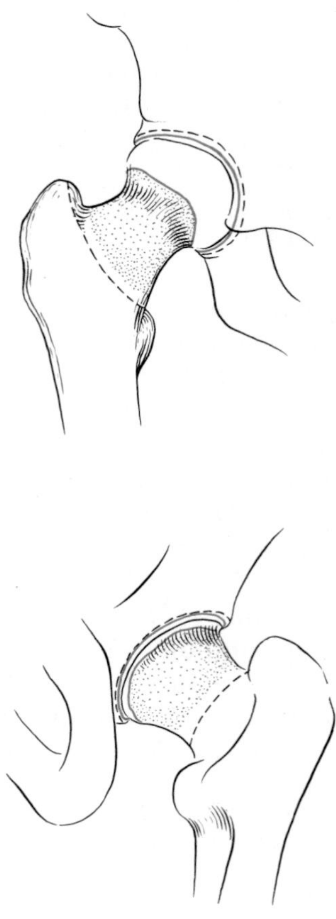

**FIG. 17-51.** Anterior and posterior views of the attachment of the capsule of the hip joint. The broken black lines represent the attachment of the fibrous capsule, and the red lines that of the synovial membrane. Red shading indicates the areas of bone over which the synovial membrane is reflected.

acetabular notch is called the **transverse acetabular ligament.** Between it and the edge of the notch is loose connective tissue through which one or more acetabular arteries enter the joint.

The **articular capsule** of the hip joint is attached proximally to the acetabular labrum (and the transverse acetabular ligament), where the fibrous membrane blends with the labrum and the synovial membrane continues over its inner surface toward the articular cartilage. From the transverse acetabular ligament, the synovial membrane passes across the acetabular notch to cover the fat and connective tissue of the acetabular fossa and be reflected as a tube around the ligament of the head.

On the femur, the capsule attaches laterally to the medial side of the greater trochanter, anteriorly to the intertrochanteric line, and medially just above the lesser trochanter—but posteriorly, it attaches to the neck of the femur a considerable distance above its base so that the lower posterior part of the neck is extracapsular (Fig. 17-51). Along the lines of attachment to the femur, the fibrous membrane of the capsule blends with the periosteum, but the synovial membrane turns proximally around the neck of the femur (often being raised here into folds by the vessels of the head and neck that run between it and the bone of the neck) and continues proximally around the neck to the edge of the articular cartilage of the head.

The fibrous membrane of the hip joint shows three prominent and one less prominent thickening, of which the important ones are described as ligaments. The ligaments are the iliofemoral, the ischiofemoral, and the pubofemoral (Fig. 17-52); these run from the pelvis to the femur. The fourth thickening is composed of circular fibers that are largely hidden by the ligaments. These fibers form the **zona orbicularis.**

The **iliofemoral ligament** is the largest and most important ligament at the hip joint and one of the strongest ligaments in the body. It is attached above to the lower part of the anterior inferior iliac spine and to most of the body of the ilium between the spine and the acetabulum. As it runs downward across the front of the hip joint toward the femur, its fibers spiral somewhat medially, and it expands to attach on the front of the greater trochanter and along the intertrochanteric line. This expanded lower part consists of two

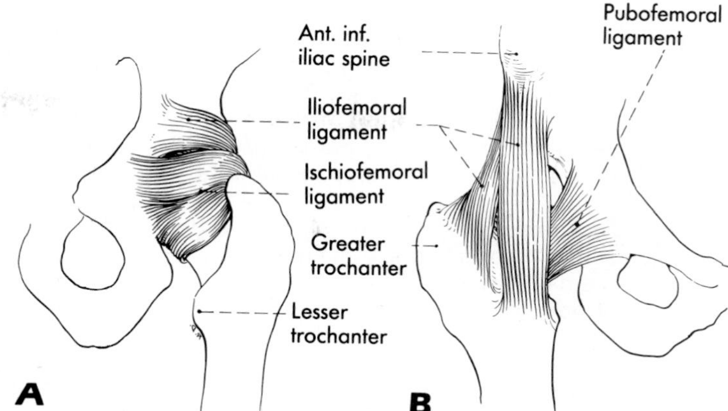


**FIG. 17-52.** Ligaments of the right hip joint seen from behind in (**A**) and from the front in (**B**).

stronger bands with an intermediate weaker portion, thus giving the ligament somewhat the form of an inverted Y; it has been called the "Y ligament" (of Bigelow).

Because it stretches across the front of the hip joint, the iliofemoral ligament strongly resists appreciable hyperextension at this joint; as the joint approaches full extension, the head of the femur is thrust ever more strongly against the ligament. Since the center of gravity normally runs slightly behind the hip joint, gravity alone tends to extend the almost extended thigh, and the iliofemoral ligament thus allows the weight of the body to be balanced on the femoral heads with no more than moderate activity of the iliacus and no activity of other muscles. Since its fibers spiral medially, the iliofemoral ligament is made more taut by medial rotation and hence tends to limit this, as do the other ligaments of the joint; however, it and the other ligaments are relaxed by lateral rotation, and this is a freer movement than is medial rotation.

The **ischiofemoral ligament,** the thinnest of the three ligaments, arises from the body of the ischium behind and below the acetabulum. Its upper fibers are directed horizontally, and its lower fibers spiral upward and laterally so that it covers most of the posterior aspect of the hip joint. It attaches primarily to the posterosuperior part of the base of the neck at its junction with the greater trochanter. Immediately below its lower fibers, therefore, the capsule of the hip joint is thin, and here circular fibers belonging to the zona orbicularis become superficially located.

The **pubofemoral ligament** is anteroinferiorly placed, arising from the body of the pubis and from an adjacent part of the superior ramus. It passes distally in front of and below the lower part of the head of the femur to blend anteriorly with the lower limb of the iliofemoral ligament and attach to the lower surface of the femoral neck. It is especially put in a position of stretch by abduction, and, therefore, presumably assists the adductor muscles in preventing excess abduction.

The hip joint is most stable when the femur is extended. This is said to result from a tightening of all three ligaments through an internal rotation and abduction that accompanies extension.

Within the cavity of the hip joint, surrounded by a tube of synovial membrane, is the **ligament of the head** (Fig. 17-50). This runs from the acetabular fossa and transverse ligament to the fovea of the head. Although it should theoretically resist adduction, it apparently does not become taut enough to function as a checking ligament in the adult. It has been said to check posterosuperior displacement of the head of the femur in the fetus and newborn, and an abnormally long ligament is said to account for the laxity of the hip joint or even frank dislocation that sometimes is present in early neonatal life.

#### Nerve and Blood Supply

The nerves to the hip joint are generally small, and there tends to be little overlap between adjacent branches. The chief **nerve supply** to the front of the joint is by one or more branches of the *femoral nerve* that may arise from that nerve directly or from its muscular branches. The anteroinferior part of the joint is usually supplied by one or more branches from the *obturator nerve* and sometimes also from the accessory obturator. The posterior aspect is supplied by a twig from *the nerve to the quadratus femoris.* Sometimes a branch of the superior gluteal nerve reaches the upper lateral part of the joint.

The heavy ligaments to the hip joint are largely avascular, so the vessels to the joint are small. They consist of small twigs that are derived from branches of **adjacent vessels,** therefore, in general, from both *femoral circumflexes,* both *gluteals,* and the *obturator.* The uppermost perforating branch of the profunda femoris may also help supply the joint. The branches from the femoral circumflex vessels that penetrate the distal part of the capsule are of appreciable size, but they are destined primarily for the femur and give off only small twigs to the capsule.

### Movements at the Hip

Movements at the hip joint are intimately concerned with posture and with walking, but the coordination of hip, knee, and ankle in these functions can best be considered after the knee and ankle have been studied. A discussion of this will be found in Chapter 18.

**Flexion** of the thigh, if the hamstrings are relaxed by flexion at the knee, is checked primarily by apposition of the soft tissues. The strongest flexor is the *iliopsoas,* but this is apparently used only when the movement must be forceful. The *tensor fasciae latae,* the *pectineus,* and the *sartorius* are probably particularly involved in the flexion demanded, for instance, in swinging the leg forward during the movement of walking, perhaps assisted also by the adductor brevis and the adductor longus. Other flexors at the hip are the anterior parts of the adductor magnus and the gluteus minimus. The gracilis flexes the thigh only if the knee is extended. (The anterior fibers of the gluteus medius, once regarded as assisting flexion, have been reported not to participate in this movement.) Of these muscles, the tensor fasciae latae, the pectineus, the adductors, and the anterior part of the gluteus minimus tend to medially rotate as they flex, but the sartorius has a tendency to rotate laterally during flexion.

**Extension** at the hip is checked by the pressure of the head of the femur against the structures lying in front of the joint—the iliofemoral ligament and the overlying tendon of the iliopsoas muscle. Of the extensors of the thigh, the *gluteus maximus* and the posterior part of the *adductor magnus* are the most important in such movements as climbing stairs or jumping, both because of their strength and the fact that they act only across the hip joint. The posterior hamstring muscles cannot act strongly in extension at the hip in these movements because in so doing they would prevent extension at the knee. The gluteus medius and piriformis are particularly poor extensors at the hip, although they will produce some extension, combined with abduction and outward rotation.

The **abductors** include, besides the *gluteus medius and minimus,* the tensor fasciae latae, the sartorius, and the piriformis, which abduct the non-weight-bearing limb very weakly. The gluteus maximus, otherwise an extensor and adductor, can assist in abduction when the thigh is flexed. However, the chief action of the abductors is not to abduct the free limb but to prevent sagging of the pelvis on the opposite side when the weight of the body is on one limb, and this is a function carried out primarily by the gluteus medius and minimus. Since these muscles (and the tensor) are innervated by the superior gluteal nerve alone, interruption of this nerve makes it impossible to stabilize the pelvis when the weight of the body is placed on the limb of the paralyzed side. In such a case, the opposite side of the pelvis drops markedly and a person so affected must sharply flex his trunk toward the side of paralysis in order to preserve his balance.

**Adduction** is usually limited by the contact of the thighs. If contact between soft tissues does not fully check the movement, then presumably upper fibers of the iliofemoral and ischiofemoral ligaments help to do so.

The adductors of the thigh are fairly numerous, but the strongest ones are those named adductor—the brevis, longus, and magnus—and the gracilis. The gluteus maximus adducts as it extends and laterally rotates; the pectineus flexes better than it adducts; and the obturator externus, the iliopsoas, and the hamstring muscles, although usually listed as adductors, presumably play little part in this movement normally. The segmental innervation of the adductor muscles is primarily from the lower lumbar spinal nerves.

**Medial rotation** is limited and is quickly checked by the increased spiral imparted to the ligaments of the hip joint by this movement. The chief medial rotators of the hip are the anterior part of the *gluteus minimus* and the much less effective tensor fasciae latae, both of which flex as they medially rotate. The gracilis, semitendinosus, semimembranosus, and posterior part of the adductor magnus are also medial rotators. The adductor muscles (including the anterior part of the magnus) and the iliopsoas have been regarded by some workers as lateral rotators, by others as medial rotators. Results of electrical stimulation support the first view, but electromyography has confirmed it only for the iliopsoas. The others act in medial rotation.

**Lateral rotation** of the thigh is presumably checked by contact of the posterior aspect of the neck of the femur with the rim of the acetabulum. The potential lateral rotators are particularly numerous, but those that are commonly used when pure rotation is desired are apparently those in the buttock, the so-called short rotators—the piriformis, the obturator internus and associated gemelli, the obturator externus, and the quadratus femoris. The gluteus maximus and the posterior fibers of the gluteus medius (but not the gluteus minimus) are also lateral rotators, the sartorius is a very weak lateral rotator, and the long head of the biceps apparently has a slight lateral rotatory action on the thigh.

Both the medial and the lateral rotators are innervated by a number of nerves. The medial rotators are supplied primarily through fibers from the lower two lumbar nerves, and the lateral rotators by lower lumbar nerves and the upper one or two sacral ones.

#### Dislocation

Dislocation of the hip can be congenital, with stretching of ligaments and sometimes maldevelopment of the acetabulum, the femoral head, or both; it requires reduction before the limb is used to bear weight.

Traumatic dislocation of the hip is not a frequent occurrence because of the deep acetabulum. When it takes place, it is usually the result of a severe blow upon the knee while the hip is flexed. The head of the femur is thus usually dislocated posteriorly, with a tearing of the posterior part of the capsule and, frequently, fracture of the acetabulum. In anterior dislocation, much rarer than posterior, the head of the femur passes around the medial edge of the iliofemoral ligament and lodges against the body of the pubis or the obturator foramen.

## GENERAL REFERENCES AND RECOMMENDED READINGS

BASMAJIAN JV, GREENLAW RK: Electromyography of iliacus and psoas with inserted fine-wire electrodes. Anat Rec 160:310, 1968

BEATON LE, ANSON BJ: The relation of the sciatic nerve and of its subdivisions to the piriformis muscle. Anat Rec 70:1, 1937

BOSCOE AR: The range of active abduction and lateral rotation at the hip joint of men. J Bone Joint Surg 14:325, 1932

CHANDLER SB: The iliopsoas bursa in man. Anat Rec 58:235, 1934

CROCK HV: A revision of the anatomy of the arteries supplying the upper end of the human femur. J Anat 99:77, 1965

EDWARDS EA, ROEBUCK JD JR: Applied anatomy of the femoral vein and its tributaries. Surg Gynecol Obstet 85:547, 1947

EVANS FG, HAYES JF, POWERS JE: "Stresscoat" deformation studies of the human femur under transverse loading. Anat Rec 116:171, 1953

EVANS FG, LISSNER HR: Studies on pelvic deformations and fractures. Anat Rec 121:141, 1955

GARDNER E: The innervation of the hip joint. Anat Rec 101:353, 1948

HOWE WW JR, LACEY T, SCHWARTZ RP: A study of the gross anatomy of the arteries supplying the proximal portion of the femur and the acetabulum. J Bone Joint Surg (Am) 32:856, 1950

INMAN VT: Functional aspects of the abductor muscles of the hip. J Bone Joint Surg 29:607, 1947

JOSEPH J: The graphic representation of movement: I. Electromyography of postural muscles. Ann Physical Med 5:185, 1960

KAPLAN EB: The iliotibial tract: Clinical and morphological significance. J Bone Joint Surg (Am) 40:817, 1958

LIEB FJ, PERRY J: Quadriceps function: An anatomical and mechanical study using amputated limbs. J Bone Joint Surg (Am) 50:1535, 1968

LIPSHUTZ BB: Studies on the blood vascular tree: I. A composite study of the femoral artery. Anat Rec 10:361, 1916

PAULY JE, SCHEVING LE: An electromyographic study of some hip and thigh muscles in man. Electromyography (Suppl 1) 8:131, 1968

ROBERTS WH: The locking mechanism of the hip joint. Anat Rec 147:321, 1963

SENIOR HD: An interpretation of the recorded arterial anomalies of the human pelvis and thigh. Am J Anat 36:1, 1925

WILLIAMS GD, MARTIN CH, MCINTYRE LR: Origin of the deep and circumflex femoral group of arteries. Anat Rec 60:189, 1934

WOODBURNE RT: The accessory obturator nerve and the innervation of the pectineus muscle. Anat Rec 136:367, 1960

# 18

# LEG AND FOOT

The leg (*crus*) and the foot (*pes*) correspond rather well to a pronated forearm and hand with the hand in extension and the arm medially rotated at the shoulder. So compared, it is obvious that the medial (tibial) side of the leg corresponds to the radial (lateral) side of the forearm, and the lateral (fibular) side of the leg corresponds to the medial or ulnar side of the forearm. Also, the anterior surface of the leg is the extensor surface, and the posterior surface, the flexor one. The plantar surface (sole of the foot) is down and the dorsum up. The big toe (*hallux*) corresponds in position to the thumb, and the little toe (*digitus minimus*) corresponds to the little finger. Many comparisons are possible between the nerves and vessels of the leg and foot and those of the forearm and hand. Some of these will be pointed out in the following descriptions.

Important differences are that although the elbow allows both flexion–extension and pronation–supination movements of the forearm, the knee allows primarily flexion–extension only, even though some rotatory movement is possible when the knee is flexed. Further, although abduction and adduction without flexion or extension are allowed at the wrist, movements at the foot are normally oblique ones: thus, abduction of the foot (movement toward the fibular side) typically also involves eversion (turning the sole of the foot out), whereas adduction of the foot (movement toward the tibial side) typically involves inversion. Another important difference between hand and foot is the rather slight mobility of the big toe as compared to the thumb.

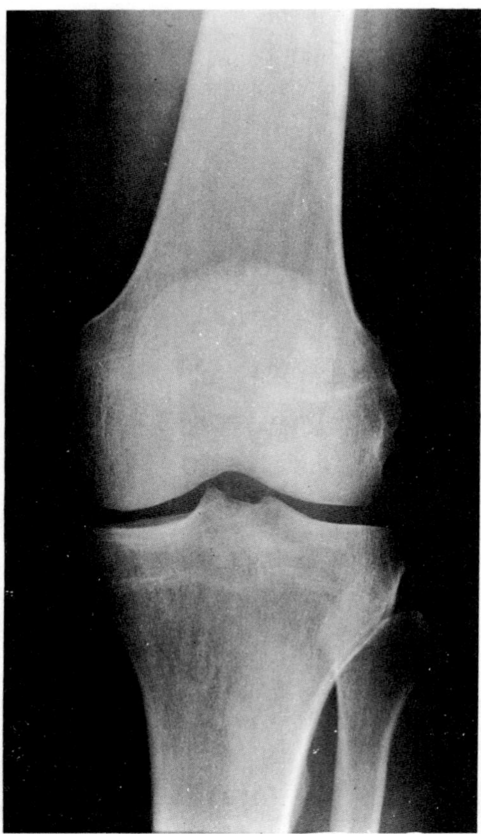

**FIG. 18-1.** Anteroposterior view of the left knee. The base of the patella, the femoral and tibial condyles (appearing particularly widely separated from each other because of the menisci), and the tibial intercondylar eminence all show clearly. Note the angle between the long axis of the femur and that of the tibia. (Courtesy of Dr. D. G. Pugh)

## SKELETON

The skeleton of the leg and foot consists of the tibia and fibula in the leg; the tarsal or ankle bones; the metatarsal bones, the more distal bones of the foot; and the phalanges, or skeleton of the digits.

### TIBIA AND FIBULA

The expanded upper end of the tibia, formed chiefly by two condyles, articulates with the femur above, and the lower surface of the lateral condyle articulates with the head of the fibula (Fig. 18-1). The synovial **tibiofibular joint** is small and merits no special comment save that sometimes it communicates with the knee joint. The bodies of the tibia and fibula are united by an **interosseous membrane,** resembling that of the forearm. A gap at its upper end allows the anterior tibial vessels to pass from posterior to anterior aspects of the leg, and a foramen in its lower part allows passage for the perforating branch of the peroneal artery. At their lower ends, the tibia and fibula are united at the **tibiofibular syndesmosis.** If this contains a diverticulum from the ankle joint, which it sometimes does, it is then properly called a *tibiofibular articulation* (synovial joint). The ligaments uniting the tibia and fibula at the syndesmosis are the **anterior** and **posterior tibiofibular ligaments,** which blend above with the interosseous membrane.

Congenitally, either the fibula or the tibia may be partially or completely absent. Of all the long bones of the limbs, the fibula is most frequently defective. In absence of the lower end of either bone, the ankle joint is malformed—the sole of the foot is turned toward the side of the defect—and toes or tarsal bones on the side of the defect may also be lacking.

#### Tibia

The upper end of the tibia consists of **tibial condyles** that bear slightly concave articular surfaces. They are separated from each other anteriorly and posteriorly by shallow sloping depressions, the anterior and posterior **intercondylar areas,** and between these by an **intercondylar eminence** with medial and lateral **intercondylar tubercles** (Figs. 18-2 and 18-3). Several ligaments of the knee attach to the intercondylar areas, but nothing attaches directly to the intercondylar eminence. Posteroinferiorly on the lateral condyle is the **fibular articular surface.**

The **body of the tibia** has anterior, medial, and interosseous **borders** and medial, lateral, and posterior **surfaces.** The anterior border begins above at the prominent **tibial tuberosity** (for the insertion of the quadriceps muscle); it and the almost flat medial surface are

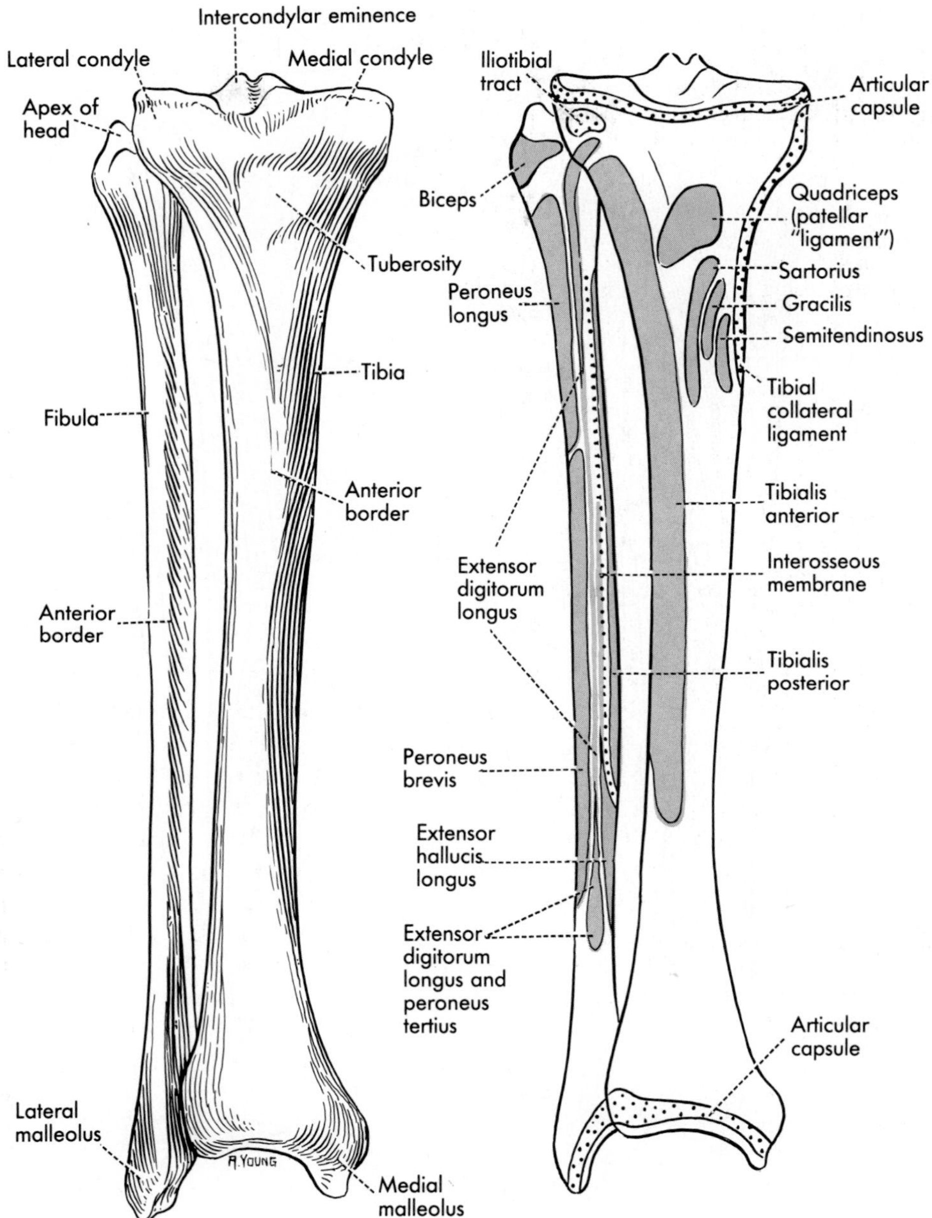


**FIG. 18-2.** Anterior views of the tibia and fibula and their muscular attachments. Origins are shown in red, insertions in blue.

largely subcutaneous and easily palpable. The medial border is somewhat rounded, and the interosseous border sharper. Neither the medial nor the lateral surface presents any particular markings, but on the upper third of the posterior surface the *line of the soleus muscle* (soleal line, oblique or popliteal line) extends obliquely medially and downward, marking the tibial origin of the soleus muscle.

At the lower end of the tibia, the **medial malleolus** projects downward. Its lateral surface and the inferior articular surface of the tibia articulate with the trochlea tali (Fig. 18-5). Behind the malleolus is the *malleolar sul-*

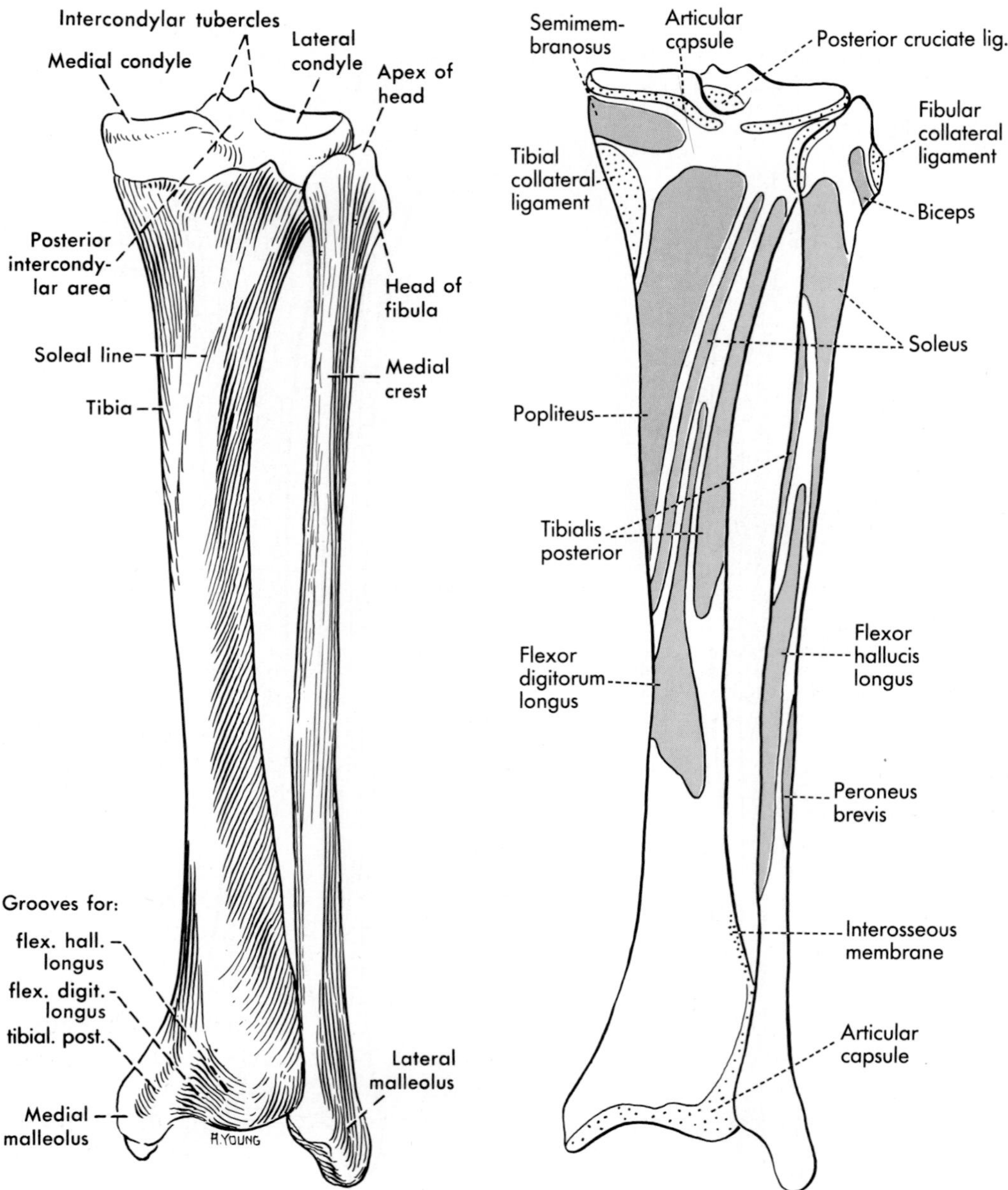


**FIG. 18-3.** Posterior views of the tibia and fibula and their muscular attachments.

cus around which tendons pass from the posterior side of the leg to the plantar aspect of the foot. On the lateral side of the lower end of the tibia is the *fibular notch* for the accommodation of that bone.

### Fibula

The expanded upper end of the fibula, the **head,** presents a pointed apex and an articular facet for its joint with the tibia (Figs. 18-2 and 18-3).

The long slender **body** is described as having anterior, interosseous, and posterior **borders** and medial, lateral, and posterior **surfaces.** However, only the interosseous border is clear-cut, and borders and surfaces spiral so that they are difficult to follow. None of them remains strictly in the position implied by its name (Fig. 18-4), and attempting to make them out on the usual bone is scarcely worth the necessary time. Perhaps most important is to note, on the upper posterior surface of the bone, the *medial crest,* which divides that surface into two parts.

The expanded lower end of the fibula is the **lateral malleolus.** It bears a facet on its medial surface for articulation with the talus. Behind and below this is a *roughened fossa* for the attachment of the *posterior talofibular ligament.*

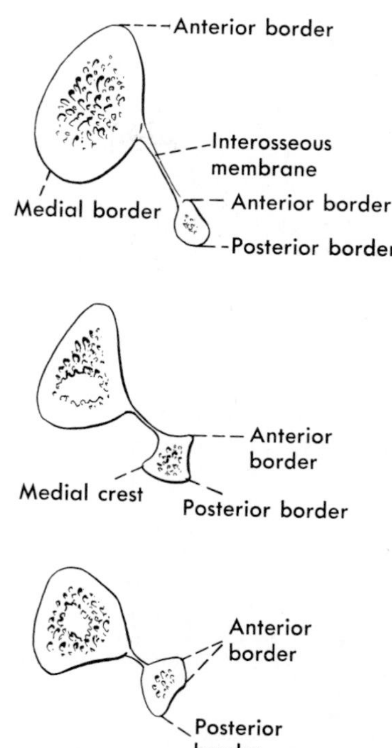


**FIG. 18-4.** Cross sections through the upper, middle, and lower parts of the bones of the leg, to illustrate especially the varying shape of the fibula.

## TARSALS, METATARSALS, AND PHALANGES

The tarsal bones form the heel (*calx*) and a proximal (posterior) part of the foot, and the metatarsal bones form the remainder of the major part of the foot. The phalanges contribute very little to the length of the foot.

### Tarsal Bones

There are seven tarsal bones (Figs. 18-5, 18-6, and 18-13). The bone that forms the heel is the calcaneus. Above it is the talus. The navicular lies in front of the talus, the cuboid lies in front of the calcaneus, and three cuneiform bones lie in front of the navicular, medial to the cuboid.

The **calcaneus** (*os calcis*), the largest of the tarsals, presents, on its upper surface, articular surfaces for the talus: the posterior one is demarcated anteriorly by a groove, the **sulcus calcanei;** anterior to this sulcus the calcaneus expands medially to form the **sustentaculum tali,** which bears the middle articular surface; and still more anteriorly, sometimes blending with the middle surface, is the anterior talar articular surface.

At the back end of the inferior surface of the calcaneus is the downward-projecting **tuber calcanei,** which bears rounded medial and lateral *tuberal processes* that support the weight upon the heel.

On the medial surface, beginning behind and extending forward below the sustentaculum tali, is the *groove for the flexor hallucis longus tendon.* On the lateral surface, there may be no particular marking, or there may be a small projection, the *peroneal trochlea,* with a slight groove behind and below it for the tendon of the peroneus longus.

The posterior end of the calcaneus is

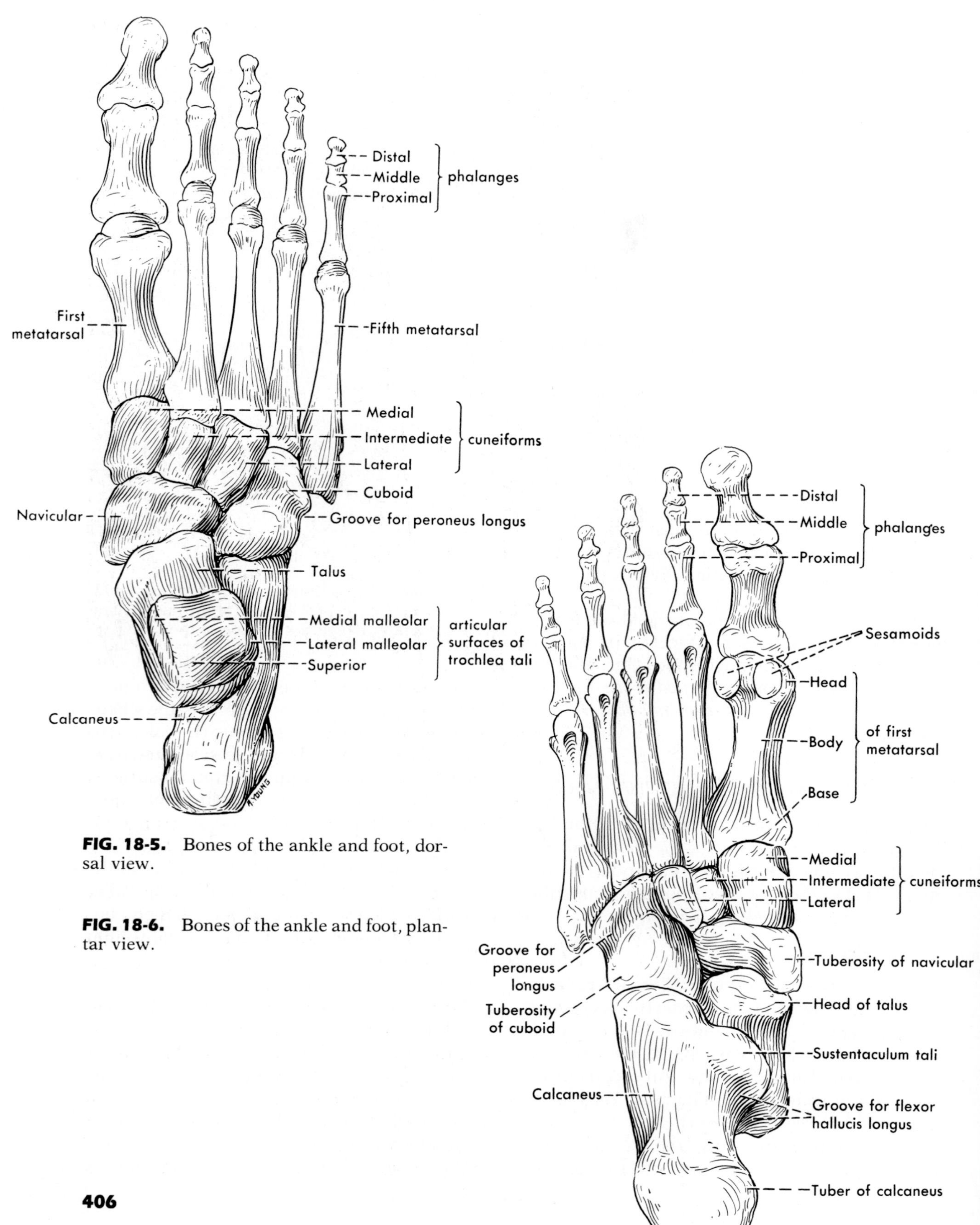


**FIG. 18-5.** Bones of the ankle and foot, dorsal view.

**FIG. 18-6.** Bones of the ankle and foot, plantar view.

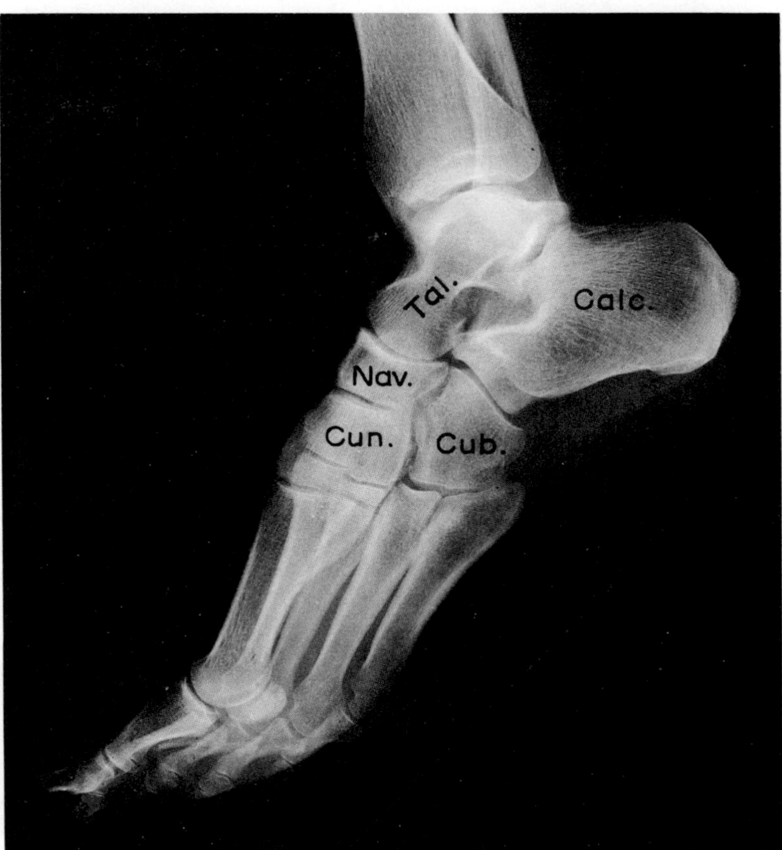


**FIG. 18-7.** A roentgenogram of the foot. **Tal.** is the talus, **Calc.** the calcaneus, **Nav.** the navicular, **Cub.** the cuboid, and **Cun.** the three cuneiforms superimposed on each other. (Courtesy of Dr. D. G. Pugh)

roughened below its middle for the attachment of the tendo calcaneus (the tendon of the important plantar flexor of the foot). The anterior end presents an articular surface for articulation with the cuboid bone.

The **talus** transmits the weight of the body to all the other weight-bearing bones of the foot (Fig. 18-7). Its posterior part, the **body,** is surmounted by the heavy, pulleylike **trochlea tali,** which presents medial and lateral articular surfaces for the malleoli and a superior articular surface for the distal end of the tibia. Below the lateral articular surface is a projection, the *lateral process,* for the attachment of a lateral ligament at the ankle, and below the medial articular surface is a rough depression for the attachment of the medial ligament. The *posterior process,* projecting backward from the body, ends in medial and lateral *tubercles* that are divided by the *sulcus for the tendon of the flexor hallucis longus muscle.* The anterior expanded end, the **head,** is separated from the body by a roughened **neck** and forms an anterior articular surface that articulates in front with the navicular bone and inferiorly rests upon an underlying ligament and the calcaneus.

On the inferior surface of the body of the talus is the oblique **sulcus tali,** which separates a large oval posterior calcaneal articular surface from a middle calcaneal articular surface that rests on the sustentaculum tali.

When the talus is in place on the calcaneus, a deep depression, the **sinus tarsi,** can be seen laterally between the two bones. This extends posteriorly and medially, deep to the neck and trochlea of the talus, to become continuous with the canal formed by the opposed sulcus calcanei and sulcus tali; the posteromedial end of the canal is just behind the sustentaculum tali. This canal is often considered a part of the sinus tarsi, but is best given the separate name of **tarsal canal.**

The **navicular bone,** so named because of its fancied resemblance to a boat, presents a concave proximal articular surface for accommodation of the head of the talus and a convex distal surface for the three cuneiform bones. Medially and inferiorly is the rough **tuberosity,** to which a major portion of the tibialis posterior tendon attaches.

The **cuboid bone** presents proximally an articular surface for the calcaneus, and distally one for the 4th and 5th metatarsals. On the upper posterior part of its medial border is a facet for the lateral cuneiform. The ridge on the lateral and inferior surfaces of the bone is the *tuberosity,* delimited anteriorly by the *sulcus for the tendon of the peroneus longus muscle.* The tuberosity may present a smooth articular facet for a seasamoid bone that frequently lies in the tendon of the peroneus longus.

Of the three **cuneiform bones,** the medial is the largest. Its lower surface is broader than its upper one, and its medial surface is rounded. This bone presents articular surfaces for the navicular, the 1st and 2nd metatarsals, and the intermediate cuneiform. The contiguous areas of the cuneiforms not occupied by articular surfaces are rough for the attachment of heavy interosseous ligaments.

The intermediate cuneiform is the smallest, and both it and the lateral cuneiform have broader dorsal surfaces and narrower plantar ones. The intermediate cuneiform has articular surfaces for the navicular, the 2nd metatarsal, and the medial and lateral cuneiforms.

The lateral cuneiform articulates mostly with the navicular, the intermediate cuneiform, the cuboid, and the 3rd metatarsal, and to a small extent with the 2nd and 4th metatarsals.

No muscles insert on the dorsal aspect of any of the tarsals, but the tibialis anterior and posterior both insert in part into the plantar surfaces of certain of the tarsals (Figs. 18-8 and 18-9).

### Metatarsals

Each of the metatarsal bones (Figs. 18-5, 18-6, and 18-10) consists of a base, a body, and a distally lying head. The **bases** all bear articular surfaces for the tarsals with which they articulate; therefore, the base of the 2nd metatarsal extends onto its sides, since it articulates with all three cuneiform bones. The 1st and 2nd metatarsals typically have no synovial cavity between them, but otherwise the sides of the bases each bear double articular facets, one above and one below, with a rough intervening area for an interosseous ligament. The *tuberosity* of the 1st metatarsal is a small projection of the plantar surface of the base, for attachment of the tendon of the peroneus longus. The *tuberosity* of the 5th metatarsal is the tapered posterior projection of the bone.

The **body** of the 1st metatarsal is heavy, but the bodies of the remainder are slender. All tend to be convex plantarward. The **heads** are covered distally, but not on their sides, by articular cartilage for the phalanges. Prominent medial and lateral **sesamoid bones** lie on the plantar surface of the head of the 1st metatarsal.

### Phalanges

The **proximal phalanges** are short. Their bases are concave for articulation with the heads of the metatarsals, their bodies slender and somewhat curved plantarward, and their heads (distal ends) pulleylike. The bases of the still shorter **middle phalanges** are somewhat saddle shaped, and their heads are pulleylike. There is no middle phalanx of the big toe, and those of the 4th and 5th toes may have practically no bodies. The **distal phalanx** of the big toe is large, but the others are tiny pieces of bone. The base of each is somewhat saddle shaped, and the distal end bears a *tuberosity* on its plantar surface.

#### OSSIFICATION

The **tibia** and **fibula** develop each from three centers of ossification—one for the body, and proximal and distal ones for the two ends (Figs. 18-11 and 18-12). The **metatarsals** and the **phalanges,**

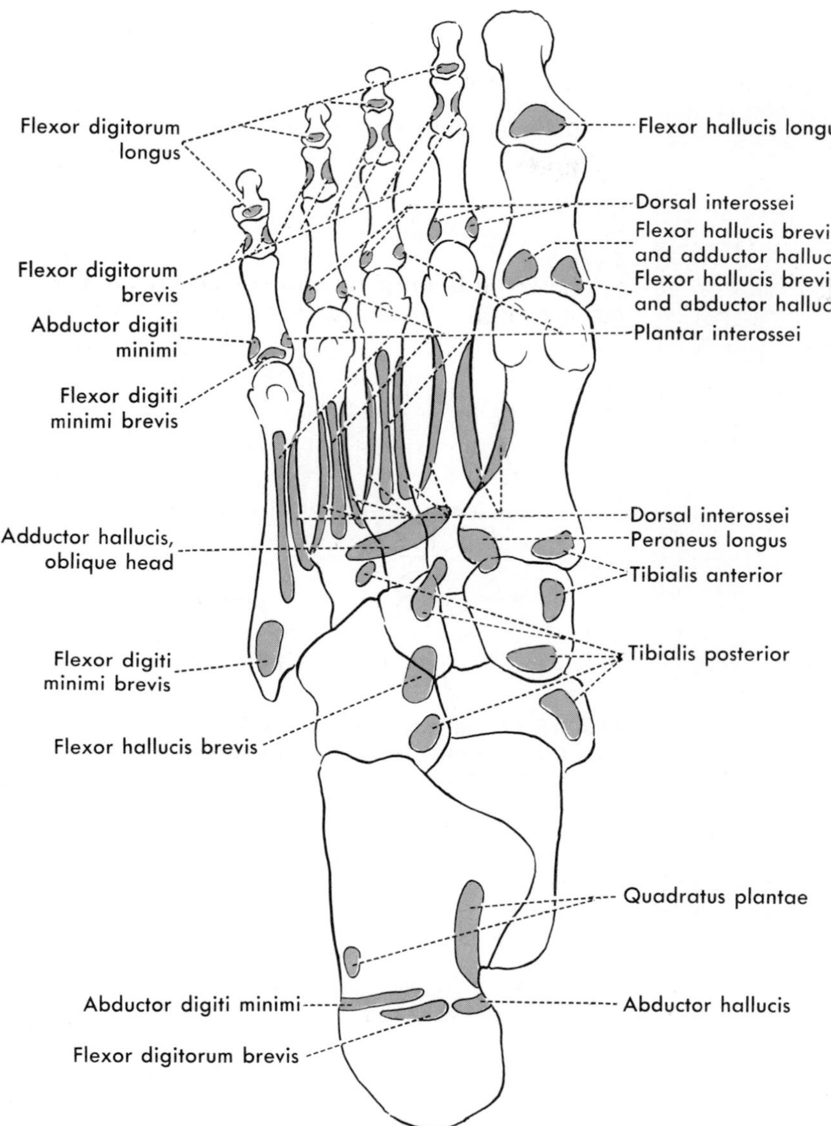


**FIG. 18-8.** Muscular attachments on the plantar surface of the foot.

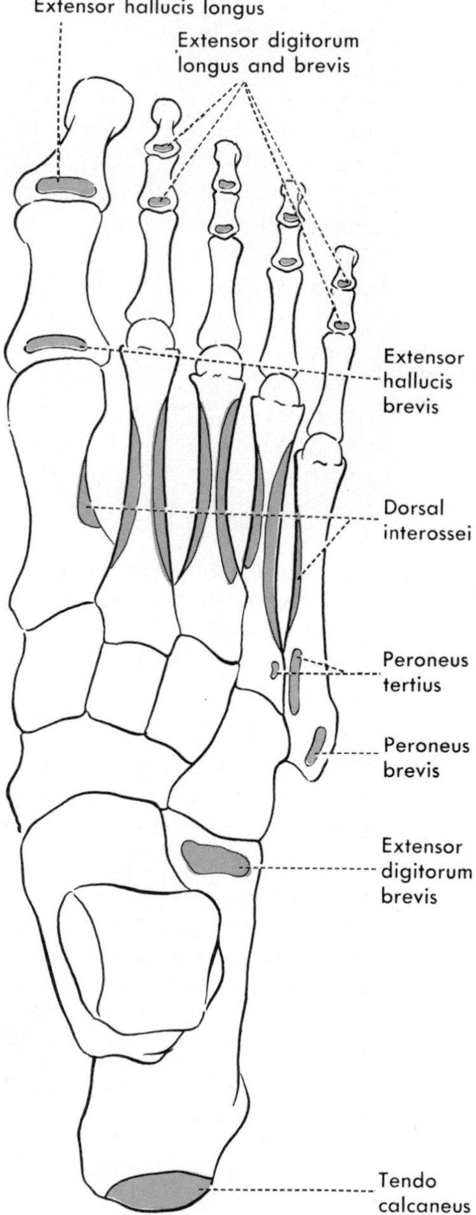


**FIG. 18-9.** Muscular attachments on the dorsum of the foot.

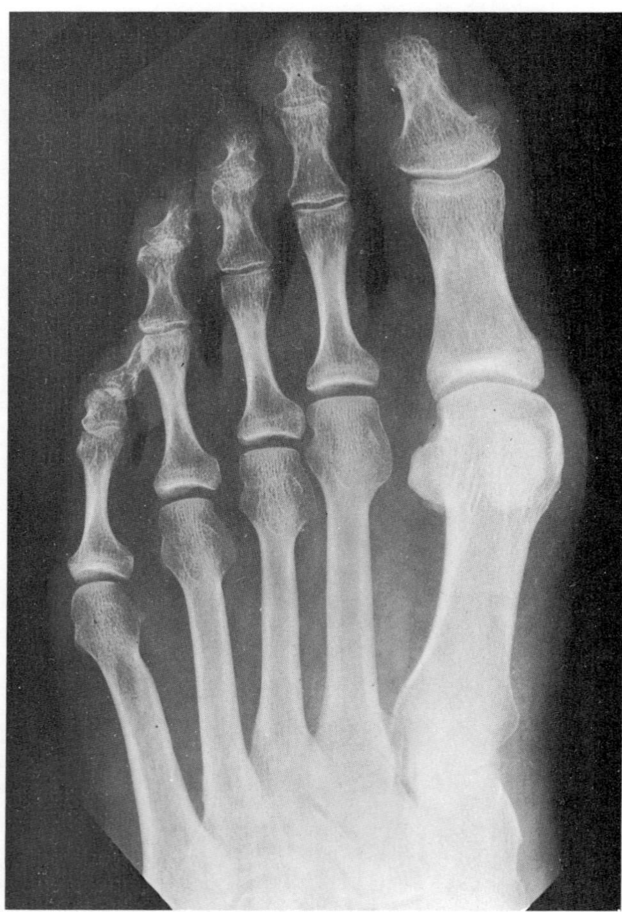

**FIG. 18-10.** The metatarsals and phalanges. The sesamoids of the big toe show plainly. (Courtesy of Dr. D. G. Pugh)

like the corresponding bones of the hand, have a center for the body but only one epiphysis for each bone; this is at the distal end of all the metatarsals except the 1st, where it is at the proximal end, and it is at the proximal end of all the phalanges. Each **tarsal** typically develops from a single center of ossification, except that the calcaneus regularly has also a small epiphysis (Fig. 18-12).

The centers of ossification for the bodies of the tibia and the metatarsals appear during the seventh and eighth weeks of fetal life, that for the body of the fibula soon thereafter, and those for the bodies of the phalanges about the tenth week. The calcaneus shows ossification during the sixth fetal month, and the talus during the seventh month.

The proximal epiphysis for the tibia appears usually during the ninth prenatal month and gives rise not only to the condyles but also to the tibial tuberosity; the center for the lower end usually appears during the first year. The epiphyseal centers for the fibula appear in the second and third years.

Normally, of the tarsals only the talus and the calcaneus show ossification at birth, but a center for the cuboid may appear shortly after birth. Centers for the other tarsals appear during the first 3 to 4 years.

Centers of ossification for the epiphyses of the metatarsals appear at approximately 2 to 3 years;

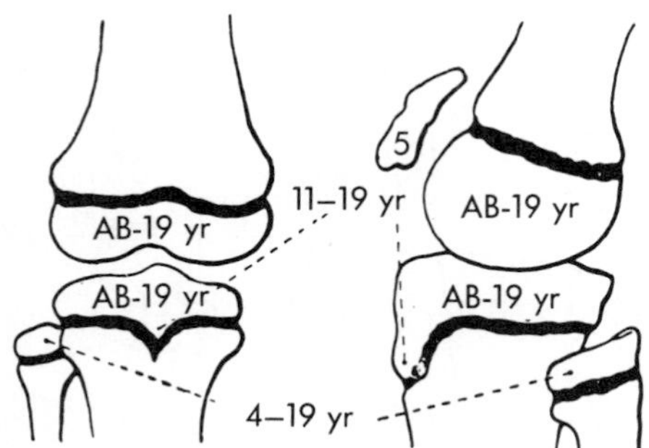


**FIG. 18-11.** Ossification at the knee—the lower femoral and upper tibial and fibular epiphyses, and the patella. **AB** indicates the presence of the epiphysis at birth. The first of two numbers indicates the time of expected appearance of an epiphysis, and the second indicates the time of expected fusion with the body. (Camp JD, Cilley EIL: Am J Roentgenol 26:905, 1931)

**FIG. 18-12.** Ossification of the ankle and foot—times of appearance and fusion of the lower epiphyses of the tibia and fibula and of the sole epiphysis of each metatarsal, each phalanx, and the calcaneus, and times of appearance of the centers for the tarsal bones. (Camp JD, Cilley EIL: Am J Roentgenol 26:905, 1931)

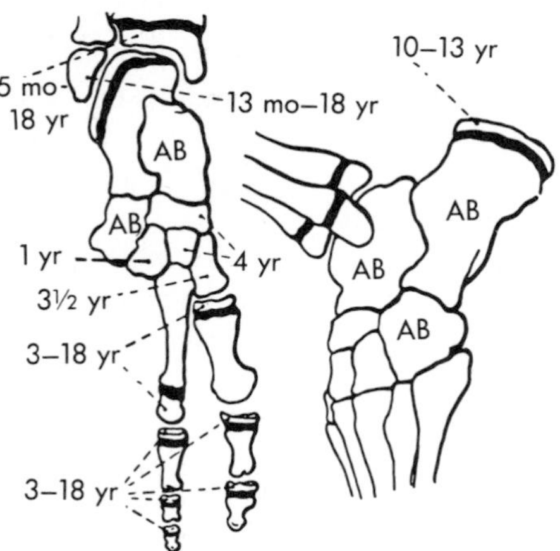

those for the epiphyses of the phalanges typically appear toward the end of the second year.

The epiphyses of the bones of the leg and foot fuse with the bodies between the ages of 14 and 19 years.

Centers of ossification of the sesamoids of the big toe are apparently constantly found after the age of 12, but do not appear earlier than the tenth year.

## THE FOOT AS A WHOLE

Details of the articulations of the foot and the role of ligaments and muscles in supporting these can best be understood after this part has been dissected and are, therefore, discussed at the end of this chapter.

### Arches

It is a matter of common observation that the normal foot is arched rather than flat on its plantar surface (Fig. 18-13). In an articulated skeletal preparation, two arches, a longitudinal and a transverse, can be seen. The *medial part* of the **longitudinal arch,** higher than the lateral, is described as beginning with the tuber calcanei and passing through the sustentaculum tali, the talus, the navicular, the three cuneiforms, and the three medial metatarsals to the heads of these bones. The *lateral side of the longitudinal arch* begins also with the tuber calcanei but passes forward through the body of the calcaneus and through the cuboid and the two lateral metatarsals to the heads of these bones. The **transverse arch** lies at the level of the distal row of tarsals and the bases of the metatarsals. It becomes flattened and disappears anteriorly, since the heads of the metatarsals all lie in the same plane. Posteriorly, it disappears as it blends with the high medial and low lateral parts of the longitudinal arch.

Although it is convenient to describe two

**FIG. 18-13.** Bones of the foot from the lateral and medial sides.

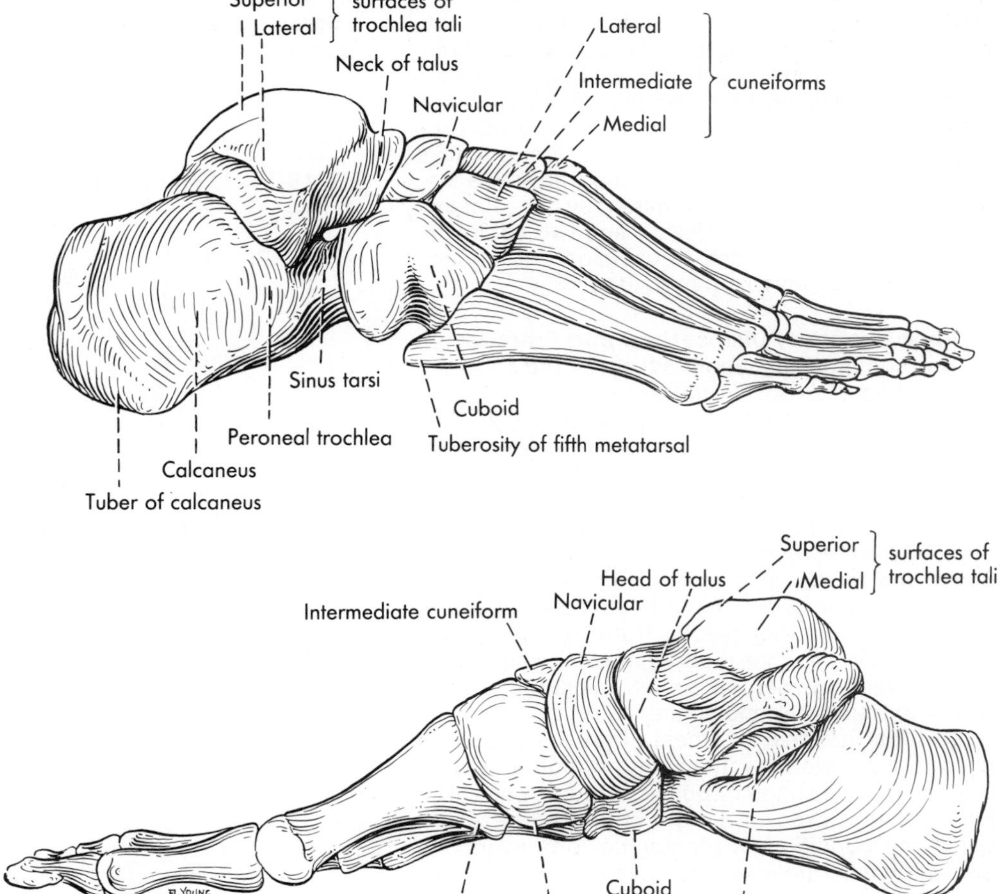

arches, they so interlock that they form a functionally single arch of complex form. Weight is distributed in this arch according to the engineering principle that the distribution of stress throughout an arch is strictly proportional to the relative heights of various parts of the arch. This is true, of course, only when the arch is loaded from its top, which is, in the normal foot, the talus. If the sole of the foot is turned outward, the lateral part of the arch supports less weight and the medial part supports more.

#### Movements of Foot

Movements of the foot, discussed in more detail later, require a short explanation here. Because the trochlea tali is grasped firmly between the medial and lateral malleoli, little movement is possible at the talocrural joint except flexion and extension. However the calcaneus, underlying the talus, can rock from side to side beneath that bone so that either its lateral or its medial surface is turned somewhat down. Thus, a certain amount of inversion and eversion of the foot is possible at the subtalar joint, the articulation between talus and calcaneus (Fig. 18-66). Movement also occurs between the posteriorly lying talus and calcaneus and the more anterior part of the foot (navicular and cuboid bones and everything anterior to them), through the transverse tarsal joint (also Fig. 18-66). Both the transverse tarsal and subtalar joints are oblique and normally move together, but the greatest amount of movement occurs at the transverse tarsal joint. The movements permitted by this joint are dorsiflexion and plantar flexion and two combinations of movements: one of eversion and abduction and one of inversion and adduction.

#### Abnormalities, Dislocation, and Fracture

The commonest abnormalities of the foot are flatfoot and clubfoot. In **flatfoot** (*pes planus*), the calcaneus so rotates that its medial side is carried downward (eversion of the plantar surface), and the talus slips downward and forward so that the head of the talus and the navicular bone, instead of forming the highest part of the medial side of the arch, come to bear weight. As they are forced down, the medial side of the arch is necessarily lengthened, and as a result, the fore part of the foot is abducted. Flatfoot results from excess weight being thrown upon the medial side of the arch, as a result of malalignment of bones, relaxation of ligaments, or muscular imbalance.

**Clubfoot** (*talipes*) is a name that has been applied to any deformity in which the foot is twisted from its natural shape or position, but nowadays is nearly always used in a more restricted sense as being the condition of plantar flexion, inversion, and adduction (talipes equinovarus). This is the most common type of distortion seen in congenital clubfoot. Its cause is not understood. One view is that the bony distortions are the primary defect, and another is that they are secondary to soft tissue defects, particularly muscle imbalance. Most postnatal clubbing of the foot can be plainly shown to be due to muscle imbalance. Conservative methods of treatment, which restore normal alignment to the foot and thus allow muscles and bones to grow in a more normal fashion, are effective in restoring a large percentage of congenital clubfeet to normal.

Other, relatively rare, anomalies of the foot include polydactylism and tarsal fusion. The latter, usually between only two bones, is not uncommon in association with partial or complete absence of the tibia or fibula, but is otherwise rare.

Approximately 30 different **accessory tarsal bones** have been described, and some sort of an accessory bone in the ankle is much more common than is one at the wrist. A separate tuberosity of the 5th metatarsal is called the *os vesalianum;* a separate lateral tubercle of the posterior process of the talus is called the *os trigonum,* although it may be the result of fracture rather than a truly accessory bone. Accessory sesamoids in connection with the toes are relatively common.

**Sprain** as the result of twisting the weight-bearing foot is the most common sprain of any joint. As

**FIG. 18-14.** Fracture of the tibia and fibula.

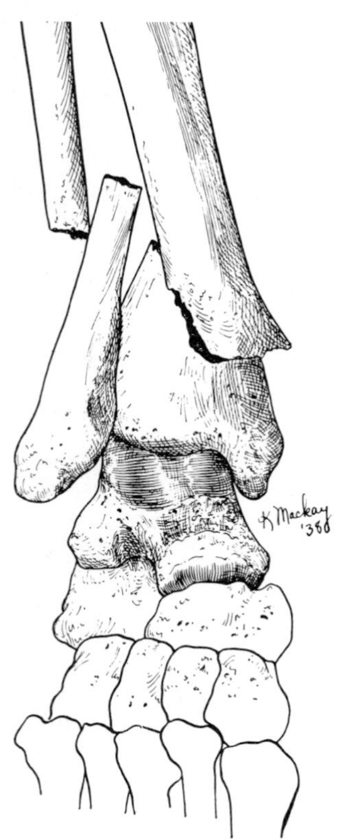

do all sprains, it involves at least partial tearing of ligaments (at the ankle) and may result in very severe disability.

**Fractures** at the ankle usually involve the lower end of the tibia or the lower ends of both tibia and fibula (Fig. 18-14) and are, like sprains, typically produced by twisting. Fracture of the talus or the calcaneus is most commonly produced by fall from a height, and fractures of the more anterior tarsals are commonly caused by a fall that twists the foot, or by something being dropped upon the foot. Fractures of the metatarsals and of the phalanges also are usually produced by direct trauma.

## FASCIA, SUPERFICIAL NERVES, AND VESSELS

The superficial fascia of the leg demands no particular consideration. It contains a variable amount of fat, and in it are embedded the major parts of the cutaneous nerves and the saphenous veins and their tributaries. On the dorsum of the foot, the superficial fascia is thin and allows mobility of the overlying skin. On the plantar surface, it is especially tough and thick and serves as a padding between the skin and the bones of the foot and anchors the skin firmly to the underlying deep fascia.

### Cutaneous Nerves

The cutaneous nerves of the leg and foot (Figs. 18-15 and 18-16) are derived partly from the femoral nerve, but mostly from the tibial and common peroneal nerves and their branches.

The **saphenous nerve,** from the femoral, becomes subcutaneous above the medial side of

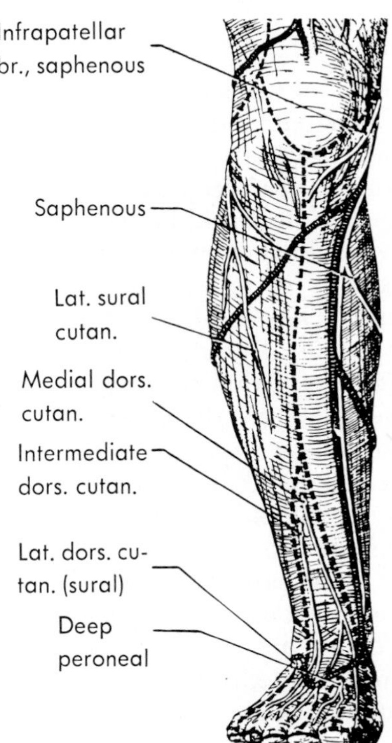


**FIG. 18-15.** Cutaneous nerves of the leg and foot, anterior view. (Corning HK: Lehrbuch der topographischen Anatomie für Studierende und Ärtze. Munich, JF Bergmann, 1923)

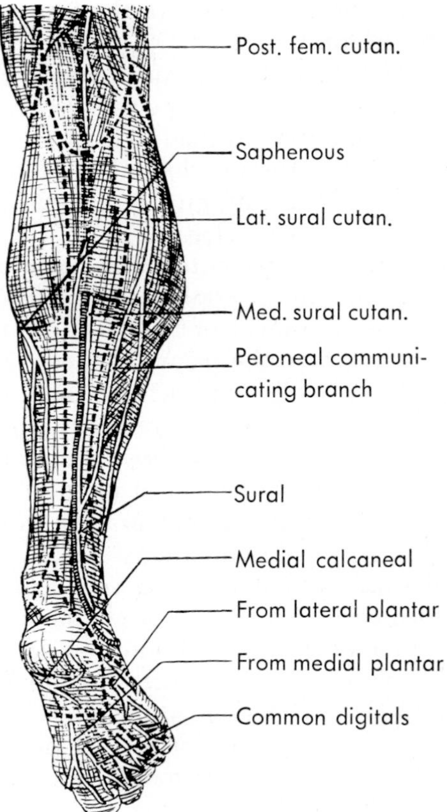


**FIG. 18-16.** Cutaneous nerves of the leg and foot, posterior view. (Corning HK: Lehrbuch der topographischen Anatomie für Studierende und Ärtze. Munich, JF Bergmann, 1923)

the knee where it emerges behind the tendon of the sartorius muscle. It gives off an *infrapatellar branch* that curves forward to supply the anteromedial part of the leg below the knee and runs downward with the great saphenous vein to give off a series of *medial crural cutaneous branches.* It then runs forward along the medial side of the foot about as far as the metatarsophalangeal joint of the big toe.

The cutaneous or terminal part of the **superficial peroneal nerve** emerges on the anterolateral side of the leg (between the lateral and the anterior muscle groups) at about the level of junction of the upper two thirds with the lower third of the leg. It gives off twigs to the leg and continues onto the foot to supply much of this, dividing into two **dorsal cutaneous nerves** as it does so. Its *medial branch* (medial dorsal cutaneous) is distributed to the medial side of the big toe and to the adjacent sides of the 2nd and 3rd toes. Its *lateral branch* (intermediate dorsal cutaneous) usually divides into two **dorsal digital** branches that supply the adjacent sides of the 3rd and 4th and the 4th and 5th toes.

The **lateral sural cutaneous nerve** typically arises from the common peroneal above the knee joint and pierces the fascia lata over the popliteal fossa. It runs down the posterolateral aspect of the calf, giving off branches to the lateral side of the leg.

The **medial sural cutaneous nerve** arises from the tibial in the popliteal fossa, usually a little below the level of the knee joint, and runs downward in the groove between the two heads of the most superficial muscle of the calf (gastrocnemius). It commonly does not penetrate the deep fascia above the middle of the leg. After emerging, it is usually joined by a **peroneal communicating branch** from the common peroneal (which may arise with the lateral sural cutaneous), and the **sural nerve** thus formed is distributed down the posterolateral side of the leg and onto the dorsal aspect of the lateral side of the foot, where it is called the **lateral dorsal cutaneous nerve.** The latter may supply only the lateral side of the little toe or may spread medially to anastomose with, or take over some of the territory supplied by, the intermediate dorsal cutaneous.

A **terminal branch of the deep peroneal nerve** emerges on the foot to supply **dorsal digital nerves** to the adjacent sides of the 1st and 2nd toes.

The **posterior femoral cutaneous nerve** is usually continued downward below the knee to supply a strip along the midline of the calf, sometimes for as much as half or two-thirds the length of the leg.

There is a good deal of variation in the amount of skin of the leg supplied by the several branches mentioned and even more variation in the branching and distribution of the nerves to the dorsum of the foot.

The plantar surface of the foot is supplied exclusively by the tibial nerve, which divides into **medial** and **lateral plantar nerves** as it passes into the foot; the lateral plantar nerve supplies a strip of skin corresponding to one and one-half toes, the medial plantar, skin corresponding to three and one-half toes.

### Vessels

The **superficial veins** of the plantar surface of the foot are small and embedded in tough subcutaneous connective tissue. They drain either directly into deep veins or around the borders of the foot and between the metatarsals into the dorsal plexus of the foot. The **dorsal plexus** is composed of relatively large veins and takes different forms, but commonly the dorsal digital veins unite to form metatarsal veins that in turn unite to form a **dorsal venous arch;** the great saphenous vein then takes origin from the medial side of this arch (see Fig. 16-13), the small saphenous vein from the lateral side (see Fig. 16-14).

From its origin along the medial side of the foot, the **great saphenous vein,** accompanied by the saphenous nerve, passes upward in front of the medial malleolus (where it is particularly convenient for intravenous infusions in infants) and then along the medial side of the leg. It receives tributaries from all surfaces of the leg, some of them originating from the small saphenous vein. It and its tributaries also send branches (communicating or perforating veins) to the deep veins of the leg.

The **small saphenous vein** passes behind the lateral malleolus and then almost straight up the middle of the calf. It receives tributaries from the posterior surface of the leg and gives off communications to the great saphenous vein and the deep veins. It may penetrate the deep fascia in the upper third of the leg or may run superficially to the popliteal fossa and penetrate the fascia there. It typically ends in the popliteal vein, but frequently, before it does so, it sends a communication around the medial side of the thigh to the great saphenous vein. Sometimes the entire small saphenous takes this course.

The saphenous system of veins is discussed also in Chapter 16 (see Figs. 16-13 and 16-14).

The superficial lymphatics of the leg have also been described in Chapter 16.

## Deep Fascia

The deep fascia of the leg (*crural fascia*) is attached around the knee to all the bony prominences, but posteriorly is directly continuous with the fascia lata across the popliteal fossa. It encircles the leg, blending with the periosteum of the subcutaneous portion of the tibia. Anteriorly and anterolaterally, in the upper part of the leg, it gives rise to some of the fibers of the muscles and is therefore closely bound to them, but it overlies more loosely the superficial posterior muscles of the calf. On the lateral side of the leg, the crural fascia gives rise to two intermuscular septa that attach to the fibula (Fig. 18-17); the **anterior intermuscular septum** passes in front of the lateral (peroneal) muscles, between them and the anterior ones, and the **posterior septum** passes behind them, between them and the muscles of the calf. The crural fascia is also continuous with a layer (unnamed in the N.A.) variously called the *transverse crural septum, transverse intermuscular septum,* or *deep transverse fascia of the leg,* that passes across the calf from one side to the other and separates the superficial muscles of the calf from the deep muscles. In its upper part, this fascia gives rise to some of the most superficial fibers of the deeper-lying muscles.

Close to the ankle, the fascia is thickened by more nearly transverse fibers that form **retinacula** about the tendons crossing the ankle. There are five retinacula here: anteriorly are the superior and inferior extensor retinacula;

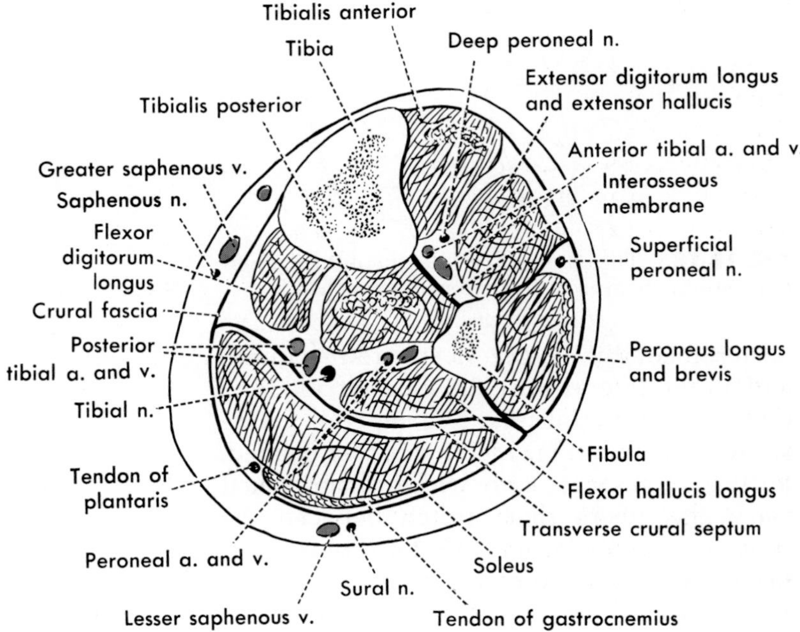


**FIG. 18-17.** Cross section through the lower part of the leg. (Redrawn from Eycleshymer AC, Schoemaker DM: A Cross-Section Anatomy. New York, Appleton, 1923)

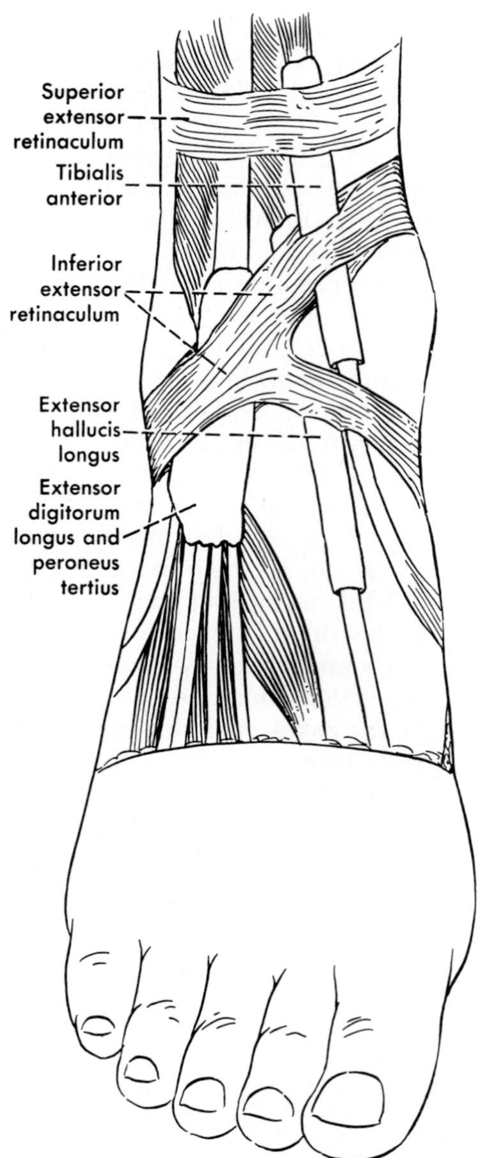


**FIG. 18-18.** The extensor retinacula of the leg and the tendon sheaths of the extensor muscles.

posteromedially is the flexor retinaculum; and posterolaterally are the superior and inferior peroneal retinacula. The **superior extensor retinaculum** (Fig. 18-18) is poorly defined, for it is represented only by a few additional transverse fibers that stretch between the tibia and the fibula. The **inferior extensor retinaculum** is more complex. It arises from the lateral side and upper surface of the calcaneus (in the sinus tarsi and the tarsal canal) and, as it crosses the front of the ankle, divides into an upper and a lower limb, so that it resembles a Y lying on its side. The upper limb attaches to the medial malleolus, but the lower one blends with the fascia of the medial side of the sole of the foot. The tendons crossing the front of the ankle run through compartments in the inferior extensor retinaculum, where they are provided with *tendon sheaths,* none of which extends any great distance onto the dorsum of the foot. There are three tendon sheaths at the front of the ankle.

The thin fascia on the dorsum of the foot is continuous above with the inferior extensor retinaculum. It encloses the tendons on the dorsum and blends on the sides with the plantar fascia.

The **flexor retinaculum** (Fig. 18-19) runs between the medial malleolus and the calcaneus. It sends three septa to the tibia and thus contains four compartments: in the most anteromedial one is the tendon of the tibialis posterior, and in the next is the tendon of the flexor digitorum longus; the third compart-

**FIG. 18-19.** The flexor retinaculum at the ankle.

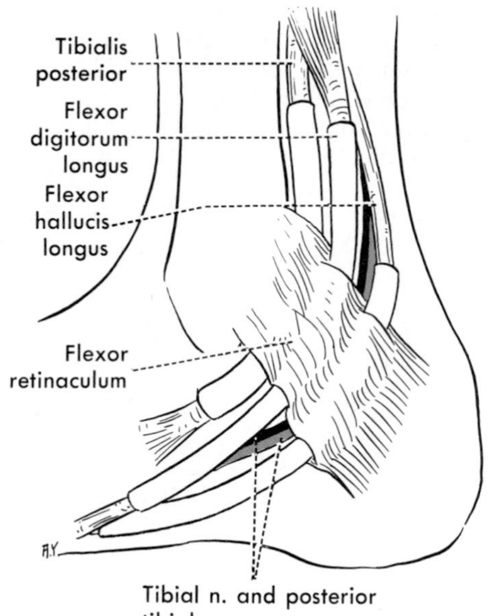

ment is occupied by the tibial nerve and the posterior tibial vessels, and the most posterior and inferior compartment is occupied by the tendon of the flexor hallucis longus. Each of the three tendons passing through the flexor retinaculum is provided with a *tendon sheath* that begins a little above the level of the retinaculum and usually extends only a short distance onto the plantar surface of the foot.

The **superior peroneal retinaculum** (Fig. 18-20) extends from the lateral malleolus to the calcaneus. Deep to it run the tendons of the peroneus longus and peroneus brevis, enclosed in a common *tendon sheath.* The **inferior peroneal retinaculum** is attached at both ends to the calcaneus, and its upper end blends with the lateral part of the inferior extensor retinaculum. From its deep surface, it sends a septum to the calcaneus that divides the common sheath and the two peroneal tendons. The part of the sheath around the peroneus brevis stops above the lateral border of the foot. The part of the sheath around the peroneus longus travels farther, approximately to the lateral border of the foot, and may continue onto the plantar surface.

**FIG. 18-20.** The peroneal retinacula.

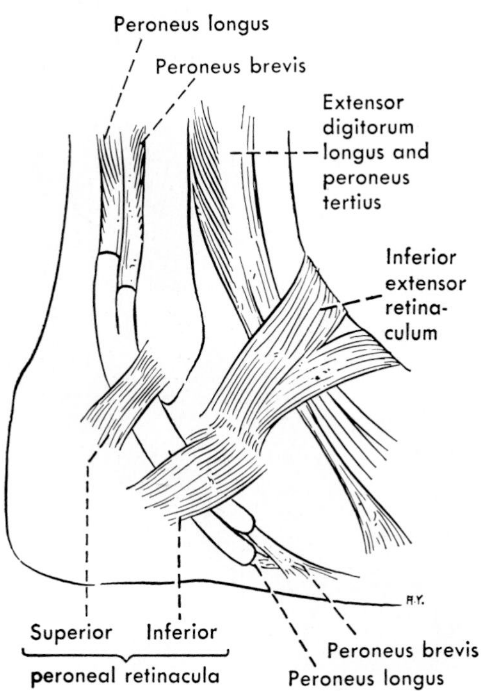


## MUSCLES AND RELATED STRUCTURES

The muscles of the leg are divisible into three groups: anterior, arising from the tibia and fibula and the intervening interosseous membrane and innervated by the deep peroneal nerve; lateral, arising from the fibula and innervated by the superficial peroneal nerve; and posterior, arising from the femur, tibia, fibula, and interosseous membrane and innervated by the tibial nerve.

Some features of the surface anatomy of the leg and ankle are shown in Figures 18-21 and 18-22, and also Figures 17-26 and 17-27.

### LATERAL MUSCLES AND RELATED STRUCTURES

The two lateral muscles of the leg are the peroneus longus and brevis (Figs. 18-23 and 18-24). These lie in a special compartment bounded by the anterior and posterior intermuscular septa. The common peroneal nerve enters the uppermost part of this compartment and divides here into deep and superficial peroneal nerves.

#### Muscles

The **peroneus longus** arises from the lateral surface of the fibula and from the adjacent deep fascia. The common peroneal nerve runs deep to it just below the fibular head. The muscle largely covers the peroneus brevis, but as its tendon approaches the ankle, it comes to lie behind the tendon of the brevis. It crosses the lateral surface of the cuboid bone and turns medially across that bone's plantar surface (where it runs in a canal roofed by the long plantar ligament) and inserts into the lateral side of the medial cuneiform and the base of the first metatarsal bone—thus, on the medial side of the foot. In its course

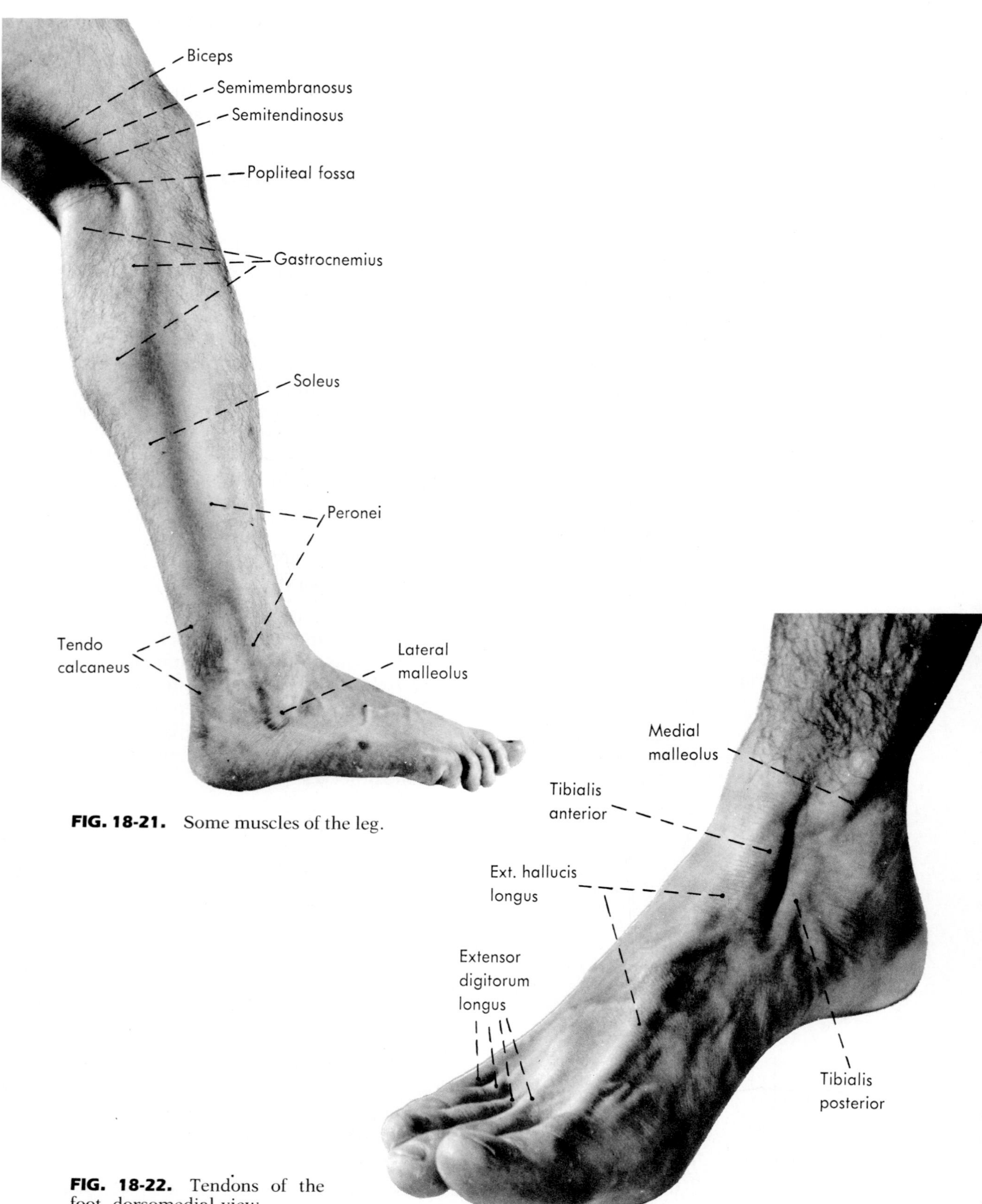


**FIG. 18-21.** Some muscles of the leg.

**FIG. 18-22.** Tendons of the foot, dorsomedial view.

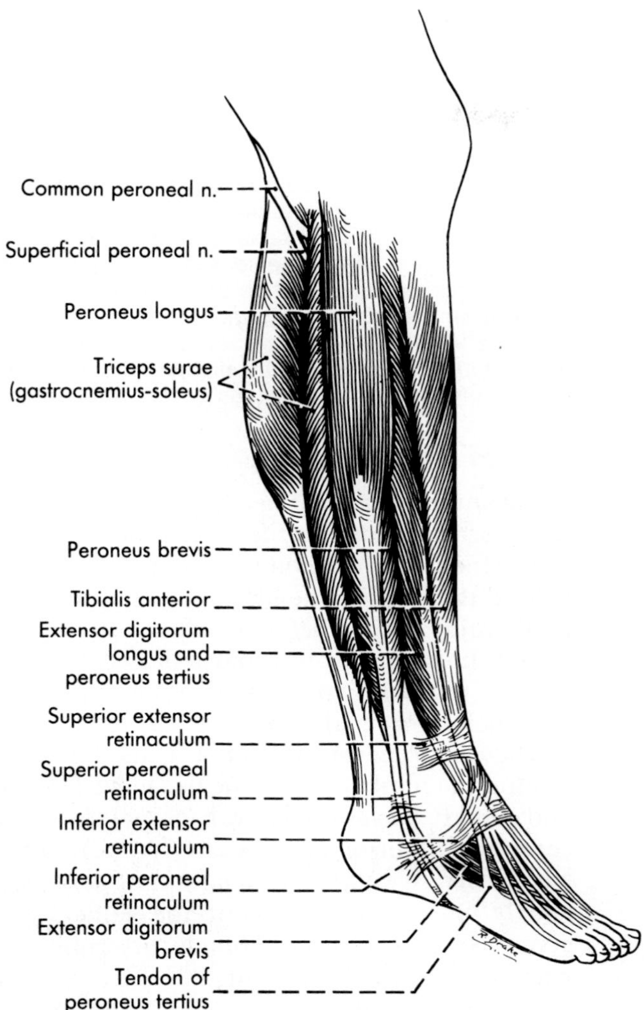


**FIG. 18-23.** Lateral view of muscles of the leg.

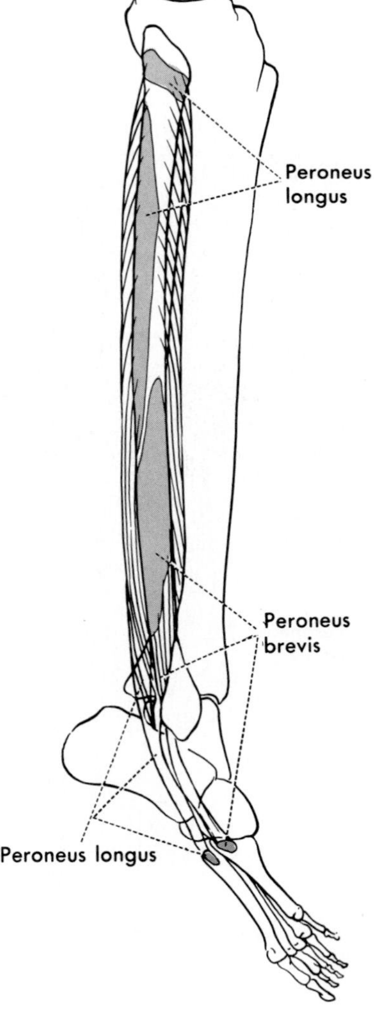


**FIG. 18-24.** Attachments of the peroneus longus and brevis.

across the foot, it is provided with a tendon sheath that may be continous around the lateral border of the foot with that at the ankle. Where it crosses the surface of the tuberosity of the cuboid, the tendon may contain a sesamoid bone.

The **peroneus brevis** arises from the lower lateral surface of the fibula and from the intermuscular septa adjacent to it. Its tendon curves forward onto the dorsum of the foot from behind the lateral malleolus and inserts on the dorsolateral surface of the base of the 5th metatarsal.

**FIG. 18-25.** Nerves in the lateral compartment of the leg.

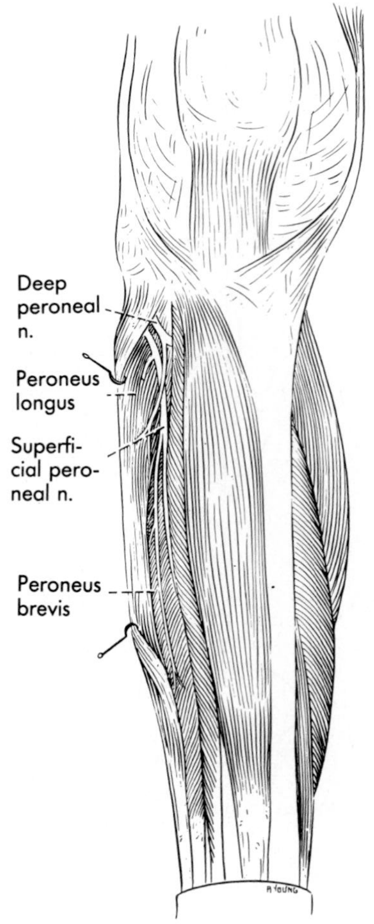


Both muscles are **innervated** by the *superficial peroneal nerve,* which descends between them. The peroneus longus frequently receives a branch from the common or the deep peroneal also. Both muscles typically receive fibers derived from the 4th and 5th lumbar and 1st sacral nerves.

The chief **action** of both muscles is to evert the foot, but since they insert anteriorly on it, they also abduct somewhat; the brevis is a better abductor than the longus. They are also weak plantar flexors and apparently act largely on the transverse tarsal rather than the talocrural joint.

Sometimes there is fusion between the muscles, or there may be an accessory slip inserting into the lateral surface of the calcaneus or into the cuboid.

### Nerves and Vessels

The **common peroneal nerve** leaves the popliteal fossa by crossing the lateral head of the gastrocnemius. It becomes subcutaneous just behind the head of the fibula, where it can be easily injured. Just before or after it has penetrated the posterior intermuscular septum, it divides into its two branches, the deep and superficial peroneal nerves.

The **deep peroneal nerve** runs forward around the fibula deep to the peroneus longus and quickly leaves the lateral compartment to enter the anterior one. Before it does so, it frequently gives off a *branch into the peroneus longus muscle* and at about the same level gives rise to a *recurrent branch* that runs upward toward the knee joint.

The **superficial peroneal nerve** (Fig. 18-25) descends in the lateral compartment, at first between the peroneus longus and the fibula and then between the peroneus longus and the peroneus brevis. After supplying both muscles, it emerges between them to supply skin of the lower part of the leg and the foot.

The only vessel of any size related to the lateral muscles is the **fibular circumflex artery.** This artery typically arises from the posterior tibial and rounds the lateral surface of the fibula a little below the peroneal nerve to end in the peroneal muscles.

## ANTERIOR MUSCLES

The four anterior muscles of the leg (Fig. 18-26) lie in a fascial compartment between the anterior intermuscular septum and the tibia. Their tendon sheaths at the ankle are shown in Figure 18-18.

The anterior tibial artery parallels the deep peroneal nerve on the front of the leg and is the chief blood supply, but branches of the posterior arteries penetrate the interosseus membrane and furnish an additional supply.

The **tibialis anterior** arises (Fig. 18-27) from the lateral surface of the tibia, from the interosseous membrane, from an upper part of its covering fascia, and from an intermuscular septum between it and the extensor digitorum longus. Its strong tendon passes across the medial side of the dorsum of the foot to insert into the medial and lower surfaces of the medial cuneiform bone and the base of the 1st metatarsal. The deep peroneal nerve and the anterior tibial artery lie deep to the muscle on its lateral side.

The upper part of the **extensor hallucis longus** is largely covered by the tibialis anterior and the extensor digitorum longus. The muscle arises (Fig. 18-28) from the anterior surface of the fibula and the adjacent interosseous membrane, emerges between the tibialis anterior and the extensor digitorum, and inserts upon the distal phalanx of the big toe, sometimes sending a slip to the proximal phalanx.

The **extensor digitorum longus** and **peroneus tertius** muscles are continuous at their origin (Fig. 18-27), although the tendon of the peroneus separates from that of the extensor digitorum before the latter divides into its subsidiary tendons.

The extensor digitorum longus has some origin from the lateral side of the lateral condyle of the tibia, the interosseous membrane, and an intermuscular septum that it shares with the tibialis anterior, but it arises mostly from the anterior surface of the fibula. The peroneus tertius arises from the fibula and the anterior intermuscular septum. The two tendons pass together across the ankle, and then the tendon of the peroneus tertius diverges laterally to insert into the dorsal surface of the base of the 5th metatarsal bone.

The tendon of the extensor digitorum longus divides into four tendons for the four lateral toes. Those to the 2nd, 3rd, and 4th toes are joined by the tendons of the extensor digitorum brevis, and all four tendons are joined by the tendons of the lumbrical muscles and, variably, by some tendinous fibers from the interossei (both are short plantar muscles of the foot). Like the extensor tendons on the fingers, they form the dorsal capsules of the joints as they pass across them. Over the proximal phalanx, each tendon complex divides into three bands, a middle and two lat-

**FIG. 18-26.** Anterior muscles of the leg; the lateral (peroneus) muscles have been removed.

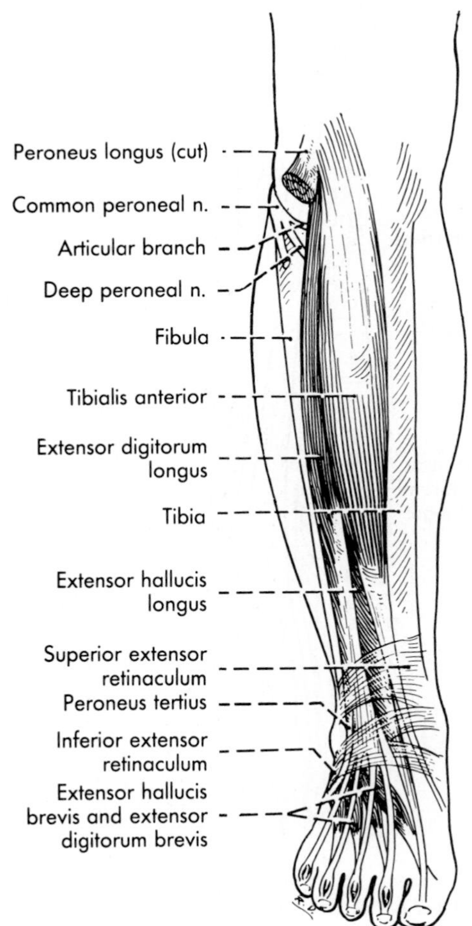

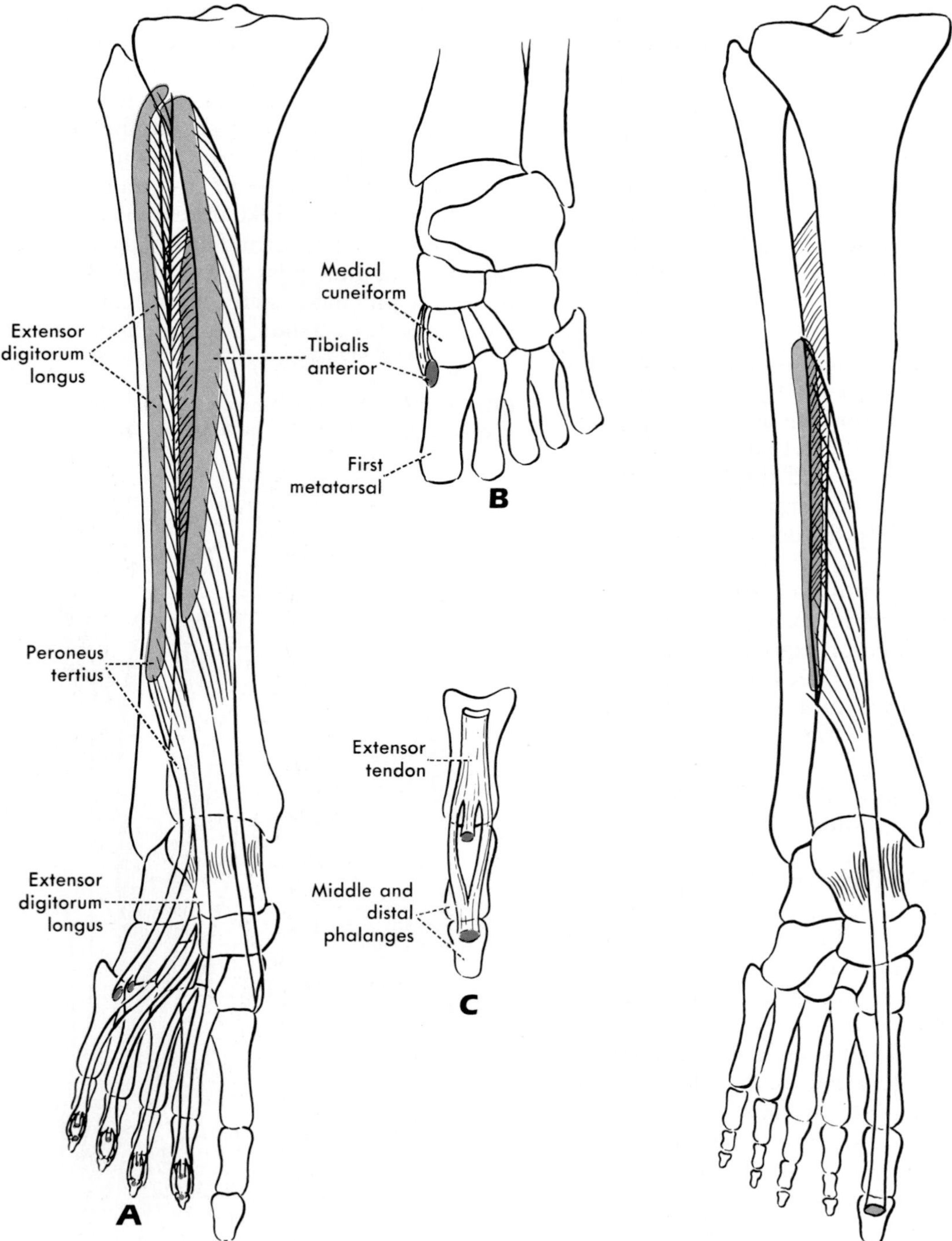


**FIG. 18-27.** Attachments of the tibialis anterior, extensor digitorum longus, and peroneus tertius muscles. **A** is an anterior view of the leg and foot, **B** a plantar view of the foot, and **C** a dorsal view of a digit to show the details of the extensor tendon here.

**FIG. 18-28.** Attachments of the extensor hallucis longus.

eral ones. The middle band inserts on the middle phalanx, and the lateral bands come together and insert on the distal phalanx (Fig. 18-27).

**Variations, Innervations, and Actions.** The peroneus tertius varies much in its size and is sometimes absent.

In some instances, the tendon of the tibialis anterior is doubled, with one part attaching to the medial cuneiform and the other to the base of the 1st metatarsal.

All four anterior muscles of the leg are **innervated** by the *deep peroneal nerve,* which brings into them fibers from the 4th and 5th lumbar and 1st sacral nerves. The nerves to most of the muscles are multiple, and one or more sometimes arises from the common peroneal nerve.

These muscles all have one **action** in common, for they all *dorsiflex the foot,* although with varying strength. The tibialis anterior is a particularly important dorsiflexor, as well as an invertor and adductor. The extensor digitorum longus and the peroneus tertius typically work together to dorsiflex, evert, and abduct the foot. The extensor digitorum longus also extends the four lateral toes, but since the interphalangeal joints of these toes are usually held in flexion by the more powerful flexors, its chief action on the toes is hyperextending at the metatarsophalangeal joints. The extensor hallucis longus is primarily an extensor of the big toe and is the weakest dorsiflexor of the foot. It is a rather weak adductor and invertor.

## Associated Nerves and Vessels

The **deep peroneal nerve,** one of the terminal branches of the common peroneal, usually arises as the latter rounds the neck of the fibula under cover of the peroneus longus muscle. It continues around the fibula and after passing deep to the extensor digitorum longus turns to run downward on the interosseous membrane in company with the anterior tibial artery. At first it lies between the extensor digitorum longus and the tibialis anterior and then between the extensor hallucis longus and the tibialis anterior. Nerve and vessels (Fig. 18-29) emerge close to the ankle to lie between the two long extensor muscles. As it enters the anterior compartment of the leg, the deep peroneal nerve gives off a series of branches into the muscles here; they vary in number and position, for several may arise by a common stem or may arise separately.

The **anterior tibial artery** and its accompanying **veins** pass through the gap at the upper end of the interosseous membrane, and the artery usually gives off the *anterior tibial recurrent artery,* which runs upward to enter

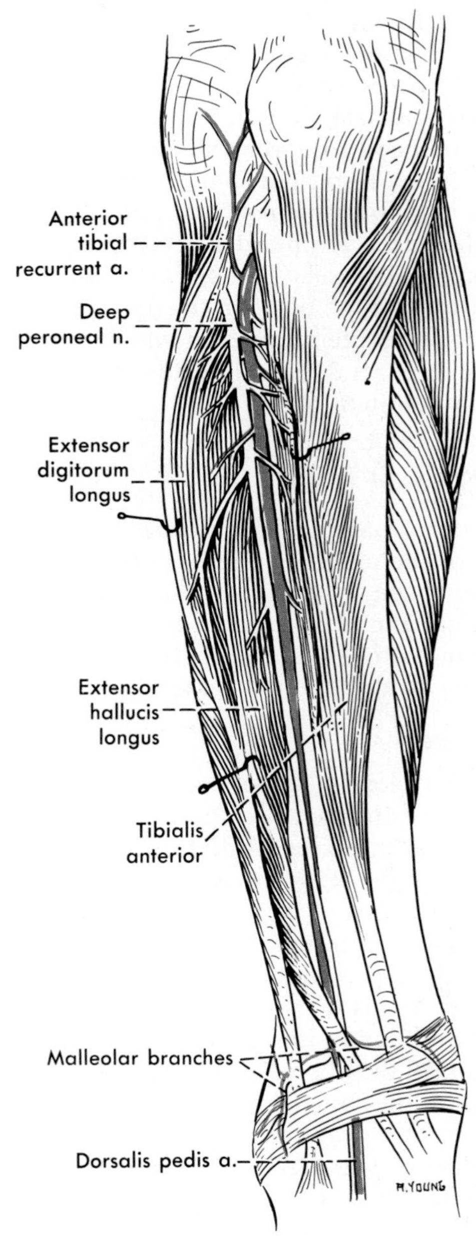


**FIG. 18-29.** The deep peroneal nerve and the anterior tibial artery.

into the anastomosis around the knee joint. As the anterior tibial descends against the interosseous membrane, it gives off branches to the surrounding muscles and also small branches that penetrate the membrane to help supply the deep muscles of the posterior side of the leg. As it crosses the ankle joint its name is changed to **dorsalis pedis.** Its last branches are usually *medial and lateral anterior malleolar arteries,* although these may arise from the dorsalis pedis. In about 3.5% of limbs, the anterior tibial artery either fails to reach the foot or is reduced to a very slender stem by the time it does so. The paired **anterior tibial veins** help form the *popliteal vein.*

A few **deep lymphatics** accompany the anterior tibial vessels. There may be a tiny anterior tibial lymph node on the upper end of the interosseous membrane, in which one of the deep lymphatics typically ends, but the others end in popliteal nodes.

## DORSUM OF FOOT

The superficial nerves and vessels and the tendons of the anterior muscles of the leg have already been described. The remaining structures are the short extensors of the toes, the continuation of the deep peroneal nerve into the foot, and the blood vessels.

### Muscles

The two closely associated muscles on the dorsum of the foot are the **extensor hallucis brevis** and the **extensor digitorum brevis** (Fig. 18-30). These arise in common from an anterolateral part of the upper surface of the calcaneus and from the deep surface of the inferior extensor retinaculum. The extensor hallucis brevis is the largest and most medial belly and sends its tendon to the proximal phalanx of the big toe. The extensor digitorum brevis divides into three muscular slips whose tendons go toward the 2nd, 3rd, and 4th toes and at approximately the level of the heads of the metatarsals join the long extensor tendons and insert with them upon the middle and distal phalanges.

These two muscles are supplied by the *deep peroneal nerve* as it passes deep to the extensor hallucis brevis.

The extensor hallucis brevis extends the proximal phalanx of the big toe. The extensor digitorum brevis aids the extensor longus in extending the other toes, but differs from the longus in that it can extend them without at the same time dorsiflexing the foot. It may send a tendon to the little toe.

### Nerves and Vessels

The **deep peroneal nerve** emerges on the dorsum of the foot between the tendons of the extensor hallucis longus and the extensor digitorum longus (Fig. 18-30). As it runs distally deep to the extensor hallucis brevis, it gives off a *lateral branch* that in part goes to the short extensor muscles and in part spreads over the dorsal surface of the foot to help supply the intertarsal joints. The remainder of the deep peroneal nerve divides into two *dorsal digital nerves* for the adjacent sides of the big and 2nd toes.

The arteries of the dorsum of the foot are variable (Fig. 18-31). The **dorsalis pedis artery** is typically the continuation of the anterior tibial and appears usually just medial to the deep peroneal nerve on the dorsum of the foot. It runs distally toward the interspace between the 1st and 2nd toes and ends by dividing into a small transversely running **arcuate artery** and a larger **deep plantar artery** that disappears between the heads of the 1st dorsal interosseous muscle. These two branches typically give off the **dorsal metatarsal arteries,** which receive communications from the plantar arch and the plantar metatarsal arteries and end as tiny **dorsal digital arteries.**

In its course, the dorsalis pedis gives off *medial and lateral tarsal arteries.* The lateral tarsal tends to anastomose with the arcuate artery, and there may be also anastomoses between the lateral tarsal, the lateral malleolar, and the perforating branch of the peroneal artery. The varying development of these anastomoses and of the dorsalis pedis accounts for most of the variations in the arterial pattern.

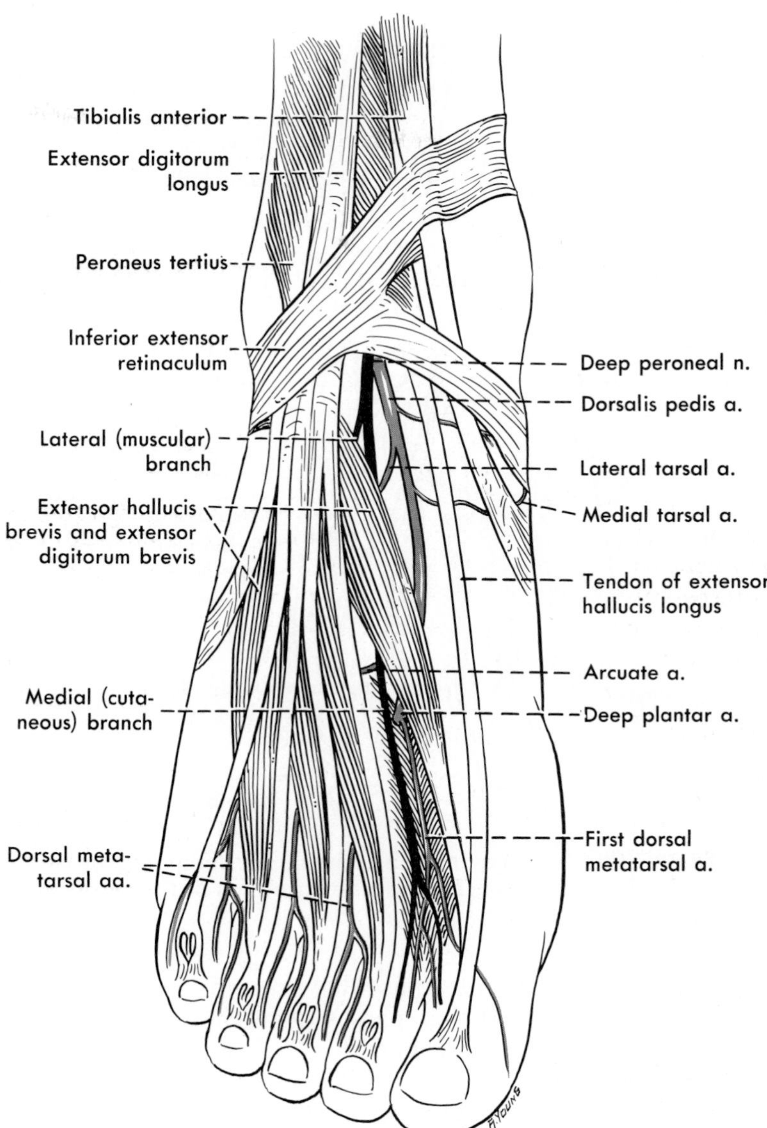


**FIG. 18-30.** Muscles and other structures of the dorsum of the foot.

When the anterior tibial artery fails to reach the foot, the *perforating branch of the peroneal,* through its anastomoses on the foot, becomes the dorsalis pedis. Similarly, if the arcuate artery is small or missing, the dorsal metatarsal branches that it usually gives off may be supplied by the lateral tarsal artery.

The **perforating branch of the peroneal artery** passes through a gap in the lower part of the interosseous membrane and descends under cover of the extensor digitorum longus and peroneus tertius muscles. It usually anastomoses with the anterior lateral malleolar branch of the tibialis anterior artery and, with this, contributes to the blood supply of the lateral side of the foot.

## POSTERIOR MUSCLES AND RELATED STRUCTURES

The muscles of the calf are divided into superficial and deep groups by the layer of fascia

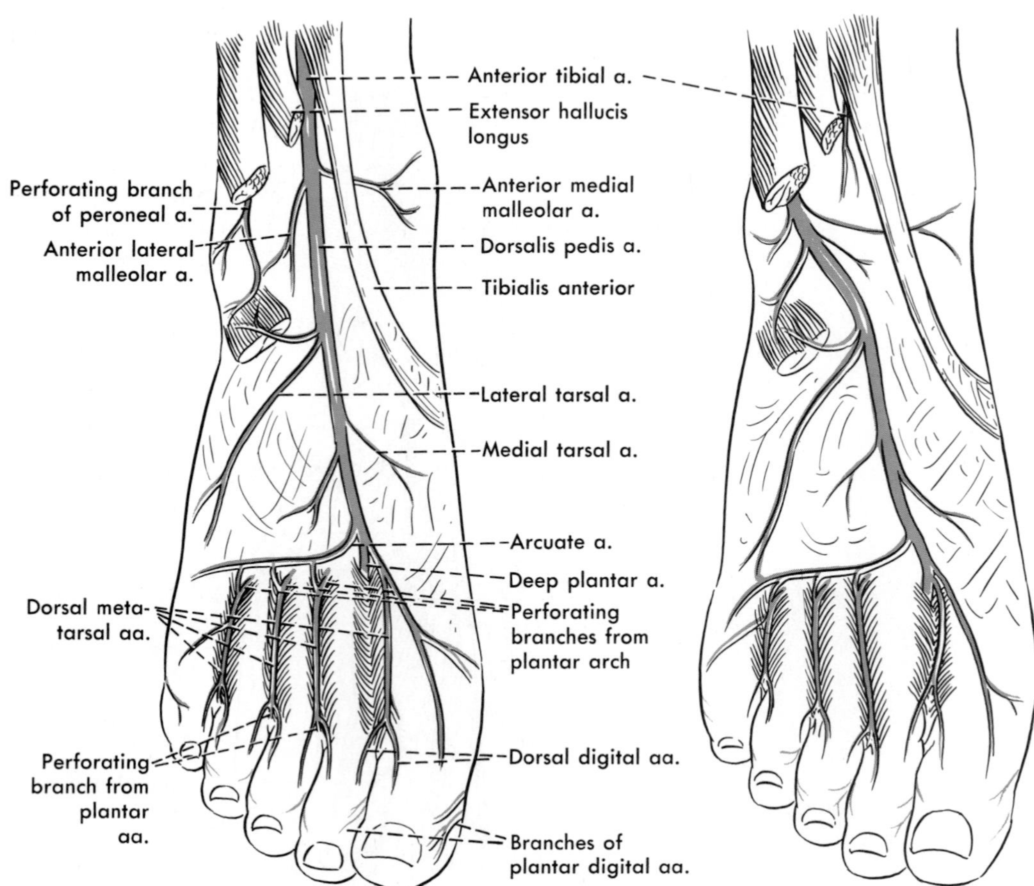


**FIG. 18-31.** Two patterns of the arteries on the dorsum of the foot; in the second, the dorsalis pedis arises from the perforating branch of the peroneal artery.

(transverse crural or intermuscular septum, deep transverse fascia) that passes from the deep fascia across the calf (Fig. 18-17). From the upper part of this fascia, the deeper muscles arise, whereas the lower part is thickened at the ankle to form the flexor retinaculum.

The tibial nerve supplies all the muscles of the calf and the plantar aspect of the foot. The popliteal artery ends in the calf by dividing into anterior and posterior tibial arteries; the latter supplies the calf and continues into the foot.

## Superficial Group

These muscles are the large and important gastrocnemius and soleus muscles, which together form the **triceps surae,** and the very small plantaris muscle (Figs. 18-32 and 18-34).

The superficial member of the triceps surae is the **gastrocnemius.** It arises (Fig. 18-33) by medial and lateral heads, from just above the medial and lateral femoral condyles, respectively. A bursa typically lies deep to each head, and that of the medial head may communicate with the cavity of the knee joint and the bursa of the semimembranosus tendon. The junction of the two heads forms the prominent upper muscular mass of the calf. Approximately halfway down the leg, the muscle gives rise to a tendon that receives on its deep surface the insertion of the soleus muscle and inserts upon the back end of the calcaneus. This combined tendon of the gastrocnemius and soleus is the **tendo calcaneus** or *tendon of Achilles.* A bursa usually lies between the tendon and the upper part of the end of the calcaneus.

The *common peroneal nerve* crosses the lateral head of the gastrocnemius close to its origin, and the *tibial nerve* and *popliteal vessels* run almost straight down in the popliteal fossa to disappear between the two heads.

The **soleus,** the second part of the triceps surae, arises from the upper part of the fibula and from the soleal line on the tibia (Fig. 18-35). Above it is largely covered by the gastrocnemius, but below the middle of the leg, it is broader than the tendon of the gastrocnemius and is, therefore, visible on either side of it. The muscle fibers of the soleus insert into the anterior surface of the tendo calcaneus, but as the tendon runs toward the heel, it twists laterally so that the part associated with the gastrocnemius inserts largely laterally, and the part belonging to the soleus inserts primarily medially. This twisting is of importance when tendon-lengthening procedures, sometimes necessary to allow the foot to be used in a normal fashion during walking, are carried out.

The *tibial nerve* and the *popliteal vessels* disappear deep (anterior) to the tendinous arch (arcus tendineus) that unites the two heads of the soleus.

Each head of the gastrocnemius receives a branch from the tibial nerve. The soleus receives one branch into its superficial surface

**FIG. 18-32.** Muscles of the calf.

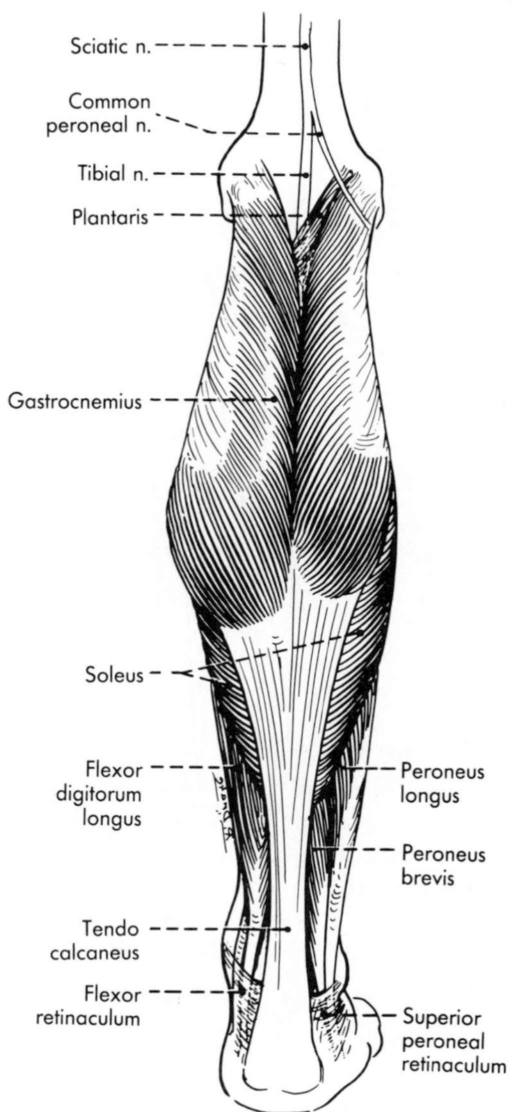


**FIG. 18-33.** Attachments of the gastrocnemius.

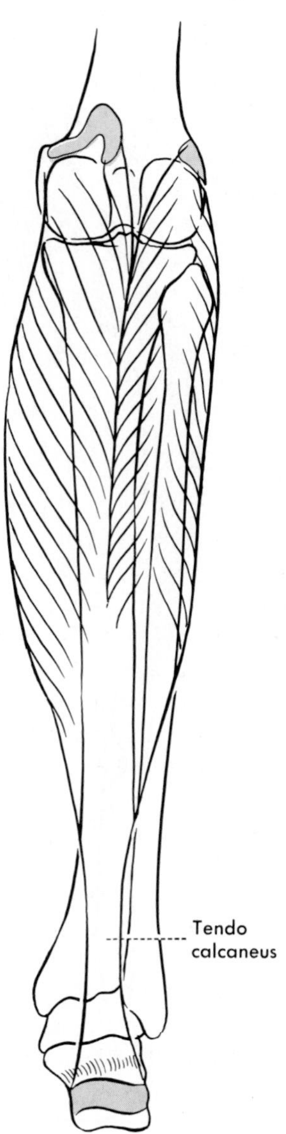

and usually a second one from the tibial after it lies deep to the muscle.

The little **plantaris** muscle, the equivalent of the palmaris longus of the forearm, arises from the femur just above the origin of the lateral head of the gastrocnemius (Fig. 18-35). It runs downward, partly covered by the gastrocnemius, to cross posterior to the tibial nerve and popliteal vessels and lie between the gastrocnemius and soleus muscles. Its muscular belly is short, commonly not more than 2 to 4 inches. Its long, slender tendon usually inserts into the calcaneus medial or anteromedial to the tendo calcaneus, but sometimes blends with it. It receives from the *tibial nerve* a small twig that often arises in common with the nerve to the lateral head of the gastrocnemius.

**Variations, Innervations, and Actions.** There are no important variations in the superficial muscles of the calf.

The **tibial nerve** supplies the gastrocnemius and

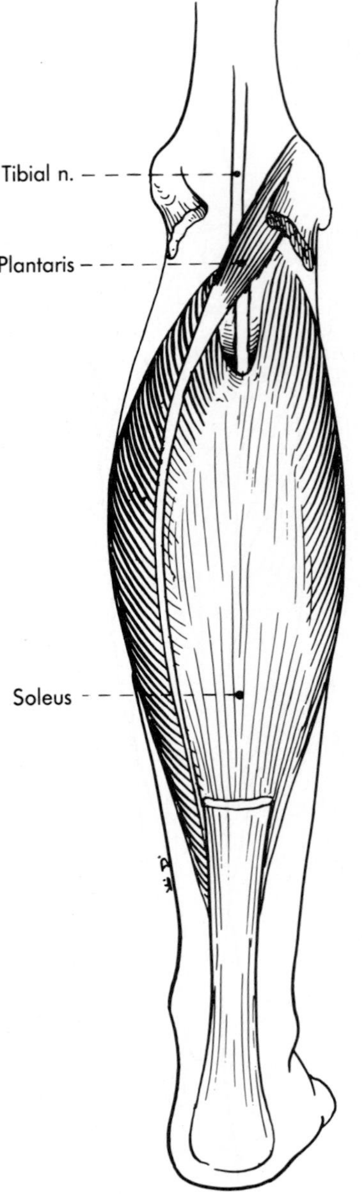


**FIG. 18-34.** The soleus and plantaris.

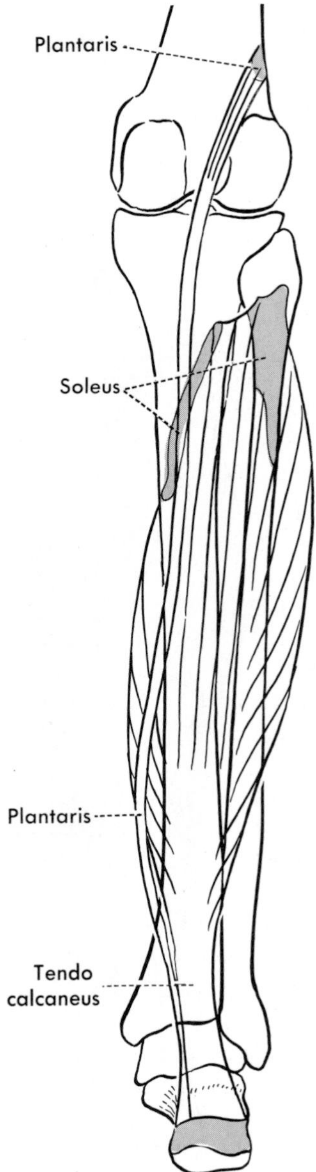


**FIG. 18-35.** Attachments of the plantaris and soleus.

the soleus with fibers derived from the first two sacral nerves, and the plantaris with fibers from the 4th and 5th lumbar and 1st sacral nerves.

The triceps surae is the important **plantar flexor** at the ankle. Although the deep calf muscles and the peroneus longus and brevis are in anatomic positions to be flexors, both their smaller bulk and their lack of leverage make them far less effective than is the triceps. Paralysis of the triceps, or rupture of the tendo calcaneus, makes walking somewhat difficult, and the patient is usually unable to raise himself on his toes. Similarly, contracture of the triceps surae maintains the foot in plantar flexion, the position known as talipes equinus (a "clubfoot" in which the weight is borne toward the distal end of the foot only, as it is in a horse's foot). Because it crosses the knee joint, the gastrocnemius will flex the knee when it acts upon a leg that is not supporting weight; when the leg is supporting weight, however, contraction of either part of the triceps resists flexion of the knee, for the muscle resists dorsiflexion at the ankle, and flexion at the knee is not possible without dorsiflexion at the ankle when an individual is standing. The small plantaris theoretically has the same function as the gastrocnemius but is of no practical importance in man.

## Deep Group

Of the four deep muscles of the calf (Fig. 18-36), the popliteus lies in the upper part of the leg, but the other three approximately parallel each other and send their tendons into the foot. The tibial nerve and popliteal vessels pass across the posterior surface of the popliteus and then onto the posterior surfaces of the other deep muscles of the leg. The popliteal artery ends at approximately the lower border of the popliteus muscle by dividing into anterior and posterior tibial arteries. The posterior tibial artery then accompanies the tibial nerve down the leg.

The **popliteus muscle** arises within the fibrous capsule of the knee joint, mostly from the lateral surface of the lateral femoral condyle, but in part from the lateral meniscus of the knee and from the arcuate ligament, part of the capsule of the knee joint (Fig. 18-43). It lies between the synovial membrane and the fibrous capsule in the joint and, as it leaves, has deep to it a diverticulum of the synovial membrane, the **subpopliteal recess.** The popliteus expands into a somewhat triangular muscle that runs downward and medially to attach into most of the posterior surface of the tibia above the soleal line (Fig. 18-37). The muscle is innervated by a twig of the *tibial nerve* that usually rounds its lower border and enters its deep surface.

The **flexor hallucis longus** arises from the posterior surface of the fibula lateral to the medial crest (Fig. 18-38), with some origin also from the covering fascia and from adjacent fascial septa that it shares with other muscles. At the ankle, its tendon lies in the most posterolateral compartment of the flexor retinaculum. In the foot it passes above the flexor digitorum longus to insert on the distal

**FIG. 18-36.** Deep muscles of the calf.

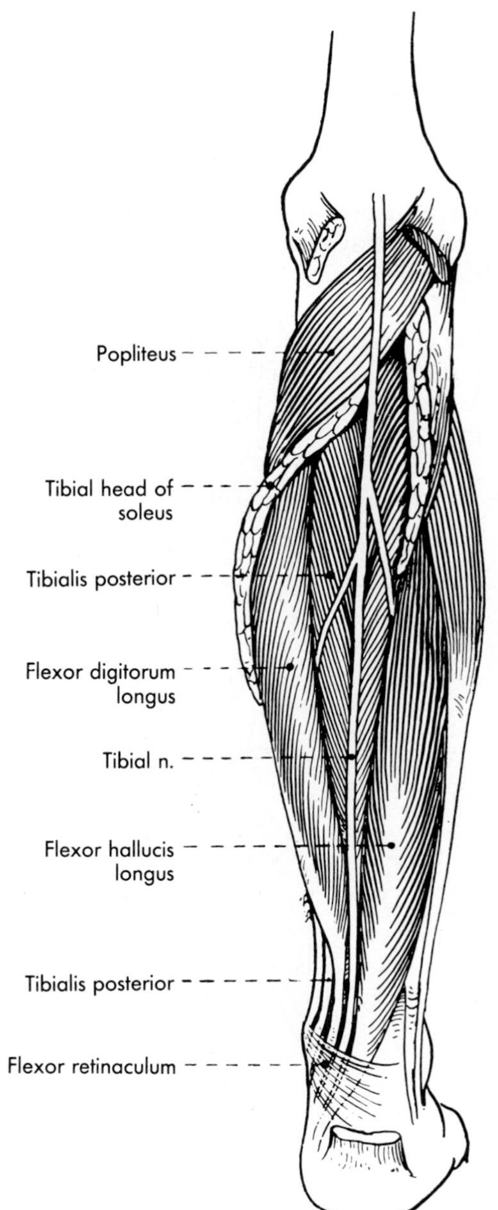

phalanx of the big toe. It receives one or more branches from the *tibial nerve* on its posterior surface.

The **tibialis posterior** lies between the flexor hallucis longus and the flexor digitorum longus and is partly overlapped by both. It arises (Fig. 18-39) from the posterior surface of the fibula between the medial crest and the interosseous border, from the interosseous membrane and a medial part of the posterior surface of the tibia, and from the covering fascia and adjacent intermuscular septa. As its tendon nears the ankle, it passes medially and forward deep to the flexor digitorum longus, and at the medial malleolus, it lies anteromedial to the tendon of the flexor digitorum longus; thus, it occupies the most anteromedial compartment in the flexor retinaculum, lying in a *tendon sheath* immediately adjacent to the medial malleolus. The tendon passes onto the plantar surface of the foot to have an extensive insertion, primarily into the tuberosity of the navicular bone but also into the plantar surfaces of the three cuneiforms, the bases of the 2nd, 3rd, and 4th metatarsals, and the cuboid bone. The *tibial nerve* and *posterior tibial vessels* have most of their deep course in the calf on the posterior surface of this muscle.

**FIG. 18-37.** Attachments of the popliteus.

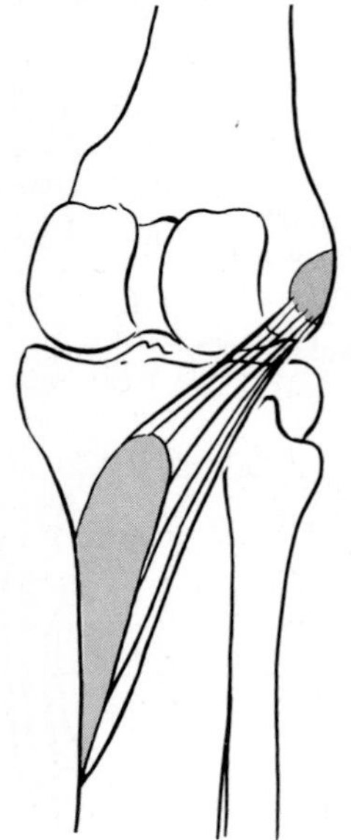

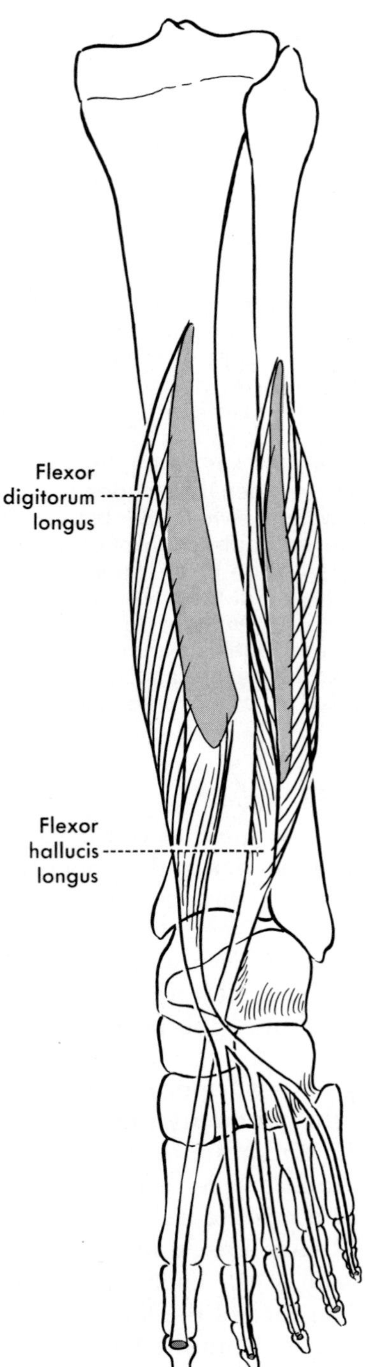


**FIG. 18-38.** Attachments of the flexor digitorum longus and flexor hallucis longus.

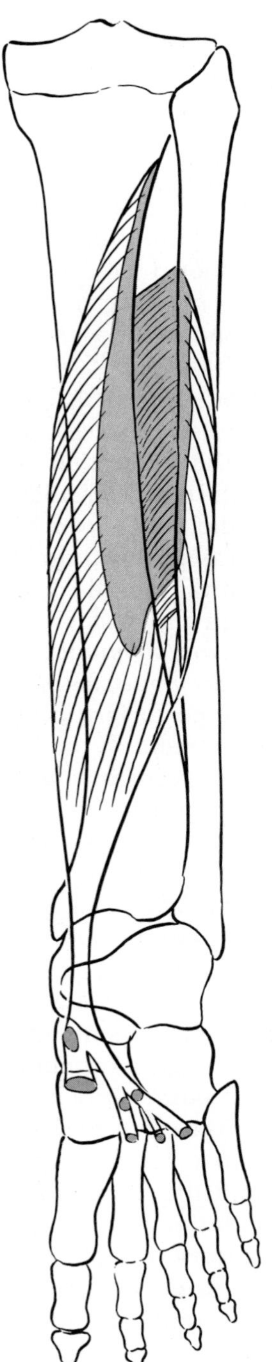

**FIG. 18-39.** Attachments of the tibialis posterior.

The **flexor digitorum longus** is, at its origin, the most medial of the deep muscles of the calf. It arises from much of the middle part of the posterior surface of the tibia (Fig. 18-38), from the covering fascia, and from the adjacent intermuscular septa. In the lower part of the leg, the tendon of this muscle is separated from the tibia by the tibialis posterior. It enters the flexor retinaculum posterolateral to the compartment for the latter tendon. On the plantar surface of the foot, the tendon passes below the tendon of the flexor hallucis longus, from which it usually receives a slip, and inserts on the distal phalanges of the four lateral toes. This muscle is the equivalent of the flexor digitorum profundus of the forearm, and in the foot, its tendons are associated in digital tendon sheaths with the tendons of the flexor digitorum brevis (the equivalent of the flexor digitorum superficialis of the forearm), which they pierce to reach their insertion.

**Variations, Innervations, and Actions.** There are apparently no important variations in the deep calf muscles. Rarely, anomalous muscles are associated with them, the most common being one arising in the leg and passing into the foot to insert into the tendons of the flexor digitorum longus or the muscle (quadratus plantae) associated with this tendon.

The **nerves** to these muscles, all from the *tibial nerve,* vary both in number and in manner of origin; two or more nerves may have a common origin or arise some distance apart. The popliteus is usually said to receive fibers from the 5th lumbar and 1st sacral nerves, as are the flexor digitorum longus and the tibialis posterior. The flexor hallucis longus is thought to be supplied by fibers from the 5th lumbar and first two sacral nerves.

Because of its oblique course across the knee, the **popliteus** medially rotates the tibia if the leg is free or, when the leg bears weight, takes its fixed point from below and therefore laterally rotates the femur. This is an important action, for during extension at the knee, there is a terminal medial rotation of the femur that needs to be undone to allow free flexion. It is itself a weak flexor; in standing with a partially flexed knee, it resists anterior displacement of the femur on the tibia. The **flexor hallucis longus** is primarily a flexor of the big toe, the **flexor digitorum longus** a flexor of the distal phalanges of the other toes, and the **tibialis posterior** primarily an invertor and adductor of the foot. All three muscles are weak plantar flexors of the foot.

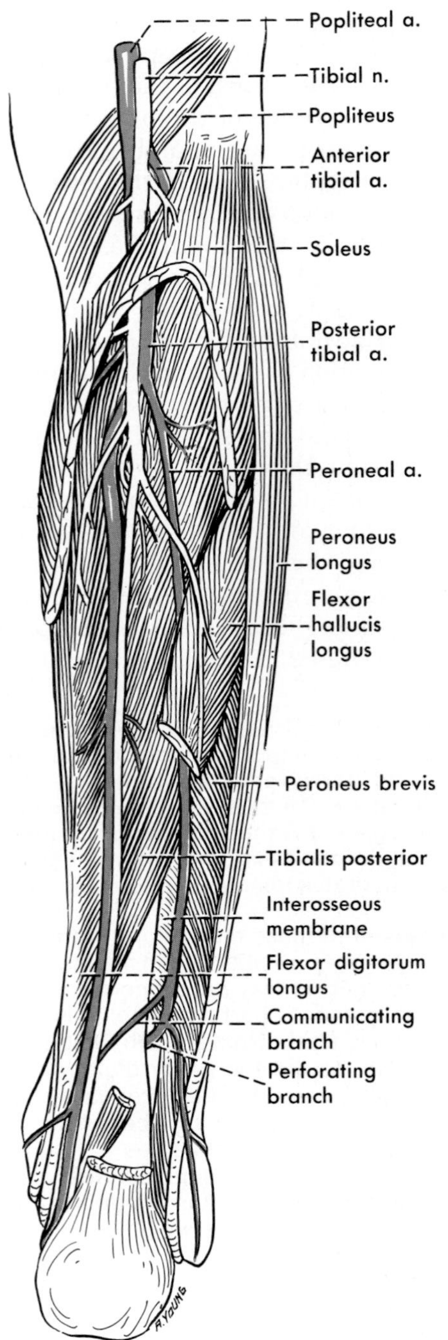


**FIG. 18-40.** Deep nerves and arteries of the calf.

## Nerves and Vessels

In the popliteal fossa, the **tibial nerve** lies posterior to the popliteal vessels (Fig. 18-40), gradually crossing from their lateral to their medial sides. Before it leaves the fossa, it gives off one or more *branches to the knee joint,* as well as the *medial sural cutaneous nerve* to skin of the leg and foot. In the lower part of the fossa, it gives off *branches to both heads* of the *gastrocnemius* and to the *plantaris* muscle and then passes with the popliteal vessels between the two heads of the gastrocnemius. Before it disappears deep to the *soleus,* it usually gives a branch into the superficial surface of that muscle and deep to it gives another branch to it and one to the *popliteus*. The tibial nerve then courses straight down the leg on the posterior surface of the tibialis posterior muscle, in company with the posterior tibial vessels, and gives off a *variable number of branches to the* three lower *deep muscles of the calf.* At the ankle, it and the posterior tibial vessels enter a special compartment in the flexor retinaculum, situated between the compartments for the flexor hallucis longus (most posterior) and the flexor digitorum longus. As it passes onto the plantar surface of the foot, it gives off *medial calcaneal branches* and divides into medial and lateral plantar nerves.

The **popliteal artery** descends through the middle of the popliteal fossa anterior to the popliteal vein and accompanies the tibial nerve deep to the gastrocnemius and soleus muscles. In its course through the fossa (see Fig. 17-49), it gives off *medial and lateral* **superior genicular arteries** that run medially and laterally around the femur to help form the collateral circulation at the knee joint; a **middle genicular artery** that pierces the posterior surface of the capsule of the joint to be distributed within the joint; and *medial and lateral* **inferior genicular arteries** that encircle the leg and also help form the anastomosis around the knee joint. Although the anastomotic channels about the knee joint (Fig. 18-41) often have been emphasized, they are rarely adequate when sudden occlusion of the popliteal artery occurs. In such cases, gangrene of the foot and leg is exceedingly common.

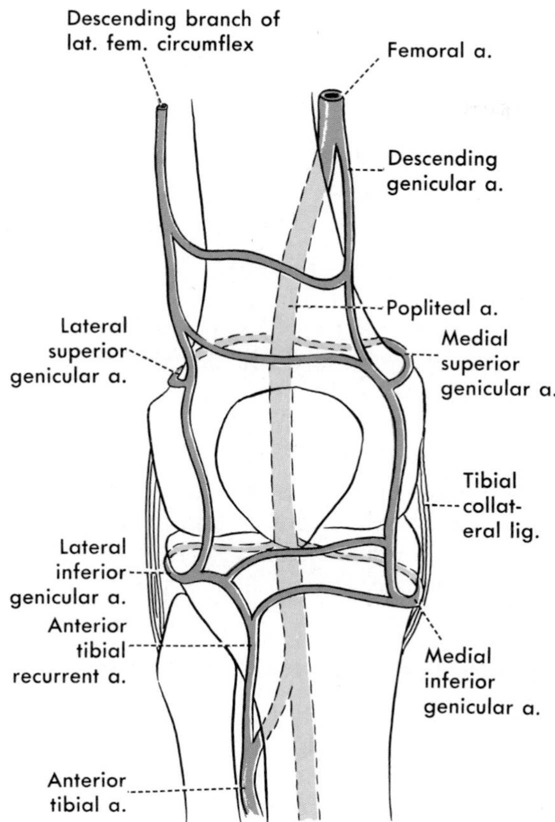


**FIG. 18-41.** Diagram of the collateral circulation around the knee joint, anterior view.

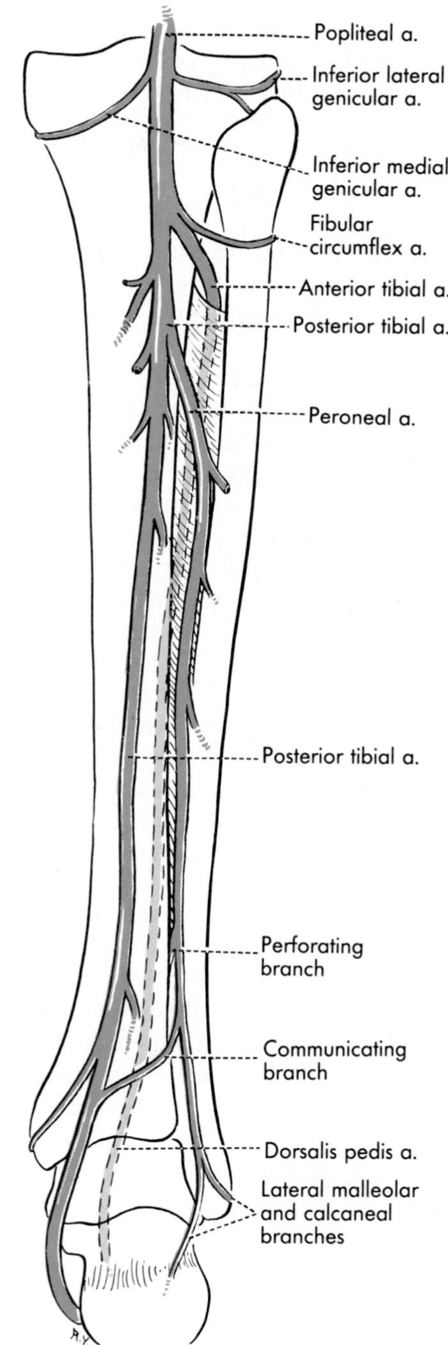


**FIG. 18-42.** Diagram of the arteries of the leg, posterior view.

The largest branches of the popliteal artery above its termination are the **sural arteries,** paired vessels that supply the gastrocnemius and soleus muscles in particular.

The popliteal artery ends by dividing into **anterior** and **posterior tibial arteries** (Fig. 18-42). This division usually occurs on the posterior surface of the popliteus muscle, and the anterior tibial then turns forward below the muscle and above the interosseous membrane to continue down the anterior aspect of the leg. Sometimes, when the division occurs higher, the anterior tibial artery passes anterior to the muscle.

The **posterior tibial artery** continues downward with the tibial nerve on the tibialis posterior muscle and runs with the nerve in a special compartment within the flexor retinaculum, dividing, as it reaches the plantar surface of the foot, into medial and lateral plantar branches. A little below its origin, it gives off a large *nutrient artery* to the tibia and usually gives off laterally the **fibular circumflex artery** that pierces the fibular origin of the soleus and winds around the neck of the fibula to supply the peroneus muscles. This artery may, however, arise from the lower end of the popliteal or from the anterior tibial. As it leaves the leg and passes into the foot, the posterior tibial artery gives off **medial malleolar** and **calcaneal** branches.

Most of the other branches of the posterior tibial artery are unnamed muscular ones, but its largest branch, the **peroneal artery,** may be larger than the continuation of the posterior tibial. It is derived from the lateral aspect of that vessel in the upper part of the leg and passes laterally across the surface of the tibialis posterior muscle to lie between the interosseous membrane and the fibula under cover of or in the flexor hallucis longus muscle. It gives off muscular branches, some of which pass through the interosseous membrane to help supply anterior muscles, and a *nutrient artery* to the fibula. A little above the ankle, it usually communicates with the posterior tibial, sends a *perforating branch* (which may arise above the communicating branch) through the interosseous membrane to reach the lateral aspect of the dorsum of the foot, and gives rise to **lateral calcaneal** and **malleolar** branches. The perforating branch of the peroneal artery sometimes gives rise to the *dorsalis pedis artery.* When the posterior tibial artery is small or deficient in the lower part of the leg, the communicating branch of the peroneal to the posterior tibial continues into the foot as the plantar arteries.

Most of the arteries of the leg are accompanied by paired veins. These unite in several different patterns to form the unpaired **popliteal vein.** Occasionally, the lower end of the popliteal vein is doubled, or it divides into two veins above to join both the femoral and the deep femoral veins.

**Deep lymphatics** in the calf accompany the posterior tibial and peroneal vessels and end in **popliteal nodes.** These few small nodes also receive both superficial lymphatics accompanying the small saphenous vein and the deep anterior tibial lymphatics. They send their efferents along the popliteal and femoral arteries.

## THE KNEE JOINT

Bones entering into the knee joint are the femur, the tibia, and the patella. Since the femur slants toward the knee, and each tibia runs almost vertically, the long axes of the femur and tibia meet at an angle of some 10° to 12° from a straight line, the tibia diverging outward from the line of the femur (Fig. 18-1). An exaggeration of this outward bend at the knee is *knock knee* or genu valgum (valgus, an adjective, means outwardly turned; the adjective for inwardly turned is varus).

### Capsule and External Ligaments

The lower part of the quadriceps tendon, the patella, and the patellar ligament replace the fibrous capsule of the knee joint anteriorly, and the iliotibial tract and patellar retinacula blend with the anterolateral and anteromedial parts of the fibrous capsule. Laterally and medially, the joint is strengthened by the fibular and tibial collateral ligaments.

The **fibular collateral ligament** (Fig. 18-43) is a rounded cord stretching between the lateral epicondyle and the head of the fibula and standing well away from the capsule of the joint. The **tibial collateral ligament,** in contrast, is a broad band attached above to the medial femoral epicondyle and below to the medial tibial condyle. It is closely applied to the capsule, although its anterior straight part is slightly separated from the capsule by connective tissue in which one or more bursae occur. Its posterior part fans out posteriorly, thus running obliquely, to blend with the capsule and through this attain a connection to the medial meniscus.

The posterior portion of the fibrous capsule of the knee joint is, for the most part, a fairly thin membrane, attached above to the inter-condylar line of the femur and below to the posterior surface of the upper end of the tibia.

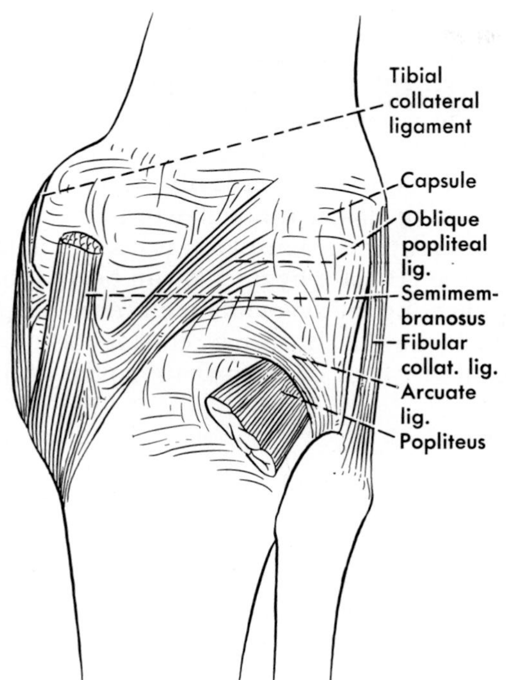


**FIG. 18-43.** Posterior view of the capsule of the knee joint.

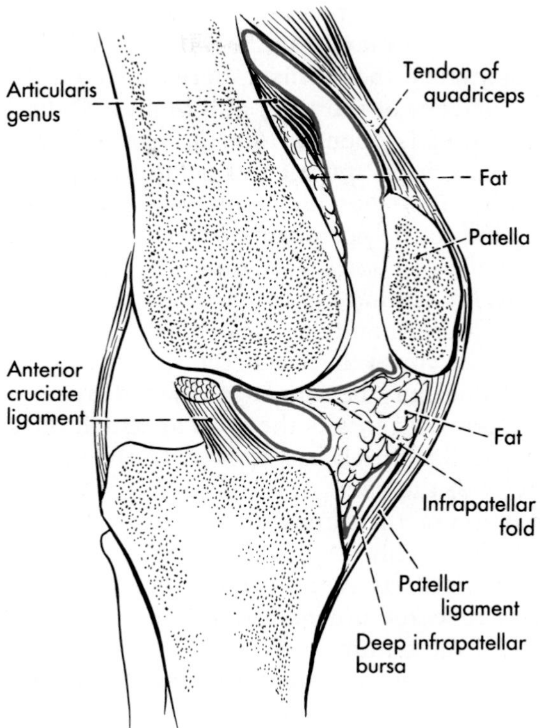


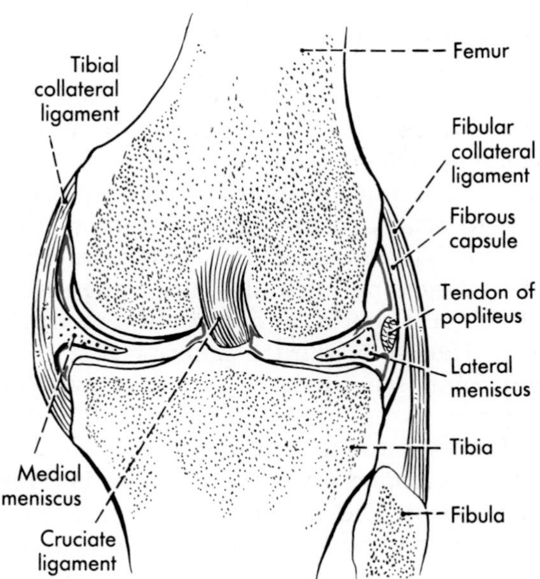


**FIG. 18-44.** Sagittal and frontal sections through the knee joint; synovial membrane is red.

Below and laterally, it presents a defect through which the popliteus muscle leaves its position from inside the capsule to run to its insertion on the tibia. One of the two special thickenings of the posterior part of the capsule is at the upper edge of this defect, that is, above the gap for the popliteus muscle. This thickening, the **arcuate ligament,** consists of fibers that are attached below to the head of the fibula but spread out above as some fibers arch upward and medially over the edge of the defect and others go more directly upward. The other thickening, the **oblique popliteal ligament,** runs upward and laterally across the posterior aspect of the knee joint. Many of its fibers are derived from the semimembranosus tendon shortly before it reaches its insertion and therefore have no independent attachment to the tibia. This ligament helps the posterior part of the capsule to resist hyperextension at the knee, a function also of the hamstring muscles.

The **cavity** of the knee joint and its synovial lining are especially complicated (Fig. 18-44), for the cavity is partially subdivided both by the synovial membrane and by fibrocartilaginous plates, the menisci, between the articular surfaces of the tibia and fibula. When the knee joint is opened anteriorly by section of the quadriceps muscle immediately above the patella, the synovial membrane can be seen applied to the posterior surface of the muscle and reflected downward upon the femur (with some fat behind) to the articular surfaces of the femoral condyles. The upper part of the synovial cavity between the quadriceps and the bone may be partially divided by a constriction that marks the level at which the **suprapatellar bursa** has fused with the synovial cavity of the knee joint; there may be no constriction or there may be a separate suprapatellar bursa above the cavity of the knee joint. The *articularis genus muscle* inserts into the layer reflected downward on the anterior surface of the femur.

As the **synovial membrane** swings forward from the sides, it covers the inner surfaces of the patellar retinacula and as it reaches the patella, ends at the articular surface of this bone. Extending downward from the patella, it forms the **infrapatellar synovial fold,** which may have subsidiary **alar folds** running out from its sides. The infrapatellar fold runs downward and backward from the patella, becoming wider as it does so, and then swings up over the anterior intercondylar area of the tibia to attach to the anterior border of the intercondylar fossa of the femur, its edges then proceeding backward along the medial and lateral borders of the intercondylar fossa. Thus although the anterior part of the knee joint is a single cavity, the lower and posterior part is paired, divided by the infrapatellar fold and its continuations around the tissue associated with the intercondylar areas of the bones. The articular surface of the medial femoral condyle, the medial meniscus, and the articular surface of the medial tibial condyle border one of the two divisions of the cavity, and the articular surface of the lateral femoral condyle, the lateral meniscus, and the articular surface of the lateral tibial condyle border the other. The paired parts of the synovial capsule are separated by the width of the intercondylar areas of the femur and tibia (Fig. 18-44B). It is to the intercondylar areas that the cruciate ligaments and the menisci are attached; thus, these ligaments and the ends of the menisci actually lie outside the cavity of the joint, although within the fibrous capsule. After the knee joint is opened anteriorly, the cruciate ligaments can be seen only by reflecting the synovial membrane, but posteriorly they can be seen by cutting the fibrous capsule without opening the synovial cavity.

## Menisci

The parts of the joint cavity between the tibial and femoral condyles are further subdivided by the menisci (Figs. 18-44 and 18-45), fairly thin, somewhat wedge-shaped semilunar pieces of fibrocartilage that are attached at their convex peripheries to the synovial membrane and in part to the fibrous capsule, but with free concave edges around which the synovial cavity continues.

The two menisci usually differ in shape; the medial meniscus is C shaped, and the lateral one is more sharply curved and approaches an incomplete O. At its ends ("horns"), each meniscus is attached by fibrous tissue to the interarticular area of the tibia (Fig. 18-45); a **transverse ligament** may unite them anteri-

**FIG. 18-45.** The menisci on the upper end of the right tibia.

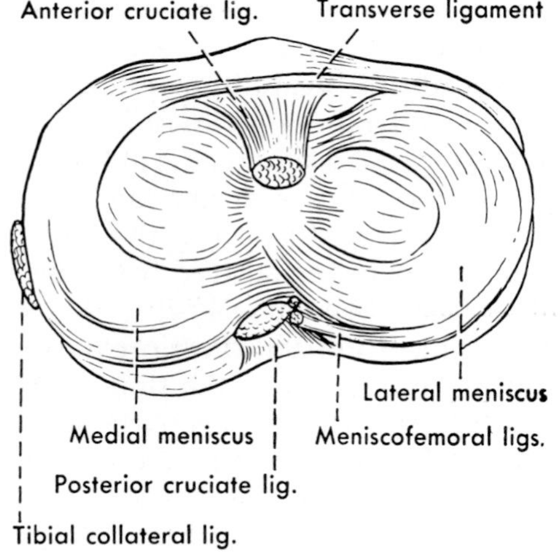

orly, but this is variable. The medial meniscus has a rather firm attachment on its convex edge to the tibial collateral ligament. The lateral meniscus is separated from much of the capsule of the joint by the tendon of origin of the popliteus muscle, which attaches in part to the meniscus. Its posterior border also often has an **anterior** or **posterior meniscofemoral ligament,** occasionally both, attached to it. This is a band of fibers that runs along the posterior cruciate ligament, in the position indicated by its name, to attach to the femur.

The chief function of the menisci is in rotation, during which they move on the tibia. The medial meniscus is moved by its attachment to the femur through the tibial collateral ligament, by the weight of the femoral condyle, and, during flexion, by the semimembranosus. The posterior part of the lateral meniscus is pulled medially and slightly anteriorly by the meniscofemoral ligament during lateral rotation of the femur. The popliteus muscle, in flexion or lateral rotation, draws the meniscus backward out of harm's way. This controlled movement of the lateral meniscus has been said to account for the fact that it is much less frequently torn in injuries to the knee than is the medial meniscus.

## Cruciate Ligaments

The cruciate ligaments are partly surrounded by the synovial cavity of the knee joint, for the synovial membrane is reflected around them anteriorly and on their sides. The longer **anterior cruciate ligament** (Fig. 18-46) arises from the anterior intercondylar area of the tibia adjacent to the medial condyle and extends obliquely upward, backward, and laterally to attach into the medial side of the lateral condyle of the femur (a part of the intercondylar fossa). The **posterior cruciate ligament** arises from the posterior intercondylar area of the tibia and passes obliquely upward, forward, and slightly medially—crossing behind the anterior cruciate ligament—to attach to the lateral surface of the medial condyle. Because both ligaments are somewhat flattened as they attach to the femur, the rolling motion of

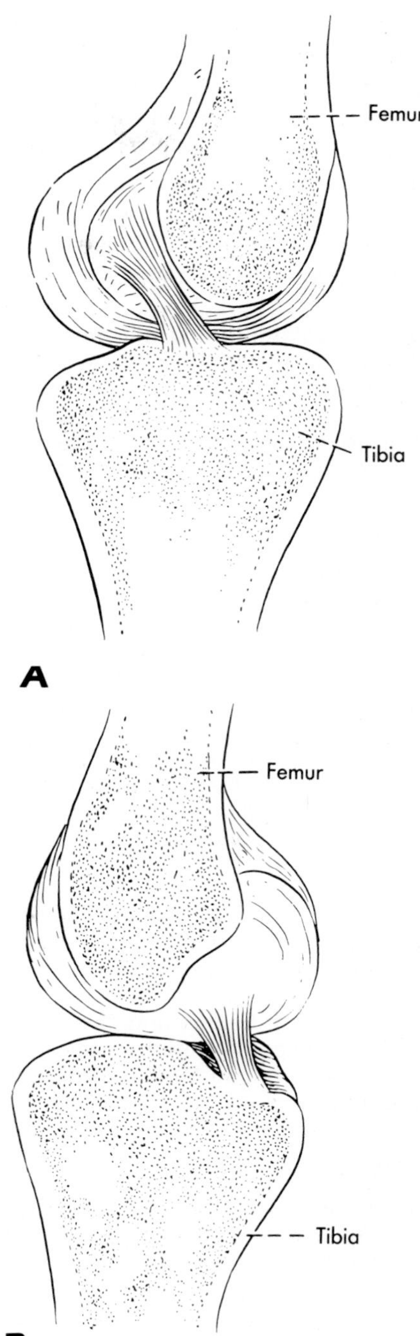


**FIG. 18-46.** **A,** a medial view of the anterior cruciate ligament; **B,** a lateral view of the posterior cruciate ligament.

the bone during flexion twists the fibers, and this twisting keeps some of the fibers under constant tension. It is now rather generally agreed that the cruciate ligaments are most tense during full extension and full flexion, but some part of each ligament is also tense during intermediate phases, so that they exert a steadying effect upon the joint throughout the entire range of movement.

**Tears** of the anterior cruciate ligament are fairly frequently associated with tears of other ligaments, particularly the tibial collateral. All the ligaments of the knee contribute to its stability, but the tibial collateral is especially important, for the major part of this strong ligament is tense in all positions of the joint and is thus a particularly valuable stabilizer.

#### Nerves and Vessels

The nerves to the knee joint are typically derived from the femoral, the obturator, and both parts of the sciatic nerve. Many of these nerves follow the arteries, but some run directly to the capsule.

The capsule of the knee joint is supplied by twigs from all the vessels that enter into the anastomosis around the knee joint (Fig. 18-41). In addition, the middle genicular artery penetrates the capsule posteriorly and is distributed especially to the tissue of the intercondylar region.

#### Movements

The chief movements at the knee joint are flexion and extension, but when the knee is partly flexed, rotation to a maximum of about 35° to 40° is also possible.

The movements of **flexion** and extension at the knee are somewhat complicated, for they consist of more than a simple rolling forward or backward of the femoral condyles on the tibial ones. Thus, as the knee is extended and the femoral condyles roll forward on the tibia, they also skid backward, so that when the anterior roll is completed, the place of contact between femur and tibia is only a little more anteriorly placed on the tibia than it was during flexion. Similarly, as the knee is flexed, the femoral condyles roll backward and skid forward on the tibia. The fact that the two femoral condyles do not have quite the same shape means also that they have to move slightly differently. The shorter, more highly curved lateral condyle exhausts its forward roll faster than does the longer and less curved articular surface of the medial condyle. In consequence, when the lateral condyle has rolled forward almost as far as it can go during extension and is being checked by the anterior cruciate ligament, the medial condyle still has some distance to go. The medial condyle, therefore, rolls forward, and at the same time skids backward, faster than does the lateral one, in order to engage its more anterior part. The greater posterior skidding of the medial condyle results, as full extension is reached, in a medial rotation of the femur on the tibia, which tightens still more the collateral ligaments of the knee. Then, when the knee is to be flexed from this extended position, a preliminary lateral rotation of the femur on the tibia (or medial rotation of the tibia on the femur) relaxes the ligaments sufficiently to allow flexion. This rotation is ordinarily brought about by the popliteus muscle.

The chief **extensor** of the knee, and the only one that directly extends it, is the quadriceps muscle. The quadriceps adds a great deal to the stability of the knee, and a strong quadriceps can often substitute satisfactorily for weakened or lax ligaments. The gluteus maximus, the gastrocnemius, and the soleus help to extend the knee by their actions at the hip and ankle joints; the gluteus maximus extends the hip joint, and the soleus and gastrocnemius flex the ankle joint (or resist dorsiflexion); they thus resist flexion of the knee, since that cannot occur without simultaneous flexion at the hip and dorsiflexion at the ankle when the limb is supporting weight.

The **flexors** at the knee are relatively numerous; the semimembranosus, semitendinosus, short head of the biceps, and gracilis are the best flexors, the sartorius a weaker one. The popliteus is said to be a weak flexor, assisting flexion more by rotating the femur on the tibia or vice versa than by its direct pull in flexion. The gastrocnemius, because it crosses behind the knee joint, flexes this joint when the limb is not supporting weight. The long head of the biceps, although assisting flexion, relaxes while flexion is still being completed by other muscles.

The **medial rotators** of the leg are the semimembranosus, semitendinosus, gracilis, sartorius, and popliteus. The **lateral rotator** of the partially flexed leg is the biceps femoris.

## PLANTAR ASPECT OF FOOT

The foot, like the hand, has a number of intrinsic muscles and also transmits the long tendons going to the digits. The nerves and vessels of the plantar surface of the foot are continuations downward of the nerves and vessels of the calf.

### FASCIA, SUPERFICIAL NERVES, AND VESSELS

The skin of the sole of the foot is supported by tough subcutaneous tissue or superficial fascia that is continuous around the sides of the

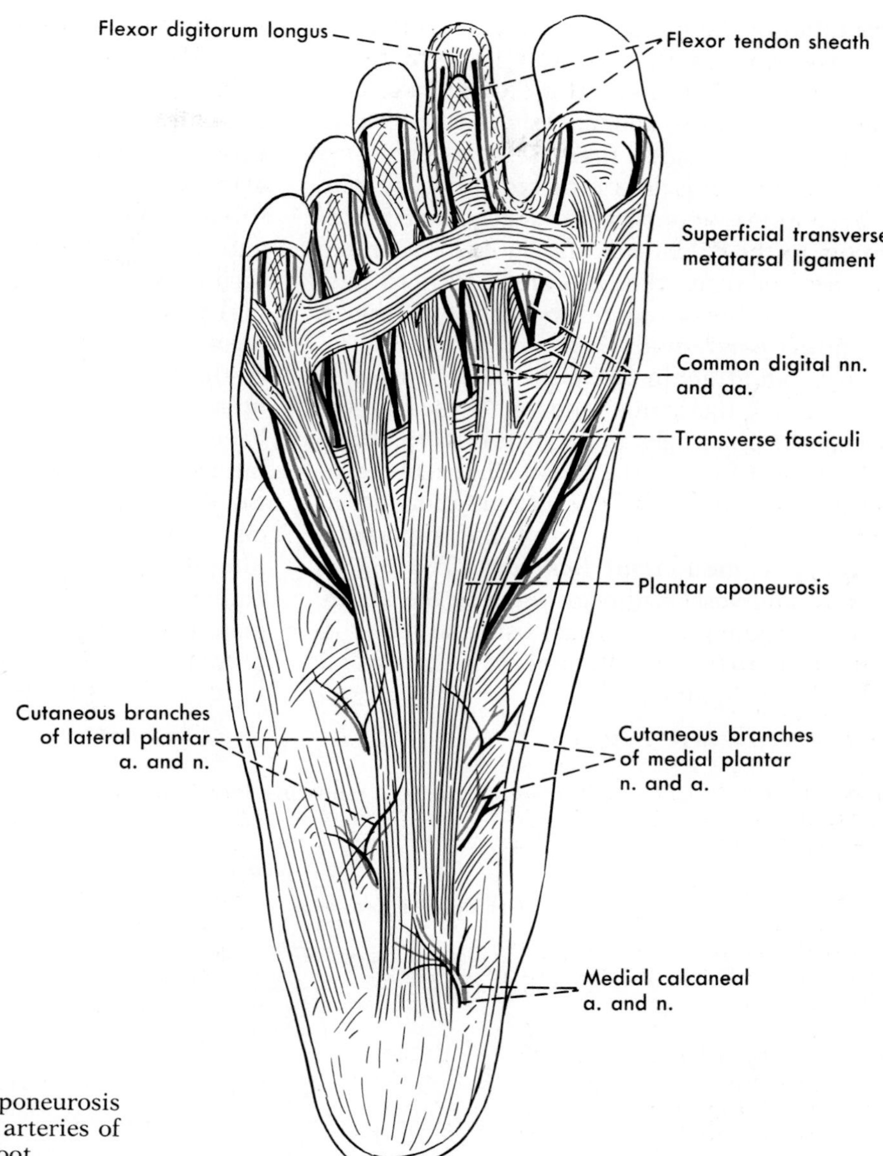


**FIG. 18-47.** The plantar aponeurosis and superficial nerves and arteries of the plantar surface of the foot.

foot with the thin subcutaneous tissue of the dorsum. Toward the toes, this tissue blends with deeper-lying tissue and continues onto the plantar surfaces of the toes to surround the digital nerves and vessels and the digital tendon sheaths in which the flexor tendons lie.

The **plantar digital nerves** and **vessels** (Figs. 18-47 and 18-49) emerge between slips of the plantar aponeurosis or deep fascia to become subcutaneous close to the bases of the digits. Of the digital nerves, those to the medial three and one-half toes are usually supplied by the medial plantar nerve, those to the lateral one and one-half by the lateral plantar nerve, a distribution similar to the distribution of median and ulnar nerves in the hand. The **digital arteries** are formed primarily by metatarsal arteries.

The skin of most of the sole is supplied by small twigs, of which the medial ones come from the medial plantar nerve and artery, the lateral ones from the lateral plantar nerve and artery.

The **deep fascia** of the foot closely resembles that of the hand. There is a central **plantar aponeurosis** that extends forward to divide into digitations that go toward the toes (Fig. 18-47). This aponeurosis is more tendinous in appearance than is the deep fascia covering the muscles of the big and little toes. The plantar aponeurosis is attached proximally to the calcaneus. Its digitations, united at first by transverse fasciculi, split to pass around the flexor tendon sheaths of the digits, blending partly with these and partly continuing to an attachment on the deep transverse metatarsal ligaments and the bases of the proximal phalanges. Superficial fibers attach into skin. Close to the heads of the metatarsals, the digitations of the plantar aponeurosis are crossed superficially by the **superficial transverse metatarsal ligament.** The digital nerves and vessels appear between the digitations of the plantar aponeurosis and pass distally deep to (above) the superficial transverse metatarsal ligament.

**Intermuscular septa** are given off from the medial and lateral edges of the plantar aponeurosis, where it continues into the less dense fascia over the muscles of the big and little toes. These septa pass dorsally and are usually described as dividing the foot into three compartments: lateral, intermediate, and medial. The **lateral septum** is usually described as attaching over the tarsals proximally, and to the 5th metatarsal distally, to form a compartment for the special muscles of the little toe. However, the **medial intermuscular septum** is more complicated. It has been described as dividing into medial and lateral leaflets as it is traced dorsally. The medial leaflet passes deep to the abductor hallucis to attach to the 1st metatarsal, and the lateral leaflet joins fascial septa running transversely in the foot (Fig. 18-48). Dyes injected into the fascial spaces containing the tendons of the flexor digitorum longus and flexor hallucis longus spread along these ten-

**FIG. 18-48.** Cross section through the foot, showing fascial layers and compartments.

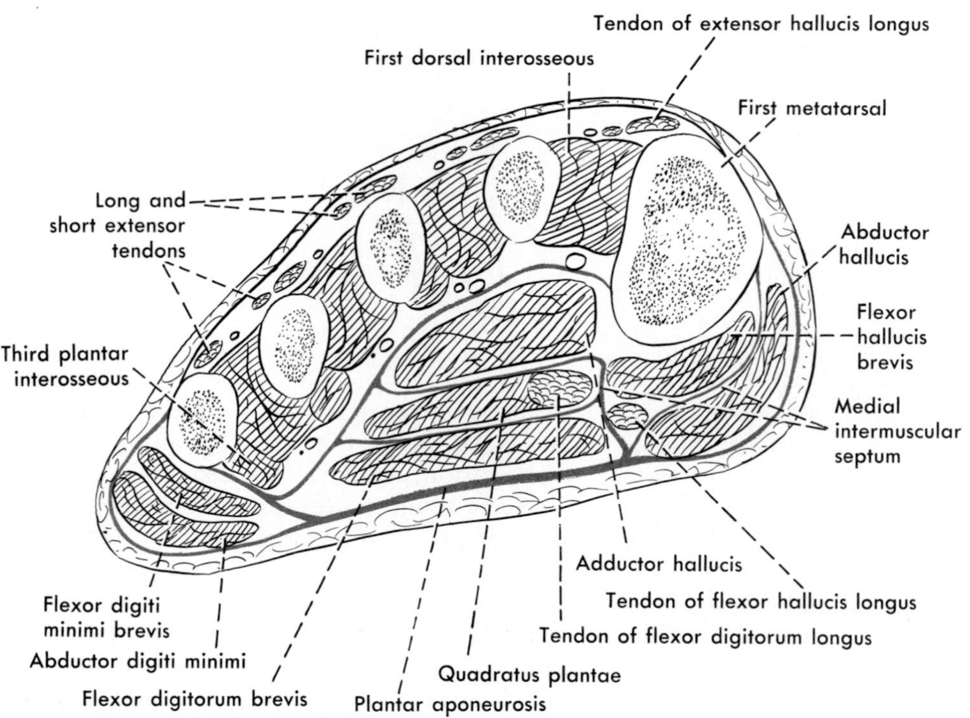

dons to the deep compartment of the leg (hence infections can also follow this route), but there was no spread of injected dye from any of the other spaces.

## MUSCLES AND RELATED STRUCTURES

Removal of the deep fascia from the sole of the foot reveals the most superficial layer of muscles and also the courses of the digital nerves and common digital arteries (Fig. 18-49). The central superficial muscle is the *flexor digitorum brevis*. Emerging on its medial side, between it and the muscles of the big toe, are the superficial branches of the *medial plantar nerve and artery*, and emerging on its lateral side, between it and the muscles of the little toe, are the superficial branches of the *lateral plantar nerve and artery*. Branches of the nerves and arteries closely parallel each other. The nerves are usually larger than the arteries, for the superficial arteries of the foot are particularly small; further, instead of medial and lateral plantar arteries forming an arch, comparable to the superficial arch of the palm, each one typically breaks up into digital arteries that accompany the digital nerves.

The **medial plantar nerve and artery** give rise to branches to the medial side of the big toe, and the nerve gives rise to *common digital branches* that divide into *proper digital nerves* for the adjacent sides of the 1st, 2nd, 3rd, and 4th toes. The first common digital nerve gives a twig to the 1st lumbrical muscle, and small *digital arteries* accompanying the nerves join the larger *plantar metatarsal arteries* from the (deep) plantar arch.

Both the **lateral plantar nerve and artery** give off a branch to the free border of the little toe, and the nerve gives rise to a *common digital branch* that runs toward the interspace between the little and 4th toes and divides to form *proper digital nerves*. The corresponding branch of the lateral plantar artery joins the 4th plantar metatarsal artery. Fairly frequently, also, the lateral plantar nerve gives off a branch that crosses the superficial surface of the flexor digitorum brevis and joins the branch of the medial plantar nerve to the interspace between the 3rd and 4th toes; a similar connection between the ulnar and median nerves of the hand is also common.

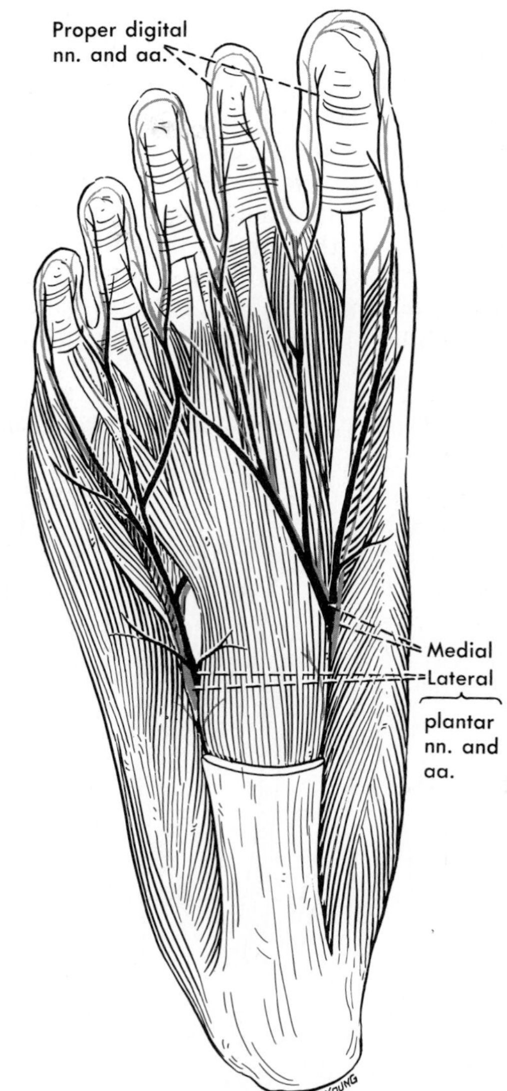


**FIG. 18-49.** Digital nerves and vessels after removal of the plantar fascia.

The muscles of the foot can most easily be described and dissected in layers. Four musculotendinous layers in the foot can be found.

### Superficial Layer

The superficial layer of the muscles of the sole (Fig. 18-50) consists of three muscles, the flexor digitorum brevis centrally, the abductor hallucis medially, and the abductor digiti minimi laterally. All three arise primarily from the calcaneus (Fig. 18-51) but take origin also from the intermuscular septa and other adjacent fascial layers.

The **abductor hallucis** arises from the medial process of the tuber calcanei, and from the lower border of the flexor retinaculum, and inserts into the medial side of the base of the proximal phalanx of the big toe. Before its insertion, it unites with the medial tendon of the flexor hallucis brevis, the combined tendon having also some insertion into the medial sesamoid bone to the big toe.

The *medial and lateral plantar nerves and vessels* enter the foot deep to this muscle side by side, for the tibial nerve and the posterior tibial vessels lie close together while they are deep to the flexor retinaculum and divide into medial and lateral plantars in this location. The lateral plantar nerve and artery turn lat-

**FIG. 18-50.** The superficial layer of plantar muscles.

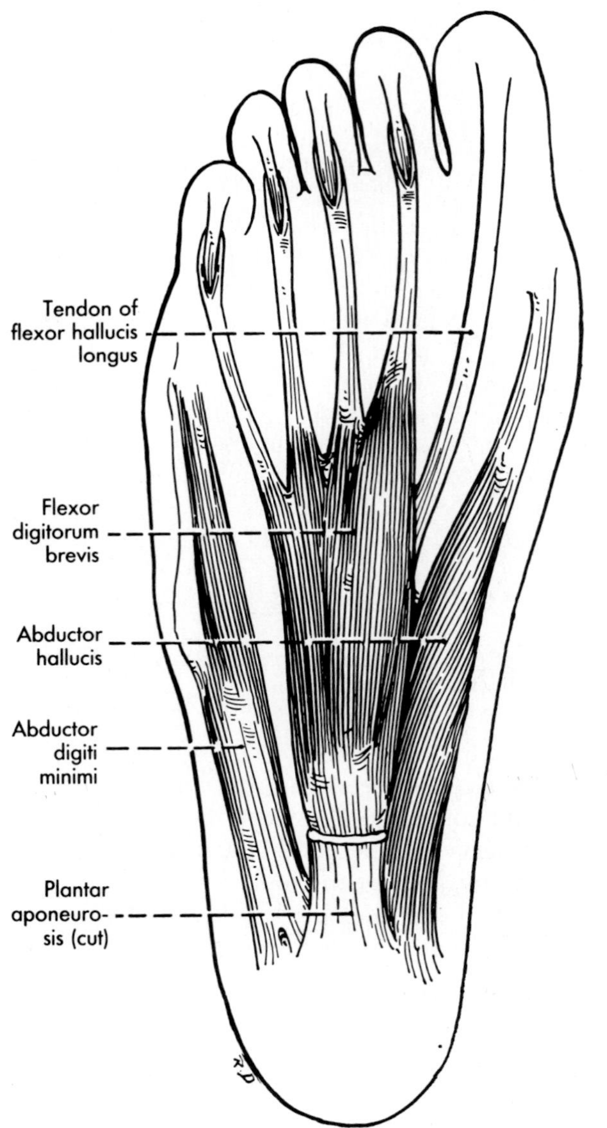


**FIG. 18-51.** Attachments of the superficial layer of plantar muscles.

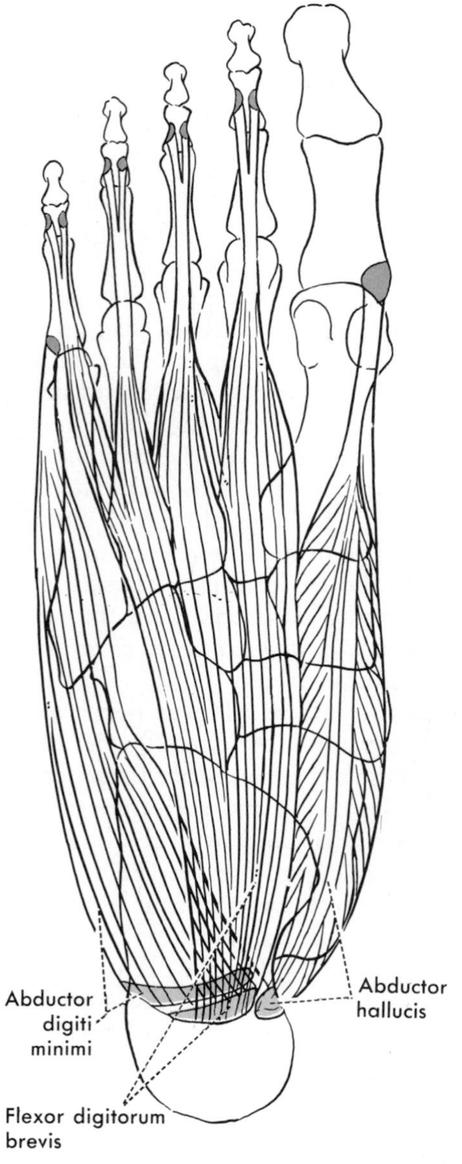

erally, but the medial plantar nerve and vessels run forward deep to the abductor hallucis. In this location, they give off branches to the abductor hallucis and supply it before emerging between the abductor hallucis and the flexor digitorum brevis.

The **flexor digitorum brevis,** the equivalent of the flexor digitorum superficialis of the forearm and hand, also arises mainly from the medial process of the tuber calcanei. It gives rise to four tendons that proceed toward the four lateral toes. These tendons lie immediately below (superficial to) the tendons of the flexor digitorum longus, and both short and long tendons enter **digital tendon sheaths** on the toes. These sheaths are essentially similar to the digital sheaths of the fingers except that they all begin blindly proximally, none normally communicating with tendon sheaths at the ankle. Within the digital tendon sheaths, the tendons of the flexor digitorum brevis divide to allow the long flexor tendons to pass through the distal phalanges; the slips then interchange fibers dorsal to the long tendons and insert on the middle phalanges. The long flexor tendons have short *vincula* over the distal interphalangeal joints, and the short flexor tendons have short vincula over the proximal interphalangeal joints, essentially similar to those in the hand; long vincula, also present, vary considerably.

The *lateral plantar nerve and artery* cross the upper (deep) surface of the flexor digitorum brevis in their course from the medial to the lateral side of the foot, but this muscle is supplied by a branch from the *medial plantar nerve.* This is comparable to the median nerve supply to the flexor digitorum superficialis in the forearm.

The lateral muscle of this group, the **abductor digiti minimi,** arises from the calcaneus and adjacent fascia and inserts on the lateral side of the proximal phalanx of the little toe. Some of its lateral fibers may attach to the tuberosity of the 5th metatarsal and constitute an accessory muscle, the **abductor ossis metatarsi quinti.** The abductor digiti minimi and the abductor of the 5th metatarsal, if present, are supplied by the *lateral plantar nerve.*

## Second Layer

The second layer of plantar structures consists of the long flexor tendons and their associated muscles (Fig. 18-52). It contains the tendon of the flexor digitorum longus, the quadratus plantae, four lumbrical muscles, and the tendon of the flexor hallucis longus.

The tendon of the **flexor digitorum longus** rounds the ankle anterior to that of the flexor hallucis longus and crosses superficial (inferior) to that tendon in the foot. It usually receives a tendinous slip from the flexor hallucis longus, and as it divides into its four tendons, it receives the insertion of the quadratus plantae muscle and gives origin to the lumbricals.

The **quadratus plantae** (accessory flexor) muscle arises by two heads, from the medial and lateral sides of the plantar surface of the calcaneus (Fig. 18-53), and inserts into the lateral edge of the tendon of the flexor digitorum longus. No equivalent of the quadratus plantae is found in the hand. The quadratus plantae receives a branch from the *lateral plantar nerve* as this crosses its superficial surface.

The **lumbrical muscles** arise from the tendons of the flexor digitorum longus, just as those in the hand arise from the profundus. The 1st (most medial) lumbrical arises from the medial side of the tendon to the 2nd toe, but each of the other three lumbricals arises from both the tendons between which it lies. As they run forward to the medial sides of the four lateral toes, they pass below the *deep transverse metatarsal ligaments* and then turn dorsally to join the extensor tendons on the proximal phalanges. The first lumbrical is supplied by a branch from the medial plantar nerve, entering its superficial surface, and the other three lumbricals are supplied by twigs from the deep branch of the lateral plantar nerve that enters their deep surfaces.

The tendon of the **flexor hallucis longus** brings with it into the foot the flexor tendon sheath that surrounds it as it lies deep to the flexor retinaculum, but this usually stops in the proximal part of the sole close to the level at which its tendon crosses above that of the flexor digitorum longus. The two tendon sheaths sometimes communicate here. The

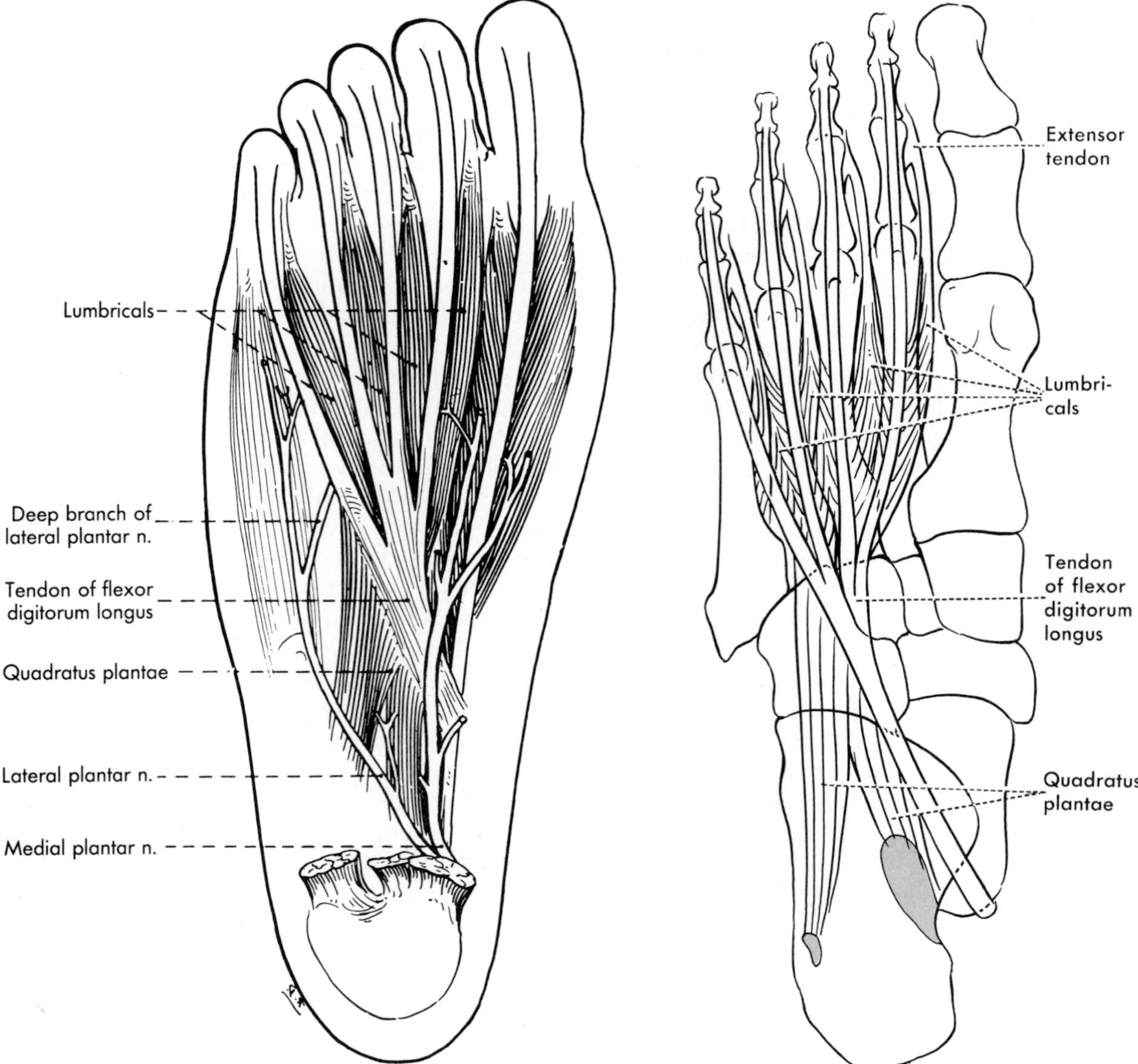


**FIG. 18-52.** The second layer of plantar muscles and the long flexor tendons in the foot.

**FIG. 18-53.** Attachments of the second layer of plantar muscles.

tendon of the flexor hallucis longus, thereafter typically bare for some distance, usually gives a slip to the tendon of the flexor digitorum longus and then runs forward on the flexor hallucis brevis to enter a digital tendon sheath in which it travels almost to its insertion on the distal phalanx of the big toe.

### Third Layer

The third layer consists of three muscles, the flexor hallucis brevis, the adductor hallucis, and the flexor digiti minimi brevis (Figs. 18-54 and 18-55).

The **flexor hallucis brevis** arises by tendinous fibers from the cuboid and lateral cuneiform bones. The muscular belly divides distally into two parts: the *medial part* blends with the insertion of the abductor hallucis, sharing the medial sesamoid of the big toe with this and inserting on the medial side of the base of the proximal phalanx; the *lateral part* blends with the two heads of the adductor hallucis, sharing the lateral sesamoid with these and inserting on the lateral side of the base of the proximal phalanx. The muscle receives a branch from the *medial plantar nerve* as this nerve runs forward on it.

The **adductor hallucis** consists (like the correspondingly named muscle in the hand) of two heads, an oblique and a transverse. The *oblique head* is usually the larger and arises from the bases of the 2nd to 4th metatarsal bones and from the long plantar ligament. The *transverse head* usually has no bony origin but, rather, one from a variable number of the plantar and deep transverse metatarsal ligaments from the 2nd to the 5th toes. The two heads converge, and as they join, they blend also with the lateral tendon of the flexor hallucis brevis, sharing the lateral sesamoid with it, and insert on the lateral side of the base of the proximal phalanx of the big toe. Each head receives a nerve from the deep branch of the *lateral plantar nerve.*

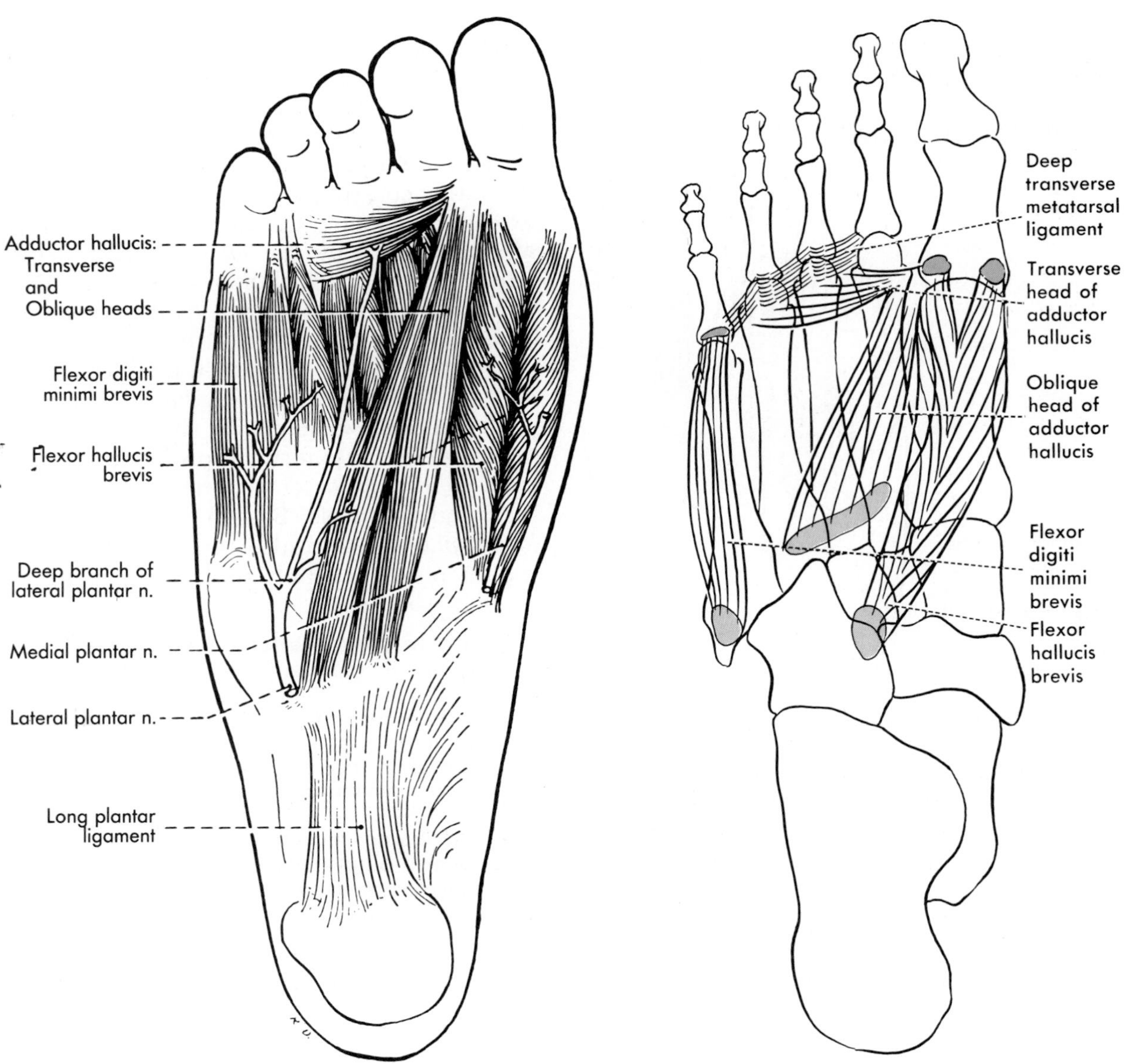


**FIG. 18-54.** The third layer of plantar muscles.

**FIG. 18-55.** Attachments of the third layer of plantar muscles.

The **flexor digiti minimi brevis** arises from the base of the 5th metatarsal bone and from the long plantar ligament as this covers the tendon of the peroneus longus. It inserts into the plantar surface of the base of the proximal phalanx of the little toe. Its tendon of insertion usually blends laterally with that of the abductor digiti minimi and often sends a slip to the extensor tendon of the little toe.

## Fourth Layer

The fourth layer consists of seven interossei (Fig. 18-56), divided into three plantar and four dorsal ones. Like the interossei of the hand, each plantar interosseous arises from a single bone, the metatarsal of the toe with which it is associated; each dorsal interosseous arises from the two metatarsals between which it lies. There is, however, a difference in arrangement of the interossei of the hand and foot, for although those of the hand are arranged around the middle digit as the midline, those of the foot are arranged around the second digit as the midline. Consequently, the **plantar interossei,** which adduct, are associated with the three lateral toes: each arises (Fig. 18-57) from the medial surface of the metatarsal and inserts on the medial side of the base of the proximal phalanx of the same digit.

**FIG. 18-56.** The interossei of the foot.

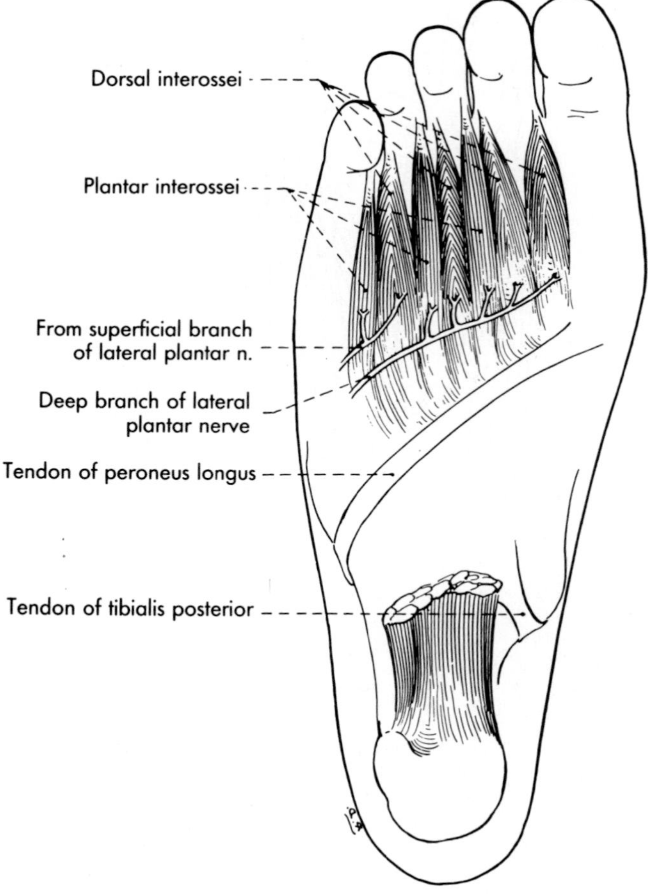


Correspondingly, the first two **dorsal interossei** attach to the 2nd toe, one on each side, and the 3rd and 4th attach to the lateral sides of the 3rd and 4th toes, respectively.

Thus, as in the hand, the plantar interossei adduct and the dorsal abduct, but this movement takes place around the second rather than the third digit. A minor difference between the interossei of hand and foot is that the latter contribute little or nothing to the extensor tendons of the toes. All the interossei are innervated by the *lateral plantar nerve,* which crosses their plantar surfaces in company with the plantar arch. As they cross the metatarsophalangeal joints, the interossei are separated from the lumbrical muscles by the *deep transverse metatarsal ligaments,* for they go dorsal to these ligaments. The *plantar metatarsal arteries* pass forward on the lower surfaces of the interossei to unite with the common digital vessels. Branches from the *deep lateral plantar nerve* also run forward across the interossei to supply some of the metatarsophalangeal joints.

### Variations, Innervations, and Actions

There are no important variations in the musculature of the sole of the foot. Most of the recorded variations are minor ones. Perhaps the most common are absence of a tendon from the flexor digitorum brevis to the little toe and partial origin of this muscle from the tendon of the flexor digitorum longus.

All the plantar muscles are **innervated** by the tibial nerve through either its medial or lateral plantar branches. The *medial plantar nerve* usually contains fibers derived from the 5th lumbar and 1st sacral nerves, and these are believed to be distributed to all the muscles that the medial plantar supplies: the abductor hallucis, the flexor hallucis brevis, the flexor digitorum brevis, and the 1st lumbrical. Similarly, the *lateral plantar nerve* usually contains fibers derived from the 1st and 2nd sacral nerves, and these are presumably distributed to all of the muscles that this nerve supplies: the adductor hallucis, the quadratus plantae, the flexor digiti minimi brevis, the abductor digiti minimi, the lateral three lumbricals, and the interossei.

The **actions** of the plantar muscles are generally indicated by the muscles' names. Thus, the abductor hallucis abducts, insofar as this is possible, the big toe at the metatarsophalangeal joint (but is a better flexor), the adductor hallucis adducts it, and the flexor hallucis brevis flexes it. The flexor digitorum brevis is primarily a flexor of the middle phalanges, and the quadratus plantae at least partially corrects the medial pull of the flexor digitorum longus and can flex the distal phalanges equally well whether the foot is being dorsiflexed or plantar flexed. The flexor digiti minimi brevis flexes, and the abductor digiti minimi abducts, the little toe. The lumbricals help to flex the metatarsophalangeal joints of the four lateral toes and, because of their insertions into the extensor tendons, can help extend the interphalangeal joints; however, the pull of the flexor tendons is usually such that the interphalangeal joints of the toes, especially the little one, are kept in permanent flexion. As already noted, the plantar interossei adduct the 3rd, 4th, and 5th toes toward the 2nd one, and the dorsal interossi abduct the 2nd, 3rd, and 4th toes. The interossei are also flexors at the metatarsophalangeal joints and are at best very weak extensors of the interphalangeal joints.

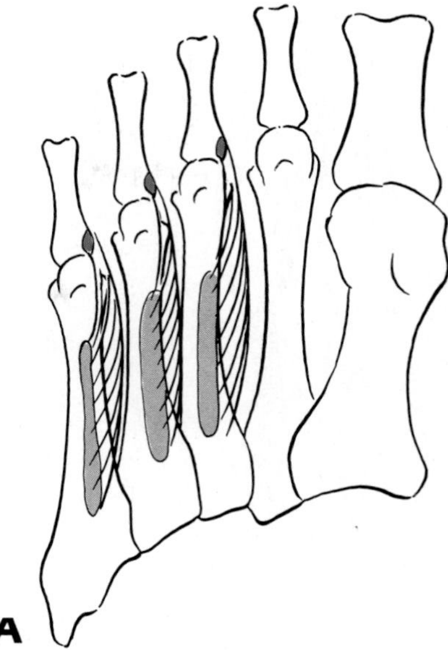


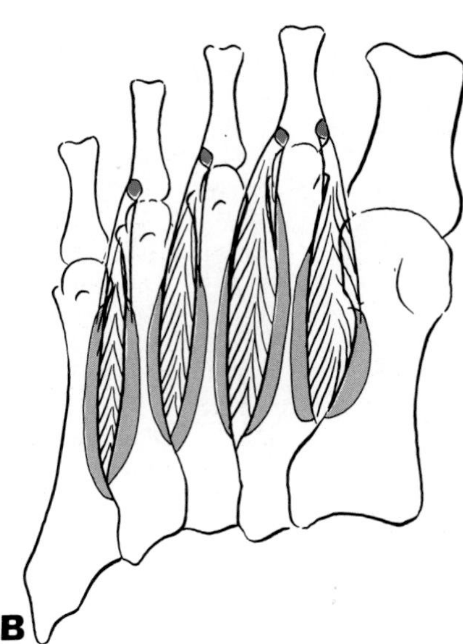


**FIG. 18-57.** Attachments of the plantar interossei, **A,** and the dorsal interossei, **B.**

## VESSELS AND NERVES

The **posterior tibial artery** and its accompanying **veins** divide under cover of the flexor retinaculum into medial and lateral plantar vessels (Fig. 18-58). These enter the foot under cover of the abductor hallucis and in company with the medial and lateral plantar nerves. The medial plantar vessels and nerves then run forward, and the lateral ones diverge laterally to pass above the flexor digitorum brevis.

The **medial plantar artery** sends its deep branch into the muscles among which it lies. Its superficial branch gives twigs to the cutaneous surface of the medial side of the foot and continues along the free side of the big toe. It also gives rise to tiny *digital arteries* that run distally and laterally across the lower surface of the flexor digitorum brevis to join the plantar metatarsal arteries and, thus, help supply the proper digital arteries on the toes (Fig. 18-49).

The **lateral plantar artery** is larger than the medial and passes obliquely forward and laterally across the foot, crossing between the flexor digitorum brevis and the quadratus plantae and associated tendon of the flexor

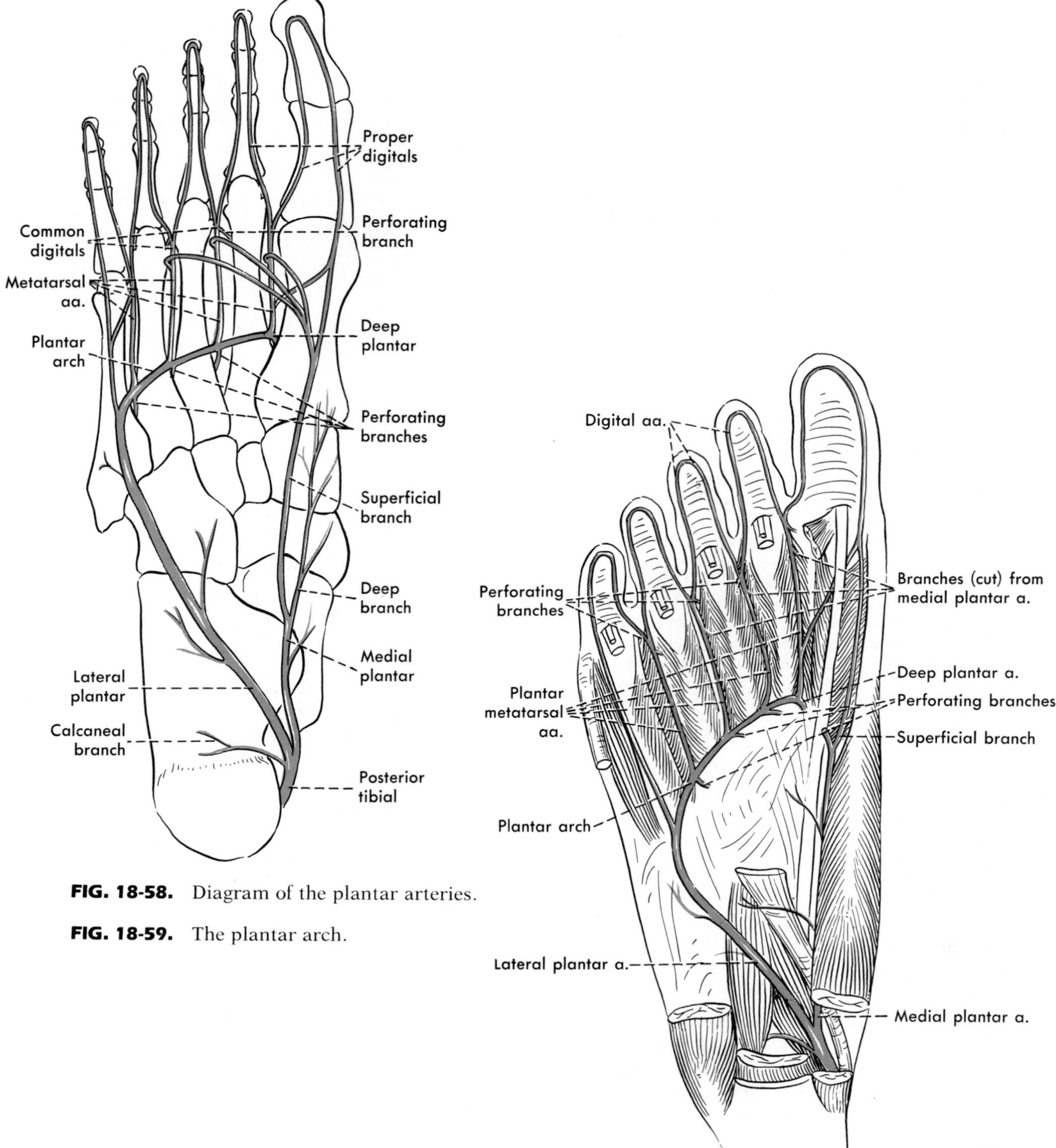


**FIG. 18-58.** Diagram of the plantar arteries.

**FIG. 18-59.** The plantar arch.

digitorum longus. It then runs forward with the lateral plantar nerve more or less between the flexor digitorum brevis and the abductor digiti minimi, giving off twigs to muscles and skin of the lateral side of the foot and *digital arteries* to the lateral one and one-half toes. It ends by turning medially across the foot on the proximal ends of the interossei, as the plantar arch (Fig. 18-59).

As the **plantar arch** crosses the foot it gives off four **plantar metatarsal arteries** that run forward on the interossei and are joined by twigs from the superficial branches of the medial and lateral plantar vessels. The short **common digital plantar arteries** then divide to form **proper digital arteries** of the toes. Proximal to the deep transverse metatarsal ligaments, the plantar metatarsal arteries give off **perforating branches** that pass dorsally to join the dorsal metatarsal arteries; distal to these ligaments they sometimes receive terminal branches of those vessels. The plantar arch also gives off **perforating branches,** three in number, that pass between the heads of the 2nd, 3rd, and 4th dorsal interossei to join the dorsal metatarsal arteries. They are sometimes the chief source of blood to the dorsal arteries.

The **deep plantar branch of the dorsalis pedis** emerges between the heads of the 1st dorsal interosseous muscle and completes the plantar arch medially. Although its relations are similar to those of the radial artery in the hand, it does not normally, as does the radial, give rise to the arch and it may be very small.

The **plantar digital veins** drain into metatarsal veins, which join the **plantar venous arch.** This drains along the medial and lateral plantar arteries to join the posterior tibial veins.

While the **tibial nerve** lies deep to the flexor retinaculum it gives off *medial calcaneal branches* to the medial side and plantar surface of the heel and divides into medial and lateral plantar nerves. These enter the sole close together and in company with the correspondingly named arteries, lying first above (deep to) the abductor hallucis. The medial plantar nerve runs forward deep to the abductor, and the lateral plantar nerve runs laterally, passing between the flexor digitorum brevis and the quadratus plantae and then turning forward.

The **medial plantar nerve** is the equivalent of the median nerve in the hand. Under cover of the abductor hallucis, it gives off a branch into this muscle, one to the flexor digitorum brevis (Fig. 18-60) and branches to the skin of

**FIG. 18-60.** Muscular distribution of the medial plantar nerve.

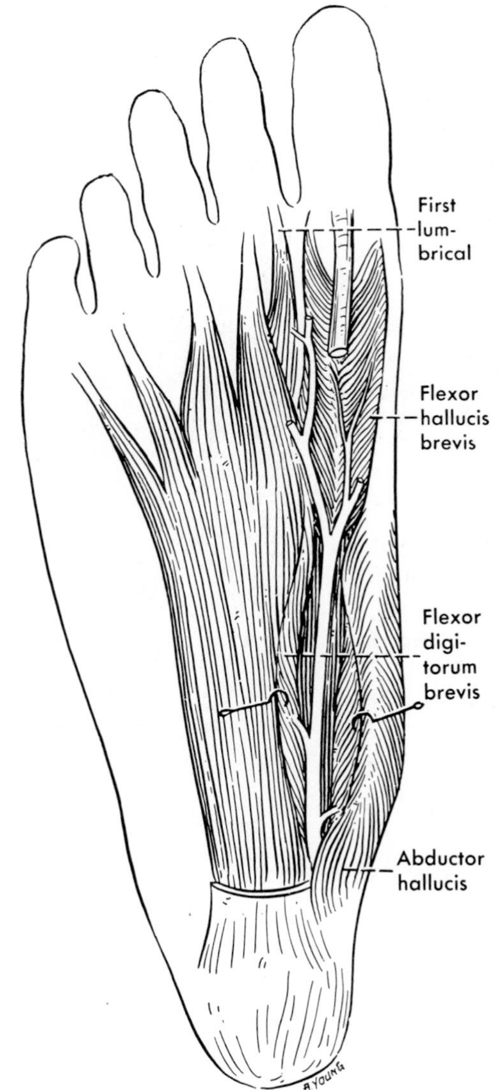

the sole of the foot. As it emerges between the abductor hallucis and the flexor digitorum brevis, it divides into four terminal **plantar digital nerves.** The most medial, the proper digital branch for the medial side of the big toe, supplies the flexor hallucis brevis and twigs to the sole of the foot before continuing on the toe. The second branch is the *common digital branch* to the adjacent sides of the big and 2nd toes; it gives a branch to the 1st lumbrical muscle and subsequently divides into proper digital nerves. The remaining two lateral branches are also common digital nerves and divide into proper digital nerves for the adjacent sides of the 2nd and 3rd and 3rd and 4th toes, respectively. The most lateral of these nerves may receive a communication from the lateral plantar nerve.

The **lateral plantar nerve** (Fig. 18-61; Fig. 18-49) is the equivalent of the ulnar nerve in the hand. As it passes laterally and forward between the flexor digitorum brevis and the quadratus plantae, it gives off a branch into the latter muscle and one that runs laterally

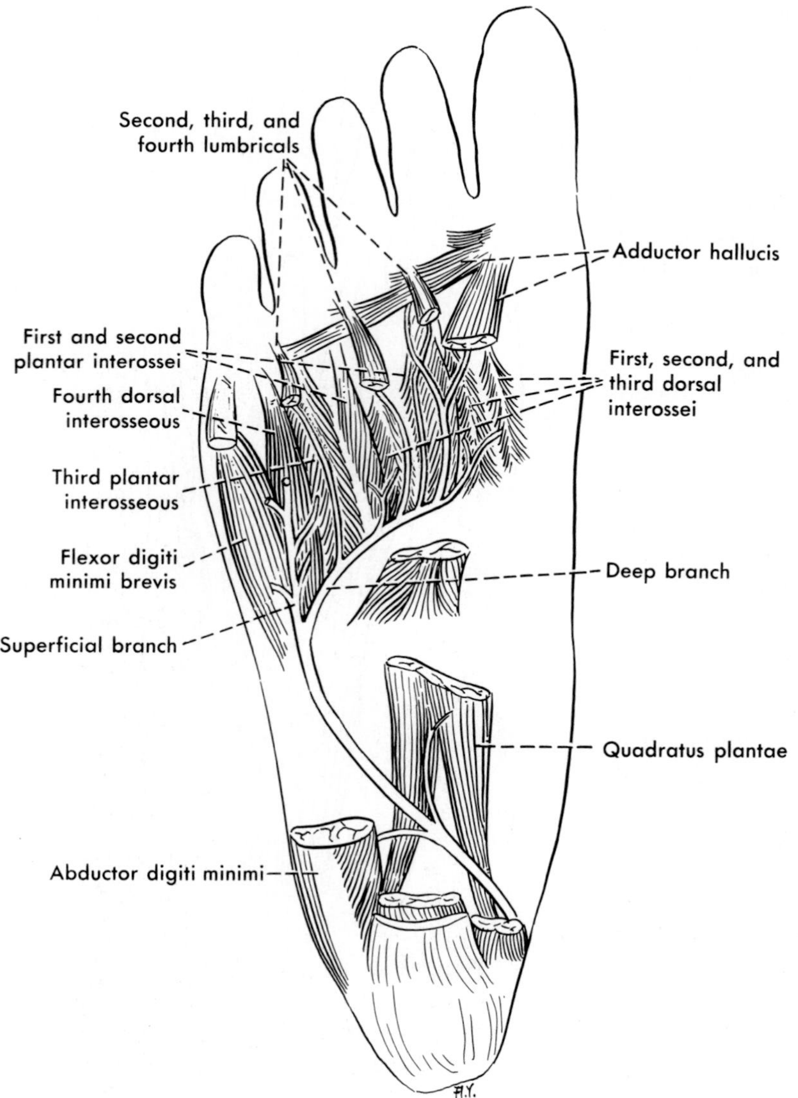


**FIG. 18-61.** Muscular distribution of the lateral plantar nerve.

to supply the abductor digiti minimi. It ends by dividing, close to the base of the 5th metatarsal bone, into deep and superficial branches. The **superficial branch** supplies the flexor digiti minimi and sometimes the 3rd plantar and 4th dorsal interossei and then emerges between the flexor digitorum brevis and the abductor digiti minimi to divide into two **digital branches,** a proper one for the free side of the little toe and a common one for the adjacent sides of the 4th and little toes. The latter, which may send a branch of communication to the adjacent common digital branch of the medial plantar nerve, divides into proper digital nerves for distribution on the toes.

The **deep branch** of the lateral plantar nerve runs medially with the plantar arch across the proximal ends of the interossei and deep to the oblique head of the adductor hallucis. In this course, it gives off muscular branches to all the interossei (except when the 3rd plantar and 4th dorsal have already been supplied by the superficial), to the lateral three lumbricals, and to both heads of the adductor hallucis.

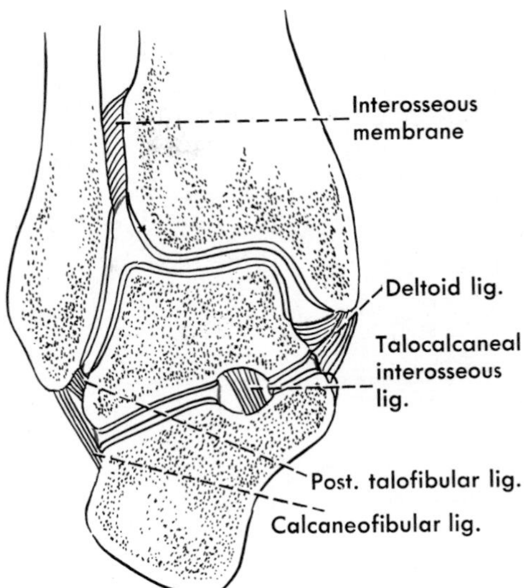


**FIG. 18-62.** Frontal section through the talocrural and subtalar joints. The interosseous ligament shown here has also been called the ligament of the tarsal canal.

**FIG. 18-63.** An anteroposterior view of the talocrural joint. (Courtesy of Dr. D. G. Pugh)

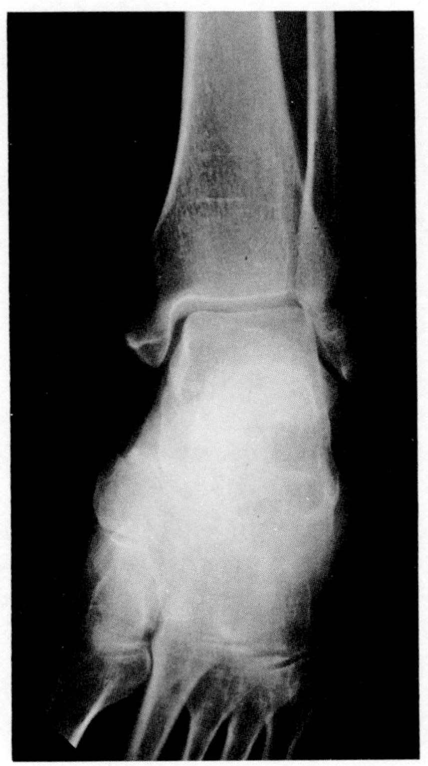

## JOINTS OF ANKLE AND FOOT

The joints of the ankle and foot consist of the talocrural joint, a number of intertarsal joints of which the subtalar and the transverse tarsal are particularly important, and tarsometatarsal, intermetatarsal, metatarsophalangeal, and interphalangeal joints.

### TALOCRURAL JOINT

The talocrural joint (Figs. 18-62 through 18-64), is formed by the trochlea tali and the lower ends of the tibia and fibula. The tibia transmits most of the weight to the talus. The lateral malleolus of the fibula and the medial malleolus of the tibia so grip the trochlea that little lateral movement is possible—there is more in plantar flexion when the narrower posterior part of the trochlea lies in the wider anterior part of the space between the malleoli, but in the weight-bearing foot in plantar

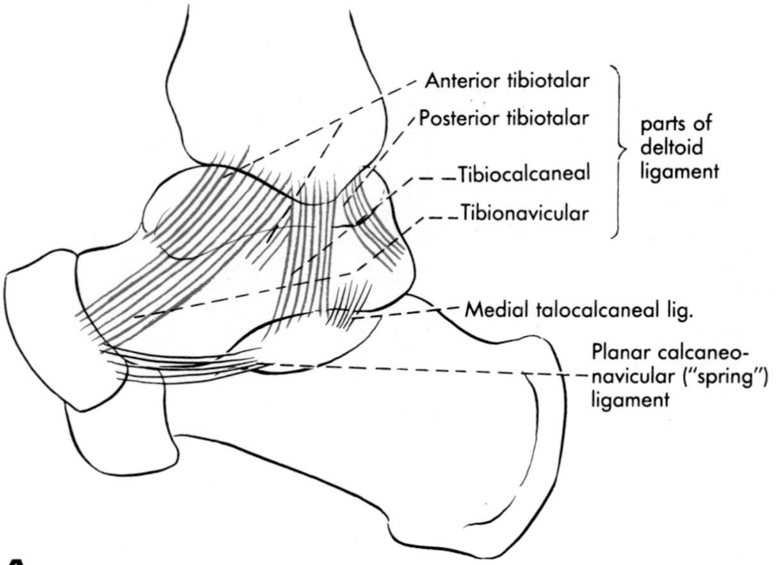

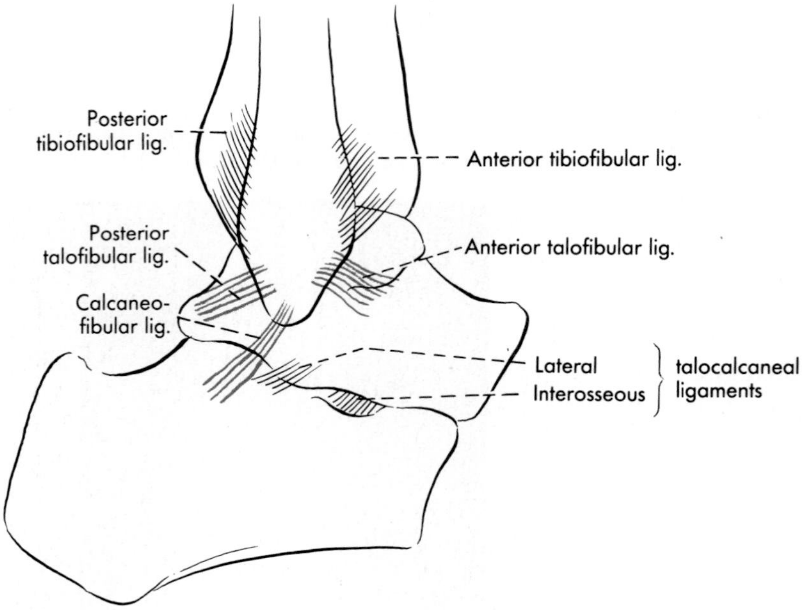

**FIG. 18-64.** Diagram of the medial ligaments at the ankle, **A**, and the lateral ones, **B**. The ligaments of the talocrural joint are red; others are black. The interosseous talocalcaneal ligament shown here is the one that has also been called the cervical ligament.

flexion the wedge shape of the joint locks it and prevents forward and downward dislocation of the leg on the talus. The tibia and fibula are held together by the *interosseous membrane* and the *anterior and posterior tibiofibular ligaments* at their lower ends. They are also held to the posterior bones of the foot by rather heavy ligaments.

On the medial side of the ankle is the heavy **deltoid (medial) ligament,** which fans out from an attachment on the medial malleolus and is described as having four parts: the *tibionavicular,* the *tibiocalcaneal,* and the *posterior tibiotalar parts* that spread out from the tibia to attach to the bones indicated by their names, and deep to the tibionavicular part, an *anterior tibiotalar part.* The deltoid ligament particularly resists eversion of the foot. Weakness of it will result in an easily turned or unstable ankle. Sprains of the ankle frequently involve tears of this ligament.

The lateral ligaments of the talocrural joint are the **anterior** and **posterior talofibular ligaments** and the **calcaneofibular ligament.** These fan out from the lateral malleolus much as do the parts of the deltoid ligament from the medial malleolus (Fig. 18-64). The lateral ligaments at the ankle, particularly the anterior talofibular, are the ones that are ruptured when an inversion sprain occurs.

Anteriorly and posteriorly, the articular capsule of the talocrural joint shows no noteworthy features.

## JOINTS OF FOOT

The intertarsal joints (Figs. 18-65 and 18-66) are the talocalcaneonavicular, the subtalar, the transverse tarsal, the calcaneocuboid, and the cuneonavicular. The cuneonavicular joint extends between the cuneiform bones as well, and usually between the lateral cuneiform and the cuboid, and may extend also between the cuboid and the navicular. The capsules of these joints are unremarkable, being short membranes that stretch from one articular surface to the next. There are, however, three sets of ligaments uniting the tarsals: dorsal tarsal ligaments, generally thin since they bear little of the duty of supporting the arch; plantar tarsal ligaments, particularly strong since they are in a position to act as bowstrings in helping to maintain the arch; and interosseous ligaments that lie deeply rather than on the surface of the joint cavity and more or less interrupt these cavities.

The ligaments of the dorsum of the foot are shown diagrammatically in Figure 18-67, and their names for the most part adequately describe their attachments. The **dorsal ligaments of the tarsus** are the *talonavicular,* the *bifurcate ligament,* consisting of the *calcaneocuboid* and the *calcaneonavicular,* the *dorsal cuboideonavicular,* the *dorsal cuneocuboid,*

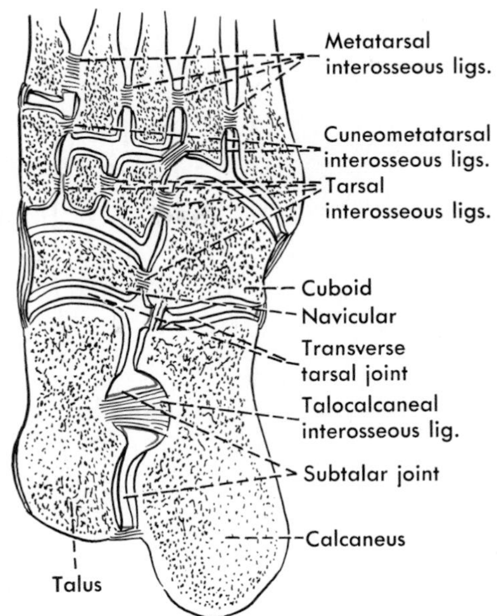


**FIG. 18-65.** An oblique section through the joints of the foot.

**FIG. 18-66.** A lateral view of some of the intertarsal joints. The arrow on the left indicates the subtalar joint; the other two arrows indicate the transverse tarsal joint. (Courtesy of Dr. D. G. Pugh)

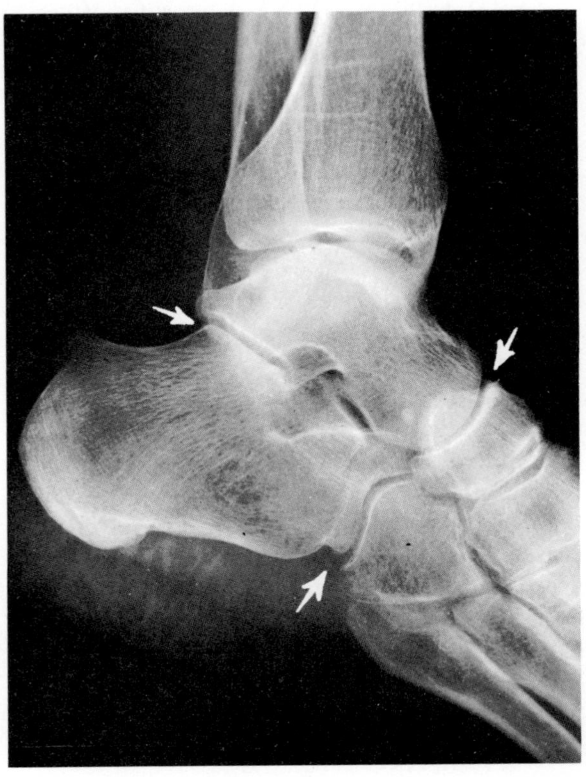

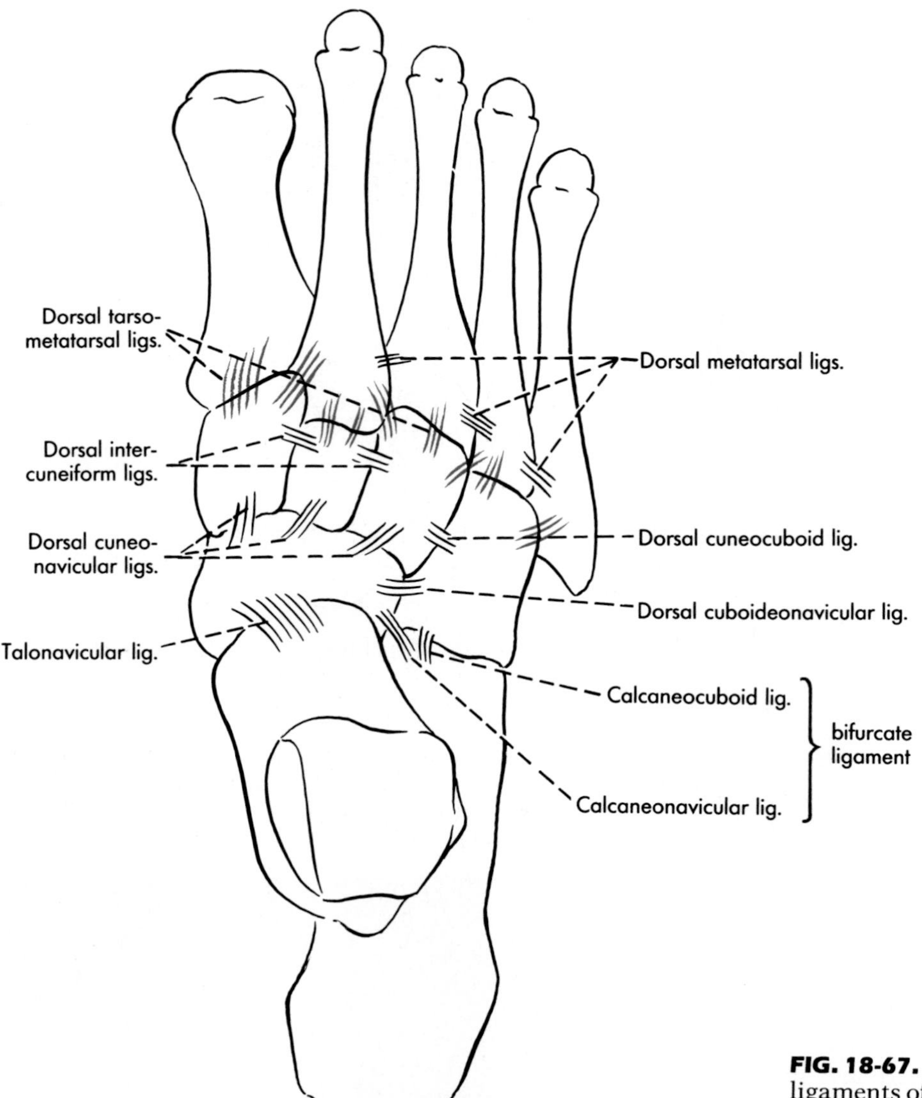


**FIG. 18-67.** Diagram of the dorsal ligaments of the ankle and foot; tarsometatarsal ligaments are red.

three *dorsal cuneonaviculars,* and two *dorsal intercuneiform.* The **dorsal tarsometatarsal ligaments** are usually arranged approximately as indicated in the figure. Three **dorsal metatarsal ligaments** connect the bases of the four lateral metatarsals.

The **plantar ligaments of the tarsus** are much stronger than the dorsal ones, and several of them cross more than one joint. They are shown diagrammatically in Figure 18-68. Of these, one of the most important ones is the **long plantar ligament,** which passes from the calcaneus as far distally as the bases of the metatarsals and in so doing forms a canal out of the sulcus for the tendon of the peroneus longus muscle on the cuboid bone (deep to it are fibers that have the same origin but attach distally to the cuboid only and form the **plantar calcaneocuboid ligament**). The other particularly important ligament is the **plantar calcaneonavicular ligament,** often called the "spring ligament," which passes from the calcaneus to the navicular (see also Fig. 18-64) and, in so doing, supports the head of the talus, which rests upon synovial membrane on the upper surface of this ligament. This ligament thus helps to support the highest part of the medial side of the arch. The other ligaments are the *plantar cuboideonavicular ligament,* three *plantar cuneonavicular ligaments,*

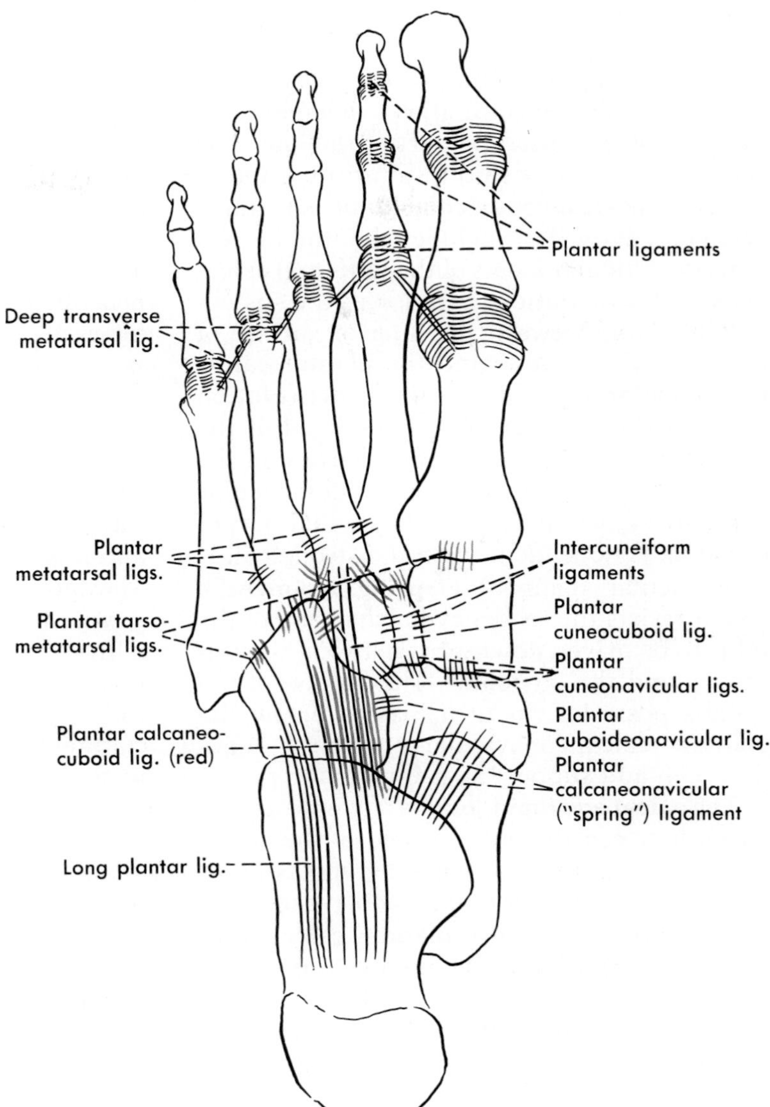


**FIG. 18-68.** Diagram of the plantar ligaments of the ankle and foot; tarsometatarsal ligaments and the deep-lying calcaneocuboid and cuneocuboid are red.

a *plantar cuneocuboid,* and two *plantar intercuneiform ligaments.*

The **plantar tarsometatarsal ligaments** are arranged approximately as indicated in Figure 18-68. There are three **plantar metatarsal ligaments** uniting the four lateral toes. Finally, the heads of the five metatarsals are connected together by **deep transverse metatarsal ligaments,** which blend also with the plantar ligaments that lie on the plantar surfaces of the metatarsophalangeal joints.

## Subtalar and Other Posterior Joints

The functional **subtalar joint** includes two joint cavities, the subtalar joint proper, the posterior articulation between the talus and the underlying calcaneus, and the talocalcaneal part of the talocalcaneonavicular joint. They are separated from each other by the tarsal canal (p. 410). In addition to its capsule, the posterior component is strengthened on its sides by the lateral and medial **talocalcaneal ligaments,** and anteriorly, its capsule blends with the strong **interosseous talocalcaneal ligament** that extends between the sulcus calcanei and the sulcus tali. This ligament has also been called the *ligament of the tarsal canal* to distinguish it from a second interosseous ligament, the **cervical ligament,** that lies anterolateral to it, arising from the floor of the sinus tarsi and attaching to the neck of the talus. The subtalar joint allows inversion and eversion of the posterior part of the foot. The cervical ligament is said to limit inversion.

The **talocalcaneonavicular joint** is a single synovial cavity that includes the middle and anterior articular facets of the talus and calcaneus; the articulation, anterior to the sustentaculum tali, between the inferior surface of the head of the talus and the **plantar calcaneonavicular ligament;** and the articulation between the anterior end of the head of the talus and the concave articular surface of the navicular. The strength of the talonavicular part, the highest portion of the medial part of the arch, is dependent upon the strong calcaneonavicular ligament **("spring ligament"),** which resists the tendency for the head of the talus to be driven downward between the calcaneus and the navicular. The tendon of the tibialis posterior, in turn, passes below the plantar calcaneonavicular ligament and, thus, can add support as it inverts the foot.

The **calcaneocuboid joint** is a simple articulation between the anterior end of the calcaneus and the cuboid. It is the highest point of the lateral part of the arch and receives its chief support from the **plantar calcaneocuboid** and **long plantar ligaments.** The tendon of the peroneus longus, passing below the cuboid bone, can also lend support to the joint, but normally contributes little to this support.

The **transverse tarsal joint** is a functional rather than an anatomic joint, for it consists medially of the *talonavicular part of the talocalcaneonavicular joint* and laterally of the *calcaneocuboid joint.* Although somewhat sinuous, the two joints together run almost transversely across the foot. The transverse tarsal joint is the line of division between the fore part of the foot and the hind part. Here, the fore part moves on the hind part in plantar flexion and dorsiflexion, inversion and adduction, and eversion and abduction.

## Other Intertarsal and Metatarsal Joints

The other intertarsal joints typically share a single synovial cavity, for the relatively large **cuneonavicular joint,** between the anterior articular surface of the navicular bone and the posterior articular surfaces of the three cuneiforms, is continuous with the *intercuneiform joints* between the cuneiform bones and usually extends also between the 3rd cuneiform and the cuboid; sometimes, however, the cuneocuboid part is separate. The three cuneiforms are connected anteriorly by two *interosseous intercuneiform ligaments,* and the lateral cuneiform and the cuboid are connected by the *interosseous cuneocuboid ligament.*

The **tarsometatarsal joint cavities** are typically three: a medial for the 1st metatarsal, an intermediate for the 2nd and 3rd metatarsals, and a lateral for the 4th and 5th metatarsals. They are separated from each other by *interosseous cuneometatarsal ligaments.* They extend a little distance proximally between the cuneiforms and between the lateral cuneiform and the cuboid, and the intermediate one often communicates, between the intermediate and medial cuneiforms, with the cuneonavicular joint. Otherwise, they are separated from this joint by interosseous intercuneiform ligaments. Between the four lateral toes, the tarsometatarsal joint cavities extend forward between the bases of the metatarsals to become continuous with the cavities of the **intermetatarsal joints.** These are sealed distally by *metatarsal interosseous ligaments,* which may extend proximally to subdivide the joint cavities into dorsal and ventral parts. Usually, there is no joint cavity between the 1st and 2nd metatarsals. The articulation here is purely through a large metatarsal interosseous ligament.

## Articulations of Digits

The articulations of the phalanges of the toes are essentially similar to those of the fingers. On the plantar surface, the capsule of each metatarsophalangeal and interphalangeal joint is thickened by a shallowly concave plate, the **plantar ligament** (Fig. 18-68), to the edges of which are attached the digital tendon sheaths as well as the lateral part of the fibrous capsule of the joint. The sides of the capsules are reinforced by **collateral ligaments,** and on the dorsal aspect, the extensor tendon largely replaces the fibrous membrane of the capsule. The plantar ligament of the metatarsophalangeal joint of the big toe contains the two sesamoids of this joint, and other ligaments may contain sesamoids.

## MOVEMENTS OF THE FOOT AND TOES

### The Foot As A Whole

In considering movements of the foot, it must be remembered that although most of the dorsiflexion and plantar flexion of the foot normally takes place at the talocrural joint, the subtalar and transverse tarsal joints allow eversion and inversion. Movements of inversion at the transverse tarsal joint are always accompanied by adduction of the fore part of the foot, and movements of eversion are always accompanied by abduction of it. Further, the fore part can be flexed and extended on the hind part at the transverse tarsal joint. Flexion obviously increases the concavity of the arch (the exaggerated condition of which is known as *pes cavus*), and dorsiflexion decreases the concavity of the arch (the exaggeration of this, flatfoot, being *pes planus*).

The triceps surae (gastrocnemius and soleus) is the only good functional **plantar flexor** at the talocrural joint. Even this, because of its short leverage arm, is at a disadvantage and must exert a pull of about 100 lb in order to exert a pressure of 50 lb on the ball of the foot. The other plantar flexors include the plantaris, of such small size that its pull is negligible, and the flexor hallucis longus, flexor digitorum longus, peroneus longus, and, from dorsiflexion, the peroneus brevis and the tibialis posterior. These muscles are weak plantar flexors, and it has been calculated that, because of their smaller size and poor leverage, they can together exert no more than about 5% of the pull that the triceps can.

The chief **dorsiflexors** of the foot are the tibialis anterior and the extensor digitorum longus and associated peroneus tertius. The extensor hallucis longus contracts weakly in dorsiflexion of the foot unless the tibialis anterior is paralyzed; then it contracts strongly in order to assist dorsiflexion but at the same time dorsiflexes the big toe.

Of the **invertors and adductors**, the best are the tibialis posterior and anterior. The flexor hallucis longus and flexor digitorum longus both also have an adductor action on the foot, but do not normally participate in this action unless the movement is strongly resisted or other invertors are paralyzed. Also, the triceps surae, in plantar flexing the foot, inverts it, apparently because of the relationship of the calcaneus to the joints of the ankle and foot, rather than of any particular direction of pull of the triceps. When the extensor hallucis longus dorsiflexes the foot, it will also invert it.

The **evertors and abductors** are the peroneus longus and brevis and the peroneus tertius with the lateral part particularly of the extensor digitorum longus.

In movements of the foot, the invertors and the evertors usually act together in combinations that permit pure plantar flexion or pure dorsiflexion. Thus, the tendency of the triceps surae to invert is counteracted during plantar flexion by the tendency of the peroneus longus to evert. In dorsiflexion the tendency of the tibialis anterior to invert is counteracted by the tendency of the peroneus tertius and lateral part of the extensor digitorum longus to evert.

### Movements of Toes

Movements of the toes are relatively simple. **Dorsiflexion** (*extension*) at the metatarsophalangeal joints is carried out by both the long and short extensors of the toes—the extensor digitorum longus and the extensor hallucis longus, and the extensor digitorum brevis and extensor hallucis brevis. Because the metatarsals slope downward from their bases to their heads, the toes are typically somewhat dorsiflexed in relation to the long axes of the metatarsals during normal standing, and the higher the heel of a shoe the more dorsiflexed they become. Increase in dorsiflexion at the metatarsophalangeal joints produces, in turn, more plantar flexion at the interphalangeal joints (for the passive pull of the long flexors overcomes the action of the extensors of these joints) so that clawing of the toes results. Because of this passive pull of the long flexors, the long extensors, except that of the big toe, largely further hyperextend the metatarsophalangeal joints instead of extending the interphalangeal ones. The lumbricals, and to a less extent the interossei, are potential extensors of the interphalangeal joints.

**Flexion** of the interphalangeal joints is brought about by the flexor digitorum longus, the quadratus plantae, the flexor digitorum brevis, and the flexor hallucis longus. The flexor digitorum longus and flexor hallucis longus act strongly only on the distal phalanges. The quadratus assists the flexor digitorum longus, perhaps particularly when the weight shifts forward with dorsiflexion of the ankle during walking, and strong contraction of the longus might interfere with dorsiflexion. It also helps to overcome the medial pull of the flexor digitorum longus. When that muscle alone is stimulated, the toes are not only flexed but also rotated so that the tips are turned medially, but when the quadratus plantae is stimulated at the same time, the rotation disappears and the toes flex in a straight direction. The flexor digitorum brevis is said to flex the middle phalanges forcefully but the proximal ones weakly.

Flexion of the metatarsophalangeal joints is brought about by all the interossei, the lumbricals, the short flexors and abductors of the big and little toes, and the adductor of the big toe. As these joints are flexed, the interphalangeal joints tend to extend, largely as a result of the passive pull of the long extensor tendons.

The usually limited amount of *abduction* and *adduction* of the toes is brought about by most of the muscles just mentioned: abduction by the dorsal interossei and the named abductors, and adduction by the planar interossei and the adductor hallucis.

### Distortion of the Foot

Although there is a considerable body of evidence to indicate that the long muscles of the leg normally contribute relatively little to the direct support of the arch, this same evidence also indicates that certain of the long muscles are particularly important in distributing weight along the medial and lateral sides of the foot, and abnormal weight bearing can much distort the foot. Similarly, there is abundant evidence that muscular imbalance, whether congenital or acquired, will gradually markedly distort the foot. Thus, distortions of the foot that result from muscle weakness or paralysis may be primary ones, produced by relative overaction of a normal muscle or by unusual shortness of a muscle when there is no paralysis; they may be secondary as a result of weight bearing; or they may be a combination of both. There are numerous possible combinations of dorsiflexion or plantar flexion and inversion or eversion, but only a few terms are necessary to describe the fundamental distortions. A "**pes calcaneus**" is a foot in which the dorsiflexors maintain the foot in a dorsiflexed condition, and the weight therefore comes on the heel; similarly, a "**pes equinus**" is one in which the chief plantar flexor, the triceps surae, maintains the foot in a varying amount of constant plantar flexion. "**Pes varus**" is one in which the sole is inverted, and since inversion and adduction go together, it is simply a foot maintained in a so called supinated position; "**pes valgus**" is the opposite, a foot that is everted and abducted and therefore maintained in the so called pronated position. Finally, as already noted, "**pes planus**" is a flat foot, and "**pes cavus**" is a highly arched one.

By considering the actions of the various muscles, the distortions of the foot that can be produced by muscle action can be fairly readily understood. For example, when the triceps is paralyzed, the foot tends to go into dorsiflexion, yet at the same time the pull of the peroneus longus will gradually plantar flex and evert the anterior part of the foot, so that a condition of pes calcaneus cavus valgus is produced. Similarly, paralysis of the peroneus longus so weakens eversion of the foot that the triceps surae and the tibialis anterior produce inversion of the resting foot; however, the peroneus longus can no longer resist the tendency of the tibialis anterior to pull upward on the medial side of the anterior end of the arch, with the result that the longitudinal arch is flattened. When weight is put on the foot, it shifts to the yielding medial side and the foot then becomes pronated and more flattened at the same time—pes planus valgus or ordinary flatfoot.

Many combinations of muscle imbalance and resulting distortion of the foot are found clinically. However, those mentioned here should be enough to indicate that there normally must be a balance between dorsiflexors and plantar flexors and between invertors and evertors. Further, there must be a balance between those muscles that plantar flex the fore part of the foot and those that dorsiflex it. If the action of the former predominates, there will be pes cavus, but if that of the latter does, there will be pes planus.

## THE LOWER LIMB IN STANDING AND WALKING

### Most Stable Position

Extension at the hip joint involves a tightening of its ligaments so that further extension is resisted. Since the center of gravity in the body normally falls behind the hip joint in standing, this joint tends to be maintained in extension by the weight of the body. Thus, quiet standing involves little muscular action. The extensors of the hip are relaxed unless one sways forward, and further extension is resisted by the ligaments, especially the iliofemoral, and apparently by slight activity of the iliacus. If the hip is slightly flexed, the posterior hamstrings become stabilizers of the joint.

The situation is much the same at the knee joint. The center of gravity normally passes slightly in front of the extended knee, and therefore, the weight of the body tends to keep this extended. Hyperextension is prevented largely by the tightness of the collateral and cruciate ligaments in extension and by a minimal contraction of the flexors of the knee. Further, in the extended position at the knee, the anterior, flatter surfaces of the femoral condyles are in contact with the tibial condyles and menisci, so this is a more stable position because of the greater area of contact. The maintenance of extension at the knee therefore involves no activity by the quadriceps muscle but, rather, a minimal activity on the part of the hamstrings to prevent hyperextension.

The situation is different at the ankle. Here the line of gravity passes in front of the joint, so there is a tendency for the foot to dorsiflex and permit the body to fall forward. This must be resisted by constant tonic action of the plantar flexors. It is chiefly the soleus that does this; the gastrocnemius usually contracts only when more powerful plantar flexion is required.

### Distribution of Weight

The tonic action of the plantar flexors also has another effect, for it distributes the weight approximately equally between the calcaneus and the ball of the foot. The talus sits so far posteriorly on the arch that, if a skeletal preparation of leg and foot is loaded, about 80% of the weight comes on the calcaneus and only 20% is transmitted through the anterior part of the arch to the heads of the metatarsals; however, in the normal standing person, the triceps surae contracts sufficiently to lessen the weight on the calcaneus and increase it on the metatarsal heads, so shifting it forward that each bears about half of the total weight.

In quiet standing on both feet, the invertors and evertors contract only enough to prevent lateral movement of the foot, but in so doing, they produce a differential distribution of weight on the ball. In a loaded cadaver foot, with the tibia vertical, the weight on the ball of the foot is distributed equally among the five metatarsals, but in the living foot, the 1st metatarsal has been said to bear about twice the weight that any of the others do and, therefore, about a third of the total weight on the anterior part of the foot. Slight extra tension of either the evertors or the invertors of the foot, not sufficient to produce demonstrable movement, will shift the weight distribution drastically—tension of the evertors throws extra weight on the 1st metatarsal, and tension of the invertors diminishes this weight and shifts it laterally.

In one study of persons with flat feet, those who had had no trouble with their arches were found to have a normal distribution of weight among the metatarsals, whereas those who had had pain were found to have more than the normal amount of weight distributed to the medial side of the arch. Apparently, therefore, the pain of flat feet results from a chronic overloading of the medial side of an abnormal arch.

### Walking and Running

In walking, the anterior swing of the thigh is brought about by flexors, apparently usually the tensor fasciae latae, the pectineus, and, perhaps, the sartorius. Although the iliopsoas is the strongest flexor at the hip, it is said that one has to make a conscious effort to use this muscle in walking, since it does not usually participate in this movement. As the hip joint is flexed, the knee also is flexed, probably both as a result of gravity and of the passive pull of the hamstrings induced by flexion at the hip, and the foot is dorsiflexed so that the toes will clear the ground. At the same time, the thrust given by the plantar flexors as the foot leaves the ground has shifted the weight to the other limb; this shift is controlled by the erector spinae of the moving side, which prevents the weight from being thrust too far laterally. The weight is supported by the gluteus medius and minimus on the supporting side, which prevent too great drooping of the pelvis on the unsupported side. There is general contraction of the muscles of the limb that is momentarily supporting the body, in order to stabilize, but not enough to prevent the forward motion of the body over the supporting limb. This forward motion is brought about by the thrust given by the moving limb as it leaves the ground, by the forward shift in the center of gravity produced by leaning forward as one walks, and by the extensor muscles at the hip. The most powerful extensor, the gluteus maximus, does not participate in extension unless a greater effort, such as climbing a steep hill or stairs, is required. As the swinging limb approaches the ground, the knee is extended by the quadriceps, and as the heel touches the ground, the dorsiflexors of the foot contract more strongly in order to control the shift of weight from the heel forward to the ball of the foot. As this forward shift occurs, the weight, which because of the usual outtoeing of the foot, falls first on the lateral border of the foot, is shifted toward the medial side of the ball. The knee usually flexes slightly and the ankle dorsiflexes, and as momentum and the hip extensors carry the center of gravity once more in front of the supporting limb, the knee extends and there is passive dorsiflexion at the ankle, followed by the plantar thrust that starts the cycle once again.

In running, much the same movements are involved, but the contraction of the muscles is faster and more powerful. Moreover, extension at the knee occurs only at the thrust from the ground, and the foot is maintained in such plantar flexion that the weight falls first on the ball of the foot, whereupon control of dorsiflexion of the foot lowers the heel almost to the ground. This diminishes the shock of landing and also permits a more powerful plantar thrust.

### Support of Arch

The forces exerted upon the arch of the foot are large, for they are the sum of the weight on the leg and the pull of the triceps surae muscle. For example, when 100 lb is to be raised, in shifting all the weight from the heel to the ball of the foot, it takes about 200 lb of pull by the triceps to accomplish this, and this 200-lb pull by the triceps plus the 100 lb of pressure on the ball amounts to a strain on the arch of 300 lb. In running, the strain on the arch obviously increases markedly because of the impact of the weight upon the ball of the foot.

There have been many conflicting concepts

concerning support of the arch. They range from the idea that the long muscles of the leg that are in a position to support the arch actually do maintain it by tonic contraction and that plantar structures have no duty in supporting the arch unless there is failure of the muscles to do so, to the idea that the support of the arch is entirely dependent upon its bony and ligamentous structure. As is often so, the truth seems to be between the two extremes.

Contraction of the muscles of the leg and foot can increase (raise) the arch in the non-weight-supporting foot, but there is good evidence that the long muscles of the leg do not normally participate appreciably in supporting the arch. Of the three long muscles often considered to be the prime support of the arch—the peroneus longus, the tibialis posterior, and the tibialis anterior—the first two are largely relaxed during quiet standing, and although they contract during plantar flexion, this contraction is by no means maximal. The third, the tibialis anterior, is also relaxed during quiet standing, but it also is necessarily relaxed during plantar flexion (since it would oppose this) and therefore cannot at all support the arch at the time support is most needed. It also has been shown that the arch is capable of supporting great weight without contraction of leg muscles, for the relaxed leg of a sitting person can be loaded with sufficient weight to produce pain on the knee without at the same time producing enough stretching of the ligaments of the arch to cause pain there.

The support of the arch, therefore, depends primarily on its intrinsic bony structure and on the soft tissues of the plantar surface of the foot. The heavy plantar ligaments and the likewise heavy plantar aponeurosis passively support the arch and during quiet standing usually do so alone, except that those muscles of the foot that can act as tie rods between the anterior and posterior ends of the arch may show slight intermittent activity. Activity increases and becomes sustained when greater stress is put on the arch through movement, particularly rising on the toes or walking.

The role of the long leg muscles in preserving the arch seems not to involve primarily any direct support but, rather, their effect in distributing the weight through the foot. Particularly, any muscular imbalance that tends to throw excess weight on the medial side of the arch, already under greater strain because it is the highest part, may produce more strain than the arch is equipped to stand. For instance, eversion of the foot not only puts greater strain upon the medial border of the foot but also slants the subtalar joint so that it inclines more medially than it usually does, thus favoring a downward and forward slipping of the head of the talus between the calcaneus and the navicular, with consequent flattening of the arch and abduction of the fore part of the foot.

## GENERAL REFERENCES AND RECOMMENDED READING

BASMAJIAN JV, STECKO G: The role of muscles in arch support of the foot: An electromyographic study. J Bone Joint Surg (Am) 45:1184, 1963

BOJSEN-MØLLER F, FLAGSTAD KE: Plantar aponeurosis and internal architecture of the ball of the foot. J Anat 121:599, 1976

BRANTIGAN OC, VOSHELL AF: The mechanics of the ligaments and menisci of the knee joint. J Bone Joint Surg 23:44, 1941

CAHILL DR: The anatomy and function of the contents of the human tarsal sinus and canal. Anat Rec 153:1, 1965

CAVE EF, ROWE CR: The patella: Its importance in derangement of the knee. J Bone Joint Surg (Am) 32:542, 1950

CUMMINS EJ, ANSON BJ, CARR BW et al: The structure of the calcaneal tendon (of Achilles) in relation to orthopedic surgery: With additional observations on the plantaris muscle. Surg Gynecol Obstet 83:107, 1946

DE PALMA AF: Diseases of the Knee: Management in Medicine and Surgery. Philadelphia, JB Lippincott, 1954

EDWARDS EA: The anatomic basis for ischemia localized in certain muscles of the lower limb. Surg Gynecol Obstet 97:87, 1953

GARDNER E: The innervation of the knee joint. Anat Rec 101:109, 1948

HALLSY JE: The muscular variations in the human foot. A quantitative study. General results of the study: 1. Muscles of the inner border of the foot and the dorsum of the great toe. Am J Anat 45:411, 1930

HELLER L, LANGMAN J: The meniscofemoral ligaments of the human knee. J Bone Joint Surg (Br) 46:307, 1964

HORWITZ MT: Normal anatomy and variations of the peripheral nerves of the leg and foot: Application in operations for vascular diseases; study of one hundred specimens. Arch Surg 36:626, 1938

HUBER JF: The arterial network supplying the dorsum of the foot. Anat Rec 80:373, 1941

JONES RL: The human foot: An experimental study of its mechanics, and the role of its muscles and ligaments in the support of the arch. Am J Anat 68:1, 1941

KAMEL R, SAKLA FB: Anatomical compartments of the sole of the human foot. Anat Rec 140:57, 1960

LAST RJ: Some anatomical details of the knee joint. J Bone Joint Surg (Br) 30:683, 1948

LOVEJOY JF JR, HARDEN TP: Popliteus muscle in man. Anat Rec 169:727, 1971

LOVELL AGH, TANNER HH: Synovial membranes, with special reference to those related to the tendons of the foot and ankle. J Anat Physiol 42:415, 1908

MANN R, INMAN VT: Phasic activity of intrinsic muscles of the foot. J Bone Joint Surg (Am) 46:469, 1964

MANTER JT: Distribution of compression forces in the joints of the human foot. Anat Rec 96:313, 1946

MARTIN BF: Observations on the muscles and tendons of the medial aspect of the sole of the foot. J Anat 98:437, 1964

MORTON DJ: The Human Foot: Its Evolution, Physiology and Functional Disorders. New York, Columbia University Press, 1935

TROTTER M: The level of termination of the popliteal artery in the white and the Negro. Am J Phys Anthropol 27:109, 1940

WILLIAMS AF: The formation of the popliteal vein. Surg Gynecol Obstet 97:769, 1953

WRIGHT DG, RENNELS DC: A study of the elastic properties of plantar fascia. J Bone Joint Surg (Am) 46:482, 1964

# PART VI:

# THORAX

# 19

# THE THORAX IN GENERAL

The thorax is the upper part of the trunk, distinguished from the abdomen by the presence of the **rib cage.** This resilient, expandable, skeletal frame is constructed of the *sternum,* the *ribs,* and the *costal cartilages* and is supported on the vertebral column, which forms the posterior wall of the **thoracic cavity,** enclosed within the rib cage. The central organs of respiration and circulation are housed within the thoracic cavity. The ribs also provide protection for some major abdominal organs.

The **thoracic wall** consists of the thoracic skeleton and associated soft tissues. Integrity of the wall is necessary for generating the subatmospheric pressure within the thoracic cavity that causes air to be sucked into the lungs during respiration.

A broad median septum, known as the **mediastinum,** partitions the thoracic cavity and separates the two hollow spaces that are filled by the right and left lungs (Fig. 19-1). The bulk of the mediastinum is made up of the **heart** enclosed in the **pericardial sac** and of the major pulmonary and systemic veins and arteries. The **trachea** and **esophagus** enter the mediastinum from the neck through the **superior thoracic aperture.** This relatively small opening between the first pair of ribs also transmits the great arteries and veins of the head, the neck, and the upper limbs. The **inferior thoracic aperture** is wide and irregular. Through it the abdominal cavity protrudes high into the chest. Abdominal and thoracic viscera are separated from one another, however, by the **diaphragm,** a musculotendinous sheet that is attached to the inner margins of the inferior thoracic aperture. The diaphragm bulges up into the chest like a dome and varies its height by rhythmic contraction and relaxation. This pistonlike action is an important factor in respiratory movements and usually contributes twice as much to pressure and capacity changes in the thorax as the movements of the ribs. The lungs follow the excursions of the thoracic walls because their surface is held apposed to the inner aspects of the walls by the **pleura.** The pleura is a serous membrane, the parietal layer of which lines the two hollow

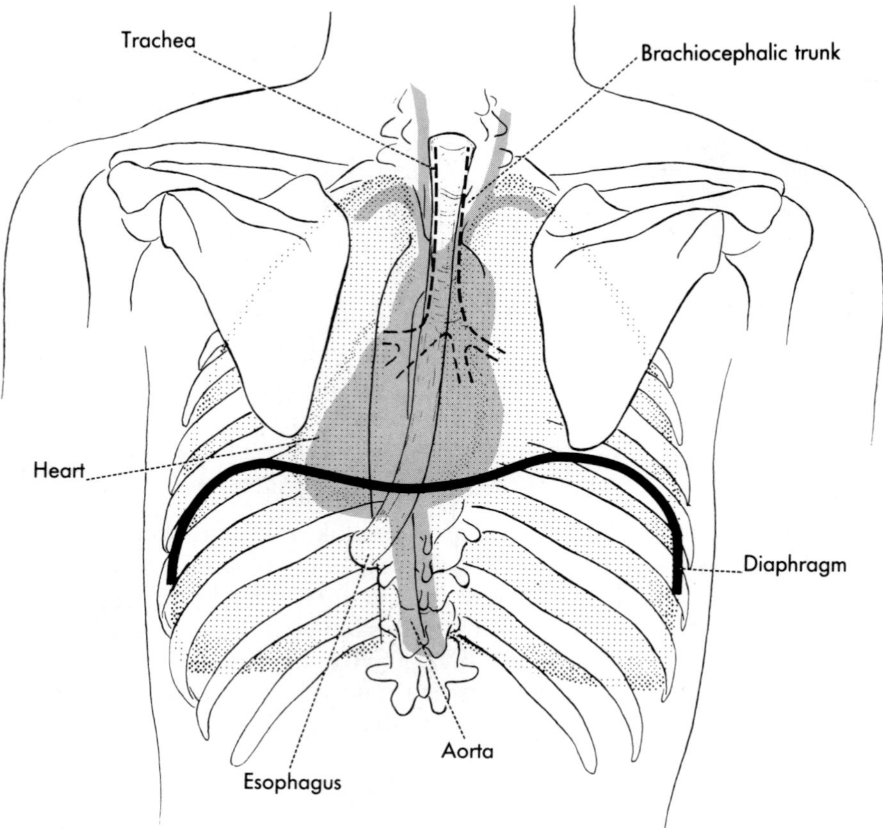


**FIG. 19-1.** The thorax seen from behind. The bulky shoulders obscure a relatively small rib cage; the diaphragm rises high up into the chest. The pleural cavities, which extend over the posterior surface of the domes of the diaphragm, are indicated by stippled areas; the heart and great arteries in the mediastinum are shown in pink.

spaces on either side of the mediastinum, whereas its visceral layer invests the expandable lungs. Thus, the thorax not only contains and protects the central organs of respiration and circulation, but its mechanisms also produce pressure changes necessary for respiration and for venous return to the heart.

## THE THORACIC WALL

Between the outer covering of skin and the inner pleural lining, the walls of the thoracic cavity are made up of more or less concentric layers of muscles and fascia that are supported by the thoracic skeleton (Fig. 19-2). The nerves, blood vessels, and lymphatics of the body wall pass among these layers, serving not only the thoracic wall but also a major part of the abdominal wall. Practically all clinical information pertaining to the thoracic viscera, and to some of the major abdominal viscera, has to be obtained by eliciting physical signs across the thoracic wall, which is possible only by knowing skeletal landmarks and the functional anatomy of the thorax.

The thoracic cage supports the bones and muscles of the pectoral girdle. The square shoulders and the bulky girdle musculature obscure the relatively small thoracic cage (Fig. 19-1). Except for a median strip of the sternum and the tips of the thoracic vertebral

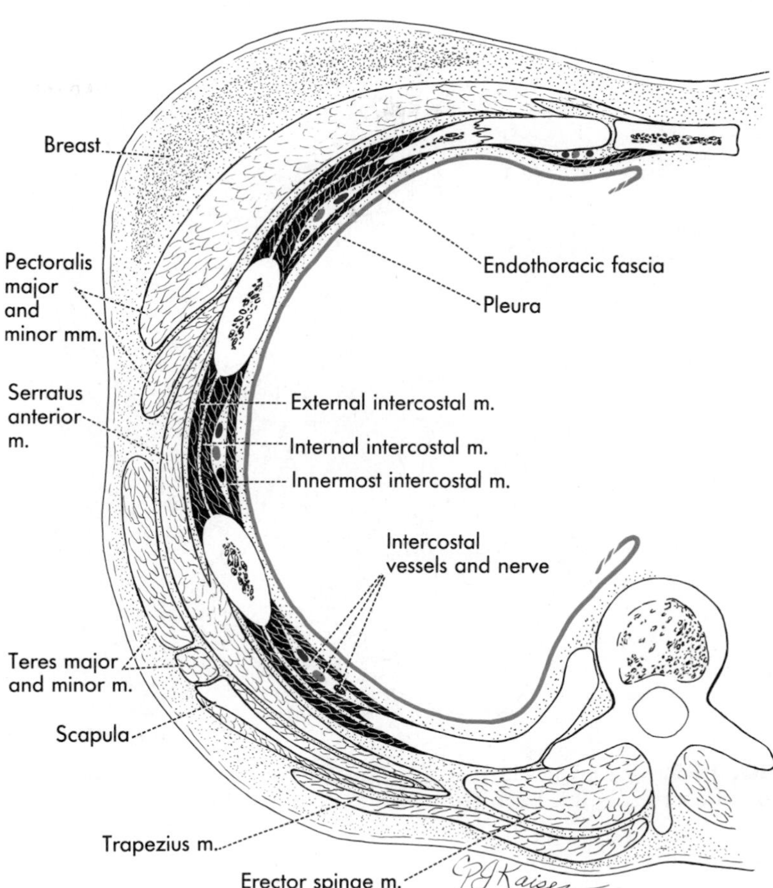


**FIG. 19-2.** Layers of the thoracic wall. The muscle layers of the thoracic wall proper are black; other muscles on the surface of the rib cage are white.

spines, the thoracic skeleton is largely covered by muscles that belong to the upper limb or the vertebral column. The more important of these are the pectoral muscles and the serratus anterior anterolaterally, the trapezius and latissimus dorsi posterolaterally, and, lying deep to the latter two, the erector spinae in the back (Fig. 19-2). Supplied by nerves other than those serving the true body wall, these muscles anatomically are not considered a component of the thoracic wall. Nevertheless, surgical and clinical access to the thoracic cavity is possible only through them. The lower part of the rib cage is covered anteriorly by the rectus abdominis and the external oblique muscles of the abdominal wall.

The female breast covers a large area of the anterior chest wall. Its adipose tissue merges with the superficial fascia along the periphery of the organ (Fig. 19-2). During palpation and percussion of the chest, the breast can readily be pushed out of the way owing to its relatively free mobility: it can be moved around on the underlying muscles (Chap. 12). Posteriorly, the scapula and its associated muscles prevent direct access to much of the chest (Fig. 19-2). However, to obviate this disadvantage in examining the lungs, the scapula may be moved out of the way by abduction of the arm and protraction of the shoulder.

## THE THORACIC SKELETON

The rib cage and the thoracic vertebral column form an irregularly shaped, truncated cone (Fig. 19-3). The cone is flattened anteroposteriorly, a peculiarly human characteristic. Ten of the 12 pairs of ribs form loops or

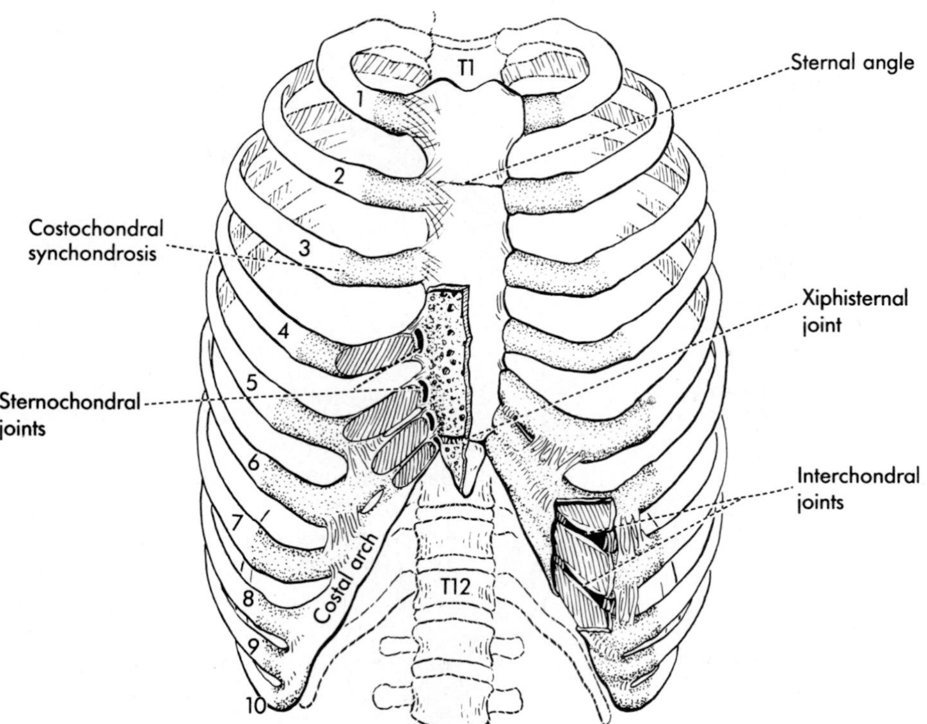


**FIG. 19-3.** Anterior view of the skeleton of the thorax. Part of the sternum and costal cartilages has been cut away to reveal some of the joints.

arches between respective vertebrae and the sternum, whereas the last two pairs of ribs float free anteriorly (**floating ribs**). All ribs slant downward from their vertebral attachment, and this slant tends to become slightly more pronounced the more caudally the ribs lie. Because the sternum anteriorly is much shorter than the length of the thoracic spine and because of the downward tilt of the shafts of the ribs, the anterior segment of each costochondral loop slants upward so that the costal cartilages may articulate with the sternum. The upward slant of the cartilages becomes gradually steeper below the 4th rib, and the 7th is the last cartilage to reach the sternum directly. The 8th, 9th, and 10th costal cartilages terminate short of the sternum and articulate with their proximal neighbor. In this manner, the **costal arch** is formed on each side. Between the costal arches in the *infrasternal angle* is the epigastrium, the uppermost region of the abdomen.

It follows from the above description that (1) the **superior thoracic aperture** is small; its plane slopes downward and forward; its boundaries are the 1st thoracic vertebra, the first pair of ribs with their cartilages, and the superior margin of the manubrium sterni; and (2) the **inferior thoracic aperture** is large and irregular; it is formed by the 12th thoracic vertebra, the 12th pair of ribs, and the costal arch made up of costal cartilages 10 to 7.

It can be readily appreciated that the thoracic cage is made both resilient and compressible by its architecture. Use is made of this property in the application of external cardiac massage during resuscitation. The costochondral arches permit rhythmic compression of the heart between the sternum and the vertebral column sufficient to effect some rhythmic pumping of blood, even though the heart itself is quiescent.

Movement of the skeletal pieces is required for respiration, and these movements are mediated by several joints. The bony shaft of each rib is directly united to its cartilage **(costochondral synchondrosis);** each typical costochondral loop articulates with the thoracic spine through two synovial joints (**costovertebral** and **costotransverse joints**), and another synovial joint joins it directly or indirectly to the sternum (**sternocostal** and **interchondral joints**). The 1st rib is an exception in that its cartilage is directly united to the manubrium, which ensures stability in the frame of the superior aperture. In addition, the three component pieces of the sternum can move in relation to one another at the **manubriosternal** and **xiphisternal joints.**

## The Sternum

The sternum, or breast bone, forms the anteromedian portion of the thoracic wall (Fig. 19-3). It is an elongated, flat bone. Its interior, filled with cancellous bone, contains hematopoietic bone marrow throughout life and has been the site for clinical bone marrow biopsies. Up to puberty the sternum consists of six segments of **sternebrae** held together by hyaline cartilage. The central four sternebrae fuse between years 14 and 21 to form the **body** (*corpus*) of the sternum; however, the superior and inferior segments remain independent as the **manubrium sterni** and the small **xiphoid** (sword-shaped) **process,** respectively. Two cartilaginous joints hold together these three parts of the sternum and permit some changes in angulation between them. This movement at the manubriosternal joint accommodates for the respiratory excursions of the ribs, which are of different lengths. In later life, however, the cartilage of the sternal joints tends to disappear, and both joints may ossify.

All three parts of the sternum are palpable. The superior margin of the manubrium forms the **jugular notch,** which can be felt between the two clavicles. It lies on one level with the 3rd thoracic vertebra. The manubriosternal joint is palpable as the **sternal angle,** a smooth ridge produced by the angulation of the bones and their slightly everted articular edges. The sternal angle is level with the 2nd costal cartilages and with the lower border of the 4th thoracic vertebra. The body of the sternum is flat, and from its lower edge, the xiphoid process inclines backward. During youth much of the process consists of cartilage and can be moved passively until its bony center has expanded and fused to the body of the sternum. Palpation of the xiphoid process in the apex of the infrasternal angle causes some discomfort; therefore, the xiphisternal junction is a more convenient landmark. Level with the xiphisternal junction is the 6th pair of costal cartilages and the 10th thoracic vertebra. The projection of the thoracic vertebrae on the sternum is telescoped owing to the anterior concavity of the thoracic vertebral column.

On either side of the jugular notch, the manubrium receives the sternal end of the clavicles in a shallow concave facet, thus forming the **sternoclavicular joints.** On the sides of the sternum are the *costal incisures,* or notches, for articulation with the ribs. Those for the first ribs lie on the sides of the manubrium; those for the second lie at the manubriosternal junction. The costal notches for the 3rd, 4th, 5th, and 6th ribs lie on the sides of the body; that for the 7th rib, the lowest to articulate with the sternum, lies at the junction of body and xiphoid process.

**Ossification.** The sternum ossifies from a number of centers that form independent of the ribs. Usually, there is a single center for the manubrium and a single one for the uppermost part of the body; each of the remaining three segments that contribute to the body may have a single or a paired center of ossification. A single center is typical for the xiphoid process. The centers, except for that of the xiphoid, which appears during the third year of life, arise (from above downward) during the late prenatal period. Fusion of the parts of the body begins below and extends upward, being completed at the 21st year.

## The Ribs

There are 12 pairs of ribs (*costae*), of which the upper seven are called **true ribs** because they form complete arches between the vertebrae and the sternum, whereas the lower five,

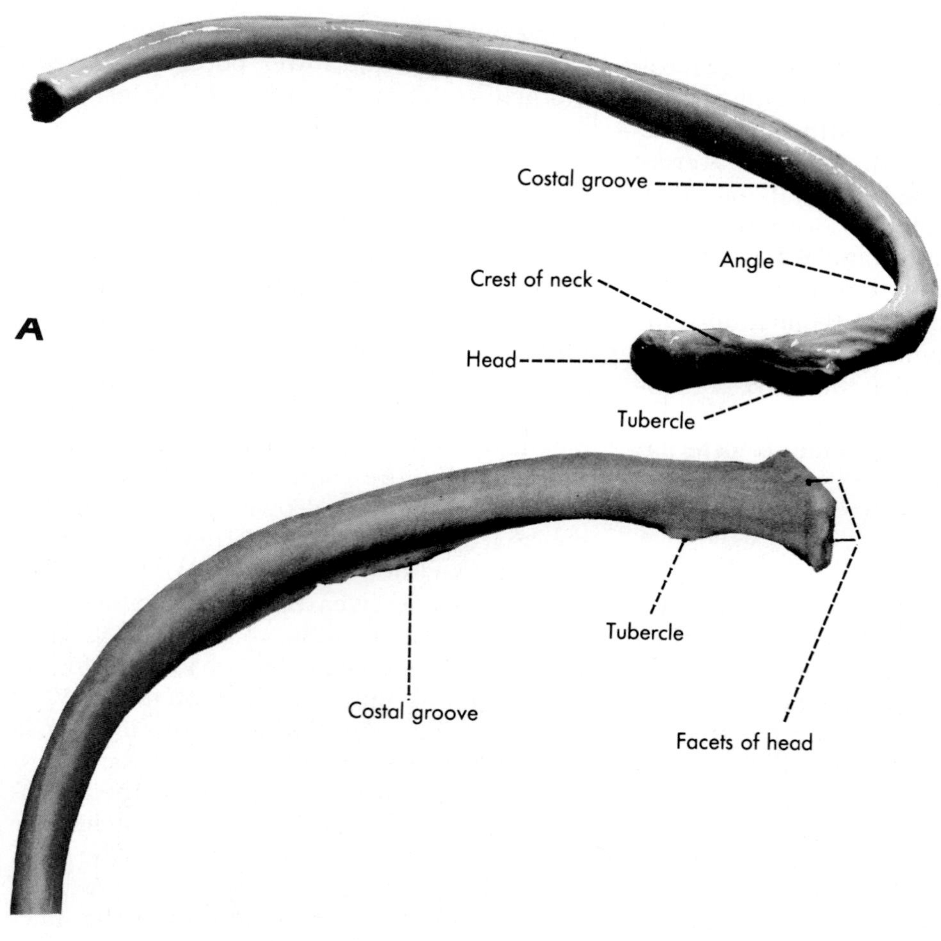


**FIG. 19-4.** The fifth right rib showing the features of a *typical* rib: **A,** viewed from the back; **B,** inner aspect of the vertebral end.

which fail to reach the sternum, are regarded as **false ribs.** The ribs are long, thin, curved bones, segmentally arranged and separated by **intercostal spaces.** Anteriorly, each rib terminates in a bar of hyaline cartilage, the **costal cartilage,** which, until it ossifies partially in old age, significantly contributes to the springiness and mobility of the rib cage. The posterior portion of the rib, made of cancellous bone, is much longer and itself has several named parts: the **head,** which abuts against the vertebral column; the relatively short **neck;** and the **body,** which has a definite bend in it known as the **costal angle** (Fig. 19-4).

**Typical Ribs.** The **head** has an upper and lower articular facet divided by a *crest* for articulation with two adjacent vertebrae in a costovertebral joint. The upper border of the narrower **neck** also displays a *crest* for the attachment of ligaments (costotransverse ligaments). The neck ends laterally at the *tubercle,* a knuckle-shaped enlargement on the outer or posterior surface of the rib. The tubercle bears an oval, convex facet for articulation with the transverse process of the similarly numbered vertebra (costotransverse joint). Lateral to the facet is a small rough area to which the lateral costotransverse ligament is attached.

The tubercle marks the junction of the neck

and body of the rib. The **body** soon turns rather sharply forward at the *costal angle* and slants downward as well as forward, reaching its lowest point just before it becomes cartilaginous. The body is twisted as well as bent and its external surface generally turns to face slightly upward, conforming to the overall cone shape of the thorax (Fig. 19-5, rib 3). The anterior end of the bony rib receives the costal cartilage in a definite pit or fossa.

The upper border of the rib is smooth and rounded, whereas the lower border projects downward as a sharp lip, most pronounced posterolaterally (Fig. 19-4). This lip shelters the *costal groove* on the internal surface of the rib in which the intercostal vessels and nerves are located.

Although ribs may **fracture** under direct violence at any point, the most frequent fractures are due to compression forces on the thorax, and these occur just anterior to the costal angle, the weakest point of the rib. The broken end tends to spring outward; however, a direct force may drive it inward, causing hemorrhage or injury to the lung, which predisposes to pneumothorax. Even though rib fractures usually heal without any splinting, a fractured costal cartilage may cause pain for a long time because the reparative capacity of cartilage is poor. Fractured cartilage heals eventually by fibrosis.

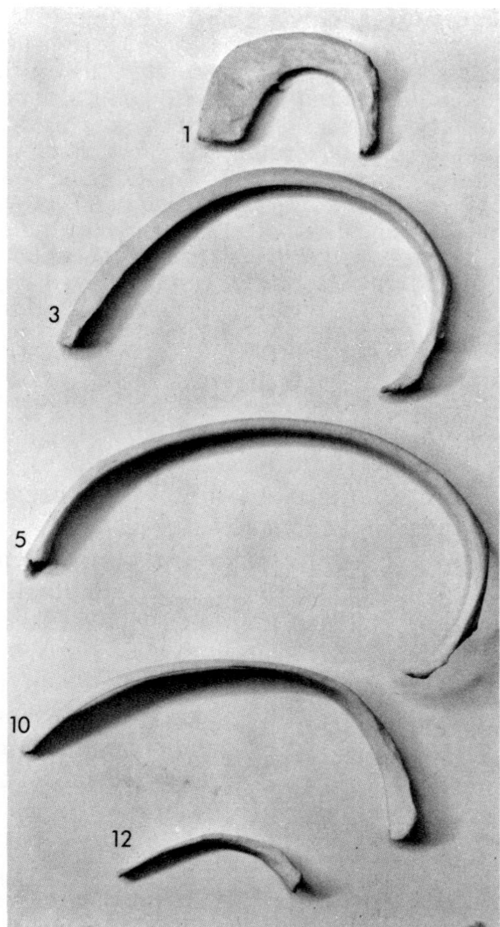


**FIG. 19-5.** The 1st, 3rd, 5th, 10th, and 12th right ribs seen from above.

**Atypical Ribs.** The above anatomic descriptions apply to all the ribs with the exception of the first two and the last three pairs, which are to some degree atypical. The **1st rib** is the most highly curved and is particularly broad (Fig. 19-5). Its inner margin forms the boundary of the superior thoracic aperture and its surfaces face primarily upward and downward. The subclavian artery exiting from the thorax marks the upper surface of the rib by a *sulcus,* as does the subclavian vein, which enters the aperture anterior to the artery. The two sulci are separated by a small eminence, the *tubercle of the anterior scalene muscle.* The head of the 1st rib bears a single articular facet, for it articulates only with the 1st thoracic vertebra. The **2nd rib** is similar in shape to the first, but larger, and its head has two facets (one each for the 1st and 2nd thoracic vertebrae). It is distinguished by a broad, rough eminence, the *tuberosity of the serratus anterior muscle.* Usually the heads of the **10th, 11th,** and **12th ribs** each articulate only with their own vertebra, and the latter two have no tubercles or angles and do not articulate with the transverse processes.

**Variations.** Ribs are present as separate bones only in the thoracic region; however, vertebrae in all regions of the spine have costal elements associated with them. Those associated with cervical and lumbar vertebrae typically fuse with their transverse processes and become parts of them. In some cases a costal element associated with either a 7th cervical or a 1st lumbar vertebra develops as a rib instead of fusing with the transverse process.

The incidence of **cervical rib** has been said to be between 0.5% and 1%. Cervical ribs are sometimes associated with signs of pressure on the brachial plexus or subclavian vessels (Chap. 12); lumbar ribs have no clinical significance except that identification of vertebral levels may be inaccurate when the ribs are counted from below. The same confusion may arise when the twelfth pair of ribs is missing or is so short as not to be palpable. There is said to be a tendency in man toward the reduction of the 12th rib. According to one survey, in more than one-fourth of cases the rib was less than 2 inches long.

**Counting of Ribs.** The ribs form important landmarks to thoracic and abdominal viscera. In most individuals, all ribs are palpable, although the first is rather inaccessible because of the overlying clavicle. Reference points for counting are the sternal angle, which identifies the 2nd rib, and the xiphisternal junction, on level with which is the 6th rib. The vertebral border of the scapula, when the arm is fully abducted, roughly corresponds in the back to this rib. The 7th costal cartilages form the apex of the infrasternal angle, and the 10th cartilages are palpable as the most inferior points of the rib cage along the costal arch.

**Ossification.** Most of the ribs are ossified from four centers: a primary center for the body and three epiphyseal centers, of which one is for the head and two are for the tubercle. The 11th and 12th ribs, however, have no separate centers for tubercles. Ossification of the body begins early in fetal life; the epiphyseal centers appear between the 16th and 20th years and unite about the 25th.

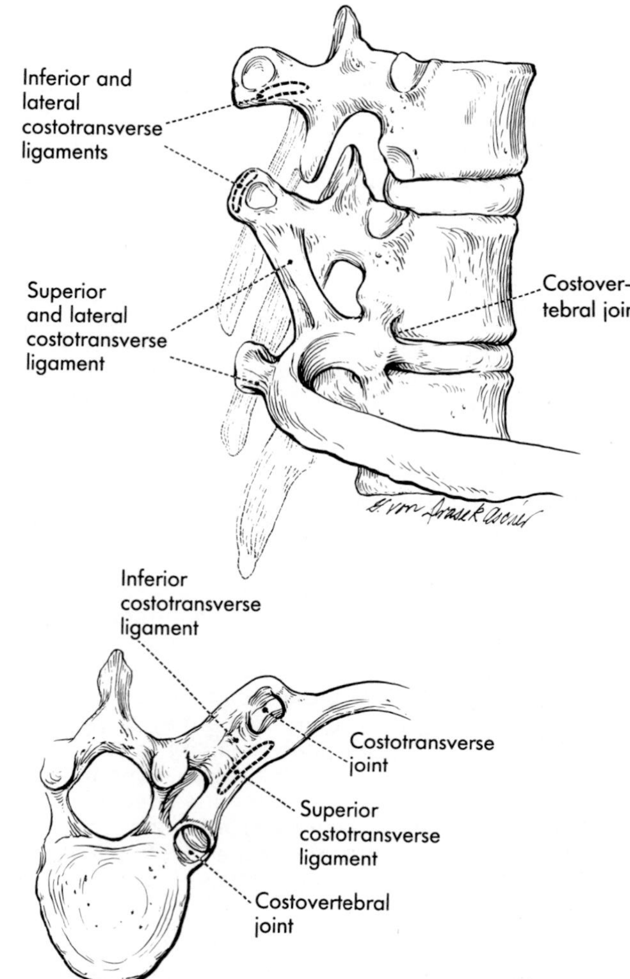


**FIG. 19-6.** Costovertebral and costotransverse joints with their associated ligaments. (Rosse C, Clawson DK: The Musculoskeletal System in Health and Disease. Hagerstown, Harper & Row, 1980)

## Joints and Movements of the Ribs

During inspiration, reduction of intrathoracic pressure is brought about by increases in the anteroposterior, lateral, and vertical diameters of the thoracic cavity. The vertical increase is due to diaphragmatic movement, whereas the anteroposterior and lateral increases depend on movements of the ribs. Most ribs articulate with the vertebral column at two joints: the *costovertebral joint,* between the head of the rib and the vertebral bodies, and the *costotransverse joint,* between the tubercle and the transverse process (Fig. 19-6). Both joints are synovial.

When the head of a rib articulates with two vertebral bodies, the cavity of the **costovertebral joint** is divided by a ligament attaching the crest of the head to the intervertebral disk. Although the articular capsule is simple, it is thickened anteriorly by a **radiate ligament,** the three components of which anchor the head to the intervertebral disk and to the bodies above and below the disk.

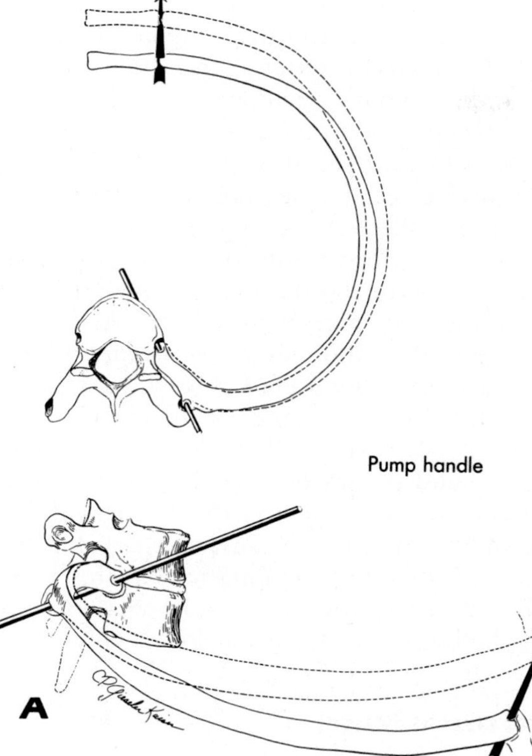


FIG. 19-7. Pump handle (A) and bucket handle (B) type movements of the ribs during respiration. The axes of the movements are indicated by solid bars.

The **costotransverse joints** have small synovial cavities that are surrounded by lax articular capsules. The joints are strengthened by three **costotransverse ligaments** (Fig. 19-6). These extend between the crest of the rib neck and the transverse process above the rib (superior ligament), between the posterior aspect of the rib neck and its own transverse process (the costotransverse ligament proper), and between the tip of the transverse process and the rough part of the costal tubercle (lateral costotransverse ligament).

The reciprocally shaped articular facets on the tubercle and transverse process determine the type of movement a rib is capable of and whether this movement increases either the anteroposterior or lateral diameter of the thorax. The tubercular facets of the upper ribs are accommodated in deep, cup-shaped sockets on the transverse processes. The tubercle is therefore compelled to roll in this fixed socket along an axis that connects the costovertebral and costotransverse joints (Fig. 19-7A). No sliding (or translation) of the tubercle is permitted.

Slight movement at these joints will be greatly amplified at the anterior end of the rib that will be raised when the tubercle, and the neck with it, rotates downward. This movement has been likened to that of a *pump handle.* The thrust is transmitted along the costal cartilages to the sternum, which becomes not only bodily elevated but also pushed forward, increasing the anteroposterior thoracic diameter. The pump-handle movement of the upper ribs is readily confirmed by laying the hands over the pectoralis major and watching them rise and fall with inspiration and expira-

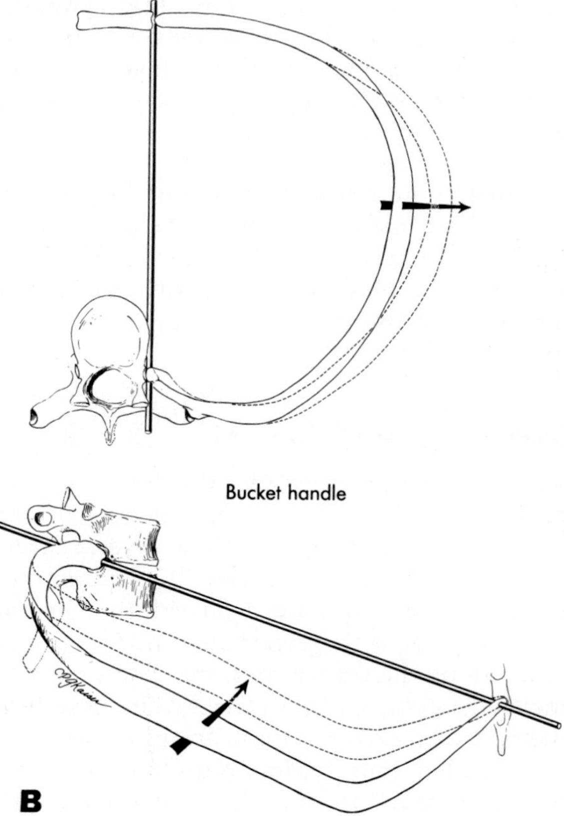

tion. Such a movement is absent over the back where the relatively fixed rib neck is permitted to roll only. Nor is the pump-handle movement demonstrable over the lower ribs. The tubercular facets of these ribs, and the articular facets of their transverse processes, are plane or flat. When the rib moves here, the tubercle rides up and down on the transverse process. The lower ribs therefore hinge on an axis that joins the costovertebral joints to the sternochondral or interchondral joints (Fig. 19-7B). Elevation of the rib will raise the entire costochondral arch as a *bucket handle* is lifted on its hinges from the side of the bucket. The lateral thoracic diameter between elevated costochondral arches will be greater than that of the same arches when they are dependent. This bucket-handle type of movement can be verified by placing the hands on the sides of the lower part of the chest. The movement will be equally pronounced whether viewed from the front or the back.

The **sternochondral** and **interchondral joints** are involved in these movements, and deformation of the costal cartilages contributes to their amplitude. The lax capsules of the interchondral joints permit some sliding of the cartilages on one another, which contributes to the visible elongation of the costal arch during inspiration and to the widening of the infrasternal angle. Limitation of rib movements due to pain or arthritic involvement of the joints of the ribs may impair respiratory function.

## MUSCLES AND FASCIA

Ventrolateral to the vertebral column and the paraspinal musculature, the muscles of the body wall are arranged in three layers: external, internal, and innermost. In the embryo, these muscles arise on each side as ventral extensions from the myotomes and meet each other along the ventral midline. These myotomal cells segregate into three layers; however, their ventral edges remain unsplit and form two parallel, longitudinal muscle masses, the *rectus muscles.* In the thoracic region, the three muscle layers derived from one myotome are separated by a rib from the muscles derived from the next myotome. These are the *intercostal muscles,* and their segmental origin remains evident throughout life. Over the abdomen, on the other hand, the consecutive myotomes fuse, although here too, the muscles retain a segmental pattern of innervation and blood supply. The rectus muscle, well developed only over the abdomen, disappears over the thorax, except in 0.5% of individuals in whom it persists as the *sternalis.* The body wall mesoderm (*somatopleur*), into which the myotomal cells migrate, provides the fascial coverings of the muscles, the lining of the body wall (endothoracic fascia and parietal pleura or endoabdominal fascia and parietal peritoneum), the superficial fascia, and the dermis. From it also develop the blood vessels and lymphatics of the body wall; however, the segmental nerves grow into the body wall from the spinal cord.

### Intercostal Muscles

There are 11 pairs of intercostal spaces and each contains an external, an internal, and an innermost intercostal muscle. All three are thin sheets of muscle whose fibers run from one rib to the next, and, although none of them spans an intercostal space from vertebra to sternum, owing to their overlap they completely seal the spaces.

Each **external intercostal muscle** begins posteriorly, just lateral to the tubercle of the rib, and extends anteriorly, slightly past the costochondral junction (Fig. 19-8). Beyond this, up to the sternum, the muscle is represented in the upper intercostal spaces by the *external intercostal membrane.* In the lower intercostal spaces, the muscular fibers of the external intercostals merge with the external oblique muscle of the anterior abdominal wall as the external oblique takes its origin from the ribs. The fibers of the external intercostal slant downward and forward from one rib to the next, and the fibers of the external oblique conform to this direction.

The **internal intercostal muscles** have a different slant; their fibers run upward and for-

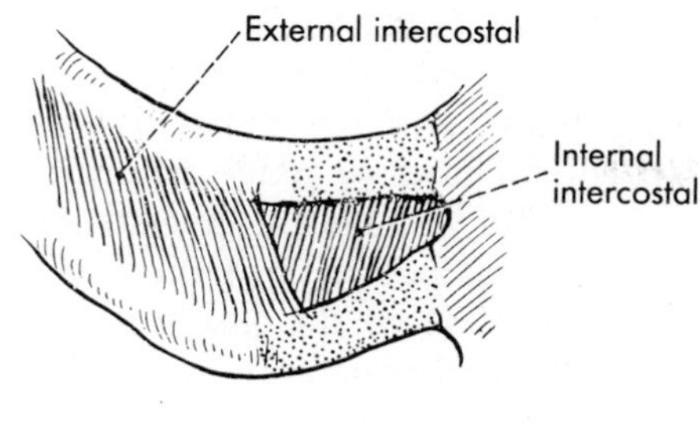


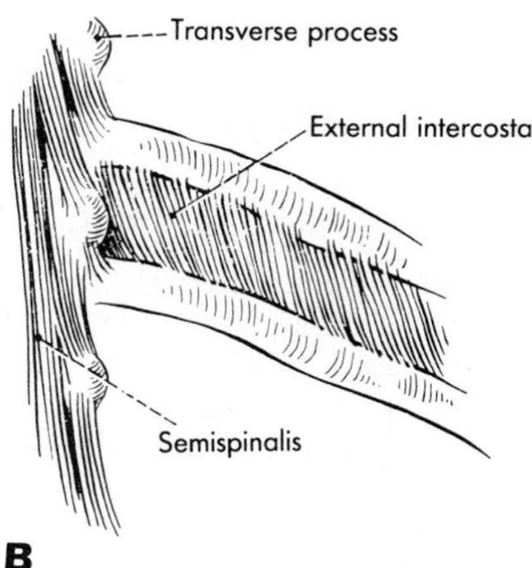


**FIG. 19-8.** Anterior (**A**) and posterior (**B**) ends of an intercostal space seen from the outside. In **A**, the anterior intercostal membrane, covering the internal intercostal muscle, is treated as if it were transparent; in **B**, some of the overlying back muscles have been cut away.

ward from the upper border of one rib to the lower border of the next above. These muscles reach the sternum anteriorly (they are visible through the external intercostal membrane, Fig. 19-9A). Posteriorly, they extend only about as far as the costal angles (Fig. 19-9B). Proximal to the angles, they are represented by *internal intercostal membranes*. The lower internal intercostal muscles merge with the internal oblique muscle of the abdomen.

The **innermost intercostal muscles** (*intercostales intimi*) lie internal to the internal intercostals. They are less well developed than the other two intercostal muscles, occupy chiefly the middle part of the length of each intercostal space, and are best distinguished by the fact that they are separated from the internal intercostals by the intercostal nerves and vessels.

The innermost muscle layer in the thorax has vestigial muscles in addition to the innermost intercostals. The **subcostal muscles** are variable fiber bundles, placed posteriorly, which span two to three intercostal spaces. The **transversus thoracis** is a thin layer of muscle whose fibers fan out from the posterior surface of the lower part of the ster-

**FIG. 19-9.** Anterior (**A**) and posterior (**B**) ends of the intercostal space seen from the inside of the thorax. In **B**, the posterior intercostal membrane, covering the external intercostal, is treated as if it were transparent.

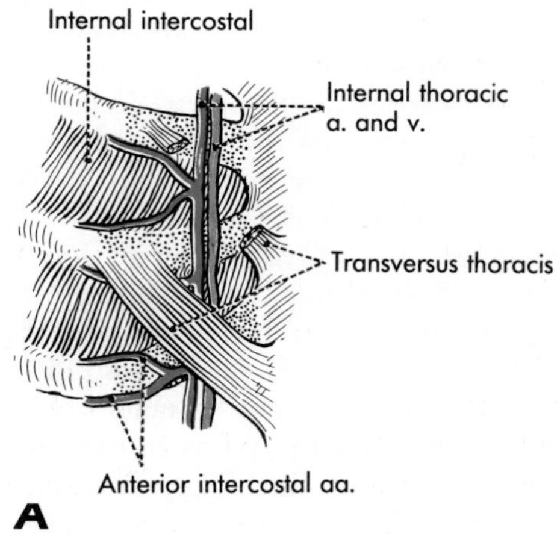


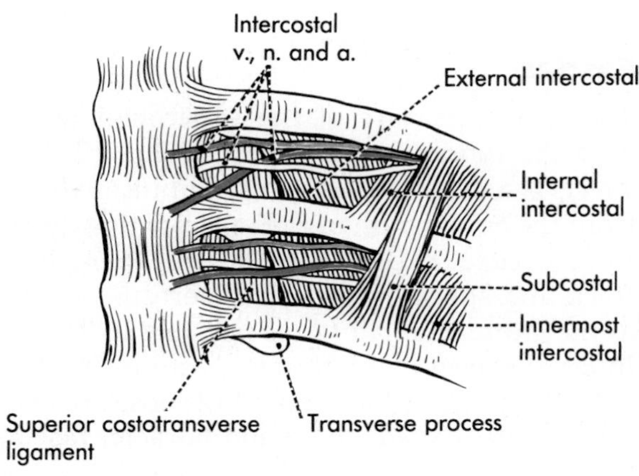

num to the neighboring costal cartilages. These two muscles, together with the innermost intercostals, are said to be equivalent to the transversus muscle of the abdomen. The peripheral fibers of the **diaphragm** are derived also from the same muscle layer; other parts of the diaphragm have different origins. The diaphragm is discussed in Chapter 25.

All the muscles of the thoracic wall are innervated by the associated intercostal nerves. Their chief function is to move the ribs, although there remains some controversy as to their specific action. Their attachments and the direction of their fibers tend to suggest that both internal and external intercostal muscles could elevate the ribs if their proximal attachment is considered as their point of origin and that both sets of muscles could depress the ribs if the inferior point of attachment is regarded as the origin. Recent electromyographic studies in man tend to confirm Galen's original teaching—that at least in some of the intercostal spaces, the external intercostals are active in inspiration and the internal intercostals, in expiration. Other studies suggest that both sets are active in forced inspiration and expiration. Intercostal muscles also contract during phonation, blowing, and sucking. When muscles of an intercostal space are paralyzed in a spare individual, paradoxical respiratory movement may be observed over that space. The tissues of the space will be sucked in during inspiration and blown out during expiration.

### Fascias

The special feature of the **superficial fascia** over the thorax is the breast (Chap. 12; Fig. 19-2). The **deep fascia** is thin and inseparable from the epimysium of the muscles that cover the rib cage, except posteriorly, where it is thick and specialized over the erector spinae as the *thoracolumbar fascia* (see Fig. 15-2). The internal aspect of the thoracic wall is lined by a hardly perceptible amount of loose connective tissue called the **endothoracic fascia,** which covers also the thoracic surface of the diaphragm (Fig. 19-2). The function of the fascia is to fix the parietal pleura to the thoracic walls and the diaphragm so that the parietal pleura moves with these during respiratory movements. The only place where the endothoracic fascia appears as a distinct layer is over the lateral portions of the superior thoracic aperture. Here the fascia forms the **suprapleural membrane** that limits bulging of the lung into the neck. The parietal pleura (and the lung with it) projects above the level of the 1st rib as the **cupula** (little dome) and is supported there by the suprapleural membrane, which is likewise dome shaped, and its highest point is fixed to the transverse process of the 7th cervical vertebra.

## INNERVATION OF THE THORACIC WALL

The body wall is innervated by the ventral rami of T1 to T12 spinal nerves. T1 to T11 are the *intercostal nerves*, and T12, analogous with them, is known as the *subcostal nerve.* These are somatic nerves and provide the motor and sensory supply to all layers of the body wall, but do not innervate the girdle musculature attached to the rib cage, even though some of their branches pass through these muscles to the skin. With the following two exceptions, thoracic ventral rami supply the skin over the entire trunk as far down as the inguinal region.

1. Over the paravertebral area of the back, the skin is supplied by the dorsal rami of the corresponding spinal nerves, which also innervate the erector spinae.
2. Anteriorly, as far down as the sternal angle, the skin is innervated by supraclavicular nerves, ventral rami of C4 and C5 (see Figs. 4-8 and 4-9).

Somatic efferent fibers in the intercostal nerves supply segmentally all the intercostal and subcostal muscles, the transversus thoracis, all three muscle layers of the abdominal wall, and the rectus abdominis. The same nerves conduct proprioceptive impulses from these muscles and also from the peripheral portions of the diaphragm, derived from the body wall. In addition to visceral afferents from the lung (Chap. 20), these somatic afferents are probably also involved in maintaining reflexly the rhythmic movements of respiration. The intercostal nerves conduct pain and other exteroceptive sensations from the skin, breast, ribs, costal cartilages, and sternum and also from the parietal pleura that lines the thoracic wall and covers the *periphery* of the diaphragm. They are sensory also to parietal peritoneum lining the entire abdominal wall and the *periphery* of the diaphragm. Pleura and peritoneum covering central regions of the diaphragm are innervated by the phrenic nerves derived from cervical segments of the cord (predominantly C4). Branches of the intercostal nerves contain sympathetic fibers that are motor to smooth muscle of the blood vessels in the body wall and of hair follicles and to sweat glands in the skin.

## Intercostal Nerves

The basic anatomy of intercostal nerves has already been described as an example of a "typical" spinal nerve (Chap. 4, Fig. 4-5). The 2nd through the 6th intercostal nerves are confined to the thorax, but the 7th to 11th intercostals and the subcostal nerve continue into the abdominal wall.

The 1st intercostal nerve represents only a small part of the ventral ramus of the 1st thoracic nerve; the major part of this ramus joins the brachial plexus. A typical intercostal nerve continues the direction of the spinal nerve as it emerges from its intervertebral foramen. After the dorsal ramus has been given off, the intercostal nerve runs, at first, outside the pleura, more or less in the middle of the intercostal space, across the internal surface of the internal intercostal membrane (Fig. 19-9B). A sympathetic ganglion is suspended from each nerve in this position. Close to the angle of the rib, the nerve enters the fascial space between the internal intercostal and the innermost intercostal muscles and also attains the shelter of the costal groove, where it accompanies the intercostal vessels. The nerve lies below the vein and the artery. As it continues forward it gives off muscular branches and a **lateral cutaneous branch** (Fig. 19-10).

The upper six nerves follow the curve of the

**FIG. 19-10.** The course and branches of an intercostal artery and an intercostal nerve shown in a schematic transverse section of the thorax.

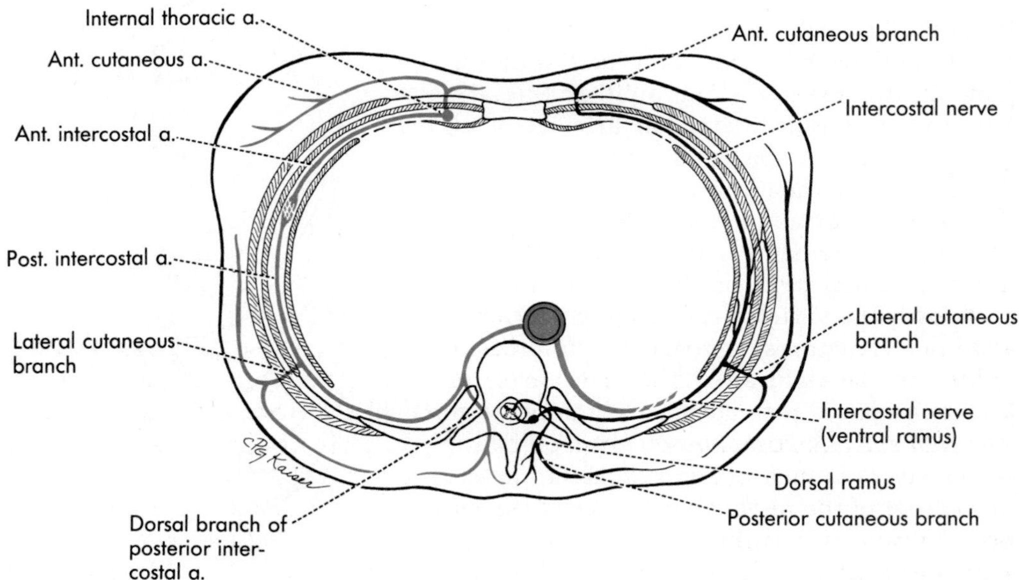

ribs and costal cartilages toward the sternum and terminate as **anterior cutaneous branches** that pierce the internal intercostal muscles, the external intercostal membrane, and the pectoralis major. Nerves 7 to 11 leave their intercostal space by crossing the internal surface of the costal arch and continue their course anteriorly between the internal oblique and transversus abdominis. The subcostal nerve running along the lower border of the 12th rib enters this fascial space without having to cross the costal arch. The termination of these nerves will be discussed with the anterior abdominal wall.

## BLOOD SUPPLY OF THE THORACIC WALL

The thoracic wall is supplied chiefly by intercostal arteries and is drained by intercostal veins. These vessels are found in each intercostal space running in the costal groove just above the respective intercostal nerves (Fig. 19-9B). Their course and branches closely conform to those of the nerve. Superficial structures over the thorax are served, in addition, by branches of the axillary and subclavian arteries and veins (Chap. 12). These vessels anastomose in the superficial fascia with branches of the intercostal vessels.

### Intercostal Arteries

In each intercostal space there are two sets of intercostal arteries, posterior and anterior, that anastomose with one another (Fig. 19-10). The **posterior intercostal arteries** of all but the first two spaces are branches of the descending thoracic aorta, which lies in the mediastinum on the left side of the vertebral column. Consequently, arteries on the right are longer than those on the left because they cross over the vertebrae, passing posterior to all other structures to reach the intercostal spaces on the right side. The arteries of the first two spaces arise from a common stem (the *supreme intercostal artery*) given off by the *costocervical branch* of the subclavian artery in the root of the neck. The arteries of the upper six spaces terminate by anastomosing with the anterior intercostal arteries of the same space; those of the lower spaces and the **subcostal artery,** like the corresponding nerves, continue to the abdominal wall.

Each posterior intercostal artery dispenses muscular branches (the largest of which, called the **collateral branch,** runs along the upper border of the rib below the space; it is accompanied sometimes by a similar branch of the nerve), a **lateral cutaneous branch,** and a **dorsal branch** that passes posteriorly with the dorsal ramus of the corresponding spinal nerve. This branch supplies the back muscles and the skin and contributes to the supply of the contents of the vertebral canal through a *spinal branch* that enters the intervertebral foramen.

The **anterior intercostal arteries** are branches of the *internal thoracic artery,* a rather slender vessel that runs down on each side of the sternum parallel with its edge and crosses the inner surface of the costal cartilages (Figs. 19-9 and 19-11). The anterior in-

**FIG. 19-11.** The course of the internal thoracic vessels: The anterior intercostal branches, usually two in each space, are not labeled. The artery is red, and the accompanying veins are black.

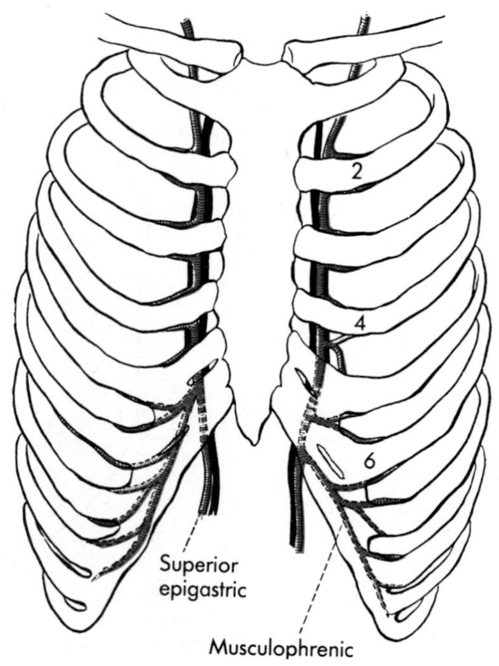

tercostal arteries are much smaller than their posterior counterparts and are somewhat variable. As a rule, they run along the lower border of each costal cartilage in the same fascial plane of the body wall as the posterior arteries. In the lower five spaces, similar branches come off the *musculophrenic artery*, one of the terminal branches of the internal thoracic, and anastomose with branches of the posterior intercostal artery rather than with the vessel itself, as is the case in the upper spaces.

### The Internal Thoracic Artery

The internal thoracic artery is given off by the subclavian artery at the base of the neck as that vessel arches over the suprapleural membrane. It descends behind the subclavian vein and the 1st rib. Its general course from this point is illustrated in Figure 19-11. The internal thoracic artery, like the intercostal vessels, runs in the neuromuscular plane of the thorax, that is, anterior to the transversus thoracis and on the internal surface of the costal cartilages and the internal intercostal muscles. Behind the 6th cartilage, it divides into its two terminal branches, the **musculophrenic** and the **superior epigastric arteries.** The former has been described; the latter descends into the abdominal wall to anastomose with the inferior epigastric artery. Apart from the anterior intercostal arteries, the internal thoracic artery gives branches to the mediastinum, the thymus, the pericardium (pericardiacophrenic artery), the sternum, and some *perforating branches.* The perforating branches pass to the skin with the anterior cutaneous branches of the intercostal nerves and, in the 2nd, 3rd, and 4th intercostal spaces, contribute to the supply of the breast; these vessels enlarge during lactation.

### Veins

All arteries described above and their branches are accompanied by veins designated by identical names. The veins anastomose in the same manner as the arteries. The **posterior intercostal veins** terminate in the *azygos* or *hemiazygos veins,* which run along the bodies of the vertebrae (Chap. 22). The veins of the upper two spaces drain into the *brachiocephalic vein,* as do the *internal thoracic veins,* which receive the **anterior intercostal veins.**

### Anastomoses

The intercostal and internal thoracic vessels participate in an important anastomotic system. Should the descending aorta or one of the venae cavae become obstructed, this anastomosis provides alternative channels for arterial and venous blood flow. The internal thoracic vessels are the cranial segment of a longitudinal, ventral anastomotic chain that links the subclavian artery and brachiocephalic vein to the external iliac vessels. The intercostal vessels connect this chain to the descending aorta and azygos system of veins. In addition, in the superficial tissues of the thorax, branches of the axillary and subclavian arteries anastomose with the intercostal arteries, providing a possible conduit for blood flow from the subclavian system to the descending aorta. Should there be an obstruction between the aortic arch and the descending aorta, as there is in coarctation of the aorta, this anastomosis assumes a great importance (p. 563). In such cases the circumscapular and intercostal arteries greatly enlarge: the former visibly pulsate around the scapula, and the latter may erode the ribs, as evidenced on x-ray films by notches along their inferior edges.

### Lymphatic Drainage

Lymphatics of the superficial tissues of the thorax, including the breast, drain primarily into **axillary lymph nodes;** some, however, follow the perforating branches of the internal thoracic vessels and terminate in **parasternal lymph nodes.** The parasternal lymph nodes are irregularly placed along the internal thoracic vessels and collect lymph from the medial half of the breast, the anterior portion of the chest wall, and the anterior mediastinum. Lymphatics running with the posterior intercostal vessels terminate in **intercostal lymph**

**nodes,** which lie along the azygos and hemiazygos veins in the posterior mediastinum. They drain lymph from the deep tissues of the chest wall; superficial tissues from the back drain to the axillary nodes.

## THE THORACIC CAVITY

### Developmental Considerations

In an early embryo, as the developing thoracic walls approximate each other ventrally, they enclose (1) a portion of the foregut, which develops into the esophagus; (2) the primitive heart tube, which has assumed a position ventral to the foregut; and (3) the rostral horseshoe-shaped portion of the intraembryonic celom, which is lined by a continuous layer of mesothelium (Fig. 19-12A). The more caudal portion of the celom, which will form the future peritoneal cavity, will be closed off from the thoracic part by the **pleuroperitoneal membranes.** These membranes later become incorporated into the diaphragm.

The heart soon sinks into the ventral portion of the celom, the future **pericardial cavity** (Fig. 19-12B). Dorsolaterally, the lungs, budding off the foregut, bulge more and more into what will become the **pleural cavity,** without breaking the mesothelial lining of the cavity (Fig. 19-12C). The pericardial cavity becomes partitioned off from the pleural cavities by bilateral septa derived from the inner mesoderm of the embryonic thoracic wall. This septum, the **pleuropericardial membrane,** is raised up and drawn across the body cavity on each side by a major vein (common cardinal vein), which drains the body wall and empties into the heart (Fig. 19-12C and D). This partition consists of "embryonic endothoracic fascia" and will develop into the **fibrous pericardium.** The mesothelial lining of the common body cavity in the thorax is thus separated into three independent blind sacs: two **pleural sacs** and the **serous pericardial sac** (Fig. 19-12E). Moreover, the definitive topography of viscera in the thoracic cavity is now established: between the two pleural sacs is the **mediastinum,** an irregular, broad, median partition of the thoracic cavity, which, for practical purposes, includes all the contents of the thoracic cavity except the lungs themselves.

The developing lungs covered with a layer of mesothelium, now designated as **visceral** or **pulmonary pleura,** grow more and more into the pleural cavity. Eventually, their pulmonary mesothelial surface will contact the **parietal pleura,** the outer wall of the sac, which is draped over the thoracic wall and the fibrous pericardium (Fig. 19-13). Parietal and visceral layers of the pleural sacs remain continuous with one another around the **root of the lung,** the pedicle that contains the pulmonary vessels and bronchi and attaches the lung to the mediastinum. The pleural cavity is thus reduced to a mere slit, which contains nothing except enough serous fluid to moisten the adjacent mesothelial surfaces. This should be borne in mind even though it is customary to speak of the lungs as if they were occupying the pleural cavity. In truth, they are outside the pleural sac.

The anterior edge of the growing lung pushes the pleura before it and peels off the fibrous pericardium more and more from the chest wall until the serous pericardial sac, with the heart in it, becomes completely enclosed in the fibrous bag (Figs. 19-12E and 19-13). This permits the two pleural sacs to approximate each other in front of the fibrous pericardium. However, they will not fuse with each other. The arrangement of pleural and pericardial sacs is shown in Figure 19-13. The serous pericardium, consisting of parietal and visceral layers, is enclosed by the fibrous pericardium in the same manner as the body wall encloses the pleural sacs. The greater part of the thoracic cavity comes to be occupied by the lungs. The boundaries and interior of this cavity are best described by considering the parietal pleura and the pleural cavity. Discussion of the lungs themselves and the contents of the mediastinum is deferred to subsequent chapters.

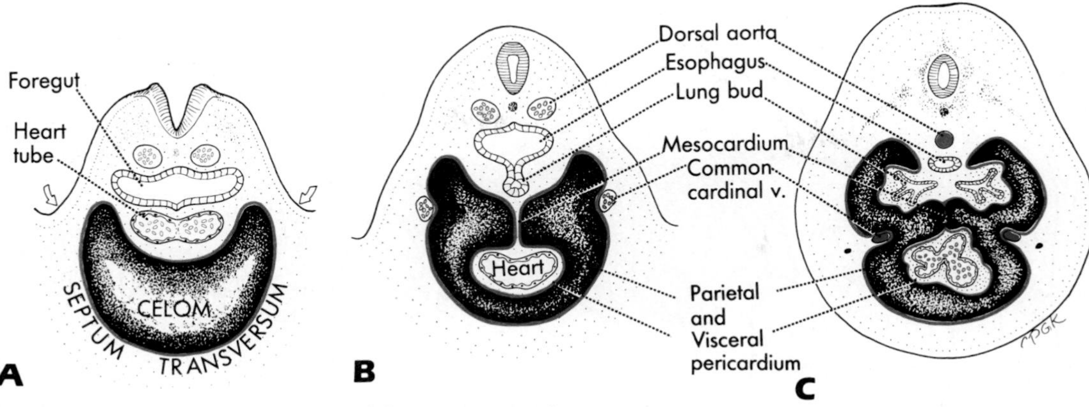


**FIG. 19-12.** The development of the serous sacs in the thoracic region of an embryo shown schematically in a series of transverse sections at successive stages of development after the completion of the head fold. The celomic cavity and its compartments are represented in heavy black stippling, and the mesothelial lining of the celom is shown in red. **A,** the flat embryonic disk is beginning to round up by the deepening of the groove between the embryonic ectoderm and the amnion (**arrows**). **B,** the primitive heart tube has sunk into the future pericardial cavity. **C,** the lung buds begin to bulge into the pleuroperitoneal canals, while the common cardinal veins elevate a ridge on the lateral wall of the celom that will become the pleuropericardial membranes. **D,** after the pleuropericardial membranes are well formed, the pleural cavities expand anteriorly (**arrows**). **E,** the formation of the fibrous pericardium completes the separation of the pericardial cavity from the pleural sacs, which continue their expansion anteriorly.

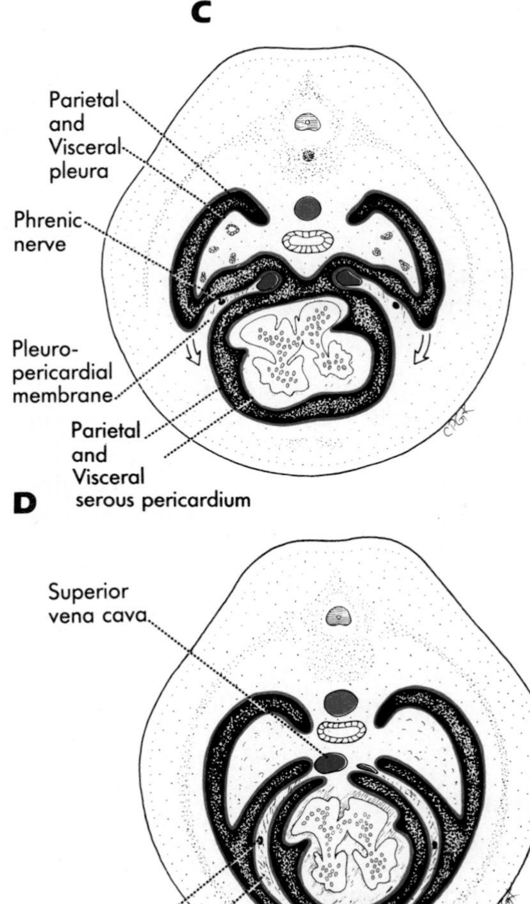


## THE PLEURA AND PLEURAL CAVITIES

The surfaces of the pleura that face into the pleural cavity are covered by mesothelium. This delicate squamous epithelium is strengthened on its abluminal surface by a substantial connective tissue membrane, an integral part of the pleura, which is sufficiently thick to support blood vessels, lymphatics, and nerves. It can be cut and sewn. The pleura can be peeled off the lung, but, because it tears readily, only with difficulty. It is easier to separate it from endothoracic fascia, especially over the fibrous pericardium, where the endothoracic fascia is sparse or laden with fat. In a dissection, it is possible to

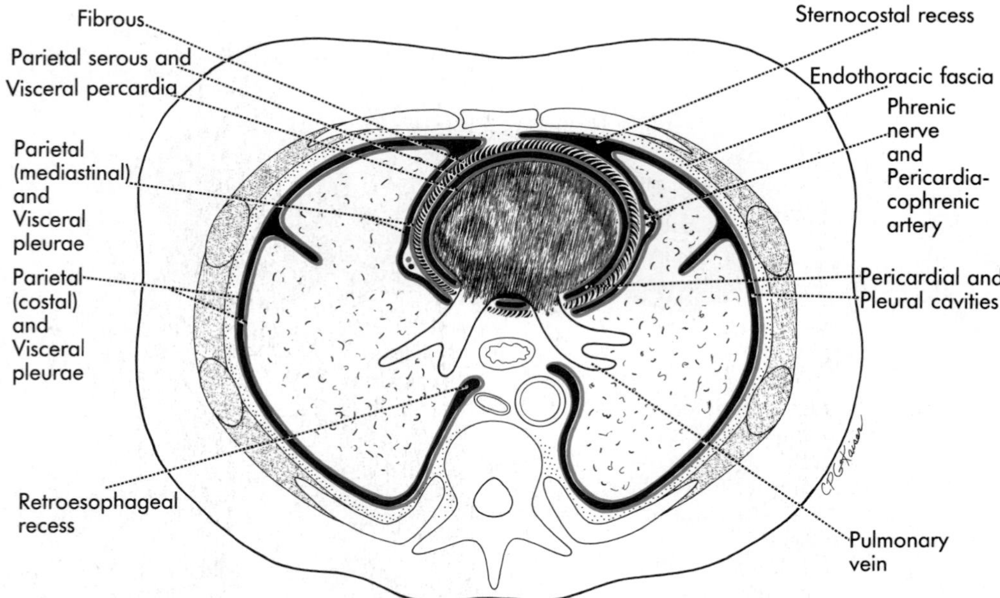


**FIG. 19-13.** The arrangement of the pleural and pericardial sacs shown in a schematic transverse section of the thorax. The serous cavities are black, and the serous membranes are red.

lift off the anterior thoracic wall, after the ribs and intercostal muscles have been cut, without opening the pleural sacs. This requires gently burrowing in the plane of the endothoracic fascia and pushing the pleura away from the wall as the wall is being lifted. It is instructive to inspect the pleural sacs from their exterior and to feel between finger and thumb the reflection of the lining off the walls onto the mediastinum or the diaphragm (Fig. 19-14).

## Parietal Pleura

The parietal pleura is divided into several named parts according to the structures it covers. The **costal pleura** lines the thoracic wall and forms the anterior, lateral, and posterior walls of the pleural cavity. Superiorly, it is continuous with the **cupula of the pleura,** which rises above the level of the 1st rib, and, inferiorly, it reflects to become the **diaphragmatic pleura,** forming the floor of the pleural and thoracic cavities. The **mediastinal pleura** is continuous anteriorly and posteriorly with the costal pleura and inferiorly with the diaphragmatic pleura. In front of the root of each lung, the phrenic nerve and the accompanying pericardiacophrenic vessels lie between the pleura and the pericardium, often surrounded by an appreciable quantity of fat. Behind the pericardial sac, the mediastinal pleura surrounds the root of the lung and, like a wide sleeve, invests it. The sleeve is very short. It is here that the parietal pleura becomes visceral pleura as the sleeve doubles back on itself, reflecting onto the lungs where the root of the lung enters the *pulmonary hilum* (Fig. 19-17; see Fig. 20-3). The wide pleural sleeve around the lung root and hilum is redundant inferiorly and, viewed from within the pulmonary cavity, hangs down as a pleural fold. This fold of pleura is the **pulmonary ligament.** After the lung is removed from the pleural cavity, the cut profiles of the two layers of the pulmonary ligament are visible, both on the lung (see Fig. 20-3) and on the mediastinal pleura (see Figs. 22-3 and 22-4).

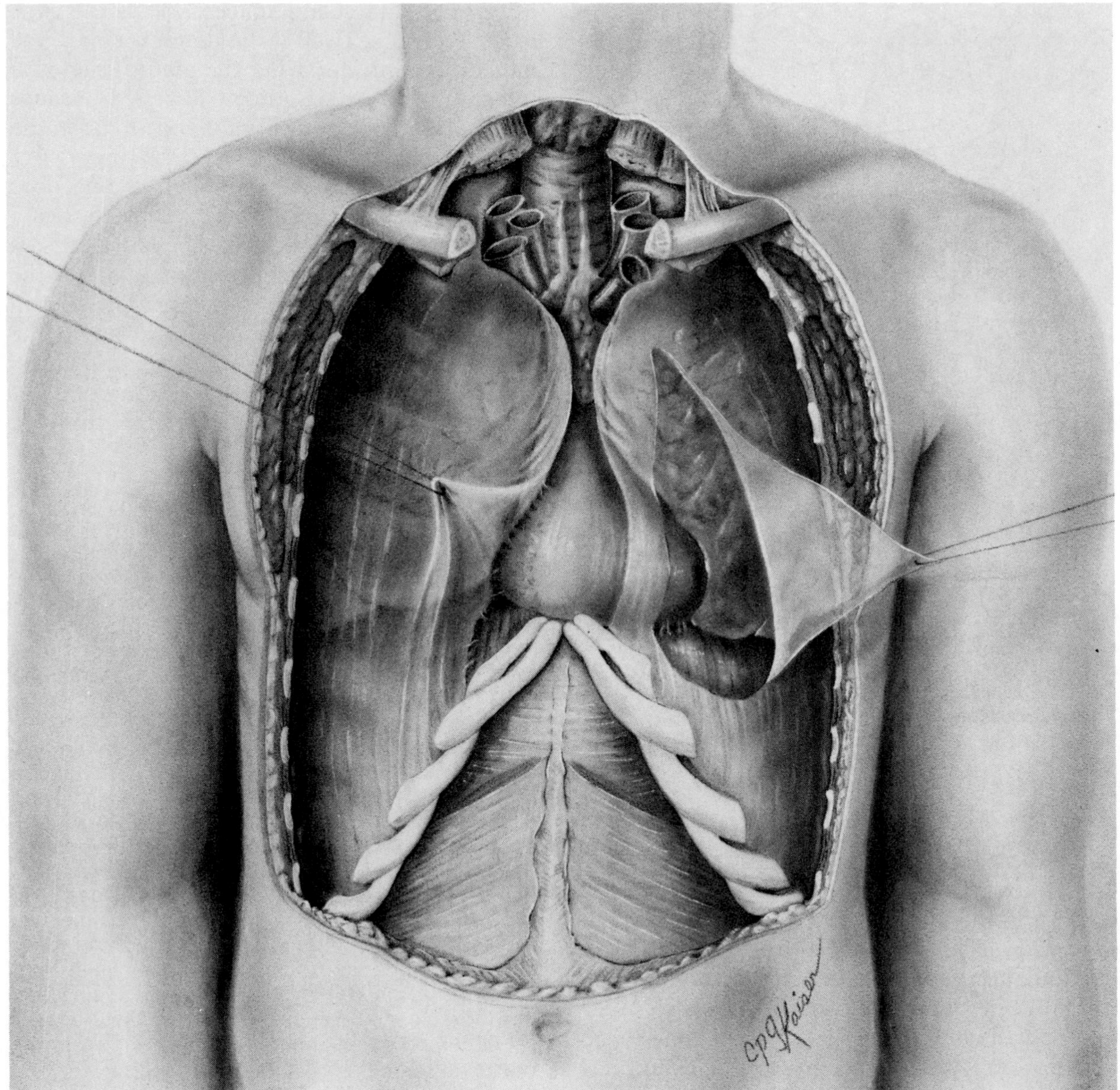

**FIG. 19-14.** A dissection of the thorax to show the arrangement of the pleural sacs. The anterior chest wall has been lifted off, the right pleural sac has been elevated from the pericardial sac and the diaphragm, and an opening has been cut in the parietal pleura on the left side to reveal the lung covered with visceral pleura.

**Lines of Pleural Reflection.** The sharp lines of reflection, where costal pleura becomes continuous with mediastinal or diaphragmatic pleura, are important because they limit the pleural cavities. These lines can be mapped on the surface of the chest in relation to bony landmarks.

The anterior lines of reflection of the parietal pleura are shown in Figure 19-15. Between the pleural cupulae, the two pleural sacs are

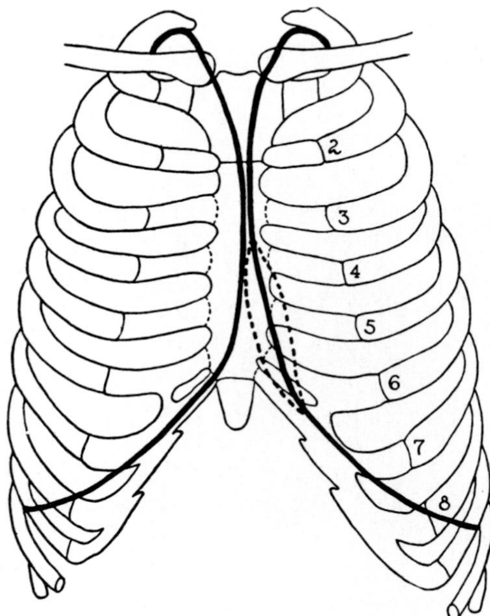


**FIG. 19-15.** Relations of the pleural reflections to the anterior thoracic wall. The two broken lines on the left pleural boundary indicate variations in the cardiac notch; within these lay 70% of Woodburne's cases. The solid line between the broken lines represents the mean found in this series. (Redrawn from Woodburne RT: Anat Rec 97:197, 1947).

far apart, but converge as they are traced down toward the sternal angle where the right and the left pleura are so close together that they almost touch at the midline. The right pleura then continues downward close to the midline of the sternum. At the lower end of this bone, it swings outward and then turns down along the 7th costal cartilage. In contrast, the anterior reflection of the left pleura typically begins to diverge laterally at about the level of the 4th rib and is usually lateral to the sternum at the level of the 5th and 6th interspaces. Thereafter, it follows the 7th costal cartilage. The deviation of the left pleura is known as the *cardiac notch.* If the notch is marked enough, it leaves sufficient room between the pleura and sternum to introduce a needle into the pericardial sac without penetrating the pleural cavity. Pericardial fluid can be obtained by inserting a needle into the 5th or 6th intercostal space close to the edge of the sternum. Even if the notch is small, the danger of contaminating the pleural sac with infected pericardial fluid is lessened because the enlarged pericardial sac tends to push the pleura laterally.

Inferiorly, the pleural reflection line does not quite coincide with the costal arch. Leaving the 7th costal cartilage, the pleura crosses the 8th rib in the midclavicular line and the 10th rib in the midaxillary line. The pleura usually reaches its lowest point at about the middle of the 11th rib and then runs posteriorly almost horizontally, swinging slightly upward as it approaches the 12th thoracic vertebra (Fig. 19-16).

The fact that the lower edge of the pleura, posteriorly, may run below the tip of a short 12th rib matters when posterior incisions, such as those for an approach to the kidney, are made. Normally, the incision is far enough lateral (because of the back muscles) to reach the 12th rib before reaching the pleura. When the 12th rib is particularly short, however, carrying the incision to this rib, or mistaking the 11th for a normal 12th, would involve opening the pleural cavity.

The posterior line of reflection between the costal and the mediastinal pleura is usually rounded rather than sharp and is therefore shown as lying either anterior to the tips of the transverse processes or immediately on each side of the vertebral bodies. However, the right pleura particularly, and the left to a lesser extent, may extend anterior to the vertebral bodies, so that sometimes the two pleural sacs almost touch on the front of the vertebral column. If they actually do so, they lie in front of the aorta, the azygos and hemiazygos veins, and the thoracic duct. These pleural pockets are located behind the esophagus and are known as the **retroesophageal recesses** (Fig. 19-13).

**Nerve and Blood Supply.** The parietal pleura shares the nerve and blood supply of the structures that the different parts cover. The costal pleura is supplied by intercostal nerves and vessels, as is the peripheral part of the diaphragmatic pleura. The phrenic nerve supplies the major, central portion of the diaphragmatic pleura and also the entire mediastinal pleura. The pain of pleurisy (pleural inflammation or irritation) is medi-

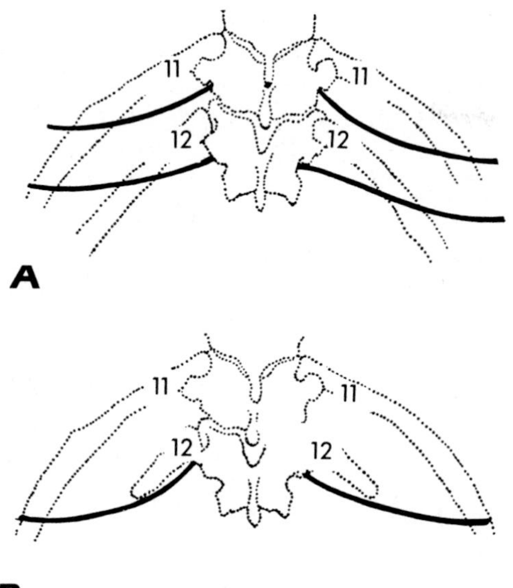


**FIG. 19-16.** The lower boundary of the pleura. **A** shows the extremes of the posterior pleural reflections in Melinkoff's series; **B** shows the usual relation of the pleura when the 12th rib is rudimentary (adapted from Melinkoff A: Arch Klin Chir 123:133, 1923)

ated along somatic afferent pathways in these nerves. By contrast, visceral pleura, supplied by autonomic nerves of the lung, is insensitive to pain stimuli.

Inflammation due to infection, irritation, emboli, or neoplasms in segments of the lung causes hyperemia of the overlying pleura, which leads to exudate formation. When the inflammation spreads across the pleural cavity and involves the parietal pleura (or when the parietal pleura is primarily inflamed due to viral infections, for instance), respiratory movements become painful because somatic pain afferents are excited. Sufficient exudate sooner or later separates the pleural layers and minimizes mechanical irritation. Resolution of the inflammation usually leaves adhesions between parietal and visceral pleura, and these, as a rule, are painless.

### Visceral Pleura and Pleural Cavities

The **visceral pleura** (also commonly called **pulmonary pleura**) is tightly attached to the outer surface of each lung and dips into the fissures of the lung (Fig. 19-13). In the fissures, the pulmonary pleura is in contact with itself, whereas over the surfaces of the lung, it touches the parietal pleura. However, only during deep inspiration is all of the parietal pleura (practically speaking) in contact with visceral pleura. In expiration and during quiet breathing, recesses exist in the pleural cavity. In these recesses, parietal pleura is in contact with parietal pleura, and the lung peels them apart as it expands into the recesses during inspiration. The pleural recesses are deepest inferiorly, where the costal pleura leaves the inner aspect of the rib cage and, in an acute angle, sweeps up onto the superior surface of the diaphragm, thus creating the **costodiaphragmatic recess** (Figs. 19-14 and 19-17). Smaller recesses exist behind the sternum, where costal pleura doubles back on itself to become mediastinal pleura, so creating the **sternocostal recesses** (Fig. 19-13). Much more shallow **retroesophageal recesses** exist where mediastinal pleura continues into costal pleura.

In quiet breathing, the inferior edge of the lungs remains about two ribs higher posterolaterally than that of the pleura (see Fig. 20-2). The surface projection of the inferior pulmonary edge is at the 6th rib in the midclavicular line, at the 8th rib in the midaxillary line, and at the 10th thoracic vertebra in the back. Thus, it is possible to thrust a large needle or a small trocar and cannula through the 8th intercostal space in the axilla and obtain a biopsy of the liver without injuring the lung, while transgressing the body wall, the pleural cavity, the diaphragm, and the peritoneal cavity.

No pleural recesses exist superiorly, and the lungs fit snugly into the pleural cavity (Fig. 19-17).

## MOVEMENTS OF RESPIRATION

The basis for the voluntary movements of respiration are the mechanical forces generated in the chest wall that are transmitted to the lung across the pleural cavity. Contrary to many accounts of respiratory movements, which explain the adherence of visceral and

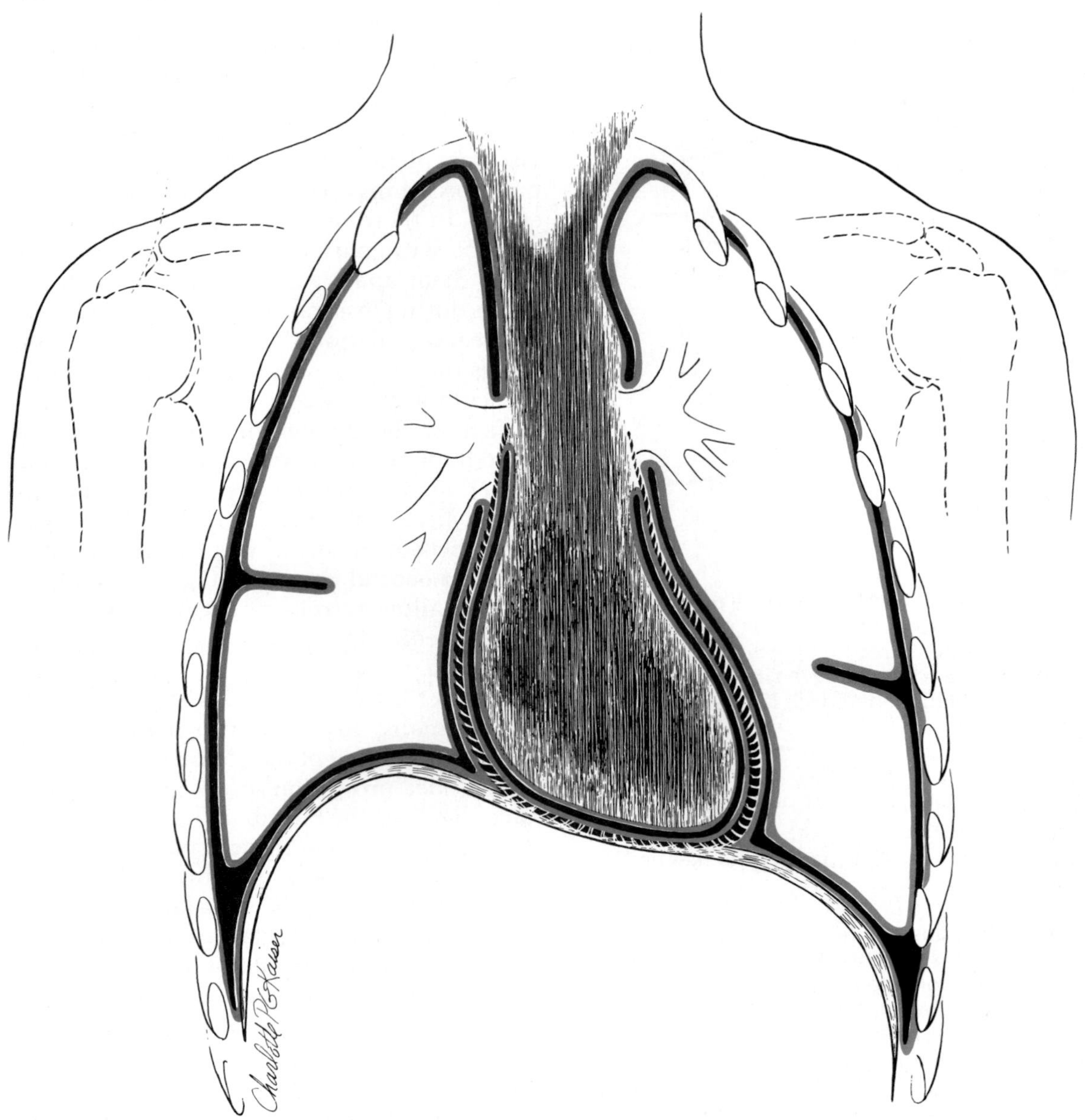

**FIG. 19-17.** The relationship of the pleural and pericardial sacs shown in a schematic coronal section of the thorax. Note the costodiaphragmatic recesses. The serous cavities are black, and the serous membranes are red.

parietal pleurae to one another by surface tension—similar to the adherence of two wet glass plates—surface force of attraction plays no part in the mechanics of respiratory movements. The forces acting across the pleural cavity are such that under normal circumstances both fluid and gas tend to be absorbed from the pleural cavity, thereby leaving the lung and the chest wall in apposition, as it were, by default.

Actually, a force is always present that, rather than keeping them together, tends to pull the pleural surfaces apart. Even at the end of a normal expiration, the pressure within the pleural cavity is 5 cm $H_2O$ below atmospheric pressure. This subatmospheric pleural pressure—often referred to as "negative" pressure—is due to retractive forces that perpetually stretch the chest wall and the lung to such an extent that they tend to pull away from each other. The lung is stretched or distended by atmospheric pressure in the alveoli, while atmospheric pressure compresses the chest wall externally. When a thoracotomy is performed, the lung collapses and the cut ends of the ribs spring outward. The lung can be made to collapse further if the pressure in the airways is lowered below the atmospheric pressure while the chest is open. In the intact chest, it is the force of recoil in the stretched elastic tissues of the lung and chest wall that is constantly pulling them apart, thereby generating the subatmospheric pleural pressure, whereas atmospheric pressure acting on the outside of the wall and on the inside of the lung is the force that keeps them together.

The balance of hydrostatic and colloid osmotic pressures in the capillaries of the visceral and parietal pleurae is such that subatmospheric pleural pressure creates a small, but definite, pressure gradient, which causes a continuous movement of fluid across the pleural cavity from the systemic capillaries to the pulmonary capillaries of the visceral pleura. This pleural fluid lubricates the pleural surfaces and facilitates their movement on one another. If the pulmonary capillary pressure becomes elevated, the visceral pleura will no longer be able to absorb the transudate, and fluid will accumulate in the pleural cavity. A rise in the protein concentration of pleural fluid, itself, will lead to the same outcome. The subatmospheric pleural pressure may cause some movement of air across the alveolar walls and the visceral pleura into the pleural cavity. The air will be absorbed promptly by the capillaries of the parietal pleura because the partial pressure of blood gases in the systemic venous capillaries is much lower than intrapleural pressure. Thus, physiological conditions prevent the accumulation of both air and fluid between the pleural layers. Therefore, as long as the muscles of respiration can overcome the resistance to air entry into the lungs, the lungs and the chest wall will not separate and will move in unison. When the lung cannot expand with the chest wall due to airway obstruction, expansion of the chest wall causes a further fall in intrapleural pressure, which leads to augmented exudation of fluid from the pleural capillaries. Thus, the space between the collapsed lung and the chest wall will always be filled with exudate as long as the surface of the lung and wall are intact.

The role of the ribs in increasing the anteroposterior and transverse diameters of the thorax by their pump-handle and bucket-handle type movements has already been discussed (p. 473). Two-thirds of the increase in thoracic capacity is due to diaphragmatic movements. This is so, even in those individuals in whom costal breathing is quite pronounced. The anatomy of the diaphragm is discussed in Chapter 25. However, it should be noted here that its domes rise from the margins of the inferior thoracic aperture as high as the 4th intercostal space on the right and the 5th space on the left. In normal, quiet breathing, rhythmic contraction of the peripherally placed muscle fibers produces an up-and-down movement of the domes without much change in their curvature. A deep inspiration calls for maximal excursion of the ribs and causes the domes to flatten as well. The inferior thoracic aperture becomes fixed by the contraction of the abdominal muscles, especially the quadratus lumborum, providing a firm base for diaphragmatic contraction. With the descent of the diaphragm, abdominal viscera are compressed, and the mediastinum is elongated. Powerful diaphragmatic contraction also elevates the lower ribs because the fibers of the diaphragm run upward from the costal arch. Diaphragmatic contraction contributes to the bucket-handle movement of the ribs.

Accessory muscles are also recruited in an increased inspiratory effort. These are the scalene muscles and the sternomastoids, which elevate the superior thoracic aperture (and the entire rib cage with it), and the muscles of the pectoral girdle, which act from their humeral insertion and elevate the ribs if the arms are fixed. For this reason, patients in respiratory distress lean on their elbows to immobilize their humeri, and they prefer the sitting to the supine position because their abdominal contents do not bulge so much into the chest.

In normal breathing, expiration is largely passive, due to the elastic recoil of lungs and rib cage and the relaxation of the diaphragm. The abdominal muscles play an important role in forced expiration. They fix the margins of the inferior thoracic aperture so that the intercostal and subcostal muscles can depress the ribs. More important, however, by raising intraabdominal pressure they push the abdominal contents into the chest, which is permitted by the relaxed diaphragm. These contractions are spasmodic in violent expiratory efforts such as coughing and sneezing, but are more refined and controlled during phonation. They may also play some role in the course of normal respiration.

## RECOMMENDED READINGS


BANYAI AL: Motion of the lung after surgically induced paralysis of the phrenic nerve. Arch Surg 37:288, 1938

BASMAJIAN JV: Muscles Alive, 2nd ed, p 287. Baltimore, Williams & Wilkins, 1974

BOYD W, BLINCOE H, HAYNER JC: Sequence of action of the diaphragm and quadratus lumborum during quiet breathing. Anat Rec 151:579, 1966

CAMPBELL EJM, AGOSTONI E, DAVIS JN: The Respiratory Muscles: Mechanics and Neural Control, 2nd ed. London, Lloyd-Luke, 1970

DAVIS PR, TROUP JDG: Human thoracic diameters at rest and during activity. J Anat 100:397, 1966

DRAPER MH, LADEFOGED P, WHITTERIDGE D: Expiratory pressures and air flow during speech. Br Med J 1:1837, 1960

GRAY DJ, GARDNER ED: The human sternochondral joints. Anat Rec 87:235, 1943

HAINES RW: Movements of the first rib. J Anat 80:94, 1946

JONES DS, BEARGIE RJ, PAULY JE: An electromyographic study of some muscles of costal respiration in man. Anat Rec 117:17, 1953

KUBIK S: Surgical Anatomy of the Thorax. Philadelphia, WB Saunders, 1970

LACHMAN E: A comparison of the posterior boundaries of the lungs and pleura as demonstrated on the cadaver and on the roentgenogram of the living. Anat Rec 83:521, 1942

MIDDLETON WS: Costodiaphragmatic adhesions and their influence on the respiratory function. Am J Med Sci 166:222, 1923

MUNRO RR, ADAMS C: Electromyography of the intercostal muscles in connected speech. Electromyography 11:365, 1971

PETERS RM: The Mechanical Basis of Respiration. Boston, Little, Brown & Co, 1969

SEIB GA: The azygos system of veins in American whites and American Negroes, including observations on the inferior caval venous system. Am J Phys Anthropol 19:39, 1934

TAYLOR A: The contribution of the intercostal muscles to the effort of respiration in man. J Physiol 151:390, 1960

WOODBURNE RT: The costomediastinal border of the left pleura in the precordial area. Anat Rec 97:197, 1947

# 20

# THE LUNGS

The lungs are paired organs specialized for the exchange of gases between atmospheric air and the blood. Their essential tissue is a squamous epithelium, a single attenuated layer of cells that forms the walls of minute spaces, the **alveoli,** and intervenes between capillaries of the pulmonary circulation and the air contained in the alveoli. The alveoli are connected to the exterior by a branching system of tubes, the **bronchial tree,** and remain filled with air even during expiration; they account for the greatest volume of the lungs by far. When the lung is normally distended with air, the chest sounds hollow to percussion, and the lungs remain afloat if immersed in water. If air has never entered the lungs, as in the case of stillbirth, the solid, glandlike lung tissue sinks in water. The bronchi and blood vessels comprise, by comparison, a small amount of tissue in the interior of this delicate, air-filled sponge, and when the alveoli collapse, the lung or one of its lobes or segments shrinks to a size many times smaller than that of its normally inflated state.

## Developmental Considerations

The tracheobronchial tree develops as a ventral diverticulum of the foregut that elongates and undergoes repeated buddings, yielding around 20 generations of endodermal tubes on both the right and the left side. These buddings entrap splanchnic mesoderm of the embryonic mediastinum. This branching establishes the anatomic pattern of the airways, the most distal of which are the alveolar sacs and the most proximal, the right and left principal bronchi. All but the main bronchi become embedded in the substance of the growing lungs, whose surfaces come to be delineated and defined by the visceral pleura as the developing lungs contact the embryonic pleural sacs, indent them, and expand with them into the thoracic cavity (see Chap. 19, Fig. 19-12). From the mesoderm entrapped among the future airways, smooth muscle and cartilage differentiate for the support of the bronchial epithelial lining. The pulmonary arterial and venous capillary beds and larger vessels, which hook up proximally with the arterial and venous ends of the developing heart, are also derived from the same mesoderm.

The internal anatomy of the lungs conforms to a segmental pattern laid down during development by the budding of the bronchi. The **bronchopulmonary segments** are the anatomic units of the lung. Clinical evaluation, as well as surgical resection, of diseased portions of the lungs relies on the anatomy of bronchopulmonary segmentation. Before dealing with the bronchial tree and the bronchopulmonary segments, the external anatomy of the lungs will be discussed.

## EXTERNAL ANATOMY OF THE LUNGS

The pliable lungs have assumed the shape of the space available to them on each side of the mediastinum in the thoracic cavity (Fig. 20-1). When they are removed from the chest in the fresh or the fixed state, each lung is roughly conical, presenting a tapered upper end, the **apex,** and a broad **base.** On each lung there are three external **surfaces** (costal, diaphragmatic, and medial) separated from each other by **borders** or margins (anterior and inferior). All surfaces are completely covered in visceral pleura, which unites with the mediastinal parietal pleura around the **root of the lung.** The root of the lung is a relatively narrow pedicle that suspends the lung from the mediastinum in the pleural cavity and enters its substance at the **pulmonary hilum. Fissures** divide the lungs into **lobes.** Typically, the left lung is divided by the **oblique fissure** into an upper and a lower lobe, whereas the right lung is divided into upper, lower, and middle lobes by the **oblique** and **horizontal fissures.** In the depths of the fissures, visceral pleura covers the *interlobar surfaces,* where it is in contact with itself; over all other surfaces visceral pleura is in contact with parietal pleura.

Even though the pleural sacs intervene between the lung and all other intrathoracic structures, a number of these structures in

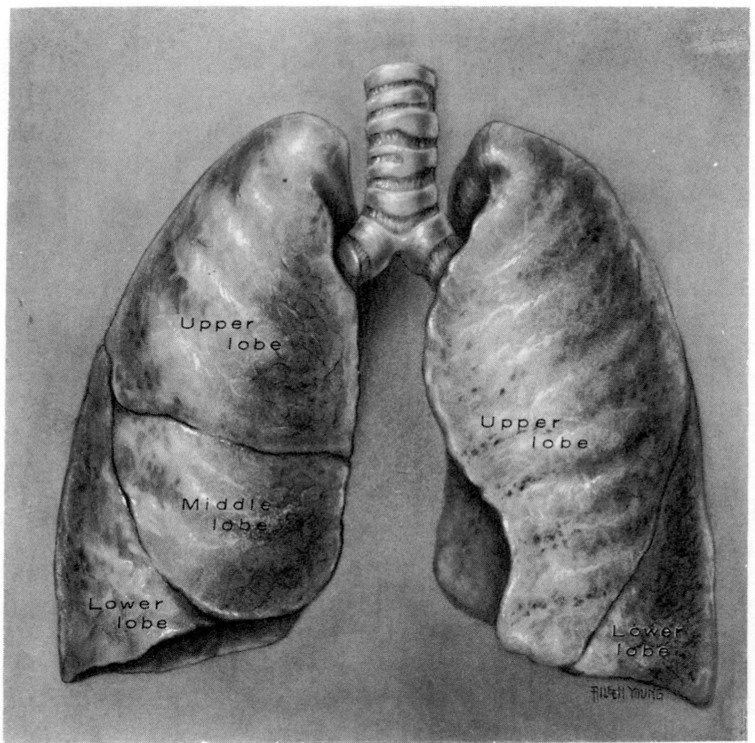


**FIG. 20-1.** Anterior view of the lungs and the lower end of the trachea.

contact with the lung through the pleura leave impressions on the pulmonary surfaces that are perceptible in the fixed, but not in the fresh, lung.

### Surfaces and Borders

The **apex** fits into the cupula of the pleura and, in the lateral portions of the superior thoracic aperture, rises into the base of the neck 3 cm above the medial third of the clavicle. It is level with the spine of the 1st thoracic vertebra. The medial surface of the right apex is in contact with the esophagus and the trachea, whereas on the left, branches of the aortic arch (common carotid and left subclavian arteries) separate the lung from the trachea. The coarse tracheal breath sounds are, therefore, transmitted to the right apex but not to the left. The subclavian arteries and veins arch over both lungs in the superior thoracic aperture, making their impressions on the lungs.

The **costal surface** is convex and extends from the vertebrae to the sternum. The **diaphragmatic surface** is concave and it is separated from the costal and medial surfaces by a sharp **inferior border,** which extends into the costodiaphragmatic recesses. On the right side, the resonant percussion note obtained over the lungs is gradually replaced by dullness as the inferior pulmonary border is approached. This is due to the presence of the solid liver under the right dome of the diaphragm. On the left, the inferior border cannot be mapped out by percussion because the fundus of the stomach under the left dome of the diaphragm is full of air and yields a resonant percussion note, like the lung itself. The surface projection of the inferior border of the lung, in relation to the pleural reflection lines, was mentioned in the previous chapter and is illustrated in Figure 20-2.

The sharp **anterior margin** occupies the sternocostal recesses and separates costal and medial surfaces of the lung. On the left, the margin is indented by the heart, creating the *cardiac notch* (Fig. 20-1). The small tonguelike process of the lung that projects below the notch is the *lingula* (Fig. 20-3B). It is part of the left upper lobe. On the **medial surface,** anterior to the hilum of the lung, the *cardiac impression* is created by the right atrium and a larger impression is created by the left ventricle, on the respective sides (Fig. 20-3). Behind the hilum, the right lung is grooved by the esophagus and the left lung is grooved by the descending aorta. Medial and costal surfaces merge with one another posteriorly, and here the lung is related to the vertebral bodies. The arch of the aorta makes an impression on the left lung as it passes above the hilum, and the azygos vein grooves the right lung as it arches over its hilum.

### The Hilum

The hilum is a somewhat wedge-shaped area at which the structures that form the root of the lung enter and leave the organ (Fig. 20-3). Most posterior in the upper part of each hilum is the bronchus; in front of it are the pulmonary artery and, in an even more anterior plane, the superior pulmonary vein. The inferior pulmonary vein is below the bronchus and occupies the space between the leaves of the *pulmonary ligament.* Usually, two bronchi are severed at the right hilum when the entire lung is removed; these are the upper lobe bronchus and the interlobar portion of the main bronchus. In the left hilum, only one bronchus is severed. The pulmonary artery is located above, rather than directly in front of, the left bronchus.

The hilum also transmits the bronchial vessels, pulmonary nerve plexuses, and lymphatics. Several bronchopulmonary lymph nodes are also located in it. The **pulmonary ligament,** the inferiorly redundant part of the pleural sleeve that surrounds the root of the lung, provides the dead space in which the root of the lung may move up and down during respiratory movements as diaphragmatic contractions pull down and release the mediastinum.

### Lobes and Fissures

As the lung grows, the spaces or fissures that separate individual bronchopulmonary buds or segments become obliterated except along two planes, evident in the fully developed

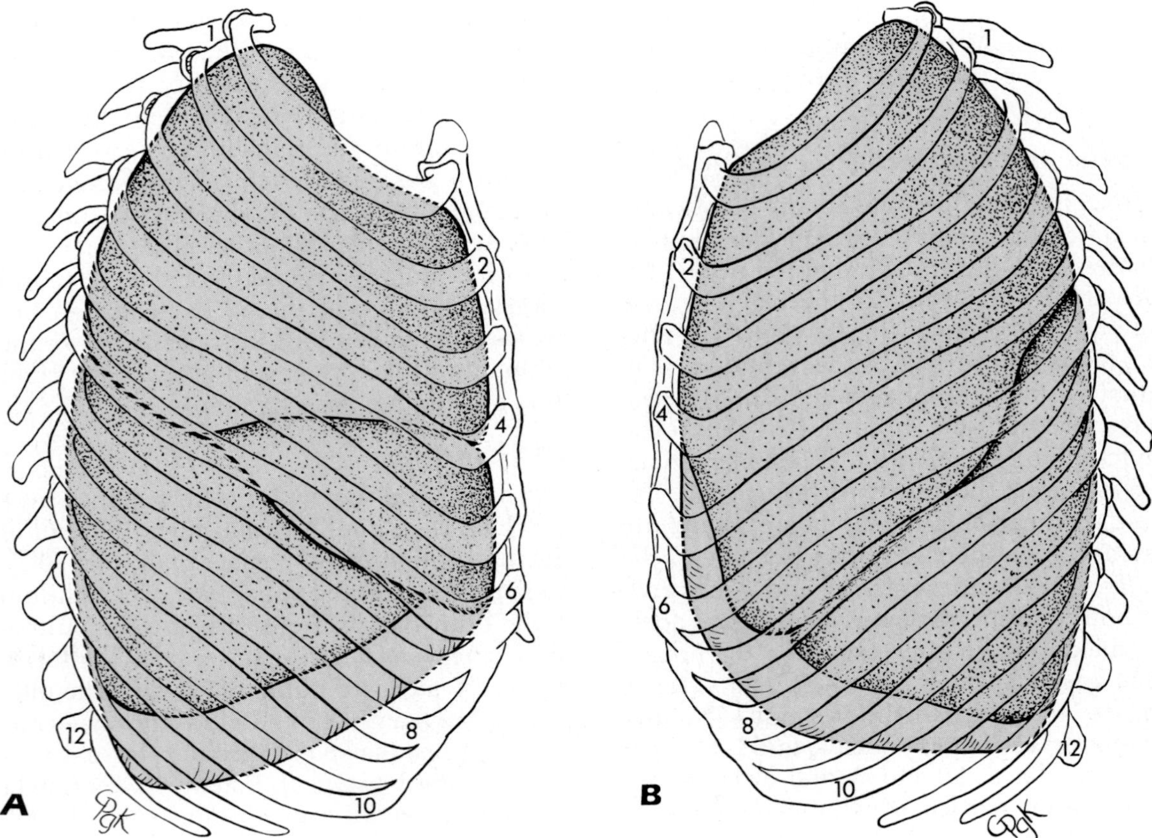


**FIG. 20-2.** The relations of the pleura, the lungs, and the fissures to the thoracic wall. The lungs are stippled, and the pleural sacs are pink. **A,** right lateral view; **B,** left lateral view; **C** *(opposite page),* posterior view. (Redrawn from Brock RC: The Anatomy of the Bronchial Tree: With Special Reference to the Surgery Abscess, 2nd ed. London, Oxford University Press, 1954)

lungs as the oblique and horizontal fissures. Although the right lung has three lobes and the left only two, the bronchopulmonary segments in right and left lungs correspond. Both anatomically and clinically, what is of significance is the underlying pattern of bronchopulmonary segmentation; the division into lobes is inconsequential. Nevertheless, the concept of lobes, and the position of fissures, is useful in locating the bronchopulmonary segments.

The fissures facilitate the movements of the lobes in relation to one another, which accommodates for greater distention and movement of the lower lobes during respiration. The lung may be divided completely by the fissures, and the lobes remain held together only at the hilum by the bronchi and pulmonary vessels. However, more often than not the fissures are incomplete or may be absent altogether, as they are between the majority of bronchopulmonary segments.

Both lungs are sliced across from their costal and mediastinal surfaces by an **oblique fissure.** This fissure cuts the vertebral border of the lung at a variable distance below the apex. However, the level of the 4th or 5th thoracic spine is an adequately accurate guide to it for mapping the fissures on the surface of the chest. The plane of the oblique fissure slopes forward and downward to intersect the diaphragmatic surface just behind the anteroin-

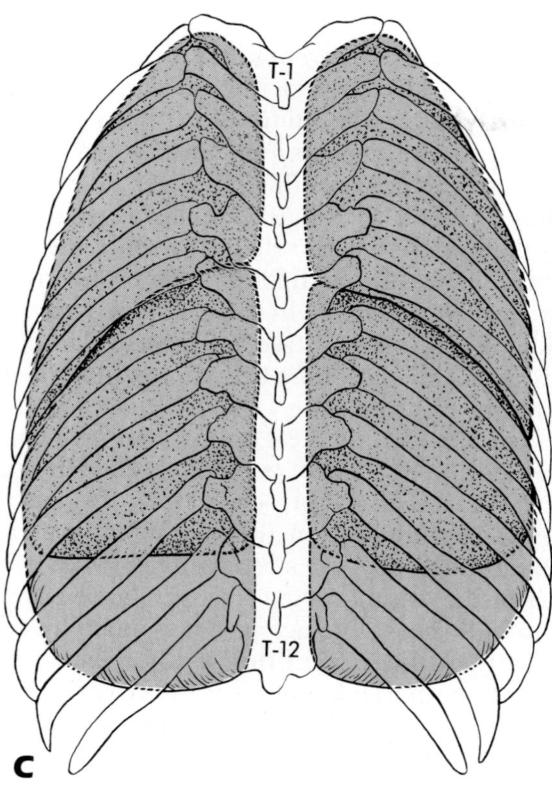


ferior border of the lung. As shown in Figure 20-2, the surface projection of the fissure follows roughly the 6th rib (or the vertebral border of the abducted and protracted scapula) and intersects the inferior border of the lung at the anterior axillary line. On the left, the fissure is somewhat lower and corresponds more closely to the 7th rib in front of the midaxillary line.

In the right lung, a **horizontal fissure** separates two bronchopulmonary segments into the **middle lobe,** the equivalents of which, on the left, remain attached to the upper lobe. The horizontal fissure commences in the oblique fissure as it crosses the midaxillary line and intersects the anterior border of the lung on level with the 4th costal cartilage (Fig. 20-2). The middle lobe is between the horizontal and oblique fissures.

Owing to the forward and downward slope of the oblique fissure, the left and right upper lobes and the middle lobe are anterior, as well as superior, to the lower lobes. Upper and middle lobes project to the anterior chest wall, where they are available for clinical evaluation, whereas none of the inferior lobe is accessible anteriorly. The converse is the case from the posterior view. The pulmonary projections posteriorly are dominated by the lower lobes; little of the upper lobes are available for examination in the back.

Sometimes, especially in the infant, fissures of varying depth can be seen in abnormal locations on the lung, delimiting anomalous lobes. Usually, the anomalous "lobes" correspond to the normal bronchopulmonary segments. The most striking abnormalities of lobation are the occurrence of a "middle lobe" in the left lung (about 8%) and the very rare *azygos lobe* in the right, produced by the azygos vein's cutting into the pleura and the apex of the lung.

## INTERNAL ANATOMY OF THE LUNGS

### THE BRONCHIAL TREE

The budding of the lung diverticulum, which lays down the pattern of the bronchial tree, does not progress in a strictly dichotomous fashion. Therefore, asymmetrical branching and trifurcation of the bronchi normally occur. Knowledge of the arrangement of the first three or four generations of bronchi has anatomic and clinical significance. The bronchi provide the framework of the lung parenchyma, and the major branches of the pulmonary arteries and veins conform to the bronchial tree in a definable manner. Not only resection of parts of the lung but also interpretation of clinical and radiologic findings have to rely on the internal anatomy of the lung.

The **bronchi** are hollow tubes kept patent by incomplete rings or plates of hyaline cartilage. They are lined by respiratory, pseudostratified, ciliated epithelium, which is rich in goblet cells. The cilia beat toward the trachea. The submucosa contains serous and mucus glands, which open into the lumen. A large amount of elastic fiber is present in the submucosa, which permits elongation and retraction of the bronchi with the respiratory movements. At the termination of the bronchi, the elastic tissue fans out into the interalveolar connective tissue septa of the lung. Within the confines of the cartilage plates, the bronchi are encircled by smooth muscle fibers arranged in two helical strands. The muscle is under neural control and is also sensitive to such stimulants in the circulation as histamine, serotonin, and norepinephrine.

## The Trachea and Principal and Lobar Bronchi

The larynx is a highly specialized sphincter located at the branching point of the alimentary and respiratory tracts (see Chap. 33). The **trachea** is the distal continuation of the larynx and it descends through the neck into the thorax, lying anterior to the esophagus. Its relations in the neck and mediastinum are described in Chapters 29 and 22, respectively. The trachea bifurcates into the **right** and **left principal bronchi,** and from each of these primary bronchi originate the secondary or **lobar bronchi.** On the right, the *superior lobar bronchus* branches off from the principal bronchus before the latter enters the hilum. The remaining main stem, sometimes called the *interlobar bronchus,* gives off, more distally, the *middle lobe bronchus,* which runs forward and downward. The main or interlobar bronchus becomes the *bronchus of the inferior lobe.*

The left principal bronchus gives off the *superior lobar bronchus* as soon as it has entered the hilum, and the remaining main stem becomes the bronchus of the inferior lobe. The counterpart of the middle lobe bronchus on the left is the *lingular bronchus,* which forms the lower division of the left upper lobe bronchus. It serves the lingula, which corresponds to the middle lobe on the right. Only the upper division of the superior lobe bronchus is equivalent to the right superior lobar bronchus. The bronchial tree is illustrated in Figure 20-4 and shows the first three or four generations of bronchi.

The division of the trachea into principal bronchi takes place behind the ascending aorta, to the right of, and below, the aortic arch. The bifurcation lies in the cadaver at about the level of the sternal angle (lower border of the 4th thoracic vertebra); however, in x-ray films of living persons, it is much lower, usually at the level of the 7th thoracic vertebra. It is marked internally by a ridge, the *carina,* that separates the openings of the two principal bronchi.

The **right principal bronchus** is wider than the left one and leaves the trachea at an angle of about 25°. It passes downward and laterally behind the superior vena cava and reaches the

**FIG. 20-3.** The medial surface of the right and left lung.

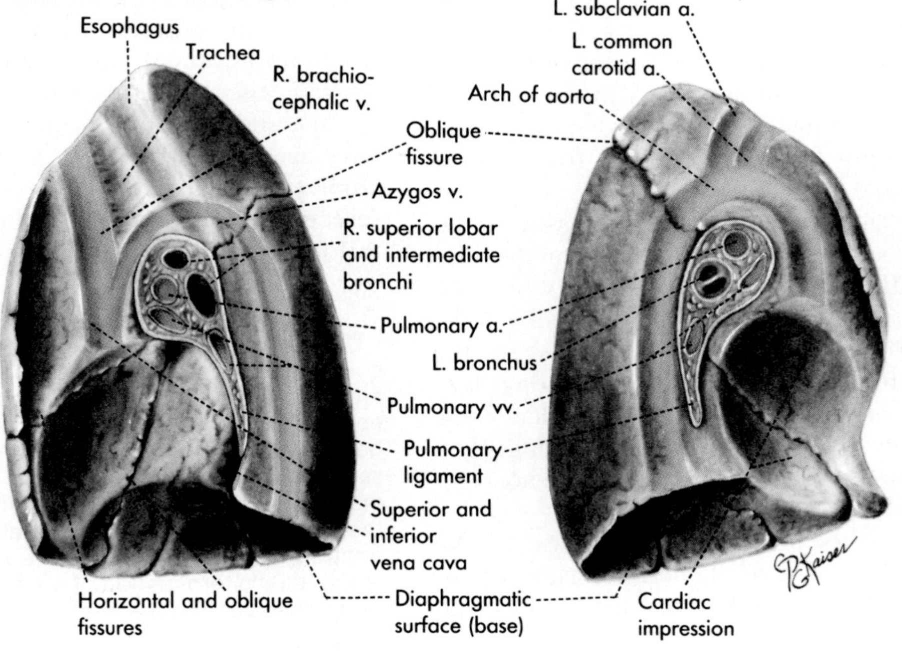

hilum of the lung after a course of only about 2.5 cm. The **left principal bronchus** is slightly smaller in diameter than the right one and almost twice as long (some 4–5 cm). It leaves the trachea at an angle of about 45° and passes below the arch of the aorta and the left pulmonary artery on its way to the hilum of the left lung.

A foreign body inhaled into the trachea is much more likely to lodge in the right bronchus because this is more directly in line with the trachea and because it is of larger caliber than the left main bronchus. The left bronchus, in spite of its greater length, is more difficult to handle surgically than the right one because of its vascular relations. Both bronchi are mobile and pliable, however, and can be easily manipulated and moved by a bronchoscope.

**Bronchoscopy.** The bronchoscope is a telescope with a built-in light source that can be passed into the bronchi through the larynx and the trachea. Bronchoscopy is usually performed under general anesthesia, although it can be done in a conscious individual if the pharyngeal, laryngeal, and tracheal mucosae are anesthetized. It provides direct inspection of the interior of the bronchi. Foreign bodies may be located, grasped by forceps, and removed from the lung through the bronchoscope. The scope is also used for obtaining biopsies of bronchial lesions and bronchial washings in which exfoliated neoplastic cells may be identified.

## Segmental Bronchi

The third generation of bronchi serve wedge-shaped districts of the lung that are defined as the bronchopulmonary segments. These tertiary bronchi are designated, therefore, as the *segmental bronchi.* The basic branching pattern is found on the right and is only slightly modified on the left. There are ten segmental bronchi on each side, and they are designated by names identical to the names of the bronchopulmonary segments they serve. The bronchi and the segments are numbered in definite sequence starting at the apex (Fig. 20-4).

The right superior lobar bronchus divides into three segmental bronchi: the **apical** (1), **posterior** (2) and **anterior** (3). The middle lobe bronchus bifurcates into the **lateral** (4) and **medial** (5) segmental bronchi. The remaining five segmental bronchi are branches of the inferior lobar bronchus. The first of these, the **superior** (6), arises high on the posterior wall of the main stem and serves the apical area of the lower lobe. The other four are called basal bronchi because they distribute to the base of the lung. They are named according to their anatomic positions: **medial basal** (7), **anterior basal** (8), **lateral basal** (9), and **posterior basal** (10).

On the left, the apical and posterior bronchi spring by a common stem, the **apicoposterior bronchus,** from the upper division of the superior lobe bronchus, which also yields the anterior bronchus. The two bronchi derived from the inferior division (lingular bronchus) are called **superior** and **inferior lingular** (4 and 5, respectively), rather than lateral and medial bronchi as on the right. The segmental bronchi in the left lower lobe are identical with those on the right, including the presence of the left medial basal bronchus, which was erroneously omitted from earlier descriptions.

**Variations.** The branching pattern described in the foregoing section is the *ideal* pattern, which serves as reference for variations that are chiefly the concern of pulmonary specialists. No two lungs are exactly alike. Many variations in the pattern of the bronchi have been described. Two segmental bronchi that usually arise separately may share a common stem as the apical and posterior bronchi on the left usually do. Additional bronchi may arise from the main stem and distribute to a segment that already has its regular bronchus. A common variation of this type is a *subsuperior bronchus,* which distributes to a zone between the superior and posterior basal segments. A portion of a segment may receive a subsegmental bronchus from a neighboring segment.

**Bronchoscopy.** It is possible to inspect the orifices of the segmental bronchi by bronchoscopy (discussed in the previous section); some even admit the bronchoscope. The directions in which the orifices of the various bronchi face are, therefore, of importance to the bronchoscopist; so is the sequence in which the orifices are encountered by the advancing bronchoscope.

**Bronchograms.** Although the bronchi are normally not visible on a plane film, they can be demonstrated radiographically by coating their mucosa with a radiopaque substance. The segmental and subsegmental bronchi can be identified on such a film (Fig. 20-5).

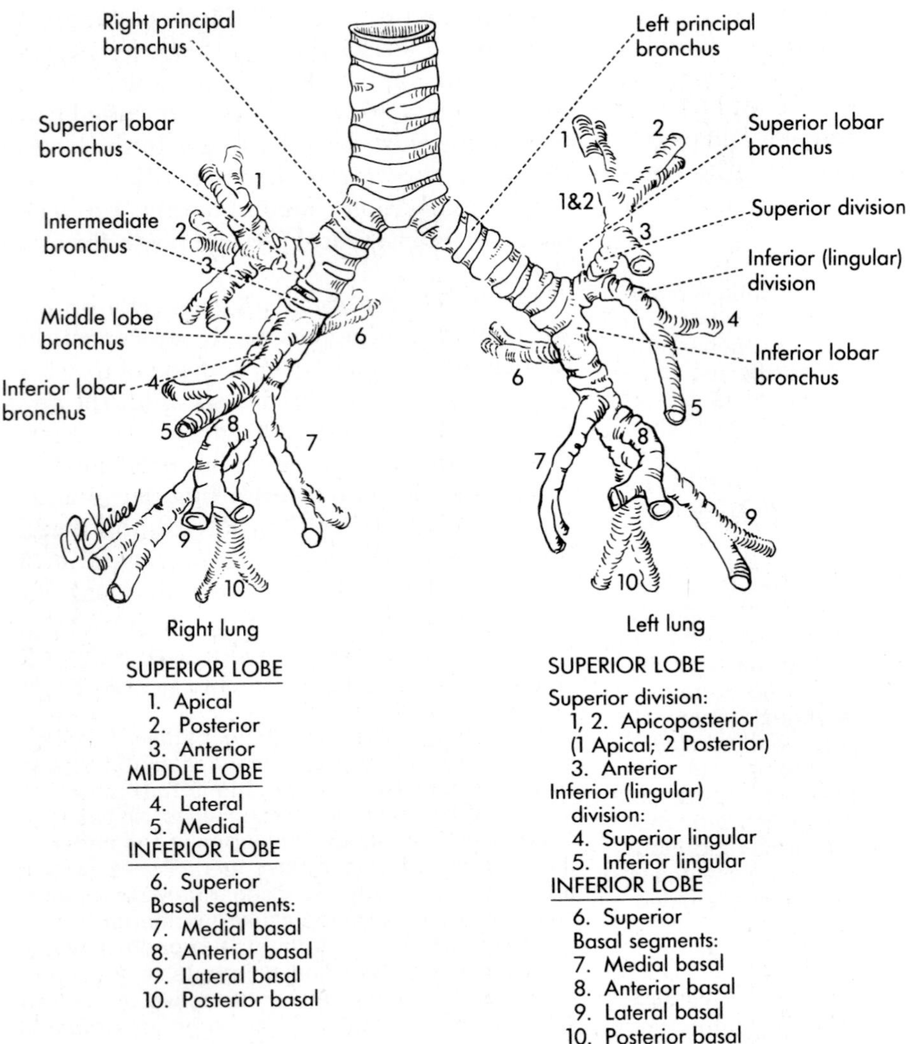


**FIG. 20-4.** The tracheobronchial tree with the segmental bronchi identified.

### The Respiratory Portion of the Bronchial Tree

Beyond the segmental bronchi, approximately ten more generations of bronchi are produced, the most distal of which are approximately 1 mm in diameter. All these tubes retain the basic structure described earlier in this section. The next two to three generations of airways are the **bronchioles,** which lack cartilage but are otherwise similar to the smallest bronchi (Fig. 20-6). The most distal of the bronchioles, the **terminal bronchiole,** forms the stem of the respiratory unit of the lung, known as the **acinus,** or lobule. As the name implies, the acinus is reminiscent of a bunch of berries, representing the alveoli. Each acinus contains over 3000 alveoli, and there are nearly 100,000 acini in the lung.

Within the acinus, each terminal bronchiole branches into several **respiratory bronchioles.** These are the most proximal respiratory passages through which gas exchange can take place. Their muscular walls are studded with a few alveoli (Fig. 20-6). The respiratory bronchioles give rise to the **alveolar ducts,** which lead into the **alveolar sacs.** The walls of the ducts and sacs are devoid of muscle and are densely crowded with alveoli. Alveolar sacs intricately interlock with their neighbors, including those of neighboring acini, and entrap between them the pulmonary capillary bed. In these most peripheral respiratory passages, the walls consist almost exclusively of the attenuated squamous epithelium of the alveolar wall, supported on a basement membrane. The sparse connective tissue in the interalveolar septa contains elastic fibers and supports the capillaries on the surface of the alveoli.

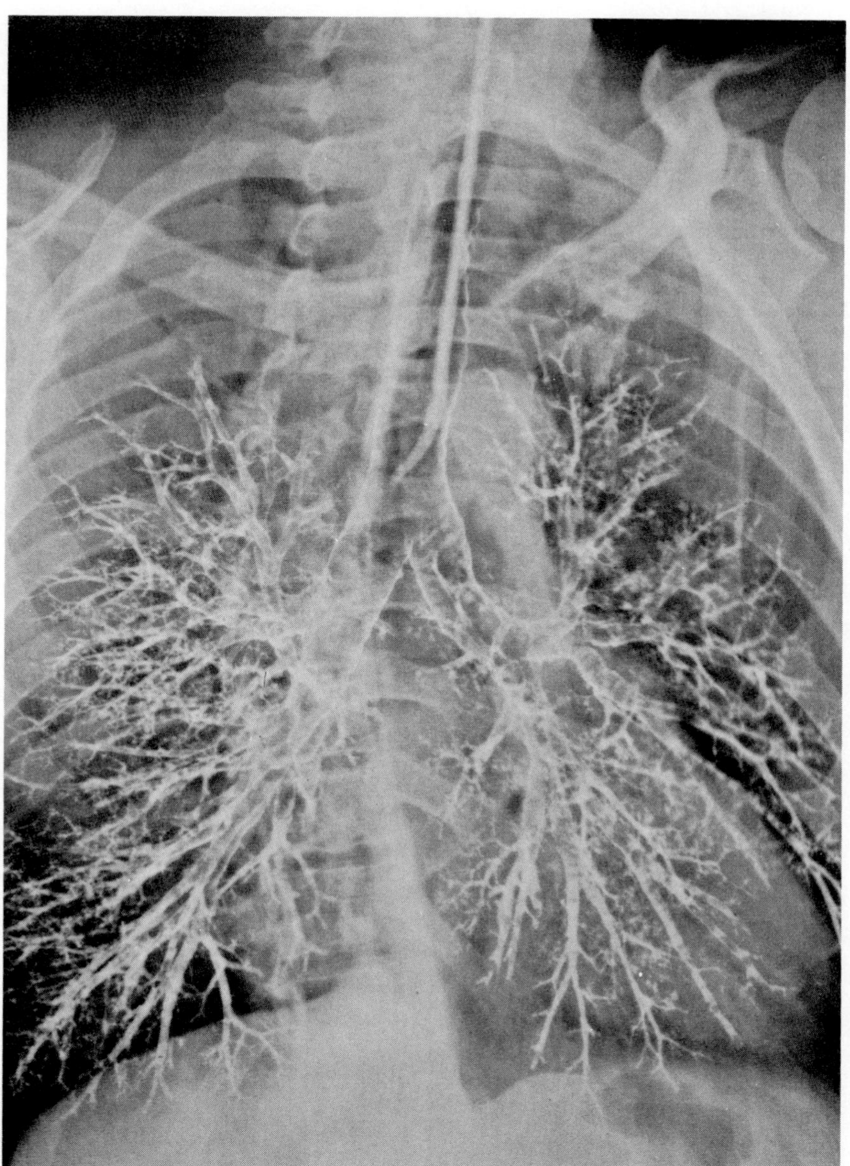

**FIG. 20-5.** A bronchogram in which both principal bronchi and most of the segmental bronchi are visualized. (Courtesy of the late Dr. E. Allen Boyden)

## BLOOD VESSELS AND LYMPHATICS OF THE LUNGS

The lung has two circulations, the pulmonary and the bronchial. The latter is part of the systemic circulation and carries blood for the nutrition of the bronchi and the connective tissue of the lung. The bronchial arteries are insignificant in size by comparison with the pulmonary arteries, which convey as much blood for gas exchange to the lungs per unit time as the aorta delivers to the entire body. Four pulmonary veins return the same volume of oxygenated blood to the left atrium; the small bronchial veins drain into the azygos and hemiazygos systems.

The pulmonary arteries end ultimately in dense capillary networks around the alveolar sacs, and the confluence of the venules that arise from these capillaries forms pulmonary veins. The capillary bed fed by the bronchial arteries is in the walls of the bronchi and in

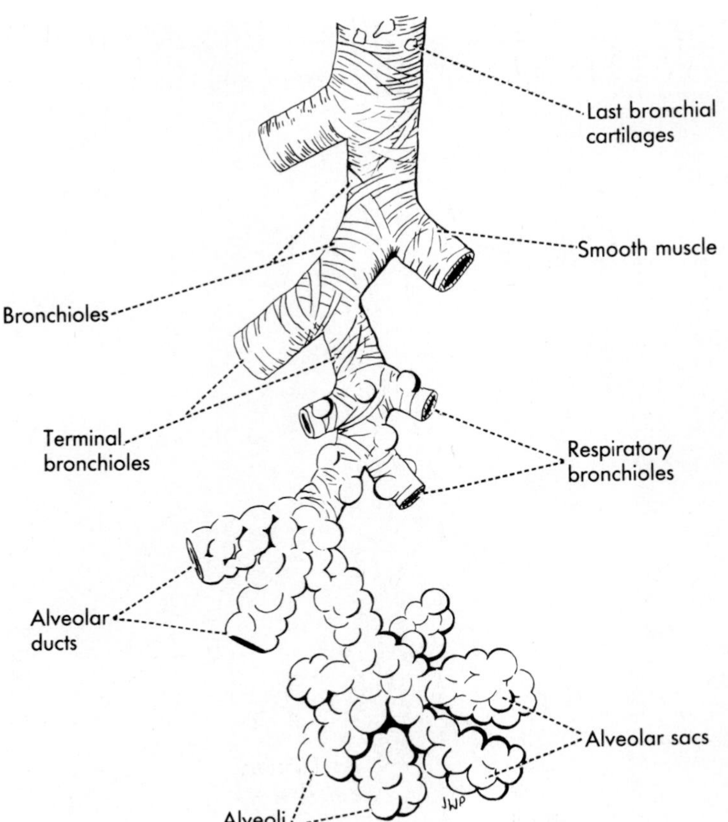


**FIG. 20-6.** The respiratory portion of the bronchial tree. (Courtesy of the late Dr. E. Allen Boyden)

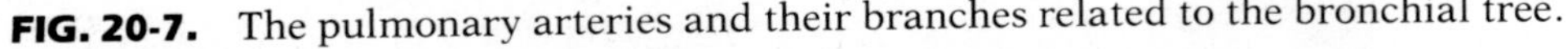

**FIG. 20-7.** The pulmonary arteries and their branches related to the bronchial tree.

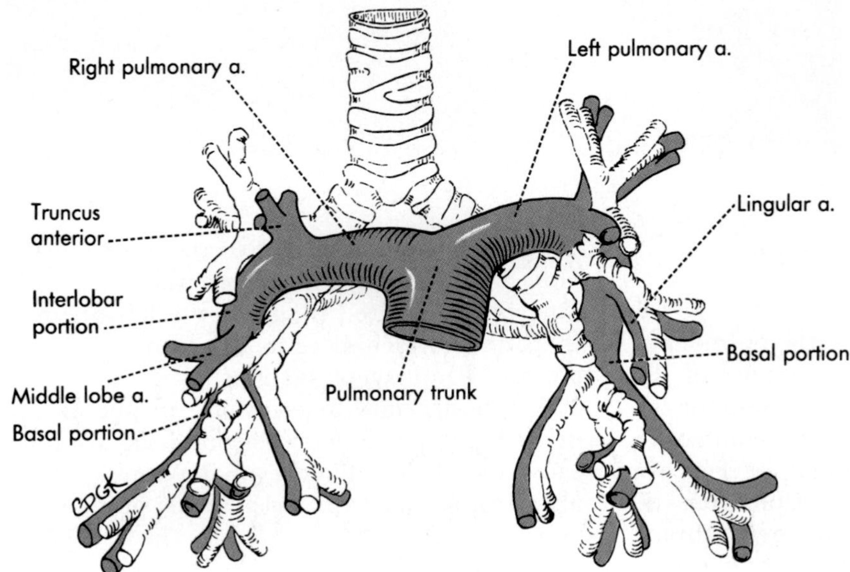

the connective tissue of the lung. Anastomoses do exist between the bronchial and pulmonary systems at the capillary level and also between some larger, precapillary branches of the bronchial and pulmonary arteries.

Since the aortic pressure is normally much higher than the pulmonary arterial pressure, the bronchial arteries deliver to the pulmonary capillaries a small amount of blood that is already oxygenated. These connections become important when there is chronic interference with the pulmonary arterial circulation. In such instances, the bronchial arteries, and their connections to the pulmonary arteries, may become very much enlarged and, in certain cases, may account for much of the arterial circulation of the lung.

### The Pulmonary Arteries

The common stem of the right and left pulmonary arteries is the **pulmonary trunk** (Fig. 20-7). The trunk arises from the right ventricle and has its entire course within the pericardial sac. Its anatomy is described with that of the heart in the next chapter.

The bifurcation of the pulmonary trunk takes place on the left of the ascending aorta in the concavity of the aortic arch. The **left pulmonary artery** immediately leaves the pericardial sac. Just outside the pericardial sac, it is connected to the aortic arch by the *ligamentum arteriosum,* the fibrous remains of the *ductus arteriosus,* which in the fetus served as a shunt between the two vessels. The **right pulmonary artery** is longer than the left, and much of it is covered by serous pericardium. It runs horizontally behind the ascending aorta and superior vena cava before it leaves the pericardial sac in the concavity of the aortic arch. The aorta emerges from the sac posterior to the superior vena cava.

Soon after leaving the pericardial sac, both right and left pulmonary arteries arch over the principal bronchi as they enter the hilum of the lung. Thereafter, both arteries descend, lying deep in the interlobar fissures lateral to the bronchi. The left pulmonary artery crosses the left principal bronchus and, at the hilum, is superior to it, whereas the right pulmonary artery crosses the interlobar portion of the right bronchus, having given off a major branch to the upper lobe (*truncus anterior*) before it enters the hilum. The branches of the pulmonary artery closely follow the segmental bronchi and receive names and numbers to

**FIG. 20-8.** The pulmonary veins and their branches related to the pulmonary arteries and the bronchial tree.

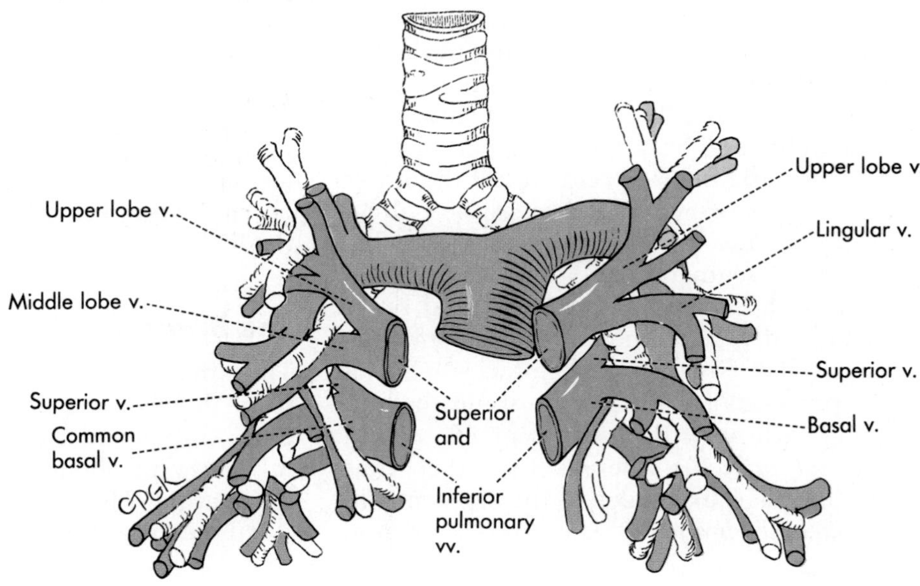

correspond with the bronchi (Fig. 20-7). These segmental arteries can be approached and dissected in the depth of the interlobar fissures. Peripheral branches of the segmental pulmonary arteries are not strictly confined to their own bronchopulmonary segments and tend to enter neighboring segments.

The above description emphasizes the basic branching pattern of the pulmonary artery. There are numerous deviations from this pattern; however, these are the concern of the thoracic surgeon. On the right, the upper lobe receives two arteries in addition to the three segmental branches of the upper lobe artery (*truncus anterior*). These additional arteries ascend to the posterior and anterior segments from the interlobar portion of the artery. The left pulmonary artery, as a rule, has no truncus anterior. The left upper lobe receives as many as four to seven independent arteries, given off from the main trunk mostly in the interlobar fissure. This renders resection of the left upper lobe, or parts of it, particularly difficult.

Anomalous pulmonary arteries sometimes arise from the aorta or its major branches at the base of the neck. The most common origin for such a vessel is from the descending aorta in the thorax; however, they may arise even in the abdomen. The abnormal arteries from the descending aorta enter the lungs through the pulmonary ligament. They may supply an otherwise normal segment of lung; often the tissue they supply is abnormal and eventually must be removed by operation.

## The Pulmonary Veins

A superior and an inferior pulmonary vein passes from the hilum of each lung to the left atrium (Fig. 20-8). The superior pulmonary veins collect blood from the right upper and middle lobes and from the left upper lobe. On both sides, the inferior veins drain the lower lobes. The primary tributaries of the pulmonary veins are related to given bronchopulmonary segments and are named and numbered according to the segmental bronchi. Typically, a pulmonary vein from each segment has two major tributaries. One of these passes along the bronchus and the pulmonary artery within the segment (*intrasegmental branch*), whereas the other runs along the inferior border of the segment (*infrasegmental branch*). The infrasegmental branches drain blood from neighboring segments, except when the intersegmental planes face into the fissures or onto the diaphragmatic surface of the lung. Here, and also in the substance of the lung, infrasegmental veins help to demarcate the boundaries of the segments and serve as guides in the dissection and resection of the segments.

The pattern of union of the segmental veins is such that, on the right, the veins of the three upper lobe segments form a large vein, the *upper lobar vein,* which receives the *middle lobe vein,* and the two form the right superior pulmonary vein. The same is the case on the left, the *lingular vein* substituting for the middle lobe vein. On both sides, the inferior pulmonary veins are formed by the confluence of two large veins, the *superior vein* and the *common basal vein.* The latter collects all the veins of the basal bronchopulmonary segments.

The left pulmonary veins empty closer together than do the right ones and, in about 25% of cases, unite to open together instead of separately into the left atrium. The two right pulmonary veins usually terminate some distance apart and rarely fuse before entering the atrium. A more striking variation is the presence of three veins, one from each lobe.

The pattern of pulmonary veins shows even more variation than that of the arteries. Truly *anomalous pulmonary veins* sometimes occur. For example, one or more pulmonary veins instead of emptying into the left atrium will join the superior vena cava or its tributaries. Otherwise, they may empty directly or indirectly into the *right* atrium. Anomalous pulmonary veins may join the azygos system or esophageal veins as the venous plexuses of the lung buds connect to veins that surround the primitive gut. The drainage of pulmonary veins into systemic veins overloads the right side of the heart. Also, since oxygenated blood must be delivered to the left side of the heart if it is to be circulated to the body as a whole, the condition is incompatible with life unless the anomalous return is only partial or an intracardiac defect allows blood to be shunted from the right to the left side.

## The Bronchial Arteries and Veins

The small bronchial arteries usually arise from the thoracic portion of the descending aorta, either directly or from a right intercostal artery at about the level of the tracheal bifurcation. They are rather easily broken in the dissecting laboratory when the lungs are mobilized for study. There may be only one bronchial artery to each lung, although multiple vessels are common. The bronchial arteries also arise fairly frequently from the arch of the aorta; sometimes one arises from a subclavian artery in the base of the neck. The bronchial arteries usually give off branches to the esophagus and then follow the bronchi into the lung, branching and rebranching with these as they supply the bronchi themselves and the adjacent connective tissue. The bronchial arteries send their branches along the interalveolar connective tissue septa to the pulmonary pleura.

The small **bronchial veins** unite along the bronchi to form a single vein that leaves the hilum and empties on the right side into the azygos vein and on the left into the hemiazygos system. Many of the bronchial veins end within the lungs in the tributaries of the pulmonary veins. The bronchial veins therefore return to the systemic veins somewhat less blood than the bronchial arteries deliver to the lung.

### Lymphatics

Plexuses of lymphatic capillaries pervade the visceral pleura, the submucosa and wall of the bronchi, the interalveolar septa and other connective tissue planes of the lungs, and the walls of the larger blood vessels. Lymph flow participates in clearing exudate from the alveoli and the pleural cavity. Inhaled minute particulate matter, such as dust and carbon particles, are also conveyed by lymphatics and filtered out by the lymph nodes. The black appearance of smokers' lungs is due to carbon particles deposited in pulmonary lymphatics and lymph nodes. Lymphatics are the primary route of spread for bronchogenic carcinoma. Lymph nodes are always involved in tuberculosis of the lung. The calcification of the nodes induced by tuberculosis will be a permanent radiographic reminder of the disease after the primary infection has been resolved.

Lymph from all the lymphatic plexuses of the lung is drained toward the hilum by lymphatic vessels that follow the bronchi. The lymph flow is interrupted by numerous lymph nodes, many of which are situated at the forking points of the bronchi. The nodes may become sufficiently enlarged to compress the bronchi and produce collapse of segments or even lobes (atelectasis; middle lobe syndrome). Enlarged hilar lymph nodes produce a characteristic shadow on x-ray film.

The most peripheral nodes embedded in the substance of the lung are the **pulmonary nodes** (Fig. 20-9). These send their efferent lymphatics to the **bronchopulmonary nodes** situated at the hilum. Several groups of **tracheobronchial nodes,** located around the bifurcation of the trachea, receive the lymph from the hila and also from several mediastinal structures, including the heart. The largest of these nodes, the **paratracheal nodes,** lie more proximally along each side of the trachea. The efferents of the paratracheal nodes unite with the parasternal lymphatics to form the *bronchomediastinal lymph trunks.* These are the chief drainage vessels of the thoracic viscera, and they empty independently on each side of the neck in the angle of junction between the subclavian and internal jugular veins. The left trunk may join the thoracic duct, and the right trunk, the right lymph duct.

Lymphatics from each lung drain chiefly to tracheobronchial and paratracheal nodes on the homolateral side. Some, but not all, of the lymphatics from the left lower lobe drain to the right side of the trachea.

When visceral pleura adheres to parietal pleura owing to inflammatory or neoplastic processes, lymph from the lungs may drain with the lymphatics of the parietal pleura to the axillary, parasternal, or intercostal nodes, thus explaining the presence of carbon particles or tumor metastases in these nodes (see Chap. 19).

## NERVE SUPPLY OF THE LUNG

Each lung is supplied by visceral efferent and visceral afferent nerve fibers through the **pulmonary plexuses.** The pulmonary plexuses enter the hilum of the lung on the surface of the

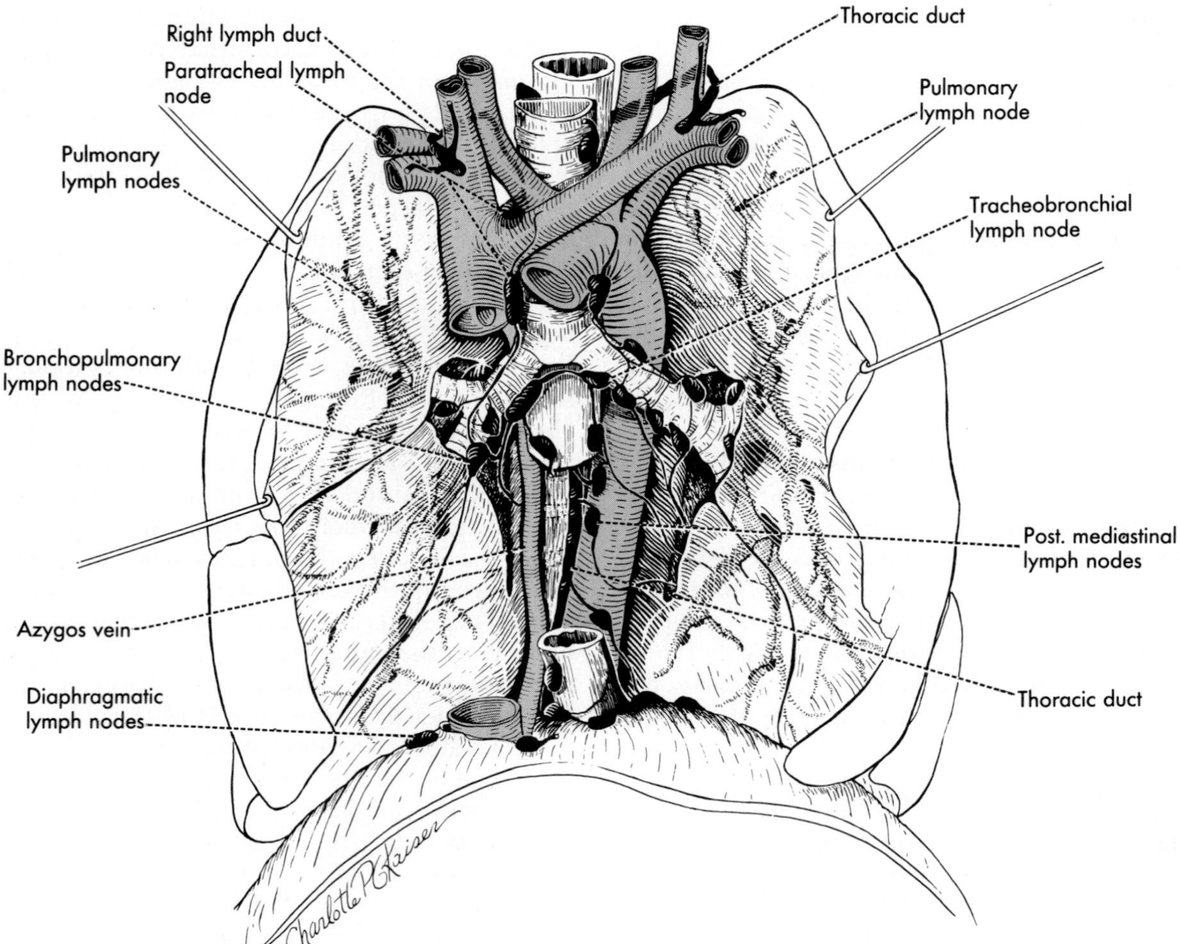


**FIG. 20-9.** Lymph nodes and lymphatics of the lungs.

bronchi and the pulmonary arteries as extensions of the *cardiac plexus,* which lies in front of the tracheal bifurcation. **Visceral efferents** are contributed to the pulmonary plexuses by the **vagus** and **thoracic sympathetic ganglia.** The sympathetic fibers in the plexus are postganglionic and the vagal fibers are preganglionic. The latter relay in small parasympathetic ganglia that are located in the peribronchial plexuses, extensions of the pulmonary plexus. **Visceral afferents** from the lung have been demonstrated only in the vagus, and their cell bodies are in the superior and inferior vagal ganglia.

The **vagus** is motor to bronchial and bronchiolar smooth muscle and produces bronchoconstriction. The vagus apparently does not innervate smooth muscle in the wall of the pulmonary vessels. **Sympathetic visceral efferents** are motor to the bronchial glands and also produce vasoconstriction of pulmonary blood vessels. It is doubtful whether such fibers terminate on bronchiolar muscles, although it is well established that sympathomimetic drugs produce bronchodilatation.

Afferents from the lung are concerned with innervation of the bronchial mucosa, sensing stretch in the alveoli, the interalveolar septa, and the pleura; monitoring pressure in the pulmonary veins; and mediating pain sensation. As stated earlier, all these impulses, including pain, travel in the vagus. They are concerned with the afferent limb of such reflexes as the cough reflex and the stretch reflex, which regulates respiration.

In summary, the anatomic evidence supported by studies on the human bronchial tree indicates that the vagus furnishes the great majority, if not all, of the afferent fibers to the lung and that it is also the constrictor of the bronchi. The sympathetic, on the other hand, is the vasoconstrictor and also furnishes secretomotor fibers to the glands of the bronchial tree.

## ANATOMIC RELATIONS IN THE ROOT OF THE LUNGS

All the structures that constitute the root of the lung have been dealt with individually in previous sections of this chapter. The relationship of these structures to one another matters because they provide principal clues to the dissection of the internal anatomy of the lungs, and they are the landmarks during the resection of the lobes and segments (Figs. 20-10 and 20-11).

The roots of the lungs are posterior to the upper part of the pericardial sac. The trachea bifurcates behind and slightly above the sac, and the principal bronchi are anchored to the heart by major blood vessels that arch over the bronchi in the obtuse angles made by the trachea and each principal bronchus (Fig. 20-10). The lower end of the trachea is displaced slightly to the right of the midline by the arch of the aorta, which occupies the angle between the trachea and the left bronchus. Before the left bronchus enters the lung, the left pulmonary artery ascends over its anterior surface and crosses it just distal to the arch of the aorta. Thus, two major blood vessels, the

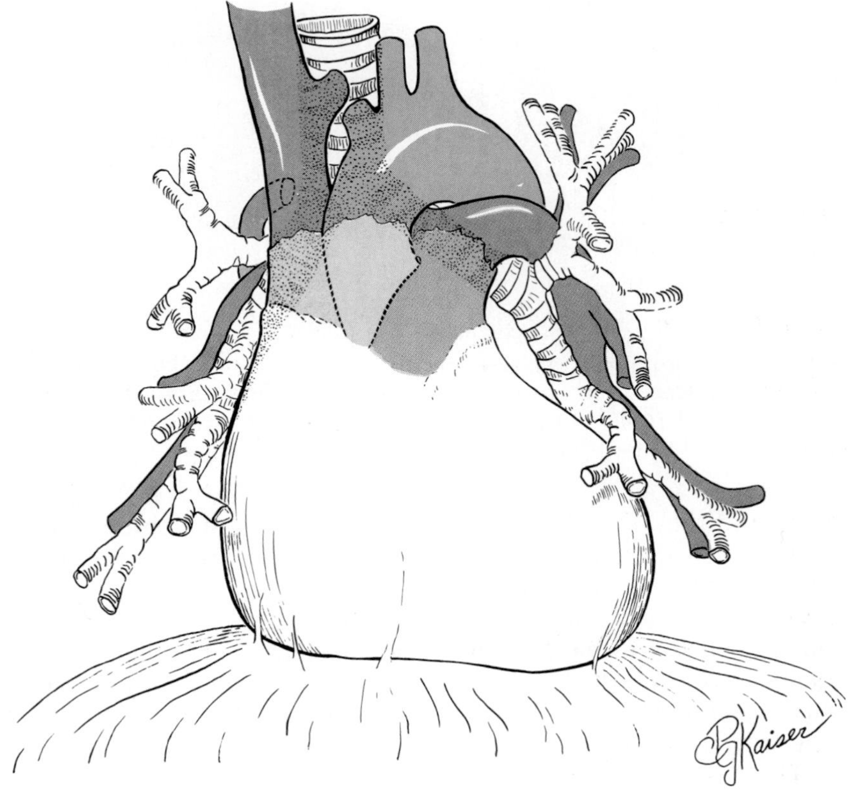

**FIG. 20-10.** The relationship of the great arteries and great veins to the principal bronchi.

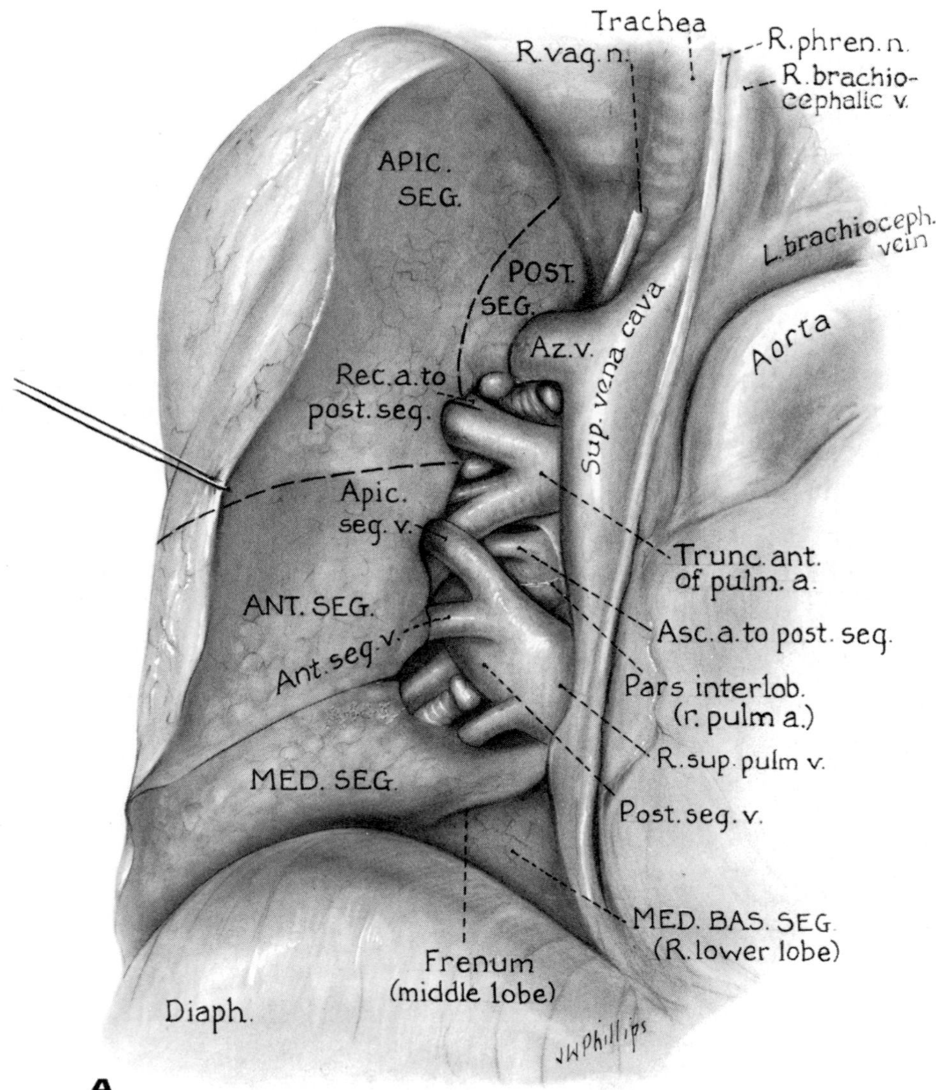


**FIG. 20-11.** Anatomic relations in the root of the right **(A)** and left **(B)** (opposite page) lungs seen in an anterior approach to the hila. (Boyden EA: The segmental anatomy of the lungs. In Myers JA (ed): Diseases of the Chest Including the Heart. Springfield, IL, Charles C Thomas, 1959)

aorta and the left pulmonary artery, come to rest on the superior surface of the left principal bronchus. On the right, only the azygos vein occupies the angle between the trachea and the bronchus because the right pulmonary artery crosses the interlobar portion of the bronchus distal to the origin of the superior lobe bronchus. The azygos vein terminates in the superior vena cava before the latter enters the pericardial sac anterior to the root of the lung.

Immediately anterior to the principal bronchi are the pulmonary arteries (Fig. 20-3). The left one ascends across its bronchus, whereas the lower right pulmonary artery passes slightly below its bronchus before it bends across it in the hilum.

The superior pulmonary vein is the most

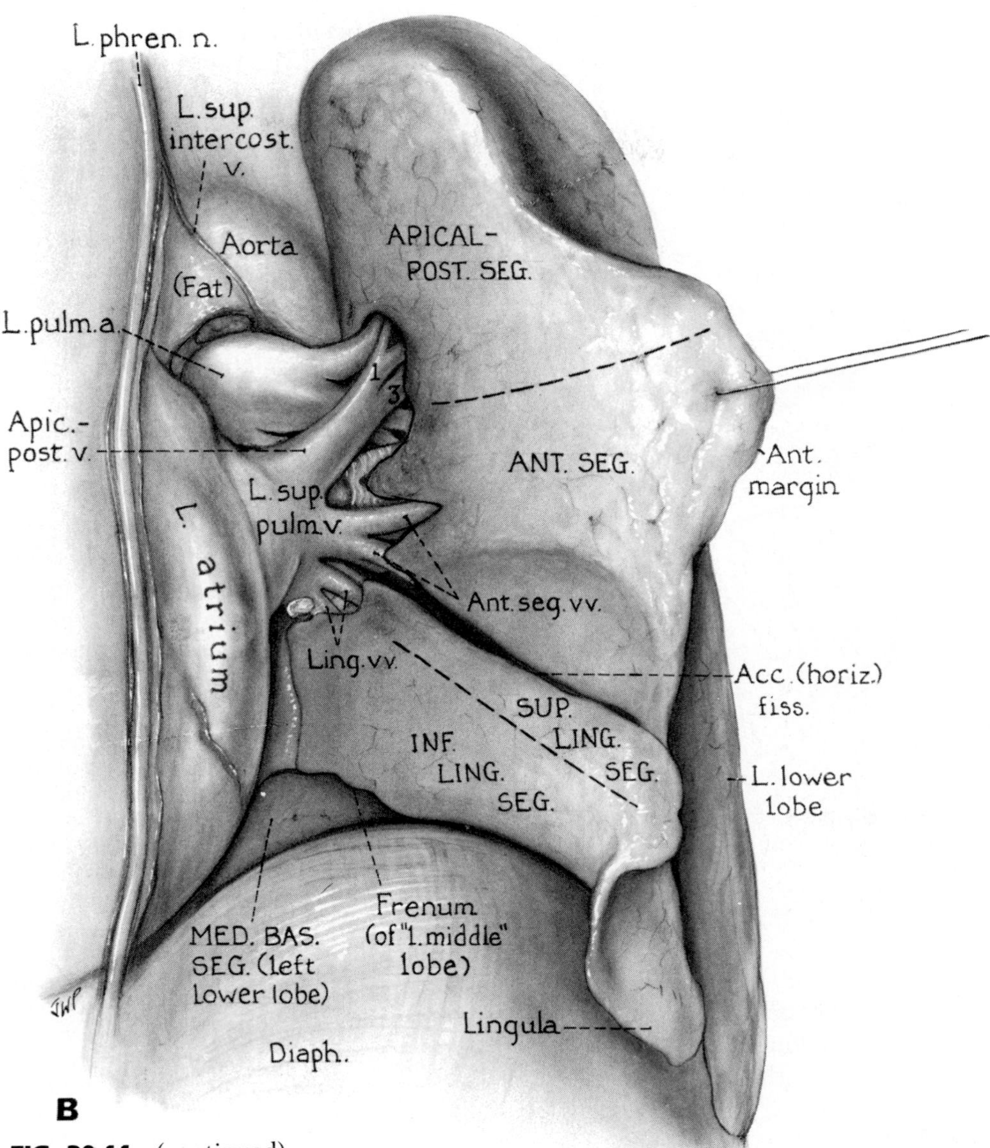


**FIG. 20-11** (continued)

anterior structure, and the inferior pulmonary vein the most inferior structure, in the root of the lung (Fig. 20-11). Thus, on both sides, the principal order of structures in an anteroposterior direction is vein, artery, bronchus. This order is reversed, of course, if the hilum is approached from the back. The inferior pulmonary veins lie below the bronchus on both sides, and below them is the pulmonary ligament into which they can expand. The inferior veins become visible in an anterior approach to the hilum only if, in addition to the anterior margins of the lungs, the middle lobe or the lingula is retracted. On the right side, the superior pulmonary vein passes behind the superior vena cava and the inferior pulmonary vein passes behind the right atrium before they pierce the pericardial sac over the left atrium. They terminate in this chamber, having practically no intrapericardial course. The left veins are shorter and are related to the lateral part of the left atrium anteriorly before they enter it.

The posterior relations of the lung roots dif-

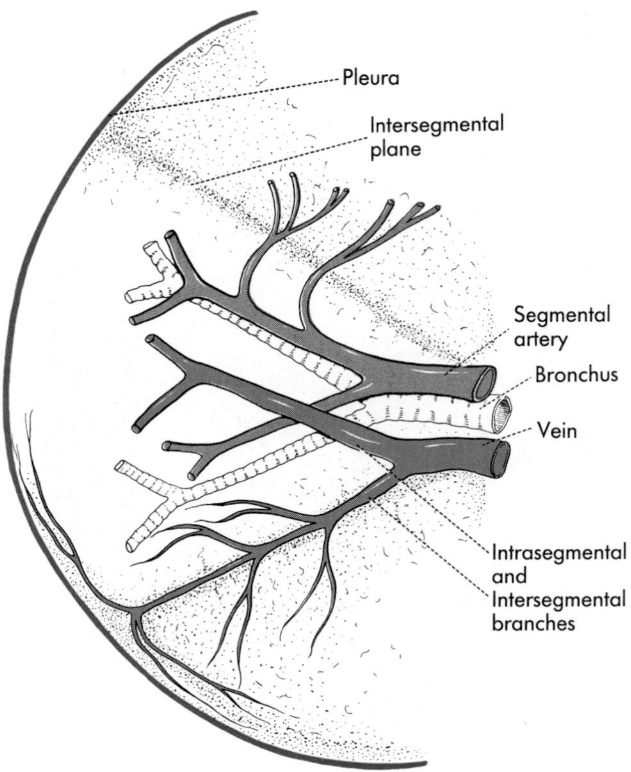


**FIG. 20-12.** A schematic representation of a bronchopulmonary segment, the bronchovascular unit of the lung.

fer on the two sides. The tracheal bifurcation and the left bronchus rest on the esophagus. More laterally, the left bronchus crosses the descending aorta, and so does the left inferior pulmonary vein at a lower level. The right inferior pulmonary vein crosses the esophagus.

On the posterior surface of the bronchi are located the delicate bronchial vessels. The vagi pass behind the hila, where they contribute numerous twigs to the bronchi and the pulmonary plexuses. Right and left vagus nerves commingle with each other behind the tracheal bifurcation, forming the *esophageal plexus.* The phrenic nerves run down on the pericardial sac and are located anterior to the hila (Fig. 20-11). The right phrenic nerve is close to the lung root as it passes from the surface of the superior vena cava onto the pericardium. The left nerve is some distance from the lung root, being pushed forward by the left ventricle.

## THE BRONCHOPULMONARY SEGMENTS AND THEIR CLINICAL IMPORTANCE

A bronchopulmonary segment is a pyramidal-shaped unit of the lung aerated by a segmental bronchus (Fig. 20-12). Each segment is served principally by a segmental branch of the pulmonary artery and a segmental pulmonary vein. Some branches of a segmental pulmonary artery do, however, enter neighboring segments, and one of the branches of a segmental pulmonary vein runs in connective tissue between neighboring segments. Thus, the bronchopulmonary segments can be conceived of as the anatomic bronchovascular units of the lungs.

If suitably prepared dyes of contrasting color are injected into the orifices of the segmental bronchi so that the most peripheral branches of the bronchial tree become filled with the dye all the way to the visceral pleura, the bronchopulmonary segments will be sharply demarcated on the surface of the lung without any mixing of colors at the boundaries of the segments. Also, if a segmental bronchus, artery, and vein are cut and clamped close to the hilum, and if traction is applied to this triad of structures, the wedge-shaped segment may be stripped away from its neighbors by blunt dissection with a gauze sponge in the intersegmental planes, and no blood vessels or bronchi of any size need be severed.

There are ten bronchopulmonary segments in each lung, and their segmental bronchi have already been described and named in the section on the bronchial tree (Fig. 20-4). The bronchopulmonary segments, as they appear on the surfaces of the lungs, are shown in Figure 20-13.

The right upper lobe consists of the **apical, posterior,** and **anterior segments,** identifiable on the costal and medial surfaces of the lung. The middle lobe has two segments, **lateral** and **medial.** Both are present on the costal surface; only the medial projects onto the me-

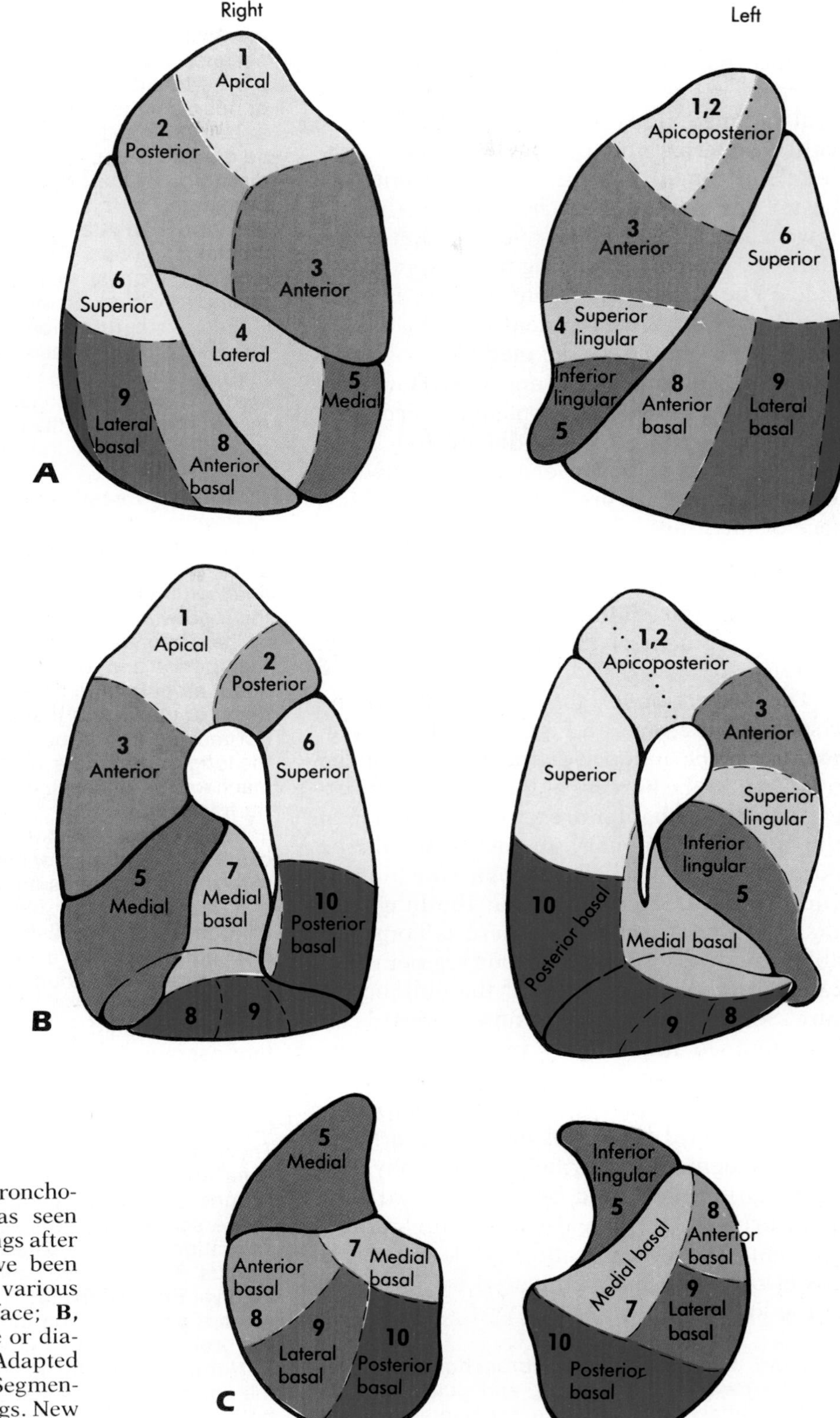


**FIG. 20-13.** The bronchopulmonary segments as seen on the surface of the lungs after segmental bronchi have been injected with dyes of various colors. **A,** lateral surface; **B,** medial surface; **C,** base or diaphragmatic surface. (Adapted from Boyden EA: The Segmental Anatomy of the Lungs. New York, McGraw-Hill, 1955)

diastinal surface. On the left, all five segments are incorporated into the superior lobe. However, the segments corresponding to the lateral and medial segments of the right lung are called **superior** and **inferior lingular** on the left. Both lingular segments project onto the costal and medial surfaces. Each of the two lower lobes consists of five segments, and these receive corresponding names on the two sides. The apex of each inferior lobe is occupied by the **superior segment,** and the base of the lobes is shared by the **medial basal, anterior basal, lateral basal,** and **posterior basal segments.** All the basal segments project onto the costal surface except the medial basal. The medial, the posterior, and the superior basal segments are present on the medial surface of the lungs.

The segments that rest on the diaphragm include all the basal segments plus the medial segment of the middle lobe and the inferior lingular segment of the left upper lobe (Fig. 20-13C).

The projection of these segments to the chest wall is shown in Figure 20-14. The segments can be mapped if the position of the fissures and the borders of the lungs described earlier in this chapter are known. This map of the bronchopulmonary segments should be borne in mind when air entry into the lungs is clinically evaluated and when the lungs are examined by x-ray films. There is considerable superimposition of various segments because of the sloping plane of the oblique fissure. Some of the segments superimposed can be examined individually in the axilla. The same principle applies to radiologic examination of the segments (Fig. 20-15). Posteroanterior and lateral exposures are required to locate a segment accurately. In a physical examination, it should be remembered that anteriorly the lung fields are dominated by segments of the upper and middle lobes and posteriorly they are dominated by segments of the lower lobes.

Some diseases, such as bronchopneumonia, lung abscess, bronchiectasis, and pulmonary infarction, affect the lung in a segmental pattern. Identification of the diseased segment is important because *postural drainage* may need to be employed to aid removal of infected exudate, mucus, or pus from the involved segments. The patient has to be positioned in such a way that the segment to be drained is uppermost so that gravity can aid the discharge of the bronchial contents into the main bronchus. For this, it is necessary to know the position of the segment and the direction of the segmental bronchus. Rhythmic and forceful percussion of the chest wall over the diseased segment facilitates dislodgement of bronchial contents.

Segmental resection may be indicated for lung abscess, bronchiectasis, and for benign, and some malignant neoplasms, if they are diagnosed sufficiently early. After the removal of one or more segments, or of a lobe, the remaining lung will expand to fill the entire pleural cavity.

#### Physical Examination of the Lungs

Physical examination of the lung essentially evaluates their functional anatomy. The physical signs employed rely on the fact that the lung is filled with air and that the air moves in and out of the lung with each respiratory cycle.

Inspection of the movements of respiration over the thorax and the abdomen gives information about air entry into the lungs. Asymmetrical movement, or the lack of movement, suggests that air is not moving in and out of a substantial portion of the lung or that there is some interference with the mechanisms of respiratory movements (paralysis, rib fracture).

The normal chest sounds hollow when percussed. Loss of a resonant percussion note over a segment or segments indicates that the portion of the lung is collapsed, filled with fluid, or solidified. Fluid in the pleural cavity also yields a dull percussion note, whereas a large amount of air in the pleural cavity will sound hyperresonant when percussed.

The movement of air in and out of the respiratory passages generates the *breath sounds,* best heard with a stethoscope. Over the lung fields the sounds are fine and are heard only during inspiration when the alveolar sacs are being distended. These are the *vesicular breath sounds.* In the trachea and the larger bronchi, much more coarse sounds are generated by the movement of air. These sounds are audible during inspiration and expiration and are known as *bronchial breath sounds.* They can be heard over the trachea, but not over the lung fields in a normal chest. However, when a portion of the lung solidifies, it transmits the bronchial breath sounds to the chest wall.

Normally, the sliding of the visceral pleura on the parietal pleura is inaudible. However, if one or both of the pleural surfaces become roughened by

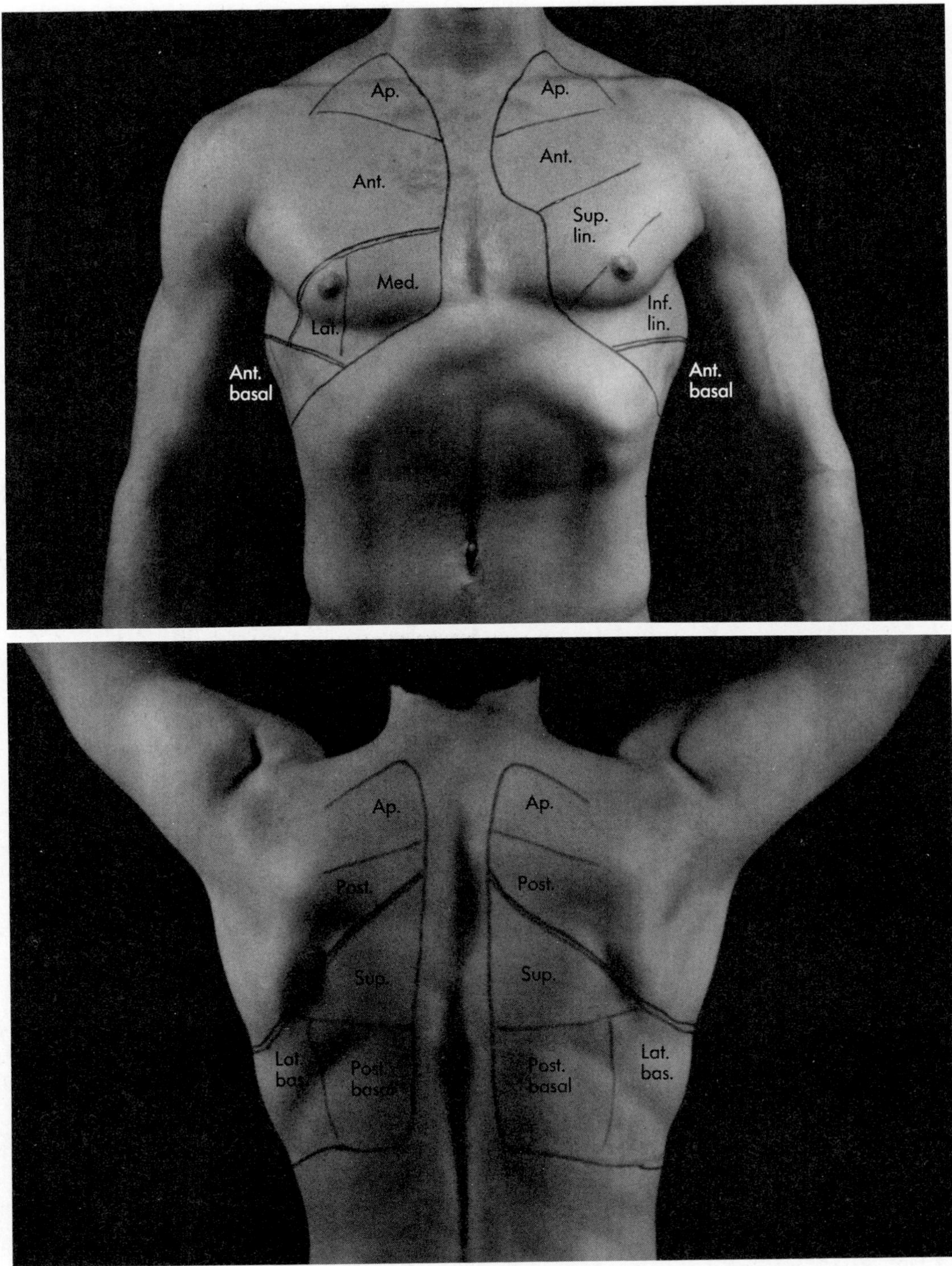


**FIG. 20-14.** Projection of the bronchopulmonary segments to the surface of the chest.

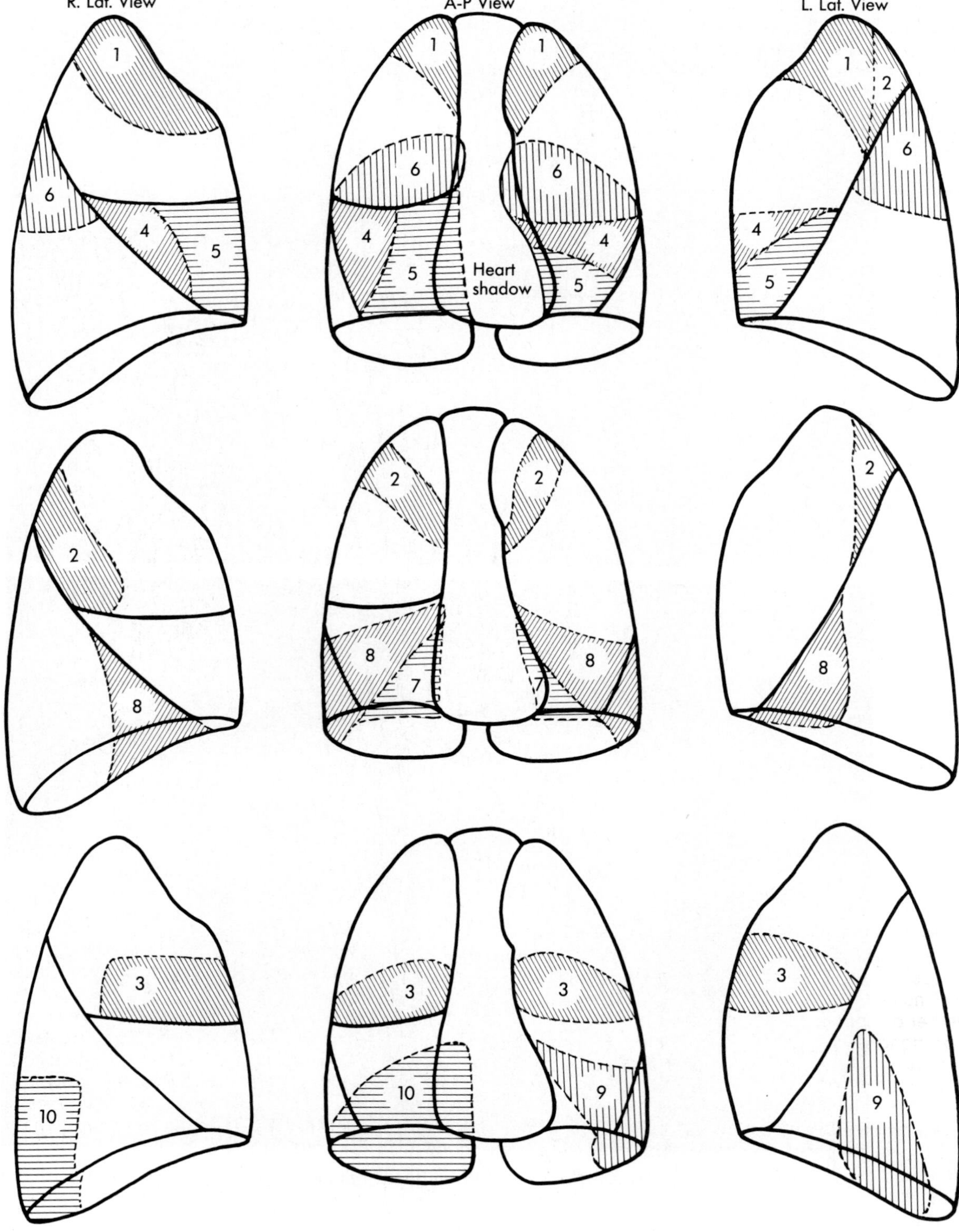

R. Lat. View
A-P View
L. Lat. View
1
1
1
1
2
6
6
6
6
4
5
4
5
Heart
shadow
4
5
4
5
2
2
2
2
8
8
7
7
8
8
3
3
3
3
10
10
9
9

**FIG. 20-15.** Diagrams illustrating the shadows of the lung segments on x-ray film. Consolidation or disease confined to individual segments create characteristic shadows that can be identified by combined lateral and anteroposterior views. The **upper row:** shadows of the apical (1), lateral (4), medial (5), and superior (6) segments; **middle row:** posterior (2), medial basal (7), and anterior basal (8) segments; **lower row:** anterior (3), lateral basal (9), and posterior basal (10) segments. (Adapted from Boyden EA: The segmental anatomy of the lungs. In Myers JA (ed): Diseases of the Chest Including the Heart. Springfield, IL: Charles C Thomas, 1959)

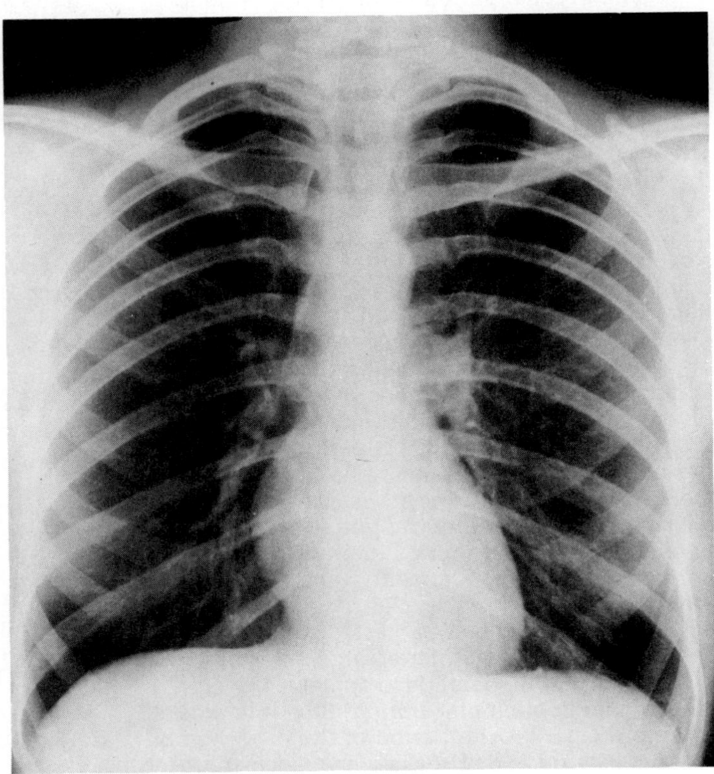

**FIG. 20-16.** An anteroposterior view of the thorax. The indistinct shadows in the lungs are caused by the pulmonary vessels. (Courtesy of Dr. D. G. Pugh)

inflammation, a *pleural rub,* which is present during both inspiration and expiration, will be heard.

Added to an understanding of pulmonary anatomy, these physical signs permit the diagnosis of whether or not air is moving normally in and out of all bronchopulmonary segments, and if some segments are not aerated, the basic cause of the condition can be deduced.

### Radiologic Examination

The lung parenchyma absorbs so little of the x-rays that pass through it that, for practical purposes, the lungs appear as transparent to x-rays as air. On a plane chest x-ray film, the mediastinal structures create a definite shadow that contrasts with the bilateral, radiolucent lung fields (Fig. 20-16). The bronchi are invisible, but the major branches of the pulmonary arteries and veins create linear shadows radiating from the hilum. Occasionally, the fissures can be identified because the small amount of pleural fluid trapped between the lobes absorbs more x-rays than the lung itself. When air is absorbed from a segment or when segments become filled with fluid or solidify, they create a shadow on the film (Fig. 20-15). The same is true of cysts, neoplasms, and enlarged lymph nodes.

## DEVELOPMENT OF THE LUNGS

In the preceding section of this chapter, repeated reference was made to the development of the lungs because the anatomy of this organ, as of any other, is most readily understood through its development. The development of the lung is divided into five periods: the embryonic, pseudoglandular, canalicular, terminal sac, and neonatal periods. The establishment of the internal gross anatomy of the lung takes place during the first two of these periods; the subsequent three phases are concerned chiefly with the differentiation and elaboration of the terminal respiratory passages for increasingly efficient gas exchange.

The **embryonic period** is the period of budding and takes place during the fourth to seventh weeks of gestation. The primordium of the lung arises as a ventral diverticulum of the foregut at the lower end of the pharynx (see Fig. 22-9). The diverticulum soon divides into right and left lung buds, which are essentially two endodermal sacs. The stalk of the lung buds (the trachea) elongates and separates from the esophagus as lobar buds appear on the two lung sacs. These, in turn, give rise by division to the bronchi of the future bronchopulmonary segments. The buds of the bronchopulmonary segments bulge on the surface of the pleura, giving it a mulberry appearance. By the end of the seventh week, growth has proceeded to several generations of subsegmental bronchi. The embryonic period ends with the closure of the pleuroperitoneal canals. During this period, branches of the pulmonary arteries and veins become associated with the bronchi. These vessels arise from the splanchnic mesoderm entrapped by the bronchial buds. Centrally, the arteries hook up with the 6th aortic arch, and the pulmonary veins, with the left atrium. All anatomic anomalies of the lung become established during this earliest of developmental stages. The bronchial arteries grow into the lungs during the next stage of development.

The **pseudoglandular period** (8th–16th weeks) is the principal growth period of the bronchi, and the requisite number of divisions in each segment are established all the way to the future respiratory bronchioles and alveolar ducts. The prolific growth by epithelial budding gives the lung a glandular appearance from which the period receives its name. The next phase is the differentiation period of the future respiratory epithelium and is called **canalicular** because the cuboidal epithelium in the region of the future alveolar sacs becomes invaded or *"canalized"* by pulmonary capillaries. Over the capillary loops, the epithelium becomes stretched and acquires the squamous character of the alveolar lining. Toward the end of this period, which lasts from the 17th to the 26th week, some of the alveolar cells begin to secrete *surfactant,* which enables the fetus, if prematurely born, to breath. During the next phase, called the **terminal sac period,** which lasts up to birth, the regions of the alveolar ducts grow and expand, forming thin-walled sacs that as yet lack true alveoli. The full-term fetus is born without alveoli. Alveolar development gets under way during the **neonatal period** and continues up to the age of seven to eight years. The peripheral respiratory zone grows by elongation, and the respiratory surface is increased by the appearance of alveoli within the expanding acini. The number of alveoli increases during the first eight years from 24 million terminal saccules to 280 million alveoli (296 million in the adult), and the air tissue interface increases from 2.8 $m^2$ to 32 $m^2$ during the same period. It is 75 $m^2$ in the adult.

## RECOMMENDED READINGS


AUER J: The development of the human pulmonary vein and its major variations. Anat Rec 101:581, 1948

BLOOMER WE, LIEBOW AA, HALES BR: Surgical Anatomy of the Bronchovascular Segments. Springfield, IL, Charles C Thomas, 1960

BOYDEN EA: Segmental Anatomy of the Lungs: A Study of the Patterns of the Segmental Bronchi and Related Pulmonary Vessels. New York, McGraw-Hill, 1955

BOYDEN EA: The segmental anatomy of the lungs. In Myers JA (ed): Diseases of the Chest Including the Heart. Springfield, IL, Charles C Thomas, 1959

BOYDEN EA: The pulmonary artery and its branches. In Luisada AA (ed): Cardiology. A loose-leaf encyclopedia sponsored by The American College of Cardiology, vol 1, p 157. New York, McGraw-Hill, 1959

BOYDEN EA: The terminal air sacs and their blood supply in a 37-day infant lung. Am J Anat 116:413, 1965

BOYDEN EA: The developing bronchial arteries in a fetus of the twelfth week. Am J Anat 129:357, 1970

BOYDEN EA: The structure of the pulmonary acinus of a child of six years and eight months. Am J Anat 132:275, 1971

BOYDEN EA: Development of the human lung. In Brennemann's Practice of Pediatrics, vol 4, chap 64. Hagerstown, Harper & Row, 1975

COLE FH, ALLEY FH, JONES RS: Aberrant systemic arteries to the lower lung. Surg Gynecol Obstet 93:589, 1951

DWINNELL FL Jr: Studies on the nerve endings of the visceral pleura. Am J Anat 118:217, 1966

FISHMAN AP, HECHT HH (eds): The Pulmonary Circulation and Interstitial Space. Chicago, University of Chicago Press, 1969

FRAZER RG, PARÉ JAP: Structure and Function of the Lung. Philadelphia, WB Saunders, 1971

HEINEMANN HO, FISHMAN AP: Nonrespiratory functions of mammalian lung. Physiol Rev 49:1, 1969

JACKSON CL, HUBER JF: Correlated applied anatomy of the bronchial tree and lungs with a system of nomenclature. Dis Chest 9:319, 1943

LARSELL O, DOW RS: The innervation of the human lung. Am J Anat 52:125, 1933

LODGE T: Anatomy of blood vessels of the human lung as applied to chest radiology. Br J Radiol 19:1, 1946

MORTON DR, KLASSEN KP, CURTIS GM: The clinical physiology of the human bronchi: II. The effect of vagus section upon pain of tracheobronchial origin. Surgery 30:800, 1951

NAGAISHI C, with the collaboration of NAGASAWA N, OKADA Y, YAMASHITA M, INABA N: Functional Anatomy and Histology of the Lung. Baltimore, University Park Press, 1972

PUMP KK: Distribution of bronchial arteries in the human lung. Chest 62:447, 1972

TOBIN CE: The bronchial arteries and their connections with other vessels in the human lung. Surg Gynecol Obstet 95:741, 1952

# 21

# THE PERICARDIUM, THE HEART, AND THE GREAT VESSELS

Basic features of the heart, and the circulation through it, have been outlined in Chapter 5. After the eighth week of embryonic development, the human heart consists of four chambers: the right and left atria and the right and left ventricles. The right and left sides of the heart become completely partitioned off from one another at birth, and on each side, blood can only flow from the atrium to the respective ventricle. The superior and inferior venae cavae return venous blood from the systemic circulation into the right atrium, and this blood is then propelled by the right ventricle through the pulmonary trunk to the lungs. Oxygenated blood from the lungs is returned by the pulmonary veins to the left atrium and is then ejected into the systemic circulation by the left ventricle through the aorta. The heart has its own blood supply: the coronary arteries and the cardiac veins. Cardiac nerves modify the rate and strength of the heart beat and also conduct visceral afferent impulses.

The heart is surrounded by its own serous cavity, the *pericardial cavity,* which is a potential space between visceral and parietal layers of the *serous pericardium.* The serous pericardial sac itself is surrounded by the *fibrous pericardium,* a substantial, dense connective tissue membrane (see Figs. 19-13 and 19-17). In addition to the heart, the pericardium encloses the roots of the major systemic and pulmonary blood vessels. The pericardium and its contents occupy the middle mediastinum. An understanding of the anatomic arrangement of the pericardium and the structures enclosed by it is aided by a preliminary consideration of development.

Sections of this chapter relating to congenital anomalies of the heart and fetal circulation were written in collaboration with **Dr. Lore Tenckhoff,** Cardiologist and Director of Cardiac Ultrasound, Children's Orthopedic Hospital and Medical Center, Seattle, Washington, and Clinical Professor of Pediatrics and Radiology, The University of Washington, Seattle, Washington.

### Developmental Considerations

**The Primitive Heart Tube.** The heart develops as two parallel, thin-walled endothelial tubes, which, with the establishment of the head fold of the embryo, come to lie ventral to the foregut and dorsal to the future pericardial cavity. The pericardial cavity at this stage of development forms the ventromedian portion of the intraembryonic celom (see Fig. 19-12A). Side-to-side fusion of the two heart tubes proceeds as they sink into the pericardial cavity, acquiring a lamina of celomic mesothelial covering, the *epicardium,* or *visceral pericardium,* on their outer surface (see Fig. 19-12B). Differential growth defines five segments of the heart tube, which, from a rostral to caudal direction, receive the following names: (1) **truncus arteriosus,** (2) **bulbus cordis,** (3) **primitive ventricle,** (4) **primitive atrium,** (5) **sinus venosus** (Fig. 21-1A and B). At the rostral end, the **aortic sac** appears as a dilatation on the truncus, and six pairs of **aortic arches** that arise from this sac skirt around the foregut and link the truncus to the bilateral *dorsal aortae* (Fig. 21-1A and B). At the caudal end, the sinus venosus receives three pairs of veins, which deliver blood to the heart from three sources: the two *umbilical veins* from the placenta, two *vitelline veins* from the yolk sac and the gut, and two *common cardinal veins* that drain blood from the body of the embryo (Figs. 21-1A and 21-22). By the completion of development, the truncus arteriosus will have given rise to the pulmonary trunk and the ascending aorta, the bulbus cordis will have given rise to the major part of the anatomic right ventricle, the primitive ventricle will have given rise to the major part of the anatomic left ventricle, and the right half of the primitive atrium and the sinus venosus will have become incorporated into the anatomic right atrium, whereas the anatomic left atrium will be formed from the left side of the primitive atrium and from the primitive common pulmonary vein that enters the embryonic atrium directly rather than through the sinus venosus.

**The Pericardium.** Soon after the heart tube sinks into the celom, the pericardial cavity becomes partitioned off from the future pleural cavity (pleuropericardial canal) by the development of the *pleuropericardial membranes* (see Fig. 19-12D). It was explained in Chapter 19, in connection with the development of divisions in the thoracic cavity, how this membrane was lifted off and dissected away from the inner mesodermal lamina of the body wall. The pleuropericardial membrane differentiates into the **fibrous pericardium** and separates the ventromedian pericardial cavity of the embryo from the dorsolateral pleural cavities. The parietal **serous pericardium** (in essence, the celomic mesothelium) lines the deep aspect of the fibrous pericardium, forms the walls of the pericardial cavity, and is continuous with the epicardium (visceral serous pericardium). An understanding of this continuity will help to explain the gross anatomy of the pericardial cavity and the existence of its sinuses.

After the dorsal mesocardium disappears (see Figs. 19-12C and 21-1C), the heart tube is suspended in the pericardial cavity only at its arterial and venous ends. The epicardium is continuous with the parietal serous pericardium around the truncus arteriosus as the truncus leaves the cavity rostrally and around the sinus venosus as this enters the cavity from the substance of the septum transversum (Fig. 21-1D). Thus, there is a single (serous) pericardial sleeve around each of the arterial and venous ends of the heart, just as there is a pleural sleeve around the root of the lung.

The heart tube grows at a faster pace than the pericardial cavity, and, as a consequence, the heart buckles, or loops, creating on the dorsal aspect a sharp infolding between the bulbus and the ventricle (Fig. 21-1B). This looping of the heart tube ap-

**FIG. 21-1.** A highly schematic representation of the primitive heart tube and the pericardial cavity. In **A,** the heart tube is seen from its ventral aspect and in **B,** from its left lateral aspect. The left and right horns of the sinus venosus (SV), with their tributaries, are embedded in the septum transversum (shaded area), and the aortic arches, given off by the truncus arteriosus (TA), skirt around the foregut to form the paired dorsal aortae (only two of the six arches are shown). In **C,** the heart tube is seen suspended in the primitive pericardial cavity. A large window has been cut in the wall of the cavity that is formed by the parietal layer of the serous pericardium. The serous pericardium is shown as a pink membrane that covers the heart tube and is continuous with the parietal serous pericardium through the *mesocardium* and through reflections at the arterial and venous ends of the heart. The mesocardium is in the process of being broken down, thus creating the transverse sinus (TS) in the pericardial cavity. Caudally, the pericardial cavity continues into the *pericardiopleural* canal, which is being pinched off from the pericardial cavity by the common cardinal veins (see also Fig. 19-12). In **D,** the heart is shown in a saggital section at the stage of development when the bulbus (B) has fused with the ventricle (V) and the atria (A) have been drawn out of the septum transversum. The continuity of the epicardium with the parietal serous pericardium can be traced now only around the truncus arteriosus and the sinus venosus.

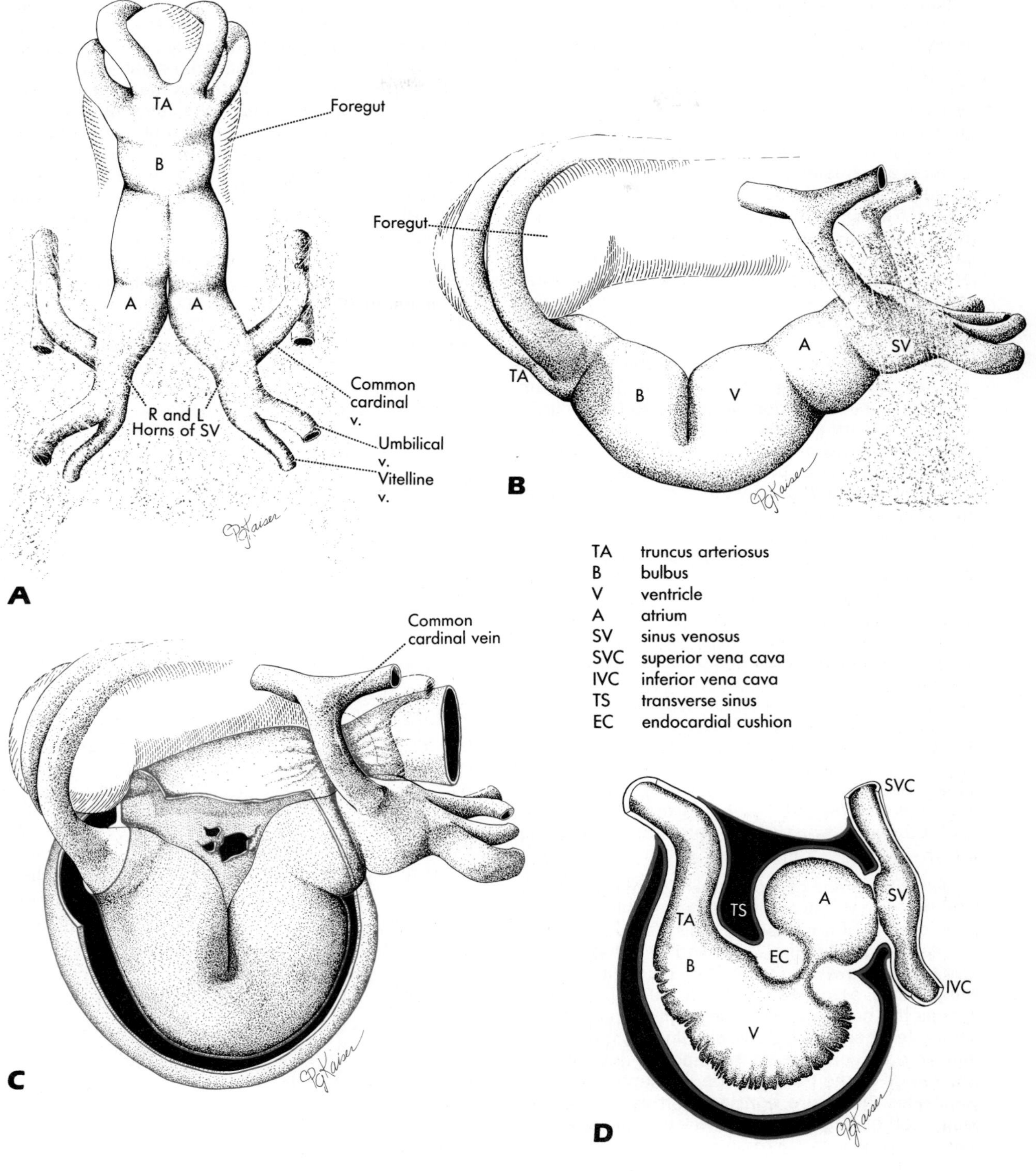


TA truncus arteriosus
B bulbus
V ventricle
A atrium
SV sinus venosus
SVC superior vena cava
IVC inferior vena cava
TS transverse sinus
EC endocardial cushion

proximates its arterial end to its venous end, an event accentuated by the emergence of the atria from the septum transversum into the pericardial cavity (Fig. 21-1D). Only a narrow channel of communication remains between the right and left portions of the pericardial cavity across the dorsal aspect of the heart where originally the mesocardium was located. This channel, limited anteriorly by the truncus and posteriorly by atria, is the future *transverse sinus* of the pericardial cavity (Fig. 21-1D).

The truncus arteriosus will become divided into two major arteries: the ascending aorta and the pulmonary trunk. However, the pericardium around them will not split, and the two great arteries remain enclosed in a single pericardial sleeve. The same is true at the venous end, although the developmental changes that occur within that sleeve are more complex. Most of the sinus venosus becomes incorporated into the right atrium; therefore, its tributaries will come to be enclosed by the pericardial sleeve. The three pairs of veins (umbilical, vitelline, and common cardinal) that drained into the primitive sinus venosus become radically modified, and by the completion of development, only two large veins, derived from the original six, enter the pericardial cavity (Fig. 21-1D). These two are the superior and inferior venae cavae. In addition, four pulmonary veins, not represented among the primitive tributaries of the sinus venosus, gain access to the left atrium through the pericardial sleeve. Thus, the six veins that enter the fully developed atria are enclosed by this common sleeve around the venous end of the heart, but these veins are not analogous with the six primitive tributaries of the sinus venosus. With the growth of the atria, the orifices of these veins are drawn away from one another, and the venous pericardial sleeve becomes distorted in a ſ shape. The *cul-de-sac* between the limbs of this inverted ſ is known as the *oblique sinus* of the pericardial cavity (Fig. 21-3).

**Establishment of Definitive Cardiac Anatomy.** The basic anatomy of the heart is established early. The looping of the heart tube displaces the bulbus not only ventrally (Fig. 21-1B and C) but also to the right (not shown in Fig. 21-1 because it is a side view). This so-called dextro loop (or "d" loop) places the bulbus to the right side of the primitive ventricle where the bulbus will develop into the definitive right ventricle. At this stage of development the flange of the bulboventricular septum partially separates the future right and left ventricles from one another (Fig. 21-1C); however, this separation is temporary because the bulboventricular septum soon becomes absorbed (Fig. 21-1D), which creates a single ventricle that will have to be divided by the development of the definitive interventricular septum.

The atria enlarge; the right atrium incorporates much of the sinus venosus, and the left takes up the primitive, common pulmonary veins. Two external features result from the atrial enlargements: (1) a deep groove develops between the atria and ventricle, known as the *coronary sulcus;* (2) the most ventral extensions of the atria, the right and left *auricles,* expand to embrace anteriorly the developing ascending aorta and pulmonary trunk. The single atrioventricular canal will be divided into right and left channels by the fusion of a dorsal and a ventral *endocardial cushion* (Fig. 21-1D). These cushions contribute to the formation of the atrioventricular valves. The development of the interatrial and interventricular septa completes the division of the heart into four chambers. These septa are marked on the exterior by shallow sulci, but the development of the septa is best discussed at the end of this chapter when embryonic equivalents of the gross structures can be defined.

## THE PERICARDIAL SAC

The pericardium is a fibrous sac that surrounds the heart and the root of the great arteries and veins as they leave or enter the heart. The outer lamina of the sac, composed of dense connective tissue, is the *fibrous pericardium.* To the inner aspect of the fibrous pericardium is closely bound the *parietal lamina of the serous pericardium,* which lines the fibrous sac and reflects onto the surface of the heart around the roots of the great vessels. Beyond these reflections, the serous membrane is known as the *visceral lamina of the pericardium* or the *epicardium*. Between visceral and parietal laminae of the serous pericardium is the *pericardial cavity,* which is completely closed and contains nothing but sufficient fluid to moisten the opposing surfaces of the serous pericardial sac. The function of the serous sac is to lubricate the moving surfaces of the heart, whereas the outer fibrous sac retains the heart in position within the thoracic cavity and limits its distention (Fig. 21-2).

### The Fibrous Pericardium

Like a cone-shaped bag, the fibrous pericardium rests with its base on the diaphragm, and its narrow opening is sealed superiorly

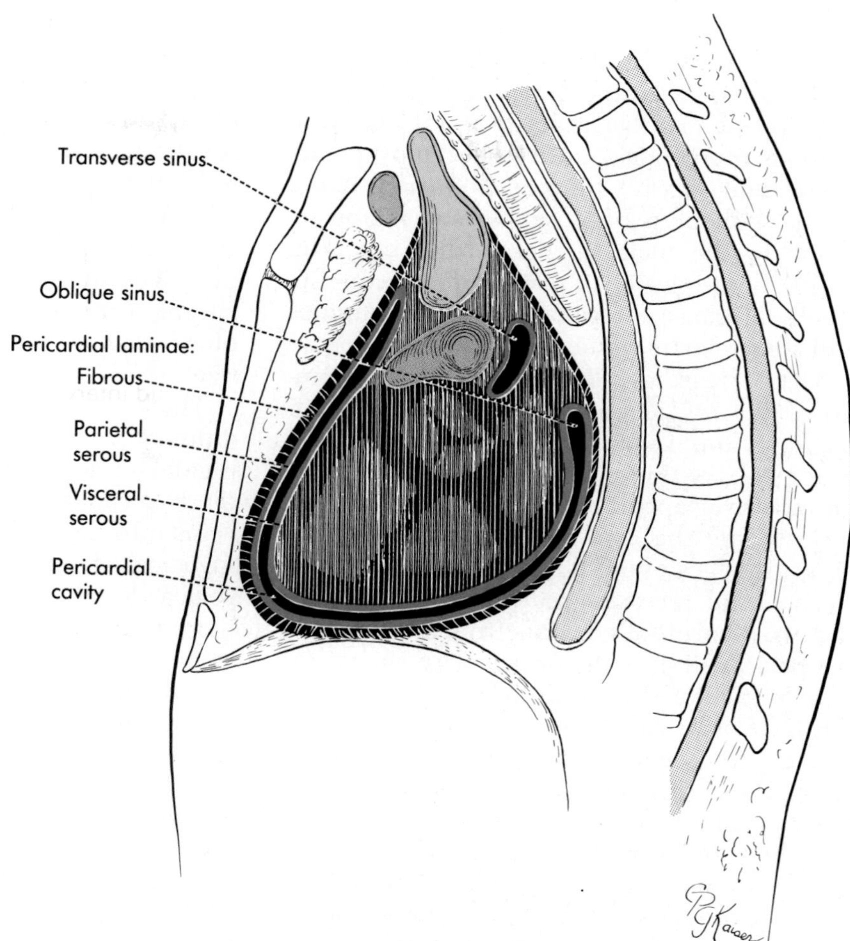


**FIG. 21-2.** A schematic representation of the pericardial sac seen in a sagittal section of the thorax. The serous pericardium is shown as a heavy red line (its thickness relative to the scale of other structures is much exaggerated), and the pericardial cavity enclosed within the serous sac is solid black (see also Figs. 19-13 and 19-17). The ascending aorta is pink; the pulmonary trunk, purple; and the left brachiocephalic vein, blue.

around the superior vena cava, the ascending aorta, and the pulmonary arteries (see Figs. 19-17 and 21-2). The base of the bag is pierced on the right by the inferior vena cava and posteriorly by the four pulmonary veins, which enter it independently.

The pericardium is fixed in the thoracic cavity because the base of the fibrous bag is inseparably fused to the anterior portion of the central tendon of the diaphragm; both structures in this region are derived from the septum transversum. Elsewhere, the pericardium is attached to the tendon by loose connective tissue. Behind the sternal angle, the fibrous pericardium blends imperceptibly with the adventitia of the superior vena cava, ascending aorta, right and left pulmonary arteries, and ligamentum arteriosum. This fusion of the connective tissues closes the pericardial sac superiorly. The anterior surface of the pericardium is attached to the sternum by two rather poorly defined and variable *sternopericardial ligaments*, which may be demonstrable in the region of the manubriosternal and xiphisternal junctions. The pretracheal fascia, which descends into the mediastinum from the neck, fuses with the anterior surface of the pericardium. Posteriorly, membranous connective tissue attaches the pericardium to the tracheal bifurcation and the principal bronchi.

With the descent of the diaphragm during inspiration, the pericardial sac is pulled downward and becomes elongated; this forces the heart into a more vertical position (see Fig. 21-19). Ascent of the diaphragm during expiration relaxes the pericardium; this permits the sac to bulge laterally and the heart to become more horizontal. These movements are evident on x-ray films as alterations in the cardiac silhouette because the heart adapts its shape to the inelastic pericardial sac (Fig. 21-19). The fibrous pericardium consists chiefly of dense interlacing collagen bundles and some elastic tissue in its deeper layers.

The pericardial sac with its contents comprises the *middle mediastinum.* The anterior mediastinum is in front of the sac and the posterior mediastinum behind it. The pericardium is overlapped and largely obscured anteriorly by the two pleural sacs and the anterior edges of the lungs, which occupy the sternocostal recesses (see Fig. 19-13). Where the two pleural sacs deviate from one another, the pericardium is in contact with the posterior surface of the sternum and the 4th and 5th left costal cartilages (see Fig. 19-14). Before adolescence, the thymus intervenes between the pericardium and the pleural sacs or the sternum, but in the adult there is little demonstrable thymic tissue in the anterior mediastinum. On each side, mediastinal pleura is draped over the lateral surface of the pericardium, with the phrenic nerve and the pericardiacophrenic vessels sandwiched between pleura and pericardium (see Fig. 19-13). The nerve and vessels are embedded in variable amounts of areolar or adipose tissue. Posteriorly, the pericardium is in contact with the esophagus, the descending thoracic aorta, and, more superiorly, both principal bronchi. On each side of these structures, the pleural sacs contact the posterior aspect of the pericardium. Occasionally, a small *infracardiac bursa* is present behind the pericardium just above the diaphragm. The bursa is a remnant of the embryonic *pneumatoenteric recess.*

## The Serous Pericardium and Pericardial Cavity

A cut made in the pericardium opens the pericardial cavity because the parietal lamina of the serous pericardium is closely adherent to the fibrous pericardium. The visceral lamina or epicardium is more loosely bound to the myocardium. The loose subepicardial connective tissue may contain a good deal of fat, especially along the blood vessels. The glistening surface of the serous pericardium is covered by a single layer of mesothelium. The mesothelium is supported by a delicate, transparent connective tissue membrane in which blood vessels, lymphatics, and nerves arborize.

The heart is completely invested in epicardium except for a posterior, narrow, and irregular area that is between the entrances of the two venae cavae and the four pulmonary veins into the atria (Fig. 21-3). Here the atrial myocardium is in contact with the fibrous pericardium. This "bare area" exists because there is an uninterrupted line of reflection between the epicardium and parietal pericardium, which encloses all six veins in a single, short, and highly distorted pericardial sleeve (Fig. 21-3). How this arrangement came about was explained earlier in this chapter, under Developmental Considerations. Of the six veins, only the superior vena cava has a significant intrapericardial portion, but it is not free in the cavity as are the ascending aorta and the pulmonary trunk. Only in the front and on its sides is the vena cava covered by serous pericardium over a length of 3.5 cm above the right atrium. Posteriorly, the vena cava contacts fibrous pericardium and the right pulmonary artery.

The superior vena cava is derived from the *right common cardinal vein* of the embryo; its left counterpart, the left common cardinal vein, normally atrophies. Sometimes the obliterated fibrous vestige of the left common cardinal vein is identifiable on the posterior wall of the pericardial cavity. It elevates a slight fold in the parietal pericardium between the left pulmonary artery and the left superior pulmonary vein. This fibrous vestige, called the *ligament of the left superior vena cava,* continues into the oblique vein of the left atrium, which drains into the coronary sinus. Sometimes both left and right common cardinal veins persist as left and right venae cavae, or the left vena cava may develop and the vessel on the right atrophy.

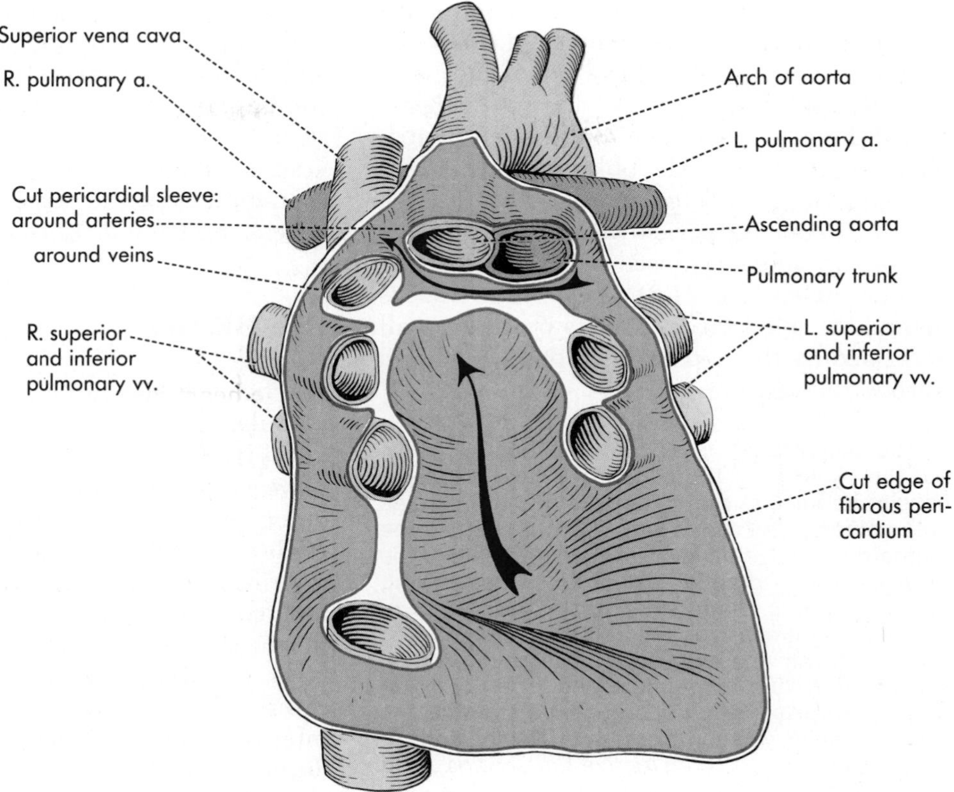


**FIG. 21-3.** The posterior wall of the pericardial sac after the heart has been removed by severing its continuity with the great arteries and veins and by cutting the two pericardial sleeves that surround the arteries and veins. The parietal serous pericardium is pink; the fibrous pericardium is white; the horizontal arrow is in the transverse sinus; the vertical arrow is in the oblique sinus of the pericardium.

The ascending aorta and pulmonary trunk may be grasped within the pericardial cavity. The two vessels can be enclosed by a ligature, or by finger and thumb, if the index finger is inserted from the left side behind the pulmonary trunk (as shown by the horizontal arrow in Fig. 21-3). This is possible because the two great arteries are enclosed by a common pericardial sleeve as explained earlier under Developmental Considerations. Anteriorly, this pericardial sleeve reflects forward from the aorta and pulmonary trunk to line the fibrous pericardium at about the level of the sternal angle where the fibrous pericardium fuses with the adventitia of these vessels (Fig. 21-2). Posteriorly, the epicardial sleeve reflects off the aorta and pulmonary trunk in a backward and downward direction to line the portion of the pericardial cavity into which the index finger was inserted in the above exercise (Figs. 21-2 and 21-3). This space is called the **transverse sinus** of the pericardium. The transverse sinus is a passage from the left to the right side of the pericardial cavity (Fig. 21-1). It is behind the great arteries. In its roof, covered by serous pericardium, runs the right pulmonary artery (as it passes to the right within the concavity of the aortic arch); its floor is formed by the left atrium. The superior vena cava (representing the venous end of the heart, Fig. 21-1D) is posterior to the sinus as the sinus opens into the main pericardial cavity to the right of the ascending aorta (Fig. 21-3).

The pericardial cavity has a second sinus called the **oblique sinus,** which is a *cul-de-sac* rather than a passage. The oblique sinus may be explored by elevating the apex of the heart and sliding the fingers behind the left ventricle and left atrium. The oblique sinus is limited superiorly and on each side by the distorted, ᒋ-shaped pericardial sleeve and the six major veins enclosed within it, as described earlier (Fig. 21-3). The oblique sinus is behind the left atrium, and its posterior wall is formed by the double layer of parietal and fibrous pericardium.

The pericardium surrounds the heart loosely, but the capacity of the pericardial cavity is small, usually about 300 ml. If there is gradual cardiac enlargement, or if fluids slowly accumulate in the pericardial cavity, the sac will accommodate to the enlargement by gradual distention. Sudden filling of the sac with blood or exudate (*cardiac tamponade*) will embarrass the action of the heart by interfering with the expansion of the atria, necessary for receiving incoming blood. The condition may be fatal. As a result of chronic pericarditis, the pericardium may become greatly thickened and adherent to the heart (*constrictive pericarditis*), and this also restricts the filling of the heart. The pericardium in these cases must be dissected away.

The most striking **congenital defect** of the pericardium results from failure of the fibrous pericardium to develop. This leaves the pleural and pericardial cavities in continuity as they are in the early embryonic stages (see Fig. 19-12). The total absence of the pericardium does not interfere with the action of the heart. In less severe cases, there is a communication between the pericardial and (usually the left) pleural cavities. This may cause problems if part of the heart herniates through the defect.

### Innervation and Blood Supply

Pericardial pain is mediated by *somatic afferents* distributed to the fibrous and parietal laminae of the pericardium by the **phrenic nerves.** The epicardium, supplied by autonomic nerves from the coronary plexuses, is insensitive to pain. Pericardial pain is felt behind the sternum and is usually due to inflammation of the pericardium caused by viral or bacterial infections or by neoplastic involvement.

The epicardium shares its blood supply with the myocardium, whereas the fibrous and parietal laminae of the pericardium are supplied by small twigs from the internal thoracic vessels, from the pericardiacophrenic vessels (themselves branches of the internal thoracic), from the aorta, from bronchial arteries, and from arteries of the diaphragm.

## THE HEART

The heart is the central organ of the cardiovascular or circulatory system. Its name in Latin is *cor* and in Greek, *kardia*. Both names serve as roots for adjectives commonly used in describing the anatomy of the heart. The walls that enclose the hollow, four-chambered interior of the heart are thick and contractile and consist of three definable layers. The thickest of these is the middle or muscular layer, the **myocardium,** covered on the heart's exterior by the **epicardium** and lined on the interior by the **endocardium.** The endocardium consists of endothelium supported by some delicate underlying connective tissue, and it is continuous with the tunica intima of the blood vessels entering and leaving the heart.

### EXTERNAL ANATOMY

### External Form

The heart somewhat resembles a short cone (Fig. 21-4) and is described as having a base and an apex as well as surfaces and borders. The borders are poorly defined on the heart itself (as one might expect from its essentially conical shape), but are useful in describing the radiologic anatomy of the heart. The division of the heart into four chambers is indicated on its surface by the coronary and interventricular sulci, which are useful landmarks for relating the chambers to external features.

**Base and Apex.** The **base** of the heart faces posteriorly and is made up largely of the **left atrium** and a more narrow portion of the right

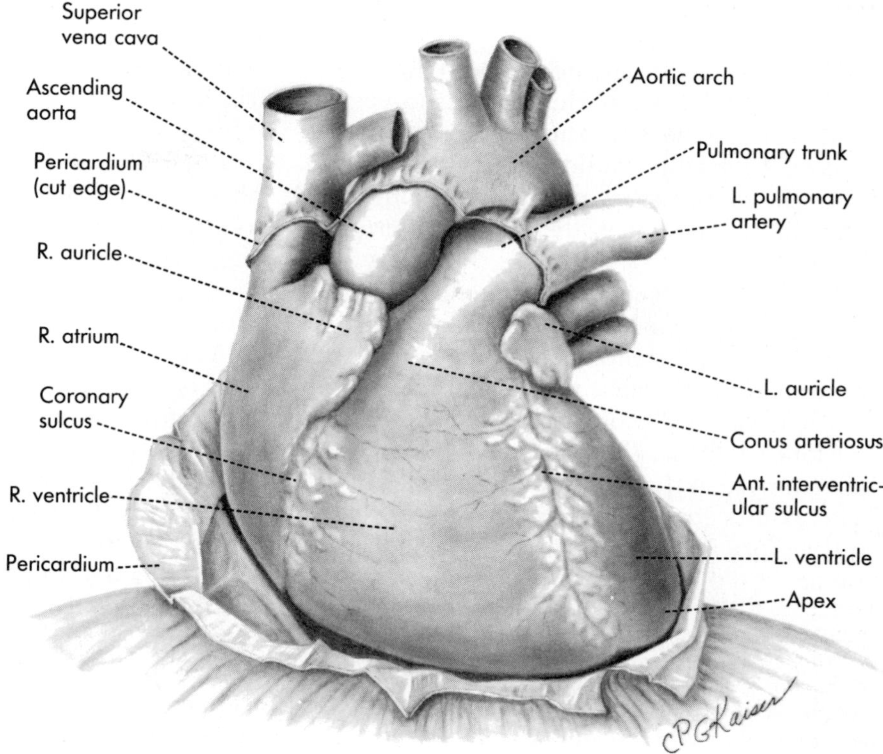


**FIG. 21-4.** Anterior view of the heart. The pericardial sac has been cut open and reflected toward the diaphragm, on which the heart is resting.

**FIG. 21-5.** Posterior view of the heart retained in the same position as shown in Figure 21-4.

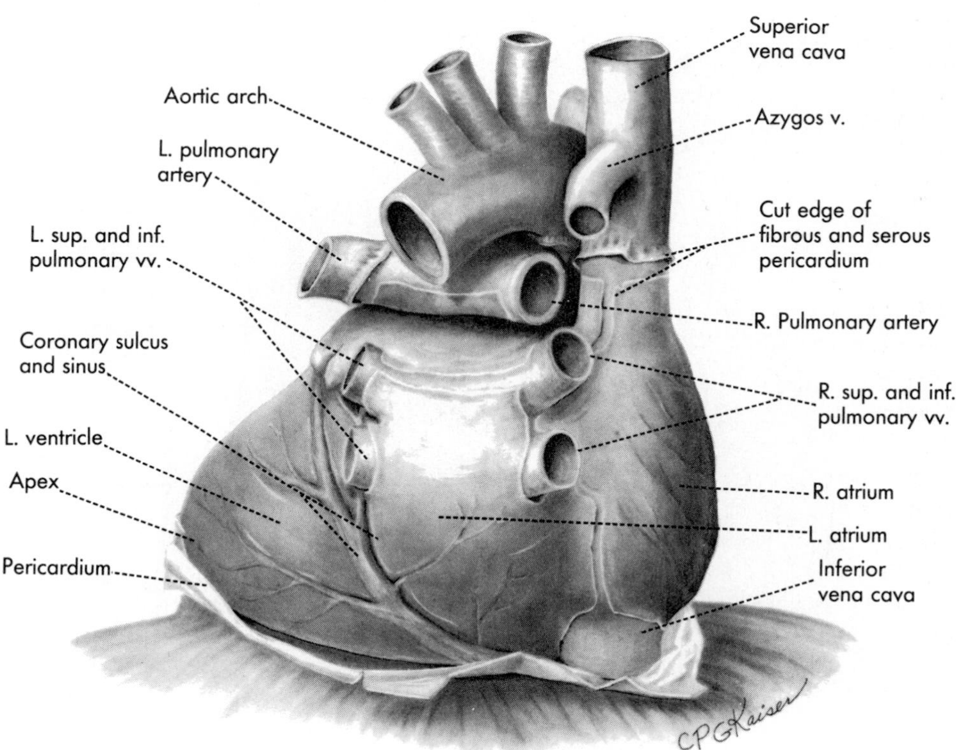

atrium (Fig. 21-5). Owing to the obliquity of the interatrial septum, the left atrium lies behind, rather than to the left, of the right atrium. A variable shallow groove situated to the right of the right pulmonary veins may indicate the position of the oblique interatrial septum.

All the great veins enter the base of the heart and fix it posteriorly to the pericardial wall. Between the entrances of the great veins into the atria, the base of the heart forms the anterior wall of the oblique pericardial sinus.

From the base, the heart projects forward to terminate in the blunt **apex.** The apex of the heart points downward and toward the left as well as forward. It is formed by the left ventricle.

**Surfaces.** In addition to the base, four surfaces are described on the heart. These are the diaphragmatic or inferior, the sternocostal or anterior, and the left and right surfaces. The **diaphragmatic surface,** formed largely by the **left ventricle** and a more narrow portion of the **right ventricle,** extends from the base to the apex, faces inferiorly, and rests on the diaphragm (Fig. 21-6). A large vein, the *coronary sinus,* lodged in the coronary sulcus, separates this surface from the base. The *posterior interventricular sulcus,* occupied by an artery and vein, runs from the coronary sulcus toward the apex, and it demarcates the ventricles from one another.

The **sternocostal surface** faces anteriorly and is dominated by the right ventricle, with the right atrium visible to the right and the left ventricle visible to the left (Fig. 21-4). Superiorly on the right, the superior vena cava leads into the **right atrium.** The free upper border of the atrium projects in front of the vena cava and extends anteriorly as the right auricle. The **auricle,** a flaplike, muscular pouch, overlaps the ascending aorta and obscures the superior portion of the coronary sulcus. The *coronary sulcus* separates the right atrium from the ventricle and runs from the root of the aorta toward the inferior vena cava (Figs. 21-4 and 21-6). The inferior vena cava is

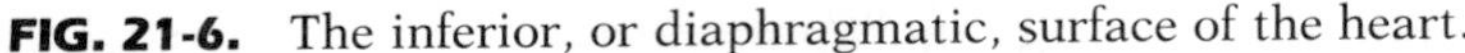

**FIG. 21-6.** The inferior, or diaphragmatic, surface of the heart.

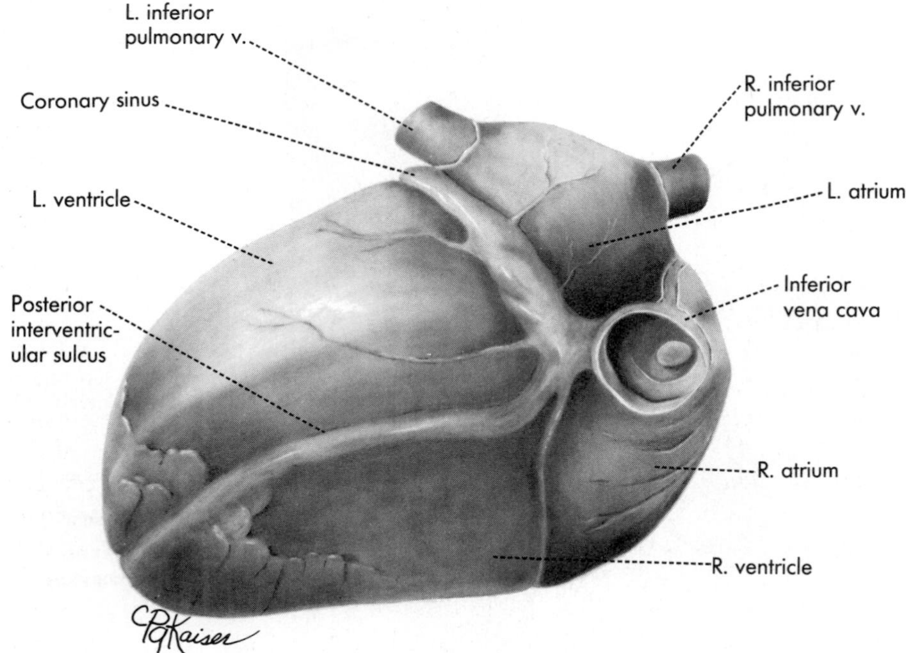

at the right inferior corner of the sternocostal surface, but is not visible from the front view. The sulcus turns posteriorly to the right inferior corner of the heart and in the back separates the base of the heart from its inferior surface. Along the inferior border of the sternocostal surface, the **right ventricle** extends from the coronary sulcus almost to the apex of the heart and is in contact with the diaphragm. Superiorly, the right ventricle tapers toward the origin of the pulmonary trunk, and this funnel-shaped portion forms the **conus arteriosus** or **infundibulum.** The infundibulum continues into the pulmonary trunk. At its root, the pulmonary trunk interrupts the continuity of the coronary sulcus. But for this interruption, the sulcus encircles the heart between the atria and the ventricles. It is identifiable to the left of the pulmonary trunk where the **left auricle** overlies the pulmonary trunk. The *anterior interventricular sulcus,* occupied by an artery and one or two veins, descends from here toward the apex and marks the position of the interventricular septum. The interventricular septum sometimes creates a slight notch (*incisura*) just to the right of the apex. The coronary sulcus proceeds to the left from the pulmonary trunk and soon turns posteriorly, where it separates the left auricle from the left ventricle.

Both the **left** and the **right surfaces** of the heart face the lungs, but only the left is described, sometimes, as the **pulmonary surface.** Both are broad and convex; the right surface consists entirely of the right atrium, and the left, of the left ventricle, with the left auricle just above.

**Margins or Borders.** One may speak of the right, left, and inferior margins of the heart, although currently, official anatomic terminology recognizes only a right margin (*margo dexter*). Truly, the right and left margins correspond to the right and left surfaces. Only along the inferior border of the heart is there a relatively sharp edge between the sternocostal and diaphragmatic surfaces of the right ventricle to warrant calling it a margin.

In a posteroanterior **chest x-ray film,** the cardiovascular shadow presents definite borders on the right and left as it abuts the radiolucent lung fields (Fig. 21-7). The inferior border is largely obscured by the liver because the heart rests on the forward slope of the diaphragm. Knowledge of the structures that constitute the borders of the cardiac silhouette is of diagnostic importance in determining the enlargement of various chambers. On a posteroanterior x-ray film the **right border** consists, from above downward, of the superior vena cava, right atrium, and inferior vena cava. The **left border** is made up of the arch of the aorta, pulmonary trunk, left auricle, and left ventricle. The right ventricle and, at the apex, the left ventricle, constitute the inferior border. Often the apex may be difficult to identify radiographically.

Radiologic examination of the heart includes lateral and oblique projections, the detailed description of which is beyond our present purpose. In the right and left lateral views, the cardiovascular shadow is made up anteriorly of the right ventricle and its conus arteriosus and posteriorly of the left atrium.

### Relations

The relationship of the cardiac chambers to one another is discussed with the surface projection and physical examination of the heart. Through the pericardium, the various chambers are related to a number of organs.

The **right atrium** is related anteriorly and laterally to the mediastinal surface of the right lung. Posterior to most of the right atrium is the left atrium. On the right, the right inferior pulmonary vein courses behind the right atrium (the superior vein is behind the superior vena cava). The sternocostal surface of the **right ventricle** is separated from the chest wall by the pericardium, the pleural sacs, and the areolar tissue in the anterior mediastinum. The **left atrium,** on the base of the heart, is in contact with the esophagus through the oblique sinus and the pericardium. When the atrium is enlarged, it displaces the esophagus, which is demonstrable radiographically by a barium swallow. Over the roof of the left atrium arch the pulmonary

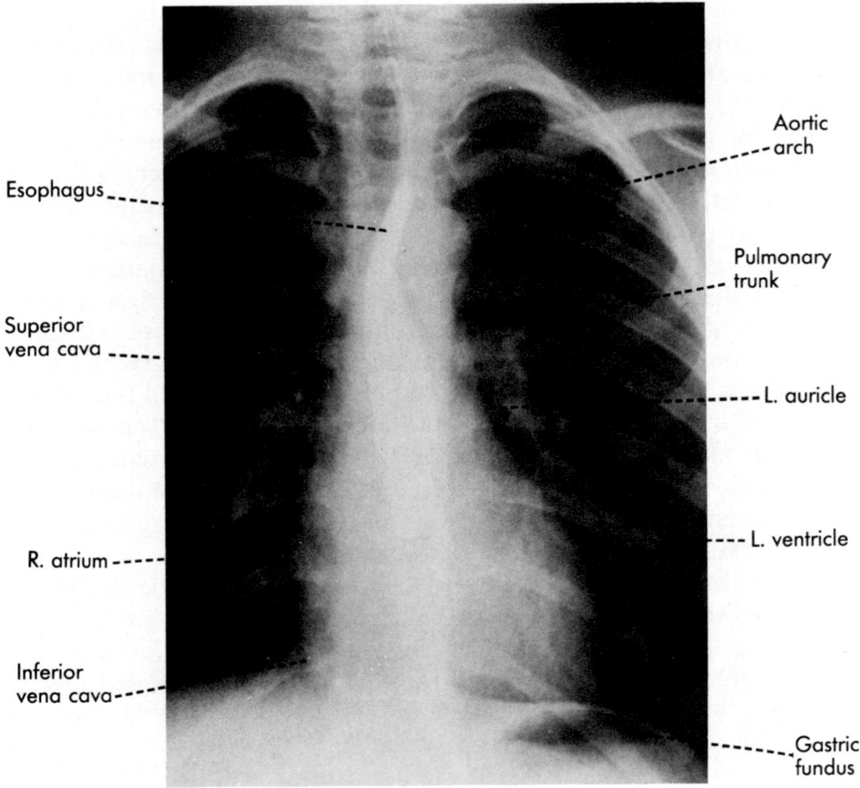


**FIG. 21-7.** A posteroanterior chest x-ray of a young man. The esophagus contains some barium, and the fundus of the stomach contains some gas.

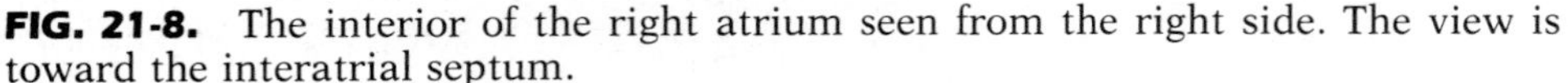

**FIG. 21-8.** The interior of the right atrium seen from the right side. The view is toward the interatrial septum.

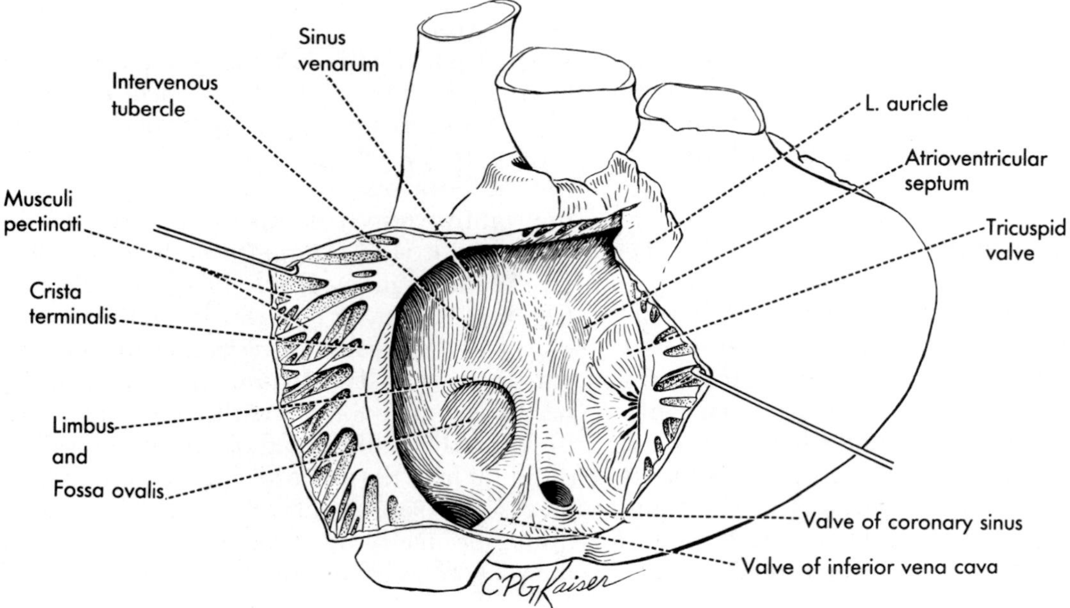

trunk and the ascending aorta. The transverse sinus intervenes between the atrium and these vessels. The **left ventricle** is related to the mediastinal surface of the left lung; the inferior surface of the ventricle is related, through the diaphragm, to the left lobe of the liver and the fundus of the stomach.

## INTERNAL ANATOMY

### The Cardiac Chambers

**The Right Atrium.** Situated on the right surface of the heart, this quadrangular chamber receives blood into its upper posterior corner from the superior vena cava, whereas the inferior vena cava and the coronary sinus open into its lower posterior corner (Fig. 21-8). From the atrium, the blood is emptied into the right ventricle through the *right atrioventricular ostium* that faces forward and medially. The ostium is guarded by the right atrioventricular or *tricuspid valve,* which will be described from its ventricular surface.

The interior of the atrium is partially divided into two main parts by the **crista terminalis,** a smooth muscular ridge that commences on the roof of the atrium just in front of the opening of the superior vena cava and extends on the lateral wall of the chamber to the anterior lip of the inferior vena cava. The cavity posterior to the crista terminalis is the **sinus venarum cavarum,** into which the two venae cavae open. The anterior half of the chamber that includes the **right auricle** is sometimes spoken of as the **atrium proper.** The walls of the sinus venarum are smooth, as is the surface of the interatrial septum, which separates the sinus from the left atrium. By contrast, the walls of the anterior half of the chamber, including the auricle, are ridged by the **musculi pectinati,** which fan out, comblike, from the crista terminalis. Only this anterior half corresponds to the primitive atrium of the embryonic heart; the sinus venarum represents the right horn of the sinus venosus, which had merged with the embryonic right atrium during development. In fact, the crista terminalis and the so-called *valve of the inferior vena cava,* a rudimentary endothelial flap on the anterior lip of the vessel, are derived from one of the valves that guarded the opening of the sinus venosus into the right atrium (Fig. 21-23). Medial to the opening of the inferior vena cava is the *ostium of the coronary sinus* bordered by a small fold, the rudimentary *valve of the coronary sinus,* also derived from the valve of the sinus venosus.

The **interatrial septum** faces both forward and to the right, for the left atrium lies, in part, behind the right atrium as well as to its left (Fig. 21-11). There is a depression in the septum just above the orifice of the inferior vena cava (Fig. 21-8). This is the **fossa ovalis,** and its prominent margin is the **limbus fossae ovalis.**

The fossa marks the location of the *foramen ovale,* an aperture through which the blood flows from the right to the left atrium before birth. The floor of the fossa is formed by the *septum primum* of the embryonic heart, which seals the foramen after birth. A small opening may remain patent superiorly in the foramen throughout life and is of no functional consequence. The limbus corresponds to the margin of the foramen ovale formed by the lower edge of the *septum secundum.* These septa are described later in the section on the development of the heart.

On the posterior wall of the sinus venarum, between the two caval openings, is a smooth eminence, the *intervenous tubercle,* which, before birth, may assist in separating the two streams of blood that enter the atrium from the two venae cavae. The tubercle may direct the stream from the superior vena cava chiefly toward the right ventricle and the stream from the inferior vena cava chiefly toward the left atrium through the foramen ovale. A number of small openings, the foramina of the smallest cardiac veins (*venae cordis minimae*), are scattered about the wall of the right atrium. Similar venous openings are found in all other chambers as well.

**The Right Ventricle.** Blood flows into the right ventricle from the right atrium in a horizontal and forward direction because the right ventricle is situated in front, as well as to the left, of the right atrioventricular opening. The cavity of the right ventricle extends almost to the apex of the heart and projects onto most of the sternocostal surface and

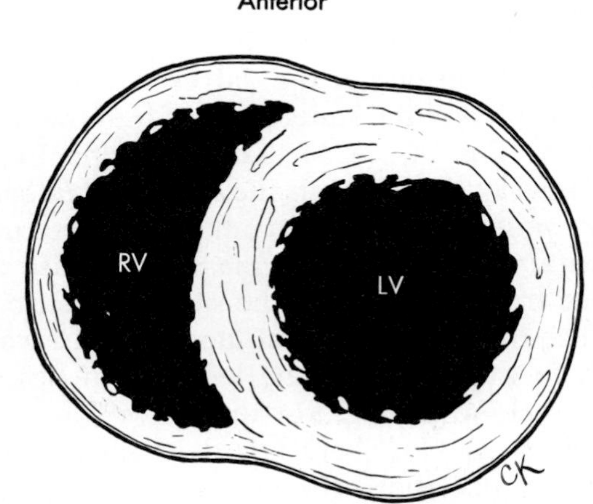


**FIG. 21-9.** A schematic representation of a cross section of the ventricles (seen from below) to illustrate the shape of their lumina and the difference in the thickness of their wall.

along the anterior edge of the diaphragmatic surface. In cross section, the cavity is C shaped (Fig. 21-9). At the base of the ventricle, blood is received from the right atrium into the posterior limb of the C; the anterior limb leads superiorly into the **conus arteriosus** or **infundibulum,** which is the outflow tract of the ventricle (Fig. 21-4). Inflow and outflow tracts are separated by a smooth muscular crest, the **crista supraventricularis,** that partially divides the interior of the ventricle into two portions (Fig. 21-10). Above, the funnel-shaped infundibulum, or conus, leads into the pulmonary trunk through the *pulmonary orifice.*

The walls of the conus arteriosus are smooth, whereas the surface of the myocardium in the rest of the ventricle has prominent fleshy ridges, the **trabeculae carneae,** and also gives rise to columnlike or nipplelike projections called *papillary muscles.*

At least one of the trabeculae carneae is elevated into a free band that forms a bridge between the interventricular septum and the anterior wall close to the apex of the ventricle. This is the *septomarginal trabecula,* formerly known as the moderator band, along which runs a branch of the atrioventricular bundle. There are usually three sets of **papillary muscles,** anterior, posterior, and septal, named according to the location of their base. From the apices of each papillary muscle several tendonlike fibrous cords, known as **chordae tendineae,** extend to the cusps of the atrioventricular valve. The largest and most constant is the anterior papillary muscle; the septal muscle is tiny, or may even be absent, and the chordae tendineae spring directly from the septum. There may be one to three posterior muscles and some chordae tendineae may, in addition, attach directly to the ventricular wall. The papillary muscles and chordae tendineae prevent the cusps of the valve from being everted into the atrium by the pressure developed in the contracting ventricle. This pressure closes the valve, and as the size of the ventricle decreases during contraction, the papillary muscles also contract so that they maintain tension on the cusps.

The **right atrioventricular valve** is also known as the **tricuspid valve** because in some hearts it consists of three cusps or leaflets. More often than not, however, there are only two cusps, and, in the literal sense, the use of the term *tricuspid* is not always apt. The bases of the cusps are secured to the fibrous ring that surrounds the atrioventricular ostium (Fig. 21-13). The chordae tendineae attach to the ventricular surface of the cusps, and when the valve is opened, the three scalloped edges of the cusps project into the ventricle. Close to their base, the cusps are continuous with one another along lines known to clinicians as *commissures.* The three cusps are named according to their position: anterior, posterior, and septal. Each receives chordae tendineae from two papillary muscles. The cusps consist of dense fibrous tissue covered by endocardium, and only along their bases are some blood vessels found in their substance.

The *orifice of the pulmonary trunk* is closed by the **pulmonary valve,** which consists of three semilunar *valvules,* or cusps (Figs. 21-10 and 21-11). Each cusp is formed by the reduplication of endothelium with little fibrous tissue between the layers, and each resembles a distended vest pocket projecting with its free edge upward into the lumen of the pulmo-

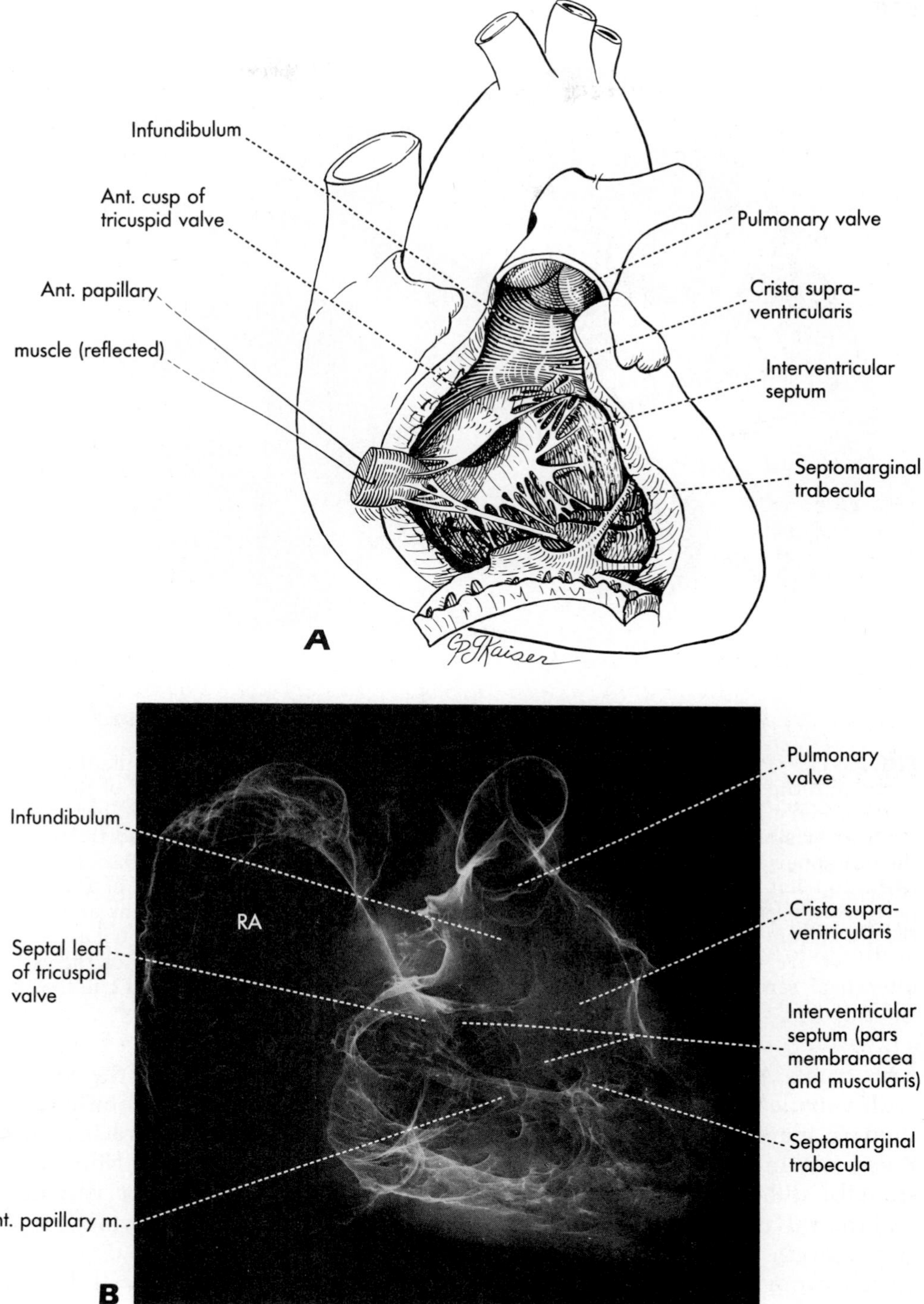


**FIG. 21-10.** The interior of the right ventricle. **A,** the drawing of a dissection, corresponds to **B,** an x-ray that was obtained after the heart was fixed, its walls impregnated with wax and the interior of the right ventricle dusted with barium powder. The crista supraventricularis separates the inflow part of the ventricle from the infundibulum, or conus arteriosus. Note the great distance between the septal leaf of the tricuspid valve and the pulmonary valve. **RA,** right atrium. (Part **B** was prepared and kindly provided by Dr. Lore Tenckhoff.)

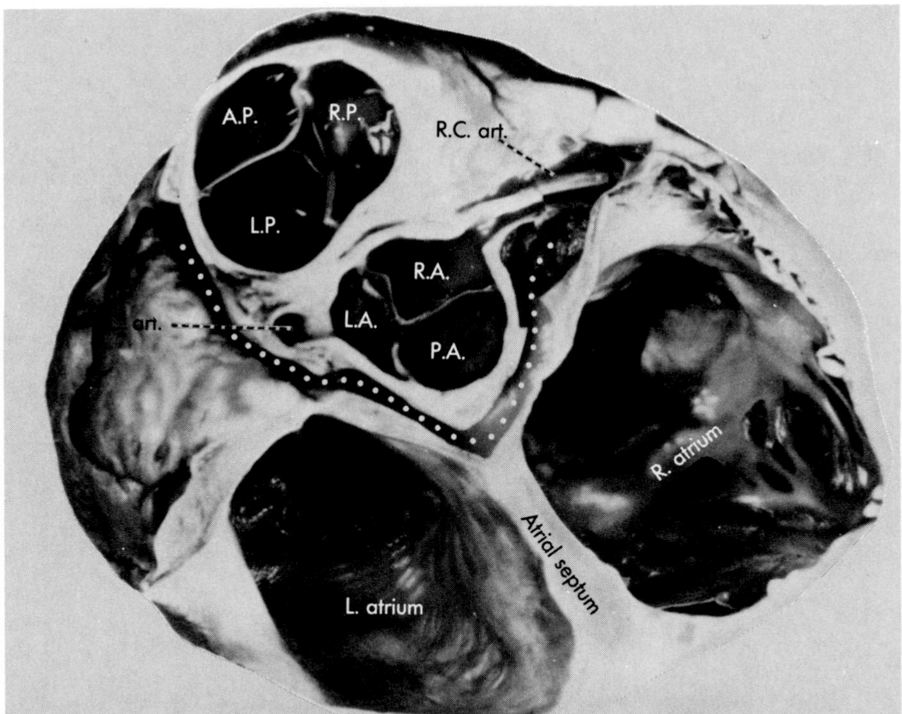


**FIG. 21-11.** Photograph of a heart cut obliquely across the pulmonary trunk, the aorta, and the two atria. The aorta is cut just above the level of the origin of the two coronary arteries (**RC art** and **LC art**), and the three valvules of the aortic valve are visible: left **(LA)**, right **(RA)**, and posterior **(PA)**. Anterior to the aorta is the pulmonary trunk, closed off from the right ventricle by the anterior **(AP)**, right **(RP)**, and left **(LP)** cusps of the pulmonary valve. The transverse sinus of the pericardium (indicated by a white dotted line) separates the pulmonary trunk and the aorta from the atria. Two cusps of the mitral valve and one cusp of the tricuspid valve are visible through the left and right atria. Note the obliquity of the interatrial septum. (Modified from Edwards JE, Burchell HB: Thorax 12:123, 1957)

nary trunk. At the middle of the free edge of each valvule is a thickening, the *nodule,* and the thin margin on each side of the nodule is the *lunula*. In the "pocket" of each valvule is a *sinus* or dilatation of the pulmonary trunk, and the valvules attach along the curved inferior margin of each sinus. After blood is ejected from the right ventricle, the nodules and lunules of each valvule are forced together tightly by the pressure of blood in the pulmonary trunk causing distention of the pulmonary sinuses. The pulmonary valve thus prevents regurgitation of blood into the ventricle. Two of the cusps of the pulmonary valve are anterior, and one is posterior. The official naming of the cusps is confusing, however, because it is based on the fetal position of the heart before it completes its rotation. Accordingly, the Nomina Anatomica recognizes one anterior valvule, one left valvule, and one right valvule (Fig. 21-11).

**The Left Atrium.** Like the right atrium, the interior of the left atrium is divided into two portions, although there is no definite line of demarcation here. The posterior half of the chamber, into which the four pulmonary veins empty, has smooth walls; this part of the atrium is derived from the proximal portions of the embryonic pulmonary veins, whose walls were incorporated into the atrium during development. The anterior portion is continuous with the **left auricle,** has *musculi pectinati* on its wall, and is derived from the embryonic atrium. The interatrial septum forms part of the anterior wall of the left atrium (Fig. 21-11). The thin area on the septum is the *valvule of the foramen ovale,* visible in the right atrium as the floor of the fossa ovalis. In the fetus, it functioned as a valve, for blood coming into the right atrium could push it aside and enter the left atrium, but blood attempting to pass in the opposite direction would force it against the rather rigid rim of the foramen. In approximately 20% of adults, the original free edge of the valvule is incompletely fused, so that the tip of a small probe can be passed between the atria.

**The Left Ventricle.** The left ventricle lies to the front of the left atrium, and blood flows toward its apex from the left atrioventricular orifice, chiefly in a forward direction. The myocardium is thickest in the wall of this chamber, which occupies the rounded lateral surface of the heart, including the apex and most of the inferior surface. The cavity of the left ventrical is conical, and the outline of its cross section is nearly circular (Fig. 21-9). Inflow and outflow tracts are not as clearly demarcated in the left ventricle as in the right one. The trabeculae carneae in the left ventricle are fine and delicate in contrast to the coarse trabeculae in the right ventricle. The septal wall is smooth and leads superiorly toward the aortic orifice. This outflow tract, sometimes designated as the **aortic vestibule,** corresponds to the conus arteriosus on the right, which lies directly anterior to it, and like the conus is derived from the embryonic bulbus cordis (Fig. 21-12).

Because the right ventricle lies largely in front of the left ventricle, the **interventricular septum** forms the anterior and right-hand wall of the left ventricle. Its thick part is the *pars muscularis.* Above, close to the atrioventricular orifices, the septum is thin and membranous; this is the *pars membranacea* (Fig. 21-12). A tiny part of the pars membranacea lies above the attachment of the septal cusp of the tricuspid valve and thus between the cavities of the left ventricle and right atrium; this is the *atrioventricular septum* (Fig. 21-8).

The *left atrioventricular ostium* opens into the posterior and right side of the upper part of the left ventricle (Fig. 21-12). The ostium is protected by the *left atrioventricular valve,* which has two cusps instead of three. The two cusps are reminiscent of a bishop's miter, hence the valve is also known as the **mitral valve.** Its cusps are anterior and posterior. Attached to the free edges of the cusps are *chordae tendineae,* which originate in *papillary muscles.* There are usually only two papillary muscles in the left ventricle, an anterior and a posterior (either may show some duplication), and the chordae tendineae of each go to both cusps of the mitral valve.

Behind the conus arteriosus of the right ventricle, the aortic vestibule gives rise to the aorta. The aortic orifice is closed by the **aortic valve,** which resembles the pulmonary valve in every respect. The three valvules, or cusps, harbor in their "pockets" the three *aortic sinuses.* The sinuses and the cusps are named right, left, and posterior (Fig. 21-11), and from the right and left sinuses originate the right and left coronary arteries, respectively. The posterior sinus and valvule are designated as the noncoronary sinus and cusp.

### Structure of the Myocardium

The cardiac muscle fibers are arranged in somewhat indistinct bundles that attach to the heavy fibrous connective tissue sometimes referred to as the **skeleton of the heart.** This connective tissue forms the foundation to which the valves—pulmonary, aortic, and atrioventricular—as well as the myocardial fiber bundles are attached. The four **fibrous rings** (*annuli fibrosi*) that support the four sets of valves are fused to one another (Fig. 21-13). Thickened areas of fusion between the aortic and left atrioventricular rings form the *fibrous trigones,* and the septum membranaceum (pars membranacea of the interventricular septum) blends with the area of fusion between the aortic and the right atrioventricular rings. The fibrous skeleton of the heart separates the musculature of the atria from that of the ventricles. Atrial bundles attach to the upper borders of the rings, and ventricular bundles attach to the lower borders. The two sets of muscle are

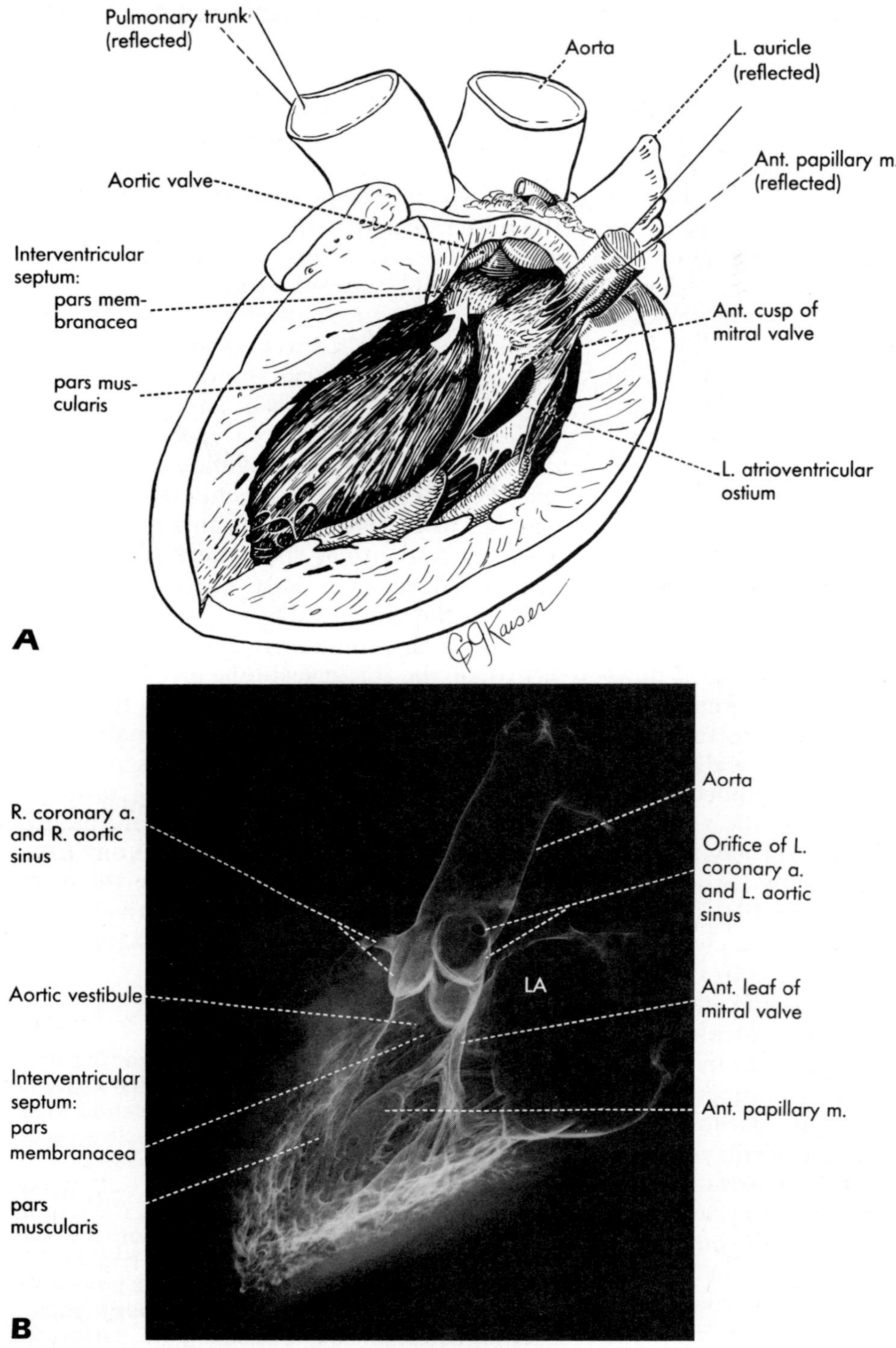


**FIG. 21-12.** The interior of the left ventricle. In **A,** a drawing of a dissection, the wall on the left side of the ventricle has been cut away. **B** is an x-ray of a heart prepared as explained in the legend of Figure 21-10. The interior of the left ventricle, left atrium **(LA),** and aorta have been dusted with barium powder. Note the proximity of the anterior cusp of the mitral valve to the aortic valve. The outflow tract of the left ventricle (aortic vestibule) is indicated in **A** by a white arrow. (Part **B** was prepared and kindly provided by Dr. Lore Tenckhoff)

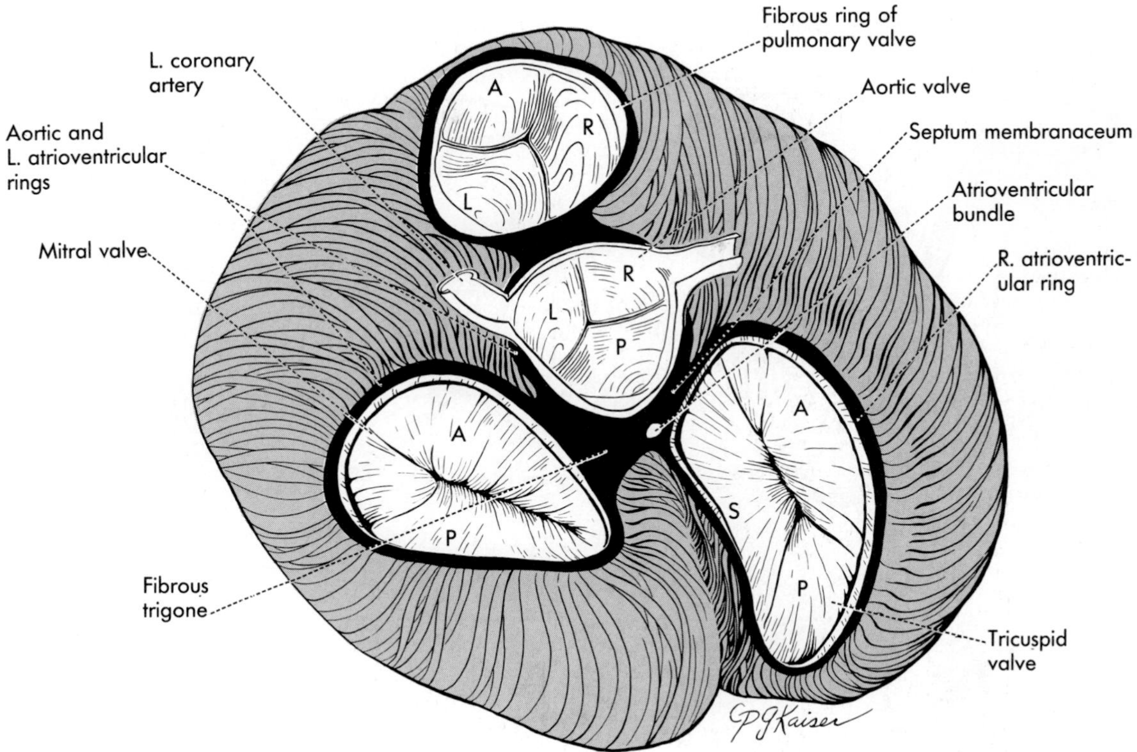


**FIG. 21-13.** The fibrous skeleton of the heart (**solid black**). The orientation of the heart is similar to that shown in Figure 21-11, but the atria have been dissected away. The venticular muscle fibers can be seen to radiate away from their attachment to the fibrous rings that support the valves. The cusps and valvules of the various valves are identified (**A**, anterior; **P**, posterior; **R**, right; **L**, left; **S**, septal).

united only by a specialized conducting bundle, the atrioventricular bundle (described under The Innervation and the Conducting System of the Heart). This bundle differs histologically from regular cardiac muscle, and it penetrates the fibrous partition.

The numerous fiber bundles of the atria are usually grouped into two systems, superficial and deep. The superficial fibers tend to run transversely across both atria and originate chiefly around the superior vena cava. Some of the deep fibers are annular and surround the several venous openings into the atria or circle the auricles. Other bundles originate from the atrioventricular fibrous ring, loop around one atrium and contribute to the interatrial septum, and then attach again to the atrioventricular ring.

Sheets of myocardial fibers encircle the ventricular chambers in a rather complex manner reminiscent of the windings of a turban. The information contained in Figure 21-14 about their anatomy is more than adequate for the purpose of this chapter.

## BLOOD SUPPLY

The arteries that supply the heart are the coronary arteries; the veins of the heart are known as the cardiac veins (venae cordis).

### Coronary Arteries

The right and left coronary arteries arise from the aorta just above its origin from the left ventricle while the aorta is still behind the pulmonary trunk (Fig. 21-15). The **right coronary artery** originates from the right aortic sinus and passes to the right, behind the pulmonary trunk, to run downward in the coronary sulcus between the right atrium and the right ventricle. Then it turns posteriorly around the right margin of the heart and continues in the coronary sulcus, supplying throughout its course the right ventricle and

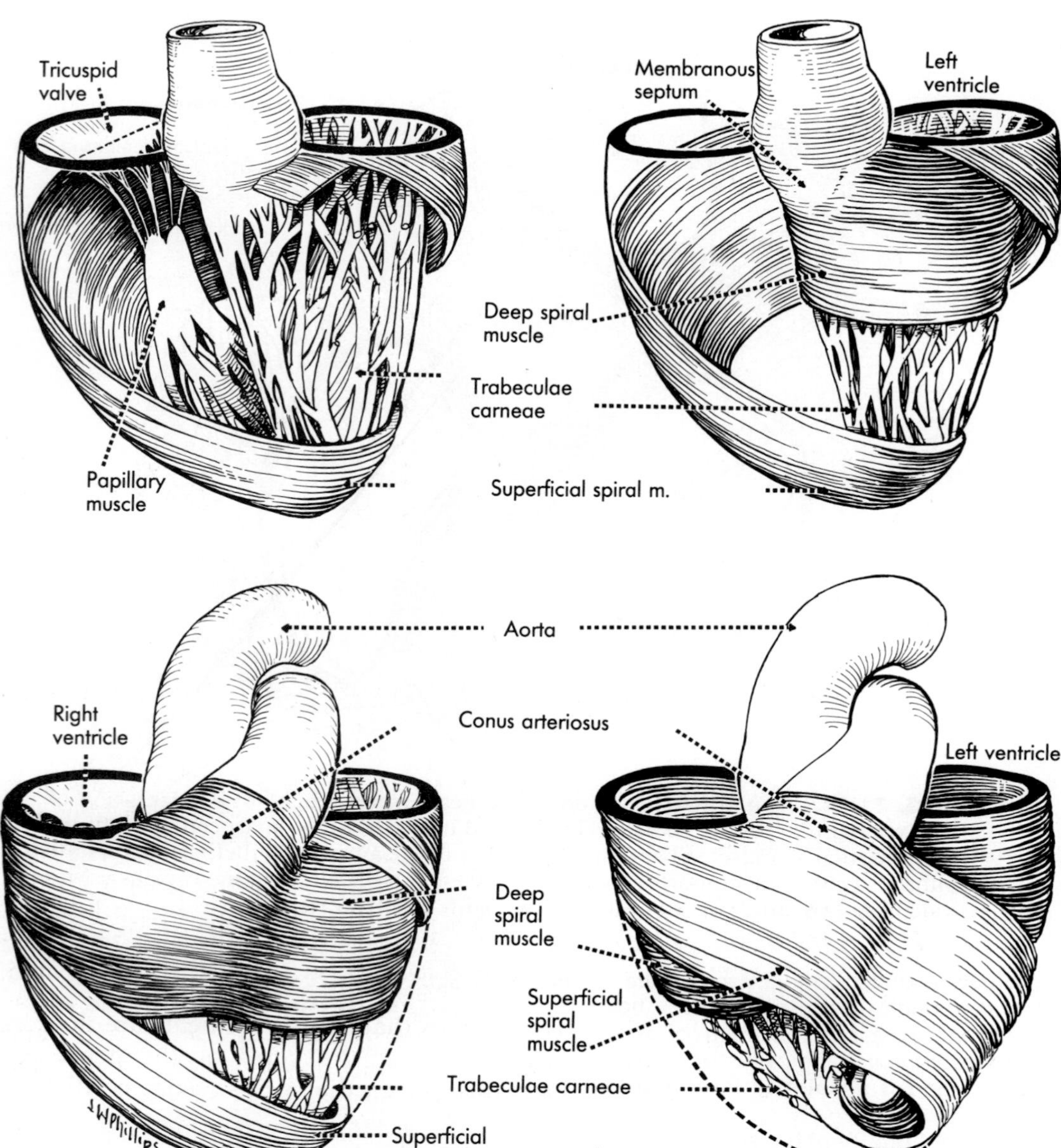


**FIG. 21-14.** Some myocardial fiber bundles of the ventricles. Superficial spiral bundles arise principally from the atrioventricular fibrous rings and form a vortex at the apex. On the inside of the ventricles, these fibers emerge from the vortex as trabeculae carneae or papillary muscles and spiral back toward the atrioventricular orifices. The deep spiral bundles are also attached to the fibrous rings; some encircle one, others both, of the ventricles. Although this anatomic arrangement could not be confirmed in canine and porcine hearts, it appears to be correct for the human heart. (After Robb JS, Robb RC: Am Heart J 23:455, 1942. Rushmer RF: Structure and Function of the Cardiovascular System, 2nd ed, p 78. Philadelphia, WB Saunders, 1976).

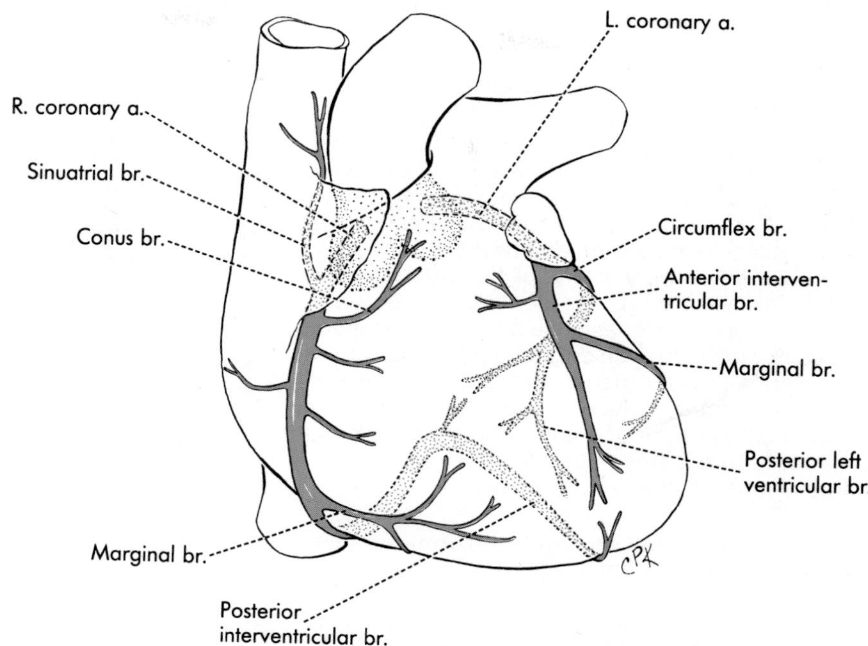


**FIG. 21-15.** The coronary arteries shown from an anterior view of the heart.

the right atrium. The largest branch of the right coronary artery is the *posterior interventricular branch,* which runs forward in the posterior interventricular sulcus toward the apex and supplies the diaphragmatic surface of both ventricles and approximately the posterior one-third of the interventricular septum. Smaller named branches include the *marginal artery* and those that are given off to the *sinuatrial* and *atrioventricular nodes* and to the *conus arteriosus.*

The **left coronary artery** arises from the left aortic sinus (Fig. 21-15). Shortly after its origin, usually while it is still behind the pulmonary trunk, the artery bifurcates into an anterior interventricular and a circumflex branch. The *anterior interventricular branch* skirts the left margin of the pulmonary trunk and descends toward the apex in the anterior interventricular sulcus. It gives branches to both ventricles, including the conus arteriosus, and to most of the interventricular septum. The *circumflex branch* runs toward the left in the coronary sulcus, first between the left auricle and left ventricle, and then it circles to the posterior aspect of the heart. It usually does not reach the posterior interventricular sulcus, but when it does, it may give rise to the posterior interventricular artery. The biggest branch of the circumflex artery, called the *posterior left ventricular branch,* supplies the diaphragmatic surface of the left ventricle. Other named branches of the circumflex artery include a *marginal,* an *intermediate,* and occasionally a *sinuatrial* and an *atrioventricular* artery.

In summary, the right coronary artery typically supplies the right ventricle, the posterior part of the left ventricle and the interventricular septum, the right atrium, and the interatrial septum, including the sinuatrial and atrioventricular nodes (parts of the conducting system of the heart). The left coronary artery regularly supplies most of the left ventricle and left atrium, and its anterior interventricular branch is the chief source of blood to the interventricular septum. This territory of supply includes the atrioventricular bundle and its branches (parts of the conducting system of the heart) as they lie in the septum. The left coronary artery may help to supply, or may be the sole supply, of the sinuatrial and atrioventricular nodes.

**Variations.** The distribution of the smaller branches of the coronary arteries is not constant, and the branching of the larger vessels may deviate from the pattern described above. As already mentioned, both interventricular branches may arise from the left coronary artery, or the anterior interventricular branch may be double. Occasionally, the two coronary arteries arise by a common stem, or the circumflex branch originates separately from the anterior interventricular branch. Either vessel may be abnormally placed on the aorta. Of more concern is the origin of a coronary artery from the pulmonary trunk, since such a vessel cannot provide oxygenated blood to the heart.

**Myocardial Infarction and Intercoronary Anastomoses.** The constant rhythmic activity of cardiac muscle makes the heart particularly dependent on a good blood supply. The density of capillaries in cardiac muscle is more than 80 times that in skeletal muscle. The coronary arteries are not end arteries. The myocardium of the normal human heart contains abundant arteriolar anastomotic channels that range in diameter from 20 $\mu$ to 200 $\mu$. Some of these anastomoses connect various branches of one coronary artery; others are intercoronary anastomoses. Despite their considerable size, the anastomoses possess only a thin wall and in healthy hearts cannot be demonstrated by arteriography because they are too small or apparently nonfunctional. However, when a major vessel is obstructed in the coronary circulation, blood will flow through the existing collateral vessels owing to the pressure gradient created between the vessels that the anastomotic channel connects. The lumen of a major vessel has to be reduced by 90% before the collateral vessels open up. The time required for the opening up of these anastomoses is not exactly known. Once flow through the collateral vessels is established, they become elongated and tortuous and may dilate 1 mm to 2 mm in diameter. The location of these collaterals has been studied by coronary angiography, and knowledge of their functional status matters in considering indications for coronary arterial surgery.

Gradual narrowing of the coronary arteries by *atheroma* leads to myocardial ischemia, which gives rise to characteristic chest pains (*angina pectoris*). When occlusion is advanced or complete, depending on the size and location of the affected artery, a varying amount of cardiac muscle becomes devitalized. The damage caused may vary from death of a small amount of muscle compatible with fibrous repair and scarring of the myocardium to such disruption of cardiac function that the "heart attack" is immediately fatal.

The number and size of anastomoses may increase with age in the coronary circulation, especially when there is progressive arterial narrowing. Many of these anastomoses develop in subepicardial fat and may reroute sufficient blood to the ischemic myocardium to prevent an infarct from occurring or reduce its size. Meticulous medical care following a nonfatal infarct often allows nature the necessary time to establish a collateral circulation.

Early surgical attempts to augment the coronary circulation included the creation of adhesions between the epicardium and such vascular structures as the greater omentum or the lung, which were approximated and affixed to the heart through an opening made through the pericardium. These procedures have now been abandoned in favor of more direct methods. The most recent of these is the *coronary bypass* operation. A localized obstruction is bypassed with a vein graft that extends from the aorta to the portion of the coronary artery distal to the obstruction.

## Cardiac Veins

Most of the venous blood is collected from the myocardium by veins that parallel the arteries (Fig. 21-16). These cardiac veins terminate in the coronary sinus, a large vein that empties into the right atrium. The rest of the blood in the coronary circulation is returned from the myocardium by small veins that open directly into the four chambers of the heart.

The **great cardiac vein** lies in the anterior interventricular sulcus and drains upward alongside the anterior interventricular branch of the left coronary artery. As it reaches the coronary sulcus, it turns to the left to run along the circumflex branch of the artery. The great cardiac vein becomes continuous with the coronary sinus at the point where the *oblique vein of the left atrium* enters it. As the **coronary sinus** continues to the right in the coronary sulcus, it is usually covered partially by superficial muscle fibers of the atrium. It ends in the posterior wall of the right atrium and, fairly close to its ending, receives the middle and small cardiac veins. The **middle cardiac vein** runs in the posterior interventricular sulcus, and the **small cardiac vein** runs in the coronary sulcus along the right coronary artery. The **posterior vein of the left**

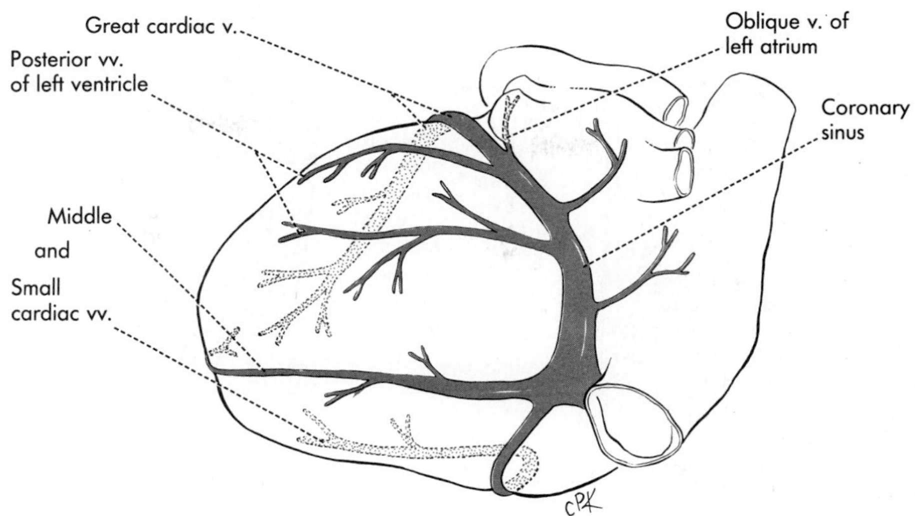


**FIG. 21-16.** The cardiac veins shown from a posterior view of the heart.

**ventricle** drains the diaphragmatic surface of the ventricle and enters the coronary sinus soon after the sinus has been formed.

The *coronary sinus* represents the left horn of the embryonic sinus venosus (Fig. 21-22); the right horn became incorporated into the right atrium (p. 518). The **oblique vein of the left atrium** is invisible or very small, but is of interest because it is the remnant of the embryonic left common cardinal vein (p. 520). When the vein is of any size, a remnant of the left superior vena cava can also usually be identified.

In addition to the tributaries of the coronary sinus, there are several small **anterior cardiac veins** that arise on the anterior surface of the right ventricle and pass across the coronary sulcus to penetrate directly the anterior wall of the right atrium. Finally, although not visible by dissection, there are, in the muscular walls of the heart, minute veins, the **least cardiac veins** (venae cordis minimae), that empty directly into the cardiac chambers. These small venous orifices are said to be most numerous in the right atrium and least numerous in the left ventricle.

### Lymphatic Drainage

The lymphatics of the heart form networks adjacent to the endocardium and the epicardium. The efferent vessels drain along the coronary arteries and empty into lymph nodes associated with the lower end of the trachea (tracheobronchial nodes; see Fig. 20-9).

## THE INNERVATION AND THE CONDUCTING SYSTEM OF THE HEART

A number of cardiac nerves derived from both the sympathetic and parasympathetic components of the autonomic nervous system commingle in the cardiac plexus, located in front of the tracheal bifurcation (Fig. 21-17). Offshoots of this plexus innervate the lungs (pulmonary plexuses described in the previous chapter) and the heart. These nerves reach the heart along the coronary arteries and serve to modify the rate and strength of the heart beat; they are not responsible for initiating or maintaining it. The atria and ventricles contract in orderly sequence before nerves contact the embryonic heart, and the same is true when the transplanted heart is severed from its nervous connections. This automatic rhythmicity is largely generated by an intrinsic "pacemaker" known as the sinuatrial node, from which impulses spread to the rest of the myocardium. The node is the first component of the heart's so-called conducting sys-

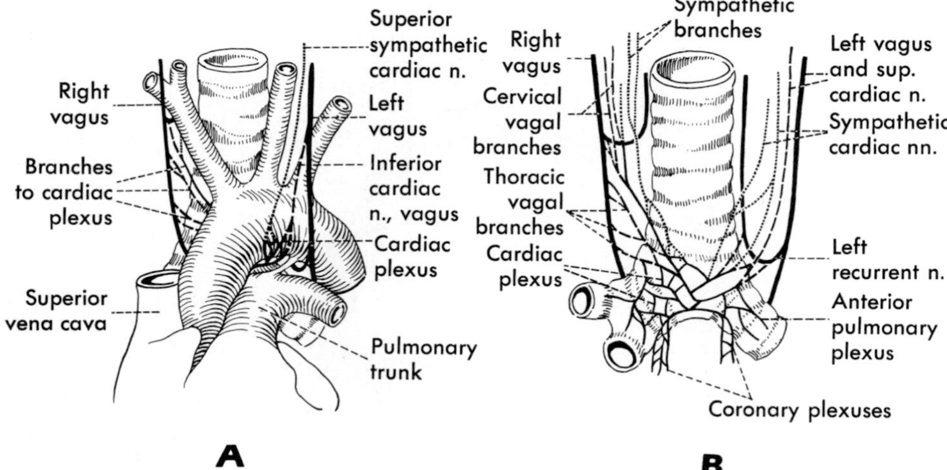


**FIG. 21-17.** Cardiac nerves and the cardiac plexus shown, in **A**, with the aorta and pulmonary trunk left *in situ* and, in **B**, with the vessels removed to expose the bifurcation of the trachea. Cardiac branches of the vagi are shown in broken lines, and cardiac branches of the cervical sympathetic ganglia are shown in dotted lines. The thoracic cardiac nerves, medial branches of the upper thoracic sympathetic ganglia, are not shown.

tem, an anatomically discrete network of myocardial tissue specialized for rapid conduction. Efferent nervous impulses primarily influence the pacemaker and coronary blood flow, whereas afferent impulses are concerned with reflex adjustments of the heart beat, circulation, and respiration, and they also mediate cardiac pain.

## The Cardiac Plexus

This diffuse network of delicate nerves extends from in front of the trachea to the aortic arch, the pulmonary trunk, and the ligamentum arteriosum, which connects the two vessels (Fig. 21-17). The **parasympathetic** input to the plexus is provided by the two vagi. Two long, slender branches (or groups of branches) arise from each vagus in the neck (*superior and inferior cardiac branches*), and a variable number of shorter nerves are given off also in the thorax, either from the vagal trunks or from their recurrent laryngeal branches (*thoracic cardiac branches*). **Sympathetic cardiac branches** likewise arise in both the neck and the thorax. The high origin of the major cardiac nerves is a reminder of the location of the developing heart, opposite cervical segments, when it first received its nerve supply. Each sympathetic trunk in the neck gives off three cardiac branches (*superior, middle, and inferior cardiac nerves*), one from the superior cervical ganglion, one from the middle, and a third from the cervicothoracic (stellate) ganglion. The small and delicate *thoracic cardiac nerves* are given off from the upper four or five ganglia of the thoracic sympathetic trunks.

**Efferent vagal fibers** that enter the cardiac plexus are preganglionic axons of neurons located in vagal nuclei of the brain stem. They synapse in minute *cardiac ganglia* found in the plexus or in the walls of the atria. Vagal stimulation slows the heart beat and constricts the coronary arteries.

**Sympathetic efferents** that enter the plexus are postganglionic. The corresponding preganglionic neurons are in the intermediolateral cell column of the first four or five thoracic cord segments, and their axons relay in ganglia of the sympathetic trunk from which the cardiac nerves issue. Sympathetic efferents increase the rapidity and strength of the heart beat and dilate the coronary vessels.

Both vagal and sympathetic nerve fibers reach the heart by the two **coronary plexuses** given off from the cardiac plexus. The nerves are distributed along the coronary vessels chiefly to the atria, the sinuatrial node, and other components of the conducting system, which are said to have a particularly rich innervation. Although numerous nerve endings have been described, the nature of the contact between nerve fibers and cardiac muscle remains unknown.

All vagal fibers to the heart can be blocked only by interruption of the individual cardiac branches of both vagi, or by interruption of the vagal trunks above the level of origin of all their cardiac branches. This is true also of the sympathetic innervation to the heart: the cervical portion of the sympathetic trunk receives no preganglionic fibers from cervical nerves, for its preganglionic fibers ascend from the upper part of the thoracic portion of the trunk (p. 58). Therefore, removal of the upper thoracic portion of the trunk over a distance of about five segments will interrupt not only the origins of the thoracic cardiac nerves but also the preganglionic fibers that synapse in sympathetic ganglia located in the neck.

**Visceral afferents** from the heart join the major part of the cardiac plexus and then pass along the sympathetic and vagal cardiac branches. The vagal afferents are concerned with cardiac reflexes, whereas afferents in the sympathetic cardiac nerves conduct pain sensation from the heart.

The vagal afferent impulses originate in various types of interoceptors that sense changes in tension or pressure (baroreceptors) or in the carbon dioxide or oxygen content of blood (chemoreceptors). Baroreceptors are located in the wall of the great veins as they enter the heart and help to regulate the cardiac output according to the amount of filling in these veins; others are in the wall of the arch of the aorta or its branches at the base of the neck where they sense changes in blood pressure and set in operation the reflexes that adjust it. Chemoreceptors (variably called aortic or paraaortic bodies, glomus aorticum, and glomus pulmonale) are located on the outside of the great vessels, primarily between the ascending aorta and the pulmonary trunk, and initiate respiratory and cardiac reflexes in response to lowered oxygen tension in aortic blood.

Whereas cardiac afferents in the sympathetic nerves may also be involved in reflexes, it is important to realize that sympathetic cardiac nerves are the sole conductors of *pain from the heart.* From the sympathetic trunks, all pain afferents eventually enter the dorsal roots of the upper four or five thoracic nerves, the same segments that give rise to preganglionic cardiac efferents. The afferent fibers in the cardiac branches of cervical sympathetic ganglia descend in the sympathetic trunks to the upper thoracic segments and join the pain fibers in the thoracic cardiac nerves. Thus, bilateral destruction of the upper five segments of the thoracic sympathetic trunks can be expected to interrupt the entire pathway for pain from the heart and has indeed been shown to abolish entirely the pain typically associated with coronary arterial disease and with aneurysms of the aortic arch. Because afferent fibers concerned with cardiac pain terminate in the dorsal horns of the same cord segments as somatic afferents contained in upper thoracic spinal nerves, the pain of *angina pectoris* usually is interpreted by the patient as being localized along the ulnar border of the upper limb (dermatomes supplied by T1 and T2) and the upper part of the thorax (dermatomes T2–T5). The pain is caused by myocardial ischemia and is sensed by pain receptors in the heart, but it is **referred** to somatic structures supplied from the same cord segments as the heart.

## The Conducting System

The conducting system of the heart consists of the sinuatrial node, the atrioventricular node, the internodal fasciculi, and the atrioventricular bundle or fasciculus, which divides into a left and a right branch, or crus (Fig. 21-18). In each ventricle, these branches, or crura, terminate in a subendocardial plexus of so-called Purkinje fibers, which become continuous with ventricular myocardial fibers. All these components of the conducting system are specially differentiated cardiac muscle fibers, separated from ordinary myocardium by delicate envelopes of connective tissue. In the human heart, the conducting system can be identified only by serial histologic sections, whereas in ungulates, the atrioventricular bundle and its branches are macroscopically visible and may be displayed by dissection or by injection of their connective tissue sheath with India ink. Histologic differences between conducting tissue and ordinary myocardium are likewise less marked in the human than in some animals.

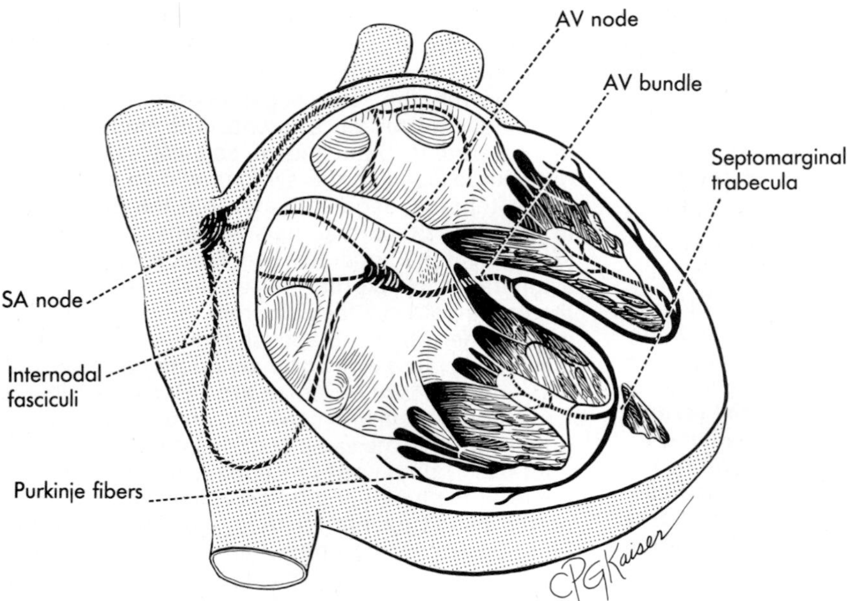


**FIG. 21-18.** The conducting system of the heart.

The **sinuatrial node** is a crescent-shaped structure, 5 mm to 8 mm in length, that occupies the whole thickness of the wall of the right atrium. It is located along the anterior circumference of the superior vena caval orifice at the upper end of the sulcus terminalis.

The **atrioventricular node** is oval and slightly smaller than the sinuatrial node. The node lies in the substance of the interatrial septum, resting with its inferior surface on the fibrous atrioventricular ring, close to the attachment of the septal cusp of the tricuspid valve. Within the septum, the node extends forward from the opening of the coronary sinus, and its cells continue anteriorly into the atrioventricular bundle, which passes through the fibrous ring and connects the node to the ventricles (Fig. 21-13).

The atrioventricular node is also connected to the sinuatrial node by three delicate fiber bundles of conducting tissue. These **internodal fasciculi,** although clearly definable anatomically, are of doubtful significance in the conduction of impulses from the sinuatrial to the atrioventricular node. According to physiological studies, impulses generated in the pacemaker spread to the atrioventricular node uniformly through the atrial myocardium, and the internodal fasciculi (as well as the interatrial fasciculus, a fourth bundle that leaves the sinuatrial node) are not involved in activating the atrioventricular node (or the myocardium of the left atrium).

The activation of the ventricles is achieved through the **atrioventricular fasciculus** or **bundle.** Disruption or block of the bundle dissociates the beat of the ventricles from that of the atria and is responsible for clinically well-defined cardiac arrhythmias. The atrioventricular bundle is a fasciculus of conducting tissue that has the thickness of a wooden matchstick. It commences as the forward continuation of the atrioventricular node. The bundle passes through the fibrous ring in a hole located at the margin of the right fibrous trigone (Fig. 21-13), and it is the only connection between the myocardium of the atria and the ventricles. On the ventricular side of the ring, the bundle runs downward and forward, skirting the posterior margin of the membranous part of the interventricular septum, and soon branches into a right and a left crus. The two **crura** straddle the upper border of the muscular part of the interventricular septum.

Each crus or branch descends toward the apex of the ventricles under the endocardium of the septum. Knowledge of these relationships matters in the surgical repair of interventricular septal defects.

The right crus of the atrioventricular bundle crosses from the septum to the ventricular wall along the *septomarginal trabecula*, whereas the left crus breaks up into branches and reaches the ventricular wall along several trabeculae carneae. These fasciculi spread out as the subendocardial **Purkinje fibers** over the ventricular wall and over the papillary muscles. Because of this arrangement, the impulses dispatched from the atrioventricular node first activate ventricular muscle in the region of the apex, assuring a "milking" action of the ventricles toward the openings of the aorta and the pulmonary trunk.

The vicinity of the sinuatrial and atrioventricular nodes and of the atrioventricular bundle is richly innervated. Nerve fibers penetrate into the node and the bundle. Ganglion cells, however, are confined to the neighborhood of the atrial portion of the conducting system.

The blood supply of the conducting system was described in the previous section with the coronary arteries. Each node is closely associated with a small artery that is named after the node. Although the myocardium that surrounds the nodes, the atrioventricular fasciculus, and its crura are highly vascular, there are no capillaries in the nodes or the bundles.

Disease conditions that interfere with conduction along the atrioventricular system necessarily interfere with the normal rhythm of ventricular contraction. Interruption of transmission along the trunk produces *heart block*, which dissociates the contraction of the atria and the ventricles, so that the atria may beat at one speed and the ventricles at a slower one. In consequence of this, the atria sometimes will attempt to force blood into an already filled or a contracting ventricle. Also, the ventricle may contract whether or not it is filled and therefore acts inefficiently. In some cases, only one branch of the atrioventricular bundle may be blocked (*bundle branch block*), which affects only one ventricle. In recent years, the condition has been alleviated by implanting in the ventricle an electrode connected to an electronic cardiac pacemaker that will deliver appropriate rhythmic shocks to the heart and produce proper ventricular contraction.

## THE GREAT VESSELS

The great vessels include the ascending aorta, the pulmonary trunk, the two venae cavae, and the four pulmonary veins. All have been encountered in this chapter, and the purpose of the present section is to summarize their anatomy. Nothing needs to be added to the description of the inferior vena cava and pulmonary veins, which have a very short intrapericardial course.

The **pulmonary trunk** and the **ascending aorta** are each about 5 cm long and 3 cm in diameter. Although the pulmonary trunk conveys venous blood from the heart to the lungs, anatomically it is an artery (as are its branches, the pulmonary arteries), and the structure of its wall is similar to that of the aorta. Both vessels run a slightly twisted or spiral course within the pericardial sac. They entwine in such a manner that the pulmonary trunk obscures a substantial portion of the ascending aorta from view. The posterior wall of the pulmonary trunk is closely applied to the anterior wall of the aorta, as both walls develop from the spiral septum that divides the embryonic truncus arteriosus and bulbus cordis into these two vessels (p. 548).

The pulmonary trunk begins at the pulmonary orifice of the right ventricle, and the aorta, at the aortic orifice of the left ventricle. Each orifice is surrounded by a fibrous ring (Fig. 21-13) and can be closed by the pulmonary and aortic semilunar valves (described earlier) that separate the lumen of each vessel from that of the respective ventricle (Fig. 21-11). The pulmonary orifice is anterior, and from it, the pulmonary trunk passes upward, backward, and to the left. The aortic orifice is posterior, and from it the aorta ascends toward the right and slightly forward. The pulmonary trunk terminates by dividing into the right and left pulmonary arteries opposite the left border of the sternum, behind the 2nd costal cartilage. The ascending aorta reaches a similar position on the right side of the sternal angle and turns backward to become the arch of the aorta. Immediately above the cusps of the aortic and pulmonary valves, there are three dilatations on the aorta and on the pulmonary trunk. These are the **aortic** and **pulmonary sinuses.** The coronary arteries arise from the left and right aortic sinuses (Fig. 21-15). The dilated base of the aorta, on which the aortic sinuses bulge, is sometimes called the *bulb of the aorta.* As the ascending aorta becomes continuous with the aortic arch, it dilates again slightly to the right. This unnamed dilatation is not present along the right margin of the cardiac outline in a posteroanterior x-ray film because to its right lies the superior vena cava. The pulmonary trunk, on the other hand, is visible along the left margin of the cardiac outline as it lies in the concavity of the aortic arch (Fig. 21-7).

As described earlier (see The Serous Pericardium and Pericardial Cavity), the pulmonary trunk and the ascending aorta are enclosed by a common sleeve of serous pericardium. The relationship of these vessels to intrapericardial structures was discussed in connection with the transverse sinus of the pericardium. The transverse sinus is posterior to the two great arteries.

Only the lower half of the **superior vena cava** is within the pericardial sac, which it enters behind the sternal angle. The vessel is formed in the superior mediastinum, and it is described in the next chapter. Its position and relations within the pericardial sac were dealt with earlier.

## THE HEART IN THE LIVING BODY

For assessing cardiac function, it is necessary to relate the general position of the heart and its specific anatomic features to landmarks on the precordium. The **precoridum** is that part of the chest wall that roughly overlies the heart and the great vessels. This section deals with the position of the heart in the chest, the cardiac outline, and the location of the valves and auscultatory areas and relates the heart sounds, briefly, to events in the cardiac cycle.

**Position.** The heart rests with its inferior surface on the central portion of the diaphragm, and its inferior border roughly corresponds to the level of the xiphisternal junction. The normal heart is somewhat larger than a person's clenched fist, and one-third of it lies to the right of the median plane and extends slightly beyond the edge of the sternum; two-thirds lie to the left of the median plane. The exact position of the heart depends on many factors, and the surface projection described below is largely an approximation that holds true for the *average* heart in an *average* individual.

In the recumbent position the diaphragm is higher and so is the heart. Due to its own weight, the heart shifts significantly toward the side the person is lying on, and it also approximates the chest wall on leaning forward. These are points to remember in palpation and auscultation of the heart. Figure 21-19 shows the changes in the same heart that are due to sustained, deep inspiration and expiration. The position of the heart is different in a narrow, slender person than in a person with a broad, stocky build (Fig. 21-20). Abdominal distention (*e.g.*, pregnancy), unequal pressure or tension in the two pleural cavities, or shift of the mediastinum due, for instance, to spinal curvatures, all have an influence on the position of the heart regardless of whether or not the heart itself is normal.

**Surface Projection.** There is only one point of the heart that can be directly identified on the precordium: the apex. A **cardiac impulse** may be visible at the apex, and palpation over it will confirm the presence of the **apex beat.** In many normal individuals, the cardiac impulse is not visible, and the apex beat is felt with difficulty or not at all. In quiet breathing and in the supine position, the apex is located in the left 5th intercostal space just medial to the midclavicular line. The outflow tracts of the ventricles are behind the sternum just below the sternal angle. The region of the pre-

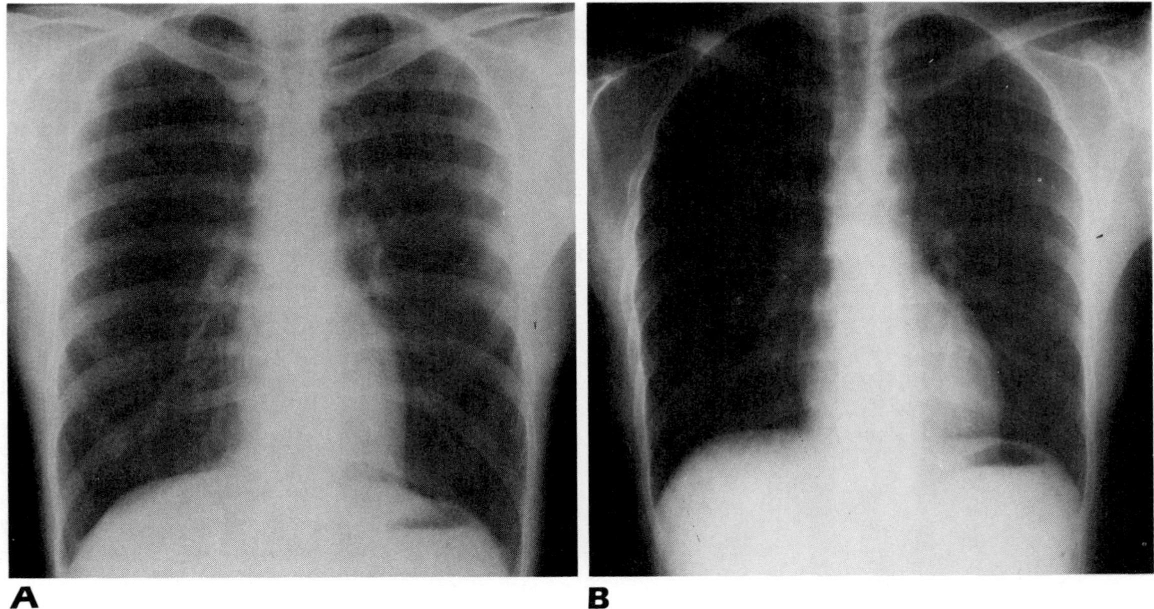


**FIG. 21-19.** Radiograph of the chest of a young man, showing changes in the shape and position of the heart during deep inspiration **(A)** and expiration **(B).**

**FIG. 21-20.** Contrasting types of hearts in persons of different body build. On the left is the slender, more vertical type of heart; on the right is a wider, more transverse type. There are numerous intergrades between these two extremes. (Courtesy of Dr. D. G. Pugh)

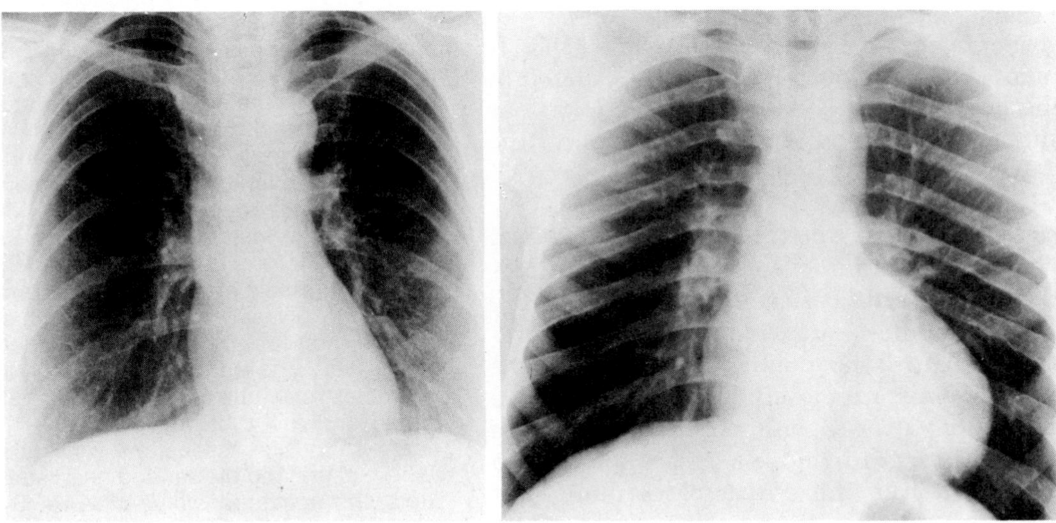

cordium that overlies the outflow tracts is known to clinicians as the **base of the heart.** It must be borne in mind that the term in this sense is different from that used by anatomists (p. 524).

Figure 21-21 shows the surface projection of the cardiac borders, chambers, and valves. The right and left borders may be confirmed by percussion of the precordium, although the overlapping lungs make it difficult to map out cardiac dullness. The coronary sulcus runs diagonally from the left upper to the right lower

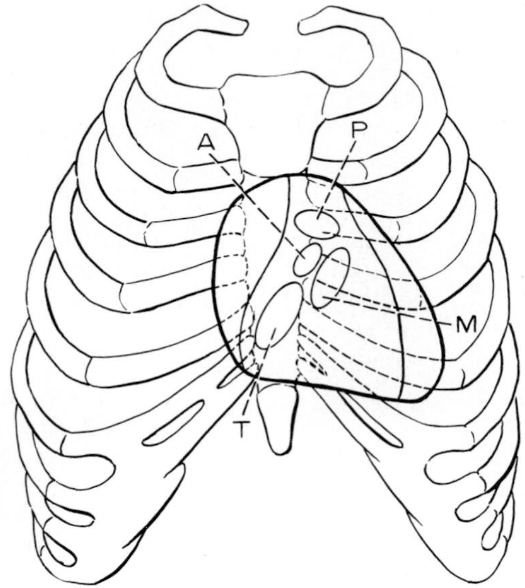


**FIG. 21-21.** Approximate projection of the heart onto the anterior thoracic wall. The large middle portion of the anterior surface of the heart is formed by the right ventricle; the right and left borders are formed by the right atrium and left ventricle, respectively. The letters **P, A, M,** and **T** identify the positions of the pulmonary, aortic, mitral, and tricuspid valves, respectively.

corner of the quadrangular cardiac outline. It marks the position of the fibrous skeleton of the heart, which is seen almost edge on but slightly from below. In it are located all the valves, which also appear edge on.

**Heart Sounds and Auscultatory Areas.** There is no unanimous agreement about precisely what physical events in the heart contribute to the generation of audible vibrations. There is no doubt, however, that the heart sounds are closely associated with the closure of valves. Opening of the valves in the normal heart is inaudible. Heart sounds can be heard over the entire precordium, but there are specific auscultation areas where sounds made by individual valves can be perceived to best advantage. For all but one valve, these areas do not coincide with the surface projection of the valves, probably because vibrations made by a valve are carried along with the blood to a point of optimal intensity.

Sounds generated by the **mitral valve** are best heard over the apex, and those of the **tricuspid valve** over the anterior wall of the right ventricle, which happens to superimpose on the projection of the valve itself. Sounds of the **aortic valve** radiate along the ascending aorta to the 2nd costal cartilage at the sternal border on the right where the aorta is closest to the chest wall. This is the aortic auscultatory area. The **pulmonary,** or pulmonic, area is along the left sternal border in the second space, where the pulmonary trunk divides.

A comprehensive discussion of the **cardiac cycle** is not the objective of this section, but correlation of the heart sounds with ventricular events serves a useful purpose in the present context.

There are two heart sounds, the first and the second, that follow each other in quick succession, and they are separated from the next doublet by a longer interval. During this interval, the ventricles are relaxed **(diastole),** whereas between the two heart sounds, the ventricles are contracting **(systole).** In diastole, the atrioventricular valves are open and the aortic and pulmonary valves are closed. Blood is pouring into the atria and flows freely into the ventricles through the open mitral and tricuspid valves. Then the ventricles begin to contract, and the rise in pressure closes the atrioventricular valves: the first heart sound is produced, signaling the beginning of ventricular systole. Simultaneously, contraction of the ventricular muscle makes the apex rise, resulting in the visible and palpable heart beat. Concurrent with the rising ventricular pressure, the aortic and pulmonary valves open silently and blood is ejected into the aorta and pulmonary trunk.

When ventricular contraction is complete, the ventricles are empty and blood under pressure in the great vessels closes the aortic and pulmonary semilunar valves, which generate the second heart sound. The second heart sound signals the end of systole and the commencement of diastole. The ventricles relax, the atrioventricular valves open silently, and the cycle is repeated. Thus, the first heart sound is associated with atrioventricular valve closure and the second heart sound is associated with semilunar valve closure. Systole is between the first and second heart sounds and diastole between the second and first heart sounds.

## EMBRYOLOGY

### NORMAL DEVELOPMENT

The heart first takes form by the fusion of two endothelial tubes that are formed in splanchnic mesoderm. The splanchnic mesoderm surrounding these tubes differentiates into the myocardium, and the outer layer of the mesodermal cells (the celomic epithelium) becomes the epicardium (see Fig. 19-12B and C). The endocardium of the fully formed heart corresponds to the primitive endothelial cardiac tubes; from this endothelium develop also the cardiac valves and most of the septa. The five divisions of the primitive heart tube (*truncus arteriosus, bulbus cordis, ventricle, atrium, sinus venosus*) were identified in the introductory section to this chapter (Fig. 21-1). The purpose of the present section is to summarize the changes that lead to the establishment of the external form of the heart, the partitioning of its cavity into four chambers, and the development of the great vessels. External and internal changes progress concomitantly, although here they are discussed under separate headings.

#### Establishment of External Form

Four major events transform the simple endothelial cardiac tube into an organ that resembles the fully developed heart: (1) flexion of the heart tube between the bulbus and the ventricle; (2) fusion of the bulbus and the ventricle; (3) dorsal migration of the atrium in relation to the ventricle, which draws out the sinus venosus from the septum transversum into the pericardial cavity; (4) enlargement of the right horn of the sinus venosus and its merging with the right side of the atrium. Only the last of these events needs discussion; the first three have already been mentioned in the introduction to this chapter.

The left horn of the sinus venosus atrophies while its right horn enlarges, owing to the development of two major anastomoses that divert venous blood from the left side of the embryo to the right side of the heart (Fig. 21-22). One anastomosis is located caudal to the heart within the liver, the other, rostral to the heart in the mediastinum. The *caudal anastomosis* is the **ductus venosus,** which shunts blood from the left umbilical and vitelline veins to the proximal segment of the right vitelline vein. The terminal portion of the inferior vena cava develops from this terminal segment of the right vitelline vein (Fig. 21-22B and C). The segments of the umbilical and vitelline veins that are excluded from the blood flow soon atrophy. The ductus venosus remains functional as long as placental blood has to reach the heart; after birth it is replaced by a fibrous cord, the *ligamentum venosum,* that is devoid of a lumen. The *proximal anastomosis* is between the left and right precardinal veins. The **precardinal anastomosis** persists as the left brachiocephalic vein, whereas the more caudal portion of the left precardinal vein atrophies, as does the left common cardinal vein. In the fully developed heart, the latter is represented by the oblique vein of the left atrium. On the right side, the precardinal vein, below the anastomosis, and the common cardinal vein together make up the superior vena cava. As a consequence of these two venous shunts, the *right horn of the sinus venosus* enlarges while its *left horn* shrinks and becomes the *coronary sinus* (Fig. 21-22). The enlarged right horn, the tributaries of which can now be called the superior and inferior venae cavae and the coronary sinus, becomes incorporated into the right side of the atrium as this chamber is being partitioned. By the sixth week of embryonic development, the heart closely resembles that of its fully developed form.

#### Establishment of Internal Form

The merging of the bulbus cordis with the primitive ventricle and the sinus venosus with the right atrium modifies and enlarges the chambers of the primitive heart. The left atrium also is augmented by the incorporation of the developing pulmonary veins into its posterior wall. The division of the interior of the heart into its four definitive chambers is the result of four major events: (1) expansion and division of the atrioventricular canal; (2) partitioning of the common atrium; (3) partitioning of the common ventricle; (4) division of the truncus arteriosus and the adjoining portion of the bulbus cordis.

The **atrioventricular canal** is the passage between the primitive atrium and the primitive ventricle, the forerunner of the anatomic left ventricle (Fig. 21-1D, p. 518). The atrioventricular canal soon expands to the right so that it opens also into the future right ventricle. The right ventricle is derived chiefly from the bulbus cordis, which is placed, by the looping of the heart, on the right side of the primitive ventricle (p. 518). Following fusion of the bulbus and the primitive ventricle, the expansion of the atrioventricular canal to the right establishes connection between the primitive atrium and the common chamber of the future left and right ventricles. This expanded atrioventricular canal is marked on the exterior of the heart by the coronary sulcus. The canal soon becomes partitioned into two orifices by swellings that are produced by the proliferation of the endocardium in the dorsal and ventral walls of the canal (Fig. 21-23). These **endocardial tubercles** or **cushions** grow toward each other and, when they meet, divide the single atrioventricular canal into left and right atrioventricular orifices, which will now connect the right and left halves of the primitive atrium with the future right and left ventricles, respectively. The tricuspid and mitral valves develop in these orifices.

The **partitioning of the atrium** is accomplished by two overlapping septa, the septum primum and the septum secundum (Fig. 21-23). The development of these septa, both produced by endothelial proliferation, is associated with three foramina: the foramen primum, the foramen secundum, and the foramen ovale. The **septum primum** descends as a

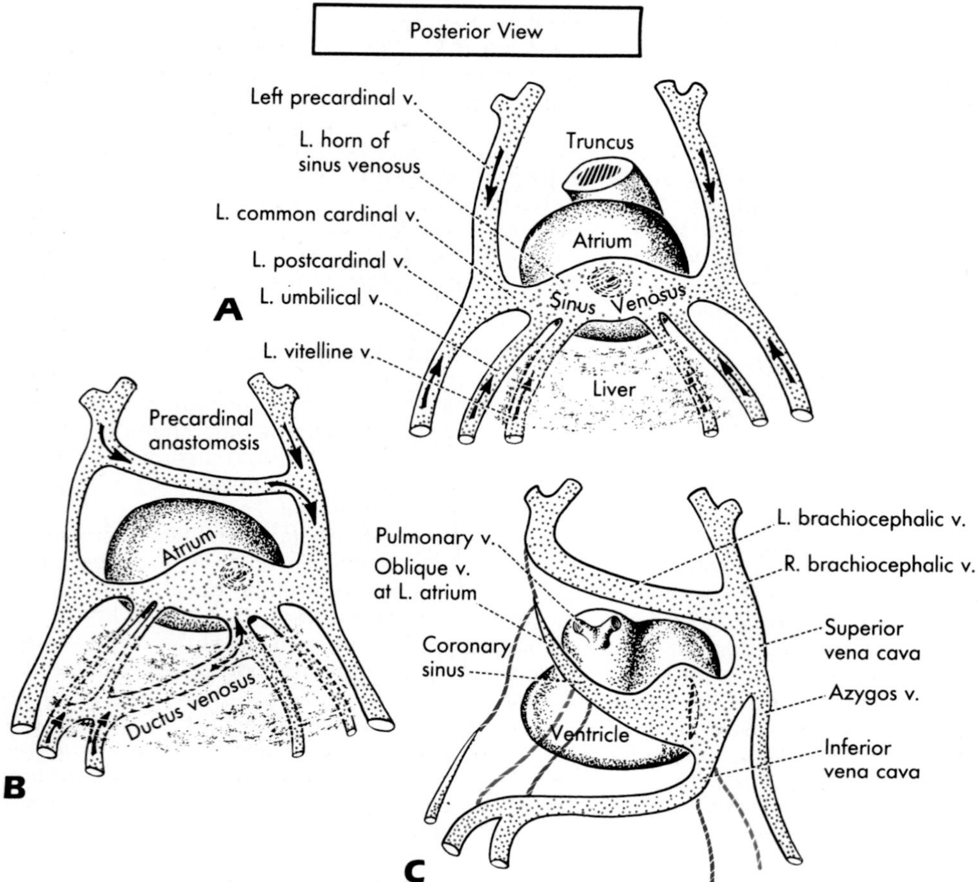


**FIG. 21-22.** Schematic representation of developmental events that modify the sinus venosus. The primitive heart tube (shown in Fig. 21-1**A** and **B**) is seen here from a caudal or posterior view. In **A**, the most primitive arrangement is shown, with the sinus venosus receiving three pairs of veins and emptying into the common atrium. Caudal and rostral anastomoses have developed between the tributaries of the sinus venosus at the stage shown in **B**, which leads to the atrophy of certain venous channels and to the enlargement and persistence of others **(C)**.

curtain from the roof of the atrium toward the bridge formed by the fused endocardial cushions, and the gap between its inferior edge and the cushions is the **foramen primum.** When the septum contacts the endocardial cushions, the foramen primum is obliterated and a second opening, the **foramen secundum,** is formed by the breakdown of an area in the septum primum close to the atrial roof.

On its right, the foramen secundum becomes covered over by the **septum secundum,** but remains patent. Like the septum primum, the septum secundum grows from the roof of the atrium toward the endocardial cushions, but most of its crescent-shaped edge (known as the *limbus fossae ovalis*) fails to reach the cushions and leaves a defect that is the **foramen ovale.** Like the foramen secundum, the foramen ovale remains patent, but is overlapped on the left by the septum primum. The septum primum acts like a flap valve (and is sometimes called the *valvule of the foramen ovale*) that permits the flow of blood from the right atrium through the foramen ovale and foramen secundum into the left atrium. After birth, the two septa become pressed against each other by the rising pressure in the left atrium, and the foramina ovale and secundum become functionally closed. The two septa may not fuse, however, for a considerable period of time, and a cardiac catheter may be passed between the apposed septa from the right to the left atrium. The foramen ovale remains probe patent up to the age of 5 years in 50% of persons and up to 20 years of age in more than 25%.

The **interventricular system** is derived from three elements: the muscular septum, the endocardial cushions, and the bulbar ridges. The *pars muscularis* of the interventricular septum commences its growth in the convex region of the cardiac loop opposite the flange of the bulboventricular septum (Fig. 21-1) even before that septum becomes absorbed. The muscular septum grows upward toward the endocardial cushions, and its crescentic upper edge progressively reduces the *interventricular foramen* (Fig. 21-23). This foramen is the communication between the right and left ventricles, which are derived from the greater part of the bulbus cordis and the primitive ventricle, respectively. The superior portion of the interventricular septum that fills in the interventricular foramen is derived from the endocardial cushions. An area in this portion of the septum remains transparent and membranous (*pars membranacea*). The anterior cusp of the mitral valve and the septal cusp of the tricuspid valve are also derived from the same endocardial tissue as the pars membranacea, and so is the lower portion of the interatrial septum. This explains the existence of the membranous atrioventricular septum (p. 531).

The part of the interventricular septum that divides the outflow tracts of the right and left ventricles below the level of the semilunar valves is situated anterosuperiorly and is formed by the bulbar ridges. The **bulbar ridges** are two longitudinal swellings that meet in the midline and divide the lumen of the distal portion of the bulbus into the *conus arteriosus* or infundibulum of the right ventricle and the *aortic vestibule* of the left ventricle. Inferiorly, these bulbar ridges fuse with the components of the interventricular septum that are derived from the endocardial cushions and the muscular septum. The *crista supraventricularis* in the right ventricle marks the lower limit of the bulbar ridges.

Above the level of the semilunar valves, the **division of the truncus arteriosus** into the pulmonary

**FIG. 21-23.** Two stages in the division of the common atrium into right and left chambers shown schematically. The plane of section through the primitive heart tube permitting this view of the interior is shown in the inset, which corresponds to Figure 21-1**D.** In **A,** only the septum primum has formed; in **B,** the septum secundum commenced its growth on the atrial roof between the septum primum and the septum spurium. The septum spurium and the right valve of the sinus venosus give rise to the crista terminalis. At the stage shown in **B,** as yet only a single (common) pulmonary vein opens into the left atrium.

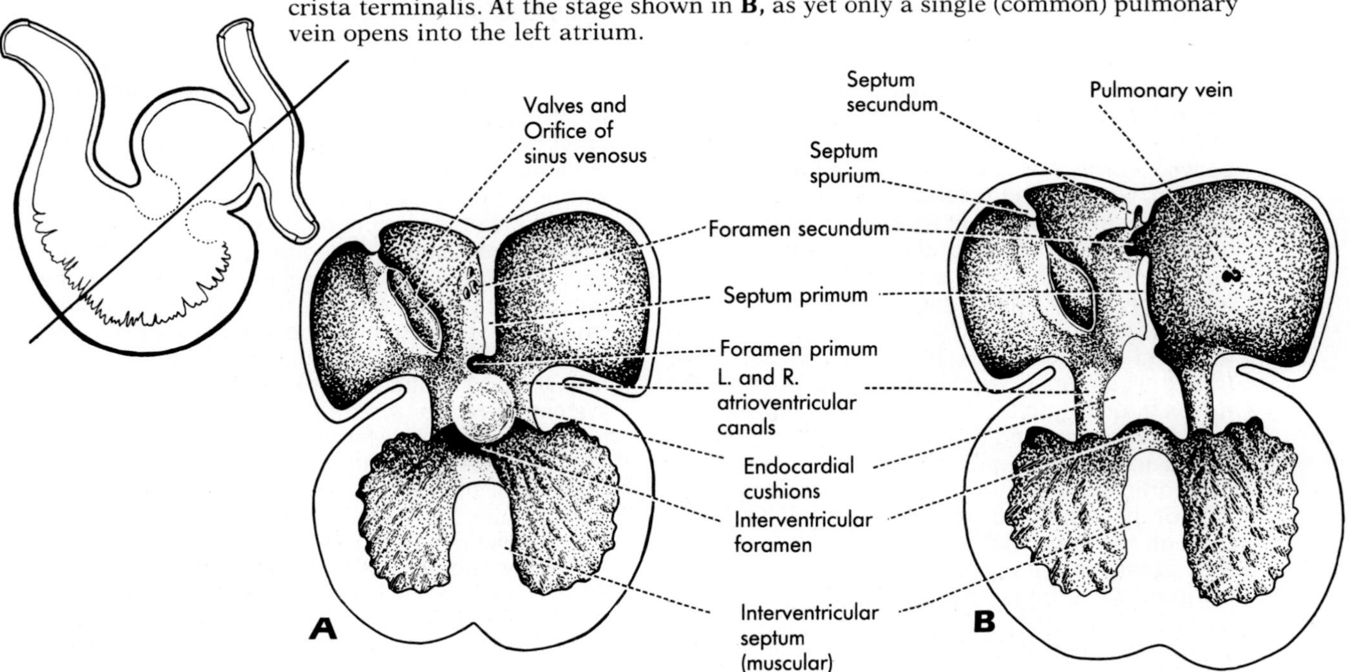

trunk and ascending aorta is accomplished by the formation of the **spiral** (or *aorticopulmonary*) **septum.** This septum is formed by two longitudinal ridges in the truncus, which are continuous with the bulbar ridges. The spiraling of the truncal ridges is such that when the aorticopulmonary septum is completed, the pulmonary trunk connects with the right ventricle and the aorta connects with the left ventricle.

After the completion of the septa, the heart rotates in such a manner that the conus arteriosus and the pulmonary trunk come to lie more in front than to the right of the aorta and the aortic vestibule. In the fully developed heart, the aortic valve is posterior and to the right of the pulmonary valve. Distally, the pulmonary trunk is in communication with the 6th pair of aortic arches (the future pulmonary artery) and the aorta is in communication with the 4th pair of aortic arches (the future aortic arch). These changes are completed by the eighth week of human development, and only after birth does the heart adapt to maintaining separate pulmonary and systemic circulatory circuits.

## FETAL CIRCULATION

The circulation in the fetus differs in several important respects from that in the adult. The major changes that adapt the circulation to the exchange of blood gases in the lungs, rather than in the placenta, take place at the time of birth and during a relatively short transitional period that extends over the first few days of neonatal life. In the adult heart, the vascular circuits fed by the right and left ventricles are maintained independently of one another, except for the minor anastomoses that exist between bronchial and pulmonary blood vessels (p. 499), whereas in the fetus, there are large intracardiac and extracardiac shunts between the two circuits. In the adult, all of the right ventricle's output is spent in the perfusion of the lungs; in the fetus, the lungs offer a high resistance to blood flow so that more than 75% of the blood pumped by the right ventricle bypasses the lungs and flows from the pulmonary trunk through the **ductus arteriosus** directly into the aorta. The ductus arteriosus is a large connecting channel between the pulmonary trunk and the aorta and is a derivative of the 6th aortic arch. In the fetus, only approximately 50% of the combined ventricular output circulates through the lower part of the body, and the other 50% passes into a pair of **umbilical arteries** given off from the terminal branches of the descending aorta. These arteries leave the body of the fetus and perfuse the placenta. Blood traversing the placenta meets very little resistance and after it has been purified of metabolites and enriched with oxygen and nutrients, it is returned to the fetus along a single **umbilical vein.** Approximately half of this blood trickles through the substance of the liver before it enters the inferior vena cava; the other half bypasses the vascular liver parenchyma and reaches the inferior vena cava more directly through a large venous channel embedded in the liver, named the **ductus venosus.**

The entrance of the inferior vena cava into the right atrium is in line with the **foramen ovale.** About two-thirds of the stream of blood entering the right atrium from the inferior vena cava is directed, by the valve of the inferior vena cava, into the left atrium through the foramina ovale and secundum. The remaining one-third of the inferior vena caval blood passes into the right ventricle, together with the venous blood that is returned to the right atrium from the superior vena cava. The right ventricle pumps this blood into the pulmonary trunk and most of it enters the aorta via the ductus arteriosus, whereas only a small fraction passes through the lung before it is returned to the left atrium. This venous blood, together with the inferior vena caval blood that entered the left atrium through the foramina ovale and secundum, passes into the left ventricle, which pumps it to the aorta. Because of the existence of the right-to-left shunt between the two atria (through the foramina ovale and secundum) and between the pulmonary trunk and aorta (through the ductus arteriosus), the two ventricles in the fetus work in parallel, pumping their combined output into the aorta, except for the small fraction that reaches the lungs from the pulmonary arteries.

Two important events change the dynamics of the circulation at the time of birth: (1) the first breath of air is taken and (2) the newborn is separated from the placenta. The factors responsible for initiating the first breath remain conjectural, but with the entry of air into the lungs and the expansion of the lungs, pulmonary vascular resistance is suddenly reduced, which greatly increases pulmonary blood flow and causes a fall of pressure in the pulmonary trunk, right ventricle, and right atrium. On the left, or systemic, side, the abrupt exclusion of the low-resistance vascular bed of the placenta from the circulation results in an increase of overall systemic vascular resistance, causing the pressure to rise in the aorta, left ventricle, and left atrium. This increase in left atrial pressure, and the fall in right atrial pressure, presses the thin septum primum against the more rigid septum secundum and functionally closes the foramen ovale.

The combined effect of the decrease in pulmonary vascular resistance and the increase in systemic vascular resistance reverses the flow through the ductus arteriosus. During the first few hours of neonatal life, blood is shunted from the aorta into the pulmonary trunk, which greatly increases blood flow through the lungs and consequently augments the amount of blood returned to the left atrium. Functional closure of the ductus begins at birth by muscular contraction; the major impetus for this is oxygen. Anatomic closure may not be complete for several months, and in addition to the contraction, it is achieved by thickening of the intima and thrombosis. The duct eventually becomes converted into a fibrous cord, the **ligamentum arteriosum.**

Changes that continue to take place during the neonatal period include the obliteration of the ductus venosus, the umbilical vein, and the umbilical arteries; an increase in the muscle mass of the left ventricle compared with that of the right; and, consistent with the pressure changes in the pulmonary and systemic circulations, an increase in smooth muscle in the systemic vascular bed compared with the relative decrease in the amount of smooth muscle associated with pulmonary blood vessels.

## ANATOMIC ABNORMALITIES OF THE HEART AND GREAT VESSELS

Anatomic abnormalities of the heart and great vessels may be *congenital* or *acquired.* The majority of congenital cardiac abnormalities are the result of interference with normal cardiac development during the first seven weeks of intrauterine life. Acquired anatomic abnormalities are caused by diseases that exert their effect on the fully developed heart after birth. In the present context, anatomic defects of the heart and great arteries and veins will be considered chiefly to illustrate crucial points of development and functional anatomy. This is particularly instructive as far as congenital abnormalities are concerned.

### Congenital Abnormalities

**The Heart.** Any of the developmental events discussed under the headings Establishment of External Form and Establishment of Internal Form may be arrested or interfered with. Discussion is limited here to a brief mention of abnormal looping of the heart and to defective development of the atrioventricular canal, the interatrial and interventricular septa, and the cardiac valves.

During **looping of the heart,** the bulbus may twist to the left side of the primitive ventricle rather than to the right, as it normally does; that is, instead of normal *dextro,* or *"d,"* looping, *levo,* or *"l,"* looping takes place. The result of *"l"* looping is that the anatomic right ventricle will lie to the left of the anatomic left ventricle. Since the atria remain unaffected by the abnormal looping, the right atrium will empty into the anatomic left ventricle, now situated on the right side of the heart. After birth, venous blood would be pumped into the aorta. This malformation, however, is usually associated with transposition of the great arteries, which assures that the hemodynamics in such a grossly abnormal heart remain essentially normal as a result of the transposition: the pulmonary trunk comes off the anatomic left ventricle (located on the right), which receives blood from the normal right atrium. In such cases of *congenitally corrected transposition,* the heart is usually located in the left side of the chest, but sometimes it may be in the right side. These cases of *isolated dextrocardia* or *dextroversion* must be distinguished from *mirror image dextrocardia,* which is found in *situs inversus.* In this condition, the left-to-right orientation of all viscera is reversed. Although there is *"l"* looping, which reverses the normal orientation of the ventricles, the atria are also reversed; the hemodynamics in mirror image dextrocardia are normal.

Several cardiac abnormalities result from defective development of the **atrioventricular canal.** The canal initially connects the primitive atrium to the future left ventricle (p. 545). If the atrioventricular canal fails to expand to connect the atrium to the future right ventricle, or bulbus, both atria will open into the left ventricle. A *double inlet left ventricle* results because the atrioventricular canal usually divides into mitral and tricuspid orifices. In such hearts, the right ventricle never develops normally: it may persist as a vestigial outflow channel or it may disappear altogether. This abnormality is also known, therefore, as *single ventricle.*

Interference with the development of the **endocardial cushions** may prevent their fusion completely or partially. If the endocardial cushions do not approximate each other, a large hole remains in the center of the heart, bordered by the defective interatrial and interventricular septa and the dorsal and ventral endocardial cushions. Such a defect is called a *complete persistent atrioventricular canal.* If the endocardial cushions approximate each other but do not fuse completely, the gaps left between the partially fused cushions remain as deficiencies in the corresponding valve leaflets: there may be a cleft in the septal cusp of the tricuspid and in the anterior cusp of the mitral valves.

When the endocardial cushions fail to approximate, the septum primum is usually also defective. Such a *septum primum defect* is one of several types of **atrial septal defects** (ASDs). Other types of ASD include the *secundum defect* and the *sinus venosus defect.* Incomplete development of the septum secundum generates an abnormally large foramen ovale, and such a rudimentary septum fails to cover the foramen secundum. The resultant uncovered hole is called an **ASD of the secundum type.** Alternatively, excessive breakdown of the upper part of the septum primum may generate an abnormally large foramen secundum, which will not be covered by a relatively normal septum secundum. Fenestration of the septum primum, as an extension of the foramen secundum, may create multiple holes in the valve of the foramen ovale that will permit communication between the two atria throughout life. Incomplete absorption of the sinus venosus into the atrium gives rise to **a sinus venosus defect.** In such a case, the superior vena caval orifice faces into the interatrial defect, which is situated close to the roof of the atria.

**Ventricular septal defects** (VSDs) likewise have several types. The defect may involve the muscular septum, the membranous septum, or the partition between the outflow tracts of the two ventricles. Muscular defects are placed low, may be single or multiple, and are usually located in the pits between trabeculae of the right ventricular myocardium. Defects in the membranous septum may give rise to communication between the left and right ventricles beneath the septal leaflet of the tricuspid valve or between the left ventricle and the right atrium above the same leaflet. These membranous defects are always below and posterior to the crista supraventricularis; the atrioventricular bundle passes along the lower margin of the defect. Defective development of the two bulbar ridges that partition the distal portion of the bulbus cordis leads to defects that are situated above the crista and permit communication between the outflow tracts of the two ventricles.

**Valvular abnormalities** may involve the atrioventricular and semilunar valves. *Atresia* of the tricuspid and mitral valves probably results from closure of the atrioventricular canals by fusion of the endocardial cushions. In the case of the semilunar valves, complete fusion along the lunules produces atresia; partial fusion results in narrowing of the opening, which is known as *stenosis.*

A complex type of congenital cardiac abnormality is the **tetralogy of Fallot,** first described by Stensen in 1673 and later by Fallot in 1888. The primary abnormality is a narrowing of the right ventricular outflow tract caused by an unequal division of the bulbus cordis (Fig. 21-24A and B). This results in the displacement of the crista supraventricularis toward the anterolateral wall of the right ventricle. The displacement narrows the outflow tract of the right ventricle, and the gap it leaves posteriorly creates a VSD. Normally, fusion of the bulbar ridges takes place at the time when the undivided outflow tract still comes off the bulbus cordis (future right ventricle); only later does the aorta line up with the primitive (left) ventricle. In the tetralogy, movement of the aorta toward the left is arrested so that the aorta continues to come off partially from the right ventricle; in other words, the aorta overrides the interventricular septum (Fig. 21-24C). Therefore, the right ventricle has to pump blood directly into the high-pressure systemic circulation through the aorta, as well as into its own abnormally narrow outflow tract. These two factors lead to right ventricular hypertrophy. Thus, the four features of the tetralogy consist of (1) infundibular narrowing (stenosis), (2) VSD, (3) overriding aorta, and (4) right ventricular hypertrophy.

**The Great Arteries and Veins.** Brief mention will be made of selected abnormalities related to the anatomy discussed in this chapter: common truncus arteriosus, transposition of the great arteries, persistent left superior vena cava, and anomalous pulmonary veins.

When the truncus arteriosus is not partitioned by the spiral septum, the single arterial trunk leaves the outflow tract of the ventricles and supplies the coronary, pulmonary, and systemic circulations. This abnormality is called a *persistent truncus arteriosus.* In complete *transposition of the great arteries,* the aorta emerges from the right ventricle and usually lies anterior to, or side by side with, the pulmonary trunk. The pulmonary trunk is given off by the left ventricle. Postnatal survival depends on the persistence of a shunt between the right and left sides of the heart (ASD, VSD, or patent ductus arteriosus). Embryological basis for the abnormality is not known. One hypothesis postulates that a mismatch occurs in the junction between the spiral ridges in the truncus and the more proximal ridges in the bulbus.

A *left superior vena cava* results from the persistence of the left common cardinal vein. Such a vessel runs down in front of the arch of the aorta and the left pulmonary artery to join a much enlarged coronary sinus. A right superior vena cava nearly always coexists with a left superior vena cava. In such cases of *double superior vena cava,* there is usually a connection in the superior mediastinum between the two vessels. A persistent left superior vena cava causes no clinical problems. Its recognition is important, however, if open heart surgery is to be undertaken.

*Anomalous drainage of pulmonary veins* (p. 500) also represents the persistence of embryonic venous channels and may occur in a variety of patterns. The venous plexus of the lungs develops as an extension of the veins that surround the foregut. The dominant veins that drain from the lung toward the foregut normally link up with the *common pulmonary vein,* which forms an outgrowth of the primitive atrium. The common pulmonary vein is eventually absorbed into the wall of the left atrium, so that in the fully formed heart, its tributaries, the superior and inferior pulmonary veins, enter the left atrium. If the pulmonary venous plexus does not link up with the common pulmonary vein, or if this vein does not develop, blood will be returned from the lungs along the existing embryonic channels. Thus, some or all of the blood from the lungs

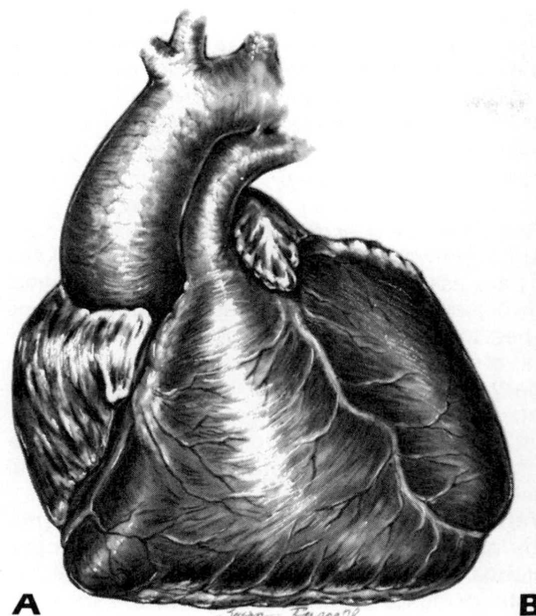


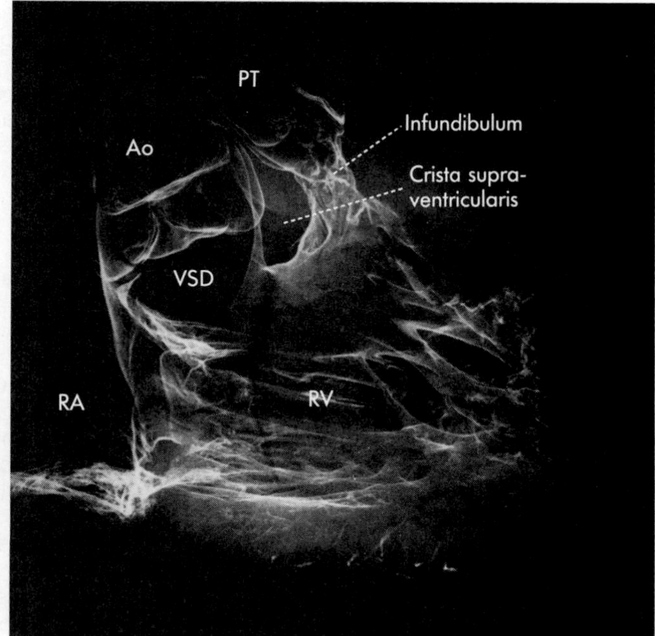


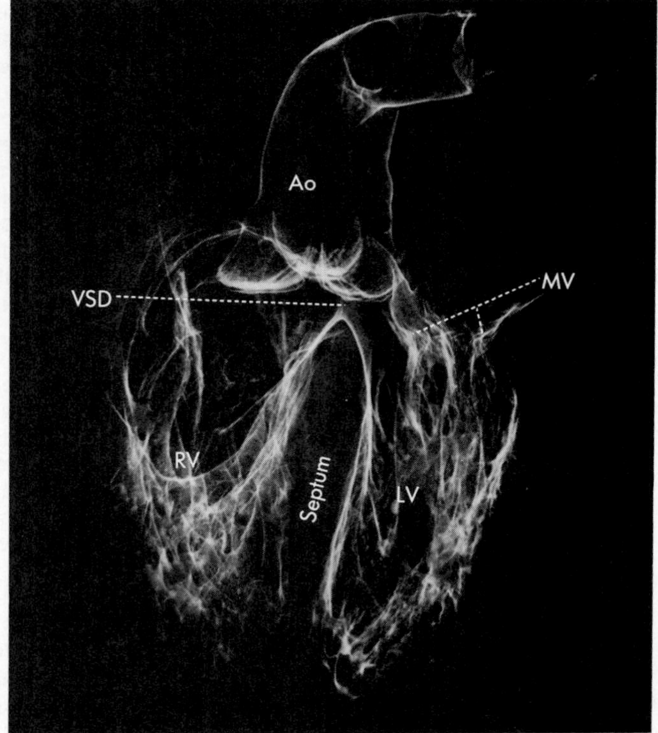


**FIG. 21-24.** Tetralogy of Fallot. **A** is a drawing of the exterior of the heart with tetralogy; **B** and **C**, x-rays taken of the heart by the method described in the legend of Fig. 21-10. **Ao,** aorta; **PT,** pulmonary trunk; **RV,** right ventricle; **LV,** left ventricle; **MV,** anterior leaf of the mitral valve; **RA,** right atrium; **VSD,** ventricular septal defect. Notice the narrow outflow tract of the right ventricle (infundibulum) in **A** and **B**; the aortic and pulmonary valves at the base of the **Ao** and **PT**; the VSD in **B** and **C**; and the overriding of the ventricular septum by the aorta in **C.** (Prepared and kindly provided by Dr. Lore Tenckhoff)

may drain into the right or left superior vena cava or into the coronary sinus. A venous channel may persist on the anterior surface of the esophagus, which can receive anomalous pulmonary veins. Such a channel drains into the portal system (left gastric vein), the ductus venosus, or the inferior vena cava. In *total anomalous pulmonary venous return,* all the blood from the lungs is conveyed by abnormal channels to veins that empty into the right side of the heart; this leads to symptoms early in life and may constitute a surgical emergency in the neonate. More commonly, the anomalous venous drainage is only partial and may remain asymptomatic into adulthood, provided the volume of blood draining from the lungs to the right atrium is not large.

**Clinical Classification of Congenital Cardiac Abnormalities.** Broadly speaking, congenital cardiac malformations are classified clinically according to the presence or absence of cyanosis. *Cyanosis* is a bluish discoloration of the skin and mucous membranes due to an increase in the amount of deoxygenated hemoglobin in the blood. In **cyanotic congenital heart disease,** cyanosis results from the mixing of venous and arterial blood, a decrease in the volume of blood that reaches the lungs, or transposition of the great arteries. Double inlet left ventricle (single ventricle), complete persistent atrioventricular canal, and persistent truncus arteriosus are examples of mixing abnormalities. Obstruction of blood flow to the lungs is found in the tetralogy of Fallot, pulmonary atresia, and tricuspid atresia. In transposition of the great arteries, cyanosis is profound, as all the venous blood is pumped out directly into the transposed aorta. Survival depends on the shunting of some venous blood into the pulmonary artery through the ductus arteriosus, an ASD, or a VSD. An ASD is created within the first days of life, when transposition is diagnosed, by tearing the interatrial septum (balloon atrial septotomy).

**Acyanotic congenital cardiac abnormalities** comprise obstructive lesions and left-to-right shunts. Stenosis of the aortic and pulmonary valves makes up the most important obstructive lesions. Left-to-right shunts occur through a persistent patent ductus arteriosus or through atrial and ventricular septal defects.

Obstructive lesions lead to hypertrophy of the chamber situated proximal to the stenosed valve. Shunts produce volume overload, which results, primarily, in dilatation and some hypertrophy of the overloaded chambers. ASDs cause right ventricular dilatation; VSDs and patent ductus lead to left ventricular dilatation.

Tremendous strides have been made in the surgical treatment of congenital heart disease over the past 20 years. Many congenital heart defects are now amenable to corrective or palliative surgery. One of the most significant advances has been in the surgery, palliative and corrective, of the critically ill neonate and infant with congenital heart disease. In the past, this age-group had the highest mortality.

### Acquired Cardiac Abnormalities

Although not all functional disturbances of the heart are associated with anatomic abnormalities, many diseases can produce gross anatomic lesions in a heart that was normal to begin with (*e.g.,* infections, rheumatic fever, hypertension, chronic pulmonary disease, myocardial infarction). Enlargement, dilatation, or hypertrophy of the cardiac chambers may occur, but from an anatomic standpoint, valvular deformities are of the greatest interest.

**Valvular abnormalities** are of two kinds: stenosis or incompetence (also known as insufficiency). In *stenosis,* narrowing is caused by fusion of the leaflets, or cusps, of the valve along the commissures, or lunules, as a consequence of scar formation during the resolution of an inflammatory process that involves the valves and the associated tissues. The valve may become scarred and distorted in such a manner that the cusps cannot occlude the orifice completely and allow regurgitation of blood into the chambers separated by the defective valve. Such a valve is *incompetent.* Thus, one may speak, for example, of mitral stenosis or mitral incompetence and of aortic stenosis or aortic incompetence. Scarring and retraction of the cusps and their chordae tendineae may render a valve both stenosed and incompetent (*e.g.,* mitral stenosis and incompetence).

The rigidity, thickening, and irregularity of deformed valves will alter the normal character of the heart sounds. Abnormal eddies will also be created as the blood passes the deformed valve. These added sounds generated by the abnormal eddies are known as *cardiac murmurs,* and they may be heard through systole, through diastole, or during both phases of the cardiac cycle. The diagnosis of a valvular abnormality is arrived at by the synthesis of several physical signs, including changes in the character of the heart sounds and the nature and timing of cardiac murmurs in relation to the cardiac cycle. Vibrations created by some valvular abnormalities may be so strong that they can be perceived not only through the stethoscope but also by a palpating hand placed on the precordium. The physical sign of such a palpable murmur is known as a *thrill.* In addition to physical signs, the shape and size of the heart on x-ray films and the electrocardiogram are routinely used in establishing the diagnosis.

Many acquired valvular abnormalities can be

corrected surgically. If the cusps of a stenosed valve are not unduly deformed, the fused commissures may be divided (commissurotomy). Most cases of valvular incompetence, however, can be corrected only by replacing the defective valve with an artificial prosthetic valve.

## RECOMMENDED READINGS

EDWARDS JE, DRY TJ, PARKER RL ET AL: An Atlas of Congenital Anomalies of the Heart and Great Vessels. Springfield, IL, Charles C Thomas, 1954

ELLISON JP, WILLIAMS TH: Sympathetic nerve pathways to the human heart and their variations. Am. J Anat 124:149, 1969

HIRSCH EF (ed): The Innervation of the Vertebrate Heart. Springfield, IL, Charles C Thomas, 1970

JAMES TN: Anatomy of the Coronary Arteries. New York, Paul B Hoeber, 1961

JAMES TN: Cardiac conduction system: Fetal and postnatal development. Am J Cardiol 25:213, 1970

JOCHEM W, SOTO B, KARP RB ET AL: Radiographic anatomy of the coronary collateral circulation. AJR 116:50, 1972

KRAMER TC: The partitioning of the truncus and conus and the formation of the membranous portion of the interventricular septum in the human heart. Am J Anat 71:343, 1942

LEVIN DC, KAUFF M, BALTAXE HA: Coronary collateral circulation. AJR 119:463, 1973

MALL FP: On the muscular architecture of the ventricles of the human heart. Am J Anat 11:211, 1911

MEREDITH J, TITUS JL: The anatomic atrial connections between sinus and A-V node. Circulation 37:566, 1968

MITCHELL GAG: Cardiovascular Innervation. Edinburgh, Livingstone, 1956

SCHAPER W: The collateral circulation of the heart. In Black DAK (ed): Clinical Studies: A North-Holland Frontiers Series, vol 1. New York, American Elsevier, 1971

SMITH RB: The occurrence and location of intrinsic cardiac ganglia and nerve plexuses in the human neonate. Anat Rec 169:33, 1971

TRUEX RC, SMYTHE MQ: Reconstruction of the human atrioventricular node. Anat Rec 158:11, 1967

TRUEX RC, SMYTHE MQ, TAYLOR MJ: Reconstruction of the human sinoatrial node. Anat Rec 159:371, 1967

WALLS EW: Dissection of the atrio-ventricular node and bundle in the human heart. J Anat 79:45, 1945

# 22

# THE MEDIASTINUM

The mediastinum is a broad, median partition, or septum, that intervenes between the two pleural sacs and extends from the sternum to the vertebral bodies. As pointed out in Chapter 19, this mass of tissue incorporates the heart, the pericardial sac, and—practically speaking—all the contents of the thoracic cavity except the lungs and the pleura. The mediastinum is a conduit for the esophagus and trachea and, closely packed around them, the blood vessels, lymph vessels, and nerves as they enter or leave the thorax on their way to or from the neck, the upper limbs, or the abdomen. Only the pericardial sac, with its contents, and the thymus are confined to the mediastinum.

For descriptive purposes, the mediastinum is arbitrarily subdivided by a transverse plane that passes through the sternal angle and the lower border of the 4th thoracic vertebra (Fig. 22-1). The **superior mediastinum** is above this plane and is limited superiorly by the superior thoracic aperture; the **inferior mediastinum** is below the plane, and the diaphragm limits it inferiorly. The inferior mediastinum is further compartmentalized based on relationships to the pericardial sac: the sac and its contents comprise the **middle mediastinum;** between the sac and the sternum is the **anterior mediastinum;** between the vertebral bodies and the pericardial sac is the **posterior mediastinum.**

This chapter is concerned chiefly with the superior and posterior mediastina, as there is little besides adipose tissue in the anterior mediastinum, and the middle mediastinum was described in the previous chapter.

## SUPERIOR MEDIASTINUM

The superior mediastinum is situated chiefly behind the manubrium sterni. Its posterior boundary, made up of the first four thoracic vertebrae, is considerably longer because the plane of the su-

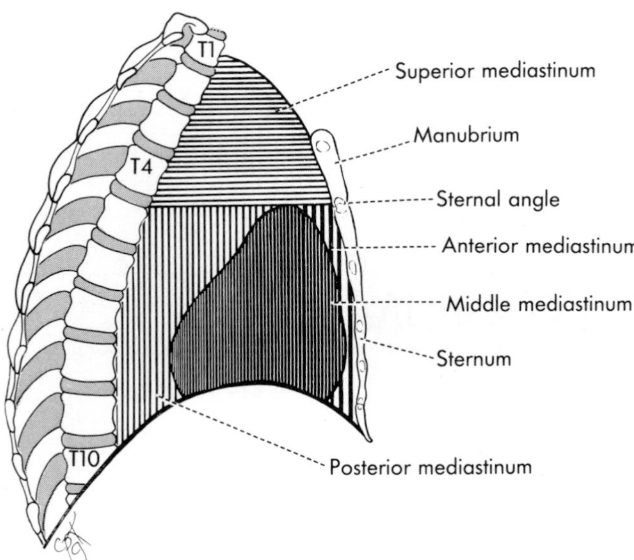


**FIG. 22-1.** Subdivisions of the mediastinum. The superior mediastinum is shaded with horizontal lines, and the compartments of the inferior mediastinum are shaded with vertical lines in various patterns.

perior thoracic aperture slopes upward from the jugular notch. Above, the superior mediastinum is continuous with the neck; below, it is continuous with both the anterior and posterior mediastina. Its central portion is directly above the pericardial sac, and laterally it is limited by parietal (mediastinal) pleura.

The main structures crowded into this small, wedge-shaped space are the trachea and esophagus, the aortic arch with its three large branches (brachiocephalic, left common carotid, and left subclavian arteries), the right and left brachiocephalic veins that form the superior vena cava, two phrenic nerves and two vagi, the recurrent laryngeal branch of the left vagus, the cardiac plexus and cardiac nerves, the thymus, the terminal portion of the thoracic duct, and a number of lymph nodes and lymph vessels. The key to understanding the relationships of these numerous structures to one another is the asymmetry of the major arteries and veins on the right and left sides of the esophagus and trachea. The trachea, with the esophagus behind it, occupies a central position in the superior mediastinum. On the left, the superior mediastinum is dominated by arteries, whereas on the right, veins predominate (Fig. 22-2).

This asymmetry results from two major rearrangements of the primitive, symmetrical pattern of embryonic vasculature:

1. On the right side, the segment of the 4th aortic arch that is incorporated into the right subclavian artery becomes displaced from the mediastinum, whereas on the left, the 4th aortic arch is retained in the mediastinum as a segment of the arch of the aorta (Fig. 22-8).
2. On the left, the common cardinal vein (left superior vena cava,) atrophies, whereas it persists on the right as the definitive superior vena cava. The anastomosis that develops between the left and right precardinal veins is represented by the more or less transverse brachiocephalic vein that lies anterior to all major structures in the superior mediastinum except the thymus (see Fig. 21–22).

The vagus and phrenic nerves enter the superior mediastinum from the neck and continue toward the diaphragm. The vagi approach each other as they pass inferiorly to form the esophageal plexus behind the root of the lungs in the posterior mediastinum. The right and left phrenic nerves, on the other hand, maintain a lateral position in the superior mediastinum and continue their course inferiorly on the lateral surfaces of the bulging pericardial sac as they head for the respective domes of the diaphragm.

## The Thymus

The most superficial structure in the superior mediastinum is the thymus (see Figs. 19-14 and 21-2). Although Nomina Anatomica lists this organ among endocrine glands, its endocrine function remains controversial and is very much outweighed by the importance of the thymus in the generation of potentially immunocompetent cells. The thymus is a *primary* or *central lymphoid organ,* its most important function being the production of

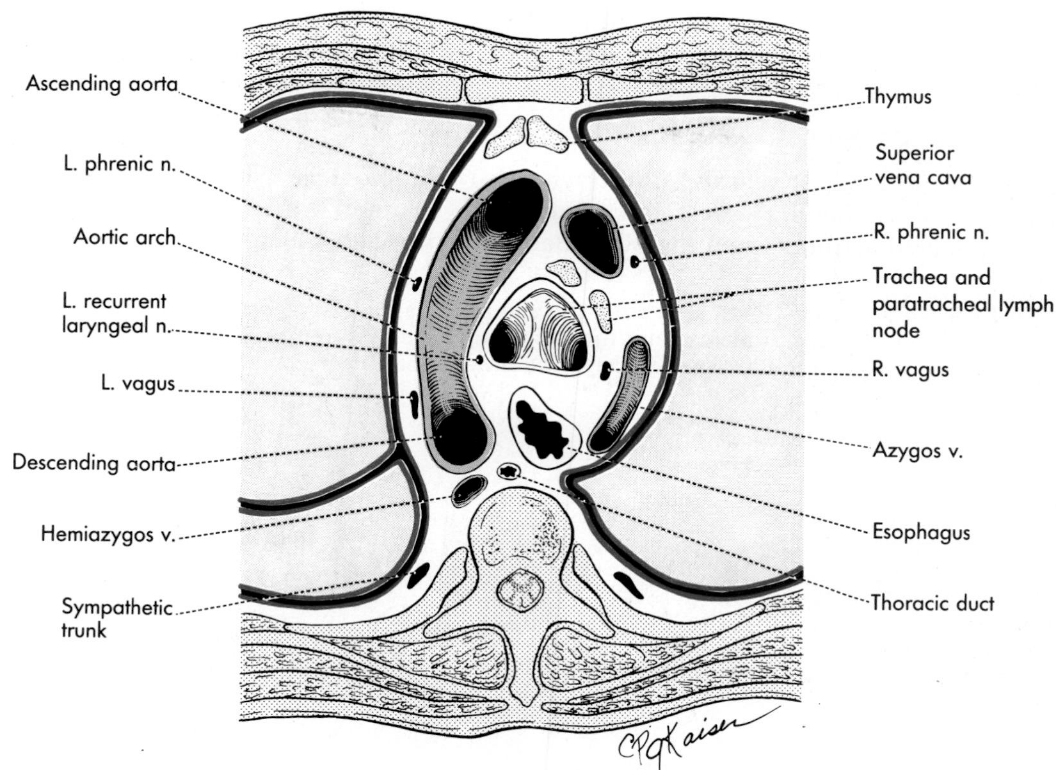


**FIG. 22-2.** A transverse section across the superior mediastinum, level with the body of the fourth thoracic vertebra. The section is being viewed from above. The pleura is indicated with solid red lines, and the pleural cavity is solid black.

lymphocytes that are discharged into the bloodstream and are seeded to the rest of the lymphatic system. The thymus consists of two small, pyramidal *lobes,* each made up of many *lobules.* The two lobes are held together by connective tissue and can be distinguished from the surrounding fat by their pinkish gray color and the glandular appearance that is imparted by the lobular structure of the organ.

Each thymic lobe develops from the 3rd pharyngeal pouch on its corresponding side. Following their separation from the pharynx, the epithelial primordia of the lobes descend into the thorax. Lymphocyte production is established during fetal development after the lobes become populated by immigrant lymphoid stem cells. The endoderm-derived epithelial elements persist into postnatal life and may be responsible for the production of humoral factors that play some role in the differentiation of lymphocytes. The thymus continues to grow, up to the age of 5 to 6 years, and progressively involutes thereafter, weighing hardly more than 10 g in the adult. It is largely replaced by fat and connective tissue, which maintain the form of the organ.

The thymus is directly behind the manubrium and may extend into the base of the neck and into the anterior mediastinum. The posterior surface of the organ is molded on the arch of the aorta, the left brachiocephalic vein, the trachea, and, more inferiorly, the pericardium. The two pleural sacs overlap the thymic lobes on each side (see Fig. 19-14). The blood supply is from branches of the internal thoracic or inferior thyroid arteries. A large thymic vein drains the gland into the left brachiocephalic vein, and smaller ones drain into the internal thoracic veins. Thymic lymphatics end in several mediastinal nodes. Small branches from cervical sympathetic ganglia and from the vagus enter the thymic lobes, and the connective tissue capsule is supplied by the phrenic nerves.

## The Trachea and Esophagus

With the neck slightly flexed, the trachea is palpable in the jugular notch as it enters the superior mediastinum from the neck. Behind it is the esophagus lying on the anterior surface of the vertebral bodies. These two tubular structures fill the superior thoracic aperture in the median plane between the apices of the lungs. Bearing witness to their common embryological origin, they remain in intimate contact with each other as they descend through the superior mediastinum. The tra-

**FIG. 22-3.** The mediastinum seen from the right pleural cavity. The chest wall, a large portion of the costal pleura, and the right lung have been removed. The structures in the mediastinum are seen through the mediastinal pleura.

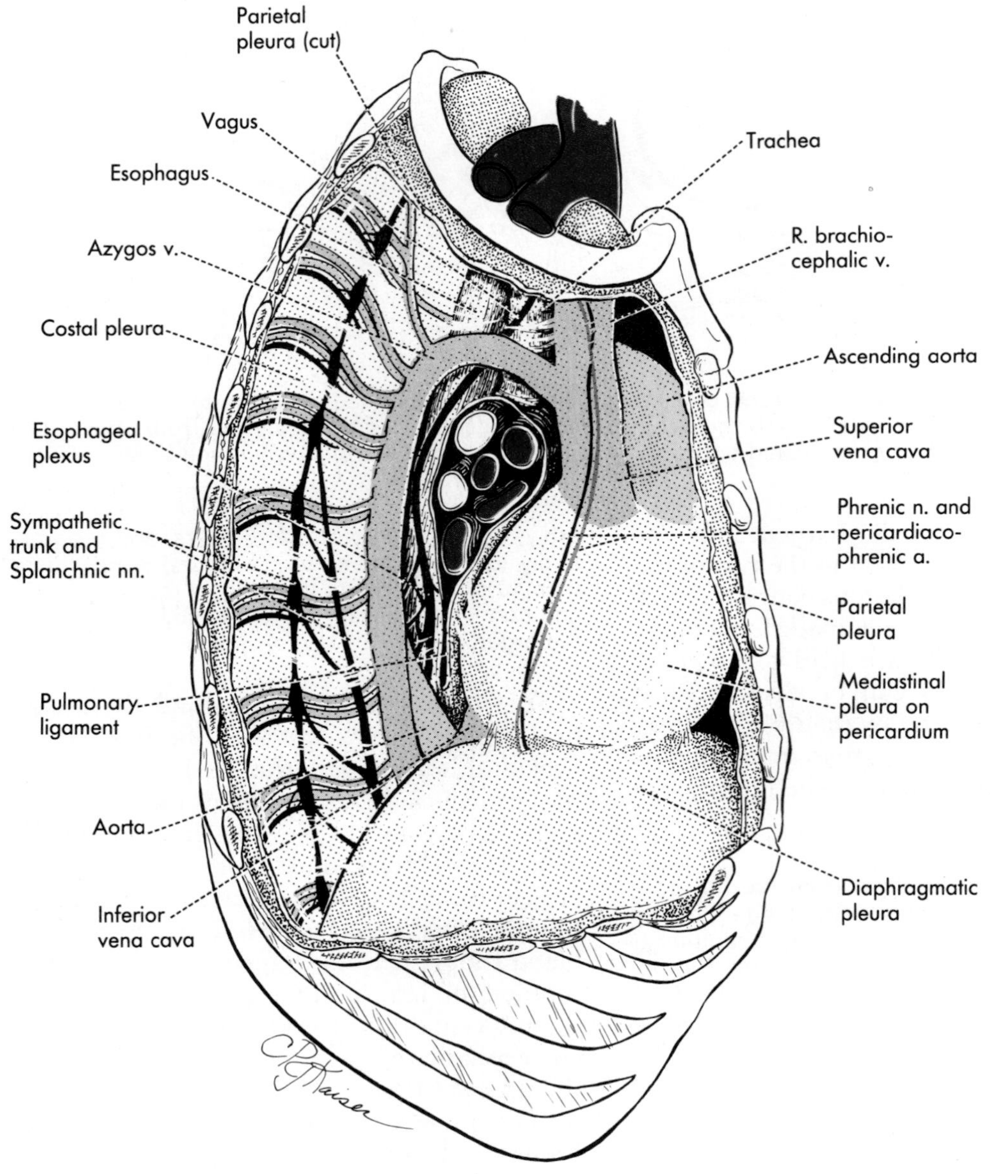

chea bifurcates into the right and left principal bronchi (p. 494) at, or slightly below, the lower boundary of the superior mediastinum, whereas the esophagus continues into the posterior mediastinum.

Above the root of the lung, both trachea and esophagus are crossed by the azygos vein on the right side (Fig. 22-3) and by the arch of the aorta on the left side (Fig. 22-4). The aortic arch indents the esophagus slightly and shifts the trachea from its median position toward the right. In its course through the superior mediastinum, the esophagus contacts the upper lobe of both lungs (with only the pleural sacs intervening), whereas the trachea becomes separated from the lung on the left side

**FIG. 22-4.** The mediastinum seen from the left pleural cavity. The chest wall, a large portion of the costal pleura, and the left lung have been removed. The structures in the mediastinum are seen through the mediastinal pleura.

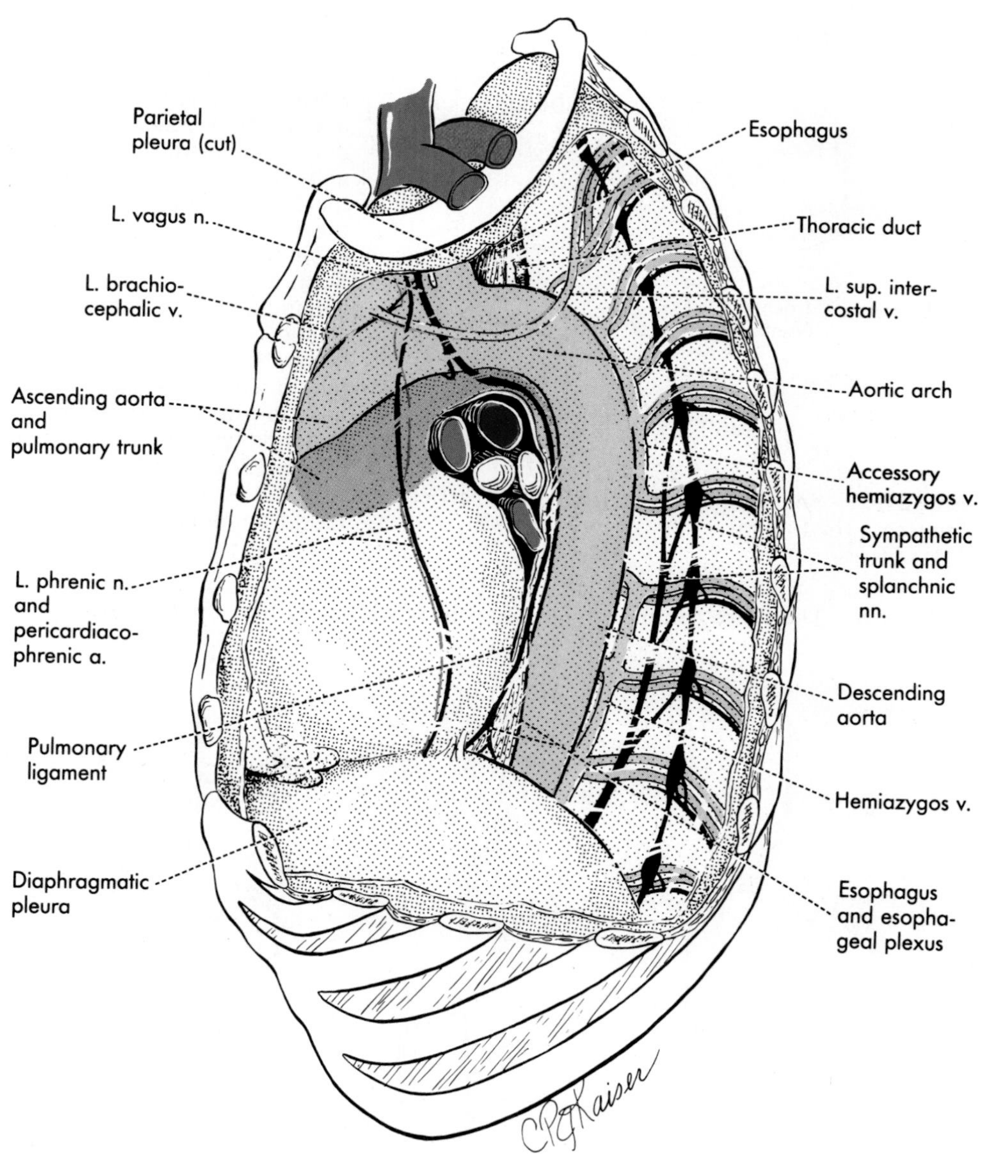

by the branches of the aortic arch (Figs. 22-2 and 22-4).

The trachea and esophagus are mobile in the mediastinum and may be shifted from their more or less median position not only by endoscopic instruments (bronchoscope, esophagoscope, gastroscope) but also by neoplasms, abscesses, diverticula, curvatures of the spine, or pressure inequality in the two pleural cavities. The trachea is elastic and becomes elongated with each inspiration. Thus, the level of its bifurcation varies over the height of several vertebrae, depending on body posture and respiratory movements.

The structure of the trachea and its blood and nerve supply are dealt with in Chapter 29, which describes the cervical portion of the trachea; the esophagus is described more fully below in the section on the posterior mediastinum.

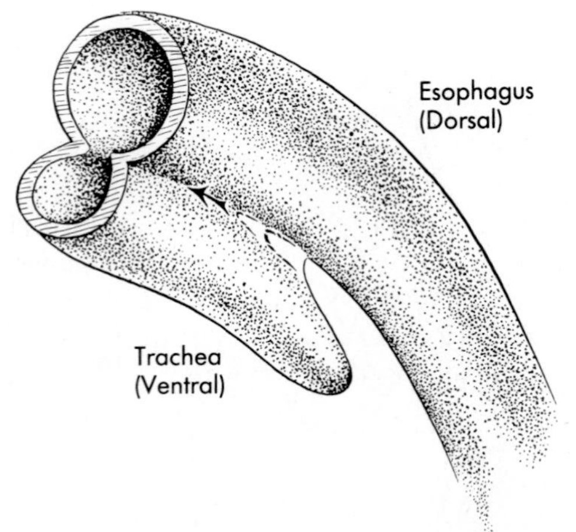


**FIG. 22-5.** Separation of the tracheal diverticulum from the foregut.

**Normal and Anomalous Development.** The trachea is formed as a ventral diverticulum of the foregut. The lumen of the foregut, located dorsal to the diverticulum, develops into the esophagus, whereas the diverticulum itself becomes pinched off and elongates (Fig. 22-5). Communication between the trachea and the esophagus is normally possible only through the pharynx and larynx. However, some type of maldevelopment is present in one of every 2000 to 3000 births. The esophagus or the trachea may become obliterated along a variable distance (**esophageal** or **tracheal atresia**), or there may be an abnormal communication between the trachea and the atretic or otherwise normal esophagus (**tracheoesophageal fistula;** Fig. 22-6). Such fistulae are usually the result of incomplete separation of the trachea and esophagus along the tracheoesophageal sulcus. Tracheal atresia is incompatible with postnatal life. However, it is important to recognize all tracheoesophageal abnormalities as soon after birth as possible. The more common types of these abnormalities are shown in Figure 22-6. To survive, the newborn must be able to swallow nourishment and saliva and prevent food from entering its lungs.

**FIG. 22-6.** The more common types of esophageal atresia and tracheoesophageal fistulae.

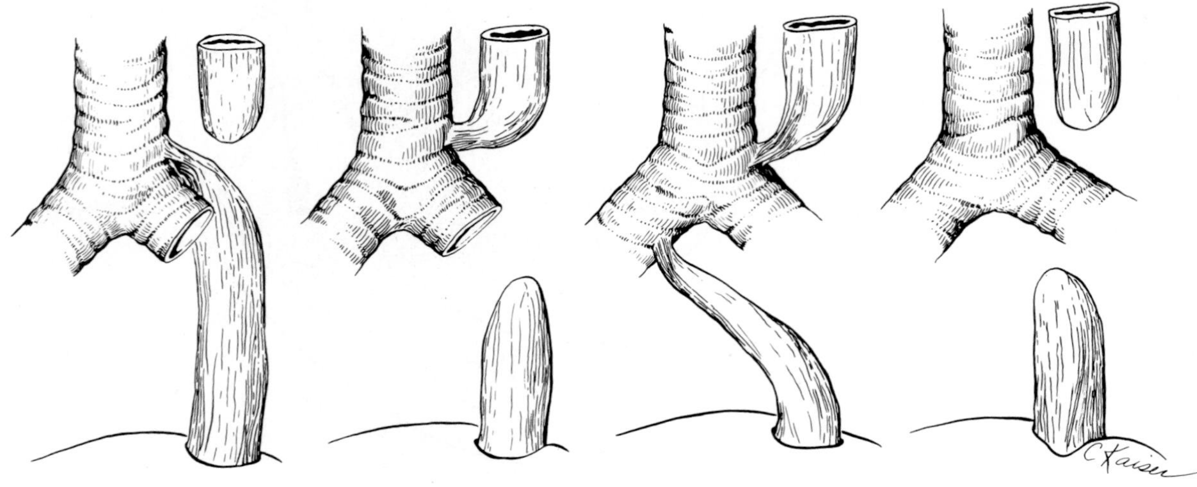

### The Arch of the Aorta and Its Branches

As soon as the ascending aorta emerges from the pericardial sac, it begins to arch backward, and the segment that runs in a nearly sagittal plane in the superior mediastinum is known as the **arch of the aorta.** On reaching the 4th thoracic vertebra, the aorta becomes vertical, and in the posterior mediastinum, it is called the **descending thoracic aorta.** The arch of the aorta is convex upward and also to the left. It accommodates the right pulmonary artery and the left bronchus in its inferior concavity; the trachea and esophagus fit into the slight concavity that faces to the right. The profile of the left curve of the aorta creates the so-called *aortic knuckle,* identifiable on a posteroanterior chest x-ray film as the upward continuation of the cardiac silhouette (see Fig. 21-7). Through the pleura, the arch indents the medial surface at the left lung just above its hilum (see Fig. 20-3B). The **ligamentum arteriosum** connects the inferior surface of the arch of the aorta to the left pulmonary artery just as the artery is given off by the pulmonary trunk. The summit of the arch reaches more than halfway up behind the manubrium and from it arise in a row three major vessels: the brachiocephalic, the left common carotid, and the left subclavian arteries (Fig. 22-7).

The **brachiocephalic artery,** the largest and most anterior of the branches, springs from the arch in the midline. Lying on the trachea, it ascends to its right side, and as the artery reaches the superior thoracic aperture, it divides into the **right subclavian** and **right common carotid arteries.**

The **left common carotid** and **left subclavian arteries** arise independently from the aortic arch in close succession beyond the brachiocephalic artery. The carotid and the subclavian ascend more or less vertically along the left side of the trachea. Near their origin all three arteries are crossed anteriorly by the left brachiocephalic vein. Between the vein and the superior thoracic aperture, their lateral surface is covered by parietal pleura (Fig. 22-3).

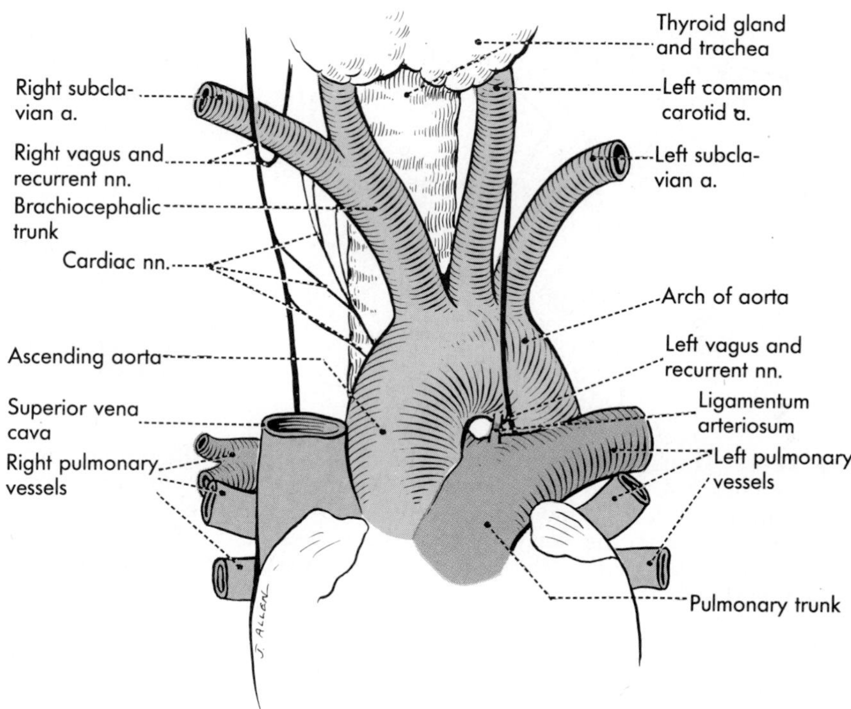


**FIG. 22-7.** The arch of the aorta and its branches with the vagus nerves and their branches.

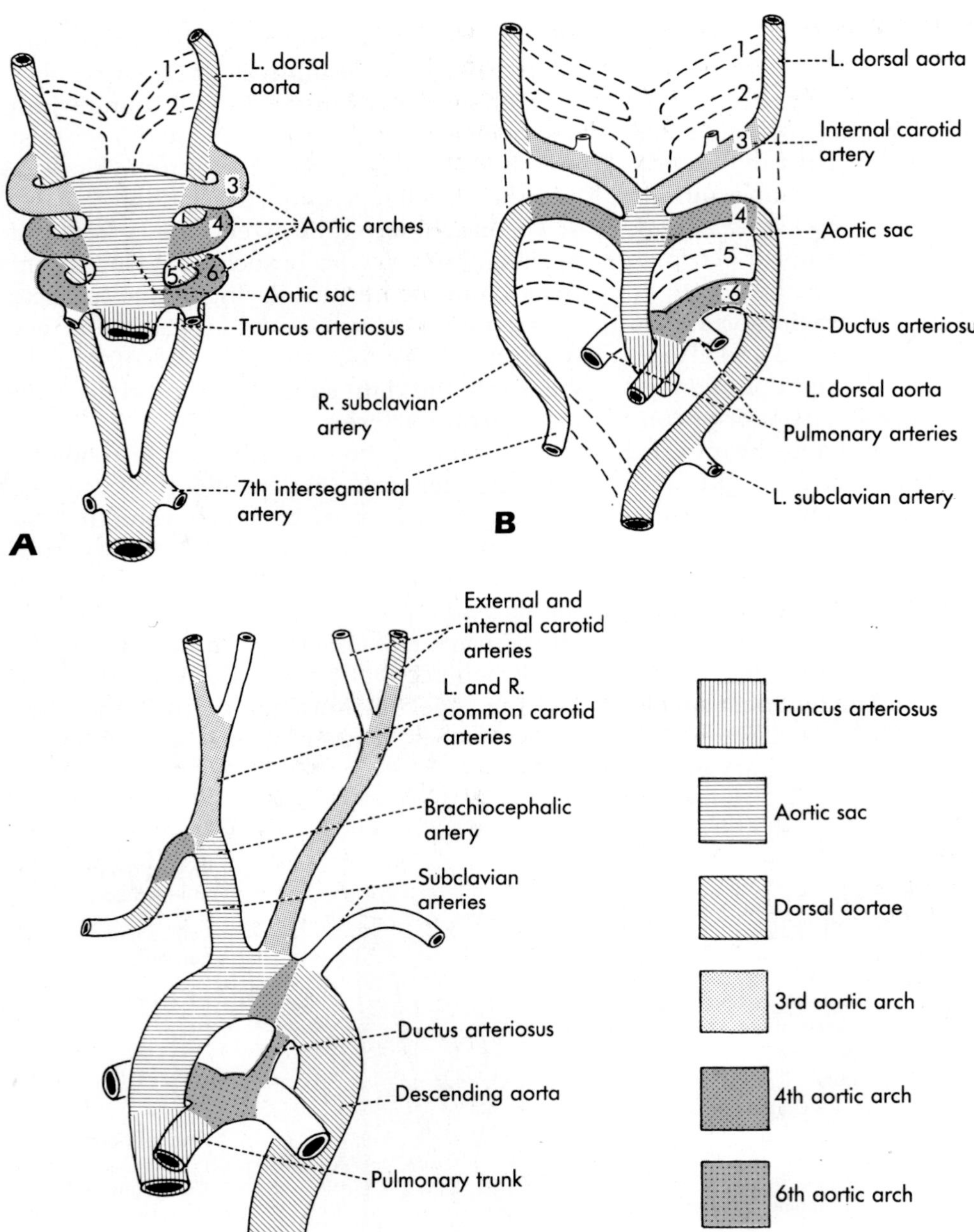


**FIG. 22-8.** Transformation of the aortic arch system in the human embryo. In **A**, aortic arches 3, 4, and 6 are seen connecting the aortic sac with the bilateral dorsal aortae; arches 1, 2, and 5 have disappeared. In **B**, the channels that disappear are still indicated. **C** shows the derivatives of the arches at completion of development. (Adapted from Moore KL: The Developing Human: Clinically Oriented Embryology, 3rd ed. Philadelphia, WB Saunders, 1982)

The three major arteries do not give off any branches in the thorax. However, one of the branches of the subclavian artery, given off at the root of the neck, reenters the superior mediastinum and continues down through the anterior mediastinum. The vessel is the **internal thoracic artery** (see Fig. 19-11). A small inconstant branch, the **thyroidea ima artery,** may arise from the aortic arch, the right common carotid, or the subclavian arteries and ascends to the thyroid gland in the neck.

**Normal and Anomalous Development.** In the human embryo, six pairs of aortic arches arise from the distal dilatation of the truncus arteriosus, which is known as the **aortic sac.** Each pair of aortic arches skirts around the foregut and connects to the dorsal aorta of the corresponding side (Fig. 22-8A; see Fig. 21-1A). In fishes, these arteries run in the branchial arches and serve the gills. In the human, normally only the arch of the aorta retains an arcuate character, and this vessel incorporates the embryonic 4th aortic arch of the left side. Other components of the aortic arch system disappear altogether or become radically modified (Fig. 22-8B and C).

Deviation from the usual set of developmental changes results in such congenital anomalies as a **double aortic arch** or a **right aortic arch** (Fig. 22-9). Both conditions predispose to compression of the esophagus and sometimes the trachea as well. Compression is due to the persistence of the vascular ring around the foregut derivatives in the case of a double arch or to the continuity of the right aortic arch with a normal descending aorta that develops on the left side. An anomalous **retroesophageal right subclavian artery** that arises from the descending aorta, rather than from the brachiocephalic artery (Fig. 22-9), may also compress the esophagus.

The process that obliterates certain segments of the aortic arch system (*e.g.*, ductus arteriosus) may involve a portion of the aorta, causing profound narrowing. The anomaly usually occurs in the vicinity of the ligamentum arteriosum and is known as **coarctation of the aorta.** The segment of the aortic arch between the left subclavian artery and the ductus arteriosus, known in the fetus as the *isthmus,* carries little blood before birth, and its narrowing may predispose to coarctation. Coarctation, however, is more common just distal to the ligamentum arteriosum (Fig. 22-10). Blood supply to the body distal to the coarctation is assured by an extensive collateral circulation that develops between branches of the aortic arch and the descending aorta.

Of less clinical importance are the **variations in the branching pattern** of the arteries given off by the arch of the aorta. The left common carotid artery may share a common trunk with the brachiocephalic or it may arise by a common trunk with the left subclavian. The left vertebral artery, normally a branch of the subclavian, may arise directly from the aortic arch.

## The Brachiocephalic Veins and the Superior Vena Cava

The brachiocephalic veins are formed by the confluence of the subclavian and internal jugular veins on each side of the root of the neck just above the superior thoracic aperture. At their commencement, the veins lie in front of the main arteries and the cervical pleura. After entering the superior mediastinum, the brachiocephalic veins become overlapped by the pleura as they descend behind the manubrium sterni. The right brachiocephalic vein is vertical and is close to the right border of the manubrium. The left vein is much longer and its course is oblique behind the manubrium as it runs in an almost transverse direction toward the right (Fig. 22-11). Behind the sternal end of the right 1st rib, the left vein unites with the right brachiocephalic vein, and their confluence forms the superior vena cava. The tributaries of the brachiocephalic veins include the vertebral and the inferior thyroid veins from the neck and the internal thoracic, superior intercostal, and thymic veins from the thorax.

The **superior vena cava** continues the vertical course of the right brachiocephalic vein, and just before it enters the pericardial sac, it receives the **azygos vein.** The azygos vein arches forward over the root of the right lung and enters the vena cava from the back. All the venous blood from the upper half of the body (except that from the heart itself) is delivered to the right atrium by the superior vena cava.

The lower half of the superior vena cava is in the pericardial sac (p. 520). In the superior mediastinum, the vena cava is in contact on its left with the commencement of the aortic arch and behind, with the trachea. On the right, through the pleura, the superior vena

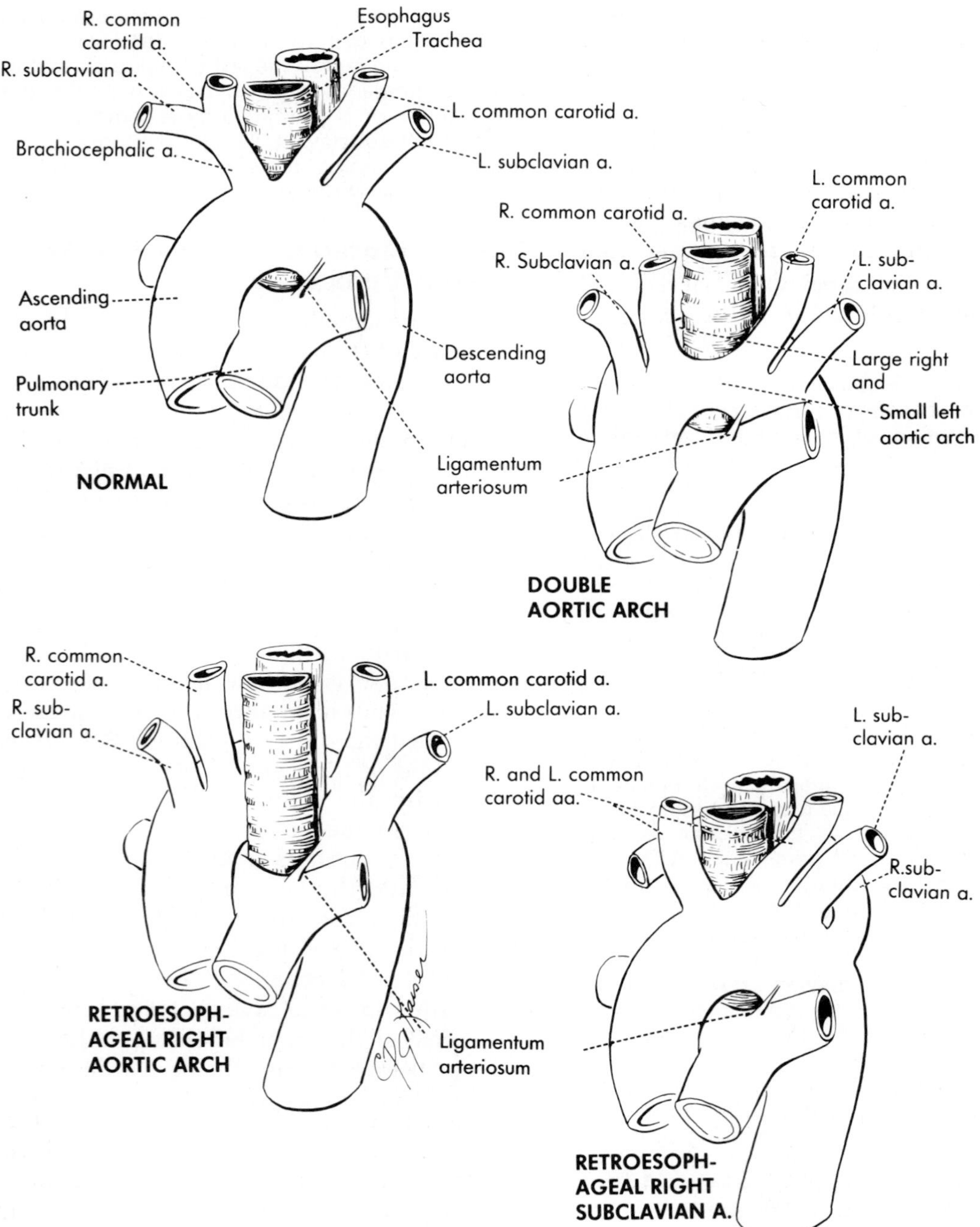


**FIG. 22-9.** Some abnormalities of the aortic arches compared with the normal arrangement.

cava contacts the upper lobe of the right lung, but it lies sufficiently forward to allow the trachea to contact the lung as well (Fig. 22-3). In the front view, the left brachiocephalic vein, lying in the same coronal plane as the vena cava, obscures the summit of the aortic arch and the roots of its three branches. Just below it, one of the tributaries of the left brachiocephalic vein, the *left superior intercostal vein,* crosses the arch of the aorta and the left vagus nerve and itself is crossed by the phrenic nerve (Fig. 22-11). Between the pleural sacs, the thymus prevents the brachiocephalic veins and the superior vena cava from coming in contact with the manubrium.

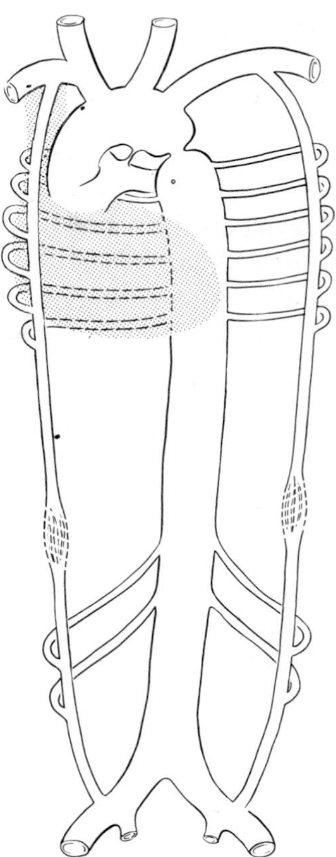

**FIG. 22-10.** Schematic representation of coarctation of the aorta with the anastomotic channels that become enlarged.

## Nerves

The main nerves that enter the superior thoracic aperture are the vagi, the phrenic nerves, and the sympathetic trunks. Cervical cardiac branches of the vagi and sympathetic ganglia also enter the superior thoracic aperture, whereas the left recurrent laryngeal nerve ascends through the aperture from the mediastinum into the neck. The cardiac nerves and the cardiac and pulmonary plexuses were discussed in Chapters 20 and 21; the sympathetic trunks will be described in the next section.

The *vagi,* the 10th pair of cranial nerves, furnish the parasympathetic innervation to all thoracic, and most abdominal, viscera (see Fig. 4-13). The *phrenic nerves,* on the other hand, are spinal nerves derived mainly from the 4th and, to some extent, from the 3rd and 5th cervical segments of the spinal cord. They are the motor nerves of the diaphragm.

**The Vagi.** At the root of the neck both right and left vagus nerves are between the internal jugular vein and the common carotid artery. Before they enter the mediastinum, they cross the first part of the subclavian artery as that vessel turns laterally. In their descent through the superior mediastinum, the vagi aim for the esophagus, on the surface of which the right and left nerves commingle and form the esophageal plexus. The structures contacted by the vagi differ on the two sides because of the asymmetry in the mediastinum. Starting behind the right brachiocephalic vein, the **right vagus** inclines posteriorly and crosses the trachea obliquely. It is easily identifiable in this position through the pleura (Fig. 22-3). Just before reaching the esophagus behind the root of the right lung, the vagus is crossed by the azygos vein. The **left vagus** does not come in contact with the trachea; rather, it descends on the surface of the left subclavian artery lying behind the left brachiocephalic vein. It becomes identifiable through the pleura as it emerges from beneath this vein and crosses the arch of the aorta (Fig. 22-4). Reaching the inferior concavity of the arch, the left vagus disappears from view as it inclines backward and medially, passing behind the root of the left lung on its way to the esophagus.

In addition to their cervical cardiac branches (p. 538), both vagi contribute **thoracic**

**cardiac branches** to the cardiac plexus (Fig. 22-7). These branches run downward and medially in the mediastinum, as do the cardiac branches that were given off in the neck. A **recurrent laryngeal branch** arises from each vagus, and these two nerves ascend to supply the larynx. The *left recurrent laryngeal nerve* arises from the left vagus at the lower border of the aortic arch (Fig. 22-11), runs medially, skirting the concavity of the arch at its junction with the ligamentum arteriosum, and then ascends on the medial side of the arch in the groove between the trachea and the esophagus. The *right recurrent laryngeal nerve* does not enter the mediastinum because it arises from the right vagus in the base of the neck. Before it commences its ascent, it hooks around the right subclavian artery (an embryonic homologue of a segment of the aortic arch [Fig. 22-8]) just above the plane of the superior thoracic aperture (Fig. 22-11).

**The Phrenic Nerves.** In addition to carrying the motor innervation to the diaphragm (C3–5), the phrenic nerves convey somatic afferents (pain sensation) from the fibrous and parietal serous pericardium, from the mediastinal and diaphragmatic portions of the parietal pleura, and from parietal peritoneum on the inferior surface of the diaphragm, as well as proprioceptive impulses from the diaphragm itself. The phrenic nerves descend to the root of the neck (Chap. 30) and enter the superior thoracic aperture just after they have crossed the subclavian vessels, passing between the artery and the vein. They are lateral to the vagus and posterolateral to the commencement of each brachiocephalic vein. The right and left phrenic nerves contact different structures as they pass through the superior mediastinum lying subjacent to the mediastinal pleura. The **right phrenic nerve** remains in contact with veins, winding its way forward on the right brachiocephalic vein and the superior vena cava (Fig. 22-3). The **left phrenic nerve** parts company with the left brachiocephalic vein and inclines forward as it descends on the arch of the aorta (Figs. 22-4 and 22-11). On the aorta, it crosses the left vagus and the left superior intercostal vein and is superficial to both. As the phrenic nerves enter the middle mediastinum, both nerves cross anterior to the root of the lung and descend along the greatest convexity of

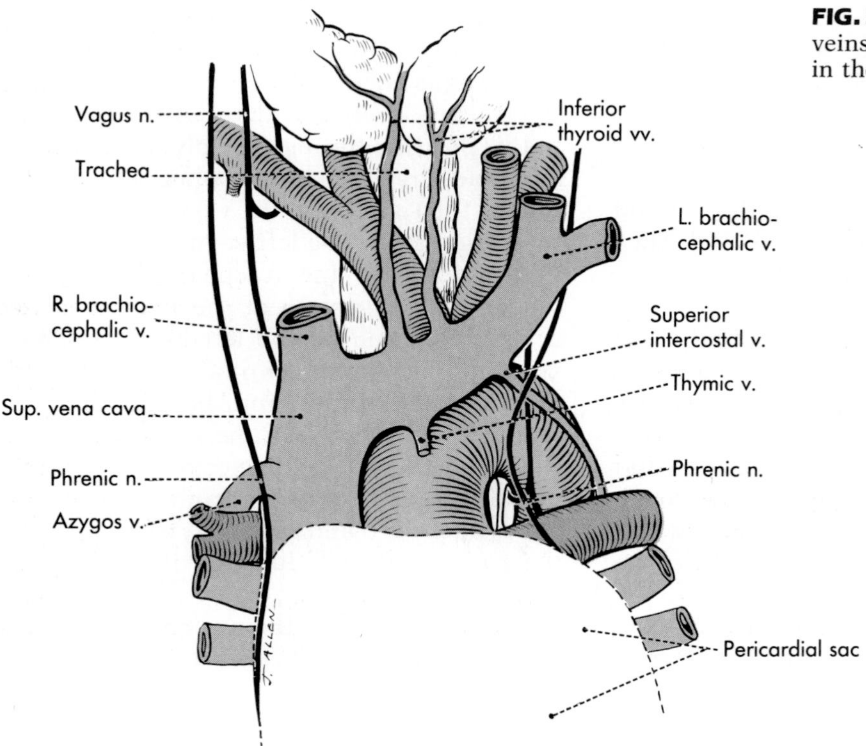


**FIG. 22-11.** The brachiocephalic veins and the superior vena cava in the superior mediastinum.

the bulging pericardial sac (Figs. 22-3 and 22-4). They are accompanied by pericardiacophrenic vessels (p. 522). The right phrenic nerve passes through the diaphragm with the inferior vena cava, and the left nerve pierces the diaphragm independently, in the vicinity of the apex of the heart.

## POSTERIOR MEDIASTINUM

The posterior mediastinum is situated chiefly behind the pericardial sac. Posteriorly, it is limited by the vertebral column, and on each side, by the pleural sacs. Superiorly, it is continuous with the posterior part of the superior mediastinum; the division between the two compartments has no anatomic basis, only a descriptive one. Several of the structures, which for convenience are described in this section, do, in fact, pass through the superior mediastinum. Inferiorly, the posterior mediastinum is limited by the diaphragm, which transmits several mediastinal structures as they enter or leave the abdominal cavity. The contents of the posterior mediastinum include the esophagus with the esophageal plexus formed by the vagi; the descending aorta and its branches; the veins of the azygos system; the thoracic duct, the largest lymphatic channel in the body; several lymph nodes; and the thoracic portions of the sympathetic trunks with their branches.

### The Esophagus

The esophagus is a muscular tube that connects the pharynx to the stomach. It is flattened by the collapse of its walls and its oval lumen is opened to diameters measuring approximately 2 cm × 3 cm by the passage of swallowed or regurgitated material. Sphincter mechanisms close off the esophagus from the pharynx and from the cardiac orifice of the stomach, which are approximately 25 cm to 35 cm apart. The esophagus is most firmly anchored at its junction with the pharynx and in the esophageal hiatus of the diaphragm. Apart from some slips of muscle that may connect it to the pleura or the left bronchus, it is relatively free along its more or less vertical descent as it follows the curvatures of the vertebral column, remaining apposed to the vertebral bodies until it approaches the diaphragm. In order to reach the esophageal hiatus, which is located in the muscular part of the diaphragm level with the 10th thoracic vertebra and to the left of the midline, the esophagus inclines forward and to the left, crossing from the right side of the aorta to the front of this vessel. In the hiatus, an extension of endoabdominal fascia, the so-called *phrenoesophageal ligament,* attaches circumferentially to the esophagus and blends, through its muscular wall, with the submucosa. This ligament secures the esophagus to the diaphragm, defines the region of the lower esophageal sphincter, and provides a seal between the abdominal and thoracic cavities.

There are four places of narrowing along the esophagus where foreign bodies are prone to lodge and where swallowed corrosives produce the greatest injury. These are also the places most commonly traumatized by the passage of instruments and are the most frequently affected by carcinoma. The first narrowing is at the beginning of the esophagus in the neck, where it is surrounded by the upper esophageal sphincter; the second is the region of contact with the aortic arch; the third is where the esophagus is crossed by the left bronchus; and the fourth is in the esophageal hiatus of the diaphragm. Some of these narrow portions can be recognized on radiologic examination of a "barium swallow" (Fig. 22-12).

**Structure and Sphincters.** The *mucosa* that lines the esophagus is formed by nonkeratinized, stratified squamous epithelium and is supported by the *lamina propria.* This loose connective tissue layer is separated from the *submucosa* by the muscularis mucosae (smooth muscle). In the submucosa are embedded numerous small mucous glands whose long ducts empty into the esophageal lumen. The inner layer of the *muscle coat* proper forms a closely wound spiral of horizontal or circular orientation, whereas in the outer layer, the muscle fasciculi are more oblique, spiraling at a steeper pitch. From the pharynx to approximately the level of the aortic arch, the muscle is striated;

below this level, it is gradually replaced by smooth muscle. The *adventitia* that surrounds the muscle coat contains many elastic fibers. Unlike other parts of the gut, there is no *serosa* (serous membrane) around the esophagus, which accounts for the difficulties in suturing the esophagus and assuring a leakproof anastomosis.

The **upper esophageal sphincter,** so-called by clinicians, is part of the inferior constrictor of the pharynx and is analogous with the *cricopharyngeus* (p. 990). The **lower esophageal sphincter** cannot be demonstrated anatomically or histologically; however, there is decisive physiological evidence for the existence of a sphincter mechanism around the lower portion of the esophagus. Many factors have been proposed to account for the resistance encountered here by swallowed material or by instruments that negotiate this region: circular esophageal muscle fibers, specialized diaphragmatic muscle (p. 628), the phrenoesophageal ligament, and intraabdominal pressure (p. 683). The most important function of this "physiological sphincter" is to prevent regurgitation of gastric contents into the esophagus. Esophageal mucosa is eroded by the gastric juice, leading to inflammation and ulceration (*reflux esophagitis*).

The **relations of the esophagus** are important because abnormalities of some of its neighboring structures may be diagnosed by the displacement or distortion of esophageal contour on x-ray films. Such conditions include osteophytes of the vertebrae, aneurysms of the aorta, distention of the left atrium, and enlarged lymph nodes associated with the trachea and bronchi. The relations of the cervical part of the esophagus are discussed in Chapter 29, and those of the upper portion of the thoracic part have been dealt with in the section on the superior mediastinum. For most of the length of the esophagus in the posterior mediastinum, the descending thoracic aorta is to its left. On the right, it is covered by mediastinal pleura (Fig. 22-3). Below the tracheal bifurcation, the following structures lie anterior to the esophagus: the right pulmonary artery, the left principal bronchus, the oblique sinus of the pericardium (see Fig. 21-1), and through it, the left atrium. The esophageal plexus formed by the vagi surrounds it (Fig. 22-13).

**Blood Supply.** The cervical part of the esophagus is supplied by branches of the inferior thyroid artery; the thoracic part is supplied by two or more *esophageal arteries* given

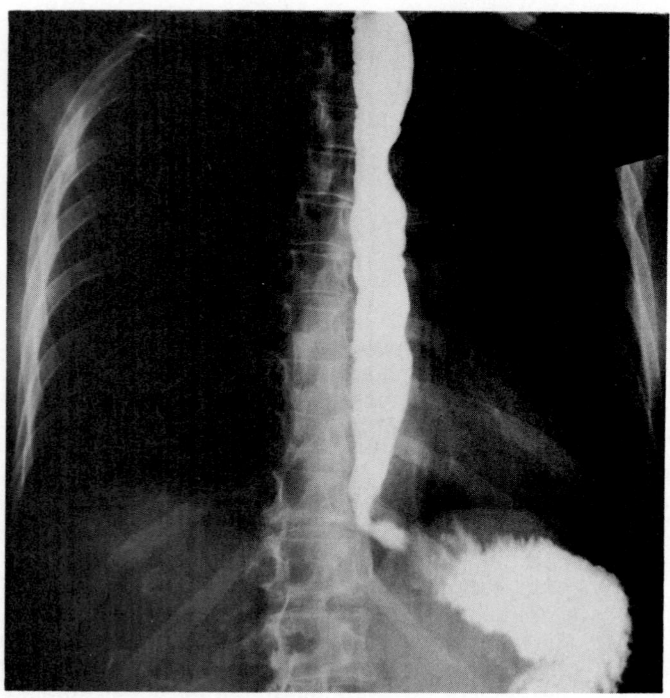

**FIG. 22-12.** Anteroposterior view of the esophagus while a barium meal was being swallowed. The sharp indentation (at the level of the upper part of vertebra T5) is caused by the arch of the aorta or the left bronchus. The esophagus appears particularly narrowed a little above the level of the stomach because barium was passing into that organ.

off by the descending aorta and by twigs from the bronchial arteries; the abdominal portion is supplied by esophageal branches of the *left gastric artery,* which ascend to the segment above the diaphragm. There are anastomoses between these territories of blood supply, but the anastomosis may be poor between the aortic and left gastric branches.

The **venous drainage** follows the arteries into the inferior thyroid, azygos, hemiazygos, and left gastric veins. It is important to remember that the latter vein belongs to the portal system, and in portal hypertension (due, for instance, to cirrhosis or scarring of the liver) the veins of the submucosa of the lower part of the esophagus become greatly engorged and distended. Tearing or rupture of these *esophageal varices* can result in a fatal hemorrhage.

**Lymph** from the esophagus drains into deep cervical nodes, posterior mediastinal nodes, and left gastric nodes.

**FIG. 22-13.** The esophageal plexus. In this example, only a few coarse branches were exchanged between the two vagi.

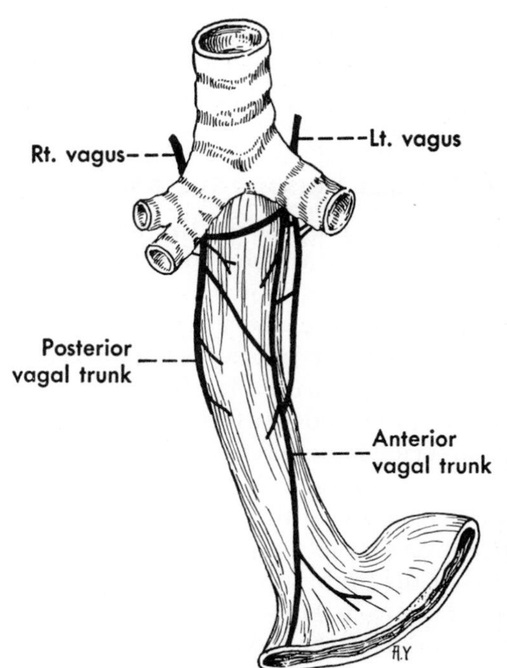


**Innervation.** The esophagus is supplied by branches of the vagus and the sympathetic trunks. Vagal branches to the striated muscle (branchial efferents) are comprised of the fibers of the *cranial accessory (11th cranial) nerve,* whereas the branches to smooth muscle are preganglionic, visceral efferents of the 10th cranial nerve, the vagus. Cervical and thoracic sympathetic ganglia contribute visceral branches to the esophagus, and some branches from the splanchnic nerves ascend from the abdomen. Their effect on the esophageal muscle is unknown. The neurons that synapse with the parasympathetic fibers of the vagus are contained in the *myenteric* and *submucosal plexuses* located in the wall of the esophagus, and these plexuses give considerable autonomy to esophageal contraction, even when incoming vagal and sympathetic fibers are severed. A peristaltic wave of contraction is initiated in the esophagus by the voluntary act of swallowing. Unlike other portions of the alimentary canal, the esophagus does not engage in spontaneous peristaltic contractions.

Passage of food along the esophagus is aided by gravity; however, esophageal contractions definitely help to transmit the bolus of food into the stomach. This is clearly demonstrated by the effective swallowing of solids or liquids while standing on one's head. Swallowing requires coordination of segmental contractions and relaxation along the esophagus and of the sphincters. *Achalasia of the esophagus* is a disorder in which the myenteric plexus and other vagal elements degenerate or are destroyed. The esophagus becomes grossly distended with food and fluid because the lower esophageal sphincter fails to relax. When the upper sphincter relaxes during sleep, the putrid contents of the esophagus may overflow into the pharynx and into the lung, causing aspiration pneumonia. In other cases, incoordination between the esophageal contraction and sphincter relaxation gives rise to so-called *diffuse spasm,* which builds up esophageal pressure and generates pain.

**Esophageal pain** is mediated along visceral afferents in the vagus and in the sympathetic nerves. The latter have their cell bodies in dorsal root ganglia T1–T10 and terminate in the corresponding segments of the spinal cord. Esophageal pain is felt substernally and may radiate to the back. Sometimes it can be confused with angina pectoris.

## The Esophageal Plexus

The two *vagus nerves,* after they have passed behind the corresponding principal bronchi and given off their branches to the lungs, converge on the esophagus and run along it into the abdomen. As they reach the esophagus, each vagus tends to divide into several trunks, and there is an interchange of branches between the nerves of the two sides to form the *esophageal plexus,* which is of varying complexity from person to person (Fig. 22-13). A little above the diaphragm, the plexus usually gives rise to two trunks. The one on the left, in which fibers of the left vagus predominate, turns forward to run on the anterior surface of the esophagus; that on the right, containing mostly right vagal fibers, turns posteriorly to run on the posterior surface of the esophagus. This change in position is consistent with the embryonic rotation of the stomach. The two newly formed nerves are called the **anterior** and **posterior vagal trunks,** and they enter the abdomen on the esophagus.

Either vagal trunk may be split into two or more parts as it passes through the diaphragm. The trunks are not necessarily placed exactly anteriorly and posteriorly as their names imply. The variations in the vagal trunks are important to the surgeon who cuts these nerves on the lower portion of the esophagus (*vagotomy*) to reduce gastric secretion in the treatment of peptic ulcer.

## The Descending Thoracic Aorta

The descending thoracic aorta is the continuation of the aortic arch, and it begins where the latter comes in contact with the vertebral column. In its descent through the posterior mediastinum, the aorta is related to the 5th to 12th thoracic vertebrae, first lying on the left side of the vertebral bodies and then gradually approaching the midline (Fig. 22-4). At the lower border of the 12th thoracic vertebra, the aorta passes through the aortic hiatus of the diaphragm, but even in this hiatus it retains its intimate relationship to the vertebrae. The portion of the descending aorta inferior to the hiatus is called the *abdominal aorta.*

Rarely, the aorta descends on the right side of the vertebral column rather than on the left. This anomaly is less frequent than a right aortic arch. Most right aortic arches continue into the descending aorta that is on the left side. When this occurs, the aortic arch passes posterior to the esophagus to link up with the descending aorta (Fig. 22-9).

**Branches.** Several small *visceral branches* are given off from the anterior surface of the descending thoracic aorta. These include two or more **bronchial** and **esophageal arteries** and pericardial and mediastinal branches. Sometimes the bronchial and esophageal arteries arise from the posterior intercostal arteries rather than from the aorta itself. A pair of *superior phrenic arteries* supply the posterior portion of the diaphragm. A series of paired *parietal branches* that supply the body wall are given off from the posterolateral surface of the aorta; they comprise nine pairs of **posterior intercostal arteries** and one pair of **subcostal arteries.** The posterior arteries of the first two intercostal spaces originate, as a rule, by a common stem (*supreme intercostal artery*) from the costocervical branch of the subclavian artery and not from the aorta. One aortic intercostal artery enters each of the intercostal spaces distal to the 3rd rib, and the subcostal artery runs along the inferior margin of the 12th rib. Their course, termination, and branches were described in Chapter 19.

## The Azygos Venous System

The primary purpose of the azygos venous system is to drain blood from the body wall. The system does, however, receive venous blood from some of the thoracic viscera and it links up with the inferior vena cava and the left renal vein in the abdomen. Blood in the azygos system normally drains upward into the superior vena cava. Because of the complex development of these veins, there is much variation in their anatomy. The azygos system has largely replaced the postcardinal veins of the embryo (p. 691), and portions of these primitive veins persist only at the commencement of the azygos system (ascending lumbar veins) and at its termination (the superior portion of the azygos vein).

The system is called *azygos* (unpaired, or lacking a mate) because the vessels of the two sides are asymmetrical, although a common pattern of organization can be readily discerned (Fig. 22-14). The azygos system consists of two longitudinal venous channels, one on each side of the vertebral column. These channels receive the intersegmental veins of the body wall and are interconnected with each other across the vertebrae at irregular intervals. The longitudinal channel on the right is a continuous vessel called the **azygos vein.** On the left, the channel usually consists of two veins: the **hemiazygos vein** inferiorly and the **accessory hemiazygos vein** superiorly. These two veins on the left are interconnected with one another, and both empty independently into the azygos vein through one or more transverse connecting veins.

The azygos and hemiazygos veins are formed by the union of two veins: the ascending lumbar vein and the subcostal vein. Inferiorly, the *ascending lumbar vein* is connected on the right to the inferior vena cava and on the left to the left renal vein (the segments of these veins are equivalent in their developmental origin). Each ascending lumbar vein receives some of the lumbar veins and passes through the diaphragm, or its aortic hiatus, to unite with the subcostal vein of the corresponding side. The azygos vein ascends in the posterior mediastinum to the level of the 4th thoracic vertebra, where it arches forward over the root of the right lung to terminate in the superior vena cava (Fig. 22-3). The hemiazygos terminates in the azygos vein at about the level of the 8th thoracic vertebra. Each of these veins receives the **posterior intercostal veins** of the spaces they pass in their course. On the right, the intercostal veins of the spaces above the azygos vein drain into the *right superior intercostal vein,* which is a tributary of the azygos. On the left, the posterior intercostal veins of the spaces above the hemiazygos are received by two veins: the *left superior intercostal vein,* which empties into

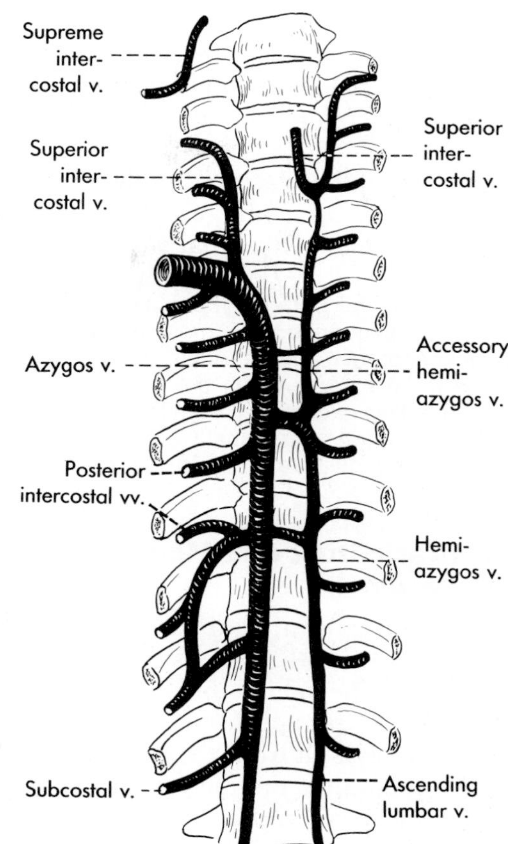


**FIG. 22-14.** The azygos system of veins. In this example, there are connections among all parts of the hemiazygos system and several connections to the azygos vein.

the left brachiocephalic vein, and the accessory hemiazygos vein, which empties into the azygos vein and is connected to both the superior intercostal and hemiazygos veins. With the exception of the first intercostal vein, which on both sides terminates in the corresponding brachiocephalic vein, the azygos vein drains venous blood from the posterolateral aspect of the thoracic and abdominal walls. Through the anastomoses of the posterior and anterior intercostal veins, the azygos system is linked to the internal thoracic veins (p. 479), and through the ascending lumbar veins, to the inferior vena cava. The azygos system is also in communication with the vertebral venous plexuses (p. 319). The azygos system, devoid of venous valves, constitutes an important channel for collateral venous circulation when the inferior or superior venae cavae are obstructed.

Tributaries of the system from thoracic viscera include the bronchial, esophageal, pericardial, and some mediastinal veins.

Variations in the pattern of the azygos system are of no consequence. The most important, although exceedingly rare, anomaly of the azygos system is for it to receive all the blood from the inferior vena cava except that from the liver, so that the azygos vein drains practically everything below the level of the diaphragm except the digestive tract.

## The Thoracic Duct and Mediastinal Lymph Nodes

Lymph draining from the greater part of the body is returned to the venous system by the thoracic duct. The duct is the upward continuation of the *cisterna chyli,* which is a saclike receptacle for lymph from the digestive tract and other abdominal and pelvic organs, from the abdominal wall and perineum, and from both lower limbs (p. 693). As soon as the thoracic duct leaves the cisterna, it enters the thorax behind the aorta through the aortic hiatus of the diaphragm. In the posterior mediastinum, the duct ascends on the front of the vertebral bodies, running between the aorta and the azygos vein (Fig. 22-15). In the midthoracic region, it inclines toward the left and leaves the thorax behind the left subclavian artery. In the base of the neck, the duct ends at the confluence of the left subclavian and left internal jugular veins. In most of its course, the thoracic duct lies behind the esophagus and is anterior to the right intercostal arteries and to the transverse, terminal portions of the hemiazygos and accessory hemiazygos veins as these vessels cross the vertebrae. On the right, it may come in contact with the retroesophageal recess of the pleura (p. 485).

The duct receives tributary vessels from several groups of lymph nodes located in the mediastinum, although most of the nodes drain first into the *bronchomediastinal lymph trunks.* The **groups of lymph nodes** in the mediastinum include the *parasternal* and *intercostal nodes* (p. 479), the large groups of *tracheobronchial lymph nodes* (p. 501), the *posterior mediastinal nodes,* and the *diaphragmatic groups of lymph nodes.* The posterior diaphragmatic, the intercostal, and the posterior mediastinal lymph nodes of the left side send their efferent vessels directly into the thoracic duct. The latter nodes, situated behind the pericardial sac and esophagus, drain adjacent structures and the diaphragmatic surface of the liver. The efferents of the parasternal and tracheobronchial nodes form the right and left bronchomediastinal lymph trunks; the left one of these may join the thoracic duct in the base of the neck or may empty independently into the confluence of the left subclavian and interjugular veins. Before its termination, the thoracic duct usually receives the left internal jugular and subclavian lymph trunks. The corresponding lymph trunks on the right side form the short **right lymphatic duct,** which has practically no intrathoracic course.

The lymph in the thoracic duct is usually white due to the presence of *chyle.* The duct is like a small vein, but is less easily seen because of its white or colorless contents. Its somewhat varicose or beaded appearance is due to the presence of valves. One set of these at the termination of the duct prevents the backflow of venous blood, which does, however, occur at the time of death. The terminal portion of the duct in the cadaver may, therefore, be confused with a vein.

Variations of the duct are common, for the lower part of the duct represents the original right member of a pair of ducts, whereas the upper part represents the original left member. Sometimes two fairly large ducts parallel each other for a distance.

Because most of the immunocompetent lymphocytes that recirculate between the blood and lymphatic tissues pass along the thoracic duct, drainage of lymph through a *thoracic duct fistula* can eliminate these cells from the body. Thoracic duct drainage may form part of the immunosuppressive regimen in recipients of organ transplants. The thoracic duct may be injured inadvertently during operations in the posterior mediastinum, and sometimes it is ruptured by violence. When such an injury is associated with a tear or cut in the pleura, lymph will pour into the pleural cavity at a rate as high as 60 ml to 190 ml/hr. The condition is known as *chylothorax,* and the accumulation of lymph (chyle) may cause collapse of

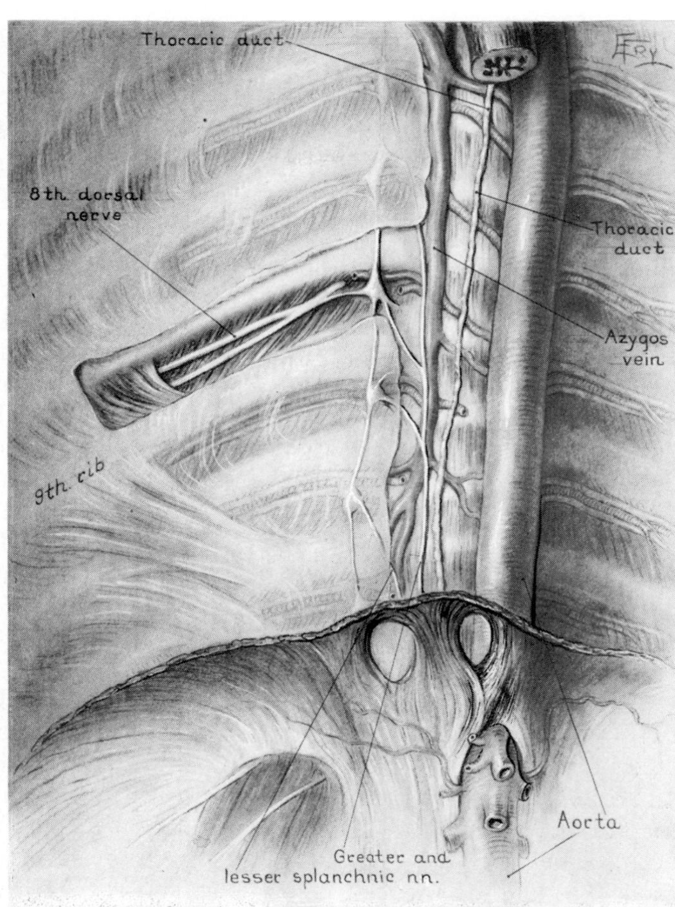


**FIG. 22-15.** Structures on the posterior thoracic wall. Some of the lower right intercostal arteries have been partially removed. (Labat G: Regional Anesthesia: Its Technique and Clinical Application. Philadelphia, WB Saunders, 1922)

the lung. Unless the defect heals, the duct may have to be ligated. The ligation usually has no noticeable consequences because numerous collateral vessels divert the lymph flow, and several communications exist between the lymphatic and venous system, although these communications have not been defined anatomically. On the other hand, when widespread malignant infiltration causes extensive blockage of the duct and other lymphatics, lymph (or chyle) may accumulate in the pleural cavity and in the peritoneal cavity (*chyle ascites*).

## The Sympathetic Trunks

The thoracic portion of the sympathetic nervous system is of particular importance because its paired trunks receive practically all of the preganglionic fibers that leave the spinal cord (p. 60). Some of these fibers synapse in the thoracic sympathetic ganglia, and the postganglionic fibers either join the thoracic spinal nerves or proceed to innervate thoracic viscera. Other preganglionic fibers either ascend in the trunks to the cervical region or descend to the lumbar and sacral portions of the trunks. Preganglionic fibers also leave the thoracic sympathetic trunks as thoracic splanchnic nerves destined for the innervation of abdominal and pelvic viscera. In addition to these preganglionic and postganglionic fibers, the trunks contain visceral afferents that find their way to the thoracic portion of the trunks, not only from thoracic organs but also from viscera of the head, neck, abdomen, and pelvis.

The right and left sympathetic trunks are essentially symmetrical. Each trunk consists of 11 to 12 ganglia linked to one another by interganglionic portions of the trunks (Fig. 22-16). Similar interganglionic branches connect the first and last thoracic ganglia to the cervi-

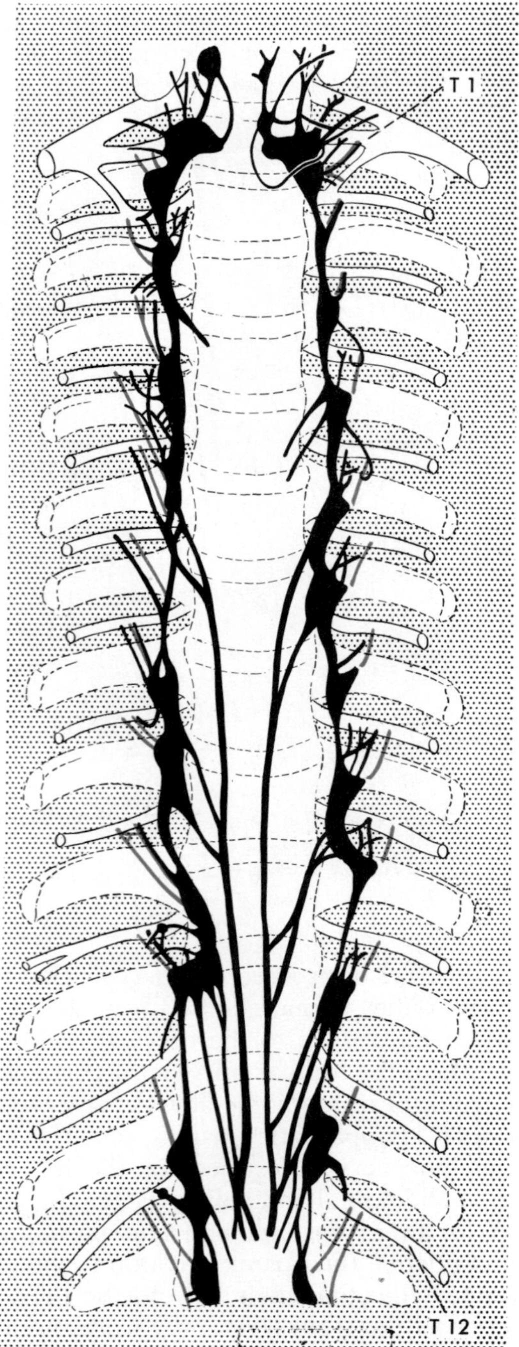


**FIG. 22-16.** The thoracic sympathetic trunks, the splanchnic nerves, and the rami communicantes. The large ganglion is the stellate; the first lumbar ganglion appears at the lower end. The white rami communicantes are shown in red, and the gray are shown in black; that they were actually white and gray rami was determined by histologic investigation. (Redrawn from Pick J, Sheehan D: J Anat 80:12, 1946)

cal and lumbar portions of the trunks, respectively. The trunks lie behind the costal pleura (Figs. 22-3 and 22-4) and, for the most part, rest on the front of the necks of the ribs. Their lower ends incline forward onto the sides of the vertebrae, so that by the time they penetrate the diaphragm, they are situated more anteriorly than laterally. Strictly speaking, therefore, only the inferior portions of the trunks are in the posterior mediastinum.

The **thoracic sympathetic ganglia** are numbered according to the intercostal nerve with which they connect. However, in 75% to 80% of people, the uppermost or first thoracic ganglion is fused with the inferior cervical ganglion and forms the *cervicothoracic,* or *stellate, ganglion* (Fig. 22-16); the last thoracic ganglion usually connects with both the 11th and the 12th thoracic nerve. Thus, there are, as a rule, 11 rather than 12 thoracic ganglia in each trunk. Each ganglion lies a little below the level of the corresponding intercostal nerve, to which it is connected by two of its branches. One of these branches consists predominantly of myelinated, preganglionic fibers and is called the *white ramus communicans;* the other consists predominantly of unmyelinated postganglionic fibers and is the *gray ramus communicans* (see also Chap. 4). Each of these rami may be split into two or more fascicles, or the two may fuse into one. Visceral afferent fibers enter the ganglion mainly by the white ramus communicans, and they continue their way to the corresponding spinal ganglion admixed with somatic afferent fibers in the intercostal nerve. In addition to the rami communicantes and the interganglionic branches, each ganglion has one or more **medial branches** that serve the viscera. The medial branches destined for thoracic viscera are small and are postganglionic (postsynaptic); those for abdominal and pelvic viscera leave the ganglia as preganglionic (presynaptic) fibers and are gathered into two or three large nerves. These are the thoracic splanchnic nerves.

There are usually three **thoracic splanchnic nerves** called the greater, lesser, and lowest splanchnic nerves (Fig. 22-17). There is much variation in their origin; among 100 cadavers,

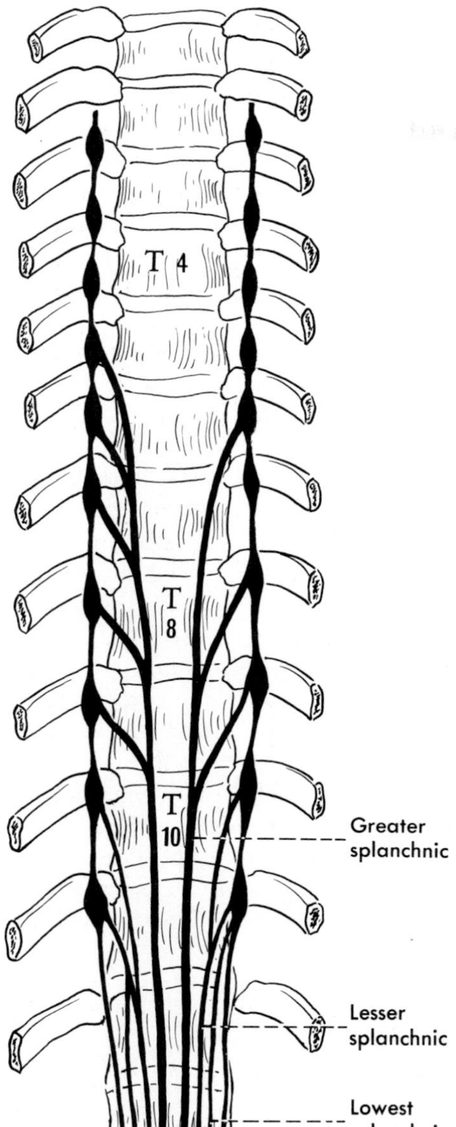


**FIG. 22-17.** Origin of the splanchnic nerves from the sympathetic trunks; they vary and are not necessarily symmetrical on the two sides.

for instance, 58 different patterns were found. The significant points to bear in mind are that these nerves are composed predominantly of preganglionic fibers that will relay in collateral ganglia of the abdomen (celiac, superior, and inferior mesenteric ganglia) and also contain afferent fibers (pain sensation) from abdominal and pelvic viscera.

The *greater splanchnic nerve* usually arises from the 5th (or 6th) to the 9th thoracic ganglia; the *lesser splanchnic nerve,* from the 9th and 10th; and the *lowest splanchnic nerve,* from the lowest thoracic ganglion. The lowest splanchnic nerve may also be represented merely by a branch of the lesser splanchnic. Nerves of considerable size are formed by the medial branches of the appropriate sets of ganglia; the greater splanchnic nerve is larger than the continuation of the sympathetic trunk into the abdomen (Fig. 22-17). All three splanchnic nerves descend toward the diaphragm in front of the vertebral column, lying medial to the sympathetic trunks. They pierce the muscular part of the diaphragm and enter the celiac ganglion or one of the associated ganglia or plexuses. The lowest splanchnic nerve, or the corresponding branch of the lesser splanchnic, usually ends in the renal plexus.

## RECOMMENDED READINGS

BARRY A: The aortic arch derivatives in the human adult. Anat Rec 111:221, 1951

BOTHA GSM: The Gastro-Oesophageal Junction. London, J & A Churchill, 1962

DAVIS HK: A statistical study of the thoracic duct in man. Am J Anat 17:211, 1915

EDWARDS LF, BAKER RC: Variations in the formation of the splanchnic nerves in man. Anat Rec 77:335, 1940

GLADSTONE RJ: Development of the inferior vena cava in the light of recent research with especial reference to certain abnormalities, and current descriptions of the ascending lumbar and azygos veins. J Anat 64:70, 1929

JACKSON RG: Anatomy of the vagus nerves in the region of the lower esophagus and the stomach. Anat Rec 103:1, 1949

JACOBSSON S-I: Clinical Anatomy and Pathology of the Thoracic Duct. Stockholm, Almquist and Wiksell, 1972

MCDONALD JJ, ANSON BJ: Variations in the origin of arteries derived from the aortic arch, in American whites and Negros. Am J Phys Anthropol 27:91, 1940

PICK J, SHEEHAN D: Sympathetic rami in man. J Anat 80:12, 1946

WRIGHT NL: Dissection study and innervation of the human aortic arch. J Anat 104:377, 1969

# PART VII

# ABDOMEN

# 23

# THE ABDOMEN IN GENERAL

The abdomen is the region of the trunk between the thorax and the pelvis. On the surface of the trunk, the abdomen is demarcated superiorly by the xiphisternal joint and the costal arches; inferiorly by the symphysis pubis, the inguinal folds, and the iliac crests; and posteriorly by the lumbar paravertebral musculature of the back. Within the trunk, the abdomen is much more extensive: the **abdominal cavity** bulges up into the rib cage and is separated from the thoracic cavity by the diaphragm, which has a highly concave abdominal surface (see Fig. 19-1); inferiorly, the abdominal cavity continues, without interruption, into the pelvis, being limited below by the pelvic floor, or pelvic diaphragm, which separates it from the perineum. The abdominal cavity is the inferior (caudal) compartment of the body cavity, or celom, and contains the major parts of the digestive and urinary systems, the internal organs of reproduction, and the peritoneum, which is the largest serous sac of the body intimately associated with the viscera.

It is customary, speaking both anatomically and clinically, to consider the abdominal cavity in two parts: the upper, larger portion is the **abdomen proper,** and the smaller part below is the **pelvic cavity.** The arbitrary plane that demarcates these two parts of the cavity coincides with the superior aperture or brim of the pelvis. The plane of the superior pelvic aperture slants forward from the sacral promontory to the upper border of the pubic symphysis. The pelvic cavity and its contents are considered separately in Chapter 27; this chapter, and the following two, deal primarily with the abdomen proper. It is important to remember, however, that through the superior pelvic aperture there is continuity of space, both within and outside the peritoneal sac, and several structures are transmitted through the aperture between the abdominal and pelvic portions of the cavity. Moreover, organs considered to be primarily abdominal (loops of small bowel, the appendix) may normally descend into the pelvis, and primarily pelvic viscera (distended bladder, pregnant uterus) normally rise up into the abdomen.

## THE ABDOMINAL WALL

Above and anteriorly, the abdomen proper is surrounded by contractile and distensible walls, whereas the posterior wall is stable and bulky by comparison. The musculotendinous diaphragm forms the mobile, dome-shaped roof, and three concentric layers of sheetlike muscles, with their aponeuroses, enclose the abdomen anterolaterally. The bodies of the lumbar vertebrae and the ala of the iliac bones account for the stability of the abdominal wall posteriorly, and the three pairs of muscles associated with these bones (quadratus lumborum, psoas major, and iliacus) increase the bulk of the posterior abdominal wall (Fig. 23-1).

It is usual to consider the diaphragm and the posterior wall separately (Chap. 25). Unless specifically qualified, the term *abdominal wall* is used by clinicians and anatomists alike to refer to the anterolateral abdominal wall only. In this sense, on the surface of the trunk, the extent of the abdominal wall corresponds to the extent of the abdomen defined in the first paragraph of this chapter. Over this region, the abdominal wall is made up of skin, superficial fascia, three layers of muscle (outer, inner, and innermost layers), and endoabdominal fascia (Fig. 23-1). Except for the lack of ribs, the arrangement of these tissues resembles the structure of the thoracic wall (Chap. 19).

The derivation and development of body wall muscles and fascias were considered in the introduction to the thoracic wall (p. 474). The ventrolateral extensions of T6–T12 (L1) myotomes into the somatopleure of the abdominal wall fuse with each other and give rise to all the muscles. Because of the absence of ribs, the only reminder of the segmental origin of the abdominal wall musculature is their segmental pattern of nerve and blood supply.

Fascial and muscular layers of the abdominal wall send extensions into the perineum. Most of these extensions provide fascial tunics for the ductus deferens in the male and the round ligament of the uterus in the female, as these structures pass through the abdominal wall. The passage through the abdominal wall, traversed by the ductus deferens or the round ligament, is known as the **inguinal canal.** Because of its clinical importance, the inguinal canal and the applied anatomy of inguinal and femoral hernia are considered separately in Chapter 26.

### MUSCLES AND FASCIAS

The outer, inner, and innermost muscle layers of the abdominal wall are formed by the external oblique, internal oblique, and transversus abdominis, respectively. These three pairs of broad, flat muscles, and their aponeuroses, form the major part of the anterolateral abdominal wall. On each side of the midline, a straplike muscle, the rectus abdominis, spans the abdominal wall between the rib cage and the symphysis pubis. The aponeuroses of the three flat muscles meet in a midline raphe, where the intertwining of their tendon fibers forms the **linea alba,** a white line of varying breadths. Before the aponeuroses meet in this raphe, they form a sheath around the rectus as they pass anteriorly or posteriorly to that muscle (Fig. 23-1).

Each muscle is covered on its deep and superficial surface by *deep fascia.* This thin connective tissue membrane is firmly attached to the muscles and their aponeuroses. There is a small amount of loose connective, or areolar, tissue between the deep fascia of the adjacent muscle layers. The **neurovascular plane** is defined by such areolar tissue between the internal oblique and the transversus abdominis. The segmental nerves and blood vessels of the body wall run in this plane and, from here, distribute their cutaneous and muscular branches that pierce the adjacent layers.

*Superficial fascia* separates the exterior of the muscular wall from the skin. On the interior, a layer of extraperitoneal, or *endoabdominal, fascia* attaches the parietal peritoneum to the transversus abdominis and merges with similar areolar tissue over the neighboring boundaries of the abdominal cavity.

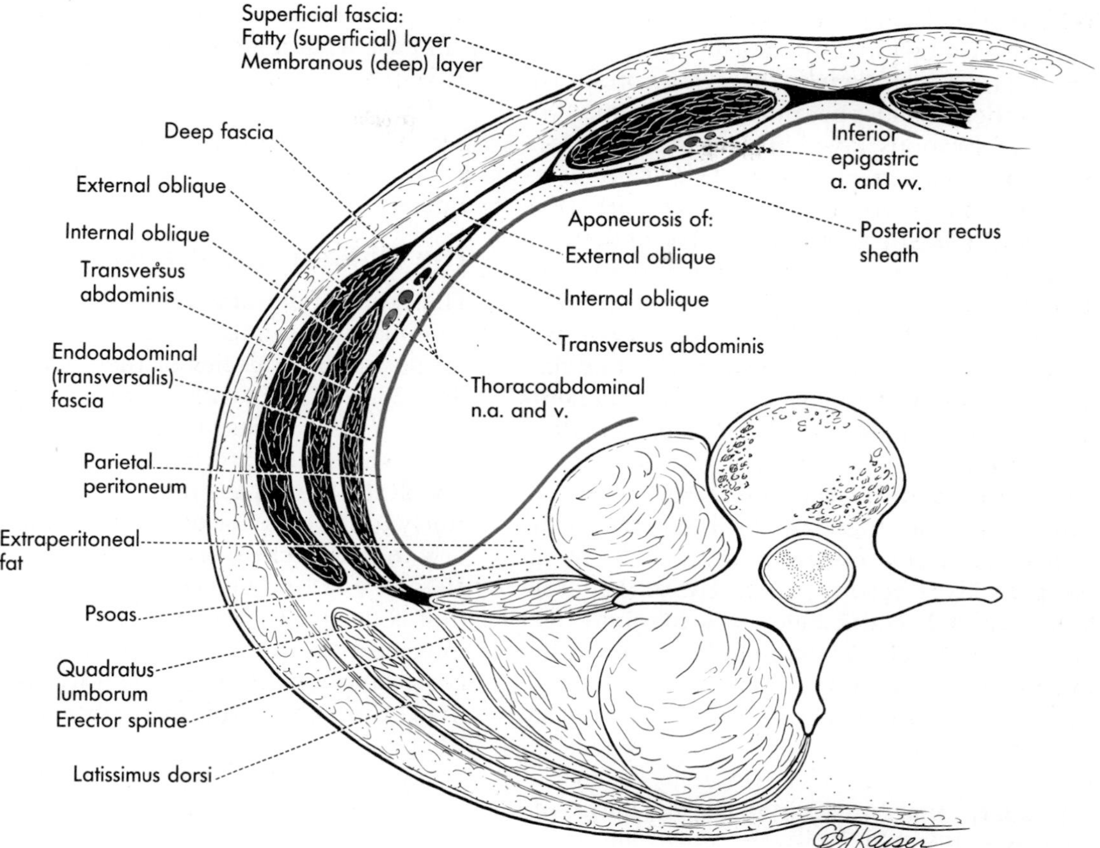


**FIG. 23-1.** A schematic transverse section of the abdomen to show the layers of the abdominal wall. The muscles of the anterolateral abdominal wall are shown in black; muscles associated with the vertebral column are white. Compare with Figure 19-2.

## Superficial Fascia

The subcutaneous tissue of the abdominal wall is a favorite repository for fat and may be several inches thick. Delicate fibrous strands, which surround the lobules of fat, are firmly anchored in the dermis, but adhere only loosely to the deep fascia of the external oblique. Consequently, the superficial fascia, and the skin with it, may be pinched up and moved around with ease, much more so than over the thigh, for instance. A cleavage plane can be readily defined by blunt dissection between the deep and the superficial fascia.

When the superficial fascia contains a moderate amount of fat, the deep surface appears membranous owing to a condensation of its overlapping, deeper fibrous strands into a definite lamella. Such a **membranous layer** can be demonstrated in most persons, and it may be dense enough to hold sutures. It can also be identified as a distinct layer by computer-assisted tomography. Thus, over the lower part of the abdomen, two layers can be defined in the superficial facia: a *superficial fatty layer* and a *deep membranous layer*. The latter is similar in structure to the deep fascia, but it is quite distinct from it anatomically.

The attachments of the membranous superficial fascia are of some importance. In the midline, the fascia is attached to the linea alba, and in the region above the pubis, it forms a midline septum that is reinforced by contributions from the deep fascia. This septum extends onto the dorsum of the penis and is known as the suspensory, or *fundiform, ligament.* To each side of the midline, the superficial fascia descends into the perineum: in the male, the fat disappears over the penis, but the membranous layer invests the penis (*superficial penile fascia*) (Fig. 23-2A) and surrounds the scrotum (*dartos fascia*) (Fig. 23-2B); in the female, both fatty and membranous layers continue into the labia majora. In both sexes, the membranous layer becomes, in effect, the *superficial fascia of the perineum,* and both superficial penile and dartos fascias can be considered as components of the superficial fascia of the perineum (Chap. 28).

Further laterally, the superficial fascia continues into the thigh: the fat blends with the superficial fascia of the thigh, and the membranous layer becomes adherent to the fascia lata a fingerbreadth below the inguinal ligament (Fig. 23-2C). Thus, in a dissection, a finger placed between the membranous superficial fascia and the deep fascia of the external oblique cannot be pushed into the thigh, but it can pass with relative ease into the penis, the wall of the scrotal sac, or the labia majora.

The superficial fascia is traversed by cutaneous nerves and blood vessels as well as by lymphatics. These nerves and vessels are considered in a separate section of this chapter.

## The Oblique and Transversus Muscles

The **external oblique** (*m. obliquus externus abdominis*) is the most superficial and the largest of the flat muscles of the abdominal wall (Fig. 23-3). Rather more than the anterior half of the muscle consists of a strong tendinous membrane, the external oblique *aponeurosis.* The fleshy part of the muscle is attached superiorly to the exterior of the rib cage and inferiorly to the outer lip of the iliac crest. The muscle has two free margins: a posterior margin that is fleshy and more or less vertical and an inferior margin that is entirely tendinous and runs obliquely between the anterior superior iliac spine and the pubic tubercle. This free

**FIG. 23-2.** Diagram showing the disposition of the superficial fascia of the lower part of the abdominal wall. The membranous layer is shown in blue. **A** is a longitudinal section of the anterior abdominal wall just to one side of the midline. **B** is a section a little more laterally. **C** is a section far enough laterally to continue into the thigh.

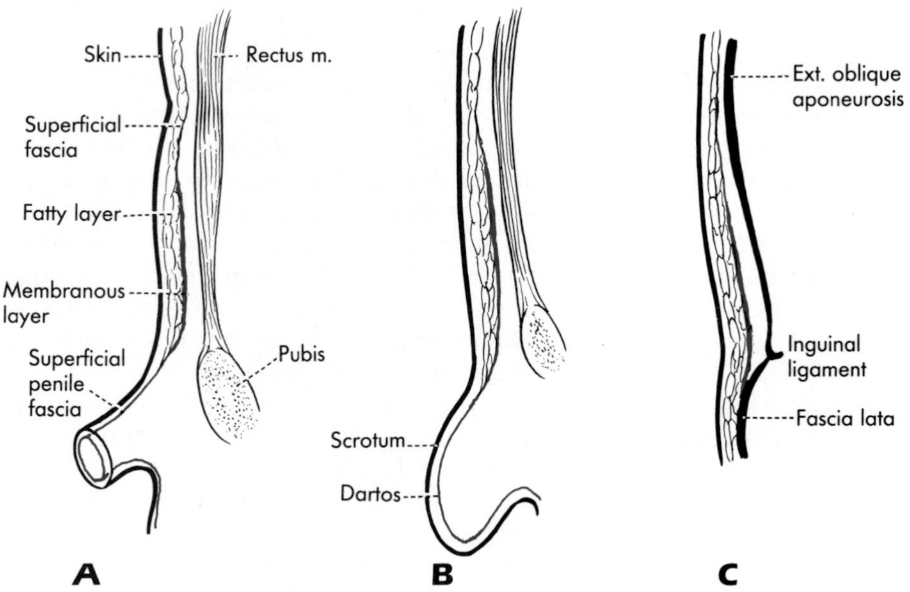

inferior margin of the external oblique aponeurosis is thickened and is known as the *inguinal ligament.*

The fleshy slips of the external oblique attach to each of the last seven or eight ribs, but fuse almost immediately to form a broad muscle with no trace of segmentation. At their costal attachments, the costal slips interdigitate with the serratus anterior and the lower slips interdigitate with the latissimus dorsi. The muscle fibers slope downward and forward from their costal attachments, the posterior fibers being the most vertical. The slips attached to the last three to four ribs remain fleshy and attach below to the anterior half of the iliac crest. The remaining slips continue as the aponeurosis. Tendon fibers in the upper portion of the aponeurosis decussate in the linea alba, whereas those in the lower portion attach to the symphysis pubis and the pubic crest. More laterally, the aponeurosis terminates by forming the inguinal ligament and by attaching to the iliac crest for a short distance beyond the anterior superior iliac spine.

The decussation in the linea alba has been shown to be rather complex: some of the tendon fibers in the external oblique aponeurosis can be traced directly into the aponeurosis of more than one of the flat muscles on the opposite side. Thus, the upper portion of each external oblique may be considered as a digastric muscle: one belly is the external oblique of one side, and the other, the internal oblique or transversus of the opposite side. Acting in unison with other flat muscles, the external obliques compress the abdomen. In this action, the costal attachment of the external oblique remains fixed and is appropriately designated as the muscle's site of origin. However, in producing movements of the trunk,

**FIG. 23-3.** The external oblique muscle.

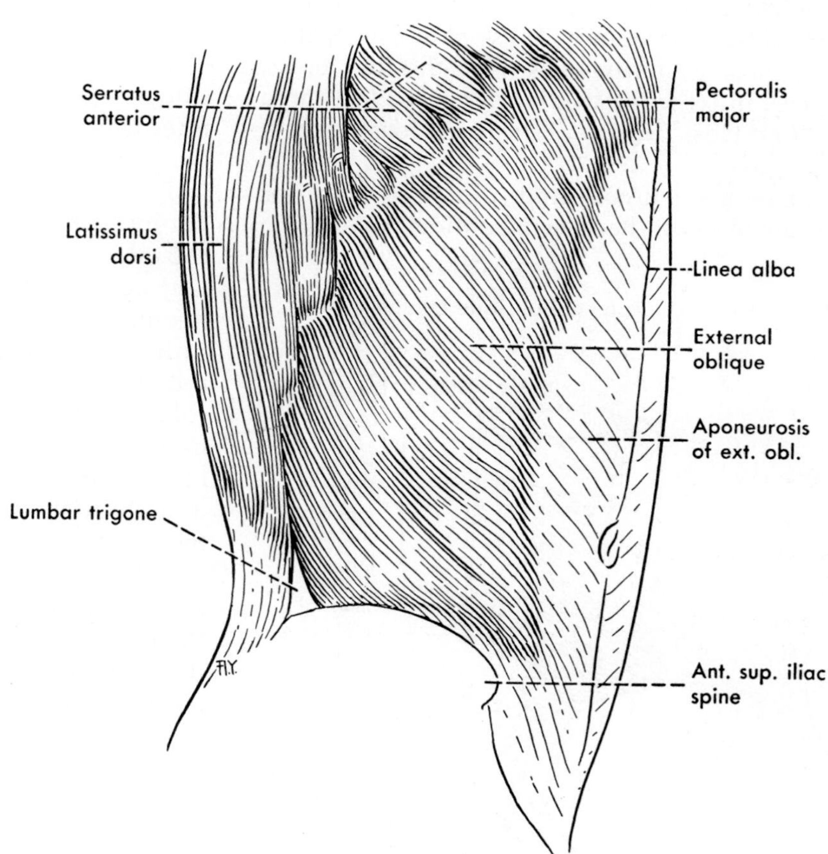

the external oblique approximates the rib cage to the stable pelvis; therefore, the costal attachment should be designated as the muscle's site of insertion.

The **internal oblique** (*m. obliquus internus abdominis*) lies immediately deep to the external oblique (Fig. 23-4). The muscle arises from the upper surface of the anterior two-thirds of the iliac crest and, in continuity with this bony attachment, from the thoracolumbar fascia behind and from the lateral third of the inguinal ligament (more precisely, from the line of fusion between the inguinal ligament and the iliopsoas fascia) in front. From this long, linear origin, fleshy fibers fan out and run mostly upward and medially, approximately at right angles to the fibers of the external oblique. The muscle does not have a free posterior border between the iliac crest and the ribs; rather, it is here that its posterior fibers arise from the thoracolumbar fascia, thereby obtaining indirect attachment to the vertebral column. These fibers, as well as the most posterior ones from the iliac crest, insert into the inferior border of the last three or four ribs. The rest of the fibers form an *aponeurosis*.

The aponeurosis and the most anterior fleshy fibers of the internal oblique have a free

**FIG. 23-4.** The internal oblique muscle.

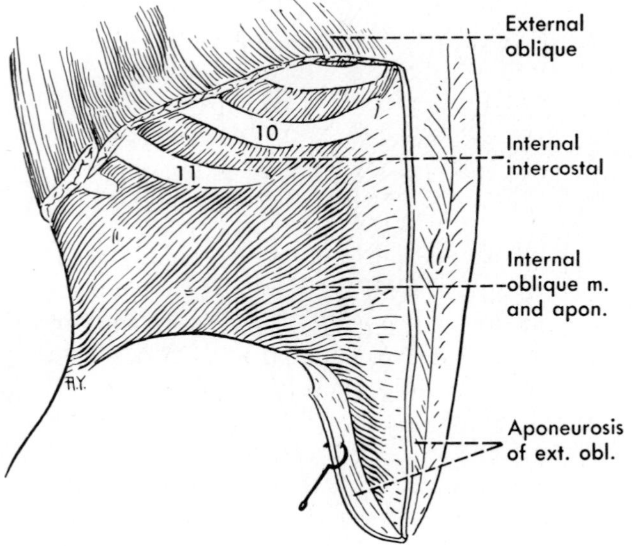


inferior border that passes downward and medially, running above and parallel with the inguinal ligament. This free border contributes to the formation of the **conjoint tendon** (*falx inguinalis*), so called because at some distance below the umbilicus, the aponeurosis of the internal oblique and the transversus abdominis fuse with each other. This fused aponeurosis passes anteriorly to the rectus, and its lateral segment, the conjoint tendon, becomes anchored to the pecten of the pubis. The rest of the fused aponeurosis decussates with the aponeuroses of the other muscles in the raphe of the linea alba.

The more superior portion of the internal oblique aponeurosis splits into an anterior and posterior layer along the lateral edge of the rectus (Fig. 23-8). The posterior layer remains fused to the transversus aponeurosis and passes with it behind the rectus; the anterior layer passes anteriorly to the rectus with the aponeurosis of the external oblique. These fused aponeurotic layers constitute the *rectus sheath* and meet their fellows of the opposite side in the linea alba. In the epigastrium, the internal oblique aponeurosis is attached to the inferior border of the costal arch.

The **transversus abdominis,** the innermost flat muscle of the abdominal wall, arises, like the internal oblique, from the iliopsoas fascia (adjacent to the lateral third of the inguinal ligament), from the iliac crest (the inner lip rather than the upper surface), and, between the iliac crest and the 12th rib, from the thoracolumbar fascia (Fig. 23-5). In addition, on the internal aspect of the lower six costal cartilages, slips of origin of the transversus interdigitate with those of the diaphragm. For the most part, the muscle fibers extend transversely and end in an *aponeurosis*. As it reaches the rectus, the aponeurosis fuses with the internal oblique aponeurosis to help form the *rectus sheath* (Fig. 23-8). Below the level of the anterior superior iliac spine, the muscle fibers tend to be directed downward and medially, paralleling the fibers of the internal oblique in this region. The lower fibers of the transversus, which arise from the iliopsoas fascia and the anterior superior iliac spine,

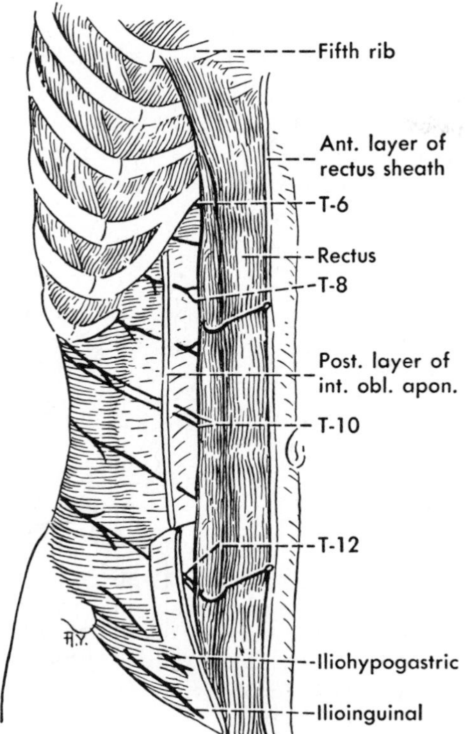


**FIG. 23-5.** The transversus abdominis muscle and nerves of the anterior abdominal wall. The rectus sheath has been opened and the rectus muscle retracted medially to show the nerves penetrating the sheath and disappearing behind the muscle; the iliohypogastric and ilioinguinal nerves are shown as they come through the internal oblique.

form the portion of the aponeurosis that, after fusing with the internal oblique aponeurosis, makes up the **conjoint tendon** (*falx inguinalis*). Like the internal oblique, the transversus abdominis also has a free musculoaponeurotic border that arches over the spermatic cord or the round ligament of the uterus as these structures enter the inguinal canal.

## The Rectus Abdominis and Its Sheath

The **rectus abdominis** arises from the pubic crest by a short tendon (Fig. 23-6). A straplike muscle belly replaces the tendon and becomes progressively broader as it ascends to the rib cage, where it inserts into the 5th, 6th, and 7th costal cartilages and into the xiphoid process. Not all the fibers of the muscle span this distance, however; *tendinous intersections*—usually three, sometimes four—divide the muscle transversely into segments. The intersections are, as a rule, not quite complete, especially posteriorly. They become prominent when the recti are contracted, and a median furrow over the linea alba appears between the two muscles (Fig. 23-7). The convex lateral border of each rectus is indicated by another shallow groove, the *linea semilunaris*. In a spare or muscular person, these three grooves are visible even when the recti are relaxed.

Overlapping the anterior surface of the rectus just above the pubis is the triangular **pyramidalis muscle** (Fig. 23-6). This little muscle arises from the body of the pubis and inserts into the linea alba. It is a tensor of the linea alba and is of no importance. The muscle may be missing on one or both sides.

The **rectus sheath** (*vagina m. recti abdominis*), formed by the aponeuroses of the three flat muscles of the abdominal wall, encloses the rectus (Fig. 23-8). The sheath has an anterior and a posterior lamina, and the two fuse with each other medially along the linea alba and laterally along the linea semilunaris. However, only the middle segment of the rectus is completely enclosed in this sheath; the posterior lamina is lacking behind both superior and inferior portions of the muscle.

The explanation for this arrangement is as follows. Along the lateral edge of the rectus, the internal oblique aponeurosis splits into an anterior and a posterior layer and these pass respectively in front of and behind the middle portion of the rectus. The transversus aponeurosis becomes fused to the posterior layer, and the external oblique aponeurosis joins, in front, the anterior layer (Fig. 23-8). The internal oblique aponeurosis does not extend up into the infrasternal angle, and the superior extent of the transversus is the costal arch. Therefore, in the infrasternal angle, the posterior lamina of the rectus sheath consists of some muscular and aponeurotic fibers of the transversus abdominis (Figs. 23-5 and 23-8), whereas above the costal margin, the rectus lies directly on the costal cartilages.

Some distance below the umbilicus, the aponeurosis of the internal oblique changes its

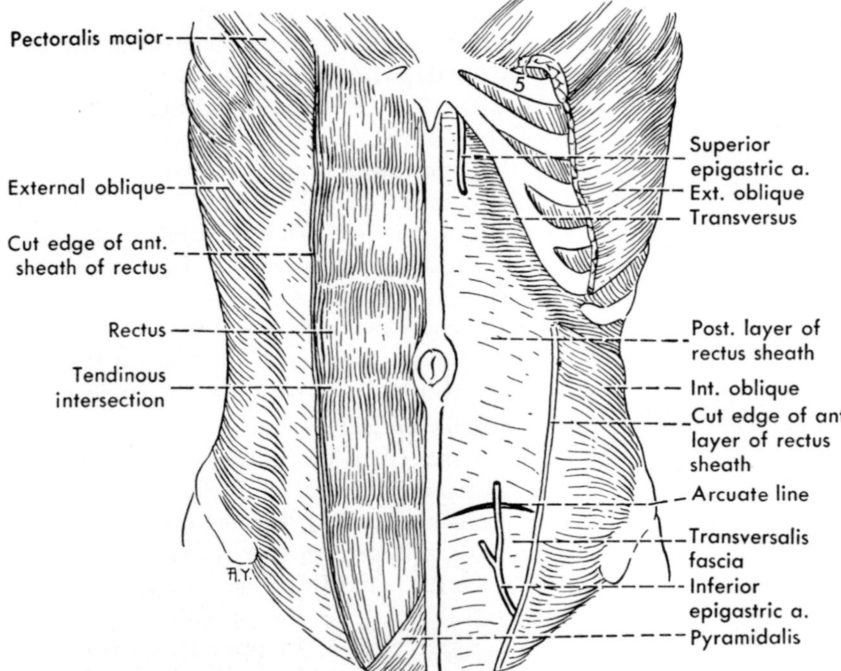


**FIG. 23-6.** Rectus abdominis and pyramidalis muscles and the posterior layer of the sheath of the rectus.

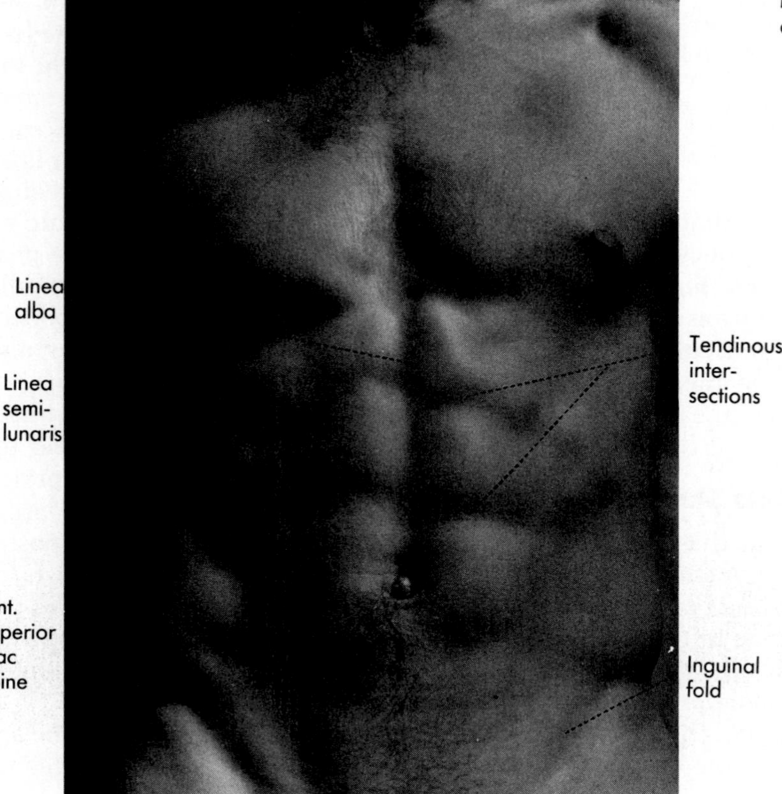


**FIG. 23-7.** The rectus abdominis contracted.

course and, rather than splitting into layers, passes unsplit in front of the rectus. Since the aponeurosis of the transversus remains fused with that of the internal oblique, it also passes to the front, creating a defect in the posterior lamina of the rectus sheath. Thus, over the inferior part of the rectus, all three aponeuroses are in the anterior lamina of the sheath, leaving the rectus in direct contact, posteriorly, with endoabdominal (transversalis) fascia (Fig. 23-8C).

If the change in the course of the tendon fibers of the internal oblique and transversus aponeuroses is abrupt, the posterior lamina of the rectus sheath ends with a sharp crescentic border somewhere between the umbilicus and the pubis. This border, concave downward, is the **arcuate line** (Fig. 23-9).

The anterior wall of the rectus sheath is intimately attached to the tendinous intersections in the rectus; elsewhere, delicate connective tissue separates the muscles from the anterior and posterior walls of the sheath. Enclosed within the sheath are also the pyramidalis muscle, anterior to the rectus, and the superior and inferior epigastric vessels, poste-

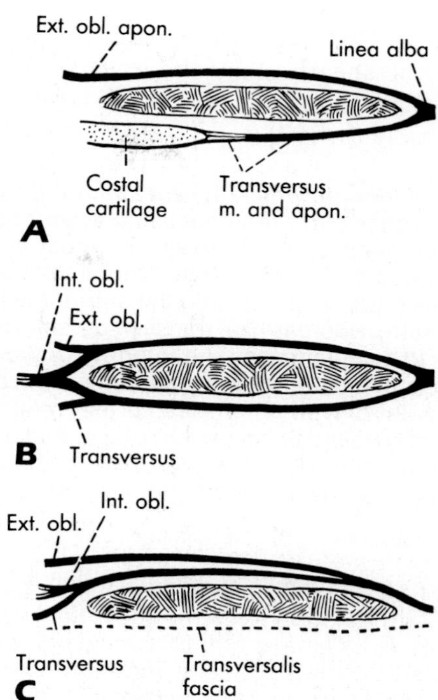


**FIG. 23-8.** Diagrams of the usually described method of formation of the rectus sheath. **A** is a transverse section through the epigastrium, **B**, one below the ribs but above the arcuate line, and **C**, one below the arcuate line.

**FIG. 23-9.** The inner surface of the anterior abdominal wall, with the transversalis fascia removed below to show the arcuate line.

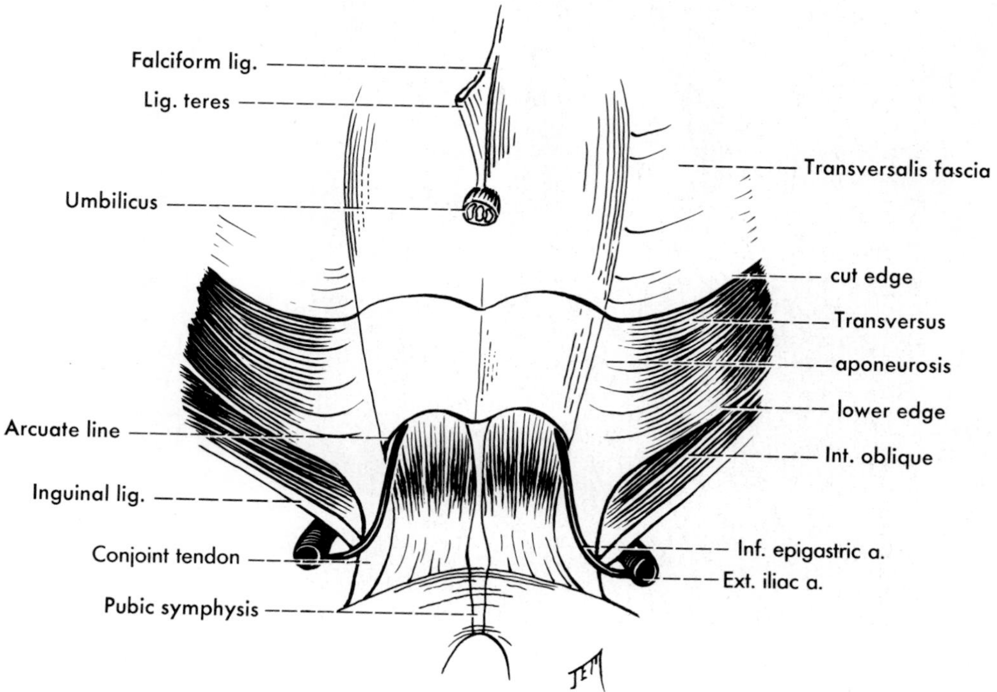

rior to it. The nerves that supply the rectus pierce the sheath along the linea semilunaris just before they sink into the lateral edge of the muscle (p. 592).

The manner in which the sheath of the rectus is formed varies more than the usual descriptions indicate. Indeed, in one survey, the most common pattern was that in which the internal oblique aponeurosis passed entirely into the anterior lamina of the sheath, whereas the transversus split to contribute to both laminae. The integrity of the rectus sheath is of considerable importance from a functional point of view. Therefore, further reference to the more detailed arrangement of the tendon fibers in the rectus sheath is included in the discussion of the actions of the anterior abdominal wall muscles.

## Innervation and Action of Abdominal Wall Muscles

The muscles of the abdominal wall are supplied segmentally by the ventral rami of T6–T12 and L1 spinal nerves. The course of these nerves in the abdominal wall is described in the next section. The rectus, external oblique, and transversus muscles with extensive fleshy attachment to the ribs receive muscular branches from the upper six of these nerves; rarely, L1 may also supply the rectus. The main belly of the internal oblique is innervated by T10–T12 and the ilioinguinal branch of L1. The iliohypogastric branch of L1 is

**FIG. 23-10.** Continuities in tendon fibers across the linea alba between the aponeuroses of the flat muscles in the abdominal wall, creating digastric arrangements. In **A**, the superficial lamina of the external oblique aponeurosis is indicated in continuous lines, and the deep lamina is indicated in interrupted lines. Fibers in the superficial lamina on the left side are continuous with fibers of the deep lamina on the right. In **B**, another set of fibers in the deep lamina of the right external oblique aponeurosis is continuous with those in the anterior lamina of the left internal oblique aponeurosis. In **C**, fibers in the anterior lamina of the right transversus aponeurosis become continuous with those in the posterior lamina of the left internal oblique aponeurosis. (Askar OM: Ann R Coll Surg Engl 59:313, 1977)

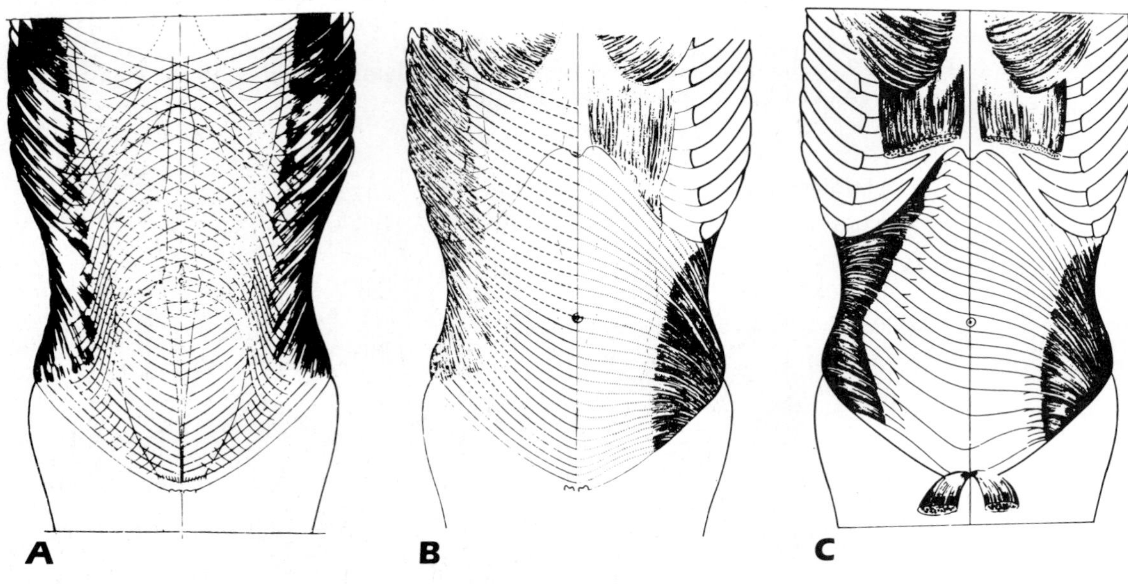

chiefly cutaneous and is motor only to the inconsequential pyramidalis muscle.

The muscles of the anterior abdominal wall perform three functions: (1) support the abdominal viscera, (2) compress the abdomen, and (3) move the trunk.

**Support.** In the supine position, the muscles are completely relaxed. During sitting and standing, it is the flat muscles and their aponeuroses that support the abdominal contents; the recti are largely passive. In the standing position, continuous electrical activity can be recorded in the lower parts of the internal oblique (and probably also in the transversus), which constantly guards the inguinal region. The importance of this function will become apparent in the discussion of inguinal hernia (Chap. 26).

The tendon fibers of the three pairs of aponeuroses are interwoven in the rectus sheath and the linea alba in such a manner that they permit the expansion of the abdomen while they support the viscera. Rather precise dissections of the aponeuroses have revealed recently that not only the internal oblique aponeurosis but also the aponeuroses of both the external oblique and the transversus split into two strata. The three sheets of tendon fibers cross each other in different directions and become continuous across the linea alba, with fibers on the opposite side that are in a stratum either superficial or deep to that of their own.

For instance, the most superficial tendon fibers of the left external oblique aponeurosis are derived from some of the deep fibers of the right external oblique, which have come to the surface at the linea alba and have continued their course laterally and downward, crossing the deep stratum of the external oblique aponeurosis at right angles (Fig. 23-10A). Other deep fibers of the right external oblique aponeurosis become continuous with the anterior lamina of the left internal oblique aponeurosis after they decussate in the linea alba (Fig. 23-10B). Similarly, there are two strata to the transversus aponeurosis. Some fibers of the anterior stratum of the right transversus aponeurosis become continuous with the posterior lamina of the left internal oblique aponeurosis (Fig. 23-10C). Other fibers in the anterior stratum of the right

**FIG. 23-11.** Plywoodlike arrangement of tendon fibers in the anterior sheath of the right rectus of a man aged 35 years. The superficial lamina (*s. ext.*), derived from the left external oblique aponeurosis, crosses at right angles the intermediate lamina (*d. ext.*), derived from the deep fibers on the right external oblique aponeurosis. The deepest lamina in the anterior sheath (*int.*) is derived from the anterior lamina of the right internal oblique, and the fiber direction coincides with that in the superficial lamina. (Rizk NN: J Anat 131:373, 1980)

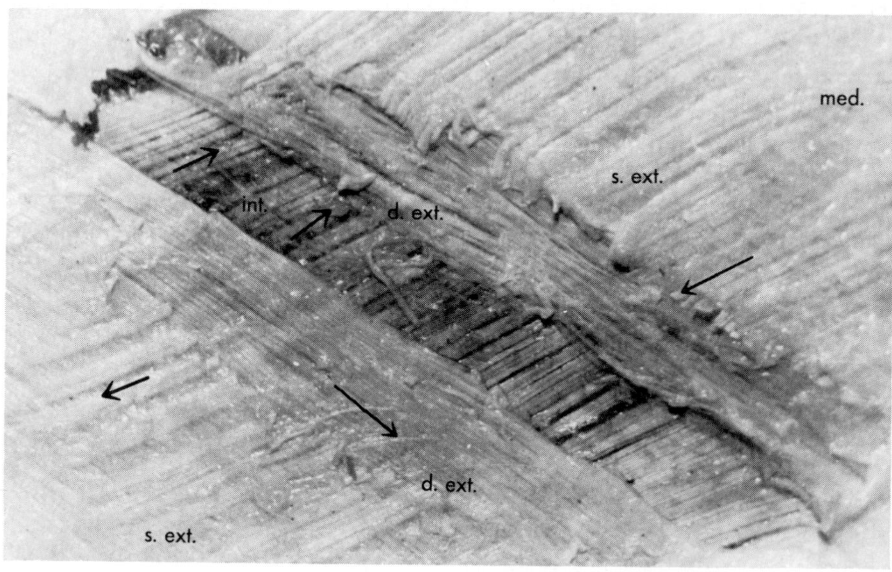

transversus aponeurosis become continuous with the posterior stratum of the left transversus aponeurosis (not illustrated). Thus, the rectus sheath, where it is completely formed, has three sets of fibers in both its anterior and posterior walls, and these fibers create a "plywoodlike" arrangement in each wall (Fig. 23-11).

This arrangement strengthens the abdominal wall considerably, owing to the perpendicular crossing of tendon fibers and the switching of tendon fibers from one stratum into another, which is chiefly responsible for holding the layers of the abdominal wall together without limiting their independent mobility. This is true also in the lower part of the abdomen, where all aponeurotic layers, with the possible exception of the posterior stratum of the transversus abdominis, pass in front of the rectus. Furthermore, the digastric arrangement created by these mechanisms ensures the unified action of the different muscle layers on the two sides.

Expansion of the abdomen is made possible, without increasing tension in the aponeuroses, by decreasing the angle of splitting in the strata, by decreasing the angles of decussation in the rectus sheath or linea alba, and by separating the parallel fiber bundles within the strata. Raising of the costal arches, which accompanies distention of the upper part of the abdomen, does, in fact, produce such an effect, since fibers in four of the six strata run parallel with the costal arch on each side.

**Compression of the Abdomen.** The obliques and the transversus, acting together bilaterally, function as a muscular girdle that exerts pressure on the abdominal contents. The digastric arrangement, established through the continuity of the tendon fibers between the aponeuroses of the two sides, ensures a uniform, bilaterally coordinated action of the muscle layers. The rectus participates little, if at all, in this function. Contraction of the flat muscles generates the increased pressure required for the forceful expulsion of air from the lungs (coughing, sneezing, singing) or of gastric contents by vomiting. The same muscles raise abdominal pressure during bearing down (necessary to empty the rectum and the bladder and to deliver the fetus from the birth canal). The obliques and the transversus are not active during quiet respiration; they are, however, the most important and only indispensable muscles of maximal voluntary expiration. They not only push the relaxed diaphragm up into the chest but also depress and compress the lower part of the rib cage.

**Trunk Movements.** The abdominal muscles can approximate the rib cage to the pelvis and thereby indirectly move the vertebral column. The type of movement each muscle is capable of producing can be deduced from its bony attachments and the direction of its fibers. However, abdominal muscles are only recruited to produce trunk movements when resistance or the force of gravity must be overcome. Thus, the recti flex the trunk when a person rises from the supine to the sitting position (they contract even when only the head is raised), but they are inactive when a person bends forward from the erect position. The external and internal obliques, acting bilaterally, can assist flexion and are being exercised, along with the rectus, during sit-ups. The right external oblique, working together with the left internal oblique, can twist the trunk to the left. Without having to overcome resistance, however, most of this movement is produced by the paravertebral muscles. Therefore, as a rule, trunk twisting exercises do not involve the abdominal wall muscles. The posterior vertical fibers of the external and internal obliques can produce lateral bending of the trunk. Their contraction will be more powerful, however, when they are called upon to raise the trunk while a person is lying on his side. The transversus contributes little to trunk movement; its chief action is in raising abdominal pressure.

An important corollary function of the abdominal wall muscles is the stabilizing effect they exert on the vertebral column by raising abdominal pressure. Strengthening the abdominal wall musculature is part of all exercise programs that aim to improve standing and sitting posture and to alleviate certain types of backache.

## Endoabdominal Fascia

In analogy with the endothoracic fascia, the loose areolar tissue that lines the entire abdominal cavity and attaches the parietal peritoneum to the abdominal walls may be designated as the *endoabdominal fascia* (see Fig. 23-1). Inferiorly, the endoabdominal fascia is continuous with the endopelvic fascia; superiorly, it lines the inferior surface of the diaphragm. This extraperitoneal connective tissue may become laden with fat, especially posteriorly, where it surrounds the kidneys. It is often referred to by radiologists as a *preperitoneal fat.* Fibrous strands of the endoabdominal fascia blend with the epimysium of the abdominal wall muscles, increasing the thickness of their deep fascia (Fig. 23-1). These membranelike fascial layers over the inner surface of the transversus abdominis and its aponeurosis, and over the psoas and iliacus muscles, are of sufficient anatomic and clinical importance to designate them by specific names: *transversalis fascia, psoas fascia, iliacus fascia.*

In the lower part of the abdomen, the **transversalis fascia** completes the posterior rectus sheath below the arcuate line when the entire transversus aponeurosis passes to the anterior surface of the rectus (Fig. 23-8). More laterally, the transversalis fascia attaches to the inferior upturned edge of the inguinal ligament (p. 714), contributes to the formation of the *femoral sheath* that surrounds the femoral vessels as they enter or leave the thigh (see p. 730, Fig. 26-9), and, by forming the *internal spermatic fascia,* provides one of the coverings of the spermatic cord or the round ligament of the uterus (see Fig. 26-7). The psoas and iliacus fascias are discussed in Chapter 25.

In the midline, the *transversalis fascia* ends by attaching to the upper border of the pubis, and the peritoneum, with some associated connective tissue, swings backward to cover the upper surface of the bladder (see Fig. 27-24). As a result, there is left between the lower part of the transversalis fascia and the pubis on the one hand, and between the peritoneum and the anterior wall of the bladder on the other, a potential *retropubic space* that is filled with a padding of extraperitoneal connective tissue, usually laden with fat.

The presence of loose endoabdominal fascia makes it possible for certain abdominal organs (*e.g.,* kidneys) and structures (*e.g.,* sympathetic trunks) to be approached surgically through abdominal incisions without entering the peritoneal cavity. For instance, resection of a segment of the lumbar sympathetic trunks (*lumbar sympathectomy*) may be performed by making an incision in each flank through the skin and the abdominal wall muscles. The sympathetic trunks lying on each side of the vertebral column can then be reached by blunt dissection in the extraperitoneal connective tissue without entering the peritoneum. Similarly, a surgeon can expose the bladder or the prostate, both lying in the pelvis, through an abdominal incision made above the pubic symphysis and then proceed inferiorly in the endoabdominal fascia of the retropubic space (*e.g.,* retropubic prostatectomy). There will be no need to cut the peritoneum. Such a dissection is possible because endoabdominal and endopelvic extraperitoneal fascias are continuous with one another.

## NERVES AND VESSELS

The innervation of the abdominal wall musculature, from the ventral rami of T6–L1 spinal nerves, has already been discussed (p. 588). The same nerves are sensory to the abdominal skin, all layers of fascia and muscle, and the parietal peritoneum associated with the abdominal walls. The arterial blood supply is from the lower posterior intercostal arteries (p. 478) and the lumbar arteries (both are branches of the aorta) and from a longitudinal, ventral anastomosis constituted by the superior and inferior epigastric arteries. These arteries are accompanied by corresponding veins. Lymph from the abdominal wall, above the umbilicus, drains to the axillary lymph nodes; below the umbilicus, it drains to the inguinal lymph nodes.

### Nerves

T6–T11 ventral rami in the abdominal wall are the continuation of the *intercostal nerves.* T12 is the *subcostal nerve,* and the two branches of L1, the most distal nerve to supply the abdominal wall, are the *iliohypogastric* and *ilioinguinal nerves* (Fig. 23-5). Maintaining their downward slope after they leave the intercostal space, the lower intercostal nerves enter the neurovascular plane of the abdominal wall by crossing the deep surface of the costal arch. Their collective name, **thoracoabdominal nerves,** aptly describes the distribution of these nerves to both thoracic and abdominal walls. These nerves, and the accompanying vessels, proceed between the internal oblique and transversus abdominis toward the linea semilunaris. They pierce the rectus sheath and penetrate the posterior surface of the rectus, which they supply. These nerves terminate as anterior cutaneous branches after piercing the rectus and emerging through the anterior lamina of its sheath. Nerve fibers may be interchanged between neighboring thoracoabdominal nerves, and sometimes a coarse plexus is formed before the nerves enter the rectus sheath, or the nerves may remain discrete and segmental.

The **subcostal, iliohypogastric,** and **ilioinguinal nerves** run in series and parallel with the thoracoabdominal nerves (Fig. 23-5). However, unlike the intercostal nerves, they enter the neurovascular plane of the abdominal wall from within the abdominal cavity. Having crossed over the surface of the posterior abdominal wall muscles (p. 694), they pierce the transversus abdominis at some point: the subcostal at the tendinous origin of the transversus from the thoracolumbar fascia, and the iliohypogastric and ilioinguinal above the iliac crest before they reach the anterior superior iliac spine. In the neurovascular plane, the iliohypogastric nerve divides into lateral and anterior cutaneous branches. Only the anterior branch supplies the abdominal wall; the lateral branch is distributed to the skin of the buttocks. The ilioinguinal nerve and the anterior branch of the iliohypogastric nerve soon pierce the internal oblique: the ilioinguinal nerve accompanies the spermatic cord in the inguinal canal, and the anterior branch of the iliohypogastric proceeds toward the medial area of the groin.

**Branches.** Muscular branches are given off by all these nerves directly from the main nerve, from a *collateral branch,* or from the cutaneous branches as these pierce the respective muscles. The innervation of the abdominal wall musculature is segmental, and the distribution of the individual nerves to the various muscles has already been discussed (p. 588). The intercostal, subcostal, and iliohypogastric nerves give off, in addition to their terminal anterior cutaneous branches, a lateral cutaneous branch.

The **lateral cutaneous branch** of the last six intercostal nerves emerges below the corresponding rib, roughly along the midaxillary line (see Fig. 12-11). Each lateral cutaneous branch divides into a posterior and anterior branch. The posterior branches supply the skin over the thorax and the back as far posteriorly as the territory of the cutaneous branches given off by the thoracic dorsal rami; the anterior branches run forward and downward supplying abdominal wall skin as far as the rectus. Skin over the recti and along the midline is supplied by the corresponding **anterior cutaneous branches** of the same nerves.

Knowledge of the segmental distribution of the cutaneous nerves over the abdomen has clinical uses. The key dermatomes to remember are T10 in the region of the umbilicus and T12 above the pubis (see Fig. 4-8). Skin between the costal arches (epigastrium) is supplied by T6, T7, and T8. Parts of L1 dermatome over the inguinal region (and buttocks) are served by the iliohypogastric nerve; other parts of L1 dermatome over the upper part of the scrotum and root of the penis or the mons pubis and the adjoining part of the labium majus are served by the ilioinguinal nerve. There is considerable overlap between neighboring dermatomes of the trunk, and no anesthesia can be demonstrated clinically in a dermatome if its spinal nerve is the only one interrupted or blocked.

In the abdominal wall, as in the trunk in general, there is a superimposition of corresponding dermatomes and myotomes. Moreover, the innervation of the parietal peritoneum also conforms to this segmental arrangement. These anatomic facts explain the abdominal reflexes and the reflex contraction of the abdominal musculature when the peritoneum is irritated.

An **abdominal reflex** is the sudden, momentary contraction of the underlying musculature when the skin of certain dermatomes is unexpectedly scratched or touched with a pointed object. Strokes are made with the object diagonally across dermatomes T7, T8, and T9 and then T10, T11, and T12, first on one side of the abdomen and then on the other. When the reflex is present, the umbilicus will be drawn toward the side of the contraction. The reflex verifies the integrity of somatic afferents and efferents in the corresponding spinal nerves and the appropriate cord segments and gives information about the intactness of suprasegmental input from the brain to the motor neurons in the segments tested. Certain types of brain damage (*e.g.*, some strokes) abolish these reflexes, even though the nerves and the spinal cord are uninjured.

Reflex contraction of the abdominal musculature protects the viscera. Blows to the abdomen rarely rupture the viscera, because the force is usually met by the reflex contraction of the muscles.

**Guarding** is the sustained involuntary contraction of segments of the abdominal musculature provoked by localized peritoneal inflammation. This may occur over an inflamed gallbladder or appendix. Somatic pain afferents excited in segments of the parietal peritoneum invoke reflex spasms of the muscles in the same segments in order to guard the underlying viscus from pressure. In cases of generalized peritonitis, the entire abdomen acquires boardlike rigidity because the entire peritoneum is inflamed.

Pain may be referred to the abdominal wall from certain abdominal viscera that send their visceral pain afferents into spinal cord segments that also receive somatic pain afferents from the abdominal wall. For instance, in the initial stages of appendicitis, before inflammation spreads to the parietal peritoneum, pain is felt over dermatome T10 around the umbilicus, because pain afferents from the appendix relay in spinal cord segment T10. Other examples of **referred pain** are discussed in the next chapter.

## Arteries and Veins

The arteries of the abdominal wall are derived from two sources: (1) the thoracic and abdominal aorta and (2) the superior and inferior epigastric arteries. The aorta contributes the **posterior intercostal arteries** of the last two intercostal spaces and the **subcostal** and the **lumbar arteries,** which are given off in series with the intercostals; the epigastric vessels have only small, unnamed branches. The aortic branches pass forward in the neurovascular plane, which they enter along with the nerves described in the previous section. They correspond to the branches of the nerves and terminate anteriorly by anastomosing with tiny branches of the epigastric arteries.

The **superior epigastric artery,** one of the terminal branches of the internal thoracic, enters first the rectus sheath and then the rectus, where it breaks up into branches (see Fig. 19-11). The **inferior epigastric artery,** a branch of the external iliac, ascends in the extraperitoneal fascia on the deep aspect of the abdominal wall, pierces the transversalis fascia behind the rectus, passes anteriorly to the arcuate line, and, roughly at the level of the umbilicus, enters the posterior surface of the rectus. Branches of the two epigastric arteries anastomose with each other within the rectus; other branches anastomose with the intercostal, subcostal, and lumbar arteries; yet others pass to the skin along with the anterior cutaneous nerves.

Minor branches of the *musculophrenic artery* (p. 479) augment the blood supply of the abdominal wall superiorly and anastomose with the superior epigastric. Inferiorly, small branches of the external iliac and femoral arteries help to supply deep and superficial tissues of the abdominal wall, respectively. The *deep circumflex iliac artery,* given off by the external iliac, and the *superficial circumflex iliac artery* (from the femoral) run toward and along the iliac crest. The *superficial epigastric artery* (from the femoral) ascends in the superficial fascia toward the umbilicus.

The superior and inferior epigastric arteries establish an **anastomosis** between the subclavian and external iliac arteries. This anastomosis enlarges and becomes an important channel for arterial blood to the lower part of the body when the aorta is obstructed. Blood will then flow posteriorly in the intercostal and lumbar vessels and inferiorly in the inferior epigastric artery. An example of aortic obstruction is *coarctation* (discussed in Chap. 21). In such cases, all anastomotic connections of abdominal wall vessels assume considerable importance.

All arteries in the abdominal wall are accompanied by corresponding veins. Blood drains to the *superior vena cava* from the **intercostal** and **subcostal veins** through the azygos and hemiazygos veins and from the **superior epigastric veins** through the internal thoracic and brachiocephalic veins. The *inferior vena cava* receives blood from the **lumbar veins** directly or through the ascending lumbar vein; from the inferior epigastric and deep circumflex iliac veins, which are tributaries of the external iliac vein; and from the **superficial epigastric** and **superficial circumflex iliac veins,** tributaries of the femoral vein. In the superficial fascia of the lateral part of the abdominal wall, there are, in addition, longitudinal venous channels, the **thoracoepigastric veins,** that do not accompany arteries. These veins drain into the femoral vein and anastomose superiorly with tributaries of the axillary vein. All venous anastomoses correspond to those described for the respective arteries.

**Anastomoses** in the abdominal wall between tributaries of the superior and inferior venae cavae provide a potential alternative route for returning blood to the heart when one of the venae cavae becomes obstructed. The veins that drain the re-

**FIG. 23-12.** Anterior view of the chief abdominal viscera shown schematically in **A** and on a plain film in **B. A** illustrates the overlap of the viscera in the supracolic compartment. In **B** (*opposite page*), the gas contained in the various viscera helps to identify them. The outline of other viscera is discernable owing to the differential radiopacity of the fat that surrounds them. The jejunum and ileum are ommited from **A.**

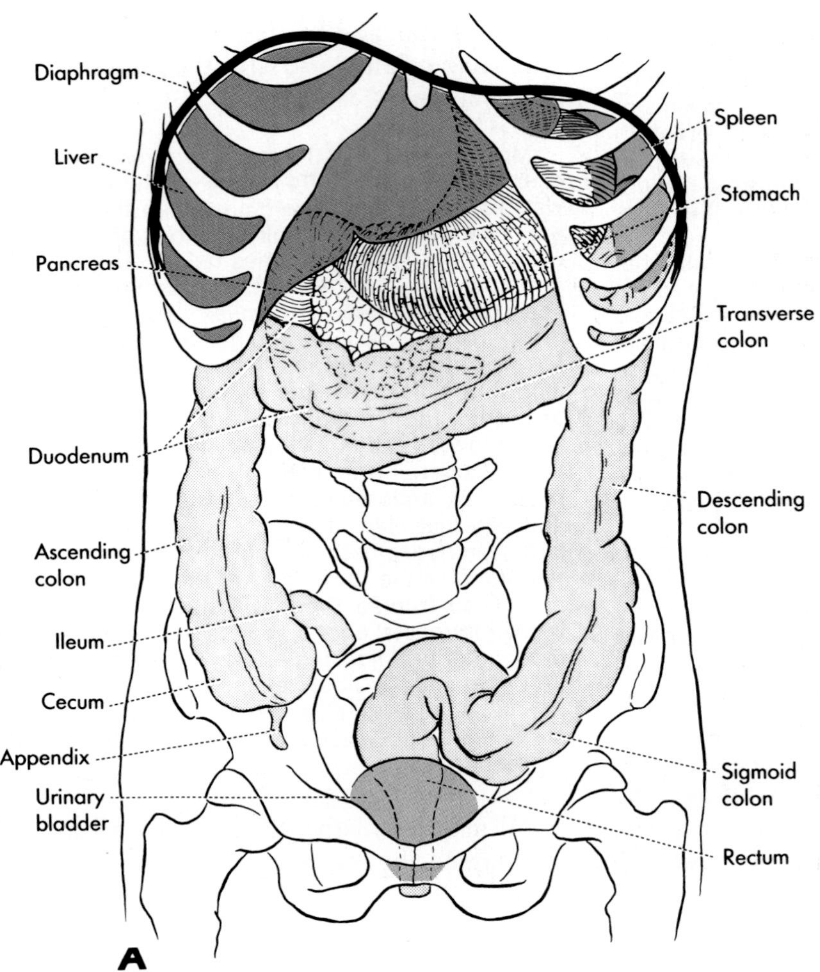

gion of the umbilicus anastomose, furthermore, with **paraumbilical veins,** located on the deep aspect of the abdominal wall. These veins drain into the portal vein. Through the enlargement of these anastomoses, portal venous blood can find its way into systemic veins when there is an increased pressure in the portal venous system (*portal hypertension;* p. 658). In such cases, the enlarged veins, radiating away from the umbilicus and visible on the surface of the abdomen, are a physical sign indicative of portal hypertension. The distended veins are known as the *caput medusae.*

### Lymphatics

Lymphatics of the superficial fascia and skin above the umbilicus drain to the **axillary nodes;** those from below the umbilicus drain to the **superficial inguinal lymph nodes.** Lymphatics from the abdominal wall muscles, the extraperitoneal connective tissue, and parietal peritoneum drain along the arteries to nodes associated with the parent vessels (lateral aortic, parasternal, external iliac, and inguinal lymph nodes).

## THE PERITONEUM AND THE PERITONEAL CAVITY

The peritoneum is a serous membrane that, like the pleura and the serous pericardium, forms a serous sac between its parietal and visceral laminae. The peritoneal sac encloses the peritoneal cavity. The **parietal peritoneum** lines the walls of the caudal compartment of the body cavity, constituted by the abdomen proper and the pelvis; the **visceral peritoneum** is applied closely to the surface of abdominal and pelvic viscera, and its redupli-

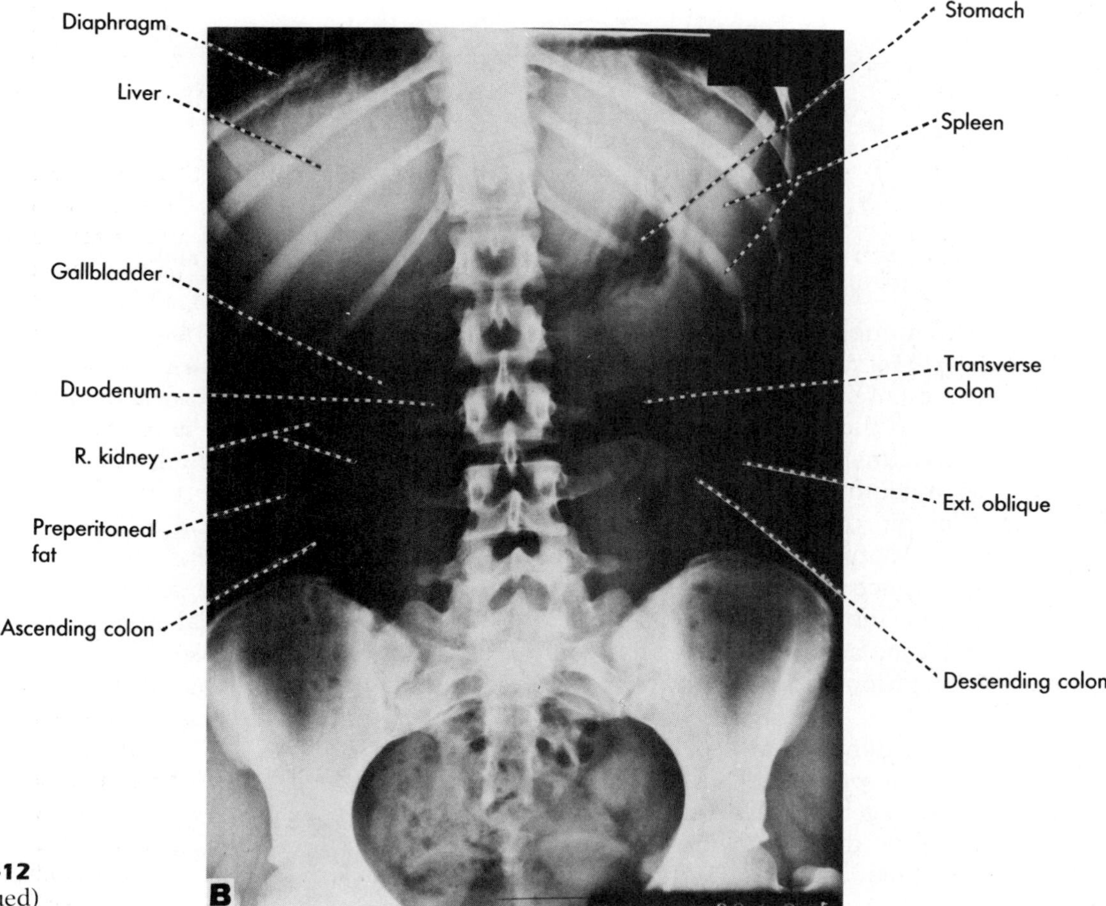


**Fig. 23-12**
(continued)

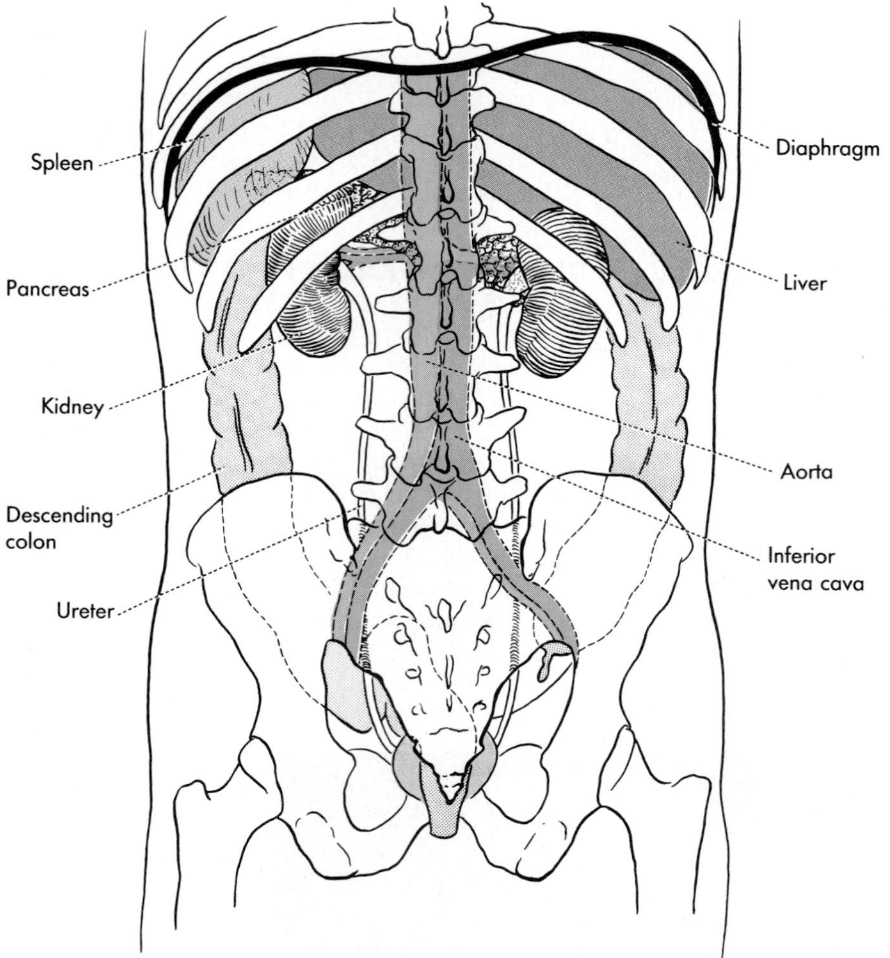


**FIG. 23-13.** Diagram of the chief abdominal viscera from behind. The stomach and transverse colon are omitted.

cations form a number of peritoneal folds that are called *mesenteries, omenta,* or *ligaments.* Where the mesenteries or ligaments reflect onto the walls of the abdominopelvic cavity, visceral peritoneum becomes continuous with parietal peritoneum. The arrangement conforms, in principle, to that of the pleura or the pericardium. Anatomically, however, the peritoneum is much more complex, owing to the secondary peritoneal attachments that the primitive gut tube acquires after it has completed its elongation and rotation during development.

The major abdominal and pelvic viscera around which the peritoneal sac is wrapped were discussed in general terms in Chapters 7 and 8 and will be dealt with in some detail again in subsequent chapters. For the purpose of orientation, Figures 23-12 and 23-13 present schematic anterior and posterior views of major abdominal organs.

The anatomy of the peritoneal cavity becomes intelligible only when the formation of the intraembryonic celom and the embryology of the gut are understood. The purpose of this section is a general anatomic survey of the peritoneal cavity along with some general developmental considerations. Further explanation of the peritoneal relationships of abdominal and pelvic viscera will be found in subsequent chapters that deal with the particular anatomy of the respective organs.

## DEVELOPMENTAL CONSIDERATIONS

**The Peritoneal Cavity and Mesenteries.** It was described in Chapter 7 how folding of the trilaminar embryonic disk defined the foregut (*proenteron*), midgut (*mesenteron*), and hindgut (*meten-*

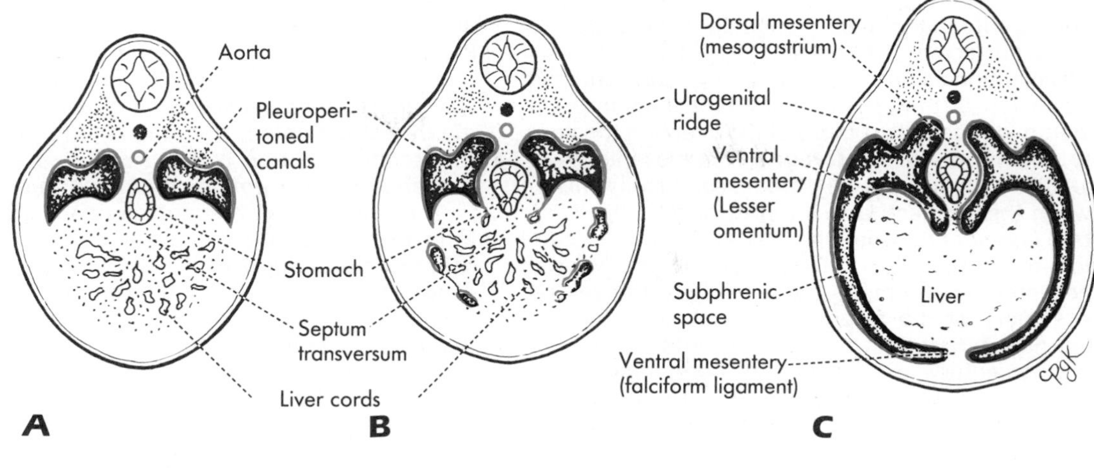


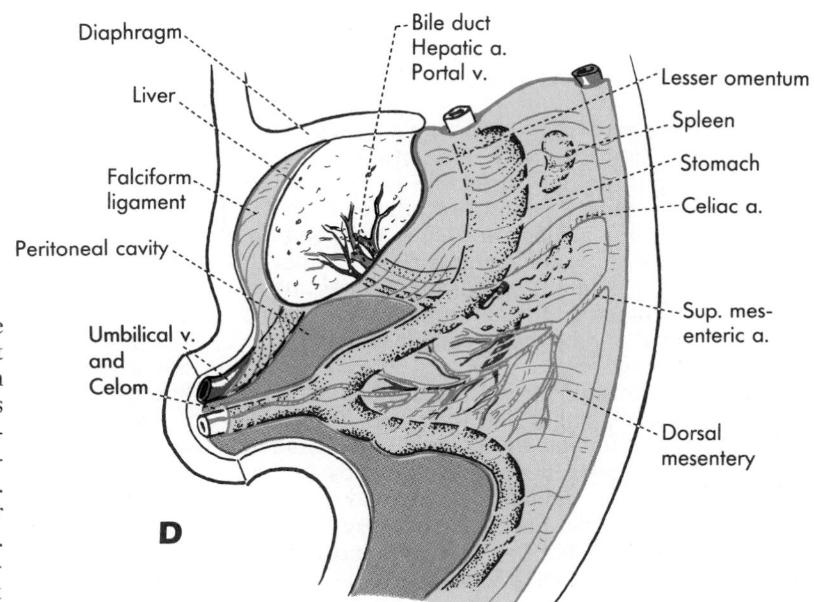


**FIG. 23-14.** Development of the ventral mesentery of the foregut from the septum transversum. In **A,** the pleuroperitoneal canals form the dorsal border of the septum transversum. **B** shows expansion of the canals into the septum. **C** shows formation of the lesser omentum and falciform ligament. In **D,** the ventral and dorsal mesenteries are viewed from the left side of the peritoneal cavity.

*teron*) and also how the caudal portions of the bilateral celomic ducts, the pleuroperitoneal canals, became related to the primitive gut (see Fig. 7-3). As a consequence of the folding, the pleuroperitoneal canals, which initially rest on the upper surface of the yolk sac, become wrapped around the gut and expand (see Fig. 7-3C). This expansion approximates the two ducts to each other along the dorsal aspect of the gut; they will be separated only by splanchnic mesoderm through which vessels and nerves reach the gut. This median sagittal septum of mesoderm that, so to speak, suspends the gut along its entire length is the **common dorsal mesentery** (see Fig. 7-3F and 7-4). Around the midgut and the hindgut, the left and right pleuroperitoneal canals approximate each other also ventrally. As their walls contact each other on the ventral aspect of the gut, they form the *ventral mesentery*. This mesentery, however, conveys no vessels or nerves and rapidly disappears, so that a single cavity, **U**-shaped in cross-sectional profile, surrounds the midgut and the hindgut (see Fig. 7-4).

Ventral fusion of the two pleuroperitoneal canals around the foregut is prevented by the presence of the septum transversum, which has grown considerably in bulk because the liver cords, sprouting from the foregut, have invaded it. The septum fills the space between the ventrolateral body wall and the foregut, which, with the two pleuroperitoneal canals alongside it, rests on the dorsal surface of the septum (Fig. 23-14).

Cavitation in the septum transversum and the expansion of the pleuroperitoneal canals into the septum define the surfaces of the liver and the **ventral mesentery of the foregut.** This mesentery will consist of two parts: (1) the *lesser omentum* between the stomach and the liver and (2) the *falciform ligament* between the liver and the ventral abdominal wall (Fig. 23-14C). Unlike the ventral mesentery of the midgut and hindgut, the ventral mesentery of the foregut conveys blood vessels and nerves that pass to the liver; both portions of this mesentery are retained as permanent structures. The falciform ligament supports the umbilical vein on its way from the umbilicus to the liver; the lesser omentum contains the derivatives of the vitelline vessels (the hepatic artery and portal vein) and the gastric and hepatic nerve plexuses, as well as the bile duct, the stalk that retains a permanent connection between the liver and the foregut (Fig. 23-14D).

A part of the septum transversum persists as the **diaphragm,** which, centrally, separates the peritoneal cavity from the pericardial cavity; more laterally, the diaphragm is interposed between the expanded pleural cavities and the peritoneum. Dorsally, the diaphragm is completed by the development of the bilateral *pleuroperitoneal membranes,* which close off the pleuroperitoneal canals from the thorax, and by the dorsal mesentery of the esophagus (Chap. 25).

It should be evident from this brief account of development that, unlike the lung and the heart, the gut does not invaginate a single serous sac to become suspended in the serous cavity by its mesentery, as is often implied by simplified explanations of the peritoneal cavity. Rather, the bilateral pleuroperitoneal canals become wrapped around the primitive gut and fuse with each other along much of its ventral aspect. By expanding, the canals create the dorsal mesentery along the entire length of the gut, and by growing into the septum transversum, the canals define the ventral mesentery along the foregut.

**Differentiation of the Gut and Its Mesenteries.** At the completion of the folding process, the gut consists of a roughly **T**-shaped tube: the vertical limb is the vitellointestinal duct connecting to the midgut; foregut and hindgut are along the horizontal bar of the **T** (see Fig. 7-3B). From the **foregut** will differentiate the esophagus (most of it located in the neck and the thorax), the *stomach,* and the proximal *half of the duodenum.* In addition, two buds appear on the caudal portion of the foregut: a *ventral diverticulum* grows into the septum transversum and gives rise to the *liver, gallbladder,* and a portion of the *pancreas;* a *dorsal diverticulum* grows into the dorsal mesentery and gives rise to the remaining and major part of the *pancreas* (see Fig. 24-21A). From the **midgut** are derived the *second half of the duodenum,* the *jejunum, ileum, cecum and appendix, ascending colon,* and most of the *transverse colon;* from the **hindgut** are derived the terminal portion of the *transverse colon,* the *descending* and *sigmoid colon,* and the *rectum.*

The midgut and hindgut elongate greatly and form a large redundant loop suspended by the dorsal mesentery (Fig. 23-14D). This loop and its mesentery become twisted and rotated (p. 611). Later, portions of the gut and their mesentery become tethered to parietal peritoneum on the dorsal wall of the cavity. These secondary peritoneal attachments are achieved by visceral peritoneum fusing with parietal peritoneum. The fusion results in the disappearance of those portions of the dorsal mesentery that suspend the duodenum and the ascending and descending colons; consequently, these structures come to lie directly on the posterior abdominal wall. The fused layers of the peritoneum disappear behind them and these portions of the gut are directly in contact with the fascia of the abdominal wall. The pancreas also becomes immobilized in this fashion across the posterior abdominal wall. The jejunum and ileum retain their mesentery, which is known as the **mesentery proper,** and so do the transverse and the sigmoid colon, which are known as the **transverse mesocolon** and the **sigmoid mesocolon,** respectively. The rectum has not had a mesentery of any size to speak of, and peritoneum is draped over only the anterior and lateral surfaces of its cranial portion.

Owing to the displacements caused by the rotation of the gut (p. 611), the secondary attachments of the persistent portions of the dorsal mesentery to the posterior abdominal wall become displaced from the dorsal midline. The dorsal mesentery of the stomach (*dorsal mesogastrium*) is modified by the development of the spleen, a collection of lymphoid tissue within the mesentery (Fig. 23-14D). In addition, the dorsal mesogastrium expands greatly into a large apronlike fold known as the **greater omentum.** Furthermore, the stomach rotates so that its original right surface faces posteriorly and its left surface anteriorly, which not only exaggerates the bulging of the greater omentum toward the left but causes the orientation of the lesser omentum to change from a sagittal to a coronal plane.

The survey of the peritoneal cavity in the fully developed state relies to some extent on these elementary developmental considerations in order to relate the complex arrangement to a rather simple underlying basic plan. The developmental events of the rotation and mesenterial fusion will be summarized after the gross anatomy of the peritoneal cavity has been surveyed.

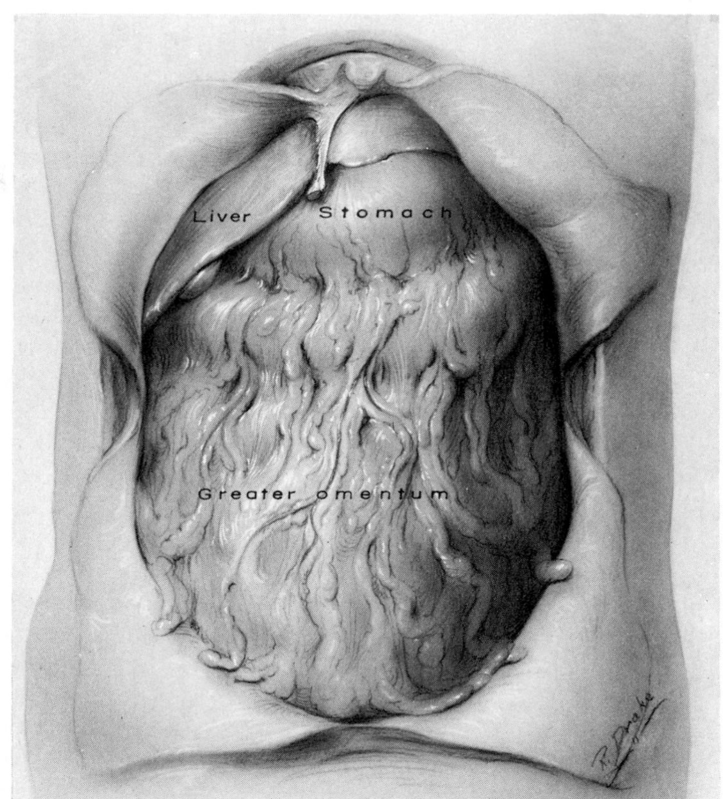


**FIG. 23-15.** The abdominal organs *in situ.* The falciform ligament, on the liver, has been largely cut away from the abdominal wall to allow the wall to be reflected. The small intestine is completely hidden by the greater omentum, as is the large intestine, but the transverse colon shows as the bulge behind the upper part of the omentum.

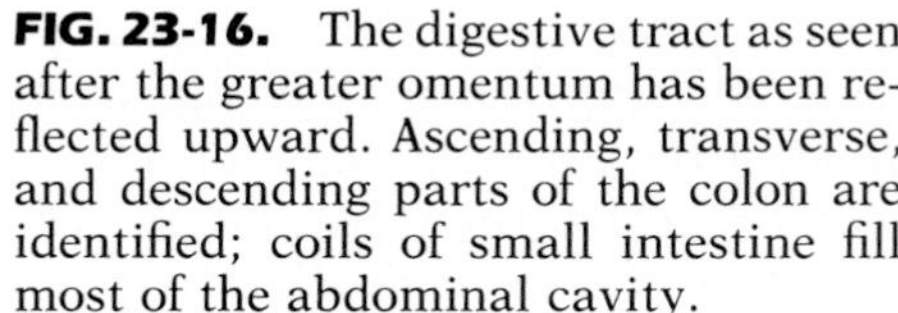

**FIG. 23-16.** The digestive tract as seen after the greater omentum has been reflected upward. Ascending, transverse, and descending parts of the colon are identified; coils of small intestine fill most of the abdominal cavity.

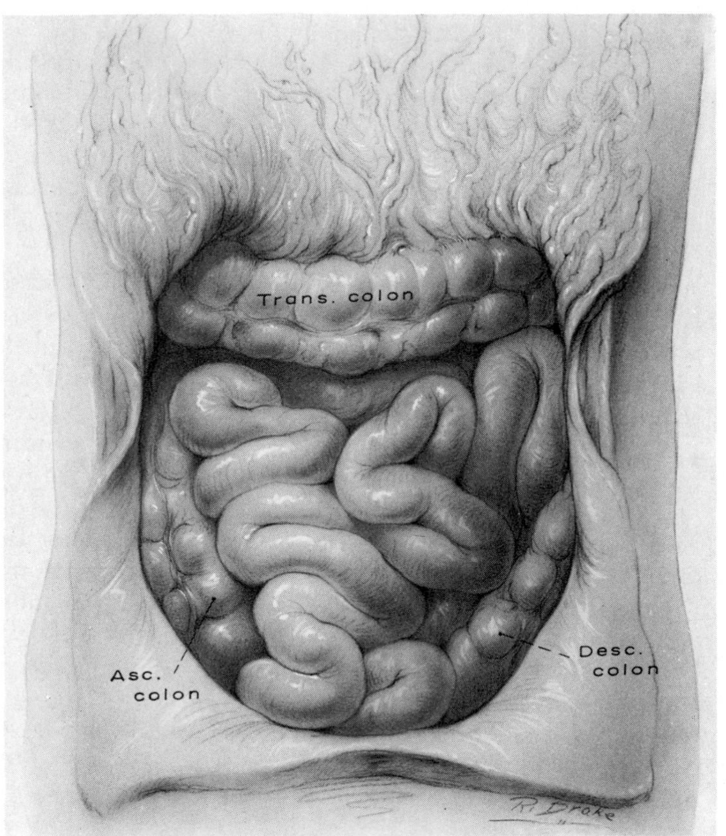

## SURVEY OF THE PERITONEAL CAVITY AND THE ABDOMINOPELVIC VISCERA

When the peritoneal sac is opened by an anterior incision of the parietal peritoneum, the first impression gained is that of a cavity filled by viscera: superiorly, the liver and the stomach, and more inferiorly, a fatty peritoneal fold, the greater omentum, and coils of intestine (Figs. 23-15 and 23-16). In truth, the peritoneal cavity contains only a small amount of serous fluid, which moistens the interior of the sac; all viscera are outside the peritoneum. Notwithstanding, certain viscera are frequently spoken of as lying in the peritoneal cavity or as being *peritoneal,* because they project into the cavity invested almost completely by visceral peritoneum; they can be reached only through the peritoneal cavity. Such organs are, for instance, the stomach, the jejunum, and the ileum. Visceral peritoneum is so closely bound to these viscera that it is considered an integral part of their wall: their *serosa*. On peritoneal organs that lack a lumen (liver, spleen), visceral peritoneum is closely bound to the organ's capsule. Other abdominal organs are spoken of as *retroperitoneal;* only their anterior surface is covered partly or completely by peritoneum. Such organs are not derived from the gut and include the kidneys and ureters, the suprarenal glands, the aorta and inferior vena cava, and the sympathetic trunks, as well as chains of lymph nodes lying along these structures. Those portions of the gut that lose their mesentery during development are sometimes described as *secondarily retroperitoneal.* Such organs are the pancreas and parts of the duodenum.

Diagrams and drawings cannot convey adequately the three-dimensional relationships of the peritoneal cavity and the mesenteries. The following description will achieve its objective only if its reading is accompanied by the exploration of a cadaver, preferably one not hardened by fixatives. The study of an embalmed body, more readily accessible, can serve the purpose adequately if reference is made to the accompanying figures for appreciating relationships that are difficult to explore in a hardened body.

### Division of the Peritoneal Cavity

The peritoneal cavity consists of the greater sac and the lesser sac. The **greater sac** (*cavum peritonei*) is the main part of the cavity that extends from the diaphragm into the pelvis. This is the sac opened by incisions of the abdominal wall, and into this sac protrude all the peritoneal organs. The **lesser sac,** also known as the *omental bursa,* is a diverticulum of the greater sac, which forms a potential space mainly behind the stomach and the lesser omentum. It is confined largely to the left side of the upper abdomen and communicates on the right with the greater sac through a passage known as the *epiploic foramen*.

Exploration of the greater sac is aided by dividing it into a *supracolic* and an *infracolic compartment,* located respectively above and below the transverse colon and its mesentery, and into a third part, the *pelvic portion* of the peritoneal cavity, which is inferior to the pelvic brim. This division of the greater sac can be demonstrated by elevating the greater omentum, which hangs down from the greater curvature of the stomach and is fused to the anterior surface of the transverse colon and its mesocolon (Fig. 23-16). The **supracolic compartment** of the greater sac is largely under cover of the costal margin and the diaphragm. In a fixed cadaver, much of it has to be examined by palpation unless the costal margin and the diaphragm are slit. The supracolic compartment contains the liver and stomach, the falciform ligament and the lesser omentum, and the greater omentum and the spleen, as well as the lesser sac. The **infracolic compartment** is filled with coils of jejunum and ileum and surrounded by the ascending, transverse, and descending colons; it leads into the pelvis inferiorly. The *pelvic portion* of the peritoneal cavity is revealed by lifting the coils of small intestine and sigmoid colon. Although its contents are the subject of a subsequent chapter, peritoneal relations in the pelvis will be studied in continuity with those of the abdomen proper.

### The Supracolic Compartment of the Greater Sac

**The Liver and Subphrenic Recesses.** The

peritoneal cavity between the liver and the diaphragm is divided anteriorly into right and left halves by the falciform ligament (Fig. 23-15).

The **falciform ligament** is a double layer of peritoneum, the ventral portion of the ventral mesentery of the foregut (Fig. 23-14). Right and left laminae of the ligament are continuous with one another around the inferior border of the ligament, which is falciform (sickle-shaped) and contains the obliterated umbilical vein, turned into a fibrous cord known as the *ligamentum teres hepatis* (round ligament of the liver). This ligament, enclosed in the free edge of the falciform ligament, spans the distance between the umbilicus and the notch on the sharp inferior border of the liver. Along the midline, right and left laminae of the falciform ligament reflect onto the deep surface of the abdominal wall and the diaphragm and become continuous with parietal peritoneum. Along their hepatic attachment, the same laminae of the falciform ligament continue as visceral peritoneum over the right and left lobes of the liver, respectively. Since this hepatic attachment is not directly opposite the line of attachment to the abdominal wall, rather to the right of the median plane, the falciform ligament itself lies in an oblique plane: its left leaf rests on the left lobe of the liver; its right leaf is in contact with parietal peritoneum of the abdominal wall and diaphragm.

On the right and left sides of the falciform ligament, the narrow peritoneal spaces confined between the liver and the diaphragm are known as the **right and left subphrenic recesses.**

These recesses may harbor infected exudate for some time after the primary abdominal pathology (cholecystitis, perforated peptic ulcer) has been treated. The resultant *subphrenic abscesses* are often difficult to detect clinically.

The subphrenic recesses are limited superiorly, just beyond the summit of the liver, by the **right and left coronary ligaments.** These are the narrow peritoneal reflections that suspend the liver from the diaphragm. Their name fancifully implies that, like a crown, albeit a much distorted one, the two ligaments encircle an area on the liver that is bare of peritoneum and is directly in contact with the diaphragm (see Figs. 24-27 and 24-28).

The part of each coronary ligament accessible in the subphrenic recess is the anterior layer (also known as the superior layer), which is continuous medially with the right and left laminae of the falciform ligament. Laterally, the left coronary ligament doubles back on itself and forms the *left triangular ligament* before it continues again medially as the posterior layer of the left coronary ligament (also known as the inferior layer) (see Fig. 24-28). The *right triangular ligament* is similarly formed, but the posterior layer of this coronary ligament diverges widely from the anterior layer, enclosing a large bare area of the right lobe of the liver and the inferior vena cava; the bare area of the left lobe is virtually nonexistent. The meeting of the posterior layers of the right and left coronary ligaments forms the lesser omentum.

**The Stomach and Lesser Omentum.** The lesser omentum and the stomach are partly covered by the left lobe of the liver. Elevation of the inferior border of the liver reveals the anterior surface of the **stomach** demarcated by the *lesser curvature,* concave toward the right, and the *greater curvature,* convex toward the left (Fig. 23-17). Above, under the left lobe of the liver, the abdominal portion of the *esophagus* enters the stomach at the *cardia;* below, under the right lobe of the liver, just to the right of the midline, the *pylorus* of the stomach leads into the first part of the *duodenum.*

The **lesser omentum** connects the inferior surface of the liver to the esophagus, the lesser curvature of the stomach, and the first part of the duodenum. Like the falciform ligament, the lesser omentum is a double layer of peritoneum, consisting of anterior and posterior (rather than left and right) laminae. It is the portion of the ventral mesentery between the liver and the foregut (Fig. 23-14). The part of the lesser omentum between the liver and the stomach is known as the *hepatogastric ligament;* this is continuous with the *hepato-*

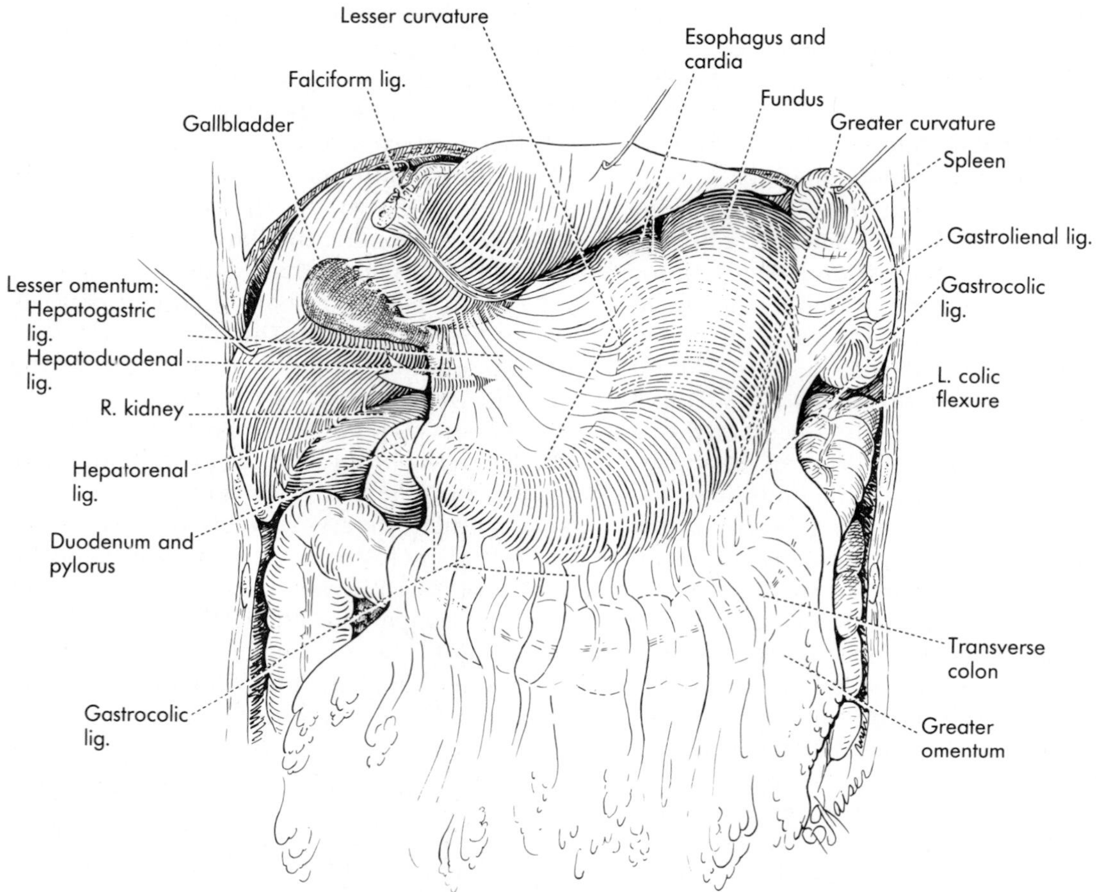


**FIG. 23-17.** The supracolic compartment showing the stomach, lesser omentum, and greater omentum. The inferior border of the liver has been elevated to reveal its inferior surface, and the anterior border of the spleen has been lifted away from the fundus of the stomach to reveal the upper part of the greater omentum. The arrow is in the epiploic foramen.

*duodenal ligament,* the portion of the omentum between the liver and the duodenum (see Fig. 23-17). The hepatoduodenal ligament terminates in a more or less vertical free border on the right. Around this border, anterior and posterior laminae of the lesser omentum are continuous with one another. The left border of the lesser omentum is not free: between the liver and the esophagus, anterior and posterior laminae reflect onto the inferior surface of the diaphragm.

Enclosed in the *free border of the hepatoduodenal ligament* are the bile duct, hepatic artery, and portal vein, the triad or pedicle that connects the liver to the foregut and its vasculature. On the inferior surface of the liver in line with this pedicle is the *gallbladder* (Fig. 23-17).

The lesser omentum and the stomach lie more or less in a coronal plane: their anterior surface faces into the greater sac; their posterior surface faces into the lesser sac. The **lesser sac,** otherwise known as the **omental bursa,** is a large irregular peritoneal *cul-de-sac,* the main portion of which is behind the stomach and the lesser omentum. The **epiploic foramen,** located behind the free edge of the lesser omentum, provides the slitlike communicating passage between the greater sac and the lesser sac. Posteriorly, the foramen is bounded by the inferior vena cava; above, by the liver; and below, by the first part of the duodenum, all covered in peritoneum.

**The Hepatorenal Recess.** The part of the greater sac into which the epiploic foramen opens is called the *hepatorenal recess* or pouch, one of the subhepatic peritoneal spaces. This recess is between the inferior surface of the right lobe of the liver and the anterior surface of the right kidney (Fig. 23-18). Superiorly, this space is limited by the posterior layer of the right coronary ligament (*hepatorenal ligament*) and inferiorly, by the right flexure of the colon. Laterally, the space leads into the right paracolic sulcus (*vide infra*).

**The Greater Omentum.** The greater omentum is a large apronlike peritoneal fold attached to the greater curvature of the stomach and the adjacent part of the duodenum (Fig. 23-15). Owing to the complex developmental changes, which will be explained later, it is difficult to appreciate that the greater omentum is the dorsal mesentery of the stomach (*dorsal mesogastrium*), composed of two peritoneal laminae folded upon themselves that connect the stomach to the posterior abdominal wall. The anterior, or outer, and the posterior, or inner, peritoneal laminae of the greater omentum are derived from the corresponding laminae of the lesser omentum after they have embraced the stomach. The anatomy of these laminae is best understood by considering the component parts of the greater omentum separately: these are the gastrocolic ligament and the gastrolienal and lienorenal ligaments. That these so-called ligaments are confluent and constitute the greater omentum would be more easily appreciated if they were designated as omenta rather than ligaments.

The **gastrocolic ligament** (omentum) is the portion of the greater omentum attached to the first part of the duodenum and the adjacent two-thirds of the greater curvature (Fig. 23-17). Its anterior and posterior peritoneal laminae are rather redundant; both double back on themselves along the inferior edge of the greater omentum before being draped over the transverse colon and its mesentery (Figs. 23-16 and 23-19), hence the name of the ligament. This name, however, is deceptive, because the laminae do not terminate on the colon; they continue to the posterior abdominal wall (Fig. 23-19). Along the inferior edge of the greater omentum, the *anterior* (*outer*) *lamina* (red in Fig. 23-19) will become the most posterior layer of the greater omentum. It ascends to the posterior abdominal wall and immediately reflects forward to become the superior lamina of the transverse mesocolon. These two adjacent laminae are usually fused to one another along a "bloodless" plane and hence may be stripped apart. The *posterior* (*inner*) *lamina* (blue in Fig. 23-19) of the gastrocolic ligament, derived from the posterior surface of the stomach, follows the anterior lamina and lines the inferior recess of the lesser sac, which extends into the greater

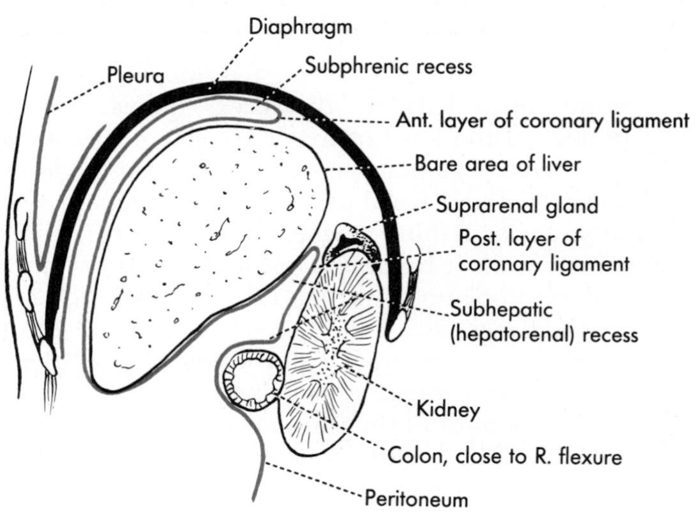


**FIG. 23-18.** Diagram of the subphrenic and subhepatic recesses.

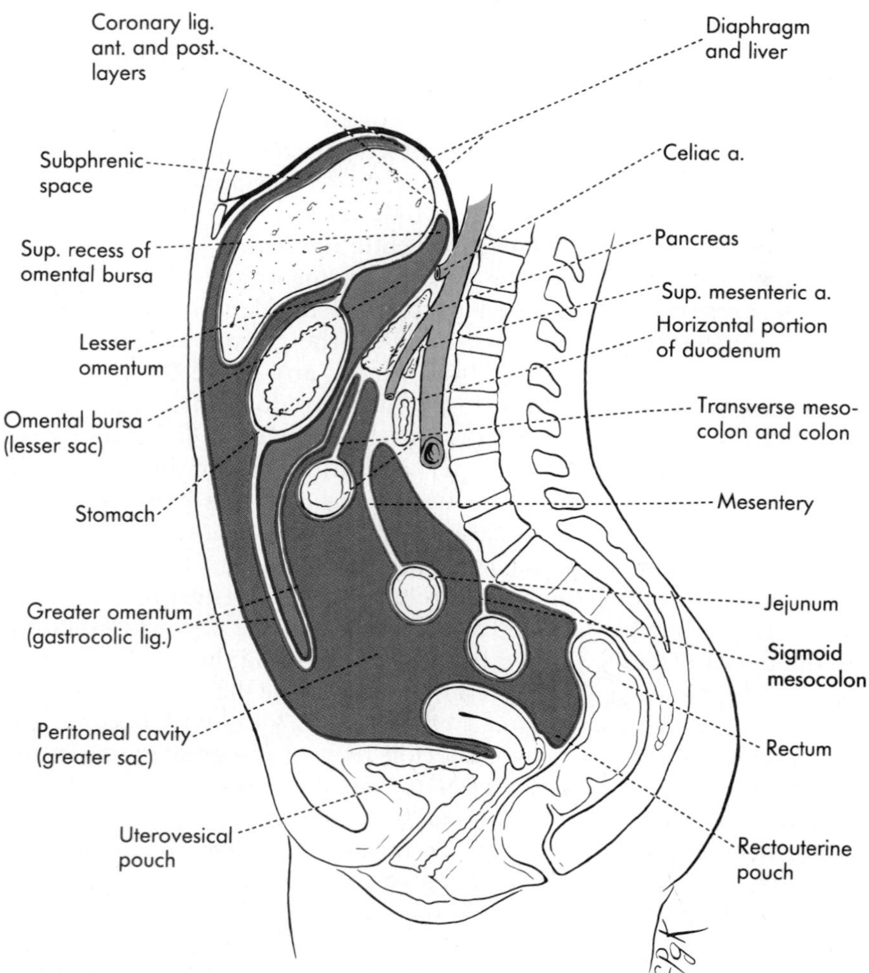


**FIG. 23-19.** Schematic sagittal section of the abdominal cavity to show the peritoneal continuity. The cut surface of the peritoneum facing into the greater sac is shown in solid red line; that facing into the omental bursa is shown in solid blue line. The cavity of the greater sac is red, that of the omental bursa, blue. The sigmoid mesocolon has been included in the section to emphasize the peritoneal reflections from the posterior abdominal wall; actually, the base of the sigmoid mesocolon is attached well to the left of the midline (see Fig. 23-20).

omentum for a variable distance. This is rarely beyond the level of the transverse colon, and below this point, the omentum usually consists of four fused peritoneal laminae. As the posterior lamina reaches the posterior abdominal wall, it becomes continuous with parietal peritoneum covering the posterior wall of the lesser sac.

The posterior attachments of the gastrocolic portion of the omentum and the posterior attachment of most of the transverse mesocolon coincide. These two mesenteries are attached to the posterior abdominal wall along a horizontal line that lies for the most part along the pancreas (Fig. 23-20).

On the right, the greater omentum has a free border created by the fusion of the anterior and posterior laminae of the gastrocolic ligament. The same is true on the left below the transverse colon (see Fig. 21-17). Above the colon, the gastrocolic portion of the omentum continues superiorly as the **gastrolienal** (gastrosplenic) and **lienorenal** (splenorenal) **ligaments** or omenta. This superior part of the

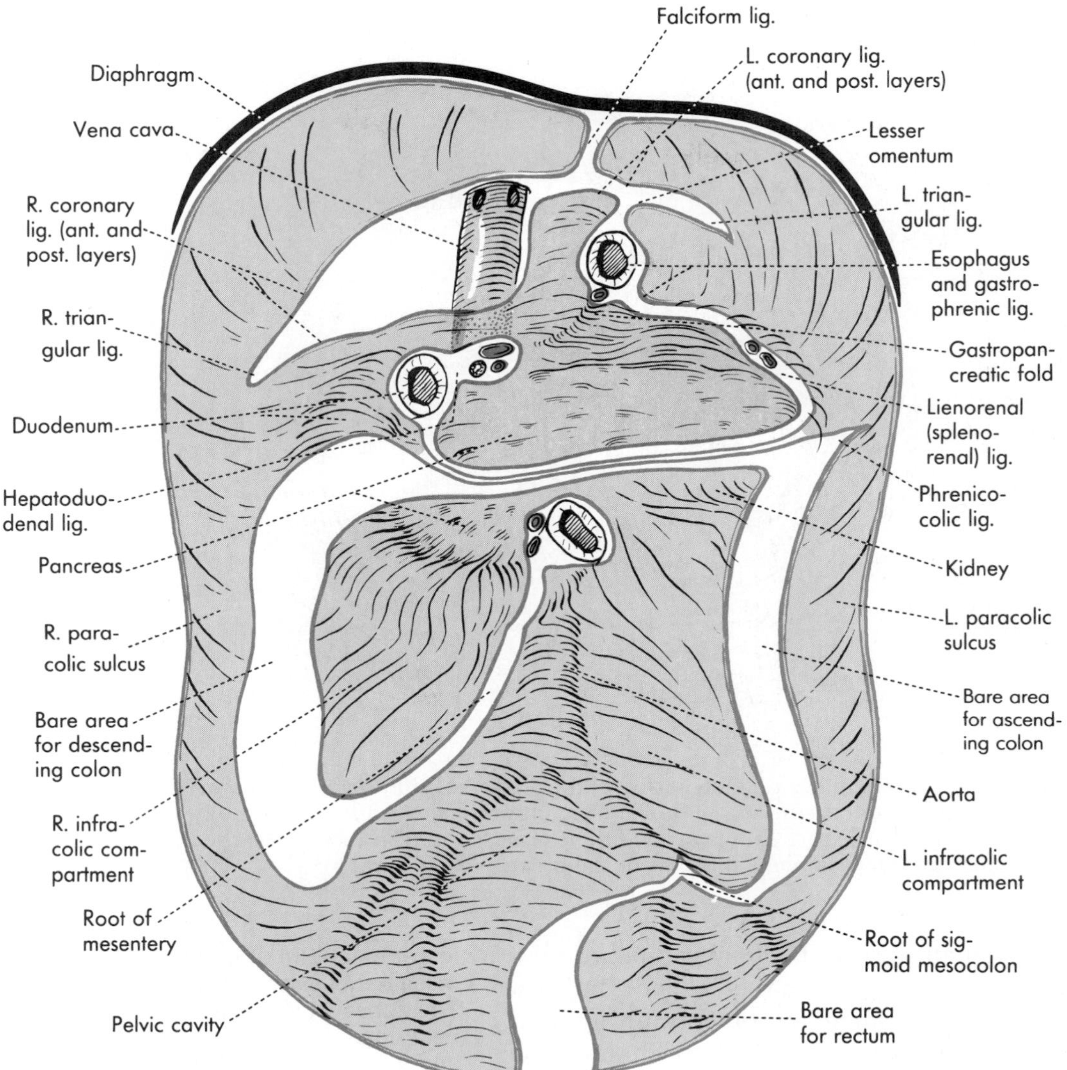


**FIG. 23-20.** Peritoneal attachments to the posterior abdominal wall and the diaphragm. The duodenum and pancreas have been left *in situ,* but all the other parts of the digestive system have been removed. The peritoneum lining the greater sac is pink, that lining the omental bursa is blue.

greater omentum connects the upper third of the greater curvature and the fundus of the stomach to the posterior abdominal wall. The parietal line of attachment is largely over the left kidney. This part of the dorsal mesogastrium is divided by the spleen into the gastrolienal and lienorenal ligaments (Fig. 23-21).

The *anterior or outer lamina of the gastrolienal ligament* (red in Fig. 23-21) passes from the stomach to the anterior lip of the hilum of the spleen, surrounds that organ, and, from the posterior lip of the splenic hilum, continues posteriorly as the outer lamina of the lienorenal ligament. The *posterior or inner lamina of the gastrolienal ligament* (blue in Fig. 23-21) is derived from the posterior aspect of the stomach, doubles back on itself at the splenic hilum, and passes posteriorly as the inner lamina of the lienorenal ligament. Lining the splenic recess of the lesser sac, this lamina only touches the hilum of the spleen; it does not cover any surface of that organ.

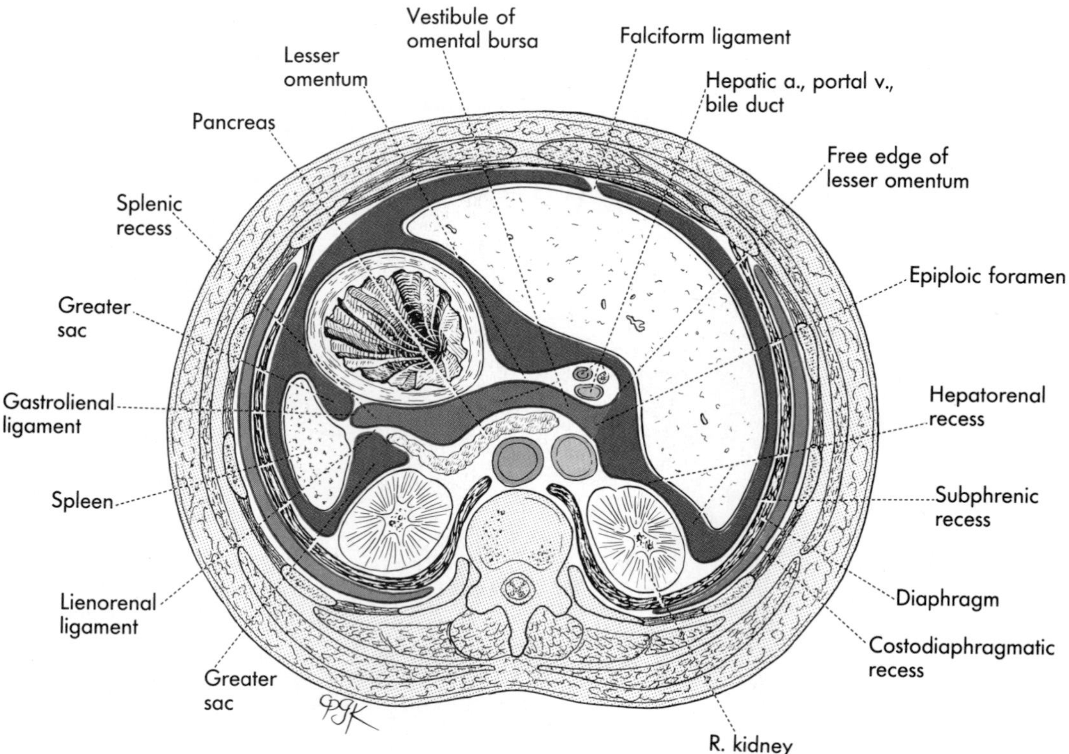


**FIG. 23-21.** A schematic representation of a transverse section across T12 vertebra seen from above to correspond with the orientation of the reader's own body. The peritoneal continuities are shown around the stomach and its mesenteries: the lining of the omental bursa is blue, and that of the greater sac is pink.

The small part of the greater omentum located above the spleen is called the **gastrophrenic ligament** (or omentum), since it connects the fundus of the stomach, and also the esophagus, to the diaphragm superior to the left kidney.

If the left hand were placed in the lesser sac and the right hand on the anterior surface of the stomach, the two hands could touch between the stomach and the spleen, with only the gastrolienal ligament intervening; between the spleen and the left kidney, with only the lienorenal ligament intervening (Fig. 23-21); and below the greater curvature of the stomach, with only the gastrocolic ligament intervening (Fig. 23-19). If the right hand were placed in the infracolic compartment, it would be separated from the left hand, still in the lesser sac, by the transverse mesocolon fused to the two laminae of the gastrocolic ligament (Fig. 23-19).

Since the epiploic foramen is too small for exploring the lesser sac through it, access to the lesser sac is afforded by creating openings in the lesser omentum or in the peritoneal ligaments and mesenteries associated with the greater omentum. The above exercise illustrates that all these peritoneal ligaments form boundaries of the lesser sac and its recesses.

When the peritoneal cavity is opened, the *greater omentum* is seldom found spread evenly in front of the intestine as anatomic illustrations suggest. The omentum is moved around, presumably by peristaltic activity of the stomach and bowel, and it may be packed into any part of the peritoneal cavity. The greater omentum can thus seal off the spread of infection. Should, for instance, the gallbladder, appendix, or other viscera become inflamed, the omentum wraps itself around the inflamed organ and may be retained there permanently by the adhesions resulting from the inflammatory process.

## The Lesser Sac (Omental Bursa)

The description of the greater and the lesser omentum has touched on much of the anatomy of the lesser sac. To comprehend the lesser sac more fully, it is necessary to consider its divisions and its posterior wall in particular (Fig. 23-20).

The epiploic foramen leads into the **vestibule** of the lesser sac, and from the vestibule diverge the three recesses of the sac: a superior, inferior, and splenic recess (Fig. 23-21).

The upward continuation of the vestibule behind the liver is the **superior recess.** In the anterior wall of the superior recess is the highly convex caudate lobe of the liver, which projects into this narrow, slitlike space (Fig. 23-27). The recess is roofed over by the diaphragm, which also forms the posterior wall (Fig. 23-19). The peritoneal lining over the diaphragm in the recess is continuous in front and on the right with the posterior layer of the right coronary ligament, and it is this ligament that limits the recess above and to the right of the caudate lobe. The posterior layer of the right coronary ligament in the vestibule and superior recess is the continuation of the hepatorenal ligament across the inferior vena cava and, beyond the epiploic foramen, upward along the left side of the vena cava (Fig. 23-20). As the vena cava passes through the diaphragm, the line of peritoneal reflection turns to the left, along the upper margin of the caudate lobe. On the left side of this lobe, the posterior layer of the right coronary ligament terminates by becoming the posterior lamina of the lesser omentum. (In other words, the lamina joins the anterior lamina of the lesser omentum, continuous with the posterior layer of the left coronary ligament.) Thus, on the left, the superior recess is closed off from the greater sac by the hepatogastric portion of the lesser omentum: the omentum runs from the deep groove of the left side of the caudate lobe to the esophagus and the lesser curvature of the stomach. As mentioned earlier, superiorly, the recess is sealed by the reflection of the posterior lamina of the lesser omentum onto the diaphragm.

Two peritoneal ridges mark the boundary between the vestibule and its inferior and splenic recesses. These ridges are the *gastropancreatic folds,* raised up from the posterior wall of the lesser sac by the branches of the celiac trunk (Fig. 23-20). The **inferior recess** is below the right gastropancreatic fold (containing the common hepatic artery) and extends behind the stomach into the greater omentum. In its posterior wall lies the pancreas and below that the posterior lamina of the greater omentum, fused with the anterior lamina to the transverse mesocolon. The **splenic recess** is to the left of the left gastropancreatic fold (containing the left gastric artery) and extends behind the stomach and in between the gastrolienal and lienorenal ligaments. In its posterior wall lie, retroperitoneally, the left suprarenal gland, the upper pole of the left kidney, and the diaphragm. The splenic artery running along the upper border of the pancreas demarcates the splenic and inferior recesses from one another. The splenic vessels and the tail of the pancreas pass forward to the hilum of the spleen, being enclosed in the lienorenal ligament (Fig. 23-21). A small portion of the abdominal aorta, from which the short celiac trunk springs, is in the posterior wall of the lesser sac where the two gastropancreatic folds converge with each other.

The lesser sac is normally empty except for the serous fluid that moistens its wall. An ulcer on the posterior wall of the stomach may perforate into the lesser sac, as may a pancreatic cyst or abscess. The gastric contents and exudate may become sealed off in the lesser sac, which may then become distended, giving rise to a *pseudocyst.* Infected material may also track its way from the lesser sac, through the epiploic foramen, into the hepatorenal space and then along the paracolic sulcus into the pelvis. Exudate or pus may also form in the hepatorenal space as a complication of gallbladder disease.

The epiploic foramen is usually too small for exploring the lesser sac through it. To do this adequately in the dissecting laboratory or at surgery, an opening has to be made in the lesser omentum or in one of the ligaments of the greater omentum. The lesser sac may also be entered from the infracolic compartment through the transverse mesocolon and the two laminae of the greater omentum fused to it.

#### Development of the Spaces and Mesenteries in the Supracolic Compartment

To explain satisfactorily the anatomy of the supracolic compartment at a time when development is complete, it is necessary to expand somewhat on the introductory remarks made in Chapter 7 and in this chapter (p. 596).

**Dorsal Mesogastrium.** To begin with, the foregut lies largely dorsal to the septum transversum, and its dorsal mesentery, the dorsal mesogastrium, is a relatively thick structure that separates the right and left pleuroperitoneal canals from one another (Fig. 23-14A through C). The attachment of the dorsal mesogastrium to the foregut identifies the greater curvature of the stomach and the future medial border of the duodenum, both of which at this stage face posteriorly. The lesser curvature of the stomach abuts against the septum transversum.

In addition to the vitelline arteries, which will be known as branches of the celiac trunk, the dorsal mesogastrium contains the spleen, the pancreas, and the vitelline veins, which in their final configuration pass as a single vessel (the portal vein) from the dorsal mesogastrium into the septum transversum (Fig. 23-14D).

**Septum Transversum.** The ventral mesentery of the foregut at this stage is the septum transversum, a wedge-shaped block of mesoderm ventral to the foregut and the pleuroperitoneal canals. The cranial surface of the septum is defined by the pericardial cavity, and the caudal surface is defined by a portion of the intraembryonic celom that extends into the umbilical cord around the vitellointestinal duct and around the loop formed by the midgut. The following structures pass forward toward the central point of the septum transversum buried just beneath the surface of the septum that faces toward the body stalk: (1) from the body stalk, the umbilical vein runs posteriorly to enter the liver; (2) from the caudal end of the foregut, the bile duct runs forward, forming the pedicle of the liver, which is expanding in the septum transversum; (3) the hepatic artery and (4) the portal vein accompany the bile duct, having entered the septum transversum from the dorsal mesogastrium as they pass along the right side of the foregut (duodenum) (Fig. 23-14D).

The peripheral cell layers of the septum transversum contribute to the formation of the *diaphragm,* especially cranially, forming its central tendon (the floor of the pericardial cavity; p. 519); the more central cells become incorporated into the liver as its connective tissue elements and the walls of its vessels (see p. 649). The *subphrenic recesses* of the peritoneal cavity develop in the cleavage plane between the peripheral cells of the septum and the liver, becoming confluent posteriorly with the pleuroperitoneal canals and caudally with the umbilical part of the celom that extends into the umbilical cord (Fig. 23-14C).

The expansions of the celomic ducts into the septum transversum define the peritoneal surfaces of the liver but stop short of surrounding it completely in three places: (1) anteriorly, where the falciform ligament remains as a mesentery supporting the umbilical vein along its inferior edge; (2) posteriorly, where the lesser omentum remains, supporting along its inferior edge the bile duct, hepatic artery, and portal vein; (3) posterosuperiorly, where the coronary ligaments surround the bare area of the liver, retaining it in contact with the portion of the diaphragm that is derived from the septum transversum.

These events transform the septum transversum into the ventral mesentery that consists of the falciform ligament and the lesser omentum, with the liver between the two.

The definitive anatomy of the ventral mesentery is established by reorientation of its component parts: (1) the free edge of the falciform ligament becomes vertical and sickle shaped owing to the elongation of the ventral body wall between the liver and the umbilicus; (2) the free edge of the lesser omentum becomes vertical owing to the descent of the stomach and duodenum; (3) the lesser omentum changes from a sagittal to a coronal plane owing to the rotation of the stomach and duodenum (*vide infra*).

It is to be emphasized that the **epiploic foramen** is created by the reorientation of the free edge of the lesser omentum into a plane that is parallel with, rather than vertical to, the posterior abdominal wall. The epiploic foramen is the slitlike space trapped behind the now vertical free edge, rather than a hole created in some existing structure.

**Omental Bursa.** The *rotation of the stomach* and duodenum, in a manner that turns their original right surface posteriorly, is probably promoted by the destabilization of the dorsal mesogastrium (Fig. 23-22). The rather bulky **dorsal mesogastrium** is invaded by the cavity of the right pleuroperitoneal canal. This cavity created in the mesogastrium thins out the dorsal mesentery and is known as the omental bursa. The location of the spleen in the far left wall of the omental bursa may contribute to the buckling of the dorsal mesogastrium upon itself, causing the spleen to bulge into the left pleuroperitoneal canal and aiding the stomach in pointing its greater curvature toward the left (Fig. 23-22). The buckling, which hinges on the spleen, divides the dorsal mesogastrium into an anterior portion between the greater curvature of the stomach and the spleen (*gastrolienal ligament*) and a posterior portion between the dorsal midline attachment of the mesentery and the spleen (*lienorenal ligament*). The latter comes to lie parallel with the surface of the left kidney and contains the pancreas and the splenic vessels.

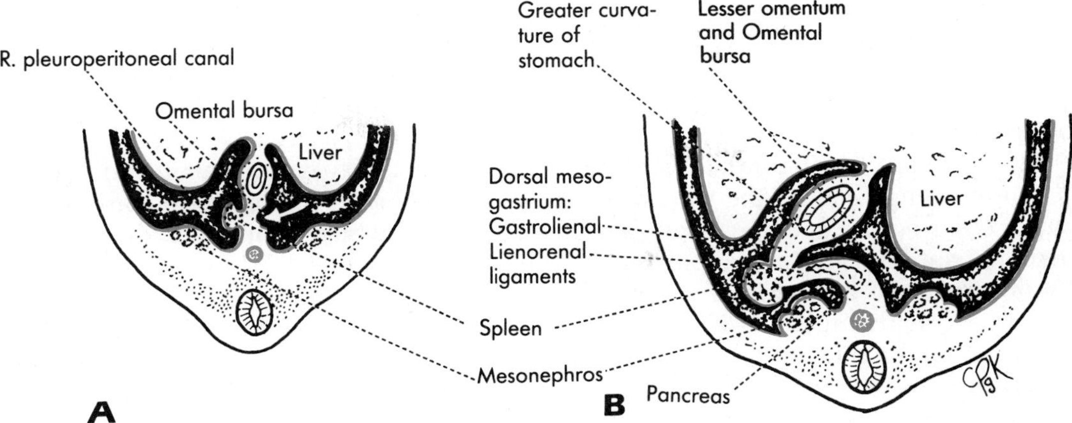


**FIG. 23-22.** Development of the omental bursa and the dorsal mesogastrium. **A** shows commencement of cavitation in the dorsal mesogastrium. (Compare with Fig. 23-14**C**.) In **B**, the stomach has rotated; fusion of the lienorenal ligament to parietal peritoneum over the kidney will establish the definitive anatomy.

This folded dorsal mesogastrium can now be called the **greater omentum,** and it becomes even more expanded to the extent that it hangs from the greater curvature of the stomach far down into the more caudal part of the peritoneal cavity. The definitive anatomy of the greater omentum is established by secondary adhesions to adjacent structures: posteriorly, it becomes tethered to the surface of the left kidney and fuses its original left lamina with the parietal peritoneum on the mesonephros (antecedent of the kidney), immobilizing the pancreas on the posterior abdominal wall (Fig. 23-22); after the midgut and hindgut obtain their definitive position in the peritoneal cavity, the greater omentum adheres to the superior surface of the transverse colon and its mesocolon.

The **cavity of the omental bursa** grows with the greater omentum, and its *splenic recess* follows the spleen toward the left; its *inferior recess* is contained in the redundant fold of the greater omentum, and part of it may become obliterated. The *vestibule* of the lesser sac is the portion of the peritoneal cavity that becomes trapped behind the lesser omentum when the stomach and the lesser omentum complete their 90° turn. The *superior recess* of the lesser sac is the cranial extension of the vestibule and is the infradiaphragmatic portion of the *pneumatoenteric recess,* for the explanation of which the literature should be consulted.

## The Infracolic Compartment of the Greater Sac

The upper limit of the infracolic compartment is the transverse colon and its mesentery (Fig. 23-16). Inferiorly, the compartment communicates with the pelvis. The attachment of the transverse mesocolon to the retroperitoneal pancreas has been discussed. To the right of the pancreas, this attachment line crosses the vertical segment of the duodenum, and the mesocolon terminates at the hepatic flexure of the colon; on the left, beyond the pancreas, the attachment line terminates at the left, or splenic, flexure of the colon (Fig. 23-20). These colonic flexures lie on the anterior surface of the kidneys and mark the junction of the transverse colon with the ascending colon on the right and the descending colon on the left. These vertical segments of the large bowel form the lateral boundaries of the infracolic compartment.

Both the ascending and the descending colon have lost their mesenteries during the development of the colon. Their posterior surface, bare of peritoneum, lies directly on the posterior abdominal wall. The visceral peritoneum on their anterior and lateral surfaces is continuous laterally with parietal peritoneum in the **paracolic sulci.** The right paracolic sulcus, or gutter, communicates above with the hepatorenal space; the left paracolic sulcus, or gutter, is limited above by a small peritoneal fold, the *phrenicocolic ligament,* pinched up between the left flexure and the diaphragm. Below, the two paracolic gutters lead into the *iliac fossae* and then into the pelvis.

Medially, peritoneum from the anterior and medial surfaces of the ascending and descending colon is continuous with parietal peritoneum on the posterior wall of the infracolic compartment.

On the posterior abdominal wall, the *duodenum* descends behind the transverse mesocolon, crossing the boundary between the supracolic and infracolic compartments (Fig. 23-23). Only the first part of the duodenum, seen in the supracolic compartment, has mesenteries; the rest lies with the pancreas across the posterior abdominal wall. In front of the 3rd lumbar vertebra, the duodenum passes horizontally from the right to the left side of the infracolic compartment. Turning upward and then abruptly forward, it becomes the *jejunum*, the union of the two creating the acute *duodenojejunal flexure*. Distally, the jejunum continues as the *ileum*. The numerous coils of the jejunum fill the upper part of the infracolic compartment; below, the coils of the ileum usually also fill the pelvis. The ileum opens into the medial side of the large intestine in the right iliac fossa. The large intestine above this *ileocecal junction* is the ascending colon; below, it is the saclike *cecum*, from which the *vermiform appendix* takes origin.

The jejunum and ileum are suspended from the posterior abdominal wall by a large continuous peritoneal fold, the **mesentery.** The mesentery is fan shaped and fluted: its intestinal border is about 40 times longer than its root, which is attached diagonally across the posterior wall of the infracolic compartment from the duodenojejunal flexure to the ileocecal junction (Fig. 23-23). Between its two peritoneal laminae, the mesentery contains the blood vessels, lymphatics, lymph nodes, and nerve plexuses that serve the jejunum and ileum. These structures are embedded in a considerable amount of fat.

Along the root of the mesentery, its right peritoneal lamina becomes continuous with peritoneum covering the pancreas and duodenum in the upper part of the infracolic compartment; below, the same lamina is continuous with parietal peritoneum, covering the ureter and the lower pole of the right kidney, as well as the muscles of the posterior abdominal wall (Fig. 23-23). The left lamina of the mesentery also reflects over the duodenum above; below, it becomes continuous with the

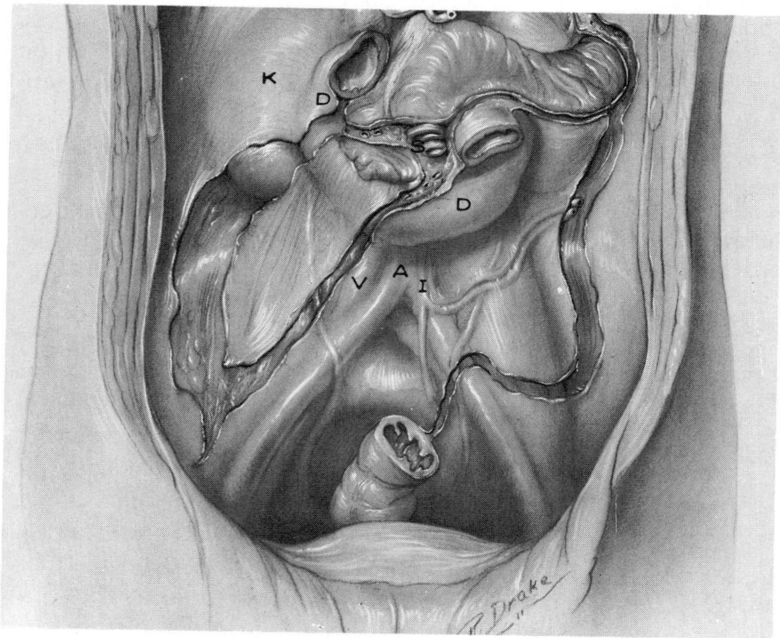


**FIG. 23-23.** The peritoneal attachments and retroperitoneal structures in the infracolic compartment. **A,** aorta; **D,** duodenum; **I,** inferior mesenteric vein; **K,** kidney; **V,** inferior vena cava. Correlate with Figure 23-20.

parietal peritoneum of an extensive quadrangular area that leads down into the pelvis. In this area, retroperitoneal structures include the lower pole of the left kidney, the left ureter, the inferior vena cava, the aorta, and both left and right common iliac vessels.

In the left iliac fossa, the infracolic compartment is limited inferiorly by the *sigmoid colon* and its mesentery, the **sigmoid mesocolon.** The sigmoid colon is the distal continuation of the descending colon. Passing into the pelvis, the sigmoid colon loses its mesentery and becomes the rectum. The sigmoid mesocolon is attached to the posterior abdominal wall along the line of an inverted **V.** The left limb of the **V** is in the iliac fossa; the right limb descends into the pelvis (Fig. 23-23).

#### Developmental Events in the Infracolic Compartment: Rotation of the Gut

The midgut and hindgut increase in length quite early in development, and although they remain suspended by the common dorsal mesentery, they soon form a loop, the apex of which is connected to the yolk sac through the umbilical cord by the vitellointestinal duct (Figs. 23-14D and 23-24). This loop will become highly convoluted as the intestine continues to grow. The intraembryonic celom fails to keep pace with this growth and the loop herniates, therefore, into the umbilical portion of the celom.

The midgut and its mesentery become twisted and turned in a particular manner during this process. When the abdominal cavity is sufficiently expanded, this physiological hernia is reduced, and the intestine returns from the umbilical cord into the abdominal cavity. The rearrangement of the intestine that results from the rotation of the midgut becomes stabilized through the fusion of extensive portions of the mesentery with parietal peritoneum on the posterior wall of the infracolic compartment.

The key to understanding the infracolic compartment is the fate of the midgut loop, which is best discussed in its simplest form (Fig. 23-24). The cranial limb of the loop is made up of the duodenum and jejunum; the future ileum is in the region of the apex of the loop and from it arises the vitellointestinal duct; on the caudal limb of the loop, the cecum is identified by a saccular dilatation, and the rest of the caudal limb continues as the colon into the hindgut.

The vitelline arteries and veins, from which the superior and inferior mesenteric vessels develop, are contained in the dorsal mesentery. The superior mesenteric artery runs to the apex of the loop (and beyond it to the yolk sac), distributing numerous branches to the cranial limb of the loop and only three branches to the caudal limb (p. 666).

For our purpose, the specific sequence of events in the rotation of the gut is immaterial; the following points are important:

1. The superior mesenteric artery and the vitellointestinal duct in line with it can be considered the axis of the rotation, which takes place counterclockwise and amounts to roughly 270°.
2. The cranial limb of the loop moves downward on the right side of the celom, and once its excursion exceeds 180°, the limb must pass *behind* (or caudal to) the superior mesenteric artery, ending in the left side of the celom (Fig. 23-24A and B). This is the component of the rotation that carries the future horizontal portion of the duodenum *behind* the superior mesenteric artery and lays it across the vertebral column, placing the duodenojejunal junction and all of the cranial limb of the loop into the left side of the infracolic compartment. The duodenum becomes fixed in this position quite early in the process of rotation.
3. The caudal limb of the loop ascends in the left side of the celom, and once its excursion exceeds 180°, the limb must cross *anterior* to the proximal end of the cranial limb and then continue its movement by descent in the right side of the celom (Fig. 23-24B and C). This is the component of the rotation that lays the future transverse colon across the descending portion of the duodenum and carries the cecum into the right side of the peritoneal cavity.

It is important to remember that during this process the mesentery does not behave as a passive sheet of tissue but as one capable of adjusting to the movement of the intestine by growth as well as by twisting and folding. Viewed from the front at the completion of the rotation, the mesentery presents an arrangement resembling a funnel: the rim of the funnel is made up of the caudal loop, (in essence, the large intestine); the coils of the cranial loop (most of the small intestine) are suspended on their mesentery within the funnel; the cavity of the funnel tapers toward the origin of the superior mesenteric artery or the duodenojejunal flexure (Fig. 23-24C and D).

The definitive arrangement is arrived at by fusion of the mesentery of the ascending and descending colons to the posterior parietal peritoneum in the infracolic compartment and the fusion of the mesentery of the jejunum and ileum to the same parietal peritoneum across a diagonal line that slopes from the duodenojejunal flexure to the ileocecal junction now located in the right iliac fossa (Fig. 23-24D). Such extensive fusion of the dorsal mesentery and portions of the intestine to the posterior abdominal wall is a feature peculiar to the human and some of the anthropoid apes.

**Abnormal Rotation.** Abnormalities of rotation are of three major types: failure of mesenterial fusion, reversed rotation, and entrapment of portions of the gut in mesenterial folds.

The most common abnormality is failure of the mesentery of the ascending colon to fuse to the posterior wall. This is usually limited to the distal part of the ascending colon but in its most severe form will leave the entire small and large intestine, up to the left colic flexure, suspended on a free mesentery. In such a case, the root of the mesentery is limited to the vicinity of the duodenojejunal flexure, and this predisposes to the twisting of the unfixed intestine and its mesentery upon itself. The twisting, called *volvulus,* obstructs the lumen of both the intestine and the blood vessels in the mesentery, constituting a major emergency.

The descending colon and its mesentery may also fail to fuse. Usually, this leaves the descending colon suspended on a short mesentery attached to the posterior abdominal wall well to the left of the midline.

The cranial limb of the midgut loop may fail to move in the appropriate direction and may cross in front of, rather than behind, the superior mesenteric artery. Such reversed rotation places the duodenum anterior, rather than posterior, to the large intestine, and if mesenterial fusion fixes the gut in this position, obstruction may result. Other types of abnormalities in rotation may produce abnormal peritoneal bands, which may be responsible for transitory or acute obstruction of either the small or the large intestine.

The rotating gut may be caught in a pocket of the mesentery, which it will distend. The most striking example of this is *paraduodenal hernias.* The small intestine becomes trapped behind the mesentery of the ascending or descending colon, and when the mesentery fuses with the posterior wall, the small intestine will be encased in a peritoneal bag on the right or left side of the duodenum.

## The Pelvic Portion of the Peritoneal Cavity

The major pelvic organs covered by peritoneum include the rectum, urinary bladder, and, in the female, uterus and its adnexae as well. The rectum is largely accommodated in the curvature of the sacrum; the bladder, when empty, is confined to the pelvis, sheltered behind the body of the two pubes and the symphysis between them. The uterus protrudes upward and forward in the space between the rectum and the bladder, and the uterine tubes extend from it laterally. Peritoneal relationships that are different in the pelves of the two sexes are examined in detail in Chapter 27. The purpose here is to understand peritoneal continuities between the abdomen proper and the pelvis.

Continuing down from the infracolic compartment, peritoneum covers the anterior and lateral surfaces of the rectum in its upper portion; bare of peritoneum are the lower third of the rectum, embedded in extraperitoneal fascia, and its entire posterior surface in contact with the sacrum.

**In the male,** the peritoneum sweeps forward from the rectum to the bladder, creating a peritoneal recess between the two organs that is the *rectovesical pouch* (Fig. 23-25A). Just before the peritoneum comes in contact with the bladder, it is draped over the upper part of the seminal vesicles. From the superior surface of the bladder, the peritoneum ascends toward the umbilicus on the deep aspect of the anterior abdominal wall. Laterally, the peritoneum of the rectum and the bladder become continuous with parietal peritoneum of the pelvic wall, which then merges across the pelvic brim with parietal peritoneum in the iliac fossae. Visible and palpable through the parietal peritoneum on each side are the ureter and ductus deferens: the ureter runs anteromedially from the pelvic wall to the bladder, and the dutus deferens runs posteromedially from the inguinal fossa in front toward the seminal vesicles in the back.

**In the female,** the peritoneum sweeps forward from the rectum onto the uterus; the peritoneal space between them is the *recto-uterine pouch* (Figs. 23-19 and 23-25B). Just before the peritoneum comes in contact with the uterus, it covers the posterior fornix of the

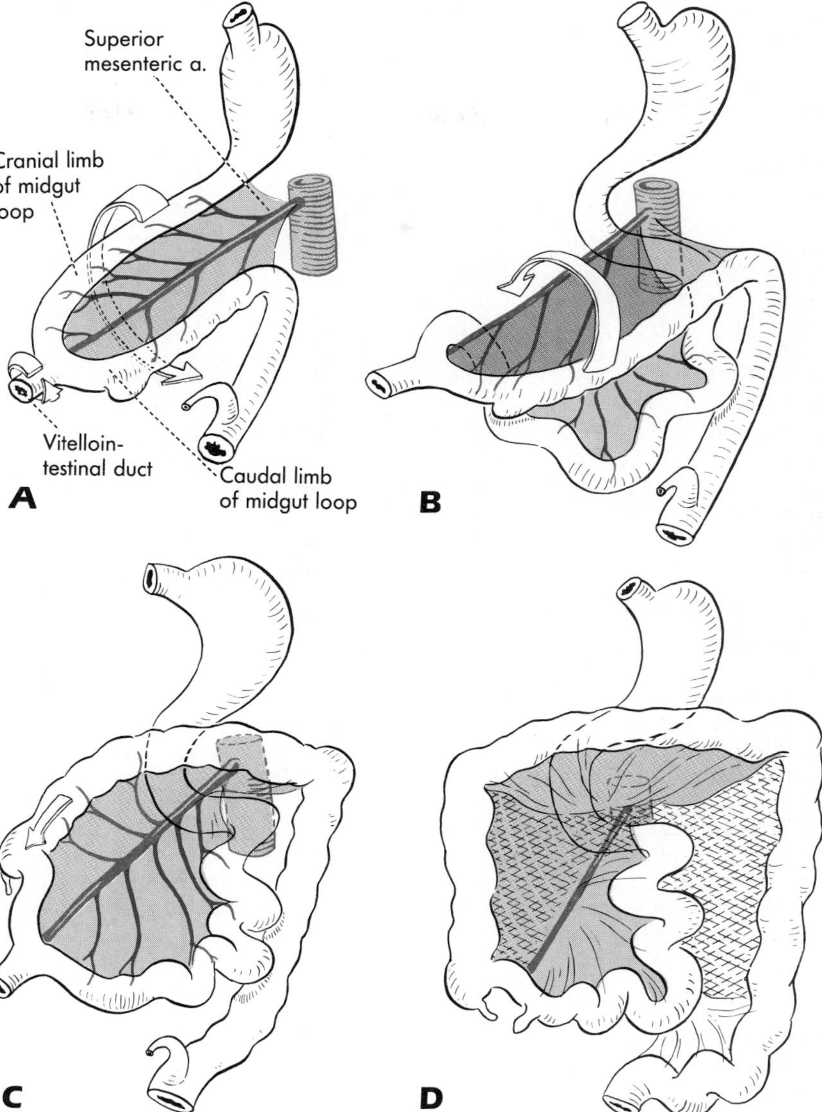


**FIG. 23-24.** Schematic representation of the rotation of the midgut. Events that occur simultaneously are arbitrarily divided into discrete steps to accord with descriptions in the text. In **A**, the primitive midgut loop is viewed from the left side. Compare with Figure 23-14**D.** The umbilical celom is not illustrated. In **B**, the cranial limb of the loop has descended into the left side of the peritoneal cavity; the duodenum is placed behind the superior mesenteric artery. In **C**, the caudal limb has ascended and is moving downward in the right side of the peritoneal cavity. In **D**, the definitive arrangement has been attained and mesenterial fusion is complete.

vagina. The peritoneum invests completely the posterior and anterior surfaces of the uterus before reflecting onto the superior surface of the bladder. The *vesicouterine pouch* is between the two organs.

Lateral to the uterus, a transverse fold of peritoneum is raised up from the floor of the pelvic cavity. The name of this fold is the **broad ligament,** which, like a mesentery, encloses the uterine tube in its superior border. Anterior and posterior laminae of the broad ligament merge with the peritoneum on respective surfaces of the uterus; reaching the lateral wall of the pelvis, the laminae reflect to become parietal peritoneum. Through the anterior lamina is visible and palpable the *round ligament of the uterus,* spanning the distance between the inguinal fossa and the uterus; through the posterior lamina, the *ligament of the ovary* can be traced from the ovary to the uterus. A reduplication of the posterior lamina suspends the ovary from the postero-

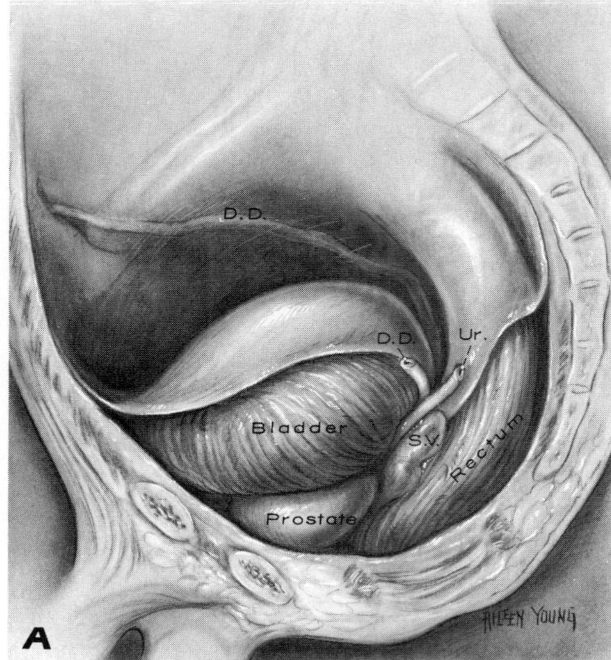


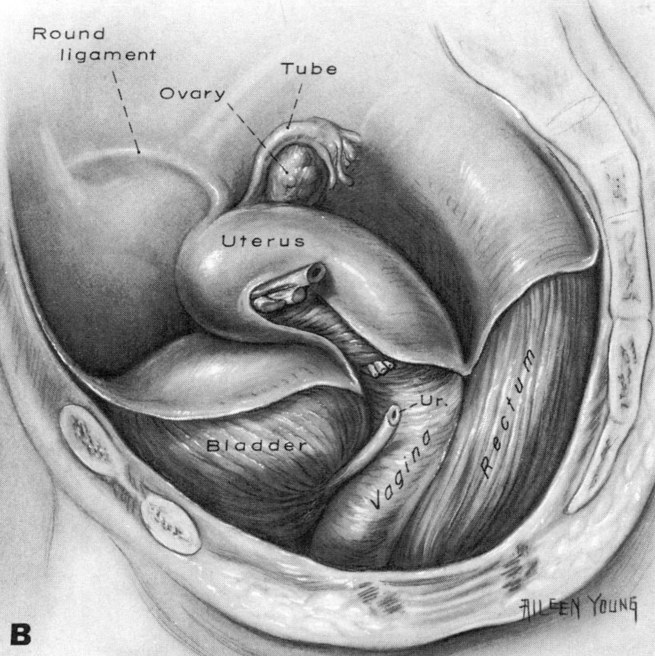


**FIG. 23-25.** The pelvic viscera with their peritoneal covering in the male **(A)** and female **(B)** pelvis. Over the rectum and bladder, the cut edges indicate the reflection of the peritoneum to the lateral pelvic wall. Over the uterus, the cut edges of the peritoneum indicate the attachment of the broad ligament to the uterus. **D.D.** identifies the right and left ductus deferentes; **Ur,** the left ureter; and **S.V.,** the left seminal vesicle. In **B,** at the base of the broad ligament, above the ureter, are the stumps of the uterine vessels as they enter the uterus. Between the cut edges of the leaves of the uppermost part of the broad ligament are stumps of the left uterine tube, of the ovarian ligament just below this, and of the round ligament just anterior to the ovarian ligament.

lateral aspect of the broad ligament. The lateral fimbriated end of the uterine tube fuses with the posterior lamina of the broad ligament in such a manner that the *ostium* of the tube surrounded by the fimbriae opens into the peritoneal cavity. Through the vagina, the uterine cavity, and the lumen of the tubes, communication is thus established between the exterior and the peritoneal cavity. In the male, the peritoneal cavity is completely closed. In the female, the ova that are shed into the peritoneal cavity can enter through the ostium into the uterine tubes, where they may be fertilized by spermatozoa deposited in the vagina.

Many spermatozoa that are ejaculated in the vagina find their way into the uterine tube and, escaping through the ostium, may accumulate in the rectouterine pouch. These spermatozoa may be aspirated from the pouch by a needle inserted into the posterior fornix of the vagina. Their presence in the aspirate rules out bilateral blockage of the uterine tubes as a cause of infertility. Infection in the female genital tract may spread along the same route as the spermatozoa and may cause peritoneal inflammation and later extensive adhesions in the pelvis.

## Deep Aspect of the Anterior Abdominal Wall

In the peritoneal tissue of the anterior abdominal wall, three ligaments extend from the region of the bladder to the umbilicus and in so doing produce more or less prominent peritoneal folds (Fig. 23-26). The **median umbilical ligament** is the remains of the *urachus,* an extension of the embryonic bladder to the umbilicus; it is attached to the apex of the bladder and produces the *median fold.* The **medial umbilical ligaments,** on each side of the median fold, are the fibrous remains of the *umbilical arteries* of the fetus, which passed just lateral to the bladder and raised up the *medial umbilical folds.* The **lateral umbilical folds,** which may not be at all prominent, are produced by the *inferior epigastric arteries.*

On each side of the median umbilical fold, just above the bladder, is a *supravesical fossa;* a *medial inguinal fossa* lies between the medial and lateral folds, and a *lateral inguinal fossa* lies lateral to each lateral fold. The medial inguinal fossa represents the *inguinal triangle,* the peritoneal area through which direct inguinal hernias occur; indirect inguinal hernias occur in the lateral inguinal fossa (Chap. 26).

At the inguinal triangles, the anterior and posterior abdominal walls meet each other. Parietal peritoneum of the iliac fossae passes from the anterior surface of the iliacus and psoas muscles onto the deep aspect of the anterior abdominal wall. Parietal peritoneum of the medial umbilical folds and the supravesical fossae is the forward continuation of the pelvic peritoneum.

### Vertical and Horizontal Continuities of the Peritoneum

Tracing peritoneal continuities in sagittal and transverse sections of the abdomen contributes greatly to understanding abdominal anatomy. Figure 23-19 illustrates that starting at the umbilicus, it is possible, in a sagittal section, to trace with a pencil the peritoneum of the greater sac into the upper abdomen, down into the pelvis, and back to the umbilicus without lifting the pencil. The same is possible in the lesser sacs, starting with the hepatic attachment of the lesser omentum, for instance. A transverse section across the epiploic foramen permits tracing of the peritoneum from the left leaf of the falciform ligament into the lesser sac and back again to the starting point (Fig. 23-21). Higher and lower sections aid in defining communication of various peritoneal spaces or the lack of such communication (Fig. 23-27).

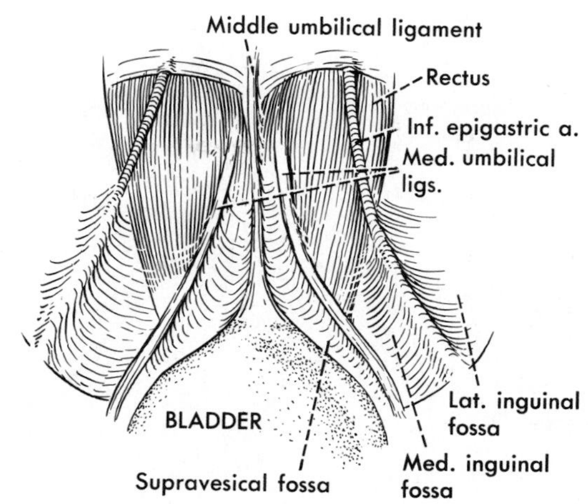


**FIG. 23-26.** View of the posterior surface of the lower part of the anterior abdominal wall, showing the structures that produce the folds and fossae related to the bladder and inguinal region.

## THE ABDOMINAL REGIONS AND LOCATION OF ABDOMINAL VISCERA

Abdominal viscera are located in reference to the surface of the body with the help of landmarks, planes that intersect the surface of the body along lines, and regions determined by some of the planes. Although in the course of a physical examination, lines are seldom drawn on the surface of the body, the student engaged in studying anatomy and acquiring the skills of physical examination benefits much from drawing these lines as an aid to learning the positions of various viscera. Such knowledge forms the basis for the examination of the abdomen. Symptoms and signs of abdominal disease are located and recorded with reference to abdominal regions mapped out on the abdominal wall.

**Landmarks.** The anatomic landmarks on the anterior abdominal wall have been described earlier in this chapter and include the xiphisternal joint, the costal arches, the umbilicus, the iliac crests and their tubercles, the anterior superior iliac spines, the inguinal folds, and the symphysis pubis (Fig. 23-28A).

**Planes and Lines.** There are sagittal and

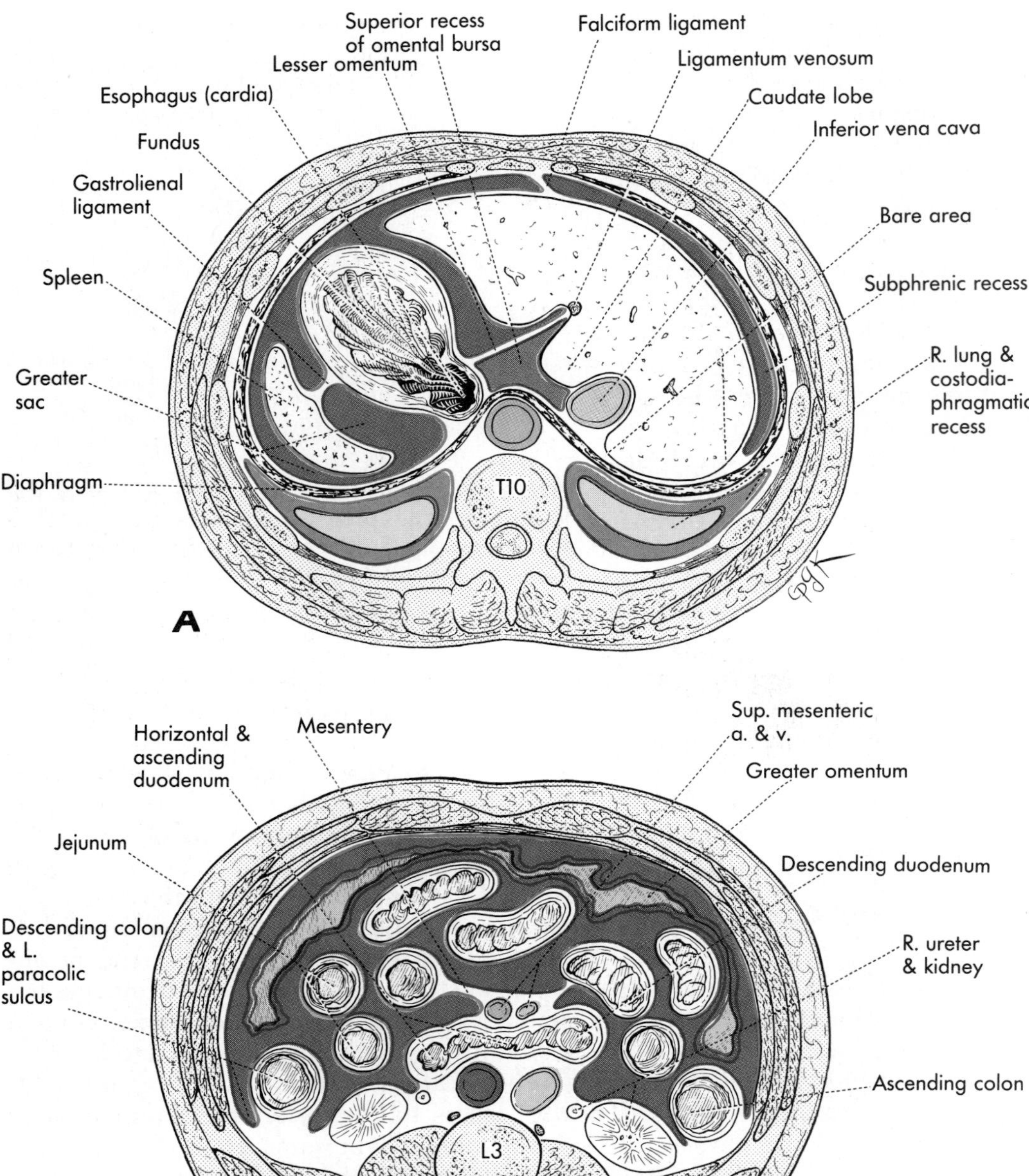


**FIG. 23-27.** Schematic transverse sections at the level of T10 **(A)** and L3 **(B).** The orientation and color scheme correspond to those of Figure 23-21.

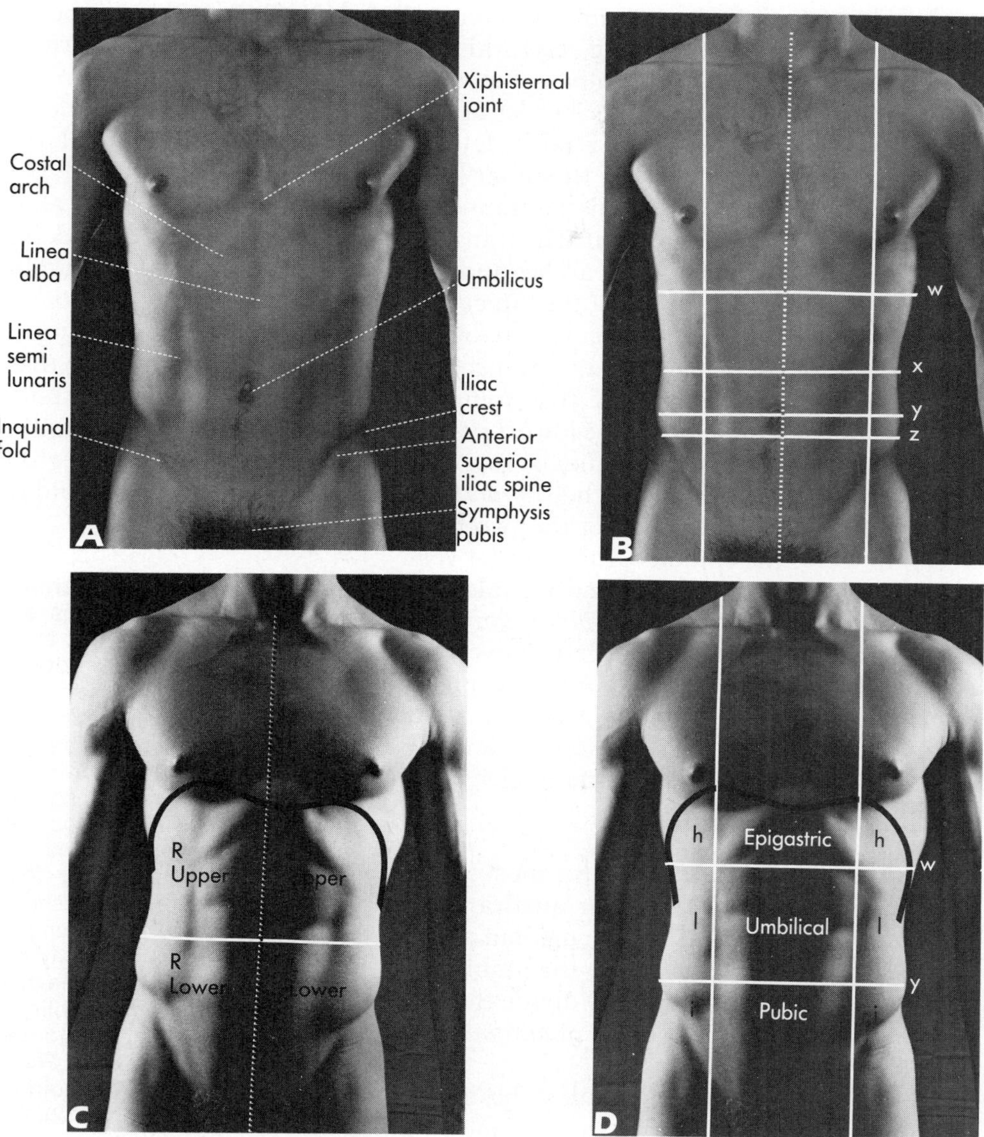


**FIG. 23-28.** Surface landmarks and regions on the anterior abdominal wall. **A,** surface landmarks. **B,** lines and planes: The median plane is indicated by a fine, interrupted line, and the midclavicular or lateral planes are indicated by solid vertical lines (**w,** transpyloric plane; **x,** subcostal plane; **y,** supracristal plane; **z,** intertubercular plane). **C,** quadrants. **D,** regions: **h,** hypochondrium; **l,** lumbar region; **i,** inguinal region.

horizontal planes used for locating viscera (Fig. 23-28B). Of the **sagittal planes,** the *median plane* coincides with the linea alba and passes through the umbilicus. The *midclavicular lines* drawn vertically from the midpoint of each clavicle define bilateral sagittal planes (often called *lateral planes*) that intersect the costal arch close to the tip of the 9th costal cartilage and the inguinal fold halfway between the anterior superior iliac spine and the symphysis pubis.

The **horizontal planes** are the transpyloric,

subcostal, supracristal, and intertubercular planes. All are useful in defining vertebral levels. The *transpyloric plane* is halfway between the jugular notch and the upper border of the symphysis pubis. It is more readily found, with sufficient accuracy, along the ulnar border of the subject's hand when the thumb is laid across the xiphisternal joint. The transpyloric plane intersects the costal arches and the midclavicular lines at the tip of the 9th costal cartilage and crosses the lower border of L1 vertebra. It is so called because it usually passes through the pylorus of the stomach. The *subcostal plane* connects the lowest points of the two costal arches and lies across the upper border of L3 vertebra. The *supracristal plane* joins the highest points of the two iliac crests; in it lies the spinous process of L4 vertebra and, in young and lean individuals, the umbilicus. The *intertubercular plane* connects the tubercles of the right and left iliac crests and passes through L5 vertebra. The plane connecting the two anterior superior iliac spines is unnamed; it passes below the level of the sacral promontory and is no longer used in mapping abdominal regions.

**Quadrants and Regions.** The most commonly used regions are the four **quadrants** of the abdomen (Fig. 23-28C). The median plane and the horizontal plane across the umbilicus (usually the supracristal plane) divide the abdomen into right and left *upper quadrants* and right and left *lower quadrants*.

An alternative method divides the abdomen into nine **regions** by two sagittal and two horizontal planes (Fig. 23-28D). The sagittal planes are in the right and left midclavicular lines, and the horizontal planes are the transpyloric and intertubercular planes.

Superiorly, sheltered by the costal cartilages, are the right and left *hypochondriac regions* extending beneath the diaphragm; between them, bounded by the costal arches and the transpyloric plane, is the *epigastrium*. The middle zone of the abdomen is made up of the *umbilical region* in the center, with the right and left *lateral regions* in the flanks. Inferiorly, the *pubic region* is in the center with each *inguinal region*, limited below by the inguinal folds, on each side. Alternative names are still in use for the regions in this lower zone: the pubic region is also known as the *hypogastrium* and the inguinal regions as the *iliac fossae*.

Named parts of the abdomen not usually recognized as clearly delimited regions are the flank (latus), loin (lumbus), and groin (inguen). In general, the *flank* is the anterolateral part of the abdominal wall, including the lateral region between the thoracic cage and the bony pelvis; the *loin* is the posterolateral part in the same area, including the paraspinal musculature; and the *groin* is the meeting point of the abdomen and the thigh along the inguinal fold.

#### Contents of the Abdominal Regions

Of the abdominal organs, only the liver, the duodenum, and the pancreas are relatively fixed. The location of most other abdominal viscera is rather variable. Nevertheless, some idea of the usual location of the major abdominal organs is valuable to both the anatomist and the clinician.

The *right hypochondrium* is filled by the liver (Fig. 23-12). For this reason, the region is dull to percussion. The *left hypochondrium* contains the fundus of the stomach, which is usually full of air (Fig. 23-12). Consequently, the percussion tone in the left hypochondrium is usually resonant. The spleen is located quite far posteriorly in the left hypochondrium, and it may be mapped out there by percussion as a dull area (p. 629, Fig. 23-13). The *epigastrium* is crossed obliquely by the liver and contains a portion of the stomach, as well as part of the pancreas (Fig. 23-12). The right and left *lateral regions* contain the ascending and descending colon, respectively (Fig. 23-12). With some experience the lower poles of the kidneys may be identified here by bimanual palpation (Fig. 23-13). The *umbilical region* is filled with coils of small intestine and usually contains also the transverse colon and the greater omentum. In the *right inguinal region* are the cecum and the root of the appendix, and in the *left inguinal region* is the sigmoid colon. Into the *pubic region* rises the full bladder and the pregnant uterus. When these organs are empty, the pubic region, or hypogastrium, contains small bowel. It is through this region that the uterus and ovaries may be felt by bimanual palpation.

In addition to giving an indication of the vertebral levels, some of the horizontal planes are useful guides to the following organs: not only the pylorus but also the fundus of the gallbladder lies in the transpyloric plane. The hilum of the right kidney is usually just inferior, and the hilum of the left kidney is just superior, to the transpyloric plane. This plane also identifies the inferior tip of the spinal cord. In the subcostal plane lies the horizontal part of the duodenum, and the supracristal plane is a good guide to the bifurcation of the abdominal aorta.

### Situs Inversus

In perhaps one of 5,000 to 20,000 persons, the disposition of the abdominal viscera is the mirror image of that just recounted: the stomach runs from right to left, the liver is largely on the left side, the cecum and appendix are on the left, and so forth. This is known as *situs inversus viscerum.* Either the thoracic or the abdominal organs alone may exhibit situs inversus, but commonly both do together. In the absence of knowledge of the presence of the condition, which can be determined by roentgenograms, faulty diagnosis and even operative approaches on the wrong side may occur, for in situs inversus, the pain from a diseased appendix or gallbladder may be localized by the patient either on the left side, where the organ is, or on the right side, where it should be.

## GENERAL REFERENCES AND RECOMMENDED READINGS

ADDISON C: On the topographical anatomy of the abdominal viscera in man, especially the gastrointestinal canal: I. J Anat 33:565; II. J Anat 34:427; III. J Anat 35:166, 277, 1899–1901

ANSON BJ, LYMAN RY, LANDER HH: The abdominal viscera *in situ:* A study of 125 consecutive cadavers. Anat Rec 67:17, 1936

BRIZON J, CASTAING J, HOURTOULLE FG: Le Peritoine. Paris, Libraire Maloine SA, 1956

CONDON RE: Surgical anatomy of the transversus abdominis and transversalis fascia. Ann Surg 173:1, 1971

DOTT NM: Anomalies of intestinal rotation: Their embryology and surgical aspects; with report of five cases. Br J Surg 11:251, 1923

FISCH AE, BRODEY PA: Computed tomography of the anterior abdominal wall: Normal anatomy and pathology. J Comput Assist Tomogr 5:728, 1981

FLINT MM, GUDGELL J: Electromyographic study of abdominal muscular activity during exercise. Res Q 36:29, 1965

FLOYD WF, SILVER PHS: Electromyographic study of patterns of activity of the anterior abdominal wall muscles in man. J Anat 84:132, 1950

FRAZER JE, ROBBINS RH: On the factors concerned in causing rotation of the intestine in man. J Anat 50:75, 1915

HUNTINGTON GS: The Anatomy of the Human Peritoneum and Abdominal Cavity Considered from the Standpoint of Development and Comparative Anatomy. Philadelphia, Lea Brothers & Co, 1903

MOODY RO: The position of the abdominal viscera in healthy, young British and American adults. J Anat 61:223, 1927

ONO K: Electromyographic studies of the abdominal wall muscles in visceroptosis: I. Analysis of patterns of activity of the abdominal wall muscles in normal adults. Tokoho J Exp Med 68;347, 1958

RIZK NN: A new description of the anterior abdominal wall in man and mammals. J Anat 131:373, 1980

TOBIN CE, BENJAMIN JA: Anatomic and clinical re-evaluation of Camper's, Scarpa's and Colles' fasciae. Surg Gynecol Obstet 88:545, 1949

WATERS RL, MORRIS JM: Effect of spinal supports on the electrical activity of muscles of the trunk. J Bone Joint Surg (Am) 52:51, 1970

# 24

# THE GUT AND ITS DERIVATIVES

This chapter deals with the stomach and intestine, the spleen, the liver, the gallbladder, and the pancreas. Before discussing the anatomy of the individual viscera, it will be useful to gain an overall appreciation of the developmental relationships between portions of the digestive tract and to understand the system of their blood supply, lymph drainage, and innervation. This can be done quite profitably as an introduction to the anatomy of these organs in view of the survey of abdominal contents presented in the preceding chapter.

## Developmental Considerations

It was explained in Chapter 23 which parts of the alimentary canal are derived from the foregut, midgut, and hindgut (p. 598). The general patterns of blood and nerve supply of the gut and its derivatives are best described by relating them to these three subdivisions of the primitive gut.

Blood is supplied to the primitive gut by a number of vitelline arteries and is drained by the vitelline veins. The **vitelline arteries** are ventral branches of the aorta. Three of the vitelline arteries persist, one for each of the main segments of the gut: the foregut is supplied by the *celiac artery* (celiac trunk), the midgut by the *superior mesenteric artery,* and the hindgut by the *inferior mesenteric artery.* **Vitelline veins** from respective segments of the gut follow the arteries in the peripheral part of their course; proximally, they all terminate in the *portal vein,* derived from a central anastomosis of the vitelline veins. The portal vein delivers the venous blood collected from all segments of the gut to the liver. After percolating through the hepatic sinusoids, the *hepatic veins* drain blood from the liver into the inferior vena cava. The vascular perfusion of the liver is augmented by the *hepatic artery,* a branch of the celiac trunk, and in the fetus also by placental blood brought to the liver by the *umbilical vein.* Most of the blood from the placenta is conducted through the liver toward the inferior vena cava by the *ductus venosus*. The ductus venosus, like the umbilical vein, is transformed into a fibrous cord known as the *ligamentum venosum,* which is embedded in the inferior surface of the liver.

Abdominal viscera derived from **intermediate mesoderm** rather than from the gut (kidneys, gonads, suprarenal glands) are supplied by paired lateral visceral branches of the aorta; their venous blood is returned to the cardinal veins of the embryo and the anastomoses that develop in association with these veins. From these venous channels is formed the *inferior vena cava*.

The inferior vena cava with all its tributaries, the abdominal aorta with its branches that supply the body wall, and the viscera derived from intermediate mesoderm are all outside the peritoneal sac. The blood vessels of the gut derived from the vitelline arteries and veins, on the other hand, are enclosed between the two laminae of the dorsal mesentery of the gut; those destined for the liver pass from the dorsal mesentery into the lesser omentum. Because of the eventual fusion of extensive portions of the dorsal mesentery to the parietal peritoneum, many of the veins and arteries that serve the gut become fixed to the posterior wall of the lesser sac and the infracolic compartment; by the completion of development these vessels appear to be retroperitoneal.

## General Pattern of Blood Supply

The **abdominal aorta** descends retroperitoneally, lying on the lumbar vertebrae. In front of the 4th lumbar vertebra, it bifurcates into right and left common iliac arteries. The abdominal aorta gives off three sets of branches: paired segmental lateral branches to the body wall; paired visceral branches to organs derived from intermediate mesoderm; and three unpaired ventral branches, the celiac, superior mesenteric, and inferior mesenteric arteries. This chapter deals only with the latter groups of vessels derived from the vitelline arteries that supply the gut and its derivatives.

The **celiac trunk** is the artery of the foregut. It originates from the aorta in front of T12 vertebra immediately below the margin of the aortic hiatus of the diaphragm. The celiac trunk is very short, and it breaks up into three branches: the *common hepatic, left gastric,* and *splenic arteries,* all of which run retroperitoneally in the posterior wall of the lesser sac. The hepatic branch of the common hepatic enters the hepatoduodenal ligament to reach the liver.

The **superior mesenteric artery** is the artery of the midgut. It arises from the aorta only a little below the celiac, in front of L1 vertebra. At its origin, it is behind the pancreas, and on its way to enter the root of the mesentery, it passes through the pancreas and crosses in front of the horizontal portion of the duodenum. Its branches to the jejunum and ileum are in the mesentery, those to the ascending colon behind the peritoneum of the infracolic compartment, and those to the transverse colon in the transverse mesocolon.

The **inferior mesenteric artery** is the artery of the hindgut. It originates from the aorta in front of L3 vertebra just behind the horizontal portion of the duodenum. At first, all its branches are retroperitoneal; those to the descending colon remain so. Branches to the transverse colon and to the sigmoid colon enter the respective mesocolons; the terminal branch supplies the rectum.

There are extensive **anastomoses** between the arteries of the gut along the stomach and the intestine. These anastomoses are of two types: those within each arterial system (*e.g.*, most branches of the celiac trunk freely anastomose with each other) and those within neighboring systems. (Branches of the celiac anastomose with those of the superior mesenteric, and branches of the superior mesenteric anastomose with those of the inferior mesenteric.)

The **superior and inferior mesenteric veins** are formed by tributaries that correspond with the branches of the respective arteries. There is no celiac vein; veins corresponding to the branches of the celiac artery enter the portal vein. The **portal vein** is formed behind the pancreas by the union of the superior mesenteric and splenic veins, and it ascends to the liver in the hepatoduodenal portion of the lesser omentum. The inferior mesenteric vein usually terminates in the splenic vein; hence, the portal vein drains all the blood from the entire intestine, the stomach, the spleen, and the pancreas. After percolating through the liver, the hepatic veins drain this blood into the inferior vena cava just before the vena cava ascends through the diaphragm.

The **inferior vena cava** drains the lower limbs, the contents of the perineum and pelvis, and the extraperitoneal abdominal organs, as well as the abdominal wall. As mentioned, the gut and its derivatives send their blood indirectly into the inferior vena cava through the liver and the hepatic veins. The inferior vena cava is formed by the union of the common iliac veins just above the sacral promontory, and it ascends along the right side of the aorta, lying in the posterior wall of the infracolic compartment, behind the pancreas and the epiploic foramen; more superiorly, it is embedded in the bare area of the liver.

There are extensive **anastomoses** between the tributaries of the portal system. Of considerable clinical importance are the potential *portasystemic anastomoses* that exist in defined locations between tributaries of the portal vein and the inferior and superior venae cavae.

### General Pattern of Lymphatic Drainage

Lymph vessels draining the gut and its derivatives accompany the blood vessels and are interrupted by a series of lymph nodes. The peripheral nodes are close to the viscera and within the mesenteries; the central nodes are in front of the aorta, around the origin of the main arteries (*preaortic lymph nodes*). Lymph from these nodes is drained by two to three large *intestinal lymph trunks* to the *cisterna chyli,* a saclike receptacle of lymph lying in front of the body of L1 vertebra. The cisterna also receives bilateral *lumbar lymph trunks,* which convey lymph to it from the lower limbs, perineum, pelvis, and abdominal wall, as well as kidneys and gonads. These lumbar lymph trunks are interrupted by a series of lymph nodes located along the vertebrae (*paraaortic or lumbar lymph nodes*).

### General Pattern of Nerve Supply

Visceral efferent and visceral afferent nerves reach and leave abdominal organs through **autonomic nerve plexuses** associated with blood vessels. These plexuses are usually named according to the artery they accompany. The *parasympathetic innervation* of the foregut and midgut is furnished by the vagi; that of the hindgut, by the pelvic splanchnic nerves derived from S2–S4 segments of the spinal cord. *Sympathetic innervation* of all abdominal and pelvic viscera is provided by the lower thoracic and upper lumbar segments of the spinal cord. Both sympathetic and parasympathetic nerves contain visceral efferents and visceral afferents.

**Synaptic relay** of parasympathetic efferents occurs in small *intramural* or *enteric ganglia* of the organs they serve; synaptic relay of sympathetic efferents occurs in *collateral ganglia* located on the aorta, chiefly at the origin of the three main vessels (celiac and superior and inferior mesenteric, as well as smaller *aortic* ganglia).

*Parasympathetic activity* stimulates glandular secretions and peristalsis and causes dilatation of the sphincters. *Sympathetic activity* constricts the sphincters and the blood vessels. Parasympathetic **visceral afferents** are primarily concerned with general visceral sensations (hunger, nausea) and afferent impulses contributing to visceral reflexes. The pseudounipolar cell bodies of parasympathetic afferents from the foregut and midgut are in the vagal ganglia at the base of the skull; those from the hindgut, in spinal (dorsal root) ganglia of S2–S4 spinal nerves. Visceral afferents in the sympathetic system are primarily concerned with pain that is induced by tension in the viscera and mesenteries. The pseudounipolar cell bodies are located in the spinal ganglia of the lower thoracic and upper lumbar dorsal roots.

The **basic plan** of the autonomic plexuses in the abdomen is best appreciated by considering the sympathetic input into them. After entering the abdomen through the diaphragm, the *thoracic splanchnic nerves* (p. 574) commingle with each other in front of the aorta, giving rise to the *aortic plexus*. This plexus is augmented by *lumbar splanchnic nerves*, medial branches of the lumbar sympathetic chain ganglia, which feed preganglionic fibers into the plexus. The *celiac and superior and inferior mesenteric plexuses* are offshoots of the aortic plexus. Before entering the celiac or superior or inferior mesenteric plexuses, preganglionic sympathetic nerve fibers relay in the celiac or superior or inferior mesenteric ganglia. The aortic plexus, below the origin of the inferior mesenteric artery, continues down into the pelvis and is known as the *superior hypogastric plexus*.

Parasympathetic fibers of vagal and pelvic splanchnic origin commingle with the sympathetic fibers in the aortic plexus and its offshoots. The vagi enter the abdomen with the esophagus, and their branches join the aortic plexus in the celiac region; the pelvic splanchnics ascend into the abdomen along branches of the inferior mesenteric artery.

Visceral afferents of both parasympathetic and sympathetic systems run along with the efferent fibers of each system. The sympathetic afferents ascend through the splanchnic nerves, the sympathetic chains, and the white rami communicantes into the spinal nerves. Parasympathetic afferents from the foregut and midgut ascend with the vagus; those from the hindgut descend to pelvic splanchnic nerves located in the pelvis.

## THE ESOPHAGUS, STOMACH, AND SPLEEN

### THE ABDOMINAL PORTION OF THE ESOPHAGUS

Of its total length of 25 cm, only the terminal 1.5 cm of the esophagus is in the abdomen. Entering the abdomen through the esophageal hiatus of the diaphragm, to the left of the midline opposite T10 vertebra, the esophagus curves to the left and joins the cardia of the stomach (Fig. 24-1). In the abdomen, the esophagus is compressed: its right margin is continuous with the lesser curvature of the stomach; its left margin creates a sharp angle with the fundus of the stomach, which is the *cardiac incisure*.

**Relations.** The esophagus, together with the most superior part of the lesser omentum attached to its right margin, forms the left boundary of the superior recess of the lesser sac (see Fig. 23-27A). The anterior lamina of the lesser omentum continues over the anterior surface of the esophagus, which faces into the most superior part of the greater sac, one of the subphrenic spaces. Limited anteriorly by the posterior lamina of the left coronary ligament, this is only a potential space (see Fig. 23-20). The esophagus is in contact with the posterior surface of the liver and may create a shallow indentation on it behind the left coronary ligament.

The posterior surface of the esophagus rests directly on the diaphragm. Peritoneum reflecting onto the diaphragm from the right and left margins of the esophagus forms the upper part of the *gastrophrenic ligament* (p. 606). Behind the esophagus, the two laminae of this ligament do not become apposed to each other; in between them the left gastric vessels pass forward from the posterior wall of the lesser sac (leaving the left gastropancreatic fold) and make contact with the esophagus and the lesser curvature of the stomach. The anterior and posterior vagal trunks, derived from the esophageal plexus, lie on respective surfaces of the esophagus (p. 570). Their position, as well as the number of the trunks, may vary.

### THE STOMACH (VENTRICULUS, GASTER)

The stomach is a saclike dilatation of the alimentary canal, closed off from the esophagus by the *cardiac sphincter* and from the duode-

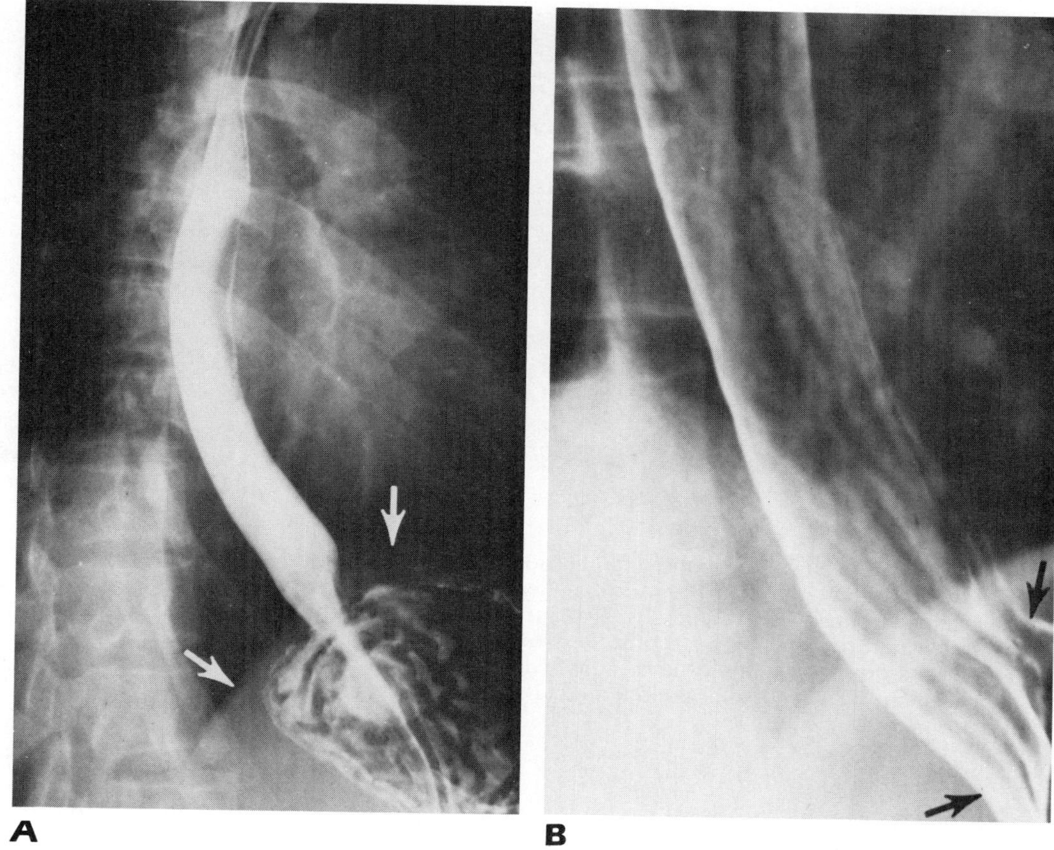


**FIG. 24-1.** X-ray films of the esophagus and cardia seen in oblique views. In **A**, the swallowed barium outlines most of the thoracic part of the esophagus, demarcated from the abdominal part by the left dome of the diaphragm (white arrows). Barium is entering the stomach through the cardia. **B** is a close-up view of the lower part of the esophagus terminating at the cardia (black arrows). When the viscera contain air, the barium adhering to their lining outlines the muscosal folds.

num by the *pyloric sphincter*. By its peristaltic activity, the stomach churns and homogenizes the swallowed foods, liquids, and saliva, adding to them its own secretions before propelling the resultant, partially digested *chyme* in aliquots into the duodenum.

The stomach lies largely under cover of the ribs and costal cartilages that form the left costal margin. It is fixed only at two points: the cardia above and the pylorus below. Between these two points, its position and shape may vary considerably, depending on the habitus (body build) of the individual, the fullness and muscle tone of the stomach itself, and the status of the surrounding viscera (Fig. 24-2).

The cardia is located behind the 7th costal cartilage, 2.5 cm to the left of the midline; the pylorus is 2.5 cm to the right of the midline in the transpyloric plane. The pylorus lies on the body of L1 vertebra, but the cardia is some distance anterior to T11.

The stomach has several **named parts**, which are not, however, demarcated by clear lines or anatomic features. Its two openings, the *cardiac and pyloric ostia*, connect it to the esophagus and the duodenum; the regions of the stomach adjacent to these openings are known as the *cardia* and the *pylorus*, respectively. Between the cardia and the pylorus, the stomach is made up of the fundus, the

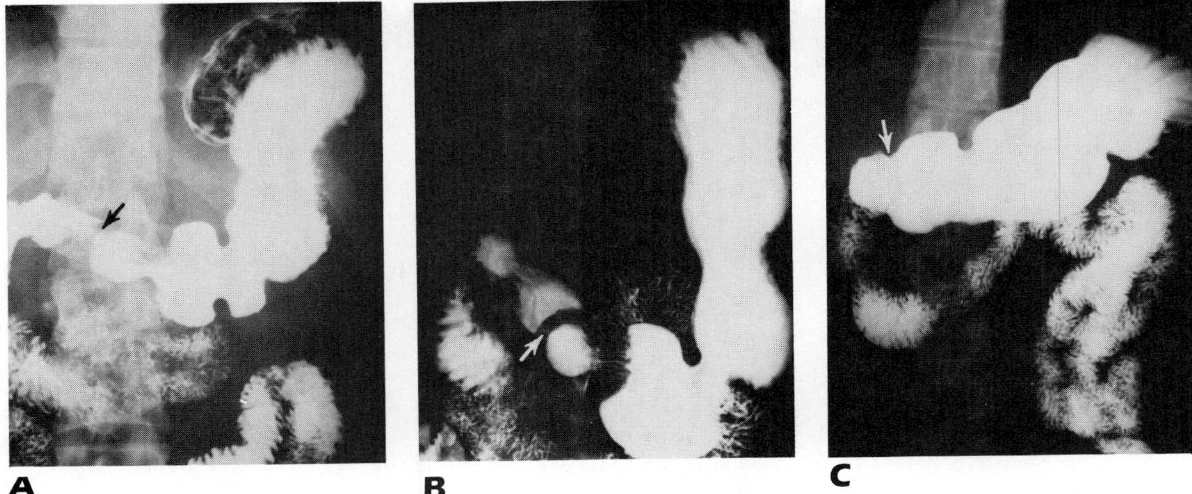


**FIG. 24-2.** Variation in the shape and position of the stomach as revealed by x-ray films following a barium meal. In all three cases illustrated, the cardia is concealed by the gastric fundus, which is filled with air; the body, pyloric antrum, pyloric canal, and duodenum are filled with barium. Peristaltic contraction distorts both greater and lesser curvatures. The pylorus is indicated by an arrow. **A,** typical configuration of the stomach; **B,** a vertical and elongated (J-shaped) stomach; **C,** a short and horizontal stomach.

body, the pyloric antrum, and the pyloric canal (Fig. 24-3). The upper left margin of the cardia is indicated by the *cardiac incisure,* the sharp angle between the left border of the esophagus and the stomach. The line drawn horizontally across the stomach, starting at the cardiac incisure, separates the *fundus* above from the *body* below. The body is continuous above also with the cardia and below with the *pyloric antrum,* which leads into the more or less tubular *pyloric canal.* The pylorus is marked on the surface of the stomach by a shallow groove in which the small prepyloric vein may be identified during surgery.

The empty stomach, especially in the fixed cadaver, presents well-defined anterior and posterior surfaces that are separated from one another by the greater and lesser curvatures. In the collapsed stomach, the curvatures form definite borders. Distention of the stomach obliterates these borders, but the curvatures remain identifiable even in the full stomach and are chiefly responsible for the characteristic shape of the organ. The *lesser curvature* is continuous with the right margin of the esophagus. It is concave to the right and faces upward. An *angular incisure* indicates the junction between the body and the pyloric antrum. To the lesser curvature is attached the hepatogastric portion of the lesser omentum; the left and right gastric arteries and veins run in the omentum closely skirting the curvature. The *greater curvature* commences at the cardiac incisure and is convex to the left as well as downward. The various named ligaments of the greater omentum are attached to it in continuity (p. 603). The right and left gastroepiploic vessels skirt the greater curvature between the two laminae of the gastrocolic ligament.

**Relations.** In addition to the omenta, the stomach has many important relations. The *anterior surface,* covered with peritoneum, faces into the greater sac and is overlapped by the left lobe of the liver; the spleen overlaps the fundus. Between the liver and the spleen, the left dome of the diaphragm separates the

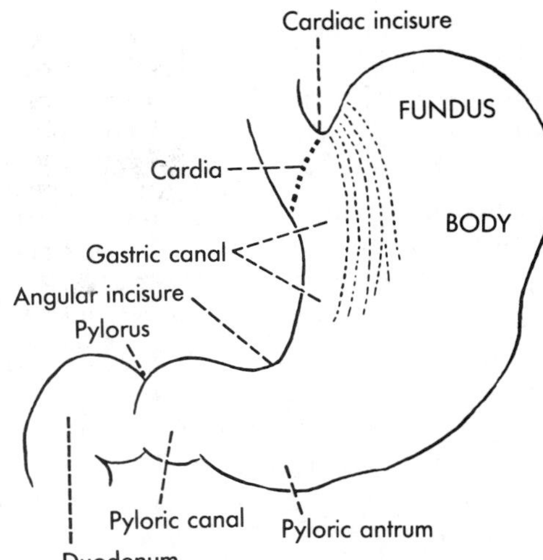


**FIG. 24-3.** Parts of the stomach. The broken lines indicate the innermost or oblique layer of muscle that bounds the gastric canal.

**FIG. 24-4.** Plain films of the left side of the chest and upper abdomen to show the relationship of the stomach to thoracic and some abdominal viscera. **A,** anteroposterior (frontal) view; **B,** left lateral view. The left dome of the diaphragm is indicated by arrows. The gas in the fundus of the stomach (*) is overlapped slightly by the shadow of gas in the left colic flexure (★); the opacity posterolateral to the gastric fundus is the spleen (⊕). Through the diaphragm, the fundus is related to the left lung, seen as a radiolucent area. The breast shadow is clearly seen in front of the left dome of the diaphragm.

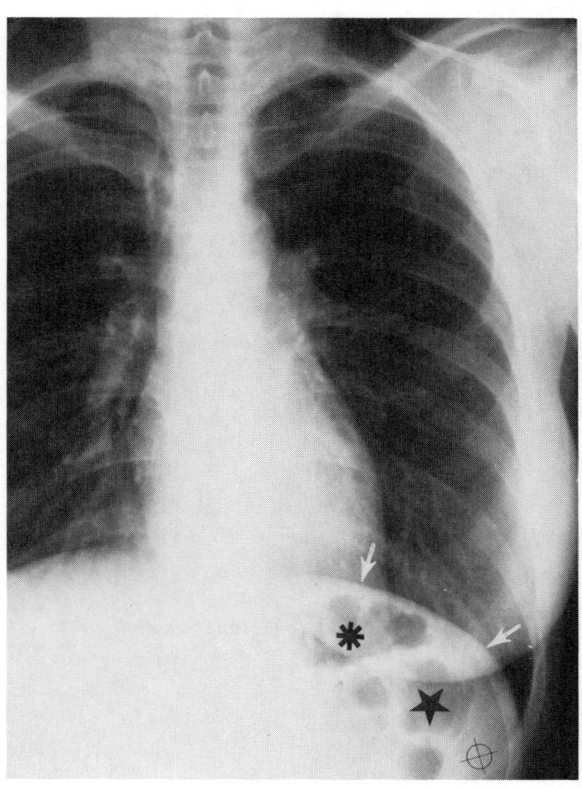

**A**

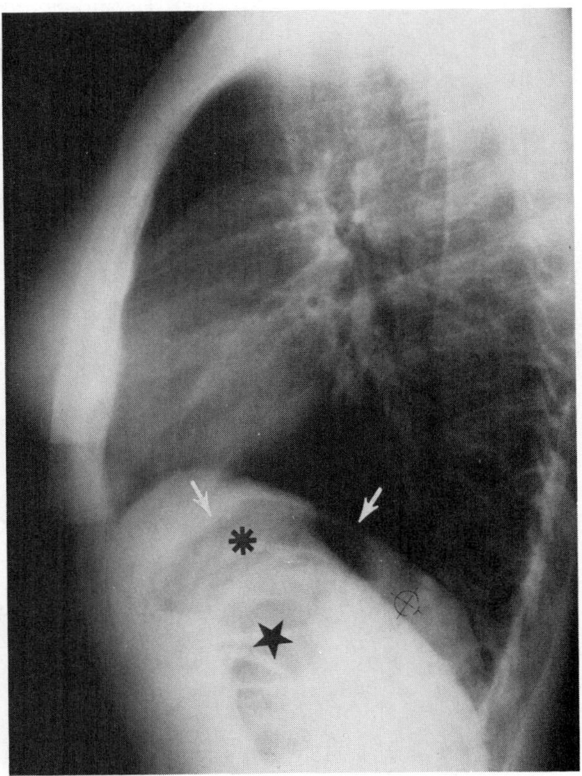

**B**

stomach from the left lung and the heart (Fig. 24-4). Below the costal margin, the stomach is in contact with the anterior abdominal wall unless the greater omentum and the transverse colon have been displaced upward and lie in front of the stomach. The *posterior surface,* covered in peritoneum, faces into the lesser sac (see Figs. 23-21 and 23-27). The so-called *stomach bed,* on which the stomach rests, is made up of retroperitoneal structures in the posterior wall of the lesser sac (pancreas, with the splenic artery along its superior border; diaphragm; left kidney and suprarenal gland), as well as the transverse mesocolon with the laminae of the greater omentum fused to it, and, outside the lesser sac, the spleen (Fig. 24-5).

**Musculature and Interior.** The form of the stomach is largely due to its musculature, which is rather thicker than in other parts of the alimentary canal. The outer, *longitudinal layer* of muscle is best developed along the curvatures and is continuous with the longitudinal muscle of the esophagus and duodenum. The inner, *circular layer* is uniform over the stomach but forms a thickened muscular ring at the pylorus; this is the *pyloric sphincter.* During gastric contractions, the sphincter is usually also contracted; its intermittent relaxation permits gradual emptying of the stomach.

There is no anatomically identifiable muscular ring or sphincter at the cardia, yet a sphincter mechanism does exist at the *esopha-*

**FIG. 24-5.** Posterior relations of the stomach shown in a computed tomogram (CT scan). The section passes through the lower part of the body of the stomach and its antrum (outlined); loops of jejunum are tucked between the lateral aspect of the stomach and the spleen (compare with Fig. 24-4). The section is seen from below, and therefore the left-to-right orientation is reversed to that in Figures 23-21 and 23-27; nevertheless, comparison with this CT scan should be instructive.

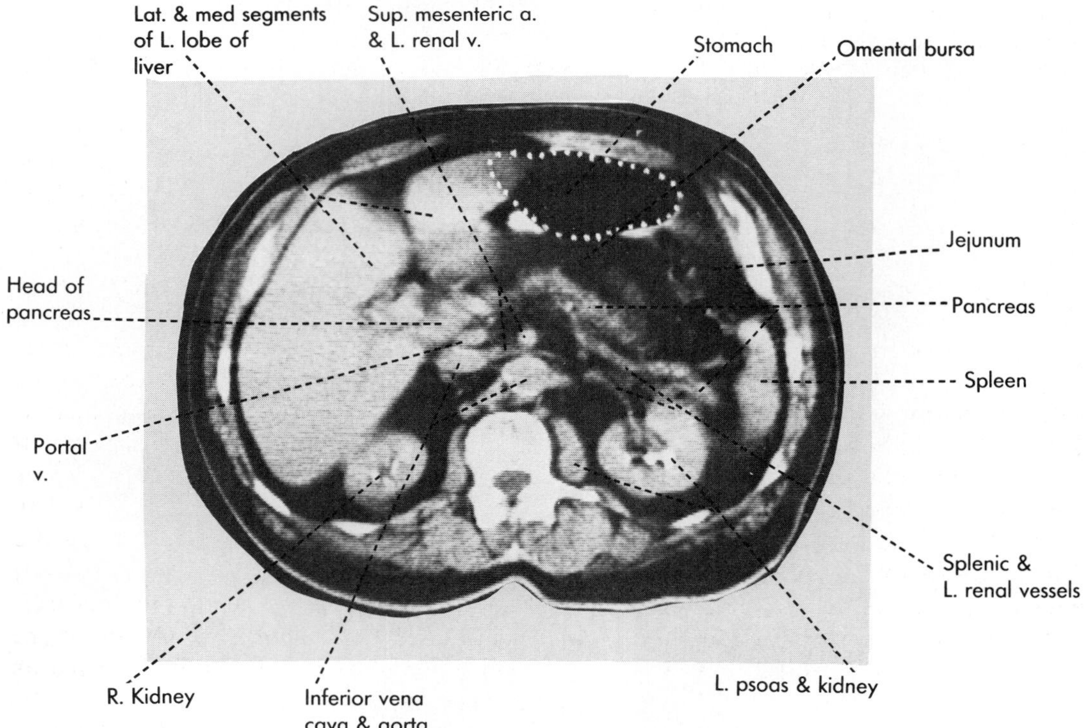

*gogastric junction,* where intraluminal pressure is higher than in either the esophagus or the stomach. Swallowed, radiopaque material can be seen to be held up momentarily before entering the stomach until relaxation of the cardiac sphincter permits its passage. Similarly, splashing sounds made by swallowed water as it is transmitted by the cardia are delayed some seconds after swallowing. These sounds can be heard by a stethoscope placed over the surface projection of the cardia. Delay or absence of the sound is a sign of obstruction at the cardia. Several mechanisms have been postulated for the functional cardiac sphincter and for the prevention of regurgitation of gastric contents into the esophagus (p. 568).

Prolonged spasm of both cardiac and pyloric sphincters may occur. In case of cardiospasm, food is held up in the thoracic part of the esophagus, which may become greatly dilated. The condition, known as *achalasia,* is thought to be due to a paucity of parasympathetic ganglion cells in the wall of the esophagus and the cardia. Spasm of the pyloric sphincter causes gastric retention and distention. *Congenital pyloric stenosis,* due to hypertrophy of the pyloric sphincter, is seen in the newborn. Division of the pyloric sphincter provides the cure. In later life, *pylorospasm* may be caused by irritation from an adjacent ulcer or by division of the vagi performed to reduce gastric secretion of acid in the treatment of peptic ulcer. The condition may require division of the sphincter (pyloroplasty), as in the treatment of congenital stenosis, or the creation of an anastomosis between the jejunum and the stomach (gastrojejunostomy) to bypass the stenosis.

Unlike other parts of the alimentary tract, the stomach possesses, internal to the circular muscle coat, an incomplete layer of *oblique muscle* fibers (Fig. 24-3). From the cardiac incisure, the oblique muscle fans out into both anterior and posterior walls and descends parallel with the lesser curvature. These oblique bundles define the *gastric canal* adjacent to the lesser curvature. Swallowed fluids first enter the gastric canal, explaining the fact that ingested corrosive materials particularly affect the lesser curvature.

The **mucosa** of the stomach is rather thick and velvety. When the stomach is empty, it is thrown into characteristic ridges or *rugae.* These rugae are most prominent in the body and fundus of the stomach (Fig. 24-1); the mucosa of the pyloric antrum and the canal presents a smoother pattern. Gastric ulcers and neoplasms distort the mucosal pattern and can be detected radiologically.

## THE SPLEEN

Although strictly speaking the spleen is not a derivative of the gut, it is best discussed along with the stomach because it is enclosed in the dorsal mesogastrium, it is anatomically closely related to the stomach, it derives its blood and nerve supply from arteries and nerves of the foregut, and its venous blood is drained into the portal vein.

The spleen is an irregular, roughly wedge-shaped, lymphoid organ. Functionally, it is a member of the immune system and is called upon primarily to eliminate senescent and damaged cells from the circulation, filter antigens and other particulate matter from the blood, and contribute to the immune response against such agents. Its removal, however, does not impair the immune response seriously.

The spleen is located in the greater sac of the peritoneum between the diaphragm and the stomach, sheltered entirely by the ribs (Fig. 24-6). The normal spleen cannot be palpated; however, it may be mapped out by percussion. The spleen is roughly as large as a clenched fist, but it varies considerably in size. It has a rather smooth, convex diaphragmatic surface that faces posterolaterally and a more irregular concave visceral surface that presents the hilum; a superior border, which may be notched, and a more rounded inferior border; and blunt upper and lower poles. The longitudinal axis of the spleen lies along the 10th rib: the upper pole is 3 cm to 4 cm from the spinous process of the 10th thoracic vertebra, and the lower pole, which is more rounded, reaches normally as far forward as the midaxillary line.

The spleen developed in the dorsal mesogastrium, so that, save for its hilum, it is com-

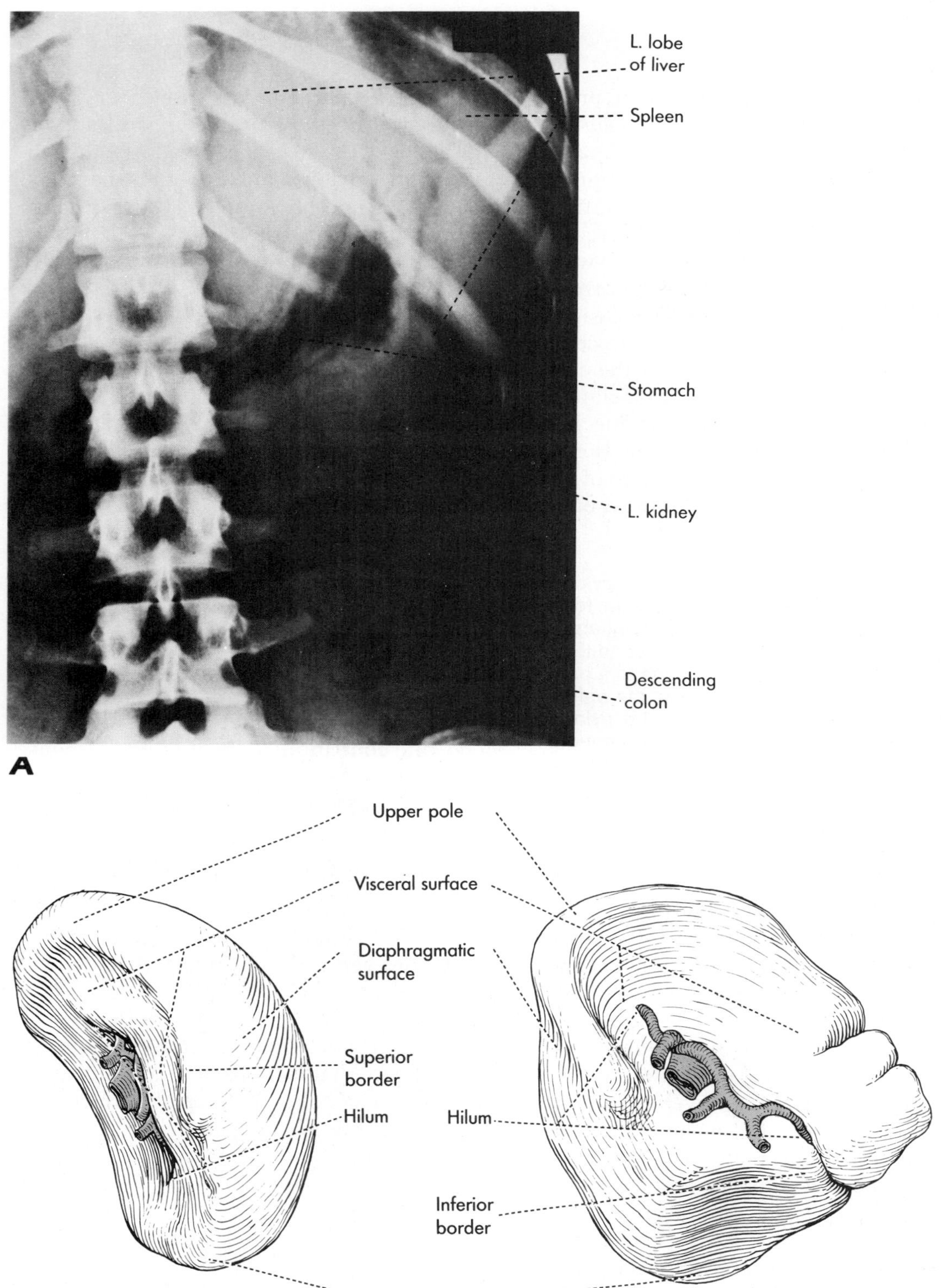


**FIG. 24-6.** The spleen. In **A,** the position of the spleen is shown in a plain anteroposterior x-ray film of the left upper abdomen. Note its relation to the stomach and the ribs. **B** shows the spleen in more or less the position it occupies in the body; **C** shows the spleen with its visceral surface turned forward.

pletely invested with peritoneum, which is fused to its fibrous capsule. The peritoneum is derived from the left lamina of the mesogastrium; the right lamina, lining the splenic recess of the lesser sac, comes in contact with the spleen only at the hilum (see Fig. 23-21). The *hilum* is a longitudinal fissure on the visceral surface. It admits the branches of the splenic vessels, which are conveyed to the spleen in the lienorenal ligament. This ligament, attached to the posterior lip of the hilum, also contains, in most cases, the tail of the pancreas, the only organ that may make direct contact with the spleen. The gastrolienal ligament, attached to the anterior lip of the hilum, conveys branches of the splenic vessels to the stomach. These two ligaments attached to the hilum make up the *splenic pedicle*.

Through the greater sac, the spleen is in contact with the stomach, the left kidney, and the splenic flexure of the colon, all of which create impressions on the visceral surface. The lower pole is in contact with the phrenicocolic ligament, which is thought, without any evidence however, to support the spleen (p. 609).

The posterolateral surface of the spleen is related through the diaphragm to the base of the left lung, the costodiaphragmatic recess of the left pleura, and the 9th, 10th, and 11th ribs (Fig. 24-6). Without necessarily tearing the diaphragm, fractures or violent displacements of these ribs may rupture the spleen, which is highly vascular and rather friable.

Owing to a variety of causes, the spleen may increase in size by as much as tenfold. When it more than doubles its size, it becomes palpable at the left costal margin, and further enlargement will extend it diagonally across the abdomen beyond the midline. There may sometimes be *accessory spleens*, located, in most cases, in the splenic pedicle. *Splenectomy* is performed when the spleen is ruptured or accidentally nicked at operation and its bleeding cannot be stopped. Splenectomy is also performed in the treatment of certain blood dyscrasias. In these cases, accessory spleens must also be removed.

## VESSELS AND NERVES

### Blood Supply

Being derived from or associated with the foregut, the abdominal esophagus, the stomach, and the spleen are supplied by branches of the celiac trunk. This artery also provides blood to the liver, gallbladder, pancreas, and a part of the duodenum, which are also derived from the foregut and are discussed later.

**Celiac Trunk.** As soon as the aorta enters the abdomen, it gives off the short celiac trunk, which juts forward from the anterior surface of the aorta and breaks up almost immediately into its three branches (Fig. 24-7). The two larger branches, the *common hepatic* and *splenic arteries,* diverge from one another along the upper border of the pancreas; the *left gastric artery* ascends toward the cardia. All these vessels are behind the posterior peritoneal wall of the lesser sac: the common hepatic in the right and the left gastric in the left gastropancreatic fold and the splenic along the upper margin of the pancreas.

The **common hepatic artery** reaches the inferior boundary of the epiploic foramen formed by the duodenum and the pancreas and passes forward into the hepatoduodenal portion of the lesser omentum. Here, it divides into a *hepatic artery proper* (which ascends in the hepatoduodenal ligament) and the *gastroduodenal artery* (which descends behind the first part of the duodenum). Two arteries arise on the right for the supply of the stomach: the *right gastric,* a branch of the hepatic artery proper, and the *right gastroepiploic,* a branch of the gastroduodenal.

On the left, the **splenic artery** reaches the hilum of the spleen along the pancreas and in the lienorenal ligament and gives off two sets of branches to the stomach: the *short gastric arteries* to the fundus and the *left gastroepiploic artery* to the body. Both sets reach the stomach in the gastrolienal ligament.

The **left gastric artery,** the smallest branch

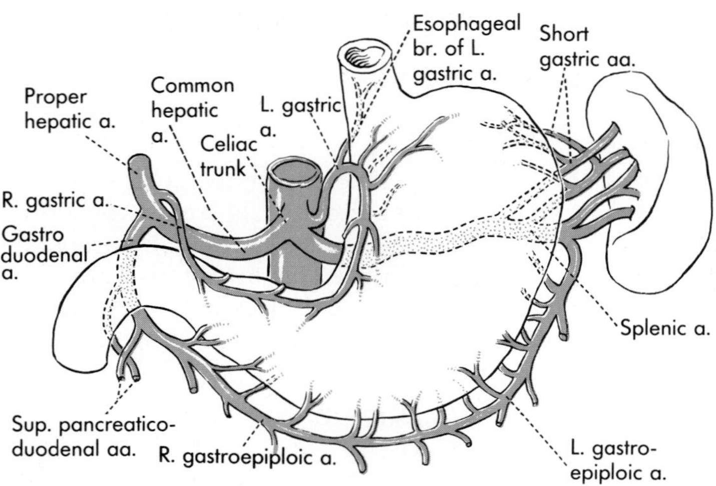


**FIG. 24-7.** Arteries of the stomach.

of the celiac trunk, on reaching the cardia, leaves the left gastropancreatic fold by turning forward onto the lesser curvature of the stomach. The artery runs toward the pylorus in the lesser omentum to meet the right gastric artery.

The **abdominal esophagus** is supplied by esophageal branches of the left gastric, given off in the gastrophrenic ligament before the left gastric enters the lesser omentum. The **stomach** receives its blood supply from two arterial arcades, one on each of its curvatures (Fig. 24-7). Each arcade is created by the anastomosis of two arteries: the right and left gastric arteries anastomose along the lesser curvature; the right and left gastroepiploic arteries anastomose along the greater curvature. Each arcade gives off numerous gastric branches to both anterior and posterior surfaces of the stomach and also epiploic branches to the lesser and greater omenta (epiploa). Anastomoses between the gastric branches link the two arcades to one another. These arteries anastomose with the short gastric arteries in the region of the fundus.

As many as 15 different branching patterns have been described for the celiac trunk. Angiography is effective in detecting these variations. In 1% of cases, the celiac trunk is completely missing and its branches arise from a common stem with the superior mesenteric (celiacomesenteric). Some branches may arise directly from the aorta.

**Veins.** All arteries are accompanied by veins of similar names. Those draining the esophagus and stomach terminate either directly in the portal vein or in the splenic vein. There is no celiac vein. The portal vein receives both *left* and *right gastric veins;* the *splenic vein* receives the *short gastric veins* and may receive both left and right *gastroepiploic veins*. The right gastroepiploic vein, however, usually drains into the superior mesenteric vein.

All the veins anastomose as freely as the arteries. The noteworthy anastomosis is in the wall of the abdominal part of the esophagus. Here, branches of the left gastric vein, part of the portal venous system, anastomose with esophageal tributaries of the accessory hemiazygos vein, part of the systemic veins. In cases of raised portal venous pressure (portal hypertension), veins subjacent to the esophageal and gastric mucosa become engorged and varicose (*esophageal and gastric varices*) and may bleed copiously.

## Lymphatics

Lymphatics in the submucosa of the esophagus and stomach form a dense anastomosing plexus. The clinical importance of the gastric lymph nodes that receive lymph from this plexus is the predilection of gastric carcinoma to spread along lymphatics. The groups of lymph nodes that first receive lymph from the stomach are named according to their location: *right* and *left gastric, right* and *left gastroepiploic, pyloric,* and *pancreaticolienal nodes.*

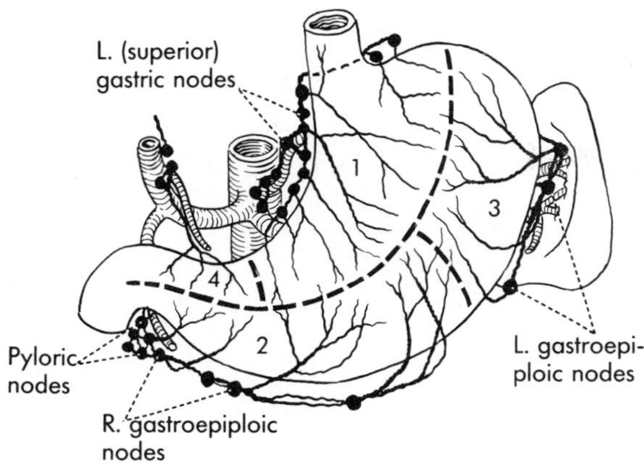


**FIG. 24-8.** The general lymphatic drainage of the stomach. The four zones described in the text are numbered, and the major groups of lymph nodes receiving this drainage are shown.

Four areas of the stomach have been defined with primary drainage to one or two of these groups of nodes (Fig. 24-8). The largest area of the stomach along its lesser curvature, including the cardia, fundus, and esophagus, drains to the *left gastric nodes* along the left gastric vessels. The second area, where carcinoma is the most frequent, includes the pyloric antrum and canal along the greater curvature and is drained by the *right gastroepiploic* and *pyloric nodes,* both located on the right part of the greater curvature. The third area of drainage is along the left part of the greater curvature to the *left gastroepiploic* and *pancreaticolienal nodes;* the latter are along the splenic artery. The fourth area, the pyloric antrum along the lesser curvature, is drained by the *right gastric nodes.* The efferent lymph from all these nodes passes through the *celiac nodes* before it enters the cisterna chyli through an intestinal lymph trunk.

The **spleen** is one of the few organs that is not pervaded by lymphatics. Tissue fluid formed in the spleen evidently freely enters the venous sinusoids. Splenic lymphatics are largely confined to the capsule and the visceral peritoneum of the organ. They drain along the splenic vessels into the pancreaticolienal nodes. It is a peculiarity of the spleen that, in comparison with the liver, cancerous metastases rarely establish themselves in it.

## Nerve Supply

Innervation of the abdominal esophagus and the stomach is provided directly by the vagi and through the subsidiary plexuses of the celiac plexus that accompany the arteries. The latter consist chiefly of sympathetic efferents and afferents.

The *anterior vagal trunk,* derived largely but not entirely from the left vagus nerve, enters the abdomen usually as a single trunk on the anterior surface of the esophagus (Fig. 24-9); sometimes, however, this trunk is double or triple. The anterior vagal trunk gives off three branches in the vicinity of the lesser curvature: the *hepatic branch* runs through the upper part of the lesser omentum and joins the plexus on the hepatic artery and portal vein; the *celiac branch* follows the left gastric artery to the celiac plexus; and the *gastric branch,* the largest of the three, follows the lesser curvature and distributes anterior gastric branches to the stomach as far as the pylorus. Innervation of the pyloric part of the stomach is reinforced by vagal fibers that run in the lesser omentum and along the hepatic artery.

The *posterior vagal trunk,* derived largely but not entirely from the right vagus nerve, enters the abdomen on the posterior surface of the esophagus and also runs along the lesser curvature (Fig. 24-10). It usually has a celiac and sometimes a hepatic branch; the continuation of the main trunk, the posterior gastric nerve, distributes its branches to the posterior surface of the stomach.

The vagus nerves largely control the secretion of acid by the parietal cells of the stomach. Since excess acid secretion is associated with the formation of peptic ulcers (found in both the stomach and the duodenum), section of the vagus trunks as they enter the abdomen is carried out to reduce the production of acid. *Vagotomy* is usually done in conjunction with resection of the ulcerated area, including the pyloric part and a portion of the body, where most of the acid-producing cells are located. More recently, a "selective vagotomy" is sometimes performed in which only the gastric branch of the vagus is cut, thus sparing the remainder of the abdominal distribution of the vagi and avoiding some of the sequelae, such as dilatation of the gallbladder, that follow total vagotomy.

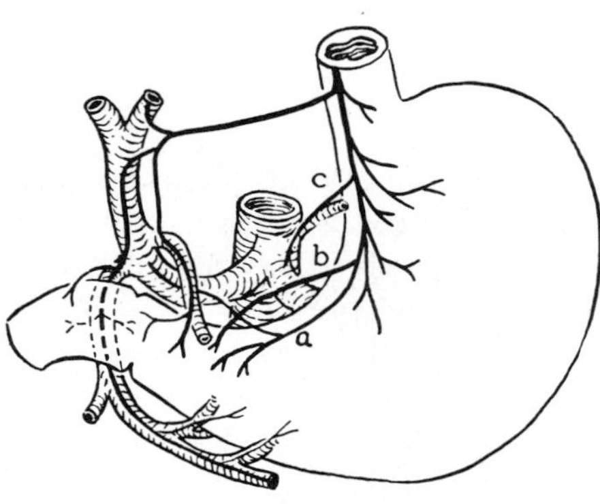


**FIG. 24-9.** Distribution of the anterior vagal trunk. **a** is its principal branch along the lesser curvature of the stomach, **b** is a branch that runs through the lesser omentum to reach the pyloric end of the stomach, and **c** is a branch running higher in the lesser omentum to join the hepatic plexus.

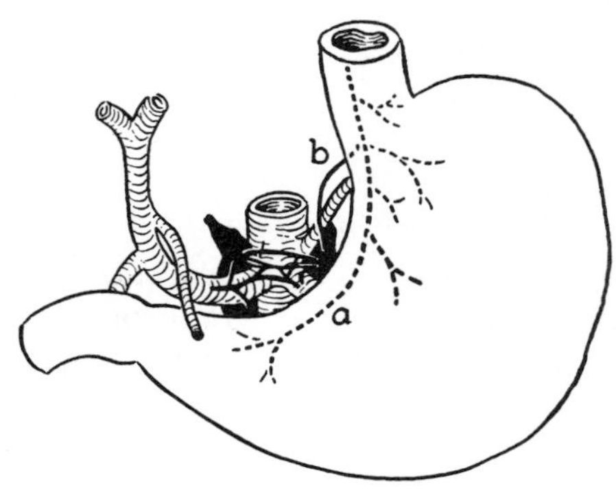


**FIG. 24-10.** Diagram of the posterior vagal trunk. **a** is its chief branch along the lesser curvature; **b** is its branch to the celiac plexus.

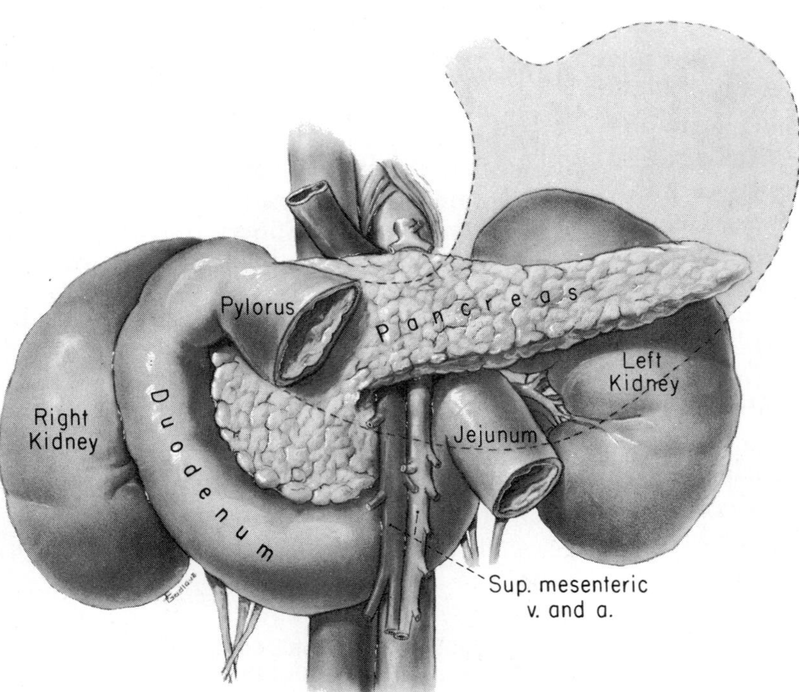


**FIG. 24-11.** Duodenum and pancreas. Except for part of the pylorus, the stomach is indicated only in outline. The uncinate process is obscured by the superior mesenteric vessels.

Both **sympathetic efferent and afferent nerves** to the stomach are derived from T6–T9 spinal cord segments. These nerve fibers are transmitted by the *greater thoracic splanchnic nerve;* preganglionic fibers relay in the celiac ganglia, and the nerves reach the stomach, esophagus, and spleen along the branches of the celiac artery. The innervation of the spleen seems to be purely sympathetic; sympathetic stimulation produces contraction of the spleen, which forces many of its red cells into the circulation.

## THE DUODENUM AND PANCREAS

### THE DUODENUM

The duodenum is the first, the shortest, and the widest part of the small intestine. It is particularly important because, in addition to contributing its own secretions (*succus entericus*) to the chyme discharged into it by the stomach, it receives bile and pancreatic juice through the common bile duct and pancreatic duct. Ancient anatomists found the length of the duodenum to be 12 fingerbreadths, hence its name, which means twelve. From the pylorus to the duodenojejunal flexure, the duodenum measures about 25 cm. Together with the pancreas, which is intimately associated with it, it is the most deeply lying portion of the alimentary tract and the least accessible to physical examination. Both organs have been thrown against the posterior abdominal wall by rotation of the gut and have become fixed there by fusion of their peritoneal covering and mesentery with parietal peritoneum (Fig. 24-11). Therefore, only the anterior surface of the duodenum and pancreas is covered in peritoneum; posteriorly, they are devoid of peritoneum.

For descriptive purposes, the duodenum is divided into four parts: the *superior, descending, horizontal,* and *ascending* portions, which are also referred to by number. A superior and inferior *duodenal flexure* separates the second, or descending, part from the first part above and the horizontal part below, respectively. Owing to the variation of the angle in the flexures, the overall shape of the duodenum varies, but it always resembles an almost complete circle that encloses the head of the pancreas (Fig. 24-12).

**FIG. 24-12.** Some variations in the shape of the duodenum; these depend in part on variations in the position of the pylorus **(P)** and of the movable superior part **(1)** of the duodenum. **2** is the descending part; **3** and **4** are the horizontal and ascending parts. These four parts are clearly distinguishable in **A**, less so in **B** and **C**, and not at all in **D**.

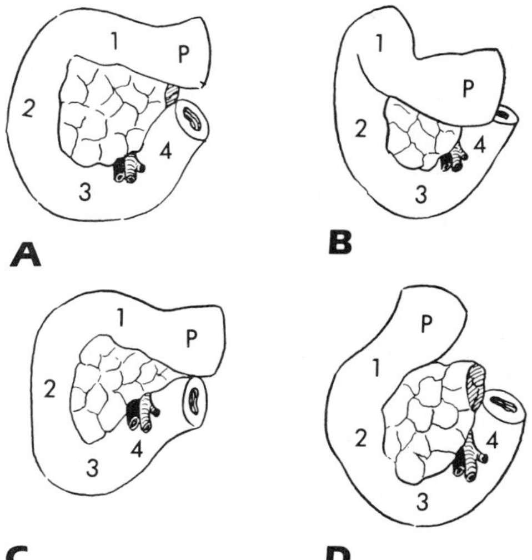


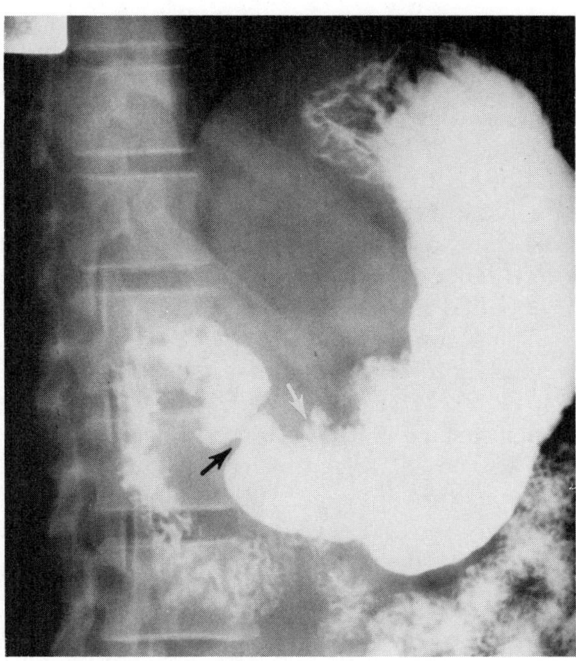

**FIG. 24-13.** The duodenum shown in an oblique view following a barium meal. The vertebrae are seen from the right side; the stomach overlaps anteriorly the duodenojejunal junction, which is located at a level above the lesser curvature. Note the direction of the four parts of the duodenum in relation to the vertebrae and also the transverse mucosal folds. **Black arrow,** pylorus; **white arrow,** duodenojejunal flexure.

At its commencement, the *first* or **superior part** of the duodenum forms the *duodenal ampulla,* or cap, into which protrudes the pylorus (Fig. 24-13). The rest slopes upward and posteriorly along the right side of L1 vertebral body. Approximately the medial half of the first part of the duodenum forms the inferior margin of the epiploic foramen, and attached to its upper margin is the free edge of the lesser omentum; to its lower margin is attached the greater omentum. These peritoneal attachments lend the first part of the duodenum a considerable degree of mobility not possessed by more distal portions. The liver overlaps the first part anteriorly, and the gallbladder is in contact with it. Posterior to it is the pancreas, the portal vein, the common bile duct, and the gastroduodenal artery, all of which cross it vertically, being partially embedded in the pancreas.

The **second part** of the duodenum descends vertically along the right side of the bodies of L1, L2, and L3 vertebrae and lies with its posterior surface on the hilum of the right kidney and its vessels. The anterior surface is covered with peritoneum except along the attachment line of the transverse mesocolon, which crosses the descending portion of the duodenum at its midpoint (Fig. 24-14). The descending duodenum is overlapped above the transverse mesocolon by the liver and below it by the transverse colon and coils of jejunum. The head of the pancreas is in direct contact with the medial surface of the descending duodenum. The common bile duct embedded in the pancreas descends parallel with the posteromedial surface of the duodenum; it is joined by the main pancreatic duct before the two pierce the duodenal wall and open into its lumen at the tip of the *major duodenal papilla* (Fig. 24-35). A *minor duodenal papilla,* located more superiorly, may mark the opening of the accessory pancreatic duct into the duodenum. These papillae are usually concealed by the circular folds of the mucosa present throughout the small intestine.

The third or **horizontal part** of the duodenum crosses to the left in front of L3 vertebra and in so doing passes over the inferior vena cava and aorta. In front of the aorta, it becomes continuous with the short **ascending** or **fourth part,** which returns to L2 vertebra on the left side of the aorta, where the sharp duodenojejunal flexure marks its junction with the jejunum. The anterior surface of both the third and fourth parts is covered by peritoneum, except where the root of the mesentery crosses them (Fig. 24-14). The superior mesenteric artery enters and the vein leaves the root of the mesentery over the third part of the duodenum. Both vessels cross the horizontal segment of the duodenum anteriorly.

**FIG. 24-14.** Some relations of the duodenum and pancreas, anterior view. Most of the vessels related to the duodenum and pancreas are not shown. The common bile duct, the hepatic artery, and the portal vein are in the hepatoduodenal ligament (cut) above the first part of the duodenum; the superior mesenteric vessels emerge between the pancreas and the duodenum and enter the root of the mesentery.

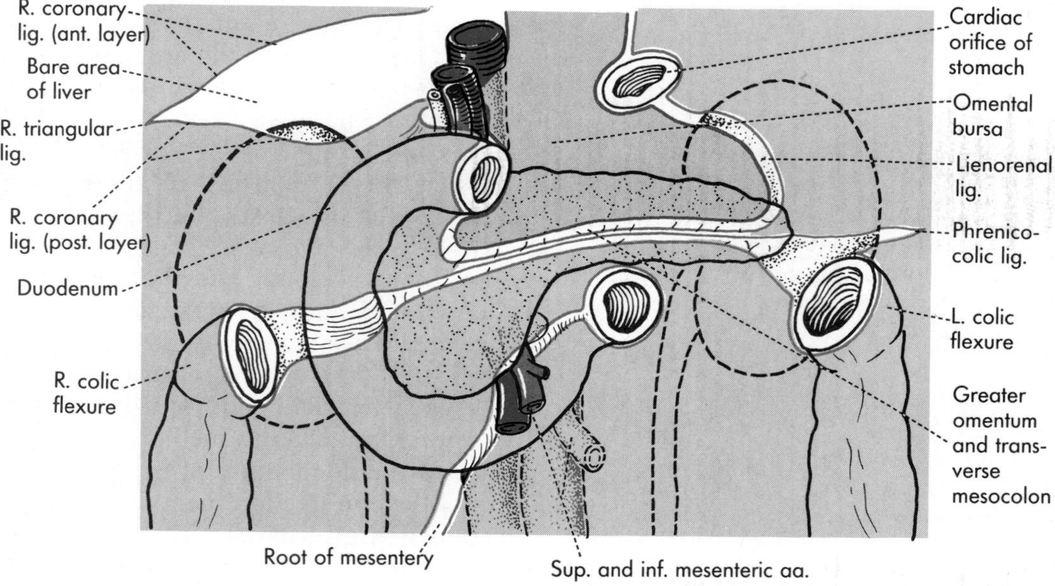

**Superior Mesenteric Artery Syndrome.** The rather unusual relationship of the superior mesenteric vessels to the horizontal part of the duodenum comes about because of the rotation of the gut. The superior mesenteric vessels form the axis around which the midgut rotates. The most proximal portion of the midgut, represented by the distal half of the duodenum, is carried behind this axis from the right side of the abdomen to the left side. This is the event that lays the horizontal portion of the duodenum across the inferior vena cava and aorta and places it posterior to the superior mesenteric vessels. These vessels may compress the duodenum, leading to distention of the proximal duodenum and the stomach (*superior mesenteric artery syndrome*). The condition is manifest by abdominal pain, nausea, and vomiting, and it is not easy to diagnose.

**Suspensory Muscle of the Duodenum.** The posterior surface of the duodenum and pancreas is attached to the posterior abdominal wall by loose connective tissue. This tissue plane may be opened up and the two organs mobilized by incising the peritoneum along the groove between the right kidney and the descending duodenum. This approach provides access to the pancreas and the portion of the bile duct embedded in it. The duodenum is secured to the posterior abdominal wall by a fibromuscular ligament known as the *suspensory muscle of the duodenum,* described in 1853 by Treitz. This ligament is variable and has two parts: one derived from the diaphragm, which contains striated muscle, and the other from the duodenal wall, which contains smooth muscle. The two parts blend with each other in the region of the celiac artery. It is thought that the presence of the suspensory muscle, which is most constant and best developed in the region of the fourth part of the duodenum, accounts for the acute angle of the duodenojejunal flexure.

**Paraduodenal Recesses.** Some peritoneal folds may be raised up on the left side of the duodenojejunal flexure (paraduodenal folds), and the pockets, or **paraduodenal recesses,** thus created present a potential danger for the entrapment of a loop of bowel (Fig. 24-15). Such *internal hernias* are quite rare, but when they have to be reduced, the surgeon must be mindful of the inferior mesenteric vein and the left colic artery, which are related to these folds.

**Development and Anomalies.** The developmental events that established the topographical relationships of the duodenum have already been dealt with (p. 635). Most positional abnormalities of the duodenum are related to malrotation of the gut. Excluding congenital pyloric obstruction and imperforate anus, the duodenum is the most common

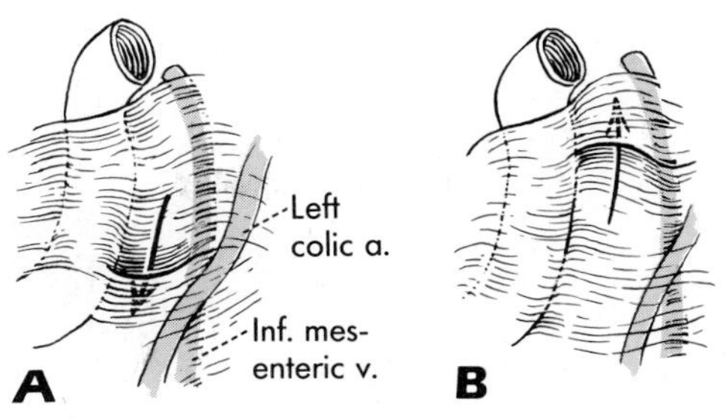


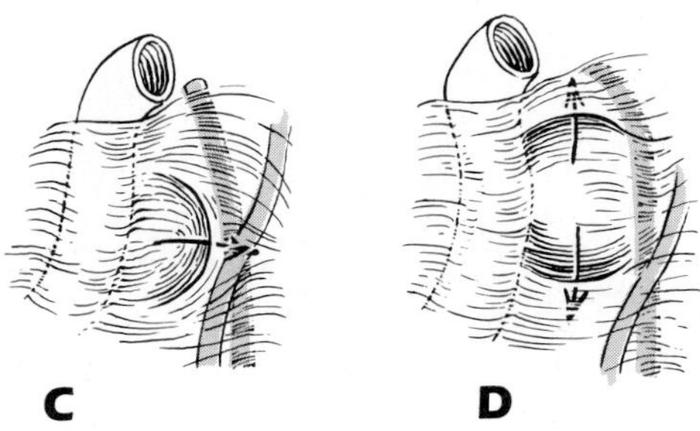


**FIG. 24-15.** Some of the folds and recesses that may be associated with the duodenum. **A** shows an inferior duodenal fold and recess; **B,** a superior fold and recess; **C,** a paraduodenal recess; and **D,** combined superior and inferior folds and recesses (these also may be associated with a paraduodenal fold and recess).

site of congenital obstruction of the digestive tract. The explanation is the complex developmental history of the duodenal lumen. Subsequent to the formation of the primitive endodermal gut tube, the lumen of the duodenum becomes occluded by epithelial proliferation. Recanalization of the duodenum is achieved by cavitation and resorption of the epithelial core, which progresses parallel with the differentiation of the surrounding splanchnic mesoderm into the muscular and connective tissue components of the duodenal wall. If a given segment remains solidly epithelial beyond a critical period, it will be replaced later by mesoderm and its derivative tissues, creating a stenosis or complete obstruction known as **atresia.** The most common site of obstruction is at the level of the major duodenal papilla. If complete, or severe, the obstruction requires surgical correction in the neonate.

**Duodenal diverticula** are diagnosed much later in life, and they may or may not be congenital. They are thought to be caused by herniation of the duodenal lining through gaps in the muscle coat where blood vessels or ducts pierce the wall. Most diverticula occur along the concavity of the second and third parts of the duodenum. The majority are

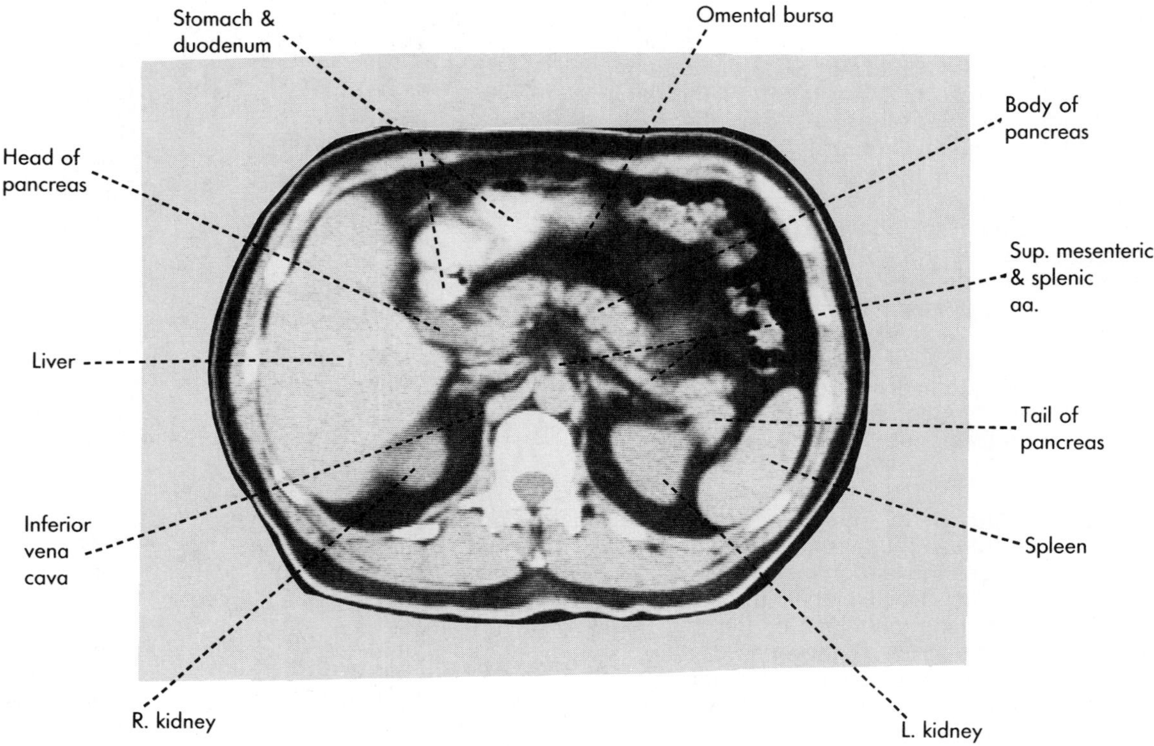


**FIG. 24-16.** The relations of the pancreas shown on a CT scan section of the abdomen. Compare with Figure 24-5.

**FIG. 24-17.** The parts of the pancreas, and some of its vascular relations, anterior view.

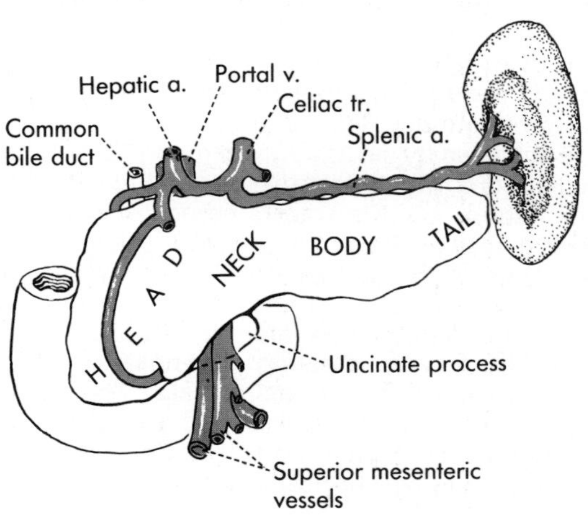


found close to the entrance of the common bile duct and chief pancreatic duct into the duodenal wall.

## THE PANCREAS

The pancreas is a large, flat, finely lobulated gland associated with the duodenum (Fig. 24-11). It produces both exocrine and endocrine secretions. The former, discharged into the duodenum, contain some of the most important enzymes for digestion; the latter, discharged into the venous system, are essential for the regulation of carbohydrate metabolism.

The pancreas is located in the upper part of the abdomen, hidden by many organs. It is not accessible by physical examination. It lies retroperitoneally, molded on the posterior abdominal wall, mostly behind the lesser sac. The pancreas extends from the right side of L1, L2, and L3 vertebrae, over the median em-

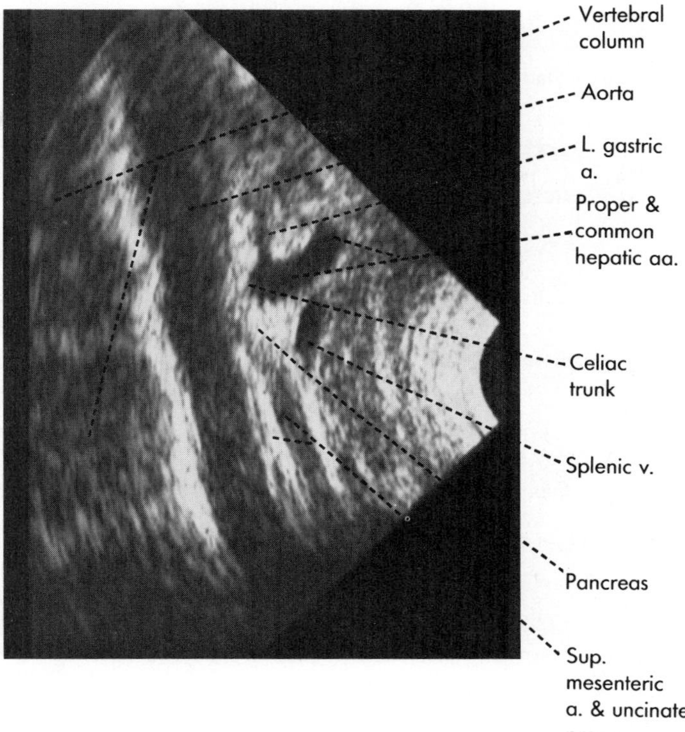


**FIG. 24-18.** Ultrasound imaging of some of the vascular relations of the pancreas. The cone of ultrasound emanates from the generator placed on the anterior abdominal wall on the right of the figure.

inence created by the bodies of these vertebrae, with the inferior vena cava and aorta in front of them, to the left as far as the hilum of the spleen. In so doing, the pancreas projects a sinuous profile in a transverse section (Fig. 24-16) that is not evident from the anterior view (Fig. 24-11).

The pancreas consists of a head, neck, body, and tail (Fig. 24-17). The broad, flat **head** fits snugly into the curve of the duodenum. Toward the left, the upper part of the head is continuous with the neck; from the lower part of the head toward the left projects the uncinate process. The **neck** is constricted by the superior mesenteric vessels, which lie in the *pancreatic incisure,* a deep groove on the posterior surface of the neck. The **uncinate process** is inferior to the neck of the pancreas and lies largely behind the superior mesenteric vessels. These vessels are trapped, so to speak, between the neck above and anteriorly and the uncinate process below and posteriorly (Fig. 24-18). The **body** lies above the duodenojejunal flexure and on the left kidney, over which it tapers into the **tail** (Fig. 24-11), which often extends into the lienorenal ligament.

**Relations.** Both anterior and posterior surfaces of the pancreas are related to many organs. The base of the transverse mesocolon attaches across the head and along the lower margin of the neck and body (Fig. 24-14). Consequently, most of the anterior surface faces into the omental bursa and, through the bursa, is related to the stomach. Above the lesser curvature, the lesser omentum, and through it the liver, may also be in contact with the pancreas (Fig. 24-11). The first part of the duodenum is either above or on the anterior surface of the head (Fig. 24-12). The lower part of the head and the narrow inferior surface of the neck and body face into the infracolic compartment (Fig. 24-14) and are in contact with loops of bowel.

The posterior surface of the head lies on the hilum of the right kidney and its vessels, the portal vein and the inferior vena cava. Close to the duodenum, the upper part of the poste-

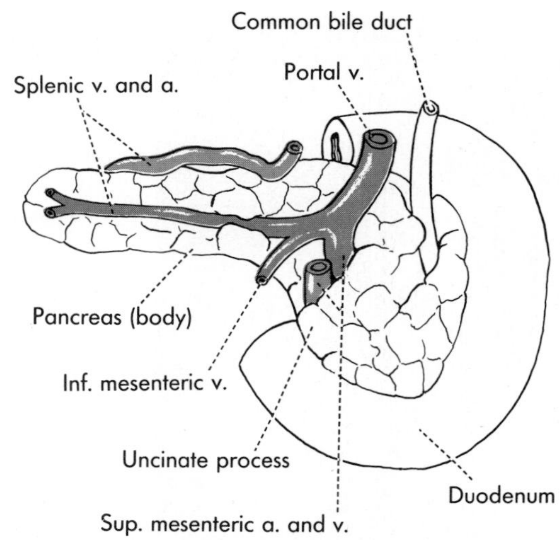


**FIG. 24-19.** Posterior view of the pancreas.

rior surface of the head is grooved or tunneled by the common bile duct (Fig. 24-19). The neck and uncinate process lie in front of the aorta; the body crosses the left kidney just above its hilum and also overlaps the left suprarenal gland and the right crus of the diaphragm.

In addition to the superior mesenteric vessels, the celiac trunk and two of its branches are also intimately associated with the pancreas (Fig. 24-17). The celiac trunk originates at the upper margin of the pancreas and may often be buried in pancreatic tissue; the common hepatic artery runs to the right along the upper margin of the neck and the head; the splenic artery runs to the left along the upper margin of the body and crosses to the front of the tail. The splenic vein is behind the pancreas and is joined there by the inferior mesenteric vein. The confluence of the splenic and superior mesenteric veins forms the portal vein on the posterior surface of the pancreas (Fig. 24-19). The tip of the tail makes contact with the spleen at its hilum in many cases and must not be damaged during splenectomy.

## The Pancreatic Ducts

The exocrine secretions of the pancreas are collected by two ducts, the chief and accessory pancreatic ducts; both drain into the second part of the duodenum (Fig. 24-20). The **chief pancreatic duct** runs the length of the pancreas, collecting radicles from the entire tail and body and from the posteroinferior part of the head, including the uncinate process. At the concave border of the duodenum, the chief duct joins the common bile duct and with it enters the duodenum (Figs. 24-32 and 24-35). The **accessory pancreatic duct** drains the anterosuperior part of the head and empties independently into the second part of the duodenum. Its opening is 2 cm above and somewhat anterior to the joint opening of the chief pancreatic and common bile duct at the major duodenal papilla.

### Development

The pancreas develops from the union of a dorsal and a ventral primordium (Fig. 24-21A). The **dorsal primordium** gives rise to the upper part of the head, the neck, body, and tail; from the **ventral primordium** develops the lower part of the head, including the uncinate process.

The dorsal pancreas arises as a bud from the dorsal side of the duodenum and grows into the dorsal mesentery toward the spleen (Fig. 24-21B). The ventral pancreas arises from the base of the liver diverticulum, which also gives rise to the gallbladder.

The two primordia become approximated to each other by differential growth limited to the left circumference of the duodenum, which transports both the common bile duct and the ventral pancreas to the posterior aspect of the duodenum. As a consequence of this migration, the ventral pancreas is swung against the right side of the dorsal mesentery and overlaps the lower border of the dorsal pancreas contained in the mesentery (Fig. 24-21C). Between the two pancreatic primordia, within the mesentery, are the vitelline arteries and veins of the midgut, from which the superior mesenteric vessels develop. The topographical relationship of these vessels to the neck and uncinate process of the pancreas thus becomes intelligible (Fig. 24-18).

Development is completed by rotation of the gut, which swings the duodenum to the right, and by the subsequent fusion of the right leaf of its mesentery to the posterior parietal peritoneum. Anastomosis between the dorsal and ventral pancreatic ducts rechannels the main stream of pancreatic secretions to the ventral pancreatic duct: although most of the chief pancreatic duct is derived from the duct of the dorsal pancreas, the ventral pancreatic duct forms the terminal portion of

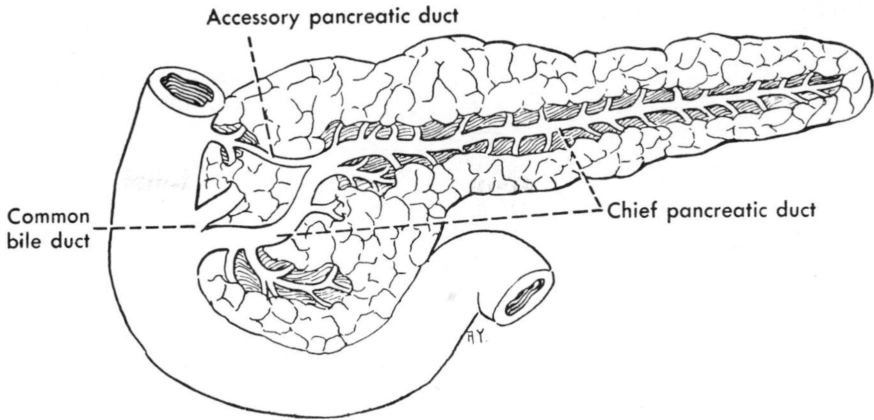


**FIG. 24-20.** The common arrangement of the pancreatic ducts.

**FIG. 24-21.** Development of the pancreas. In **A**, the ventral diverticulum of the duodenum has differentiated into the liver, gallbladder, and ventral pancreas (unlabeled); the dorsal diverticulum enclosed in the dorsal mesentery forms the dorsal pancreas. In **B**, the ducts for bile and pancreatic secretions have been defined. The position of the portal vein and the superior mesenteric vessels is indicated. In **C**, differential growth involving the left half of the circumference of the duodenum displaces the ventral pancreas and the common bile duct posteriorly and to the left, trapping the superior mesenteric vessels between the dorsal and ventral pancreas. In **D**, the two pancreatic primordia fuse; the ducts of the dorsal and ventral pancreas (shown in black and white, respectively) anastomose, and the chief duct will empty its contents into the duodenum through the duct of the ventral pancreas. The proximal portion of the dorsal pancreatic duct becomes the accessory pancreatic duct.

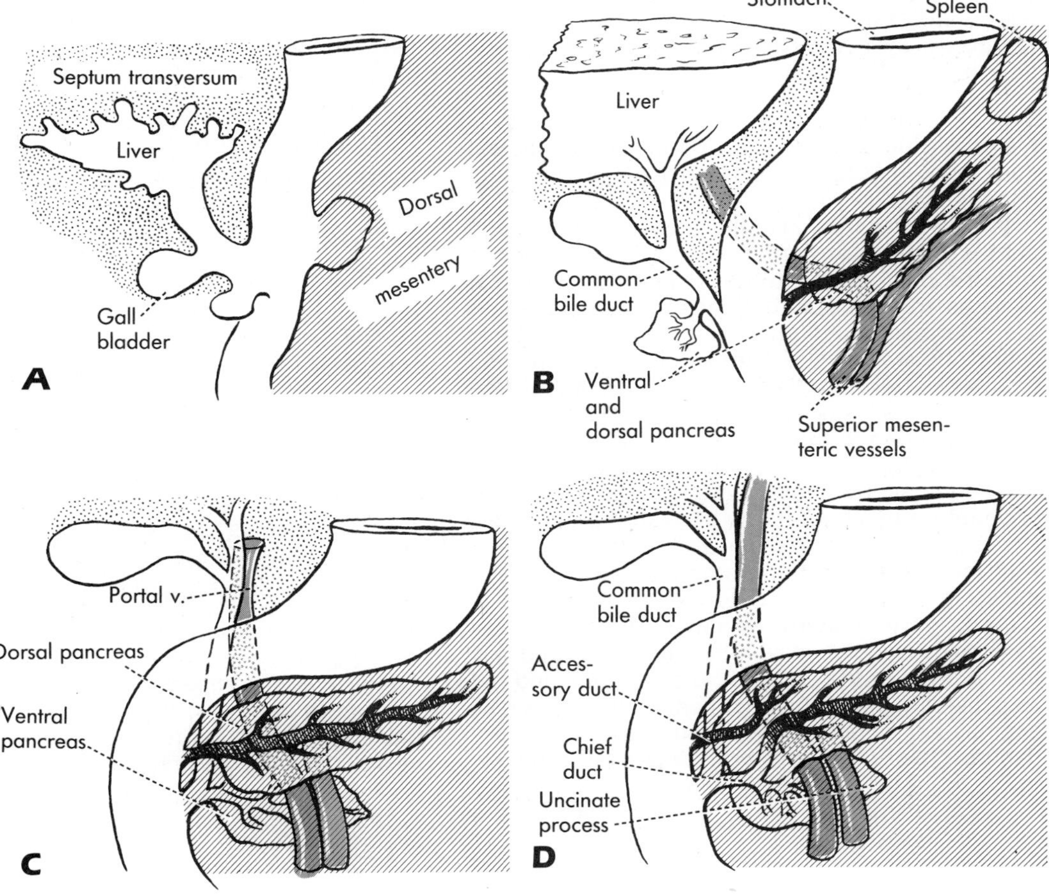

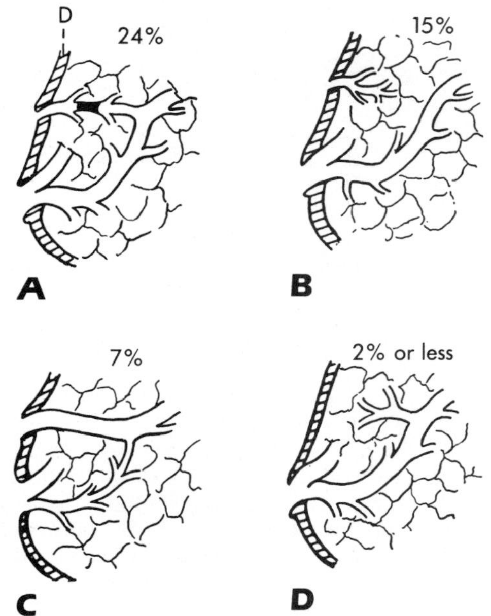


**FIG. 24-22.** Variations in the duct system of the pancreas. The duodenal wall is shaded and the duodenal lumen is on the left of each figure. **A** shows occlusion of the accessory duct; **B** shows discontinuity between the two ducts; **C** shows so-called inversion of the ducts, where the accessory duct is larger; and **D** shows absence of the duodenal end of the accessory duct. The percentages indicate the approximate incidence of each of these conditions.

the chief duct; the terminal portion of the dorsal duct gives rise to the accessory pancreatic duct (Fig. 24-21D).

A common variation in the accessory pancreatic duct is whether or not it has a patent connection with the chief pancreatic duct and therefore can serve as an accessory drainage to the pancreas if the chief pancreatic duct is occluded close to or within the wall of the duodenum. The combined results of several studies indicate that in about 40% of adults the accessory pancreatic duct has no patent connection to the chief duct; in about 7%, the accessory duct is as large as, or larger than, the chief duct and therefore appears as the direct continuation of this duct in the body of the pancreas. Occasionally, the accessory duct does not connect with the duodenum but drains only into the chief duct (Fig. 24-22).

**Accessory** or **aberrant pancreatic tissue** may develop in association with the duodenum and less frequently with the stomach, the jejunum, and the ileal diverticulum. A band of pancreatic tissue may encircle and constrict the second part of the duodenum. The cause of this congenital anomaly, known as *annular pancreas,* remains unknown, although several theories of its embryogenesis have been proposed.

The endocrine cells of the pancreas, which form the *pancreatic islets,* are believed to differentiate and sequestrate from the acinar cells of the organ quite early in development. Failure of some of the acini of the exocrine pancreas to connect during development with the duct system may give rise to *congenital pancreatic cysts.*

## VESSELS AND NERVES

### Blood Supply

The duodenum and pancreas are supplied by branches of the celiac trunk and the superior mesenteric artery. The first part of the duodenum has a poor blood supply furnished by small branches of the gastroduodenal artery; the second, third, and fourth parts of the duodenum and the head of the pancreas are served by two parallel arterial arcades located in the concavity of the duodenum on the surface of or embedded in the head of the pancreas. The arcades are fed from above by branches of the gastroduodenal artery and from below by the superior mesenteric artery. The neck, body, and tail of the pancreas are supplied by the splenic artery.

The **first part of the duodenum** is supplied by the *supraduodenal* and *retroduodenal arteries,* branches of the gastroduodenal (Fig. 24-23). These vessels may be given off by the gastroduodenal artery directly or by one or other of its named branches.

The **arterial arcades,** on the head of the pancreas, are made up of four freely anastomosing *pancreaticoduodenal arteries;* anterior and posterior vessels above, given off by the gastroduodenal artery, and anterior and posterior vessels below, given off by the superior mesenteric (Fig. 24-24). Either the anterior or the posterior arcades may be incomplete, but for such small channels, they are remarkably constant.

The *anterior superior pancreaticoduodenal artery* arises as one of the terminal branches of the gastroduodenal artery after this artery has descended behind the first part of the duodenum. The

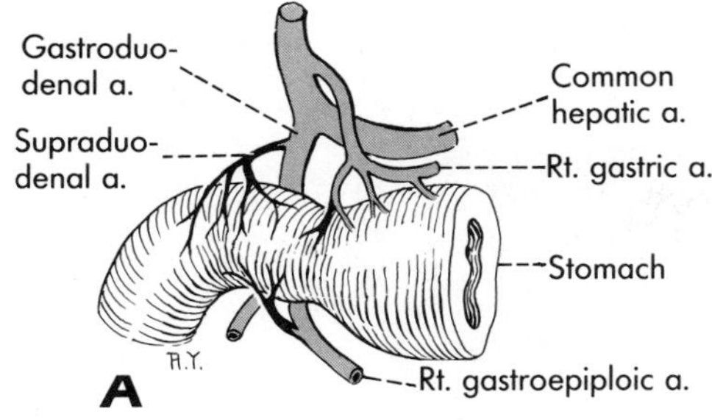


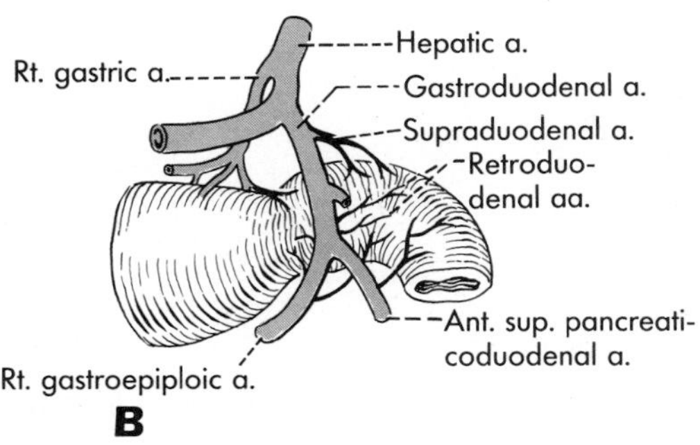


**FIG. 24-23.** Anterior **(A)** and posterior **(B)** views of the blood supply to the first part of the duodenum.

**FIG. 24-24.** The pancreaticoduodenal arcades and the blood supply to the duodenum and the head of the pancreas. The first part of the duodenum has been elevated and turned to the left, revealing its posterior surface.

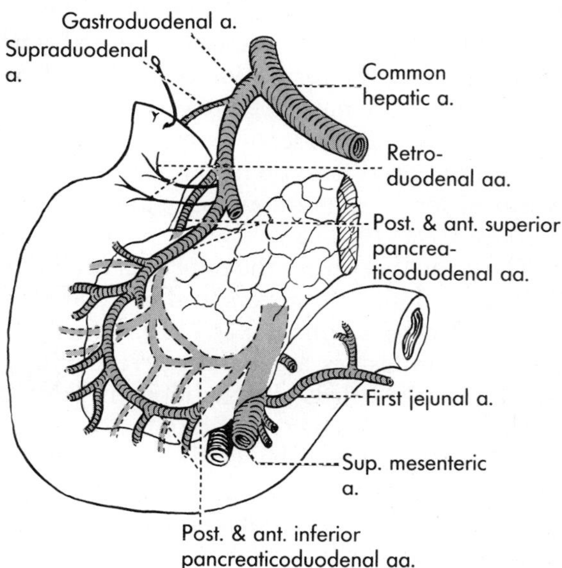


other terminal branch, the right gastroepiploic, has already been traced to the greater curvature of the stomach. The *posterior superior pancreaticoduodenal artery* is given off by the gastroduodenal before it divides into its terminal branches. The posterior superior pancreaticoduodenal artery runs inferiorly along the left side of the common bile duct.

The two *inferior pancreaticoduodenal arteries* arise by a common stem (the first branch given off by the superior mesenteric artery), which divides into an anterior and a posterior vessel. These ascend on or in the head of the pancreas to anastomose with the respective arteries derived from the gastroduodenal. Both anterior and posterior arcades give off branches into the head of the pancreas and a series of straight vessels to the second, third, and fourth parts of the duodenum. The blood supply of the fourth part and that of the duodenojejunal flexure is augmented by duodenal branches of the *superior mesenteric artery* and the first *jejunal artery*.

The rest of the pancreas beyond the head is supplied by a series of **pancreatic branches** given off by the splenic artery (Fig. 24-25). The most proximal of these vessels anastomose with the arteries that supply the head. The named vessels are large and constant: the *dorsal pancreatic artery* is usually given off by the splenic close to its origin from the celiac trunk, or it may arise from the trunk itself or from the common hepatic artery; the *great pancreatic artery* (*a. pancreatica magna*) enters the middle of the body of the pancreas; the *caudal pancreatic artery*, or arteries to the tail, is from a branch of the splenic or from the left gastroepiploic artery. Numerous other splenic branches to the pancreas anastomose freely with other pancreatic vessels. One of these anastomotic arteries is sometimes designated as the *inferior pancreatic*.

The **venous drainage** of the duodenum and pancreas follows, in general, the arterial supply. There are anterior and posterior venous arcades that parallel the arterial arcades and a number of small twigs from the first part of the duodenum that empty into pancreaticoduodenal veins or the right gastroepiploic or the portal vein. One of these veins, the *prepyloric vein*, was discussed earlier, as a landmark for the pylorus (p. 626).

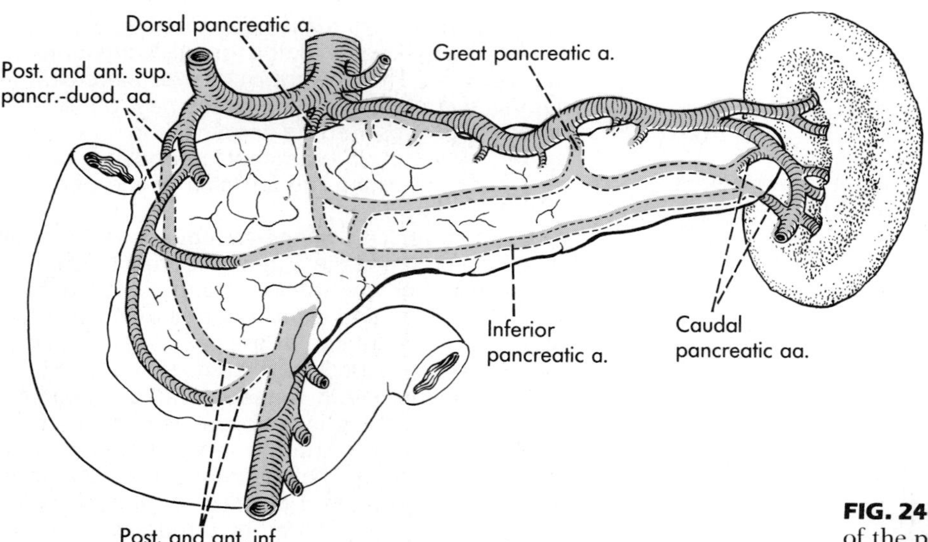


**FIG. 24-25.** The chief arteries of the pancreas and their anastomoses.

The venous drainage of the body and tail of the pancreas is by a variable number of veins that empty into the splenic vein as this lies embedded in the posterior surface of the pancreas.

## Lymphatics

The lymphatics of the duodenum and pancreas tend to follow the blood vessels. Those from the duodenum and the head of the pancreas drain, in part, into celiac and superior mesenteric lymph nodes. Some lymphatics also terminate in lumbar lymph nodes. Anterior lymphatics drain also into the pyloric nodes already mentioned in connection with the lymphatic drainage of the stomach. From the body and tail of the pancreas, there is drainage into the pancreaticolienal nodes along the splenic vessels, which send their efferents to the celiac nodes. Carcinoma of the head of the pancreas or the duodenum may, therefore, involve three major groups of lymph nodes: celiac, superior mesenteric, and upper lumbar.

## Nerve Supply

The nerves to the duodenum and pancreas are derived from the **celiac** and **superior mesenteric plexuses,** which are continuous with each other. The nerves to the duodenum follow the vessels, and those to the pancreas (*pancreatic plexus*) arise directly from the two major plexuses and enter the posterior surface of the pancreas.

It has been claimed that the sympathetic fibers to the pancreas end exclusively on the blood vessels but that the parasympathetic vagal fibers end in connection with both the acinar cells and the cells of the islets. The physiology of the vagal innervation of the pancreas is not well understood, for although vagal fibers are presumably concerned with some phase of the formation or release of the pancreatic enzymes, the pancreas is largely controlled by a duodenal hormone, secretin. Vagotomy produces no clear effect on the composition or the amount of pancreatic secretions.

The pain fibers from the pancreas run in the thoracic splanchnic nerves and are conveyed to spinal cord segments T6–T10. Pancreatic pain is manifest as a constant severe pain in the upper two-thirds of the abdomen. Pancreatic pain may also be referred to the back over T10–L2 vertebrae. The explanation of this may be related to the somatic innervation by the spinal cord segments of the structures upon which the pancreas lies or to the excitation of the somatic receptor neurons in the spinal cord by relay of the visceral pain afferents.

## THE LIVER, GALLBLADDER, AND BILIARY DUCTS

### THE LIVER

The liver (*hepar*) is the largest and most vascular organ in the body. In addition to being the chief site of intermediary metabolism, the liver secretes bile and a number of hormones, synthesizes serum proteins and lipids, and processes not only the products of digestion but most endogenous and exogenous substances, including toxins and drugs, that enter the circulation. It participates in the elimination of senescent cells and particulate matter from the bloodstream and during much of fetal life produces hematopoietic cells of all types.

**Position.** The liver lies largely under cover of the costal cartilages, occupying much of the upper abdomen, especially on the right. It extends from the right hypochondrium, which it fills almost completely, across the epigastrium into the left hypochondrium as far as the left lateral line (Fig. 24-26).

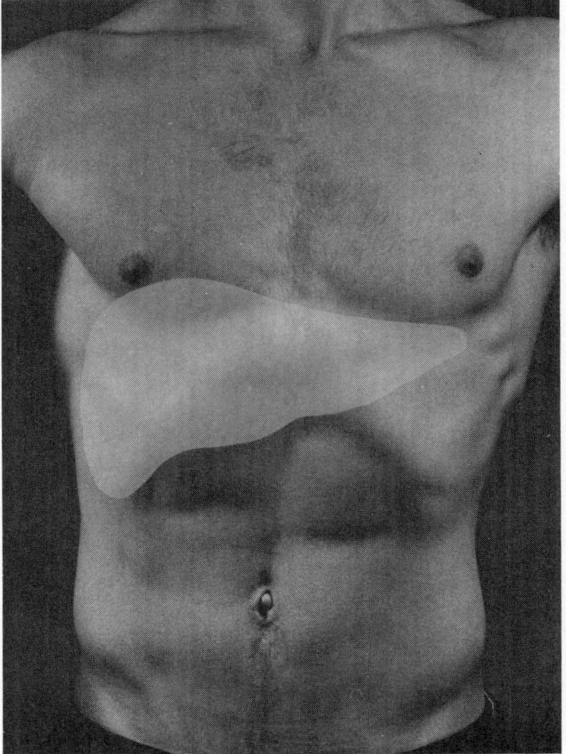

**FIG. 24-26.** The surface markings of the liver.

The upper extent of the liver under the right dome of the diaphragm is readily demonstrated by the transition of a resonant percussion tone into a dull one. This occurs just below the nipple in the 4th intercostal space when the subject is supine and about one intercostal space lower in the standing or sitting position. In the midline, the upper limit of the liver is behind the xiphisternal joint and along the left lateral line, one or two intercostal spaces lower than on the right side. The liver is difficult to percuss on the left owing to the gas in the gastric fundus, which the liver overlaps.

The lower limit of the liver is sheltered by the right costal margin as far medially as the tip of the 9th costal cartilage, then it crosses the epigastrium obliquely, less than a hand's breadth below the xiphisternal joint; beyond the left costal margin, the lower border of the liver tapers upward into the tongue-shaped left lobe.

The liver follows the excursions of the diaphragm: ascent and descent of its lower edge during respiratory movements may be detected by an experienced examining hand provided the abdominal musculature is relaxed. In the newborn and the young infant, the lower border of the liver extends below the costal margin.

**Size, Shape, and Surfaces.** The liver weighs 1 kg to 2 kg, contributing close to one-fortieth of the total body weight. In the newborn and infant, its relative size is considerably greater.

The liver is an irregular, wedge-shaped organ (Fig. 24-27), on which only two surfaces and one margin can be defined distinctly: a diaphragmatic and a visceral surface and an inferior margin.

The **diaphragmatic surface** includes smooth *peritoneal areas* that face upward, anteriorly and to the right (Fig. 24-27) and an irregular *bare area* devoid of peritoneum facing posteriorly (Fig. 24-28).

The most notable features on the diaphragmatic surface are the inferior vena cava and the peritoneal ligaments that connect the liver to the diaphragm. The *inferior vena cava* is embedded in the liver in a deep *sulcus* located in the left portion of the bare area (Fig. 24-28). This sulcus is roofed over in most cases by fibrous tissue, called the *ligament of the inferior vena cava,* which may contain hepatic tissue converting the sulcus into a tunnel. The *peritoneal ligaments* are the falciform ligament and the left and right coronary and triangular ligaments, which have already been described in Chapter 23 (Figs. 24-27 and 24-28).

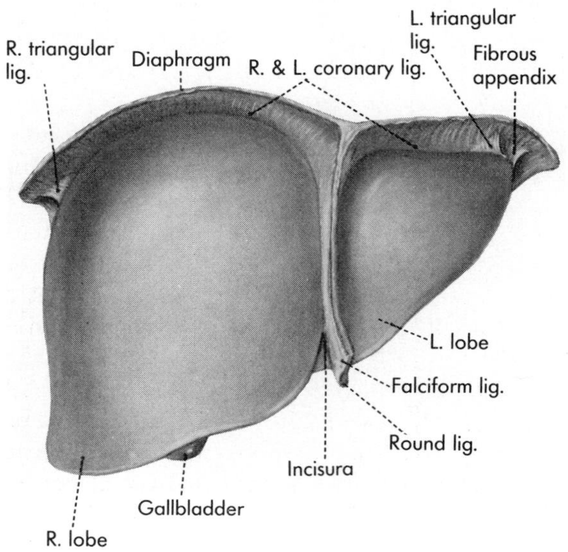


**FIG. 24-27.** Anterior view of the liver and its peritoneal ligaments. (Popper H, Schaffner F: Liver: Structure and Function. New York, McGraw-Hill, 1957)

The relatively flat **visceral surface,** also covered by peritoneum, is divided into several areas by deep fissures and impressions adjacent viscera have made on it (Fig. 24-29). This surface faces downward as well as posteriorly and is separated in front from the diaphragmatic surface by the sharp *inferior margin* and in the back by the posterior lamina of the coronary ligament (p. 601; Fig. 24-28).

The most notable features of the visceral surface are the gallbladder, the fissure for the ligamentum teres hepatis (round ligament), the fissure for the ligamentum venosum, and the porta hepatis (Fig. 24-29). The *gallbladder* lies in an elongated *fossa* that runs from the inferior margin of the liver in front toward the inferior vena cava in the bare area and leads into the porta hepatis. The gallbladder is retained in the fossa partly by the continuity of the hepatic peritoneum across the inferior surface of it. The *ligamentum teres* continues from the free edge of the falciform ligament toward the porta hepatis, buried in its fissure,

**FIG. 24-28.** Posterior view of the liver and its peritoneal attachments, illustrated schematically. **A** shows the posterior diaphragmatic and visceral surfaces, with lines of reflection of peritoneum and ligaments indicated; **B** shows attachments to the diaphragm.

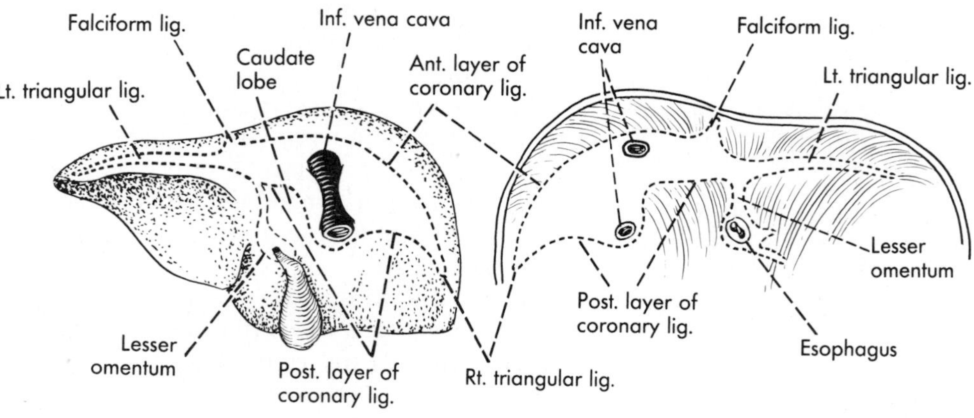

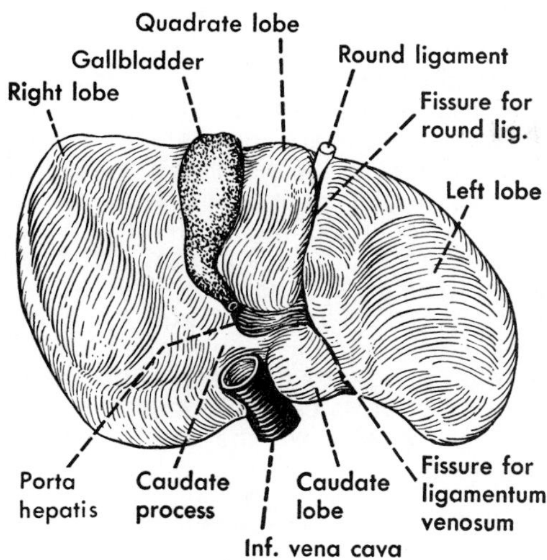


**FIG. 24-29.** The visceral surface of the liver continuous with the posterior diaphragmatic surface.

which is more or less parallel with the gallbladder. Beyond the porta hepatis, in line with the fissure for the ligamentum teres, is a second fissure in which is buried the *ligamentum venosum.*

The **porta hepatis** is a transverse fissure located between the neck of the gallbladder and the junction of the fissures for the ligamenta teres and venosum. Through the porta hepatis, the hepatic artery and portal vein enter the liver and the hepatic duct leave it. Just below the porta, the hepatic ducts are joined by the cystic duct from the gallbladder. Right and left leaves of the lesser omentum, attached in the fissure for the ligamentum venosum, prolong their hepatic attachment to the anterior and posterior lips of the porta hepatis, respectively. At the right limit of the porta, the two leaves become continuous with one another as they form the uppermost portion of the free edge of the lesser omentum (hepatoduodenal ligament), embracing in it the structures that enter and leave the porta hepatis. These structures in the hepatoduodenal ligament are often referred to as the *hepatic pedicle.*

**Lobes and Segments.** On the diaphragmatic surface, the attachment of the falciform ligament of the liver demarcates the right lobe from the left (Fig. 24-27). The right lobe, which forms the base of the wedge-shaped liver, is approximately six times the size of the tongue-shaped left lobe. On the visceral surface, the two fissures, the porta hepatis, and the gallbladder define four lobes: the **right lobe** to the right of the gallbladder; the **left lobe** to the left of the fissures of the ligamenta teres and venosum; the **quadrate lobe** between the gallbladder and the fissure for the ligamentum teres in front of the porta hepatis; and, behind the porta, the **caudate lobe** between the fissure for the ligamentum venosum and the inferior vena cava (Fig. 24-29).

These lobes, demarcated by surface features, are useful landmarks, but they do not correspond to the structural units or hepatic segments that are established by the intrahepatic branching of the bile ducts, to which the branches of the hepatic artery and the portal vein conform. Although not as well defined as the bronchopulmonary segments, the hepatic segments can be demonstrated by injection techniques and corrosion casts of the bile ducts and hepatic blood vessels.

There are **four hepatic segments,** *anterior, posterior, lateral,* and *medial,* and each segment is divided into an *upper* and a *lower area* (Fig. 24-30). The *anterior and posterior segments* are on the right; they are drained by the right hepatic duct and served by the right branches of the hepatic artery and portal vein and, therefore, constitute the "true" right lobe. The *medial and lateral segments* constitute the "true" left lobe. These two true lobes are of equal size and weight, and the junction between them falls in a nearly sagittal plane, passing through the fossa of the gallbladder and the sulcus for the inferior vena cava. This plane is some distance to the right of the falciform ligament.

There is no identifiable demarcation between the anterior and posterior segments of the right lobe; the fissures of the ligamenta teres and venosum mark the junction between the lateral and medial segments of the left lobe. The caudate and quadrate lobes are incorporated into the upper and lower areas of

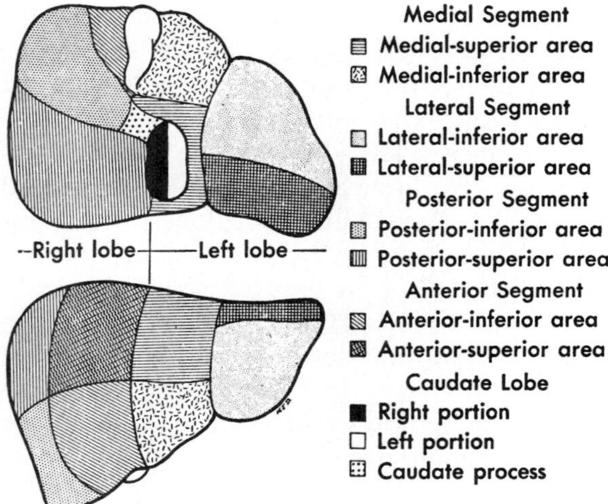


**FIG. 24-30.** Divisions of the liver, based upon the biliary drainage of the organ. The upper figure is an inferior view of the liver, and the lower figure is an anterior view. (Healey JE Jr, Schroy PC, Sorenson RJ: J Int Coll Surg 20:133, 1953)

the medial segments respectively. The caudate lobe, in fact, could be regarded as a special segment; it is supplied by an independent branch of the right and left hepatic arteries and the right and left radicles of the portal vein and is drained into both right and left hepatic ducts.

The anatomy of hepatic segmentation is still of controversial usefulness in partial resection of the liver: a true lobe, rather than a segment, must be resected in most instances of partial hepatectomy.

**Relations.** The peritoneal areas on the **diaphragmatic surface** of the liver are related to the diaphragm through the *subphrenic spaces* described in Chapter 23. Through the diaphragm, the costodiaphragmatic recesses and the base of the right and, to a lesser extent, the left lung are related to the liver. The central tendon of the diaphragm separates the liver from the pericardial cavity and the heart, which makes a shallow impression on the upper surface of the liver.

The clinical importance of these relationships can be emphasized by some examples: A liver biopsy is obtained by thrusting a small trocar and cannula into the right lobe of the liver. After withdrawal of the trocar, a small core of liver parenchyma is aspirated. The trocar and cannula are usually inserted into the 7th or 8th intercostal space in the midaxillary line. Before reaching the hepatic peritoneum, the instrument has to traverse the tissues of the intercostal space, the costal pleura, the costodiaphragmatic recess, the diaphragmatic pleura, the diaphragm, the diaphragmatic peritoneum, and the subphrenic recess. Another example is the discharge of a hepatic abscess: it may burst, not only into a subphrenic space, but through the diaphragm into the pleural cavity or even into a basal bronchus when the inflammatory process fixes the base of the lung to the diaphragm.

Most of the **bare area** of the liver is in direct contact with the diaphragm and, in addition, with the inferior vena cava and, just to its right, with the right suprarenal gland and a small area of the right kidney (Fig. 24-31). The potential anastomoses of venous capillaries between the liver and the diaphragm that exist in the bare area will open up and become functional under certain pathologic conditions.

To the left of the inferior vena cava, the liver is indented by the vertebral column and the aorta; both are separated from it by the diaphragm.

The **visceral surface** makes contact with many viscera, all, except for the gallbladder, through the peritoneal cavity (Fig. 24-31). To the right of the gallbladder, the duodenum, the right kidney, and, further anteriorly, the right flexure of the colon and the transverse colon make impressions on the *right lobe*. Between the kidney and the liver is the hepatorenal recess, which communicates with the omental bursa (p. 602). To the left of the gallbladder, the *quadrate lobe* is in contact with the lesser omentum and the pyloric part of the stomach. Behind and above the porta hepatis, the *caudate lobe* faces into the slitlike superior recess of the lesser sac (see Fig. 23-27). The narrow *caudate process* between the porta hepatis and the inferior vena cava forms the roof of the epiploic foramen. The inferior surface of the *left lobe* is in contact with the fundus and body of the stomach. Along the blunt posterior border of the liver, the esophagus makes

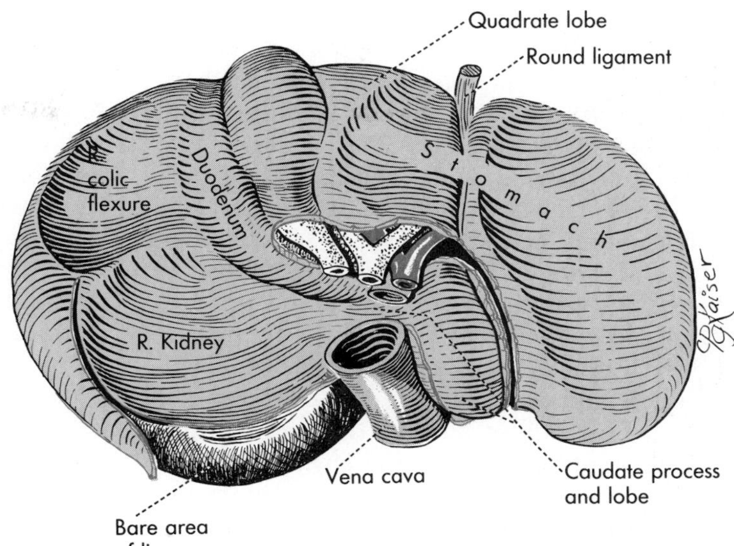


**FIG. 24-31.** Relations of the visceral surface of the liver.

an impression just to the left of the lesser omentum.

#### Development

The liver develops from the interaction of endodermal cells derived from the rostral lip of the archenteron (the region of the junction between the yolk sac and the primitive gut) and prochordal splanchnic mesoderm, which, after completion of the head fold, becomes the septum transversum. The mesodermal cells induce the endoderm to proliferate and form the hepatic diverticulum, which grows into the septum transversum; the endodermal cells induce the mesoderm to form the hepatic sinusoids.

The septum transversum contains the vitelline veins and the umbilical veins before the hepatic diverticulum invades it. These vessels subdivide to form the sinusoids, and the sinusoids invade the hepatic diverticulum, breaking it up into cords of hepatocytes, which later become rearranged to create the radially disposed sheets of liver cells in the hepatic lobules. Some cells lose connection with the hepatic diverticulum and develop independently.

Bile canaliculi and ductules are formed in the substance of the liver, and these tubes establish connections with the extrahepatic bile ducts as a secondary event at a later stage. This process of intrahepatic bile duct formation is responsible for establishing the hepatic segments. Failure of union between some of these bile ducts with the biliary tree may be the cause of cyst formation in the liver.

It has been proposed that some hepatocytes arise from the mesothelium of the celomic lining and become commingled with the endoderm-derived hepatocytes, being indistinguishable from them in every respect. The mesoderm of the septum transversum contributes to the liver all its connective tissue, peritoneal coverings, blood vessels, and, except for the epithelial lining, walls of all bile ducts.

The rapidly growing liver distends the septum transversum and acquires peritoneal surfaces, except for the bare area, where the original relationship with the septum transversum is retained. Extrahepatic portions of the septum transversum form a part of the diaphragm and the ventral mesentery (p. 597). After the ninth week, the growth rate of the left lobe of the liver regresses, and some of its hepatocytes degenerate. Such degeneration affects mainly the left lobe and may be so complete as to leave a *fibrous appendage* at the left extremity of the lobe (Fig. 24-27; p. 657).

There are numerous **variations** in the segmental division of the liver, as well as of the branchings of its ducts and vessels. Few abnormalities of lobulation exist. *Riedel's lobe* is an extension of normal hepatic tissue from the inferior margin of the liver, usually from the right lobe. Its significance is that it may be mistaken for an abnormal abdominal mass. Rarely, there may be an anomalous extension of hepatic tissue through the diaphragm into the chest.

## THE GALLBLADDER AND THE BILIARY DUCTS

Bile is produced by hepatocytes and collected in the tiny canaliculi bordered by the hepatocytes themselves. These *bile canaliculi* drain at the periphery of the hepatic lobules into thin-walled *bile ductules* that run toward the porta hepatis along the branches of the portal vein and hepatic artery. They coalesce into larger and larger ducts, eventually forming the *segmental bile ducts* and the ducts of the true right and left lobes (Fig. 24-36). The *right and left hepatic ducts* emerge from the liver in the fissure of the porta hepatis and unite to form the *common hepatic duct*. This duct is joined by the cystic duct, the duct of the gallbladder, and the two form the common bile duct. As already recounted, the common bile duct is joined in turn by the chief pancreatic duct before the two open together into the duodenum (Fig. 24-32).

## The Common Hepatic Duct

The usual length of the common hepatic duct varies from 2.5 cm to 5 cm, since the right and left hepatic ducts may join deep in the porta hepatis or may descend into the lesser omentum before joining. On the other hand, there may be no common hepatic duct at all, for the cystic duct may join the right hepatic duct before the latter joins the left hepatic duct. Normally, the cystic duct joins the right side of the common hepatic duct at an acute angle in front of the portal vein. In the porta hepatis and the lesser omentum, the common hepatic duct lies on the right side of the hepatic artery in front of the portal vein.

In about one-fifth of all bodies, a hepatic duct that normally joins the duct system within the substance of the liver emerges independently to join one of the extrahepatic ducts, most often the common bile duct. Such a duct is called an *aberrant* or *accessory bile duct*.

**FIG. 24-32.** The extrahepatic biliary ducts and the gallbladder.

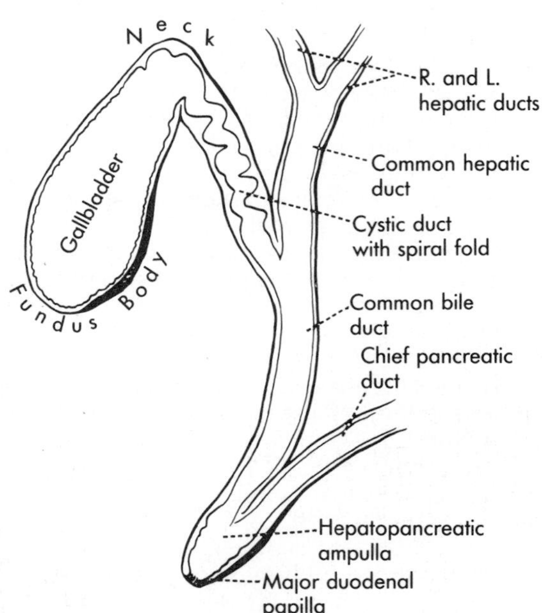


## The Gallbladder and the Cystic Duct

The gallbladder (*vesica fellea;* "fel" meaning bile or gall) is an elongated pear-shaped sac in which bile is stored and concentrated. Bile enters and leaves the gallbladder through the cystic duct. The gallbladder lies on the visceral surface of the liver (Fig. 24-31); its position and peritoneal covering on this hepatic surface have already been discussed. The nonperitoneal upper surface of the gallbladder is attached by connective tissue to a shallow fossa on the liver located between the right lobe and the quadrate lobe. Sometimes the gallbladder is invested almost completely by peritoneum and may be suspended from the liver from a mesentery (floating gallbladder).

The gallbladder is rarely congenitally absent; it may be intrahepatic or may not have its normal position on the visceral surface of the liver. Rarely, the fundus of the gallbladder is bifid, or there may be two gallbladders.

The gallbladder consists of the fundus, body, and neck (Fig. 24-32). The **fundus** is the expanded blind anterior end of the organ projecting beyond the inferior margin of the liver. It is covered completely in peritoneum and is in contact with the anterior abdominal wall just below the tip of the 9th right costal cartilage. The **body** tapers toward the neck, which lies in the porta hepatis. The junction of the body and neck is sometimes straight, more often angular. The **neck** may show a pouchlike dilatation toward the right (Hartmann's pouch); it has been shown to be a pathologic feature. The neck turns sharply downward as it becomes continuous with the cystic duct.

The **cystic duct** is up to 5 cm long and runs backward and downward from the neck of the gallbladder. The junction with the common hepatic duct usually takes place immediately below the porta hepatis; however, the two ducts may parallel each other for some distance and, on occasion, may not join until they have almost reached the duodenum.

The mucous membrane of the cystic duct is raised up into a *spiral fold* that consists of five to ten irregular turns (Figs. 24-32 and 24-33); it is continuous with a similar fold in the neck of the gallbladder. The spiral fold (sometimes called a valve) is believed to serve the purpose of keeping the duct open so that bile can pass through it both in and out of the gallbladder. When the common bile duct is closed at its inferior end, bile secreted by the liver fills the duct and passes along the cystic duct into the gallbladder. When the common bile duct is open, bile flows into it from the common hepatic and cystic ducts. The flow of bile is augmented by the contraction of the gallbladder, which is coordinated with the relaxation of the sphincter of the common bile duct through the action of cholecystokinin, a hormone released by the duodenal mucosa.

**Relations.** The relations of the gallbladder and biliary ducts have clinical relevance for the diagnosis of gallbladder disease and for its surgical treatment.

The inferior surface of the gallbladder, close to its neck, is in contact with the first part of the duodenum; further anteriorly, the body rests on the descending part of the duodenum and on the tranverse colon. In the cadaver, green stain is usually present on the organs with which the gallbladder is in contact. Inflammation of the gallbladder (*cholecystitis*) may cause it to adhere to its neighboring organs and also to the greater omentum. Inflamed and suppurating tissue may break down, creating a fistula between the gallbladder and the duodenum or the transverse colon.

The region of union of the **cystic** and **common hepatic ducts** is of particular interest in the common operative procedure of *cholescystectomy* (removal of the gallbladder). The cystic duct and common hepatic duct define two

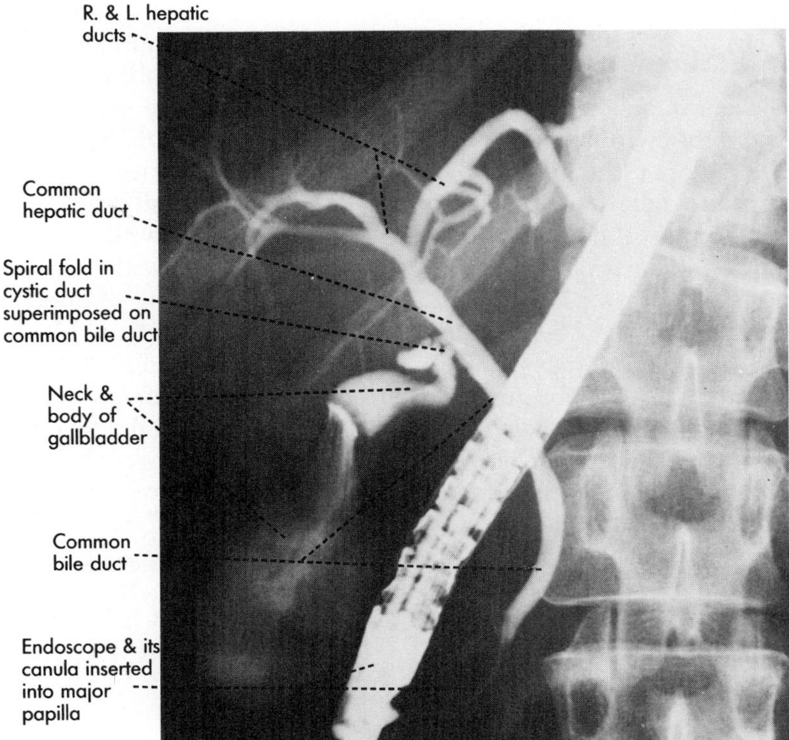


**FIG. 24-33.** The gallbladder, extrahepatic ducts, and some intrahepatic biliary ducts revealed by endoscopic retrograde cholecystography. For explanation of the procedure, see page 653.

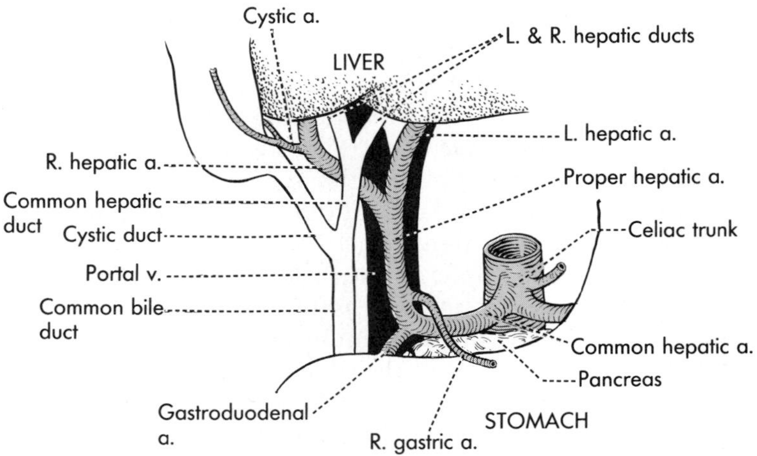


**FIG. 24-34.** Normal arrangement of the biliary duct system, the hepatic artery, and the portal vein as they run in the free edge of the lesser omentum (hepatoduodenal ligament).

sides of a triangle, the base of which is formed by the liver (Fig. 24-34). This *cystohepatic triangle* (of Calot) contains the right hepatic artery, the cystic artery, and most aberrant or accessory bile ducts that may be present. There is, however, considerable variation in the position of structures in this region, and successful cholecystectomy demands a meticulous dissection and careful identification of all ducts and vessels. Damage to the common hepatic or common bile duct usually results in stricture from scar formation, and this may also impede biliary flow to the extent that it will threaten the life of the individual.

## The Common Bile Duct

The common bile duct (*ductus choledochus*) drains bile from both the liver and the gallbladder. The duct descends almost vertically from just below the porta hepatis in the right free border of the lesser omentum. It parallels the hepatic artery and portal vein, lying to the right of both structures and in front of the portal vein (Fig. 24-34). Artery, vein, and duct can be grasped together between finger and thumb with the forefinger inserted into the epiploic foramen. Leaving the lesser omentum, the bile duct descends behind the first part of the duodenum along the gastroduodenal artery and comes to lie on, or embedded in, the posterior surface of the head of the pancreas (Fig. 24-19). Here it lies anterior to the inferior vena cava and a variable distance to the right of the posteromedial wall of the descending duodenum. In the head of the pancreas, the duct turns right and is joined by the chief pancreatic duct, and the two enter the posterior aspect of the duodenal wall obliquely at the midpoint of the descending part (Fig. 24-20).

Embedded in the duodenal wall, the two ducts empty into the **hepatopancreatic ampulla** (of Vater), a short dilated chamber protruding from the wall into the duodenal lumen as the *major duodenal papilla* (Fig. 24-35). The ampulla, as well as the intramural parts of the bile duct and pancreatic ducts, is surrounded by smooth-muscle **sphincters.** The sphincter around the ampulla has long been known as the *sphincter of Oddi*. All three sphincters are apparently independent of the muscle coat of the duodenum. The longitudinal and circular muscles of the duodenum swing around the ducts, creating potential weaknesses in the duodenal wall, predisposing to duodenal diverticula. These sphincters remain closed until gastric contents enter the duodenum, stimulating its mucosa to release cholecystokinin. This hormone, in addition to causing contraction of the gallbladder, relaxes the sphincters, permitting bile and pancreatic secretions to enter the duodenum.

#### Clinical and Radiologic Examination

The position of the fundus of the gallbladder can be located at the tip of the right 9th costal cartilage

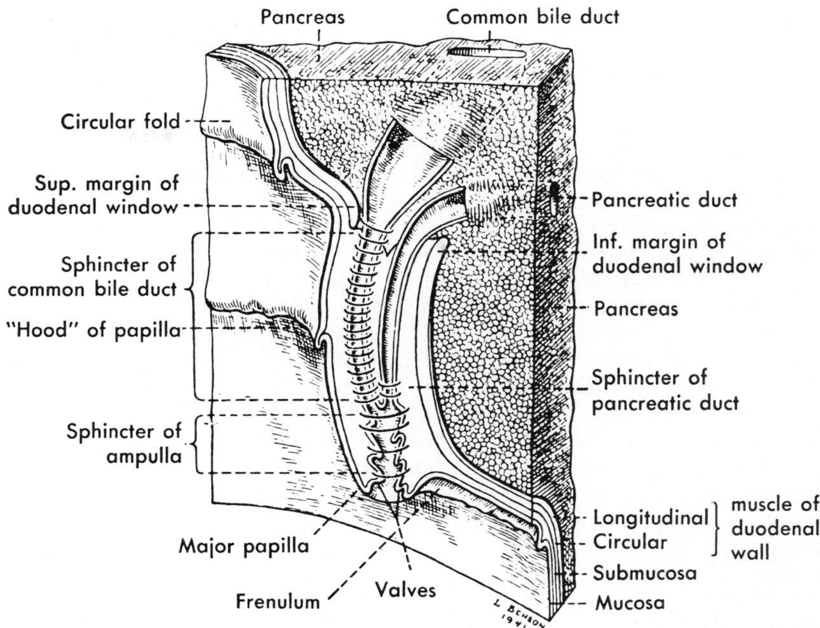


**FIG. 24-35.** Diagrammatic representation of the sphincter of the hepatopancreatic ampulla. Note that much of it, represented by the encircling lines, actually is around the common bile duct rather than the ampulla, and that there is also a small sphincter of the pancreatic duct. (Boyden EA: Surgery 10:567, 1941)

by the intersection of the right costal arch and the transpyloric plane or by extrapolating the line that joins the left anterior superior iliac spine to the umbilicus as far as the costal arch. During the movements of respiration, the gallbladder moves up and down with the inferior edge of the liver. Although the position of the gallbladder is known to vary with body type of the individual, the landmarks discussed here are clinically useful.

In the normal subject, the gallbladder cannot be felt by palpation. When the gallbladder is inflamed or distended, the patient will experience pain and will catch his breath as the gallbladder descends with an inspiration and meets the pressure of the palpating hand placed just below the right costal margin.

The gallbladder may be visualized radiographically following the ingestion of iodinated lipid-soluble substances, which, consequent to absorption, are secreted in the bile. Concentration of the substance in the gallbladder renders the organ radiopaque (*oral cholecystography*). The emptying of such a radiopaque gallbladder can be demonstrated by feeding the patient substances that effectively stimulate the release of cholecystokinin (*e.g.*, egg yolk and cream). The rate and extent of emptying can be monitored on serial x-ray films.

Some iodinated compounds are excreted rapidly in the bile following their intravenous administration (*intravenous cholangiography*) and are therefore suitable for radiologic examination of the major bile ducts. Concentration of the bile in the gallbladder is not required.

A small catheter containing fiber optics for endoscopy can be swallowed by the patient and advanced into the duodenum. Under direct vision through the endoscope, the catheter may be inserted into the hepatopancreatic ampulla, and the bile duct or the pancreatic duct or both may be injected with x-ray-opaque contrast medium (*endoscopic retrograde cholangiography* [Figs. 24-33 and 24-36]). The retrograde injection outlines the extrahepatic and intrahepatic bile ducts for radiologic examination. *Percutaneous transhepatic cholangiography* is performed by inserting a flexible metal needle through the skin into the liver and injecting contrast medium into a bile duct.

**Biliary Obstruction and Gallstones.** An obstruction to flow of bile into the duodenum causes distention of the bile ducts and the gallbladder, leading eventually to absorption of the yellow bile pigments into the circulation; this becomes manifest as a type of *jaundice*. Obstruction of bile flow may occur anywhere along the biliary tree and may be caused by spasm of the sphincters, gallstones, or external pressure on the ducts by an adjacent neoplasm. Biliary stasis predisposes to infection and to the formation of gallstones.

Not all the factors leading to the formation of gallstones are understood, but one of them seems to be sluggish flow of bile with the resultant absorption of water from it and precipitation of some of the contents in the form of stones. Gallstones can be formed within the hepatic duct, within the radicles in the liver, or in the gallbladder. Many gallstones are radiolucent and cannot be seen on x-ray film. They may be revealed as filling defects on cholangiograms or on cholecystograms.

Stones that pass into the cystic duct may obstruct this duct, or if they enter the common bile duct, they may be arrested anywhere along its length, causing obstruction and spasms. The common bile duct is narrowest at the point of its entry into the duodenal wall; stones may get held up here or at the ostium of the hepatopancreatic ampulla. There may be excruciating pain as a result of stretching of the duct or spasm of its wall and the sphincters. The intensity of the pain would warrant the use of morphine; however, this drug is known to induce spasm of the ampullary sphincter and therefore cannot be used to relieve the pain of biliary colic.

It remains controversial as to what extent backflow of bile into the pancreatic ducts contributes to *pancreatitis* when the ampulla is occluded. Spasm of the ampullary sphincter has been held responsible for certain cases of pain originating in the biliary system and for some cases of pancreatitis. The sphincter has, in fact, been sectioned to relieve these conditions.

## VESSELS AND NERVES

### Blood Supply

The **liver** is perfused with blood from two sources: the hepatic artery proper and the portal vein. In a recumbent normal adult, nearly one-third of the cardiac output passes through the liver. Roughly 80% of this is delivered through the portal vein; the 20% delivered through the hepatic artery furnishes up to 80% of the oxygen requirements of the liver. After the arterial and portal blood percolates through the sinusoids of the hepatic lobules, it is collected at their center by radicles of the hepatic veins, which convey it to the inferior vena cava.

The **gallbladder** and some of the extrahepatic biliary ducts receive their arterial supply from branches of the hepatic artery proper and, lower down, from the gastroduodenal ar-

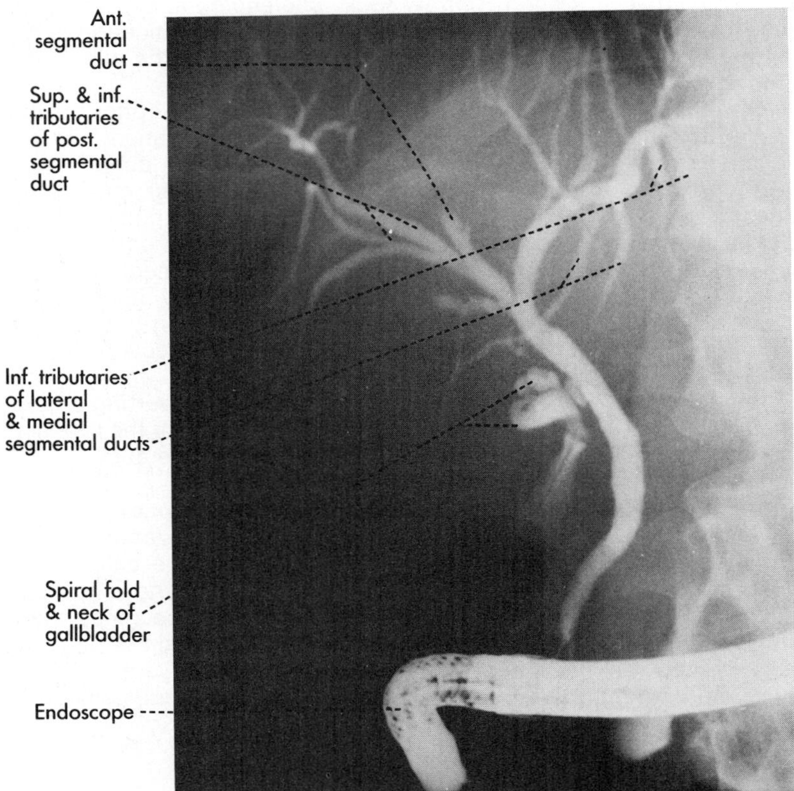


**FIG. 24-36.** Some of the segmental bile ducts revealed by endoscopic retrograde cholangiography. See also Figure 24-33.

teries; their venous blood is returned for the most part to the liver or the portal vein.

In contrast to the arteries of the stomach, duodenum, pancreas, and the rest of the digestive tract, the hepatic artery does not form any anastomoses, and its branches are end arteries. Furthermore, the segmental arteries within the liver can also be considered end arteries, as the only anastomoses between the territories of distribution are through tiny vessels of the subcapsular region. The few arterial twigs that enter the liver over its bare area from the phrenic vessels in the diaphragm have no functional importance.

**The Hepatic Artery Proper.** On reaching the inferior margin of the epiploic foramen, the *common hepatic artery,* one of the three branches of the celiac trunk, divides into two vessels: the gastroduodenal artery descends behind the first part of the duodenum; the hepatic artery proper ascends in the hepatoduodenal ligament toward the porta hepatis (Fig. 24-34).

In the lesser omentum, the *proper hepatic artery* lies to the left of the common bile duct and in front of the portal vein. As it nears the liver, it divides into right and left hepatic arteries. This division typically takes place to the left of the common hepatic duct. The left hepatic artery retains this relationship to the hepatic duct as they disappear into the porta hepatis. The right hepatic artery runs upward and turns to the right, crossing behind the common bile duct in approximately 85% of cases (Fig. 24-34); in 15% of cases, it crosses in front of the duct.

If the bifurcation of the proper hepatic artery is low, the right hepatic artery may lie in front of the common bile duct or may cross in front of both the common bile duct and the cystic duct. In any event, the right hepatic artery will be found in the *cystohepatic triangle* (Fig. 24-34). In this triangle, lying close to the cystic duct and the neck of the gallbladder, the right hepatic artery gives off the cystic artery.

As the **cystic artery** reaches the gallbladder, it divides into two branches, one of which runs on the serous surface of the gallbladder, the other on its hepatic surface between the gallbladder and hepatic substance.

The **blood supply to the extrahepatic biliary ducts** is somewhat variable, but the major supply typically comes from the *posterior superior pancreaticoduodenal artery,* a branch of the gastroduodenal, and is supplemented above by branches from the right or left hepatic arteries or cystic artery.

The **venous drainage of the gallbladder** is, in part, through veins that pass directly into the substance of the liver, joining branches of the portal vein, and, in part, through veins that cross the neck to enter the liver or join the ascending veins of the common bile duct that follow the hepatic ducts into the liver. Only rarely does a cystic vein enter the portal vein directly. The major venous drainage of the common bile duct is upward, but veins from its lower end may communicate with veins of the duodenum and pancreas and also with the portal vein.

#### Variations

Variations in the arterial supply of the liver and gallbladder are extremely common, particularly if the relations of the arteries to the duct system are taken into consideration as well.

In approximately a third of all bodies, an artery of abnormal origin (not derived from the proper hepatic) can be found entering the liver. Such arteries are called **aberrant hepatic arteries.** These arteries usually replace a segmental branch of the hepatic artery that would normally be given off in the substance of the liver.

Aberrant arteries may go to either side or both sides of the liver. Those that go only into the left side are more commonly derived from the left gastric artery. An aberrant right hepatic artery usually arises from the superior mesenteric artery or the aorta. There may be two aberrant arteries of different origin. In about 4% of bodies, the entire common or proper hepatic artery is aberrant, arising from the superior mesenteric, the aorta, or the left gastric artery.

Aberrant arteries of superior mesenteric and aortic origin may run behind, instead of in front of, the portal vein. Aberrant arteries always present a hazard in biliary surgery because of the varied relationships that they may present. This is also true of arteries of normal origin that pursue unusual courses.

The **cystic artery** also may show a number of variations, both in its manner of origin and course.

It should be obvious by now that when variations, both in the duct system and in the hepatic artery and its branches, are considered, what can be regarded as an absolutely normal set of ducts, arteries, and interrelationships is actually not common—one estimate has been that this occurs in only about one-third of bodies.

**The Portal Vein.** The portal vein is formed behind the pancreas by the union of the splenic and superior mesenteric veins (Fig. 24-37). It drains, indirectly, the inferior mesenteric vein, which enters the terminal part of the splenic vein, and receives the veins from the duodenum and stomach. The portal vein thus collects blood from the entire digestive tract except the liver, to which it carries its blood. On entering the liver, the portal vein breaks up into branches that end in the hepatic sinusoids, where portal blood is brought into intimate contact with the hepatocytes.

After its formation, the portal vein emerges above the upper border of the pancreas and enters the lesser omentum, usually being crossed anteriorly before it does so by the hepatic artery; it then lies behind and to the left of the common bile duct and behind the proper hepatic artery (Fig. 24-34). Between its level of formation and its disappearance into the liver, the portal vein typically receives tributaries from the stomach and from the upper part of the pancreas and duodenum. The *left gastric vein* usually joins it low, and the *right gastric vein* often runs up some distance in the lesser omentum to reach it. It also receives the pancreaticoduodenal vein, other veins from the pancreas, and the *paraumbilical veins* running along the ligamentum teres in the falciform ligament. Rarely, the portal vein receives the cystic vein before ending in the porta hepatis and dividing into right and left branches. The left branch of the portal vein is joined by the *round ligament of the liver,* and the same branch gives off the *ligamentum venosum,* which connects it to the left hepatic vein (Fig. 24-38).

The **ligamenta teres hepatis and venosum** represent two segments of the major venous channel through which the placental blood reaches the fetal heart. The *round ligament of the liver* represents the obliterated left umbilical vein (the right umbilical vein having disappeared early in development), and the ligamentum venosum represents the ductus venosus. The *umbilical vein* runs from the umbilicus in the free border of the falciform ligament to the porta hepatis, creating a furrow on the visceral surface of the liver. Only a portion of the nutrient- and oxygen-rich placental blood delivered by the umbilical vein is spent in the liver, mixing in the hepatic sinuses with blood from both the hepatic

**FIG. 24-37.** Formation of the portal vein and the most frequent sites of terminations of its tributaries in it. (Modified from Douglass BE, Baggenstoss AH, Hollinshead WH: Surg Gynecol Obstet 91:562, 1950)

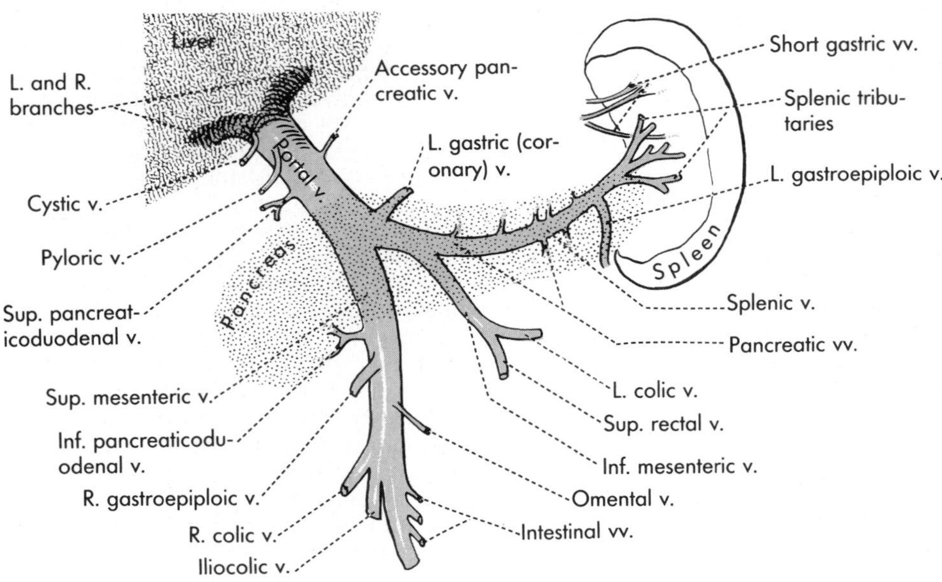

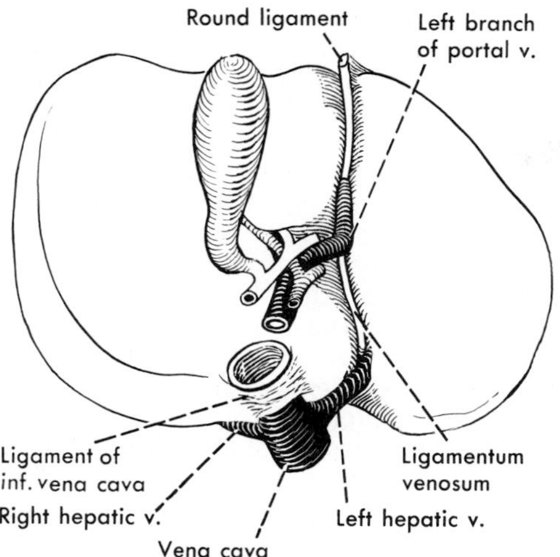


**FIG. 24-38.** Relations of the round ligament (ligamentum teres) and the ligamentum venosum. The left hepatic artery and duct have been cut away, and the fissures for the ligaments have been spread to show their contents.

artery and the vitelline veins, which give rise to the portal vein. Much of the placental blood is diverted from the liver parenchyma by the **ductus venosus.** This large venous channel is embedded in the liver and shunts blood from the umbilical vein to the proximal end of the vitelline vein, from which the hepatic veins and the proximal segment of the inferior vena cava develop.

After birth, both the umbilical vein and the ductus venosus become nonfunctional, occluded, and fibrosed, but a lumen can be opened up in the umbilical vein for a considerable time, permitting the withdrawal or transfusion of blood during the neonatal period. In the hepatic end of the round ligament, a lumen may persist into adulthood, and the "ligament" can be used to inject substances into the portal vein or to withdraw blood from it.

The left branch of the portal vein makes a sharp angle with the umbilical vein, whereas the right branch lies more directly in line with it. In the fetus, this arrangement places the left lobe at a circulatory disadvantage, causing the growth rate of the left lobe to lag behind that of the right. As a consequence, hepatocytes may disappear from the tip of the tongue-shaped left lobe, leaving behind the fibrous capsule and the hepatic peritoneum as the *fibrous appendix of the liver* at the left extremity of the left lobe (Fig. 24-27).

**Hepatic Veins.** The blood in the sinusoids of the liver derived from the arterial and portal venous circulation is collected into radicles of the hepatic veins. These veins, however, do not accompany their portal triads made up of branches of the bile duct, hepatic artery, and portal vein; rather, they run between them. The major stems of the hepatic veins lie between subsegments and segments rather than in the segments; they are intersegmental. The largest hepatic veins are three: the *left hepatic vein* between the medial and lateral segments of the true left lobe, the *middle hepatic* between the true right and left lobes, and the *right hepatic* between the anterior and posterior segments of the right lobe (Fig. 24-39). These three veins may enter the inferior vena cava independently, but the left and middle veins usually join, so that only two major hepatic veins enter the vena cava. They enter while the vena cava is in its sulcus on the bare area of the liver and immediately before it pierces the diaphragm to end in the heart.

#### Cirrhosis of the Liver and Portal Hypertension

The liver consists of polyhedral *lobules,* each of roughly 1 mm diameter and made up of anastomosing sheets or laminae of hepatocytes radiating away from a *central canal* containing a radicle of the hepatic vein. At the periphery of the lobules, in the corners of the polyhedron, run the *portal triads,*

**FIG. 24-39.** The hepatic veins. (Healey JE Jr: J Int Coll Surg 22:542, 1954)

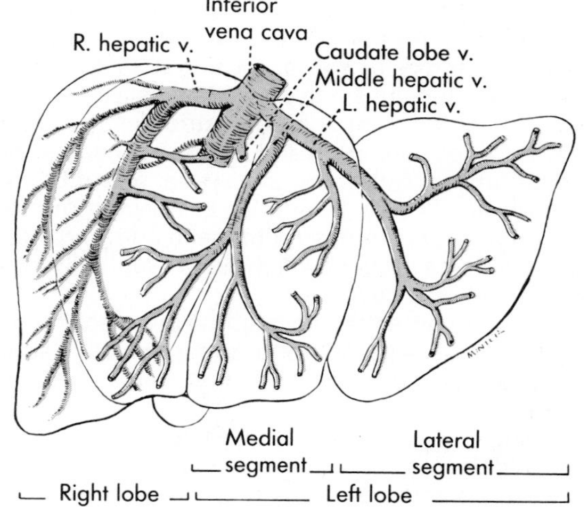

consisting of a bile duct and branches of the hepatic artery and the portal vein. The radially arranged venous sinusoids occupy the lacunae between the sheets of liver cells within a lobule and connect the portal vein of the triad to the central vein of the lobule. The sinusoids also receive minute branches given off by the hepatic arteries of the triads. The permeable sinusoidal wall, consisting of endothelial cells and some macrophages (Kupffer cells), separates the blood from the lacunar surfaces of hepatocytes. The portal triads are embedded in perilobular connective tissue, which pervades the liver and is continuous with the capsule of the organ lying subjacent to the hepatic peritoneum.

Following injury and death of the hepatocytes (caused by infection, toxins, alcohol, poisons), their regeneration is usually prevented by excessive scar-tissue formation produced by proliferation of the perilobular connective tissue. The resultant hepatic fibrosis exhibits a variety of patterns and is known as *cirrhosis.* Resistance to blood flow through the fibrosed or cirrhotic liver is increased, causing a build up of pressure in the radicles of the portal vein in the triads. Since no valves exist in the portal vein and its tributaries, the increased venous pressure will affect the entire portal venous system, causing engorgement and distention of all its tributaries, as well as of the spleen. This condition is known as *portal hypertension.*

Some relief of high venous pressure is obtained by opening up anastomoses between veins that drain into the portal system and those that drain into tributaries of the two venae cavae. There are three specific locations where such **portasystemic anastomoses** produce grossly dilated venous varicosities: *esophageal varices* at the lower end of the esophagus, the *caput medusae* around the umbilicus, and *hemorrhoids* or piles in the anal canal and lower end of the rectum. In the wall of the esophagus, the tributaries of the left gastric vein anastomose with those of the azygos system (p. 570); around the umbilicus, the paraumbilical veins connect to tributaries of the epigastric veins, which drain into both superior and inferior venae cavae (p. 595); in the wall of the anal canal and rectum, superior rectal veins, tributaries of the inferior mesenteric vein, anastomose with the middle and inferior rectal veins, both of which drain eventually into the inferior vena cava.

It is well to note that the majority of hemorrhoids have other causes than portal hypertension. Caput medusae may occur also when the superior or inferior vena cava is obstructed. The sole cause of esophageal varices is portal hypertension.

In addition to the three specific sites of portacaval anastomoses, anastomoses develop between the portal and systemic venous systems wherever nonperitoneal areas of the intestine, liver, and pancreas are in contact with the body wall. These include the posterior surface of the pancreas, the duodenum, and the ascending and descending colon, as well as the bare area of the liver.

The common method of reducing portal pressure is to divert blood from the portal to the caval system by an operative anastomosis. This usually is done by creating a fistula between the portal vein and the inferior vena cava as they lie close together below the liver (*portacaval anastomosis*) or by anastomosing the splenic vein to the left renal vein (*splenorenal anastomosis*) after removal of the spleen. The effectiveness of the latter procedure is a consequence of the absence of valves in the portal system whereby blood can run retrogradely through the splenic vein into the renal vein more easily than it can run through the engorged esophageal veins.

## Lymphatics

A network of superficial lymphatics exists in the capsule of the liver underneath its peritoneum; lymphatics accompanying the portal triads constitute the deep lymphatics. Most of the **superficial lymphatics** from the posterior aspect of both the diaphragmatic and visceral surfaces of the liver converge toward the bare area and pass with the inferior vena cava through the diaphragm to terminate in the *posterior mediastinal lymph nodes.* These lymph nodes drain into the thoracic duct. The posterior part of the left lobe drains to the *left gastric nodes.* Superficial lymphatics from the lower part of the anterior aspect of the diaphragmatic surface run around the inferior edge of the liver, join with those of the anterior portion of the visceral surface, and drain into *hepatic nodes* located at the porta hepatis and, lower down, along the hepatic artery. These hepatic nodes also receive lymph from the gallbladder and the extrahepatic biliary ducts. They drain their lymph to the celiac nodes and through an intestinal lymph trunk into the cisterna chyli. Many superficial lymphatics from the upper part of the diaphragmatic surface of the liver run through the falciform ligament, turn upward along the superior epigastric vessels, and terminate in *parasternal lymph nodes.*

**Deep lymphatics** of the liver form ascending and descending lymph trunks. The ascending trunk follows the hepatic veins and the inferior vena cava to the posterior mediastinal lymph nodes; the descending trunk passes through the porta hepatis to the hepatic lymph nodes.

### Nerve Supply

The nerves to the liver, gallbladder, and extrahepatic ducts run in the **hepatic plexus,** which, for the most part, originates from the celiac plexus and follows the hepatic arteries and the portal vein to the liver. Close to the liver, the plexus is usually joined by one or more hepatic branches from the anterior vagal trunk and sometimes from the posterior vagal trunk. These vagal fibers reach the hepatic plexus through the lesser omentum. A part of the plexus surrounds the hepatic arteries and gives off twigs to the duct system and to the gallbladder. Both sympathetic and vagal fibers are said to end on the gallbladder and the extrahepatic and intrahepatic ducts; blood vessels receive sympathetic fibers only.

Sympathetic stimulation causes constriction of the branches of both the hepatic artery and the portal vein. The nerves in the liver apparently do not affect the rate of bile formation. In man, sympathectomy does not appear to affect the gallbladder, but vagotomy leads to its enlargement and seems to slow its emptying. Vagotomy has been suspected to increase the incidence of gallstones.

Among the fibers of the hepatic plexus are *visceral afferents* concerned with visceral pain. These fibers belong to the sympathetic components of the autonomic nervous system and reach the sympathetic trunks by passing through the celiac plexus and the splanchnic nerves. They enter the spinal cord through dorsal roots of the 6th to 9th thoracic nerves. Pain from the gallbladder tends to be referred to the right side of the thoracic wall in the region of the 6th to 9th ribs and extends back toward the inferior angle of the scapula. Most of the pain fibers seem to be in the right splanchnic nerves, but some may also reach the cord via the left splanchnic nerves.

As already mentioned, pain from the spasm of the sphincters and the muscle of the wall of the biliary ducts can be excruciating. Distention of the hepatic capsule and hepatic peritoneum by swelling and inflammation of the liver, as in hepatitis, is also painful, felt in the epigastrium and sometimes referred to the shoulder.

## THE SMALL AND LARGE INTESTINE

### THE JEJUNUM AND ILEUM

The small intestine (*intestinum tenue*) is made up of the rather short and thick duodenum, discussed earlier in this chapter, and of the much longer and more mobile jejunum and ileum. The jejunum and ileum are specialized for the absorption of digested foodstuffs, vitamins, and electrolytes, and this specialization is reflected in the large surface of their mucosa. Their combined length, when excised from an unembalmed body, varies from 5 m to 10 m, with an average length of approximately 7 m. By definition, the upper two-fifths of the intestine between the duodenojejunal junction and the cecum are designated as the jejunum and the distal three-fifths are designated as the ileum.

In the abdominal cavity, this long tube is disposed in a series of coils and loops that fill almost completely the infracolic compartment and the pelvic portion of the peritoneal cavity (Fig. 24-40). Visceral peritoneum completely invests the jejunum and ileum except along a narrow strip where it becomes continuous with the two peritoneal laminae of the *mesentery,* which suspends the coils of the small bowel from the posterior abdominal wall.

The jejunum begins at the **duodenojejunal flexure** on the left side of L2 vertebra; its first loops lie in the upper part of the left infracolic compartment. Although some of its coils may extend into the pelvis, it generally occupies the umbilical region (Fig. 24-40).

There are histologic differences between the upper end of the jejunum and the lower end of the ileum, but the transition between

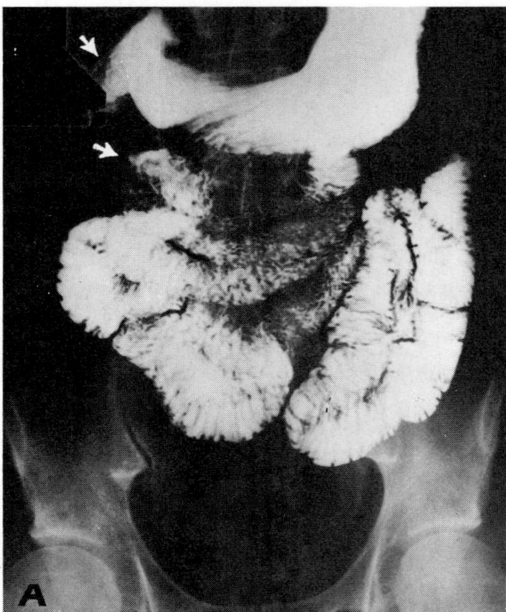


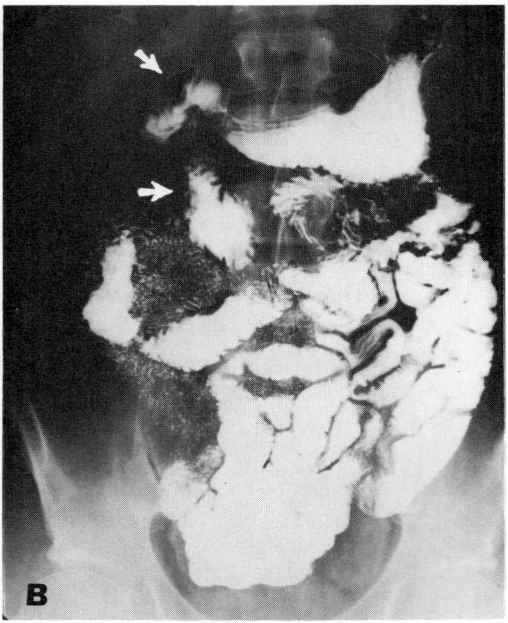


**FIG. 24-40.** X-ray films of the digestive tract after a barium meal. In **A**, the stomach, parts of the duodenum (arrows), and the jejunum are well visualized; in **B**, taken 1 hour and 45 minutes later, some material still remains in the stomach, duodenum, and jejunum, but most of it is now in the ileum. The upper arrow in each figure points to the "duodenal cap," partly hidden in the first figure by the pyloric end of the stomach; the lower arrow points to the lower part of the descending limb of the duodenum. (Courtesy of Dr. D. G. Pugh)

them is gradual. Loops of ileum are found mainly in the hypogastrium and the pelvic cavity, from which the terminal portion of the ileum ascends to open into the posteromedial side of the junction between the cecum and the ascending colon, located in the right iliac fossa.

The jejunum is slightly wider than the ileum and has a thicker wall on account of its thick mucosa. The mucosa is thrown into *circular folds* in the jejunum, whereas circular folds in the ileum are small and sparse. Distinctive *lymphoid follicles* (Peyer's patches), formed in the submucosa of the ileum, are visible through the epithelium and may be several centimeters long; such large follicles are absent in the jejunum.

The lumen of the jejunum and ileum is filled with liquid and gas. In the recumbent position, the gas rises to the surface, and for this reason, percussion of the abdomen in the umbilical region yields a resonant sound. Peristaltic activity in the small intestine is responsible for generating most of the *bowel sounds,* which are often audible to the unaided ear. In the normal abdomen, bowel sounds can always be detected with a stethoscope. In the cadaver, the jejunum is usually empty and collapsed (jejunum meaning empty; ileum meaning coiled).

The absorptive surface of the mucosa is increased not only by the circular folds but also by the characteristic *villi,* which give the fresh mucosal surface a lush, pink, velvetlike appearance. Since most of the absorption takes place in the jejunum and ileum, removal of large segments of the jejunum and ileum leads to grave nutritional problems.

The mobility of the jejunum and ileum is enhanced by the **mesentery,** described in Chapter 23 (p. 610). Its attachment to the posterior abdominal wall, known as the *root* or *base of the mesentery,* is about 15 cm long.

During development, the midline attachment of the mesentery has shifted to become

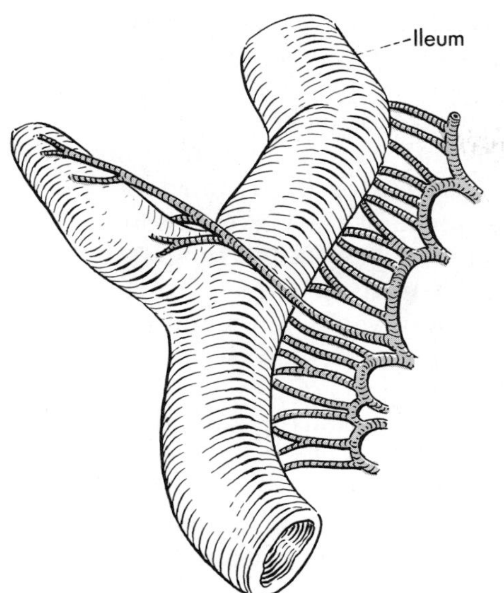


**FIG. 24-41.** Ileal or Meckel's diverticulum. It has an independent blood supply from an arcade of the intestinal arteries.

oblique. Starting at the duodenojejunal junction, the root of the mesentery descends into the right iliac fossa; in so doing, it crosses the horizontal part of the duodenum, the aorta, the inferior vena cava, the right ureter, and the right psoas muscle, terminating at the junction of the ileum with the large intestine. The height of the mesentery from root to intestinal border varies between 12 cm and 25 cm, being the broadest in the central portion. Although the root of the mesentery is essentially straight, the intestinal border is very much folded as it follows the coils of the jejunum and ileum.

#### The Ileal Diverticulum

The most common anomaly of the small intestine is the ileal diverticulum or *Meckel's diverticulum*, a protrusion from the antimesenteric border of the ileum (Fig. 24-41). It represents a persistence of part of the vitelline duct that joined the midgut loop to the yolk sac. The diverticulum has been found in 1% to 2.5% of persons in whom it was sought. It usually is located 10 cm to 15 cm from the ileocecal junction, and it may be 2 cm to 5 cm long. Occasionally, it is attached to the umbilicus by a fibrous cord, and, very rarely, a patency in the cord persists, in which case the diverticulum opens to the exterior at the umbilicus. The ileal diverticulum is particularly prone to pathologic change; therefore, if it is discovered during an abdominal operation, it is usually removed.

## THE LARGE INTESTINE

**General Anatomy.** The large intestine (*intestinum crassum*) begins at the ileocecal junction and ends at the anus. It is approximately 1.5 m long and consists of the cecum and appendix, the ascending, transverse, descending, and sigmoid colons, the rectum, and the anal canal. The function of the large intestine is to convert the liquid contents of the ileum into semisolid feces by the time the sigmoid colon is reached. This is accomplished by the absorption of fluid and electrolytes.

The ascending and descending colons are located in the flanks on the right and left side of the abdominal cavity respectively. With the transverse colon above and the sigmoid colon below, they surround a quadrangular space in the peritoneal cavity (Fig. 24-42) that is filled with the coils of small intestine. The transverse and sigmoid colons possess considerable

**FIG. 24-42.** The large intestine.

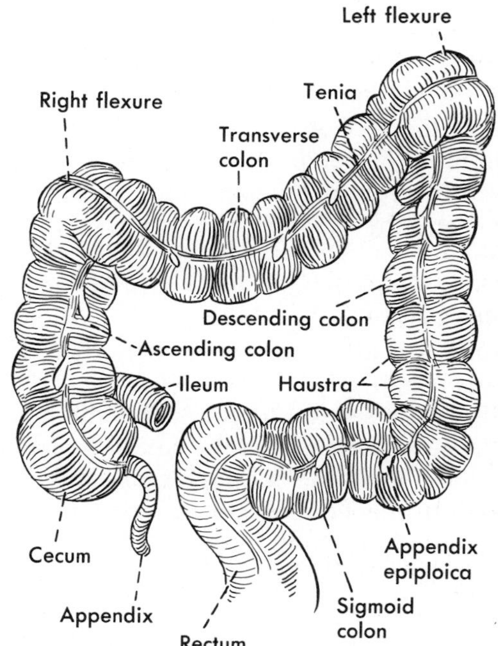

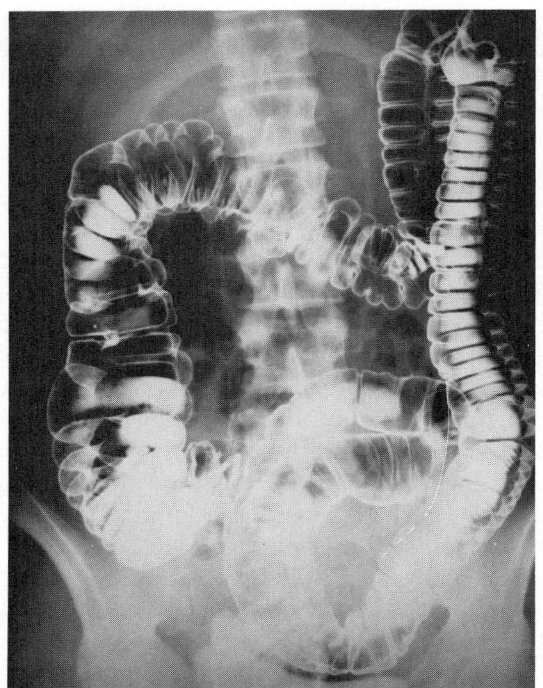

**FIG. 24-43.** An x-ray film of the large intestine using "double contrast." Following a barium enema, air was introduced into the bowel. Note the haustra, the different levels of the right and left colic flexures, the narrow appendix related to the terminal ileum (seen end on), and the sigmoid colon, which forms a loop larger than usual.

mobility, as each is suspended on a mesentery; the ascending and descending colons, as well as the rectum, are fixed to the posterior abdominal or pelvic wall. The cecum and appendix are completely peritoneal as they hang free from the inferior end of the ascending colon.

This chapter is concerned with the large intestine as far distally as the sigmoid colon; the rectum is dealt with along with other pelvic organs (Chap. 27).

There are several features that distinguish the large intestine from the small. In general, the large intestine has a larger caliber, although it may contract to a diameter smaller than that of the small intestine. The longitudinal muscle of the cecum and colon, rather than forming a continuous coat on the exterior of the circular muscle, as in most parts of the digestive tube, is gathered into three narrow ribbonlike bands called the *teniae* (taeniae) *coli*. One tenia is at the antimesenteric border or on the anterior surface of the colon and cecum, and the other two are equidistance, each one-third around the circumference of the bowel. Because the teniae are shorter than the gut tube itself, or because some fibers of the teniae stray from the main band and invaginate the gut wall, the cecum and colon present a series of sacculations called *haustra coli* (Figs. 24-42 and 24-43). These haustrations involve the circular muscle, the submucosa, and the mucosa. Another typical feature of the cecum and colon are the *appendices epiploicae,* pendant-shaped bodies of fat enclosed by peritoneum, hanging from the teniae.

## The Cecum

The cecum (*intestinum caecum;* blind intestine) is the saccular commencement of the large intestine located in the right iliac fossa (Fig. 24-44A). At its base, the cecum is continuous superiorly with the ascending colon; their junction is marked on the interior by the *ileocecal valve.* The frenula of the valve encircle approximately the posteromedial third of the cecocolic junction and guard the opening of the ileum (Fig. 24-44B). The *vermiform* (wormlike) *appendix* is attached to the posteromedial surface of the cecum, some 2 cm inferior to the ileocecal opening, and its narrow lumen communicates with the spacious cecum.

**Development.** The cecum and appendix commence their development as the cone-shaped, blind end of the large intestine, the apex of which is the tip of the appendix. Growth of the appendix becomes retarded, while that of the cecum proceeds, creating an abrupt demarcation between the two. Moreover, in the majority of cases, the right anterior haustrum of the cecum expands more than the remaining two haustra, which displaces the base of the appendix from the inferior tip of the cecum to the posteromedial wall and converts the anterior haustrum into the spacious blind inferior end of the cecum. The three teniae of the cecum, however, retain their fetal position and meet each other at the root of the appendix.

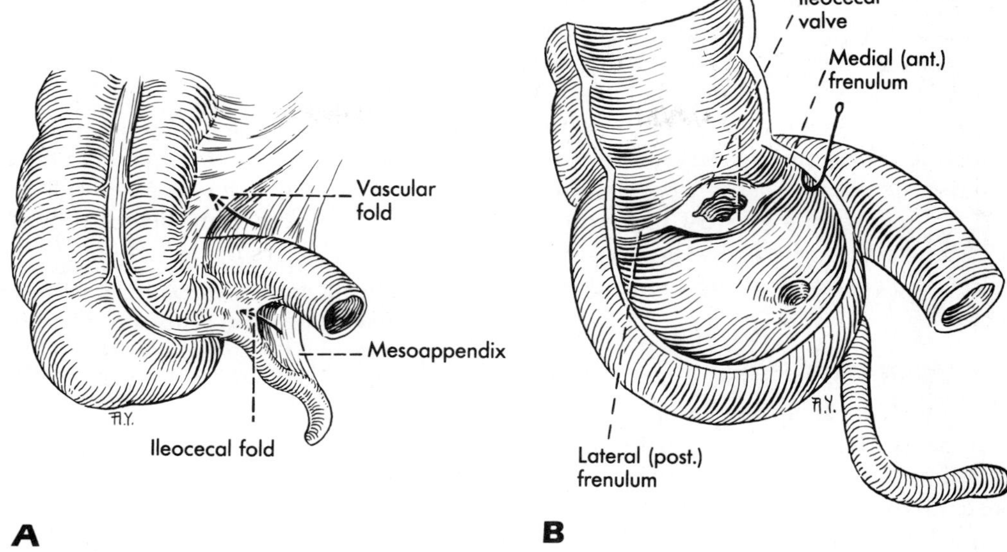


**FIG. 24-44.** **A** shows peritoneal folds associated with the terminal portion of the ileum and with the cecum. The arrows indicate the superior and inferior ileocecal recesses. **B** shows the cecum opened by removal of parts of its anterior wall.

In 78% to 90% of cases, development proceeds as described; in the remaining cases, it may be arrested at some stage, explaining the variations in the shape of the cecum and the location of the base of the appendix.

**Peritoneal Folds and Recesses.** The cecum has no mesentery; it projects from the antimesenteric side of the gut and therefore is covered on all sides with serosa. However, fusion of the ascending colon to the posterior abdominal wall may extend and involve the cecum for a variable distance, reducing its mobility.

The cecum is connected to the parietal peritoneum of the iliac fossa, usually by two **cecal folds** of peritoneum that limit, on each side, a peritoneal space behind the cecum called the **retrocecal recess.** When the fusion of the ascending colon to the posterior wall is halted, the retrocecal recess will continue as a *retrocolic recess* behind the ascending colon. Additional cecal folds may divide these recesses into compartments.

There are two other peritoneal recesses associated with the medial side of the cecum that are formed by two small peritoneal folds. The **superior ileocecal recess** is behind a peritoneal fold raised up by the anterior cecal artery as it approaches the medial side of the cecum, passing anterior to the termination of the ileum (Fig. 24-44A); this fold is called, therefore, the *vascular fold of the cecum.* The **inferior ileocecal recess** is behind the *ileocecal fold,* which is a bloodless fold of the cecum that joins the antimesenteric border of the terminal ileum to the cecum and the base of the appendix (Fig. 24-44A).

**Relations.** The cecum is located just above the lateral third of the inguinal ligament. Its anterior surface is in contact with the parietal peritoneum of the anterior abdominal wall; posteriorly, the retrocecal recess separates it from the iliacus and the psoas muscle. It is accessible to palpation in the right iliac fossa; its liquid contents do not offer resistance to the palpating fingers, which can readily reach the underlying firm iliacus. Consequently, the right iliac fossa feels empty to palpation.

**The Ileocecal Ostium.** On the interior of the cecum, the ileocecal ostium is surrounded by lips or flaps that protrude into the lumen of the cecocolic junction and contain some circular muscle derived partly from ileal and partly from cecal musculature: this is the *ileocecal valve.* Upper and lower lips unite and continue circumferentially at the frenula of the ileocecal valve. This valve is not particularly efficient, since x-ray films often reveal backflow through it into the ileum when the colon and cecum are filled with contrast material.

## The Appendix

The length of the appendix is quite variable; it may be as much as 20 cm, although usually it is half as long. In addition to being fixed by its base to the posteromedial surface of the cecum, the appendix is attached by its own mesentery, the *mesoappendix,* to the inferior border of the terminal portion of the ileum (Fig. 24-44A). The *appendicular artery* and *vein* enter the mesoappendix from behind the terminal ileum and run in the mesentery to the tip of the appendix. Usually, the appendicular artery is an end artery and its thrombosis, caused by appendicitis, results in gangrene of the appendix.

The appendix may lie in several positions. Most often it is hidden in the retrocecal recess. It may hang down into the pelvis or curve along the inferior margin of the cecum, or it may lie on the anterior surface of the terminal ileum or be tucked behind it. The physical signs that permit diagnosis of appendicitis are greatly influenced by the position of the appendix.

**Appendicitis.** The appendix is a specialized part of the digestive tract; its function is not completely understood. It has thick walls because its submucosa is filled with numerous lymphoid follicles. Its narrow lumen becomes readily occluded, and water may be absorbed from its contents to the extent that a *fecolith* (calcified bolus of feces) may form, which is often visible on x-ray films. Occlusion and stasis predispose to infection and inflammation, presenting the clinical picture of appendicitis. Until surgical treatment of appendicitis was introduced at the end of the last century, most cases were fatal.

Inflammation of the appendix induces visceral pain owing to distention of the organ and its covering serosa. This pain is perceived in the central abdomen and is poorly localized. When the inflammation spreads to parietal peritoneum, the pain shifts to the right iliac fossa. The precise location of the pain and the nature of tenderness are greatly influenced by the location of the appendix. A subcecal or preileal appendix will produce inflammation of the parietal peritoneum on the anterior abdominal wall with the attendant localized spasm in the anterior abdominal wall muscles and hypersensitivity of the overlying skin. This makes diagnosis relatively straightforward. The physical sign of *rebound tenderness,* that is, eliciting pain by slowly and progressively pressing on the abdominal wall in the iliac fossa and then suddenly letting it go, is explained by the sudden stretch experienced by the parietal peritoneum when the palpating hand lets go and the abdominal muscles jump back into position.

A pelvic appendix may cause signs and symptoms suggestive of pelvic disease (inflammation of the uterine tubes or ovary) and may be palpated rectally. Least definite are the physical signs of an inflamed retrocecal or retroileal appendix, since the appendix is inaccessible to palpation in these locations. Inflamed parietal peritoneum in the iliac fossa may cause spasm of the iliopsoas or may irritate the right ureter.

The tip of the appendix may be in variable positions, but its base is relatively constant at the junction of the middle and lateral thirds of a line that joins the right anterior superior iliac spine to the umbilicus (McBurney's point). In cases of *situs inversus,* the appendix, together with the cecum, is located on the left rather than on the right side.

## The Colon

The **ascending colon** begins at the upper border of the ileocecal junction and continues up the posterior body wall until just below the liver. In front of the right kidney, it makes a sharp bend to the left (Figs. 24-42 and 24-43). This bend creates the *right colic flexure* (sometimes called the hepatic flexure), beyond which the large intestine is called the transverse colon. Traced from the root of the mesentery of the small intestine, the peritoneum of the right infracolic compartment passes over the anterior surface of the ascending colon and swings back a little before continuing on the posterior abdominal wall. This creates the right *paracolic sulcus,* or gutter, lateral to the ascending colon.

The **paracolic sulcus** tends to conduct infectious material originating in the region of the appendix to the hepatorenal recess or, in the reverse direction, from a subphrenic or subhepatic abscess into the pelvis. It is also along the paracolic sulcus that the surgeon incises peritoneum when it is necessary to mobilize the ascending colon and its blood vessels. The blood vessels and lymphatics lie in the retroperitoneal connective tissue between the root of the mesentery of the small intestine and the left border of the colon, but this is tissue that once composed the mesentery of the ascending colon. By mobilizing the large intestine on its lateral avascular border and loosening it gently toward the midline, the connective tissue, containing the blood vessels and lymphatics, can be split from endoabdominal fascia and lifted as if it were still part of the mesentery.

The **transverse colon** begins at the right colic flexure and runs across in front of the coils of the small intestine to the left side, where it ends in the *left colic flexure* (also called the splenic flexure). The left colic flexure is situated higher and further posteriorly than the right colic flexure. An empty transverse colon may run obliquely upward from right to left, but a full one, especially when the person is standing, usually loops down a variable distance in front of the small intestine (Fig. 24-43). The *transverse mesocolon* has already been described with its relations to the omental bursa and the infracolic compartment (p. 604). Its attachments to the posterior abdominal wall have also been recounted (Fig. 24-14). The blood vessels and lymphatics of the transverse colon course through the transverse mesocolon. The first part of the transverse colon lies against the liver and gallbladder. Sometimes the lesser omentum extends farther to the right than usual to form the *hepatocolic ligament,* which connects the liver and gallbladder to the right colic flexure.

The **descending colon** begins at the left colic flexure, where the large intestine loses its mesentery in front of the left kidney. It passes down on the left side with a *paracolic sulcus,* like that of the ascending colon, related to it laterally. At or below the crest of the ilium, the colon acquires a mesentery once again and its name changes to sigmoid colon.

There is great variation in the length of the **sigmoid colon** (so called because it frequently takes the form of a Greek letter sigma). A loop of it may be so long as to be susceptible to *volvulus,* twisting on itself to produce obstruction. The attachment of the *sigmoid mesocolon* to the posterior body wall varies. It may run obliquely downward and medially across the pelvic brim; more often, the attachment resembles an inverted **V** that encloses between its limbs an *intersigmoid recess* of the peritoneum at the pelvic brim. The external and common iliac vessels are in the floor of the recess, and if the recess is deep, the ureter frequently passes across the vessels here on its way from the abdomen to pelvis. The sigmoid colon becomes the rectum when it loses its mesentery; this usually takes place in front of the 3rd sacral vertebra.

**Diverticulitis.** Diverticulitis is a common affliction of the colon, particularly on the left side. It is thought that pockets of colic mucosa herniate through the gaps created in the muscle coat of the colon by the entry of blood vessels along the mesenteric border or what was the mesenteric border. Fecal material may become impacted in these diverticula, which subsequently become inflamed; this is *diverticulitis.* Diverticulitis may present a clinical picture similar to that of appendicitis and may be confused with a left-sided appendix.

## VESSELS AND NERVES

### Blood Supply

The jejunum, ileum, cecum, ascending colon, and most of the transverse colon belong to the midgut; their arterial supply is furnished by the *superior mesenteric artery*. The hindgut, consisting of the left portion of the transverse colon, the descending and sigmoid colon, and the rectum, is supplied by the *inferior mesenteric artery.* The territories of drainage of the *superior* and *inferior mesenteric veins* conform to this basic plan of arterial distribution, and tributaries of the veins receive names that correspond with those of the arterial branches.

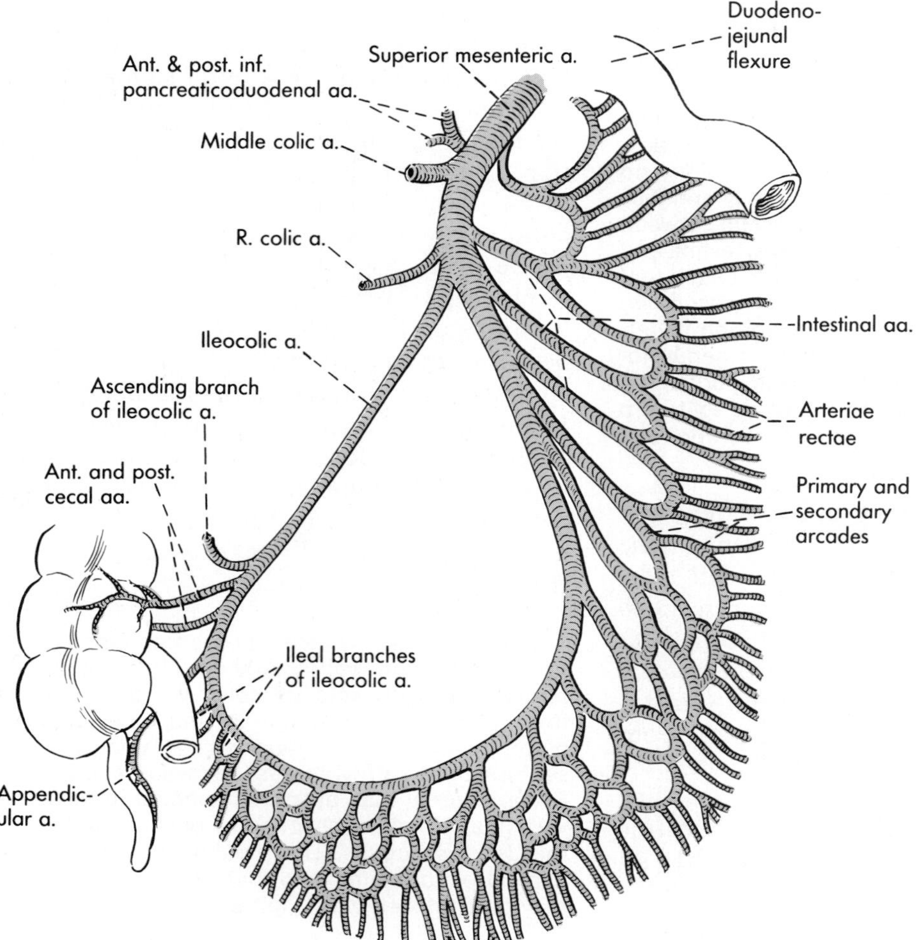


**FIG. 24-45.** Diagram of the blood supply to the jejunum and ileum: the ileocolic artery lies at the base of the mesentery of the intestine; the stem of the superior mesenteric artery swings into the mesentery.

**The Superior Mesenteric Artery.** The origin and initial course of the superior mesenteric artery (Fig. 24-45) behind and through the pancreas have already been encountered (p. 622), but some recapitulation may not be redundant here. It is the second *ventral branch* of the aorta, given off slightly below the celiac trunk, opposite the lower border of L1 vertebra. The artery descends in a groove on the posterior surface of the neck of the pancreas. Immediately below its origin, it crosses the left renal vein, which lies between it and the aorta. Below the inferior margin of the neck of the pancreas, it crosses anteriorly to the uncinate process and the horizontal portion of the duodenum; these also separate it from the aorta (Fig. 24-18). As it passes over the duodenum, it enters the root of the mesentery (Fig. 24-14). Within the mesentery, the main arterial stem describes an arc that spans the distance between the horizontal duodenum and the ileocecal junction, where the superior mesenteric artery terminates by anastomosing with one of its own branches, the ileocolic artery (Fig. 24-45).

The superior mesenteric artery gives off three sets of **branches:** (1) several small arteries before it enters the root of the mesentery;

(2) three large arteries for the supply of the large bowel from the right side of the proximal part of its course; and (3) an uninterrupted series of arteries for the jejunum and ileum from the left side of the arc it describes in the mesentery (Fig. 24-45).

The first group of small arteries include the **inferior pancreaticoduodenal stem,** which divides into anterior and posterior vessels (p. 643), and the first few of the jejunal arteries. Of the arteries to the large intestine, the **middle colic artery** is the first to arise from the right side of the superior mesenteric. This branch is usually given off at the inferior margin of the neck of the pancreas before the superior mesenteric artery enters the mesentery. The middle colic artery passes into the transverse mesocolon. On approaching the colon, it divides into *right and left branches* (Fig. 24-46). They contribute to the *marginal artery* by anastomosing with the adjoining vessels along the inner border of the colon (Fig. 24-47). The second branch given off from the right side is the **right colic artery.** It crosses the right half of the infracolic compartment retroperitoneally and, nearing the ascending colon, divides into an *ascending and descending branch,* which also contribute to the formation of the marginal artery (Figs. 24-46 and 24-47). The third artery given off from the right side of the superior mesenteric is the **ileocolic artery.** It runs a more or less straight course toward the ileocecal junction along the root of the mesentery or retroperitoneally, along the right side of the root of the mesentery. Near the ileocecal junction, it gives off several branches (Figs. 24-45 and 24-46): the *ascending colic artery,* which passes up along the ascending colon and contributes to the formation of the marginal artery; anterior and posterior *cecal arteries* (the anterior one contained in the vascular fold of the cecum, the posterior one passing retroperitoneally); the *appendicular artery,* which descends behind the ileum and enters the mesoappendix; and the *ileal branch,* which anastomoses with the terminal portion of the superior mesenteric artery.

**FIG. 24-46.** The blood supply of the colon.

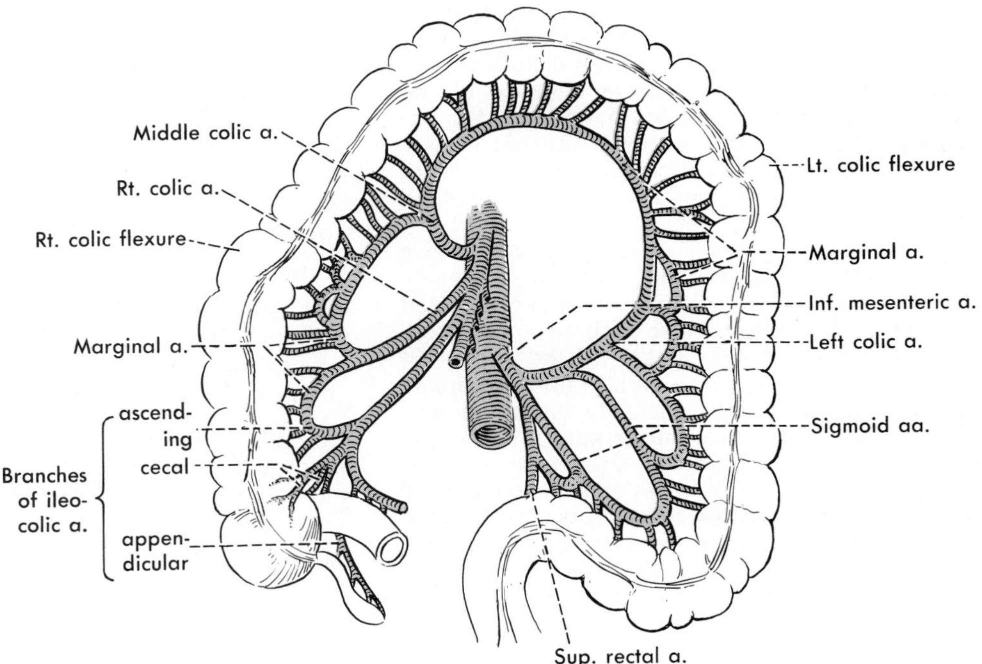

The **jejunal and ileal arteries,** given off in the mesentery from the convex left side of the superior mesenteric artery, may number up to 20. Each of these parallel vessels terminates in two branches that anastomose with the branch of their neighbors, forming a row of *arterial arcades.* The secondary branches that spring from these arcades reduplicate this pattern, adding tiers of arcades as the height of the mesentery increases (Fig. 24-45). From the last tier of arcades, straight vessels (*arteriae rectae*) approach the jejunum and ileum and enter their wall without anastomosing again along the mesenteric border of the gut.

In summary, the superior mesenteric artery supplies the distal half of the duodenum, part of the head of the pancreas, and the entire jejunum, ileum, cecum, appendix, and ascending colon and also most of the transverse colon. The branches to the small intestine form anastomosing arcades in the mesentery; the branches to the colon anastomose to form the marginal artery.

**Intestinal Strangulation and Gangrene.** Upon reaching the small intestine, the arteriae rectae pass deep to the peritoneum of one side or the other; sometimes an artery may branch and supply both sides simultaneously. These arteries anastomose at the antimesenteric border with those from the other side of the gut. Thus, although normal twisting of the mesentery does not disturb the circulation of the small intestine because of the collateral circulation afforded by the arcades, occlusion of a series of arcades or straight vessels seriously interferes with circulation and may lead to surgical emergency because of strangulation of the bowel.

When a loop of small intestine is caught in a peritoneal recess or in a hole in one of the mesenteries, not only may its lumen become obstructed but its blood vessels will also become compressed. Such compression first closes off the veins. The artery continues to pump blood into the trapped coil of bowel with no possibility for return flow through the veins. This causes edema and swelling, gradually obstructing the arteries. Thus, intestinal obstruction in such cases of so-called **internal hernias** will be complicated by *strangulation* of the bowel, and this, in turn, leads to *gangrene.* The same sequence of events may ensue when a part of the small or large intestine enters the inguinal or femoral canal in the form of a hernia (Chap. 26).

**The Inferior Mesenteric Artery.** The third ventral branch of the aorta arises about 4 cm above the aortic bifurcation behind or immediately below the horizontal part of the duodenum at the level of the 3rd lumbar vertebra. The inferior mesenteric artery descends in front and then along the left side of the aorta, lying beneath the peritoneal floor of the left infracolic compartment (Fig. 24-46). Reaching the left common iliac vessels, the artery enters the sigmoid mesocolon and continues into the pelvis as the *superior rectal artery.*

The **branches** of the inferior mesenteric artery are the left colic artery, two to four sigmoid arteries, and its terminal branch, the superior rectal artery. The **left colic artery** runs retroperitoneally toward the left and divides into *ascending* and *descending branches* that contribute to the marginal artery along the descending colon, the left colic flexure, and the transverse colon. Often before the left colic divides, it gives off a sigmoid artery that anastomoses with the **sigmoid arteries** proper given off directly by the inferior mesenteric. Sigmoid branches are also usually contributed by the superior rectal artery.

**The Marginal Artery.** Although not recognized by Nomina Anatomica as being worthy of a name of its own, a continuous arterial channel can be identified and dissected in many bodies that skirts the inner margin of the large intestine from the cecocolic junction to the rectosigmoid junction: surgeons call it the *marginal artery* (Fig. 24-47). This anastomotic channel consists of the ascending branch of the ileocolic artery; the descending and ascending branches of the right colic; the right and left branches of the middle colic; the ascending, descending, and sigmoid branches of the left colic; the sigmoid branches of the inferior mesenteric; and the superior rectal. When well developed, the marginal artery can serve as a good source of collateral circulation to a part of the colon whose chief arterial stem has been obstructed or ligated.

The collateral circulation to the left side of the colon is frequently made use of when, for

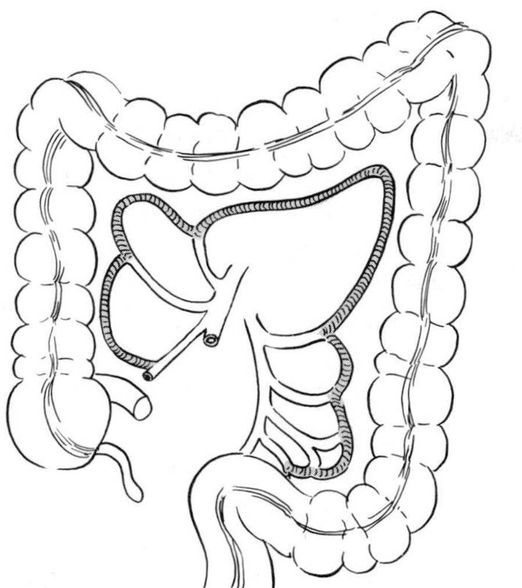

**FIG. 24-47.** Diagram of the marginal artery (colored).

instance, some of the sigmoid colon or rectum must be removed. In certain of these operations, the inferior mesenteric artery is ligated close to its origin, and in these instances, an adequate collateral circulation downward from the middle colic and upward from the middle rectal arteries usually can be obtained.

Although anastomoses in the rectum have traditionally been regarded as inadequate, and the anastomoses at the left colic flexure sometimes so, injection experiments have shown that both the middle colic and middle rectal arteries may be filled through the marginal artery by injection into the other vessel. The weakest part of the marginal artery is often between the ileocolic and right colic arteries, where there may be no anastomosis at all.

The marginal artery may be close to the wall of the bowel or some distance away from it; in the latter case, there may be more than one arcade formed by the anastomosing branches. From the arcades straight arteries (*arteriae rectae*) are given off that pass to the colon, some penetrating the wall of the colon at the mesenteric tenia, others running anteriorly and posteriorly around its surfaces to enter the other teniae.

There is some variation in the distribution in all the vessels to the colon. Perhaps the most variable is the distribution of the middle colic artery. This typically supplies the major part of the transverse colon but not the colic flexures, which are supplied by the right and left colic arteries, respectively. Sometimes, however, the middle colic supplies one or both flexures. Ligation of the middle colic artery predisposes more often to ischemia of a part of the colon than does ligation of any other colic artery. Occasionally, a colic vessel or one of its branches may be entirely absent.

**Veins.** All the major veins of the digestive tract in the abdomen (except the hepatic veins) are part of the portal system.

The **superior mesenteric vein** receives the drainage from a part of the head of the pancreas and duodenum, from the entire length of the jejunum and ileum, from the cecum and appendix, and from the ascending and most of the transverse colon; it thus parallels the artery. The veins of the jejunum and ileum follow essentially the same pattern as the arteries and unite to form the *superior mesenteric vein.* There are typically *ileocolic* and *right colic veins* joining the superior mesenteric vein. There may be more than one *middle colic vein,* and it may join the superior mesenteric or unite with pancreatic veins or veins from the stomach to enter the portal vein instead of the superior mesenteric.

At the upper part of the root of the mesentery, the superior mesenteric vein lies in front and slightly to the right of the superior mesenteric artery and passes in front of the duodenum to end by uniting behind the pancreas with the splenic vein, thereby forming the portal vein.

The superior rectal, sigmoid, and left colic veins unite to form the **inferior mesenteric vein,** which drains the descending colon, sigmoid colon, and much of the rectum, plus the left part of the transverse colon. It ends at a higher level than the artery originates, passing upward on the left of the aorta and the duodenojejunal flexure (Fig. 24-15) to disappear behind the pancreas, where it joins the splenic vein. Sometimes, instead, it curves medially at the lower border of the pancreas to join the superior mesenteric vein or the angle between the splenic and superior mesenteric veins.

Since all venous drainage of the digestive tract, including that of the stomach and upper duodenum, reach the liver, venous spread of gastrointestinal neoplasms almost always results in metastases to the liver.

## Lymphatics

As in other parts of the digestive tract, lymphatics of the small and large intestine follow the blood vessels and are filtered through several sets of lymph nodes. Lymph nodes are situated both close to the digestive tract and more centrally. Drainage is, in general, through a converging series of lymphatics and nodes, of which the chief and terminal ones are the *superior mesenteric* and *inferior mesenteric lymph nodes,* lying in front of the aorta. Lymph leaving the digestive tract does not necessarily run through each node along its pathway, for only a few lymphatics end in any one node; others bypass it and go to neighboring or more proximal nodes. In general, however, by the time lymph from the digestive tract has reached the superior or inferior mesenteric nodes, it has passed through several sets of regional lymph nodes.

It is for this reason that resection of a part of the gut containing a cancerous growth must be coupled with extensive excision of the mesenteries and all the draining lymph nodes.

Lymphatics of the small intestine perform a special function, the absorption of fat. Because of the emulsified fat in the lymph, the lymphatics of the mesentery are milky white and are therefore called *lacteals.* Their lymph is the *chyle.* Lacteals can be readily demonstrated in experimental animals by laparotomy following a fatty meal.

The lymphatics from most of the small intestine and the ascending and transverse colons drain into **superior mesenteric lymph**

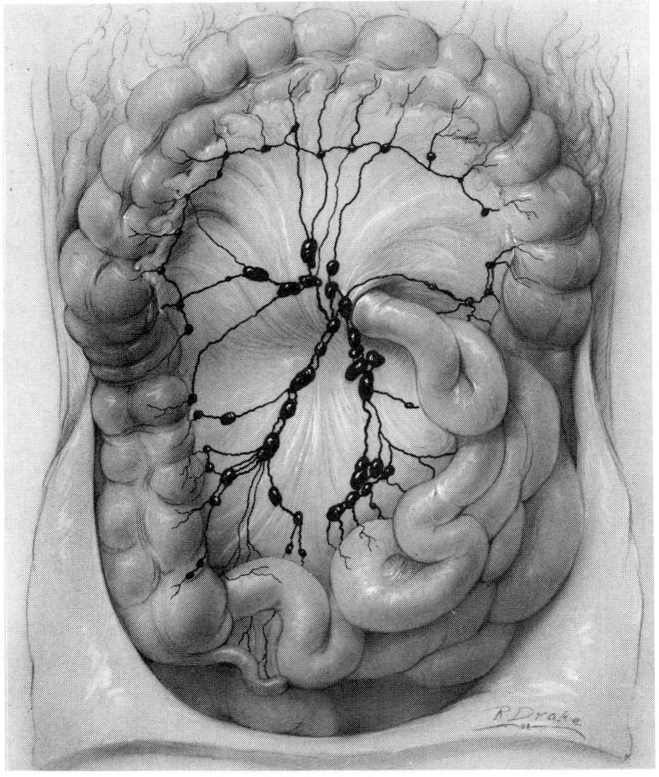

**FIG. 24-48.** Lymphatics and nodes associated with the superior mesenteric artery and its branches and draining the ascending and transverse parts of the colon and most of the small intestine. Nodes along the marginal artery, the middle and right colics, the ileocolic, and the stem of the superior mesenteric artery in the mesentery are all recognizable. (Desjardins AU: Arch Surg 38:714, 1939)

**nodes** (Fig. 24-48). The main nodes of this group are large, lying around the origin of the artery. They communicate with the adjacent celiac and upper lumbar nodes. Efferent lymphatics of the celiac and superior mesenteric lymph nodes together form an *intestinal lymph trunk* that terminates in the cisterna chyli (see Fig. 25-14).

Subsidiary nodes of the superior mesenteric group are the well over 100 *mesenteric nodes* situated in the mesentery of the small intestine along the arcades and branches of the superior mesenteric artery; *ileocolic nodes* along that artery, receiving the drainage of the cecum and appendix; and *right and middle colic nodes,* along the vessels of the same name. Since the superior mesenteric artery helps supply the pancreas and duodenum, a part of the drainage of these organs is also into the superior mesenteric nodes.

The lymphatics accompanying the inferior mesenteric vessels (Fig. 24-49) drain the major part of the rectum, the sigmoid and the descending colon, the left colic flexure, and the left end of the transverse colon. The lymph passes through one or more *left colic nodes* associated with the branches of the inferior mesenteric vessels, and the lymphatics converge on the large **inferior mesenteric lymph nodes** associated with the stem of the inferior mesenteric artery. Lymphatics that leave the inferior mesenteric nodes enter **lumbar nodes** situated along the aorta and the inferior vena cava.

The lymphatics from the left colic flexure and approximately the left third of the transverse colon drain downward along the ascending branch of the left colic artery. Only some of them join the inferior mesenteric nodes; others diverge from them at the duodenojejunal flexure and follow the terminal course of the inferior mesenteric vein superiorly and terminate in the superior mesenteric nodes. Thus, the left colic flexure and the left third of the transverse colon have dual lymphatic drainage: along branches of the

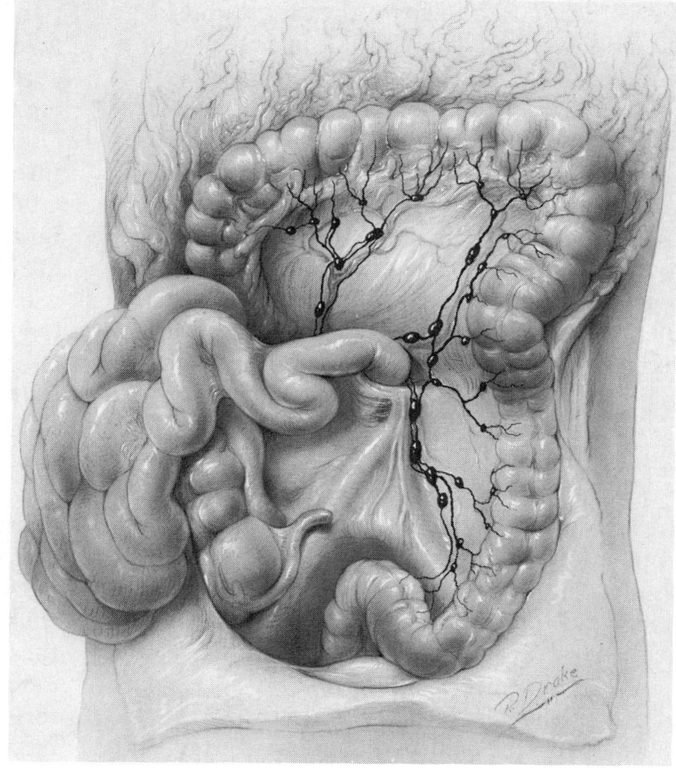

**FIG. 24-49.** Lymphatics and lymph nodes associated with the inferior mesenteric and middle colic arteries and draining the transverse, descending, and sigmoid parts of the colon (the rectal drainage, Fig. 27-20, is not shown here). Note that lymphatics from the region of the left colic flexure diverge close to the duodenojejunal junction, so that lymph can pass either above this junction to superior mesenteric nodes or continue downward to inferior mesenteric nodes. (Desjardins AU: Arch Surg 38:714, 1939)

middle colic artery to the superior mesenteric node, and along branches of the left colic artery to both superior and inferior mesenteric nodes.

Efferent lymphatics of the celiac and superior mesenteric nodes together form the **intestinal lymph trunk** (sometimes more than one). That trunk delivers lymph into the cisterna chyli from the visceral surface of the liver, the spleen, stomach, pancreas, and duodenum and from all the small intestine and the right side of the colon, including the major portion of the transverse colon. The *lumbar nodes* receive the drainage of the left side of the colon via the inferior mesenteric nodes and empty into the cisterna chyli. The lumbar nodes also receive the drainage from the kidneys, suprarenal glands, gonads, lower limbs, lower parts of the abdominal wall, perineum, and most of the pelvic viscera; thus, they are not primarily associated with the digestive tract and are discussed later.

## Nerve Supply

Nerve fibers to and from the small and large intestine are conveyed in the *superior* and *inferior mesenteric plexuses,* which are offshoots of the *aortic plexus* along the arteries of corresponding name and along subsidiary plexuses that follow branches of the two mesenteric vessels. The superior mesenteric plexus is continuous above with the celiac plexus and below through the *intermesenteric segment* of the *aortic plexus* with the inferior mesenteric plexus. Below the origin of the inferior mesenteric artery, the aortic plexus continues inferiorly and, beyond the aortic bifurcation, becomes the *superior hypogastric plexus,* which descends into the pelvis.

All the plexuses contain sympathetic collateral ganglia in which sympathetic preganglionic efferent fibers synapse. The largest of these ganglia are the celiac, superior mesenteric, and inferior mesenteric ganglia, located at the root of the respective vessels. There are, however, numerous other smaller aortic ganglia, the majority of microscopic size.

The **superior mesenteric plexus** and its subsidiary plexuses are composed of *vagal* parasympathetic fibers of both *efferent* and *afferent* functional types and of visceral efferents and afferents that connect the plexus to the sympathetic chain, passing through the thoracic splanchnic nerves.

The **vagal fibers** reach and leave the celiac plexus through the celiac branches of the vagal trunks, given off as they lie on the abdominal esophagus (Figs. 24-9, 24-10, and 24-50). Separate branches of the vagal trunks can be traced to the liver and stomach (p. 633; Fig. 24-9), but the vagal fibers for the rest of the foregut and midgut are intermingled with sympathetic fibers in the celiac and superior mesenteric plexuses. Vagal fibers running along the middle colic and marginal arteries reach possibly as far distally in the digestive tract as the left colic flexure. *Vagal efferents* pass through the aortic ganglia without a synapse; their postganglionic cell bodies are located in the *enteric ganglia.* Vagal efferents, in general, increase peristaltic activity and, at least in part, secretory activity; they inhibit the ileocecal sphincter. However, the vagi apparently have their chief effect on the stomach; the main results of vagotomy are decrease in the acid secretion and rate of emptying of the stomach, with no apparent change in the function of the small or large intestine.

The function of the *vagal visceral afferents* is largely unknown; they are believed to mediate the feelings of nausea and distention and are probably also involved in visceral reflexes (*e.g.,* gastrocecal reflex that activates the discharge of ileal contents into the cecum upon food entering the stomach).

The sympathetic components of the superior mesenteric plexus reach it through the *thoracic splanchnic nerves,* which, having passed through the diaphragm, enter the celiac plexus and, through it, descend to the superior mesenteric plexus (Fig. 24-50). Sympathetic *visceral efferents* synapse in the superior mesenteric and neighboring smaller aortic ganglia. The postsynaptic fibers reach the enteric plexus along the offshoots of the superior mesenteric plexuses that run with the

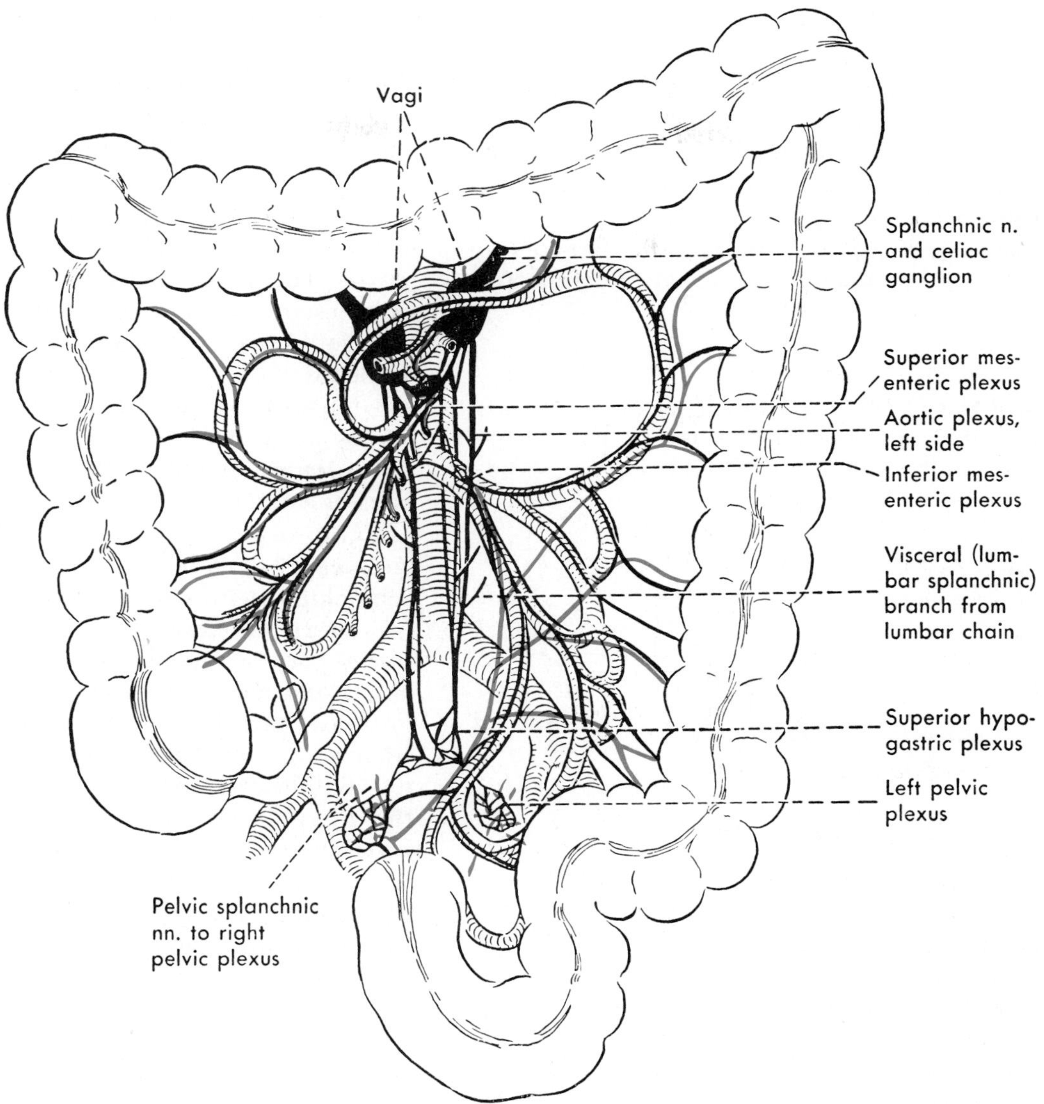


**FIG. 24-50.** Nerves of the large intestine. Sympathetic efferents and the visceral afferents that accompany them are represented by black lines; vagal parasympathetic fibers and sacral parasympathetic fibers are represented by red lines. In this diagram, the vagi are shown distributed as far as the transverse colon; parasympathetic fibers of sacral origin are shown supplying the descending colon.

branches of the superior mesenteric artery. Their function is to inhibit the smooth muscle of the digestive tract and possibly its secretory activity. Their most marked effect is, however, on blood vessels, which they constrict.

*Thoracolumbar sympathectomy* produces dilatation of abdominal blood vessels, and the vasodilatation in the extensive abdominal vascular bed contributes to lowering the blood pressure. Sympathectomy does not, however, produce any marked or constant change in the activity of the digestive tract.

**Visceral afferents** in the sympathetic nerves are chiefly concerned with the mediation of pain. They ascend through the plexuses, the ganglia, and the splanchnic nerves; pass through the sympathetic trunks and their ganglia; and, via their white rami communicantes, join the spinal nerve and its dorsal root. Their cell bodies are located in the spinal ganglia. The spinal segments that receive afferent input from abdominal viscera are, in general, the same as those that contain the visceral efferent neurons for the appropriate organs (Fig. 24-51).

**FIG. 24-51.** The approximate segmental distribution of sympathetic efferent and visceral afferent fibers to the abdominal viscera.

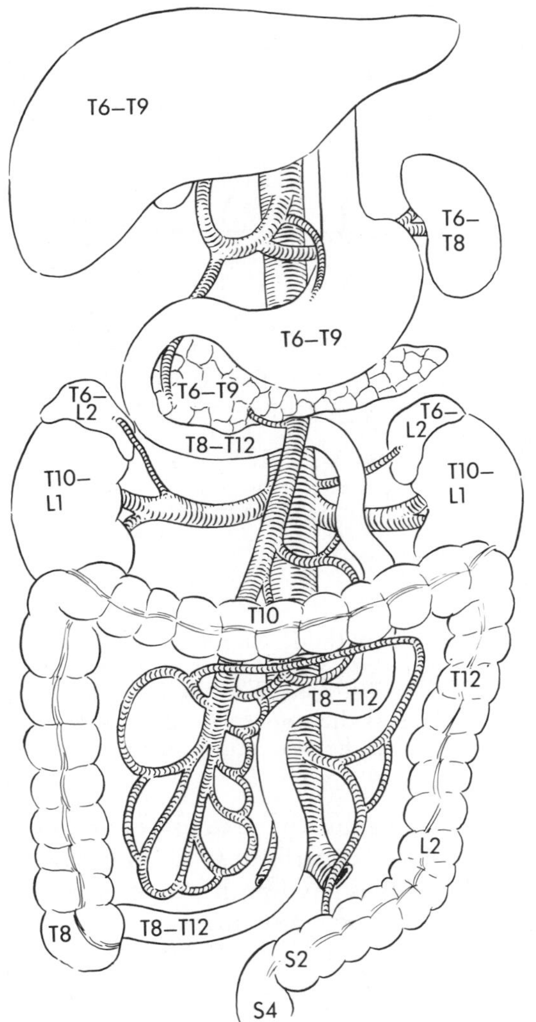


The **inferior mesenteric plexus** receives sympathetic fibers from the superior mesenteric plexus through the intermesenteric plexus (Fig. 24-50). The connections and distribution of both efferent and afferent nerves belonging to the sympathetic system conform to the same principles and pattern as those just described for the superior mesenteric plexus. The sympathetic input into the inferior mesenteric plexus is reinforced, however, by the *lumbar splanchnic nerves*, visceral branches of the upper lumbar ganglia, which are in series with the thoracic splanchnic nerves.

The *parasympathetic components* of the inferior mesenteric plexus are not derived from the vagi but rather from pelvic splanchnic nerves. *Visceral efferents* of the **pelvic splanchnic nerves** have their cell bodies in the intermediolateral cell column of S2–S4 spinal cord segments. Their axons enter the pelvis and ascend through the pelvic plexus into the mesentery of the sigmoid colon, where they commingle along the marginal artery with the sympathetic fibers derived from the inferior mesenteric plexus. Like the vagal fibers in the foregut and midgut, they synapse in enteric ganglia in the wall of the hindgut (descending and sigmoid colon and the rectum). It is uncertain whether any *visceral afferents* of the descending and sigmoid colons pass along the pelvic splanchnic nerves, but pain afferents definitely follow the sympathetic pathway in the inferior mesenteric plexus. Exactly where the vagal parasympathetic and sacral parasympathetic fibers meet is not known, but it is believed to be no farther distally than the left colic flexure. Although pelvic splanchnic nerves certainly reach the sigmoid and the descending colon, it seems that parasympathetic fibers are absent from the plexus around the stem of the inferior mesenteric artery and also from the aortic plexus between the artery and the aortic bifurcation.

### Visceral Pain

Most accounts indicate that afferent fibers accompany the vagus from abdominal organs; however, the presence of sacral parasympathetic afferents from the descending and sigmoid colon is controversial. On the other hand, it has been clearly shown that all pain from abdominal viscera is conducted by fibers that run with the sympathetic system and enter the spinal cord through dorsal roots of the same nerves that give rise to the preganglionic sympathetic fibers to the viscera (Fig. 24-51). Although pain arising in the abdominal viscera can be alleviated by sectioning the splanchnic nerves or removing the thoracolumbar parts of the sympathetic trunks, the pain will not be eliminated if disease also involves parietal peritoneum, which is innervated through somatic nerves of the body wall. A sympathectomy may be carried out for the relief of pain, for instance, in severe pancreatitis or, in some cases, a painful and inoperable carcinoma. The sympathectomy will be effective in the relief of pain only as long as the disease is confined to the organs. Once invasion of somatic structures occurs, the pain will return and be of a different character.

It should be remembered that afferent pathways from the viscera that run in the sympathetic system pass through the dorsal roots of the spinal nerves, where they join somatic afferents from the body wall. Moreover, within the spinal cord, visceral afferents may end on the same group of interneurons as somatic afferents that have entered the cord in the same dorsal root. Thus, some pain fibers from the viscera and from the body wall may share the same neurons in the posterior horn of the spinal cord. This sharing of a common interneuronal pool is believed to be the chief explanation for "**referred pain.**" Referred pain is pain that actually originates in a viscus but is perceived as if located in a somatic structure such as the skin or the underlying musculature. This somatic pain is rather well localized and quite distinct from visceral pain, which is deep seated and poorly localized. Because of the shared interneuron pool, the sufferer cannot distinguish whether the pain stimulus reaches the spinal cord along somatic or visceral pain afferents. He will allocate the pain to the somatic region, since this has good cortical representation.

## GENERAL REFERENCES AND RECOMMENDED READINGS


BARCLAY AE: The Digestive Tract. London, Cambridge University Press, 1936

BOYDEN EA: The anatomy of the choledochoduodenal junction in man. Surg Gynecol Obstet 104:641, 1957

BOYDEN EA, COPE JG, BILL AH JR: Anatomy and embryology of congenital intrinsic obstruction of the duodenum. Am J Surg 114:190, 1967

BOYDEN EA, VAN BUSKIRK C: Rate of emptying of biliary tract following section of vagi or of all extrinsic nerves. Proc Soc Exp Biol Med 53:174, 1943

CAULDWELL EW, ANSON BJ: The visceral branches of the abdominal aorta: Topographical relationships. Am J Anat 73:27, 1943

DAVIES F, HARDING HE: Pouch of Hartmann. Lancet 1:193, 1942

DOUGLASS BE, BAGGENSTOSS AH, HOLLINSHEAD WH: The anatomy of the portal vein and its tributaries. Surg Gynecol Obstet 91:562, 1950

ELIAS H: Origin and early development of the liver in various vertebrates. Acta Hepatol 3:1, 1955

FLEISCHNER FG, SAYEGH V: Assessment of the size of the liver: Roentgenologic considerations. N Engl J Med 259:271, 1958

HAYWARD J: The lower end of the oesophagus. Thorax 16:36, 1961

HEALEY JE JR, SCHROY PC: Anatomy of the biliary ducts within the human liver: Analysis of the prevailing pattern of branchings and the major variations of the biliary ducts. Arch Surg 66:599, 1953

HUNT JN, KNOX MT: Regulation of gastric emptying. In Code EF (ed): Handbook of Physiology, Sect 6, Alimentary Canal, vol 4, p 1917. Washington, American Physiological Society, 1968

JAMIESON JK, DOBSON JF: The lymphatic system of the liver. Lancet 1:1161, 1907

JEFFERSON G: The human stomach and the canalis gastricus (Lewis). J Anat 49:165, 1915

KLEITSCH WP: Anatomy of the pancreas: A study with special reference to the duct system. Arch Surg 71:795, 1955

LEWIS FT: The form of the stomach in human embryos with notes upon the nomenclature of the stomach. Am J Anat 13:477, 1912

MAISEL H: The position of the human vermiform appendix in fetal and adult age groups. Anat Rec 136:385, 1960

MICHELS NA: The variational anatomy of the spleen and splenic artery. Am J Anat 70:21, 1942

MICHELS NA: Blood Supply and Anatomy of the Upper Abdominal Organs: With a Descriptive Atlas. Philadelphia, JB Lippincott, 1955

MICHELS NA, SIDDHARTH P, KORNBLITH PL ET AL: The variant blood supply to the small and large intestines: Its import in regional resections. A new anatomic study based on four hundred dissections, with a complete review of the literature. J Int Coll Surg 39:127, 1963

POPPER H, SCHAFFNER F: Liver: Structure and Function. New York, McGraw-Hill, 1957

RAY BS, NEILL CL: Abdominal visceral sensation in man. Ann Surg 126:709, 1947

ROSENBERG JC, DIDIO LJA: *In vivo* appearance and function of the termination of the ileum as observed directly through a cecostomy. Am J Gastroenterol 52:411, 1969

THOMPSON IM: On arteries and ducts in hepatic pedicle: Study in statistical anatomy. Univ Calif Publ Anat 1:55, 1933

TREVES F: The Anatomy of the Intestinal Canal and Peritoneum in Man. London, HK Lewis, 1885

UNDERHILL BML: Intestinal length in man. Br Med J 2:1243, 1955

WAKELEY CPG: The position of the vermiform appendix as ascertained by an analysis of 10,000 cases. J Anat 67:277, 1933

WHILLIS J: Lower end of the esophagus. J Anat 66:132, 1931

WOODBURNE RT, OLSEN LL: The arteries of the pancreas. Anat Rec 111:255, 1951

# 25

# THE POSTERIOR ABDOMINAL WALL AND ASSOCIATED ORGANS

As noted in Chapter 23, the posterior abdominal wall is constructed on a different plan than the distensible anterolateral walls of the abdomen (see Fig. 23-1). The posterior abdominal wall is bulky and stable because of the lumbar vertebral column and the iliac bones, to which are attached the quadratus lumborum, psoas, and iliacus muscles (Fig. 25-1). These muscles, rather than supporting the abdominal viscera and controlling abdominal pressure as the muscles of the anterior abdominal wall do, act on the vertebral column and the hip joint.

The posterior part of the abdominal wall is formed, in the midline, by the lumbar portion of the *vertebral column;* laterally, on each side of this, by the *psoas major* and *quadratus lumborum muscles;* and in the iliac region, by the *iliacus muscle,* which covers the internal surface of the wing of the ilium. Lateral to the quadratus lumborum is the *transversus abdominis.* The *diaphragm* may be considered along with the posterior abdominal wall, because it not only forms the roof of the abdominal cavity but it completes the posterior wall superiorly.

Embedded in the psoas major is the *lumbar plexus,* formed by ventral rami of the lumbar spinal nerves. From the plexus issue major nerves for the supply of the lower limb and the inguinal region. These nerves descend on the abdominal surface of the muscles of the posterior wall.

On the anterior surface of the lumbar vertebral bodies lie the *abdominal aorta,* with its associated *autonomic nerve plexuses,* and the *inferior vena cava* (Fig. 25-1). On the lateral aspects of the same vertebrae are the *sympathetic trunks* and chains of *lymph nodes* and *lymphatics*. The organs that are developmentally associated with the posterior abdominal wall are derived from *intermediate mesoderm* and the overlying celomic epithelium. They are the *kidneys*, the *suprarenal glands*, and the *gonads*. In the female, the gonads have descended, with their ducts, into the pelvis; in the male, they have descended into the perineum. They will be considered in the appropriate chapters. The kidneys, ureters, and suprarenal glands remain on the posterior abdominal wall outside the peritoneal sac and will be dealt with in the last section of this chapter.

## THE MUSCLES AND FASCIAS

### The Quadratus Lumborum

A roughly quadrilateral muscle, the quadratus lumborum, fills the medial half of the gap between the last rib, the iliac crest, and the tips of the lumbar transverse processes (Fig. 25-1). It is attached to all of these bones and to the iliolumbar ligament (see Fig. 17-36).

The **actions** of the quadratus lumborum include depression and stabilization of the 12th rib and lateral bending of the trunk. In forced expiration, when the diaphragm is relaxed, the quadratus lumborum depresses the 12th rib, providing the base for intercostal muscle action that can then depress the other ribs. Fixation of the 12th rib provides the stable base for diaphragmatic contraction, a chief factor in inspiration. In lateral bending of the trunk, the quadratus lumborum approximates the rib cage to the iliac crest on its own side. Acting together, the muscles of the two sides may extend the lumbar spine, increasing its lordosis.

The quadratus lumborum is supplied by the ventral rami of T12, L1–4.

**Relations.** The most medial portion of the quadratus lumborum is overlapped anteriorly by the psoas major (Fig. 25-1). Along its lateral edge, the transversus abdominis and internal oblique arise from the anterior layer of the thoracolumbar fascia, lying behind the muscle (see Fig. 15-2). Close to the 12th rib, the quadratus lumborum is crossed anteriorly by the *lateral arcuate ligament* (Fig. 25-3). The *subcostal nerve* enters the abdomen by passing behind the ligament and then crosses the quadratus lumborum. It descends on the anterior surface of the muscle, more or less parallel with branches of the lumbar plexus that emerge along the lateral border of the psoas (Fig. 25-2). The *subcostal artery* and the anterior branches of the *lumbar arteries*, however, pass posterior to the muscle. In addition to the nerves, the anterior relations of the quadratus lumborum on each side are the kidneys and the colon.

### The Psoas and Iliacus

The psoas and iliacus are two separate muscles in the abdomen, but they exert their flexor action on the hip through a common tendon inserted into the lesser trochanter. In this context, they were described in Chapter 17 (see Figs. 17-32, 17-36, and 17-38).

The **psoas major** is shaped like an elongated cone; its apex, represented by the tendon of insertion, points into the thigh. The psoas covers the anterolateral surface of the lumbar vertebral bodies, largely filling the space between the transverse processes and the vertebral bodies (Fig. 25-23; see Fig. 23-1). The muscle arises by several sets of slips: from the anterior margin of the lower edge of all lumbar transverse processes; from the anterolateral circumference of the intervertebral disks between vertebrae T12–L5 and the adjacent margins of the vertebral bodies; and from five small, tendinous arches that span the slight concavity of each vertebral body between its upper and lower margin (see Fig. 17-36).

The lumbar plexus is enclosed within the muscle between the fibers that arise from the transverse process and those attached to the disks and vertebral bodies. The lumbar arteries and veins and the rami communicantes of lumbar sympathetic ganglia skirt the vertebral bodies running underneath the five tendinous arches of the psoas origin.

From this wide vertebral attachment, the

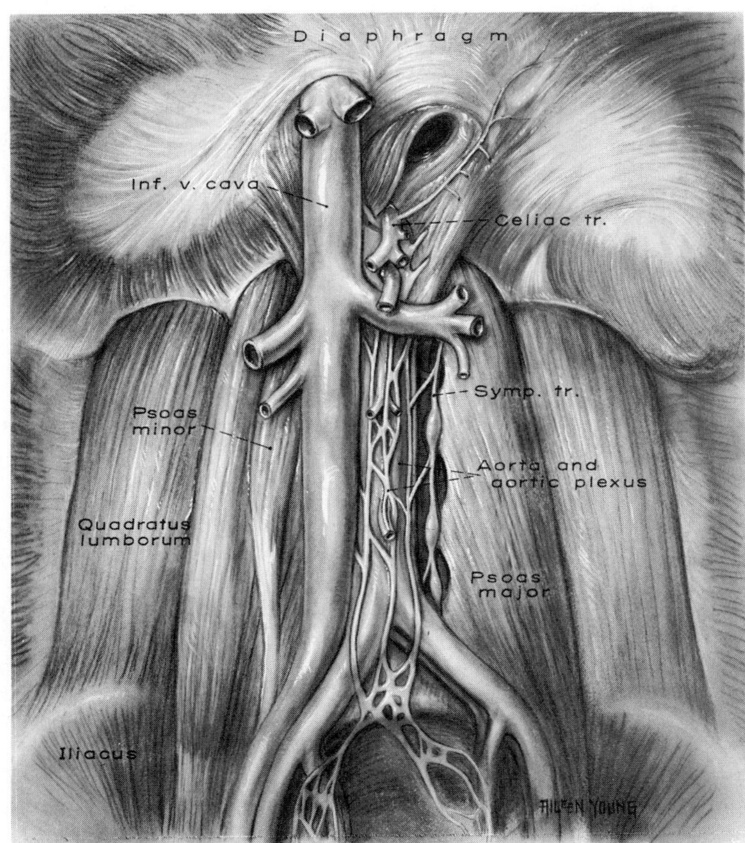


**FIG. 25-1.** The great vessels and the autonomic nerves on the posterior abdominal wall. Around the celiac and superior mesenteric stems as they arise from the aorta are ganglia of the celiac plexus, and below these, the aortic plexus lies on the front of the aorta and then, crossing the aortic bifurcation, proceeds into the pelvis. The left lumbar sympathetic trunk is visible alongside the aorta; the right one is hidden by the inferior vena cava.

psoas tapers downward, crosses in front of the ala of the sacrum and the sacroiliac joint, and then runs along the pelvic brim where the iliacus is immediately lateral to it.

The fan-shaped **iliacus** arises from most of the inner surface of the wing of the ilum (iliac fossa) and inserts many of its fibers into the tendon of the psoas as the two muscles leave the abdomen between the inguinal ligament and the superior ramus of the pubis (see Fig. 17-36).

A slender muscle bundle, the **psoas minor,** is present on the anterior surface of the psoas major in less than half of the bodies. It arises with the highest fibers of the psoas major and its long, narrow tendon inserts at or near the iliopubic eminence.

**Action.** In addition to flexing the thigh on a stabilized trunk, the iliopsoas is an important flexor of the trunk at the hip. In the erect position, the force for this movement is provided by gravity. However, when the trunk is flexed against gravity, as during sit-ups, most of the required force is generated by the iliopsoas, especially when the movement is performed with the knees straight on the ground. The psoas major is not an effective flexor of the lumbar spine itself, in view of its rather posterior attachment to the vertebrae in relation to the lumbar curvature. The contribution of the psoas to vertebral movements and of the iliopsoas to rotation at the hip remains controversial.

The psoas is innervated by ventral rami L1–L3, and the iliacus by branches of the femoral nerve (L2,3).

**Relations.** The psoas is crossed superiorly by the *medial arcuate ligament,* and on its an-

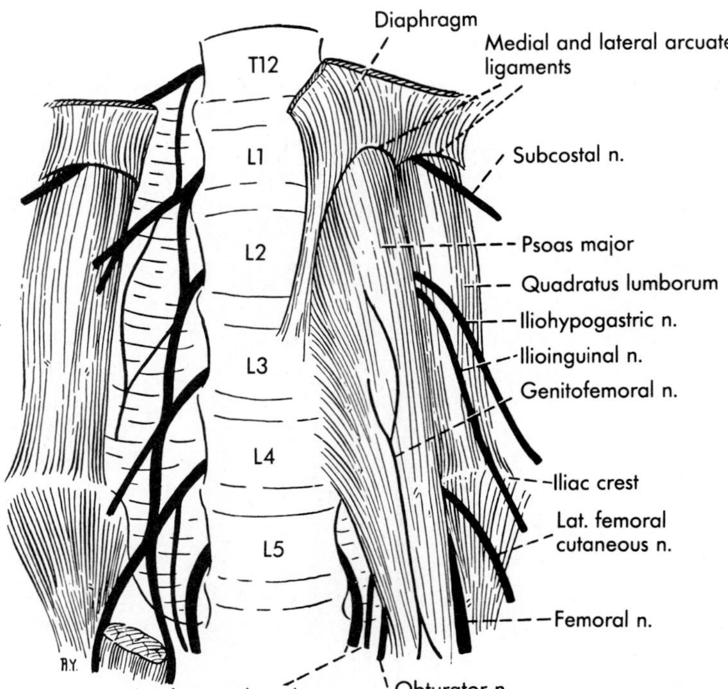


**FIG. 25-2.** The lumbar plexus. The right psoas muscle and the associated part of the diaphragm have been removed.

terior surface lie the kidneys (Fig. 25-1). Lower down, the muscles are crossed anteriorly by the ureters, the testicular or ovarian vessels, and the vessels of the ascending and descending colon. The right psoas is crossed also by the root of the mesentery of the small intestine and the left psoas by the root of the sigmoid mesocolon. The lumbar portions of the two sympathetic trunks lie on the front of the psoas major and on the vertebral column just at the origin of the muscle. The lower and medial part of the psoas is covered by the common iliac and external iliac vessels as these run to and along the pelvic brim.

The *lumbar plexus* takes form within the substance of the psoas major. The plexus can be displayed only by tearing the muscle, for it does not divide the muscle into planes. Its branches emerge through the psoas anteriorly, along its lateral border, or along the medial border (Fig. 25-2). The *genitofemoral nerve,* a branch of the lumbar plexus, runs downward on the front of the psoas major and may be mistaken for the tendon of the psoas minor. The genitofemoral nerve divides on the lower part of the iliopsoas into a genital and a femoral branch; the former leaves the abdomen through the deep inguinal ring, and the latter by passing along the external iliac artery. The branches that emerge through the lateral border of the psoas are, in order from above downward, the iliohypogastric and ilioinguinal nerves, the lateral femoral cutaneous nerve, and the femoral nerve.

The *femoral nerve* emerges from the muscle close to the iliac crest, runs down in the groove between the psoas and iliacus muscles, and enters the thigh with these muscles. The *lateral femoral cutaneous nerve,* emerging slightly higher, passes across the lower part of the quadratus lumborum and across the iliac crest to run on the surface of the iliacus muscle. It makes its exit from the abdomen behind the lateral part of the inguinal ligament, or it may pierce through the ligament. The *iliohypogastric* and *ilioinguinal nerves* emerge separately or together through the upper part of the muscle and run laterally and downward across the quadratus lumborum before turning forward on the internal surface of the transversus abdominis. They enter the transversus only slightly above the iliac crest, some 2 cm to 3 cm posterior to the anterior superior iliac spine.

On the wing of the sacrum, concealed by the common iliac vessels, three important structures lie on the medial side of the psoas: (1) the *lumbosacral trunk,* made up of fibers from L4 and L5 ventral rami, which connect the lumbar plexus to the sacral plexus; (2) the *obturator nerve,* a branch of the lumbar plexus, which leaves the pelvis through the obturator foramen to supply the adductor compartment of the thigh; and (3) the *iliolumbar artery,* a branch of the internal iliac, which ascends from the pelvis and gives off a branch that, along the lateral side of the psoas, fans out into the iliacus.

The iliacus muscle is covered with peritoneum, except where the cecum and ascending colon or the descending colon lies against its upper part.

### Fascias of the Posterior Abdominal Muscles

Although the fascia on the abdominal surface of the quadratus lumborum, psoas, and iliacus can be considered analogous with the transversalis fascia, there are sufficient specializations to merit mention (p. 591). The membranous fascia on the surface of these muscles is distinct from the extraperitoneal fat that may be quite voluminous in this region and will be considered with the kidneys.

The transversalis fascia, which clothes the inner surface of the transversus muscle, is continued onto the lower surface of the diaphragm as the *diaphragmatic fascia;* it also continues over the anterior surfaces of the quadratus lumborum and psoas major muscles as the fascia of these muscles. After attachment to the iliac crest, the transversalis fascia clothes the iliacus muscle as the iliacus fascia.

The fascia of the quadratus lumborum and psoas major muscles is thickened above where the diaphragm crosses the muscles. The thickenings, called the *lateral* and *medial arcuate ligaments* (lumbocostal aches), give rise to fibers of the diaphragm and are considered a part of this muscle. Through the ligaments, the diaphragm is tightly attached to the muscles, thus sealing any potential aperture behind the diaphragm in this location.

Branches of the lumbar plexus are beneath the psoas and quadratus lumborum fascia, but the common and external iliac vessels lie on the surface of the psoas fascia. The psoas fascia descends into the thigh with the muscle and its tendon. An abscess of one of the lumbar vertebral bodies may discharge its contents into the psoas fascia and track down into the groin, where the resultant swelling (*psoas abscess*) may be confused with a hernia. Behind the external iliac vessels, the iliacus fascia contributes to the formation of the posterior wall of the femoral sheath (p. 730), through which a femoral hernia protrudes into the groin.

### The Diaphragm

The diaphragm is a curved, musculotendinous sheet intervening between the thoracic and abdominal cavities. Those structures that pass between the thorax and the abdomen necessarily penetrate the diaphragm or pass behind it.

**Parts.** The fibers of the diaphragm are arranged radially around a **central tendon** and are divisible into sternal, costal, and lumbar portions (Fig. 25-3).

The small **sternal portion** of the diaphragm arises from the posterior surface of the xiphoid process and runs upward and backward to insert into the central tendon. The extensive **costal portion** of the muscle arises from the inner surface of the 7th, 8th, and 9th costal cartilages and the distal ends of the last three ribs. Between the origin from the 12th rib and the origin from the vertebral column, the muscular fibers arise from the arcuate ligaments.

The *lateral arcuate ligament* extends across the quadratus lumborum from the 12th rib to the transverse process of the first lumbar vertebra. The *medial arcuate ligament* extends

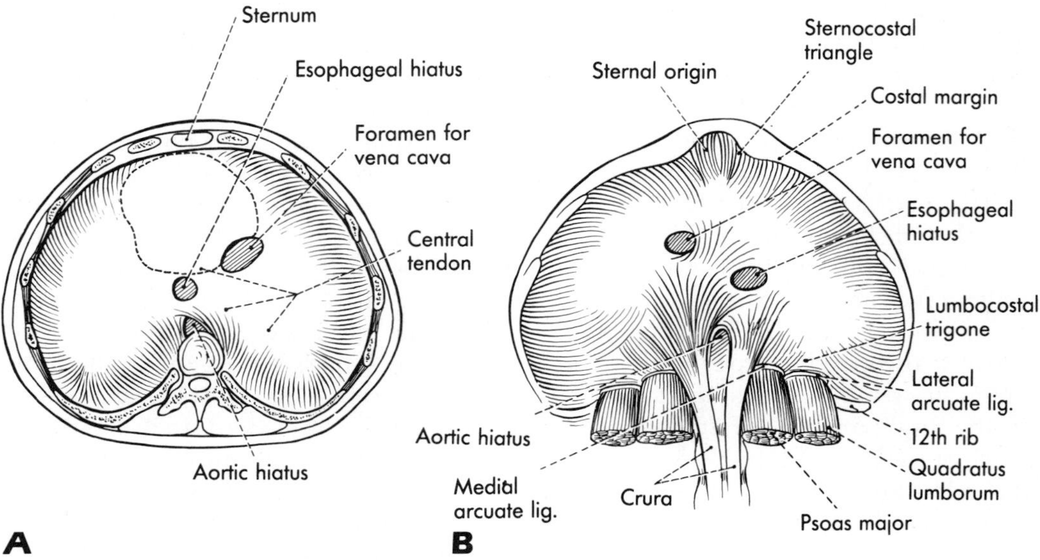


**FIG. 25-3.** Superior, **A,** and inferior, **B,** views of the diaphragm.

from this transverse process across the psoas major to the side of the body of L1 or L2 vertebra to blend here with the corresponding crus of the lumbar part of the diaphragm. That part arising from the arcuate ligaments, therefore, represents the region of transition between the costal and lumbar part of the diaphragm.

Sometimes the diaphragm is thin and nonmuscular above the lateral arcuate ligament, apparently as a result of degeneration of the muscle that should be here. This weak place is the *lumbocostal trigone* and is more commonly present on the left side than on the right. It is of clinical importance because the kidney lies directly in front of it, and in operations on the kidney, inadvertent breakthrough of the thin lumbocostal triangle into the pleural cavity may occur since the pleura of the costodiaphragmatic recess is immediately posterior.

The **lumbar part** of the diaphragm arises by two **crura,** tendinous in their lower parts where they are firmly attached to the front of the vertebral column and becoming muscular as they curve forward into the diaphragm. The crura are separated by a gap, the *aortic hiatus.* The *right crus* of the diaphragm is usually wider and about one vertebral segment longer than the *left crus* and is attached to the front of the upper three or four lumbar vertebrae. At the level of the 12th thoracic vertebra, the two crura are united over the front of the aortic hiatus by a tendinous arch, the *median arcuate ligament.* Beyond this, the muscular fibers derived from the right crus spread out so that they regularly pass on both sides of the esophageal hiatus. In contrast, the left crus typically sends few or no fibers to the right of the esophageal hiatus, its fibers curving primarily to the left of this.

**Apertures.** Structures that pass between the thoracic and abdominal cavities do so through three large, named apertures and through several unnamed, smaller openings or gaps. The three major apertures are the aortic and esophageal hiatuses and the foramen for the inferior vena cava.

The **aortic hiatus** is located behind, rather than within, the diaphragm. It is enclosed by the 12th thoracic vertebra, the two crura, and the median arcuate ligament. Through the hiatus, the *aorta* descends, changing its name as it does so from the descending thoracic aorta to the descending abdominal aorta. The *thoracic duct* ascends through the hiatus as does, sometimes, the *azygos vein.* Because these structures are located behind the diaphragm, they are not compressed by its contraction.

The **esophageal hiatus** lies slightly to the left of the midline and is enclosed by the insertion of the medial fibers of the right crus into the central tendon. Since it is situated farther forward on the curve of the diaphragm than is the aortic hiatus, it also lies higher, approximately at the level of the 10th thoracic vertebra in the relaxed diaphragm. Through the hiatus pass, with the *esophagus,* the *vagal trunks* and the *esophageal branches of the left gastric vessels.*

Communication between the thoracic and abdominal cavities through the hiatus is sealed off by the *phrenoesophageal ligament.* This ligament is a lamina of transversalis (diaphragmatic) fascia that surrounds the esophagus as it lies in the hiatus and attaches to its muscular wall some distance above the gastroesophageal junction, penetrating as far as the submucosa of the esophagus. Enlargement of the hiatus involves stretching the diaphragmatic muscle and the phrenoesophageal ligament. Protrusion of abdominal contents into the thorax through such an enlarged opening is known as a *hiatus hernia.*

It is controversial whether or not diaphragmatic contraction, especially that of the fibers of the right crus, provide a sphincter action around the esophageal hiatus. Factors implicated in the sphincter mechanisms at the cardia were discussed in Chapter 24.

The **foramen for the inferior vena cava** lies still farther forward and correspondingly higher than the esophageal hiatus at the height of the right dome of the diaphragm. In the cadaver, the foramen lies at the level of the 8th thoracic vertebra. It is within the tendinous part of the diaphragm, and the inferior vena cava is firmly attached to the margins of the foramen as it passes through the diaphragm. Thus, contraction of the muscular parts of the diaphragm stretches the caval opening, aiding venous flow from the abdomen toward the right atrium. Branches of the right phrenic nerve pass through the opening along with the inferior vena cava.

Of the **smaller structures** passing between the thorax and abdomen, the superior epigastric vessels are anterior, and all others pass behind the posterior part of the diaphragm. The *sternocostal triangle* is a small gap between the sternal and costal fibers of the diaphragm (Fig. 25-3). Through it, on each side of the xiphisternum, pass the *superior epigastric vessels* (the continuation of internal thoracic vessels to the anterior abdominal wall). These vessels are accompanied by some lymphatics.

The path of the posterior structures is not constant; they simply pierce through the muscular fibers or pass behind the arcuate ligaments without producing a real hiatus in the diaphragm. The *sympathetic trunks* descend behind the medial end of the medial arcuate ligaments; the *subcostal nerves* descend behind the lateral arcuate ligaments. In addition, three *splanchnic nerves* pierce each crus; the abdominal beginnings of the *azygos and hemiazygos veins* pass behind the crura. The *phrenic nerves* pierce the central tendon of the diaphragm or the muscle fibers close to the tendon, the left phrenic at the apex of the pericardial sac and the right near the foramen of the inferior vena cava.

**Relations.** The upper surface of the diaphragm is largely covered by pleura and pericardium and, through these, is in contact with the base of the *lungs* and with the *heart.* The lower surface is, in part, covered by adherent peritoneum, continuous with the peritoneum of the falciform, coronary, and triangular ligaments of the liver and, between the liver and esophagus, the lesser omentum. Through this peritoneum, the diaphragm is in contact with the *liver* on the right side and with the *stomach* and *spleen* on the left. The peritoneal spaces between the diaphragm and these organs are the *subphrenic spaces.* The lower posterior part of the diaphragm, facing mostly forward, is widely separated from the peritoneum by the *kidneys,* with their adipose capsules and surrounding fat, and those retroperitoneal structures that lie in front of the kidneys. Except for intervening fat, the *suprarenal glands* and the upper poles of the kidneys rest directly on the diaphragm.

**Blood Supply.** The chief blood supply to the diaphragm reaches its abdominal surface from the *inferior phrenic arteries.* These paired arteries, as a rule, arise from the aorta just as it enters the abdomen. A branch of each pierces the diaphragm to help supply the pericardial sac and becomes continuous with the *pericardiacophrenic arteries* (branches of the internal thoracics). Other vessels to the diaphragm are the small *superior phrenic arteries* from the thoracic aorta, and branches from the *musculophrenic,* the terminal branch of the internal thoracic that runs along the costal attachment of the diaphragm.

Of the *veins,* the most important are the inferior phrenic veins. The right inferior phrenic vein ends in the upper part of the inferior vena cava; the left may do so, but it usually descends and ends in the left suprarenal vein or the left renal vein.

**Innervation and Action.** The **motor innervation** of the diaphragm appears to be entirely from the two phrenic nerves (C3–C5), which descend through the thorax between the pericardium and pleura and enter the diaphragm. The **sensory innervation** is from two sources. A large central area of the diaphragm is supplied with sensory fibers from the phrenic nerves, but its peripheral part is supplied by twigs of the intercostal nerves that run into it where it is attached to the ribs. In consequence of this sensory innervation, pain from the central part of the diaphragm will be referred to the base of the neck and to the shoulder, the cutaneous area innervated by the 3rd to 5th cervical nerves; pain from the periphery of the diaphragm is referred to the costal area or the anterior abdominal wall.

The diaphragm is the most important **muscle of respiration.** Its contribution to respiratory movements was described in Chapter 19. Together with the anterior abdominal muscles, the diaphragm is also important in raising abdominal pressure necessary for the evacuation of abdominal contents: defecation, micturition, and parturition.

**FIG. 25-4.** Development of the diaphragm. Parts derived from the septum transversum are shown in pink; those from the mesoesophagus are uncolored. Parts derived from the pleuroperitoneal membranes are blue; those from the body wall are stippled.

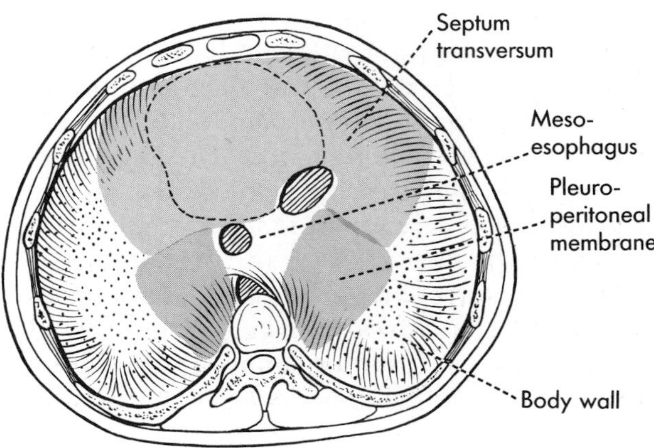


**Development.** The diaphragm develops from tissue derived from four sources: the septum transversum, the dorsal mesentery of the esophagus or mesoesophagus, the pleuroperitoneal membranes, and the somatic mesoderm of the body wall (Fig. 25-4).

The **septum transversum,** which has been discussed before in several contexts, forms the central tendon and most of the muscular part of the diaphragm anterior to it. The **mesoesophagus** contributes the median portion of the diaphragm behind the central tendon, including the crura and the parts that surround the esophageal hiatus. The **pleuroperitoneal membranes** grow from the walls of the pleuroperitoneal canals toward the posterolateral edge of the septum transversum and the mesoesophagus and thereby close off the communication between the pleural and peritoneal parts of the celomic ducts. The tissue they contribute to the diaphragm is in the posterolateral portion of the central tendon. The most peripheral muscular parts of the diaphragm, forming the medial walls of the costodiaphragmatic recesses, are derived from **body wall.** The expansion of the pleural sacs in a caudal direction splits off an inner lamina of somatic mesoderm, with which the septum transversum is continuous, and pushes it medially, creating the costodiaphragmatic recesses. This process is similar to the one that forms the fibrous pericardium from the same lamina of somatic mesoderm (see Fig. 19-12).

**Congenital and Acquired Diaphragmatic Hernias.** Probably because of the presence of the liver, the right pleuroperitoneal canal closes early. The left one persists longer, and if it has not closed by the time the intestine is returned to the peritoneal cavity from the umbilical celom, abdominal contents will herniate up into the thorax. The most common type of *congenital diaphragmatic hernia* is one in which the stomach or loops of small or large intestine have passed through the left pleuroperitoneal canal to lie in the left pleural cavity. Both cardiac and pyloric ends of the stomach remain below the diaphragm, but the organ is pushed "upside down" into the pleural cavity.

Much more rarely, a similar herniation occurs after the diaphragm has been completed but before the thin pleuroperitoneal membrane has been strengthened by muscle. In these cases, the herniated intestines lie in a sac consisting of peritoneum internally and pleura externally that projects up into the thoracic cavity (*eventration of the diaphragm*).

The two other locations in which diaphragmatic hernias ordinarily occur are the esophageal hiatus and the sternocostal triangle. Hernia in either location may be congenital; however, esophageal hiatal hernia, the more common, is far more frequently acquired. In the common type of **hiatal hernia,** the cardia and fundus of the stomach slide upward through an enlarged esophageal hiatus, somewhat as if the esophagus had pulled them up. In such *sliding hernias,* the cardiac notch is obliterated, and the upper end of the stomach seems to be simply a widening of the esophagus (Fig. 25-5A). In the other type of hiatal hernia, called *paraesophageal hernia,* the gastroesophageal opening remains in the abdomen, but the fundus and perhaps a part of the body of the stomach balloon up through the hiatus into the thorax (Fig. 25-5B).

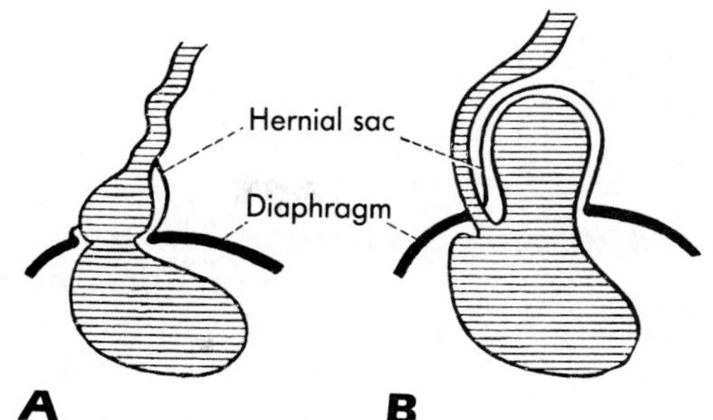


**FIG. 25-5.** Schema of the two chief types of hernia through the esophageal hiatus, viewed from the right side. **A** is the sliding type, shown here with some redundancy of the esophagus; **B** is the paraesophageal type. Note the difference both in the location of and in the angle formed by the esophagogastric junction.

## THE VESSELS OF THE POSTERIOR WALL

### THE ABDOMINAL AORTA

The aorta enters the abdomen through the aortic hiatus and passes downward on the front of the vertebral column to end by bifurcating in front of the lower part of the 4th lumbar vertebra (Figs. 25-1, 25-6, and 25-7). On the surface of the body, the abdominal aorta can be represented as a midline structure extending from about 3 cm above the transpyloric plane to the supracristal plane (usually the umbilicus). Its pulsations may be palpated and in lean individuals can also be seen.

**Relations.** A complex *plexus of autonomic nerves* is closely applied to the anterior surface of the aorta from the aortic hiatus to the bifurcation. Several large structures cross the anterior surface of the aorta: just below the origin of the celiac trunk, the *pancreas;* just below the origin of the superior mesenteric artery, the *left renal vein;* and slightly lower, the horizontal portion of the *duodenum.* The terminal segment of the aorta and its bifurcation are covered by *parietal peritoneum* of the left infracolic compartment. The *inferior vena cava* is on the right of the abdominal aorta throughout its course; only the right crus of the diaphragm intervenes partially between the two parallel vessels. Below the pancreas, on the left of the aorta, the fourth part of the duodenum ascends to the duodenojejunal flexure, and lower down, the *inferior mesenteric artery* remains in contact with the left side of the aorta for some distance.

#### Branches

The abdominal aorta gives off four types of branches: unpaired ventral visceral branches to the gut and its derivatives, paired visceral branches to organs derived from intermediate mesoderm, dorsal branches to the body wall, and terminal branches.

There are three **ventral visceral branches:** the celiac trunk and the superior and inferior mesenteric arteries, all of which have been described in detail in Chapter 24.

**Paired Visceral Branches.** The paired visceral branches are the inferior phrenic and middle suprarenal arteries, both of which supply the suprarenal glands, and the renal and gonadal (testicular or ovarian) arteries.

The **inferior phrenic arteries** may arise separately or by a common stem from the aorta in the aortic hiatus. Sometimes they are given off by the celiac trunk. They run upward and laterally, passing medial to the suprarenal glands, and furnish these glands with a major part of their blood supply before they disappear into the diaphragm (p. 684).

The small and variable **middle suprarenal arteries,** given off from the lateral aspect of the aorta just above the origin of the renal artery, help supply the suprarenal glands and anastomose with branches of the inferior phrenic and with the *inferior suprarenal arteries,* branches of the renal arteries.

The **renal arteries** are given off at right angles from the lateral aspect of the aorta at the level of L1 vertebra, only a little below the origin of the superior mesenteric. The right artery passes behind the inferior vena cava; the left is behind the left renal vein.

The origin and much of the abdominal course of the **testicular** and **ovarian arteries** are similar. The paired vessels usually arise from the front of the aorta, below the origin of the renal arteries, either at the same or at dif-

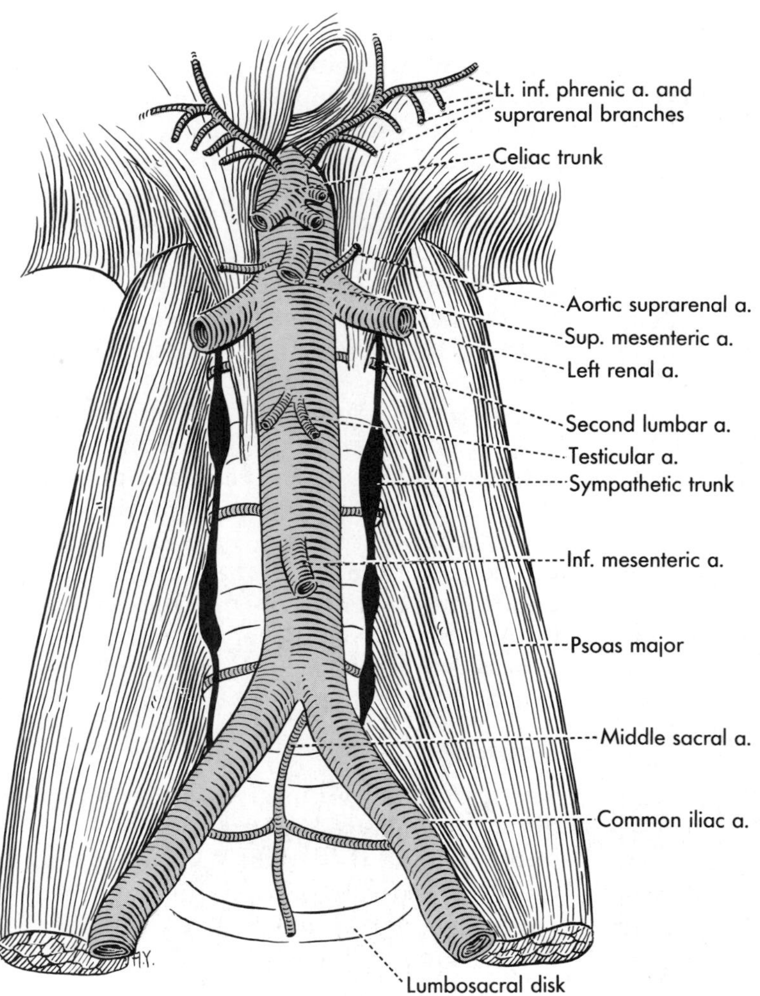


**FIG. 25-6.** The abdominal part of the aorta and its branches. The inferior vena cava is omitted from the drawing.

ferent levels or even by a common stem. The left artery may loop upward to pass behind and above the renal vein before turning down. Otherwise, both vessels run downward and laterally on the anterior surface of the psoas muscle (Fig. 25-19). Throughout most of its course, each artery is accompanied by the corresponding vein. The arteries usually give off a branch to the adipose capsule of the kidney and, as they pass in front of the ureter, contribute to its blood supply.

After crossing the ureters, the vessels pursue a different course in the two sexes. *In the male,* the vessels cross the iliac fossa in a continuation of their downward and forward course and then enter the deep inguinal ring. They become incorporated into the spermatic cord and reach the testis with it. *In the female,* they approach the pelvic brim and turn downward and medially as they enter the pelvis.

The high origin of the vessels indicates the embryonic position of the gonads, which have migrated caudally in the female as well as in the male.

**Dorsal Branches.** There are usually four pairs of lumbar arteries given off from the dorsal aspect of the aorta and a single median sacral artery.

The first two pairs of **lumbar arteries** run behind or through the crura of the diaphragm. All four pairs are closely applied to the front of the vertebral bodies. As they pass laterally, they lie behind the lumbar lymphatic chains and the lumbar sympathetic trunks on both sides; on the right, they lie behind the inferior vena cava. They disappear between the psoas major muscle and the vertebral bodies through the tendinous arches from which the psoas arises (Fig. 25-6).

**FIG. 25-7.** An abdominal arteriogram. Contrast medium was injected into the aorta just above the origin of the celiac artery through a catheter. The catheter, barely visible along the left margin of the aorta, was inserted into the right femoral artery and advanced superiorly in the lumen of the external and common iliac arteries into the aorta. Compare with Figure 24-7**B**.

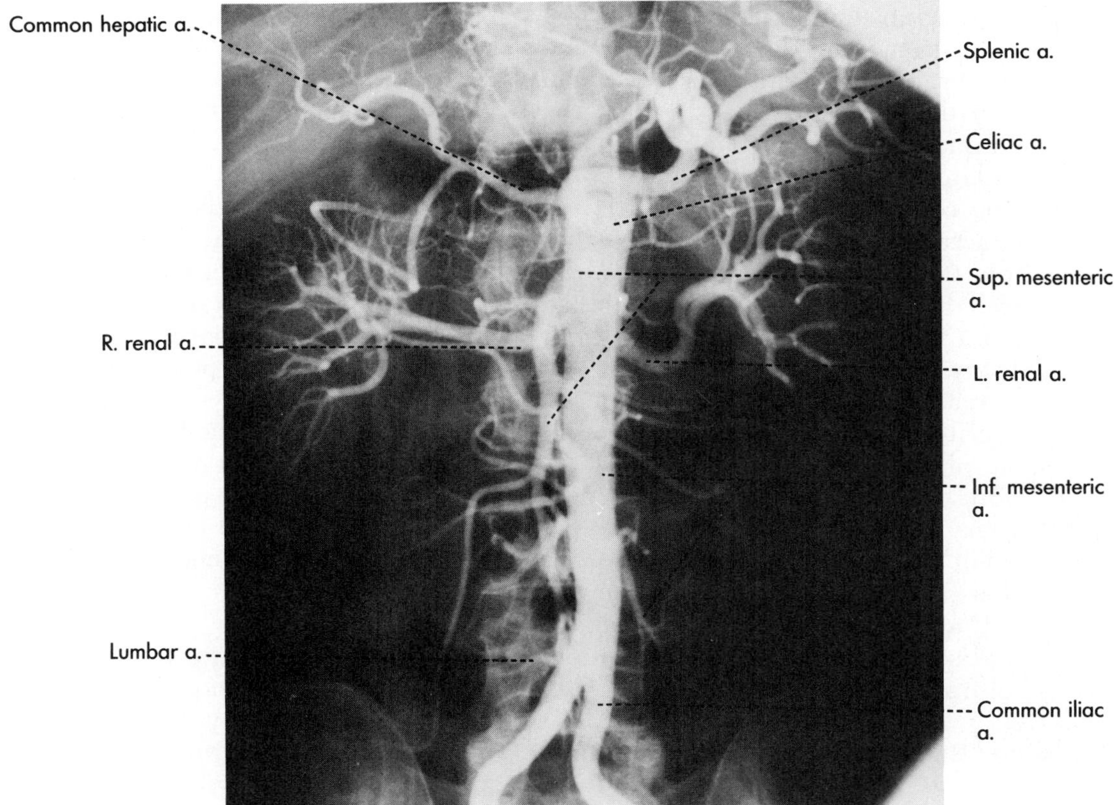

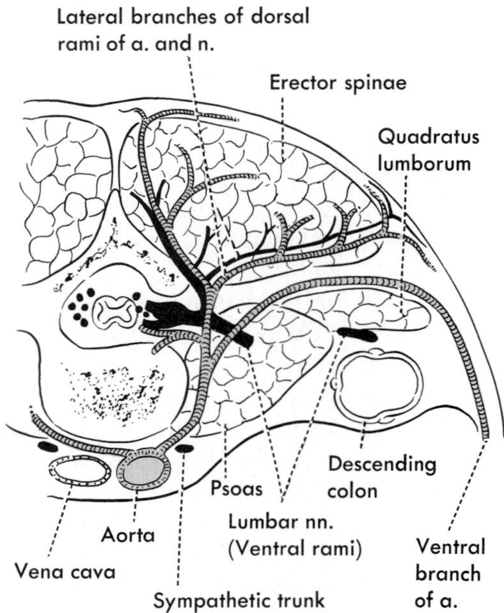


**FIG. 25-8.** Diagram of a lumbar artery.

Like the posterior intercostal arteries, each lumbar artery gives off a *dorsal branch* that turns backward to supply the musculature of the back and, in turn, gives off a *spinal branch* to the vertebral column and the nerve roots. The rest of the artery continues behind the psoas and quadratus lumborum muscles into the anterolateral abdominal wall (Fig. 25-8).

The true continuation of the aorta is the **median sacral artery,** a small vessel that arises from the posterior surface of the aorta just above the level of its bifurcation (Fig. 25-6). This vessel emerges between the common iliac arteries, closely applied to the front of the vertebral column and covered anteriorly by peritoneum and the continuation of the aortic plexus in the retroperitoneal tissues. Before it reaches the pelvis, the median sacral artery may give off a 5th, or lowest, pair of small lumbar arteries. Sometimes, the 4th lumbar arteries may arise from the median sacral artery as well, or the median sacral may arise from one of the lumbar arteries instead of from the aorta.

**Terminal Branches.** The large terminal branches of the aorta are the right and left **common iliac arteries.** These diverge and run downward and laterally, with the corresponding common iliac veins (Fig. 25-19). The arteries follow the medial border of the psoas major muscle to the pelvic brim, where each common iliac artery divides into an internal and an external iliac artery. The *internal iliac artery* enters the pelvis. Its course and branches are described in Chapter 27. The *external iliac artery* continues to follow the iliopsoas and leaves the abdomen between the inguinal ligament and the superior ramus of the pubis.

Usually, there are no branches of the common iliac arteries other than their terminal branches. The branches of the external iliac are described in Chapter 26.

## THE INFERIOR VENA CAVA AND ITS TRIBUTARIES

The inferior vena cava is formed by the junction of the two common iliac veins on the right anterior surface of the 5th lumbar vertebra and conveys venous blood to the right atrium from all parts of the body below the diaphragm. It ascends in front of the lumbar vertebrae on the right of and parallel with the abdominal aorta (Fig. 25-9). Passing behind the right lobe of the liver, it inclines forward with the curvature of the diaphragm to pierce its central tendon, where it is on level with, but some distance anterior to, the 8th thoracic vertebra. In the thorax, the inferior vena cava has a short intrapericardial course before it opens into the lower posterior corner of the right atrium.

On the surface of the body, the position of the inferior vena cava may be indicated by a band extending from a point in the intertubercular plane 2 cm to 3 cm to the right of the midline to the sternal end of the right 6th costal cartilage.

**Relations.** The inferior vena cava begins behind the *right common iliac artery,* above which it is covered by *parietal peritoneum* of the infracolic compartment and is crossed by the *root of the mesentery*. Retroperitoneally, the right *ureter* may overlap it, and the right *testicular* or *ovarian artery* passes over it. The anterior surface of the vena cava is crossed by both the horizontal and superior parts of the *duodenum*, and in between the two, the head of the *pancreas*, the *common bile duct*, and the *portal vein* lie in direct contact with it. Above the duodenum, the inferior vena cava is again covered by peritoneum as it forms the posterior boundary of the *epiploic foramen*. Beyond the posterior layer of the right coronary ligament, the vena cava lies in a deep groove, or tunnel, of the *liver*, which forms the boundary between the right and caudate lobes.

The more important structures upon which the inferior vena cava lies include some of the *lumbar arteries;* the *right renal, suprarenal,* and *inferior phrenic arteries;* and the *right sympathetic trunk*. On its right, the inferior vena cava is in contact with the right suprarenal gland, the right kidney and ureter, and the descending part of the duodenum; on the left, below the aortic hiatus, lies the abdominal aorta.

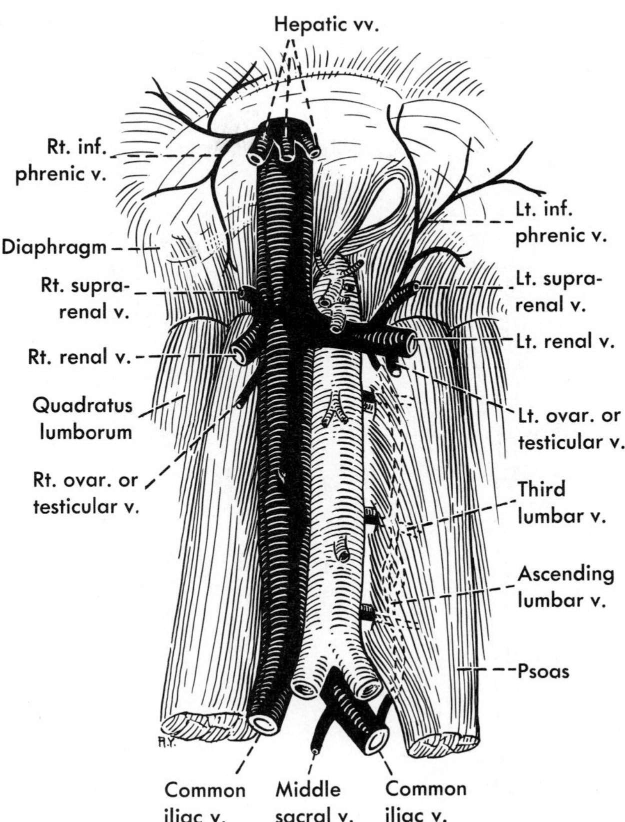


**FIG. 25-9.** The inferior vena cava and its tributaries.

**Tributaries.** Through the *common iliac veins,* the inferior vena cava collects all the blood from the pelvis, the perineum, and the lower limbs. The pelvis and perineum are drained by the *internal iliac vein,* and the lower limbs by the *external iliac vein;* the two unite to form the *common iliac vein* on each side. Blood from the digestive tract, having been collected by the portal system and passed through the liver, enters the vena cava through two or three *hepatic veins* that open into the vena cava just below the diaphragm (see Fig. 24-39). Between the hepatic veins and common iliac veins, the inferior vena cava receives the *right inferior phrenic, right suprarenal,* and *right testicular* or *ovarian veins,* both *renal veins,* and a variable number of the four pairs of *lumbar veins*. The left inferior phrenic, the left suprarenal, and the left testicular or ovarian veins drain into the left renal vein, a developmental homologue of a segment of the inferior vena cava. Only the lumbar veins need to be described here, as all other tributaries of the inferior vena cava are dealt with in the sections on the organs they drain.

The **lumbar veins** follow the distribution of the corresponding arteries, as do, in general, the other tributaries of the vena cava. The lumbar veins enter the inferior vena cava in an irregular pattern; the second left lumbar vein frequently enters the left renal vein. In the abdominal wall, the tributaries of the lumbar veins communicate with those of the epigastric veins. In front of the transverse processes, the lumbar veins of each side are connected to one another, just before they enter the inferior vena cava, by a vertical anastomotic venous channel, the *ascending lumbar vein*. These veins are the abdominal counterparts of the azygos and hemiazygos veins of the posterior mediastinum (see Fig. 22-14). Inferiorly, the ascending lumbar vein connects

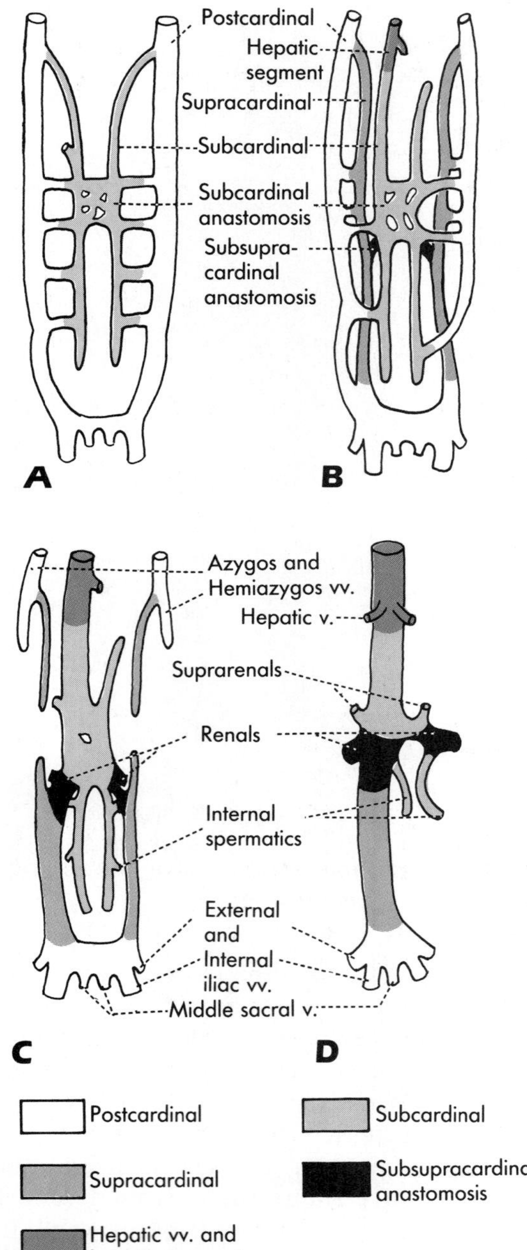


**FIG. 25-10.** Development of the inferior vena cava and its tributaries. In **A**, the postcardinals are the chief drainage of the caudal part of the body; in **B**, the supracardinals have developed and are taking over this drainage; in **C**, the lower portions of the supracardinals are draining into the subcardinal system; and in **D**, the definitive condition is shown.

the lumbar veins with the common or internal iliac veins and the iliolumbar vein. Superiorly, the ascending lumbar veins pass through the crura of the diaphragm or through the aortic hiatus and, uniting with the subcostal veins, form the azygos vein on the right and the hemiazygos vein on the left. Through a small vein, the right ascending lumbar vein connects with the inferior vena cava and the left ascending lumbar vein connects with the left renal vein.

The ascending lumbar vein is, therefore, a potential source of collateral circulation from the lower part of the body when the inferior vena cava is occluded. Since the lumbar veins communicate with the internal vertebral venous plexuses as well, the plexuses around the spinal cord form an important part of this collateral drainage. Other channels that may be involved are the epigastric veins, the superficial veins of the abdomen, and the gonadal veins. Sudden obstruction of the vena cava above the renal level, however, interferes too much with renal function to be tolerated.

**Development.** The inferior vena cava is a composite of several generations of longitudinal venous channels and anastomoses that interlink these channels. These vessels appear at different stages and disappear partially or completely in the course of development. The tributaries of the vena cava are persistent segments of some of these embryonic veins. A description of these veins is not our purpose here, and mention of them is made only to indicate that the explanation of the anatomy of the inferior vena cava and its tributaries may be found in the manner in which these veins unite to form a single vessel. Errors or arrests in this process provide the clues for the congenital anomalies associated with the inferior vena cava.

The embryonic veins are concerned with draining the body wall, the lower limb buds, and the mesonephric ridges. Some of the veins are located behind the mesonephric ridges, others within them. The so-called *postrenal segment* of the inferior vena cava, lying caudal to the renal vessels, develops from veins that are most dorsally placed, whereas the so-called *prerenal segment,* cranial to the renal veins, develops from more ventrally lying veins. This explains the forward inclination of the vena cava in its course above the level of the renal veins and also why the right inferior phrenic, suprarenal, and renal arteries pass behind the vena cava, whereas the right testicular or ovarian artery passes in front of it.

The first set of bilateral veins is the **postcardinal veins,** which disappear except at the cranial end, where they form the azygos vein and the left superior intercostal vein, and at the caudal end, where the anastomosis between the two postcardinal veins persists as the left common iliac vein (white in Fig. 25-10A). The postcardinal veins are replaced by the **supracardinal veins;** the right one of the pair forms the postrenal segment of the inferior vena cava (blue in Fig. 25-10B and C). The third set, called the **subcardinal veins,** is formed more ventrally within the mesonephros, which it drains through its tributaries. The right subcardinal vein becomes incorporated into the prerenal segment of the inferior vena cava below the liver (red in Fig. 25-10).

The subcardinal veins establish connections with both the supracardinal veins (*the subsupracardinal anastomosis,* black in Fig. 25-10) and, ventral to the aorta, wih eath other (*subcardinal anastomosis*). The left renal vein is derived from the subcardinal anastomosis, explaining its position anterior to the aorta. The continuity between postrenal and prerenal segments of the inferior vena cava is established by a portion derived from the subsupracardinal anastomosis. The tributaries of the inferior vena cava are all derived from the subcardinal system.

The most cranial portion of the prerenal segment of the vena cava is formed by the *right vitelline vein,* which also gives rise to the hepatic veins, and by a special vessel that grows from the hepatic veins to link up with the subcardinal portion of the inferior vena cava; the latter vein constitutes the *hepatic segment* of the fully formed vessel (purple in Fig. 25-10).

**Anomalies.** Most anomalies of the inferior vena cava involve the postrenal segment. The usual ones are *double vena cava* and *left inferior vena cava,* caused by persistence on the left side of veins that normally disappear (Fig. 25-11). If the left vena cava is the smaller of a pair, it may seem to join the left renal vein. If it is larger or the only one, it crosses to the right side in front of the aorta at the renal level and receives the left renal, suprarenal, and gonadal veins. If the right subcardinal rather than the right supracardinal, vein persists to form the postrenal segment of the vena cava, the whole vessel is displaced ventrally, causing the ureter to pass behind it (retrocaval ureter).

Very rarely, also, the lower end of the inferior vena cava is so formed that it lies anterior, rather than posterior, to the right common iliac artery.

The hepatic portion of the prerenal segment may fail to develop, in which case the bilateral supracardinal veins that form the azygos and hemiazygos systems are the channels through which blood can return from below the diaphragm to the heart.

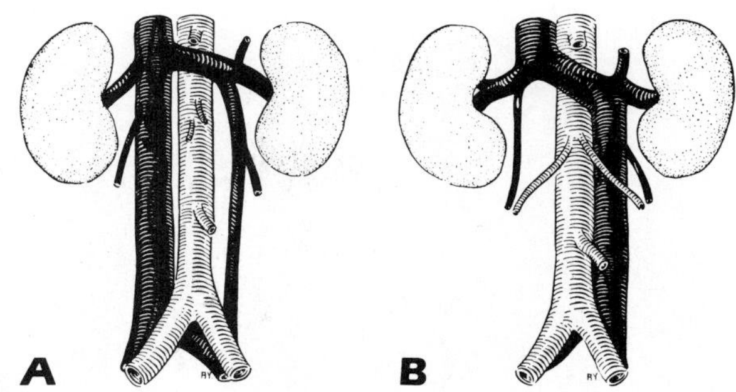


**FIG. 25-11.** Double inferior vena cava, **A,** and left inferior vena cava, **B.** Note that above the renal level, however, the inferior vena cava is single and on the right side in both cases.

## THE LYMPHATICS

Lymph from all parts of the body below the diaphragm is directed toward lymph vessels and chains of lymph nodes that ascend around the great vessels of the posterior abdominal wall. The lymphatics from the lower limb, the perineum and buttocks, and the lower part of the anterior abdominal wall converge upon the *inguinal lymph nodes,* which, in turn, drain upward to a series of *external* and *common iliac nodes* lying along and around the vessels of the same name. Being located along the pelvic brim, these nodes receive also the drainage from the *internal iliac nodes,* which, in turn, drain most of the urinary and genital organs in the pelvis and a part of the rectum, the chief drainage of which is along the inferior mesenteric artery. Right and left sets of common iliac nodes meet and exchange lymphatics at the level of the aortic bifurcation.

Above the aortic bifurcation, lymphatics continue upward along the aorta and inferior vena cava (Figs. 25-12 and 25-13). These chains of nodes are known as the **lumbar nodes** (also called *aortic* or *caval* nodes). Although they interchange lymphatics across

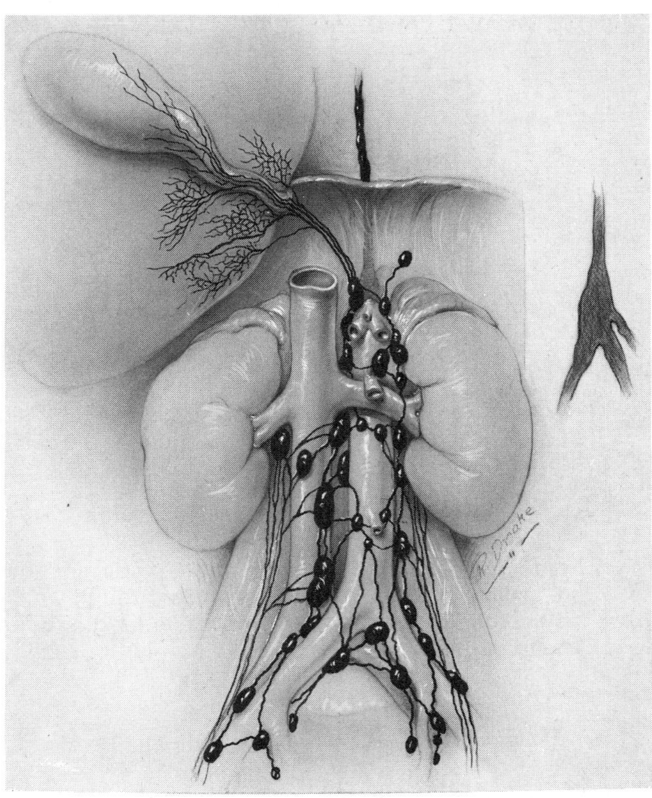

**FIG. 25-12.** Iliac and lumbar lymph nodes. Some of the nodes above the left renal vein are superior mesenteric and celiac nodes, and lymph vessels from the liver and gallbladder can be seen joining the latter. The lower part of the thoracic duct is indicated, and the small inset shows one type of cisterna chyli. (Desjardins AU: Arch Surg 38:714, 1939)

**FIG. 25-13.** Iliac and lower lumbar nodes injected through lymphatics of the feet. (Courtesy of Dr. W. E. Miller)

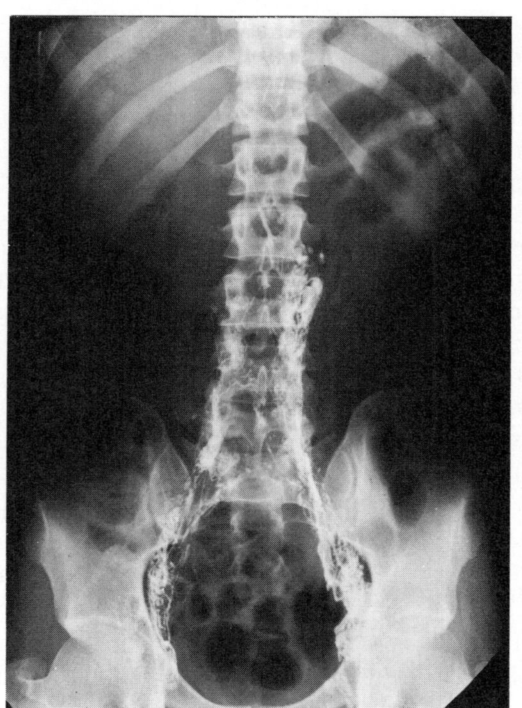

the midline, these nodes are best thought of as a left chain and a right chain lying along the aorta and the inferior vena cava, respectively. The lumbar nodes receive lymphatics not only from the common iliac nodes but also directly from the body wall, kidneys, and suprarenals, as well as from the testes or ovaries.

At the level of origin of the inferior mesenteric artery, the lumbar nodes also receive the lymphatic drainage from the left side of the colon; at the level of origin of the superior mesenteric and celiac arteries, they exchange lymphatics with the superior mesenteric and celiac nodes. In this region, lumbar nodes are so mingled with these nodes that it is impossible to say exactly which is a lumbar node and which are nodes connected more directly with the digestive tract.

The chief efferent vessels from the upper lumbar nodes on each side unite to form the lumbar lymphatic trunk, and the two **lumbar trunks** are joined by the intestinal trunk in the formation of the cisterna chyli. The **intestinal trunk** (there may be more than one) is a short stem formed by efferent lymphatics from the

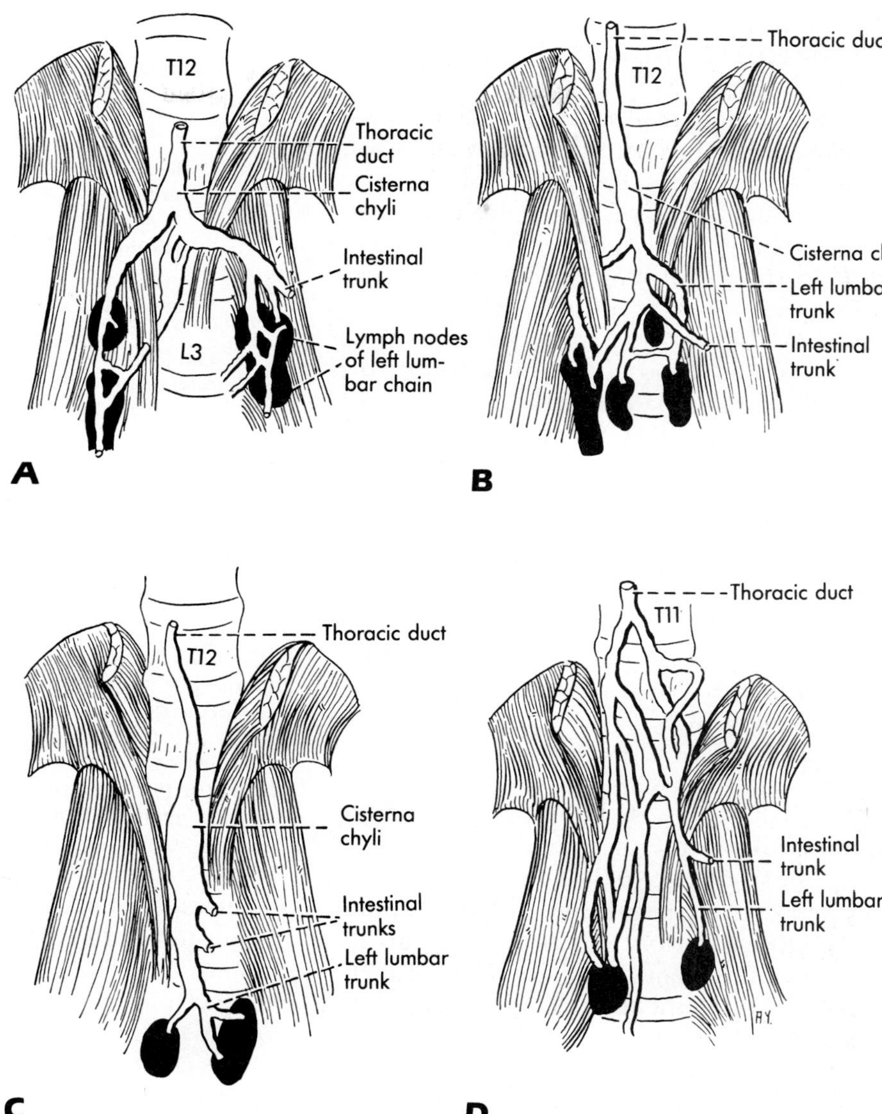


**FIG. 25-14.** Some variations in the formation of the thoracic duct. (Adapted from Jossifow GM: Arch Anat Physiol Anat Abt, p 68, 1906)

superior mesenteric and celiac nodes. It usually joins the left lumbar trunk before the two lumbar trunks unite.

The **cisterna chyli** is the lower expanded end of the thoracic duct. It lies in the upper part of the abdomen behind the right side of the aorta, in front of the upper right lumbar vessels, on the bodies of L1–L2 vertebrae (Fig. 25-14). Its continuation, the thoracic duct, passes through the aortic hiatus and ascends through the thorax as previously described.

The cisterna chyli varies considerably in its mode of formation, its size, and its placement (Fig. 25-14). There may be no visible enlargement, and the lymph trunks may unite in highly variable patterns. However, regardless of its method of formation, the important point is that the thoracic duct receives, at its lower end, practically all the lymphatic drainage from the body below the level of the diaphragm, including that from the digestive tract.

## THE NERVE PLEXUSES AND THE SYMPATHETIC TRUNKS

Two qualitatively distinct nerve plexuses are associated with the posterior abdominal wall: the lumbar plexus, composed of somatic nerves, and the aortic plexus, composed of autonomic nerves. The lumbar plexus is enclosed in the psoas major, and its branches have already been identified earlier in this chapter as they emerge from the psoas. The aortic plexus is disposed on the anterior surface of the abdominal aorta, and its subsidiary plexuses have been described in the previous chapter in relation with the innervation of the gut and its derivatives. The two sympathetic trunks concerned with relaying sympathetic efferents to the lumbar plexus (*i.e.*, to the lower limbs) and, to a lesser extent, to abdominal viscera lie lateral to the aorta and inferior vena cava on the sides of the lumbar vertebral bodies.

### LUMBAR PLEXUS

The lumbar plexus is the upper part of the lumbosacral plexus, the nerve plexus of the lower limbs (see Fig. 16-7). The lumbar plexus is formed by the ventral rami of L1–L3 spinal nerves, with contributions from T12 and L4 ventral rami. Except for a cutaneous branch, which extends to the foot, nerves given off by the lumbar plexus are distributed only as far distally as the knee. The lumbar plexus is joined to the sacral plexus by the *lumbosacral trunk,* which is composed of part of L4 and the whole of L5 ventral ramus. The sacral part of the lumbosacral plexus, described in the chapter on the pelvis, supplies all the parts of the lower limb not innervated by the branches of the lumbar plexus.

The ventral rami of T12–L4 may be thought of as the *roots* of the plexus, and, as they emerge from the intervertebral foramina, they lie within the substance of the psoas muscle. Each of the L1–L4 roots splits within the muscle into an *anterior* and a *posterior division.* The branches of the lumbar plexus, like those of the brachial and sacral

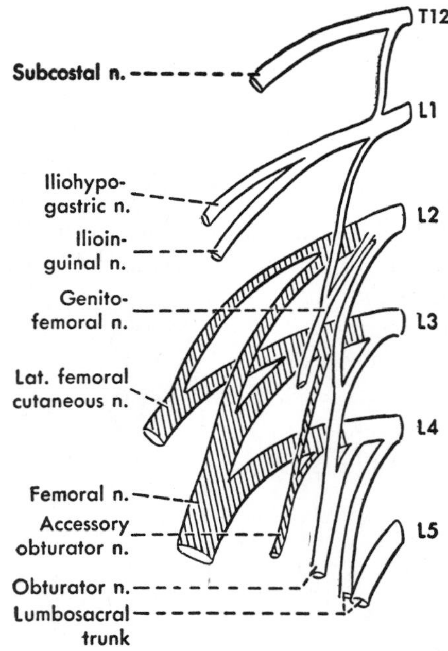


**FIG. 25-15.** A common form of the lumbar plexus. The ventral rami of the nerves, identified by the lettering, and the anterior divisions of the plexus are unshaded; the posterior divisions and branches of the plexus are shaded. The accessory obturator nerve is sometimes regarded as being an anterior branch, and sometimes a posterior one.

plexuses, are formed by either the anterior or the posterior divisions of certain ventral rami and are then distributed to the skin and musculature of the original anterior and posterior compartments of the limb (Fig. 25-15).

The common method of formation of the lumbar plexus is that the first lumbar root, which may or may not receive fibers from the 12th thoracic, gives rise to a larger stem that subsequently divides into the *iliohypogastric* and the *ilioinguinal nerves.* The smaller stem of L1 root descends to unite with the small part of the anterior division of the 2nd lumbar root, to form the *genitofemoral nerve;* the rest of the L2 anterior division contributes to the formation of the *obturator nerve.* The larger, posterior division of the 2nd lumbar ventral ramus gives origin to parts of both the *lateral femoral cutaneous* and *femoral nerves.* The anterior and posterior divisions of the 3rd lumbar root similarly contribute to the obturator, the lateral femoral cutaneous, and the femoral nerves. The 4th lumbar root divides before it enters the plexus and sends its major portion into the lumbar plexus and the smaller one into the *lumbosacral trunk*. The posterior division of L4 contributes to the femoral nerve, and the anterior division, to the obturator. The 2nd to 4th or the 3rd and 4th lumbar roots sometimes also give rise to an *accessory obturator nerve* that arises between the femoral and obturator nerves and is not definitely assignable to either the posterior or the anterior division.

The part of L4 root that contributes to the lumbosacral trunk passes downward to join the ventral ramus of L5. The lumbosacral trunk carries L4 and L5 fibers to the sacral plexus. The distributions of the obturator, femoral, and lateral femoral cutaneous nerves to the lower limb are described in chapters concerned with the lower limb. The iliohypogastric and ilioinguinal nerves are distributed mostly to the skin of the lowest part of the abdomen and upper part of the thigh and buttock. The iliohypogastric nerve may also innervate the pyramidalis muscle. The femoral branch of the genitofemoral goes to the skin of the thigh, and the genital branch goes to the cremaster muscle and to the skin of the scrotum and labia majus.

**Variations.** The chief variation of the lumbar plexus is in its lower boundary. The lumbar plexus is regarded as normal if it receives more than half the fibers of the 4th lumbar ventral ramus, all of those from the 3rd, and none from the 5th. If, in contrast, the plexus receives a minor contribution from the 4th lumbar root, or only some of the fibers from the 3rd lumbar root with or without those of the 4th, the plexus is regarded as *prefixed* (that is, moved somewhat cranially). If, on the other hand, almost all or all of the 4th root joins the lumbar plexus, and some of the 5th root does likewise, the plexus is regarded as *postfixed* (that is, having a caudal border that is lower than usual). The cranial border of the plexus (the highest level from which it receives fibers) also may vary, and this variation is apparently independent of the level of the caudal border.

Variations in the composition of the lumbar plexus may affect the composition of the individual branches of the plexus. Although the reported variations are numerous, the basic pattern described is a useful guide to clinical evaluation of the plexus. For instance, the femoral nerve usually receives fibers from the 2nd, 3rd, and 4th lumbar nerves, and the noted variations from this are that it may receive fibers from L1 or from L5 or from both. Thus, the shift in the composition of the femoral nerve usually is not more than a segment, which, clinically, is not critical.

## THE LUMBAR SYMPATHETIC TRUNKS

The lumbar parts of the paired sympathetic trunks lie on the sides of the vertebral bodies at the origin of the psoas major muscle from these bodies (Fig. 25-6). Each is a continuation of the thoracic part of the trunk on its own side. The interganglionic branch from the lowest thoracic to the uppermost lumbar ganglion is typically slender and usually runs behind the medial arcuate ligament to enter the abdomen. The upper part of the lumbar trunk lies somewhat behind the crus. Each trunk is continued over the sacral promontory as the sacral part of the sympathetic trunk. In the abdomen, the left trunk lies to the left of, but slightly behind, the aorta; the right trunk lies behind the inferior vena cava. They are located, however, in front of the lumbar arteries and veins.

Although five *lumbar ganglia* would be expected, because there are five lumbar segments of the body, their number and placement on each side vary (Fig. 25-16). The definitive ganglia are the result of fusion and splitting of the primitive ganglia. Because of the simultaneous variations in number and connections of the ganglia, it is difficult to number them logically or to be sure that similarly numbered ganglia are similar in their connections.

Owing to the variability of the lumbar ganglia, there is considerable variation in their *rami communicantes*. As a rule, however, only the upper lumbar ganglia receive *white rami*, which bring to the sympathetic trunk preganglionic fibers, but they may receive several of these because of the fusion of the ganglia (Fig. 25-17). All ganglia give off *gray rami communicantes*, some of them more than one, to the

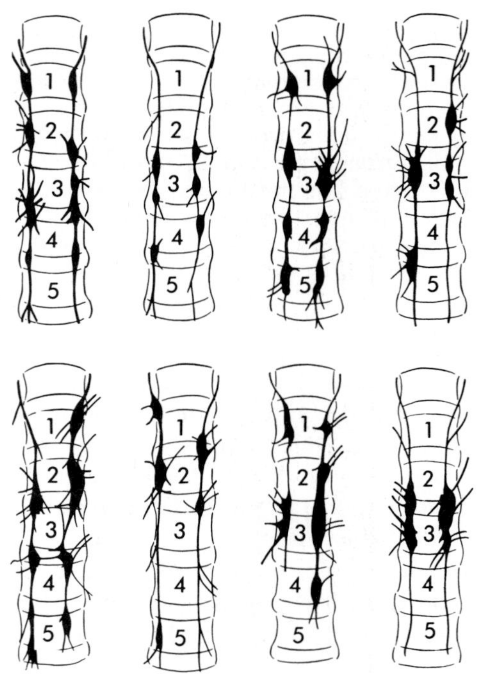


**FIG. 25-16.** Variations in the lumbar sympathetic trunks. The two sides of the same body are rarely symmetrical in regard to both number and placement of the lumbar ganglia. (Redrawn from Yeager GH, Cowley RA: Ann Surg 127:953, 1948)

ventral rami of the lumbar nerves. Also, each lumbar nerve usually receives gray rami from more than one ganglion. Some of these postganglionic sympathetic efferents find their way into the dorsal rami of these nerves as well, running medially to the point of division of the spinal nerve.

The upper lumbar ganglia give medial branches that are the *lumbar splanchnic nerves* (Fig. 25-17). These nerves contain preganglionic visceral efferent fibers that relay in the aortic ganglia after the lumbar splanchnic nerves join the aortic plexus, which they help to form. The lumbar splanchnic nerves also convey visceral afferents from the aortic plexus to the lumbar ganglia and, through them and their white rami communicantes, to the upper lumbar spinal nerves.

### Sympathectomies

Segments of the sympathetic trunks have been removed (sympathectomy), mainly with two objectives: (1) improving the circulation in the lower limbs by reducing vasospasms caused by increased sympathetic activity and (2) lowering the blood pressure by vasodilatation in the large vascular bed of abdominal viscera. Because of the availability of a variety of suitable drugs, sympathectomies are now done less frequently.

For the sympathetic innervation of the lower limb, which receives all its postganglionic fibers through the lumbar and sacral nerves by way of the gray rami communicantes, the standard operation is to remove the upper segment of the lumbar trunk, which receives all the white rami communicantes. For hypertension due to sympathetic overactivity, cutting the splanchnic nerves and removing the upper parts of the lumbar trunks, often together with the lower parts of the thoracic sympathetic trunks (thoracolumbar sympathectomy), has been the operation of choice. The effect of de-

**FIG. 25-17.** The rami communicantes of the lumbar sympathetic trunk in a specimen in which the rami were both dissected and examined histologically. The white rami, as determined by histologic examinations, are shown in red; the gray rami, in black. (Redrawn from Pick J, Sheehan D: J Anat 80:12, 1946)

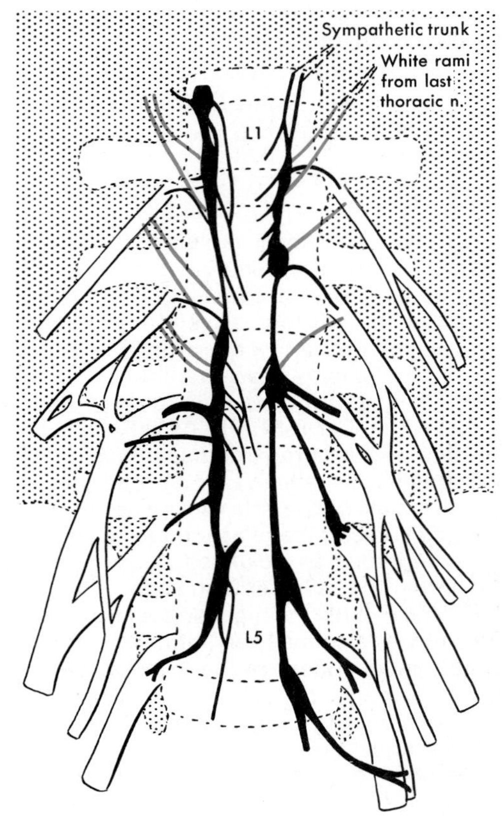

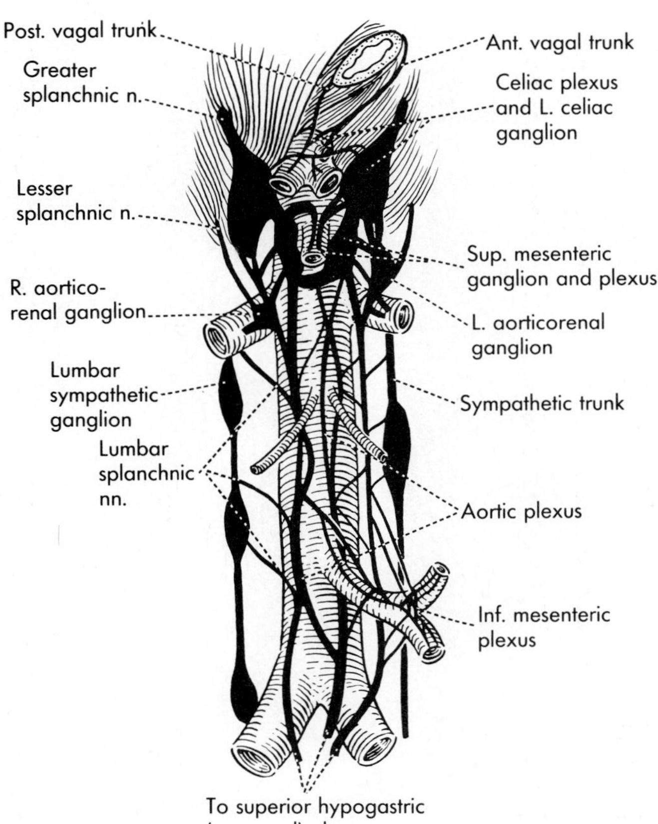


**FIG. 25-18.** The autonomic nerve plexuses and ganglia associated with the aorta. The aortic plexus is below the level of the superior mesenteric artery.

nervation on the abdominal and pelvic organs is generalized vasodilatation. In the male, a complicating, undesirable effect is interference with ejaculation (p. 774).

## THE AORTIC PLEXUS

The functional components and the anatomic subdivisions of the aortic plexus have been mentioned in Chapter 24 with relation to the innervation of the gut and its derivatives. The aortic plexus is a continuous network of nerves extending on the surface of the aorta from the aortic hiatus to the bifurcation and, below that, into the pelvis (Fig. 25-18). It has several parts named according to the major branches of the aorta.

The superior and most dense portion of the plexus is the **celiac plexus,** continuous below with the **superior mesenteric plexus.** The part of the aortic plexus intervening between the superior and inferior mesenteric arteries is called the **intermesenteric plexus.** Some of the fibers of the intermesenteric plexus continue along the inferior mesenteric artery as the **inferior mesenteric plexus;** others descend toward the sacral promontory and, beyond the bifurcation, are designated as the **superior hypogastric plexus.**

The aortic plexus receives input from the following sources: (1) thoracic splanchnic nerves (greater, lesser, and lowest), composed of the visceral branches of the lower thoracic sympathetic ganglia; (2) lumbar splanchnic nerves, visceral branches of lumbar sympathetic ganglia; (3) celiac branches of the vagi. The plexus also contains visceral afferents from all abdominal and most pelvic viscera, which leave the plexus along the vagi and the thoracic and lumbar splanchnic nerves. The pelvic splanchnic nerves, which furnish parasympathetic fibers to the descending colon and the rest of the hindgut, do not enter the aortic plexus; they follow the sigmoid arteries on their ascent from the pelvis (see Fig. 24-50).

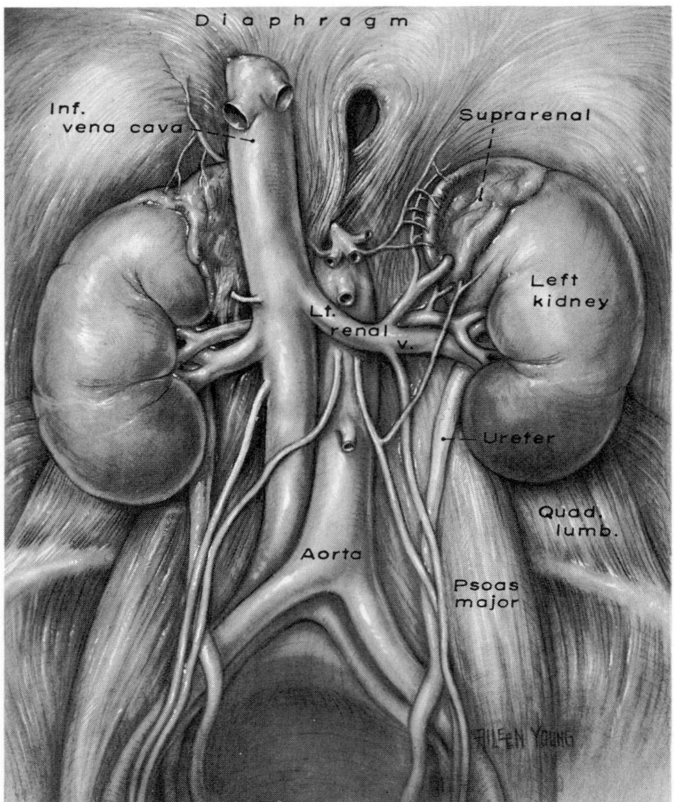


**FIG. 25-19.** The kidneys and suprarenal glands.

Thus, the aortic plexus is composed chiefly of sympathetic fibers with some vagal fibers admixed in its superior parts.

The branches of the aortic plexus are the subsidiary plexuses that accompany the branches of the aorta (*e.g.*, hepatic, renal, inferior mesenteric).

Throughout its length, from the uppermost part of the abdomen down into the pelvis, the aortic plexus contains numerous minute ganglia within which preganglionic sympathetic fibers synapse. It also contains a few large ganglia. The **celiac ganglia,** the largest ones, are paired structures lying in the lateral parts of the celiac plexus on the sides of the aorta and the crura of the diaphragm at about the level of the origin of the celiac trunk. They regularly receive the greater splanchnic nerves and may receive the lesser; through them, therefore, come most of the fibers that descend from the thorax. More or less continuous with the celiac ganglia are the **superior mesenteric ganglia,** which are sometimes identifiable as separate ganglia and sometimes not. These lie on each side of the origin of the superior mesenteric artery and may unite below the artery to form an unpaired ganglion. A smaller ganglion, the **aorticorenal ganglion,** frequently is identifiable a little lower, lying close to the origin of the renal artery. The **inferior mesenteric ganglion** often is so embedded in the inferior mesenteric plexus or so subdivided into smaller ganglia that it is difficult or impossible to find.

## THE KIDNEYS, URETERS, AND SUPRARENAL GLANDS

The kidneys, ureters, and suprarenal glands are located in the retroperitoneal connective tissue of the posterior abdominal wall (Fig.

25-19). Their close anatomic association is a consequence of their developmental history. The three structures commence their development far away from each other. The close association of the kidneys with the ureters, their excretory ducts, is essential for function; the kidneys and suprarenal glands have nothing in common functionally, although both are essential to life. Following the separate anatomic descriptions of these three paired organs, their relations, blood supply, lymph drainage, innervation, and development will be considered together.

## THE KIDNEYS

The kidneys (*renes*) are two somewhat bean-shaped, reddish-brown organs whose function is the maintenance of the body's fluid and electrolyte balance. They regulate the volume and composition of the urine, which they excrete and discharge through the ureters into the urinary bladder. The kidneys also secrete substances into the circulation that regulate blood pressure and certain processes of hematopoiesis.

Each kidney is about 11 cm long, 6 cm wide, and 3 cm thick.

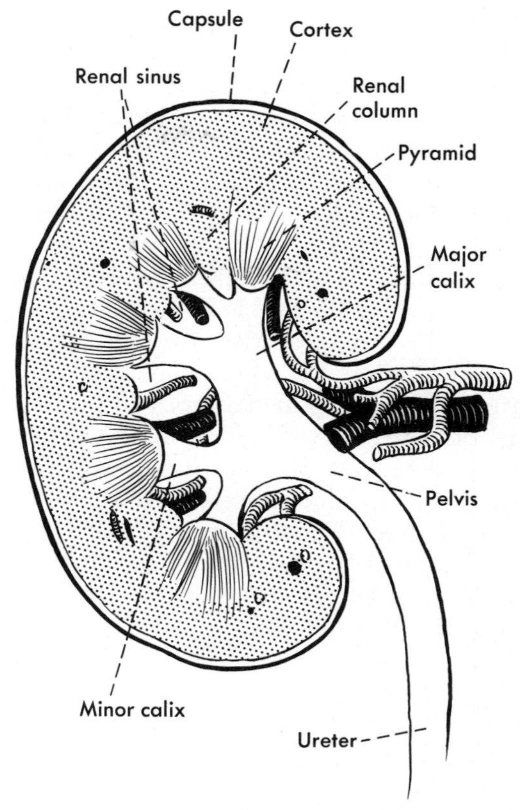


**FIG. 25-20.** The kidney. The renal vessels anterior to the renal pelvis have been cut in removing the anterior half of the kidney; the fat in the renal sinus has also been removed.

**Parts and Structure.** Each kidney has a smooth anterior and posterior *surface*, separated by the lateral and medial *margins*. The two margins become confluent with each other around the rather blunt superior and inferior extremities or *poles* of the kidneys. Much of the medial margin is occupied by the *hilum*, through which the renal vessels, lymphatics, and nerves enter or leave the *renal sinus*, the space enclosed by the renal tissue. The sinus and hilum contain also, along with some fat, the *renal pelvis*, which is the expanded upper end of the ureter, shaped like a complex funnel and distinct from the renal *parenchyma*.

A section through the hilum in a plane parallel with the renal surfaces reveals the renal sinus with its contents and the structure of the renal parenchyma (Fig. 25-20). The parenchyma is enclosed by the *fibrous capsule*, which fits the kidney tightly but is not bound to it; once incised, it can be easily stripped from the kidney. The renal parenchyma consists of two parts: the outer *cortex*, which forms a continuous broad band of tissue subjacent to the capsule, and the inner *medulla*, which is discontinuous owing to the projections of the cortex toward the renal sinus. These projections are the *renal columns*, and the individual portions of the medulla between them are known as *renal pyramids*.

The renal cortex, which is rather pale, dense, and homogeneous on macroscopic examination, contains mainly the *renal corpuscles* (about 1 million in each kidney) and the convoluted portions of the *renal tubules*, whereas the renal pyramids, distinguishable from the cortex by their darker color and longitudinal striations, contain the descending and ascending limbs of the renal tubules and the *collecting tubules*. With the aid of a hand lens, it is possible to see the extension of the striations from the base of the pyramids into the cortex. These striations are the *medullary rays* and, like the medulla, contain collecting tubules.

The nipplelike apex of each pyramid points into the renal sinus and is known as the *renal papilla*. The renal papillae are perforated by the termination of collecting tubules and drain the urine into the *minor calices,* subdivisions of the renal pelvis. A minor calix may receive several papillae, as there are, in each kidney, 5 to 18 renal papillae and only up to 13 minor calices. Each pyramid, with the peripheral cortex between its base and the capsule, constitutes a *lobe* of the kidney; there are 5 to 18 lobes in each kidney.

**Microscopic Structure.** The renal parenchyma consists of a mass of *uriniferous tubules* and blood vessels. Each uriniferous tubule has two component parts, distinct both functionally and developmentally: (1) the *nephron,* composed of a *renal corpuscle* and a *renal tubule,* and (2) *collecting tubules,* in which several renal tubules terminate. Urine is produced by the nephron and is then conducted to the minor calices by the collecting tubules.

The renal corpuscle consists of the *glomerulus,* a tuft of capillaries that produces a filtrate of blood plasma discharged into the *glomerular capsule,* the expanded end of the renal tubule around the glomerulus. The rest of the renal tubule has several named segments that are concerned with the modification of the glomerular filtrate, converting it into urine by the time the collecting tubule is reached.

**Position.** The kidneys lie in the paravertebral gutters, the depth of which is reduced by the psoas major. The vertebral levels of the kidney vary somewhat with body build and

**FIG. 25-21.** Retrograde pyelograms of the kidneys. Contrast medium is introduced into the renal pelvis through a catheter (visible in the film on the right) that was inserted into the ureter from the bladder. One of the kidneys has two and the other has three major calices. Only one of the minor calices shows, but their pattern is obviously different in the two kidneys. (Braasch WF, Emmett JL: Clinical Urography. Philadelphia, WB Saunders, 1951).

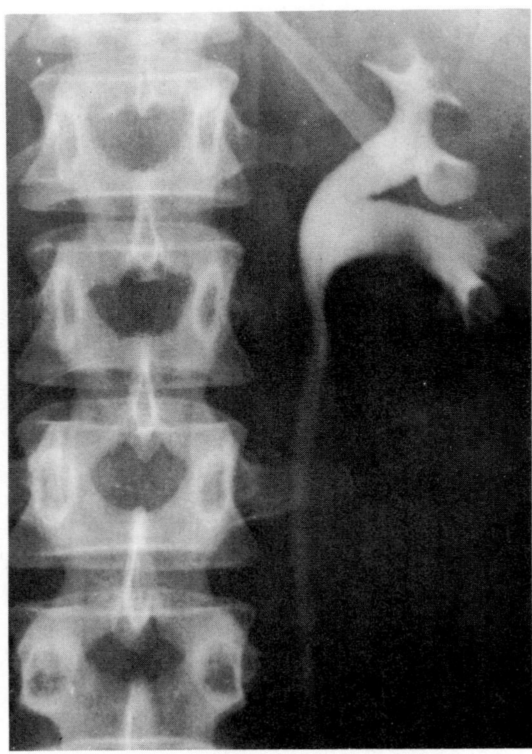

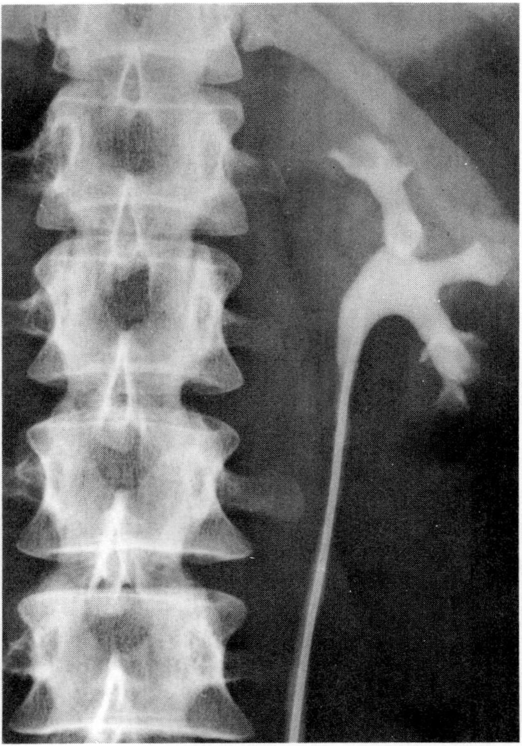

posture and also with the excursions of the diaphragm. However, when a person is in the supine position, the kidneys tend to extend from the 12th thoracic to the 3rd lumbar vertebra, with the right kidney usually a little lower than the left. The transpyloric plane passes through the upper part of the hilum of the right kidney and through the lower part of the hilum of the left. The upper poles of the kidneys are closer to one another than the lower poles; thus, both their long and transverse axes are oblique. The hila face anteriorly as well as medially, the lateral margin being more posterior.

The kidneys are difficult to palpate; the rib cage and the bulky paravertebral muscles make them inaccessible. When the abdominal wall is relaxed in a supine subject, the lower poles of the kidneys may be caught in the lumbar region of the abdomen between a hand placed just below the costal margin and the other posteriorly between the last rib and the iliac crest. Projected to the anterior abdominal wall, the renal hila are just medial to the point where the transpyloric plane intersects the costal margin and, on the back, 5 cm from the spinous process of L1 vertebra. The upper poles are in the epigastrium, each 2.5 cm from the midline, approximately 5 cm above the hilum. The lower poles are 7.5 cm from the midline, slightly above the supracristal plane.

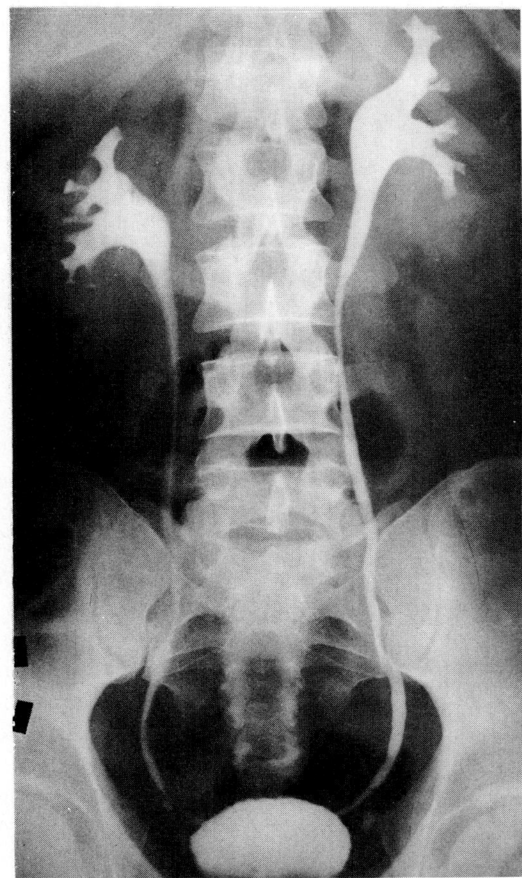

**FIG. 25-22.** An ascending or retrograde pyelogram showing the renal pelves and calices and the courses of the ureters. The urinary bladder also is visualized. (Braasch WF, Emmett JL: Clinical Urography. Philadelphia, WB Saunders, 1951)

## THE RENAL PELVIS AND URETER

In the sinus of the kidney, the renal pelvis divides into two or three **major calices,** which, in turn, divide into **minor calices** (Fig. 25-21). Into the minor calices, urine is discharged through the pores of the *cribriform plate,* which caps the tip of each renal papilla.

The renal pelvis extends through the hilum and, as it tapers to a narrow tube outside the kidney, becomes the ureter. The ureteropelvic junction is rather indefinite and is located approximately opposite the lower pole of the kidney. From here the ureter descends more or less vertically in the extraperitoneal fascia of the posterior abdominal wall (Fig. 25-22). At the pelvic brim, where the ureters cross the common iliac vessels, they incline medially and continue their intrapelvic course to the urinary bladder (Chap. 27).

**Hydronephrosis.** After its discharge from the renal tubule into the collecting ducts, the urine is not changed until it is voided from the bladder. Its passage through the renal pelvis and ureters is aided by peristaltic activity. There is constant production of urine by the kidneys, and obstruction of its flow along the ureters causes pressure to build up proximal to the obstruction, producing distention of the renal pelvis and its calices. This condition is known as hydronephrosis and also involves the ureter above the obstruction. Persistent and severe hydronephrosis results in damage to and eventually atrophy of the kidney.

The obstruction may be caused by stenosis of the ureteropelvic junction, by compression of the ureter, or by a urinary calculus (stone) lodged anywhere along the ureter. Hydronephrosis may result also from abnormalities in ureteric peristalsis or from abnormalities at the junction of the ureter with the bladder.

**Urinary Calculi.** Urinary stasis in the renal pelvis predisposes to the formation of stones, especially if certain substances are excreted in abnormally high concentrations in the urine. There are several types of renal calculi. If they are small, they are passed down the ureter and are voided. Larger ones may cause spasm and obstruction of the ureter, announced by excruciating pain. Still larger ones may not leave the renal pelvis, where they may form a cast of one or more of the calices. As they increase in size, they may eventually fragment.

## THE SUPRARENAL GLAND

The two suprarenal glands are roughly triangular, compact bodies, closely applied to the upper poles of the kidney (Fig. 25-19). They consist of an outer, pale cortex and an inner, almost black, medulla; the two are quite distinct, not only macroscopically and histologically but also with respect to development and function. The composition and function of both cortex and medulla are adequately described in Chapter 9.

The right suprarenal gland is rather pyramidal in shape; the left one is semilunar, larger, and flatter. The largest dimension of the glands does not exceed 5 cm. Their base, molded on the upper pole of the kidneys, is concave, and each gland has an anterior surface, covered partly with peritoneum on the right and completely on the left, and a posterior surface that rests against the diaphragm. Each gland has a small hilum on its anterior surface; a single vein issues from it. Many arterial twigs enter the glands around their periphery.

**FIG. 25-23.** Diagram of the renal fascia in a cross section through the posterior body wall.

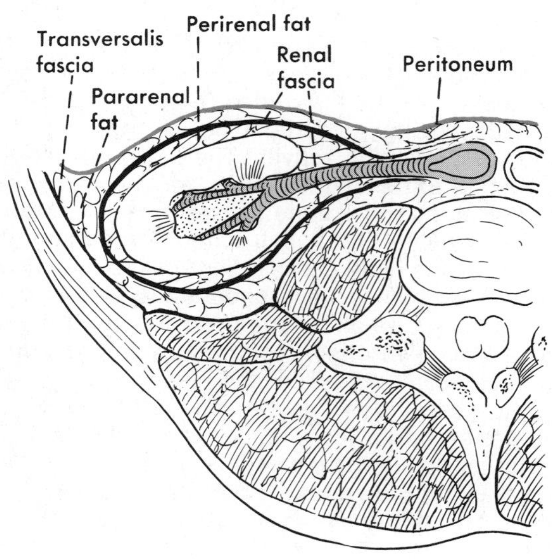


## RELATED FASCIAS AND ORGANS

The retroperitoneal connective tissue that surrounds the kidneys, ureters, and suprarenal glands is organized in a particular manner to form the *renal fascia* and the *adipose capsule* of the kidney. Other organs and structures are related through this connective tissue to the kidneys, ureters, and suprarenal glands.

### Adipose Capsule and Renal Fascia

The *perirenal fat,* in immediate contact with the kidney, ureter, and suprarenal glands, is regarded as the *adipose capsule.* This perirenal fat extends through the hilum, and in the renal sinus, the renal vessels, lymphatics, nerves, and calices of the renal pelvis are embedded in it. A membranous condensation of connective tissue surrounds this perirenal fat and separates it from the general adipose tissue of the posterior abdominal wall. The membranous component is the *renal fascia,* and the fat outside this fascia is known as the *pararenal fat* (Fig. 25-23).

It is the **renal fascia,** rather than the adipose capsule, that is of importance. The renal fascia surrounds each kidney as a separate sheath, and it must be incised in any operation on the kidney, whether from an anterior or a posterior approach. Descriptions of the fascia and its continuities vary, but it is usually described as a complete covering of both the kidney and suprarenal gland, sometimes sending a septum between them.

The renal fascia on the anterior surface of the kidney passes around the lateral border to become continuous with that on the posterior surface (Fig. 25-23). Traced medially, the two layers are continuous in a similar way over the medial margin of the kidney, except around the renal vessels, where they have been described as being traceable across the vertebral column, both in front of and behind the great vessels, to the corresponding layers around the other kidneys. It is generally granted, however, that passage of material other than injected gas does not occur across the midline. Traced upward, the renal fascia passes over the suprarenal gland, and its anterior layer merges with that of the posterior layer above the gland. The layers have been described as meeting also below the kidney, except where the ureter leaves. Because the renal fascia is open below, around the ureter, infections around the kidney may descend into the pelvis, and infections in the pelvis ascend to the kidney. The fat within the fascia forms a potential space in which fluid or injected air can accumulate and be more or less retained by the fascia.

Gas is sometimes injected into the extraperitoneal tissue of the pelvis and is permitted to ascend into the adipose capsule so as to outline the kidney and suprarenal gland for roentgenographic examination, particularly when a tumor is suspected.

**Radiology.** The outline of the kidneys can often be made out on a plain x-ray film of the abdomen because of contrast in the density of the radiolucent pararenal and perirenal fat and the more radiopaque renal parenchyma (see Fig. 23-12B). *Excretory urography* is a procedure that makes use of the ability of the kidneys to concentrate and excrete rather rapidly certain substances injected intravenously. When such substances are tagged with radiopaque molecules (*e.g.*, iodine), the kidneys become radiopaque soon after the injection, and their anatomy, as well as that of the renal pelvis, can be examined on x-ray films (Fig. 25-24). Such an x-ray is called an *intravenous pyelogram* (IVP). The anatomy of the ureters and renal pelves can be studied with greater accuracy if they are filled with contrast medium directly through a catheter inserted through the bladder into the orifice of the ureter (Figs. 25-21 and 25-22; *ascending* or *retrograde urography; retrograde pyelogram*).

## Related Structures and Organs

The posterior relations of the kidneys, ureters, and suprarenal glands are quite similar on the two sides, whereas anteriorly, there are major differences.

**Posterior Relations.** Both suprarenal glands and roughly the upper third of both kidneys lie on the diaphragm. Through the costal parts and the lumbocostal trigones of

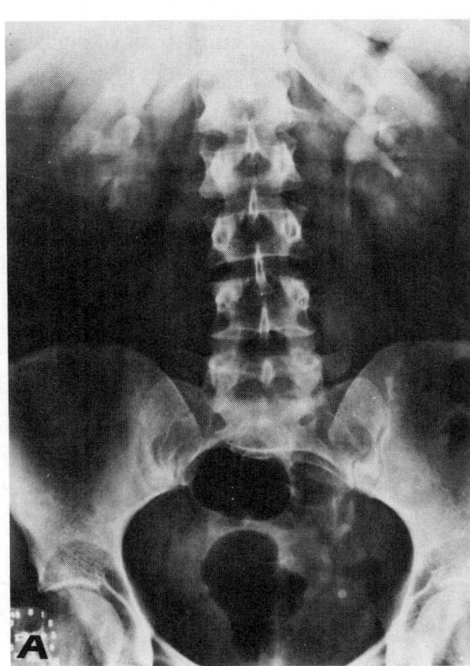


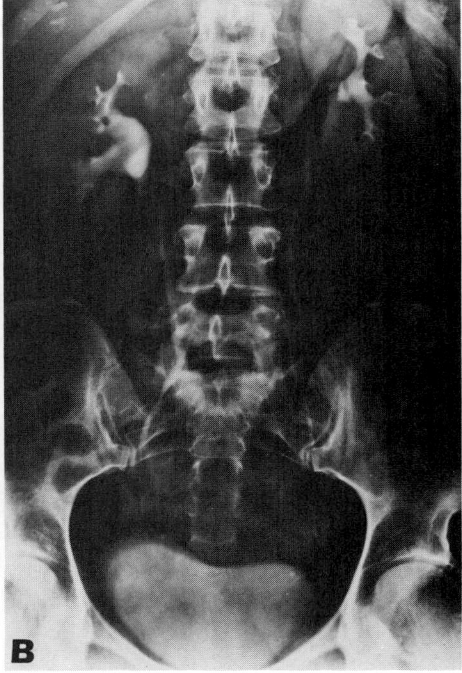


**FIG. 25-24.** Two intravenous pyelograms (IVPs) in two different subjects. In **A,** exposed shortly after intravenous injection of the contrast medium, the renal parenchyma is radiopaque owing to the concentration of the contrast medium by the kidneys, which is beginning to outline also the calices and the renal pelvis. In **B,** exposed much later, the contrast medium fills the renal pelves, the ureters, and the bladder.

the diaphragm, these organs are related to the pleura, quite intimately if there are few or no muscle fibers present in the trigones. The 11th and 12th ribs are behind the suprarenal glands and the left kidney. The costodiaphragmatic recess and the diaphragm intervene between the 11th rib and the renal fascia, whereas the 12th rib is in direct contact with the fascia of both kidneys. Fractures or displacements of this rib may cause injury to the kidney. The pleural sacs extend below the lower border of the last rib and are present behind the kidneys and the diaphragm in the triangular area bounded by the 12th rib, the medial arcuate line, and L1 transverse process.

Both lateral and medial lumbocostal arches (arcuate lines) cross behind the kidneys. Below the medial arch, the medial margin of the kidney rests on the psoas major, as does the renal pelvis and, lower down, the ureter. The posterior branches of the renal artery enter the renal hilum, running posteriorly to the upper part of the renal pelvis. Each ureter is separated from the tips of the lumbar transverse processes by the psoas major muscle. As noted before, the ureter lies on the psoas as far down as the pelvic brim, where the ureters cross anterior to the common iliac vessels.

Below the lateral arcuate line, the renal fascia is in contact with the quadratus lumborum. Near their lateral margins, the kidneys lie on the transversus abdominis. The subcostal and iliohypogastric nerves cross these two muscles behind the kidney.

**FIG. 25-25.** Peritoneal relationships of the kidneys. The stippled areas are those in which the kidneys are directly related to the peritoneum through the fat and connective tissue over them. Peritoneum of the greater sac is pink; that of the omental bursa is blue.

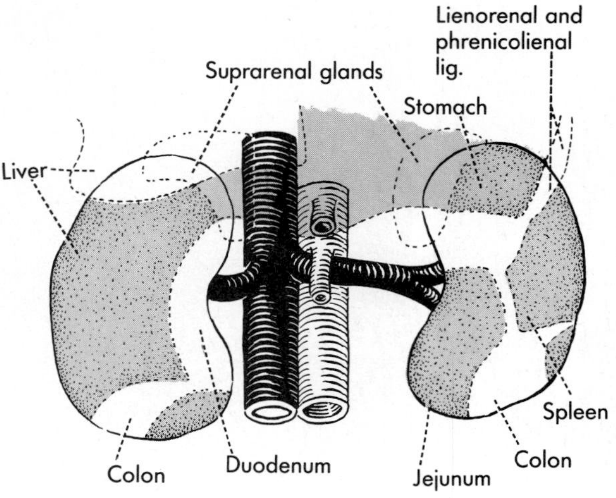


**Anterior Relations on the Right.** The upper half of the right suprarenal gland and, to its right, the upper pole of the right kidney lie behind the bare area of the liver. Medially, the suprarenal gland is overlapped by the inferior vena cava (Figs. 25-19 and 25-25). The kidney and the suprarenal gland are crossed by the posterior layer of the right coronary ligament (hepatorenal ligament), and below this ligament, both are covered by parietal peritoneum. Their anterior surface here faces into the hepatorenal recess (compare Figs. 25-25 and 23-18). Through this recess, the kidney and suprarenal gland are in contact with the inferior surface of the right lobe of the liver (see Fig. 24-31). The medial margin of the kidney, in the region of the hilum, is nonperitoneal; the descending part of the duodenum lies on it. Just above the inferior pole, the right colic flexure is in direct contact with the renal fascia (Fig. 25-25). The lower pole itself, however, is covered by parietal peritoneum of the right infracolic compartment.

The right *renal pelvis* is crossed anteriorly by the anterior branches of the renal vessels, which, together with the pelvis, are behind the duodenum. Below the duodenum, the right ureter descends behind the peritoneum of the right infracolic compartment, where it is crossed by the testicular or ovarian vessels, the right colic and iliocolic vessels, and the root of the mesentery.

**Anterior Relations on the Left.** The left suprarenal gland and the upper pole of the left kidney face into the omental bursa and are covered by parietal peritoneum (Fig. 25-25). Through the cavity of the bursa, the stomach is in contact with this area, limited toward the left by the attachment of lienorenal ligament to the kidney. On the left side of the ligament, parietal peritoneum of the greater sac covers the kidney, and through the cavity of this sac, the spleen is in contact with the anterior surface and the lateral margin of the kidney. Below the gastric area, the tail of the pancreas rests directly on the renal fascia, and it overlaps much of the hilum and its vessels. The lower pole and the adjoining anteromedial surface are covered by parietal peritoneum of the left infracolic compartment; loops of jejunum are in contact with it. Lateral to the lower pole, the left colic flexure and descending colon are in direct contact with the renal fascia.

The *renal pelvis* is crossed anteriorly by branches of the renal vessels and the tail of the pancreas. The left ureter is retroperitoneal in the left infracolic compartment and is crossed by the left colic vessel and, just above the pelvic brim, by the root of the sigmoid mesocolon.

## VESSELS AND NERVES

### Arterial Supply and Venous Drainage

**Suprarenal Glands.** The arterial supply to the suprarenal gland is derived from three or four sources: the inferior phrenic arteries, the aorta, and the renal and, on occasion, the gonadal arteries (Fig. 25-26).

As the *inferior phrenic arteries* pass just above and medial to the suprarenal glands, each artery usually gives off a series of branches into the gland of its own side before it supplies the diaphragm. These arteries constitute the **superior suprarenal arteries,** and they have long been named as though there were only a single artery. They enter the upper part of the gland over a considerable expanse of its anterior and posterior surface.

There usually is at least one artery to each gland given off by the aorta just above the origin of the renal arteries. This artery (or arteries) is called the **middle suprarenal artery.** Also, one or more arteries reach the suprarenal gland from the adjacent renal artery and are called the **inferior suprarenal artery.** In addition to these three regular sources of blood supply, other vessels running close by may also supply branches to the suprarenal gland. Most constant of these are the arteries to the gonads.

Since any of the arteries approaching the gland may branch or rebranch, the number of vessels entering it may be quite numerous—as many as 50 individual stems have been counted. There is no regular position in which these branches enter the gland; they enter over much of the surface.

In contrast to the arteries, the main drainage of the gland is into a single **suprarenal vein** that leaves the gland through its hilum,

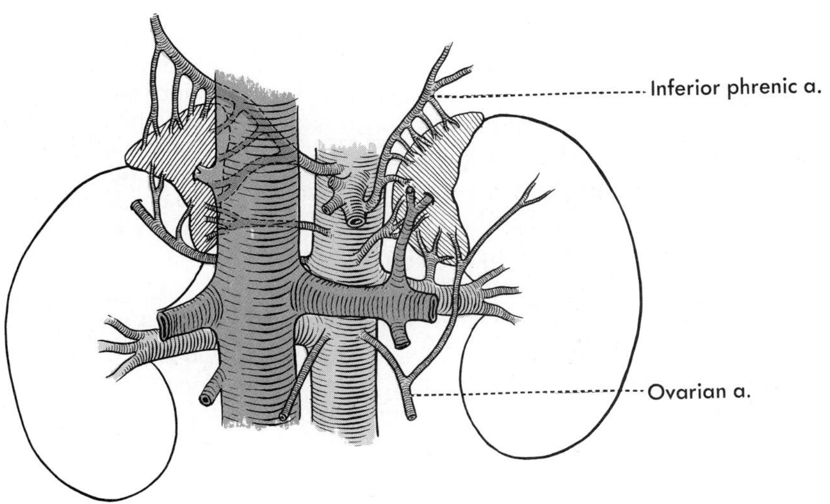


**FIG. 25-26.** Vessels of the suprarenal glands; note the single vein from each gland, and the multiple arteries. (Modified from Hollinshead WH: Surg Clin North Am 32:1115, 1952)

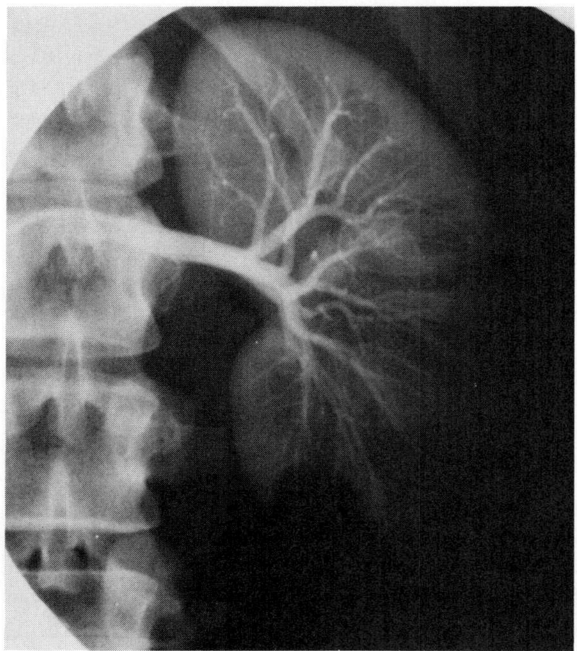

**FIG. 25-27.** A renal arteriogram. Contrast medium was injected into the left renal artery through a catheter just visible in the aorta close to the origin of the artery.

although there are also several tiny venous twigs that connect with veins in the adjacent connective tissue. On the left side, the suprarenal vein is usually joined by the inferior phrenic vein of that side, and separately or together, they empty into the left renal vein. On the right side, the suprarenal vein joins the inferior vena cava. Usually the suprarenal vein is very short, emerging from the gland close to the vena cava.

**Kidneys.** The two renal arteries arise from the lateral aspects of the aorta, only a little below the origin of the superior mesenteric artery, most commonly at the level of the lower third of the 1st lumbar vertebra to the upper third of the 2nd. The right artery tends to arise a little higher than the left one (Fig. 25-27) and passes behind the inferior vena cava. The renal veins lie in front of the renal arteries, largely hiding them (Fig. 25-19). On their way to the renal hilum, the renal arteries give off their nonrenal branches and, just before reaching the hilum, divide into *segmental renal arteries,* which pass around the renal pelvis into the renal sinus.

The renal artery gives off its *suprarenal branch,* sends one or more *ureteric arteries* downward and one or more *capsular branches* into the adipose capsule of the kidney. Usually, fairly close to the renal pelvis, the renal artery divides into *anterior* and *posterior rami* that pass on respective sides of the renal pelvis and the major calices. Each of these arteries supplies a vascular segment of the kidney and each is known as a segmental artery (Fig. 25-28).

**Vascular Segments.** Within the renal sinus, anterior and posterior arterial rami rebranch. Although the pattern is quite variable, the distribution is constant enough to allow the division of the kidney into vascular segments that correspond to the prevailing vascular pattern.

The anterior rami of the renal artery are three in number: the **superior, anterior,** and **inferior segmental arteries;** they supply segments of the kidney named correspondingly (Fig. 25-28). There is usually a **posterior ramus** of the renal artery supplying the posterior segment. Superior and inferior segments occupying the region of the upper and lower poles are represented on both anterior and posterior surfaces of the kidney; the anterior and posterior segments are limited to the corresponding surfaces. Vascular patterns that depart markedly from the one described here have also been reported.

As the branches of the renal artery run through the renal sinus, they give off twigs to the connective tissue of the sinus and to the renal pelvis and calices. Their major branches penetrate the renal parenchyma as **interlobar arteries.** These become the **arcuate arteries** after a sharp turn reorients them across the base of the pyramids. In this position, they give rise to the **interlobular arteries** from which the smaller vessels of the kidney are derived.

The segmental arteries do not anastomose with each other; obstruction of a segmental artery leads to cessation of function and death of that segment of the kidney served by the artery.

In contrast to the arteries, the **veins** within the kidney anastomose with each other and have no segmental distribution. All tributaries of the renal vein usually pass in front of the renal pelvis, but sometimes one may pass behind it.

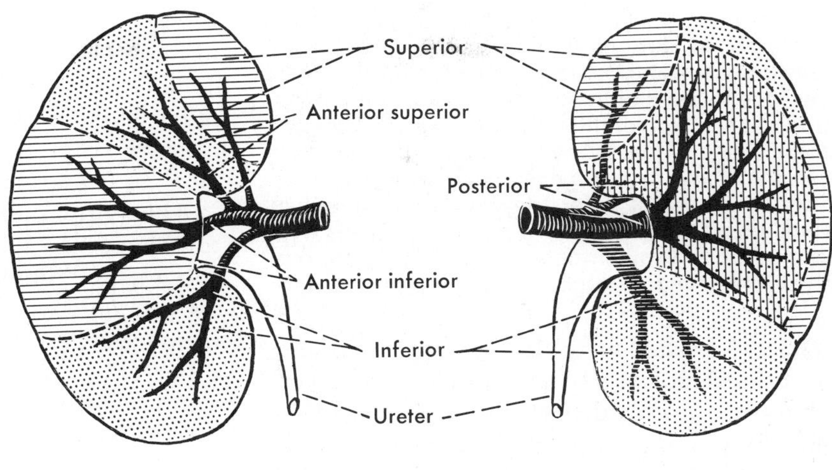


**FIG. 25-28.** The segmental renal arteries and the renal segments.

The **left renal vein** crosses in front of the aorta and behind the descending stem of the superior mesenteric artery to join the inferior vena cava. The short **right renal vein** joins the inferior vena cava at about the same level.

The right renal vein usually receives no tributaries other than those from the kidney; the left renal vein usually receives the *upper lumbar vein,* a communication from the *ascending lumbar vein,* the combined stem of the *left inferior phrenic* and *suprarenal veins,* and the vein from the left gonad (*ovarian or testicular*). The right ovarian or testicular vein typically enters the inferior vena cava below the entrance of the right renal vein, whereas the right suprarenal vein and the right inferior phrenic enter the vena cava separately above the renal vein.

**Variations.** The renal, inferior phrenic, suprarenal, and gonadal arteries are all members of a series of segmental *mesonephric arteries* that supplied the urogenital ridges (p. 708). These ridges contain the suprarenal gland, the developing kidneys, and the gonads. In approximately one-third of bodies, more than one artery persists to an otherwise normal kidney. Almost a third of such kidneys receive two or more of these so-called accessory renal arteries. Two or three renal arteries occur as often on one side as the other.

Whether these accessory renal arteries pass behind or in front of the inferior vena cava depends on their level of origin from the aorta. When they arise in the vicinity of normal renal arteries or above them, they will pass behind the prerenal segment of the inferior vena cava, along with the inferior phrenic or suprarenal branches of the aorta. Should they arise from the part of the aorta that parallels the postrenal segment of the inferior vena cava, they will pass in front of it and also in front of the renal pelvis or ureter. These inferior accessory arteries may compress the ureter or the renal pelvis, predisposing to hydronephrosis.

Although most of the branches of the renal artery and most of the tributaries of the renal vein pass through the hilum of the kidney, arterial branches that do not do so are fairly common. Such extrahilar arteries penetrate the parenchyma of the kidney on its external surface. They may arise from the renal artery or the aorta, or they may have an aberrant origin from a different vessel. Their unexpected presence is a hazard in operations on the kidney. They are particularly frequent to the upper pole.

None of the so-called accessory renal arteries, be they hilar or extrahilar, can be regarded as truly *accessory* to the normal renal vasculature. They do not anastomose with the segmental arteries. Although these arteries are sometimes called by surgeons "aberrant," this term should be applied to them only when they do not arise from the renal artery or the aorta.

Multiple **renal veins** are almost as frequent as multiple arteries on the right side but are rare on the left. However, the left renal vein may bifurcate and pass both in front of and behind the aorta, thus forming a *circumaortic venous ring,* or it may have no anterior connection at all and will pass behind, rather than in front of, the aorta.

Unlike the arteries, the venous tributaries anastomose within the kidney, and there are also minor anastomoses through the fibrous capsule, with veins that are not tributaries of the renal vein. However, due to the inefficiency of these anastomoses, sudden occlusion of the right renal vein causes necrosis of the whole kidney. The left renal vein, on the other hand, has larger connections to the extrarenal veins that enter the caval system. It has been ligated and divided in a number of cases with no evidence of impairment of renal function, owing, it is believed, to larger anastomoses on the left.

**Ureter.** In its course through the abdomen, the ureter receives twigs from adjacent vessels but in no fixed pattern. The *renal artery* rather regularly supplies the upper end. Ureteric branches can often be found from the aorta, the testicular or ovarian, and the iliac arteries. After it has entered the pelvis, the ureter receives one or more branches from pelvic arteries. Typically, all the vessels reaching the ureter divide into ascending and descending branches that form a longitudinal anastomosis along it. This anastomosis may be poor in places because of the uneven spacing of the vessels.

The **veins** of the ureter conform to the pattern of arterial supply.

### Lymphatics

The suprarenal glands, kidneys, and ureters drain their lymphatics into the lumbar nodes (p. 691).

### Innervation

**Suprarenal Glands.** As the greater splanchnic nerves pass the suprarenal glands on their way to the celiac plexus, they give off fibers directly to the glands, which mingle with fibers derived from the celiac plexus. The *suprarenal plexus* formed by these nerves consists of *preganglionic* sympathetic fibers that terminate on the cells of the suprarenal medulla. These cells are equivalent to the autonomic ganglia and release their neurotransmitter substances directly into the venous sinusoids of the medulla (see Chap. 9). The suprarenal cortex has no nerve supply.

**Kidneys.** Extensions of the celiac plexus form a *renal plexus* around the renal artery. This plexus is joined by the lowest splanchnic nerve or by the renal branch of the lesser splanchnic nerve. The plexus contains small *renal ganglia,* one of which, the *aorticorenal ganglion,* is of macroscopic size and is located close to each renal artery.

The sympathetic fibers in the renal plexus are concerned with the control of blood vessels to the kidney. Section of them increases the blood flow through the kidney and thereby produces diuresis, but apparently has no other effect on renal function and no effect upon the pressure of urine in the renal calices. The renal plexus also contains visceral afferent fibers concerned with pain, which terminate in T10–L1 spinal cord segments. Renal pain is usually referred to the back in the region of the costal angle.

**Ureters.** The *ureteric plexus* is formed by delicate nerve filaments derived from various parts of the aortic plexus. The nerves contain sympathetic efferents and afferents. Stimulation of the renal plexus, which contributes to the ureteric plexus, increases peristaltic activity of the ureter, but section of the plexus does not interfere with ureteric peristalsis. The visceral afferents concerned with pain relay in segments T11–L2. Ureteric pain, mentioned in relation to urinary calculi, is referred to a wide area, including parts of the abdominal wall above the iliac crest, the suprapubic region, the genitals, and the medial aspects of the thigh and leg.

## DEVELOPMENT AND CONGENITAL ANOMALIES

### Normal Development

**The Urogenital Ridges.** The kidneys, gonads, and suprarenal cortex develop from intermediate mesoderm. The longitudinal, bilateral mass of **intermediate mesoderm** becomes defined in the trilaminar embryo lateral to the somites. Once the intraembryonic celom is formed, the lateral surface of this bar of mesoderm comes to face into each celomic duct (see Figs. 7-2 and 7-3). Growth and differentiation of the intermediate mesoderm produce two longitudinal ridges that bulge into the celomic ducts on either side of the dorsal mesentery of the primitive gut (see Fig. 7-3). These

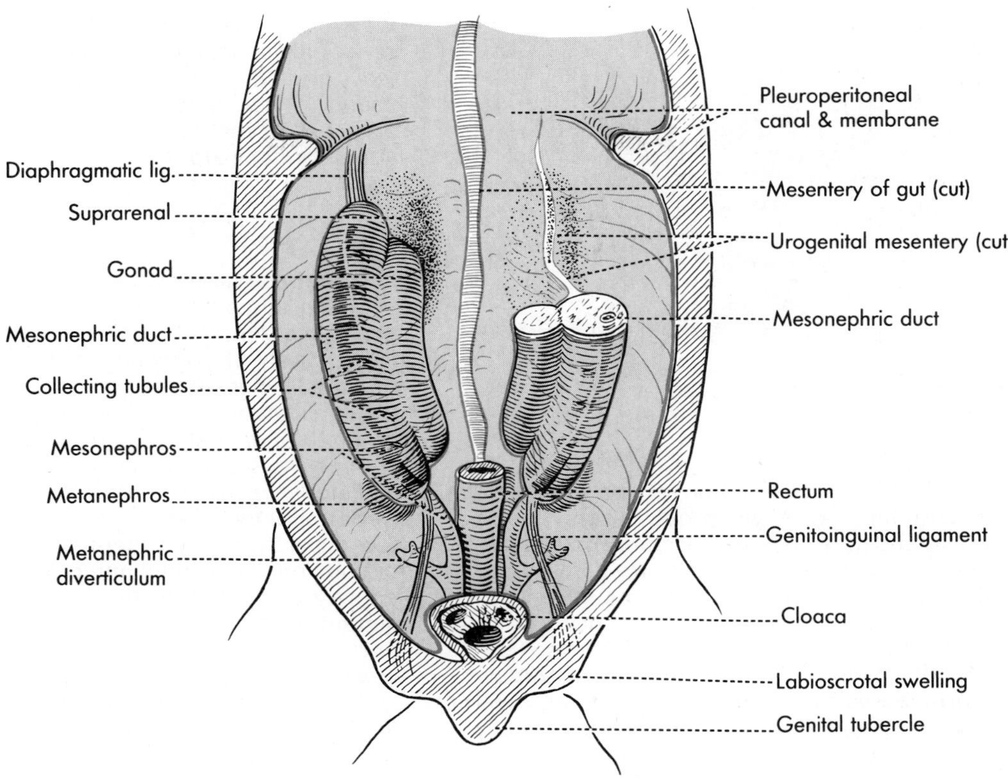


**FIG. 25-29.** The urogenital ridges, their mesentery and associated ligaments, illustrated diagrammatically. Except for the rectum, the gut has been excised with its mesentery. The mesothelial lining of the celom is colored pink. The paramesonephric ducts are omitted from this scheme, although they develop in both sexes. For the relationship of the paramesonephric ducts to the mesonephros and the mesonephric duct, see Figure 27-16**A**.

so-called *urogenital ridges* extend from the lower cervical to the caudal regions of the embryo. The celomic or abdominal surface of the urogenital ridges is covered with the celomic mesothelium derived from cells of the intermediate mesoderm itself. This mesothelium is destined to become parietal peritoneum in this region and to contribute to the gonads and suprarenal gland. The attachment of the ridges to the dorsal wall of the celom is spoken of as the *urogenital mesentery.*

The anterolateral part of the urogenital ridges is occupied by *nephrogenic tissue,* from which develop three successive and overlapping generations of kidneys: *pronephros, mesonephros,* and *metanephros.* The anteromedial portion of the ridges is occupied by the primitive gonads, and behind them is the suprarenal gland (Fig. 25-29).

**Gonads and Suprarenal Glands.** Proliferation of the celomic mesothelium that overlies the gonadal and suprarenal surfaces of the urogenital ridges contributes to the formation of both the gonads (ovaries and testes) and the suprarenal cortex. In addition, each organ becomes invaded by immigrant cells: the gonads by the *primitive germ cells* from the yolk sac and the suprarenal glands by cells derived from the *neural crest.* The primitive germ cells are the ancestors of oogonia and spermatogonia; from the neural crest cells develops the suprarenal medulla. All autonomic ganglia, whose function is similar to the suprarenal medulla, are also derived from neural crest cells.

**Kidneys.** The **pronephros** and **mesonephros** are segmental organs located in the lower cervical and thoracolumbar regions of the embryo. They receive a series of segmental branches from the aorta. Early in development, urine is discharged directly into the celom; later, it is discharged into bilateral longitudinal ducts called at first the pronephric ducts and later the *mesonephric ducts,* which empty into the cloaca. The **metanephros,**

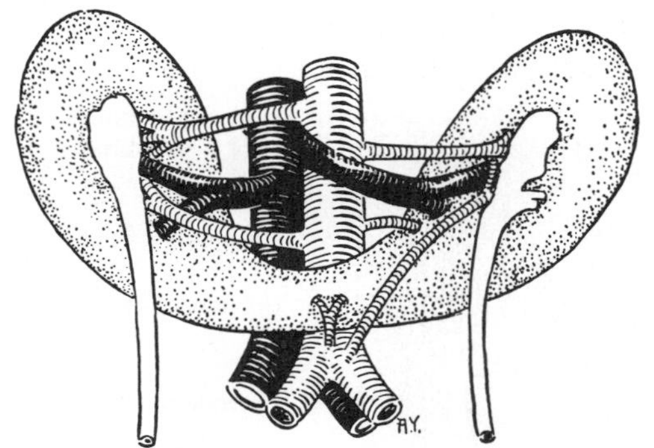


**FIG. 25-30.** Horseshoe kidney.

located in the sacral region of the embryo, becomes functional as its predecessors, the pronephros and mesonephros, decline and disappear.

Although the mesonephric duct persists in the male (p. 757), urine produced by the metanephros will be conveyed to the cloaca by a new duct that sprouts from the caudal end of the mesonephric duct in the form of the **metanephric diverticulum,** known also as the *ureteric bud* (see Fig. 27-15). This diverticulum elongates and grows cranially to invade the metanephros, located at the caudal end of the urogenital ridges (Fig. 25-29). Contact between the metanephric diverticulum and the metanephros induces differentiation of the renal corpuscles and renal tubule from metanephric tissue and differentiation of the collecting tubules from the cranial end of the diverticulum. This establishes the definitive kidney. More caudal parts of the metanephric diverticulum become the renal pelvis and the ureter.

**Changes in Relationships.** Disappearance of the pronephros and mesonephros leaves the ovary or testis, the mesonephric duct, and an additional duct called the paramesonephric duct (explained in Chap. 27) suspended in the urogenital mesentery. This mesentery with its contents slides down the posterior abdominal wall into the pelvis (further discussed in Chap. 27), leaving behind the suprarenal glands in a retroperitoneal position.

The definitive kidneys, after they have been formed in the pelvis, ascend on the posterior abdominal wall, but the mechanisms of this migration are unknown. It is probable that differential growth of the various vertebral regions and the elongation of the ureteric bud are responsible for the relative and absolute cranial migration of the definitive kidneys. This takes place behind the peritoneum and brings the kidneys in contact with the suprarenal glands, which, up to the time of birth, are larger than the kidneys themselves. As the kidney ascends, its connections to the segmental arteries that originally supply the mesonephros also change. The reduction of these mesonephric branches of the aorta normally results in a single artery for each kidney.

### Renal Abnormalities

Gross abnormalities of the kidneys or ureters occur in at least 3% to 4% of persons. There are a large variety of congenital abnormalities that can be broadly classed as (1) agenesis and hypoplasia, (2) duplications, (3) malposition and malrotation, and (4) congenital cystic disease. By far the most common are abnormalities of position.

Fetuses in whom both kidneys fail to develop **(renal agenesis)** survive to term but die within a few days after their separation from the placental circulation. One functioning kidney, however, is entirely compatible with life. A single kidney, resulting from unilateral renal agenesis, may be abnormally located, as may be one or both of an otherwise normal pair of kidneys.

Splitting of the ureteric bud during its growth leads to partial or complete **duplication** of the kidney, owing to reduplication of the induction process in the metanephros. The renal duplication is usually incomplete, and the kidney consists of two fused masses, one above the other, with separate renal pelves that empty into a bifid ureter or into two separate ureters. The extra ureter may open into the normal ureter (bifid ureter), in which case the ureteric bud probably split close to the metanephros. If the split was close to the mesonephric duct, the extra ureter may end independently in the bladder (into which the distal portion of the mesonephric duct, bearing the metanephric diverticulum, becomes absorbed), or it may end ectopically in the urethra. It is interesting that usually the upper part of a "double" kidney and the upper ureter are the abnormal structures. In such cases, usually the upper ureter terminates ectopically and the lower one has a normal course into the bladder.

Arrests or errors in the ascent of the kidney result in **ectopia.** In *simple renal ectopia,* one or both kidneys are placed lower than usual, owing to the arrest of a normal process. A large proportion of ectopic kidneys are in the pelvis, and they typically receive their blood supply from whatever artery lies close to them. Many such kidneys have multiple blood supplies.

It is a general rule that the lower the kidneys lie, the closer together they are. Thus, ectopic kidneys placed low in the abdomen tend to fuse together, usually at their lower poles, to form a somewhat horseshoe-shaped mass called *horseshoe kidneys* (Fig. 25-30).

In *crossed ectopia,* a kidney crosses to the opposite side during its ascent. Usually, when this occurs, the kidney that is crossed fuses with the kidney on the normal side, so that there is a single renal mass. However, the fact that one ureter descends on one side and the other crosses the vertebral column to descend on the other distinguishes such a mass from that of a partial duplication of the kidney.

Among the most common abnormalities of position of the kidneys is **abnormal rotation.** The definitive kidneys begin their development with their hila facing forward. Subsequently, they turn 90° and point their hila medially. This rotation may fail to occur, go too far, or be in the reverse direction—thus, a renal pelvis may face posteriorly or laterally. Abnormal rotation is regularly associated with abnormally low kidneys, but may occur in normally placed ones.

A rare *abnormality of the right ureter* is for it to run behind the inferior vena cava. This anomaly, called **retrocaval** or *postcaval ureter,* is a consequence of abnormal formation of the postrenal segment of the inferior vena cava (p. 690). Retrocaval ureter usually calls attention to itself eventually because of obstruction by the vena cava.

Several types of **renal cysts** have congenital origin. The most widely accepted explanation of *polycystic kidneys,* a condition in which the organs are riddled with multiple cysts, is the widespread failure of the renal tubules, derived from metanephric tissue, to join up with the collecting tubules derived from the ureteric bud. The cysts arise because the urine produced by the nephrons cannot be drained into the calices.

## GENERAL REFERENCE AND RECOMMENDED READINGS


ALLISON PR: Reflux esophagitis, sliding hiatal hernia and anatomy of repair. Surg Gynecol Obstet 92:419, 1951

ANSON BJ, CAULDWELL EW, PICK JW ET AL: The blood supply of the kidney, suprarenal gland, and associated structures. Surg Gynecol Obstet 84:313, 1947

ARVIS G: Considerations anatomiques sur le hile et le sinus du rein. Ann Radiol 12:75, 1969

BOYARSKY L, LABAY P, GLENN JF: More evidence for ureteral nerve function and its clinical implications. J Urol 99:533, 1968

BOYD W, BLINCOE H, HAYNER JC: Sequence of action of the diaphragm and quadratus lumborum during quiet breathing. Anat Rec 151:579, 1965

CAREY JM, HOLLINSHEAD WH: Anatomic study of the esophageal hiatus. Surg Gynecol Obstet 100:196, 1955

GRAVES FT: Anatomy of the intrarenal arteries and its application to segmental resection of the kidney. Br J Surg 42:132, 1954

INKE G, SCHNEIDER W, SCHNEIDER U: Anzahl der Papillen und der Vasi uriniferi der menschlichen Niere. Anat Anz 118:241, 1966

LISTERUD MB, HARKINS HN: Anatomy of the esophageal hiatus. Arch Surg 76:835, 1958

MCKIBBIN B: The action of the iliopsoas in the newborn. J Bone Joint Surg (Br) 50:161, 1968

PICK J: The Autonomic Nervous System: Morphological, Comparative, Clinical, and Surgical Aspects. Philadelphia, JB Lippincott, 1970

TARNAY TJ: Diaphragmatic hernia. Ann Thorac Surg 5:66, 1968

TOBIN CE: The renal fascia and its relation to the transversalis fascia. Anat Rec 89:295, 1944

WELLS LJ: Development of the human diaphragm and pleural sacs. Contrib Embryol 35:107, 1954

# 26

# THE INGUINAL AND FEMORAL CANALS; THE SCROTUM; THE APPLIED ANATOMY OF HERNIA

On the surface of the body, the inguinal region is the meeting place of the anterior abdominal wall and the thigh; in the interior of the body, the anterior and posterior walls of the abdomen meet each other in the inguinal region. This area has particular clinical importance because it is here that abdominal contents most commonly protrude through the abdominal wall and present themselves as hernias.

The key to the anatomy of this region is the **inguinal ligament,** already encountered in Chapter 23. Structures that pass between the abdominal cavity and the scrotum or the labia majora do so *above* the inguinal ligament; the slitlike passage in the abdominal wall that transmits them is the **inguinal canal.** In the female, only a fibromuscular band, the *round ligament of the uterus,* passes through the inguinal canal to attach to the skin of the labia majora. In the male, on the other hand, the canal transmits the duct of the testis (*ductus deferens*) with its associated vessels and nerves; wrapped in fascial tunics, together they form the *spermatic cord.* Consequently, the inguinal canal is of greater significance in the male than in the female.

In both sexes, abdominal contents that should normally remain in the abdomen may descend through the inguinal canal to form an **inguinal hernia.** Inguinal hernias emerge as swellings from the inguinal canal and will eventually distend the scrotum or the labium majus.

Structures that pass between the abdominal cavity and the thigh

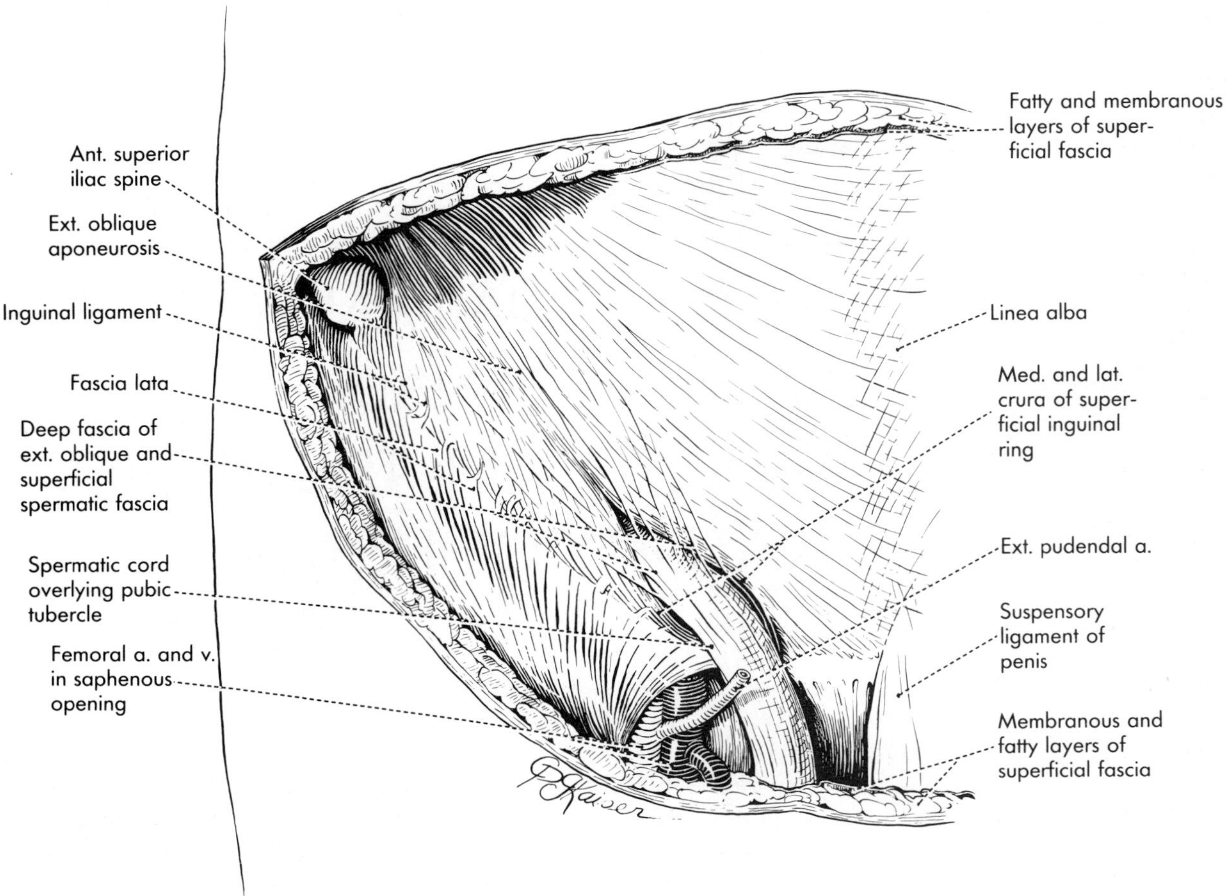


**FIG. 26-1.** A dissection of the inguinal region showing the anterior view of the inguinal ligament, the related fascias, the superficial inguinal ring, and the saphenous opening.

do so *beneath* the inguinal ligament. These are the external iliac artery and vein, wrapped in a fascial tunic called the *femoral sheath*. The potential space within the femoral sheath is the **femoral canal.** This canal may become distended by abdominal contents that should normally remain in the abdomen, forming a **femoral hernia**, which descends beneath the inguinal ligament into the thigh; it cannot enter the scrotum or the labia majora.

To explain the anatomy of the inguinal and femoral canals, a requisite for understanding inguinal and femoral hernia, the inguinal ligament will be described first. This chapter also deals with the scrotum, because it is anatomically continuous with the inguinal canal, despite the fact that the scrotum is located in the perineum. The description of the labia majora, however, is deferred to Chapter 28, where it is included with other parts of the vulva.

## THE INGUINAL LIGAMENT

The inguinal ligament is not a true ligament; rather, it is the inferior free border of the external oblique aponeurosis, thickened and reinforced by collagen fiber bundles that run from the anterior superior iliac spine to the pubic tubercle (p. 582, Fig. 26-1). When viewed from the front, the external oblique aponeurosis is rounded because its free border is turned posteriorly; anteriorly, there is a smooth transition between the fibers of the ligament and those of the aponeurosis proper. Thus, the inguinal ligament has an upper and a lower surface.

The inguinal ligament is not visible readily on either the superficial or the deep aspect of the anterior abdominal wall. It is concealed by the fascias attached to it. Anteriorly, the **fascia lata** is attached to the rounded lower border of the external oblique aponeurosis; posteriorly, the attachment of the **transversalis fascia** along the free edge of the ligament obscures that edge to some degree.

Because of the pull of the fascia lata on the inguinal ligament, the ligament is convex downward when the lower limbs are in line with the trunk; flexion of the hip relaxes and straightens the ligament.

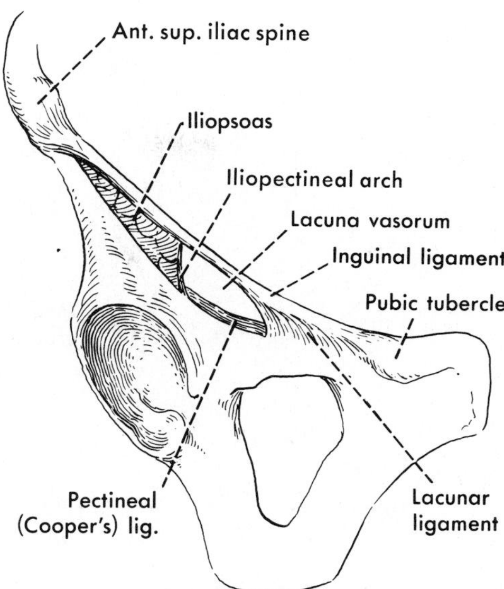


**FIG. 26-2.** Inguinal, lacunar, and pectineal ligaments and the muscular and vascular compartments behind the inguinal ligament, seen from below.

**Attachments.** The inguinal ligament is best exposed for the study of its attachments by incising the external oblique aponeurosis just above and parallel with the inguinal ligament. Laterally, the inguinal ligament is attached to the *anterior superior iliac spine* and medially, to the *pubic tubercle* (Fig. 26-2). Before the ligament reaches the pubic tubercle, some fibers diverge from its posterior free edge and attach to the medial 2 cm to 3 cm of the *pecten* of the superior ramus of the pubis (see Fig. 17-4). This crescent-shaped expansion is called the **lacunar ligament,** and it fills in a triangular gap between the medial end of the inguinal ligament and the pecten pubis (Fig. 26-2). At the apex of this triangle is the pubic tubercle; the base is formed by the sharp, free edge of the lacunar ligament, which faces laterally and forms the medial boundary of the *lacuna vasorum,* the space through which the external iliac vessels leave and enter the thigh (Fig. 26-2). Some fibers of the lacunar ligament continue laterally along the pecten pubis and contribute to the formation of a strong fibrous band, the **pectineal ligament,** which is fused with the periosteum of the superior ramus of the pubis.

As the inguinal ligament ends on the pubic tubercle, it gives off a few fibers that reflect upward and run through the aponeurosis into the linea alba. These fibers constitute the **reflected part** of the inguinal ligament (Fig. 26-3). This ligament lies in the posterior wall of the *superficial inguinal ring.*

**Relations.** The space between the inguinal ligament and the superior ramus of the pubis is divided into two compartments by a fascial septum, the **iliopectineal arch,** continuous both with the inguinal ligament and with the iliopsoas fascia (Fig. 26-2). The lateral compartment, the *lacuna musculorum,* contains the iliopsoas muscle with the femoral nerve lying on it; the medial compartment, the *lacuna vasorum,* transmits the external iliac vessels, as mentioned earlier. The medial half of the inguinal ligament has a free surface facing into the lacuna vasorum; its lateral half is bound to the iliopsoas fascia. The most inferior fibers of the internal oblique and transversus abdominis arise from the line of fusion between the inguinal ligament and the iliopsoas fascia. The upper surface of the medial

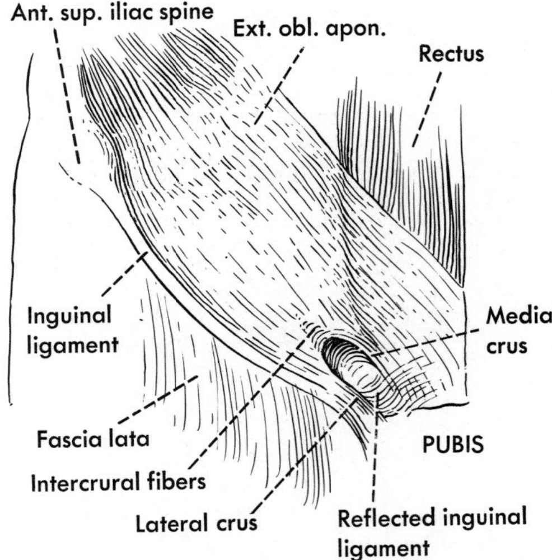


**FIG. 26-3.** The superficial inguinal ring.

half of the ligament forms the floor of the inguinal canal. Just medial to the anterior superior iliac spine, the inguinal ligament is usually pierced by the *lateral femoral cutaneous nerve.* The nerve may be compressed by the ligament, giving rise to pain in the area of the nerve's distribution (see Fig. 16-2). The medial end of the ligament, attached to the pubic tubercle, forms the inferior boundary of the *superficial inguinal ring* and is designated as the lateral crus of the ring.

## THE INGUINAL CANAL

The inguinal canal is an oblique passage through the anterior abdominal wall connecting the extraperitoneal space of the abdomen to the scrotum or the labia majora. The canal commences at the *deep inguinal ring* and terminates at the *superficial inguinal ring.* It is 4 cm to 5 cm long and slopes medially and downward.

### BOUNDARIES

#### The Deep Inguinal Ring

The deep inguinal ring appears as an oval defect in the transversalis fascia when the deep aspect of the fascia is exposed by reflecting the parietal peritoneum from it. In truth, the transversalis fascia, rather than being defective, is prolonged anteriorly from the margins of the deep ring into the inguinal canal. It forms a fascial sleeve that surrounds the ductus deferens or the round ligament of the uterus, both of which enter the deep ring from the abdomen with their accompanying vessels and nerves. This fascial sleeve is the **internal spermatic fascia,** and the structures enclosed in it constitute the **spermatic cord** in the male or the **round ligament** in the female. The deep ring is best defined, therefore, as the junction of the internal spermatic fascia with the transversalis fascia proper.

The deep ring is located at the *midinguinal point,* midway between the anterior superior iliac spine and the pubic symphysis, slightly more than 1 cm above the inguinal ligament. Immediately anterior to the deep inguinal ring is the lower border of the **transversus abdominis,** and the spermatic cord enters the inguinal canal by passing below the inferior margin of this muscle. The fibers constituting this part of the transversus arise farther laterally from the line of fusion between the inguinal ligament and the iliopsoas fascia. As the fibers arch medially, they cross in front of the ring, so that only part of the ring is below the inferior border of the muscle (Fig. 26-4). When the transversus contracts, as it always does when abdominal pressure is raised, its lower border moves downward, providing a firm covering for the deep inguinal ring. On its deep surface, the ring is covered by extraperitoneal fat and parietal peritoneum.

#### The Superficial Inguinal Ring

The superficial inguinal ring is a more or less triangular defect in the aponeurosis of the external oblique, formed by the separation of its fibers (Fig. 26-1). The base of this triangular hiatus is the pubic crest, and its sides are the *crura* of the ring. The **lateral crus** lies really inferiorly and is the inguinal ligament attached to the pubic tubercle; the **medial crus** is superior and is formed by that part of the external oblique aponeurosis that attaches to the symphysis and the body of the pubis. Laterally, at the apex of the triangle, the two crura are bound together by the more or less obvious *intercrural fibers* of the aponeurosis. If these are well defined, they give the lateral margin of the superficial ring a rounded rather than an angular contour (Fig. 26-3).

The superficial inguinal ring, like the deep inguinal ring, is not a true opening. A thin layer of connective tissue continues from the margins of the superficial ring toward the scrotum or the labium majus and invests the spermatic cord or the round ligament as these exit from the inguinal canal (Fig. 26-1). This connective tissue sleeve is the **superficial spermatic fascia** and is derived from the deep fascia that covers both superficial and deep surfaces of the external oblique aponeurosis. Therefore, although the margins of the superficial ring are much better defined than those of the deep ring, they are not clearly visible until the superficial spermatic fascia is cut away from the edges of the crura.

## Walls

The floor of the inguinal canal is formed by the medial half of the inguinal ligament and the lacunar ligament. The roof is open; the canal communicates superiorly with the tissue spaces that separate the external and internal obliques and the transversus.

The entire length of the **anterior wall** of the inguinal canal is formed by the external oblique aponeurosis. As already mentioned, in front of the deep ring, the transversus abdominis reinforces the anterior wall of the canal and so does the internal oblique (Fig. 26-4). This inferior part of the **internal oblique** arises just in front of the transversus on the lateral half of the inguinal ligament, and as its fibers run medially, they lie in front of the deep ring. Here, the internal oblique splits to allow the spermatic cord or the round ligament to pass through it. Beyond this hiatus in the muscle, the internal oblique becomes rather thin and aponeurotic, and as it continues medially, now in the posterior wall of the inguinal canal, its inferior free border rests on the posterior edge of the inguinal ligament. The internal oblique terminates by fusing with the aponeurosis of the transversus; the two form the conjoint tendon and contribute to the inferior part of the anterior rectus sheath (p. 584).

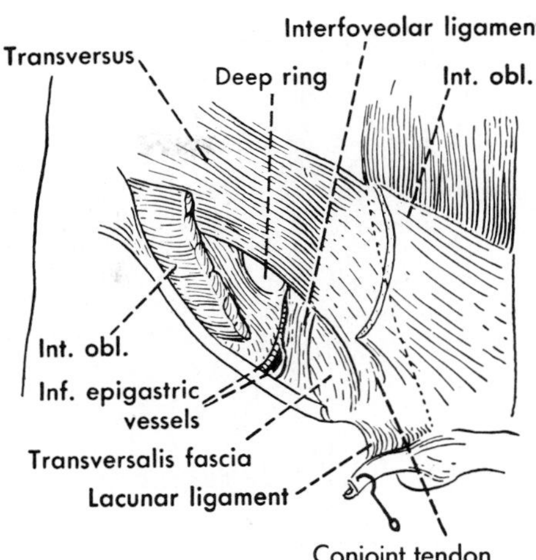


**FIG. 26-4.** Anterior view of the deep inguinal ring. The inguinal region deep to the internal oblique; the internal spermatic fascia, derived from the transversalis fascia, has been cut away from this fascia at the deep inguinal ring. The sloping fibers bordering the deep ring inferiorly constitute the iliopubic tract.

The **posterior wall** of the inguinal canal is rather complex and important because it is here that weaknesses predispose to the formation of inguinal hernia. The posterior wall is made up of the transversalis fascia, the conjoint tendon, and the inferior fibers of the internal oblique just described.

The **transversalis fascia** fills the hiatus between the inferior border of the transversus and the inguinal ligament. The fascia is thickened along the ligament and forms a more or less distinct band that runs from the iliopsoas fascia to the pubis. When this band is recognizable, it is called the *iliopubic tract* (Fig. 26-4). At its lateral end, the tract forms the inferior boundary of the deep inguinal ring. The medial boundary of the deep ring is also reinforced by a thickening of the transversalis fascia; this is the *interfoveolar ligament.* When present, this ligament extends from the lower border of the transversus to the inguinal ligament, more or less in front of the inferior epi-

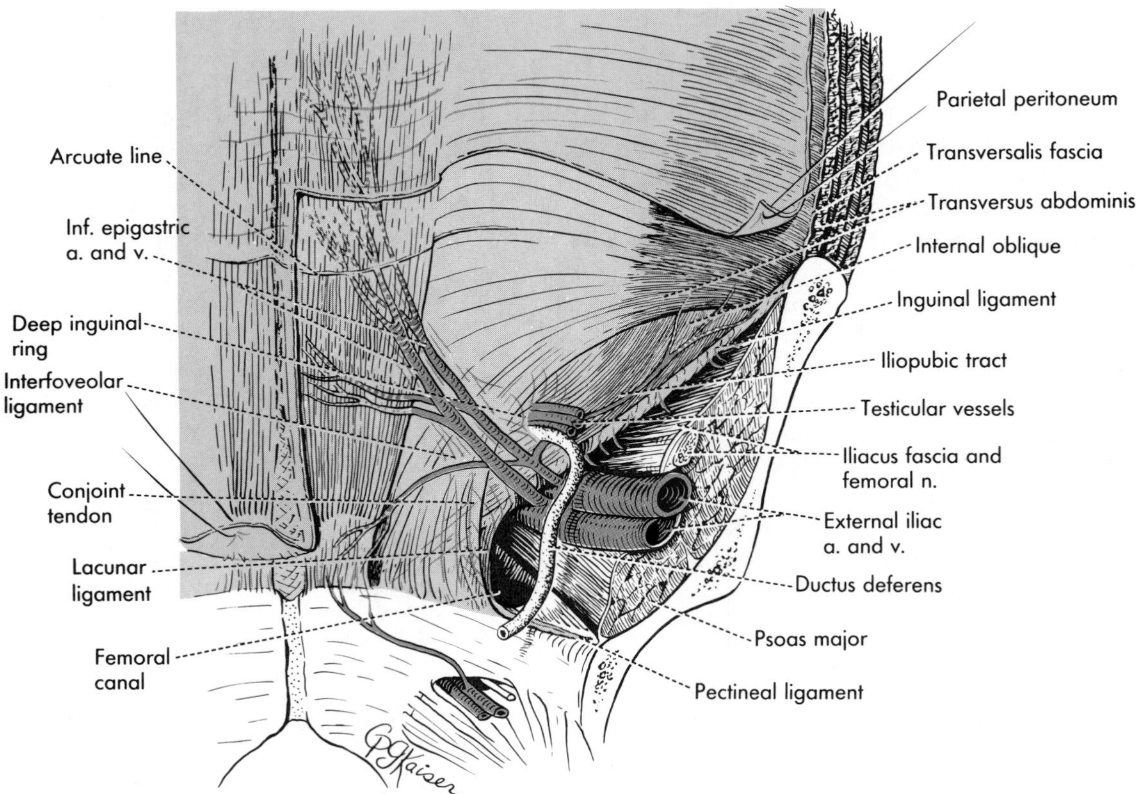


**FIG. 26-5.** The posterior aspect of the anterior abdominal wall. The peritoneum (pink) has been cut away to reveal the features of the transversalis fascia (blue) and the tendons and ligaments (black) that contribute to the posterior wall of the inguinal canal. The femoral canal and the structures that pass inferior to the inguinal ligament are also illustrated. The inguinal triangle is bordered by the rectus (unlabeled), the inferior epigastric artery, and the medial portion of the inguinal ligament.

gastric vessels (Fig. 26-4). Medially, the interfoveolar ligament blends into the conjoint tendon.

The **conjoint tendon** (*falx inguinalis*) is the combined aponeurosis of the transversus and internal oblique inserted into the pectineal ligament along the superior ramus of the pubis (Fig. 26-5). Most of the fused aponeurosis passes in front of the rectus as part of its sheath and inserts into the pubic crest; only the part lateral to the rectus is designated as the conjoint tendon. The conjoint tendon forms the posterior wall of the most medial portion of the inguinal canal, more or less directly behind the superficial inguinal ring. It varies how far laterally the conjoint tendon extends, as does how strong it is and how distinct it is from the interfoveolar ligament with which its lateral border merges.

The deep aspect of the posterior wall of the inguinal canal is covered by parietal peritoneum of the **lateral** and **medial inguinal fossae** described in Chapter 23 (see Fig. 23-26). The deep inguinal ring is located in the lateral inguinal fossa just lateral to the inferior epigastric vessels. The medial inguinal fossa overlies a triangular area bordered by the rectus, the inguinal ligament, and the inferior epigastric vessels: it is known as the *inguinal triangle* (Fig. 26-5).

A summary of the posterior wall of the inguinal canal according to these landmarks is relevant to the applied anatomy of inguinal hernia. **In the inguinal triangle,** the medial portion of the posterior wall consists of the *conjoint tendon* (with transversalis fascia on its deep surface) and, in front of it, the inconsequential *reflected part of the inguinal ligament* (Fig. 26-3). Lateral to the crescentic edge of the conjoint tendon, *transversalis fascia,* incorporating the interfoveolar ligament and the iliopubic tract, completes the posterior wall as far as the inferior epigastric vessels. Directly in front of this fascia, the *hiatus* in the internal oblique transmits the spermatic cord and a few attenuated fibers of the *internal oblique* and its aponeurosis, paralleling the inguinal ligament. **Lateral to the inferior epigastric vessels** is the deep inguinal ring, the lower margin of which is formed, as noted earlier, by the iliopubic tract.

### CONTENTS

The inguinal canal contains the spermatic cord or the round ligament and the ilioinguinal nerve. The structures that constitute the spermatic cord and the round ligament pass through both deep and superficial inguinal rings; the ilioinguinal nerve enters the canal laterally and leaves it through the superficial ring.

The **ilioinguinal nerve,** a branch of the lumbar plexus (see Fig. 25-15), enters the abdominal wall by piercing the deep surface of the transversus abdominis just above the anterior superior iliac spine and soon after that passes through the internal oblique. It supplies both muscles and descends into the inguinal canal between the internal and external obliques. In the canal, the nerve runs along the inferior aspect of the spermatic cord to the superficial inguinal ring, where it pierces the external spermatic fascia and distributes its branches to the skin of the upper part of the thigh, scrotum, and penis, or the labia majora and mons pubis.

## The Spermatic Cord

The spermatic cord is the pedicle of the testis, and it also connects the scrotum to the abdomen. It is composed of the structures that pass through the deep inguinal ring and fascial coverings contributed to it by the layers of the abdominal wall.

In the male, the principal structures that pass through the deep inguinal ring include the ductus deferens, the testicular artery and veins, lymphatics, and autonomic nerve plexuses around the testicular artery and the ductus deferens. Less important structures that enter the deep ring are the genital branch of the genitofemoral nerve and small arteries that supply the ductus itself and the cremaster muscle.

**The Ductus Deferens.** The ductus deferens, also known as the *vas deferens,* conveys spermatozoa and secretions produced by the testis to the ejaculatory ducts. The deferent duct commences behind the lower pole of the testis as the continuation of the duct of the epididymis (Fig. 26-6) and terminates in the pelvis

**FIG. 26-6.** Blood supply of the testis.

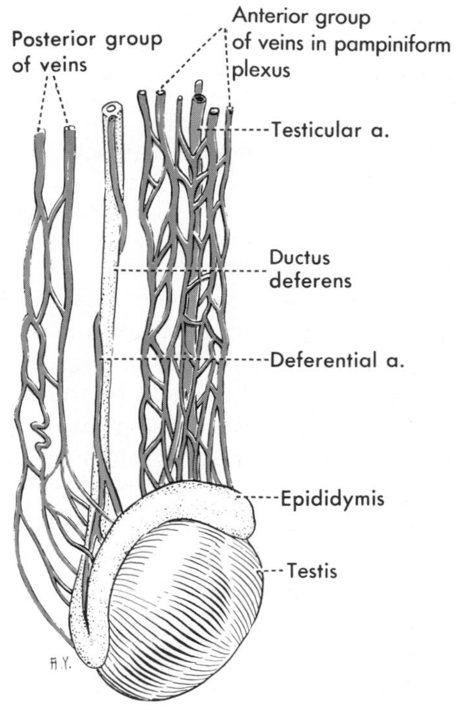

just above the prostate by forming the ejaculatory duct (see Fig. 27-29). The deferent duct ascends in the scrotum behind the testis and then in the spermatic cord, entering the inguinal canal through the superficial ring and leaving it through the deep ring. At the deep ring, the duct bends sharply medially and, embedded in pelvic fascia, pursues its intrapelvic course toward the prostate (p. 773).

The deferent duct can be identified as a 2-mm to 3-mm thick palpable cord when the neck of the scrotum is rolled between finger and thumb. The lumen of the duct is quite narrow, and its thickness is due to the smooth muscle in its walls. The duct is usually filled with spermatozoa; contraction of its walls, induced by sympathetic activity, discharges the spermatozoa into the ejaculate.

The ductus (or vas) deferens may be exposed under local anesthetic by an incision in the skin in the neck of the scrotum before the vas enters the superficial inguinal ring. Bilateral ligation and division of the vas is called **vasectomy** and renders an individual permanently sterile unless the duct is surgically reanastomosed.

The vasectomized individual will retain full sexual function and will continue to produce an ejaculate, but spermatozoa will be absent from it. Although vasectomy may seem an ideal method of contraception in the male, it has not been resolved to what degree absorption of the sperm (which continue to be produced at an apparently normal rate) contributes to certain manifestations of autoimmunity sometimes observed in vasectomized individuals.

**The Testicular Artery.** The testicular artery, a branch of the abdominal aorta, descends on the posterior abdominal wall to the deep inguinal ring, where it joins the ductus deferens (p. 687). Enclosed in the spermatic cord, the artery enters the scrotum where it lies anterior to the ductus and surrounded by the *pampiniform plexus,* formed by the veins that drain the testis (Fig. 26-6). These veins emerge from the back of the testis and epididymis and give rise to some 10 to 12 veins, which, as they ascend, anastomose with each other, mainly around the testicular artery. This is the **pampiniform** (tendrillike) **plexus,** drained superiorly by three to four veins that pass through the inguinal canal and terminate on each side in two **testicular veins** at the deep inguinal ring. These veins ascend with the testicular artery to end in the inferior vena cava on the right and in the left renal vein on the left (Chap. 25).

There is experimental evidence, obtained mainly in the dog and ram, that the cooler venous blood in the pampiniform plexus lowers the temperature of the blood in the testicular artery. This contributes to providing the temperature optimal for spermatogenesis, which has to be below that of the abdominal cavity.

The testicular vein and the pampiniform plexus are devoid of valves except for an occasional one at their termination in the inferior vena cava or the left renal vein. Engorgement and distention of these veins leads to varicosities in the pampiniform plexus that are palpable through the scrotum. The condition, known as **varicocele,** is much more common on the left than on the right, presumably related, in many cases, to the fact that the left renal artery crosses, and partially compresses, the left testicular vein just before its termination. In the supine position, both the varicosities and the discomfort caused by the distention of the veins disappear because the venous pressure is relieved from the pampiniform plexus.

The two small arteries that enter the deep ring are the **artery of the ductus deferens,** a branch of the internal iliac, and the **cremasteric artery,** a branch of the inferior epigastric. The latter supplies only the tunics of the spermatic cord, but the artery of the ductus supplies the ductus and the epididymis and anastomoses with the testicular artery. These anastomoses, however, are not large enough to provide an adequate blood supply to the testis when the testicular artery becomes occluded. Such occlusion most often results from **torsion of the testis.** Twisting of the testis, with the spermatic cord as its pedicle, occludes the veins and the testicular artery, and unless this is relieved, the testis will die. Indeed, this has been the outcome in the majority of cases.

**The Spermatic Fascias.** The fascial coverings contributed to the spermatic cord by lay-

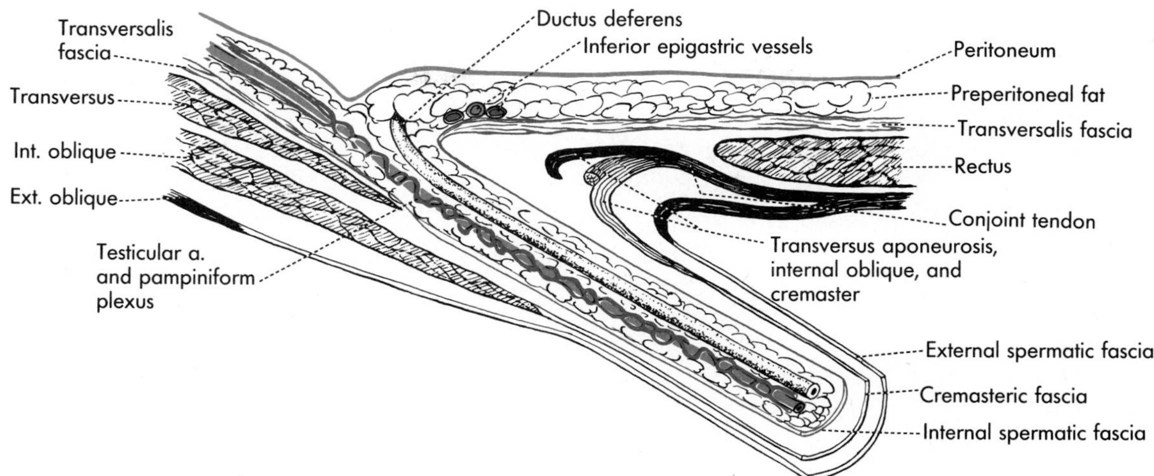


**FIG. 26-7.** The continuity between the layers of the abdominal wall and the tunics of the spermatic cord in a hypothetical longitudinal section through the upper part of the spermatic cord. As in Figure 26-5, the peritoneum is red and the transversalis fascia and its continuation, the internal spermatic fascia, are blue.

ers of the abdominal wall are the internal and external spermatic and the cremasteric fascias (Fig. 26-7). As explained already with the description of the deep and superficial inguinal rings, the **internal spermatic fascia** is derived from the transversalis fascia and the **external spermatic fascia** is derived from the fascias of the external oblique. The **cremasteric fascia** is derived chiefly from the internal oblique (Fig. 26-8) and, in the inguinal canal, is wrapped around the internal spermatic fascia, forming the outer covering of the spermatic cord. However, inferior to the superficial ring, the cremasteric fascia is inside the external spermatic fascia. All these fascias become continuous with layers of the scrotal sac (Fig. 26-9).

The cremasteric fascia differs from the other fascial tunics of the spermatic cord in that it contains loops of muscle fasciculi constituting the **cremaster muscle.** As the lower fibers of the internal oblique part to let the spermatic cord pass through, not only the deep fascia but also some of the muscle fibers of the internal oblique are carried along the cord in the form of loops that reach down into the scrotum (Fig. 26-8). Although most of the loops of the cremaster are traceable to the in-

**FIG. 26-8.** Course of the spermatic cord through the internal oblique muscle; in **B**, the cremaster muscle and the cord have been removed to show the defect in the internal oblique.

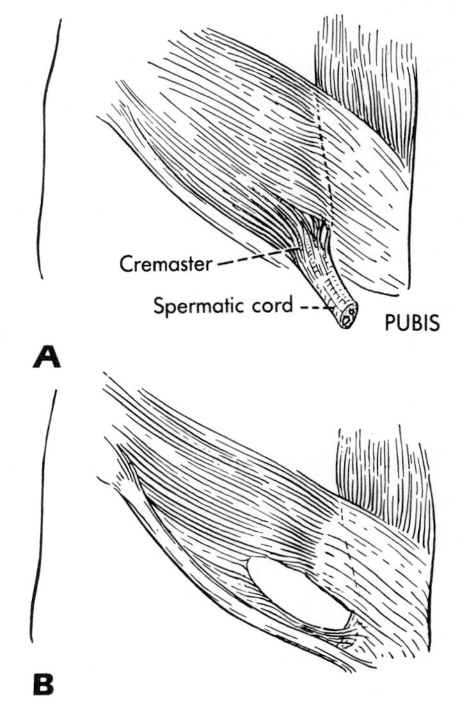

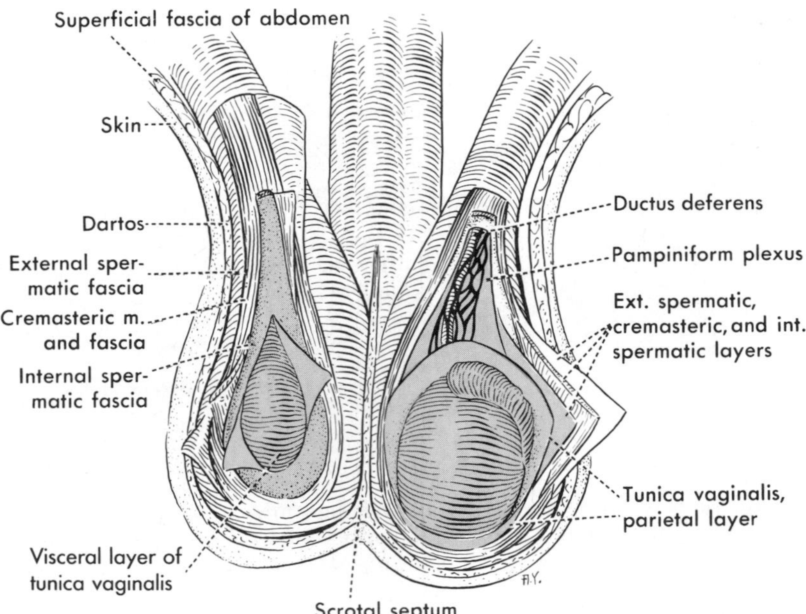


**FIG. 26-9.** The scrotum and the coverings of the spermatic cord and testis from the front. On the reader's right, the tunica vaginalis has been opened to show the testis and the epididymis, which are here covered by the visceral layer of the tunic. Parietal and visceral layers of the tunica vaginalis are pink, and the internal spermatic fascia is blue.

ternal oblique, some of the loops may be derived from the lower border of the transversus and others may have an independent origin on the inguinal ligament.

Contraction of the cremaster elevates the testis. Both testes are elevated just prior to ejaculation. The cremaster is innervated by the genital branch of the *genitofemoral nerve* (L1), which enters the inguinal canal through the deep inguinal ring. Sudden stroking of the skin on the medial side of the thigh provokes unilateral reflex contraction of the cremaster, manifest by momentary elevation of the testis, particularly well seen when the scrotal sac is relaxed. This phenomenon is known as the **cremasteric reflex;** it tests L1 segment of the spinal cord.

### The Round Ligament

The *ligamentum teres uteri* is a fibromuscular band that passes retroperitoneally from the uterus, located in the pelvis, to the deep inguinal ring. After traversing the inguinal canal, the ligament breaks up into fibrous strands that merge with the connective tissue of the labium majus.

In the inguinal canal, the round ligament acquires the same fascial tunics as the spermatic cord. However, apart from the ligament itself, these fascias contain no structure of significance and no genital duct or major vessel; they usually fuse with the round ligament almost as soon as they are formed. Therefore, the portion of the ligament surrounded by the fascial tunics is known by the same name as its intrapelvic portion devoid of fascial coverings.

## THE SCROTUM AND ITS CONTENTS

### THE SCROTUM

The scrotum is a sac formed by the continuation of the skin and superficial facia of the abdominal wall into the perineum. It hangs below the pubis and the root of the penis and, together with the penis, constitutes the *external genitalia* of the male. Its interior is divided into right and left halves, each containing a testis with its associated ducts, a serous sac called the *tunica vaginalis* applied to the surface of the testis, and the inferior portion of the spermatic cord, including its fascial tunics.

The appearance of the dark, sparsely hairy **scrotal skin** is influenced by the smooth muscle contained in the dermis and underlying fascia. The skin may be smooth and flaccid or wrinkled and drawn closely around the testis. A median *scrotal raphe* indicates the fusion of the bilateral labioscrotal swellings that formed the scrotum. The raphe extends onto the inferior surface of the penis and posteriorly in the perineum as far as the anus.

Beneath the skin, and closely bound to it, is the **dartos** (*tunica dartos*), the continuation of the membranous layer of the superficial fascia from the abdominal wall into the scrotum (see Fig. 23-2). This fascia in the scrotum is devoid of fat and contains a significant amount of smooth muscle, the *dartos muscle.* Deep to the scrotal raphe, the dartos extends inward as the *scrotal septum,* forming a partition between the right and left halves of the scrotum (Fig. 26-9). In addition to its connection with the abdominal superficial fascia, the dartos is continuous with the superficial penile fascia and, posteriorly, with the superficial fascia of the perineum.

Within each scrotal cavity, the **tunics of the spermatic cord** end as a trilaminar sac enclosing the tunica vaginalis and the testis (Fig. 26-9). The *external spermatic, cremasteric,* and *internal spermatic fascias* are all present in the scrotum, although their fusion with each other may make it difficult to dissect them separately.

The **tunica vaginalis testis** is an extension of the peritoneal sac into the scrotum. Beyond the neonatal period, however, the tunica vaginalis normally does not communicate with the peritoneal cavity. The *processus vaginalis,* which connected the two in the fetus, becomes obliterated along the inguinal canal and the upper part of the scrotum and either disappears altogether or persists as a thin fibrous band. The tunica vaginalis has a *parietal layer,* fused by its nonserous external surface with the internal spermatic fascia, and a *visceral layer,* bound to the anterolateral surface of the testis and epididymis (Fig. 26-9). The cavity of the tunica vaginalis contains a small amount of serous fluid that moistens the apposed, inner mesothelial surfaces of the visceral and parietal laminae of the tunica.

The cavity of the tunica vaginalis may become distended with serous fluid, forming what is known as a **hydrocele.** Unless the anteroposterior orientation of the testis is reversed (anteversion of the testis), such a swelling, quite translucent when a beam of light is shone through it, is always located anterior to the testis.

Hydrocele exists in many varieties. *Congenital hydrocele* is present at birth and is associated with the persistence of a communication between the peritoneal cavity and the tunica vaginalis through a patent *processus vaginalis.* Even when there is no communication, a hydrocele may suddenly appear in young children or at any age caused by inflammation of the tunica vaginalis. When the hydrocele is chronic, the swelling may attain quite an enormous size and need to be drained periodically.

The **blood supply, lymph drainage,** and **innervation of the scrotum** are quite distinct from that of the testis and the spermatic cord. The main scrotal vessels and nerves belong to the perineum and are discussed in Chapter 28.

## THE TESTIS AND ITS DUCTS

The testis is the male gonad: its function is the production of spermatozoa and the secretion of testosterone (or dihydrotestosterone), a hormone responsible for the development and maintenance of the secondary sex characteristics of maleness. The spermatozoa leave the testis through the *ductuli efferentes,* which unite to form the *duct of the epididymis.* This highly coiled duct forms a compact body, the *epididymis,* and gives rise to the *ductus deferens.*

### The Testis

Each testis is a firm, ellipsoid organ, measuring approximately 4 × 3 × 2.5 cm. The longest diameter is between the rounded superior and inferior poles, and this axis is tilted slightly anteriorly and laterally. The testis, convex and smooth on all surfaces, is slightly flattened from side to side, presenting rounded anterior and posterior borders and more extensive lateral and medial surfaces. The epididymis is applied to the posterior border and protrudes from it laterally; it is in contact with the testis from its superior to its inferior pole (Fig. 26-6).

The testis and the epididymis invaginate the **tunica vaginalis** from behind. Therefore, the visceral lamina of the tunica covers the testis and epididymis everywhere except along their posterior border and along the posterior part of the area of contact between the two. Anteriorly, a deep groove, the *sinus epididymidis,* intervenes between the testis and the epididymis and is lined by the tunica vaginalis. The tunica vaginalis also covers the anterior surface of the spermatic cord for a variable distance above the testis.

The tunica vaginalis is identical in appearance and structure to the peritoneum. It has been conjectured that, like the ovary, the testis is devoid of visceral peritoneum because the celomic epithelium covering the surface of the embryonic gonads (*germinal epithelium*) becomes incorporated into both ovary and testis to form their cortex. Whether it is the remnant of the germinal epithelium or the celomic lining, mesothelium covers the surfaces of the testis that face into the cavity of the tunica vaginalis, and this mesothelium is continuous with that of the parietal lamina of the tunica vaginalis.

When the tunica vaginalis is particularly extensive, the "bare area" along its posterior border will be narrow, and the testis will appear as if suspended on a short mesentery. Such a mesentery may be designated the *mesorchium* (orchid is one of the Greek names for the testis). A well-defined mesorchium predisposes to torsion of the testis (p. 720).

**FIG. 26-10.** Diagram of the duct system of the testis.

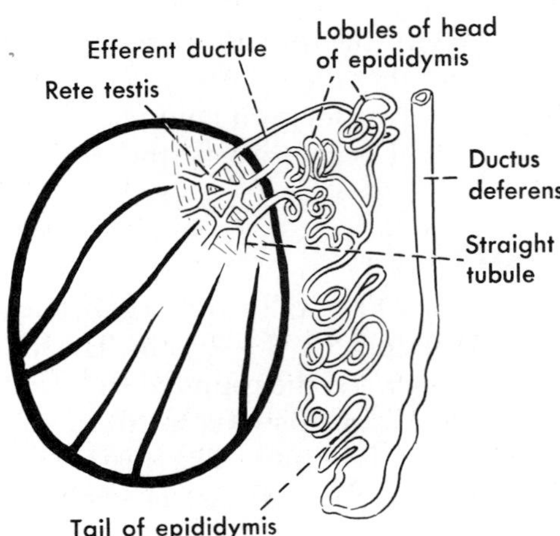


**Mediastinum Testis.** Vessels, nerves, and lymphatics enter and leave the testis and epididymis along their posterior border, which is devoid of the tunica vaginalis (Fig. 26-6). The **seminiferous tubules** in the interior of the testis also converge toward the posterior border and discharge their contents into a duct system, the **rete testis,** which is, in turn, connected to the epididymis by the **ductuli efferentes,** located near the superior pole of the testis (Fig. 26-10). The posterior segment of the testis, containing the rete, is known as the *mediastinum testis.*

**Internal Structure.** Subjacent to the tunica vaginalis, the testis is invested by a dense layer of fibrous tissue, the **tunica albuginea,** which is thinner over the mediastinum. Small septa (*septula testis*) extend from the deep surface of the tunica albuginea into the testis and subdivide it into 200 to 300 roughly pyramid-shaped compartments or **lobules.** The apex of each lobule points toward the mediastinum, and within the lobule lie one to three **seminiferous tubules.** These tubules are 0.1 mm to 0.3 mm in diameter and may measure nearly 1 m in length. Most of the tubule packed into the lobule is highly convoluted (*tubuli seminiferi contorti*), but one or both ends of all tubules pointing toward the mediastinum become straight (*tubuli seminiferi recti*). In the mediastinum these straight seminiferous tubules terminate in a labyrinth of intercommunicating channels that form the **rete testes.** As the rete passes through the tunica albuginea, it links up with 10 to 20 efferent ductules.

## The Epididymis

*Didymos* is the Greek equivalent of the Latin word *testis.* The epididymis is a comma-shaped, compact body formed by tortuous tubules bound together by areolar tissue. The epididymis consists of a head (*caput*) overlying the superior pole of the testis, a tapering body (*corpus*), and a tail (*cauda*). The body lies along the posterolateral aspect of the testis, and the tail reaches below the inferior pole of the testis.

The head of the epididymis is made up of 10 to 20 lobules, each consisting of an efferent ductule that becomes highly convoluted after leaving the testis. These ductules unite with each other in the epididymis to form the **duct of the epididymis** (Fig. 26-10). The body and tail are made up of convolutions of this duct. Beyond the tail, the duct is called the **ductus deferens** and becomes more and more straight as it ascends behind the testis into the spermatic cord.

Although the epididymis itself measures barely more than 4 cm, the length of its duct exceeds 6 m. During their passage through the epididymis, the spermatozoa undergo maturation and acquire motility and the ability to swim in one direction. It is not yet known to what extent this maturational process is influenced by the secretions of the epididymal epithelium.

**Vestigial Structures.** There are vestigial structures associated with the epididymis and the upper pole of the testis. These are the *appendices of the testis and the epididymis*, the *aberrant ductules*, and the *paradidymis* (Fig. 26-11). All these vestiges are remnants of embryonic structures associated with the mesonephros and the developing gonads, discussed in the next section. Their clinical significance is that either appendix may undergo torsion, giving rise to intense pain. The condition must be distinguished clinically from torsion of the testis itself (p. 720). The aberrant ductules or the paradidymis may give rise to cysts within the scrotum. Such cysts must be distinguished from hydrocele and from cysts of the processus vaginalis.

## Blood Supply, Lymph Drainage, and Innervation

**Blood Supply.** The testicular artery and vein and the pampiniform plexus were discussed with the spermatic cord earlier in this chapter. The distribution of the intratesticular branches of these vessels follows a connective tissue lamina (*tunica vasculosa*) that is

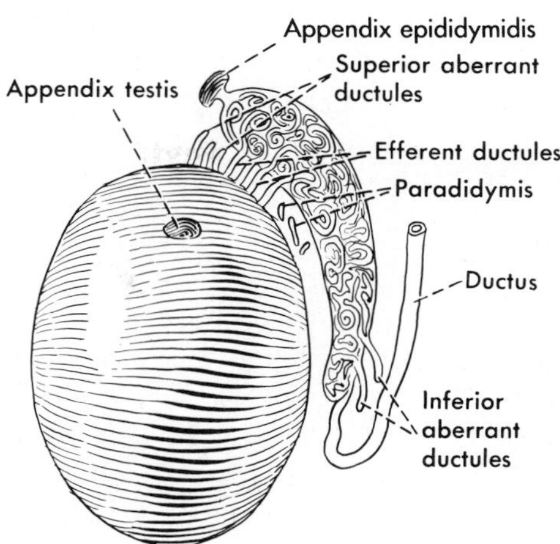


**FIG. 26-11.** Embryonic remains sometimes associated with the testis and its ducts. The appendix testis is a remnant of the paramesonephric duct, and the others are remains of the mesonephric system.

subjacent to the tunica albuginea and is continuous with the mediastinum testis. Extensions of this vascular lamina along the septula testis supply blood to the lobules. However, capillaries do not penetrate the seminiferous tubules. There is evidence that a barrier exists between the blood and the lumen of the duct systems in which spermatozoa are produced and transported.

**Lymphatics.** The lymphatics of the testis and epididymis drain upward as components of the spermatic cord. After they have passed through the inguinal canal, they run upward in the abdomen along the testicular vessels and join the upper lumbar nodes. In consequence of this drainage, metastatic carcinoma from the testis first involves the lymph nodes close to the renal level rather than the inguinal nodes, although they are nearest topographically.

**Nerve Supply.** Autonomic nerves accompany both the testicular vessels and the ductus deferens. The **testicular plexus** is derived from the aortic and renal plexuses, and the sympathetic efferent and afferent fibers connect chiefly to T10–T11 segments of the spinal

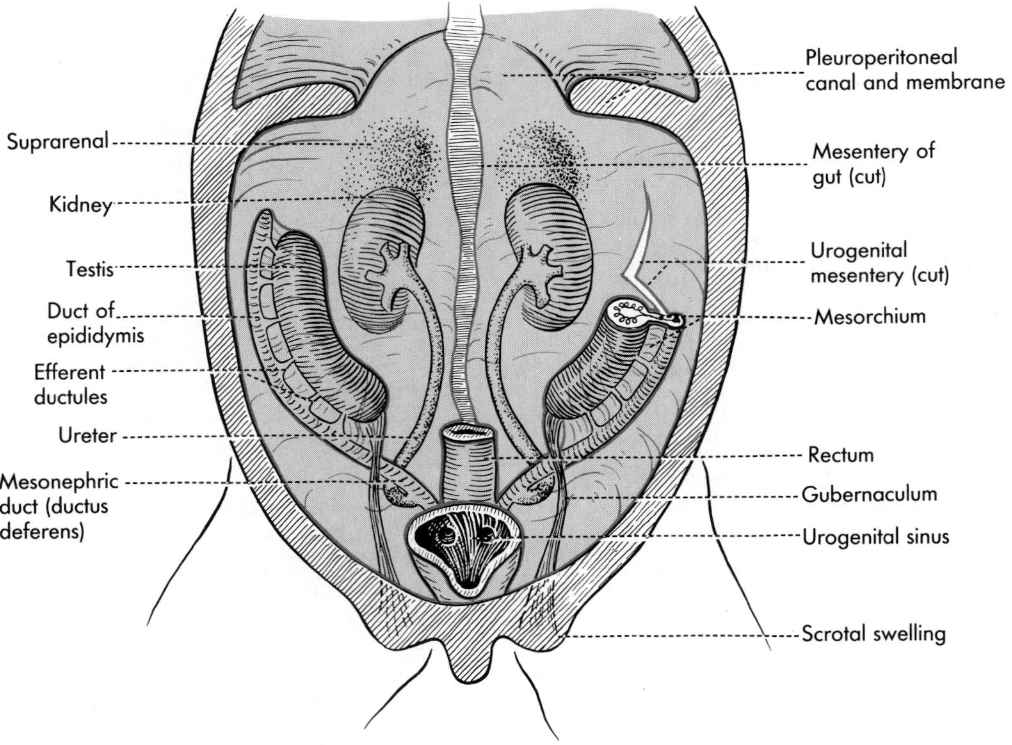


**FIG. 26-12.** Development of the testis and its duct system shown diagrammatically. This drawing represents a stage later than that in Figure 25-29. The mesothelial lining of the celom is pink. The definitive kidneys have ascended the posterior abdominal wall; the testis, with its ligaments and ducts, is still highly placed.

cord through the thoracic splanchnic nerves. The efferent fibers are evidently vasomotor; the visceral afferents are concerned with pain. The testis is quite sensitive to pressure, and the pain of it is localized in the testis as long as the cutaneous nerves of the scrotum are intact. Testicular pain may also be referred to the lower thoracic segments.

The **deferential plexus** becomes associated with the ductus during its intrapelvic course and is derived from branches of the superior and inferior hypogastric plexuses. The exact origins and pathways of the nerves are not known. Sympathetic efferents in this plexus are responsible for contraction of the muscle wall of the epididymis and ductus deferens, and there is some evidence that parasympathetic fibers derived from the pelvic splanchnics relax the smooth muscle of these ducts. Pain afferents seem to end in the same cord segments as those in the testicular plexus. Epididymal pain cannot be distinguished from testicular pain.

## DEVELOPMENT AND DESCENT OF THE TESTIS

### The Testis and Its Ducts

The testis develops in the urogenital ridges lying along the medial side of the mesonephros (see Fig. 25-29). It is connected to the mesonephros by a peritoneal fold, the *mesorchium.*

The primitive testis consists of a *cortex,* formed chiefly by the proliferation of celomic mesothelium (*germinal epithelium*), and a *medulla,* adjacent to the mesonephros and derived from the mesenchyme of intermediate mesoderm. Both the cortex and medulla form cords of cells: from the former develop the seminiferous tubules, from the latter, the rete testis. The seminiferous tubules incorporate the primitive germ cells that migrate to the testis from the yolk sac, and these tubules inosculate with the rete.

Although the mesonephros atrophies and disappears, its collecting tubules and its chief duct persist. Some of the collecting tubules establish connections with the rete testis and become the ductuli efferentes. The mesonephric duct is thus put in communication with the seminiferous tubules (Fig. 26-12). The definitive pathway for spermatozoa is established by the transformation of the urinary duct of the mesonephros into ducts that transport spermatozoa. From the mesonephric duct differentiates the duct of the epididymis, the ductus deferens, and the ejaculatory duct. Its blind cranial end persists as the appendix of the epididymis. Some mesonephric collecting tubules do not connect with the rete and become the *aberrant ductules* and the *paradidymis,* associated with the epididymis (Fig. 26-11); those that do connect with the rete form not only the ductuli efferentes but the lobules of the caput epididymidis.

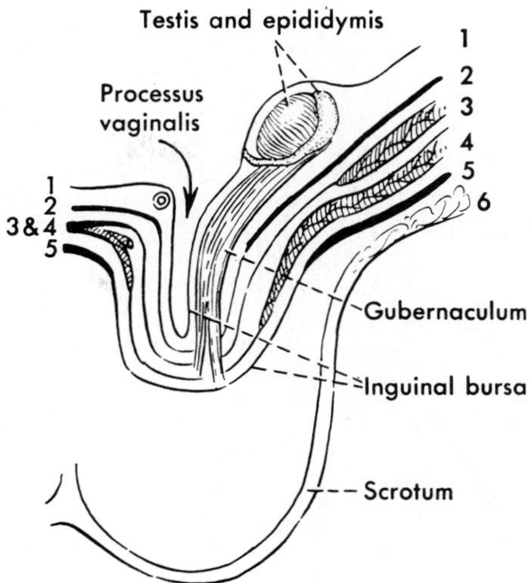


**FIG. 26-13.** Diagram of the testis, inguinal bursa, and scrotum before descent of the testis. The numbered layers are **1,** peritoneum; **2,** transversalis fascia; **3,** transversus muscle and tendon; **4,** internal oblique muscle and tendon; **5,** aponeurosis of the external oblique; **6,** skin and subcutaneous tissue (superficial fascia). Note that the wall of the inguinal bursa consists of all the layers through the derivative (external spermatic fascia) of the external oblique; the considerable space existing at this stage between the bursa and the scrotum proper is occupied by loose connective tissue. The gubernaculum, along the course of which the testis descends, is shown here somewhat smaller than it really is; at the time of descent, it is as large in diameter as the combined testis and epididymis.

### Descent of the Testis

The caudal pole of the urogenital ridges is connected to the skin of the labioscrotal swellings by the *genitoinguinal ligament* (see Fig. 25-26). Once the mesonephros disappears, this ligament spans the distance between the lower pole of the testis and the future scrotal skin; it is renamed in the male the **gubernaculum testis.**

Concomitant with the differentiation of the abdominal wall musculature, the fascial coverings of the spermatic cord develop around the gubernaculum in the region of the inguinal canal. At this stage, the future inguinal canal is essentially vertical.

Like the testis, the gubernaculum is retroperitoneal on the posterior abdominal wall and in the lateral inguinal fossa. From the lateral inguinal fossa a tubular extension of the peritoneal sac grows into the mesenchyme of the gubernaculum and forms the **processus vaginalis.** The processus vaginalis extends through the inguinal canal into the scrotum, forming, with the layers of the spermatic cord, what is called the *inguinal bursa* (Fig. 26-13). Some unknown factors initiate the descent of the testis after the inguinal bursa has been formed.

Presumably guided by the gubernaculum, the testis slides down from the posterior abdominal wall, through the inguinal canal, to pass retroperitoneally outside and behind the processus vaginalis but within the sleeve of the internal spermatic fascia. The gubernaculum shortens as the testis progresses and pulls its ducts with it into the inguinal canal. After the testis has reached the scrotum, the processus vaginalis becomes obliterated and persists only in the scrotum as the tunica vaginalis. The gubernaculum cannot be identified as a discrete structure in the fully developed scrotum. Descent of the testis occurs during the eighth month of fetal life but may be delayed for a year or so after birth.

In the female, the **round ligament of the uterus** is the equivalent of the caudal portion of the gubernaculum. Although the ovary does not descend into the inguinal canal, the inguinal bursa is formed the same way as in the male.

**Abnormalities of Descent.** Failure of the testis to descend properly results in **cryptorchidism** or **ectopia testis.** In the former, the testis remains in the abdominal cavity or lodges in the inguinal canal rather than emerging through the superficial inguinal ring; in the latter, the testis may emerge through the ring but will lie between the superficial abdominal fascia and the abdominal muscles above the ring or will assume various anomalous positions (Fig. 26-14). Testes that do not descend into the scrotum usually are sterile. An undescended testis is kept at too high a temperature to

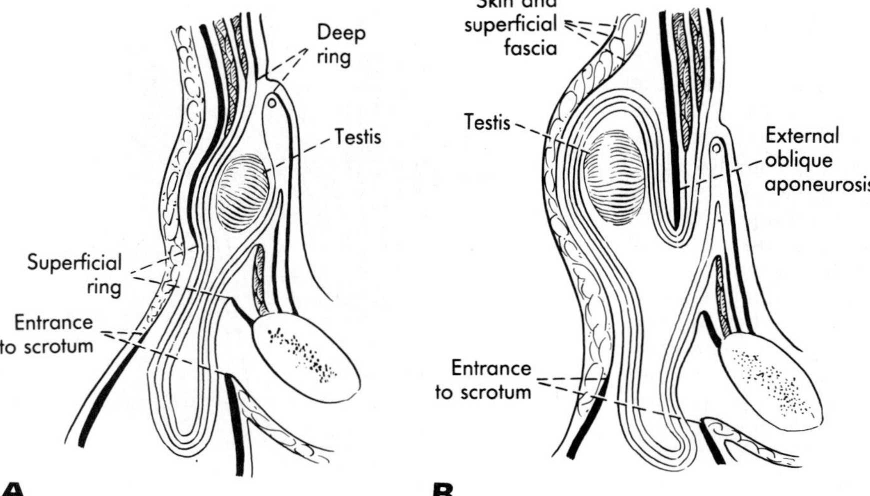


**FIG. 26-14.** Two types of maldescent of the testis. In **A,** it is lodged in the inguinal canal, whereas in **B,** it has emerged through the superficial ring but lies deep to skin and superficial fascia of the inguinal region.

permit spermatogenesis. It begins to show increasing evidence of damage after the age of 5 years; therefore, if neither testis is in the scrotum by that time and the descent cannot be induced by appropriate administration of hormones, at least one testis should be placed in the scrotum by surgical means.

## INGUINAL HERNIA

### Hernias In General

The protrusion of any structure through an opening or passage that it normally does not traverse is called a hernia or "rupture." The most well known types of hernias are protrusions of abdominal contents through the anterolateral abdominal wall. There are many hernias, however, that are entirely internal and may not involve abdominal contents. For instance, the heart may herniate through a defect in the pericardium or a muscle through a defect in its deep fascia. Previous chapters have already mentioned several types of internal abdominal hernia: hiatus hernia, a protrusion of the stomach through the diaphragm, and paraduodenal and other internal intestinal hernias through holes in the mesentery or behind peritoneal folds.

Hernias through the abdominal wall may occur through the inguinal and femoral canals, through or along the umbilicus, through tears in the linea alba or rectus sheath, or through the scar of an incision that has been made in the anterolateral abdominal wall, as well as in some other rather unusual places.

In the planning of abdominal incisions, consideration is given to factors that will prevent the formation of hernias through the weakness created by the scar tissue. For instance, the appendectomy incision takes advantage of the fact that the internal and external oblique muscles run at right angles to one another, and the incision separates rather than transects the fibers of each muscle. Paramedian incisions often are made through the anterior sheath of the rectus, and after retraction of the muscle laterally, the posterior rectus sheath is incised, so that when the muscle is replaced, it guards the scars created.

### Direct and Indirect Inguinal Hernias

Abdominal contents may enter the inguinal canal directly through its posterior wall or indirectly through the deep inguinal ring. Consequently, two types of inguinal hernias are distinguished: *direct* and *indirect.* Both types emerge through the superficial inguinal ring and present themselves as swellings in the groin, and both types may eventually descend into the scrotum. The hernias that enter the inguinal canal directly through its posterior wall bulge more or less directly into the superficial inguinal ring, especially if the ring is rather large; this is another reason for designating them direct hernias.

Both direct and indirect inguinal hernias are usually contained in a peritoneal sac that surrounds a herniating loop of small or large intestine or a portion of the greater omentum. It is possible, however, for a nonperitoneal structure, most commonly the bladder, to form an inguinal hernia without any peritoneal covering around it.

By definition, an inguinal hernia whose neck is located lateral to the inferior epigastric vessels is called *indirect;* one that commences medial to the inferior epigastric vessels is designated as *direct.*

Indirect inguinal hernias may be *congenital* or *acquired;* essentially all direct inguinal hernias are acquired.

A **congenital indirect hernia** is the result of a *patent processus vaginalis* that connects the peritoneal cavity with the tunica vaginalis (Fig. 26-15). The hernia may not appear until adulthood. In **acquired indirect hernias,** the peritoneal sac of the hernia does not communicate with the tunica vaginalis. It is suspected, however, that the majority of acquired indirect inguinal hernias occur because a variable length of the proximal part of the processus vaginalis remains patent. Such hernias are within the spermatic cord, and in the inguinal canal, their peritoneal sac is surrounded by the internal spermatic and cremasteric fascias. Beyond the superficial ring, they are also covered by the external spermatic fascia.

The effective shutter mechanisms of the transversus and internal oblique over the deep inguinal ring prevent such hernias from occurring even when the proximal portion of the processus is patent (p. 716; Fig. 26-4). The precipitating factors of hernias remain speculative.

After an indirect inguinal hernia has been reduced, its reappearance may be prevented by firm pressure applied, even by one finger, over the deep inguinal ring. If the hernia reappears when the patient increases abdominal pressure, while the deep inguinal ring is controlled by the finger placed over it, the hernia is likely to be a direct one.

Because of its narrow neck and the obliquity of its passage through the inguinal canal, an indirect hernia is liable to obstruct and strangulate. Strangulation implies that the blood supply to the contents of the hernia has been cut off and gangrene is likely to set in. Either of these complications is a surgical emergency. Therefore, the surgical repair of indirect hernias is indicated once the diagnosis is established.

The neck of a **direct inguinal hernia** is wide. The hernial sac is formed by parietal peritoneum of the medial inguinal or supravesical fossae (see Fig. 23-26). Pushing transversalis fascia in front of it, the hernial sac may slip under the crescentic edge of the conjoint tendon and under the inferior fibers of the internal oblique, which lie parallel with the inguinal ligament, or it may stretch and push the conjoint tendon and the internal oblique in front of it. Until such a hernia emerges through the superficial inguinal ring, it will not be inside the spermatic cord (Fig. 26-16); beyond the superficial ring, it will lie between the superficial spermatic and cremasteric fascias. A direct inguinal hernia may also enter the inguinal canal through the hiatus of the internal oblique, through which the spermatic cord passes, especially if this hiatus is large (Fig. 26-8). In such a case, the hernia will be within the cremasteric fascia. Because of their wide neck, direct inguinal hernias are more readily reduced than indirect inguinal hernias, and they are less likely to obstruct or strangulate.

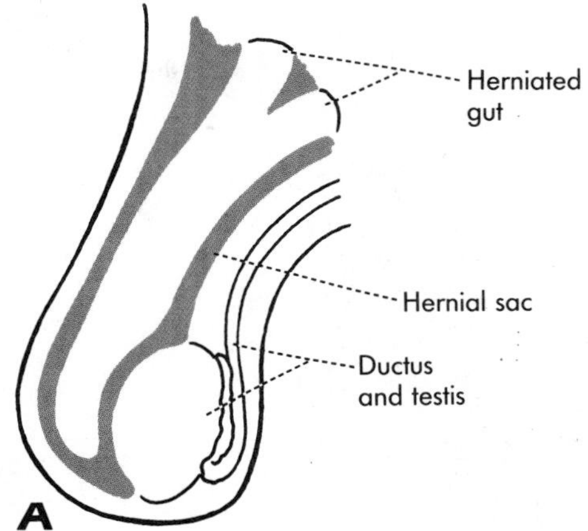


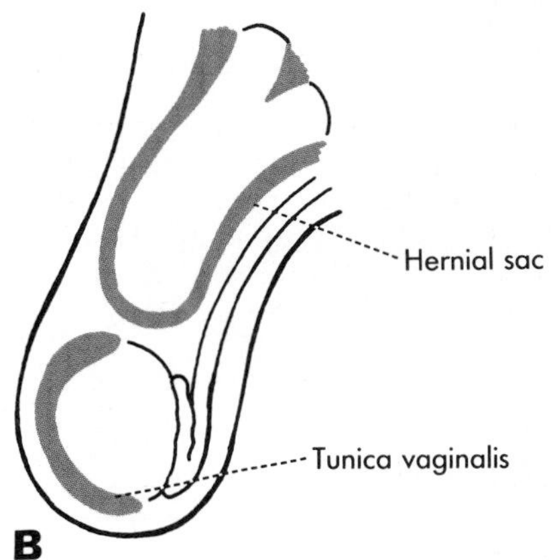


**FIG. 26-15.** Two types of indirect inguinal hernia. **A** is the type usually called congenital; **B** is usually called acquired.

**Hernial Repair.** Repair of any abdominal hernia includes reducing it, ligating the neck of the peritoneal hernial sac so that the peritoneum possesses a smooth surface rather than a funnel-shaped diverticulum that invites further herniation, closing the enlarged opening through which the hernia has escaped, and strengthening the region in order to mitigate a recurrence.

In the female, both the deep and superficial inguinal rings may be closed. In the male, however, they can be narrowed only to the extent that they allow passage of the spermatic cord. The specific technique used to strengthen the walls of the inguinal canal varies with the size of the hernia, its type,

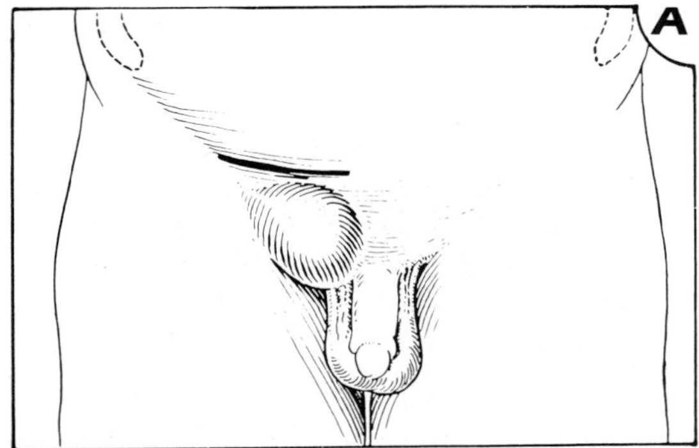

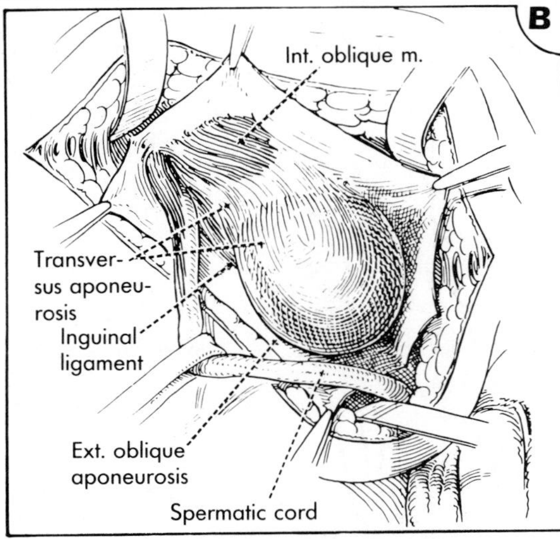


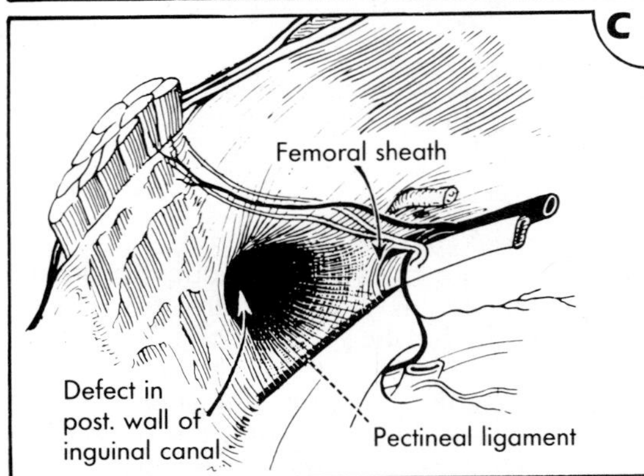


**FIG. 26-16.** The appearance of a direct inguinal hernia in an anterior view, in **A,** with the line of incision marked on the skin, and in **B,** with the external oblique and internal oblique aponeuroses divided, opening up the inguinal canal. **C** shows the deep or posterior aspect of the inguinal canal with the forward herniation of the transversalis fascia and the conjoint tendon into the inguinal canal. (Anson BJ, McVay CB: Surgical Anatomy, 5th ed, vol 1. Philadelphia, WB Saunders, 1971)

and the preference of the surgeon. In indirect hernia, the technique usually includes (1) narrowing the enlarged superficial and deep rings so that they accommodate the cord with little room to spare and (2) reinforcing the wall of the inguinal canal in one manner or another. The best-known techniques involve bringing down the borders of the internal oblique and transversus muscles and suturing them to the inguinal ligament or the pectineal ligament.

## THE FEMORAL SHEATH AND FEMORAL HERNIA

In the upright anatomic position, the pelvis is so oriented that the inguinal ligament lies anterior and somewhat below the level of the superior ramus of the pubis. The psoas and iliacus muscles pass vertically behind the lateral part of the inguinal ligament into the thigh (see Fig. 17-32) and fill the *lacuna musculorum* (Fig. 26-17). Between the two muscles is the femoral nerve. The transversalis fascia on the deep surface of the transversus meets and becomes continuous with the iliopsoas fascia along the lateral part of the inguinal ligament.

Behind the medial part of the inguinal ligament, in the lacuna vasorum, the external iliac artery descends vertically into the thigh to become the femoral artery; the external iliac vein, the continuation of the femoral vein, ascends on the medial side of the artery (see Fig. 17-43; Fig. 26-17). The space behind the inguinal ligament, through which these vessels pass, is bounded medially by the free edge of the lacunar ligament and the conjoint tendon, posteriorly by the pectineal ligament and the superior ramus of the pubis, and laterally by the iliopectineal arch, which separates the artery from the psoas and the femoral nerve.

Just above the inguinal ligament, the transversalis fascia forms a reinforced band, the *iliopubic tract;* below the inguinal ligament, the iliopubic tract continues into the lacuna vasorum and forms a sheath around the femoral vessels (Fig. 26-18). This funnel-shaped diverticulum of the transversalis fascia, called the **femoral sheath,** extends around the femo-

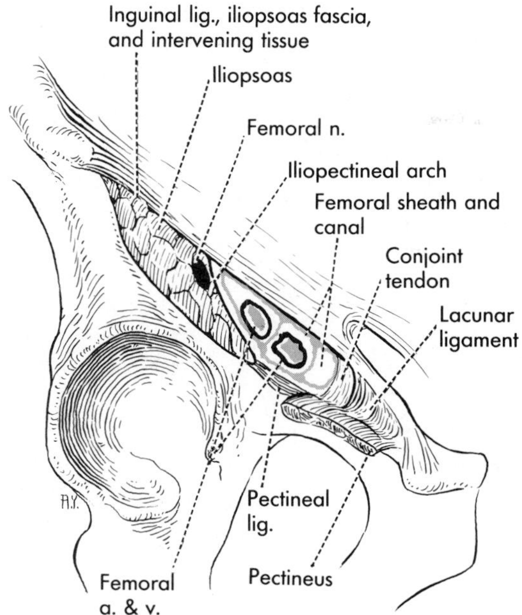


**FIG. 26-17.** The space between the inguinal ligament and the coxal bone. The iliopectineal arch divides this into the laterally lying lacuna musculorum, occupied by the iliopsoas muscle and the femoral nerve, and the medially lying lacuna vasorum, occupied by the femoral vessels and the femoral canal enclosed in the femoral sheath, shown in blue.

ral vessels into the femoral triangle as if the vessels had drawn it out into the thigh (see Fig. 17-43). It has been described in Chapter 17.

Two septa divide the femoral sheath into three compartments: the lateral compartment contains the femoral artery; the intermediate compartment contains the femoral vein; and the medial compartment is essentially empty except for some areolar tissue, lymphatics, and a lymph node. The medial compartment of the femoral sheath is the **femoral canal,** a potential space into which the femoral vein can expand when the venous return through it increases during walking and running. The femoral canal is cone shaped, with its base, the **femoral ring,** facing into the abdomen. The medial margin of the femoral ring is formed by the lacunar ligament and the conjoint tendon (Fig. 26-19); its anterior lip is formed by the iliopubic tract along the inguinal ligament; the pectineal ligament borders it posteriorly.

The ring, measuring just over 1 cm in diameter, is open above except for a filmy layer of extraperitoneal tissue that stretches across it and is known as the *femoral septum.* Only extraperitoneal connective tissue intervenes between the femoral ring and the parietal peritoneum. The ring represents an area of potential weakness through which femoral hernias descend into the femoral canal. The apex of the 1 cm to 1.5 cm long canal points into the femoral triangle and is formed by the fusion of the femoral sheath with the adventitia of the femoral vein behind the saphenous opening (see Fig. 17-43).

#### Femoral Hernia

Femoral hernias are always acquired; a preformed peritoneal sac is never present in the femoral canal. A hernia is likely to occur when the femoral ring is large, but the exact predisposing causes are variable and largely unknown. The peritoneal sac forced into the canal by the hernia is covered

**FIG. 26-18.** The transversalis fascia in the inguinofemoral region seen from the front, showing the continuity with the femoral sheath and the transverse thickening ("iliopubic tract") across the femoral ring. The internal oblique is elevated; the spermatic cord and the inguinal ligament are cut and displaced. (Redrawn from Clark JH, Hashimoto EI: Surg Gynecol Obstet 82:840, 1946)

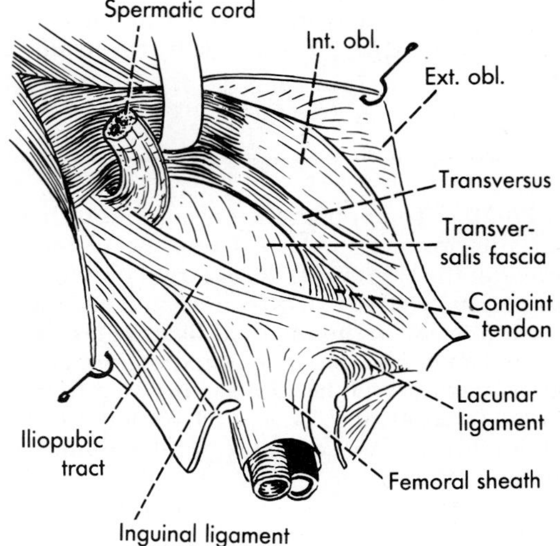

by extraperitoneal fat. The hernia dilates the femoral sheath and protrudes into the subcutaneous tissue of the upper part of the thigh, through the *saphenous hiatus*, stretching before it the *fascia cribosa* (p. 373; Fig. 26-20).

The femoral hernia is always inferior to the inguinal ligament and can be further distinguished from an inguinal hernia by verifying that it does not emerge through the superficial inguinal ring. After

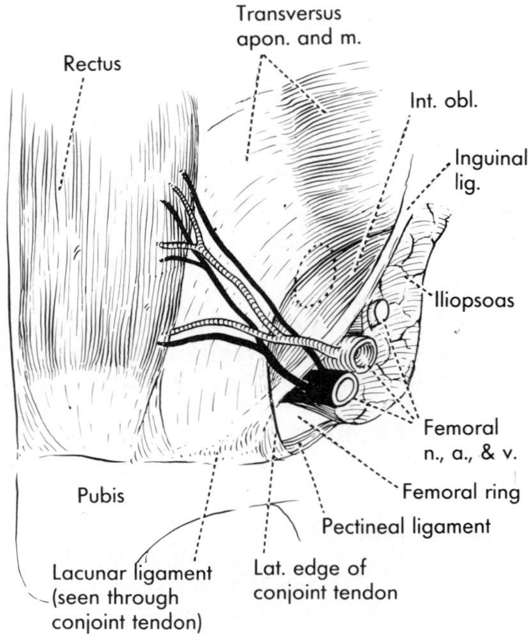


**FIG. 26-19.** The inguinal region and the upper end of the femoral canal from the inner aspect of the anterior abdominal wall, after removal of the transversalis fascia. The location of the deep inguinal ring is indicated by the broken oval. Figure 26-5 shows the same view but with the transversalis fascia intact.

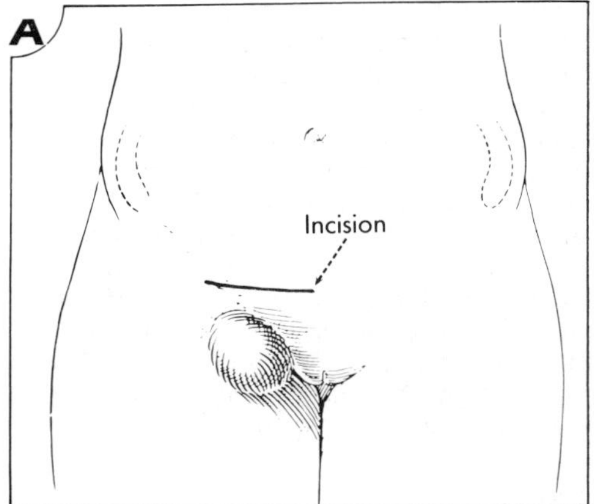


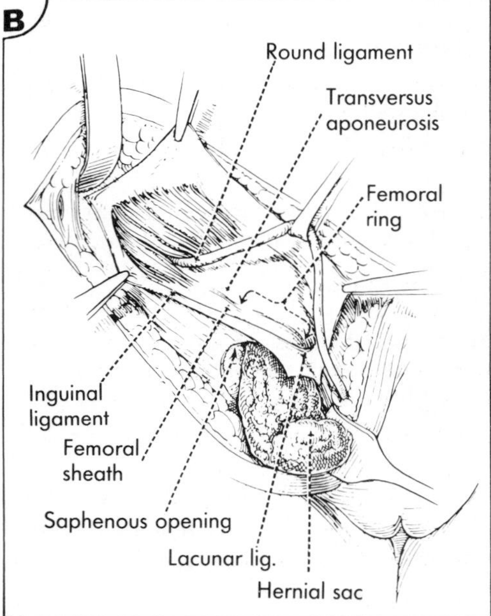


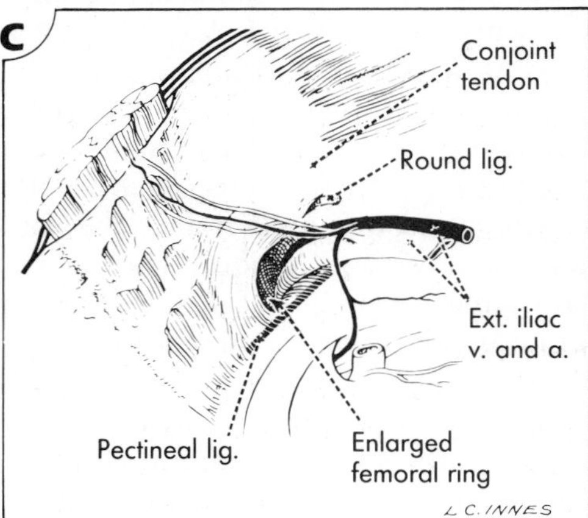


**FIG. 26-20.** The appearance of a femoral hernia in an anterior view **(A)** with the line of incision marked on the skin. In **B,** the external oblique aponeurosis has been divided. The hernial sac pushes forward inferior to the inguinal ligament. **C** shows the posterior aspect of the inguinal region with the parts named and the hernial sac omitted from the enlarged femoral ring. (Anson BJ, McVay CB: Surgical Anatomy, 5th ed, vol 1. Philadelphia, WB Saunders, 1971)

a hernia has been reduced, a finger (usually the little finger) may be inserted into the superficial inguinal ring by invaginating the skin of the scrotum. When abdominal pressure is raised, an inguinal hernia can be felt pushing against the finger in the superficial ring; a femoral hernia will appear below the ring without contacting the finger.

Because of the sharp ligamentous lips of the femoral ring, a femoral hernia is particularly likely to obstruct and strangulate. Repair of the hernia follows the same principles as that of an inguinal hernia. The femoral ring may be closed completely by suturing the conjoint tendon or the inguinal ligament to the pectineal ligament.

### Vascular Relationships

Just before it leaves the abdomen, the **external iliac artery** gives rise to two named branches, the inferior epigastric and the deep iliac circumflex.

The **inferior epigastric artery** arises from the front of the external iliac artery and turns forward and upward on the anterior abdominal wall between the peritoneum and the transversalis fascia (Fig. 26-19). As it runs upward, it is directed also somewhat medially; it passes just on the medial side of the deep inguinal ring and in its course raises the lateral umbilical fold of peritoneum (see Fig. 23-26). It pierces the transversalis fascia below the arcuate line to reach the deep surface of the rectus abdominis and, shortly thereafter, enters the muscle to run upward in its substance. Close to its origin, the inferior epigastric artery gives off, in the male, the **cremasteric artery,** which follows the spermatic cord and supplies the tunics of the cord and testis, or in the female, a small **artery of the round ligament** that follows the round ligament of the uterus. The inferior epigastric also gives rise to a **pubic branch** that runs medially and downward, passing close to the femoral ring. It sends anastomotic twigs to the obturator artery and, behind the pubic symphysis, anastomoses with its fellow of the opposite side and with vessels on the adjacent surface of the bladder. The connection to the obturator artery may become the anomalous stem of this pubic branch. The obturator artery leaves the pelvis through the obturator foramen just below the superior ramus of the pubis.

The deep **iliac circumflex artery** arises from the lateral side of the external iliac and runs laterally and upward across the iliopsoas toward the anterior superior iliac spine. It pierces the transversus to run above the iliac crest, between the transversus and the internal oblique.

## GENERAL REFERENCE AND RECOMMENDED READINGS


ANSON BJ, MORGAN EH, MCVAY CB: Surgical anatomy of the inguinal region based upon a study of 500 body-halves. Surg Gynecol Obstet 111:707, 1960

BACKHOUSE KM, BUTLER H: The development of the coverings of the testis and cord. J Anat 92:645, 1958

BACKHOUSE KM, BUTLER H: The gubernaculum testis of the pig. J Anat 94:107, 1960

BARRETT WC: A note on the internal cremaster muscle. Anat Rec 109:392, 1951

BLUNT MJ: Posterior wall of the inguinal canal. Br J Surg 39:230, 1951

DOYLE JF: The superficial inguinal arch: A reassessment of what has been called the inguinal ligament. J Anat 108:297, 1971

GUNN SA, GOULD TC: Vasculature of the testes and adnexa. In Handbook of Physiology, Sect 7, vol II, p 117. Washington DC, American Physiological Society, 1975

HARRISON RG, WEINER JS: Vascular pattern of the mammalian testis and their functional significance. J Exp Biol 26:304, 1949

HILL EC: The vascularization of the human testis. Am J Anat 9:463, 1909

HODSON N: The nerves of the testis, epididymis, and scrotum. In Johnson AD, Gomes WR, Vandemark NL (eds): The Testis, vol 1, p 47. New York, Academic Press, 1970

HOLSTEIN AF, ROOSEN-RUNGE EC: Atlas of Human Spermatogenesis. Berlin, Grosse, 1981

JOHNSON AD: The influence of cadmium on the testis. In Johnson AD, Gomes WR (eds): The Testis, vol IV, p 565. New York, Academic Press, 1977

JOHNSON FP: Dissections of human seminiferous tubules. Anat Rec 59:187, 1934

KOHLER FP: On the etiology of varicocele. J Urol 97:741, 1967

KORMANO M, SUORANTA H: Microvascular organization of the adult human testis. Anat Rec 170:31, 1971

LYTLE WJ: Inguinal anatomy. J Anat 128:581, 1979

MCVAY CB: The normal and pathologic anatomy of the transversus abdominis muscle in the inguinal and femoral hernia. Surg Clin North Am 51:1251, 1971

MCVAY CB, ANSON BF: Aponeurotic and fascial continuities in the abdomen, pelvis and thigh. Anat Rec 76:213, 1940

MADDEN JL, HAKIM S, AGOROGIANNIS B: The anatomy and repair of inguinal hernias. Surg Clin North Am 51:1269, 1971

MITCHELL GAG: The innervation of the kidney, ureter, testicle and epididymis. J Anat 70:10, 1935

MITCHELL GAG: The innervation of the ovary, uterine tube, testis, and epididymis. J Anat 72:508, 1938

SHAFIK A: The cremasteric muscle. In Johnson AD, Gomes WR (eds): The Testis, vol IV, p 481. New York, Academic Press, 1977

WAITES GMH: Temperature regulation and the testis. In Johnson AD, Gomes WR, Vandemark NL (eds): The Testis, vol I, p 241. New York, Academic Press, 1970

WAITES GMH, MOULE GR: Relation of vascular heat exchange to temperature regulation in the testis of the ram. J Reprod Fertil 2:213, 1961

WELLS LJ: Descent of the testis: Anatomical and hormonal considerations. Surgery 14:436, 1943

WENDLER D: Histologisch-histochemische Befunde an der Schleimhaut des Ducture deferens (pars funicularis) beim geschlechtsreifen Mann. Acta Histochem 31:48, 1968

PART VIII

# THE PELVIS AND PERINEUM

# 27

# THE PELVIS

The primary function of the **bony pelvis,** made up of the two *ossa coxae* and the *sacrum,* is the transmission of forces between the lower limbs and the axial skeleton. The anatomy of these bones was described in Chapters 14 and 17 (see Figs. 14-7, 17-1, and 17-9). The osseoligamentous ring, formed by the bony pelvis, encloses the **pelvic cavity,** the inferior portion of the body cavity introduced in Chapter 23.

The word "pelvis" means "basin" and is used for referring to both the bony pelvis and the pelvic cavity. In this chapter, the term is used chiefly in the latter sense, since the purpose is to describe the pelvic cavity with its apertures, walls, and contents.

Insofar as the pelvis contains most of the internal genitalia of the female and some of those of the male, pelvic contents differ in the two sexes. In the female, the pelvic internal genitalia include the ovaries, the uterine tubes, and the uterus, with their associated peritoneal and fibromuscular ligaments, and the upper part of the vagina; the male pelvic internal genitalia are the ductus deferentes, the seminal vesicles and ejaculatory ducts, and the prostate. Notwithstanding these sexual differences, and those evident in its skeleton, the basic anatomy of the pelvis is the same in the two sexes: this is true of the boundaries, fascias, vessels, and nerves of the pelvis, as well as of its nongenital viscera, the bladder and the rectum. Except for the ovaries and, to some extent, the rectum, pelvic viscera are supplied by branches of the internal iliac artery; the internal iliac vein drains them. Pelvic lymphatics generally follow the blood vessels but do not all empty into the internal iliac nodes. The pelvis contains the *sacral plexus,* the largest plexus of somatic nerves in the body, from which issue some of the major nerves of the lower limb. The autonomic plexus for the supply of pelvic viscera is the *inferior hypogastric,* or *pelvic, plexus;* its offshoots reach the viscera along the arteries.

The first section of this chapter discusses the boundaries of the pelvic cavity, including its apertures and the bony, ligamentous, and muscular elements in its walls. These features have a bearing on the anatomy of the birth canal, the passage through which the fetus is delivered. The sections dealing with pelvic viscera are preceded by developmental considerations that explain not only the embryological but also the topographical relationships between pelvic viscera.

## BOUNDARIES OF THE PELVIC CAVITY

The boundaries of the pelvic cavity are formed by the pelvic skeleton and the pelvic musculature. The pelvis has a wide, more or less oval, *superior aperture,* or *inlet,* and a more narrow, somewhat rectangular, *inferior aperture,* or *outlet.* Through the superior aperture, the pelvic cavity communicates freely with the abdominal cavity proper (p. 579); inferiorly, the pelvic cavity is limited by a muscular floor, the *pelvic diaphragm,* which is attached to the pelvic walls. The *inferior pelvic aperture* is below the pelvic diaphragm and, therefore, outside the pelvic cavity; it provides the osseoligamentous frame of the perineum. The walls of the pelvic cavity taper downward like those of a basin and are formed anteriorly by the body of the two pubes and the symphysis between them; posteriorly, they are formed by the sacrum, and anterolaterally, by the obturator internus muscle, covering most of the lower half of the internal surface of the coxal bone (see Fig. 17-6). Posterolaterally, between the greater sciatic notch of the coxal bone and the sacrum, the greater sciatic foramen provides exit for nerves and vessels from the pelvis to the gluteal region and the perineum. The foramen is largely filled by the piriformis muscle (see Fig. 17-21).

**FIG. 27-1.** Commonly measured diameters of the pelvic inlet of the female.

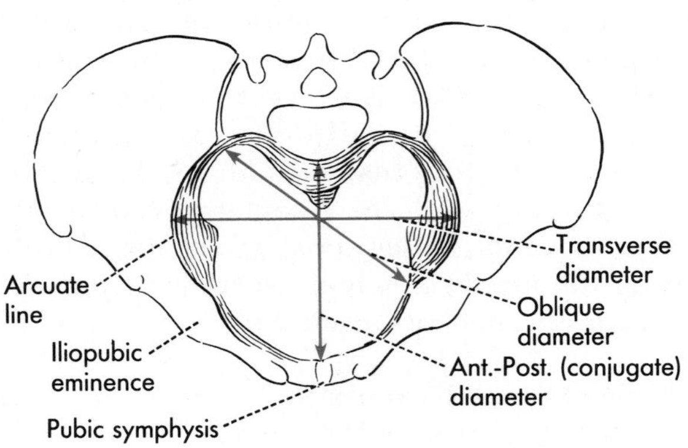


### THE PELVIC SKELETON

The size of the pelvic cavity is determined by the pelvic skeleton. The limits imposed by the bony pelvis are so important in obstetric considerations that the muscles and fascias contributing to the pelvic walls are not taken into account at all in assessments of the size of the pelvis. The pelvic apertures are bordered largely by bones, and, as already noted, the anterior and posterior walls of the pelvic cavity are formed by bare bone.

#### The Pelvic Apertures

The **pelvic inlet** is bordered by the pelvic brim, formed on each side by the *pubic crest,* the *linea terminalis* (consisting of the pecten pubis and the arcuate line on the ilium; see Fig. 17-4), and the *ala of the sacrum* (Fig. 27-1). Posteriorly, the sacral promontory, perched on a level higher than the linea terminalis, juts forward into the pelvic inlet, rendering it somewhat heart shaped rather than oval. The two sacroiliac joints and the symphysis pubis firmly unite these bones and permit essentially no change in the dimensions of the pelvic inlet, although during pregnancy, the ligaments of the joints may slacken and there may be some play between the bones.

During much of its growth and development, the fetus enclosed in the enlarged uterus is located in the abdominal cavity above the pelvic brim. For delivery to take place, the fetal head has to be admitted into the pelvic cavity through the pelvic inlet. This "engagement" of the fetal head usually occurs sometime before the commencement of labor. Clearly, the shape and dimensions of the pelvic inlet have obstetric consequences, and these are assessed by physical and radiologic measurements of various pelvic diameters.

The commonly measured **diameters of the superior aperture** are the conjugate and transverse diameters (Fig. 27-1). The *conjugate,* or anteroposterior, *diameter* is measured from the upper border of the symphysis pubis to the sacral promontory (*true conjugate diameter;* this can only be done radiologically), or from the lower border of the symphysis pubis to the promontory (*diagonal conjugate;* this can be obtained by vaginal examination, since the fingers of the examiner can reach the sacrum). The *transverse diameter* is the greatest distance obtainable between bilateral symmetrical points on the linea terminalis. Apart from radiologic measurements, this can be inferred from the external dimensions of the pelvis or from the transverse diameter of the pelvic cavity measured between the ischial spines, palpable *per vaginam.* A number of other diameters may be measured for obstetric diagnosis and prognosis in the pelvic inlet, outlet, and cavity; however, because the superior pelvic aperture is the most variable in shape, pelves are classified according to the ratio of the conjugate and transverse diameters in this aperture.

Four major **types of pelves** are recognized: anthropoid, android, gynecoid, and platypelloid pelves. In the first two, the conjugate diameter is longer than the transverse; in the latter two, the reverse is the case. Anthropoid and android pelves predominate in males; most women have gynecoid or android pelves. Platypelloid pelvis is rare and shows a pronounced anteroposterior flattening; an anthropoid pelvis is flattened from side to side. Both types, as well as an android pelvis, may make engagement of the fetal head difficult.

The **inferior pelvic aperture** is formed anteriorly by the *pubic arch,* made up of the inferior margin of the pubic symphysis, and the conjoint rami of the pubis and ischium, terminating behind in the ischial tuberosity. Posteriorly, the aperture is bordered by the *sacrotuberous ligaments,* which connect the ischial tuberosities to the dorsal aspect of the sacrum and the coccyx. The coccyx juts forward into the inferior aperture (see Fig. 28-1).

After the fetal head progresses beyond the pelvic floor into the perineum, it is delivered through the inferior pelvic aperture, its occiput usually passing in the pubic arch. A narrow pubic arch, characteristic of anthropoid and android pelves, may displace the fetal head posteriorly to the extent that the soft tissues of the perineum will be torn.

## Orientation of the Bony Pelvis

Correct orientation of the pelvis is necessary for appreciating the topographical relationships of pelvic and perineal viscera and the mechanisms that support them. In the upright anatomic position, the pelvis is so oriented that the right and left anterior superior iliac spines and pubic tubercles are in the same vertical plane. Indeed, it is possible for a lean person facing a wall to bring all four bony points in contact with the wall. If a transverse plane is placed at right angles to the vertical plane across the superior border of the symphysis pubis, it will pass through the ischial spines and, in the female, also through the tip of the coccyx (Fig. 27-2). The plane of the pelvic inlet makes approximately a 60° angle with such a horizontal plane, whereas the plane of the pelvic outlet lies almost parallel with it. The consideration of these planes makes it clear that (1) the superior pelvic aperture faces more anteriorly than upward; (2) the anterior wall of the pelvic cavity is much shorter than its posterior wall; and (3) the axis of the pelvic cavity, which runs through the central point of the inlet and the outlet, is curved, almost paralleling the sacral curvature (Fig. 27-2).

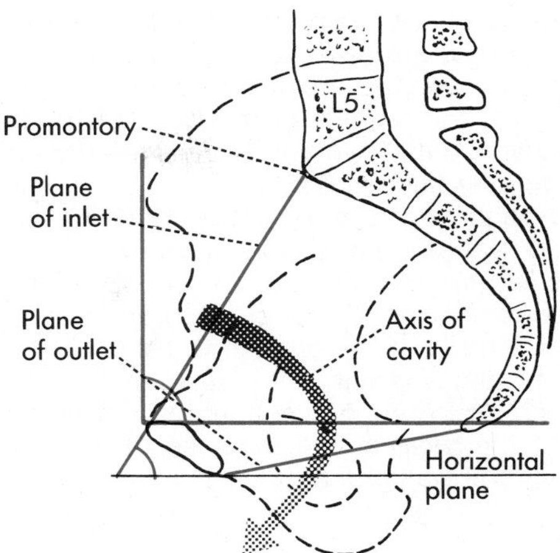


**FIG. 27-2.** Orientation of the pelvis: the red vertical line shows the alignment of the anterior superior iliac spine with the pubic tubercle; the red horizontal line shows the alignment of the pubic symphysis, ischial spine, and tip of the coccyx. The planes of the pelvic inlet and outlet are shown in blue.

### Skeletal Sex Differences

Adaptations of the female pelvis to childbearing

are reflected in several features of the pelvic skeleton. Although the absolute measurements of any part of the pelvis may be greater in the male than in the female, the relative proportions of the female pelvis give it a more roomy cavity, wider apertures, and a lighter skeletal frame.

The pelvic cavity of the male tends to be more conical; that of the female is more cylindrical. The female *sacrum* is shorter and wider than the male, and its concavity is deeper. In the female, more than two-thirds of the base of the sacrum is made up by the ala, whereas in the male, the width of the first sacral vertebral body occupies more than one-third of the base at the expense of the ala. The anterolateral wall of the pelvis is relatively wider in the female, as expressed by several features: the pubic tubercles are farther apart; the distance between the symphysis and the anterior lip of the acetabulum is greater in the female than the diameter of the acetabulum, whereas in the male, these measurements are approximately equal. This gives a triangular shape to the obturator foramen in the female; in the male, the foramen is more or less round.

The greater sciatic notch and pubic arch are wider in the female than in the male. In the male, the pubic arch makes an acute angle considerably less than 90°. In the female, the arch is smooth and rounded; the line of the two inferior pubic rami, if extrapolated forward, would make an angle close to 90°. In the male, the ischiopubic rami are rather robust and everted to give attachment to the crura of the penis; in the female, the rami are rather delicate.

## THE PELVIC MUSCULATURE

Although the pelvic cavity is part of the body cavity derived from the intraembryonic celom, the muscles in its walls bear no resemblance to those of the thorax and abdomen proper. The *piriformis* and *obturator internus* are both muscles of the lower limb that lie on the interior of the pelvic skeleton and, consequently, line part of the pelvic cavity. The several named parts of the levator ani muscle and the coccygeus arise in continuity from the wall of the pelvic cavity and, fusing with their counterparts of the opposite side along the midline of the gutter-shaped pelvic floor, constitute the *pelvic diaphragm*. This diaphragm separates the pelvic cavity from the perineum. The levator ani and coccygeus have evolved from the musculature of the tail, supplying a diaphragm across the pelvic outlet,

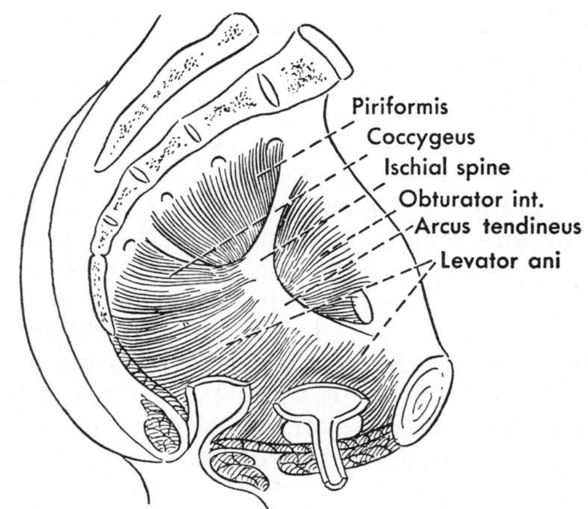


**FIG. 27-3.** Muscles of the pelvic wall and floor. Stumps of the hemisected rectum and the bladder and prostate have been left in place, and the sphincter ani externus and the urogenital diaphragm are shown, unlabeled, below the levator ani.

necessitated by an erect posture and wide bony pelvis.

### The Piriformis and Obturator Internus

Both the piriformis and the obturator internus are lateral rotators of the thigh; both were encountered in the gluteal region (see Figs. 17-19 through 17-23).

The **piriformis** arises on the posterior wall of the pelvis from the area of bone in between and lateral to the pelvic foramina of the sacrum. Between the two muscles, the bodies of the 2nd, 3rd, and 4th sacral vertebrae are bare, covered only with periosteum, as are the 1st sacral vertebra and the ala of the sacrum above the origin of the piriformis. The muscle completes the posterior wall laterally and leaves the pelvis through the greater sciatic foramen, which it nearly fills (Fig. 27-3). The ventral rami of sacral spinal nerves emerging from the pelvic foramina of the sacrum unite on the pelvic surface of the piriformis to form the sacral plexus, and the major branches of the plexus leave the pelvis with the muscle.

The **obturator internus** covers a large area on the internal surface of the coxal bone below the pelvic brim. It arises from the medial surface of the obturator membrane, from the adjoining margins of the obturator foramen (except the superior pubic ramus, which is bare), and from the broad strip of bone above and behind the obturator foramen formed by the fused bodies of the ilium and the ischium (see Fig. 17-6). The fan-shaped muscle tapers to a narrow tendon lodged in the lesser sciatic notch, where it turns abruptly laterally and heads with the tendon of the piriformis toward the greater trochanter. Less than the upper half of the muscle faces into the pelvic cavity; the lower part forms the lateral wall of the perineum. The medial surface of the muscle is covered by the dense *obturator fascia,* and to this fascia is attached the major part of the pelvic diaphragm.

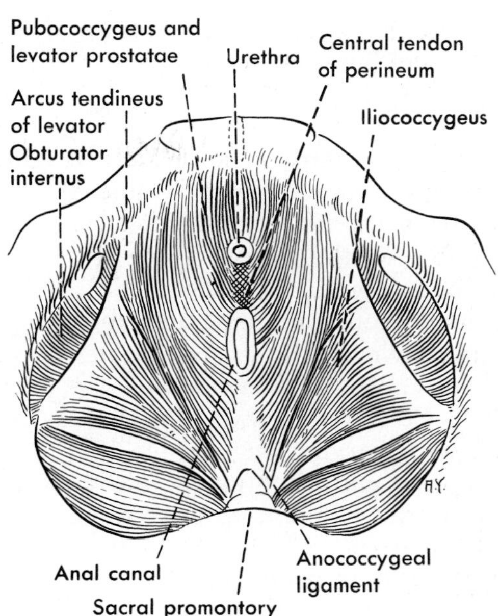


**FIG. 27-4.** The levator ani from above.

## The Pelvic Diaphragm

The pelvic diaphragm consists of the *coccygeus muscle* posteriorly and the more extensive and complex *levator ani* anterolaterally (Figs. 27-3 and 27-4). The diaphragm is a thin sheet of muscle and its halves form the sloping floor of the pelvis, through which the pelvic effluents, the urethra, vagina, and anal canal, pass into the perineum. The diaphragm is sometimes compared to, and may be demonstrated as, a funnel slotted into the pelvic cavity. The rim of the funnel fits snugly against the body of the pubes, the obturator internus, the ischial spine, and the sacrum. The stem of the funnel may represent the urethra, the vagina, or the anal canal. Unlike the funnel, however, the diaphragm is incomplete both posteriorly and anteriorly: posteriorly, the two coccygeus muscles leave the coccyx and sacrum bare between them; anteriorly, there is a **U**-shaped deficiency between the two levator ani muscles, through which the urethra, vagina, and anal canal pass. The anterior deficiency is known as the **urogenital hiatus.** It is closed completely, largely by the pelvic effluents accommodated in it and by fibers of the levator ani that attach to the urethra, vagina, and anal canal. The pelvic fascia that covers half of the pelvic diaphragm fuses with that of the opposite side and with the fascias that sur-

**FIG. 27-5.** The pelvic diaphragm of the male from below.

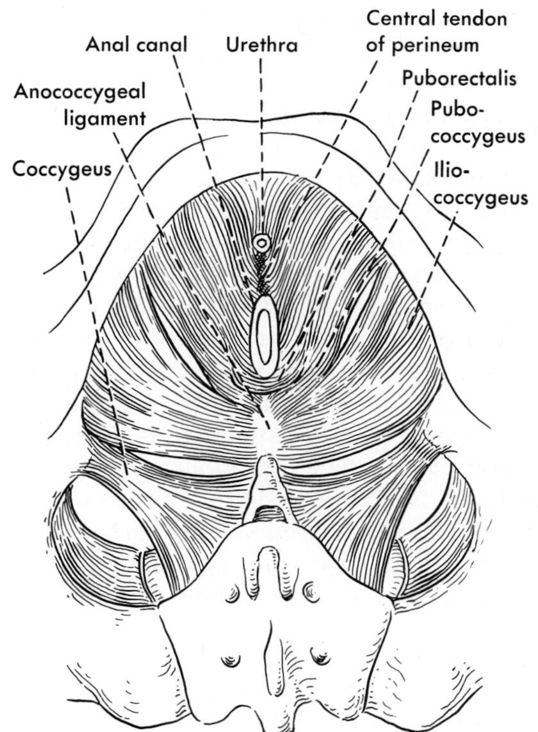

round the pelvic effluents, sealing any communication between the pelvis and the perineum through the hiatus.

**The Levator Ani.** The muscle originates along a semicircular line that skirts the pelvic walls from the pelvic surface of the body of the pubis to the ischial spine. In between these bony points, the levator ani is attached to a bandlike reinforcement in the obturator fascia known as the *arcus tendineus* (or the *tendinous arch of the levator ani*). The fibers of the muscle sweep backward as well as downward with varying degrees of obliquity and insert, as already mentioned, into the walls of the urethra, vagina, and anal canal and beyond the urogenital hiatus into the *anococcygeal ligament,* or raphe, and the coccyx.

The named parts of the levator ani are the pubococcygeus (levator prostatae or pubovaginalis), the puborectalis, and the iliococcygeus. None of these parts is a distinct entity, but their fibers have different directions and relationships.

The **pubococcygeus muscle** runs posteriorly from the body of the pubis and the anterior part of the tendinous arch to the anococcygeal ligament and the coccyx (Figs. 27-4 and 27-5). The medial fasciculi of the two muscles border the urogenital hiatus, and some of their fibers terminate by inserting into the structures that fill the hiatus between the pubic symphysis and the anococcygeal ligament. The most anterior of these fibers insert into the urethra, or sweep behind the prostate or the vagina, and end in the central tendon of the perineum, a fibromuscular body (*perineal body*) projecting from the perineum into the urogenital hiatus. In the male, this part of the muscle is called the **levator prostatae;** in the female, it is called the **pubovaginalis.** Some fibers of the pubovaginalis blend with the wall of the vagina and will be surrounded in the perineum by the sphincter vaginae; others, bypassing the vagina, can act as an additional sphincter around it.

The **puborectalis** is a relatively thick bundle in the levator ani, best defined on the perineal surface of the pubococcygeus (Fig. 27-5). On each side, the muscle sweeps backward from the pubis and fuses with its fellow of the opposite side behind the junction of the rectum with the anal canal, forming a sling around the anorectal junction. Some fibers blend with the longitudinal muscle coat of the anal canal. The puborectalis is chiefly responsible for the angulation at the perineal flexure between the rectum and the anal canal. As it pulls the anorectal junction forward, the muscle has a sphincterlike action and contributes to anal continence.

The **iliococcygeus** arises from the posterior part of the arcus tendineus and the ischial spine. It inserts into the anococcygeal raphe and the coccyx, just below the insertion of the coccygeus.

**The Coccygeus.** In conformity with the parts of the levator ani, the coccygeus could be called the ischiococcygeus. The small triangular muscle arises from the ischial spine and expands to insert on the lateral borders of the lower two sacral and upper two coccygeal segments. On its external surface, it is blended with the *sacrospinous ligament,* which forms the inferior limit of the greater sciatic foramen.

### Innervation and Actions

The **piriformis** and **obturator internus** are supplied chiefly by S1 segment through small branches of the sacral plexus (Chap. 17). The **levator ani** is innervated on its pelvic surface by twigs from the 4th (sometimes also the 3rd) sacral ventral ramus. The anterior part of the muscle, particularly the puborectalis, usually receives a branch from the pudendal nerve after this nerve has left the pelvis and lies on the inferior surface of the muscle. The **coccygeus** is supplied by the 4th and 5th sacral ventral rami.

The levator ani has important functions in the regulation of abdominal and pelvic pressure and contracts whenever abdominal pressure is raised. The muscle is particularly involved with the voluntary control of micturition (p. 773) and the support of the uterus (p. 784). The pelvic diaphragm supports all the pelvic viscera, and insertion of some of its fibers into the central tendon of the perineum is an important factor in this respect. Its contraction raises the entire pelvic floor. Prenatal exercises are aimed at conscious relaxation of the pelvic diaphragm, whereas the strengthening of the muscle has been advocated for preventing the prolapse of pelvic viscera.

## PELVIC FASCIA, VESSELS, AND NERVES

The major pelvic organs protrude into the cavity of the pelvis from its floor, lined up between the anterior and the posterior wall along the midline; only the ovaries, uterine tubes, and deferent ducts are laterally placed. The pelvic peritoneum is draped over these organs, and, although it descends into the recesses between the viscera, it does not make contact with the pelvic floor. The voluminous, irregular space between the pelvic peritoneum and the pelvic floor and walls is filled with *pelvic fascia*. Although much of this fascia is loose areolar tissue, capable of accommodating to the changing dimensions of the distensible pelvic viscera, specialized condensations in this fascia form neurovascular sheaths, whereas others support and separate the pelvic organs (Fig. 27-6).

Blood vessels and nerves are arranged in the pelvis more or less in layers concentric with the posterolateral pelvic wall. They are embedded in fascial laminae that can be defined in the plane of these vessels and nerves. The most exterior lamina in contact with the

**FIG. 27-6.** Highly diagrammatic schema of the fascia of the pelvis in a coronal section. The closely related fascia on the obturator internus muscles and the outer surface of the levator ani is indicated by broken lines.

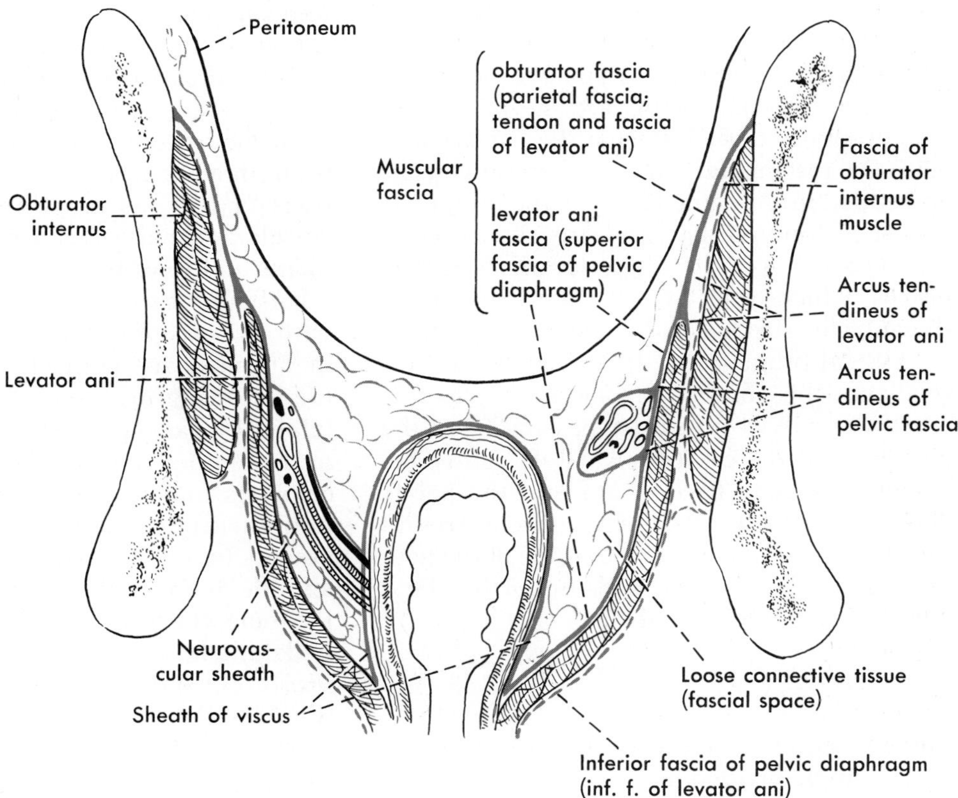

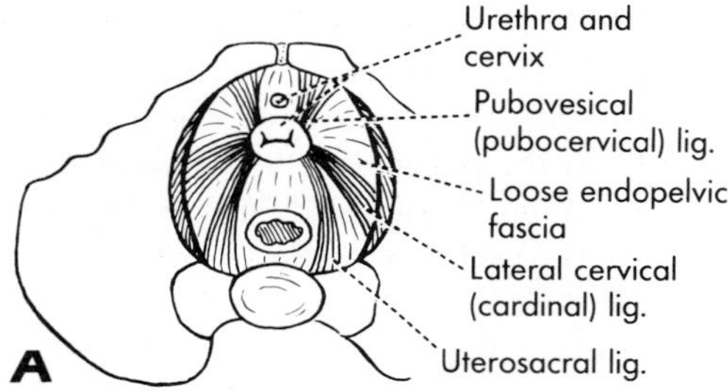


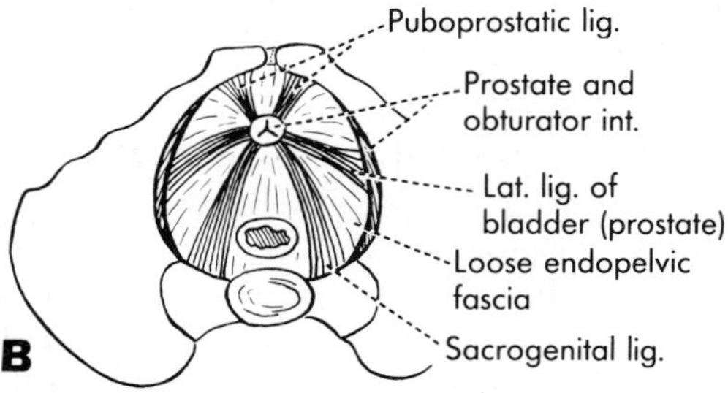


**FIG. 27-7.** Schematic representation of condensations of the subperitoneal pelvic connective tissue into so-called ligaments in the female **(A)** and male **(B)** seen from a superior view after removal of the pelvic viscera.

muscular walls is that of the somatic nerves and contains the sacral plexus with its branches. Internal to this are the blood vessels, branches of the internal iliac artery and vein. On the medial surface of these vessels, embedded in the same fascia as the ureters, is the inferior hypogastric plexus of nerves, a continuation of the superior hypogastric plexus, which descends into the pelvis from the anterior surface of the abdominal aorta.

Visceral branches of the internal iliac vessels and the inferior hypogastric plexuses reach the pelvic organs by passing from the periphery of the cavity toward its center between the pelvic peritoneum and the pelvic floor. Condensations of pelvic fascia around these vessels and nerves form some of the neurovascular sheaths and the supporting ligaments of the pelvic organs.

## PELVIC FASCIA

The connective tissue continuum within the pelvis can be divided into *parietal* and *visceral pelvic fascias* and *subperitoneal pelvic connective tissue*.

**Parietal pelvic fascia** forms more or less dense membranes on the pelvic surface of the muscles and blends with the periosteum of the bony pelvic boundaries. Particularly well defined is the *obturator fascia* (Fig. 27-6). Indeed, it is thought that this fascia between the linea terminalis and the arcus tendineus of the levator ani is a vestige of the levator ani itself that, in the course of evolution, lowered its origin from the pelvic brim to the tendinous arch. Much thinner layers of fascias exist over the piriformis and the pelvic diaphragm. The latter, called the *superior fascia of the pelvic diaphragm,* is continuous across the pelvic floor and blends with the obturator fascia laterally and with the visceral pelvic fascias at the urogenital hiatus.

The **visceral pelvic fascia** invests the bladder, prostate, vagina, uterus, and rectum in sheaths, or sleeves, of connective tissue in a manner that allows their distention. In the male, the heaviest of these fascias is the *prostatic fascia,* or sheath, which is nondistensible and surrounds not only the prostate but also the prostatic venous plexus.

The **subperitoneal pelvic connective tissue** is the continuation of extraperitoneal fascia from the abdomen into the pelvis. There has been much controversy whether condensations of this fascia, which ensheaths blood vessels and nerves, can be regarded as ligaments and to what extent these putative ligaments support the pelvic contents. Although, with the exception of the puboprostatic ligaments, none are named in *Nomina Anatomica,* the following ligaments have been described (albeit by various names) and are considered in the applied anatomy of the pelvis by gynecologists and urologists: in the female, the uterosacral ligaments, the lateral cervical ligaments (or cardinal ligaments of the uterus), and the pubovesical ligaments (Fig. 27-7A); in the male, the fibrous tissue of the sacrogenital folds, the lateral ligaments of the bladder (or prostate), and the puboprostatic ligaments (Fig. 27-7B).

The puboprostatic and pubovesical ligaments, equivalent in the two sexes, are sometimes considered part of the superior fascia of the pelvic diaphragm; they do not contain any blood vessels. All ligaments blend medially with the visceral fascia of either the prostate, bladder, vagina, or cervix and laterally with the superior fascia of the pelvic diaphragm. The latter junction is often evident as the *arcus tendineus of the pelvic fascia,* distinct from that of the levator ani, located at a higher level (Fig. 27-6). Many of these ligaments incorporate some smooth muscle. Their contents and relationships are discussed with the respective vessels, nerves, and organs.

There are a number of **fascial septa** in the subperitoneal connective tissue of the pelvis, some of which at times are considered part of the visceral fascial sheaths. These septa include the *vesicovaginal septum* in front of the vagina and the uterine cervix and the *peritoneoperineal fascia,* called the *rectovesical septum* in the male and the *rectovaginal septum* in the female. It has been both asserted and denied that this septum represents the peritoneum of the rectovesical or rectouterine pouch, which during development receded from its original contact with the pelvic floor. More pronounced in the male, the rectovesical septum is attached above to the peritoneum of the rectovesical pouch and blends on each side with the parietal fascia on the lateral pelvic walls and below with the fascia of the pelvic floor (Fig. 27-24). It may be a strong white membrane or it may be quite thin. It divides, in essence, the subperitoneal connective tissue space of the pelvis into anterior and posterior compartments.

## BLOOD VESSELS AND LYMPHATICS

Although the internal iliac vessels supply and drain all pelvic structures, there are other vessels that significantly contribute to the circulation in the pelvis. These smaller vessels are the ovarian artery and vein, the superior rectal artery and vein, and, much less important, the median sacral artery and vein.

### Internal Iliac Artery

The internal iliac artery is one of the terminal branches of the common iliac artery, arising at the pelvic brim in front of the sacroiliac joint. The other terminal branch, the external iliac artery, continues in the direction of the common iliac artery along the pelvic brim, whereas the internal iliac passes downward into the pelvis across the common or external iliac vein.

In front of the greater sciatic foramen, the artery breaks up into a number of branches that supply the pelvic viscera as well as the perineum and the proximal parts of the lower limbs (Fig. 27-8).

In the fetus, blood was returned to the placental circulation through the umbilical artery, the first branch of the internal iliac. The umbilical artery, after passing forward just below the pelvic brim, ascended to the umbilicus. After birth, the extrapelvic portion of the umbilical artery becomes obliterated, and the intrapelvic portion decreases in relative size and supplies the bladder.

There is great variation in the precise branching pattern of the internal iliac artery. Nine major types of branching and 49 subtypes have been described. The structures and regions supplied by the branches of the artery, however, are quite constant (Fig. 27-9). Before the artery breaks up into its named branches, it usually divides into an **anterior** and a **posterior trunk.** It is common for all the pelvic visceral branches to arise from the anterior trunk, along with the artery of the perineum called the internal pudendal.

**Branches.** The **visceral branches** of the internal iliac artery include the following:

1. The *umbilical artery,* which gives off the *superior vesical arteries* to the upper part of the bladder and the *artery of the ductus deferens,* continues forward as the medial umbilical ligament. The artery of the ductus supplies, in addition to the deferent duct, the ureter, the seminal vesicles, and part of the bladder.
2. The *inferior vesical artery* is present in the male. It reaches the bladder and the prostate along the lateral ligament of the bladder and supplies both.
3. The *middle rectal artery* may arise from a

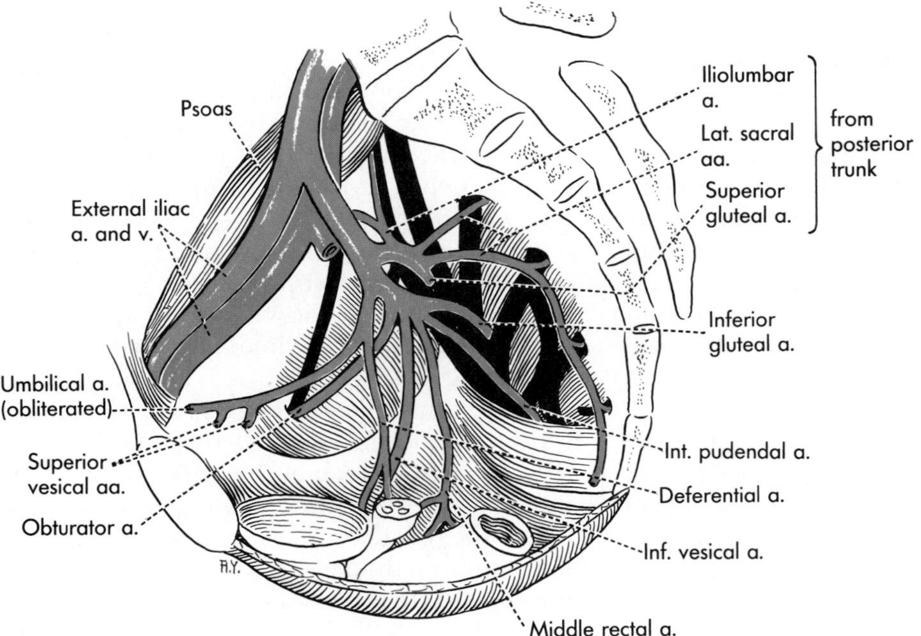


**FIG. 27-8.** Branches of the internal iliac artery in the male. The pattern seen here is a common one, but varies greatly among bodies. The obturator nerve and nerves of the sacral plexus are shown in black.

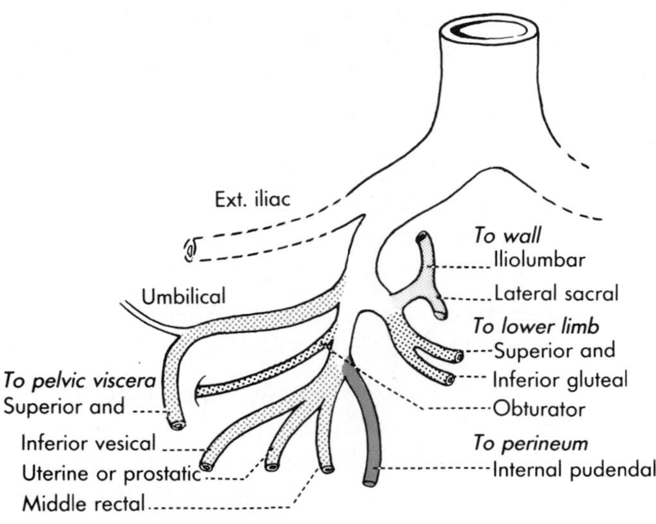


**FIG. 27-9.** Groups of branches of the internal iliac artery given off for the supply of the pelvic wall, the pelvic viscera, the perineum, and the lower limb, identified by different colors and patterns of shading.

common stem with the inferior vesical; it enters the rectum just above the pelvic floor and supplies chiefly its muscle coat. It anastomoses in the rectal wall with both superior and inferior rectal arteries.

4. The *uterine artery* descends on the pelvic wall and turns medially along the lateral cervical ligament. It passes above and in front of the ureter, just lateral to the cervix. In addition to supplying the uterus, it gives rise to the vaginal arter-

ies and terminates in its tubal branch. It also helps to supply the ovary. The artery replaces the inferior vesical artery of the male.

5. One or more *vaginal arteries* may arise from the internal iliac with the uterine artery or independently. They run to the side of the vagina in the pelvis and descend in its wall to the perineum. In the pelvis, they also supply the bladder and the rectum. Some anatomists regard the vaginal arteries as replacing the inferior vesical arteries of the male.

All five visceral branches arise from the anterior trunk of the internal iliac (Figs. 27-8 and 27-9).

**Branches to the pelvic wall** are given off by the posterior trunk of the internal iliac artery. The *iliolumbar artery* leaves the pelvis by ascending along the lumbosacral trunk in front of the ala of the sacrum. It divides into an iliac branch, which supplies the iliacus (p. 681), and a lumbar branch, which replaces the 5th lumbar arteries.

There are usually two *lateral sacral arteries.* The upper artery disappears into the 1st sacral foramen; the other descends close to the pelvic sacral foramina and sends a branch into the foramina. If there is only one artery, it gives a large branch into the first sacral foramen before descending. The lateral sacral arteries correspond essentially to the dorsal branches of the intercostal and lumbar arteries. They give rise to branches that supply the bone of the sacrum and the dural sheaths of the nerve roots. Some twigs emerge from the dorsal sacral foramina to supply the musculature and skin on the dorsal surface of the sacrum.

**Branches to the lower limb** include the obturator artery and the superior and inferior gluteal arteries. The *obturator artery* is a branch of the anterior trunk and runs parallel with the umbilical artery along the pelvic wall on the surface of the obturator internus muscle, to which it gives branches. The obturator nerve runs above the artery and the obturator vein runs below it. They all converge upon the obturator canal and disappear into the thigh. The obturator artery thereafter is distributed primarily to the obturator externus muscle and the hip joint. Fairly commonly (in about 25%–30% of sides) the obturator artery arises from the inferior epigastric artery or the external iliac rather than from the internal iliac. A large anastomotic vessel between the pubic branch of the inferior epigastric and the obturator artery may function as a second origin of the obturator artery. Such an anastomotic vessel or an anomalous obturator artery runs close to or across the femoral ring to reach the obturator canal. Because of its close relationship to the neck of a femoral hernia, such an artery is in danger of being divided when a strangulated or obstructed femoral hernia is relieved.

Of the two *gluteal arteries,* the inferior is usually given off by the anterior trunk of the internal iliac and the superior gluteal artery is usually given off by the posterior trunk. Both arteries leave the pelvis through the greater sciatic foramen, the superior artery above and the inferior artery below the piriformis.

The **internal pudendal artery,** a large branch of the anterior trunk, leaves the pelvis between the piriformis and coccygeus and descends vertically on the exterior of the levator ani into the perineum (p. 794).

### The Internal Iliac Vein and Its Tributaries

The internal iliac vein lies, for the most part, between the lateral pelvic wall and the internal iliac artery. It joins the external iliac vein to form the common iliac vein. Its tributaries largely parallel the branches of the artery except that there is no umbilical vein in the pelvis.

The tributaries of the internal iliac vein communicate freely with each other and also with veins lying outside their territory of drainage. Most important of these venous communications are those with the vertebral venous plexus and with the veins of the portal system. Veins passing through the pelvic foramina of the sacrum link the tributaries of the internal iliac vein to the vertebral venous plexus. Metastases from neoplasms of the pelvic viscera or infected emboli from the same organs may pass through these veins and lodge in the cancellous bone of the vertebrae or may even reach the cranial cavity.

In the wall of the rectum, the middle and inferior rectal veins, tributaries of the internal iliac vein, anastomose with the superior rectal vein, part of the portal venous system. As a result of portal hypertension, blood may be shunted from the superior rectal veins into the tributaries of the internal iliac vein (p. 657). On the other hand, since there are no valves in the veins of the pelvis, venous blood may leave the pelvis along the route of least resistance. This includes the internal iliac veins, the vertebral venous plexus, or the anastomosis in the wall of the rectum.

### Other Vessels

The **ovarian artery** and its two accompanying veins cross the pelvic brim some distance anterior to the internal iliac vessels. Their course, on the posterior abdominal wall, was traced in Chapter 25. Below the pelvic brim, the vessels run in the suspensory ligament of the ovary and in the broad ligament before reaching the ovary and anastomosing with tubal branches of the uterine vessels.

The **superior rectal vessels** cross the pelvic brim of the left side in the sigmoid mesocolon. They were described in Chapter 24, and their pelvic distribution is discussed with the rectum.

Although the **median sacral artery** was originally the terminal part of the aorta, it arises from that vessel's posterior aspect a little above the aortic bifurcation. It descends over the sacral promontory close against the bone and behind the superior hypogastric plexus. It runs downward for the length of the sacrum, embedded in connective tissue on the anterior surface of this bone, and gives off twigs into the bone and lateral twigs that anastomose with the lateral sacral vessels. Just beyond the tip of the coccyx, it ends and is connected with the median sacral veins by a series of arteriovenous anastomoses that form the *coccygeal body*. Two **median sacral veins** parallel the artery and unite before ending in the left common iliac vein as this crosses to the right to help form the inferior vena cava.

### Lymphatics and Lymph Nodes

The lymph nodes in the pelvis are divided into two general groups, the *internal iliac nodes*, which are associated with various branches of the internal iliac vessels, and *sacral nodes*, which lie on the front of the sacrum. These nodes are embedded in the general extraperitoneal connective tissue associated with the pelvic wall and floor and receive much of the lymphatic drainage from the pelvic viscera. Their removal, part of the treatment of some pelvic neoplasms, involves dissection of the nerves and vessels of the lateral pelvic wall, cleaning as much connective tissue as possible from around the nodes. The lymphatics from the viscera pass, in general, along the blood vessels. Not all, however, end in internal iliac or sacral nodes. Some may reach nodes along the brim of the pelvis (external and common iliac nodes); others from the ovary, the uterine tube, and the fundus of the uterus drain upward into lumbar nodes. The chief drainage from the rectum is upward along the superior rectal vessels to the inferior mesenteric nodes, and the drainage from the lowest part of the anal canal and the vagina is downward and forward into superficial inguinal nodes. These nodes also receive some lymphatics from the uterus, which accompany the round ligament.

## NERVES

Nerves enter the pelvis through the superior pelvic aperture or through the pelvic foramina of the sacrum. They leave it by passing over the brim of the urogenital diaphragm or through the urogenital hiatus; some ascend out of the pelvis through the superior aperture. The entering **somatic nerves** include the lumbosacral trunk and the obturator nerve on the ala of the sacrum (Fig. 27-10) and the ventral rami of sacral nerves through the pelvic sacral foramina (the minute coccygeal nerves are inconsequential). Entering **autonomic nerves** include the two sympathetic trunks

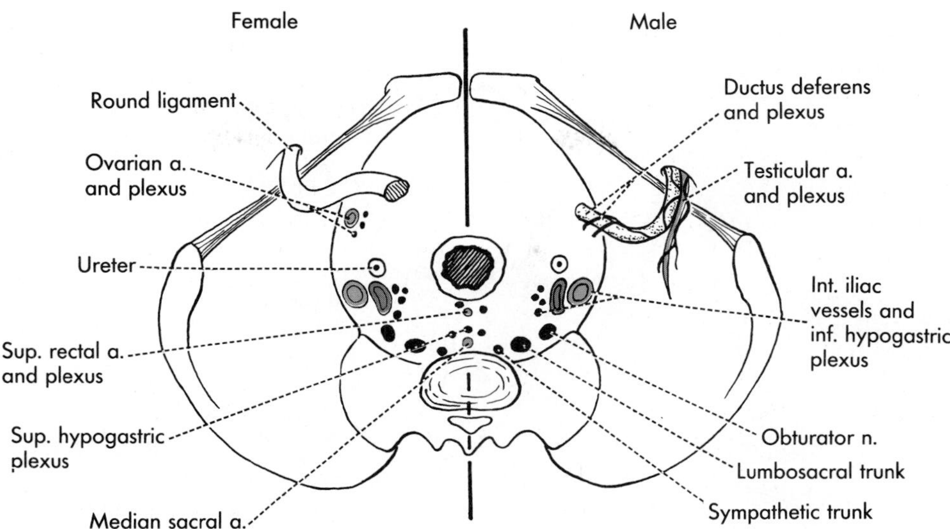


**FIG. 27-10.** A schematic representation of nerves, vessels, and other structures that pass through the pelvic inlet. The left side shows the female pelvis, the right side, the male. Although the superior and inferior hypogastric plexuses are at different levels, both are included in the diagram.

and the hypogastric nerves over the sacral promontory; the ovarian plexus around the ovarian vessels as they cross the pelvic brim; the superior rectal nerve plexus, descending with the superior rectal artery in the sigmoid mesocolon (Fig. 27-10); and the pelvic splanchnic nerves, entering the pelvis with the ventral rami of the sacral nerves.

The branches of the sacral plexus leave the pelvis through the greater sciatic foramen, and the obturator nerve leaves through the obturator canal. Exiting autonomic nerves include the sympathetic fibers for the lower limb and perineum, running with the branches of the sacral plexus, and the *cavernous nerves,* extensions of the inferior hypogastric plexus through the urogenital hiatus for the supply of erectile tissue in the perineum. Autonomic nerve plexuses ascend through the superior pelvic aperture around the ductus deferens (deferential plexus) and along the superior rectal artery for the supply of the colon distal to the left flexure.

### The Sacral Plexus

The sacral plexus is the lower part of the lumbosacral plexus, the nerve plexus of the lower limbs (see Fig. 16-7). The plexus takes form on the posterior wall of the pelvis, just lateral to the pelvic foramina of the sacrum, and most of it disappears in the buttock just as it gives rise to its branches (Fig. 27-11). The major part of the plexus lies on the anterior surface of the piriformis muscle, and all the larger branches pass through the greater sciatic foramen, most of them below the piriformis, to appear in the buttock.

The sacral plexus is formed by the union of the lumbosacral trunk, containing some of the fibers from the 4th lumbar ventral ramus, all those from the 5th lumbar ventral ramus, and the ventral rami of the first three or four sacral nerves. The nerves formed by the plexus supply the musculature and skin of the buttock, the posterior compartment of the thigh, and the entire leg and foot below the knee, except for the cutaneous area medially, which is supplied by the saphenous nerve, a branch of the femoral nerve derived from the lumbar plexus. The sacral plexus gives off also the chief somatic nerve of the perineum called the pudendal nerve and sends branches to the pelvic diaphragm as well.

The lumbosacral trunk and the sacral ven-

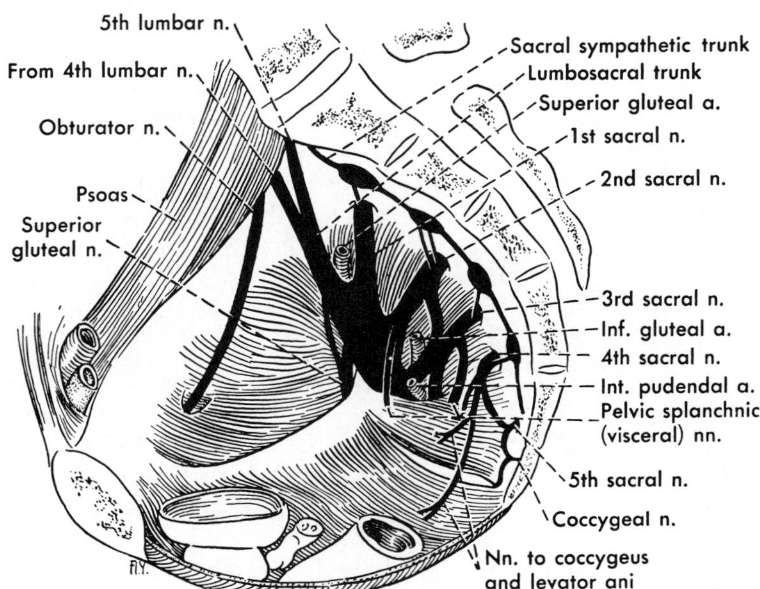


**FIG. 27-11.** The sacral plexus in the pelvis. The muscle visible between the sacral nerves is the piriformis. The obturator nerve is a branch of the lumbar plexus, but traverses the pelvis.

tral rami can be considered the *roots* of the plexus. Conforming to the general plan of the limb plexus, these roots divide into *anterior* and *posterior divisions.* Such a division, however, is more difficult to demonstrate by dissection than are the divisions in the brachial or lumbar plexuses. The *branches* of the sacral plexus, like those of the brachial and lumbar plexuses, are formed either by the anterior or by the posterior divisions of certain ventral rami and are then distributed to the skin and musculature over the original anterior and posterior compartments of the limb, respectively (Fig. 27-12).

After the sacral ventral rami have emerged from the pelvic sacral foramina and before they enter into the formation of the plexus, each root of the plexus receives a ramus communicans from the sacral sympathetic trunk; this brings into the nerves postganglionic sympathetic fibers to be distributed to the lower limb and the perineum.

**Branches.** The branches of the sacral plexus are difficult to recognize in a dissection of the pelvis, because many of them arise as the plexus leaves the pelvis. The largest branch of the plexus is the **sciatic nerve,** the largest nerve in the body. The sciatic nerve, however, is a composite nerve consisting of the *common peroneal nerve,* formed by posterior divisions, and the *tibial nerve,* formed by anterior divisions. These two nerves separate from each other some distance above the knee (see Fig. 17-48). The **common peroneal nerve** is formed by the posterior divisions of L4–S2, and the other nerves derived from the posterior divisions are as follows (Fig. 27-12): the *superior gluteal nerve* from L4, L5, and S1; the *inferior gluteal* from L5, S1, and S2; a lateral part of the *posterior femoral cutaneous* from S1 and S2; one or more nerves from S1 and S2 *to the piriformis muscle;* and often a *perforating cutaneous branch* from S2 and S3 to the skin of the buttock.

The nerves formed from the anterior divisions of the plexus are the **tibial nerve** from L4–S3; two small nerves, the *nerve to the quadratus femoris* and *inferior gemellus* from L4, L5, and S1 and the *nerve to the obturator internus* and *superior gemellus* from L5, S1, and S2; the medial part of the *posterior femoral cutaneous* from S2 and S3; and the **pudendal nerve** from S2 and S3 or from S2, S3, and S4. Twigs from the 4th sacral nerve also supply the coccygeus and levator ani muscles. The

5th sacral and coccygeal nerves, not considered a part of the sacral plexus, unite to form the *anococcygeal nerves,* which contribute to the innervation of the skin between the anus and the tip of the coccyx.

Branches of the sacral ventral rami not considered part of the sacral plexus are the **pelvic splanchnic nerves.** These nerves given off by S3 and S4 ventral rami (and sometimes S2 as well) contribute to the formation of the *inferior hypogastric,* or *pelvic, plexus,* taking into that plexus preganglionic parasympathetic fibers and conveying visceral afferents from the pelvic plexus to sacral segments of the spinal cord.

The superior gluteal nerve makes its exit from the pelvis above the upper border of the piriformis muscle in company with the superior gluteal vessels; all the other branches of the sacral plexus that leave the pelvis do so below the piriformis muscle. The plexus, as a whole, is covered on its pelvic surface by the internal iliac vessels. The superior gluteal vessels always, the inferior gluteal and internal pudendal vessels often, pass through the plexus as they leave the pelvis. The precise distribution of the branches of the sacral plexus is described, for the most part, in chapters on the lower limb. The distribution of the pudendal nerve to the perineum is described in Chapter 28.

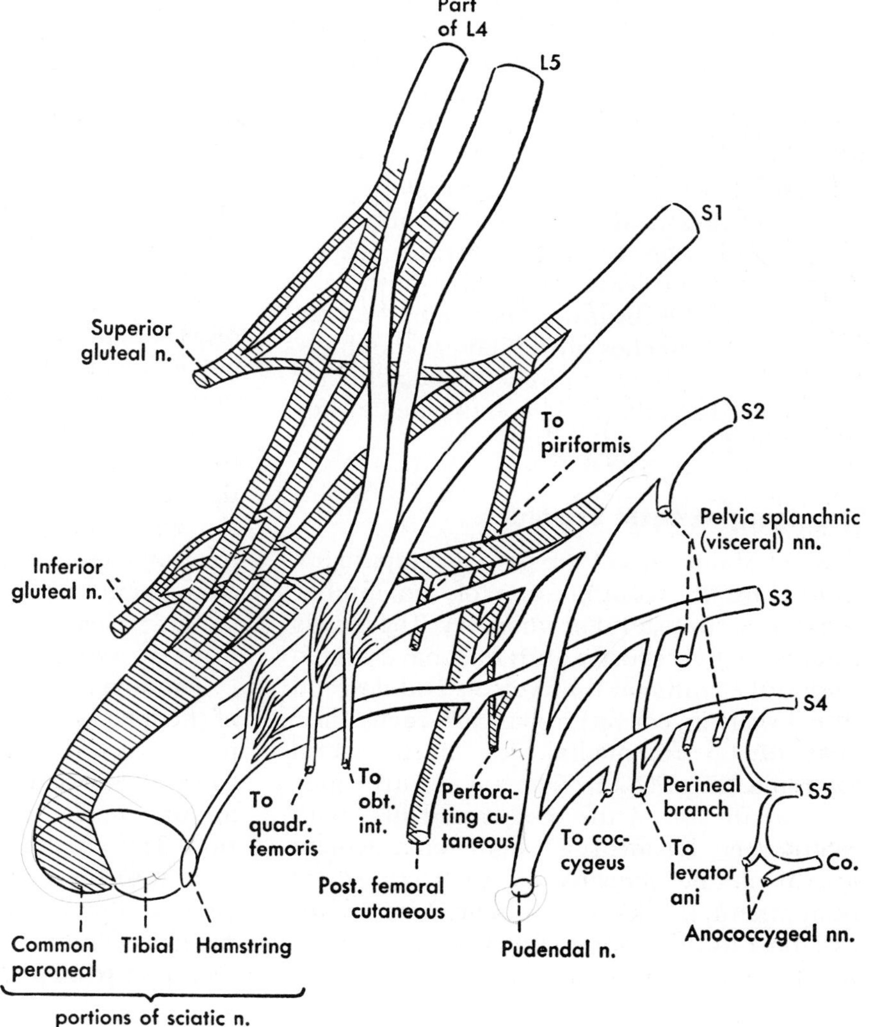


**FIG. 27-12.** A diagram of the sacral plexus. The roots of the plexus, the anterior divisions of the roots, and the branches of the plexus composed of the anterior divisions are shown in white (unshaded); the posterior divisions of the roots and their branches are shaded.

**Variations.** Of the many variations in the formation and branching of the sacral plexus, most important are prefixation and postfixation of the plexus. The prefixed type may receive all the fibers from L4, L5, and S1 and sometimes fibers from L3; it may or may not receive fibers from S2 and S3. In a postfixed plexus, fibers from S4 enter the sciatic nerve, but its important characteristic is that it receives fewer fibers than normal from L4 or that the highest nerve contributing to it may be L5. The composition of various branches of the sacral plexus necessarily varies with the segments contributing to the plexus.

## The Obturator Nerve

The formation of the obturator nerve was described in Chapter 25 and its distribution in Chapters 16 and 17. Having been formed by the ventral rami of L2, L3, and L4, the obturator nerve emerges from the medial border of the psoas near the pelvic brim and descends into the pelvis on the ala of the sacrum, lying lateral to the lumbosacral trunks; lower down, it is on the lateral side of the internal iliac vessels (Figs. 27-8 and 27-11). Keeping to the lateral pelvic wall, the nerve slopes forward and downward to meet the obturator artery and vein at the obturator canal, through which it leaves the pelvis. In the pelvis, it lies on the obturator fascia and does not give off any branches until it enters the thigh. Here it is distributed to some of the adductor muscles and an area of skin on the medial side of the thigh.

## The Sympathetic Trunks

The sacral portions of the paired sympathetic trunks are the continuation of the lumbar trunks over the sacral promontory. The trunks enter the pelvis shortly after they have passed behind the common iliac vessels and descend on the sacrum, converging toward each other. They lie on the medial side of the sacral foramina (Fig. 27-11). The two trunks may meet at the tip of the coccyx and fuse with each other to form a slight enlargement known as the *ganglion impar.* Each sacral sympathetic trunk tends to bear three or four ganglia, but the number may vary from one to six. The interganglionic portions of the trunk consist primarily of descending fibers that are mostly preganglionic. These fibers have entered the lumbar portion of the trunks through white rami communicantes given off by upper lumbar nerves. Most of them synapse in the sacral sympathetic ganglia and give off *gray rami communicantes* to the sacral nerves, thus furnishing these nerves with the postganglionic fibers they carry to the lower limbs and perineum.

The ganglia also have slender visceral branches (one might call them *sacral splanchnic nerves*) consisting presumably of preganglionic fibers that join the inferior hypogastric plexus. The precise function of these nerves is not known. They may also convey some visceral afferent fibers to the trunks from the inferior hypogastric plexus that presumably terminate in lumbar segments of the spinal cord. There may be transverse or oblique connections between the two sympathetic trunks across the front of the sacrum.

## The Autonomic Plexuses

The chief autonomic plexus of the pelvis is the **inferior hypogastric plexus,** also known as the **pelvic plexus.** All pelvic viscera receive visceral efferent and afferent nerves through this plexus. In addition to the nerves that form the pelvic plexus, the superior rectal and ovarian plexuses bring autonomic nerves to the rectum and ovary, respectively, but do not pass through the inferior hypogastric plexus. Both the superior rectal and ovarian plexuses consist chiefly of sympathetic fibers, but in the inferior hypogastric plexus, parasympathetic components predominate, although the plexus receives substantial contributions also from the sympathetic system.

The **superior rectal plexus** is a continuation of the inferior mesenteric plexus along the superior rectal vessels. As such, it is derived from the aortic plexus. Its sympathetic fibers are presumably concerned with vasoconstriction. The visceral afferents in the plexus ascend into the inferior mesenteric plexus and contain afferents from the sigmoid colon but not the rectum.

The **ovarian plexus** is derived from the aortic and renal plexuses. This plexus contains sympathetic visceral efferents from T10–T11 segments and is distributed to the ovary and the uterine tube. In the pelvis, the plexus communicates with the inferior hypogastric plexus. Apart from vasoconstriction, its effects on the ovary and uterine tube are unknown.

The **inferior hypogastric plexus** is formed by lateral extensions of the superior hypogastric plexus, known as the *hypogastric nerves,* and the *pelvic splanchnic nerves.* Each of these merit description before discussing the inferior hypogastric plexus itself.

**The Superior Hypogastric Plexus and Hypogastric Nerves.** The superior hypogastric plexus (presacral nerve) is the direct extension of the aortic plexus below the aortic bifurcation (Fig. 27-13). It lies immediately behind the peritoneum and descends over the anterior surface of the 5th lumbar vertebra in the retroperitoneal tissue that continues downward in front of the sacrum and is often referred to as "presacral fascia." Like the aortic plexus, the superior hypogastric plexus consists of a mixture of preganglionic and postganglionic sympathetic fibers, small ganglia, and visceral afferents that mediate pain sensation from the fundus and upper part of the uterus and follow the route of sympathetic efferents. The visceral efferents are from T10–L2 segments; the afferents also terminate in the same segments. While in front of the lum-

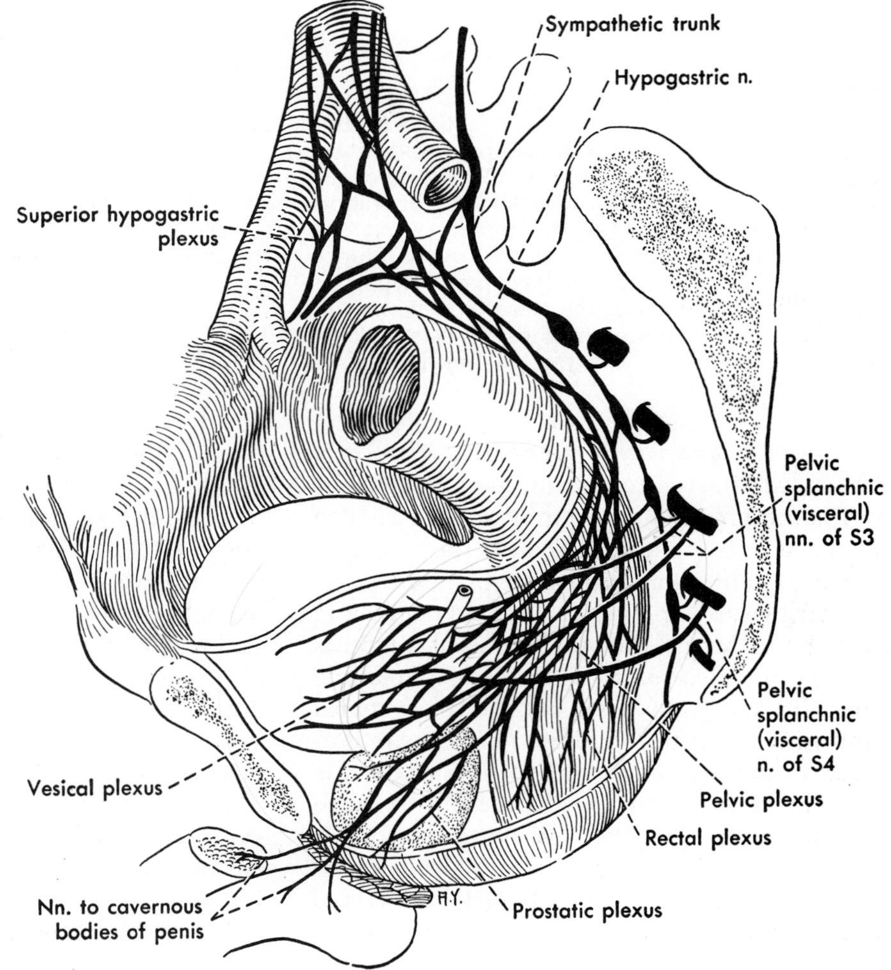


**FIG. 27-13.** The superior and inferior hypogastric plexuses and the sympathetic trunks in the male pelvis seen from the left side. The subsidiary plexuses of the inferior hypogastric plexus (the vesical, prostatic, and rectal in the male; the vesical, uterovaginal, and rectal in the female) run medially through the connective tissue below the peritoneal level to reach the viscera that they supply. The fibers that ascend in the sigmoid mesentery to reach the descending colon are not shown here.

bar vertebra, these small fibers may form a true plexus or they may become condensed into one or two nerve trunks.

The superior hypogastric plexus ends by bifurcating into **right** and **left hypogastric nerves** (Fig. 27-13). The two hypogastric nerves (in truth, plexuses) diverge from each other at about the level of the sacral promontory and run down and forward along the walls of the pelvis in the lamina of the pelvic fascia closest to the peritoneum. Slightly more anteriorly, the ureters enter the pelvis in the same lamina and, in a dissection, serve as a guide to the hypogastric nerves and the inferior hypogastric plexus, in which the hypogastric nerves terminate well below the level of the pelvic peritoneum.

The hypogastric nerves and the superior hypogastric plexus apparently do not contain any parasympathetic fibers. The sympathetic efferents are vasomotor.

**Presacral Neurectomy.** Resection of the superior hypogastric plexus has been practiced for the relief of *dysmenorrhea* (excessive pain associated with menstruation) because of the presence of uterine pain afferents in the plexus. This operation has no detectable effect on the control of micturition or defecation nor on uterine or ovarian function. It has been claimed that the superior hypogastric plexus conducts some sensation from the bladder, but the important sensory nerves from both the bladder and the rectum travel with the pelvic parasympathetic system.

**The Pelvic Splanchnic Nerves.** The pelvic splanchnic nerves represent the sacral parasympathetic outflow, and they are known also as the **nervi erigentes,** because they are the nerves capable of causing erection of the penis and clitoris. The pelvic splanchnic nerves are also the major pathway for visceral afferents for most pelvic organs. The pelvic splanchnic nerves take origin from two or three sacral ventral rami just after these nerves have emerged from the pelvic sacral foramina (Fig. 27-13). The largest contribution is from S3, with a smaller one from S4 and occasionally S2. The pelvic splanchnic nerves run forward and medially through the branches of the internal iliac vessels and the lamina of the pelvic fascia that contains them, and on the medial surface of these vessels, they mingle with the hypogastric nerves, contributing to the formation of the inferior hypogastric plexus (Fig. 27-13).

The **visceral efferents** in these nerves have their cell bodies in the intermediolateral column of S2–S4 segments. These preganglionic fibers leave the spinal cord along the ventral root, and without passing through the ganglia of the sacral sympathetic trunk, they synapse either in ganglia located in the inferior hypogastric plexus or in the walls of the viscera they innervate (Fig. 27-14). They are motor to the muscle wall of the bladder and the rectum. **Visceral afferents** ascend through the dorsal root to a cell body in the spinal ganglia of S2–S4 spinal nerves. It is generally stated that they mediate not only general visceral afferents but also sensations of pain from all pelvic organs, which is noteworthy since pain sensation from thoracic and abdominal viscera in general follows the sympathetic pathway. There is evidence, however, that the uterus is an exception to this generalization (p. 787).

**The Inferior Hypogastric Plexus.** This large, dense plexus is formed by the commingling of the hypogastric and pelvic splanchnic nerves (Fig. 27-13). The small visceral branches of the sacral sympathetic ganglia (sacral splanchnics) also feed into this plexus, though their precise contribution is uncertain.

The inferior hypogastric plexus is 2.5 cm high and 3 cm to 5 cm long. It lies against the posterolateral pelvic wall, internal to the branches of the internal iliac vessels and lateral to the rectum, the vagina, and the base of the bladder. Its subsidiary plexuses, embedded in some of the ligaments formed by the subperitoneal connective tissue, extend medially toward the viscera.

Posteriorly, the inferior hypogastric plexus gives off the *middle rectal plexus* (often known simply as the rectal plexus); farther forward, in front of the rectum, it gives rise, in the female, to the *uterovaginal plexus;* anteriorly, in

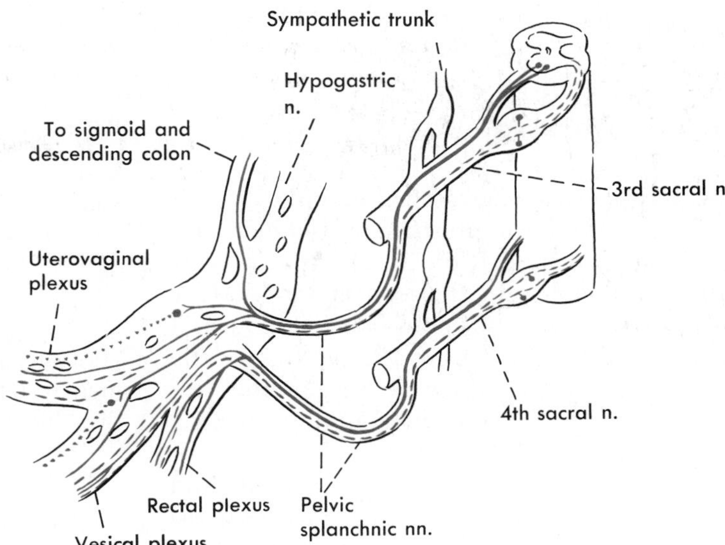


**FIG. 27-14.** Schema of the pelvic splanchnic nerves and their contribution to the pelvic plexus. Visceral afferents are indicated by broken lines, preganglionic fibers by solid lines, and postganglionic fibers by dotted lines. Many of the postganglionic parasympathetic cell bodies are in the wall of the viscera.

both sexes, it gives rise to the *vesical plexus.* Fibers from the lower part of the uterovaginal plexus form the vaginal nerves. In the male, a few fibers from the vesical plexus follow the ductus deferens and form the *deferential plexus,* and other fibers from the lower part of the vesical plexus form the *prostatic plexus.* Finally, fibers from the prostatic plexus in the male and the vesical plexus in the female follow the urethra through the urogenital hiatus to be distributed to the corpora cavernosa of the penis or the clitoris and to other erectile tissue. These fibers constitute the *cavernous nerves.* Except for the cavernous nerves, the various plexuses, in general, reach the viscera in company with the vessels to the viscera.

In addition to these branches, both inferior hypogastric plexuses give rise to ascending fibers that join behind the upper end of the rectum and run through the sigmoid mesocolon to be distributed to the sigmoid and the descending colon. These fibers have never been named, although they have been described by several investigators.

The pelvic plexus contains **pelvic ganglia,** in which both sympathetic and parasympathetic preganglionic fibers synapse. Usually, sympathetic and parasympathetic ganglia are separate. Thus, the plexus consists of preganglionic and postganglionic sympathetic and parasympathetic fibers and of visceral afferents. Presumably, most of the branches of the pelvic plexus transmit sympathetic, parasympathetic, and visceral afferent fibers. However, it is not known whether any sympathetic efferents pass into the nerves to the cavernous bodies of the penis and clitoris or whether any parasympathetic efferents pass into the uterovaginal plexus. Further, the ascending pathway to the sigmoid and the descending colon apparently consists entirely of preganglionic parasympathetic fibers. They synapse in the enteric ganglia that lie in the subserous, myenteric, and submucosal plexuses. There is clinical evidence that the sensory fibers in the pelvic splanchnic nerves reach no higher than the rectum and that pain afferents from the descending and the sigmoid colon accompany the sympathetic system and run through the lumbar splanchnic nerves and the sympathetic trunks before reaching the spinal ganglia.

Paucity or absence of the enteric ganglia of the rectum or lower part of the colon creates difficulty in the passage of feces and usually leads to gross dilatation of the large intestine above the inactive segment. This condition is known as **megacolon.**

Because the pelvic splanchnic nerves and the inferior hypogastric plexus long have been known to be responsible for contraction of the bladder, it has often been assumed that transient retention of urine following operations upon the rectum (a common complication) is a result of injury to the inferior hypogastric plexus or the pelvic splanchnic nerves. There is evidence, however, that the pelvic diaphragm is the chief controller of bladder emptying, and spasm resulting from damage to this diaphragm, rather than to the inferior hypogastric plexus, may be the cause of urinary retention (p. 773).

## PELVIC VISCERA

The pelvic viscera developed equally in both sexes are the rectum, the urinary bladder, and the ureters. Although quite separate in their fully developed state, the rectum and bladder are derived from a common endodermal chamber called the *cloaca*. The reasons for the differences in the pelvic viscera of the two sexes are twofold: (1) the descent of the female gonad is arrested in the pelvis, whereas that of the male gonad proceeds to the perineum, and (2) the genital ducts that transport gametes develop from two distinct duct systems in the male and the female.

Both genital duct systems appear in association with the mesonephros and terminate in the dorsal wall of the ventral division of the cloaca, called the *urogenital sinus*. This interposes the genital ducts and their derivatives between the urogenital sinus, from which the bladder and urethra develop, and the *rectum*, derived from the dorsal division of the cloaca.

The purpose of the following section is to consider the development of pelvic viscera in sufficient detail to make their particular anatomy and topographical relationship intelligible.

### DEVELOPMENTAL CONSIDERATIONS

#### The Cloaca and Its Divisions

The cloaca is formed in the tail of the embryo by the union of the hindgut with the allantois. The **allantois** appears very early as a diverticulum of the yolk sac into the body stalk. Its proximal segment becomes incorporated into the embryo as a consequence of the tail fold and forms the ventral portion of the cloaca (see Fig. 7-3). The ventral portion of the cloaca is broader than its dorsal portion, derived from the hindgut, and the posterolateral wall of the ventral portion admits the two mesonephric ducts descending from the urogenital ridges (Fig. 27-15; see Fig. 25-29).

The cloaca is surrounded by intraembryonic mesoderm unsplit by the celomic ducts, and its floor, surfacing on the ventral aspect of the tail, is the *cloacal membrane.* After its expansion with the growth of the tail, the cloaca becomes partitioned into a ventral and a dorsal compartment, the urogenital sinus and the rectum, respectively. The partitioning is effected by the **urorectal septum,** formed by the proliferation of the mesoderm, filling the sulcus between and above the broad ventral and narrow dorsal parts of the cloaca. The growth of the urorectal septum progresses toward the cloacal membrane, and when the septum contacts the membrane, it divides it into a posterior *anal* and an anterior *urogenital membrane* (Fig. 27-15A through C). With the perforation of the anal membrane, the definitive anatomy of the rectum and anal canal is essentially established; however, there are several complex and rather poorly understood developmental steps required for transforming the urogenital sinus into the bladder, the various segments of the urethra, the prostate, and the vestibule of the vagina.

**The Urogenital Sinus.** The cranial, or **vesical, part** of the urogenital sinus and the adjoining allantois expand to form much of the urinary bladder. The rest of the allantois becomes the **urachus,** later to be transformed into the median umbilical ligament. Caudally, the urogenital sinus is divided into two additional parts: a pelvic and a phallic portion. From the **pelvic portion** develops most of the urethra of the female or the prostatic urethra of the male, which also gives rise to buds that form the prostate. The phallic portion of the sinus, to be discussed in Chapter 28, extends into the phallus to form the vestibule of the vagina in the female and the penile urethra in the male (Fig. 27-15D and E).

After the metanephric diverticula (ureteric buds) have appeared on the mesonephric ducts (see Fig. 25-29), each ureter acquires a direct and independent opening into the bladder, it is thought by the absorption of the terminal segments of the mesonephric ducts into the posterior wall of the primitive bladder. Subsequently, the new openings of the mesonephric ducts are transferred to the pelvic portion of the urogenital sinus by an extension of this absorption process. It is believed that as the mesonephric ducts loop caudally between the ureteric openings, they contact the dorsal wall of the urogenital sinus and become incorporated into it (Fig. 27-15F through I).

There are two consequences of this absorption process. (1) The triangular area (trigone) in the posterior wall of the urogenital sinus between the openings of the ureters and the mesonephric ducts is formed by mesoderm derived from the ducts, whereas the rest of the sinus lining is derived from endoderm. It is believed, however, that endoderm later overgrows this mesodermal component of the bladder and urethral wall. (2) The mesonephric ducts, which function as the male genital ducts, open into the pelvic portion of the urogenital sinus along with the glands of the prostate. The female

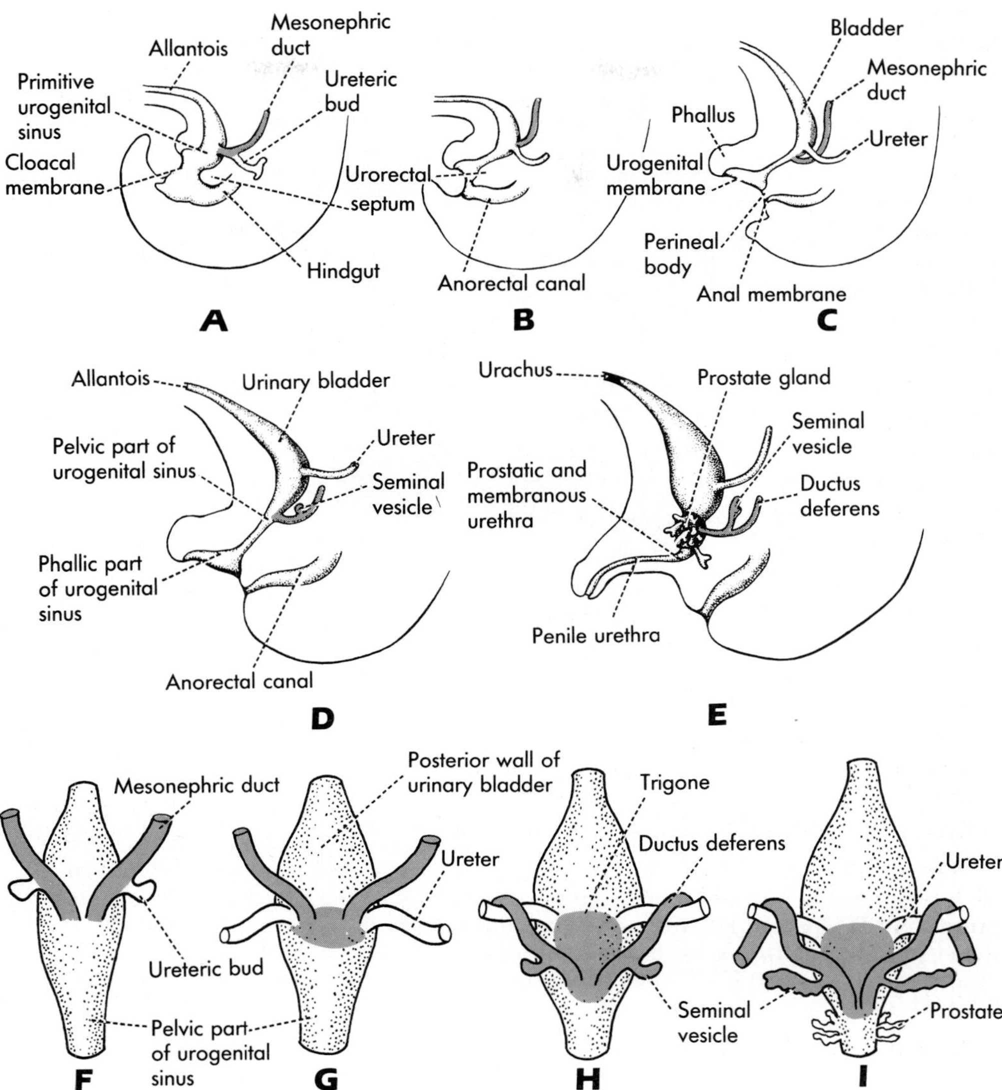


**FIG. 27-15.** Stages of embryonic development illustrated schematically to show the division of the cloaca into the urogenital sinus and the anorectal canal (**A** through **C**), the development of the urogenital sinus (**D** through **E**), and the absorption of the mesonephric ducts into the posterior wall of the urogenital sinus (**F** through **I**). Note the development of the prostate and the seminal vesicles. (Adpated from Langman J: Medical Embryology, 3rd ed. Baltimore, Williams & Wilkins, 1975)

genital ducts also contact the pelvic part of the urogenital sinus, and, at their point of contact, the posterior sinus wall gives rise to the vagina.

### The Genital Ducts

As explained in Chapter 26, the male gonad connects its seminiferous tubules with the **mesonephric duct,** and along this duct, spermatozoa pass to the pelvic portion of the urogenital sinus. The female gonad does not connect directly with any duct system; the ova are discharged into the peritoneal cavity and are picked up and transported by a highly specialized duct system developed from the paramesonephric ducts.

The **paramesonephric ducts** develop in the urogenital ridges parallel with the mesonephric ducts and accompany them to the pelvic portion of the urogenital sinus (Fig. 27-16). In addition to trans-

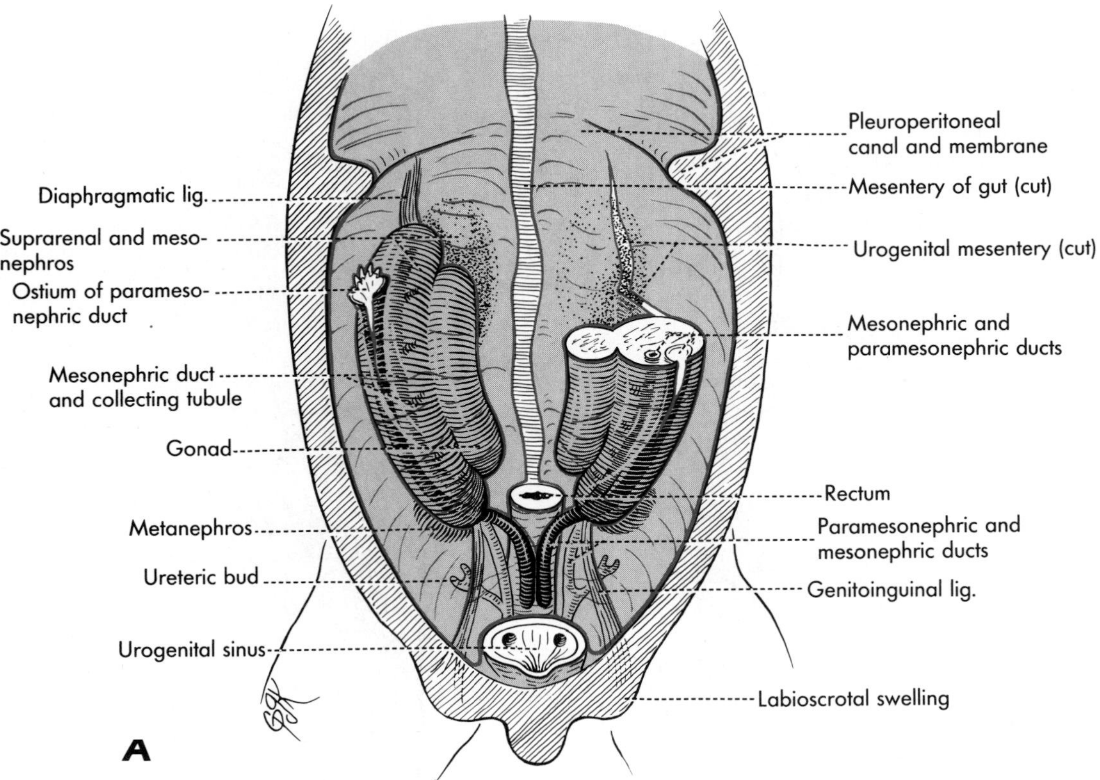


**FIG. 27-16.** The development of the paramesonephric ducts in the urogenital ridges illustrated diagrammatically. Except for the rectum, the gut has been excised with its mesentery. The mesothelial lining of the celom is colored pink. Two successive stages of development are shown: In **A**, the mesonephros and the mesonephric ducts are present; in **B** (opposite), the mesonephros has regressed and its duct atrophied. On the right side, the ridges have been transected to reveal the urogenital mesentery. Compare with Figures 25-29 and 26-12, from which the paramesonephric ducts have been omitted.

mitting gametes, the uterus and uterine tubes, derived from the paramesonephric ducts, provide the conditions necessary for fertilization and the development of the zygote into a viable offspring. The different functional requirements of the genital duct systems in the two sexes are instructively contrasted by comparing their development and later their anatomy.

The mesonephric duct is formed by the longitudinal anastomosis of the nephric ducts, originally located in the pronephros; it conducts urine to the cloaca. The paramesonephric ducts are formed by a longitudinal invagination of the celomic mesothelium on the urogenital ridges; they never convey urine. Cranially, the mesonephric ducts end blindly; the paramesonephric ducts are open because the fusion of the lips of the groove that forms the ducts stops short of sealing their cranial end, leaving an ostium through which the ducts communicate with the celom from the time of their inception.

Once the mesonephros disappears, both sets of ducts are suspended in the urogenital mesentery along the lateral side of the gonad. Because the female gonad remains in the pelvis, this mesentery persists as the broad ligament, but disappears in the male as the testis descends into the inguinal canal.

Each duct converges medially to approximate its counterpart of the opposite side. In so doing, both sets of ducts cross the caudal pole of the gonads, their genitoinguinal ligament, and the ureter. The paramesonephric ducts fuse with each other in the median plane, and their fused caudal tip makes contact with the urogenital sinus. The unfused, lateral parts of the ducts become the uterine tubes,

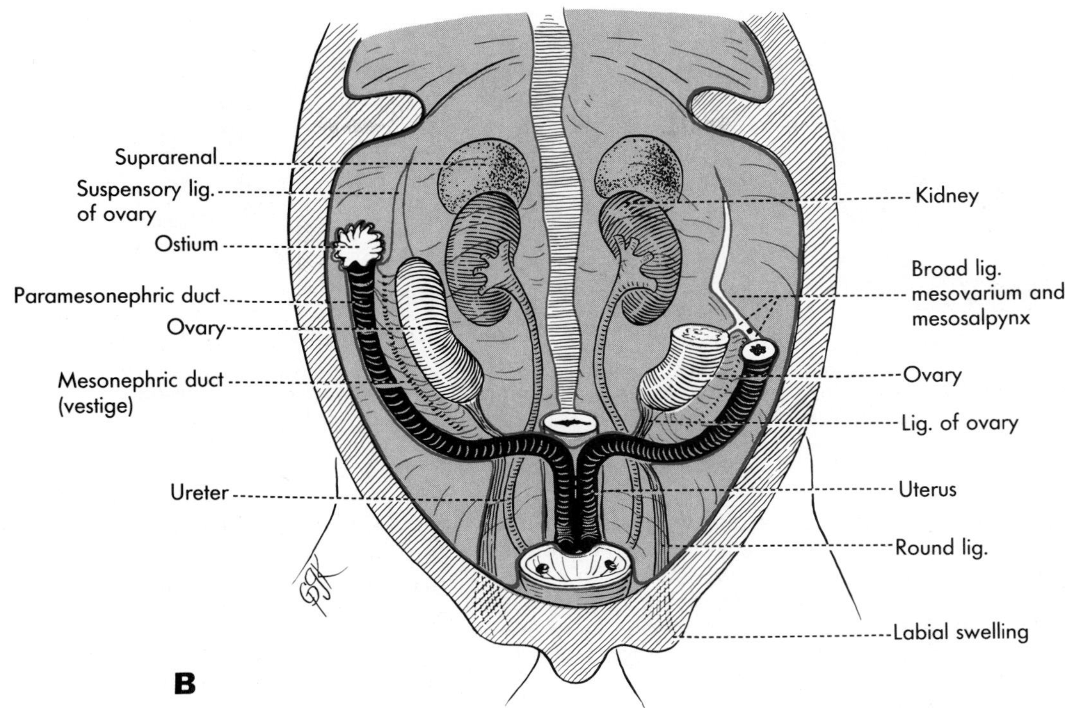


**FIG. 27-16** (continued)

their fused portions form the uterus, and from the tip of the tubes develops the uterine cervix. The mesonephric ducts open into the urogenital sinus on each side of the midline as the ejaculatory ducts; they remain unfused.

Both mesonephric and paramesonephric ducts become tethered to the genitoinguinal ligament as they cross it; this remains evident as the attachment of the gubernaculum to the cauda epididymidis as well as to the testis (a factor probably responsible for the alignment of the epididymis along the testis) and as the attachment of the ligament of the ovary and round ligament (together representing the genitoinguinal ligament) to the uterus. The attachment of the genitoinguinal ligament to the paramesonephric ducts, which persists in the female pelvis, is probably the chief factor for the retention of the ovaries in the pelvic cavity.

The mesonephric ducts regress and the paramesonephric ducts persist unless a male gonad is formed in the embryo. The testis secretes factors that suppress the paramesonephric ducts (which commence to atrophy at the point where they cross the testis) and promote the differentiation of the mesonephric ducts. The latter concerns the events of linking the seminiferous tubules to the mesonephric duct (p. 727; see Fig. 26-12) and, at the caudal end of the duct, the development of the seminal vesicles. The *seminal vesicles* appear as diverticula similar to the ureteric buds before the ducts enter the pelvic part of the urogenital sinus. The same factors promote the appearance of the prostatic buds on the wall of the pelvic urogenital sinus; the buds come to surround not only the urethra but the ejaculatory ducts as well. The suppressed paramesonephric duct system is represented in the male by the appendix testis and probably also the prostatic utricle. The vestigial mesonephric duct system in the female is represented by blind tubules in the mesentery of the ovary (paraoophoron and epoophoron) and a vestigial duct (of Gärtner) along the wall of the uterus.

#### Formation of the Vagina

As the tips of the fused paramesonephric ducts contact the urogenital sinus, they form an eminence known as the *sinus tubercle,* which evidently induces the endoderm on the dorsal wall of the pelvic part of the sinus to proliferate. This proliferation produces the **sinuvaginal bulb** or *"vaginal plate,"* a flattened cylinder of tissue composed of sinus endoderm, which, as it grows, causes the sinus tubercle to recede from the sinus lumen further and further in a craniodorsal direction, into the urorectal septum (Fig. 27-17). When the sinuvaginal bulb becomes canalized, the vagina is formed

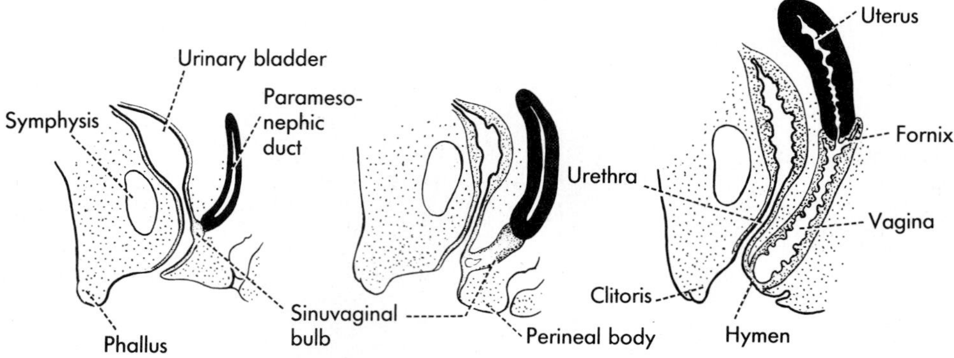


**FIG. 27-17.** Schematic sagittal sections showing the formation of the vagina. (Langman J: Medical Embryology, 3rd ed. Baltimore, Williams & Wilkins, 1975)

with a lumen. The tubercle differentiates into the cervix, and its canal links the vagina to the uterine cavity; the sinus wall separating the lumen of the vagina from that of the sinus forms the *hymen.* It is controversial whether any contributions are made by the paramesonephric ducts to the formation of the sinuvaginal bulb.

The phallic part of the urogenital sinus, caudal and anterior to the hymen, becomes the vestibule of the vagina into which the urethra also opens.

## THE RECTUM

Located in the pelvis, the rectum is the inferior segment of the large intestine, in which feces accumulate until their evacuation through the anal canal. The rectum begins at the rectosigmoid junction, in front of the 3rd sacral vertebra, as the continuation of the sigmoid colon; it ends at the anorectal junction, in front of the tip of the coccyx, as the rectum, passing through the pelvic floor, leads into the anal canal. The rectum is about 12 cm long; its upper limit cannot be reached by an index finger inserted through the anus.

The **sigmoid colon** descends into the pelvis across the left sacroiliac joint, the attachment of its mesocolon crossing the left common iliac or external iliac artery and vein at the pelvic brim and sometimes the left ureter. The sigmoid mesocolon becomes shorter as it descends in front of the sacrum and is lost at the rectosigmoid junction. Although the transition between the sigmoid colon and the rectum is a gradual one, several external features identify the **rectosigmoid junction.** At this junction, the bowel loses its mesentery and becomes closely applied to the curve of the sacrum, so that it presents an anterior concavity, the *sacral flexure,* the commencement of which may make a definite angle with the sigmoid colon (Fig. 27-18). There is a gradual broadening of the teniae coli of the sigmoid colon to form broad anterior and posterior bands that meet laterally and form a complete layer of longitudinal muscle around the rectum. With this change, there is a disappearance of the haustra, sacculations that characterize most of the colon. The appendices epiploicae, another characteristic of the colon, are lost from the sigmoid colon after it enters the pelvis.

Contrary to what is implied by its name, the rectum is not straight; it presents not only the anteroposterior sacral flexure but also three lateral curves or bends that give it a sinuous profile in the anterior view, although both the beginning and the end of the rectum are retained in the midline. The upper and lower curves are convex to the right, the middle one to the left (Fig. 27-19). The part of the rectum in the region of the middle and lower curves is called the *rectal ampulla,* because it is somewhat dilated and is especially distensible. At its inferior end, the *perineal flexure* markedly angulates the anorectal junction; the anal canal bends posteriorly at the level of the puborectalis muscle (p. 790).

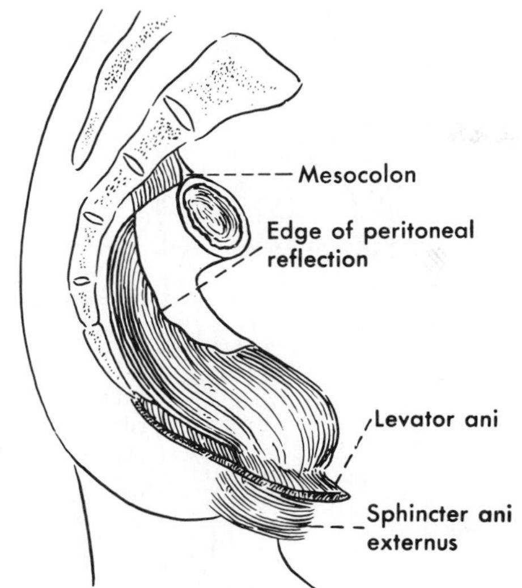


**FIG. 27-18.** Lateral view of the rectum and the lower end of the sigmoid colon. Both sacral and perineal flexures are clear.

As the rectum reaches the pelvic diaphragm, most of its longitudinal muscle continues downward along the anal canal, but a few fibers may reflect from it anteriorly and posteriorly on the upper surface of the diaphragm. In the male, the anterior fibers pass to the urethra and are known as the *recto-urethral muscle;* the slips that pass backward to the coccyx form the *rectococcygeus muscle*.

On the **interior,** there is a change in the character of the mucosal lining in the region of the rectosigmoid junction. The fluffy, rugose mucosa of the colon gives rise to the smooth mucosa of the rectum. Three permanent semilunar folds project into the lumen of the rectum from its lateral walls, located in the depths of each of the lateral curvatures. These are the *transverse rectal folds,* formed by the reduplication of the mucosa, submucosa, and circular muscle coat.

**Relations.** *Peritoneum* covers the front and sides of the rectum in its upper third, only the front in the middle third, and none of it in the lower third (Fig. 27-22). The peritoneum sweeps forward from the rectum to the base of the bladder in the male or the posterior surface of the vagina and uterus in the female, creating the *rectovesical* and *rectouterine pouches,* respectively. In these peritoneal spaces, the rectum is related to loops of small bowel or sigmoid colon.

The *pararectal fossae* of the peritoneal cavity surround the rectum laterally and are limited on the sides by the sacrogenital folds in the male and the rectouterine folds in the female.

Except for its peritoneal surfaces, the rectum is surrounded by pelvic fascia. The *rectal fascia* is a filmy layer of areolar tissue that forms a tubular sheath for the rectum loose enough to allow distention of the ampulla. The condensations of subperitoneal pelvic fascia around the middle rectal vessels form the so-called *lateral ligaments of the rectum.* The anorectal junction is connected to the sacrum by an avascular condensation of fascia called the fascia of Waldeyer. Anteriorly, the *rectovesical septum* (p. 745) separates the rectum from the prostate, seminal vesicles, and bladder or the *rectovaginal septum* from the vagina. The deferent ducts and the terminal part of the ureters are also anterior to this peritoneoperineal fascia. The posterior surface of the rectum rests on the lower sacral vertebrae, the median sacral and superior rectal vessels, the pelvic sacral foramina filled by the sacral ventral rami, and the piriformis muscles. The inferior hypogastric plexuses are related to the lateral sides of the lower part of the rectum.

## Blood Supply and Lymph Drainage

The rectum is supplied primarily by the **superior rectal artery.** This vessel, the continuation of the inferior mesenteric artery, reaches the rectum in the sigmoid mesocolon. At about the rectosigmoid junction, the superior rectal artery divides into right and left branches that run forward around the sides of the rectum in the rectal fascia, give off branches into the bowel, and disappear into its walls (Fig. 27-19A). They continue into the anal canal.

The lower part of the rectum also receives the *middle rectal arteries,* branches of the inter-

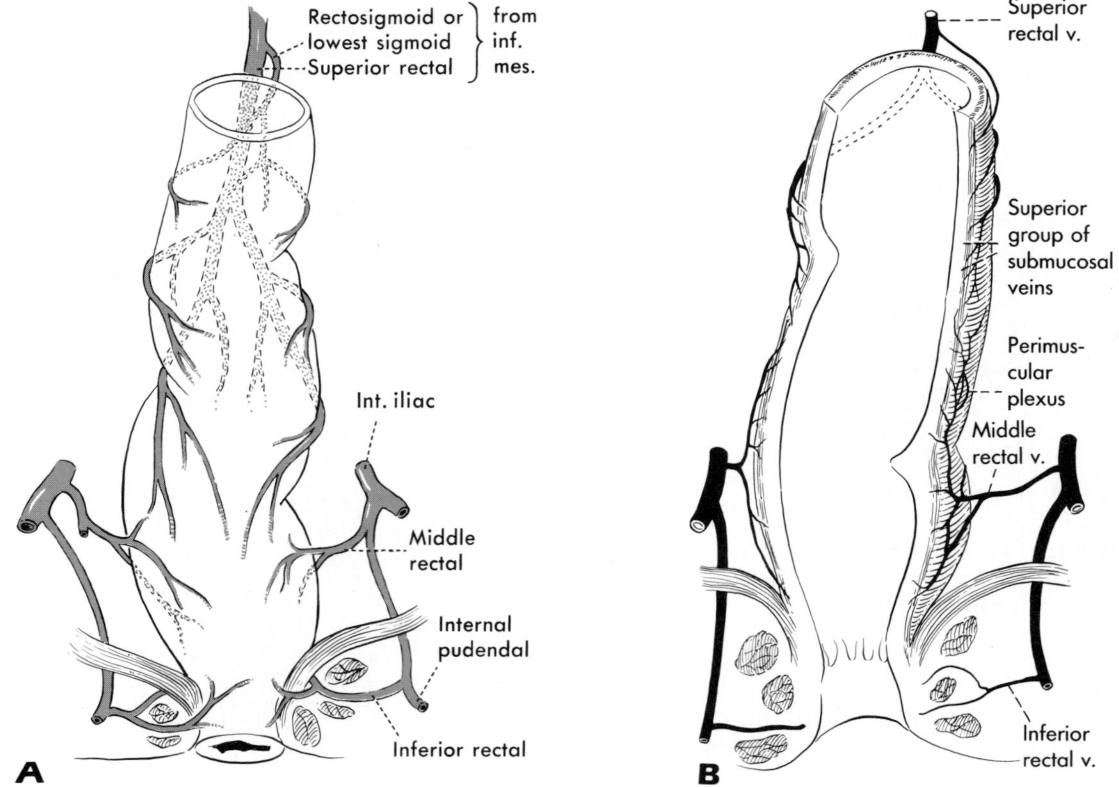


**FIG. 27-19.** Blood supply of the rectum and anal canal. **A,** arteries; **B,** veins.

nal iliac arteries. These arteries are variable in size and in their place of origin; they anastomose freely with the superior rectals. The *inferior rectal arteries* supply the anal canal rather than the rectum.

The **venous drainage** of the rectum follows the arterial supply. The chief drainage from the *rectal plexus* is into the superior rectal veins. These are at first paired, as is the lower end of the artery, but unite at about the rectosigmoid junction into a single *superior rectal vein. Middle rectal veins* drain the musculature more than they do the more abundant vessels in the mucosa (Fig. 27-19B).

The **lymphatic drainage** of the rectum is also primarily upward, paralleling the superior rectal blood vessels (Fig. 27-20). Numerous lymph nodes are associated with these lymphatics. The superior rectal lymphatics enter into the *inferior mesenteric nodes* around the origin of the inferior mesenteric artery. The lymphatics that parallel the middle rectal vessels on the upper surface of the pelvic diaphragm empty into the *internal iliac nodes.* Finally, the lymphatic plexus of the lower part of the rectum is directly continuous with that of the upper part of the anal canal and joins the lymphatics that accompany the inferior rectal and internal pudendal blood vessels below the pelvic diaphragm. Since this route follows the internal pudendal vessels, it passes with them through the buttock into the pelvis, where it ends in *internal iliac nodes* as the middle route.

Carcinomas of the rectum metastasize along the lymphatics. Those situated particularly low in the rectum or in the anal canal may metastasize by still another route: although the connections between

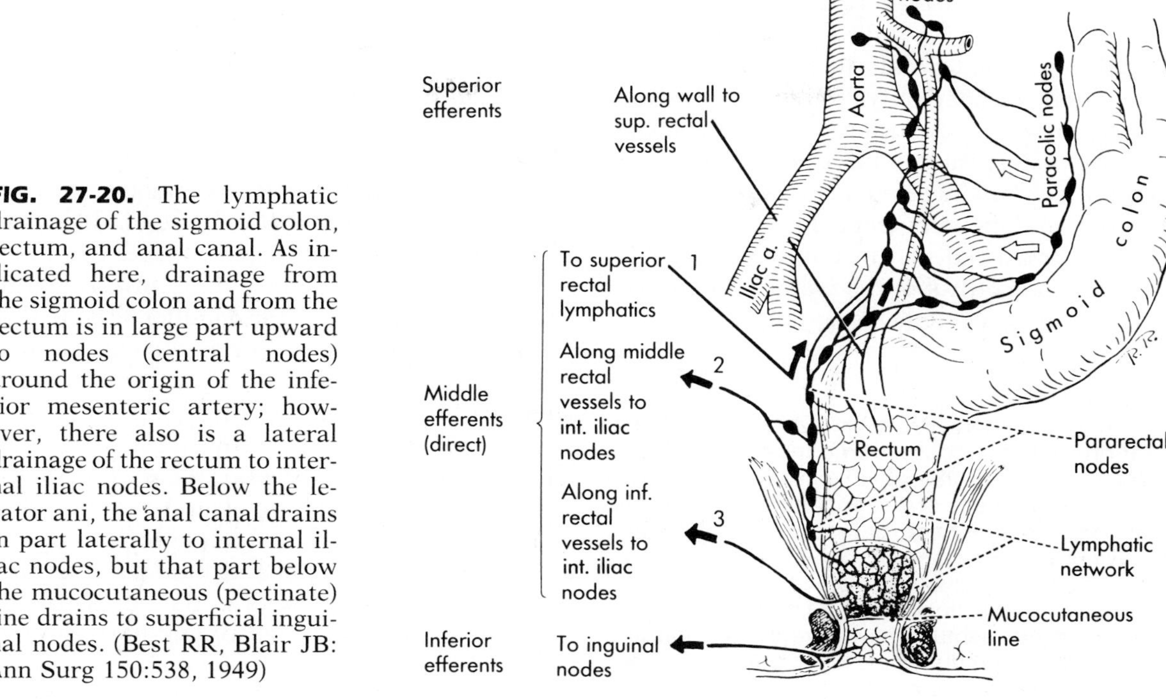


**FIG. 27-20.** The lymphatic drainage of the sigmoid colon, rectum, and anal canal. As indicated here, drainage from the sigmoid colon and from the rectum is in large part upward to nodes (central nodes) around the origin of the inferior mesenteric artery; however, there also is a lateral drainage of the rectum to internal iliac nodes. Below the levator ani, the anal canal drains in part laterally to internal iliac nodes, but that part below the mucocutaneous (pectinate) line drains to superficial inguinal nodes. (Best RR, Blair JB: Ann Surg 150:538, 1949)

the lymphatic plexus in the upper part of the anal canal and that in the lower part are few, a low-lying carcinoma may spread through these connections when other lymphatic pathways are blocked. The carcinoma may also grow across the "lymphatic divide" and involve the lymphatics of the lower part of the anal canal. The drainage of this part is into the *superficial inguinal nodes.*

## Innervation

The **rectal plexus** receives its nerves from two plexuses: the superior rectal and the middle rectal. The *superior rectal plexus* descends along the artery of the same name and may supply rectal blood vessels, but it is not important in the physiology of the rectum. The motor fibers to the rectum appear to be entirely parasympathetic and are conveyed in the *middle rectal plexus,* derived from the inferior hypogastric plexus. Moreover, the afferent supply to the rectum, both the fibers concerned with pain and those that sense the presence of feces or gas in the rectum, belong to the parasympathetic system. Thus, the rectum receives both its afferent and efferent innervation through the *pelvic splanchnic nerves* in the rectal plexus.

**Emptying of the bowel** occurs as the result of the activity of the pelvic splanchnic nerves, which increase the peristaltic activity of the rectum. This, with the help of increased abdominal pressure, moves the feces through the anal canal. Reflex restraint of a bowel movement is brought about through the activity of the voluntary muscle around the anal canal. As peristaltic activity increases the pressure in the rectum, the afferent fibers in the lower part of the rectum are stimulated and bring about a reflex contraction of the voluntary musculature. This contraction lasts just long enough to resist the increasing pressure caused by peristalsis, disappearing as the peristaltic activity abates. The initiation of the reflex activity depends upon the presence of a definite part of the rectum. If the entire rectum is removed and the colon anastomosed to the anal canal, the passage of gas or feces can be resisted only by the rather incompetent voluntary contraction of the external anal sphincter; thus, a large majority of persons with complete removal of the rectum suffer from anal incompetence. In contrast, if the lower 2 cm or so of the rectum can be retained, good control of the passage of gas and feces can be expected.

#### Anomalies

Congenital anomalies of the rectum include **imperforate anus** (discussed in connection with the anal canal in Chap. 28), which may involve an arrest in the development of the rectum as well as the anal canal, and **fistulae** between the rectum and the viscera located anterior to it. These fistulous connections are readily explained on the basis of the division of the cloaca (p. 756). Defective growth of the urorectal septum may leave the bladder and the rectum in continuity below through a *persistent cloaca.* More common are narrow apertures between the rectum and the derivatives of the urogenital sinus, which constitute *rectovesical* or *rectourethral fistulae* (Fig. 27-21A and B). These communications persist owing to gaps in the urorectal septum. A *rectovaginal fistula* can be explained by the development of the vagina in the urorectal septum with a defect in the mesoderm between the vagina and the rectum (Fig. 27-21C and D). Fistulae of the rectum are often associated with imperforate anus.

## THE URINARY BLADDER, URETERS, URETHRA, AND PROSTATE

### The Urinary Bladder

The urinary bladder (*vesica urinaria*) accumulates urine more or less continuously discharged into it by the ureters, until the bladder walls are sufficiently distended to activate the reflexes for micturition. The bladder is located behind the two pubes and the pubic symphysis; when empty, it is entirely within the pelvic cavity. As the bladder fills, it rises above the pelvic brim, coming in contact with the posterior surface of the anterior abdominal wall (Fig. 27-22); if fully distended, it may reach as high as the umbilicus. During infancy and early childhood, the dimensions of the pelvis are small and even the empty bladder is largely above the pelvic brim.

Although the distended bladder rises as a dome, the empty viscus is flat, presenting a roughly triangular superior and posterior surface, the latter called the *base,* or *fundus,* and two triangular inferolateral surfaces. The anterior angle of this tetrahedron is the *apex* with the median umbilical ligament attached to it; the inferior angle leads into the urethra and is called the *neck* of the bladder. The two posterolateral angles admit the ureters and are unnamed. The superior and inferolateral surfaces make up the *body* of the bladder.

**FIG. 27-21.** Some types of rectal fistulae associated with imperforate anus in the male (**A** and **B**) and female (**C** and **D**). (Ladd WE, Gross RE: Am J Surg 23:167, 1934)

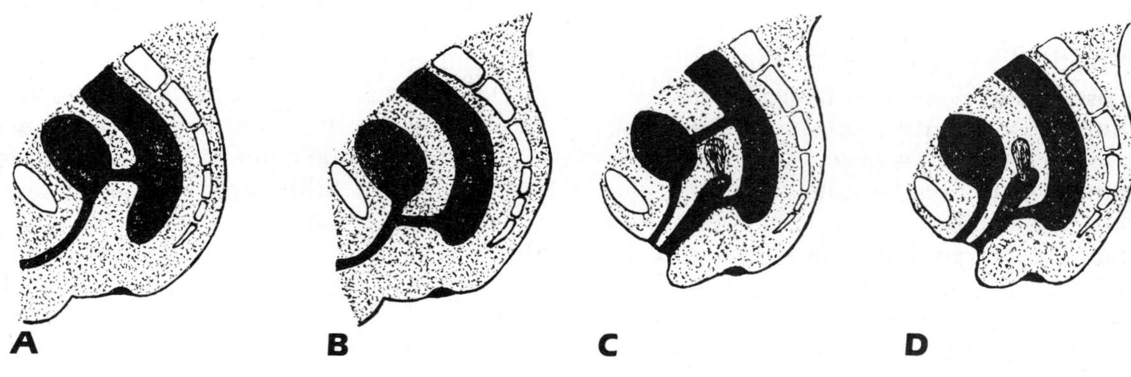

**Interior and Structure.** The pink vascular mucosa of the bladder is loosely attached to the underlying muscular wall on the interior of the body, but is closely adherent over the fundus. Therefore, in the empty contracted bladder, the mucosa of the bladder base is conspicuously smooth, contrasting with the small irregular rugae over the body.

On the interior, the base of the bladder is defined by the two *ureteric ostia* and the *internal urethral orifice*. The smooth triangular area outlined by these three openings is called the **vesical trigone** (Fig. 27-23). The *interureteric fold* is a ridge in the mucosa between the two ureteric ostia; the *uvula* is a smooth and small eminence at the inferior corner of the trigone just above the internal urethral orifice. With advancing age, the uvula becomes exaggerated, owing to the enlargement of the underlying median lobe of the prostate. The trigonal submucosa, especially that over the uvula, contains mucous glands or glands similar to those of the prostate. Hypertrophy of these glands may contribute to the factors causing urinary obstruction at the bladder neck.

The **musculature** of the bladder (*tunica muscularis*) as a whole is referred to as the *detrusor muscle*. It consists of an interlacing network of smooth-muscle bundles that run longitudinally, transversely, and obliquely and that change their course from one direction to another. Many of the longitudinal fibers, both on the interior and exterior of the bladder, continue into the urethra. Some of the outer fibers that descend to the bladder neck reflect forward as the *pubovesical muscle* and mingle with the fibrous tissue of the puboprostatic and pubovesical ligaments. Similar muscle slips that diverge posteriorly from the bladder neck blend with the sacrogenital or rectouterine ligaments and constitute the *rectovesical muscle*.

Over the trigone, an innermost lamina of muscle is formed by the longitudinal muscle of the ureters, which passes with the ureters obliquely through the bladder wall. This *trigonal muscle*, spread out between the trigonal mucosa and the detrusor, connects the ureteric ostia to each other (beneath the interureteric fold) and each ostium to the internal urethral orifice. Many of the fibers descend into the posterior wall of the urethra. The function of the muscle has been controversial. There is evidence to suggest that it maintains the obliquity of the ureters through the bladder wall and helps to close their ostia. At the same time, the contraction of the trigonal muscle contributes to opening the urethral orifice. Should the mechanism that closes the intramural part of the ureters fail, as the bladder contracts, urine will be forced retrogradely up the ureters during voiding. If there is infection in the bladder, it can reach the renal pelvis, eventually causing pyelonephritis.

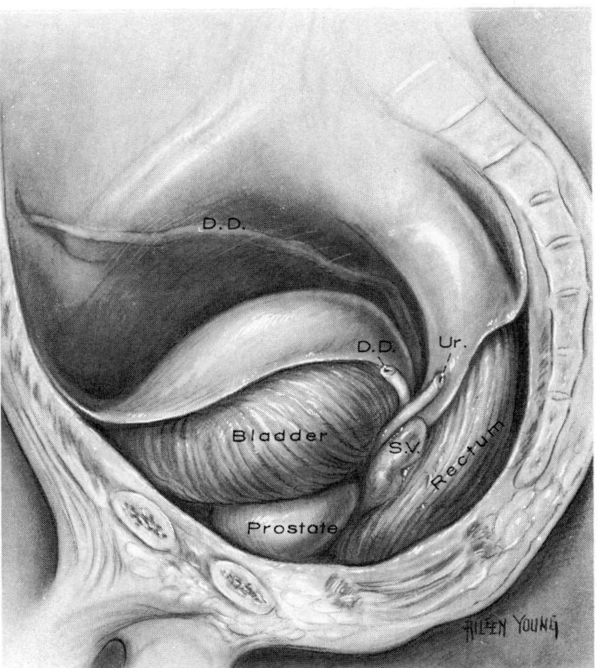


**FIG. 27-22.** Pelvic viscera in the male. The peritoneum over the bladder and rectum was cut where it was reflected upon the lateral pelvic wall. **D.D.** identifies the right and left ductus deferentes; **Ur**, the left ureter; and **S.V.**, the left seminal vesicle. The ureter has here been displaced somewhat forward from where it was running retroperitoneally on the lateral pelvic wall, and the lower end of the sigmoid colon is so close to the sacrum that the flexure between it and the rectum is almost obliterated.

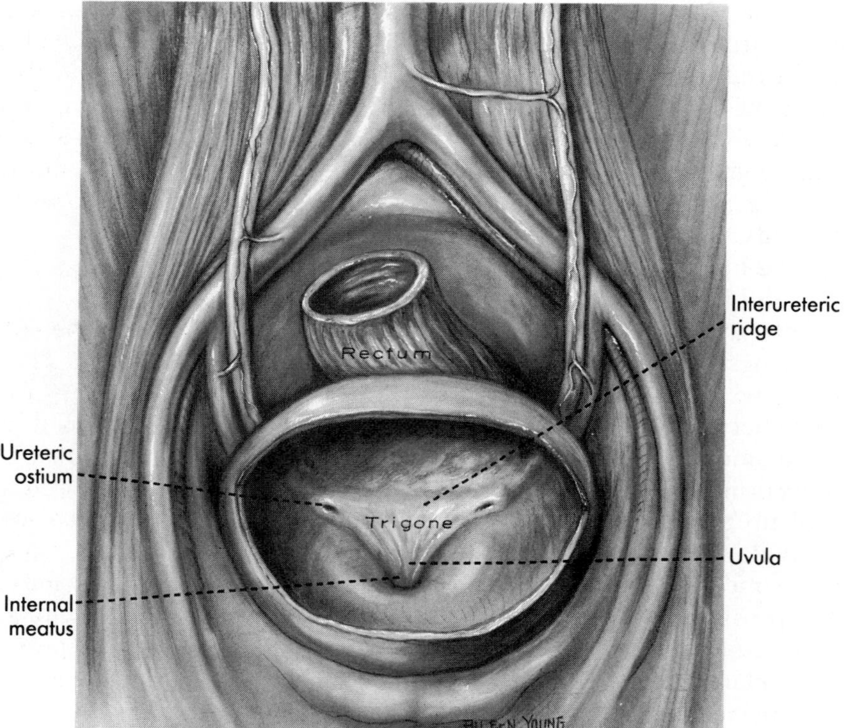


**FIG. 27-23.** The trigone of the urinary bladder, lying between the openings of the two ureters and that of the urethra.

Unlike the intestine, the bladder has no continuous circular muscle. Furthermore, several surveys have verified that there is no anatomically demonstrable *vesical sphincter* at the junction of the bladder and the urethra. The *"internal sphincter of the bladder"* is a functional entity that prevents urine from entering the urethra and the ejaculate from entering the bladder, but the mechanisms responsible for its maintenance are not understood completely.

**Relations.** Peritoneum covers the bladder on its superior surface and is continuous anteriorly with the median and medial umbilical folds and with the parietal peritoneum of the supravesical fossae (Fig. 27-22). Posteriorly, the peritoneal continuities are different in the two sexes. *In the male,* the peritoneum descends onto the base of the bladder, lining the rectovesical pouch, the peritoneal recess that intervenes between the bladder and the rectum (Fig. 27-22); posterolaterally, it continues as the sacrogenital folds. *In the female,* the peritoneum reflects onto the uterus without covering the base and lines the shallow vesicouterine pouch (Fig. 27-36).

In the male, subjacent to the perineum, the ductus deferentes descend in direct contact with the base of the bladder, having crossed to the medial side of the ureters (Fig. 27-29). The bladder neck rests on the upper surface of the prostate. Posteriorly, between the prostate and the deferent ducts, the seminal vesicles rest against the base of the bladder.

In the female, the base of the bladder is bound to the supravaginal part of the cervix and the anterior wall of the vagina by the fusion of the fascial sheaths of the organs (Fig. 27-36). Inferolaterally, the empty bladder rests on the pubic bones and the pelvic diaphragm with the areolar tissue and ligaments of the pelvic fascia intervening.

The neck and base of the bladder are relatively firmly fixed by their fascias and ligaments and move up and down with the contraction of the pelvic diaphragm. Their relations are not altered by distention of the bladder. As the body of the bladder distends, its superior surface rises, elevating the peritoneum and bringing its inferolateral surface more and more in contact with fascia on the anterior abdominal wall (Fig. 27-24).

**Ligaments and Fascias.** The neck of the bladder is connected to the pubis on the upper surface of the pelvic diaphragm by the puboprostatic ligaments in the male and the pubovesical ligaments in the female, and to the lateral wall by the lateral ligaments of the bladder (Fig. 27-6). The latter blend posteriorly with the tissue of the sacrogenital folds of the male and the lateral cervical ligaments of the female. All these ligaments contain smooth muscle reflected into them from the bladder neck. Vessels, nerves, and the ureters reach the bladder embedded in the lateral and posterior ligaments. Posterolaterally, these ligaments form the floor of the perivesical spaces, which are roofed over by the peritoneum and contain loose connective tissue continuous with these ligaments. Anteriorly, the retropubic space is similarly limited below by the puboprostatic and pubovesical ligaments (Fig. 27-24).

The *median umbilical ligament,* attached to the apex of the bladder, is the remnant of the embryonic urachus; at its base, the ligament usually retains a lumen communicating with the bladder. Rarely, urine will be discharged from the umbilicus because the patency of the urachus persists along its entire length.

Below the floor of the rectovesical pouch, the bladder base is separated from the rectal fascia by the rectovesical septum (Fig. 27-25). The connective tissue space anterior to the bladder is divided into two compartments by a similar lamina of fascia called the *umbilical prevesical fascia,* which connects the two medial umbilical ligaments to the median umbilical ligament and separates the immediate perivesical space from the retropubic space (Fig. 27-24). This space provides an extraperitoneal approach not only to the bladder and prostate, as mentioned earlier (p. 591), but also to the pregnant uterus, which, like the distended bladder, elevates the peritoneum and contacts the anterior abdominal wall above the bladder. In a cesarean section, the uterus is opened through the retropubic space without incising the peritoneum.

## The Ureters

The ureters have been described from the kidney to the pelvic brim in Chapter 25. They cross the pelvic brim at approximately the level of the bifurcation of the common iliac artery. The left ureter is related to the base of the sigmoid colon. In the female, both ureters may also be closely related to the ovarian vessels, for these often cross the pelvic brim just lateral to the ureters. In surgical procedures on the ovaries, the ureter is liable to damage in securing the ovarian vessels at the pelvic brim, one of the common sites of iatrogenic injury to the ureter in the female.

In both sexes, after it crosses the pelvic brim, the ureter lies at first immediately deep to the peritoneum, medial to the internal iliac artery, passing downward and forward along the lateral pelvic wall. In the female, it tends to run just behind the ovary, forming the posterior boundary of a shallow fossa, against which the ovary usually lies. As they reach the level of the peritoneal floor of the pelvis, the two ureters converge toward each other and enter the connective tissue of the sacrogenital folds or the rectouterine folds of the female, in which they continue medially with the vessels and nerves contained in these ligaments.

As it approaches the bladder, the ureter runs more deeply and, in the male, comes in contact with the ductus deferens. The ureter passes below the ductus and then in front of the seminal vesicles (Fig. 27-29). It enters the posterolateral aspect of the bladder and courses obliquely through its wall.

In numerous gynecologic procedures, the relations of the ureter as it approaches the bladder are particularly important in the female pelvis. In the lateral cervical ligaments, the ureter is closely associated with the blood

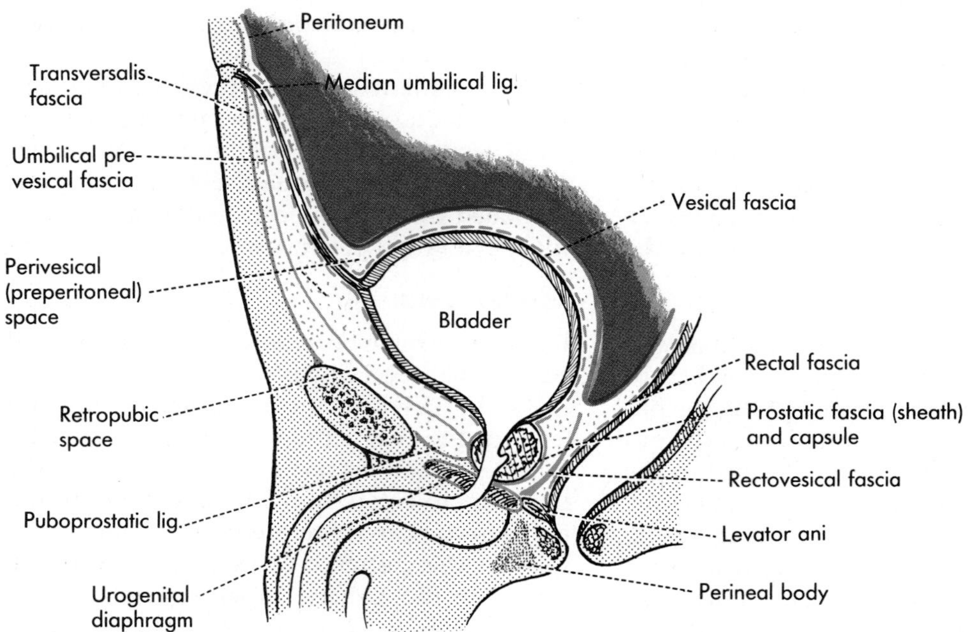


**FIG. 27-24.** Fascial spaces and laminae associated with the bladder, prostate, and rectum. The bladder is shown distended.

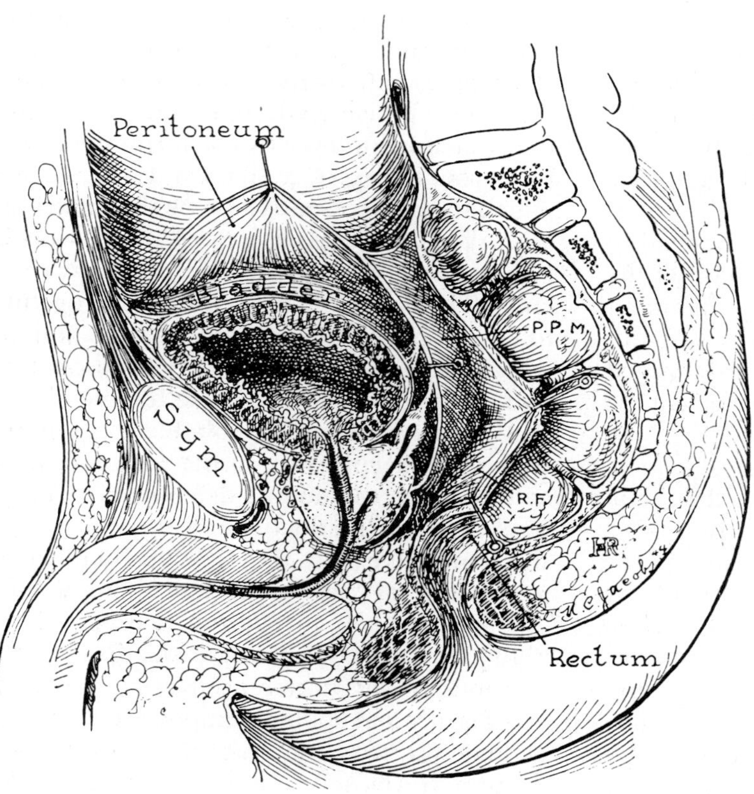


**FIG. 27-25.** A dissection of the rectovesical septum (**P.P.M.,** or peritoneoperineal membrane). **R.F.** is the rectal fascia, here retracted backward with the rectum. (Tobin CE, Benjamin JA: Surg Gynecol Obstet 80:373, 1945)

vessels and nerves of the uterus and vagina. It passes forward and medially at the base of the broad ligament and, in so doing, is crossed above and in front by the uterine artery as this vessel runs a more direct transverse course toward the uterus. The crossing of the ureter by the uterine artery occurs about 1.5 cm lateral to the uterus, but can vary markedly when pathologic conditions have distorted relationships. This crossing is oblique; the two structures are in contact for approximately a centimeter or more. This is another common site of surgical injury to the ureter. The ureter may be injured by a clamp, or even ligated and divided, when the uterine vessels are clamped to control uterine bleeding, or it may be ligated and divided in the process of removing the uterus (hysterectomy).

After crossing behind and below the uterine artery, the ureter continues its course forward and medially and passes to the front of the vagina surrounded by the upper parts of the vesical nerve plexus. It enters the posterolateral aspect of the bladder just as it does in the male.

**Ectopic Ureter.** Interference with the normal process of the absorption of the mesonephric duct into the dorsal wall of the urogenital sinus may displace one or both ureteric ostia from their normal location (p. 756). This condition is known as ectopic ureter.

The most common ectopia is one in which the ureter opens distal to the bladder into the prostatic urethra in the male or into the urethra or vaginal vestibule in the female. Openings into the ductus deferens or seminal vesicle or the vagina also occur sometimes. These ectopias can be explained by the embryonic derivation of these structures either from the mesonephric ducts or from portions of the urogenital sinus. Rarely, an ectopic ureter opens into the rectum, the explanation for which is not so obvious.

## The Urethra

The urethra connects the bladder to the exterior. It commences at the internal urethral orifice of the bladder and terminates at the external orifice located at the tip of the penis or in the vestibule of the vagina. In the female, the urethra transmits only urine; in the male, it is the final passage for both urine and semen.

The **female urethra** is a simple tube, measuring barely 4 cm in length. It is closely fused to the anterior vaginal wall, runs with the vagina through the urogenital hiatus, and opens just anterior to the vagina into the vestibule (see Fig. 28-18). Its walls are simple, containing much elastic tissue as well as smooth muscle continuous with that of the bladder and voluntary muscle derived from the urogenital diaphragm.

The *urogenital diaphragm,* described with the perineum (Chap. 28), consists of muscles that span the triangular space between the conjoint rami of the pubis and ischium and includes the sphincter urethrae.

The **male urethra** is approximately 20 cm long and is divisible into three portions: the part that leaves the bladder, called the *prostatic part,* because it is surrounded by the prostate; a part that passes through the urogenital diaphragm, called the *pars membranacea* or membranous urethra, because it is between the superior and inferior connective tissue membranes of the urogenital diaphragm; and the third, most distal part, enclosed by the corpus spongiosum of the penis and therefore called the *pars spongiosa.* The membranous and spongy parts of the urethra are located in the perineum and are described in Chapter 28. Only the prostatic urethra is in the pelvis.

The average length of the **prostatic urethra** is about 2.5 cm, and its lumen is spindle shaped (Fig. 27-26). Its walls are formed by the prostate itself, for the ducts and glands that compose the prostate are outgrowths from the urethra (p. 756), and the muscular fibrous tissue of the prostate is part of the original urethral wall.

On the posterior wall of the prostatic urethra is a longitudinal ridge, the *urethral crest,* raised up largely by the continuation of the trigonal muscle into the urethra. The crest is continuous above with the uvula of the bladder (Fig. 27-26). A similar crest exists in the female urethra. In the male, the crest widens to form a smooth eminence, the *colliculus*

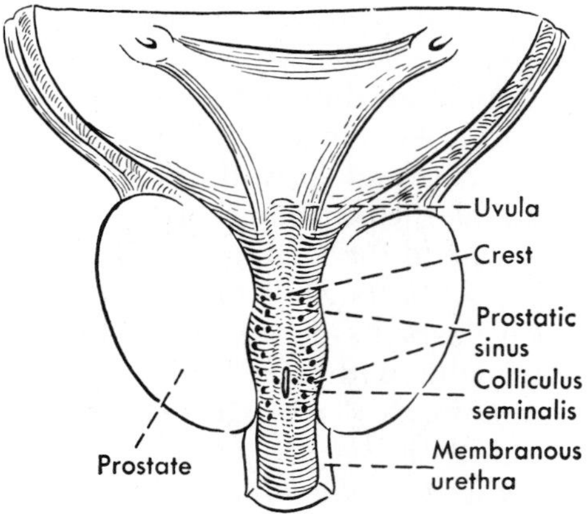


**FIG. 27-26.** The posterior wall of the prostatic urethra. The relatively large midline opening on the colliculus seminalis is the utriculus; on each side of it is the small opening of an ejaculatory duct. The similarly small openings in the prostatic sinuses are those of prostatic ducts.

*seminalis,* which urologists call the *verumontanum.* The colliculus is 2 mm to 4 mm long, and its center lies about two-thirds of the distance down the prostatic urethra. In the midline and usually just distal to its center, the *prostatic utricle* opens on the colliculus by a rounded or slitlike aperture. The utricle is believed by some to be the remnant of the fused paramesonephric ducts, which form the uterus, and is sometimes called the *uterus masculinus*. It is more likely that the utricle is a homologue of the vagina and should rather be called the *vagina masculina*.

The *ejaculatory ducts* also open on the colliculus on each side of the utricle. Their openings are tiny and often cannot be seen in the cadaver, but in living persons it is possible to catheterize them and inject radiopaque contrast medium into the genital ducts (Fig. 27-30).

On each side of the colliculus and the urethral crest is a sulcus called the *prostatic sinus*. In it, the 12 to 20 tiny orifices are the openings of the prostatic ducts (Fig. 27-26).

## The Prostate

The prostate is an encapsulated gland developed only in the male around the urethral lumen between the neck of the bladder and the pelvic floor. It is developmentally and actually the thickened wall of this part of the urethra. Its firm consistency is due to the significant amount of smooth muscle present in its stroma, which is continuous with the musculature of the bladder. Secretions of the glandular follicles of the prostate account for much of the seminal plasma.

The prostate is deeply placed, in contact with the gutter-shaped pelvic floor, directly behind the lower border of the symphysis pubis (Figs. 27-22, 27-24, and 27-25). It is shaped like an asymmetrical cone: its **apex** points inferiorly, resting on the pubococcygeus, and through the urogenital hiatus is in contact with the superior fascia of the urogenital diaphragm. The oval **base,** fused to the bladder neck, measures approximately 2 and 4 cm along its anteroposterior and lateral diameters, respectively. The vertical extent of the posterior surface is greater than that of the anterior surface, and the gland bulges posterolaterally, creating a shallow median sulcus on the posterior surface, which can be palpated by rectal examination. The two inferolateral surfaces are broad posteriorly and converge on the narrow **isthmus** that connects the two sides anteriorly in front of the urethra and may consist of only fibromuscular tissue.

**Lobes.** Behind the urethra, the prostate is traversed obliquely by the two ejaculatory ducts, which enter its posterosuperior margin behind the bladder and slope medially toward their orifices on the colliculus seminalis (Figs. 27-26 and 27-27). The two ducts, with the urethra, define a more or less cone-shaped core of the gland, which is regarded as its **median lobe.** The base of the median lobe underlies the uvula of the bladder, and the blind prostatic utricle extends into it between the ejaculatory ducts.

Although no definite lobulation is evident on either the interior or the exterior of the prostate, the isthmus is designated as the **anterior lobe** and the two bulging posterolateral portions as the **lateral lobes.** The tissue joining the latter in the posteromedian sulcus is the **posterior lobe.**

**Prostatic Enlargement.** Benign hypertrophy (*prostatic hyperplasia*), common after middle age, affects preferentially the median lobe. The lobe may bulge into the prostatic urethra and obstruct it, or its base may bulge into the bladder and, acting as a ball valve, be forced into the internal urethral orifice by the vesical pressure developed for emptying the bladder and obstruct the flow of urine. Hyperplasia of the uvular submucosal glands contributes to this obstruction. The benign tumor may be pared away as it bulges into the prostatic urethra through a cytoscope (*transurethral resection*), or it may be shelled out after the urethra has been opened by transecting the isthmus of the prostate (*retropubic prostatectomy*). The prostate may also be approached surgically through the bladder or from the perineum. Malignant change (carcinoma of the prostate) most often affects the posterior lobe, and its treatment usually combines radiation and hormone therapy with some type of resection of the cancerous tissue.

**Capsule and Ligaments.** Outside its firm, white, shiny capsule, continuous with the fibromuscular trabeculae of the gland, the prostate is surrounded by a rather dense fascial sheath, with the rich prostatic plexus embedded in it. This prostatic fascia also contains nerves and lymphatics and is continuous above the level of the pelvic floor with the puboprostatic ligaments, the lateral ligaments of the bladder, and the connective tissue of the sacrogenital folds (Fig. 27-6). Directly behind the prostate, the rectovesical fascia separates the prostatic sheath from the rectal fascia (Fig. 27-25).

The fibers of the pubococcygeus insert into the prostatic fascia and the prostatic capsule, and these fibers are called the *levator prostatae muscle*.

**Internal Structure.** Inside the capsule, the major part of the prostate consists of numerous glands that have grown out from the posterolateral aspect of the urethra and expanded to form the glandular substance that is mixed with fibromuscular tissue of the urethral wall. An internal and peripheral zone have been defined histologically in the gland. It is the glands of the peripheral zone posteriorly that are prone to carcinomatous transformation and those of the internal zone that are prone to benign hypertrophy. The ducts of the glands combine to form the 12 to 20 prostatic ducts opening into the prostatic sinus of the urethra.

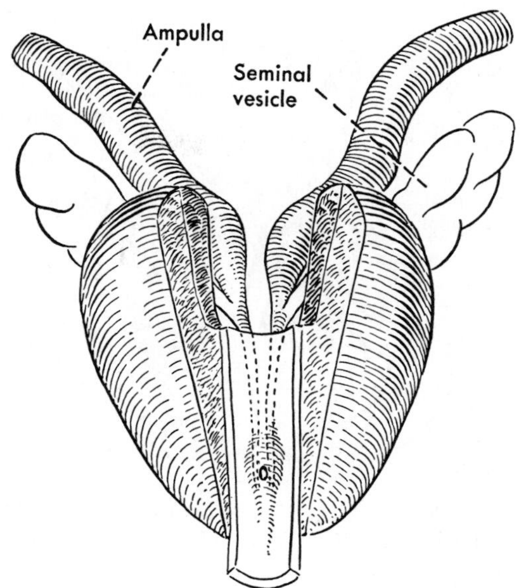


**FIG. 27-27.** The ejaculatory ducts seen from the front. A midsegment of the upper part of the prostate has been completely removed to the level at which the ejaculatory ducts enter the gland, and a corresponding segment of the anterior urethral wall and the associated part of the prostate has been removed below. The intraglandular portions of the duct are indicated by broken lines.

## Blood Supply and Lymph Drainage

The **arteries of the urinary bladder** vary in number and have been variously named. Usually, two or three arteries arise from the umbilical artery as it runs anterolateral to the apex of the bladder and are distributed to the apex and the upper part of the body; these are the *superior vesical arteries* (Fig. 27-28). Close to its origin, the umbilical artery usually gives off a vessel, the *artery of the ductus deferens* (deferential artery), which has also been called the *middle vesical artery*. It is distributed to the body of the bladder as well as to the ductus and seminal vesicles. In the female, the middle vesical artery is a branch of the uterine artery.

There is, in addition, an *inferior vesical artery* on each side. This is inconstant in its ori-

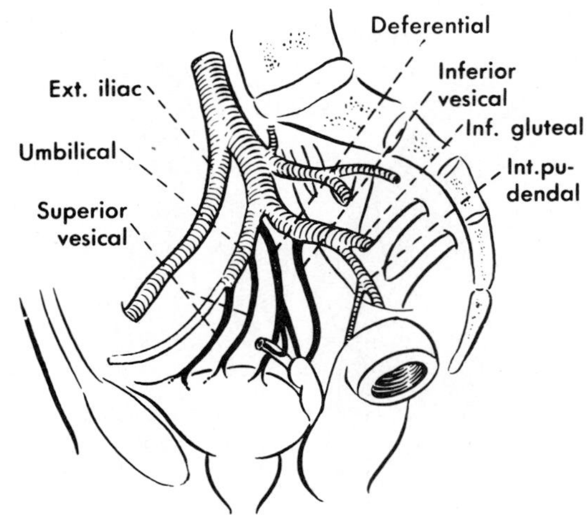


**FIG. 27-28.** Common origins of the chief arteries (black) that supply the bladder. (Redrawn from Braithwaite JL: Br J Urol 24:64, 1952)

gin, for it may arise from many different parts of the internal iliac system. It runs through the connective tissue close to the floor of the pelvis and is distributed to an inferior part of the bladder, including the vesical neck, and also sends branches to the prostate. In the female, the inferior vesical artery is a branch of the vaginal artery.

In the pelvis, the **ureters** receive one or more arterial branches. There may be a branch from the common iliac artery, as the ureter crosses the pelvic brim, or from the internal iliac artery within the pelvis. A rather regular branch to the ureter is given off by the artery to the ductus deferens or by a vesical artery. This branch enters the ureter close to the bladder.

The **arteries of the prostate** are derived from the inferior vesical, middle rectal, and internal pudendal arteries.

The veins of the bladder, urethra, and prostate form fairly dense venous plexuses in the fascial sheaths. The **prostatic venous plexus** lies lateral to the gland and receives the dorsal vein of the penis. Superiorly, the prostatic plexus joins the **vesical venous plexus** around the neck of the bladder. The venous drainage from the plexuses is, in general, laterally and posteriorly along the inferior vesical arteries into the internal iliac veins. There is usually also some venous drainage anterolaterally into the lower ends of the external iliac veins. The internal iliac veins communicate freely with the veins draining the vertebral column and the coxal bone, explaining the metastases of prostatic carcinoma to the coxal bone and the rest of the vertebral column as mentioned earlier (p. 747).

The female lacks a prostatic plexus, and the dorsal vein of the clitoris drains into the vesical plexus. This plexus drains, as in the male, into the internal iliac vein directly, but is also united to the vaginal plexus; therefore, it is drained, in part, by the uterine veins. Pelvic cancer in the female spreads primarily along lymphatics.

The **lymphatic drainage** of much **of the bladder** is laterally and upward across the pelvic brim into *external* and *common iliac nodes*. Some of the drainage is more posterior, into *internal iliac nodes* situated along the branches of these vessels, and these, in turn, drain into the common iliac nodes. In the female, some of the lymphatics of the bladder unite with vaginal and uterine lymphatics and drain to the internal iliac and sacral nodes, but others drain laterally into the iliac nodes as in the male.

The **lymphatic drainage of the prostate** is similar to that of the bladder, with which its lymphatics communicate. The prostatic lymphatics begin largely or entirely in the capsule of the prostate, not in the prostatic tissue. This may explain why metastases from the prostate are more likely to pass along veins than along lymphatics.

## Innervation

The **vesical** and **prostatic nerve plexuses** are forward extensions of the inferior hypogastric plexus. In the female, the *vesical plexus* is less distinct, because it merges with the larger, *uterovaginal plexus* (p. 787).

The parasympathetic fibers in the vesical plexus are known to be involved in reflex emptying of the bladder, but the role and even the distribution of the sympathetic fibers are disputed. It has been both claimed and denied that sympathetic fibers end only on the blood vessels and the trigonal muscle rather than on the detrusor muscle. However, certain experimental evidence, including pharmacologic data, seems to indicate that the old and generally discarded concept that sympathetic innervation relaxes the detrusor may be true after all.

The prostatic musculature, as that of other parts of the genital duct system, is apparently innervated by sympathetic and not by parasympathetic efferents. Sympathetic excitation produces contraction of the smooth muscle of the gland, emptying the prostatic secretions into the urethra during ejaculation.

The **deferential plexus** around the ductus deferens is an offshoot of the vesical plexus. It leaves the pelvis along the deferent duct.

Some fibers of the prostatic plexus accompany the urethra through the urogenital hiatus into the perineum to the erectile bodies of the penis; these are the cavernous nerves. Similar nerves to the clitoris and bulb of the vestibule are derived from the vesical plexus in the female.

**Visceral afferents** from the bladder and the prostate are believed to reach the sacral spinal cord segments through the pelvic splanchnic nerves.

**Neural Control of Micturition.** Urine will flow from the bladder through the urethra when the vesical pressure is higher than urethral resistance. Contraction of the detrusor muscle is the usual way in which vesical pressure is raised. Although a few workers insist that the once widely held concept of voluntary control over the autonomic system innervating the bladder is valid, much evidence indicates that it is the somatic nerves that supply the voluntary muscles related to the bladder that start and stop micturition. The part of the pelvic diaphragm that directly supports the vesical neck (pubococcygeus) has been shown to relax so that the vesical neck moves downward. This apparently decreases the resistance in the urethra and at the same time reflexly increases the activity of the detrusor, allowing urine to flow freely if the vesical pressure is high enough. If it is not, it can be increased by voluntary contraction of the abdominal muscles and the diaphragm. Cessation of urination is then brought about by contraction of the pubococcygeus. The voluntary muscle around the urethra (sphincter urethrae) is thought to relax and contract with the pelvic diaphragm, but is not generally considered to be essential to voluntary control.

In support of this concept, it has been claimed that malfunction of the pelvic diaphragm is a cause of loss of urinary control. Spasm of the diaphragm, for instance, has been said to be responsible for the urinary retention that complicates rectal operations. Similarly, it has been shown in women that strengthening the muscles of the pelvic outlet by exercise will often cure urinary incontinence, indicating that it is voluntary rather than involuntary muscle that primarily controls the emptying of the bladder.

## THE MALE GENITAL DUCTS

The ductuli efferentes and the epididymis and its duct have been described in Chapter 26, and the ductus deferens was traced as far as the deep inguinal ring. The male genital ducts in the pelvis include the pelvic portion of the deferent ducts, the seminal vesicles, and the ejaculatory ducts.

**The Ductus Deferens.** The ductus, more commonly known by clinicians as the *vas deferens,* enters the abdominal cavity through the deep inguinal ring. At this ring, it leaves the testicular vessels by turning medially, soon passing the brim of the pelvis and running downward and backward along the lateral pelvic wall (Fig. 27-22). In so doing, it enters the lateral ligaments of the bladder, crosses above and medial to the ureter, and turns medially to converge with the other ductus on the posterior aspect of the bladder (Figs. 27-29 and 27-30). It is retroperitoneal throughout its course.

Beyond the point of crossing the ureter, the ductus deferens enlarges to form the **ampulla.** The ampullae of the two ducts come close together in the midline, then each suddenly narrows, before uniting with the duct of the semi-

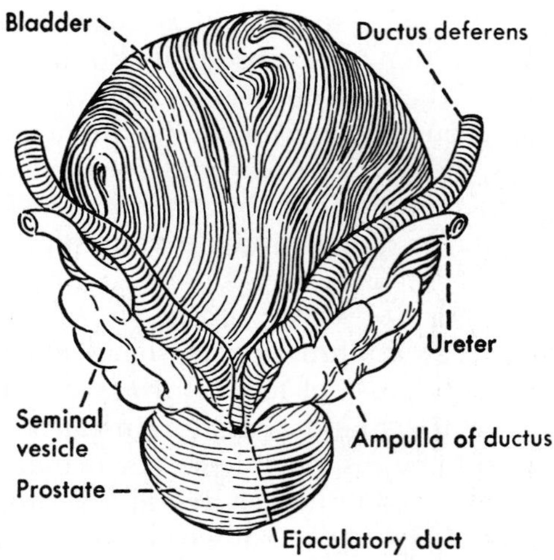


**FIG. 27-29.** Posterior view of the bladder, prostate, and seminal vesicles.

nal vesicle to form the ejaculatory duct (Figs. 27-27 and 27-29).

**The Seminal Vesicles.** Each seminal vesicle is a nearly 15-cm long blind tube, folded upon itself. The coils of the tube are held together by connective tissue, and therefore each seminal vesicle appears as a compact, elongated, oval body less than a third of the length of its tube. Its coils give the vesicle a lobulated appearance (Figs. 27-27, 27-29, and 27-30). The distal end of the duct of the seminal vesicle becomes narrow and, together with the ductus deferens, forms the ejaculatory duct. The secretions of the epithelium of the vesicles are discharged into the ejaculate by the contraction of the muscular walls of the vesicle. This adds fructose to the seminal plasma, necessary for maintaining the motility of the spermatozoa. Spermatozoa are stored in the ampulla, the ductus, and the epididymis, but not in the seminal vesicle.

The seminal vesicles lie below and lateral to the ampullae of the ductus, resting against the base of the bladder (Fig. 27-29). Posteriorly, both they and the ampullae of the ductus are separated from the rectum by the rectovesical fascia (Fig. 27-4).

**The Ejaculatory Ducts.** The thin-walled ejaculatory ducts are about 2 cm long. They taper from their origins to their termination and are about 0.5 mm in diameter when they end. They converge toward each other as they run through the prostate between the median and lateral lobes, so that they open close together on the colliculus seminalis just lateral to the utriculus (Fig. 27-27).

**Blood Vessels, Lymphatics, and Nerves.** Close to its lower end, the ductus deferens receives the *deferential artery*, which is derived from the umbilical artery. It follows the duct to the testis. The seminal vesicles receive their blood supply from the deferential and inferior vesical arteries and often receive also branches from the middle rectal arteries.

The **veins** of the seminal vesicles join the vesical plexus of veins, and the **lymphatics** join those of the prostate and bladder.

The **innervation** of the smooth muscle of the seminal vesicles, ductus deferentes, and ejaculatory ducts, like that of the prostate, is apparently exclusively by sympathetic fibers. Ejaculation is a function of the sympathetic nervous system. Unlike the testis, which receives its nerve supply from a higher level along the testicular artery (p. 725), the nerve fibers in the *deferential plexus* are derived from the pelvic plexus and the superior hypogastric plexus.

Lumbar sympathectomy may interfere with ejaculation, either because it eliminates the efferent input into the plexus or because it affects the functional vesical sphincter, permitting the semen to be discharged from the prostatic urethra into the bladder rather than into the membranous and spongy urethra (retrograde ejaculation).

## THE OVARY AND THE DERIVATIVES OF THE FEMALE GENITAL DUCTS

The ovary is the female gonad: its function is the production of ova and the secretion of estrogen and progesterone. Through these hormones, the ovary influences the cyclic maturation and discharge of the ova and the development and maintenance of the second-

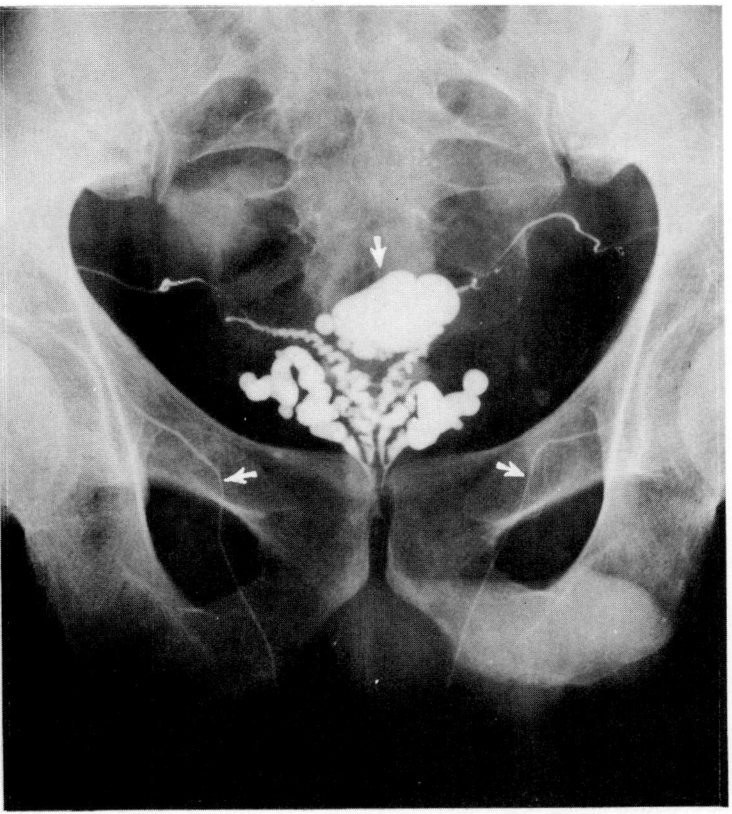

**FIG. 27-30.** The genital duct system in the male. Both ductus deferentes were injected with radiopaque material and show particularly well where they cross the pubes (lateral arrows) and their pelvic course. The larger convoluted lumina, as the two ducts converge, are the ampullary parts of the ducts; the still larger convoluted structures lateral to these are the seminal vesicles. On the reader's right, the junction of the ductus deferens and the seminal vesicle, and the origin of the ejaculatory duct at this junction, can be seen. Some of the injected medium has entered the bladder (middle arrow). (Braasch WF, Emmett JL: Clinical Urography. Philadelphia, WB Saunders, 1951)

ary female sex organs and secondary somatic sex characteristics of the female phenotype. Furthermore, these hormones establish the requirements for implantation of the embryo and maintain those necessary for retention of the embryo and fetus in the uterine cavity until the time of birth. As already recounted in Chapter 23, the ova are transported to the uterus by the uterine tubes, which communicate with the peritoneal cavity. The uterine cavity communicates, in turn, with the vagina through the cervical canal. The uterus, its tubes, and the ovaries are associated with the *broad ligament,* which, acting as a mesentery, suspends the uterine tubes and the ovaries from the pelvic floor (Fig. 27-31).

## The Ovary

Each ovary (*ovarium*) is a firm, almond-shaped organ, about 3 cm long, 1.5 cm wide, and 1 cm thick. Its surface, devoid of peritoneum, is smooth until puberty; thereafter, the scars left by the degenerating corpora lutea (*vide infra*) render it somewhat irregular.

The ovary is suspended from the posterior lamina of the broad ligament by its own mesentery, the *mesovarium,* and is embraced anteriorly and laterally by the uterine tube (Fig. 27-32). It is quite mobile, and its position in the pelvic portion of the peritoneal cavity may vary. However, in the young, nulliparous woman, the ovary lies against the lateral pelvic wall (Fig. 27-36), held there by its suspensory ligament (Fig. 27-31). It has a lateral and medial surface, a mesovarian and a free border, and a tubal and uterine extremity, or pole.

The **lateral surface** lies against the parietal peritoneum of the pelvic wall in the *ovarian fossa,* a depression bordered by the internal iliac artery and the ureter. In the fossa, the ovary is related to the obturator nerve and

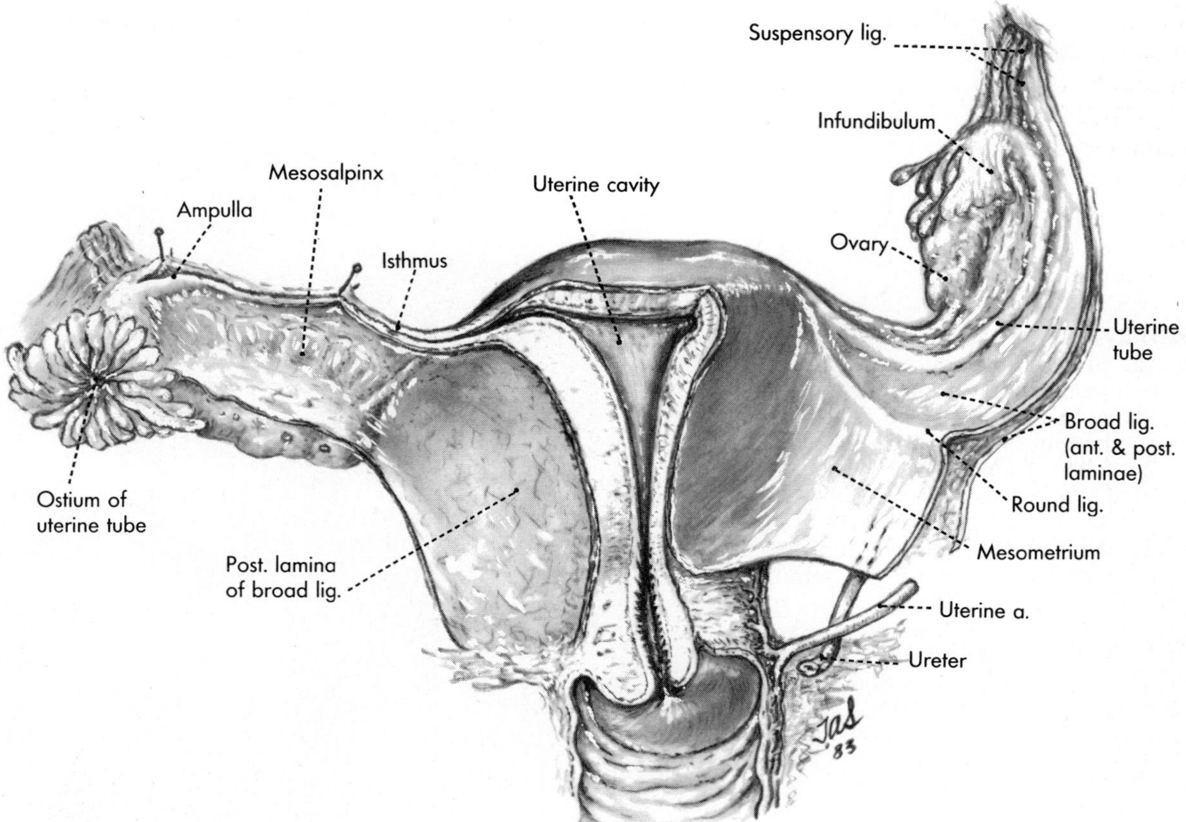


**FIG. 27-31.** The uterus and its adnexa viewed from the front. The anterior leaf of the right broad ligament has been resected with the anterior wall of the uterus and uterine tube. Adapted from Blandau RJ: The female reproductive system. In Weiss L (ed): Histology: Cell and Tissue Biology, 5th ed. New York, Elsevier Biomedical, 1983

**FIG. 27-32.** The right ovary seen from behind.

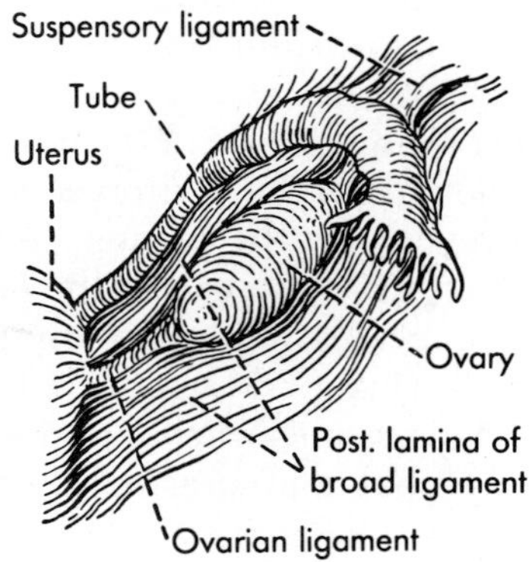


vessels. Its **medial surface** faces toward the pararectal fossae and the rectouterine pouch and is in contact with loops of bowel.

To appreciate the peritoneal and ligamentous attachments of the ovary, it needs to be lifted from its fossa and gently pulled upon. The fimbriated lateral end of the uterine tube is tethered to the pole of the ovary that points laterally and upward; hence, this pole is called its *tubal extremity* (Figs. 27-31 and 27-32). This extremity is connected to the pelvic brim by the **suspensory ligament of the ovary,** a peritoneal fold draped over the ovarian vessels and nerves, representing the lateral continuation of the broad ligament beyond the uterine tube (Figs. 27-31 and 27-32). The opposite pole of the ovary, pointing toward the uterus and downward, is its *uterine extremity.*

This is attached to the uterus in the inferior angle of the uterotubal junction by a fibromuscular band, the **ligament of the ovary** (Fig. 27-32). This ligament is within the broad ligament and raises a ridge in its posterior lamina.

The **mesovarium,** a reduplication of the posterior lamina of the broad ligament, is attached along the anterior, or *mesovarian, border* of the ovary (Fig. 27-33). The mesovarium contains the vessels and nerves of the ovary, which enter and leave the organ through its hilum, to which the mesovarium is attached.

**Development and Structure.** The ovary, like the testis, develops in the urogenital ridges; during their initial phase of development, the two gonads are indistinguishable (p. 709; see Fig. 25-29). The proliferation of celomic mesothelium forms the cortex of the ovary; its more vascular medulla is derived from the mesenchyme of the intermediate mesoderm. Primordial germ cells, derived from the yolk sac, invade the cortex. There they will develop into oogonia and oocytes, surrounded by cells of the cortex, which form a follicle around each developing egg.

The outer surface of the ovary is devoid of peritoneum. The so-called *germinal epithelium* (which does not give rise to oocytes) merges with the mesovarian mesothelium along the mesovarian border of the ovary. Subjacent to the germinal epithelium is the *tunica albuginea,* a collagenous stratum that surrounds the cortex and is breached by the *ovarian follicles* as they mature, enlarge, and burst on the surface of the ovary. The cortex contains such follicles in various stages of maturation. After the discharge of the ovum from the follicle (ovulation), the follicular cells are transformed into the *corpus luteum,* a yellow, glandular body that persists through most of pregnancy if the ovum has been fertilized and implantation has occurred. Otherwise, the corpus luteum degenerates before the next ovulation. The degenerated corpora lutea form small fibrosed bodies called *corpora albicans,* also located in the ovarian cortex.

## The Uterine Tube

The uterine tubes (*tuba uterina, salpinx uterina*), also known as fallopian tubes, are the bilateral ducts that extend from the uterus to the ovary and connect the uterine cavity to the peritoneal cavity (Figs. 27-34 and 27-35). Each tube is about 10 cm long and is almost completely surrounded by peritoneum along the superior margin of the broad ligament as the anterior and posterior laminae of the ligament become continuous with one another around the tube (Fig. 27-33). The part of the broad ligament between the tube and the base of the mesovarium is called the **mesosalpinx.** The lateral part of the tube arches over the lateral pole of the ovary and turns posteriorly (Figs. 27-32 and 27-36). Its trumpet-shaped, expanded open end becomes closely applied to the ovary.

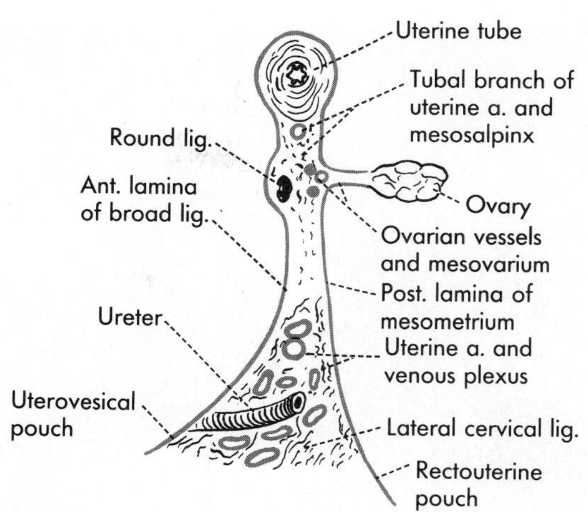


**FIG. 27-33.** A schematic sagittal section across the broad ligament, uterine tube, and ovary.

**FIG. 27-34.** The right uterine tube, opened from behind.

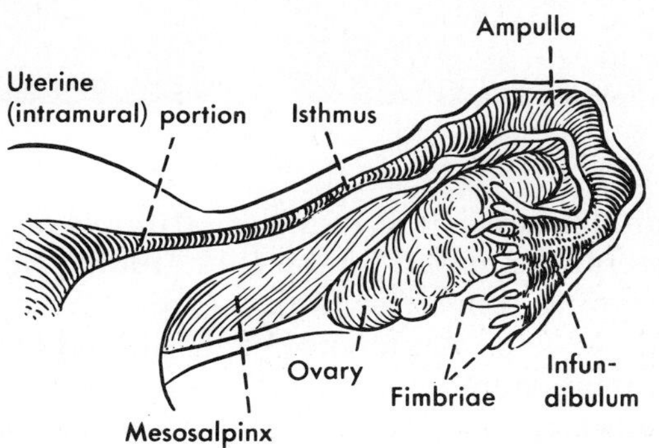


**Parts.** Four parts are recognizable on each tube: the infundibulum, the ampulla, the isthmus, and the uterine part (Fig. 27-34). The **in-**

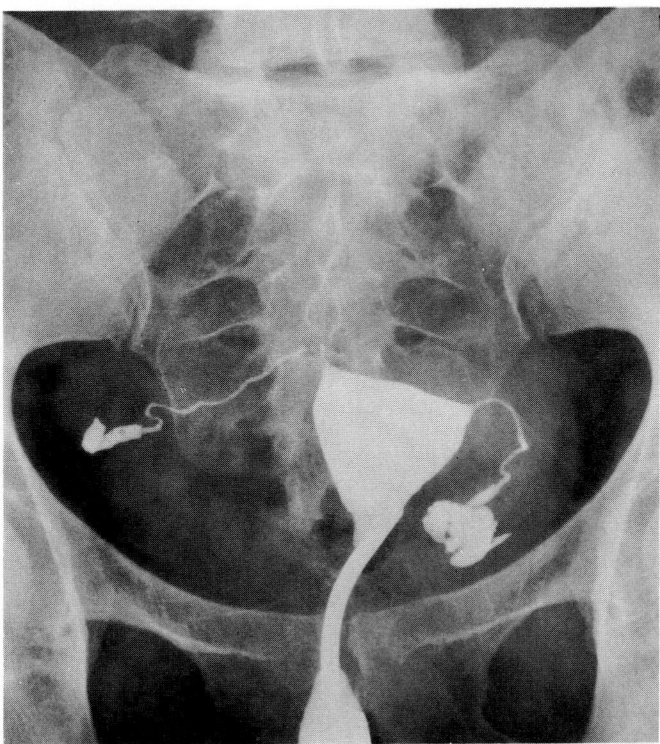

**FIG. 27-35.** A uterosalpingogram. The catheter through which the radiopaque material has been injected extends through the vagina into the cavity of the body of the uterus; to the left of the upper part of the catheter can be seen part of the cavity of the isthmus and cervix. Both uterine tubes are clearly visualized, and from that on the reader's right, radiopaque material has entered the peritoneal cavity. (Courtesy of Dr. R. B. Wilson)

**fundibulum** is the funnel or trumpet-shaped lateral expansion of the tube (infundibulum meaning funnel; salpinx meaning trumpet). The wide circumference of the infundibulum presents numerous fingerlike processes called **fimbriae,** one of which, the *ovarian fimbria,* is attached to the ovary. The exterior of the infundibulum is completely peritoneal, but the internal surface of the fimbriae and the tapering cavity of the tube are covered by highly ciliated columnar epithelium thrown into irregular longitudinal folds that converge toward a small opening, the *abdominal ostium of the tube,* located at the bottom of the infundibulum. The **ampulla,** succeeding the infundibulum, is wide, thin walled, and tortuous. It leads into the more narrow **isthmus.** The **uterine part of the tube,** also called the *intramural part,* traverses the thick uterine wall and, through the *uterine ostium* (or *uterotubal junction*), opens into the uterine cavity.

The uterine tube has a narrow lumen that is filled by the highly complex folds of its ciliated mucosa. Nevertheless, the patency of the tubes can be tested, as it is done in investigations of infertility, by filling the uterine cavity with radiopaque contrast medium through the cervix and observing on x-ray film the spillage of the medium into the peritoneal cavity through the tubes (Fig. 27-35). This procedure is known as salpingography. In addition to the mucosa, the wall of the tubes has the usual layers of submucosa, muscularis, and serosa.

**Gamete Transport.** The function of the uterine tubes in coordinating egg and sperm transport for fertilization to take place is highly complex and incompletely understood. Although there is evidence that an ovum discharged from one ovary can be transported through the peritoneal cavity and picked up by the uterine tube of the opposite side, under normal circumstances, ovum transport is less hazardous. The fimbriae are usually closely applied to the surface of the ovary. Aided by muscular contractions of the tube and the ligaments (all of which contain smooth muscle), the fimbriated infundibulum appears to sweep the surface of the ovary. The ovum discharged from the ripe follicle is surrounded by the *cumulus ovaricus,* an aggregation of several thousand follicular cells held together around the egg rather tenaciously. The cilia of the fimbria can effectively grasp the cumulus, and it may, in fact, be difficult to pull the cumulus away from the fimbria with a pair of forceps. Thus, it is an irregular mass of cells, rather than a naked egg, that is swept into the abdominal ostium of the uterine tube. The cilia carry the cumulus to the ampulla, where the egg is liberated from the cumulus and its progress along the tube is held up for some time awaiting fertilization.

The passage of spermatozoa along the uterine and isthmic parts of the tube is believed to be aided not only by their own motility but also by the activity of the tube. Normally, fertilization takes place in the ampulla, and the descent of the fertilized egg into the uterus is delayed for 6 to 7 days. Thus, not only fertilization but also cleavage and blastocyst formation take place in the uterine tube. The factors responsible for this delay, necessary for the uterine mucosa to become receptive to the conceptus, remain unknown.

Interference with the timing of this mechanism may result in the implantation of the conceptus in the tube **(tubal ectopic pregnancy),** inevitably leading to rupture of the tube, a major emergency. Should the fertilized egg be transported into the peritoneal cavity rather than the uterus, or should the egg be fertilized in the peritoneal cavity, an **abdominal ectopic pregnancy** may result. The embryo and fetus may develop outside the uterus and have to be delivered by laparotomy.

The most common cause of **infertility** in women is blockage of the uterine tubes. This usually results from pelvic inflammatory disease, often a consequence of the spread of infection along the initially patent tubes into the pelvis. Surgical interruption of the tubes is the procedure for permanent sterilization, a not infrequent method of contraception. The tubes are doubly ligated and divided between the two ligatures. Reanastomosis of the

**FIG. 27-36.** **A.** Pelvic viscera in the female. The cut edges of peritoneum indicate the reflection to the lateral pelvic wall. **Ur.** is the left ureter. At the base of the broad ligament, above the ureter, are the stumps of the uterine vessels as they enter the uterus. Between the cut edges of the leaves of the uppermost part of the broad ligament are stumps of the left uterine tube, of the ovarian ligament just below this, and of the round ligament just anterior to the ovarian ligament. **B.** A diagrammatic sagittal section of the female pelvis.

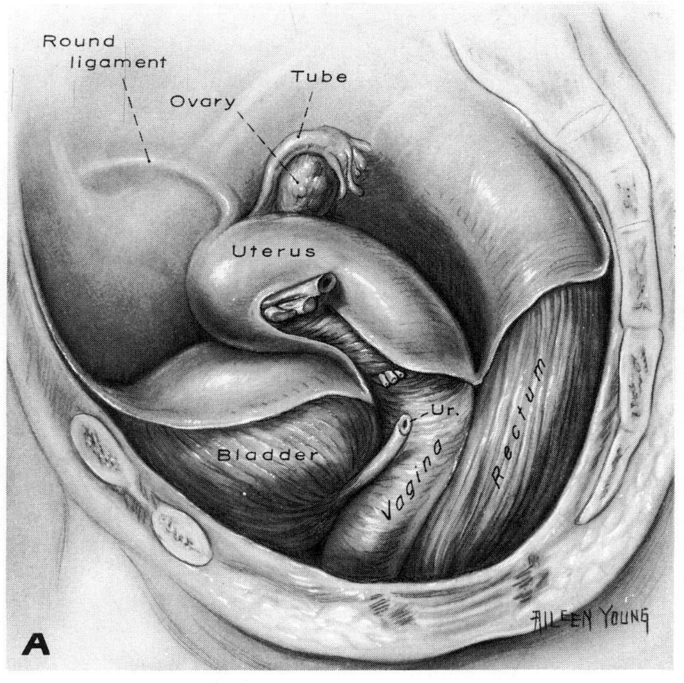


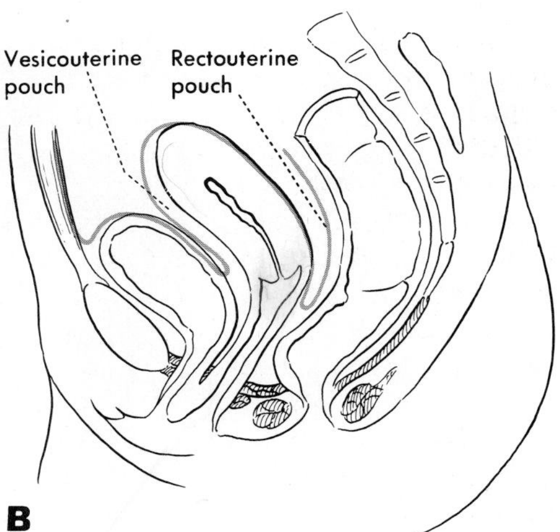

tubes and reestablishment of fertility has been achieved with microsurgical techniques.

## The Uterus

The uterus, or womb, is the organ of gestation. It is a somewhat pear-shaped, hollow organ, slightly flattened anteroposteriorly (Fig. 27-37). Its thick muscular walls enclose a slitlike cavity, the triangular anterior and posterior walls of which are essentially in contact with one another (Fig. 27-35).

The uterus is located behind the bladder and in front of the rectum, entirely below the plane of the pelvic brim. The vertical length of the uterus is about 8 cm. Its broadest part, between the attachment of the uterine tubes, measures 5 cm, and its anteroposterior thickness is about half that much. During pregnancy, the uterus undergoes an astonishing degree of enlargement. This involves hypertrophy of its walls, ligaments, and peritoneal coverings, as well as the growth of the vessels and nerves associated with it. No less astonishing is the involution of the organ after parturition.

**Parts.** The upper, broad, piriform part of the uterus is its **corpus,** or *body;* a narrow, more cylindrical inferior portion is the *neck,* or **cervix.** The dome-shaped part of the body, projecting above the level of the entrance of the uterine tubes, is the *fundus.* The tapering inferior part of the body leads into the cervix; the cervix projects into the vagina through the anterior wall of its vault, which demarcates the *supravaginal* and *vaginal parts of the cervix.* The vaginal lumen surrounding the cervix is called the **fornices of the vagina.** The lower part of the body adjoining the supravaginal cervix is designated as the **isthmus,** or the *lower uterine segment* (Fig. 27-38).

The posterior surface of the body and supravaginal cervix is covered with peritoneum and is convex (Figs. 27-36 and 27-39). It is in contact with coils of intestine and is hence called its intestinal surface. The concave anterior surface is peritoneal only as far down as the cervix; both peritoneal and bare parts are related to the bladder, and this surface is therefore called the *vesical surface of the uterus* (Fig. 27-36). The rounded right and left borders give attachment to the two laminae of the broad ligament and superiorly receive the uterine tubes (Figs. 27-36 and 27-39). The part of the broad ligament adjacent to the borders of the uterus is called the **mesometrium.**

**Uterine Cavity.** The triangular cavity of the uterus, which, in comparison to the size of the organ, is surprisingly small, receives the uterine ostia of the uterine tubes and leads inferiorly through the **internal os** into the cervical canal (Figs. 27-35 and 27-38). The **cervical ca-**

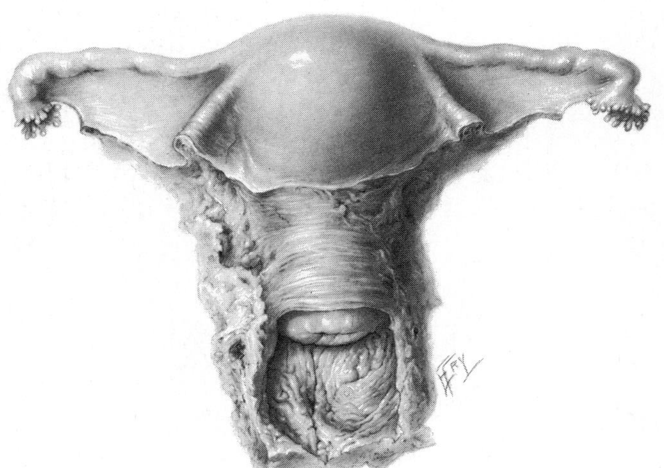

**FIG. 27-37.** Anterior view of the uterus and upper part of the vagina. Parts of the broad ligaments, enclosing the tubes and the upper ends of the round ligaments, have been left on the sides, and the vagina has been opened to show the lower end of the cervix uteri. Note the subperitoneal pelvic connective tissue (parametrium) along the side of the uterus and in between the laminae of the broad ligament.

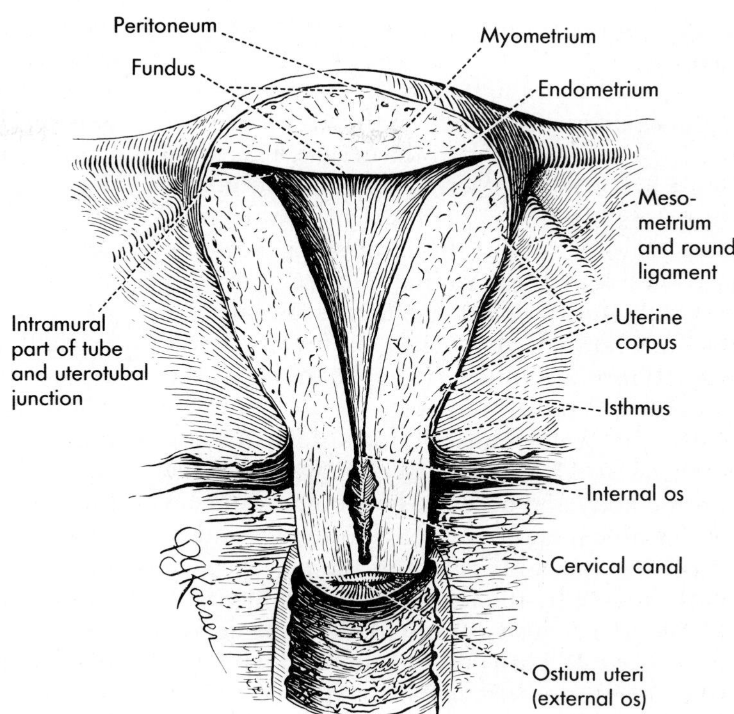


**FIG. 27-38.** A coronal section of the uterus to show its parts.

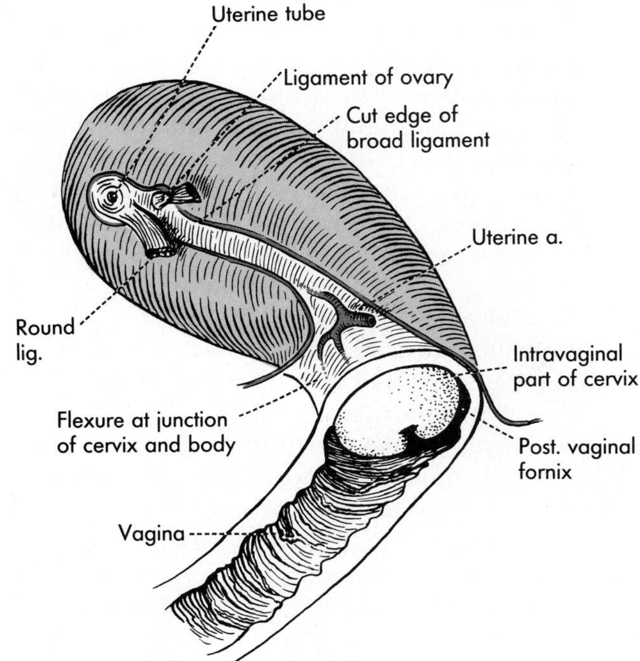


**FIG. 27-39.** The uterus seen from the left side to show the relationship of the cervix to the vagina. Note the relationship of the peritoneum to the cervix and vagina anteriorly and posteriorly.

**nal** is somewhat spindle shaped; it opens through the **external os** (*ostium uteri*) into the superior part of the vagina. The external os is a transverse slit guarded by the broad anterior and posterior lips, or the *labia of the cervix*.

The lumen of the cervical canal is normally filled by a viscous mucus plug, the consistency of which changes during the menstrual cycle. The strands of mucus are stabilized in the canal by the *palmate folds* of the cervical mucosa, which interlock as they project into the canal from its anterior and posterior walls. The cervical canal remains closed during pregnancy until the second stage of labor, and its premature dilatation is a sign of imminent abortion. The isthmus, on the other hand, becomes dilated rather early in pregnancy and is incorporated into the expanding uterine cavity.

**Orientation.** The body of the uterus is normally bent anteriorly on the cervix. This relationship is spoken of as *anteflexion*. The axis of the cervix is likewise bent forward on the axis of the vagina, so that the external os faces the posterior wall of the vagina. This relationship is called *anteversion* of the uterus. Because of the anteversion, the depth of the posterior vaginal fornix is much greater than that of the anterior or lateral fornices.

The factors responsible for anteflexion are intrinsic to the fibromuscular walls of the uterine body and cervix. The factors responsible for anteversion are not completely understood, but probably relate to the pull exerted on the cervix by the rectouterine ligaments. A full bladder tends to eliminate to some degree the angles of both anteflexion and anteversion. Permanent reversal of these angles, called retroflexion and retroversion, has been suspected as a cause of infertility.

**Structure.** The mucous membrane lining the uterine cavity, called **endometrium,** is a cuboidal, largely nonciliated epithelium. Its invaginations form simple glands extending into the cellular and vascular *submucosa*. The cervical canal is lined by columnar cells, ciliated in the upper part of the canal; however, the exterior of the vaginal cervix facing into the fornices is covered by stratified squamous epithelium. The cervical glands secrete the cervical mucus.

During each *menstrual cycle,* the endometrium undergoes cyclic changes and is shed with a variable quantity of blood at the end of each cycle; this constitutes the menstrual flow. The mucosa of the uterine tubes and the cervical canal are not shed.

The tunica muscularis, or **myometrium,** is more than a centimeter thick over most of the organ. It consists of interlacing bundles of smooth muscle, divisible into external, intermediate, and internal laminae, but these are rather indistinct. The smooth muscle is intermixed with fibrous and elastic tissue; fibrous tissue predominates in the cervix. The visceral pelvic fascia on the exterior of the body and cervix is called the **parametrium** and **paracervix,** respectively (*metra* is a Greek word meaning uterus). The paracervix is dense and continuous with the supporting ligaments of the uterus (Fig. 27-37). The peritoneal covering of the uterus constitutes its serosa and reflects from its lateral borders as the **mesometrium,** the medial part of the broad ligament. The cervix is devoid of serosa, except for the supravaginal part posteriorly, where it forms the anterior wall of the rectouterine pouch.

**Ligaments.** Three types of ligaments are associated with the uterus: a peritoneal ligament, two fibromuscular ligaments derived from the genitoinguinal ligament, and several ligaments formed by condensations of subperitoneal pelvic connective tissue.

The peritoneal ligament of the uterus is the **broad ligament,** consisting of three named parts: mesometrium, mesosalpinx, and mesovarium (Figs. 27-31 and 27-33). Each has been described in preceding sections of this chapter.

The broad ligament is derived from the *urogenital mesentery,* which, after the regression of the mesonephros and its duct in the female, supports the paramesonephric ducts and the ovary (p. 709; see Fig. 25-29). It contains the vestiges of the mesonephric ducts, the epoophoron and its duct, the paraoophoron, and the duct of Gärtner, as well as the derivatives of the genitoinguinal ligament and blood vessels and nerves that supply the ovary, uterus, and uterine tubes.

The two broad ligaments, with the uterus between them, form a transverse septum in the pelvic part of the peritoneal cavity. The anterior lamina becomes continuous at the base of the ligament with the parietal peritoneum of the paravesical fossae; these fossae communicate with each other through the uterovesical pouch. The posterior lamina becomes continuous, at the base of the ligament, with the parietal peritoneum of the rectouterine folds and the pararectal fossae,

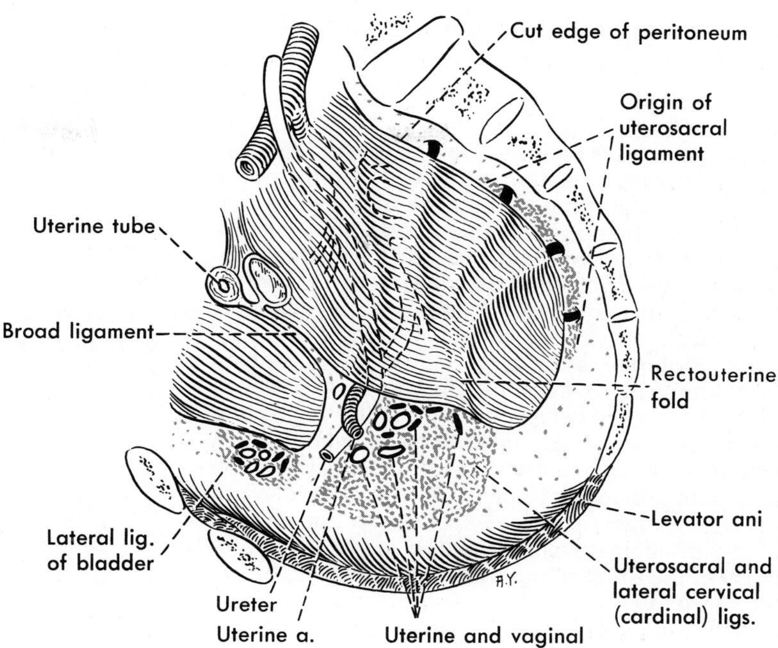


**FIG. 27-40.** Schema of the lateral cervical, or cardinal, and uterosacral ligaments in a section of the pelvis lateral to the uterus. The subperitoneal tissue that constitutes the ligaments is in blue. The course of the ureter and of some branches of the internal iliac artery is indicated by broken lines.

which communicate with each other through the rectouterine pouch (Fig. 27-40). The base of the ligament is some distance above the pelvic diaphragm, and just below it run the lateral cervical ligaments with the uterine vessels and the ureter embedded in them.

A transverse mesenterial septum, such as is formed by the broad ligament, is lacking in the male pelvis, since the descent of the male gonad into the scrotum and the regression of the paramesonephric ducts eliminate the urogenital mesentery in the male.

The **derivatives of the genitoinguinal ligament** in the female are the *ligament of the ovary* and the *round ligament of the uterus.* Both attach to the uterus (p. 758). As already described, the ovarian ligament runs in the broad ligament from the uterine pole of the ovary to attach to the inferior angle of the uterotubal junction (Fig. 27-32). The forward continuation of the ovarian ligament from its uterine attachment is the **round ligament of the uterus** (*ligamentum teres uteri;* Figs. 27-31 and 27-39). The ligament raises a ridge in the anterior lamina of the broad ligament and in the floor of the paravesical fossa as it runs retroperitoneally from the uterotubal junction to enter the deep inguinal ring (Fig. 27-36). Its fate in the inguinal canal has already been described (p. 722).

Uterine **ligaments formed by pelvic fascia** occupy a pyramidal space on the lateral side of the cervix, bordered above by pelvic peritoneum and below by the sloping muscular pelvic floor (Fig. 27-40). The ligaments that can be defined in this mass of connective tissue are continuous medially with the *paracervix,* the dense fascial sheath around the supravaginal part of the cervix, and the upper part of the vagina. Peripherally, they attach to the pelvic walls (Fig. 27-7).

The posterior ligament of this mass of connective tissue is contained in the uterosacral folds and is called the *uterosacral ligament,* attached posteriorly more to the sacrum than to the rectum. Smooth muscle contained in these ligaments makes up the *rectouterine muscle.* The dense connective tissue below the base of the broad ligament is described as the *lateral cervical ligament,* also called by gynecologists Mackenrodt's ligaments or *cardinal ligaments of the uterus,* implying their importance in support of the uterus. Anteriorly, the *pubocervical ligaments* connect the cervix to the posterior surface of the pubes.

**Support.** It is believed that the tension of these ligaments keeps the uterus suspended in the pelvic cavity. It is also thought that the uterosacral ligaments are responsible for anteversion of the uterus. Although both pelvic and urogenital diaphragms are involved in the support of the uterus, stretching or tearing of the ligaments of the cervix is probably an important factor in producing prolapse. **Prolapse** is the protrusion of pelvic viscera through the pelvic floor into the vagina. In prolapse of the uterus, the cervix descends in the vagina and may appear at the vaginal orifice or even outside it, everting the vagina. The base of the bladder similarly may prolapse or bulge into the anterior vaginal wall, forming a *cystocele,* or the rectum into the posterior vaginal wall, forming a *rectocele.* A prolapse of the urethra into the anterior vaginal wall is a *urethrocele.*

The treatment of prolapse involves tightening the ligaments of the cervix, especially the lateral cervical ligaments, and repairing other damage, followed by strengthening of the muscles of the pelvic and urogenital diaphragms.

## The Vagina

The vagina is the female organ of copulation; the word vagina means sheath. The entrance into the vagina (*ostium vaginae*) is in the vestibule between the labia minora and is described with the external genitalia in Chapter 28. The vagina passes upward with a posterior inclination through the urogenital diaphragm and the urogenital hiatus (Fig. 27-36). Its major part is located in the pelvis, where it terminates by fusing around the cervix of the uterus.

The vagina is about 10 cm long and is greatly distensible. Normally, it is flattened anteroposteriorly, its anterior and posterior walls lying in contact with each other. At the upper end of the vagina, often called the *vaginal vault,* the lumen forms recesses, or *fornices,* around the vaginal part of the cervix. Because of the angle made between the vagina and the cervix, the posterior vaginal wall is longer than the anterior wall and the posterior fornix is deeper than the anterior and lateral fornices.

The anterior wall of the vagina is related to the urethra and the base of the bladder; the urethra is, in essence, embedded in the vaginal wall. The posterior wall of the fornix is covered by peritoneum of the rectouterine pouch, and below that, the vagina is separated from the rectum by the rectovaginal septum (peritoneoperineal septum). Its perineal portion is separated from the anal canal by the *perineal body* (p. 796).

**Structure.** The bulk of the wall of the vagina consists of a tunica muscularis in which smooth muscle and dense connective tissue are mixed. There are many elastic fibers in this connective tissue. The outer part of the vaginal wall is largely connective tissue and contains the vaginal plexus of veins, the vaginal arteries, and the vaginal nerves. The mucosa is thrown into transverse folds (vaginal rugae) and a longitudinal fold called the *vaginal column,* on both the anterior and posterior walls. Its surface epithelium is stratified squamous, which undergoes cyclic changes that can be correlated with the ovarian cycle but are much less pronounced than in many mammals. Nevertheless, a vaginal smear technique can give some information regarding the stage of the menstrual cycle. It was the study of vaginal smears that led to the technique of diagnosing early carcinomatous change of the cervix uteri through cervical smears.

#### Congenital Abnormalities of the Uterus and Vagina

Abnormalities of the paramesonephric ducts and their derivatives may be due to agenesis, noncanalization, and failure of fusion (Fig. 27-41). Failure of fusion of the paramesonephric ducts, the most common group of abnormalities, can result in complete duplication of the uterus and the vagina, or the body of the uterus may be split into two equal parts (bicornate uterus) with a single or double cervix. The terminal part of one or both of the paramesonephric ducts may fail to canalize and consequently atrophy, resulting in a uterus unicornus or in the absence of the uterus or the uterine tubes or both.

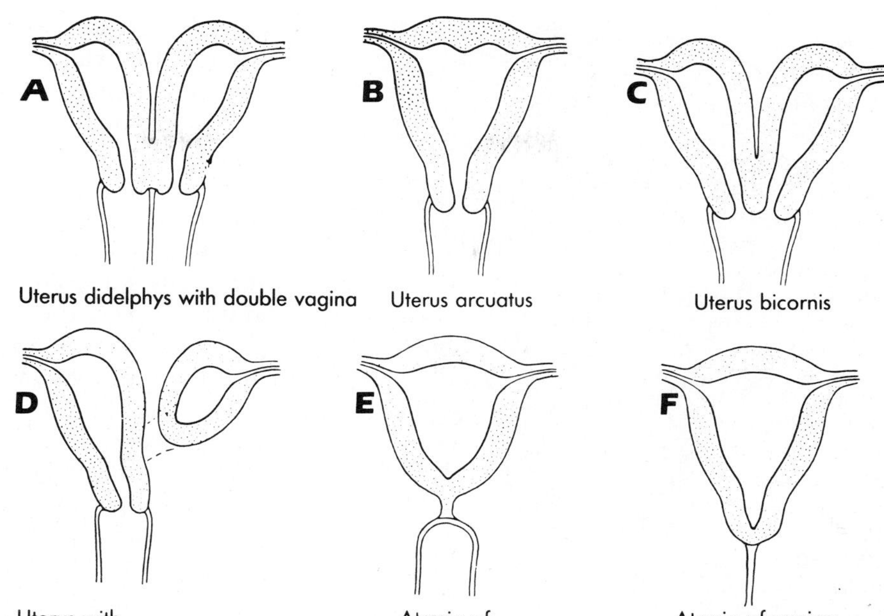


**FIG. 27-41.** Schematic representation of the main abnormalities of the uterus and vagina, caused by persistence of the uterine septum or obliteration of the lumen of the uterine canal. (Langman J: Medical Embryology, 3rd ed. Baltimore, Williams & Wilkins, 1975)

The vagina may be absent because the sinuvaginal bulb failed to develop or the vaginal plate failed to canalize. The external genitalia in such cases may be normal. Incomplete canalization may leave septa in the vagina or result in an abnormally thick hymen. If a normal vagina is absent, an artificial vagina can be constructed surgically by creating a cleavage plane along the rectovesical fascia and lining it with skin.

Faulty development of the urorectal septum may result in rectovaginal fistulae (Fig. 27-21).

## Blood Supply and Lymph Drainage

The blood supply of the uterus and uterine tubes is derived from the uterine artery. This artery also contributes to the supply of the ovary and the vagina, although each of these relies to a large extent on the ovarian and vaginal arteries, respectively. Veins and lymphatics of the female genital tract largely follow the arteries.

**Arteries.** The blood supply of the uterus is through paired **uterine arteries.** These are branches of the internal iliac artery and run medially toward the uterus close to the upper surface of the lateral ligaments of the cervix along the base of the broad ligament. This connective tissue also contains the uterine veins and a uterovaginal plexus of nerves. The ureter also passes through this tissue, coursing posteroinferior to the uterine artery on its way to the bladder (p. 767). Just above the lateral vaginal fornix, the uterine artery approaches the supravaginal part of the cervix as soon as it has crossed the ureter (Fig. 27-42). Here it gives off branches to the cervix and also one or more to the vagina and then turns upward on the lateral aspect of the uterus, either between the leaves of the broad ligament or in the substance of the uterus. A little below the uterine fundus, the artery gives off tubal and ovarian branches, and the remainder of the vessel ends in the fundus.

The **tubal branch** of the uterine artery leaves the uterus close to the attachment of the uterotubal junction and runs laterally in the mesosalpinx to supply the whole length of the tube (Fig. 27-43). The **ovarian branch** enters the uterine pole of the ovary through the mesovarium and supplements the chief supply delivered through the ovarian artery. The **ovarian artery,** described in Chapter 25 as far as the pelvic brim, enters the mesovarium through the suspensory ligament of the ovary.

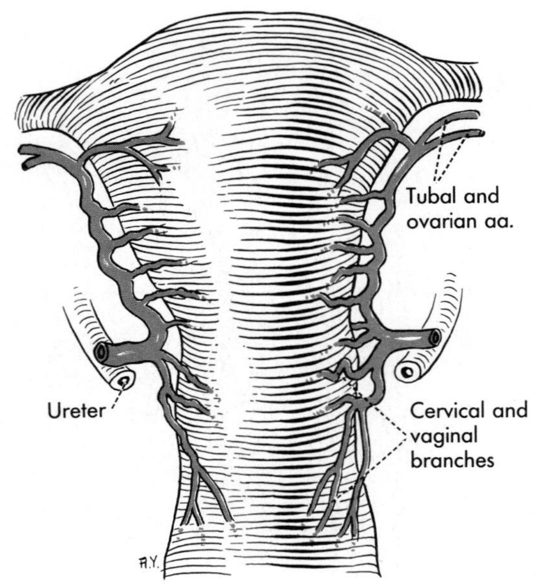


**FIG. 27-42.** The uterine artery.

**FIG. 27-43.** The blood supply of the ovary.

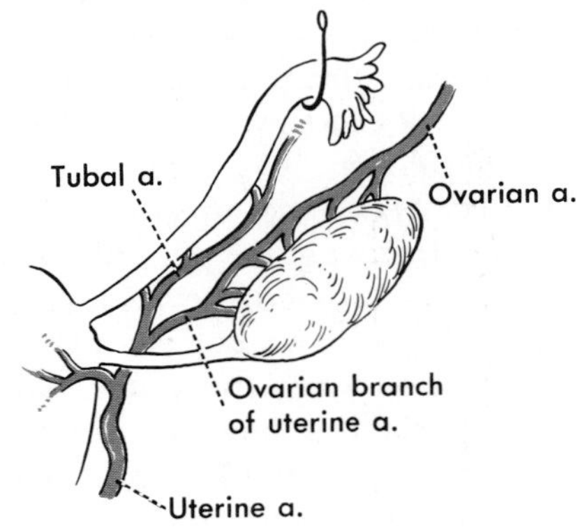


During pregnancy, all these vessels become very much enlarged.

The chief blood supply of the vagina is through **paired vaginal arteries** (Fig. 27-44). These arteries arise from the internal iliac independently or are branches of the uterine arteries. They may be multiple. Branches of the uterine arteries to the cervix uteri help supply the upper part of the vagina. The lower part of the vagina is supplied by branches of the internal pudendal artery.

**Veins.** A venous plexus surrounds the vagina and the uterus. These plexuses are located in the fascial sheaths of the organs. Although the **vaginal plexus** has connections with the vesical plexus, for the most part it drains upward into the **uterine plexus.** The uterine plexus also receives the veins from the uterine tubes. The **uterine veins** begin in the uterine plexus and drain laterally, usually as two trunks on each side, and join the internal iliac vein. The **ovarian veins** leave the mesovarium through the suspensory ligaments and ascend to the inferior vena cava on the right and the renal vein on the left.

**Lymphatics.** The distribution of lymphatics from the **uterus** is particularly important because of the relative frequency of uterine carcinoma, which spreads preferentially along lymphatics. The lymphatics diverge widely and empty into a number of differently situated nodes. A few from the **fundus** and upper part of the body drain along the round ligament into *superficial inguinal nodes.* But most of the upper lymphatics pass laterally in the upper parts of the broad ligaments. They unite with lymphatics from the **uterine tube** and **ovary** and pass with the ovarian vessels over the pelvic brim. Instead of ending in iliac nodes at this level, they continue upward along the ovarian vessels and empty into nodes of the lumbar chain close to the level of the renal vessels. The drainage of the **lower part of the body** of the uterus and **cervix,** where carcinoma most frequently originates, is largely to nodes within the pelvis. The lymphatics of the lower part of the body and of the cervix pass laterally in the lower part of the broad ligament. Many end in *internal iliac nodes;* others reach the pelvic brim and end in *external iliac nodes;* still others run posteriorly and medially to empty into *sacral nodes* or nodes associated with the common iliac vessels.

Because of this wide area of lymph drainage from the cervix, successful treatment of cervical carcinoma usually requires resection of the entire uterus (hysterectomy; *ystera,* another Greek name for the uterus) plus thorough dissection of lymph nodes on the pelvic walls and in the inguinal region.

The **lymphatic drainage of the vagina** is in

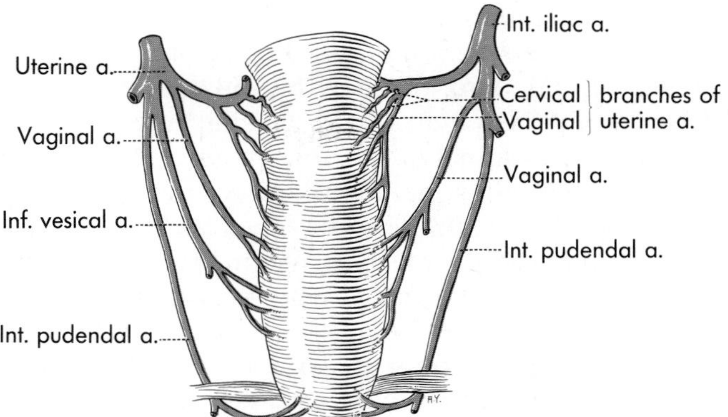


**FIG. 27-44.** Arteries to the vagina. On the reader's right, the vaginal artery arises from a stem common to it and other branches of the internal iliac; on the left, a vaginal artery arising from the uterine artery is supplemented by a branch arising from the inferior vesical artery, which is distributed mostly to the bladder.

two directions. From about the upper three-fourths, the lymphatics run upward and laterally, joining those of the cervix uteri to empty into *internal iliac nodes;* the lower fourth drains downward to the perineum and, hence, into *superficial inguinal lymph nodes.*

## Innervation

A dense **uterovaginal plexus** of nerves passes to the uterus with the uterine vessels through the lateral cervical ligament. The plexus reaches the uterus at about the junction of the cervix and body. The plexus gives off *vaginal nerves* and a particularly abundant supply to the cervix before it turns upward with the uterine artery to supply the rest of the uterus.

The uterovaginal plexus consists primarily of visceral afferent and sympathetic efferent fibers and contains few, if any, parasympathetic efferents. The physiology of the motor supply to the uterus is not understood. Innervation is not necessary to the functions of the uterus, not excepting its contractions during labor.

In contrast to the afferent innervation of most of the pelvic organs, that of the body of the uterus travels from the inferior hypogastric plexus with the sympathetic system through the superior hypogastric and aortic plexuses. Pain fibers from the body of the uterus evidently enter the spinal cord through the last two thoracic nerves. These are the nerves that mediate pain sensation during the first stage of labor, and this pain is referred to the lower thoracic and lumbar region in the back. The elimination of this pain requires that anesthetic be injected into the epidural space or into the thecal sac. This procedure blocks all nerves that leave and enter the lumbar and sacral parts of the spinal cord.

It has long been assumed that pain fibers from the cervix uteri do not follow those from the body; rather they were believed to pass from the inferior hypogastric plexus with the pelvic splanchnic nerves into S2–S4 segments of the cord. Several clinical studies cast doubt on this assumption: paravertebral block of the 11th and 12th thoracic nerves, paravertebral block of the upper lumbar sympathetic trunks, or presacral neurectomy all eliminate pain caused by dilatation of the cervix during the first stage of labor. This indicates that pain afferents from the cervix ascend from the inferior hypogastric plexus with pain afferents from the uterine body. Pain caused by dilatation of the cervix may also be blocked by injecting anesthetic around the *paracervical plexus of nerves* (the uterovaginal plexus) through the lateral fornices of the vagina. Such a paracervical block is effective also when the cervix has to be dilated artificially for scraping out the uterus. The cervix, like the vagina, is insensitive to cutting and burning; pain is caused only by its dilatation.

Nothing particularly is known of **nerves to the uterine tubes.** Presumably, they are derived from the uterovaginal plexus of nerves and from the ovarian plexus.

The **vaginal nerves** are mostly from the uterovaginal plexus, but little is known concerning their physiology. Presumably, most of the pain fibers from the vagina travel with the sacral parasympathetic fibers, as do the pain fibers from the cervix uteri, and therefore enter the spinal cord through S2–S4 nerves. These nerves respond only to tension in the vagina. Large vaginal lacerations may remain painless unless they include the lower part of the vagina. Approximately the lower 2 cm to 3 cm of the vagina receives its innervation from the pudendal nerves, which, although originating from the same sacral segments, are somatic rather than visceral nerves and convey somatic rather than visceral pain afferents.

The **nerves to the ovary** probably include sympathetic efferent and visceral afferent fibers, but ovarian function is not at all dependent on motor nerve supply. These nerves form a delicate plexus, not grossly visible except at the level of origin of the ovarian artery, where it can be seen to be derived from the aortic plexus.

## GENERAL REFERENCES AND RECOMMENDED READINGS

ASHLEY GL, ANSON BJ: The hypogastric artery in American whites and Negroes. Am J Phys Anthropol 28:381, 1941

BERGLAS B, RUBIN IC: Study of the supportive structures of the uterus by levator myography. Surg Gynecol Obstet 97:677, 1953

BONICA JJ: Principles and Practice of Obstetric Analgesia and Anesthesia, vol 1. Philadelphia, FA Davis, 1967

BORELL U, FERNSTRÖM I: Radiologic pelvimetry. Acta Radiol (Suppl) 191:3, 1960

BOXALL TA, SMART PJG, GRIFFITHS JD: The blood-supply of the distal segment of the rectum in anterior resection. Br J Surg 50:399, 1963

BOYARSKY S (ed): Neurogenic Bladder. Baltimore, Williams & Wilkins, 1967

BRAITHWAITE JL: The arterial supply of the male urinary bladder. Br J Urol 24:64, 1952

BRAITHWAITE JL: Variations in origin of the parietal branches of the internal iliac artery. J Anat 86:423, 1952

CLEGG EJ: The arterial supply of the human prostate and seminal vesicles. J Anat 89:209, 1955

CURTIS AH, ANSON BJ, BEATON LE: The anatomy of the subperitoneal tissues and ligamentous structures in relation to surgery of the female pelvic viscera. Surg Gynecol Obstet 70:643, 1940

CURTIS AH, ANSON BJ, McVAY CB: The anatomy of the pelvic and urogenital diaphragms, in relation to urethrocele and cystocele. Surg Gynecol Obstet 68:161, 1939

FRANCIS CC: The Human Pelvis. St Louis, CV Mosby, 1952

GASTON EA: The physiology of fecal continence. Surg Gynecol Obstet 87:280, 1948

HUTCH JA, RAMBO ON JR: A study of the anatomy of the prostate, prostatic urethra and urinary sphincter systems. J Urol 104:443, 1970

KIMMEL DL, McCREA LE: The development of the pelvic plexuses and the distribution of the pelvic splanchnic nerves in the human embryo and fetus. J Comp Neurol 110:271, 1958

KOFF AK: Development of the vagina in the human fetus. Contrib Embryol 24:59, 1933

KURU M: Nervous control of micturition. Physiol Rev 45:425, 1965

LANGWORTHY OR, MURPHY EL: Nerve endings in the urinary bladder. J Comp Neurol 71:487, 1939

LOWSLEY OS: The development of the human prostate gland with reference to the development of other structures at the neck of the urinary bladder. Am J Anat 13:299, 1912

MILLEY PS, NICHOLS DH: A correlative investigation of the human rectovaginal septum. Anat Rec 163:443, 1969

MUELLNER SR: The voluntary control of micturition in man. J Urol 80:473, 1958

PARKS AG, PORTER NH, MELZAK J: Experimental study of the reflex mechanism controlling the muscles of the pelvic floor. Dis Colon Rectum 5:407, 1962

REYNOLDS SRM: Physiology of the Uterus, 2nd ed. New York, Paul B Hoeber, 1949

RICCI JV, LISA JR, THOM CH et al: The relationship of the vagina to adjacent organs in reconstructive surgery: A histologic study. Am J Surg 74:387, 1947

ROBERTS WH, HABENICHT J, KRISHINGNER G: The pelvis fasciae and their neural and vascular relationships. Anat Rec 149:707, 1964

TOBIN CE, BENJAMIN JA: Anatomical and surgical restudy of Denonvilliers' fascia. Surg Gynecol Obstet 80:373, 1945

UHLENHUTH E, HUNTER DWT, LOECHEL WE: Problems in the Anatomy of the Pelvis. Philadelphia, JB Lippincott, 1952

WOODBURNE RT: Anatomy of the bladder and bladder outlet. J Urol 100:474, 1968

# 28

# THE PERINEUM

The perineum is the most inferior region of the trunk, located between the thighs and the buttocks. It contains the anal canal, with its external opening, the anus, and the associated sphincters, and the external genitalia, with the fascias and muscles that support and surround them. Superiorly, the perineum is limited by the pelvic diaphragm; inferiorly, it presents a free surface covered by skin. The lateral walls of the perineum are formed by the medial surface of the inferior pubic and ischial rami, the obturator internus below the attachment of the levator ani, and, posterolaterally, the medial surface of the sacrotuberous ligaments overlapped by the gluteus maximus.

The osseoligamentous frame of the perineum is the **inferior pelvic aperture,** described in Chapter 27 (Fig. 28-1). The area enclosed by this rhomboidal or diamond-shaped frame can be divided into two triangular regions by a line connecting the two ischial tuberosities: anteriorly, the *urogenital region* and posteriorly, the *anal region.* These regions are often spoken of as the urogenital and anal *triangles.* The **anal region** contains the anal canal and the ischiorectal fossae on each side of the anal canal. In the **urogenital region,** a muscular shelf stretches between the conjoint ischiopubic rami of the two sides; this is the *urogenital diaphragm.* It is pierced by the urethra and, in the female, also by the vagina. The diaphragm serves as a foundation for the attachment of the external genitalia. The external genitalia in the male include the penis, made up of three cavernous erectile bodies (two corpora cavernosa and the corpus spongiosum), and the scrotum. The female external genitalia include the labia majora, the labia minora, the vestibule of the vagina in between them, the bilateral erectile bulbs of the vestibule, the clitoris, and the mons pubis. Fascias attached to the urogenital diaphragm define two spaces of considerable anatomic and clinical importance; these are the *superficial* and *deep perineal spaces,* or pouches. The superficial perineal space contains all the external genitalia and the superficial perineal muscles associated with them;

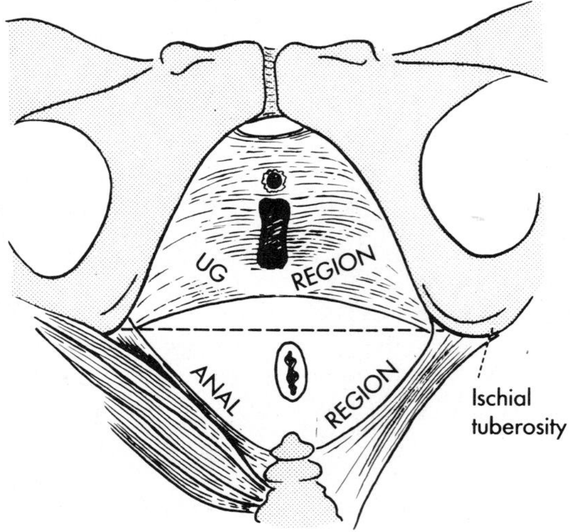


**FIG. 28-1.** Schema of the regions of the perineum. The arbitrary division between the urogenital and anal regions (triangles) is indicated by an interrupted line.

the deep space is filled by the deep muscles of the perineum, which constitute the urogenital diaphragm.

The chief blood supply of the perineum is provided by the *internal pudendal artery*. Blood is drained from the perineum, in general, by the internal pudendal veins, but the blood from some of the erectile bodies of the external genitalia is returned to the prostatic or vesical plexuses by the *deep dorsal vein* of the penis or clitoris, respectively. Most of the perineal lymphatics do not follow the internal pudendal vessels; perineal structures drain primarily to the inguinal lymph nodes. The motor and sensory supply to somatic structures in the perineum is provided chiefly by the *pudendal nerve,* a branch of the sacral plexus; parasympathetic efferents responsible for erection enter the perineum through the urogenital hiatus in the *cavernous nerves,* branches of the prostatic or vesical plexuses; sympathetic nerves reach perineal structures in the pudendal nerves.

This chapter will first describe the anal region, followed by the urogenital region. The blood supply, lymphatic drainage, and innervation will then be considered for the perineum as a whole.

# THE ANAL REGION

## The Anal Canal

The anal canal is the terminal part of the alimentary tract. It commences at the *perineal flexure,* where it is continuous with the rectum, and terminates at the anus, its opening to the exterior. The anorectal junction at the perineal flexure is marked by the narrowing of the rectal ampulla and by the change of the forward inclination of the rectum to the backward inclination of the anal canal (Fig. 28-2). The flexure between the rectum and anal canal is due chiefly to the forward pull of the puborectalis as it swings around the anorectal junction in the posterior boundary of the urogenital hiatus. The anal canal, approximately 4 cm long, is thus directed posteriorly as well as downward. As the longitudinal muscle coat of the rectum continues down on the anal canal, it becomes gradually replaced by fibroelastic tissue. Fibromuscular contributions from the puborectalis reinforce this outer layer of the anal canal, which is then surrounded by the *sphincter ani externus,* formed by voluntary striated muscle. Fibromuscular

**FIG. 28-2.** The lower part of the rectum and the anal canal hemisected and viewed from the left.

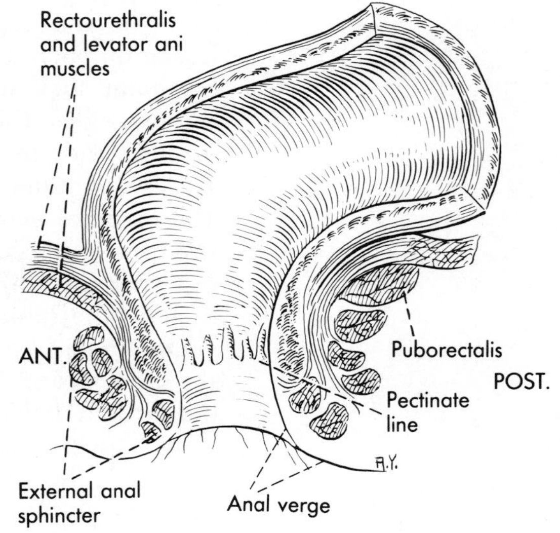

and fatty bodies are related to the anal canal outside its external sphincter on all sides: anteriorly, the perineal body, or central tendon of the perineum (to be described later); posteriorly, the anococcygeal body, or ligament, which separates the anal canal from the coccyx; and laterally, the mass of fat in the ischiorectal fossae.

**Sphincters.** The circular muscle coat of the rectum becomes thicker as it continues below the anorectal junction to form the **internal anal sphincter,** which surrounds the upper two-thirds of the anal canal. Delicate septa from the outer longitudinal fibromuscular coat of the anal canal penetrate the internal sphincter and spread out in the submucosa; others fan out into the dermis of the perianal skin and the fibrous tissue of the ischiorectal fat pads. The thickest of these septa limits the internal sphincter inferiorly and is identifiable on the interior of the anal canal as the *intersphincteric groove*.

The **external sphincter** (*sphincter ani externus*) surrounds the entire length of the anal canal, and it has been described as consisting of three parts: subcutaneous, superficial, and deep (Fig. 28-2). It is difficult to distinguish these three parts, and, in the opinion of some, only two parts can be recognized. The most inferior, **subcutaneous part** of the sphincter surrounds the lower third of the canal, below the internal sphincter; the **superficial** and **deep parts** overlap the internal sphincter, and the deep part fuses above with the puborectalis muscle.

There are truly annular fibers in the subcutaneous and deep parts of the sphincter; many other fibers, as well as most of those in the superficial part of the sphincter, run in parallel bundles along the sides of the anal canal and decussate in front and behind it as they attach to the anococcygeal ligament behind and the *central tendon of the perineum* (perineal body) in front of the anus (Fig. 28-3).

The tone of the external and internal sphincters keeps the lateral walls of the canal apposed to each other except when the sphincters relax during defecation. Probably as a consequence of the anteroposterior direc-

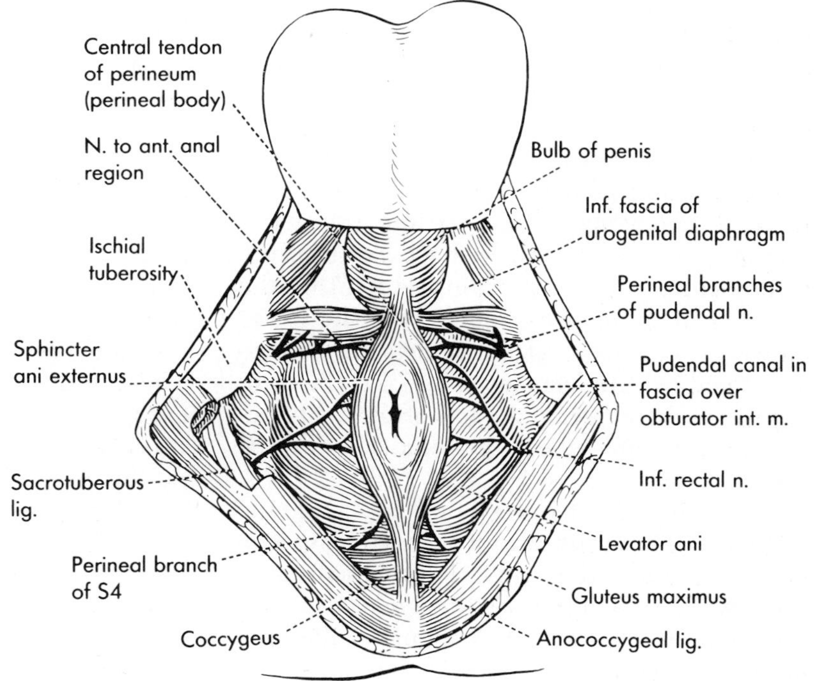


**FIG. 28-3.** Muscles and nerves of the anal region. Also shown is the posterior part of the urogenital region. The internal pudendal artery and its inferior rectal and perineal branches, although not shown here, accompany the pudendal nerve and its branches.

tion of many of the fibers in the external sphincter, the anus is a longitudinal slitlike opening rather than a circular one. The anal skin around the opening is puckered by the *corrugater cutis ani,* which consists of smooth muscle and elastic fibers derived from the septa of the external fibromuscular coat that have traversed the internal sphincter and submucosa of the canal. The same strands of tissue create tiny, circumscribed compartments in the perianal subcutaneous tissue. Distention of these compartments by perianal abscesses or hemorrhage is particularly painful.

On rectal examination, the tone of the sphincters offers a definite resistance to the finger inserted into the anal canal. The intersphincteric groove can be palpated between the subcutaneous part of the external sphincter and the internal sphincter. The so-called *anorectal ring,* palpable above the intersphincteric groove, is formed by the fused fibers of the deep parts of the external sphincter and the puborectalis. On account of the slinglike puborectalis, the ring is strongest posteriorly and its tearing or division results in rectal incontinence.

The external sphincter is innervated by the inferior rectal branch of the pudendal nerve and the internal sphincter by autonomic nerves that descend from the superior rectal plexus.

**Interior.** For clinical reasons, it is useful to relate the internal features of the anal canal to its dual **embryological derivation.** The upper part of the anal canal, like the rectum, develops from the dorsal compartment of the cloaca and is lined by endoderm. The lower part is derived from the *proctodeum,* a depression lined by ectoderm (see Fig. 7-2). The ectodermal and endodermal parts of the canal become confluent when the anal membrane breaks down.

Approximately halfway up the canal, the mucosa is raised up into a transverse row of six to ten folds that encircle the anal canal (Fig. 28-4). These folds are the *anal valves,* and the serrated line formed by them is called the *pectinate line* (dentate line). At the meeting of

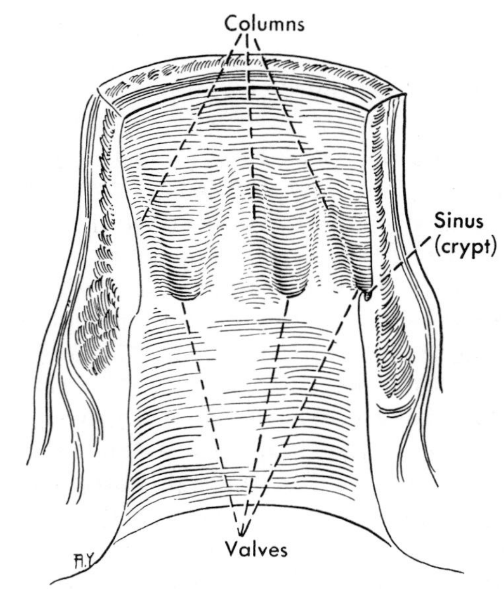


**FIG. 28-4.** The anal columns.

adjacent valves, the mucosa is raised up into longitudinal folds, the *anal columns,* that extend into the upper part of the anal canal. Below the pectinate line, a narrow band of mucosa, the so-called *transitional zone,* also known as the *pecten,* encircles the anal canal and is limited below by the intersphincteric groove. The intersphincteric groove has long been known as the *white line,* described in 1863 by Hilton (despite the fact that it does not appear white in either the living or the cadaver).

There is no agreement with respect to these anatomic features on the precise location of the anal membrane, which marks the junction of the endoderm- and the ectoderm-derived portions of the anal canal. The indiscriminate use of a number of terms for the description of this junction has further confused the issue. The ectodermal–endodermal junction, however, is somewhere in the region of the pecten; it is certainly not above the pectineal line and not below the "white line." The pecten, therefore, represents the "watershed" or "divide" between the type of epithelium lining the canal and also between the two sources of blood and nerve supply and the territory of lymphatic drainage.

The *mucous membrane* above the pectineal

line is similar to that of the rectum, whereas below the pecten, it is nonkeratinized, stratified squamous epithelium devoid of hairs. There is a transition between these two types of epithelium across the pecten. Carcinomas arising in either region are distinct in their nature and clinical history. Above the pecten, the mucosa is supplied by branches of the superior rectal artery and drained by the superior rectal vein. The radicles of these vessels raise up the anal columns, and transverse anastomotic vessels run between them at the bases of the anal valves. This submucous venous anastomosis is the **internal "rectal" venous plexus.** The veins in three of these six to ten columns (two on the left and one on the right) are liable to distention, together with the plexus behind the valves. This causes them to bulge into the anal canal, forming *internal hemorrhoids,* or piles. The anal canal below the pecten is supplied and drained by the inferior rectal vessels, and the **inferior "rectal" venous plexus,** formed in the submucosa, can, when distended, give rise to *external hemorrhoids.* The mucosa above the pecten is innervated by autonomic nerves and is, therefore, insensitive to touching, pricking, and cutting. It is the site of choice for injecting hemorrhoids to promote their thrombosis and fibrosis. The mucosa over and below the pecten is extremely sensitive, being supplied by somatic nerves contained in the inferior rectal branch of the pudendal nerve. Fissures and tears in the lining of the anal canal, often commencing at the valves, are very painful.

Each of the valves shelters a small pocket of space, the *anal sinus.* Into the sinus open the *anal glands,* which extend into the submucosa and sometimes into the muscle coat. Suppuration in these glands results in abscesses and may lead to the formation of fistulae.

The lymphatic drainage of the anal canal above the pecten is upward along the superior rectal vessels and below the pecten, with the inferior rectal vessels into the superficial inguinal nodes.

**Developmental Anomalies.** The common congenital anomaly of the rectum and anal canal is failure of the bowel to open to the outside in the normal fashion. Regardless of whether the anomaly is primarily in the rectum or the anal canal, the condition is referred to as **imperforate anus.** Imperforate anus has been estimated to occur in 1 in every 1500 to 5000 newborns. Imperforate anus is of several different types (Fig. 28-5). It may consist only of a marked stenosis rather than a true lack of any opening, there may be little tissue between the bowel and the outside, or the bowel may end high on the inside. There may be no sign of an anus, or there may be a normal anus that leads into a blind anal canal or a blind rectum. Imperforate anus is often associated with fistulous connections between the rectum and the derivatives of the urogenital sinus (see Fig. 27-21).

## The Ischiorectal Fossae

The ischiorectal fossae are roughly wedge-shaped spaces on each side of the anal canal (Fig. 28-6). The anal canal and the perineal body separate the fossae of the two sides; the only communication between them is poste-

**FIG. 28-5.** Some congenital anorectal anomalies. Type I is a stenosis, here both at the anus and above; type II, a simple type of imperforate anus; type III, imperforate anus, in which the bowel ends blindly a considerable distance above the perineum; type IV, an atresia of the rectum. (Ladd WE, Gross RE: Am J Surg 23:167, 1934)

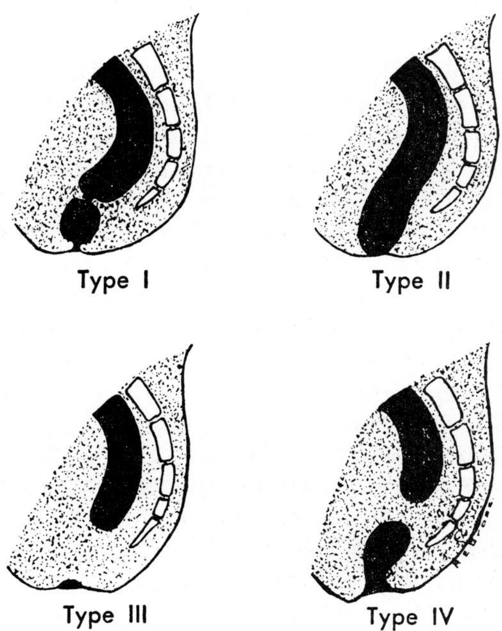

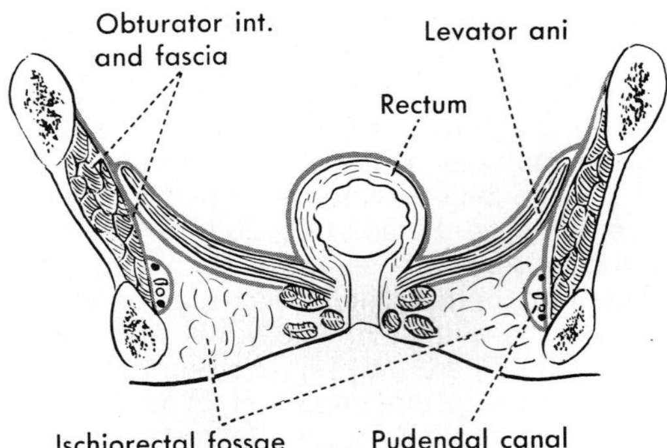


**FIG. 28-6.** Coronal section through the ischiorectal fossae. The obturator fascia and the superior and inferior fascias of the levator ani are shown in blue.

riorly through a narrow potential space located deep to the fibers of the sphincter ani externus that attach to the coccyx.

The more or less vertical **lateral wall** of each fossa is formed by the obturator internus, which covers the medial side of the obturator foramen and the fused bodies of the ilium and ischium above the ischial tuberosity; the **base** of the fossa is perineal skin, and its sloping **superomedial wall** is the levator ani. The fossa is sealed off above by the fusion of the inferior fascia of the pelvic diaphragm to the obturator fascia and medially by the fusion of the same fascia to the external anal sphincters, the outer longitudinal fibromuscular coat of the anal canal, and the perineal body. Anteriorly, the fossa is limited by the fusion of the perineal fascia to the posterior margin of the urogenital diaphragm (Fig. 28-12B). Above the urogenital diaphragm, the *anterior recess* of the fossa extends forward. The **anterior recesses** are narrow spaces between the pelvic and urogenital diaphragm, limited laterally by the conjoint rami of the ischium and pubis and medially by the fusion of the inferior fascia of the pelvic diaphragm to the superior fascia of the urogenital diaphragm (Fig. 28-8A). Posterolaterally, the ischiorectal fossae are continuous with a potential space in the buttocks located deep to the gluteus maximus, the inferior border of which overlaps the fossae posteriorly.

The subcutaneous connective tissue of the anal region expands to fill the fossa, forming an **adipose body** that adapts its shape to the fossa and permits distention of the anal canal during defecation. The adipose body is permeated by tough fibrous strands that do not form well-defined compartments and, therefore, permit the expansion and spread of abscesses in the fat without generating tension and pain.

In addition to the fat, the ischiorectal fossa contains the *pudendal nerve* and *internal pudendal vessels* in a fascial canal that runs along the lateral wall of the fossa, and the *inferior rectal nerve and vessels,* which cross from the lateral wall toward the anal canal.

**The Pudendal Canal.** The pudendal canal is a fascial sheath in which the pudendal nerve and the internal pudendal artery reach the perineum and the internal pudendal veins leave it. The **pudendal nerve,** a branch of the sacral plexus, and the **internal pudendal artery,** a branch of the internal iliac, leave the pelvis between the piriformis and coccygeus, passing over the upper margin of the pelvic diaphragm and entering the gluteal region through the greater sciatic foramen. Destined for the perineum, the nerve and artery run vertically downward and, in so doing, skirt the lateral side of the ischial spine. They attain the medial surface of the obturator internus just below the spine by passing through the lesser sciatic foramen, enclosed by the sacrospinous and sacrotuberous ligaments. Clinging to the medial surface of the obturator internus, the vessels are in the lateral wall of the ischiorectal fossa. Some distance above the ischial tuberosity, they change their course forward, heading for the urogenital region. The fascial sheath of the nerve and vessels is fused to the obturator fascia and is called the *pudendal canal.* The canal extends forward, sheltered by the *falciform process* of the sacrotuberous ligament along the ischiopubic ramus. It terminates at the posterior border of the urogenital diaphragm, which the nerve and vessels penetrate. As they continue forward along the pubic ramus, they are surrounded by the attachment of the muscles of the urogenital diaphragm to that ramus.

The branches of the pudendal nerve and internal pudendal artery to the anal canal are given off just above the ischial tuberosity, and they run medially in the fat of the ischiorectal fossa; they are called the inferior rectal nerve and artery. The inferior rectal vein runs with the nerve and artery to join the internal pudendal veins in the pudendal canal.

## THE UROGENITAL REGION

### THE UROGENITAL DIAPHRAGM, THE PERINEAL FASCIAS AND SPACES

#### The Urogenital Diaphragm

The urogenital diaphragm is a continuous sheet of muscle spanning the triangular space bordered on each side by the conjoint rami of the ischium and pubis (Fig. 28-7). The fascias that cover the superior and inferior surfaces of the muscle are part of the diaphragm (Fig. 28-8). These fascias, called the **superior** and **inferior fascia of the urogenital diaphragm,** fuse with each other along the relatively short anterior and much longer posterior margins of the muscle. Along the anterior margin of the diaphragm, the fusion of its fascias creates the *transverse perineal ligament,* which is a short distance posterior to the *arcuate pubic ligament* lying along the inferior border of the symphysis pubis. The gap between the two ligaments transmits the deep dorsal vein of the penis or clitoris from the perineum into the pelvis. At the midpoint of the posterior margin of the diaphragm is a fibromuscular tendinous mass called the *central tendon of the perineum* or **perineal body,** described in the next section. The urogenital diaphragm lies below the anterior part of the pelvic diaphragm, and its central region forms the floor of the urogenital hiatus. In the female, the diaphragm is less distinct than in the male, because the vagina passing through it more or less splits the diaphragm into right and left halves. Its function as the foundation of the

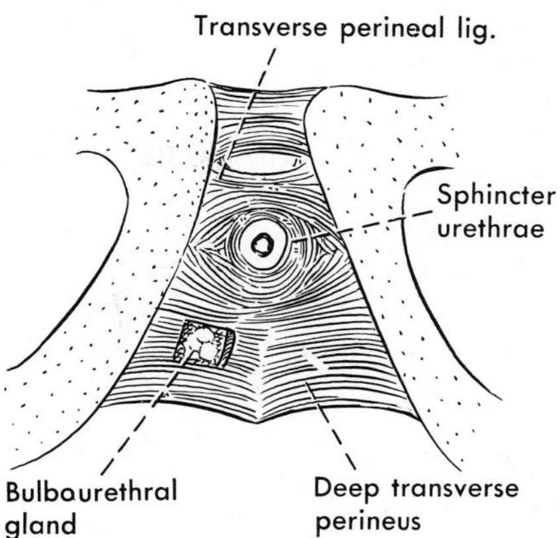


**FIG. 28-7.** The muscle of the urogenital diaphragm of the male seen from below after removal of the inferior layer of fascia (perineal membrane). A piece has been cut from the muscle on the left to show the bulbourethral gland.

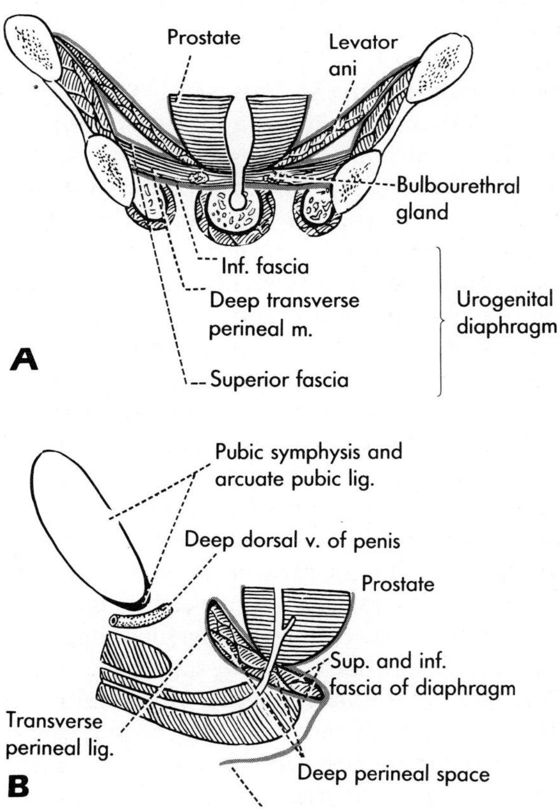


**FIG. 28-8.** The urogenital diaphragm and the deep perineal space in coronal section through its posterior part, **A,** and in sagittal section, **B.** The fascial layers are shown in blue.

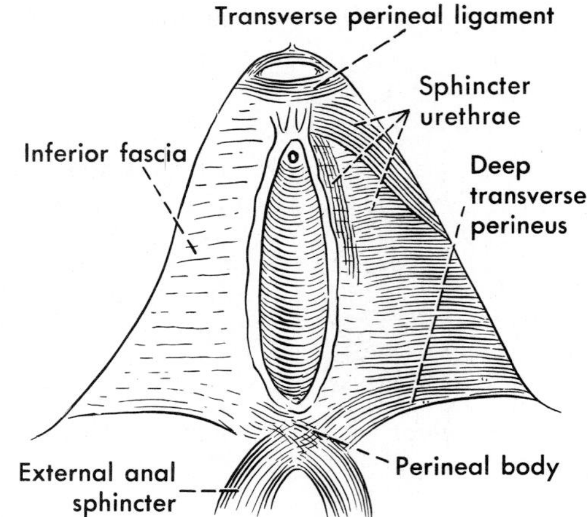


**FIG. 28-9.** The urogenital diaphragm of the female, from below. On the reader's left, the inferior fascia is intact; on the right, it has been removed to show the musculature.

urogenital region is more obvious in the male, because it is uninterrupted except for the passage of the urethra.

**The Deep Perineal Muscles.** The muscle sheet of the urogenital diaphragm consists of the deep muscles of the perineum. For descriptive purposes, two main parts are distinguished in this muscle: anteriorly, the *sphincter urethrae,* and posteriorly, the *deep transversus perinei* (Figs. 28-7 and 28-9). The fibers of both muscles run, in general, transversely from each ischiopubic ramus toward the midline. Those in front arch around the urethra and make up the **sphincter urethrae;** those further posteriorly meet in the central tendon of the perineum and are called the **deep transverse perineal muscle.** It may be possible to define some intrinsic circular fibers around the urethra in the sphincter urethrae and others that run on each side of the urethra from the transverse perineal ligament to the perineal body.

In the female, the urethra is embedded in the anterior wall of the vagina, the sphincter urethrae arches around the vagina as well as the urethra, and some of its fibers blend with the vaginal wall (Fig. 28-9). As in the male, the deep transversus perinei inserts into the perineal body, located behind the vagina.

Both deep muscles of the perineum are innervated by perineal branches of the pudendal nerve. Their contraction puts tension on the perineal body and elevates it; thereby, they presumably contribute to the support this body provides for the pelvic viscera. The sphincter urethrae compresses the urethra, especially when there is urine in the bladder. It can also interrupt the stream of urine.

## The Perineal Body

More recently called the *central tendon of the perineum,* the perineal body is a pyramid-shaped mass of fibromuscular tissue, rather than a flat or round tendon, located between the anal canal and the vagina (or the bulb of the penis). A number of perineal muscles terminate in it. The perineal body is larger in the female than in the male, filling the space between the divergent vagina and anal canal (Fig. 28-10). It is of considerable importance in obstetrics and gynecology.

The base of the perineal body rests against the perineal skin between the anus and the vestibule of the vagina; its apex points into the urogenital hiatus, where it is continuous with the rectovaginal septum.

The muscles whose fibers are interlaced in it include the pubovaginalis and pubococcygeus (parts of the pelvic diaphragm), both deep perineal muscles (the sphincter urethrae and deep transversus perinei), the sphincter ani externus, and, of the superficial perineal muscles, the bulbospongiosus and the superficial transversus perinei, which are described below.

The perineal body permits a remarkable degree of stretching of the perineum as the presenting fetal head distends the vagina and the entire perineum. Overstretching may tear the perineal body, and such tears may or may not involve the perineal skin.

Deliberate division of the perineal body during delivery is called an **episiotomy.** A cut is made with a pair of scissors toward the anus, starting at the midpoint of the posterior margin of the vaginal orifice (the introitus). This is done to prevent an uncontrolled or concealed tear and to make suturing

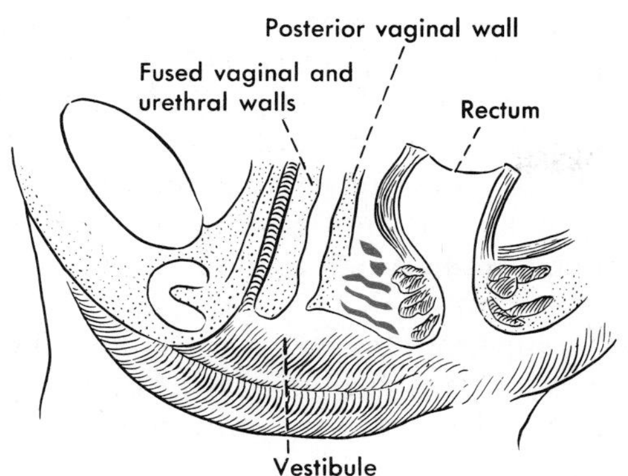


**FIG. 28-10.** The central tendon or perineal body (red) in saggital section.

of the perineum easier than it would be to repair a tear. When the perineal body is torn, contraction of the perineal muscles attached to it will widen the gap in the tear and, rather than strengthening, will weaken the support of pelvic viscera.

The support provided by the perineal body for pelvic viscera is illustrated by the sequelae that develop should its integrity not be restored. These sequelae include various types of *prolapse*, discussed in Chapter 27.

## The Deep Perineal Space

The superior and inferior fascias of the urogenital diaphragm enclose what is called the deep perineal space, or pouch. In truth, this is not a space, not even a potential one, since it is completely filled by the deep perineal muscles, whose fibers attach to both fascial membranes that cover the surfaces of the diaphragm.

The deep perineal space is completely closed, and it does not communicate with other perineal or pelvic spaces. The rather thin **superior fascia of the diaphragm** forms the floor of the anterior recesses of the ischiorectal fossa and medially fuses with the inferior fascia of the pelvic diaphragm (Fig. 28-8). No communication exists between the pelvic cavity and the perineum around the urethra or vagina, which pass through the deep space, because the urogenital diaphragm is fused to the wall of these structures. The **inferior fascia of the urogenital diaphragm** is significantly thicker and is often called the **perineal membrane.** This fascia forms the roof of the superficial perineal space and to it are attached the external genitalia.

The deep perineal space contains, in addition to the deep perineal muscles and the urethra and vagina, the bulbourethral glands in the male. The branches of the pudendal nerve and internal pudendal artery entering the deep space from the pudendal canal are buried among the fibers of the deep perineal muscles as they attach along the inferior pubic ramus.

**The Membranous Urethra.** The part of the urethra between the superior and inferior fascias of the urogenital diaphragm is called the membranous urethra (*pars membranecea urethrae*). The sphincter urethrae that surrounds the membranous urethra extends above the diaphragm, especially in the anterior urethral wall, into the prostatic part of the male urethra or the pelvic part of the female urethra.

**In the male,** the membranous urethra is the shortest segment of the urethra and its walls are the thinnest. It connects the prostatic and spongy parts of the urethra, and as it penetrates the urogenital diaphragm, it is surrounded by the urethral sphincter. It is approximately a centimeter long, passing from the apex of the prostate into the bulb of the penis (Fig. 28-11). It enters not the back end but the upper surface of the bulb and, as it does so, turns forward in the corpus spongiosum.

The **female urethra** terminates at the urethral orifice in the vestibule as soon as it has passed through the urogenital diaphragm. The **paraurethral glands,** considered by some to be the homologues of the prostate, are embedded in the wall of the urethra and their ducts open along the urethra into the vestibule. There are simple, small mucous glands opening into the lumen of both the male and female urethra.

The membranous part of the urethra in the male is especially subject to damage through physical violence. It may be perforated by ill-advised attempts to force a probe through it, or it may be

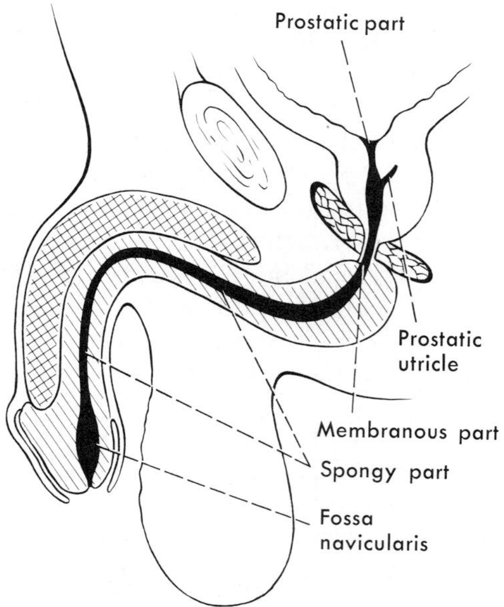


**FIG. 28-11.** Subdivisions of the urethra.

ruptured below the diaphragm as a result of a fall astride an object. Rupture above the diaphragm usually results from a fracture of the pelvis with displacement of the prostatic part of the urethra. In crushing injuries of the pelvis, common in automobile accidents, the bladder and prostate are displaced upward and backward with such force that the urethra ruptures between the apex of the prostate and the pelvic diaphragm.

**The Bulbourethral Glands.** A small gland located on each side of the membranous urethra is embedded in the muscle of the urogenital diaphragm (Fig. 28-16). The ducts of these bulbourethral glands descend through the diaphragm, piercing the perineal membrane, and terminate in the spongy urethra. The clear mucus they secrete during sexual excitation lubricates the urethral orifice and the glans penis. The female homologues of the glands are the greater vestibular glands, located below the inferior fascia of the urogenital diaphragm.

## The Superficial Perineal Space

Deep to the skin of the urogenital region, a potential space surrounds the external genitalia; this space, called the superficial perineal space, or pouch, is limited by the superficial perineal fascia and the inferior fascia of the urogenital diaphragm. It is more extensive and clinically more important in the male than in the female, because, in the male, it surrounds the scrotum and penis, whereas in the female, it is split by the vestibule and is confined on each side to the labia majora and minora.

The **superficial perineal fascia** is a continuation of the membranous layer of the superficial fascia from the anterior abdominal wall into the perineum (see Fig. 23-2). Frequently known as Colles' fascia, the superficial perineal fascia is a thin membrane; the *superficial penile fascia* and the *tunica dartos* are two named parts of it. At the symphysis pubis, the membranous superficial fascia of the abdomen becomes the superficial penile fascia, which invests the penis. This fascia, doubling back on itself in the prepuce, attaches to the penis around the neck of the glans (Fig. 28-12). Below the penis, the superficial penile fascia invests the scrotum as the tunica dartos, acquiring a significant amount of smooth muscle (p. 723). Posterior to the scrotum, the dartos continues as a fibrous membrane called the *superficial perineal fascia* and attaches to the posterior margin of the urogenital diaphragm. Lateral to the penis and scrotum, the superficial perineal fascia is attached to the ischiopubic rami, thus sealing the superficial perineal space everywhere except anterosuperiorly. The lateral and posterior attachments of the superficial perineal fascia are similar in the female perineum, but medially, the fascia fuses with the margins of the vaginal orifice.

Another layer of membranous fascia can be defined in the urogenital region that is deep to the superficial perineal fascia. It surrounds more intimately the cavernous bodies that constitute the penis and clitoris and the superficial perineal muscles associated with them. This is the **deep perineal fascia,** which essentially divides the superficial perineal space into a superficial and deep compartment. The deep perineal fascia over the penis becomes the *deep penile fascia* and does not descend into the scrotum (Fig. 28-12). At the root of the penis, it attaches laterally to the ischiopubic rami and posteriorly to the mar-

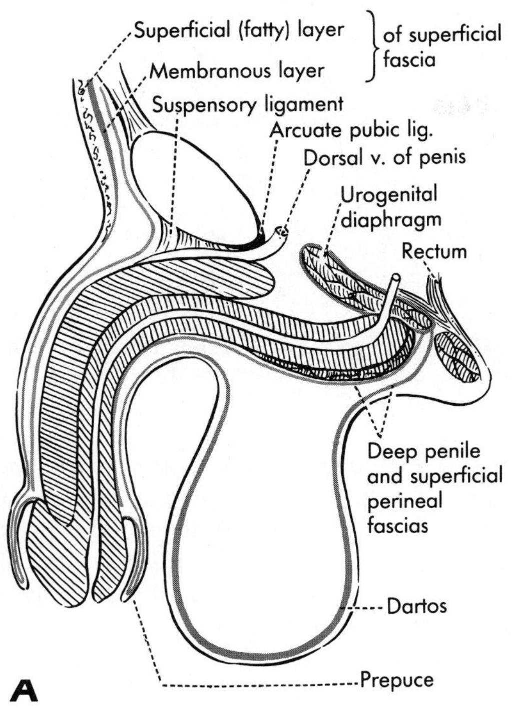


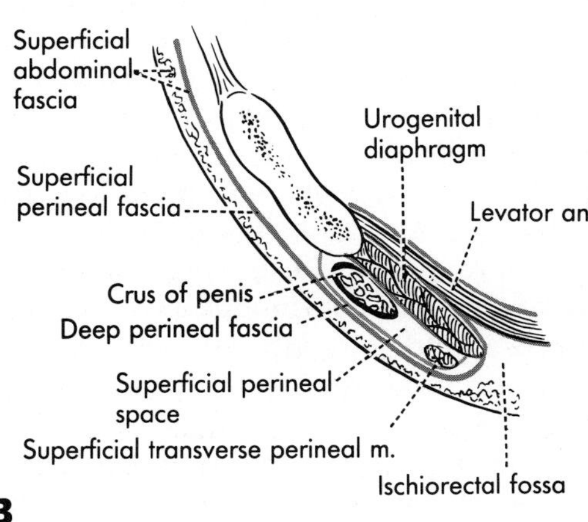


**FIG. 28-12.** Continuities of the subcutaneous connective tissue (superficial fascia) and deep fascia of the abdominal wall with fascia of the penis, scrotum, and perineum. **A** is a sagittal section close to the midline; **B**, one lateral to the penis and scrotum. The fascias of the urogenital diaphragm and the levator ani are blue, as in previous illustrations; the membranous layer of the superficial abdominal fascia is also blue, as is its continuation, the superficial penile and superficial perineal fascias. The deep perineal fascia and its continuation, the deep penile fascia, are red.

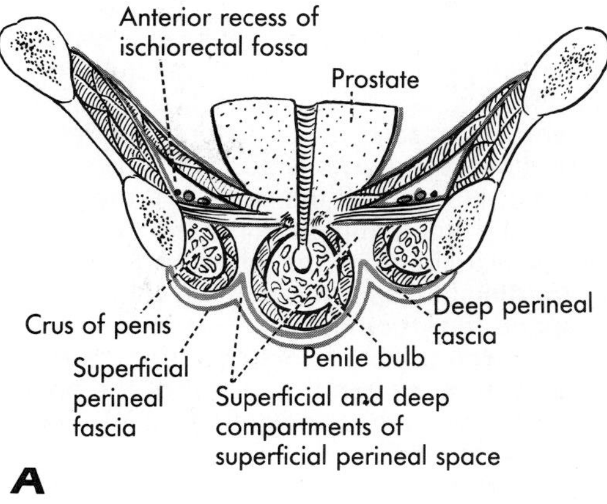


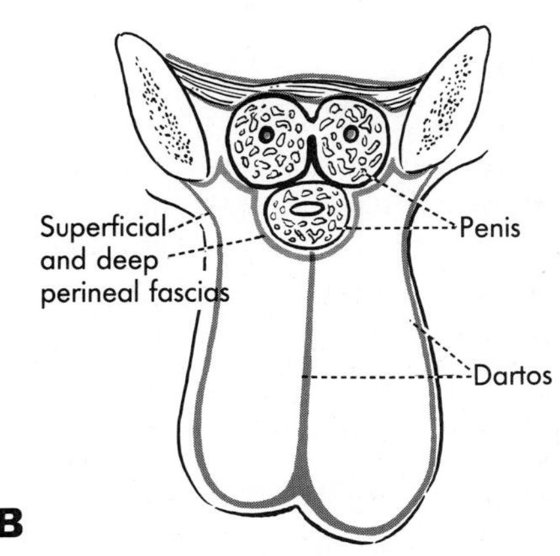


**FIG. 28-13.** Diagram of the superficial and deep perineal fascias and the dartos in coronal section through the level of the penile bulb, **A**, and the scrotum, **B.** The nerve and vessels lying in the "anterior recess" of the ischiorectal fossa are the dorsal nerve of the penis and the pudendal vessels. The color designation of the fascias corresponds to that in Figure 28-12.

gin of the urogenital diaphragm, just deep to the attachments of the superficial perineal fascia (Figs. 28-12 and 28-13). Anteriorly, the deep perineal fascia fuses to the symphysis pubis. Thus, the deep compartment of the superficial perineal space is closed anterosuperiorly, whereas the superficial compartment is continuous with the tissue space between the deep fascia of the external oblique and the membranous layer of the superficial fascia of the anterior abdominal wall.

Injuries to the male urethra below the urogenital diaphragm are frequent and cause extravasation of urine in the perineum. The extravasated urine will be confined to the perineum if it is limited by the deep perineal fascia; it will distend the deep compartment of the superficial perineal space, causing swelling of the shaft of the penis and the perineum between the ischiopubic rami. If the deep perineal fascia is also injured, the urine will escape into the superficial compartment of the superficial perineal space, distending not only the perineum between the ischiopubic rami in front of the anus but also the scrotal sac and the shaft of the penis, including the prepuce, and will ascend into the anterior abdominal wall.

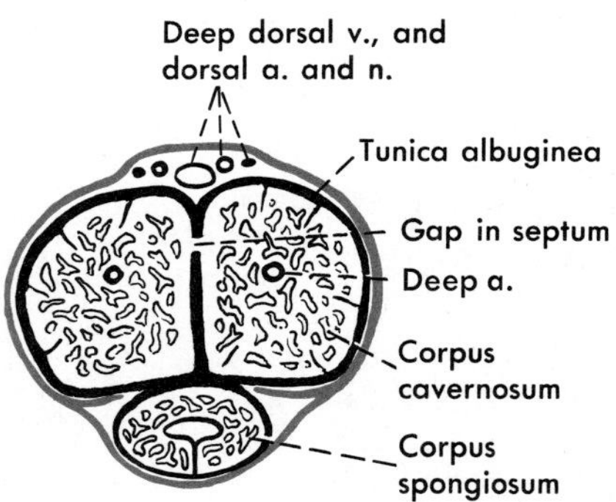


**FIG. 28-15.** Cross section through the body of the penis. The deep penile fascia is shown in red.

## THE MALE GENITALIA AND THE SUPERFICIAL PERINEAL MUSCLES

### The Penis

The penis is the male organ of copulation, capable of becoming hard and erect due to its engorgement with blood, a requirement for its intromission into the vagina. The penis consists of two parts: the *body* (*corpus*, or shaft) and the *root* (*radix*).

**FIG. 28-14.** The urethral surface of the penis. The muscles in the superficial perineal space have been removed to show the crura and the penile bulb.

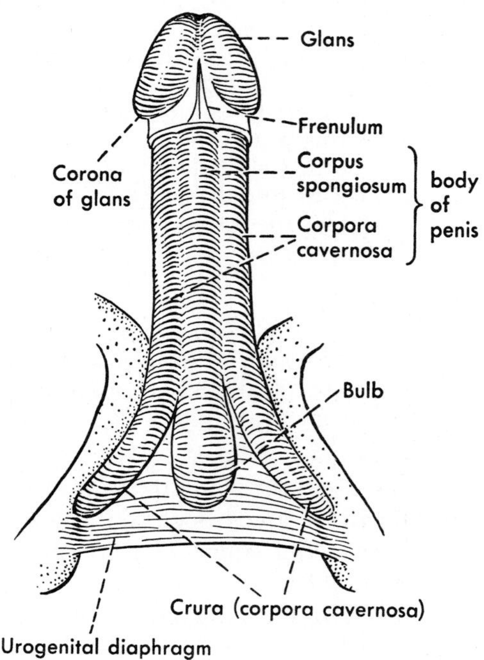


In the flaccid state of the organ, the **body** hangs free below the symphysis pubis, anterior to the scrotum, and terminates in an acornlike enlargement, the *glans penis*. The body has a so-called *dorsal surface*, which faces anterosuperiorly when the penis is flaccid, and a *urethral surface*, which faces the scrotum (Fig. 28-12). In the body of the penis, three parallel, cylindrical masses of erectile tissue are held together by the penile fascias and by the fusion of their capsules: along the midline of the urethral surface is the *corpus spongiosum*, containing the urethra, and along the dorsal surface are the two *corpora cavernosa* (Figs. 28-14 and 28-15).

The **root of the penis** is deep to the scrotum, attached to the inferior fascia of the urogenital diaphragm. It consists of the divergent *crura of the penis*, continuous with the corpora cavernosa, and the *bulb of the penis* between them, continuous with the corpus spongiosum (Fig. 28-14).

Each of the three corpora and their posterior extensions into the root of the penis are composed of spongy connective tissue with vascular sinuses, or *cavernae*, between the fibrous trabecular network. Their *tunica albuginea*, continuous with the trabeculae, forms the capsule of these cavernous bodies. The arteries of the penis penetrate the tunica albuginea, arborize in the trabeculae, and terminate in spiral branches called *helicine arteries*, which, in response to excitation of the cavernous nerves (derived from the "nervi erigentes"), pour arterial blood into the cavernae at a faster rate than it can leave through the cavernous veins. Erection is achieved by tumescence of the corpora with arterial blood. Thus, compression of veins or contraction of muscles plays no important part in erection; it is, in essence, an arterial phenomenon. The commingling of elastic fibers and smooth muscle with the predominantly collagenous trabeculae and tunics of the corpora permits the enlargement of the penis as it becomes erect and accounts also for the return to its smaller, flaccid state after excitation of the cavernous nerves ceases.

**The Corpora Cavernosa.** Each corpus cavernosum commences in the root of the penis as the **crus** attached to the everted medial surface of the conjoint ischiopubic ramus and to the perineal membrane along the bone. Reaching the inferior margin of the pubic symphysis, the two crura approximate each other, their tunicae albugineae fuse, and they continue forward into the body of the penis as the corpora cavernosa (Fig. 28-14). The median **septum** (*septum penis*) that separates the corpora is penetrated by blood vessels and, posteriorly, is fenestrated, permitting communication between the vascular spaces in the two corpora (Fig. 28-15). The corpus spongiosum is fused to the groove on the urethral surface of the corpora cavernosa without significant vascular connections. The blunt, distal ends of the corpora cavernosa stop short of the tip of the penis and are capped by the glans penis (Fig. 28-16).

**The Corpus Spongiosum.** Smaller in diameter along the body of the penis than the corpora cavernosa, the corpus spongiosum expands anteriorly as the glans and posteriorly as the bulb of the penis. It encloses along its entire length the spongy part of the urethra.

The base of the **glans** is wider than the circumference of the corpora cavernosa; its projecting margin, the *corona glandis,* shelters a groove at the junction of the glans with the corpora cavernosa called the *neck of the glans* (Fig. 28-14). Extending from the tip of the glans to its urethral surface is the slitlike opening of the urethra, the *external urethral ostium.*

The posterior expanded end of the corpus spongiosum, the **bulb of the penis,** is tightly attached to the inferior fascia of the urogenital diaphragm between the crura (Figs. 28-14 and 28-16). Its adherent upper surface, rather than its convex posterior end, is penetrated by the urethra, which turns forward as soon as it enters the bulb (Fig. 28-11). The ducts of the bulbourethral glands, descending from the deep perineal space, also penetrate the bulb on each side of the urethra and terminate in it after traversing the bulb (Fig. 28-16).

**The Spongy Urethra.** The terminal part of the urethra, called the spongy urethra (*pars spongiosa urethrae*), is continuous proximally with the membranous urethra (p. 797). It extends within the corpus spongiosum from the perineal membrane through the bulb to the external urethral ostium on the glans penis (Fig. 28-11). Within the glans, the urethra dilates to form the **fossa navicularis,** but the external ostium is the narrowest and the least

**FIG. 28-16.** Diagrammatic sagittal section through the penis. The bulbourethral gland, shown here, actually lies lateral to the urethra (Figs. 28-7 and 28-8) and therefore would not show in a section close to the midline.

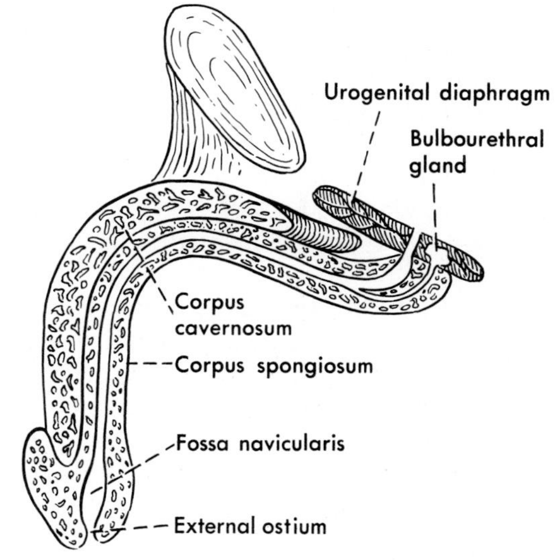

dilatable part of the urethra. A small mucosal valve may be present at the proximal end of the fossa, and a mucosal pit, the *lacuna magna,* is in the roof of the fossa. There are numerous smaller lacunae in the mucosal lining of the spongy urethra, and a number of mucous glands also open into it.

**Penile Fascias and Ligaments.** The superficial fascia of the perineum over the penis is called the **superficial penile fascia.** It is subjacent to the skin without any fat intervening between the two. It doubles back with the skin of the prepuce and is attached around the neck of the glans.

The **deep penile fascia** is continuous with the deep perineal fascia. It binds the corpora cavernosa and corpus spongiosum together in the body of the penis and also attaches around the neck of the glans. The apposition of these two fascias accounts for the mobility of the skin over the shaft of the penis. The superficial vessels and nerves of the penis run from the dorsum of the penis between superficial and deep penile fascias; they supply the skin. The deep dorsal vein, dorsal arteries, and dorsal nerves of the penis are enclosed by the deep penile fascia, and they penetrate the tunica albuginea of the corpora cavernosa and spongiosum (Fig. 28-15).

The proximal end of the body is attached to the pubic symphysis by the triangular **suspensory ligament** of the penis, which fuses with the deep penile fascia (Fig. 28-12). The **fundiform ligament** of the penis is a continuation of the median septum of the membranous layer of the superficial fascia attached to the linea alba (p. 580). The ligament encircles the body of the penis below the pubic symphysis and blends with the superficial penile fascia.

**The Penile Skin and Prepuce.** The skin over the body of the penis is thin, hairless, and very mobile. On the urethral surface, it presents a midline raphe continuous with the raphe of the scrotum. Over the glans, the penile skin forms a redundant hood, called the **prepuce** (*preputium penis*), by doubling back on itself and attaching around the neck of the glans (Fig. 28-12). The prepuce covers the glans to a variable extent and can be retracted from it completely. Over the urethral surface, the prepuce forms a small median fold called the *frenulum,* which leads to the posterior end of the external urethral ostium. Preputial glands are located along the frenulum, the corona, and the neck of the glans. They secrete *smegma,* a white sebaceous material that accumulates in the preputial sac. The skin over the glans is so thin that it is semitransparent and firmly bound to the tunica albuginea of the glans. Over the lips of the urethral meatus, it is continuous with the urethral mucosa.

**Circumcision.** Resection of the prepuce, or circumcision, has been a ritual practiced by certain races and in certain cultures since time immemorial. It has been adopted for putative hygienic reasons in some modern societies. Many generations of Americans have been routinely circumcised during the newborn period. The adoption of this practice was largely based on apparent evidence for the pathogenic effects of smegma: the incidence of cervical carcinoma among Jewish women, whose consorts have for generations been circumcised, was found to be lower than in uncircumcised populations. The validity of the interpretation of the findings has now been called into question. Furthermore, the availability of modern amenities of personal hygiene argues against the necessity or advisability of routine circumcision, even if smegma should be shown to be carcinogenic. Circumcision may be indicated when the preputial sac becomes chronically inflamed (*balanitis*) and causes fibrosis and stricture of the preputial skin (*phimosis*), preventing its retraction from the glans. Retraction of the prepuce is necessary for successful coitus.

### The Scrotum

The anatomy of the scrotum and its contents was discussed in Chapter 26.

### The Superficial Perineal Muscles

The muscles of the superficial perineal space are mostly associated with the root of the penis. They are rather insubstantial sheets of voluntary muscle wrapped around the crura and the bulb of the penis and a small transverse muscle along the posterior edge of the urogenital diaphragm (Fig. 28-17).

The **superficial transverse perineal muscle** passes from the front end of the ischial tuberosity to meet its fellow at the midline in the perineal body. Both perineal fascias turn around the posterior borders of these muscles to fuse with the posterior border of the urogenital diaphragm.

The **ischiocavernosus muscle,** on each side, surrounds the free surface of each crus of the penis and, like the penile crus, is attached to the ischiopubic ramus. The muscles end by a tendinous insertion into the corpus cavernosum of their own side, just as the two parts come together.

The **bulbospongiosus** arises, in part, from the central tendon of the perineum and raphe of the penis and wraps around the bulb and the posterior end of the corpus spongiosum. The muscle inserts into the upper surface of the bulb and the corpus spongiosum. The most anterior fibers of the muscle pass around the entire body of the penis to end on its dorsal aspect deep to the deep penile fascia. Through the central tendon, there may be a variable amount of continuity with the sphincter ani externus.

All these muscles are covered by the deep perineal fascia and are supplied by twigs from the perineal branches of the pudendal nerve. The bulbospongiosus aids the emptying of the urethra at the end of micturition and during ejaculation. The functions of the ischiocavernosus and superficial transverse muscles are not readily obvious.

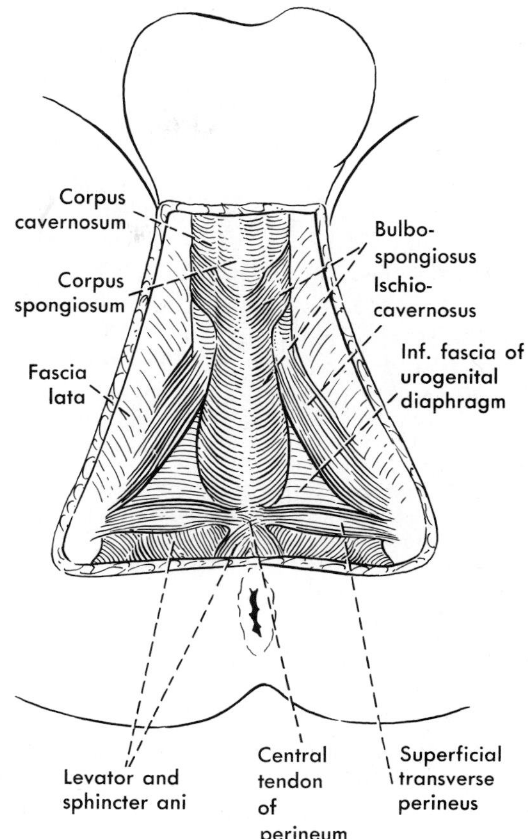


**FIG. 28-17.** Muscles of the superficial perineal space in the male. The superficial perineal fascia has been removed.

## THE FEMALE EXTERNAL GENITALIA

The female external genitalia are collectively referred to as the **vulva.** The vulva consists of the mons pubis, the labia majora and minora, the clitoris, and the bulb of the vestibule and the vestibule of the vagina, into which open the orifices of the vagina, the urethra, and the ducts of the paraurethral and vestibular glands (Fig. 28-18).

The *mons pubis* is a conspicuous, subcutaneous fat pad over the pubic bones and symphysis. It is covered by pubic hair and is largely absent in the male. The other components of the vulva are homologues of the male external genitalia. Their anatomy will be discussed before considering the developmental relationships between the respective parts.

**The Labia Majora and Minora.** Each **labium majus** is a broad, longitudinal fold of skin filled with subcutaneous fat and fibrous tissue, continuous anteriorly with the subcutaneous tissue of the mons pubis and posteriorly with ischiorectal fat. The lateral surface of the labia facing the thighs is hairy; their smooth medial surface is studded with sebaceous glands and encloses the *pudendal cleft* (*rima pudendi*). The junction of the two labia anteriorly and posteriorly are the rather indistinct *anterior* and *posterior commissures.*

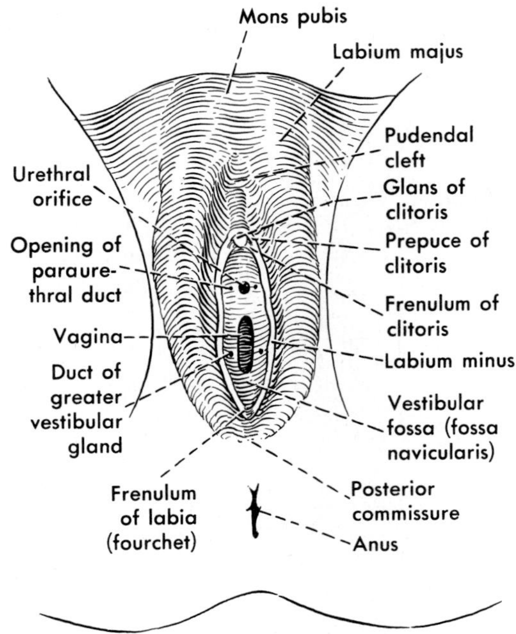


**FIG. 28-18.** External genitalia of the female.

The labia majora are homologues of the two halves of the scrotal sac, and some subcutaneous smooth muscle in them, the homologue of the dartos. The fibromuscular strands of the round ligament of the uterus are interlaced in the connective tissue of the labia, and a patent processus vaginalis may extend into them.

Each **labium minus** is a smaller fold of skin in the pudendal cleft and, unlike the labia majora, contains no fat. Their lateral surfaces are in contact with the smooth, inner surface of the labia majora, and their medial surfaces are in contact with each other. Between them is the vestibule of the vagina, which is opened up by separation of the labia minora. Posteriorly, the labia minora are united by a fold, the *frenulum of the labia,* also called the "fourchette." Anteriorly, they approach the clitoris, and each labium divides into two tiny folds that fuse with their fellow of the opposite side around the clitoris. The two upper folds unite over the clitoris to form the *prepuce of the clitoris;* the two lower folds meet each other on the undersurface of the clitoris as the *frenulum of the clitoris*. Glands located along these folds secrete white sebaceous material.

The labia minora are homologues of the skin that covers part of the penis. Along the base, and partly in the substance of each labium minus, is located an oval-shaped mass of erectile tissue called the **bulb of the vestibule.** Each bulb is a homologue of half of the bulb of the penis and the posterior part of the corpus spongiosum. The bulb of the vestibule, however, does not surround the urethra. Each bulb is attached to the inferior fascia of the urogenital diaphragm and is enclosed by the deep perineal fascia. As they taper toward the clitoris, the bulbs are joined to one another and to the undersurface of the clitoris by the *pars intermedia* and the *commissura of the bulb* (Fig. 28-19).

**The Clitoris.** The clitoris is the homologue of the penis, but it consists of only two erectile bodies, the *corpora cavernosa clitoridis,* and is not traversed by the urethra. The corpora cavernosa commence as the *crura of the clitoris,* attached to the ischiopubic rami and the inferior fascia of the urogenital diaphragm. They unite in the midline to form the body of the clitoris, which is connected to the symphysis pubis by the *suspensory ligament.* The body ends in the tiny *glans* (Fig. 28-19).

Like the penis, the clitoris and the bulb of the vestibule become tumescent during sexual excitation. The clitoris, richly supplied with sensory nerve endings, plays a dominant role in the excitatory phase of the sexual response.

**The Vestibule of the Vagina.** The space bordered by the labia minora and their frenulum is the vestibule of the vagina. The anterior part of the vestibule receives the **external urethral ostium,** the margins of which are rather raised and puckered. On each side of this ostium are the tiny openings of the *paraurethral glands* (of Skene). A short distance posterior to the urethral orifice is the **orifice of the vagina,** also known as the *introitus.* Although above the urogenital diaphragm the lumen of the vagina is a transverse slit, the vaginal orifice is elongated in the sagittal plane (Fig. 28-19). It may be partially closed by the hymen. The **hymen** is a thin fold of mucous membrane of variable extent and shape; it is usually crescentic, covering the posterior margin of the vaginal orifice. If it is imperforate, it must be incised at the time of puberty to provide exit for the menstrual flow. After the hymen has been ruptured, by coitus or by other means, it is recognizable only as small tags of mucous membrane, the *carunculae hymenalis*.

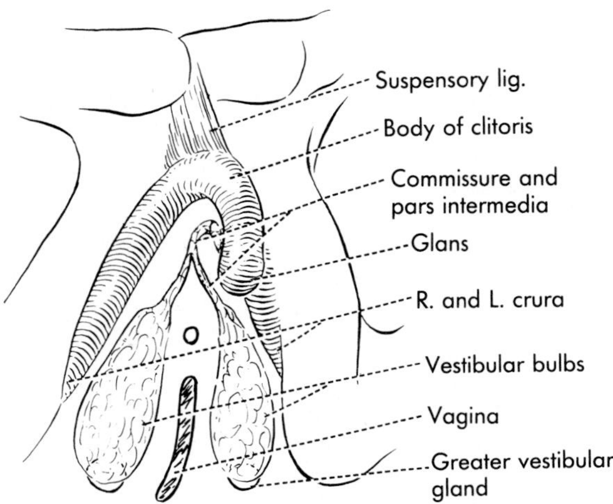


**FIG. 28-19.** The clitoris. The connections of the vestibular bulbs with the glans clitoridis are shown.

The space in the vestibule posterior to the vaginal opening is the *vestibular fossa*. Into it open, on each side, the ducts of the greater vestibular glands. The **greater vestibular glands,** associated for a long time with the name of Bartholin, are roughly pea-sized, lobulated structures located at the posterior pole of the bulbs of the vestibule. They are the homologues of the bulbourethral glands. They, like the numerous *lesser vestibular glands,* which also open into the vestibule, secrete mucus and lubricate the vulva.

## The Superficial Perineal Muscles

The superficial perineal space of the female contains essentially the same structures as that of the male. The muscles associated with the erectile bodies of the female external genitalia correspond to those of the male (Fig. 28-20). The **ischiocavernosus muscles** surround the crura of the clitoris; the **bulbospongiosus muscles** remain separate on the two sides, and, as they enclose the bulb of the vestibule, they surround the orifice of the vagina. Posteriorly, these are attached to the perineal body, as are the two **superficial transverse perineal**

**FIG. 28-20.** Muscles of the superficial perineal space in the female.

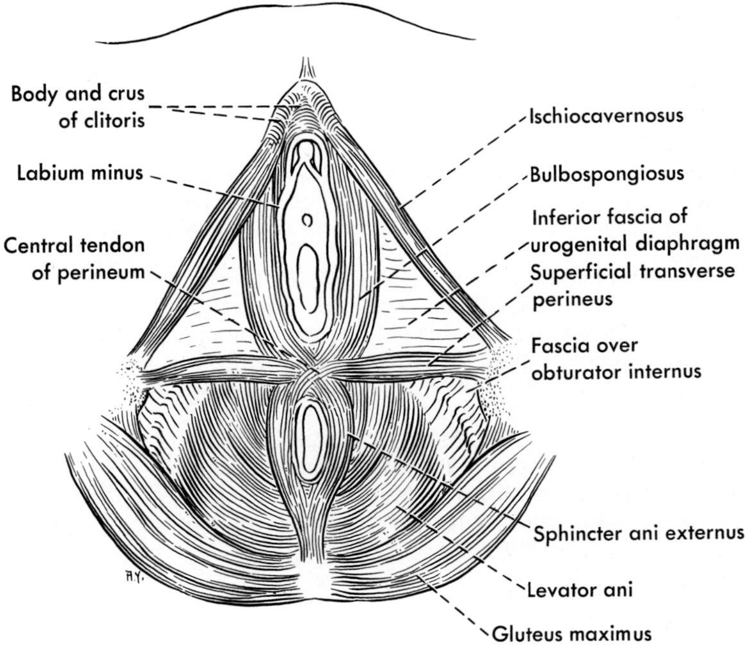

**muscles.** All these muscles are less well developed than in the male.

### DEVELOPMENT OF THE EXTERNAL GENITALIA

As the urorectal septum reaches the ventral surface of the tail, it divides the cloacal membrane into the anal membrane posteriorly and the urogenital membrane anteriorly (p. 756; see Fig. 27-15). The caudal edge of the septum contacting the cloacal membrane will form the perineal body. The primordia of the external genitalia develop as swellings around the cranial and lateral margins of the urogenital membrane: the *genital tubercle* in the midline cranially, and the *urogenital folds* and *labioscrotal swellings* laterally. The genital tubercle soon elongates into the *phallus* and along its caudal surface draws forward with it a narrow prolongation of the urogenital membrane, deep to which is the *phallic part of the urogenital sinus.* The margins of the urogenital membranes soon become prominent, forming the *urogenital folds* and causing the urogenital membrane to lie in a *sulcus.* Just lateral to the urogenital folds, a pair of swellings appear called the *labioscrotal swellings* (Fig. 28-21). The membrane in the urogenital sulcus breaks down, opening the phallic part of the urogenital sinus to the amniotic cavity and permitting urine formed by the mesonephros and metanephros to be voided. The urogenital folds, united by a commissure posteriorly, form the lips of the open phallic urogenital sinus, continuous above with the pelvic part of the urogenital sinus (p. 756).

These events establish, in essence, the definitive anatomy of the female external genitalia. From the phallus develop the cavernous bodies of the corpora cavernosa and glans clitoridis; from the urogenital folds, the labia minora with the bulb of the vestibule within their substance; and from the labioscrotal swellings, the labia majora. The phallic urogenital sinus becomes the vestibule of the vagina into which open the urethra, representing the pelvic part of the urogenital sinus, and the vagina (p. 760).

The establishment of the definitive anatomy of the male external genitalia requires three additional events that modify this simple arrangement. (1) The urogenital folds fuse along the phallus, resealing the phallic urogenital sinus, uniting the two parts of the bulb of the penis, which develop in the urogenital folds, and enclosing the spongy urethra. (2) The navicular fossa of the urethra is formed by the canalization of the glans penis. (3) The labioscrotal swellings approximate each other and fuse on the ventral surface of the root of the phallus and form the scrotal sac.

**Anomalies.** Congenital anomalies of the external genitalia are common. They may be minor, or they may make sex determination difficult and are often associated with maldevelopment of the gonads, the mesonephric and paramesonephric duct systems, and the kidneys.

Fusion of the urogenital folds and labioscrotal swellings may be arrested at various stages in the male. In its most severe form, the perineum resembles the vulva. When the abnormality is limited to incomplete fusion of the urogenital folds, the urethra does not reach the glans and opens somewhere along the urethral surface of the penis. This is known as **hypospadias.** When the urethral opening is close to the glans, this imposes no disability. However, when the opening is more proximal on the body of the penis, the corpus spongiosum does not develop properly and, distal to the opening, is replaced by a fibrous band that causes a marked ventral flexure of the penis. A little boy so affected cannot urinate standing up, and in an adult, the penile flexure prevents coitus. In such cases, the penis must be straightened surgically and the urethra reconstructed.

Rarely, there may be a urethral opening on the dorsum of the penis. This is called **epispadias,** and the abnormality is related to **extrophy of the bladder.** When somatic mesoderm fails to migrate into the ventral body wall distal to the umbilicus, the surface ectoderm apposed to the endoderm of the vesical part of the urogenital sinus breaks down, and the trigone of the bladder becomes exposed on the anterior abdominal wall. This is extrophy of the bladder. The defect in the anterior abdominal wall may extend to the dorsum of the phallus, or it may be limited to the phallus. The latter condition is epispadias. Both extrophy and epispadias may occur in the male and in the female.

Congenital anomalies of the female external genitalia consist of various degrees of fusion between the urogenital folds (labia minora) and labioscrotal swelling (labia majora) usually associated with hypertrophy of the clitoris.

## BLOOD SUPPLY, LYMPHATIC DRAINAGE, AND INNERVATION

The introduction to this chapter has already identified the internal pudendal artery and pudendal nerve as the chief sources of blood and nerve supply to the perineum. The course of these structures has been traced from the pelvis, through the pudendal canal, into the urogenital region (p. 794). The purpose of this section is to deal, in particular, with the branches of distribution of this artery and nerve, as well as with others that supplement the supply of the perineum, and to enlarge on

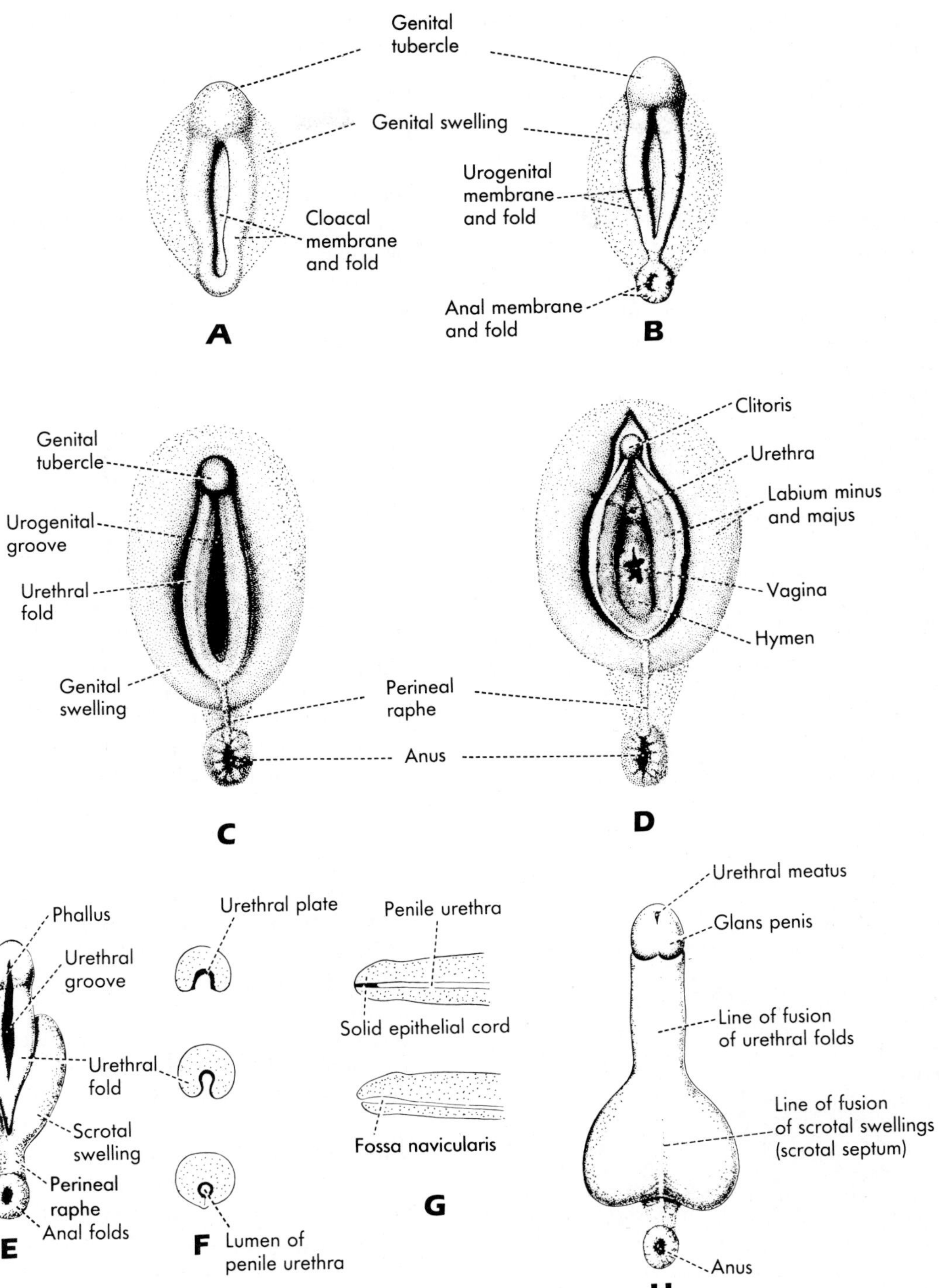


**FIG. 28-21.** Development of the external genitalia. **A** and **B**, indifferent stage; **C** and **D**, female; **E** through **H**, male. **F** and **G** show the development of the penile urethra. (Langman J: Medical Embryology, 3rd ed. Baltimore, Williams & Wilkins, 1975)

the general statements made in the chapter with reference to the venous and lymphatic drainage of the perineum.

## ARTERIAL SUPPLY

### The Internal Pudendal Artery

A branch of the anterior division of the internal iliac artery in the pelvis, the internal pudendal artery, enters the perineum from the buttock through the lesser sciatic foramen and passes downward and forward in the lateral wall of the ischiorectal fossa enclosed with the pudendal nerve in the pudendal canal. Reaching the urogenital diaphragm, it passes along the ischiopubic ramus in the deep perineal space and terminates by dividing into a *deep* and *dorsal artery of the penis* or *clitoris* before it reaches the transverse perineal ligament (Fig. 28-22).

**FIG. 28-22.** The internal pudendal artery. The perineal artery and its branches are cut away on the right, as is most of the corpus cavernosum of the penis, to show the deep branches; the course of the artery through the urogenital diaphragm and the anterior recess of the ischiorectal fossa is indicated by dotted outline.

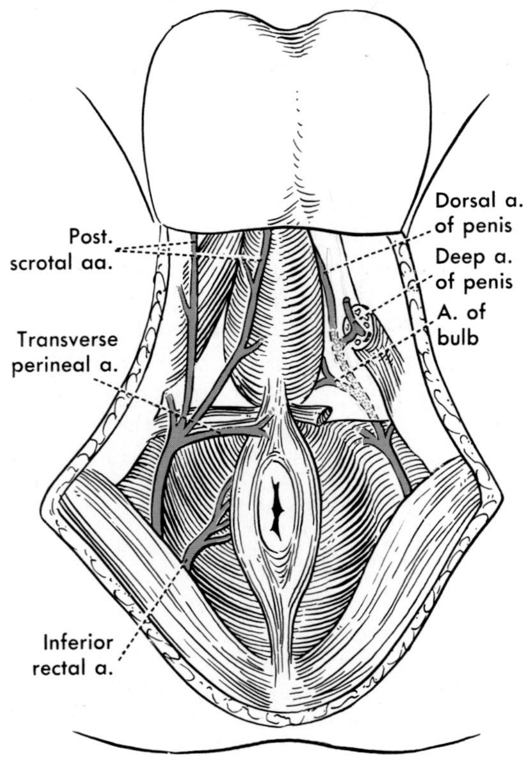


In addition to its terminal branches and a number of small muscular branches distributed to the pelvic and urogenital diaphragms, the internal pudendal artery gives off a number of named branches: the inferior rectal, the perineal, and the urethral arteries and the artery of the bulb of the penis or the bulb of the vestibule.

The first branch of the artery, the **inferior rectal,** is given off just above the ischial tuberosity. It runs medially across the ischiorectal fossa to supply the muscles and the lining of the anal canal and the perianal skin. In the wall of the canal, it anastomoses with branches of the middle and superior rectal arteries and with those of the inferior rectal artery from the opposite side.

The **perineal artery** is given off in the vicinity of the posterior margin of the urogenital diaphragm. It almost immediately divides into *transverse perineal* and *posterior scrotal* or *posterior labial* branches (Fig. 28-23). These run through the superficial perineal space and supply perineal subcutaneous tissue in the scrotum or the labia, as implied by their name. They also give twigs to the lower part of the vagina along with twigs from other branches of the internal pudendal.

The **artery of the bulb** arises within the deep perineal space just lateral to the bulb and runs medially to pierce the inferior fascia of the urogenital diaphragm and enter the bulb. It also supplies the bulbourethral gland. In the female, it is distributed to the bulb of the vestibule and the greater vestibular gland.

The **urethral artery** arises distal to, or sometimes with, the artery of the bulb; it also pierces the inferior fascia and, after entering the corpus spongiosum, runs distally in this body to anastomose with branches of the dorsal artery. In the female, this branch is absent or inconspicuous.

The **deep artery of the penis** or **clitoris,** one of the terminal branches of the internal pudendal, leaves the deep perineal space by piercing the perineal membrane and entering

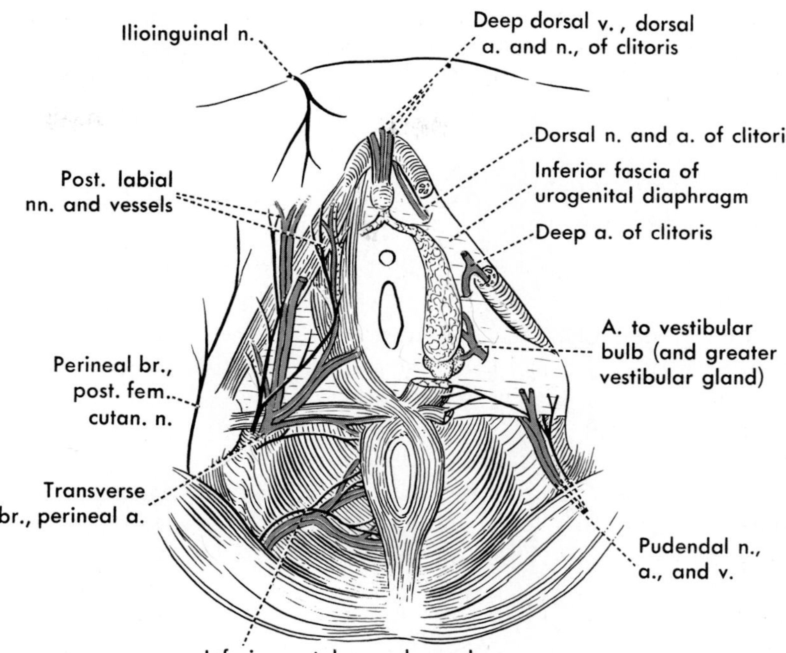


**FIG. 28-23.** Nerves and vessels of the female perineum.

the crus of the penis or the clitoris attached to that membrane. It continues along the axis of the corpus cavernosum, supplying the erectile tissue and anastomosing with branches of the dorsal artery of the penis or clitoris that pierce the tunica albuginea of the corpora cavernosa. The **dorsal artery of the penis** or **clitoris** pierces the inferior fascia of the diaphragm farther forward, in company with the dorsal nerve, medial to the crus. The two arteries and nerves reach the dorsal surface by passing between the corpora cavernosa and corpus spongiosum. The arteries run on each side of the centrally placed deep dorsal vein of the penis, and along the lateral side of the arteries run the dorsal nerves (Fig. 28-24). Over the body of the penis, the vessels and nerves are deep to the deep penile fascia, in contact with the tunica albuginea of the corpora cavernosa (Fig. 28-15). As the arteries are traced forward on the penis, the vessels and nerves give off branches that run around the body of the penis. Their deep branches penetrate the corpora cavernosa and spongiosum. The artery terminates in the glans penis. The dorsal artery of the clitoris pursues a similar course (Fig. 28-23). In the corpora cavernosa and in the glans, these arteries anastomose with the deep artery of the penis or clitoris.

### The External Pudendal Arteries

The blood supply to the skin and superficial fascia of the external genitalia is provided largely by the external pudendal arteries (Fig. 28-25). There are usually two on each side, a *deep* and a *superficial external pudendal artery*. They are branches of the femoral artery that emerge through the cribriform fascia and pass in the fatty subcutaneous tissue medially across the thigh and the spermatic cord or the round ligament. Their branches enter the scrotum or labia majora and the dorsum of the penis. They anastomose with the posterior scrotal or labial branches derived from the internal pudendal artery and with the branches of the dorsal artery of the penis.

## VENOUS DRAINAGE

Venous tributaries accompany all the branches of the internal and external pudendal arteries and receive names corresponding

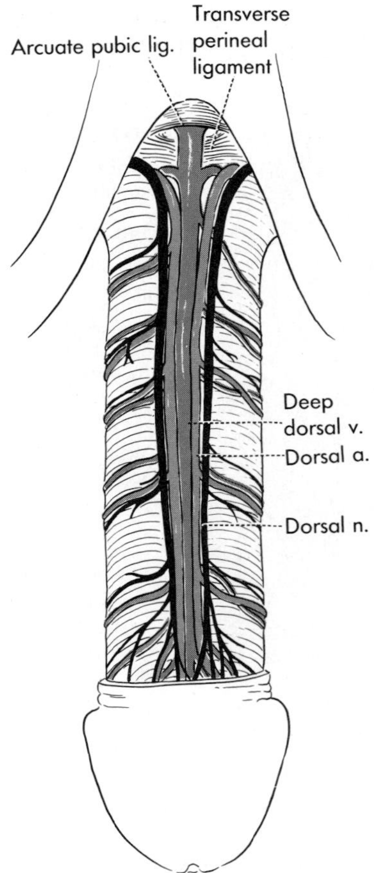


**FIG. 28-24.** Dorsal nerves and vessels of the penis, after removal of the penile fascia.

to their arterial branches (Figs. 28-23 and 28-25). Veins from the bulb of the penis or vestibule, the scrotum or labia majora, and the anal canal join to form two venae comitantes of the internal pudendal artery, called the **internal pudendal veins,** which terminate in the internal iliac vein on reaching the pelvis through the pudendal canal and the buttock. The **external pudendal veins** are formed on each side by tributaries from the skin of the penis, prepuce (*superficial dorsal vein of the penis*), and scrotum or labia majora and terminate in the great saphenous vein. (There is no vein along the deep artery of the penis or clitoris, and most of the blood from the cavernae of the corpora cavernosa and the glans penis or clitoris is drained by tributaries of the **deep dorsal vein of the penis** or **clitoris;** Figs. 28-23 and 28-24). The deep dorsal vein is an unpaired vein that runs between the two dorsal arteries of the penis or clitoris deep to the deep fascia. It leaves the perineum through the gap between the transverse perineal ligament and the arcuate pubic ligament and joins the prostatic plexus in the male and the vesical plexus in the female. It communicates with tributaries of the internal pudendal veins.

## LYMPHATIC DRAINAGE

Most perineal structures send their lymphatics along the branches of the external pudendal vessels to the superficial inguinal nodes. A few lymphatics from the deep structures of the perineum pass with the internal pudendal vessels into the pelvis to end in the internal iliac nodes. Thus, lymphatics of the lower part of the anal canal, the perineal skin (including the scrotum as well as the glans penis), the spongy urethra, and the entire vulva drain to the inguinal nodes. Enlargement of these nodes may be the first sign of an infective or neoplastic lesion in the superficial tissues of the perineum. Lymphatics from the deep perineal space, the membranous urethra, and the vagina just above the hymen, as well as some part of the anal canal, drain along the internal pudendal vessels in the internal iliac nodes. The upper part of the anal canal drains upward along superior rectal vessels (see Fig. 27-20), and the upper part of the vagina has the same lymphatic drainage as the cervix uteri (p. 786).

## INNERVATION

### The Pudendal Nerve

Formed by S2–S4 ventral rami, the pudendal nerve is a branch of the sacral plexus. It reaches the perineum with the internal pudendal artery and gives off branches that correspond closely to those of the artery (Figs. 28-23 and 28-26). Approaching the posterior edge

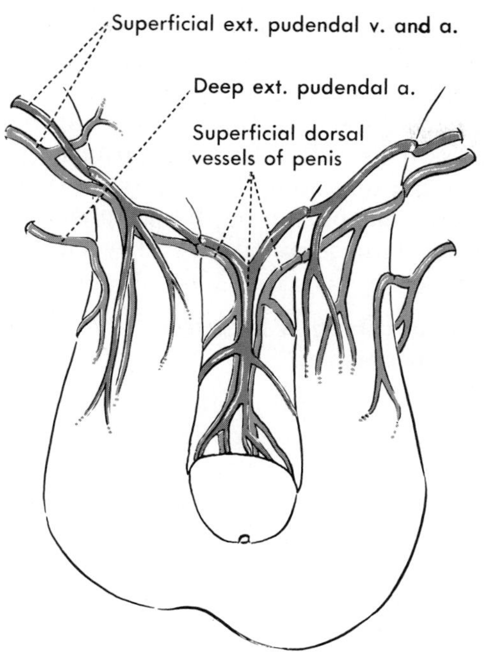


**FIG. 28-25.** The external pudendal vessels. Those on the penis supply the skin and subcutaneous tissue, not the erectile tissue.

of the urogenital diaphragm, the nerve divides into its terminal branches, the *perineal nerve* and the *dorsal nerve of the penis or clitoris,* while it is within the canal. The pudendal nerve gives off the inferior rectal nerve in the posterior part of the ischiorectal fossa before it divides into its terminal branches.

The **inferior rectal nerve** crosses the ischiorectal fossa from its lateral wall to the anal canal, running with the vessels of similar name. It supplies the external anal sphincters, the lining of the lower part of the anal canal, and the perianal skin.

The **perineal nerve** breaks up into numerous muscular branches and two long cutaneous nerves, the *posterior scrotal* or *posterior labial* branches. The muscular branches are distributed to all the deep and superficial muscles of the perineum, including the external anal sphincter and a part of the levator ani. One of the branches enters the bulb of the penis or vestibule. The posterior scrotal or labial branches pass forward in the superficial perineal space and are distributed to the scrotum or labia majora.

The **dorsal nerve of the penis** or **clitoris** continues forward along the ischiopubic rami with the dorsal artery of the penis or clitoris, running either in the deep perineal space or in the anterior recess of the ischiorectal fossa. The nerve gives a branch to the corpus cavernosum and then reaches the dorsum of the penis or clitoris running on the lateral side of the dorsal artery of the penis or clitoris (Figs. 28-23 and 28-24). It terminates in the glans penis or clitoris.

Thus, in essence, the pudendal nerve is the

**FIG. 28-26.** Nerves in the superficial perineal space in the male. On the reader's right, the posterior scrotal nerves (superficial branches of the perineal nerve) have been cut away. In the ischiorectal fossa, the inferior rectal nerve, a branch of the pudendal nerve, is shown, but the dorsal nerve of the penis, lying above the perineal nerve, is omitted.

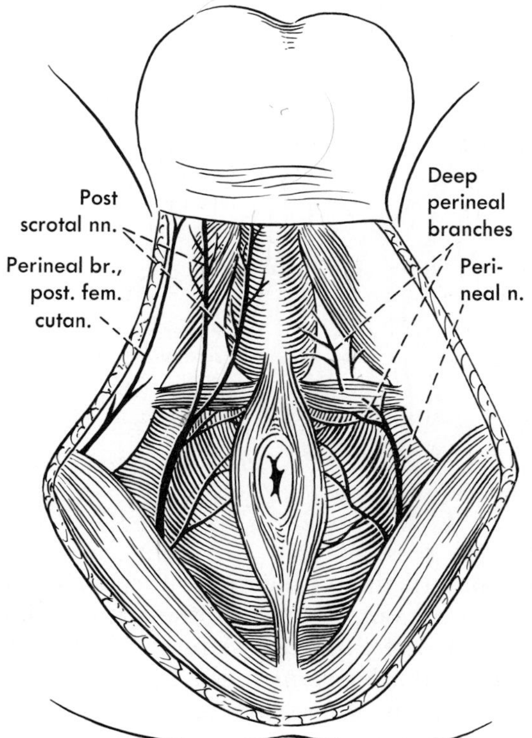

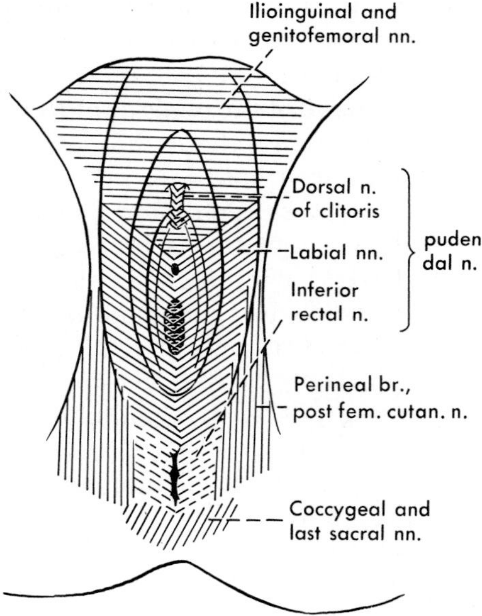


**FIG. 28-27.** General cutaneous distribution of nerves to the female perineum.

**FIG. 28-28.** Cutaneous nerves of the scrotum; anterior view.

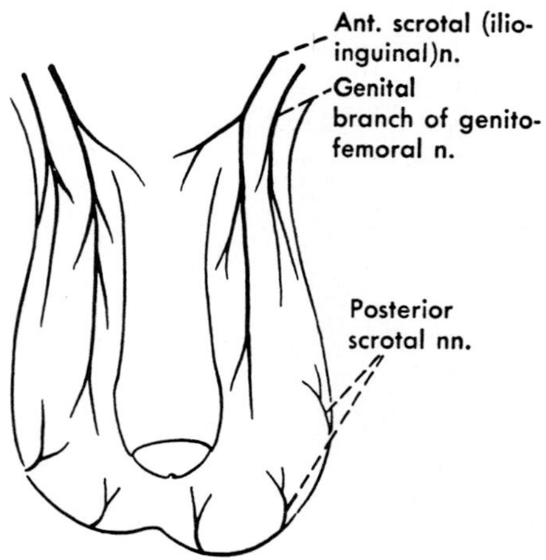


sole somatic motor nerve of the perineum, and it is sensory to most of the perineal skin. Knowledge of the sensory innervation of the perineum is important in the female because of the frequent gynecologic and obstetric procedures that may require local anesthetics. The pudendal nerve conducts sensations from the prepuce and the penis; from the glans penis or clitoridis; from the vestibule of the vagina, including the lower part of the vagina; from part of the anal canal; and from the perianal skin, as well as from the posterior parts of the labia majora and the scrotum (Fig. 28-27). The areas of skin on the periphery of the perineum are supplied by other cutaneous nerves.

## Other Cutaneous Nerves of the Perineum

Skin over the lateral portions of much of the ischiorectal fossa is supplied by the *perineal branches* of the **posterior femoral cutaneous nerves** (derived from the sacral plexus) that course forward superficially at about the level of the ischial rami. The mons pubis and the anterior portion of the labia are supplied by **anterior labial nerves** from the *ilioinguinal nerve* and by the *genitofemoral nerve.* Both these nerves are derived from the lumbar plexus and contain L1 segments (Fig. 28-27).

The *anterior scrotal nerves* are likewise branches of the ilioinguinal and genitofemoral nerves, and they supply the skin over the anterior surface of the scrotum and the root of the penis (Fig. 28-28).

#### Perineal Anesthesia

During some gynecologic and obstetric procedures, local anesthesia of the perineum can obviate a general anesthetic. The most important nerve to block is the pudendal nerve. A **pudendal block** is performed by palpating the ischial spine through the vagina and injecting local anesthetic around it through a needle inserted either through the perineal skin or through the vaginal wall. In addition, the anterior parts of the labia and the mons pubis may also need to be infiltrated with anesthetic. Since somatic afferents do not reach above the lower part of the vagina, the female genital tract may be cut, cauterized, and sutured after a pudendal block. The cervix is only sensitive to stretching and not to other stimuli. To eliminate the pain caused by dilatation of the cervix, a paracervical block is required (p. 787).

## Pelvic Splanchnic Nerves

The visceral efferents required for erection of the penis or clitoris are derived from the pelvic splanchnic nerves. They reach the perineum along the urethra, passing through the urogenital hiatus and diaphragm. Fractures of the pelvis may damage these nerves and cause impotence.

## GENERAL REFERENCE AND RECOMMENDED READINGS

ALVAREZ-MORUJO A: Terminal arteries of the penis. Acta Anat 67:387, 1967

CURTIS AH, ANSON BJ, ASHLEY FL: Further studies in gynecological anatomy and related clinical problems. Surg Gynecol Obstet 74:709, 1942

CURTIS AH, ANSON BJ, MCVAY CB: The anatomy of the pelvic and urogenital diaphragms, in relation to urethrocele and cystocele. Surg Gynecol Obstet 68:161, 1939

DUTHIE HL, GAIRNS FW: Sensory nerve-endings and sensation in the anal region of man. Br J Surg 47:585, 1960

GOLIGHER JC, LEACOCK AG, BROSSY JJ: The surgical anatomy of the anal canal. Br J Surg 43:51, 1955

JOHNSON FP: The development of the rectum in the human embryo. Am J Anat 16:1, 1914

LADD WE, GROSS RE: Congenital malformations of anus and rectum: Report of 162 cases. Am J Surg 23:167, 1934

MASTERS WH, JOHNSON VE: Human Sexual Response. Boston, Little, Brown, 1966

MILLIGAN ETC, MORGAN CN: Surgical anatomy of the anal canal: With special reference to anorectal fistulae. Lancet 2:1150, 1934

NEWMAN HF: Tonus of the voluntary anal and urethral sphincters. Arch Neurol Psychiatry 61:445, 1949

ROBERTS WH, HABENICHT J, KRISHINGNER G: The pelvic and perineal fasciae and their neural and vascular relationships. Anat Rec 149:707, 1964

TOBIN CE, BENJAMIN JA: Anatomic and clinical re-evaluation of Camper's, Scarpa's, and Colles' fasciae. Surg Gynecol Obstet 88:545, 1949

UHLENHUTH E, SMITH RD, DAY ED ET AL: A re-investigation of Colles' and Buck's fasciae in the male. J Urol 62:542, 1949

WALLS EW: Observations on the microscopic anatomy of the human anal canal. Br J Surg 45:504, 1958

WOODBURNE RT: Anatomy of the bladder and bladder outlet. J Urol 100:474, 1968

PART IX

# HEAD AND NECK

# 29

# HEAD AND NECK IN GENERAL

The head and neck are particularly complex and difficult to dissect, the former because of the bone of the skull and lower jaw, and the latter because so many important structures are crowded together in the front of the neck.

The head (*caput*) is roughly divisible into cranium, or brain case, and face. By definition, a cranial nerve is one that emerges through the cranium, and therefore all 12 cranial nerves must be sought in the head. Most of them are also distributed there, but two, the vagus and the accessory, have particularly long courses in the neck, and others appear where neck and head blend. Except for the skeleton, there is no clear demarcation between the head and neck; the floor of the mouth, for instance, can be regarded as part of the head or part of the neck. Similarly, muscles continue from one to the other, as do veins and arteries, so that it is only as the head is dissected that the upper parts of many of the soft structures of the neck can be seen. Thus, the pharynx (Fig. 29-1), into which both nose and mouth open, begins in the head but extends to the lower border of the larynx in the neck before giving way to the esophagus; of the large vessels in the neck, the internal jugular vein receives most of its blood from the head, and the common carotid artery sends most of its blood to the head.

The neck (*collum*) consists of an anterior region, the *cervix,* and a posterior region, the *nucha*. The nucha consists primarily of the vertebral column and its associated muscles and therefore has been described in the section on the back. It may be helpful to recall that although there are only seven cervical vertebrae, there are eight cervical nerves. The trapezius and levator scapulae muscles, also attaching to the cervical vertebral column, have been described with other muscles of the shoulder, the group to which they belong. In this section of the text, therefore, we are primarily interested in

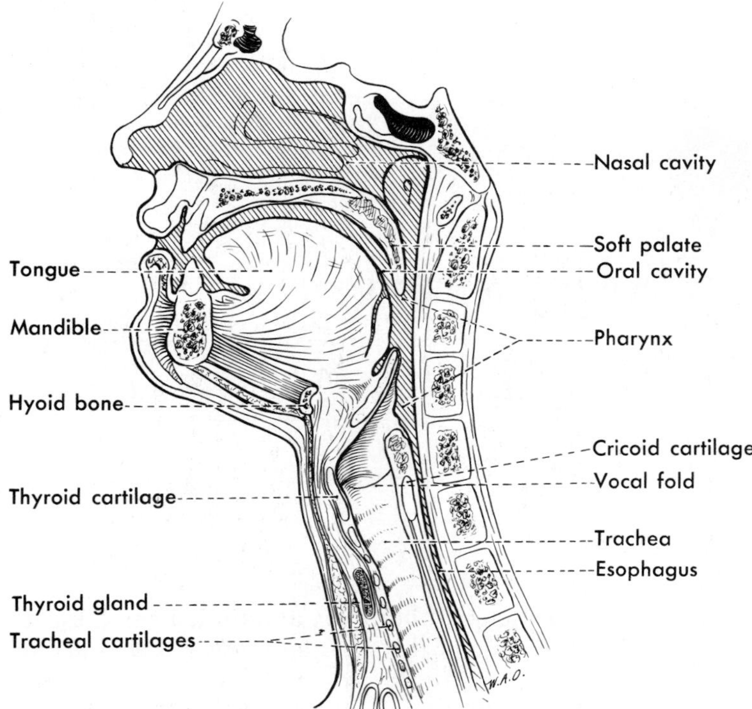


**FIG. 29-1.** The respiratory and digestive systems in a sagittal section of the head and neck.

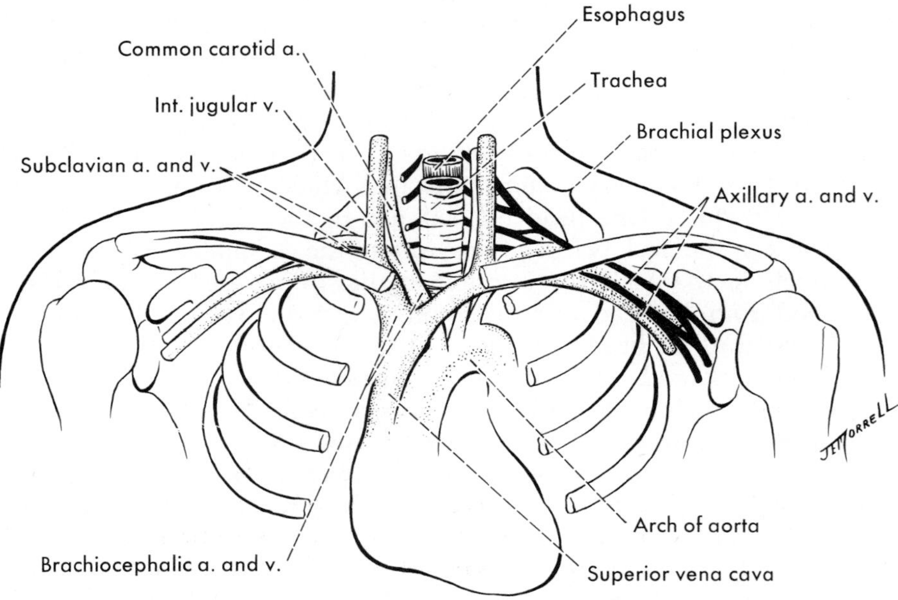


**FIG. 29-2.** Diagram of some of the more important continuities between structures in the base of the neck and those of the thorax and upper limb.

the cervix rather than the neck as a whole. For practical purposes, the cervix (which also means neck) can be thought of as extending from about the lower border of the mandible (lower jaw) to the level of the first rib and as extending posterolaterally behind the clavicular insertion of the trapezius.

Brief mention already has been made of the continuities between neck and head. Equally important are the continuities at the base of the neck between neck and thorax and neck and upper limb (Fig. 29-2). These continuities mean that in a regional dissection of the neck, neither the origin nor the destination of many of the most important structures can be seen, yet the student must know something of these matters if their functional significance is to be grasped.

The most obvious continuity at the base of the neck is the extension of the *trachea,* or windpipe, from the neck into the thorax, where it is distributed to the lungs. Immediately behind it is the *esophagus,* which must pass through the thorax to reach the abdomen. Originating in the thorax from the arch of the aorta (just above the heart) are the great arterial stems found at the base of the neck. On the right side, a single stem, the *brachiocephalic artery,* enters the neck, but in the base of the neck, it divides into two arteries, the *right subclavian* and the *right common carotid.* The *left subclavian* and *left common carotid arteries* have separate origins from the arch of the aorta and, therefore, appear separately in the neck. On both sides, the common carotid runs upward to be distributed mostly to the head, and the subclavian runs laterally at the base of the neck. The subclavian gives off branches to the neck and shoulder and to the thorax, but ends by crossing the first rib behind the clavicle, where, as the axillary artery, it is the chief artery of the upper limb.

In the same fashion, the two *subclavian veins* are continuations of the two axillary veins, the chief veins of the upper limb. On both sides, they unite with the *internal jugular vein,* from the head, to form a brachiocephalic vein. These two veins enter the thorax and then unite to form the superior vena cava, a short trunk that enters the heart.

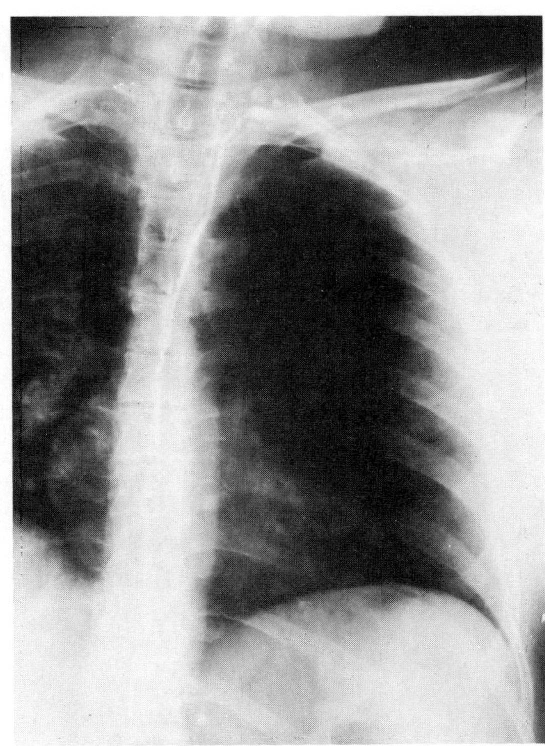

**FIG. 29-3.** Lymphangiogram of the upper end of the thoracic duct as it leaves the thorax and enters the neck. (Courtesy of Dr. W. E. Miller)

The *brachial plexus,* which supplies the nerves to all except a few muscles of the upper limb, does originate in the neck. However, it extends into the axilla around the axillary artery and gives off most of its branches here, so that even its general form is not apparent in a dissection confined to the neck.

Besides smaller vessels that run into the thorax from the neck, and lymphatics, usually not dissectible, that enter the neck from the axilla, two other important structures should be mentioned. The *vagus nerves,* which descend through the length of the neck, continue into the thorax and thence to the abdomen. On the left side, the *thoracic duct,* the largest lymphatic channel in the body, joins the venous system close to the junction of the internal jugular and subclavian veins. The thoracic duct begins in the abdomen, where it receives lymph from the lower limbs and the abdomen and pelvis, and traverses the thorax before reaching the neck (Fig. 29-3).

# 30

# THE NECK

The anterior part of the neck, or cervix, has two important landmarks: in the anterior midline is the *laryngeal prominence* (the hyoid bone is palpable just above the larynx), and on each side, extending obliquely upward and backward between the sternum and clavicle and the prominence (mastoid process) behind the ear is the *sternocleidomastoid muscle* (Fig. 30-1). The sternocleidomastoid divides the cervical region of its side into an anterior and a lateral part, usually called the *anterior* and *posterior triangles* (Fig. 30-2). Of these, the anterior triangle is bordered above by the mandible, medially by the cervical midline, and laterally by the anterior border of the sternocleidomastoid muscle. The posterior triangle is bordered by the clavicle below and by the posterior and anterior borders, respectively, of the sternocleidomastoid and trapezius muscles. Subsidiary triangles within the two major triangles also are described, as indicated in the figure, but become apparent only after reflection of overlying skin and fascia.

## THE SKELETON AND SUPERFICIAL STRUCTURES

### Skeletal Structures

The **cervical vertebrae** have been described previously. Their bodies, largely covered by prevertebral muscles, are the deepest structures in the front of the neck, and their transverse processes give attachment to anterior and anterolateral muscles of the neck as well as to muscles of the back and shoulder. Of these processes, those of the atlas (1st cervical) are palpable below the mastoid processes, but those of the others are too deeply buried in muscles to be felt distinctly. The cervical nerves run laterally and downward in the grooves between the anterior and posterior tubercles of the transverse processes and emerge between or through the muscles that attach to the tubercles.

The **thyroid cartilage** (shield-shaped cartilage) is a part of the

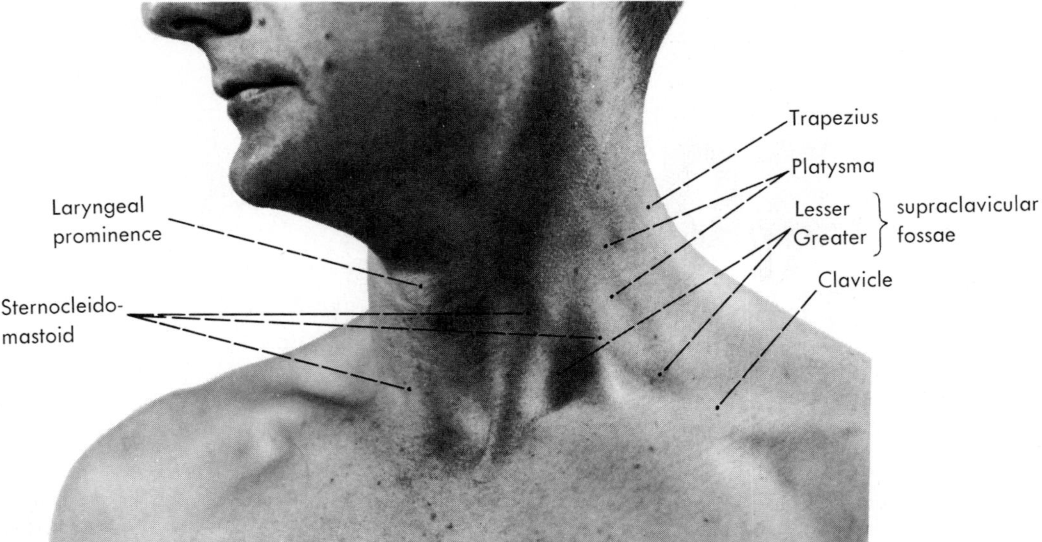


**FIG. 30-1.** Surface anatomy of the neck.

skeleton of the larynx and therefore is described with that. It also serves, however, for attachment of certain muscles of the neck. Its two laminae, facing anterolaterally, meet in the anterior midline at an angle that is especially prominent above and is surmounted by the superior thyroid notch.

The **cricoid cartilage,** also a part of the laryngeal skeleton, lies below the thyroid cartilage; unlike the latter, it completely encircles the larynx. Its anterior part, the arch, is palpable immediately below the thyroid cartilage. Below the arch are the smaller cartilages of the trachea.

The **hyoid bone,** above the larynx, is U shaped. Its front part is its body, and its posterolateral free extremities are its **greater cornua** (horns). These can be felt easily in the living person if the hyoid bone is displaced manually to the side that is to be palpated. The junction of the body with each greater horn is marked on the upper surface by a small projection, the lesser cornu. The hyoid bone receives the attachments of anterior neck muscles on both its upper and lower borders and is highly mobile (note its movement during swallowing). It also gives attachment to certain muscles of the tongue, providing a stable or a mobile base, as necessary, for that organ.

Ligaments and membranes fill the spaces between the thyroid cartilage and the hyoid bone, the thyroid and cricoid cartilages, and the latter and the upper tracheal cartilage. Of these, the **thyrohyoid membrane** is most extensive; nerves and vessels penetrate this membrane to enter the larynx.

The skin and the tela subcutanea (superficial fascia) of the cervix usually are rather thin, but there may be an appreciable amount of subcutaneous fat. Within the superficial fascia is an exceedingly thin muscle, the **platysma muscle,** that begins in the tela subcutanea over the upper part of the thorax, passes over the clavicle, and runs upward and somewhat medially in the neck and across the mandible to blend with superficially located facial muscles (see Fig. 31-16). Descending on its deep surface close to the junction of the posterior and inferior borders of the lower jaw (the angle of the mandible) is its nerve, the cervical branch of the facial nerve. Also deep to it, and then piercing it, are cutaneous nerves of the neck. The platysma muscle has no very important action, but will wrinkle transversely the skin of the neck and help to

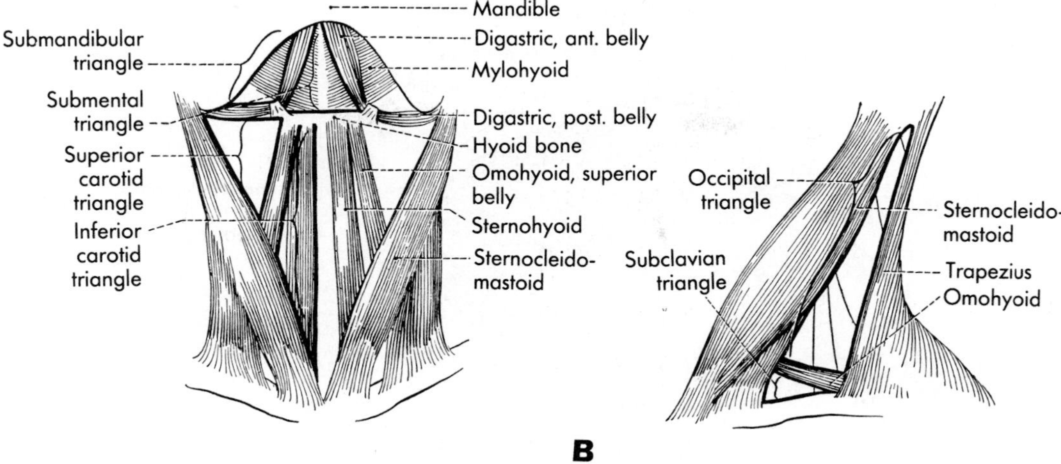


**FIG. 30-2.** The anterior triangle of the neck **(A)**; the posterior triangle **(B)**; and subsidiary triangles into which they can be divided. The superior carotid triangle often is called "the carotid triangle," and the inferior one "the muscular triangle."

open the mouth. It is the cervical equivalent of the facial muscles. Certain superficial veins of the neck lie immediately deep to the platysma in the superficial fascia.

## Cervical Fascia

Removal of the tela subcutanea and the platysma does not clearly reveal the musculature and other structures of the cervix, for these are covered by the cervical fascia (often called deep cervical fascia). The cervical fascia is rather complicated, and the more complete descriptions of it include layers that are not named in the N.A. Further, the looser connective tissue between layers is described as constituting fascial spaces of the neck (see also p. 826), none of which is named in the N.A. Opinions differ much among investigators as to how some of the fascias are arranged and which fascial spaces communicate with each other and which do not. However, there has been so much discussion of the fascia and fascial spaces of the neck that the student will undoubtedly be expected to be at least vaguely familiar with the question at the time he enters his clinical work.

The cervical fascia is in three layers—a *superficial,* a *pretracheal,* and a *prevertebral* (Fig. 30-3). In the anterolateral part of the neck, where these layers are fairly closely associated with each other, they form the *carotid sheath* around the great vessels of the neck, the internal jugular vein and common carotid artery. A simple concept is that the superficial layer surrounds all the important structures in the neck; the prevertebral layer surrounds the vertebral column and the muscles closely connected with it; and the pretracheal layer helps form a visceral compartment around the chief viscera, trachea and esophagus, with the carotid sheath forming a neurovascular compartment.

The **superficial layer** of the cervical fascia can be traced posteriorly to the cervical spinous processes and the ligamentum nuchae. It passes both superficial and deep to the trapezius muscle and then, coming together to form a single layer, passes forward across the posterior triangle of the neck to the posterior border of the sternocleidomastoid muscle. Upon reaching this muscle, the fascia again divides to pass on both sides of it and continue across the front of the neck, joining the layer of the other side in the anterior midline. It also gives off from its deep surface a special investment around the omohyoid muscle that runs from the scapula to the hyoid bone, but takes an abrupt turn in the anterior part of the

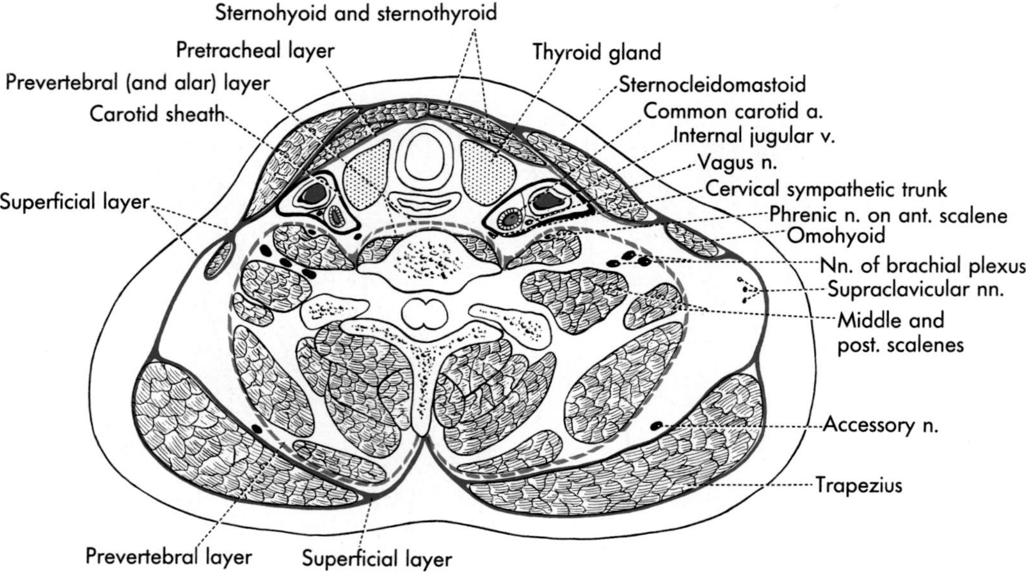


**FIG. 30-3.** Diagram of the fascial layers of the neck, in a cross section below the level of the larynx.

neck (Fig. 30-10). This fascial investment holds the muscle to the clavicle, so that it pulls the hyoid bone primarily downward rather than laterally. Medial and deep to the sternocleidomastoid, the superficial layer gives off delicate laminae that pass between and behind the muscles in front of the trachea, and it blends with the connective tissue (carotid sheath) around the great vessels lateral to the trachea and mostly behind the sternocleidomastoid.

Traced vertically, the superficial layer of cervical fascia over and posterior to the sternocleidomastoid simply blends with fascia of the head, forming a part of the scalp. Anteriorly, this layer is attached below to the clavicle and above to the hyoid bone (Fig. 30-4), after which it continues from the hyoid bone to the lower border of the mandible. Above and behind the hyoid bone, it runs upward to envelop muscles connected with the mandible, and between the mandible and the ear, it splits to envelop the parotid gland. Thereafter, it attaches to the skull. Between the sternocleidomastoid muscles, the superficial lamina attaches, below, to the anterior and posterior surfaces of the sternum; in consequence, there is, between these two attachments, a small blind *suprasternal space.*

The cutaneous nerves of the neck (see the following section) enter the superficial layer of cervical fascia at the posterior border of the sternocleidomastoid and thereafter run, in part, in it before entering the tela subcutanea. In cleaning the fascia from the external surface of the sternocleidomastoid muscle, care must be taken to isolate the nerves as they run around the posterior border of the muscle. Otherwise, removing the fascia will leave their stumps only. The major superficial veins of the neck lie largely external to the fascia, and there often is a communication between the two anterior jugular veins in the suprasternal space.

The **pretracheal lamina** of cervical fascia stretches across the front of the neck immediately behind the infrahyoid muscles and usually is described as continuous on both sides with the part of the superficial layer deep to the sternocleidomastoid muscles (Fig. 30-3). (It also is described as surrounding the trachea, esophagus, and thyroid gland, or the trachea and esophagus but not the thyroid gland.) It is attached above to the thyroid car-

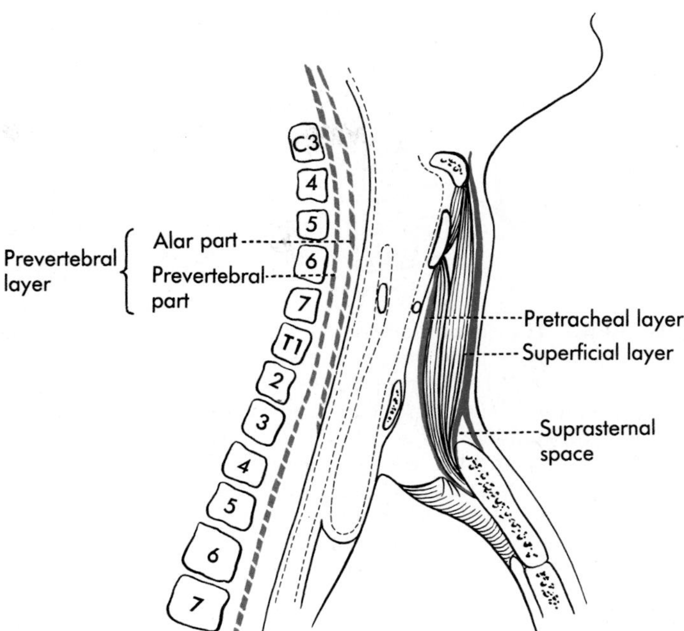


**FIG. 30-4.** Diagram of the fascial layers of the neck in a longitudinal section. (Redrawn from Grodinsky M, Holyoke EA: Am J Anat 63:367, 1938)

tilage (Fig. 30-4) and runs downward behind the infrahyoid or "strap" muscles to their origins from the posterior aspect of the sternum. Here it blends with connective tissue between the pericardial sac and the sternum and with the adventitia of the great vessels as they leave or enter the pericardial sac. On its deep surface, the pretracheal layer blends laterally with the carotid sheath, just as does the superficial layer. The pretracheal layer of the cervical fascia is generally thin and probably of no great importance in itself; however, it is supported anteriorly by the infrahyoid muscles, so it and these muscles, and the stronger carotid sheath laterally, bound the area in which the trachea, esophagus, and thyroid gland lie in loose connective tissue in front of the vertebral column. The space so enclosed often is known as the *visceral compartment*.

The **prevertebral lamina,** the third layer of cervical fascia, begins, as does the superficial layer, in the posterior midline (Fig. 30-3). It is here closely adjacent to the superficial layer, but it covers the outer surfaces of the muscles of the back and as it proceeds forward forms the floor of the posterior triangle of the neck. Particularly at the base of the neck, there is loose connective tissue between the superficial and prevertebral layers through which run nerves and vessels to the upper limb. As the prevertebral layer continues forward from the superficial surface of the muscles of the back, it covers anterolateral muscles, the scalenes, connected with the vertebral column, and this part is often called **scalene fascia**. It then attaches to the transverse processes of the cervical vertebrae and splits into two layers that continue across the front of the vertebral column to fuse again on the other side as they attach to the transverse processes.

Only this most anterior part is truly prevertebral in position; sometimes the two layers here are given different names (Fig. 30-4). Between them there is some loose connective tissue, constituting one of the fascial spaces of the neck. The two parts of the prevertebral layer in front of the vertebral column behave differently. Although both attach above to the skull, the *anterior lamina* ends below by blending with the fascia on the posterior wall of the esophagus in the upper part of the thorax, thus obliterating the posterior part of the visceral compartment, but the *posterior lamina* continues downward along the front of

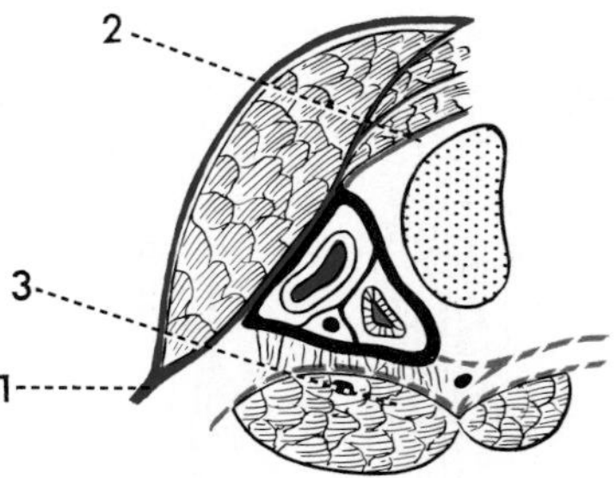


**FIG. 30-5.** The carotid sheath. **1** is the part of the superficial layer that passes deep to the sternocleidomastoid, **2** is the pretracheal layer, and **3** is the prevertebral layer on the anterior scalene muscle. The internal jugular vein laterally, the common carotid artery, and the vagus nerve appear within the sheath; the phrenic nerve lies behind the prevertebral layer, and the sympathetic trunk lies in front of it, between it and the attachment it receives from the carotid sheath.

the thoracic portion of the vertebral column.

The cutaneous nerves of the neck penetrate the prevertebral fascia in their courses to a superficial position. However, the lower cervical nerves and the accompanying artery (subclavian), instead of penetrating this fascia, carry with them a sleevelike diverticulum of the fascia as they emerge from behind it. This accompanies them to the axilla, where it contributes to the *axillary sheath*.

The **carotid sheath** is a condensation of connective tissue around the great vessels of the neck and is particularly prominent because these vessels are so large. Superficially, it blends with or is formed by the superficial layer of the cervical fascia deep to the sternocleidomastoid muscle and the adjacent part of the pretracheal layer. The anteromedial and posterior layers of the carotid sheath come together medially to complete the sheath, and here looser connective tissue unites the posterior layer of the sheath to the prevertebral fascia (Fig. 30-5). The carotid sheath contains the internal jugular vein anterolaterally, the common carotid artery medially, and the main trunk of the vagus nerve posteriorly. The cervical sympathetic trunk lies behind the carotid sheath, between it and the prevertebral fascia.

**Fascial Spaces.** Accounts of the *fascial spaces* of the neck (the looser connective tissue between fascial layers) vary markedly, and there is little agreement among clinicians as to whether infections spread through spaces that are continuous with each other, thus making them of clinical importance, or whether they do not. None of the spaces have official names.

The potential space in front of the trachea and behind the infrahyoid muscles and pretracheal fascia (part of the visceral compartment) usually is called the **pretracheal space.** It is limited above by the attachment of fascia and muscles to the thyroid cartilage and below by the blending of the pretracheal fascia with connective tissue in the anterior mediastinum (Fig. 30-6). The space behind the esophagus, between it and the prevertebral fascia, is the **retrovisceral** or **retropharyngeal space.** It extends above to the base of the skull, behind the pharynx as well as the esophagus, and is limited below in the posterior mediastinum by the attachment of the anterior layer of the prevertebral fascia to the esophagus. According to most, but not all, accounts, the two spaces are continuous with each other around the thyroid gland; thus, infections originating close to the pharynx have been described as descending in the retrovisceral space to

**FIG. 30-6.** The visceral compartment (horizontally lined) in a sagittal section of the neck. The "danger space" also is indicated. (Adapted from Grodinsky M, Holyoke EA: Am J Anat 63:367, 1938)

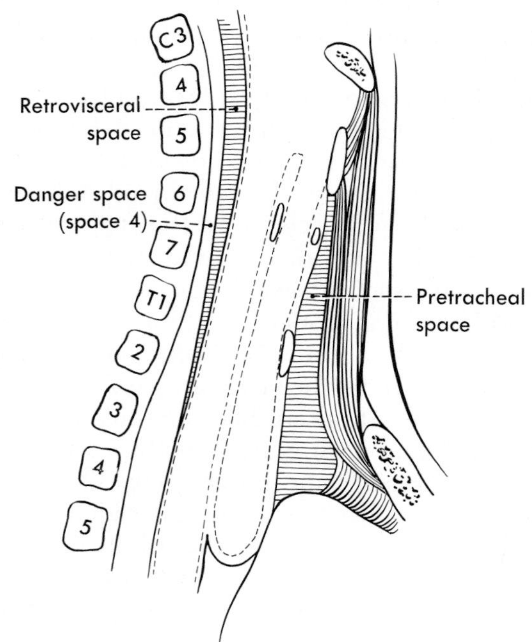

the posterior mediastinum, or sometimes passing forward into the pretracheal space and involving the anterior mediastinum. The potential **space between the two layers of prevertebral fascia** in front of the vertebral column has been called the "danger space," because although it also begins above at the base of the skull, it extends downward throughout the thorax. The **space of the carotid sheath** sometimes has been described as reaching the base of the skull and sometimes as being obliterated at the level of branching of the common carotid artery.

## Superficial Nerves

The cutaneous nerves of the neck are all branches of the cervical plexus, and the upper three appear at the posterior edge of the sternocleidomastoid muscle (Fig. 30-8). The highest, the **lesser occipital nerve,** appears at the upper part of the posterior border of the sternocleidomastoid and runs vertically upward toward the mastoid process, to be distributed to skin and scalp behind the ear. Below it, close to the middle of the muscle, is the **great auricular nerve.** This also runs vertically upward, usually accompanied by the external jugular vein, about in line with the front of the ear. It supplies much of the external ear and some skin of the face below and in front of the ear. A little below it is the **transversus colli** or transverse cervical **nerve.** As or soon after it rounds the posterior border of the sternocleidomastoid muscle, it divides into two major branches. One of these runs upward and forward, and one downward and forward, so that between them they supply most of the skin of the anterior part of the neck.

The lowest set of cutaneous nerves, not so intimately related to the posterior border of the sternocleidomastoid, consists of three **supraclavicular nerves.** They enter the superficial lamina of the cervical fascia in the posterior triangle, and the *medial supraclavicular nerve* then runs downward and anteriorly across the lower part of the sternocleidomastoid to supply anterior skin at the base of the neck and over approximately the upper two intercostal spaces. The *intermediate* nerve descends through about the middle of the base of the posterior triangle and supplies skin of the base of the neck and the upper part of the thorax. The *lateral* (posterior) *supraclavicular nerve* runs over the anterior border of the trapezius muscle to supply skin toward and over the tip of the shoulder.

## Superficial Veins

The superficial veins of the neck are the external and anterior jugulars (Fig. 30-7). The **external jugular** begins on the superficial sur-

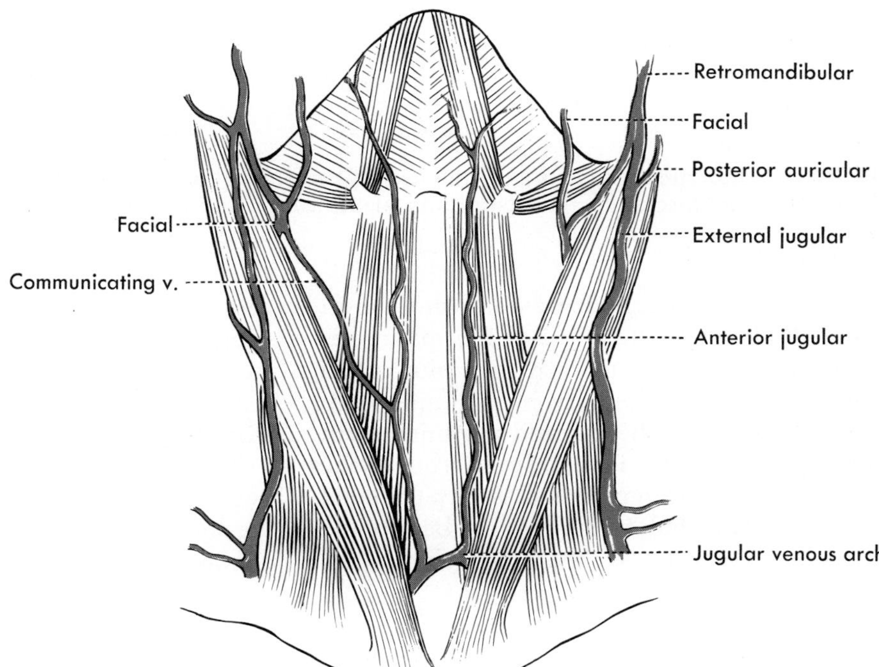


**FIG. 30-7.** Superficial veins of the neck.

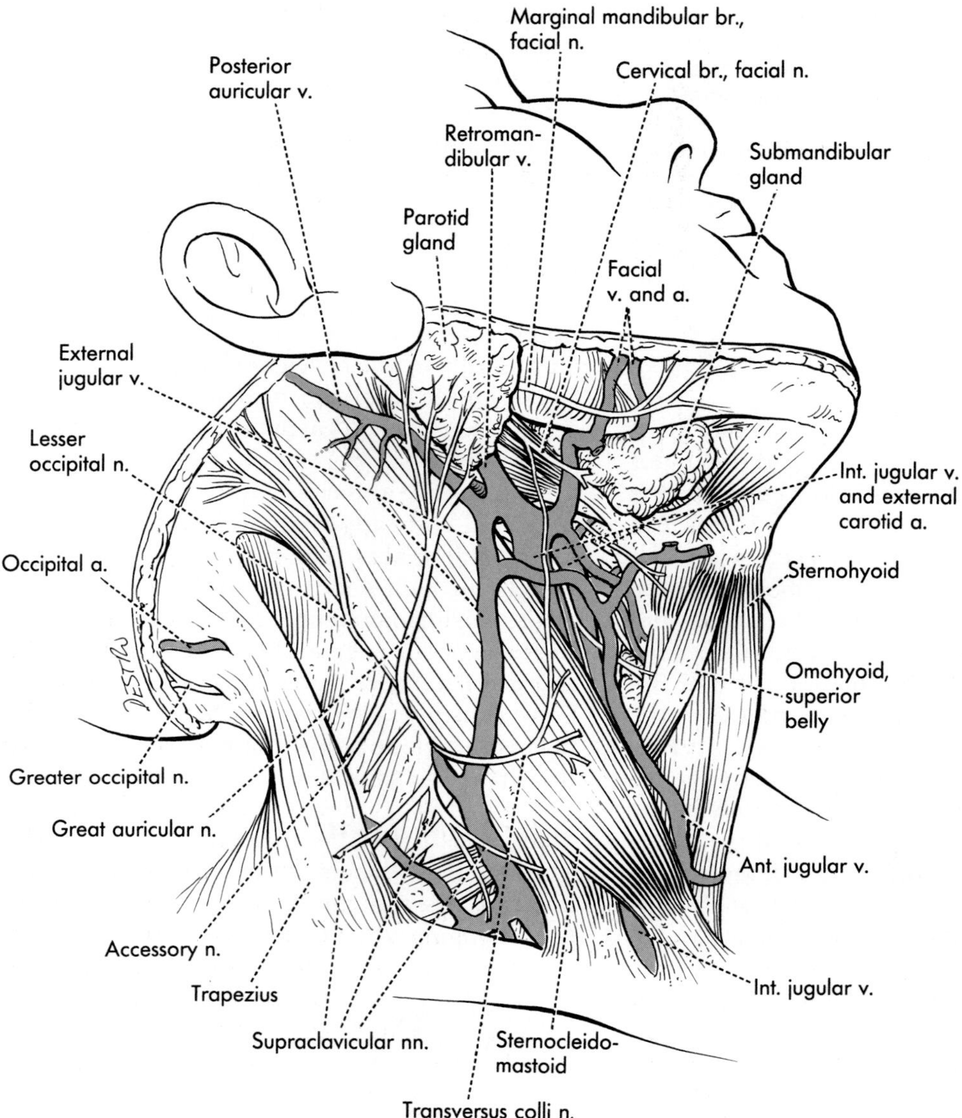


**FIG. 30-8.** Superficial structures of the neck after removal of the platysma muscle and the cervical fascia. Further detail and identification can be found in following figures.

face of the sternocleidomastoid muscle at about the level of the angle of the mandible, where it usually is formed by the union of the posterior auricular vein, from skin and scalp behind and above the ear and a branch from the retromandibular vein, from in front of the ear. The external jugular is closely paralleled by the great auricular nerve. At the base of the neck, just posterior to the sternocleidomastoid muscle, the external jugular turns more deeply to pierce the deep fascia and end in the subclavian vein. Close to its termination, it receives the anterior jugular vein and veins from the shoulder.

The variable **anterior jugular vein** typically begins by the union of small veins between the hyoid bone and the chin and may receive connections from the external jugular or the

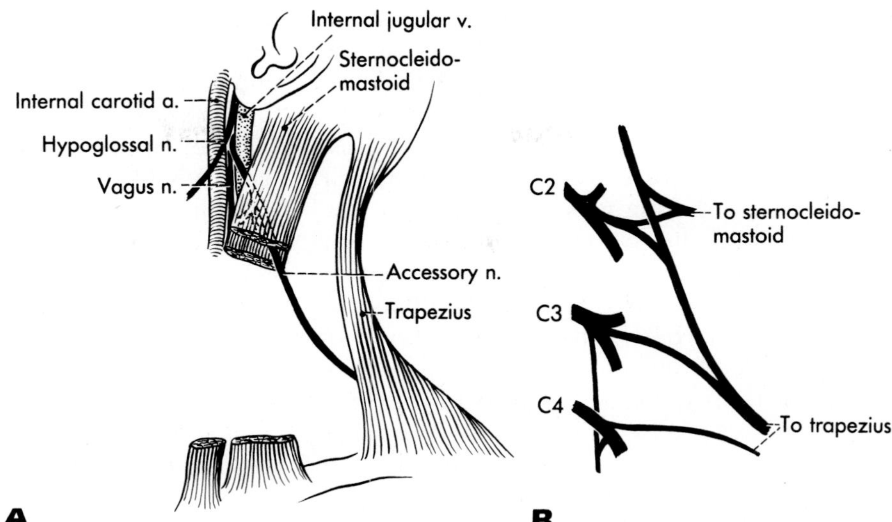


**FIG. 30-9.** Usual course and relations **(A)** of the accessory nerve, and its connections **(B)** with the cervical plexus.

fascial vein; either of these connections may form its upper end, or it may be absent. It often sends a communication, the **jugular venous arch,** to the vein of the opposite side through the suprasternal space. It ends by passing deep to the sternocleidomastoid muscle and entering the external jugular vein.

## STERNOCLEIDOMASTOID MUSCLE

The sternocleidomastoid muscle arises by a tendinous head from the sternum and by a broader but thin muscular head from the medial part of the clavicle. The two heads unite, and the muscle extends obliquely upward and laterally across the neck to insert into the mastoid process behind the ear.

The **spinal accessory nerve** (Figs. 30-8 and 30-9) runs posteriorly and downward to enter the deep surface of the sternocleidomastoid. It usually runs through the muscle but sometimes runs deep to it. After supplying the sternocleidomastoid, the nerve runs downward and laterally through the posterior triangle to disappear deep to the trapezius and supply that muscle also. The accessory nerve is joined by cervical nerve fibers, typically from C2, that enter the muscle as a part of the accessory nerve and contribute afferent fibers.

The muscle receives two fairly prominent vascular twigs. The upper one is from the *occipital artery,* and the lower one is from the *superior thyroid artery.* It also receives a lesser blood supply from vessels at the base of the neck and from vessels related to its upper end.

**Action.** The sternocleidomastoid, acting alone, laterally flexes the neck and rotates the face to the opposite side. The two muscles acting together flex the head and neck forcibly. Spasm of the muscle, usually of unknown origin but sometimes congenital, is one cause of a flexion deformity of the neck known as *wryneck* or *torticollis;* other muscles that rotate and flex the neck also may contribute to torticollis.

# THE ANTERIOR TRIANGLE

Removal of the superficial lamina of the cervical fascia between the two sternocleidomastoid muscles allows identification of the subsidiary triangles in the anterior triangle. These (Fig. 30-2) consist of two important ones, the submandibular triangle and the carotid (superior carotid) triangle, and two others that are not in the N.A. but usually are called the submental triangle and the inferior carotid or muscular triangle.

The **submandibular triangle** lies below the

border of the mandible and above the hyoid bone. It is bordered anteroinferiorly by the anterior belly of the diagastric muscle and posteroinferiorly by the posterior belly of the digastric and the associated stylohyoid muscle. It is largely filled by the submandibular gland, one of the major salivary glands. The **carotid triangle** lies below the submandibular triangle, and its upper border is the stylohyoid and the posterior belly of the digastric. Its posterior border is the anterior border of the sternocleidomastoid muscle. Its anteroinferior border is the superior belly of the omohyoid.

The **submental triangle** lies above the hyoid bone between the submandibular triangle and the anterior midline, and the **muscular** or inferior carotid **triangle** lies between the (superior) carotid triangle and the anterior midline below the hyoid bone. Both these triangles have muscular floors.

## MUSCLES, VESSELS, AND NERVES

### Carotid Sheath and Contents

The major structures in the carotid sheath are the internal jugular vein, the common carotid artery, and the vagus nerve. The carotid sheath begins at the base of the neck, where it is continuous with the connective tissue around the great vessels here, and usually is described as ending at about the upper border of the thyroid cartilage in the same manner, although some authorities claim to have traced it higher. In its lower part, it is covered by the sternocleidomastoid muscle, and it also is crossed anteriorly by the omohyoid muscle. Care must be taken in removing it from about the vascular structures, for nerves to the infrahyoid (strap) muscles traverse it; one part of the loop that supplies these muscles (*ansa cervicalis*) may run lateral and anterior to the vein in the carotid sheath to join the other part that runs between the artery and vein, or both may run deep to the vein (Fig. 30-10). Many of the deep lymph nodes of the neck lie in the carotid sheath, along the internal jugular vein and between it and the common carotid artery. These nodes may be small and scattered and not easily visible in the usual dissection, but they are important when there is carcinoma of the mouth, larynx, or other structures of the head and neck.

The **internal jugular vein** (Fig. 30-10) emerges from deep to the posterior belly of the digastric muscle and receives one or more veins at about the level of the hyoid bone. The one vein regularly entering the internal jugular at this level is the facial, usually after it has been joined by part of the retromandibular vein. The *facial vein* runs downward superficially across the submandibular gland, but the *retromandibular vein* usually divides at the anterior border of the sternocleidomastoid muscle to contribute partly to the external jugular and partly to the facial. Other veins entering the internal jugular at about this level are from deeper structures and are difficult to identify until they have been traced to the region from which they come. Further, they may join together in various combinations to empty into the internal jugular or may join the facial vein and empty with it into the internal jugular. The largest vein here is typically the *lingual vein*, but there also may be *pharyngeal veins*, and there is a *superior thyroid vein*. Below this level, the internal jugular vein usually receives no tributaries until it is at the level of the middle of the thyroid gland, where it often receives a *middle thyroid vein*. Below this, it receives no other tributaries and ends by joining the subclavian vein at the base of the neck. The union of these two veins forms the brachiocephalic vein.

The dilated upper end of the internal jugular vein, the *superior bulb*, is located in the jugular fossa of the skull. It is primarily a continuation of the sigmoid venous sinus. The lower end of the vein, also dilated, is the *inferior bulb*.

The *thoracic duct*, the great lymphatic channel of the thorax, joins the angle of junction of the internal jugular and subclavian veins on the left side (Fig. 30-27) or either of these veins close to their junction. The smaller *right lymphatic duct* or the several channels that may represent it, identifiable with more difficulty, empty in a corresponding position on the right side.

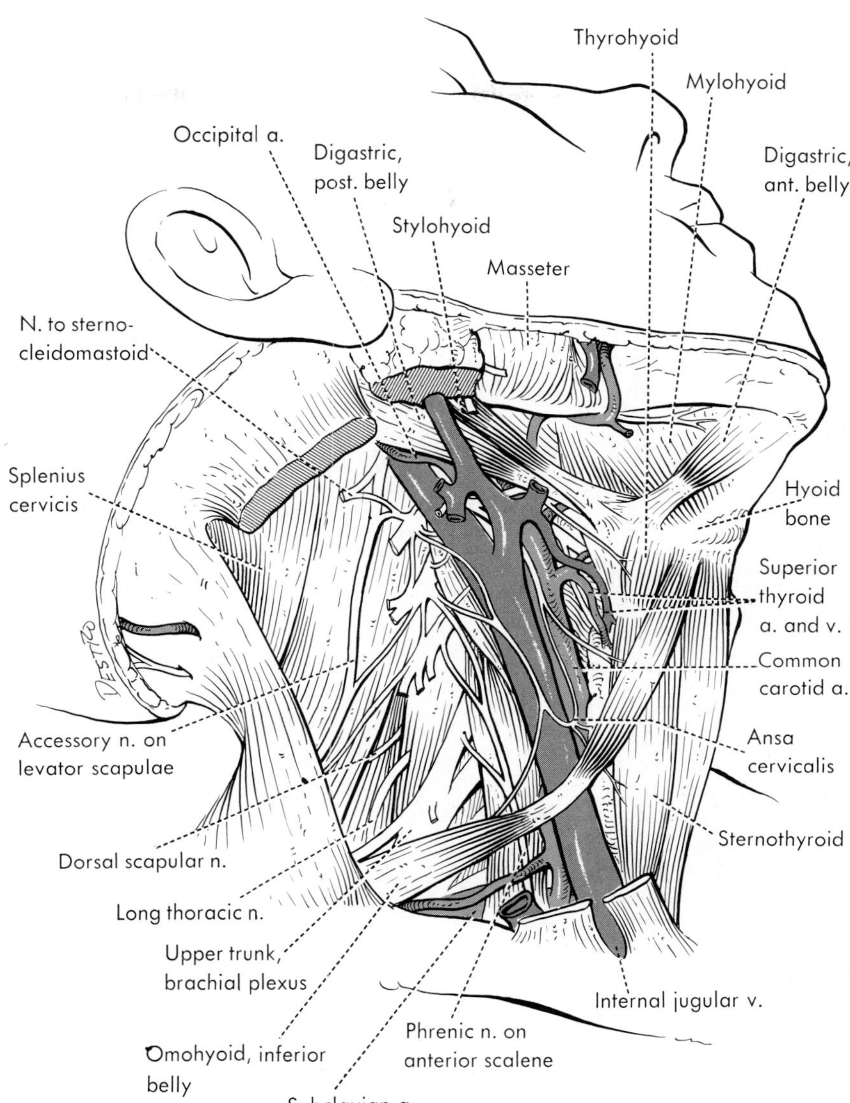


**FIG. 30-10.** Structures in a deeper dissection of the neck. The superficial veins and the sternocleidomastoid muscle have been removed, as have the submandibular gland and a segment of the facial vein, and the cutaneous nerves have been cut down to short stumps arising from the second, third, and fourth cervical nerves.

The internal jugular vein largely covers the **common carotid artery.** The right common carotid and the right subclavian separate from each other at the base of the neck, arising as the terminal branches of the brachiocephalic trunk, so the origin of this carotid can be seen in a dissection of the neck. The left common carotid, however, is an independent branch from the arch of the aorta, and it enters the neck separate from and a little anteromedial to the left subclavian artery. The two common carotids run upward, one on each side of the trachea (where the pulse can easily be felt) to about the upper border of the thyroid cartilage before they give off any branches. Here each common carotid divides into its terminal branches, the internal and external carotids.

Each **vagus nerve** (Figs. 30-11 and 30-12) lies behind and somewhat between the common carotid artery and the internal jugular vein. Of its upper branches, given off while it is related to the internal carotid artery and the internal jugular vein, the *superior laryn-*

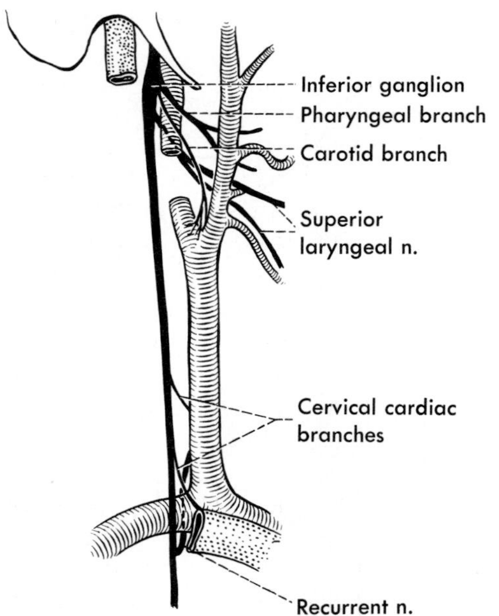


**FIG. 30-11.** Course of the right vagus nerve in the neck.

*geal nerve* descends into the neck (Fig. 30-12), dividing into external and internal branches that are both distributed to the larynx. In the lower part of the neck each vagus gives off slender *superior and inferior cardiac branches* that run downward and medially, usually passing behind the subclavian artery to disappear into the thorax and join the cardiac plexus. At the base of the neck each nerve passes in front of the subclavian artery (but behind the vein) before disappearing into the thorax. At the lower border of the artery, the right vagus gives off the **right recurrent laryngeal nerve,** which loops below and behind the artery to run upward and medially in the groove between the trachea and esophagus.

**FIG. 30-12.** A still deeper dissection of the neck. The internal jugular vein has now been removed, as have the long infrahyoid muscle and parts of the stylohyoid muscle, of both bellies of the digastric, and of the clavicle. The only remaining part of the ansa cervicalis is the stump of its superior root projecting downward from the hypoglossal nerve. The upper and middle trunks of the brachial plexus are recognizable; the nerve to the subclavius has been largely removed.

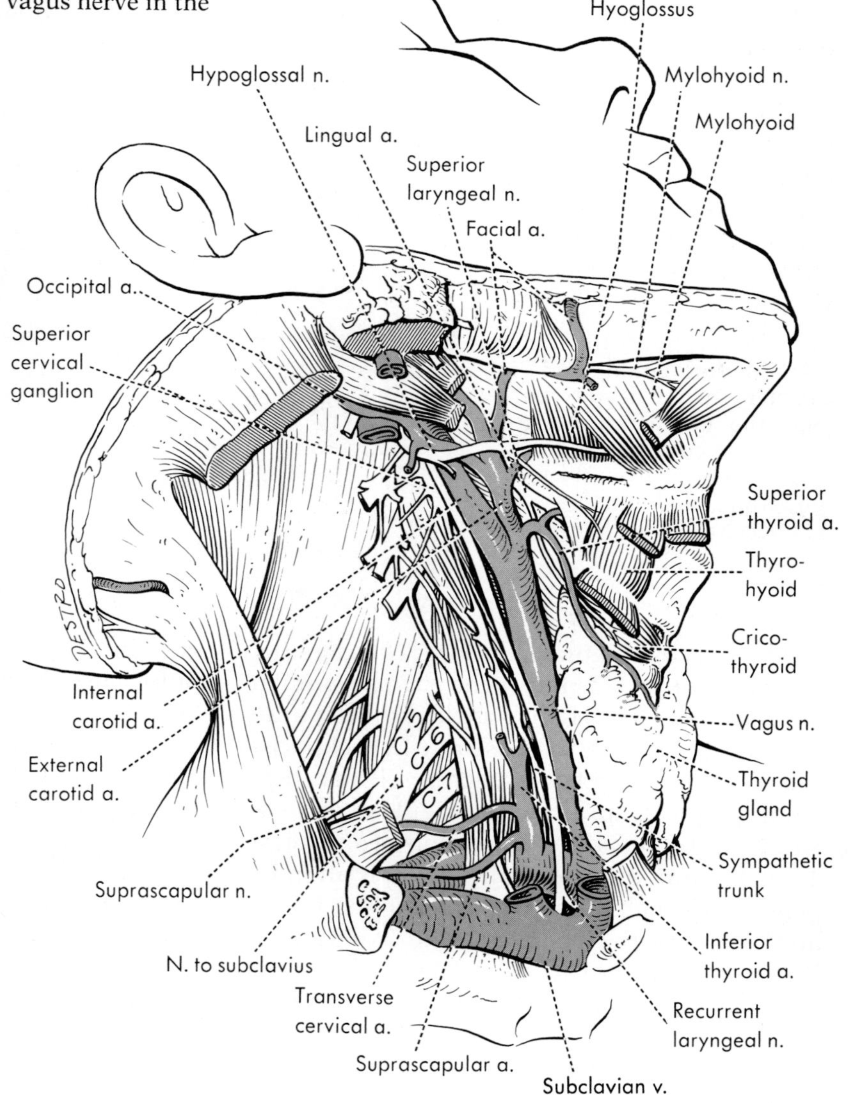

The left recurrent laryngeal nerve originates in the thorax and runs below and behind the arch of the aorta. In the neck it follows a course similar to that of the right nerve. Each recurrent nerve gives twigs to the esophagus and trachea and ends as an *inferior laryngeal nerve.*

### Infrahyoid Muscles

The sternohyoid, omohyoid, sternothyroid, and thyrohyoid muscles are often referred to as "strap muscles" (Figs. 30-8, 30-10, and 30-13). They cover the front and much of the sides of the larynx, trachea, and thyroid gland and, with a deeper-lying suprahyoid muscle (geniohyoid), represent a cervical continuation of the muscle mass that, in the abdomen, forms the rectus abdominis.

Most superficially and on each side of the midline, ascending from an origin on the posterior surface of the manubrium sterni and the sternal end of the clavicle, are the **sternohyoid muscles.** These thin, flat muscles attach above to the body of the hyoid bone. Lateral to the sternohyoid muscle is the *superior belly* of the **omohyoid muscle,** which attaches to the hyoid bone just lateral to the attachment of the sternohyoid. In the upper part of its course, this muscle almost parallels the sternohyoid, but it diverges somewhat laterally as it runs downward (passing in front of the carotid sheath and its contents). In the lower part of the neck, its muscle fibers usually give way to a tendon through which it is joined to its inferior belly; the tendinous junction may or may not be obvious. The *inferior belly* of the omohyoid runs laterally, inferiorly, and posteriorly across the posterior triangle of the neck, dividing this triangle into an upper occipital and a lower subclavian triangle (Fig. 30-2), and disappears deep to the trapezius. It attaches to the superior border of the scapula (omo meaning shoulder) just medial to the scapular notch.

Deep to the sternohyoid are the sternothyroid and the thyrohyoid. The **sternothyroid** takes origin from the posterior surface of the manubrium and attaches above to an oblique line on the thyroid cartilage. The **thyrohyoid** runs from this line to the hyoid bone. Its lateral border usually appears posterolateral to the omohyoid.

The infrahyoid muscles are *innervated* by fibers from upper cervical nerves, which reach them in a somewhat peculiar fashion (Fig. 30-13). The nerves of the lower part of these muscles are given off from a loop, the **ansa cervicalis.** The inferior root of the loop is a direct branch from the cervical plexus, typically containing fibers from the 2nd and 3rd cervical nerves. It descends either deep or superficial to the internal jugular vein and passes medially to form a loop (ansa meaning loop) with the terminal part of the superior root

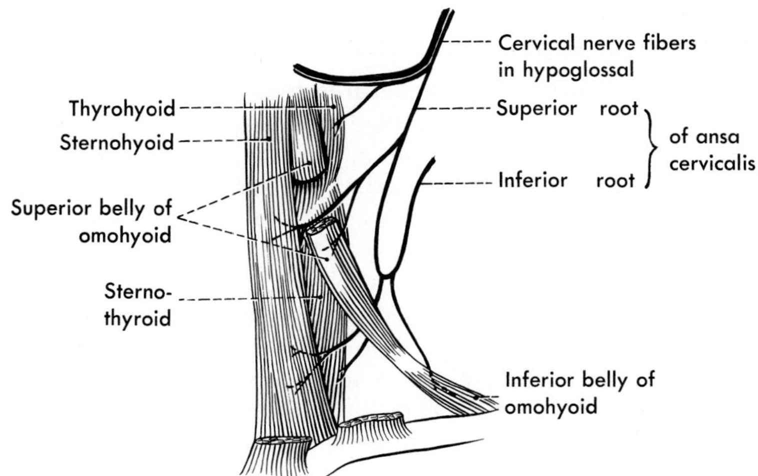


**FIG. 30-13.** The ansa cervicalis.

from which branches are given off to the inferior belly of the omohyoid and to the lower parts of the sternohyoid and sternothyroid. Depending upon the relation of the inferior root, the loop may be either around the internal jugular vein or deep to it. The superior root of the ansa cervicalis appears to take origin from the hypoglossal (12th cranial) nerve, usually just visible at about the level of the hyoid bone. It descends between the common carotid artery and the internal jugular vein and sends a branch to the superior belly of the omohyoid and one to the sternohyoid and sternothyroid muscles close to the level of the thyroid cartilage; it then joins the inferior root to complete the loop. Actually, the superior root contains not hypoglossal but cervical nerve fibers that leave the 1st and 2nd cervical nerves or the latter alone and join the hypoglossal a little before they leave it via the superior root. Other cervical nerve fibers in the hypoglossal continue farther forward and give rise to a thyrohyoid branch that descends to enter the thyrohyoid muscle. (Still other cervical nerve fibers pass even farther forward with the hypoglossal and supply the geniohyoid.)

The infrahyoid muscles lower the larynx and the hyoid bone, a motion occurring in singing a low note and following elevation of these structures during swallowing. With suprahyoid muscles, they fix the hyoid bone, thus providing a firm base upon which the tongue can move.

## Superficial Suprahyoid Structures

Most of the suprahyoid structures related to the mandible are best examined during or after dissection of that part, but certain superficial ones should be noted early even though they must be examined more carefully later (Figs. 30-8, 30-10, and 30-12; see also p. 903).

Extending downward and forward from deep to the sternocleidomastoid muscle, the lower end of the parotid gland, and the angle of the jaw are two closely associated muscles, the **stylohyoid** and the **posterior belly of the digastric.** Their origins are too deep to be observed at this time, but can be seen in Figure 31-36. A little above the hyoid bone, the deeper-lying stylohyoid divides to pass on both sides of the tendon of the digastric and attach to the hyoid bone. The tendon from the posterior belly of the digastric, after passing through the stylohyoid, is held to the hyoid bone by a fascial sling or by tendinous fibers; thereafter, it gives place to muscle fibers which, as the **anterior belly of the digastric,** run forward and slightly upward to attach to the inner surface of the lower border of the mandible just lateral to the midline.

Immediately above the anterior belly of the digastric is the **mylohyoid.** This arises from the inner surface of the mandible and inserts with its fellow of the opposite side into a midline raphe and, posteriorly, into the hyoid bone; it thus fills the space between the mandible and the hyoid bone and forms a movable floor for the mouth. It and the anterior belly of the digastric muscle are innervated by a branch of the trigeminal or 5th cranial nerve that lies on the lower surface of the mylohyoid muscle. A small artery accompanies the nerve.

The two bellies of the digastric may be overlapped superficially by the **submandibular gland,** a salivary gland that lies partly below and partly medial to the mandible and both below and behind the mylohyoid muscle. The duct of this gland lies above the mylohyoid muscle. Associated with the gland are **submandibular lymph nodes** that receive some of the lymphatic drainage from the face, jaws, and tongue. The **facial vein** leaves the face by crossing the mandible at the level of the submandibular gland and runs superficially across this gland to reach the internal jugular. The **facial artery,** a branch of the external carotid, crosses the lower border of the mandible just in front of the facial vein; however, it first runs deep to the submandibular gland and appears between it and the mandible. A branch (marginal mandibular) of the **facial nerve** runs forward toward muscles of the lower lip and chin at about the lower border of the mandible and typically crosses the facial vessels superficially. Sometimes this nerve is above the lower border of the mandible; sometimes it passes across the submandibular gland.

Posteriorly, the external and internal carotid arteries, the internal jugular vein, and, more deeply, the vagus nerve can be traced upward until they disappear deep to the digastric. Running downward and forward lateral to the vessels and the occipital artery (the branch of the external carotid directed upward and backward) is the **hypoglossal** or *12th cranial nerve* (Fig. 30-14). This nerve usually turns forward lateral to the occipital artery and below its sternocleidomastoid branch, continuing forward a little above the level of the hyoid bone to disappear deep to the suprahyoid muscles. It gives off the superior root of the ansa cervicalis at about the point at which it makes its turn around the occipital artery.

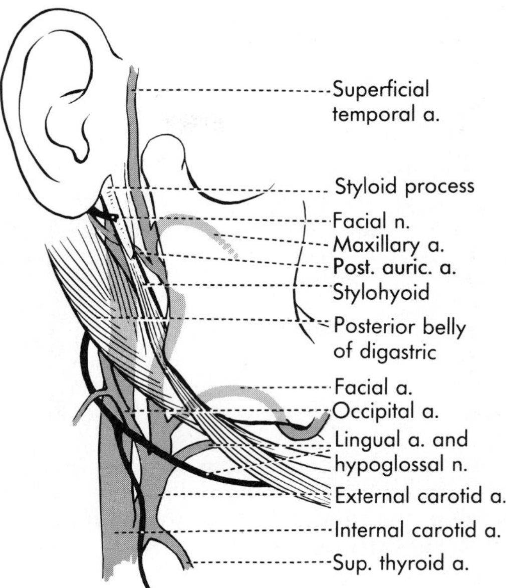


**FIG. 30-14.** Relations of the carotid arteries and the external carotid branches in the upper part of the neck and the retromandibular region. The ascending pharyngeal artery is not shown.

## Internal and External Carotids

Only parts of these vessels appear in the neck, but branches of the external carotid spread so widely to the neck and head that portions of them will be seen in various regions.

The **carotid bifurcation** lies at approximately the level of the upper border of the thyroid cartilage (Fig. 30-12). At the bifurcation, the **internal carotid** usually is posterior or posterolateral to the external carotid, from which it always can be distinguished by the fact that it gives off no branches in the neck, whereas the external does.

The common carotid at its bifurcation and the first part of the internal carotid are often somewhat dilated, and form the **carotid sinus.** This is a segment provided with nerve endings that are sensitive to pressure, and the sinus normally responds to increases in blood pressure within it by initiating impulses that reflexly lower the pressure. Behind the upper end of the common carotid, with its upper pole usually projecting into the angle between the vessels, is a flattened, somewhat ovoid body, the **glomus caroticum** (carotid body). This is a highly vascular epithelial body that also contains special nerve fibers. It responds to chemical changes in the composition of the blood, particularly to reduced oxygen, and reflexly increases the depth and rapidity of breathing. The pulse rate and blood pressure also rise. Both carotid sinus and body are innervated by nerve fibers (carotid sinus branch) that descend between the two carotid vessels to the bifurcation, primarily from the 9th (glossopharyngeal) cranial nerve.

Beyond its origin, the **internal carotid artery** is inclined somewhat more medially than the external carotid and therefore comes to lie more medial than posterior to that vessel. After the internal carotid disappears deep to the stylohyoid and the posterior belly of the digastric, it runs upward, deep to the mandible, to enter the skull and gives off no branches until it has done so.

The **external carotid artery** gives off some of its branches in the neck below or close to the level of the hyoid bone and some during its further course upward medial to the jaw. There are eight named branches, and most of those rebranch extensively, for the external carotid supplies most of the structures of the head except the brain and the contents of the orbit and helps supply structures in the neck. The branches of the external carotid are the superior thyroid, lingual, ascending pharyngeal, occipital, facial, posterior auricular, maxillary, and superficial temporal (Figs. 30-14 and 30-15).

Ligation of the external carotid artery, sometimes carried out to help control bleeding from a branch that is relatively inaccessible, cuts down the

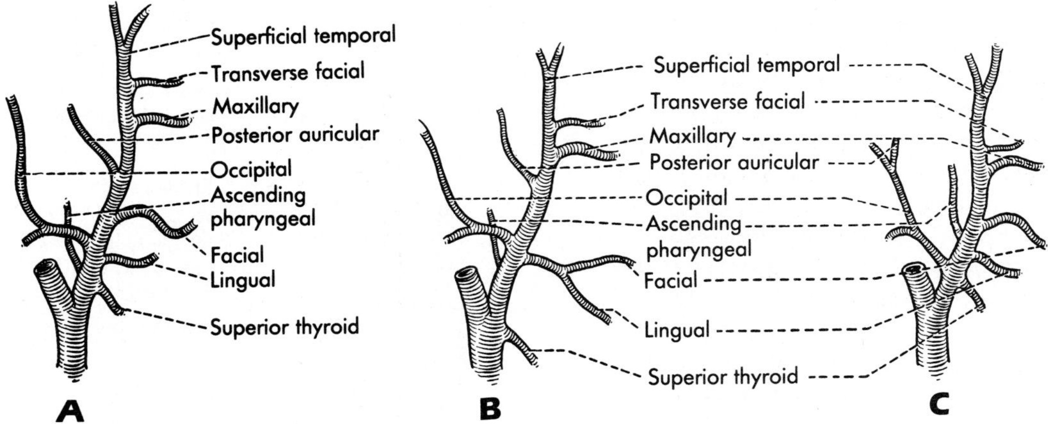


**FIG. 30-15.** Some variations in the branching of the external carotid artery. **A** shows an approximately normal pattern. In **B** the facial and lingual arteries arise by a common stem, and in **C** the posterior auricular arises from the occipital rather than the external carotid directly.

blood flow through the artery and its branches but does not abolish it. Rather, blood flows retrogradely into the carotid from the other side through such anastomoses as those of the face and the scalp.

The first branch of the external carotid, arising anteriorly close to or even at the carotid bifurcation, usually is the **superior thyroid artery.** This descends obliquely downward and forward, disappearing deep to the infrahyoid muscles to reach the upper pole of the thyroid gland.

Above the superior thyroid artery, the external carotid gives off, also anteriorly, the **lingual artery.** This originates at about the level of the hyoid bone and runs forward or loops upward and forward to disappear deep to the suprahyoid muscles and the submandibular gland. If the lingual artery branches into two major stems before it disappears, one of the two stems probably is the facial artery, for sometimes lingual and facial arteries arise by a common **linguofacial trunk.** Sometimes the lingual artery is not visible because it arises deep to or above the posterior belly of the digastric.

The small **ascending pharyngeal artery** arises from the deep surface of the external carotid, usually between the levels of origin of the superior thyroid and lingual arteries. It has a deep course upward, on the posterolateral wall of the pharynx, to which it is largely distributed.

The **occipital artery** arises from the posterior aspect of the external carotid, usually a little below the level of the hyoid bone. It runs upward and posteriorly, gives off a branch to the sternocleidomastoid muscle (below which the hypoglossal nerve turns forward), and disappears deep to the posterior belly of the digastric. It later appears in the upper part of the posterior triangle of the neck.

The remaining branches of the carotid, with the possible exception of the facial (which may arise with a lingual of normal origin), arise from the external carotid after that artery has disappeared deep to the posterior belly of the digastric and the associated stylohyoid muscle. After doing so, the external carotid runs deep to the angle of the mandible and then just deep to the posterior border of the jaw (ramus of the mandible), therefore deep to or through the parotid gland, which lies between the ear and the jaw. This gives it a course almost straight up in front of the ear.

Under cover of the suprahyoid muscles or of the mandible, the external carotid gives off its third anterior branch, the **facial artery.** This loops upward and forward and then downward over the submandibular gland. After giving off certain branches, it emerges between the gland and the lower border of the mandible and turns upward on the face.

The small **posterior auricular artery** arises below the ear from the posterior aspect of the external carotid and runs upward and backward to be distributed primarily behind the ear.

The **maxillary artery** is given off medial to the upper part of the jaw and runs forward. It is distributed particularly to the jaws, the palate, and the inside of the nose.

The **superficial temporal artery** is the upward continuation of the external carotid artery after the maxillary has been given off. It is distributed primarily to the side of the head (temple) and is easily palpable in front of the upper part of the ear.

## Cervical Plexus

The cutaneous branches of the cervical plexus (Fig. 30-8) are its largest ones. After reflection of the sternocleidomastoid muscle, under cover of which they lie, they can be traced to their origins, and the formation of the cervical plexus can be examined. The nerves contributing to the plexus are primarily the ventral rami of the 2nd, 3rd, and 4th cervical nerves (Figs. 30-12 and 30-16); the 1st cervical nerve frequently but not always reaches the plexus by forming a loop with some of the fibers of the 2nd cervical nerve. From this loop, or from the 2nd cervical nerve directly, fibers are given off to the hypoglossal nerve. It is these fibers that form the superior root of the ansa cervicalis, the nerve to the thyrohyoid muscle, and that to the geniohyoid.

In addition to the branch of the hypoglossal nerve, the 2nd cervical also sends a branch to the spinal accessory nerve, to be distributed with it to the sternocleidomastoid and trapezius muscles (or this branch may enter the sternocleidomastoid independently). It contains afferent fibers to the muscle. Still another branch from the 2nd cervical runs downward to join a branch from the 3rd and form the inferior root of the ansa cervicalis.

The small muscular branches of the cervical plexus include twigs to the adjacent muscles of the neck—the longus muscles on the front of the vertebral column, the scalenus medius, more laterally,

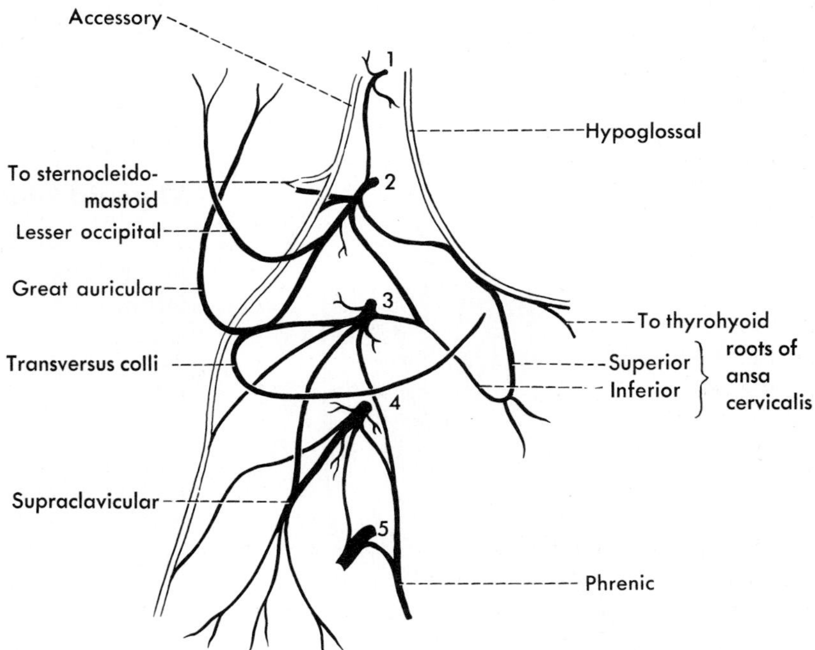


**FIG. 30-16.** Diagram of the cervical plexus. The small unlabeled branches are muscular ones.

and the levator scapulae, which arises from the transverse processes of cervical vertebrae posterior to the scalene muscles.

An important muscular branch of the cervical plexus is the **phrenic nerve,** although this does not arise exclusively from the cervical plexus. It takes origin from the 3rd, 4th, and 5th cervical nerves (the 5th cervical regularly contributes to the brachial plexus) and descends on the anterior surface of the anterior scalene muscle (Fig. 30-17), behind the fascia on this muscle. The roots contributing to the phrenic nerve may unite after only a short course, or they may be long. In some instances, therefore, two parts of the phrenic nerve parallel each other for a variable distance on the anterior scalene, in which case one of them, usually the lower, is called an *accessory phrenic nerve*. They join either low in the neck or in the thorax. Typically the phrenic nerve enters the thorax by passing behind the subclavian vein, but sometimes it or an accessory phrenic will run in front of the vein.

The nerves of the cervical plexus receive (close to the level at which they appear) rami communicantes, most of which descend from the superior cervical ganglion, located high in the neck, but these may run through muscles and be difficult to identify.

**FIG. 30-17.** The right phrenic nerve in the neck. The arteries crossing it are the transverse cervical, above, and the suprascapular, below.

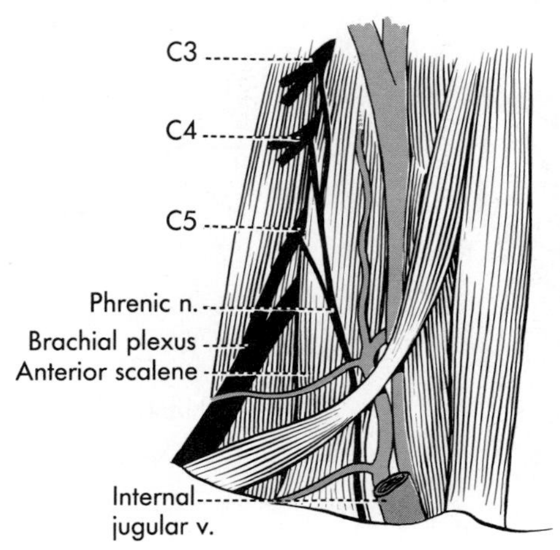


The cutaneous branches of the cervical plexus arise in the following manner. The **lesser occipital nerve** usually is a direct branch from the main stem of the 2nd cervical nerve. The larger remaining part of this stem then unites with a part of the 3rd cervical nerve to form a trunk from which arise the **great auricular** and the **transversus colli** (transverse cervical) **nerves.** Another part of the 3rd cervical nerve runs downward to unite with a major part of the 4th and form a common supraclavicular trunk, which then divides into the three **supraclavicular nerves.**

Finally, both the 3rd and 4th cervical nerves typically send a branch to the accessory nerve, or directly into the deep surface of the trapezius, to furnish sensory fibers to this muscle. The 4th cervical nerve may send a branch downward to join the 5th cervical and participate in the formation of the brachial plexus.

## THYROID AND PARATHYROID GLANDS

The thyroid and parathyroid glands are endocrine or ductless glands (Chap. 9) that originate from the pharynx but "migrate" caudally (have their relative positions changed by growth) to assume their definitive position in the neck. The thyroid gland is the largest endocrine gland in the body and is unpaired. The parathyroid glands are small, and there are typically four of them.

### Thyroid Gland

The two large **lobes** of the thyroid gland lie anterolateral to the trachea and larynx. The **isthmus** unites the lobes across the front of the trachea, immediately below the larynx. Sometimes a pointed process, the pyramidal lobe, projects upward from the isthmus, usually close to the anterior midline but occasionally asymmetrically. This may be attached by connective tissue to the front of the thyroid cartilage or may even extend to an attachment on the hyoid bone.

**Developmentally,** the thyroid gland grows out from the floor of the pharynx in the region in which the tongue (lingua, glossa) later develops, and its duct of origin is the **thyroglossal duct.** The duct normally disappears early in development, but a part of it may persist, usually close to the hyoid bone, as a *thyroglossal duct cyst;* or the duct may retain its connection to the pit on the tongue (the foramen cecum) where the gland originated and form a fistula. Sometimes, also, part or all of the thyroid gland fails to "migrate" into the neck and remains in close association with the tongue. This is called a "lingual thyroid."

The normally placed thyroid gland lies in the visceral compartment of the neck, in a space bordered anteriorly by the pretracheal fascia and the infrahyoid muscles, laterally by the two carotid sheaths, and posteriorly by the prevertebral layer of the cervical fascia.

The thyroid gland receives two pairs of arteries (Figs. 30-18 and 30-19) and is drained by two or three pairs of veins. Close to its origin, the **superior thyroid artery** gives off an infrahyoid branch that runs along the lower border of the hyoid bone, a branch into the sternocleidomastoid muscle, and the transversely running superior laryngeal artery that disappears deep to the thyrohyoid muscle (with the internal branch of the superior laryngeal nerve) to enter the larynx. It then descends, at first in close company with the external branch of the superior laryngeal nerve, gives off a cricothyroid branch that runs forward between the thyroid and cricoid cartilages, and, as it reaches the upper pole of the thyroid gland, divides into anterior and posterior branches. The posterior branch disappears into the posterior surface of the gland, but the anterior one usually runs along the upper medial border of the lobe and may anastomose with the vessel of the opposite side along the upper border of the isthmus.

The superior thyroid artery is accompanied by the **superior thyroid vein,** which joins the internal jugular either independently or after union with other veins that enter the same general region of that vessel. Venous anastomoses on the anterior surface of the thyroid gland are frequently fairly prominent and usually can be seen to unite the superior and inferior thyroid veins. There may or may not be a **middle thyroid vein** arising at about the midlevel of the gland and passing directly laterally to perforate the carotid sheath and end in the internal jugular vein.

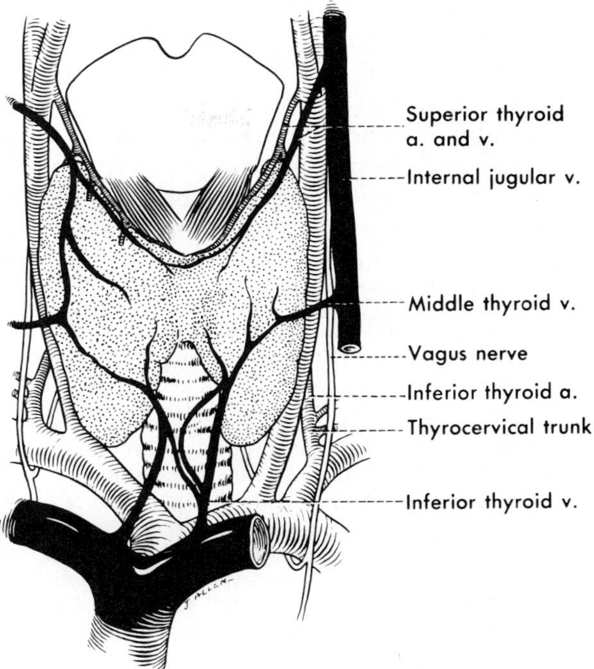


**FIG. 30-18.** Anterior view of the thyroid gland and its vessels. (Hollinshead WH: Surg Clin North Am, Aug 1952, p 1115)

**FIG. 30-19.** Posterior view of the thyroid and parathyroid glands. Superior and inferior thyroid arteries are shown. (Hollinshead WH: Surg Clin North Am, Aug 1952, p 1115)

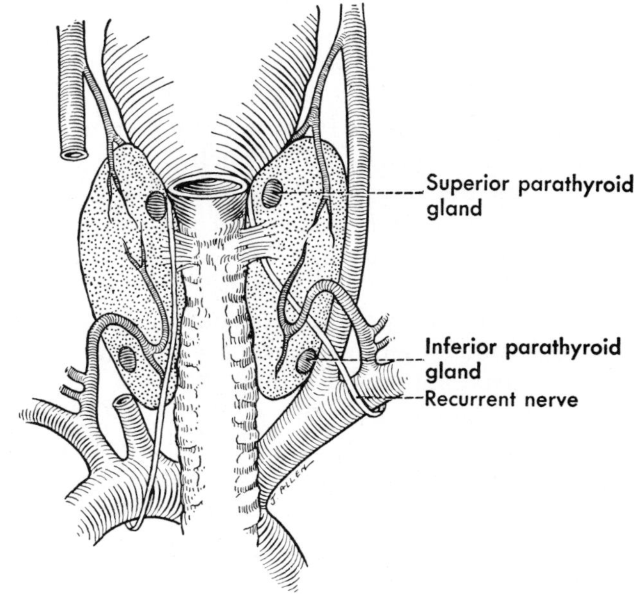

The **inferior thyroid veins** are more obvious than are the inferior thyroid arteries, for they take form in part on the anterior surface of the gland (although they drain posterior parts also) and descend in front of the trachea. They may unite to form a single stem, in which case this typically ends in the left brachiocephalic vein, or the left vein may end in the left brachiocephalic and the right one in the right brachiocephalic. The two inferior thyroid veins may interchange numerous connections to form a **plexus thyroideus impar** in front of the trachea.

**Tracheostomies** are commonly done below the isthmus of the thyroid gland, by cutting approximately the third and fourth tracheal cartilages. In this approach, the inferior thyroid veins and, if the incision is carried too far toward the sternum, the jugular venous arch (and even, occasionally, an anomalous artery) are potential sources of bleeding. Some workers prefer to displace the isthmus and cut the trachea just below the cricoid cartilage.

The **inferior thyroid artery** is a branch of the thyrocervical trunk, which arises from the subclavian artery. It runs upward and medially behind the carotid sheath, pierces the prevertebral layer of fascia behind the thyroid gland, and then usually loops downward, commonly dividing into two or more branches as it nears the gland. While it lies behind the prevertebral fascia, it gives off the **ascending cervical artery,** which ascends on the anterior and lateral muscles of the vertebral column, gives branches into them, and also gives off spinal branches that pass through the intervertebral foramina to help supply the vertebral column and the spinal cord and its meninges. The inferior thyroid artery crosses the cervical sympathetic trunk behind the carotid sheath, usually behind this trunk, but sometimes in front of it or between two parts of a double trunk.

After penetrating the prevertebral fascia, the inferior thyroid artery gives off pharyngeal branches to the lower part of the pharynx and esophageal and tracheal branches to the upper parts of these structures. As it passes medially behind the thyroid gland, it crosses the recurrent laryngeal nerve, which runs more vertically; it may cross in front of the nerve or behind it or send branches on both sides of it. The small inferior laryngeal artery, from the inferior thyroid, runs upward with the recurrent nerve to help supply the larynx.

The lobes of the thyroid gland are attached to the cricoid cartilage and several of the upper tracheal rings by some fairly dense connective tissue, the adherent zone or **suspensory ligament of the gland,** and the recurrent laryngeal nerve passes close to or even through this on its way to the larynx. The nerve or one of its branches sometimes is injured at this level in thyroidectomy, or it may be injured at the level at which it crosses the superior thyroid artery.

Sometimes an abnormal artery to the thyroid gland (usually on only one side) arises from the arch of the aorta, the brachiocephalic trunk, or the lower end of a common carotid and supplements or replaces the inferior thyroid artery. This is called a *thyroidea ima* (lowest thyroid) *artery.*

*Nerves* to the thyroid gland accompany the thyroid vessels, lying in their adventitia. They probably are entirely vasomotor; neither the thyroid nor the parathyroid glands depend upon the nervous system to govern their secretory activity.

## Parathyroids

The four parathyroid glands are small and particularly difficult to recognize in the cadaver because they are about the same color as the thyroid gland and as lymph nodes; however, in the living person their yellowish color contrasts fairly well with the deep red of the thyroid gland. Each slightly oval parathyroid tends to be approximately 4 mm to 6 mm in diameter and 1 mm to 2 mm thick. Normally, they all lie on the posterior and posteromedial surfaces of the lateral lobes of the thyroid (Fig. 30-19). The two glands of one side usually are known as the superior and inferior parathyroid glands, or sometimes, in reference to their origin from branchial pouches, as parathyroid IV (the upper gland) and parathyroid III (the lower one). However, either parathyroid may not be in its proper position—for instance, one may lie on the superior thyroid artery or even as high as the carotid bifurcation, or one may lie on the inferior thyroid artery or even in the mediastinum. Also, one may lie laterally or anteriorly on the thyroid instead of posteriorly.

It is not always possible to distinguish between superior and inferior parathyroid glands or, indeed, to locate four glands. Removal of all parathyroid tissue has serious consequences (Chap. 9): surgeons, therefore, take care to preserve at least some of the glands during thyroidectomy. The parathyroid glands typically receive their blood supply from the inferior thyroid arteries; when these are ligated, however, the glands can receive blood through the anastomoses that the thyroid arteries make with other vessels that supply the pharynx, larynx, and esophagus.

### TRACHEA AND ESOPHAGUS

The tubular **trachea** begins immediately below the larynx and is supported by a series of C-shaped cartilages united by elastic *annular ligaments,* with the incomplete part of the cartilaginous ring facing posteriorly. The posterior wall of the trachea is entirely membranous and muscular (*musculus trachealis*). Immediately behind it is the esophagus. The trachea descends in the anterior midline. In front of it lie the infrahyoid muscles, the isthmus of the thyroid gland, the inferior thyroid veins, and usually some pretracheal lymph nodes; on each side lie the lateral lobes of the thyroid gland and the carotid sheaths. The trachea disappears at the base of the neck behind the great vessels and the sternum. In the lateral grooves, between the upper part of the trachea and the esophagus, are the recurrent laryngeal nerves. They give off branches to the trachea and esophagus before they disappear into the larynx. The larynx is discussed in a later chapter.

The **pharynx** and the upper end of the **esophagus** are discussed also in a later chapter. The pharynx is somewhat funnel shaped, and the anterior wall of its small lower end is also the posterior wall of the larynx. Concerning the esophagus, it need only be noted here that it is a tubelike continuation of the pharynx arising at the lower border of the larynx and that, rather than remaining constantly open, it is collapsed anteroposteriorly except when something is being swallowed. Its cervical part is supplied by twigs from the recurrent laryngeal nerves and the inferior thyroid arteries. The muscle of the esophagus (outer longitudinal and inner circular) in the neck is typically striated muscle, being supplied by branchial efferent rather than visceral efferent elements of the vagus nerves.

## THE SCALENES AND POSTERIOR TRIANGLE

The scalene (scalenus) muscles lie in part deep to the sternocleidomastoid but appear also in the posterior triangle. There are three of them, anterior, middle, and posterior (Fig. 30-20). Each anterior scalene has the phrenic nerve on its anterior surface; each also is crossed anteriorly by arteries to the shoulder that arise from the thyrocervical trunk and by the subclavian vein of its side. The subclavian artery, however, passes behind the anterior scalene.

On the left side, the thoracic duct turns laterally in front of the anterior scalene to enter approximately the angle of junction between internal jugular and subclavian veins. The subclavian artery and most of the elements of the brachial plexus emerge between the anterior and middle scalene muscles and therefore in part hide the latter muscle. Two nerves, the dorsal scapular nerve (to the rhomboids) and the long thoracic nerve (to the serratus anterior), usually run through the middle scalene.

**FIG. 30-20.** The scalene muscles.

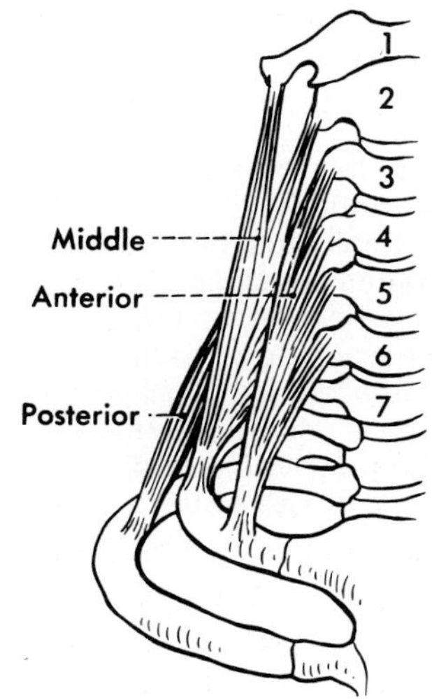

The **anterior scalene** muscle usually arises from the anterior tubercles of the transverse processes of approximately the 3rd or 4th to the 6th cervical vertebra. It runs downward and somewhat laterally. The phrenic nerve on its anterior surface runs downward and slightly medially, at an angle to the fibers of the muscle. The muscle ends in a tendon that attaches to the first rib.

The **middle scalene** muscle usually is the largest of the three, having an origin from the transverse processes of all the cervical vertebrae, all except the 1st or the 1st and 2nd, or sometimes all except the 7th. It also runs downward and laterally and inserts on the 1st rib. The gap between the anterior and middle scalenes is traversed by the brachial plexus and the subclavian artery, and those structures sometimes suffer damage here as they cross the 1st rib (p. 187).

The **posterior scalene** usually is a small muscle. It arises from the posterior tubercles of the transverse processes of about the 5th and 6th cervical vertebrae and descends between the middle scalene and the levator scapulae to cross the 1st rib and insert upon the 2nd or 3rd.

The scalene muscles are innervated by twigs from nerves that contribute to the cervical and brachial plexuses. Because the particular segmental nerves concerned are those most intimately related to the muscle, they vary with the origin of the muscle. They are flexors and rotators of the neck and head, but also are respiratory muscles, for they take their fixed point from above, usually, and elevate the thorax.

In addition to the three scalene muscles named, there is fairly frequently, between the anterior and middle one, a small muscular bundle known as the **scalenus minimus.** This usually arises from the anterior tubercle of either the 6th or the 7th cervical vertebra and passes downward between the anterior and middle scalenes. It may attach entirely to the 1st rib or pass behind the rib to attach into the fascia over the lung, the suprapleural membrane, or it may attach to both structures. In attaching to the 1st rib, it may separate some elements of the brachial plexus from other parts or the subclavian artery from the brachial plexus. Presumably, a scalenus minimus sometimes is responsible for signs of compression of the brachial plexus.

Immediately behind the posterior scalene muscle is the **levator scapulae,** a muscle of the shoulder that arises from cervical transverse processes. The accessory nerve runs downward on it across the posterior triangle. Above and behind the levator scapulae, the splenius muscles, muscles of the back, form part of the floor of the triangle. The occipital artery usually appears in the uppermost part of the triangle as it passes from under cover of the sternocleidomastoid to disappear deep to the trapezius muscle, which it then pierces to turn upward in the scalp.

## Brachial Plexus

Only a part of the brachial plexus lies in the neck, and almost all its branches arise in the axilla (after it has crossed the 1st rib). A more nearly complete description of this plexus will be found beginning with page 183; see also Figure 12-21 for some relations in the neck. The ventral rami contributing to the plexus are those of the 5th, 6th, 7th, and 8th cervical nerves and the 1st thoracic; there may or may not be a branch descending from the 4th cervical nerve to join the 5th and thus contribute to the plexus. A small part of the ventral ramus of the 1st thoracic nerve, instead of ascending across the 1st rib to join the brachial plexus, continues below the rib as the 1st intercostal nerve.

The first named parts of the brachial plexus are three **trunks** (Fig. 30-21). In the usual brachial plexus, the 5th and 6th cervical nerves, with any contribution received from the 4th, emerge from between the anterior and middle scalene muscles and join on the anterolateral surface of the middle scalene to form the *superior trunk* of the plexus. From this superior trunk there usually is given off a tiny **nerve to the subclavius** (Fig. 30-12), which runs down in front of the plexus and behind the clavicle to end in the subclavius muscle, and a much larger **suprascapular nerve,** which runs downward, laterally, and posteriorly to disappear deep to the trapezius and join the suprascapular artery in a course toward the shoulder.

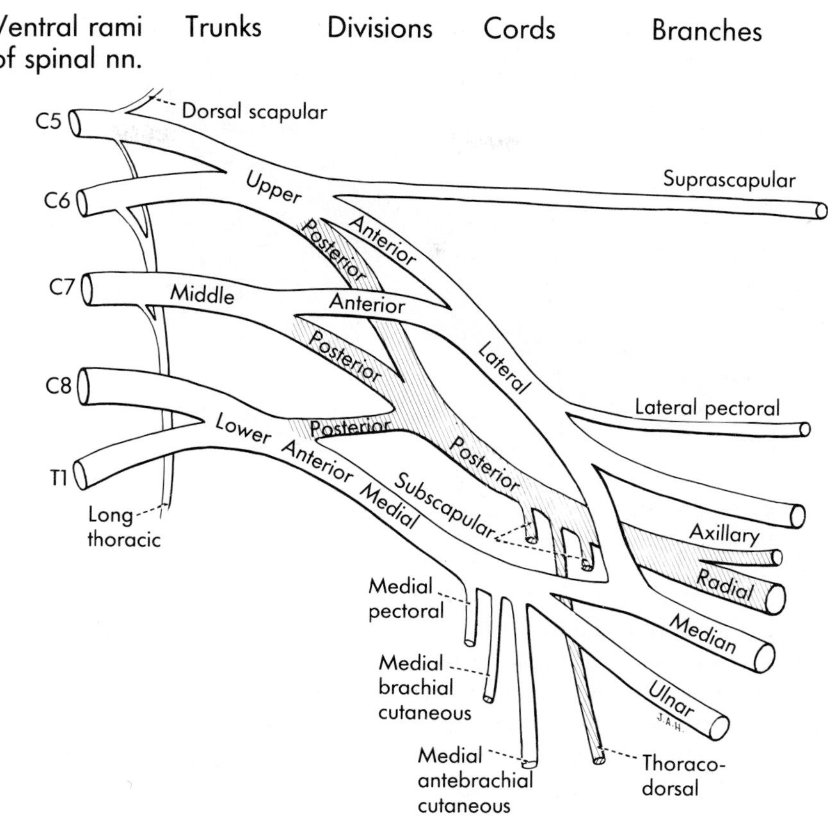


**FIG. 30-21.** Schema of the left brachial plexus; the tiny nerve to the subclavius, from the upper trunk, is here omitted. The posterior divisions and the posterior cord and its branches are shaded.

The *middle trunk* of the brachial plexus is the ventral ramus of the 7th cervical nerve and runs downward and laterally on the middle scalene. The *inferior trunk* is formed by the union of the ventral ramus of the 8th cervical with that of the 1st thoracic (after the small 1st intercostal has been given off); the union of these nerves may be behind the anterior scalene muscle or lateral to it. Neither the middle nor the inferior trunk gives off any branches comparable to those of the upper trunk.

The trunks are in turn succeeded by **divisions.** A variable distance from their origin, each trunk divides into an anterior and a posterior division. These divisions represent a sorting of fibers for the nerves that go to the original anterior and original posterior aspects of the limb, respectively. They usually give off no branches, but are in turn succeeded by three **cords.** The anterior divisions of the upper and middle trunk unite soon after their origin to form the **lateral cord** of the brachial plexus. The posterior divisions of the upper and middle trunks also typically unite soon after their origin to form a major part of the **posterior cord,** which is then joined by the small posterior division of the lower trunk. The large anterior division of the lower trunk continues as the **medial cord.** The lower trunk may cross the 1st rib and thus leave the neck before it divides; in this case, the medial cord is formed in the axilla and the posterior cord is completed in the axilla. In leaving the neck, the lateral and posterior cords, or the major part of the posterior cord, lie above and therefore lateral to the subclavian artery (Fig. 30-22), but the lower trunk or the medial cord lies behind it, against the 1st rib. The cords usually give off their first branches after they

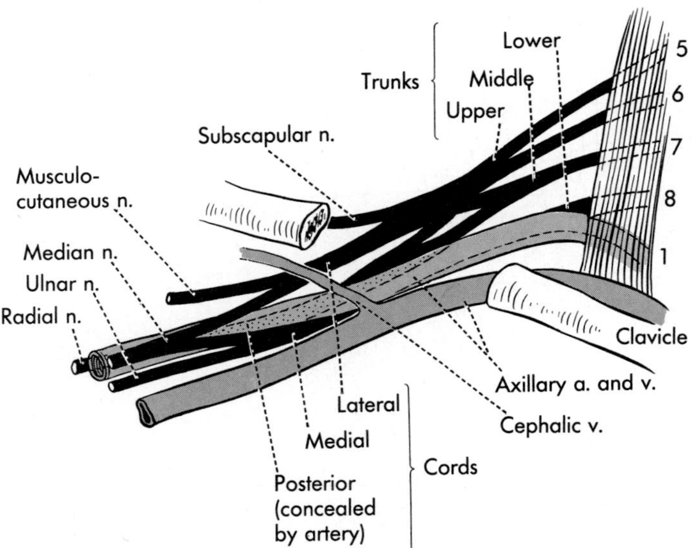


**FIG. 30-22.** The relation of the brachial plexus to the subclavian and axillary arteries. The posterior cord is exaggeratedly prolonged in order to show its position.

reach the axilla. Their disposition and their branches are described with that part.

In addition to the nerve to the subclavius and the suprascapular nerve, other branches arising in the neck include the dorsal scapular and long thoracic nerves, twigs to the scalene and the longus cervicis muscles, and the root of the phrenic nerve that is contributed by the 5th cervical. The **dorsal scapular nerve** arises from the ventral ramus of the 5th cervical nerve behind the anterior scalene muscle and passes downward and backward through the middle scalene to course to the shoulder deep to the trapezius. The **long thoracic nerve** arises typically from the ventral rami of the 5th, 6th, and 7th cervical nerves. These roots may come together in the substance of the middle scalene, or they may run separately through, anterior or posterior to the muscle. After the roots come together, the nerve descends almost vertically on the middle scalene to reach the upper slip of origin of the serratus anterior muscle and descends on that muscle along the lateral thoracic wall.

It already has been noted (Chap. 12) that the common variations in the brachial plexus are those in the level of origin of its branches and that this sometimes is in part an artifact due to the manner in which the connective tissue surrounding the nerves is dissected. The length of plexus in the neck is so short that most variations are found in the axilla; however, sometimes the nerves themselves, rather than the trunks, divide into anterior and posterior divisions, so that the trunks are not clear. Similarly, the suprascapular nerve may arise from the posterior sides of the 5th and 6th cervical nerves by separate roots instead of arising from the upper trunk, or a branch such as a pectoral nerve may arise higher than usual. These variations do not, of course, change the composition of the plexus or its branches. Real variations in the formation of its cords and branches, such as a contribution from C8 to the lateral cord, are rare.

## Vessels and Nerves

**Arteries and Veins.** Besides the occipital artery, already mentioned, the vessels that appear in the posterior triangle are largely the subclavian artery and its branches and the subclavian vein and its tributaries; lymphatics and lymph nodes, not very obvious, are arranged along these vessels.

The two subclavian arteries have different origins, but similar courses in the neck (Fig. 30-23). The **left subclavian,** a branch of the arch of the aorta, arises in the thorax behind the left border of the manubrium sterni. It ascends almost vertically on the left side of the trachea and adjacent to the parietal pleura of the left lung. At the base of the neck, it arches

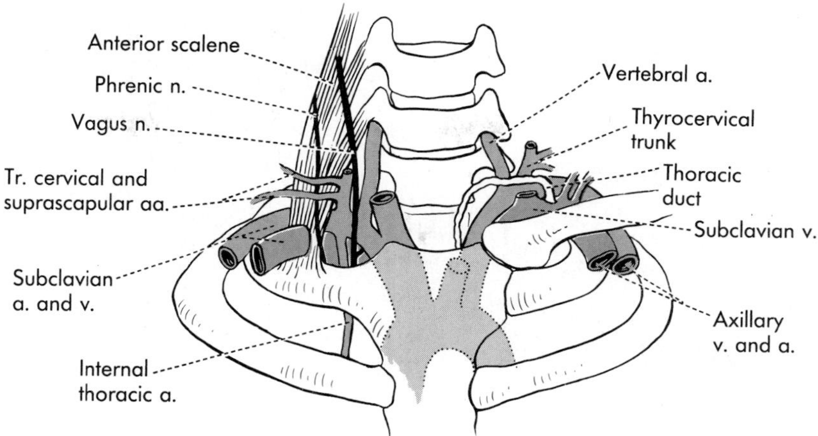


**FIG. 30-23.** Vascular relations at the base of the neck. The left common carotid is indicated as a stump behind the sternum, and the left anterior scalene is removed, so the fact that the thoracic duct runs behind the common carotid but in front of the scalene is not apparent here.

upward and laterally in front of the apex of the lung to disappear behind the left anterior scalene muscle. The **right subclavian artery,** in contrast, originates at the base of the neck behind the right sternoclavicular joint, where it and the right common carotid are the terminal branches of the brachiocephalic trunk. It then curves laterally across the anterior surface of the apex of the right lung and disappears behind the right anterior scalene muscle. The part of each subclavian artery between its origin and the medial edge of the anterior scalene muscle frequently is known as the first part of the artery; the second part lies behind the anterior scalene, and the third part lies between the lateral edge of the muscle and the 1st rib. At the latter level the subclavian artery enters the axilla and its name changes to axillary artery.

The anterior scalene muscle separates the subclavian artery from the subclavian vein. The third part of the artery lies in front of the lower trunk of the brachial plexus and immediately behind the subclavian vein. The upper and middle trunks of the plexus and their continuations lie mostly above the level of the artery but become closely related to its upper (lateral) aspect close to the level of the 1st rib.

The usual description of the subclavian artery has been that it has four branches, all of which arise from the first part of the artery (medial to the anterior scalene), but frequently a fifth branch arises from the second or third part of the subclavian. The branches of the subclavian artery that typically arise from the first part of the subclavian are the vertebral artery, the thyrocervical trunk, the internal thoracic artery, and the costocervical trunk. A branch arising from the second or third part of the subclavian may be the transverse cervical or the suprascapular artery, both of which are said to have a "normal" origin when they arise from the thyrocervical trunk; most frequently, it is an abnormally arising branch of the transverse cervical artery. An artery arising from the second or third part of the subclavian artery usually passes through rather than in front of the brachial plexus.

The **vertebral artery** (Fig. 30-24) typically arises from the posterosuperior aspect of the subclavian artery, although occasionally the left one may arise from the arch of the aorta. It runs upward and slightly posteriorly and usually enters the transverse foramen of the 6th cervical vertebra, sometimes higher but very rarely lower. In this first part of its

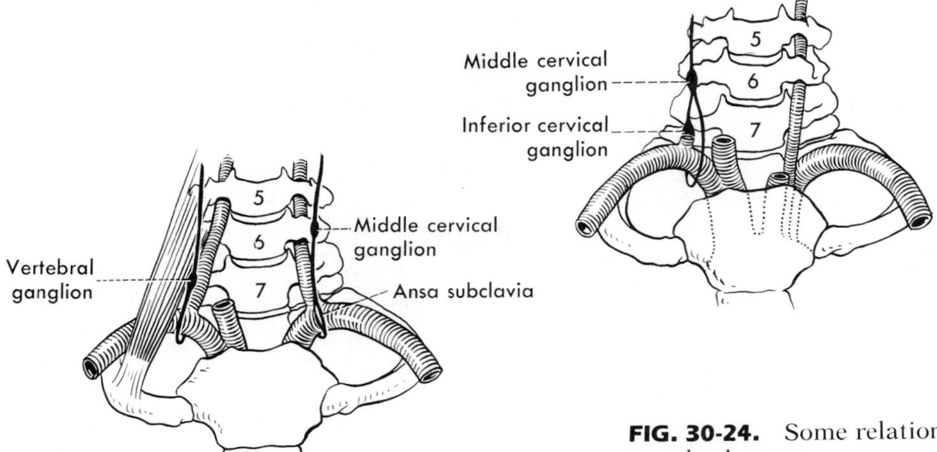


**FIG. 30-24.** Some relations and variations of the vertebral artery.

**FIG. 30-25.** The thyrocervical trunk.

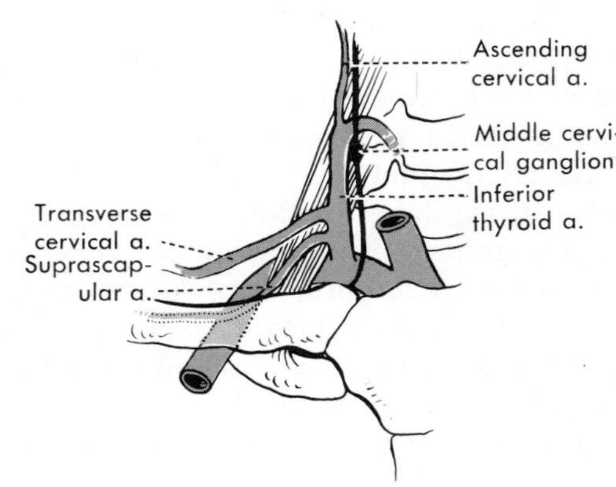


course, it passes in front of the cervical transverse processes below its level of entrance. The upper end of the cervicothoracic or stellate ganglion lies behind or posteromedial to it at its origin from the subclavian artery, and the vertebral artery is accompanied by rather heavy nerve filaments derived from this ganglion. A vertebral ganglion may lie on the anterior surface of the artery at its origin and be connected to the upper end of the stellate by fibers that pass on both sides of the artery. The lower part of the vertebral artery is accompanied by a **vertebral vein,** which usually forms a plexus around its upper part. Just before this vein ends in the brachiocephalic vein, it receives the **anterior vertebral vein,** a small vein descending in front of the transverse processes parallel to the ascending cervical artery.

After entering a transverse foramen, the vertebral artery runs upward through successive transverse foramina to the base of the skull, accompanied by the vertebral vein. After emerging through the vertebral foramen of the 1st cervical vertebra, it turns dorsally and medially on the upper surface of this vertebra and enters the vertebral canal by penetrating the posterior atlantooccipital membrane. Its major distribution, to the brain, is described in Chapter 32. Its branches in the neck are small. They consist of some that supply adjacent deep muscles of the neck and spinal branches that pass through intervertebral foramina to enter the vertebral canal. In the suboccipital triangle, some of its branches anastomose with branches of the occipital and deep cervical arteries. The vertebral vein does not accompany the artery into the vertebral canal, but participates in the formation of the **suboccipital venous plexus.** This plexus receives the occipital veins, may receive emissary veins (from inside the skull), and gives rise to the vertebral and deep cervical veins.

The **thyrocervical trunk** (Fig. 30-25) arises from the upper anterior part of the subclavian artery, close to the medial border of the anterior scalene. After giving off the transverse cervical (transversa colli) and suprascapular arteries, it ends as the inferior thyroid. The inferior thyroid artery ascends with a somewhat medial course and already has been described. The **transverse cervical artery,** the upper of the two branches to the shoulder, runs laterally and posteriorly across the front of the anterior scalene muscle and the brachial plexus. At the posterior border of the posterior triangle, it disappears deep to the trapezius muscle. Its further course has been described (p. 203); it need only be noted here that it may divide into a superficial and a deep branch, which are distributed deep to the trapezius and deep to the rhomboid muscles, respectively. Commonly, the vessel corresponding to the superficial branch arises independently from the thyrocervical trunk and is then known as the **superficial cervical artery.** The vessel corresponding to the deep branch then arises from the subclavian artery and is known as the **dorsal** or **descending scapular artery.**

The other branch of the thyrocervical trunk, the **suprascapular artery,** is usually the first branch from the trunk. It runs transversely across the neck parallel to but below the transverse cervical artery and behind the clavicle. It passes in front of the anterior scalene muscle and the brachial plexus and, as it reaches the lateral border of the plexus, is joined in its course by the suprascapular nerve. Nerve and artery disappear together deep to the trapezius, and their further course is described with the shoulder. The suprascapular artery has been reported to arise directly from the subclavian in more than 20% of instances.

Single or paired **transverse cervical** and **suprascapular veins** accompany the arteries. These veins do not join the inferior thyroid vein, however, but enter the lower end of the external jugular vein or the proximal part of the subclavian.

The **internal thoracic artery** (formerly internal mammary) arises from the anteroinferior aspect of the subclavian at about the same level as the thyrocervical trunk. It passes downward, forward, and somewhat medially, lying on the anterior aspect of the parietal pleura covering the apex of the lung, and behind the brachiocephalic vein and the sternal end of the clavicle. It enters the thorax behind the cartilage of the 1st rib. The **internal thoracic veins** do not quite parallel the uppermost parts of the arteries, for they enter the brachiocephalic veins rather than the subclavians.

The **costocervical trunk** arises from the posterior aspect of the subclavian artery just medial to or behind the medial border of the anterior scalene muscle. It runs posterosuperiorly to divide into deep cervical and supreme intercostal arteries. The **supreme intercostal artery** supplies approximately the upper two intercostal spaces with posterior intercostal arteries. The **deep cervical artery** passes backward between the transverse process of the 7th cervical vertebra and the neck of the 1st rib, and below the 8th cervical nerve. It then ascends between two muscles of the back (semispinalis capitis and semispinalis cervicis), anastomosing as it does so with branches of the ascending pharyngeal and vertebral arteries and ending above by anastomosing with a descending branch of the occipital artery. Sometimes the deep cervical and the supreme intercostal arise separately from the subclavian, or the supreme intercostal may be lacking and supplanted by branches from the aorta.

Each **subclavian vein** is the direct continuation of the axillary vein on its side and therefore begins at the upper border of the 1st rib. At the base of the neck, the vein runs at first in front of and partly below the subclavian artery and then is separated from that by the anterior scalene muscle and the phrenic nerve. In front of the medial border of the muscle, the vein ends by joining the internal jugular vein, thereby forming the upper end of the brachiocephalic vein. The two **brachiocephalic veins,** therefore, begin behind the medial ends of the clavicles and pass downward to enter the thorax.

The subclavian vein has only one constant named tributary, the **external jugular.** This enters the subclavian close to the posterior border of the sternocleidomastoid muscle. Through the external jugular, the subclavian receives blood from the suprascapular and transverse cervical veins. The veins corresponding to other branches of the subclavian artery empty into the brachiocephalic: the in-

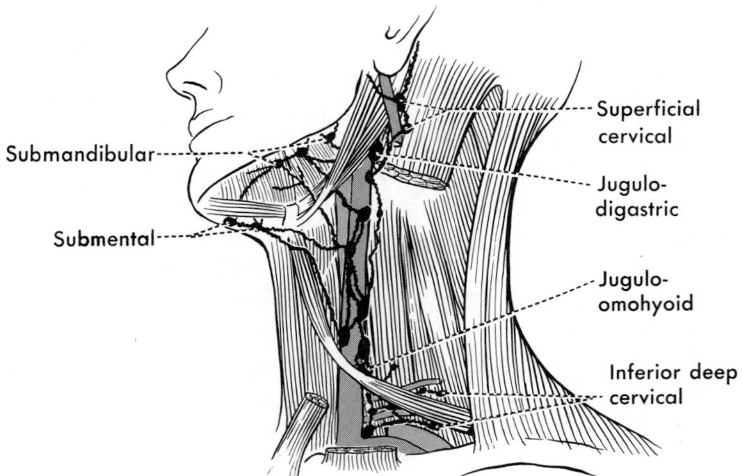


**FIG. 30-26.** Superficial and deep lymph nodes of the submandibular region and neck.

ternal thoracic veins enter the brachiocephalic veins in the thorax; the vertebral veins empty into the upper ends of the brachiocephalics; and the supreme intercostal and deep cervical veins empty separately into either the vertebral or the upper end of the brachiocephalic.

**Lymph Nodes and Lymphatics.** Note already has been made that **deep cervical lymph nodes** lie in the connective tissue of the carotid sheath, particularly closely related to the internal jugular vein. Sometimes they are divided roughly into a superior and an inferior group, and two nodes usually are named (Fig. 30-26): the **jugulodigastric node** lies at about the level at which the digastric muscle crosses the internal jugular vein, and the **juguloomohyoid node** lies at the level of crossing of the omohyoid muscle and the vein. The deep cervical lymph nodes receive, directly or indirectly, most of the lymphatics from the head and neck. Among the nodes draining into them are the *submental nodes,* in the submental region; *submandibular nodes,* associated with the submandibular salivary gland; *retropharyngeal nodes* that lie posterolateral to the pharynx; and nodes that lie along the course of the facial vessels. Lymphatics from the face, tongue, and other parts of the head also bypass the regional nodes along their course and enter deep cervical nodes directly. In addition to the nodes mentioned, *occipital, retroauricular,* and *superficial* and *deep parotid nodes* drain into the deep cervical nodes. They also drain, in part, into a few superficial cervical nodes that lie along the course of the external jugular vein. There also are nodes and lymphatics along the accessory nerve and the suprascapular and transverse cervical vessels (the latter two sets usually included among the inferior deep nodes) in the posterior triangle and deep to the trapezius that drain into the nodes along the internal jugular vein.

In the operation known as "radical neck dissection," carried out when carcinoma has reached some of the deep cervical nodes, the tissue containing them is removed as completely as possible and in one piece as far as that can be managed. The sternocleidomastoid muscle and the part of the omohyoid lying deep and medial to it, the internal jugular vein and the connective tissue around it and the carotid artery and the vagus nerve, and the connective tissue in the posterior triangle of the neck are dissected out from the base of the neck to the lower border of the jaw. The dissection includes the submandibular gland and the lower end of the parotid gland, both bellies of the digastric muscle, and the stylohyoid muscle. Often, the accessory nerve is resected and tissue is dissected from deep to the trapezius muscle. The major arteries, the vagus nerve, the brachial plexus, and the phrenic nerve are spared, but the cutaneous branches of the cervical plexus are sacrificed. The attempt is to remove in one block all the node-bearing tissue of one side of the neck.

The formation and thoracic course of the **thoracic duct** have been described. It enters the neck slightly to the left of the midline, lying on the vertebral column and behind the esophagus. As it diverges farther to the left, it passes behind the carotid sheath and its contents, but in front of the parietal pleura over the apex of the left lung and in front of the branches of the subclavian arising medial to the anterior scalene. As it runs laterally or arches laterally and downward to enter approximately the angle of union of internal jugular and left subclavian veins, it passes also in front of the anterior scalene muscle and the phrenic nerve (Fig. 30-27). It may run upward only high enough to reach the angle of junction of the two veins, or it may run much too high and farther laterally than necessary and then loop downward and medially, thereby being exposed to injury in almost any operation at the base of the neck. Close to its ending, it usually receives a **jugular trunk** from the deep cervical nodes and a **subclavian trunk** from the axillary nodes on its side and may receive the **left bronchomediastinal trunk;** or these vessels may enter the venous system independently or in various combinations. These smaller lymphatic trunks are difficult to recognize in the usual dissection; indeed, even the thoracic duct may not be easily recognized, because its proximal end is frequently filled with blood in the cadaver.

A **right lymphatic** (right thoracic) **duct,** representing a vessel formed by a combination of jugular, subclavian, and bronchomediastinal trunks, may enter the angle of junction of the right subclavian and internal jugular veins, but frequently the vessels on the right do not combine to form a single trunk.

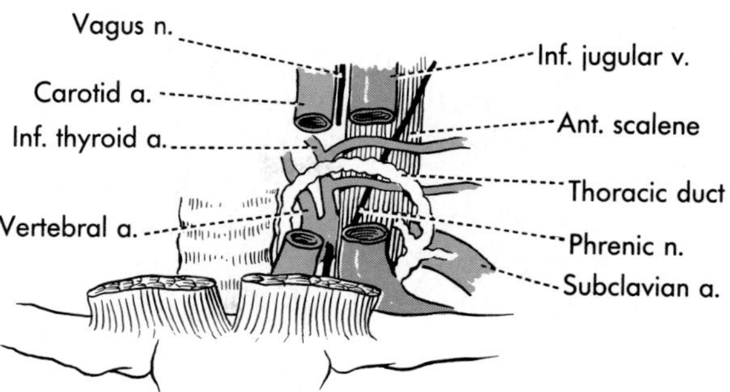


**FIG. 30-27.** The thoracic duct in the neck; it passes behind the structures within the carotid sheath, and these are shown with a segment of each removed.

**FIG. 30-28.** The cervical sympathetic trunks; the two trunks are not necessarily symmetrical, and there may be both a middle cervical and a vertebral ganglion on one side.

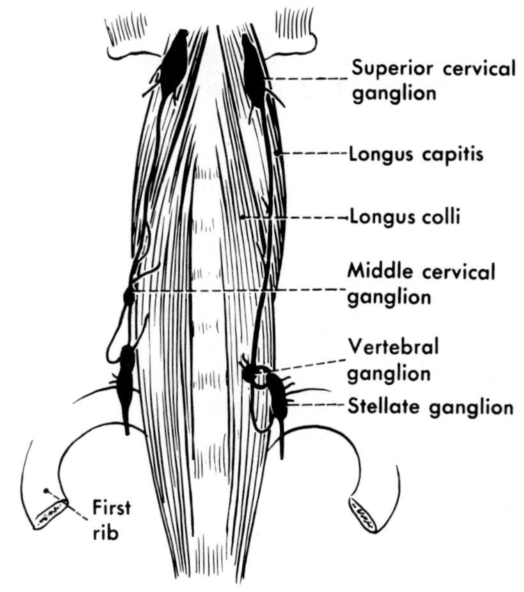


**Sympathetic Trunks and Ganglia.** The lowest ganglion of the **cervical sympathetic trunk** (Fig. 30-28) is the **stellate** or **cervicothoracic ganglion,** formed by fusion of the lowest cervical and 1st thoracic ganglion; sometimes these are separate, and there is then an **inferior cervical ganglion.** The ganglion is particularly large if it is truly cervicothoracic. It lies in front of the neck of the 1st rib and the transverse process of the 7th cervical vertebra, behind the upper end of the first part of the subclavian artery and the origin of the vertebral artery, and just above or partly behind the dome of the pleura. It receives one or more white rami communicantes (preganglionic fibers) from the 1st thoracic nerve and may receive one from the 2nd. It gives off gray rami communicantes to the 1st thoracic and lower cervical nerves, thus contributing postganglionic fibers to the brachial plexus, and sends other rami (grouped to form the *vertebral nerve*) along the vertebral artery to the upper nerves of the brachial plexus.

The cervical sympathetic trunk passes upward on the front of the longus colli and longus capitis muscles, behind the common carotid artery while that is in the carotid sheath, and then behind the internal carotid. It is a rather slender strand of fibers and ends above in the **superior cervical ganglion.** This elongated ganglion lies in front of the upper two cervical vertebrae. It sends rami communicantes to the upper three or four cervical nerves and a small **external carotid nerve,** difficult to find by dissection, downward to the external carotid to form a delicate external carotid plexus that follows branches of that vessel. Its largest branch is a broad band of fibers from its upper end, the **internal carotid nerve.** This breaks up into the internal carotid plexus, which follows the artery into the skull.

Between the cervicothoracic and the superior cervical ganglia, there are one or two other ganglia. One, the **middle cervical ganglion,** lies on the trunk at the level of the transverse process of the 6th cervical vertebra. It typically is connected to the cervicothoracic ganglion both by fibers that pass in the trunk behind the subclavian artery and by fibers, constituting the **ansa subclavia** (Fig. 30-24), that loop down in front of the subclavian artery and then turn below it to reach the cervicothoracic ganglion. A second ganglion, the **vertebral ganglion,** may lie anterior or anteromedial to the vertebral artery at its origin and when present may receive the ansa subclavia, in addition to being connected, behind the subclavian artery, to the cervicothoracic ganglion by fibers that usually pass on both sides of the vertebral artery.

The vertebral or middle cervical ganglia typically supply rami communicantes to about the 4th and 5th cervical nerves. There is considerable variation as to exactly which cervical nerves receive rami from which ganglia, and it is often difficult to determine this without extreme care in dissection, since many of the rami communicantes disappear into the longus muscles on their way to the spinal nerves. Superior, middle, and inferior **cardiac nerves** (the superior cardiac nerves are very tiny) arise from or close to the similarly named ganglia and run downward to join the cardiac plexus in the thorax (p. 540).

**The Prevertebral Muscles.** The **longus colli** and **capitis** have been mentioned, but no complete view of them can be obtained until a great deal of dissection has been done on the neck and head. It should suffice to say here that they are complex in structure, closely bound to the front of the vertebral bodies and transverse processes, and consist of two overlapping muscles, the lower being the longus colli and the upper the longus capitis (Fig. 30-28). They are primarily flexors of the neck and head.

Lying partly behind the upper end of the longus capitis is the **rectus capitis anterior,** which arises from the lateral mass of the atlas and inserts on the basilar part of the occipital bone. Lateral to it is the **rectus capitis lateralis,** arising from the transverse process of the atlas and inserting more laterally on the occipital bone.

## RECOMMENDED READINGS


ARMSTRONG WG, HINTON JW: Multiple divisions of the recurrent laryngeal nerve: An anatomic study. Arch Surg 62:532, 1951

BACHHUBER CA: Complications of thyroid surgery: Anatomy of the recurrent laryngeal nerve, middle thyroid vein and inferior thyroid artery. Am J Surg 60:96, 1943

BECK AL: Deep neck infection. Ann Otol Rhinol Laryngol 56:439, 1947

BROWN S: The external jugular vein in American whites and Negroes. Am J Phys Anthropol 28:213, 1941

COLLER FA, YGLESIAS L: The relation of the spread of infection to fascial planes in the neck and thorax. Surgery 1:323, 1937

DASELER EH, ANSON BJ: Surgical anatomy of the subclavian artery and its branches. Surg Gynecol Obstet 108:149, 1959

DOW DR: The anatomy of rudimentary first thoracic

ribs, with special reference to the arrangement of the brachial plexus. J Anat 59:166, 1925

GILMOUR JR, MARTIN WJ: The weight of the parathyroid glands. J Pathol Bacteriol 44:431, 1937

GRODINSKY M, HOLYOKE EA: The fasciae and fascial spaces of the head, neck and adjacent regions. Am J Anat 63:367, 1938

HARRIS W: The true form of the brachial plexus, and its motor distribution. J Anat Physiol 38:399, 1904

HUELKE DF: A study of the transverse cervical and dorsal scapular arteries. Anat Rec 132:233, 1958

KELLY WO: Phrenic nerve paralysis: Special consideration of the accessory phrenic nerve. J Thorac Surg 19:923, 1950

KERR AT: The brachial plexus of nerves in man, the variations in its formation and branches. Am J Anat 23:285, 1918

KIMMEL DL: The cervical sympathetic rami and the vertebral plexus in the human fetus. J Comp Neurol 112:141, 1959

MILLZNER RJ: The normal variations in the position of the human parathyroid glands. Anat Rec 48:399, 1931

NORRIS EH: The parathyroid glands and the lateral thyroid in man: Their morphogenesis, histogenesis, topographic anatomy and prenatal growth. Contrib Embryol 26:247, 1937

PIKKIEFF E: On subcutaneous veins of the neck. J Anat 72:119, 1937

REED AF: The relations of the inferior laryngeal nerve to the inferior thyroid artery. Anat Rec 85:17, 1943

SCHMIDT CF, COMROE JH JR: Functions of the carotid and aortic bodies. Physiol Rev 20:115, 1940

SINGER E: Human brachial plexus united into a single cord: Description and interpretation. Anat Rec 55:411, 1933

SIWE SA: The cervical part of the gangliated cord, with special reference to its connections with the spinal nerves and certain cerebral nerves. Am J Anat 48:479, 1931

SMITH JR: Lymphatic cannulation of the head and neck. Plast Reconstr Surg 32:607, 1963

SUNDERLAND S, BEDBROOK GM: Narrowing of the second part of the subclavian artery. Anat Rec 104:299, 1949

# 31

# SKULL, FACE, AND JAWS

Intelligent study of the soft tissues of the head is impossible without knowledge of the skull, and a clean skull with a removable top should therefore be studied before dissection of the head is undertaken.

## THE SKULL

The skull consists of 22 bones, 21 of which are firmly bound together. One, the mandible, or bone of the lower jaw, is movable and articulates with the remainder of the skull through paired synovial joints. The terms "skull" and "cranium" often are used as synonyms, but the bones of the skull as a whole are sometimes subdivided into cranial and facial ones. In the N.A. classification, the bones of the cranium are the unpaired occipital, sphenoid, frontal, ethmoid, and vomer and the paired parietals, temporals, inferior nasal conchae, lacrimals, and nasals. Of these, however, the vomer, the inferior nasal conchae, and the lacrimal and nasal bones could more logically be regarded as bones of the facial skeleton, since they do not help form the brain case. The bones of the face, in the N.A. classification, are the paired maxillae, palatines, and zygomatic bones and the unpaired mandible.

Knowledge of the skull as a whole, of the way in which its bones fit together, and of the chief foramina through which structures enter or leave the cranial cavity is much more important than are details of each bone. If the latter are to be studied, individual bones rather than the skull as a whole will prove more useful. In addition to face and cranium, a few general terms that are applied to the skull should be understood. The skull contains five large cavities, four of which, the paired nasal cavities and the paired orbits (so called because the eyeballs rotate in them), open freely to the outside. The fifth and largest cavity, the cranial cavity, houses the brain and, except at the base of the skull where brain and spinal cord are continuous, is a closed cavity during life.

The roof of the cranial cavity is referred to as the **calvaria.** The outer periosteum of the cranium is known as the **pericranium.** Most of the cranial bones have dense external and internal **laminae** (tables) separated by spongy bone known as the **diploë.** Relatively large venous channels, the *diploic veins,* run within canals in the diploë and empty their blood in some instances into veins outside the skull, in others into cranial venous sinuses. Some of the bones of the skull—particularly the frontal, ethmoid, sphenoid, and maxillary bones—have their diploë in part replaced by air-filled cavities, the **paranasal sinuses,** that grow from the nasal cavity into adjacent bones. Similarly, the paired temporal bones largely surround the middle ear cavity, from which other air-filled extensions (mastoid air cells) grow into adjacent bone.

The highest part of the skull when it is held in an approximately normal position (with the floor of the orbit at the level of the opening of the external ear) is the **vertex. Frontal** refers to the forehead; **occipital** to the back of the head; and **temporal** to the side of the head. The internal surface of the base of the skull (*basis cranii interna*) is the floor of the cranial cavity. It and the external base deserve particularly careful study. Although most of the bones of the skull are bound together by sutures, in a few locations the gap between bones is greater and the articulation is a synchondrosis rather than a suture.

**FIG. 31-1.** Lateral view of the skull. **MAX., OCC., SPH., GR. WING,** and **ZYGO.** identify the maxillary and occipital bones, the greater wing of the sphenoid, and the zygomatic bone.

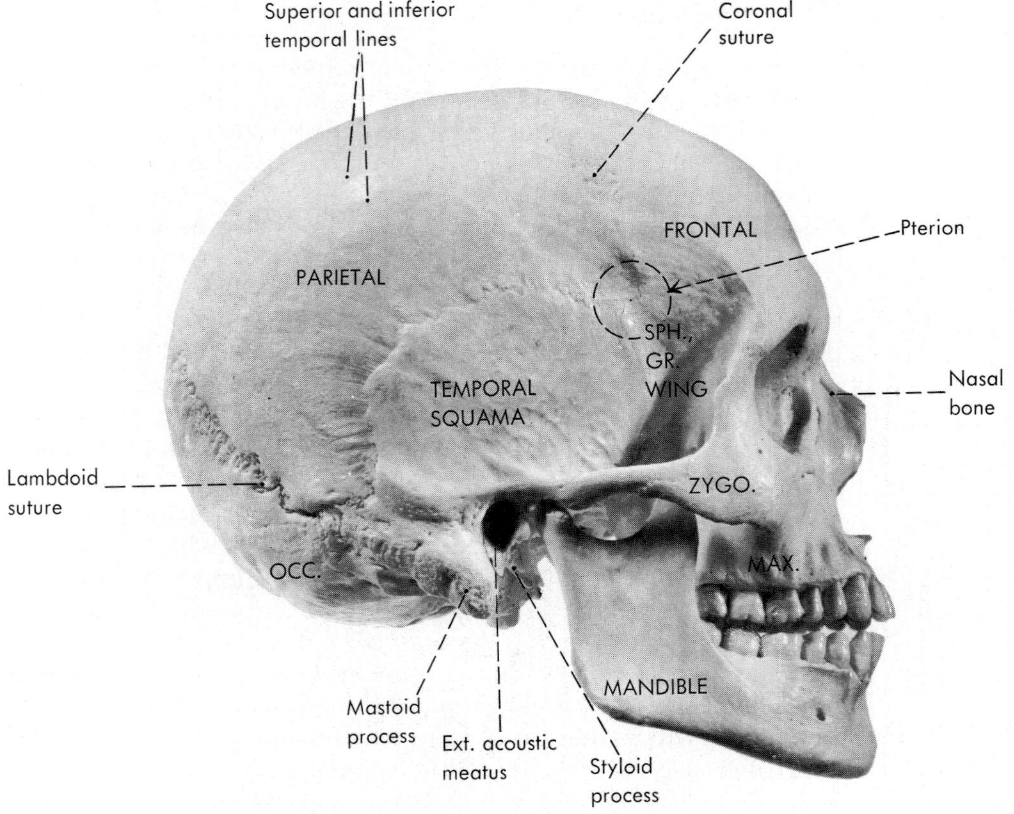

The names of many of the sutures indicate the bones between which the suture in question lies—for instance, the frontozygomatic suture is between the frontal and zygomatic bones, the sphenoparietal between sphenoid and parietal bones—and in these cases there is no necessity of learning their names. In other cases, however, the terms are not so obvious: one must learn, for instance, where the lambdoid suture is and that the suture between the temporal and parietal bones is regarded as two sutures, of which one is named the squamous suture and the other the parietomastoid suture.

## EXTERIOR OF THE SKULL

The calvaria (Figs. 31-1, 31-2, and 31-6) is formed anteriorly by the unpaired **frontal bone,** behind this by the paired **parietal bones,** and posteriorly by the unpaired **occipital bone,** which, however, participates only a little in the calvaria although it forms most of the back and extends also onto the base of the skull. The suture between the frontal bone and the two parietal bones is the **coronal suture** (hence the coronal plane of the body is the same as the frontal plane); that between the two parietal bones is the **sagittal suture** (hence the planes of the body through or parallel to it are sagittal planes). The sagittal suture is so called because in the infant, before the bones of the skull are firmly united, it and fontanels associated with it make it somewhat resemble an arrow (sagitta). Fontanels or fonticuli are the soft places in an infant's skull, where the membrane in which the bones of the calvaria are formed has not yet been replaced by bone. The inverted V-shaped suture between the two parietal bones and the occipital bone is the **lambdoid suture** (Fig. 31-2), so called because of its resemblance to the Greek letter lambda (λ).

**FIG. 31-2.** Posterior view of the skull.

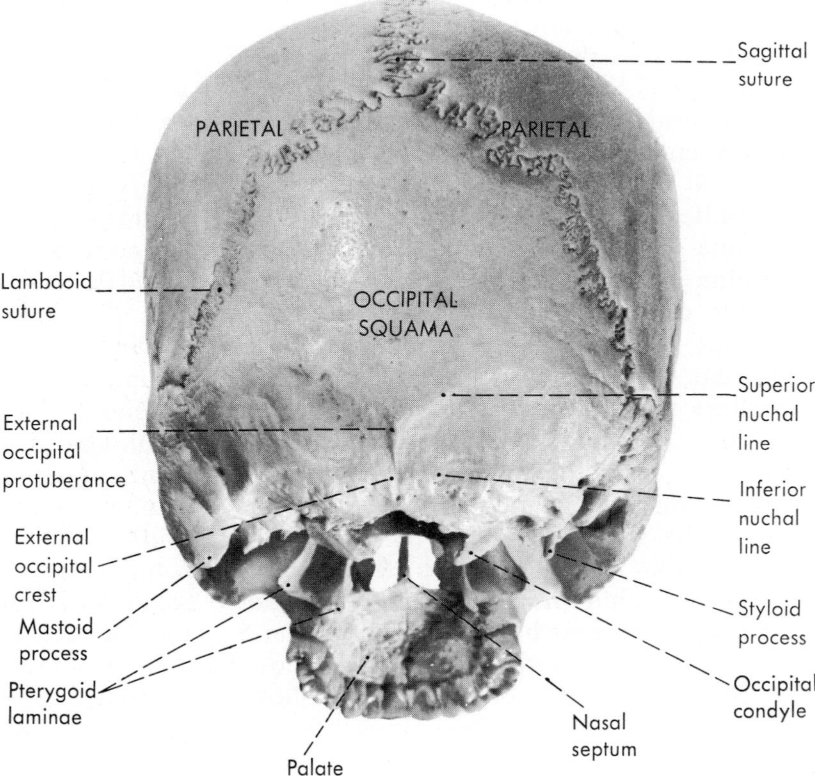

There may be small bones, sutural bones, in the sagittal or lambdoid sutures.

## Lateral View

In a lateral view of the skull, in which the bones of the calvaria also can be seen, most of the bones visible (Fig. 31-1) are those of the cranium; however, the bone forming the lateral rim of the orbit and the bony prominence of the cheek is the zygomatic bone, and below and medial to it is the maxilla, the tooth-bearing bone of the upper jaw. The mandible is described later; it is best removed while studying the rest of the skull.

The orbital surface of the **zygomatic bone** forms a part of the lateral wall of the orbit and is not visible from the side; the rounded lateral surface forms the prominence of the cheek; and the temporal surface is posterior and medial. The zygomatic bone sends a **frontal process** upward along the lateral border of the orbit to articulate with the frontal bone close to the roof of the orbit and a freely projecting **temporal process** almost horizontally backward to articulate with a similar forward projection from the temporal bone; together the two processes form the **zygomatic arch.** On its lateral surface, the zygomatic bone usually presents a small **zygomaticofacial foramen,** and on its temporal surface, a small **zygomaticotemporal foramen.** These two foramina communicate with one or two foramina in the orbital surface of the bone, and the canals through the bone transmit branches of the maxillary nerve. Sutures unite the zygomatic bone medially with the maxillary bone; above, at the lateral rim of the orbit, with the frontal; posteriorly, below the frontal, with the sphenoid bone (greater wing of this bone); and with the temporal through the zygomatic arch.

Below the zygomatic, and extending in front of and behind it, is the **maxilla.** Many of its features can best be seen in anterior and basal views of the skull, described in following sections. The **body** of the bone is hollow, containing the large **maxillary sinus.** There is a **zygomatic process** that extends upward to articulate with the zygomatic bone, and the tooth-bearing part of the maxilla is the **alveolar process.** (Alveolus means pit, and the term refers to the sockets or dental alveoli in which the teeth are set.) On the back end of the alveolar process is a small projection, the **maxillary tuber.** Behind and above this is a thin plate of bone, the lateral lamina of the pterygoid process of the sphenoid bone, described on a following page.

Just behind the zygomatic bone and largely below the frontal, forming part of the lateral wall of both the cranial cavity and the orbit, is the **greater wing of the sphenoid bone.** In addition to articulating with the zygomatic, frontal, and parietal bones, the greater wing shares a sphenosquamous suture with the squamous (flat) part of the temporal bone. The area where the frontal, parietal, sphenoid, and temporal bones are all close together is known as the **pterion.**

The somewhat concave outer surface of the greater wing of the sphenoid and the squamous part of the temporal bone together form the concavity on the side of the skull, deep to the zygomatic arch, known as the **temporal fossa.** At the lower border of the temporal fossa, the greater wing of the sphenoid presents a sharp ridge, the **infratemporal crest,** below which the bone is more horizontal, forming a part of the floor of the cranial cavity. The concavity below the crest, thus behind the maxilla and below the sphenoid and temporal bones, is the **infratemporal fossa** (p. 860). The lateral lamina of the pterygoid process of the sphenoid bone forms much of the medial wall of this fossa.

Immediately behind the greater wing of the sphenoid bone, the part of that bone best studied in lateral view, is the **temporal bone.** This forms much of the lower lateral part of the skull, contributes to the base, and houses the middle and internal ears. It is particularly complicated. The **squamous part** (*pars squamosa*) of this bone has been identified as the flat plate articulating anteriorly with the greater wing of the sphenoid. Behind the *sphenosquamous suture* it shares the curved *squamous suture* with the parietal bone. The pars squamosa participates with the sphenoid in forming the medial wall of the temporal fossa and the roof of the infratemporal fossa. From its lower part, the **zygomatic process** projects laterally and forward to complete the zygomatic arch. Behind the zygomatic process is the **external acoustic** (auditory) **meatus,** or ear canal. The thin part of the temporal bone forming the anterior wall, floor, and part of the posterior wall of this canal is the **pars tympanica;** details of the pars tympanica can best be examined when the base of the skull is studied. Just posterosuperior to the canal, there is often a sharp crest, the *suprameatal spine.* Above this, leading toward the root of the zygomatic arch, is a triangle that marks a surgical approach to both the tympanic cavity and the mastoid air cells.

The bony prominence behind the external acoustic meatus is the **mastoid process,** the posterior end of the **petrous portion of the temporal bone** (petrous, because much of this part of the bone, with the exception of the mastoid process, is particularly dense, hence somewhat rocklike—the meaning of petrous). Most of the petrous part of the temporal appears at the base of the skull and therefore is described later. The mastoid part articulates with the parietal bone through a *parietomastoid suture* and with the occipital bone through an *occipitomastoid suture.* These two sutures are continuous with each other, and the parietomastoid is in turn continuous anteriorly with the squamous suture. On the lateral surface of the bone above the mastoid process proper, there usually is a *mastoid foramen* through which a vein (emissary vein) leaves the skull. Projecting downward from the lower surface of the petrous portion of the temporal bone, in front of the mastoid process, is the more slender **styloid process;** this can be examined better in an external view of the base.

Only a little of the occipital bone can be seen from the lateral side; this is best examined in posterior and inferior views.

## Posterior View

The chief bone of the posterior wall of the skull is the **occipital bone.** It consists of *basilar and lateral parts and the squama,* the part best seen in posterior view (Fig. 31-2). The *lambdoid sutures* through which it articulates with the parietal bones, and at the back end of the sagittal suture, are both visible. The upper part of the squama may persist as a separate bone here, the *interparietal.* At its lower end, which may be indistinct, the lambdoid suture joins the parietomastoid and occipitomastoid sutures. The projection on the posterior surface of the squama is the **external occipital protuberance,** easily palpable in the living person. Its highest point is called the **inion.** There may be a ridge, the *external occipital crest,* extending downward in the midline from the protuberance. Extending laterally from it is the curved **superior nuchal line,** which may have a *supreme nuchal line* just above it. The *inferior nuchal line* is much lower, barely visible in a posterior view of the skull.

In a posterior view, also, the mastoid process again can be recognized. The deep groove on its inferomedial surface is the **mastoid incisure** or notch, the origin of the posterior belly of the digastric muscle. The styloid process may be visible anteromedial to the mastoid process.

## Basal View

In external views of the base of the skull (Figs. 31-3 and 31-4), the **occipital bone** is the prominent posterior element. The large foramen that it surrounds is the **foramen magnum,** through which brain and spinal cord are continuous. Anterolateral to the foramen are the **occipital condyles** for articulation with the atlas (1st cervical vertebra). Behind each occipital condyle is a depression, the **condylar fossa,** into which there usually opens an oblique **condylar canal** that transmits a vein from a cranial venous sinus inside the skull (sigmoid sinus) to the suboccipital plexus of veins outside the skull. Anteriorly, under cover of about the middle of each condyle and running almost horizontally, is the **hypoglossal canal** for the emergence of the nerve of that name (cranial nerve XII).

There are no distinct boundaries between the squama, the lateral parts, and the basilar part of the occipital bone, but in general the **squama** lies behind and above the foramen

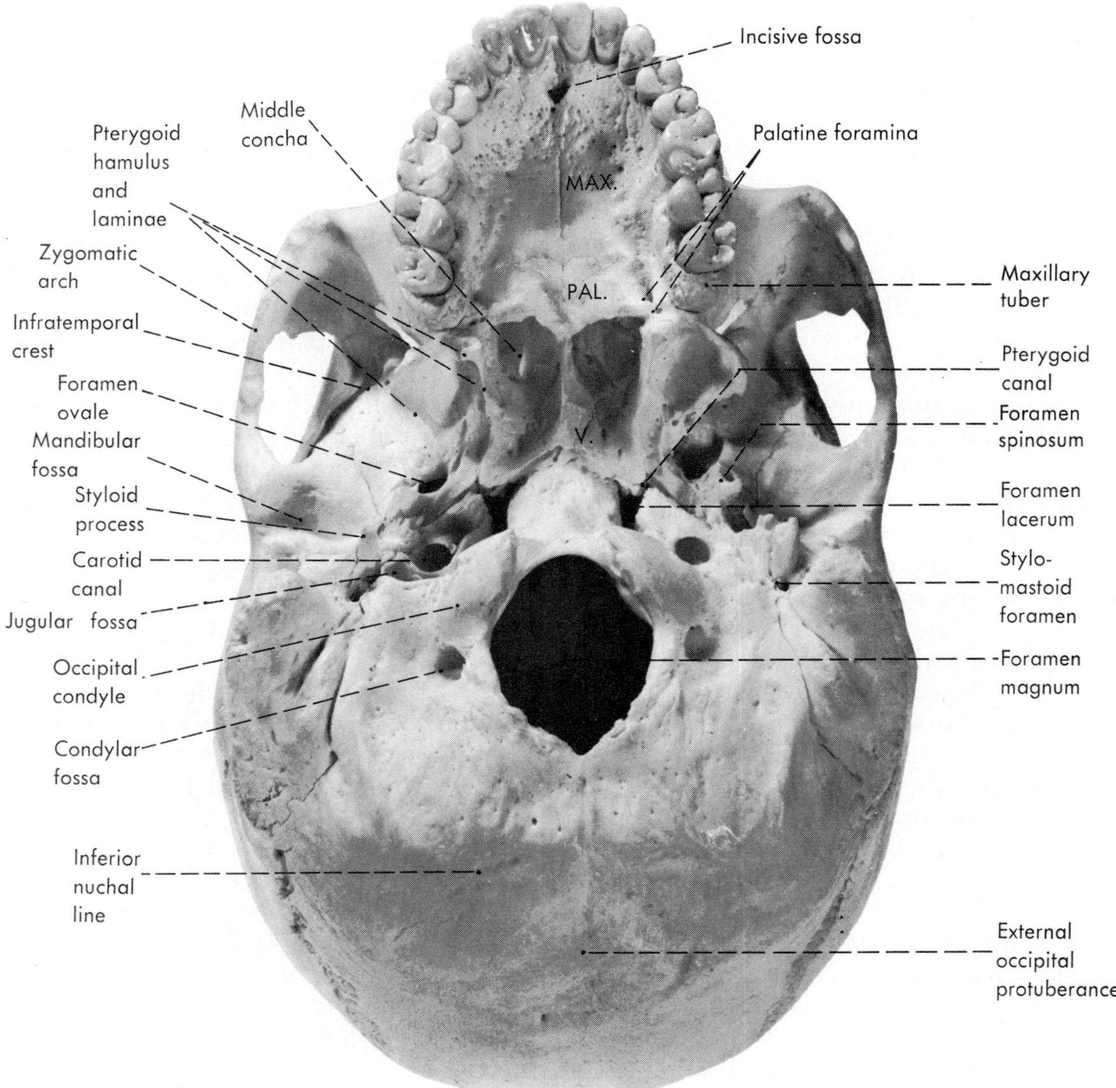


**FIG. 31-3.** A view of the external surface of the base of the skull with the mandible removed. Here the view is at the same time slightly forward. See also the following figure. **MAX.** and **PAL.** identify the parts of the maxillary and palatine bones forming the hard palate; **V.** is the vomer.

magnum. On it, about halfway between the foramen and the external occipital protuberance, is the *inferior nuchal line.* Muscles of the back of the neck attach both below this line and between it and the superior nuchal line. Each **pars lateralis** lies lateral to the foramen magnum and blends with the squama and basal part, and the **pars basilaris** extends forward as a midline bar of bone about an inch wide. Lateral to the lateral and basal parts are the petrous parts of the temporal bones, which include the mastoid processes.

Anteriorly, the basilar part of the occipital bone is fused, often indistinguishably, to the slanting posterior part of the body of the sphenoid bone; these parts together form the **clivus.** Laterally, there is a gap, the **foramen lacerum,** between the occipital, the sphenoid,

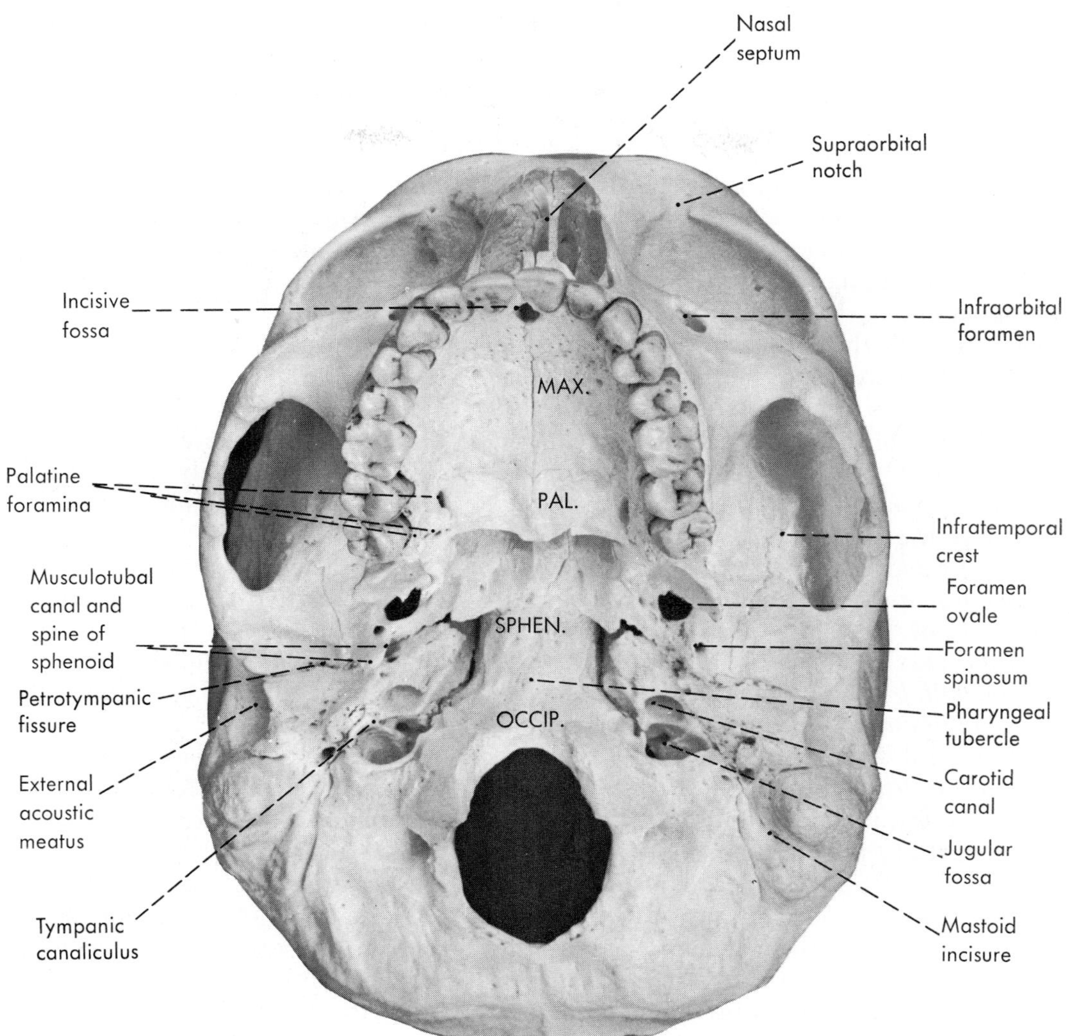


**FIG. 31-4.** Another view of the external surface of the base of the skull, this time with the skull so tipped that one looks also slightly backward. Certain features are recognizable in both views, but each also shows structures not recognizable in the other: for instance, the jugular fossa in the present figure and the foramen lacerum in the preceding figure. **MAX.** and **PAL.** are as in the preceding figure; **SPHEN.** and **OCCIP.** identify the bodies of the sphenoid and occipital bones.

and the petrous part of the temporal. This is occupied in life by cartilage and ligamentous tissue that unite the three bones and therefore constitute three *synchondroses*. The basal part of the occipital bone presents no other features of particular interest except at about its middle, where there is a small **pharyngeal tubercle** to which the superior constrictor muscle of the pharynx is attached. In front of this tubercle, the mucous membrane of the upper part of the pharynx lies against the occipital bone.

The **petrous part of the temporal bone** forms the floor of the posterior part of the cranial cavity lateral to the occipital bone. Its mastoid process and notch already have been noted. Anteromedial to the mastoid process is the **styloid process,** of varying thickness and length and often broken in the prepared skull. Just behind the base of the styloid process is a prominent foramen, the **stylomastoid foramen,** through which the facial (7th cranial) nerve leaves the skull. Medial to the styloid process, between it and the occipital condyle, the temporal bone presents a pronounced depression, the **jugular fossa,** bounded posteriorly by the **jugular process.** The fossa houses the upper end of the internal jugular vein and opens through the **jugular foramen** into the interior of the skull. In the lateral wall of the jugular fossa is the **mastoid canaliculus** for the auricular branch of the vagus nerve. In the ridge anterior to the fossa is the opening of the **tympanic canaliculus,** for the tympanic branch of the 9th nerve. The opening may be in a slight depression, the *fossula petrosa.* Anteromedial to the jugular foramen, the petrous part of the temporal bone is separated from the pars basilaris of the occipital by the *petrooccipital fissure,* which meets the *petrosphenoid fissure* (lateral to the petrous part of the temporal) at the anterior end of the pars petrosa to form the opening with jagged edges, the **foramen lacerum,** already noted.

The part of the petrous portion of the temporal bone in front of the ear is known as the **petrous apex.** It ends anteriorly at the foramen lacerum. On the external base of the skull, it is marked by the external opening of the **carotid canal,** which lies immediately in front of the jugular foramen. The carotid canal extends, first, vertically upward in the temporal bone, but then turns abruptly to run anteriorly and medially throughout the length of the petrous apex, emerging just above the foramen lacerum.

In front of the styloid process, closely fused to the petrous part of the temporal bone, is the **tympanic part.** This can be seen to form the anterior wall and floor of the external acoustic meatus and some of its posterior wall. It extends forward and medially, giving off an expansion or *sheath* to the base of the styloid process, to a level a little in front of the external opening of the carotid canal. In front of the lateral part of the pars tympanica is a small slit or fissure, the **tympanosquamous fissure,** that, traced medially, divides into two parallel and closely adjacent fissures. The posterior one, the **petrotympanic fissure,** important because the chorda tympani, a branch of the facial nerve that runs through the middle ear cavity, emerges through it. The fissure in front of the pars tympanica also lies just behind the **mandibular fossa,** the concavity of the inferior surface of the pars squamosa for articulation with the mandible. Anterior to the mandibular fossa is the rounded **articular tubercle.**

At the anteromedial end of the tympanic part, between it and the petrous part, is an opening that actually represents the bony anterior ends of two canals, an upper for the tensor tympani muscle (a muscle extending into the middle ear cavity) and a lower for the auditory tube that connects the middle ear cavity to the pharynx. This bony tube is the **musculotubal canal** and in life is divided into a small upper and a larger lower part by a thin bony septum.

Anterior to the base of the occipital bone, and anterior and lateral to the petrous apex, the **sphenoid bone** forms the major part of the floor of the skull. The body of the bone, in line with the basilar part of the occipital bone, contains paired **sphenoid sinuses,** but is largely covered in inferior view by a bone, the **vomer,** that forms a part of the septum of the nasal cavity and by the **pterygoid processes** that form the lateral walls of the posterior openings of the nasal cavities. Each pterygoid process is also attached to a **greater wing** of the sphenoid bone. This wing, already seen in lateral view, extends backward medial to the temporal squama to the angle between that and the petrous part of the temporal. The fourth part of the sphenoid bone, the **lesser wing,** can best be seen in an inner view of the base of the skull (Fig. 31-9).

The flattened inferior surface of the greater wing of the sphenoid bone is the **infratemporal fossa.** Laterally, it is separated from the temporal fossa on the side of the skull by the *infratemporal crest.* Posteromedially, between it and the adjacent part of the temporal bone, there is a groove that housed the carti-

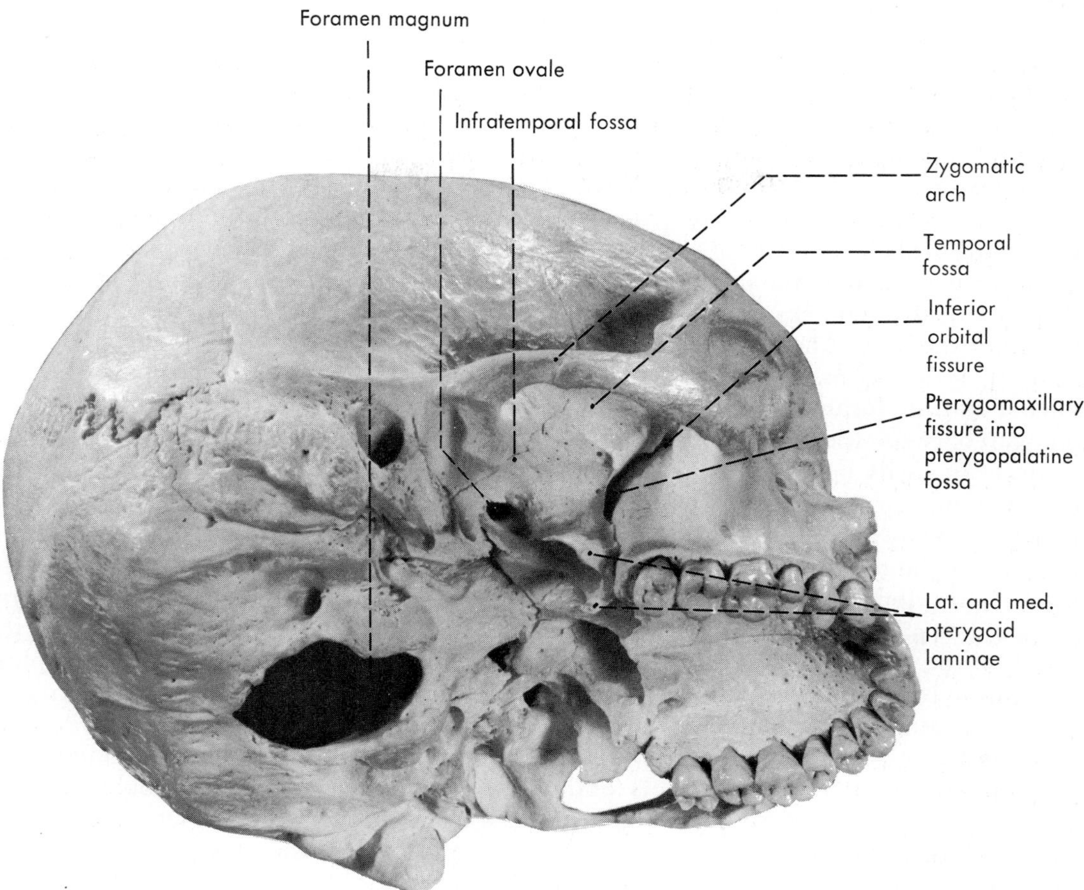


**FIG. 31-5.** An oblique view of the base and lateral aspect of the skull, to show the pterygopalatine fossa; it and the inferior orbital fissure opening into it are seen through the pterygomaxillary fissure.

laginous part of the *auditory tube,* continued forward from the bony part of the tube in the musculotubal canal. The prominent **foramen ovale,** through which the mandibular branch of the trigeminal (5th cranial) nerve makes its exit from the skull, opens into the posterior part of the infratemporal fossa. Posterolateral to the foramen ovale is the small round **foramen spinosum.** Through this the middle meningeal artery, the largest of the vessels to the coverings of the brain, enters the cranial cavity. The foramen spinosum obtains its name from the **sphenoid spine,** a small projection on the posterior tip of the greater wing of the sphenoid posterior to the foramen.

Anteriorly, the infratemporal fossa extends to the posterior (infratemporal) surface of the body of the maxilla and here is bounded medially by a thin plate of bone, the **lateral pterygoid lamina.** Upon tilting the skull to obtain a view from below and laterally (Fig. 31-5), it can be seen that although the lateral pterygoid lamina is fused to the maxilla below, it is separated above from that bone by a fissure, the **pterygomaxillary fissure,** that opens into the infratemporal fossa. The pterygomaxillary fissure is continuous anteriorly with a fissure (inferior orbital) that opens above into the orbit, but it also opens above and medially into a deep recess, the **pterygopalatine fossa.** This important fossa extends posteriorly, behind the maxilla, between the pterygoid part of the sphenoid bone and the perpendicular lamina of the palatine bone, which is here part of the lateral wall of the nasal cavity. Part of the maxillary artery enters the pterygopalatine fossa through the pterygomaxillary fissure, and the maxillary branch of the trigeminal nerve enters the fossa as it leaves the skull through the foramen rotundum, which opens into the posterior end of the fossa. From the fossa, both artery and nerve send small branches downward to the posterior surface of the maxilla, on which there are two or more **alveolar foramina** through which they enter the bone. They send large branches medially into the nasal cavity through the **sphenopalatine foramen**, which lies in the thin wall between the fossa and the nasal cavity, and can easily be seen through the posterior openings of the nose. They also send branches downward to appear on the posterior part of the bony palate through the several palatine foramina located here. Anteriorly, a major part of the maxillary nerve and a branch of the artery pass into the floor of the orbit through the inferior orbital fissure. Posteriorly, a small **palatovaginal canal** and a larger and more important **pterygoid canal**, both transmitting nerves and vessels (and described in a following paragraph), open into the pterygopalatine fossa.

The **pterygoid processes** of the sphenoid bone, the lateral laminae of which already have been noted, have several important relations. Each pterygoid process, projecting downward from the body and greater wing of the sphenoid bone, has two laminae (plates), the *lateral and a medial lamina.* These are continuous with each other above and seem also to be continuous below. Here, however, the apparent continuity is brought about by the *pyramidal process of the palatine bone* (the small bone that forms the posterior edge of the bony palate). This process is fused to the maxilla and to both laminae of the pterygoid process. The concavity between the two pterygoid plates is the **pterygoid fossa.** At the lower end of the medial plate is a posteriorly projecting hook of bone, the **hamulus.** The tendon of one of the muscles of the palate (tensor veli palatini) passes downward lateral to the hamulus and then turns medially in a notch, the **pterygoid incisure,** on its lower surface. At its upper end, the thin edge of the medial pterygoid plate expands into two slight ridges that enclose between them a triangular flat area, the **scaphoid fossa;** the tensor of the palate, just mentioned, arises in part from this.

The process extending medially from the medial pterygoid plate over the inferior surface of the body of the sphenoid is the *processus vaginalis.* It meets the expanded upper end, or **ala, of the vomer** (the bone between the two nasal cavities). On the lower surface of the vaginal process there is frequently a small but clear groove that leads forward into a foramen. This foramen lies between the vaginal process and the perpendicular lamina of the palatine bone (which forms the lateral wall of the nose immediately in front of the medial pterygoid process) and is therefore called the **palatovaginal canal.** It transmits tiny nerve and arterial twigs from the pterygopalatine fossa to the roof of the pharynx. At the posterior end of the vaginal process, where it is joined by the medial side of the pterygoid plate, there is a small bulge, and immediately above that, opening into the anterior wall of the foramen lacerum, is the posterior end of the **pterygoid** (*vidian*) **canal.** The pterygoid canal transmits an important but rather slender nerve (nerve of the pterygoid canal) that typically leaves the interior of the skull at the foramen lacerum, enters the canal, and, through this, runs forward into the pterygopalatine fossa where it ends in an autonomic ganglion.

The posterior openings of the nose, already mentioned, are the **choanae.** Each is about twice as broad in the vertical diameter as it is in the transverse one. The bony septum extending between them is the **vomer.** The articulation of its expanded upper end, the **ala,** with the body of the sphenoid and the vaginal parts of the pterygoid processes has been noted. Inferiorly, the vomer articulates with the upper surface of the bony palate; anteriorly, it articulates with other bone and cartilage of the nasal septum (see Fig. 33-29).

The thin curved plates of bone projecting inward from the lateral wall of the nose are the **conchae;** three of these can be seen through the choana. The **inferior nasal concha** really is a separate bone, but is fused laterally to the palatine and maxillary bones (which form the lateral walls of the nasal cavity at the level of the concha). Above the inferior nasal concha is the **middle nasal concha.** The middle concha is almost as long as the inferior concha, but frequently appears thicker at its base owing to the invasion of air cells from the ethmoid sinus. This is a part of the *ethmoid labyrinth,* a lateral part of the ethmoid bone that forms the upper lateral wall of the nasal cavity. Above the middle nasal concha, but primarily at its posterior end and therefore most easily visible posteriorly, is the **superior nasal concha,** also a part of the ethmoid labyrinth. The nasal cavities are described in more detail in Chapter 33. Concerning the bony wall, only one thing needs to be added at this time: just above the back end of the middle nasal concha is the **sphenopalatine foramen** that transmits important nerves and vessels from the pterygopalatine fossa to the nasal cavity.

The floor of the nasal cavities, the **bony palate** (*hard palate* when soft tissues cover the bone), is formed posteriorly by the **horizontal laminae** of the palatine bones. Each **palatine bone** is somewhat L shaped: its horizontal lamina meets that of the other side in the midline, and its **perpendicular lamina** or plate runs upward as a posterior part of the lateral wall of the nasal cavity. The perpendicular lamina is fused posteriorly, through its pyramidal process, to the pterygoid process of the sphenoid bone, anterolaterally to the maxilla, and is almost hidden by the two. The sphenopalatine foramen, already noted, lies above an upper part of this lamina, between it and the body of the sphenoid. Behind the *sphenopalatine foramen,* the perpendicular plate articulates with the body of the sphenoid and the pterygoid process, and in front of the foramen, it extends into the floor of the orbit.

Anterolaterally, in the suture between the horizontal laminae of the palatine bones and the processes in front of them, are the paired **greater palatine foramina;** behind each of these, in the palatine bone, are one or two **lesser palatine foramina.** These foramina are the lower ends of the **greater** and **lesser palatine canals** through which nerves and vessels reach both the hard and the soft palate (the soft palate is attached to the posterior border of the hard palate). If bristles are passed up these canals, they will be found to converge above in the pteryopalatine fossa. Where the horizontal laminae meet in the midline, they form a posteriorly projecting **posterior nasal spine;** on their upper surfaces they send up a **nasal crest** for articulation with the vomer; and on their lower surfaces, a variable distance from the posterior edges, there may be a distinct ridge, the **palatine crest.**

The bony palate in front of the horizontal laminae of the palatine bone is formed by the fusion of the **palatine processes of the two maxillae.** Each blends laterally and anteriorly with the body of the maxilla and the alveolar process (the downward-projecting, tooth-bearing part) of the maxilla. Posterolaterally, there usually is a groove leading forward from the greater palatine foramen in the osseous palate. This transmits the large nerves and vessels of the hard palate. Sometimes there is a bony ridge, the *torus palatinus,* along the intermaxillary suture. In the anterior midline, where the palatine processes blend with the alveolar processes, there is a deep pit, the **incisive fossa.** In its walls, sometimes visible and sometimes not, are two to four **incisive foramina,** which are the lower openings of **incisive canals** that open above into the nasal cavity. There are most commonly four canals, two for blood vessels (one on each side) and two for nerves.

## Anterior View

In an anterior view of the skull (Fig. 31-6), it is obvious that the two maxillae form a large part of the border of the anterior opening of the nasal cavities (**piriform aperture** in the bony skull) and also much of the inferior and medial border of the entrance to the orbit.

Upon inspection of the **nasal cavity,** the fused **nasal crests,** extending upward from the palatine processes to form the lowest part of

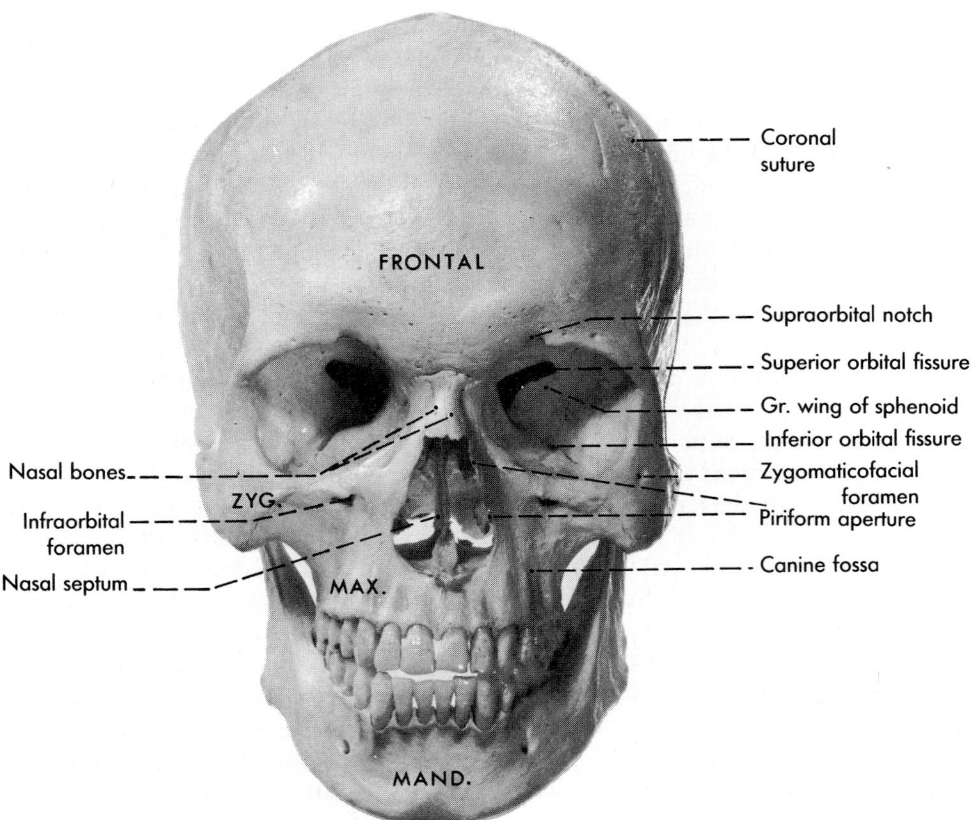


**FIG. 31-6.** Anterior view of the skull. **ZYG.** is the zygomatic bone; **MAX.** and **MAND.** are the maxilla and mandible.

the nasal septum, can be recognized. The crests end anteriorly in an **anterior nasal spine.** Posteriorly, they articulate with the vomer, but anteriorly, in the dry skull, their upper free borders are partly separated, thus presenting a cleft. Into this cleft, during life, there fits a cartilage that largely completes the nasal septum. The upper part of the bony septum is the **perpendicular plate** of the **ethmoid bone.** This bone also forms the narrow upper roof of the nose, the **cribriform plate,** best seen from inside the cranial cavity. It sends downward, on each side of the cribriform plate, the **ethmoid labyrinth** that separates the nose from the orbit. The ethmoid labyrinth appears on the medial orbital wall as the **orbital lamina** of the ethmoid bone and projects into the nasal cavity as the superior and middle nasal conchae, already seen through the choanae. The front ends of the inferior nasal conchae also are easily recognized in an anterior view of the nasal cavity. The middle conchae can be seen somewhat indistinctly above them, but the superior conchae are typically too far back and placed too high to be visible. Other features of the bony walls of the nasal cavity are best appreciated during dissection of this part.

Returning to the **maxilla,** the **alveolar process,** bearing the teeth, can be recognized again as it extends downward below the level of the palatine process. The body of the maxilla lies lateral to the lower part of the nasal cavity. Rather than being convex, as are its lateral and posterior surfaces, the anterior surface of the maxilla presents a concavity, the **canine fossa,** just below the prominence of the cheek and above the alveolar process.

Above the canine fossa, a little below the lower border of the orbit, is the large **infraorbital foramen.** This transmits the largest branch of the maxillary nerve (a part of the trigeminal) as it comes out onto the face to supply skin of the cheek, upper lip, and side of the nose. Laterally and above, the body of the maxilla articulates with the zygomatic bone and medial to and behind this forms a large part of the floor of the orbit (and the roof of the maxillary sinus). Most of the details of the bony orbit can be studied more profitably just preceding study of the soft tissues it contains and therefore are to be found in Chapter 33. It need only be noted that the **frontal process** of the maxilla extends upward between the nose and the orbit, forming a part of the lateral wall of the nasal cavity, to articulate with the frontal and nasal bones.

The two **nasal bones** articulate with each other in the midline, articulate laterally with the frontal processes of the maxillae, and articulate above with the frontal bone. Their lower free edges form the uppermost part of the piriform aperture. In life, they articulate with nasal cartilages that support the lower part of the nose. Posteroinferiorly, they articulate with the nasal septum. On the internal surface of each nasal bone there may be a slight groove, the *ethmoidal sulcus;* this accommodates a branch of the ophthalmic division of the trigeminal nerve, which leaves the nasal cavity to pass between the nasal bone and the nasal cartilage to a subcutaneous position.

The **frontal bone** has already been seen in superior and lateral views. The part of the frontal bone that forms the forehead is the **squama.** Just above and paralleling each supraorbital margin, the squama presents a ridge, the **superciliary arch.** Between the two superciliary arches, there is, in the midline, a slight protuberance, the **glabella,** in which some remains of the *frontal* or *metopic suture* (between the originally paired frontal bones) may be visible. The swellings of the squama that form the prominence of the forehead on each side are the *frontal tubers* or eminences. The supraorbital margin of the frontal bone is marked on its inner third by a **supraorbital foramen,** or supraorbital notch, that transmits the supraorbital nerve and vessels as these leave the orbit and turn up on the forehead. Sometimes, medial to the supraorbital foramen, there is a second notch or foramen, the **frontal notch** (foramen). It accommodates a medial branch of the supraorbital nerve that once was known as the frontal nerve.

Lateral to the orbit, the frontal bone sends a zygomatic process downward to articulate with the zygomatic bone and complete the lateral wall of the orbit. Between the two orbits, it also sends a short nasal process downward to articulate with the nasal bones and the frontal processes of the maxillae. Finally, the frontal bone has almost horizontal parts, each called a **pars orbitalis,** that extend backward from the supraorbital margins as major parts of the roofs of the two orbits.

The squama of the frontal bone is thick. It presents hard inner and outer laminae, between which lies diploë except in that part of the bone that is invaded by the **frontal sinuses.** These are paired sinuses, but vary greatly in size and often are asymmetrical. They usually extend into both the squama and the orbital part of the frontal bone. The orbital part contrasts greatly with the squama in strength, for it is a thin plate of bone or, where it contains a part of the frontal sinus, two very thin plates.

## Mandible

The mandible (Figs. 31-7 and 31-8), the unpaired bone of the lower jaw, consists of the tooth-bearing **body** and the more vertically disposed **ramus** that receives the insertions of the chief muscles of the jaw and articulates with the temporal bone. Ramus and body meet posteriorly at the **angle.** The *body* of the mandible is divided into two parts; the lower is the *base,* and the upper part, bearing the teeth, is the *pars alveolaris* or alveolar process.

The base of the mandible shows a swelling on its anteroinferior surface where the two sides come together; this is the **mental protuberance.** The lower, lateral, part of the mental protuberance, the **mental tubercle,** is fre-

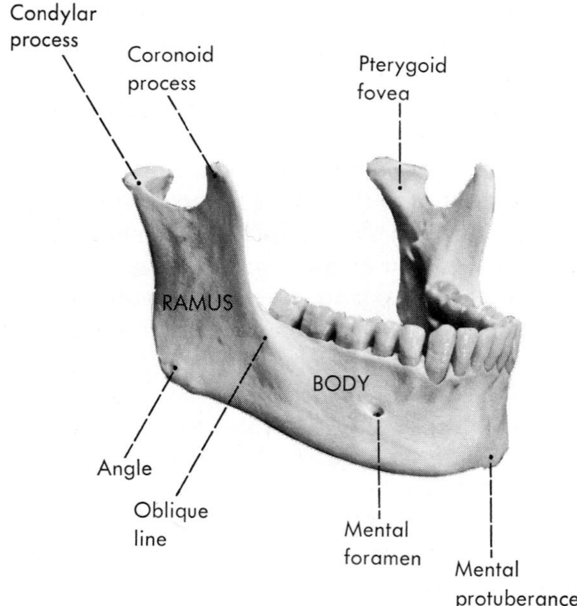


**FIG. 31-7.** Anterolateral view of the mandible.

**FIG. 31-8.** Inner surface of the mandible.

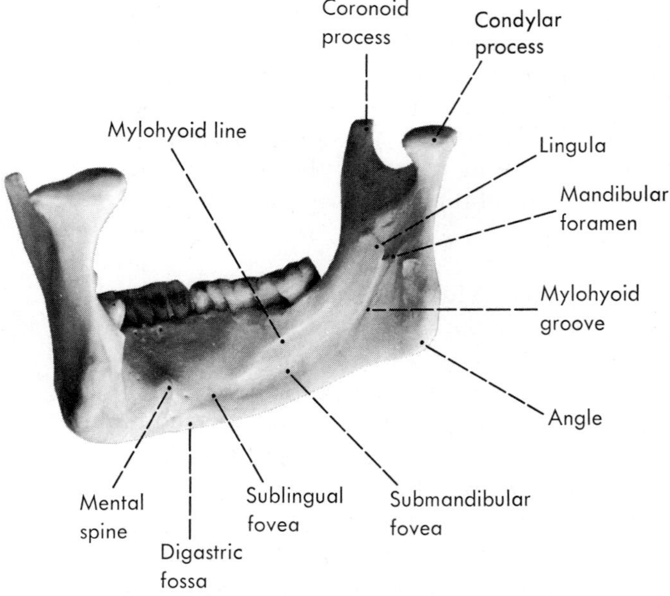


quently more pronounced. Anterolaterally, about halfway between the upper border of the alveolar process and the lower border of the base, is the **mental foramen.** A large branch of the mandibular nerve emerges through this foramen, after having supplied the teeth, and supplies mucous membrane and skin of the lower lip and chin. The ridge running down from the front of the ramus onto the body of the mandible is the *oblique line.* It usually is traceable to the mental tubercle, but is not well marked except close to the ramus.

On the inner surface of the mandible, there is, in the midline, a roughened projection, the **mental spine.** This may be bilateral instead of in the midline, or it may be divided into an upper and lower spine, for one pair of muscles attaches to the lower part of the spine and another pair attaches to the upper part. Lateral to the mental spine on each side is a slight concavity, the **sublingual fovea,** against which the anterior end of the sublingual salivary gland fits. Below each sublingual fovea, on the posteroinferior border of the mandible, is another slightly concave area, the **digastric fossa,** where the anterior belly of the digastric muscle is attached. Starting between the sublingual and digastric impressions, but not particularly marked until it is posterior to the former, is the **mylohyoid line.** This marks the attachment of the mylohyoid muscle, the muscle that forms the floor of the mouth. The mylohyoid line extends upward and backward for the entire length of the alveolar process. The long concave area below the major part of the mylohyoid line is the **submandibular fovea,** accommodating the salivary gland of the same name.

The almost perpendicular **ramus** of the mandible is continuous anteriorly with the alveolar process and base. Posteriorly, it ends below at the angle and above it presents two processes. Of these, the anterior, sharper **coronoid process** serves for the attachment of a muscle, and the *head* of the posterior **condylar process** helps form the temporomandibular joint. Below the head is the *neck,* and anteromedially, at the junction of the head and neck, is the **pterygoid fovea,** representing the attachment of part of the lateral pterygoid muscle. The concavity between the coronoid and condylar processes is the **mandibular notch** (*incisure*).

On the inner surface of the ramus of the

mandible is the **mandibular foramen.** It is the entrance to the **mandibular canal,** which runs forward in the mandible deep to the roots of the teeth and carries to them their nerves and vessels. Parts of these nerves and vessels emerge on the outer surface of the mandible through the mental foramen. Anterior to and above the mandibular foramen is a thin projection of bone, the mandibular **lingula.** It somewhat overlaps the foramen and is a landmark for injection of the nerve here. Leading downward from the mandibular foramen is a small groove, the **mylohyoid groove,** which indicates the course of the mylohyoid nerve and vessels. The mylohyoid nerve and vessels leave the nerve and vessels that enter the mandible and run downward on the inner surface of the bone.

Not only the size but the shape of the mandible varies much with age and the condition of the dentition. At birth, the ramus of the mandible makes an obtuse angle with the body, because the condylar process projects more posteriorly than upward (Fig. 31-13). As the teeth appear and the child uses them for chewing, the body of the mandible increases in size; the ramus grows particularly fast posteriorly and becomes more nearly vertical, so the angle becomes less obtuse. As the teeth are lost in old age, the alveolar part of the mandible is absorbed, and if dentures are not regularly worn, the altered muscular pull on attempted occlusion results in the condylar process being gradually displaced backward, so that the angle returns toward the infantile condition.

## INTERIOR OF THE SKULL

### Roof

The inner surface of the calvaria presents little of interest. The only structures worthy of note are the grooves that blood vessels leave here. In the sagittal plane, most marked in the region of the sagittal suture between the parietal bones but extending also downward on the inner surface of the occipital and on the inner surface of the frontal, is the **sulcus of the superior sagittal sinus.** This great venous sinus becomes larger as it is traced backward, because it receives the veins from the upper surface of the adjacent brain. On each side of the sulcus of the superior sagittal sinus, usually on the inner surface of the parietal bone but sometimes also on the frontal, there may be two or three depressions, the *foveolae granulares* (p. 926). Laterally, extending upward from the cut lower edge of the calvaria, are a variable number of grooves that accommodate the meningeal vessels; almost all these markings are caused by the middle meningeal artery and its accompanying veins and can be seen better when the base of the skull is examined.

### Base of the Skull

The interior of the base of the skull, the **cranial floor,** presents three levels (Fig. 31-9). The anterior part, the highest, is largely also the roof of the two orbits and of the nasal cavities and is the **anterior cranial fossa.** The middle part, the **middle cranial fossa,** lies at a lower level. Its deeper lateral parts are separated from the anterior fossa by sharp ridges of bone, often known as the "sphenoid ridges," but are united across the midline by a narrower and higher portion (the *sella turcica*). The whole fossa roughly resembles a butterfly. Finally, the most posterior part of the floor, surrounding the foramen magnum and largely separated from the middle fossa by the ridges ("petrous ridges") formed by the converging petrous parts of the two temporal bones, is the **posterior cranial fossa.** The internal anatomy of the skull is easier to learn if it is studied in terms of the cranial fossae.

**Anterior Cranial Fossa.** The **crista galli** (cock's comb), a part of the **ethmoid bone,** is the sharp projection of bone anteriorly in the midline of the floor of the anterior cranial fossa. Attached to the crista is the anterior end of a sheet of dura (*falx cerebri*) that partly separates the two cerebral hemispheres. In front of the crista, on the inner surface of the frontal bone, is the **frontal crest** (also for the attachment of the falx). This crest, sharp below, widens above and gives place to the *sulcus for the superior sagittal sinus.* There may or may not

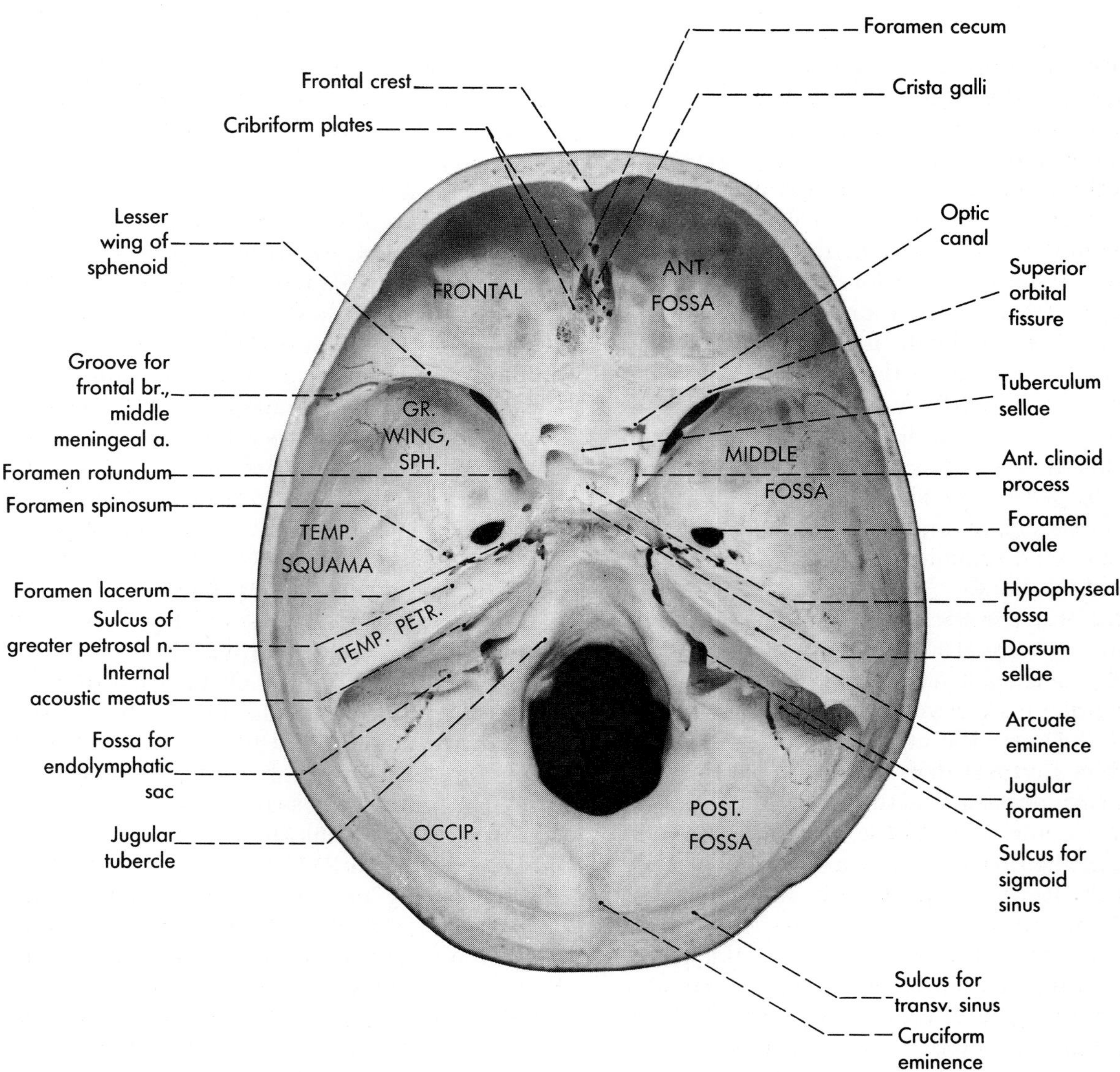


**FIG. 31-9.** Interior of the base of the skull. **GR. WING, SPH.** is the greater wing of the sphenoid bone; **TEMP. SQUAMA** and **TEMP. PETR.** are the squamous and petrous parts of the temporal bone; and **OCCIP.** is the occipital bone.

be, between the frontal crest and the crista galli, the small **foramen cecum** that leads into the nasal cavity. A small vein, often lacking in the adult, runs through this foramen and connects the veins of the nose with the intracranial sinuses. On each side of and behind the crista galli, the lowest part of the floor of the anterior fossa is formed by a part of the ethmoid bone that presents numerous holes that lead from the nasal cavity (transmitting filaments of the olfactory nerve); this is the **cribriform plate** (*lamina cribrosa*) of the ethmoid. Lateral to the cribriform plates and arching gently upward are the orbital parts of the frontal bone, which form major parts of the roofs of the orbits.

On the sides of the cribriform plates, between these and the frontal bone, are the medial ends of the small *anterior* and *posterior ethmoidal canals.* These medial ends are often difficult to observe but their other ends, the *ethmoidal foramina,* can be seen on the medial wall of the orbit. Through each of these canals, a blood vessel enters the cranial cavity, runs forward along the lateral border of the cribriform plate, and then turns downward into the nose. A nerve accompanies the anterior artery.

Behind the cribriform plates and the orbital parts of the frontal bone, the floor of the anterior fossa is formed by the **body** and the **lesser wings of the sphenoid bone.** The anterior border of a groove, the *chiasmatic sulcus,* across the body of the sphenoid, demarcates the anterior fossa from the middle part of the middle fossa. As the lesser wings project laterally, they form the sharp boundaries between the lateral parts of the anterior and middle cranial fossae, and it is these that often are referred to as the "sphenoid ridges." At their medial ends, the lesser wings project posteromedially as free processes, the **anterior clinoid processes,** that serve for the attachment of a part of the cranial dura mater.

**Middle Cranial Fossa.** The middle cranial fossa is particularly complex, because of its shape and because of the numerous foramina and grooves related to it. Between the body and the lesser wings of the sphenoid, anterior to the anterior clinoid processes, are the **optic canals,** opening anteriorly into the orbits and each transmitting an optic nerve and an accompanying (ophthalmic) artery. The **chiasmatic sulcus** unites the two canals and is bordered posteriorly in the midline by a small rounded elevation, the **tuberculum sellae.** The remaining small midline part of the middle cranial fossa, behind the optic canals and the chiasmatic sulci, is the **sella turcica** (Turk's saddle); it is higher than the lateral parts and is formed by the body of the sphenoid bone. The anterior part of the sella turcica is the tuberculum. Posteriorly, the sella is bounded by the **dorsum sellae,** a ridge of bone that projects upward like the back of a saddle. Projecting forward from each side of the dorsum is a **posterior clinoid process,** usually not so marked as the anterior clinoid process. The deepest part of the sella is the **hypophyseal fossa,** so called because it houses the hypophysis. Tumors of the hypophysis, or long-standing increased intracranial pressure, produces decalcification, erosion, and expansion of the sella turcica, which can be detected in roentgenograms.

On each side, posterolateral to the dorsum sellae, is a jagged foramen, the **foramen lacerum,** already noted on the external surface of the skull. Here the tip of the petrous portion of the temporal bone fits in between the body and the greater wing of the sphenoid and presents the anterior end of the carotid canal. Leading upward from just above the foramen lacerum is the **carotid sulcus.** The internal carotid artery leaves the petrous tip to run in this sulcus to the undersurface of the anterior clinoid process, where it turns medially toward the tuberculum sellae and then leaves the cranial floor by turning superiorly to penetrate the dura and distribute branches to the brain. Just at the point where the carotid artery turns superiorly on the side of the sella turcica, there may be a *middle clinoid process.* Sometimes this is so marked that it meets or almost meets the anterior clinoid process, thus converting the terminal part of the carotid sulcus into a foramen.

Much of the floor of each lateral expanded part of the middle cranial fossa is formed by a **greater wing of the sphenoid bone** (see also p. 856). Anteriorly, lateral to the anterior clinoid process, this wing is separated from the lesser wing by the **superior orbital fissure.** Lateral to this fissure the greater and lesser wings fuse. Behind the base of the superior orbital fissure, and at about the level of the middle of the sella turcica, there is a rounded foramen that is directed anteriorly and slightly laterally. This is the **foramen rotundum.** It opens into the pterygopalatine fossa and transmits the maxillary nerve, the second branch of the trigeminal nerve (the first or ophthalmic branch makes its exit through the superior orbital fissure). Behind the foramen rotundum is a larger oval foramen that is directed downward. This is the **foramen ovale,** which opens into the infratemporal fossa and transmits the third or mandibular branch of the trigeminal nerve. The trigeminal and other nerves in the middle cranial fossa are shown in Figure 31-10.

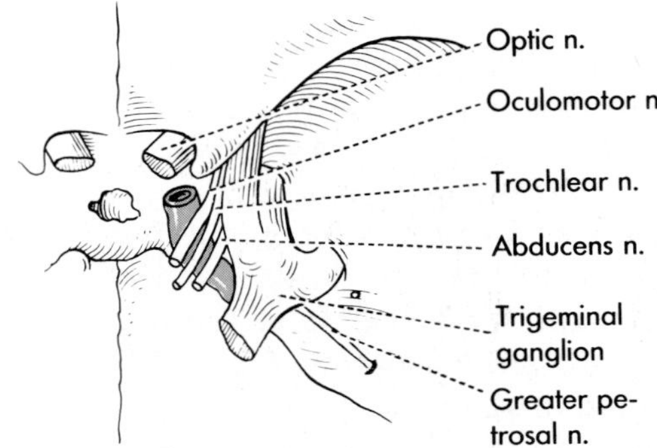


**FIG. 31-10.** The trigeminal and other nerves in the middle cranial fossa.

Posterolateral to the foramen ovale is the **foramen spinosum,** also opening below into the infratemporal fossa. This foramen transmits the middle meningeal artery, the largest of the arteries supplying the cranial dura and adjacent skull. Leading laterally from the foramen spinosum is a groove that marks the course of the middle meningeal artery. At a variable distance lateral to the foramen, this groove divides into two, to accommodate the frontal and parietal branches of the artery.

The upwardly bulging posteromedial walls of the middle cranial fossa are formed by the petrous parts of the temporal bones (*pars petrosa;* see also p. 859). The anterior part or **petrous apex** (in front of the part containing the internal ear) usually presents, immediately above the foramen lacerum, a **trigeminal impression** that marks the position of the ganglion of the trigeminal nerve in the middle fossa.

Extending backward and laterally from the foramen lacerum, there usually is visible a slight groove that enters a foramen on the anterior surface of the petrous part. This groove is the **sulcus of the greater petrosal nerve,** and the foramen is the *hiatus of the canal of the greater petrosal nerve.* Anterolateral to the sulcus and usually ending in a small foramen just medial to the foramen spinosum, there may be visible a second groove, the **sulcus of the lesser petrosal nerve,** transmitting the nerve of that name. The tiny opening at the upper end of this groove is the *hiatus of the canal of the lesser petrosal nerve.* Both petrosal nerves arise in the temporal bone and make their exits through its anterior surface to appear inside the cranial cavity before they leave the skull. The most lateral part of the petrous part of the temporal bone on the anterior surface is the **tegmen tympani,** the roof of the middle ear cavity; the bone here is fairly thin. Anteromedial to the tegmen tympani is a protrusion of the bone, the **arcuate eminence,** that marks the position of the anterior semicircular canal of the internal ear.

The upper edge (superior margin) of the pars petrosa separates the middle cranial fossa from the posterior fossa; it may show a groove, the **sulcus for the superior petrosal sinus.**

**Posterior Cranial Fossa.** The anterior wall of the posterior cranial fossa is the dorsum sellae in the midline and the petrous parts of the temporal bones ("petrous ridges") laterally. The lateral parts of the fossa are, during life, roofed by the *tentorium cerebelli,* a layer of dura mater that is attached anterolaterally to the superior margins of the two petrous bones and posteriorly to the occipital bone. The tentorium presents a central notch extending backward from the two posterior clinoid processes. On the posterior surface of the petrous temporal is the opening of a canal, the **internal acoustic meatus,** that passes laterally into the bone. This opening is the *porus acusticus internus,* more often called the internal acoustic or auditory foramen. The 7th and 8th (facial and vestibulocochlear) nerves enter the internal acoustic meatus (Fig. 31-11).

Medial to, behind, and below the internal acoustic meatus, the temporal bone is separated from the occipital by the elongated **jugular foramen.** The deep, curved groove at the junction of temporal and occipital bones, extending downward and then medially and forward to the back end of this foramen, is the **sulcus for the sigmoid sinus.** This large venous sinus empties through the posterolateral part of the foramen into the upper end of the internal jugular vein. A variably developed projection, or *intrajugular process,* partially separates this part of the jugular foramen from the middle part, which transmits the vagus and accessory nerves (cranial nerves X and XI). The anteromedial part of the jugular foramen transmits the glossopharyngeal nerve (cranial nerve IX) and the inferior petrosal sinus.

On the posterior surface of the petrous part of the temporal bone, just lateral to and above the opening of the internal acoustic meatus, is a shallow depression, the *subarcuate fossa,* with one or two small vascular foramina opening into it. Below and lateral to that, extending almost to the groove for the sigmoid sinus, is a larger shallow depression that houses a flat sac (*endolymphatic sac*) from the internal ear. On the lower border of the bulge between these two depressions, at about the level of the internal meatus but often hidden by the bulge or a projecting scale of bone, is a crevice that represents the **external aperture of the vestibular aqueduct.** This aqueduct transmits the endolymphatic duct, which connects the endolymphatic sac to the internal ear.

In the anterior midline, the body of the sphenoid bone behind the dorsum sellae is fused to the basal portion of the occipital bone. This part of the floor of the posterior cranial fossa that slants downward and backward is the **clivus** (that is, the declining part). It is limited posteriorly by the foramen magnum and has already been seen in an external view. The *sulci for the inferior petrosal sinuses,* ending posteriorly at the jugular foramina, are along the sides of the clivus. Just medial to each jugular foramen, the occipital bone presents a rounded eminence called the **jugular tubercle.** Under cover of this tubercle, directed laterally and somewhat anteriorly, is the **hypoglossal canal** through which the hypoglossal nerve leaves the cranial cavity. The **condylar canal** opens posterolateral to the jugular tubercle into the floor of the last part of the groove for the sigmoid sinus, or into the posterior wall of the jugular foramen.

Behind and lateral to the foramen magnum, and posteromedial to the sigmoid sinuses, the floor of the posterior cranial fossa is concave where it accommodates the cerebellar hemispheres. The two concavities are separated from each other by a ridge placed in the sagittal plane. This ridge intersects a transverse ridge that bounds the posterior cranial fossa above, and the cross so formed is the **cruciform eminence.** The most prominent portion of the cruciform eminence is the **internal occipital protuberance.** It usually lies at about the level of the external occipital protuberance. The upper border of the transverse part of the cruciform eminence is the lower border of the **sulci for the transverse sinuses;** these two sulci lead laterally into the **sulci of the sigmoid sinuses,** for transverse and sigmoid sinuses are two parts of the same vascular channel. The tentorium cerebelli, the fold of dura covering the cerebellum and roofing the posterior fossa, is attached on both sides of the grooves for the transverse sinuses. The part of the skull above these sulci is occupied by the occipital lobes of the cerebral hemispheres of the brain.

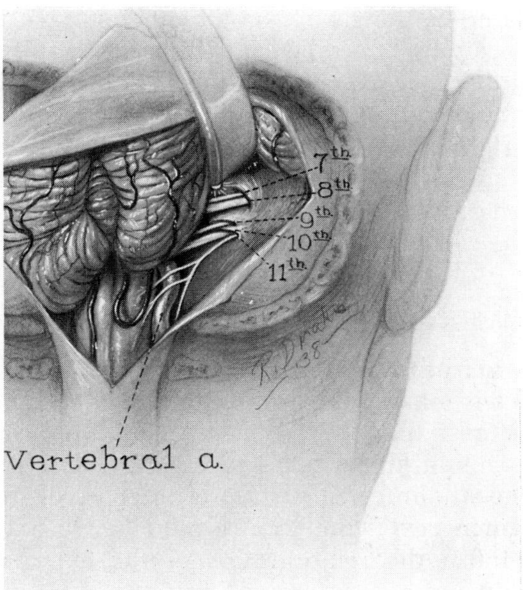


**FIG. 31-11.** The seventh through the 11th nerves as they leave the posterior cranial fossa, with the cerebellum retracted. An enlarged vertebral artery here bulges laterally against them. (Craig WM: Am Heart J 17:40, 1939)

Running downward on the inner surface of the occipital bone above the internal occipital protuberance, usually to the right of the midline and joining particularly the groove for the right transverse sinus, is the back end of the **sulcus for the superior sagittal sinus,** much of which is visible on the inner surface of the calvaria.

## SUMMARY OF FORAMINA OF THE SKULL

The foramina of the skull are so numerous that a summary of the important ones and the structures that they transmit should be useful. This summary is based upon the description of the internal surface of the base of the skull. In reviewing it, it would be wise to note also the relationships of the external openings.

### In Anterior Fossa

The important openings into the anterior cranial fossa are those of the **cribriform plates.** The numerous foramina in each plate lead from the cranial cavity into the uppermost part of the nose, and most of them transmit filaments of the olfactory or 1st cranial nerve, the nerve of smell. On the sides of the cribriform plates are the openings of the two ethmoidal canals.

### In Middle Fossa

The **optic canals,** situated in the most anterior part of the middle cranial fossa just anteromedial to the anterior clinoid processes, transmit two structures each. One is the optic or 2nd cranial nerve, the nerve of sight, and the other, the ophthalmic branch of the internal carotid artery, which supplies structures within the orbit.

Lateral to the optic canal, situated lateral to the body of the sphenoid between that bone's greater and lesser wings, is the somewhat triangular gap of the **superior orbital fissure.** This, like the optic canal, leads into the back of the orbit. The structures traversing this fissure are the ophthalmic vein, the ophthalmic or first division of the trigeminal nerve, and all three nerves that supply muscles of the orbit: the oculomotor or 3rd cranial nerve, the trochlear or 4th, and the abducens or 6th.

The **foramen rotundum,** in the floor of the middle cranial fossa behind the superior orbital fissure, opens into the upper posterior part of the pterygopalatine fossa. It transmits only one structure, the maxillary or second branch of the trigeminal nerve. Behind it is the **foramen ovale,** opening into the infratemporal fossa and transmitting the mandibular or third branch of the trigeminal nerve. A small artery, the accessory middle meningeal artery, may enter the skull by running retrogradely along the nerve. The ganglion of the trigeminal nerve, from which spring all three of its major branches, is located posteromedial to these three openings, lying against the anterior surface of the petrous portion of the temporal bone.

In the floor of the carotid groove, below the anterior end of the carotid canal in the petrous part of the temporal bone, can be seen the **foramen lacerum.** It should be obvious that the internal carotid artery, since it enters the middle fossa above the foramen, *does not* run through the foramen lacerum, which during life is almost completely filled by cartilage. (Sometimes the foramen lacerum is defined as including the upper border of the front end of the carotid canal, so the upper part of the foramen is then described as containing the internal carotid.) The groove for the greater petrosal nerve usually leads to the lateral part of the foramen lacerum. This nerve, together with a twig from the nerve plexus around the carotid artery, leaves the cranial cavity at the anterior lip of the foramen, where the posterior end of the pterygoid canal presents itself. The only structures actually passing through the foramen between the outside of the skull and the middle cranial fossa, or vice versa, are a meningeal twig of the ascending pharyngeal artery and some small veins and meningeal lymphatics.

Just posterolateral to the foramen ovale is the **foramen spinosum.** This transmits the middle meningeal artery and accompanying veins.

### In Posterior Fossa

The largest foramen here is, of course, the **foramen magnum,** through which the brain and the spinal cord are continuous with each other. The two vertebral arteries ascend through this foramen to reach the brain; the venous channels of the cranial dura communicate through it with the vertebral venous plexuses around the spinal cord; and the spinal roots of the accessory nerves, which arise from the spinal cord, ascend through it.

Although all the cranial nerves except the first two are connected with those parts of the brain that lie in the posterior cranial fossa, the three to the muscles of the orbit (3rd, 4th, and 6th) run forward in the dura to the superior orbital fissure and therefore need no foramina for exit from the posterior fossa. Similarly, the 5th crosses the depression on the anterior part of the petrous apex, and the foramina for exit of its three branches have been noted in the description of the middle cranial fossa. Thus, the nerves that need foramina for leaving the posterior cranial fossa are the 7th through the 12th. The **internal acoustic meatus,** on the posterior surface of the petrous portion of the temporal bone, serves for the exit of both the 7th and 8th nerves (facial and vestibulocochlear). Both the vestibular (balance) and the cochlear (hearing) parts of the 8th nerve end in the internal ear, which is entirely enclosed within the temporal bone, and therefore they need no external openings. The facial nerve, however, leaves the temporal bone by turning downward to make its exit through the stylomastoid foramen. The 7th and 8th nerves are accompanied in the internal acoustic meatus by the labyrinthine artery, a branch to the internal ear from one of the arteries to the brain.

The **jugular foramen** transmits veins and nerves and usually also small meningeal branches of arteries. Anteriorly, the glossopharyngeal (9th cranial) nerve leaves the skull, and the inferior petrosal sinus leaves to join the internal jugular vein; posteriorly, the sigmoid sinus leaves to form the internal jugular, and meningeal arteries enter; between these the 10th (vagus) and 11th (accessory) cranial nerves make their exists from the skull. Finally, the **hypoglossal canal,** under cover of the jugular tubercle, transmits the hypoglossal (12th cranial) nerve and opens externally just anterolateral to and above the occipital condyle.

These, then, are the particularly important foramina of the skull and the structures that they transmit. Other foramina, such as the condylar foramina and the mastoid foramina, can be grouped together as *emissary foramina.* They transmit connections between the venous sinuses inside the skull and the veins outside. These connections are known as emissary veins.

## DEVELOPMENT AND GROWTH OF THE SKULL

#### Development of the Skull as a Whole

The bones of the skull develop in part as cartilage bones, in part as membrane bones (p. 27). Formation of cartilage begins during the second month at the base of the skull and is almost complete by the end of the third month of fetal life. In general, the **chondrocranium** (cartilaginous cranium) supports the brain and protects the internal ear and the nose. Cartilage surrounds the foramen magnum, giving rise to most of the occipital bone; it surrounds the internal ear on each side, giving rise to the petrous parts of the temporal bones; it extends forward below the brain, giving rise to most of the body, the lesser wing, and a small part of the greater wing of the sphenoid; and it surrounds the nasal cavities, giving rise to the ethmoid bone, the inferior nasal conchae, and a front part of the body of the sphenoid bone. Parts of the nasal chondrocranium persist as the nasal cartilages of the adult. The nasal bones and the maxillae develop in the membrane adjacent to the cartilaginous nasal capsule.

Besides these cartilages, those of the upper two **pharyngeal** or **branchial arches** also are connected with the skull (Fig. 31-12). The upper end of the first or mandibular branchial cartilage becomes enclosed in the developing middle ear cavity and gives rise to parts of two bones in this cavity (malleus and incus). The lower part does not give rise to the mandible, but disappears as membrane bone is formed around it. Similarly, the upper end of the hyoid or 2nd cartilage also presents itself in the middle ear cavity and subsequently forms the stapes. Another part fuses with the temporal bone to form the styloid process. The lower part of the hyoid cartilage unites with the 3rd branchial carti-

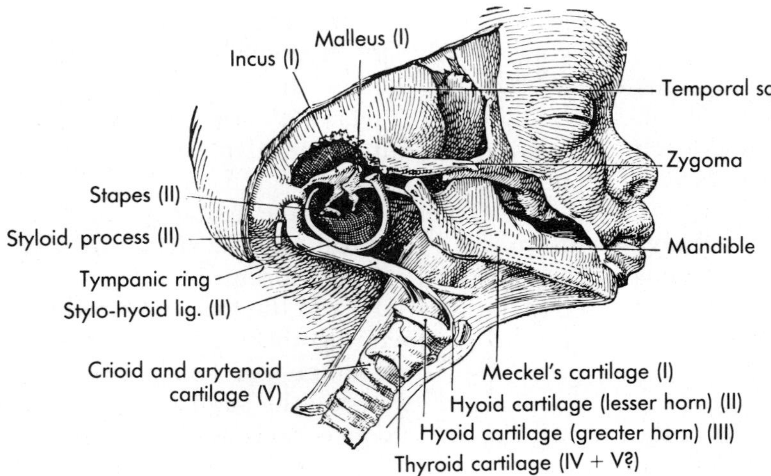


**FIG. 31-12.** Skeletal derivatives of the branchial arches (identified by Roman numerals) as seen in a lateral dissection of the fetal head. (After Kollmann. Arey LB: Developmental Anatomy, 6th ed. Philadelphia, WB Saunders, 1954)

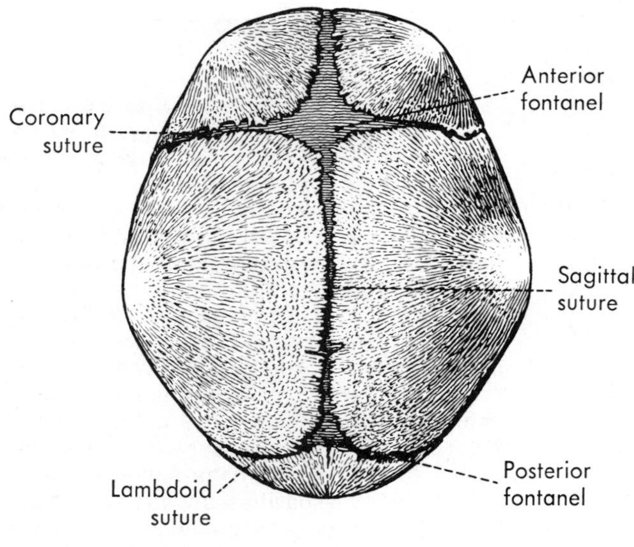


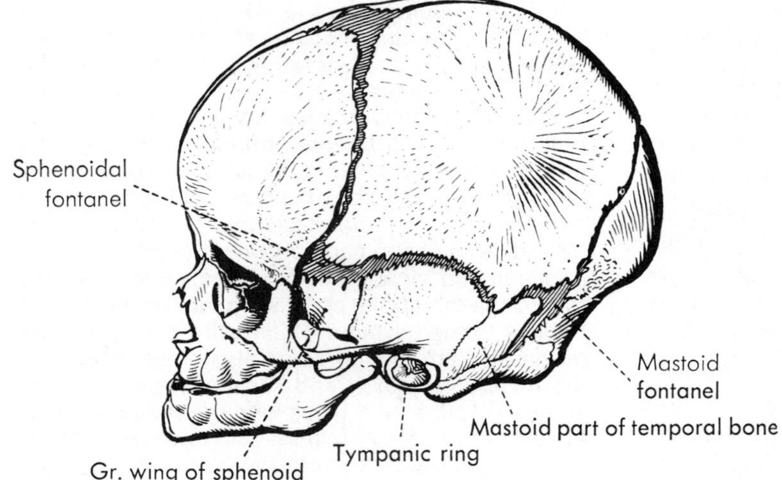


**FIG. 31-13.** Superior and lateral views of the skull of a newborn, showing the sutures and fontanels. (Benninghoff A: Lehrbuch der Anatomie des Menschen, vol 1. Berlin and Munich, Urban & Schwarzenberg, 1949)

lage to form the hyoid bone, and the part connecting the styloid process and hyoid bone typically is replaced by the stylohyoid ligament.

From this brief account, it should be clear that many of the bones of the skull are formed in part or entirely as membrane bones. Details of ossification are too complicated to be entered into at any length here, but in general the earliest centers of ossification both in cartilage bones (for example, the occipital) and in membrane ones (for example, the squama of the temporal) appear during the sixth and seventh weeks. Ossification is not complete at the time of birth; the bones forming the sides and roof of the skull still are united by membrane; and some of those of the base of the skull are united by cartilage (both of these permitting further growth of the skull).

The largest areas of membrane at birth are (1) at the junction of frontal and parietal bones and the sagittal suture, this forming the **anterior fontanel** (fonticulus anterior, Fig. 31-13); (2) posteroinferiorly, between the parietal, temporal, and occipital bones, this forming the **mastoid fontanel** (fonticulus); (3) posteriorly, in the angle between the two parietal bones and the occipital one at the sagittal suture, this forming the **posterior fontanel;** and (4) laterally, in the angle between sphenoid, parietal, and frontal bones, this forming the **sphenoidal fontanel.** The mastoid and sphenoidal fontanels lie deep to muscle, but the anterior and posterior lie deep to only the scalp and constitute the "soft spots" of an infant's head. As the bones continue to grow in the membrane, they meet; the posterior fontanel closes approximately 2 months after birth, the sphenoidal and mastoid ones at approximately 3 months and 1 year, respectively, and the anterior one sometime during the second year.

Growth of the cranium is particularly rapid during the first year, and growth of the skull as a whole continues fairly rapidly to about the age of 7. Thereafter, the skull grows more slowly until the age of puberty, at which time growth again is accelerated. The rapid growth of the face is especially associated with a rapid enlargement of the paranasal sinuses. Until the bones of the skull begin to interlock, their growth occurs in the membrane separating them. Thereafter, growth probably is through absorption on the inner surface of the skull and addition to the outer surface. Although most of the cartilage of the skull is replaced by bone, cartilage persists in connection with the nasal cavity, as already noted, to form a part of the adult skeleton. Cartilage also persists between the sphenoid and the ethmoid bones and at the foramen lacerum between the occipital bone, the sphenoid, and the petrous portion of the temporal.

### Development of Individual Bones

Following is a brief resumé of the development of the bones of the skull. The **occipital bone** ossifies in cartilage from five centers, one for the base of the bone, one each for the lateral (condylar) parts, and two (that quickly fuse) for the squama below the superior nuchal line. The part of the squama above the superior nuchal line develops in membrane from two centers that soon fuse with each other and subsequently with the lower part of the squama. At birth, there usually is some incomplete fusion laterally between upper and lower parts.

Each **parietal bone** is developed from two centers that appear in the membrane and spread to form a single continuous mass of bone. The **frontal bone,** originally paired, develops likewise in membrane. There is one chief center for each half of the bone, but secondary centers for smaller parts of it appear later. These centers fuse together several months before birth, but the frontal bone at birth is still paired, and fusion in the midline is not complete until the fifth or sixth year. Traces of the frontal or metopic suture may remain in the adult skull, especially in the region of the glabella.

The **temporal bone** has a particularly complicated developmental history, for, like the occipital bone, it is formed partly as a membrane bone and partly as a cartilage bone. The squama, including the zygomatic process, is a membrane bone and ossifies from a single center; the petrous portion of the temporal bone, including the mastoid process, is ossified about the internal ear from four centers that fuse together (but at birth, although most of the petrous part of the temporal bone is well formed, the mastoid process is a mere nubbin; it develops as an inferior projection at the time of puberty, when the mastoid air cells develop). The styloid process also develops in cartilage (of the hyoid arch) from two centers, one above the other; sometimes these fail to fuse. The tympanic part, like the squama developed in membrane, arises from a single center and at birth consists of a ring of bone (*annulus tympanicus*, or tympanic ring) incomplete above (Fig. 31-12). After birth, the lower part of the ring expands medially, laterally, and downward to form the tympanic plate visible on the base of the skull.

The **sphenoid bone** also presents a number of centers of ossification, most of which are in cartilage and form the body, the lesser wing, and the pterygoid plates; most of the greater wings are formed in membrane. At least seven pairs of centers are described as contributing to the ossification of the sphenoid bone. At birth, this bone has three unfused parts; the middle one consists of the body, lesser wings, and medial pterygoid plates; the lateral ones consist of the greater wings and lateral pterygoid plates.

The **ethmoid bone** and the **inferior nasal conchae** ossify from the cartilage of the nasal capsule.

Each inferior concha is ossified from a single center. Several centers contribute to the ethmoid, whose ossification is not completed until after birth.

The remaining bones—the lacrimals, vomer, nasals, maxillae, palatines, zygomatics, and mandible—are ossified in membrane. Each **lacrimal bone** is ossified from a single center, the **vomer** is ossified from two centers, and each **nasal bone** is ossified from a single center. All these appear in membrane associated with the cartilaginous nasal capsule. The major part of each **maxilla** is ossified from a single center that spreads to form all the bone except that lying in front of the incisive fossa. The latter part, bearing the incisor teeth, is, in many animals, a separate bone (the premaxilla) and even in man develops as a separate one. Each premaxillary portion develops from at least two centers that usually fuse together. Traces of fusion between the premaxillary and maxillary portions, in the form of an indistinct suture that runs from the posterior border of the incisive fossa to the interval between the canine and lateral incisor teeth, are common in young adults. It is along this line that the cleft of unilateral cleft palate lies.

Each **palatine bone** usually is ossified from a single center, as is each **zygomatic bone;** apparently, either bone may sometimes have a second center. The **mandible** is ossified from paired centers of ossification, one for each half. It develops primarily in the membrane overlying the cartilage of the mandibular arch **(Meckel's cartilage),** but a small part of it is sometimes said to develop from this cartilage.

**FIG. 31-14.** The layers of the scalp.

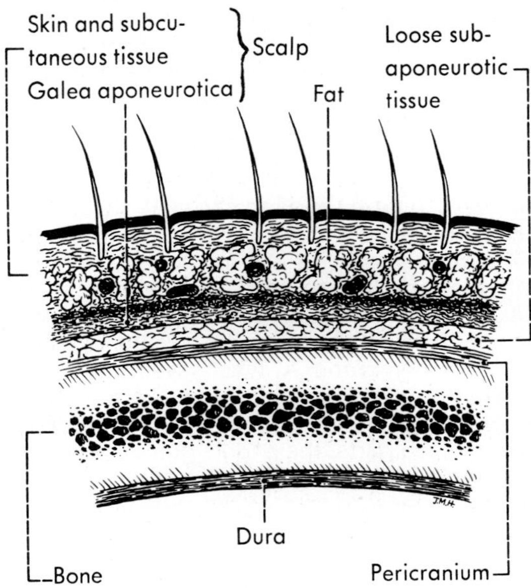


## THE FACE AND SCALP

The subcutaneous connective tissue of the **face** blends with the deeper fascia surrounding the muscles, but the skin is particularly mobile because most of the muscles insert into it. Over the lips (labia), the external openings of the nose (nares), and the margins of the eyelids (palpebrae), the skin becomes continuous with the mucous membranes that line the oral and nasal cavities and (as the conjunctiva) the inner surfaces of the eyelids. The skin of the face varies considerably in thickness and is exceedingly thin in the eyelids; because of this, and the insertion of facial muscles into it, considerable care must be exercised in removing the skin if the underlying muscles are to remain relatively intact.

The **scalp** consists of three layers, firmly united (Fig. 31-14). The outer layer is the skin proper, normally provided with abundant hairs; the deepest layer is the strong **galea aponeurotica** (epicranial aponeurosis), a tendinous layer that covers the calvaria; and the intermediate or subcutaneous layer is composed of rather dense connective tissue that binds the skin to the galea and contains also an appreciable amount of fat. The nerves and blood vessels run in this intermediate layer, between the galea and the skin. The arteries of the scalp anastomose freely with each other both on the same side and across the midline, and they are so held by the dense connective tissue around them that they tend to remain open after they are cut; for both these reasons, bleeding from wounds of the scalp is particularly free.

In contrast to its firm attachment to the skin, the galea aponeurotica is separated from the **pericranium** (the periosteum on the outer side of the calvaria) by a layer of loose connective tissue that allows movement of the scalp over the skull.

The pericranium, not a part of the scalp, possesses little osteogenic capacity as compared to most periosteum. Except over the sutures, it is rather loosely attached to the bones of the calvaria.

A few additional terms must be understood

in studying the face. The upper and lower **lips** (labium superius, labium inferius) and the angle of the mouth (angulus oris) are easily understandable and need no comment. The depression in the center of the upper lip that extends down from the nose is the *philtrum*. The **nose** will be studied in some detail later; at the moment, it should be noted only that the *nares* (the anterior openings) are separated from each other by the midline *septum nasi* and that the flared part of the nose lateral to each naris is the *ala*. The tip of the nose is the *apex*, the *root* is the attachment of the nose to the forehead, and the *dorsum* is the free border between. Extending downward on each side from the ala toward the corner of the mouth is a groove, the *nasolabial sulcus*.

The **external ear** also will be studied more fully later. All that need be noted now is that the part of the external ear that projects from the side of the head is the *auricle*, that the dependent soft portion of the auricle (devoid of cartilage) is the *lobule*, and that the little part projecting backward over the external opening of the external acoustic meatus or auditory canal is the *tragus*.

Most of the names applied to the **regions** of the face are self-explanatory: the *nasal* and *oral regions* are those about the nose and mouth; the *mental region* is that of the chin; the *orbital and infraorbital* regions are about and below the orbits, respectively; the *buccal region* is that of the soft part of the cheek; the *zygomatic region* is that of the prominence of the cheek—of the zygomatic bone; and the *masseteric* and *parotid regions* are those of the ramus of the mandible (covered by the masseter muscle) and the parotid gland that lies between this ramus and the external ear. None of these regions is sharply defined.

The **innervation** of the skin of the face is largely through the three branches (ophthalmic, maxillary, mandibular) of the trigeminal nerve (Fig. 31-15), although some skin over the angle of the mandible and the back part of the ramus is supplied through ascending branches of cervical nerves. The cutaneous branches of the trigeminal nerve necessarily must pierce facial muscles or fascia to reach the skin, and since only their terminal twigs do so, they cannot be traced in the subcutaneous tissue; after the muscles of the face have been dissected, the nerves must be sought close to the bones of the skull in their regions of emergence. In the case of the face, therefore, its innervation cannot be determined by dissection only; rather, this has been determined largely by clinical means. The branches of the facial nerve, which is the motor supply to all the facial muscles, also run deep to the muscles.

The cutaneous nerves of the scalp include some of those that supply the face, some branches of the cervical plexus (greater auricular and lesser occipital), and, posteriorly, dorsal rami of upper cervical nerves that include the large greater occipital nerve. Most of them are closely associated with blood vessels.

**Cutaneous vessels** to the face, like cutaneous nerves, are tiny; the larger vessels run deep to the facial muscles. Therefore, if care is taken to remove only the skin and fascia from the outer surface of the superficial facial muscles and not to undercut the muscles, the nerves and vessels of the face can be left intact

**FIG. 31-15.** The distribution of cutaneous nerves to the face and scalp; the distributions of each of the three great divisions of the trigeminal nerves—the ophthalmic, maxillary, and mandibular—are shown as a whole rather than in terms of the distribution of their several named branches.

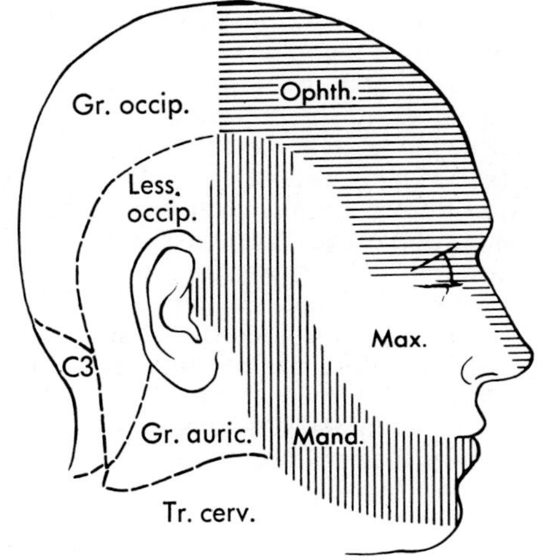

for study after the muscles have been studied. The vessels of the scalp are superficially placed, but the layer in which they are embedded is so tough that they are difficult to dissect. They are continuations of vessels, mostly branches of the external carotid, that supply the face or upper structures in the neck. It is the pulsating distention of the arteries of the scalp that accounts for most of the pain in migraine headache.

## FACIAL MUSCLES

All the facial (mimetic) muscles are differentiated from a premuscle mass that originates from the hyoid arch, as do also the platysma muscle in the neck and a few deeper-lying muscles. They vary a good deal in development from one person to the next, and there may be some blending between muscles. Most of them are thin and flat and need little description other than a reference to their shapes and attachments. They are named primarily from their actions.

Damage to the facial nerve or its branches produces variable amounts of weakness or paralysis of facial muscles, sometimes called "Bell's palsy." Weakness is, of course, particularly noticeable about the mobile mouth, where it may be evidenced even in repose by a sagging corner and becomes obvious as a result of the asymmetry attending an attempt to show the teeth or to smile. Upper facial weakness can be similarly brought out by having a patient attempt to frown, raise his eyebrows, or close his eyes tightly. Paralysis of an entire side of the face is an indication that the facial nerve as a whole has been damaged, the level of injury being either in the brain at the level of origin of the nerve, along the course of the nerve in the skull (for instance, as it runs through the temporal bone), or between its exit from the skull and its dispersion into diverging branches as it emerges from the parotid gland. Weakness rather than complete paralysis of a group of muscles typically results from injury to facial nerve branches, because of the overlap in their distribution. A peculiarity of the voluntary control over the facial muscles is that only a lower, rather than a complete, unilateral facial paralysis results from the usual cerebral stroke that paralyzes, in general, half the body; the part of the facial nucleus controlling the muscles of the forehead is bilaterally controlled from the cerebral cortex, and these muscles are therefore spared in a unilateral lesion above the level of the nerve's origin.

### Muscles about the Mouth

Beginning below, on the mandible, there are several muscles that are attached to it and to skin of the chin or to skin and mucosa of the lower lip (Fig. 31-16). The somewhat triangular muscle arising close to the lower border of the mandible at about the level of the corner of the mouth is the **depressor anguli oris.** From a broad base, where it is partly blended with fibers of the platysma, it tapers as it proceeds toward the corner of the mouth to blend with other muscles here. Medial to it, and in part covered by it, is the **depressor labii inferioris** muscle. This somewhat quadrilateral muscle runs upward to blend with the muscle about the mouth (orbicularis oris) and insert also into the lower lip. It and the depressor anguli oris cover the mental foramen. The fibers of the depressor labii inferioris almost meet those of the opposite side as they reach their insertion, but the origins of the two muscles are some distance from the midline. Appearing between these, arising also from the mandible but running downward to insert into the skin of the chin, are the two **mentalis** muscles, one on each side of the midline.

The muscle encircling the mouth and forming the muscular substance of the lips is the **orbicularis oris.** Some of the fibers of this rather complex muscle arise from the maxilla above the incisor teeth, but many of them are continuations from adjacent muscles, particularly from the buccinator (in the cheek), the depressor of the angle of the mouth, and the levator of the angle of the mouth. Some of the fibers of the upper and lower lips decussate at the angle of the lips. The muscle fibers insert into the skin and mucous membrane of the lips, particularly medially.

Extending medially and usually somewhat upward to its insertion into the angle of the mouth is the **risorius** (risus meaning "laughter" or "grin"), a rather straplike muscle that arises from the fascia over the parotid gland but blends in its lower part with the platysma. Above the risorius, and covered in part

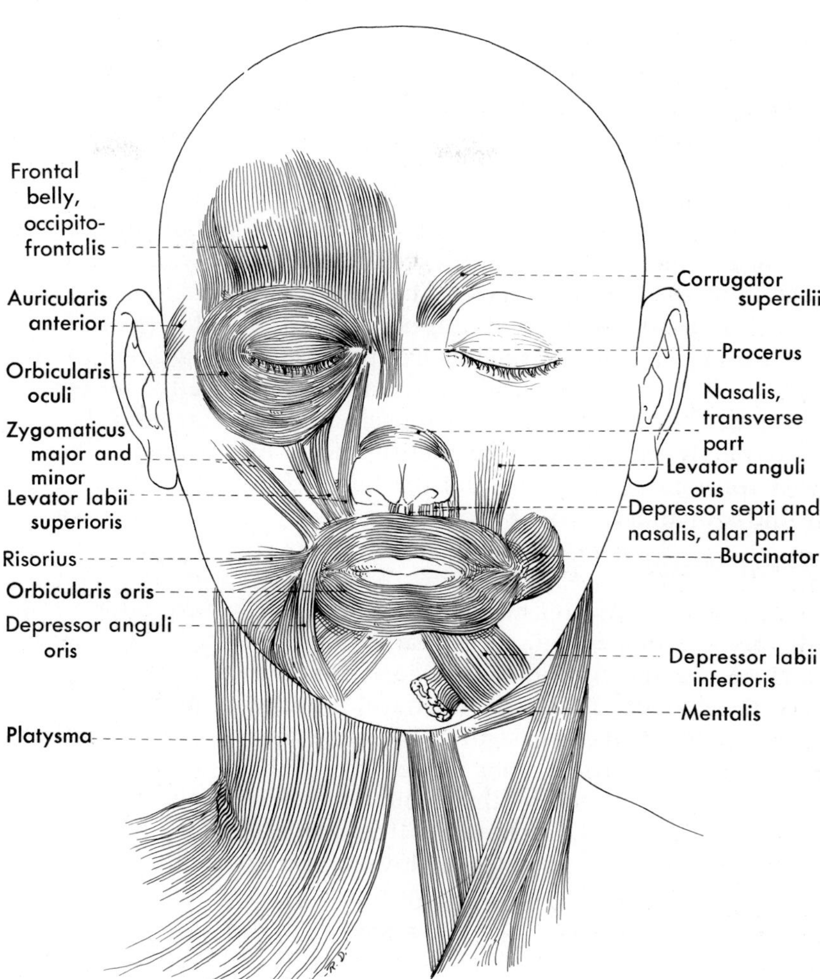


**FIG. 31-16.** Muscles of the face, and the platysma muscle. The facial muscles shown on the reader's left are the more superficial ones; on the right, most of these have been removed to show the deeper-lying muscles. The medial muscle labeled "levator labii superioris" is the levator labii superioris alaeque nasi.

by the fat pad that lies in the cheek, is the **buccinator** muscle (the bugler's muscle). This is the muscular part of the cheek. It arises in part from both the maxilla and the mandible, and posteriorly, deep to the mandible, it arises along the pterygomandibular raphe, a line that extends from the medial pterygoid plate to the mandible (see Fig. 34-3). The duct of the parotid gland pierces the buccinator to reach the oral cavity. As the muscle passes forward to the angle of the mouth, it is partly covered by the other muscles inserting here. Some of its fibers run into the upper lip and some into the lower lip without crossing, becoming continuous here with the orbicularis oris. Others cross each other to reach the upper and lower lips, respectively.

Extending downward and forward toward the angle of the mouth from an origin on the zygomatic arch or the fascia over the upper part of the parotid gland is the **zygomaticus major.** This inserts into skin at the angle of the mouth and becomes in part continuous with the orbicularis oris. It pulls the angle of the mouth laterally and upward, as in a smile.

There are three muscles running downward into the upper lip, and a fourth, deeper-lying, related to the angle of the mouth. The **zygomaticus minor** arises from the zygomatic bone (where it is often continuous with the muscle around the eye, the orbicularis oculi) and runs downward and forward to insert into the upper lip; the **levator labii superioris** arises from the maxilla just above the infraorbital foramen and inserts into the upper lip; and the **levator labii superioris alaeque nasi** (levator of the upper lip and of the ala of the nose) arises from the frontal process of the maxilla alongside the nose and descends with a slightly lateral course. The labial part of the levator labii superioris alaeque nasi inserts into the skin of the lip, but a medial part inserts into the ala of the nose. The last of the muscles connected with the upper lip is the **levator anguli oris.** This arises from the maxilla below the infraorbital foramen and under cover of the three superficial muscles. It inserts partly into skin of the mouth and is in part continuous with fibers of the lower part of the orbicularis oris.

## Nasal Muscles

The muscles connected with the nose are usually poorly developed. As already noted, a part of the levator labii superioris alaeque nasi inserts into the ala. Of the nasal muscles proper, the **depressor septi** is a somewhat quadrilateral muscle that arises from the maxilla under cover of the orbicularis oris and passes upward, in contact with its fellow at the midline, to insert into the lowest part of the septum of the nose. It draws the septum downward. Just lateral to it is the **nasalis,** also arising from the maxilla. One part of this muscle inserts into the lower margin of the ala of the nose, and another part crosses the dorsum of the nose to be united to its fellow of the opposite side by a tendon.

The **procerus,** a thin, bandlike muscle blended with the one on the other side, arises from the nasal bone and runs upward to insert into skin of the forehead. The two muscles pull the skin down, forming horizontal wrinkles between the eyebrows.

## Muscles of the Eyelids, Ear, and Scalp

The large muscle of the eyelids is the **orbicularis oculi.** It surrounds the orbit and extends into both lids. Of the three parts of this muscle, one is placed deeply and can be seen only when the orbit is dissected. The parts visible in a superficial dissection are the *pars orbitalis* and the *pars palpebralis;* the latter, thinner and paler, is the portion in the lids. Both parts work together in closing the lids, but the orbital part is much heavier and therefore closes them more forcibly. These fibers arise at the medial side of the orbit, in part from a heavy ligament (*medial palpebral ligament*) that holds parts of the eyelids against the medial wall, and in part from the bone above and below this ligament. From these origins, the fibers originating above the medial palpebral ligament and those originating below it usually become continuous with each other around the lateral border of the orbit with no obvious intersection; some of the fibers here may be continuous with the zygomaticus minor. The upper and lower palpebral fibers, in contrast, usually intersect in a more or less distinct line, the *lateral palpebral raphe,* at the lateral angle of the lids.

The other muscle of the lid, which raises it, is not really a facial muscle but an orbital one and is described with the orbit.

Originating deep to the upper part of the orbicularis oculi, from the frontal bone just above the nose, are two rather stout bundles of fibers that run laterally to insert into the skin of the eyebrows. These are the **corrugator supercilii** muscles. They draw the eyebrows together so as to produce vertical wrinkles in the forehead above the nose.

There are three extrinsic muscles of the ear: anterior, superior, and posterior (Fig. 31-17). They are variably developed. The **anterior auricular** muscle arises from the temporal fascia in front of and above the auricle and runs downward and backward to insert into it. The **superior auricular,** the best developed of the three, also arises from the temporal fascia above the ear and converges, so that it is fan shaped, to insert into the upper medial side of the auricle. The **posterior auricular,** a small muscle, arises behind the ear from the mastoid process of the temporal bone and inserts into the medial side of the auricle, running approximately horizontally.

The paired muscles of the scalp are referred to as the **epicranius muscles.** Their parts insert into the tendinous galea aponeurotica that forms the deepest layer of the scalp and is

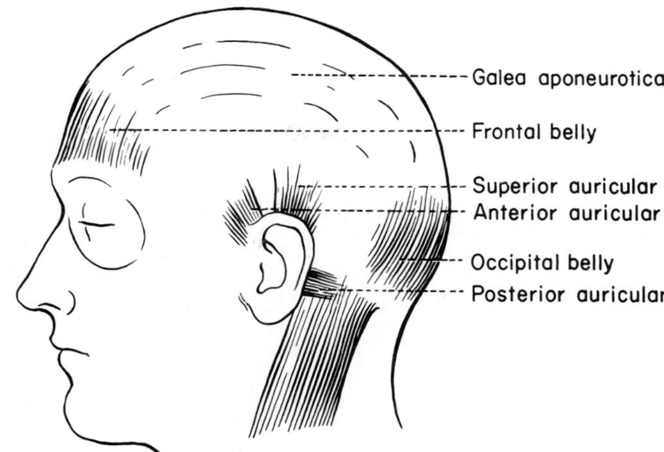


**FIG. 31-17.** The occipitofrontalis and auricular muscles.

separated from the calvaria by looser connective tissue so that the scalp as a whole is movable. The chief components of the epicranius muscles are the **occipitofrontalis** muscles with frontal and occipital bellies. The frontal belly begins above the orbit where it is attached to the skin of the eyebrow (and can thus draw this upward to produce horizontal wrinkles in the forehead). It blends with the orbicularis oculi muscle, extending up through most of the height of the forehead to attach above into the galea aponeurotica. The two muscles together usually cover the entire forehead, meeting in the midline. The occipital belly is smaller, forming a flat band that arises from the lateral half or more of the superior nuchal line on the occipital bone. Its fibers run upward to end, like the frontalis, in the galea. Alternating contraction of the occipitofrontalis bellies produces forward and backward movement of the scalp. It is the tense, involuntary, sustained contraction of the epicranius muscles that accounts for most of the pain of tension headache, probably the most common headache seen in clinical practice.

## PAROTID GLAND AND FACIAL NERVE

### Parotid Gland

The parotid gland, largest of the three most important paired salivary glands (the others being the submandibular and the sublingual), is an outgrowth from the mouth. Its duct empties through the cheek, piercing the buccinator muscle. The gland itself is lodged partly superficial to and partly behind the ramus of the mandible and the masseter muscle that covers it (Figs. 31-18 and 31-19). The gland extends from about the inferior border of the mandible to the level of the zygomatic arch, but varies considerably in size and shape. Superficially, it overlaps a posterior part of the masseter muscle, and it largely fills the space between the ramus of the mandible and the anterior border of the sternocleidomastoid muscle. Deeply, it extends between the ramus of the mandible anteriorly and the sternocleidomastoid muscle, mastoid process, and external acoustic meatus posteriorly. Because of this relationship, the deep part of the gland is compressed between the mandible and the posteriorly lying bone when the mouth is opened; hence, this movement may be painful when the gland is swollen by stored secretion and may be highly painful or impossible when it is greatly swollen, as in mumps.

A number of structures traverse or lie just deep to the parotid gland. Most important of these is the **facial nerve,** the branches of which emerge at the anterior, upper, and lower borders of the gland. The facial nerve enters the deep surface of the gland as a single stem, passing posterolateral to the styloid process as it does so. Within the substance of

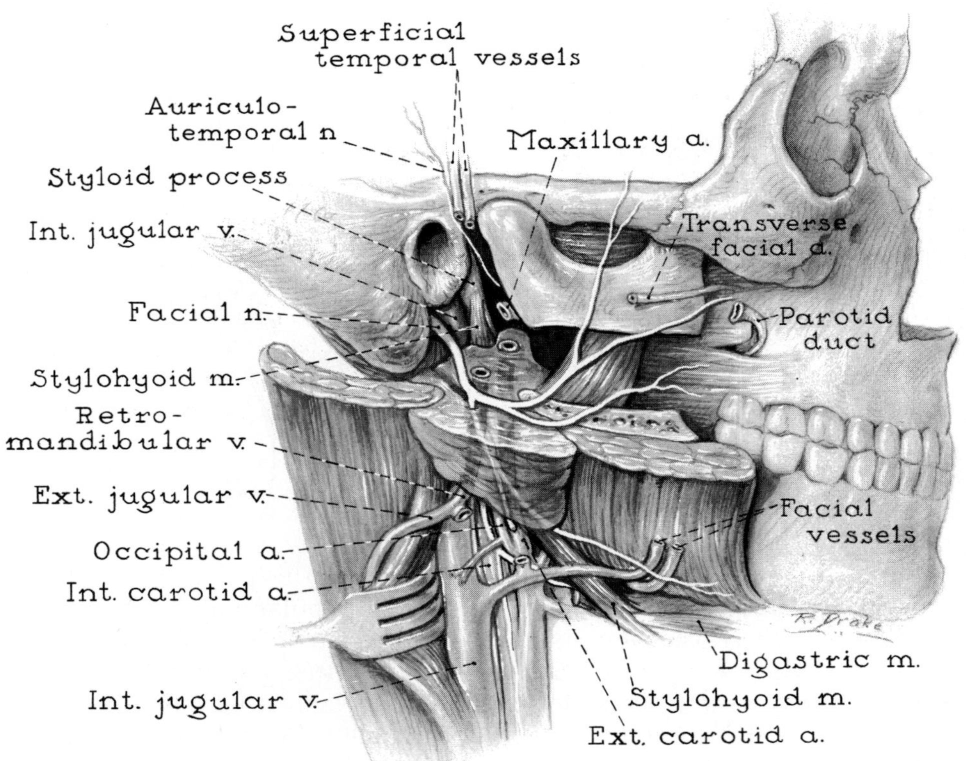


**FIG. 31-18.** Relations of the parotid gland. Upper parts of the sternocleidomastoid muscle, the parotid gland, and the mandible with the overlying masseter muscle have been removed to show deep relations of the gland and the courses of the facial nerve, external carotid artery, and retromandibular vein through it. (Beahrs OH, Adson MA: Am J Surg 95:885, 1958)

the gland, it may at first divide into two stems or it may rebranch and anastomose to form a **parotid plexus.** Regardless of the presence or absence of a parotid plexus in the gland, the facial nerve usually leaves the shelter of the gland as five or more branches.

In its course through the parotid gland, the facial nerve runs superficial to those chief blood vessels that traverse the gland, but is interwoven with the glandular tissue and its ducts. Thus, removal of part or all of the parotid gland demands the most meticulous dissection if the nerve is to be spared.

At the upper pole of the gland, the **superficial temporal vein and artery** appear immediately in front of the ear. The vein runs deep to the upper pole of the gland and then through the deeper part of the gland, where it unites with the maxillary vein (which runs transversely medial to the mandible) to form the **retromandibular vein.** This descends through the substance of the gland and usually divides into two parts before or just after emerging at its lower pole. The posterior part joins the posterior auricular vein to form the external jugular, and the anterior part joins the facial vein and empties with it into the internal jugular. The retromandibular vein may be small, or it may empty exclusively into the external or internal jugular or the facial vein.

The upper end of the **external carotid artery** ascends deep to or through the deep part of the parotid gland, although it is separated from the lower part of the gland by the digastric muscle and other muscles attached to the hyoid bone. Above this level, the external carotid runs along the posterior aspect of the ramus of the mandible deep to the retromandibular vein. It gives off twigs to the gland and a little below the ear divides into maxillary and superficial temporal branches. The maxillary artery passes almost horizontally forward deep to the mandible, and the superficial temporal artery ascends in a groove on the deep surface of the gland. One of its branches, the **transverse facial artery,** emerges from under cover of the gland and runs forward across the face a little below the zygomatic arch. Accompanying the superficial temporal vessels above the parotid gland is the **auriculotemporal nerve,** a branch of the mandibular.

The *posterior belly of the digastric muscle,* close to its origin, lies posterior to the parotid gland and as it runs downward comes to lie medial to the gland. Also, the *styloid process* lies just deep to the gland, the muscles originating from this process being fairly intimately related to it; the large *internal jugular vein* lies close to or against the deep part of the gland. Of these various structures, it is the facial nerve that is most likely to be injured in removing a part or all of the gland (as when tumors arise in its substance). The possibility of injury to the nerve usually is minimized by identifying the nerve before it enters the gland and following it or its branches forward through the gland, dissecting glandular tissue away from its superficial surface. However, it is often easier in the cadaver to trace branches of the facial nerve as they emerge from the parotid gland back into the substance of the gland to their union with other branches.

The **duct of the parotid gland** is formed within the glandular substance and emerges at its anterior border. It runs almost horizontally forward across the superficial surface of the masseter muscle, at about the level of the tip of the lobule of the ear (therefore below the transverse facial artery), and at the anterior border of the masseter turns deeply to penetrate the buccinator muscle and open on the inside of the cheek at about the level of the crown of the second molar tooth of the upper jaw. Sometimes the part of the duct closest to the gland has along its upper border a bit of glandular tissue known as the **accessory parotid gland.**

Sensory twigs from the great auricular and the auriculotemporal nerves end in the parotid gland. The auriculotemporal branches also contain secretory fibers to the gland, derived from the otic ganglion through that ganglion's *communicating branch* with the auriculotemporal.

## Facial Nerve

Before the facial nerve enters the parotid gland, it gives off several branches, and in the gland it gives rise to five named branches or sets of branches that appear on the face: temporal, zygomatic, buccal, marginal mandibular, and cervical (Fig. 31-19). These are named according to their general course and distribution. They all are distributed to voluntary muscle and consist largely of motor fibers. They do contain sensory fibers, however, which have been regarded as proprioceptive to the muscles of the face (just as the motor fibers are motor to these muscles) or as being concerned with pain. Although there is no evidence that the facial nerve is concerned with cutaneous pain, many authors have attributed deep pain from the face to conduction by the facial nerve.

Of the branches of the facial nerve, the temporal, zygomatic, and buccal branches often are double or triple as they leave the parotid gland, or divide soon thereafter; the marginal mandibular branch and the cervical branch (ramus colli) usually are single, although sometimes the former is double. As these diverging branches are traced farther distally, they subdivide further and there may be communicating loops between similarly named branches or adjacent, differently named branches; thus, the pattern of the facial nerve on the face is variable.

In general, the **temporal branches** supply the frontal belly of the occipitofrontal muscle, the corrugator supercilii, the orbicularis oculi, and the anterior and superior auricular muscles; a temporal branch to the auricular muscles may join the auriculotemporal branch of the mandibular nerve and be distributed with that. The **zygomatic branches,** similarly, typically help supply the orbicularis oculi and supply all the muscles connected with the upper lip and the external aperture of the nose, as well as a part of the

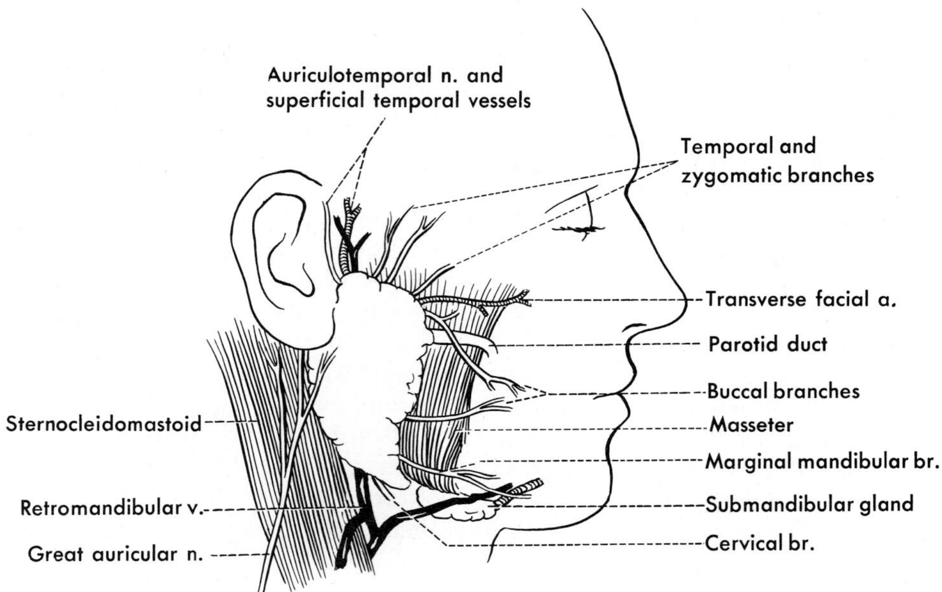


**FIG. 31-19.** Superficial relations of the parotid gland and the branches of the facial nerve on the face and in the neck.

buccinator muscle; some of these branches anastomose with the cutaneous branches of the infraorbital nerve (maxillary division of the trigeminal) as this emerges from the infraorbital foramen.

**Buccal branches** of the nerve may at first run with zygomatic ones and then diverge downward toward the angle of the mouth, or they may come off the lower part of the facial nerve and run approximately horizontally across the masseter toward the angle. They supply the muscles converging on the angle of the mouth, including the buccinator and the levators of the upper lip and angle of the mouth that also are innervated by zygomatic branches and the depressors of the lower lip and angle of the mouth that also are innervated by the marginal mandibular branch. They anastomose with the buccal branch of the mandibular nerve that supplies skin and mucous membrane of the cheek. The **marginal mandibular branch** passes across the masseter and the external surface of the mandible close to the lower border of the mandible, hence its name. Occasionally, it may run lower, across the superficial surface of the submandibular gland. Like the other branches of the facial nerve, it runs deep to most of the superficial facial muscles, and it is also deep to the platysma muscle. However, it typically crosses superficial to the facial vein and artery, being most easily identified by the surgeon as it crosses the vein. It supplies muscles connected with the lower lip and those that run upward to the corner of the mouth. The marginal mandibular branch anastomoses with the mental nerve (a branch of the trigeminal) as this emerges to go to skin of the chin. Finally, the **cervical branch** of the facial nerve leaves the parotid gland close to its lower end and runs downward and then forward just behind and below the angle of the mandible to enter the deep surface of the platysma muscle, which it supplies. It usually anastomoses with the transversus colli (transverse cervical) nerve, the cutaneous nerve of this region.

## VESSELS

The arteries of the face and scalp, except the forehead, are branches of the external carotid. The veins drain into the jugular system.

## Facial Vessels

The chief artery of the face is the **facial artery.** This appears at the lower border of the mandible between that bone and the submandibular gland, just in front of the masseter muscle, for after its origin from the external carotid, it at first lies deep to some of the suprahyoid muscles. However, the **facial vein,** which lies immediately behind the artery at the lower border of the mandible, has a different course below that; it passes downward over the superficial surface of the submandibular gland to enter the internal jugular, usually being joined before it does so by that part of the retromandibular vein that does not go to the external jugular and, variably, by pharyngeal, lingual, and superior thyroid veins. Where the facial artery and vein lie against the submandibular gland, they are connected to it by small twigs. The artery gives off a **submental artery** that runs forward beneath the chin, and the vein receives a corresponding tributary.

At the lower border of the mandible, the facial vessels lie deep to the platysma muscle and, at about the same level, are crossed superficially by the marginal mandibular branch of the facial nerve. Above this, they lie deep to most of the facial muscles along their course and deep to the branches of the facial nerve that supply these muscles. The facial artery may be tortuous or fairly straight, but in any case, the course of the facial vessels is an oblique one past the corner of the mouth and along the side of the nose to the angle between the eye and the nose (Fig. 31-20).

Before the artery reaches the corner of the lip, it gives off an **inferior labial artery** that runs medially in the lower lip and anastomoses with its fellow of the opposite side; this artery may be double. It next gives off the **superior labial artery** to the upper lip. Thereafter, it runs along the side of the nose toward the medial angle of the eye, as the **angular artery.** Instead of ending by breaking up into small branches, the angular artery may anastomose with a terminal branch of the ophthalmic artery (the dorsalis nasi) that leaves the orbit and runs downward to supply the nose. Sometimes, also, the angular artery is very

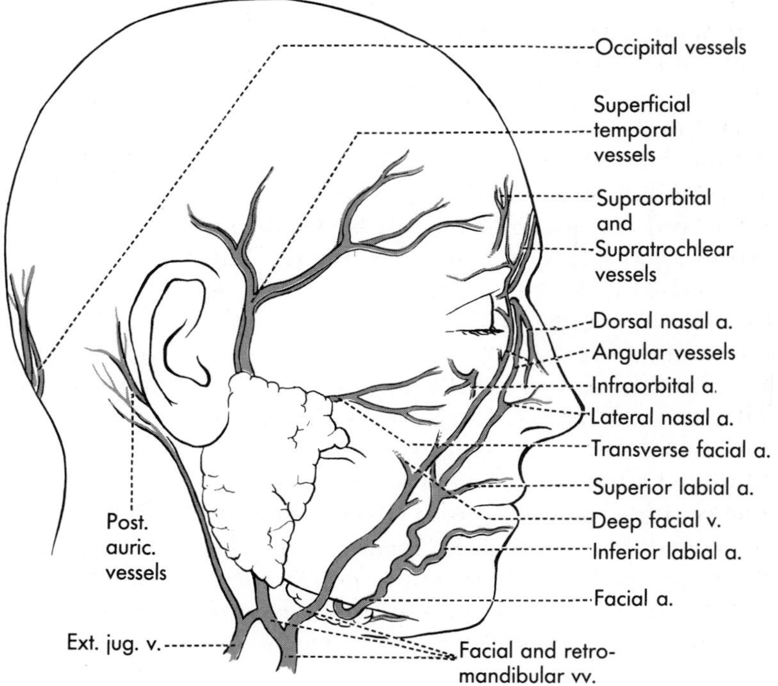


**FIG. 31-20.** Blood vessels of the face and scalp.

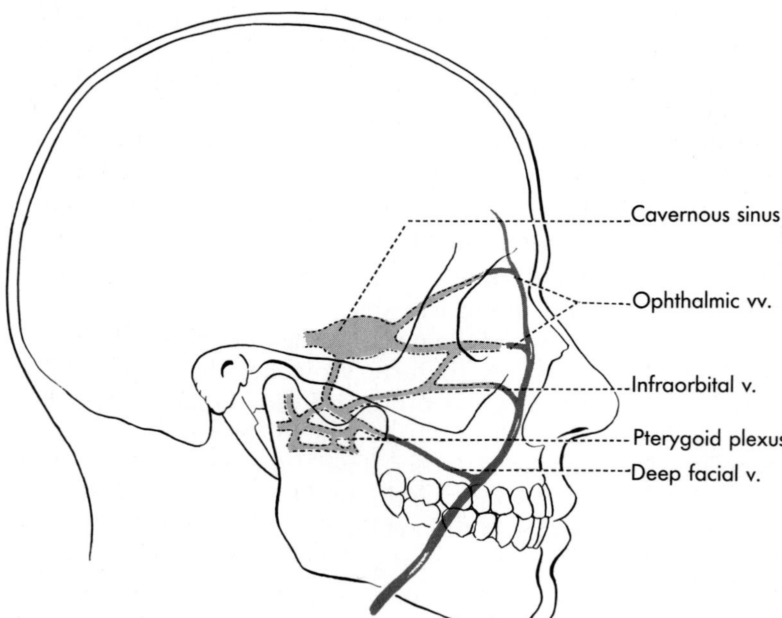


**FIG. 31-21.** Chief connections of the facial vein to the cavernous sinus.

poorly developed or is absent. The **angular vein,** the upper end of the **facial vein,** lies immediately adjacent to the angular artery. It often has no obvious beginning, however, for when traced upward, it is usually continuous with veins of the forehead, usually sends a connection through the upper lid to enter the orbit and join the ophthalmic vein there, and may also send a communication through the lower lid.

Since the veins in the orbit drain blood into intracranial venous sinuses, the angular vein, when well developed, is one of the important extracranial veins having anastomotic connections with the intracranial venous system. As it lies beside the nose, the angular vein receives small veins from the nose and from both eyelids; as the facial vein descends past the mouth, it receives superior and inferior labial veins. In addition to tributaries corresponding to the branches of the artery, the facial vein also receives, as it lies on the buccinator muscle, the **deep facial vein.** This emerges from deep to the ramus of the mandible and its covering muscles in company with a small artery and nerve (buccal artery and buccal branch of the mandibular nerve). Above this, it also may receive the front end of the infraorbital vein as this emerges from the infraorbital foramen.

Through the deep facial vein and veins in the orbit, and also often through the infraorbital vein, the facial vein is connected to the cavernous sinus within the cranial cavity (Fig. 31-21). Since none of these veins is provided with valves, blood from the face either may drain down through the facial vein or, when conditions of pressure are different, may run through the connections of this vein to the cavernous sinus. If infectious material from the face is carried along the latter course, thrombosis of the cavernous sinus and ensuing meningitis are likely to occur; serious neurologic damage or even death may result. Infection and thrombosis of the cavernous sinus commonly are initiated by squeezing pustules around the upper lip or side of the nose and even by careful surgical opening of such pustules; thus, this region is often known as the "danger area" of the face.

## Branches of the Ophthalmic Artery

In addition to the facial vessels, branches of the ophthalmic vessels (the vessels within the orbit) also appear close to the medial angle of the eye, where they pierce the upper lid. The

**dorsalis nasi** artery emerges through the upper lid and runs downward on the side of the nose to supply that, often anastomosing with the upper end of the angular artery. Another branch of the ophthalmic, the **supratrochlear artery,** appears just above or with the dorsal nasal branch, to turn upward on the forehead. Here it runs close to the midline, in company with the supratrochlear vein and a correspondingly named nerve (a branch of the ophthalmic division of the trigeminal). Lateral to the supratrochlear vessels and nerve are the **supraorbital vessels** and nerve. They round the upper border of the orbit by passing through the supraorbital foramen or notch. The nerve is the largest cutaneous branch of the ophthalmic division of the trigeminal nerve. Both supratrochlear and supraorbital vessels and nerves round the upper rim of the orbit immediately against the bone, under cover of the orbicularis oculi and frontalis muscles. They then enter the scalp to lie in its tough subcutaneous or middle layer. Here the vessels anastomose freely with the other vessels supplying the scalp.

The other branches of the ophthalmic artery to appear on the face are small **palpebral arteries** that supply the eyelids and are described with these parts.

### Other Blood Vessels

The remaining arteries, all branches of the external carotid, supply the scalp chiefly, although the superficial temporal also helps to supply the face. The **superficial temporal vessels** appear above the parotid gland as they cross the zygomatic arch. Each soon divides into a *frontal* and a *parietal branch.* Both branches enter the dense connective tissue of the scalp where they anastomose with each other, with vessels of the opposite side, and with the supraorbital and other vessels that supply the scalp. The auriculotemporal nerve runs with the superficial temporal vessels as they lie in front of the ear. Before dividing into its terminal branches, the superficial temporal artery gives off small *anterior auricular* and *parotid branches,* the *transverse facial,* a *zygomaticoorbital artery,* and, usually just below the level of the zygomatic arch, a *middle temporal branch.* The zygomaticoorbital artery, which may arise from the middle temporal, runs forward toward the orbit along the upper border of the zygomatic arch enclosed in the temporal fascia. The middle temporal crosses the arch deep to the superficial temporal and then pierces the temporal fascia to reach the deep surface of the temporal muscle.

The tributaries of the **superficial temporal vein** accompany the branches of the artery. The vein ends by joining the maxillary vein thus forming the retromandibular vein.

The small **posterior auricular artery** arises from the posterior aspect of the external carotid at about the point at which that artery emerges from deep to the posterior belly of the digastric, or sometimes it arises from the occipital artery. It runs upward and backward behind the ear to help supply the external ear and scalp behind the ear. It also gives off *stylomastoid* and *posterior tympanic arteries,* which contribute to the supply of the temporal bone and the middle ear cavity. The **posterior auricular vein,** usually much larger than the artery, leaves the artery at the upper border of the sternocleidomastoid muscle and runs superficial to this muscle to join the external jugular vein.

The **occipital artery** emerges from under cover of the sternocleidomastoid to cross the uppermost part of the posterior triangle of the neck and turn upward, piercing the deep fascia and usually the trapezius muscle, into the scalp. Here it is accompanied by the greater occipital nerve and the occipital vein, and it anastomoses with the posterior auricular and superficial temporal arteries. Just before it turns upward, it gives off a descending branch that anastomoses with the transverse cervical, deep cervical, and vertebral arteries. The **occipital vein** accompanies the artery only on the scalp. Where the artery turns upward, the vein ends in the suboccipital venous plexus, drained by the deep cervical and vertebral veins, but it may communicate with the posterior auricular vein and help form the external jugular.

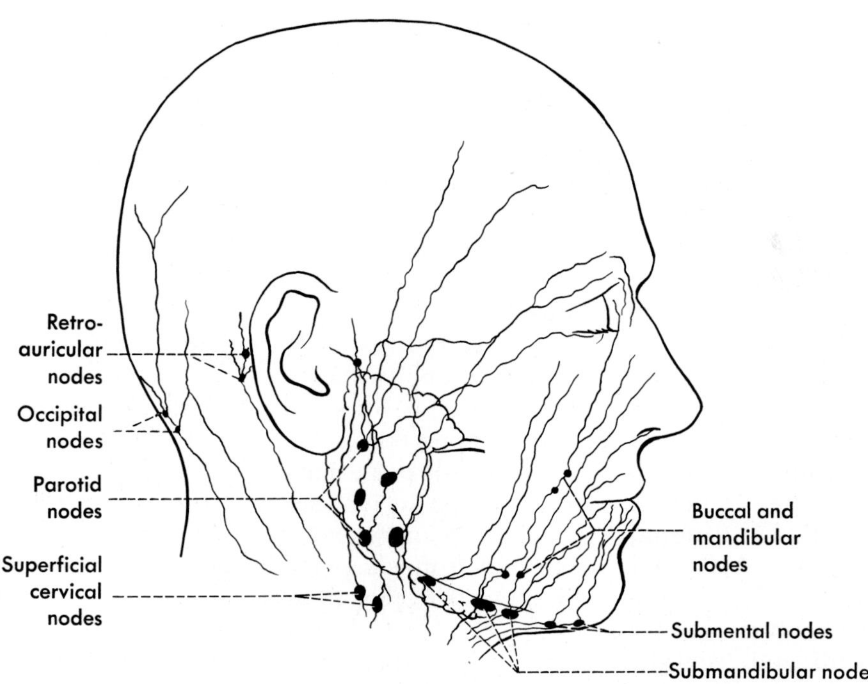


**FIG. 31-22.** Lymphatics of the face and scalp.

## Lymphatics

The lymphatics of the face and scalp are fairly simple (Fig. 31-22). Those from the anterior part of the face accompany the facial blood vessels or, in the case of the lower lip, drain directly downward. A few tiny lymph nodes (*buccal* and *mandibular*) occur along the facial vessels, but most of the anterior lymphatics drain into nodes associated with the **submandibular gland,** into **submental nodes,** or, beyond these, into upper nodes of the **deep cervical set.** Lymphatics from more laterally on the face, including parts of the eyelids, drain diagonally downward and posteriorly toward the parotid gland, as do the lymphatics from the frontal region of the scalp. Associated with this gland, some lying superficially and some more deeply, are **parotid nodes.** These, in turn, drain downward along the retromandibular vein to empty in part into the superficial lymphatics and nodes along the outer surface of the sternocleidomastoid muscle and in part into upper nodes of the deep cervical chain. Lymphatics from the parietal region of the scalp drain, in part, into the parotid nodes in front of the ear and in part into **retroauricular nodes** in the back of the ear that in turn drain into upper deep cervical nodes. The largest retroauricular node frequently is palpable as it lies over the mastoid process, especially in children. Lymphatics from the occipital region end, in part, in **occipital nodes** (which in turn drain into upper deep cervical ones) and, in part, directly in upper deep cervical nodes.

# CUTANEOUS NERVES

The face and scalp are supplied largely by the trigeminal nerve and the upper cervical nerves (Fig. 31-23). The chief exception to this is a bit of skin behind the ear that is supplied mostly by a small twig from the vagus or 10th cranial nerve.

## Trigeminal Distribution

The larger cutaneous nerves of the face already have been noted, for they anastomose with branches of the facial nerve, and those on the forehead accompany the arteries. The trigeminal or fifth cranial nerve provides the

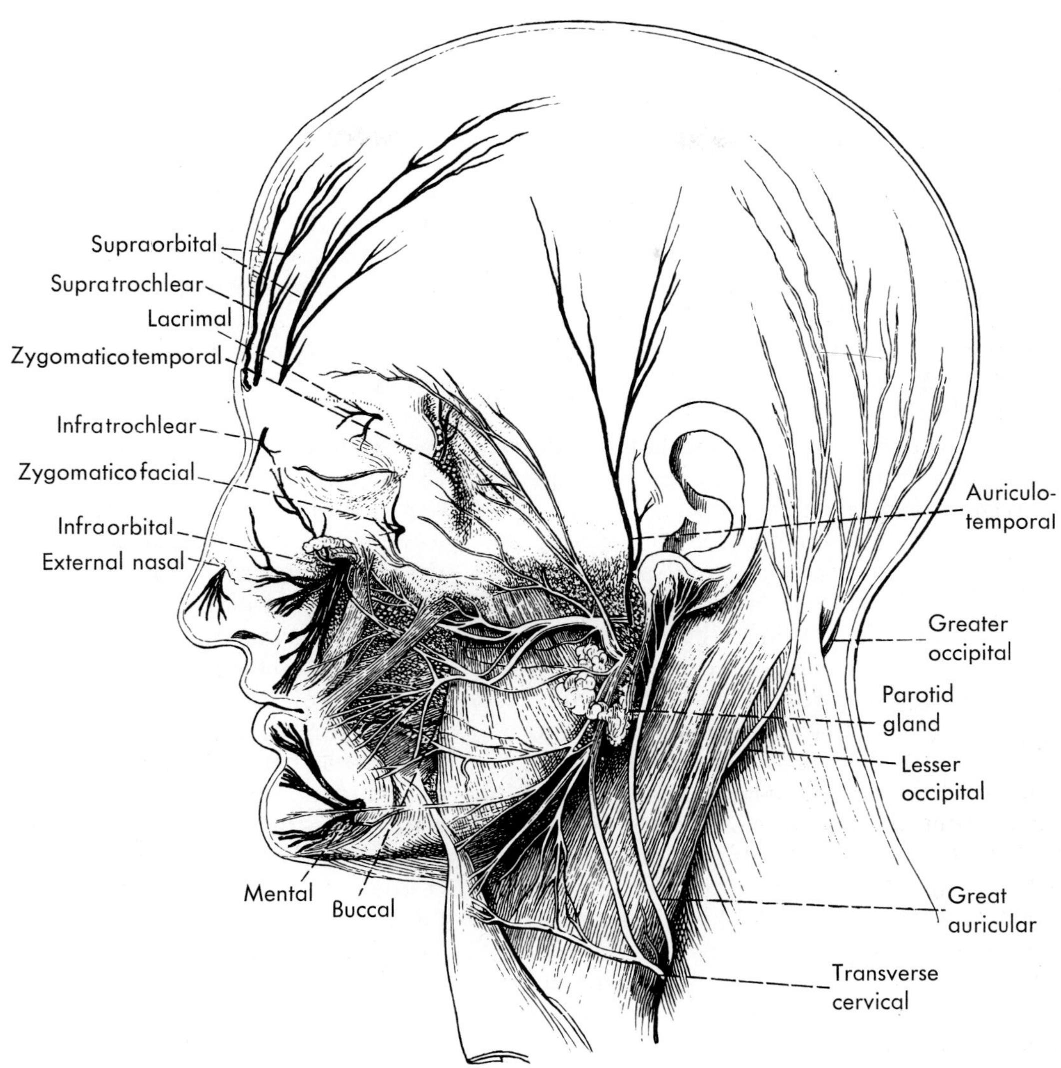


**FIG. 31-23.** The cutaneous branches of the trigeminal nerve, black. Shown in outline are upper branches of the cervical plexus and branches of the facial nerve. The parotid gland has been partly removed. (Henle J: Handbuch der systematischen Anatomie des Menschen, vol 3, pt 2. Braunschweig, Vieweg und Sohn, 1868)

sensory supply to the skin of most of the face, except for much of the ear, some skin immediately in front of this, and that over the angle of the jaw, which are innervated by fibers from upper cervical nerves.

All three divisions of the trigeminal nerve supply the face. In general, the **ophthalmic** or upper division supplies the upper eyelid, the entire dorsum and the upper parts of the sides of the nose, the forehead, and the scalp about as far back as the interauricular line (Fig. 31-15). The **maxillary** or second division supplies the lower eyelid, the side of the lower part of the nose, the skin on and above the prominence of the cheek, and that of the upper lip. The **mandibular** nerve supplies an anterior part of the lateral surface of the external ear, as well as much of the skin of the external acoustic canal and the eardrum or tympanic membrane; skin extending up in front of the ear onto the scalp; the skin of the cheek lateral to the lips; and the lower lip and chin. Each division of the trigeminal nerve is represented on the face by several branches, and most of these branches appear on the face by running close against the skull or emerging from foramina in the skull. They, therefore, are deep-lying branches and are most easily found in their deep positions. Traced superficially, they anastomose with branches of the facial nerve and penetrate the overlying facial muscles and therefore cannot really be traced accurately to their distribution.

Five **branches of the ophthalmic nerve** reach the skin of the face. The largest of these is the **supraorbital nerve,** running through the supraorbital foramen or notch and turning up on the forehead, where it usually divides into a medial and a lateral branch and is distributed to the forehead and to the scalp about as far back as the interauricular line. Medial to it is the smaller **supratrochlear nerve,** distributed to a medial part of the skin of the forehead. Each of these accompanies the similarly named blood vessels. Also leaving the orbit close to the medial angle of the eye, but lying below the supratrochlear nerve and therefore not against the upper bony rim of the orbit, is the **infratrochlear nerve.** This gives off twigs to the lateral side of the nose and to the upper eyelid. About halfway down the dorsum of the nose there emerges from between the nasal bone and the nasal cartilage a small branch of the ophthalmic, the **external nasal branch of the anterior ethmoidal nerve** that then runs subcutaneously down the nose to the tip, supplying the distal part of the dorsum. The fifth cutaneous branch of the ophthalmic is a twig that is difficult or often impossible to find by dissection; it is the **lacrimal nerve,** which pierces the lateral part of the upper eyelid and is distributed to this lid and the immediately adjacent skin.

Three **branches of the maxillary nerve** reach the skin of the face. The **infraorbital,** the largest, emerges from the infraorbital foramen with a small artery and vein. It anastomoses with zygomatic branches of the facial nerve and spreads out toward skin of the ala of the nose, the upper lip, and the lower eyelid. A small branch, the **zygomaticofacial,** emerges from the zygomatic bone on the anterior surface of the prominence of the cheek and supplies skin here and on up toward the lateral angle of the eye. The third cutaneous branch, the **zygomaticotemporal,** emerges from the skull through the anterior wall of the temporal fossa, but then penetrates the temporal fascia anterior to the temporal muscle to supply skin of the anterior part of the temple just lateral to the eye.

There also are three cutaneous **branches of the mandibular nerve.** The largest of these is the mental, a continuation of the nerve (inferior alveolar) that supplies the teeth of the lower jaw. The **mental nerve** emerges from the mental foramen, anastomoses with the facial nerve, and distributes branches to the chin and lower lip. The second cutaneous branch is the **buccal nerve.** This small nerve emerges from deep to the ramus of the mandible and appears on the external surface of the buccinator muscle, where it anastomoses with buccal branches of the facial nerve. It sends branches through the buccinator muscle to supply the mucous membrane on the inside of the cheek, and its superficial branches supply the skin of the outer surface. The third cutaneous branch is the **auriculotemporal.** This emerges from deep to the upper end of the parotid gland and crosses the back end of the zygomatic arch immediately in front of the ear, in company with the superficial temporal vessels. As the nerve passes behind the temporomandibular joint, it gives a twig to this; as it passes the external acoustic meatus, it gives off a nerve of the meatus that supplies most of this and most of the tympanic membrane; it gives twigs to the parotid gland; and it supplies a small anterior part of the upper lateral surface of the ear before continuing to the skin and scalp of the posterior part of the temporal region. The auriculotemporal nerve often receives the facial nerve fibers to the anterior and superior auricular muscles.

That small part of the skin of the face that is not supplied by the trigeminal nerve—some of that over the angle of the jaw and extending up to include much of the skin of the external ear—is supplied from the cervical plexus through the **transverse cervical** (transversus colli) and **great auricular nerves** (Fig. 31-15). Skin of the upper part of the posterior or medial surface of the ear and skin and scalp just behind the ear are supplied largely by the **lesser occipital nerve** from the cervical plexus. Finally, the scalp posterior to the interauricular line, and behind the distributions of right and left auriculotemporal and lesser occipital nerves, is supplied by dorsal branches of cervical nerves. The **greater occipital nerve,** primarily the cutaneous branch of the dorsal ramus of C2, runs with the occipital artery and reaches the posterior distribution of the supraorbital branch of the ophthalmic. A branch from the dorsal ramus of C3, the **3rd occipital nerve,** may run upward to supply scalp close to the midline about as high as the external occipital protuberance.

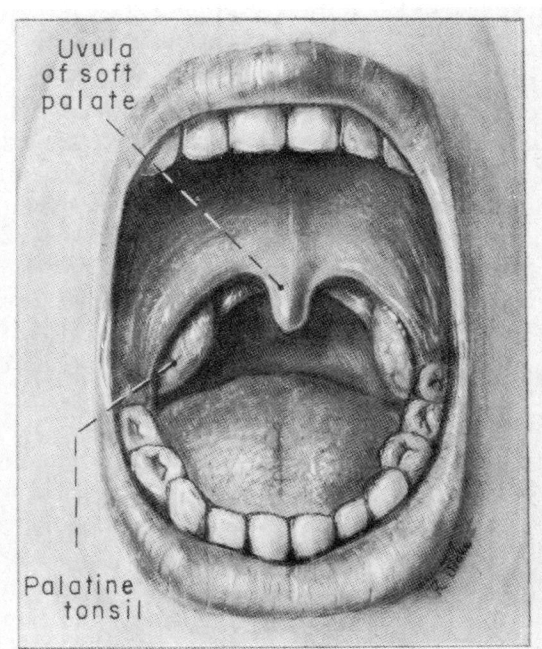


**FIG. 31-24.** The oral cavity, soft palate, and palatine tonsils. The palatoglossal arch partly hides the tonsil; the palatopharyngeal arch, visible above, is hidden by the tonsil below.

## THE MOUTH AND JAWS

### Oral Cavity

The mouth or oral cavity (Fig. 31-24) is bounded externally by the cheeks (*buccae*) and the lips (*labia*), above by the palate, and below by mucous membrane that connects the floor of the mouth to the tongue. The major part of the tongue projects into the oral cavity, as do the teeth. The **rima oris** is the opening between the lips, and the cavity is divided into two parts: the **vestibule,** the part lying between the cheeks and lips on the one hand and the teeth and gums on the other, and the **oral cavity proper,** that part within the arches of the teeth. In the vestibule, a small midline fold, the **frenulum,** can be seen connecting the upper lip to the gum; there usually is a similar frenulum of the lower lip. Except at the frenula, the transition between the mucosa of the lips and cheek and that of the gums is generally smooth.

The parotid gland, the largest of the salivary glands, opens through the substance of the cheek, its point of opening being indicated by a **parotid papilla** that is situated at about the level of the upper second molar (next to last or last) tooth. The parotid papilla, like most of the other features of the soft tissues of the mouth, is best seen in the living individual. Adequate examination of the mouth of the cadaver is difficult.

Numerous small glands also open into the vestibule. Sebaceous glands of the lips are especially concentrated at approximately the mucocutaneous junction. They are superficial and often appear as a series of small yellowish bodies close to the free borders of the lips. Fewer sebaceous glands occur in the cheeks close to the teeth.

Mixed or seromucous labial glands are closely packed in the submucosa of the lips, where they can be palpated easily. Buccal glands are less numerous; most lie in the submucosa of the cheek, but in the region of the molar teeth a few usually lie on the outer surface of the buccinator muscle and are called molar glands.

The muscles, nerves, and vessels of the lips and cheek have been described with the face.

Within the oral cavity proper there is a **frenulum** on the lower surface of the tongue (*lingua*) in the anterior midline. On each side of this frenulum is a projection of the mucous membrane, the **sublingual caruncle** (papilla), that contains the termination of the duct of the submandibular salivary gland as this opens into the mouth. Running backward from the caruncle, there may be in the floor of the mouth a bulge, the **sublingual fold,** produced by the sublingual salivary gland that lies immediately below the mucosa here. Small sublingual ducts open through the mucosa in this position, and a sublingual duct usually joins the submandibular duct to open on the sublingual caruncle.

The roof of the mouth is formed largely by the hard palate, which consists of the bony palate and the mucosa and glands associated with this. The smaller posterior part of the roof is formed by the soft palate, which also projects downward to separate the oral cavity partially from the nasal part of the pharynx. The palate and the tonsillar region, where the oral cavity opens into the pharynx, are described in Chapter 34. The margins of the hard palate are, of course, continuous with the alveolar processes of the maxillae, which bear the upper teeth.

**Teeth.** The anatomy of the gums and teeth should be examined in preparation for the dissection of the jaws. The **gums** (*gingivae*) are layers of mucous membrane that cover the alveolar processes of the upper and lower jaws. The inner and outer gums are continuous with each other in the spaces between the teeth, and those of the upper jaw are continuous with the mucosa of the lips and cheek and with that of the hard palate, respectively. Those of the lower jaw are similarly continuous with mucosa of the lips and cheek and the floor of the mouth lateral to and below the tongue.

The teeth (dentes) of the upper and lower jaws form **superior** and **inferior dental arches.** The **permanent teeth,** beginning in the anterior midline and proceeding laterally and posteriorly to the end of the arch, are similarly named on each side and in both upper and lower jaws. In this order, there are two incisive teeth (**incisors**), one **canine,** two **premolars,** and three **molars.** The last of these is slow to erupt, often does so imperfectly, and is a "wisdom tooth" (*dens serotinus*). A complete set of permanent teeth consists of 32, eight on each side of each jaw (Fig. 31-25; see also Figs. 31-3 to 31-8); however, many adults do not have their third molars, either because these failed to erupt or because they were removed. (They may push against the second molar and

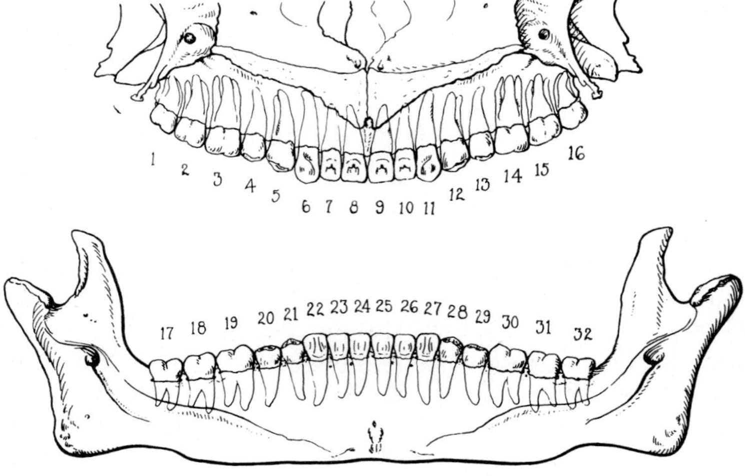


**FIG. 31-25.** Lingual aspect of the upper and lower teeth, with both dental arches widely spread. The method used here for numbering the teeth is only one of several.

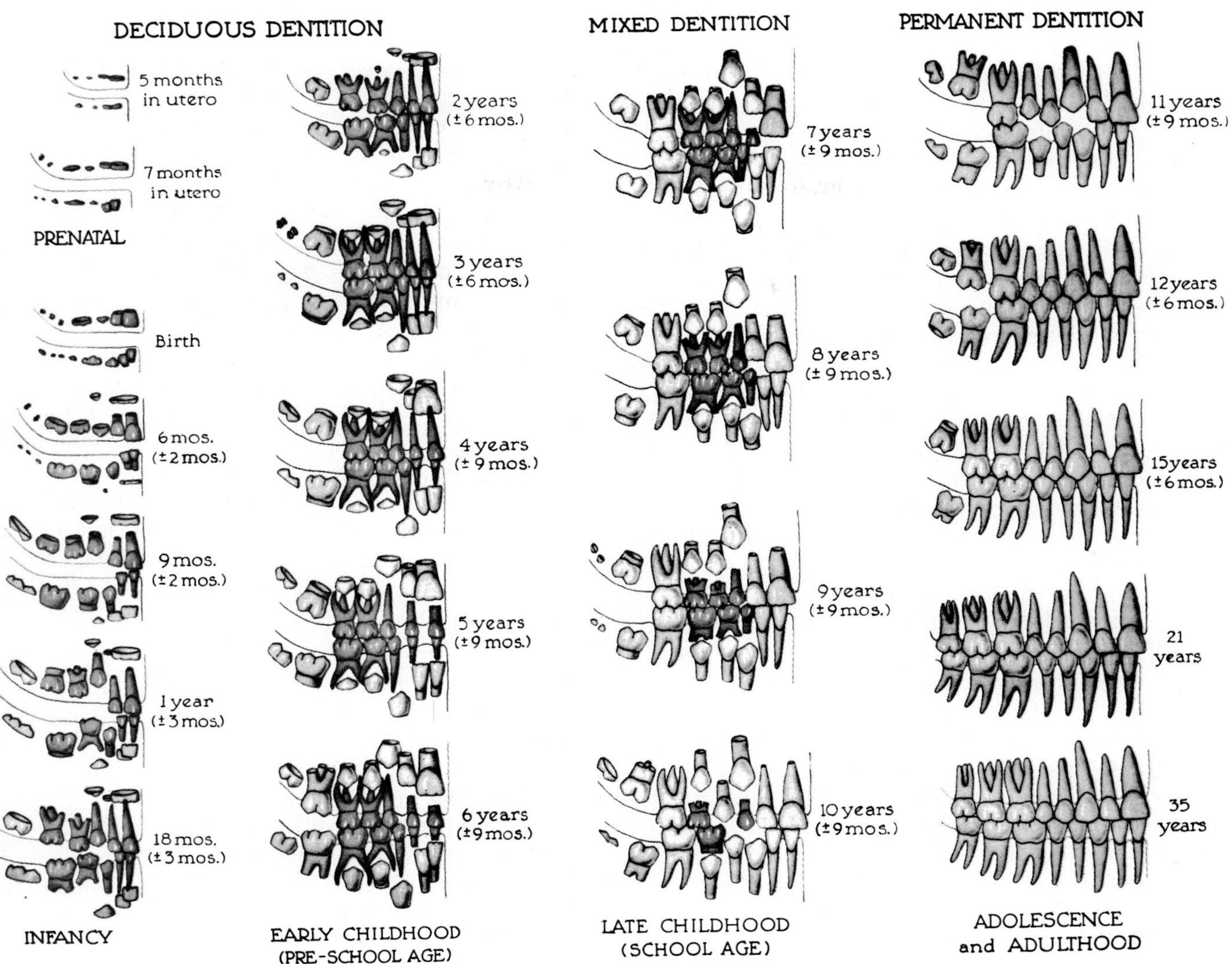


**FIG. 31-26.** Development of dentition in the human. The deciduous teeth are dark, and the permanent ones are lighter. Note that the full quota of deciduous teeth is shown present and erupted at the ages of 2 and 3 years. (Schours I, Massler M: A chart prepared for and distributed by the American Dental Association, 2nd ed, 1944)

lie at such angles that they cannot erupt.) In contrast, the **deciduous** (baby) **teeth** are only 20. There are two incisors and a single canine on each side, as is the case of the permanent teeth; the canines are followed by two molars that lie in the positions of the permanent premolar teeth (Fig. 31-26).

There are certain obvious differences among the teeth, but all of them have the same basic structure. A tooth is largely formed of modified bone known as *dentine,* but the dentine of the **crown** (*corona dentis*) is covered by *enamel.* This thins out and disappears at the **neck** (*collum*), and the dentine of the root or roots is covered instead by *cementum,* which anchors the root in its alveolus or socket. Normally, the gum attaches at the neck of the tooth so that the crown is exposed and the root is enclosed in bone. If the gingivae and bone recede to uncover part of the bony root, that and the enamel-covered crown then are known as the **clinical crown;** the remainder of the root is known as the **clinical root.** The **surfaces** of the teeth are described as *occlusal* (masticatory), those that meet during chewing: *vestibular* or *facial* (outer), and *lingual* (inner); and surfaces of contact, which are designated as *mesial* (toward the center of the dental arch) and *distal* (towards the ends of the arch).

Each **incisor tooth,** designed for biting and cutting, presents an *incisal edge* (margo incisalis). The upper incisors, and particularly the upper canines, often show a marked thickening, the **cingulum,** on their lingual surfaces close to the gum. The canines end in blunted points instead of incisal edges. The **premolar** and **molar teeth,** designed for grinding and chewing, have their occlusal surfaces roughened by raised *cusps* (tubercles) that assist in the grinding process. There are typically two cusps on the premolars (hence these are frequently called "bicuspids"), either three or four on the upper molars, and four or five (often the latter) on the lower molars. In the posterior teeth, a ridge, or *crista triangularis,* extends toward the center of the occlusal surface from the apex of each cusp, and if two crests meet, they form a *crista transversalis.*

The **cavum dentis,** or cavity within a tooth, is divided into the cavity of the crown (cavum coronale) and a **root canal,** both filled with **dental pulp.** When there are several roots to a tooth, the root canals fuse to form the cavum of the crown. The nerves and vessels in the dental pulp enter through an **apical foramen** at the tip of each dental root.

The **roots** of the teeth are most easily examined in roentgenograms (Fig. 31-27). Their great density makes them stand out plainly against the bone of the alveolar process. In such roentgenograms, the periosteum lining the alveolus, the **periodontium** (periodontal ligament or membrane), also is visible, as a dark line just outside the tooth, and the cavum dentis is similarly visible. The incisor and canine teeth regularly have each a single root. The premolar teeth do also, but the root sometimes is grooved, and that of the upper first premolar, particularly, may be bifid. The molar teeth of the lower jaw usually have two roots each, and those of the upper, three roots, but the third molar (wisdom tooth) in either jaw may show fusion between its roots.

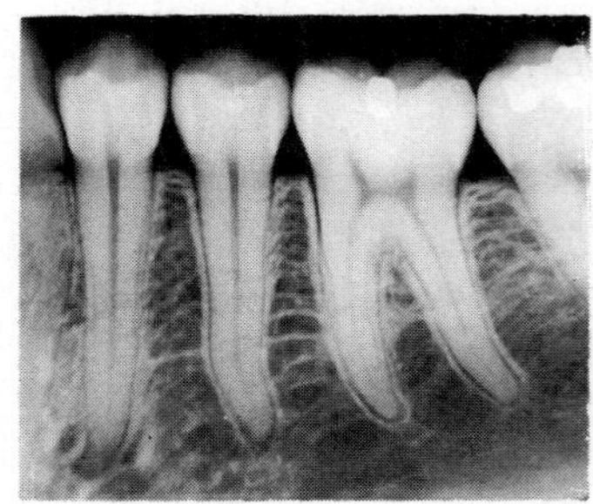

**FIG. 31-27.** A roentgenogram of the two premolar teeth and the first molar tooth of the lower jaw. The root canals within the teeth show clearly. The periodontium (periodontal membrane) is the thin dark line immediately adjacent to the root or roots of each tooth. The thin layer of dense bone that lines the alveolus is known to dentists as the "lamina dura." The crown of the first molar contains a filling, as does the part of the second molar that is visible. (Stafne EC: Oral Roentgenographic Diagnosis. Philadelphia, WB Saunders, 1958)

A few nerve twigs and blood vessels reach the outside of the tooth by penetrating the bone of the alveolar process. They are derived from the vessels and nerves of the gingivae. However, the major nerves and vessels of the teeth run in the bone of the jaws; at the apex of each root some branches enter the root canal to ramify in the pulp, but others extend into the periodontium to end there or pass through the bone to reach the gingivae.

The nerves and blood vessels of the gingival mucosa are largely those that supply the teeth; however, the labial and buccal gingivae also receive twigs from the nerves that supply the lips and cheek. The lymphatics from the lingual gingiva drain between the teeth to join those of the vestibular gingiva; these drain into buccal, mandibular, and submandibular glands.

## Muscles, Nerves, and Vessels of the Jaws

The muscles concerned primarily with chewing attach to the mandible and often are known as the muscles of mastication or as mandibular muscles. They are responsible not only for the bite but also for the side-to-side movement of chewing. They all are innervated by the mandibular nerve, the lowest division of the trigeminal nerve and the only branch of the nerve that contains voluntary motor fibers. The mandibular nerve also innervates the teeth of the lower jaw, but the maxillary nerve, the second or middle division of the trigeminal, innervates those of the upper jaw. The blood supply to both the upper and the lower jaw is through the maxillary artery, which with the superficial temporal is one of the terminal branches of the external carotid.

The suprahyoid muscles attaching to the mandible usually are not regarded as muscles of mastication, although they may assist in movements of the jaw. Some of them may be identified without further dissection (Fig. 31-28), but they can be examined more thoroughly as the mandible is removed (Figs. 31-36 and 31-40). The four muscles of mastication are the masseter, the temporal, and two pterygoid muscles. The masseter and temporal muscles (Fig. 31-29) are superficial ones, are covered on their superficial surfaces by deep fascia. The masseteric fascia is thin and is continued posteriorly as the thicker parotid fascia around that gland. The temporal fascia is much thicker and hides that muscle. The temporal fascia blends above with the tendinous galea aponeurotica. Below, it attaches to the zygomatic arch, dividing as it does so into two layers, a superficial and a deep one.

**The Masseter.** The muscle is largely covered by the parotid gland posteriorly and by facial muscles anteriorly. Deep to the facial muscles, it is crossed superficially by the parotid duct, above this by the transverse facial artery, and both above and below the duct by branches of the facial nerve (Fig. 31-19). The facial vessels lie just in front of its lower end as they cross the lower border of the mandible.

The masseter muscle largely covers the ramus of the mandible. It consists of two blended parts, of which the superficial arises from about the anterior two-thirds of the lower border of the zygomatic arch and extends downward and somewhat posteriorly. The deep part arises from the whole deep surface of the zygomatic arch and extends vertically downward. The muscle inserts on almost the entire lateral surface of the ramus of the mandible, including a lower part of the coronoid process; there is, however, no insertion on the condylar process. The *masseteric nerve and vessels* enter the deep surface of the masseter by passing through the mandibular notch.

**The Temporalis.** The large, fan-shaped temporal muscle covers much of the side of the head, arising from the temporal fossa and the deep surface of the temporal fascia and converging onto the coronoid process. The anterior fibers run almost vertically downward;

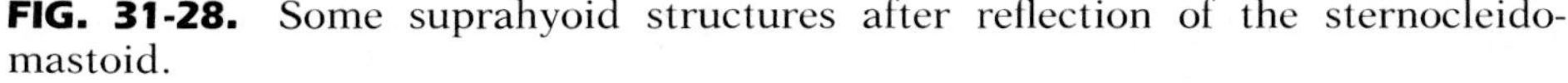
**FIG. 31-28.** Some suprahyoid structures after reflection of the sternocleidomastoid.

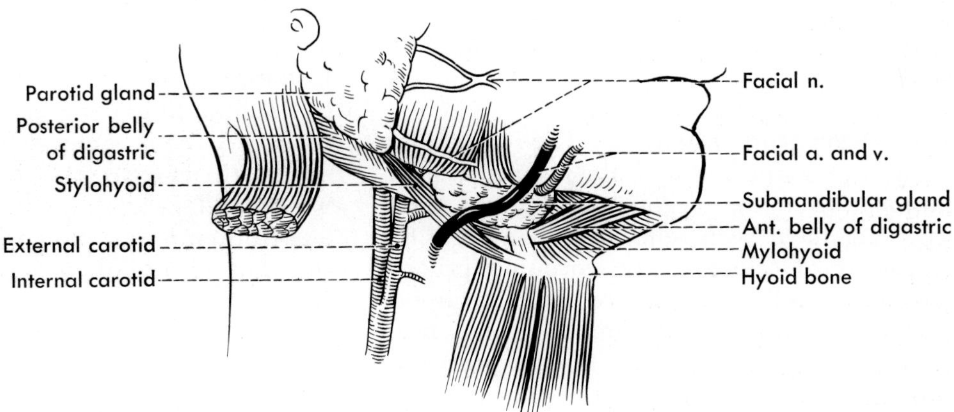

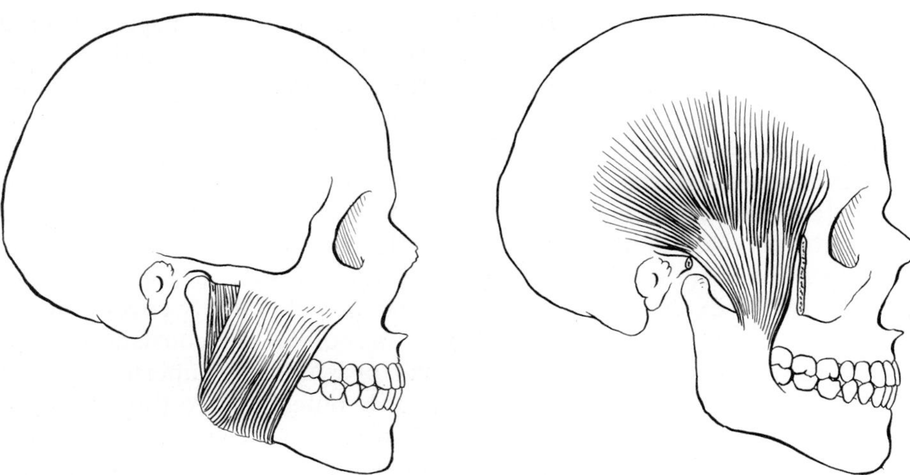

**FIG. 31-29.** Masseter and temporal muscles.

the posterior fibers run both forward and downward. The muscle is thick as it reaches the coronoid process. The insertion is into the upper and anterior border of the coronoid process externally and along the whole internal surface of that process and the ramus of the mandible below it almost as far as the alveolar process.

The masseteric nerve and vessels pass behind the muscle to reach the masseter muscle. The buccal nerve and vessels pass downward and forward medial to the temporal, or sometimes through some of its fibers, to reach the cheek. Its nerve supply usually is two deep *temporal nerves,* the posterior commonly being larger, that run upward between the muscle and the bone of the temporal fossa. Each nerve typically is accompanied by a deep temporal artery, from the maxillary. These supply both the muscle and the adjacent bone of the skull.

**The Pterygoid Muscles.** The remaining structures connected with the jaw are best examined after the zygomatic arch and the masseter have been reflected downward, the temporal muscle and the coronoid process have been reflected upward, and the upper part of the ramus of the mandible has been removed (Fig. 31-30). The **lateral pterygoid muscle** runs almost horizontally backward to its insertion on the mandible. It arises by two heads, a small upper one from the infratemporal fossa and a larger lower one from the lateral surface of the lateral pterygoid plate. It inserts into the neck of the mandible, the capsule of the temporomandibular joint, and the articular disk that lies within the joint. The two heads are immediately adjacent and blend as they are traced toward the mandible. They are largely covered by a dense pterygoid plexus of veins that surrounds them and the maxillary artery.

Since the *mandibular nerve* lies deep to the lateral pterygoid, most of the branches of the nerve are intimately related to the muscle (Fig. 31-31). Thus, the anterior deep temporal and the buccal nerves usually pass between the two heads of the muscle. The posterior deep temporal and the masseteric nerve usually pass between the upper head and the infratemporal fossa. Two of the largest branches of the mandibular nerve, the lingual and the inferior alveolar, emerge below the lower border of the lateral pterygoid and run downward and forward; the third large branch, the auriculotemporal, passes posteriorly deep to the muscle and then continues deep to the neck of the mandible before turning upward just behind the temporomandibular joint.

The **medial pterygoid muscle** partly covers, and is partly covered by, the inferior fibers of

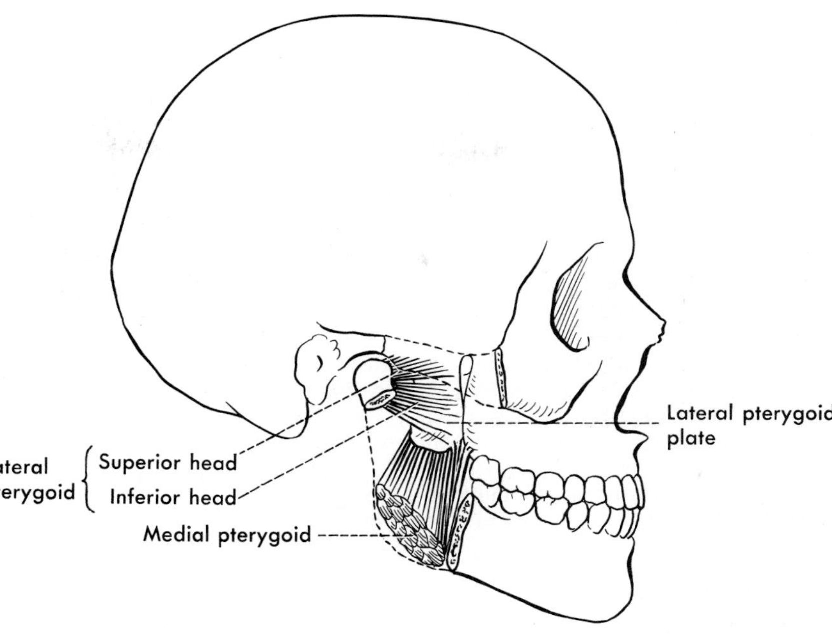


**FIG. 31-30.** The pterygoid muscles after removal of the temporal and masseter muscles, the zygomatic arch, and most of the posterior part of the mandible.

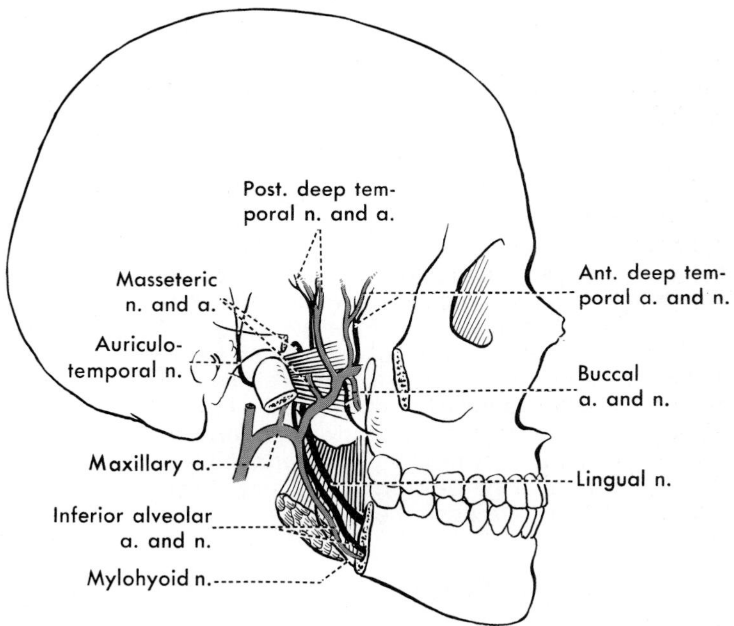


**FIG. 31-31.** Relations of nerves and vessels to the pterygoid muscles. The middle meningeal artery arises just above the leader to the maxillary artery.

the lateral pterygoid. A small part of the muscle arises from the tuber of the maxilla and overlaps the lowest fibers of the lateral pterygoid at their origin. The major part arises from the medial side of the lateral pterygoid plate and from the pyramidal process of the palatine bone (in the lower part of the anterior wall of the pterygoid fossa) and therefore appears at the lower border of the lateral pterygoid. The fibers of the medial pterygoid muscle run downward, posteriorly, and laterally to insert on the inner side of the ramus of the mandible below and behind the mylohyoid groove down to the lower border of the angle. It is innervated by a branch into its deep surface that leaves the mandibular nerve before that nerve gives off most of its other branches.

The masseter, temporal, and medial pterygoid muscles are powerful closers of the jaw, accounting for the strength of the bite. The temporal muscle abducts (deviates) the jaw to the same side, but the masseter and pterygoid abduct to the opposite side. With proper synchronization, therefore, these muscles can produce the grinding movement of chewing. The tongue and the buccinator muscle also aid mastication, but in a different way; the tongue positions the food on the teeth, and the buccinator muscle helps to maintain it there during chewing. The posterior fibers of the temporal muscle are the chief retractor of the mandible (pulling it back after it has been jutted forward) and also are primarily responsible for maintaining the resting position of closure of the mouth. The two heads of the lateral pterygoid have always been considered to have the same actions, but it has now been reported that they do not. Although both apparently aid in protracting (pulling forward) the mandible and thus deviating it to the opposite side when they act unilaterally, the superior head is said to become active only when the mouth is closed or being closed. The inferior head, however, pulls the condyle forward and downward and is active in opening the mouth—the function once assigned to both heads. When the mandible is protracted, the **sphenomandibular ligament** and a thickening of the deep cervical fascia called the **stylomandibular ligament** (extending from the styloid process to the posterior border of the angle of the jaw) have been said to act to keep the angle of the mandible from sliding as far forward as the condyles. This permits the jaw to rotate around a line that joins the centers of right and left rami: as the condyles go forward, the chin thus goes downward. The anterior belly of the digastric, and the geniohyoid and mylohyoid muscles, also help to open the mouth.

**The Masticator Fascial Space.** The loose connective tissue deep to the ramus of the mandible and around the lower part of the temporal muscle is described as forming the masticator fascial space. It is this space that contains the pterygoid muscles, most of the branches of the mandibular nerve, and the branches of the maxillary artery. Although the masticator space lies lateral to the lateral pharyngeal space, there is no communication between the two. As the *superficial lamina* of the **cervical** (deep cervical) **fascia** reaches the lower border of the mandible, a part of it attaches to the mandible and then continues upward superficial to the masseter and the parotid gland, attaching also to the zygomatic arch and then blending with the fascia over the temporal muscle. A second, *deep layer* of the cervical fascia splits off, however, and runs deep to both pterygoid muscles to reach the skull, thus separating the masticator and lateral pharyngeal spaces (Fig. 31-32). The masticator space is subdivided into a number of subsidiary spaces by fascia around the various muscles. It also extends forward superficially to include the **buccal fat pad** (overlying the buccinator muscle) and deeply to include the connective tissue and neurovascular structures in the pterygopalatine fossa.

In the masticator space the medial pterygoid muscle is crossed externally by the lingual and inferior alveolar nerves, which emerge below the lateral pterygoid muscle. The inferior alveolar nerve runs to the mandibular foramen on the medial side of the ramus. The lingual nerve passes farther forward, to the tongue. Inferior alveolar vessels, the artery a branch of the maxillary and the vein joining the pterygoid plexus, accompany the inferior alveolar nerve. The back part of the muscle is also crossed superficially by the **sphenomandibular ligament,** a thin band that runs from the sphenoid spine and the tympanosquamous fissure to the lingula and the adjacent inner surface of the mandible. The inferior alveolar nerve and vessels run between it and the mandible. The ligament represents remains of the cartilage of the mandibular arch (Meckel's cartilage) between the mandible and the skull.

**Temporomandibular Joint.** The temporomandibular articulation is fairly simple but differs from many joints in that there are two synovial cavities, separated by an articular disk. Although it is reinforced externally by a **lateral ligament,** the capsule generally is lax between the disk and the temporal bone and much stronger both medially and laterally between the disk and the mandible. The **articu-**

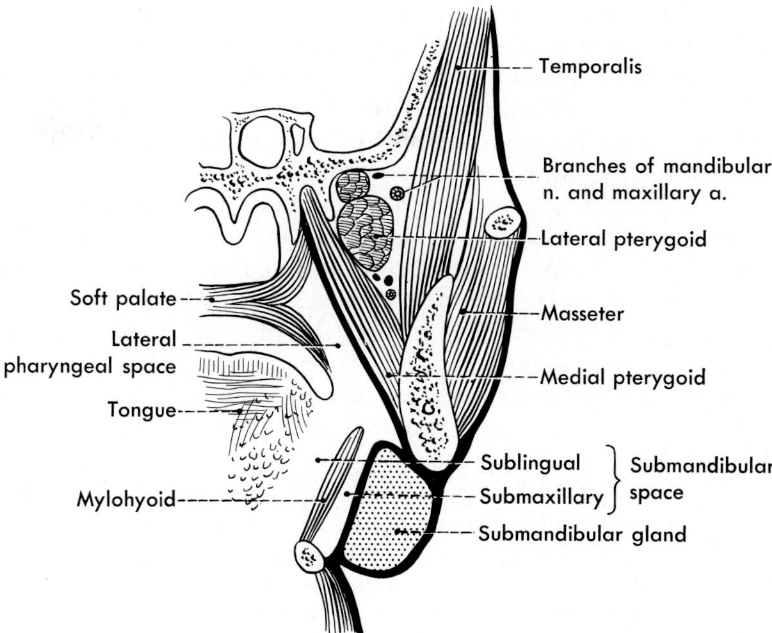


**FIG. 31-32.** Schema of half of a frontal section close to the angle of the jaw, showing the disposition of the superficial layer of the deep fascia (heavy lines) above the level of the hyoid bone. After splitting to surround the submandibular gland, it attaches to the mandible and thence is reflected up both external and internal to the muscles of mastication. The spaces among the muscles of mastication together form the masticator space; also shown here is the submandibular space, lying both above and below the mylohyoid muscle, and its posterior communication with the lateral pharyngeal space, lying lateral to the pharynx and in part medial to the medial pterygoid muscle.

**lar disk,** partly fibrocartilage but largely dense fibrous connective tissue, is loosely attached to the capsule posteriorly but strongly attached anteriorly where part of the tendon of the lateral pterygoid muscle blends with it. It moves forward with protraction of the mandible, but hinge movements—opening and closing—of the jaw take place between the disk and the mandible.

Because of the protraction when the mouth is opened widely, a great yawn or other excessively wide opening of the mouth will in some persons be accompanied by so much forward movement that the disk and condyle slide across the articular tubercle into the infratemporal fossa. The muscular pull is so altered by this dislocation that the jaw then is held open by the masseter and medial pterygoid. The dislocation must be reduced by forcing the condyle downward and then slipping it back into place.

The pressure produced by the closers of the jaw is normally borne almost entirely by the molar teeth and the synovial membrane of the joint, which extends in part over the articular surfaces and contains numerous nerves and vessels. Thus, malocclusion, or any factor that leads to spastic contraction of the muscles (*trismus*), may bring pressure to bear on the sensitive synovial membrane and cause pain. The spasm of the muscles may also be a cause of pain. Further, degenerative changes in the joint may be produced by pressure on the blood vessels of the membrane.

**Vessels and Nerves Related to Jaws.** The **inferior alveolar nerve and vessels**—the nerve from the mandibular, the artery from the maxillary—appear below the lateral pterygoid muscle and run downward across the medial pterygoid, passing between the mandible and the sphenomandibular ligament to reach the mandibular foramen. Just before they enter the foramen, each gives off a **mylohyoid branch** that runs downward on the medial side of the mandible, in the mylohyoid groove, and onto the lower surface of the mylohyoid muscle. Within the mandibular canal, nerve and vessels give off branches to the teeth until they reach the level of the mental foramen. Here a large branch of the nerve, and twigs of the artery and its accompanying vein, emerge as the **mental nerve and vessels.** The remaining parts of the nerves and vessels continue forward to supply teeth anterior to the mental foramen. The incisors, the canine, and the first premolar lie anterior to this foramen in about 50% of sides, for the most common location of the foramen is at the level of the second premolar tooth. Within the mandibular canal, the branches of the inferior alveolar nerve form an **inferior dental plexus,** which gives rise to both dental and gingival branches.

Some of the branches of the maxillary artery and most of those of the mandibular nerve already have been described.

The **maxillary artery** arises as one of the terminal branches of the external carotid artery at the posterior border of the ramus of the mandible, adjacent to or in the deep part of the parotid gland. It passes forward almost horizontally medial to the ramus of the mandible, at or a little below the level of the neck (see both Figs. 30-14 and 31-31). Its branches are particularly numerous (Fig. 31-33). Close to its origin, it gives off a small **deep auricular artery** that passes upward in the parotid gland to supply the temporomandibular joint and the external acoustic meatus, and it may give off a tiny **anterior tympanic artery** that helps supply the tympanic cavity and the inner surface of the tympanic membrane.

Farther forward, the artery gives rise to two larger branches, the middle meningeal artery and the inferior alveolar. The inferior alveolar and its terminal branch, the mental, already have been noted. The **middle meningeal artery** runs upward deep to the lateral pterygoid muscle, passes between the two roots of origin of the auriculotemporal nerve, and enters the skull through the foramen spinosum. It may give off, or there may be given off directly from the maxillary, an **accessory meningeal artery.** This artery gives off branches to extracranial structures adjacent to it and often enters the skull through the foramen ovale (by which the mandibular nerve leaves the skull) to reach the meninges around the trigeminal ganglion.

As the maxillary artery reaches the lower border of the lateral pterygoid muscle, it may run deep or superficial to it. At about this level

**FIG. 31-33.** Branches of the maxillary artery.

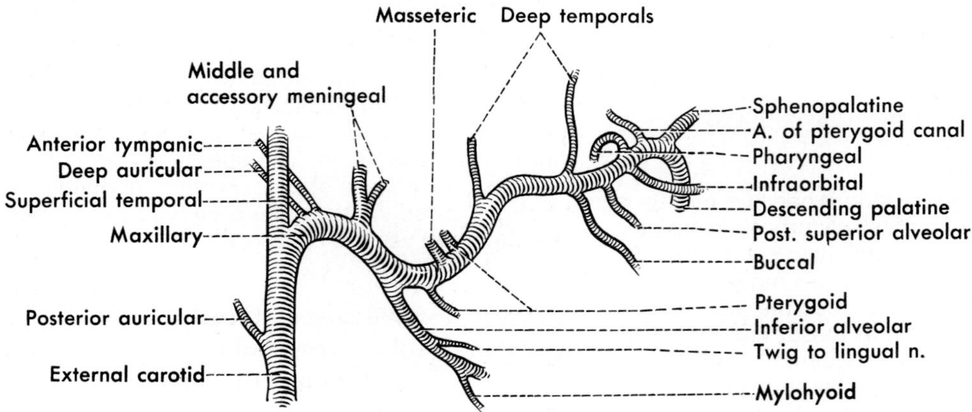

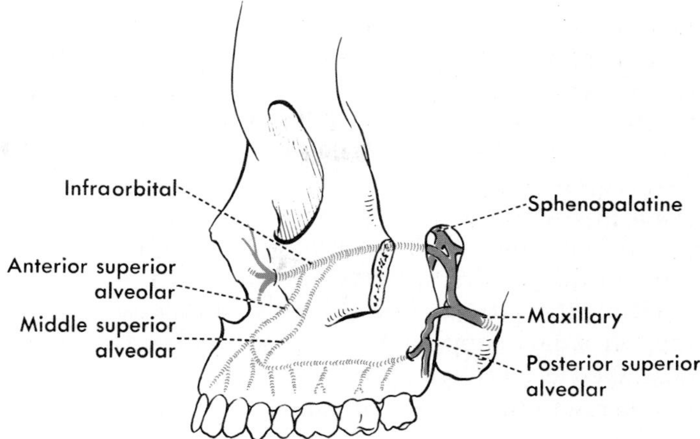


**FIG. 31-34.** Blood supply of the upper teeth.

it gives off the **masseteric artery,** and if it runs deep to the lateral pterygoid, it then emerges between the two heads of the muscle to give off the **deep temporal** and **buccal arteries.** There are also branches to the pterygoid muscles.

The last part of the maxillary artery passes deeply, through the pterygomaxillary fissure into the pterygopalatine fossa. As it does so, it gives off one or more **posterior superior alveolar arteries** that run downward on the posterior surface of the maxilla, accompanied by posterior superior alveolar nerves. These arteries enter foramina on the posterior surface of the maxilla and run downward and forward in the alveolar process, supplying dental branches to the more posterior upper teeth (Fig. 31-34). Other terminal branches of the maxillary artery can be identified only with difficulty, if at all, at this time. Most of them will be seen in deeper dissections. They are a tiny **artery of the pterygoid canal** and a **pharyngeal branch** (entering the palatovaginal canal), to the pharynx; a larger **infraorbital artery,** a part of which emerges at the infraorbital foramen but which supplies through **anterior superior alveolar** branches (one of which is sometimes called the middle) the anterior teeth of the upper jaw; a large **descending palatine** that appears on the palate as greater and lesser palatine arteries; and a **sphenopalatine** that passes straight medially from the pterygopalatine fossa into the nasal cavity.

The **pterygoid plexus** of veins is a venous plexus around the maxillary artery and both superficial and deep to the lateral pterygoid muscle. It is drained posteriorly by the maxillary vein, but also has other connections: in addition to receiving veins corresponding to the branches of the maxillary artery, it is connected to the facial vein by the deep facial vein, accompanying the buccal branch of the maxillary artery; to the cavernous sinus in the skull by veins that travel with the mandibular nerve through the foramen ovale; and to the middle meningeal veins inside the skull through veins that travel through the foramen spinosum with the middle meningeal artery.

As the **mandibular nerve** (see both Figs. 31-31 and 32-22) emerges from the foramen ovale, it consists of two roots, a large superficial (lateral) *sensory root* and a small, deeper-lying *motor root.* These join almost immediately outside the foramen ovale to form the mixed nerve. In this position, deep to the lateral pterygoid muscle, it gives off a tiny **meningeal branch** that runs backward to join the middle meningeal artery and accompany it into the skull, and it gives rise to the **medial pterygoid nerve.**

On the medial side of the mandibular nerve, sometimes closely applied to it, sometimes slightly farther out on a nerve stem to the medial pterygoid and tensor veli palatini muscles, is the **otic ganglion.** This ganglion is one of the four parasympathetic ganglia of the head. It receives its preganglionic fibers through a nerve that reaches it from posteriorly, the *lesser petrosal nerve,* a branch of the glossopharyngeal nerve. It gives off its postganglionic fibers into the auriculotemporal branch of the mandibular nerve (through its *ramus communicans with the auriculotemporal*); whence they pass into that nerve's parotid branches as the secretory pathway to the parotid gland. The **nerve of the tensor veli palatini,** the latter being the only muscle of the soft palate supplied by the mandibular, is described as a branch of the ganglion, as is the tiny **nerve to the tensor tympani** that parallels the lesser petrosal, proceeding posteriorly toward the middle ear. Actually, of course, these nerves contain no fibers from the otic ganglion, for they go to voluntary muscles; it is simply that they are so small that as they pass by the ganglion they seem to be branches of it.

After giving off these branches, the mandibular nerve usually divides into a smaller anterior part and a larger posterior part. From the *anterior part* arise the masseteric, the two deep temporal, the lateral pterygoid, and the buccal nerves, so that this part of the nerve is largely to muscles. From the *posterior part* arise the auriculotemporal, the inferior alveolar, and the lingual. The **masseteric nerve** usually passes between the upper head of the lateral pterygoid muscle and the infratemporal fossa and reaches the muscle through the mandibular notch. The **deep temporal nerves** (there may be more than two) may both turn upward deep to the temporal muscle by passing between the lateral pterygoid muscle and the infratemporal fossa, or the posterior nerve may take this course, and the anterior emerge between the two heads of the lateral pterygoid to cross the muscle. The **lateral pterygoid nerve** usually arises with the buccal and enters the deep surface of the muscle. The **buccal nerve** emerges between the two heads of the lateral pterygoid muscle and runs forward and downward to supply skin and mucous membrane of the cheek.

The **auriculotemporal nerve** arises by two roots, between which the middle meningeal artery passes. These roots come together behind the artery, and the nerve passes posteriorly just deep to the mandible, turning upward behind the temporomandibular joint and deep to the parotid gland and joining the superficial temporal vessels in their course above the parotid gland.

The inferior alveolar nerve already has been described. The **lingual nerve** at first almost parallels it, running downward above and in front of it but gradually diverging from it. After the lingual nerve leaves the lateral surface of the medial pterygoid muscle, it runs between the mandible and the uppermost muscle of the pharynx, the superior pharyngeal constrictor; its further course to the tongue is described with that organ.

Close to its origin, the lingual nerve receives the **chorda tympani,** a branch of the facial nerve that brings into it parasympathetic preganglionic fibers for the **submandibular ganglion** with which this nerve is connected, and taste fibers that this nerve conducts to the anterior two-thirds of the tongue. The chorda tympani emerges from the petrotympanic fissure, just behind and medial to the temporomandibular articulation. It runs downward and forward, passes deep to the middle meningeal artery and the inferior alveolar branch of the mandibular nerve, and joins the posterior border of the lingual nerve. It receives a twig (*ramus communicans with the chorda tympani*) from the otic ganglion.

Branches of the **maxillary nerve,** the **posterior superior alveolar nerves,** leave the nerve

**FIG. 31-35.** The superior alveolar nerves. A middle superior alveolar appears in Figure 32-21.

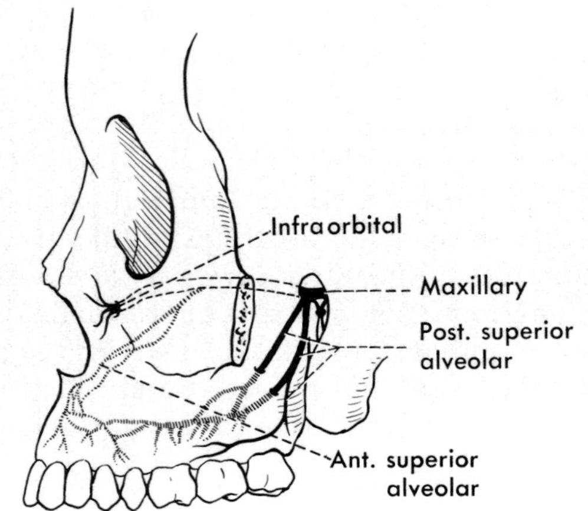

while it is in the pterygopalatine fossa and descend along the posterior surface of the maxilla with the corresponding vessels to run forward and supply the posterior teeth (Fig. 31-35). There usually are three of these nerves. One usually simply supplies the gums rather than entering the bone, and the other two enter the bone. Apparently, they never run any farther forward than the canine tooth and often supply only the molar teeth. There also are **anterior superior alveolar nerves** that are given off farther forward from the maxillary nerves. They run downward in the bone of the anterior wall of the maxilla to supply the more anterior teeth and usually spread back as far as the second premolar; apparently, they frequently cross the midline to help supply the medial incisor of the other side. There also is a less constant **middle superior alveolar nerve** to the premolars, so that these teeth apparently can be supplied by any combination of posterior, middle, and anterior nerves. The superior alveolar nerves form a **superior dental plexus** above the roots of the teeth; this gives rise to both dental and gingival branches. Because of their minute size, it is difficult to trace any of the superior alveolar nerves through the bone.

## Suprahyoid Structures

The suprahyoid muscles are the digastric, the stylohyoid, the mylohyoid, and the geniohyoid. They are classified as muscles of the neck, but the mylohyoid is also regarded as forming the floor of the mouth, since it fills much of the space between the two sides of the body of the mandible. The digastric and the stylohyoid lie below the mylohyoid (Fig. 31-36), but the geniohyoid lies above it.

A little connective tissue and some lymph nodes lie between the mylohyoid muscle and the fascia that covers it superficially (the superficial lamina of the cervical fascia, attached below to the hyoid bone and above to the mandible). The potential space here communicates around the posterior border of the mylohyoid muscle with the greater amount of loose connective tissue deep to the tongue (sometimes called the sublingual space), so the spaces above and below the mylohyoid are conveniently grouped together as the **submandibular fascial space** (Fig. 31-32).

The *mylohyoid muscle* and the superficial relations of the **submandibular gland** are described on page 834. The gland largely fills the triangle between the two bellies of the digastric and the lower border of the mandible and extends upward deep to the mandible. It lies partly on the lower surface of the mylohyoid and partly behind that muscle against the lateral surface of a muscle of the tongue, the hyoglossus. Its duct, and often a process of glandular tissue, run forward above the mylohyoid (Figs. 31-38 and 31-39). The hypoglossal nerve lies deep to it or below it,

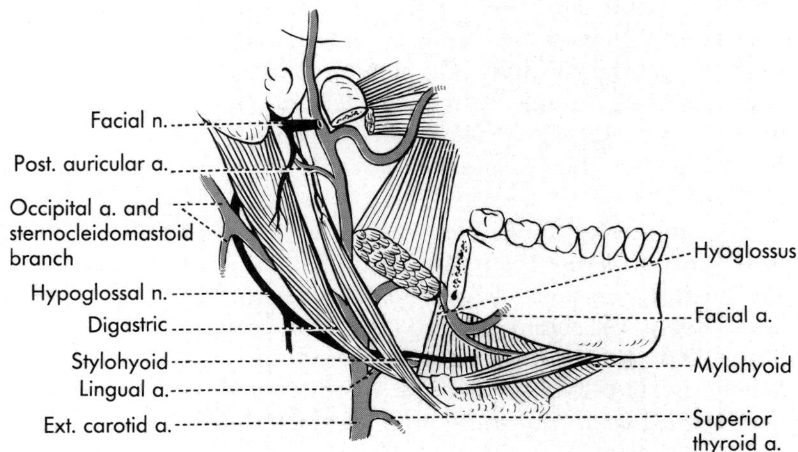


**FIG. 31-36.** The more superficial suprahyoid muscles after removal of the sternocleidomastoid muscle, the parotid and submandibular glands, and the posterior part of the mandible.

against the hyoglossus muscle and close to the hyoid bone, before disappearing above the mylohyoid. Posteriorly, the lingual artery lies deep to it for a short distance. The artery then goes deep to the hyoglossus muscle (Fig. 31-36).

The **digastric** and **stylohyoid** muscles can be seen more fully after the parotid gland and part of the mandible have been removed. The **posterior belly of the digastric,** innervated by a branch of the facial nerve, arises from the mastoid notch on the medial side of the mastoid process. The *tendon* connecting the two bellies passes through the stylohyoid and is held to the hyoid bone by fascia or tendon or both. The **anterior belly of the digastric,** proceeding forward and medially from the connecting tendon, attaches on the inner surface of the mandible just lateral to the midline. The mylohyoid nerve, a branch of the inferior alveolar that reaches the lower surface of the muscle, also supplies the anterior belly of the digastric. The **stylohyoid,** the most lateral muscle arising from the styloid process, is innervated by a branch of the facial nerve. It runs downward and forward, superficial to the external carotid, facial, and lingual arteries, splits to pass on both sides of the digastric tendon, and inserts into the hyoid bone.

After the mylohyoid muscle is reflected, the **geniohyoid** can be seen from below and can be examined more completely after various structures lying lateral to it and the tongue muscles have been reflected (Fig. 31-40). It extends from the lower part of the mental spine backward and downward to the body of the hyoid bone, the muscles of the two sides being adjacent at the midline. The geniohyoid is a suprahyoid strap muscle and is innervated by fibers from the 1st cervical nerve. These fibers join the hypoglossal nerve and travel with it.

The suprahyoid muscles raise the hyoid bone, and therefore the tongue and the floor of the mouth, or, acting with the infrahyoid ones, steady the hyoid bone so that it can provide a firm base for movements of the tongue. When the hyoid bone is fixed by the infrahyoid muscles, the suprahyoid ones can assist in opening the mouth. During swallowing, these muscles move the hyoid bone up. When high notes are sung, they also move the hyoid bone and, therefore, the larynx up, but when low notes are sung the infrahyoid muscles lower the hyoid bone and larynx.

## THE TONGUE

The attached part of the tongue, through which muscles reach it deep to the mucous membrane, is its **root,** and the upper surface is its **dorsum.** The dorsum bounds part of the oral cavity, but the most posterior part faces posteriorly and is part of the anterior wall of the epiglottic valleculae and adjacent parts of the pharynx. The major part of the tongue is the **body,** extending to the **apex** or tip. Anterior to the root, the body and tip have also an

**FIG. 31-37.** The dorsum of the tongue. The folds forming the lateral boundaries of the epiglottic valleculae are the lateral glossoepiglottic folds. (Rankin FW, Crisp NW: West J Surg Obstet Gynecol 40:105, 1932)

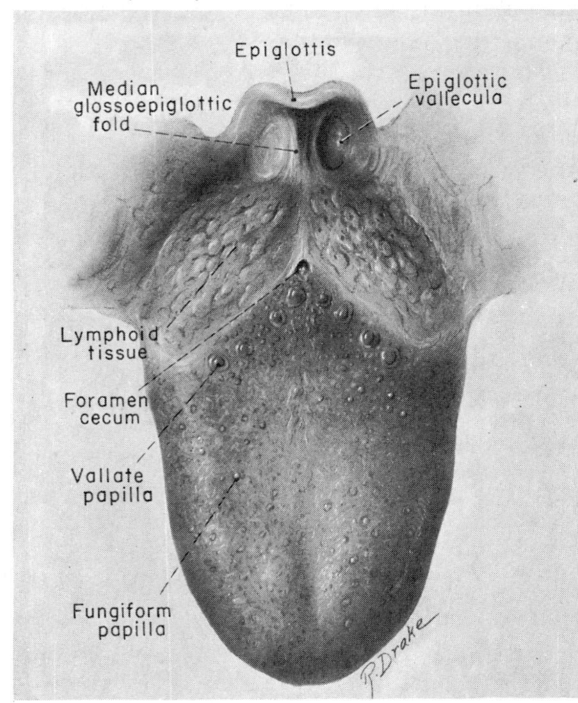

inferior surface, and it is to this surface that the frenulum linguae is attached. On the lower surface of the tongue on each side of the frenulum, the deep lingual vein usually is obvious through the mucosa.

The dorsum of the oral part of the tongue is velvety because it is covered by numerous small **papillae,** the *filiform papillae,* among which can be seen an occasional larger *fungiform papillae* (Fig. 31-37). The latter are provided with a few taste buds; the former have none. At the back end of the oral part, arranged in a V with the apex pointed posteriorly, are large *vallate papillae* that are studded with numerous taste buds. Each papilla is deeply encircled by a groove. Immediately behind the vallate papillae is a V-shaped groove, the **sulcus terminalis.** This usually is described as separating the oral from the pharyngeal part of the dorsum. At the apex of the sulcus lies the **foramen cecum,** a tiny blind pit that indicates the point of origin of the thryoglossal duct (p. 839). A shallow median sulcus that extends from the apex of the tongue toward the foramen cecum marks the attachment of the septum linguae. The pharyngeal part of the dorsum linguae is best examined in connection with the pharynx (Chap. 34).

The musculature of the tongue lies above the hyoid bone and largely medial to the mandible, and therefore it and the associated nerves and vessels can best be examined by removing most of the body of the mandible almost as far forward as the midline.

### Structures Lateral to the Root

The **sublingual gland,** the smallest of the three chief salivary glands, lies between the mucous membrane of the floor of the mouth above, the mylohyoid muscle below, the mandible laterally, and muscles of the tongue medially. It varies in size but usually is some 35 mm to 45 mm long. It is flattened mediolaterally, and its posterior end is slender, but it expands vertically at its anterior end (Fig. 31-38). From its upper surface, a series of small ducts, the **minor sublingual ducts,** empty through the mucous membrane of the floor of the mouth immediately above the gland. There are commonly about a dozen of these. A duct at the anterior end of the gland may open into the submandibular duct. Called the **major sublingual duct,** it may be lacking or no larger than the minor sublingual ducts.

The **duct of the submandibular gland,** frequently accompanied by a process of the gland, runs above the mylohyoid muscle to disappear deep to the sublingual gland. The lingual nerve, descending and running forward, crosses lateral to the submandibular duct and also disappears deep to the sublingual gland; and the hypoglossal nerve, lying at a lower level, likewise passes forward between the gland and the muscles of the tongue before dividing into branches and extending deeper into the tongue. When the sublingual gland is elevated or removed, the submandibular duct can be traced forward to its ending on the sublingual caruncle, and the course of the lingual nerve into the tongue can be seen: after crossing the submandibular duct laterally, the nerve passes under the duct and then turns up medial to it into the tongue, thus making a loop around the duct (Fig. 31-39).

The **lingual nerve** has already been described as leaving the mandibular nerve, re-

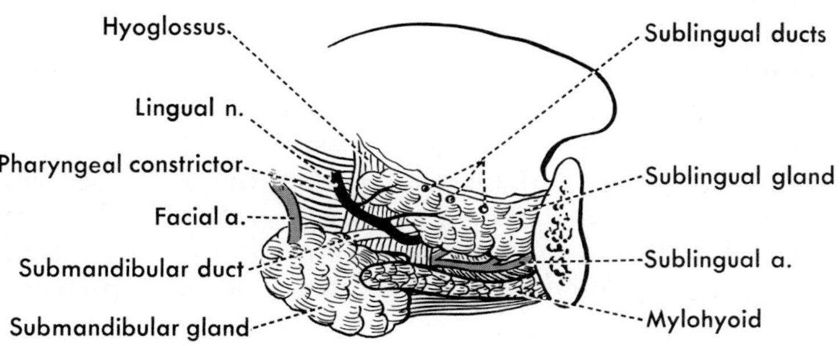


**FIG. 31-38.** Submandibular and sublingual glands from the lateral aspect after removal of the mandible.

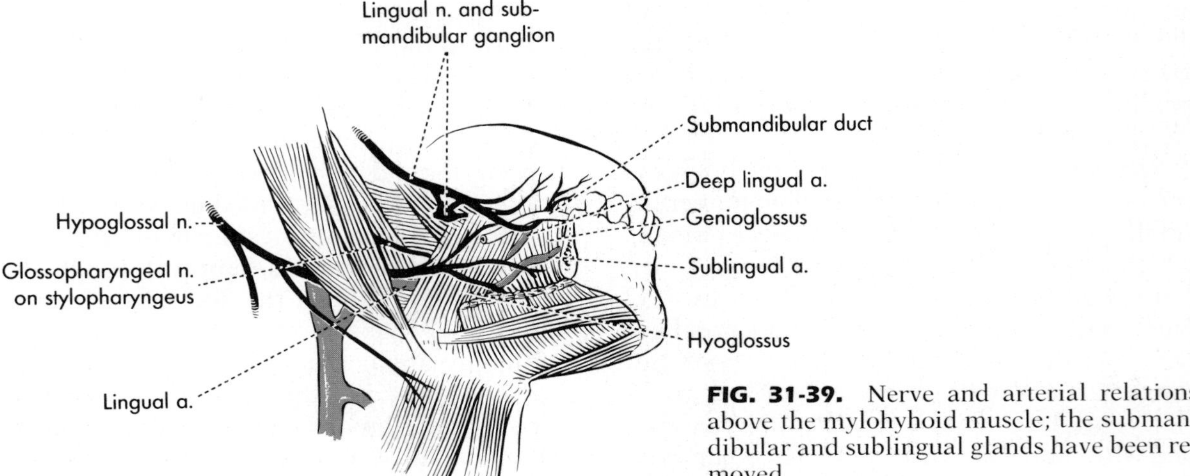


**FIG. 31-39.** Nerve and arterial relations above the mylohyoid muscle; the submandibular and sublingual glands have been removed.

ceiving the chorda tympani, and running downward and forward across the lateral surface of the medial pterygoid muscle. Beyond this, it runs adjacent to the superior constrictor muscle of the pharynx, medial to the mandible, and deep to the mucous membrane of the floor of the mouth lateral to the tongue. As it curves forward on the side of the tongue, it bears the **submandibular ganglion** (Fig. 31-39). This parasympathetic ganglion is connected to the lingual nerve by rami communicantes and gives off into the submandibular gland one or more nerves that are formed of postganglionic fibers from the ganglion. The preganglionic fibers are furnished by the *chorda tympani,* a branch of the facial. The submandibular ganglion also sends fibers to the lingual nerve to be distributed to the sublingual gland and small glands of the oral cavity. As the nerve runs forward deep to the sublingual gland, it gives off fibers to the floor of the mouth and to the sublingual gland and breaks up into branches to the mucous membrane of the anterior two-thirds of the tongue. Postganglionic, secretomotor fibers from the submandibular ganglion are distributed to glands; the fibers of trigeminal origin, representing a great majority of its fibers, furnish general sensation to the tongue—touch, pain, heat and cold—and the fibers of facial origin are for taste alone.

The **hypoglossal nerve** runs forward above the hyoid bone, crossing lateral to the hyoglossus muscle, supplying this muscle, and then branching as it continues forward deep to the sublingual gland. It supplies all the muscles of the tongue, and one branch, which contains cervical nerve fibers, supplies the geniohyoid muscle.

The largest part of the **lingual vein,** here called the *vena comitans of the hypoglossal nerve,* runs with the hypoglossal nerve, but the lingual artery runs deep to the hyoglossus muscle. Much of it turns up into the tongue, but the **sublingual artery** continues forward and appears anterior to the hyoglossus in the floor of the mouth, medial to the sublingual gland.

## Muscles

The tongue consists largely of muscle fibers. Its **intrinsic muscles** are so arranged that they can, by appropriate action, change the shape of the tongue—flatten it, curl it, point it, and the like. Superior and inferior *longitudinal muscles,* and *transverse* and *vertical muscles,* are named, although both of the latter are somewhat oblique. The muscles of the two sides are separated, except at the tip, by a fascial **septum linguae,** and this also almost completely separates the branches of the two lingual arteries; in midline sectioning of the tongue, therefore, there is little bleeding. The "transverse" and "vertical" muscles form the major part of the tongue. The longitudinal muscles are arranged in two relatively narrow bands, one on the dorsum of the tongue immediately beneath the mucous membrane and one toward the lower surface.

The **extrinsic muscles** of the tongue are three on each side: the hyoglossus (with its subdivision, the chondroglossus), the styloglossus, and the genioglossus (Fig. 31-40).

The **hyoglossus** is a flat, quadrilateral muscle that arises from the body and the greater cornu of the hyoid bone, partly above and partly behind the mylohyoid muscle, and passes upward and forward into the tongue. The lingual nerve, the submandibular duct and sublingual gland, and the hypoglossal nerve with its accompanying vein lie lateral to the muscle, and the lingual artery runs deep (medial) to it. The muscle ends in the tongue by becoming interlaced with the other muscle fibers there. The **chondroglossus** is a small bit of muscle that arises from the lesser cornu of the hyoid bone, passing into the tongue with the hyoglossus. It really is a slip of the latter muscle and is not always present.

The **styloglossus** arises from the anterior border of the styloid process and from the *stylohyoid ligament,* a slender band extending from the tip of the styloid process to the lesser cornu of the hyoid bone. The muscle runs forward, downward, and medially, to insert into the side of the tongue, mingling with fibers of the other muscles.

The **genioglossus** arises from the mental spine immediately above the geniohyoid. From this origin, it fans out as it runs backward: the lowest fibers insert into the body of the hyoid bone; the greater number of fibers run obliquely upward and posteriorly to blend with other muscles throughout the whole body of the tongue; and the most anterior fibers curve up and then forward to extend to the tip of the tongue.

The three extrinsic muscles, like the intrinsic muscles, are **innervated** by the hypoglossal nerve. They act with the intrinsic muscles in moving the tongue, and their chief actions are obvious from their positions and directions: the hyoglossus flattens the tongue, approximating the dorsum to the hyoid bone; the styloglossus pulls the tongue upward and backward; and the genioglossus pulls the body of the tongue forward and downward and the hyoid bone forward, thus helping to protrude the tongue (or, through action of its anterior fibers, can retract the tip of the protruded tongue). Since the tongue can be protruded in the midline only if the muscles of both sides work together, a hypoglossal nerve paralysis can be easily diagnosed by having the patient protrude the tongue as far as possible; the tongue is protruded only by the action of the muscles on the sound side and therefore deviates to the paralyzed side.

Another muscle, smaller than those just described, also attaches to the tongue. This, the **palatoglossus,** is described with the muscles of the soft palate. It enters the tongue anterior to and above the styloglossus, but it is improbable that it contributes to movements of the tongue.

## Lingual Vessels

After its origin from the external carotid, the **lingual artery** runs forward above the hyoid bone and deep to the hyoglossus muscle (Fig.

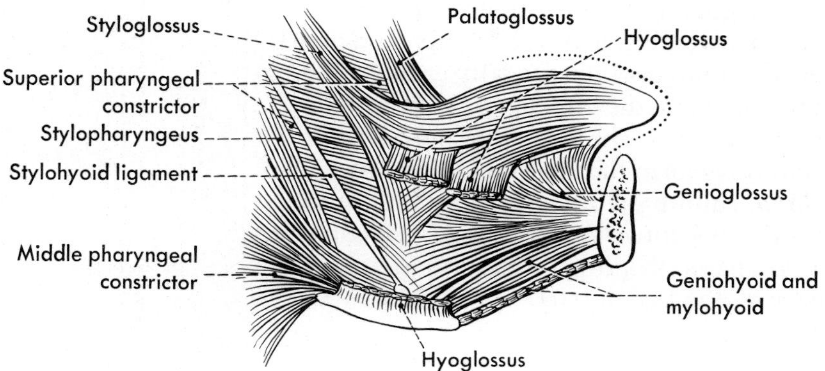


**FIG. 31-40.** Muscles of the tongue.

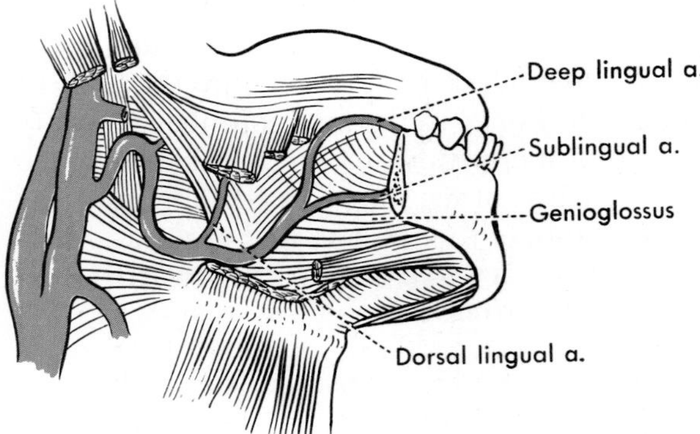


**FIG. 31-41.** The lingual artery.

31-41). It has a small suprahyoid branch that runs along the upper border of that bone, superficial to the hyoglossus muscle, but its other branches arise while it lies deep to that muscle.

In addition to a twig to the tonsillar region, the lingual artery gives off one or two **dorsalis linguae** rami, which pass upward between the hyoglossus and genioglossus muscles to the dorsum of the tongue. They supply the posterior part of the tongue, especially its pharyngeal part. The **sublingual artery** arises close to the anterior border of the hyoglossus muscle and continues forward between the mylohyoid and genioglossus muscles to supply these, the geniohyoid, and the sublingual gland. The remainder of the lingual artery, larger than the sublingual, is the **deep lingual** (*profunda linguae*). This runs forward on the lower surface of the tongue, between the inferior longitudinal muscle and the genioglossus, and when it reaches the free lower surface of the tongue, it is immediately adjacent to the mucous membrane. It is accompanied here by the **deep lingual vein,** usually easily recognizable through the mucous membrane in the living person. The deep lingual and sublingual veins form the **vena comitans nervi hypoglossi,** running with that nerve lateral to the hyoglossus muscle. The dorsal lingual veins accompany the lingual artery in its course deep to the hyoglossus.

The **lymphatics** of the tongue are of particular importance because of the relative frequency of carcinoma of that organ. Those from the *anterior part* of the tongue run downward among the lingual muscles, penetrate the mylohyoid, and end in part in *submental nodes* (see Fig. 30-26), in part in *submandibular nodes,* and in part, bypassing these nodes, in nodes of the *deep cervical chain* as low as the juguloomohyoid one (where the omohyoid muscle crosses the internal jugular vein). Lymphatics from a more *posterior part* of the tongue run behind the edge of the mylohyoid and join *deep cervical nodes,* and those from about the posterior third of the tongue penetrate the lateral pharyngeal wall to end in deep cervical nodes. Further, lymphatics from the *central portion* of the tongue, in contrast to those from the margin, drain both to the same and to the opposite side.

In consequence of this drainage, metastatic carcinoma from the tongue may be widely disseminated through the submental and submandibular regions and along the internal jugular vein. The operation designed to remove such metastatic lesions is called "radical neck dissection" or "block dissection of the neck" (p. 848).

## Nerves

The courses of the **lingual** and **hypoglossal** nerves to the tongue have already been described. The hypoglossal nerve supplies all the lingual muscles, extrinsic and intrinsic. The lingual nerve is distributed to the mucous membrane in front of the vallate papillae, thus to about the anterior two-thirds of the tongue (Fig. 31-42). The afferent fibers that the

**FIG. 31-42.** Sensory innervation of the tongue.

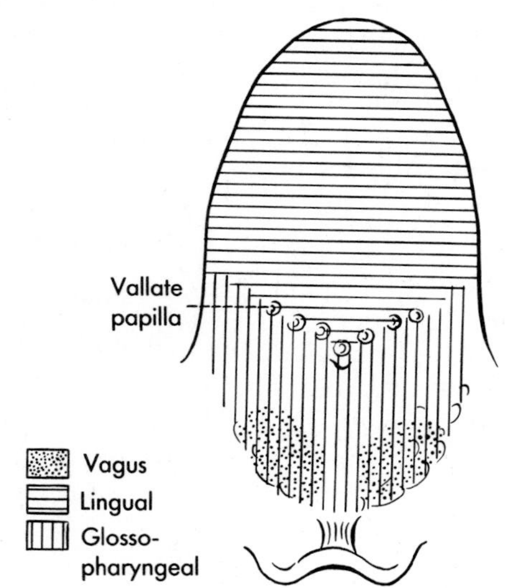

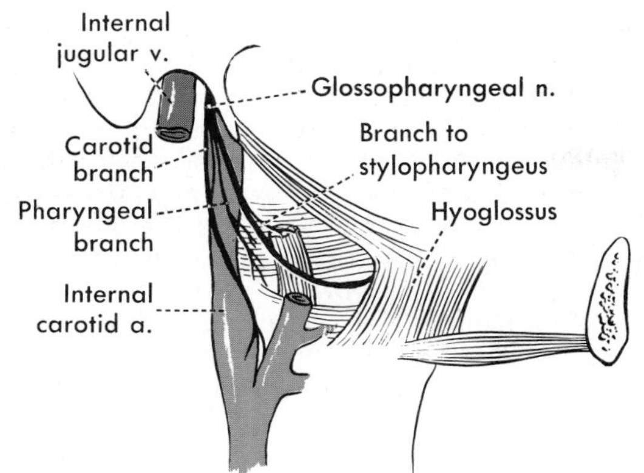


**FIG. 31-43.** The glossopharyngeal nerve after it leaves the jugular fossa.

lingual nerve receives from the chorda tympani branch of the facial nerve are distributed to the taste buds of the anterior two-thirds of the tongue and mediate only taste. The fibers in the lingual nerve that are of trigeminal origin mediate only general sensation from the mucous membrane.

The **glossopharyngeal nerve** (Fig. 31-43) supplies most of the mucous membrane of the posterior third of the tongue both with fibers for taste, ending especially in the vallate papillae, and with fibers of general sensation. This innervation is supplemented on the most posterior part of the tongue by twigs from the *vagus nerves.* The glossopharyngeal nerve runs downward and forward between the external and internal carotid arteries, giving off a carotid branch (*ramus sinus carotici*), which supplies both the carotid sinus and the carotid body, and one or more pharyngeal branches that join the vagus in forming the pharyngeal plexus. It then runs downward and forward, lying first below and then lateral to the most medial muscle (stylopharyngeus) that arises from the styloid process and giving off to that muscle its only *muscular branch.* In its course to the tongue, it runs along the lateral pharyngeal wall immediately lateral to the tonsil, giving off one or more *tonsillar branches* to the pharynx. It then turns upward into the posterior part of the tongue, coursing deep to the styloglossus and hyoglossus muscles, and breaks up into its terminal branches.

## RECOMMENDED READINGS


BENNETT GA, HUTCHINSON RC: Experimental studies on the movements of the mammalian tongue: II. The protrusion mechanism of the tongue (dog). Anat Rec 94:57, 1946

BERNICK S: Innervation of teeth and periodontium after enzymatic removal of collagenous elements. Oral Surg Oral Med Oral Path 10:323, 1957

BURCH JG: The cranial attachment of the sphenomandibular (tympanomandibular) ligament. Anat Rec 156:433, 1966

CASTELLI W: Vascular architecture of the human adult mandible. J Dent Res 42:786, 1963

CHANDLER SB, DEREZINSKI CF: The variations of the middle meningeal artery within the middle cranial fossa. Anat Rec 62:309, 1935

CORBIN KB, HARRISON F: The sensory innervation of the spinal accessory and tongue musculature in the Rhesus monkey. Brain 62:191, 1939

CUSHING H: The sensory distribution of the fifth cranial nerve. Bull Johns Hopkins Hosp 15:213, 1904

GAUGHRAN GRL: Mylohyoid boutonniere and sublingual bouton. J Anat 97:565, 1963

HARPMAN JA, WOOLLARD HH: The tendon of the lateral pterygoid muscle. J Anat 73:112, 1938

HARROWER G: Variations in the region of the foramen magnum. J Anat 57:178, 1923

HAYES ER, ELLIOTT R: Distribution of the taste buds on the tongue of the kitten, with particular reference to those innervated by the chorda tympani branch of the facial nerve. J Comp Neurol 76:227, 1942

HUNT JR: The sensory field of the facial nerve: A further contribution to the symptomatology of the geniculate ganglion. Brain 38:418, 1915

JAMIESON JK, DOBSON JF: The lymphatics of the tongue: With particular reference to the removal of lymphatic glands in cancer of the tongue. Br J Surg 8:80, 1920

LEWIS D, DANDY WE: The course of the nerve fibers transmitting sensation of taste. Arch Surg 21:249, 1930

LEWIS WH: The cartilaginous skull of a human embryo 21 millimeters in length. Contrib Embryol 9:299, 1920

MAES U: Infections of the "dangerous area" of the face. Surgery 2:789, 1937

MCCORMACK LJ, CAULDWELL EW, ANSON BJ: The surgical anatomy of the facial nerve with special reference to the parotid gland. Surg Gynecol Obstet 80:620, 1945

MCKENZIE J: The parotid gland in relation to the facial nerve. J Anat 82:183, 1948

MOFFETT BC JR, JOHNSON LC, MCCABE JB ET AL: Articular remodeling in the adult human temporomandibular joint. Am J Anat 115:119, 1964

SAUNDERS RL DE CH: Microradiographic studies of human adult and fetal dental pulp vessels. In: X-Ray Microscopy and Microradiography, p 561. New York, Academic Press, 1957

SICHER H: Structural and functional basis for disorders of the temporomandibular articulation. J Oral Surg 13:275, 1955

STAFNE EC, HOLLINSHEAD WH: Roentgenographic observations on the stylohyoid chain. Oral Surg Oral Med Oral Path 15:1195, 1962

STARKIE C, STEWARD D: The intramandibular course of the inferior dental nerve. J Anat 65:319, 1931

WARD GE, CANTRELL JR, ALLAN WB: The surgical treatment of lingual thyroid. Ann Surg 139:537, 1954

WEDDELL G, HARPMAN JA, LAMBLEY DG ET AL: The innervation of the musculature of the tongue. J Anat 74:255, 1940

# 32

# THE CRANIAL PARTS OF THE NERVOUS SYSTEM

The cranial parts of the nervous system consist of the brain, lodged within the cranial cavity; the 12 pairs of nerves that leave the cranial cavity; and the parts of the autonomic system, sympathetic and parasympathetic, that reach the head or arise there.

The interior of the bony cranium has already been described. It should be studied again in connection with the following description, where the importance of the various markings on the interior of the skull, and of the foramina at the base of the skull, will become clearer. The contents of the cranial cavity are the brain and its meninges, associated blood vessels, and parts of the cranial nerves.

## BRAIN AND MENINGES

After the calvaria has been cut by a circular incision, it can be pried loose from the underlying membrane, the dura mater, with no great difficulty, for it is only in the midline that there is a strong attachment between the dura and the overlying bone. The outer part of the dura actually is periosteum, but this periosteum is, like the pericranium, not very potent in forming bone. Within the dura mater, the brain is surrounded by the arachnoid and the pia. Both of these membranes are continuous with and structurally similar to the arachnoid and pia that surround the spinal cord.

Some basic knowledge of the anatomy of the brain is necessary to an understanding of the meningeal relationships and of the relationships of various parts of the brain to the bony walls of the cranial cavity. This basic knowledge is best obtained by study of a whole or half brain removed from the cranial cavity and properly hardened, and the following description is based on such brains. Since neuroanatomy, which is usually studied separately, is beyond the province of this book, this description is a general one designed merely to introduce the student to gross aspects of the subject.

## CEREBRAL HEMISPHERES

The great bulk of the brain, and all of that visible after removal of the skull cap, is formed by the two large paired cerebral hemispheres. Each hemisphere is described as having three surfaces: a convex one; a base, the slightly concave inferior surface; and a flat medial surface. The cleft between the medial surfaces of the two hemispheres is the **longitudinal cerebral fissure.** At the bottom of this fissure, the hemispheres are connected by a large bundle of transverse fibers, the **corpus callosum.**

**Lobes.** Each hemisphere is divided into four **lobes,** which are named from their chief bony relations (the frontal lobe next to the frontal bone, the temporal lobe next to the temporal bone, and so forth). The lobes are divided from each other by certain arbitrarily determined lines based upon the folds and grooves (*gyri* and *sulci*) of the hemisphere.

The **frontal lobe** extends from the *frontal pole* of the brain to the **central sulcus** (Figs. 32-1 and 32-2). Since it lies mostly in the anterior cranial fossa, its lower surface is shallowly concave to fit the orbital roof. Some distance behind the frontal pole a prominent fissure, the **lateral sulcus,** begins below the frontal lobe and runs backward and upward into the substance of the hemisphere. The part of the hemisphere below this sulcus is the **temporal lobe.** Its convex anterior end, the *temporal pole,* fits into the large lateral part of the middle cranial fossa. The back end of the cerebral hemisphere, the **occipital lobe,** is continuous with the temporal lobe and, like the posterior part of that lobe, lies on a shelf of dura, the *tentorium cerebelli,* that separates it from the posterior cranial fossa. The tentorium lies in the **transverse cerebral fissure,** below the cerebral hemispheres and above the cerebellum. The part of the cerebral hemisphere that lies between frontal, temporal, and occipital lobes is the **parietal lobe.** It is separated from the frontal lobe by the central sulcus. Similarly, it is largely separated from the temporal lobe by the lateral sulcus, but at the back end of this sulcus, parietal, temporal, and occipital lobes are confluent on the lateral

**FIG. 32-1.** Lobes, gyri, and sulci of the lateral surface of the cerebral hemisphere. (DeJong RN: The Neurological Examination, 2nd ed. New York, Paul B Hoeber, 1958)

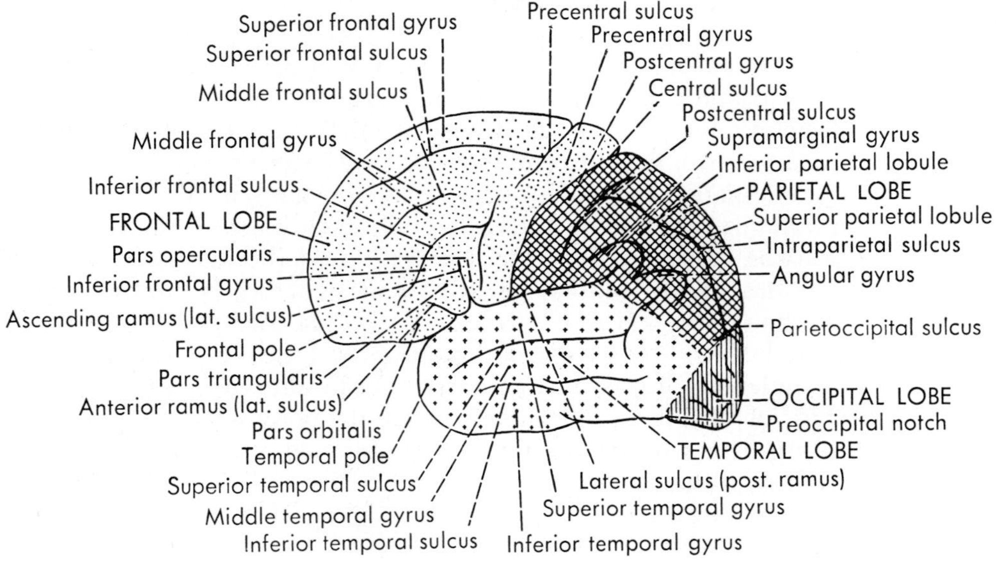

surface of the hemisphere. The boundary between the parietal and occipital lobes, and the occipital and the temporal lobes, is best seen on the medial surface of the hemisphere. Here the **parietooccipital sulcus** is its upper part, and the division is completed by a line drawn from that sulcus to the *preoccipital notch* on the base of the hemisphere.

Although the gyri and sulci vary somewhat in different cerebral hemispheres, many are constant enough to be named, and most of them usually are recognizable in all cerebral hemispheres. Some knowledge of them is necessary if important craniocerebral relations are to be grasped.

**Sulci and Gyri.** Of the sulci visible **on the convex surface** of the brain, the *central* and *lateral sulci* are the most important. Immediately in front of the central sulcus is the *precentral gyrus,* and the frontal lobe anterior to this gyrus is composed of superior, middle, and inferior *frontal gyri.* Behind the central sulcus and above the transverse one is the *postcentral gyrus,* part of the parietal lobe. This lobe also has superior and inferior *parietal lobules.* The *supramarginal* and *angular gyri,* at the level of transition between parietal, temporal, and occipital lobes, are regarded as parts of the parietal lobe. There are three gyri, superior, middle, and inferior *temporal,* on the lateral surface of the temporal lobe. No individual gyri are named on the lateral surface of the occipital lobe.

**On the medial side** of the hemisphere, the *gyrus cinguli* lies immediately above the corpus callosum. Above it, proceeding from anterior to posterior, are the medial or superior frontal gyrus, the *paracentral lobule* (a fusion of the precentral and postcentral gyri), and the *precuneus.* The *cuneus,* behind and below the precuneus, is separated from that by the parietooccipital sulcus and is bounded below by the *calcarine sulcus.* Below the calcarine sulcus is the *lingual gyrus.* Gyri of the medial and inferior parts of the temporal lobe include the *parahippocampal gyrus,* an anterior continuation of the lingual gyrus that bears the

**FIG. 32-2.** Lobes, gyri, and sulci of the medial surface of the cerebral hemisphere. The "limbic lobe," indicated by the oblique parallel lines, is not an anatomic lobe in the sense that the other lobes are, but is thought to constitute a functional unit; its parts frequently are assigned to the frontal, parietal, occipital, and temporal lobes, according to their positions. (DeJong RN: The Neurological Examination, 2nd ed. New York, Paul B Hoeber, 1958)

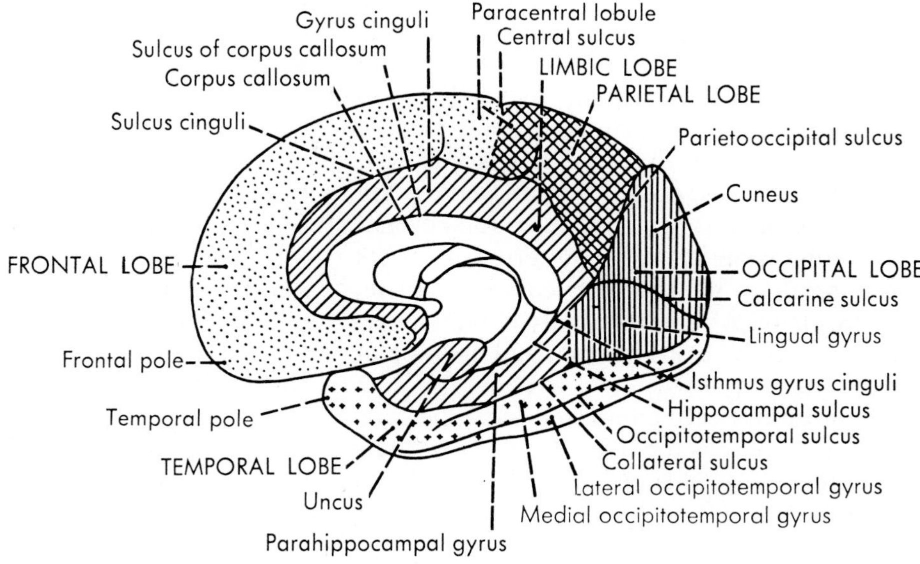

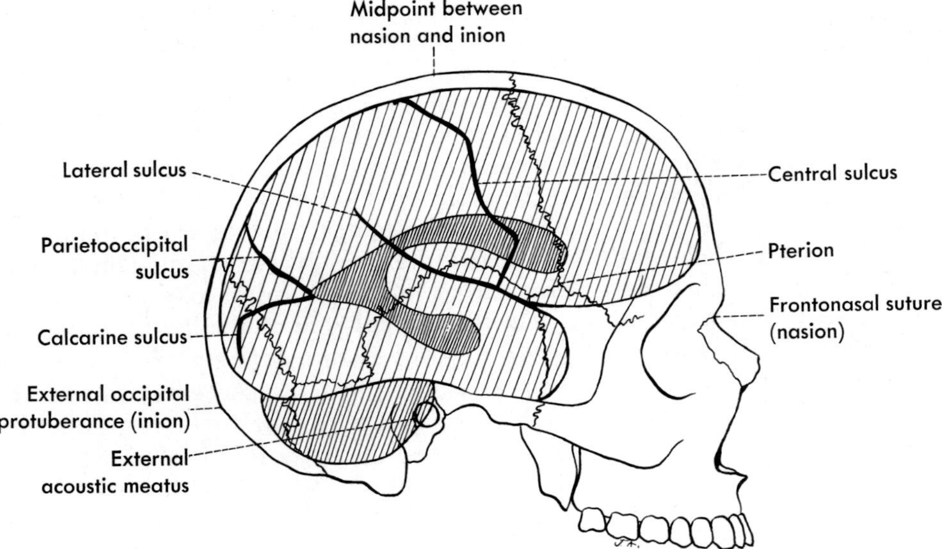


**FIG. 32-3.** Some relations of the brain to the skull. The lateral ventricle is indicated by the heavily shaded area in the cerebral hemisphere. The relations of the sulci to the skull are those found in a roentgenogram of a hemisected head in which the chief sulci had been marked with lead.

hooklike *uncus* anteriorly; the *medial occipitotemporal gyrus* lateral to the lingual and parahippocampal gyri; and the *lateral occipitotemporal gyrus,* continuous with the inferior temporal gyrus on the lateral surface of the hemisphere.

The approximate relations of the lobes and important sulci of the hemispheres to some of the sutures of the skull are shown in Figure 32-3. These relations vary somewhat among heads, both because of variations in skulls and because of variations in hemispheres. It is perhaps particularly important to note that the central sulcus, the division between frontal and parietal lobes, does not at all correspond to the coronal suture. It is not in the coronal plane, and even its most anterior part (its lower end) is behind the coronal suture, whereas its upper end is usually behind a vertical line erected from the external opening of the ear.

The cranial nerve connected with the cerebral hemisphere is the *olfactory,* or 1st cranial nerve. It consists of such tiny filaments that it is not visible after the brain has been removed. These filaments end in the **olfactory bulb,** an elongated mass of gray matter lying in a groove on the lower surface of the frontal lobe and connected to the base of the hemisphere by the slender olfactory tract. The olfactory bulb lies in the anterior cranial fossa on the cribriform plate. Apertures in the cribriform plate transmit the olfactory nerve.

**Interior.** The cerebral hemispheres are outgrowths from the upper end of the hollow neural tube (p. 39) and are themselves hollow: each contains a cavity, called a **lateral ventricle** (Fig. 32-7), which has its *anterior horn* in the frontal lobe, continues back through the frontal and parietal lobes as the *body,* and sends a *posterior horn* toward the occipital lobe and an *inferior horn* curving downward and forward into the temporal lobe. The lateral ventricles do not communicate with each other, but each opens into a centrally placed **third ventricle** that lies between the right and left thalamus.

Most of the thickness of the wall of a cerebral hemisphere is composed of white matter (nerve fibers), but on the inner surface of the wall, close to the lateral ventricle, are some deeply placed masses of gray matter (nerve cells), the largest of which forms the **corpus striatum.** The corpus striatum consists of the **caudate** and the **lenticular** (lentiform) **nucleus** and is traversed by a large fiber bundle, the **internal capsule** (Fig. 32-4).

**Cortex.** In contrast to the spinal cord, many nerve cells of the cerebral hemisphere have migrated to the outer surface where they completely cover the white matter with a layer of gray matter. This gray matter is the *pallium* or **cerebral cortex,** and its thickness varies from a little more than 1 mm to as much as 4.5 mm. The cortex covers the outer surfaces of the gyri and dips down into the sulci. About two-thirds of the cortex is in the sulci.

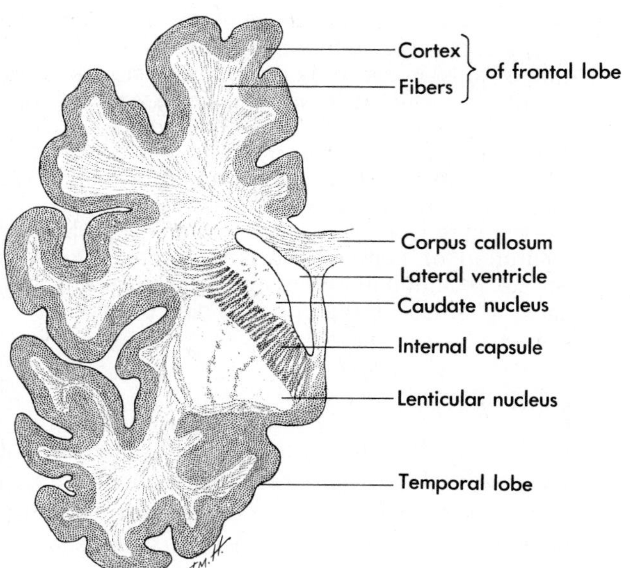


**FIG. 32-4.** A frontal section through the front part of an entire cerebral hemisphere of a child.

**Functional Aspects.** The fibers in the walls of the cerebral hemispheres connect different parts of the same hemisphere with each other, including not only adjacent gyri but also, for instance, the frontal and temporal lobes. They connect the two hemispheres, by way of the corpus callosum, and they connect the cerebral hemispheres to lower parts of the brain and to the spinal cord. These fibers include both fibers coming into a hemisphere and ones originating in the hemisphere and then leaving it to run downward.

The corpus striatum and certain associated masses, often grouped together as the **basal ganglia,** are reflex centers concerned particularly with voluntary muscle. Injury to them may result in changes in muscle tone and in the appearance of unwanted repetitive movements that interfere with the carrying out of desired movements.

Different parts of the cerebral cortex have specific functions, and within an area of given function, there may be smaller areas that are concerned with different parts of the body. Thus, the precentral gyrus is largely concerned with initiating voluntary movement (hence is known as the **motor area**). The adjacent postcentral gyrus is mostly concerned with the recognition of sensations from skin, muscles, and joints and, therefore, is known as the **sensory area.** In both motor and sensory areas, the lower limb is represented toward the top and onto the medial side of the hemisphere (paracentral lobule), the upper limb is represented lower on the lateral surface, and the face, tongue, and larynx are represented on the lowest part. The body is thus represented essentially upside down. It is necessary to remember also that each cortex controls primarily the opposite side of the body, for most ascending impulses cross before they reach the hemisphere, and most descending ones cross after they leave it.

Other important sensory areas include that of sight, centered around the calcarine sulcus, and that of hearing, below the lateral sulcus on the upper surface of the temporal lobe. More posterior parts of the parietal lobe have to do with particularly complex judgments and syntheses, such as recognition of an object placed in the hand through its shape, weight, and texture, or understanding language spoken or written. In one hemisphere, usually the left in right-handed persons, there is, at the back end of the inferior frontal gyrus, a "motor speech area" whose function is to coordinate the muscles used in speaking. When it is injured, there is no paralysis of muscles, yet the patient says words only with great difficulty (ataxic or motor aphasia).

Parts of the frontal and parietal lobes assist the motor area in controlling voluntary movement, but the larger anterior part of the frontal lobe (called prefrontal cortex), much of the temporal lobe, and much of the medial surface of the hemisphere anterior to the visual area have to do largely with mental activity rather than movement or sensation.

Because of the specialization in the brain, injuries to it, resulting, for instance, from wounds, tumors, or occlusion of blood vessels, may produce a great variety of symptoms. Sometimes it is possible to localize the area of injury accurately from the symptoms, but in other cases the lesion is diffuse or involves a part of the brain the function of which is not well understood. Since functional areas of the cerebral cortex are relatively large, lesions of it may affect only one function primarily, or the function of only a part of the body. For instance, there may be a partial loss of sight or paralysis of one limb only. However, many of the fibers going to and from the broad expanse of cortex are collected in the **internal capsule,** and damage to this bundle may affect many motor and sensory activities of a hemisphere. Damage to the internal capsule from infarction or hemorrhage is a common form of stroke, resulting in loss of or decrease in sensation and movement of the opposite side of the body.

## CEREBELLUM

Immediately below the posterior portions of the cerebral hemispheres, and separated from them, when *in situ,* by the tentorium cerebelli, is the cerebellum (little brain). It consists of paired lateral parts, the **cerebellar hemispheres,** and a smaller midline portion, the **vermis.** The vermis is continuous with both hemispheres but is so much smaller that it occupies a notch between them posteriorly and inferiorly. The cerebellum lies in the posterior cranial fossa. The convexity of the hemisphere fits into the concavity of the occipital bone and the adjacent petrous part of the temporal bone.

The cerebellum resembles the cerebral hemisphere in having a cortex that covers a much larger center of white matter. It also has certain masses of gray matter, the **cerebellar nuclei,** embedded deeply in it. The **cerebellar cortex** presents folia and fissures, names that correspond to the sulci and gyri of the cerebral hemisphere. Unlike the cerebral hemisphere, the cerebellum is solid rather than hollow. It forms, however, a part of the roof of the cavity in the brain at this level (called the fourth ventricle).

No cranial nerve is directly attached to the cerebellum, although the 8th (vestibulocochlear) nerve, which is attached to the brain close to the cerebellum, has intimate connections with this part.

**FIG. 32-5.** The medial side of a hemisected brain. The cerebral hemispheres have been separated from each other along the longitudinal cerebral fissure; the corpus callosum, brain stem, and cerebellum have been cut. (Henle J: Handbuch der systematischen Anatomie des Menschen, vol 3, pt 2. Braunschweig, Vieweg und Sohn, 1868)

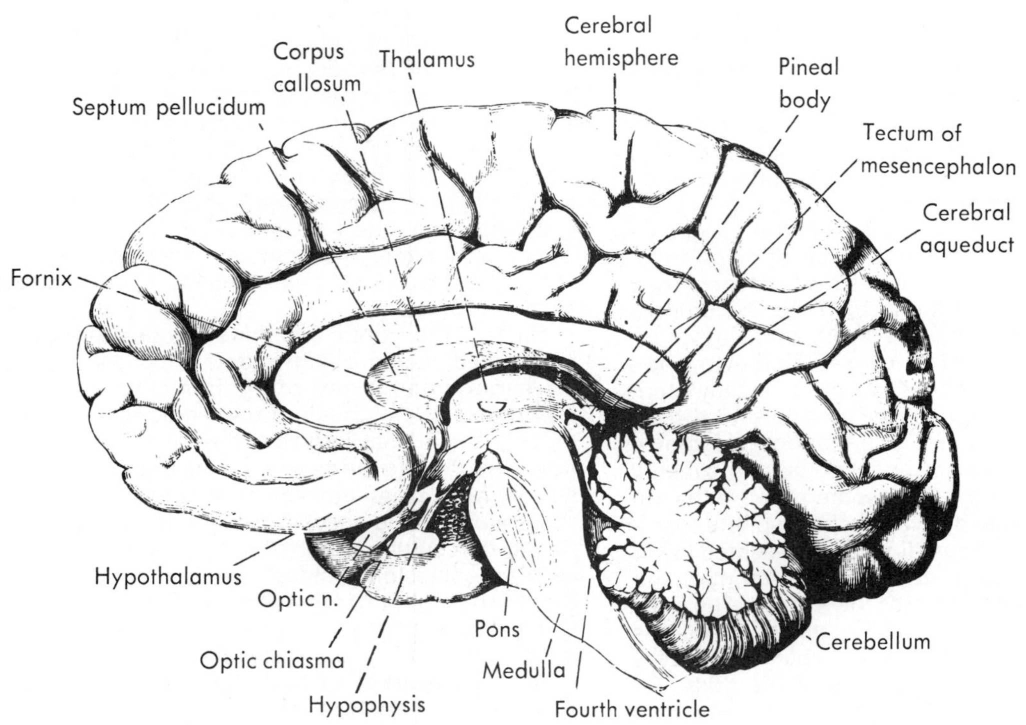

The cerebellum has to do, in part, with the maintenance of balance, and injuries to it of the type sometimes seen in children from tumors may primarily affect balance. However, the largest portion of the cerebellum, in conjunction with the cerebral hemispheres and other parts of the brain, is concerned with helping to control the contraction of voluntary muscles and especially with controlling the time and strength of contraction of various muscles so that a desired movement is carried out smoothly and accurately. Damage to the cerebellum, therefore, may result in disturbances of voluntary movement.

## BRAIN STEM

The remaining parts of the brain are so hidden by the cerebral hemispheres and cerebellum that, unless dissected specimens are available, they must be examined largely from below and in sagittal sections of the brain (Figs. 32-5 and 32-6). These parts are collectively called the brain stem and contain the third and fourth ventricles and the aqueduct, cavities derived from that of the neural tube. As in the spinal cord, the gray matter of the brain stem is, for the most part, placed centrally, close to the ventricles and aqueduct, whereas the white matter lies largely on the exterior. There is, however, more mixing of gray and white matter in the brain stem than there is in the spinal cord, and the gray matter, instead of forming continuous masses similar to the gray columns of the spinal cord,

**FIG. 32-6.** View of the base of the brain and the ventral surface of the brain stem and an upper part of the spinal cord. (Henle J: Handbuch der systematischen Anatomie des Menschen, vol 3, pt 2. Braunschweig, Vieweg und Sohn, 1868)

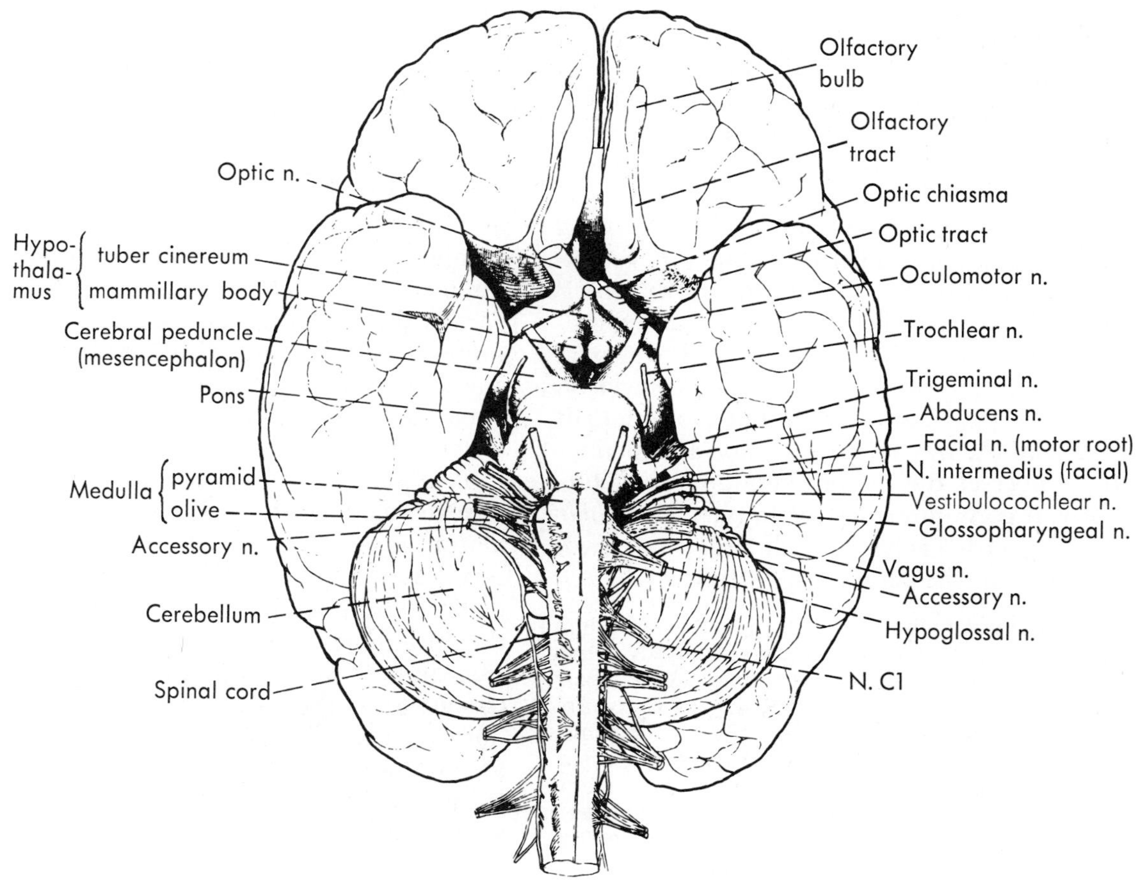

is separated into numerous smaller masses known as **nuclei.** Many of the nuclei are motor or sensory nuclei of the cranial nerves.

The uppermost part of the brain stem is the **diencephalon.** Only a little of this can be seen from the ventral surface, for the mesencephalon and cerebral hemispheres are close together here. The part visible ventrally is the **hypothalamus,** the region above and posterior to the **optic chiasma** (a prominent cross band of fibers composed of some of the fibers of the optic or 2nd cranial nerve that cross to the opposite side). The nerves coming into the optic chiasma are the **optic nerves.** Leaving the chiasma and running posterolaterally are the two **optic tracts.** They also are part of the diencephalon. Behind the optic chiasma is a grayish protuberance, the **tuber cinereum,** that gives rise to a funnellike **infundibulum.** The **hypophysis,** which lies in the concavity of the sella turcica, normally is attached to the infundibulum. The little rounded eminences (*mammillary bodies*) behind the infundibulum also belong to the hypothalamus.

The **thalamus** is the largest part of the diencephalon, but much of it is buried in the cerebral hemispheres and cannot be seen, except in a hemisected brain. The thick right and left thalami are separated from each other by a slitlike cavity, the **third ventricle.** The ventricle's roof is thin, like that of the fourth ventricle, and usually has been torn away in prepared specimens. Its floor and the lower part of its walls are the hypothalamus. Posteriorly, the third ventricle is connected to the fourth ventricle by the narrow cerebral aqueduct that traverses the mesencephalon. Anterolaterally, it communicates by an *interventricular foramen* with each of the lateral (first two) ventricles in the cerebral hemispheres.

The diencephalon lies anterior to the tentorium cerebelli, and the hypothalamus lies directly above the narrow midline part of the middle cranial fossa, the sella turcica, that contains the hypophysis. Because the hypothalamus and the optic chiasma, nerves, and tracts are so close to the hypophysis, tumors of this gland may affect the visual system, causing partial blindness (commonly in both eyes), or may affect the functions of the hypothalamus. Operations on tumors of the gland are particularly delicate because of the vascular relations here, the presence of the optic system, and the hypothalamus. The hypothalamus is concerned with many basic functions of the body, including temperature regulation, appetite, and the control of the hypophysis, the "master gland" of the endocrine system (p. 130). Operations in its region are, therefore, sometimes attended by alarming or even fatal disturbances of these basic mechanisms—for instance, a transient period of extremely high fever often follows removal of tumors here.

The only nerves attaching to the diencephalon are the optic nerves, already noted. These, originating in the eyeballs, enter the cranial cavity through the optic canals, just at the anterior end of the sella turcica.

The diencephalon is followed by the **midbrain** or **mesencephalon.** On its ventral surface are two heavy fiber bundles, the **cerebral peduncles,** that converge as they run inferiorly and disappear into the pons. On the dorsal surface of the midbrain are two swellings on each side, the superior and inferior **colliculi** (little hills). Running through it is a cavity, usually somewhat smaller than the lead of a pencil, that is the **cerebral aqueduct,** the mesencephalic part of the ventricular system.

The mesencephalon lies at the junction of the posterior and middle cranial fossae, partly in both. Since the central part of the middle cranial fossa above the concavity of the sella turcica is almost horizontal, and the central part of the posterior cranial fossa, the clivus, slopes markedly downward, the brain stem is angulated here; thus, the mesencephalon is longer dorsally, where the colliculi are, than it is ventrally at the acute angle of the bend.

Two pairs of cranial nerves arise from the mesencephalon. The **3rd cranial nerve (oculomotor)** leaves it close to the medial edge of the cerebral peduncle, at the upper border of the pons. It runs laterally and forward to enter the dura mater just anterolateral to the posterior clinoid process. The **4th cranial nerve (trochlear)** is the only one that attaches to the dorsal aspect of the brain. It leaves the brain at the lower border of the inferior colliculus. It runs at first laterally and then bends forward around the colliculus, passing through the posterior cranial fossa immediately below the tentorium cerebelli to enter the lower surface of this dural shelf just behind the dorsum sellae.

The **pons** or **metencephalon** succeeds the mesencephalon. All that is visible from the surface is a band of fibers, of considerable superoinferior width when viewed anteriorly but narrower and more cordlike when viewed from the side. This band looks like a bridge between the two cerebellar hemispheres and gives the part its name (pons meaning bridge). Actually, the pontine fibers do not connect the two cerebellar hemispheres; rather, each side of the pons is part of the connection between the opposite cerebral hemisphere and the cerebellar hemisphere into which it can be traced. Dorsal to the transverse pontine fibers is a mixture of nuclei and fibers that makes this part of the pons essentially similar in structure to the upper part of the medulla. The upper part of the fourth ventricle lies in the pons, largely covered by the cerebellum.

The pons lies in the most anterior part of the posterior cranial fossa, against the upper part of the clivus and the posterior wall of the dorsum sellae. Only one cranial nerve, the **5th** or **trigeminal,** attaches to it. This leaves laterally, through the narrower lateral part, and has two roots: a large *sensory root* and a small *motor root.* The two roots run anterolaterally from the posterior into the middle cranial fossa by passing across a notch on the upper border of the anterior end of the petrous part of the temporal bone. This notch is converted into a foramen by the attachment of the tentorium cerebelli above it. The *trigeminal ganglion* lies in the middle cranial fossa, and its branches leave the skull by foramina in this fossa.

The last part of the brain stem, obviously an upward continuation from the spinal cord, is the **medulla oblongata (myelencephalon).** Although the upper end of the medulla differs markedly in both size and structure from the cord, the lower end is very much like the cord, and there is no sharp transition between the two. The medulla usually is described as beginning at the level of the foramen magnum or at the uppermost rootlet of the 1st spinal nerve (these being at about the same level). The medulla expands, particularly laterally, as it is traced upward. Its upper border is the level at which it disappears deep to a heavy bundle of transversely placed fibers, the pons. The lowermost end of the medulla contains a tiny central canal, the upward continuation of the central canal of the spinal cord. As the medulla expands laterally, however, the central canal widens rapidly into the **fourth ventricle,** which is eccentrically placed in the medulla. Instead of lying centrally, it lies dorsally, and much of its roof is a thin membrane that stretches between the thicker lateral walls. It is frequently torn away from the specimen so that one can look into the fourth ventricle. The dorsal surface of the medulla lies against the lower and anterior surface of the cerebellum, which forms a major part of the roof of the fourth ventricle. At its upper end, above the cerebellum, the fourth ventricle tapers again to a small canal, the cerebral aqueduct.

The medulla lies in the posterior cranial fossa, as does the cerebellum, with its lower surface against the clivus. The last seven cranial nerves are attached to the medulla. Of these, the 6th nerve arises ventrally (anteriorly), at the caudal border of the pons; the 7th and 8th arise laterally just caudal to the lateral part of the pons and partly under cover of the cerebellar hemisphere; the 9th, 10th, and a part of the 11th arise by a series of small filaments from the lateral side of the medulla; and the 12th leaves the medulla anteriorly or ventrally, as does the 6th, arising by a number of rootlets that lie in a groove just medial to a smooth ovoid enlargement (the olive) that is present anterolaterally on the medulla.

The **6th (abducens) nerve** has a short course downward and forward from the medulla to the dura over the clivus. After entering the dura, it runs forward in it as far as the superior orbital fissure, through which it enters the orbit.

The **7th (facial) and 8th (vestibulocochlear) nerves** run laterally together from their origin, and both enter the internal acoustic meatus, on the posterior surface of the petrous part of the temporal bone.

The **9th (glossopharyngeal), 10th (vagus),** and **11th (cranial accessory) nerves** run laterally and slightly forward to enter the anterior and middle compartments in the jugular foramen. The cranial roots of the accessory nerve are joined by spinal roots that leave the spinal cord and ascend alongside it to enter through the foramen magnum.

The last cranial nerve, the **12th (hypoglossal),** runs laterally to the hypoglossal canal.

### VENTRICLES AND CEREBROSPINAL FLUID

The cerebrospinal fluid fills the subarachnoid space and, therefore, surrounds the brain and spinal cord. Most of this fluid is formed in the ventricles of the brain where plexuses of blood vessels, the choroid plexuses, project into the ventricles by invaginating the medial walls of the lateral ventricles and the roofs of the third and fourth ventricles. The fluid formed within the lateral ventricles passes through the interventricular foramina into the third ventricle (Fig. 32-7), which adds additional fluid. It then passes through the narrow aqueduct into the fourth ventricle, where more fluid is added, and escapes from the fourth ventricle into the subarachnoid space through paired *lateral apertures* on the sides of the roof, which open close to the cerebellum and the 9th nerve, and through a *median aperture* in the caudal part of the roof. Cerebrospinal fluid continues to be formed in the lateral ventricles, sometimes at an unreduced rate, after complete ablation of the choroid plexuses. The source of its formation under these circumstances is unknown.

*Meningitis* (inflammation of the membranes of the brain) may result in the formation of scar tissue that closes the foramina of the fourth ventricle, or a tumor in the mesencephalon may occlude the aqueduct, preventing the escape of cerebrospinal fluid into the subarachnoid space. The result of this occlusion is *hydrocephalus.* The pressure in hydrocephalus dilates the ventricles of the brain and exerts pressure upon the cerebral hemispheres, which are damaged in an adult by being squeezed between the fluid inside and the unyielding skull outside. In infants and young children, in whom the bones of the cranium are not yet joined, the internal pressure enlarges both the brain and the skull, widening the sutures and the fontanels. When the condition is not relieved, the cranium may become exceedingly large and the brain substance so stretched that it may be no more than a thin membrane. Such extreme conditions necessarily result in idiocy. It is, however, possible in some cases to remove the obstruction by operation and in others to devise a drainage system that allows escape of the fluid.

Air can be injected into one or both lateral ventricles to demonstrate them for roentgenologic examination (Fig. 32-8). Proper views of an air ventriculogram may show distortion or displacement of a ventricle (Fig. 32-9) and thus aid in diagnosing or localizing a tumor within the cranial cavity.

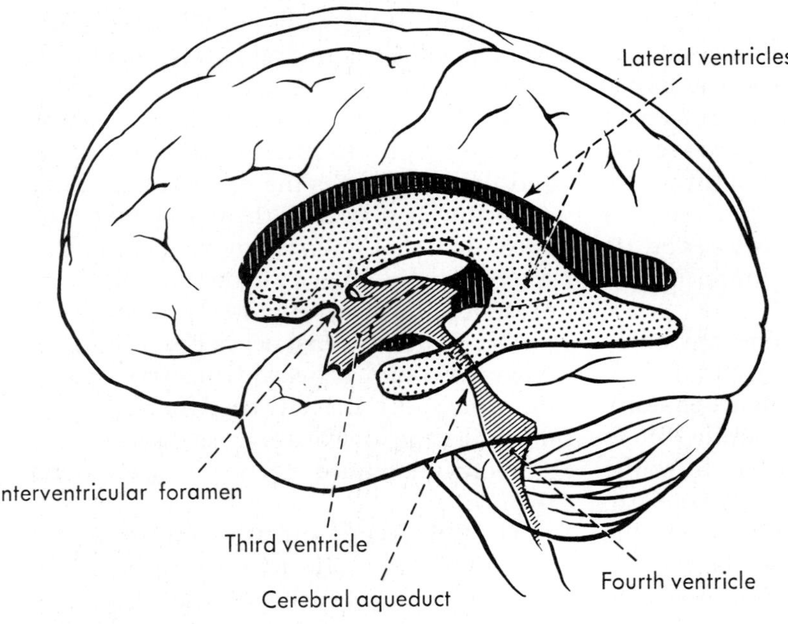


**FIG. 32-7.** Diagram of the ventricular system of the brain. (Rushton JD: Neurology for Nurses. Minneapolis, Burgess Publishing Co, copyright Mayo Association, 1959)

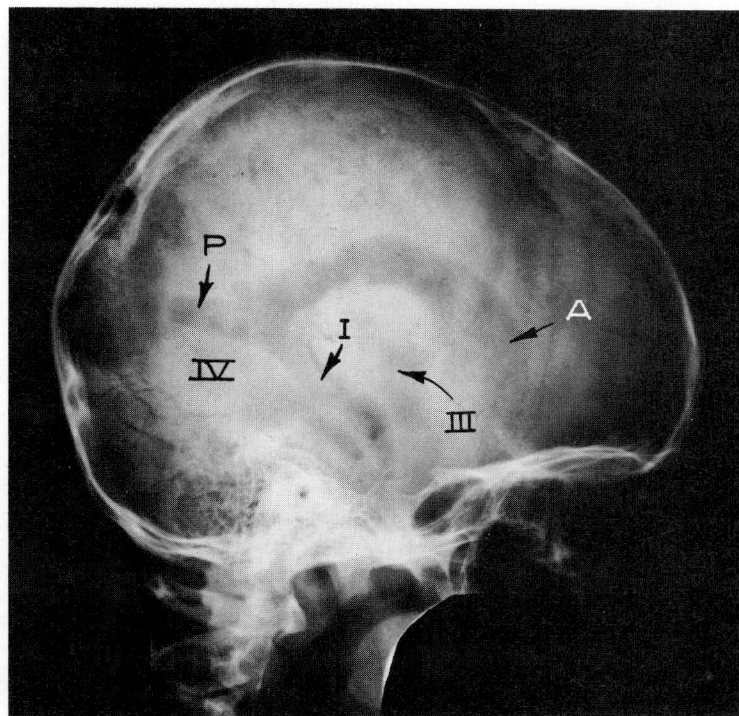


**FIG. 32-8.** A lateral ventriculogram in which the ventricular system has been fairly well filled with air. **A, P,** and **I** identify the anterior, posterior, and inferior horns, respectively, of the lateral ventricle, while **III** and **IV** identify the third and fourth ventricles. Some of the ventricular system of the other side also shows faintly. (Courtesy of Dr. D. G. Pugh)

After the cerebrospinal fluid has escaped into the subarachnoid space, some of it circulates downward around the spinal cord, but that which remains in the cranial cavity around the brain circulates upward to the dorsal midline. Here **arachnoid villi** project into the lateral lacunae alongside the superior sagittal sinus, and through these thin-walled villi, the cerebrospinal fluid is returned to the bloodstream.

Small amounts of cerebrospinal fluid are apparently added from the tissue spaces, which drain to the surface of the brain along minute perivascular spaces. These apparently act in place of lymphatics in the central nervous system, which has no true lymphatics.

## CRANIAL MENINGES

The meninges of the brain (Fig. 32-10) are continuous with those about the spinal cord, receive the same names—dura mater, arachnoid, and pia mater—and are generally similar in structure and arrangement save in one respect: the cranial dura mater, instead of being separated from the bone and its covering periosteum by an extradural (epidural) space, as is that of the spinal cord, is fused to the periosteum on the inner surface of the skull. The periosteal portion of the dura, like

**FIG. 32-9.** An anteroposterior ventriculogram showing displacement of the left lateral ventricle to the right by a tumor. The round areas above the ventricular system are the burr holes through which air was introduced into the ventricles. (Courtesy of Dr. D. G. Pugh)

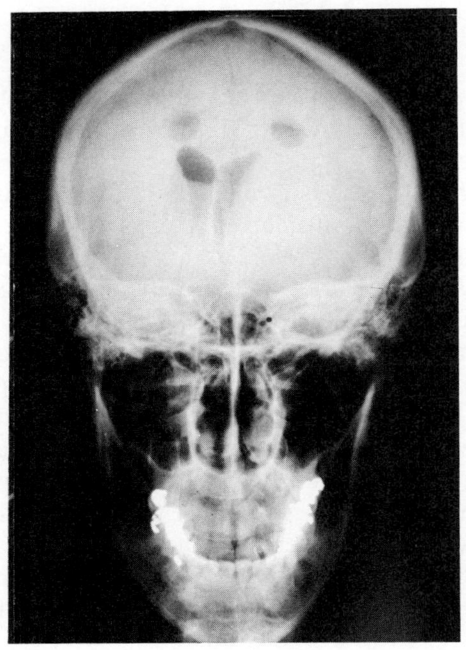

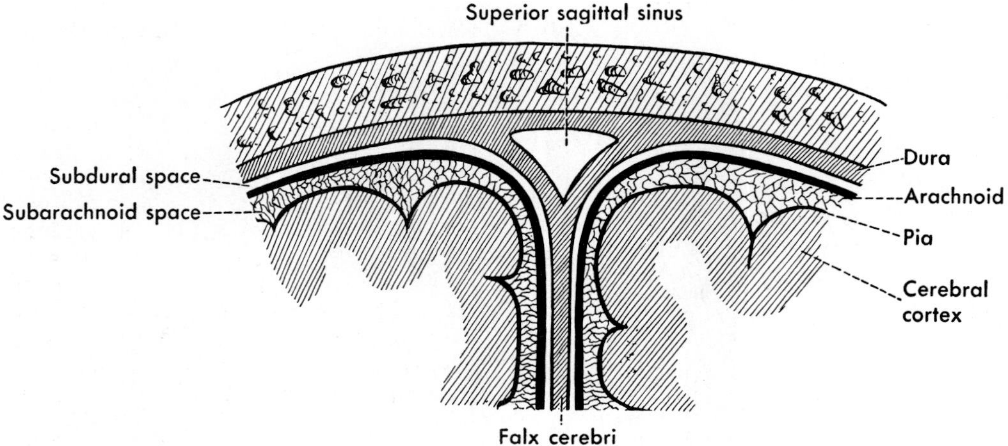


**FIG. 32-10.** A small part of the skull and the meninges close to the upper medial surface of the cerebral hemispheres. The slitlike subdural space is here much exaggerated.

**FIG. 32-11.** Difference between the spinal and cranial epidural spaces. The cranial dura has been elevated to form an epidural space.

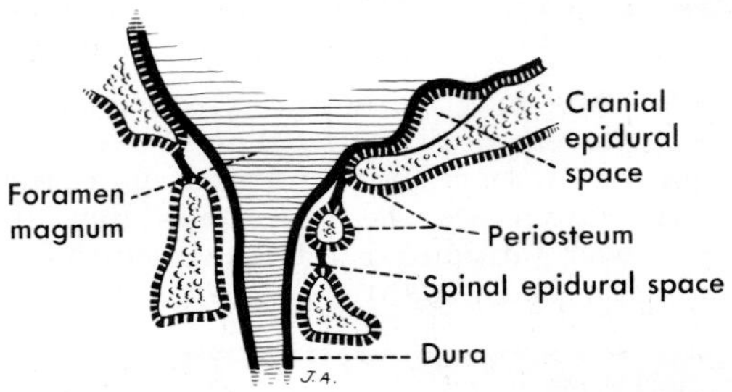


the periosteum on the outer surface of the skull, is not, however, tightly attached to the bones of the skull; the strongest attachments are along venous sinuses and, at the base of the skull, to projecting ridges of bone and about the foramina. Except over the venous sinuses, therefore, at which levels one would not usually want to remove bone anyway, the calvaria is readily separated from the dura, thus facilitating removal of bone for an operation upon the brain. The so-called **epidural space** (Fig. 32-11) in the cranial cavity is actually subperiosteal, between the periosteal layer of the dura and the bone of the skull. It, therefore, is a potential space, which becomes real when, for instance, there is an accumulation of blood between the dura and the bone as a result of a fractured skull.

## Dura Mater and Venous Sinuses

Except for certain shelflike projections, the dura mater conforms accurately to the shape of the cranial cavity. The projections that partially subdivide the cranial cavity (Fig. 32-12) are the falx cerebri (cerebral fold) and the tentorium cerebelli (tent of the cerebellum); there also are two smaller projections, the falx cerebelli and the diaphragma sellae.

The **falx cerebri** is a crescentic fold that projects downward between the two cerebral hemispheres. At its attachment to the convexity of the dura, the falx is divided into right and left layers, between which lies the *superior sagittal sinus*. In its free lower border, above the corpus callosum, is the small *inferior sagittal sinus*. At its front end, the falx is attached below to the crista galli and frontal crest, and at its back end, it is attached to and blends with the upper surface of the tentorium cerebelli.

The **tentorium cerebelli** is a layer of dura that intervenes between the lower surfaces of the cerebral hemispheres and the upper surface of the cerebellum and roofs the posterior cranial fossa. It is attached posteriorly to the occipital bone along the grooves for the *transverse sinuses,* which it partly encloses, and laterally to the uppermost part of the petrous portions of the temporal bones and to the posterior clinoid processes. The center of the tentorium, where it meets the falx cerebri, is higher than its sides, giving it a tentlike appearance. Its anterior free border is likewise highest in the midline and from this midline curves downward and forward to the posterior clinoid process on each side. Thus, a somewhat oval aperture is left between the free edge of the falx and the posterior surface of the dorsum sellae. This aperture, the *tentorial incisure* or notch, is occupied by the upper end of the mesencephalon.

The **falx cerebelli** is a small fold in the posterior midline of the posterior cranial fossa, running upward to the lower surface of the tentorium cerebelli. Its free edge projects between the two cerebellar hemispheres.

The **diaphragma sellae** is a layer of dura that largely covers the hypophyseal fossa in the sella turcica, but has in its middle an aperture, of variable size, through which passes the hypophyseal stalk connecting the hypothalamus and the hypophysis. Hypophyseal tumors, to affect the hypothalamic region or the optic system above them, must either expand through the opening in the diaphragm or bulge it upward.

The tentorial incisure is a little larger than is necessary to accommodate the brain stem; hence, when the pressure above the tentorium is markedly greater than that below (as may be true when there is a tumor), a part of the adjacent temporal lobe may herniate through the incisure. Tentorial herniation may lacerate the temporal lobe against the tough edge of the tentorium or, more seriously, stretch or compress various cranial nerves and compress blood vessels and the brain stem.

The dura mater has relatively little blood supply, for the middle meningeal arteries, the largest, supply the skull more than they do the dura. If these arteries are injured, as they may be in skull fracture, they can give rise to serious bleeding either epidurally or into the subarachnoid space. Tiny branches from several other arteries, including the internal carotid, also usually help supply the dura of the middle fossa at the base of the skull.

After the **middle meningeal artery** enters the skull through the foramen spinosum, it runs laterally and divides into a *frontal* and a *parietal branch,* the courses of which are indicated by grooves on the inner surface of the skull. Smaller branches of

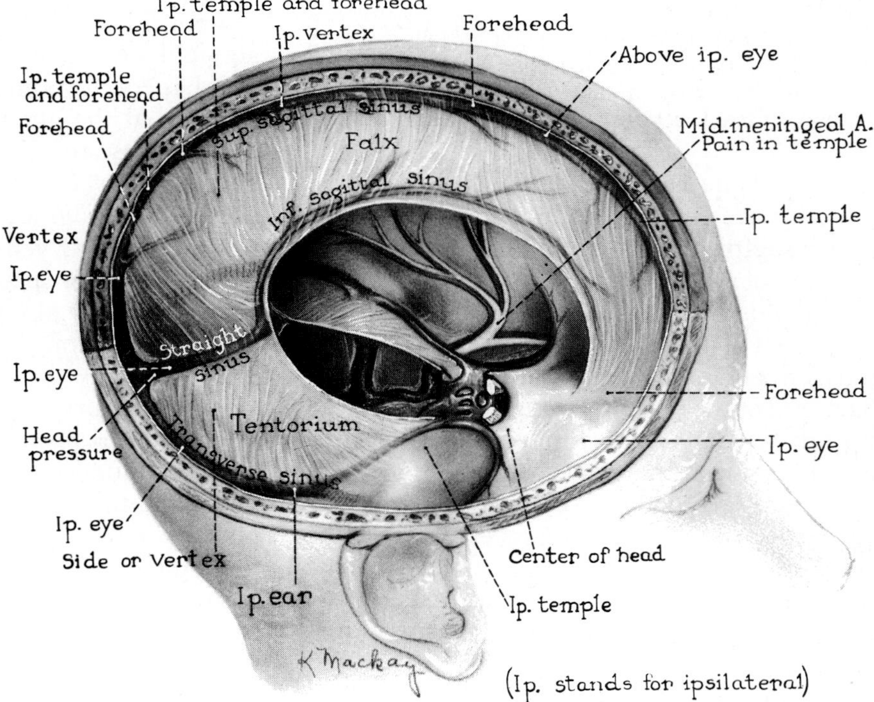


**FIG. 32-12.** The falx cerebri and tentorium cerebelli, the venous sinuses related to them, and the areas of the head to which pain is referred by the patient when the dura is stimulated. Behind the cut ends of the optic nerves is the diaphragma sellae with its central aperture. (Baker GS, Adson AW: Minn Med 26:282, 1943)

the middle meningeal include a *petrosal branch* that enters the canal of the greater petrosal nerve, a *superior tympanic artery* that enters the tympanic cavity from above, and a *communicating branch* with the lacrimal artery in the orbit.

**Anterior meningeal arteries** are small twigs from the ethmoidal arteries to the floor of the anterior cranial fossa. **Posterior meningeal arteries** usually are two, a branch of the ascending pharyngeal and one of the occipital artery, which enter the skull through the jugular foramen and supply dura in the posterior fossa.

Sensory nerve filaments are distributed to the dura, especially along the courses of the venous sinuses and the middle meningeal artery. Although the dura is relatively insensitive, stimulation along the vessels, particularly, gives rise to pain that is "referred to" (localized by a conscious patient as coming from) various parts of the head. The **meningeal nerves** to the anterior and middle cranial fossae and to the supratentorial dura are all derived from the trigeminal nerve: the anterior ethmoidal branch of the ophthalmic division gives twigs to the floor of the anterior cranial fossa, and just before it leaves the middle cranial fossa, the stem of the ophthalmic gives off a tentorial branch that sweeps backward and upward over the tentorium cerebelli and the falx cerebri; the maxillary and mandibular divisions both give rise to meningeal branches that supply the floor of the middle fossa and otherwise largely follow the middle meningeal artery. In the posterior fossa are meningeal branches of the vagus and hypoglossal nerves. Both branches apparently consist largely of fibers from upper cervical nerves that have joined the main nerves extracranially, and the hypoglossal branch, at least, consists entirely of such fibers.

Most of the venous sinuses that lie within the dura are rather simply arranged. The **superior sagittal sinus** lies in the midline at the junction of the upper border of the falx cerebri with the dura over the convexity of the hemispheres (Figs. 32-10 and 32-12). It begins at the front end of the crista galli and ends at the back end of the junction of the upper border of the falx with the tentorium cerebelli, where it is the largest of the several sinuses that come together here to form the **confluence of the sinuses.** Lateral to the main channel of the superior sagittal sinus, there are a variable number of expansions, the **lateral lacunae,** of otherwise tiny meningeal veins that enter the sinus.

The **inferior sagittal sinus** runs in the free edge of the falx cerebri. As the falx joins the tentorium, the sinus and the *great cerebral vein,* draining deeper parts of the brain, unite to form the **straight sinus** (sinus rectus). This runs posteriorly, still in the midline, at the junction of falx cerebri and tentorium cerebelli and also ends in the confluence of the sinuses. The **occipital sinus** often is small but sometimes is very large. It begins in paired parts, usually called **marginal sinuses,** around the foramen magnum and runs upward, at the attached border of the falx cerebelli, to join the confluence of the sinuses (Fig. 32-13). It and the marginal sinuses communicate with the internal vertebral venous plexuses.

From the confluence of the sinuses, there proceed laterally, at the posterior border of the attachment of the tentorium to the skull, two **transverse sinuses** (Fig. 32-13). Usually the right one is larger, for the superior sagittal sinus often turns primarily to the right, giving only a small communication to the left transverse sinus. When this occurs, the occipital and straight sinuses usually join the left transverse sinus. The transverse sinuses run laterally in the attached edge of the tentorium to the posterior ends of the petrous parts of the temporal bones. Here their names change to **sigmoid sinuses,** each of which courses downward, medially, and forward, in a deep groove situated at the junction of occipital and temporal bones, and leaves the skull through the posterior part of the jugular foramen. Outside the skull, the sigmoid sinus is continued as the internal jugular vein.

The **superior petrosal sinus** is relatively small and runs along the upper border of the petrous part of the temporal bone. Posteriorly, it joins the upper end of the sigmoid sinus. Anteriorly, it joins the cavernous sinus. The **inferior petrosal sinus,** usually larger than the superior, runs along the sphenoid and occipital bones at their junction with the petrous portion of the temporal. It originates from the cavernous sinus and leaves the skull through the anterior part of the jugular foramen, joining the internal jugular vein just outside the foramen. Between the two inferior petrosal sinuses extending downward on the clivus to the foramen magnum is a plexus of small veins, the **basilar plexus.**

The **cavernous sinuses** lie on each side of the sella turcica, where each surrounds the internal carotid artery of its side and is traversed also by the nerves that go to the orbit. Instead of containing a single large cavity, as do most of the cranial venous sinuses, the cavity of the cavernous sinus is broken up into numerous communicating chambers, and blood flow through it is, therefore, slow. The cavernous sinuses of the two sides communicate with each other through an *intercavernous sinus* situated in the diaphragma sellae. They drain posteriorly through the superior and inferior petrosal sinuses. Anteriorly, each receives the ophthalmic vein or veins and also the small **sphenoparietal sinus** that lies along the posterior edge of the lesser wing of the sphenoid bone.

Although most of the blood from the brain leaves the skull by way of the internal jugular veins, into which the sigmoid and inferior petrosal sinuses empty, some can leave by other routes. The basilar plexus and the occipital sinus anastomose through the foramen magnum with the internal vertebral venous plexuses, and since neither these nor the cranial venous sinuses have any valves, blood can leave by this route. The superior or both ophthalmic veins, through their connections with both the cavernous sinus and the angular vein, likewise can conduct blood in either direction. There also are a number of fairly constant **emissary veins** through which blood can escape from the skull.

There are approximately six paired communications between vessels inside and those outside the skull that can be termed emissary veins, and there may be unpaired ones. The three pairs that are named emissary veins are the *parietal emissary veins,* which penetrate the parietal bones posterior to their middles, only a little on each side of the midline, and connect the superior sagittal sinus to tributaries of the two occipital veins; one or more *mastoid emissary veins,* which emerge through the mastoid processes to connect the sigmoid sinuses to the occipital or posterior auricular veins; and the *condylar emissary veins,* not always present, which pass through the condylar canals and connect the lower ends of the sigmoid sinuses with the suboccipital plexus just below the base of the skull. There may also be an unpaired *occipital emissary vein*

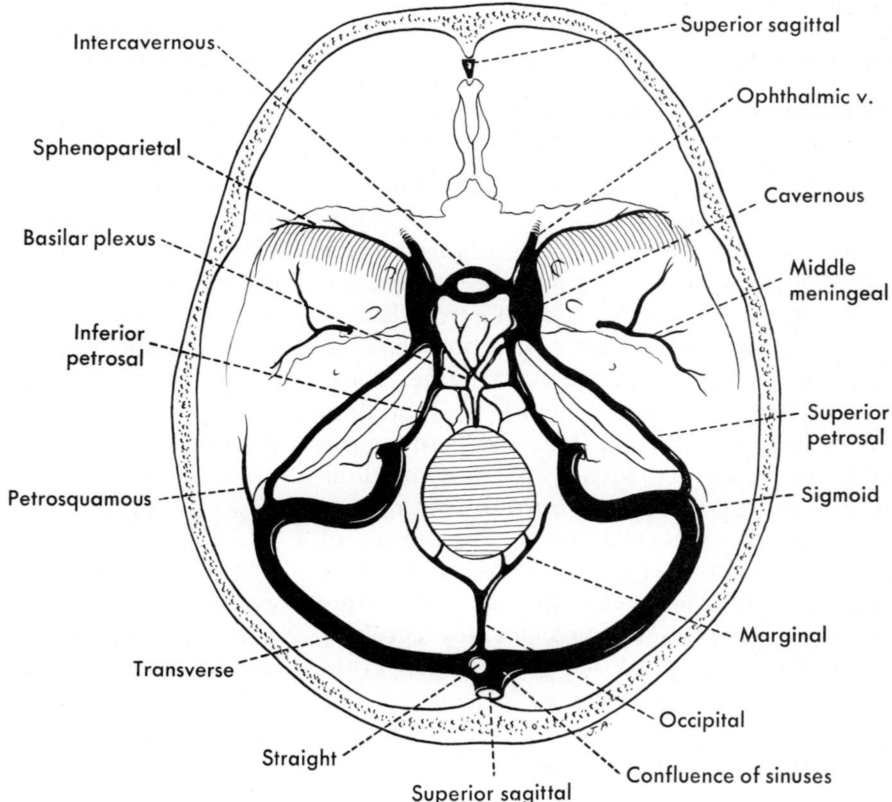


**FIG. 32-13.** The chief venous sinuses at the base of the skull. The petrosquamous sinus, shown here on the left side only, is variable in its presence.

passing through the occipital protuberance and connecting the confluence of the sinuses to a tributary of one of the occipital veins. The front end of the superior sagittal sinus in the child, although not often in the adult, is connected to veins of the nose through emissary veins in the foramen cecum.

There also are three plexuses that are really emissary veins: an *internal carotid venous plexus* around the artery connects the cavernous sinus with the internal jugular vein or the pharyngeal plexus (which empties into the internal jugular); a *venous plexus of the foramen ovale,* accompanying the mandibular branch of the 5th nerve, connects the cavernous sinus to the pterygoid plexus of veins; and a *venous plexus of the hypoglossal canal,* accompanying the hypoglossal nerve, connects the lower part of the occipital sinus with the upper end of the internal jugular vein or the terminal part of the inferior petrosal sinus.

The emissary veins cannot generally be much enlarged because their size is limited by the surrounding bone; however, they, and particularly the connections to the vertebral venous plexuses, apparently can carry a great deal of blood: it was long held that simultaneous occlusion of both internal jugular veins would necessarily be fatal, since they receive by far the major amount of blood leaving the cranial cavity; however, it has been shown that it is possible to ligate and remove both internal jugular veins (in a radical dissection of the neck for carcinoma) in one operation.

### Arachnoid, Pia, and Subarachnoid Space

The smooth outer surface of the **arachnoid** lies against the inner surface of the dura, separated from it only by a slitlike subdural space that contains enough fluid to keep the adjacent surfaces moist. From the inner surface of the arachnoid extend the cobwebby trabeculae that give the membrane its name. These cross the subarachnoid space and become continuous with the pia (Fig. 32-10). Along the superior sagittal sinus, the arachnoid is tightly attached to the dura by *arachnoid granulations* (macroscopic collections of arachnoid villi) that it sends through the dura to project into the lateral lacunae of the sinus and that provide drainage for the cerebrospinal fluid. The positions of the largest granulations often are marked on the inner surface of the skull by *foveolae granulares* that result from absorption of the adjacent bone.

The **pia mater** consists of a thin layer of connective tissue covered by flattened cells continuous with those of the arachnoid. It is closely attached to the substance of the brain and follows its every contour. Thus, in contrast to the arachnoid, it dips into the sulci of the cerebral cortex, and it closely invests the roots of the cranial nerves as they leave the brain. Because the pia follows the contours of the brain, whereas the arachnoid follows those of the dura, the subarachnoid space is larger at some levels than at others. For instance, it is larger over the sulci than the gyri of the hemispheres.

The larger expansions of the subarachnoid space are known as **cisterns,** since they contain the largest accumulations of cerebrospinal fluid (Fig. 32-14). Those named by anatomists are the *cisterna cerebellomedullaris,* often called the cisterna magna, in the angle between the inferior surface of the cerebellum and the posterior surface of the medulla; the *cistern of the lateral cerebral fossa,* above the temporal pole; the *cisterna chiasmatis,* anterior to and above the optic chiasma; and the *cisterna interpeduncularis,* between the cerebral peduncles. The boundaries of none of these are really distinct, and the last-named two have been grouped together as the *cisterna basalis*. Clinicians use a variety of names to identify not only these but many others, including dorsolateral extensions from the cisterns at the base of the brain, extensions from the cisterna cerebellomedullaris over the cerebellum and midbrain, and a cistern between the two cerebral hemispheres just above the corpus callosum. They can be studied by injecting air or radiopaque material into them (*encephalography*).

## BLOOD SUPPLY OF THE BRAIN

There are two sets of arteries to the brain, the paired internal carotids and the paired vertebrals (Fig. 32-16). From its origin in the neck, the **internal carotid** ascends posterolateral to the wall of the pharynx to enter the carotid canal in the lower surface of the petrous portion of the temporal bone. This canal turns

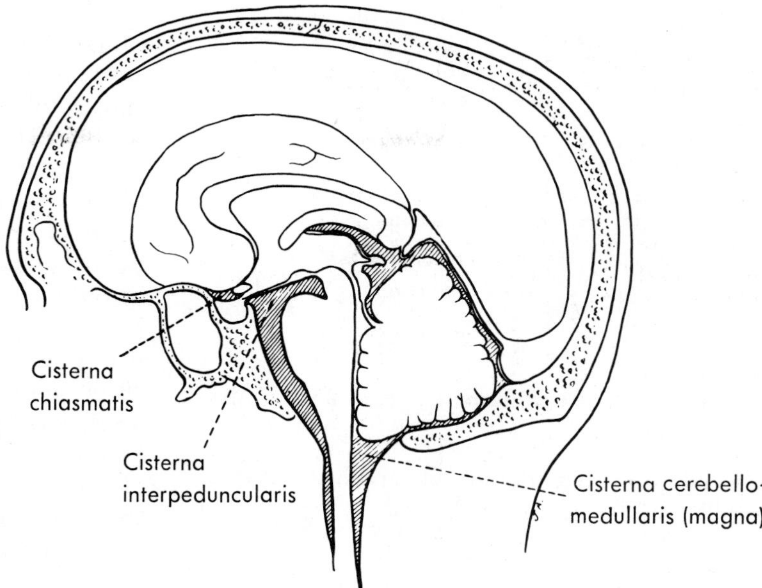


**FIG. 32-14.** Subarachnoid cisterns in a median sagittal section of the head. The labeled ones are those recognized in the official terminology, but those over the cerebellum and midbrain, and one ventral to the pons, also can be seen.

forward in the bone, and the internal carotid leaves the canal at the apex of the petrous part, just above the foramen lacerum. It runs upward and forward on the side of the sella turcica, through the cavernous sinus. It then turns upward and medially just medial to the anterior clinoid process, penetrating the dura and giving off a branch, the **ophthalmic artery,** that follows the lower surface of the optic nerve to reach the orbit. The part of the carotid artery close to the sella is often referred to as the "carotid siphon." The chief branches of the carotid are to the brain and, therefore, are best examined on a brain that has been removed from the skull, as are also the branches of the vertebral arteries.

### Vertebral Arteries

The vertebral arteries reach the interior of the skull by ascending through the transverse foramina of the cervical vertebrae, turning medially along the upper surface of the posterior arch of the atlas and then penetrating the posterior atlantooccipital membrane and the underlying dura to enter the subarachnoid space and pass through the foramen magnum into the cranial cavity.

While the vertebral arteries lie somewhat lateral to the lower end of the medulla, each gives off a small **posterior spinal artery** that runs down the spinal cord at about the point of attachment of the dorsal nerve roots. Before the vertebral arteries come together on the ventral surface of the medulla, each gives off a root of the **anterior spinal artery** (Fig. 32-15). These two roots unite to form an artery that runs down in the anterior median fissure of the cord. These spinal branches are small and by no means sufficient to supply the length of the spinal cord. They are reinforced at irregular intervals by segmental arteries that pierce the dura and run along the spinal nerve roots to join them. Each vertebral artery also gives off branches to the medulla and a **posterior inferior cerebellar artery** to the posteroinferior surface of the cerebellum. The vertebral arteries end by fusing to form the **basilar artery.**

The basilar artery runs forward on the lower surface of the pons, gives off branches to that, and gives rise to two pairs of arteries to the cerebellum. Each **anterior inferior cerebellar artery** arises at about the level of the 8th nerve and often loops laterally along the nerve. It usually gives off the **labyrinthine artery** (to the internal ear), which follows the

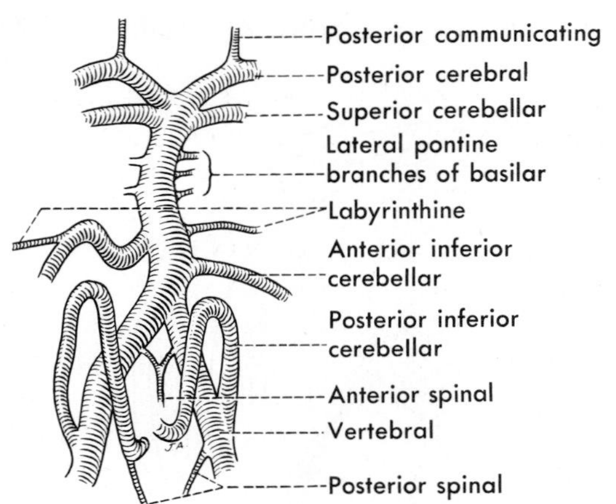


**FIG. 32-15.** The vertebral arteries and their intracranial branches.

**FIG. 32-16.** Formation of the circulus arteriosus cerebri.

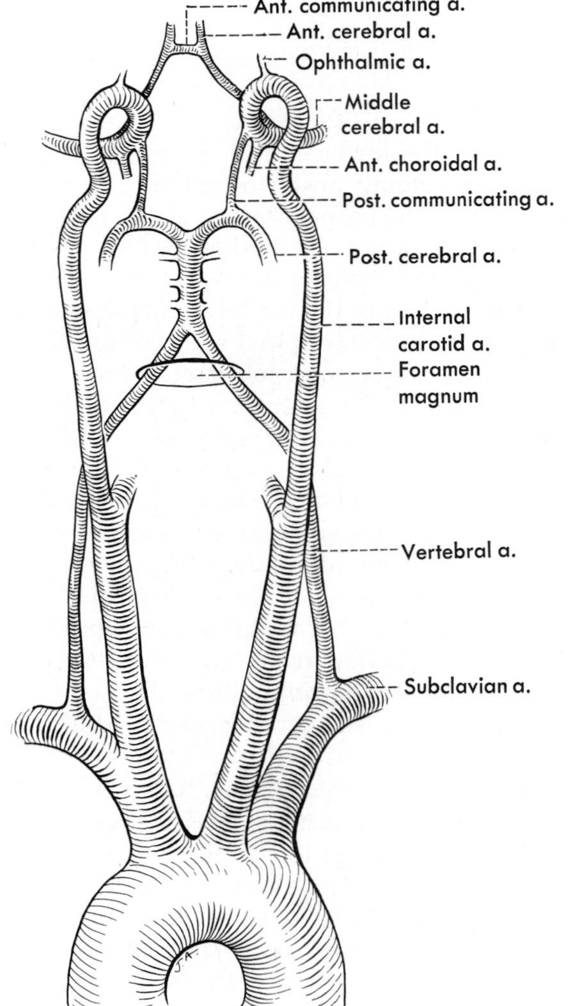


7th and 8th nerves into the internal acoustic meatus, and it is distributed to the anterior or ventral surface of the cerebellum. The **superior cerebellar arteries** arise just before the basilar artery ends. They run laterally around the brain stem, which they help to supply, to reach the superior surface of the cerebellum.

Just beyond the origin of the superior cerebellar arteries, the basilar artery bifurcates into the **posterior cerebral arteries.** These also run laterally around and help to supply the brain stem. They then pass above the tentorium cerebelli and are distributed to the lower surface of the back part of the temporal lobe and to the occipital lobe. On the surface of the cortex, the cerebral arteries (there are three on each hemisphere) anastomose with each other; their central branches that pass into the brain do not anastomose, however.

## Arterial Circle

The *circulus arteriosus cerebri* (circle of Willis) is a somewhat hexagonal formation of blood vessels at the base of the brain close to the sella turcica. It connects the vertebral and internal carotid arteries to each other and to the vessels of the opposite side (Fig. 32-16). Anteriorly, the circle is formed by the *internal carotid arteries,* their *anterior cerebral* branches, and an *anterior communicating artery* that connects the two anterior cerebrals. Each internal carotid also sends a *posterior communicating artery* to connect with the *posterior cerebral* of its side, and the origins of the posterior cerebrals from the basilar complete the circle posteriorly. Parts of the circle, particularly the posterior communicating arteries, vary much in size. Depending upon this, blood may or may not be easily shunted from one side of the brain to the other, or from the carotid to the basilar system, or vice versa.

## Internal Carotid

The internal carotid gives rise to an **anterior choroidal** (the chief blood supply to the choroid plexus in the lateral ventricle, but also supplying brain tissue) and the **posterior communicating artery** before dividing into the anterior and middle cerebral arteries. The **mid-**

**dle cerebral artery** runs upward and laterally in the lateral cerebral fissure (between the temporal and frontal lobes) and spreads out so as to supply the lower and lateral surface of the cerebral hemisphere as far back as the distribution of the posterior cerebral artery on the lateral surface (Fig. 32-17). In addition to cortical branches, it and other branches of the internal carotid give off, close to that vessel, central or perforating branches that pierce the base of the brain. Some of these supply the important internal capsule (the bundle of fibers leading from and to the cerebral hemisphere) and part of the adjacent corpus striatum and are called **striate arteries.** Occlusion of or hemorrhage from some of these vessels is a common type of stroke.

The **anterior cerebral arteries,** after communicating with each other, proceed forward, upward, and then backward in the longitudinal fissure, each supplying the medial surface of its cerebral hemisphere back to the distri-

**FIG. 32-17.** Distribution of the three cerebral arteries in lateral, medial, and inferior views of the cerebral hemisphere. The cortical branches are here named in more detail than in official anatomic terminology. (Mettler FA: Neuroanatomy, 2nd ed. St Louis, CV Mosby, 1948)

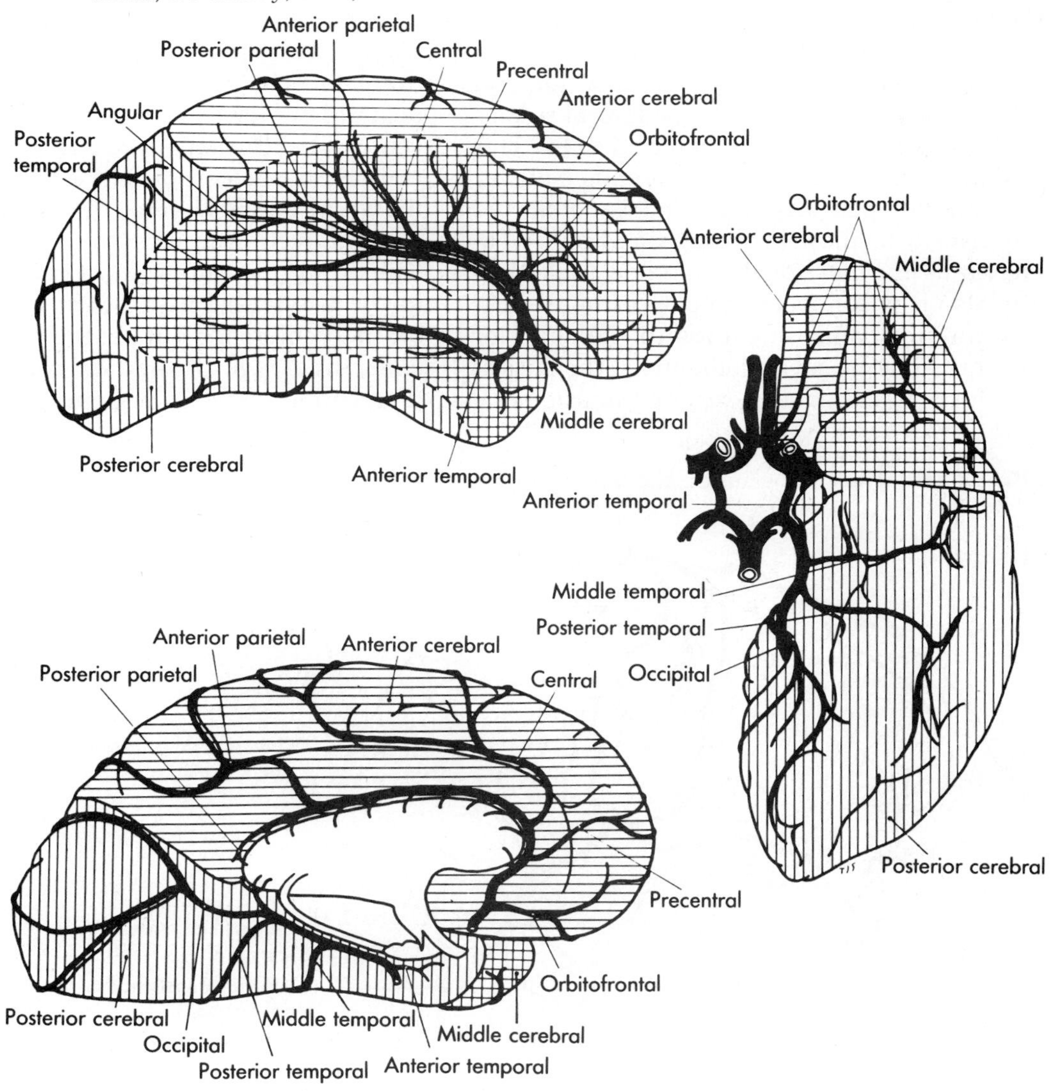

bution of the posterior cerebral. Branches of the anterior cerebral run onto the upper lateral part of the cerebral hemisphere, helping to supply this and anastomosing with branches of the middle and posterior cerebrals.

## Venous Drainage

The venous drainage of the brain is through the venous sinuses in the dura mater. The drainage from the convex and most of the medial surface of the cerebral hemisphere is into superficial veins. **Superior cerebral veins,** from the upper part of the hemisphere, run upward to empty into the superior sagittal sinus (Fig. 32-18). **Inferior cerebral veins,** from the lower part of the hemisphere, join venous sinuses situated at the base of the skull. Running horizontally between the superior and inferior cerebral veins, over the lateral sulcus, is the **superficial middle cerebral vein.** This drains anteriorly into the cavernous sinus or the superior petrosal sinus and is connected posteriorly to the transverse sinus by a vein known as the *inferior anastomotic vein.* Usually also it is connected to one or more superior cerebral veins and thus to the superior sagittal sinus by several channels, the largest of which is called the *superior anastomotic vein.*

The **deep middle cerebral vein** lies in the depth of the lateral sulcus, on the hidden cortex (the insula) that here connects the frontal, parietal, and temporal lobes. This vein unites lateral to the hypothalamus with veins from the inferior and medial surfaces of the anterior part of the hemisphere and with striatal veins to form the **basal vein** (vein of Rosenthal). The basal vein then runs laterally and dorsally around the cerebral peduncle to end in the **great cerebral vein.**

Veins from the deeper parts of each cerebral hemisphere begin as the **thalamostriate vein** in the floor of the lateral ventricle, the **vein of the septum pellucidum,** on part of the medial wall of the ventricle, and the **choroidal vein** that drains the choroid plexus of the ventricle. These unite to form the **internal cerebral vein.** The two internal cerebral veins run posteriorly side by side in the roof of the third ventricle and at about the level of the posterior end of the corpus callosum unite to form the **great cerebral vein** (of Galen). The great cerebral vein receives the basal veins and veins from the midbrain as it runs backward in the transverse fissure just above the midbrain. It enters the dura at the point where the lower free edge of the falx cerebri joins the tentorium cerebelli and here joins the inferior sagittal sinus to form the straight sinus.

**FIG. 32-18.** The chief veins on the lateral surface of the cerebral hemisphere.

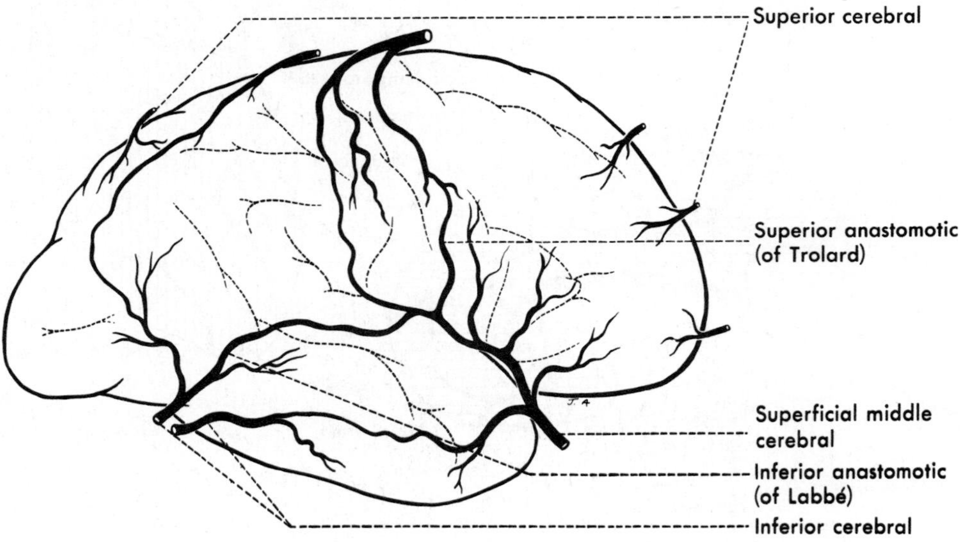

**Superior cerebellar veins** enter the great cerebral vein and the transverse and superior petrosal sinuses. **Inferior cerebellar veins** enter the straight or a transverse sinus and the inferior petrosal and occipital sinuses.

## CRANIAL NERVES

The extracranial courses and branches of most of the cranial nerves are described later; only their intracranial courses and the composition and distribution of each nerve are summarized here. Their attachments to the brain are shown in Figure 32-6, and their positions at the base of the skull and certain vascular relations are shown in Figure 32-19.

Among the 12 pairs of cranial nerves, some are sensory only, some are voluntary motor only, some are mixed nerves, and some contain autonomic fibers in addition to voluntary motor fibers (see also p. 54). The cranial nerves that contain sensory fibers typically have ganglia, collections of nerve cells located outside the central nervous system, that give rise to these sensory fibers and are the cranial equivalent of the dorsal root (sensory) ganglia of spinal nerves. The voluntary motor and preganglionic autonomic fibers arise from cell bodies, grouped together as cranial nuclei, that lie inside the central nervous system and correspond collectively to the anterior and lateral gray columns of the spinal cord. Unlike spinal nerves, however, the motor and sensory fibers of the cranial nerves do not necessarily form different roots, but may be mixed together in a single root.

### Olfactory Nerve

The olfactory nerve, or nerve of smell, is purely sensory. Its cell bodies are located in the mucosa of

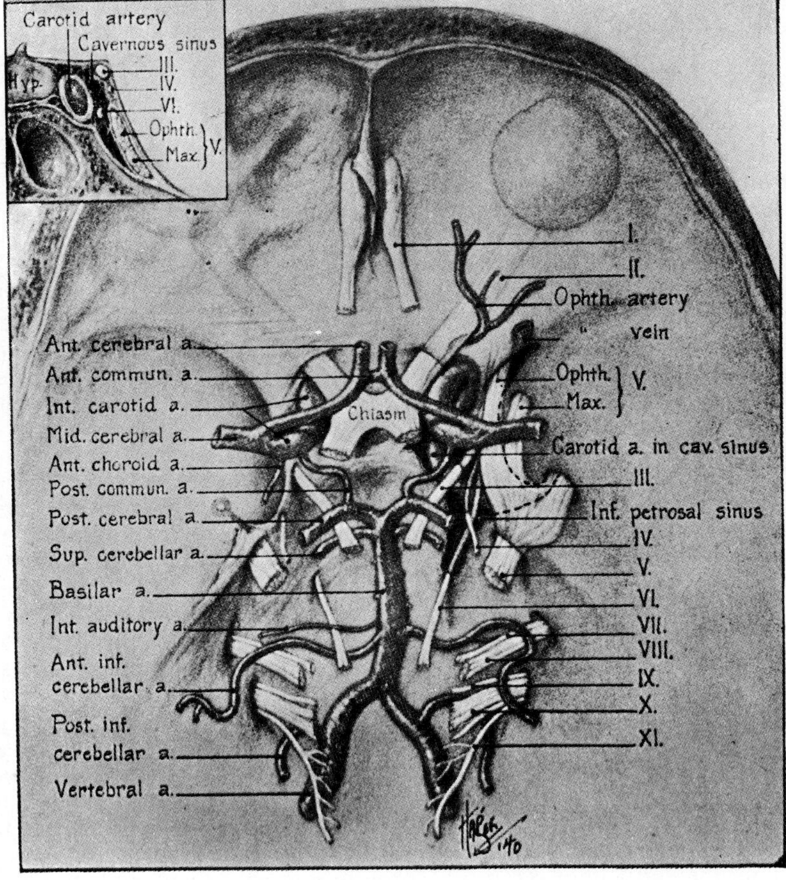


**FIG. 32-19.** Nerves and arteries at the base of the skull. On the right side, the floor of the anterior cranial fossa is shown as semitransparent, so that within the orbit the eyeball, optic nerve, and ophthalmic artery can be seen. The broken line over the ganglion and branches of the fifth nerve indicates the approximate posterior and lateral extent of the cavernous sinus in relation to this nerve, and the small inset is a frontal section to show relationships within this sinus. **I.** is the olfactory bulb and tract; the other Roman numerals identify the cranial nerves. "**Int. auditory a.**" is the name by which the artery now called the labyrinthine has long been known. (Dandy WE: Intracranial Arterial Aneurysms. Ithaca, New York, Comstock, 1944)

the uppermost part of the nasal cavity. The fibers from these cells unite to form a number of small filaments that enter the skull through the cribriform plate to end in the olfactory bulb.

### Optic Nerve

The optic or 2nd cranial nerve, the nerve of sight, usually is said to be sensory only, but may contain efferent fibers of unknown origin and function. However, it actually is neither developmentally nor anatomically a nerve: the ganglion cells giving rise to the fibers of the nerve constitute a layer of the retina of the eyeball, and the retina is an outgrowth from the central nervous system. Further, the optic nerve is surrounded by meninges, as is the central nervous system, and contains neuroglia throughout its length. The optic "nerve" really is a tract of the brain, connecting two parts.

After it leaves the back of the eyeball, the optic nerve has a slightly sinuous course backward through the orbit, thus allowing for movement of the eyeball. It enters the cranial cavity through the optic canal, and the two optic nerves converge toward each other. They join to form the **optic chiasma,** attached to the hypothalamus, and in the chiasma exchange fibers: the fibers from the medial half of each retina cross to the opposite side; those from the lateral half of each retina remain uncrossed. The fibers behind the chiasma are a continuation of those in the optic nerves and the chiasma, but are called the **optic tracts.** Because of the chiasma, an optic tract differs fundamentally from an optic nerve in composition; an optic nerve contains all the fibers from one eye, whereas an optic tract contains fibers from the lateral half of the eye on its own side and from the medial half of the other eye. Since these halves of the two eyes receive impulses from the opposite side (for instance, the right half of each retina receives light that originates to the left of the body), the optic system behaves like most other sensory systems: impulses originating on one side of the body cross to end in the cerebral cortex of the opposite side. The right optic tract, transmitting impulses evoked in the right half of each retina by light waves from the left, relays these to the right cerebral cortex (Fig. 32-20).

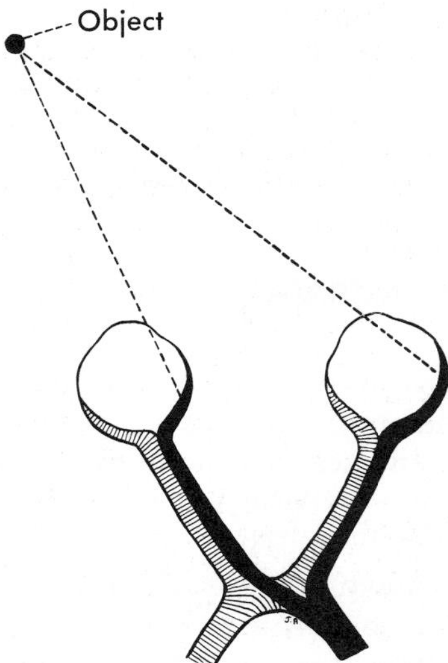


**FIG. 32-20.** The manner in which visual stimuli on one side are conducted from both eyes to the opposite cerebral hemisphere.

### Oculomotor and Trochlear Nerves

The 3rd and 4th cranial nerves are two of the three nerves that supply the voluntary muscles in the orbit, which are concerned with moving the eyeball and raising the upper eyelid. The **oculomotor** supplies five of the seven muscles (all except the superior oblique and the lateral rectus) and is composed largely of voluntary motor fibers. It also contains, however, preganglionic autonomic fibers that synapse in a small ganglion, the ciliary ganglion, located within the orbit. Through this ganglion, the oculomotor nerve controls the smooth muscle that is responsible for constriction of the pupil of the eye and for accommodation of the lens to close vision.

The oculomotor nerve leaves the floor of the midbrain just in front of the pons, through the medial part of the cerebral peduncle. It runs forward through the subarachnoid space and pierces the dura over the cavernous sinus just anterolateral to the posterior clinoid process. It then runs forward in the wall of the sinus and enters the orbit through the superior orbital fissure.

The **trochlear nerve** supplies only one voluntary muscle (superior oblique) in the orbit, and no smooth muscle. It leaves the dorsal surface of the brain stem just below the inferior colliculus and runs forward against the lower surface of the tentorium cerebelli. Just behind the posterior clinoid process, it pierces the lower surface of the tentorium close to its free edge and runs forward in the dura of the lateral wall of the cavernous sinus, leaving the skull through the superior orbital fissure.

Muscle spindles (sensory organs, p. 106) are in those voluntary muscles supplied by the oculomotor and trochlear nerves, and the sensory fibers ending in the spindles are in these nerves for at least a part of their course. However, neither nerve has a sensory ganglion (corresponding to a dorsal root ganglion of a spinal nerve), and the cell bodies giving rise to these sensory fibers have never been certainly identified. They are believed either to lie within the central nervous system, in contrast to cell bodies of most sensory fibers, or to lie in the trigeminal ganglion.

### Trigeminal Nerve

The trigeminal (5th cranial) nerve is so named because it has three major branches. It contains both sensory and motor fibers. The sensory fibers are distributed particularly to the skin of the face and to the upper and lower teeth, and the motor fibers to muscles associated with the jaws. It is attached to the side of the pontine region of the brain stem and has two roots, a large sensory root and a small so-called motor root that does contain all the motor fibers but also proprioceptive (afferent or sensory) fibers. The two roots run close together across the upper surface of the tip of the petrous portion of the temporal bone to enter the middle cranial fossa, where the sensory root bears the **trigeminal ganglion.** This ganglion is covered by the dura of the middle cranial fossa, but its proximal part is surrounded by cerebrospinal fluid: the subarachnoid space around the pons continues across the petrous bone around the roots of the nerve and expands over the proximal part of the ganglion. This expansion into the middle fossa is the *trigeminal* (Meckel's) *cave*. When the sensory root of the 5th nerve is cut to alleviate the painful *trigeminal neuralgia* (tic douloureux), this often is done in the subarachnoid space of the trigeminal cave.

From the trigeminal ganglion arise the three branches of the trigeminal nerve, the ophthalmic, maxillary, and mandibular. These leave the skull through three separate apertures. The **ophthalmic nerve** runs forward in the dura of the lateral wall of the cavernous sinus and enters the orbit through the superior orbital fissure. It contributes branches to the eyeball and the upper part of the nasal cavity, but for the most part leaves the orbit to supply skin of the upper eyelid, the dorsum of the nose, the forehead, and the scalp about as far back as the interauricular line (see Figs. 31-15 and 31-23).

The **maxillary nerve** (Fig. 32-21) leaves the skull by way of the foramen rotundum, which opens into the pterygopalatine fossa. Suspended from it at this point is the pterygopalatine ganglion, which sends fibers into the branches of the maxillary nerve that go to the interior of the nose and to the palate. While it is in the pterygopalatine fossa, the maxillary nerve gives off palatine and nasal branches, branches to the posterior upper teeth, and the zygomatic nerve, which runs with the re-

**FIG. 32-21.** Diagram of the maxillary nerve. (Henle J: Handbuch der systematischen Anatomie des Menschen, vol 3, pt 2. Braunschweig, Vieweg und Sohn, 1868)

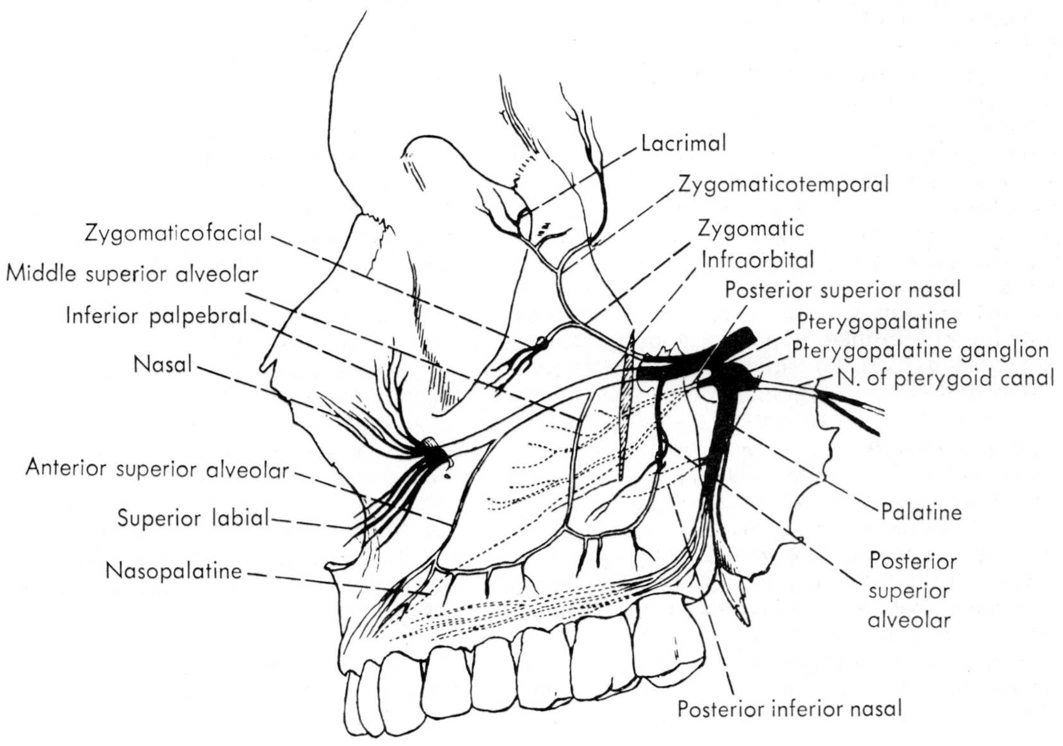

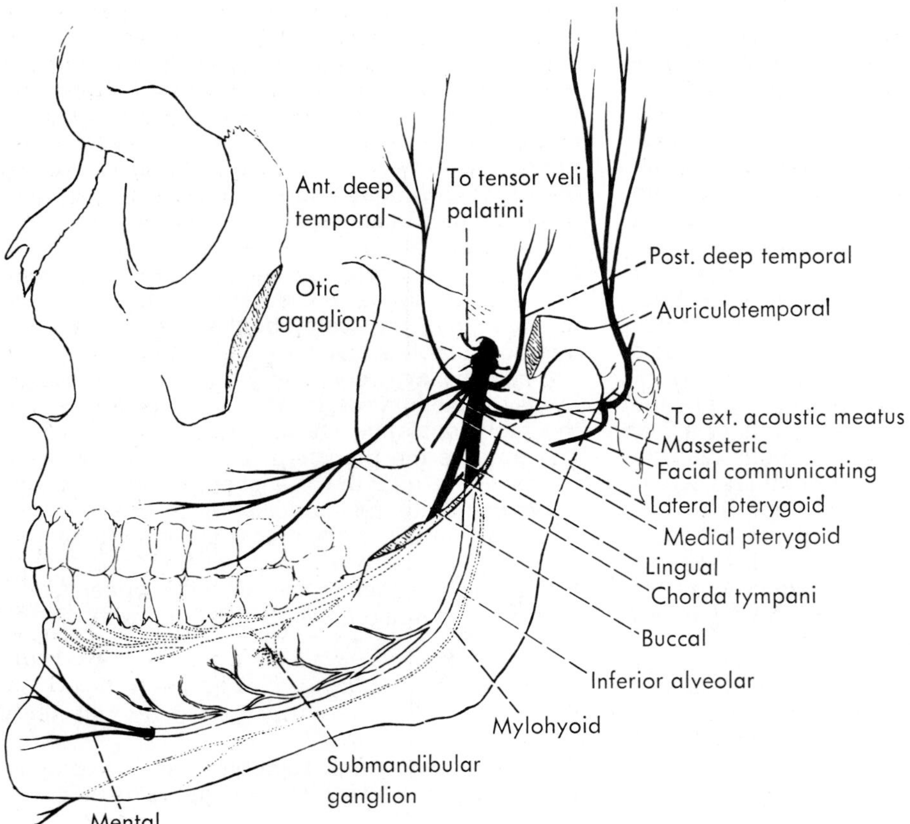


**FIG. 32-22.** Diagram of the mandibular nerve. (Henle J: Handbuch der systematischen Anatomie des Menschen, vol 3, pt 2. Braunschweig, Vieweg und Sohn, 1868)

maining part of the maxillary into the orbit. The major continuation of the maxillary nerve, the infraorbital, leaves the pterygopalatine fossa by passing through the inferior orbital fissure into the orbit. Here it runs first in a groove and then in a canal in the floor of the orbit. It gives off superior alveolar nerves and then emerges to go to skin of the lower lid, side of the nose, upper lip, and upper part of the cheek (Fig. 31-23). The zygomatic nerve divides within the orbit into zygomaticotemporal and zygomaticofacial branches, which also emerge to become cutaneous.

The **mandibular nerve** leaves the skull through the foramen ovale as two roots, a sensory root from the ganglion and the motor root, which join just outside the skull. The mandibular is the only branch of the trigeminal containing voluntary motor fibers. They are distributed to the muscles of mastication and to the mylohyoid, the anterior belly of the digastric, the tensor veli palatini, and the tensor tympani muscles (Fig. 32-22). In addition, the mandibular nerve has large sensory branches: the lingual nerve, to the anterior two-thirds of the tongue; the inferior alveolar, to the lower teeth and the skin of the chin; the auriculotemporal nerve; and the buccal nerve. The otic ganglion is on the medial surface of the mandibular nerve immediately outside the foramen ovale. The submandibular ganglion, another parasympathetic ganglion of the head, is suspended from the lingual branch of the mandibular nerve.

Whereas all four of the cranial parasympathetic ganglia are located close to or on some branch of the trigeminal nerve, the preganglionic fibers to these ganglia, although they may run in a branch of the trigeminal, actually come from other nerves—the 3rd, 7th, or 9th. The roots of the trigeminal nerve contain no autonomic fibers. The postganglionic fibers from the ganglia, however, typically join and are distributed with peripheral branches of the trigeminal.

### Abducens Nerve

The abducens or 6th cranial nerve is like the trochlear in supplying only one voluntary muscle of the orbit (the lateral rectus). Like the 3rd and 4th nerves, it contains afferent fibers to muscle spindles but has no ganglion, and the locus of the cells of origin of these fibers is unknown; all three of the eye-muscle nerves frequently are classed as purely motor ones.

The abducens leaves the floor of the brain stem just behind the pons and has a short course downward and forward into the dura of the floor of the posterior cranial fossa. After entering the dura, it runs forward through the cavernous sinus and the superior orbital fissure into the orbit.

### Facial Nerve

The facial or 7th cranial nerve (Fig. 32-23) is so known because it supplies the muscles of the face. It leaves the lateral surface of the medulla just caudal to the pons and anterior to the cerebellum, in company with the 8th nerve, and runs laterally with this nerve to the internal acoustic meatus. At the distal end of the meatus the facial nerve enters the facial canal in the petrous part of the temporal bone. This canal lies partly in the wall between the internal and middle ears and opens below at the stylomastoid foramen on the base of the skull. The ganglion of the nerve is located where the nerve makes a sharp turn in the facial canal, and because it is located at this bend (geniculum), it is called the **geniculate ganglion.**

Outside the stylomastoid foramen, the facial nerve passes lateral to the styloid process and into the parotid gland. Before passing into the gland, it gives off a branch to the posterior belly of the digastic and to the stylohyoid muscle and a *posterior auricular* branch that runs upward behind the ear to supply the posterior auricular muscle and the occipital belly of the occipitofrontal muscle. Its distribution on the face has already been described.

**FIG. 32-23.** Diagram of the facial nerve. Voluntary motor fibers are indicated by solid lines; parasympathetic ones, both preganglionic and postganglionic, by dotted lines; and sensory fibers by broken lines. The only sensory fibers shown here are the best-known ones, those of taste that are distributed to the tongue. (Sections of Neurology and Section of Physiology, Mayo Clinic and Mayo Foundation: Clinical Examinations in Neurology, 2nd ed. Philadelphia, WB Saunders, 1963. After Strong OS, Elwyn A: Human Neuroanatomy, 3rd ed. Baltimore, Williams & Wilkins, 1953.)

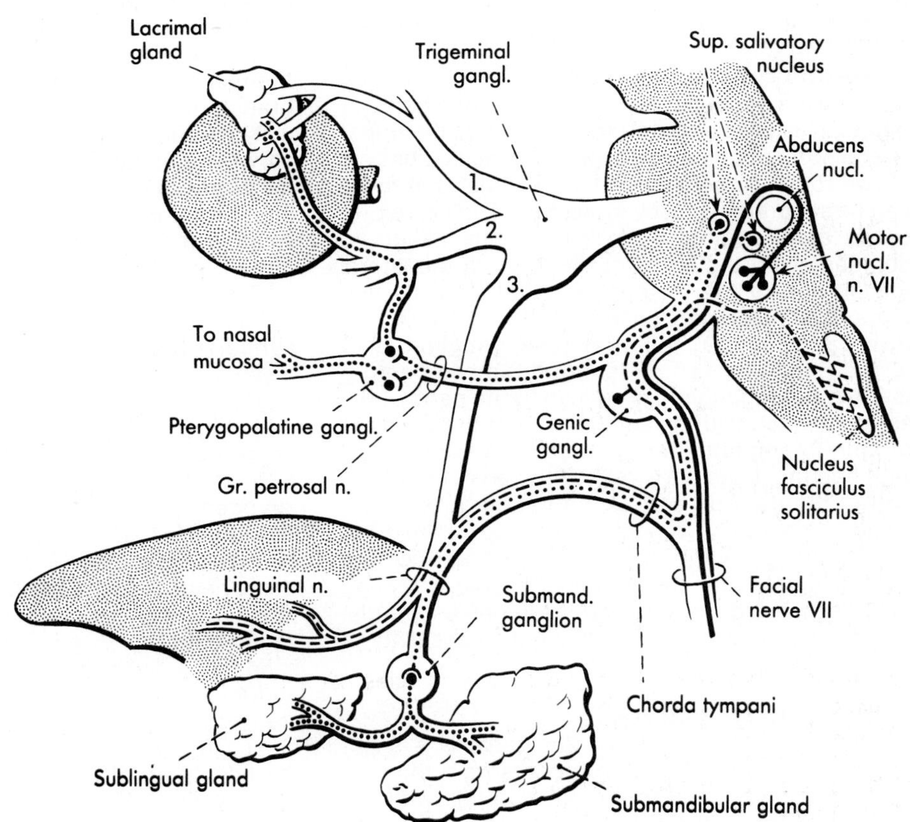

The **nervus intermedius** is the sensory and parasympathetic motor root of the facial nerve; the large root of the facial is voluntary motor. It is called nervus intermedius because it often lies between the 7th and 8th cranial nerves. It joins the facial nerve proximal to the geniculate ganglion and contributes fibers to both branches of the nerve that contain parasympathetic fibers (greater petrosal and chorda tympani).

The voluntary motor fibers of the facial nerve go to all the muscles of the face, including those of the scalp, and to the platysma, the posterior belly of the digastric, the stylohyoid, and the stapedius muscle. The autonomic fibers go to the pterygopalatine and submandibular ganglia through the greater petrosal and chorda tympani nerves. The exact distribution and function of many of the sensory fibers of the facial nerve are not known, but some are thought to be concerned with deep pain from the face; some of them apparently are distributed to a small part of the soft palate; a few may reach the middle ear cavity; and the few cutaneous fibers that the nerve contains are distributed to skin on the posterior surface of the ear, along with similar fibers from the 9th and 10th nerves. The best-known sensory fibers in the facial nerve are those for taste on the anterior two-thirds of the tongue. These reach the tongue through the chorda tympani branch of the facial and the lingual branch of the mandibular, which the chorda tympani joins.

### Vestibulocochlear Nerve

The 8th cranial nerve (n. vestibulocochlearis, or octavus) has two parts. At its origin from the upper lateral surface of the brain stem, the nerve consists of two roots, a cochlear and a vestibular, but these can be separated only by dissection. The nerve proceeds laterally into the internal acoustic meatus, in company with the facial nerve. Toward the lateral (distal) end of the meatus, it divides into its two parts, vestibular and cochlear. Each part has its own ganglion, and the **vestibular ganglion** is in turn divided into two parts. From the two parts of the vestibular ganglion, fibers go to the parts of the ear connected with balance, but from the cochlear or **spiral ganglion** fibers go to the part connected with hearing. Both parts of the 8th nerve, therefore, end in the petrous part of the temporal bone, instead of emerging from the skull as do all the other cranial nerves. Efferent fibers have been demonstrated in both branches of the 8th nerve. They end in the sense organs to which these nerves are distributed and apparently influence the sensitivity of these organs.

### Glossopharyngeal Nerve

The glossopharyngeal or 9th nerve is a small one, arising by two or more rootlets from the side of the medulla dorsal to the olive, in line with the rootlets of the vagus nerve. It runs laterally to the front part of the jugular foramen and penetrates the dura separate from the vagus nerve. As it emerges through the jugular foramen, it bears two ganglia close together; the **superior ganglion** is particularly small, but the **inferior ganglion** is a little larger. Outside the jugular foramen, the glossopharyngeal nerve gives off a tympanic branch, sends a branch of communication to the auricular branch of the vagus, gives rise to a large carotid sinus branch that also supplies the carotid body, and sends one or more pharyngeal branches to the pharynx (see Fig. 31-43). These branches are all sensory. The only voluntary motor branch of the 9th nerve is to the stylopharyngeus muscle. Beyond these branches, the 9th nerve runs forward along the side of the pharynx, to which it gives off one or more tonsillar branches, and then continues into the tongue to end in the mucosa and taste buds of the posterior third, supplying both general sensation and taste. The preganglionic fibers that lie in the 9th nerve end in the otic ganglion and are described with that.

### Vagus Nerve

The vagus or 10th cranial nerve arises by a number of rootlets emerging from the lateral border of the medulla dorsal to the olive. As these run laterally, they join to form the vagus nerve, which passes through the middle part of the jugular foramen, with the accessory nerve just behind it. At this point, the accessory nerve divides into two branches, an internal and an external. The internal branch then joins the vagus and is distributed with it. It is properly a part of the vagus, not of the accessory. The vagus, like the glossopharyngeal, typically bears a **superior** and an **inferior ganglion;** the inferior ganglion is large, and the superior one small. It is the wide distribution of the vagus that gives it its name, which means wandering.

The vagus sends a small meningeal branch to the dura of the posterior cranial fossa and a small auricular branch to a part of the external acoustic meatus and tympanic membrane and to skin behind the ear. The branches of the vagus in the neck (see Fig. 30-11) are one or more pharyngeal branches, motor to the muscles of the pharynx and most of the muscles of the soft palate; the superior laryngeal nerve, sensory to the larynx and motor to one muscle, an external one on the larynx (cricothyroid); superior and inferior cardiac branches, to the heart; and the right recurrent laryngeal nerve (the left one arises in the thorax). In the thorax, the vagus contributes fibers to the heart, lungs, and esophagus and descends on the esophagus to the abdomen. In the abdomen, it is distributed to the stomach and the liver and, after joining the celiac plexus, to the digestive tube about as far distally as the beginning of the descending colon. Branches of the vagus nerve in the neck and thorax contain voluntary motor fibers, autonomic fibers, and sensory fibers including those for pain. The abdominal part of the vagus consists of preganglionic parasympathetic fibers that end in the enteric ganglia located in the walls of the intestinal tract, and afferent fibers of largely unknown reflex functions.

### Accessory Nerve

The accessory nerve arises by two sets of rootlets: **cranial** rootlets that are in line with the vagal rootlets on the side of the medulla, and **spinal** rootlets that ascend alongside the spinal cord, between the dorsal and the ventral roots. These may come from as low as the 5th cervical segment and join each other as the nerve ascends. The cranial and spinal roots come together briefly as the nerve passes through the jugular foramen immediately behind the vagus, but then the fibers of the cranial root separate as the *internal ramus* and join the vagus (Fig. 32-24) to be distributed with that. The *external ramus* of the nerve, the spinal accessory nerve, then supplies only two voluntary muscles, the sternocleidomastoid and the trapezius. Afferent fibers in the accessory nerve are from cervical nerves.

### Hypoglossal Nerve

The hypoglossal or 12th cranial nerve apparently is a purely motor nerve, containing no afferent fibers. It arises from the anterior surface of the medulla, just ventral to the olive. Its rootlets come together and it pierces the dura mater at the hypoglossal canal to leave the skull, sometimes as two filaments, sometimes as a single nerve. After the nerve leaves the hypoglossal canal and runs downward, it receives a communication from upper cervical nerves that helps form the ansa cervicalis. The hypoglossal nerve itself is motor to the extrinsic and intrinsic muscles of the tongue.

## AUTONOMIC SYSTEM IN THE HEAD

The autonomic system in the head consists of both sympathetic and parasympathetic fibers, but the main autonomic ganglia in the head are parasympathetic. Two of these, the otic and the submandibular, have been described with the jaw; two others yet to be described lie within the orbit or adjacent to the nose.

### CRANIAL SYMPATHETIC SYSTEM

Most of the postganglionic sympathetic fibers to the face and head are derived from the **superior cervical ganglion** in the upper part of the neck. This ganglion sends fibers along both external and internal carotid arteries. An **external carotid nerve** or several nerves run downward from the ganglion to the base of the external carotid artery. These form a plexus along the external carotid and its branches, sending a smaller plexus downward along the common carotid. The **common carotid plexus** ends on the artery. The **external carotid plexus** ends in part on the branches of this artery, but also in the salivary glands and in the sweat glands of the lower part of the face. A tiny **jugular nerve** from the superior cervical ganglion divides to join the glossopharyngeal and vagus nerves.

The largest branch from the superior cervical ganglion to the head is the **internal carotid nerve,** which passes onto the internal carotid artery and forms an internal carotid plexus about this. The **internal carotid plexus** follows the carotid artery through its extracranial course and onto the part of the artery lying in the cavernous sinus. Here, many of the sympathetic fibers leave the internal carotid, but some of them follow the vessel and its cerebral branches (although autonomic fibers exert relatively little effect on cerebral blood vessels). It is believed also that sensory fibers run with the motor fibers in the carotid plexuses, just as they do in many other sympathetic branches.

While it is close to the middle ear cavity, the internal carotid plexus gives off twigs that join the tympanic plexus on the medial wall of the cavity, but the two main groups of sympathetic fibers leaving the plexus do so in the cavernous sinus. One, the *sympathetic branch to the ciliary ganglion,*

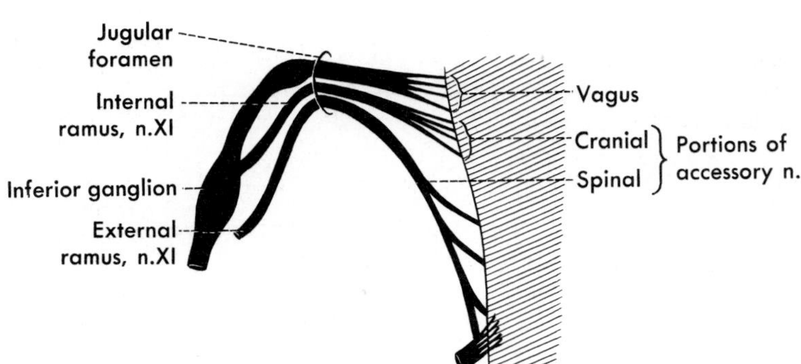


**FIG. 32-24.** The grouping of the cranial rootlets of the accessory nerve as the internal ramus that joins the vagus, and the continuation of the spinal rootlets as the external ramus, or accessory nerve of gross anatomy.

runs forward through the cavernous sinus to the superior orbital fissure to enter the orbit and join the ciliary ganglion. However, its postganglionic sympathetic fibers do not end in this parasympathetic ganglion, but pass through it and help form the short ciliary nerves that the ganglion sends to the eyeball.

The other large group of sympathetic fibers forms the **deep petrosal nerve.** This leaves the internal carotid artery while that vessel lies on the side of the sella turcica and passes to the anterior lip of the foramen lacerum, as does the greater petrosal, to join the greater petrosal nerve and form the nerve of the pterygoid canal (Fig. 32-25). This runs forward to the pterygopalatine fossa, where it ends in the pterygopalatine ganglion. This ganglion is attached to the lower border of the maxillary nerve, and its apparent branches of distribution (primarily to the nose and the palate) contain more 5th nerve fibers than they do sympathetic or parasympathetic fibers. The sympathetic fibers contributed by the deep petrosal nerve do not synapse within the pterygopalatine ganglion but simply pass through it to be distributed with its branches.

Since the preganglionic sympathetic fibers to the face and head arise from thoracic nerves and traverse the cervical sympathetic trunk to reach the superior cervical ganglion, interruption of this trunk denervates the glands and smooth muscle supplied by the ganglion. The resultant ptosis (drooping) of the eyelid and pupillary constriction (the smooth muscle of the upper lid and the dilator pupillae are both supplied by the sympathetic system) are known as *Horner's syndrome.*

## CRANIAL PARASYMPATHETIC SYSTEM

The parasympathetic system consists of both cranial and sacral fibers (p. 62), but only the cranial ones are to be considered here; the sacral parasympathetic system is described in connection with the pelvis.

There are four major pairs of parasympathetic ganglia in the head; they are the ciliary, the pterygopalatine, the submandibular, and the otic. There also are four cranial nerves that contain preganglionic fibers; these are the oculomotor, the facial, the glossopharyngeal, and the vagus. However, the last-named nerve sends its parasympathetic fibers to structures in the thorax and abdomen, those of

**FIG. 32-25.** Diagram of the connection of the pterygopalatine ganglion. Preganglionic parasympathetic fibers (which synapse in the ganglion) are indicated by long broken lines, postganglionic parasympathetic fibers arising from the ganglion by short broken lines, and postganglionic sympathetic fibers by dotted lines. The only afferent fibers (solid lines) shown are those of the fifth nerve, although the deep petrosal nerve is said to contain a few (of spinal origin) and the greater petrosal has afferent fibers of facial origin that are distributed to the back part of the soft palate and probably also, sometimes, fibers for taste from the tongue.

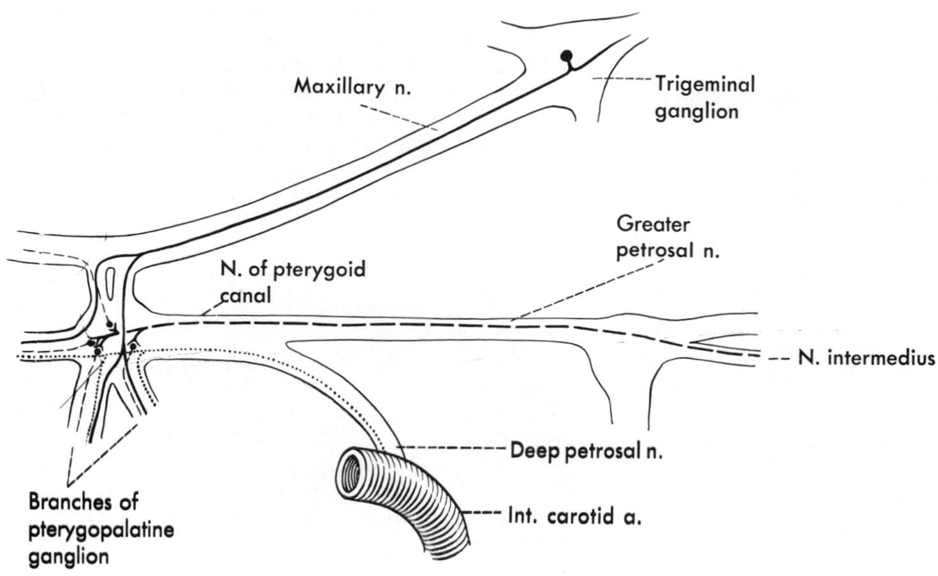

**Table 32-1. Autonomic Ganglia of the Head**

| Ganglion | Preganglionic Fibers Via: | Postganglionic Fibers Distributed To: |
|---|---|---|
| Ciliary | Oculomotor nerve and its inferior division | Ciliary muscle, sphincter pupillae |
| Pterygopalatine | Nervus intermedius part of facial, greater petrosal, and nerve of pterygoid canal | Nose, palate, and lacrimal gland |
| Submandibular | Nervus intermedius part of facial, chorda tympani, and lingual nerve | Submandibular and sublingual salivary glands and smaller glands of oral cavity |
| Otic | Glossopharyngeal nerve, its tympanic branch, and lesser petrosal | Parotid salivary gland |

the thorax ending in small ganglia in the pulmonary and cardiac plexuses, and those of the abdomen ending upon the enteric ganglia in the walls of the digestive tract. Thus, although the vagal parasympathetic fibers belong to the cranial system, they are not connected with cranial parasympathetic ganglia.

Of the four ganglia, the oculomotor nerve supplies preganglionic fibers to the ciliary only, the facial supplies fibers to both the pterygopalatine and the submandibular, and the glossopharyngeal supplies fibers to the otic (Table 32-1). Most of the ganglia also have sympathetic and sensory fibers closely associated with them. The sympathetic fibers are derived from the superior cervical ganglion and pass through or close against the ganglion; similarly, the sensory fibers pass through or adjacent to the ganglion and are for the most part derived from the branches of the trigeminal nerve with which the cranial parasympathetic ganglia are closely connected.

### Ciliary Ganglion

The ciliary ganglion is a small ganglion situated toward the back of the orbit on the lateral side of the optic nerve. It receives its preganglionic fibers through a short stout *oculomotor root* that arises from the oculomotor nerve; it also receives a *communicating branch* from the nasociliary branch of the ophthalmic nerve, which carries sensory fibers; and it usually receives a *sympathetic branch* from the internal carotid plexus. The postganglionic fibers from the ciliary ganglion, mixed with sensory and sympathetic fibers that bypass the ganglion, form short ciliary nerves that enter the eyeball to control the smooth muscle there.

### Pterygopalatine Ganglion

The pterygopalatine ganglion lies in the pterygopalatine fossa, immediately adjacent to the lateral nasal wall at the level of the back end of the middle nasal concha. In the pterygopalatine fossa the ganglion lies immediately below the maxillary branch of the trigeminal nerve and is attached to it by two *pterygopalatine nerves;* most of the fibers of these nerves run past the ganglion and form a major part of the so-called branches of the ganglion (orbital, posterior nasal, nasopalatine, palatine, and pharyngeal). The pterygopalatine ganglion receives its preganglionic parasympathetic fibers from the facial nerve; it is joined also by sympathetic fibers (that run past it but are distributed with its branches) from the internal carotid plexus. The sympathetic and parasympathetic fibers reach the ganglion in the same nerve, the *nerve of the pterygoid canal;* this is formed by the union of the *deep petrosal nerve,* carrying sympathetic fibers from the internal carotid plexus, and the greater petrosal nerve from the facial (Fig. 32-25). As the greater and the deep petrosal nerves reach the anterior lip of the foramen lacerum, they unite and enter the pterygoid canal, which lies in the floor of the sphenoid sinus and opens anteriorly into the pterygopalatine fossa. (The canal is also known as the vidian canal, whence the nerve gets its alternative name, *vidian nerve.*)

The *greater petrosal nerve* leaves the facial nerve in the pars petrosa of the temporal bone at the level of the geniculate ganglion; it runs forward and medially through the pars petrosa in a small canal, escaping through the hiatus of this canal to run in the middle cranial fossa to the foramen lacerum, when it is joined by the deep petrosal. The greater petrosal nerve carries not only parasympathetic fibers but also sensory ones, both running in the nervus intermedius part of the facial nerve.

The postganglionic fibers derived from the pterygopalatine ganglion run for the most part in the nasal and palatine nerves; although these are listed as branches of the ganglion, it should be emphasized again that they consist of a mixture of sensory fibers derived from the maxillary nerve, sympathetic fibers brought in by the deep petrosal, and only in part of postganglionic parasympathetic fibers. The sensory fibers derived from the facial nerve probably run with the palatine nerves to reach a small part of the soft palate.

The postganglionic parasympathetic fibers to the nose and soft palate are both vasodilatory and secretory; their activity accounts for some of the speed with which the nose can become stuffed up. The pterygopalatine ganglion also sends postganglionic fibers to the maxillary nerve by way of the pterygopalatine nerves; these travel with the maxillary nerve and then with its zygomatic and zygomaticotemporal branches to reach the lacrimal gland, to which they are secretory.

#### Otic Ganglion

The otic ganglion lies on the medial side of the mandibular branch of the trigeminal nerve, just outside the foramen ovale and deep in the infratemporal fossa. It may be adherent to the medial side of the mandibular nerve or may be on some of its motor branches. Indeed, the nerves to the tensor veli palatini and to the tensor tympani muscles are listed as branches of the otic ganglion, although so far as is known they contain no fibers from this ganglion but consist, rather, of fibers derived from the mandibular nerve. The otic ganglion receives its preganglionic fibers from the *lesser petrosal nerve.* This is a derivative of the glossopharyngeal, but the fibers have a somewhat circuitous course. The glossopharyngeal nerve gives off a *tympanic branch* that enters the middle ear cavity and there, with sympathetic twigs from the internal carotid plexus, and usually a twig from the facial nerve, forms a tympanic plexus; the preganglionic fibers of the glossopharyngeal that are destined for the otic ganglion run through the plexus, however, and as they leave it form the lesser petrosal nerve (Fig. 32-26). This emerges into the middle cranial fossa, runs just lateral to the greater petrosal nerve, and leaves the skull close to or through the foramen ovale to join the otic ganglion. There also may be in the lesser petrosal nerve some fibers derived from the facial nerve that reach it through communications between the greater and lesser petrosal nerves; these are too tiny to be found in the usual dissection. The ganglion has a *communicating ramus* with the *chorda tympani* nerve through which it also can receive facial fibers or send glossopharyngeal fibers toward the submandibular ganglion. The postganglionic fibers of the otic ganglion are believed to go entirely into the auriculotemporal branch of the mandibular nerve and to be distributed, with sensory fibers of this nerve, to the parotid gland.

#### Submandibular Ganglion

The submandibular ganglion lies medial to the mandible, suspended from the lingual nerve and close to the upper border of the submandibular gland. It is attached to the lingual nerve by *communicating branches* that represent both preganglionic fibers coming into the ganglion and postganglionic fibers that leave it. Its preganglionic fibers reach the lingual branch of the trigeminal through the *chorda tympani* (Fig. 32-23). This leaves the facial nerve in the petrous part of the temporal bone,

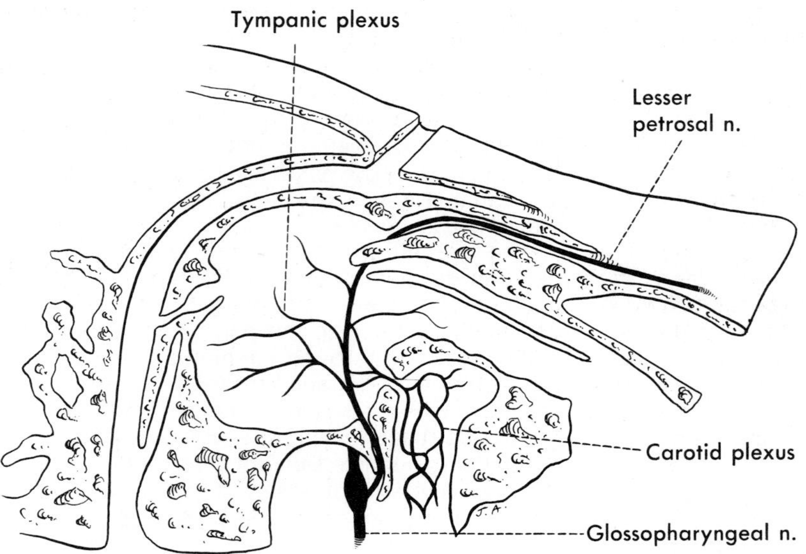


**FIG. 32-26.** The tympanic plexus and the origin of the lesser petrosal nerve.

traverses the middle ear cavity, and leaves the skull through the petrotympanic fissure. It soon joins the posterior border of the lingual nerve. The preganglionic fibers then form a part of the lingual nerve until they leave it to end in the submandibular ganglion. Some authorities believe that the ganglion receives some preganglionic fibers from the glossopharyngeal nerve also, through the ramus communicans with the otic ganglion.

The submandibular ganglion gives off some of its postganglionic fibers as secretory fibers to the submandibular gland. Other postganglionic fibers return to the lingual nerve and are distributed through branches of this nerve to the sublingual salivary gland and to smaller glands of the oral cavity.

#### Nervus Terminalis

The nervus terminalis, seldom seen in the dissecting laboratory, now is regarded as containing both autonomic and sensory fibers, although little is known concerning it. It consists of a few fibers that lie in the pia mater on the lower surface of the frontal lobe of the brain, just medial to the olfactory tract. Its fibers join olfactory nerve fibers and are distributed to the nasal mucosa. Along the course of the nerves there are small groups of ganglion cells, constituting the terminal ganglia; thus, these nerves seem to contain both preganglionic and postganglionic fibers, as well as afferent ones. Their function is unknown.

## GENERAL REFERENCES AND RECOMMENDED READINGS


BEEVOR CE: On the distribution of the different arteries supplying the human brain. Philos Trans R Soc sB200:1, 1909

BURR HS, ROBINSON GB: An anatomical study of the gasserian ganglion, with particular reference to the nature and extent of Meckel's cave. Anat Rec 29:269, 1925

CAMPBELL EH: The cavernous sinus: Anatomical and clinical considerations. Ann Otol Rhinol Laryngol 42:51, 1933

DUBOIS FS, FOLEY JO: Experimental studies on the vagus and spinal accessory nerves in the cat. Anat Rec 64:285, 1936

EVERETT NB: Functional Neuroanatomy, 6th ed. Philadelphia, Lea & Febiger, 1971

FORBES HS: Physiologic regulation of the cerebral circulation. Arch Neurol Psychiatry 43:804, 1940

HAYNER JC: Variations of the torcular Herophili and transverse sinuses. Anat Rec 103:542, 1949

KERR FWL: The divisional organization of afferent fibres of the trigeminal nerve. Brain 86:721, 1963

KULLMAN GL, DYCK PJ, CODY DTR: Anatomy of the mastoid portion of the facial nerve. Arch Otolaryngol 93:29, 1971

KUNTZ A: The Autonomic Nervous System, 4th ed. Philadelphia, Lea & Febiger, 1953

MCCULLOUGH AW: Some anomalies of the cerebral arterial circle (of Willis) and related vessels. Anat Rec 142:537, 1962

MITCHELL GAG: Anatomy of the Autonomic Nervous System. Edinburgh, Livingstone, 1953

PENFIELD W, MCNAUGHTON F: Dural headache and innervation of the dura mater. Arch Neurol Psychiatry 44:43, 1940

RANSON SW, CLARK SL: The Anatomy of the Nervous System, 10th ed. Philadelphia, WB Saunders, 1959

RAY BS, HINSEY JC, GEOHEGAN WA: Observations on the distribution of the sympathetic nerves to the pupil and upper extremity as determined by stimulation of the anterior roots in man. Ann Surg 118:647, 1943

STOPFORD JSB: The arteries of the pons and medulla. J Anat 50:131; 255, 1916

STREETER GL: The development of the venous sinuses of the dura mater in the human embryo. Am J Anat 18:145, 1915

SUNDERLAND S: Neurovascular relations and anomalies at the base of the brain. J Neurol Neurosurg Psychiatry ns11:243, 1948

TARLOV IM: Sensory and motor roots of the glossopharyngeal nerve and the vagus-spinal accessory complex. Arch Neurol Psychiatry 44:1018, 1940

TORKILDSEN A: The gross anatomy of the lateral ventricles. J Anat 68:480, 1934

WALTNER JG: Anatomic variations of the lateral and sigmoid sinuses. Arch Otolaryngol 39:307, 1944

# 33

# THE EAR, ORBIT, AND NOSE

The ear, orbit, and nose comprise or contain the three major organs of special sense—the ear that of hearing, the orbit that of sight, and the nose that of smell. Taste, also a special sense, is of course mediated through taste buds that are primarily on the tongue. Unlike the nerve fibers from the larger organs of special sense, fibers for taste are incorporated in several of the cranial nerves—the facial and glossopharyngeal primarily, and in small degree, the vagus—along with fibers of other functions. In contrast, the ear, the eye, and the olfactory part of the nose have their own nerves. Associated with the organs of sense are other parts, such as a transmission system by which sound reaches the inner part of the ear, muscles and nerves that are responsible for movements of the eye, the nerves and vessels to parts of the nose other than the small area that is concerned with smell.

## EAR

The ear is divided for purposes of description into the external ear, which includes not only the part projecting from the side of the head but also the canal leading inward (Fig. 33-1); the middle ear, separated from the external ear canal by a membrane but opening into the pharynx through a narrow tube; and the internal ear, which receives sound vibrations transmitted to it through the middle ear and has to do also with the balance of the head and body.

### EXTERNAL EAR

The two parts of the external ear are the auricle (auricula), the projecting part of the ear, and the external acoustic meatus, the ear canal.

The chief named parts of the **auricle,** as seen in a lateral view, are shown in Figure 33-2. As can be determined by palpation, the *lobule,* consisting of skin and intervening connective tissue, is the only part of the auricle that is not supported by cartilage. Elsewhere, the

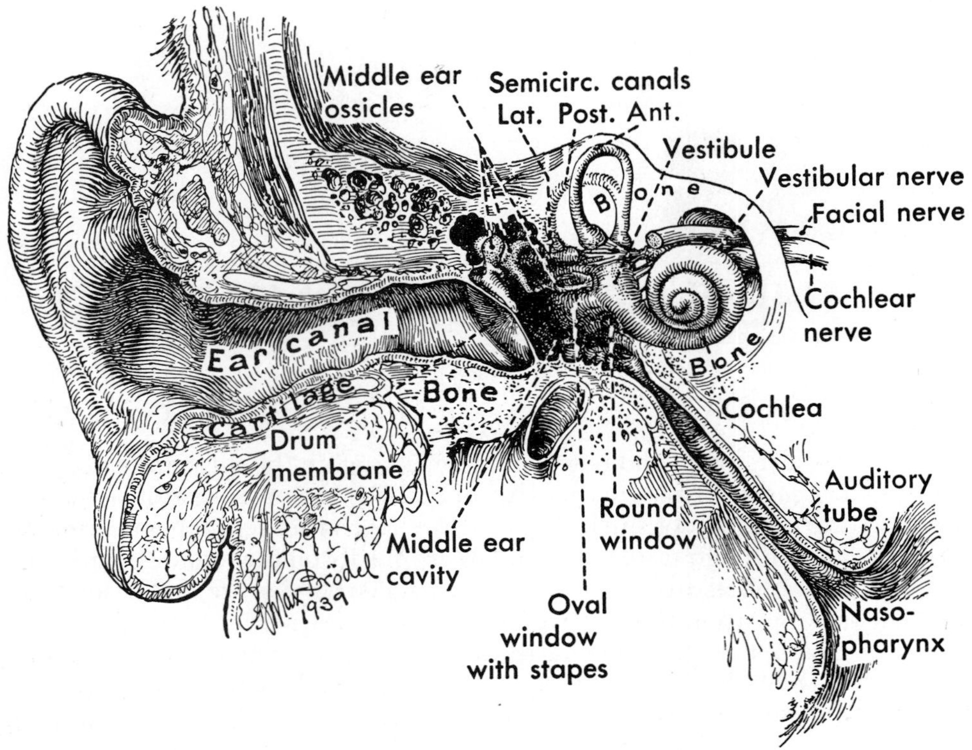


**FIG. 33-1.** The external, middle, and internal ear. (Brödel M: Three Unpublished Drawings of the Anatomy of the Human Ear. Philadelphia, WB Saunders, 1946)

**FIG. 33-2.** Parts of the auricle.

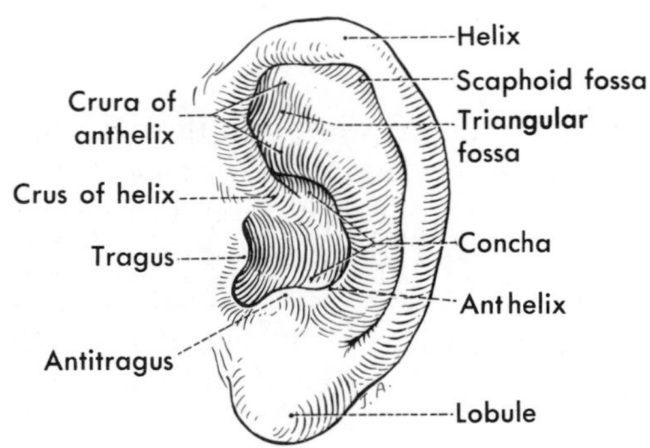


skin of the auricle is firmly attached to the perichondrium that covers the auricular cartilage, and the shape of the cartilage can be largely seen in the intact ear. The one feature of this cartilage that cannot be so appreciated is that a part of it projects inward, from the *tragus* and the notch between it and the *antitragus,* to form the anterior and inferior wall of a part of the external acoustic meatus. The medial side of the ear has few named parts, and these are eminences corresponding to the depressions on its lateral surface—the eminence of the *concha,* for instance. The auricular muscles have already been described with the facial ones.

The **external acoustic meatus** (external auditory meatus, external ear canal) leads from the deepest part of the concha to the tympanic membrane or eardrum. It is not straight, nor is it of uniform diameter throughout; as it is traced inward, a first part is directed somewhat forward and upward, a second succeeding part is directed slightly backward, and the third and longest part turns again to run forward and slightly downward.

A lateral part of the meatus, about 8 mm in length, has for its walls the troughlike cartilage of the meatus that extends inward from the cartilage of the auricle. The deficiency in the roof is completed by dense fibrous tissue, which converts the trough into a canal. Anteriorly and inferiorly, the cartilage extends farther medially, so that there is a part of the canal that has a bony roof and posterior wall, but a cartilaginous floor and anterior wall. The cartilage is firmly attached to the bony part of the canal by dense connective tissue. In the newborn, the bony canal has almost no length, for it is formed primarily by the small piece of bone known as the **tympanic ring** (see Fig. 31-12), to which the eardrum is attached. The term "tympanic ring" sometimes is used to designate that part of the temporal bone to which the eardrum is attached in the adult. Since the tympanic ring and the tympanic part of the temporal bone in the adult are incomplete above (where there is a gap called the *tympanic incisure*), the roof of the bony external meatus is completed here by the petrous part of the temporal bone.

The **tympanic membrane** or eardrum membrane (Figs. 33-1 and 33-3) is at the medial end of the acoustic meatus and separates this from the middle ear or tympanic cavity. In the newborn, it faces almost inferiorly and only slightly laterally, but with growth of the tympanic ring, its position changes. Even in the adult, however, it is still oblique, sloping medially from top to bottom and medially from posteriorly to anteriorly, so that the anterior wall and floor of the external meatus are longer than its roof and posterior wall. The tympanic membrane consists of a central fibrous core, composed of radial and circular fibers, interposed between a thin layer of skin on the side of the acoustic meatus and one of mucosa on the side of the tympanic cavity. At its periphery, it is attached by a fibrocartilaginous ring into a groove, the *sulcus tympanicus,* in the tympanic part of the temporal bone, and at its center, it is attached to the handle of the malleus (a bone of the middle ear). At this attachment, the *umbo,* the membrane is drawn inward so that it is somewhat cone shaped. Above and slightly anterior to the umbo, another part of the malleus (lateral process) likewise is attached to the inner surface of the membrane. Extending forward and backward from this *mallear prominence* are anterior and posterior *mallear folds* on the inner surface of the membrane. The part of the membrane above the mallear prominence and folds is attached peripherally at the tympanic incisure and is devoid of the central layer of fibrous tissue that forms the major part of the eardrum. This thin part is the *pars flaccida,* and the major remaining part is the *pars tensa.*

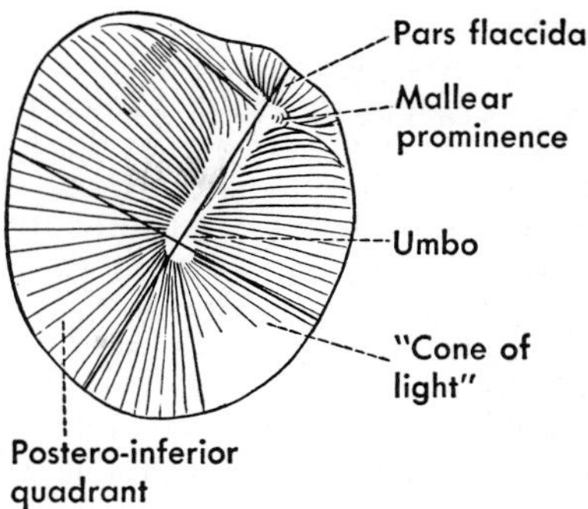


**FIG. 33-3.** The right tympanic membrane as it might appear through an otoscope, but with its quadrants indicated on it. Since the quadrants are determined by a line through the handle (manubrium) of the malleus and one at right angles to this, these lines are not perpendicular and horizontal. The "cone of light" is a reflection of light from the membrane.

The tympanic membrane responds to air vibrations that are collected by the auricle and concentrated upon it through the acoustic canal. As it moves in response to such vibrations, it necessarily also moves the malleus, which is connected to it. The malleus in turn articulates with a second bone, and this articulates with a third (Fig. 33-6). Through the movement of this chain of bones vibrations are transmitted across the middle ear cavity to the internal ear.

The **blood supply** of the lateral surface of the auricle is from twigs (anterior auricular branches) of the superficial temporal artery. These help supply the meatus, which is supplied also by the deep auricular artery (a branch of the maxillary that runs through the parotid gland to the meatus and continues to the external surface of the eardrum). The medial side of the auricle is supplied by the auricular branch of the posterior auricular artery, a branch of the external carotid.

The **nerve supply** to most of the auricle is through the great auricular nerve. The auriculotemporal branch of the mandibular also contributes to the lateral surface of the auricle, as does the lesser occipital nerve to the medial surface. The major part of the external meatus and outer surface of the tympanic membrane is innervated through the auriculotemporal, but a posterior and lower part of both meatus and membrane is supplied by the tiny auricular branch of the vagus. This may contain fibers from the glossopharyngeal and facial nerves in addition to vagal fibers and supplies also a small bit of skin where the ear is attached over the mastoid process.

## MIDDLE EAR

The middle ear or **tympanic cavity** (Figs. 33-1 and 33-4 through 33-7) lies in the temporal bone and is separated from the external acoustic meatus by the tympanic membrane. It has a greater height than the meatus and the tympanic membrane. Its floor is a little below the level of the inferior border of the membrane, and it extends well above the upper border. This upper extension of the tympanic cavity is the **epitympanic recess,** often referred to by clinicians as the "attic."

The tympanic cavity is wider above than it is below, and the epitympanic recess is still wider; however, the inward slant of the tympanic membrane to the umbo considerably narrows the cavity at the umbo's level. The floor of the cavity, called its *jugular wall* because it is also the roof of the jugular fossa seen on the outside of the skull, is exceedingly thin unless it is invaded by mastoid air cells. It presents a tiny **tympanic canaliculus** through which the *tympanic branch of the glossopharyngeal nerve* enters the middle ear. The anterior or *carotid wall* of the tympanic cavity is incomplete. Below, a thin layer of bone intervenes between the cavity and the internal carotid artery. The bone presents small *caroticotympanic foramina* through which caroticotympanic rami of the internal carotid artery, and one or more nerve twigs from the internal carotid plexus, enter the tympanic cavity. Above this partial wall is the relatively large *opening of the auditory tube.* The *roof* of the cavity is the **tegmen tympani** or tegmental wall; it also is thin and forms a small posterolateral part of the floor of the middle cranial fossa.

The posterior or *mastoid wall* of the tympanic cavity is, like the anterior wall, incomplete above. Below, there is a bony wall between the mastoid air cells and the tympanic cavity, but above this, the cavity extends back, as the **aditus** (*aditus ad antrum,* or entrance to the antrum), into the mastoid part of the bone. As it reaches the mastoid part, the aditus gives way to the larger cavity of the **mastoid antrum.** From the mastoid antrum a series of connecting air-filled and mucosa-lined spaces extend through most of the mastoid process; these are the **mastoid air cells.** They may extend forward into the roof, floor, or walls of the tympanic cavity and may come to partly surround some of the particularly hard bone that encloses the internal ear.

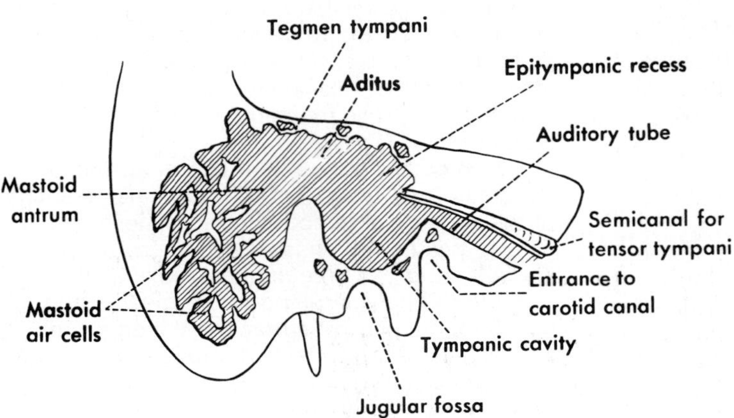


**FIG. 33-4.** Diagram of the middle ear cavity and its connecting air spaces, in a sagittal section.

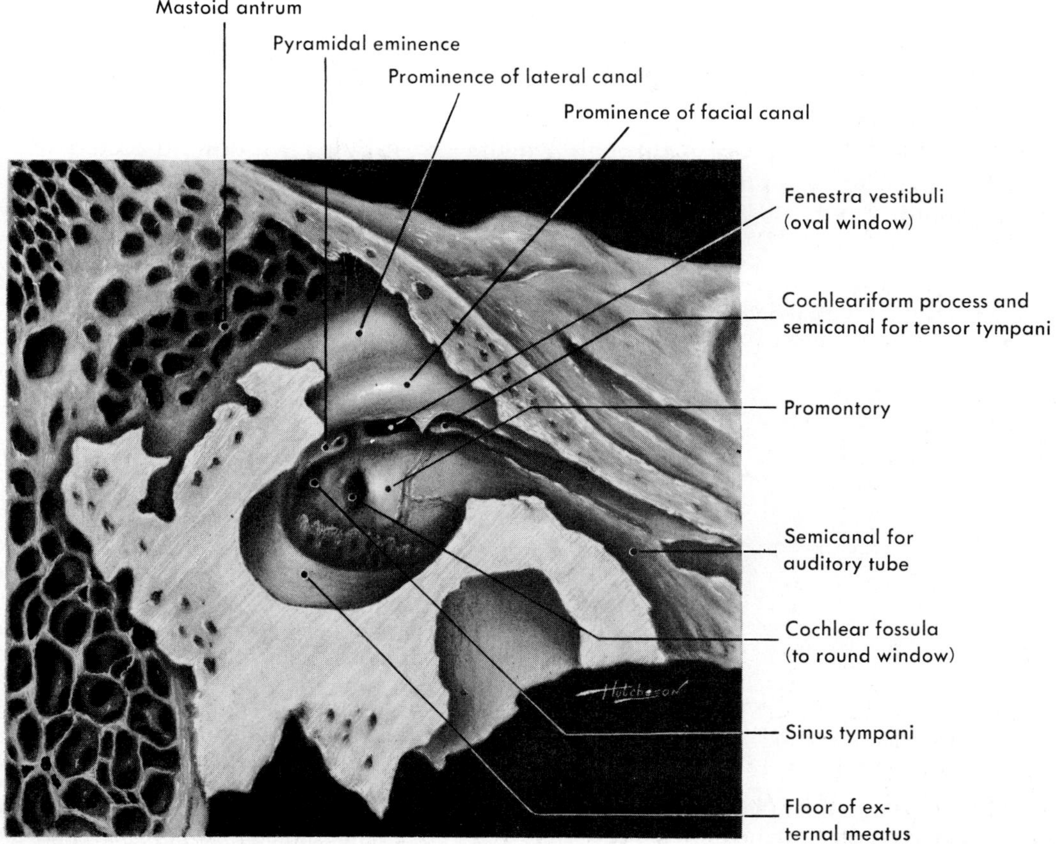


**FIG. 33-5.** The medial wall of the tympanic cavity.

It is because of the sinuosity of the connections among the mastoid air cells that mastoiditis, before the advent of antibiotics, was regularly treated by exenteration (removal) of the mastoid cells. Necrosis of bone as a result of mastoiditis easily can lead to fatal infection, for the internal jugular vein is adjacent to the floor of the tympanic cavity, and the sigmoid sinus is adjacent to the mastoid air cells posteromedially and may be separated from them by only a thin lamina of bone.

Because of the close relations of the middle ear to great vessels and to the brain, severe bleeding or the escape of cerebrospinal fluid through a ruptured tympanic membrane may occur following an accident or blow on the head. Either is, of course, evidence of a dangerous fracture.

The medial or *labyrinthine wall* of the tympanic cavity (Fig. 33-5) is of special interest. The prominent bulge of this medial wall is the **promontory,** formed by the large basal coil of the cochlea, the part of the internal ear concerned with hearing. On the promontory there is a delicate **tympanic plexus** (see Fig. 32-26), contributed to mostly by the *tympanic branch of the glossopharyngeal nerve* but joined also by twigs from the internal carotid plexus (*caroticotympanic nerves*) and usually by a *twig from the facial nerve* or its greater petrosal branch. Extensions from this plexus supply the mucous membrane of the tympanic cavity, the tympanic antrum and mastoid air cells, and the auditory tube. In addition, a part of the tympanic branch of the 9th nerve leaves the promontory to travel forward as the *lesser petrosal nerve* to the otic ganglion. Above the promontory in a depression, the *fossula fenestrae vestibuli,* is the opening into which the base of the stapes fits. This is the **fenestra vestibuli** (vestibular window, or oval window because of its shape). Below this, posteroinferior to the promontory, is another depression, the *fossula fenestrae cochleae* or cochlear fossula, that leads to a second opening into the internal ear, the **fenestra cochleae** (cochlear window, also called round window).

This opening is sealed during life by the **secondary tympanic membrane.**

Above and anterior to the fenestra vestibuli is the **cochleariform process** through which the tendon of the *tensor tympani muscle* enters the tympanic cavity. A little posterior to this fenestra is the tiny **pyramidal eminence** through which the tendon of the *stapedius muscle* enters the tympanic cavity. Above the vestibular window, directed posteriorly and somewhat downward, is the **prominence of the facial canal;** the *facial nerve* produces this, bulging the bone here after it has left the geniculate ganglion. Above and somewhat posterior to the prominence of the facial canal, on the medial wall of the aditus, is a broader bulge, the **prominence of the lateral semicircular canal,** caused by the enlarged anterior or ampullary end of this canal (and the adjacent ampullary end of the superior canal). These canals are part of the portion of the ear having to do with balance.

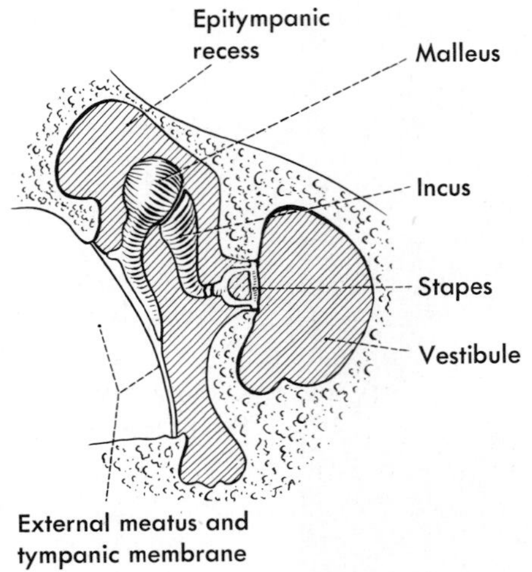


**FIG. 33-6.** Diagram of the three bones of the middle ear cavity as they extend between the tympanic membrane and the vestibule of the internal ear. In order to show its form, the stapes has here been rotated 90°—its crura actually are anterior and posterior rather than superior and inferior.

Recognition of the prominence of the facial canal and of the prominence of the lateral semicircular canal is critical in attempts to restore hearing by making an additional opening into the internal ear (fenestration), because the opening must be made into the lateral semicircular canal at its ampullary end. This operation is done to afford a second movable window between the middle and internal ear when, because of ossification of the membrane by which it is held in the fenestra vestibuli, the stapes becomes immovable (*otosclerosis*). It is also particularly important, in operations for mastoiditis, to recognize the course of the facial nerve.

A second way of restoring hearing after otosclerosis is to mobilize or replace the stapes. Stapes operations have certain advantages over fenestration in restoring hearing and are currently more often employed as the primary operation for otosclerotic deafness.

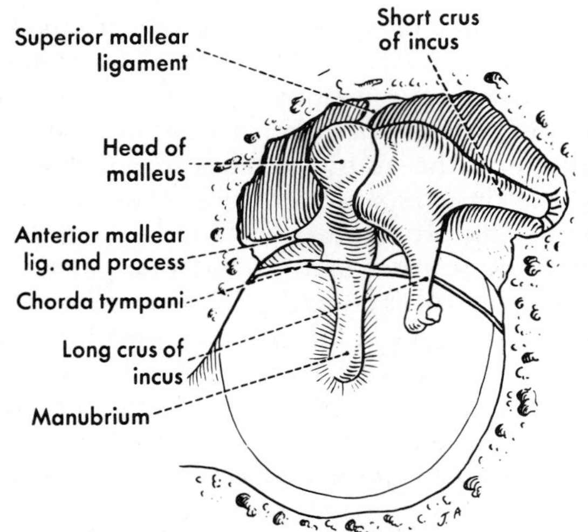


**FIG. 33.7** The malleus, incus, and tympanic membrane seen from the medial side.

The **auditory tube** (pharyngotympanic tube, eustachian tube) opens through the anterior wall of the tympanic cavity. It originally was an outgrowth from the pharynx, representing the first pharyngeal pouch, and the middle ear cavity and mastoid air cells are expansions from it. From the ear, it extends forward, medially, and downward in a bony tube, called a **semicanal,** that lies in the temporal bone and opens on the external base of the skull between the foramen spinosum and the carotid canal. During life, however, the tube has also a cartilaginous part; the medial wall and the upper part of the lateral wall are composed of a folded piece of cartilage. The remainder of the tube is completed by membrane. The cartilage extends to the wall of the pharynx, where the opening of the tube and the elevation (torus) above and medial to this opening can be seen on the interior of the pharynx.

Immediately above the auditory tube, opening into the tympanic cavity through the cochleariform process, is the tube or "semicanal" for the **tensor tympani muscle.** The muscle arises in part from this canal and in part from the cartilage of the auditory tube and the greater wing of the sphenoid dorsolateral to the tube. As it enters the tympanic cavity, the tendon of the muscle turns sharply laterally to attach to the malleus.

The tympanic cavity and inner surface of the tympanic membrane, the mastoid antrum and cells, and the auditory tube are **innervated** by the tympanic plexus, therefore primarily through the tympanic branch of the glossopharyngeal nerve. There are a number of **arteries,** all of them too tiny to be recognizable in an ordinary dissection. They include caroticotympanic branches from the internal carotid, parts of the stylomastoid and posterior tympanic arteries from the posterior auricular, the inferior tympanic artery from the ascending pharyngeal, the superior tympanic artery from the middle meningeal, and the anterior tympanic artery from the maxillary.

The auditory tube is normally closed at its pharyngeal end, but must on occasion be opened if the pressure in the tympanic cavity is to be equalized with atmospheric pressure. The **tensor veli palatini,** which arises in part from the cartilaginous part of the tube, is primarily responsible for opening it. The fact that this muscle contracts when one yawns or swallows explains the effect these actions have in alleviating the discomfort arising from unequalized pressure.

## Auditory Ossicles

The three bones of the middle ear cavity (Figs. 33-6 and 33-7) are the malleus (hammer), incus (anvil), and stapes (stirrup). They extend across the cavity (covered by mucous membrane) from the tympanic membrane to the internal ear.

The round upper end of the **malleus** is its *head* (caput). This fits into the concavity of the body of the incus, which lies behind it; both bones lie in part in the epitympanic recess. Below a slight neck, the malleus has a short *anterior process* and a more prominent *lateral process* that is attached to the inner surface of the tympanic membrane. The long lower part of the malleus is the *manubrium* (handle), attached at its lower end to the umbo. The tendon of the **tensor tympani muscle** attaches to the upper part of the manubrium. The tensor tympani muscle is innervated by a branch of the mandibular nerve and by its reflex contraction tends to check too great movement of the eardrum in response to loud noises. The malleus is attached to the eardrum by both the manubrium and the lateral process; it is suspended by ligaments, including one that comes down from the tegmen tympani; and it has a synovial joint, provided with ligaments of elastic tissue, between it and the incus.

Below the posterior mallear fold of the tympanic membrane, and passing upward and forward medial to the malleus so as to blend with the anterior fold, is a fold of mucosa that contains the *chorda tympani nerve.* This slender nerve leaves the facial nerve in the posterior wall of the tympanic cavity and traverses the *canaliculus of the chorda tympani* between the facial canal and the cavity. It then arches upward and forward between the incus and the malleus and leaves the tympanic cavity through a second canaliculus to emerge at the petrotympanic fissure. Beyond this, it joins the lingual branch of the mandibular nerve.

The body of the **incus** is concave anteriorly, where it articulates with the head of the malleus. From the body extend two *crura,* of which the upper, the short crus, extends almost horizontally backward to be attached by a ligament to the upper part of the posterior wall of the tympanic cavity. The long crus projects downward into the tympanic cavity, almost paralleling the manubrium mallei but lying posteromedial to it. At its lower end, this crus suddenly bends medially to articulate with the third bone, the stapes, through a nodule of cartilage called the *lenticular process.*

The **stapes,** shaped like a stirrup, has a head that receives the articulation of the long process of the incus. From the head proceed two limbs (crura), which join the *base* of the stapes, the footplate of the stirrup. The base is approximately oval and fits into the vestibular window on the medial wall of the tympanic cavity. Its inner surface is in contact with fluid in the inner ear, so when the stapes is removed one can look from the middle into the inner ear. The stapes is held in place by an *annular ligament* that not only seals the fluid within the inner ear but allows the stapes to move back and forth as vibratory impulses are transmitted to it from the eardrum via the malleus and the incus. The tendon of the small **stapedius muscle** emerges from the posterior aspect of the tympanum through an aperture at the end of the pyramidal eminence and attaches to the head of the stapes. This little muscle, innervated by a twig of the facial nerve, reflexly contracts to prevent too great oscillation of the stapes.

## Facial Nerve

As evidenced by the prominence of the facial canal, the facial nerve is closely related to the middle and internal ears during a part of its course. The relations and branches of this part of the nerve, therefore, are best examined during dissection of the ear.

The facial nerve enters the petrous part of the temporal bone through the internal acoustic meatus (Fig. 33-8), in company with the 8th (vestibulocochlear) nerve and the labyrinthine artery. It lies above the 8th nerve and has two roots, a large voluntary motor root and a smaller nervus intermedius that consists of sensory and parasympathetic fibers and lies between the motor root and the 8th nerve or is attached to the latter. The two roots join in the acoustic canal. At the lateral end of the acoustic canal, the facial nerve enters the **facial canal,** in which it runs to the stylomastoid foramen. In this course, it first passes laterally between the upper parts of the bony vestibule and the bony cochlea, where it bears its sensory ganglion, the **geniculate ganglion,** and gives off the greater petrosal nerve. This nerve enters the middle cranial fossa through the hiatus of the canal of the greater petrosal nerve. At the geniculate ganglion, the facial nerve makes an abrupt turn (the *geniculum*) posteriorly, and, as it passes across the lateral wall of the vestibule (medial wall of the tympanic cavity) just below the lateral semicircular canal, it raises the prominence of the facial canal (Fig. 33-5). Behind the vestibule, the nerve turns somewhat more gradually to run downward in the bony wall between the middle ear cavity and the mastoid antrum and air cells.

Before it emerges at the stylomastoid foramen, the facial nerve gives off three small branches: a twig to the stapedius as it passes that muscle; the chorda tympani, already described; and a tiny branch that joins the auricular branch of the vagus.

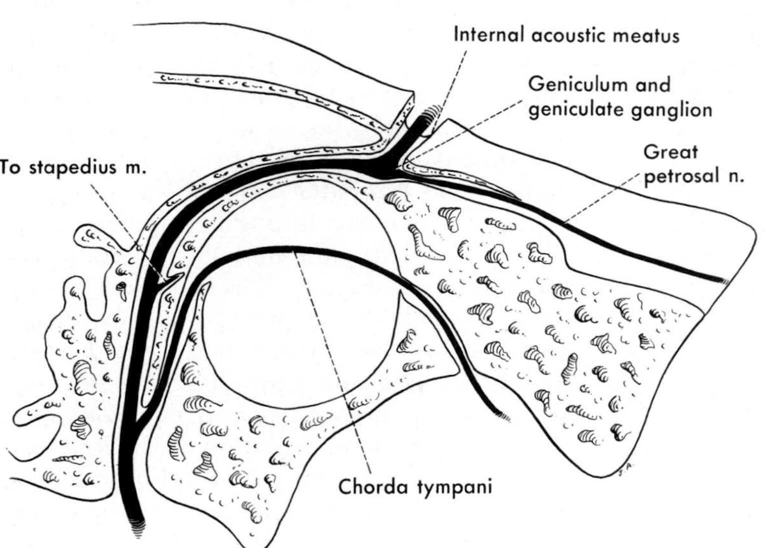


**FIG. 33-8.** The facial nerve in the temporal bone. The relation of the chorda tympani to the malleus and incus is shown in Figure 33-7.

In the lower part of its course in the facial canal, the facial nerve is accompanied by the **stylomastoid artery,** a branch of the posterior auricular that runs retrogradely along the nerve, helping to supply it and the middle ear cavity.

## INTERNAL EAR

The internal ear consists of a series of cavities in the petrous part of the temporal bone (these have particularly dense bony walls and give the name petrous—rocklike—to this part of the bone) and a series of membranous ducts and sacs that lie within the bony cavities. The bony part of the internal ear is known as the **osseous labyrinth,** and the membranous ducts and sacs are known as the **membranous labyrinth.** The membranous labyrinth does not fill the bony one (Fig. 33-9), although it is in places closely attached to the inner wall of this and in other places is attached by delicate trabeculae. The space between the membranous and bony labyrinths is filled with **perilymph** or perilymphatic fluid. Similarly, the membranous labyrinth is filled with **endolymph** or endolymphatic fluid. The bony labyrinth forms a continuous mass, and the cavity within it is one continuous cavity. Similarly, all parts of the membranous labyrinth are connected together. However, both bony and membranous labyrinths are divisible into anterior and posterior parts, with different functions. The anterior part, the **cochlea,** is concerned with hearing, and the bony cochlea contains a membranous **cochlear duct.** The posterior part is concerned with balance. Its osseous parts are the **vestibule** and three **semicircular canals.** The vestibule contains two membranous parts, the **utriculus** and the **sacculus,** and each bony semicircular canal contains a membranous canal, called a **semicircular duct.**

Descriptions of the internal ear are easier to follow if emphasis is placed on its functional divisions rather than the differences between the bony and membranous parts. The following description is based necessarily in part upon histologic studies, since in gross dissection of the ear little more can be done than to ascertain the position of the chief bony parts and display the perilymphatic cavity by removing a part of the wall of the bony labyrinth. Because of the small size of the internal ear, this dissection is a tedious and time-consuming one.

### Vestibule and Semicircular Canals

The **vestibule** is the central chamber of the bony labyrinth. The bony cochlea is continuous with it anteriorly, and the bony semicircular canals, lying largely above and lateral to it, each are continuous with it at both their ends. Its relationship to the middle ear cavity can best be observed by noting that the fenestra vestibuli, into which the stapes fits, opens from the tympanic cavity into the vestibule when the stapes is removed and that a cavity within the promontory of the middle ear (produced by the large basal turn of the cochlea) also opens into the vestibule. (Another part of the perilymphatic cavity in the basal turn of the cochlea is separated from the middle ear cavity by the secondary tympanic membrane of the round or cochlear window.)

The vestibule (Figs. 33-9 and 33-10) is a small oval chamber approximately 6 mm long anteroposteriorly, a millimeter or two less in vertical diameter, and perhaps 3 mm wide. The facial nerve lies first anterior and dorsal to the vestibule and then in the lateral wall of the vestibule. The medial wall of the vestibule, between this cavity and the internal acoustic meatus, presents three small areas with minute perforations that transmit branches of the vestibular nerve from the meatus to the sense organs in the vestibule.

The sense organs are within two membranous sacs: the posterior sac, somewhat oval, is the **utriculus,** and the anterior sac, more rounded, is the **sacculus.** Each makes an impression on the medial wall of the vestibule, and they are united only by a tiny **utriculosaccular duct.** From the utriculosaccular duct, the **endolymphatic duct** runs posteriorly to emerge through the bone of the posterior cranial fossa (posterior surface of the petrous bone) and expand into a blind pouch, the **endolymphatic sac,** in the dura just above the sigmoid sinus. The utricule and saccule usu-

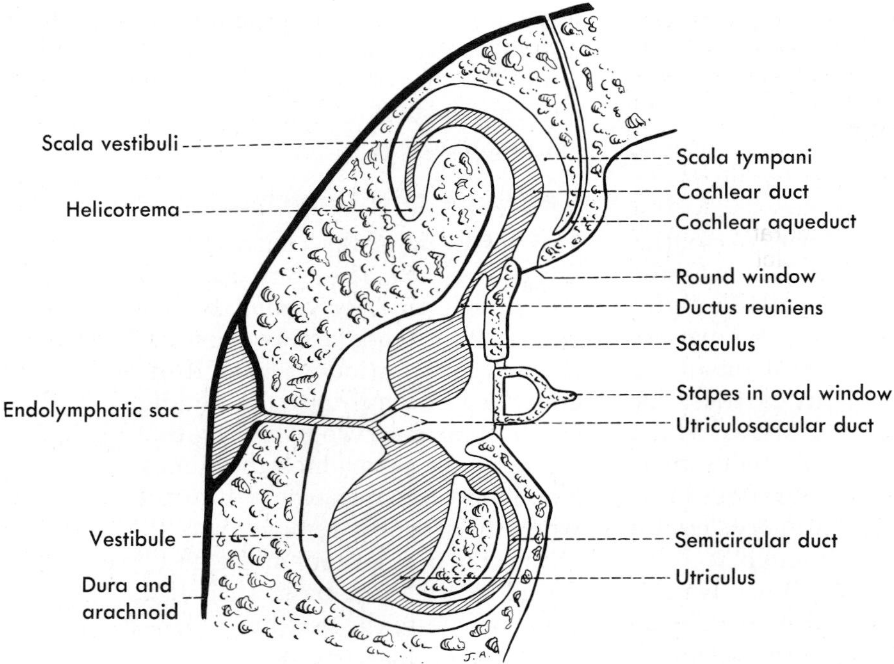


**FIG. 33-9.** Diagram of the perilymphatic and endolymphatic spaces; the latter, in the membranous labyrinth, is shaded.

**FIG. 33-10.** Casts of the bony labyrinth (perilymphatic space). **A** is a lateral view of the left labyrinth, **B** a medial view of the right one, and **C** a view of the left one from above. (Henle J: Handbuch der systematischen Anatomie des Menschen, vol 2. Braunschweig, Vieweg und Sohn, 1868)

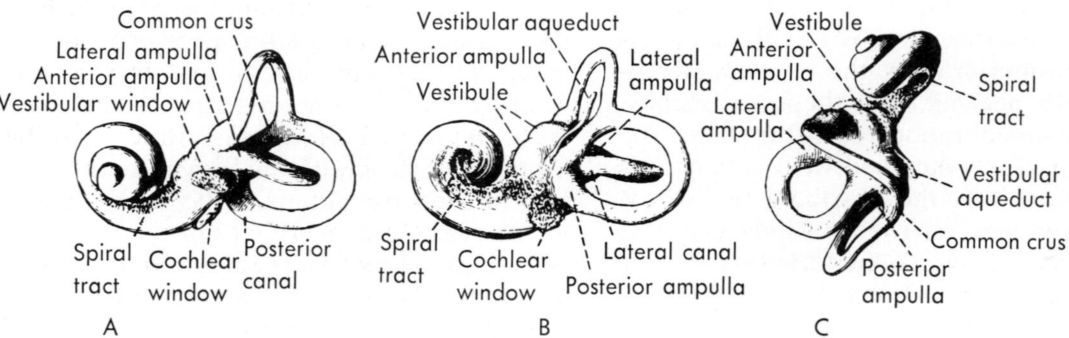


ally are destroyed in gross dissection; however, the endolymphatic sac can be found and opened by incising the dura in the fovea for this sac (p. 871). The bony canal through which the endolymphatic duct runs necessarily opens at its proximal end into the vestibule and is called the **vestibular aqueduct.**

Abnormal increase in the endolymph, termed hydrops of the internal ear or Ménière's disease, produces distressing symptoms and may eventually destroy the sense organs of the ear. An operation to alleviate this condition takes advantage of the position of the endolymphatic sac to create an opening between it and the subarachnoid space and thus relieve the endolymphatic pressure.

The sense organs within the utricle and saccule are essentially similar. They are called **maculae,** and there is a utricular and a saccular macula. A macula consists of a thick layer of epithelium, called neuroepithelium because its important functional component consists of "hair cells" that receive the stimulation. These cells have tiny processes projecting from their free surfaces, and the processes are embedded in a membrane in which are also embedded tiny crystals of calcium carbonate, called **statoconia** (otoconia) or otoliths. The neuroepithelial cells are innervated by fibers of the vestibular division of the 8th nerve. They are stimulated by the statoconia acting on their hairs.

The utriculus and sacculus are generally believed to be concerned entirely with balance (although some authorities think the sacculus may also have something to do with hearing) and are known to respond to gravity—that is, they initiate reflexes in response to the position of the head (the relation of the maculae to gravity), the statoconia either pulling or pushing upon the hairs. As a simple example of some of the complicated reflexes related to the position of the head, lowering the nose of a cat properly prepared to show these reflexes changes the relationship of the maculae of the utriculus and sacculus to gravity and brings about a reflex response consisting of flexion of the anterior limbs and extension of the hind limbs—an appropriate reflex if one imagines that the cat, in lowering its head, wants to get its nose and mouth closer to a rat hole.

The **semicircular canals and ducts** have a somewhat different function from the utriculus and sacculus, for although they also are connected with balance, it is apparently movement *per se,* not position, that stimulates them. Each bony semicircular canal is much larger than the membranous canal (semicircular duct) that it contains, but one end of each duct is dilated to form a structure known as the **ampulla,** and the bony canal is somewhat dilated here also.

The bony canals necessarily move with the head, of which they are a part, and the ducts are so attached to the canals that they also move. The canals and ducts are so arranged that movement in any direction will necessarily move one or more somewhat along its long axis, with the result that fluid within the duct will exert a pressure in the direction opposite from the movement (just as water in a drinking glass first stands still when the glass is rotated). In the ampulla of each semicircular duct there is a large projection, the **ampullary crest,** that contains neuroepithelial cells whose hairs project into a gelatinous mass (*cupula*) that occludes the lumen of the membranous ampulla. Apparently, the pressure of the endolymph bends the cupula and this in turn stimulates the hair cells.

An important reflex connection of the semicircular ducts is with the muscles that move the eyes; when the head is moving, the eyes can move in the opposite direction to allow fixation of gaze—as in watching objects from a moving train. Each canal apparently controls certain muscles of the eyeball, so that stimulation of one will produce movement of the eyes in one direction and stimulation of another canal will produce movement in another direction. This movement, relatively slow, is followed by a quick movement in the opposite direction, and this combination of slow and quick movements is **nystagmus.** Clinically, the semicircular ducts can be tested by putting cold or warm water in the external ear canal or by rapidly revolving the patient with the head in different positions. The nystagmus that normally results is then described according to the direction of the quick movement (because it is easier to observe than the slow movement).

In the dissecting room one rarely sees the semicircular ducts, for they are of such size as to be barely visible grossly, are almost transparent, and frequently are destroyed in opening the bony semicircular canals. However, a good concept of the arrangement of the ducts can be obtained by examining a dissection of the canals. The three semicircular canals are called anterior, posterior, and lateral. Each canal opens at both its ends into the cavity of the vestibule, and each semicircular duct opens at both its ends into the utriculus; each has at one end an enlargement that identifies the ampullary end.

The **anterior semicircular canal** is placed vertically, and its position is indicated on the upper surface of the petrous part of the temporal bone (in the posterior part of the floor of the middle cranial fossa) by the arcuate eminence. Its anterior end, which bears the ampulla, is both anterior and lateral, for this canal lies at an angle of about 45° to the sagittal plane. The ampulla lies at the lower part of the curve at the anterior end, just before the canal joins the vestibule. The posterior end of the anterior canal curves downward and unites with the anterior end of the posterior canal, the two forming a **common crus** that opens into the upper and medial part of the vestibule. The **posterior semicircular canal** also is vertical, like the anterior one, but from the common crus that it shares with the anterior canal, it is directed posteriorly and laterally, so that although it also makes an angle of about 45° with the sagittal plane, it makes one of about 90° with the anterior canal. From the common crus, the posterior canal curves posteriorly, laterally, and downward, and its ampulla is on its posteroinferior end. Beyond the ampulla, the canal opens into the lower posterior part of the vestibule. The **lateral semicircular canal,** shorter than the anterior and posterior canals, has its ampullary end situated just below the ampulla of the anterior canal and opens into the vestibule immediately above the fenestra vestibuli. The lateral canal arches laterally, backward, and slightly downward (it is not quite horizontal) to pass through the loop of the posterior canal and open into the vestibule between the openings of the common crus and the ampullary end of the posterior canal.

The **semicircular ducts** need little further description. Since they lie within the canals, they have the same orientation; the anterior ends of the anterior and lateral ducts, and the posterior end of the posterior duct, have the membranous ampullae, and the posterior end of the anterior duct unites with the anterior end of the posterior duct to form a common crus. In short, just as there are only five openings for the three semicircular canals into the vestibule, there are only five openings of the semicircular ducts into the utriculus. Further, each duct of one ear lies in a plane at almost right angles to each other duct, but in the two ears, the lateral semicircular ducts are approximately parallel, and the anterior duct of one ear is approximately parallel to the posterior one of the other.

## Cochlea

The cochlea, concerned with hearing, consists of a bony cochlea and a membranous cochlear duct. Endolymph fills the cochlear duct, and perilymph largely surrounds this and fills the rest of the bony cochlea. The cochlea somewhat resembles the shell of a snail and presents about two and three-fourths turns, or, disregarding the turns, it takes the form of a short cone. The wide base of the cone is the **base** of the cochlea, and the apex is the **cupula.** Part of the basal turn of the cochlea protrudes into the middle ear cavity as the promontory and bears the cochlear window, and the base lies against both the vestibule and the distal end of the internal acoustic meatus. Where it abuts against the vestibule, part of the cavity within it is continuous with the vestibule. The part of the base lying against the meatus presents apertures through which the cochlear division of the 8th nerve leaves the cochlea.

The cochlea is most easily described as if it so sat upon its base that the line between the center of the base and the cupula were vertical. Actually, this axis of the cochlea (usually called its long axis, although it is only about 5 mm long and the base measures almost twice that across) runs anteriorly, laterally, and slightly upward.

Within the center of the cochlea is a central piece of bone extending from its base toward the cupula. This, the **modiolus,** forms a bony core around which the cochlear turns are arranged (Fig. 33-11). Through its hollow center, it transmits the branches of the cochlear nerve to the internal acoustic meatus. Between the turns of the cochlea, the modiolus is continuous with the outer bone of the cochlear wall. Projecting laterally from the modiolus at about the center of each turn, like the threads of a screw, is a thin lamina of bone, the **spiral osseous lamina.** At its upper end, instead of attaching the modiolus to the cupula, the spiral lamina ends freely in a little hook, the **hamulus.** The lamina contains tiny canals for nerve fibers. At the attached end of the lamina, these canals unite to form a larger one, the **spiral canal** of the cochlea, which houses the **spiral** (cochlear) **ganglion** and opens into the hollow of the modiolus.

The spiral bony lamina only partly subdivides the perilymphatic space of the cochlea into two passages, but this division is completed by the **cochlear duct.** This duct is somewhat triangular in cross section and is attached centrally at its apex to the osseous spiral lamina and peripherally to the outer wall of the bony canal. The cochlear duct will be described in more detail shortly. At the moment, the point to be made is merely that the osseous spiral lamina and the cochlear duct together divide the perilymphatic space within a single turn into two parts (Fig. 33-12). Of these parts, one (the upper one when the cochlea is stood upon its base) opens below into the vestibule and hence is known as the **scala vestibuli.** Through its opening, peri-

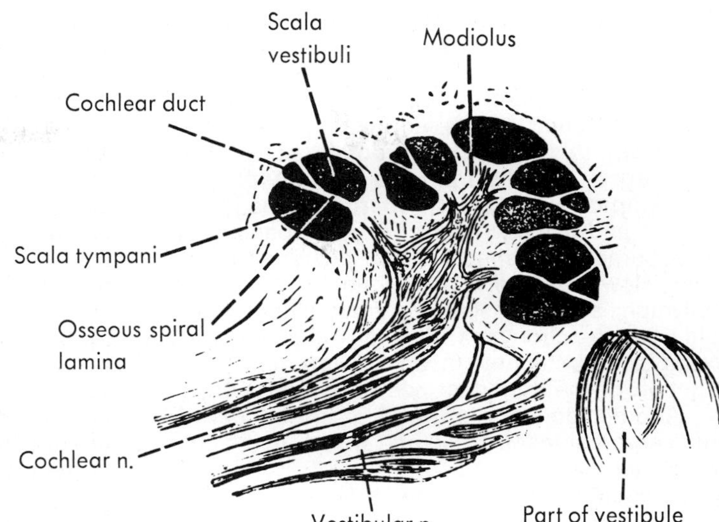


**FIG. 33-11.** A longitudinal section through the cochlea. (Modified from Henle J: Handbuch der systematischen Anatomie des Menschen, vol 2. Braunschweig, Vieweg und Sohn, 1868)

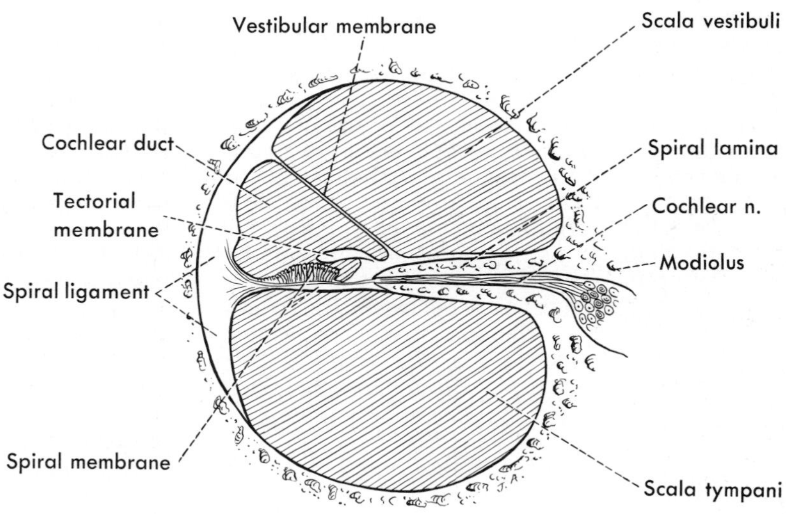


**FIG. 33-12.** Section through a single turn of the cochlea.

lymph can be freely interchanged between the vestibule and the cochlea. The other, the **scala tympani,** opens below into the tympanic cavity through the fenestra cochleae in the dried condition. During life, the perilymph is prevented from escaping here by the **secondary tympanic membrane** that occludes this window.

Although separate elsewhere, the scala vestibuli and the scala tympani are continuous with each other at the cupula (their junction is called the *helicotrema*). Thus, each turn of the cochlea contains a perilymphatic space, the scala vestibuli, that is coiling upward from the base to the cupula and a continuation of this space, the scala tympani, that is coiling downward from the cupula to the base. Because they are continuous, movement of perilymph in the vestibule, induced by movement of the stapes, can induce movement of the fluid up the scala vestibuli to the cupula, then down the scala tympani to the cochlear fenestra (or pressure can be transmitted from one scala to the other through the membranous cochlear duct). Since fluid is essentially incompressible, an inward movement of the stapes is accompanied by an outward movement of the secondary tympanic membrane and *vice versa,* and mobility of the two windows is thought to be essential for proper function of the ear. Although sound waves can enter the internal ear through the cochlear window, they can have little effect if the stapes or some other part of the wall cannot move. This is the rationale upon which fenestration, providing a second opening into the perilymphatic space and sealing it with a thin and movable membrane, is done.

The origin of the perilymph, the few drops of fluid that fill the perilymphatic spaces of the vestibule, semicircular canals, and cochlear scalae, is not known. A tiny **cochlear aqueduct** (*perilymphatic duct*) extends from the lower end of the scala tympani to the anterior border of the jugular foramen and usually is described as opening into the subarachnoid space of the posterior cranial fossa. However, the much-argued concept that perilymph is derived entirely from cerebrospinal fluid, which circulates through the aqueduct, seems to be untenable in view of reports that destruction of the aqueduct has no effect upon the ear. Another concept is that it is in part derived from cerebrospinal fluid, in part from blood, and in part by secretion.

The **cochlear duct,** the membranous part of the cochlea, is attached to the saccule by a tiny canal, the *ductus reuniens* (Fig. 33-9), below which the cochlear duct ends blindly. After winding around the modiolus attached to the spiral lamina, the cochlear duct ends at the cupula, and the helicotrema, connecting the two scala, passes around this blind end.

The peripheral or *external wall* of the cochlear duct is the lining of the bony cochlea (periosteum, to which is attached the lining epithelium of the duct), greatly thickened to form the **spiral ligament.** Deep to the epithelium over the upper part of the spiral ligament are numerous small blood vessels forming the **stria vascularis.** This stria and its associated epithelial cells are believed to be largely responsible for the formation of the endolymph, and it has been suggested that its part in the scala vestibuli may be concerned with the formation and reabsorption of the perilymph. Other markings on this wall are of less importance.

The thin *roof* of the cochlear duct intervening between the endolymphatic and the perilymphatic cavities of the scala vestibuli is the **vestibular membrane.** It has a thin core of connective tissue between two very thin layers of flattened epithelium, and it stretches between the upper part of the spiral ligament and the upper surface of the bony spiral lamina. Perilymph and endolymph differ markedly in their chemical makeup, and the vestibular membrane allows no free mixing between the two. Endolymph is believed to be absorbed largely through the endolymphatic sac, a part of which is very vascular. Both perilymph and endolymph are, according to some authorities, resorbed as well as formed at the *stria vascularis.*

The third or *tympanic wall* of the cochlear duct (between the duct and the scala tympani) is in part formed by the upper surface of the bony spiral lamina; the remainder is formed by the **spiral membrane.** The spiral membrane consists of a fibrous base of connective tissue, the **basilar lamina,** more often known as the basilar membrane, and the *spiral organ,* epithelium of the cochlear duct that is supported by the basilar lamina. The fibers of the basilar lamina or membrane are attached peripherally to a prominent projection (*basilar crest*) of the spiral ligament and centrally to the lower (tympanic) lip of the free edge of the osseous spiral lamina. The anatomy and vibratory properties of the basilar membrane have been extensively investigated.

The histology of the complicated **spiral organ** (or Corti) need not be described in detail here. Suffice it to say that among its tall epithelial cells are some neuroepithelial hair cells, from the outer surface of which project tiny hairlike processes. Nerve fibers end around these epithelial cells, and some of their hairlike processes are attached, in the living condition, to a fibrogelatinous mass, the **tectorial membrane,** that lies against the otherwise free surface of the spiral organ. This membrane is attached to the upper or vestibular lip of the spiral lamina, and its free edge extends just beyond the outermost neuroepithelial cells.

The exact way in which nerve impulses are set up in the cochlear nerve as a result of vibration of the perilymph or endolymph is not known. Most theories have held that the basilar lamina (membrane) vibrates, but others have held that it is the tectorial membrane, or both membranes. Whatever the case, movement of either membrane (or of the endolymph in the case of those cells not attached to the tectorial membrane) would produce a bending of the processes or "hairs," which apparently activates the neuroepithelial cells. These obviously are rather crude concepts, and there is evidence that the mechanism by which vibratory energy is translated into nerve impulses is much more complicated than this.

Although both the scala tympani and the scala vestibuli are larger in the basal than the apical coil, the basilar lamina is shorter because the spiral bony lamina projects a greater proportion of the width in the basal coil; thus, the basilar lamina gradually becomes broader as the apex is approached. Similarly, the tectorial membrane becomes larger as it is traced from base to apex. This has been interpreted to mean that there is a localization in which higher tones are appreciated along the basal turn of the cochlea, and lower tones toward the apex, but present opinion is that this is only approximate; high tones do appear to be localized, but any one low tone appears to be appreciated over a wide upper part of the cochlea.

### Blood and Nerve Supply

The bony labyrinth is supplied by the same vessels that supply adjacent parts of the temporal bone (for instance, the anterior tympanic branch of the maxillary, the stylomastoid branch of the posterior auricular, and the petrosal branch derived from the middle meningeal), but the membranous labyrinth has its own blood supply, distinct from this. The artery supplying it is the **labyrinthine artery.** It arises from the anterior inferior cerebellar artery or from the basilar and travels the length of the internal acoustic meatus with the facial and 8th nerves. The pattern of branching of the labyrinthine artery apparently varies among ears. However, there is regularly a *cochlear branch,* running in the modiolus and distributed to the cochlear duct and especially the stria vascularis, except a proximal part of the basal coil, and there usually are two *vestibular branches,* one distributed partly to the basal cochlear coil but primarily to the utriculus, sacculus, and semicircular ducts, and the other supplying only these parts.

The veins include vestibular veins and a spiral vein in the modiolus. The chief drainage is through a **labyrinthine vein** formed at the lateral end of the internal acoustic meatus and opening into either the inferior petrosal or the sigmoid sinus. There are two other small veins, a **vein of the vestibular aqueduct** that passes through that channel to enter the superior petrosal sinus, and a **vein of the cochlear aqueduct** (caniculus) that parallels the aqueduct and opens into the inferior petrosal sinus or the internal jugular vein.

The **8th nerve** has already been largely described, both in the preceding description of the internal ear and on page 936. Its distribution to the ear (Fig. 33-13) therefore can be summarized briefly. At the distal (lateral) end of the internal acoustic meatus, the 8th nerve divides into a vestibular and a cochlear part. The **vestibular part,** in turn, divides into a superior and an inferior part, each of which bears a part of the **vestibular ganglion.** From the superior part, a short stem divides into branches that penetrate the bony wall between the end of the acoustic meatus and the cavity of the vestibule, ending in the ampullae of the anterior and lateral ducts, the macula of the utriculus, and a part of the macula of the saccule. The inferior part of the ganglion gives a branch to the major part of the macula of the saccule and one to the ampulla of the posterior semicircular duct. (The branch that passes from the inferior division of the vestibular nerve to the spiral ganglion of the cochlea apparently contains efferent or olivocochlear fibers that influence the sensitivity of the cochlea; the function of a twig from the cochlear ganglion to the saccule is not known.)

The ganglion cells of the **cochlear part** of the 8th nerve form the **spiral ganglion,** lying within the spiral canal of the cochlea. These cells send their fibers distally through the canals of the bony spiral lamina to reach the spiral organ. They send their fibers proximally into the modiolus, those from the upper (apical) part running through a central canal in the modiolus, those from lower parts through a series of spiral canals, and all of them opening through foramina at the lateral end of the internal acoustic meatus, where they unite to form the cochlear division of the nerve. As this is traced proximally, it unites with the vestibular division.

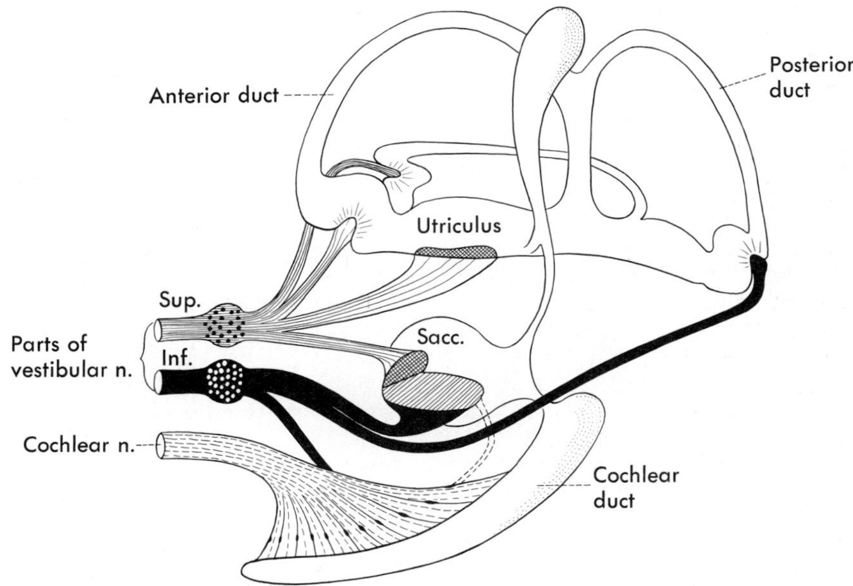


**FIG. 33-13.** Schema of the distribution of the eighth nerve. (Redrawn from Hardy M: Anat Rec 59:412, 1934)

## ORBIT AND EYE

The orbit largely surrounds and protects the eyeball and the nerves and vessels concerned with it. It lies below the anterior cranial fossa and mostly in front of the middle cranial fossa. Branches from the ophthalmic artery (from the internal carotid within the cranial cavity) supply the structures in the orbit, and some of them continue to the face and forehead, in company with similarly named branches of the ophthalmic nerve (a branch of the trigeminal). The optic nerve, originating from the eyeball, has its chief course within the orbit; it ends soon after it enters the cranial cavity. There are seven voluntary muscles, six of which move the eyeball. The remaining one is the levator of the upper lid. These seven muscles are supplied by three nerves (oculomotor, trochlear, abducens), none of which has a cutaneous distribution and all of which end in the orbit.

### Bony Orbit

The orbit is somewhat pyramidal in shape and has superior, medial, inferior, and lateral walls. However, its long axis does not parallel the sagittal plane; rather, the medial walls of the two orbits are approximately parallel (Fig. 33-19), and their lateral walls slope medially as they run posteriorly. The base of the pyramid is the anterior opening or aditus of the orbit. It has prominent supraorbital and infraorbital margins that meet laterally to give it a well-defined lateral border but are separated from each other medially by the more gradual transition between the medial wall of the orbit and the lateral side of the nose. At the back end, or *apex,* of the orbit there is a large, somewhat triangular aperture, the **superior orbital fissure,** that opens into the front end of the middle cranial fossa and just medial to this, visible when the orbit is inspected somewhat from the lateral side, a rounded aperture, the **optic canal,** that also opens into the middle cranial fossa (Fig. 33-14).

In the lateral part of the floor (inferior wall) of the orbit, separating it from the lateral wall, is a long fissure, the **inferior orbital fissure.** Its back end opens into the pterygopalatine fossa, its front end into the temporal fossa, and its middle into the most anterior part of the infratemporal fossa. Proceeding forward in the floor of the orbit from the junction of about the anterior third and the posterior two-thirds of the inferior orbital fissure is the **infraorbital groove.** This houses the largest branch of the maxillary nerve, which at the front end of the groove enters the infraorbital canal that opens onto the face below the infraorbital rim.

The **roof** of the orbit is composed almost entirely of the frontal bone. At the very back end, a little of the roof, medial to the upper part of the superior orbital fissure, is formed

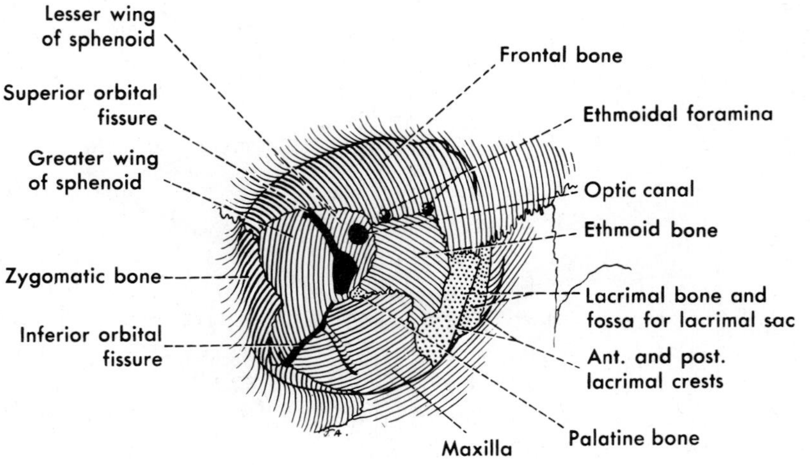


**FIG. 33-14.** The bony orbit, in anterolateral view.

by the lesser wing of the sphenoid bone, which, with the body of the sphenoid medially, surrounds the optic canal. The roof intervenes between the orbit and the anterior cranial fossa. It is thin enough to be translucent and is readily penetrated, so that even such a thing as a pencil brought forcibly in contact with it may penetrate the brain. A part of the frontal sinus usually lies in the roof of the orbit, particularly anteromedially; the extent of the sinus in the roof varies. Anterolaterally, in the roof and lateral wall, is a large smooth depression, the **fossa for the lacrimal gland** (the tear gland). Anteromedially, at the junction of roof and medial wall, there is a small depression on the frontal bone, or a small spine, or both. These are, respectively, the **trochlear fovea** or **spine** and represent the attachment of a pulley (trochlea) through which one of the muscles to the eyeball runs. On the medial side of the supraorbital margin is the prominent **supraorbital foramen** or **notch** for transmission of the supraorbital nerve and vessels to the forehead, and medial to that there may be a less prominent **frontal notch** or **foramen** for a medial branch of the nerve.

The **medial wall** of the orbit is formed largely by the orbital lamina of the ethmoid bone. The orbital lamina contains ethmoid air cells and is so thin that the intercellular walls and, therefore, the pattern of some of the air cells usually can be seen without difficulty upon examination of a dry skull. At about the junction of roof and medial wall, on or close to the frontoethmoid suture, are two small **ethmoidal foramina,** the orbital ends of the anterior and posterior ethmoidal canals; through them arteries and nerves leave the orbit.

Anterior to the ethmoid bone is the small lacrimal bone. This also is usually invaded by ethmoid air cells. Toward its anterior border is a fairly sharp margin, the *posterior lacrimal crest,* in front of which is a concavity, the **fossa for the lacrimal sac.** The frontal process of the maxilla extends upward in front of the lacrimal bone and forms the *anterior lacrimal crest* that bounds the fossa anteriorly. The fossa for the lacrimal sac becomes deeper as it is traced downward and ends in the **nasolacrimal canal,** which opens below into the nasal cavity and houses the nasolacrimal duct.

The **floor** of inferior wall of the orbit is formed largely by the maxilla, which articulates medially with the lacrimal and ethmoid bones and, except anteriorly, is separated from the lateral wall by the inferior orbital fissure. The floor here also is the roof of the maxillary sinus. Anterior to the inferior orbital fissure, the zygomatic bone, which forms part of the lateral wall of the orbit, also contributes to its floor. Posteriorly, just in front of the back end of the inferior orbital fissure, at the point at which the maxilla, the ethmoid, and the lesser wing of the sphenoid seem to come together, the orbital process of the palatine bone forms a tiny part of the floor; this is not recognizable as a separate bone in some skulls.

A major part of the **lateral wall** of the orbit anteriorly is formed by the zygomatic bone, but the frontal bone curves downward to meet it. Behind the zygomatic bone, lateral to the inferior orbital fissure, is the greater wing of the sphenoid bone. The anterior part of the lateral wall separates the orbit from the temporal fossa (through which the orbit can be approached at operation), and a small posterior part separates it from the middle cranial fossa.

The orbit is lined by periosteum that is given the special name **periorbita.** It is continuous over the rim of the orbit and through the inferior orbital fissure with the periosteum of the outer surface of the skull, and through the superior orbital fissure and the optic canal with the periosteal or outer layer of the dura mater.

### Eyelids: Lacrimal Apparatus

The skin of the eyelids or **palpebrae** (Fig. 33-15) is very delicate, and careless removal of it will remove also the thin layer of muscle deep to it. There is only a little connective tissue between the skin and muscle and deep to the muscle. Since it is particularly loose tissue, it allows ready accumulation of fluid—as, for instance, the minor hemorrhage resulting in a "black eye."

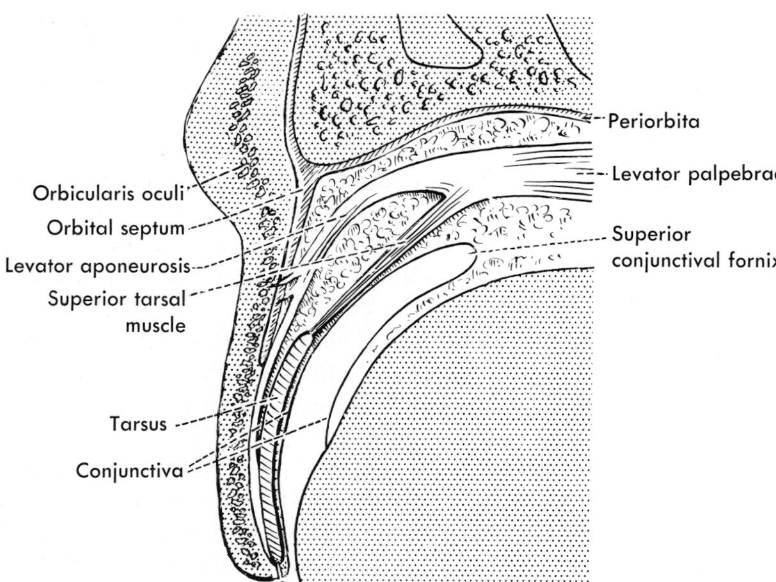


**FIG. 33-15.** Structure of the upper eyelid; the space around the levator and its continuations into the lid is greatly exaggerated here.

The muscle in the lid is the **palpebral portion of the orbicularis oculi;** the *orbital* part surrounds the orbit, and both these parts have already been described with the facial muscles (p. 880). It need only be noted here that the palpebral portion is delicate and extends to the margin of the lid and that at the medial corner of the eye, many of the fibers of both palpebral and orbital parts attach to a ligamentous structure, the **medial palpebral ligament.** This ligament is attached medially to the anterior lacrimal crest. Laterally, under cover of the orbicularis, it bifurcates and attaches to the heavy structures (tarsi) that produce the curved shape of the upper and lower lids. Lateral to the corner of the eye, a **lateral palpebral raphe** may mark the intersection of fibers in the upper and lower lids, but this is not a palpebral ligament—the **lateral palpebral ligament** lies behind the muscle and gives no attachment to it.

A third part of the muscle is also demonstrable on careful dissection, for some of the fibers close to the free borders of the lids pass deeply as they near the medial corner of the lids, run behind the lacrimal sac, and attach to the posterior lacrimal crest. These form the **pars lacrimalis of the orbicularis oculi.** This part of the muscle surrounds two tiny tubes (the lacrimal canaliculi) that lead from the medial ends of the eyelids to the lacrimal sac (Fig. 33-16), carrying into it the tears that accumulate between the lids and the eyeball.

On the posterior surface of each lid is the thin **conjunctiva,** continuous at the margin of the lid with the skin. In the upper lid, it is reflected upward on the posterior surface of the lid and then turns down to run in front of the eyeball and attach to it close to the periphery of the cornea (the transparent front of the eyeball). Thus, there is a *palpebral* and a *bulbar* (eyeball) *layer* of conjunctiva; the angle that they make with each other is the **superior conjunctival fornix.** Similarly, the lower part of the conjunctiva is divisible into palpebral and bulbar parts and forms an **inferior conjunctival fornix.**

Ducts of the lacrimal gland open into the superior conjunctival fornix, primarily its lateral part, and blinking movements keep the tears spread evenly over the eyeball. In the medial corner of the lids, on the posterior surface of each, is a little projection, the **lacrimal papilla,** that contains the opening or punctum of the **lacrimal canaliculus.** This drains excess tears. The two lacrimal canaliculi run medi-

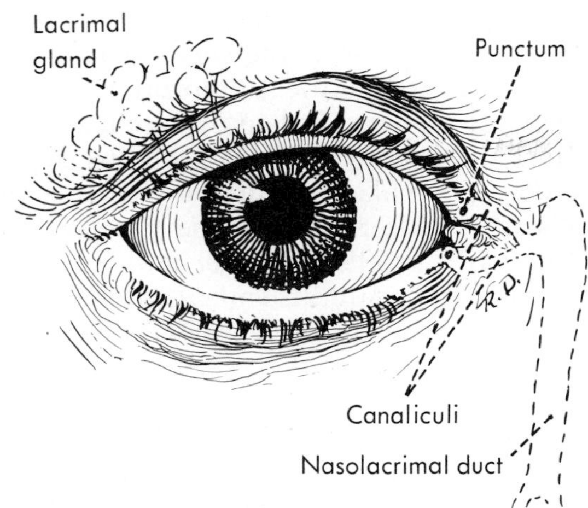


**FIG. 33-16.** The lacrimal apparatus. (Wakefield EG: Clinical Diagnosis. New York, Appleton, 1955)

ally, converging toward each other and usually joining, to empty into the lacrimal sac. The **lacrimal sac** is blind above (Fig. 33-16), but below, it opens into the **nasolacrimal duct,** the lower opening of which can be seen when the nasal cavity is examined. The sac as a whole is lodged in the fossa for the lacrimal sac, between the anterior and posterior lacrimal crests (on the maxillary and lacrimal bones, respectively). It lies largely behind the medial palpebral ligament (although its upper end protrudes above that) but in front of the lacrimal part of the orbicularis oculi muscle. It also is surrounded by a special layer of **lacrimal fascia** that blends at the lacrimal crests with the periosteum in the fossa of the sac.

The major part of the thickness of each lid is formed by a number of **tarsal glands** that are embedded in dense connective tissue and form a plate, the **tarsus,** that is almost cartilaginous in appearance and maintains the shape of the lid (Figs. 33-15 and 33-17). The tarsi extend to the free margin of the lids, where their glands open. They are attached to the sides of the orbit by the palpebral ligaments. The medial palpebral ligament has already been noted. The lateral *palpebral ligament* lies behind both the orbicularis oculi muscle and some connective tissue and blends with other connective tissue behind it. Each palpebral ligament is bifid at its palpebral end and attaches both to the superior and to the inferior tarsus. The superior tarsus is larger than the inferior one, and its upper border is much more arched than is the lower border of the inferior tarsus. Until they are dissected free, however, neither has a free edge except at the edges of the lids, for they are overlapped by connective tissue that extends from the margin of the orbit into the lids and blends with the front of the tarsi.

The connective tissue extending over the front of both tarsi is derived from the periosteum at the margin of the orbit and simply projects down or up into the lid, as the case may be. It is known as the **orbital septum** and at its origin forms the most posterior part of the lid; its posterior surface is adjacent to the fat and connective tissue within the orbit.

In the upper lid, but not the lower, the tissue in front of the tarsus is also contributed to by tendinous fibers from the **levator palpebrae superioris** (p. 966) that run down behind the orbital septum to blend with it on the front of the tarsus and, according to some accounts, pass in part through the septum toward the skin. Behind the levator, a voluntary muscle, is a layer of smooth muscle that arises from its inferior surface and attaches below into the upper border of the superior tarsus; this is the **superior tarsal muscle** (Figs. 33-15 and 33-17, and p. 966). An **inferior tarsal muscle** attaching to the lower edge of the inferior tarsus cannot be demonstrated by dissection.

**FIG. 33-17.** The tarsi and the palpebral ligaments.

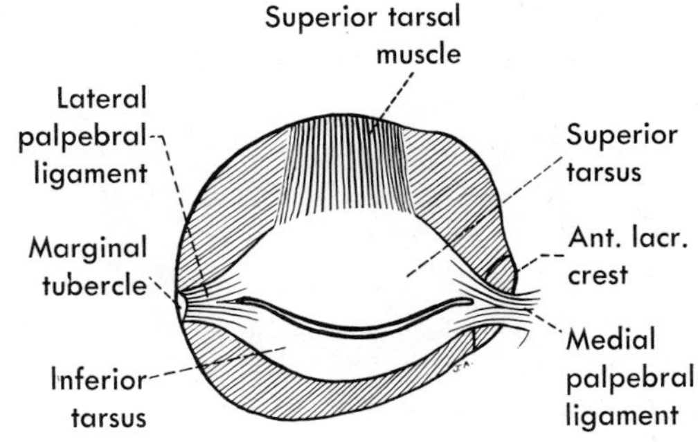

Two **fascial spaces** are described in the lid: one, the *preseptal space,* lies in front of the orbital septum, between that and the orbicularis oculi; the other, the *pretarsal space,* lies behind the tendon of the levator palpebrae and is bounded posteriorly by the tarsus and the superior tarsal muscle and superiorly by the origin of that muscle from the levator palpebrae.

The **lacrimal gland** is located in the lateral part of the orbit behind the orbital septum. Its anterior part is deeply indented, and thus divided into two parts, by the tendon of the levator. The larger part, lying above the levator in the fossa for the gland, is the *pars orbitalis.* A slender *palpebral part* of the gland extends behind the levator into the upper lid, where it lies just deep to the conjunctiva on the posterior surface of the lid. The nerves and vessels of the lacrimal gland are described later.

The **cutaneous innervation of the lids** is through tiny branches of the large nerves that appear close to the rim of the orbit. The lower lid is supplied by branches from the *infraorbital nerve.* At the lateral corner of the eye, tiny twigs of the *lacrimal nerve* emerge to supply skin here but are difficult to find, for this nerve ends mostly in the lacrimal gland. In the upper medial corner of the orbit, three nerves, branches of the ophthalmic nerve, as is the lacrimal, penetrate the orbital septum and give twigs to the upper lid before continuing to the forehead and the side of the nose. The largest of them is the *supraorbital nerve,* which runs through the supraorbital foramen or notch to turn upward on the forehead close against the frontal bone. The smaller *supratrochlear nerve* likewise turns upward on the forehead a little medial to the supraorbital, and the still smaller *infratrochlear nerve* appears below the supratrochlear, running medially and downward to supply the upper part of the nose.

Associated with the nerves as they emerge from the orbit are **branches of the ophthalmic artery** (Fig. 33-18). The *supraorbital branch* turns up on the forehead with the supraorbital nerve, and the *supratrochlear* runs with the supratrochlear nerve. Similarly, the *dorsal nasal artery,* much larger than the infratrochlear nerve, turns downward to supply the upper part of the external nose. At the medial corner of the eye, between it and the nose, and in front of the medial palpebral ligament, are the angular artery and vein, upper parts of the *facial vessels.* The artery may inosculate with the dorsal nasal, so that it is impossible to say at what point the vessels join. Part of the *angular vein* enters the lid, above the medial palpebral ligament, to help form the superior ophthalmic vein. Above this level, the angular vein is formed by the junction of the supratrochlear and a part of the supraorbital vein. Lateral to and below the orbit are the *transverse facial vessels,* and lateral to and

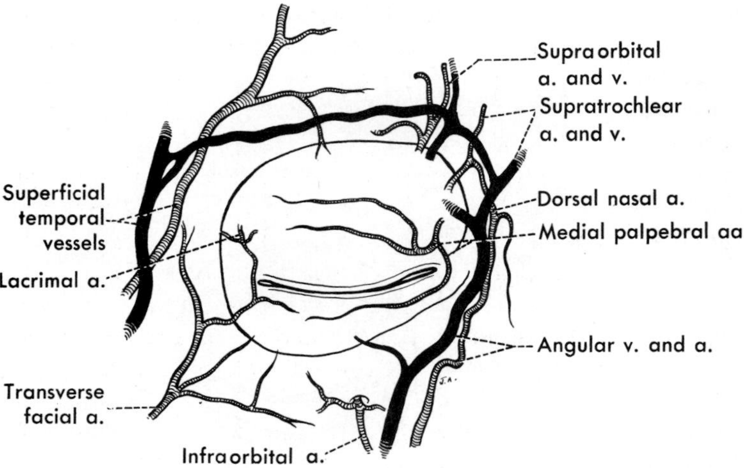


**FIG. 33-18.** Chief vessels around the margin of the orbit and in the lids.

above the orbit are the frontal branches of the *superficial temporal vessels,* which may anastomose broadly with the supraorbital vessels. Part of the supraorbital vein joins the supratrochlear vein to form the *angular vein,* but another part turns into the orbit to help form the *superior ophthalmic vein.*

All these vessels about the orbit have twigs to or from the lids, but these supply primarily the skin of the lids and the orbicularis oculi, and the blood supply to the deeper structures of the lids, particularly the tarsi, is primarily through the **medial palpebral arteries.** These, like the supraorbital and supratrochlear, are branches of the ophthalmic artery. The **superior medial palpebral** appears above the medial palpebral ligament and runs laterally, usually dividing into an upper and lower branch as it does so. The lower branch runs at first on the front of the tarsus and usually disappears into it, and the upper branch runs above the tarsus. The lower branch is described as anastomosing with a *lateral palpebral branch of the lacrimal artery* (also a branch of the ophthalmic) that appears at the lateral corner of the eye, to form the **superior palpebral arch.**

The *inferior medial palpebral artery* passes downward behind the medial palpebral ligament and then turns laterally to run on the anterior surface of the inferior tarsus. It usually disappears into this tarsus, but is described as anastomosing with a *lateral palpebral branch of the lacrimal artery* to form the **inferior palpebral arch.** The comparable **medial palpebral veins** drain into the angular vein.

Most of the lymphatic drainage of the lids is downward and backward toward the parotid nodes, but some of it, particularly from the medial corner of the eye, is downward along the angular and facial vessels to the submandibular lymph nodes.

## MUSCLES AND ASSOCIATED STRUCTURES

The contents of the orbit are best studied by removing its roof, that is, much of the floor of the anterior cranial fossa. In doing this, the frontal sinus usually is opened, for it commonly sends an extension of varying size into the roof of the orbit, and some of the mucosa of the ethmoid cells may be exposed medially. The bone of the roof of the orbit strips off easily from the periorbita (periosteum lining the orbit), and the latter can be incised carefully without damaging the underlying structures. When this is done, the muscles will be seen to be partly concealed by fascia and connected to each other by intervening connective tissue that has a smooth outer surface where it was in contact with the orbital roof and walls. Since most of the muscles arise close together at the posterior end of the orbit and diverge to surround the eyeball as they are traced forward, they and the connective tissue connecting them form a cone. Most of the nerves and vessels of the eyeball lie within this muscle cone, but several appear peripheral to it and can be seen as soon as the periorbita of the roof of the orbit has been reflected.

### Peripheral Nerves and Vessels

The largest peripheral structure is the **frontal nerve,** which enters the orbit through the superior orbital fissure above the origin of the bulbar muscles and, therefore, never lies within the muscle cone (Fig. 33-19). It is the largest branch of the ophthalmic nerve and is given off by this nerve in the cavernous sinus. As it runs forward, it divides into the lateral supraorbital and medial supratrochlear nerves, which round the upper rim of the orbit. Before the supraorbital nerve leaves the orbit, it is joined by the supraorbital artery, which emerges from the connective tissue and fat among the muscles to run forward in a superficial position.

Lateral to the frontal nerve, running along the upper part of the lateral wall of the orbit, is the **lacrimal nerve.** It likewise is given off from the ophthalmic within the cavernous sinus and passes, with the frontal nerve, through the superior orbital fissure above the origins of the bulbar muscles. As it runs forward, it is joined by lacrimal vessels that emerge from the muscle cone. Finally, the **trochlear nerve,** which, like the ophthalmic,

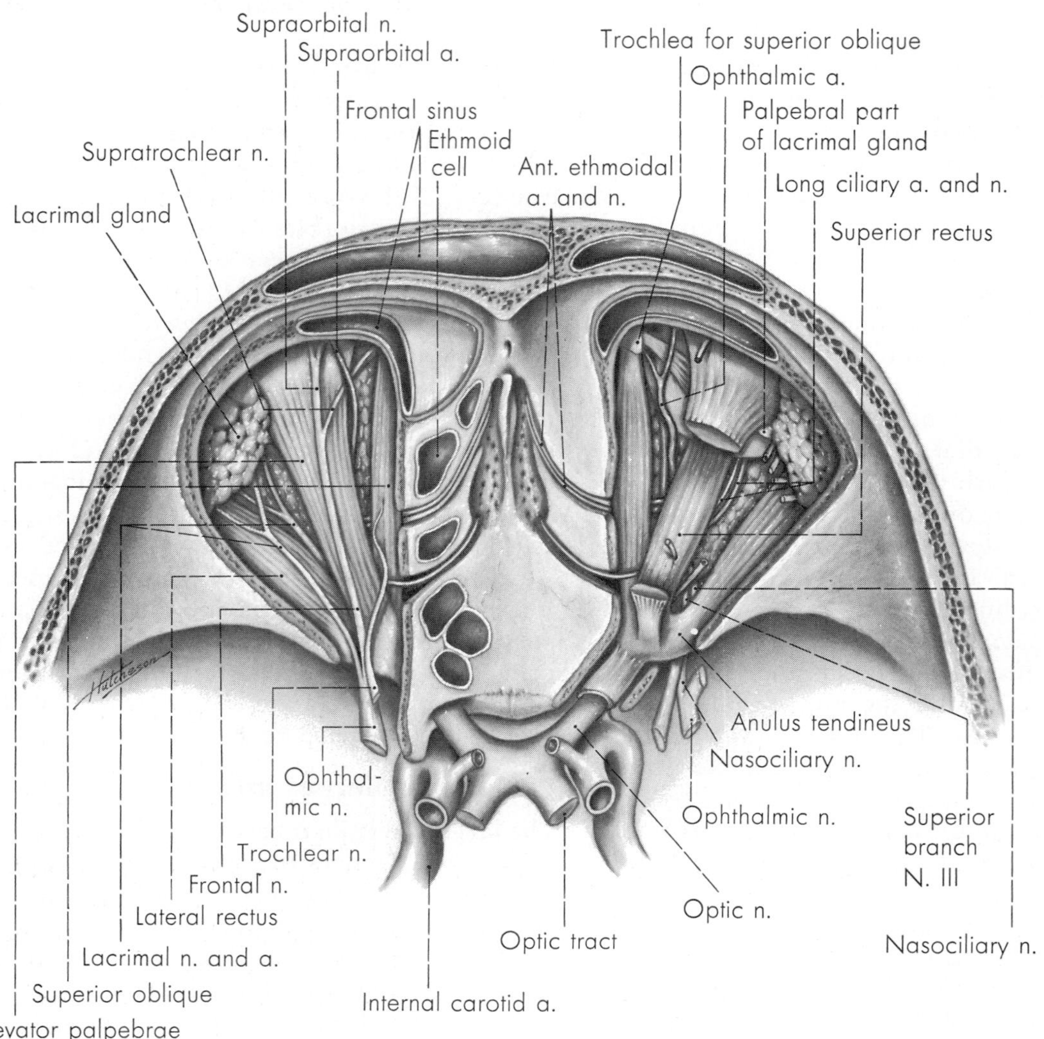


**FIG. 33-19.** General anatomy of the orbit. On the left, bone and periorbita have been removed; on the right, the more superficially lying nerves and vessels have been removed, as has a part of the levator palpebrae, and the proximal part of the nasociliary nerve within the orbit has been moved laterally and anteriorly to bring it into view.

traverses the cavernous sinus, enters the orbit with the frontal and lacrimal nerves and turns medially across the uppermost muscle here (the levator palpebrae) to reach the upper border of the muscle that lies highest on the medial wall of the orbit; this is the superior oblique muscle, which the trochlear nerve supplies.

The remaining vessels and nerves of the orbit lie within the cone formed by the muscles, embedded in the fat that occupies most of the space between the muscles. They are best seen, therefore, after the upper muscles have been reflected and the fat and connective tissue carefully dissected out.

## Annulus Tendineus

Six of the seven voluntary muscles of the orbit arise from its posterior end. Four are rectus muscles and run forward to insert into the eyeball in the positions implied by their names: **superior rectus, medial rectus, inferior rectus,** and **lateral rectus.** These four muscles are blended at their origin to form a com-

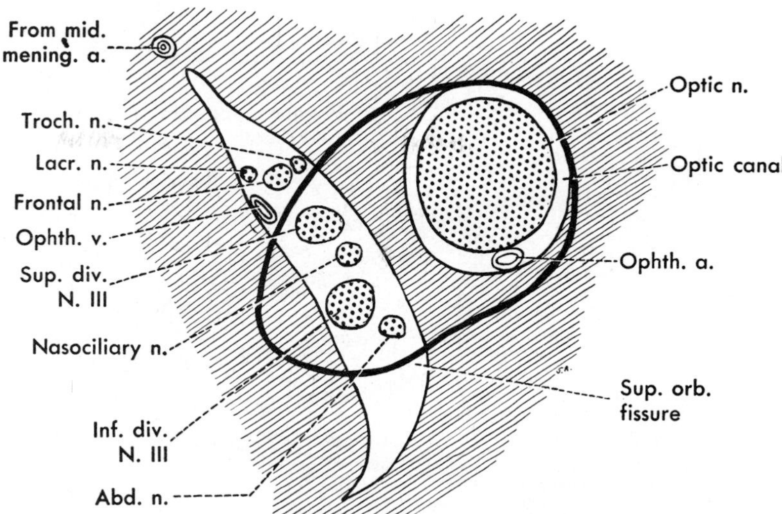


**FIG. 33-20.** Diagram of the annulus tendineus and relation of nerves and vessels of the orbit to it.

mon tendon in the form of a ring, the **annulus tendineus communis,** that is attached to the body and lesser wing of the sphenoid bone medial to, above, and below the optic canal and laterally runs across the superior orbital fissure so as to enclose a part of this within the ring (Fig. 33-20). The nerves that lie outside the muscle cone traverse the superior orbital fissure above the annulus, but the remaining nerves enter the orbit through that part of the superior orbital fissure enclosed by the annulus. The optic nerve and the ophthalmic artery also lie within the annulus, since this encloses the optic canal. The ophthalmic vein passes through the superior orbital fissure in a variable position, either above, through, or below the annulus.

Arising immediately adjacent to the annulus, but not forming a part of it, are two other muscles. One is the **levator palpebrae superioris,** the other the **superior oblique.** The seventh muscle of the orbit, the **inferior oblique,** arises from the floor of the front of the orbit and also runs obliquely to the eyeball. It, therefore, has no particular relation to the annulus.

## Fascia

All the muscles of the orbit are surrounded by fascial sheaths that become thicker as they are traced forward toward the eyeball. Where the four rectus and the two oblique muscles become applied to the eyeball, the fascia on them becomes continuous with a layer of fascia about the eyeball (Fig. 33-21). This layer ensheathes a major part of the eyeball, being attached to the sclera (the white part of the eyeball) posteriorly, close to the optic nerve, and anteriorly, just behind the cornea or transparent part of the eyeball. This is the sheath of the eyeball (**vagina bulbi** or **bulbar sheath**), also known as "Tenon's capsule." As the sheaths of the muscles blend with it, their tendons pass between it and the eyeball to reach their points of insertion. Here, the bulbar sheath is, therefore, separated from the sclera, although elsewhere it is bound to that by rather delicate connective tissue trabeculae, so that there is, between the sheath and the eyeball, an **episcleral space.**

Just before the fascial sheaths of the four rectus muscles blend with the bulbar sheath, they expand laterally to fuse with each other, thus forming what is usually called the **intermuscular membrane.** Tumors or other masses lying internal to this membrane and the rectus muscles may not be visible unless the space among the muscles is explored, and therefore the orbit is frequently described as being subdivided into two spaces, one inside and one outside the muscle cone.

## Muscles and Their Nerves

Of the seven voluntary muscles of the orbit (Figs. 33-19, 33-21, and 33-22), the superior

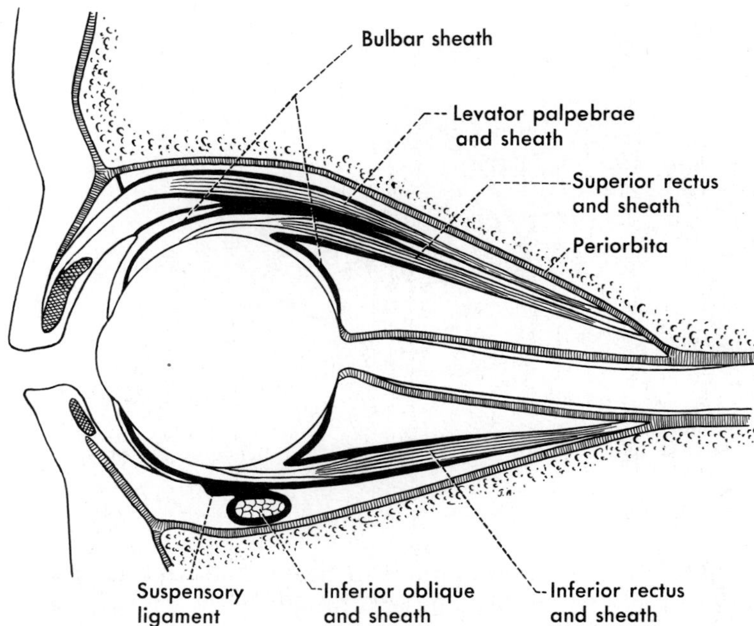


**FIG. 33-21.** Fascia of the orbit in sagittal section.

oblique is innervated by the trochlear nerve, and the lateral rectus by the abducens; all the others are innervated by the oculomotor.

The **levator palpebrae superioris** arises just above the superior rectus and follows that muscle forward, but instead of attaching to the eyeball, it continues into the lid. As it approaches the lid, its tendon widens rapidly and sends expansions ("horns") to attach to the medial and lateral walls of the orbit, but its major part attaches to the front of the superior tarsus (Fig. 33-15). From the fascia between the levator and the superior rectus, there arises a thin lamina that attaches to the superior conjunctival fornix, so that these muscles pull this up also as they elevate the lid and turn the eyeball up. Fascia from the upper surface of the levator turns upward to attach to the superior orbital margin behind the orbital septum and often is described as forming a check ligament for the muscle. The levator palpebrae is innervated by the *superior branch of the oculomotor nerve* (the nerve divides into superior and inferior branches just before it enters the orbit), which first innervates the underlying superior rectus and then passes through this muscle to end in the levator palpebrae.

The **superior tarsal muscle**, often called "Mueller's muscle" by clinicians, arises from the lower surface of the levator palpebrae and inserts into the upper border of the superior tarsus. It is smooth muscle, innervated by *sympathetic fibers from the superior cervical ganglion,* and its denervation produces the ptosis or drooping of the lid that is characteristic of *Horner's syndrome* (p. 938).

The **superior oblique muscle** arises medial to and a little below the origin of the levator palpebrae and, like it, just outside the annulus. Although it is straight at first, it runs obliquely to the eyeball. As it runs forward, it lies along the upper medial part of the wall of the orbit, above the upper edge of the medial rectus muscle. It becomes tendinous before it reaches the front of the orbit and thereafter is tendinous to its insertion on the eyeball. At the upper medial corner of the front of the orbit, its fascial sheath thickens and blends with a loop of dense fibrous connective tissue that contains a bit of cartilage and is attached at both ends to the frontal bone so as to form a pulley (*trochlea*). At the trochlea, the tendon turns sharply backward, downward, and laterally, surrounded by a synovial sheath. Beyond the trochlea, the fascial sheath is thick.

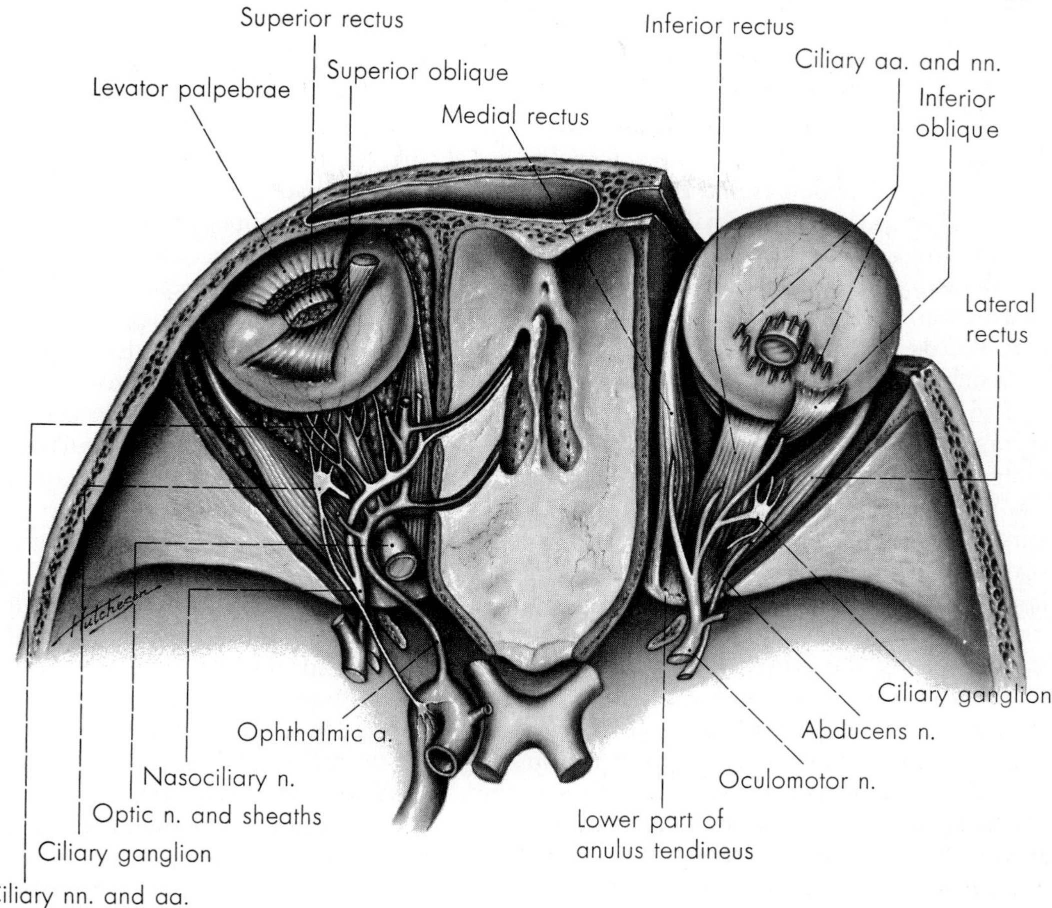


**FIG. 33-22.** Further anatomy of the orbit. On the left, the levator palpebrae, superior rectus, and superior oblique have largely been removed and a segment has been taken from the optic nerve; on the right, most of the nerves and vessels of the orbit have been removed to show the abducens nerve and the inferior branch of the oculomotor nerve.

It fuses with the bulbar sheath medial to the entrance of the superior rectus into this sheath, but the superior oblique tendon continues posteriorly, laterally, and downward close against the upper and posterior surface of the eyeball. In this part of its passageway through the episcleral space, it passes deep to the superior rectus, between that muscle and the eyeball. As it nears its insertion lateral to and above the optic nerve (therefore, in the upper lateral quadrant of the posterior half of the eyeball), the tendon of the superior oblique expands from a cord to a thin membrane, and the insertion of the tendon is, therefore, into a variably curved line on the sclera.

The superior oblique muscle is innervated by the *trochlear nerve*, which enters its upper border. When it contracts, it turns the eyeball so that the pupil is directed downward and outward.

The **superior rectus muscle** arises from the lesser wing of the sphenoid immediately above the optic canal, and its tendon, therefore, forms the highest part of the annulus. As it runs forward, it is directed also somewhat laterally, as it must be to remain in the center of the orbit, and just before it reaches the eyeball, its fascial sheath separates from the sheath of the overlying levator. After its sheath has fused with the bulbar sheath, the muscle continues forward in the episcleral space, passing above the tendon of the superior oblique and inserting into the sclera above and behind the sclerocorneal junction. It is supplied by the *superior branch of the oculomotor nerve,* which is the uppermost nerve passing through the superior orbital fissure within the annulus. Shortly after it has entered the orbit, this stout branch turns upward to enter the proximal part of the superior rectus and continue into the overlying levator palpebrae.

The tendon of origin of the **lateral rectus** forms a lateral part of the annulus, particularly the upper and lateral limb of that part of the ring that crosses the superior orbital fissure. The nerves entering the orbit through this part of the fissure, therefore, are particularly intimately related to the lateral rectus muscle. The muscle runs forward against the lateral wall of the orbit. Just before its fascial sheath fuses with the bulbar sheath, it gives off an expansion that attaches to the lateral wall of the orbit to form a *check ligament* (*lacertus of the lateral rectus*) that limits the action of the muscle. After a course within the bulbar sheath, the muscle inserts into the sclera on the lateral side of the eyeball. The *abducens nerve* enters the orbit closely applied to the medial side of the lateral rectus and runs forward in this position only a short distance before entering the muscle.

The **medial rectus muscle** contributes the upper medial part of the annulus. Like the superior rectus, it is closely applied to the sheath of the optic nerve. It runs forward along the medial wall of the orbit, below the levator, and enters the episcleral space to attach to the medial side of the sclera. Its sheath, like that of the lateral rectus, gives off a *check ligament* that attaches to the medial wall of the orbit. The medial rectus is supplied by the *inferior branch of the oculomotor nerve;* this nerve lies just above the inferior rectus muscle, and its branch to the medial rectus runs medially below the optic nerve.

The **inferior rectus** contributes the inferior part of the annulus. It runs forward and laterally, as does the superior rectus. After it penetrates the bulbar sheath, it passes above the inferior oblique and inserts anteriorly on the eyeball. It is innervated by a branch from the *inferior division of the oculomotor nerve.*

The remaining voluntary muscle of the orbit, the **inferior oblique,** arises from the floor of the orbit anteromedially, just behind the orbital septum and often with some attachment to the fascia covering the lacrimal sac. It runs laterally and posteriorly to curve around the lower surface of the eyeball and insert into the lower lateral quadrant of the posterior half of the sclera, therefore below the insertion of the superior oblique. As the muscle reaches the eyeball, its surrounding fascia contributes to the bulbar sheath, which is particularly thick below (where it is contributed to by the fascia of both the inferior rectus and the inferior oblique, Fig. 33-21), and forms a sort of hammock that stretches upward to the fascia contributed by the medial and lateral rectus muscles. (This lower thickened part of the bulbar sheath is known to clinicians as the *suspensory,* or Lockwood's, *ligament.*) The inferior oblique passes below the inferior rectus muscle as it enters the bulbar sheath. It, like the medial and inferior recti, is innervated by the *inferior branch of the oculomotor nerve.* This muscle directs the pupil upward and laterally.

All four rectus muscles insert on the anterior half of the eyeball, and their actions can be deduced from this. The medial and lateral rectus muscles direct the pupil medially and laterally, respectively. The superior rectus directs the pupil upward, and the inferior rectus directs it downward, but because neither pulls in a direction parallel to the long axis of the eye, both muscles also tend to direct the pupil medially. It is this medial pull of the superior and inferior recti that normally is overcome by the pull of the obliques. The inferior oblique directs the pupil laterally and upward, and therefore, when it and the superior rectus work together, a pure upward movement of the pupil can be obtained. Similarly, the superior oblique directs the pupil downward and laterally, and therefore, when it and the inferior rectus work together, a pure downward movement can be obtained. Because of their attachments to the eyeball, the superior oblique and the superior rectus have long been described as rotating the eyeball medially, and the inferior oblique and inferior rectus as rotating it laterally. It has now been shown, however, that rotation is not a normal movement and that the superior and inferior recti actually prevent the rotation that the obliques produce when acting alone.

## Other Nerves and Vessels

The frontal and lacrimal branches of the ophthalmic nerve, and the motor nerves to the orbit, now have been largely described. However, there is an additional branch from the inferior division of the oculomotor nerve that has not yet been mentioned; this is an *oculomotor root of the ciliary ganglion* that brings into it preganglionic fibers. It is a short, stout trunk, arising from the inferior branch just before that ends by branching into the nerves to the inferior rectus and inferior oblique, or arising from the first part of the nerve to the inferior oblique.

The **ciliary ganglion** (Figs. 33-22 and 33-23) lies well back in the orbit, just lateral to the optic nerve and between that and the lateral rectus muscle. It is a small ganglion, not much larger than the head of an ordinary pin. Sometimes it is double. In addition to its oculomotor root, the ganglion also receives a slender *ramus communicans from the nasociliary nerve* (a branch of the ophthalmic). This may leave the nasociliary nerve while that is in the cavernous sinus or after it has entered the orbit, but, in any case, runs forward just lateral to the optic nerve to end in the upper part of the ciliary ganglion. There also may be a third delicate root to the ciliary ganglion, a *branch from the sympathetic plexus on the internal carotid artery*, or the sympathetic fibers may reach the ganglion with the branch of the nasociliary nerve or follow the ophthalmic artery.

The branches given off by the ciliary ganglion are **short ciliary nerves** and consist of postganglionic parasympathetic fibers originating in the ganglion, afferent fibers from the nasociliary nerve that run through the ganglion, and postganglionic sympathetic fibers that do the same thing. They proceed forward toward the eyeball, accompanied by a number of small branches of the ophthalmic artery, and, although they originate lateral to the optic nerve, they spread out and branch as

**FIG. 33-23.** The ciliary ganglion and nerves. Parasympathetic fibers are red, the preganglionic fibers (in the oculomotor nerve) being indicated by solid lines, the postganglionic ones (in the short ciliary nerves) by broken lines. Trigeminal sensory fibers are indicated by the black dotted lines. Sympathetic fibers are also black, with preganglionic fibers (from the thoracic part of the spinal cord) indicated by solid lines and postganglionic fibers (from the superior cervical ganglion) indicated by broken lines.

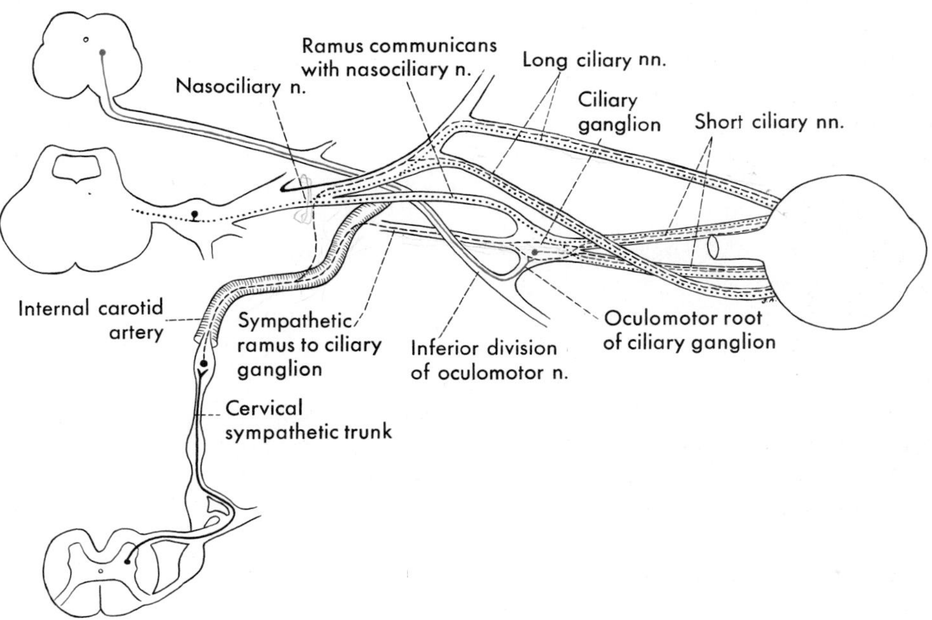

they run forward so that as they enter the eyeball close to the optic nerve, they almost surround that nerve.

The **nasociliary nerve** (Figs. 33-19 and 33-22) leaves the rest of the ophthalmic nerve in the cavernous sinus and, unlike the frontal and lacrimal branches of that nerve, enters the orbit through the annulus, passing medially between the superior and inferior divisions of the oculomotor nerve as it does so. It runs above the optic nerve toward the medial wall of the orbit, giving off two delicate **long ciliary nerves** that run forward to enter the eyeball medial and lateral to the optic nerve, outside the general circle formed by the short ciliary nerves. They contain afferent fibers for the eyeball and, usually, sympathetic fibers derived from the internal carotid plexus.

When the nasociliary nerve reaches the medial wall of the orbit, it runs forward between the superior oblique and medial rectus muscles, giving off either one or two ethmoidal nerves. The small **posterior ethmoidal nerve,** if present, enters the posterior ethmoidal canal and supplies sensory twigs to the sphenoid sinus and perhaps posterior ethmoid cells. The **anterior ethmoidal nerve,** larger and constant, enters the anterior ethmoidal foramen or canal, which opens into the anterior cranial fossa just at the lateral edge of the cribriform plate. The nerve then runs forward on the lateral edge of the plate and at the anterior end turns downward to enter the nose. Its nasal branches are described in a following section. The ethmoidal nerves are accompanied by ethmoidal arteries.

**FIG. 33-24.** Diagram of the ophthalmic artery.

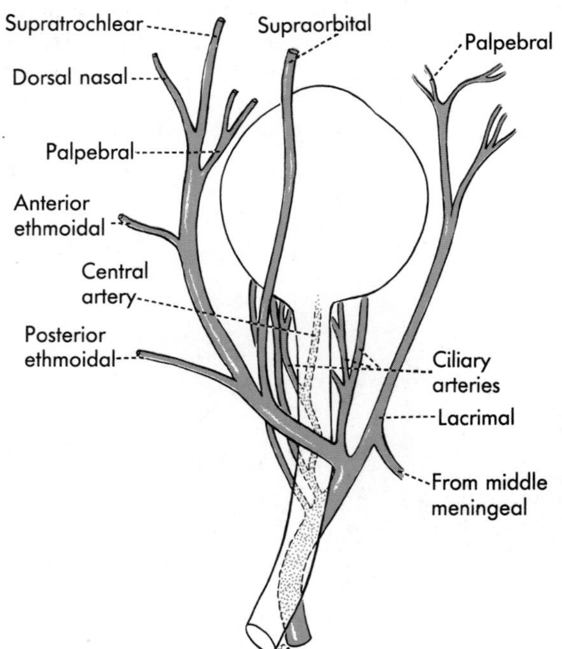


After giving off its anterior ethmoidal branch, the nasociliary continues as the **infratrochlear nerve,** which passes just below the trochlea of the superior oblique muscle, penetrates the orbital septum, and gives branches to the side of the nose and the medial corner of the lids.

The largest nerve in the orbit, and the central structure in the muscle cone, is the **optic nerve.** This leaves the back of the eyeball and takes a somewhat sinuous course to the optic canal. In the orbit, the optic nerve is surrounded by a heavy sheath, the *external sheath* of the nerve, that is continuous with the dura and arachnoid within the skull. On the surface of the nerve is a thin layer of pia mater, the *internal sheath,* and between these two layers is an *intervaginal space* that is a continuation of the subarachnoid space. The optic nerve is the only nerve that is related to meninges throughout its length, and that is because it really is not a nerve but a tract of the brain.

The **ophthalmic artery** (Figs. 33-19, 33-22, and 33-24) enters the orbit through the optic canal, where it lies below the optic nerve. As it emerges into the orbit, it deviates to lie lateral to the nerve and here gives off the **lacrimal artery.** This runs forward at first within the muscle cone but then emerges to run in the upper lateral part of the orbit with the lacrimal nerve and supply the lacrimal gland. Small twigs, the *lateral palpebral arteries,* continue into the lid to complete the superior and inferior palpebral arches. The lacrimal artery usually receives a tiny *anastomotic branch* from the middle meningeal artery that enters the orbit through or just lateral to the upper end of the superior orbital fissure.

Sometimes this branch is larger, in which case it may form the origin of the lacrimal, and, very rarely, it gives rise to the whole ophthalmic artery.

After the ophthalmic artery has given off the lacrimal artery, it turns medially, usually above the optic nerve. As it does so, it gives off several branches that in turn subdivide as they run forward and become mingled with the short ciliary nerves from the ciliary ganglion. Most of these branches are **short posterior ciliary arteries** and enter the eyeball close to the optic nerve, along with the short ciliary nerves. One branch on each side is a **long posterior ciliary artery** that enters the eyeball medial or lateral to, respectively, the entrance of the short arteries.

Among the ciliary vessels is the **central artery of the retina.** This passes onto the inferior surface of the optic nerve, penetrates its sheaths, and passes to the center of the nerve, in which position it enters the eyeball. It divides into four retinal arterioles, one to each quadrant, which run in the superficial or vitreous surface of the retina. It is accompanied by a central vein that, after leaving the optic nerve, enters the cavernous sinus. The retinal vessels can be clearly seen with the aid of an ophthalmoscope, and engorgement of the veins, swelling of the disk of the optic nerve (papilledema), or both commonly indicate arterial disease or increased intracranial pressure.

After the ophthalmic artery has crossed the optic nerve, it gives off its **supraorbital** branch, which runs forward at first within the muscle cone but then emerges to join the supraorbital nerve and appear with that on the forehead. The artery then runs forward in company with the nasociliary nerve along the medial wall of the orbit and gives off **posterior** and **anterior ethmoidal arteries.** Both arteries enter the anterior cranial fossa, and the anterior ethmoidal usually gives rise to an **anterior meningeal artery.** Both also enter the nose and supply an upper part of it.

Beyond the anterior ethmoidal artery, the ophthalmic artery leaves the orbit, usually first giving off a **medial palpebral artery** that subdivides into superior and inferior medial palpebrals, and then dividing into terminal **dorsal nasal** and **supratrochlear arteries.** The palpebral arches that the medial and lateral palpebral arteries form supply the deeper structures of the lids and most of the bulbar conjunctiva. It is worth noting again that, because of the anastomoses between the branches of the ophthalmic artery and the branches of the external carotid on the face and forehead, the ophthalmic artery can conduct blood of external carotid origin to orbital structures.

In addition to the branches already mentioned, the ophthalmic artery also gives rise to muscular branches that enter all the muscles arising from the posterior part of the orbit, usually close to the entrance of the nerve into the muscle. These arteries run forward in the muscles, and usually two tiny twigs emerge on the surface of the tendons of the rectus muscles. These are the **anterior ciliary arteries,** which supply the sclera and conjunctiva closest to the cornea and then penetrate the sclera and anastomose inside the eyeball with posterior ciliary vessels.

There are two ophthalmic veins, superior and inferior (Fig. 33-25). As a rule, dissection of these is unsatisfactory because they are fragile and the nerves and arteries are more important. The **superior ophthalmic vein** begins within the orbit by the union of a stem from the supraorbital vein with one from the upper end of the angular vein. As the superior ophthalmic runs back in the upper medial

**FIG. 33-25.** Diagram of the ophthalmic veins.

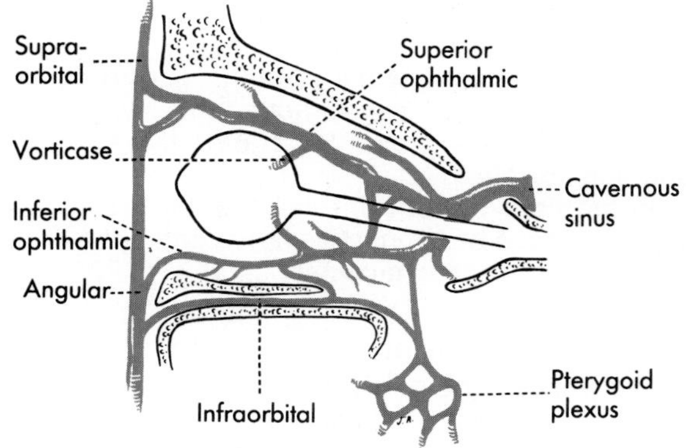

part of the orbit, it receives branches corresponding to most of those of the ophthalmic artery—**anterior** and **posterior ethmoidal, lacrimal, muscular,** and a part of the **central vein of the retina.** Behind the eyeball, it also receives the upper two of the four **vorticose veins,** which are veins that leave the posterior aspect of the eyeball some distance from the optic nerve and penetrate the sheath of the eyeball to end in the ophthalmic veins. The superior ophthalmic vein usually also receives the inferior ophthalmic vein just before the venous drainage leaves the orbit.

The **inferior ophthalmic vein,** smaller than the superior, begins by the union of small veins in the floor of the orbit. It may receive a communication from the angular vein through the lower lid, and, as it runs back, has communications also with the superior ophthalmic vein. It receives the veins from the lower bulbar muscles and the two inferior vorticose veins, which are placed below and on each side of the optic nerve in positions corresponding to those occupied by the superior vorticose veins above the nerve. The inferior ophthalmic vein also communicates through the inferior orbital fissure with the pterygoid plexus of veins, and often with the infraorbital vein.

The common stem formed by the two ophthalmic veins usually leaves the orbit through the part of the superior orbital fissure that lies above the annular tendon, but it may pass through the annulus, or occasionally even below, to enter the cavernous sinus. When the two veins do not join, but empty independently into the sinus, one may have one position, the other another.

The **central vein of the retina** is said to empty independently into the cavernous sinus always, but usually to have a branch that joins the superior ophthalmic vein.

In the floor of the orbit, the inferior orbital fissure is bridged by connective tissue and by smooth muscle, the **orbitalis muscle.**

Just deep to the orbitalis muscle in the posterior part of the orbit is the **maxillary nerve** as it runs forward in the pterygopalatine fossa. As the nerve enters the orbit to run along its floor in the infraorbital groove, it gives off the **zygomatic nerve.** The zygomatic nerve runs upward and forward in the periosteum of the lateral wall of the orbit and divides into **zygomaticofacial** and **zygomaticotemporal** branches. These both penetrate the bony lateral wall of the orbit to become subcutaneous, but before the zygomaticotemporal branch does so, it sends a twig upward to join the lacrimal nerve. It is this branch that brings secretory fibers to the lacrimal gland. The secretory fibers are postganglionic and arise from the pterygopalatine ganglion (whose preganglionic fibers are from the 7th nerve). They join the maxillary nerve, immediately below which the pterygopalatine ganglion lies, and run forward with it into its zygomatic and zygomaticotemporal branches.

Accompanying the infraorbital nerve, the continuation of the maxillary, in the floor of the orbit are the infraorbital artery and vein. These may help supply structures of the floor.

## EYEBALL

The eyeball (*bulbus oculi*) usually cannot be satisfactorily examined in the dissecting room, and it is described here only in enough detail to give a general concept of its structure and function. Histologic and more advanced texts should be consulted for details.

### Chief Layers

The major part of the eyeball consists of three layers (Fig. 33-26): an outer **fibrous tunic,** the *sclera* (white of the eye) and *cornea;* a **vascular tunic,** the *choroid* and its anterior extensions that form most of the *ciliary body* and *iris;* and a **tunica interna,** the sensory part or *retina.* The inner surface of the back part of the eyeball usually is called the **fundus.**

The **sclera** is composed of connective tissue and forms about five-sixths of the surface of the eyeball. It is continuous anteriorly with the transparent **cornea,** which allows light to enter the eyeball and is commented upon further in a following section. The cornea is part of a smaller sphere than is the sclera and, be-

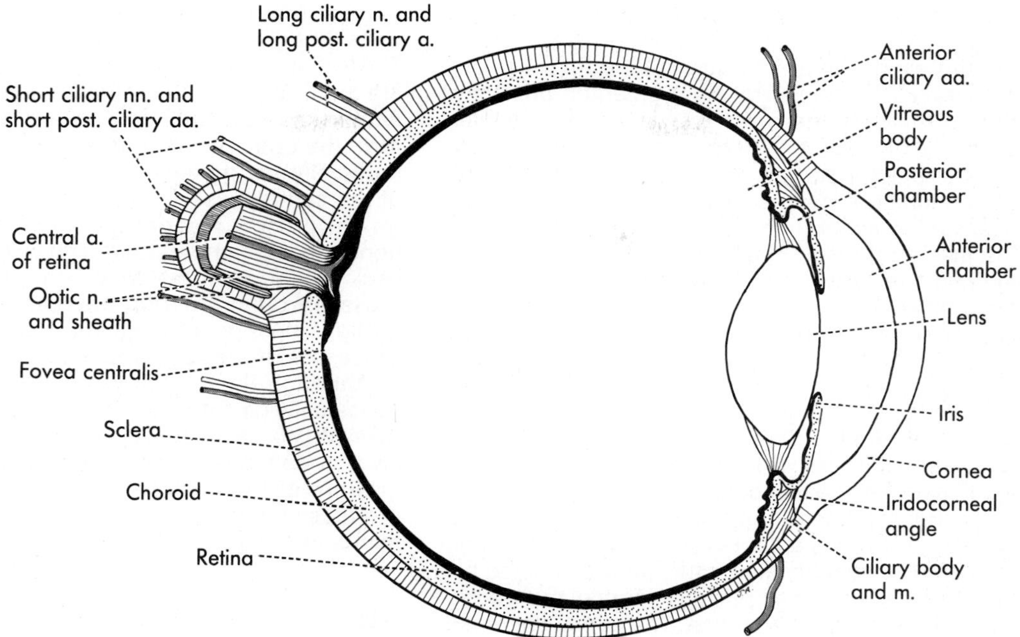


**FIG. 33-26.** Schema of the eyeball in horizontal section.

cause of its greater curvature, protrudes in front of the sclera, so that the anteroposterior axis of the eye is its longest one.

The **choroid** consists primarily of blood vessels that are fed by the short posterior ciliary arteries and recurrent arteries from the front of the eyeball and drained by the four vorticose veins. Its smallest blood vessels form a choroidocapillary layer adjacent to the retina, but the largest ones lie peripherally; the long ciliary arteries, which supply more anterior structures, run between the choroid and the sclera. The choroid is nutrient to the retina, primarily to the outer part that is adjacent to the choroid.

For some 6 mm to 7 mm behind the sclerocorneal junction (*limbus* of the cornea) the choroid is thickened and tightly attached to the sclera. This part is the **ciliary body** (so called because of the numerous fine folds on its inner surface). To the posterior part of the ciliary body are attached fibers that suspend the lens (zonular fibers, often referred to collectively as the *suspensory ligament of the lens*), and anteriorly, a continuation of the choroid leaves the sclera to extend almost transversely across the eye and form most of the iris. Much of the bulk of the ciliary body is formed by the vascular layer of the choroid, but this body also contains smooth muscle fibers that form the **ciliary muscle.** Most of these muscle fibers are longitudinal (meridional) and radial fibers, but some of them are circular. They all presumably function to narrow the diameter of the ring formed by the ciliary body, thus releasing tension on the zonular fibers (suspensory ligament). The ciliary muscle is innervated by parasympathetic fibers through the oculomotor nerve and the ciliary ganglion.

The thick part of the **retina,** the innermost layer of the eyeball, can be easily stripped off as far forward as the posterior border of the ciliary body; here it tears away in an irregular line (*ora serrata*). Anterior to the retina is a thin layer closely attached to the posterior surfaces of the ciliary body and the iris. The *ciliary and iridial parts* of the retina are not sensitive to light. The part that is sensitive to light is called the *pars optica.*

The optic part of the retina consists of two layers, an outer pigmented layer and an inner nervous or cerebral stratum. The *pigmented layer* is a single layer of heavily pigmented cuboidal cells. It is this layer only that is continued as the ciliary and iridial parts of the retina, and it is tightly adherent to the choroid everywhere. The *cerebral stratum* or nervous layer is attached to the pigmented one only around the optic nerve as that leaves the eyeball and at the ora serrata. Therefore, it is this layer only that can be stripped off the eyeball, and when retinal detachment occurs as a result of accident or disease, the separation is between the two layers. The layers represent the outer and inner layers of an optic cup developed by invagination of one side of a hollow, ball-like optic vesicle and therefore are adjacent to each other rather than closely attached.

A little medial to the posterior pole of the eye there is, in the retina, a whitish circle about 1.5 mm in diameter; this, produced by the optic nerve as it leaves the retina, is the **optic disk,** or disk of the optic nerve. (It also has been called the "nerve head" and the "papilla of the optic nerve.") It is the thickest part of the retina except where it presents, close to its middle, a depression, the excavation of the disk. The sensory elements that are affected by light are lacking in the optic disk; hence, this represents a **blind spot** in the retina. The blind spot must be differentiated from other blind spots or areas that may be brought about by disease. The central artery of the retina branches as it enters the retina at the optic disk, and the central vein is formed here by the convergence of its tributaries. Examination of the vessels is an important part of examination of the fundus through the ophthalmoscope.

Lateral to the optic disk, almost exactly at the posterior pole of the eye, is a region that is slightly yellow and is known as the **macula** (formerly macula lutea, or yellow spot). In its center is a depression, the **fovea centralis,** where the nervous part of the retina is thinnest and vision is most accurate. The fovea centralis and macula represent the area of central vision, and most movements of the eyes are attempts to bring light to focus on the macula.

The retina has been described as having from eight to ten layers including the pigmented part, but there actually are only three layers of nerve cells. The outermost or neuroepithelial layer, adjacent to the pigmented stratum, is composed of cells whose processes are shaped somewhat like rods and cones and, therefore, are so named. The **rods** have associated with them visual purple and are sensitive only to varying degrees of light (usually described as being for "night vision"); **cones,** employed in more accurate vision ("day vision," including color vision), are most numerous at the macula, where there are no rods. They become less numerous as the retina is traced toward the ora serrata, and the more peripheral parts of the retina contain none at all. Next is a layer formed primarily but not exclusively by **bipolar cells** that connect the rods and cones to the third layer of cells. This third or innermost layer is made up of the **ganglion cells** that give rise to the fibers of the optic nerve. These fibers, lying most superficially (that is, they are the first part of the retina to be reached by light entering the eyeball), converge at the optic disk to form the optic nerve. In this layer also, located superficially among the nerve fibers, are the branches of the central vessels of the retina.

The precise manner in which the retina functions is not yet understood. The layers of the retina, back to the level at which the visual purple is present around the rods, are perfectly transparent, for light must penetrate them in order to reach the processes of the rods and cones.

Although the central vessels of the retina send delicate twigs into the deeper part of the retina, these do not reach the layer of rods and cones. This layer is dependent upon the closer-lying capillary network of the choroid. Thus, retinal detachment produces blindness by depriving the receptive elements of adequate nutrition.

## Iris

The iris, with its central aperture, the **pupil,** projects from the anteromedial surface of the ciliary body as a thin perforated disk, similar to the diaphragm of a camera. Its major bulk is a continuation of the choroid, and the anterior ciliary vessels enter it to anastomose with the posterior ones. In the white infant, all the pigment of the iris is in its posterior part at birth, and although the pigment actually is dark, it appears blue when seen through the other layers. In nonwhites the pigment is deposited more generally through the iris; hence, the eye does not appear blue. During infancy the more anterior part of the iris may become heavily pigmented, and the blue eyes turn to brown; if the anterior pigmentation is less, the eyes turn to gray or remain blue.

The most important feature of the iris is its ability to change the diameter of the pupil, in a fashion somewhat similar to the change in the aperture of a camera diaphragm. This change is brought about by smooth muscle lying in the iris. Most of the muscle is circularly arranged and constitutes the **sphincter pupillae.** When the sphincter pupillae contracts, it reduces the size of the pupil, thus protecting the retina from excessive or unneeded light. It is innervated through the oculomotor nerve by way of the ciliary ganglion and the short ciliary nerves. Radially arranged fibers constitute a **dilator pupillae** innervated by the sympathetic system. The preganglionic fibers leave the spinal cord through the upper thoracic nerves, especially T1, and run upward in the cervical sympathetic trunk to synapse in the superior cervical ganglion. The postganglionic fibers from the superior cervical ganglion join the internal carotid plexus and reach the iris either by joining the ophthalmic nerve and its long ciliary branches or by joining the sympathetic ramus to the ciliary ganglion and mingling with other fibers of the short ciliary nerves. Among the fibers to the iris there also are many sensory ones, particularly concerned with pain; fibers for pain also end in the ciliary body.

## Refracting Media

In reaching the retina, light must travel through several transparent substances intervening between the retina and the outside air. Since all these have a tendency to bend or refract light rays that pass through them, they are grouped collectively as the refracting media of the eye. They consist of the cornea, aqueous humor, lens, and vitreous body.

The **cornea** is the anterior transparent continuation of the sclera, although its curvature is sharper. It consists mostly of connective tissue. Epithelium on its anterior surface is continuous with the epithelium of the bulbar conjunctiva; epithelium on its posterior surface is called the *endothelium of the anterior chamber.*

The **anterior chamber** of the eye lies behind the cornea, between that and the iris and lens. It is filled with fluid, the **aqueous humor,** that apparently is formed, probably from the epithelium of the ciliary body, in the small **posterior chamber,** behind the attachment of the iris. After passing through the pupil into the anterior chamber, the aqueous humor, which constantly is being produced, is drained off through spaces that permeate a structure at the iridocorneal angle called the *pectinate ligament* and open into a circular venous channel, the **sinus venosus sclerae,** or *canal of Schlemm* (Fig. 33-27). If the rates of formation and absorption get out of balance, the amount of aqueous humor, and therefore its pressure, may be increased above normal. This is known as *glaucoma* and, when it is severe, results in degeneration of the retina and blindness.

The **lens** lies immediately behind the iris, attached to the ciliary body by the gelatinous **ciliary zonule,** which consists of **zonular fibers** and intervening spaces. Between the lens and zonule and the iris is the posterior chamber of the eye.

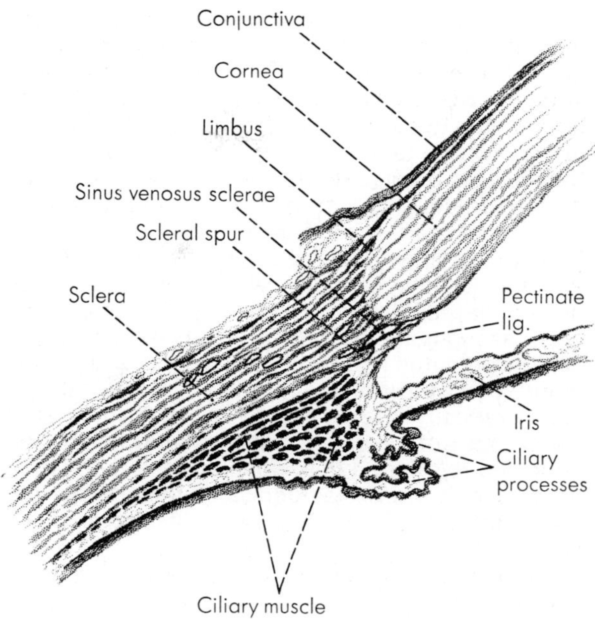


**FIG. 33-27.** The iridocorneal angle.

The lens is a biconvex disk measuring about 9 mm to 10 mm in diameter and about 4 mm to 5 mm in anteroposterior thickness. Its posterior surface is more highly curved than its anterior one. It is made up of anucleate cells, which are normally perfectly transparent; an opacity of the lens is known as a *cataract.* The lens is the only refracting medium of the eye whose refracting ability can be varied from moment to moment to maintain visual acuity: the lens is elastic but is so held under tension by the zonular fibers that it is somewhat flattened. When the ciliary muscle contracts and thereby relaxes the zonular fibers, the lens becomes more convex.

The large **vitreous chamber** behind the lens and the posterior chamber is filled with a gelatinous transparent mass known as the **vitreous body,** frequently referred to as the vitreous humor. In contrast to the aqueous humor, the vitreous body does not undergo constant replacement; it is formed early in development and cannot be replaced thereafter.

### Some Functional Aspects

Certain functional aspects of bringing light rays to a focus on the retina can now be briefly considered. In the resting normal eye (with the ciliary muscle relaxed), the refraction of the flattened lens is just enough to bring into sharp focus on the retina an object seen at a distance, from which the light rays are traveling approximately parallel by the time they reach the eye. Because it is in the resting eye that the lens is most flattened, the resting eye is adapted for distance vision, not near vision. The closer an object is to the eye, the more the light rays from that object diverge as they approach the eye and, therefore, the more refraction there must be to bring these rays to a focus on the retina. This greater refraction is brought about by change in the shape of the lens (*accommodation*). For near vision, the ciliary muscle contracts, releasing the tension on the zonular fibers and allowing the lens to become more convex. Although the increased thickness of the lens is a passive process, brought about by the elasticity of this body, it should be noted that near vision requires muscular effort, that of the ciliary muscle. In middle age, the lens gradually loses elasticity, so that, in spite of contraction of the ciliary muscle, it may remain too flattened to afford a good focus on the retina for a near object; this condition, which makes near objects difficult to see clearly, is known as *presbyopia.*

Certain other abnormalities of vision have nothing particularly to do with the ability of the lens to adjust itself for near and far vision. The cornea, the first refracting medium that the light strikes as it enters the eye, may have slight irregularities in its curve; this accounts for the indistinctness of vision known as *astigmatism.* Further, the eyeball may have grown out of proportion to its refracting system, to a point beyond which the lens can adjust by changes in its shape. Thus, if the eyeball is too long for its refracting system, objects will be brought to a focus in front of rather than on the retina, whereas if it is too short, the focus will be behind the retina. Either condition, obviously, results in poor vision. Too little refracting power for the length of the eyeball, resulting in a focus point behind the retina, is known as *hyperopia* or *hypermetropia,* or farsightedness; distance vision is the best here, since less refracting power is needed for distant objects. Too great refractive power for the length of the eyeball, with the result that the focus lies in front of the retina, is *myopia* or near-sightedness; here vision is better for close objects, where greater refractive power is needed anyway.

## NOSE AND PARANASAL SINUSES

The development and some general anatomic features of the nose are mentioned in the discussion of the respiratory system (Chap. 7). The paranasal sinuses were encountered during the description of the skull.

### External Nose

The external nose sometimes is compared to a three-sided pyramid (and its skeleton is sometimes referred to as the "nasal pyramid"). The smallest side of the pyramid faces downward and surrounds the nares before it becomes continuous with the upper lip; the other two sides are the sides of the nose. The margin along which these two sides meet is the **dorsum nasi.** This extends from the **root** of the nose, the part continuous with the forehead, to the apex or tip. The flared lower part of the side of the nose is the **ala** (wing). The external openings of the nose are the **nares** (nostrils). They are separated from each other by the lower border of the **nasal septum.**

Over the movable alae and tip of the nose, the skin is tightly attached to the supporting cartilage, but over the rest of the nose it is movable on the underlying skeleton. The blood supply, from both angular and dorsal nasal vessels, and its innervation, mostly from the ophthalmic nerve, have been described in connection with the face.

The **skeleton of the external nose** (Fig. 33-28) is contributed to laterally by the fron-

**FIG. 33-28.** The nasal cartilages, somewhat spread apart.

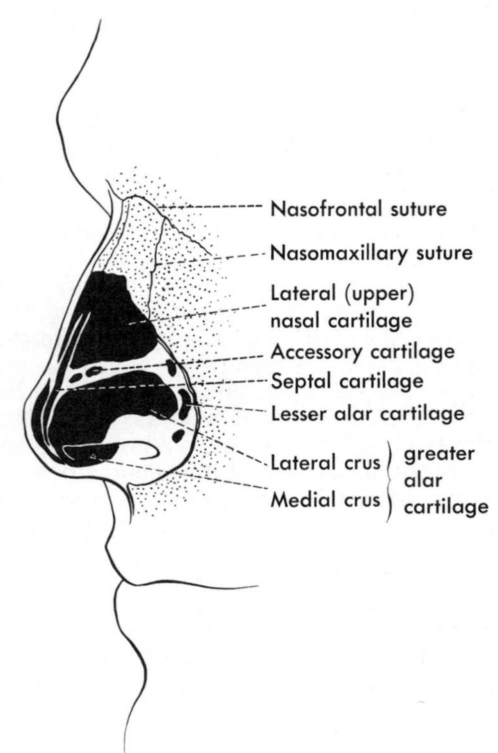

tal processes of the maxillas and superiorly by the nasal part of the frontal bone. Articulating with both of these, and completing the bony part of the skeleton, are the two small **nasal bones.** The remainder of the skeleton of the external nose is cartilaginous.

Articulating with the nasal and maxillary bones are the **lateral nasal cartilages,** often referred to as the upper cartilages. In their upper parts, these cartilages are continuous with each other through the **septal cartilage,** which passes backward and downward from them to articulate with the bony septum of the nose. They and the septal cartilage actually are a single piece of cartilage, although named as if they were separate. In their lower parts, the lateral nasal cartilages are separated from each other by a cleft in which the septal cartilage appears on the dorsum of the nose. The septal cartilage extends farther downward toward the apex of the nose than do the lateral nasal cartilages. Below the lateral nasal and septal cartilages are the **greater alar cartilages** (often referred to as the lower nasal cartilages). These are separate cartilages, and each one has a relatively large lateral crus that supports the ala (and overlaps the lower border of the lateral nasal cartilage) and a smaller medial crus that lies close to its fellow of the opposite side and forms a lower part of the nasal septum. In the angle between the lateral nasal cartilage, the lateral crus of the greater alar cartilage, and the septal cartilage, there commonly are one or more small **accessory nasal** (sesamoid) **cartilages.** Posterior to the lateral nasal cartilage and the lateral crus of the major alar cartilage, there are usually several **lesser alar cartilages.** These various cartilages are united by dense connective tissue, and this tissue, with fat added, forms a part of the ala.

## NASAL CAVITY

Each nasal cavity is narrow above and wider below, and the two cavities are separated from each other by the nasal septum. Each cavity begins at the **naris** and extends back to the **choana,** the opening into the pharynx. The first part of the cavity is the **vestibule** (Fig. 33-30), the lower part of which is lined with skin and provided with hairs. The vestibule is bounded laterally by the greater alar cartilage, and its upper limit, the *limen nasi,* is a crescentic infolding of the lateral wall at the lower border of the lateral nasal cartilage; the medial wall of the vestibule, formed by the nasal septum, has no marking to delimit it from the remainder of the nasal cavity.

The entire medial and lateral nasal walls posterior to the vestibule, except in the uppermost narrow part of the nasal cavity, are covered by a thick glandular and vascular mucous membrane and constitute the **respiratory region** (warming and humidifying the inspired air). In the narrow roof of the nose, extending downward a little on the medial and lateral sides, is the thinner mucous membrane of the **olfactory region** in which the olfactory nerves arise. The mucous membrane of the respiratory region is provided with cilia that have in general a beat backward toward the choanae so as to carry mucus into the throat where it can be swallowed. The mucous membrane of the conchae (Fig. 33-30) is particularly vascular, presenting a so-called *cavernous plexus* that resembles erectile tissue. Because of its vascularity, the nasal mucous membrane easily becomes swollen, and when this is combined with increased secretion of mucus a "stopped-up nose" results. Most nose drops contain some epinephrinelike substance that decreases the secretion of mucus and produces vasoconstriction of the arterioles, thus relieving the nasal congestion.

### Nasal Septum

The nasal septum presents little in the way of markings and can be studied in part to advantage in a dried skull. As seen through the piriform apertures, the large posterior part of the septum is bony. The upper part of the bony septum is formed by the **perpendicular plate of the ethmoid bone** (Fig. 33-29). Its lower part is formed by the **vomer,** with some contribution from the nasal crests of the maxillary and palatine bones. Most of the remainder of the septum, the anterior and lower part, is cartilaginous and is formed chiefly by the septal cartilage.

The **septal cartilage** has a tongue-and-groove articulation with the edges of the bony septum. Alongside its lower edge is a small cartilage, the **vomeronasal cartilage;** this is connected with a little blind pouch from the nasal cavity, the **vomeronasal organ,** that extends backward between the two cartilages. The septal cartilage does not reach the distal part of the nasal septum; this is formed by the medial crura of the two greater alar cartilages and, therefore, is a double wall. The two medial crura are not closely bound together, and the groove between them can be felt distinctly at the tip of the nose. They are bound to the lower edge of the septal cartilage by a sufficient amount

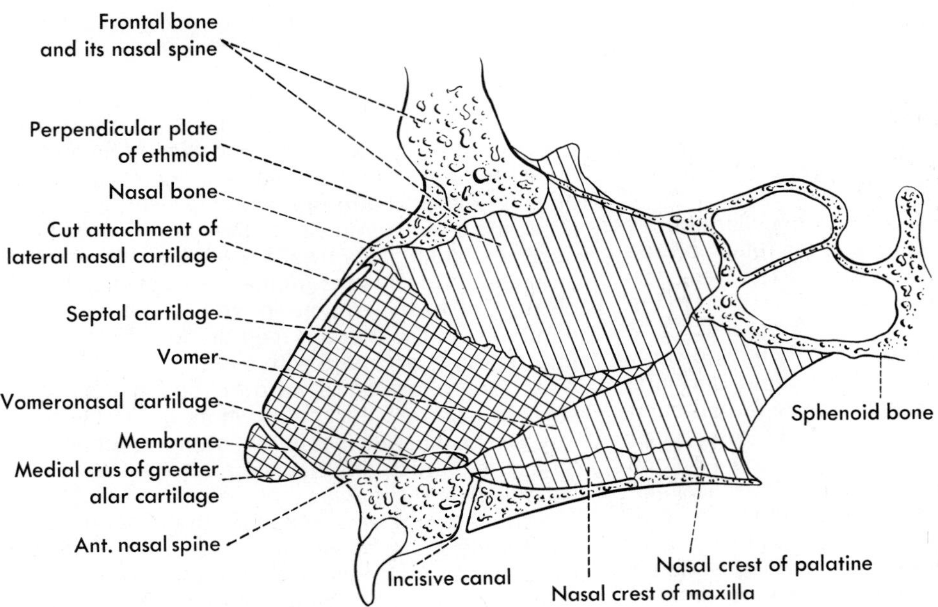


**FIG. 33-29.** Components of the nasal septum. Cartilage is crosshatched.

of connective tissue to allow them to be freely movable, and the part of the septum that they form is known as the mobile part.

The nasal septum is covered by mucous membrane, of which an upper posterior part, just below the cribriform plate of the ethmoid, is olfactory. The vomeronasal organ, rudimentary and not always discernible in man, is in some mammals large and lined by olfactory epithelium.

Frequently, the nasal septum is deviated to one side or the other and, in conjunction with deviation, may present spurs that project still farther into one nasal cavity. A deviated septum and spur may come in contact with the projecting conchae from the lateral nasal wall, resulting in partial occlusion of the nasal cavity or in discomfort as a result of the contact.

The **nerves and vessels** of the nasal septum are branches of those that also supply the lateral nasal wall and are described in a following section.

## Lateral Nasal Wall

The lateral nasal wall presents a number of features deserving study (Figs. 33-30 and 33-31). Most prominent are the **conchae** (also called "turbinates"), curved shelves of bone covered by mucosa that project from the lateral nasal wall and greatly increase the respiratory surface of the nose. They are named inferior, middle, and superior nasal conchae, according to their position. The **inferior concha** is the longest, with the **middle** being almost as long although it does not come quite as far forward. The **superior concha** is much smaller, only about half the length of the middle concha, and lies above the posterior half of this concha. Above the back end of the superior concha there may be an inconspicuous bulge, the supreme concha. The air passageways deep to the conchae are known as the inferior, middle, and superior **nasal meatuses,** respectively.

Only a few other markings on the lateral wall of the nose can be seen until the conchae are reflected. At the anterior border of the middle concha, a depressed area, the *atrium of the middle meatus,* leads upward into the middle meatus. Anterior to and above the atrium is a variably sized projection, the *agger nasi,* that contains one or more ethmoid air cells. Anterior to and above this is a slight groove, the **olfactory sulcus,** that leads upward to the olfactory region of the nose. Relatively little air normally flows over the sulcus or the agger nasi to the uppermost part of the nose during quiet respiration, so that, when one wants to smell the inspired air better, one sniffs to set up air currents that will carry the air to the olfactory region.

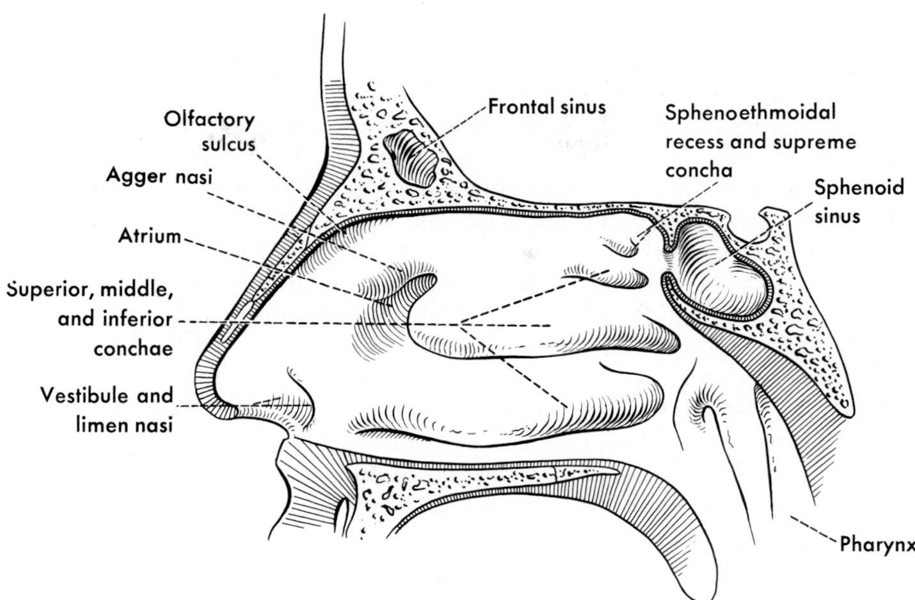


**FIG. 33-30.** The lateral nasal wall.

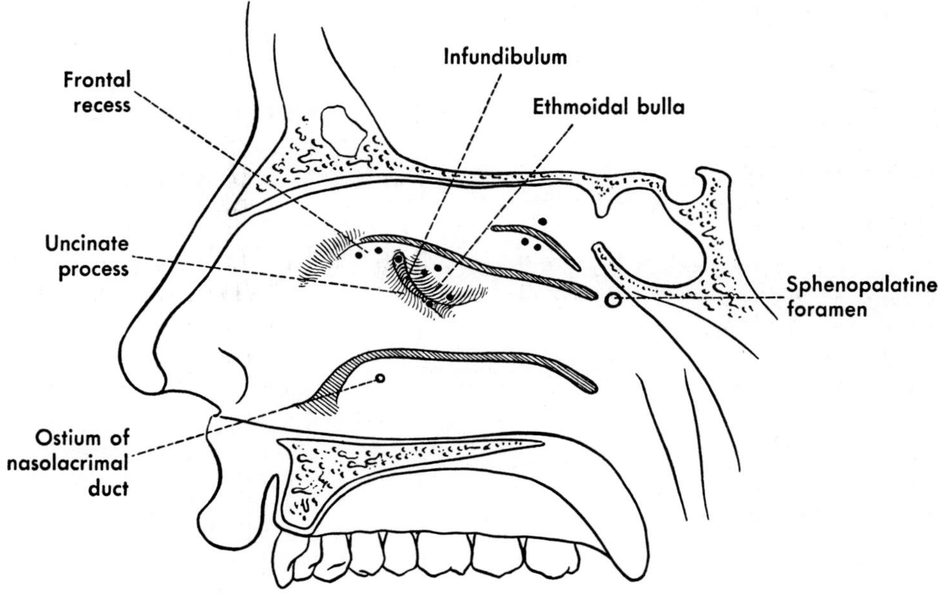


**FIG. 33-31.** The lateral nasal wall after removal of the conchae; their cut bases are represented by appropriate shaded areas. The unlabeled black dots are openings of ethmoid cells.

At the uppermost posterior part of the nose, the nasal cavity is bounded posteriorly by a part of the body of the sphenoid bone known as the **sphenoid concha** (this does not at all resemble the nasal conchae). The sphenoid sinus opens through the sphenoid concha; the part of the nasal cavity here is known as the **sphenoethmoidal recess.** The part of the nasal cavity behind the middle and inferior conchae is the **nasopharyngeal meatus;** it ends at the choana where the nasal cavity opens into the nasal part of the pharynx.

The superior and middle nasal conchae are parts of the **ethmoid labyrinth,** the paired parts of the ethmoid bones that form much of the lateral walls of the nasal cavities and medial walls of the orbits. In the bases of the bones and the ethmoid labyrinth, are the air cavities or ethmoid cells that together constitute the ethmoid or **ethmoidal sinus** (Fig. 33-37). Some of the ethmoid cells open into the superior meatus, and sometimes one or more may open above the superior concha. Other ethmoid cells open under cover of the middle concha. After the middle concha is removed by cutting it close to its base (Fig. 33-31), the openings of some of these cells and additional modeling of the labyrinth can be observed. A rounded bulge projecting downward and forward below the root of attachment of the middle concha is the **ethmoid bulla,** formed by ethmoid air cells that open into the middle meatus. It varies in size with the size and number of the cells present. Anterior to and below it and projecting upward toward it is a thin process, the **uncinate process.** The gap between the bulla and the uncinate process is the **semilunar hiatus,** which leads downward and forward into a curved channel, the **ethmoid infundibulum,** lying lateral to the uncinate process. The maxillary sinus usually opens into the infundibulum under cover of the uncinate process. At its anterior and upper end, the infundibulum expands and regularly receives the openings of ethmoid cells (which may also open into the middle meatus in front of the infundibulum). In many cases, the infundibulum also receives the opening of the frontal sinus.

The only opening under cover of the inferior concha is that of the **nasolacrimal duct.** The exact position of this opening varies, but it usually is fairly high under the curved anterior end of the concha. The opening also varies in size and shape; it frequently is partly covered by a fold of mucous membrane, the lacrimal fold.

**FIG. 33-32.** Distribution of nerves to the lateral nasal wall.

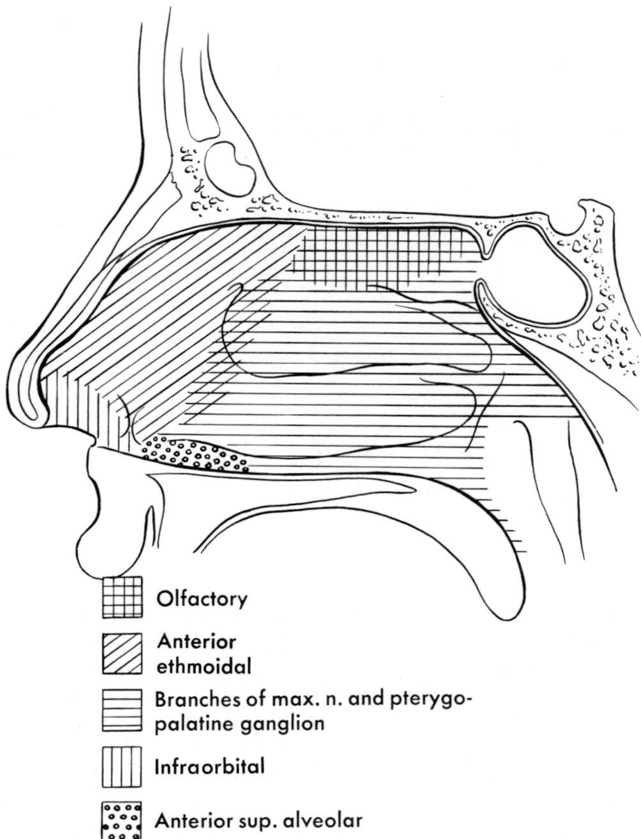


## Nerves and Vessels

The nerves and blood vessels of the nasal cavity are fairly simple, although difficult to demonstrate on the average cadaver. The general areas of distribution of various nerves to the lateral nasal wall are shown in Figure 33-32.

The **olfactory nerves,** concerned with smell only, arise from cells in the uppermost part of the nasal mucosa, largely that over the superior concha and a similar area on the septum. The central processes of these cells form the delicate filaments that constitute the olfactory nerve. These penetrate the cribriform plate and end in the overlying olfactory bulb.

The largest nerves and blood vessels of the nasal cavity enter it through the sphenopalatine foramen, in the lateral nasal wall just behind the posterior end of the middle nasal concha. It connects the nasal cavity and the pterygopalatine fossa. In the pterygopalatine fossa are the maxillary nerve, the pterygopalatine ganglion, and branches of the maxillary artery. Sensory fibers from the maxillary nerve combine with postganglionic parasympathetic fibers from the pterygopalatine ganglion and with sympathetic fibers that have reached the ganglion from the deep petrosal nerve to form lateral

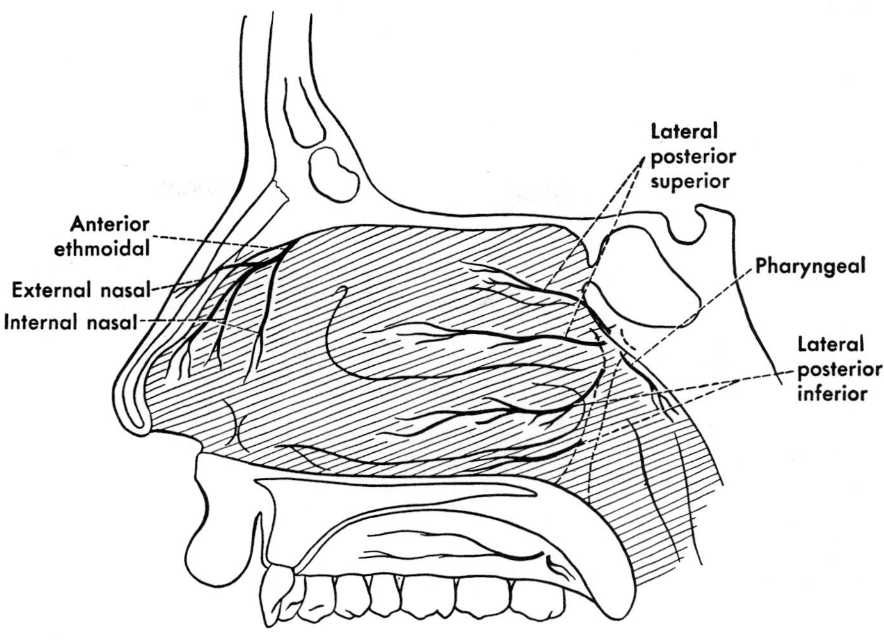


**FIG. 33-33.** Nerves of the respiratory part of the lateral nasal wall.

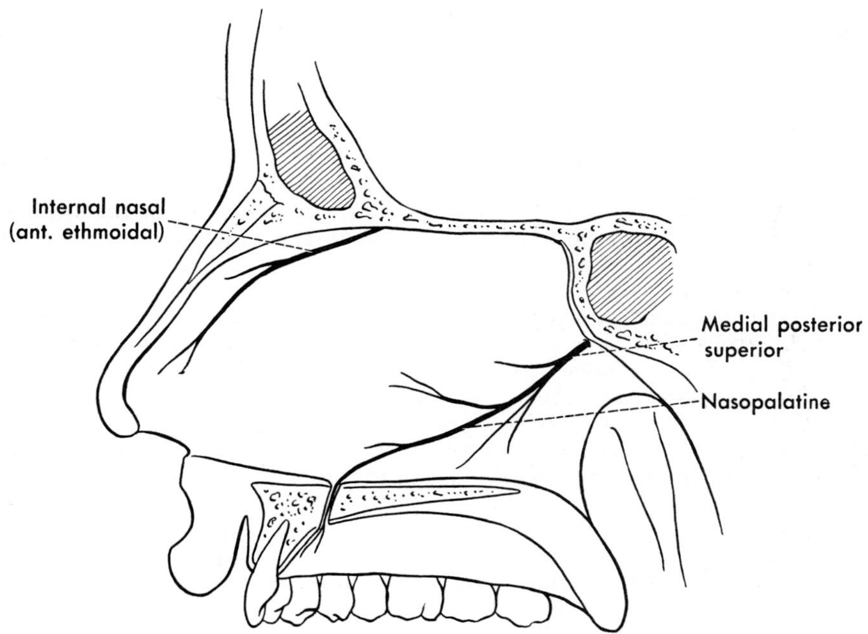


**FIG. 33-34.** Nerves of the respiratory part of the nasal septum.

and medial posterior superior nasal branches (Figs. 33-33 and 33-34). The lateral branches turn forward after traversing the sphenopalatine foramen, and the medial branches cross the body of the sphenoid bone to reach the septum. One of the latter, the **nasopalatine,** continues downward and forward on the septum, eventually passes through an incisive canal between the nasal cavity and the mouth, and ends by innervating mucosa of the hard palate just behind the incisor teeth. The other posterior nerves of the nasal cavity, the **lateral posterior inferior nasals,** are distributed to the posterior parts of the middle and inferior meatuses and the inferior concha. They descend from the maxillary nerve and the pterygopalatine ganglion as part of the greater palatine nerve, then leave it to penetrate the lateral nasal wall.

The anterior part of the nasal cavity is inner-

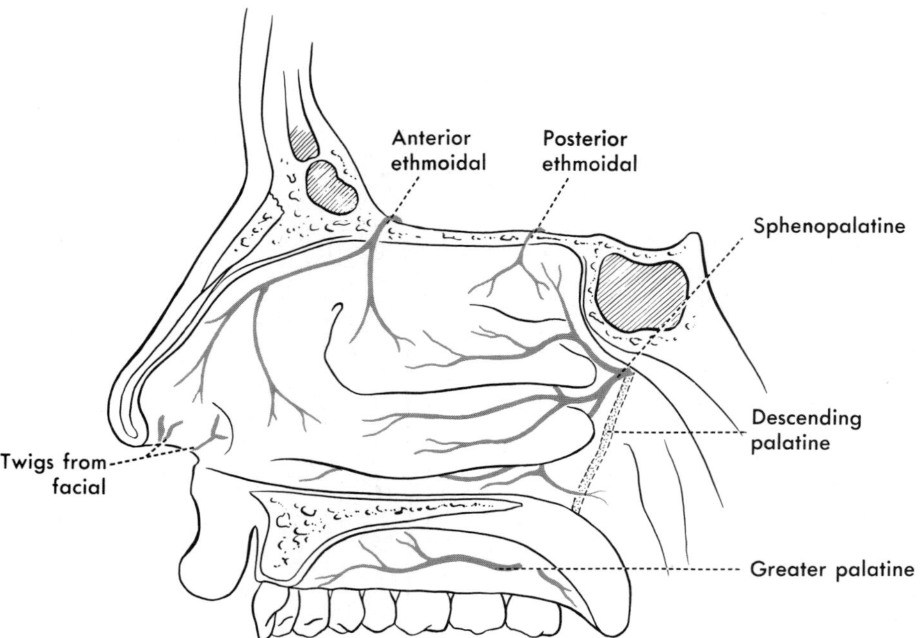


**FIG. 33-35.** Arteries of the lateral nasal wall.

vated largely through the *ophthalmic nerve,* although twigs from the maxillary nerve supply parts of the vestibule and of the inferior meatus. Above, the innervation is through the **anterior ethmoidal nerve,** a branch of the nasociliary. It enters the nasal cavity through a slit in its roof on the side of the crista galli and, besides giving twigs to the anterior ethmoidal cells and to the frontal sinus, distributes internal lateral and medial nasal branches to the anterior parts of the nasal cavity. The external nasal branch continues downward and forward to leave the nasal cavity between the nasal bone and the lateral nasal cartilage and supply skin of the dorsum of the nose.

The **arteries** of the nose (Fig. 33-35) run in general with the larger nerves. The *sphenopalatine artery,* a branch of the maxillary, enters the nasal cavity through the sphenopalatine foramen with the posterior superior nasal nerves and divides into *lateral* and *septal* (medial) *posterior nasal arteries.* In addition, the posterior and anterior ethmoidal branches of the ophthalmic artery enter the nasal cavity through its roof, alongside the cribriform plate, and divide into medial and lateral branches that are distributed to an upper but particularly an anterior part of the nasal cavity. Finally, the *superior labial artery* (from the facial) usually sends a branch into the nasal cavity along the septum, and the *greater palatine artery* (from the maxillary) sends a branch upward through an incisive canal to reach an anterior lower part of the nasal septum. Here there are rather broad anastomoses between the major arteries of the nose, and this area often is involved when there is nosebleed.

More serious bleeding from the nose may arise from the large vessels, just after they have entered the nasal cavity. Bleeding close to the back end of the middle concha, from the sphenopalatine artery, often can be checked by packs placed in this position. The sphenopalatine artery can be ligated if the severity of the bleeding warrants it, or the external carotid artery can be ligated in the neck, to decrease the amount of blood delivered to the sphenopalatine. Bleeding from the roof of the nasal cavity originates from one of the ethmoidal arteries. Severe bleeding from these arteries has been halted by ligating them intraorbitally.

Most of the **lymphatics** of the nasal cavity join those of the pharynx.

## PARANASAL SINUSES

The paranasal sinuses (Figs. 33-36 through 33-38) are diverticula of the nasal cavity, which grow from this cavity into neighboring bones and replace the diploë there. Of the four sinuses, the frontal, the maxillary, and the sphenoid are typically large and paired, but

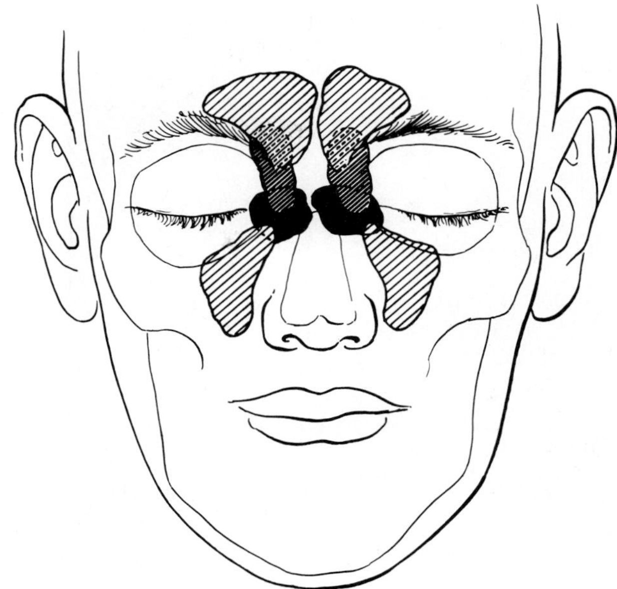

**FIG. 33-36.** Diagram of the paranasal sinuses. The frontal and maxillary sinuses are shaded by widely spaced lines, the ethmoid sinuses are shaded by closely spaced ones, and the sphenoid sinuses are black. (Wakefield EG: Clinical Diagnosis. New York, Appleton, 1955)

**FIG. 33-37.** The ethmoid air cells, together constituting the ethmoid sinus. **b.e.** is the ethmoid bulla; **Sin. sphen.** is the sphenoid sinus. The posterior ethmoid cells are here more heavily shaded than are the anterior and middle ones. "Pharyngeal tube" is an older name for the auditory tube. (Corning HK: Lehrbuch der topographischen Anatomie für Studierende Und Ärtze. Munich, JF Bergmann, 1923)

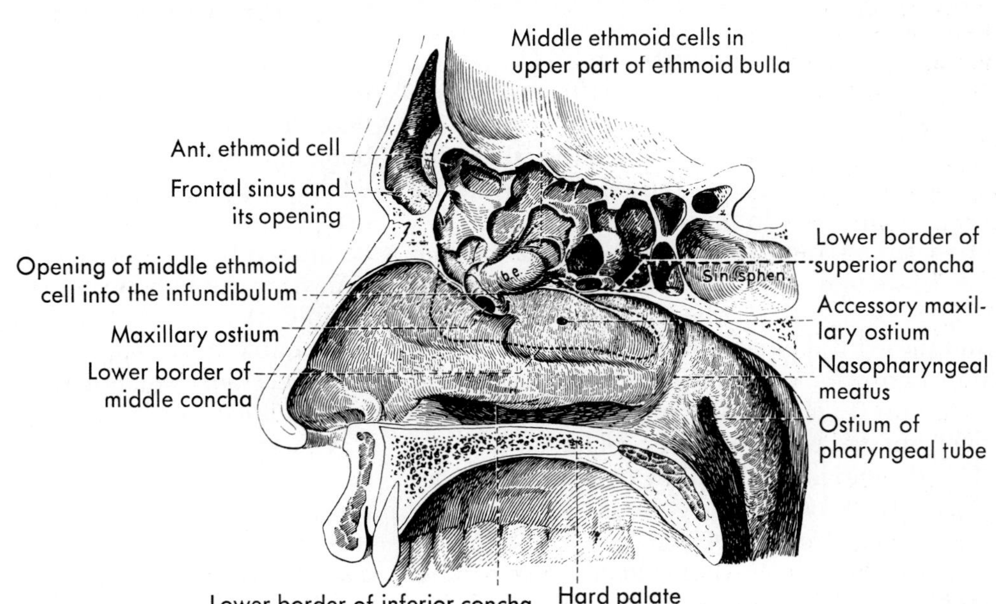

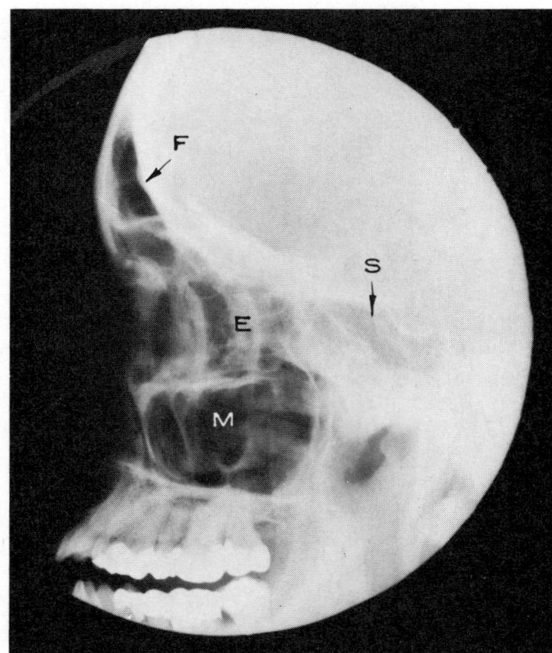


**FIG. 33-38.** A lateral view of the paranasal sinuses. (Courtesy of Dr. D. G. Pugh)

the ethmoid sinus of each side consists of a varying number of smaller air-filled cavities known as the ethmoid cells.

The **frontal sinuses** lie in the frontal bone, as their name implies. They vary much in size; one may be considerably larger than the other and grow beyond the midline, overlapping or pushing to one side the sinus of the other side. Usually a frontal sinus has both a perpendicular and a horizontal extension (Fig. 33-38), so that it lies in both the squama and the orbital part of the frontal bone. Sometimes anterior ethmoid cells protrude markedly into it. The frontal sinus opens beneath the front end of the middle concha, either into the ethmoid infundibulum or anterior to or above this groove in a region known to clinicians as the frontal recess.

The ethmoid cells that collectively form the **ethmoid sinus** vary in number; there may be as few as three or as many as 18. The *anterior ethmoid cells,* which tend to be more numerous but also usually are smaller than the middle and posterior ones, open into the middle meatus, either into the ethmoid infundibulum or the frontal recess. Some of these cells are closely related developmentally to the frontal sinus, and it is one of these that may bulge into that sinus. The *middle ethmoid cells* (also known as bullar cells, since they form the ethmoid bulla) average about three in number, as compared to an average of five or six for the anterior cells. They also open into the middle meatus, either on the surface of the bulla or immediately above it beneath the attached edge of the middle concha. (Many clinicians include the middle cells with the anterior ones, describing as anterior cells all those that open into the middle meatus.) The *posterior ethmoid cells* are reported to vary from none to six; they typically open into the superior meatus, but occasionally one or more open above the superior concha.

Sometimes a posterior ethmoid cell bulges into the sphenoid sinus, or sometimes it becomes intimately related to the optic canal, growing partly or even entirely around the canal so that the optic nerve is separated from the mucosa of the air cell by only a thin lamina of bone. Total blindness of an eye has been reported from careless exploration of such a posterior ethmoid cell.

The **maxillary sinus** lies in the prominence of the cheek, where its roof, lateral wall, and floor are composed primarily of the maxillary bone and its medial wall is the lateral nasal wall. Its opening is high on its medial wall, usually into the infundibulum where it may be hidden by the uncinate process, but sometimes far enough posteriorly to be into the middle meatus itself and therefore more approachable for irrigation when there is infection. Accessory openings into the middle meatus are fairly common.

The **sphenoid sinuses** are placed close together in the body of the sphenoid bone, and therefore the hypophysis, lying in the sella turcica, is above them. On each side are the carotid artery, cavernous sinus, and ophthalmic and maxillary branches of the trigeminal nerve; the nerve of the pterygoid canal runs through the anterior part of the floor. In addition, if a sinus extends forward and upward sufficiently, it becomes closely related to the optic canal. The two sinuses vary much in size and rarely are symmetrical; the bony septum between them usually is deviated to one side or the other. Each sphenoid sinus opens through its anterior wall, some distance above its floor, into the sphenoethmoidal recess of the nasal cavity.

From the positions of their openings, it is obvious that, of the large paranasal sinuses, only the frontal sinus has gravity drainage in the erect posture; the maxillary and sphenoid sinuses never do, and the openings of ethmoid cells vary so much that there is no good position for gravity drainage of all of them. Gravity drainage from the maxillary sinuses is best when one is lying on the side opposite the affected sinus. Similarly, best gravity drainage from the sphenoid sinuses can be secured when one lies face down. However, all the paranasal sinuses are provided with *ciliated epithelium*, and the beat of this cilia is toward the normal ostium; thus, under normal circumstances the secretions are carried to the ostia and discharged into the nose. In sinusitis, the ciliary action is not sufficient to empty the sinuses, the ostia of which often are partly or completely closed by swelling of the mucous membrane; hence, lavage of the sinuses and shrinkage of the membrane to allow them to become once again aerated may be necessary to relieve the condition. Accessory openings for drainage are sometimes made surgically.

The ethmoid cells begin to develop during fetal life, and it is because of this earlier development that they may occupy positions where they bulge into the later-developing frontal and sphenoid sinuses. The frontal and sphenoid sinuses are represented only by rudimentary diverticula at birth; the maxillary sinus is present but small at birth. The maxillary, frontal, and sphenoid sinuses develop fairly rapidly during the first 7 to 8 years of life, but the maxillary sinuses do not reach their full development until after the second or permanent dentition has been acquired, and the frontal ones do not reach theirs until after puberty. The invasion of the frontal and maxillary bones by the sinuses apparently is an important cause of the change in features occurring between babyhood and puberty.

The **blood and nerve supply** of the sphenoid, ethmoid, and frontal sinuses is largely from the nerves and vessels that supply the nose—the posterior superior nasal vessels and nerves, the posterior ethmoidal artery and nerve (when this is present) to the sphenoid sinus and more posteriorly lying ethmoid cells, and anterior ethmoidal arteries and nerves to anteriorly lying ethmoid cells and the frontal sinus. The supraorbital artery and nerve are the chief supply to the frontal sinus and also may supply anterior ethmoid cells. The innervation and blood supply to the maxillary sinus is from twigs that leave the posterior superior alveolar and infraorbital nerves and arteries as they run in the wall of the maxillary sinus in close relationship to the mucosa.

## GENERAL REFERENCES AND RECOMMENDED READINGS


ALDRED P, HALLPIKE CS, LEDOUX A: Observations on the osmotic pressure of the endolymph. J Physiol 98:446, 1940

BAST TH, ANSON BJ: The Temporal Bone and the Ear. Springfield, IL, Charles C Thomas, 1949

BEATIE JC, STILWELL DL JR: Innervation of the eye. Anat Rec 141:45, 1961

BLANTON PL, BIGGS NL: Eighteen hundred years of controversy: The paranasal sinuses. Am J Anat 124:135, 1969

CHRISTENSEN K: The innervation of the nasal mucosa, with special reference to its afferent supply. Ann Otol Rhinol Laryngol 43:1066, 1934

COOPER S, DANIEL PM: Muscle spindles in human extrinsic eye muscles. Brain 72:1, 1949

CROWE SJ, HUGHSON W, WITTING EG: Function of the tensor tympani muscle: An experimental study. Arch Otolaryngol 14:575, 1931

DAVIS H: Biophysics and physiology of the inner ear. Physiol Rev 37:1, 1957

DOHLMAN G: Investigations in the function of the semicircular canals. Acta Otolaryngol (Suppl) 51:211, 1944

FINE BS, TOUSIMIS AJ: The structure of the vitreous body and the suspensory ligaments of the lens. Arch Ophthalmol 65:95, 1961

GRAVES GO, EDWARDS LF: The Eustachian tube: A review of its descriptive, microscopic, topographic and clinical anatomy. Arch Otolaryngol 39:359, 1944

GUILD SR: The circulation of the endolymph. Am J Anat 39:359, 1944

HARDESTY I: On the proportions, development and attachment of the tectorial membrane. Am J Anat 18:1, 1915

HILDING A: The physiology of drainage of nasal mucus: I. The flow of the mucus currents through the drainage system of the nasal mucosa and its relation to ciliary activity. Arch Otolaryngol 15:92, 1932

ISAKSSON I: Studies on congenital genuine blepharoptosis: Morphological and functional investigation of the upper eyelid. Acta Ophthalmol, Suppl 72, 1962

KASPER KA: Nasofrontal connections: A study based

on one hundred consecutive dissections. Arch Otolaryngol 23:322, 1936

PEARSON AA: The development of the nervus terminalis in man. J Comp Neurol 75:39, 1941

RITTER FN, LAWRENCE M: A histological and experimental study of cochlear aqueduct patency in the adult human. Laryngoscope 75:1224, 1965

SCHAEFFER JP: The Nose, Paranasal Sinuses, Nasolacrimal Passageways, and Olfactory Organ in Man. Philadelphia, Blakiston, 1920

VAN ALYEA OE: Ethmoid labyrinth: Anatomic study, with consideration of the clinical significance of its structural characteristics. Arch Otolaryngol 29:881, 1939

VAN ALYEA OE: Sphenoid sinus: Anatomic study, with consideration of the clinical significance of the structural characteristics of the sphenoid sinus. Arch Otolaryngol 34:225, 1941

WEVER EG: The mechanics of hair-cell stimulation. Ann Otol Rhinol Laryngol 80:786, 1971

WOLFF E: The Anatomy of the Eye and Orbit: Including the Central Connections, Development, and Comparative Anatomy of the Visual Apparatus, ed 3rd. Philadelphia, Blakiston, 1948

YOUNG MW: Fluid channels in the temporal bone. Anat Rec 112:136, 1952

# 34

# PHARYNX AND LARYNX

The pharynx is the continuation of the digestive cavity from the mouth, but also receives in its upper part the posterior openings of the nasal cavities, the choanae. That part above the level of the soft palate is the **pars nasalis** (nasal pharynx); that between the level of the soft palate and the entrance into the larynx is the **pars oralis;** and the part posterior to the larynx down to the beginning of the esophagus is the **laryngeal part** (Fig. 34-4). Because of the openings of the nose, mouth, and larynx into the pharynx, the musculature of the pharyngeal wall is largely lateral and posterior; even below the opening from the mouth, there is no anterior muscular wall to the pharynx, for the back part of the tongue at first provides this wall, and below this and the laryngeal opening, the posterior wall of the larynx is the anterior pharyngeal wall.

A certain amount of loose connective tissue lies between the two pterygoid muscles and the lateral wall of the pharynx, and this is described as forming the **lateral pharyngeal fascial space.** This space is bounded above by the base of the skull and below by the attachment of the cervical fascia to the hyoid bone, submandibular gland, and mandible (see Fig. 31-32). Anteriorly, however, it becomes continuous with the potential spaces above and below the mylohyoid muscles ("submandibular space"), and posteriorly, it is continuous with the retropharyngeal space.

The **retropharyngeal space** is the loose connective tissue between the pharynx and the vertebral column. It is the uppermost part of the retrovisceral space as that extends upward to the base of the skull, and really, therefore, should not be separately named.

## Exterior of the Pharynx

The external surface of the pharynx consists of voluntary muscle covered by a thin **buccopharyngeal fascia** continuous with that on the outer surface of the buccinator muscle. It can best be examined after the posterior part of the occipital bone and the vertebral column have been separated from the pharynx, so that it can be seen from behind as well as laterally. In close relationship posterolaterally are the 9th, 10th, and 11th cranial nerves as they emerge through the jugular foramen; just behind these is the upper end of the internal jugular vein as it leaves the jugular fossa; crossing behind the structures from the jugular foramen and running downward, laterally, and forward is the hypoglossal nerve; and anteromedial to them is the upper end of the internal carotid artery as this enters the back end of the carotid canal.

The **blood supply** of the upper part of the pharynx is from the **ascending pharyngeal artery,** which runs upward along the posterolateral wall, and from small descending branches of the palatine arteries; twigs from the superior and inferior thyroid arteries supply the lower part (Fig. 34-1). The **veins** form a *pharyngeal plexus* on the posterior surface of the pharynx. This plexus drains laterally at irregular intervals into the pterygoid plexus, the superior and inferior thyroid veins, and often by separate pharyngeal veins that enter the lower end of the facial vein or the internal jugular close to this ending. Posterolaterally on the pharyngeal wall, not much above the level of the carotid bifurcation, there usually is a large **retropharyngeal lymph node;** there may be smaller nodes present also. These receive most of the lymphatic drainage of the pharynx, the nasal cavity, and a posterior part of the tongue.

The **nerve supply** to the pharynx is from the *pharyngeal plexus,* formed by the union of the pharyngeal branches of the glossopharyngeal and vagus nerves. The vagal contribution to the pharyngeal plexus supplies all the muscles of the pharynx except the stylopharyngeus and most of those of the soft palate (tensor veli palatini excepted), but the lowermost muscle of the pharynx, the inferior constrictor, may receive a twig from another branch of the vagus (external laryngeal nerve) that goes primarily to the larynx. It usually is stated that the glossopharyngeal contribution to the pharyngeal plexus is sensory to all the pharynx between the levels of the opening of the auditory tube and the larynx, but the nerve may not reach as high as the auditory tube, leaving much of the pars nasalis to be

**FIG. 34-1.** Nerves and arteries of the pharynx.

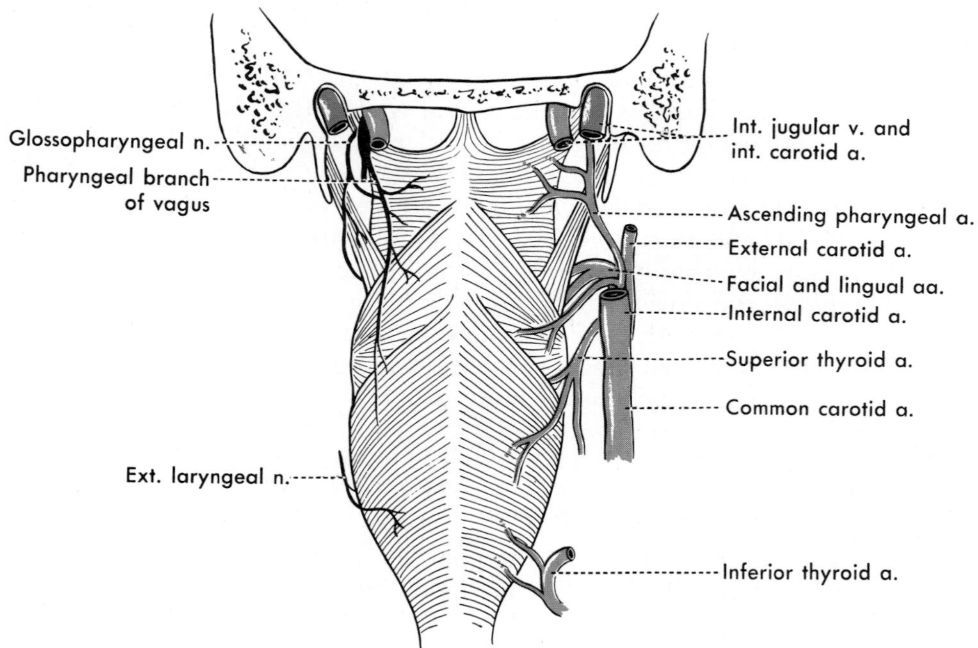

supplied by the 5th nerve; the vagus nerve apparently may supply sensation to more than the laryngeal part of the pharynx.

The four paired **muscles of the pharynx** visible from the outside (Figs. 34-2 and 34-3) are rather simply arranged. There are three constrictors, each overlapping the one above, and there is a more longitudinally arranged muscle, the stylopharyngeus, that runs downward and disappears between the superior and middle constrictors. The muscle fibers of each pair of pharyngeal constrictors meet in the posterior midline to form a *pharyngeal raphe.*

The **superior constrictor** is a broad muscle whose upper part does not completely cover the wall of the pharynx. It has a free upper edge that originates from the posterior border of the medial pterygoid plate, runs posteriorly and somewhat downward, and then loops up to meet the similar component from the other side. The two parts are then attached above by a midline band to the pharyngeal tubercle of the occipital bone. The gap above the muscle on each side is traversed by the auditory tube, but otherwise it is sealed by a heavy **pharyngobasilar fascia** attached above to the base of the skull and to the cartilaginous portion of the auditory tube and anteriorly to the posterior border of the medial pterygoid plate. The pharyngobasilar fascia supports the mucous membrane of most of the nasal part of the pharynx, but becomes much thinner below at about the level of the soft palate. Two muscles of the palate lie adjacent to it, and when the fascia is removed, the gap above the superior constrictor is largely hidden by these (Fig. 34-3).

In addition to its origin from the medial pterygoid plate, the superior constrictor arises also from the *pterygomandibular raphe,* which extends downward from the hamulus of the plate to the mandible and gives origin anteriorly to the buccinator muscle in the cheek. Below this, it takes origin from the mylohyoid line of the mandible, and below this, from the lateral surface of the muscles of the tongue. Four parts corresponding to these four origins are named (Fig. 34-3), but these terms are seldom used. The lower part of the superior constrictor is covered posteriorly by

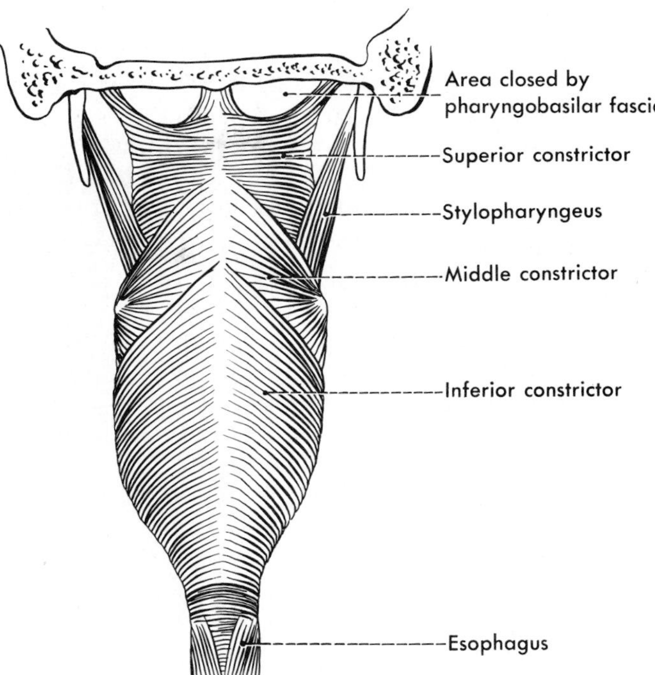


**FIG. 34-2.** The pharyngeal muscles seen from behind.

the middle constrictor, but anterolaterally, there is a gap between the two through which the stylopharyngeus muscle passes.

The **middle constrictor** arises from both the greater and lesser horns of the hyoid bone and from the lower part of the stylohyoid ligament above the lesser horn. From this relatively narrow origin, it spreads out in a fan-shaped manner so that its upper fibers overlap the superior constrictor posteriorly, and its lower fibers are in turn overlapped by the inferior constrictor.

The **inferior constrictor** arises from both thyroid and cricoid cartilages. The fibers of thyroid origin run backward, the upper ones, especially, running upward at the same time, to insert into the pharyngeal raphe and overlap the middle constrictor. The lowest fibers of thyroid origin have only a little upward direction, and the uppermost fibers of cricoid origin run still less upward. The lower fibers run almost directly transversely and blend below with the circular fibers of the esophagus, which are here exposed.

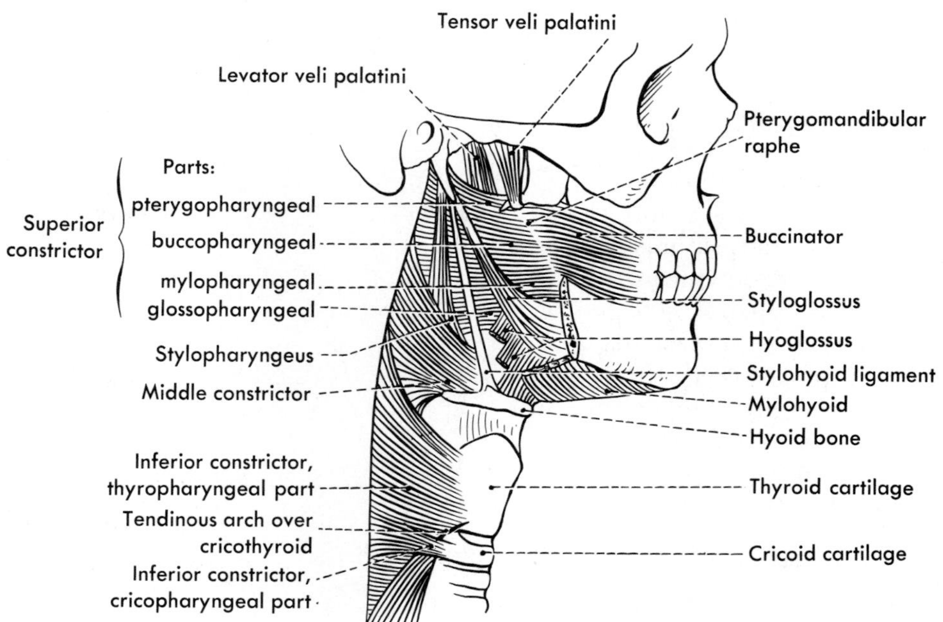


**FIG. 34-3.** Lateral view of the pharyngeal muscles.

The cricopharyngeal part of the inferior constrictor, often called the **cricopharyngeus muscle,** is of considerable importance. In contrast to the other pharyngeal constrictor fibers, it maintains a tonic contraction until swallowing is started and thus serves as the sphincter between the pharynx and the esophagus. This normally prevents regurgitation to the laryngeal level of material passing retrogradely from the stomach into the esophagus, unless there is active vomiting. Also, spasm of the more transverse fibers of the muscle may be a cause of obstruction here and allow the development of a diverticulum (usually called a hypopharyngeal diverticulum). Such diverticula are directed downward and may become very large sacs that interfere with the nutrition of the individual and must be treated surgically.

The **stylopharyngeus muscle** arises from the medial side of the styloid process and passes downward and medially between the external and internal carotid arteries. It enters the pharynx through the gap between the superior and middle constrictors and spreads out on the inner surface of the latter to blend with it and insert in part on the upper and posterior borders of the thyroid cartilage.

The stylopharyngeus is the only muscle innervated by the *glossopharyngeal nerve.* This nerve runs downward medial to the stylopharyngeus and then turns forward toward the tongue around the posterior border and outer surface of the muscle, giving off its lone motor branch at about this point. The stylopharyngeus helps to raise the pharynx, but paralysis of it has no noticeable effect on swallowing.

## INTERIOR OF THE PHARYNX

Of the parts of the cavity of the pharynx (Fig. 34-4), the nasal part often is called nasopharynx and the laryngeal part, the hypopharynx.

### Nasal Part

The nasal part of the pharynx has a roof and lateral and posterior walls. There is, essentially, no anterior wall, since the choanae open here. The roof, called the **fornix,** consists of mucous membrane closely applied to the basal portions of the sphenoid and occipital bones. The lateral and posterior walls consist of the superior constrictors, the pharyngobasilar fascia that lines their internal surfaces, and the mucosa.

The soft palate forms the floor of the anterior part of the nasal pharynx and is the only really mobile wall of this part of the pharynx; thus, the nasal pharynx remains constantly

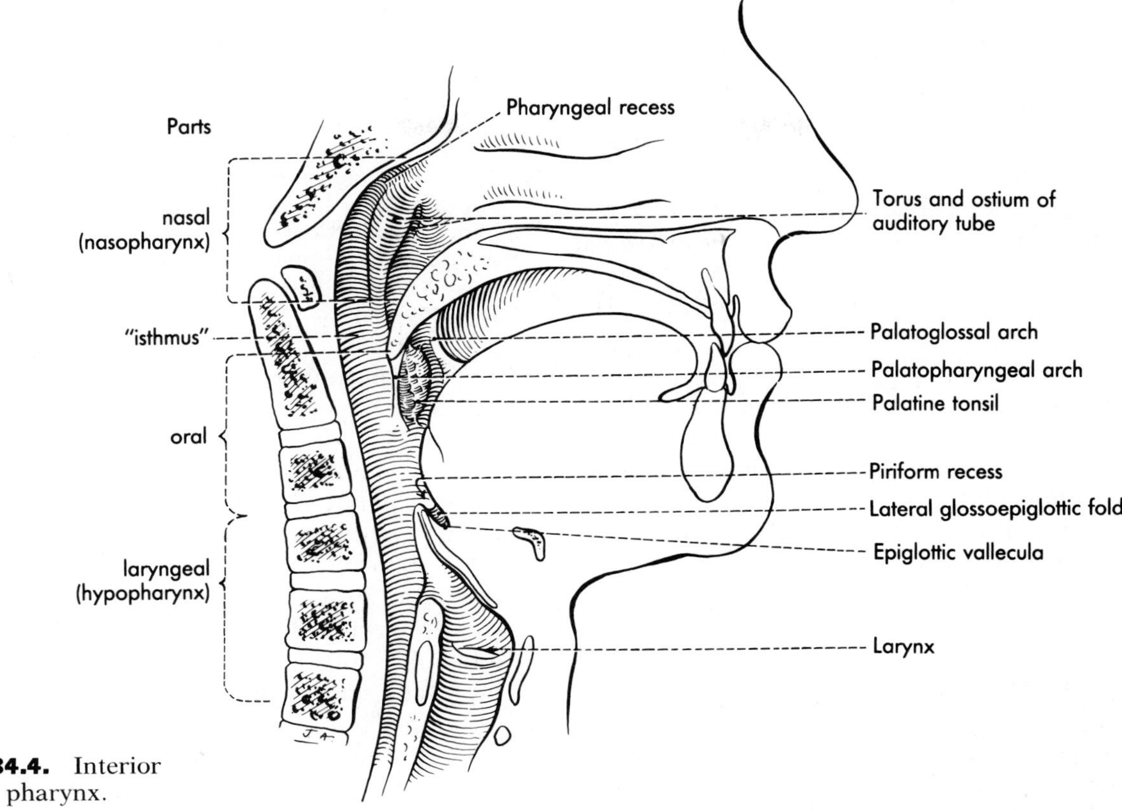


**FIG. 34.4.** Interior of the pharynx.

open. On its lateral wall, above the soft palate, is the **pharyngeal ostium of the auditory tube,** the tube to the middle ear cavity.

Above and posterior to the ostium is an elevation, the **torus tubarius,** or tubal torus, produced by the cartilage of the tube. Proceeding down from the torus is a slight fold, the **salpingopharyngeal fold** (salpinx meaning tube). Behind the torus and the salpingopharyngeal fold is a slitlike lateral projection of the pharynx, the **pharyngeal recess.** A slight **salpingopalatine fold** runs from the anterior border of the torus tubarius toward the palate. Below the torus tubarius, in front of the salpingopharyngeal fold, another fold or bulge, the **torus levatorius,** or levator torus, is formed by the levator veli palatini and is especially evident when the muscle contracts.

The opening of the nasopharynx behind the soft palate into the oral pharynx is usually called the **pharyngeal isthmus.** At this level the nasal part of the pharynx can be completely closed off from the oral part, for the soft palate can be pulled backward and upward by the levator veli palatini to meet the posterior wall. The meeting of soft palate and posterior pharyngeal wall is necessary for proper phonation, especially of consonants, and also is necessary if fluid swallowed under pressure is to be kept from running into the nose (for instance, one can drink bending over, or standing on one's head, because of the complete closure here).

In the posterior part of the roof and the upper part of the posterior wall of the nasal part of the pharynx is an accumulation of lymphoid tissue that may be prominent in children but that becomes indistinct or disappears by adulthood; this is the **pharyngeal tonsil.** Similar accumulations of lymphoid tissue are in children associated with the posterior lip of the ostium of the auditory tube and are called the **tubal tonsil.** When the pharyngeal and tubal tonsils are enlarged, they are referred to as **adenoids.** These may cause difficulty in nasal breathing because of obstruction of the nasal pharnyx, and if the ostium of the tube is occluded, there may be hearing loss because of gradual absorption of the air in the middle ear cavity.

## Oral Part

The oral part of the pharynx opens above, behind the soft palate, into the nasal part. It receives anteriorly the opening from the mouth and below this is bordered by the posterior part of the dorsum of the tongue. Behind the tongue, the oral part of the pharynx extends downward lateral and posterior to the upwardly projecting epiglottis, a portion of the larynx, to become continuous with the laryngeal part.

The **fauces** (meaning throat), the lateral boundaries of the *faucial isthmus,* or the opening of the mouth into the pharynx, deserves special attention. Each consists of two folds ("pillars of the fauces") between which the palatine tonsil, usually known simply as "the tonsil," lies (see both Figs. 31-24 and 34-5). The anterior fold, the **palatoglossal arch,** curves downward and forward from the soft palate to the tongue. The posterior fold, the **palatopharyngeal arch,** extends downward from the posterolateral border of the soft palate along the wall of the pharynx. Each arch contains a muscle, similarly named—that is, palatoglossus and palatopharyngeus muscles. Between these two arches is the somewhat almond-shaped mass of lymphoid tissue that constitutes the palatine tonsil and that largely fills the space between the folds, the **tonsillar fossa.** The roof and floor of the faucial isthmus, not well defined, are the soft palate and the dorsum of the tongue.

**FIG. 34-5.** The palatine tonsil and its relations.

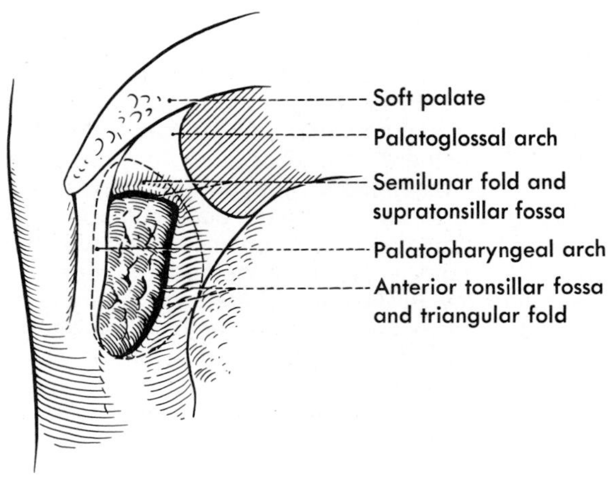


A fold of mucous membrane, the **semilunar fold,** extends between the palatoglossal and palatopharyngeal arches across the upper part of the tonsil, and between it and the tonsil is a slit, the **supratonsillar fossa.** A second fold, the **triangular fold,** projects backward from the palatoglossal arch and either may be fused to the tonsil or may leave a slit that usually is called the anterior tonsillar fossa.

The **palatine tonsil** varies considerably in size and shape, for it may bulge markedly between the folds or may be flat and almost hidden in the tonsillar fossa. It often extends into the soft palate deep to the semilunar fold. Its surface is covered with epithelium that has pits, the *tonsillar crypts,* passing into the lymphoid substance. Between it and the superior pharyngeal constrictor, which forms most of its muscular bed, is the pharyngobasilar fascia; the part adjacent to the tonsil sends septa into it and often is described as the *tonsillar capsule.* Loose connective tissue between the "capsule" and the superior constrictor muscle forms a line of cleavage that facilitates removal of the tonsil.

The largest **artery** of the tonsil, the *tonsillar branch of the facial,* enters its lower pole. The ascending pharyngeal, lingual, descending palatine, and ascending palatine branch of the facial usually also are described as having small tonsillar branches. The **glossopharyngeal nerve** runs forward to the tongue medial to the hyoglossus muscle after passing across the gap between the superior and middle pharyngeal constrictors close to the lower pole of the tonsil. Because of this relationship, edema about the nerve may result from tonsillectomy, and some patients complain of temporary loss of taste following this operation. The nerve gives off a tonsillar branch which, along with branches of the 9th from the pharyngeal plexus, supplies the region of the tonsil.

Below the fauces, the pharynx is bounded anteriorly by the posterior part of the dorsum of the tongue. This, in addition to presenting the vallate taste buds, also has an accumulation of lymphoid tissue beneath its mucosa. This lymphoid tissue on the pharyngeal surface of the tongue constitutes the **lingual tonsil** and may be sufficiently enlarged to need removal when other tonsillar tissue is removed.

It may be noted that the pharyngeal tonsils posteriorly and above, the palatine tonsils laterally, and the lingual tonsil anteriorly and below form an oblique ring of lymphoid tissue around the pharynx. This apparently has the function of tending to halt infection at this level, but when it becomes enlarged as a result of disease, it is no longer of use as a defense mechanism, and its enlargement may, of course, cause obstruction.

The **epiglottis,** behind the tongue, is united to that structure by a midline and two lateral folds, the **median** and the **lateral glossoepiglottic folds,** respectively. The paired depressions between the median and lateral glossoepiglottic folds are the **epiglottic valleculae** (see both Figs. 31-37 and 34-4).

## Laryngeal Part

The laryngeal part of the pharynx, continuous with the oral part at the level of the upper border of the epiglottis, is wide above but narrows rapidly below, at the level of the cricoid cartilage of the larynx (Fig. 34-6), to become continuous with the esophagus at the lower border of the cartilage. The anterior wall of this part of the pharynx is the larynx: above is the posterior surface of the epiglottis and below the opening into the larynx are certain muscles of the larynx and the lamina or expanded posterior part of the cricoid cartilage, covered posteriorly by pharyngeal mucous membrame.

The pharynx also extends lateral to the larynx, being separated from that by the **aryepiglottic folds** that run from the upper posterior border of the larynx to the sides of the epiglottis. These lateral extensions are the **piriform recesses** (also called sinuses). As the pharynx narrows at the cricoid level, the piriform recesses are obliterated; they, thus, are blind forward extensions of the pharynx.

Since each epiglottic vallecula is a shallow basin, an object such as a safety pin that is thought to have been swallowed or inspired may sometimes lodge in it, or such an object may lodge in a piriform recess. Both the valleculae and the recesses, therefore, usually are examined for objects thought to be inhaled, before a child is subjected to bronchoscopy.

**Swallowing** (deglutition) is a complex act typically involving contraction of the pharyngeal constrictors and the esophagus from above downward. As the bolus is passed into the oral part of the pharynx, the soft palate is raised to come in contact with the posterior pharyngeal wall, and the pharynx as a whole is raised by the action of its longitudinal muscle. Contraction of the superior pharyngeal constrictor passes the bolus into the region of the relaxed middle constrictor, which in turn contracts, followed by the inferior constrictor. The cricopharyngeal part of the inferior constrictor ("cricoesophageal sphincter") relaxes as swallowing is

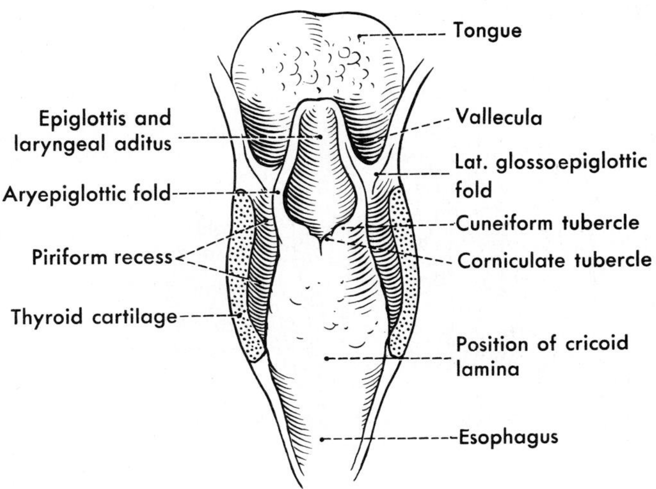


**FIG. 34-6.** The larynx from behind, after removal of the posterior pharyngeal wall.

started, and a descending contraction of the esophagus passes the bolus into the stomach. The larynx is protected in part by contraction of its sphincteric muscles and in part by the fact that the food or liquid tends to pass to the sides of the epiglottis rather than directly over the laryngeal aditus.

## ATTACHMENT OF THE ESOPHAGUS

The mucosa of the esophagus is continuous with that of the pharynx at the lower border of the cricoid cartilage, but the musculature of the two structures is continuous only posteriorly. Here the longitudinal (external) layer of esophageal muscle separates into two bands that diverge anteriorly around the sides of the esophagus, thereby exposing the inner circular muscle posteriorly (Fig. 34-2). After all the longitudinal muscle is concentrated in the anterior esophageal wall, the two bands unite to form a **cricoesophageal tendon** that is attached to the posterior surface of the cricoid lamina (Fig. 34-7). The posteriorly exposed circular fibers at the upper end of the esophagus blend with the lowest fibers of the inferior pharyngeal constrictor (cricopharyngeal part); anteriorly, however, the fibers largely encircle the esophagus, but a few uppermost ones join the sides of the cricoesophageal tendon. Since the inferior constrictor attaches to the sides of the cricoid cartilage, this leaves a small gap laterally between the lower edge of the cricopharyngeal part of the inferior constrictor and the esophageal muscle. The inferior laryngeal nerve and vessels enter the larynx here (Fig. 34-20).

**FIG. 34-7.** Attachment of the esophagus to the larynx.

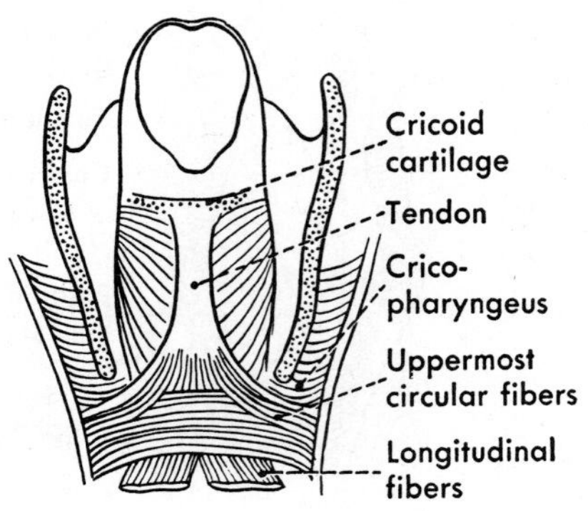


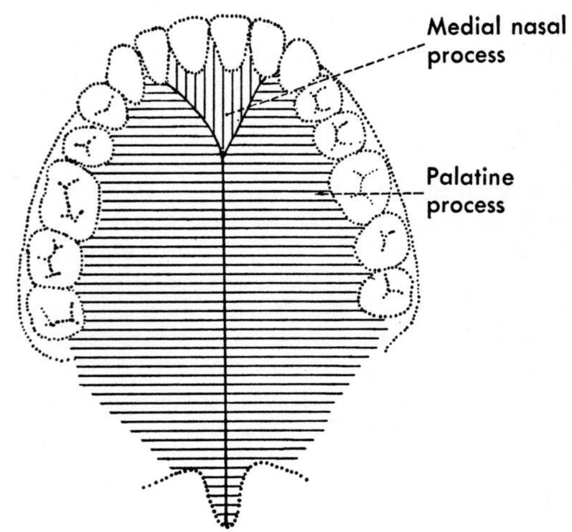


**FIG. 34-8.** Elements entering into the formation of the palate and anterior alveolar region.

## PALATE

The palate separates the nose from the mouth and partially separates the nasal and oral parts of the pharynx. Its major anterior part, the **hard palate,** consists of the bony palate with a covering of mucosa and numerous mucous glands.

**Development.** The palate is formed by two palatine (lateral palatine) processes that develop from the maxilla and fuse with each other in the midline and with the fused medial nasal or medial palatine processes (also called the premaxillary or primary palate) anteriorly (Fig. 34-8). Failure of the nasal process to develop or to fuse with one or both lateral processes, or failure of the lateral processes to fuse, produces **cleft palate.** Associated with cleft palate, or sometimes occurring alone, there may be a **cleft upper lip.** Whether the cleft is bilateral, unilateral, or a large midline defect, it often has been called harelip. (Hares and rabbits normally have a cleft upper lip, but their cleft is a midline one. The abnormal cleft in man is rarely in the midline, occurring more commonly along the line of fusion between medial nasal and palatine processes.) Clefts that involve the posterior part of the palate interfere with swallowing and with phonation. Orthodontists, plastic surgeons, and speech specialists often work together as a team in determin-

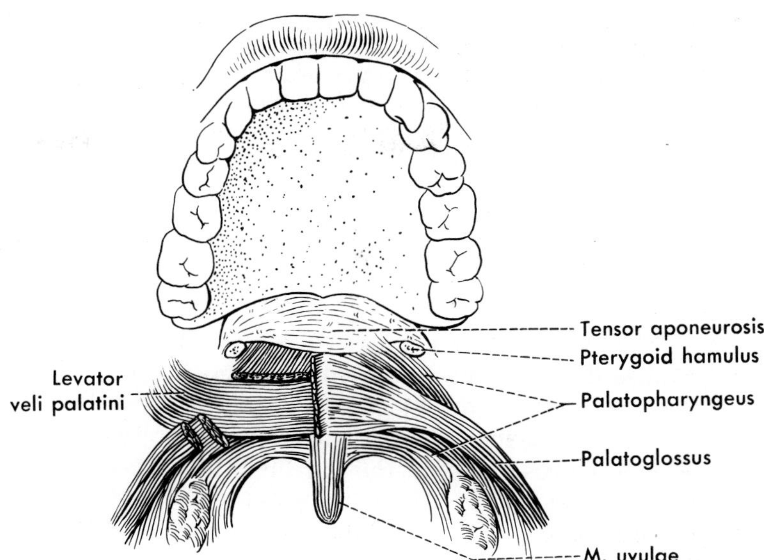


**FIG. 34-9.** Muscles of the soft palate.

ing at what age an operation should be done to produce maximum benefit.

The **soft palate** (palatinum molle) is attached anteriorly to the hard palate and blends laterally with the pharynx. Its posteroinferior part, the **velum palatini** (a term used also to mean the entire soft palate), is more in the coronal than the horizontal plane, and from its free edge, a nipplelike projection, the **uvula**, hangs down. The soft palate is mainly muscular in structure (Fig. 34-9), but numerous mucous glands lie between the mucosa and the muscles. With the exception of the tensor veli palatini, the muscles are innervated through the vagus nerve by way of the pharyngeal plexus; the tensor is innervated by a branch from the mandibular nerve.

## Muscles

The **levator veli palatini** (Fig. 34-3) arises from the inferior surface of the petrous portion of the temporal bone and from a part of the medial surface of the cartilaginous portion of the auditory tube, just inside the attachment of the pharyngobasilar fascia to the tube and the temporal bone. It runs obliquely downward and medially into the posterior part of the soft palate to become continuous with the muscle of the opposite side. As it passes to the palate below the ostium of the auditory tube, it is responsible for the *levator torus,* which may be obvious only when the muscle contracts. The insertion end of this muscle forms much of the bulk of the soft palate.

The **tensor veli palatini** (Figs. 34-3 and 34-9) lies lateral and in part anterior to the levator. It arises from the scaphoid fossa (at the base of the medial pterygoid process), the spine of the sphenoid bone, and the lateral surface of the cartilaginous portion of the auditory tube (outside the pharyngobasilar fascia, which separates it from the levator) and extends vertically downward lateral to the medial pterygoid process. Its posterior border runs forward, but its anterior border runs straight down, so that at the level of the pterygoid hamulus, it has become narrow. Here it gives rise to a tendon that runs medially below the pterygoid hamulus and spreads out as the more aponeurotic anterior part of the soft palate. Anteriorly, the aponeurosis is attached to the posterior edge and lower surface of the bony palate, and it helps give origin to muscles of the soft palate. While it is related to the medial pterygoid plate, therefore deep in the infratemporal fossa, the tensor receives its nerve supply from the mandibular nerve, which lies in the infratemporal fossa lateral to the muscle.

These two muscles have different actions on the soft palate. The levator veli palatini raises the soft palate and in so doing also pulls it backward so as to make it approach the posterior wall of the pharynx. The tensor veli palatini, by reason of the sharp turn that it makes around the pterygoid hamulus, cannot raise the soft palate but, rather, pulls laterally and, therefore, tenses it. Both muscles have been said to open the auditory tube (as in yawning, to allow air to enter the middle ear cavity), but the role of the levator in this process is less certain than that of the tensor.

The **musculus uvulae** arises by paired slips from the posterior border of the hard palate just on each side of the midline, and from the palatine aponeurosis behind the hard palate. The slips run backward on each side of the midline into the uvula, blending as they do so.

The **palatoglossus** is a thin sheet of muscle that begins on the lower surface of the soft palate where it is continuous with its fellow of the opposite side. It runs laterally, downward, and forward in front of the tonsil, forming the palatoglossal arch, and inserts into the dorsum and side of the tongue.

The **palatopharyngeus** is attached to the posterior border of the hard palate and to the aponeurotic layer of the soft palate, and is, in part, continuous with its fellow of the opposite side. It is split into a superior (posterior) and an inferior (anterior) layer, between which lie the musculus uvulae and the palatine part of the levator. At the lateral border of the palate, the two layers blend and descend internal to the superior pharyngeal constrictor. Some of the anterior fibers form a part of the bed of the tonsil, and some of the posterior ones form the palatopharyngeal arch. The posterior fibers are joined by the small **salpingopharyngeus muscle,** a muscle of the pharynx that takes origin from the pharyngeal end of the auditory tube and raises the salpingopharyngeal fold in the posterolateral wall of the pharynx. The muscle fibers of the two palatopharyngeal muscles then spread toward the posterior midline. They end by attaching, in part, into the thyroid cartilage but largely into the pharyngobasilar fascia.

The palatopharyngeus muscle is a levator of the pharynx. The salpingopharyngeus also is potentially a pharyngeal levator, but typically is a tiny muscle and may be lacking; it is of no real importance. The palatopharyngeus and the palatoglossus can both depress the palate, and when they do so, they narrow the faucial isthmus.

## Nerves and Vessels

The chief nerves and vessels of the palate descend through greater and lesser palatine canals that open above into the pterygopalatine fossa and below by a major and usually two minor palatine foramina on the lower surface

**FIG. 34-10.** The palatine nerves.

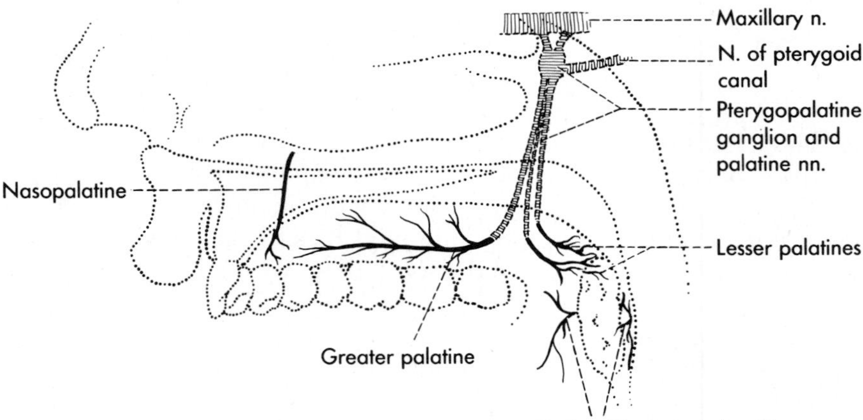

of the hard palate. The nerves descending in the canals are listed as branches of the pterygopalatine ganglion. Actually, the fibers composing them are largely from the maxillary, but as these fibers pass the pterygopalatine ganglion, they probably are joined by some postganglionic parasympathetic fibers from it, sympathetic fibers from the internal carotid plexus, and sensory fibers from the facial nerve, the last two brought to the ganglion by the nerve of the pterygoid canal. As the palatine nerves run downward through the palatine canals, they give off posterior inferior nasal branches to the lateral nasal wall and usually emerge on the palate as three nerves (Fig. 34-10).

The **greater palatine nerve,** the largest, emerges from the greater palatine foramen and turns forward to supply the hard palate and the inner surface of the gums of the teeth as far forward as the canine tooth. A small area immediately behind the incisor teeth is supplied by the **nasopalatine nerve,** which descends through an incisive canal from the nose. The **lesser palatine nerves,** emerging through lesser palatine foramina, supply the soft palate and the upper tonsillar region. It presumably is one of these nerves that contains sensory fibers from the facial nerve to a posterior part of the soft palate.

The **descending palatine artery,** derived from the terminal portion of the maxillary artery in the pterygopalatine fossa, is a short trunk that gives off lesser palatine arteries and continues as the **greater palatine artery.** The latter runs, with the nerve of the same name, through the greater palatine canal and emerges at the greater palatine foramen to run forward on the hard palate with the nerve. Instead of stopping short of the incisor teeth, it sends a terminal branch through an incisive canal to reach a lower part of the nasal septum. The **lesser palatine arteries** descend with the lesser palatine nerves to supply the soft palate and the upper tonsillar region. The small **ascending palatine branch** of the facial artery runs over the upper border of the superior constrictor and along the levator veli palatini to anastomose with them.

## LARYNX

The larynx, part of the respiratory system, opens posteriorly and above into the pharynx; inferiorly, it is continued by the trachea. Although it is supported by cartilages, its aperture can be varied, for it is provided with muscles that can bring folds together, as in holding the breath or in vocalizing (in which the vocal folds are close together and set in vibration by the expired air), or separate them, as when one gasps for breath.

### LARYNGEAL SKELETON

The major cartilages of the larynx are the thyroid, cricoid, arytenoid, and epiglottic (Figs. 34-11 and 34-12). The largest of these is the **thyroid cartilage,** which forms the laryngeal prominence in the neck. The thyroid cartilage, as its name (shield shaped) indicates, does not extend around the larynx. It consists of *right and left laminae* that meet in the anterior midline. Projecting upward from the posterior border of each lamina is a *superior cornu,* and projecting downward is an *inferior cornu.* The upper border and superior cornua of the thyroid cartilage are united to the hyoid bone above by the *thyrohyoid membrane,* the thickened lateral edges of which are called the **thyrohyoid ligaments** and may contain small triticeal cartilages. The thickened midline part is called the median thyrohyoid ligament. Each side of the thyrohyoid membrane presents, posterolaterally, an aperture through which the superior laryngeal nerve (internal branch) and vessels enter the larynx. The thyroid cartilage articulates below, through its inferior horns, with the cricoid cartilage and, besides these synovial articulations, is attached to this same cartilage by the **cricothyroid ligament.**

Only a part of the **cricoid cartilage** can be seen anteriorly; this is mostly its *arch,* united above to the thyroid cartilage by the *cricothyroid ligament* (in the anterior midline) and below to the uppermost ring of the trachea by a *cricotracheal ligament.* Posteriorly, the cricoid cartilage expands to form a *lamina,* on the

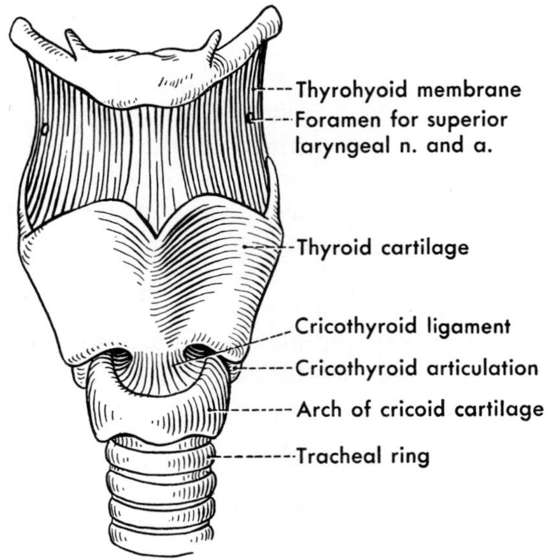


**FIG. 34-11.** Cartilages of the larynx, anterior view.

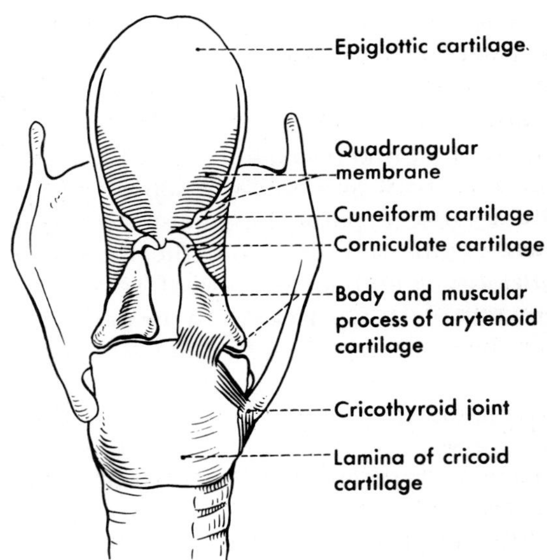


**FIG. 34-12.** Cartilages of the larynx, posterior view.

sides of which are the cricothyroid synovial articulations.

Articulating with the upper lateral borders of the cricoid lamina are the **arytenoid cartilages.** These are particularly important since they are movable upon the cricoid cartilage and have the *vocal folds* attached to them. Each arytenoid cartilage somewhat resembles a three-sided pyramid. The base presents a synovial articular surface on which the arytenoid cartilage can slide laterally and medially, forward and backward, or rotate upon the cricoid cartilage. Projecting laterally from the base is a short blunt *muscular process* (two important muscles insert upon it); projecting anteriorly is a thinner process, the *vocal process,* to which the vocal cords and folds are attached. The posterior surface of the arytenoid cartilage is somewhat concave and gives attachment to a muscle that runs between the two cartilages; the medial surface is small, faces the medial surface of the other cartilage, and is covered by mucous membrane of the larynx; and the anterolateral surface, the largest, gives attachment to thin muscles that line the larynx and also has a large group of mucous glands between it and the mucous membrane. The apex of the cartilage is surmounted by a slender curved **corniculate cartilage** that is directed posteromedially but does not move independently of the arytenoid cartilage. Its upper end raises the mucosa to produce a small *corniculate tubercle.*

Extending upward, laterally, and forward from the upper part of each arytenoid cartilage and from its surmounting corniculate cartilage to the lateral border of the epiglottic cartilage, there is, in the intact condition, a fold of mucous membrane, the **aryepiglottic fold,** supported by a thin lamina of connective tissue, the **quadrangular membrane** (Fig. 34-12). This membrane separates the piriform recess from the entrance into the larynx, and in its free border a little above and anterior to the corniculate tubercle, there may be a small **cuneiform cartilage** that produces a *cuneiform tubercle* (Fig. 34-13).

The **epiglottic cartilage** is a thin unpaired cartilage supporting the epiglottis, which projects upward in the pharynx behind the tongue. The cartilage is broad above but narrows below to a thin stalk or *petiolus.* The upper part of its anterior surface, clothed with mucous membrane, borders the epiglottic valleculae posteriorly. Its posterior surface, also clothed with mucous membrane, forms the anterior wall of the laryngeal vestibule, the first part of the larynx. Below the valleculae, the anterior surface of the epiglottic cartilage is attached to the posterior surface of the hyoid bone by a *hypoepiglottic ligament,* and the petiolus is attached to the posterior surface of the thyroid cartilage by a *thyroepiglottic ligament.* From the edges of the epiglottis, the *aryepiglottic folds,* already mentioned, extend downward to the corniculate and arytenoid cartilages.

## INTERIOR OF LARYNX

In a hemisection of the larynx (Fig. 34-13), the general anatomy of its walls can be studied; relatively little of the cavity can be seen from above (Fig. 34-14). The aditus or entrance is bounded above by the superior border of the epiglottis, laterally by the aryepiglottic folds, and inferiorly and posteriorly by the interarytenoid fold (enclosing the arytenoideus muscles). The cavity of the larynx is divided into three parts: the **vestibule,** extending from the aditus to the vestibular folds, prominent transverse folds of mucous membrane; the **ventricle,** lying between the vestibular and vocal folds; and the **infraglottic** (subglottic) **cavity,** lying below the vocal folds and extending downward to the lower border of the cricoid cartilage. The **glottis** consists of the vocal folds and the slit between these.

The bulky **vestibular folds** extend transversely on either side of the larynx and enclose between them a slit, the **rima vestibuli.** Underlying these folds are muscle fibers that can, by their contraction, bring the folds together, and it is this apposition of folds that is essential to holding the breath against pressure in the thoracic cavity when one is, for instance, exerting strain as in lifting a weight.

The **laryngeal ventricles,** lateral expansions of the laryngeal cavity between the vestibular folds and the sharper-edged vocal folds, undercut somewhat the vestibular folds, thereby helping these to resist the pressure of outgoing air when they are brought together. Toward the anterior end of the laryngeal ventricle there is a small blind sac protruding upward, the **laryngeal saccule.**

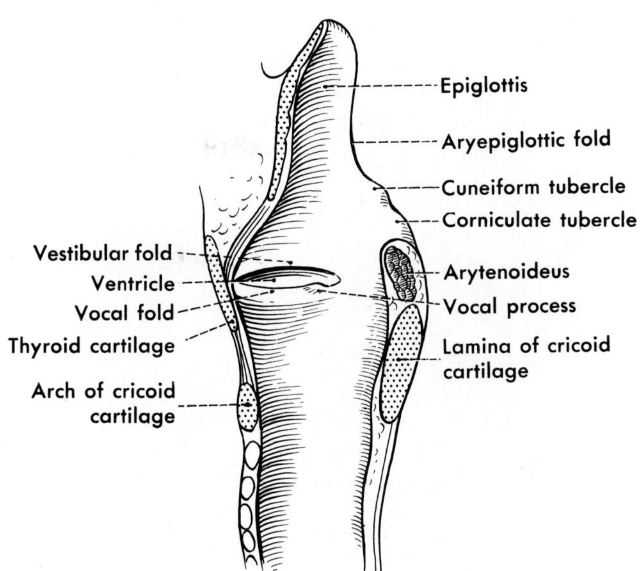


**FIG. 34-13.** Interior of the larynx.

Rarely, the laryngeal saccule is congenitally enlarged and protrudes out through the thyrohyoid membrane to lie deep to the strap muscles of the neck (as it does in certain monkeys normally). It can also be enlarged by frequent increases in thoracic pressure. Thus, its enlargement can be considered an occupational hazard for such persons as glassblowers.

The **vocal folds** below the ventricles project medially and somewhat upward. Because of their direction, they readily are pushed aside by expired air (this occurs in phonation) but tend to resist inspiration of air when they are together. The slit between the two vocal folds is the **rima glottidis.** The anterior part of each fold is formed by the vocal ligament (these border the pars intermembranacea of the rima). The posterior part of each fold is formed by the vocal process of the arytenoid cartilage and borders the pars intercartilagina.

The membranous vocal folds, commonly called the vocal cords, are formed by thickened, upwardly projecting edges of a fibroelastic membrane, the **conus elasticus,** that lies between the laryngeal muscles and the mucous membrane (Fig. 34-15). The conus elasticus arises below from the entire upper border of the arch of the cricoid cartilage; the

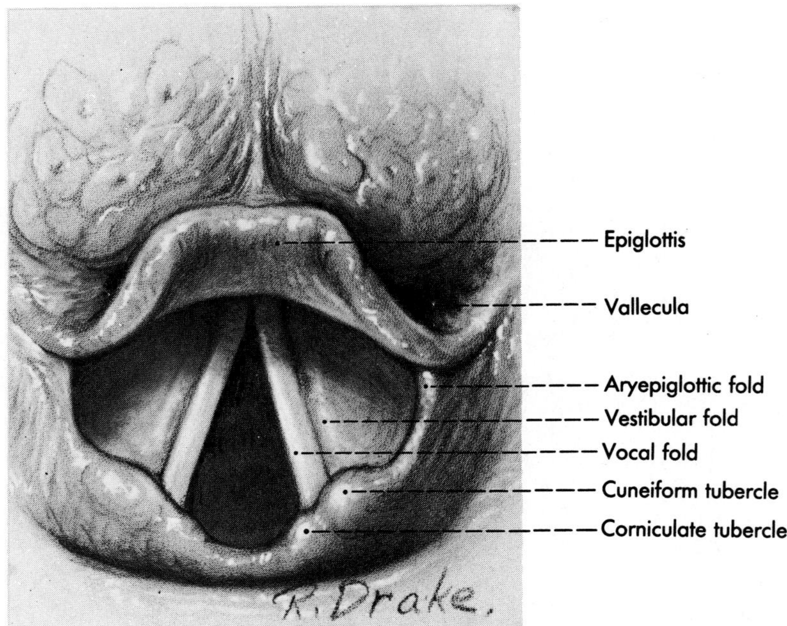


**FIG. 34-14.** View of the larynx from above with the vocal cords abducted.

cricothyroid ligament, visible externally, is the thickened anterior part of the conus. Its upper borders, the **vocal ligaments,** are attached posteriorly to the vocal processes of the two arytenoid cartilages; anteriorly they are attached almost together to the posterior surface of the thyroid cartilage (Fig. 34-16). Apposition of the two vocal folds is necessary for normal phonation, since this allows the setting up of vibrations by them as the air passes between them. Similarly, their abduction is necessary to widen the passageway and allow the utmost in respiratory activity. Since their anterior ends are attached to the thyroid cartilage, it is their posterior ends that are moved apart or brought together by movements of the arytenoid cartilages.

**FIG. 34-15.** Diagram of the conus elasticus.

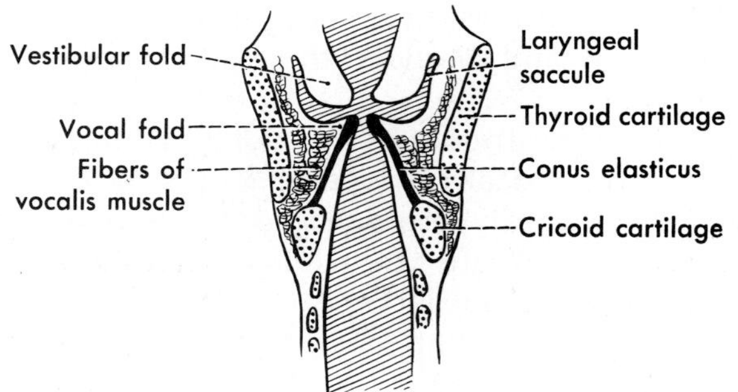


The **infraglottic portion** of the larynx presents nothing of particular importance. It is wider anteroposteriorly than laterally. It narrows anteroposteriorly as it is traced downward toward the trachea, and its narrower lateral dimension widens. Anteriorly, the mucosa of the infraglottic portion of the larynx rests against a lower part of the thyroid cartilage, the cricothyroid ligament, and the arch of the cricoid cartilage. Posteriorly, it rests against the anterior surface of the lamina of the cricoid cartilage. At the lower border of the cricoid cartilage, the larynx is continuous with the trachea.

#### Elastic and Epithelial Layers

The larynx is lined by a fibroelastic membrane. The mucosa of the larynx is closely attached to the inner surface of this membrane, and many of the muscles are closely attached to the outer surface. The heaviest part of the fibroelastic membrane is the lowest part, or *conus elasticus,* ending above in the vocal ligaments, and the part above is the *quadrangular membrane.* This is particularly thin over the ventricles, but thickens in the vestibular folds to form vestibular ligaments. Above this, it runs up-

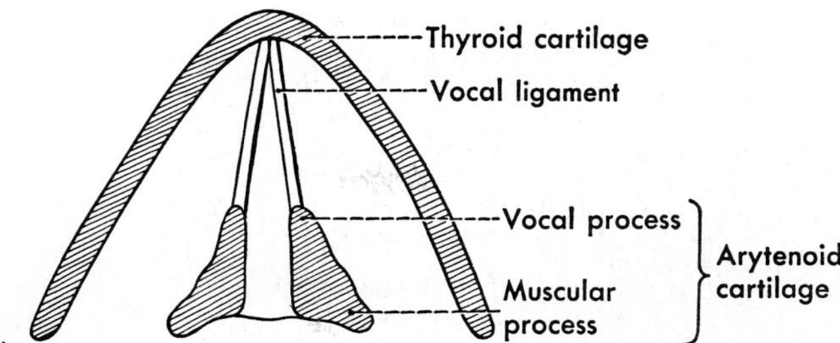


**FIG. 34-16.** Attachments of the vocal ligaments.

ward and forward from the arytenoid and corniculate cartilages to the sides of the epiglottic cartilage, thus supporting the aryepiglottic folds.

The vocal ligaments are covered by stratified squamous epithelium, as is most of the wall of the vestibule, but columnar ciliated epithelium, similar to that of the trachea, is present in the laryngeal ventricle and below the glottis.

## Muscles

The muscles listed as laryngeal ones are confined to the larynx and are divisible into two general groups: one group, composed of the aryepiglotticus, the thyroepiglotticus, the thyroarytenoideus, and the oblique arytenoideus, is primarily concerned with protection of the larynx and is sphincteric in action on the laryngeal aditus and vestibule; the other group, composed of the cricothyroideus, the cricoarytenoideus, the vocalis, and the transverse arytenoideus, is concerned with adjustments of the larynx and vocal cords during phonation and respiration. The larynx as a whole is also acted upon by the infrahyoid muscles, and indeed by all muscles attaching to the hyoid bone, since it moves with the hyoid bone.

The laryngeal muscles are innervated by the vagus nerves (the fibers reaching these nerves through the so-called cranial rootlets and internal ramus of the accessory nerve). The cricothyroid muscle, the only external muscle of the larynx, is the only one innervated by the superior laryngeal nerve (which is otherwise only sensory and secretory to the larynx). The remaining muscles are innervated through the recurrent laryngeal branch of the vagus, which enters the larynx from below.

The **cricothyroid muscle** (Fig. 34-20) arises from the arch of the cricoid cartilage and runs upward and backward, fanning out as it does so, to insert on the lower border and lower medial surface of the thyroid cartilage. The two muscles cover the lateral part of the cricothyroid ligament. When they act together, they tilt the cricoid cartilage on the thyroid. In consequence, they lengthen the distance between the posterior surface of the thyroid cartilage and the vocal processes of the arytenoid cartilages and thus lengthen the vocal cords (Fig. 34-17). Further, since in the relaxed condition the vocal cords are bowed, this movement also tends to straighten the vocal cords and thus bring them closer to the midline.

The **external branch of the superior laryngeal nerve** descends on the lateral pharyngeal wall just deep to the superior thyroid artery to reach the cricothyroid muscle and innervate it.

After the pharyngeal mucosa has been removed from the posterior and lateral surfaces of the cricoid and arytenoid cartilages, and a lateral part of the thyroid cartilage has been cut away, most of the remaining muscles of

**FIG. 34-17.** Action of the cricothyroid in lengthening and thinning the vocal cord by tilting the cricoid and thyroid cartilages on each other.

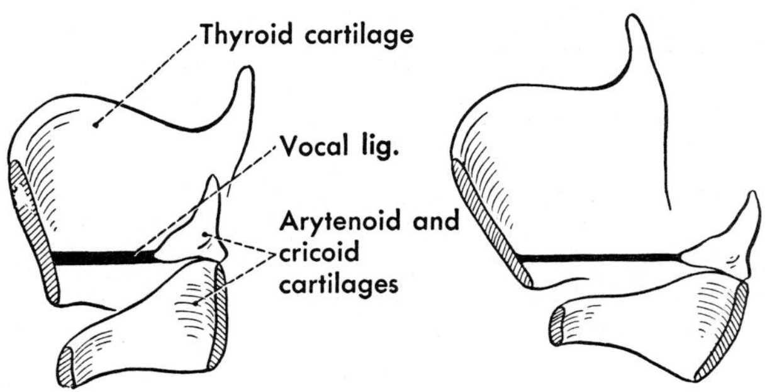

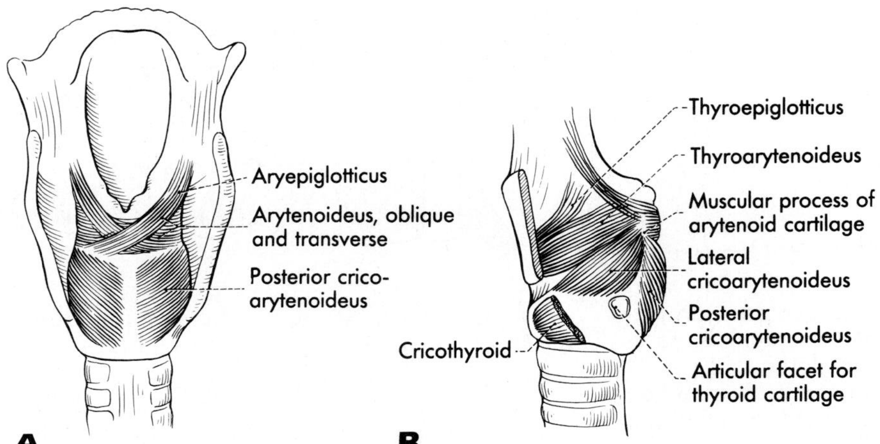


**FIG. 34-18.** Muscles of the larynx in posterior view, **A,** and lateral view, **B,** after removal of the thyroid cartilage.

the larynx can be seen (Fig. 34-18). The posterior surface of the lamina of the cricoid cartilage is largely covered on each side by the two **posterior cricoarytenoideus muscles,** which arise from the lamina. The fibers of each muscle converge to an insertion on the muscular process of the arytenoid cartilage. By their contraction, these muscles pull the muscular processes of the arytenoid cartilages medially and backward (Fig. 34-19A), thus producing a rotation of the arytenoid cartilages at the cricoarytenoid joints and swinging the vocal processes laterally. At the same time, they pull the arytenoid cartilages downward on the obliquely sloping cricoarytenoid joints, so that the two cartilages slide away from each other. Thus, both intercartilaginous and intermembranous parts of the rima glottidis are widened; indeed, the posterior cricoarytenoid muscles are the only effective abductors of the vocal cords and, therefore, are particularly important in the preservation of an airway at the rima glottidis.

The **lateral cricoarytenoideus** is smaller than the posterior one. It arises from the upper border of the posterior part of the arch of the cricoid and runs upward and backward to insert on the muscular process of the arytenoid cartilage. It pulls the muscular process of the arytenoid cartilage forward and laterally (Fig. 34-19B), rotating and sliding the arytenoid cartilage so that it and its vocal process are brought closer to the midline; thus, the lateral cricoarytenoidei are adductors of the vocal cords.

The **transverse arytenoideus** is an unpaired muscle that stretches between the posterior surfaces of the two arytenoid cartilages, uniting them below the interarytenoid notch. It draws the two cartilages together when it contracts, thus tending to close the rima glottidis.

Posterior to the heavy transverse fibers of the transverse arytenoideus are two slender bands, the **oblique arytenoidei,** that arise from the muscular processes of the two arytenoid cartilages. The fibers swing upward across the midline, crossing those of the other side, to reach the apex of the arytenoid cartilage, where many of them insert. The remainder, joined by other fibers arising here from the arytenoid cartilage, extend upward in the aryepiglottic folds and form the **aryepiglottic muscle.** The oblique arytenoids and the aryepiglottic muscles together constitute a sphincter of the laryngeal aditus. By their contraction, they help to prevent solids and liquids from entering the vestibule.

On the side of the larynx deep to the thyroid cartilage is a thin lamina of muscle, best developed at the level of the ventricular fold, that is the **thyroarytenoideus muscle.** This arises from the inner surface of the lower part of the thyroid cartilage close to the angle formed by the two laminae. It runs backward and somewhat upward to an insertion on the lateral border of the arytenoid cartilage. It is a

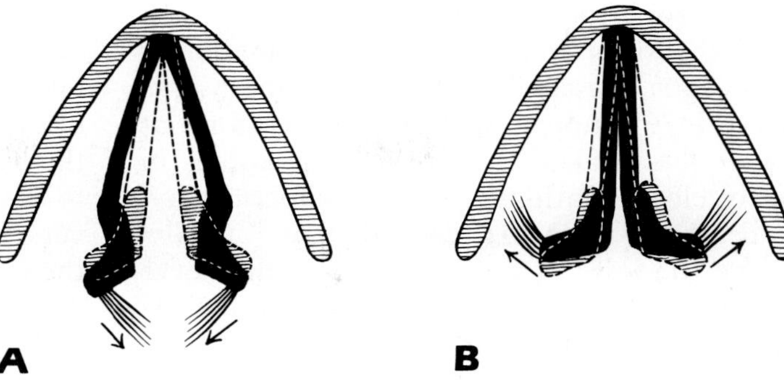


**FIG. 34-19.** **A** shows the action of the posterior cricoarytenoid in adducting the vocal cord; **B** shows the action of the lateral cricoarytenoid in adducting the cord. The final position of the vocal cord and arytenoid cartilage is shown in black, and the starting position is shown in outline.

sphincter of the vestibule, drawing the arytenoid cartilages closer to the thyroid cartilage and approximating the vestibular folds.

Associated with the upper border of the thyroarytenoideus muscle are other fibers, the **thyroepiglotticus,** that arise in common with the thyroarytenoideus but leave it to sweep upward into the aryepiglottic fold. Here they blend with the aryepiglotticus and insert in part into the quadrangular membrane and in part into the epiglottic cartilage. The thyroepiglotticus is a part of the sphincteric mechanism of aditus and vestibule.

The **vocalis muscle** is closely associated with the conus elasticus and the vocal ligament (Fig. 34-15). It lies largely below and deep to the thyroarytenoideus. It is thicker than the thyroarytenoideus and somewhat triangular in cross section. It arises from the inferior part of the angle between the two laminae of the thyroid cartilage and runs backward to an insertion on the lateral surface of the vocal process and the adjacent anterolateral surface of the arytenoid cartilage, but between these two extremes, it is usually said to be attached along the entire length of the vocal ligament or the conus elasticus below the ligament (the details of its insertion are not agreed upon). By its contraction, the vocalis muscle would tend to shorten the vocal cord if it acted alone; however, it probably acts when the length of the cord already is fixed by the cricothyroid muscle. By varying the degree of its contraction, it varies the tension of the vocal fold, and it also can hold a posterior part of the fold firmly against the other fold, allowing air to escape only between anterior parts of the folds. Exactly how it does the latter is not agreed upon. It has been said variously to pull the conus elasticus upward so that the vocal folds are brought together over a greater area than that provided by their thin edges and to do the opposite, either abducting slightly a variable anterior segment of the cord or so pulling the cord forward that an anterior part of the two adducted cords is more easily separated by the air stream than is the posterior part.

## Nerves and Vessels

The nerves and arteries of the larynx consist of the superior and inferior laryngeal nerves and arteries (Fig. 34-20).

**FIG. 34-20.** Nerves and arteries to the larynx.

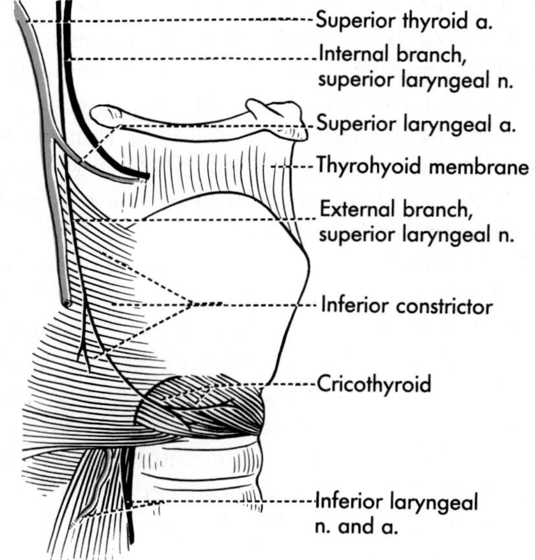

The **superior laryngeal nerve,** a branch from the lower end of the inferior ganglion of the vagus, descends anterior to the vagus nerve in the upper part of the neck and, at about the level of the hyoid bone, comes in fairly close relationship to the superior thyroid artery, lying just deep to that. Usually, above the level of the hyoid bone, the nerve divides into external and internal branches. The **external branch** descends behind the larynx on the inferior constrictor and with the superior thyroid artery until it curves forward below the thyroid cartilage to end in the cricothyroid muscle, frequently giving off a branch into the inferior constrictor before it does so.

The **internal branch** of the superior laryngeal nerve curves forward below the hyoid bone and accompanies the superior laryngeal artery, a branch of the superior thyroid, in a course across the posterior part of the thyrohyoid membrane. Soon, however, nerve and artery penetrate the membrane and run in the wall or floor of the piriform recess (where the nerve can be anesthetized). The nerve breaks up into branches, most of which penetrate the thyroarytenoideus muscle to be distributed both upward and downward to the mucosal surface of the larynx, largely above the vocal folds (and also to the piriform recess, the epiglottic vallecula, and a small posterior part of the dorsum of the tongue). One or more branches of the superior laryngeal nerve descend to reach the mucous membrane on the pharyngeal surface of the larynx and penetrate the transverse arytenoideus muscle to reach the interior of the larynx. One of these descending branches often anastomoses with the inferior laryngeal nerve.

The **inferior laryngeal nerve** is the terminal part of the recurrent laryngeal, a branch of the vagus. The courses of the recurrent nerves in the neck have already been described. Each nerve enters the larynx just posterior to the cricothyroid articulation in the lateral gap between the lower edge of the inferior pharyngeal constrictor and the esophagus. Fairly often before it enters the larynx, but if not before then after it has entered, the inferior laryngeal nerve divides into two or more branches. The muscular distribution of the branches is variable, but if there are only two, the **posterior branch** is likely to be distributed to the posterior cricoarytenoid and the transverse and oblique arytenoidei muscles, and the **anterior** branch to the remaining internal muscles of the larynx. Twigs also supply the laryngeal mucosa below the vocal folds. Any of the branches may communicate with the internal laryngeal nerve.

The **inferior laryngeal artery** is a small branch of the inferior thyroid artery that accompanies the inferior laryngeal nerve into the larynx. It anastomoses there with the larger **superior laryngeal artery,** from the superior thyroid, and the two supply the larynx and adjacent portions of the pharynx. The laryngeal arteries are accompanied by laryngeal veins. The **superior laryngeal vein** typically joins the superior thyroid vein and through this empties into the internal jugular. The **inferior laryngeal vein** joins the inferior thyroid vein of its side or anastomosing channels between the thyroid veins across the front of the trachea.

The **lymphatics** of the larynx drain along the vessels and, therefore, both upward and downward. Some of them end in very small lymph nodes that lie on the thyrohyoid membrane and cricotracheal ligament or on the upper end of the trachea, but these in turn drain to the deep cervical nodes, so that all the drainage of the larynx ends here.

#### Some Functional Aspects

The functions of the individual muscles of the larynx already have been noted, and functional aspects of the larynx as a whole can be fairly briefly summarized. The vestibule of the larynx is particularly sensitive to foreign objects, so that when these come in contact with its epithelium, violent coughing ensues in order to expel them. This involves first a marked abduction of the vocal cords, brought about particularly by the posterior cricoarytenoid muscles, in order to enlarge the airway as much as possible and allow a gasp for breath; then a closure of the larynx, primarily brought about by the vestibular folds and therefore by the thyroarytenoideus muscle; and when this closure is suddenly released following increase in pressure in the thorax, the air erupts as a cough.

In quiet inspiration, the vocal cords are slightly abducted and somewhat curved, and the arytenoid cartilages also are apart, so that there is a fair-sized slit at the rima glottidis. During forced inspiration, the rima glottidis is widened as much as possible, which means that the arytenoid cartilages are pulled laterally and the vocal processes rotated laterally by the posterior cricoarytenoid muscles. During phonation, except for the lowest tones, the vocal cords become straight and move together to meet in the midline. This apparently is brought about by the action of several muscles simultaneously. The cricothyroids adduct and lengthen the vocal cord and fold because they tilt the thyroid and cricoid cartilages upon each other; the lateral cricoarytenoids adduct the vocal cords by swinging the vocal processes together; and the vocalis muscles apparently act differentially on the vocal folds so that a variable anterior segment vibrates.

Although it commonly is the escape of air between anterior segments of the two vocal folds that produces the vibrations that then are modified by the tongue, lips, palate, and so forth, to produce speech, other vibrating mechanisms can be used. For instance, the vestibular folds can be used for vocalization, although the voice then is more hoarse than usual. Also, after the larynx has been removed, as it may have to be for carcinoma, a patient may produce a certain amount of intelligible sound by learning to govern the escape of swallowed air from the stomach and the esophagus.

It perhaps cannot be claimed that the mechanism of vocalization is completely understood, but it seems clear that under normal circumstances, control of tone (pitch) is brought about by variations in the length of the vibrating segments of the vocal folds. It is the anterior ends of the vocal folds that vibrate for the highest tones, and as progressively lower tones are produced, longer and longer anterior segments of the folds vibrate. Men generally have both longer and heavier vocal folds than women, and this is the reason the male voice is deeper. The muscles that particularly affect the length of the vocal folds and of their vibratory portions are the cricothyroideus, which lengthens the fold as a whole, and the vocalis, which apparently governs the length of the vibratory portion of the fold. Along with changes in the vibratory portions of the folds, the larynx as a whole also is raised or lowered—raised for high tones, lowered for low tones—so that the pharynx, a part of the resonating chamber above the larynx, is altered in length for different tones, somewhat as organ pipes of different lengths are used for the production of different tones. Insofar as the vocal cords themselves are concerned, it seems clear that changes in pitch are not produced simply by altering the tenseness of the vocal ligaments, as was once believed. The tenseness of the vocal folds is controlled by the vocalis muscles, not by stretch of the vocal ligaments, so that these muscles control both the tenseness and the mass of the vibrating segments of the vocal folds and are responsible for the fine variations in pitch of which the human larynx is capable.

Since the inferior laryngeal nerve innervates all the muscles of the larynx except the cricothyroid, and therefore all the muscles concerned with active movement of the vocal cords, paralysis of that nerve produces paralysis of the vocal cord on its side. A cord so paralyzed is at first bowed outward and can be neither abducted nor adducted. Since the normal vocal cord cannot meet it, the voice is poor (but in longer-standing paralysis, described below, a paralyzed cord may gradually move toward the midline, with consequent improvement of the voice to a point where it seems normal). Paralysis of the left nerve may occur as a result of a lesion in the mediastinum, since this nerve turns upward around the arch of the aorta, or either nerve may be interrupted in the neck. Each nerve is closely related to the posterior aspect of the thyroid gland just before it enters the larynx, and below that crosses the inferior thyroid artery (it may run behind the artery, between its branches, or in front of it); great care, therefore, must be taken to avoid injuring the nerves during thyroidectomy.

In bilateral paralysis of the vocal cords, the voice is almost lost, since the cords cannot be moved together. Often, however, bilaterally paralyzed cords gradually become less bowed and, therefore, move toward each other, as a result of the pull of the unparalyzed cricothyroid muscles. The voice improves as adduction increases, but since the cords cannot be abducted, the airway is simultaneously narrowed. The narrowing may be so severe that operative intervention, usually involving removal of the arytenoid cartilage, is necessary.

If a superior laryngeal nerve or its external branch is interrupted (as they may be in securing the superior thyroid artery in a thyroidectomy), the cricothyroid muscle cannot lengthen the vocal cord. Although such a cord will move as readily to the midline as will its mate on the unaffected side, it is not as straight and tense, for the vocalis muscle is not as effective in acting on the cord as a whole as is the cricothyroid. Consequently, because of the lack of tenseness of one cord, there is a tendency to hoarseness and easy tiring of the voice.

Variations in the position and tenseness of the vocal folds after nerve injury are impossible to quantitate by laryngoscopy, so clinicians commonly classify vocal cords simply as bowed or straight and as movable or immovable.

## GENERAL REFERENCES AND RECOMMENDED READINGS


BASMAJIAN JV, DUTTA CR: Electromyography of the pharyngeal constrictors and levator palati in man. Anat Rec 139:561, 1961

DORRANCE GM, BRANSFIELD JW: Studies in the anatomy and repair of cleft palate. Surg Gynecol Obstet 84:878, 1947

FINK BR: Tensor mechanism of the vocal folds. Ann Otol Rhinol Laryngol 71:591, 1962

FINK BR: The Human Larynx: A functional study. New York, Raven Press, 1975

LEMERE F: Innervation of the larynx: I. Innervation of laryngeal muscles. Am J Anat 51:417, 1932

LEMERE F: Innervation of the larynx: III. Experimental paralysis of the laryngeal nerves. Arch Otolaryngol 18:413, 1933

MITCHINSON AGH, YOFFEY JM: Changes in the vocal folds in humming low and high notes: A radiographic study. J Anat 82:88, 1948

NEGUS VE: The Mechanism of the Larynx. St Louis, CV Mosby, 1929

PRESSMAN JJ: Sphincter action of the larynx. Arch Otolaryngol 33:351, 1941

PRESSMAN JJ, KELEMEN G: Physiology of the larynx. Physiol Rev 35:506, 1955

RAMSEY GH, WATSON JS, GRAMIAK R et al: Cinefluorographic analysis of the mechanism of swallowing. Radiology 64:498, 1955

RAVEN RW: Pouches of the pharynx and oesophagus: With special reference to the embryological and morphological aspects. Br J Surg 21:235, 1933

RICH AR: The innervation of the tensor veli palatini and levator veli palatini muscles. Bull Johns Hopkins Hosp 31:305, 1920

SPRAGUE JM: The innervation of the pharynx in the Rhesus monkey, and the formation of the pharyngeal plexus in primates. Anat Rec 90:197, 1944

TURNER WA: On the innervation of the muscles of the soft palate. J Anat Physiol 23:523, 1889

WARDILL WEM, WHILLIS J: Movements of the soft palate: With special reference to the function of the tensor palati muscle. Surg Gynecol Obstet 62:836, 1936

# GLOSSARY OF SOME SYNONYMS AND EPONYMS

This relatively short list includes B.N.A. and other terms that are in fairly common use but differ considerably from the N.A. and therefore need definition. With the occasional exception of useful terms that the N.A. omits entirely, the continued use of these terms should be avoided, but the student will encounter them and need to understand them. Eponyms, such as those listed here and numerous others, should be discouraged, since they are often vague, do not necessarily indicate the person who first described the structure, and are, of course, totally nondescriptive.

**Alcock's canal:** pudendal canal
**Ansa hypoglossi:** ansa cervicalis
**Antrum:** maxillary sinus
  **tympanic:** mastoid antrum
**Appendage, auricular:** auricle (of heart)
**Arches, lumbocostal:** arcuate ligaments
  **volar arterial:** palmar arterial arches
**Artery, auditory, internal:** labyrinthine
  **buccinator:** buccal
  **coronary of stomach:** left gastric, *or* left and right gastrics
  **dental:** alveolar
  **descending, of heart:** interventricular
  **digital, volar:** palmar digital
  **dorsal interosseous, of forearm:** posterior interosseous
  **frontal:** supratrochlear
  **genicular, supreme:** descending genicular
  **hemorrhoidal:** rectal
  **hypogastric:** internal iliac
  **innominate:** brachiocephalic
  **interosseous, dorsal and volar:** posterior and anterior interosseous
  **of ligamentum teres femoris:** a. of the ligament of the femoral head
  **mammary, internal:** internal thoracic
  **maxillary, external:** facial

**Artery**
**internal:** maxillary
**metacarpal, volar:** palmar metacarpal
**spermatic, internal:** testicular *or* ovarian
**transverse scapular:** suprascapular
**vesiculodeferential:** a. of ductus deferens
**Auditory:** acoustic
**Auerbach's plexus and ganglia:** myenteric portion of the enteric nerve plexus
**Bartholin's duct:** major sublingual duct
**gland:** greater vestibular gland
**Bell's muscle:** lateral edges of the trigonal muscle of the urinary bladder
**Bigelow, "Y" ligament of:** iliofemoral ligament
**Bochdalek's gap:** lumbocostal trigone
**Bone, astragalus:** talus
**cuneiform, first, second, and third:** medial, intermediate, and lateral cuneiforms
**innominate:** os coxae
**malar:** zygomatic
**multangular, greater and lesser:** trapezium and trapezoid
**navicular, of hand:** scaphoid
**scaphoid, of foot:** navicular
**turbinate:** nasal concha
**Botallo's duct and ligament:** ductus arteriosus and ligamentum arteriosum
**Broca's convolution:** posterior end of left inferior frontal gyrus
**Buck's fascia:** deep penile fascia
**Burn's space:** suprasternal fascial space
**Calot, triangle of:** cystohepatic triangle; *see* Triangle
**Camper's fascia:** fatty part of superficial fascia (tela subcutanea) on lower abdomen
**Canal, pterygopalatine:** palatine canals
**of Schlemm:** sinus venosus sclerae
**semicircular, membranous:** semicircular duct
**subsartorial:** adductor canal
**Canthi (canthus):** the palpebral commissures, or angles at which the eyelids meet
**Cartilage, lower nasal:** greater alar
**sesamoid, nasal:** accessory
**upper nasal:** lateral
**Chain, sympathetic:** sympathetic trunk
**Chopart's joint:** transverse tarsal joint
**Cisterna magna:** cerebellomedullary cistern
**Cloquet's septum; node:** femoral septum; lymph node at the femoral ring
**Colles' fascia:** superficial perineal fascia (membranous, deep to the tela subcutanea)
**Cooper's ligaments:** a. pectineal ligament, b. suspensory ligaments of breast
**Corpus cavernosum urethrae:** corpus spongiosum penis
**Cortex, of cerebral hemisphere:** pallium
**Corti, ganglion and organ of:** spiral ganglion and spiral organ of ear
**Cowper's glands:** bulbourethral glands
**Cuvier, duct of:** common cardinal vein
**Denonvillier's fascia:** rectovesical septum (peritoneoperineal membrane)
**Dorsal:** posterior, usually, except in hand and for nerve roots
**Douglas, cavity or pouch of:** rectouterine excavation or pouch
**fold of:** rectouterine fold
**line or fold of:** arcuate line of sheath of rectus abdominis
**Duct, submaxillary:** submandibular duct
**Eminence, iliopectineal:** iliopubic eminence
**Eustachian tube:** auditory tube
**valve:** valve of inferior vena cava
**Fallopian aqueduct or canal:** facial canal
**foramen:** hiatus of canal of greater petrosal nerve
**tube:** uterine tube
**Fascia bulbi:** vagina bulbi
**lumbodorsal:** thoracolumbar
**renal:** fibrous portion of the adipose renal capsule
**Sibson's:** suprapleural membrane
**Fenestra ovalis:** fenestra vestibuli
**rotunda:** fenestra cochleae
**Filum terminale, internal:** filum terminale (of cord)
**external:** filum of the spinal dura mater
**Fissure, of cerebral pallium:** sulcus
**pterygopalatine:** pterygomaxillary fissure
**sphenoidal:** superior orbital fissure
**sphenoidal, inferior,** *or* **sphenomaxillary:** inferior orbital fissure
**Flexure, hepatic:** right colic
**lienal:** left colic
**Fold, ileocecal, inferior:** ileocecal
**superior:** vascular fold of cecum
**malleolar:** mallear
**ventricular:** vestibular
**Foramen, optic:** optic canal
**Fossa, antecubital:** cubital
**navicularis:** vestibular fossa
**ovalis femoris:** saphenous hiatus

**Fourchette:** frenulum of labia minora

**Foveae, inguinal and supravesical:** inguinal and supravesical fossae

**Frankenhauser's ganglion:** uterovaginal nerve plexus

**Gärtner's duct:** longitudinal duct of epoöphoron, remains of mesonephric duct in female

**Galen, vein of:** great cerebral vein

**Ganglion, cervical, inferior:** unfortunately not included in the N.A. This is the lowest cervical sympathetic ganglion, frequently but not always fused with the first thoracic to form the cervicothoracic or stellate ganglion

**jugular:** superior, of vagus

**nodose:** inferior, of vagus

**petrous:** inferior, of glossopharyngeal

**semilunar:** trigeminal

**sphenopalatine:** pterygopalatine

**submaxillary:** submandibular

**Gasserian ganglion:** trigeminal ganglion

**Genu, external, of facial nerve:** geniculum

**Gerota's capsule or fascia:** renal fascia; *see* Fascia

**Gimbernat's ligament:** lacunar ligament

**Gland, adrenal:** suprarenal

**lymph:** lymph node

**submaxillary:** submandibular

**Glaserian fissure:** petrotympanic fissure

**Glisson's capsule:** perivascular fibrous capsule of liver

**Groove, bicipital, of humerus:** intertubercular groove

**Gyrus, fusiform:** medial occipitotemporal gyrus

**hippocampal:** parahippocampal gyrus

**Hasner, fold or valve of:** lacrimal fold

**Haversian canals:** central canal of osteon

**Heister, spiral valve of:** spiral fold of cystic duct

**Henle's ligament:** part of conjoint tendon

**Hensen's duct:** ductus reuniens

**Hering, nerve of:** carotid sinus branch of cranial nerve IX

**Herophilus, torcular of:** confluence of the (cranial venous) sinuses

**Hesselbach's ligament:** interfoveolar ligament

**triangle:** inguinal triangle

**Hiatus, of facial canal:** of canal of greater petrosal nerve

**Highmore, antrum of:** maxillary sinus

**His, bundle of:** atrioventricular bundle

**Horner's muscle:** pars lacriminalis of orbicularis oculi

**Houston's valves or folds:** transverse rectal folds

**Humphrey, ligament of:** anterior meniscofemoral ligament

**Hunter's canal:** adductor canal

**Inscription, tendinous:** tendinous intersection

**Ischiadic:** sciatic

**Jacobson's cartilage, organ:** vomeronasal cartilage and organ

**nerve:** tympanic branch, cranial nerve IX

**Labbé, vein of:** inferior anastomotic vein of cerebrum

**Labrum, glenoid (of hip):** acetabular labrum

**Lacertus fibrosus:** bicipital aponeurosis

**Lamina papyracea:** orbital lamina of ethmoid bone

**Langer's lines:** cleavage lines of skin

**Ligament, cruciate crural:** inferior extensor retinaculum of ankle

**dorsal carpal:** extensor retinaculum at wrist

**iliopectineal:** iliopectineal arch

**laciniate:** flexor retinaculum at ankle

**periodontal:** same as periodontal membrane

**supensory of eyeball:** thickened lower part of bulbar sheath

**transverse carpal:** flexor retinaculum at wrist

**transverse crural:** superior extensor retinaculum of leg and ankle

**umbilical, lateral:** medial umbilical ligament

**Ligamentum teres (of femur):** ligament of the head of the femur

**Line, pectinate or mucocutaneous:** a line connecting the anal valves at the bases of the anal columns

**Linea semicircularis:** arcuate line of sheath of rectus abdominis

**Lister's tubercle:** dorsal radial tubercle; *see* Tubercle

**Littré, glands of:** urethral glands

**Lockwood's ligament:** thickened lower part of the bulbar sheath

**Louis (Ludovici), angle of:** sternal angle

**Luschka's body or glomus:** coccygeal body

**foramina:** lateral apertures of fourth ventricle

**Mackenrodt's ligament:** cardinal (lateral cervical) ligament of uterus

**Magendie, foramen of:** median aperture of fourth ventricle

**Marshall, ligament or fold of:** fold of the left superior vena cava in the pericardial sac

**vein of:** oblique vein of left atrium

**Mayo, vein of:** prepyloric vein

**Meatus, urethral:** external urethral ostium in male

**Meckel's cartilage:** the cartilage of the first branchial arch
  **cave:** trigeminal cavum, the subarachnoid space around cranial nerve V as it lies in the middle cranial fossa
  **diverticulum:** ileal diverticulum, a persistent proximal part of the vitellointestinal duct
  **ganglion:** pterygopalatine, *sometimes* submandibular ganglion
**Meibomian glands:** tarsal glands
**Membrane, basilar (of ear):** lamina basilaris
  **periodontal:** the dental term for the periodontium; in dentistry, periodontium includes also the surrounding bone and the gums
**Monro, foramina of:** interventricular foramina
**Morgagni, appendix of hydatid of:** appendix testis *or* vesicular appendix of epoöphoron
  **columns of:** anal columns
  **foramen of:** a. foramen cecum linguae, b. sternocostal triangle; *see* Triangle
  **fossa of:** fossa navicularis urethrae
  **sinus of:** a. laryngeal ventricle, b. anal sinus, c. space between superior constrictor and base of skull, d. pharyngeal recess
**Morison, pouch or space of:** hepatorenal recess
**Müller's duct:** paramesonephric duct
  **muscle:** any of four bits of smooth muscle related to orbit, but most commonly the superior tarsal muscle
**Muscle, bulbocavernosus:** bulbospongiosus
  **caninus:** levator anguli oris
  **compressor naris:** transverse part of nasalis
  **dilator naris:** alar part of nasalis
  **flexor digitorum sublimis:** flexor digitorum superficialis
  **quadratus labii inferioris:** depressor labii inferioris
  **quadratus labii superioris:** zygomaticus minor, levator labii superioris, *and* levator labii superioris alaeque nasi
  **thyroarytenoideus externus:** thyroarytenoideus
    **internus:** vocalis
  **triangularis:** depressor anguli oris
**Nerve, acoustic:** vestibulocochlear
  **anterior thoracic:** pectoral nerves
  **buccinator (N. V):** buccal branch of trigeminal nerve
  **cutaneous colli:** transversus colli
  **dorsal, in forearm:** posterior
  **hemorrhoidal:** rectal
  **of Hering:** carotid sinus nerve
  **petrosal, superficial, greater and lesser:** greater and lesser petrosals
**Nerve**
  **statoacoustic:** vestibulocochlear
  **vidian:** nerve of pterygoid canal
  **volar:** in forearm, anterior; in hand, palmar
**Nuck, canal of:** persistent processur vaginalis in female
**Oddi, sphincter of:** sphincter of hepatopancreatic ampulla
**Olive, inferior:** olive
**Os, external:** uterine ostium
**Otoliths:** statoconia
**Pacchionian bodies or granulations:** arachnoidal granulations
**Passavant, fold or ridge of:** a fold developing on the posterior pharyngeal wall to help close the nasopharynx
**Petit, triangle of:** lumbar triangle
**Plexus, hypogastric:** pelvic plexus
**Portio major, of trigeminal nerve:** sensory root
  **minor, of trigeminal nerve:** motor root
**Pouches, perineal:** perineal spaces
**Poupart's ligament:** inguinal ligament
**Process, odontoid:** dens of axis
**Purkinje fibers:** modified muscle fibers of the conduction system of the heart
**Ranvier, nodes of:** constrictions of the myelin of nerve fibers
**Rathke's pouch:** hypophyseal, craniobuccal, *or* neurobuccal pouch, from which the anterior lobe of the hypophysis develops
**Reil, island of:** insula of cerebral hemisphere
**Reissner's membrane:** vestibular membrane
**Retzius, cave of:** retropubic space
  **veins of:** retroperitoneal veins connecting portal and caval systems
**Riolan, arc of:** usually an anastomosis between the left and middle colic arteries at the base of the mesocolon
**Rivinus, ducts of:** lesser sublingual ducts
**Rolandic fissure:** central sulcus of cerebral hemisphere
**Rosenmüller, fossa of:** pharyngeal recess
**Rosenthal, vein of:** basal vein of cerebrum
**Sac, lesser:** omental bursa
**Santorini, cartilage of:** corniculate cartilage
  **duct of:** accessory pancreatic duct
**Sappey's veins:** small veins in the falciform ligament
**Scala media:** cochlear duct
**Scarpa's fascia:** membranous part of superficial fascia of lower abdomen
  **ganglion:** vestibular ganglion

**Scarpa's**
**triangle:** femoral triangle
**Schlemm, canal of:** sinus venosus sclerae
**Schwann, cells of:** neurilemma cells
**Sibson's fascia:** suprapleural membrane
**Sinus, piriform:** piriform recess
**rectal:** anal sinus
**Skene's ducts or glands:** paraurethral ducts or glands
**Spieghel's (spigelian) line:** semilunar line of abdomen
**lobe:** caudate lobe of liver
**Stensen, canal of:** incisive canal
**duct of:** parotid duct
**Sylvian aqueduct:** cerebral aqueduct
**fissure:** lateral sulcus of cerebral hemisphere
**Tables, of skull:** laminae
**Tenon's capsule or fascia:** the fascial sheath of the eyeball
**Thebesian valve:** valve of coronary sinus
**veins:** least cardiac veins
**Treitz, muscle or ligament of:** suspensory muscle of duodenum
**Treves, bloodless fold of:** ileocecal fold
**Triangle, cystohepatic:** triangle between cystic duct, common hepatic duct, and liver
**sternocostal:** triangle between sternal and costal origins of diaphragm, transmitting superior epigastric vessels
**Trolard, vein of:** superior anastomotic vein of cerebrum
**Tuberosity, bicipital:** radial
**of humerus:** tubercle
**Tunnel, carpal:** carpal canal
**Turbinate:** concha
**Urethra, anterior:** spongy, or spongy and membranous, part of male urethra
**posterior:** prostatic, or prostatic and membranous, part of male urethra
**Uvea, or uveal tract:** the choroid, ciliary body, and iris of the eyeball
**Valsalva, sinus of:** aortic sinus
**Vas deferens:** ductus deferens
**Vasa efferentia testis:** ductuli efferentes
**Vater, ampulla of:** hepatopancreatic ampulla
**Vein, coronary:** usually left gastric
**facial, anterior:** facial
**common:** lower end of facial vein after it is joined by retromandibular vein
**posterior:** retromandibular
**hemorrhoidal:** rectal
**hypogastric:** internal iliac
**innominate:** brachiocephalic
**maxillary, internal:** maxillary
**pyloric:** usually right gastric, sometimes prepyloric vein
**Verumontanum:** colliculus seminalis
**Vidian canal, nerve:** peterygoid canal, nerve of pterygoid canal
**Volar:** in forearm, anterior; in hand, palmar
**Waldeyer's ring:** lymphatic ring in the pharynx
**Wharton's duct:** submandibular duct
**Willis, circle of:** circulus arteriosus cerebri
**Window:** *see* Fenestra
**Winslow, foramen of:** epiploic foramen
**Wirsung, duct of:** pancreatic (chief) duct
**Wolffian body and duct:** mesonephros and mesonephric duct
**Wormian bones:** sutural bones of skull
**Wrisberg, cartilage of:** cuneiform cartilage
**ganglion of:** a cardiac ganglion
**ligament of:** posterior meniscofemoral ligament
**Zinn, annulus of:** annulus tendineus communis
**Zuckerkandl, organs of:** paired chromaffin-cell bodies on the aorta near origin of inferior mesenteric artery

# INDEX

With few exceptions, the structures and parts listed in this index must first be sought under the noun that identifies them, rather than under any of their descriptive adjectives; the latter are used for the alphabetical arrangement under the noun. Thus, all arteries are listed under *A*, and the superficial temporal artery should be sought under the subheading "temporal" under "Artery(ies)"; similarly, all muscles are listed under *M*, and the medial pterygoid muscle should be sought under the subheading "pterygoid" under "Muscle(s)." With few exceptions, also, the terms in this index are N.A. ones. The Glossary identifies many of the major synonyms.

The index is also a regional one, listing under the specific part of the body concerned the page numbers on which discussions of the nerves, blood vessels, and so forth of that part will be found—except where such discussions form such an integral part of a short description that they can readily be located in that description. Page references in *italics* identify illustrations or tables.